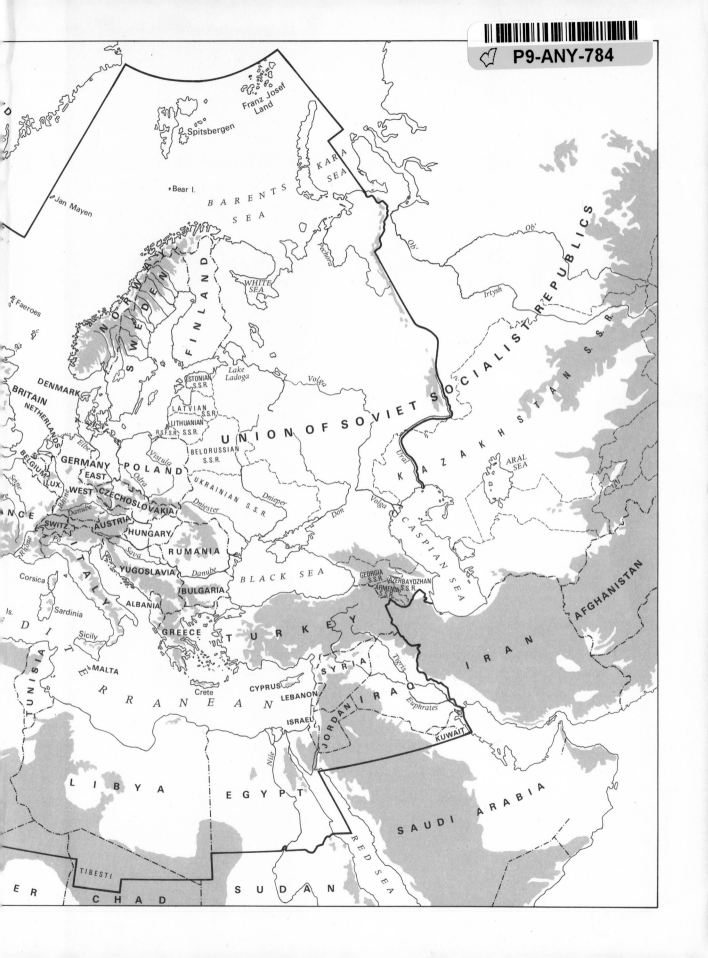

Franz Josef
Land

Spitsbergen

Jan Mayen

Bear I.

*B A R E N T S*

*SEA*

KARA
SEA

Ob'

Ob'

Faeroes

WHITE
SEA

Irtysh

U N I O N   O F   S O V I E T   S O C I A L I S T   R E P U B L I C S

Pechora

Volga

S.
S.
R.

DENMARK

BRITAIN

NETHERLANDS

ESTONIAN
S.S.R.

Lake
Ladoga

LATVIAN
S.S.R.

R.S.F.S.R.

LITHUANIAN
S.S.R.

K A Z A K H S T A N

ARAL
SEA

Elbe

BELGIUM

LUX.

GERMANY

EAST

WEST

POLAND

Vistula

Odra

BELORUSSIAN
S.S.R.

UKRAINIAN  S.S.R.

Dnieper

Don

Ural

Volga

Seine

CZECHOSLOVAKIA

Dniester

ANCE

Rhine

SWITZ.

AUSTRIA

HUNGARY

RUMANIA

Danube

Sava

*CASPIAN  SEA*

Rhône

ITALY

YUGOSLAVIA

Danube

*B L A C K   S E A*

GEORGIA
S.S.R.

AZERBAYDZHAN

Corsica

BULGARIA

ARMENIA S.S.R.

Is.

Sardinia

ALBANIA

GREECE

T U R K E Y

I R A N

AFGHANISTAN

Sicily

MALTA

*M E D I T E R*

CYPRUS

Crete

SYRIA

Tigris

LEBANON

*R A N E A N*

ISRAEL

JORDAN

IRAQ

Euphrates

TUNISIA

Nile

KUWAIT

L I B Y A

E G Y P T

S A U D I   A R A B I A

RED
SEA

TIBESTI

ER

C H A D

S U D A N

Handbook of the
Birds of Europe
the Middle East and
North Africa

The Birds of the
Western Palearctic

Volume I

# Handbook of the
# Birds of Europe
# the Middle East and
# North Africa

## The Birds of the Western Palearctic

## Volume I · Ostrich to Ducks

Stanley Cramp *Chief Editor*    K E L Simmons

I J Ferguson-Lees    Robert Gillmor    P A D Hollom

Robert Hudson    E M Nicholson    M A Ogilvie

P J S Olney    K H Voous    Jan Wattel

OXFORD   LONDON   NEW YORK

OXFORD UNIVERSITY PRESS

*Oxford University Press, Walton Street, Oxford OX2 6DP*

*Oxford New York Toronto*
*Delhi Bombay Calcutta Madras Karachi*
*Petaling Jaya Singapore Hong Kong Tokyo*
*Nairobi Dar es Salaam Cape Town*
*Melbourne Auckland*

*and associated companies*
*Beirut Berlin Ibadan Nicosia*

© *Oxford University Press, 1977*

*Published in the United States by*
*Oxford University Press, New York*

*First published 1977*
*Reprinted 1978, 1980, 1982, 1984, 1986*

DEDICATED TO THE MEMORY OF

H F WITHERBY

(1873–1943)

EDITOR OF *THE HANDBOOK OF BRITISH BIRDS*

(1938–1941)

**British Library Cataloguing in Publication Data**

Handbook of the birds of Europe, the Middle East, and North Africa:
 the birds of the Western Palearctic.
 Vol. 1: Ostrich-Ducks.
 Bibl. – Index.
 ISBN 0-19-857358-8
 1. Cramp, Stanley and Simmons, KEL
 598.2′94
 598.2′91′822    QL690.A1
 Birds – Europe, Middle East and North Africa

*Printed in Hong Kong*

# CONTENTS

Contents

Contents

# INTRODUCTION

*The Handbook of British Birds* (Witherby *et al.* 1938–41) represented a pioneer effort to review and present clearly the entire up-to-date knowledge of birds of Britain and Ireland. Its scholarly standard and concise presentation made it a firm basis for countless and diverse advances in ornithological research. Due in no small part to the stimulus it provided, ornithological knowledge has shown remarkable advances since it appeared. There has been a clear need for some years for a work which would incorporate the mass of new knowledge, now scattered and often difficult of access in the journals of many countries, to provide a work of reference both for the professional scientist and the ever-growing body of amateurs whose range of interest continues to widen. Such a work should also focus attention on the still considerable gaps in our information, even of common species, and on the correlations between different aspects of our knowledge, so providing a stimulus for further studies.

However, the *Handbook of British Birds* reflected the tendency, prevalent as late as the middle of this century, for serious interest in birds to be concentrated largely within the national boundaries of a country, leaving the rest to be dealt with peripherally. Parallel with this, there was a widespread reluctance to regard birds as an essential and major part of biological science. Since then ornithology has formed a main growing point in studies of speciation and evolution, population dynamics, migration and dispersal, navigation, and ethology, while the impact of the *Handbook* impressed on ornithologists the need for, and the feasibility of, a more comprehensive approach. Thus *Ptitsy Sovetskogo Soyuza* (*Birds of the Soviet Union*) (Dementiev and Gladkov 1951–4), while constructed on similar general lines, covered a major part of a zoogeographical region, and this important Russian initiative was followed by Palmer in the *Handbook of North American Birds* (1962, etc., and in progress) and more recently by the *Handbuch der Vögel Mitteleuropas* (Bauer and Glutz 1966, etc., and in progress) dealing primarily with a compact group of central European countries. In the light of such progress, the mere revision of the *Handbook of British Birds*, focused on a small group of islands off western Europe, could not be justified. It became clear that any new approach must deal with the zoogeographical province of which Britain and Ireland form a part—the western Palearctic. Our original intention was to define this so as to fill the obvious gap between the Russian and North American works, but after hearing the comments of many ornithologists (including those of the USSR, who were anxious for us to extend the boundaries as far east as possible) we decided to include the whole of European Russia.

The Palearctic, like all faunal regions, is not susceptible to precise definition. Charles Vaurie, the author of the most recent authoritative work on this region (*The Birds of the Palearctic Fauna* 1959–65), adopted a more liberal interpretation than that of his great predecessor Ernst Hartert (*Die Vögel der paläarktischen Fauna* 1903–22). We have been somewhat more restrictive than Vaurie in delimiting the western and southern boundaries of the Palearctic. In the west, we have excluded Greenland (east Greenland was included by Vaurie in 1959 and the whole in 1965) but have followed him in covering all the eastern Atlantic islands south to the Cape Verde Islands, to which we have added the Banc d'Arguin group (but not the adjoining mainland) where the extensive recent researches by de Naurois (1969*a*) have clearly demonstrated the Palearctic character of the avifauna. The important habitats of the eastern North Atlantic and Arctic Oceans are also included. In the south, we have, like Vaurie, covered the Sahara south to the northern borders of the Sahel region, but, while including the mountain massif of Tibesti, we have excluded those of Aïr and Ennedi, where in our view the Ethiopian element predominates; farther east, in Egypt and Saudi Arabia, our boundary runs to the north of his. If the boundary of the Palearctic region eludes any logically unquestionable demarcation, the determination of the eastern limits of the western part is clearly largely arbitrary. For the reasons given earlier, we have included all European Russia, using the limits internationally agreed for the *Flora Europaea* (Tutin *et al.* 1964), and, further south, we have taken in the remaining regions of the USSR between the Caspian Sea and the Black Sea, Turkey, Iraq, and Kuwait, but have excluded the whole of Iran.

There have been considerable differences of opinion and practice regarding the sequence of families, genera, and species (see, e.g. Lack 1966*b*). In view of the existing confusion in this respect, Voous (1969) undertook to prepare a list of Holarctic bird species, which would promote a badly needed uniformity and (at least) short-term stability without being inflexible or interfering with taxonomic research. The first part of this list, treating non-passerines, has since appeared (Voous 1973) and has met with wide approval as a means for stabilizing nomenclature and sequence in lists of Holarctic birds. The second part, covering the passerines, is now in an advanced stage of preparation. We decided to follow this list and to deviate from it only in those cases where fresh research presented us with overwhelming reasons to do so. These cases (if any) will be fully discussed and documented.

The main body of the work consists of the species accounts, but there are also general summaries covering the main taxa (taxonomic categories) above the level of the genus—order, family, and, if necessary, subfamily and tribe—compiled on a world basis with special reference to west Palearctic forms. The summary for each order is brief, usually consisting of (1) a characterization of the type of birds concerned together with indication of size, (2) a list of the constituent families, and (3) a comment on its relationship, if any, to other orders; where there is just one family in an order, the treatment is transferred to that for the family. The summary for each family usually briefly covers (1) the type and size of bird; (2) the number of species, genera, and, if necessary, other groupings —both formal (tribes) and informal (superspecies, species-groups); (3) representation in the west Palearctic; (4) certain, mainly external, morphological characters (for more detailed taxonomic diagnoses, see such works as Witherby *et. al.* 1938–41); (5) world distribution; (6) habitat; (7) population trends in the west Palearctic; (8) movements; (9) food and feeding habits; (10) social pattern; (11) social behaviour, especially display; (12) voice; (13) other behaviour characters; and (14) breeding biology. In the case of other behavioural characters (not usually treated in the species accounts), the topics covered include special features of *comfort behaviour* (bathing, oiling, preening, comfort-movements, etc.), *resting posture, thermoregulatory behaviour* (especially heat responses such as sunning), and manner of *head-scratching*—whether *indirect* ('overwing') or *direct* ('underwing'); see, e.g., Simmons (1961, 1974*b*). These summaries have been abbreviated in the case of groups only marginally represented in the west Palearctic or consisting of one or a few species. Where appropriate, information usually given under the family is transferred to the summary for the subfamily or tribe. The tribal category, coming between the genus and family or subfamily, has been used sparingly but is useful in the case of certain large or taxonomically complex groups, e.g. the Anatidae (wildfowl). No separate summary is given for the genus because this taxon is generally recognized to be the most arbitrary in the system of classification and the most subject to controversy and change.

The scientific names of the species treated are in accordance with the International Code of Zoological Nomenclature (2nd edition 1964) and with subsequent rulings of the International Commission on Zoological Nomenclature. The Code and rulings cover only the formal side of nomenclature and leave room for independent taxonomic judgement on the boundaries of genera and species, which may affect the actual names that are used. So, in some species a synonym of the species name is given; this is an acceptable alternative dependent upon different generic or specific allocation of these species. Many species show geographical variation to such an extent that more than one *subspecies* is recognized; these have been termed *polytypic*. Species in which no subspecies are recognized are said to be *monotypic*. Subspecies

are groups of similar populations, belonging to a single species, inhabiting a geographic subdivision of the species range, and recognizably differing from other populations of the same species (Mayr 1969). We have tried to avoid naming stages in a cline, or populations differing from others in averages only but overlapping largely in range of measurements or colour shades. Our subspecies concept is a wide one. The word *race* is to be taken as a full synonym of subspecies.

For polytypic species, a list of west Palearctic subspecies is given, those not breeding in the area being marked as *migrant* or *accidental*. A separate sentence mentions *extralimital* races giving names where there are one or two, but the total number only where there are more. In a polytypic species, one of the subspecies has the same name as the species itself; this is termed the *nominate* subspecies, e.g. under *Anser fabalis* (Bean Goose), *A. fabalis fabalis* is referred to as 'nominate *fabalis*'. It is not necessarily the most 'typical', central, or most widespread subspecies but merely the one first named.

Vernacular names in the various languages have been taken from national check-lists, from Jørgensen (1958), from *A Field Guide to the Birds of Britain and Europe* (Peterson *et al.* 1974), or have been suggested by country correspondents.

The species accounts are then divided into the following sections, each the special responsibility of the editor named:

*Field Characters* I J Ferguson-Lees
*Habitat* E M Nicholson
*Distribution* and *Population* S Cramp
*Movements* R Hudson
*Food* P J S Olney
*Social Pattern and Behaviour* Dr K E L Simmons
*Voice* E M Nicholson
*Breeding* M A Ogilvie
*Plumages, Bare Parts, Moults, Measurements, Weights, Structure,* and *Geographical Variation* Dr J Wattel

The scope of each section, together with the terms and conventions used, and a discussion of special problems is outlined in the following pages. There are, at the time of writing, 601 breeding species in the western Palearctic (including introduced species with viable feral populations and species extinct in historical times) and for these the species accounts include all the above sections. For regular non-breeding migrants to the western Palearctic (11), the sections on Population, Social Pattern and Behaviour, and Breeding are omitted, and for accidental visitors which have occurred in the area since 1900 (131, and increasing annually), the section on Food also.

All the above categories (breeding species, regular migrants, and accidentals since 1900) are illustrated by paintings showing every plumage which is identifiable in the field, including, for example, breeding, non-breeding, immature, juvenile, and (except for regular migrants) nestling or downy young. Flight patterns are illustrated where appropriate. The paintings in Volume 1 are the work of the following artists (whose initials appear at the end of the caption for each plate): Paul Barruel, Dr C J F Coombs, Dr N W Cusa, Robert Gillmor, Peter Hayman, and Sir Peter Scott. Robert Gillmor, the editor with general responsibility for these illustrations, has also prepared all the line drawings in the text except those otherwise credited. R J Connor and A C Parker have provided photographs of eggs and down.

Finally, there are a number of species which do not merit full treatment or any illustration. They fall into two main categories: (1) those for which valid records exist but which have not been recorded in the west Palearctic since 1900 or have occurred only on ships at sea (chiefly Nearctic species in the east Atlantic); and (2) those with only a dubious western Palearctic status, such as specimens of doubtful origin, disputed identifications, and escapes from captivity which have not yet established viable feral populations. Species in both categories are each treated in a single short paragraph concerned with distribution and migration only; those in the second category are distinguished by being in smaller type. Fossil species are not covered in this work.

We have not attempted to make independent judgements on the validity of records of vagrants, but have followed national checklists and the advice of our country correspondents (see Distribution), except in a few instances where authoritative criticisms have been published, e.g. Bourne (1967) for the Procellariiformes.

# FIELD CHARACTERS

This section summarizes information on size, shape, and plumage as they appear in the field, differentiating where relevant between the sexes, breeding and non-breeding plumages, and immature stages, and making comparisons as necessary with similar or other confusion species; it also includes brief notes on general habits and behaviour where these aid field identification. The presentation is normally in four paragraphs.

The first paragraph opens with ranges of total length and wing-span, both in centimetres. These require some explanation. Previous European reference books have given a single median figure for total length, but such medians are often taken too precisely and used even in distinguishing birds side by side on size alone. In fact, birds vary as significantly in size as many other animals do and, in extreme cases, may be as much as 10% or more on either side of the median. The measurement of total length is the conventional one of the dead bird or skin on its back measured from tip of bill to tip of tail; one American field guide (Robbins *et al.* 1966) has used measurements of length based on field measurements, from tip of bill to tip of tail, of live birds 'hand-held in natural positions', but comparable data are not available in Europe and, in any case, this method results in reducing the size of, for example, a Mallard *Anas platyrhynchos* (with neck retracted) to only 2·5 cm larger than a Sandwich Tern *Sterna sandvicensis*, whereas the difference by the conventional method is 18 cm; the latter is thus a better indication of bulk. Where necessary, however, allowance must still be made for unusual lengths of bill, neck, or tail; attention is always drawn to these in the first paragraph. Wing-spans are more difficult to define, because museum skins (with wings set in the folded position) are not measurable in this way. Nevertheless, as they are an important indication of size in flight, it has been thought desirable to make an attempt at including them. Figures have been freely taken from such works as Palmer (1962) and Robbins *et al.* (1966), backed up by data from live birds and skins, and rough formulae for calculating wing-spans in each family have thus been devised. The results are only approximate and intended as a general guide. The remainder of the first paragraph includes a brief summary of character, shape, and proportions, and of the colours of the adult (or adult male in breeding plumage), together with an indication of the extent and separability in the field of sexual, seasonal, and immature differences.

The second paragraph gives field descriptions, under separate headings of male and female, of breeding and non-breeding plumages (with intermediates where appropriate), as many immature stages as necessary and, where appropriate, of morphs. Characters which are common to male and female, or adult and immature, are not usually repeated after the first mention; this applies particularly to bare parts. Sometimes the requisite degree of detail can be noted only after close observation under favourable conditions, but one of the aims of this paragraph is to encourage observers to distinguish sexual, seasonal, and immature plumages as far as possible in the field. Too often a specific identification is considered adequate and there is little attempt to distinguish more than the most obvious plumages, but important biological data on migration patterns, age of first breeding, and social behaviour are obtainable with fuller sexing and ageing. Information on moulting times is included where this is relevant to sexing or ageing.

The third paragraph draws attention to similar species which may be a source of confusion, also noting briefly their most important features. The fourth and last paragraph summarizes general characters and behavioural aspects which aid identification, including habitat, gait on land, swimming or diving where appropriate, feeding behaviour, flight, gregariousness, and voice. It is intended only as a general pointer and, for fuller information on some of these aspects, the reader is referred to the sections on Habitat, Voice, Food, and Social Pattern and Behaviour.

References are not normally given in this section, except

where there have been special studies of little-known species or where a disputed character is mentioned.

# HABITAT

Observers who have trained themselves to note in the field precise, uniformly defined features of avian form and plumage have not yet generally accustomed themselves to paying similarly critical attention to describing the components and characteristics of habitat. Casual observation has long since shown the probability of encountering particular bird species within certain recognizable habitats. Most of these have loose popular names, often confused and ambiguous, which the majority of observers have been content to use. Similar problems have arisen throughout the environmental sciences. A scientific approach to habitat is slowly emerging as precise standard terms gain currency in the description of, for example, vegetation, local climates, hydrobiology, and landforms. It is evident that ornithologists cannot continue to use terminology and concepts in the description of environment which botanists or other specialists concerned would find unacceptably loose and misleading. On the other hand, field ornithologists cannot be expected to master all these subjects, or to acquire more than a superficial knowledge of the ecological principles and techniques by which they are harmonized for use in the biosphere. A compromise approach is therefore required, which will assist progressively to eliminate unscientific terminology and attitudes and to substitute more exactly defined and systematically organized working methods, understandable to those specializing in the other relevant life and earth sciences.

Recent advances in these sciences have largely been reviewed and digested through the International Biological Programme. Its results would now enable a radically new format and terminology to be adopted here which would be widely accepted among interested branches of the scientific community, but would be strange and difficult for ornithologists.

After careful consideration, such an approach has not been adopted for three main reasons. First, it would fail to take account of the extent to which field observers have empirically, however loosely and unscientifically, evolved their own practical and reliable way of making instant judgements on the suitability of whatever terrain may be before them to one or other species of bird. Given this largely subconscious capacity for taking a bird's-eye view, birdwatchers might justly feel that they were being asked to abandon quick, proved, and convenient methods for slower, untested, and awkward ones.

Secondly, a great volume of imprecise data gathered over many decades on this basis would prove unusable in a more precise system, because it would be impossible to reconstruct with any accuracy the old data in the fresh terms.

Thirdly, the replacement of anecdotal and traditional by scientifically tested and standardized terms presupposes that a complete and agreed basis for such terms is available from the environmental sciences. That, unfortunately, is not yet the case. The task is complicated not only by international and other differences of approach but by the rapidity with which important recognizable habitats are being drastically modified, especially by changing land use and development, and by the spread of blended or degraded variants which are difficult to diagnose and define.

So, in a work of this character, the best course is an intermediate one, conserving so far as possible familiar terms but assigning to them reasonably precise meanings consistent with modern scientific usage, and treating the habitat of each species in a regular order of presentation and on a more comprehensive basis. Underlying the species and family treatments therefore is a systematic framework, derived from the World Checksheet Survey of the Conservation Terrestrial Section of the International Biological Programme, and a Glossary of Terms newly designed to bring the description of habitats into line with international scientific practice so far as is consistent with meeting ornithological requirements and with a common-sense degree of simplicity. As the framework provided by the International Biological Programme does not cover unvegetated aquatic habitats or human artefacts it has been used here in the fully comprehensive expanded form developed in the Geogram system (see Nicholson 1973). It is of course impossible to relate such an approach fully to the many overlapping systems which have been proposed within particular branches of study or countries, but as many as possible of these have been taken into account. Those familiar with other methods, for example in the treatment of vegetation, should find no insuperable obstacle in achieving some degree of reconciliation.

A further parallel is offered by the effective systematization and adoption of precise methods and terms in making sight identifications. Descriptions of habitat have hitherto tended to dwell on such conspicuous points as obvious land-forms (including water) and the height, spacing, and general character of vegetation, to the neglect of past and present human influences unless crudely obvious, and of ecological factors. Use of airspace and underwater zones has also been widely neglected. Even in careful and detailed studies, results have not usually dealt fully with the varying attraction to different species of birds of specific habitat situations, or the precise linkages and distinctions in requirements which govern the habitat selection of different species of birds at different stages of their life-cycle. Not much has yet been done to build up a picture of total use, season by season, of a total range of habitats.

To define where each species should not be found is no less essential than to forecast where it should. No other class of living organisms can match birds in extent of occupancy of the entire range of habitats on earth, or in serving to monitor the capacity of these for supporting life. The study of bird habitat is therefore not merely an aid to finding and recognition of each species, but provides valuable indicators for the state of the biosphere.

In many cases attempts at full standard presentation are frustrated by gaps in material, for example in relation to altitudinal range and to moulting habits and displacements. Over the west Palearctic it often happens that regional variations can make a precise statement partially misleading unless it is lengthily qualified or explained, and discretion has had to be used in omitting details of trivial or local significance, in order not to swamp or confuse the presentation of what is normal.

On the other hand, habitat descriptions of many species are less complete and more superficial than they should be for lack of studies in depth. This is well illustrated by the studies made by Fjeldså (1973*a*, *b*) of the Slavonian Grebe *Podiceps auritus*. In order to test correlations between subspecific and other genetic distinctions and habitat selection he made thorough investigations of a number of isolated populations in north Norway and Iceland for comparison with Swedish and Finnish stocks. He visited 966 sites in brackish bays, slow-flowing rivers, pools, and lakes, besides using several thousand literature sources. In the lakes, primary factors were morphometry (such as extent, depth, and altitude, and the profile under and above water); topography, including islands; geology; vegetation (marginal, emergent, floating, and submerged); water quality (eutrophic, mixotrophic, dystrophic and oligotrophic, and hard or soft); and biological productivity. Distance from the sea or from other lakes, marshes, and small water bodies was also relevant. Analysis of emergent vegetation revealed the high importance for nest siting of distance from shore and from open water; vegetation type; density and cover; anchorage; and water depth. In all cases records were made of water colour, sight depth, plankton turbidity, and bottom sediment. The percentage of water area carrying emergent vegetation was a further relevant factor. The significance of depth of water from surface to bed might be less than the depth from surface to top of submerged vegetation. The age of birds occupying territories of different types also proved significant, marginal sites tending to be left to the youngest. Account had to be taken of correlations with neighbourhood of other birds and specific plants. With all this wealth of detail the investigation was still incomplete. It ignored precise reactions to human presence or influence, and frequency and height of flight movements, for example. Such a case serves to emphasize how inadequate is our knowledge of factors determining habitat for almost all species of birds, and to show what opportunities exist for comparable studies in depth in countless other instances.

In the absence of such up-to-date comprehensive studies, reliance has to be placed largely on the literature, most of which consists of sketchy or fragmentary contributions, few of which are sufficiently precise and well-informed about habitat factors to enable specific conclusions to be drawn. Many are not, or are only partly, primary sources. It is necessary to cross-check and condense a very large number of references, few of which stand independently as sole or principal sources for any particular fact. This creates difficulties in deciding which out of countless sources consulted to select for citation. There is no problem for the small number of contributions tackling in depth the habitat aspects of a single species or group, or where some significant fact comes directly from a particular observer, especially in a source not much quoted and liable to be overlooked.

From the nature of the Habitat section, however, it is more frequently a case of producing valid generalizations from many different records, which cannot all be cited and no one of which stands out as of major significance or authenticity. Attempts to distil such generalizations from the literature have been previously made at different times and with differing degrees of success by many authorities, whose judgement and perception has often made our task easier. Some, such as Bauer and Glutz (1966–9), have included copious lists of references, and those who wish to pursue habitat records in such detail would do well to consult them.

With certain exceptions in the interests of convenience, the Habitat sections for every family and species follow a common pattern. Normally they begin with the biogeographical setting of the distribution of the species in terms of its latitudinal and altitudinal range and climatic limits and preferences, and the broad terrestrial and aquatic groupings in which it occurs, Then follow more details (including land-form and vegetation) of the habitats used for breeding, foraging, roosting, assembling to moult, stopping on migration, wintering, and other purposes, of the zones used in or under water or in the air, and of relations with man.

Care has been taken to minimize duplication with other sections of this work, but in certain cases some overlap is inherent in the plan. For example, where conservation problems stem from or must be considered in relation to habitat, they are treated in this section. Where, however, they are more or less self-contained (e.g. shooting pressure or mortality from toxic chemicals) they are handled only under Population. Although specific recommendations have not normally been included, efforts have been made to bring out opportunities and needs for further observation and research, which are plentiful and important.

The following Glossary of Terms and Systematic List of Basic Habitat Types aim to relate the wording in descriptions to ecological and geographic usage.

## GLOSSARY

In attempting to attain greater precision and completeness in describing the habitat preference of each bird species, a major obstacle is presented by the vague and ambiguous terms in common use, even in scientific records, for many of the elements in the environment. That environment is composed of a multiplicity of components, each of which conforms to ecological principles and can be scientifically described, not necessarily in some new jargon but by giving coherent, clearly defined meanings to many words in everyday use. This approach has been adopted over several decades by leading plant ecologists whose definitions for such words as bog and fen are largely followed here. It is hoped that in this way readers will be enabled to check more accurately the characteristics and limits of observed habitats and to acquire a more discriminating attitude to the semantics of the various terms employed, thus gradually bringing about a greater measure of clarification and standardization in their use. Owing to the great variations of these elements in the field, and the different, often conflicting practice in the use of terms, complete standardization must be recognized as unattainable.

The function of this glossary is simply to define principal terms, leaving their refinements and subdivisions to be treated in the text as they arise. For those less familiar with the English language, the glossary should prove more enlightening than reference to normal dictionaries, which inevitably suffer from this lack of a precision adequate to the study of bird habitats. Wherever possible, widely used equivalent terms in other European languages are included.

The guiding principle has been to conform in all important respects to the best recent authorities in each field, especially ecologists, geographers, and hydrologists, but not to hesitate to define more fully and to set quantitative criteria wherever advisable to facilitate accurate recording and comparison of ornithological data. Special care has been taken to minimize ambiguity and unconscious overlapping of terms. For convenient reference, words defined in this glossary are printed in capitals wherever else they occur in it.

AQUATIC VEGETATION. Grouped into four classes (1) MARGINAL for emergent plants forming a more or less narrow zone along the edge of standing or flowing open water (2) EMERGENT for plants other than MARGINAL which are rooted on the bottom and project above the surface (3) FLOATING for plants forming mats, streamers or other patterns on the water surface and (4) BOTTOM for plants of which no part reaches or comes near to the surface.

ARCTIC–ALPINE. The climatic zone, above the TREELINE, situated at increasing altitude towards lower latitudes, which carries many dwarf woody plants forming a HEATH cover below the snowline. The term TUNDRA is sometimes misleadingly extended to this zone. In exceptional areas, plants characteristic of this zone descend locally much lower, even to sea level in upper mid-latitudes.

ARCTIC and SUBARCTIC. The boundary between these in the west Palearctic is taken as passing through southern Iceland, extreme northern Norway and the entry to the White Sea, thence ranging down to as low as 65°N where it crosses the Urals. The southern boundary of the SUBARCTIC runs from the intersection of 30°W and 60°N in the Atlantic to the Norway coast just above the Arctic Circle, thence down near Bergen, Oslo, north of Stockholm, through Helsinki and near Leningrad to the Urals about 50°N.

BARRAGE. Artificial body of open fresh water normally of more than 1 km² extent and often much larger, formed by a major dam across a river, usually in steep terrain, and frequently much deeper than 10 m. 'Man-made lakes' is a term justifiably applied to some very large BARRAGES. *See also* RESERVOIR.

BOG. Peatland which receives its water supply only from precipitation. *See* MIRE and MARSH.

BOREAL. Used principally in relation to a large group of plants and animals whose distribution pattern can be matched with northerly climatic factors. It overlaps the ARCTIC and SUBARCTIC regions, but extends into the north temperate deciduous forest zone. It is unconfined to any one biome or group of communities within its climatic range. The species concerned all stop short at a southern limit within the coolest temperate zone.

BRECK. Type of undulating or flattish lowland country, often sandy and marginal for traditional forms of agriculture, which accordingly has relapsed to HEATH or a kind of WOODLAND SAVANNA between intermittent and well-separated periods of cultivation.

BUSH. More or less dense shrub, usually 1–3 m tall, with many woody branches and seasonally ample foliage providing good cover; sometimes a group of shrubs forming a single mass. (Not used here in the sense, probably borrowed from the Dutch *bosch*, of wild forest or scrub country.)

CANAL. Artificial channel of fresh or saline water, from 5 m to 50 m or occasionally much wider, sometimes on one level, otherwise on different levels linked by locks or lifts, with no consistent flow. Disused canals tend to become blocked with AQUATIC VEGETATION and their banks become overgrown to constitute an entirely different habitat.

CANOPY. The more or less continuous network of branches and foliage forming the top layer of FOREST, any gaps in which are normally narrower than the crowns of the separate trees.

CANYON. *See* RAVINE.

CARR. Preferred use implies more or less mature tree cover, either fringing or often succeeding a FEN or BOG after drainage or other local change eventually forming WOODLAND on wet, usually flooded site; also applicable to wet alluvial woods (cf. German *Bruch*).

CLEARING. The counterpart of a GLADE, but produced by tree-cutting or other human intervention and tending to be larger, up to 10 ha or more.

CLOSED VEGETATION. Where plants are in contact or overlap. *See* SPARSE VEGETATION.

CONTINENTAL SHELF. The underwater plateau extending from the coast to *c.* 100 fathoms (*c.* 180 m) deep, beyond which the CONTINENTAL SLOPE falls steeply toward the ocean bottom. *See* OFFSHORE and PELAGIC.

COPSE. Small stand of trees, generally not larger than 5 ha, usually on level or gently sloping ground or in lowland valley, not obviously planted, and mainly deciduous with substantial undergrowth.

COPPICE. Cutting woody stems on a rotational basis, often in association with a dispersed silviculture of taller trees, is coppicing and produces coppice with standards, usually in small woodlands. Accordingly a counterpart of COPSE, into which a COPPICE may become converted if it ceases to be managed as above.

CRAG. An abrupt, precipitous rock face, not necessarily high, characteristically from glacial pressures which produce a gentle slope behind it in the opposite direction. Commonly used of inland precipices forming a local and limited counterpart to a sea cliff.

CREEPER. *See* VINE.

DELTA. A complex river mouth or stream mouth where much silt is deposited, splitting the flow into numerous channels separated by emerging banks and reedbeds, and often giving rise to separate LAGOONS.

DESERT. A large tract of waterless land in mid or low latitudes where vegetation is either absent or virtually insignificant.

DIKE or DYKE. Artificial watercourse formed by excavation, commonly for drainage or water supply, at least 2 m wide and often regulated by a sluice or spillway. If less than 2 m wide it is a DITCH or TRENCH; if used for navigation or irrigation and over 5 m wide it is a CANAL. (Not used here in the derived sense of a bank of earth produced by excavation, although the bank may be regarded as part of the DIKE.)

DITCH. See DIKE.

DYSTROPHIC. Applied to somewhat acid and shallow water, especially in north, with few aquatic organisms but much marginal vegetation. See LAKE.

ESTUARY. Formed where a river entering the sea broadens out into a single INLET.

EUTROPHIC. See LAKE.

FEN. Peatland receiving its water supply from outside. See also MIRE.

FLOODLAND. Land so sited in relation to normal run-off of water that it becomes inundated, usually only to shallow depth, at regular or frequent intervals, but never long enough wholly to destroy terrestrial vegetation cover (including farm crops where appropriate) and terrestrial animals.

FOREST. Area of natural or semi-natural tall tree cover, mainly or wholly with closed CANOPY above 5 m and typically above 20 m, continuous over a substantial area, not less than 1000 ha and typically much more.

FOREST TUNDRA or WOODED TUNDRA. The zone, several or sometimes many kilometres wide, of stunted forest mixed with lichens and grassland and, in favoured valleys, with stands of well-grown trees, beyond the limit of continuous TAIGA, but by definition incompatible with true TUNDRA, which begins strictly beyond the TREELINE, and not merely beyond the forest edge.

GARIGUE. Open, low-growing calcareous scrub on lowland, often rocky slopes or cliffs, in MEDITERRANEAN CLIMATE, and due to grazing, burning, former cultivation or other human interference with earlier FOREST. Includes more herbage, grasses, and bare patches and fewer woody plants than MAQUIS, but may be transitional towards that type or towards colonization by woodland trees such as Aleppo pine *Pinus halepensis*.

GEYSER. A hot SPRING projecting water into the air.

GLADE. Small, open, but sheltered and usually elongated space within a FOREST or other WOODLAND, often with a relatively benign microclimate, and either natural or resembling a natural opening, vegetated with grass, herbage or BUSHES. See also CLEARING.

GORGE. A steep-sided, narrow, deep valley, often flanked by CRAGS. See also RAVINE.

GRASSLAND. Broad general term for areas of low CLOSED VEGETATION, predominantly of grasses and excluding woody plants, aquatics, mosses or lichens. See also STEPPE.

GRAVEL PIT. An area, commonly 5–50 m deeper than surrounding land surface, exploited for extraction of a loose, detrital sediment containing a large proportion of pebbly material, of 2–50 mm diameter and therefore much coarser than sand. A worked pit may remain unvegetated or may become overgrown or inundated with water or both. See also OPENCAST, QUARRY.

GROVE. More or less close stand of mature trees, natural or planted, with little or no undergrowth, usually applied to a feature in level, open landscape and of maximum extent 1 ha.

GULLY or GULLEY. See Ravine.

HEATH. Dry, normally lowland area of usually flat, sandy or gravelly soil having little peat element and carrying evergreen dwarf scrub plants such as heather *Erica*, normally under 0·5 m high; sometimes interspersed with scattered pines *Pinus*, birches *Betula* or other trees. Often traceable to earlier deforestation or to abandoned cultivation.

HEDGE or HEDGEROW. Artificial, narrow, dense, often straight linear barrier of woody vegetation, normally kept within 1–3 m high, especially for separating farm fields; variant type interspersed with shade trees.

ICE. Subject of standardized definition in Ice Glossary of the Scott Polar Research Institute at Cambridge (*Polar Record* 1956 and 1958), and elsewhere by Dunbar (1955). The most relevant types here are: Pack-ice: more or less hummocky ice, always formed at sea and unattached, varying from open to fully consolidated, but often divided by channels termed leads; Icesheet: layer of ice covering large area of land, continuously moving along defined channel from higher to lower ground, and usually eventually to sea, and Iceberg, always consisting of mass of ice broken away from a glacier, at least 5 m above sea surface, and thus distinct from much thinner but often more extended Ice floes or Icefields, always of marine origin. Ice on fresh or sheltered marine water rarely assumes forms differing much from associated water surface.

INLET. Used as a general term to cover any bay, gulf, fiord, outer estuary, or smaller indentation in a coastline of sea or large lake, which produces a more sheltered or richer habitat than neighbouring unindented coast and thus may attract concentrations of birds.

INSHORE. Ideally would be defined in terms of complex of factors such as depth, temperature gradient, and LITTORAL influences, but it is more practical to adopt a standard distance relevant to observed movements of land-based birds and to the convenience of land-based observation. This may be arbitrarily set at 5 km from low water mark, rather than at 'maximum of 4 or 5 miles out to sea' in Witherby *et al.* (1941). Where there are islands or islets well in sight of the shore, all intervening water and zone up to 5 km beyond them should be included as INSHORE, as should seas of depth less than 6 m.

ISLET. A very small island, typically under 1 km².

KARST. Area of weathered limestone marked by abrupt ridges, caverns, sinks, and underground water flows causing solution. The bare pavement surface is broken by deep clefts and interspersed with grassland or woody vegetation, and locally by cultivation.

KRUMMHOLZ. Twisted, gnarled and distorted woody vegetation, especially coniferous, produced within marginal zone for tree growth, normally on mountains or coast.

LAGG. The fringe of a MIRE, originally simply drainage channel, but by extension area of its influence, often distinguished from rest of mire by different water quality and growth of moisture-loving trees and shrubs, usually at scrub height.

LAGOON. Enclosed shallow coastal lake, pool, or landlocked marine inlet contained by sand, shingle or earth banks or coral reefs; strictly natural, but often extended to similar water bodies partly or wholly man-made.

LAKE. Natural body of standing fresh water arbitrarily defined as having over 100 ha of open water. Subject to wind turbulence and wave action on margins. If very deep (over 60 m) or deep (8–60 m), tends to have little marginal vegetation and to be OLIGOTROPHIC (infertile). If shallow, especially under 8 m, more favourable to aquatic vegetation and likely to be EUTROPHIC (fertile), sometimes to excess and involving algal proliferation. If more acid and shallow, especially in north, more akin to BOGS with fewer aquatic organisms, but much marginal vegetation, and classed as MESOTROPHIC or DYSTROPHIC. *See also* POOL.

LITTORAL. Intertidal zone of a tidal SEA or OCEAN, extended or alternatively employed for non-tidal waters, to include the underwater limit to which the more abundant attached plants can grow, sometimes to a depth of 30 m or more, corresponding to light penetration. May thus overlap INSHORE zone at sub-surface levels.

MANGROVE. Although applied by some to trees or shrubs of a certain family, used ecologically (often under variant MANGROVE SWAMP) to mean any population of trees adapted to live along muddy LITTORALS in the tropics, where they can resist strong wave action and exhibit a closely intertwined habit of growth, seldom reaching any great height (normally well below 20 m).

MAQUIS, MACCHIA. Type of degenerate forest forming a dense scrub community of prickly or spiny plants 1–3 m tall, and of others such as pistachio *Pistacia lentiscus*, often contrasted with GARIGUE as typical of non-calcareous soils. While distinction is valuable for habitat description, countless variations and blends occur in practice, and edge-effects attract a wider range of bird species, for example where patches of cultivation occur. Both types are characterized by a large element of broad-leaved sclerophyllous plants. Chaparral of western USA is equivalent.

MARSH. Wetland area in which standing water does not usually cover the surface, apart from scattered POOLS or fringes of LAKES, RIVERS, and other water bodies. Usually excludes peatland (see MIRES) and is associated with mineral soil, coastal levels, and depressions in river basins, usually at low or moderate altitude. May be of any size, but often small.

MEADOW. More or less compact and often enclosed stand of usually permanent grassland and mixed herbage, normally free of trees or bushes, not currently subjected to grazing and in traditional farming practice usually cut for hay. Applied also to floating vegetation and to high Alpine grasslands which may be seasonally grazed, and therefore strictly pastures.

MEDITERRANEAN CLIMATE (or Western Margin Warm Temperate). Characterized by hot summers with high sunshine amounts and almost complete drought, as well as mild winters, usually with no month below 6 °C mean, but with substantial rainfall creating overall subhumid state.

MEGAPHYLL. Describes plants with leaves 50 cm or more long, and at least 5 cm wide.

MERSE. *See* SALT-MARSH.

MESOPHYLL. Describes plants with leaves intermediate between MEGAPHYLL and MYCROPHYLL.

MESOTROPHIC. *See* DYSTROPHIC and LAKE.

MICROPHYLL. Describes plants with leaves; if a tree, under 25 mm in greatest dimension; if a shrub, of 10 mm or less, and if a dwarf shrub, of 5 mm or less.

MIRE. An arbitrary redefinition by modern botanists of a loose country word, now used to cover all peatlands classified according to whether their water supply comes from outside, bringing in adequate nutrients, or is derived from precipitation on them, and is accordingly poor in nutrients, with a low pH value; and transition types. In habitat terms the first group are FENS or FEN MIRES, which may be sloping or flat; in the latter case varying according to whether they are flooded permanently or for long or brief periods, giving rise to sedge, scrub, alder, or other vegetation, while some in the TUNDRA zone are permanently frozen at subsoil level (see PERMAFROST). The second group includes blanket, raised, and flat BOGS, coastal or upland, continental or oceanic. These are often called MOSSES in northern Britain. The common factor of all types of MIRE is the presence of peat.

MOOR. A general term applied to acid, peaty, usually upland

tracts, including BOGS, which are unenclosed, typically covered with heather *Calluna*, moorgrass *Molinia*, bogmoss *Sphagnum* etc., and not used primarily as pasture. *See also* HEATH.

MOUND. Any artificial pile or rounded bank of earth or other material not exceeding 1 ha and rising more than 1 m; including large pitheaps, spoil heaps, refuse heaps, archaeological or landscaping eminences, and similar natural eminences.

MUDFLAT. Often an extensive area, wholly or mainly unvegetated, level but sometimes dissected by a network of small channels, exposed either regularly by tides or seasonally or periodically by ceasing to be submerged with water, formed of fine clayey particles often containing much water and having sticky and slippery properties.

MUSKEG. Clear low-lying area, usually within a coniferous FOREST, often formerly open water which has become filled with peat moss, over which fringing tree stands are gradually advancing. Muskegs are typically fairly narrow and small, often appearing from the air as a mosaic of glade-like sheltered spaces.

OCEAN. The main worldwide body of interconnected salt waters, with a normal salinity range of 3·3 to 3·7%.

OCEANIC. Used in two senses: (1) relating to OCEAN, as distinct from SEA, and (2) relating to a climate or region significantly influenced by the neighbourhood of the OCEAN. *See also* SUBOCEANIC and PELAGIC.

OFFSHORE. The zone extending from the limits of INSHORE waters to the edge of the CONTINENTAL SHELF, or CONTINENTAL SLOPE.

OLIGOTROPHIC. *See* LAKE.

OPENCAST or OPENPIT MINING. Extractive operation in which an 'overburden' of up to 100–200 m in depth (exceptionally more) is stripped off mechanically and stored for use in eventual restoration, while coal, or other required mineral is exposed and removed. Differs from QUARRY in preponderance of soft materials and suitability for site restoration. Formerly sites were often abandoned at this stage, but modern practice calls for restoration to agriculture or some other use, after normally 3–10 years. Opencast sites often exceed 100 ha in area and may extend to over 1000 ha. Their successive phases involve drastic alterations of habitat, often including new abrupt land-forms and new water bodies. Called Stripmining in North America.

OPEN VEGETATION. Plants separated by less than their diameter. *See also* SPARSE VEGETATION.

ORCHARD. Enclosure planted with fruit trees which are, or will become when fully grown, at least 2–3 m tall, excluding such stands of planted shrubs as coffee and tea, as are without tree cover. Here understood to include widely spaced stands, e.g. GROVES of olives in certain Mediterranean countries which are similarly cultivated but may be undersown with cereal or other ground crops, and which might not be called orchards in common usage.

ORNAMENTAL WATERS or PONDS. Purposely created for amenity in parks and other landscapes, these rarely extend over 100 ha, but are often loosely called LAKES.

OUED. *See* WADI.

OVERHEAD LINE. Used for any cable, wire or other continuous artificial line for telephone, telegraphy, supply of electric power or traction, including its supports (whether metal pylons, wooden posts, or concrete or other standards) but not electrified fences or other structures less than 3 m high.

PALM. Tree-like monocotyledonous plant. *See also* TREE.

PARK. Originally any enclosed, small or medium tract of land reserved for hunting, but has acquired a confusing range of meanings. For habitat and landscape most relevant (also PARKLAND) is as temperate, often artificial counterpart of SAVANNA, blending scattered trees and bushes singly or in clumps or SPINNEYS with predominantly grassland cover. Frequently form small managed units in humid climates especially on undulating landforms. Many areas managed as National Parks or city parks contain no such habitats, while car parks, amusement parks, safari parks, and other modern uses of the term are irrelevant here.

PELAGIC. The most characteristic zone of the deep OCEAN beyond the CONTINENTAL SLOPE.

PERMAFROST. Permanently frozen subsoil, and hence the zone over which this condition persists. *See also* TUNDRA.

pH. Standard measure of acidity and alkalinity, e.g. in water and soil, taken in terms of the number (on inverse logarithmic scale) of hydrogen ions or positively charged hydrogen particles in a solution. A pH value of 7 indicates neutrality; higher values, up to 14, indicate alkalinity, and lower values acidity.

PLANTATION. As FOREST, WOOD, SPINNEY, and COPSE are defined as consisting of natural, semi-natural or spontaneous tree cover, it follows that all tree stands planted by man during the past 400 years (other than ORCHARDS and minor distinctive forms such as roadside avenues and narrow shelterbelts alongside fields and gardens) should be classed as PLANTATIONS. They may conveniently be divided into timber plantations, ornamental plantations, sporting plantations, treecrop plantations (such as COPPICE), and protective plantations, but this degree of breakdown is not normally needed for ornithological purposes.

POND. Small body of fresh standing water, having less than 10 ha of open water formed artificially by hollowing out or impoundment. (By analogy or confusion sometimes wrongly used for similar pools of natural origin.)

POOL. Natural body of standing fresh water with less than 100 ha of open water. *See* LAKE. Also used for a small body of still water in a watercourse, often seasonally cut off. *See also* RIVER.

QUARRY. Site of open-air extraction usually of deposits of hard rock so placed that they can be excavated in quantity out of a slope or cliff by working from a lower level, generally by blasting or cutting. Consequently a worked quarry produces a steep rock face or faces incapable in most cases of remedial treatment, but often offering ledges and holes attractive to predators and other birds. Quarries do not often exceed 1–2 ha. *See also* OPENCAST.

RAVINE or GORGE. Narrow valley, deep in proportion to its width, usually with rocky precipitous sides but sometimes between abrupt grassy slopes, often carrying a TORRENT below. A miniature RAVINE is a GULLY or GULLEY; a very large one is a CANYON.

REEDBED. Level area of sluggish drainage, usually with shallow standing water during much of the year, carrying dense tall uniform stands of emergent aquatic plants such as reeds *Phragmites*, reedmace *Typha*, reed meadowgrass *Glyceria*, and *Papyrus*. Applicable not only to extensive tracts but to fragments down to a fraction of a hectare, and to sites where small areas of open water are enclosed by such vegetation.

REEF. Typically, narrow or very narrow ridge of rock or coral, just above or below the surface of the water, often forming a natural breakwater enclosing a relatively calm LAGOON of protected water. Also applied to shorter and less linear rock structures nearly or just emerging above the water surface.

RESERVOIR. Artificial body of open fresh water, normally above 10 ha in area, formed by excavation or impoundment in a depression or by enlargement of a natural pool or lake. *See also* BARRAGE.

RIVER. Major flowing watercourse, more than 5 m broad between banks, having either a single channel or multiple channels which may be braided into a network of interlaced streams between wide banks of sand, earth or gravel. Seasonal flow fluctuates, even ceasing entirely for periods more or less extended and regular, leaving occasional standing POOLS. Where the banks are low there may be ancillary backwaters, oxbows, and small wetlands, in a riverain zone. *See* STREAM.

SAHEL. Zone of SAVANNA south of the Sahara in western Africa.

SALINA. Saltpan, salt pond or LAGOON, or other seasonally or generally flooded area of high enough salinity for economic exploitation, usually applied in Spain to those in which the water area is dissected by artificial banks.

SALT LAKE. Natural body of standing water, usually without an outflow, containing over 300 parts per million (ppm) dissolved solids. In extreme cases up to 100 000 ppm. High evaporation and fluctuating inflow are common features. Some of these lakes may be more specifically defined as soda lakes or alkali lakes. Properly the latter term refers to those containing alkali salts of pH 8·5 and higher, but shallow, muddy, seasonal lakes drying out to barren hard mud with saline surface are often called 'alkali flats' or sometimes 'white alkali'. The term 'salt lake' is best reserved for those which at least seasonally attain 100 ha.

SALTING. *See* SALT-MARSH.

SALT-MARSH. Low-lying coastal or estuarine land regularly or intermittently overflowed by the tide, and covered with halophytic vegetation; also called SALTING, saltings being distinct from the intertidal zone occupied by mudflats and seaweeds.

SAND-DUNE. Ridge or complex of ridges of sand piled up by wind action, usually on sea coasts or in deserts, sometimes higher than 100 m, sometimes crescentic; either mobile or fixed, juvenile or senile. Old dunes may become vegetated and even forest-clad.

SAVANNA. A controversial term best used only in a broad sense for landscapes influenced by alternating dry and rainy seasons in lower and mid latitudes. Generally these have closed cover of grassland, mainly level or undulating, with scattered well-spaced trees and bushes, singly or in small fairly open groups permitting distant views in all directions. SAVANNA tends to be semi-arid, with no or infrequent wetland elements, and should be at least semi-natural and in sizeable tracts. Man-made landscapes of similar appearance are PARKLANDS.

SCRUB. Characterized by low trees, often below 5 m, whose growth is checked or much retarded by adverse natural or human influences, or which are unsuited to create tall forest, at least in the particular climatic, soil, or other conditions. MAQUIS and THICKET are types of SCRUB, which is also often found for long periods after fire or clear-felling and, for example, along margins of BOGS and ARCTIC-ALPINE HEATHS. BUSHES are often interspersed among SCRUB.

SCLEROPHYLL. Applies to plants, with hard or stiff leaves and similar organs.

SEA. One of the smaller divisions of an OCEAN, adjoining a coastline and sometimes nearly land-locked. Very large bodies of inland salt water not communicating with the OCEAN, such as the Caspian, are also in common usage loosely called SEAS. SEAS may have a much higher or lower salinity than that of the OCEAN, ranging in the Red Sea up to 4·1%.

SKERRY (usually grouped as SKERRIES). Cluster or archipelago of rocky islets over which the sea breaks in stormy weather.

SPARSE VEGETATION. Used of plants spaced more than twice as far apart as their diameter, distinguished from OPEN VEGETATION, where the plants are on average separated by less than their diameter, and CLOSED VEGETATION, where they are in contact or overlap. These terms relate to whichever layer of vegetation shows the most complete coverage.

SPINNEY. Small, often elongated, stand of trees with undergrowth, sometimes a relict of former forest on a steep slope or in a valley head, not normally larger than 1 ha and typically of deciduous species.

SPIT. Narrow, low ridge or bank of sediment, formed of any material, attached to the land at one end and terminating in open water at the other.

SPRING. Site where enough water issues through a natural opening to make an appreciable but not necessarily continuous current. If the temperature is more than 10 °C above the yearly mean of the surrounding air, it is a HOT SPRING; if the water is projected into the air, a HOT SPRING becomes a GEYSER.

STEPPE. Treeless, and often waterless, extensive level lowland grassland of warm lower mid-latitudes in continental Palearctic. Sometimes misleadingly applied to less dry grasslands such as North American prairie, sometimes at much higher altitudes than true STEPPE, and to narrow transitional zones, e.g. between Mediterranean forest and desert.

STREAM. Perennial watercourse, or one whose bed seasonally contains appreciable water, mostly or wholly below 5 m wide, having a sloping profile sufficient to produce flow but not to create chronic turbulence or rapids. *See also* TORRENT.

STRIPMINING. *See* OPENCAST.

SUBOCEANIC WATERS. Taken to include any beyond the CONTINENTAL SLOPE which are within 500 km of the nearest land, and therefore subject to significant terrestrial influences in terms of bird distribution and of pollution. The broad term OCEANIC includes those waters, except insofar as they form a distinct subdivision for such purposes as distribution. *See also* PELAGIC.

SWAMP. Differs from a MARSH in that its saturated surface is permanently, or at least throughout summer, below water level; nevertheless richly vegetated, typically with reed *Phragmites*, *Papyrus* or other dense-growing tall aquatic plants, and frequently with moisture-loving trees. Applicable in such conditions to MIRES as well as non-peatland habitats. May be very large.

TAIGA (or TAYGA). Dense, predominantly coniferous forest, little disturbed or dissected and often containing swampy areas such as MUSKEGS. Occupies the broad belt south of the TUNDRA in upper mid latitudes; frequently buffered from it by the intermediate, more open and fragmented transition zone, sometimes quite broad, which is often contradictorily referred to as WOODED TUNDRA or FOREST TUNDRA.

TARN. Natural POOL, often at some altitude, normally small and at most not much above 1 ha with low pH value and no inflow stream of significant volume or length, often on tableland or undulating ground without developed drainage pattern, or alternatively in a glacial mountain cirque or coomb.

TEMPERATE. Mid-latitude climatic zone including *Cool temperate* and *Warm temperate*. In west Palearctic, humid cool temperate climate with short cold season (fewer than 6 months with mean temperature below 6°C) is confined to Britain, Ireland, and band from north Spain through France and Low Countries to Scandinavia below about 62°N. Extension with longer winter cold season covers most of remaining land between 40°N and 60°N except Mediterranean coast and part of USSR between the Black Sea and southern Urals.

THERMAL. Rising current of warmer air, strong enough to provide lift for soaring birds.

THICKET. Dense stand of low trees, usually under 5 m high, forming close scrub in which deciduous and thorny species are often blended with shrubs, bracken, and other plants tolerant of shade and moisture. Accordingly a temperate counterpart to

MAQUIS, and a successional stage.

TORRENT. Perennial watercourse, in mountainous or upland regions, with rocky bed and banks characterized by steep gradient, sudden fluctuations in flow, and usually many emerging rocks and gravel banks. Breadth variable, mainly below 10 m. Rapids or waterfalls may be frequent.

TREE. Large non-climbing woody plant, usually not less than 5 m tall, with complex underground root system and one or more hard bark-clad stems carrying a network of branches and leaves. PALMS Palmaceae are not true TREES. They should be referred to separately as palms, and their groupings as clusters or PLANTATIONS. Bamboos are canes or woody grasses, and are also excluded, even when of forest height.

TREELINE or TREELIMIT (Timberline in USA). The limit of tolerable exposure to cold, wind, and other adverse climatic elements, either at higher altitudes or at higher latitudes, beyond which trees are unable to grow, and TUNDRA or ARCTIC-ALPINE communities take over. Towards the TREELINE there may be a zone of stunted and distorted tree growth known as KRUMMHOLZ.

TRENCH. *See* DIKE.

TUNDRA. Loosely associated, like the arctic region, with the 10 °C isotherm, north of which the mean temperature of the warmest month falls below that value and tree growth ceases to be possible. The zone of treeless TUNDRA within the western Palearctic is mainly between the Kara and White Seas north of the Arctic Circle, with small outliers along the Urals and the Murmansk coast, as well as on islands from Novaya Zemlya to Iceland. This TUNDRA is mainly lowland and near coasts of the Arctic Ocean; it contains a wide variety of vegetation types, all low-growing, and is typically conditioned by PERMAFROST.

UPWELLING. Displacement of warmer surface water, usually by offshore winds, causing cooler water rich in nutrients to rise to the surface in well-marked zones, often generating much increased biological productivity and thus attracting substantial population of marine birds.

VINE or CREEPER. Climbing plant, which may be woody (i.e. a liana) and ascend to substantial height on host trees, sometimes stretching from tree to tree.

WADI or OUED. Watercourse, sometimes flowing only once a year or less, in arid countries. The term is customarily extended to include the immediate valley which it occupies. When flow does occur through intense rainfall, it is liable to be sudden, copious, and destructive, often briefly becoming a braided RIVER, and leaving surface POOLS and sumps of groundwater which support substantial vegetation, including TREES. Nullah, not used here, is largely a synonym.

WEED. Rapidly growing pioneer plant, usually colonizing a recently cleared area.

WETLAND. Generic term used broadly to cover all non-marine aquatic habitats, large or small, whether permanently under standing or flowing water, or intermittently inundated, or brought into a moist state for significant period, either by natural or artificial means. The Ramsar Convention extends it to cover marine waters to depths of 6 m at low tide.

WOOD. Area of natural or spontaneous tree cover, smaller than a FOREST but larger than a COPSE, and not necessarily having a closed canopy.

WOODLAND. Used as a broad general term to cover all types and scales of tree cover where density of stands exceeds those falling within PARKLAND and SAVANNA. This corresponds to the French *terrains boisés*, and, in international professional usage, to the clumsy 'forested land'; it does not correspond to North American practice which distinguishes WOODLAND from FOREST as an open stand of trees, usually below 15 m tall, with an intervening good growth of grasses or shrubs. This is classed here as WOODLAND SAVANNA.

WOODLAND SAVANNA. Intermediate between SAVANNA and FOREST, consisting of grassland or shrubs interspersed with frequent, usually fairly low trees, singly or in groups, too widely spaced to form anything like a continuous canopy or to attract strictly forest birds.

ZONE. An elongated belt or region distinguished by some common character indicated by its name, which differentiates it from other adjoining ZONES, e.g. Steppe zone.

# DISTRIBUTION

The basic information on the status and distribution of species in the western Palearctic has been supplied by correspondents for each country (except Albania). Their names are given on p. 35, and their initials are shown after any data in the text furnished by them. For all breeding species and regular migrants, the maps are intended to provide the essential information on distribution, and the text (which gives supplementary information on changes, occasional breeding, etc.) should be read in conjunction with them. Species of accidental occurrence in the west Palearctic are not mapped, and the text briefly summarizes their world distribution. For all species the paragraph in the text headed Accidental refers only to the west Palearctic, though some information on vagrancy elsewhere is often given in Movements. Again, maps of the western Palearctic are in most cases largely based on data supplied by the country correspondents. Knowledge of breeding distribution (shown in RED on all maps) varies enormously in different parts of the western Palearctic. At one extreme are those countries which have completed Atlas schemes, notably Britain and Ireland and France. At the other extreme are countries, especially in the south and east, where so few detailed studies have been made by ornithologists (in some cases virtually none in recent years) that any map prepared at the present time must either underestimate or exaggerate the true state of affairs. In between there is almost every possible variation in the information available, which again differs between the various species. Careful consideration has been given to the possibility of distinguishing in some way on the maps the degree of accuracy in the basic information, but the very extent of the variation in knowledge makes it impossible to devise any feasible system. The maps therefore represent, as accurately as possible, the state of our current knowledge—in some cases detailed and up-to-date, in others deficient to a greater or lesser degree. The heartening spread of the European Atlas schemes means that for a growing number of countries much greater accuracy should be possible in future volumes. For those where no Atlas schemes are at present envisaged, it is hoped that the maps will encourage ornithologists to visit little-known areas or make available their unpublished data and so help to correct deficiencies.

On the maps, GREY is used to indicate winter distribution, i.e. those areas where the species concerned is regularly present in normal winters; in mild or severe winters, the distributions may differ considerably. For pelagic species—in Volume 1 petrels, gadfly-petrels, and shearwaters (Procellariidae); storm-petrels (Hydrobatidae); tropicbirds (Phaethontidae); frigatebirds (Fregatidae); and boobies and gannets (Sulidae)—however it represents their marine distribution throughout the year (seasonal changes are described under Movements). Our knowledge of marine distribution for most pelagic species, especially away from the regular shipping routes, is still inadequate, despite the considerable advances in knowledge in recent years, especially from the summaries published e.g. by the Royal Naval Bird-watching Society in *Sea Swallow*, King (1974), and Watson *et al.* (1971).

The maps of world distribution, using the same basic colours of red for breeding and grey for winter or marine distribution, are on a much smaller scale and are necessarily more diagrammatic. For large areas of the world, moreover, detailed information is simply not available to map with precision most species, especially those with a patchy distribution. The world maps, therefore, in some cases indicate only the general picture using the standard published sources. However, for the Ethiopian region, Dr D W Snow has generously made available the draft maps for his forthcoming *Atlas of Speciation of African Non-passerine Birds*, and C W Benson has kindly assisted with distribution for Madagascar and neighbouring islands.

# POPULATION

Despite the growing interest in census work, detailed counts or even estimates of breeding numbers are available for only a limited number of species in few countries. Regional or national estimates are given where available or local counts where information is scanty, together with any data which illustrate trends. Apart from the references to published work, many of the country correspondents listed on p. 35 have helped; here again, this is indicated by the use of their initials. In those species where winter counts throw light on total populations or population trends (notably in the Anatidae), summaries of these are included. This section concludes with details of any post-fledging survival or mortality studies (breeding success is covered in the Breeding section), and the age of the oldest known wild-ringed bird of the species concerned.

# MOVEMENTS

This section relates to the subject of migratory movements and should not be confused with the use of this word as a behaviour term. The summaries for each species deal essentially with migration patterns and seasonality of movement, and largely ignore the still speculative subject of orientation; they are thus concerned with 'where' and 'when' rather than 'how'. For modern statements of orientation theory, see Griffin (1965), Matthews (1968), and Evans (1966, 1968).

Each species account begins with a brief statement of whether the species is *migratory* (all or most individuals make regular seasonal movements between breeding and wintering ranges); *partially migratory* (populations contain substantial migratory and non-migratory elements); *dispersive* (makes more or less random movements, lacking direction bias other than topographical limitations); *resident* (basically non-migratory though some individuals may move long distances); or *sedentary* (individuals not normally moving more than 50 km). In specialized habitats, there may also be *nomadism* (irregular movements by desert and semi-desert species in response to rainfall or drought) or *vertical displacement* (altitudinal movements by mountain breeders which descend to foothills or valleys in winter). In the Palearctic, migration seasons are governed by temperature and day-length and their effects on food supply, but in Ethiopian Africa, from where several species straggle north into the west Palearctic, migration seasons are less well defined and are for some species governed to a large extent by seasonality of rainfall; such species are termed *rains migrants*.

In the interval between the end of a breeding season and the onset of autumn migration, birds may make random movements, sometimes over long distances, the short duration distinguishing them from normal dispersal as defined above; these movements are termed *post-fledging dispersal* in the case of juveniles, and *post-breeding dispersal* in the case of adults. True migrations involve *emigration* (departures) and *immigration* (arrivals), and in Europe these tend to be basically north-south in autumn and south-north in spring (with longitudinal displacements also); some species show evidence of *loop migration*, which is a spring return by a different route from that used in autumn. Though each migratory population tends to have a *standard direction* (the mean direction of movement between breeding and wintering areas), there may be deflections along *leading lines* (visible and roughly linear topographical features followed on occasion by migrants), even causing *retromigration* (movements temporarily in an inappropriate direction for the season). These latter should not be confused with *reversed migration*, which is general and sustained movement in a seasonally inappropriate direction. Migrants may also be deflected by *drift* (displacement from normal route by cross-winds), leading to *re-determined migration* (movements of drifted migrants back towards the normal route or quarters); related terms are *one-direction navigation* (flights on constant headings without compensation for drift or artificial displacement), and *goal-orientation* (flights that do compensate for displacement, however caused, so that birds arrive in normal quarters).

Special types of movement believed to apply only to limited groups of species are: *abmigration* (a spring migration by a bird that had wintered in its natal area, best known among ducks (Anatinae) where due to early pairing); *aberrant migration* (a winter visitor migrating in spring to a breeding area other than that from which it came the previous autumn, also best demonstrated by ducks); *eruptions*, which are irregular (not annual) departures from breeding range, sometimes *en masse*, these being *irruptions* in the areas they go to; *moult migration*, which is a special summer movement between breeding grounds and restricted moulting areas, also typical of waterfowl (Anatidae) but now being found in other non-passerines; and *wrecks*, which are abnormally large inshore and inland visitations by seabirds, often with high mortality, usually after severe storms at sea. Two other special terms used are *Ortstreue* (recurrence in the same wintering or breeding area in successive years) and *Zugunruhe* (restlessness preceding migration departures); both are German words for which there are no exact English equivalents.

The African winter quarters of European breeding birds (trans-Saharan migrants) are now better known than 30 years ago, mainly through the work of Elgood *et al.* (1966), Moreau (1972), and Morel and Roux (1966, 1973); one of the more important developments has been the ornithological exploration of the Banc d'Arguin, Mauritania, which receives frequent mention in the main text. Unfortunately, there is still a paucity of information on the nature of migration or dispersal in the south-east of our area, and in respect of west Palearctic breeders migrating through and wintering in Iran and Arabia.

Generally, where wide-ranging species are concerned, only those populations which spend some part of their annual cycle in the west Palearctic are dealt with in detail; populations which are the origins of stragglers to our area are treated more briefly, while extralimital populations are for the most part ignored although breeding and wintering areas are all shown on the distribution maps.

The subject of the timing of movements is only briefly covered. In such a huge area there are, inevitably, major variations and it would take much more space than is available to give comprehensive coverage. Only broad outlines are given, and readers with particular interest in this aspect should consult national avifaunas. In the ANNUAL CYCLE DIAGRAM for each species, the data on timing of movements refer to birds from a specific part of our area, always the same as that named under Breeding; where there are 2 bands in the Movements circle the outer is labelled P (passage migrants) or M (moult migration) and the inner band always refers to spring and autumn migrations of the local breeding birds.

Information on migratory behaviour and characteristics (e.g. degree of flocking, flight formations, whether movements are diurnal or nocturnal) is usually treated in the family summaries.

# FOOD

For many species, useful information on foods and feeding behaviour is sparse and scattered, and may be difficult to evaluate. There are, for instance, numerous observations of feeding behaviour, though they are rarely related statistically to changes in diet, availability of foods, locality or season. Anecdotal observations, particularly of unusual or unidentified foods, are widespread, but their importance can seldom be assessed. Quantitative data based on analyses of the contents of the digestive tract are obviously more reliable as guides to what the normal diet may be. However, the methods of assessing the contents vary considerably, and have been well reviewed by Hartley (1948) and Bartonek (1968). Most commonly the results are expressed as a percentage of the total volume of food eaten, and as a percentage of the total frequency (i.e. frequency of occurrence) of food eaten. Less commonly, the results are given as a percentage of weight (wet or dry) of food eaten, or as based on a scale of importance. Each method has inherent faults and virtues, which vary with species and composition of diet. No one method is fault-proof. Much work has been based mainly on the contents of the gizzard, though these may reflect to a large extent the digestibility of different food items (Dillon 1959; Bartonek and Hickey 1969; Swanson and Bartonek 1970; Bengtson 1971). For example, an over-emphasis of the importance of items which take longer to be digested, such as seeds, is common. Where possible, only information on the contents of the oesophagus has been used, since this is thought to provide a more accurate picture of true diet composition (Sugden 1973; Bengtson 1975). The data may also be influenced by the time which elapses between death and removal of the contents of the digestive tract, owing to post-mortem digestion and bacterial action (Koersveld 1951; Dillery 1965).

Information on the variations in diet with changes in season, sex, and age are available for very few species. Even where such information has been published, it is often based on an inadequate sample size and is limited to a particular period, e.g. the shooting season. In many studies, practical considerations rather than statistical requirements dictate sample size. It is equally rare for quantitative correlation to be shown between changes in diet and changes in locality, or consideration given to the main ecological factors which may influence the availability and abundance of preferred foods. Few attempts have been made to evaluate daily intake, or the nutritional and calorific values of food items. Ideally, the relative importance of different foods should be expressed in terms of metabolizable energy contributed, but these figures are rarely available.

Comparative studies have seldom been made on species living in the same area, particularly in relation to availability of foods and possible competition between species, especially closely related species (Olney 1965).

Each species is treated as a whole, though where there are known diet differences between subspecies or races, they have been included. The main emphasis has been given to data from the western Palearctic, and extralimital data are given only when of special interest or when western Palearctic information is sparse. A brief statement of the main foods taken is followed by a description of feeding methods, which where possible are related to the type of food being taken and to any significant correlation with anatomical and physiological adaptations. For most species, a list of major food items recorded is then given with, when available, their size range. This is followed by a précis, and where possible comparative assessment, of detailed quantitative studies, with particular emphasis on changes in diet and behaviour related to season and locality. Where available, an estimate of daily food intake, and nutritional and calorific values, are given. The final paragraph deals with the food of young, though the data are few and they are rarely quantitative.

Scientific and common names of food items are given, the nomenclature being based on the following references.

### Plants

Tutin, T G *et al.* (1964, 1968, 1972, and in preparation). *Flora Europaea*. Cambridge.

Komarov, V L *et al.* (1934–64). *Flora USSR*. Leningrad and Moscow.

Czerepanov, S K (1973). *Flora USSR: Supplements and corrections*. Moscow.

Where not covered by the above, the basic and standard floras listed in Tutin *et al.* (1972).

### Vertebrates

Ellerman, J R and Morrison-Scott, T C S (1966). *Checklist of Palearctic and Indian Mammals 1758–1946*. London.

Mertens, R and Wermuth, H (1966). *Die Amphibien und Reptilien Europas*. Frankfurt-am-Main.

Wheeler, A (1969). *Fishes of the British Isles and Northwest Europe*. London.

Bailey, R M *et al.* (1960). A list of common and scientific names for fishes from the United States and Canada. *Spec. Publs. Am. Fish. Soc.* No. 2, 1–102.

### Invertebrates

Illies, J (ed.) (1967). *Limnofauna Europaea*. Stuttgart.

Nordsieck, F (1968). *Die europäischen Meeres-Gehäuseschnecken (Prosobranchia) vom Eismeer bis Kapverden und Mittelmeer*. Stuttgart.

Nordsieck, F (1968), *Die europäischen Meeres-(Bivalvia), vom Eismeer bis Kapverden, Mittelmeer und Schwarzes Meer*. Stuttgart.

# SOCIAL PATTERN AND BEHAVIOUR

This section is divided into two parts: (1) social pattern, and (2) social behaviour, each in its own numbered paragraph. The term *social* is used here mainly in its more restricted senses of indicating, respectively: (1) aspects of society, and (2) mutual relationships including interaction or communication between two or more individuals (see, e.g. Tinbergen 1953); it is not equivalent to words such as 'gregarious' or 'sociable'. The accounts are largely factual, but, in some cases, interpretative comments have been included, particularly on the influence of ecological factors—though causal analyses of motivational factors involved in displays have generally been considered to be beyond our chosen scope. By giving the facts where possible, the aim has been to provide source material for use in future analyses of the behavioural ecology of species and larger taxa, bearing in mind that various, apparently disparate aspects of a bird's biology—involving its habitat selection, movements, feeding, social pattern, social behaviour, breeding, etc.—are often inter-related; such 'co-adaptations' enable it to survive and reproduce in the face of the exigencies and vicissitudes of its environment (see, e.g. Crook 1964, 1965; Simmons 1970a, 1974a). When possible, the accounts have been prepared in consultation with the ornithologists having specialist knowledge of the species concerned; in many cases, they have also provided information on calls which have been used both in this section and that on Voice.

1. The aim in the first paragraph is to outline the types of association and dispersion of the species (its social pattern or organization); for general reviews of this topic, see especially Lack (1954, 1968), Wynne-Edwards (1962), Crook (1965, 1970), McKinney (1973). The term *dispersion* is used in the special sense of the 'state of spatial separation' between individuals or larger assemblages of birds within their habitat; it should not be confused with 'dispersal' (see Movements). Dispersion is of 3 types: (1) *random*, (2) *aggregated* (individuals, pairs, or family groups less dispersed than they would be if distributed at random), and (3) *over-dispersed* (more dispersed and more evenly distributed than if at random); see Salt and Hollick (1946), Hinde (1956). The treatments deal principally with adults but also, where relevant, with family groups and dependent young; they concentrate too on intraspecific aspects of social pattern, though interspecific ones are covered when of particular importance—especially in the case of interspecific territorialism (Simmons 1951; Orians and Willson 1964; Murray 1971). The paragraph starts with a brief general characterization of the kind of association typical of the species through all or most of the year, i.e. gregarious (in family parties, flocks), in pairs, or solitary. There follows a summary of the kinds of association and spatial dispersion found outside the

breeding season, with details of flocking, winter territories, etc. The account then proceeds under three sub-headings, as follows. BONDS. The type of pairing and mating system is treated first; see, especially, Lack (1940), Selander (1965), Orians (1969, 1971), Kunkel (1974). It should be noted that *pairing* here refers to the process of pair-formation, and *mating* to copulation. Pair-bonds are classified as (1) monogamous, (2) polygynous, or (3) polyandrous. If *monogamous*, the bond is a firm one between one male and one female: for part or all of one breeding cycle only (*seasonal*, a new partner usually being obtained next time), or for a longer period (*sustained* or *long-term*, with or without a break in association during the non-breeding or winter period); in some cases, the bond may be *life-long*. Except in the case of species which usually practice only seasonal monogamy, it is customary to speak of *divorce* when at least one member of a previous pair, both having survived, pairs with a third bird (Richdale 1951). If *polygynous*, the bond is between a male and two or more females (*bigamy* and *polygamy* respectively), either simultaneously or successively during one breeding season (see von Haartman 1951); when the male and a group of females associate together, this is usually known as *harem polygamy*. If *polyandrous*, the bond is between one female and two or more males, either simultaneously or, more usually, successively during one breeding season. In some cases, the mating association between males and females may be *promiscuous* without the establishment of any true pair-bond; this may be the only type of heterosexual relationship in the species, or males or females may (e.g.) have promiscuous matings with other individuals of the opposite sex while maintaining a stable pair-bond with a regular partner (the mate)—such promiscuity being either frequent and widespread (species-characteristic) or essentially casual (perhaps on the individual level only). Other information given, when available, includes sex-ratios, age at which pairing first takes place, and stage in the annual cycle in which pair-formation starts and the pair-bond ends. Finally, the type of parental care is indicated together with (e.g.) an outline of any well-developed social organization within the family group affecting parent–young relations and relations between siblings (other aspects being treated under Breeding); also included are associations outside the family group itself, such as *crèching* (regular formations of groups of unfledged young). BREEDING DISPERSION. This sub-section indicates solitary or colonial nesting, and whether there is a territorial or non-territorial system (the two not being mutually exclusive as is sometimes supposed); intraspecific variation in dispersion pattern, correlated with variations in local features of the habitat, is also covered—but details of sites and nests will be found in the Breeding section. In the case of colonial (aggregated) nesters, the distribution of the nests—loose or dense— is indicated where possible; in the case of solitary (over-dispersed) nesters, the distances between neighbouring nests—usually in general terms only. The topic of territory has attracted much attention in the ornithological literature; see, e.g., Lack and Lack (1933), Nice (1941), Hinde (1956), Tinbergen (1957). In the case of the territorial species, the type of territory held is indicated plus, where possible, the size; also the various activities conducted, more or less exclusively, within it. A *territory* may be defined as 'any defended area' (Noble 1939; but see Emlen 1957) occupied exclusively by a single bird, pair, or larger social unit. Its essential characteristics are that it has a fixed topographical location and that there are social conventions associated with its maintenance—involving aggression and advertisement by the owners and avoidance and non-aggression by others (except, of course, when there are territorial disputes). Types of territory include: (1) *all-purpose territory*, a large and usually well-defined area used for all or most maintenance and reproductive activities, including the raising of the young (termed a *permanent territory* when also occupied outside the breeding season); (2) *breeding-only territory*, a large or fairly large area, used much as in type 1 but not for feeding to any great extent; (3) *nest-area territory*, a small area round the nest, not used regularly for feeding; (4) *nest-site territory*, restricted to the nest; (5) *pairing territory*, small area used only for pair-formation; (6) *mating territory*, used only for copulation (types 5 and 6 are sometimes combined); (7) *brood territory*, an area of variable size, used for raising the young; (8) *feeding territory*, an area of variable size, separate from the territory where breeding takes place, used for obtaining food; (9) *loafing territory*, a small area, away from the immediate vicinity of the nest-site (but often adjoining or near), used for resting, preening, etc. (also termed a *waiting territory* when used by the non-incubating or off-duty bird, or by the pair as a meeting place). Territories used by single birds, such as pre-pairing or feeding territories, are sometimes called *individual territories*; feeding territories occupied outside the breeding season are also termed *winter territories*. In many cases, of course, precise information is not available and more generalized categories (such as *breeding territory*) have to be used. Separate territories may be held by the same birds at the same or different stages in the cycle, e.g. separate pairing and breeding-only territories, separate nest-site and feeding territories. Nest-area and nest-site territories are particularly common in colonial birds, but are also found in species with over-dispersed nests—in some cases, as part of a larger home-range, The concept of home-range has not been extensively studied by ornithologists, as pointed out by Dzubin (1955). A *home-range* may be described as the area over which an individual, pair, or larger social unit is most active for part or all of the annual cycle; the area as a whole is undefended (or ineffectually defended), has no precise boundaries, and may overlap with the home-ranges of other birds of the same species. As we have seen, the home-range may contain a defended nest-area or nest-site territory and also

defended loafing and waiting territories—as, e.g., in the Anatinae (ducks). If the whole area frequented is defended, it should be termed a territory and not a home-range. ROOSTING. The main aim of this sub-section is to characterize the roosting (sleeping) patterns of the species, both outside and during the breeding season where possible—though information on this whole topic is often most difficult to come by. Roosting is classified as (1) communal or solitary, etc., and (2) nocturnal or diurnal. Any diurnal 'loafing' (resting and preening) is also indicated, though usually only for gregarious species. Information is given on typical sites (classified as protected, concealed, traditional, temporary, etc.), times of arrival and departure, and any other relevant aspects. For recent discussion on roosting patterns and their adaptive significance, see Simmons (1965), Ward and Zahavi (1973).

2. The main aim of the second paragraph is to describe objectively the major aspects of the interactions between conspecific individuals—flock members, mates, parents and young, siblings—and the situations in which they occur. The paragraph sometimes starts with a general statement pointing out (e.g.) any special features of the social behaviour of the particular species, any problems of interpretation, etc. In general, however, interpretation has been limited to the allocation of behaviour to certain categories and use of certain terms (see below) without attempting to establish correlations between the species social behaviour, social pattern, and other features of its life-history and ecology. Particular attention is paid to *displays* (species-characteristic behaviour patterns functioning in communication as social signals) both visual and vocal—though the actual descriptions of calls and other auditory signals are given separately in the Voice section. Visual displays consist of postures and movements derived from more basic, everyday, non-signal behaviour patterns, such as feeding behaviour, nest-building movements, locomotory intention-movements, thermoregulatory plumage postures, comfort behaviour, etc. (see, e.g. Daanje 1950; Tinbergen 1952; Morris 1956; McKinney 1965*b*). These have usually been modified (*ritualized*) to a greater or lesser extent in the course of evolution to serve their new signal function, becoming stereotyped in form sometimes to such an extent that their origins may be wholly obscured. In some cases, however, the displays still so closely resemble their precursors that they can be recognized as signals only by their regular and formal occurrence in certain social situations; their identity as displays may also sometimes be established by comparison with homologous but more highly ritualized displays in closely related species. Displays may be wholly visual or also have vocal components, or they may be more or less wholly vocal. It is not uncommon, however, to refer just to the visual performances as 'displays' and to distinguish the vocal ones as 'vocalizations'. As far as possible, all displays

(both visual and vocal) have been given distinctive names (with capital initial letters) that are non-interpretative, using the same or similar names for homologous displays in related species; these have been taken or modified from the literature or consultants' drafts in most cases but, where necessary, new terms have been introduced. It should be noted that the use of the same or similar names for the displays of species in unrelated taxa does not imply homology. Though the approach is basically *ethological* (concerned with biological study of bird behaviour), the more technical language of that discipline has in general been avoided; in particular, we have not concerned ourselves with interpretative concepts such as 'displacement activities', 'releasers', etc., many of which are at present under re-evaluation and are, in some cases, outmoded and of little practical use to the field observer. Treatments vary between taxonomic groups. Where the behaviour is simple or little known, no sub-headings are used. Where there are complex behaviour features, special headings may be needed: e.g. WATER-COURTSHIP in the Podicipedidae (grebes), PURSUIT-FLIGHTS in the Anatinae (ducks), LEK BEHAVIOUR in the Tetraonidae (grouse). In certain cases, behaviour serving to integrate members of the same flock is treated under FLOCK BEHAVIOUR. In general, however, the information is organized under three standard sub-headings, as follows. ANTAGONISTIC BEHAVIOUR. This deals with the whole complex of aggressive (hostile) and escape responses involving birds, often of the same sex, which have no mutual bonds. The term *antagonistic behaviour* was coined to replace in part the motivational term 'agonistic behaviour' of both comparative psychology and ethology (see, e.g., Morris 1954); this is also used—as 'antagonistic behaviour' is here not—to apply to similar elements which sometimes occur within the heterosexual responses of paired birds. In general, the term is applied to the behaviour involved in disputes over food or mates, during the establishment and maintenance of territory, etc. It can also be related to the phenomenon of *individual distance* (see Conder 1949)—that small, exclusive area of space that individuals of many species maintain around themselves by showing aggression towards or avoiding other individuals—save those, perhaps, with which they have strong bonds. This area has no topographical restriction (as has a territory) but moves with the bird; in some cases, similar areas may be defended around the mate or young. The forms of antagonistic behaviour include (1) *threat-displays* (functioning to induce withdrawal by a rival); (2) *advertising displays*, such as song and song-flights (to draw the attention of rivals to the displaying bird, often the occupant of a territory); (3) *appeasing displays* (to forestall attack well in advance); (4) *submissive displays* (to inhibit or stop attack that is imminent or in progress)—as well as (5) *attack* and *fighting* (including all significant components from *hostile approach* to full, mutual *combat*), and (6) *escape behaviour* (fleeing). The term 'fighting' is sometimes used to cover all overtly

aggressive aspects of antagonistic behaviour, but is used here in its more usual sense of involving actual physical contact between birds. Some displays, including vocal ones such as song, have a dual function of repelling rivals and attracting mates; termed '*Imponiergehaben*' in German and 'self-advertising behaviour'—or, less satisfactorily 'advertising behaviour' in English—such phenomena are usually first treated under the present sub-heading and mentioned again under the next. The concept of *social dominance* and the formation of *hierarchies* or *peck-orders* (see, e.g., Noble 1939) has been applied here with caution for it has been much abused in animal studies (see critiques by Gartlan, e.g. 1968); such organization frequently arises only in conditions of artificial crowding, e.g. highly constrained captivity, and may have no reality in many natural populations. On the other hand, some wild populations are so structured, at least at certain phases of the annual cycle; also much valuable information is forthcoming from studies on the behaviour of captive birds, especially in the description of displays. Such information is therefore included where appropriate—though only when it is clear that there have been no distortions. HETEROSEXUAL BEHAVIOUR. This subsection deals with those interactions between birds of the opposite sex that are involved in the formation and maintenance of pair-bonds, with copulation, etc. It covers a much wider field than mating behaviour, including those more disruptive elements, arising from hostility and fear, that may also occur during (e.g.) pair-formation, especially in the early stages. Treatment usually proceeds chronologically through the breeding cycle, starting with pair-formation and ending with functional copulation and any nest-relief ceremonies during incubation or later. Special attention is given to pair-formation, indicating if possible where and when it starts, how it progresses, the roles of the sexes, manner of sex identification, etc.; where necessary, distinction is made between initial mate-selection and the engagement-period (that part of the cycle between pairing and nesting). Other situations considered include site-selection, meeting (especially after separation, when displays can be elaborate), and copulation. The behaviour involved includes: (1) *advertising displays*, e.g. by lone birds to attract mates—these may or may not be different from those used in antagonistic situations; (2) *pair-courtship*; (3) *communal courtship*; (4) *greeting displays*; (5) *allopreening*; (6) *courtship-feeding*; (7) *nest-site showing*; (8) *pre-copulatory and post-copulatory displays*. The term *courtship* is used broadly to cover those displays involved in both the establishment and maintenance of the pair-bond, though different forms may be used at different stages in the breeding cycle. Together with both types of advertising display, they tend to be highly species-characteristic, especially when two or more closely related sympatric species occur together, thus preventing interspecific pairings. Courtship displays may be *mutual*—with the sexes performing similar or different (but often interchangeable) roles, simultaneously or reciprocally—or *unilateral*, usually by the male. Mutual displays are often termed *ceremonies*, particularly when elaborate. Courtship is characterized as *communal* when a number of birds gather expressly to display; the displays then performed may be the same as between isolated males and females (*pair-courtship*), or quite distinct. *Greeting ceremonies* (and mutual courtship generally) are particularly prevalent in species in which male and female are alike in plumage and behaviour, especially if they meet solely or mainly at the nest, as in many colonial birds; joint activity of this sort may help the participants to identify one another individually (often by voice) and suppress the hostility that the site occupant shows to any close-approaching bird of the same species. A special form of mutual greeting display, termed the *Triumph Ceremony* occurs in some groups—notably the Anserini (swans and geese)—during or after antagonistic encounters, especially in species with strong pair-bonds. *Allopreening* (Cullen 1963), which may or may not be ritualized, is the name now given to what used to be called 'mutual preening'; such preening may be *simultaneous* (the birds preen each other at the same time), *reciprocal* (they take it in turns), or *unilateral* (non-reciprocal, i.e. by one bird only), and may involve special soliciting postures—see, especially, Harrison (1965). In some cases, the allopreening is so highly ritualized that no contact between bill and feathers is made; at the extreme, non-vocal sounds may be produced, as in some Ardeinae (herons). In *courtship-feeding*, which may continue after egg-laying where only one sex incubates, one partner (nearly always the male) gives food to the other which usually solicits it in the manner of food-begging young (see, e.g., Armstrong 1947); if the behaviour of neither sex is ritualized, however, it is better to term it *food-presentation*. As pointed out by Royama (1966), such feeding of the female by the male provides her with a highly significant amount of food during critical periods of egg-formation and incubation. The bonding function of such behaviour, especially when ritualized, should not be overlooked; in extreme cases, which are far from uncommon, it has become so ritualized that food is seldom or never passed to the mate. Finally, it should be noted that copulation and the displays associated with it often occur—as does mate-selection—well in advance of ovulation and egg-laying in some species, as a form of pair-courtship. RELATIONS WITHIN FAMILY GROUP. This deals with communication and other social interactions between adults and young and between siblings. These include (e.g.) behaviour associated with feeding and food-begging, appeasement displays, parental hostility towards the young, parental warning behaviour (but not anti-predator reactions such as distraction-display), displays and fighting, etc. between the young, and other relevant information not covered in the Breeding section.

# VOICE

Exceptionally complex and difficult problems are met in attempting to develop a scientific method of describing and interpreting the vocal utterances of birds. Primarily, bird vocalizations are conditioned by anatomical variations in the syrinx, the functional mechanism of which is still imperfectly understood. Certain species have evolved an elaborate communication system of distinct calls, each having a specific signal function within an observable situation. Others use voice much less frequently, and often on a more primitive level. The task of description and interpretation is complicated not only by the wide scope for variation offered by the parameters of sound, pitch, loudness, duration, and tone quality but by differences linked to sex, age, emotional state, individuality, and region. Voice accordingly must be considered in the light of Social Pattern and Behaviour.

Songs are vocal displays. The human eye, when well trained, is capable of resolving and fixing much of the visual information afforded by birds and their activities, except the fastest. The human ear is much more limited and fallible. It is also much harder to train in promptly resolving and accurately recalling bird sounds. Birds themselves can resolve bird sounds at least ten times faster than men. Moreover, at the highest frequencies in use by birds, the hearing of most observers, although reaching similar levels, becomes undependable. Even within lower frequency bands, there are often inexplicable gaps or weaknesses, commonly unrecognized by the observer and not necessarily constant. Moreover, apart from musicians and linguists within their special fields, probably far fewer people have well-trained ears than well-trained eyes.

While the risks of distortion of a visual image due to weather, topography, distance, or poor light are fairly familiar, corresponding interferences with sound reception are less well known. An example is the Doppler Effect which, when the sound source or the hearer are moving towards or away from each other, leads to an apparent rise or decline in the pitch of the sound. Bird utterances exist only within a moment of time, and for practical purposes only in so far as they are, at that moment, received, acted upon, memorized, or recorded by an interested listener. Few items of ornithological information are so likely to be lost or subject to error. Further confusion results from the lack of progress so far in assigning to descriptive words in ordinary use scientifically clear and distinct meanings. From the standpoint of semantics the entire range of words suitable for describing bird utterances is in all languages more or less strongly influenced by onomatopoeic echoes often carrying a misleading message.

Attempts to express sound in letters and syllables, supported by accents or other indications of emphasis, pitch, and tempo, may lead to a bewildering range of alternative renderings. Such is the fallibility of the human ear that it persists in 'hearing' birds voicing consonants which objective methods do not confirm, although it is perhaps too late to rename the Chiffchaff *Phylloscopus collybita* the 'iffaff' or, in German, the 'ilpalp'. Where an utterance happens to approximate to a human voice or a familiar instrument, as in the traditional 'cuckoo', 'coucou', or 'kuckuck', passable results are obtainable by use of musical notation. Proposals for standard methods of bird voice recording in the field, such as those of Saunders (1951), North (1950), or Hold (1970), have unfortunately not so far led to sufficient agreement or widespread adoption to become established. They have at least helped to clarify the problems of accurate description. Any competent descriptions accompanied by syllabic renderings can tell us something useful about the pattern, pitch, and tone of a call.

In studying bird utterances, the hopes of many have recently become centred on sound recordings. These have great advantages. They can be stored and replayed at will. They can be analysed at convenient speeds with sophisticated equipment enabling their components and characteristics to be more clearly perceived, and they can be converted into visible sonagrams or melograms (see below). Much progress has recently been made in recording on tape a wide range of utterances of many species of birds in various countries. The recordists deserve praise for their dedication and skill, and no less for co-operating so amicably and helpfully in an international network for exchange and dissemination of results. The immediate effect, however, in view of the youthful state of bio-acoustic studies, is to throw up a bewildering array of new queries and unresolved problems, which can only be treated tentatively in the present state of knowledge.

Such problems, which directly impinged on the preparation of the Voice section, are aggravated by external complications. The rapid growth of bio-acoustic studies, of which research into bird vocalizations forms part, gives rise to a need for co-ordinating techniques, standards, and terminology, and for establishing new concepts and methods, going beyond the concerns of ornithology. Such needs have been taken into account only indirectly in the present work, but they will doubtless become more important in future.

## REQUIREMENTS FOR ADEQUATE TREATMENT

A more serious immediate challenge is the emerging awareness in behavioural studies of the significance of vocal signals and communications, which have hitherto tended to be assigned no more than an ancillary role in interpreting many performances in visual terms. In field identification also, the potential usefulness of recognition through utterances has often been underrated. Treatment of voice as a self-contained branch of study is accordingly inadmissible. Every utterance has to be evaluated in relation to its role in behaviour, communication, and location, to its relevance and reliability in specific identification and individual recognition, to its significance for geographically isolated populations and for taxonomy, and to the anatomically specialized organs which govern it, as well as to its linkage with sex, age, physiological state, and surrounding circumstances.

Many obstacles hinder the integration of bird-voice analysis with such other branches, having their own incompatible requirements, practices, and vocabulary. The task of producing a comprehensive, fully digested and standardized scientific treatment of bird voice, as well as non-vocal signals, can be broken down into stages, at each of which serious difficulties arise. They are:

1. Understanding the actual production of sound as determined by the anatomy and physiology of the bird.
2. Variations in reception of the resulting utterance due to the position of the bird in relation to the observer or recorder, and to topography, weather, and background noise.
3. Variations in reception due to differing capabilities, age, education, and training of observers.
4. Limitations inherent in different types of recording equipment and ancillary devices such as parabolic reflectors and microphones, and in capacity to discriminate between noise and signal, which may distort or confuse material available for analysis.
5. Limitations in media and equipment for playback of recorded sound.
6. Limitations of conversion of sound into a visual image by, for example, a sonagraph or melograph, or by use of a computer.
7. Use of techniques and discretion in editing a sonagram or melogram to distinguish signal from background noise.
8. Scrutiny of available descriptions, discs, and tapes and preparation of a critical list of hitherto recognized utterances.
9. Critical comparison of data from description with data from tapes and other recordings, and with relevant behavioural studies.
10. Preparation of a digested comprehensive account, with sonagram or melogram illustrations, after analysis and reconciliation of inconsistencies.
11. Summation of general voice characteristics of the species or group, and role of voice in its life-history, indicating points or areas needing further study.

## ORGANIZATION OF THE VOICE SECTION

For each of these stages data are partly lacking and techniques are capable of improvement. To bring together all the varied types of expertise required, and to co-ordinate and progress their inputs, has proved a formidable task. Some of the difficulties and points for decision should be briefly outlined. Not only is the number of observers interested in bird voice unduly small in relation to those active in other fields, but the use of their results is compromised by the continuing lack of any generally accepted and readily understood method of comparative verbal description of bird utterances, partly owing to widespread ignorance of the use of phonetics.

For many species, available data are limited to a few descriptions, based perhaps on a single individual and even on a single utterance. Comprehensive and critical studies exist as yet only for a small minority. Most authors have felt compelled by this situation to confine themselves to piecemeal assembly of odd scraps of information, too often merely copied, with variations, in one work after another. While subject to similar constraints, we have had the advantage of access to a large selection of documented utterances recorded on tape. By analysing these in comparison with a newly ordered abstract of conventional verbal descriptions and renderings in syllabic form, a confrontation has for the first time been achieved within the west Palearctic between the experience of field observers and the evidence of recorded sound, itself largely from the field.

Although no rigid formal structure has been maintained in the text, it usually begins with a brief general outline of the voice of the species, any special features, its relation to allied forms, its significance in the life-pattern, and the adequacy or otherwise of study and data available. The various calls are then listed, wherever appropriate separately for male, female, and young, bearing in mind the principal aspects. These include, objectively, the duration, frequency, amplitude, and the 'timbre' or tonal quality (which covers, e.g., distribution and relative strength of harmonics, presence or absence of transients, modulation rate in frequency and amplitude, etc.) and, subjectively, time, pitch, loudness and range of audibility, and tone quality which covers character of utterance and circumstances conditioning the utterance as heard. Account is also taken of influences on and limits of the production mechanism, and of the social background and signal value of the sound pattern, detailed treatment of which occurs in Social Pattern and Behaviour.

Even where one or more of these aspects may be inapplicable, they need to be taken into account in evaluating the vocabulary of the species. Uniform utterances on the one hand, and variable utterances on the other need to be clearly distinguished. Attention must also be given to individual, local, and regional variations in utterances, or absence of particular utterances in one or other sex, or at certain seasons or ages, or in certain parts of the range. Fully reliable generalized accounts of the language of a species have in most cases yet to be produced. Analysis of data shows that certain species and groups, and certain areas, have been badly neglected, and that even where primary data about voice appear fairly complete

there may be shortcomings in standards of description and analysis or recording.

The attainment of 'word-consistency' in descriptions has presented severe difficulties. In face of the ambiguities, confusions, and vaguenesses often enshrined in the literature, we have found it acceptable neither to continue on the same lines, not to attempt at a single leap to adopt a new set of rigorously scientific descriptive terms unfamiliar to readers and even to workers in this field. As in the Habitat section, however, we look forward to a time when all descriptive terminology, however loosely the words may be employed in normal usage, will be screened against clearly formulated scientific definitions, together affording a comprehensive coverage of ornithological requirements for accurate communication.

For the present we have had to be content to begin by eliminating clearly misleading or unserviceable expressions in order to build up some approximation to an acceptable vocabulary. In order to help readers, especially those for whom English is not their first language, we have tried wherever possible to include a number of synonyms or partial equivalents, which may afford fuller clues to the nature of each utterance. We have also, wherever possible, included complementary physical indications, such as duration, intervals and numbers and rates of repetitions, and comparisons with other sounds, as well as sonagrams and melograms, to assist in what can at best only be a groping towards communication. Often even this has been frustrated by uncertainty over the intention of the writer of an original description, and whether his chosen words accurately conveyed his intention. Where translation from one language to another has been involved the difficulties are compounded.

The greatest recent advance in the study of bird voice has been the widespread success in obtaining tape recordings, both in the wild and in captivity, for comparison with verbal descriptions which they may for some purposes replace. The reconciliation of the two has been accorded the highest priority in preparing the present work. In planning it much thought was given to the inclusion of reproducible sound material, for instance on gramophone records, but the disadvantages outweighed the advantages. We have therefore relied upon visual processing in the form of sonagrams and melograms.

## VOICE ILLUSTRATIONS

Two forms of diagrammatic illustration are used, both derived from precise measurements made by electronic instrumentation: (1) sound spectrograms (sometimes called audio-spectrograms), usually abbreviated to 'sonagrams', and (2) melograms.

Sonagrams (produced by the Kay Sona-Graph, designed and made by the Kay Electric Company, New Jersey, USA) are basically graphs showing the distribution of sound energy in three parameters: (a) frequency (pitch), vertical scale, (b) time (duration), horizontal scale, (c) amplitude (loudness), depth of shading from pale grey to dense black. Frequency is measured in kilocycles, now usually referred to as kiloHertz (kHz) and it is important to note that the frequency scale is linear, that is, it proceeds by evenly spaced divisions from 50 Hz to 8 kHz ($\frac{1}{2}$ inch per 1000 Hertz on the original graphs except in the lowest division 0–1 kHz, where, with the wide-band filter in operation, there is a slight downward shift resulting in a narrower band). Since the frequency ratio of the octave is 2:1 the vertical scale expands by a corresponding extent for each ascending octave:

1–2 kHz = 1 octave occupying $\frac{1}{2}$ inch on the original graphs

2–4 kHz = 1 octave occupying 1 inch on the original graphs

4–8 kHz = 1 octave occupying 2 inch on the original graphs

Everything below 1 kHz, about 5 octaves in man's auditory range, is compressed within the lowest division 50 Hz–1 kHz. Since many—probably most—avian vocalizations fall within the range 1–8 kHz, the coarseness of the analysis within the first 1 kHz is of little consequence. The best frequency resolution of the Sona-Graph is 45 Hz; this, too, is coarse in the lower reaches of the frequency spectrum where the semitone, the smallest musical interval *easily* perceived by all humans, may measure only 2 Hz or less; but in the region of 1 kHz the semitone measures about 60 Hz, and so a fair degree of accuracy may be obtained in the upper octaves. Either of 2 band-width filters may be used, 45 Hz or 300 Hz, the former giving better frequency resolution and the latter better temporal resolution (Figs. 1, 2a, b). Time resolution excellent at 0·0015 second. This is important in the analysis of bird sounds since it is now accepted that the temporal resolution of birds is much finer than that of man, probably ten times better, though higher figures have been suggested. Thus the avian ear may distinguish 200 discrete sounds per second while, to the human ear, sounds fuse at about 20 per second.

The maximum duration of a single sonagram is about 2·5 s. However, for the purpose of illustrating sections of continuous song exceeding 2·5 s it is a simple matter to make several overlapping sonagrams and thereafter match and join them. A sonagram of 1·2 s can also be made; this gives better temporal detail.

Flexibility of the analytical process may be obtained by varying the playback speed of recordings fed to the Sona-Graph (Fig. 3). Half-speed playback results in a clearer picture of temporal detail (the duration of a complete sonagram being reduced to 1·25 s) and may also be used to halve the frequency of sounds exceeding 8 kHz, the maximum frequency shown on the usual sonagram. Contrary to some expectations, such high-frequency sounds occur rather rarely in bird songs and calls. Doubling the playback speed of a recording obscures some of the temporal detail and extends the horizontal scale to 5 s but is useful in obtaining a clearer picture of low-frequency sounds.

The delicate shading showing intensity of sound is lost in reproduction but, when change of intensity is important (for purposes of description or recognition of vocalizations), a running amplitude display can be added to a sonagram. This correlates exactly with the time axis but the vertical scale of the subsidiary graph is in decibels (Fig. 4).

Sonagrams show the acoustic spectrum of each sound so that much information about the nature of the sound and 'timbre'—or tone quality—of a tone may be deduced. 'White noise', containing virtually all frequencies, spreads irregularly over the entire spectrum (Fig. 5), while the thin, pure tones from a signal generator appear as a narrow horizontal line (Fig. 1). Tones produced by the voice and musical instruments are usually rich in harmonics (Fig. 6). The average ear does not analyse harmonics into discrete entities as it does—or can usually be trained to analyse—the several simultaneously sounding notes in a chord. Instead of hearing harmonics as they appear on a sonagram we are aware of a particular tone quality which differs from one instrument to another according to the relative strength of the harmonics. Tone quality is also much affected by starting transients—the brief initiating sounds which precede, and may follow, the steady state of a tone. Other factors which affect the quality of a tone and which may be read from a sonagram are frequency modulation and amplitude modulation. The former is akin to the vibrato of the string player or singer but the modulation rate is generally much faster in bird sounds; it appears on the sonagram as rapid and regular frequency alternation of fairly uniform range (Fig. 7). Amplitude modulation is a waxing and waning of the amplitude of a sound (Fig. 8); it may be coupled to frequency modulation.

Noises, i.e. those sounds lacking a sufficient concentration of energy within narrow margins of a principal frequency to give tones of identifiable pitch, appear as broad band patches of energy. The frequency range is often wide and may encompass the available spectrum, as in the hiss of the Mute Swan *Cygnus olor* (Fig. 9) or the noise made by the key depression of an electric typewriter (Fig. 10). Many bird songs comprise a mixture of tones and noises (Fig. 11).

Melograms are produced by a new instrument (the Melograph) designed and constructed by members of the Electronic Department, Institute of Physics, University of Uppsala, Sweden. It analyses both frequency and amplitude and prints, by means of an ultraviolet or an ink-jet writer, separate but simultaneous curves at various writing speeds. It differs from the Sona-Graph in that it selects only the fundamental frequency. Thus no information about acoustic structure is displayed, but the frequency analysis is much more precise. Frequency is

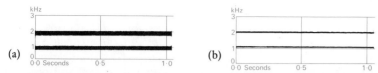

Fig. 1. Pure tones from a signal generator at 1 kHz and 2 kHz: Sound spectrographic analysis, (a) with wide band filter (300 Hz), (b) with narrow band filter (45 Hz).

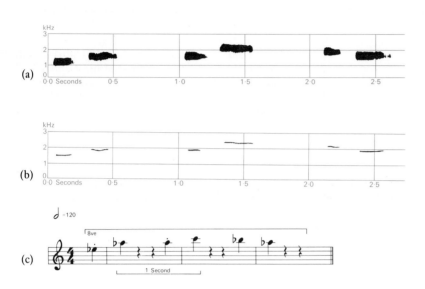

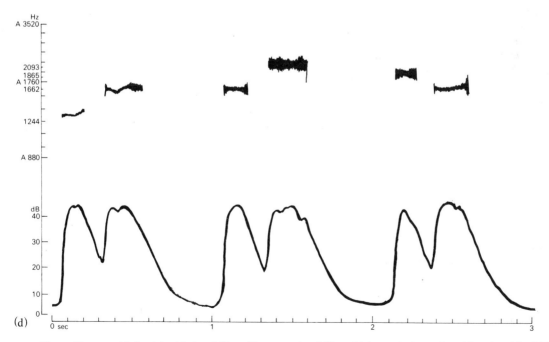

Fig. 2. Human whistle: (a) wide-band filter, (b) narrow-band filter, (c) in musical notation, (d) analysed by Melograph (horizontal scale in seconds, upper vertical scale in Hz and whole tones, lower vertical scale relative intensity in dB).

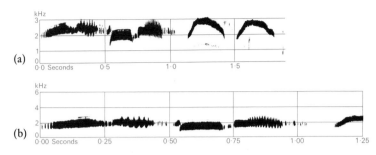

(a)

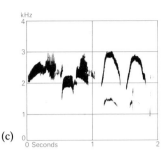

(b)

(c)

Fig. 3. Wide-band sonagrams of song phrase of Blackbird *Turdus merula*: (*a*) normal playback speed, (*b*) half-speed playback of first part, (*c*) double-speed playback.

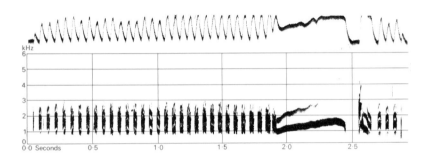

Fig. 4. Wide-band sonagram of 'churring' of Leach's Storm-petrel *Oceanodroma leucorhoa*. The upper curve shows relative loudness.

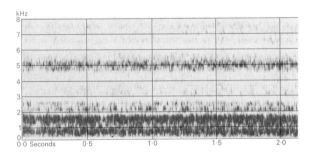

Fig. 5. Wide-band sonagram of white noise, which embraces all frequencies of the available spectrum.

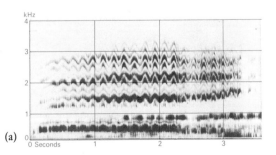

(a)

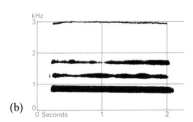

(b)

Fig. 6. Illustrations of harmonic contents of sound: (*a*) human ♂ voice singing the vowel 'ay' (fundamental frequency 246 Hz), (*b*) sopranino recorder tone at A one octave above tuning A (fundamental frequency 880 Hz).

(a)

Fig. 7. frequency modulation in Blackbird *Turdus merula*; unit 2 uttered at *c.* 50 per second and unit 4 at *c.* 100 per second. (*a*) Sonagram, (*b*) melogram, (*c*) melogram of 3 units from 7(*b*) analysed at 40 cm/s.

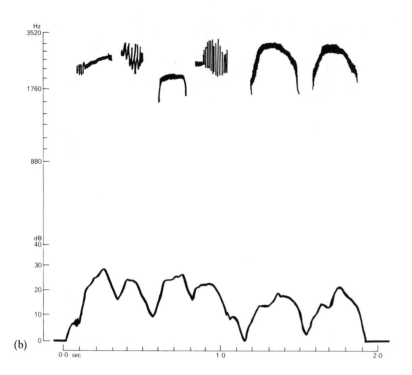

(b)

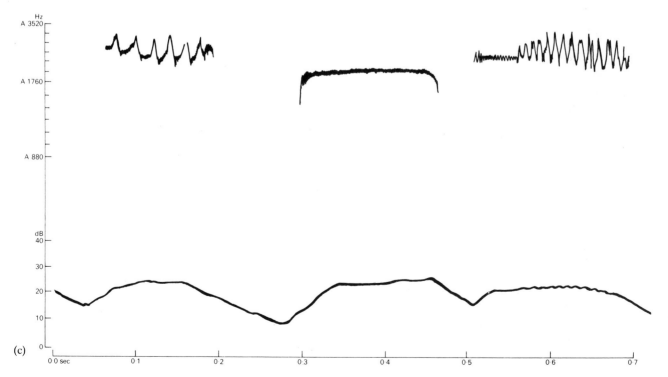

(c)

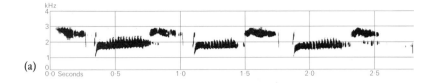

(a)

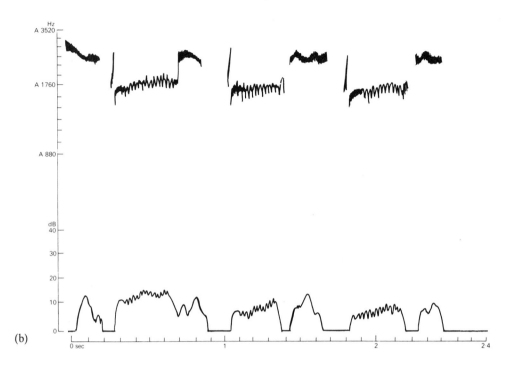

(b)

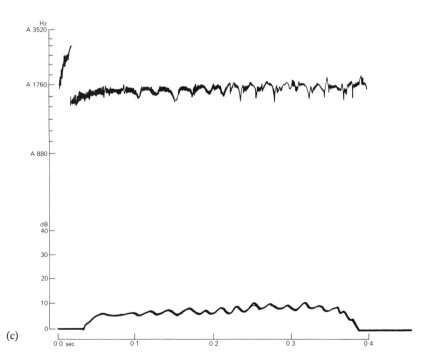

(c)

Fig. 8. Frequency modulation with amplitude modulation in Blackbird *Turdus merula*: (*a*) sonagram, (*b*) melogram, (*c*) melogram of single unit from 8(*b*) analysed at 40 cm/s.

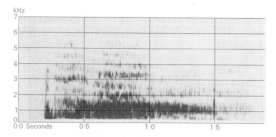

Fig. 9. Hiss of Mute Swan *Cygnus olor* (compare Fig. 5).

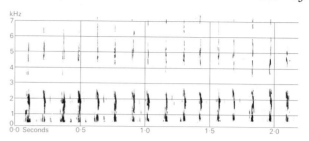

Fig. 10. Click-like sounds made by depressing keys of electric typewriter (wide-band sonagram).

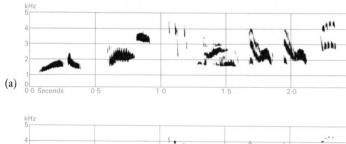

(a)

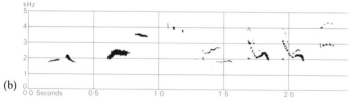

(b)

Fig. 11. Song phrases of Blackbird *Turdus merula* illustrating tones (units 1–4); clicks (units 5–7) and noises (units 8 to end). (*a*) wide-band sonagram (*b*) narrow-band sonagram.

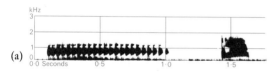

(a)

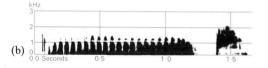

(b)

Fig. 12. (*a*) Sonagram of Inciting-call of Ruddy Shelduck *Tadorna ferruginea*. (*b*) Human imitation of same call with initial consonant 'k' (initiating consonants in bird vocalizations are often imagined, probably because man cannot start vowel sounds as quickly as birds).

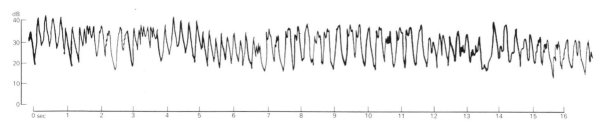

Fig. 13. Melographic amplitude curve of part of song of Reed Warbler *Acrocephalus scirpaceus*.

measured in Hertz and whole tones on a logarithmic scale so that all octaves are equidistant, as in musical notation. Starting from 55 Hz, any octave, 2 octaves, or 5 octaves may be selected from the frequency spectrum and the vertical axis of the graph can be restricted to the vocal range of the species under discussion. The ultraviolet writer used in conjunction with the Melograph—Bell and Howell's Direct Writing Recording Oscillograph—permits 12 writing speeds ranging from 3 cm/min to 180 cm/s. For the diagrams, a speed of 8 cm (3·2 in)/s is generally used (Figs. 2d, 7b, 8b) but higher speeds are helpful where particular attention to detail is necessary (Figs. 7c, 8c). Low writing speeds are useful for long-term analyses, such as those required for measuring time intervals between calls or song in a long series (Fig. 13). The amplitude trace correlates exactly with the frequency trace and ranges from 0–40 dB (Figs 7b, c, 8b, c).

The limitations of the instrument, as set against the Sona-Graph, are that it will not register unvoiced sounds, neither can it analyse sounds comprising a mixture of unrelated frequencies or display acoustic spectra. It is essentially an instrument for registering tones as distinct from noises and in this it is more efficient than the Sona-Graph. Its flexible horizontal and vertical scales greatly facilitate analysis and permit certain aspects of sounds to be highlighted. Long recordings can be analysed continuously with a compromise between running speed and the total length of the paper strip: 30 metres in this instance.

As a result of the different attributes of the two instruments, the Sona-Graph is used to analyse any calls and songs incorporating noise component, and when spectral analysis is desirable. The Melograph is used for analysing predominantly musical vocalizations and, in particular, those of relatively low frequency such as the calls of doves *Columba*, cuckoos *Cuculus*, bittern *Botaurus*, etc. Reproductions of sonagrams and melograms are made from photographic prints of the original diagrams; where background noise interferes with the signal representation, it has been painted out.

The figures will help in the mental realization of sonagrams and melograms. With practice, it becomes possible to read and interpret the diagrams and gain a very fair impression of the sounds portrayed. In many cases, they provide at a glance much information which could only be conveyed, if at all, by lengthy and complex verbal description. It is best to begin by looking at temporal factors, establishing the time pattern in mind, and then to note the most obvious pitch movements—ascending, descending, steady or variable pitch. Distinguishing between tones and noises presents no problems (Figs. 1, 2, 5, 6, 9, 10, 11) but it is often difficult to imagine the quality of the sound, for example, whether the tone is 'reedy' or 'flute-like', or the noise is growling, grunting, screeching, clucking, etc. In such instances much can be learned from

studying the diagrams of sounds of well-known species and noting similarities and differences between these and unfamiliar sounds.

## PHONETIC RENDERINGS

As an aid to realizing the diagrams in this manner, phonetic renderings are given in the text and sometimes added to the illustrations. Phonetic, or onomatopoeic, representation is beset by pitfalls on account of its subjective nature; no major studies have been made of the phonetic content of bird sounds (except a few with reference to the imitative abilities of talking birds) and it is readily apparent that even within a group of ornithologists experienced in recording and describing avian sounds, there is considerable disagreement in this sphere, especially in connection with the supposed consonantal content of sounds. The rapid initiation of many bird sounds gives a strong impression of the presence of a consonant but it is often impossible to decide which consonant. Hence where consonants are given they should be treated as no more than suggestive of the kind of sound that may be heard (Fig. 12).

It is generally much easier to reach agreement about vowel sounds, probably on account of their longer duration but here, too, there are problems which may often be attributed to individual variation of voice within a species; thus one member of a flock may be heard to pronounce 'a' as in cat and another 'e' as in pet. Study of the recordings suggests that this may, at times, be attributable to sexual dimorphism of voice, but little work has been done in this field and it could provide a fruitful project for research. A simple but functionally adequate system is provided by the 16 vowel sounds used in Pitman's shorthand. These are readily memorized from three short sentences. Long vowels: 'Pa may we all go too?' rendered 'ah, ay, ee, aw, oh, oo'. Short vowels: 'That pen is not much good', rendered 'a, e, i, o, u, oo'. Dipthongs: 'I now enjoy music', rendered 'ī, ow, oy, ū'. That these vowels suffice for general communication in English is obvious from the success of the system. There is little point in attempting to distinguish between, for example, 'ah' and 'ar' when individual variation may extend to 'aw' for the same call. It must also be emphasized that in using material from a variety of sources and languages, full consistency is unobtainable.

Many requirements for further study emerge from what has been said above. As the study of bird voice lags so far behind some other facets of ornithological research, a few lines must be devoted to explaining its requirements. These include:

1.  Systematic appreciation of the extent of the field and of the descriptive and experimental techniques for exploring it, including all necessary standardization of terms and methods.

2. Further development of existing laboratory facilities, archives, and networks of specialist communication, with a view to expanding resources and filling in gaps in structure, including closer relations with other branches of bioacoustics and improved international provision for co-operation and publication.

3. Increased facilities for training and opportunities for research by young scientists in this area of bioacoustics.

4. Compilation of an authoritative newly worked compendium of the vocabulary of as many species as possible on an international basis, to lay a foundation for future work, on lines sketched in the present Voice section.

5. Initiation of a series of co-operative studies designed to bring together researches on voice, behaviour, and communications theory, and to demonstrate the potential of a modern scientific effort in bioacoustics.

By way of illustration we include a few possible topics for specific studies which have come to our attention in preparing this volume:

1. Intraspecific variation of vocalization (a) Regional, (b) Group, (c) Individual.
2. Relation between habitat type and vocalization (e.g. it is claimed that species calling from a forest floor use lower frequencies than those in the canopy and that inhabitants of coniferous woods produce quieter sounds than those of deciduous woodland).
3. Relation between territory size and vocalization type.
4. The study of racial characteristics of voice as an aid in taxonomy.
5. An investigation of female song and the possibilities of more duetting in species of temperate regions.
6. Diurnal variation in song.
7. A study of song outside the breeding season.
8. Development of song in the individual.
9. Interspecific imitation.
10. The extent of imitation within different environments.

## GLOSSARY

AMPLITUDE. (*See* INTENSITY, LOUDNESS, VOLUME). The increase or decrease of air pressure at a given point during a sound. The maximum value of the displacement in an oscillatory motion (waveform).

ANTIPHONAL SINGING. (*See* DUETTING). Precisely timed alternating singing or calling normally restricted to the performance of male and female for the formation and maintenance of the pair-bond.

CALL NOTES (CALLS). Sounds of brief duration used mainly to alert to danger and to coordinate group behaviour (e.g. flight-call, alarm-call, flock-coordination call, etc). Most of these calls do not vary with the season and, in contrast to song, are not strongly influenced by the state of seasonal development of the sex hormones.

COMMUNAL SINGING. Tendency of social groups or adjacent territory-holders to sing together in response to the same stimulus or as a result of social facilitation.

COUNTER SINGING. Alternation of song which may occur when two or more territory holders are responding to one another.

CYCLE. *See* HERTZ. When a waveform repeats itself each complete repetition is called a cycle; e.g. a string or other sound source vibrating at 440 cycles per second (Hertz) produces tuning A.

DECIBEL. A measure comparing the intensity or loudness of two or more sounds. The decibel scale is logarithmic. It ranges from the threshold of hearing, reaching an intensity that is painful to the human ear at about 130 dB.

DIALECT. Produced when there is a tendency among local populations of a species to approximate their songs to a common pattern as a result of mutual imitation.

DOPPLER EFFECT. The apparent change in the pitch of a sound when either the sound source or the listener are moving towards or away from each other. When the distance between the two is decreasing the apparent pitch is raised and *vice versa*. In practice this is especially noticeable in the sounds made by wing-beats.

DUETTING. Duetting normally occurs when two members of a pair sing simultaneously or antiphonally as part of the courtship display, to maintain the pair-bond or to maintain contact.

FREQUENCY. (*See* PITCH). The number of cycles occurring per unit of time. The unit of time is normally 1 second, and frequency is then measured in hertz.

FUNDAMENTAL FREQUENCY. The lowest component frequency of a periodic wave; the basic factor that determines the pitch of a sound.

GLISSANDO. An extremely rapid succession of tones either ascending or descending in pitch.

HARMONIC. (*See* OVERTONE or UPPER PARTIAL). A harmonic is a component of a periodic wave having a frequency which is a simple multiple of the fundamental, e.g. the component whose frequency is twice that of the fundamental is called the second harmonic (*See also* TONAL QUALITY).

HERTZ. Synonym for cycles per second.

INSTRUMENTAL SOUNDS OR MECHANICAL SOUNDS. Non-vocal sounds used for communication, e.g. drumming, bill-clappering, wing-clapping, etc.

KILOHERTZ or KILOCYCLE. 1000 Hertz (i.e. 1000 cycles per second).

LEGATO. Sounds performed in a smooth or connected manner. The opposite of staccato performance.

LOCATORY SOUND. Direction giving. A sound which contains cues by means of which the position of its source can be determined.

MELODY. Change of pitch in time.

MELOGRAM. A graph showing synchronous curves of principal frequencies and their relative amplitudes in real time.

MELOGRAPH. An instrument for measuring and displaying changes in principal frequencies and amplitude in time.

MNEMONICS. Words or sentences designed to aid the recall of sound patterns.

MODULATION. Any periodic alternation of frequency, intensity or phase of a sound wave.

NOISE. Non-periodic sound waves.

NOISE, WHITE. A sound with an approximately equal amount of power at every component frequency over the audible range.

NOTE. A single tone of definite pitch.

OCTAVE. The interval between two frequencies having the ratio 2:1. The second harmonic is one octave above the fundamental.

OVERTONE. (*See* HARMONIC). A partial having a frequency higher than that of the basic frequency. Often used as a synonym for harmonic.

PARABOLIC REFLECTOR. A device used for reflecting sound waves into a reversed microphone in order to give increased directionality by collecting more sound energy from a particular direction than would the microphone alone. At the same time it helps to exclude sounds from other directions.

PHRASE. Short musical passage forming part of a longer passage. In bird song a number of song-phrases (or songs) comprise a song repertoire.

PITCH (*See* FREQUENCY). The pitch of a note is determined by the frequency of vibration of the particles in the sound wave, which is equal to the frequency of the source: the greater the number of vibrations (cycles) per second (or Hertz), the higher the pitch. Pitch refers to the auditory sensation, frequency to the physical measurement

POLYPHONIC SINGING. A type of duet singing in which the contributions from the 2 (or more) birds synchronize or overlap, each part having its own temporal and pitch patterning.

PRIMARY SONG. Sometimes used to refer to the normal loud specific song. Also called territorial or advertising song.

REACTION TIME, AUDITORY. The briefest period of time between stimulus and response.

REPERTOIRE. The total vocal and instrumental output of any individual. Often applied only to song but should, strictly, include all vocalizations and mechanical (instrumental) sounds used in communication.

RHYTHM. In musical usage the term is often restricted to temporal patterning. However, a time pattern—with or without pitch patterning—may or may not be performed rhythmically. Rhythm embraces those minute gradations of time and intensity which give to abstract sound patterns a forward impetus to points of climax and rest. Its expression does not require a regularly recurring strong beat and many bird songs are profoundly rhythmic.

SEMITONE. The smallest musical interval in common use in Western music. There are 12 semitones in an octave. Bird songs often incorporate smaller intervals than the semitone but it is not known whether birds are able to distinguish much smaller pitch intervals than can man.

SONAGRAM (Abbreviation of Sound Spectrogram). A precise visual representation of a sound pattern. Distribution of sound energy is shown as a function of time. Loudness (intensity) is discernible in the depth of shading. Much of the information on a sonagram is not immediately apparent to the ear, e.g. harmonics are not heard as discrete sounds occurring simultaneously but as a single tone with a quality (timbre) which is largely dependent upon the relative strength and distribution of the harmonics (See Introduction to Voice Section).

SONG. A relatively complex pattern of sounds in time which may be repeated exactly and is, consequently, recognizable not only at the specific level but often at the group and individual levels. Song has both tonal and temporal form. PRIMARY SONG (q.v.) is usually under the control of the sex hormones and its utterance is therefore largely confined to the breeding season.

SOUND SPECTROGRAPHIC ANALYSIS. The sound spectrograph is an instrument for making spectral analyses of single sounds or series of sounds. Sounds are recorded magnetically on the edge of a 12-in metal turntable which then revolves repeatedly. At each revolution the signal is scanned by a 45-Hz or 300-Hz filter, beginning with the low frequencies and working upwards. The output is recorded on dry facsimile paper attached to a drum which revolves synchronously with the turntable. A recording stylus shifts gradually up the frequency scale in step with the scanning oscillator, thereby recording the frequency components at any given instant. Amplitude variations are shown by fluctuations of intensity at the output of the filter, the darker regions on the paper indicating the louder sounds and *vice versa*. Precise measurement of the relative amplitudes of the components of a sound may be obtained by means of a sectioner. This gives a graph of any required instant (about 1/24 s) of the relative amplitude against frequency in cycles per second (Hertz).

SOUND SPECTRUM (Acoustic structure). An acoustic spectrum shows the distribution of energy among the component frequencies, intensities, and phases of a sound at any given moment in time.

SUBSONG. A useful general term to denote forms of quiet song. More specifically it refers to a quiet, extended warbling in which fragments of the full (or primary) song may be heard, often with imitations of other species. A bird uttering subsong is usually perched low and in an inconspicuous place; the beak may be closed while singing. Subsong is not confined to out-of-season utterances and may readily be heard during the period of full song.

SYLLABLE. This has been used to describe various successive sections of vocalizations. By analogy with verbal use: if the term is used to describe a section of the unit 'word', it may—in avian vocalizations—be used to describe a section of the discrete unit as shown on the sound spectrogram (sonagram).

SYRINX. The avian organ of voice or song. Unlike the human larynx it is situated at the lower end of the trachea and the junction of the two bronchi. Although variable in structure and complexity between groups of birds, it is generally a bony and cartilaginous chamber containing membranes which are activated by the passage of air from the lungs.

TEMPORAL RESOLUTION. A measure of the minimal time lapse between successive sounds which are still perceptible as discrete.

TIMBRE. Tonal quality, q.v.

TONAL QUALITY. That aspect of a sound which distinguishes it from other sounds of the same pitch. Physically, tone quality is determined by the presence or absence of upper partials (*see* OVERTONE) and their relative intensity and distribution, by transients, and by the rate and range of frequency and/or amplitude modulation.

TONE. In general use a tone is a musical sound (i.e. pitched) as distinct from a noise. In musical theory the term means a definite interval between two notes, that is, 1/6th of an octave or 2 semitones.

TRANSCRIPTION. The representation of avian vocalizations in symbolic form e.g. words, notation, onomatopoeia, etc.

TRANSIENT. The brief sounds which precede and follow the steady state of a tone. The frequency of the transient vibrations is the natural period of the system; the frequency of the steady state vibrations is the frequency of the driving force.

TRANSPOSITION. The exact repetition of a sound pattern at a higher or lower pitch while the pitch and time relationships within the pattern remain constant.

TREMOLO. Rapid reiteration of one note. A very rapid tremolo may have the appearance of amplitude modulation.

UNISON and UNISON SINGING. Sounding at the same pitch and in the same time pattern. Synchronous singing by 2 or more birds in identical time and pitch patterns.

UNIT. A discrete section of a figure. A unit may be simple or complex; simple units comprise a single continuous sound sometimes divisible into syllables. Complex units comprise 2 or more discrete sub-units and are distinguished by exact repetition of the entire complex.

VELOCITY OF SOUND. The rate at which sound waves travel

through a given medium. In air this is 1129 feet per second at 20 °C. Velocity increases with a rise in air temperature by about 1 foot per second for a rise of 1 °C.

VIBRATO. Rapid and continuous perodic rise and fall in pitch. A form of frequency modulation.

VOCALIZATION. An umbrella term for all avian sounds produced by the voice; song, calls, distress cries, etc.

VOLUME. The subjective experience of intensity.

WAVELENGTH. $\lambda = v/f$. The length of a sound wave is determined by dividing the velocity of sound by the frequency of the sound in question; thus tuning A at 440 cycles per second (Hertz) = (1129/440) ft = 2·57 ft.

WHISTLE. A term often used to describe those sounds which most nearly approach pure tones; that is, having a fundamental but no overtones.

# BREEDING

The breeding season is taken as that period within which the species lays and incubates its eggs and rears its young to the flying stage. Some authorities would also include pair-formation and nest-building but these activities cannot be pinned down as accurately in time as the laying of an egg and may take place over a prolonged period that would be misleading to define as the breeding season.

The ANNUAL CYCLE DIAGRAMS show the normal season for the occurrence of eggs (E) and unfledged young (Y) with margins for early eggs and late broods. The diagram is in each case for a specified geographical area which is usually a large one taking in a considerable proportion of the species' range. Good data on variations in breeding season in different parts of the range are rarely available, although a south to north, or perhaps south-west to north-east, cline undoubtedly exists for many species. For some species the information on season is insufficient to compile a meaningful diagram; exceptionally all that is known is a handful of dates when eggs or young have been recorded.

The timing of the breeding season is clearly affected by such factors as climate, latitude, food availability, and the physiological condition of the bird. The timing of the onset of breeding is also affected by the proximate factors of spring temperatures and rainfall, and the dependent conditions of vegetation growth and existing food supply. While this general assumption is probably true for the majority of species, comments have been restricted to those species for which actual studies on this aspect have been carried out.

THE NEST: its site, construction, and building. The description of the nest site is restricted to its precise location; the broader aspects being covered in Habitat. The emphasis has been placed on the normal situation, omitting lists of exceptional, often aberrant sites. Also included here is the re-use of nests in subsequent years.

The materials used in the construction of the nest are given fairly precisely in those relatively few cases where they have been examined in detail. For the majority of species, however, the information is rather generalized. As with nest site, the exceptions have not been given undue prominence. Average dimensions of nests are usually quite easy to take but published accounts deal with mostly small samples, if these are stated at all. Detailed observations are still required on the building of nests of most species; in particular information is lacking on the role of the sexes, the time taken, and the time of day.

EGGS. These are described briefly in terms of shape, texture, and eolour. The first and last of these should be read in conjunction with the colour plates of eggs (reproduced actual size). The principal terms used in describing shape are oval (or ovate), elliptical, pointed (or pyriform), and rounded (or sub-elliptical), which should be self-explanatory. These may be qualified by adjectives such as blunt, broad, short, or long.

Measurements and weights of eggs are given to the nearest whole millimetre and gram respectively. Range and sample size are included wherever possible, and a sample size of 100 has been regarded as acceptable though by no means always achieved. The most important single source of information on egg sizes is the work of Schönwetter (1967, and in preparation), which also includes information on eggshell weights, and geographical and subspecific variation.

CLUTCH. The ideal information under this heading is a large sample of clutches (100 as a reasonable minimum) set out giving the percentage distribution of the different clutch sizes, together with the mean. This has been achieved only for a minority of species, however, and for the remainder it has been possible to give only the limits of normal clutch sizes, plus extremes. Sufficiently large samples from different parts of the range of a species to allow regional comparisons are given where available, as are examples of seasonal variation in clutch size.

The extremes of clutch size must be treated with great caution. At the lower end of the scale there is always a strong possibility that one or more eggs may have disappeared before the observer recorded the size, while at the upper end there is the complication of more than one female laying in the same nest. Both these occurrences can, of course, alter the true percentage distribution as well as the mean, yet neither can be completely allowed for. For a few species, for example among the ducks, where laying by two or more females in the same nest is quite regularly recorded, information is given separately for these *dump* nests.

The number of broods normally reared is given together with any known regional variation. The occurrence of a replacement for a lost clutch is mentioned though only as a probability in the many instances where it is likely but has not been proved. Most single-brooded species seem capable of laying a replacement clutch when their first is lost, particularly when this happens early in the incubation period. In a small number of species the replacement clutch is known to be smaller than the first, but for many more information is lacking. The normal interval between laying eggs is stated where known.

INCUBATION. The incubation period is given as accurately as has been recorded, though this is usually a range of days. The period refers always to that taken for one egg to hatch from the time incubation starts, rather than for the full clutch, though some authors do not make clear which they are giving. The division of incubation between the parents is stated where known, as is the stage of laying at which incubation begins, and its subsequent effect on the hatching pattern. These items tend to follow family, or at least generic lines and consequently where it has not been reported it can usually be inferred. Similarly disposal of eggshells after hatching is in general a family characteristic.

YOUNG. The first terms under this heading refer to the state of maturity of the young at hatching, as follows: (1) *Precocial*: well developed at hatching—covered in down, eyes open, and capable of coordinated movements, including a measure of terrestrial or aquatic locomotion; may feed themselves, or be fed by their parents. (2) *Altricial*: poorly developed at hatching—little or no down, eyes closed, capable of little movement; always fed by parents. (3) *Semi-altricial*: much as (2) but with a better covering of down, eyes open, or closed for only a short time after hatching, and rather more active.

These three terms are coupled with three more: (4) *Nidifugous*: leaves nest and its vicinity soon after hatching. (5) *Nidicolous*: stays in nest for some time after hatching. (6) *Semi-nidifugous*: stays in or near the nest for some time after hatching although physically capable of leaving altogether.

The parts played by the parents in feeding, caring for, and defending the young are dealt with next. Where the parents regurgitate food for the young this is divided into *complete* (food deposited on the nest or ground), and *incomplete* (food retained in parent's throat or mouth, and obtained by the young there). Distraction-displays and other anti-predator reactions are mentioned but not described in detail.

FLEDGING TO MATURITY. The fledging period is given as precisely as possible. It is not recorded for a surprising number of species, and authorities differ about others. Equally the dependence of the young on the parents after fledging is largely unknown, being particularly difficult to establish in those species which embark on a migration together but arrive at their destination separately. The age of first breeding is stated where known.

BREEDING SUCCESS. This can be given only for a minority of species. A study lasting several years may not be sufficient to give a reliable figure in the case of long-lived species.

# PLUMAGES AND RELATED TOPICS

PLUMAGES. This section is primarily intended for use with the bird in the hand, although it should also be of value when studying close-up photographs or detailed field notes. Usually one subspecies is described, the characters of others being given in the final section on Geographical Variation. The description of a plumage is presented in the sequence: head, neck, upperside and underside of body, tail, and wings (for names of plumage areas see figures on pp. 32–33). The plumages are discussed in the following sequence: adult breeding, adult non-breeding, nestling or downy young, juvenile, and immatures. The descriptions refer to fully developed plumages, but it should be kept in mind that elements of several plumages may be found in the same bird, e.g. during active moult or when moult is arrested.

The *breeding plumage* is defined as the plumage worn during part or all of the nesting season, but sometimes acquired long before. In some species of ducks (Anatinae), the females nest in a plumage which is homologous to non-breeding, but display and pair-up in a plumage which is homologous to adult breeding in other birds. Such aberrant cases are indicated in the text by inserting the words 'do not nest in this plumage' after the sub-heading 'Adult breeding'. In many species, the breeding plumage regularly alternates with a *non-breeding plumage*, acquired during the post-breeding moult (see below); in others the same plumage is worn during the whole year and simply termed *adult*. Occasionally, other plumages may be recognized in the course of the annual cycle; these are named *supplementary*. A non-breeding or supplementary plumage which is cryptic, worn for a comparatively short period, and which alternates with a colourful breeding plumage is called *eclipse*, as in some Anatinae.

The first full plumage following the nestling or downy young stage is called *juvenile* in all species, even in those, such as some seabirds, in which it is identical with adult. The juvenile plumage may be followed by one or more recognizable *immature* plumages. If the plumage following the juvenile is identical with the adult, there are no named immature plumages, but this does not imply that the bird breeds in this first adult plumage for sexual maturity may be deferred several years, as in petrels (Procellariidae). The precise sequence of plumages will be clear from the descriptions. Some species, e.g. certain herons (Ardeidae), breed before they are in completely adult plumage; for these the terms *sub-adult breeding* and *sub-adult non-breeding* are reserved.

A new terminology for plumages and moults was proposed by Humphrey and Parkes (1959). In this, the breeding plumage is termed 'alternate' and the non-breeding 'basic'. Stresemann (1963a) criticized these terms because it is not always clear whether the non-breeding plumage is the basic one. Nevertheless, the system of names is widely used in North America. To us, the terms given above seem more straightforward and easier to understand. They link the various plumages with the phases in the life-cycle for which they have been evolved.

BARE PARTS. These are described for the same race as the plumages in the sequence: iris, bill, bare skin on head (if present), leg, and foot. Bare-part colours cannot be studied satisfactorily in museum skins for they often fade after death; we have therefore relied on notes on specimen labels, descriptions of live and fresh dead birds, and occasionally (as supplementary evidence) colour photographs.

MOULTS. The following moults are recognized: adult post-breeding, adult pre-breeding, post-juvenile, and a variable number of moults of immature plumages. In species in which the adult wears a supplementary plumage, there is an additional moult for which no special terminology is used. The main moult of the annual cycle is almost invariably the *post-breeding moult*. This is usually *complete*, involving not only the body plumage, but also the flight-feathers and tail-feathers. The *pre-breeding moult* is normally *partial*, the flight-feathers, tail, and other parts of the plumage being retained. The replacement of nestling down by juvenile plumage could be termed *pre-juvenile moult*; it is not described in the Moults section, but information is presented in the description of the nestling or downy young. The *post-juvenile moult* is the first moult in which contour feathers are replaced. It is variable in timing and extent, being complete or partial (sometimes involving only a few feathers), ranging in timing from late summer of the first calendar year to autumn of the second.

The description of a moult contains information about timing, sequence of primary replacement, and for partial moults, about the parts of the plumage which are replaced. Timing of moult in adults is compared, where appropriate, with timing of breeding and migration in the ANNUAL CYCLE DIAGRAM, where P = primaries, B = body. Sequence of primary replacement may be *descendant* (from the carpal joint outwards), rarely *ascendant* (from the outermost primary inwards), *irregular*, or *simultaneous*. In a special type of descendant moult, a new moult series starts before the preceding one has reached the outermost primary, so that several moult centres are simultaneously active. We have called this type of moult *serially descendant*; Stresemann and Stresemann (1966) use the word *Staffelmauser*.

MEASUREMENTS. In most species, data are given for length of wing, tail, bill, tarsus, and middle toe. All measurements are given in millimetres (unit of measurement being omitted from the text). The wing is measured by pressing it against a rule and stretching it fully; *wing length* is the distance from carpal joint to the tip of the longest primary. The tail is measured with dividers, *tail length* being the distance from the point where the central tail feathers emerge from the skin to the tip of the longest feather. Bill, tarsus, and middle toe are measured with vernier calipers. *Bill length* is the chord of culmen from implantation of the feathers to the tip of the upper mandible. In some groups, it is preferable to measure from the nasal-frontal hinge in the skull or from the rim of cere to the tip of the upper mandible. Such cases will be indicated in the family summaries. *Tarsal length* is measured from the middle point of the joint between tibia and tarsus behind to the middle point of the joint between tarsus and middle toe in front of the leg. *Middle toe* is measured from this point to the tip of the middle claw. In some groups it is customary to measure the middle toe without the claw; this will be indicated in the family summaries of such groups. In the tables of measurements length of the middle toe with claw appears simply as 'toe'.

For each measurement, the following parameters are given: mean (m), standard deviation (s), number of specimens measured (n), lowest value of series (l) and highest value of series (h). These are presented in a standard form: m (s; n) l–h. If any of these values is unknown, the sequence is shortened to m (n) l–h, to m (l–h), or to l–h (n). Means over 100 are given to the nearest whole number, means under 100 to one decimal place. Standard deviations over 10 are given to one decimal place, those under 10 to two decimal places. Range is given in whole numbers, except when the values are under 20. The statistical significance of the difference between two means is tested with a *t*-test (see, e.g., Sokal and Rohlf 1969). Where values differ markedly from those given elsewhere, this is discussed.

At the head of each table, the sources of the material are specified, using the following abbreviations:
BMNH, British Museum (Natural History), Tring
BTO, British Trust for Ornithology, Tring
MNHN, Muséum National d'Histoire Naturelle, Paris
NMW, Naturhistorisches Museum, Wien
RIN, Rijksinstituut voor Natuurbeheer, Arnhem
RMNH, Rijksmuseum van Natuurlijke Historie, Leiden
RSME, Royal Scottish Museum, Edinburgh
ZFMK, Zoologisches Forschungsinstitut und Museum Alexander Koenig, Bonn
ZMA, Zoölogisch Museum (Instituut voor Taxonomische Zoölogie), Amsterdam
ZMK, Universitetets Zoologiske Museum, København
ZMO, Zoologisk Museum, Universitetet i Oslo

WEIGHTS. These are taken from notes on specimen labels, from published or unpublished data on birds captured for ringing, and from other literature sources. They are presented in the same way as measurements. All weights are given in grams (unit of measurement being omitted from text). If information exists on biological relevance of seasonal or daily weight variations, this is briefly summarized.

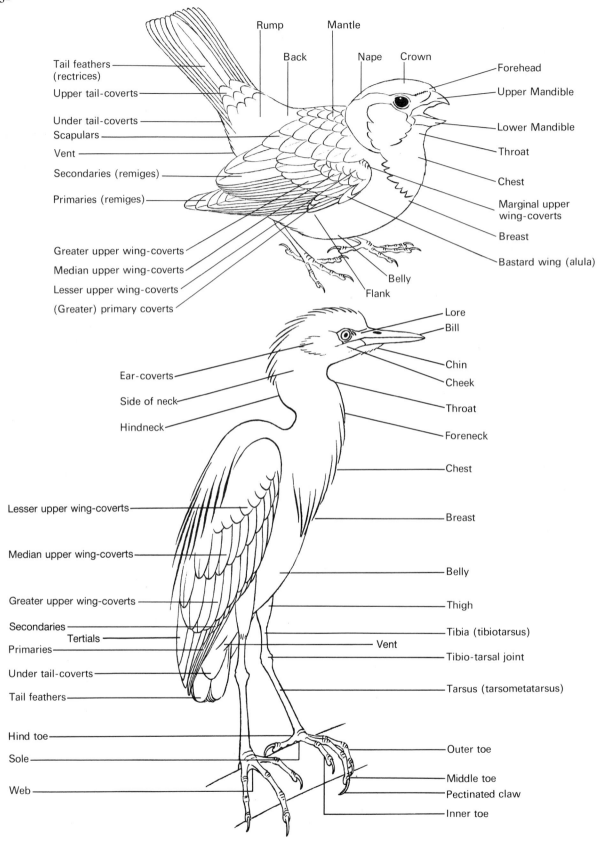

Rump

Mantle

Back

Nape Crown

Forehead

Tail feathers (rectrices)

Upper Mandible

Upper tail-coverts

Lower Mandible

Under tail-coverts

Throat

Scapulars

Vent

Chest

Secondaries (remiges)

Marginal upper wing-coverts

Primaries (remiges)

Breast

Greater upper wing-coverts

Bastard wing (alula)

Median upper wing-coverts

Lesser upper wing-coverts

Belly

(Greater) primary coverts

Flank

Lore

Bill

Ear-coverts

Chin

Side of neck

Cheek

Hindneck

Throat

Foreneck

Chest

Lesser upper wing-coverts

Breast

Median upper wing-coverts

Belly

Greater upper wing-coverts

Thigh

Secondaries

Tibia (tibiotarsus)

Tertials

Primaries

Vent

Tibio-tarsal joint

Under tail-coverts

Tarsus (tarsometatarsus)

Tail feathers

Hind toe

Sole

Outer toe

Middle toe

Web

Pectinated claw

Inner toe

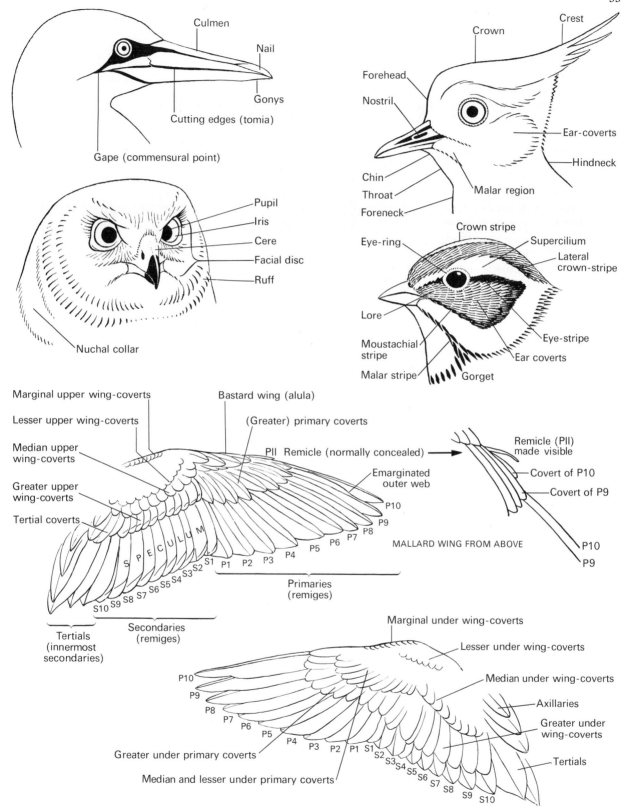

Culmen

Nail

Gonys

Cutting edges (tomia)

Gape (commensural point)

Crown

Crest

Forehead

Nostril

Ear-coverts

Hindneck

Chin

Throat

Malar region

Foreneck

Pupil

Iris

Cere

Facial disc

Ruff

Nuchal collar

Crown stripe

Eye-ring

Supercilium

Lateral
crown-stripe

Lore

Eye-stripe

Moustachial
stripe

Ear coverts

Malar stripe

Gorget

Marginal upper wing-coverts

Bastard wing (alula)

Lesser upper wing-coverts

(Greater) primary coverts

Median upper
wing-coverts

PII Remicle (normally concealed)

Remicle (PII)
made visible

Greater upper
wing-coverts

Emarginated
outer web

Covert of P10

Tertial coverts

Covert of P9

P10

P9

P8

S P E C U L U M

P7

P6

MALLARD WING FROM ABOVE

P10

P5

S1  P1  P2  P3  P4

P9

S2
S3
S4
S5
S6

S7

S8

S9

S10

Primaries
(remiges)

Tertials
(innermost
secondaries)

Secondaries
(remiges)

Marginal under wing-coverts

Lesser under wing-coverts

Median under wing-coverts

P10

Axillaries

P9

P8

Greater under
wing-coverts

P7

P6

P5

Tertials

P4

P3  P2  P1  S1

S2
S3
S4
S5
S6

S7

Greater under primary coverts

S8

Median and lesser under primary coverts

S9

S10

MALLARD WING FROM BELOW

STRUCTURE. The following points are covered: shape of wing; number of primaries and secondaries; wing formula (expressed as differences from tips of primaries to tip of wing); shape of tail, bill and leg; proportions of toes; and other structural peculiarities. Primaries are numbered from the carpal joint outwards (*descendantly*) as has become common practice. Individual primaries are indicated by p and a number, p1 being the innermost, p10 the outermost functional primary in most species. The vestigial remicle is found distally from p10 and is interpreted as a reduced p11 (p12 in some groups). Sometimes, it is considered a modified covert (Stresemann 1963*b*). The secondaries are numbered from the carpal joint inwards, abbreviated as s1, s2, etc. The tail-feathers are numbered from the central one outwards: t1, t2 etc.

GEOGRAPHICAL VARIATION. The general nature of the geographical variation is indicated (even where no formal subspecies are recognized) and differences between the recognized subspecies are summarized. Reasons are given for rejection of subspecies accepted by other authors.

## GLOSSARY

Many of the terms used in the Plumages and associated sections are shown in the diagrams on pp. 32–33; others are given below.

ADULT. In practice, defined as birds in plumage no longer changing with age.

AIGRETTE. Loose, fine ornamental feather with free barbs.

BAR. Feature of colour pattern oriented transversely on feather.

BARBULES. Tiny branches of barbs of feather.

BODY FEATHERS. All contour feathers with exception of tail-feathers and feathers of wing.

BOOTED TARSUS. Covered with single horny sheath.

BRISTLE. Stiff, hairlike feather usually with a few barbs at base of shaft.

BULLA. Tymphanic element of the SYRINX when it has a bubble-like appearance (plural: bullae).

CARPAL JOINT. Wrist joint.

CLINE. Series of populations of a species showing gradation in one or more characters from one end of its range to the other (hence CLINAL VARIATION).

CONSPECIFIC. Belonging to same species.

CONTOUR FEATHERS. All feathers with well defined webs belonging to outer cover of body, including flight-feathers and tail-feathers (opposite: DOWN FEATHERS).

DIASTATAXIS. Arrangement of feathers in wing in which 5th greater upper covert has no corresponding secondary (opposite EUTAXIS).

DIMORPHISM. The existence of 2 differing morphs.

DISTAL. Pertaining to part of feather, leg, wing, etc. which is farthest from body.

EUTAXIS. Arrangement of feathers in wing in which secondary corresponding to 5th greater upper covert is present (opposite DIASTATAXIS).

FENESTRA. Membranaceous 'window' in tracheal BULLA (plural: fenestrae).

FILOPLUMES. Fine, hairlike feathers with small tuft of barbs at tip (occasionally at other places along shaft).

FORM. Neutral term indicating an individual variant or taxonomic unit (e.g. species or subspecies).

GONYS. Ventral ridge of lower mandible from tip to forking of rami.

LAMELLAE. Small plates or scales.

LAPPET. Wattle at corner of mouth.

MORPH. One of 2 or more well-defined forms belonging to same population.

PAPILLA. Small, conical eminence.

PLUME. Type of ornamental feather.

POWDER-DOWN. Soft and friable down producing fine dust particles used in plumage care.

PROXIMAL. Pertaining to part of feather, leg, wing, etc. which is closest to body (opposite: DISTAL).

PTERYLOSIS. The way in which feathers are arranged on the skin, usually in feather tracts wearing contour feathers (pterylae), and featherless spaces (apteria) wearing at most only down.

RAMI, MANDIBULAR. Halves of lower jaw.

REMICLE. Vestigial outermost primary.

RETICULATE TARSUS. With numerous small scales, randomly arranged.

SCUTELLATE TARSUS. With one or more distinct rows of relatively large scales.

SHAFT-STREAK. Narrow streak of contrasting colour along shaft of feather.

SIGNIFICANT. Shown by statistical test as unlikely to be due to accident (said of difference between means of 2 or more samples).

STANDARD DEVIATION (S.D.) Statistical parameter estimating scatter around mean in a sample of biometric data. Theoretically 99% of sample lies within $2 \cdot 58 \times$ S.D. from mean, 95% within $1 \cdot 96 \times$ S.D.

STREAK. Feature of colour pattern oriented longitudinally on feather.

SYRINX. 'Voice box' of birds, lying at lower end of trachea.

TRACHEA. Windpipe.

VARIATION. Differences in any character between animals of the same species. The following broad types of variation may be recognized: individual (between individuals belonging to same population, same sex and age and studied at same season); seasonal; sexual; age; and geographical.

WING FORMULA. Configuration of tips of primaries relative to each other, expressed as distances of tip of each primary to tip of longest.

# ACKNOWLEDGEMENTS

The preparation of this volume has involved a major commitment of both money and time. The heavy financial burden, which has been vastly aggravated by the rampant inflation of recent years, could not have been contemplated without the generous help and understanding shown by the Delegates of the Oxford University Press, while, both in the early stages and during a later critical period, the assistance and helpful encouragement provided by the Pilgrim Trust were crucial. We are indebted also to the British Ornithologists' Union and the Ernest Kleinwort Foundation for grants at the beginning, and more recently to the Commission of the European Economic Communities, The Dulverton Trust, the Stichting P A Hens Memorial Fund, Fondation Européenne de la Culture, Sir A Landsborough Thomson, the Radcliffe Trust, the Royal Society for the Protection of Birds, and the Science Research Council.

Time has been given generously by many ornithologists throughout the world. First, by members of the Editorial Board, and we are grateful to the British Trust for Ornithology, the Wildfowl Trust, the University of Amsterdam, the University of Leicester, and the Zoological Society of London for facilitating their labours.

Secondly, valuable assistance was given to the Editors by D I M Wallace and C S Roselaar who prepared just under one-third of the sections on Field Characters and more than half of those on Plumages and related topics respectively, while Sir A Landsborough Thomson prepared some sections on Movements; these contributions are indicated by their intials at the end of the relevant sections. Treatment of all the complex aspects of the Voice sections was organized through a team of specialists. The input and handling of tape recordings obtained from the British Library of Wildlife Sounds and from collaborators in many countries (especially Sture Palmér and Sveriges Radio) was the responsibility of P J Sellar. Processing and analysis of the tapes in the form of sonagrams or melograms, and their comparative evaluation, were undertaken by Joan Hall-Craggs under the scientific direction of Professor W H Thorpe. Dr K E L Simmons advised on the functions and context of the particular utterances and provided many drafts and much essential material. Dr R W Warner handled the critical task of preparing voice descriptions and tying them in with the sonagrams.

Thirdly, by the distinguished panel of advisory editors, including, in addition to Professor W H Thorpe, Dr K Curry-Lindahl, Professor Jean Dorst, Professor Yu A Isakov, Professor Finn Salomonsen, and Sir A Landsborough Thomson.

Fourthly, by the correspondents who, as stated earlier, so generously provided much of the basic data on status, distribution, and population in respect of the various countries, as well as giving help and advice in many other ways:

ALGERIA  E D H Johnson
AUSTRIA  Dr H Schifter
BELGIUM AND LUXEMBOURG  H Wille, Comte L Lippens
BULGARIA  Dr S Dontchev
CHAD  J Vielliard
CZECHOSLOVAKIA  Dr K Hudec
DENMARK  Prof. F Salomonsen, T Dybbro
EGYPT  Dr G E Watson
FINLAND  Prof. L von Haartman
FRANCE  Dr C Erard, C Jouanin, L Yeatman, Dr F Roux, Dr J-Y Monnat
GERMANY (EAST) DDR  Prof. E. Rutschke
GERMANY (WEST) FDR  Dr E Bezzel
GREECE  W Bauer
HUNGARY  Dr I Sterbetz, Dr A Keve, Dr L Horváth
ICELAND  Dr F Gudmundsson
IRAQ  P V Georg Kainady
IRELAND  Major R F Ruttledge
ISRAEL  Prof. H Mendelssohn
ITALY  Prof. S Frugis
JORDAN  I J Ferguson-Lees and P A D Hollom
KUWAIT  S Howe
LEBANON  Dr H Kumerloeve
LIBYA  G Bundy and J Morgan
MALI  Dr J M Thiollay
MALTA  J Sultana, C Gauci
MAURITANIA  Abbé R de Naurois
MOROCCO  The late K D Smith, Dr P Robin, M Giraud-Audine, J Pineau, J D R Vernon
NETHERLANDS  Dr J Wattel, C S Roselaar
NIGER  Dr J M Thiollay
NORWAY  Prof. Svein Haftorn
POLAND  Dr A Dyrcz and L Tomiałojć
PORTUGAL  M D England, J F Bugalhao
RIO DE ORO  Dr J A Valverde
RUMANIA  M Talpeanu
SPAIN  Prof. F Bernis
SWEDEN  Dr S Matthiasson
SWITZERLAND  W Thönen
SYRIA  Dr H Kumerloeve
TUNISIA  M Smart
TURKEY  Ornithological Society of Turkey
YUGOSLAVIA  V F Vasić
USSR  Prof. Yu A Isakov

Fifthly, we are deeply grateful to Dr H M S Blair, Miss D M Keating, and M Wilson who gave us valuable help with translations.

Sixthly, we thank those who made available photographs, sketches and published material on which the drawings by Robert Gillmor illustrating the Social Pattern and Behaviour sections were based. Their names are given at the end of the relative sections. We also thank Cornell University Press for permission to use for this purpose the illustrations in their publication *Handbook of Waterfowl Behavior* (1965) by P A Johnsgard.

Finally, we are indebted to the following ornithologists, professional and amateur throughout the world, who have assisted us in ways too diverse to specify in detail, though credits are given in the text as appropriate:— A Aasgaard, E Allsopp, A B Amerson, A Anderson, J Arkell, G L Atkinson-Willes, Dr R S Bailey, Dr H Bandorf, Dr D A Bannerman, Dr E K Barth, M A S Beaman, Ms G Bell, Dr S-A Bengtson, C W Benson, H Berry, D Blaker, A A Blok, Dr Z Bochenski, Dr P Bondesen, J Boswall, Dr W R P Bourne, H Boyd, Col C L Boyle, P L Britton, R Broad, J Brock, Dr L H Brown, the late Prof. E Brun, Dr R L Bruggers, Prof. H Burns, Dr J Cadbury, Dr L H Campbell, E Carp, Dr C Chappuis, D A Christie, G Collis, J Cooper, P E Davis, Dr S R Derrickson, J Dufour, Prof. G M Dunnet, A Dupuy, N Elkins, Dr C C H Elliott, Sir H F I Elliott, Dr S K Eltringham, A J Erskine, R D Etchécopar, Miss M E Evans, Dr E Fabricius, Dr J Fjeldså, Dr L Ferdinand, Dr C Ferry, Dr H Fischer, Prof. V E Flint, Miss E Forster, Dr C H Fry, C Fuller, R Furness, Major M D Gallagher, Dr V M Galushin, Prof. A Gardarsson, R W Genever, Dr P Géroudet, Dr F B Gill, Dr F Goethe, Dr F G Gooch, D Goodwin, J Gordon, W W H Gunn, J A Hancock, E Hardy, Dr M P Harris, Drs L Hartsuijker, Dr F Haverschmidt, Dr H Hays, T Hedley Bell, U Hirsch, D T Holyoak, Prof. C E Huntington, I Inglis, A Ingolfsson, H Insley, R J J Iribarren, J Jack, Prof. P A Johnsgard, A R Johnson, E T Jones, S Jonson, Dr M P Kahl, Dr J Kear, B King, J Kirby, P J Knight, Dr L Kooh, Drs P P A M Kop, D E Ladhams, Dr D A Lancaster, Dr G Lid, Miss P Lind, M A Macdonald, D McChesney, J A McGeoch, Prof. F McKinney, L B McPherson, W M Makins, Ms L S Maltby-Prevett, Dr S Marchant, J R Mather, Prof. G V T Matthews, Miss V Maxse, Dr G F Mees, O J Merne, Prof. A J Meyerriecks, Dr H Milne, P J Morgan, A Mrugasiewicz, Dr D Munteanu, Dr R K Murton, Prof. M T Myres, R Nakatsubo, Dr J B Nelson, J Neuschwander, Dr I Newton, the late Dr G Niethammer, Dr L Nilsson, Prof. D F Owen, Dr M Owen, Dr Ralph S Palmer, Sture Palmér, Dr K Paludan, Dr I J Patterson, Dr C J Pennycuick, F W Perowne, Dr C M Perrins, Dr A Petersen, Drs E P R Poorter, R F Porter, R J Prytherch, Dr P Radford, Miss W Reade, R Ridgway, H Ringleben, J-C Roché, P Round, Prof. R A Ryder, L Saari, A K M St Joseph, D I Sales, Dr E G F Sauer, Dr D A Saunders, C D W Savage, Miss M Schommer, Th Schmidt, Prof. E Schüz, D A Scott, Sir Peter Scott, P Shelton, L C Shove, Prof. W R Siegfried, J B Sigurdsson, E Simms, M Sinclair, Dr S Sjölander, J J Smit, Dr R I Smith, Mrs B K Snow, Dr D W Snow, A Soper, A Stagg, Miss J Stannard, Prof. R W Storer, Dr A Studer-Thiersch, M K Swales, R L Swann, Dr A Tamisier, Dr L Tickell, O Ungev, Dr E K Urban, Dr G F van Tets, Dr G Vauk, Z Veselovský, Dr W J M Vestjens, J Vielliard, A Vittery, S Wahlström, A Walker, Dr J Warham, Dr W E Waters, Dr U Weidmann, Dr M W Weller, Col J D Wellings, A Wetmore, C M N White, S White, J P Williams, Dr A Zahavi, A Zino.

As so many have given their time and skill so generously, it seems appropriate to repeat here that the entire project has been planned on a non-profit basis. It is intended that any surplus of royalty revenue will be devoted to further ornithological research in the west Palearctic region, especially in those countries with limited finances available for such a purpose. In due course, it is proposed to set up a special and independent trust fund for this purpose.

**CITATION**
The editors recommend that for references to this work in scientific publications the following citation should be used: Cramp, S and Simmons, K E L (eds.) (1977) *The Birds of the Western Palearctic*, Vol. 1.

# Order STRUTHIONIFORMES

Huge, flightless, cursorial birds. 3 living families Casuariidae, Rheidae, and Struthionidae, placed in separate suborders by some (Sibley and Ahlquist 1972), but also combined with Tinamiformes and Apterygiformes in single order Palaeognathiformes (Cracraft 1974), or split into 3 separate orders (Storer 1971). Only Struthionidae in west Palearctic. Formerly resemblances between these groups often interpreted as due to convergence, but currently considered—from morphological (Bock 1963; Cracraft 1974), biochemical (Sibley and Frelin 1972), behavioural, and parasitological evidence—to have had common origin. Relationships to other birds unknown, but generally thought to be derived from flying ancestors.

# Family STRUTHIONIDAE ostriches

One living species in single genus *Struthio*. Now only marginally west Palearctic, where almost extinct owing to human persecution. Body broad and heavy, neck long, head small, thigh musculature strongly developed. Wings reduced, consisting of many loosely structured feathers, functioning in display; flightless. No oil gland. Aftershaft absent. (For further details see *Struthio camelus*.) General behaviour characters include: direct head-scratching; true yawning; dissipation of heat by panting with open mouth, exposing bare skin below bill; resting with neck upright and locked in sigmoid position, as well as lying with neck flat on ground; elaborate, often communal dusting in sand, etc. (no bathing in water).

## *Struthio camelus* Ostrich

Du. Struisvogel    Fr. Autruche    Ge. Strauss
Ru. Африканский страус    Sp. Avestruz    Sw. Struts

PLATE I
[after page 38]

*Strutio Camelus* Linnaeus, 1758

Polytypic. Nominate *camelus* Linnaeus, 1758, west and south Sahara and adjacent Sahel region from Senegal to Sudan, formerly north to Morocco, Algeria, Tunisia, Libya, and Egypt; *syriacus* Rothschild, 1919, probably extinct (see Distribution), formerly Syrian Desert and north Arabia. 3 extralimital, including *australis* Gurney, 1868, South Africa south of Cunene and Zambesi Rivers.

**Field characters.** Largest living bird, taller than man and $1\frac{1}{2}$–2 times heavier: height of ♂ 210–275 cm to top of head (up to 130–140 cm to back); ♀ 175–190 cm. Flightless. Small head and long, mostly bare neck; long, thick, and powerful legs with bare thighs and only two toes; dark body; and whitish curled plumes forming ends of wings and tail. Sexes easily separable; no seasonal differences except ♂ bill and tarsus redder in breeding season. ADULT MALE. Apart from larger size, has jet black body, white curling primaries and tail and, in Saharan and some other races, reddish neck and thighs. ADULT FEMALE. Brown body with pale feather edgings, off-white or brownish-white wings and tail, and more grey-brown down on neck. JUVENILE. Resembles ♀ but more uniformly grey-brown, with less distinctly coloured wing tips and tail and still more down on neck.

Unmistakable, even a chick distinguishable by its thin, unfeathered neck from Arabian Bustard *Ardeotis arabs* of similar size.

In west Palearctic, confined to broken semi-desert and wadis where nomadic over considerable areas. Usually wild, difficult to approach closer than 100 m and, although not attempting concealment, surprisingly easy to overlook; unable to fly, but evades danger by running at speeds of up to 50–70 km/h (even month-old chick can run at 50 km/h).

**Habitat.** In west Palearctic, remnant in hot fringing desert-steppes of west Sahara, independent of forest or any other continuous plant cover, and avoiding savanna where trees and shrubs frequent. Mainly lowland, below about 100 m except where following up dry wadi beds towards mountains. Habitat generally unvaried and continuously although nomadically occupied at low density throughout year; shared with specialized African semi-desert fauna and flora, wholly unvegetated tracts being avoided. Valverde (1957) distinguished 2 favoured habitats, often neighbouring: (1) dry beds of wadis within broad valley bottoms of uneven terrain (where can withdraw inconspicuously from approaching danger), with sparse cover of such favoured plants as *Nucularia*

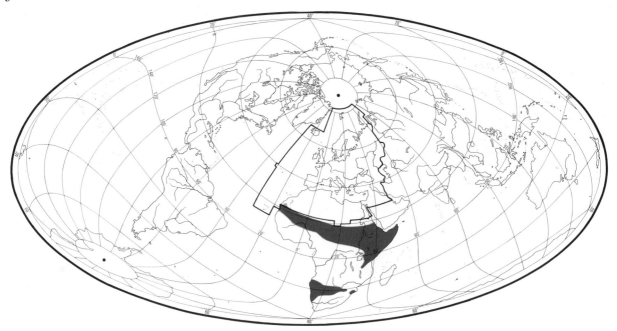

*perrini*; and (2) zone of Sahelian desert-savanna plains flanking southern edge of bare desert, where gentle slopes carry thistles and coarse grass *Andropogon foveolatus*, with such associates as *Panicum*, *Crotalaria*, and *Citrullus*. Latter habitat shared with local representatives of larks *Mirafra*, *Eremophila*, and *Eremopterix*. In Tunisia, eggshell remains indicate decidedly richer steppe-like habitats used before recent pressures brought local extinction. Avoids regions modified or frequently disturbed by man. Unique among west Palearctic birds in having no aerial, aquatic, forest, mountain, or man-made component in habitat.

**Distribution.** Marked decline throughout west Palearctic range in last 120 years or so. Nominate *camelus* once widespread in North Africa. Eggshells, not fossilized, found in Tunisia and many parts of Sahara, while species numerous in high plateau of Algeria and Morocco to mid-19th century, surviving in some areas to 1900 (Heim de Balsac and Mayaud 1952). Fossil eggshells found Lanzarote, Canary Islands (Sauer and Rothe 1972). Eggshells also found over much of desert and sub-desert region of northern Libya (Toschi 1969) where probably, as in Egypt, extinct 19th century (Meinertzhagen 1930; Vaurie 1965). In western Sahara, small numbers still breeding in 1950s (see full survey in Valverde 1957); 2 pairs nested 1963 and odd pairs may have bred 1970–2 in Rio de Oro and near borders of Algeria and Mauritania (J A Valverde). In Middle East, *S.c. syriacus* fairly common to 1914 in desert areas north to *c*. 33½°N and east to deserts west of Kuwait and Bahrain, but extinct by 1941 and exterminated in Empty Quarter of Arabia by *c*. 1900

(Meinertzhagen 1954); however, dying bird reported brought down by floods south-west Jordan, February 1966 (IUCN *Red Data Book*, 1967).

Accidental. In recent years occasional Tibesti, northern Chad (Malbrant 1954; Malzy 1962), and perhaps northern Niger (JMT).

**Population.** Now almost extinct in western Sahara, although a few individuals probably still exist (J A Valverde). Hunting pressures (and to lesser extent collection of eggs), which have caused extinction over large areas, make survival of *S. c. camelus* unlikely without rigorous protection.

**Movements.** Nomadic. Arid biotope necessitates wanderings for food or water, most pronounced in areas affected by drought; such movement unlikely to be seasonal in view of irregular rainfall over desert and subdesert regions. In Rio de Oro, occasionally wanders north to Draa region, while 82 seen at Negyir in April 1955 after virtual disappearance from Sbayera (over 200 km to north-east) where common 2 months earlier (Valverde 1957). In Sahel zone of Mali and Chad, generally restricted to wadis, but more northward penetration into Sahara after rain with retreat on return of drought (Malbrant 1954; Malzy 1962). No information on movements of former Arabian population (*S. c. syriacus*).

**Food.** Omnivorous, though chiefly daytime browser on plant materials including shrubs, succulents, and seeds. Will travel great distances in search of food and water (Bouet 1955; Valverde 1957; Sauer 1971).

PLATE 1. *Struthio camelus* Ostrich (p. 37): **1** ad ♂, **2** ad ♀, **3** juv, **4** downy young. (PB)

PLATE 2. *Phoenicopterus ruber* Greater Flamingo (p. 359): **1** ad ♂, **2** ad ♀, **3** imm, **4** juv, **5** nestling (10–15 days). *Phoenicopterus minor* Lesser Flamingo (p. 366): **6** ad ♂, **7** ad ♀, **8** imm, **9** juv. (PS)

PLATE 3. *Gavia stellata* Red-throated Diver (p. 43): 1 ad breeding, 2 ad pre-breeding moult, 3 imm, 4 juv, 5 downy young 1st down. (PB)

PLATE 4. *Gavia arctica* Black-throated Diver (p. 50): 1 ad breeding, 2 imm, 3 juv, 4 downy young 1st down, 5 downy young 2nd down. (PB)

In Spanish Sahara, based on observations and 3 stomach analyses, mainly seeds of *Crotalaria saharae*, roots, leaves, and flowers of an unidentified thistle, larvae of locust *Schistocerca gregaria*, and fruits of *Citrullus colocynthis*; less important were plants *Astericus graveolens*, *Andropogon foveolatus*, *Salvia aegyptiaca*, and *Zilla spinosa* (Valverde 1957). Order of preference in Nigeria: small shrub *Cassia*, juicy leaves of *Oxystelma bornouense* (parasitic creeper on *Acacia*), leaves of ground creeping gourd *Cucumis* and of small tree *Moerua rigida* (Buchanan 1921). Fruits of fig *Ficus carica* (Archer and Godman 1937) and of *C. colocynthis* and shoots of *Mimosa* (Bouet 1955) also recorded. In south-west Africa, desert grass *Stipagrostis uniplumus* and succulent *Bohenia* taken (Louw 1972). Animal materials include insects, especially termites and locusts, small tortoises, and lizards (Archer and Godman 1937; Dekeyser 1957; Meinertzhagen 1954; Sauer 1971). Most water needs, especially of young, derived from succulent plants, but free water normally essential and, when possible, drinks daily and copiously (Sauer 1971). Predilection for picking up small bright objects well known: one stomach analysis (captive bird) revealed nearly 120 g of scrap iron, wire pieces, belt buckles, coins, nails, and even the enclosure padlock key (Dekeyser 1947).

**Social pattern and behaviour.** Virtually unknown in Palearctic races. Present account based on material supplied by E G F Sauer from extralimital observations on *S. c. australis* (e.g. Sauer 1971, 1972a; Sauer and Sauer 1959, 1966), mainly in Namib Desert (South West Africa). Social life probably basically similar to that of indigenous races, as indicated by casual records in Barth (1857–8), Greenway (1958), Heim de Balsac and Mayaud (1962), Meinertzhagen (1954), Niethammer (1971), Schiffers (1967).

1. Highly flexible opportunist breeder, modifying organization according to varying and unpredictable conditions in largely arid habitat. Gregarious; local population made up of distinct social units based on family, from which super-families (i.e. adults plus young of various origin), large herds, mating groups, and flocks of immatures derive. For most of year in family or super-family groups usually, where common, up to several dozen and occasionally to *c.* 100 of both sexes. Social rank-order in each group, with adult ♂♂ and major ♀♀ (see below) dominant. Neighbouring flocks assemble, often daily, in large herds up to several hundred in communal areas centred round water. Each flock occupies own undefended home-range, with roost sites and feeding grounds, and avoids other flocks except at communal centres; radius of daily activity 10–20 km or more depending on availability of food and water. Nomadism and migration often induced by drought. BONDS. Individuals maintain social contact with others at all times of year; associations, in smaller groups at least, based on individual recognition. During breeding, monogamous pair-bond may prevail in isolated populations with only a few adults; otherwise commonly polygynous with one ♂ maintaining harem of 2–5 (often 3) ♀♀. Pair-formation frequently initiated by major ♀; ♂ often remains passive when approached by minor ♀♀ leaving major ♀ to accept into group or not; these usually driven off once contribution made to joint clutch. Bond, especially between ♂ and major ♀, sustained by pronounced nest-site tenacity and, in older birds, often

maintained outside breeding season. Minor ♀♀ that have left harem sometimes entice ♂ which may break bond with major ♀; if latter successfully hatches chicks on own, may later associate with another ♂ who then fosters brood. Eggs and young normally cared for, however, by father and major ♀, bond between parents and young usually lasting up to 1 year or until young abandoned or driven off at onset of new reproductive period. Under exceptional environmental conditions producing sudden rich food supply, family group may break up when chicks only 2–3 weeks old; adults become sexually active again and highly aggressive towards all juveniles. Abandoned and lost chicks frequently adopted; multiple adoptions lead to formation of large, stable super-family flocks of up to several hundred juveniles of various ages. These often led by 2 ♂♂, though larger number of adults of both sexes and also sub-adults often join flock as it grows, while original parents (especially ♀) sometimes leave. When abandoned by guards, juveniles remain in roving parties up to several hundred strong and seek to attach themselves to any sub-adult or family party encountered. Large assemblies of unguarded immatures tend to break into smaller units. Immatures usually maintain close contact among themselves until pair-formation, which may lead to heterosexual engagement groups, polygamous in composition, as early as 1st or 2nd year – although do not breed until 3 or 4 years old. Chicks fledged in small numbers outside current breeding season frequently treated as outcasts and live solitarily. BREEDING DISPERSION. At start of new breeding period, sexes segregate into temporary flocks, up to about 50 ♂♂ frequenting neighbourhood of equal or often smaller numbers of ♀♀. Subsequently, pairs of polygamous mating groups leave assemblies to set up nest-territories; later still, solitary adults, temporarily isolated from mates while departing from and arriving at communal area, territory or nest, also numerous. Nests commonly several km apart, usually at maximum calling-distance of ♂♂. Territory restricted to nest-site and surroundings; to 50 m in sheltered terrain but beyond 800 m in open plains. Prior to incubation, trespassers chased from territory, but later tolerated by sitting bird even when passing close to nest. Undefended home-range surrounds or adjoins territory and extends to communal area. Family leaves territory when chicks some 3–5 days old. Local population also contains many non-breeders throughout season. ROOSTING. Loosely communal at traditional roosts, preferably in sandy places, sleeping within sight and voice contact of others of flock at distances of from several to over 30 m. Strictly diurnal; cease activity with brief preening bout some 40 min after sunset, rising about 40 min before sunrise. Commonly orient downwind with head raised during light sleep but on ground during relaxed sleep, either straight forward or bent to side of body. During incubation, off-duty adult spends night mainly outside territory, occasionally within 10–20 m of nest. For nocturnal booming see Voice. Small chicks sleep in contact with parent.

2. Much social interaction within flock; vestigial wings and bushy tail play important part in display. Adult ♂♂ and major ♀♀ lead, e.g. by initiating united movement or communal dust-bathing; at danger, ♂ herds mates as well as others of flock in own fleeing direction by elaborate Flash-displays, zigzagging towards, away from, and from behind them while running at full speed, throwing wings high, and often calling. ANTAGONISTIC BEHAVIOUR. Aggression mainly means of self-advertisement of dominant individuals within group and expression of social rank; also due to sexual rivalry at pair-formation among ♀♀ as well as ♂♂, while disinterested individual will respond with threat or attack during courtship (see below). In community centres, adult guards frequently defend chicks against aggressive conspecifics.

A

Hostile encounters between members of same or different flocks usually brief, consisting of threats in form of calls—Hissing, Snorting, or Boo-calling—or Upright-display (see A). Aggressor will also peck or, with forward and downward stroke, kick rival; attacks can lead to short chases. Threatened birds run off or assume Subordinate-posture in appeasement, holding head low on U-shaped neck with wings and tail pointing downwards (see B). More prolonged encounters usually have sexual overtones and consist of Hiss-snort-and-kick displays. In homosexual encounters between ♂♂, hostility may be blocked by sudden show of courtship (see below) given as appeasement. GROUP-DISPLAY. In all-♂, pre-nesting flock (see above), ♂♂ perform ritualized hostile behaviour as group in front of ♀♀ during pair-formation. They chase and circle one another. Hiss, Snort, and Boo-call, Wing-flap by raising one or both wings high above back, and Tail-raise. Position of tail indicates mood and social rank; highest, and even bent over back, in dominant ♂.

COURTSHIP. Pair-formation occurs in communal area; often initiated by major ♀ who usually completes moult and comes into breeding condition ahead of ♂. ♀ postures in front of ♂, very erect and raising closed wings at shoulder while urinating and defaecating. Major ♀♀ will also incite ♂♂ to fight and perform group-display. ♂ also initiates pair-formation at times, but response to ♀'s overtures usually delayed until bright colours fully developed on bare parts. ♂ then behaves similarly, and also engages in other displays while stationary or circling partner. These include Wing-flagging and Tail-raising, as in group-display, also Wing-sweeping and Wing-beating with alternate and simultaneous movements of wings respectively, and Penial-display in which fully extruded, erect and enlarged penis (which, like cloacal wall, becomes bright red) exhibited and swung from side to side while ♂ ceremonially urinates and defaecates.
MATING BEHAVIOUR. Pre-copulatory courtship often initiated by ♂'s singling out one of mates and driving her to secluded area where they perform synchronized Ceremonial-feeding; with heads usually close together near ground, they make rapid but superficial, nervous pecking movements in unison. This leads to Rocking-display when ♂ makes symbolic nest-building movements in sandy spot, dropping to ground and beating wings while rocking from side to side, twisting neck in continuous 'corkscrew' action and repeatedly uttering four-syllable Courtship-song. ♀ remains near or walks round ♂ with slow gait, wings and tail drooping and head low (O-posture); when she drops to ground, ♂ mounts and copulates. Short, synchronized bout of self-feeding by both follows copulation. RELATIONS WITHIN FAMILY GROUP. Social contact between chicks and adults, and among chicks themselves, begins with Pre-hatching calls by chicks while still in unpipped eggs; first given one or more days before hatching, triggering and synchronizing hatching efforts of siblings and alerting adults. When dry, chicks venture from under parent into nest surrounds, learning to recognize one another, while flocking and crouching together and becoming socially imprinted on parents. Social contact enforced by Distress- and Contact-calls. When endangered by carnivores or raptors, adult guards protect chicks by attack or Distraction-display (see Breeding), while chicks scatter, crouch, and freeze until danger passed or guard signals all-clear; meanwhile, ♀ may decoy chicks.
(Figs A–B from photographs by E G F Sauer.)

**Voice.** Based on material supplied by E G F Sauer for extralimital *S. c. australis*. Wide range of vocalizations, some minutely differentiated, still at preliminary stage of analysis; include hoarse and guttural, occasionally clear, calls of adults, adult hissing, snorting, and booming calls, and complex melodious calls of chicks. Non-vocal sounds, produced in antagonistic situations, include beak-snapping and stomach-rumbling.

Most common call of adults muted 'boo', varying from hoarse to clear, uttered in antagonistic situations—threat, attack, escape—and during distraction-display.

Booming 'song' of adult ♂♂ mostly 4-syllabic 'boo boo booh hoo', often repeated several times at short intervals, varying in length and temporal pattern between individuals. Serves at least 3 purposes: establishment of territory, courtship, and maintaining distance from other animals, particularly when disturbed at night. When establishing territories, ♂♂ boom day and night, with peaks 1–2 hours after sunrise and sunset. Booming carries

B

several km, ♂♂ frequently responding to each other. Nocturnal booming also triggered by predators, and by noises (e.g. thunder).

CALLS OF YOUNG. Melodious Contact-call of chick first uttered 1 to several days before hatching. Establishes and maintains contact between parents and siblings. Chicks also produce distress and fear notes.

**Breeding.** SEASON. Most eggs October–November in north-west Africa, but laying can extend to February, with some dependence on timing of rains (Valverde 1957; Heim de Balsac and Mayaud 1962). Eggs found October in Sudan (Dekeyser 1957). SITE. On ground in sandy places; in open or sometimes in natural shelter such as dry river bed. Re-used in successive years. Nest: shallow depression, often with low wall formed from excavated material; *c.* 3 m overall diameter. Building: mostly by ♂, sometimes aided by ♀♀, scraping with feet then moulding with body. EGGS. Ovate, smooth and shiny, though can be rough and pitted; creamy white. 159 × 131 mm (142–175 × 120–145), sample 48; calculated weight 1600 g (*camelus*) (Schönwetter 1967); 755–1618 g, sample not given (*australis*) (Sauer and Sauer 1966a). Clutch: each ♀ lays 4–8 eggs; usual number of ♀♀ laying in nest 3 or 4; total clutch size 20–25 (16–60) (Sauer and Sauer 1966a). One brood. Eggs laid at 2-day intervals. INCUBATION. 39–42 days. Begins on completion of laying. Normally by one ('major') ♀ during day and ♂ by night, but exceptionally other ('minor') ♀♀ incubate for short spells (Sauer and Sauer 1966a). Hatching synchronous. Eggshells left in nest. YOUNG. Precocial and nidifugous; self-feeding. Cared for by 2 parents, usually but not always ♂ and 'major' ♀. Both parents, but especially ♂, have elaborate Distraction-display running to and fro and dropping to ground, flapping wings. FLEDGING TO MATURITY. Flightless, but full grown at *c.* 18 months. Parents accompany young for *c.* 12 months or until next breeding season. Mature at 3–4 years. BREEDING SUCCESS. No data.

**Plumages** (nominate *camelus*). ADULT MALE. Skin of head with stiff, white, hairlike feathers eyelashes dusky; neck with sparse white down. Ring of white feathers round lower neck, rest of plumage black. Tail and wings white. ADULT FEMALE. Neck more densely feathered than ♂, with grey down. Plumage brown-grey with white feather-edges; occasionally some white feathers in wing or tail DOWNY YOUNG. Down of crown sandy brown, neck paler, throat off-white. Pattern of 8 longitudinal black stripes, continuous on crown and hind-neck, broken into series of black

spots on sides and front of neck. Upperparts of body with mixture of sandy and dark grey down feathers, tipped by several prolonged, flattened and twisted barbs, either black, orange-brown or whitish, giving appearance of pile of straw (Jehl 1971). Breast pale sandy, belly off-white. At *c.* 1 month, juvenile feathers push out down. Juvenile feathers on back full-grown at 11 weeks, but remains of down retained much longer (Brinckmann and Haefelfinger 1954). JUVENILE. Like ♀, but down on neck denser, and some remains of down present elsewhere.

**Bare parts.** ADULT MALE. Iris brown. Bill horn-yellow, lower mandible tinged red. Skin of head, neck, flanks and thighs pink-flesh; bald patch on crown pale brown. Tarsus with flesh-coloured, yellow-margined shields (Salvadori 1895; Bannerman 1930). Neck in breeding season fiery red, and rest of bare skin brighter (Reichenow 1900–1). In *australis*, head and neck blue-grey; as courtship developes, red pigment appears on face; dorsal surface of shins, and toes; at about same time, cloacal wall and penis turn from pale to bright red (Sauer and Sauer 1966b). ADULT FEMALE. Bare skin of head, neck, flanks, and thighs pale brown; bald patch on crown as ♂; tarsus with horn-coloured shields, darker on toes. DOWNY YOUNG. Iris dark brown. Bill flesh, darker on upper mandible. Leg dirty flesh (Reichenow 1900–1).

**Moults.** Little information for wild birds. Adult (Rio de Oro) in full moult April–June, after breeding in winter and early spring (Valverde 1957). Captive juveniles moulted into adult plumage at 9–11 months (Brinckmann and Haefelfinger 1954).

**Measurements.** Height of standing bird *c.* 2·75 m. Bill from cere 62–84, from gape 120–143; tarsus 450–530, average of 7 adults 490. *S. c. syriacus* smaller: tarsus 390–465, average of 6 adults 420. (Hartert 1921–2; Bannerman 1930; Vaurie 1965.)

**Weights.** Range of 20 semi-domesticated adult *australis* 63–104 kg (Schmidt-Nielsen *et al.* 1969).

**Structure.** Largest living bird. Flightless. Feathers, including tail and wing, of loose structure without definite webs. Bill broad and flat, head small, neck elongated. Legs long and powerful, clad with horny shields in front; 2 toes, lateral much smaller and without claw.

**Geographical variation.** Involves colour of bare skin, occurrence and form of bald patch on crown, size, and structure of egg-shell. *S. c. syriacus* similar to nominate *camelus* but smaller (Rothschild 1919). Population from Rio de Oro separated as distinct subspecies *spatzi* by Stresemann (1926) and considered to be characterized by small size and pore pattern of egg-shells. Pores in nominate *camelus* solitary and diffusely scattered, in *spatzi* clustered and mixed with irregular grooves (Schönwetter 1927). However, pore patterns may be individually variable (Sauer 1972b), so *spatzi* treated as synonymous with nominate *camelus* (see also Vaurie 1965).

# Order GAVIIFORMES

## Family GAVIIDAE  divers

Medium to large, foot-propelled diving birds. Sole family in order; 4 closely related species in single genus *Gavia*. Relationships with other avian orders remote and not fully established. Gaviiformes formerly considered close to Podicipediformes (grebes), but similarities due to convergence (Stresemann 1927–34). Skeletal features (Storer 1956) and egg-white proteins (Sibley and Ahlquist 1972) suggest affinities with Charadriiformes (see also Storer 1971). All 4 species breeding in west Palaearctic. Body elongated, legs set far back, neck short. Wings narrow, relatively small, and strongly pointed. 11 primaries, attenuated towards tip; p10 longest, p11 minute. At least 23 secondaries; diastataxic. Tail short, 16–20 feathers, largely hidden by tail-coverts. Bill long, straight, and pointed; nostrils narrow and slit-like. Tarsus laterally flattened, 3 front toes connected by webs; hind toe small and elevated; nails small. Oil gland feathered. Feathers with aftershaft. Sexes similar in plumage, ♂ averaging larger. Breeding plumage strikingly patterned; non-breeding dull, grey and white. Plumage dense and compact, generally harsher than in Podicipedidae, less downy; soft and velvety on head and neck. 2 moults per year; flight-feathers shed simultaneously. Young precocial and nidifugous. 2 downy plumages, dusky above, paler below; both develop from same follicles as later contour feathers, adhering on top of these. Juvenile and immature plumages generally similar to adult non-breeding. Fully developed adult plumage acquired in 3rd calendar year.

Northern Holarctic; circumpolar. Ecological counterparts in higher northern latitudes of Podicipedidae (grebes). Both make minimal use of terrestrial habitats, breed almost exclusively on fresh waters (usually standing), and are governed in habitat by availability of small or medium fish or other aquatic life, using varying expanses of water whence such food can be readily obtained by shallow or medium dives. However, Gaviidae show substantial ecological distinctions. Largely arctic and subarctic breeding range often forces vacation of summer habitat by early autumn, and may prevent re-occupation until following June, thus involving long transit stays on neighbouring unfrozen water, often marine. More readily use inshore waters of exposed coasts, or even move offshore, in winter and spring. Whereas grebes rarely breed on waters incapable of fully sustaining adults and young, some divers fly long distances for food to more prolific waters than those where they nest. Although southern breeding range overlaps northern limits of certain grebes in some areas, members of both groups only rarely share same water. Breeding habitats of divers more seasonally changeable, often deeper and less eutrophic, and much less often in forest or woodland. Unlike grebes, commonly avoid emergent vegetation. Good, all-round visibility for approaching danger from land appears to be a critical factor in many cases. Fly much more frequently and regularly, not only for escape and feeding but also during social interactions. Distribution in more northern areas often inadequately known, especially overlap between Great Northern *G. immer* and White-billed *G. adamsii*. Little evidence of recent changes in ranges or numbers within west Palearctic. Migratory and dispersive there, making some major overland passages. So far as known, movements mainly nocturnal.

Feed principally on fish, caught mainly by underwater pursuit. Not highly gregarious; occur singly, in pairs, or in small flocks outside breeding season, mostly on sea. Breed mainly in dispersed pairs, but will gather in small flocks in breeding area, especially in spring. Strictly territorial throughout summer; in larger species, especially, pairs mostly solitary but smaller species prone to form communities of adjacent territory holders. Red-throated Diver *G. stellata* habitually feeds away from breeding water, ferrying in food for young; same system sometimes adopted by *G. arctica* but never by 2 largest species which find much, if not all food within territory. Monogamous pair-bond; probably life-long and maintained over winter, pair using same water, territory, and nest-site annually. Advertisement and defence of territory mainly by calls. Significance of elaborate mutual water-displays and social gatherings during breeding season disputed. According to, e.g. Dunker (1974) promote pair-formation and maintenance of pair-bond, but displays interpreted by Sjölander and Ågren (1972) as mainly territorial; shown to intruding birds, including those feeding in flocks, with new pairs formed in territory, not in winter or spring flocks. Territorial displays elaborate in *G. stellata*, less so in *G. arctica*, and least well developed in *G. immer* and *G. adamsii*; actual fights rare but vicious. Courtship simple and apparently performed mainly by new pairs; older pairs display little and tend to start nesting immediately after occupation of territory. Copulation ashore at nest-site or platform; no conspicuous displays at or near site. Voice highly distinctive; of remarkable carrying power and often of musical quality. Uttered on water and in flight, often by 2 or more birds responding to one another, dominating all other sounds over tundra or taiga, both by day and night. Silent outside breeding season. Comfort behaviour apparently much as in other waterbirds; often preen ventrally by rolling over on side in water, exposing underparts.

Bathing spectacular at times, involving rolling, somer-saulting, and diving; interpreted as display by some early writers. Other general behavioural characters include direct head-scratching and resting with bill on back over wing (not inserted behind).

Seasonal breeders with limited laying period. Nest-sites on land or sometimes in shallow water; on land, rarely more than 1 m from shore, in open situation or screened by low vegetation. Nests typically low mounds of aquatic vegetation with shallow cup; occasionally little or none. Built by both sexes and often re-used in subsequent years. 'False' nests, or mating platforms, built by at least one species. Clutches normally 2 eggs, laid at interval of 1–3 days. Probably determinate layers. Single brood; replacements after egg loss possible in some species. Incubation periods 24–30 days. Both parents incubate, starting with first egg; each has single median brood-patch. Eggs not covered in absence of adults. Eggshells removed or left in nest. Young hatch asynchronously; food-dependent and reared by both parents. Carried on back when small. No bond between siblings. Fledging periods 55–75 days. Young become independent after fledging, probably during or after first autumn migration. Age of maturity, where known, 2–3 years.

## *Gavia stellata* Red-throated Diver

PLATES 3 and 6
[facing page 39 and after page 62]

Du. Roodkeelduiker    Fr. Plongeon catmarin    Ge. Sterntaucher
Ru. Краснозобая гагара    Sp. Colimbo chico    Sw. Smålom    N.Am. Red-throated Loon

*Colymbus Stellatus* Pontoppidan, 1763

Monotypic

**Field characters.** 53–69 cm (body 32–43 cm), ♂ averaging slightly larger with heavier head and bill (not distinctive unless pair together); wingspan 106–116 cm. Smallest diver (though not noticeably smaller than Black-throated *G. arctica* in field), with slender bill appearing uptilted, partly because of inclination of head and partly because of usually angled lower mandible and straight or slightly concave culmen. Sexes similar, but breeding and non-breeding plumages quite different; juvenile distinguishable.

ADULT BREEDING. Head and sides of neck dark grey; rear crown and hindneck narrowly and indistinctly striped black and white, extending to bolder striping on sides of breast; lower foreneck dull vinaceous red, appearing black at distance. Rest of upperparts almost uniform dark grey-brown (glossy green tinge when freshly moulted), with a few pale spots on wing-coverts visible only at close range. Underparts white, except for predominantly brown flanks and under tail-coverts. Bill dull blue-grey with black, grey or buff stripe along culmen; eyes wine red; legs grey-black. ADULT NON-BREEDING. Whole upperparts much paler, head particularly so with noticeably white face, eyering, and in some cases forecrown (combining to give look of surprise); neck mainly white and, with only narrow strip of grey at rear between crown and mantle, often looking extremely thin. Rest of upperparts grey, finely spotted with white and noticeably paler on mantle than wing-coverts. Underparts white with less brown on flanks and under tail-coverts. Bill dark slate-grey, sometimes looking black. JUVENILE. Similar to adult non-breeding, but mantle duller and browner with fewer spots and thus often darker than wing-coverts; neck darker due to extensive speckling on sides and throat; forehead also darker. Underparts tinged brown, noticeably on flanks.

Bill pale grey, rarely showing still paler tone on basal part; eyes red-brown. FIRST WINTER. Some retain dark juvenile plumage until January, rarely to March, but many quickly become indistinguishable from adult non-breeding except

Fig. I Divers (Gaviidae) in winter plumage. From top to bottom, Red-throated *G. stellata*, White-billed *G. adamsii*, Great Northern *G. immer*, and Black-throated *G. arctica*. 1st winter birds on right; adults on left. Note how upper 2 species habitually hold their heads up. See species accounts for other diagnostic characters (bill colours and shapes, and patterns of face, neck, and upper parts).

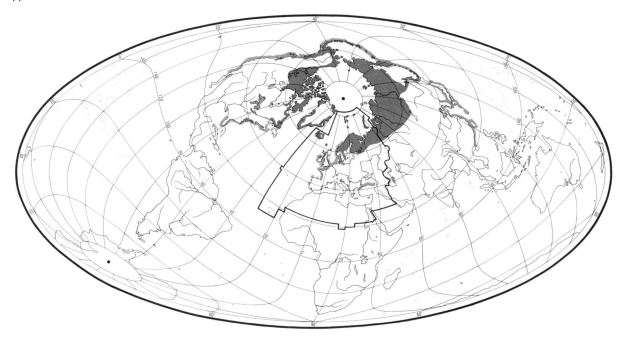

for retention of noticeably pale bill, even though some juvenile feathers may remain and flight-feathers are not moulted.

As red throat of *G. stellata* looks black at distance, breeding adults sometimes misidentified as *G. arctica* by the unwary, but that species has heavier, straight bill, larger head, bold striping behind black throat-patch, and conspicuous spotting on upperparts. Confusion with other divers most likely in non-breeding plumages, but speckled upperparts, paler appearance, thin neck, small head, and always more uptilted bill even when lower mandible straight (see Fig. I) aid identification at reasonable ranges. In flight, grey along back of neck appears as almost straight continuation of lower edge of grey back, whereas in other divers it curves downwards.

Sometimes floats high on water, but more often swims low; often raises itself to flap wings and, when preening or bathing, turns over on side or even completely on back. Dives quietly by putting head down and thrusting under water, sometimes with forward spring. Ungainly on land; usually heaves itself forward by thrusts of feet, sometimes aided by wings. When alarmed, may submerge whole body until head or only bill show above surface, but often takes flight. Rises much more easily from water than other divers and can take off from much smaller area. Alone among divers, recorded alighting on and taking off from land. Flight swift and direct with rapid wing-beats. More sociable than other divers, sometimes nesting in loose colonies; in winter, often singly, though sometimes in scattered concentrations of 50–200 and in even larger numbers on migration.

**Habitat.** Free-ranging between small or tiny, shallow, freshwater pools and large, deep lakes or sheltered marine inlets. Mobility in air and on water, and unusual hardiness, permit occupancy of breeding habitat up to 83°N in polar lands, and to altitudes of up to 200 m (frequently) and 700 m (exceptionally) in subarctic and boreal zones, extending marginally into temperate regions. Therefore most northerly and often most densely distributed of divers, occupying in breeding season majority of favourable waters in suitable regions, down to pools covering fractions of hectare, mainly along coastal fringes and through lower and middle river basins. Prefers shallow pools with well-vegetated banks, promontories and islets sparsely fringed with cotton grass *Eriophorum*, sedge *Carex*, and similar plants. Avoids dense floating or emergent vegetation and steep rocks above water, and strongly prefers treeless terrain, although sometimes using pools in taiga zone beyond fringe of tundra. Tolerates peat-stained oligotrophic water, and occasionally slow-flowing rivers. Entirely aquatic and aerial, normally on land only for nesting. Inadequacy of food on small breeding pools requires frequent flighting to better fishing lakes or sea, involving dual aquatic habitat and much use of airspace up to at least 1000 m. Mobility aided by ability to rise readily from small water surfaces, and exceptionally from land.

After breeding, shifts to ice-free warmer regions where numbers concentrate loosely along shallow inshore or coastal waters, even diving in waves breaking on beach; also visits lakes, pools, reservoirs and even large rivers, but flies less than in breeding season and often low over water.

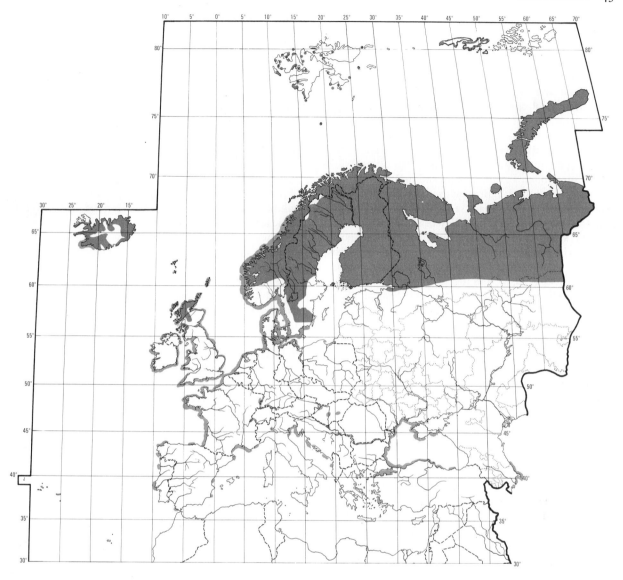

Rarely in close contact with man or artefacts, but fairly tolerant of human presence and limited disturbance, breeding within sight of highways, and living conspicuously. Flexibility, opportunism, and adaptability to varying conditions and climates more marked than in other divers.

**Distribution.** BRITAIN: after decrease Scotland 19th century, reported somewhat increasing (Alexander and Lack 1944); increase has continued in Shetland, Orkney, and Hebrides, but no permanent extension of range (Parslow 1967). NORWAY: numbers greatly diminished due to human persecution (Haftorn 1971). USSR. Estonia: bred once (Onno 1970).

Accidental. Austria, Greece, Rumania, Malta, Morocco.

**Population.** FINLAND: *c.* 2000 pairs (Merikallio 1958). ICELAND: at least 1000 pairs, perhaps considerably more (Gardarsson 1975). IRELAND: at least 2 pairs breeding regularly, with 1–2 pairs irregularly (RFR).

Survival. Oldest ringed bird 23 years 8 months (Rydzewski 1973).

**Movements.** Migratory and dispersive. Freshwater breeding lakes deserted in autumn; winters on tidal water, mainly inshore. From Spitsbergen, most depart September, virtually all by early October (Løvenskiold 1964). In Lapland and USSR, autumn movements begin when young capable of strong flight (usually late August), continuing until final exodus when fresh waters freeze about October. Has occurred off Kola Peninsula in December. Arrives Black Sea September–November, though some still passing through Ukraine December (Dementiev and Gladkov 1951). Spring passage away from winter quarters during April and May, or March on western seaboard, when moves to northern fjords, estuaries, and bays awaiting inland thaw; reappears on Greenland and Spitsbergen lakes May and early June; present off Murmansk coast from late April but May or June before return to tundra lakes (Dementiev and Gladkov 1951). British breeding lochs occupied mainly April to September–October (Baxter and Rintoul 1953).

Principal wintering areas in Europe: Atlantic and North Sea coasts around Britain and Ireland and from western Norway at least south to Brittany, with smaller numbers in Kattegat and Baltic, Black and Caspian Seas; small minority penetrate south into Bay of Biscay and, in cold winters, Mediterranean. Migration routes not well known. 4 recoveries birds ringed Greenland (west and east coasts) all from west Europe: south Norway (Kristiansand), England (Essex, Kent), and inland France (Dordogne). 2 Swedish chicks found midwinter in France (Seine Maritime) and Netherlands; and 2 chicks from Scotland (Shetland) also found French coast (Finistère, Vendée). 2 full-grown birds ringed Gulf of Bothnia, Finland, recovered Kent and north Russia (Tumen); latter suggests that, as in better-known case of Black-throated Diver *G. arctica*, part of USSR population migrates to or through Baltic area, though presumed Russian and west Siberian birds also winter Black and Caspian Seas. Possibly some from island populations disperse pelagically without long-distance migration: one Icelandic breeder recovered Icelandic coastal waters January. Occasional summer records from North and Baltic Seas presumably non-breeders. A few winter on larger lakes of central Europe (e.g. Switzerland), but many inland winter records probably sickly or storm-blown birds.

Migrates by day and night, usually in ones and twos, occasionally small parties. Large numbers may occur in loose associations on favoured staging waters, but no Palearctic records comparable to that of 1200 on Lake Ontario, Canada, in October (Palmer 1962). See also Bannerman (1959) for further details.

**Food.** Principally fish, captured by seizing (not spearing) with bill in underwater pursuit after dive from surface, usually without forward spring; normally feet used for propulsion, but occasionally wings. Depths recorded from 2–9 m and submergence times of up to $1\frac{1}{2}$ min, with average *c*.1 min.

From marine areas, most frequently recorded fish cod *Gadus morhua*, herrings *Clupea harengus* (van Damme 1968 and others), and sprats *Sprattus sprattus*; followed by gobies Gobiidae, sticklebacks *Gasterosteus* and *Spinachia*, sand-eels Ammodytidae, flounders *Platichthys*, coalfish *Pollachius virens*, butterfish *Pholis gunellus*, sculpin *Leptocottus armatus*, tomcod *Microgadus tomcod* and, in arctic, such species as polar-cod *Boreogadus saida* (Peder-

sen 1942) and capelin *Mallotus villosus*. Freshwater fish included trout and salmon *Salmo*, char *Salvelinus alpinus*, brook trout *Salvelinus fontinalis*, roach *Rutilus* and dace *Leuciscus*, bleak *Alburnus*, and perch *Perca*. In 173 stomach contents collected October–February from coastal areas and fjords of Denmark, only fish found, mostly small, but a few up to 25 cm long; *Gadus morhua* present in 71% of stomachs, forming 54% of total volume, and only fish in 38%; gobies (mainly *Chaparrudo flavescens* and *Pomatoschistus minutus*), sticklebacks (mainly *Gasterosteus aculeatus* and less frequently *Pungitius pungitius*), and *Clupea harengus* found respectively in 24%, 19%, and 18% of stomachs, forming 14%, 12%, and 10% of total volume; 11 other fish recorded formed 10% of total (Madsen 1957).

Other items include fish spawn, frogs, invertebrates such as crustaceans (especially crabs, shrimps, and prawns), molluscs (e.g. mussels *Mytilus* and snails *Valvata* and *Limnaea*), water insects, and annelids (e.g. nereids and leeches); many of these may have been already within fish. Plant material usually listed as possibly accidental traces, though Cottam (1936) recorded 2 instances where moss (Hypnaceae) occupied 38% of total stomach contents. Salomonsen (1950a) noted that in high arctic springs, when waters still ice-covered, plant material may be only food available and taken.

Young fed on crustaceans and aquatic insects for first 3–4 days, then fish (Pedersen 1942) small enough to be eaten whole (Hall and Arnold 1966); this contrasts with Rankin (1947) where fish brought to young in nest all too large and eventually eaten by parent. After 28 days young can eat like adults (van Braun *et al.* 1968).

**Social pattern and behaviour.** Based mainly on material supplied by S Sjölander; see especially Huxley (1923) and Bylin (1971).

1. Rather more gregarious than other divers. Usually small, scattered flocks on sea in winter, though solitary individuals and couples not infrequent and larger numbers occasionally at good feeding spots. In northern part of range especially, larger, loose flocks form and wait for nesting pools to re-open, frequenting edge of sea-ice, open lakes, or rivers. BONDS. Monogamous pair-bond, probably sustained over winter and life-long. Both parents tend young at least until fledging when family leaves nesting water; some pairs continue to feed young on sea in autumn (Sjölander 1973). Siblings show no social attachment to each other. BREEDING DISPERSION. Unlike other divers, nests mostly on small waters down to 10–20 m long, seldom using pools over 5 ha; food obtained wholly or largely elsewhere, on sea, bigger lakes, or rivers, so territory small and used only for breeding. No firm evidence of separate feeding territories though aggressive behaviour may occur between feeding birds (e.g. Keith 1937; Sjölander 1973). Where many small pools available, only 1 pair nests on each; this appears to be preferred dispersion, but in some areas loose colonies occur on larger waters, with several pairs on same lake, and rarely up to 50 pairs nesting within a few metres of each other (Collett 1877). Pair spends summer in territory, except for feeding. Same territory and even same nest often used from year to year, sometimes at least by same pair (Bylin 1971), but if

severely disturbed may move to another pool even when with young (see below). ROOSTING. During breeding season sleeps in territory, mainly morning, late afternoon, and, when dark, on water, usually in pairs; rarely on land. At other times apparently mainly at sea, singly or in small flocks.

2. Some aggressive behaviour in larger winter flocks and, together with calling, in pre-nesting gatherings (Keith 1937). At end of winter and in early spring, pairs (presumably formed in previous seasons) make daily reconnaissance flights inland and, as soon as enough open water, occupy nesting waters, often starting courtship and copulation immediately (e.g. Zedlitz 1913). Absence of courtship in winter and spring flocks, occurrence of lone, calling birds on previously unoccupied waters in early spring, disturbance of established pairs by roving individuals, and ability of ♂ to obtain new mate within a few days of death of old, indicate new pairs formed in summer quarters rather than in wintering areas (Sjölander 1973). Young ♂ probably first occupies territory and waits for ♀ to visit and eventually join him; such couples perform more courtship than older pairs which often omit first stage (see below). Use of small breeding waters, need to feed outside territory, and often considerable interaction with neighbours on some lakes, have all influenced social behaviour patterns: thus territorial behaviour more complex than in other divers and, probably because of stronger selection pressure to maintain pair-bond, often performed as ritualized, mutual ceremony by pair. For latter reason, and because members of isolated pair tend to show territorial displays to each other, especially when one returns to territory after excursion, such behaviour previously interpreted as courtship (e.g. Huxley 1923). True courtship distinct, however, and much simpler (as in other divers). TERRITORIAL BEHAVIOUR. Diver flying over occupied territory elicits Wailing-call and often also Long-call; both also given in response to similar calls from neighbours as well as to intruder on water, though latter then induces further responses by one, or more usually, both of pair. They first swim up to intruder in Alert-posture (see A) which differs from equivalent in other divers in that neck held straight and bill pointed upwards, probably because main field of binocular vision below bill, though posture also exposes red throat-patch to adversary. Pair next start Bill-dipping and later Splash-diving with upward kicks of feet on submergence. Croaking-call sometimes uttered during encounters up to this stage, e.g. briefly during Splash-dive; more common, however, during disturbances not involving conspecifics. Subsequently, pair likely to perform mutual Snake Ceremony (Huxley 1923), swimming slowly side by side with body deeply sunk and neck obliquely outstretched while Long-calling in duet (see B), and then even more intense but silent Plesiosaur-race (Huxley 1923), in which intruder tends to participate; similar to Snake Ceremony but with bodies steeply raised in Penguin-posture, bills lifted (see C). One of pair will also sometimes chase intruder in Zig-zag race in Snake-posture, as in Snake Ceremony, but with wings lifted alternately on side nearest other. May also assume Penguin-posture on own, standing upright, trampling water, and often turning round; may be equivalent of Fencing-posture of other divers, also occurring at higher stages of aggression, as well as Rushes over water with wings either flapping or held in V-position (Sjölander 1973). When actually fighting, rivals stab at head or seize neck; spearing from below water and underwater fights also recorded (Ruthke 1938; Rankin 1947). Aggression also shown to other waterbirds, especially Anatidae, and consists mostly of rushing at, or diving and emerging close to intruder. HETEROSEXUAL BEHAVIOUR. First stage of pre-copulatory courtship consists of mutual Bill-dipping and Splash-diving: with necks straight and stiff at angle of *c*. 45° forward, ♂ and ♀ lie opposed on water and begin to Bill-dip synchronously with accelerating tempo; then dive past each other, sometimes making fast rushes under water, before they turn and repeat sequence one or more times. One will sometimes perform such behaviour unilaterally when partner unresponsive (Bylin 1971). Second stage initiated by ♀ who starts Search-swimming, swimming slowly close to shore in Crouch-posture with neck retracted and bill drawn in to breast, hiding throat-patch, eventually jumping ashore at suitable spot; here she Invites in similar posture while lying still (see D) or, especially near egg-laying, making nest-building movements, with or without material (Bylin 1971; Sjölander 1973). Pre-copulatory courtship often repeated several times before each copulation; when responsive, ♂ follows ♀ ashore and immediately mounts, standing almost vertical with feet on her back and head bent down (e.g. Zedlitz 1913). Cloacal contact established for 10–20 s, during which ♂ often tramples with feet and sometimes moans; afterwards, drops back behind ♀ and immediately returns to water. ♀ remains crouched, sometimes extending neck horizontally, and after ♂ has departed may stay ashore for few minutes, often making incipient nest-building movements. Copulations

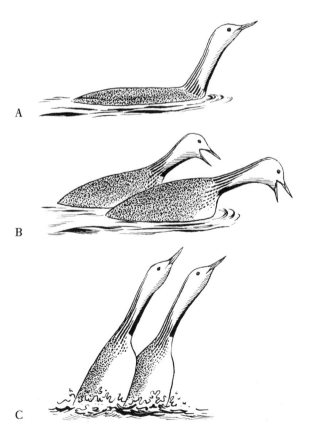

A

B

C

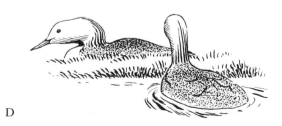

D

up to 6 times daily from arrival until egg-laying or occasionally just after, with normally 1 or 2 around sunrise; often incomplete, especially at beginning of cycle. Repeated invitations and copulations at same spot may create apparent mating-platforms but, unlike grebes (Podicipedidae), latter not essential to successful copulation and mainly consequence of scarcity of suitable landing places. Rarely, couples may copulate at water not used eventually for nesting. RELATIONS WITHIN FAMILY GROUP. Both parents provide food from 1st day, brood young ashore, at first at nest then at any suitable spot, for longer period than other divers, but carrying on back rare. Vocal contact by low moaning (adults) and shrill chirping (young); when fully separated, all use Long-call (von Braun *et al.* 1968). Chick begs by swimming in half circle round front of parent, chirping and pecking at breast or bill, continuing up to several hours if adult unresponsive. Young often left alone during fishing excursions. Adults seldom defend young but usually swim away from source of disturbance, croaking, Splash-diving, and sometimes rushing; when alarmed, chick does not seek refuge with parent but flattens and swims towards shore, hiding by lying still on water. Young can move for long distances on land by 'frog-jumping' and this enables parents to change pools when severely disturbed; chicks then crawl, led by incessant Long-calls of parents to which they answer with same call (von Braun *et al.* 1968). Siblings mutually highly aggressive especially when hungry, fighting almost continuously when, as often, parental feeding-rate low. In such cases one often becomes dominant, taking all food or preventing other from begging, and other may die of starvation within 3–4 days; this probably main reason why 2 fully fledged siblings rather uncommon.

(Figs A–D after original drawings by S Sjölander.)

**Voice.** Calls same for both sexes, except for Long-call.

(1) Croaking-call. Short barking, like Carrion Crow *Corvus corone*, uttered as warning-call when shore disturbances, e.g. by man. (2) Flight-cackle. Rhythmically repeated (about 5 per second), goose-like cackle (see I); territorial, used in flight when passing conspecifics on water or in air, and when circling own pond before landing or after take-off. (3) Wailing-call. Around 1 s duration, of descending pitch, and with fundamental around 1 kHz accompanied by strong harmonics that impart characteristic tonal quality, rather resembling meowing of cat (see II).

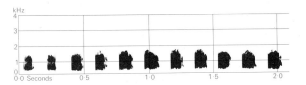

I P J Sellar Shetland August 1970

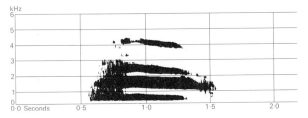

II M Sinclair Shetland August 1964

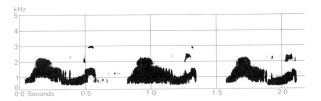

III M Sinclair Shetland August 1964

(4) Long-call. Like cooing of Collared Dove *Streptopelia decaocto* but harsher, louder, and audibly pulsed, with units of ascending and descending pitch repeated up to 10 times (see III). Sexes differ: ♀ version longer and softer. Calls often given in duet, with considerable overlapping. Territorial and bond-strengthening; in territorial encounters, contacts between mates, and during 'frog-jumping' (see Social Pattern and Behaviour). (5) Moaning-call. Low-pitched, uttered with bill closed. Mainly contact-call between mates and between parents and young; also given during copulation.

CALLS OF YOUNG. (1) Chirping-call: shrill, indistinct, uttered with bill closed; used to contact parents and in begging. (2) Long-call: used in reply to similar call of adults.

**Breeding.** SEASON. Diagram: southern parts of range; dependent in north on thaw. SITE. In shallow water up to 10 m from shore, or on bank usually within 0·5 m of water, though falling level can increase this. Can be in emergent vegetation but often completely exposed. Sites often re-used in successive years. Nest: heap of moss or aquatic vegetation, averaging 27 cm diameter and with cup of 4–5·5 cm depth (Gudmundsson 1952). Sometimes scrape or flattened hummock with no added material. Building: probably by both sexes. EGGS. Ovate, glossy; dark olive with blackish-brown spots and blotches. 75 × 48 mm (65–83 × 42–49), sample 100 (Witherby *et al.* 1940). Weight 83 gm (73–90), sample not given (Dementiev and Gladkov 1951). Clutch: 2 (1–3). Of 39 clutches, Shetland: 1 egg, 21%; 2, 79%; mean 1·79 (G Bundy). One brood. Replacement clutches laid after egg loss. Of 14 2nd clutches, Shetland: 1 egg, 50%, 2, 50%; mean 1·5 (G Bundy). Eggs laid at intervals of 1–2 days, though up to 8 recorded (Brandt 1940). INCUBATION. 26–28 days (24–29) (Dementiev and Gladkov 1951; Witherby *et al.* 1940), Keith (1937) gave record of one egg taking 26 days, while Brandt (1940) and Johnson (1935) stated between 36–40 days—which seems long, especially in comparison with other Gaviidae. Mean of 27·05 days (range 24–29), sample 19 (Shetland, G Bundy). Begins with 1st egg; hatching asynchronous. Both parents incubate, though ♀ probably does larger share. Eggshells usually removed from nest (Keith 1937; Hall and Arnold 1966), but sometimes left (Nethersole-Thompson 1942). YOUNG. Precocial and nidifugous. Fed by both parents, with young leaving nest at 24-hours old and swimming to meet parent bringing food (Hall and Arnold 1966). FLEDGING TO MATURITY. Fledging period 43 days (range 38–48), sample 18 (Shetland, G Bundy); estimated at *c.* 8 weeks (Witherby *et al.* 1940). Young independent probably after leaving breeding area. Age of first breeding unknown; probably 2–3 years as other Gaviidae. BREEDING SUCCESS. Of 91 eggs laid, Shetland 1974, 29·7% hatched and 19·8% fledged; predation and disturbance main adverse factors (G Bundy).

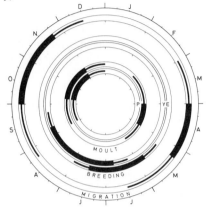

**Plumages.** ADULT BREEDING. Crown grey, feathers with darker centres. Hindneck striped black and white; similar shorter stripes on upper mantle and sides of chest. Rest of upperparts shiny dark brown, with small pale spots. Chin and sides of head and neck pure grey; deep chestnut triangle on throat. Breast and belly white with narrow grey band across vent, under tail-coverts mixed white and brown. Flanks dark brown and white. Tail dark brown. Primaries, primary coverts, and secondaries dark brown, paler below; secondaries with a little white at tips, almost disappearing with wear. Upper wing-coverts brown; marginal near body without spots, lesser with small white spots, median and greater with white margins near tips. Under wing-coverts white, outer on primaries with pale grey outer webs, greater near body grey with white margins; axillaries white with dark brown shaft streaks. ADULT NON-BREEDING. Crown and hindneck finely streaked dark grey-brown and white. Rest of upperparts dark grey-brown with small white spots. Underparts white with narrow, broken dusky band across vent and some dark under tail-coverts. Sides of head and neck mottled grey and white, sometimes only white. Flanks mixed dark brown and white. Tail-feathers dark brown with white tips, paler below. Wing as breeding, but white spots on upper coverts brighter and feathers less worn. DOWNY YOUNG. 1st down dark brown to dark grey above, slightly paler sides of head and neck, flanks, throat, and breast; lower breast and belly light grey. 2nd down paler, pushed out by developing juvenile feathers (Palmer 1962). JUVENILE. Like adult non-breeding but colour of upperparts browner, spots less pure white, elongate, forming V-marks on shoulders and scapulars; spots on lower back, rump, and upper tail-coverts vague. Feathers of vent narrowly edged grey-brown; under tail-coverts similar, longer ones grey-brown with white subterminal band. Tail-feathers tipped dusky. IMMATURE. Juvenile plumage

gradually replaced by adult non-breeding; recognizable only as long as some juvenile feathers remain on back, wing-coverts, throat, or vent. Median and greater upper wing-coverts not moulted and appear strongly worn. IMMATURE SUMMER. Like adult breeding but mixed with retained immature winter and probably some juvenile feathers, throat patch paler, primaries strongly worn (Palmer 1962).

**Bare parts.** ADULT BREEDING. Iris brown-red. Bill black. Foot black on outer side, pale on inner side; webs fleshy with dark margins. ADULT NON-BREEDING. Bill pale grey with dark grey culmen. DOWNY YOUNG. Iris brown; bill dusky. JUVENILE. As adult non-breeding, but iris browner and bill paler.

**Moults.** ADULT POST-BREEDING. Complete; flight-feathers simultaneous. Timing variable, starting late September to December. ADULT PRE-BREEDING. Partial; body, tail, and some lesser wing-coverts. Starts February–early April; first in full breeding dress mid-April. POST-JUVENILE. Partial; body, tail, and some wing-coverts. From December (perhaps earlier in some) adult non-breeding feathers present on throat and back. Late birds start moulting February. IMMATURE SPRING. May–late July. Begins before end of juvenile moult. Variable, but rather large number of winter feathers replaced; juvenile flight-feathers retained. IMMATURE AUTUMN. Presumably like adult post-breeding, but flight-feathers moulted earlier, in summer, not autumn, of 2nd calendar year.

**Measurements.** Adult. Skins (RMNH, ZMA); additional data on range from published sources.

| | | | | | |
|---|---|---|---|---|---|
| WING | ♂ 292 | (10·1 ; 12) | 265–310 | ♀ 281 | (4·6 ; 10) | 257–308 |
| TAIL | 53·4 | (1·65; 10) | 42–57 | 50·1 | (2·51; 10) | 47–54 |
| BILL | 55·1 | ( 2·98; 11) | 48–61 | 51·3 | (1·16; 10) | 46–55 |
| TARSUS | 75·1 | ( 4·15; 10) | 66–82 | 70·8 | (3·29; 10) | 65–77 |

Sex difference significant. Variability in ♀ underestimated in measured sample, so ranges given from all sources.

**Weights.** Lean winter ZMA; summer, Bauer and Glutz (1966), De Korte (1972).

| | | | | | |
|---|---|---|---|---|---|
| winter | ♂ 1341 | (118·4; 6) | 1170–1456 | ♀1144 | (117·2; 9) | 988–1302 |
| summer | 1729 | (175·8; 7) | 1370–1900 | 1477 | ( 88·8; 5) | 1410–1613 |

Russian birds heavier, 4 ♂♂ mean 1888, maximum 2460; 11 ♀♀ mean 1652, maximum 2255 (Dementiev and Gladkov 1951). D J Stevenson gives data on increase in weight of growing chicks, Iceland, in relation to increase of wing length (figures in brackets show number of birds weighed):

| WING | under 100 (3) | 100–149 (5) | 150–199 (3) | 200–249 (7) | over 249 (4) |
|---|---|---|---|---|---|
| WEIGHT | 154–450 | 710–870 | 920–1080 | 1100–1410 | 1180–1460 |

**Structure.** 11 primaries: p10 and p9 longest, each may be 2–4 shorter than other; p8 *c.*10 shorter, p7 *c.*30, p1 *c.*125. 23 secondaries. Tail short, 18 feathers. Bill slender, culmen very slightly downcurved, lower mandible sloping upwards, giving bill upturned aspect.

**Geographical variation.** Birds from Spitsbergen and Franz Josef Land said to differ from other populations in having greyish edges to mantle feathers in breeding plumage, but character variable and does not warrant recognition of subspecies *squamata* (De Korte 1972).

# *Gavia arctica* **Black-throated Diver**

Du. Parelduiker     Fr. Plongeon arctique     Ge. Prachttaucher
Ru. Чернозобая гагара     Sp. Colimbo ártico     Sw. Storlom     N.Am. Arctic Loon

*Colymbus arcticus* Linnaeus, 1758

Polytypic. Nominate *arctica* (Linnaeus 1758), west Palearctic east to river Lena, Siberia; *viridigularis* Dwight, 1918, accidental, Palearctic east from river Lena, possibly also Cape Prince of Wales, Alaska. Extralimital: *pacifica* (Lawrence, 1858), North America, and coastal region of north-east Siberia (sometimes considered separate species).

**Field characters.** 58–73 cm (body 36–46 cm); wing-span 110–130 cm; averages slightly bulkier and longer-bodied than Red-throated Diver *G. stellata*, with noticeably heavier head. Straight-billed, medium-sized, evenly proportioned diver, lacking obvious structural characters. Summer plumage uniquely combines grey head with chequered black and white upperparts. In other plumages, darker upperparts than those of other divers contributing to impression of low carriage in water, common to all *Gavia*, most marked in this species. Sexes similar; marked seasonal variation; juveniles separable at close range.

ADULT BREEDING. Crown, upper face, and hindneck grey, lower face dark grey, shading into black chin and throat. Sides of neck narrowly striped black and white; narrow half necklace of white streaks divides throat from large black patch on foreneck, latter widening towards base; neck-side stripes extend over sides of breast. Mantle and scapulars basically blue-black, relieved by 4 areas of regular transverse white spots and checks (those on scapulars with wider marks and more prominent); wing and upper tail-coverts black, spotted white. Quills black, underwing white, merging with brown-black undersides

of quills. Underparts white but top flanks black or partly so. Bill black; legs and feet grey-black; eyes red. ADULT WINTER. Plumage darkens more than any other diver. Forehead, mantle, and scapulars uniform dark brown; darker than crown and hindneck which show greyish cast and appear paler (reverse of pattern in Great Northern Diver *G. immer*). Underparts silky-white except for brown spots or streaks across upper throat, on sides of chest, and along top flanks. Bill grey, with darker culmen and tip. JUVENILE. Like winter adult, but upperparts browner (except for blackish forehead) and less uniform, with round (not square) feather-margins grey and appearing as pale scales in good light. Bill pale grey, sometimes bluish-white with dark culmen and tip even more obvious; eyes browner than adult.

Confusion with other divers in breeding plumage unlikely but winter adults and immatures can be difficult to identify since, of the 3 common species, *G. arctica* intermediate. Distinction from atypical, dark *G. stellata* best based on larger size, straight (sometimes apparently down-turned) bill, dark forehead, and level head carriage; from small *G. immer* by slender bill, sloping (not steep)

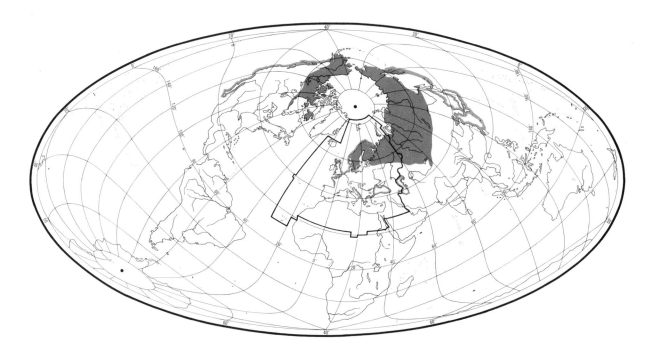

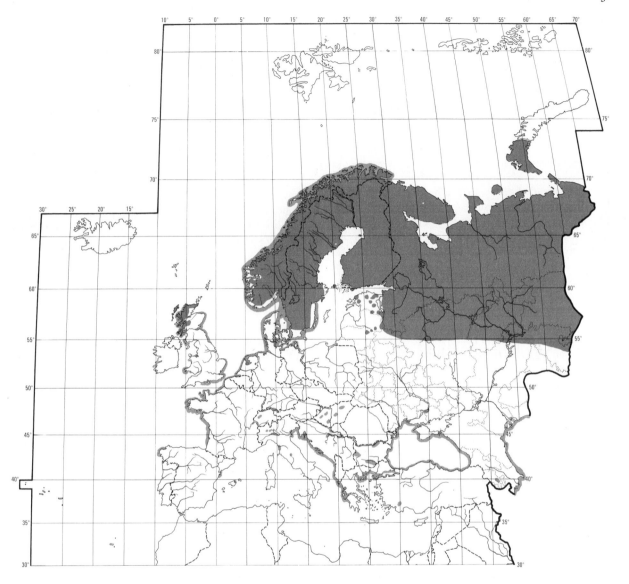

forehead, thinner neck with no indented pale half-collar, and uniform upperparts darker, not paler than hindneck (see Fig I, page 43). Flight as *G. stellata* but wings not raised as high on upstroke and action a little slower.

General behaviour as *G. stellata*, though prefers deeper, large hill lakes for nesting. Winters almost exclusively in coastal waters; less sociable at sea than *G. stellata*, rarely forming even loose assemblies. Breeding birds vocal; call notes rarely heard in winter but lower, less quacking than *G. stellata*.                                                                    DIMW

**Habitat.** Differs from Red-throated Diver *G. stellata* in preferring boreal to arctic tundra habitats and fairly large, deep lakes to small, shallow pools. Typically selects open, deep, productive lakes or extensive pools in tundra, or not uncommonly in coniferous forest such as taiga; usually

somewhat sheltered, mature, and undisturbed, with choice of small islets, peninsulas, or other inaccessible and inconspicuous nest-sites immediately beside deep water. Sometimes uses shallows protected by dense growth e.g. of willow *Salix*; infrequently on quiet backwaters or oxbows of large slow rivers, or sheltered non-tidal marine inlets. Fringing trees and emergent vegetation tolerated, provided ample open water for security from surprise. Avoids sharing with other divers, normally requiring 50–150 ha for single pair. Concentrates breeding season activities on one large fairly deep water, but on lakes among sphagnum bogs in taiga where fishing poor will go for food to other lakes or rivers, sometimes 5–10 km distant (Y A Isakov). Infrequently resorts to marine fishing waters while breeding. Lakes liable to strongly fluctuating water level avoided for breeding (Rankin 1947). Human disturbance, especially from boats or noisy parties, easily causes desertion.

Outside breeding season remains conservative, sensitive to disturbance, and almost entirely aquatic or aerial in habitat. Great majority resort to sheltered coastal marine waters, not often recorded far from land. Wintering on natural lakes or barrages, lagoons or large rivers relatively infrequent.

**Distribution.** BRITAIN: prior to 1944 some decrease in Scotland owing to human persecution (Alexander and Lack 1944); probably no marked change in recent years, but has extended range to south-west Scotland where breeding since 1951 (Parslow 1967; British Ornithologists' Union 1971). SWEDEN: recently extended range and increased (Curry-Lindahl 1959). EAST GERMANY: probably bred 1968 (*Beitr. Vogelk.* 1973, **19**, 170–4). POLAND: bred 5 localities Koszalin province in 19th century, and 2 districts elsewhere up to 1938 (Tomiałojć 1972).

Accidental. Faeroes, Bear Island, Spitsbergen, Luxembourg, Austria, Spain, Algeria, possibly Morocco, Israel.

**Population.** FINLAND: *c.* 2000 pairs (Merikallio 1958).

Survival. Migrants ringed east Baltic 1931–41 produced 18 recoveries over 12 years old (Vogelwarte Rossitten). Oldest ringed bird 27 years (*Orn. Fenn.* 1962, **39**, 132).

**Movements.** Migratory and dispersive. Some Scottish breeding lochs deserted August, but movement to salt water chiefly September–October; return about April (Baxter and Rintoul 1953). In Lapland and USSR, timing of movements to and from breeding lakes associated with thawing and freezing of fresh waters; spring return faster. Some reach Baltic by September and most at sea by mid-October, but November record from Taimyr and December ones from Smolensk (Dementiev and Gladkov 1951) indicate protracted nature of autumn passage. Main departures from Baltic and Black Sea winter quarters from mid-April to mid-May; reaches Murmansk and Lapland coasts from late April, moving inland early May; but breeding lakes not reoccupied until late May or early June on northern tundras. Non-breeders occasionally summer well south of breeding range. Wintering area of Scottish breeders unknown (no relevant recoveries). Evidently some immigration into or through British coastal waters (scarce Ireland), since winter numbers believed to exceed size of breeding population; winter visitors in North Sea probably Norwegian breeders as no recoveries there despite large-scale ringing in Baltic. Movements of Continental populations generally better understood; Baltic important wintering area and migration route for birds breeding over vast region from Scandinavia to at least Taimyr Peninsula, latter travelling over 6000 km (Bodenstein and Schüz 1944; Schüz 1954, 1957c). Tundra breeders said to migrate north to arctic seas before turning west towards Scandinavia (Dementiev and Gladkov 1951),

but subsequent movements unknown and concept perhaps speculative. Most important routes from USSR appear to be those of forest-zone breeders (at least) which move west to south-west overland; doubtless some fly direct to Black Sea winter quarters, which others reach flying overland across east Europe from Baltic, mainly October and November. Apparently major northward movement in spring from Black Sea to Baltic, as large numbers passing Kaliningrad in May, thence east into USSR. Some Fenno-Scandian birds among those reaching Black Sea, where 2 recoveries from Sweden and one from Finland. Small numbers occur Mediterranean (singles ringed Baltic recovered Aegean and Adriatic) and larger lakes of central Europe; one ringed Switzerland found in May in Komi ASSR. Breeders from south-west Siberia thought to winter in Turkestan and on Caspian Sea (Dementiev and Gladkov 1951).

East Siberian race, *G. a. viridigularis*, migrates east to south-east to Pacific and Bering Sea, but possibly straggles to Europe; several in summer plumage ringed Kaliningrad in May believed *viridigularis* (or intergrades) and one from there recovered at 129°E on Lena delta (Stresemann 1936; Schüz 1954), but 2 Dutch records no longer acceptable (Voous 1961).

**Food.** Chiefly fish obtained by diving from surface, generally using only feet for propulsion, occasionally wings. Depths recorded 3–6 m, and submergence times up to 2 minutes (average 45 s). Longest dive in 4 hours observation 63 s, shortest 5; out of 201 dives, 96 between 48–50 s (Joyce and Joyce 1959).

In winter, probably almost entirely fish, and other foods recorded may have been taken accidentally or eaten within stomachs of fish. Crustaceans, molluscs, and aquatic insects sometimes taken in appreciable quantities, more especially during breeding season. Frogs (Naumann 1903) and leeches (Grote 1950) also recorded. Plant materials in stomachs usually mere traces, though seaweeds said to form bulk of food in some instances (Madsen 1957); 1 stomach contained 95% seeds and fibres of mare's-tails *Hippuris*, pondweeds *Potamogeton*, and bulrushes *Scirpus* (Palmer 1962). Foods from marine habitats principally small shoal fish, such as gobies Gobidae, sticklebacks Gasterosteidae, herrings *Clupea harengus* and sprats *Sprattus sprattus*, sand-smelts *Atherina*, sand-eels Ammodytidae and larger fish such as cod *Gadus morhua*. Crustaceans and annelids 23·4% and molluscs 15·5% of volume in 7 birds (Collinge 1924–7). Fish records from freshwater areas include perch *Perca*, trout *Salmo trutta*, bleak *Alburnus*, dace *Leuciscus* and roach *Rutilus*, carp *Cyprinus* and according to Beamish (1945), char *Salvelinus*. Invertebrates include water-beetles (Coleoptera), water-boatmen (Hemiptera), caddisfly (Trichoptera), and dragonfly (Odonata) larvae, water snails (Mollusca), shrimps (Crustacea), and occasionally, as only item, crayfish *Astacus fluviatilis*. Records of small ducklings (Lesser Scaup *Aythya*

A

*affinis* and Tufted Duck *A. fuligula*) eaten (Lensink 1967; Dementiev and Gladkov 1951) may be due to extreme territorial hostility. 123 stomach contents from Danish marine areas October–February all contained fish remains, a few traces of crustaceans, polychaetes, and molluscs, probably derived from fish; cod *Gadus morhua*, gobies (mainly *Pomatoschistus minutus*, and *Chaparrudo flavescens*), and sticklebacks (chiefly *Gasterosteus aculeatus*) formed 90% of total and only fish in 80% of birds examined; cod and gobies qualitatively of about equal importance, about ⅓ of total diet, and sticklebacks about ⅕; 12 other species of fish normally found near the bottom formed less than 10% of total (Madsen 1957).

**Social pattern and behaviour.** Based mainly on material supplied by S Sjölander; see especially Lehtonen (1970), Sjölander (1968), Dunker (1974, 1975).

1. Breeds in isolated pairs and occurs singly, in apparent pairs or small flocks in winter, usually on sea. Winter flocks closely follow thaw on nesting grounds. BONDS. Evidence suggests monogamous, life-long pair-bond (see below). Both parents care for young until family migrates at end of season; parent–young bond does not appear strong, young often left alone for hours during fishing excursions. Fledged young highly self-reliant, spending much time apart from parents though remaining in territory. Almost no bond between siblings. BREEDING DISPERSION. In southern part of west Palearctic breeding range, found on large clear-water lakes with abundant fish where occupies large, all-purpose territory as Great Northern Diver *G. immer*. Normally 1 pair on open waters of 50–150 ha; if peninsulas, inlets, and islands provide several nest-sites, more pairs occur but rarely closer than 200–300 m. As in other divers, territorial density much influenced by food supply and lakes rich in fish may be entirely divided up into territories of 5–10 ha. In more arctic part of range, on much smaller pools down to less than 1 ha; then, as with Red-throated Diver *G. stellata*, territory used for breeding only and food obtained from rivers, large lakes, or (less commonly) sea. Same territory, though not necessarily same nest, used from year to year; pair remains within territory until migration except during fishing excursions. Breeders and non-breeders often congregate to feed in small flocks, especially in early morning, often at same spot daily and from afar, fishing in line (Sjölander 1968). ROOSTING. As in *G. stellata*.

2. At start of breeding season, pairs make daily flights from sea, larger lakes, and rivers, and settle in territory as soon as possible. Most arrive already paired, start courtship and copulation immediately, and often lay within week of return. Much behaviour closely similar to other divers, though territorial behaviour simpler than in *G. stellata*, but more ritualized than in other black-and-white divers. TERRITORIAL BEHAVIOUR. ♂

especially highly vocal, uttering Wails and Long-calls mainly in response to calling of territorial neighbours and sight of flying bird. Pair lack Long-call Duet of *G. stellata*; also mutual Snake Ceremony and Plesiosaur-race. Intruder in territory approached by one of pair, or often both, in Alert-posture which differs from corresponding display of *G. stellata* in that neck has S-shaped bend; usually followed by Circle-dance (see A) in which all manoeuvre closely in Alert-posture while exposing throat-patch and necklace (Sjölander 1968). Intruder often then leaves without further interaction; if stays, owners intensify their Bill-dipping (see A) which leads to Splash-diving, with or without several upward kicks of feet, accompanied by Short-call peculiar to present species. Croaking also common in this situation, often as alternating duet between pair; Long-calls also occasionally uttered. Participants also engage in Rushes in wide arcs over water often for long distances while flapping wings (see B) or occasionally holding them up in V. When highly aggressive (e.g. immediately before fighting), adopts Fencing-posture, standing upright, trampling water, with neck strongly arched backwards and bill on breast, and then sometimes moves slowly forwards in series of Bow-jumps, often with spread wings, alternating between Bowing and Bill-dipping in water and raising head and body (Lehtonen 1970). Fights rare and similar to those of *G. stellata*, i.e. with stabbing, wing-beating, and spearing; occasionally participant gets killed, either from spearing or drowning (Sjölander 1968). Members of summer feeding flocks spend much time performing Circle-dances and other antagonistic behaviour. HETEROSEXUAL BEHAVIOUR. As in other divers, initial courtship on water consists of Bill-dipping and Splash-diving by pair; seen best in late breeders (probably newly formed pairs) when sometimes intermingled with territorial behaviour, but often omitted especially by old-established pairs. Splash-diving not accompanied by call (as in territorial encounters) and tends to be followed by underwater Rushes. ♀ then Search-swims and Crouches before getting ashore and Inviting, often repeating sequence again and again before ♂ follows. Copulation occurs normally near sunrise from day of arrival until egg-laying or later, and may be repeated 3–5 times daily; mainly in territory, exceptionally on lakes where pair awaits spring thaw (Sjölander 1968). RELATIONS WITHIN FAMILY GROUP. Essentially similar to *G. stellata* including vocal communication (though young of present species not so mobile on land). Siblings highly aggressive towards each other; fighting may lead to death of weaker, though both not uncommonly survive (Lehtonen 1970).

(Figs. A–B after original drawings by S Sjölander.)

B

**Voice**. Both sexes, especially ♂, vocal in breeding season; mainly silent in winter. Following based on material supplied by S Sjölander; see also Lehtonen (1970), Dunker (1975).

(1) Croaking-call. Long, pulsed and hoarse, like Raven *Corvus corax*, repeated several times, sometimes in dual calling. Given as warning (e.g. when disturbed), and also in territorial encounters (see I). (2) Wailing-call. 3 brief notes of increasing loudness and pitch (see II), given in territorial encounters and between neighbouring pairs. (3) Long-call. Resembles low-pitched whistle; composed of 2 notes of definite pitch with marked pitch increase in longer second note (see III). Yodel-like breaks occur between calls, repeated 3–5 times, sometimes with short, low-intensity calls between. Main territorial call. (4) Short-call. Short, intense, lasting *c.* 0·2 s with sharply rising pitch (see IV); given before Splash-dive towards intruding conspecifics; also towards predators, aircraft, etc. (5) Moaning-call. Contact-call in seeking absent mate or young; rises in pitch in middle syllable.

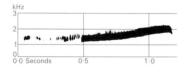

I Sture Palmér/Sveriges Radio Sweden May 1959

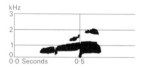

II S Sjölander Sweden

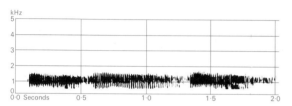

III Sture Palmér/Sveriges Radio Sweden May 1959

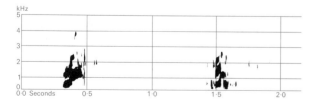

IV S Sjölander Sweden

CALLS OF YOUNG. Chirping-call: shrill indistinct contact-call used by chicks and juveniles; difficult to locate, and resembles that of Red-throated Diver *G. stellata*.

**Breeding**. SEASON. For southern parts of west Palearctic see diagram; in northern Europe and USSR dependent on thaw of breeding lakes. SITE. In shallow water up to 10 m from shore, or on land, not more than 120 cm from water (most 30–70 cm), most frequently on islands 90% of 85 Finnish nests on north shores (Lehtonen 1970). Nest-site always dry at laying. Rarely, nest floating (Dementiev and Gladkov 1951). Old nest-site re-used (Lindberg 1968). Nest: heap of moss or water-weed, usually with distinct cup, less often merely scrape or flattened mound. Occasionally larger structure, with base of twigs (England 1957). Average dimensions of 25 nests, 44–58 cm by 37–52 cm, up to 10 cm high, and with rim raised from 0·3–2·2 cm to form shallow cup of diameter 20–25 cm (Lehtonen 1970; Dementiev and Gladkov 1951). 10 nests average 17·5 cm above water (range 8–34 cm); if wave action threatens completed nest, ♀ may raise further 20–40 cm. Up to 3 false nests sometimes built by ♂ simultaneously or after true nest, often 1–2 m apart and 5–10 m from true nest, occasionally up to 200 m (Lehtonen 1970). Building: true nest mostly by ♂, with ♀ helping; false nests apparently always by ♂. EGGS. Ovate, slightly glossy; greenish olive to dark umber with black spots and blotches. 84 × 52 mm (76–95 × 46–56), sample 100 (Schönwetter 1967). Weight 120 gm (92–134), sample 15 (Lehtonen 1970). Clutch: 2 (1–3). Of 78 Finnish first clutches: 1 egg, 13%; 2, 86%; 3, 1%; of 7 replacement clutches, 1 egg, 2; 2 eggs, 4; and 3 eggs, 1 (Lehtonen 1970). One brood. Replacement clutches laid after egg loss. Eggs laid at 1–2-day intervals, though up to 4 recorded (Lehtonen 1970). INCUBATION. 28–30 days. By both sexes, though mainly ♀. Begins with first egg; hatching asynchronous. Eggshells generally removed on hatching (Lindberg 1968); sometimes left in nest (Nethersole-Thompson 1942). YOUNG. Precocial and nidifugous. Fed by both parents. Spend first 2–3 nights on nest, then sleep on water under parents' wings (Lindberg 1968). Begin to catch fish at 60–70 days (Lehtonen 1970). FLEDGING TO MATURITY. Fledging period *c.*60–65 days. Become independent after leaving breeding area. Mature at 2–3 years. BREEDING SUCCESS. Of 159 eggs laid in Finland, 1962–8, 72 (45%) hatched and 34 young (21%) fledged (Lehtonen 1970).

**Plumages** (nominate *arctica*). ADULT BREEDING. Crown and hindneck grey, sides of head shading into black of chin. Rest of upperparts glossy black with 4 elongated groups of regularly arranged, squarish white spots at shoulders and on scapulars. Chin and throat black; throat with purple gloss, separated from chin by transverse row of 10–15 short white streaks; sides of neck and chest striped black and white. Flanks black with some white streaks; rest of underparts white, brown-grey band over anal region, long under tail-coverts blackish-grey, some with white tips. Tail brown-black. Primaries glossy black, paler below and at bases of inner webs; secondaries black at tips, pale grey on lower halves, margins of inner webs white. Upper wing-coverts, except outermost, black with subterminal white spots; under wing-coverts white, greater with pale grey outer webs, axillaries white, mostly with dark shafts widening at tips. ADULT NON-BREEDING. Upperparts dark brown, paler on crown and hindneck, scapulars sometimes with pale subterminal spots. Underparts and sides of head white; often brown-grey band over anal region; sides of chest streaked grey; flanks as back. Tail dark grey with narrow white tip, paler below; long under tail-coverts grey, tipped white. Wing as in breeding, though some white-spotted upper coverts may be replaced by dark brown. DOWNY YOUNG. First down sooty brown, paler below. Second down paler, centre of belly whitish; attached to tips of first. Short juvenile down develops among second down; growing body feathers push second down out. JUVENILE. As adult non-breeding but feathers of upperparts rounded, not square, edged light grey giving scaly appearance; tips of under tail-coverts grey; white tips of tail-feathers narrower; no white spots on upper wing-coverts. IMMATURE. During 2nd calendar year juvenile plumage replaced by another distinguishable from adult non-breeding only by unspotted wing-coverts (see Moults).

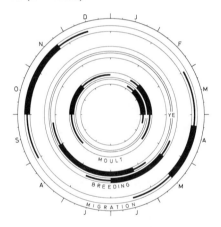

**Bare parts.** ADULT BREEDING. Iris red with narrow silvery inner ring. Bill black. Tarsus black on outer side, light grey on inner; toes grey, webs mixed flesh and grey. ADULT NON-BREEDING. Bill grey, ridge of culmen dark. JUVENILE. Iris brown. Bill blue-white. Foot paler than adult.

**Moults.** ADULT POST-BREEDING. Partial; body, tail, and some upper wing-coverts; starts round base of bill and continues on head and body. Timing variable, about September in some, others not before December. Some breeding feathers on back and scapulars often retained; sometimes no post-breeding moult in autumn, old breeding plumage replaced by new from January onwards. Presence of white-spotted coverts on upper wing separates adult non-breeding from immature. ADULT PRE-BREEDING. Complete; primaries simultaneous. Starts round base of bill January or later; flight-feathers February–April. POST-JUVENILE. Apparently 2nd calendar year birds moult slowly and

continuously during most of year. Different plumages and moults hard to separate. Body feathers gradually replaced, January (exceptional December)–summer, by plumage like adult non-breeding, slightly blacker on mantle with faint gloss and without white-spotted upper wing-coverts. In autumn this may be replaced partly or wholly by another plumage like adult non-breeding but still with unspotted wings. Juvenile flight-feathers July–August. FIRST PRE-BREEDING. February–May of 3rd calendar year, apparently later than adult. Flight-feathers April–May. When spotted wing-coverts acquired indistinguishable from adults. (Schüz 1936; Palmer 1962).

**Measurements.** Adult, breeding plumage. Skins (RMNH, ZMK); *viridigularis* from Austin and Kuroda (1953).

nominate *arctica*

| | | | | | |
|---|---|---|---|---|---|
| WING | ♂ 324 (17·4; 10) | 294–343 | ♀ 309 (16·3; 0) | 282–337 |
| TAIL | 58·7 (4·40; 10) | 53–67 | 57·8 (3·15; 8) | 51–61 |
| BILL | 60·7 (5·57; 9) | 52–68 | 60·2 (5·14; 10) | 52–68 |
| TARSUS | 82·2 (6·06; 11) | 72–89 | 78·8 (5·41; 10) | 71–87 |

| | ♂ and ♀ *viridigularis* | | ♂ and ♀ *pacifica* | |
|---|---|---|---|---|
| WING | 320 (—; 15) | 286–356 | 299·9(12·1; 21) | 279–326 |
| BILL | 64·5 (—; 15) | 53–71 | 55·6( 5·08; 23) | 48–66 |

Sex difference not significant, but ♂ tends to be larger. Individual variation very great; Edberg (1957) gave range of wing ♂ *arctica* 283–356 (see also Voous 1961).

**Weights.** Data scanty. RMNH, ZMK, ZMA, Bauer and Glutz (1966), and Dementiev and Gladkov (1951).

| Nominate *arctica*, winter | | | summer | | |
|---|---|---|---|---|---|
| 3 | ♂♂ | 1316–2607 | 2 | ♂♂ | 3310–3400 |
| 1 | ♀ | 1688 | 3 | ♀♀ | 2037–2471 |
| 13 | ♂♀ | 1300–2673 | 5 | ♂♀ | 2480–3000 |

Bauer and Glutz (1966) supposed young lighter than adults, but not confirmed by our data: 2 first winter ♂♂ in February 1316 and 2673. Apparently great individual variation as in other measurements. *G. a. viridigularis* ranges 2255–3792.

**Structure.** 11 primaries: p10 longest, p9 1–7 shorter, p8 15–22, p7 30–45, p1 130–160. 22–24 secondaries. Tail short, 16–18 feathers. Bill strong, but shorter and less heavy than *G. immer*; tip of upper mandible slightly downcurved.

**Geographical variation.** Insufficiently known and obscured by great individual variation. *G. a. viridigularis* averages slightly larger, especially bill and foot, and has green gloss on throat in breeding plumage; *G. a. pacifica* smaller with pale grey nape and hindneck and purple gloss on throat. *G. a. viridigularis* and *pacifica* found breeding sympatrically in Anadyr, north-east Siberia, and on Cape Prince of Wales (Vaurie 1965). Further study may prove them specifically distinct, but winter plumage indistinguishable (Godfrey 1966).

**Recognition.** Large adult non-breeding and juvenile *G. arctica* may resemble small *G. immer*. Differentiating characters: less heavy bill, hindneck paler than back (darker than back in *G. immer*), narrow shaft streak on axillaries, no light half-collar on side of neck; in adult non-breeding no light spots on mantle and scapulars (apart from pure white spots on retained breeding feathers); in juvenile no pale spots on lesser upper wing-coverts.

## *Gavia immer* **Great Northern Diver**

PLATES 5 and 6
[after page 62]

Du. IJsduiker    Fr. Plongeon imbrin    Ge. Eistaucher
Ru. Черноклювая гагара    Sp. Colimbo grande    Sw. Islom    N.Am. Common Loon

*Colymbus Immer* Brünnich, 1764

Monotypic

**Field characters.** 69–91 cm (body 43–56 cm), wingspan 127–147 cm, averaging smaller than White-billed Diver *G. adamsii* and smallest overlapping with largest Black-throated *G. arctica*. Spear-billed, heavy, large, bulky diver, with steep forehead and thick neck contributing to stockiness and weight of foreparts. Summer plumage uniquely combines wholly dark bill and head with boldly spotted and chequered upperparts. Other plumages, essentially dark above and whitish below, not distinctive and separation from both *G. adamsii* and large *G. arctica* can be difficult. Sexes similar; marked seasonal variation; juveniles separable at close range.

ADULT BREEDING. Head and neck black, glossed mainly green, with incomplete, broken white necklace on lower throat and broad half-collar of black and white lines on sides of mid-neck; stripes on breast sides black, not joining neck lines. Mantle and scapulars black, with bold, transverse lines of white spots and checks, enlarging towards rear of scapulars; wing- and upper tail-coverts and flanks dull black, with many white spots. Quills black; underwing white, with some brown marks, fading into brown-black of undersides of quills. Rest of underparts white. Bill black, very short legs and feet grey-black, iris wine-red. ADULT NON-BREEDING. Upperparts dark, often

oily brown, with upper head and hindneck darker than back (the reverse of pattern in non-breeding *G. arctica*), dark sides of neck with obvious white crescent-shaped indentation halfway down, incomplete dark collar at neck base, and back feathers showing paler margins or bands, most evident on edges of scapulars. Underparts generally white, with brown marks producing mottled or spotted pattern on sides of chest and top flanks. Bill grey, with blackish culmen diagnostic (even if tip pale). JUVENILE. Like non-breeding adult but feather margins of back much more distinct, being grey and shaped as scallops; separation of dark and pale plumage occluded by fine streaking along border, particularly on foreneck. Bill bluish-white, even cream, with dusky patches at base, but culmen again always dark; eyes browner than adult.

Uniformly black bill and head of breeding adult prevents confusion with same-aged *G. adamsii* and *G. arctica* but separation of non-breeding and immature birds from those species more difficult. Confusion of pale-billed birds with *G. adamsii* clarified by Burn and Mather (1974) who listed following points as crucial to identification of this species: culmen always dark (even when rest of bill pale), dark cheeks, more uniform upperparts, and (in exceptional field circumstances or in the hand) dark brown primary shafts, less pointed maxilla feathering reaching

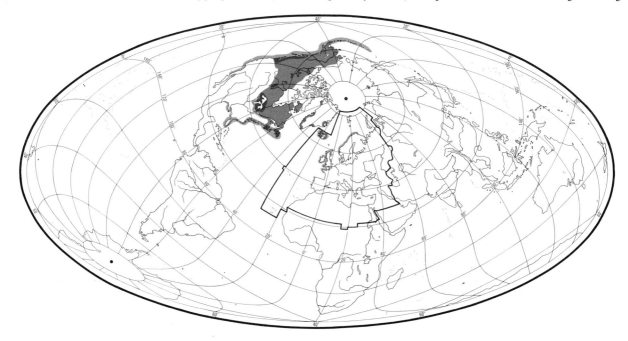

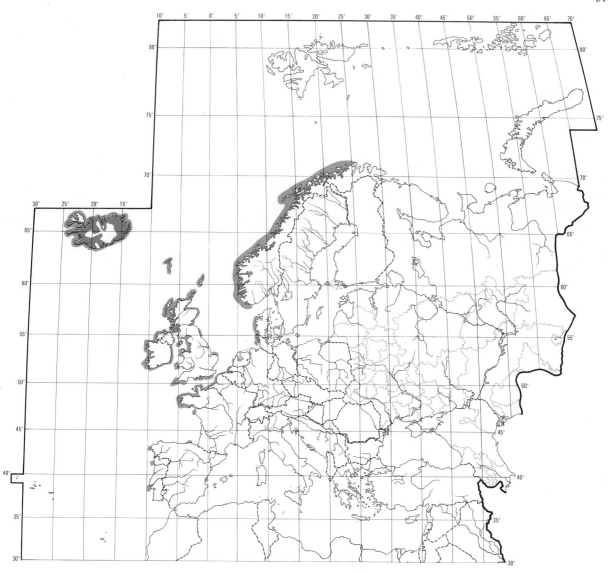

but not covering upper edge of nostril, and convex side to bulky bill. Compared to *G. arctica* even small birds distinguished by heavier bill, steeper forehead, thicker, more patterned neck, and upperparts paler than hindneck and less uniform. Confusion with Red-throated Diver *G. stellata* unlikely, though at distance worn or discoloured birds of that species can show similar pattern and some *G. immer* have apparently up-tilted bills. Attention to general bulk, weight of foreparts (particularly neck), and head carriage should prevent mistake. (See Fig I, page 43).

Flight character as other divers, but wing action noticeably slower and more powerful than smaller species (recalling that of goose *Anser* or cormorant *Phalacrocorax*) and noticeably shallower than *G. stellata*; flies less than smaller species, taking off laboriously and landing with greater impact.

Prefers large, deep lakes for nesting, tolerating surroundings varying from coniferous forest to scrub tundra. Winters mainly in coastal waters but more regular inland than *G. arctica*; less sociable at sea than *G. stellata*, rarely forming even small groups except on migration. Highly vocal in breeding season; mainly silent in winter, though moaning-call heard from flocks in autumn.    DIMW

**Habitat.** Basically Nearctic and mainly subarctic breeding range concentrated largely within northern coniferous forest zone, overlapping fairly freely on to tundra and using suitable waters of open, treeless regions in south Greenland, Iceland, and Bear Island. Though sometimes on smaller pools or shallow bays, selects mainly medium-sized or large, remote, undisturbed lakes with extensive, fairly deep and open water, establishing spacious territories well stocked with fish and containing safe nesting islets. Requires ample room for withdrawal on remote approach of danger, by swimming, diving, or taking flight with long take-off run so tree-fringed margins normally avoided. Habitat preferences thus similar to Black-throated Diver *G. arctica* but demands even more space. Uses sites far inland much more often than White-billed Diver *G. adamsii*, but, when breeding near coast, flights freely to fish in tidal waters. Avoids floating or emergent vegetation and tends to keep clear of ice.

Before closure by ice, population makes long-distance shift to south, usually to marine waters, often off rocky and exposed coasts but also commonly in bays, channels, and sheltered inlets even along low-lying, shallow coasts. Sometimes on inland lakes and reservoirs, but rarely rivers. Occupies most western winter quarters of divers in east North Atlantic, especially off north-west Scotland, Ireland, and south-west England. Normally flies low over sea but higher, up to 500 m or more, over land; makes frequent, wide, circling, noisy, aerial excursions over breeding territory. Tolerates only more or less distant human presence and, apart from reservoirs, does not use artefacts.

**Distribution.** In west Palearctic, breeds regularly only Iceland and perhaps Bear Island. Nested Scotland 1970 (British Ornithologists' Union 1971) and possibly earlier (see Movements); may have bred Faeroes in 18th century (Williamson 1970). Seen in summer Jan Mayen Island, but no proof of breeding (Marshall 1952). In North America, more data required on northern limits, especially overlap with *G. adamsii* (Godfrey 1966, and compare Palmer 1962).

Accidental. Belgium, Luxembourg, Netherlands, Spitsbergen, Sweden, East Germany, Poland, USSR, Bulgaria, Czechoslovakia, Hungary, Austria, Switzerland, Yugoslavia, Greece, Rumania, Turkey, Italy, Algeria, Azores.

**Population.** ICELAND: probably 100–300 pairs (Gardarsson 1975). BEAR ISLAND: perhaps 1 pair only (Williams 1971).

Survival. Oldest ringed bird 7 years 10 months (Kennard 1975).

**Movements.** From Iceland and Greenland main departures, singly or in small parties, September–October, though some winter southern coasts (Salomonsen 1950a; Bannerman 1959). Minority reach winter quarters (e.g. Scotland) from mid-August, probably mainly immatures or failed breeders. Infrequent records of family parties on north Scottish coasts, August (Venables and Venables 1955; Macmillan 1970) perhaps from local breeding, since by then juveniles hardly capable of sustained flight for minimum 800 km crossing from Iceland, while possibly family bond broken before autumn migration (Bent 1919; Palmer 1962). Freezing of freshwater lakes over most of breeding range results in maritime winter distribution. Spring return early May to mid-June according to latitude and weather, remaining in bays and fjords until thaw of inland ice. Pre-breeders summer chiefly in northern coastal waters (Palmer 1962), some regularly in Shetland (Venables and Venables 1955), but seldom North Sea.

Migrations little understood in absence of European ringing recoveries. West Palearctic winterers may be mainly of Icelandic origin, but some probably from Greenland or even Canadian Arctic: one Faeroe Islands *c.* 1860 contained Eskimo arrow (Salomonsen 1935), and others collected Iceland in winter had ingested grit not of Icelandic origin (Gudmundsson 1972). Winters regularly, though numbers relatively small, around Britain and Ireland and in North Sea; irregularly east to Varanger Fjord and Latvia, and south to Azores, west Mediterranean, and Madeira. Casual to inland European waters; Black Sea reports erroneous (Dementiev and Gladkov 1951).

Nearctic birds winter Atlantic and Pacific coasts of Canada and USA, with coastal migration and broad-front passage overland to and from breeding range; important stopping area on Great Lakes, but up to 300 may gather temporarily on small lakes in adverse weather (Palmer 1962).

**Food.** Primarily fish up to 28 cm, but varies with locality and season, and diet can include crustaceans, molluscs, annelids, insects, and amphibia. In some breeding areas where no fish occur, crustaceans, insects, and molluscs taken instead (Munro 1945). When searching, regularly dips bill and forehead underwater before diving silently from surface. Feet used in propulsion but, particularly in spurts and turning, also wings. Small fish swallowed under water, larger or spiny fish and crabs brought to surface and mutilated before swallowing. Though diving depths of up to 70 m quoted, based mainly on birds in fish nets, more normal depths probably 4–10 m (Roberts 1932; Schorger 1947). Maximum diving times probably 1 min (Wolton *et al.* 1950; Stewart 1967); durations above this (up to 8 mins), most likely caused by factors other than food-finding, e.g. escaping (Schorger 1947; Salomonsen 1950a).

Fish include haddock *Melanogrammus aeglefinus*, cod *Gadus morhua*, whiting *Merlangius merlangus*, herring *Clupea harengus*, sprat *Sprattus sprattus*, gurnard *Eutrigla gurnardus*, bull rout *Myoxocephalus scorpius*, sand-eel (Ammodytidae), pipefish (Syngathidae), goby (Gobiidae), flat-fish (Pleuronectidae), eel-pout *Zoarces*, eel *Anguilla anguilla*, stickleback (Gasterosteidae), trout *Salmo trutta*, perch *Perca fluviatilis*, roach *Rutilus rutilus*, and char *Salvelinus alpinus*. Crustaceans include crabs (*Portunus, Carcinus*), shrimps, and prawns; molluscs, razorshell (*Solen*), *Planorbis*, and small cephalopods; annelids, polychaet worms, and leeches; aquatic insects, caddisfly larvae (Trichoptera), waterboatmen (Corixidae), and dragonfly nymphs (Odonata). Young birds occasionally recorded (Meinertzhagen 1941), but may be due to territorial hostility rather than feeding urge. Plant material occasionally found in some quantity, mainly roots, seeds, moss, and shoots of willow *Salix* (Collinge 1924–27; Madsen 1957; Salomonsen 1950). For American records, see Palmer (1962).

In 38 stomachs collected during winter in British Isles 55·3% fish, 24·0% crustaceans, 18·5% molluscs, and 2·1% unidentified materials (Collinge 1924–27). 3 collected December and January in Denmark, contained *Platichthys flesus* in one; *Gadus morhua*, *Myoxocephalus scorpius*, and *Zoarces viviparus* in another; and *M. scorpius*, *Z. viviparus*, *Anguilla anguilla*, and *Gasterosteus aculeatus* in 3rd (Madsen 1957).

**Social pattern and behaviour.** Based mainly on material supplied by S Sjölander; see especially Olson and Marshall (1952), Sjölander and Ågren (1972).

1. In isolated pairs when breeding, and singly or in small flocks on sea in winter, though little known about organization outside breeding season. BONDS. Monogamous pair-bond, probably life-long. Both parents tend chicks until time of migration; parent-young bond seems somewhat stronger than in Red-throated Diver *G. stellata* and Black-throated Diver *G. arctica*. No social attachment between siblings. BREEDING DISPERSION. On medium or large clear-water lakes occupies large all-purpose territories. Numbers of pairs on any lake depend on local topography, but pairs usually nest out of sight of one another and rarely closer than 400–500 m (Munro 1945; Olson and Marshall 1952). Breeders and non-breeders spend entire summer in territory; feeding on other lakes and on sea rare, though feeding flocks reported. ROOSTING. As in *G. stellata*.

2. Little is known about pair-formation. Presence of lone, calling individuals on breeding water where later joined by second bird, with more display than in apparent older pairs, suggests initial pairing in summer quarters (Sjölander and Ågren 1972). Spring occupation of traditional territory by already paired birds, with little display and tendency to start nesting almost immediately, indicates established pair remain together in winter. TERRITORIAL BEHAVIOUR. Highly aggressive in breeding territory but behaviour even less ritualized than that of *G. arctica*. Territorial defence and advertisement effected by calls, especially Long-call. Tremolo corresponds in function to Croaking in *G. arctica* and *G. stellata*, but performed more often and frequently given by pair in duet combined with Long-calls; answered by nearest neighbours, producing well-known 'call-concerts' on still nights or mornings (Olsen and Marshall 1952). In boundary encounters especially, Long-call preceded and accompanied by Forward-posture with body crouched low in water, neck and head extended along water, and bill tipped slightly up and open when calling (see Rummel and Goetzinger 1975). In response to intrusion, one or both of pair approach in Alert-posture and silently perform Circle-dances, Bill-dipping, and Splash-dives; also Rushes interspersed with Bow-jumping during which Fencing-posture (see A), mostly with spread wings, adopted (Sjölander and Ågren 1972). Latter behaviour also occurs while bird remains stationary and, compared with *G. arctica*, S-bend neck-posture less pronounced though feathers of forehead more conspicuous; also Circle-dance much less ritualized and Bow-jumping shorter and less elaborate. Fights rare but vigorous (Munro 1945; Yeates 1950). Circle-dances and other antagonistic

A

behaviour also occur in summer feeding flocks, though less conspicuous than in *G. arctica*. HETEROSEXUAL BEHAVIOUR. Courtship. Much as described for *G. arctica* and *G. stellata*, consisting of Bill-dipping and Splash-dive, Search-swimming, and Inviting. Many sequences daily, especially when ♂ unresponsive, and first phase (Bill-dipping and Splash-dive) often omitted (Sjölander and Ågren 1972). Copulation as *G. stellata* several times daily from arrival to egg-laying (Tate 1969). RELATIONS WITHIN FAMILY GROUP. Much as in other divers but young spend more time (sometimes greater part of day) riding on parent's back during first 2 weeks, probably as adaptation to windy conditions on preferred large lakes. Adults more prone to defend young on disturbance, using Rushes, Bow-jumps, Fencing-posture, and Tremolo-calls.

(Fig A after drawing by S Sjölander.)

**Voice.** Mainly silent in winter quarters, though calls reported from flocks especially in autumn.

Wide variety of loud, sustained, rising or falling, mostly low-pitched, wailing or howling calls given by both sexes in breeding area (see particularly Olson and Marshall 1952). Following distinguished by S Sjölander, of which all, except Moaning-call, connected in some way with territorial advertisement and defence, and often uttered in response to similar vocalizations by neighbours, producing outbreaks of (often prolonged) concerted calling, mostly between dusk and dawn. (1) Tremolo-call. Call of 3–10, usually 4–5 notes running together and rapidly repeated (see I). Often strident or screaming, gives impression of uncontrolled, idiotic laughter; said to be due to even intensity throughout (Palmer 1962) but S. Sjölander found first 2 syllables lower pitched than rest. Often performed in duet with mate, sometimes combined with Long-calls. Typical response to disturbance or conspecifics flying over territory. (2) Wailing-call. Abruptly rising wail with yodel-like breaks between notes; also likened to wolf-like howling by Olsen and Marshall (1952) who give usual rendering as 'ahaa-ooo-oooo-oooo-ooo-ahh'. Sample (see II) lasts 1–7 secs with a two-step rise in pitch. Typically given during low-intensity territorial encounters; see also Olson and Marshall (1952). (3) Long-call. 'Yodel' of Olson and Marshall (1952). Prolonged, slowly rising, throaty whistle with yodel-like breaks between notes, last 2 of which being rhythmically alternated usually 3–6 times, at end of rising phase. Extremely variable, weird and wild-sounding; described as 'a-a-whoo-queee-queee-wheoooo-queee' repeated up to 3–6 times, with pitch rising on 3rd syllable and undulating on rest (Olson and Marshall 1952). Sample (see III) begins with 3 units of increasing pitch, rising abruptly but without separation, followed by 2 units rhythmically alternated 5 times with yodel-like breaks between. Main territorial call, typically uttered in response to intruders. (4) Moaning-call. Man-like, moaning sound rising slightly in pitch. Main contact-call when seeking absent mate or young. May be 'Hoot' of Olson and Marshall (1952): short, abrupt, medium-pitched. 1-syllable call uttered at irregular intervals from 1–10 secs; far-carrying, though not loud. Most often heard from flocks especially in autumn, usually as 'talking' call by one bird at a time.

CALLS OF YOUNG. Shrill, indistinct chirping, resembling equivalent call of Red-throated Diver *G. stellata* (S Sjölander). Likened to peeping of domestic chicks (Olson and Marshall 1952); constantly given when tiny, less frequently when older. Thin, prolonged, high-pitched 'heeeuer' also heard from 3-week old, captive chick. Voice becomes more adult-like in autumn but still high and thin, and infrequently used.

**Breeding.** SEASON. See diagram. Onset of breeding in northern parts of range dependent upon thaw. SITE. Always close to water, preferably of diving depth; on islands or promontories, and usually screened by vegetation. Previous year's sites often re-used. Nest: mound of vegetation, occasionally only a scrape. Average diameter 37–47 cm and height 4–10 cm (Gudmundsson 1952a). Building: probably by both sexes. Material often added during incubation by sitting bird gathering vegetation within reach (Palmer 1962). Nest shaped by movements of bill, feet, and body. EGGS. Elongate ovate, slightly glossy; olive-green to dark brown with small dark brown spots, occasionally a few larger black blotches. $90 \times 58$ mm ($84–102 \times 54–63$), sample 100; calculated weight 167 g (Schönwetter 1967). Clutch: 2 (1–3). One brood. Replacement clutches laid 5–14 days after egg loss (Olson and Marshall 1952). Egg-laying interval at least 2 days. INCUBATION. 24–25 days. By both sexes though ♀ does larger share. Begins with 1st egg; hatching asynchronous. Eggshells may be removed on hatching or left in nest (Palmer 1962; R S Palmer). YOUNG. Precocial and nidifugous. Cared for and fed by both parents. Ride on parents' back for up to 14 days, occasionally 21. Can dive at 2 days old, but still fed by parents at 42 days (Olson and Marshall 1952). ♂ defends brood against predators by rushing over surface of water with head and neck outstretched. FLEDGING TO MATURITY. Fledging period not certain; estimated 70–77 days (Olson and Marshall 1952), though no flights seen until young older than this. Become independent after fledging, though some stay with parents up to and possibly including migration. Mature at 2 years. BREEDING SUCCESS. No data from west Palearctic. In Minnesota, 42 breeding pairs laid 63 eggs, lost 32, hatched 27, and reared 21 young to beginning of September (Olson and Marshall 1952).

I  S Sjölander  Alaska

II  S. Sjölander Alaska

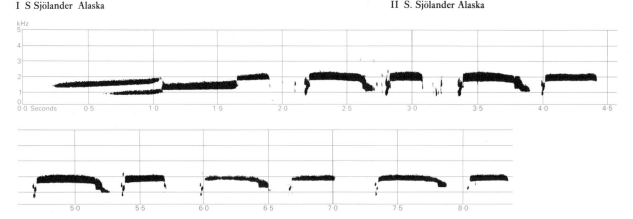

III Sture Palmér/Sveriges Radio Iceland June 1967

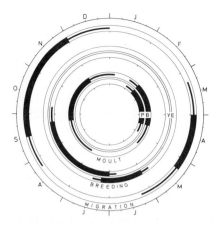

**Plumages.** ADULT BREEDING. Head and neck black, with green and purple metallic gloss; chin, throat, and lower neck strongly glossed green; transverse row of 6–8 short white streaks on throat; half-collar of *c*.15 similar streaks on either side of neck. Rest of upperparts black with white spots, small and round on centre of mantle and upper wing-coverts, large and square on scapulars, tiny on rump. Breast and belly white, with band of white-spotted black feathers across vent; under tail-coverts dull black, shorter ones with white tips or spots. Sides of chest streaked black and white, flanks black with white spots. Tail black, paler below. Primaries black with dark brown shafts; secondaries similar, inner ones with 1–2 white spots. Under wing-coverts white, greater with grey shaft-streaks; under primary coverts grey with white margins to inner webs; axillaries white with dark shaft-streaks. ADULT NON-BREEDING. Upperparts dark brown-grey, mantle and scapulars with pale markings suggestive of white spots in breeding; hindneck darker than mantle. Underparts white with dark band across vent, sometimes obscure; under tail-coverts white. White crescent on side of neck, protruding into dark grey of dorsal part; dark half-collar lower down with ends sometimes almost meeting on foreneck (Richter *et al.* 1970). Flanks mottled dark brown and white. Tail-feathers tipped white. Wing as breeding, with bright white spots on upper wing-coverts. DOWNY YOUNG. 1st down dark brown-grey with white spots on chin, throat, and breast; centre of belly white. 2nd down similar but paler. Juvenile feather growth starts *c*.3 weeks after hatching; fully fledged *c*. 8 weeks. JUVENILE. Like adult non-breeding, but upperparts browner; feathers of mantle, scapulars, rump, and flanks rounded with wide pale margins; throat and sides of neck with fine brown specks; under tail-coverts dusky; tail-feathers with dusky tips; no white spots on inner secondaries and upper wing-coverts. IMMATURE. In spring of 2nd calendar year, juvenile plumage gradually replaced by one resembling adult non-breeding, but lacking white-spotted upper wing-coverts; distinguished from juvenile by square feathers on back, with pale spots on both webs, and by white tips to tail-feathers. Often many juvenile feathers retained. Another plumage may appear in summer (Palmer 1962) with crown and hindneck

darker; mantle and scapulars dark brown, slightly glossy, with pale spots; lesser upper wing-coverts with round dirty-white spots. Tail-feathers black or narrowly tipped white. Not known whether all individuals acquire this immature summer plumage; great individual variation and juvenile and immature winter feathers often retained. Birds in 2nd winter differ from 1st winter by absence of juvenile feathers and brown specks on throat; from adult non-breeding by absence of bright white spots on upper wing-coverts, but indistinguishable when these appear during pre-breeding moult in spring of 3rd calendar year.

**Bare parts** ADULT BREEDING. Iris red to brown-red. Bill black. Foot outside black, inside pale grey, webs with flesh-coloured centres. ADULT NON-BREEDING. Culmen brown to dark brown, rest of bill variably grey. DOWNY YOUNG. Iris morocco-red. Bill grey, dark at base, light at tip. Foot blackish (Olson and Marshall 1952). JUVENILE. Like adult non-breeding; iris browner; bill paler, more blue-grey, but ridge of culmen always dark.

**Moults.** ADULT POST-BREEDING. Partial; body, tail, and part of wing-coverts; begins late July–early October, completed October–January. ADULT PRE-BREEDING. Complete; primaries simultaneously February–April. Rest of plumage starting February–March, complete April–May. Birds moulting late may be in 3rd calendar year acquiring adult breeding for first time. POST-JUVENILE AND SUBSEQUENT PLUMAGES. Birds in 2nd calendar year apparently almost continuously moulting. Tail and part of body moulted February to late spring. Not sharply separated from summer moult so not clear whether 2 separate plumages may be distinguished (Palmer 1962) or only one (Witherby 1940). Juvenile flight-feathers moulted simultaneously in summer of 2nd calendar year. Moult into 2nd immature winter in autumn; timing variable, almost completed or just starting December. Next moult, first pre-breeding, complete; parallel to adult pre-breeding but probably later.

**Measurements.** Adult breeding; ♂♂ and ♀♀ combined. Skins. (RMNH, ZMA, ZMK).

| | | | | | |
|---|---|---|---|---|---|
| WING | 366 | (13·4 ; 29) 331–400 | BILL | 80·2 ( 3·33; 27) | 72–89 |
| TAIL | 66·8 ( 3·37; 21) | 58–75 | TARSUS | 91·4 ( 3·39; 25) | 83–100 |

♂♂ average slightly larger than ♀♀, difference not significant. Hartert (1912–21) and Witherby *et al.* (1940) gave extreme wing lengths 405 and 408.

**Weights.** Limited data. 2 ♂♂ 3600 and 3990; 3 unsexed, winter 4250, March 4350, May 4480 (Bauer and Glutz 1966); juvenile ♀, winter 2780 (RMNH). Range 1600–8000, given by Palmer (1962), seems improbable.

**Structure.** Wing narrow, pointed. 11 primaries: p10 longest, p9 0–5 shorter, p8 12–20, p7 30–45, p1 *c*.150. P11 minute and hidden. Tail short; 20 feathers. Bill heavy; cutting edge of upper mandible slightly curved, sometimes almost straight. Juvenile bill often less heavy than adult. Mandibular rami usually not completely fused at angle of gonys. Nostril elongate with small tubercle on dorsal rim. Feathering at base of upper mandible falls short of this tubercle (Burn and Mather 1974).

## *Gavia adamsii* **White-billed Diver**

PLATES 5 and 6
[after page 62]

Du. Geelsnavelduiker    Fr. Plongeon à bec blanc    Ge. Gelbschnabel-Eistaucher
Ru. Белоклювая гагара    Sp. Colimbo de Adams    Sw. Vitnäbbad islom    N.Am. Yellow-billed Loon

*Colymbus adamsii* Gray, 1859

Monotypic

**Field characters.** 76–91 cm (body 46–56 cm), wingspan 137–152 cm; largest diver, bulkier than Great Northern *G. immer* and with longer bill and wings. Scissor-billed, large-headed, huge diver, often with bump on forecrown exaggerating angle of forehead and adding to diagnostic uptilt of both head and bill. Breeding plumage essentially as *G. immer* but pale cream bill affords immediate recognition. Other plumages also similar to *G. immer* and, when bill dusky, separation can be difficult. Bill shape now shown to be untrustworthy (Burn and Mather 1974) though typical birds have straight culmen and lower mandible distinctly angled upwards. Sexes similar; marked seasonal variation; juvenile separable at close range.

ADULT BREEDING. At close range, plumage differs from *G. immer* in mainly purple gloss on throat and neck (looking blacker), fewer, broader streaks on necklace and half-collars, and larger checks on upperparts, especially on scapulars. Bill ivory to pale straw (depending partly on intensity of light); legs and feet dark grey; eyes red-brown. ADULT WINTER. Similar to *G. immer* but typically with paler face and sides of head (with whitish feathers extending from cheeks to round eye, recalling pattern of Red-throated Diver *G. stellata*), large spots on sides of nape whitish and made more obvious by dark surround, less demarcation between colour of upper and underparts, and more complete pale brown neck collar. Bill pale cream, with variable dusty cast over bases and sides of both mandibles near head; pale culmen remains diagnostic. JUVENILE. Again similar to *G. immer* but typically with less definite pattern on head and neck, greyer-brown upperparts, with noticeably wider and paler tips to feathers of mantle, scapulars, and wing-coverts. Bill shorter than adult, showing similar colour variation though pale culmen again diagnostic; eyes browner than adult.

White bill of breeding adult unmistakable but separation of winter and immature birds from *G. immer* not easy, requiring greatest concentration on bill colour and structure, head shape and carriage, and upperparts pattern. Burn and Mather (1974) list following points as crucial to identification of this species: culmen uniform with or paler than rest of bill, pale cheeks, and (in exceptional field circumstances or in the hand) pale white to yellow-brown primary shafts, extremely pointed maxilla feathering almost covering upper edge of nostril, and flat sides to rather narrower bill. At distance, bill shape and carriage combined with pale face may suggest *G. stellata* but large head, general bulk (especially thick neck), and scaled, not spotted upperparts should allow diagnosis. (See Fig I, page 43.)

Flight and general behaviour essentially as *G. immer*. Voice as *G. immer* but louder and harsher. DIMW

**Habitat.** High arctic breeder along littoral of Arctic Ocean and on flowing and standing water of tundra zone, almost entirely beyond tree-limit. Prefers low-rimmed fresh waters, occurring at least up to 425 m in foothills in Alaska (Sage 1971), but in west Palearctic avoids mountainous terrain. Requires adequate ice-free period and immunity from sudden changes in water level while breeding. Unlike Great Northern Diver *G. immer*, shows no marked preference for large lakes or lagoons, nor for nesting on islets or other safe sites concealed from ground predators by tall plants. In Alaska, Sage (1971) found nests on smallish pools of 17 and 43 ha, with maximum depths 3 m or less and apparent pH 6–7. Evidence contradictory on use or avoidance of coastal belt for breeding and on use of river sites; apparently both favoured in European Russia, but neither in North America. Distribution implies some marginal breeding sites in wooded tundra or taiga zones, but data on habitat still inadequate and possibly unrepresentative. Flies readily to feed at distance from breeding waters. Outside breeding season, lives almost wholly at sea, mainly in closest ice-free waters especially inshore and up inlets. Rarely and briefly on fresh waters in winter and only accidentally on or close to land, so rarely observed by ornithologists, and even less known then than on breeding grounds. Apparent low numbers and restricted choice of suitable habitats indicate problems for conservation with opening up of high Arctic.

**Distribution.** Precise distribution still unclear because some areas not fully surveyed, while breeding often sporadic in USSR and confusion has occurred with Great Northern Diver *G. immer* in North America (Dementiev and Gladkov 1951; Godfrey 1966).

Accidental. Britain, Netherlands, West Germany, Denmark, Sweden, East Germany, Poland, Czechoslovakia, Austria, Yugoslavia.

**Population.** No census information, but rare west Palearctic, being fairly common in USSR only in north-east Siberia (Dementiev and Gladkov 1951).

**Movements.** Migratory. Like Great Northern Diver *G. immer*, breeds on high latitude fresh waters but winters on

PLATE 5. *Gavia immer* Great Northern Diver (p. 56): **1** ad breeding, **2** imm, **3** juv, **4** downy young 1st down. *G. adamsii* White-billed Diver (p. 62): **5** ad breeding, **6** imm, **7** juv. (PB)

PLATE 6. *Gavia immer* (p. 56): **1** ad breeding, **2** juv. *G. adamsii* (p. 62): **3** ad breeding, **4** imm. *G. arctica* (p. 50): **5** ad breeding, **6** juv. *G. stellata* (p. 43): **7** ad breeding, **8** ad non-breeding. (PB)

PLATE 7. *Podilymbus podiceps* Pied-billed Grebe (p. 67): **1** ad breeding, **2** ad non-breeding, **3** juv. (PB)

PLATE 8. *Tachybaptus ruficollis* Little Grebe (p. 69): **1** ad breeding, **2** ad non-breeding, **3** juv, **4** downy young. (PB)

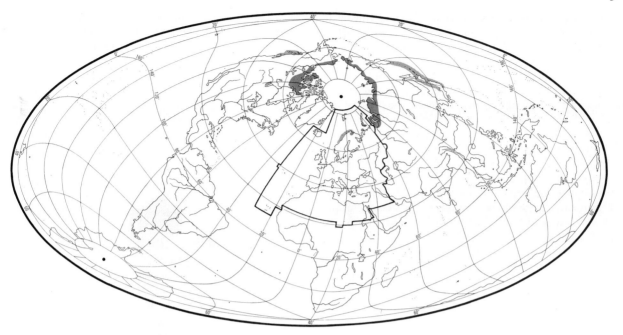

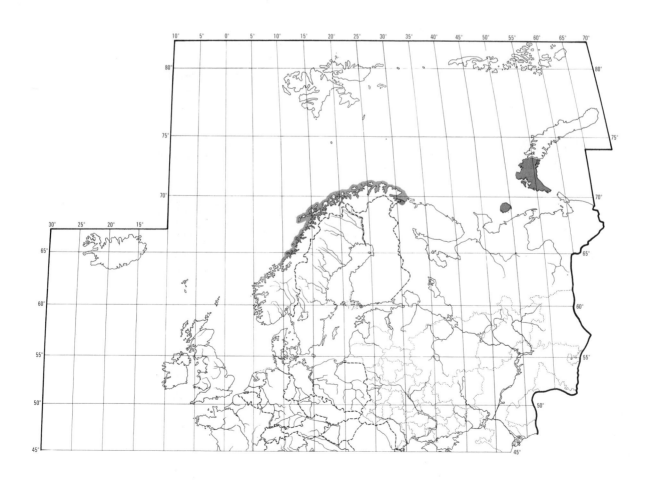

salt water. Freshwater arrivals and departures linked to thawing and freezing; at sea mainly October–May; when late thaw breeding lakes not re-occupied until late June (Sage 1971). No ringing recoveries for Eurasia, but principal wintering areas thought to be Bering Sea and adjacent parts of north Pacific (Palmer 1962), and, in west Palearctic, off north Norway (Collet 1894a; Haftorn 1971). Thus main migrations east and west along arctic coasts of Eurasia. Smaller numbers winter off west Norway, and stragglers reach North Sea where identified almost annually in recent years. Infrequent Baltic, though observations in 1960s indicate that may pass regularly in small numbers along Gulfs of Finland and Bothnia (L von Haartman). Exceptional inalnd in western Europe. Presumably west Palearctic winterers originate European Russia, but some may come from farther east.

Nearctic breeders appear to winter in restricted area off coasts of British Columbia and south-east Alaska; apparently follow coast around Alaskan peninsula, though possibly some overland movement from Pacific coast to Great Slave Lake region; only vagrant to west Atlantic. In spring, remains offshore until ice breaks on rivers and later on tundra lakes and pools (Palmer 1962; Sage 1971).

Migrates singly or in small parties, possibly family groups in autumn.

**Food.** Mainly fish caught by underwater pursuit after dive from surface. Crustaceans, molluscs, and marine annelids recorded, though possibly mainly taken accidentally or via fish stomachs. 3 birds from Alaska coast contained sculpin *Leptocottus armatus* 19·7%, great sculpin *Myoxocephalus joak* 13%, undetermined Cottidae 56·7%, tomcod *Microgadus proximus* 10%, crustaceans 0·3%, and molluscs 0·3%; all fish species readily available in area (Cottam and Knappen 1939). Other fish recorded include bull rout *Myoxocephalus scorpius* (Collett 1894a) and cod *Gadus morhua* (Bailey 1922).

**Social pattern and behaviour.** Based mainly on material supplied by S Sjölander.
1. Similar to closely related Great Northern Diver *G. immer*, occurring in isolated pairs when nesting, and singly, in couples, or small parties in winter when, however, even less known. BONDS. Evidence suggests monogamous, life-long pair-bond: same nesting waters and sites used annually; paired birds arrived together in spring and show same simple courtship as other divers. Bonds between parents and young much as in *G. immer*; almost no bond between siblings. BREEDING DISPERSION. Similar to *G. immer*. Where pools abundant, prefers those 10–20 ha, one pair to a pool. On larger waters, prefers inlet or adjacent pool probably seeking protection for nest and young from wave action. Territory size much as in *G. immer*, but pairs not closer than 600 m. Pair spends summer in territory, apart from fishing excursions, especially to rivers. ROOSTING. As Red-throated Diver *G. stellata*.
2. TERRITORIAL BEHAVIOUR. Defence and advertisement mainly by calls, similar to same of *G. immer* but lower pitched and shorter. Behaviour at territorial intrusions as in latter species:

approach in Alert-posture, Circle-dances, Bill-dipping, and Splash-dives (silent), often close to shore. Rushes, Jumps, and Fencing-posture (often with Bow-jumping) as in *G. immer*—though Fencing-posture, with widely spread wings (see A) and quick turns to left or right, seems more common. Such antagonistic behaviour also seen in summer feeding flocks, especially when flocking on occupied territory. HETEROSEXUAL BEHAVIOUR. Courtship similar to other divers: Bill-dipping and Splash-dive followed by Search-swimming and Inviting. ♂ often unresponsive. Copulation (see B) as in other divers. RELATIONS WITHIN FAMILY GROUP. Similar to *G. immer*, though more prone to brood young on land (due to windy habitat).

(Figs A–B after original drawings by S Sjölander.)

A

B

**Voice.** Resembles that of Great Northern Diver *G. immer*, but louder and harsher (Bent 1919); also less frequently used. Tremolo-call, Wailing-call, Long-call, and Moaning-call distinguished by S Sjölander in breeding area. All closely resemble corresponding calls of *G. immer*, and uttered in similar circumstances, but Tremolo pitched higher (see I) and Long-call (see II) lower, while Long-call also has shorter ending. Flight-cackle version of Tremolo described as evenly pitched 'ha-ha-ha-ha-ha-ha-ha' (Palmer 1962).

CALLS OF YOUNG. Utter shrill, indistinct chirping, similar to *G. immer* and Red-throated Diver *G. stellata*.

I S Sjölander Alaska

II S Sjölander Alaska

**Breeding.** SEASON. Eggs from early June, dependent upon thaw. SITE. Similar to Great Northern Diver *G. immer*. Nest: similar to *G. immer* but sometimes turf cut and overturned to form mud platform (Sage 1971). Building: probably by both sexes. EGGS. Elliptical ovate, slightly glossy; olive to umber, spotted and blotched with black. 89 × 57 mm (80–100 × 54–62), sample 20; calculated weight 160 g (Schönwetter 1967). Weight of 2 fresh eggs 161 and 146 g (Kretchmar and Leonovitch 1965). Clutch: 2. One brood. No information on replacement clutches. INCUBATION. Period not recorded but probably similar to *G. immer*. By both sexes. Begins with 1st egg; hatching asynchronous. YOUNG. Precocial and nidifugous. Cared for and fed by both parents who defend young vigorously. FLEDGING TO MATURITY. Fledging period not recorded. Young may stay with parents until after migration from breeding place (Sage 1971). Age of maturity not known. BREEDING SUCCESS. No data.

**Plumages.** ADULT BREEDING. Like Great Northern Diver *G. immer* but gloss on throat purple, not green; white streaks on throat and collar wider and fewer (*c*.10 on half of collar against *c*.15 in *G. immer*). White spots on sides of mantle and scapulars larger and more rectangular. Fewer spots on rump, none on lower back. Shafts of primaries in all plumages pale cream to yellow-brown with darker tips, not dark brown. ADULT NON-BREEDING. Like *G. immer*, but sides of head paler, pale area extending over eye. Pale spots on mantle and scapulars larger. DOWNY YOUNG. Down probably as *G. immer*, but paler. JUVENILE. Sides of head more extensively pale, like adult non-breeding; upperparts and flanks paler with wider and paler grey feather margins. IMMATURE. Sequence of plumages similar to *G. immer*, but plumages tend to be paler.

**Bare parts.** ADULT BREEDING. Iris red-brown. Bill pale yellow. Foot on outer surface dark brown-grey, on inner surface pale grey. ADULT NON-BREEDING. Bill pale yellow, darker at base. JUVENILE. Iris brown. Bill pale, darker at base; culmen always much paler than *G. immer*.

**Moults.** Apparently similar to *G. immer*. Primaries of adult simultaneously renewed February to April (first described but wrongly interpreted by Collett 1894*a*).

**Measurements.** Adult, skins (RMNH, ZMA, ZMK). Sexes similar; combined.

| | | | |
|---|---|---|---|
| WING | 381 (4·90; 7) 376–402 | BILL | 90·4 (3·20; 10) 83–96 |
| TAIL | 65·3 (3·47; 10) 62–72 | TARSUS | 91·9 (2·74; 11) 89–97 |

Range of wing greater in larger samples: 364–405 (sample 22) (Dementiev and Gladkov 1951); 392 (11·8; 16) 370–419 (Collett 1894*a*). Juveniles have shorter wings and bills than adults. Bill of juvenile and 1st winter: 79·2 (5·26; 5) 71–84. 1st summer and older: 92·4 (3·62; 31) 85–102 (Burn and Mather 1974). *G. adamsii* differs from *G. immer* in proportions; wing and bill averaging significantly longer, but tail and tarsus same size. Juvenile bill not fully developed does not differ from *G. immer*.

**Weights.** 2 summer ♂♂ 5700, 5800; September ♂ 5000; 5 summer ♀♀ 4400–6400 (Palmer 1962). 3 summer ♀♀ 4050, 4715, 6300; ♂ 5700; unsexed: 4250 (Bauer and Glutz 1966). Lean 1st winter ♂ *c*.3700 (ZMA).

**Structure.** Differences from *G. immer*: 18 tail-feathers; sides of bill flatter, bill more strongly compressed laterally; ridge of culmen usually straight, lower mandible usually angled at gonys. Characteristic bill shape develops in spring of 2nd calendar year: juvenile like *G. immer*, but has small knob at angle of gonys (hardly seen but easily felt). Mandibular rami always completely fused at gonys. Feathering at base of upper mandible extends beyond tubercle in dorsal rim of nostril (Burn and Mather 1974).

# Order PODICIPEDIFORMES

# Family PODICIPEDIDAE grebes

Small to medium-large, foot-propelled diving birds. Sole family in order. Morphology and egg-white protein analysis suggest no ties with any other particular group of waterbirds (Storer 1971; Sibley and Ahlquist 1972). Formerly considered closely related to Gaviidae (divers) but similarities due to convergence (Stolpe 1935; Storer 1960, 1971). About 20 species in 6 genera (R W Storer; see also Simmons 1962; Storer 1963): *Rollandia*, *Tachybaptus*, *Podilymbus*, *Poliocephalus*, *Podiceps*, and *Aechmophorus*. 1 *Tachybaptus* and 4 *Podiceps* breeding, west Palearctic; 1 *Podilymbus* accidental. Body elongated or rotund; feet placed far back. Neck relatively longer than in Gaviidae. Wings small and narrow; flight-feathers curved and, when folded, fit closely to body, hidden under contour plumage.

12 primaries, p10 longest, p12 minute. 17–22 secondaries; usually diastataxic. 3 species permanently flightless; all confined to large lakes in Neotropical region. Tail-tuft short, downy; not composed of normal remiges. Bill highly variable: from long, stout, and diver-like, long and pointed, to short and sturdy. Nostrils narrow slits. Narrow strip of skin on lores bare. Tarsi markedly flattened laterally, serrated along rear edge. Toes broadly lobed, front 3 connected by small webs at base; hind toe raised, flattened, with small lobe. Nails broad and flat; those of middle toe laterally serrated. Oil gland feathered. Feathers with aftershaft. Sexes similar in plumage (♀ often somewhat duller), ♂ averaging larger. Breeding plumage characterized by display markings on head, particularly in *Podiceps*; non-breeding plumage plain, dark above, white below. Plumage loose on upperparts, downy on back and rump; dense and satiny on breast and belly. Moult of flight feathers simultaneous, causing temporary flightlessness. Young precocial and aberrantly semi-nidifugous (leave nest but at first carried by parents). 1 downy plumage with pattern of stripes in all but one species; often conspicuous, especially on head and neck. Definitive feathers develop from same follicles, pushing out down. Loral patches large and at first bare. These, and inflatable outgrowth of bare, wrinkled skin on crown (e.g. in *Podiceps*) may function in thermo-regulation (Borodulina 1971) or location of optimum warmth, but also act as social signals to parent when infused with blood (Simmons 1970c). Juvenile resembles adult non-breeding but often retains traces of striping on head; reaches fully adult plumage in 2nd calendar year.

Nearly cosmopolitan, main centre of adaptive radiation in New World (especially Neotropical region). Highly specialized, aquatic group avoiding vicinity of ice or snow or intensely cold water; also oligotrophic, deep, or turbulent waters, or exposed rocks or gravels. Prefer muddy, clay, or sandy bottoms, with emergent, floating, or submerged vegetation, especially for breeding. Opportunists, readily colonizing newly flooded or excavated areas, especially with depths of 2–3 m and of small to moderate size, or presenting sheltered inlets or channels. Basic environmental conditions determining siting of nests include vegetation density, depth of water, and distance from open water, specific composition of plant communities playing minor role (Gotman 1965); safety from predation also paramount (Simmons 1974a). When migrating or wintering, tolerate more open waters (fresh or saline) where food ample, but avoid exposed coasts or offshore and oceanic waters, preferring sheltered bays or inlets or vicinity of inshore sandbanks or mudbanks. Habitat requirements involve localized distribution, but aerial reconnaissance (often nocturnal) effective in ensuring use of most suitable places. See also Gaviidae for comparisons with that group. Complex changes in distribution within west Palearctic in last 100 years. Most marked in Great Crested Grebe *Podiceps cristatus* (increase

and spread) and Black-necked Grebe *P. nigricollis* (large range expansion in north and west). All species vulnerable to habitat changes and water pollution. Dispersive and, within northern part of west Palearctic range, largely migratory, many populations moving from smaller, inland fresh waters to larger lakes or sea. Travel nocturnally by overland flight, diurnally by coastal swimming. In many areas, only Little Grebe *T. ruficollis* remains essentially freshwater bird during winter. Moult movement demonstrated for *P. cristatus* and may be expected in other grebes. Compared with divers, more winter inland where may occur in large concentrations; even on tidal water, small flocks more usual than in divers.

Feed mainly on aquatic arthropods and fish, obtained primarily by underwater pursuit with wings held close to body. Other methods used to varying extent depending on availability of food items, but specialized surface-feeding characteristic of *P. nigricollis*. Proportion of arthropods in diet varies with season, locality, and feeding method. *T. ruficollis* probably takes relatively more molluscs than other Palearctic species; *P. cristatus* generally more fish. Feathers usually found in stomachs of both adults and young, relatively more and more often in fish-eating species; break down into felt-like mush, often forming ball. Habitually eat own feathers, mainly during self-preening; given to young from day of hatching. Significance disputed (see, e.g. Hanzák 1952; Simmons 1956; Storer 1969; Fjeldså 1973c). Pellets reported from Pied-billed Grebe *Podilymbus podiceps*, Slavonian Grebe *Podiceps auritus*, and *P. cristatus* (Storer 1961, 1969; Simmons 1973), those of last species, at least, containing feathers.

Though some species highly sociable more or less throughout year (breeding colonially), most not strongly gregarious and often solitary, and all usually disperse to feed. Monogamous pair-bond, often of seasonal duration though some overwinter in pairs, especially *T. ruficollis*. Territorial at least during breeding season; territory size, utilization for feeding, and duration of occupancy variable. Establishment of breeding territory follows pair-formation, often well in advance of nesting. Tend then to be particularly aggressive. Highly complex, mutual displays (water-courtship) accompany pair-formation and continue into post-pairing (or engagement) period. Different displays associated with platform-behaviour (building of special mating structures, soliciting, and copulation); as both soliciting and mounting by both sexes occur well before need to fertilize eggs, also termed platform-courtship. Copulation typically out of water, on platform or nest. True Courtship-feeding recorded only in Western Grebe *A. occidentalis* but more or less casual presentation of food (by ♂ to ♀ early in cycle, by either sex later) occurs in most other species. No elaborate ceremonies at nest-relief. Not so vocally conspicuous or distinctive as divers, with variety of trilling, whistling, wailing, and barking calls; more cover-haunting species, however, often highly vocal. In *Podiceps* and *Aechmophorus* particularly, species-characteristic Advertising-calls of lone birds important in pair-formation and re-establishing contact with mate or young; usually show appreciable variation, facilitating individual identification. Except *T. ruficollis*, west Palearctic species mostly silent outside breeding season; seldom or never call in flight. Comfort behaviour much as in other waterbirds, with frequent oiling and preening; preen ventrally in conspicuous and characteristic manner by rolling over in water. Bathe and then oil plumage after each feeding session; bathing at times involves splash-diving, swimming with rear-end submerged and body vertical in water, and wallowing. As well as unique habit of feather-eating (see above), general behavioural characters include: direct head-scratching; both true yawning and jaw-stretching movements; throat-touching to drain water from bill; resting with bill inserted frontally into side of neck, and foot on other side stowed away in flank pocket.

Extended laying period in most species; start often dependent on growth of emergent vegetation and, in northern parts of range, on spring thaw. Nests placed in water, anchored to vegetation, built up from bottom, or partly floating; exceptionally on land, close to water. Except in habitually colonial species, usually well dispersed in concealing or protective sites, though sometimes more grouped where sites short or one particularly safe place available. Shallow heaps of aquatic vegetation, often substantial below waterline; built by both sexes. Eggs elongate, white or cream, with chalky covering rapidly becoming stained. Clutches 2–7 (1–10); 1 or 2 broods, occasionally 3. Replacements laid after loss of both eggs or young. Indeterminate layers. Eggs laid at intervals of 1–2 days. Incubation periods 20–28 days, beginning with 1st or 2nd egg. By both sexes equally; each has single large median brood patch. Eggs covered with material if parent disturbed from nest. Eggshells mostly removed. Young hatch asynchronously. Cared for and fed by both parents, riding on back while small. Brooded on back in nest during hatching period; later on water, or on nest or specially built platform. In some species, brood often divided between parents later; see especially *P. cristatus*. Fledging periods, where known, 44–79 days. Young independent before or after. Mature at 1 or 2 years.

## *Podilymbus podiceps* Pied-billed Grebe

PLATE 7
[facing page 63]

Du. Dikbekfuut    Fr. Grèbe à bec cerclé    Ge. Bindentaucher
Ru. Пестроклювая поганка    Sp. Macá picopinto    Sw. Svartvitnäbbad dopping

*Colymbus Podiceps* Linnaeus, 1758

Polytypic. Nominate *podiceps* (Linnaeus, 1758), North America; *antillarum* Bangs, 1913, Antilles; *antarcticus* (Lesson, 1842), South America. Stragglers to west Palearctic probably nominate *podiceps*, but subspecies not determined.

**Field characters.** 31–38 cm, of which body 18–22 cm; wingspan 56–64 cm. Small, stocky grebe with short, thick, chicken-like bill, large head, and blunt-ended stern; breeding adult has whitish and black bill, white orbital ring, blackish forehead, black bib, and otherwise brownish plumage except for white undertail. Sexes nearly alike, but marked seasonal differences; juvenile distinctive.

Adult Breeding. Dominant feature is heavy, parti-coloured bill, more than half as deep as it is long with decurved ridge and blunt end, mainly bluish-white with broad, black band in middle; white orbital ring round dark brown-red eyes also conspicuous. Forehead brown-black shading to dark brown on crown and green-brown on nape; sides of head and neck grey-brown contrasting with clearly defined black throat patch bordered with whitish line at sides (this pattern usually more clear-cut in ♂). Upperparts dark brown, often with hoary look in fresh plumage; breast and flanks barred and blotched with dark brown on whitish ground, contrasting with usually conspicuous white under tail-coverts. In flight, wings mostly dark brown without any continuous patch of white as in *Podiceps*. Adult Non-breeding. Distinctive black markings lost. Bill yellowish, to dusky grey with dull greenish tinge, but no black band. Forehead and crown dark grey-brown; throat whitish, sometimes with traces of black; rest of head grey-buff with slightly darker line through eye. Neck, breast, and flanks grey-buff tinged reddish, to rich reddish tawny. Otherwise as in breeding plumage, though white rear often more conspicuous. Juvenile. Bill yellow-brown; no orbital ring. Sides of head and neck variably dark to grey-brown, blotched and streaked irregularly with white; throat white; rest of neck brown-buff. Upperparts black-brown; breast and flanks rather dark brown mixed with buff; belly and under tail-coverts white, though latter not as conspicuous as in adult. Late juveniles migrate in this plumage. By 1st winter, which may not be attained until near end of year, becomes like adult non-breeding, but bill greyish, white orbital ring still missing, and neck, breast, and flanks yellow-brown to grey-buff instead of reddish tawny.

Breeding adult unmistakable. Non-breeding adult and immature, being generally brown and blunt-sterned, likely only to be overlooked as Little Grebe *Tachybaptus ruficollis*, which is smaller and dumpier with darker and less stout bill, smaller head, shorter neck, and brown-buff (not white) throat and grey-brown under tail-coverts.

As catholic as *T. ruficollis* in choice of habitats, may be seen on passage and in winter in wide variety of usually shallow waters. Generally wary and often skulking, particularly in breeding season, though this has also been characteristic of some vagrants to Britain; in America usually out in open in autumn and winter. Swims less buoyantly than *T. ruficollis*, sometimes with head similarly sunk low, sometimes with neck extended; except when alert or alarmed, feathers fluffed out and white undertail conspicuous. Dives neatly with downward thrust of head. When alarmed, slowly sinks beneath surface until only head or sometimes only bill remains, then either moves away in this submerged position or dives and swims away before breaking surface again with head. Rests and preens on water, only occasionally coming ashore at water's edge or perching on partially submerged objects, such as branches. Except on migration, rarely flies and then usually only for short distances low over water with rapid wing-beats. Often rather solitary and vagrants in Europe likely to occur singly. Generally silent outside breeding season, but vagrants to Britain in spring and summer have used the loud, resonant and distinctive song of slurred, gulping notes, increasing in speed and loudness and rising in pitch before falling away again, or often drawn out into wailing finale.

**Habitat.** Well-vegetated, shallow, fresh waters, standing or slow moving, with ample marginal emergent and bottom growth of plants. Requires some open water, if necessary in form of channels intersecting marsh, or occasionally ponds in dense swamp growth. Compact all-purpose territory provides for food and cover, nest-sites, and escape from danger. Migrates from parts of breeding range frozen during winter, when most seek identical habitat in milder zones; some shift to brackish estuarine waters, or infrequently to sheltered inlets of sea or open lakes. Sometimes winters on floodlands. Like other grebes, comes to land infrequently except for nesting or resting at water's edge. Averse to observation or disturbance, using floating or emergent vegetation to watch intruder unperceived while almost wholly submerged. Dives freely and can fly strongly on migration, usually at night.

**Distribution.** Nominate *podiceps* breeds from central Mexico over whole of North America north to southern British Columbia, south Mackenzie, most of Alberta, Saskatchewan and Manitoba, central Ontario, south-west Quebec, southern New Brunswick, Prince Edward Island, and Nova Scotia. 2 other races resident over Central America, Caribbean, and South America, except southern tip (Palmer 1962; Godfrey 1966).

Accidental. Britain and Ireland. 7 records, perhaps concerning only 3 individuals: in Somerset, December 1963, August–October 1965, May and July–November 1966, May–October 1967, and May–July 1968; Yorkshire, June–November 1965; Norfolk, November 1968 (British

Ornithologists' Union 1971). Azores: singles October 1927 (♀), October 1954, October 1964, January 1969 (Bannerman and Bannerman 1966; *Bull. BOC* 1969, **89**, 110).

**Movements.** Migratory, dispersive, and sedentary. Breeding birds from much of central, north, and north-east North America migrate south for winter, apparently chiefly to south USA (California to Florida and Carolinas) and Mexico, where overlap largely sedentary nesting population. Autumn passage begins August, with main body of migrants moving over protracted period September–November. Return movements begin March; re-occupation of breeding waters dependent on thaw, mostly April and early May. Nominate race regular Bermuda in winter; has straggled to Panama, Cuba, Newfoundland, Labrador, south Baffin Island, and west Palearctic (see Distribution).

Migrates nocturnally. Not particularly gregarious, though some large concentrations reported at times, chiefly at migration staging points; exceptionally, up to 20 000 on inland Salton Sea (California), November. Basically freshwater species at all seasons, but will move to sheltered bays and estuaries in hard weather. See Bent (1919), Palmer (1962).

**Voice.** See Field Characters.

**Plumages** (nominate *podiceps*). ADULT BREEDING. Crown glossy brown-black, finely streaked tawny, hindneck slightly lighter. Chin and upper throat black, narrowly bordered white; foreneck and sides of head and neck pale grey-brown. Upperparts glossy brown-black. Feathers of chest and flanks with brown-grey subterminal bars and golden-buff hair-like fringes; rest of underparts white, mottled black towards vent and flanks; under tail-coverts white. Tail dark brown. Primaries grey-brown, silver-grey below; secondaries similar, but each with individually variable and sometimes absent white spot near tip (lacking on tertials). Upper wing-coverts brown with darker edges; under wing-coverts white, longer ones washed grey; axillaries white with dark shaft-streaks. ADULT NON-BREEDING. Crown and hindneck brown, paler than breeding; chin and upper throat white, mottled black during moult; foreneck and sides of head and neck pale cinnamon-buff; fringes of feathers on chest and flanks tawny, not golden-buff; rest of plumage like breeding. JUVENILE. Head and neck with pattern of white and rufous or dark brown spots and streaks; lower foreneck rufous-buff; upperparts browner than adult with long, hairy remnants of down to tips of feathers; otherwise like adult winter but chest and flanks greyer, less tawny. Downy pattern of head and neck disappears in 1st autumn; afterwards differs from adult winter only in paler, more ochre-brown neck, breast, and flanks (Palmer 1962; Prytherch 1965).

**Bare parts.** ADULT BREEDING. Iris brown to dark brown, sometimes flecked with gold. Eyering white. Bill white to pale blue, darker along culmen, with prominent transverse black band. Leg black to dark grey, paler on inner surface, sometimes slightly tinged green. ADULT NON-BREEDING. Like breeding, but bill brown tinged yellow, lower mandible paler; black band absent or faintly indicated. JUVENILE. Iris dark brown. Upper mandible dark brown or dark yellow-brown, lower pink-buff. Leg dark olive-grey with black markings. IMMATURE. Iris yellow-brown to dark brown, sometimes with yellow spots or yellow outer ring. Eyering pale dull green. Upper mandible grey, lower buff, sometimes with traces of cross-bar. (Munro 1941*b*; Roberts 1955; Palmer 1962.)

**Moults.** ADULT POST-BREEDING. Complete; August to mid-November or later, individually variable. Starts with simultaneous loss of flight-feathers. When these finished, body moult begins, completed in 1–2 months. ADULT PRE-BREEDING. Partial; spring. Poorly known; straggler in Britain still in non-breeding 15 May 1966 (Ladhams *et al.* 1967) but in full breeding 14 May 1968 (Simmons 1969*b*). Extent of moult of body feathers not known, but includes at least chin and throat. POST-JUVENILE. Partial; autumn to December. Unknown proportion of juvenile feathers slowly and gradually replaced by immature. Subsequent moults as adult.

**Measurements.** Data from Wetmore (1965).

Nominate *podiceps*

| | | | | | | | |
|---|---|---|---|---|---|---|---|
| WING | ♂ 130 | (13) | 124–135 | ♀120 | (18) | 115–126 |
| BILL | 22·7 | (13) | 20–24 | 19·3 | (18) | 17–21 |
| TARSUS | 42·3 | (13) | 40–44 | 38·5 | (18) | 35–41 |
| TOE | 56·2 | (7) | 55–57 | 51·7 | (14) | 47–54 |
| *antarcticus* | | | | | | |
| WING | ♂ 132 | (13) | 130–138 | 124 | (8) | 119–131 |
| BILL | 24·3 | (12) | 23–27 | 21·9 | (8) | 20–24 |

*P. p. antillarum* has shorter wing; ♂ 122–126, ♀ 112–115. Additional data for nominate *podiceps*: bill ♂♂ 21·8 (1·78; 8), ♀♀ 19·4 (2·61; 9); tarsus ♂♂ 42·0 (1·45; 10), ♀♀ 37·8 (1·01; 10) (Zusi and Storer 1969). Sex difference significant.

**Weights.** Adult ♂♂ 485–559 (5), ♀♀ 281–435 (3); immature ♂♂ 282–566 (20), ♀♀ 189–389 (13) (Palmer 1962).

**Structure.** Wings short and rounded. 12 primaries: p10 longest, p9 0–2 shorter, p11 1–3, p8 3–5, p7 10–15, p1 40–50; p12 minute. P8–p10 emarginated on outer webs, p9–p11 on inner. Tail downy tuft, but more prominent than in other grebes. Bill high and short, depth *c.*½ of length; tip strongly decurved; in juvenile more laterally compressed than in adult. Feathers on forehead with stiff hair-like tips. Tarsus and toes long; toes broadly lobed and connected by webs up to base of 2nd joint (Storer 1960); foot relatively larger than in other grebes; outer toe *c.*105% of middle, inner *c.* 82%, hind *c.* 27%.

**Geographical variation.** Slight and not well understood. *P. p. antarcticus* averages larger, particularly bill. Crown and hindneck blacker, rest of upperparts greyer, underparts more extensively mottled grey. Individuals from South America may match North American ones (Chapman 1926). *P. p. antillarum* said to be smaller and darker, especially in juvenile plumage; deserves further study (Hellmayr and Conover 1948). Central American birds apparently variable in size; have been assigned both to *podiceps* and *antillarum*.

## *Tachybaptus ruficollis* Little Grebe

PLATE 8
[facing page 63]

DU. Dodaars    FR. Grèbe castagneux    GE. Zwergtaucher
RU. Малая поганка    SP. Zampullín chico    SW. Smådopping

*Colymbus ruficollis* Pallas, 1764

Polytypic. Nominate *ruficollis* (Pallas, 1764), Europe, north-west Africa, Turkey, and Israel; *capensis* (Salvadori, 1884), Caucasus, Transcaucasia, Armenia, and Egypt (west Palearctic), also rest of Africa, and south Asia east to Burma; *iraquensis* (Ticehurst, 1923), Iraq and south-west Iran. Extralimital: 6 or more races in other parts of Old World (Vaurie 1965).

**Field characters.** 25–29 cm, of which body 14–16 cm; wingspan 40–45 cm. Smallest Palearctic grebe, with short, relatively stout bill, short neck, and dumpy, blunt-ended body, and only indigenous species without ornamental head feathers or conspicuous patch of white in wing; breeding adult mainly rich, dark brown above and paler brown below, with bright chestnut throat and cheeks, and conspicuous, pale gape patch. Sexes alike, but marked seasonal differences; juvenile distinctive.

ADULT BREEDING. Crown, hindneck, upperparts, and upper breast largely black-brown and slightly glossy; lores, chin, and upper throat black; cheeks, throat, and foreneck chestnut, forming clearly defined panel. Flanks slightly paler black-brown diffused by white feather bases; belly variable mixture of silky white and black-brown, which becomes blacker through abrasion; under tail-coverts grey-brown. Bill black with whitish tip, and patch of bright yellow-green at gape forming distinctive light mark on otherwise dark head; eyes red-brown. In flight, wings generally brown above, usually with hardly any white showing on secondaries (although sometimes indistinct whitish area), and underwings white. ADULT NON-BREEDING. Whole plumage much paler. Crown, hindneck, and upperparts dull brown; chin white; cheeks, throat, and foreneck brown-buff or paler buff variably mixed with dull chestnut. Slightly darker buff-brown band on upper breast, often broken in middle; flanks paler buff; rest of breast and belly largely silky white and conspicuous in flight; under tail-coverts pale grey-brown. Light patch on gape usually invisible, being much reduced and often absent. JUVENILE. Crown and upperparts dark brown; sides of head and neck variably dark to buff-brown, often tinged rufous on neck, with irregular stripy pattern of white patches and streaks, extending to some white feathers on back of neck. Upper breast and flanks buff-brown; rest of breast and belly silky white variably mixed with brown; under tail-coverts black-grey. Bill horn-brown. In 1st-winter plumage, assumed by any

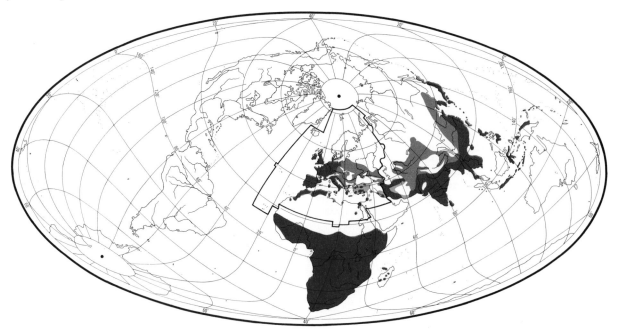

time September–January, generally paler than adult, but not reliably distinguishable owing to individual variation.

Breeding adult unmistakable, and pale gape patch stands out at considerable distances. In winter, pale-cheeked individuals sometimes confused with Slavonian Grebe *Podiceps auritus* and, more particularly, Black-necked Grebe *P. nigricollis*, but both larger and more contrasted, blacker above and whiter below, with finer bills, longer necks, more pointed sterns, conspicuous pink eyes, and white secondaries in flight. Greater possibility of confusion with vagrant Pied-billed Grebe *Podilymbus podiceps*, also dumpy with stout bill and blunt stern, and similarly lacks white patch in wing, but larger, with much heavier, paler bill, bigger head, longer neck, whiter throat, and white undertail.

Found in wide range of aquatic habitats, breeding on large lakes, small ponds, quiet rivers, or even marshland ditches, and wintering also on bare reservoirs and sheltered estuaries. Swims buoyantly with feathers often fluffed out at rear of body, thus accentuating dumpy shape and making blunt stern look remarkably broad. Dives by sub-merging head first or with distinct jump and splash, occasionally several birds in unison. When alarmed, sometimes submerges until only head above surface and may remain so concealed in vegetation for several minutes. Seldom seen on land, but stands and walks more easily than most grebes, with body almost erect; can even run; sometimes perches on partially submerged branches and railings. Flies more readily, with rapid wing-beats, than most grebes, particularly if disturbed, but usually only for short distances and low over water, often diving as soon as it touches surface. Usually singly or in small parties, but also at times in concentrations of several tens or even

hundreds. For much of year easily visible, but fre-quently shy and skulking in breeding season, though attracting attention by distinctive, loud, whinnying trill.

**Habitat.** Spectrum broadest of family, but mostly at lower end of range in water depth (often less than 1 m) and extent (often less than 1 ha). Spans greatest altitudinal range, commonly to 600 m and exceptionally to 1500 m, and attains highest breeding densities of non-colonial species (commonly 1 and up to 5 pairs per ha) (Bauer and Glutz 1966). Adapted to wide range of habitats from boreal to tropical zones, both in open country and in forest with dense stands of trees growing even into water, with clutter of emerging branches. Does not, however, normally tolerate unlimited invasion of open water by thick stands of emergent plants such as reed *Phragmites* or rush *Scirpus*. Likes muddy bottoms and margins, often with dense growth of submerged aquatic plants; tolerates surfaces much encumbered with floating vegetation so long as diving and swimming not inhibited. On large lakes and reservoirs, favours most sheltered and shallowest bays or channels, but also uses small waters, including artificial ponds down to fractions of 1 ha, farm and woodland ponds, ponds by sewage farms and factories or power stations, depressions through mining, gravel pits, and sometimes ornamental waters in town parks. Also on canals and slow-moving rivers, deltas, streams, and oxbows or backwaters; exceptionally on deep torrents (Witherby *et al.* 1940). Outside breeding season, on more open and exposed waters and on coasts and estuaries free from strong waves. Often occupies waters disturbed by man, but expert in concealment or avoidance of close contact, and can thus maintain itself on waters too disturbed by boat or

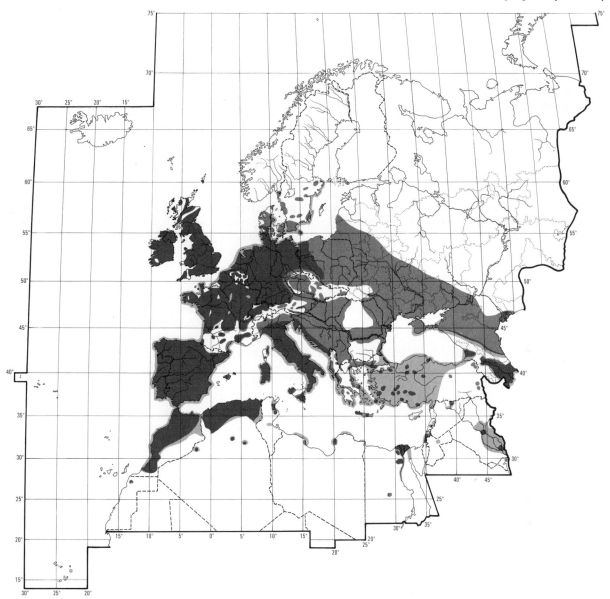

other traffic for other grebes. Pattern of living requires little use of airspace except for seasonal movement to fresh habitat, when flies readily although not recorded above low or medium heights. Frequent appearances on distant or unlikely waters indicate intensity of aerial reconnaissance of opportunities for new colonization.

**Distribution.** NORWAY: first bred 1973 (*Sterna* 1974, **13**, 116–20). SWEDEN: bred in south since 1905, but unstable in northern areas due to hard winters (Bauer and Glutz 1966; Voous 1960). FINLAND: nested once 19th century and perhaps 1953–5 (Merikallio 1958; LvH). CYPRUS: bred 19th century and early 20th; now occasional,

breeding 1967, 1969, and 1970 (Stewart and Christiansen 1971).

Accidental. Faeroes, Malta, Azores, Canary Islands.

**Population.** BRITAIN: some evidence of increase since end 19th century but probably little recent change (Parslow 1967). BELGIUM: *c.* 160 pairs; declined by half since 1952 due to pollution and drainage (Lippens and Wille 1972). DENMARK: enquiry in 1960–7 suggested 500–2000 pairs, but survey incomplete (Preuss 1969). NETHERLANDS: perhaps 1000–2000 pairs, with recent declines some areas (van Herp and Dekker 1966).

Survival. Oldest ringed bird 13 years (Rydzewski 1973).

**Movements.** Resident, dispersive, and migratory. Less marked shift to coasts or large lakes for winter than in other European grebes. Some move to sheltered estuaries and harbours, accentuated in severe winters, but not normally seen in significant numbers on sea. Extent of movement in different parts of breeding range linked to winter temperatures. In northern parts (notably south Sweden and Baltic States) and east Europe, regular winter freezing of smaller rivers and lakes compels movement which may be southward migration or dispersal to nearest open water. In Britain, as elsewhere in temperate and south Europe, many basically resident though others (perhaps mainly juveniles) disperse. Autumn flocking noted on some large reservoirs in south-east England; exceptionally, concentrations up to 300 in Essex in August–September (Hudson and Pyman 1968). Movement begins late July, normally continuing into November (Bauer and Glutz 1966); in Russia, migration noted on lower Don in October, with winter arrivals in Crimea in November (Dementiev and Gladkov 1951). Cold weather movements can occur at any time during winter. Breeding territories, if deserted, re-occupied mainly March to early May. Autumn and spring records on and near British east coasts, especially at lighthouses, indicate immigration from continent (Bannerman 1959); extent unknown, but one ringed Denmark in September recovered in Hertfordshire, January. Ringing recoveries throw limited light on origins of migrants since most ringed and recovered autumn or winter, when perhaps already well away from natal areas. General indications: breeders from north-west Europe migrate or disperse in south to south-west directions; those from central and west-central Europe move (if at all) WNW to south-west; while those from east-central and south-east Europe tend to move west to south. One ringed as chick Czechoslovakia shot Italy, November; only July adult from Belgium found north France, March. 3 ringed Denmark late August and early September recovered England (January, see above), north Germany (December) and Netherlands (November). Individuals ringed Switzerland midwinter found Czechoslovakia in August and Hungary in July, perhaps near natal areas. Others, both ringed and recovered between late September and April, show movements between north-west Germany, Low Countries, and west France; between lakes of Switzerland, Austria, south Germany, and France; and between central Europe and Italy. Winter recoveries not fitting above: ringed England September–November, found France (December, March) and Netherlands ('spring'); and one ringed Switzerland November found Poland subsequent November. More breeding season ringing necessary to elucidate complex movements.

Degree of movement between Europe and north Africa uncertain; but flock 25–30 flying north between Tunisia and Sardinia, 6 April 1967 (Schenk 1970). Common winter visitor to south Caspian September to March–April, augmenting resident breeders (D A Scott); these presumably *T. r. capensis* which breeds south and east of Caucasus. No evidence of any long-distance movement by local race *T. r. iraquensis*. Vagrant to Arabia (Bahrein, Oman, Aden); the only specimen, taken Oman, October, was *capensis* (Smith 1969).

**Food.** Chiefly insects and larvae, molluscs (more so than other west Palearctic grebes), crustaceans, amphibian larvae, and small fish. Obtained (1) chiefly by diving (up to 1 m, rarely 2 m depths, average submergence times 10–25 s), (2) rather less by swimming with head and neck immersed, (3) from surface of water, vegetation and banks, and (4) when swimming rapidly to and fro and snatching insects from air, often with neck extended and body almost upright, even occasionally leaping out of water (Bandorf 1970). Smaller prey swallowed under water, larger prey (e.g. fish over 3 cm long) brought to surface. Daylight feeder, usually singly or, especially in summer, in pairs. Will congregate where prey concentrated (Bandorf 1970).

Composition of diet determined by season and environment (see tables 2 and 3 in Bandorf 1970). Insects and larvae predominate in breeding season, and in some areas also in winter: chiefly mayflies (Ephemeroptera), stoneflies (Plecoptera), dragonfly larvae (Odonata), waterbugs (*Notonecta*, *Plea*, *Naucoris*, *Gerris*, *Corixa*), beetles (Carabidae, *Brychius*, *Cnemidotus*, *Haliplus*, *Calathus*, *Pterostichus*, *Hyphydrus*, *Hydroporus*, *Agabus*, *Colymbetes*, *Dytiscus*, *Spercheus*, Curculionidae), flies (*Bibio*, Chironomidae, *Atherix*, *Eristalis*), caddisflies (Rhyacophilidae, *Hydropsyche*, Phryganidae). Molluscs include *Lymnaea*, *Physa*, *Planorbis*, *Bithynia*, *Valvata*, *Paludina*, *Rissoa*, and *Lacuna*. Crustaceans mainly waterlice *Asellus*, and shrimps *Gammarus* and, in summer, waterfleas Entomostraca; leeches (3–5 cm long) rarely, amphibian tadpoles, occasionally small frogs and salamanders. Fish (mostly 5–7 cm long) form considerable proportion (40–50%) of diet in winter (Bandorf 1970), but rarely eaten in summer, though sticklebacks (Gasterosteidae) taken in places by breeding pairs (Steenhuizen 1934). Fish also include carp *Cyprinus carpio*, gudgeon *Gobio gobio*, minnow *Phoxinus phoxinus*, roach *Rutilus rutilus*, dace *Leuciscus leuciscus*, rudd *Scardinius erythrophthalmus*, bream *Abramis brama*, bleak *Alburnus alburnus*, perch *Perca fluviatilis*, miller's thumb *Cottus gobio*. Plant materials, mainly taken accidentally, but occasionally on purpose, include algae, leaves, stems, seeds of aquatic species, moss, stoneworts (Characea) (Naumann 1903; Madon 1931; Noll and Schmalz 1935; Witherby *et al.* 1940; Waentig 1953; Bandorf 1970).

Of 66 stomachs from Bodensee, Germany, in winter, 87% contained insects (chiefly caddisfly larvae, less stonefly and beetles) and molluscs (especially *Lymnaea* and *Planorbis*), 67% fish (mainly *C. gobio*) (Noll and Schmalz 1935). 7 from Danish freshwater (1 in May, 1 September, 5 November–December) all contained insects, chiefly *Corixa*, *Notonecta*, and caddisfly larvae; 3 contained fish of which 10-spined stickleback *P. pungitius* identified in 1 (Madsen 1957). 1 stomach, Rio de Oro, May, contained larvae of aquatic beetles and dragonflies (Valverde 1957).

Young eat mainly insects and larvae and some plant materials (Bandorf 1970).

**Social pattern and behaviour.** Based mainly on material supplied by J Fjeldså and H Bandorf; see especially Bandorf (1968, 1970) and, for earlier studies, Selous (1915), Huxley (1919), Hartley (1933, 1937).

1. Usually not highly gregarious. Solitary outside breeding season, or in scattered pairs or small, loose parties of 5–30; sometimes up to 700. Normally flocks only when loafing; disperses to feed, when sometimes associates commensally with other waterbirds—see, e.g. King (1963), Ashmole *et al.* (1956). Juveniles often gather after fledging; also adults during post-breeding wing moult, in rich feeding areas, and later those without winter territories. Many pairs hold winter territories, comprising both feeding and loafing areas; often maintained for several successive winters. If possible, pair will occupy same territory throughout year. Other winter territories sometimes temporary, e.g. those held for while after initial pair-formation. BONDS. Monogamous pair-bond maintained for whole breeding cycle or longer; probably often life-long in case of pairs with permanent territories. Pair-formation starts in late summer moulting flocks, or in autumn and winter flocks. Engagement period thus often prolonged; though bonds initially tend to be unstable, most seem to be in pairs at arrival on breeding waters in February–April. Both parents tend young, family usually keeping together until young independent, but evidently at least a measure of brood-division. Larger chicks, if tolerated after hatching of 2nd brood, may help feed new siblings (J Fjeldså). BREEDING DISPERSION. Highly territorial, with cryptic nests usually well dispersed; if not held during winter, territory established 4–10 days after spring arrival. On small waters, whole area filled with territories; used for courtship, self-feeding, loafing, copulation, nesting, and rearing young. On larger waters, adults often obtain food outside territory and desert it entirely after young *c.* 3 weeks old. Territory size less variable than in *Podiceps*, averaging 1000 m² on Fetcham Pond, England (Hartley 1933); mean 44 territories, Denmark, 1520 m² (J Fjeldså); 32, Germany, 1600 m² (Bandorf 1970). Defence most intense prior to egg-laying, with 2nd peak just after hatching. Occasionally, 2 pairs may nest less than 10 m apart despite extensive vegetation permitting wider dispersal; loose colonialism also reported infrequently, pairs defending only immediately vicinity of nest. Pair sometimes changes territory when re-nesting, or in case of 2nd (or even 3rd) broods, and single-brooded pairs may abandon territory and move to moulting areas from late June when young independent. No information on any non-breeding population surplus, but all 44 pairs in Danish study area established territories (J Fjeldså). ROOSTING. In breeding area, adults and, later, parents carrying young rest and roost on or besides platform or nest at intervals during day and throughout night; in pairs or family parties otherwise, well hidden in flooded shore vegetation or in sheltered bays. At end of season, non-breeders, juveniles, and transients roost communally in small groups on water within reed peninsulas, under bushes, etc. On moulting and wintering waters, though pairs often roost within own territories, many birds spend night in groups of 10–50 in sheltered spots on edge of well-flooded littoral vegetation; occasionally use air spaces below ice edge of frozen shore. On some waters (e.g. Bodensee), flocks of 100–700 roost offshore; do

not usually associate with other waterfowl. In winter, start to arrive at roost site *c.*½ hour before sunset and all assembled by dark; begin to depart 60–70 min before sunrise and all gone 25–35 min later. Roosting by spring migrants confined to short periods during day as night spent travelling. For further information, see Bandorf (1970). Often active on moonlit nights, especially in spring when nocturnal courtship occurs.

2. Much more vocal than most *Podiceps*, characteristically giving far-carrying Trills, singly or in duet with mate, in many social situations. Heterosexual displays mutual as in *Podiceps* (Bandorf 1968, 1970; J Fjeldså) but less ritualized, while dichotomy between water and platform-courtship not so distinct. ANTAGONISTIC BEHAVIOUR. In winter flock, bird with fish often pursued by others. Trespassers into both winter and breeding territory driven off by one or both of pair together, often after initial approach flight accompanied by Attack-calls, though actual fights rare in winter. Hostility against sexual or territorial rivals particularly intense during establishment and maintenance of nest-territory. Encounters then more complex. 3 main forms of threat: (1) Attack-upright, with erect neck, ruffled head feathers, bill sometimes slightly lowered, rear part of body somewhat lifted, tail-tuft cocked; (2) Forward-display, with head and neck stretched low over water, usually while Attack-calls given; (3) Hunched-display, with neck in, bill pointing slightly down, entire plumage ruffled, folded wings lifted on back (see A) or tilted forward somewhat. Hunched-display less likely to lead to attack than other threat; usually accompanied by Threat-trills, and also component of Triumph Ceremony between pair (see below). During mutual, close-range threat, rivals may raise feet laterally above surface, causing splash (Water-kicking); violent Splash-diving and springing dives tend to follow, as in distraction-display to man near nest (see Gyllin 1965). May also alternate between close, face-to-face threat and formal swimming together side-by-side (Parallel-swimming), while inhibited approach dives and more ritualized Token-diving (see Great Crested Grebe *P. cristatus*) also occur. In attack, one mainly patters or skids across water at other with neck extended, or dives and bites it from under water. If opponent flees, chased by rapid swimming or skidding, or by underwater pursuit-dive. In one-sided fights, aggressor tries to throw itself upon other and force it below surface, biting and pecking head and neck; in full mutual combat, rise breast to breast in vertical posture, kicking with feet and stabbing with bills. As encounter abates, Alarm-calls, Water-kicking, and body-shakes occur. Appeasing behaviour includes: (1) Prone-posture when defeated, bird remaining immobile, crouched on water with bill partly submerged; (2) Submissive-posture, with curved back, withdrawn neck, raised bill; (3) Furtive-upright, with elongated neck inclined slightly forward, bill somewhat raised, plumage sleeked. May retreat in Alarm-upright, with neck vertical and rear-end of body sunk deep in water. Escapes by pattering, swimming, or diving. WATER-COURTSHIP. Occurs any time of year, but especially during pair-formation and engagement period prior to nesting; peaks September–October, February–March. In flock, where new pair-bonds apparently formed gradually, Hunched-display with Trilling most frequent courtship. At this and other times, individuals or small groups may engage in prolonged wing-waving in upright pose, often then pattering over surface (Wilson 1959; Ladhams 1968; J Fjeldså)—but this seems not to be display (Simmons 1968b). Pairing may follow Advertising by single bird of either sex which swims slowly or floats on spot, calling (often in long series) with neck erect, bill slightly raised. However, initial pairing far less well studied and understood than in *Podiceps*. Unlike latter, elaborate ceremony when ♂ and ♀ meet after Advertising poorly developed. In primitive Discovery

Ceremony observed by J Fjeldså: (1) bird Advertised, (2) assumed Forward-display when other dived towards it, (3) turned in anxious manner, then changed to posture somewhat like Cat-display of *Podiceps* as other emerged in Ghostly-penguin display (see C); (4) Triumph Ceremony with Trilling followed (see A). More usually, however, couple (1) assume Hunched-display, (2) dive or swim towards each other (or, rarely, approach in Forward-display), (3) face for Triumph Ceremony, giving Courtship-trills in duet. Hunched-display then usually less intense than when performed independently, especially in ♀, and birds frequently turn parallel or pivot (moving rear-ends from side to side) while calling. During antagonistic encounters, often alternate between turning towards intruder in threat (see B) and facing each other in Triumph Ceremony (see A), noisily Trilling; frequently, thus elicit reciprocal Triumph Ceremonies from neighbouring pairs in own territories. In simpler Trill Ceremony, pair (1) mainly face and Trill, often in duet at rate of up to 80 per hour, while (2) pivoting, and (3) performing Slow Head-turns—pointing bill now to one side, now to other. Trill Ceremony begins, accompanies, or follows other ceremonies.

Head-shaking Ceremony often follows or replaces Triumph and Trill Ceremonies: (1) birds erect necks and begin a few, rapid lateral Head-waggles with bills slightly lowered, (2) continue with Slow Head-turns. Rarely, preceded by pair patterning side by side over water in Parallel-flight. After both Triumph and Head-shaking Ceremonies, may turn and Parallel-swim together for up to 10 m, ending with pecking at water, body-shakes, or fetching weed. May also make series of Parallel-dives or Parallel-flights, these grading into unritualized diving or pattering one after the other, ♂ probably chasing ♀. When highly excited, e.g. after antagonistic encounters, sometimes (1) meet breast to breast, (2) rise in water with vigorous rapid foot-paddling into type of mutual Penguin-dance; this, however, less elevated than in *Podiceps* and soft weeds or other material never held. In well-established pairs, meeting after separation, advertising less elaborate, with low-intensity postures, pecking at water, pivoting, and body-shakes. Other water-courtship behaviour includes Food-presentation and apparently related Bill-touching. PLATFORM-COURTSHIP. Unlike *Podiceps* grebes, carries out both types of courtship to much greater extent in cover within

A

B

C

D

E

territory, water-courtship often occuring immediately before building, soliciting, and copulation at 1 or more special mating-platforms. Triumph, Trill, and Head-shaking Ceremonies all lead at times to repeated mutual or unilateral Weed-presentation as pair swim about excitedly or join and face each other while holding weed, with necks erect and head feathers raised; at extreme, in Weed-trick Ceremony, face thus or swim side by side, Trilling, after diving for weed. Eventually, perform Ceremonial-building, facing across platform and alternately picking up and dropping same item of weed in front of one another. This, and normal building, lead to soliciting by either sex. When Inviting, on water or platform: (1) remains largely immobile with sharply kinked neck extended forward, (2) utters Inviting-call, (3) frequently dips bill in water. While Inviting on platform (see D), if mate unresponsive: (1) rises periodically, (2) makes intention-movements of arranging material, (3) 'freezes' for *c*. 1 s, standing high up on toes, and performs Rearing-display with neck bent down, bill pointing towards floor, dorsal plumage ruffled, while Wing-quivering (see *P. cristatus*); (4) arranges material, (5) finally sits down and Invites again. Other bird may show no response, bring weed, or even get up on platform and Invite briefly by side of partner. In early part of cycle, events may proceed no further, pair just taking turns in soliciting. Later, but still well before egg-laying, copulation added to sequence. Preliminary behaviour of bird in water largely unritualized. Immediately before mounting, however, repeatedly Slow Head-turns in exaggerated manner while simultaneously pivoting in opposite direction and starts to give Buzzing-calls. Latter may continue during and after copulation, usually becoming more intense as bird mounts and ending in post-copulatory Trill. Mounting similar to *Podiceps* but lasts longer; during copulation, other raises head (see E) and rubs nape rhythmically against mate's belly and · breast, or sometimes just Head-waggles. Copulating bird dismounts quickly over other's head into water at edge of nest, Water-treads in upright position, and stands erect with head feathers raised. Other stands similarly behind it on nest; they remain thus, facing in same direction, for several seconds, sometimes facing away from one another or making Slow Head-turns, often finally perform Weed-trick Ceremony. Both sexes take active role in copulation, often mounting reciprocally at intervals of $2\frac{1}{2}$–5 min, with 2nd copulation often

more intense than 1st. No special ceremony at change-over during incubation, though pair sometimes then perform Triumph Ceremony. RELATIONS WITHIN FAMILY GROUP. Brood stays mostly in nest for week after hatching of first chick; then gradually start moving about, mainly in cover within territory. Though young tend to follow any moving object for first 3–4 days, parents carry them on backs at times until *c*. 12 days old. Chicks board adult unaided, initially at any point but soon learn to climb up from behind. At least in early stages, parent may at times make definitive boarding-invitation while calling: deliberately backs up to chick and presents rear-end with wings parted and feathers on lower back fluffed open conspicuously, exposing oil gland (M D England; see photo in Beven 1972). Carrying said to be less developed than in *Podiceps*, but brood makes more use of nest and auxilliary platforms for loafing out of water. In early stages, parents take turn in carrying, on water or on platform, while mate brings food. Chicks initially peck at white tip of adult's bill, or coloured patch at base, but soon learn to take food direct. Young shed from back by body-shake; when carrying period over, adult avoids those that try to board, shaking them off, splash-diving, or pattering away. By then, chicks may accompany feeding adult, and family may separate (at least temporarily) into 2 sub-groups (Selous 1905, 1915; Ahlén 1966) as (e.g.) in *P. cristatus*, but extent of full brood-division uncertain. Chicks communicate with parents by calls, including Peeping-call, much as in *P. cristatus*. When approaching, e.g. for food, Peep intensely while drawing in neck and slightly lifting bill; will follow other objects thus, even swimming grass snake *Natrix natrix* (Ahlén 1966). When begging, however, stretch neck forward with bill raised, body partly or wholly submerged, while Foot-splashing and giving series of Peeps. Also seen to turn away while performing peculiar, side-to-side swaying of head (K E L Simmons). When attacked by parent, take up Prone-posture on water (see above), with or without Foot-splashing, and Bill-hide. Triumph Ceremony, Head-shaking, carrying weed, Parallel-swimming, and chasing frequent between siblings; may lead to temporary partnerships. Platform building, Inviting, and even full copulation sequences reported in juveniles, some of which begin to form initial pair-bonds at 3 months.

(Figs A–E after original drawings by J Fjeldså.)

**Voice.** Consists of several mainly twittering sounds not unlike Slavonian Grebe *P. auritus* (J Fjeldså), but more varied and frequent. Trilling main utterance (see I): wild-sounding, shrill, rippling peal of notes like high-pitched whinny of horse, with certain quality of laughter (Selous 1915; Huxley 1919); termed 'tittering' by Hartley (1937). Often loud and clear (Coward 1920), but considerable variation in pitch, loudness, and duration. Sometimes just single phrase, but higher-pitched version longer, oscillating up and down in series of short phrases, with periodic crescendos; may start suddenly and run down scale. Used in variety of social situations by individual or often by pair in duet. Calls of sexes identical. Following based mainly on Bandorf (1968) who recognized several distinct calls; also short, soft, rattling noise (rendered 'dig-dig-dig . . .') made by rapid, repeated movements of mandibles.

(1) Courtship-trill. Main component 'bi-i-i-i-i-i . . .', often introduced by 'düg e-düg e-düg . . .', 'e-di-edi edidi', or other variants. Often falls in pitch towards end, or sometimes rises and falls (Voigt 1950), and may end with more stuttering 'dyg-edyg-edyg-edyg . . .' (J Fjeldså). Given by mates mainly during Triumph and Trill Ceremonies (see I, 2 birds). Occasionally uttered in flight (Ruthke 1935; van Lynden 1962), latter describing it then as 'tit lululululu . . .'. (2) Threat-trill. Harsher, resonant, slightly lower pitched, whirring 'kükükükü . . .' or 'dütütütü . . .'. Also described as harder, more chittering 'tytytytyty . . .' or 'brie-bie-bie-bie-bie-bibibi . . .' (J Fjeldså). Often prolonged, lasting up to 5 min with only brief intervals. Typical territorial call; often precedes hostile encounter, accompanying Hunched-display of threatening individual, or Triumph Ceremony (etc.) of pair. (3) Buzzing. Sustained, low, vibrant 'bir(i) r(i) r(i) . . .', uttered before, during, and after copulation. Described as subdued chattering by Selous (1915); ends in trilling 'i-i-i-i-i-i-i-i' (J Fjeldså), and usually gets louder as bird mounts. (4) Inviting-call. Low-pitched, repeated 'djack', uttered by soliciting bird on mating-platform or water (J Fjeldså). (5) Contact-calls. Short, quick, high-pitched, distant-sounding 'bii-ib', 'ididüd', 'bü-ag', etc. Typically consists of conversational undertones between mates in 2–5 note phrases (J Fjeldså). Often answered by 'bee-äb' from partner. Uttered both by day and night, especially when going to roost in evening; occasionally in flight. (6) Greeting-call. Low 'bib' or 'bib-bib'. Uttered when 2 or more birds meet. Probably much the same as (5).

(7) Alarm-calls. Sharp, hard, metallic 'pit', 'pic', 'plit', or 'whit', less frequently 'tick-tick-tick' (described as short, sharp, piercing 'ik ik ik' by Selous 1915). Usually loud, but sometimes soft and subdued. Given (e.g.) near nest when disturbed, during distraction-display, and at end of antagonistic encounters. (8) Anxiety-call. Series of 1 or more uniform wailing notes: 'brieh', 'brieh . . .', 'quibiebiebieb . . . wäg-wäg-wäg, 'dühgühgühgük . . .', etc. Often answered by mate: e.g. 'wähgähgähgäh (1st bird)— 'brieh-brieh . . .' (2nd). Typically given when surprised. Preceded by soft, low trills, also serves as Advertising-call by lone bird, paired or unpaired, seeking mate (J Fjeldsåå). (9) Attack-call. Subdued but sharp, resonant 'düdüdüdüd', 'üdüdü-düd', and other variants. Given during aggressive skirmishing, or when other species approach nest or brood. (10) Submissive-call. Low, slow 'diag' or 'düg düg düg'. May be same as (4). (11) Parental calls. Short, soft, low 'bib' given before feeding young; quiet 'bib', 'bibib', or 'bibibib' (also rendered 'biäg', 'didü', and 'düd') in rallying young. Probably variants of Contact and other calls.

CALLS OF YOUNG. Various high-pitched piping calls much as in *Podiceps*, with 3 main types—Peeping, Squeaking, and Wheedling (see Great Crested Grebe *P. cristatus*). From 1st day, give 'wid' or 'bid' notes. From 3rd day, these develop into Food-call—'sid-sid-sidsidsidsid . . .'—which varies individually and according to age (Bandorf 1968); described by Ahlén (1966) as whistling 'bi-vi-vi-vi . . .' given almost continuously throughout day, with increased intensity when parent surfaces from dive or when chick approaches it. Contact-note: 'wid wid wid', developing into adult version by 53rd day; adult-like Alarm-call first heard on 32nd day (Bandorf 1968). Trill of juvenile: 'dredredredre', with more wooden tone than adult.

**Breeding.** SEASON. See diagram for north-west and central Europe. No information from Mediterranean area. Dependent upon growth of emergent vegetation and on water level. SITE. On water, among emergent vegetation or beneath overhanging branches. Nest: floating platform of water weeds anchored to submerged vegetation, often twigs or branches of bushes growing in water. Diameter at base up to 60 cm, average diameter of nest cup 4 cm, height of nest above water 4·5 cm (Ahlén 1966). Several platforms built before one completed as nest. Building: by

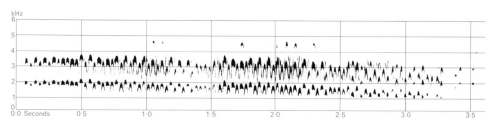

I C Chappuis/Sveriges Radio France June 1968

both sexes. Aquatic plants collected and heaped on submerged vegetation, then shorter pieces of plant material collected from bottom and laid crosswise in all directions. Cup formed with bill, feet, and belly (Ahlén 1966). EGGS. Elongate or pointed ovate, smooth, dull; white or cream becoming stained red-brown during incubation. 38 × 26 mm (33–43 × 24–28), sample 100; calculated weight 14 g (Schönwetter 1967). Clutch: 4–6 (2–7, up to 10 recorded). 2 broods, possibly 3 occasionally. Replacement clutches laid after egg loss. 2nd clutch laid when young 15 days old or older (Bauer and Glutz 1966). Eggs usually laid daily, occasionally 2-day interval. INCUBATION. 20–1 days. By both sexes. Begins with 1st or 2nd egg, sometimes later; hatching asynchronous. Eggshells mostly removed. Eggs covered with nest material if incubating bird leaves without being relieved. YOUNG. Precocial and semi-nidifugous. Cared for and fed by both parents. Carried on parent's back, under wings, when small but not as commonly as with Great Crested Grebe *Podiceps cristatus* as more use made of nest and supplementary platforms. Young fed by parents, bill to bill; as young grow, prey sometimes released in front of them to be caught (Ahlén 1966). Also given feathers. Parents active in defence of eggs and young against predators, including use of threat postures and splash-diving with sharp call. FLEDGING TO MATURITY. Fledging period 44–48 days. Become independent at 30–40 days but may beg from parents after that time (Bandorf 1970). BREEDING SUCCESS. Mean brood size at hatching 4·6 (7 broods); mean brood size at fledging 1·8 (31 broods) giving survival of 40% (Ahlén 1966). Eggs lost to avian predators and to weather and water level factors. Young taken by avian predators and by pike *Esox lucius*.

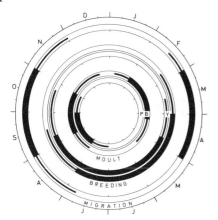

**Plumages** (nominate *ruficollis*). ADULT BREEDING. Head and hindneck glossy black, darkest on nape. Ear-coverts, throat, and sides of neck chestnut; lower throat dark brown. Upperparts glossy black, sides of rump with pale tawny patches. Underparts black, mottled silver-white, lightest on belly; vent grey. Chest and flanks washed light brown. Tail-tuft black. Primaries and outer webs of secondaries brown-grey, inner webs of secondaries with variable amount of white; tertials dark brown. Upper wing-coverts dark brown, under wing-coverts and axillaries white.

ADULT NON-BREEDING. Crown, hindneck, and rest of upperparts olive-brown, sometimes with black shaft-streaks. Tawny patches on sides of rump much paler than breeding. Cheeks, throat, sides of neck, chest, and flanks pale sandy-buff; flanks mottled grey and white. Breast and belly silver-white, vent pale grey. Wings as breeding. DOWNY YOUNG. Crown, hindneck and back black-brown with pattern of light stripes; forehead silver-white. On centre of crown patch of rufous-brown down with black tips; no bare patch on head. Lores bare. Narrow tawny or whitish line above eye; streak behind eye and spot on ear-coverts white. On back 5 tawny lines, central one (sometimes hardly visible) continuation of 2 white stripes on hindneck joining on upper mantle. On side of neck, white to brown stripe ending at chest. Chin and throat with median black line and pale brown line on either side, whitish on chin, but much individual variation. Breast and belly white, vent grey; flanks dark brown with 1–2 pale lines. Pattern of stripes on back disappears at *c.* 4 weeks; flight-feathers then begin to grow. Down pushed out by juvenile feathers and remains adhered to their tips (for details see Bandorf 1970). JUVENILE. Crown, hindneck, mantle, and back dark brown; remains of nestling pattern on sides of head and neck, often some warm rufous feathers on side of neck. Chin white with some brown mottlings, throat buff, grading into grey-buff chest. Centre of breast and belly white, flanks grey-brown, mottled dark brown on lower flanks, vent grey. Wing as adult. During autumn, pattern on head and neck disappears and olive-brown feathers replace dark brown on upperside. At this stage, juveniles may be confused with adults moulting from black breeding plumage to olive-brown non-breeding; only means of recognition, remains of down on feathers of back. IMMATURE WINTER. Like adult non-breeding; not distinguishable after remnants of down worn off.

**Bare parts.** ADULT. Iris brown-red. Bill black with white tip, base of lower mandible and adjacent patch of skin yellow-green; in winter brown-black, lower mandible more extensively yellow. Foot green-black, inside tarsus and inner and middle toes dull olive. DOWNY YOUNG. Iris dark brown. Bill at hatching yellow with faint red hue, darker at base of upper mandible; at *c.* 3 weeks with irregular black cross-bar over nostrils. Foot black. JUVENILE. Bill horn-brown, lower mandible paler. (Witherby *et al.* 1940; Bandorf 1970.)

**Moults.** ADULT POST-BREEDING. Complete; flight-feathers simultaneous. Much variation in timing, dependent on end of breeding cycle. Body moult gradual, head and neck first; August–November, rarely December. Wings July–October, main period August–September; flightless 3–4 weeks. ADULT PRE-BREEDING. Partial; body, tail, some tertials and inner wing coverts. January–April; much variation in timing, later and slower in cold springs. Single feathers growing until July. POST-JUVENILE. Partial; body and tail. Not sharply separated from pre-juvenile moult. Variable with time of hatching, early birds finishing September, some still moulting December. (Witherby *et al* 1940; Bandorf 1970; skins ZMA.)

**Measurements.** Nominate *ruficollis*, full-grown; skins (ZMA).

| | | | | | |
|---|---|---|---|---|---|
| WING | ♂ 101·4 (2·98; 20) | 95–106 | ♀ 99·0 (2·88; 20) | 90–102 |
| BILL | 18·9 (1·24; 21) | 16–21 | 17·0 (0·92; 20) | 15–19 |
| TARSUS | 36·0 (1·23; 20) | 34–38 | 34·6 (1·73; 20) | 30–38 |
| TOE | 47·0 (1·26; 11) | 44–48 | 45·4 (2·80; 10) | 40–50 |

*capensis*, South Africa.

| | | | | | |
|---|---|---|---|---|---|
| WING | ♂ 103·8 (2·99; 4) | 100–107 | ♀ 99·2 (2·93; 6) | 95–102 |
| BILL | 20·2 (0·96; 4) | 19–21 | 18·5 (0·55; 6) | 18–19 |

Sex differences significant, except for middle toe.

**Weights.** Netherlands, autumn, winter, spring (ZMA); (1) without subcutaneous fat, (2) with fat.

(1)     ♂    140 ( 4·9; 5)   133–146   ♀   130 ( 7·7; 8)   117–142
(2)            193 (27·8; 17)   131–236       187 (25·1; 14)   147–235

Sex difference in group (1) significant. Maximum weight autumn, Germany ♂ 265, ♀ 211; winter, Switzerland ♂ 300, ♀ 315 (Bauer and Glutz 1966).

**Structure.** Wing short; 12 primaries: p10 longest, p11 nearly equal to 4 shorter, p9 equal to 2 shorter, p8 2–6 shorter, p7 6–10, p6 10–14, p1 28–32; p12 minute. P9 and p10 emarginated outer web, p8 slightly; p10 and p11 inner web. Bill short and stout, high at base. Outer toe about as long as middle, inner c.80%, hind c.25%.

**Geographical variation.** Clinal. Affects size and amount of white in secondaries. *T. r. capensis* more white in secondaries, particularly in Asiatic part of range; *iraquensis* smaller (average wing ♂ 93, ♀ 87·3) and intermediate in amount of white in wing (Vaurie 1965). *T. r. capensis* on Madagascar hybridizing with indigenous form *rufolavatus*, usually considered distinct species (Voous and Payne 1965). *T. ruficollis* and Australian *T. novaehollandiae* form superspecies.

## *Podiceps major* (Boddaert, 1783) Great Grebe

Fr. Grand grèbe     Ge. Südamerika-Rothalstaucher

Breeds temperate South America from Chile, Paraguay, and extreme south Brazil southwards to Tierra del Fuego; apparently non-migratory on published evidence, though stragglers have occurred Peru and Falkland Islands. Thus unlikely to reach Europe unaided, and there must be doubts about origins of 2 old specimens in Spanish museums said to have been taken Prat de Llobregat, Barcelona, February 1908 (*Ardeola* 1954, 1, 124–7) and Collera, Valencia, some time between 1900–10 (*Ardeola* 1969, 13, 233–5).

## *Podiceps cristatus* Great Crested Grebe

PLATE 9
[after page 110]

Du. Fuut     Fr. Grèbe huppé     Ge. Haubentaucher
Ru. Чомга     Sp. Somormujo lavanco     Sw. Skäggdopping

*Colymbus cristatus* Linnaeus, 1758

Polytypic. Nominate *cristatus* (Linnaeus, 1758), Palearctic. Extralimital: *infuscatus* Salvadori, 1884, Africa; *australis* Gould, 1844, Australia and New Zealand.

**Field characters.** 46–51 cm, of which body 29–32 cm, ♂ averaging larger; wingspan 85–90 cm. Largest Palearctic grebe, with long neck and dagger-like bill; breeding adult has pink bill, black crest, chestnut and black tippets, and otherwise grey-brown upperparts and white face and underparts. Sexes alike (♂ with longer, heavier bill and, usually, longer crest and larger tippets), but marked seasonal differences; juvenile and immature distinctive.

ADULT BREEDING. Long bill carmine with dark brown ridge and pale tip; eyes red. Crown black extending into double crest at rear; distinctive tippets, erected in characteristic displays, chestnut with black tips; in contrast, facial area and narrow supercilium white, with black line from base of bill to eye. Back of neck black-grey; sides of neck variably tinged rufous-buff; rest of upperparts brown-black. Flanks also brown-black, with rufous tinge; foreneck and rest of underparts silky white and, when bird rolls on side to preen, whole body looks white. In flight, appearance surprising with white patch on secondaries and white border to front of inner wing; also white underwing. ADULT NON-BREEDING. Tippets lost and crest reduced. Crown browner black, separated from black line from bill to eye by clear white supercilium; rest of head largely white with variable blackish and chestnut marks on side. Remaining plumage much as breeding, but less rufous to sides of neck and flanks (neck looks largely white with black-grey line down rear). JUVENILE. Whole head and neck largely white with broken stripes of black-brown mixed, on sides of neck, with some pale rufous. Rest of body much as adult, but upperparts browner. Bill paler pink with bluish tinge to base. FIRST WINTER. Juvenile plumage moulted from July onwards, but often not completely lost until end of year. Then much like adult winter, but still whiter sides to head without blackish and chestnut marks (sometimes tinged dusky) and much shorter crest facilitate identification.

Breeding adult unmistakable. Even in winter, can be confused only with Red-necked Grebe *P. grisegena*, which is smaller with thicker neck and stockier build, black bill with yellow base, no white supercilium, and grey neck.

In breeding season, found on fairly shallow lakes, gravel pits, and reservoirs, but in winter many resort to estuaries and sea coasts, while others congregate on large sheets of fresh water, often devoid of emergent vegetation. Not at all skulking and always easily visible on open water. Swims with body low and neck erect or floats at rest with neck sunk on shoulders and bill often buried at side of neck. Dives from stationary position by sinking lower on surface while depressing feathers of head and body, then submerging bill-first with forward and downward swing of neck, body following, or occasionally with forward jump. Comparatively seldom seen in flight, when extended neck and feet depressed below line of body give hump-backed appearance; rapid wing-beats look laboured. Seldom seen on land, when walks awkwardly with body inclined forward or flops along on breast. Usually singly or in pairs, sometimes in scattered parties, and occasionally in loose parties in winter; exceptionally breeds in colonies.

**Habitat.** In west Palearctic, mainly within temperate and Mediterranean zone, preferring cool to cold standing fresh or brackish waters, natural or artificial, usually 0·5–5 m deep, with flat or sloping banks and muddy or sandy bottoms, affording considerable open areas, normally of at least 1 ha per pair. Prefers ample, but not too dense, emergent aquatic vegetation, especially fringing, with some submerged bottom cover and limited floating growth, in non-acidic water of medium to high biological productivity. On large sheets of water, prefers shallow sheltered bays with islets or fronting reedbeds; avoids deep, rocky, or unduly narrow water bodies but tolerates dense fringing stands of forest trees, or banks with sparse or low vegetation cover. Occasionally on soda lakes or other saline inland waters. Quickly finds and colonizes new sites including ponds, especially fishponds, gravel pits, reservoirs, lagoons, ornamental waters in country or occasionally in town parks, canals, and drainage channels; also quiet reaches of rivers with backwaters, oxbows or broader pools, where rate of flow approximates to standing waters and sufficient area compactly available. Fish essential but, within wide limits, amount of fish said not to influence population density of species (Hanzák 1952); does not normally feed elsewhere than on breeding water.

Outside breeding season commonly, but otherwise infrequently, on deltas, brackish estuaries, and tidal channels or lagoons, and on shallow, relatively sheltered, marine inshore waters; then also often on large exposed inland lakes and reservoirs. Avoids ice and rough wave action.

Pragmatic concerning occupancy of artificial waters and acceptance of human presence, developing marked tolerance in absence of destruction or direct injury; cutting or destruction of emergent vegetation and drainage of fishponds, however, greatly affects carrying capacity. Often active aerially on moonlight nights, visiting neighbouring waters in search of mate or territory. Otherwise uses airspace somewhat sparingly, although strong flier once airborne, not often at much height. Movement on land restricted to water's edge. Mainly lowland species, although exceptionally breeding (outside west Palearctic) to well above 4000 m, infrequently up to 1000 m and mostly below 300 m. In tropics, however, habitually on high lakes.

**Distribution.** BRITAIN AND IRELAND. Scarce by 2nd half 18th century; major decline from 1851 due to shooting for plumage trade, then increase and spread from 1880 (Harrisson and Hollom 1931). Little major alteration in range since 1930s (some colonization in north Scotland and south-west England proved only temporary), except in Ireland where increase in range and numbers this century, especially since *c.* 1920, while parts of south-east colonized in last 20 years (Parslow 1967). NETHERLANDS: probably unknown between 1500 and *c.* 1600, but from 1700 to *c.* 1900 apparently more common, though large numbers killed for plumage and by fishermen; probably increased after 1925 (Leys and de Wilde 1971). NORWAY: first bred 1904; some spread, colonizing Oslo Fjord area 1943 (Haftorn 1971). FINLAND and SWEDEN: spread north (Kalela 1949; Curry-Lindahl 1960). CENTRAL EUROPE: some parts of breeding range settled recently, breeding Westfalen and many lakes edge of Alps in e.g. Austria and Switzerland in 20th century, probably due to increasing eutrophication (Bauer and Glutz 1966). PORTUGAL: breeding proved in south 1973 (MDE). ISRAEL: regular breeder, Huleh, until drained; sporadic in recent years (HM). Formerly bred Cyprus, earlier this century (Stewart and Christensen 1971), and Sicily, where now winter visitor (GS). Spread northwards in Fenno-Scandia linked with climatic amelioration; for other possible factors, see Population.

Accidental. Iceland, Jordan, Azores.

**Population.** BRITAIN AND IRELAND. After decline to only 42 pairs in England 1860, increased steadily to *c.*1150 breeding pairs in England and Wales 1931, when incomplete census showed *c.*80 pairs in Scotland (Harrisson and Hollom 1931). Later sample censuses to 1955 suggested further general increase, perhaps with decline in hard winters. Full census in Britain 1965 showed rise to 4132–4734 birds compared with *c.* 2800 in 1931, apparently due to fairly steady increase from 1949 onwards; greatest increases in counties with many new reservoirs and gravel pits (Prestt and Mills 1966). In Scotland 333–345 adults in 1973, little change since 1965 (Smith 1974). FRANCE: small extension of range (Yeatman 1971). NETHERLANDS. Increasing: estimated 300 pairs or less 1932; probably 3300–3500 pairs 1966; 3600–3700 pairs 1967 (Leys and de Wilde 1971). BELGIUM: increase since 1900; *c.* 40 pairs, 1953–54; *c.* 50 pairs, 1959; 60–70 pairs, 1966 (Lippens and Wille 1972; Leys and de Wilde 1971). NORWAY. First bred Jaeren 1904; several pairs, 1905; 20 pairs 1919; 30 pairs 1968 (Haftorn 1971). Whole country *c.* 50 pairs (Leys and de Wilde 1971). DENMARK: survey 1960–67 suggested between 2200 and 2500 pairs (Preuss 1969). SWEDEN: *c.* 3000 pairs (Leys and de Wilde 1971), FINLAND: *c.* 5000 pairs (Merikallio 1958). USSR: Estonia, *c.* 775 pairs 1951–7; *c.* 1200 pairs 1967–9 (Onno

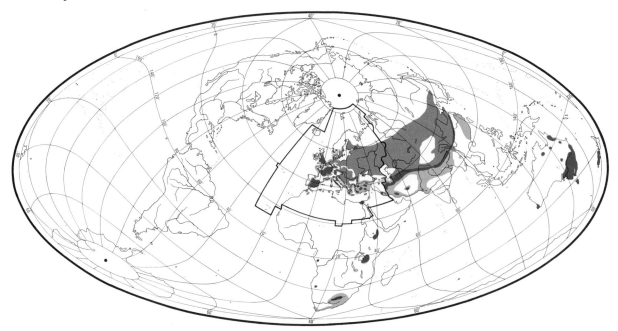

1960–71). WEST GERMANY: Baden-Württemberg, at least 1250 pairs, 1968 (Hölzinger 1970); Bayern, c. 800 pairs, 1968–70 (EB); Hessen, 54–62 pairs, 1964–66, decreasing (Berg–Schlosser 1968); Nordrhein–Westfalen, 250 pairs 1967, decreasing (Mebs 1972). Generally increasing in protected areas, but decreasing where bathing and water sports allowed (EB). EAST GERMANY: for many local counts, some suggesting recent increases, see Melde (1973). SWITZERLAND: large increase since early 20th century; very common lowland lakes with e.g. 300–400 pairs on 4 km of shore, Lake Geneva (Glutz 1962; WT). FRANCE: c. 80 pairs, Jura (Leys and de Wilde 1971); numbers increased (Yeatman 1971). SPAIN: c. 6–12 pairs in Marismas, 1960s; probably increased since (Ree 1973). TUNISIA: Lake Kelbia, c. 60 pairs, 1968, when water-level high (Jarry 1969).

After decreases in several countries in middle 19th century, partly due to slaughter for plumage trade, increased with protection towards end of century, though in Britain recovery began before protection could have any marked effect (Harrisson and Hollom 1931). Population increases may have played a part in northward expansion in Fenno-Scandia, though climatic amelioration usually given as main cause. More recently increases assisted by spread of man-made habitats, including new reservoirs, gravel pits, and (in Netherlands) enclosure of IJsselmeer and estuaries, and increasing eutrophication of these and other waters. Decreases in some areas due to disturbance, e.g. from water sports. Numbers affected by hard winters in many parts of range and susceptible to water pollution. In Britain, pesticide residues sometimes high but apparently no widespread effect on breeding success (Prestt and Jefferies 1969).

Survival. Oldest ringed bird 9 years 8 months (Rydzewski 1973).

**Movements.** Migratory and dispersive; unlikely that any west Palearctic population truly sedentary though some individuals may be. In west Europe, at least, many make short moult movements to large lakes and reservoirs; ♀♀ tend to depart first, ♂♂ to moult on breeding waters (Simmons 1970c). Movements away from nesting areas begin July–August; moulting concentrations build up on certain waters then, and in September are joined by birds that moulted elsewhere. In autumn, over 600 Chew Valley Lake, England, (K E L Simmons), and up to 20 000 (never over 25% juveniles) IJsselmeer, Netherlands (Vlug 1974). Many small waters deserted at this time. Marked shifts from inland waters to sea in winter, though an exception in central Europe where up to 22 000 overwinter Switzerland, especially Bodensee and Lakes Geneva and Neuchâtel (Bauer and Glutz 1966; Géroudet 1974); on Bodensee September–October numbers (usually 3000–4000) no higher than in winter, and lower than in March–April when usually 4000–6000 (Jacoby et al. 1970). In west Europe, dispersal to coast gradual after moult, many remaining inland October–November, though fewer by January; these remain unless or until water freezes. In north and east Europe, more strictly summer migrant as lakes more often frozen in winter; in Russia, most emigrate October, though in Ukraine some linger inland into early November. Return movements gradual, from mid-February in Britain but not until April in USSR; most territories taken up by early May though presumed non-breeders arrive into June.

As few ringed, little known of migration routes, though

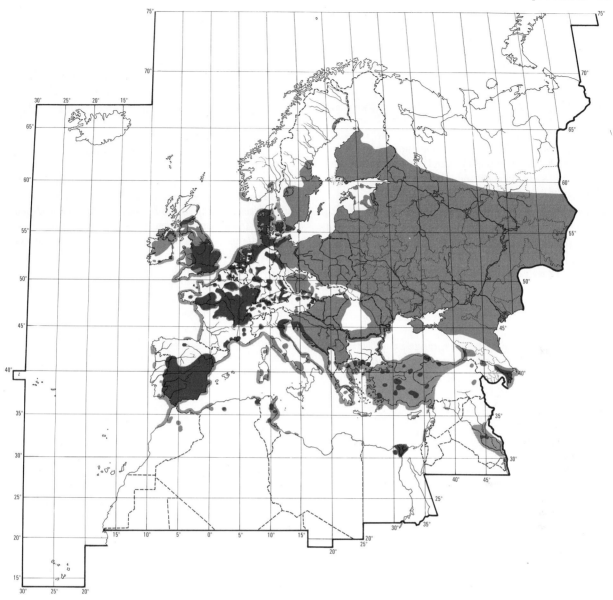

apparently broad-front overland as well as coastal. Some autumn and winter immigration into British waters, where single recoveries Kent and Suffolk from Denmark and Netherlands, but no evidence of significant increase then. Large numbers overwintering Switzerland include some from west-central Europe, with recoveries to or from Bavaria, Schleswig-Holstein, Netherlands, and north France. Unfledged young ringed Swiss lakes found November–March in France (including Mediterranean and Atlantic coasts), Italy, Austria, and Bavaria. Czechoslovakian birds found Bulgaria (October) and Italy (December). Mid-winter recoveries of one from Finland to inland West Germany (Köln), and from Sweden and Norway to Netherlands; yet many Baltic region birds believed to migrate south-east in autumn, joining those of east Europe which winter in Bosporus, Sea of Marmara, and Black Sea (Bannerman 1959). Single recoveries Sweden to Bulgaria (January), and Finland to Black Sea (May) and north Ukraine (October); also autumn recoveries in Ukraine from East Germany, Poland, and Kaliningrad; one ringed Crimea (April) found that summer Poland; and individuals ringed June in Denmark and East Germany (Oberlausitz) recovered USSR in, respectively, east Black Sea (38°E) and Voronezh (40°E), displacements of 1800–2000 km. Black Sea major wintering area, with over 20000 in one concentration off Turkish coast (Yesilirmak delta) January 1969, while mid-winter gatherings up to 4000 on some inland lakes of Anatolia and Central Plateau (*Rep. Orn. Soc. Turkey* 1968–9). Also winter visitor to south Caspian in tens of thousands (D A

Scott), presumably mostly from USSR. Much smaller numbers winter Iberia and west Mediterranean, and unknown whether these include long-distance migrants; highest counts 300 on inland Barrage de Caia, Portugal, January 1973 (Biber and Hoffman 1973) and 120 Mar Chica, Morocco, January 1963 (Smith 1965). Limited southward coastal movement possible; straggler seen Sénégal delta January 1971 (Morel and Roux 1973) probably of Palearctic origin, since *P. c. infuscatus*, breeding east and south Africa, does not occur in west Africa north of Gabon (Cawkell and Moreau 1963; White 1965). See also Dementiev and Gladkov (1951); Bauer and Glutz (1966); Melde (1973).

**Food.** In contrast to other Palearctic grebes, chiefly fish; also to lesser extent, aquatic invertebrates. Fish, especially larger specimens, obtained mainly by diving in water clear of dense vegetation: from stationary position on surface, sinks low in water and, with forward and downward swing of neck, usually submerges smoothly without noticeable leap; under water, searches for and pursues prey by foot-propelled swimming with neck extended and wings folded away. Dives generally last less than 30 s (see Simmons 1955; Ladhams 1968). Mean 26, range 15–41 (Harrisson and Hollom 1932); mean 24, maximum 56 (Hanzák 1952). Mean of 450 dives on one pool 19·5 (range 5–30, but usually 10–25); elsewhere, in smaller samples, 26 (22–30) and 31 (24–37) (K E L Simmons). Other recent records include: 18 (5–45) (Ladhams 1968); 22 (8–39) (Hancock and Bacon 1970). Duration of feeding dives varies with different factors; although means of samples correlated with water depth, length of individual dives determined by hunting success as most prey chased in water layer between surface and bottom (Simmons 1970b). Usually in depths of 2–4 m or less (Harrisson and Hollom 1932), but, judging from 161 birds entangled in fishing nets on Sempacher See, Switzerland, apparently can descend as much as 30 m; there, dives deeper in winter than summer with no correlation between transparency of water and depth of dive (Hofer 1969). ♀♀ also said to dive deeper than ♂♂ but this not substantiated elsewhere, ♀♀ tending more to dive in shallower water and in cover than ♂♂ (Simmons 1970c). Large fish—and sticklebacks (Gasterosteidae)—always brought to surface; held at or just below head, mandibulated in beak, and swallowed whole, alive and head first, preparation of defensively spined sticklebacks often being particularly prolonged; other, smaller prey items eaten under water (Simmons 1955). Other feeding methods include: (1) hunting from surface, usually over submerged vegetation, for small fish and invertebrates by peering under water, neck-dipping, up-ending, and plunging to re-appear tail-first; (2) picking up prey, mainly insects, from water surface or aquatic vegetation; (3) snatching insects from air above head. Last 2 such methods only of minor importance, and 1st favoured rather more by ♀♀ than ♂♂ (Simmons 1970c). Daytime feeder, normally solitary or in pairs, though will congregate at times over shoals of small fish. In summer, tends to be most active in early morning and evening (Harrisson and Hollom 1932; Hanzák 1952), though, particularly in winter and when feeding young, more often alternates between spells of feeding and loafing throughout day (Simmons 1955).

Fish include roach *Rutilus rutilus*, bleak *Alburnus alburnus*, gudgeon *Gobio gobio*, dace *Leuciscus leuciscus*, rudd *Scardinius erythrophthalmus*, tench *Tinca tinca*, goldfish *Carassius auratus*, minnow *Phoxinus phoxinus*, perch *Perca fluviatilis*, bream *Abramis*, pike *Esox lucius*, trout *Salmo trutta*, char *Salvelinus alpinus*, sticklebacks *Gasterosteus*, and eel *Anguilla anguilla*; spawn may also be taken. Insects and larvae include dragonflies (Odonata), beetles (Coleoptera), caddisflies (Trichoptera), ants (Hymenoptera), waterbugs (Hemiptera), flies (Diptera), moths (Lepidoptera), and stoneflies (Plecoptera). Crustaceans include crayfish *Astacus*, shrimps *Gammarus* and *Pandalus*. Molluscs include snails *Limnaea* and *Valvata*, but mollusc remains in stomachs probably mainly result of eating mollusc-feeding fish (Geiger 1957). Spiders (Araneae), amphibia (frogs, tadpoles), and occasionally newts, and grass-snakes *Natrix natrix* also recorded. Plant remains found in small amounts in stomachs include seeds (willow *Salix*, sedge *Carex elata*), parts of reed *Phragmites*, pondweeds, and algae. (Hofer 1915; Maag 1917; Widmer 1928; Madon 1926, 1931; Witherby *et al.* 1940; Poncy 1941, 1943; Buxton 1952; Dementiev and Gladkov 1951; Hanzák 1952; Geiger 1957; Madsen 1957; Burkhard 1961.) Fish range from 3–21 cm (Heuscher 1907; Madon 1931; Hanzák 1952); mean 13 cm (Geiger 1957), but larval fry and invertebrates down to *c.* 5 mm also taken, at times in large numbers, more often by ♀♀ than ♂♂, mainly by surface-hunting (Simmons 1970c). Fish taken throughout year; insects and their larvae chiefly from spring to autumn.

77 adults and 30 juveniles, June and July, Bieler See, Switzerland, contained predominantly fish (Geiger 1957). In adults, in order of frequency: roach, bleak, gudgeon, and perch, with insect remains in 70 (91%) though volume small and mainly beetle *Donacia clavipes* and ant *Formica rufa* (latter taken from water surface). In juveniles: chiefly small perch and insects. Concluded that fisheries on Bieler See not radically affected, as mainly surplus perch taken, and few cyprinoids or pike. Change of diet as year progresses indicated at Zürich See, Switzerland: in April and May, 94 stomachs had fish in 61 (24 species, chiefly roach and bleak), insect remains in 87 (mainly beetles, dragonfly nymphs, and ants), and seeds in 86 (Hofer 1915); in May in 104 adults and 10 juveniles, insects, molluscs and seeds occurred but fish (mainly bleak, roach, dace, rudd, and perch) predominated, occurring in 78 (Maag 1917); in October, 17 of 20 stomachs contained fish (chiefly roach, bleak, and perch), and only 1 contained insects (Knopfli 1935). Denmark: 25 stomachs from

freshwater areas contained fish (mostly cyprinoids) and less insects; 29 stomachs from brackish and saltwater areas contained fish (most frequently *Gobius*, and less herring *Clupea harengus*, sticklebacks *Gasterosteus*, cod *Gadus morhua*, and cyprinoids) with 31% crustaceans (Crangonidae and Palaemonidae) and less molluscs, polychaetes, insects, and plant remains (Madsen 1957). Insects can predominate, especially in years of local abundance, e.g. 1915 when 34 (42%) of 66 stomachs contained cockchafer *Melolontha vulgaris*, 13 being crammed full (Knopfli 1956).

Daily food intake, based on captive birds, amounts to about one-fifth of body weight, i.e. 150–250 g per day (Heinroth and Heinroth 1927–8; Knopfli 1935; Geiger 1957).

Young, which depend on parents for food for at least 10 weeks, given mainly small fish and insects, possibly insects mainly at beginning (Geiger 1957). For parental feeding frequency, see Breeding.

Habitual feather eating, by adults and young; ball of whole and decomposed feathers often nearly fills stomach, and feathers given to chicks from hatching (see, e.g. Harrisson and Hollom 1932; Rankin 1947; Geiger 1957; Madsen 1957). Link with pellet formation suggested (Hanzák 1952; Simmons 1956), and pellets consisting of feathers, vegetable debris, and fish remains recently described (Simmons 1973). Adults eat own intestinal worms (almost certainly cestodes), and feed these to young (Simmons 1975*b*).

**Social pattern and behaviour.** Most fully studied of all grebes; for general accounts see, e.g., Harrisson and Hollom (1932), Hanzák (1952), Simmons (1955), Bannerman (1959), Bauer and Glutz (1966), Melde (1973).

1. Not highly gregarious. Outside breeding season, often solitary, especially when feeding, though occasionally forms temporary congregations of up to 100 or more (Simmons 1974*a*). Will also gather at times in loose groups, mainly when loafing. Numbers then often small, but larger on certain big lakes, particularly late summer to autumn during flightless period of wing moult. Later in winter and spring, some birds associate in small flocks, usually of *c.* 10–20 (P P A M Kop), in which pair-formation gradually takes place. Though most non-territorial when not nesting, a few individuals hold small, often temporary, winter feeding territories while some pairs remain year round in breeding territory if weather permits (Simmons 1974*a*). On independence, juveniles tend to form small, loose parties for a time in late summer and autumn. BONDS. Monogamous pair-bond. Typically, pair-formation initiated from mid-winter before selection and occupation of territory. Engagement period often long and some pairs unstable, in part due to competition for sites. Pair remain together for all or part of cycle, at least until brood divided; many pairs split up around time young leave, often when ♀ also departs before post-breeding wing moult, though some bonds re-established in late summer or autumn and a few pairs remain intact into early winter or occasionally next season (Simmons 1974*a*). Both parents tend young; social structure of family complex, and affinity of each adult lies increasingly with certain young of divided brood rather than mate (Simmons 1970*c*, 1974*a*). When family of 2 or more chicks,

brood often divided by 6th week into 2 groups, each in charge of single parent. Each adult rarely or never feeds young in other group and often hostile to them. Sometimes, groups operate more or less independently, adults showing hostility to each other; may separate totally. Within each group, 1 favoured or 'in-chick' given feeding priority over 1 or more 'out-chicks'. Adult may be hostile to young in own group, but mainly out-chick. Before brood-division, age-hierarchy between siblings gives advantage to larger young. Further details in Simmons (1970*c*); for discussion of adaptive aspects, see Simmons (1974*a*). Young of 1st brood reported helping to feed siblings in 2nd (J Fjeldså), even carrying them on back (P P A M Kop). BREEDING DISPERSION. Typically territorial (Venables and Lack 1934, 1936; McCartan and Simmons 1956), with well-dispersed nests usually concealed at vulnerable sites, though often clearly visible at safer ones (Simmons 1974*a*). Territory often established well before nesting, usually from mid-winter (if mild) to early spring; used subsequently at least for mating and nesting, though cover may at first be absent and breeding delayed until growth of vegetation or drop in water level. Great variation in size and later usage, depending in part on local features of habitat (see also Simmons 1955, 1959). Some pairs include up to or over 1 ha of water, others merely defend nest and immediate surrounds. Colonies sometimes formed when dispersed sites scarce or particularly safe area (such as large reedbed) available; usually loose, but occasionally dense. When territory large, pair often obtain most or all food and raise young within; when small, both solitary and colonial pairs feed and rear young elsewhere, some establishing separate brood-territory while others more nomadic. When breeding water deficient in food, exceptionally will fly to feeding water elsewhere and return with food for young, as on certain pools bordering dykes on IJsselmeer (Leys *et al.* 1969); here territories 300–600 m², with nests sometimes only a few metres apart, whereas those on other local waters of 5000 m² mean size, each including extensive area of open water and part of reed fringe (P P A M Kop). Non-breeders remain mostly in loose group away from occupied territories; may include suspected homosexual pairs, probably often ♂♂, of 1st-year birds (Kop 1971*b*). ROOSTING. Singly, or in pairs, family groups, or (when not breeding) in loose parties; on water, both while loafing periodically during day and at night, when activity tends to cease well after sunset. In open, or amongst surface vegetation, or in edges of well-flooded cover, often in shelter of lee shore. Loafs and roosts in territory while nesting. Sometimes active on moonlit nights, especially during initial period of pairing and later during territory establishment—when nocturnal courtship and other social behaviour occur.

2. In post-breeding and winter flocks, all social activity much reduced (Simmons 1968*b*) until water-courtship and antagonistic encounters associated with pair-formation begin, usually from December or January. Following account based mainly on Simmons (1955, 1970*c*, 1975*a*), supplemented by information from P P A M Kop and J Fjeldså; for earlier studies, see especially Selous (1901, 1920–1), Huxley (1914, 1924*a*). ANTAGONISTIC BEHAVIOUR. Hostility shown to sexual and territorial rivals, especially during initial pairing, territory establishment, and engagement period; also in defence of young. By both sexes, but ♂♂ usually more aggressive. Forward-display most common threat: head and neck lowered and stretched over water, tippets moderately to well spread; directed with Barking-call at more distant birds, but silent or with Growling-call at closer quarters. At higher intensity, back feathers ruffled and neck awash; during close encounters, may also lift folded wings over back, or occasionally assume Wing-spread posture (as in Cat-

display), partly opening and rotating them forward so upper surfaces shown frontally. Threatening bird may float on spot or drift slowly (Threat-lurking) or swim purposefully along (Threat-cruising), e.g. when patrolling border of territory or searching for intruder. Also stalks or chases rival in series of Sinister-dives. These evidently partially ritualized: bird submerges from Forward-display and swims rapidly just below surface, course often indicated by ripple; may partly surface or emerge head only on way. Attacks by surfacing bill first under rival, or by patter-flying over water with neck extended. In one-sided fight, seizes rival by head or neck, attempting to force it under water. In full combat, both leap forward and upward to clash breast to breast; paddle feet, stab and grapple with bills, and loosely beat with open wings. In close confrontations, one or both may adopt Defensive-upright with neck erect, tippets full spread, and crest vertical; bill often pointed down with mandibles widely open as loud Growling or Snarling-calls given. Such behaviour occurs mostly during disputes at territorial boundaries, but Threat-lurking even more characteristic; interspersed by Token-dives in which bird submerges as if to approach under water but surfaces instead at or near original spot. Apparent appeasing behaviour includes: unilateral Head-waggling, Habit-preening, and, by ♀♀ especially, full Cat-display (see A)—all more typical of water-courtship. Self-preening frequent after encounters; in winter flock, adoption of resting posture seems to reduce likelihood of aggressive contact (Simmons 1968b). When moderately alert, holds head and neck erect with crest up, tippets relaxed (Alarm-upright); when frightened, elongates and sleeks neck even more, crest and tippets tightly depressed (Furtive-upright)—in both cases, may utter Clicking-call. Escapes usually by diving and swimming away under water, or first patters or skids over surface, then dives; sometimes flies. WATER-COURTSHIP. Characteristic of pre-egg stage of cycle, especially initial pairing, territory establishment, and earlier part of engagement period; with partial exception of Head-shaking, not shown later in cycle as often claimed. Starts in open water, among solitary birds or those in loose flock, continuing later in territory. Consists largely of 4 highly ritualized, mutual ceremonies. Either sex takes initiative, often drawing attention by vocal Advertising (Simmons 1954): floats on spot or swims slowly while giving special Croaking-call (or sometimes Growling) with neck erect and throat distended. Advertising also performed by paired bird separated from mate. Discovery Ceremony most characteristic of initial pairing phase; often follows Advertising. Roles of sexes different but fully interchangeable: (1) one dives and, periodically surfacing head only, swims towards other just under surface so that course indicated by wake (Ripple-approach); (2) other watches, adopting Cat-display (see A); (3) first bird then emerges suddenly not far from second with back to it in Ghostly-penguin display (see B), surging into vertical position with bill down, before rotating to face it; (4) couple start Head-shaking. Unlike other west Palearctic *Podiceps*, never rise in Penguin-dance at climax of Discovery Ceremony. Head-shaking Ceremony itself also performed independently; most common ceremony throughout engagement period, though at decreasing mean intensity. Often occurs spontaneously as greeting when pair join after separation, or periodically when pair together; also induced by activity of others, frequently during and after antagonistic encounters. Roles of sexes identical, movements performed simultaneously or reciprocally while birds face closely, necks upright (♂'s head higher than ♀'s), tail-tufts cocked. 3 main phases in full performances, but each may also occur on own: (1) with bill open and lowered, crest erected vertically, and tippets fully spread, each rapidly performs Low Head-waggles from side

to side in narrow arc while uttering Ticking-call (see C); (2) with bills closed and tippets partly closed, change to repeated, silent bursts of High Head-waggling with elevated bills interspersed with Slow Head-turning in horizontal plane; (3) one or both now also Habit-preen, dipping neck to flick up scapular feathers quickly with bill. Particularly in early phase of pairing and start of engagement period, Head-shaking interrupted at times by Retreat Ceremony (Simmons 1959), often after encounter with intruders or rivals. Roles of sexes different and interchangeable, though less so than in Discovery Ceremony. 2–4 phases: (1) during phase-2 Head-shaking, one of pair suddenly performs Ceremonial-flight past or away from other, pattering over water for several metres before subsiding with back to it in Cat-display; (2) then turns thus and faces other which remains defensively in

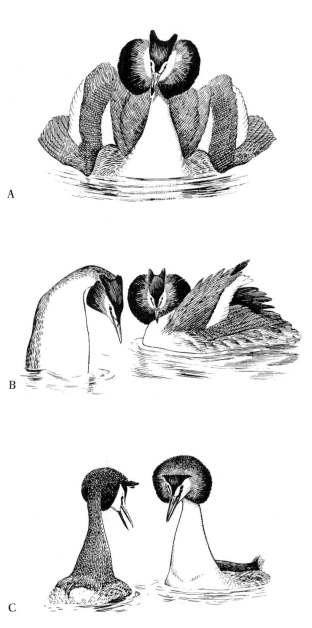

A

B

C

D                          E                          F

original position with ruffled back feathers, or briefly in Cat-display; (3) pair may then link up again and Head-shake, sometimes then (4) performing 2nd Retreat, often with roles reversed. In majority of cases, first Ceremonial-flight performed by ♀. Last and most complex ritual, Weed Ceremony, first occurs during engagement period just before appearance of platform-courtship, typically in territory off potential mating or nesting sites. Roles of sexes identical, in 5 main phases (see also Simmons 1968c). (1) During intense Head-shaking, birds slowly withdraw from one another while Habit-preening (see D), moving apart in Statue-upright (see E) with neck erect, tippets spread in full circle with marked forward slant, crest depressed, giving Twanging-call (Ceremonial Turning-and-Sailing-away). (2) When well apart, submerge in slow, deliberate Weed-dive, collect material (usually soft weed) under water, and surface (Weed-trick). (3) Swim quickly towards one another with heads low (Weed-approach) and (4) meet and rise abruptly in vertical posture, usually breast to breast, in Weed-dance. During latter feet trample rapidly as both Weed-swing laterally in unison with a few or many Fast Head-turns, ♂'s head higher than ♀'s (see F). Then subside and (5) Head-shake, dropping weed as High Head-waggling begins. During engagement period, ♂ sometimes offers fish to ♀; such Food-presentation irregular and largely unritualized, though at times associated with Head-shaking. Water-courtship declines when birds enter main platform-courtship phase, ceasing as egg-laying nears except for some (mainly low-intensity) Head-shaking; latter may persist thus as greeting throughout period pair-bond maintained, but bouts usually induced (see above). PLATFORM-COURTSHIP. Early in territory-occupation phase of engagement period, when pair seek mating sites, either bird (or both simultaneously) may Invite-on-water with back to mate, floating on spot in essentially same posture and giving same call as later when Inviting on platform. Not associated with other platform-behaviour though sometimes occurs at or near site where platform or nest built later when vegetation grown up. Pair later construct 1 or more rudimentary mating-platforms; for details of building methods, see Simmons (1955). Unlike other west Palearctic *Podiceps*, may also at times engage in Ceremonial-building, much as in Little Grebe *Tachybaptus ruficollis* but more formally (J Fjeldså). Now take turns in soliciting on platform. When Inviting there, bird extends kinked neck forward with head plumes depressed, remaining immobile thus while intermittently giving Twanging-call. Also periodically performs Rearing-display—but only when mate close behind: suddenly stands with neck arched down, bill near

floor and tippets partly open, and Wing-quivers rapidly in bursts of 3 shakes, flashing white marks while inclining head sideways and squinting back at mate. Other brings material, drifts near, or self-preens; may High Head-waggle or show intention of mounting. Early in engagement period, Mating Ceremony does not usually go beyond this. Copulation added subsequently; at first, still well in advance of laying, performed by both sexes as part of platform-courtship but later mainly by ♂ for insemination when ♀ eventually ovulates. Before mounting, bird on water gives Mooing-call, expands tippets fully, and raises crest; then leaps up on to other's back, uttering Rattling-call, copulates quickly, dismounts over other's head, re-enters water in upright position while vigorously stamping feet (Post-copulatory Water-treading), and starts High Head-waggling with back to mate. Latter, after staying in full Inviting-display throughout, now raises head and also High Head-waggles; ensuing Head-shaking bout brief or fairly prolonged as birds face or one in water swims slowly away. During incubation and hatching period, display at nest usually minimal, though pair will sometimes Head-shake a little at change-over when some relieved birds hold out and loosely vibrate wings; this Wing-shuffling display not known from any other grebe. RELATIONS WITHIN FAMILY GROUP. Young carried on parents' back for *c.* first 3 weeks. Climb aboard largely unaided, but tiny chicks encouraged by adult raising wings and back feathers; particularly at moments of stress, such boarding-invitation re-inforced by Growling or Croaking-calls, and assistance given by lifting both feet sideways along surface. Young shed from back by body-shake or wing-flap. During first weeks, follow any moving object, even boat. Apparently do not recognize parents until beginning of 6th week though parents recognize own young from early stage, possibly through voice (P P A M Kop). Young communicate with parents by piping calls of 3 main types. Adults have no definite 'food-call'; give Advertising Croaks when cannot find young. For 3–4 weeks after hatching, both parents tend all young impartially until brood division begins. Individual identification of young now assisted by variation in patterns of facial marking within brood (Simmons 1970c). When attacking young of other group, or own out-chick, adult swims or ploughs after them, though sometimes chick seized by head and ducked. Young flee or remain near—Facing-away, Bill-hiding, etc; also show more active appeasing behaviour including (1) Wheedling and Peeping-calls, (2) sinking in water, (3) Foot-splashing; (4) Gaping; (5) Head-pointing; (6) Head-tossing; (7) Chin-bracing, and (8) Skin-flushing—during which bare patches on face and crown become bright red (full details in Simmons 1970c). Such behaviour also shown when chicks join adult, especially when begging for food. Interaction between adult and in-chick particularly complex, and association may continue at times well beyond fledging. Fighting between siblings not uncommon from later part of carrying period; mostly head pecking by larger, smaller usually moving away or Bill-hiding. Disputes rarer among older young, but can take form of persistent pursuits. Adult-like water-courtship between siblings occurs at times from 3rd week; largely unilateral or mutual High Head-waggling and Fast Head-turning with weed, though incomplete versions of all ceremonies and, rarely, full Weed Ceremonies also recorded. Inviting-on-water occasional from 6th week. Older chicks sometimes solicit on raised sites, but no copulations seen; may build platforms, even depositing material on back of sibling or parent. Head-shaking and soliciting not uncommon between independent juveniles.

(Figs A–F after drawings by R J Prytherch in Simmons 1975a).

**Voice.** Adults have basic repertory of c. 10 distinctive calls (Simmons 1970c), mostly harsh, nasal, open-throated, or guttural; those of ♂ and ♀ identical. Not so vocal as some grebes.

(1) Barking. Loud, far-carrying 'rah-rah-rah . . .' (see I), given in phrases of 1–12 or more notes. Typical long-distance threat or self-assertive call ('song'); sometimes also used as approach-call when joining mate. Quiet grunting version given at times. (2) Growling. Loud, rather drawn-out 'g(h)ar' or 'gorrr' (see II), given singly or repeated slowly. More a close-quarter, intense threat-call; also anxiety-call, e.g. when contact lost with mate or chick. Occurs at times during heterosexual encounters, especially in early stages of pairing. (3) Snarling. Loud, prolonged 'gaaaa . . .' Most often uttered during close, antagonistic encounters involving defensive threat and fighting. Also accompanies intense defensive or offensive reactions against predators. (4) Twanging. Quiet, metallic 'gung'; given in repeated short phrases, often of 2–3 notes. Uttered during close-quarter, antagonistic encounters when bird not inclined to attack. Also regular contact-call when pair together, and uttered during Sailing-away phase of Weed Ceremony and during Inviting-display on water or platform. (5) Clicking. Quiet, metallic 'kek' given repeatedly. Alarm-call when disturbed. (6) Clucking. Quiet, hen-like 'cluk . . .' given slowly and repeatedly during pair movement about territory; also low-intensity meeting-call during approach phase of Head-shaking Ceremony and incipient Low Head-waggling. (7) Ticking. Brief 'ktic-ktic . . .' uttered continuously during 1st main phase of Head-shaking Ceremony. Recording (see III) shows pair Ticking, one bird calling at higher pitch. High-intensity meeting-call not heard at any other time. (8) Croaking (or Crowing). Loud, far-carrying, resonant 'grr-owp', 'ah-rrrrrrr-c' (see IV), etc.; followed by much quieter, moaning 'row-ah' (probably largely due to intake of air and not part of vocal signal). Advertising-call, given singly or slowly repeated up to 6 times, typically in characteristic posture with neck upright by lone

bird—either unpaired prior to pair-formation, or paired and seeking own mate or young. Considerable variation in tone and phrasing of main component, thus providing means of individual identification. Sometimes accompanied or replaced by Growling. May also Advertise with neck stretched forward horizontally giving 'kek-kek-kek' notes (P P A M Kop). (9) Mooing. Loud, cow-like 'eu-eu-eu . . .' uttered in long crescendo. Pre-copulatory call. (10) Rattling. Loud, harsh series of 3 notes in continuous accelerating phrase. Copulation call, following on from (9).

CALLS OF DOWNY YOUNG. Chicks highly vocal, with 3 main calls, each variants of basic piping. (1) Squeaking. Barely disyllabic 'wee(-u)p', 'quee(-u)p', or 'pee(-ee)p' given singly but often repeatedly; call of unhatched chick and uttered in moments of stress (e.g. when lost) by hatchlings and older young. (2) Wheedling. Rhythmic, rising and falling 'quee EEEP peep-peep . . .', etc., delivered continuously and persistently in presence of

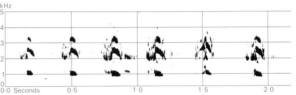

I Sture Palmér Sweden April 1966

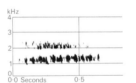

II Sture Palmér Sweden April 1966

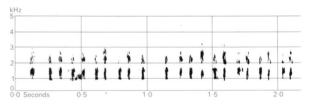

III Sture Palmér Sweden April 1966

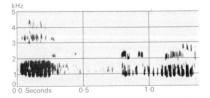

IV Sture Palmér Sweden April 1966

parent, especially during loafing; provides stimulus for adult to provide food. (3) Peeping. Repeated, often shrill 'pee-pee-pee . . .', becoming more vociferous and almost disyllabic at high intensity; common approach-call of young, especially when being fed. Both (2) and (3) also vocal components of appeasing displays to parent.

CALLS OF JUVENILES. Chick-like calls lost after *c.* 14 weeks. Chattering only specific juvenile call: repeated, harsh, almost duck-like 'er-er-er . . .' with, at various times, buzzing, rattling, braying, or rasping character; given in similar situations as (2) and (3) above by young maintaining prolonged, post-fledging association with parent. From *c.* 17th week, juvenile versions of adult Barking, Twanging, Clucking, and Ticking appear.

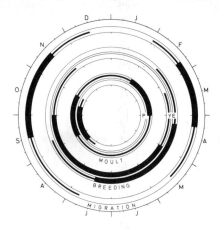

**Breeding.** Main references: Simmons (1970*c*, 1974*a*), England; Harrisson and Hollom (1932), Britain; Prestt and Jefferies (1969), Britain; Leys and De Wilde (1968, 1971), Leys *et al.* (1969), Netherlands. SEASON. See diagram for Britain; laying in January recently confirmed. Netherlands: March–October, mainly May–June. Extended season also elsewhere in range; determined mainly by availability of safe sites, and thus on growth of emergent vegetation. SITE. Usually in water, well concealed within reeds or other aquatic plants. 758 nests, Netherlands, in 22 different types of vegetation, including *c.* 70% in *Phragmites communis* (48%), *Scirpus lacustris*, and *Typha angustifolia*; further 13% in flooded or trailing willow *Salix*. Sometimes with little or no cover, or out in open on emerging mud or reservoir masonry; occasionally on dry land, mainly when stranded by drop in water level. Usually well dispersed, but sometimes in colonies; these usually small and loose, but nests reported as close as 2 m. New nests built for 2nd brood. Nest: substantial heap of aquatic vegetation, lying mostly below water; floating, tethered, or built up from bottom. In Netherlands, at least 40 plant species reported, plus *c.*10 other types of material (including mud and rubbish such as paper); plants include *Phragmites* in 83% of 758 nests, *Typha* (29%), *Scirpus* (23%), various roots (17%), rotten weed (11·5%), twigs (4·5%). Up to 45 cm overall diameter, 30 cm deep; above

surface, 28–45 cm diameter, 4–7 cm high; shallow cup *c.*10 cm diameter, 2 cm deep. No special lining, but finer material usually placed on rim and used to cover eggs. 1 or more separate and usually less substantial mating-platforms often built, but no auxiliary platforms for young. Building: by both sexes after choosing site together. Though sometimes taking only 1–2 days, usual period 6–8 with laying starting 48 hours later (Bauer and Glutz 1966); in Holland, 62% of 263 nests built in less than 10 days, rest taking up to 28 though mostly not more than 14. Throughout day, especially in early stages, but usually inhibited by rain or cold. Materials gathered mostly as floating or submerged debris; obtained in vicinity of site, when nest in thick cover may have area cleared round it, or from all over territory. Some material added during incubation, especially if water rises. EGGS. Elongated oval, dull chalky white, quickly staining orange-red to brown. 54 × 37 mm (47–64 × 32–40), sample 273; weight 42 g (34–50), sample 55 (Netherlands). Clutch: 1–6 (–9); 7 or more usually by 2 ♀♀, but clutches of 7–9 exceptionally by same ♀, Netherlands. Britain: usually 3–4 (66%), mean 3·5. Netherlands: mean 3·5, sample 700; no significant seasonal variation May–July, but substantial decrease later when most clutches of 1–2 laid (probably mostly by 1st-year birds). Europe: 1 egg, 2%; 2, 6%; 3, 20%; 4, 42%; 5, 22%; 6, 6%; 7 or 8, less than 2%; mean 4·0, sample 818 (calculated from Melde 1973). Mean clutch-size often varies significantly from water to water in same area. Normally 1 brood, but 2 in up to 11% of pairs, Britain. 2nd clutch usually laid when first young 6–7 weeks old and still unfledged. Replacement clutches, sometimes smaller than first, laid when both eggs and young lost. Eggs usually laid at 48-hour intervals, mostly in morning. INCUBATION. 28 (27–29) days for each egg. Starts with first (on day of laying or next); hatching asynchronous, usually at 48-hour intervals (sometimes only 24 hours between 1st and 2nd chick). By both sexes equally in spells of 10–492 min; mean 108, sample 181, increasing during incubation period (England). Eggs often covered with nest material when incubating bird disturbed from nest. Eggshells dropped from nest or carried a short distance away. YOUNG. Precocial and semi-nidifugous. Cared for and fed by both parents equally. Nest abandoned after end of hatching period; not normally used for loafing or roosting subsequently. Young carried on back by parents in turn while other brings food, persistently up to day 14, intermittently to day 21; thereafter, guarded until *c.* day 30 though left alone earlier if food scarce. Main diet whole fish, though sometimes mainly insects in first few days; also given feathers throughout dependence period but mainly in weeks 1–2. Food items and feathers given singly in tip of bill. Feeding rate variable depending on brood-size and size of prey; up to 95 feeds per hour, mean 12 (England, sample 8000). If only 1 chick, single parent may assume sole responsibility; broods of more than 1 divided in weeks 4–6, each parent tending own group only and favouring one of chicks in particular (see Social Pattern and Behaviour). Adults aggressively defend eggs and young against avian predators; warn young to dive by splash-diving. Young can feed themselves from week 8 but food-dependent until at least week 10; mean duration of parental feeding 13 weeks, but favoured chicks usually fed longer than others—up to week 15 (exceptionally week 23) as against weeks 10–11, sample 12 (England). FLEDGING TO MATURITY. Fledging period 71–79 days, sample 12. Become independent then or later, with wide variation within same family, favoured chicks leaving later (mean week 17) than others (mean week 12). First breeding probably often delayed until 2nd year, but may pair and hold territory in 1st; sometimes breed then, mostly late in season. BREEDING SUCCESS. Britain: 431 pairs reared 589 young, mean 1·3; 169 pairs, 250 young, mean 1·5. England: 98 pairs, 115 young, mean 1·2. Broods of 1–5 reared, but mean brood-size: 2·3, sample 169 (Britain); 2·1, sample 124 (England); 2·1, sample 535 (Europe, Melde 1973). Egg-fertility high and, though viable eggs occasionally abandoned, main losses due to predation of young (especially by pike *Esox lucius* and other large fish), starvation, and bad weather.

**Plumages.** ADULT BREEDING. Crown glossy black with short transverse crest ending in 2 long tufts at sides of nape. Nape feathers and strongly developed tippet (or frill) chestnut at base, otherwise black. Crests and tippets, average longer in ♂. Cheeks and narrow line over eye white or pale buff; chin pure white. Upper hindneck glossy black, lower dark grey; rest of upperparts black with slight brown tinge and grey feather edges. Some lower scapulars mainly white. Foreneck white, separated from chin by black and chestnut of tippets; rest of underparts shiny white. Flanks mixed rufous and dusky. Tail-tuft black above, white and buff below. Primaries dark brown-grey, paler below, white at base, inner tipped white. Secondaries white, outermost with mainly brown-grey outer web, inner brown-black. Marginal upper wing-coverts white, rest dark brown-grey, but inner webs of greater coverts in outer part of wing mainly white. Under wing-coverts white, some near base of primaries grey. Axillaries white, often with buff or grey spots at tips. ADULT NON-BREEDING. Crown dark grey, crest short, tippets absent or indicated by some rufous or black feathers. Sides of head, chin, and throat white. Upperparts with wider grey feather edges than breeding. Underparts white; flanks mottled dusky. Tail-tuft without buff. Wings like breeding. DOWNY YOUNG. Head, neck, and upper-parts with pattern of black and pale buff or white lines and spots, ligher than Red-necked Grebe *P. grisegena*; breast and belly white, vent dark grey. Forehead and sides of crown off-white, bordered ventrally by black line running over eye to side of nape. Crown black with central bare spot and large white patch on nape (larger than *P. grisegena*). Circular bare spots on lores. Sides of head with pattern of white and black lines, heavy black streaks over eye and on lower cheek most prominent; narrow white line straight behind eye characteristic. Chin white with variable rows of black spots at sides. Neck off-white with dorsal, ventral, and on each side 2 lateral black lines. Ventral line split in two and ending near chest; dorsal and lateral lines continuing on upperparts and flanks in pattern of 9 black and 8 grey-buff lines. Black lines on flanks dissolved into rows of dark spots. Lower flanks grey-buff. Contour feathers start to grow soon after hatching (P P A M

Kop); down remains adhered to tips and young retain downy appearance until well grown. Pale stripes on upperparts gradually disappear. Feathers on underside full-grown and free of down first, followed by those on back and lastly on head. Flight-feathers develop at same time as back. Bare spots on lores covered by white feathers at 6–7 weeks, bare crown spot by black feathers at 10. New body feathers continue to grow slowly during 1st autumn. (Niethammer 1964; Storer 1967; Kop 1971a Fjeldså 1973e; ZMA). JUVENILE. Similar to adult non-breeding. Black and white pattern on sides of head may be retained until late autumn of 1st calendar year. Narrow white edges to crown feathers and hairy remnants of down may persist into 1st winter. Region of tippet marked with pale rufous and black spots. Wings like adult, but usually more grey on secondaries, and inner webs of outer greater coverts dark grey, only occasionally white. IMMATURE WINTER. Differs from adult non-breeding by virtual absence of white on inner webs of greater coverts and by hind rim of tarsus, which is strongly serrated and double in adult, slightly serrated and hardly double in immature (Kop 1971a). Specimens with remnants of down to feathers of upperparts easily recognized as immature, but other characters not always fully reliable. SUB-ADULT BREEDING. As adult breeding, but tippets probably on average less strongly developed. Birds with some hairs to tips of scapulars, slightly serrated hind rim of tarsus, and hardly any white in greater coverts are 2nd calendar year.

**Bare parts.** ADULT. Iris crimson with narrow inner ring of orange. Culmen and cutting edges of bill dark horn; sides grey, strongly washed carmine. Bare skin of lores reddish-black. Outside of tarsus and sole of foot olive-green, inside and upperside of toes green-yellow; nails tinged blue. DOWNY YOUNG. Iris pale yellow. Bill reddish at base, white at tip, with transverse black bar in middle. Bare triangle on crown and circular bare spots on lores pale flesh, becoming red by vascularization. Leg and foot slate-grey with pink edges to lobes of toes. JUVENILE. Iris orange with narrow inner ring yellow. Bill paler on sides than in adult. Leg and foot mottled dark brown.

**Moults.** ADULT POST-BREEDING. Complete; July–December. Primaries and secondaries simultaneous; flightless for 3–4 weeks; August–October; ♂♂ tend to moult 2–3 weeks before ♀♀ (Simmons 1970c, 1974a). ADULT PRE-BREEDING. Partial; December to late spring. Traces of tippets first of breeding plumage to appear. Pre-breeding and post-breeding moult not sharply separated, moulting almost continuous; most of plumage renewed twice a year, but white underparts once. Moult of scapulars starts in front, proceeds backward, and begins anew when completed. Moult slows in winter, stops entirely on underside. POST-JUVENILE. Partial; involving many body feathers and apparently some wing-coverts, but no longer scapulars and white underparts. January to late spring of 2nd calendar year; leads to sub-adult breeding plumage. (Kop 1971a; P P A M Kop.)

**Measurements.** Adult, skins (BMNH, ZMA); ranges from various literature sources.

| | ♂ | | | ♀ | | |
|---|---|---|---|---|---|---|
| WING | 195 | (4·2; 28) | 175–209 | 184 | (5·6; 19) | 168–199 |
| BILL | 51·8 | (1·9; 22) | 41–55 | 46·1 | (1·9; 23) | 38–50 |
| TARSUS | 63·6 | (2·6; 23) | 59–71 | 61·6 | (2·4; 23) | 57–65 |
| TOE | 69·9 | (2·4; 10) | 65–73 | 67·3 | (1·6; 10) | 64–70 |

Additional wing-lengths from adults in flesh. Netherlands (ZMA): ♂ 195 (4·8; 25); ♀ 188 (5·1; 25). Bodensee (Jacoby and Schuster 1972): ♂ 194 (13) 183–200; ♀ 185 (26) 174–194. Sex differences significant.

**Weights.** (1) Netherlands (ZMA), (2) Bodensee (Jacoby and Schuster 1972).

(1) Lean ♂ 738 (79; 7) 596–813 ♀ 609 (52; 4) 568–686
(1) Fat 920 (144; 15) 687–1180 830 (169; 12) 619–1125
(2) Fat 1325 (13) 1060–1490 1198 (26) 1020–1380

Range of weights (Bauer and Glutz 1966): ♂♂ 750–1400; ♀♀ 642–1180; sex unknown 590–1310.

**Structure.** Wings short and narrow. 12 primaries: p10 longest, p11 often as long or up to 5 shorter, p9 0–5 shorter, p8 9–16, p7 18–28, p6 26–34, p1 63–76; p12 minute. p9 and p10 emarginated on outer webs, p10 and p11 on inner. 21–22 secondaries (Stephan 1970; Stübs 1972). Bill straight and stout, pointed; relatively longer than in *P. grisegena*; nostrils narrowly oblong in shallow groove near ridge of culmen. Strip of bare skin between base of upper mandible and eye. Outer toe *c*.106% of middle, inner *c*.83%, hind *c*.25%.

**Geographical variation.** None in west Palearctic. *P. c. infuscatus* and *australis* both darker, without white line over eye, flanks dark brown without rufous tinges (Hartert 1912–21); *infuscatus* slightly shorter wing on average than nominate *cristatus*; *australis* appears sturdier. Most populations of *infuscatus* and *australis* do not develop recognizable non-breeding plumage, but South African *infuscatus* may have one (Benson and Irwin 1964). Australian populations formerly separated as *christiani*, but considered inseparable from *australis* by Vaurie (1965).

## *Podiceps grisegena* Red-necked Grebe

PLATE 10
[after page 110]

Du. Roodhalsfuut    Fr. Grèbe jougris    Ge. Rothalstaucher
Ru. Серощекая поганка    Sp. Somormujo cuellirrojo    Sw. Gråhakedopping

*Colymbus grisegena* Boddaert, 1783

Polytypic. Nominate *grisegena* (Boddaert, 1783), west Palearctic; *holboellii* Reinhardt, 1854, east Palearctic and west Nearctic, accidental.

**Field characters.** 40–50 cm (body 22–30 cm), wingspan 77–85 cm; medium-large, rather stocky and thick-necked, dark-crowned grebe, showing at all seasons pale cheeks contrasting with crown and neck colours and dark bill with striking yellow base. Sexes alike, but marked seasonal variation; juvenile separable in 1st autumn.

ADULT BREEDING. Whole crown (down to eye level) and short crest tufts at rear dull black, lower face and cheeks pale grey, outlined in white; front and sides of neck and breast chestnut. Hindneck and upperparts, including all but leading wing-coverts, dark brown; underparts dull white, with mottled brown flanks. Primaries brown-black, secondaries white. Bill strikingly patterned, with black culmen and tip contrasting with rich yellow base. Eyes warm brown. Legs and feet dark grey. ADULT NON-BREEDING. Basic plumage pattern retained, though neck becomes dusky grey on sides and dull white on front and cheeks whiten; lower chest mottled grey-brown and flank marks duller. Bill paler than breeding but dark tip usually obvious. JUVENILE. Plumage pattern as breeding adult, including red tinge to neck; distinguished by striped sides to crown and face. Bill pattern present but colours less intense than in adult.

Breeding adults unmistakable and birds in other plumages unlikely to be confused with slightly larger and longer-necked Great Crested Grebe *P. cristatus* at close range (though at longer distances or in poor light, differentiation requires attention to character as well as plumage). *P. grisegena* distinct, with shorter, stouter bill, deeper head shape (often showing round crown in winter), thicker, less elongated neck, and bulkier body—all contributing to stocky silhouette and lack of serpentine appearance often suggested by *P. cristatus*. Except in juvenile, crown totally black (lacking white above eye of *P. cristatus*) and giving a frowning expression to face. Relative bill size rules out confusion with smaller grebes. Flight silhouette less attenuated than *P. cristatus* and less variegated upperwing pattern makes flight action look more regular.    DIMW

**Habitat.** Overlaps with Slavonian Grebe *P. auritus*, displacing it from sites suited to both (Fjeldså 1973*a*). Markedly more continental, however, flourishing best away from oceanic fringing lands and arctic-alpine terrain, and thinning out towards boreal regions and warmer temperate zone. Predominantly in lowland basins and great plains below 100 m; exceptionally up to 1800 m (Turkey). Compared with Great Crested Grebe *P. cristatus*, favours smaller waters, often down to 3 ha or in Denmark even under 1 ha, with less fish, higher ratio of emergent vegetation, and often closer surrounding tall forest. Needs shallow water, normally diving less than 2 m (Simmons 1970*b*), although up to 10 m recorded. Nests farther within reedbeds than grebes feeding more on open water, such as *P. cristatus* (Gotzman 1965), seeking easy underwater approach with sufficient density of stems to assure nest stability. Often associated with breeding colonies of *Larus* gulls or other waterbirds. Sometimes breeds on backwaters of large rivers or estuaries and on pools cut off from sea, and can adapt to using fish-ponds filled with water only from March–October (Gotzman 1965), but range of tolerance for breeding habitat fairly limited.

After breeding, moves to more open, especially

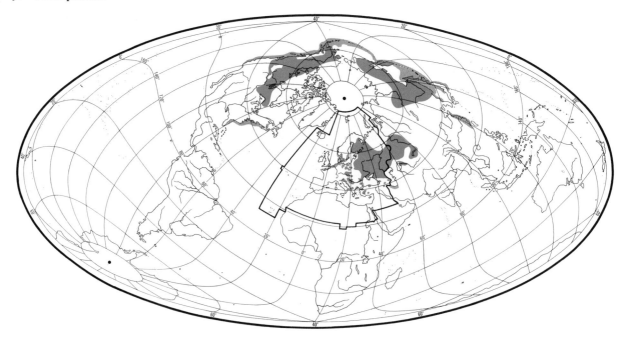

estuarine or coastal waters, where fish accessible. Like other grebes, little on land except at nest. Flies somewhat infrequently, taking off clumsily, at low or medium altitude, making extended journeys mainly at night. Fairly tolerant of human presence.

**Distribution.** Map shows main areas of known regular breeding, but often sporadic and irregular within these, e.g. USSR (Dementiev and Gladkov 1951), as well as on edge of range where signs of retreat in some parts in 20th century. EAST AND WEST GERMANY: marked decline in Mecklenburg and Saxony (especially in west, though still numerous Oberlausitz) and breeding irregular further west (Bauer and Glutz 1966). CZECHOSLOVAKIA: regular breeder 19th century and to some extent until 1920s; now single pairs nest occasionally (K Hudec). AUSTRIA: bred in east first half 19th century (Neusiedlersee and Danube valley) and young found 1955; now accidental (HS). FINLAND: apparent recession in extreme south (Merikallio 1958). NETHERLANDS: occasional breeder 1918, 1927, and 1966 (JW). FRANCE: probably bred Lorraine before 1919 (CE). GREECE: now extinct after draining of Nestos delta where formerly *c.* 10 pairs (WB). MOROCCO: evidence of breeding 19th century unsatisfactory (KDS).

Accidental. Iceland, Faeroes, Spitsbergen, Spain, Albania, Algeria, Tunisia, Lebanon, Egypt.

**Population.** DENMARK: 350–400 pairs (Preuss 1969). FINLAND: *c.* 2000 pairs (Merikallio 1958). HUNGARY: 50–60 pairs (L Horváth). WEST GERMANY: *c.* 120 pairs Schleswig-Holstein (Schmidt 1967). POLAND: at least 540 pairs, and perhaps up to 1000 pairs (AD, LT). USSR: *c.* 120 pairs Estonia, probably some increase last few decades (Onno 1970). In absence of regular and extensive counts, impossible to assess population trends. In north-west Europe (Thomasson 1953), considered rare in 17th and 18th century becoming increasingly common from middle 19th century. No evidence of increase this century; numbers apparently fluctuating and probably some decline on edges of range.

**Movements.** Migratory and dispersive. After breeding, most migrate or disperse to tidal water. In west Palearctic principal wintering areas western seaboard (Norway and Britain to Bay of Biscay), Baltic and Caspian Sea; lesser numbers also Black Sea (Dementiev and Gladkov 1951), and apparently Adriatic and Aegean. 19th century reports of large numbers Morocco and Tunisia, but no modern confirmation and seems now scarce to rare winter visitor to south side Mediterranean (Bannerman 1959). Some winter inland waters, especially larger ones such as Swiss lakes (Bauer and Glutz 1966). Present British coastal waters chiefly October–March; scarce Ireland. Southward passage noted Åland Channel (entrance Gulf of Bothnia) August–October (Andersson 1954); departures from Schleswig–Holstein breeding lakes commence August. In USSR, autumn movements at peak October to early November (Dementiev and Gladkov 1951). Occasional large midwinter influxes (associated with severe weather) occur, as Britain 1865, 1891, 1895, 1922, and 1937 (Witherby *et al.* 1940). Main spring departures from western seaboard March, from Caspian Sea April; throughout range of nominate *grisegena*, breeding waters reoccupied late March to early May, averaging earlier in west.

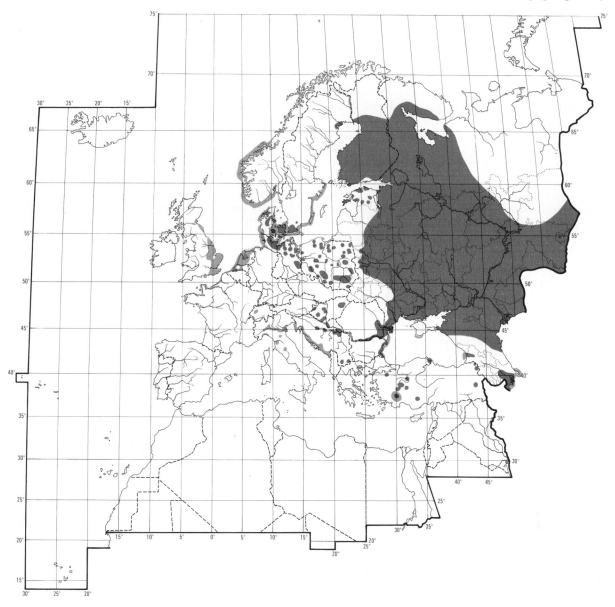

Migrates singly or in small parties. In Åland Channel migration chiefly nocturnal, though continued by swimming by day (Andersson 1954); in North America (*holboellii*) passage nocturnal overland, but flies by day in coastal movements (Palmer 1962). Movements imperfectly known. Presumably north and north-west Europe breeders winter North Sea, Atlantic coasts and Baltic, perhaps with some from European Russia (west to Urals), though others of unknown origins winter Black Sea. Caspian winterers believed breed Volga basin and west Siberia (Dementiev and Gladkov 1951). Only 3 long-distance ringing recoveries: juvenile from Oberlausitz (East Germany) found north-west France, February; one ringed Denmark (October) found Finland, May; one

ringed Switzerland (December) recovered Hungary, June.

*P. g. holboellii*, breeding east Asia and North America, winters coasts on both sides of Pacific and in west Atlantic, latter reached via Great Lakes. Gregarious at migratory staging points in North America, in concentrations up to 500 (Palmer 1962). Vagrant to west Europe.

**Food.** Mainly invertebrates, especially aquatic and terrestrial insects and their larvae, and to lesser extent, fish; obtained by diving and by swimming on surface with head submerged. Insects also taken off aquatic vegetation, and from or just above surface (Madsen 1957; Wobus 1964). Prey caught underwater brought to surface if large and difficult and, when fish, manoeuvred into place to be swallowed head first; occasionally even dropped back into water and seized again in quick dive (Wobus 1964). Fish killed either by crushing head with beak or by severe shaking, though sometimes swallowed alive (Shelley 1930). Dives last up to 1 minute in larger race *holboellii* (Bent 1919; Palmer 1962) though generally less than 30 s in nominate race (Madsen 1957). 155 dives at one locality, depth 3–4 m, averaged 25 s (Hancock and Bacon 1970); 59 and 130 at others 24·8 and 29·3 s respectively, depending mainly on depths of water (Simmons 1970b). Normally daytime feeder with periods of inactivity at mid-day and in afternoon (Wobus 1964).

Nominate race, notwithstanding large size, has diet similar to Slavonian Grebe *P. auritus* (see especially Fjeldså 1973c). Chiefly invertebrates, particularly from marshes and small lakes during summer months; primarily insects and their larvae, including dragonflies, waterbugs (*Corixa, Notonecta, Ranatra, Naucoris*), aquatic and terrestrial beetles (*Amara, Harpalus, Dytiscus, Agabus, Colymbetes, Hydroporus, Noterus, Rhantus, Donacia, Haemonia, Helophorus, Hydrobius, Lixus, Cetonia*). Less fish, though local variations, particularly when feeding from large lakes and coast, especially in winter when fish (up to 25 cm) may predominate. From marine areas: young herring *Clupea*, pipefish *Syngnathus*, goldsinny *Ctenolabrus*, sculpin (Cottidae), sticklebacks *Gasterosteus* and *Spinachia*, blenny *Zoarces*, sand-eel (Ammodytidae), butterfish *Pholis*. From freshwater areas: stickleback, *Gasterosteus*, perch *Perca*, roach *Rutilus*, eel *Anguilla*. Also recorded: crustaceans (*Triops, Palaemon*, small crabs), frogs mainly tadpoles, molluscs, aquatic worms and plants (probably mainly digested accidentally) including algae, mosses, seeds, and aquatic plant portions. (Naumann 1905; Eckstein 1970; Madon 1931; Witherby *et al.* 1940; Dementiev and Gladkov 1951; Madsen 1957; Onno 1958; Markuze 1965; Fjeldså 1973c).

34 stomachs from different parts of Europe contained insects in 28 (82·4%), fish in 8 (23·6%), molluscs in 3 (88%), crustaceans in 1 (Madon 1931). 25 stomachs from Danish salt and brackish water areas October–January all contained fish, mainly *Gobius* (52%) and *Gadus* (44%), and 18 (72%) crustaceans, chiefly shrimps and prawns (Crangonidae and Palaemonidae); 5 from Danish freshwater areas, same period, contained fish, mainly *Gasterosteus*, and insects, though in less quantity (Madsen 1957). 73 stomachs from fish ponds near Astrakhan, USSR, showed 92–100% invertebrates, mainly *Triops, Esteria*, Hemipterans (especially *Ranatra linearis* and *Naucoris cimicoides*); waterbeetles (dytiscids and hydrophilids) and larvae; other beetles *Donacia* and *Lixus*; and some amphibians, mainly tadpoles, and few fish (Markuze 1965).

See Wetmore (1924) and Palmer (1962) for American studies: suggests *holboellii* mainly piscivorous, i.e. ecological equivalent of Great Crested Grebe *P. cristatus*.

**Social pattern and behaviour.** Based mainly on material supplied by J Fjeldså and R W Storer.

1. Least gregarious of west Palearctic grebes. Outside breeding season, nominate race most frequently single or in twos, or in small, loose assemblages—e.g. in late summer before and during wing moult; only exceptionally in large flocks. Usually feeds singly, but sometimes in mixed groups with other waterbirds, especially when food abundant. BONDS. Monogamous pair-bond of at least seasonal duration, maintained through most of growth period of young or longer. Pair-formation starts on migration or on breeding waters. Both parents tend young. Social structure of family group not fully studied; no information on brood-division but, in *holboellii*, chicks often accompanied by both parents. Young from 1st brood, if still present, may assist in feeding of 2nd. BREEDING DISPERSION. Pairs typically territorial with well-dispersed, cryptic nests. Sometimes also in loose colonies of up to *c.* 20 pairs, each defending 70–110 m of shoreline; nests rarely as close as 10 m (see, e.g., Wobus 1964). ROOSTING. On water, during day as well as at night.

2. Highly vocal in breeding season, including loud, far-carrying Song. Appears to serve general purpose of territorial advertisement as well as accompanying displays of water-courtship; delivered in characteristic posture with head raised and feathers of crests and hindneck erected. Following based on Wobus (1964); supplemented by information from J Fjeldså (on nominate *grisegena*), incorporating material from Th Schmidt, and from R W Storer (*holboellii*). See also Fjeldså (1973b), Storer (1969). ANTAGONISTIC BEHAVIOUR. Occurs during halts on migration as well as in breeding area; by both sexes, but especially ♂♂, by night as well as day. Threat includes: (1) Hunched-display with head withdrawn into shoulders, dorsal plumage raised, and wings partly lifted on back; (2) Forward-display much as in *P. cristatus*. In latter display, most of body often submerged with neck awash and bill somewhat raised (Threat-lurking); in Hunched-display, bill usually horizontal or lowered, sometimes touching water, but may be raised. From account and photograph in Sage (1973), probably also uses (3) Defensive-upright with crests raised as threat at times. Token-dives, underwater pursuit, supplanting attacks from below surface or by pattering over water, and breast-to-breast fighting all much as in *P. cristatus*, though full Sinister-dives not reported. When attacked, escapes by pattering across water (with or without use of wings), diving, or flying. In extreme Alarm or Furtive-upright, all feathers depressed, neck stretched up and somewhat back, bill horizontal; as bird swims away, rear-end of body may be low in water, front-end high (K E L Simmons). WATER-COURTSHIP. Though similar in some features to that of better-known *P. cristatus* and Slavonian Grebe *P. auritus*, apparently lacks any form of Habit-preening or Retreat Ceremony, and no casual Food-presentation recorded. Also much more vocal, giving Display-call (Song) during most ceremonies, and greater emphasis placed on Penguin-dance. Advertising appears to be less specialized and frequent than in other *Podiceps*. Performed by lone bird in posture similar to that adopted when singing; Quacking-call may be given but probably replaced, to some extent at least, by Song. Advertising by one or both birds may precede Discovery Ceremony. Latter much as in *P. auritus*, with similar type of Cat-display—i.e. in Wing-spread component, primaries not so fully unfolded nor wing tilted forward as found in *P. cristatus* and Black-necked Grebe *P. nigricollis* (see A). (1) During approach dive, first bird emerges at times in Bouncy-posture, with head resting on mid-back area, breast puffed out, tail-tuft cocked, or shows head only; (2) performs Ghostly-penguin display 3–4 m away from bird in Cat-display, emerging

A

slowly with back to it while uttering Display-call loudly (see A); (3) still calling, joins other; (4) both rise breast to breast in Penguin-dance (see B); (5) subside and Head-shake (see C). Discovery Ceremony ends as ♂ and ♀ (6) move formally apart (Ceremonial Turning-away), or they may sometimes turn during Penguin-dance and swim slowly side by side through water in Penguin-posture (Parallel-barging) before subsiding and Turning-away; in *holboellii*, Penguin-dance occasionally replaced by Parallel-rush over water. In typical Penguin-dance, both give Display-call and one bird (almost certainly ♂) always rises much higher than other; as well as Slow Head-turning (see Little Grebe *T. ruficollis*), both also open mandibles and either or both may grasp at other's bill or make short stabs towards its breast. In terminal Head-shaking, Slow Head-turning only movement. Both Head-shaking and Penguin-dance also occur as separate ceremonies. In Head-shaking Ceremony, face and perform (1) Low Head-waggles (see *P. cristatus*) while uttering Display-call and (2) Slow Head-turns silently. Penguin-dance Ceremony particularly common (at least in nominate *grisegena*), probably largely taking over role of Head-shaking during engagement period; pair rarely rise high in water, and Parallel-barging and Parallel-swimming variants frequent. Weed-trick and Weed-dance, much as in *P. cristatus*, also reported but full details of whole Weed Ceremony lacking; Weed-dance itself

accompanied by loud calling. In *holboellii* at least, after pair-bond well established, Discovery Ceremony replaced by pair meeting and simply engaging in Display-call duet; may then also Head-shake, sometimes rising in Penguin-dance. In both races, pair may link up and perform Triumph Ceremony, assuming Hunched-display and duetting, especially after antagonistic encounters. Both Penguin-dance and Parallel-barging persist into later phases of breeding cycle in latter situation, while bird disturbed by man at nest may rise on water in Penguin-posture. PLATFORM-COURTSHIP. Similar to other *Podiceps* (see especially Schmidt 1970). Includes (1) Inviting (see D), (2) Rearing (see E), (3) copulation (see F), and (4) Post-copulatory Water-treading (see G); last more ritualized than in other *Podiceps*. Mounting preceded and accompanied by calls (see D and F). Inviting bird may slightly raise head during copulation (see F) and afterwards rises in so-called Ecstatic-posture (see G) as mate in water turns and faces it, initiating bout of Slow Head-turning (see H). Reversed mounting observed early in season. Inviting-on-water also not uncommon, especially prior to platform building. Apparent weed-presentation and attempted copulation on water reported by Hemming (1968); see also Johnston (1953). RELATIONS WITHIN FAMILY GROUP. No detailed information.

(Figs A–H after original drawings by J Fjeldså.)

B

C

D

E

F

G

H

**Voice.** Varied and, in breeding season, used to much greater extent than in other Palearctic *Podiceps*. Similar in both sexes. Mostly silent during rest of year. Best account: Wobus (1964); see also Sim (1904), Bent (1919), White (1931) for *holboellii*.

Though difficult to identify all calls in literature from descriptions in Wobus (1964), following based on his treatment, supplemented by information from R W Storer (*holboellii*) and other sources. (1) Song and display-call. Remarkable, loud, diver-like sound: roaring, wailing, howling, or hooting 'uööh . . .' (see I), with considerable individual variation. Recalls, especially when given from cover, dabchicks *Tachybaptus* and particularly pied-billed grebes *Podilymbus*, rather than typical *Podiceps*. Uttered by one bird or by pair in duet, during night as well as day, often from cover. Surprisingly loud sequences of up to 60 consecutive notes delivered during singing encounters between birds in different territories. Phrases during water-courtship displays usually shorter: 4–10 (spring), 1–4 (summer), rising in intensity and pitch. Variability probably facilitates individual identification, and, in duets, call of one bird (almost certainly ♀) noticeably weaker than other's. These factors, plus likelihood that other calls often combined with 'uööh' notes, seem largely to account for great variation in published accounts; also, earlier descriptions of 2-toned call—like combination of pig-like squealing and foal-like whinnying (see Witherby *et al.* 1940)–did not allow for duetting. Calls of presumed ♂ screaming or squealing in character, strongly nasal, distinctly higher pitched and audible at greater distance; those of presumed ♀ shorter, tremulous, and even more whinnying, neighing, or braying in character. Other renderings include, for nominate *grisegena*, 'ee(-)er EE(-)ER-EE(-ER) ee(-)er . . .' (K E L Simmons); for *holboellii*, 'ah-oo ah-oo ah-ooo ah-ah-ah-ah' sometimes ending in more staccato, trilling 'whaa whaa whaa whaa whaa chitter—r-r-r-r-r', 'quonk . . .' (coarse, nasal braying), and harshly raucous 'carrack awwack caawwrrak' (see Bent 1919). (2) Quacking-call. Quacking 'wack . . .' or 'ack . . .' like Decrescendo-call of Mallard ♀ *Anas platyrhynchos*. Normally 4–10 notes, but occasionally up to 100 — e.g. 'aakaak-aakaak-aakaakaak-aakaak-aakaak-aakaakaak' (see Bauer and Glutz 1966). Variable: often loud and emphatic, uttered in short or long sequences, slowly or fast, but sometimes quieter; often combined with song-notes. Typically given by lone bird. Though no Advertising-call specified by Wobus (1964), probably used at times as such and also as Anxiety-call; in latter case, described as duck-like 'quek-quek' (see II) (K E L Simmons). In *holboellii*, loud, nasal 'a a a n', somewhat similar to comparable call of Slavonian Grebe *P. auritus*, used as Advertising-call; uttered by either ♂ or ♀, or both, prior to Discovery Ceremony and also in combination with duet calls and, possibly, other notes (R W Storer). (3) Clucking-call. Fast, soft 'ga . . .' in series of up to 10 notes, usually 4–5. Harsher version rendered 'keck . . .'; also

'cuk' and 'teck' (see Sage 1973). Given by both members of pair during pair-formation and later near platform or nest during platform-courtship and mating phases. Often alternates with Quacking-call. Probably homologous to Platform-call of *P. auritus* (R W Storer), and same as 'whit-tah' of *holboellii* (Sim 1904). (4) Hissing-call, etc. Hissing, lightly snarling, throaty sound. Snake-like (Sage 1973). Given when disturbed, e.g. at nest. Bird in encounter with Coot *Fulica atra* near nest gave long, drawn-out, triple 'airrrr . . .—airrrr . . .—airrrr . . .' like Snarling of Great Crested Grebe *P. cristatus* (K E L Simmons). In general, threat-calls not well documented. Rather loud, grating, nasal scream (somewhat like song but no so loud) described for *holboellii* when angry (Sim 1904; White 1931). (5) Alarm-call. Short, sharp 'äck' ('tjäck', 'tjäp', 'teck', etc.). Uttered at intervals of ½–1 s, e.g. when disturbed by man. (6) Purring-call. Drawn-out, largely unarticulated 'ääääää . . .'. Given most frequently by bird performing Inviting-display while soliciting on platform. (7) Rattling-call. Deep, trilling 'u' sound turning immediately into rapid series of higher 'r' sounds, e.g. 'uer rrrrrrrrrrrrrrrr'. Repeated 3–6 times by copulating bird.

CALLS OF YOUNG. Short, sharp 'jig' ('iejp' or 'pieh') first given, at intervals of 1–2 s, up to 2 days before and during hatching; homologous to Squeak of *P. cristatus*. Begging and contact-call: 'bibibibibibibibi' similar in structure and rhythm to song of Chiffchaff *Phylloscopus collybita*. When 6–7 weeks old, but still dependent, also begins to utter soft 'ga' which later develops into 'gagaga' with only slight 'jip' component.

I Sture Palmér/Sveriges Radio Sweden June 1965

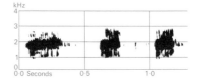

II Sture Palmér/Sveriges Radio Sweden June 1965

**Breeding.** SEASON. See diagram for central Europe. Dependent on thaw in north of range. SITE. In water, usually among emergent vegetation but sometimes in open though with thick cover near. Usually solitary but sometimes in loose colonies. Nest: floating heap of aquatic weed anchored to submerged vegetation. Diameter at base 40–90 cm, diameter at water level 18–50 cm, height of nest 18–90 cm, height above water 2·7–10·5 cm, diameter of cup 5–18 cm and depth 1·5–7 cm (Wobus 1964). Several platforms built before actual nest. Building: by both sexes gathering plants from under water and piling on fixed vegetation; cup formed by bill, feet, and belly. EGGS. Ovoid, not glossy; dull white becoming scratched and stained during incubation. 51 × 34 mm (47–58 × 30–37), sample 100; calculated weight 31 g (Schönwetter 1967). Clutch: 4–5 (2–6); rare larger clutches probably always by 2 ♀♀. Normally one brood, rarely 2. Replacement clutches laid after egg loss or desertion, up to 4 times (Palmer 1962). Eggs laid daily. INCUBATION. 20–23 days. By both sexes. Begins with 1st egg, occasionally later (Dementiev and Gladkov 1951); hatching asynchronous. Eggs covered with nest material when incubating bird disturbed from nest. Eggshells mostly removed. YOUNG. Precocial and semi-nidifugous. Cared for and fed by both parents. When small, carried by one parent in turn and fed by other or sometimes by one carrying. One parent may care for first hatched young while other incubates but sometimes last eggs abandoned (Palmer 1962). Young brooded near nest on platforms which may be early nest-building attempts. Fed bill to bill by parents; also given feathers by them. Parents aggressive in defence of young. FLEDGING TO MATURITY. Fledging period not recorded, but over 72 days in one case (Palmer 1962). Time of independence not recorded. Mature at 2 years. BREEDING SUCCESS. No data.

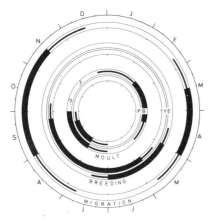

**Plumages.** ADULT BREEDING. Crown, hindneck, and 2 crests of elongated feathers at back and sides of crown glossy black. Rest of upperparts brown-black with narrow brown-grey feather edges. Chin, throat, and sides of head mouse-grey, conspicuously bordered white. Foreneck and sides of neck chestnut, chest paler with bronze gloss. Breast and belly white, often mottled grey; vent pale grey; sides of breast mottled grey, rufous, and white; flanks mottled grey and white. Tail-tuft of downy, black feathers; partly white below. Primaries brown-grey with black shafts; outermost secondary similar, next 2 or 3 increasingly spotted white, followed by c. 10 mainly white with shafts basally black, and c. 3 brown-grey with white or rufous at base; innermost brown-black like mantle. Upper wing-coverts brown-grey, near body brown-black, but marginal and part of lesser white. Under wing-coverts and axillaries white. ADULT NON-BREEDING. Upperparts dark grey, feathers of mantle and scapulars with paler edges. Chin, throat, and sides of head white; mouse-grey patch behind eye, followed by narrow white crescent. Foreneck and sides of neck pale grey. Rest of underparts white, chest and flanks

mottled grey, sometimes breast and belly also; vent washed pale grey. Wing as breeding, but innermost secondaries and their coverts greyer. DOWNY YOUNG. Crown black with central bare spot and white patch on nape. Hindneck black with narrow white stripe on either side of centre. Forehead and sides of crown white, continued as buff-white line on side of neck; bordered ventrally by black line over eye, which broadens to irregular spot, interrupted or bent sharply downwards on side of upper neck. Sides of head, chin and throat white with pattern of black streaks, short black line under eye most prominent. Lores bare. Pattern on foreneck variable, central black line which splits near chest or 2 separate lines meeting on lower neck. Back and flanks with pattern of brown-black and pale brown stripes; dark stripes broken into rows of spots on flanks. Breast and belly white, vent washed pale grey (Storer 1967). JUVENILE. Crown dark brown, hindneck brown-grey, rest of upperparts dark brown with pale feather fringes and remains of down to long feathers. Chin and throat white, sides of head with remains of downy pattern. Foreneck, sides of neck and chest tawny-buff, paler than in adult breeding. Rest of underparts white, flanks and vent grey. Wings as adult, but usually with less white in secondaries. IMMATURE WINTER. Like adult non-breeding, not certainly recognizable after remains of down have disappeared. West Palearctic populations do not seem to develop grey-headed immature winter plumage as illustrated for holboellii in Palmer (1962). IMMATURE SUMMER. Doubtfully distinct from adult breeding. Said to be duller with whiter chin and throat, browner crown, mottled cheeks, and dusky mottling in chestnut of neck (Bent 1919), but this may refer to specimens in first pre-breeding moult.

**Bare parts.** ADULT BREEDING. Iris dark brown. Bill black, bright yellow at base. Foot black to dark grey-green on outside and underside; inside and upperside of toes pale yellow to greenish-grey, sometimes mottled brown. ADULT NON-BREEDING. Bill paler than breeding, blue-grey to grey-black, yellow at base. DOWNY YOUNG. Iris olive-brown. Bare spots on lores and crown paler or brighter red. Bill buff with black spots. JUVENILE. Iris yellow, apparently not becoming brown before summer of 2nd calendar year (Palmer 1962). Bill averages paler and more extensively yellow than adult non-breeding; yellow at base, dark greenish-brown at tip and along edge of culmen, greenish-yellow in between. IMMATURE WINTER. May differ from adult non-breeding by yellow iris and lighter bill, but confirmation required.

**Moults.** Poorly known. ADULT POST-BREEDING. Complete; primaries simultaneous; July–September. Failed breeders may moult early (Palmer 1962). ADULT PRE-BREEDING. Partial; presumably entire body, tail, innermost secondaries and wing-coverts (Witherby et al. 1940); December–May. Late moulters may be in 2nd calendar year, indistinguishable from older birds. Many adults in complete non-breeding plumage at least until February, others in complete breeding mid-March (Wobus 1964). POST-JUVENILE. Partial; head, neck, rest of upperparts, chest and flanks; September to late January. Some nearly finished November, others just starting.

**Measurements.** Full-grown (wing older than juvenile only). Entire range of nominate grisegena, but mainly Britain and Netherlands winter; skins (BMNH, RMNH, ZMA).

| | ♂ | | | ♀ | | |
|---|---|---|---|---|---|---|
| WING | 175 | (8·09; 21) | 164–193 | 169 | (9·79; 23) | 153–182 |
| BILL | 40·0 | (2·47; 25) | 34–44 | 37·1 | (2·25; 30) | 33–42 |
| TARSUS | 55·9 | (3·36; 17) | 51–63 | 53·8 | (2·48; 26) | 49–60 |
| TOE | 65·2 | (2·75; 4) | 62–68 | 62·9 | (2·51; 11) | 59–67 |

Sex difference significant, except for middle toe. Wings of recognizable juveniles average shorter: ♂ 169 (5·00; 11) 160–177, ♀ 162 (8·24; 13) 153–177, but no difference in bill and tarsus. Swedish specimens longer wings (mean 16 ♂♂ 178, 10 ♀♀ 176), but no difference in other measurements (Ahlén 1961). *P. g. holboellii* much larger, particularly bill. Mean wing: 20 ♂♂ 197, range 183–212; 20 ♀♀ 190, range 173–204. Mean bill: 20 ♂♂ 50·2, range 47–56; 14 ♀♀ 46·7, range 45–55. (Mean wing from Vaurie 1965, bill from Palmer 1962, ranges from various sources.)

**Weights.** Germany: breeding ♂♂ 806, 823, 900, 925; breeding ♀♀ 692, 823, 873; juvenile ♂ 582, ♀ 620 (Bauer and Glutz 1966). Netherlands, beached birds: non-breeding ♂ 742; mean 10 non-breeding ♀♀ 476, range 316–764; juvenile ♂ 378 (ZMA). *P. g. holboellii* heavier, range 743–1270 (Palmer 1962).

**Structure.** 12 primaries: p10 (rarely p11) longest, p11 equal—2 shorter, p9 1–4 shorter, p8 9–13, p7 15–20, p6 23–29, p1 56–67. P12 minute, hidden by coverts. P9–p11 emarginated inner webs, p9 and p10 outer. Outer web of p11 narrow. 20 secondaries (Snow 1967). Bill straight, shorter and sturdier than *P. cristatus*. Outer toe *c.* 108% of middle, inner *c.* 83%, hind *c.* 25%.

**Geographical variation.** *P. g. holboellii* larger than nominate *grisegena*. Long-winged population in south-east Kazakhstan considered separate subspecies *balchaschensis* by some authors (Dolgushin 1960), but rather outpost of *holboellii*.

## *Podiceps auritus* Slavonian Grebe

PLATE 11
[facing page 111]

Du. Kuifduiker    Fr. Grèbe esclavon    Ge. Ohrentaucher
Ru. Красношейная поганка    Sp. Zampullín cuellirrojo    Sw. Svarthakedopping    N.Am. Horned Grebe

*Colymbus auritus* Linnaeus, 1758

Monotypic

**Field characters.** 31–38 cm (body 18–23 cm), wingspan 59–65 cm; distinctly smaller than Great Crested Grebe *P. cristatus* and Red-necked Grebe *P. grisegena*, but bulkiest of 3 small grebes regular in west Palearctic. Small, dumpy, rather stub-billed grebe, with relatively larger, flatter head and thicker neck than Black-necked *P. nigricollis*. Breeding adult combines black facial tippets with golden lateral crest tufts and chestnut neck, but birds in all other plumages essentially black above and white below. Sexes alike though ♂ larger and brighter when breeding; marked seasonal variation. Juvenile separable.

ADULT BREEDING. Head black, with golden crest running back from and slightly upwards from each eye. Neck, upper breast, and upper flanks chestnut. Hindneck and upperparts glossy brown-black; rest of underparts white except for grey vent. Dark brown primaries and wing-coverts contrast with white secondaries and shoulder patch. Bill black, white tip. Eyes red. Legs and feet grey. ADULT NON-BREEDING. Becomes black above and white below. Crown and loral stripe black, narrow strip of dark grey down hindneck (often broken with pale feathers), rest of upperparts drab black. Upper lores greyish-white; rest of face, sides of nape, foreneck, and rest of underparts white, marked only by dusty wash or mottling on sides of lower neck and breast, upper flanks and vent. Bill colour less intense and pale base to mandibles enlarged. JUVENILE. Pattern as winter adult, but upperparts browner, with sides of nape and face also mottled brown.

In summer, black head with golden horns and, in winter, strong contrast between upper- and underparts preclude confusion with Little Grebe *Tachybaptus ruficollis*. In summer, flat crown and chestnut neck differentiate from *P. nigricollis* but confusion possible at all other seasons and ages especially at longer ranges in poor light. *P. nigricollis* slightly smaller, differs in bill shape, and black of crown extends lower on face. *P. auritus* has stubby, straight, evenly profiled bill (with mandible tips gently curving together, not bent upwards as in *P. nigricollis*) and white of face abuts eye and extends over whole cheek, round on to sides of nape (and even rear crown) to form high pale collar around throat. White also extends down front of neck. In some immatures, these characters ill-marked and birds with unusually thin bills can be confusing, separable only by careful observation of general structure and forehead pattern. When afloat, exhibits notable buoyancy like *P. nigricollis* (though less so than *T. ruficollis*). When alert, neck of *P. auritus* usually more erect and bill carriage horizontal. Flight silhouette virtually identical to *P. nigricollis* but action stronger with rapid wing beats; track noticeably direct. Swims expertly, coping with rougher water than other small grebes; dives for food, often jumping first.

Resorts to inland waters to breed, but most birds winter in inshore waters of sea. Breeding call a low, rippling trill; not very vocal in winter.    DIMW

**Habitat.** Mainly in boreal climatic zone, markedly more northerly than congeners, with frost-free season around or below 90 days. Fjeldså (1973a,b) pointed out that in Finland and Sweden somewhat restricted to shallow eutrophic lakes and pools, mainly 1–10 ha, with rich surrounding vegetation, while in north Norway and Iceland additionally uses oligotrophic open lakes with sterile exposed shores and generally more opportunist, probably reflecting enforced gradual adaptation to climatic cooling and growing acidity of lake types, where free from competition by congeners. Among these Black-necked Grebe *P. nigricollis* apparently complementary species

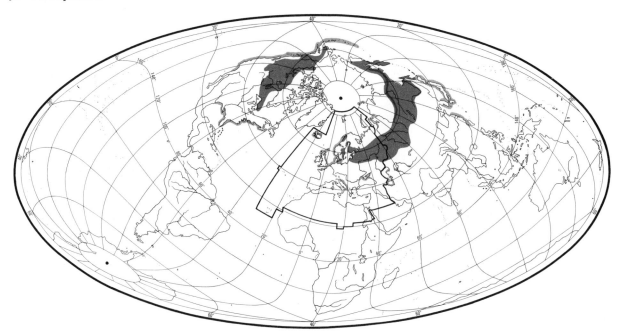

occupying corresponding shallow waters in steppe-deciduous forest zone, with little overlap geographically but more colonial breeding pattern. More frequently overlaps with Red-necked Grebe *P. grisegena* and then rarely maintains itself in interspecific competition, being left with only smaller suitable waters lacking dense emergent vegetation. Little overlap with Great Crested Grebe *P. cristatus*.

Uses more upland sites than congeners, but infrequently above 300 m; often with bare shores but readily accepts surrounding trees. Normal depth no more than 2·5 m, preferably less, requiring submerged vegetation and avoiding gravel, mud, or silt bottoms. Chooses for nest somewhat sparse emergent vegetation such as sedge *Carex*, horsetail *Equisetum*, shrub willow *Salix*, and sometimes reed *Phragmites*, at least 10 m from margin, at depths up to 1¼ m with ready access to open water but giving shelter from wave action. Early availability of cover after thaw often significant. Locally breeding on brackish sounds between rocky islands. When nesting on small pools down to 0·10 ha (Lake Mývatn, Iceland), will flight to forage in larger waters; flies readily, visiting neighbouring waters but observations of high flight lacking, and little known from marginal tundra, steppe-desert, and plateau habitats. Some stay throughout year in breeding quarters, but many move along river systems or larger lakes to sheltered coastal waters and estuaries, diving well and accepting moderate depths. Tolerates fairly close human presence.

**Distribution.** Changes in north-west Europe described by Fjeldså (1973*a*), who suggested distribution once much wider, then decline due to formation of more humous and acid lakes; more recent expansion influenced by eutrophication due to human activities. BRITAIN: first bred Scotland 1908; still spreading and increasing slowly (Parslow 1967). FAEROES: bred occasionally since 1935 (Williamson 1970). NORWAY: range apparently extended both north and south, with numerous local increases (Haftorn 1971). SWEDEN: rapid spread 20th century over lowlands of centre and south; these areas now fusing, with further increases in last 10 years, though marked decline Scania (Fjeldså 1973*a*). DENMARK: formerly bred north-west up to *c.* 1860 (Fjeldså 1973*a*). FINLAND: recent decline in south-west (Fjeldså 1973*a*). POLAND: bred 1972 (A D).

Accidental. Jan Mayen, Spitsbergen, Spain, Hungary, Switzerland, Greece, Rumania, Turkey (perhaps regular winter in small numbers), Bulgaria, Israel, Cyprus, Tunisia, Azores, Madeira.

**Population.** ICELAND: *c.* 500–750 pairs (Gardarsson 1975). SCOTLAND: *c.* 50 pairs (Dennis 1973); 57–62 pairs (Sharrock *et al.* 1975). NORWAY: *c.* 500 pairs. SWEDEN: *c.* 1000 pairs. FINLAND: estimate of 3000 pairs (Merikallio 1958) perhaps too high (Fjeldså 1973*a*). ESTONIA: *c.* 500 pairs (Onno 1970).

**Movements.** Migratory. Small breeding waters deserted for winter, birds moving mainly to inshore seas, but some to large lakes (Voous 1960). On western edge of breeding range may be dispersive, moving only to nearest coasts. Autumn movements begin late August, peak October–November. Main departures from winter quarters March–April; reach Iceland, south Sweden and south Finland in April, but breeding waters further east may not

be reoccupied until late May, dependent on thaw. Scottish lochs occupied from March or early April to September, failed breeders leaving first (Bannerman 1959). Records for May to early June from North Sea may include late migrants, but later must be non-breeders.

Isolated, large-billed populations breeding Iceland, Faeroes (occasionally), Scotland and north Norway, appear from museum speciments to winter around Faeroes (few), Scotland, Atlantic coast Ireland, and on west coast Norway from Troms (69½°N) south to Jaeren, while some straggle to Greenland (Fjeldså 1973a). Depart Iceland, though a few may occasionally overwinter south-west coast. One ringed Iceland recovered Faeroes in November.

Small-billed birds (breeding Sweden eastwards) apparently winter further south and east: west Baltic, Kattegat, Skagerrak, west coast Denmark to southern North Sea and Brittany; small numbers across west-central Europe (notably lakes in France and Switzerland) to Mediterranean and Adriatic; also on Black, Caspian and Aral Seas. Data from skins suggest contact with large-billed birds only east Scotland and south-west Norway (Fjeldså 1973a). Migration routes little known; one ringed Sweden recovered Netherlands (December); while one ringed Vologda, USSR, found Yorkshire, England (April) shows that some Russian birds reach as far west as North Sea.

In North America migrates overland at night, but often diurnally also along coasts (Palmer 1962); no comparable information from Palearctic.

**Food.** Chiefly arthropods (especially insects and larvae), and fish. Obtained mainly by diving, bird usually gliding into water with no splash though in rough conditions will make more violent spring so that body clears water entirely (Fjeldså 1973c). Dives with wings folded against body and using feet for propulsion (Storer 1969). Diving times 1–73 s, mostly in range 7–25 s, with pauses mainly 6–15 s (Fjeldså 1973c). Vary considerably with different localized conditions. Up to 3 min alleged (Eaton 1910) but generally less, with average 30–35 s (Bent 1919; Madsen 1957). Mean of 11 and 25 dives, 19·2 and 17·4 s respectively, range 8·2–25·8 s (Heintzelman and Newbury 1964). Mean of 120 and 82 dives at 2 localities 19·5 and 17·5 s, variation due at least partly to differences in depth (Ladhams 1968). As must seek, chase, and capture individual prey under water, duration of dive less dependent on water depth than in bottom-feeding species but more on density of prey: at depth of c. 1 m, diving times of parents feeding young averaged 22 ± 10, 13 ± 7·5, and 7·4 s at densities of less than 1000, 1000–4000, and over 4000 animals per m² bottom respectively (see Fjeldså 1973c for further details). Usually dives in shallow water, but up to 25 m reported (Palmer 1962). Other feeding methods include snatching insects from air or from aquatic vegetation; picking up floating items from surface, or skimming with bill if prey density high; peering under water and submerging completely (Fjeldså 1973c). Feeds mainly by daylight; usually solitary, even members of flock dispersing, but will congregate over large shoals of fish, while synchronized diving of large flocks into swell recorded (Bergman 1936). Group feeding by c. 200 individuals reported (Palmer 1962), and commensal feeding with Surf Scoter *Melanitta perspicillata* in USA (Paulsen 1969).

Diet composition depends mainly on feeding behaviour and habitat. Fishes include stickleback, adults and eggs *Gasterosteus aculeatus*, eel *Anguilla anguilla*, and trout *Salmo trutta* fry. Insects include larvae and adults (sometimes pupae) of mayflies (Ephemeroptera), damsel flies (Zygoptera), stoneflies (Plecoptera), terrestrial and aquatic beetles (Coleoptera), lacewings (Neuroptera), caddisflies (Trichoptera), flies (Diptera), aquatic bugs (Hemiptera). Also eats molluscs, crustaceans and annelids. Vegetable material occasionally recorded, including grasses and seeds, water plants and algae (for details see Witherby *et al.* 1940; Madsen 1957; Fjeldså 1973c).

126 stomachs from Holarctic, mainly in winter, contained fish in 51, crustaceans in 32, unknown quantity of insects, and one mollusc (Madon 1931). One stomach from Denmark, October, contained 15-spined stickleback *Spinachia spinachia*, 10–20 small gobies *Gobius*, insect remains, and insect eggs (Madsen 1957). 56 stomachs from Estonia contained mainly amphipods (*Gammarus*), aerial insects, and some beetles *Donacia* and Zygoptera, but no fish (Onno 1958). In 49 stomachs from 7 breeding localities north Norway and Lake Mývatn area, Iceland, mainly in July, 2864 individual prey recorded; by numbers, diet dominated by crustaceans (Cladocera) 17·7%, various aerial insects which dropped on to surface 17·6%, larvae 15·2%, and pupae 7·3% of chironomid midges, small fish 14·1%, aquatic coleopterans 12·3% and their larvae 5·1%. However, by weight fish dominated 69·1%, and cladocerans only 0·5% (Fjeldså 1973c). In birds with heavy bills from Atlantic part of range (see Geographical Variation), feeding appears to be unspecialized and flexible; although normal food is nektonic, including both arthropods and fish, may shift to what is in abundant supply and easily available e.g. cladocerans, *Eurycercus lamellatus* and *Daphnia longispina*, while smaller-billed birds elsewhere apparently more specialized for diet of arthropods (Fjeldså 1973c).

At sea, winter food mainly fish (Wetmore 1924), especially sculpins (Icelidae and Cottidae), and some crustaceans (mysids, amphipods, and decapods).

9 stomachs from small chicks, mainly from pot-holes near Lake Mývatn, contained 5% crustaceans, 6% corixids, 37·2% coleopterans, 14·1% trichopterans, 28·4% chironomids, 4·9% terrestrial insects, and 4·4% stickle-backs *G. aculeatus*. In some localities however chicks seen to be fed mainly with fish. Appears to be no selection for particular foods for chicks (Fjeldså 1973c).

Feathers normally found in all stomachs; for significance and possible function, see Podicipedidae.

For North American studies, see McAtee and Beal (1912), Wetmore (1924), and Palmer (1962).

**Social pattern and behaviour.** Based mainly on material supplied by J Fjeldså and R W Storer.

1. Not highly gregarious. Outside breeding season, Palearctic birds mainly single, in twos, or small parties, but in flocks of up to 60 on spring passage and also after arrival, especially if weather cold. Gathers mainly when loafing, usually dispersing to feed; will, however, congregate over large shoals of small fish (Fjeldså 1973c), and synchronized diving into swell or on flood-tide reported (Bergman 1936; King 1971). Winter feeding territories not known. Birds of same age tend to associate, often in temporary pairs. American birds more often in loose flocks, up to 500 (Helmuth 1929); may forage alone, or in groups, or in mixed flocks with small *Larus* gulls. BONDS. Monogamous pair-bond the rule, lasting at least through major part of growth period of young. Nesting trio (♂, 2 ♀♀) once found in Iceland (J Fjeldså), and occasional simultaneous bigamy also suggested in American birds from large clutches (see Bent 1919). Pair-formation starts in winter quarters or during migration, before establishment of territory; in Mývatn area, Iceland, and various waters in north Norway, most paired on arrival, rest soon after (Fjeldså 1973d). Evidently first pairs and often breeds in 1st year. Both parents tend young. Brood sometimes more or less permanently divided after 2nd week (Fjeldså 1973d); young from 1st brood, if still present, may assist in feeding of 2nd. After departure of parents, mainly August or early September, siblings and juveniles from other broods may associate in small groups. BREEDING DISPER-

SION. Territorial, typically with well-dispersed, cryptic nests. Although temporary territory sometimes established during halts on migration, and mating platforms built, true nest-territory established soon after arrival on eventual breeding water. Usually 1 pair per small pool, with 2 pairs only if feeding area 5–10 times minimum needed for successful raising of one brood. On larger waters, usually 1 pair to each bay or sedgebed. Modal size of 550 territories, Mývatn area and north Norway, *c.* 0·4 ha, range 0·01–2·2 ha. Nest-building and copulation in territory. Some pairs, on smaller pools and well-secluded bays on larger waters, also feed in territory (mean size 0·54 ha, range 0·07–2·2 ha) but others mostly elsewhere; at extreme, on poor waters, travel to feed elsewhere (like some divers *Gavia*). Especially where food supply rich, territory maintained almost to autumn passage, but usually abandoned a few days after hatching on poor lakes or where nests more crowded; also kept for any 2nd brood but new one often established for re-nesting after failure (especially if due to predation). Small, loose colonies of a few territorial pairs reported on some waters, usually where cover limited, but denser colonies (up to 15 pairs) rare. Non-breeding 'surplus' reported from some waters (*c.* 10% of total population), but these mainly birds prevented from nesting by lack of sites and many probably attempt to breed later in season. For further information, including discussion on relationship between nesting density, food supply, and level of inter-pair aggression, see Fjeldså (1973*d*). ROOSTING. On water, both by day and night, probably much as in Great Crested Grebe *P. cristatus*.

2. Like Red-necked Grebe *P. grisegena*, to which most similar, vocally conspicuous in breeding season though to much lesser extent. Recent comprehensive studies by Storer (1969), North America, and Fjeldså (1973*b, d*), north Norway and Mývatn area of Iceland. ANTAGONISTIC BEHAVIOUR. 3 main types of threat: (1) Forward-display (see A, or with neck more outstretched, dorsal plumage raised, wings partly lifted over back); (2) Hunched-display (see B) with Chittering-call; (3) Attack-upright (see C, or with wings raised as in other version of Forward-display). In high-intensity Forward-display, may almost disappear at times below water or, in north European and Icelandic birds at least, perform Token-dives. In flock, mainly low-intensity Hunched-display with chin resting on breast and Chittering. After arrival on fresh water in spring, paired ♂♂ especially hostile towards birds that Advertise or engage in mutual water-courtship; threaten mainly with Attack-upright at close quarters, but, at longer range, respond with Triumph Ceremony and much self-preening. Territorial defence by both sexes but ♂ more pugnacious. Forward-display then more typical or, in defensive situations, Hunched-display. Threat generally most intense just prior to egg-laying. Encounters usually end with much self-preening from all participants and Triumph Ceremony by pairs. Particularly during border encounters, sustained calling (Chittering or Trilling) by 2 or more pairs often stimulates further pairs to join in while performing induced Triumph Ceremonies themselves. Attack-upright may be followed by patter-flying along water or flying at rival, or by diving and pecking from under water. Fighting common, especially during pair-formation and in transition period between flocking and solitary dispersion in territories. Combatants run together, standing vertical almost out of water, using wings for balance and striking; also collide breast to breast, each attempting to push other off balance—peck at hindneck if rival falls and sometimes apparently kick. In pauses in fighting, occasionally adopt Penguin or Cat postures (see below). When retreating, or trespassing, assumes Furtive-upright with all plumage sleeked except small black tuft on back of crown (see D); when beaten in combat, or indicating submission, adopts

Prone-posture on water (see Little Grebe *T. ruficollis*). Escapes by flying, pattering, or diving away. In Alarm-upright, even crown-tuft depressed and Alarm-call given. WATER-COURTSHIP. Occurs in winter quarters (from about January), on spring migration, and on breeding water, during and (to lesser extent) after pair-formation. Activity minimal amongst birds in flocks, pairing taking place gradually with little display; tend to disperse when others Advertise or perform water-courtship ceremonies. Among solitary birds, pairing based on Discovery Ceremony followed by Weed Ceremony involving unique Weed-rush. Unpaired birds attracted by Advertising of lone birds, latter assuming typical upright posture giving loud call. Advertising by 2 birds alternately often leads to Discovery Ceremony when one

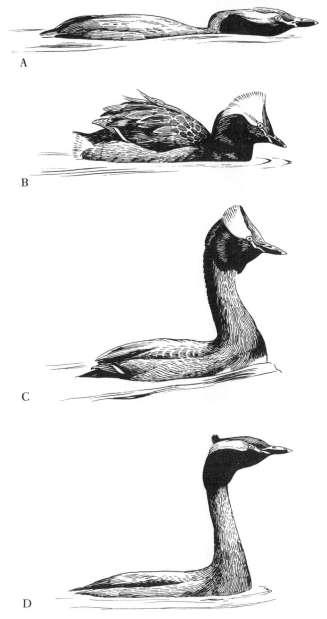

A

B

C

D

dives and other assumes Cat-display. Discovery Ceremony of similar type to those of *P. grisegena* and Black-necked Grebe *P. nigricollis*: (1) approach dive involves emergences in bathing-like Bouncy-posture (see E); (2) Cat-display (see F) most like that of *P. grisegena*; (3) Ghostly-penguin display (see F) performed with back to mate, *c*.1 m away; (4) mutual Penguin-dance (see G), with Fast Head-turns, forms climax of ceremony; (5) this periodically interrupted as birds subside and Habit-preen, on back or flank, before rising again; (6) sequence usually terminated when pair Ceremonially Turn-away from each other. However, (7) they occasionally swim slowly side by side in Penguin-posture (Parallel-barging), or diverge thus. At times, gradual transition between ritualized travelling side by side (Parallel-barging, diving, swimming, or pattering) and less differentiated swimming, diving, and pattering after one another, with ♂ mainly chasing ♀ (much as in *T. ruficollis*). After

Discovery Ceremony, often feed and separate, but, in early stages of pair-formation especially, mutual diving after Turning-away may lead to Weed Ceremony. Then (1) both complete Weed-trick (see *P. cristatus*); (2) come together carrying weed (Weed-approach); (3) rise breast to breast in Weed-dance but, instead of remaining thus, immediately turn parallel and move quickly side by side for 5–10 m across water (Weed-rush; see H); (4) subside, move apart, and come together again before (5) performing another Weed-rush in a different direction; (6) continue thus up to 15 times. In newly formed or temporary pairs, Discovery Ceremony performed as greeting after separation. Occasionally then followed by Retreat Ceremony as (1) one bird, after Turning-away, makes Ceremonial-flight from other, (2) assumes Cat-display and faces back towards it; (3) another Discovery Ceremony then initiated. Weed Ceremonies (with Weed-rush) also occur frequently on own, mostly before pairing; infrequent in territory later but then separate Penguin-dance Ceremony sometimes performed, especially after territorial encounters or when disturbed. Weed Ceremony with more prolonged Weed-dance (much as in *P. cristatus* and *P. grisegena*) and no Weed-rushes occasionally reported in Old World, but more details needed. In North America and over much of Old World range, separate Head-shaking Ceremony may replace Discovery Ceremony as greeting for a time immediately after birds pair-up: (1) they face with tippets and ear tufts spread; (2) perform a series of rapid, lateral Head-waggles of variable duration. At times,

E

F

G

H

however, Head-waggling gains in intensity and leads to (3) Penguin-dance, usually interrupted by (4) spells of Habit-preening. Sequence may also end with (5) Ceremonial Turning-away followed by Weed Ceremony (with Weed-rush). In north Norway and Iceland, Head-shaking as separate ceremony relatively unimportant. Occurs in post-pairing period, mainly after territorial encounter or Triumph Ceremony. Involves (1) lateral Head-waggles, which appear hardly ritualized, (2) Slow Head-turning, (3) Throat-touching, (4) Habit-preening; (5) terminates in Turning-away or Pivoting. In Triumph Ceremony, pair swim side by side in Hunched-display while Trilling in duet; often follows antagonistic encounters but, in north Norway and Iceland, also mainly replaces Head-shaking and other ceremonies as pair-bond strengthens. In pairs already formed on arrival, occasional Triumph Ceremonies often only water-courtship shown. Occurs spontaneously and as meeting ceremony throughout rest of season; often no calling and much self-preening. Other infrequent interactions between pair include Food-presentation by ♂ to ♀ of much the same undifferentiated type found in some other *Podiceps* and in *Tachybaptus*. PLATFORM-COURTSHIP. Site-selection mainly by pair together, uttering low calls with plumage relaxed; also associated with Inviting-on-water and Advertising-calls. Building initiated mainly by ♂, although both sexes sometimes construct own rudimentary platforms (mainly new-formed pairs). Weed-presentation by ♂ to ♀ occurs at times. When platform ready, one bird may lead other to it with neck stretched obliquely forward while giving Advertising-call; often preceded by Inviting-on-water. At platform, take turns in soliciting, performing typical Inviting and Rearing displays. Pre-mounting behaviour of bird in water variable; includes rapid lateral Head-waggling, self-preening, or utterance of Advertise-call in oblique posture. Copulation of typical *Podiceps* type except that, in north Norway and Iceland, some Inviting birds will raise head and even touch breast of other. Mounting preceded by pecking at rear of Inviting bird and Copulation-trills which continue with increasing intensity until just before dismounting. Followed by Post-copulatory Water-treading by active bird, which may then briefly Head-turn with back to other or facing it. Latter may Head-turn reciprocally or, occasionally, rise in Ecstatic-posture much as in *P. grisegena*. During platform-courtship stage, reversed mounting by ♀ not infrequent, especially early in season, but, just prior to egg-laying, ♀ does most soliciting and only ♂ mounts. At nest-relief, activities include Food-presentation, giving of Advertising-calls with neck held horizontally forward over nest, Inviting, and Duet-trilling. RELATIONS WITHIN FAMILY GROUP. Generally much as in other *Podiceps* (see *P. cristatus*). At least in north Norway and Iceland, early peck-order between young, brood-division, and associated features of family social organization; but less pronounced than in *P. cristatus*. Parental hostility, mainly in response to begging or attempts at boarding, develops when young over 2 weeks old. When begging, chicks stretch neck forward, Foot-splash, and Peep loudly. Parents warn young with Alarm or Anxiety-calls; also give Advertising-calls when contact lost with young or mate. Interactions between siblings much as in *P. cristatus* but less studied. Head-shaking seen between two aged 1 month.

(Figs A–H after originals by J Fjeldså; see also Fjeldså 1973*b*.)

**Voice.** In some respects rather similar to both Red-necked Grebe *P. grisegena* and Little Grebe *Tachybaptus ruficollis*, but repertoire (basically same in both sexes) includes harsher and more guttural notes than latter while Display-call of former lacking. Mostly silent outside breeding season and in flight. Studied in most detail by Storer (1969) and Fjeldså (1973*b*).

Following main calls in breeding area; probably cover most or all those mentioned by Dubois (1919), Gordon and Gordon (1928), Ruthke (1938*b*), Hosking (1939), Witherby *et al.* (1940), Täcklind (1954). Loud splashing noise of feet, e.g. during distraction-display, only non-vocal signal. (1) Advertising-call (see I). Loud, nasal 'aaanrrh', 'jaorrrrh', 'ij-arrrrr', 'arrrr', etc., usually descending in pitch and ending in rattle (or harsh trill). Variable, probably facilitating individual identification. Given by lone birds in typical Advertising-posture during pair-formation and when separated from mate or young. Similar calls uttered near and from platform, before copulation, or on nest (e.g. at nest-relief). (2) Alarm and Anxiety-calls. Closely similar to last, differing mainly in nuance (usually shriller). A 'ko-wee kowee' also serves as an Alarm-call (R W Storer); likened to call described by Dubois (1919) as resembling double squeak of dry wheelbarrow, but this also ascribed to Duet-trilling by Fjeldså (1973*b*). See also (7) and (8) below. (3) Duet-trilling. Loud, accelerating trill, e.g. 'dji-ji-ji-ji-ji- . . . ji-ji-jrrh' (but highly variable) rather like similar call of *T. ruficollis*. Given by pair together, mostly during Triumph Ceremony on meeting or after antagonistic encounter. Silent during main ceremonies of water-courtship. (4) Copulation-trill. Intense form of trilling closely similar to last. Uttered several times in succession by active bird before and during mounting, rising in pitch and intensity. (5) Threat-chittering. Similar to (3), but more staccato and stuttering, with less continuous cadence and shorter phrases of 2–3 notes, e.g. low 'kru-vu, kru-vu, kru-vu . . .', 'djii-ji dji-ji dji-ji . . .'; also more intense, squeaky version, 'dji-ji—JOARRH dji-ji-JOARRH . . .'. Version heard in Scotland: 'kitta' repeated 1–3 times followed by 'kee-a' (K E L Simmons). Threat-call typically given in Hunched-posture, often by pair together. Intergrades completely with Duet-trilling; when 2 or more pairs answer one another (Chittering or Trilling), produce characteristic, sustained chorus or 'calling concert'. (6) Contact-chittering. Low 'kru-uuck . . .' or 'uck uck . . .'. Uttered by pair, e.g. when investigating nest-sites; audible only at close range. Pairs near nests in Scotland gave repeated, conversational clucking notes similar to Clucking-call of Great Crested Grebe *P. cristatus*; uncertain whether this low-intensity (5), (6), or (7). (7) Platform-call. Soft, 2–3 syllable 'ga- . . .' or 'gjack' by one

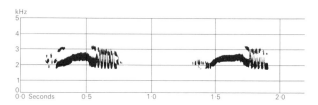

I Sture Palmér/Sveriges Radio Sweden May 1969

or both of pair (most often ♂) at or in vicinity of platform, e.g. when building and soliciting (including Inviting-on-water). Intensity variable: may grade into goose-like honking note which may also serve, at least partly, as Alarm-call. (8) Other calls. Include (a) low call, audible only at close range, uttered by bird in Furtive-upright posture; (b) variety of notes, often associated with Splash-diving, given when disturbed with chicks or by some bold birds disturbed at nest—hissing or crying sounds, shrill 'i-jrrrrh', lower 'grrr', short 'djeck', 'nje'; (c) soft 'wheck, between a squeak and grunt, by ♂ (alone or in presence of ♀) before diving. Conversational undertone from flocks.

CALLS OF YOUNG. Much as in other grebes. Main utterance peeping 'pee-a', similar to *T. ruficollis*, and not unlike domestic chick; used mainly during food-begging. Gradually changes into more adult-like chittering.

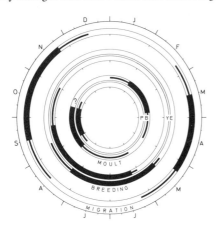

**Breeding.** Based largely on Fjeldså (1973*d*); mainly Iceland. SEASON. See diagram. Dependent on thaw in northern part of range. SITE. In water, in sheltered bays of lakes and pools; usually in cover of emergent vegetation. Of 113 Icelandic nests, 82 in dense sedge *Carex* beds, 17 in open water or among rocks, 11 under overhanging willows *Salix*, 2 in sedge beds and under willow, 1 in sparse sedge. Of 45 nests, 35 within 2 m of open water, 10 between 2–4 m, 2 at 5 m, and 1 at 7 m. Generally solitary, but sometimes in loose colonies; Iceland colony had 36 nests, with smallest distance between nests 3 m (Clase *et al.* 1960). Nest: heap of water weed, anchored to aquatic vegetation, or sometimes to rocks or bottom. Of 113 Icelandic nests, 97 were floating, anchored to vegetation; 11 built on rocks at or just below water level; and 5 built up from bottom (Clase *et al.* 1960). Mean diameter at base 52 cm (20–125), depth below water 32 cm (4–75), height above water 3–5 cm (0·5–14). Platforms sometimes built before nest. Building: by both sexes, most often begun by ♂. Usually built 4–7 days before egg laying, occasionally taking only 3–4 hours. Vegetation brought up from under water and placed on anchorage; cup formed with bill, feet, and belly. EGGS. Elongated oval, not glossy; chalky white, soon becoming heavily stained red or brown. 46 × 31 mm

(39–50 × 28–34), sample 499; calculated weight 23 g (Schönwetter 1967). Clutch: 4–5 (1–7). Mean of 507 Icelandic clutches 3·75. Usually one brood, occasionally 2. Repeat clutches laid after egg loss; of similar size to 1st clutches. Eggs usually laid daily, but 2-day intervals recorded (Dubois 1919). INCUBATION. 22–25 days. By both sexes. Begins gradually, sometimes with first egg, rarely not until clutch complete. Eggs covered with material when sitting bird disturbed from nest. Hatching asynchronous. Eggshells removed. YOUNG. Precocial and seminidifugous. Cared for and fed by both parents. Young carried on parents' backs from 2nd day. Can dive at 10 days. Fed bill to bill; also given feathers. Parents defend young with aggressive behaviour and splash-diving distraction-display. FLEDGING TO MATURITY. Fledging period 55–60 days. Young become independent at 45 days or less. Age of maturity probably 2. BREEDING SUCCESS. Of 721 Icelandic nests, 24·5% complete clutches totally lost; successful nests lost 15·3% of eggs. Of 1332 eggs in 339 nests, 63% hatched, 13·6% taken by man, 7·7% predated, 2·8% deserted, 2·0% infertile, remainder lost from miscellaneous causes. From average brood of 3·58 at hatching, 62·5% young survived to 10 days, 53% to 20 days, then no significant decline to fledging.

**Plumages.** ADULT BREEDING. Crown glossy black. Lores chestnut; over and behind eye, broad golden brown lateral tuft or 'horn' of elongated feathers. Lower down on side of head thick tippet (or frill) of elongated black feathers. Hindneck grey-brown; rest of upperparts glossy black with grey feather edges. Chin and throat black; foreneck, sides of neck, chest, and flanks chestnut; rest of underparts white, vent and under tail-coverts pale grey. Tail-tuft of downy black feathers. Primaries grey; secondaries mainly white, outer 2 partly and innermost wholly grey. Upper wing-coverts grey, innermost with variable amount of white, marginal near leading edge white. Under wing-coverts and axillaries white. ADULT NON-BREEDING. Crown glossy black. Hindneck grey-black, sides of neck down to lower throat grey. Mantle and scapulars glossy black with grey feather edges; back and rump duller. Chin, throat, cheeks up to level of eye, and sides of upper neck white. Pale grey spot in front of eye. Foreneck very light mouse-grey. Rest of underparts white, vent and under tail-coverts pale grey. Flanks mottled grey and white. Tail and wing as breeding. DOWNY YOUNG. Very similar to *P. grisegena*. White patch on nape larger, fewer black spots on sides of head, mid-lateral neck-stripe sometimes doubled (Fjeldså 1973*a*), light areas on head and neck more buff, light stripes on back whiter, dark stripes on flanks less strongly broken into spots (Storer 1967). JUVENILE. Similar to adult non-breeding. Remains of downy pattern on head, border between black crown and white cheeks diffuse. Upperparts with brown tinge and without pale feather edges. Flanks white, with hairy, grey feather tips. Usually some remnants of down on feathers of back and scapulars. IMMATURE WINTER. Indistinguishable from adult non-breeding, except by less worn mantle feathers and shorter primaries (Fjeldså 1973*a*).

**Bare parts.** ADULT BREEDING. Iris red, narrow silvery ring round pupil. Line of bare skin on lores dusky pink. Bill black, white at tip, flesh at base. Leg and foot blue to grey-green, outside and underside brown-black; toes edged yellow. ADULT NON-BREEDING. Iris pink. Bill dusky. DOWNY YOUNG. Iris grey. Bare lores and crown patch pink. Bill pale pink, with two black bars on upper mandible. Foot dark grey. JUVENILE. Bill paler than adult.

**Moults.** ADULT POST-BREEDING. Complete; primaries simultaneously, August–September in ♂♂, probably October in ♀♀. Body plumage starts early June to mid-July with loss of ornamental feathers on head; much individual variation in sequence and timing, ♀♀ averaging *c*.1 month earlier than ♂♂. Wings start with moult of lesser coverts. ADULT PRE-BREEDING. March. POST-JUVENILE. Partial. Head, neck, chest, flanks, some feathers of upperparts, and belly, and some lesser wing-coverts; mid-winter to late March. POST-JUVENILE. Partial. Head, neck, rest of upperparts, flanks, and narrow strips of white feathers on sides of belly; starting mid-October to late November, completed late January. FIRST PRE-BREEDING. Later than adult pre-breeding, early March to mid-May or later, starting soon after completion of post-juvenile; thus first winter birds moulting almost continually. (Fjeldså 1973*a*).

**Measurements.** Full-grown, skins (BMNH, ZMA, ZMFK).

| | | | | | |
|---|---|---|---|---|---|
| WING | ♂ 144 | (5·28; 26) | 136–158 | ♀141 (4·98; 22) | 131–153 |
| BILL | 23·3 | (1·07; 19) | 21–25 | 22·5 (1·36; 15) | 20–25 |
| TARSUS | 45·9 | (2·67; 7) | 42–50 | 44·2 (2·06; 4) | 42–46 |
| TOE | 53·8 | (3·66; 6) | 49–58 | 50·0 (1·83; 4) | 48–52 |

Iceland and Norway, all ages (Fjeldså 1973*a*): wing ♂ 146 (5·3; 59) 132–159, ♀ 140 (4·2; 52) 131–151; bill (average) ♂ 23·6, ♀ 22·2; tarsus (average) ♂ 45·0, ♀ 42·2; toe (average) ♂ 54·6, ♀ 52·8. ♂♂ larger than ♀♀, but differences not significant. Old birds have longer wings than young, average 148·9 against 142·1 for

Icelandic and 141·4 against 138 for Swedish and Finnish specimens (see Fjeldså 1973*a* for further details).

**Weights.** Iceland and Norway, summer: 449 (14) 415–470. Baltic area, summer: 375 (5) 300–450. Germany and Switzerland, winter: ♂♂ 424 (8) 320–470; ♀♀ 364 (4) 350–376. Records of 600 and 720 for ♀♀, Iceland, exceptional or erroneous. (Fjeldså 1973*a*; Bauer and Glutz 1966.)

**Structure.** Wing short. 12 primaries: p10 longest, p11 0–3 shorter, p9 1–4 shorter, p8 9–13, p7 16–20, p6 22–28, p1 48–58; p12 minute, *c*. 10 shorter than primary coverts. P9 to p11 emarginated on both webs. Bill short, sturdy, straight. Outer toe *c*. 107% of middle, inner *c*. 82%, hind *c*. 25%.

**Geographical variation.** Populations in Atlantic part of range large, with heavy bill, average height 9·3; in Baltic area 7·9; in Russia, Kazakhstan, and Altai 7·2; in North America 7·9. Similar variation in width of bill. Occasional intermediates occur in north Scandinavia (Fjeldså 1973*a*). American populations greyer above, particularly on crown, more pronounced in non-breeding plumage; grey feather edges on mantle and back wider; lateral tuft on head paler in adult breeding (Parkes 1952). Cline of decreasing colour saturation runs eastward through east Palearctic; east Asian populations intermediate between west Palearctic and American. For Atlantic birds name *arcticus* Boie, 1822 available; for American *cornutus* (Gmelin, 1788); but naming of subspecies considered unnecessary as differences slight and partly clinal, intermediates occur, and single birds rarely identifiable.

## *Podiceps nigricollis* Black-necked Grebe

PLATE 12
[facing page 111]

DU. Geoorde Fuut    FR. Grèbe à cou noir    GE. Schwarzhalstaucher
RU. Черношейная поганка    SP. Zampullín cuellinegro    SW. Svarthalsad dopping    N.AM. Eared Grebe

*Podiceps nigricollis* C L Brehm, 1831

Polytypic. Nominate *nigricollis* Brehm, 1831, Palearctic. Extralimital: *gurneyi* (Roberts, 1919), Africa; *californicus* Heermann, 1854, North America.

**Field characters.** 28–34 cm (body 18–20 cm); wingspan 56–60 cm; marginally smaller and slighter than Slavonian Grebe *P. auritus*. Small, square-headed, round-backed grebe, with greater extent of black on head and neck than other small grebes and slim, uptilted bill. Breeding adult uniquely combines black head, neck, and upperparts with spray of golden feathers falling from eye over cheeks to nape. Birds of all other ages essentially black above and white below, requiring careful separation from *P. auritus*. Sexes alike; marked seasonal variation. Juvenile separable.

ADULT BREEDING. Head, with markedly steep forecrown and elongated crown, glossy black, relieved by brilliant sprays of narrow, golden feathers springing from behind eyes and cuving downwards over rear ear-coverts; neck, upper breast, and whole upperparts black, becoming brown on rump. Rest of underparts mainly white, with flanks mottled chestnut and black; vent pale brown. Upperwing pattern similar to *P. auritus* but inner primaries as well as most secondaries white or partially so. Eyes warm pink. Bill black; legs and feet grey. ADULT

NON-BREEDING. Loses dark foreparts of breeding dress, becoming dusky or black above and white below. Crown, lores, upper ear-coverts (particularly central feathers), centre of nape, hindneck and rest of upperparts all dusky or brown-black. Lower face and throat white, usually contrasting strongly with dusky wash or smudges over foreneck; rest of underparts white, mottled dusky along upper flanks. Bill duller, grey with pale pink base and often whitish tip. JUVENILE. Pattern as adult non-breeding but browner, with duskier cheeks, buff tinge to sides of upper neck and broad dusky collar above breast.

Breeding adult unmistakable but, in winter, adult and young can be difficult to separate from *P. auritus*. When visible, bill shape diagnostic but some immatures show little uptilt, necessitating concentration on line of division of plumage tones on head and neck. In *P. nigricollis*, black of crown extends lower on face, particularly on rear cheeks; high pale collar below head visible from in front but not from behind as in *P. auritus*. Postures afloat similar to *P. auritus* but neck usually slightly coiled and head as

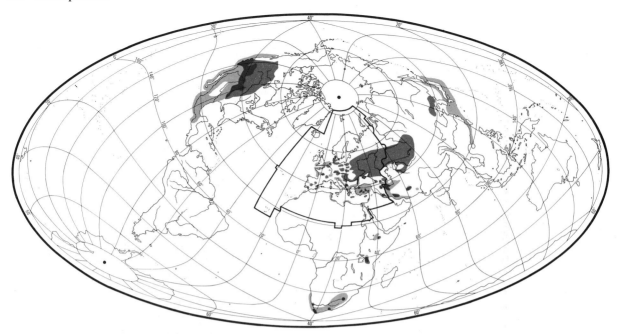

well as bill often uptilted. Flight silhouette as Little Grebe *Tachybaptus ruficollis* but, in winter, can also recall small auk (Alcidae); flight action rapid and quite strong (but more reluctant to take wing than other grebes). Swims well; feeds by diving usually without initial jump, and also, characteristically, by skimming insects off water surface.

Resorts to inland waters to breed; winters on wide variety of such waters and inshore seas. Quieter than other grebes when breeding; slurred trill and peeping call most frequent notes.                    DIMW

**Habitat.** Marked by exceptional instability, both ecologically and in terms of continuity and density of occupation. Feeding less on fish and more on insects and molluscs than other Palearctic grebes (except Little Grebe *Tachybaptus ruficollis*), relies for suitable habitat on small, shallow, plentifully vegetated and highly productive waters, which over much of range tend to multiply in wet and to disappear in dry seasons. Such dessication especially prevalent in Mediterranean, North African, and south-west Asian sectors of west Palearctic, and in parts of eastern Europe where population densest. More oceanic westerly regions and boreal or arctic zone sparsely and erratically occupied, corresponding niches here being taken by Slavonian Grebe *P. auritus*. Generally lowland, but to 2000 m in Transcaucasia (Dementiev and Gladkov 1951) and 660 m West Germany (Bauer and Glutz 1966). Inhabits high lakes in tropical Africa. Requirements more eutrophic than *P. auritus*; less associated with dense stands of emergent vegetation than Great Crested Grebe *P. cristatus*. Freely uses large newly flooded areas, where large colonies build up at short notice. Requires easy access to shallow bottoms rich in plant and animal life, preferring

warmer, drier, and more sheltered locations, including fishponds, reservoirs, sewage farms, and quiet, well-vegetated reaches or backwaters of rivers. Exploits other waters where suitable opportunities arise either spontaneously or through human intervention. Sites often abandoned after one or few years. Often associates with colonies of gulls *Larus*, terns *Chlidonias* and other aquatic birds.

Outside breeding season shifts to open standing waters or to sheltered estuaries, arms of the sea, and inshore shallows in bays or channels. Although diving expertly, gathers more food from surface than other grebes; flies readily at no great height, and usually independently. Somewhat wary, although so freely using human artefacts. Frequency of appearance where previously absent indicates active widespread aerial reconnaissance. For full discussion, see Frieling (1933).

**Distribution.** Breeding sites changed frequently, especially on peripheries of range, influenced by changes in water level and possibly other factors (e.g. disturbance). Map of west Palearctic shows sites occupied since 1966 in areas where regular information available; elsewhere, especially in centre and east of range, often no precise data. On world map, breeding sites often restricted to parts of areas shown. Tendency to increase on western and northern edges of range in last 100 years often ascribed to dessication of lakes, especially in steppe areas of Caspian region, causing invasions in years of greatest aridity (Frieling 1933; Kalela 1949). Existing data hardly support any firm correlation; for some recent changes, at least, conditions in breeding areas probably more significant. In some southern parts of range (Spain, Italy, North Africa,

and Cyprus) appears to have declined recently.

BRITAIN AND IRELAND. First recorded breeding Wales 1904, Ireland 1915, England 1918, Scotland 1930. Breeding in many counties of England and some in Wales and south Scotland has been sporadic; now confined to central Scottish lowlands. In Ireland, bred regularly 1915–57 in Roscommon, with up to 250 pairs in one colony 1929–32, and in 3 or 4 other counties; some may still nest but no proof since 1966 (Witherby *et al.* 1940; British Ornithologists' Union 1971; Ruttledge 1966; RFR). NETHERLANDS: bred twice 19th century, then regularly since 1918 in small numbers (JW). BELGIUM: first proved breeding 1931, though may have bred 1921 and 1924 (Lippens and Wille 1972). DENMARK: first bred 1876

(Bauer and Glutz 1966); some spread, but fluctuating (see Population). SWEDEN: first proved breeding 1927; permanent breeding areas restricted to few localities in extreme south (Curry-Lindahl 1957). FINLAND: reported breeding 1932 and 1954–5 (Merikallio 1958, LvH). POLAND: first reported breeding in most provinces 1880–90, but scarce (Tomiałojć 1972). CZECHOSLOVAKIA: considerable increase and spread at the beginning of 20th century; now most common grebe in lowland areas (Hudec and Černý 1972). SWITZERLAND: first proved breeding 1942; since 1960 a few pairs have nested almost annually (WT). ITALY: although apparently widely distributed formerly (Voous 1960), no longer breeds Sicily; may nest elsewhere in limited areas but breeding not proved in recent years (GS, SF). GREECE: no proved breeding records (WB). IRAQ: reported resident in small numbers (Allouse 1953); present status uncertain (PVGK). CYPRUS: bred earlier this century. NORTH AFRICA: formerly bred Algeria and Morocco and probably Tunisia (Heim de Balsac and Mayaud 1962), but not found nesting in Morocco in last 10 years (PR) and present status Algeria and Tunisia uncertain (EDHJ, MS).

Accidental. Norway, Kuwait, Madeira, Canary Islands.

**Population.** SCOTLAND: *c.* 25 pairs (Dennis 1973); *c.* 15 pairs, 1974 (Sharrock *et al.* 1975). NETHERLANDS: marked fluctuations since regular breeding from 1918, peak years 1918–21, 1925–8, and 1942–7; low in 1960, but slight increase since (JW). BELGIUM: up to 10 pairs annually since 1952; fewer in dry years (Lippens and Wille 1972). WEST GERMANY. Considerable numbers only in east Schleswig-Holstein and Bavaria. Bavaria: *c.* 200 pairs 1968, but decreased 1964–6. Baden–Württemberg: *c.* 50–100 pairs, recent decrease in some parts. Hessen: after 1945, 4–6 pairs annually; perhaps now absent (EB). DENMARK: fluctuating population, 1960–7, of 124–200 pairs (Preuss 1969). SWEDEN: no census data, but breeding numbers show marked fluctuations (SM). CZECHOSLO-VAKIA: marked fluctuations near Bino 1885–1973, varying from a few to 470 pairs (Fiala 1974).

**Movements.** Migratory and dispersive. Winters on lakes, reservoirs, and coastal waters; compared with Slavonian Grebe *P. auritus*, much smaller proportion on salt water (Voous 1960). Autumn movements protracted; dispersal begins mid-August, at height in October, lasting to late November. Majority in winter quarters November–March. Return begins March; through North Sea countries, central Europe, and Russia in April; most breeding lakes re-occupied by early May. Migrant Britain and Ireland mid-July to mid-May (Witherby *et al.* 1940), extremes presumably failed and non-breeders. Possibly rather sedentary in southern parts of range, e.g. Spain (Bernis 1966; Voous 1960), and reported as such on Lake Sevan, Armenia (Dementiev and Gladkov 1951).

In west Palearctic, winters chiefly coasts from British Isles to Iberia, on ice-free lakes (especially France, Switzerland, Turkey), and on and around Mediterranean, Adriatic, Black, and Caspian Seas. Three-figure concentrations exceptional west Europe, but larger numbers in east: 18600 on Lake Burdur, Turkey, December 1970 (Koning and Dijksen 1971), while hundreds of thousands winter Iran, mainly south Caspian (D A Scott). The few recoveries indicate south-west and south-east autumn movements; 7 from Denmark recovered Netherlands, west France (4) and Italy (2); 9 Czechoslovakia found Switzerland (3), Italy (2), Sicily, Corsica, Ukraine, and Caucasus; while one ringed Switzerland in August found north-west Spain January. Unknown whether any Russian breeders reach west Europe, but one ringed Czechoslo-vakia, May, found Tumen (69 °E) 2 years later (Formánek

1970). Winters Iran (large numbers), also Turkestan, Pakistan, and north India; presumably mainly USSR breeders.

More gregarious than Slavonian Grebe in winter and on migration staging waters; in North America, where much more numerous than in Europe, concentrations up to 20000 reported Oregon on passage, while perhaps half million winter inland Salton Sea, California (Palmer 1962). Migration principally nocturnal in North America, but in Palearctic also diurnal (Dementiev and Gladkov 1951).

**Food.** Chiefly insects and larvae, and, depending on locality and time of year, molluscs, crustaceans, amphibians, and small fish. Taken in shallow water or on surface. Food often collected from surface by skimming with slightly recurved and dorso-ventrally flattened bill (Fjeldså 1973*a*); with quick right and left movements of head; and when swimming with head and neck immersed (van IJzendoorn 1944). Also catches flying insects from just above surface. Dives in depths up to $5\frac{1}{2}$ m; duration 9–50 s, though mostly 23–35 s (Witherby *et al.* 1940; Madsen 1957). Normally daytime feeder, singly or in small parties.

Insects (adults and larvae) predominate during breeding season; include aquatic and terrestrial beetles (*Dytiscus, Badister, Noterus, Haliplus, Agonum, Amara, Harpalus, Notiophilus, Helophorus, Aphodius, Donacia, Cassida, Drilus, Phylan, Apion, Gymnetron, Sitona*); caddisflies and larvae (*Hydropsyche, Phryganea, Rhyacophila*); waterbugs (*Corixa, Notonecta, Velia*); dragonfly larvae (*Libellula, Gomphus*); earwigs; mayflies; ants; and flies (Chirono-midae, Simuliidae, *Muscaria*). Also takes molluscs (small bivalves and univalves); crustaceans (*Gammarus, Crangon, Mysis, Asellus*), especially from coastal areas, watermites (Hydrachnida) and spiders (Araneidae); amphibians (small frogs and tadpoles); and fish (small perch *Perca*, gobies *Gobius*). Plant materials, probably ingested accidentally, include algae, stoneworts (*Chara*), mosses and grasses. (Naumann 1905; Madon 1931; Witherby *et al.* 1940; Dementiev and Gladkov 1951; Madsen 1957.)

From coast of Denmark in January, one contained *Gobius, Mysis*, and undetermined insect adults and eggs. One from freshwater in May contained mainly insects: mosquito larvae (Culicidae), caddisfly larvae, beetles including *Haemonia* and Haliplidae, flies and waterboat-men *Corixa*); also some spiders and a watermite (Madsen 1957). In 79 stomachs from various European sources, 81·0% contained insects and larvae, 12·7% molluscs, 11·4% crustaceans, 7·6% fish, 27·8% plant remains, and 82·3% feathers (Madon 1931).

Juveniles fed chiefly on insects (Dementiev and Gladkov 1951).

**Social pattern and behaviour.** Based mainly on material supplied by J Fjeldså and R W Storer.

1. Most gregarious of Palearctic grebes. Outside breeding season, though sometimes solitary, often in flocks. These

frequently large, especially during halts on migration and in winter quarters, though aggregations on scale observed in USA (see Movements) not reported from west Palearctic. Flock formation often loose, with scattered singles and groups, but dense rafts also occur on salt water (Meinertzhagen 1950; McLachlan and Liversidge 1957). In flocks for a time after spring return to breeding waters; these sometimes large, though again apparently not on scale observed in USA. BONDS. Monogamous pair-bond of at least seasonal duration. Pair-formation usually starts during halts on spring migration, continuing after arrival; some changes in composition of pairs noted before start of nest-building (McAllister 1958). Sometimes in pairs during winter (King 1967; J Fjeldså). Both sexes tend young. Little published information on social structure of family groups, but evidence of crèching system not found in other Palearctic grebes (J Fjeldså): though period of dependency apparently shorter than in *Podiceps* of comparable size (see van IJzendoorn 1944), young sometimes assemble in multi-family groups after first 2–3 weeks, with loose family bonds, no brood-division, and feeding of foreign young. Such crèches may be attended by only a few adults while others start 2nd broods or depart; also, young of 1st brood sometimes remain with parents and assist in feeding of 2nd brood siblings. In some cases, also permanent brood-division (see van IJzendoorn 1944; Prinzinger 1974). BREEDING DISPERSION. Typically colonial, often densely in large numbers at more or less exposed but inaccessible sites; nests also sometimes cryptic and more widely scattered, partly depending on availability of sites. Territory restricted to immediate vicinity of nest and usually abandoned after hatching, though some families continue to use nest for loafing for up to 2 weeks after (J Fjeldså). ROOSTING. Except when nest used, on water both by day and night, probably much as in Great Crested Grebe *P. cristatus*.

2. Less well known than in other Palearctic grebes from which differs in some features. Following based largely on information on nominate *nigricollis* from J Fjeldså (see also Wittgen 1962; Prinzinger 1974) and on *californicus* by R W Storer (see also Wetmore 1920; McAllister 1958). ANTAGONISTIC BEHAVIOUR.

Hostility shown mainly during pairing period, e.g. when Advertising bird (see below) approaches courting pair, and during site-establishment, when intruders within *c.* 0·6 m of platform driven away (McAllister 1958). By both sexes, but ♂♂ more aggressive. Threat much as in other grebes, often by pair together. Normally Forward-display (Wittgen 1962; Fjeldså 1973*b*) with bill often open, neck arched with head above water (see A) or extended, awash, along water; crown feathers raised and lowered at frequent intervals. Version with neck diagonally forward, illustrated by McAllister (1958) and Prinzinger (1974), not typical. Hunched-display, much as in Slavonian Grebe *P. auritus*, also frequent but Chittering-call not so loud. Attacks by swimming towards rival or pattering across water, uttering Chittering-call; may then dive and 'torpedo' it from below surface. Breast-to-breast fights, much like those of *P. auritus*, also occur but infrequent and brief compared with other Palearctic grebes. In Alarm-upright, bill horizontal and feathers of head and neck depressed. Escapes by swimming away, pattering over water, crash-diving from patter, or diving breast first from floating position on surface. WATER-COURTSHIP. In contrast to west Palearctic congeners, but like more closely related species in South America, lacks any form of true Weed Ceremony. Otherwise, most like *P. auritus*. Performed mutually by both sexes, typically in flock; starts during halts on spring passage but most intense immediately after arrival on breeding water, ending more or less abruptly as sites taken up. Advertising common, especially in period after arrival when many pair-bonds still unstable (though continues throughout cycle when bird separated from mate or young); distinctive call given in typical upright posture with crest raised (see B). Discovery Ceremony of much the same type as in *P. auritus* and Red-necked Grebe *P. grisegena*: (1) approach of diving bird sometimes punctuated by 1–3 emergences in Bouncy-posture; (2) Ghostly-penguin display (see C) occurs rather further away from bird in Cat-display than in *P. auritus* but not so far as in *P. grisegena*; (3) Cat-display (see C) of type found in *P. cristatus*; (4) couples often call loudly as they move towards each other; (5) mutual Penguin-dance (see D)

A

B

C

D

E

at climax of ceremony reportedly accompanied by same call as birds Head-shake (Slow Head-turns in precise tempo, sometimes alternating with 2 rapid lateral Head-waggles). End of sequence variable: pair may perform (6) mutual Parallel-rush in Penguin-posture across water (see E); (7) Head-turning while swimming normally; (8) self-preening; (9) combination of 2 or more of these (usually in above sequence); (10) Retreat Ceremony much as in *P. auritus* leading to another Discovery Ceremony; (11) Ceremonial Turning-away. Last infrequent and poorly developed, probably ·because of absence of Weed Ceremony. Parallel-rush may sometimes end (12) in a series of unison dives (Wittgen 1962). Such Rushing-and-diving sequences also occur as a separate ceremony, mainly after aggressive encounters. Penguin-dance (with Head-shaking) and Head-shaking on own also performed as independent ceremonies. Though in *californicus* pair said to swim side by side while Head-shaking (McAllister 1958), in nominate *nigricollis* typically face closely (see F); latter race also Habit-preens much as in *P. cristatus*. Just

after initial pairing, Rushing and Rushing-and-diving Ceremonies frequent: Parallel-swimming, with necks erect and crest raised, grades into Parallel-rushes in Penguin-posture, latter often interrupted by unison Parallel-dives. Head-shaking Ceremony replaces Discovery but often transition between Head-shaking and Triumph Ceremony, latter becoming most frequent display after pair-formation, at least in nominate *nigricollis*. Triumph Ceremony much as in *P. auritus* (see G); ends with Pivoting and much preening of a more or less stereotyped kind (see H), though not perfunctory as in true Habit-preening. No Food-presentation to mate recorded, but Bill-touching evidently occurs at times. Significance, and extent of ritualization, of some other putative water-courtship activities (see, e.g., Wetmore 1920) uncertain. PLATFORM-COURTSHIP. Much as in other *Podiceps* (see especially Franke 1969). Includes: (1) Inviting-on-water before platform built, (2) Inviting, (3) Rearing with Wing-quivering, and (4) copulation on platform—both sexes sharing roles. Abortive Weed-presentation reported by Askey and Boyd (1944). Mounting preceded by Copulation-trills from active bird; these continue during copulation when accompanied by different call from other (Copulation-duet). Before active bird dismounts, Inviting one may exceptionally raise head and rub nape against former's breast (Th Schmidt). Post-copulatory behaviour includes Water-treading by active bird, Slow Head-turning by both (one with back to other), and self-preening. RELATIONS WITHIN FAMILY GROUP. Little detailed information. Parental hostility noted after brood-division; chick approaches adult for food, calling with head low, immediately swimming away from it afterwards (van IJzendoorn 1944). Head-shaking between young seen from about 8th week (J Fjeldså).

(Figs A–H after original drawings by J Fjeldså.)

F

G

H

PLATE 9. *Podiceps cristatus* Great Crested Grebe (p. 78): **1** ad breeding, **2** ad non-breeding, **3** ad post-breeding moult, **4** imm winter, **5** juv, **6** downy young. (PB)

PLATE 10. *Podiceps grisegena* Red-necked Grebe (p. 89): **1** ad breeding, **2** ad non-breeding, **3** ad post-breeding moult, **4** imm winter, **5** juv, **6** downy young. (PB)

PLATE 11. *Podiceps auritus* Slavonian Grebe (p. 97): **1** ad breeding, **2** ad non-breeding, **3** ad post-breeding moult, **4** juv, **5** downy young. (PB)

PLATE 12. *Podiceps nigricollis* Black-necked Grebe (p. 105): **1** ad breeding, **2** ad non-breeding, **3** ad post-breeding moult, **4** juv, **5** downy young. (PB)

**Voice.** Rather similar in some cases to Slavonian Grebe *P. auritus* but generally quieter and less harsh. Full repertoire inadequately known but probably same in both sexes as in other grebes, i.e. attribution of certain calls to ♂ only dubious. Most vocal in breeding season, especially during pairing and site-establishment. Apparently does not call in flight.

Following based on outline provided by R W Storer; see also McAllister (1958) and Prinzinger (1974). (1) Advertising-call. 'Poo(–)eee(–chk)'. Ascending, plaintive, flute-like and slurred 'poo-eee' followed by a soft, guttural 'chk' which can be heard only at close range (see I). Also rendered 'huiick-ted', 'puu-ii-(ch)', etc; variable. First given by lone, unmated birds early in season (so-called 'bachelor-call'), later by paired birds out of contact with mate or young. Used more frequently than equivalent calls of other Palearctic grebes. (2) Threat-chittering. Loud, often prolonged chittering; rapid alternation of 2 notes; each note individually accented. See also next call. (3) Display-trill. Shrill chittering, somewhat like (2) but more musical. Resembles 'poo-eee' of (1) but uttered much more quickly. Given before and probably during mutual Penguin-dance display. Also probably at other times judging from general accounts in literature of rippling or trilling calls not unlike those of *P. auritus* or of Little Grebe *Tachybaptus ruficollis*, though evidently not so loud. These rendered 'bidder-vidder-vidder-vidder' in Witherby *et al.* (1940); and as 'bidvidevidevide' by Söderberg (1950) who described them as clear and mellow whistling heard mostly at night. Probable that calls (2) and (3) often confused, so last descriptions could apply to either. (4) Copulation-trill. Intense, eerie trill. Given by active bird before and during copulation. Rendered 'sih sih sih si-si-urrrrrr' by Franke (1969). Also said (McAllister 1958) to accompany soliciting in Inviting-posture (but see next call). (5) Inviting-call. Low, rapid 'goah . . . .'. Uttered by soliciting bird on water or platform. Given simultaneously with trill of active bird, produces characteristic Copulation-duet (see especially Franke 1969). (6) Alarm-calls. Most usually 'whit', repeated 2–3 times. Like *T. ruficollis*, but not so loud and sharp. Also warbler-like 'tschak' (Prinzinger 1974).

CALLS OF YOUNG. Contact-call: 'bibib' (or 'quiefieb' while still in egg). Food-call: whimpering 'piie'.

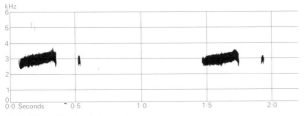

I E Allsop England August 1971

**Breeding.** SEASON. See diagram for central Europe. Factors governing onset of breeding include temperature

(McAllister 1958). SITE. In shallow water, usually in cover of reed *Phragmites* and sedge *Carex* beds. Normally colonial, from under 10 nests to several hundreds with nests as close as 1 m; in North America, much larger colonies up to thousands (Palmer 1962). Often in deliberate association with small gulls *Larus*, terns *Chlidonias*, etc. (see e.g. Durango 1954; Gauckler and Krauss 1968). Nest: floating heap of water weed anchored to aquatic vegetation. Average outer diameter 26 cm (20–34), internal diameter 3·0 cm (1·5–4), sample 21 nests (Bochenski 1961). Several nests built before final one; not just mating platforms as eggs may be laid in each (Palmer 1962). Building: by both sexes. Shaped by bird sitting on nest and piling weed around it. EGGS. Ovate, chalky white, not glossy, soon becoming stained light to dark brown. $44 \times 30$ mm ($39$–$49 \times 27$–$34$), sample 250; calculated weight 21 g (Schönwetter 1967). Clutch: 3–4 (1–6). Mean of 21 clutches 3·2 (range 2–5) (Bochenski 1961). Of 293 American clutches: 1 egg, 0·3%; 2, 3·1%; 3, 54·2%; 4, 35·5%; 5, 4·8%; 6, 2·1%; mean 3·48 (McAllister 1958). Eggs laid daily but one or more days missed during most laying sequences. Most eggs laid in morning. Indeterminate layer (McAllister 1958). One brood, sometimes 2. Replacement clutches laid after egg loss. INCUBATION. 20–2 days. By both sexes. Starts with 1st egg. Eggs usually covered when incubating bird disturbed off nest. When clutch not complete, 37% covered eggs, when clutch complete 75% covered; sample 267 (Broekhuysen and Frost 1968a). Hatching asynchronous. Eggshells removed. YOUNG. Precocial and semi-nidifugous. Cared for and fed by both parents, or in crèche. Ride on parents' backs when small, being fed there. Fed bill to bill; also given feathers. In Netherlands, each parent took part of brood to separate feeding area and did not reunite (van IJzendoorn 1944). Parents defend eggs and young against predators. FLEDGING TO MATURITY. Fledging period and age at independence not fully recorded, but said to be fully independent in as little as 21 days (van IJzendoorn 1944). Age of maturity probably 2, but possibly 1. BREEDING SUCCESS. No data from west Palearctic. Of 223 eggs laid in

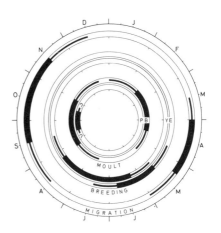

colony in South Africa, only 12 young left nest but this probably not typical; major causes of loss were predators and wave action in high winds (Broekhuysen and Frost 1968*b*).

**Plumages.** ADULT BREEDING. Head and neck to upper chest black; feathers of crown elongated; tuft of long, golden-brown feathers on side of head behind eye. Mantle and scapulars black, back and rump dark brown. Breast and belly white, mottled along border with black chest, vent brown-grey; flanks chestnut mottled black and silver-grey. Tail-tuft of downy feathers; upper half black, lower half dirty white. Outer primaries dark brown-grey, p6 often with small white spot at tip, p5 with inner web mottled white, p4–p1 increasingly white. Primary coverts dark grey. About 11 outer secondaries white, inner increasingly dark grey. Upper wing-coverts dark brown, under wing-coverts white, axillaries partly grey. ADULT NON-BREEDING. Upperparts grey-black. On side of head, grey reaches down below eye; ear-coverts grey. On upper neck, white half-collar with ends nearly meeting on hindneck. Chin and upper throat white, lower throat and foreneck mouse-grey. Rest of underside white, vent brown-grey, flanks grey. Tail and wing as breeding. DOWNY YOUNG. Upperparts brown-black with variable pattern of paler stripes, darker than other species of *Podiceps*. Lores and spot on centre of crown bare. Forehead dirty white; 2 pale grey lines on each side of crown converge on forehead in irregular double V, sometimes reduced to number of pale spots. Hindneck with 2 pale lines. Back with several pale stripes, sometimes hardly visible. Sides of head and neck white with black-brown spots, roughly arranged in 2 rows. Underparts white. On chin and throat, 2 rows of tiny spots; foreneck pale grey; vent and flanks brown-black. Pattern soon disappears; older chicks brown-black above and white below. JUVENILE. Similar to adult winter. Crown and hindneck with browner hue; rest of upperparts paler, tinged olive, feathers with remains of down to tips, soon mixed with black immature winter feathers. Sides of head mottled, of upper neck tinged buff; flanks paler than adult, pale olive grey. IMMATURE WINTER. Like adult non-breeding.

**Bare parts.** ADULT BREEDING. Iris bright orange-pink with narrow silver-white ring round pupil, eyelid orange-red. Bill black. Foot black on outside, lighter, with blue or green hue on inside and upperside of toes. ADULT NON-BREEDING. Bill lighter than breeding, blue-grey, culmen black, gonys pale. Foot sepia on outside, grey-green to grey-blue on inside, lobes paler. DOWNY YOUNG. Iris dark brown. Bare spot on crown flesh and lores pink, becoming scarlet by vascularization. Bill fleshy white with 2 black bars and tawny base. Foot grey, edges of lobes scarlet. JUVENILE. As adult non-breeding.

**Moults.** ADULT POST-BREEDING. Complete; primaries simultaneous. From mid-June, birds with young July to mid-September. Many in winter plumage September, others probably not before November (Gauckler and Kraus 1968). Timing of wing moult unknown, presumably August–September. ADULT PRE-BREEDING. Partial; body and tail; February–April. Many in full breeding plumage April, but some (possibly in 2nd calendar year) still moulting May. POST-JUVENILE. Partial; body and tail; summer to December. Apparently starts soon after completion of juvenile plumage. After moult indistinguishable from adult non-breeding.

**Measurements.** Nominate *nigricollis*, older than juvenile; skins (BMNH, RMNH, ZMA).

| | ♂ | | ♀ | |
|---|---|---|---|---|
| WING | 134 (3·77; 20) | 127–139 | 130 (2·82; 19) | 124–136 |
| BILL | 23·4 (1·04; 18) | 21–26 | 21·5 (1·25; 15) | 20–24 |
| TARSUS | 43·2 (1·73; 18) | 40–45 | 42·1 (1·64; 18) | 39–45 |
| TOE | 50·2 (1·99; 13) | 47–54 | 49·9 (1·76; 9) | 47–53 |

Sex difference significant for wing and bill. *P. n. gurneyi* small, mean wing ♂♀ 125, bill 21·7 (McLachlan and Liversidge 1957). *P. n. californicus* shorter wing, longer bill; mean wing ♂ 133, ♀ 127; bill ♂ 26·5, ♀ 23·7 (Palmer 1962).

**Weights.** Nominate *nigricollis* (Bauer and Glutz 1966; BTO; ZMA). Breeding: 3 ♂♂ 265–402, 2 ♀♀ 213, 298. Non-breeding: 4 ♂♂ 366–450; ♀ 280; 2 unsexed 230, 297. Juveniles: emaciated ♂ 144; fat ♂ 366; fat ♀ 354. Downy young at hatching 13–14.

**Structure.** 12 primaries: p11 and p10 longest, either may be 1 or 2 shorter than other; p9 2–4 shorter, p8 10–12, p7 17–20, p6 23–27, p1 51–56; p12 minute, concealed. P11 emarginated inner web, p10 slightly; p10 emarginated outer web, p9 slightly; outer web of p11 narrow. About 18–19 secondaries. Bill wide at base, compressed at tip; culmen slightly upturned, gonys sharply sloping upwards, giving markedly tilted aspect. Outer toe *c.* 108% of middle, inner *c.* 82%, hind *c.* 26%.

**Geographical variation.** *P. n. gurneyi* smaller; greyer on head, neck, and upperparts; tufts paler, straw-yellow; and lesser upper wing-coverts variably tinged rufous in adult. No separate winter plumage. *P. n. californicus* longer bill on average; inner primaries brown-grey, not white. *P. nigricollis* forms superspecies with *P. andinus* from east Andes of Colombia (Simmons 1962; Mayr and Short 1970).

**Nomenclature.** Until recently sometimes named *Colymbus caspicus* Hablizl, 1783 or *Podiceps caspicus*. Both names *Colymbus* Linnaeus, 1758 (Int. Comm. Zool. Nomencl. 1956*a*) and *caspicus* Hablizl, 1783 (Int. Comm. Zool. Nomencl. 1956*b*) suppressed by the International Commission on Zoological Nomenclature.

# Order PROCELLARIIFORMES

Huge to small tube-nosed seabirds. 4 families, all but Pelecanoididae (diving petrels) represented in west Palearctic, though Diomedeidae only as stragglers. Clearly defined order, not closely related to any other, but some morphological and biochemical characters suggest distant affinities to Sphenisciformes (penguins) and probably Pelecaniformes (pelicans, etc.). Strictly marine, coming to land only to breed.

Nasal, salt-secreting glands large. Nasal olfactory organ better developed than in most birds, including all other seabirds, indicating highly adaptive sense of smell, possibly connected with food seeking and social interaction; most species possess musty body odour. With exception of Pelecanoididae, all store large quantities of oily fluid in proventriculus at certain times of year. Recent research shows this stomach-oil not a secretion of glandular cells but dietary residue, permitting adults to forage at long distances from colony and feed young, at least partly, on converted and highly nutritious liquid food (see Serventy *et al.* 1971). In some open-nesting species, ejection of stomach-oil used in territorial defence and against predators by both adults and young.

# Family DIOMEDEIDAE albatrosses

Huge seabirds relying on dynamic sailing flight. 13 species in 2 genera: *Diomedea* and *Phoebetria*. Smaller members of *Diomedea* known as 'mollymawks', but, apart from size, differences from larger species no more important than those within species groups. Body short and broad, head relatively large, neck short. Wings long and narrow with long arm; fold in 3, about equal, parts when closed. 11 primaries; p10 longest, p11 minute. Many secondaries, up to 37 or 40 in Wandering Albatross *D. exulans*; diastataxic. Tail short and square in *Diomedea*, longer and wedge-shaped in *Phoebetria*; 12 feathers. Bill heavy, horny sheath divided in conspicuous plates (arrangement important in separating species), hook at tip, nostrils in lateral tubes. Legs strong, stance upright (most other Procellariiformes crouch on tarsi), 3 front toes connected by webs, hind toe lacking (or represented by vestigial bump). Oil gland feathered. Aftershaft absent. Sexes similar in plumage (except *D. exulans*), ♂ averaging larger. Mainly white with some black on wings, white with upperwing and mantle dark, or mainly dark grey. Single moult per year, or once every other year in larger species; sequence of primary moult unknown. Juveniles resemble adults, except in *D. exulans*, but generally separable; reach fully adult plumage only after several years—up to 30 or more in *D. exulans*.

Oceanic; southern hemisphere and north Pacific. Rare vagrants in west Palearctic. Food mainly fish and squid taken while floating on sea by seizing from surface or just below by up-ending or shallow diving; will feed on faeces of cetaceans. Also scavengers on antarctic whaling industry and follow ships, taking all kinds of edible refuse but especially fat.

## *Diomedea melanophris* Black-browed Albatross

PLATE 13
[after page 134]

Du. Wenkbrauwalbatros    Fr. Albatros à sourcil noir    Ge. Mollymauk
Ru. Чернобровый альбатрос    Sp. Albatros ojeroso    Sw. Svartbrynad albatross

*Diomedea melanophris* Temminck, 1828

Polytypic. Nominate *melanophris* Temminck, 1828, islands in Southern Ocean. Extralimital: *impavida* (Mathews, 1912), Campbell and Antipodes Islands, New Zealand.

**Field characters.** 80–95 cm, ♂ averaging larger with longer bill; wingspan 213–246 cm. One of larger mollymawks, or medium-sized albatrosses, with white underbody and rump, grey tail (sometimes almost invisible against sea) and grey mantle between sooty-brown wings (general pattern typical of this group), but with heavy head and thick neck giving humped appearance like outsize Fulmar *Fulmarus glacialis*. Sexes alike and no seasonal differences, but distinct immature plumages.

ADULT. Head wholly white except for dark 'eyebrows' (present to varying extent in most mollymawks, but so marked in this species as to give frowning look); mantle slightly darker grey than in other species. Small to quite large white central stripe along underwing with broad, somewhat ill-defined, brownish-black margins, wider in front; stripe tends to merge with dark edges most towards wing tip. Bill yellow, orange-yellow, or pale flesh with brighter pink tip and thin black vertical line where joins head; swollen at base and tapering towards tip so does not appear disproportionate (see below). Feet usually bluish-grey, but ranging from purplish to whitish. IMMATURE. Duller and drabber, with top of head and neck grey at first, although soon changing to white, leaving lower neck bluish-grey. Underwing at first shows little white, which tends to merge with dark edges and tip, though central white stripe soon becomes prominent. Bill blackish-grey with darker tip, but quickly develops horn-coloured patches on lateral plates, later becoming increasingly yellow at base until sub-adult has yellow bill with dark tip (last sign of immaturity).

Albatrosses distinguished from other seabirds by large size (considerably bigger than Gannet *Sula bassana*), extremely long narrow wings, stout bodies, long necks, large heads, long heavy beaks hooked at tip of upper mandible and covered with horny lateral plates, short tails, short legs placed far back, and continuous powerful gliding flight. Specific identification points are distribution of white on underwing and whether clearly demarcated from dark wing margins; colour and shape of bill; amount of grey on head; extent and shape of dark patch round eye; presence or absence of white mark behind and below eye; colour of mantle; and general shape and build. *D. melanophris* in west Palearctic needs to be distinguished from 3 other mollymawks which might occur (see Warham *et al.* 1966). (1) Grey-headed Albatross *D. chrysostoma*. Similar in plumage and underwing pattern (though white normally more extensive with less well-defined margins), but adult has glossy black bill with orange-pink tip and strips of brilliant yellow or orange along ridges of both mandibles; grey head and nape with white mark just below and behind eye; dark feathers in eye socket not enough to give frowning look; and slightly paler mantle. Immatures, especially when young, not easily separable from *D. melanophris*, but darker (often browner) grey of head more sharply demarcated from white breast, and blackish-grey bill soon developing narrow, horn-coloured strips along ridges of both mandibles, which gradually brighten into yellow of adult. (2) Yellow-nosed Albatross *D. chlororhynchos*. Smaller and rather lightly built. Adult much more white on underwing, though dark anterior margin still fairly broad; black bill with yellow strip along ridge and pinkish tip, but with relatively little taper giving top-heavy appearance; extremely pale grey head looking white at distance, with dark feathers round eye in shape of downward-pointing triangle. Immature closely similar to adult, but with pure white head and black bill. (3) Shy Albatross *D. cauta*. Largest mollymawk. Adult has underwing almost entirely white except for thin dark edges and black tip; deep-based bill with yellow line above and below shading into orange tip and with blue-grey sides running up in front of eyes, set off by dark line extending down and back in horseshoe behind nostrils. Immature closely similar to adult, but whole bill grey or olive.

Most sociable of albatrosses and most fearless of man while at sea, approaching nearest to ships and shore. Like other albatrosses, spends much time in air and, except when little wind, glides to and fro on stiff, almost motionless wings with large, webbed feet open as rudders or balancers on each side of tail; builds up speed in rapid downward sweep to leeward on flexed wings, banks along wave troughs on upward currents from crests and then soars into wind on straight wings to heights of 15–20 m, banks again in half-circle, and repeats process again and again. Flaps wings only occasionally, usually when stalling momentarily in turn, and normally without raising them above level of body, though flexes and twists them at carpal joints to take advantage of air currents; in gales may rake wings steeply back. Will also hang in air over sterns of ships. When wind drops, may flap more often, but in calms prefers to settle on sea, riding buoyantly. Usually alights gently with wings raised, but sometimes plunges in from height of several metres with wings half-open and momentarily submerges; also dives from surface. To take flight again, patters along with outstretched wings until

sufficient impetus obtained, usually flapping until clear of waves. Gait on land awkward and waddling, but, unlike Procellariidae and Hydrobatidae, holds tarsi straight; stretches head out low in front and moves it from side to side at each step, raising it on stopping; if hurried, may spread wings and half-run, but then frequently trips and falls forward on breast; owing to great wing spread, cannot rise from level ground unless fair wind blowing, and then not under about 50–100 m. Generally silent when solitary. Active nocturnally and follows ships on moonlit nights.

**Habitat.** Oceanic, from pack-ice edge into temperate and to some extent subtropical seas of substantial biological productivity. Ranges nomadically over vast marine areas at unlimited distance from land; also uses fjords and channels between islands, especially in stormy weather, and often fishes inshore, sometimes near breakwaters and quays, even among flocks of gulls (Laridae). Often accompanies ships, making close approach and taking discarded or offered food (Murphy 1936). Tireless on wing, at low or medium height above waves; swims well, especially when feeding. On land, seeks undisturbed natural sites for breeding, usually on remote oceanic islands, on steep slopes at top of cliffs and also on flattish tablelands, permitting ready take-off, up to 300–400 m. Stragglers in west Palearctic recorded sitting on sea-cliffs with Gannets *Sula bassana*.

**Distribution.** Breeds southern oceans between 60 °S and Tropic of Capricorn; on islands off Cape Horn, Staten Island, Ildefonso Island, South Georgia, Falkland Islands, Kerguelen, Heard Island, Macquarie Island, Campbell Island and Antipodes Island, and possibly Crozet Islands (Watson *et al.* 1971).

Accidental. Britain and Ireland: 13 recorded to 1968, all but 2 during 1963–8, one frequenting Bass Rock gannetry during summers 1967–8 (British Ornithologists' Union 1971). Since then, single birds Bass Rock 1969, Orkney 1969, Shetland and Fife 1972, Sussex and Yorkshire 1974, and one frequenting gannetry, Unst, Shetland, 1974 and 1975. Spitsbergen: one killed June 1878 at sea to west (Løvenskiold 1964). Iceland: Westmann Islands, one July 1966 (Bourne 1967a). Faeroes: adult ♀ summered for 34 years up to 1894; adult ♀ shot at sea 64 km south-west, May 1900 (Bourne 1967a). Norway: young ♂ shot May 1943 (Bourne 1967a). For indeterminate sight records, some of which may refer to this species, see Bourne (1967a).

**Movements.** Migratory and dispersive. Immatures dispersed pelagically until begin visiting colonies at 3–5 years; adults at sea except when on duty at nest. Ranges south to pack-ice edge, north to about Tropic of Capricorn, occasionally beyond. Movements away from colonies studied intensively in South Atlantic, area of probable origin of North Atlantic stragglers. Of birds ringed Falkland Islands, recovered to 1966 (mainly 1–3 years old), 88% recovered eastern South America (Brazil, Uruguay, Argentina) and 12% South Africa; of those ringed South Georgia, 95·6% recovered southern Africa (including Angola), 1·6% South America, and 3·8% Australia (Tickell 1967). Thus strong evidence of separate winter ranges for young of these populations, seemingly with little circumpolar movement; more information needed for adults, but probably similar. Many occur in Australian waters and South Pacific, presumably mainly from colonies in Indian Ocean and New Zealand area. Vagrant to northern hemisphere, but most often identified albatross in eastern North Atlantic and adjacent seas, where individuals may remain several years (see Distribution); in western North Atlantic, recorded Massachusetts (2), North Carolina (2), Martinique, and Greenland (DuMont 1973), fewer than for Yellow-nosed Albatross *D. chlororhynchos*.

**Voice.** Mainly silent at sea, but may utter harsh cries or peevish croaking during disputes over food (Palmer 1962; Serventy *et al.* 1971).

**Plumages** (both subspecies). ADULT. Head, neck, rump, upper tail-coverts, and underparts white. Black eye-streak. Mantle grey, back darker, scapulars and upper wing-coverts dark slate. Tail slate with pale shafts. Primaries and secondaries black, paler towards base particularly on inner web. Under wing-coverts white, lightly shaded grey, those at leading edge dark slate like upperwing; axillaries white. JUVENILE. Crown, hindneck, and sides of neck grey. Narrow collar over lower throat mottled grey. Greater and median under wing-coverts white with grey tips, lesser and marginal slate; axillaries white tipped grey. Otherwise like adult. Grey on head and neck mostly disappears with wear; replaced, at least on neck, at next moult by similar grey-tipped feathers of more ashy tinge (Falla 1937; Witherby 1922). White neck of adults apparently acquired only after several years.

**Bare parts.** ADULT. Iris brown or yellow-brown. Bill yellow with orange tip, narrow black line between horny sheath and feathers at base. Foot pale flesh, varying from almost white to purple-flesh. JUVENILE. Bill grey-black soon developing horny patches on sides; later yellow at base. Yellow with dark tip in subadult (Warham *et al.* 1966).

**Moults.** Poorly known. Moulting birds (adults and juveniles) collected in November and January–May. Apparently one complete moult each year, primaries either descendant or irregular. Juvenile primaries retained for *c.* 1 year, becoming very worn (Murphy 1936).

**Measurements.** Birds at sea or beached (Condon 1936; Holgersen 1945; Lowe and Kinnear 1930); sexes combined.

| | Australian and New Zealand sector | | | Rest of oceans | | |
|---|---|---|---|---|---|---|
| WING | 508 | (14·7; 23) | 480–540 | 516 | (15·9; 26) | 462–543 |
| TAIL | | | | 200 | (8·5; 7) | 189–215 |
| BILL | 112 | (4·59; 32) | 99–120 | 118 | (3·27; 29) | 110–124 |
| TARSUS | 82·2 | (3·83; 30) | 75–94 | 85 | (2·40; 27) | 80–90 |
| TOE | 131 | (6·11; 6) | 124–140 | | | — |

Small and large measurements occur in both groups, but preponderance of small in Australia–New Zealand sector; subspecies *impavida* presumably averaging smaller than nominate. Due to wide-ranging habits, separation of 2 populations at sea not absolute. Differences significant only for bill and tarsus.

**Weights.** Live birds caught at sea: 3515 (10) 3175–3855 (Dixon 1933). 2 ♂♂ washed ashore South Australia 2322 and 2491; ♀ 2210 (Condon 1936). ♂ Tristan 3940; ♀ 3250 (Hagen 1952). ♂ South Georgia 3950 (Holgersen 1945).

**Structure.** Wings long and narrow; wingspan 2130–2280 (Dixon 1933), maximum 2460 (Murphy 1936). 11 primaries: p10 longest, p9 5–20 mm shorter, p8 30–45, p7 60–90; p11 small, hidden by primary coverts (Witherby *et al.* 1940). Tail short, slightly rounded; 12 feathers. Bill heavy, nasal tubes short, no fleshy membrane between dorsal and lateral horny plates as in other small albatrosses, dorsal plate broad and round at base, lateral plate widest at base (Condon 1936). No hind toe.

**Geographical variation.** Populations in New Zealand region apparently average smaller; iris yellow or golden brown (dark brown in other populations); eye-mark more strongly developed, particularly in front of eye; underwing darker, but too variable to be useful in identification.

**Recognition.** Bill structure diagnostic.

## *Diomedea chlororhynchos* Gmelin, 1789 Yellow-nosed Albatross

FR. Albatros à bec jaune    GE. Gelbnasenalbatros

## *Diomedea chrysostoma* Forster, 1785 Grey-headed Albatross

FR. Albatros à tête grise    GE. Graukopfalbatros

Two rather similar species, often confused in older literature. *D. chlororhynchos* occurs mainly temperate and subtropical South Atlantic and Indian Oceans, but ranges east to New Zealand and wanders north to tropics (Palmer 1962). *D. chrysostoma* breeds Cape Horn, South Georgia, Marion Island, Crozets, Kerguelen, Macquarie Island and Campbell Island (not Falklands, as often stated) in Southern Ocean and rarely identified north of 40 °S; perhaps makes circumpolar movements in Southern Ocean since several ringed South Georgia recovered off New Zealand and southern Australia (Tickell 1967).

*D. chlororhynchos* reliably identified 13 times on Atlantic coasts North America (*Amer. Birds*, **27**, 1973, 563–5), but *D. chrysostoma* unlikely to occur naturally in Northern Hemisphere. Preserved skeleton of albatross obtained Westmann Islands, Iceland, c.1844, appears to be *D. chlororhynchos* or *D. chrysostoma*; one *D. chlororhynchos* reported obtained Basses Pyrénées, France, August 1889, but skin now in Bayonne Museum is *D. chrysostoma*

and appears to be substitute; 2 *D. chrystostoma* said obtained (one preserved) near Oslo, Norway, 1834, but data vague and importation from Southern Hemisphere likely (Bourne 1967a).

## *Diomedea nigripes* Audubon, 1839 **Black-footed Albatross**

FR. Albatros à pattes noires     GE. Schwarzfussalbatros

Essentially North Pacific species breeding several of Leeward Islands (Hawaii), and formerly various Volcano, Bonin and Micronesian islands and Seven Islands of Izu. Pelagic range extends to both sides of North Pacific and from equator to southern Bering Sea (exceptionally to 65 °N); only one reliable

southern hemisphere record, from New Zealand (Palmer 1962). One killed near Messina, Sicily, 10 November 1971, seems unlikely to have reached Mediterranean unaided, and release by sailors thought probable (Sorci *et al.* 1972).

## *Diomedea exulans* **Wandering Albatross**

PLATE 13
[after page 134]

DU. Grote Albatros     FR. Albatros hurleur     GE. Wanderalbatros
RU. Странствующий альбатрос     SP. Albatros común     SW. Jättealbatross

*Diomedea exulans* Linnaeus, 1758

Polytypic. Nominate *exulans* Linnaeus, 1758, Tristan da Cunha and Gough Island; *chionoptera* Salvin, 1896, South Georgia, islands in south Indian Ocean, and Macquarie Island; subspecific identity of populations in New Zealand region not fully clear.

**Field characters.** 110–135 cm, wingspan 275–305 cm, ♂ averaging larger with longer bill. Large albatross with mainly white adult plumage, short tail, and massive, extremely pale bill. Sexes show some differences when fully adult and apparently minor seasonal changes, but distinct immature plumages.

ADULT MALE. Wholly white except for black wing-tips and dark line along rear edge of underwing; bill pale pink (appearing almost white), eyelids greenish-white to bright blue or pink, and feet pink to mauve; salmon-pink areas sometimes present on ear-coverts and round neck, and bills of some nesting individuals or populations tend to become yellower. ADULT FEMALE. Brown flecks on crown, forming patch or cap, and blackish mottling on upperwing, breast and tail, but many younger ♂♂ of breeding age also show some of these characters. JUVENILE. Brown body and upperwing, but face white and underwing and bill as adult; through successive stages, body gradually becomes mottled and then white, and basal part of upperwing paler, until dark patch on breast and dark feathers in tail usually last immature features to be lost in ♂.

While immature feathers remain, easily separable from Royal Albatross *D. epomophora* which has no distinct juvenile plumage, but which at all ages closely resembles adult ♂ *D. exulans*: distinguishing characters are former's black cutting edge to upper mandible and, in northern race *sanfordi*, almost entirely dark upperwing which with white body provides combination not found in *D. exulans*. Individuals known to follow ships for up to 6 consecutive days and nights. (See Jameson 1958a, b; Warham *et al.* 1966.)

**Habitat.** Principally aerial and marine, preferring fresh or strong prevailing winds and cool or cold water, though rarely in contact with ice. Comes to land only for nesting, usually on ridges or near slopes and banks where can become readily airborne, particularly on windward ends of islands or promontories. Flight closely linked with upward thrust of air derived from reaction of swell or waves upon wind, rising rarely above 15 m. Exceptionally recorded soaring in circles in hundreds up to at least 1500 m in sunny, calm weather of early antarctic summer (Murphy 1936). Although freely alighting and swimming well on ocean, meets severe difficulty in taking flight in calm, foggy, or rainy conditions, requiring long take-off run, sometimes without success. Expert at following ships, even after repeatedly falling far out of sight astern. Takes advantage of whaling activities, but not otherwise given to seeking human neighbourhood or artefacts.

**Distribution.** Breeds Tristan da Cunha, Gough Island, South Georgia, Marion Island, Prince Edward Island, Crozet Islands, Kerguelen, Antipodes Island, Campbell Island (few), Auckland Islands, Macquarie Island, and occasionally Amsterdam Island.

Accidental. Immature ♂, of small subantarctic type, killed Palermo, Sicily, October 1957, and immature seen 37°40′N 9°45′W about 80 km off Portugal, October 1963 (Bourne 1967a, where other, doubtful records discussed).

**Movements.** Migratory and dispersive. Pelagic distribution mainly between 30°S and 60°S, within west-wind zone; small numbers reach high Antarctic latitudes in southern summer; straggles north to equator, occasionally beyond. Migration may be circumpolar; South Georgia birds occur off New South Wales June–September, while numerous ringing recoveries to and from Australian seas of birds natal to colonies in southern Indian Ocean, involving adults and immatures. Summer ranges of pre-breeders not adequately defined (except from South Georgia where recoveries off Argentina), but concentrations in cold currents off Chile, Argentina, and South Africa. See Dixon (1933), Tickell (1968), Tickell and Gibson (1968). Movements of Tristan da Cunha and Gough Island populations not known.

**Voice.** Mainly silent at sea, but throaty, gargling sounds uttered during squabbles over food (Serventy *et al.* 1971).

**Plumages** (both subspecies). ADULT MALE. Head, neck, body, and tail pure white. Pink or orange stain over ear-coverts in many individuals, probably exudate from ear channel (Murphy 1936). Primaries and primary-coverts black, white at base; inner secondaries white, remainder with black tips, increasing towards carpal joint. Upper wing-coverts white, mixed with dark brown on outer wing, greater on inner wing spotted dark brown; sometimes upper wing and back marked with characteristic fine dark vermiculations. Under wing-coverts and axillaries white. This largely white plumage worn only by about half breeding ♂♂ in any population, rest having dark mottling on head, back, wings, or tail. ADULT FEMALE. Like ♂, but less advanced towards white plumage, always with some dark mottling; some, however, have completely white head, back, wings, or tail. Sexing by plumage impossible, but ♂ of breeding pair usually whiter than ♀. JUVENILE. Dark brown except forehead, sides of face, chin, upper throat, and under wing-coverts. Belly grey-white in birds on Ile de la Possession, Crozets (Voisin 1969). SUBSEQUENT PLUMAGES. Characterized by gradual increase of white. In early stages crown and hindneck brown-grey mottled white; mantle, back, and rump brown-grey; scapulars and upper wing-coverts black-brown; chin and throat white, wide grey band across chest; centre of breast and belly white, flanks grey; lower belly white freckled grey; tail dark brown. Later grey band on chest disappears, whole upperside increasingly white with dark mottlings and vermiculations ('leopard stage'); large white spot on upper wing near elbow developing to mostly white upper wing. Complete sequence of plumage may take 20–30 years or more. For details see Murphy (1936), Gibson (1967), and Tickell (1968).

**Bare parts.** ADULT AND JUVENILE. Iris brown; eyelids white tinged green or blue, sometimes described as scarlet but thought by Falla (1937) due to infusion with blood. Bill pale pink acquiring yellow-buff hue in breeding season. Leg and foot pale grey with blue, pink, or fleshy tinges; grey with pink spot at heel in breeding season.

**Moults.** Little known. ADULT POST-BREEDING. Complete; primaries either descendant or irregular. Probably one moult

every other year, alternating with year of successful rearing of young (Stresemann and Stresemann 1966). Specimens renewing outer primaries collected September (Wilson 1907) and October (Falla 1937). IMMATURE. One ♀ *exulans* (December, Cape Seas) has most primaries new, p10 not fully grown; another ('spring') shows irregular moult of primaries (ZMA), suggesting immatures moult about same time as adults.

**Measurements.** Breeding adults Bird Island, South Georgia, measured in flesh (Tickell 1968).

| | ♂ | | | ♀ | | |
|---|---|---|---|---|---|---|
| WING | 679 | (14·4; 21) | 655–710 | 657 | (13·7; 22) | 630–680 |
| TAIL | 227 | (7·4; 21) | 215–246 | 215 | (6·2; 23) | 206–227 |
| BILL | 169 | (4·03; 21) | 163–180 | 164 | (3·62; 23) | 155–171 |
| TARSUS | 118 | (4·61; 21) | 110–127 | 113 | (4·27; 23) | 106–123 |
| TOE | 184 | (6·93; 21) | 172–193 | 175 | (4·73; 22) | 165–181 |

Difference between sexes statistically significant. Figures for bill and tarsus correspond closely with those for Indian Ocean birds: mean 10 ♂♂, 171 and 123, 4 ♀♀, 161 and 114 (Murphy 1936); wing and tail measurements taken differently, not comparable. Nominate *exulans* smaller (data from various sources).

| | ♂ | | | ♀ | | |
|---|---|---|---|---|---|---|
| wing | 635 | (8·7; 5) | 627–648 | 610 | | |
| BILL | 147 | (4·72; 7) | 138–151 | 144 | (4·76; 5) | 139–148 |

Bill ♂, ♀, and unsexed 145 (5·05; 15) 136–51; wing of unsexed specimen small, 600 (Dabbene 1926).

**Weights.** Adults breeding South Georgia: ♂♂ 9768 (875; 20) 8193–11 907; ♀♀ 7686 (559; 22) 6719–8703 (Tickell 1968). Ile de la Possession: unsexed 9575 (1138; 40) 6800–11 500 (Voisin 1968). 108 unsexed at sea off New South Wales: 5900–11 300 mean 8300; not including one ringed South Georgia 12 200 (Gibson 1967). Nominate *exulans* apparently lighter than *chionoptera*: ♂♂ 7170, 7264; ♀ 6810.

**Structure.** Wing long and narrow, wingspan of *chionoptera* 2720–3454 (Murphy 1936; Gibson 1967). 11 primaries: p10 longest, p9 slightly shorter, p8 about 30–40, others regularly shorter in steps of about 40–60. Tail short and rounded; 12 feathers. Head large. Bill long and heavy, nasal tubes short, nostrils oval, directed slightly upwards, feathers at side of lower mandible projecting beyond base of nasal tube. No hind toe.

**Geographical variation.** South Atlantic populations (nominate *exulans*) smaller than *chionoptera*. Sub-antarctic populations in New Zealand area also small-sized. Serventy *et al.* (1971) restricted name *exulans* to these, but this not in keeping with rules of nomenclature as type locality of *exulans* Cape of Good Hope. Immature ♂ Sicily identified as nominate *exulans*, supported by small measurements: wing 630, bill 150 (Orlando 1958). Murphy (1936) named southern subspecies *exulans*, using the name *dabbenena* for the Tristan and Gough population, but seems not justified in view of type locality of nominate race in sub-tropical south Atlantic (Bierman and Voous 1950; W R P Bourne).

**Recognition.** Adult *D. exulans* may be confused with *D. epomophora*, but latter mainly white from fledging, with more prominent nasal tubes, circular nostrils, black line along cutting edge of upper mandible, and black eyelids (Murphy 1936; Tickell 1968).

## *Diomedea epomophora* Lesson, 1825 **Royal Albatross**

FR. Albatros royal    GE. Königsalbatros

Breeds locally New Zealand (Otago Peninsula) and outliers (Campbell, Enderby and Chatham Islands). Winters extensively cool currents off west and east coasts of South America, some therefore passing through Drake Passage (Robertson and Kinsky 1972). Not known to range so far north as Wandering Albatross

*D. exulans*, and not yet identified in South African waters. Heim de Balsac and Mayaud (1962) include on basis of skull reputedly found on west coast Morocco, *c.* 1885; but this has also appeared in literature as *D. exulans* and, in any case, origin of specimen doubtful (Bourne 1967a).

# Family PROCELLARIIDAE fulmars, prions, petrels, shearwaters

Huge to medium-sized pelagic seabirds. Most diverse of Procellariiformes. About 55 species in 12 genera, which may be divided in 4 groups: (1) *Macronectes, Fulmarus, Thalassoica*, and *Daption* (fulmars), to which monotypic *Pagodroma* (Snow Petrel) also best assigned; (2) *Halobaena* (Blue Petrel) and *Pachyptila* (prions); (3) *Pterodroma* and *Bulweria* (gadfly-petrels); (4) *Procellaria, Calonectris*, and *Puffinus* (shearwaters) (Alexander *et al.* 1965). 6 species breeding in west Palearctic, 2 others regular migrants, 1 accidental. Body ovate, more elongated in shearwaters. Wings long and narrow. 11 primaries: p10 longest, p11 minute. 20–29 secondaries; short, diastataxic. Low gliding flight on stiff wings; shearwaters also use wings for underwater propulsion. Tail short; 12 feathers (14 in *Fulmarus* and *Daption*, 16 in *Macronectes*). Bill heavy, more slender in shearwaters, broad in prions; hooked at tip; horny sheath divided in plates; nostrils in dorsal tubes of varying length. Legs set far back, bird crouching on tarsi when on land, gait shuffling, except *Macronectes*. Tarsus laterally flattened, round in gadfly-petrels; lower part of tibia bare. 3 anterior toes webbed, hind toe rudimentary, elevated; nails sharp and curved. Oil gland feathered. Aftershaft small. Peculiar musky odour, evident even in old skins. Sexes similar in plumage, ♂ usually larger than ♀. Plumage mostly black or grey above and white below or all-dark; pure white in *Pagodroma*, boldly patterned in *Thalassoica* and *Daption*. Several species occur in both light and dark morphs. 1 moult per year, primaries moulting descendantly. Young semi-altricial and nidicolous, development slow. 2 downs, both long and fluffy, juvenile feathers growing immediately under 2nd down (see Godman 1907–10). Juvenile like adult, except in *Macronectes*.

Occur in all oceans. Essentially pelagic; though less specialized than Diomedeidae, highly adapted for living out of contact with land and inshore waters. Most abundant in cool or cold waters, rich in plankton and mainly clear of floating ice. Although good swimmers, more often aerial except when resting and feeding; capable of flying for sustained periods close to or between waves. With partial exception of gadfly-petrels, rarely climb into medium or higher airspace except when approaching some breeding places. Nest ashore, on sea cliffs, high on slopes or escarpments, or lofty plateaux or mountains, sometimes inland. Clumsy and feeble on land, avoiding open ground as much as possible. Most hole-nesting species strictly nocturnal when visiting land, though Cory's Shearwater *C. diomedea* partly diurnal at some stations; open-nesting species largely diurnal at colonies. Mostly indifferent to human presence or direct influence unless persecuted, but highly selective of breeding localities to secure immunity from predators and disturbance and to reduce difficulties of landing and access to sites. Hole nesters especially susceptible to ground pests, such as rats, deliberately or accidentally introduced. In west Palearctic, breeding distribution of nocturnal species inadequately known in some areas, and few reliable estimates of numbers. For marine distribution, information still more limited. Major increase and spread of boreal population of Fulmar *F. glacialis* for over 200 years and continuing; local declines of some other species, due to human predation. Migratory and dispersive. Some species, e.g. larger shearwaters *Puffinus*, long-distance migrants; in others, of which *F. glacialis* typical, birds of breeding age mostly stay within feeding range of colony and visit it much of year, leaving long-distance dispersal to immatures. Pre-breeders visit colonies from age of 2 years, arriving later and departing earlier—thus extending migration seasons. Immatures may summer an ocean's width from colony of birth. Passage nocturnal and diurnal, more usually in flocks. Feeding movements often of some days' duration, covering long distances. A few species vulnerable to being driven ashore in gales, but not normally in mass 'wrecks' suffered by certain Hydrobatidae (storm-petrels).

Food chiefly fish, cephalopods, and crustaceans, often as plankton; also offal and carrion. Obtained, sometimes by scavenging, in variety of ways—(1) flight-feeding, mainly swooping to surface and pattering; (2) plunge-diving, mainly pursuit-plunging; (3) surface-feeding, mainly by seizing and filtering prey; (4) surface-diving and underwater pursuit. Flight-feeding employed by minority of species, including gadfly-petrels *Pterodroma*; surface-feeding mainly by *Bulweria*, prions, and *Fulmarus* and allies, in which distensible gular pouches and muscular system for creating suction association with filtering plankton from water; plunge-diving and surface-diving mainly by shearwaters *Puffinus* and allies. Some species follow ships (and whales) and live on offal from fishing vessels or whalers. Mostly gregarious at sea, though some, e.g. Bulwer's Petrel *B. bulwerii*, typically solitary. Colonial breeders; moderately territorial so far as known, defending nest at least. Non-migratory species in particular, frequent breeding area well in advance of laying. Long-term monogamous pair-bonds evident, but maintained on seasonal basis with no indication of association between mates (or parents and fledged young) at sea. Roost at or near site when ashore, though after egg-laying mostly at sea except when incubating or feeding young; at sea, apparently rest mainly on water (at least in non-tropical forms). Little known about antagonistic and heterosexual behaviour of nocturnal, hole-nesting species, but these show little specialization for visual display; tactile interaction (e.g. by Allopreening and Billing), and probably olfactory, important while vocal communication highly developed. Copulation ashore, on ground or in nest-hole. Loud, complex, and individually variable calls given largely during flight approach in darkness to nesting site; also from ground and within burrow or hole. Aerial activity over colony less complex than in Hydrobatidae. Unlike latter, flight and ground calls evidently similar in most or all species though largely diurnal, open-nesting species differ from hole-nesting ones in many respects. Mostly silent at sea, notable exception being *F. glacialis*. Comfort behaviour not closely studied; disputed whether stomach-oil, rather than preen-gland oil in normal way, used to dress plumage—though this may explain musty odour. Other behavioural characters include direct (so far as known) head-scratching.

In west Palearctic, seasonal breeders with relatively restricted laying period. Nest on ledges, when little more than scrape with minimum of added material, or in holes, both natural or excavated by both sexes, up to 1 m long; chamber can be bare or lined with stones, grass, roots, etc; some extralimital species nest on surface. Eggs ovate, white, not glossy. Clutches always 1; single broods and no replacements recorded from those species studied. Incubation periods long, 45–55 days, and variable; sexes take roughly equal shares in spells of 1–11 days, both developing single median brood-patch. Eggshells probably always left in nest. Young cared for and fed, by incomplete regurgitation, by both parents; initially on regurgitated stomach-oil, later on invertebrates and small fish. Rarely left alone for first 1–2 weeks, then fed only at night (hole-nesters) or during day (cliff-ledge nesters); little communication then between parents and young except briefly during feeding visits. Fledging periods shorter on average in cliff-ledge nesters than in hole nesters. Young deserted at fledging (cliff-ledge nesters) or up to 10 days before (hole nesters). Age of maturity not less than 3–4 years, and up to 6–12.

## *Macronectes giganteus* Southern Giant Petrel

PLATE 14
[facing page 135]

Du. Reuzenstormvogel    Fr. Pétrel géant    Ge. Riesensturmvogel
Ru. Гигантский буревестник    Sp. Abanto marino    Sw. Jättestormfågel

*Procellaria gigantea* Gmelin, 1789

Monotypic

**Field characters.** 75–90 cm, ♂ slightly but distinctly larger with heavier bill; wingspan 180–210 cm. Huge, grey-brown and white fulmar, near size of small albatrosses (Diomedeidae), but with markedly stouter body (so that long wings appear disproportionately small and narrow), massive pale bill with hooked tip and raised nasal tube extending for three-fifths of length, and short, rounded tail. Sexes alike (apart from size) and no seasonal differences, but polymorphic; only dark morph has distinct immature plumages.

ADULT. Much commoner dark morph has mottled dark and pale grey head, including crown, nape, and neck down to upper breast, as well as leading edge to wings, and these areas become increasingly pale until whole head almost pure white in oldest birds; rest of body, wings, and tail all brown-grey to dark grey, variably mottled on lower breast and flanks, and with blacker grey flight-feathers when freshly moulted. Scarce white morph entirely white apart from asymmetrical scattering of dark feathers on body, but not on wings or tail. Note that, in both morphs, whole head grey to white with variable mottling. Bill appears uniformly pale at distance and deeper than those of albatrosses, but at closer ranges shows as pale yellow-green to pink-ochre with nails of both mandibles forming lucent green tip. Eyes brown, grey-brown or pale grey; legs black to grey-brown, or dull blue-grey in some light individuals. JUVENILE. Dark morph sooty black, soon fading to dull black-brown or chocolate with some greyish on face and throat; during successive moults over next 7 years or more, becomes gradually paler, and more mottled on head, foreneck, upper breast, and leading edge to wings. Light morph white from first plumage, with variable scattering of dark feathers, like adult. Bare parts apparently like adult.

Huge size makes confusion with Diomedeidae more likely than with other Procellariidae, and dark immature could be mistaken for Sooty Albatross *Phoebetria fusca* or Light-mantled Sooty Albatross *P. palpebrata*, but these have much slimmer bodies, longer and more wedge-shaped tails, longer wings, smaller heads, and black bills. Disproportionately stout body of *Macronectes*, with short, rounded tail, smaller-looking wings, large head, and massive, pale, tube-nosed bill, should make distinction easy. Separation from closely similar Northern Giant Petrel *M. halli* much more difficult: though never white and generally darker than the much more variable *M. giganteus*, *M. halli* has nearly identical immature plumages and subadult *M. giganteus* may pass through a stage which is exactly like adult *M. halli*. In *M. halli*, however, only feathers of face, throat and upper breast (not those of crown, back of head, or leading edge of wings) become paler mottled grey in older birds and only those in facial area around base of bill (chin, cheeks, and narrow strip over forehead) actually become white; thus, *M. halli* never 'white-headed' and, as adult, shows marked contrast between dark crown/hindneck and pale face/chin. As *M. halli* always has dark leading edge to wing, the pale leading edge of adult *M. giganteus*, often conspicuous in flight, is diagnostic. *M. halli* also shows much greater tendency to grey eyes in adults, with some so pale as to appear white at distance. Colour of bill tip only distinguishing feature for all ages; end plates of both mandibles, but particularly upper, being dull pink to dark pinkish-red (sometimes tending to brown or yellow-brown) in *M. halli* and essentially green in *M. giganteus*.

Gliding flight on stiff wings distinctive, but without effortless quality or characteristic upwind and downwind pattern of albatrosses; on calm days, flaps more often. Alighting on water, looks clumsy with webbed feet widely spread; feet also used as rudders in free flight. Take-off equally clumsy; often flounders into waves when failing to get necessary lift. Regularly follows ships. Dives well. Stands, walks, and even runs more easily than other petrels, with waddling gait and head upright. Typically, scavenger and predator which gathers gregariously on dead whales and seals and kills other seabirds and their young. Silent away from breeding grounds. Account based on Warham (1962); Bourne and Warham (1966); Warham *et al.* (1966); Voisin (1968); Johnstone (1974); Serventy *et al.* (1971); Conroy (1972); Simpson (1972); Warham and Bourne (1974).

**Habitat.** Cold or cool pelagic waters from antarctic ice edge, and leads amid pack ice, to southern temperate oceanic zone, preferring windy and troubled seas, but also penetrating channels between land masses, e.g. in Magellan region, Patagonia. Largely a scavenger, particularly where whales, sea-lions, or seals slaughtered or processed, but also commonly fishes, and takes live prey on land, even hunting introduced rabbits and rats on breeding islands (Murphy 1936). Like Wandering Albatross *Diomedea exulans*, a frequent neighbour, seeks high ground, often *c.* 100–300 m, with open grassy or stony sites for colonial nesting, on islands or promontories, usually near water and often facing prevailing wind. In this, differs from Northern Giant Petrel *M. halli*, which nests usually scattered in secluded sheltered sites with heavy vegetation, or in broken coastal terrain affording concealment and also feeds more regularly in coastal waters (Watson *et al.* 1971). Fairly mobile on land, walking and running clumsily for some distance, especially when topography or calm inhibit aerial take-off. Will climb out of water on to, e.g., dead whale.

**Distribution.** Breeds Antarctic continent (Enderby Land, Wilkes Land, Adélie Land), Antarctic Peninsula, South Shetlands, South Orkneys, South Sandwich Islands, South Georgia, Falkland Islands (Bourne and Warham 1966), Diego Ramirez Islands south of Cape Horn (needs confirmation), Marion Island (Van Zinderen

Bakker 1971, where Prince Edward Island implied, but not expressly mentioned), Crozets, Heard, and Macquarie Island (Watson *et al.* 1971; Conroy 1972), and possibly Kerguelen. Recorded Bouvet but no evidence of breeding.

Accidental. Adult seen 48°23′N 5°37′W off Ushant, France, November 1967 (*Ardea* 1969, **57**, 92).

**Movements.** Migratory, dispersive or perhaps resident, according to age. Wide pelagic distribution Southern Ocean, from pack-ice edge north to sub-tropics and occasionally beyond. During first 2 years, juveniles circumnavigate globe in trade winds and may wander north to tropics: ringing at main colonies has provided numerous recoveries in South Africa, central Indian Ocean, Australia, New Zealand, southern Polynesia, and South America. Apparently older age groups remain in higher latitudes, perhaps in seas around colonies, or in far south, in nearest open water, with rather few individuals aged 3 or more years making transoceanic flights. (See Stonehouse 1958; Tickell and Scotland 1961; Sladen *et al.* 1968.) Apart from record off Ushant, only other valid northern hemisphere report of *Macronectes* from Midway Atoll, Hawaii, where one (species not determined) seen in December 1959, 1961, and 1962 (Fisher 1965).

**Voice.** Normally silent at sea, except for braying sound when disputing over food (J Warham).

**Plumages.** Polymorphic, with white and dark morph. ADULT. WHITE MORPH. White, with scattered dark feathers (occasional leucistic individuals lack these, Conroy *et al.* 1976); wings and tail white. DARK MORPH. Head, neck, and chest pale grey or nearly white, slightly mottled darker grey, sometimes pure white. Rest of upperparts and upper wing-coverts grey-brown, newly moulted feathers dark brown-grey with dark ashy margin, soon fading by wear (Falla 1937). Underparts like back or paler, occasionally wholly pale grey. Tail grey-brown. Flight-feathers black-brown when fresh, fading to grey-brown; under wing-coverts and axillaries as rest of underparts; leading edge of wing pale (Johnstone 1974).

JUVENILE. WHITE MORPH. White like adult from 1st plumage on (Lowe and Kinnear 1930). DARK MORPH. Sooty black all over, later fading to brown. Subsequent plumages become gradually like adult, some white beginning to appear on throat, slowly spreading over head and neck which may, however, still be quite dark at 13 years (Carrick and Ingham 1970).

**Bare parts.** ADULT. Iris varying from dark brown through grey-brown to pale grey; in white morph usually brown. Bill pale yellow-green, pink-ochre, or pale horn; nails at tips of both mandibles lucent green. Foot dark grey-brown to black; dull blue-grey to pink-grey in some white birds. Leucistic birds have uniformly horn-coloured bill and pink foot (Conroy *et al.* 1976). Bare parts of juveniles and immatures apparently like adult. (Falla 1937; Voisin 1968; Johnstone 1974.)

**Moults.** ADULT POST-BREEDING. Complete; primaries descendant. Body moult starts on abdomen about time of egg-laying and continues with breast, head, neck, and back. Primaries begin after hatching of chick. In northern colonies (Macquarie Island, Warham 1962), breeding 1 month earlier than in southern

(Signy Island, Conroy 1972), but primary moult 2 months earlier (December against February). Moult may not be completed when adults leave colony, last stages possibly slow, as in Fulmar *Fulmarus glacialis*. Some records of winter moult: May and August (Lowe and Kinnear 1930), June (Falla 1937). Finished at colony on return (Conroy 1972). Tail starts when p6 or p7 growing, probably finishing at same time as wing (Stresemann and Stresemann 1966). Failed breeders start earlier than successful; immatures earlier than adults.

**Measurements.** Data from Bourne and Warham (1966), Falkland Island birds and obviously mis-sexed specimens excluded; Voisin (1968); skins (ZMA).

| | | | | | | |
|---|---|---|---|---|---|---|
| WING | ♂ 537 | (13·9; 10) | 520–565 | ♀ 501 | (16·4; 16) | 460–526 |
| TAIL | 182 | (5·64; 11) | 171–189 | 173 | (6·57; 14) | 163–184 |
| BILL | 99·9 | (2·69; 13) | 96–105 | 89·0 | (4·15; 17) | 81–98 |
| TARSUS | 96·7 | (4·12; 12) | 89–103 | 88·8 | (4·93; 16) | 84–99 |
| TOE | 134 | (5·05; 11) | 126–143 | 126 | (3·83; 14) | 121–131 |

♂ significantly larger in all measurements than ♀.
Falkland Islands birds smaller:

| | | | | | | |
|---|---|---|---|---|---|---|
| WING | ♂ 511 | (12·2; 4) | 500–528 | ♀480 | (11·0; 5) | 462–490 |
| BILL | 98·2 | (3·51; 4) | 94–102 | 83·4 | (4·15; 5) | 78–87 |

Live birds, Signy Island (South Orkneys, Conroy 1972).

| | | | | | | |
|---|---|---|---|---|---|---|
| WING | ♂ 554 | (10·7; 13) | 534–571 | ♀516 | (15·2; 13) | 498–541 |
| BILL | 101 | (2·45; 66) | 97–109 | 87·6 | (3·00; 73) | 82–97 |

**Weights.** Possession Island (Crozets): ♂♂ 5190 (9) 4650–5300, ♀♀ 3944 (7) 3350–4900. Adélie Land: ♂♂ 4986 (3) 4400–5898, ♀

3350 (Voisin 1968). Signy Island: ♂♂ 4940 (410; 37) 4100–5800, ♀♀ 3850 (370; 37) 3000–4800, some seasonal variation (Conroy 1972). At sea ♂ 5000, ♀ 4150 (Bierman and Voous 1950).

**Structure.** 11 primaries: p10 longest, p9 slightly shorter, p8 *c.* 25, others regularly shorter in steps of 30–40. Tail strongly rounded, 16 feathers: t8 *c.* 40 shorter than t1. Bill long and massive, nasal tubes $\frac{1}{2}$–$\frac{3}{5}$ of length, wide, nostrils directed forwards. Tarsus heavy, somewhat laterally compressed.

**Geographical variation.** Larger body size, but smaller extremities, towards south; difference not conspicuous (Voisin 1968). Falkland Islands population has shorter wing than others. Proportion of white morph variable, under 10% in entire population. Absent or scarce in antarctic colonies; proportion on antarctic and subantarctic islands varying between 0 and 15·1%, without relation to climatic conditions or geographical latitude (Shaughnessy 1971). In subantarctic, *M. giganteus* replaced by similar Northern Giant Petrel *M. halli*. Latter dark grey with pale forehead, cheeks, and throat in adult plumage, nails at tip of bill dark pink tending to ochre or brown; no white morph. Wing of *M. halli* averages shorter and bill longer, but differences slight. At least in Macquarie Island, Crozets, and South Georgia (where *M. halli* discovered nesting in 1971, Brit. Antarctic Survey unpubl.), *M. giganteus* and *M. halli* behave as distinct species. Apparently also sympatric on Marion Island and probably Kerguelen. In Falkland Islands, intermediate population occurs and specific separation seems more doubtful (Bourne and Warham 1966; Voisin 1968; Watson *et al.* 1971; Johnstone 1974.)

## *Fulmarus glacialis* Fulmar

PLATE 14
[facing page 135]

DU. Noordse Stormvogel     FR. Pétrel fulmar     GE. Eissturmvogel
RU. Глупыш     SP. Fulmar     SW. Stormfågel     N.AM. Northern Fulmar

*Procellaria glacialis* Linnaeus, 1761

Polytypic. Nominate *glacialis* (Linnaeus, 1761), north Atlantic. Extralimital: *rodgersii* Cassin, 1862, north Pacific.

**Field characters.** 45–50 cm, wingspan 102–112 cm; slightly larger than large shearwaters, Cory's *Calonectris diomedea* and Great *Puffinus gravis*, and noticeably larger than Kittiwake *Rissa tridactyla*. Strong-billed, bull-headed, robust, large petrel, with commonest plumage patterns recalling gulls *Larus*, not shearwaters. Sexes alike although ♂♂ usually with bigger bill; polymorphic (see Geographical Variation) but no seasonal variation (except through wear or moult); juvenile inseparable.

PALE MORPHS. Head, neck, and underparts white, often tinged faintly yellow; upperparts grey, mottled and with silvery to brown tone (dependent on wear); rump and tail pale grey, latter with whitish rim. Inner wing coverts and inner flight-feathers uniform with back but carpal feathers and primaries darker and often noticeably variegate, with blackish tips and vanes contrasting with paler, even whitish bases and inner webs; underwing whitish with irregular dusky border, most marked at carpal joint and on primaries. At distance, pale patch on primaries obvious (particularly in worn plumage). DARK MORPHS. More

uniformly coloured, with head, neck, and underparts blue or smoky grey and darker parts of other plumage generally more intense, though pale bases to primaries still obvious. Bill of all morphs strong and stubby, with tubed nostrils prominent, raised above basal half of upper mandible and blunt hook at tip; tips of both mandibles usually yellow but rest varies from olive-green to bluish-grey, tubes dark grey to pale brown (at distance bill appears yellowish-green). Short legs and long feet yellow to green, most often bluish pink. Eye dark, emphasized by small patch of dark grey feathers in front of socket.

At long range and in difficult light, confusion possible with large shearwaters but when adequately seen, immediately separable by stubby bill, uniformly white (or pale) head, essential greyness of upperparts (or whole body), rather broad wings, and stocky shape. Flight action in all winds strong and confident, with bursts of stiff wing strokes alternating with accomplished gliding; in higher winds, skilled in banking and sliding through troughs, over crests of waves. Flight action recalls albatrosses *Diomedea*

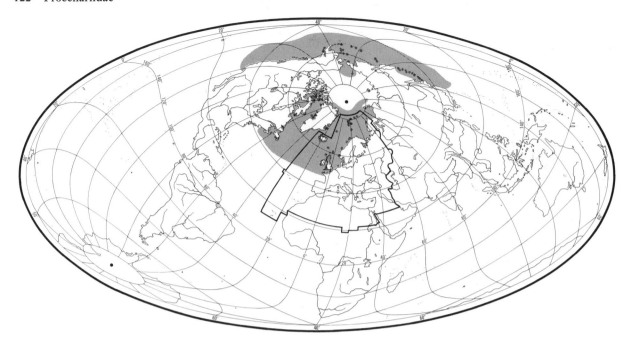

more than any shearwater and, like both, uses kicks to assist take-off from sea. Patrols breeding cliffs effortlessly, combining long passes with precise turns and astonishing stalls in strong upcurrents, yet may find difficulty in landing directly at nest site.

Essentially gregarious at larger breeding sites but forerunners of expanding population occur in tiny colonies. Attracted to waste food provided by man, large numbers attending trawlers at sea, harbour landings, and other sources including sewage outfalls. At sea, generally tolerant of other seabirds but normally dominates food competitors; at breeding sites, adults and young defend themselves by spitting foul-smelling oil. Courting birds croon and cackle loudly, flocks at sea also vocal, but single birds quiet. Chicks make variety of buzzing and shrill cackling sounds.                                    DIMW

**Habitat.** In west Palearctic, covers entire marine sector south from polar pack-ice through arctic, sub-arctic, and Gulf Stream waters to *c.* 50°N with slight shift in winter out of most northerly seas into less cold waters down to 40°–42°N. Only exceptionally on brackish or fresh water, using less shallow waters to continental slope, but especially fishing banks, upwellings, current rips, and near offshore reefs. Wherever open water occurs, even along leads through pack ice, faces rigorous conditions of storm and cold, but adapted to cope with varying weather, temperature, and currents, where biological productivity, or indirect food supply through trawling offal, adequate. Warmer waters avoided, apparently because of insufficient food; shows little discomfort at high air temperatures in normal habitat. Predominantly aerial, with full oceanic

capability, although rarely flies over 100 m above sea. Rarely alights on land other than at actual or prospective breeding sites, normally within 2 km of sea and mostly immediately facing it. Breeds mainly on narrow ledge or precipitous sea cliffs or crags, at all heights from under 10 m to 300 m, and even 1000 m on inland crags in Spitsbergen (Fisher 1952). Usually on turf-clad or bare soil, less frequently naked rock, or at mouth of burrows or crevices, or on unused stone buildings; less commonly on flat surfaces without immediate aerial take-off. Fearless of human approach on land when breeding and will feed near vessels at sea. Readily exploits human activities and artefacts.

**Distribution.** Boreal population of warmer waters of eastern North Atlantic spreading and increasing markedly for over 200 years. Began Iceland 1753 or earlier, expanding from one colony all round coasts; colonized Faeroes *c.* 1839; coasts of Britain and Ireland (excluding St Kilda, where long resident) 1878; first proved breeding Norway 1924, northern France 1960, and West Germany 1972. Spread and increase described in detail by Fisher (1952, 1966) who argued correlation with increase of offal, first from Arctic whaling, then from fish-trawlers. Wynne-Edwards (1962) noted no detectable increase or expansion in other parts of its extensive range and considered correlation with increased offal poor; he suggested new genotype might have arisen in Iceland which was able to spread into lower latitudes and nest in small, straggling colonies. Salomonsen (1965) stressed that expansion coincided with increase in area of warmer waters in eastern North Atlantic due to climatic change in previous hundred

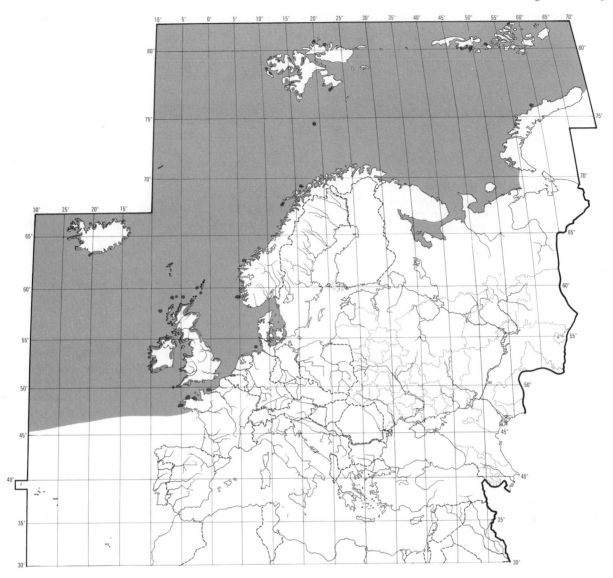

years, while Brown (1970) noted that in western North Atlantic, feeding distribution more closely linked with cool water than with supply of fish offal.

Accidental. Finland 7, Poland 2, Czechoslovakia 5, Yugoslavia 1, Spain 1, Portugal 1.

**Population.** ICELAND: increased from one colony in 1640 to 155 in 1949; population estimates for some colonies in Fisher (1952), Bannerman (1959). BRITAIN AND IRELAND. In 1969–70 some 305 000 occupied sites, over four-fifths in north and north-west Scotland. Numbers in St Kilda group may have reached ceiling, fluctuating around 40 000 sites; elsewhere numbers rose from 24 sites in 1879 to 2039 in 1909; 35 223 in 1939; 70 472 in 1949; 97 039 in 1959, and

269 039 in 1969–70. Slower rate of increase 1949–59 probably misleading due to incomplete coverage. Total increase 1949–70 *c*. 280% compared with *c*. 420% in period 1929–49, showing some slackening, especially in Outer Hebrides, but still *c*. 7% a year compound (Cramp *et al.* 1974). NORWAY: rather slow growth after first colonization, to about 850 pairs 1972, and *c*. 1100 pairs 1974 (E Brun). SPITSBERGEN: decrease since cessation of whaling (Løvenskiold 1964). BEAR ISLAND: *c*. 200 000 sites 1932 (Bertram and Lack 1933). FRANCE. 3 breeding areas Brittany and one Normandy; present elsewhere in nesting season. Numbers still small, *c*. 30 pairs (J-Y Monnat). WEST GERMANY: first bred Heligoland 1972; at least 2 pairs successful 1973 (G Vauk). USSR: estimated 25 000 birds in Barents Sea area (Uspenski 1959).

Survival. Mean annual survival rate 94·48. Adult average life-span 17·63±2·30 years. 2 birds ringed as adults 1950 still breeding 1974, so probably some 34 years old (G M Dunnet and A Anderson).

**Movements.** Migratory and dispersive. In temperate and sub-arctic latitudes, breeding adults present at sea within feeding range (up to 320 km, Palmer 1962) of colonies for most of year. Breeding ledges often visited during autumn and winter; inshore waters mostly deserted during moulting period August–October (Fisher 1952), though in recent years some present all year at older colonies, those arriving July–August newly moulted and presumably immatures or failed breeders (Dott 1973). Ringing evidence suggests adults do not truly migrate; such recoveries as Scotland to Faeroes and Denmark, or Norway to Faeroes and Netherlands, represent pelagic dispersal. In high arctic, mainly absent from frozen seas November–February; begins visiting colonies March, often earlier Spitsbergen, penetrating far up leads of open water, but retreating before unbroken sea-ice (Dementiev and Gladkov 1951; Løvenskiold 1964). Juveniles fledge August–September, dispersing and migrating over great distances; probably immatures constitute majority seen mid-ocean.

In west Palearctic, occurs all year in Atlantic, polar and North Seas, English Channel, and Bay of Biscay; less often Kattegat and rarely Baltic. Recorded north to 85°05′N (79°E) (Dementiev and Gladkov 1951); normal pelagic range extends from ice-edge (seasonally variable) south to about 48°N or further (to about 41°N) on Newfoundland Banks. Straggles south to New England and Madeira; has also occurred far inland in Europe (see Distribution). Most young have moved well away from colonies by early October; a few British-ringed reach Bay of Biscay, though most move west to north-east into Atlantic and northern seas. Major year-round feeding concentrations (high proportions immatures) on Grand Banks off Newfoundland and in Varanger Fjord–Barents Sea area, regions of frontal turbulence and high productivity. Generally assumed that Canadian and Greenland birds feed on Grand Banks, while those from Spitsbergen and Soviet arctic islands feed Barents Sea area, but situation more complex as both major feeding grounds have much higher proportions of light-phase southern morphs than occur in nearest breeding localities. Several ringed Greenland recovered Newfoundland area but also 2 Bay of Biscay and Portugal; one from Spitsbergen in Scotland (Orkney), while British-ringed birds recovered Canada (mainly Newfoundland), south and west Greenland, Iceland, north Norway, and Barents Sea to 72°30′N. Small-billed and dark-phase birds of high arctic morphs not uncommon off Atlantic coast of Canada (Palmer 1962; Salomonsen 1971); generally rare in west Europe though many in major North Sea wreck in February–March 1962 (Mathiasson 1963c; Pashby and Cudworth 1969). Ringing de-monstrates fidelity of return to breeding sites (Dunnet et al. 1963; Dunnet and Anderson 1965).

**Food.** Mainly crustaceans, cephalopods, fish, fish offal, and carrion (especially whale, walrus, and seal), mostly by surface-seizing whilst floating or swimming. Also, infrequently, by pursuit-plunging, reaching depths of over 4 m using legs and half-opened wings (see Fisher 1952). Competitive or neutral feeder, rarely if ever co-operative probably because live fish form only small part of diet (J B Nelson), but will congregate in large numbers (see Social Pattern). Wide ranging, by night and day. Voracious feeders, so much so that at times unable to rise from surface.

In Greenland waters, mainly planktonic crustaceans (especially amphipods *Hyperia*, *Gammarus*, *Themisto libellula*) and schizopods, isopods, and cumaceans. Also polychaetes, especially nereids; occasionally fish, coalfish *Pollachius virens*, ling *Urophycis* (though possibly only as dead and drifting, as these species do not occur in area), and offal from whalers and fisheries (Cottam and Hanson 1938; Deichmann 1909; Salomonsen 1950a). 50 stomachs from Davis Strait contained pelagic snails, ctenophores *Beroe*, lenses and beaks of cephalopods (probably via whale faeces), and red oil from crustacean plankton (Hagerup 1926). 32 stomachs from West Spitsbergen in summer contained mainly crustaceans *Thysanoessa inermis* (23), and fewer *Myis oculata*, *Themisto libellula*, *Pseudalibrotus littoralis*, *Sagitta elegans artica* (chaetognath), cephalopod beaks, and fish offal. As *Thysanoessa* swarms disappear in September, then feed chiefly on floating refuse, jellyfish, and other plankton (Hartley 1934; Hartley and Fisher 1936). On Bear Island in 23 stomachs, cephalopods, polychaetes, and fish (Duffey and Sergeant 1950). From Jan Mayen, principally shrimp *Hymenodora glacialis*, fish remains, herring *Clupea harengus*, sand-eel Ammodytidae, and cuttlefish *Sepia*; from Barents Sea, mainly pteropods *Clione limacina* and *Limacina limacina* (Palmer 1962; Voous 1949). Floating refuse, fish offal, oil, and blubber can form substantial part of diet (Dementiev and Gladkov 1951; Witherby et al. 1940; Boswall 1960; Brown 1970; Løvenskiold 1964). For food data for *F. g. rodgersii* see Palmer (1962), Dementiev and Gladkov (1951), Preble and McAtee (1923).

Young fed once or twice a day by regurgitation; at least at beginning, only on liquid of fish offal and plankton (Cramp et al. 1974).

**Social pattern and behaviour.** Compiled largely from material supplied by J B Nelson (some unpublished); see also Fisher (1952) and other references cited.

1. Highly gregarious both when breeding and when feeding at sea. Outside breeding season, though often in ones or twos, frequently in flocks or as dispersed aggregations; nomadic in some areas, concentrations constantly shifting, more or less evenly distributed in others. Often thousands in feeding flocks;

such concentrations due to largely opportunistic feeding habits. Spectacular congregations at whaler and trawler operations, while large numbers also attracted by clearly demarcated feeding zones, often glacier faces. BONDS. Monogamous pair-bond, sustained at least over several seasons; renewed each breeding season after separation in summer and autumn. Once having bred successfully, change of mate or nest-site unusual, though after breeding failure marked birds known to move, as pairs, to new sites (Dunnet *et al.* 1963). Age at first breeding 6–12 years. Both parents tend single young until fledging, after which juvenile wholly independent. BREEDING DISPERSION. Nests in loose or moderately dense colonies varying greatly in size. Pairs not spaced regularly nor usually congregated densely but, where terrain relatively even, often 0·8–1·7 m apart; rarely within contact distance of neighbours and often at greater distances, with some pairs isolated. Little tendency for new pairs to nest near existing ones; young colonies often sparse rather than clustered at optimal density in one small area. Site attachment slow to form, pairs spending part or all of season (possibly longer) establishing site before first breeding. Highly territorial at colony, defending nest-site territory used for courtship, mating, incubation, and care of young; as far as known, no extra sites established for pairing or other activities. Most British colonies attended for at least 10 months of year, birds absent only for part of September and October (M A Macdonald); then sporadically into breeding season. Pairs absent 2–3 weeks prior to egg-laying (so-called 'honeymoon-period') but, as laying extends over *c.*3 weeks, colony never wholly deserted. ROOSTING. Roosts at or near site only when breeding. Otherwise (including at night during most of pre-breeding period) on sea, easily riding out gales and breaking water. Off-duty birds often gather in flocks on water during breeding season and sometimes sleep there.

2. At sea, little positive communication between individuals; those feeding on plankton zones, where dispersed or in dense packs, do not squabble over food but at carcasses, etc., often aggressive to each other and to competing species such as gulls (Laridae), uttering variety of harsh grunting sounds. Though largely diurnal at colony and nesting in exposed sites (unlike other west Palearctic Procellariiformes), behaviour relatively little studied and difficult to interpret; sexes closely similar in plumage and habits. As often nests fairly densely in huge colonies, spends much of year there, faithful to site and mate, evolution of rich repertory of ritualized displays might be expected; in fact, apparent paucity of stereotyped behaviour. In marked contrast to nocturnal Procellariiformes, vocally relatively unspecialized; in particular, lacks conspicuous flight and site-advertising calls. AERIAL ACTIVITY. Spends much time merely flying silently round cliffs in nesting area, frequently soaring using upcurrents (see Pennycuick and Webbe 1959). These often 'prospecting flights', involving investigation from air of sites, rivals, potential mates, and others (both pairs and singles). Bird repeatedly sails up, stalls or hovers, looking in at site (see A), elicits response from occupant, and then glides away. Such investigation at times followed by landing and interaction with occupant. On other occasions, mainly when flapping steadily along in vicinity of colony, gives quiet, repeated Flight-calls (Pennycuick and Webbe 1959); these uttered with open bill, swollen throat, and slight head movement and appear to be low-intensity form of Cackling-call more usually heard from birds sitting at sites (see below) or settled in groups on sea near colony. Also gives quiet, distinctive call on wing when surprised or disturbed by man, e.g. when gliding near cliff-top or while hovering near intruder in vicinity of nest (Pennycuick and Webbe 1959); such vocalizations not yet recorded in intraspecific

A

encounters. BEHAVIOUR AT SITE. On ledges, fighting rare and simple in form: merely grips rival, say by wing, perhaps till both fall from site, still locked together. More often, disputes settled by 'spitting' (see, e.g. Duffey 1951; Armstrong 1951; Pennycuick and Webbe 1959): when intruder approaches, bird at site may make convulsive, retching movements accompanied by squeaky note superimposed on low, growling one; forceful ejection of stomach-oil at opponent may follow but latter usually flees before this necessary, especially if spitting bird makes decisive approach movement first. Spitting behaviour also performed with Head-waving display (see below) when accompanied by quick upward snap of head and usually followed by vigorous champing of mandibles and swallowing. Commonest and most conspicuous display between rivals and mates, vocal cackling with associated Head-waving; this Cackling-display occurs confusingly in many contexts at site (when directed to distant, even overflying bird or one close), and elsewhere. Throat and neck variably, sometimes grossly distended and bill usually opened widely (see B), while head sways from side to side and up and down; head may be stretched far forward or bent right back as displaying bird orientates its Head-waving towards one eliciting it, frequently Head-shaking. Between bouts of Head-waving, swallows conspicuously and tail-wags. Head-waving to partner often followed by pecks, usually directed to head, neck, or breast, evoking slight but quick withdrawal movements. Other behaviour between mated birds at site includes Nibbling, Billing, and copulation. Nibbling unilateral activity of bill aimed at head and neck of partner; still only vaguely described, may best be considered form of allopreening. When Nibbling directed at mate's bill, difficult to know whether food passed. Nibbling can follow or grade into more vigorous interaction, Billing, in which birds jab open bills at each other with quick, irregular head movements. Early in season, pairs frequently jab and evade each other thus, lightly pecking but avoiding reciprocating jabs at same time; in some cases, ♀ initiates Billing more frequently than ♂. Mutual Billing also occurs as simple greeting after mates meet at site after absence. During copulation, ♀ passive and ♂ arches neck over ♀'s head, Head-shaking rapidly (M A Macdonald). Little or no preliminaries and no post-copulatory display. Once incubation started rarely together at site. Nest-relief accompanied by little or no ceremony (Williamson 1965): incomer

B

sometimes Nibbles other before and both sometimes perform brief Cackling-display at change-over itself. RELATIONSHIP WITHIN FAMILY GROUP. Based mainly on observations by Lennox Campbell (per J B Nelson); see also Duffey (1951), Pennycuick and Webbe (1959). Chick attended constantly for first 2 weeks, after which attendance gradually declines until, from start of 5th week, parents simply visit ledge briefly at times, mainly for feeding. When first left alone, chick reacts to sudden approach of any adult by spitting, squatting back on tail and ejecting stream of oil (and sometimes food) while making spasmodic forward movements of head and retching noises. Parents usually land at some distance, approach gradually, cackling frequently, and eventually appease chick; then readily accepted. After 2nd week, chick recognizes own parents, but will still not tolerate intruder. Spitting also directed at other species (e.g. Herring Gull *Larus argentatus* and Kittiwake *Rissa tridactyla*), but later elicited less readily; as up to 50 g of oil may be ejected, reduction of response would seem adaptive once chick large and less vulnerable. During 2nd week (possibly earlier), chick begs actively by lunging upwards at parent's bill; later, chick sways head from side to side, interspersing fencing movements of bill, and utters harsh, repetitive Food-call when begging. Parent typically precedes feeding of young chick by touching bill and head; later, starts to feed mainly on hearing chick's Food-call, making thin, whining sounds itself while retching. Parents also Allopreen chick, particularly on head and neck.

(Figs A–B from photographs by C J Pennycuick and J B Nelson.)

**Voice.** Silent when alone at sea, though feeding flocks often noisy, uttering wide variety of grunts and cackles (Fisher 1952).

Main call ashore on breeding cliffs, given in many different social situations, best described as 'cackling', but also has small repertoire of other distinctive calls (Pennycuick and Webbe 1959). (1) Cackling. Highly variable; numerous descriptions listed by Fisher (1952) encompassing 'every pitch of intensity—from a quiet crooning to an orgasm of vocal excitement—and with every variation that the combination of the consonants c, g, h, k, q, and r can produce'. Performed, mainly at site, by pair during courtship and in territorial aggression against intruders; also by parent when approaching own chick, by single birds apparently without provocation, and sometimes by groups on water near colony (Pennycuick and Webbe 1959). In captivity, given only in company of conspecifics, isolated birds being silent (Kritzler 1948). 2 forms distinguished by Pennycuick and Webbe (1959). In 'slow cackling', braying 'aaark' note given at rate of usually 3 per second, each interspersed with rasping, breathy inspiration ('aawww'), e.g. 'AAARK - aaww - AAARK - aawww - AAARK-. . .' or 'aark-aaww, AARK-aaww' (see I). In less emphatic 'fast cackling', 'cock' note repeated at rate of up to 10 per second, with same periodically interjected inspiration, e.g. 'cock-cock-cock-cock-cock-aawww-cock-cock-cock-cock-. . .' or (see II) 'aaw-kak-kak-kak' (P J Sellar). Both forms normally given in discrete outbursts, typically starting with series of slow 'cock' notes, then developing into crescendo often containing

'aaark' notes, finally dying away; but calling more continuous when birds excited or provoked. Active pair called at rate of 54–60 outbursts of cackling per hour, each lasting 10–20 s. Highly explosive 'queck-queck-queck' version (see III) given as bird arrives at ledge; usually followed by long burst of fast cackling (P J Sellar). (2) Flight-calls. Of at least 2 types. First variable, repeated, soft coo or croon (Fisher 1952) given, usually in steady, flapping flight near cliff, at rate of 1–3 units per second and interspersed with usual rasping 'aawww' inspiration (Pennycuick and Webbe 1959); may be low-intensity form of aerial cackling, but full cackling also evidently performed in flight at times (Fisher 1952). Other flight-call more distinctive: sharp, quiet sneeze given either as apparent alarm-call when bird surprised, e.g. by sudden appearance of man on cliff, or as apparent mobbing-call, e.g. while hovering above man near nest (Pennycuick and Webbe 1959). (3) Spitting-call. Squeaky 'f-chee, f-chee' superimposed on low, growling note; associated with preparations for aggressive ejection of stomach oil, and extremely efficacious in inducing prompt withdrawal by trespassers of own species. (4) Feeding-call. Thin, whining sound like small puppy uttered while parent retches to produce food for chick; follows series of cackles as adult approaches chick.

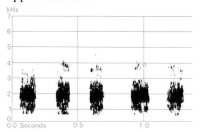

I P J Sellar Scotland July 1970

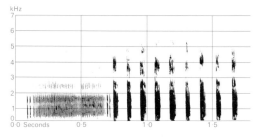

II P J Sellar Scotland July 1970

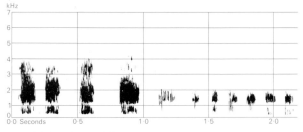

III P J Sellar Scotland June 1974

CALLS OF YOUNG. Small chicks said to produce soft, quacking note which later develops into continuous, monosyllabic buzzing (Fisher 1952). Continuously repeated, shrill note, given during feeding and apparently not developing until between 3rd and 5th week (Duffey 1951), may be same. 2 distinctive calls described by Pennycuick and Webbe (1959): (1) squeaky Spitting-call, given at rate of up to 3 units per second (like adult version, but lacks growling note); (2) low, regular, monotonous, cawing Food-call, given at rate of $1\frac{1}{2}$–3 units per second, each syllable lasting $c. \frac{1}{4}$ s.

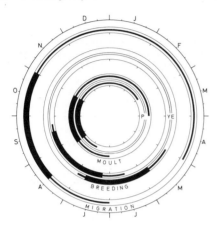

**Breeding.** SEASON. Outside Arctic, laying begins $c.$ 10th May; last eggs hatch mid-July (G M Dunnet and A Anderson). First young hatch end June, most early July. Last fledge mid-September. In Arctic, laying unusual before mid-May with peak early to mid-June. SITE. Cliff ledges, hollows, and recesses in banks, old buildings, grass and earth slopes, sand dunes and, rarely, inland sites. Colonial but rarely dense; on even terrain nests $c.$ 1–2 m apart. Nest: no material on bare rock; shallow depression on soft soil, occasionally few small flat stones as lining, Building: scrape formed by turning on breast, and probably by use of feet. EGGS. Pointed oval, thin-shelled, 0·2 mm (Dunnet *et al.* 1963), rough; white. 74 × 51 mm (68–82 × 43–54), sample 150 (Schönwetter 1967). Weight 101 g (65–122), sample 34 (Mougin 1967). Clutch: 1; 2 sometimes, but probably always produced by 2 ♀♀. Surprising report of up to 15% nests with 2 eggs, Westmann Islands (Lockley 1936). One brood. No replacement layings reported. INCUBATION. 52–53 days (41–57) (Fisher 1952). 49 days (47–51), sample 56 (Mougin 1967). By both sexes with ♀ taking first spell, usually $c.$ 24 hours, rarely up to 3 days; then ♂ has spell of $c.$ 7 days (range 1–11), followed by alternate, roughly equal spells averaging 3·5 days, until last 7–10 days of incubation, when average only 1 day (Williamson 1952; Dunnet *et al.* 1963; Mougin 1967). Eggshells probably not deliberately removed. YOUNG. Semi-altricial and nidicolous. Cared for and fed by both parents. One parent nearly always present at nest for first 2 weeks or so of chick's life, defending nest

and young against intruders (avian or mammal) by spitting oil. Young can also spit oil and regurgitate food as defence. Young fed by placing head inside parent's bill and parent regurgitating. Feeding rate declines up to 14 days before fledging. FLEDGING TO MATURITY. Fledging period 46 days (range 41–57, most 46–51) (Fisher 1952). Independent at or just before fledging. Age of first breeding 6–12 years, mean 9·2, sample 19 (G M Dunnet and A Anderson). BREEDING SUCCESS. $c.$ 45% of eggs lost, most in first 3 days after laying, and around hatching (Mougin 1967; Dunnet *et al.* 1963). Most of 14·5% loss of young also due to predation by large gulls *Larus* and crows *Corvus*, especially during first 2 weeks after hatching (G M Dunnet and A Anderson).

**Plumages.** ADULT. Ranges from white with grey back, tail, and wings to wholly dark grey, slightly lighter on underparts. Lightest birds have head, neck, upper back, and underparts white, often tinged faintly yellow; blackish spot in front of eye. Mantle and upper wing-coverts blue-grey, tinged brown when worn; rump and upper tail-coverts lighter. Tail light grey, narrowly tipped white, paler below. Primaries and primary-coverts dark grey, shading to white at base of inner web, paler below; shafts off-white, brownish-grey at tip; secondaries similar, but lighter. Under wing-coverts white, grey at distal part of leading edge; axillaries white. Darkest birds have crown, hind-neck, mantle, and upper wing grey, darker than mantle of light birds; sides of head and underparts lighter grey; tail without white tip; under wing-coverts and axillaries grey. Fisher (1952) divided whole range of plumages into 4 morphs: 'double light'—head and underparts wholly white; 'light'—crown and hindneck light grey, belly white or light grey, breast white; 'dark'—crown grey, not as dark as back, breast light grey; 'double dark'—almost uniformly dark blue-grey, wings as dark as their tips. NESTLING. Shows variability corresponding to adult (skins ZFMK). 1st down of lightest birds white with light grey mantle and small grey vent patch. 1st down of darkest grey with dirty white chin and lores; crown light grey, rest of upperparts darker, underparts lighter. Others intermediate. 2nd down slightly darker than first. Wing quills appear $c.$ 12 days after hatching, tail quills few days later, body quills $c.$ 20 days. Webs of flight-feathers protrude from sheaths $c.$ 18 days. Down shed from neck at 25 days, from head at 29, from most of body except vent at 40. Fully feathered at 50 days (Mougin 1967). JUVENILE. Like adult, mostly without yellow tinge on head, but not certainly identifiable.

**Bare parts.** ADULT and JUVENILE. Iris dark brown. Bill usually yellow at tip; rest variable, olive-green to blue-grey; tubes darker, yellow-brown to grey-black, sometimes mottled blackish. Leg varies from yellow or greenish to livid-flesh. NESTLING. Bill and leg grey at hatching, changing to black in 1st week.

**Moults.** Variation in timing insufficiently studied. ADULT POST-BREEDING. Complete. Primary moult normally starts after departure from breeding colony; July or earlier after nesting failure, late August after successful breeding, though Carrick and Dunnet (1954) recorded wing moult beginning mid-August while still at site with nestling. Primaries lost and replaced in rapid succession, but in final stage outer ones grow slowly, not reaching full length before late February. Body moult begins before wings and continues until at least mid-February, with feathers of 2 successive generations present for most of year. Tail renewed simultaneously with outer primaries (Stresemann and Stresemann 1966). POST-JUVENILE. Early, May–September, probably becoming progressively later in subsequent years.

**Measurements.** Full grown; skins (BMNH, ZFMK, ZMA). Scottish data: Dunnet and Anderson (1961).

Spitsbergen

| | | | | | | | |
|---|---|---|---|---|---|---|---|
| WING | ♂ | 339 | (8·30; 11) | 325–353 | ♀323 | (5·13; 20) | 312–332 |
| TAIL | | 125 | (4·37; 10) | 118–131 | 122 | (2·47; 13) | 118–128 |
| BILL | | 37·5 | (1·26; 17) | 36–40 | 35·7 | (1·45; 22) | 33–38 |
| TARSUS | | 53·7 | (1·42; 10) | 51–55 | 50·2 | (0·89; 14) | 49–52 |
| TOE | | 67·9 | (2·79; 7) | 65–73 | 63·0 | (1·65; 12) | 60–65 |

Bear Island

| | | | | | | | |
|---|---|---|---|---|---|---|---|
| WING | ♂ | 339 | (6·64; 25) | 329–356 | ♀328 | (5·94; 7) | 319–337 |
| BILL | | 38·8 | (1·49; 27) | 36–41 | 37·1 | (1·01; 8) | 35–38 |

Jan Mayen

| | | | | | | | |
|---|---|---|---|---|---|---|---|
| BILL | ♂ | 39·2 | (1·61; 6) | 37–42 | ♀ 36·0 | (1·73; 10) | 33–39 |

Scotland

| | | | | | | | |
|---|---|---|---|---|---|---|---|
| WING | ♂ | 340 | (9·46; 18) | 324–355 | ♀ 323 | (8·13; 17) | 309–336 |
| BILL | | 40·9 | (1·34; 23) | 38–44 | 37·7 | (0·97; 26) | 36–39 |

Sex differences significant. Additional data on measurements in Salomonsen (1950b), Wynne-Edwards (1952b), and Mathiasson (1963c). Young about to leave breeding colony have shorter wings and tails than adults: mean Scotland 290 and 110 (Mougin 1967).

**Weights.** ZMA, Dunnet and Anderson (1961).

Spitsbergen, summer

| | | | | | |
|---|---|---|---|---|---|
| | ♂ 795 | (57·2; 8) | 730–895 | ♀ 628 | (60·3; 12) | 535–757 |

Scotland, museum

| | | | | | |
|---|---|---|---|---|---|
| | 835 | (81·2; 13) | 689–957 | 700 | (29·2; 10) | 661–760 |

Scotland, live breeding

| | | | | | |
|---|---|---|---|---|---|
| | 884 | (80·8; 17) | 760–1000 | 706 | (57·7; 18) | 610–855 |

Sex difference significant. Summer birds Scotland significantly heavier than those from Spitsbergen. In emaciated condition, ♂ not heavier than ♀; minimum of those washed ashore Netherlands in winter ♂ 438, ♀ 436. Nestlings Scotland at maximum weight, mean 1120 (970–1440) $1\frac{1}{2}$–$3\frac{1}{2}$ weeks before fledging, decreasing to mean 860 (650–1230) at fledging, still heavier than adults (Mougin 1967).

**Structure.** 11 primaries: p10 longest, p9 2–15 shorter, p8 14–33, p7 34–55, p6 55–79, p1 164–210. 20 secondaries. Tail rounded, 14 feathers: t7 12–20 shorter than t1. Bill heavy; nasal tubes $\frac{1}{3}$–$\frac{2}{5}$ of its length, wide, nostrils directed forwards. Tarsus stout, somewhat laterally compressed.

**Geographical variation.** Pacific *F. g. rodgersii* has more slender bill; lightest colour morph lighter, darkest darker than in nominate *glacialis*. Geographical variation in Atlantic extensively studied by Salomonsen (1965). Both proportion of dark birds in population and size of bill vary, but variation not correlated. In many high-arctic localities (Baffin Island, north-east Greenland, Spitsbergen), dark and intermediate morphs (Fisher's 'double dark', 'dark', and 'light') form great majority of population. All populations of southern, boreal part of Atlantic range overwhelmingly 'double light'. In west Greenland and at Jan Mayen, 99% of breeding birds 'double light' or 'light' (but 18 Jan Mayen, BMNH, 39% 'light', 55% 'dark', 6% 'double dark'). On Bear Island, 'dark' birds slightly less predominant than in Spitsbergen, 'double light' apparently amounting to 10% of population. Variation in bill length of breeding birds insufficiently known. Small bills found in Baffin and Devon Islands (♂♂, average 36·2). Data for Greenland scattered and conflicting; apparently various populations differ considerably, but generally bills slightly longer than in north Canada. At Spitsbergen, bill longer (♂♂, average 37·5) than in Baffin Island. Again, larger bills found at Bear Island (38·8), Jan Mayen (39·2), and Iceland, Faeroes, and British Isles (combined average of 34 ♂♂ 40·8): see Wynne-Edwards (1952b). Salomonsen divided Atlantic Fulmar into 2 subspecies: *glacialis* comprising high-arctic, dark, short-billed populations and *auduboni* Bonaparte, 1857, comprising all light populations (mainly boreal). Division between light and dark morphs not rigid, however, as grade imperceptibly into one another. Variation in bill appears less closely correlated with water temperature than Salomonsen supposed (Spitsbergen intermediate between Baffin Island and west Europe) and shows considerable overlap among populations. Moreover range of mainly dark birds split by light populations in west Greenland. Differences described by Salomonsen represent general trends, but not sufficient for acceptance of subspecies.

## *Daption capense* (Linnaeus, 1758)  **Cape Pigeon**

FR. Pétrel damier    GE. Kapsturmvogel

Abundant in Southern Ocean, breeding many antarctic and sub-antarctic localities between latitudes $68\frac{1}{2}$°S (Vestfold Hills, Wilkes Land) and $46\frac{1}{2}$°S (Crozet Islands). Strongly migratory; ranges north into sub-tropical seas, exceptionally to equator, and circumpolar movements indicated by recoveries of South Orkney birds in New Hebrides, New Zealand, and Chile. Has occurred Northern Hemisphere in west Mexico, California, and south India. Several reports from European seas of which 5 possibly genuine. Single specimens shot Mediterranean France, October 1844; Wales, 1879; Ireland, October 1881; and skull found on Netherlands beach, 1930 (Bourné 1967a). Also immature captured off Sciacca (Sicily) in September 1964 (Massa 1974). None currently accepted in national lists due to possibility of importation by sailors, but seems as likely to occur naturally as albatrosses *Diomedea exulans* and *D. melanophris*.

## *Pterodroma neglecta* (Schlegel, 1863) **Kermadec Petrel**

FR. Pétrel de Kermadec    GE. Kermadek-Sturmvogel

## *Pterodroma leucoptera* (Gould, 1844) **Collared Petrel**

FR. Pétrel à ailes blanches    GE. Brustbandsturmvogel

One *P. neglecta* reported found dead inland Cheshire, England, April 1908 (Newstead and Coward 1908), and one *P. leucoptera* said shot on Cardigan coast, Wales, November or December 1889 (Salvin 1891). Neither record now accepted in view of improbability of these tropical and sub-tropical Pacific Ocean petrels occurring naturally in North Atlantic. Sight record of *P.* *neglecta* from east USA October 1959, after hurricane, seems more likely to have been very similar Trinidade Petrel *P. arminjoniana*, which breeds in South Atlantic (Bourne 1967a); indeed, Iredale (1914) had earlier expressed belief that Cheshire specimen was *P. arminjoniana*.

## *Pterodroma hasitata* (Kuhl, 1820) **Black-capped Petrel**

FR Pétrel diablotin    GE. Teufelssturmvogel

Rather few modern sightings of *P. h. hasitata*, and breeding sites on inland cliffs of Hispaniola but recently discovered (Wingate 1964). Only recent pelagic records of numbers from Caribbean towards end of winter breeding season (Mörzer Bruyns 1967). Away from colonies, appears to disperse at sea, often far from land, in Caribbean and to north in west Atlantic. Occurs as vagrant to USA in late summer (including several after hurricanes), at least 2, taken August, being in moult, while 2 seen 38°N 69°30'W on 16 August 1966; hence possibly shifts north in sub-tropics for summer moult (Bourne 1967a, 1970; Palmer 1962). One found inland Norfolk, England, March or April 1850 (Newton 1852), only accepted Palearctic record (*P. h. hasitata*).

## *Pterodroma mollis* **Soft-plumaged Petrel**

PLATES 15 and 16
[after page 158]

DU. Atlantische Grijze Stormvogel    FR. Pétrel de Madère    GE. Weichfedersturmvogel
RU. Серохвостый тайфунник    SP. Petrel de Madeira    SW. Sydlig stormfågel

*Procellaria mollis* Gould, 1844

Polytypic. *P. m. feae* (Salvadori 1899), Bugio Island, Desertas, and Cape Verde Islands; *madeira* Mathews, 1934, mainland of Madeira. Extralimital: nominate *mollis* (Gould 1844), South Atlantic and Indian Oceans.

**Field characters.** 32–37 cm, wingspan 83–95 cm; about size of Manx Shearwater *Puffinus puffinus*. Largely grey above, and white below except for grey underwing and grey band across breast. Sexes alike and no seasonal differences; juvenile inseparable.

Entirely slate-grey above, but for scaly-looking forehead, browner head, broad but inconspicuous blackish band across wing-coverts, and paler tail. White cheeks, underbody and undertail, except for more or less well-defined grey band across upper breast; dark grey underwing with darker leading edge and inconspicuous narrow white strip in centre. Bill black; legs and feet pink, becoming blackish-brown on toes and distal parts of webs. Rare dark morph entirely blue-grey above and paler grey below, but known only from South Atlantic and Indian Oceans.

Normal light morph distinguished from other petrels and shearwaters by combination of dark (but not black) upperparts, dark underwing, and grey breast-band, but dark morph more difficult and virtually indistinguishable in field from Kerguelen Petrel *P. brevirostris*. Like other gadfly-petrels, flies fast with erratic careening action, towering into sky. Settles on sea. Seen singly or in parties or flocks. Often occurs in coastal waters; sometimes follows ships. (See Alexander 1955; Elliott 1957; Serventy *et al.* 1971.)

**Habitat.** Restricted within south-west oceanic sector of west Palearctic to few small islands used for breeding and to lowest airspace over neighbouring marine waters. Makes nesting burrows where plateau areas above cliffs provide suitable soft soil overgrown with herbage, flying

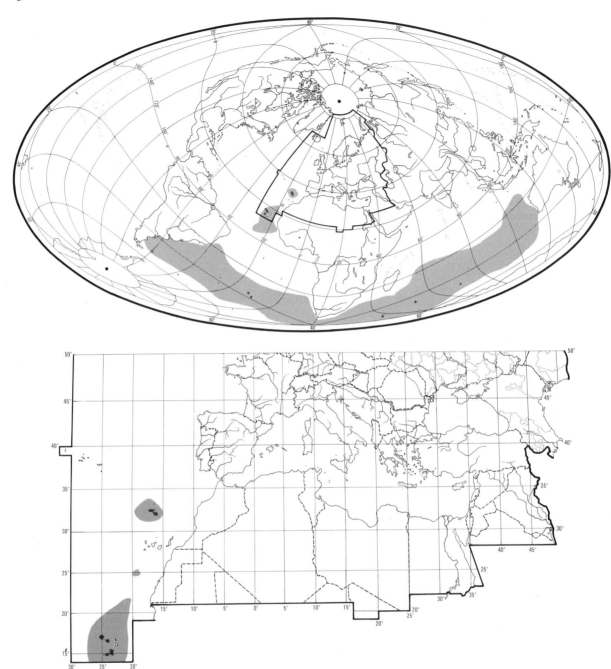

directly to them (Gruanin *et al.* 1969); sometimes also at 500–1000 m on precipices (Bannerman and Bannerman 1965, 1968) and to 1500 m at foot of cliffs on Madeira (R de Naurois).

**Distribution.** MADEIRA: *P. m. madeira* breeds only main island, on inland cliffs (Bannerman and Bannerman 1965); breeding June 1969, Curral das Freiras (R de Naurois). PORTO SANTO group: *P. m. feae* no longer breeding (Jouanin *et al.* 1969). Marine range little known (see Movements).

Accidental. Israel: one found dead west shore of Dead Sea, February 1963 (HM).

**Population.** MADEIRA: 2 small colonies, perhaps 50 pairs, 1969 (R de Naurois). DESERTAS: restricted to Bugio, numbers small (45–50 sites found 1967–8, but probably more), and future precarious (Jouanin *et al.* 1969). CAPE VERDE ISLANDS: population does not exceed some hundreds of pairs; considerable human predation, as fat used medicinally (de Naurois 1969*b*).

**Movements.** Little evidence of movement, but one of least known seabirds in North Atlantic. With staggered breeding seasons, present around sub-tropical North Atlantic islands during much of year. Extent of marine range unknown. Type specimen of *feae* collected October at 06°50′N 23°46′W (Bourne 1957); still most southern record of species in North Atlantic. Murphy (1924) reported present September 1912 north-west of Cape Verdes to 23°17′N 28°19′W and south to 9°N, though later (Murphy 1967) stated species not yet encountered away from vicinities of North Atlantic colonies. Only 3 recent pelagic reports: 2 at 25°N 19°W (south-west of Canary Islands) September 1951 (Bourne 1955*b*), and singles at 12°N 17½°W (off Guinea) April 1963 (Bourne 1965), and 19°N 17°W (off Mauritania) September 1966 (Bourne and Dixon 1973); yet old report of large numbers at 11°10′N (off Guinea) on 27 May (Bannerman 1914). More numerous southern hemisphere, where other breeding populations. Widely reported in west-wind zone of South Atlantic and Indian Oceans, though some remain around Tristan da Cunha group colonies for much of year, apparently rare June–July only (Elliott 1957).

**Food.** Little specific information on diet or feeding behaviour; only squid and small fish recorded, species unidentified (Bannerman 1968; Hagen 1952).

**Social pattern and behaviour.** Compiled mainly from material supplied by Frank B Gill.
1. Gregarious, in small flocks at sea at all seasons; also when breeding. BONDS. Nothing definitely known. As in other petrels, pair-bond presumably monogamous with both parents tending young until about time of fledging. BREEDING DISPERSION. In small colonies, but inventory of occupancy and density difficult as burrows inaccessible. ROOSTING. Nothing recorded.
2. Burrows occupied only on seasonal basis. Ashore very shy and strictly nocturnal, any disturbance at night causing birds to fly to sea (Jouanin *et al.* 1969). Calling in flight begins after sunset with peak then and again before sunrise; usually silent on moonlit nights (Jouanin and Roux 1966). Aerial flights and chases, with extended calls observed at Gough and Antipodes Islands (Serventy *et al.* 1971). Nothing known of behaviour in burrow.

**Voice.** Silent at sea. At breeding places, adult's nocturnal Flight-call variously described. In *P. m. feae*, as 'pleasant, high pitched tittering call' (Bourne 1955*b*, Cape Verde Islands), but also as long wail terminated by hiccup, latter inaudible unless close (Jouanin *et al.* 1969, Desertas). In nominate *mollis*, recording of Flight-call on Gough Island (see 1) agrees with description of Jouanin *et al.* (1969) for *feae*, but in Tristan da Cunha group described as shrill whistle 'tree-pi-pee' (Elliott 1957) and melancholy, flute-like 'uuuuuuu-hi' with last syllable rising noticeably in pitch (Hagen 1952). In *feae*, continuous, cackling 'gon-gon' call reported from birds on breeding ledges (Bourne 1955*b*), 'Gon-Gon' being local name for *P. mollis*. In nominate *mollis*, birds in burrows on Inaccessible gave calls similar to those heard in flight (Elliott 1957).

CALLS OF YOUNG. Nestlings of nominate *mollis* utter piping 'pee-chee', repeated several times (M K Swales, recording).

**Breeding.** SEASON. Desertas: main laying period July–August, with young in nest to at least December (Jouanin *et al.* 1969). Madeira: May (R de Naurois). Cape Verde Islands: main laying period December–January; one record of egg laid February (de Naurois 1969*b*). SITE. In boulder screes, on flat or steeply sloping ground. Colonial, with nests 0·3–2 m apart (Bannerman and Bannerman 1965). Nest: burrow in ground or between rocks, or natural crevices between rocks. Building: no details. EGGS. Elliptical ovate; white. 44 × 58 mm (57–60 × 42–45), sample 4; weight of 2 eggs, 59 and 60 g (Jouanin *et al.* 1969). Clutch: 1. One brood. No information on replacements. INCUBATION. Period not recorded. YOUNG. Semi-altricial and nidicolous. FLEDGING TO MATURITY. Fledging period not recorded. Age of maturity unknown. BREEDING SUCCESS. No data.

**Plumages** (subspecies *feae*). ADULT AND JUVENILE. Crown dark grey, feathers of forehead edged white, hindneck lighter than crown. Rest of upperparts grey, tinged dark brown on scapulars and upper wing-coverts; upper tail-coverts much lighter than rump. Sides of head mottled grey and white, black spot in front of, below, and behind eye. Underside white; sides of chest pale grey, sometimes forming interrupted band across; flanks vermiculated grey. Tail grey, paler below, outer feathers with inner webs mottled grey and white. Primaries black on outer web, dark grey on inner; secondaries paler. Under wing-coverts grey, white bases shining through; axillaries mottled grey and white. NESTLING. First down apparently undescribed. Second down grey, bare patch on throat. Second down of nominate *mollis* nearly uniform dark leaden grey with brownish hue; little brighter on ventral surface; half naked patch round base of bill (Hagen 1952); other races unknown.

**Bare parts.** ADULT AND JUVENILE. Iris dark brown; bill black; tarsus, inner toe, and base of webs flesh pink, remainder of foot black.

**Moults.** Unknown for North Atlantic races. Nominate *mollis*, breeding November–April, found moulting winter and spring (July, August, and November); various age classes in population may moult asynchronously.

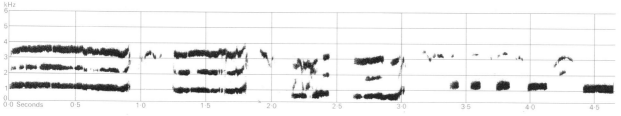

I M K Swales/BBC Gough Island 1955/6

**Measurements.** Full grown; skins (BMNH, MNHN, RMNH, and ZMA).

| *feae* ♂, ♀ and unsexed | | | *mollis* ♂, ♀ and unsexed | |
|---|---|---|---|---|
| WING | 268·4 (3·43; 18) | 263–273 | 254·1 (6·70; 15) | 239–263 |
| TAIL | 113·3 (2·66; 6) | 108–115 | 113·4 (3·25; 15) | 108–120 |
| BILL | 28·4 (0·88; 19) | 26–30 | 27·6 (0·96; 24) | 26–31 |
| TARSUS | 35·0 (1·59; 20) | 32–38 | 35·2 (0·86; 16) | 34–37 |
| TOE | 45·0 (1·52; 20) | 41–47 | 44·8 (1·21; 15) | 43–46 |

| *madeira* | ♂ | ♀ | mean |
|---|---|---|---|
| WING | 247, 248 | 251, 254 | 250 |
| TAIL | 104, 106 | 105, 110 | 106 |
| BILL | 24·5, 25·6 | 25·7, 26·5 | 25·6 |
| TARSUS | 30, 31 | 31, 31 | 31 |
| TOE | 39, 41 | 40, 41 | 40 |

Sexes not significantly different, so measurements combined. Wing and bill of nominate *mollis* significantly shorter than *feae*. *P. m. madeira* smaller in all measurements.

**Weights.** Adults nesting Bugio: 295, 330, 335, 340 and 355 (Jouanin *et al.* 1969). Adult (nominate *mollis*) Inaccessible, Tristan da Cunha: 257. Nearly full-grown chicks, Tristan da Cunha: 298 and 370 (Hagen 1952). Average of 146 adults Gough: 254 (202–342) (Swales 1965b).

**Structure** (subspecies *feae*). Wing long and narrow. 11 primaries: p10 longest, p9 0–5 shorter, p8 10–17 shorter, p7 27–35, p6 43–57, p1 157–163. P11 minute. Tail wedge-shaped; 12 feathers, t6 25–35 shorter than t1. Bill stout, nasal tubes $\frac{1}{4}$–$\frac{1}{3}$ of length, nostrils directed forwards. Tarsus slender, round. In *P. m. madeira*, p1 141–150 shorter than p10, bill much more slender. Bill in nominate *mollis* intermediate. Distal part of wing less pointed in nominate *mollis* than in North Atlantic races, p8 7–12 shorter than p10, p7 22–26, p6 37–46.

**Geographical variation.** *P. m. feae* largest race; nominate *mollis* shorter wing and slightly shorter bill; *madeira* small in all dimensions and distinguished by short, slender bill and short tarsus and toes (see Measurements). All 3 races very similar in plumage. Nominate *mollis* usually has well-developed grey band across chest, under wing-coverts with white or pale grey edges, but white bases concealed; in addition, some very dark with underparts wholly slate grey, while others intermediate, with grey wash encroaching upon belly from flanks. *P. m. madeira* more heavily mottled on flanks than *feae*. (See also Bourne 1957.)

**Recognition.** Differs from all other *Pterodroma* in combination of white underside with grey patches on side of chest, dusky underwing, and black eyespot (Alexander 1955). Rare dark morph very similar to *P. brevirostris*, but latter has narrower primaries with pale centres and shafts (Bourne 1957; but see Bourne 1966b).

## *Bulweria bulwerii* Bulwer's Petrel

PLATES 15 and 16
[after page 158]

Du. Bulwers Stormvogel     Fr. Pétrel de Bulwer     Ge. Bulwersturmvogel
Ru. Длиннохвостый тайфунник     Sp. Petrel de Bulwer     Sw. Spetssjärtad stormfågel

*Procellaria Bulwerii* Jardine and Selby, 1828

Monotypic

**Field characters.** 26–28 cm, wingspan 68–73 cm; intermediate in size between storm-petrels (Hydrobatidae) and other gadfly-petrels and small shearwaters (Procellariidae). All-dark gadfly-petrel (no white rump) and so more like dark-rumped storm-petrel than *Pterodroma* or *Puffinus*, with rather long, wedge-shaped tail giving 'tail-heavy' look. Sexes alike and no seasonal differences; juvenile inseparable.

Entirely sooty-brown above except for grey-brown greater coverts forming pale band across wing. Dull sooty-brown below, paler and greyer on cheeks and upper throat. Bill deep and black; rather short legs pale flesh, becoming dusky on ends and outsides of feet.

Most likely to be confused in west Palearctic with Sooty Shearwater *Puffinus griseus* or Swinhoe's Storm-petrel *Oceanodroma monorhis* (or dark-rumped race of Leach's Storm-petrel *O. leucorhoa*), but distinguished by flight, by size where direct comparison possible, and by length and shape of tail and bill. Bill much shorter than in *Puffinus*, but deeper than in Hydrobatidae, giving stubby appearance. Jouanin's Petrel *B. fallax* considerably larger; various other dark gadfly-petrels and shearwaters need to be distinguished in southern hemisphere.

Wings long and slender and, like other gadfly-petrels, flies fast with swooping, careening action, towering on the upswing, quite different from the fluttering action of storm-petrels. Stated rarely to flock, and generally feeds singly well out to sea. See Alexander (1955); Bourne (1960); and Sharrock (1968).

**Habitat.** Within warm oceanic sector in west Palearctic. Breeds often near sea-level in small holes and crevices under large boulders or talus, but also at higher elevations. Apparently not in burrows, but sometimes in holes in walls. Intense competition often restricts it to openings too small for Cory's Shearwater *Calonectris diomedea* (Lockley 1952). Occasionally sits on open ground or on roads. Nocturnal on land or near colonies (Jespersen 1930; Bannerman 1959); generally occurs singly, flying low over surface. Swimming and feeding habits still uncertain.

**Distribution.** In North Atlantic, most favoured breeding sites Selvagens, Porto Santo group, and Desertas; also nests Canary Islands (Tenerife and off Lanzarote) and Cape Verde Islands (Cima and Razo), and once Santa Maria, most southerly of Azores (Bannerman 1963; Bannerman and Bannerman 1965, 1966, 1968).

Accidental. Britain: May 1837, February 1908. Ireland:

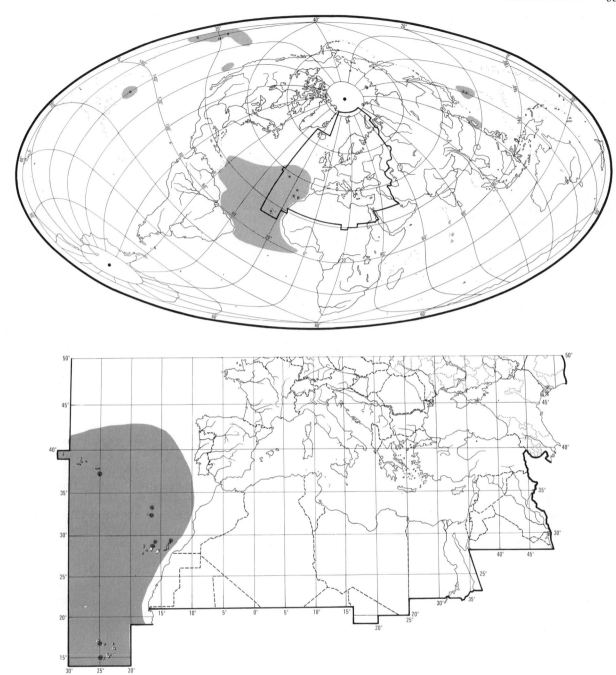

August 1965. France/Italy: on lightship between Corsica and Genoa 1898. (Bourne 1967*a*.)

**Population.** CAPE VERDE ISLANDS: probably tens of pairs (de Naurois 1969*b*). CANARY ISLANDS: *c*.100 pairs breed Montana Clara, off Lanzarote (Lovegrove 1971).

**Movements.** Migratory. Absent from vicinities of North Atlantic colonies in winter, when apparently majority have moved south-west (but some south) into tropical Atlantic.

Timing of movements uncertain, though loose flocks noted 8°N along 29°W to 44°W between 7–10 October (Bourne and Dixon 1973). Pelagic distribution not well known; scattered observations November–March mainly 10°N to 7°S and 20°W to 50°W, chiefly in mid-ocean, though the few South Atlantic sightings concentrated off north-east Brazil in December–January (van Oordt and Kruijt 1953; Bourne and Dixon 1973). Evidently a few penetrate west to Sargasso Sea and off Lesser Antilles, since singles seen

18°51′N 54°12′W in December 1951 (van Oordt and Kruijt 1953), 10°N 60°W in October 1966, and 19½°N 63°W in August 1968 (Bourne and Dixon 1973), and one found dead Trinidad, January 1961 (ffrench 1963). Surprisingly few records off bulge of West Africa, though presumed migrants noted offshore Gambia to Liberia April, October, and December (Mayo 1948; van Oordt and Kruijt 1953), and once (April) inner Gulf of Guinea at 1°36′N 3°25′E (Wallace 1973).

Singles seen 5°55′N 35°50′W in May 1961 and 9½°N 54°W in July 1967 (Bourne 1962; Bourne and Dixon 1973) indicate some, presumed immatures, summer in wintering area. One, straggler on present knowledge, noted 21°S 10°W January 1952 (van Oordt and Kruijt 1953). Summer range extends to north of Azores, since several May–June records 37°N to 42°N and 11½°W to 24½°W, and seen at 39½°N off Portugal in September (Bourne 1967b). See also *Sea Swallow* 1962–70.

**Food.** Limited data available. Mainly plankton such as fish eggs, ctenophores, and polychaete worms (Bent 1922; Watson 1966). Almost entirely nocturnal feeder.

**Social pattern and behaviour.** Compiled mainly from material supplied by F B Gill and D T Holyoak, with Pacific Ocean Biological Survey Program data from Sand Island, central Pacific (A B Amerson Jr, P Shelton, A Wetmore).

1. Gregarious when breeding but at sea usually occurs singly at all seasons, though loosely associating groups occasionally reported (van Oordt and Kruijt 1953; Bourne and Dixon 1973). BONDS. Strong, monogamous pair-bond indicated ashore, renewed each season after absence at sea. At Sand Island, though a few birds may have changed mates or taken new ones when former partner lost, probably most of 96 birds ringed bred together each year in 1964–69 (at least 2 pairs proved to do so). Initial pair-formation (and site-selection) presumably mainly in 4th year: at Sand Island, birds ringed as nestlings first returned to colony, in more or less same area where hatched, in 3rd (1 record) and 4th years (4 records), 2 of latter developing brood-patches and courting though not proved to breed. Both parents tend chick until several days before fledging. BREEDING DISPERSION. Colonial, in groups varying from a few pairs to many thousands. Density depends largely on distribution of natural cavities; latter often close together with common entrance to many, though nest-chambers separate. Nothing known about territorial behaviour, but calls persistently from burrow. ROOSTING. Little known; apparently feeds chiefly at night, and commonly seen sitting on water during day.

2. Highly pelagic, but ashore for *c*.5 months each year to breed (or, in case of pre-breeders, to establish pair-bond and nest-site). Strictly nocturnal at colony, coming in from sea at dusk or after dark and departing shortly before dawn; when caught and released during day, searches for dark place and does not usually fly out to sea (Bannerman 1914; Lockley 1952). Starts frequenting colony *c*.6 weeks before laying (pre-breeders at least 2 months later). Highly localized circling flight, often brief and apparently mainly silent, at dusk after arrival from sea (Lockley 1952; Fisher and Lockley 1954). Barking-call also given on wing. Otherwise breeders most secretive and seldom located except in nest-hole where presence indicated by frequently uttered Barking-calls; at Sand Island, easily made to leave hole by imitation of same vocalization. Antagonistic, courtship, and other

behaviour virtually unknown. Pair together in hole, Selvagens, uttered quickened and continuous version of Barking-call (Lockley 1952). On Sand Island, adults noted resting on breast a little apart and uttering Barking-call, with extended heads and swelling throats; when quiet, lay prostrate on breast or shuffled along while barely raising body free of ground for brief movement at a time. After laying, only single adult found in burrow at any one time. Incubating birds give Barking-call constantly during night, occasionally during day (Lockley 1952). While breeders with eggs or young, non-breeders at Sand Island (pairs and singles) continue to occupy sites and to call. At Laysan Island, 2 groups of at least 1000 each found sitting around on ground at night in June; such 'clubs' not recorded elsewhere (Ely and Clapp 1973), though solitary birds reported calling from open ground at Sand Island late in season. RELATIONS WITHIN FAMILY GROUP. Little known. At Sand Island, chick guarded only for first few days; thereafter most unattended, even at night, visiting adult staying only long enough to feed it. Chick remains in burrow by day, but at Sand Island sometimes seen in open at night; utters characteristic call (A Wetmore).

**Voice.** Silent at sea. Usually noisy during nocturnal visits to breeding colony.

(1) Barking-call. Low, barking 'chuff', often repeated (*c*. 4 per s), resembling steam engine pulling away (see I). Uttered in flight, on ground, and from nest-hole. Also described as quiet but deep, frog-like 'whok' or 'whow', rather like faint barking of hound (Lockley 1942, 1952), and comical little barking call, common to both sexes (A Wetmore). Fast, excited version, producing stream of 'whok' notes, given non-stop for over 1 min by pair in nest-hole; not unlike deep call of the Great Black-backed Gull *Larus marinus* during coition, but much fainter (Lockley 1952). (2) Song. Loud and cheerful, 5-note call, consisting of 4 higher notes and a lower, more prolonged note, the whole repeated several (usually 3) times; heard frequently at night on Selvagens (Ogilvie-Grant 1896), but not reported by Lockley (1952) who thought Barking-call only vocalization. Further study needed. (3) Alarm-call. Dovelike, penetrating but low moan, resembling 'who who'; uttered when disturbed (see Bent 1922); could be Barking-call or variant.

CALL OF YOUNG. Low, whistling wheeze, somewhat like note of young pigeon *Columba*, reported from Sand Island, central Pacific (A Wetmore).

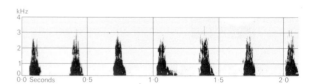

I W and J Ward Hawaiian Islands August 1961

PLATE 13. *(facing) Diomedea exulans* Wandering Albatross (p. 116): **1a, 1b** ad ♂, **2** ♀ breeding (not fully ad), **3** older imm, **4** imm with last vestige of juvenile body plumage, **5** juv. *Diomedea melanophris* Black-browed Albatross (p. 113): **6a, 6b** ad, **7** imm, **8** juv. (PJH)

**Breeding.** SEASON. Madeira, Desertas, Selvagens, Canary Islands: eggs end May and June, first young early July, most fledging by end October (Bannerman 1959, 1963; Lockley 1952). Cape Verde Islands: most incubating end January to end March, but eggs also found May (de Naurois 1969b). SITE. In rock scree, cliffs, or stone walls. Usually colonial; nests from 1 to several metres apart, but solitary nests also found (de Naurois 1969b). Nest: natural hole in rock or scree, up to 1 m deep. Building: no excavation. EGGS. Ovate to elliptical, not glossy; pure white. 42 × 31 mm (40–46 × 29–33), sample 80; calculated weight 21 g (Schönwetter 1967). Clutch: 1. One brood. No data on replacement eggs. INCUBATION. 44 and 47 days recorded on Johnston Atoll, Pacific Ocean (P Shelton). By both sexes, in spells of 1–5 days (Lockley 1952). YOUNG. Semi-atricial and nidicolous. Fed by parents at night. Abandoned several days before fledging. FLEDGING TO MATURITY. Fledging period on Johnston Atoll 62 days (57–67), sample 8 (P Shelton). Become independent some days before fledging. Age of maturity not certainly known, but chicks hatched 1965 on Johnston Atoll present 1969 as adults with brood patches. BREEDING SUCCEESS. No data from west Palearctic. Of 71 eggs on Johnston Atoll, 67% hatched, losses due mostly to high tides; 85% of hatched young fledged (P Shelton).

**Plumages.** ADULT AND JUVENILE. Sooty brown, glossier above, duller below; crown slightly darker than mantle, chin grey. Tail-feathers sooty-brown; primaries and secondaries sooty-brown, darker on outer webs, pale near base of inner webs; greater and median upper wing-coverts pale brown-grey. NESTLING. 1st down long and dense, sooty, slightly paler below; sparse on lores, round eyes, and on chin. 2nd down apparently similar.

**Bare parts.** ADULT AND JUVENILE. Iris dark brown. Bill black. Tarsus, inner joints of inner toes, and base of inner web pale flesh; outside of tarsus sometimes darker, rest of toes and webs dusky.

**Moults.** ADULT POST-BREEDING. No precise data. Apparently during winter, as May specimens (Cape Verde Islands) in fresh plumage (Murphy 1924).

**Measurements.** ADULT AND JUVENILE. North Atlantic. Skins (BMNH, MNHN, ZFMK).

| | | | | | | | |
|---|---|---|---|---|---|---|---|
| WING | ♂ 200 | (3·90; 29) | 191–207 | ♀ 199 | (4·07; 24) | 193–209 |
| TAIL | 108 | (3·10; 29) | 103–114 | 109 | (3·83; 24) | 102–116 |
| BILL | 21·6 | (0·78; 29) | 19–23 | 21·1 | (0·69; 24) | 20–22 |
| TARSUS | 27·8 | (1·24; 29) | 24–29 | 27·1 | (1·14; 24) | 25–30 |
| TOE | 29·1 | (1·44; 19) | 25–32 | 28·6 | (1·33; 14) | 26–32 |

Sex difference significant for bill and tarsus.

**Weights.** Breeding ♀♀ Deserta Grande, July: 87, 89, 95, 98, 98. ♀ found dying Porto Santo, July: 73 (MNHN).

**Structure.** Wing long and narrow. 11 primaries: p10 longest, p9 0–4 shorter, p8 6–10, p7 17–21, p6 32–38, p1 107–119. Tail long, strongly wedge-shaped. 12 feathers: t6 *c.* 40 shorter than t1. Bill stout, nasal tubes $\frac{1}{4}$–$\frac{1}{3}$ of length; nostrils small, circular, directed forward. Tarsus slender, round.

**Geographical variation.** North Atlantic and North Pacific populations similar. *B. fallax* Jouanin, 1955, and *B. macgillivrayi* (Gray, 1859) sometimes considered conspecific. *B. fallax*, being much larger, with less pronounced wing bar, may actually be less closely related (CSR). *B. macgillivrayi*, known only from unique specimen (Fiji), probably all-black member of some species of *Pterodroma*.

## *Bulweria fallax* Jouanin, 1955 **Jouanin's Petrel**

FR. Pétrel de Jouanin  GE. Jouanins Bulwersturmvogel

Tropical western Indian Ocean; breeding sites unknown, but most likely islands off Kuria Muria Islands, south-east Arabia, where concentrations reported March–August. Movements not clear, but apparently common offshore there all year, others dispersing west to Gulf of Aden (once to Kenya), east to Gulf of Oman and into Arabian Sea, south to equator and sometimes beyond, southernmost 8°30′S 58°E (Bailey 1966, 1968; Gill 1967). Unknown to what extent penetrates Red Sea; several dark petrels tentatively recorded as *B. fallax* seen May at Elat, Gulf of Aqaba (Safriel 1968). 3 seen (one obtained) Cinadolmo (Treviso) Italy, 2 November 1953 (Giol 1957); identity certain but record so extraordinary that unnatural origin, such as release by sailor, must be regarded as possibility (C Jouanin). Nevertheless, vagrancy potential shown by 1967 record from Hawaii (*Condor* 1971, 73, 490). Differs from *B. bulwerii* chiefly in size, *fallax* being larger, and they may be conspecific (Bourne 1960); several species of dark-plumaged petrels, difficult to separate at sea, occur in Indian Ocean area—see Bourne (1960) for comparisons.

PLATE 14. (*facing*) *Fulmarus glacialis* Fulmar (p. 121): 1 ad double light morph, 2 dark morph, 3 double dark morph, 4 nestling. *Macronectes giganteus* Southern Giant Petrel (p. 119): 5 ad, 6 ad white morph, 7 imm. *M. halli* Northern Giant Petrel (p. 119): 8 ad, 9 juv (included for comparison). (PJH)

## *Calonectris diomedea* **Cory's Shearwater**

Du. Kuhls Pijlstormvogel    Fr. Puffin cendré    Ge. Gelbschnabel-Sturmtaucher
Ru. Желтоклювый буревестник    Sp. Pardela cenicienta    Sw. Kuhls lira

*Procellaria diomedea* Scopoli, 1769

Polytypic. Nominate *diomedea* (Scopoli, 1769), Mediterranean; *borealis* (Cory, 1881), subtropical east Atlantic; *edwardsii* (Oustalet, 1883), Cape Verde Islands.

**Field characters.** 45–46 cm, wingspan 100–125 cm; longer-winged than Great Shearwater *Puffinus gravis* and less compact than Fulmar *Fulmarus glacialis*. Pale-billed, large-headed, full-bodied, and long-winged shearwater; brown above and white below, lacking any obvious marks and closely recalling (in North Atlantic) only Great Shearwater. Sexes alike; no seasonal variation; juvenile inseparable. 3 races in west Palearctic; nominate *diomedea* and *borealis* indistinguishable in field but *edwardsii* distinctive (Bourne 1955).

Nominate *diomedea* and east Atlantic race *borealis*. Head and hindneck grey-brown, faintly mottled; rest of upperparts sooty brown, with most feathers distinctly fringed paler brown or grey, longest upper tail-coverts tipped whitish (occasionally forming pale horseshoe mark); tail dark brown. Scapulars and inner wing dark brown, with less distinct pale fringes to feathers, carpal feathers, secondaries, and primaries blackish-brown (inner webs pale brownish-white, but rarely show). Underwing almost uniformly white, merging into dusky brown border, most obvious on trailing edge. Sides of neck and chest pale grey-brown, mottled and merging with centre of throat; rest of underparts all white except for faint brown-grey barring on flanks and sides of under tail-coverts. Bill, particularly of Mediterranean race (nominate *diomedea*), strong and long, with tubed nostrils inconspicuous (over basal fifth of upper mandible) but hook obvious; yellow-horn with dusky marks behind nail (at distance bill appears browner). Legs and feet pale flesh, dusky on sides of tarsal joints and webs. Cape Verde Islands race *edwardsii* smaller, shorter bill and wings (by 10–15%) and noticeably darker above.

Imperfect observation may cause confusion with *P. gravis*, *F. glacialis*, and east Mediterranean race of Manx Shearwater *P. puffinus yelkouan*, but separable from first by heavier and looser-winged appearance, more uniform plumage pattern with grey hood (not blackish cap), completely unmarked belly and vent, and thicker, pale bill; from second by longer, more flexible wings, dusky head, brown upperparts, dark tail, and longer bill; and from third by considerably larger size (by more than a quarter), pale foreparts, totally white underparts, and thick, pale bill.

Flight action strong and graceful, including in light breezes distinctive, regular alternation of deep, lazy flaps and low glides (with wings depressed and angled back); and in high winds prolonged gliding and banking on bowed wings, wheeling, and even soaring. At distance can recall *F. glacialis* but silhouette more attenuated and wings much less stiffly beaten with tips bending down noticeably. Flies well over breeding sites, forming wheeling assemblies before dusk. Gait an awkward, waddling shuffle, with frequent, often unsuccessful use of wings. Highly gregarious when breeding and numerous at sea round colonies, but over deep ocean markedly less gregarious than *P. gravis* (though concentrations do occur during migrations). Attends fishing boats (even following them close to land) but most food obtained naturally by plunging, often in association with other Procellariiformes or gulls *Larus*. Adults vocal at breeding grounds but apparently silent at sea.    DIMW

**Habitat.** Mainly in warm marine waters from temperate to sub-tropical zones of North Atlantic and in Mediterranean, avoiding colder seas but equally satisfied with pelagic, offshore, and inshore waters. Tolerant of high winds and rough seas, only exceptionally becoming storm-driven over mainland. Varies characteristic shearwater flight by rising higher above water than related species on upwind tack; aerial activity much interrupted by alighting on water, often in groups. Sometimes forms large rafts on open sea; once so dense off Guinea-Bissau that flock mistaken for land (Palmer 1962). Although scavenging, does not turn aside to follow ships. Concentrates from South Atlantic wintering area on breeding sites, nearly all on uninhabited islands or on mountainous terrain up to 30 km or more from coast, including volcanic craters and lava flows, boulder fields, caves, rabbit burrows, or excavated burrows in sand or even hard earth; also sometimes, where farmed, in artificial nest-holes from which large, fat nestlings removed. Nests often near other Procellariiformes, probably through similar requirements. On land, usually nocturnal wherever man or other predators threaten. See also Bannerman (1963) and Bannerman and Bannerman (1965).

**Distribution.** May also breed elsewhere in Mediterranean, e.g. islands off Turkey and Islas Chafarinas, Morocco (RP, KDS).

Accidental. Faeroes, Belgium, Netherlands, Denmark, West Germany, East Germany, Poland, Czechoslovakia, Austria, Switzerland, Bulgaria, Cyprus, Israel, Egypt.

**Population.** SELVAGENS: colony probably largest in Atlantic, with, until recently, estimated minimum 22000

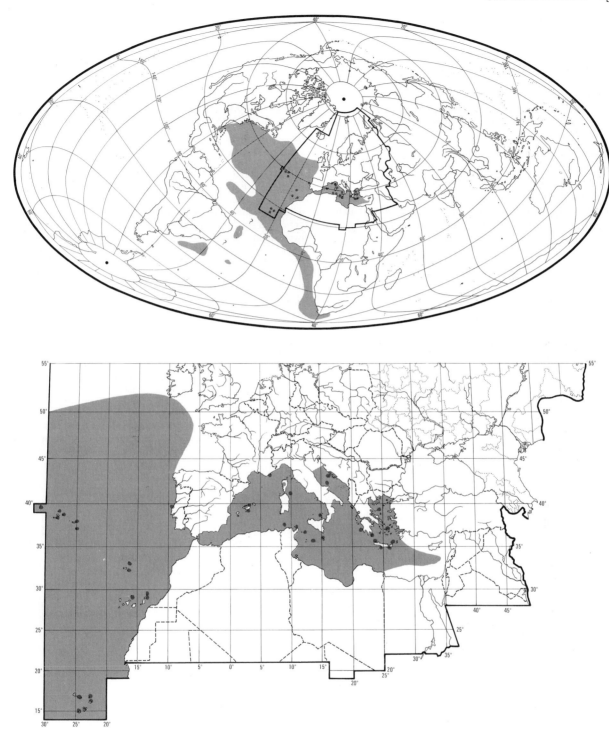

pairs. For over century up to 24 000 young taken annually for food, down, and oil; now fishermen also taking adults and eggs, so far fewer young collected in recent years. DESERTAS and PORTO SANTO group: numbers also reduced (Roux and Jouanin 1968). CANARY ISLANDS, MADEIRA, and AZORES: by far most numerous shearwater (Bannerman 1963). CAPE VERDE ISLANDS: probably several thousands (de Naurois 1969*b*). TUNISIA: some 10 000 present at dusk Ile de Zembra, May 1968 (Jarry 1969).

**Movements.** Migratory. Nominate race and *borealis* both strongly migratory; no detailed information for *edwardsii*, though absent from vicinity of colonies outside breeding season. Nominate *diomedea* and *borealis* both collected off New England in northern summer and autumn and off South Africa in northern winter and spring (Palmer 1962; Clancey 1965), so both probably have rather similar migrations, though *borealis* identified more often in New World. Present in breeding areas late February (when often marked eastward movement in Mediterranean) to October–November; additional influxes, doubtless pre-breeders, about May–June when non-breeders become widespread across temperate North Atlantic from Bay of Biscay to New England coast (Palmer 1962). Occasionally large numbers off south-west Ireland between July and September (Wright *et al.* 1964); otherwise scarce visitor to north-west Europe, mainly off Atlantic coasts, irregularly English Channel and North Sea.

Non-breeders move south in autumn, joined by birds from colonies in October–November, making rapid transequatorial passage to main pelagic wintering area now known to lie off South Africa (Bierman and Voous 1950; van Oordt and Kruijt 1953). Largest numbers off Cape Province between 30°S–40°S, 10°E–20°E, some reach Natal seas (*Ostrich* 1973, **44**, 85), and others fan out during southward passage to become thinly distributed over sub-tropical South Atlantic. Ringed immature *borealis* among those reaching Uruguay and Brazil, may be derived, at least in part, from pre-breeders that summered off New England; *borealis* identified occasionally in Caribbean, and once New Zealand (January 1934). Largely absent from temperate North Atlantic after November, but isolated exceptions include report of 43 in Aegean Sea, December 1957 (Horváth 1959*a*), and specimen of *borealis* in England in January; birds of uncertain origin occur off west Africa in winter in small numbers (Bourne 1965) and may be *edwardsii* whose wintering range unknown. Does not enter Indian Ocean further than Natal; Kerguelen reports erroneous (Bourne 1955*a*). Some pre-breeders remain south, notably Cape seas, until May, but main return movement, largely up eastern side of Atlantic, between late January and March; several atypical June records from tropics of small numbers migrating north in flocks of Great Shearwater *P. gravis* (Bourne and Dixon 1973). See also annual seabird reports in *Sea Swallow* 1961–73.

**Food.** Chiefly fish, fish spawn, cephalopods, and crustaceans; North Atlantic race *borealis* also scavenges offal thrown from fishing vessels, particularly livers of cod *Gadus morhua* and other oily substances. Feeds mainly at night, chiefly by skimming over surface and surface-feeding. Though Audubon (1835) stated that *borealis* 'frequently dives', apparently seldom does so; probably mainly shallow plunging, from 5 m and with up to 2 s immersion (Bauer and Glutz 1966) in pursuit of surface prey. Birds near Azores in August seen hovering, and then surface-plunging to take fish (Beven 1946*a*). Follows whales and schools of predatory fish to pick up food scraps (Bent 1922) and small fish driven to surface (Watson 1966).

Seen feeding on schools of herring *Clupea harengus* off Massachusetts (Baird 1887). Fish, crab remains, sea-weeds, and oily substances found in stomachs (Audubon 1835). Food of Mediterranean race (nominate *diomedea*) mainly cephalopods, of which some probably *Loligo vulgaris* (Witherby *et al.* 1940).

**Social pattern and behaviour.** Compiled mainly from material supplied by Frank B Gill.

1. Gregarious when breeding and at sea, though extent of winter flocking poorly known. Non-breeding period spent entirely at sea, birds moving quickly far from land once breeding over (Bannerman 1965). BONDS. Monogamous pair-bond of unknown duration. Both sexes incubate and tend young until just before fledging. BREEDING DISPERSION. Often nests in huge, dense colonies though some islands occupied by only few breeders. No evidence of competition for sites; no territorial behaviour described. ROOSTING. Little known; presumably rests on sea during winter and when not incubating.

2. Arrives suddenly in great rafts on sea off breeding station, assembling just offshore and then moving inland in large flocks at sunset (Fisher and Lockley 1954; Zino 1971). Otherwise partly diurnal, with off-duty and unemployed birds returning from sea about 2 hours before sunset; molestation by man, however, causes more strictly nocturnal behaviour at some colonies (Lockley 1952). Calling in flight reaches peak at sunset; colony then silent all night but calling resumed at dawn when many birds can again be seen flying (Jouanin and Roux 1966). Incubation and other reproductive activities during day done silently. In courtship, mates sit facing and Allopreen, nibbling at each other's head and bill; during copulation, they utter harsh calls and afterwards ♀ nibbles ♂'s tail and stern (Lockley 1952).

**Voice.** Silent at sea, extremely noisy at breeding grounds.

Following descriptions (mainly from Lockley 1942, 1952) refer to *C. d. borealis*. Nocturnal Flight-call wild, harsh, rasping, sobbing scream or wail like man assembling phlegm in throat: 'kaa-ough' or sometimes 'koo-ough' (but hard to render satisfactorily), uttered 3–4 times with distinct sighing, audible only at close range, after each double note probably due to indrawing of breath. Portuguese name for species ('Cagarra') onomatopoeic rendering of Flight-call. Copulation-call of ♂ harsh, continuous 'ka-ka-ka-ka'. At least some calls of ♂ and ♀ different. When handled, screams of ♂♂ louder than those of ♀♀, latter uttering only short 'ka-ka-ka' (Lockley 1952). Call of ♀ described as rasping 'ia-ia-ia', of ♂ guttural, sobbing wail 'ia-gow-a-gow-a-gow' resembling crying of human infant; sonagram (see I) illustrates similar call recorded in Portugal like crow of domestic chicken after laying egg, followed by 3 shorter 'ow ow ow' notes. Birds on Selvagens mainly silent by day, although sometimes call even in bright sunlight; calling most intense at dusk and dawn (Zino 1971).

Cape Verde Islands race *edwardsii* evidently differs in voice from *borealis* (Bannerman and Bannerman 1968) but

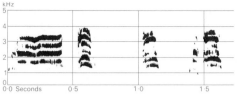

I Jacques Dufour/BBC Portugal

accounts vary. Described as 'deep laughing note and high-pitched reply' (Bourne 1955a, of birds calling on sea cliffs); as weird call like whistling of Wigeon *Anas penelope* (Alexander 1898); and as ringing 'ha-oo' given in flight (Murphy 1924). Calls of nominate *diomedea* apparently not adequately described.

CALLS OF YOUNG. Chicks utter excited squeaking noise while pecking at parent's bill, apparently in attempt to solicit food (Zino 1971).

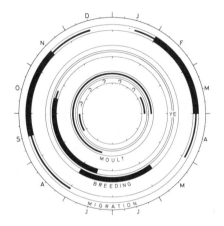

**Breeding.** SEASON. See diagram for Atlantic islands. Little data Mediterranean. High degree of synchronization, 49 eggs laid in 52 nests observed in Selvagens colony in period 27 May–6 June (Zino 1971). SITE. On ground, or among rocks, cliffs, and scree. Colonial; nests often almost touching and, in small caves, with common entrance hole. Nest: in cavity at end of tunnel in soft soil, or natural hole in rocks; sometimes above ground sheltered only by thick vegetation. Tunnel up to 1 m long with entrance hole *c.*20–30 cm diameter; passage often winding or with right-angle bend (Bannerman 1963). Nest can be built up into platform of stones; shells and other material sometimes added (Roux and Jouanin 1968). Building: excavation, where made, probably by both sexes. EGGS. Ovate to elliptical, not glossy; white. 69 × 45 mm (63–74 × 42–53), sample 85; calculated weight 74 g (*C. d. diomedea*). 75 × 50 mm (66–83 × 45–55), sample 80 (Schönwetter 1967); weight 104 g (95–115), sample 46 (Zino 1971) (*C. d. borealis*). CLUTCH. 1. One brood. No information on replacements. INCUBATION. Mean 53·8 days (52–55), sample 25; once 56 days (Zino 1971). By both sexes, ♂ taking 1st spell of mean 6·4 days (range 1–10); later spells average 6 days (Zino 1971). YOUNG. Semi-altricial and nidicolous. Brooded by parent continuously at first, for average of 4·2 days (range 2–8), sample 24, then left and fed only at night; eventually deserted and makes own way to sea (Roux and Jouanin 1968; Zino 1971). FLEDGING TO MATURITY. Fledging period not precisely known, but some young in nest at *c.* 85 days and others fledged before 97 days (Zino 1971). Independent some days before fledging. Age of maturity unknown. BREEDING SUCCESS. Of 42 eggs laid on Selvagens, 71% hatched; 7 failures due to predation by gulls (Laridae), 3 addled, 1 dead in shell, 1 cause unknown; of 30 young hatched, 29 still alive at *c.* 9 weeks (Zino 1971).

**Plumages** (subspecies *borealis*; for others, see Geographical Variation). ADULT AND JUVENILE. Crown and hindneck grey-brown. Mantle, back, and rump light brown, feathers widely edged grey with narrow white fringes in fresh plumage; scapulars darker, warmer brown. Tips of long upper tail-coverts white, mottled brown. Sides of head mottled grey and white, grading into white of chin; chin and throat white with some grey streaks and specks. Rest of underparts white, with some grey mottling on under tail-coverts and sides of chest. Tail dark brown, paler below. Primaries dark brown, paler below, basal half of inner webs pale grey; primary coverts dark brown. Secondaries as primaries with white spot on basal part of inner web; innermost like scapulars. Upper wing-coverts brown; under wing-coverts white, those at leading edge brown with pale edges; axillaries white. NESTLING: 1st down brown, slightly paler below; forehead, chin, and throat nearly bare, with short white down. 2nd down longer and slightly darker, centre of chest dirty white. 2nd down begins to grow 10 days after hatching; tail-feathers 2½ weeks; secondaries 3½; primaries 4 weeks. Body feathers appear first on head and upperparts, predominate on back when primaries emerge (Mallet and Coghlan 1964).

**Bare parts.** ADULT and JUVENILE. Iris brown, Bill yellow. Inner surface of tarsus and inner toes pale flesh, outer surface and outer toe and web dusky.

**Moults.** ADULT POST-BREEDING. Complete; primaries descendant. Timing unknown as relevant specimens scarce, but some in moult September (Witherby *et al.* 1940) when majority not yet started (Mayaud 1949–50); those collected in March at breeding places have new wing-feathers, some *edwardsii* showing p10 in last stage of growth. Body feathers July–March, starting when eggs hatch (Jouanin 1964); moult slow, first feathers already much worn when last grow. IMMATURE. No data; birds moulting early (see above) may be immature.

**Measurements.** Skins (BMNH, RMNH, ZFMK, ZMA) with additional data from Murphy (1924) and Murphy and Chapin (1929).

*borealis*

| | | | | | | |
|---|---|---|---|---|---|---|
| WING | ♂ | 361, 367 | | ♀358 | (4·50; 11) | 347–363 |
| TAIL | | 138, 145 | | 136 | (3·01; 11) | 131–141 |
| BILL | | 55·5 (1·73; 52) | 51–59 | 52·8 | (1·85; 60) | 49–57 |
| TARSUS | | 57·0 (1·24; 52) | 54–59 | 54·6 | (1·32; 61) | 51–57 |
| TOE | | 71·5 (1·92; 52) | 67–74 | 68·8 | (1·92; 61) | 64–73 |

*diomedea*

| | | | | | | |
|---|---|---|---|---|---|---|
| WING | ♂ | 346 (4·68; 9) | 339–351 | ♀339 | (6·83; 5) | 330–347 |
| BILL | | 51·2 (1·51; 17) | 49–55 | 47·3 | (1·78; 16) | 45–50 |

*edwardsii*

| | | | | | | |
|---|---|---|---|---|---|---|
| WING | ♂ | 313 (6·51; 9) | 298–321 | ♀308 | (6·21; 7) | 302–319 |
| BILL | | 44·6 (1·52; 52) | 41–46 | 42·5 | (1·54; 52) | 39–46 |

Sex differences significant, except wing *edwardsii*; wing and tail *borealis* not tested. Wing: ♂ *borealis* 363 (20) 349–372; ♂ *diomedea* 346 (20) 335–361; ♂ *edwardsii* 309 (20) 300–325 (Vaurie 1965).

**Weights.** Breeding *borealis*, Selvagens (Zino 1971): ♂ 1014 (55·9; 26) 940–1130; ♀ 877 (61·6; 26) 730–1000. Breeding *borealis*, Azores (de Chavigny and Mayaud 1932): ♂ 857–962; ♀ 755–813. Breeding *diomedea*, Malta (BTO): sexes combined 654 (62·5; 12) 560–730. Maximum nestlings Azores: ♂ 1100, ♀ 800–875 (De Chavigny and Mayaud 1932), unsexed 960 (Mallett and Coghlan 1964). Average of 21 at *c.* 53 days Selvagens: 1092 (Zino 1971).

**Structure.** Wing long and narrow. 11 primaries: p10 longest, p9 0–10 shorter, p8 14–26, p7 38–51, p6 62–74, p1 180–201; p11 small, *c.* 15–25 shorter than primary coverts. Primaries evenly tapering, p10 with narrow outer web. Tail short, strongly rounded. 12 feathers: t6 26–36 shorter than t1. Bill long and stout with strong terminal hook, nasal tubes *c.* ⅕ of its length; nostrils directed obliquely upwards. Foot strong, middle and outer toes *c.* 25% longer than tarsus, inner slightly shorter; hind toe raised, reduced to short claw.

**Geographical variation.** Nominate *diomedea* smaller than *borealis*; bill less massive; head and mantle paler; chin and throat less marked grey; prominent white wedge on inner web of primaries, projecting well beyond under wing-coverts. *C. d. edwardsii* much smaller; bill more slender, black instead of yellow; tail relatively longer, but (contrary to Bourne 1955*a*) not more square (t1–t6 in 2 ♂♂ 34 and 40, in 2 ♀♀ 29 and 39); crown and neck darker brown; sides of head more contrasting; mantle darker, but lighter than lower back, latter as dark as upper wing-coverts; no white on distal part of primaries.

## *Puffinus gravis* Great Shearwater

PLATES 15 and 16
[after page 158]

Du. Grote Pijlstormvogel    Fr. Puffin majeur    Ge. Grosser Sturmtaucher
Ru. Пестробрюхий большой буревестник    Sp. Pardela capirotada    Sw. Större lira

*Procellaria gravis* O'Reilly, 1818

Monotypic

**Field characters.** 43–51 cm, wingspan 100–118 cm; shorter-winged than Cory's Shearwater *Calonectris diomedea* but a third larger than Atlantic and west Mediterranean races of Manx Shearwater *P. puffinus* (nominate *puffinus* and *mauretanicus*). Dark-billed, rather thickset, large, and long-winged shearwater, dark brown above and white below, with distinct headcap; confusable only with *C. diomedea* or east Mediterranean race of Manx *P. p. yelkouan*. Sexes alike; no seasonal variation (apart from wear or moult); juveniles inseparable.

ALL PLUMAGES. Headcap (compromising crown, upper cheeks, and nape) sooty brown, hindneck, pale grey-brown (in some lights merging with white of neck to form collar behind dark headcap). Rest of upperparts sooty-brown but appearing paler due to most feathers being fringed lighter brown (greyish when worn); distal upper tail-coverts tipped and fringed white, usually forming inverted horseshoe mark over tail. Scapulars and wing-coverts as back, but tail, carpal, and flight-feathers brown-black (inner webs of primaries white at base and often showing as blotches when plumage worn). Underwing white, but heavily blotched with dark brown from wing-pit to carpal joint and bordered by black-brown, especially on trailing edge. Rest of underparts also mostly white, but centre of belly, flanks, and undertail blotched dark brown; in fresh and little worn plumage, belly marks form long dark patch visible when bird tilts over in flight. Bill strong with tubed nostrils inconspicuous (over basal quarter of upper mandible) but hook obvious; dark blackish-horn, base of lower mandible grey (looking all black at distance). Legs and feet pale flesh to bright pink, browner on outer tarsus and toes.

Distinctive shearwater visiting North Atlantic during southern winter; best separated from *C. diomedea* or *P. p. yelkouan* by obvious headcap, blotched underwing, and belly patch; pale horseshoe mark over tail not diagnostic since some *C. diomedea* also show it.

Flight action urgent and powerful, with, in all winds, rapid beats of stiff wings interspersed with glides and banking, recalling *P. puffinus* not *C. diomedea*; only occasionally soars but flocks on migration pass in characteristic lines of steeply looping birds. Flight silhouette similar to *C. diomedea* but wing tips rounder and held straighter. Gait shuffling but effective. Abundant and gregarious in North Atlantic, concentrating at times over fishing areas, where fearlessly exploits offal sources. Squabbling birds squawk raucously, or utter harsh cries and screams.                                          DIMW

**Habitat.** Oceanic, in cool to fairly cold pelagic waters of South and North Atlantic, migrating quickly without stopping across intervening tropical and sub-tropical seas, and usually avoiding both mid-ocean and inshore zones, preferring (in non-breeding season) chilly climates, even with scattered icebergs, to sunny, warmer currents sought by Cory's Shearwater *Calonectris diomedea*. Stops short of Arctic pack-ice, and occupies west Palearctic habitats almost exclusively during third quarter of year, sometimes well into October in east Atlantic. Less frequently inshore, or in narrow seas such as English Channel, than Sooty Shearwater *Puffinus griseus*. In flight, except at breeding stations, confined to lowest belt of airspace, but less aerial than other shearwaters and more given to resting and feeding for long periods on water, often in rafts, making shallow dives below surface. Indifferent to shipping but

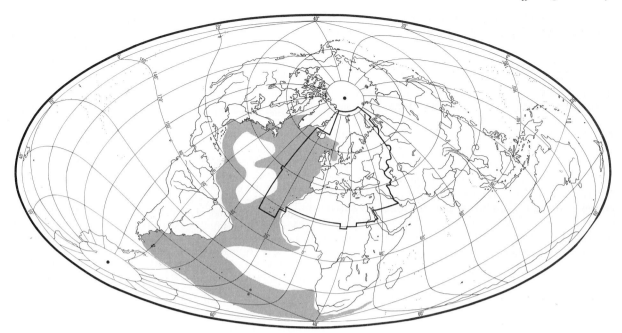

will approach fishing vessels for food, or follow whales and porpoises. Unaffected by stormy weather. Sometimes overlaps with other shearwaters, but does not normally associate with them. Breeding habitat in South Atlantic up to *c.* 400 m on any kind of terrain permitting excavation of burrows, including slopes under tussock grass or woodland (Rowan 1952).

**Distribution.** Only 3 main breeding sites—Nightingale Island and Inaccessible Island, in Tristan da Cunha group, and Gough Island, although single pair found breeding Kidney Island, Falkland Islands, 1961 (Woods 1970).

Estimated 2 million pairs Nightingale Island and at least 150 000 pairs Inaccessible Island, and minimum of 600 000 pairs Gough Island (Rowan 1952; Elliott 1970).

Accidental. Norway (possibly annual), Sweden, Denmark, Netherlands, Germany, Italy (Sardinia), Algeria, Morocco.

**Movements.** Migratory. Absent from main breeding stations chiefly April (adults) and May (juveniles) until late August or early September (Rowan 1952), when undertakes transequatorial migration into North Atlantic (Voous and Wattel 1963, on which this account based). Apparently most depart breeding colonies in north-west direction on narrow front, passing horn of Brazil (Metcalfe 1966; Bourne 1966a), reaching Atlantic coasts of North America in May and early June; northward movement occurs east Atlantic, but numbers small. Present off Newfoundland Grand Banks all through northern summer, and later also off Greenland, adults undergoing rapid moult. Widely distributed western half of North Atlantic in June when northward passage out of tropics still continuing, e.g.

estimated 25 000 flying north 12 June over east Sargasso about 17°N 53°W; these presumably juveniles, which fledge several weeks after adult migration begun. Virtually all in winter quarters by July, when present north to 65°–66°N in Davis and Denmark Straits; maximum abundance high latitudes early August. During July and August, spreads east across North Atlantic, reaching Rockall Bank and European coasts, mainly well offshore Scotland to Iberian Peninsula; largest numbers occur from August–October when occasionally, in adverse weather, visits Irish inshore waters in numbers (Newell 1968); rare North Sea; has straggled to Heligoland Bight, Skagerrak, and Kattegat. Because of relative lateness, those in European waters presumed non-breeders (Rowan 1952). Rapid return passage by adults starts August; some back at colonies by end August, and most breeders return by mid-September; but surprisingly few August mid-ocean records from tropical North Atlantic. As yet uncertain whether main southward passage concentrated in western Atlantic (as in northward movement) or, as more likely, on broad front across whole ocean (Rowan 1952; Bourne 1970); usually scarce off west African coasts. In northern hemisphere marked reduction by November, despite one report in 1954 of thousands off Nova Scotia on 11th (Voous and Wattel 1963); only stragglers north of equator December–April. Presumed immatures thinly dispersed over South Atlantic during breeding season, with numbers off South Africa, especially October–January (Liversidge 1959); and concentration in cold current between Argentina and Scotia Arc islands (South Georgia, South Sandwich); moulting immatures off Tierra del Fuego in January and February (Watson 1971) may mostly be Tristan da Cunha or Gough birds, but numbers breeding

Falkland Islands unknown. See also Palmer (1962), Lambert (1971), Stresemann and Stresemann (1970).

**Food.** Mainly fish and cephalopods, taken on and under water mostly by pursuit-plunging, pursuit-diving, and by surface-seizing. Surfaces to swallow prey. Typically daytime feeder, and often, though not always, gregarious.

Follows and feeds on shoals of surface-swimming fish, and on squids which come to surface when little light available (Murphy 1936); also scavenges extensively on fish entrails thrown overboard from fishing boats (Collins 1884; Palmer 1962), and, especially in north, takes capelin *Mallotus villosus* (Rees 1961). Often found near whales and porpoises, probably mainly to feed on fish which they drive to surface; early records showed many squid beaks in stomachs (Townsend 1905), but Wynne-Edwards (1935) suggested these obtained indirectly from feeding on cetacean faeces. Bent (1922) quoted sand-eels Ammodytidae as prey, and Watson (1966) mentioned crustaceans taken, chiefly at night. Of 10 stomachs from Tristan da Cunha, 5 contained cephalopod remains, and 1 each fish remains, a few amphipods *Themisto gaudichaudi*, decapod pieces (possibly *Palinurus*), and grass leaves; 7 also included indeterminate mosses (Hagen 1952). Of 38 stomachs from Newfoundland Grand Banks, 24% contained squid including *Illex illecebrosus*, which occurs in large shoals at surface at night, and 63% fish remains mainly scavenged from fishing vessels, including capelin and fish livers; also euphausiid crustaceans *Thysanoessa inermis* probably derived from fish (Rees 1961).

Over 20 chicks from Nightingale Island near Tristan da Cunha had stomachs filled with squid remains and occasional crustaceans; after desertion by parents, nibbled grass and ate soil (Rowan 1952).

**Voice.** See Field Characters.

**Plumages.** ADULT AND JUVENILE. Top of head to below eye dark brown, sharply outlined against white chin and throat; some white mottling below eye. White crescent on side of neck forms collar with pale hindneck; rest of upperparts grey-brown, feathers edged pale brown to off-white, longest scapulars dark brown with narrow white margins. Long upper tail-coverts white, irregularly barred pale brown away from tips. Underparts white, sides of chest grey, flanks spotted brown, lower flanks and under tail-coverts grey-brown with pale feather edges; on centre of belly grey patch of variable extent. Tail dark brown. Primaries dark brown, inner webs paler with white wedge on inner two-thirds; outer webs of secondaries grey, inner white with grey spot at tips. Upper wing-coverts darker than back, greater tinged grey like secondaries; underwing white spotted and streaked brown, axillaries white with brown tips.

**Bare parts.** ADULT AND JUVENILE. Iris brown. Bill dark horn, grey at base of lower mandible, white spot on terminal hook. Foot flesh; varies from white to (particularly in juvenile) bright pink; outer side of tarsus and outer toe black.

**Moults.** (See Stresemann 1970 for details). ADULT POST-BREEDING. Complete; primaries descendant. From end of May to mid-August, when birds in winter quarters, primaries moulted in rapid sequence, up to 6 growing simultaneously; take-off may be hindered occasionally, but never flightless. In early stage of moult, white band on upper wing where shedding of coverts exposes white bases of secondaries and greater primary coverts. Body feathers moulted at same time as wing, tail somewhat later. Sequence of secondary moult as in Manx Shearwater *P. puffinus* (Mayaud 1949–50). POST-JUVENILE. Partial, said to moult body plumage only; July–August of 1st calendar year. SUBSEQUENT MOULTS. In 2nd calendar year, moulting earlier than adult, beginning late March or April, when still in South Atlantic. Birds having shed p1–p4 in January off Tierra del Fuego have been in 2nd year (Watson 1970). Many in wing moult present in North Atlantic when adults already at breeding grounds (Mayaud 1949–50; Voous 1970) probably older immatures, but more data needed. Also not known whether all early moulters 2nd year birds or whether some older.

**Measurements.** FULL GROWN. Skins BMNH, RMNH:

| | ♂ | | | ♀ | | |
|---|---|---|---|---|---|---|
| WING | 332 | (8·61; 10) | 318–348 | 318 | (11·3; 6) | 301–334 |
| TAIL | 117 | (3·66; 10) | 113–126 | 113 | (4·37; 8) | 109–120 |
| BILL | 46·0 | (1·71; 12) | 43–50 | 44·0 | (1·14; 8) | 43–47 |
| TARSUS | 59·8 | (1·48; 12) | 58–63 | 59·0 | (1·20; 8) | 57–60 |
| TOE | 71·6 | (2·67; 10) | 68–77 | 69·4 | (1·34; 5) | 68–71 |

Sex difference significant for wing and bill. Long wings may be found in newly-fledged birds recognizable by remains of down (Elliott 1957); maxima in table (♂ 348, ♀ 334, Nightingale Island, mid-May) pertain to this category.

**Weights.** Breeding adults 834 (59·1; 14) 715–950, sex-difference negligible (Hagen 1952).

**Structure.** Wing long and narrow. 11 primaries: p10 longest, p9 2–9 mm shorter, p8 16–26, p7 31–50, p6 50–75, p1 162–193; p11 minute, pointed, hidden by coverts. Primaries evenly tapering, p10 with narrow outer web. *c.* 20 secondaries. Tail short, strongly rounded, t6 *c.* 25 shorter than t1. Bill powerful with strong terminal hook, but more slender than Cory's Shearwater *Calonectris diomedea*; nasal tubes $\frac{1}{5}-\frac{1}{4}$ of bill-length, nostrils directed obliquely upwards. Tarsus stout, laterally compressed; middle and outer toes equal, inner slightly shorter, hind toe rudimentary.

**Geographical variation.** None within species. Sometimes considered to form subantarctic circumpolar superspecies with *P. carneipes* in south Indian Ocean and south-west Pacific and *P. creatopus* in south-east Pacific (Palmer 1962).

## *Puffinus griseus* Sooty Shearwater

DU. Grauwe Pijlstormvogel    FR. Puffin fuligineux    GE. Dunkler Sturmtaucher
RU. Серый буревестник    SP. Pardela sombría    SW. Grå lira

*Procellaria grisea* Gmelin, 1789

Monotypic

**Field characters.** 40–51 cm, wingspan 94–109 cm; larger than Balearic race of Manx Shearwater *P. puffinus mauretanicus*, with longer, heavier body and greater wingspan, but smaller than Great Shearwater *P. gravis*, particularly in head size and wing breadth. Long-billed, fairly small-headed, large shearwater, with noticeably bulky body, narrow wings set well back and almost uniform, dark plumage, recalling, in east Atlantic, only *P. p. mauretanicus*. Sexes alike, no seasonal variation; juvenile inseparable.

ALL PLUMAGES. Almost totally dark grey or sooty brown; in good light, faint buff or grey fringes to feathers give paler (yellowish) cast to scapulars and wing-coverts (causing contrast with blackish flight-feathers and tail) while paler, slightly warmer underparts can then be seen. Normally, however, dark plumage relieved only by strikingly pale buffish or greyish-white wing linings (from wing-pit to basal parts of quills), showing up at surprisingly long distances and often suddenly shining as bird tilts over in flight. Bill black; long and fairly strong, with tubed nostrils raised above basal quarter of upper mandible. Legs and feet slate, tinged purple, particularly on webs.

Combination of uniform dark body and pale wing linings unique in shearwaters of east Atlantic; in poor light, however, much care necessary to distinguish from darkest *P. p. mauretanicus* but these smaller, slighter in body, and much shorter in wing-length.

Flight recalls *P. puffinus* but action more powerful with scuttling, often backswept wingbeats appearing almost mechanical and flight tracks more direct (even in high winds); rarely high above waves. Gait on land as other shearwaters. Less common in North Atlantic (and less gregarious) than *P. gravis*, being rarely seen in large groups, but much more regular within sight of land than *P. gravis*. Most food obtained from surface or by purposeful diving, and underwater flying. Quieter at sea than *P. gravis*, but raucous calls and screams reported (Palmer 1962). DIMW

**Habitat.** Basically cold pelagic waters of southern hemisphere where concentrates in vast numbers at high densities. Outside breeding season, many migrate towards arctic waters, spending least possible time in warmer tropical and sub-tropical seas on way and tending even to avoid the Gulf Stream, in which numbers fewer than in northern Pacific or colder west Atlantic currents. In east Atlantic, mostly over continental shelf but fairly frequently offshore in small numbers and occasionally inshore,

although never normally coming to land in west Palearctic. In flight, as other shearwaters, usually confined to lowest airspace and able to cope with severe storms; dives freely but shallowly, swimming strongly under water, on which often rests. Strongly colonial at southern hemisphere breeding stations, burrowing to nest even in fairly hard ground and at heights apparently up to 1500 m, although normally much nearer sea and often under thick, even woody vegetation.

**Distribution.** Passage visitor to North Atlantic in non-breeding season (see Movements).

Accidental: West Germany, Norway, Sweden, Bear Island, Netherlands, Belgium, Italy, Algeria, Azores.

**Movements.** Migratory. Cold-water species, breeding sub-antarctic; most migrating rapidly across equatorial seas to winter in temperate zones of North Atlantic and North Pacific where evidence of clockwise movements with prevailing winds (Phillips 1963a, 1963b, on which following account largely based). Adults present or near colonies early October to April (some earlier); chick fledged by late May. During late southern summer (chiefly February–March) large feeding movements into Southern Ocean mainly between Kerguelen and Gaussberg Ridge (70°E) and Macquarie and Balleny Ridge (150°E), south to pack-ice; must involve non-breeders, since this high-latitude feeding zone beyond range of breeding adults, which have off-duty spells of only 4–6 days. No evidence of large-scale southward movements in western hemisphere by immatures of South American origins, but regular southern summer reports off southern Africa (exceptionally north to Fernando Po, where once collected January) presumably of non-breeders.

Northward movements begin late March but mostly April–May. Few ringing recoveries, but these indicate migration from Antipodes into Pacific and from Falkland Islands into Atlantic; lack of subspeciation suggests gene-flow between populations. Presumably most Cape Horn and Tierra del Fuego breeders migrate into Pacific since species common off Pacific coasts of Americas, while far fewer winter in Atlantic than in Pacific. Some present off South Africa in July during southern winter (Stanford 1953). Has been observed off east coasts South America most months but chiefly April–May and late August to September (Friedmann 1927), coinciding with main movements from and to colonies. Following rapid northward, transequatorial migration up west Atlantic,

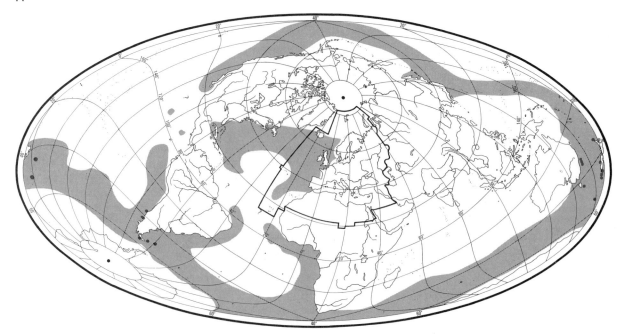

first arrivals off North American coasts in April and early May, exceptionally March (Rees 1964); still largely concentrated off New England and Newfoundland area in June (very few European records then); begins spreading across temperate North Atlantic in July. Does not normally penetrate so far north as Great Shearwater *P. gravis*, but moves further east, small numbers reaching Norwegian Sea.

Numbers begin decreasing in western North Atlantic in August, when return passage has begun. Others, however, apparently move east since more mid-ocean records in August–September; and then becomes more widespread off north-west European coasts (usually well offshore) including concentrations on Rockall Bank and Faeroese fishing grounds. At same time, southerly movements occur off east Atlantic seaboard south to Morocco, and to lesser extent in North Sea and English Channel (Oliver 1971). Many fewer North Atlantic flocks in October, and only stragglers November–March. Those crossing to eastern North Atlantic probably mainly non-breeders, since large-scale return passage occurring off South America by late August, and birds remaining in west Palearctic waters by September–October probably could not return to colonies in time for successful breeding. Only small-scale migration off west Africa, mainly September–November (see, e.g. Bourne and Dixon 1973), so presumably also non-breeders.

Food. Mainly cephalopods, crustaceans, and fish. Feeds largely from surface or by short, shallow dives, often close inshore during day, but also on wanderings in open sea. In Far East, often follows whales and feeds on organisms in their vomit or faeces, especially small Entomostraca and squids; also apparently takes small fish and crustaceans brought to surface by their movements. In Soviet waters, often observed where small planktonic Entomostraca and shoals of capelin concentrate, but rarely approach fishing vessels (Dementiev and Gladkov 1952).

Fish mostly small shoal species, especially capelin *Mallotus villosus*, sand-eels Ammodytidae, and anchovies *Engraulis*. Fond of oily foods, especially livers of cod *Gadus morhua* and hake *Merluccius* (Collins 1884); in many hundreds examined, squid remains (Cephalopoda) always present and, in North Atlantic, flocks seen feeding on dead and mutilated squids. Small fish found in birds from near Cape Horn (Murphy 1936), though Oustalet (1891) recorded only crustaceans in others collected from that area. Birds at various seasons along coast of Peru contained anchovies, squid remains, and larvae of several crabs, including *Pinnaxia transversalis* and *Cancer polydon* (Murphy 1936).

**Voice.** See Field Characters.

**Plumages.** ADULT AND JUVENILE. Upperparts, tail, and wings sooty brown, feathers edged paler. Underparts brown-grey, sometimes as dark as upperparts but mostly paler, feathers narrowly edged off-white; chin pale grey to white. Primaries black-brown, grey-brown on inner webs, grey underneath; secondaries similar. Under wing-coverts grey to dirty white, with dark shaft-streaks, much lighter than rest of underparts; marginal ones darker; axillaries dark brown.

**Bare parts.** ADULT. Iris dark brown. Bill grey-black. Tarsus black-brown outer surface, purple-flesh inner, webs purple-flesh (Serventy *et al.* 1971).

**Moults.** ADULT POST-BREEDING. Complete; primaries descendant, May–August. IMMATURES. Presumably earlier, but records of early moult (from February onward) may also pertain to unsuccessful breeders. Occasional birds in moult as late as November (Mayaud 1949–50). Pattern of moult similar to *P. gravis*, but fewer details known. Sequence of secondary moult as in *P. puffinus*.

**Measurements.** Full-grown; skins (BMNH, RMNH, ZFMK). Additional data for live birds, Whero Island, New Zealand (Richdale 1963).

| skins | | | live adults | | |
|---|---|---|---|---|---|
| WING | 300 | (9·87; 18) | 283–315 | 304 | (6·9; 100) | 287–322 |
| TAIL | 89·9 | (2·35; 18) | 86–96 | 86·6 | (2·7; 100) | 84–93 |
| BILL | 41·4 | (1·43; 19) | 38–44 | 41·9 | (1·6; 100) | 38–46 |
| TARSUS | 56·7 | (2·16; 19) | 53–60 | — | | |
| TOE | 64·1 | (2·97; 14) | 59–70 | 67·6 | (2·2; 100) | 62–74 |

Differences between sexes not significant; data combined.

**Weights.** Adults, December–January, Whero Island 787 (64·0; 100) 666–978.

**Structure.** Wing long and narrow. 11 primaries: p10 longest, p9 3–8 shorter, p8 15–25, p7 32–50, p6 54–75. p1 156–175; p11 minute. Primaries evenly tapering, p10 with narrow outer web. 20 secondaries (Mayaud 1949–50). Tail short, strongly rounded. 12 feathers: t6 15–30 shorter than t1. Bill much more slender than in Great *P. gravis* and Sooty Shearwaters *Calonectris diomedea* with less powerful hook; nasal tubes *c.* $\frac{1}{4}$ of bill length, nostrils directed obliquely upwards. Structure of foot as *P. gravis*.

  **Recognition.** Closely similar to Australian *P. tenuirostris*, which may at times be as light on underwing, but always distinguished by longer bill, 38–46 against 29–34 (Eisenmann and Serventy 1962), and almost always by longer tail, 84–96 against 75–89 (Palmer 1962).

## *Puffinus puffinus* Manx Shearwater

PLATE 15
[after page 158]

DU. Noordse Pijlstormvogel   FR. Puffin des Anglais   GE. Schwarzschnabel-Sturmtaucher
RU. Обыкновенный буревестник   SP. Pardela pichoneta   SW. Mindre lira

*Procellaria puffinus* Brünnich, 1764

Polytypic. Nominate *puffinus* (Brünnich, 1764), east temperate North Atlantic; *mauretanicus* Lowe, 1921, Balearics; *yelkouan* (Acerbi, 1827) other Mediterranean islands. 5 extralimital in Pacific; sometimes considered separate species.

**Field characters.** 30–38 cm, wingspan 76–89 cm; size variable, with a few runt individuals of nominate race almost as small as Little Shearwater *P. assimilis*, and largest birds of Balearic race *mauretanicus* approaching size of smaller Sooty Shearwaters *P. griseus*. Long-billed, fairly small-headed, lithe, medium-sized shearwater, dark above and paler below in all forms. Sexes alike; no seasonal variation; juvenile inseparable. 3 races in west Palearctic distinguishable in the field.
  ATLANTIC RACE *puffinus*. Whole upperparts, including lores, ear-coverts, and flight-feathers, black (wearing brown-black), whole underparts and underwing white, former faintly marked dusky-brown on flanks and outer under tail-coverts, and latter with blackish marks in wing-pit and obvious blackish border. WEST MEDITERRANEAN RACE *mauretanicus*. Slightly larger and most distinct in plumage: upperparts noticeably paler and browner; underparts, including wing lining, much duskier or browner (so appear more uniform and in dull light sometimes all dark). Lack of sharp contrast between upper and underparts quite distinct from *puffinus* or next race. EAST MEDITERRANEAN RACE *yelkouan*. Close to nominate *puffinus* in size and plumage but in most birds upperparts brownish-black, under tail-coverts brownish-grey, underwing, flanks, and sides of neck all washed dirty brown. In all races, bill black with grey at base of lower mandible; long and thin (compared with larger shearwaters) with tubed nostrils short and inconspicuously raised over base of upper mandible. Legs and feet pale greyish-pink marked blackish.
  Commonest shearwater in region, but variations in size and plumage complex. Special care required to separate small individuals of nominate *puffinus* or *yelkouan* from Little *P. assimilis* and Audubon's Shearwaters *P. lherminieri*; larger *yelkouan* need to be distinguished from Great Shearwater *P. gravis* and, especially, dark *mauretanicus* from Sooty Shearwater *P. griseus* (for distinctions, see other species). Partial albinos have been recorded (M P Harris).
  Flight action typical of family: rapid and stiff-winged, alternating between periods of gliding and banking and bursts of flapping, either of whole wing or, with shallower, flicking action, of primaries only, as birds hurry between waves, often changing direction. In high winds, flight modified by almost continual banking and gliding with wingbeats reduced to merest ripple of primaries, birds occasionally rising as high as 10 metres above wave crests. At long range, flocks appear as lines of small black or white cruciform shapes wheeling regularly along horizon. Flies well over land when returning to colonies at night, manoeuvring competently over slopes and up cliffs. Swims freely with head carried above chest and tail and wing tips clear of surface, often resting on water in same attitude. Gait shuffling but fairly free, on bent tarsi; adults heavy with food, and young, both use wings and beaks to assist themselves over rough ground or up launch points. Highly gregarious when breeding, forming huge pre-dusk assemblies near colonies; in smaller groups at sea except in favoured feeding areas. Food taken from surface of sea while swimming, and from under water by plunging and diving. Breeding adults utter variety of crowing, cooing, and screaming noises; mainly silent at sea.

DIMW

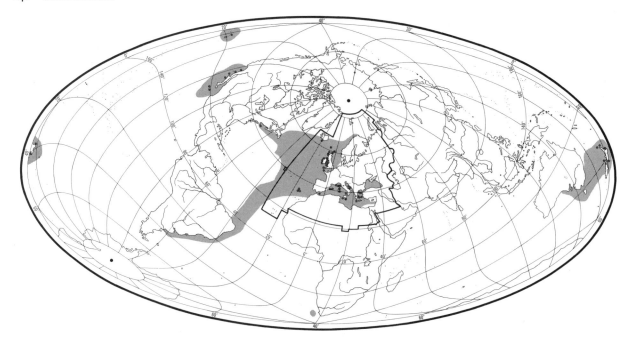

**Habitat.** Marine and aerial within sub-oceanic waters of eastern North Atlantic from subarctic fringe to subtropical fringe of north Canary Islands, and in winter into tropical seas of eastern South America. Nominate *puffinus* prefers continental shelf to deep ocean, and usually avoids inshore waters unless visiting nest. (Mediterranean races *mauretanicus* and *yelkouan* have to pass through narrows regularly.) Breeds normally on inshore islands from sea-level to *c.* 700 m. Visits only in darkness, often with minimum time on ground surface for alighting or take-off, but sometimes, especially non-breeders, may spend hours on surface. Land habitat otherwise subterranean, in dark burrows or crevices, mainly on flat tops or slopes of islands or promontories with fairly deep soil and rough herbage, often shared with rabbits *Oryctolagus cuniculus* and Puffins *Fratercula arctica*; sometimes on mountains near sea (Rhum, west Scotland). Exceptionally will nest under buildings (Skomer and Skokholm, Wales), but normally avoids human contacts. On wing, keeps usually within 10 m of water surface, flight style being wholly adapted to this zone of airspace, deriving lift from airflow over sea. Vulnerable to disturbance of breeding sites and to introductions of predators.

**Distribution.** Inadequately known in many parts of range. In North Africa, breeding suspected on Islas Chafarinas (Morocco) and Zembra and Zembretta (Tunisia) (Heim de Balsac and Mayaud 1962; Jarry 1969). Perhaps several undiscovered in Aegean; enormous numbers of *yelkouan* passing daily through Bosporus and Dardanelles presumably engaged in feeding movements from Aegean colonies, or possibly undiscovered ones in Sea of Marmara. Bred islet in Black Sea, off Bulgaria 1963

(SD). In Azores, known breeding only Flores 1865, and more recently Corvo (Bannerman and Bannerman 1966); no valid breeding record for Canary Islands (Bannerman 1963). In North America, first recorded nesting 1973; one pair, island off Massachusetts, USA (J Baird). Formerly bred Bermuda.

Accidental. Sweden, West Germany, Austria, Switzerland, Rumania, Egypt.

**Population.** FAEROES: estimated 10 000 pairs 1969–70 (W R P Bourne). BRITAIN and IRELAND. Several island colonies reduced or became extinct, especially in last century, often through accidental introduction of brown rat *Rattus norvegicus* (Parslow 1967). Few reliable population figures, but on Skokholm, Wales, estimated 30 000–40 000 pairs by ringing/recapture techniques (Harris 1966; Perrins 1967); Skomer, Wales, *c.* 95 000 pairs 1971 (Corkhill 1973); and Rhum, Scotland, *c.* 70 000 pairs using sample counts (P Wormell). Total population Britain and Ireland almost certainly over 175 000 pairs and may well exceed 300 000 pairs (Cramp *et al.* 1974). FRANCE. Major decrease of nominate *puffinus*. Banneg, Molène archipelago, Brittany, now only breeding place; 30 pairs 1960, now under 10; abundant there last century and early this, and present elsewhere in archipelago and nearby in 1950s (J-Y M).

Survival. Ringing results indicate 27% of young ringed prior to fledging on Skokholm, Wales, survived to age of 2 years or more; survival rate positively correlated with weight at time of fledging. For breeding adults (normally 4 years or older), mean annual survival rate 90%, indicating average life span from first breeding of *c.* 10 years (Perrins *et al.* 1973). On Skomer, Wales, predation by Great Black-

backed Gulls *Larus marinus* estimated to account each year for less than 2% of adult shearwaters and smaller number of fledglings (Corkhill 1973). Oldest ringed bird 27 years 2 months ringed as full-grown, so at least 29 years old (C M Perrins).

**Movements.** Different populations migratory or dispersive. Nominate race almost total migrant to northern breeding areas, where rarely recorded November–January; majority perform long journeys, including regular post-breeding migration from Europe to South America by juveniles and adults. Southward passage begins July, juveniles following September; return movements to breeding areas late February to early April, with influx of older immatures about May. Details of movements mainly from ringing in Wales. Breeding birds make feeding movements of some days duration, but no confirmation yet for theory that these reach Bay of Biscay (Lockley 1953), though some suggestion from recoveries that may do so in pre-laying exodus; could not do so after hatching of young when absences too short. Large winter concentration, especially October–December, off east coast of South America, 10°S–40°S, but most recoveries between 20°S–30°S (Thomson 1965; Cook and Mills 1972). Dearth of recoveries anywhere between January and early February suggests return by different route, possible counter-clockwise movement in South Atlantic, for which some support from recent sightings of small numbers off southern Africa in January–February (Lambert 1971). Immatures do not come ashore at colonies in 2nd calendar year but degree of return to European seas uncertain (Harris 1966). Small numbers occur off New England–Newfoundland area (vagrant to south Greenland) late March to mid-September (Post 1967); the few recoveries indicate these include high proportion of 2nd calendar year birds which may have migrated north up west Atlantic from winter quarters (Perrins et al. 1973). Rare in North Sea; has occurred as vagrant in several European countries with exceptional records far inland. In Britain, a few juveniles found inland after gales in most autumns. Atypical recoveries of Welsh-ringed birds (mostly recently fledged) in Norway and Denmark (September), Mediterranean France (February), Canary Islands (September), Rio de Oro (September), Ghana (October), Trinidad and Grenada (November), and South Australia (November). One of unknown origin found New Zealand June 1972 (Kinsky and Fowler 1973). Nominate race also breeds subtropical North Atlantic islands; movements unknown, but reported absent from Madeira November–February.

After breeding, *P. p. mauretanicus* more widely dispersed in Mediterranean; many or all emerge into Atlantic and occur off west Morocco, some then moving north off coasts of west Europe to Britain, exceptionally to Denmark and south Norway; regular in English Channel June or July to October, and not infrequent in (mainly western) North Sea (Nicholson 1952); but winter range inadequately known. *P. p. yelkouan* likewise more widely dispersed in Mediterranean area after breeding, some moving west to Straits of Gibraltar but largest numbers in Adriatic, Aegean and Black Seas, and off Levant coast (Frank 1952).

Like other shearwaters, markedly gregarious, moving both by night and day; commonly seen in small flocks flying low over sea. Some nestlings ringed Wales reach southern limit of tropics within 5–6 weeks, and one fledgling reached Brazil within 17 days. During this southward migration birds do not spend time in the tropics (Harris and Hansen 1974) and the young may be able almost to reach Brazil without having to feed (Perrins et al. 1973). Breeding adult used in homing experiment crossed North Atlantic (5150 km) in $12\frac{1}{2}$ days (Mazzeo 1953). Many ringing records of accurate return to breeding place.

ALT

**Food.** Fish, mostly small; also cephalopods, small crustaceans, and surface floating offal. Feeds by day by pursuit-plunging, pursuit-diving, and by surface-seizing; in varying numbers from single birds to small flocks.

Fish, mostly Clupeidae and especially herrings *Clupea harengus*, sprats *Sprattus sprattus*, pilchards or sardines *Sardina pilchardus*, and anchovies *Engraulis encrasicolus*.

Remains of fish and small cephalopods found in stomachs from south-west Irish seas (Jespersen 1930),

whereas mostly anchovies (as many as 43 small ones in single stomach) found in birds from Crimean waters (Dementiev and Gladkov 1952). Large flocks congregate to feed on huge anchovy shoals that gather on surface near eastern shore of Crimea October–December (Yu A Isakov). Lockley (1953) suggested that British populations appeared to follow migratory shoals of sardines as they moved north; these may, however, not have been breeding birds (W R P Bourne).

Young fed partly predigested fish remains, particularly sprats and sand-eels Ammodytidae (Lockley 1942); also young herrings (Witherby et al. 1940).

**Social pattern and behaviour.** Compiled mainly from material supplied by M P Harris and C M Perrins, based on observations on Skokholm, Wales; see also Lockley (1942, 1959).

1. Gregarious, especially when breeding. Feeds in flocks at all times of year (little winter information) but aggregations may be caused by patchy food supply. In breeding season, often assembles off shore in late afternoon in large rafts before visiting colonies after dark. BONDS. Life-long, monogamous pair-bond the rule though divorces do occur, i.e. both partners survive between seasons but then mate with other individuals; rarely, new adult pair will breed in first season, but more usual to associate in burrow for season before breeding. No evidence that paired birds associate at sea. Immatures do not normally return to natal sub-colony when 1 year old but in 2nd, 3rd, and 4th years of life do so in growing numbers and increasingly early in season, usually pairing and occupying burrow for at least part of one season before nesting (Harris 1966b; Perrins et al. 1973). Both parents tend young until 8–9 days before fledging. BREEDING DISPERSION. Nests in dense, often huge colonies, almost certainly through choice as apparently suitable areas and islands nearby often not used. Nest-burrows often grouped into discrete sub-colonies, perhaps due to lie of land; breeding adults remain faithful to these sub-colonies. Colonial nesting may be advantageous in reducing predation, though disadvantageous in facilitating interference between pairs (causing egg losses) and spread of epizootics of puffinosis. Breeders minimally territorial, defending only nest-chamber, not entrance or ground round site; several pairs may share single entrance but not normally same chamber. Competition for nest-sites, both intraspecific and interspecific (especially with Puffins *Fratercula arctica*), occurs at some colonies. ROOSTING. Presumably rests mostly on water as does not visit land except for breeding and prospecting. Only duty bird normally sleeps in burrow when ashore, but before laying and after failure of nesting attempt both spend some days there; this, however, probably connected with territorial protection of burrow and with courtship.

2. Social activity in offshore gatherings usually confined to periodic flights when part or all of flock rises and flies round in circles or figures-of-eight (Lockley 1942). Entirely nocturnal ashore at breeding grounds, activity reduced on moonlit nights (Harris 1966a). Display and related behaviour virtually unknown due to nocturnal and subterranean breeding habits, but highly vocal at colony during night—in air, on ground, and from within burrow. Calling of all sorts inhibited on moonlit nights, when birds which do come ashore usually silent. Calls also heard rarely from burrows by day, or from rafts at sea. AERIAL ACTIVITY. Incomers utter characteristic Flight-calls at night as they fly in from sea and circle colonies; others flying over and possibly prospecting places where do not breed also call. Nocturnal flight-

calling most intense early in season when breeding adults return, then again much later when numerous non-breeders present; latter return to colony later than adults, at Skokholm commonest during new moons of June and July (Perrins *et al.* 1973). GROUND ACTIVITY. Breeding pairs occupy burrows several weeks before laying, spending about quarter of available days ashore (see Roosting). Underground, pitch of Burrow-calls given by lone bird or pair varies with excitement, e.g. disturbance of pair often followed by increase in intensity and volume of calling, perhaps to warn off intruders. Birds already in burrows answer those flight-calling, and mates possibly recognize each other's voices individually. Similar but quieter, more conversational notes uttered by courting pair in burrow; as well as calling, some reciprocal Allopreening occurs and Courtship-feeding of one of pair by other, accompanied by much calling, also once recorded. Copulation probably normally in burrow, though also reported on surface nearby at night (Lockley 1942); ♂ grips ♀ by feathers of neck and both call. RELATIONSHIP WITHIN FAMILY GROUP. Little information. As with other petrels, chick largely left alone in burrow after brooding period over, and visited only for feeding. At entry of parent into burrow, chick starts Begging-call (Lockley 1942).

**Voice.** Nominate *puffinus*. Mainly silent at sea even when assembled in flocks off breeding places (Lockley 1942), though calls, of type heard at night ashore, sometimes given (see Witherby *et al.* 1940).

Variety of rapidly repeated, raucous, often piercing crows, coos, croons, howls, and screams uttered over land, on ground, and in burrow, often by pair in duet, especially during period before eggs laid (Witherby *et al.* 1940; Lockley 1942, 1959). Such calling largely nocturnal; inhibited by moonlight, though occasionally heard from burrow during day. Considerable individual variation, evidently mainly in pitch (Lockley 1942, 1959). Generally begins with staccato cackle, then crooning caw, followed by brief, harsh indrawing of breath, e.g. 'cack-cack-cack-carr-hoo' (Lockley 1959); from recordings, also evident that inspiration of breath may sometimes come at beginning of phrase instead of at end (J Hall-Craggs). Other renderings include following adapted from Witherby *et al.* (1940): guttural 'goch-och-aarka-a', commonly with inflections resembling 'it-i-corka' or 'it-IS-yor-folt' (see I).

CALLS OF YOUNG. Little known, but 'excited squeaks' and 'piping cry' heard during feeding, when parent also makes 'muffled chortling' sound (Lockley 1942, 1959). 'Piping cry' generally used, often increasing in intensity when humans walked over burrow; nearly full-grown juveniles occasionally uttered adult 'cack-cack-cack-carr-hoo' cry when handled (P J Conder).

**Breeding.** Based mainly on Harris (1966*a*, *b*); Skokholm, Wales. SEASON. See diagram for British Isles. Eastern Mediterranean: eggs April–May. Western Mediterranean: eggs mid-March onwards (Palmer 1962). SITE. On flat or sloping ground usually close to sea, but sometimes 1–2 km inland. Colonial; burrows close together. Burrows re-used in successive years. Nest: burrow excavated in soft ground with chamber at end. Usually 1–2 m long but sometimes only 50 cm; diameter of hole *c.* 15 cm. Rabbit *Oryctolagus cuniculus* holes also used. Nest chamber usually lined with small pieces of grass, bracken, and roots. Some green vegetation taken into chamber after hatching. Building: burrows excavated by both sexes using bill and feet. EGGS. Broad ovate, not glossy; white. 61 × 42 mm (56–68 × 39–45), sample 100 (Schönwetter 1967). Weight 58 g (52–63), sample 10. Clutch: 1. One brood. Replacement egg recorded only rarely and in one case may have followed mating of ♀ with different ♂. INCUBATION. 51 days (47–55), sample 43. Can be prolonged by chilling, egg being remarkably resistant at all stages of incubation; up to 63 days recorded (Matthews 1954; Ralphs 1956; C M Perrins). By both sexes, ♂ usually taking 1st spell. Spells average 5·9 days (1–26), sample 288, with 6–12 spells per incubation period. YOUNG. Semi-altricial and nidicolous. Cared for and fed by both parents. One parent stays with young for first few days and feeds it at intervals; thereafter only night visits made to nest. 2613 visits gave mean interval between feeds of 1·69 days (range of mean intervals 1·47–2·21); only 4·5% of intervals greater than 3 days. Visiting slightly inhibited during full moon though incubation spells not affected (Ralphs 1956). Young deserted by parents at *c.* 60 days, average 8–9 days (2–15, sample 68) before leaving nest. FLEDGING TO MATURITY. Fledging period 70 days (62–76), sample 53. Become independent when deserted by parents (see above). Age of first breeding usually 5 or 6; single records of 3 and 4, with a few not until 8 or older. BREEDING SUCCESS. Of 56 eggs laid, 78% hatched, most eggs being lost during burrow disputes between adults. Of 44 chicks hatched, 95·5% fledged; losses from starvation, diseases, trampling, flooding of burrows, etc.

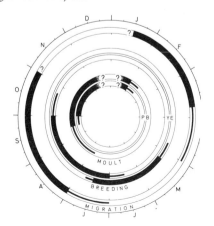

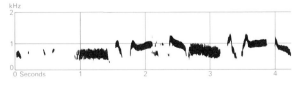

I A Soper/BBC Wales June 1954

**Plumages** (nominate *puffinus*). ADULT AND JUVENILE. Upperside black, extending below eye on side of head, sharply contrasting with white chin. Underparts white; sides of neck mottled black and white, with white crescent behind ear coverts; sides of breast mottled, occasionally a few dark spots on centre of chest; lower flanks usually with some dark spots; lateral under tail-coverts variably mottled black near tip; thighs black. Tail black. Flight-feathers black, paler below; under wing-coverts white, greater tipped grey, lesser near leading edge grey; axillaries white with tips variably mottled or suffused dark grey. Much variation in dark mottling on sides, but belly always pure white. NESTLING. First down long and soft, short round base of bill and eyes; grey-brown above, whitish below, centre of belly grey-brown. Second down slightly darker and denser. The 2 downy plumages sometimes difficult to separate (Witherby *et al.* 1940). Second down begins to grow at 6–12 days after hatching; wing quills appear at 15 days, body quills at 17, tail quills at 28. Primaries protruding from sheaths at 25–30 days (Ralphs 1955, 1956).

**Bare parts**. ADULT AND JUVENILE. Iris black-brown. Bill black, lower mandible paler, cutting edge tinged green. Tarsus, inner toe, and base of middle toe pale flesh, brighter tinge in juvenile; back of tarsus, rest of toes, and claws black; webs pale pink with black markings.

**Moults**. ADULT POST-BREEDING. Complete; primaries descendant. Main moult of nominate *puffinus* in South American winter quarters; poorly known. Body feathers start July–August, followed by wing feathers; moult completed February–April (Witherby *et al.* 1940). Non-breeding immatures may moult earlier (Murphy 1936; Palmer 1962). Subspecies *mauretanicus* moults June–October when present Atlantic coast of Europe. Primaries, secondaries, and body feathers start about same time, tail later. Secondaries moulted in four groups, s1–s4, s5–s12, 13, or 14, and three innermost ascendantly, s18 or 19–s14, 13 or 12 descendantly. Tail irregular (Mayaud 1931).

**Measurements**. Full-grown; skins (BMNH, MNHN, RMNH, ZFMK, and ZMA).

nominate *puffinus*

| | ♂ | | | ♀ | | |
|---|---|---|---|---|---|---|
| WING | 239 | (3·66; 19) | 231–243 | 235 | (4·50; 14) | 226–242 |
| TAIL | 74·7 | (2·85; 15) | 70–79 | 74·9 | (2·89; 9) | 70–79 |
| BILL | 34·9 | (1·30; 16) | 33–38 | 34·3 | (1·26; 15) | 31–36 |
| TARSUS | 45·6 | (1·46; 16) | 43–48 | 44·4 | (0·85; 14) | 43–46 |
| TOE | 49·5 | (1·31; 8) | 48–52 | 47·4 | (1·60; 8) | 44–49 |

| | *mauretanicus* ♂ and ♀ | | | *yelkouan* ♂ and ♀ | | |
|---|---|---|---|---|---|---|
| WING | 246 | (6·02; 19) | 235–256 | 235 | (5·16; 23) | 224–244 |
| TAIL | 74·8 | (2·22; 19) | 70–79 | 69·9 | (3·14; 23) | 64–76 |
| BILL | 38·4 | (1·69; 25) | 36–42 | 35·7 | (1·61; 23) | 32–38 |
| TARSUS | 48·1 | (1·39; 24) | 46–51 | 45·5 | (1·20; 23) | 43–47 |
| TOE | 54·6 | (1·33; 9) | 52–56 | 50·6 | (2·28; 20) | 47–55 |

Sex difference significant for wing, tarsus and middle toe of nominate *puffinus*; not significant for other subspecies, hence sexes combined. Average wing of 10 live British breeding birds 236 (s.d. 3·74), of 23 *yelkouan* from Malta 229 (s.d. 6·22); bill of 14 Malta birds 36·9 (s.d. 1·45) (BTO). 4 adult *yelkouan* with exceptionally short wings (209, 211, 214, 216) not included in tables. Mayaud (1932) mentioned nominate *puffinus* with wing ♂ 216, ♀ 214.

**Weights**. British breeding birds: March 478 (29·92; 67) 430–545, June 424 (42·43; 97) 350–535; difference significant (BTO). Averages of individual birds in breeding season Skokholm ranged ♂♂ 395–459, ♀♀ 375–447. Maximum nestling weights in 2 years averaged 645 and 610, fledging weights 465 and 444 (Harris 1966a), so fledging juveniles as heavy as adults in good condition. Adult ♂♂ *mauretanicus* 537 (25·4; 6) 490–565, ♀♀ 506 (26·0; 7) 472–550 (Mayaud 1932). Adult *yelkouan* breeding Malta: March 349 (18·5; 6), April and May 416 (29·3; 15); difference between these groups significant.

**Structure**. Wing long and narrow. 11 primaries: p10 longest, p9 0–5 shorter (exceptionally 8 in *yelkouan*), p8 10–20, p7 22–34, p6 36–49, p1 114–135. P11 minute. 19–22 secondaries. Tail slightly rounded, 12 feathers: t6 5–10 shorter than t1. Bill long, slender; nasal tubes about $\frac{1}{3}$ of length, slightly raised above line of culmen, nostrils directed upwards. Tarsus laterally compressed.

**Geographical variation**. *P. p. mauretanicus* slightly larger than nominate *puffinus*; *yelkouan* about as large as *puffinus*, but tail shorter; *mauretanicus* also relatively short-tailed (see Measurements). In contrast to nominate *puffinus*, both Mediterranean races have upperparts tinged brown. Otherwise, *yelkouan* similar to *puffinus*, but normally more dusky on axillaries and under tail-coverts. White crescent on side of neck often absent, but sometimes well-developed. Brown tinge of upperparts only diagnostic character in such birds, but occasionally upperside black with slight tinge of brown-grey approaching nominate *puffinus* still more (Nisbet and Smout 1957). *P. m. mauretanicus* has underparts variably tinged brown-grey; in some only throat, sides of breast, flanks, lower belly, and under tail-coverts, in others whole underside. No white crescent on side of neck, no sharp contrast between brown and white on side of head, axillaries entirely grey. For other subspecies see Murphy (1952) and Palmer (1962).

# *Puffinus assimilis* Little Shearwater

PLATE 15
[after page 158]

Du. Kleine Pijlstormvogel    Fr. Petit Puffin    Ge. Kleiner Sturmtaucher
Ru. Буревестник-крошка    Sp. Pardela chica    Sw. Dvärglira

*Puffinus assimilis* Gould, 1838

Polytypic. *P. a. baroli* (Bonaparte, 1857), small islands off Madeira, Selvagens, Canaries, and Azores; *boydi* Mathews, 1912, Cape Verde Islands. 7 extralimital in southern oceans, including nominate *assimilis* Gould, 1838, Lord Howe and Norfolk Islands, and *elegans* Giglioli and Salvadori, 1869, Tristan da Cunha and Gough Island.

**Field characters.** 25–30 cm, wingspan 58–67 cm; much smaller (*c.* two-thirds) than normal-sized Manx Shearwater *P. puffinus* and shorter-winged than any race of that species; similar to Puffin *Fratercula arctica* in total length. A small-billed, short, small, and rather compact shearwater, black above and white below recalling (in east Atlantic) only nominate race of *P. puffinus*. Sexes alike; no seasonal variation, juveniles inseparable. 2 races in west Palearctic rarely distinguishable in the field.

Madeiran and Canary Islands race *baroli*. Crown, hindneck, upper body, flight-feathers, and tail slaty-black (wearing duller but not brown); sides of face and sides of neck mottled white and black, appearing whitish at distance; underparts, including central undertail coverts, white. Underwing, including basal primary area, white with smudges on axillaries and distinct slate border, particularly on trailing edge. Bill short (three-quarters that of *P. puffinus*) with tubed nostrils short and inconspicuously raised over base of upper mandible; black with blue-grey cast on basal sections of both mandibles. Legs and feet of adults chalk-blue, centre of webs and edges dusky grey; of immatures livid flesh. Cape Verde Islands race *boydi*. Undertail coverts mostly black, upperparts duller and browner, and basal area of undersurface of primaries dusky; bare parts apparently bluer.

Difficult to separate from small, particularly runt individuals of *P. puffinus* (both nominate *puffinus* and *yelkouan*), but when seen in circumstances allowing comparison, smaller, shorter bill, restricted headcap, white face and side of upper neck, flight action, and (in *boydi*) dark undertail allow certain identification. Separation from Audubon's Shearwater *P. lherminieri* even more difficult but again smaller bill, fully black (not brownish black) upperparts, cleaner cheeks and sides of neck, blue legs (in adults), white undertail (in *baroli*), and combination of shorter wings and longer tail (in both races) may permit identification in ideal conditions.

Flight action distinct from larger shearwaters, fluttering with many more wingbeats and less gliding or banking over given distance (recalling auks *Alca* or *Fratercula*) but, occasionally, in high winds, modified by inclusion of banking and gliding in manner of *P. puffinus*; progress in flight no less rapid than *P. puffinus*, just less free. Outer wing looks short compared to *P. puffinus* and *P. lherminieri*, heightening auk-like appearance. Gait essentially as *P. puffinus* but somewhat nimbler. Sociable at

breeding colonies but concentrations in mid-ocean unknown, and all vagrants occur singly or in small parties. Feeds from surface or by shallow plunge-diving. Chorus of breeding adults sounds like feeble, high-pitched *P. puffinus*; silent at sea.                DIMW

**Habitat.** Unlike other west Palearctic shearwaters, largely confined to tropical, sub-tropical, and other relatively warm waters close to breeding islands. Like relatives, normally keeps to lowest airspace over sea, flying fast and diving adroitly; seldom, however, needs to ride out storms. Apparently less aerial and more aquatic than most Procellariiformes; somewhat sociable at sea but scarcity of observations and nocturnal habits prevent firm assessment. Pays frequent visits outside breeding period to nesting sites, up to 15 km inland and to 500 m altitude, in clefts, crevices, and other holes (sometimes unoccupied burrows) in rocky or talus slopes, or among dense tussocks; thus one of few pelagic birds habitually coming to land at any season of year. Compared with Manx Shearwater *P. puffinus* spends much more time on surface of ground when at colony. Vulnerable to predation from ground-living hunting creatures, including skinks *Ocypoda cursor*. No evidence of group or flock movement far from breeding stations, or of other than casual long-distance dispersal by individuals. See Bannerman and Bannerman (1968).

**Distribution.** Azores: only one proved case of breeding (São Miguel 1951), but probably nests Graciosa and Santa Maria. Desertas: record of breeding should be treated with caution (Bannerman and Bannerman 1965, 1968). Marine distribution generally inadequately known (see Movements).

Accidental. Britain and Ireland over 40, most 1964–68, (see Movements); France 5; Denmark 1; West Germany (1 inland, Bodensee); Italy 2; Morocco 1. Where specimens obtained, all *baroli*.

**Population.** Cape Verde Islands: probably several thousands; decrease on Razo due to human predation (de Naurois 1969*b*). Canary Islands: small numbers breeding Allegranza, Montaña Clara, and Graciosa, north of Lanzarote (Lovegrove 1971).

**Movements.** Present around colonies east Atlantic all or most of year, partly due to varying nesting seasons (see

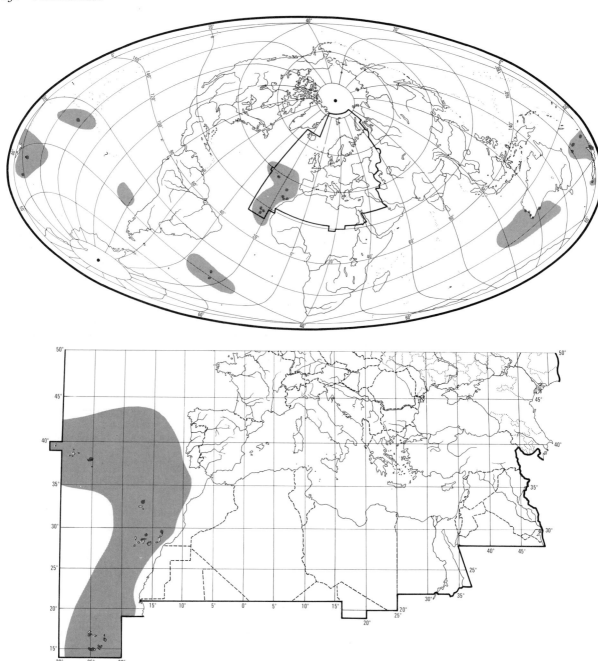

Breeding). All populations dispersive to some extent after breeding, but moulting birds collected ashore Cape Verde Islands late August and apparently small numbers visit nest sites there throughout year, as at Tristan da Cunha (Bourne 1955*b*; Rowan 1952). No long-distance migration; comparatively sedentary tendency in all oceans, with resulting lack of gene-flow, has presumably made possible high degree of subspeciation (Warham 1958*b*). Dispersal largely within tropics and sub-tropics, perhaps mainly by immatures; no clear pattern of dates. Occurs in small numbers off Atlantic coasts of Iberian Peninsula, Morocco, Rio de Oro, and Mauritania, and (fewer) southwards off west African coasts to about 9°N (e.g. Bourne 1963, 1965); also recorded Gulf of Guinea, but possibly from undiscovered colony in region (W R P Bourne).

Northern limits of pelagic dispersal not yet defined; rare visitor to Britain (specimens all *baroli*), but sightings of 26 individuals off west Ireland 1964–68 (mainly August–October) (Smith *et al.* 1969) suggest may be more than accidental visitor to Western Approaches. At least 5 reports of small shearwaters (*P. assimilis* or Audubon's Shearwater *P. lherminieri*, or both) in mid North Atlantic in region of 42°N–50°N, 33½°W–49°W August, November and December (Post 1967; Bourne 1970), nearest breeding station being in Azores (*baroli*). This race also vagrant (specimen records) to North America (South Carolina, Nova Scotia) and various countries in west Palearctic (see Distribution).

**Food.** Information limited. Chiefly fish and cephalopods caught by surface-seizing, also by pursuit-plunging and pursuit-diving just below surface. 6 stomachs of race *boydi* contained fish and cephalopod remains up to 10 cm (Bourne 1955); also in 2 cases an amorphous yellow-green mass probably due to bile-staining (W R P Bourne). In other races, small fish, cephalopods, and crustaceans recorded (Palmer 1962; Warham 1958*b*), but no data for race *baroli*.

**Social pattern and behaviour.** Based mainly on material supplied by J Warham.

1. Gregarious, especially ashore. At sea, though sometimes singly, small flocks more frequent in waters round island colonies. Disperses at sea after breeding but, like number of other non-migratory petrels, later pays nocturnal visits to nesting places well before next breeding season. Thus recorded Selvagens in July, 6 months before laying (Jouanin 1964), while some present ashore on Eclipse Island, Western Australia, during 10 months of year, and plentiful and noisy at night at least 4 months before laying (Warham 1957). At least some then visiting sites previous breeders; those on Selvagens, July, in full moult and ♀♀ examined had laid eggs in previous breeding seasons (Jouanin 1964). Timetable of visits by other age-classes unknown. Term 'protogamic' proposed to describe such returns to breeding place between 2 reproductive periods (Jouanin 1964). BONDS. Obviously in pairs at breeding station; however, little known about duration of pair-bond in absence of large-scale ringing but sustained monogamy likely as in other, better-known shearwaters. Both parents tend chick until some days before fledging. BREEDING DISPERSION. Forms small to large colonies on islands. Nests close together in large colonies where ground suits tunnelling, but more scattered on rocky substrates where niches not plentiful; no precise data on densities. Territory, consisting of burrow and immediate vicinity, defended from intruders up to 4 months before first layings of season; birds also defend own individual-distances on ground within colony. ROOSTING. Presumably rests and sleeps in or near nest-cavity when ashore; otherwise on sea.

2. Strictly nocturnal at breeding station, arriving after dark and departing before dawn. Little recorded of behaviour there, except for extralimital race *tunneyi* on Eclipse Island (Warham 1955, 1958*b*). AERIAL ACTIVITY. At least during protogamic period (see above), chases between individuals apparently of opposite sex occur over colony, with much calling; such flights seem less frequent during breeding season, at least in Western Australia, those occurring then probably largely due to younger, pre-breeding element of local population. These chases differ from incoming flights of birds locating own burrows, when several trial passes made before each flutters wings through small amplitude, stalls, and drops to ground, with little calling throughout (Warham 1958*b*). GROUND ACTIVITY. Nocturnal fighting recorded on Eclipse Island between birds of unknown

sex, apparently for possession of burrow: individual at entrance held on to wing of another partly within tunnel and protracted struggle ensued to accompaniment of raucous growling cries; eventually, attacked bird forced passage from burrow and staggered off, leaving attacker free to enter without further opposition. In threat against human intruder, some attacked with bill and lifted half-open wings from body, but others docile when handled. During protogamic visits, those present during day remain silent out of sight in burrows, but can sometimes be stimulated to call by close footfalls. At night, birds heard and seen calling when 2 or more together on ground; calls also heard occasionally from lone ones perched in trees. On Gough Island, nocturnal calling delivered from rock or other prominent perch (as well as in flight), but never heard from burrow (Swales 1965*a*). On Cape Verde Islands, up to 7 noted gathering on exposed mounds to display (Bourne 1955*b*); on Eclipse Island, pairs or, at most, trios call, Allopreen, and Bill-fence on ground while vocal activity from flying birds also at high pitch; much social stimulation apparent at such times, though no copulations observed. Pair engage in calling duets together, interspersed with Allopreening; latter unilateral, reciprocal, or simultaneous, bird working tip of bill among feathers of head and nape of partner. Such behaviour may well lead at some stage to copulation, but no first-hand accounts. Apparently no differences between mutual displays in breeding season and in earlier protogamic period, but all displays more frequent during latter. Pair illustrated calling in duet after one had fed chick (Warham 1958*b*); both had arched napes, swollen throats, and somewhat lowered bills. RELATIONSHIP WITHIN FAMILY GROUP. Parent-chick behaviour described by Warham (1958*b*) at burrows fitted with false roofs and lit with red light. To initiate feeding, chick begged by giving chirruping calls and driving bill at parent's head. Adult meanwhile often Allopreened chick's head, this leading to mutual Bill-fencing, chick still calling vigorously, and eventual feeding in silence during which chick usually beat wings. Series of such meals, usually given soon after parent's evening return, generally completed within 30 minutes, after which both adult and chick tended to sleep.

**Voice.** Silent at sea. Highly vocal at night while ashore, especially when no moonlight and during period of first landfall. Descriptions of main calls at breeding grounds vary considerably; difficult to reconcile, both within same race and between different races.

On Selvagens, nocturnal calling of adult *P. a. baroli* rendered 'phwee-her-her-wher' but with many variants; resembles typical scream of Manx Shearwater *P. puffinus*, but higher and weaker, 'more of a squeal than a scream', rising to excited, wheezing finish (Lockley 1942, 1952). Basically same calls heard both in flight and from birds courting in crevice. From recordings, calls may be described as consisting of variable, 5–8 syllable, laughing phrase, e.g. 'ha HA ha-ha-ha HOOOO' (see I), 'ha HA ha ha-ha-ha HOOOOOO', 'ha HA ha-ha-ha-ha-ha HOOOOO', etc., all with stress on short 2nd note and prolonged final one; sound high-pitched, but fundamentals low (J Hall-Craggs). Also extremely low-pitched, 5-syllable, growling call uttered in same time pattern as 'ha HA-ha-ha HOO' variant of laughing call, apparently given simultaneously by second bird. Flight-call of *P. a. boydi* (Cape Verde Islands) described as high-pitched, crowing

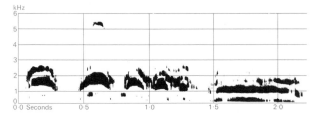

I Alexander Zino/Sveriges Radio Selvagens September 1969

'karki-karroo', birds disturbed or courting in burrows utter variety of squalling notes, rising to typical crowing phrase with increasing excitement (Palmer 1962). 2 birds of race *elegans* recorded while fighting by M K Swales on Gough Island gave variety of long and short, shrill and harsh cries; quite different from laughing call, and in tone quality resembling call of chick of this race described below (J Hall-Craggs). In extralimital race *tunneyi*, raucous growls reported to accompany fighting; also uttered by birds withdrawn from burrows (Warham 1955; Serventy *et al.* 1971).

Main calls of extralimital races *elegans* and *tunneyi* described by Elliott (1957) and Palmer (1962) and by Warham (1955) and Serventy *et al.* (1971) respectively. In *tunneyi*, calls given in flight and from burrow said to be same: throaty, asthmatical 'wah-i-wah-i-wah-i-wah-ooo', with sobbing intake and spluttering termination. In *elegans*, croaks and moans of birds courting on ground contrast strongly with highly variable and curiously unpetrel-like Flight-call—a sharp, plaintive, melodious, haunting, and rather drawn-out whistle ('preep-preep'), though this sometimes develops into shrill scream or lower, dove-like coo. In *tunneyi*, nocturnal choruses of duetting pairs common.

CALLS OF YOUNG. Unknown in west Palearctic races. In *tunneyi*, chicks utter liquid chirrups when soliciting food (Warham 1955); in *elegans*, recording indicates long, shrill, somewhat metallic cries of chick, rather like sound produced by blowing through clarinet reed (J Hall-Craggs).

Breeding. SEASON. Canary Islands: eggs February, earliest young mid-March. Cape Verde Islands: eggs February and March; fledging from June (Bannerman 1959; Bannerman and Bannerman 1963; de Naurois 1969*b*). SITE. In rocky ground, caves, cliffs. Generally colonial, but occasionally scattered nests. Nest: tunnel in soft soil, hole between rocks, and under fallen boulders; also uses old tunnels of other Procellariidae and Hydrobatidae. Building: tunnel excavated in soft ground probably by both sexes. EGGS. Variable, blunt or pointed ovate, not glossy; white. 50 × 35 mm (45–54 × 33–38), sample 55; calculated weight 31 g (Schönwetter 1967). Clutch: 1. One brood. No information on replacements. INCUBATION. Period not recorded in west Palearctic, but 52–58 days for *P. a. tunneyi* in Australia (Glauert 1946). By

both sexes, changing every 48 hours, *P. a. tunneyi* (Glauert 1946). YOUNG. Semi-altricial and nidicolous. Fed by both parents at night. Deserted 8–11 days before fledging (Glauert 1946). FLEDGING TO MATURITY. Fledging period not recorded in west Palearctic, but 70–75 days for *P. a. tunneyi* (Glauert 1946). Becomes independent 8–11 days before fledging. Age of maturity not known. BREEDING SUCCESS. No data.

Plumages (subspecies *baroli*; for others see Geographical Variation). ADULT. Upperparts slate-black, distal greater coverts with narrow white tips. Underparts white, including most under tail-coverts, but lateral and long ones variably marked slate. Sides of face in front of and round eye, sides of neck, and sides of chest mottled dark grey and white, mottling rarely extending across upper breast; flanks with a few grey spots or streaks. Tail black. Primaries and secondaries black, paler on inner webs towards white base. Under wing-coverts white, marginal mottled dark grey; axillaries white, variably marked with grey tips or shaft-streaks. NESTLING. 1st down on upperparts long, grey with brown wash; shorter and paler on underparts, grading to white in middle, narrow grey-brown patch on vent. Chin nearly bare. 2nd down denser, darker, white-tipped; white on underside more extensive. JUVENILE. Newly-fledged birds probably recognizable by pale tips of feathers of upperparts, apparently more pronounced than in freshly moulted adults; otherwise like adult.

Bare parts. ADULT and JUVENILE. Iris dark brown, Bill lead-grey, darker on ridge of culmen. Foot blue, outer side of tarsus and outer toe black. NESTLING. Back of tarsus, outer toe, and sole black; rest of tarsus, 2 inner toes and webs livid flesh (labels on speciments ZFMK, collected by D A Bannerman).

Moults. ADULT POST-BREEDING: Complete. Primaries descendant. About May–August in *P. a. baroli*, but body feathers may start earlier. Moulting birds on Selvagens in July (Jouanin 1964; see also Mayaud 1931). *P. a. boydi* on Cape Verde Islands in full body moult late August, but not apparently moulting primaries (Bourne 1955*b*). Precise relation of breeding season to timing of moult unknown, as also moult of other age-classes.

Measurements. Adult and juvenile (BMNH, MNHN).

|  |  | *baroli* | | | *boydi* | |
|---|---|---|---|---|---|---|
| WING | ♂ 184 | (4·93; 7) | 176–190 | ♂ ♀188 | (3·91; 15) | 180–193 |
|  | ♀ 179 | (5·01; 6) | 170–185 |  | — |  |
| TAIL | 71·8 | (3·66; 14) | 67–78 | 77·6 | (3·96; 14) | 71–84 |
| BILL | ♂ 26·1 | (0·97; 8) | 24–28 | ♂ ♀ 25·2 | (0·95; 14) | 23–28 |
|  | ♀ 25·0 | (0·55; 6) | 24–26 |  | — |  |
| TARSUS | 37·2 | (1·12; 14) | 36–39 | 37·4 | (1·31; 16) | 35–39 |
| TOE | 40·9 | (1·80; 12) | 37–44 | 41·5 | (1·36; 15) | 39–44 |

Although sample small, *baroli* ♂♂ appear to have longer wing and bill than ♀♀, but no difference apparent in other measurements. Witherby *et al.* (1940) gave ranges for wing and bill of ♂ *baroli* 170–187 and 25–28, ♀ 174–182 and 24–25. Longer tail of *boydi* significant; other apparent differences between subspecies need corroboration.

Weights. None recorded for North Atlantic races. Full-grown *elegans* Gough Island: 226 (91), 170–275 (Swales 1965).

Structure. Wing long and narrow. 11 primaries: p10 longest, p9 0–2 shorter, p8 5–10, p7 14–19, p6 25–31, p1 170–192; p11 minute, concealed under coverts. Tail moderately rounded, 12 feathers: t6 5–10 shorter than t1. Bill small and slender; nasal tubes $\frac{1}{4}$–$\frac{1}{3}$ of bill-length, hardly emerging above line of culmen. Tarsus slender, laterally compressed.

**Geographical variation.** Southernmost populations (e.g. *elegans*) tend to be larger, more blue-black above, with pronounced white feather edges, and purer white underwing and under tail-coverts. Populations of lower latitudes resemble North Atlantic *baroli*. Tropical Cape Verde population (*boydi*) differs from *baroli* in brown tinge of upperparts, mostly dark brown under tail-coverts, and pale brown basal parts of inner webs of primaries. In these characters, *boydi* intermediate between other races of *P. assimilis* and Audubon's Shearwater *P.*

*lherminieri*. Several authors (e.g. Vaurie 1965) have united these in single, world-wide species of small shearwaters. In Atlantic, however, *P. assimilis* apparently of southern origin and Audubon's Shearwater *P. lherminieri* of Pacific origin, so for purpose of this work, separation seems warranted. *P. lherminieri* differs from *P. assimilis* by browner upperparts, wholly dark under tail-coverts, darker bases of primaries, and flesh-coloured instead of blue foot; also nestling more extensively white underneath, especially on throat. In addition, nominate *lherminieri* from tropical west Atlantic larger than *baroli* and *boydi*, wing averaging 201, bill 29. In colour of foot, nestling, size, voice, and behaviour, *boydi* similar to other races of *P. assimilis*, *P. assimilis*, so included in that species (see Bourne 1955*b*).

## *Puffinus lherminieri* Lesson 1839 Audubon's Shearwater

FR. Puffin obscur    GE. Audubon-Sturmtaucher

Only west Palearctic record, from Sussex, England, January 1936 (Witherby *et al.* 1940), no longer included in official list (British Ornithologists' Union 1971). *P. l. lherminieri* breeds Caribbean and disperses over sub-tropical western North Atlantic; seldom

in waters far from colonies, though a few wander north in Gulf Stream as far as New England (Post 1967). *P. l. persicus* of Persian Gulf may be recorded sooner or later in Kuwait waters.

# Family HYDROBATIDAE storm-petrels

Small seabirds. 18–21 species in 8 genera. Fall into 2 groups representing main radiations, each originating in different hemispheres, though overlapping in tropics: (1) southern species (*Oceanites*, *Garrodia*, *Pelagodroma*, *Fregetta*, and *Nesofregetta*), and (2) northern species (*Hydrobates*, *Halocyptena*, and *Oceanodroma*). In west Palearctic only *Pelagodroma* (1 species breeding), *Hydrobates* (1 species breeding), *Oceanodroma* (2 species breeding, 1 accidental) and *Oceanites* (1 off-season visitor). 1st group characterized by long and slender legs, rather short and rounded wings with 10–11 secondaries and 2nd by short legs, long and pointed wings with *c.* 14 secondaries. Usually all-black or black plumage with white rump. Flight light, erratic, usually low over water, bouncing, fluttering, swooping, or skimming. 11 primaries: p9 longest, p11 minute. Secondaries rather short, diastataxic. Tail mostly long, often forked, but square or rounded in some; 12 feathers. Bill fairly small, but with heavily hooked tip; nostrils fused to single tube over basal part of culmen, opening often directed upwards. Tarsus rounded; lower part of tibia bare. 3 anterior toes webbed, hind toe rudimentary; nails sharp, flattened in some species. Crouch on tarsi when on land, gait shuffling. Oil gland feathered. Aftershaft small. Peculiar musky odour. Sexes similar in plumage, ♀ slightly larger than ♂. 1 moult per year; primaries replaced in descendant order. Young semi-altricial and nidicolous, development slow. 2 downs, but at least in some long-legged species only 1; long and fluffy. Juveniles similar to adults.

    Cosmopolitan in all oceans. Resort to inshore waters and

land solely for breeding, usually on inaccessible and remote coasts and islands, seeking concealment in holes, crevices, or burrows; usually come ashore only at night. Usually fly at low height above land or water. Interspecific differences of habitat mainly related to preferences for warmer or cooler temperate seas and to extent of dispersal at long distances from breeding quarters. In case of Wilson's Storm-petrel *Oceanites oceanicus*, this leads to temporary seasonal inflow of Antarctic population into regions of North Atlantic temporarily vacated by local species when breeding. By contrast, those breeding on island groups in subtropical and warm temperate Atlantic appear to remain largely within easy distance of nesting sites throughout year. Possibly owing to lower biological productivity, extensive areas of mid-ocean seem to be relatively unfrequented—though this situation may be more apparent than real as Madeiran Storm-petrel *Oceanodroma castro* found there even in poor tropical blue waters. Perhaps habitually nocturnal as well as diurnal at sea. Small size and elusive pattern of life have seriously restricted collection of comprehensive data. In west Palearctic, breeding distribution still not fully known and limited data on marine distribution in many areas. Little information on changes in ranges or numbers. West Palearctic species apparently all long-distance migrants, wintering sub-tropical and tropical seas, though only Storm Petrel *Hydrobates pelagicus* and *O. oceanicus* known to be consistently trans-equatorial. Pelagic movements both nocturnal and diurnal; more often attracted to ship lights and lighthouses than Procellariidae. Also much more prone to wrecks after gales (see especially Leach's Storm-petrel *Oceanodroma leucorhoa*).

Principal foods planktonic crustaceans, molluscs, small fish; also oily and fatty substances. Some species habitually follow ships, taking scraps and natural prey in wake. Mainly flight-feeders, taking prey at surface while hovering, fluttering, pattering, stepping, and to lesser extent, swooping—often following erratic path; to much lesser extent, also surface-feed by seizing prey while swimming. Seldom plunge-dive or surface-dive. Gregarious or solitary at sea. Colonial when breeding, colonies containing many unemployed birds including pre-breeders. Long-term, monogamous pair-bonds the rule, often maintained from season to season though unlikely that mates (or parents and fledged young) associate at sea in any season. Probable that pair-bond continues from year to year largely because of site-tenacity rather than vice-versa (but see White-faced Storm-petrel *Pelagodroma marina*). When ashore, roost in nest-hole but largely absent at sea after egg-laying except when taking turns at incubating, brooding, or feeding young; at sea, presumably rest at times on water, but may do so mainly on wing. Limited information on display and related behaviour. Seem highly tolerant of conspecifics near or actually within nest-hole; in some species, notably *H. pelagicus*, visitors not uncommonly enter occupied burrows, particularly before egg-laying. Due to nocturnal and subterranean nesting habits, little specialization for visual displays; heterosexual communication tactile (e.g. Allopreening) and probably olfactory. Great emphasis on vocal display. Song (or Burrow-call) given as advertisement from hole by pair or single bird, mainly at night. Prolonged, often slow, more or less continuous churring, purring, or crooning with distinctive temporal pattern; often relatively far-carrying. Nocturnal Flight-circuiting over colony, with loud Flight-calls, another notable feature, starting well in advance of laying; significance not fully established. Little known on vocalization at sea, nocturnal calling while feeding gregariously probably more common and widespread than generally believed. Little known of comfort behaviour or general behavioural characters.

Protracted breeding seasons except in sub-arctic where much more strictly seasonal. Shortage of secure breeding cavities may sometimes involve successive occupancy by different species with consequential adjustments of breeding seasons. Nest in holes, both natural or excavated by both sexes, often with tunnel up to 1 m long; chamber left bare or lined with grasses, roots, etc. Eggs ovate, not glossy; white. Clutch 1; single brood. Replacements possible but rarely recorded. Incubation period long, 40–50 days; both sexes, each with single median brood-patch, take approximately equal shares in spells of up to 6 days. Eggshells usually left in nest. Young cared for and fed, by incomplete regurgitation, by both parents. Not left alone for first 5–7 days, then visited nightly for feeding with little association otherwise; feeding frequency drops towards fledging, but not abandoned until fledging at 59–73 days when become independent. Mature at 4–5 years where known.

## *Oceanites oceanicus* Wilson's Storm-petrel

PLATES 17 and 18
[facing page 159 and after page 182]

Du. Wilsons Stormvogeltje  Fr. Pétrel océanite  Ge. Buntfüssige Sturmschwalbe
Ru. Темнобрюхий океанник  Sp. Paíño de Wilson  Sw. Havslöpare  N.Am. Wilson's Storm Petrel

*Procellaria oceanica* Kuhl, 1820

Polytypic. Nominate *oceanicus* (Kuhl, 1820), subantarctic South Atlantic and Indian Ocean: *exasperatus* Mathews, 1912, antarctic; both migrate into west Palearctic. Extralimital: *maorianus* (Mathews, 1932), New Zealand seas.

**Field characters.** 15–19 cm, wingspan 38–42 cm; slightly larger than Storm Petrel *Hydrobates pelagicus*. Black with white rump, long legs usually projecting well beyond square tail, but sometimes drawn in. Sexes alike and no seasonal differences; juvenile not separable.

Mainly sooty-black above and slightly browner below, darkest on flight-feathers and tail, apart from white rump patch extending to lower flanks and lateral under tail-coverts and, particularly in fresh plumage, narrow light diagonal across inner wing (caused by white tips and fringes to greyish greater coverts). Tail noticeably square, rounded at corners. Bill and long, spindly legs black, feet with variable yellow centres to webs (often invisible, and rather inconspicuous even at close range and when legs dangling).

Distinguished from other white-rumped, black storm-petrels by combination of long legs (with yellow on webs when visible), square tail, and in fresh plumage, narrow pale wing diagonal. Also differs from *H. pelagicus*, which it most closely resembles, in uniformly dark underwing and generally darker appearance, rounded wings and robust shape, and flight compared with swallows (Hirundinidae) rather than bats (Chiroptera).

Like *H. pelagicus*, flight alternation of fluttering and gliding, but fluttering stronger and interspersed glides longer; also regularly patters on surface like other storm petrels, quickly dipping feet 3–4 times between short-glides as it hangs on extended wings to pick up food. In addition, 'walks' (or, rather, hops) on surface of water, planing down with rather short and broad wings level or

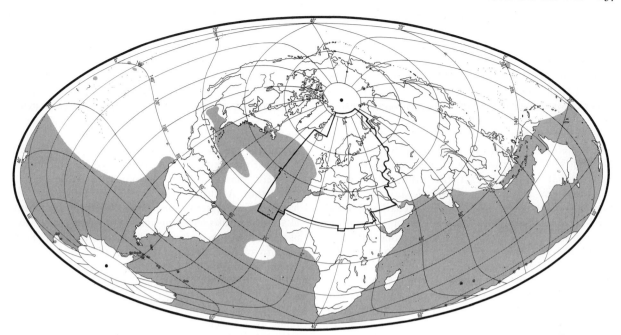

slightly raised, tail fanned, and bill angled down, stretching forward with legs until toes dip in water and then pushing back with them as it rises; will also 'stand' on water, facing into wind with wings rigid and webbed feet acting as anchors, while it drifts backwards in breeze. Sometimes flies up to 2–3 metres above sea in more purposeful, less fluttering manner with feet retracted, looking like wader; or darts high in air, like feeding House Martin *Delichon urbica*; or, towards sunset, dashes to and fro, shooting up in air and plunging down in excited manner. Feeds mainly in flight, but recorded diving after food. Sometimes settles on water, occasionally in groups. Less markedly pelagic than some other storm-petrels, frequently found in shallower waters of continental shelves; often follows ships and fishing boats, circling in wake. Singly or in parties up to hundreds and, in Antarctic, even thousands round whaling stations. Generally silent at sea like other storm-petrels, but feeding flocks sometimes repeatedly utter peeping or chattering sounds, even after dark. See Alexander (1955), Witherby *et al.* (1940), Palmer (1962), Serventy *et al.* (1971).

**Habitat.** In west Palearctic, marginal and atypical, small minority of non-breeding birds from southern latitudes temporarily occupying warm waters in Cape Verdes and Madeira seas and up to Bay of Biscay during northern summer. Despite scientific name, avoids central tracts of ocean, preferring offshore and even inshore waters, especially where cool currents bring concentration of plankton; often follows ships closely. Infrequently seen swimming and rarely diving; predominantly in air immediately over water, with habit of pattering feet on surface. When catching insects or displaying over land sometimes rises to some height. In west Palearctic hardly ever touches land, being relatively immune from storm-driven catastrophes; crowded concentrations typical in other parts of range not encountered, and normally distributed singly or in small parties over wide expanses of sea. Like other storm-petrels, frequents mainly rocky, more or less remote islands for breeding.

**Distribution.** Known breeding areas shown on map (based on Watson *et al.* 1971), but probably nests on almost every suitable exposed rocky area of coast of Antarctic continent and may breed on majority of offshore islands (Palmer 1962).

Accidental. Britain and Ireland: at least 7, possibly others 19th century, with 3 recent—1967, 1969, 1970 (British Ornithologists' Union 1971). France: singles 1872, 1883 (CE). West Germany: one 1963 (Bauer and Glutz 1966). Italy: singles Sardinia 1863, 1956 (SF).

**Movements.** Migratory. Present at antarctic and sub-antarctic colonies November to mid-April, when pelagic distribution mainly south of 50°S, but some immatures spend southern summer in tropics off South America and South Africa. Off-duty adults and non-breeders then widely distributed in higher latitudes of Southern Ocean, sometimes in thousands around whaling operations (Falla 1937); recorded south to $75\frac{1}{2}$°S in Weddell Sea. Migrates north into all oceans, beginning late February to early March when observed flying west in antarctic waters from Ross Sea to Atlantic longitudes (van Oordt and Kruijt 1953); this, with known scarcity in Pacific sector of Antarctica (Holgersen 1957), indicates importance of Atlantic and Indian Oceans as wintering areas. Large northward movements Weddell Sea mid-March (Bierman and Voous 1950) and major emergence into subantarctic waters April.

In Atlantic, main migration up west side, between St Paul's Rocks and Brazilian coast; rapid passage across tropics, first arrivals reaching USA coasts third week April. Smaller numbers move north off Africa: said to advance slower than on west side (Roberts 1940), but some at least present off north-west Africa by late April. Northward movements continue throughout May, but still virtually absent from central parts of North and South Atlantic. Most have crossed equator by mid-June; then concentrations in Gulf Stream off North America, where numbers vast all through northern summer. Fairly common in Canary Current, off Iberia and in Bay of Biscay late May–September; these have probably moved up east Atlantic since seldom recorded in any numbers between 30°W and 60°W, but crossings from Brazil to west Africa may occur. Normally north to 45°N or 47°N on both sides of North Atlantic, and therefore rare off north-west Europe, though reported at sea north to Rockall; up to 300 around weather ship at 52½°N 20°W September–October 1966 (Tuck 1967), and large numbers at 50°N 15°W in late July 1938 (Roberts 1940) exceptional. A few enter Mediterranean, exceptionally east to Sardinia; c. 100 at 37°16′N 0°56′E 9 August 1960 (Bourne 1961). Surprisingly seldom blown ashore on either side of Atlantic (see Distribution); presumably adapted to bad weather in antarctic environment.

Return movements begin September, when already scattered records off South America and South Africa, and in full swing October; disappears from North American waters earlier than from eastern North Atlantic, and some west to east movement suggested (Roberts 1940) but unconfirmed; stragglers into November, and exceptional December records from Cape Verde Islands and St Paul's Rocks. Main southward routes again offshore from South America (chiefly) and Africa; infrequently seen in mid-ocean though apparently some cross from 'bulge' of West Africa to 'horn' of Brazil via St Paul's Rocks. Return to Antarctic seas dependent on pack-ice conditions, but mainly November; by December all but some immatures confined again to Southern Ocean and higher latitudes.

In Indian Ocean, where winters in tropics, first arrive off east Africa in mid-April, becoming widely distributed and common May–July when moves north all across, east to Australia, north to Bay of Bengal, Ceylon, Arabia and to c. 20°N in Red Sea (occasionally as far as southern Egypt); concentrates in August–October off Arabia prior to rapid withdrawal in early November when large numbers noted off Cape Comorin; only stragglers thereafter (MacLaren 1946; Phillips 1947, 1955). (See also Roberts 1940; Palmer 1962; and annual seabird reports in *Sea Swallow* 1961–70.)

**Food.** Usually feeds from surface while skimming over water, but also hovers and 'stands on water' to pull at floating food and pick up oil globules (Roberts 1940). On occasions will dive, and sometimes swims; commonly follows ships in most parts of its range.

In antarctic waters, probably chiefly planktonic crustaceans especially euphausids (Falla 1937), including *Euphausia superba* (Roberts 1940) and on Kerguelen an amphipod *Euthemisto*. (Falla 1937). Also feeds extensively on offal from whaling stations, especially oil globules and fat particles (Murphy 1936). Other food items rarely specifically recorded though cephalopod remains found in 2 birds from Cape Denison (Falla 1937) and in 2 from southern Atlantic (Bierman and Voous 1950). Few specific data from Atlantic, but particularly associated with fishing industry, feeding on its waste products. Probably to lesser extent also feeds on planktonic crustaceans, small fish, and fish eggs, molluscs and cetacean faeces (Witherby *et al.* 1940; Roberts 1940; Watson 1966).

**Voice.** See Field Characters.

**Plumages.** ADULT. Body mainly black-brown; chin and belly duller, crown and sides of head tinged grey. Feathers of lower rump black-brown tipped white, upper tail-coverts white without dark spots on tips, forming sharply defined white patch. Feathers of lower flanks and lateral under tail-coverts white, some with black subterminal spots. Tail-feathers sooty black, white at base of inner web. Flight-feathers sooty black, shorter edged white, especially inner secondaries; greater upper wing-coverts grey, median dark grey, with white tips and fringes, lesser sooty black; underwing sooty brown. In worn plumage, body browner, no grey tinge on head; white edges in wing feathers disappear. JUVENILE. As adult, but narrow white edges on feathers of belly and white feathers on lores; soon wear off (Murphy 1936). Outer primary apparently pointed as in juvenile *Hydrobates* (C S Roselaer; see also Mayaud 1941).

**Bare parts.** ADULT AND JUVENILE. Iris brown; bill, tarsus, and toes black, webs variable amount of orange-yellow with black margins.

**Moults.** ADULT POST-BREEDING. Complete; primaries descendant. In Atlantic and Pacific wintering areas, moult of body feathers and primaries starts after arrival in May, lasting till about end August. In tropical Indian Ocean, moult apparently somewhat later. For details of replacement of flight-feathers see Mayaud (1949–50). POST-JUVENILE. Complete; in tropical areas, October–May.

**Measurements.** ADULT BREEDING. Nominate *oceanicus*, Crozet Islands, Falkland Islands, Tierra del Fuego, and Kerguelen

PLATE 15. (*facing*) *Calonectris diomedea* Cory's Shearwater (p. 136): 1 ad, 2 nestling. *Puffinus gravis* Great Shearwater (p. 140): 3 ad. *Puffinus griseus* Sooty Shearwater (p. 143): 4 ad. *Puffinus puffinus* Manx Shearwater (p. 145). *P. p. puffinus*: 5 ad, 5a wing length 240 mm, 5b flight and swimming positions, 6 nestling. *P. p. mauretanicus*: 7 ad, 7a wing length 256 mm. *P. p. yelkouan*: 8 ad, 8a wing length 244 mm, 8b wing length 224 mm, 9 nestling. *Puffinus assimilis* Little Shearwater (p. 151). *P. a. baroli*: 10 ad. *P. a. boydi*: 11 ad, 12 nestling. *Bulweria bulwerii* Bulwer's Petrel (p. 132): 13 ad, 14 nestling. *Pterodroma mollis* Soft-plumage Petrel (p. 129): 15 ad, 16 nestling. (PJH)

Island; *O. o. exasperatus* from South Shetland Islands and Antarctic continent. Skins (BNMH, MNHN, RMNH, ZMA, ZMO).

| | *O. o. oceanicus* | | *O. o. exasperatus* | |
|---|---|---|---|---|
| WING | 142 (5·48; 21) | 133–150 | 154 (3·68; 30) | 147–163 |
| TAIL | 60·1 (3·60; 21) | 52–67 | 67·9 (3·16; 30) | 63–75 |
| BILL | 11·6 (0·49; 21) | 10–13 | 12·2 (0·42; 31) | 11–14 |
| TARSUS | 34·7 (0·75; 20) | 33–37 | 34·4 (0·88; 31) | 32–36 |
| TOE | 27·1 (0·74; 21) | 25–29 | 27·2 (1·10; 31) | 25–30 |

♂♂ tend to be a little smaller than ♀♀, but differences not significant.

**Weights.** ADULT. *O. o. exasperatus* Signy I., South Orkney Islands, South Atlantic (Beck and Brown 1972).

| | | |
|---|---|---|
| On arrival breeding area, November | 39·5 (2·41; 61) | 34–45 |
| Laying and start breeding, January | 37·6 (2·29; 75) | 33–43 |
| On departure, April | 46·4 (2·38; 14) | 42–50 |

Sex differences slight. Arrive moderately fat November; weight lower December–March during courtship, laying, breeding, and feeding of nestlings; quick weight increase after desertion of nestlings prior to migration, when much fat stored (average of 5 with fat extracted 30·7). Moulting ♀, July, off Surinam, 27·5 (RMNH). Nominate *oceanicus*. ♀, February, Kerguelen, 25

(MNHN); July, Arabian Sea, *c.* 24 (Bailey 1966); moulting ♂, November, off north-east Australia, 25 (ZMO).

**Structure.** 11 primaries: p9 longest, p10 5–10, p8 3–8, p7 10–19, p6 21–30, p1 75–87 shorter. Tail only slightly forked; t6 2–8 longer than t1 in *exasperatus*, 0–5 in *oceanicus*. Bill rather short, broad at base; nasal tube *c.* 40% of length of bill. Tarsus and toes long and slender, smooth; outer toe about 1 and inner toe 4 mm shorter than middle; nails somewhat flattened, sharp.

**Geographical variation.** No variation in plumage. Considerable variation in size, birds of Kerguelen (sometimes separated as *parvus* Falla, 1937) somewhat larger than birds of Tierra del Fuego, Falklands, and Crozet Islands, but much overlap (for use of name *oceanicus* for smaller subspecies instead of *parvus* see Bourne 1964). Hardly any overlap between smaller Kerguelen and larger Antarctic populations, but *oceanicus* of South Georgia somewhat larger than Kerguelen birds, showing more overlap with *exasperatus*. *O. o. maorianus*, a white-breasted form known only from 3 speciments taken off New Zealand long ago, agrees in size with South Georgian birds; may belong to all or partly white-breasted population breeding (or formerly breeding) on islands in New Zealand sector of high latitude subantarctic zone, or be a rare individual variation (Murphy and Snyder 1952). CSR

## *Pelagodroma marina* White-faced Storm-petrel (Frigate Petrel)

PLATES 17 and 18
[facing page 159 and after page 182]

Du. Bont Stormvogeltje    Fr. Pétrel frégate    Ge. Weissgesichtsturmschwalbe
Ru. Морская качурка    Sp. Paíño pechialbo    Sw. Fregatstormsvala    N.Am. White-faced Storm Petrel

*Procellaria marina* Latham, 1790

Polytypic. *P. m. hypoleuca* (Webb, Berthelot, and Moquin-Tandon, 1841), Selvagens; *eadesi* Bourne, 1953, Cape Verde Islands. 4 extralimital, of which *dulciae* Mathews, 1912, South Australia and Indian Ocean, may reach Red Sea.

**Field characters.** 20–21 cm, wingspan 41–43 cm; near size of Leach's Storm-petrel *Oceanodroma leucorhoa*, but shorter-winged. Only Palearctic storm-petrel with almost entirely white underparts; dark grey above with light grey rump, feet projecting well beyond almost square tail. Sexes alike and no seasonal differences; juvenile separable at close range.

ADULT. Crown dark grey shading to brown-grey on back and upper wing-coverts, latter edged with dark curve from carpal joint to trailing edge. Rump light grey, sharply demarcated from slightly forked but often rather square-looking tail. Distinctive pattern on sides of head, with white extending back as broad eyebrow to merge at rear with another white band reaching back from chin, thus enclosing dark grey patch below eye to ear-coverts. Underparts, including under wing-coverts, wholly white apart from darker remiges and slight grey wash on sides of

PLATE 17. (*facing*) *Oceanodroma monorhis* Swinhoe's Storm-petrel (p. 173): **1** ad. *Oceanodroma leucorhoa* Leach's Storm-petrel (p. 168): **2** ad, **3** nestling. *Oceanites oceanicus* Wilson's Storm-petrel (p. 156): **4** ad. *Oceanodroma castro* Madeiran Storm-petrel (p. 174): **5** ad, **6** nestling. *Hydrobates pelagicus* Storm Petrel (p. 163): **7** ad, **8** nestling. *Pelagodroma marina eadesi* White-faced Storm-petrel (p. 159): **9** ad, **10** nestling. (PJH)

breast and flanks. Bill and extraordinarily long legs black, feet with yellow webs (rather inconspicuous except when legs dangling). JUVENILE. Paler head, whiter rump, whitish-edged wing-coverts, and whiter-tipped scapulars, secondaries, and inner primaries.

Diagnostic combination of white underparts and greyish upperparts, with dark cheek patch on otherwise white face and dark curve across wings, precludes confusion with any other storm-petrel, but general pattern not unlike Grey Phalarope *Phalaropus fulicarius* in winter plumage, especially when floating high on water.

Flight erratic, but stronger than other storm-petrels with more banking and less fluttering; recognizable at long range by habit of dancing from side to side like pendulum at intervals of about 2 s; does not really patter on water (see Wilson's *Oceanites oceanicus*), but sometimes dangles long legs to touch surface with toes (or ground on way to burrow) giving impression of rapid hops or bouncing along on springs; also sails on stiff wings, skidding along close to surface, and can fly quite fast. Feeds by lowering legs and splashing into water with body to pick up surface plankton. Markedly pelagic; singly or in small parties. See Alexander (1955); Witherby *et al.* (1940); Warham (1958); Palmer (1962); Serventy *et al.* (1971).

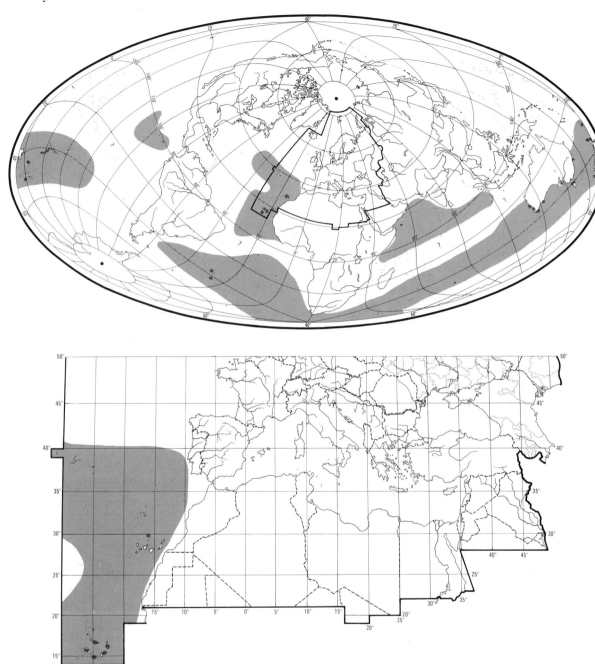

**Habitat.** Information limited. Localized in west Palearctic within warm oceanic waters, apparently shifting mainly west and south over deep waters when not breeding, but some coastal visitation off Africa (Rankin and Duffey 1948), and susceptible to storm displacements. Seldom, if ever, follows ships. Aerial, in lowest airspace above sea; relative time spent swimming unknown. Comes to land only to breed colonially on arid, undisturbed, small islands, sometimes volcanic and rising 100 m or more; nesting areas usually flat with enough soil to permit burrowing, often among low vegetation. Nest burrows used by other species in succession. Habitat choice and awkwardness on land lead to heavy predation by cats, various bird predators, and, on Cape Verde Islands, by crabs. (Bannerman 1963; Bannerman and Bannerman 1965, 1968.)

**Distribution.** Canary Islands: occurs irregularly, but no evidence of breeding except on nearby Selvagens (Bannerman 1963). Cape Verde Islands: nests locally on

Cima, Branco, Larje Branco and dos Passaros (de Naurois 1969*b*).

Accidental. Britain: Inner Hebrides, January 1897; off Cornwall, August 1963 (Buckley and Wurster 1970). France: Bay of Biscay, August 1963. Netherlands: November, 1974.

**Population.** SELVAGENS: in 1963 breeding population estimated to be of order of half million individuals (Jouanin and Roux 1965). CAPE VERDE ISLANDS: in 1951 on Cima 10000–50000 storm-petrel burrows apparently mainly this species (Bourne 1955*b*). Oldest ringed bird 10 years 1 month (Rydzewski 1974).

**Movements.** Migratory and dispersive. Movements of North Atlantic populations little known. Two races involved: *hypoleuca* breeds Selvagens in northern summer, returning to colonies from February, departing end September (Jouanin and Roux 1965); *eadesi* breeds Cape Verde Islands in northern winter, returning about November. Occurs around Canaries and Madeiran group, presumably feeding in Canary Current upwelling. Surprisingly few pelagic reports, most offshore from west and north-west Africa from 9°N–35°N and 13°W–25°W, especially off Morocco and Mauritania. Few scattered records far to west and north of colonies, mostly August–October, suggest post-breeding dispersal into central North Atlantic; single specimens of *hypoleuca* collected 320 km east of Nantucket, USA, September 1885 and in Azores May 1912; and single *eadesi* 160 km off Long Island, USA, August 1953 (Bourne 1966*a*) and at *c*. 42°N 66°W September 1959 (C S Roselaar). Other pelagic records, not subspecifically identified, span North Atlantic longitudinally, and include 6 more in Gulf Stream off New England (Buckley and Wurster 1970; Barnhill and DuMont 1973); mostly singles, but small numbers 45°N–46°N, 37°–45°W October 1944 ahead of major storm front (Rankin and Duffey 1948), 5 at 45½°N 43½°W in August 1957, and several at 40°N 9°52′W (west of Portugal) August 1962 (Buckley and Wurster 1970).

Southern hemisphere populations absent from vicinities of colonies chiefly May–September when dispersed at sea. Many *dulciae* from Australia migrate into tropical Indian Ocean where recorded all round north periphery west to Mombasa (e.g. Voous 1965). Nominate *marina* of Tristan da Cunha and Gough Island winters within tropical South Atlantic, dispersing west to offshore South America, east almost to Africa; while limited northward dispersal indicated by scattered records at sea between Ascension Island and Angola (north to 6½°S) April–October. (See also Lambert 1971; annual seabird reports in *Sea Swallow* 1961–70.)

**Food.** Little information, mostly from southern hemisphere, but apparently chiefly planktonic crustaceans and, to lesser extent, squids. Feeds mainly by pattering, snatching food as it runs along surface, and by lowering legs and splashing in with body. Seldom follows ships but will follow schools of whales and dolphins (Bailey 1966). Specimen of nominate race on Tristan da Cunha, January, contained numerous copepods and a few euphausids (Hagen 1952); stomach of one from south-east coast of Arabia, small otoliths of fishes (Bailey 1966). Regurgitations from New Zealand (race *maoriana*) mainly euphausids, with some barnacle larvae and cephalopod remains (Richdale 1943). In Australia, parents nightly fed chicks with fishy, oily paste comprised chiefly of whale-food, probably *Euphausia* (Campbell and Mattingley 1906); young bird in burrow contained pasty substance, including small shrimp (Mathews 1912).

**Social pattern and behaviour.** Based mainly on material supplied by J Warham.

1. Gregarious while breeding, feeding in small flocks considerable distances from nesting islands. Usually singly at other seasons. BONDS. Evidence for monogamous pair-bond, sustained from one season to next. In extralimital race *maoriana*, average duration of pair-bond at Whero Island, New Zealand, 1·4 years; 36% of breeders re-united with same mate in following season and some pairs intact for at least 4 years, most changes due to disappearance of at least one partner, with little divorce (Richdale 1965). If mate lost, another soon acquired. Maintenance of pair-bond said to keep pair in same burrow, rather than vice versa (but see other Hydrobatidae). At Whero, pre-breeders constituting up to two-thirds of adult population wandered over surface investigating possible nest-sites; probably formed pairs at this time before first breeding in subsequent seasons. Both adults evidently tend chick, latter independent on fledging. BREEDING DISPERSION. Colonial, often in large numbers. Burrows at times densely grouped in suitable soft ground, up to one hole per square metre (Murphy 1936). On Whero, population of *c*. 2000 in 0·2 ha of soil also extensively burrowed by other petrel species. Nest-sites on Whero retained from year to year; 90% of 184 breeders had not shifted in following season, 80% after 2, and 60% after 3. Apparently territorial only in defending burrow. ROOSTING. No specific information; presumably as in other nocturnal petrels.

2. Strictly nocturnal in visiting land. At New Zealand colonies, activity fluctuates with phase of moon and few ashore on bright nights, although breeders still tend chicks. Little information on display and related behaviour. AERIAL ACTIVITY. Flighting over breeding grounds after dark well-marked feature, birds moving about 'like huge long-legged flies' (Lockley 1952) and appearing to engage in much erratic chasing; though said to be silent on wing (Murphy 1936), Flight-calling frequent at least at some stations. High numbers overhead at chick-rearing stage thought to be due to visits from unemployed birds which appear to keep away on moonlit nights. GROUND ACTIVITY. Owing to inaccessibility of nest-chambers, behaviour underground unknown apart from frequent calling which starts as soon as bird enters burrow at night (Murphy 1936) and which may occasionally be heard during day (Warham 1958*a*). Calling also common on surface near burrows. Fighting recorded during apparent disputes over ownership of burrows. Later in season, some unemployed birds remain ashore by day in burrows not in use by breeders. RELATIONS WITHIN FAMILY GROUP. Little known.

**Voice.** Poorly known; various descriptions from widely dispersed breeding stations (mainly Australian) do not tally, but full range of vocalizations evidently not yet described from any (J Warham). Apparently silent at sea; also mainly silent on wing even over breeding colony (Murphy 1936; Witherby *et al.* 1940), at least when with eggs and young, though extent of aerial calling earlier in cycle still not adequately investigated.

In Australian race *dulciae*, main call heard at night from birds on or below ground, mournful 'woo' repeated *c.* once per second; also develops into siren-like moaning 'oooo-aaaooo' (Warham 1958*a*). Same call sometimes given from burrows during day. Bird utters mouse-like squeaking immediately on entering burrow (Littler 1910); apparently same sound ('loud squeaking calls') also given by parent long after entering burrow to attend chick (Falla 1934). Other calls or renderings from Australian sites include low, frisky 'chee-ur' (Campbell 1933) and low rasping note (see Witherby *et al.* 1940); also low twittering note, given by parent while alighting at burrow (Falla 1934). No adequate description of Flight-calls from Australian sources, though 'low cries' on taking wing mentioned in Witherby *et al.* (1940). In race *maoriana*, evident from Richdale (1965) that considerable aerial calling occurs at some New Zealand breeding islands when large numbers in flight over colony on dark nights; this presumably involves 'canary-like' notes also mentioned in same paper and which give rise to great chorus of squeaking and chirping sounds after dark.

In west Palearctic, call of birds at night on Selvagens (race *hypoleuca*) described as 'very faint, and not unlike that of Redshank *Tringa totanus* heard at some distance' (Lockley 1952); at least at times, this call given by birds in flight (Lockley 1942). Also repeated 'ooo' call, apparently similar to mournful call of *dulciae*, but at nearly 5 per second (see I). On Cape Verde Islands (race *eadesi*), utters 'grating noises like those of a pair of rusty springs set in motion' (Alexander 1898). Otherwise only other vocalizations mentioned: slight groaning noise from burrows, and short grunting note heard from birds caught while incubating (see Witherby *et al.* 1940).

CALLS OF YOUNG. Unknown in west Palearctic. In race *dulciae*, chick makes purring noise when adult enters nest-chamber (see Witherby *et al.* 1940; Serventy *et al.* 1971).

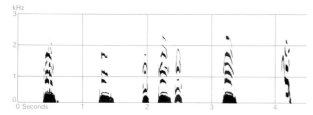

I Alexander Zino Selvagens

**Breeding.** SEASON. Selvagens: eggs laid March–April; young in nest to 3rd week July (Bannerman 1963). Cape Verde Islands: eggs laid end January–March (de Naurois 1969*b*). SITE. In sandy ground, usually among thick vegetation which conceals entrance hole. Colonial, with holes *c.* 1 m apart. Nest: excavated hole in soft ground, 5–20 cm deep, then horizontal tunnel up to 1 m long; often winding, ending in nest chamber *c.* 15 cm diameter. No material (Bannerman 1959). Building: holes dug by both sexes. EGGS. Ovate; white, more or less finely spotted with purplish or dark red spots, often grouped at large end. 36 × 27 mm (34–38 × 25–27), sample 26; calculated weight 14 g (Schönwetter 1967). Clutch: 1. One brood. No data on replacements. INCUBATION. Period not known. By both sexes. YOUNG. Semi-altricial and nidicolous. Fed by parents at night; abandoned shortly before fledging. FLEDGING TO MATURITY. Fledging period not recorded. Becomes independent when abandoned by parents. Age of maturity not known. BREEDING SUCCESS. No data.

**Plumages.** *P. m. hypoleuca.* ADULT. Forehead white with some grey mottling; crown and hindneck dark grey. Mantle, back, and scapulars grey, rump and upper tail-coverts much lighter, ash-grey. Tail black, grey at base. Sides of face white, spot round eye extending in streak to nape dark grey contrasting with white streak over eye, and with white cheeks and throat. Sides of chest grey, rest of underparts white; some ash-grey mottling on long under tail-coverts, variable amount of grey streaks along shafts of some flank feathers. Flight feathers black, grading to white on base of inner web, paler below. Upper wing-coverts dark brown-grey, greater with pale tips, under wing-coverts and axillaries white, marginal coverts grey with white tips. In fresh plumage, feathers of mantle, back and rump, tail-coverts, greater upper wing-coverts, and secondaries tipped white; in worn plumage, white tips disappear and grey of feathers becomes brownish. NESTLING. One down only; pale mouse-grey, paler still below; long and thick on upperparts, concealing bald spot on crown, shorter below; chin, upper-throat, and lores nearly bare. JUVENILE. In fresh plumage, white tips of feathers of upperparts, wing-coverts, and secondaries apparently more prominent than in adults. When newly grown, primaries also narrowly edged white at tips. Tip of outer primary pointed, against rounded in adults, but intermediates occur. Grey feathers of upper tail-coverts and lower flanks often more extensively barred white.

**Bare parts.** ADULT AND JUVENILE. Iris brown; bill, tarsus, and toes black; webs yellow, but at least in some juveniles flesh-coloured, with narrow black margin.

**Moults.** ADULT POST-BREEDING. Insufficiently studied for North Atlantic populations. Southern hemisphere subspecies moult some body feathers just before end of breeding season, finishing in winter quarters. Wing moult starts soon after arrival winter quarters; inner primaries and secondaries replaced quickly, but outer primaries grow more slowly. Tail moult after major wing moult. POST-JUVENILE. No data.

**Measurements.** Adults, breeding; skins (BMNH, MNHN, RMNH, ZMA). Fork: difference between t6 and t1.

| | *hypoleuca* | | *eadesi* | |
|---|---|---|---|---|
| WING | 161 | (5·50; 16) | 153–171 | 162 | (4·08; 11) | 154–169 |
| TAIL | 75·1 | (3·95; 16) | 68–84 | 72·6 | (3·85; 11) | 68–79 |
| FORK | 7·2 | (2·17; 16) | 4–12 | 4·5 | (0·93; 11) | 3–6 |
| BILL | 17·3 | (0·45; 15) | 16·6–18·2 | 18·6 | (0·91; 11) | 17·5–20·0 |
| TARSUS | 44·2 | (1·32; 16) | 41·6–46·3 | 44·9 | (1·24; 11) | 42·4–46·2 |
| TOE | 36·4 | (1·25; 16) | 34·0–38·5 | 36·3 | (1·25; 11) | 33·6–38·9 |

Difference between subspecies significant for fork and bill.

**Weights.** 17 breeding adults *hypoleuca* from Selvagens: 42–60, mean 48·9 (Jouanin and Roux 1965). Single ♀ *eadesi* off Cape Verde Islands: 52 (ZMA). No data on nestlings. Extralimital *maoriana* (New Zealand). Adult: 47·2 (4·0; 100) 40–62. Nestlings become heavier: mean 71·6, maximum 107 (8 days before departure), sample 32. On leaving nests still significantly heavier than adults: 53·7 (6·0; 40) 42–68 (Richdale 1965).

**Structure.** 11 primaries: p9 longest, p10 7–15 shorter, p8 2–6, p7 13–20, p6 25–33, p1 72–90. Tail slightly forked. Bill long and slender, nasal tubes *c.* 30% bill-length. Tarsus and toes long and slender, scutellated; middle toe *c.* 80% tarsus, outer toe almost equal to middle, inner *c.* 15% shorter.

**Geographical variation.** Cape Verde birds (*P. m. eadesi*) differ from *hypoleuca* in having forehead whiter, more sharply contrasting with crown; back paler grey; more white on sides of neck, sometimes forming almost complete white collar; longer bill. Subspecies of southern hemisphere with shorter tarsi: those breeding north of subtropical convergence much like *hypoleuca*, but greyer on sides of chest; southern ones more heavily pigmented, with shorter bill and toes and longer, more deeply forked tail (for details see Murphy and Irving 1951; Bourne 1953).

CSR

## *Hydrobates pelagicus* Storm Petrel

PLATES 17 and 18
[facing page 159 and after page 182]

Du. Stormvogeltje     Fr. Pétrel tempête     Ge. Sturmschwalbe
Ru. Прямохвостая качурка     Sp. Paíño común     Sw. Stormsvala

*Procellaria pelagica* Linnaeus, 1758

Monotypic

**Field characters.** 14–18 cm, wingspan 36–39 cm; smallest west Palearctic petrel. Black with white rump, feet not projecting beyond slightly rounded tail. Sexes alike and no seasonal differences; juvenile similar.

ADULT. Sooty-black above and slightly browner below, darkest on flight-feathers and tail, apart from white rump patch extending to lower flanks and lateral under tail-coverts, small irregular whitish patch on underwing-coverts (sometimes white and quite prominent) and, in fresh plumage, narrow whitish line along edge of greater upper wing-coverts. Tail often looks square, but actually slightly rounded. Bill, legs, and feet black. JUVENILE. In fresh plumage, light edgings to feathers of upperparts.

Smaller and darker than other west Palearctic storm-petrels, distinguished from all by variable white bar on underwing; from Leach's *Oceanodroma leucorhoa* and Madeiran *O. castro* by size, blacker coloration, and, more particularly, shorter wings without broad diagonal bar, squared tail, and different flight; from Wilson's *Oceanites oceanicus* chiefly by short black legs and feet not extending beyond tail and, again, different flight.

Flight weak-looking, almost continuous, bat-like fluttering interspersed with short glides. At intervals, patters with feet on surface as it hangs with raised wings to pick up food, but does not walk on water like *O. oceanicus*. Sometimes settles on sea. Markedly pelagic. Sometimes follows ships, fluttering back and forth across wake. Usually singly or in small parties.

**Habitat.** Distinctively north-east Atlantic and west Mediterranean range within lowest 10 m band of airspace over pelagic, offshore, and to less extent inshore marine waters. Found especially in intermediate offshore and sub-oceanic zones between littoral and deep ocean, from 10 °C isotherm (barely overlapping sub-arctic) down to 25°C isotherm; overlaps tropics in winter. Finer habitat components governing seasonal range not yet worked out, but doubtless linked with varying availability of food. Able to live in severe wind and wave conditions, remote from land, but rarely goes near floating ice. Alights and swims freely on sea in all weathers, but much more aerial than aquatic. Indifferent to human presence and artefacts; takes advantage to feed in disturbance in wake of ships. Comes to land solely for breeding and by night only, on unsheltered and undisturbed islands, islets, or, more rarely, promontories of mainland. For breeding, prefers rocky outcrops, narrow crevices between stones, even in walls, ruins, or burrows of other species, seeking maximum freedom from ground and winged predators and human disturbance; often on low undulating surfaces in hinterland beyond cliff-tops.

**Distribution.** Current breeding distribution imperfectly known in many areas. NORWAY: birds with brood patches and others which laid eggs on capture recently caught Lofoten Islands (Myrberget *et al.* 1969; E Brun). GREECE: summer records may indicate breeding on islands, but no proof (WB). NORTH AFRICA: formerly bred Ile de la Galite, Tunisia; but no recent records on Mediterranean coast (Heim de Balsac and Mayaud 1962). MADEIRA: reported breeding Desertas, 1849, suspect (Bannerman and Bannerman 1965). CANARY ISLANDS: apparently rarely breeds, though nesting proved 1931 (Bannerman 1963).

Accidental. Belgium, Netherlands, Denmark, Spitsbergen, West Germany, Sweden, Poland, Czechoslovakia, USSR, Austria, Switzerland, Greece, Cyprus, Libya.

**Population.** BRITAIN AND IRELAND. Disappearance from some Scottish islands and marked decrease Isles of Scilly

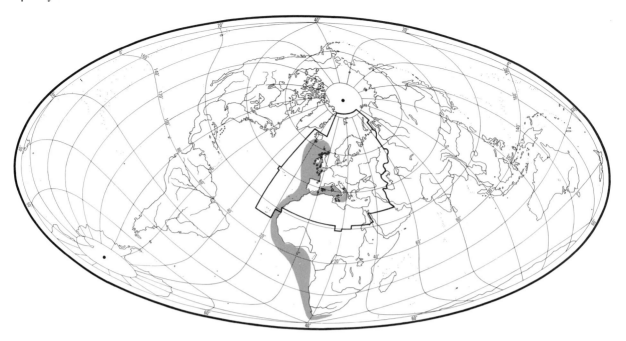

indicate possible decline towards end 19th century; in 20th century, 3 Irish islands abandoned, and decrease reported Cornwall (Parslow 1967). History of known colonies, with some estimates of number 1969–70, in Cramp *et al.* (1974). FRANCE: Brittany, 414–451 pairs, 1970–71; Biarritz, 50 pairs, 1947 (J-Y M).

Survival. Annual survival of adults 87–88% (Scott 1970). Oldest ringed bird: one ringed 1950 as full-grown last recaptured 1969, so probably over 20 years old (Scott 1970).

**Movements.** Migratory and dispersive. Restricted to east Atlantic and Mediterranean; winters in strength off South Africa as discovered by van Oordt and Kruijt (1953). Dispersed at sea when not breeding: adults pelagic mainly October–April (Davis 1957). Degree of movement by Mediterranean breeders uncertain. Occurs in autumn in east Mediterranean, stragglers entering Black Sea and Sea of Azov in October (Dementiev and Gladkov 1951); reported resident around Malta (Roberts 1954), and individuals ringed Filfla recovered nearby in January and in Tunisia in late December; no conclusive evidence of Mediterranean breeders emerging into Atlantic, but few winter records in Mediterranean. Marked southward migration by Atlantic breeders, though very small numbers (possibly late migrants) noted in north-temperate waters in early winter. Main departures from British and Irish waters September–November; migrants noted in Bay of Biscay and off Iberia up to December, though arrivals off west Africa from mid-November. Some occur in winter north to offshore Rio de Oro and Mauritania (Irish breeder recovered 19°N on 28 January); but most

transequatorial migrants winter (December–April) in cool waters off South West and South Africa, some south to 38°S, 300 km south of Cape Agulhas (van Oordt and Kruijt 1953, Lambert 1971). Dutch museums' winter specimens of juveniles all from west Africa, adults all South Africa (J Wattel). Scottish breeder recovered False Bay, Cape Town, February; another on south coast of Natal on late date of 2 May. Old reports from Zambesi mouth (Mozambique) and Red Sea unsubstantiated (W R P Bourne).

Substantial northward passage offshore from west Africa in March and April; April and May records from Cape seas and May ones in tropics probably mainly pre-breeders making more leisurely return towards colonies, which they visit after adults established there. Numerous recoveries demonstrating accuracy of return to colonies by adults; also many recorded inter-colony movements between those of English Channel, Wales and Ireland, and between Scotland and Faeroe Islands, mostly by presumed pre-breeders. Present far out to sea in north-east Atlantic in latitudes of colonies during breeding season as 'off-duty' adults and non-breeders; infrequently found west of 20°W, but scattering of records out to 30°W, exceptionally to 40°W (Wynne-Edwards 1935; Rankin and Duffey 1948). One authentic North American record: Sable Island, off Nova Scotia, in August 1970 (McNeil and Burton 1971); oft-quoted reports from Canada during 1882–1901 due to misidentifications (Godfrey 1966). Small numbers occur in North Sea, and number obtained late in year at Danish lights. Inland only after storms, when recorded many European countries. (See also annual seabird reports in *Sea Swallow* 1961–70.)

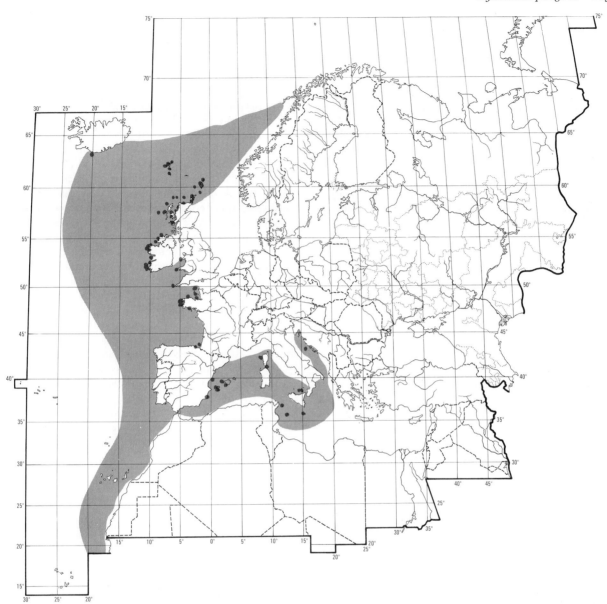

**Food.** Mainly surface crustaceans, small fish, medusae, cephalopods, and oily and fatty materials. Feeds during day from surface by pattering, hovering, and snatching without alighting; alone or in loose groups. Sometimes, but not often, follows ships or other birds (W R P Bourne).

Cephalopod remains found in 5 birds from north Europe; in 1 also small fish remains, and wings of aphids presumably blown on to water (Voous 1948, 1954; Bierman and Voous 1950). Also feeds at carcasses of whales (Godman 1907–10), ship's garbage (Macgillivray 1852), and once floating blood of shot Great Black-backed Gull *Larus marinus* (W R P Bourne). Regurgitations of breeding birds contained headless fish about 5 cm long; nesting season may be timed to take advantage of late summer abundance of sprats *Sprattus sprattus* and small fry.

Young fed regurgitated, predigested grey pulp (Davis 1957). Almost whole diet, July–October, small herrings *Clupea harengus* and sprats, with infrequently crustaceans (Scott 1970).

**Social pattern and behaviour.** Based mainly on material supplied by P E Davis, from studies on Skokholm, Wales (Davis 1957). See also Scott (1970).

1. Gregarious, especially when nesting. Even in breeding season, often forages in scattered groups at sea, with wide individual spacing, and converges at rich food supplies. Non-breeding period spent entirely at sea, apparently frequently in loose groups of greatly variable size, though often also solitary.

BONDS. Monogamous pair-bond, sustained or life-long in duration. Nest-hole appears to be prime bond between members of pair, ♂ and ♀ returning to same one each spring. Immatures do not come back to colony until 2nd year, some not until 3rd; pairs often formed during these pre-breeding visits, occupying hole for one season and returning next year to nest for first time, mostly in 4th or 5th year. Experienced breeders return some 2 weeks earlier than new ones, latter earlier than immatures; this reduces likelihood of divorce, and re-matings rare unless one of pair has died. Divorces occur exceptionally in other situations, however, i.e. after early nesting failure, excessive disturbance or destruction of nest-hole. Both sexes care for chick until fledging or just before. Most improbable that pairs, or parents and young, associate at sea. BREEDING DISPERSION. Nests in colonies, a few to several hundreds; occasional solitary pairs. Nest-holes frequently close, at times sharing common entrance, though nest-chambers then in separate ramifications of hole. Apparently non-territorial; nest-chambers not defended against intruders during pre-egg stage (usually 4–8 weeks) and intruder often found with one bird of pair, rarely with both, by night or day. Non-breeders and other unemployed birds may number up to half of those visiting holes during breeding season; colonies probably have many other visitors which do not even enter holes, including in late summer large number of non-breeders which do not land. Most birds breed on natal island though ringing shows that many individuals visit 2 or more colonies in successive years, even within single season; one record of breeding in another colony. ROOSTING. Except when occupying nest-hole, likely to roost on surface of sea. Being active both by day, and at least in summer months, by night, probably sleeps only for brief and irregular periods.

2. Entirely nocturnal at breeding station, earliest birds coming to land while still some daylight, latest departing less than 1 hour before sunrise. Visits not affected by stage of moon, but reduced in periods of strong winds (Scott 1970). As behaviour at colony occurs at night or within nest-hole, observation limited largely to what may be seen by twilight or moonlight, artificial white light tending to cause disturbance. Relatively little known, therefore, about display and related behaviour. Highly vocal at colony, both in air and at nest-site. AERIAL ACTIVITY. In late May and early June, engages in Display-flights over breeding grounds at any hour of night. Usually 2 birds, which appear to be paired, involved; circuit in fairly limited area above nest-hole, probably starting and usually ending there. One closely chases other, and flight more rapid and direct than normal; course usually quite irregular, though often circular. At times, couple separate, together form figure-of-eight, and then resume chase when meeting again at centre; occasionally collide. Third bird sometimes joins pair. Though silent at sea and during normal flight over land, those circuiting give loud and variable Flight-call at intervals; becomes louder and more frequent as excitement mounts, particularly when 3rd bird involved. Short snatches of song and, rarely, Wick-calls, also uttered on wing. BEHAVIOUR IN BURROW. Little direct evidence of antagonistic behaviour between adults, and visitors appear to be received amicably even in nest-chamber, at least until egg laid. Visitors apparently attracted by distinctive Purr-call (or song) of occupant, both singing and visiting virtually ceasing after egg laid. Persistent Purr-calls appear to be given by bird alone in hole and comparatively seldom heard where 2 present, so evidently not performed as duet by pair (as in Leach's Storm-petrel *Oceanodroma leucorhoa*). Probably both sexes sing, though confirmation needed. Song mainly nocturnal, but not uncommon during day, particularly in afternoon and early evening. Main function apparently to advertise occupation of hole to other adults. Non-breeders continue singing throughout season. Most copulations presumably in nest-hole; observed only twice (both in day), ♂ Allopreening feathers of ♀'s nape, both keeping wings closed and silent (Scott 1970). RELATIONS WITHIN FAMILY GROUP. See Davis (1957).

**Voice.** Silent at sea. Following calls given at breeding places (based mainly on Davis 1957).

(1) Purr-call (Song). Harsh, purring 'arrr-r-r-r-r . . .' (see I) sustained and penetrating though not loud, ending with abrupt, guttural hiccoughing 'chikka' (see II). Duration usually short, though sometimes prolonged for several hours. Unit-rate in churring 30–40 notes per sec (faster than in Leach's Storm-petrel *Oceanodroma leucorhoa*); intervals between onset of breath-notes 1·4–1·8 s (shorter); duration of breath-notes 0·20–0·28 s (shorter) (J Hall-Craggs). Uttered in burrow, when frequently interjected with abrupt, guttural, hiccoughing 'chikka' or 'cuch-ah' (see II); mainly at night. Also, rarely, given in flight. (See also Social Pattern and Behaviour.) (2) Flight-call Resembles 'terr-CHICK'; uttered during aerial courtship chases, becoming louder and more frequent with increasing excitement, particularly if original 'pair' joined by 3rd bird. Variable, with syllables transposed or 1st omitted, and sometimes given with snatches of Purr-call. (3) Wick-call. Rapid 'wick-wick-wick' (see III), also occasionally heard during aerial chases. (4) Alarm-call. Resembles 'up-CHERRK'; rather similar to Flight-call. Used when bird senses danger outside burrow. (5) Distress-call. Rapidly repeated 'pee-pee-pee' uttered by frightened birds; also, exceptionally during squabbling in nest-holes.

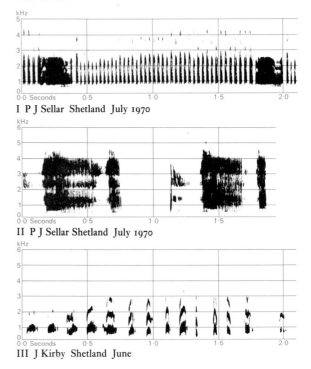

I  P J Sellar  Shetland  July 1970

II  P J Sellar  Shetland  July 1970

III  J Kirby  Shetland  June

CALLS OF YOUNG. Prolonged, sibilant 'pee-pee-pee' uttered at feeding-time. Quieter version of adult Distress-call given by chicks not accustomed to being handled.

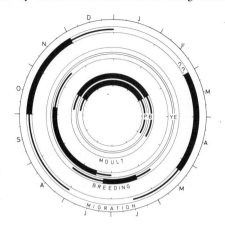

**Breeding.** Based mainly on Davis (1957) and Scott (1970); Skokholm, Wales. SEASON. See diagram for Britain and Ireland. On Skokholm, extreme laying dates 28 May and 20 August. 80% eggs laid 4 weeks mid-June to mid-July; 88% eggs hatched between 3rd week June and 2nd week August. Laying period slightly later in more northern colonies, with (late) young present until end of year. Malta: eggs laid end May to early June. Balearics: laying just started 12 May. Canary Islands: single downy young found September. SITE. In natural crevices, among boulders just above high water mark, under stones, and in fissures in rocks; also in soil on flat ground, less often in hedge banks, stone walls, and old burrows of rabbit *Oryctolagus cuniculus*. Old or occupied burrows of Manx Shearwater *Puffinus puffinus* and Puffin *Fratercula arctica* also used, sometimes sharing common entrance; rarely under buildings or other artefacts. Usually colonial; some pairs share common entrance hole, with separate chambers. Sites used in successive years, especially by successful breeders. Nest: tunnel 5–8 cm diameter, entrance a little smaller; 10 cm to 3 m or more long, average *c.* 75 cm; shallow depression *c.* 7 cm diameter at end of tunnel. Usually no material, occasionally small pieces of grass, bracken *Pteridium aquilinum*, or seaweed; rarely built up into substantial nest. Building: both sexes excavate with bill and feet in soft soil. EGG. Blunt ovate; pure white, unmarked or with a few red-brown specks concentrated at large end, but these soon lost. 28 × 21 mm (25–30 × 19–23), sample 130 (Schönwetter 1967). Weight 7 g (6–8), sample 15. Clutch: 1. One brood. Replacement eggs occasionally laid after egg loss, but some cases at least due to re-mating after loss of first mate; of 5 known cases, from 15–30 days elapsed between loss and re-laying. Eggs laid at night. INCUBATION. 40·6 ± 2·0 days (38–50), sample 36; periods of over 40 days usually indicate egg chilled during incubation. By both sexes, with ♂ taking first spell after laying, then roughly equal shares with average spell 3

days (1–5). Eggshells left in nest, rarely removed. YOUNG. Semi-altricial and nidicolous. Hatched blind, eyes open at about day 5. Cared for and fed by both parents. Left alone after 6–7 days, then fed on *c.* 82% of nights until *c.* 50 days old. Fed by bill-to-bill partial regurgitation. No apparent reduction in feeding visits during full moon. Visits drop off towards fledging time but no complete desertion; mean interval between last feed and fledging 2 days (0–7), sample 32. FLEDGING TO MATURITY. Fledging period 62·8 ± 3·5 days (56–73), sample 32 (Davis 1957); 70 days (61–86) sample 53 (Scott 1970). Becomes independent at about same time. Age of first breeding 4 or 5 years. BREEDING SUCCESS. Of 214 eggs laid, 62% hatched with losses from chilling and infertility; 66% fledged of 133 young hatched. Overall success varied from 0·27–0·49 young fledged per egg laid.

**Plumages.** ADULT. Sooty black, paler and greyer on forehead and chin, and browner on belly. Lower rump sooty black with white feather-tips; upper tail-coverts white, longer black-tipped, often with dark brown shafts; white of lower rump and upper tail-coverts forms rectangular white patch. Under tail-coverts white with black tips, but white mostly concealed, except on lateral feathers where forms white patches at sides of tail-base. Tail feathers sooty black, basally white, except middle pair. Upper wing sooty black, secondaries white at base of inner webs; when fresh, narrow white edges to outer webs of greater upper wing-coverts. Underwing sooty brown; white tips and edges to axillaries and greater under wing-coverts form irregular bar. NESTLING. 1st down long, soft, silvery grey; paler below; lores, round eye, and chin bare, bald spot on crown. 2nd down similar. JUVENILE. As adult, but in fresh plumage broader white edges to greater wing-coverts; black tips of some longer upper tail-coverts often partly edged white. Until 1st wing moult, tip of outer primary pointed (rounded in adults), sometimes difficult to see when worn; intermediates may be 2nd year birds (C S Roselaar; see also Mayaud 1941).

**Bare parts.** ADULT AND JUVENILE. Iris dark brown; bill, tarsus, and toes black. NESTLING. Leg, foot, and bill pinkish-grey, except for black bill tip; becoming darker during growth (Davis 1957).

**Moults.** ADULT POST-BREEDING. Complete; primaries descendant. Wing and tail start during 2nd half of incubation, *c.* 70% show moult of primaries on peak fledging date. Body starts during incubation. Non-breeders, and possibly also failed breeders, start earlier. Moult of flight-feathers and body continues slowly in winter-quarters; a number growing outer primaries on arrival on breeding ground, about half still in secondary moult then. Moult may be interrupted by breeding season. POST-JUVENILE. First signs of body moult during mid-winter. Flight-feathers start about June, finished October–November of 2nd calendar year. (Mayaud 1949–50; Scott 1970).

**Measurements.** Full-grown. Skins, Britain (breeding) and at sea Europe to South Africa (winter) (ZMA, RMNH).

| | ♂ | | | ♀ | | |
|---|---|---|---|---|---|---|
| WING | 120 | (2·95; 19) | 116–127 | 121 | (2·93; 25) | 116–127 |
| TAIL | 52·6 | (2·47; 20) | 47–56 | 54·5 | (3·72; 25) | 46–61* |
| BILL | 10·8 | (0·49; 20) | 9·5–11·5 | 11·0 | (0·41; 24) | 10·8–11·8 |
| TARSUS | 21·5 | (0·68; 20) | 19·6–22·3 | 22·1 | (0·87; 25) | 20·2–24·3* |
| TOE | 19·6 | (0·86; 19) | 17·8–21·0 | 19·9 | (0·69; 23) | 18·9–21·2 |

* Sex differences significant.

Live breeding adults: Britain, wing 120 (3·32; 129) 111–128; Malta, wing 123 (2·54; 73) 118–128, bill 12·5 (0·42; 10) 12–13 (BTO).

**Weights.** ADULT. St Kilda: June to mid-July, 25·3 (1·7; 27) 23–29; August, 23·7 (1·2; 35) 22–27 (Waters 1964). Skokholm: 50 summer, average 28 (Davis 1957). Britain: July, 25·4 (2·10; 62) 18·4–31·0 (BTO). Malta: June, 28·9 (2·96; 41) 24·1–38·5 (BTO). JUVENILE. Blown inland after gales, only 14·5–17 (ZMA). NESTLING. Becomes heavier than adult, reaching peak weight of 50·3±3·3 (55) c. 2 weeks before departure; at fledging (63rd day) average 34·3±3·0 (55) (Davis 1957; Scott 1970).

**Structure.** 11 primaries: p9 longest, p10 6–13, p8 1–3, p7 7–12, p6 16–19, p1 56–66 shorter. Tail square; length of white rump in middle 10·5–18 (14·7) in adults. Bill slender, decurved; nasal tubes c. 40% bill-length. Tarsus and toes slender, scutellate; inner toe 18% and outer toe 8% shorter than middle.

**Geographical variation.** None. Formerly recognized subspecies *melitensis* (Schembri, 1843), Mediterranean, said to be darker and larger, but differences from British birds too small to warrant subspecies recognition (apparently larger bill of Malta birds, cited in Measurements, possibly due to different measuring method and shrinking of bill in skins).

## *Oceanodroma leucorhoa* Leach's Storm-petrel

PLATES 17 and 18
[facing page 159 and after page 182]

DU. Vaal Stormvogeltje   FR. Pétrel culblanc   GE. Wellenläufer
RU. Северная качурка   SP. Paiño de Leach   SW. Klykstjärtad stormsvala   N.AM. Leach's Storm Petrel

*Procellaria leucorhoa* Vieillot, 1817

Polytypic. Nominate *leucorhoa* (Vieillot, 1817), North Atlantic, North Pacific from Hokkaido to Alaska. 3 or 4 extralimital subspecies west coast of North America, south of range of nominate.

**Field characters.** 19–22 cm, wingspan 45–48 cm; marginally largest west Palearctic storm-petrel. Blackish-brown with white rump, long wings, feet hardly projecting beyond forked tail. Sexes alike and no seasonal differences; juvenile similar.

ADULT. Blackish-brown above and browner below, darkest on flight-feathers and tail, apart from white rump patch divided down middle by grey strip (not easy to see) and just extending to lower flanks, broad pale diagonal across inner wings, and paler scapulars and underwing; tail well forked; bill, legs, and feet black. JUVENILE. Sometimes has light edges to body-feathers.

Difficult to separate from Madeiran Storm-petrel *O. castro* in field, but distinguished from Storm Petrel *Hydrobates pelagicus* by larger size, slenderer shape, longer wings with broad diagonal bar, paler and browner plumage, different flight and, in close view, longer forked tail; from Wilson's *Oceanites oceanicus* by same features (latter has narrow wing diagonal) and by shorter, all-dark legs.

Flight distinctive, leaping and darting through air with constant changes of speed, direction and action, sometimes gliding and banking like shearwater *Puffinus*, sometimes hovering like gull *Larus*, sometimes flapping buoyantly like tern *Sterna*, with unexpectedly deep wingbeat. Does not patter with feet on surface like *O. oceanicus* and most other storm-petrels, or walk on water. Sometimes settles briefly on sea. Less markedly pelagic than some other storm-petrels, frequently found throughout year in shallower waters of continental shelves; does not ordinarily follow ships, though will come close to boats when fish offal thrown overboard. Usually singly or in small parties.

**Habitat.** Cool or cold waters fringing sub-arctic, or linked with oceanic convergences or upwellings in lower latitudes; mainly offshore or over continental shelf, but unlike most of similarly specialized relatives also dispersed in non-breeding season right across tropical and sub-tropical Atlantic, relatively few continuing south towards colder sub-antarctic seas. Otherwise comes ashore exclusively to nest on undisturbed offshore islands, preferring slopes or plateau areas to precipices, and using natural rock crevices in talus or among outcrops, burrows in soft soil and holes in old walls or ruins or, in North America, among tree roots. Clumsy and feeble on land, spending minimum time on open ground. Wholly nocturnal, yet remains vulnerable to predation. At sea, flies invariably in lowest airspace immediately above surface but, although swimming well, not often seen on water, and apparently does not dive.

**Distribution.** BRITAIN AND IRELAND: now only 4 certain colonies, all north Scotland, but found nesting Sule Skerry 1933 and suspected elsewhere; in Ireland has bred Kerry and Mayo in past, but no recent proof (see Cramp *et al.* 1974 for details). Breeding proved Foula 1974 (*Scott. Birds* 1975, **8**, 321–3). FAEROES: discovered nesting 1934, Mykines (Williamson 1970). NORWAY: small colony Lofotens group from 1961 and perhaps earlier (Haftorn 1971); birds found with brood patches, but breeding not confirmed (Myrberget *et al.* 1969).

Accidental. Belgium, occurred in 25 years 1900–70 (Lippens and Wille 1972); also West Germany, East Germany, Sweden, Finland, Poland, Austria, Switzerland, Italy, Malta, Israel, Morocco.

**Population.** BRITAIN AND IRELAND: census data limited, but total population probably under 10 000 pairs (Cramp *et al.* 1974); possible increase North Rona and Flannans (Parslow 1967). FAEROES: 4–5 pairs Mykines 1934, but several thousand pairs in 1940s; uncertain whether major increase or previously overlooked (Williamson 1970). Estimated 500–1000 pairs 1969 (W R P Bourne). East Atlantic colonies form only minute proportion of world population; in west Atlantic, population in Newfoundland colonies alone estimated in millions, despite general recent decline perhaps linked with decrease in food supply (Huntington 1963).

Survival. Observed survival rate for breeding birds (North America) *c.* 70%, but actual rate must be higher in order to maintain the species (Huntington and Burtt 1972). Oldest ringed bird 24 years (Kennard 1975).

**Movements.** Migratory. Nominate *leucorhoa*, breeding North Atlantic and North Pacific, migrates south to winter in regions of tropical convergences. Stragglers remain in cooler waters of North Atlantic during northern winter, though old January–March reports from Greenland seas lack conviction (W R P Bourne). In September, adults desert chicks, pre-breeders have departed much earlier. Numbers build up at sea on both sides of Atlantic approximately in latitudes of colonies September–October; thereafter decline rapidly in west though peak numbers in east October–November; numbers, occasionally large, noted October in mid-Atlantic (45°N–48°N, 45°W–22°W) (Rankin and Duffey 1948; Philipson and Doncaster 1951). Thus likely that, as shown for Arctic Tern *Sterna paradisaea*, some North American *O. leucorhoa* cross to European side before moving into tropics: one ringed in Newfoundland colony recovered Huelva, Spain, January, while big wreck in north-west Europe, autumn 1952 (see below), involved too many for European population alone. Nevertheless migration occurs down west Atlantic, passing Bermuda September–November, some entering Caribbean (Voous 1967), majority continuing south to offshore Brazil where rather common in equatorial waters December–January. Major autumn passage down east Atlantic (regular Bay of Biscay but scarce North Sea); common in Canaries Current on 11 November (Bierman and Voous 1950).

Main arrivals in tropics off west Africa from late November, some fanning out then, though main concentrations over upwellings (typically not mid-oceanic). Fairly numerous outer Gulf of Guinea January. Uncertain, but probably rather small, proportion may pass beyond tropics, a few reaching Cape seas (Liversidge 1959). One seen Southern Ocean at 57°40′S 5°00′E in January 1948 (Bierman and Voous 1950) must (if identified correctly) have emerged from South Atlantic.

Little known of return passage: northward migration both sides of Atlantic, March; most adults presumed to have left tropics by mid-April, as first arrivals Newfoundland in 1st week May and British colonies reoccupied from late April. Some, presumably 1st year birds from state of moult, linger in tropics into northern summer (e.g. Ritchie and Bourne 1966). Colonies visited at least by 2nd

year birds as pre-breeders. No evidence of longitudinal concentration on northward passage.

Essentially oceanic, but occasionally blown ashore or inland by storms at sea; generally few involved but two major wrecks in last 85 years: 26 September–10 October 1891 and 21 October–8 November 1952, both with greatest mortality Britain and Ireland (Evans 1892; Macpherson 1892; Boyd 1954); both followed persistent severe westerly gales in Atlantic. In 1891, mainly adults (young not fledged at time of 26–28 September storms), but juveniles included in 1952 wreck; generally emaciated, a 1952 sample *c.* 35% underweight. In 1952, *c.* 2600 in Bridgwater Bay, Somerset, alone, and probably over 7000 died Britain and Ireland, some numbers inland. Many fewer died elsewhere, but 70 Belgium compared with only 10 during 1928–1951 (Lippens 1953), and one as far inland as Switzerland, while several in south France (Herault) in early November thought driven overland from Bay of Biscay (Boyd 1954). 2 Indian Ocean vagrants in Kenya and Sharjah (Lapthorn *et al.* 1970) more likely to have originated from Pacific than Atlantic.

In homing experiment 2 released Sussex, England, returned 4800 km (Great Circle distance) to their colony in Canada (New Brunswick) in 13·7 days, daily average of at least 350 km (Billings 1968).

**Food.** Chiefly planktonic crustaceans, molluscs, small fish; also oily and fatty substances, especially from fish offal thrown overboard. Feeds mainly from surface while skimming or hovering and snatching; rarely settles on water to feed (Miller 1937) and apparently never dives. Follows feeding whales and seals for waste left on surface (Bent 1922; Gross 1935), for their faeces (Van Kammen 1916) and, if wounded, for traces of oil released (Bent 1922). Though thought to be largely diurnal feeder (Wynne-Edwards 1935), a number of planktonic crustaceans on which it feeds surface only after dark.

Planktonic crustaceans identified include euphausid shrimp *Meganctiphanes norvegica* and copepod *Temora longicornis* in British waters (Witherby *et al.* 1940); hyperiid amphipods from 3 stomachs collected Forrester Island, Alaska (Heath 1915); and larval stage of spiny lobster *Palinurus* in spring and summer off south California (Anthony 1898). Cephalopods (squid) important part of diet in Bay of Fundy (Gross 1935). Small fish and fish spawn said taken, but little direct evidence apart from Palmer (1962) and Miller (1937).

**Social pattern and behaviour.** Compiled mainly from material supplied by C E Huntington and W E Waters.

1. Gregarious at breeding grounds; mainly solitary at sea. Pelagic and nomadic outside breeding season so far as known; occasionally in small, loose flocks, apparently when attracted by abundant food, when also associates freely with other petrels. BONDS. Monogamous pair-bond; of life-long duration (shown by ringing) but renewed each breeding season. Probably based on strong site-fidelity rather than individual recognition; even

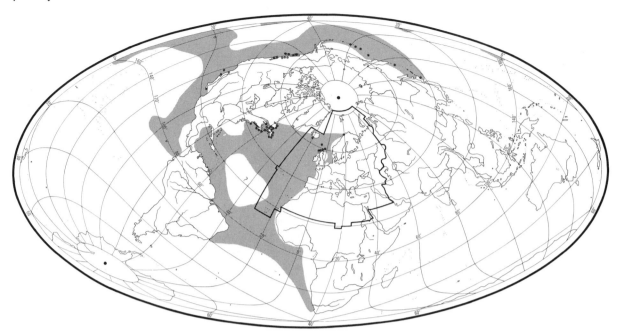

relatively uncommon cases where pair moves from one nest-site to another may result from pair-formation at previous season's site before move made (Huntington 1966; Wilbur 1969). No evidence of any association between pair at sea. Though 1st year birds twice reported from burrows (Gross 1947), at Kent Island, New Brunswick, pre-breeders normally return to colony and probably pair-up one year before first breeding at 5 (occasionally 4) years (Huntington and Burtt 1972), pairs with small gonads being found in burrows in July (Palmer 1962). Both adults tend chick until just before fledging when fully independent. BREEDING DISPERSION. Nests colonially in burrows; often grouped, even where population small, and densely concentrated at some stations, e.g. mean density wooded part of Gull Island, Newfoundland, July 1960, 1 occupied burrow per m² (Huntington 1963). Occasionally one burrow entrance serves 2 or more nest-chambers, but more frequently single nest has 2 entrances. Largely non-territorial, sharing feeding grounds, air-space over colony, and ground surface near burrows; but nest-chamber itself evidently defended. ROOSTING. No specific information; presumably as in other nocturnal petrels.

2. Normally comes ashore for breeding, and then only at night. First returns St Kilda, April (Waters 1964). After spending day at sea, usually well away from breeding grounds, starts arriving over land c.1–2 hours after sunset; earliest arrivals on shortest summer nights when, in latitude of west Palearctic colonies, darkness incomplete. Arrives rather later on light nights and when windy or in thick fog, while during gale colony seems practically deserted for whole night. Time of departure for sea variable; some leave burrows up to 20 min before first in-comers, while others remain until less than 90 min before sun-rise. AERIAL ACTIVITY. Repeated nocturnal Display-flight of single birds, pairs, and even larger groups above or near colony marked feature of short summer nights. Reaches first peak May and early June, then declines somewhat until second peak July and August, latter probably associated with pre-breeders then newly arrived (Waters 1964). Much less activity over land September; occasionally recorded well into October (North Rona). Display-flight occurs over irregular course and may be communal display, sex-advertisement, or sexual chase, but

probable that olfactory location of individual burrows also involved (Grubb 1974). Flighting birds hard to follow in darkness, but evidently perform 2 types of Display-flight (Ainslie and Atkinson 1937). (1) Rapid, erratic flight up to 10 m above ground commoner; when circular or orbital, may be centred on particular burrow and bird will make series of concentric circles, at least 3 or 4 in number, each rather tighter than one before, until dropping suddenly to ground and disappearing into burrow—although sometimes taking-off again (Waters 1964). Such burrow-locating flights not associated with calling by second bird underground. (2) Slower, and apparently more purposeful flight low over ground or canopy; bird rapidly vibrates wings and hovers with legs trailing, or even stands with raised wings apparently picking up things (possibly insects or vegetable matter). Aerial activity of both types continues most of night, with hundreds dashing overhead, flying fast, often brushing against other birds, trees, and stones; apparent frenzy slowly declines after midnight, though some flighting continues almost to first light. If colony dense, concentration so high that collisions occur, birds sometimes tumbling to ground; after falling, they may immediately disappear underground together as if incident deliberate (Williamson 1948). While Display-flights in progress, other birds give Purr-calls from underground, especially in May and June and regularly until August. Intensity of flighting varies during night at any given spot over colony, and flying birds tend to gather near others Purr-calling (Darling 1940), being also attracted when recording played back (Grubb 1973). Though those arriving first after dark often silent until joined by others, Display-flight typically accompanied by Chatter-call which shows considerable individual variation, probably sufficient for mates to recognize one another (Williamson 1948). Calls repeated several times a minute, but frequency variable and reduced during slower, low-level version of Display-flight. When Chatter-calls given in close succession, bird sometimes breaks into short burst of Purr-call, though this ordinarily heard only from burrow. As night advances, volume of noise gradually grows until peak reached some time after midnight. Little reduction in calling on moonlit nights at British colonies, contrasting with noticeable reduction at Kent Island, New Brunswick, further

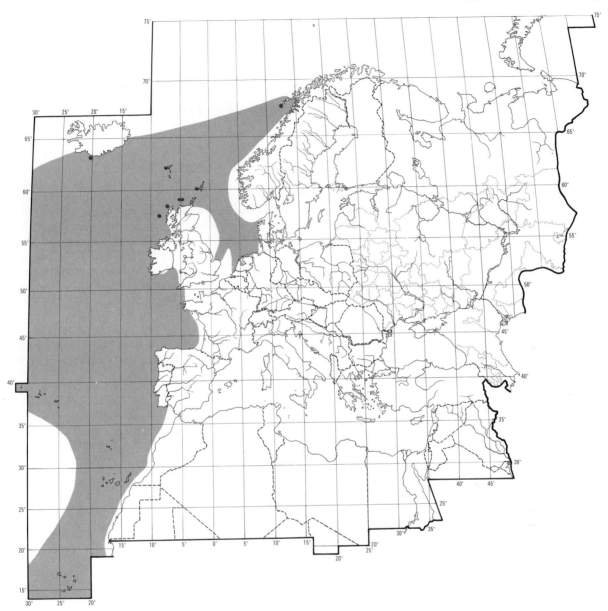

south and with many Herring Gulls *Larus argentatus* near.
BEHAVIOUR IN BURROW. Little known about display and related
behaviour underground. No visual observations made on
antagonistic behaviour between adults, but excited Screech-
calling (such as occurs when strange bird experimentally
introduced into occupied burrow and occasionally during
handling) sometimes heard from underground and may result
from real intrusions, strangers being found in nest-burrows at
times. Though Chatter-calling often given from surface of
ground at night, final approach to and entry of burrow after
flighting usually silent, bird often then immediately Chatter-
calling from within (Grubb 1974). Because pair remains inside
burrow most of time while ashore, communication between ♂
and ♀ evidently mostly vocal and perhaps tactile and olfactory. As
Purr-calling usually uttered only when pair together in burrow,
this probably main heterosexual activity, ♂ and ♀ engaging in
frequent and prolonged duets; such calling mainly nocturnal but

occasionally heard after sunrise, even at mid-day. ♂ in burrow
said to attract ♀ by calling (Gross 1935) but this still not
demonstrated, though recordings of Chatter-call and Purr-call
will induce approach by flying birds (Grubb 1973). Burrow
normally excavated by ♂. Copulation undescribed; presumably
occurs underground after nest completed, burrow later being
deserted for 1 day and egg laid following night (Gross 1935).
Unfinished burrows dug spasmodically as late as August and
occupants, probably pre-breeders, call from within. RELATIONS
WITHIN FAMILY GROUP. No information.

**Voice.** Usually silent at sea, though sometimes utters same
Chatter-call as over land (Palmer 1962).

2 main calls given nocturnally at colony (Ainslie and
Atkinson 1937; Grubb 1973). (1) Purr-call (Song or
Burrow-call). Almost continuous series of musical croon-

ing, churring, or purring notes, each sequence rising in pitch, then, with short 'whee' note caused by indrawing of breath, starting again at lower pitch. Differences in calling rate, time interval between breath-notes, and duration of breath-notes separate this species from Storm Petrel *Hydrobates pelagicus*. Unit rate in churring 15–22 notes per s (slower than in *H. pelagicus*); interval between onset of breath-notes 3·06–4·8 s (longer); duration of breath-notes 0·38–0·65 s (longer) (J Hall-Craggs). Often interrupted, preceded, or followed by Chatter-call (see below). Best rendering: prolonged '. . . r-r-r-r-r-r-r' followed by soft, high-pitched 'ooee-' or 'whee' breath-note (see I). Given mainly from within burrow, often as duet by pair; occasionally heard on surface or in flight. (2) Chatter-call. Often repeated, rapid sequence of 8 or more musical, staccato notes with descending cadence, and ending in slurred trill. Highly variable in pitch, emphasis, and duration, with sex differences; sometimes almost screamed. Resembles 'pā-parakĕpāktō—quā quă qua qua quă' (see II) (P J Sellar); sometimes abbreviated. Given freely on wing, especially during Display-flight over colony, almost throughout pepriod birds visit land; harsher, shorter, and much less frequent in September. Sometimes also uttered on ground, but particularly common just after bird has entered burrow.

Other calls include soft, high-intensity squeals or wheezing sounds (see III), produced during pauses in Purr-calling (J Hall-Craggs), clicking note uttered from burrow towards intruder (conspecific or man), by both parents and young, probably as warning or challenge; also described as repeated, half-purred 'choo shoo'. Harsh Screech-call also reported when intruder enters burrow; same call occasionally given when birds removed from hole, although more usually several squeaking notes sometimes followed by Chatter-call.

CALLS OF YOUNG. Small chicks utter plaintive, piping 'peeee-peeee-peeee' with beak closed (Witherby *et al.*

1940); softer during feeding, more prolonged in older chicks (Palmer 1962).

**Breeding.** SEASON. See diagram for British Isles. No information for Faeroes and Iceland. SITE. In banks, under boulders, under tree stumps, in grassed-over ruins of buildings (North Rona), and occasionally in piles of stones (Ainslie and Atkinson 1937; Gross 1935). Colonial; tunnel entrances often close together, up to 7 in sq m recorded (Bent 1922); also up to 5 birds recorded nesting in side branches of one tunnel (Bent 1922). Nests re-used in subsequent years (Palmer 1962). Nest: shallow depression in enlarged chamber at end or in side of tunnel; average dimensions of chamber 16 × 16 × 9 cm; average depth below ground of 10 chambers 25 cm (12–40); length of 10 tunnels 51 cm (28–87) though up to 180 cm recorded; size of entrance hole 8 × 7 cm (6–14 × 6–9) (Ainslie and Atkinson 1937; Gross 1935). Nest material varies from nil to substantial amount; depression usually lined with small twigs, grass stems, and leaves. Building: by ♂ only (Gross 1935). Bird starting excavation, rolled sideways on one foot and metatarsal and threw soil back with other; changed feet after 10–12 strokes. Used this technique for centre of depression; for edges rested on breast and kicked with both feet alternately 15–20 times, throwing soil back up to 1 m. Bill sometimes used to loosen soil (Grubb 1970). EGGS. Elongated or blunt ovate; dull white, some with band of minute lilac spots towards large end. 33 × 24 mm (30–36 × 22–23), sample 100 (Schönwetter 1967). Weight 9 g (8–10), sample 10 (Gross 1935). Clutch: 1. One brood. Replacement eggs occasionally laid, but may involve change of mate. INCUBATION. 41–42 days. By both sexes, in approximately equal spells of *c.* 3 days (1–6), with ♂ taking 1st spell after laying (Wilbur 1969). Eggshells left in nest. YOUNG. Semi-altricial and nidicolous. Cared for and fed by both parents. Brooded continuously for first 5 days, then fed nightly. Fed by bill-to-bill, partial regurgitation.

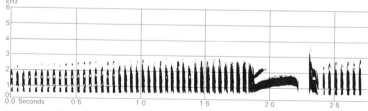

I  P J Sellar  Iceland  June 1968

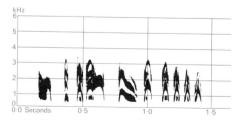

II  P J Sellar  Iceland  June 1968

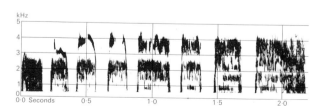

III  P J Sellar  Iceland  June 1968

Feeding rate drops off towards fledging. No desertion period. FLEDGING TO MATURITY. Fledging period 63–70 days. Become independent about same time. Age of first breeding, Kent Island, New Brunswick, 5 or occasionally 4 years (Huntington and Burtt 1972). BREEDING SUCCESS. No data.

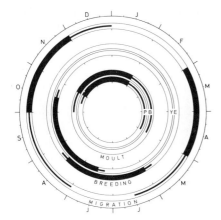

**Plumages** (nominate *leucorhoa*). ADULT. Mostly black-brown, paler ashy on chin and forehead, browner below, with strong grey or slate-blue tinge on upper parts and chest in fresh plumage, black-brown when worn. Patch on lower rump and upper tail-coverts white, some longer feathers with black tips and shafts, and central ones often all dark; shorter feathers sometimes irregularly tipped black-brown, with brown shafts and bases. White rump not so sharply defined as in other white-rumped storm-petrels, white often divided by dark in middle. Shorter lateral under tail-coverts white, shafts brown; longer ones white, broadly tipped sooty-brown. Tail feathers black with variable amount of white on base of outer ones. Upper wing black-brown, but greater and median upper wing-coverts and tertials grey-brown with white edges, most prominent on tertials and inner coverts. Underwing black-brown. Flight-feathers black. NESTLING. 1st down bluish grey, base tinged brown; bare around neck, face and chin. 2nd down longer, darker (Gross 1935). JUVENILE. Light edges of upper wing-coverts and scapulars somewhat broader than in adult. Outer primary pointed at tip (rounded in adult), but sometimes difficult to see when worn, and intermediates occur.

**Bare parts.** ADULT AND JUVENILE. Iris brown; bill, tarsus, and foot black. NESTLING. At hatching, tip of bill black, rest pale horn; tarsus and foot light-grey, sometimes flesh-colour;

pigmentation increases during growth. Bare skin around eyes grey-blue at hatching; later darker (Gross 1935).

**Moults.** ADULT POST-BREEDING. Complete; primaries descendant. Moult of some body feathers starts at end of breeding cycle, tail and wing soon after, mainly October when often still near breeding grounds; adults blown ashore Europe, October–November, show first stage of tail and primary moult. Main moult in winter quarters, November–February; outer primaries grow more slowly, all renewed by April. Non-breeders may start primary moult in August–September. POST-JUVENILE. Complete. Flight-feathers start about April of 2nd calendar year, completed by October–December. Wing moult starts gradually later in following years; in non-breeders visiting colonies from July or later.

**Measurements.** Full-grown, west Europe and Atlantic to South Africa; skins (RMNH, ZMA).

| | | | | | |
|---|---|---|---|---|---|
| WING | ♂ 158 | (3·76; 47) | 148–165 | ♀158 (3·56; 54) | 152–166 |
| TAIL | 80·8 | (4·15; 47) | 74–91 | 80·5 (3·47; 53) | 73–87 |
| CULMEN | 15·7 | (0·46; 50) | 14·2–16·6 | 15·7 (0·50; 56) | 14·7–16·9 |
| TARSUS | 24·0 | (0·58; 50) | 22·9–25·5 | 24·1 (0·70; 55) | 22·3–25·5 |
| TOE | 24·2 | (0·68; 48) | 22·5–25·6 | 24·4 (0·92; 51) | 22·6–26·5 |

Sex difference not significant.
Live adults, breeding, Britain: wing 160 (3·41; 39) 154–166 (BTO).

**Weights.** ADULT. St Kilda: May, 44·9 (2·0; 8) 41–48; end June to mid-August, 45·0 (3·5; 28) 39–55 (Waters 1964). Britain: July, 44·1 (2·86; 37) 40–50 (BTO). Wintering (including juveniles), at sea Azores to South Africa, August–April, 39·9 (4·89; 14) 33–52 (RMNH, ZMA). JUVENILE. Stranded after gale in Netherlands, only 29·5 (4·39; 6) 26–37. NESTLING. New Brunswick, Canada; average 1st day 6·4; 34th and 39th day, 69·5 (Gross 1935); September, 66·8 (39) 44–90, exceptionally 23 and 30 (Palmer 1962).

**Structure.** 11 primaries: p9 longest, p10 5–12, p8 1–5, p7 9–15, p6 20–28, p1 76–92 shorter. Tail forked, 12 feathers: outer 3 elongated, t6 19·0 (2·1; 50) 16–26 longer than t1. Bill rather slender, decurved, firm hook, nasal tubes *c.* 40% bill-length. Tarsus and toes slender, scutulated; outer toe almost equals middle toe, inner toe *c.* 20% shorter.

**Geographical variation.** Difference in measurements between North Atlantic and North Pacific nominate not significant. Along west coast of North America from south-east Alaska to Baja California, birds become smaller and browner, with reduction of white on rump. (See Austin 1952.)                    CSR

## *Oceanodroma monorhis* Swinhoe's Storm-petrel

PLATES 17 and 18
[facing page 159 and after page 182]

DU. Chinees Stormvogeltje     FR. Pétrel tempête de Swinhoe     GE. Swinhoes Wellenläufer
RU. Малая качурка     SP. Paiño de Swinhoe     SW. Svart stormsvala

*Thalassidroma monorhis* Swinhoe, 1867

Monotypic

**Field characters.** 19–20 cm, wingspan 44–46 cm; slightly smaller than Leach's Storm-petrel *Oceanodroma leucorhoa*, with which sometimes regarded as conspecific,

though appearance quite distinct from Atlantic population of that species. Dark brown with no white rump, feet hardly projecting beyond forked tail. Sexes alike and no seasonal differences; juvenile inseparable.

Generally dark brown, browner than *O. leucorhoa*, with slightly paler forehead and upper wing-coverts, and greyer brown underparts including under wing-coverts. Black bill and feet. No detailed descriptions available of flight action or behaviour at sea. See Alexander (1955).

**Habitat.** Compared with Leach's Storm-petrel *O. leucorhoa*, prefers warmer currents in breeding season, but similarly resorts for nesting to small, uninhabited islets near mainland or larger islands, usually in immediate vicinity of open ocean. Little known about marine off-season habitats, but occurs in tropical and even equatorial waters reaching 30°C (Bailey 1966), straying exceptionally into land-locked seas (see Distribution).

**Distribution.** Breeds islands off Honshu, northern Kyushu, southern and western Korea, and off Shantung. Perhaps also in Ryu Kyu Islands south to small islands off north-east Formosa (Vaurie 1965).

In west Palearctic, one 14 January 1958, Elat, Gulf of Aqaba, Israel; specimen in Tel-Aviv University Museum (HM).

**Movements.** Migratory. From breeding stations disperses over west tropical Pacific and around East Indies, where most common storm-petrel, since Pacific populations of *O. leucorhoa* winter in central and eastern parts. Ranges regularly west to Singapore. Relatively small numbers pass through Straits of Malacca into tropical Indian Ocean and Arabian Sea (most records), west to offshore Somalia; present in these regions at least February–August, probably non-breeders (Gibson-Hill 1953; Bailey *et al.* 1968).

**Voice.** Nothing known.

**Plumages.** ADULT. Entirely black-brown, including rump; blue-grey tinge on crown, mantle, sides of head and breast in fresh plumage; somewhat duller below. Tail and flight-feathers black, basal parts of inner webs paler; inner secondaries edged grey when freshly moulted. Greater and median upper wing-coverts light sooty, greater narrowly edged grey. JUVENILE. Like adult, but edges of greater coverts wider; median coverts narrowly edged pale grey.

**Bare parts.** ADULT AND JUVENILE. Iris brown; bill, tarsus, and toes black.

**Moults.** ADULT POST-BREEDING. Complete; primaries descendant. Some start moult of primaries and tail on arrival winter quarters November, others still not moulting December. One winter quarters March in full body and wing moult (Bailey *et al.* 1968; RMHN). POST-JUVENILE. No information.

**Measurements.** Full grown. Japan, China, and Java Sea (BMNH, RMNH).

| | | | | |
|---|---|---|---|---|
| WING | ♂ 152 | (3·58; 12) | 146–157 | ♀158 (5·27; 5) 150–165 |
| TAIL | | 70·8 (3·44; 12) | 65–77 | 75·0 (1·79; 5) 73–78 |
| BILL | | 14·5 (0·44; 12) | 13·7–15·2 | 14·6 (0·31; 5) 14·1–15·0 |
| TARSUS | | 23·8 (0·89; 12) | 22·3–25·2 | 23·9 (0·42; 5) 22·3–24·6 |
| TOE | | 23·2 (0·55; 12) | 22·1–23·8 | 23·8 (0·91; 5) 22·5–25·1 |

Sex differences significant for wing and tail.

**Weights.** 2 north-west Indian Ocean *c.* 40 and 38–40; one exhausted Elat, Israel, 23 (Bailey *et al.* 1968; RMHN); one off China *c.* 40 (BMNH).

**Structure.** 11 primaries: p9 longest, p10 4–9 shorter, p8 2–5, p6 21–27, p4 49–55, p1 79–88 shorter, p11 minute. Tail forked: t6 11–19, t3 0–3 longer than t1. Bill short, with strong terminal hook, nasal tubes *c.*35% bill-length. Outer toe slightly shorter than middle and inner toe *c.* 20% shorter.

**Geographical variation.** Sometimes treated as subspecies of Leach's Storm-petrel *O. leucorhoa* (see Palmer 1962), being superficially similar to all-dark *O. l. chapmani*. Latter, however, connected with nominate *leucorhoa* along coast of east North Pacific by populations with intermediate size and rump-coloration, while in west North Pacific no intergradation between nominate *leucorhoa* and *O. monorhis*, which replace each other abruptly between Hokkaido and Honshu. *O. monorhis* differs from *O. leucorhoa* in proportions; culmen of former relatively short and outer tail-feathers not so prolonged. Wintering areas also differ. *O. monorhis* can be considered as early offshoot of *O. leucorhoa* ancestor or vice versa, but future studies on reproductive behaviour will have to elucidate whether *O. monorhis* still conspecific with present *O. leucorhoa* populations. Until then, safe to treat *O. monorhis* and *O. leucorhoa*, together with *O. homochroa* (Coues) from west California, as one superspecies (see also Austin 1952 and Vaurie 1965). CSR

## *Oceanodroma castro* Madeiran Storm-petrel

PLATES 17 and 18
[facing page 159 and after page 182]

DU. Madeirastormvogeltje FR. Pétrel de Castro GE. Madeira-Wellenläufer
RU. Мадейрская качурка SP. Paíño de Madeira SW. Oceanlöpare

*Thalassidroma castro* Harcourt, 1851

Monotypic

**Field characters.** 19–21 cm, wingspan 44–46 cm; almost same size as Leach's Storm-petrel *Oceanodroma leucorhoa*. Blackish-brown with white rump, long wings, feet hardly projecting beyond forked tail, bill short and stubby. Sexes alike and no seasonal differences; juvenile similar.

ADULT. Brownish-black above and browner below,

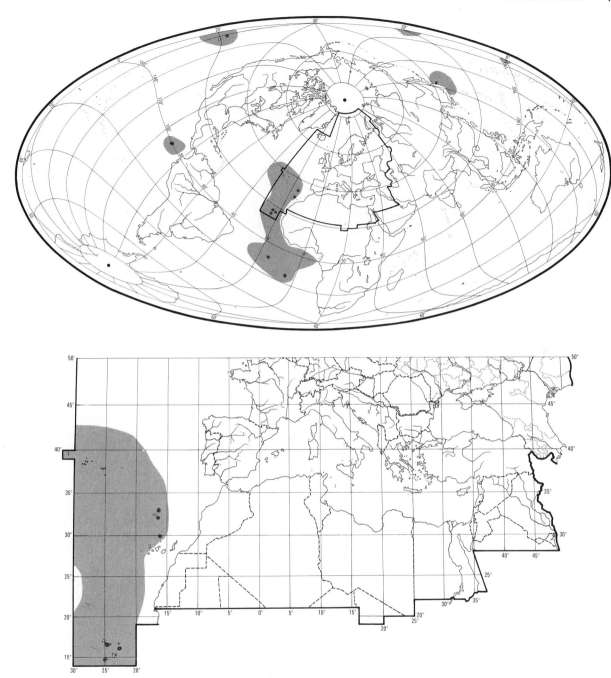

darkest on flight-feathers and tail, apart from white rump extending to lower flanks and lateral under tail-coverts, and slightly browner upper wing-coverts forming indistinct panel. Tail with shallow fork, but often looks square. Bill, legs, and feet black. JUVENILE. In fresh plumage, white tips and edges to wing-coverts, scapulars, and inner secondaries, but these soon abraded.

Difficult to distinguish at sea from *O. leucorhoa*, which it closely resembles but for less marked panel on inner wing (lacking pale rear edge), less forked tail (though this varies with amount of tail spread), prominent all-white rump patch (though grey central feathers of *O. leucorhoa* visible only in good light and at close range), more even distal edge to rump patch, and more white on lower flanks and under tail-coverts. More easily distinguished from Storm Petrel *Hydrobates pelagicus* by larger size, more slender shape, longer wing, browner plumage and, in close view, longer forked tail; also from Wilson's Storm-petrel *Oceanites oceanicus* by same characters together with shorter, all-dark legs and absence of narrow whitish line across wing. Flight action best distinction: intermediate between rapid, rather bounding and erratic flight of *O. leucorhoa*, and fluttering flight of *H. pelagicus* (M P Harris), being typically fairly steady with even, rather high beats interspersed with glides on wings bowed below horizontal (K E L Simmons).

Does not patter with feet on surface as freely as *O. oceanicus* and most other storm-petrels. Markedly pelagic, staying well out to sea; unlike *O. leucorhoa* stated to follow ships regularly (Watson 1966), but this disputed (Simmons 1969a; M P Harris). Generally rather solitary.

**Habitat.** Ecological counterpart of Leach's Storm-petrel *O. leucorhoa* in tropical and adjoining temperate sector of west Palearctic seas (Fisher and Lockley 1954), occupying warm deep pelagic waters throughout year, without making significant seasonal shift in habitat. Unlike *O. leucorhoa*, North Atlantic stock concentrated within west Palearctic, at lower density corresponding to lower biological productivity of ecological niche, partly shared with Leach's during northern winter. Only small, uninhabited islets used for breeding; shared in places with much more numerous White-faced Storm-petrel *Pelagodroma marina*, which migrates after breeding to more productive seas (Bannerman and Bannerman 1968). Nests in rock crevices, caverns, and covered ledges as well as in burrows in soil, often on flat low-lying areas, but also in cliffs up to at least 400 m. Takes readily to nest boxes and clefts in old walls (Allan 1962). Like other storm-petrels habitually flies in lowest airspace over sea. Habit of coming to land during most months renders it susceptible to ground predators and human interference.

**Distribution.** AZORES: probably breeding islets off Graciosa and Santa Maria, but never thoroughly explored, and no proof (Bannerman and Bannerman 1966). Marine range little known (see Movements).

Accidental. Britain and Ireland: Hampshire, November 1911, and Mayo, October 1931 (British Ornithologists' Union 1971). Spain: Badajoz, February 1970 (*Ardeola* 1971, **15**, 101).

**Population.** Little information. CAPE VERDE ISLANDS: numbers 1% or less of White-faced Storm-petrel *Pelagodroma marina* (Bannerman and Bannerman 1968).

Survival. Suspected annual mortality not above 5–7% on Galapagos, but no reliable estimates (Harris 1969a). Oldest ringed bird 10 years 11 months (Rydzewski 1974).

**Movements.** Dispersive. Little known of marine range due to difficulty of separating from other storm-petrels at sea. Apparently highly pelagic and seldom seen inshore near major land masses (Bourne 1967a), but no evidence yet of true migration. Probably North Atlantic breeders occur regularly well offshore from bulge of west Africa, since various records south to 10°N in east Atlantic (Bierman and Voous 1950; Bourne 1963, 1965, 1966a; Lambert 1971). In Gulf of Guinea, scattered records

include: moderate numbers 4°N in February (Bourne 1962), a few at 2°N 3½°E April and off Sierra Leone July (Wallace 1973), and specimens collected São Tomé; but these could be from Ascension, St Helena, or Cape Verde Islands (see also Geographical Variation). A few wanderers or storm-blown individuals recorded far from normal range, including 7 in east USA (Maryland, Pennsylvania, Indiana, Florida), and singles in Brazil, Cuba, and Canada (Ontario) (Palmer 1962; Watson 1966; Garrido and Montaña 1968; Baxter 1970). These suggest pelagic range may extend into mid-Atlantic; confirmation lacking, though Berndt *et al.* (1966) saw storm-petrels believed to be this species in mid-ocean 43½°–48½°N and west to 42½°W in June and July, which might represent dispersal from Azores. Seen by Harris and Hansen (1974) only close to known breeding stations in October–November trans-atlantic voyages.

**Food.** Largely crustaceans, fish, oily scraps and refuse, taken from surface (King 1967; Watson 1966). Mainly solitary; apparently does not follow ships regularly. Appears to feed well away from land, presumably by day (Harris 1969a).

15 stomachs examined Galapagos; 14 had fish-eye lenses (up to 19) or otoliths, 4 cephalopod remains. One cephalopod beak identified as belonging to a myopsid of estimated weight 3–4 g. 2 adults mist-netted April regurgitated fish, length 50 and 37 mm. Droppings of chick, August, contained single fish otolith and smashed cephalopod beak. Evidence indicated main food small fish (probably mainly size of regurgitated sample to judge from otolith size) and squids, all caught on or near the surface. (Harris 1969a).

**Social pattern and behaviour.** Based mainly on material supplied by D W Snow and M P Harris. Studied in most detail extralimitally, mainly at Ascension (Allan 1962) and in Galapagos (Harris 1969a).

1. Gregarious when nesting; pelagic and probably solitary in off-season, dispersing widely. BONDS. No information on pair-bond from west Palearctic, but evidence elsewhere suggests sustained monogamy the rule, with pairs faithful to same nest-site from season to season; in Galapagos same pairs maintained from one season to next, while at Ascension no known changes in partners in one season even when one of pair disappeared. Both parents care for single nestling until just before fledging. BREEDING DISPERSION. Usually loosely colonial, with burrows reaching maximum density at Ascension (Boatswainbird Islet) of one per m². No indication of defence of burrow by owners. At Ascension, numbers of non-breeding adults and pre-breeders present during breeding season. In Galapagos, youngest bird (not breeding) recorded at colony was in its 5th year. Immatures visit colony for one or more seasons before breeding, and, unlike many petrels, return early in season before and during egg-laying period, one record indicating that they may then establish burrow-site used for subsequent nesting. ROOSTING. Like other petrels presumably rests and sleeps mostly on surface of sea. In Galapagos, burrow occupied up to 72 days before egg laid (most visits 3–4 weeks before); birds can spend up to 5 days at a time there, both sexes commonly being present even during day. ♀♀ mostly absent during 2 weeks prior to laying but ♂♂ often

continue to occupy burrow, presumably to maintain ownership; after laying, burrow usually occupied only by duty adult.

2. Individuals occasionally over land by day, but largely nocturnal at breeding stations and activity much reduced on moonlit nights. Gathers over colony at dusk at beginning of breeding season, Flight-calling and flying round, mostly engaged in characteristic Flight-circuiting in which each bird follows its own course; ♀ will also fly together on roughly constant path and aerial 'leap-frog.' observed in which they passed alternately one above the other in gliding flight. Aerial collisions not uncommon during courtship period. In Display-flight, bird calls and flies with shallow flapping, or glides with wings held stiff and straight (occasionally high above back) in wide irregular circles, planing repeatedly past chosen nest-site and eventually landing at it. Function of this flighting apparently to select and occupy nest-hole and to establish relationships with neighbouring pairs; uncertain whether pair-formation takes place during Display-flights or in holes. Birds underground at nest-sites utter characteristic, far-carrying Song or Burrow-call which is quite different from Purr-calls of Leach's Storm-petrel *O. leucorhoa*. Copulation probably occurs only in burrow, not in air. Subdued call uttered by one of pair, with fanned tail, may be prelude to mating but single copulation observed in Galapagos, 33 days before egg laid, silent throughout; this lasted at least 3 minutes, ♂ gently pecked ♀'s head, moving bill from side to side across head with special emphasis at base of upper mandible.

**Voice.** No known calls at sea. Following uttered at breeding places (Allan 1962).

(1) Flight-call. Irregular repetition of short phrase: relatively low-pitched, prolonged initial note, followed by slurred whistling chatter, and ending with 2 staccato notes of higher pitch 'kair chuch-a-chuck chuk chuk . . .' (see I). Also described as 'soft, squeaky note exactly like that of a finger rubbed hard on a windowpane' (Lockley 1952). Uttered over colony after arrival from sea, from dusk onwards; faster and harsher during Display-flight. Inhibited by moonlight. Quieter version sometimes heard from burrow. (2) Burrow-call. Penetrating, guttural purring 'urr-rrr-rrr-rrr-rrr' interspersed with occasional sharp 'wicka' breath-notes (see II), uttered by lone bird in nest-hole, intensified at entry of intruder. (3) Pre-copulation-call. Subdued 'wick-ick-ick', probably preceding mounting.

CALLS OF YOUNG. High-pitched, mellow 'tueep-tueep-tueep', slow and continuous; given as adult enters burrow, and during begging and feeding.

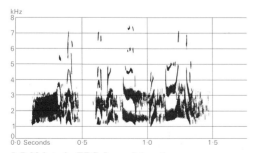

I  R Nakatsubo/BBC  Japan July 1960

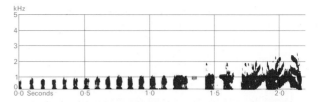

II  A Zino/Sveriges Radio  Selvagens June 1969

**Breeding.** Based mainly on Allan (1962), Ascension; Harris (1969*a*), Galapagos. SEASON. Madeira: eggs found every month of year except May in 2 main periods, June–September and October–December in each year (Bannerman 1954); perhaps involves different individuals, but may be one season plus out-of-season nesting (as Ascension), or 2 populations, each breeding annually, but 6 months out of phase (as Galapagos). Cape Verde Islands: eggs mid-January to May, perhaps not same period each year (de Naurois 1969*b*). Selvagens: eggs late June or early July to September (Bannerman 1963). SITE. In rocky ground, in cliffs in soft soil or guano. Colonial, but density rarely high. Pairs return to same sites each year on Galapagos. Nest: hole in ground or between rocks. Burrow 5 cm diameter, 12–30 cm deep, extending laterally 65 cm or more (Bannerman 1959). Sometimes only under projecting rock, concealing bird from sight but not from light (Galapagos). Also uses nest-boxes on Boatswainbird Island, Ascension. Building: both sexes excavate in soft ground sites. EGGS. Blunt oval, no gloss; white, occasionally zone of pale red spots at larger end, soon disappearing. 32 × 24 mm (30–35 × 21–25), sample 44 (Ascension) (Stonehouse 1963). weight 9·5 g (6–11), sample 28 (Galapagos). Clutch: 1. One brood. Replacement laying possible (Galapagos). Eggs laid at night. INCUBATION. 42 days (39–51) (Galapagos). By both sexes; spells 4–7 days, extremes 2–11 (Galapagos). Eggshells not removed, but occasionally kicked out. YOUNG. Semi-altricial and nidicolous. Cared for and fed by both parents. Brooded continuously for up to 7 days, then fed on 69% of nights (Galapagos). Fed by partial regurgitation from back of parent's throat. Abandoned 1–2 nights before fledging. FLEDGING TO MATURITY. Fledging period 69 days (60–72), exceptionally 107 (Galapagos); 64 days (59–72) (Ascension). Becomes independent when deserted by parents. Age of maturity not known; unlikely to be before 3 or 4 years. BREEDING SUCCESS. Galapagos: in 3 seasons, 60% of 229 eggs hatched and 50% of young reared, so 0·3 young per pair; nesting success showed marked decline within each season. Ascension: 49% of 154 eggs hatched and 67% of young fledged.

**Plumages.** ADULT. Mainly brown-black, with blue-grey tinge hindcrown to back, sides of head, and chest; chin, belly, and median under tail-coverts black-brown. Lower rump-coverts brown-black with white tips, upper tail-coverts white, longer ones tipped black, forming prominent white band without dark shafts or spots; white patches at sides of tailbase. Tail-feathers black, broad white base on outer, decreasing amount of white on inner. Flight-feathers black, more brown-black below. Upper wing-coverts brown-black, but median and greater dark brown, longer feathers with narrow buff margins. Underwing black-brown. NESTLING. 1st down dense, pale grey; 2nd down darker (Allan 1962). JUVENILE. As adult, but more prominent light edges to upper wing-coverts; white edges to black tips of some longer upper tail-coverts (Witherby *et al.* 1940); pointed outer primary (rounded in adult) apparently as in *Hydrobates*.

**Bare parts.** ADULT AND JUVENILE. Iris dark brown. Bill, tarsus, and foot black. NESTLING. Bill and foot light flesh-coloured, dark pigmentation spreading during growth from tip of bill and joints of tarsus and foot (Allan 1962).

**Moults.** ADULT POST-BREEDING. Complete; primaries descendant. Starts with inner primary and body feathers during last stage of breeding; wing moult just completed on arrival for next breeding cycle. Non- and failed breeders may start earlier. Tail moult irregular. POST-JUVENILE. Probably much earlier than adults, as in related species. (Allan 1962; Harris 1969*a*.)

**Measurements.** Adults from Selvagens, Madeira, Cape Verde Islands (MNHN, RMNH, J G van Marle).

| | | | | | | |
|---|---|---|---|---|---|---|
| WING | ♂ 148 | (3·51; 18) | 142–154 | ♀156 | (4·41; 8) | 149–161 |
| TAIL | | 67·2 (2·97; 18) | 61–74 | 70·1 (2·71; 8) | | 67–75 |
| BILL | | 14·4 (0·67; 18) | 13·0–15·5 | 14·8 (0·40; 8) | | 14·5–15·7 |
| TARSUS | | 22·2 (0·56; 18) | 21·0–23·0 | 22·5 (0·42; 8) | | 22·1–23·5 |
| TOE | | 21·9 (0·99; 16) | 20·3–23·1 | 22·5 (1·02; 8) | | 21·5–23·5 |

Sex differences significant for wing and tail.

**Weights.** ADULT. Galapagos, 28·5–56·5, mean 41·7, sample 376 (Harris 1969*a*). Ascension Island mean 43·5 ±5·0, sample 12 (Allan 1962). NESTLING AND JUVENILE. 1–5th day, average 11·1 ±3·4, sample 8; maximum on 46–50th day, 74·8 ±7·9, sample 28; at departure 48·6 ±1·2, sample 7 (Allan 1962).

**Structure.** 11 primaries: p9 longest, p10 5–9, p8 1–4, p7 8–13, p6 18–26 and p1 88–96 shorter. Tail forked, 12 feathers: t6 4–12 (mean 5·7) shorter than t1. Length of white rump in middle 5–18 (mean 13·9) in adults. Bill short, strongly hooked; nasal tube *c.* 40% bill-length. Tarsus and toes slender, scutulate; outer toe equals middle, inner *c.* 20% shorter.

**Geographical variation.** Several subspecies described, based on differences in bill structure, amount of white on tail and rump, wing length, etc, but much overlap and differences not statistically significant. As a rule, populations breeding in warmer surface water have longer extremities and less white in rump compared with those of colder waters. 4 specimens of unknown breeding population, collected Gulf of Guinea, large with little white in rump; possibly distinct subspecies, as significantly different from all other known populations (Harris 1969*a*).

# Order PELECANIFORMES

Medium-sized to huge aquatic birds, both marine and freshwater. Totipalmate (all 4 toes connected by webs). 6 families all breeding in west Palearctic, though Phaethontidae, Anhingidae, and Fregatidae only marginally. Phaethontidae and Fregatidae most aberrant of order; not closely related to each other, or, probably, to rest of order as presently constituted. Relationships of Pelecaniformes as a whole to other orders not clear, but morphological and behavioural characters suggest distant affinities to Procellariiformes and Ciconiiformes (van Tets 1965).

# Family PHAETHONTIDAE  tropicbirds

Medium-sized, highly aerial seabirds. 3 closely related species in single genus *Phaethon*. 1 species in west Palearctic. Body elongated, neck short, wings long and narrow. 11 primaries; p10 longest, p11 minute. Secondaries diastataxic. Flight sustained, with powerful wing-beats. Tail with 12–14 feathers, central pair elongated to long streamers. Bill powerful, decurved, tip not hooked, cutting edges serrated; nostrils well-developed slits. No bare skin in gular area. Tarsus extremely short; totipalmate, nail of middle toe without comb but with horny flange in Red-billed Tropicbird *P. aethereus*. Stance crouching, gait shuffling. Oil gland feathered. Aftershaft absent. Sexes similar in plumage and size. Plumage white with black markings on head and upperparts (tail streamers pink in Red-tailed Tropicbird *P. rubricauda*); white sometimes tinged salmon. Bare parts vividly coloured. 1 moult per cycle. Primaries replaced in serially descendant order. Young semi-altricial and nidicolous; clad in white to pale grey down at hatching. Contour feathers do not develop from same follicles as natal down. Juveniles resemble adult but distinguishable; reach adult plumage in 2nd calendar year.

Occur in all oceans, tropical and subtropical. Essentially pelagic; ranging at great distances from land, entering water briefly to catch food, and coming to land only for nesting. Only limited extensions from tropical and sub-tropical waters, beyond which wandering does not occur. In contrast to Procellariidae and Hydrobatidae, do not habitually fly low over water, and unlike Diomedeidae do not rely on lift from winds or aircurrents, being equally efficient in still air with exceptional flight endurance. Barely penetrate west Palearctic where declining in range and numbers.

Principal foods fish (particularly flying-fish) and cephalopods (especially squid); caught during day mainly by plunge-diving from good height but often with little deep penetration under water (surface-plunging and shallow-plunging), though will sometimes descend for several metres. Often attracted by ships but do not appear to follow or forage in wake. At sea, mostly solitary but also regularly in twos; will join feeding congregations of other seabirds. At breeding station, often nest in loose groups if terrain permits and engage in communal display-flights (2 or more birds) before laying. Monogamous pair-bond often of long-term duration, but successful pairs more likely to reunite in subsequent season than failed breeders (see, e.g., Fleet 1974); unlike in most other seabirds, possible that some pairs maintain contact at sea—both when breeding and at other times. Defend nest-sites vigorously, showing marked site-tenacity from season to season, especially if same mate present. Roost at sea, on water or (probably) in air, when not breeding. Because of restricted nature of nest-site and awkwardness of locomotion on land, no specialization for visual display ashore; pair usually sit side by side at site for long periods with little activity. Voice relatively unspecialized. Calls given mainly on wing, both in vicinity of colony and at sea, and to lesser extent, from site. Comfort behaviour and general behavioural characters not closely studied.

Breeding seasons often prolonged, with individuals laying at intervals of 9–12 months. Nest a scrape with little or no added material; in cavity or large hole, or under shelter of boulder or bush. Eggs quite unlike those of other Pelecaniformes: variable in shape and colour, fawn to rich purple-brown, mottled or blotched darker especially at ends; not glossy. Clutch 1: single brood. Replacements laid after egg loss, but proportion varies depending on current feeding conditions. Incubation periods 40–46 days; by both sexes, each with single median brood-patch, in variable spells of 3–16 days. Young cared for and fed by both parents by regurgitation; in *P. rubricauda* at least, adult inserts bill into gullet of young and disgorges food directly (unlike in all other Pelecaniformes). In earlier stages, at least 1 feed per day but feeding intervals longer (up to 2–3 days) as chick grows; left alone for long periods before this, sometimes after only a few days. Fledging periods highly variable (65–90 days) and dependent on current food supply. Become independent at same time; no desertion period. Age of maturity unknown.

## *Phaethon aethereus* Red-billed Tropicbird

PLATE 19
[between pages 182 and 183]

Du. Roodsnavelkeerkringvogel    Fr. Paille en queue éthérée    Ge. Rotschnabeltropikvogel
Ru. Красноклювый фаэтон    Sp. Rabijunco común    Sw. Rödnäbbad tropikfågel

*Phaëthon aethereus* Linnaeus, 1758

Polytypic. *P. a. mesonauta* Peters, 1930, Cape Verde Islands, islets off Sénégal, West Indies, east Pacific; *indicus* Hume, 1876, Red Sea, Persian Gulf, Arabian Sea. Extralimital: nominate *aethereus* Linnaeus, 1758, Ascension Island, St Helena, and Fernando Noronha.

**Field characters.** 90–105 cm, of which half tail-streamers; wingspan 100–110 cm. Medium-sized, plump, predominantly white, tern-like seabird with narrow, pointed wings, conspicuous red bill, and greatly elongated tail-streamers in adult stages. Sexes alike in pattern (streamers of ♂ some 12 cm longer) and no seasonal differences, but Atlantic and Red Sea/Persian Gulf populations distinct in field; juvenile distinguishable.

ADULT. Tropical North Atlantic race *P. a. mesonauta* silky white (suffused rosy pink in fresh plumage) apart from broad black stripe through eye (extending as broken band across nape in worn plumage), narrow black barring on upperparts, black marks on lesser coverts, and black on upper surfaces of outer primaries and inner secondaries; bright red bill, grey legs, black feet. Red Sea/Indian Ocean race *P. a. indicus* has black eye-stripe greatly reduced (often missing behind eye) and reddish-orange bill with blackish cutting edges. JUVENILE. Similar (both races) but black eye-stripes usually joined across nape, black barring on upperparts broader and closer, tail black-tipped and lacking streamers.

Combination of mainly white plumage with characteristic black markings, narrow pointed wings, tail-streamers, and distinctive pigeon-like flight makes adult unmistakable among west Palearctic birds, and even moulting adult without streamers or juvenile easily identified. Both other species of tropicbirds breed tropical Atlantic or Indian Oceans and might occur as vagrants. *P. aethereus* largest of the three and distinguished by barred back, red bill, white streamers, and, when present, black across nape. Yellow-billed or White-tailed Tropicbird *P. lepturus* has white back, black band on wing, yellow to orange bill, and

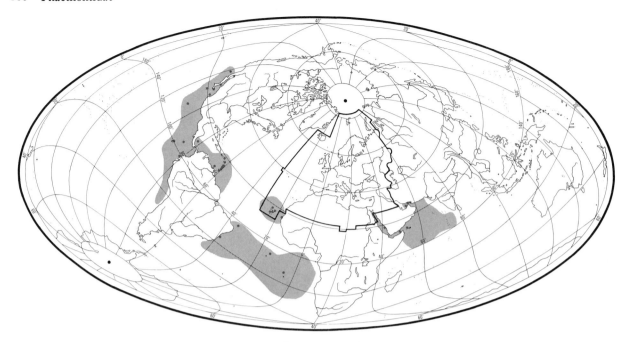

broader white streamers. Red-tailed *P. rubricauda* has white back, less black in wing, and red streamers with black shafts. Juveniles of these other 2 species less heavily barred than young *P. aethereus* on hindneck and back, and lack black band across nape.

Purposeful flight direct with rapid, pigeon-like wing-beats, but at other times fluttering alternated with soaring glides. Feeds by plunging vertically, often after hovering. Pelagic and may be seen several hundred km from land. Usually solitary except during breeding, when gregarious rather than colonial, but sometimes seen in twos and, more rarely, small parties out to sea. Legs short and so cannot walk upright, but shuffles along with belly touching ground. Occasionally settles on sea, but seldom remains long on surface, when floats high with tail raised, streamers forming arc. Although sometimes attracted to ships, normally circles and departs, rather than follows. Tame when nesting.

**Habitat.** Mainly in airspace, cruising commonly at 30 m or more above sea, but heights not accurately recorded; probably reaches levels at least comparable with highest breeding places at *c*. 500 m. Frequently deviates to circle ships; on emerging from steep plunge-dive after prey will sit on water, taking off quickly and easily. Not ordinarily so exposed to storms as pelagic species outside tropical seas. Radius of foraging around breeding stations not recorded; otherwise rarely inshore and less frequently offshore than pelagic, covering great distances. Nests in rocky crevices at whatever altitude these occur, from rocks on beach within 2 m of high tide and upwards (Nelson 1899), so long as inaccessible to predators and, so far as possible, to man. Prefers uninhabited smaller islands, or high crags well inland on larger islands, permitting take-off by immediate launching over brink; mobility on ground extremely limited. Makes repeated approaches to nest before finally alighting; often seen flying over intermediate forest cover. At sea, more often singly or in pairs than in groups.

**Distribution.** Limited and decreasing area occupied in west Palearctic. CAPE VERDE ISLANDS: declined; believed bred formerly on São Vicente; probably still nests Santo Antão; few pairs on Razo (Bannerman and Bannerman 1968, de Naurois 1969*b*). EGYPT: on islands Gulf of Suez; egg said to have been taken Shadwan Island (Meinertzhagen 1930) and immature collected from nest, West Ashafri Island, May 1926 (Borman 1929) but no recent information. KUWAIT: according to Meinertzhagen (1954), certainly bred Um-al-Maradim and Kubbar, but not included as breeding there by Ticehurst (1925) and no recent evidence of nesting (SH).

Accidental. Madeira: one August 1893; 2 in September 1966 (Bannerman and Bannerman 1965, 1968). See also Vielliard (*Alauda*, 1972, **40**, 399–400) for records of *Phaethon* not specifically identified Banc d'Arguin and Morocco.

**Population.** CAPE VERDE ISLANDS: alarming decline, as many taken by fishermen (Bannerman and Bannerman 1968); incomplete survey (Rombos, Brava, São Tiago, Boa Vista) suggested total population under 1000 (de Naurois 1969*b*). Oldest ringed bird 6 years 11 months (Rydzewski 1974).

**Movements.** Dispersive and resident. No evidence of regular seasonal movements; some adults near colonies all year though few ashore August–October (Bannerman and

Bannerman 1968). Dispersal at times far from land; perhaps young disperse further than breeding adults (Palmer 1962) but Galapagos adults move up to 1500 km (Harris 1969*b*).

In absence of ringing, movements away from Cape Verde Islands and small Sénégal colony indistinguishable; apparently dispersal chiefly east and south-east into upwellings off west Africa, from Sénégal to Guinea-Bissau, though few recent records (Bourne 1966; Lambert 1971). One seen west of Cape Verde Islands at 16°N 34°W in December 1963 (Bannerman 1969); 3 at 20°14′N 18°19′W (west of Mauritania) in January 1941 (Mayo 1948); also 2 Madeiran records (see Distribution). Reported in Gulf of Guinea and off north Angola, but these most likely from Ascension Island or St Helena. Some Caribbean breeders disperse into western Sargasso Sea; one vagrant reached Long Island, New York, June 1963 after tropical storm (Bull 1964).

Disperses widely over Gulf of Suez, Red Sea, Gulf of Aden, Arabian Sea, and Persian Gulf from colonies within these areas; has occurred western India, Laccadive Islands, Ceylon; Bay of Bengal, and Madagascar (Meinertzhagen 1954; Ali and Ripley 1968).

**Food.** Chiefly fish and squids caught mainly by plunge-diving vertically into water where remains only a few seconds; also takes flying fish on wing. Prey never speared or stabbed; caught and held crossways in beak, and swallowed under water or as emerging (Gifford 1913; Murphy 1936; Stonehouse 1962*b*).

Stonehouse (1962*a, b*) examined regurgitated food, Ascension Island, and found most frequently small (10–20 cm) flying fish *Exocoetus volitans* and *Oxyporhamphus micropterus*. Other fish included small (10–25 cm) *Ophioblennius webbii*, *Holocentrus ascensionis*, and unidentified immature forms; also larger flying fish (over 25 cm), probably *Cypselurus* and *Hirundichthyes*. Remains of squid *Hyaloteuthis pelagicus* also reported.

Squid and small fish fed to small chicks (Murphy 1936). Stonehouse (1962*b*) found that young fed within 12–16 hours of hatching; food varied with age of the chick: small chicks fed on squid or fish of 4–6 cm and partially digested larger fish, larger chicks received flying fish up to 30 cm long.

**Social pattern and behaviour.** 1. Gregarious when nesting; pelagic during off-season, dispersing widely, solitary or at times apparently in pairs but rarely in flocks, even at fish shoals. BONDS. Monogamous; pair-bond maintained from one season to next at Ascension (Stonehouse 1962*b*) and in Galapagos (Harris 1969*b*). Both parents tend single nestling until fledging. BREEDING DISPERSION. Colonial, usually in loose groups depending on distribution of holes and other cavities. ROOSTING. Little information. Said always to roost at night on ledges and holes in rocky cliffs (Palmer 1962), but as comes to land only to breed, when occupies own nest cavity, doubtful if roosts ashore at other times (M P Harris). When absent at sea, presumably rests on water or in air.

2. Little known about behaviour at sea, apart from calling. ANTAGONISTIC BEHAVIOUR. Intense competition for limited nest-sites appears to be usual at many stations (Stonehouse 1962*b*; Snow 1965, and others), contestants fighting with beaks and uttering rasping screams apparently similar to those given when disturbed by human intruder. At Ascension, groups of 2–3 gather round occupied sites, settled on breasts; may remain thus, inactive, often for long periods before resident starts pecking vigorously at any intruder within reach (Stonehouse 1960). Curiously static encounters also recorded between 2 contestants, 1 in nest, other outside: both sit for up to half hour staring fixedly at ground, sky, or other point but rarely at each other, one occasionally shuffling a few cm towards other, calling softly; then suddenly leap on one another, interlocking bills, remaining still thus for up to further half hour before one retreats slowly. Sometimes winner also flies off. Struggles may cause bleeding wounds, and most birds scarred on head, neck, or flanks. HETEROSEXUAL BEHAVIOUR. Like all tropicbirds, performs aerial displays (van Tets 1965). At Ascension, appears over colonies by day for some 4–6 weeks at beginning of breeding season and engages in group flights (Stonehouse 1960, 1962*b*). Parties of 4 or more birds, often in pairs, fly together in close formation (see A): alternate periods of rapid wing-beats, while wheeling or climbing, with long, swooping glides; during latter, call loudly and tails of uppermost birds held downwards and flicked rapidly from side to side, central streamers, trailing and waving, often producing shimmering effect. At Revilla Gigedos Island, off Pacific coast of Mexico, small, loose groups of 5–10 recorded flying in wide circles, at least 1 or 2 in each group calling repeatedly (Brattstrom and Howell 1956). Pairs leave the group at times: ♂ and ♀ glide together, one about 30 cm above other, for 100–300 m, descending to within 6 m of sea; then separate and rejoin same or different group. During glide, upper bird holds wings bent downwards and lower holds wings up, so that wing-tips are very close but apparently not actually touching. 2 birds also seen flying zigzag course, changing direction every 5–10 wing-beats. At Ascension, display-flight evidently ended with one of party going to nest cavity where joined by one or more others. Before laying,

A

pair occupies nest-site together; apparently perform no complex rituals there, pair-formation and courtship being largely aerial (van Tets 1965). Copulation occurs at nest, with few preliminaries apart from some 'caressing', probably Allopreening or Billing; apparently similar to Yellow-billed Tropicbird *P. lepturus* in which ♂ mounts extremely briefly, with wings extended and tail depressed to right or left of ♀'s, and pair immediately leave site together after ejaculation (Stonehouse 1960, 1962*b*). RELATIONS WITHIN FAMILY GROUP. No detailed information. For calls of parents and young, see Voice.

(Fig A after illustrations in Stonehouse 1960, 1962*b*.)

**Voice.** Highly vocal; especially noisy at colony. Calls uttered both in air and at nest; described as shrill grating, rattling, or clicking, also as short high rasping (Gifford 1913), although this does not convey loud piercing screaming or whistling quality stressed by others.

(1) Flight-call. Shrill, electrifying whistle uttered by single birds and pairs at sea (Murphy 1936). Probably similar to extended series of frenzied, shrill screams and cackles, like trilling of boatswain's pipe, given during aerial display (Stonehouse 1960, 1962*b*). Series of harsh screeching notes, uttered continuously during such flights, likened to those of Common Tern *Sterna hirundo* (Alexander 1898). (2) Ground-call. Similar to Flight-call, uttered at nest when disturbed (Gifford 1913). Persistent piercing yells, shrieks, and screams ('sustained caterwauling') uttered from or near nest-hole during disputes and when disturbed (Stonehouse 1960, 1962*b*) probably more intense version of same call. (3) Parental-call. Guttural clicks and 'chucks' uttered by adults while feeding young (Stonehouse 1962*b*).

CALLS OF YOUNG. Persistent, vigorous, hoarse, guttural chirping uttered whilst food-begging (Stonehouse 1962*b*; Gross 1912). Young also said to utter short, high-pitched, rasping notes like adult; heard first from chick only day or so old when handled (Gifford 1913).

**Breeding.** Based mainly on Stonehouse (1962*b*), Ascension; Snow (1965) and Harris (1969*b*), Galapagos. SEASON. Cape Verde Islands: eggs laid December–March; prolonged season, but appears to breed annually (de Naurois 1969*b*). Red Sea: eggs found April, newly fledged young late November (Palmer 1962). Ascension: eggs laid every month, but individuals lay at 9–12 month intervals; successful breeders at intervals of 335 days (306–358), sample 13; unsuccessful breeders at 290 days (259–336) after egg loss, sample 7, and 322 days (294–357) after chick loss, sample 6. Galapagos: at one colony eggs laid every month with much re-laying after egg loss obscuring any seasonal peaks; at another, distinct, apparently annual season. SITE. In shallow cavity; on ledge, flat ground, or cliff; under boulder. Colonial; nests as close as 1 metre, availability of sites mainly influencing dispersion. Nest: scrape with little or no material. Building: by both parents, forming scrape with body and feet. EGGS. Ovate, not glossy; white, well covered with dark brown streaks and smears. 65 × 45 mm (55–76 × 33–50), sample 245 (Ascen-

sion); weight 67 g (32–84), sample 32 (Stonehouse 1963). 56 × 42 (52–63 × 37–46), sample not given; calculated weight 54 g (*P. a. mesonauta*). 60 × 44 mm (55–64 × 41–48), sample 6; calculated weight 62 g (*P. a. indicus*) (Schönwetter 1967). Clutch: 1. One brood. Re-lays after egg loss, period between layings 42–112 days, sample 7 (Ascension). INCUBATION. 43 days (42–44) (Ascension). By both sexes. YOUNG. Semi-altricial and nidicolous. Cared for and fed by both parents. Fed by partial regurgitation; on entirely predigested food while small, then with whole food items. Parental visits probably daily at start, reducing slowly until *c.* 70th day when fall off rapidly to one per 2–3 days (Ascension). FLEDGING TO MATURITY. Fledging period 80–90 days, extreme over 110 days (Ascension). Becomes independent at same time. Age of maturity not known. BREEDING SUCCESS. Of 328 eggs laid at Ascension, 30·8% lost before hatching, 17·7% lost as chicks, 51·5% reared; almost all eggs and chicks lost as results of fighting by adults of same species and also with Yellow-billed Tropicbird *P. lepturus*. On Galapagos, availability of food and competition for nest sites, main factors influencing breeding success.

**Plumages** (subspecies *mesonauta*). ADULT. Feathers of crown white with black bases. Black spot in front of eye continued as black line through eye, bordering crown, more or less interrupted on nape. Rest of upperparts white, densely and narrowly barred black or dark grey. Sides of head and underparts white; long flank feathers mottled dark grey, feathers at sides of under tail-coverts barred dark grey. Tail-feathers white with basally black shafts; central ones strongly elongated. Outermost primaries black with white tips and white margins to inner webs; rest of primaries progressively more white, innermost white with black shaft-streak on basal half. Under surface of primaries white. Outer primary coverts black, central tipped white, inner white with black shaft; marginal coverts of primaries mainly white. Outer secondaries white; inner black with white margins. Upper wing-coverts white, lesser and marginal with subterminal black V-marks. Under wing-coverts and axillaries white. NESTLING. In long, fluffy down from hatching; delicate grey above, white below. First feathers (scapulars) appear at 13–15 days, primaries at 24–27 days, tail-feathers at 30–35 days; fully feathered at 55 days (data for nominate *aethereus*, Stonehouse 1962*b*). JUVENILE. Like adult, but black on nape more solid, tail-feathers with subterminal black spot, and tail streamers lacking.

**Bare parts.** ADULT. Iris black or brown-black. Bill red. Tarsus, outer and inner toes, and basal parts of middle toe and webs pale yellow or pale buff; distal parts of middle toe and webs black.

PLATE 18. (*facing*) *Oceanodroma monorhis* Swinhoe's Storm-petrel ▶ (p. 173): 1 ad. *Pelagodroma marina eadesi* White-faced Storm-petrel (p. 159): 2 ad, 3 ad in worn plumage showing pale wing patch, 4 juv or freshly moulted ad. *Oceanodroma leucorhoa* Leach's Storm-petrel (p. 168): 5 ad. *Oceanodroma castro* Madeiran Storm-petrel (p. 174): 6 ad. *Oceanites oceanicus* Wilson's Storm-petrel (p. 156): 7 ad in fresh plumage, 7a (in steady flight), 7b and 7c (in feeding flight), 7d ad in worn plumage. *Hydrobates pelagicus* Storm Petrel (p. 163): 8 juv, 8a juv (on migration flight), 8b ad (below). (PJH)

1                    2                    3                    4

5                    6                    7                    8

7a          7b          7c          7d                    8a          8b

NESTLING. Bill cream. Foot 'entirely light' (Murphy 1936). JUVENILE. Bill yellow.

**Moults.** ADULT POST-BREEDING. Complete; primaries serially descendant (Stresemann and Stresemann 1966). Starts after breeding and lasts 19–29 weeks, average 24 weeks. Moult of elongated tail-feathers apparently irregular (Stonehouse 1962b; see also Chasen 1933). POST-JUVENILE. No data.

**Measurements.** Adult, skins (BMNH, RMNH, ZMA).

| | | | | | | |
|---|---|---|---|---|---|---|
| WING | ♂ 310 | (9·44; 9) | 296–330 | ♀ 313 | (4·31; 6) | 307–319 |
| TAIL | 114 | (4·50; 7) | 109–120 | 110 | (6·02; 5) | 105–120 |
| BILL | 62·1 | (1·73; 10) | 59–65 | 61·4 | (2·07; 7) | 60–66 |
| TARSUS | 29·5 | (0·97; 10) | 28–31 | 28·1 | (0·90; 7) | 27–29 |
| TOE | 45·4 | (0·97; 10) | 44–47 | 44·1 | (1·21; 7) | 42–46 |

Difference between sexes significant only for tarsus and toe. Tail measured without central streamers: these strongly variable through wear and moult, average 10 ♂♂ 628, 10 ♀♀ 510 (Palmer 1962). *P. a. indicus* significantly smaller: wing ♂♀ 295 (13·6; 6) 281–315, bill ♂♀ 55·7 (4·23; 6) 52–62.

**Weights.** Emaciated ♂ 427 (ZMA). Young of nominate *aethereus* about to fledge 730 (Stonehouse 1962b).

**Structure.** Wing long and narrow. 11 primaries: p10 longest, p9 *c.* 10 shorter, p8 *c.* 25, p7 *c.* 45, p6 *c.* 70, p5 *c.* 100; p11 *c.* 10 shorter than primary coverts. Primaries tapering, inner web of p10 slightly emarginated. Tail wedge-shaped, 14 feathers; central pair projects far beyond rest, tapering to narrow point, shaft almost bare at tip. Bill strong, heavy at base, decurved, serrated along cutting edges. Tarsus short and slender; outer toe *c.* 90% of middle, inner *c.* 85%, hind *c.* 37%. Horny flange on medial side of middle claw, lacking in other species of *Phaethon* (Palmer 1962).

**Geographical variation.** Slight. Nominate *aethereus* has dark bars on upperside and dark areas of primaries and secondaries paler, more grey; primary coverts with wider white tips and margins; probably slightly longer winged. *P. a. indicus* averages smaller; has black streak through eye less well developed; and bill more orange with black cutting edges. (Palmer 1962; Vaurie 1965.)

# Family SULIDAE boobies, gannets

Medium-sized to large seabirds. 9 species falling into 2 groups (1) boobies (6 species), (2) gannets (3 species forming single superspecies). Though all assigned here to single genus *Sula* many authorities separate boobies *Sula* (sens strict.) from gannets *Morus*. 1 booby and 1 gannet breed in west Palearctic. Body elongated; neck thick, of medium length; head large. ♂ and ♀ equally large in gannets, ♀ larger than ♂ in boobies. Wings long, narrow, and pointed. 11 primaries: p9 or p10 longest. About 28 secondaries; short, diastataxic. Flight strong, alternately flapping and gliding. Tail long, strongly wedgeshaped; 12–16 pointed feathers. Bill long and stout, gradually tapering towards curved nail at tip; not hooked except in Abbott's Booby *S. abbotti*. Cutting edges serrated. Nostrils closed. Gular and facial skin bare, more extensively in boobies than in gannets. Thus in former: eyes of most set well within facial skin with clearly visible, thick, fleshy orbital ring; line of demarcation between facial skin and feathers often comes well above eyes, producing bare forehead; skin on side of inner part of lower mandible bare, also all of gular area and inter-ramal space – all results in characteristic, more elongated and human-look to face than in gannets. Leg short and stout, totipalmate; nail of middle toe medially with comb. Stance upright, slightly tilted backward; gait waddling. Oil gland feathered. Aftershaft absent. Plumage predominantly white with black on wings, or white variegated with darker tinges on upperparts. Red-footed Booby *S. sula* polymorphic with white, grey, and brown morphs. Brown Booby *S. leucogaster* brown with white belly. Bare parts often brightly coloured. Sexes similar in plumage, sometimes differing in bare-part colours, which in ♂ may also change markedly during season of display and breeding. 1 moult per cycle; primaries replaced in serially descendant order. Young altricial and nidicolous, naked at hatching. Down at first sparse, later denser; white. Contour feathers do not grow from same follicles as down. Juveniles differ from adults, usually with less white in plumage. Reach full adult plumage in 2–4 years.

Occur in all oceans. Boobies tropical and sub-tropical; 3 species (*S. leucogaster*, Masked Booby *S. dactylatra*, *S. sula*) sympatrically pan-tropical, others more restricted. Gannets typically temperate, extending into sub-tropical and tropical only in off-season; species allopatric, based on North Atlantic (Gannet *S. bassana*), southern Africa (Cape Gannet *S. capensis*), and Australia–New Zealand (Australasian Gannet *S. serrator*). Strictly marine, inshore and offshore rather than pelagic, apart from some boobies, but often making long daily journeys between fishing areas and breeding stations. Sites usually in open on islands or islets, on cliffs or flat ground; some boobies also breed and perch habitually in trees. Differ from Phalacrocoracidae (cormorants) in more aerial habits, especially plunging from air after food, and in more marine habitat, frequently out of

◀ PLATE 20. *(facing)* *Sula leucogaster* Brown Booby (p. 185). *S. l. plotus*: 1 ad ♂, 2 ad ♀ (facial pattern), 3 ad ♀ (in flight), 4 post-juvenile moult, 5 juv (facial pattern), 6 juv, 7 juv (with worn shorter tail), 8 nestling. *Sula bassana* Gannet (p. 191): 9 ad breeding, 10 sub-adults, 11 post-juvenile moult, 12 juv, 13 nestling. (PJH)

sight of land. In forms with white plumage especially, conspicuous at a distance owing to large size, diurnal habits, and distinctly higher level of normal flight above water than chosen by most seabirds. Wanderings involve little change of habitat. In west Palearctic, *S. leucogaster* restricted to small numbers in extreme south, while *S. bassana*, after marked declines in 19th century, has shown notable increase and spread, especially in recent years, probably mainly due to protection. Migratory and dispersive, but some booby populations highly sedentary.

Feed mainly on large fish, especially shoaling species (*S. bassana*) and flying-fish (boobies); also (mainly boobies) cephalopods, especially squid, and crustaceans, especially prawns—though latter probably only of minor importance, mostly taken indirectly via main prey. Food captured chiefly by deep plunge-diving, though other methods also used (see *S. leucogaster*). In blue waters, pan-tropical boobies, like most other tropical seabirds, particularly dependent on schools of large predatory fish and cetaceans to flush their prey up into vicinity of surface. Food-piracy reported from *S. leucogaster*. Solitary or gregarious at sea, often hunting singly but congregating where prey densely abundant; often return to colony or roost in flight formations. Monogamous pair-bond, often of long-term duration especially in boobies; maintained principally at nest-site with no evidence of association between mates (or parents or fledged young) away from immediate vicinity of breeding station. Defend nest-site vigorously, showing marked site-tenacity between seasons. Though gannets roost on water at sea when not breeding, boobies tend to return to land at night, some populations roosting at or near nest-sites throughout year. Correlated with large size, conspicuousness, and use of open nest-sites in dense or loose colonies, visual displays moderately to well developed; unilateral and mutual, relating mainly to site-ownership and heterosexual advertisement (chiefly ♂♂) and to greeting and individual identification (both sexes, especially when one joins other at nest). Most display occurs at or near site, or when flying in to site, but some of smaller boobies also display in air. Pronounced dichotomy between behaviour of boobies and gannets, displays of latter generally being more ritualized, with evolution of quite different signal meaning to distinctive Sky-pointing display (see especially Nelson 1970). Voice harsh and relatively unspecialized in gannets; similar in both sexes, used chiefly in aggression, landing displays, and greeting ceremony with conclusive evidence of individual variation permitting mutual identification between mates (and chicks and parents). Used freely on wing over and near colony, but mainly silent at sea. In most

boobies marked difference in voices of ♂♂ and ♀♀, correlated with difference in tracheal anatomy. Comfort behaviour generally similar to that of other waterbirds. Bathing at times spectacular, involving shallow plunging, beating water with both wings, and rolling on side with one wing raised; followed by oiling and preening. Heat dissipated by characteristic gular-fluttering; boobies also sun themselves with wings extended at rear. Other behaviour characters include: direct head-scratching (with use of pectinated middle claw); true yawning and jaw-stretching; resting with head on back, bill inserted into scapulars.

Breeding annual and strictly seasonal in gannets. Though generally more protracted, also annual in some species and populations of boobies but less-than-annual and non-seasonal in others. Success or failure in breeding, length of successful cycles, and common response to favourable food supply among factors determining periodicity of breeding, peaks in laying, and nesting of some pairs out of phase with rest (see, e.g. Simmons 1967b). In *S. abbotti*, cycle lasts over 1 year and successful pairs evidently nest only every other year. Nest on ground, cliff ledges, or in trees; use shallow depressions in ground without lining, assemble simple rims or drums of solidified guano, or build loosely woven nests of twigs or flotsam, or substantial heaps of vegetation cemented together with excreta. Construction by both sexes, but typically ♂ brings material. Colonial; often densely in gannets, but usually much more loosely in boobies some of which sometimes also solitary. Eggs ovate, pale blue, green, or white staining brown; covered with thick chalky coat. Clutches 1–4; laid at intervals of up to 5 days, incubation starting with 1st egg. Replacements laid after egg loss; in some boobies, also sometimes after loss of young. Incubation periods 40–57 days, by both sexes in approximately equal spells; eggs cupped between webs (no brood-patches). Eggshells left in nest or removed. Hatching asynchronous. Young cared for and fed by both parents, usually by incomplete regurgitation; 1–3 times daily in gannets but often at longer intervals in boobies. Brooded continuously for first 2–3 weeks, then guarded by 1 or both parents throughout cycle (*S. bassana*) or for as long as possible (boobies). Typically only single surviving chick when 2 hatch in *S. leucogaster* and *S. dactylatra*, younger evicted by older. Fledging periods 86–*c*. 170 days; much variation in boobies especially depending on current food supply. Gannets become independent at fledging (no desertion period); variable and often prolonged period of post-fledging care, based on nest-site, in boobies. Age of maturity from 4–6 years.

## Sula dactylatra Lesson, 1831 Masked Booby

Fʀ. Fou masqué    Gᴇ. Maskentölpel

Nominate race now rare in Caribbean (Palmer 1962), but larger colonies Ascension Island and islands off Brazil. Confined to tropical and sub-tropical seas. Immature Gannet *Sula bassana* often misidentified as this species off West Africa (W R P Bourne). Small numbers claimed seen off Rio de Oro and Cape Verde Islands, May 1947 (Bierman and Voous 1950); these, too, probably misidentified, and therefore no valid west Palearctic records (Oreel and Voous 1974). Another race, *melanops*, breeds central and western Indian Ocean, and in dispersal has penetrated up Red Sea at least to Port Sudan (20°N) (Meinertzhagen 1954).

## Sula leucogaster Brown Booby

PLATE 20
[facing page 183]

Dᴜ. Bruine Gent    Fʀ. Fou brun    Gᴇ. Brauntölpel
Rᴜ. Желтоногая олуша    Sᴘ. Alcatraz pardo    Sᴡ. Brun sula

*Pelecanus Leucogaster* Boddaert, 1783

Polytypic. Nominate *leucogaster* (Boddaert, 1783), Atlantic Ocean; *plotus* (Forster, 1844), Indian and Pacific Oceans. Extralimital: 1–3 races in extreme east Pacific.

**Field characters.** 64–74 cm, wingspan 132–150 cm; ♂ noticeably smaller than ♀, and both considerably smaller (by up to 30%) than Gannet *S. bassana*. Medium-sized, slim, dapper booby with long, lanceolate tail (proportionately longer in ♂). Dark brown upperparts and frontal aspect contrast strikingly with white of lower breast, rest of underparts, and lining of inner wing; line of demarcation across chest sharply defined and diagnostic. Sexes alike in plumage type, with no seasonal variation; juvenile separable. Sexes differ in facial, gular, and leg colour, only ♂ showing marked variation in relation to stages of breeding cycle; also considerable variation between ♂♂, and to lesser extent ♀♀, of different, discrete populations (whether of same race or not). 2 races in west Palearctic, diverging slightly in plumage characters.
    Aᴅᴜʟᴛ. Plumage of head, neck, upperparts, upperwing, tail, and upper breast chocolate-brown; of uniform tone in race *plotus* but darker on head, neck, upper mantle, and upper breast in nominate *leucogaster*, forming distinctive 'cowl'. Lower breast, flanks, belly, and under tail-coverts white; also wing-pit and central area of inner wing, white contrasting with dark leading and trailing edges and distal section. In all populations, so far as known, ♀♀ characteristically show conspicuous dark patch (or 'false eye') on facial skin in front of eye and ♂♂, at least when in immediate pre-breeding colour, bluish orbital skin. Jᴜᴠᴇɴɪʟᴇ. Wholly dark (including bill and facial skin) but with essentially same plumage pattern as adult, including characteristic line of demarcation on chest. Brown of upperparts, etc., less deep and duller than adult, that of lower breast, etc., and wing-lining slightly paler than rest of plumage; no spots or mottling. Iᴍᴍᴀᴛᴜʀᴇ ᴀɴᴅ Sᴜʙ-ᴀᴅᴜʟᴛ. Initial changes involve bare-part colours, which progressively resemble adult, young ♀ developing dark mark in front of eye, young ♂ darkish eye-ring. Paler brown areas of plumage become progressively whiter, passing through mottled and then more uniform buffish white stages, and rest darker.
    Gannet-like shape and behaviour make confusion unlikely with any seabirds in area except other Sulidae. Juvenile Masked Booby *S. dactylatra* often misidentified in tropics as adult of present species. Juveniles and immatures of both these boobies quite different from one another, though more chance of confusion between juvenile *S. leucogaster* and all dark plumages of polymorphic Red-footed Booby *S. sula*. Confusion unlikely, however, between young *S. leucogaster* and young Gannets *S. bassana*: former much smaller and, at all times, more uniformly brown on areas brown in adult, with definite suggestion of adult-type pattern (see above) at least at closer range; also lack extensive spotting or, later, increasingly black and white spangling of *S. bassana*, mottled effect in immature being confined to lower breast, rest of underparts, and wing-lining. When seen well, this and other boobies also differ from true gannets by distinctive facial characters (see under Sulidae).
    Altogether more agile on wing than proportionately heavier and more ponderous *S. bassana*, flight action (though gannet-like) noticeably lighter and quicker. Usually seen singly or in small parties at sea, but will congregate in large numbers at fish shoals, sometimes travelling back to roost or colony in discrete groups, moving in irregular line or shallow, asymmetrical V-formation, usually low over water. Plunge-dives for food, usually from low height above water and at steep angles, specializing in feeding in shallow water and in aerial pursuit of flying-fish. Will settle on sea but usually only briefly, preferring to perch on rocks and artefacts (see Habitat), otherwise remaining on wing while at sea. Far less pelagic and more coastal than most other Sulidae,

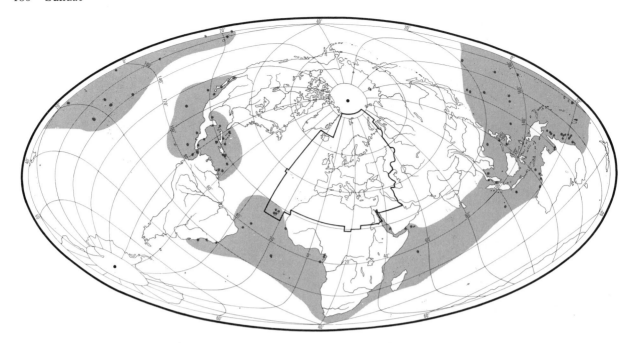

coming freely to land where perches adroitly. Walks fairly well with waddling action and frequent hops. Highly vocal at colony, but generally silent at sea; much louder honking of ♀ and immatures predominate over weak whistling of ♂.

**Habitat.** Tropical, aerial, and marine; sometimes pelagic but mostly in coastal waters, often maintaining daily contact throughout year with breeding site or roost on islands (both oceanic and continental), islets, or coral reefs. Breeds freely at steep sites, such as cliffs, even, when not persecuted, nesting on bare rock; also on slopes and flat ground, even sandy beaches, sometimes flanked or shaded by dense vegetation. Roosts in similar places, and in trees in some parts of range (not west Palearctic); will also use artefacts such as buoys or ship's rigging, or even backs of swimming turtles (Murphy 1958). Perches competently where necessary at restricted or precarious sites. Where nesting or roosting on flat surfaces, prefers ready access to slope or precipice for easy take-off; nevertheless, smaller ♂ especially, much more agile in taking wing than certain congeners, e.g. Gannet *Sula bassana*.

Fishes as close to breeding station as resources permit, often inshore and in extremely shallow water (Simmons 1967b), even when muddy and discoloured. Where necessary will journey up to *c*. 100 km to feeding grounds (Palmer 1962). Little affected by storms. Flies low, also *c*. 15 m or more above water. Active in air in vicinity of nest-site, especially on windy days or when engaging in aerial displays. Swims strongly. Largely indifferent to human presence, except where persecuted. In Cape Verde Islands, exploitation of breeding colonies for food and failure to protect has resulted in severe diminution and extirpation of colonies (Bannerman and Bannerman 1968). Though little contact with man at many stations, at some disastrously affected—particularly by introduction of cats.

**Distribution.** *S. l. plotus* recorded breeding on islets in northern Red Sea and Gulf of Suez (Meinertzhagen 1930, 1954); no recent information. *S. l. leucogaster* nests on several of Cape Verde Islands (Bannerman and Bannerman 1968).

Accidental. Azores: one obtained off São Miguel, summer 1966 (Bannerman and Bannerman 1966). Mauritania: one juvenile captured Banc d'Arguin July 1960 (Heim de Balsac and Mayaud 1962).

**Population.** Numbers considerably reduced on Cape Verde Islands due, at least in part, to human predation (Bannerman and Bannerman 1968); recent surveys, though incomplete, suggest total population under 2000 pairs (de Naurois 1969b).

Survival. At 2 colonies of nominate *leucogaster* Ascension Island, studied 1962–64 with subsequent visits, annual mortality in population of *c*. 100 adults calculated, from repeated censuses and individual identification, as probably no more than 3·3% and possibly only 2–3% (assuming no sex difference), giving average life-expectancy of at least 30 and possibly up to 50 years (Simmons 1967b). 4 birds ringed as adults 1958 still present early 1973 (K E L Simmons).

**Movements.** Dispersive and resident. Present near breeding stations all year, though some (possibly mainly immatures) disperse within tropical and sub-tropical seas; basically inshore species. In east Atlantic, nominate race has restricted range in up-wellings off Sénégal and Mauritania and around Cape Verde Islands and Bissagos Islands, Guinea-Bissau (Vaurie 1965); but picture obscured by frequency with which immature Gannets *Sula bassana* misidentified off west Africa (W R P Bourne). Those recorded in tropical east Atlantic presumably from more southerly colonies (Gulf of Guinea, Ascension Island). In west Atlantic, stragglers from Caribbean colonies to Bermuda and off USA coasts to New England and even Nova Scotia (Palmer 1962), so Azores bird (see Distribution) as likely to have come from Caribbean as Cape Verde Islands. Other Palearctic race, *plotus*, occurs warmer parts of Indian Ocean and adjacent seas, mainly in vicinities of colonies, and restricted movements indicated by its absence from east Arabian Sea and Persian Gulf; occurs throughout Red Sea, also in Gulfs of Suez and Aqaba, with small numbers present Aqaba at least March–September (Safriel 1968); common off Aden only May–June (Meinertzhagen 1954).

**Food.** Mostly fish; also squid and prawns. Obtained mainly by versatile plunge-diving from air, usually obliquely from low height or from more or less level flight close to surface; will also plunge more steeply and with deeper penetration from *c.* 15 m and engage in aerial pursuit of flying-fish (Exocoetidae), either seizing prey in air or, more usually, following it under water in pursuit-plunge. Less frequently feeds by contact-swooping, hovering, plunging from perch, or surface-seizing; also habitually practices food-piracy against own species, congeners, and even frigatebirds *Fregata* (see Simmons 1967b, 1972). When plunging, tail spread and wings held partly open until moment of impact with water; then wings fully extended backwards in typical manner of Sulidae. Smaller ♂ more manoeuvrable than larger and heavier ♀. Typically inshore feeder, often plunging into rough and shallow water: into surf at Ascension Island (Dorward 1962a; Simmons 1967b); into depths of 1·5–1·8 m at Christmas Island, Indian Ocean, from heights of 9–12 m and submerging for 25–40 s (Gibson-Hill 1947). Often solitary feeder but will congregate, frequently in large numbers, at big shoals of fish, in tropics especially when these driven to surface—or into shallow water—by large predatory fish and cetaceans.

Fish include: flying-fish (*Exocoetus* and *Cypselurus*); halfbeak *Hemiramphus*; mullet *Mugil*; sea catfish *Galeichthys*; garfish *Belone ardeola; Cottus;* parrot-fish (Coridae); and flatfish (Clark 1903; Murphy 1936; Gibson-Hill 1951; Voous 1957a; Palmer 1962). In Dahlak Archipelago, Red Sea, takes many sardines *Harengula punctata* when these make seasonal appearance in large numbers (Lewinsohn and Fishelson 1967). At Ascension Island, while nominate *leucogaster* takes some squid, main diet fish: of 10 species of pelagic fish recorded, flying-fish (mostly *Exocoetus volitans*), *Ophioblennius webbii*, and anchovy *Engraulis* main prey, but garfish *Scomberesox*

*saurus* and pelagic carangid *Selar crumenophthalmus* taken inshore, latter (and other unknown species) often as fry when driven in by predatory fish; often in size-range 5–7 cm, but some up to 37 cm (Stonehouse 1962; Dorward 1962a; Simmons 1967b). While many flying-fish taken at Ascension, and more or less same range of other pelagic fish and squid, diet more varied than that of larger Masked Booby *S. dactylatra* at same station, more often to exclusion of flying-fish (Dorward 1962a) and with significantly more inshore fishing; *S. dactylatra* also main victim of food-piracy by *S. leucogaster* (Simmons 1967b, 1972).

**Social pattern and behaviour.** Limited information west Palearctic, mainly from Cape Verde Islands (Murphy 1924). Extralimital studies chiefly Ascension Island: Boatswainbird Islet (Dorward 1962a, b) and Georgetown Stacks (Simmons 1967a, b, 1970a).

1. Gregarious, but less than most boobies. Usually singly or in small parties at sea, at times in larger congregation at fish shoals; often returns to colony or roost in small formations. Fledged juvenile returns to birth site for weeks or months. BONDS. Pair-bond basically monogamous, often sustained from year to year and probably often life-long. At Georgetown Stacks, 83% of pairs remained intact over 2-year study, each pair living together at site even when not breeding. Pairing begins from end 2nd year (Ascension), but little definite information. Both parents tend young; latter fed after fledging, especially by ♀, often for prolonged period. Bond between juvenile and parents maintained chiefly by persistent begging and return to natal site by former; no bond between siblings, elder almost invariably evicting younger from site. BREEDING DISPERSION. Usually small, loose colonies; also isolated nests. Strictly territorial, both sexes defending small area, including nest or potential site and adjacent perches; also often 1 or more separate perching sites; some, mainly younger birds, occupying only perching-territory in colony. Sites occupied, often daily, throughout year at some colonies: at Georgetown Stacks over 2 years, nearly 90% of pairs retained same favourable site and territory; some ringed individuals still at same site or near 11 years later. Older immatures and young adults establish sites during 2nd and 3rd year (Ascension). ROOSTING. Most populations maintain daily contact with land and sleep ashore; not known to sleep regularly on sea at night. At Ascension, usually spends part or all day at sea; when long feeding trips necessary often leaves soon after sunrise and returns late, sometimes landing after dark. Established adults and dependent juveniles roost in territory or on elevated perch in vicinity; others at variety of sites unoccupied for breeding, including communal roosts on predator-free headlands, offshore stacks, etc. After successful feeding during day, may assemble in rafts on sea or day-roosts along shore.

2. Information mostly from Georgetown Stacks, Ascension. ANTAGONISTIC BEHAVIOUR. Most disputes in colony due to territorial and sexual rivalry. Behaviour of sexes mostly similar, but larger ♀ more aggressive, with much louder, far-carrying calls. On return to territory or perch, invariably performs Landing-ritual, alighting conspicuously giving Landing-calls. Sometimes followed by Bowing at any rival beyond pecking range: site-bird while Threat-calling looks at other, stretching neck up, then lowers bill towards ground and finally raises head; may repeatedly raise and lower head rhythmically while pecking at and raising nest-material. Bowing-display shown more often in other situations, especially in response to Bowing and calling by

neighbours and to rival flying over territory. In mild encounters, will also stretch neck forward and persistently nibble at substrate. Established birds engage in flight-circuits from territory, often making reconnaissances over adjoining territories, sometimes in low-level Provoking-flight with Chin-bracing movement on passing over rival's head. When responding to rival prior to take-off, Parades (see below) to vantage point in Pre-flight Upright with neck fully stretched and erect, bill often raised sharply; sometimes Wing-shakes before taking wing. In disputes at close quarters, Hunches-and-ruffles defensively, crouching with neck withdrawn, feathers erected; tail may be cocked and fanned. Bill-thrusts thus at rival, with mutual Bill-jabbing if it responds; usually accompanied by lateral Bill-shaking and sometimes by submissive Bill-tuck and Facing-away (see below). Attacking behaviour may consist merely of partial move towards rival or, especially when rival alights close by unexpectedly, of Wings-up and Darting in which site-bird suddenly flicks up both wings and jumps forward, bill-thrusting persistently. When fighting, rival seized by bill or neck and pushed back over edge or down slope; sometimes prolonged (especially between ♀♀), with participants interlocked, wings and tails spread-eagled. Dive-attacks on trespasser made from air or, more frequently, latter supplanted by site-bird landing on or near place it occupied, giving intense Landing-calls. Rival also chased in flight over water, with loud Threat-calls; may end ceremonially in Rise-and-fall when participants fly up close together, stall and then separate.
HETEROSEXUAL BEHAVIOUR. ♀ dominant in pair interactions, tending to show more overt hostility than ♂. When mate joined at site from flight, pair engage in Greeting Ceremony (see A): returning bird lands near other or jumps in close; both Bill-point and Bill-touch opposing each other's bill frontally, calling. Often followed by further Billing; or pair Bill-spar, poking and fencing at each other in more hostile manner (but distinct from Bill-jabbing of rivals). Billing and sparring, after Bill-pointing, occur frequently when pair close together at site. Symbolic-feeding occasionally follows Bill-pointing: ♂ inserts bill inside that of ♀, or encases ♀'s bill in his. ♂ occasionally Allopreens ♀'s head. Especially during sparring, pair often Bill-shake and perform various submissive postures in which bill averted; includes ritualized Bill-tucking (see B), with neck withdrawn and bill pressed down, and Face-away with bill turned away so that side or back of head towards mate. Facing-away also shown as bird Parades ceremonially from partner with deliberate gait and tail cocked; bill raised (Bill-up-Face-away; see C) or lowered (Bill-down-Face-away)—former, especially, often followed by flight. As mate moves away, stationary bird may quietly Bow. ♂ only often then performs one or more Sky-pointing Salutes, quickly throwing up head with neck vertical (see D), uttering characteristic Wheeze-whistle. ♂ often prefaces both Bowing and Saluting with Stretch-up-and-stare, slowly raising neck while gazing at ♀, usually Saluting only if gaze returned. Bowing by either sex or Saluting by ♂ also typically performed when mate perched outside contact range. ♂ Salutes ♀, often persistently, if she flies over territory; less frequently if perched close. Pair often fly out from and back to territory in joint flight-circuit, with ♂ almost invariably close behind ♀; ♂ sometimes then performs aerial Salutes, throwing head up while soaring in graceful swallow-glide, giving Wheeze-whistle. After Billing at site, one or both often Bows away from mate, sequence Billing-and-Bowing often initiating Nest-activity, involving various nest-building movements during which pair also raise, 'show', and present items to one another and give persistent Nest-calls, often crossing necks in reaching to peck material. Nest-activity also follows return of mate to site, especially ♂ with material, and pair may change

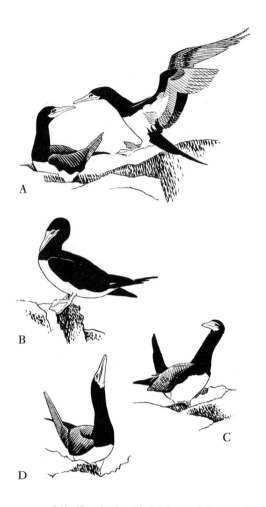

places ceremonially (Symbolic-relief). Nest-activity may develop into Mating Ceremony in which ♀ solicits by remaining still in nest, leaning down and Quiver-nibbling material repeatedly, suddenly ceasing to give Nest-calls. ♂ mounts, rhythmically dipping bill towards back of ♀'s head while positioning (but never seizing it as in Gannet *S. bassana*), and copulates. ♀ may, however unseat ♂ at any stage, often by sparring, and full sequences occur only just prior to and during egg-laying. After copulation, ♂ and ♀ often Bill and Bill-spar, and ♀ may parade away with bill averted then rejoin ♀ in arranging nest-material. RELATIONS WITHIN FAMILY GROUP. Parents take turns to brood, shield, or guard young first 5–6 weeks; guarding persists during entire nestling period only if food situation favourable. Elder sibling evicts other by persistent bill-thrusting; parents take no part in eviction, but do not recover or feed displaced chick which soon disappears (one case of rearing of twin young at Georgetown Stacks). Adults frequently Allopreen chick, especially when smaller. Chick often plays with parent's bill and plumage, usually giving Chuckling-calls; sometimes joins in Nest-activities with parents. ♂ seen to present material to older chick; also occasionally Sky-points at it. On return of parent, especially after long absence, chick sometimes shows defensive, submissive, or even hostile behaviour, including Bill-tuck, Facing-away and lunging while giving Roar, with mouth wide open, arms lifted and glare. Such behaviour more usual with intruders in absence of parents, when also Bill-hides if attacked. Main interaction between parents and young, feeding and begging. When begging directly, hatchling merely lifts head and bill vertically with wobbling action, but by 3rd week also touches adult's bill; later chick attempts to prise open parent's bill or repeatedly prods it. Also begs indirectly by looking at parent with neck withdrawn, while either Head-nodding continuously up and down (smaller young especially) or Head-wagging in figure-of-eight movement. Chick also raises elbows and flexes wrist, but full Wing-waving performed mainly by full-grown chick and juvenile. Begging in smaller chick sometimes accompanied by ordinary Chuckling-calls, but true Food-call given by older chick and, particularly returning juvenile. Chick stays at nest or immediate vicinity until well-grown. In first weeks after fledging, juvenile usually waits at site, leaving mainly to bathe and drink; later spends increasingly more time at sea during day. Parent-young contact declines after fledging; mainly confined to brief meetings for feeding, adults tending to avoid juvenile. Persistent begging by juveniles of higher intensity and much less restrained than in flightless chick. (Figs A–D from Simmons 1967a, 1970a.)

**Voice.** Studied in detail only at Ascension Island (K E L Simmons). Moderately vocal, but few distinctive calls; majority seem basically aggressive, uttered in wide variety of antagonistic and heterosexual situations. Voices of ♂ and ♀ distinct; that of ♂ high-pitched, sibilant, surprisingly quiet whistling (mostly variations of the soft, repeated 'schwee' note); that of ♀ lower-pitched, harsh, strident, much louder honking and grunting (mostly variations of hard, repeated 'arr' note), far-carrying and dominating colony. Voice of young like ♀'s. In each sex, qualitative differences between calls in different situations small, though quantitative differences evident, especially in ♀, e.g. calls usually louder, longer, and more emphatic when aggressive. Differences less apparent in ♂ because voice quieter and less penetrating, but, during hostilities, calls tend to change in tone and delivery, becoming more emphatic and slurred. Both sexes most vocal at colony or roost, but will call at sea during disputes over food. Calls of ♂ usually uttered with mandibles more or less closed at tip; ♀ usually has bill open. Both sexes produce mechanical bill-rattling noises, but these not loud and seem to have no signal function. Following vocalizations most readily recognized at colony.

(1) Landing-call. Series of thin whistles (♂) and harsh honks or grunts (♀). Uttered on return to own nest site or perch, whether or not other birds in vicinity. Calling starts in flight, and reaches peak at landing; may cease after touch-down, especially if not directed at rival. (2) Threat-call. More vehement and prolonged. Directed at neighbour or intruder. May start as Landing-call, continuing long after alighting; also uttered during Bowing-display, supplanting, flight-pursuit, and in other antagonistic situations. In ♀, louder than ordinary Landing-calls; composed of up to 12 or more notes, usually in 2 phrases. In ♂, series of slurred notes, often coughed out rhythmically. (3) Roar. Sudden, loud, abbreviated, intimidating 'shout', given by ♀ when intruder attempts to alight close by; usually induces instant flight. ♂ equivalent, if any, not

heard. (4) Greeting-call. ♂ whistles and ♀ honks or, more usually grunts in friendly fashion during Greeting-Ceremony and other heterosexual encounters involving bills. (5) Wheeze-whistle. Exclusively ♂, uttered only when performing Sky-pointing Salutes to ♀; disyllabic 'swee-oo', given as head thrown back. Considerable individual variation, mainly in quality and stress of component syllables, but also in speed and phrasing; Wheeze-call of each ♂ in colony probably distinct, aiding individual identification. (6) Nest-calls. Distinctive variants of whistling (♂) and grunting (♀) uttered by both sexes more or less simultaneously during Nest-activity. Long series of single, spaced, conversational notes: quiet 'swoo' (♂) and monotonous 'uh' (♀). Nest-calls of ♀ cease when she solicits copulation.

CALLS OF YOUNG. Unfledged young utter various chuckling, clucking, or gosling-like calls, both when alone and especially when with parent. Emphatic guttural version given as Food-call while Head-nodding, Head-waggling, bill-prodding, and (in older chicks) Wing-waving. Develops into repeated persistent ticking, clicking, or 'tucking' Food-call of free-flying, dependent juvenile; uttered incessantly at site if parent present, and in air while flying in to site. Young utter loud, flight-inducing 'Roar' at intruders, and even to parents after long absences; rasping version heard from small young, while Roar of older chicks resembles that of adult ♀. Nestlings also utter defensive, crow-like 'craa-craa-craa' calls towards persistent intruders at nest-site.

**Breeding.** Main reference: Simmons 1967b; Ascension Island. SEASON. Cape Verde Islands: eggs and young every month of year (Bannerman 1914), peak laying February and May (Murphy 1924), January–March (Bourne 1955b), May (de Naurois 1969b); cycle probably not annual, and different colonies may breed at different times. Little information from Red Sea, but laying recorded June–July islands in mouth of Gulf of Suez (Moreau 1950); in Dahlak Archipelago, Ethiopia, though breeding extended (Clapham 1964), laying peak mainly October–November, coinciding with influx of sardines *Harengula punctata* (Lewinsohn and Fishelson 1967). Elsewhere consistently stated to breed throughout year, usually with annual peak, though 2 peaks indicated on St Giles Island, Tobago (Dinsmore and ffrench 1969) and Christmas Island, Pacific Ocean (Schreiber and Ashmole 1970). At Ascension (only station where studied in detail), laying peak approximately every 8 months tending to follow influx of prey-fish inshore. 8-month periodicity due mainly to unsuccessful breeders. Cycles of successful pairs, 1962–4, 43 weeks (25–70), sample 20; did not breed again until young fully independent, nesting late in next period or missing it. Interval between (1) successive clutches of successful pairs 49 weeks (25–78), sample 18; (2) end of successful cycle (independence of young) and next laying 6 weeks (most 1–14), sample 26. SITE. On ground, ledges of cliffs, niches among rocks. Colonial; mostly in small loose groups (occasionally thousands); nests usually well beyond contact range; isolated nests not rare, At Ascension, most sites occupied by same pairs throughout year, whether breeding or not, and used for successive nestings. Nest: hollow in ground or rocks; with or without material depending on site. Amount of material varies from a few scraps to quite large, loosely interwoven but often compacted structure; grass, twigs, etc. used and wide variety of flotsam and jetsam; mostly from sea or stolen from other nests. Building: by both sexes, but material brought almost exclusively by ♂. Cup shaped by sitting bird rotating. In softer ground, may make or enlarge hollow by alternate backward pawing action of feet. EGGS. Oval; pale blue or greenish with chalky white covering, staining brown. 60 × 41 mm (57–66 × 35–43), sample 48 (Schönwetter 1967). At Ascension, 59 × 40 mm (50–67 × 36–44), sample 130; weight 52 g (39–65), sample 82 (Stonehouse 1963). Clutch: 1–2; rarely 3. On Christmas Island, Indian Ocean: 1 egg, 57%; 2, 42%; 3, 1%; (Nelson 1970). 2nd egg laid 5 days (4–6) after 1st. One brood. At Ascension, replacement clutches laid after total loss of eggs or young. Up to 4 replacements recorded. Replacement rate variable, up to 91%. Average interval between loss and re-laying, 5 weeks; significantly shorter on average when food abundant inshore. INCUBATION. 43–47 days Ascension (Dorward 1962d); 40–43 days Christmas Island, Indian Ocean (Gibson-Hill 1947), mean 43 (Nelson 1970). Begins with 1st egg; hatching asynchrous. By both sexes, about equally. At Ascension in 1957–9, average spells 12 hours (1–37) (Dorward 1962a); in 1962–4, 5–7 hours. At Christmas Island, Indian Ocean, spells c. 24–30 hours (Nelson 1970). No brood-patch, eggs incubated cupped against parents' webs, or merely shielded from sun, lying on top of webs. Displaced eggs rolled back in with bill. Eggshells usually discarded from nest. YOUNG. Altricial and nidicolous. Tended and fed by both parents. Brooded for first 21 days mainly on adults' webs, then guarded for further 15 days or longer; throughout nestling period if food supply favourable, otherwise left alone while parents forage. Fed by partial regurgitation, at least once daily, mostly afternoon or evening. Young fed equally by both parents when small, later mainly by ♀. Young remain at or near site throughout nestling period. Adults make no attempt to retrieve displaced young and do not feed them outside nest-territory. FLEDGING TO MATURITY. Fledging period variable, depending on food supply. At Ascension 1957–9, 119 days, sample 5 (Dorward 1962a); in 1962–4, 91 days (84–105), sample 24. At Christmas Island, Indian Ocean, 91–105 days (Gibson-Hill 1947), mean 98 (Nelson 1971). Juvenile returns to nest-site for variable, but usually long, period after fledging. Up to 4 weeks only Ascension 1957–9, young begging though rarely fed (Dorward 1962a); in 1962–4, juveniles returned, begged, and were fed for average 23 weeks (7–51), sample 22, period varying with food supply. Independent thereafter, most leaving colony immediately. Age of first breeding unknown, but adult plumage attained by 4th year and, at Ascension, some birds ringed 1958 breeding 1962. BREEDING SUCCESS. Of 185 clutches at Ascension in one breeding period, when severe food shortage, 71 hatched at least 1 young, but only 18 young fledged (Dorward 1962a) In 1962–4, only 12% clutches successful, sample 149; but (with replacement layings) 35% of pairs successfully raised young in one breeding period, 26% in another, and not more than 10% in a third; survival rate of young estimated as 87%. Losses due to action of sea, and breakage of eggs and displacement of surviving young; but egg desertion and death of young by starvation (both due to food shortage) important during some breeding periods. 2 eggs often laid, but typically only 1 young (at most) survives long after hatching; first-hatched chick invariably evicts smaller sibling from site. Adults unable to raise more than 1 when artificially given 2 young (Dorward 1962a). Exceptional record of natural twins being reared (Ascension 1962–3). 2nd egg acts as 'insurance' if first egg or chick lost (Dorward 1962a); one of series of adaptations enabling at least some young to be reared when feeding conditions poor. At Christmas Island, Indian Ocean, 2-egg clutches gave rise to significantly more fledged young than single-egg ones (Nelson 1970).

**Plumages** (nominate *leucogaster*). ADULT. Head, neck, upper mantle, and chest dark brown; rest of upperparts including upper wing-coverts brown. Lower breast and rest of underparts white, sharply separated from brown chest. Tail brown. Primaries dark brown, slightly glossy on outer webs, paler below. Secondaries brown. Greater under wing-coverts and median in inner part of wing white; marginal, lesser, median in outer part of wing, and under primary coverts dark grey-brown. Axillaries white. NESTLING. Almost naked at hatching. Down develops 9th–26th day (Dorward 1962a); white, long, and dense; sparse on sides of face; absent on chin and throat. JUVENILE. Head, neck, upperparts, and chest brown, lighter than adult; lower breast and rest of underparts pale grey-brown, feathers fringed white. Demarcation between chest and lower breast sharp, often conspicuous; some, however, have breast and belly browner, less grey, making separation from chest less obvious. Tail-feathers and flight-feathers brown, frosted grey; outer webs of primaries darker. Under wing-coverts grey-brown, greater and median in inner part of wing paler. IMMATURE. Head, neck, upperparts, and chest darker brown than juvenile; feathers of lower breast and belly pale grey with white fringes and white bases.

**Bare parts.** ADULT. Nominate *leucogaster*. Iris silver-white to pale grey. Bill yellow at base, pale flesh or pale blue-grey at tip. Insufficiently documented geographical, sexual, and, at least in ♂♂, cyclical variation in colour of bare skin on head and foot. ♂ at Ascension Island in pre-breeding colour: facial skin, gular pouch, and foot bright yellow, eyering bright blue; during incubation and care of young, facial and gular skin faintly greenish-yellow, foot pale green to deep lime-green, eyering dark blue; non-breeding similar with sea-green blaze on forehead, foot dull yellow to green. ♀ at Ascension: facial and gular skin, eyering, and foot at all times pale yellow and black spot in front of eye (Simmons 1967b). Cape Verde Islands, few data: face of ♂ breeding bright yellow, eyering blue, gular pouch slate-blue; face of ♀ pale yellow with lead-grey spot before eye (skins BMNH). St Paul's Rocks: bill and facial skin of ♀ breeding occasionally blue (skins BMNH). Subspecies *plotus*. Facial skin apparently mostly darker than nominate *leucogaster*. Red Sea: iris off-white to straw-yellow; bill of ♂ pale green, lower mandible tinged pink; facial skin bright turquoise; foot pale green (not known whether these colours retained all year); bill and facial skin of ♀ pale yellow or ashy grey (perhaps seasonal variation); foot pale yellow or grey-green (skins BMNH). In ♂♂ from other localities, bill sometimes blue, facial skin yellow-green. Facial skin of ♂, Christmas Island, dull purple or dark purple-grey, of ♀ light green-yellow with dark blue spot in front of eye; eyelid in both sexes dull blue (Gibson-Hill 1947); iris of ♀ sometimes red (Voous 1964). NESTLING. Iris pale grey. Bill and face black. Foot dark grey. When feathers develop, bill, face, and foot become paler, dark grey; foot orange-pink at fledging (Simmons 1967*b*). JUVENILE. Iris pale grey. Bill dark grey; later becoming purple-flesh, greenish or yellow-green at base. Facial skin grey, flesh-brown at 5–8 weeks after fledging, lemon-yellow at 11–15 weeks. Foot orange, becoming dull yellow at 20–30 weeks. IMMATURE. Data scanty. Gular skin greenish (Voous 1957*b*, skins ZMA). In *plotus*, bill described as blue-grey or pale olive buff, yellow at base; facial and gular skin blue-grey, sometimes with yellow tinge. Foot pale yellow; sometimes tinged pink, or olive-buff (skins BMNH, ZMA).

**Moults.** ADULT. Primaries serially descendant; 1 or 2 active centres in wing. Moult usually suspended during breeding. Not known how long completion of one series lasts. Timing of body moult not recorded, presumably occurs slowly during non-breeding period. POST-JUVENILE. 1st series of primaries starts at *c*.10 months, 2nd series at *c*.15 months, when 1st has proceeded to p8 (Dorward 1962*a*). Analogous to *S. dactylatra*, body starts with 1st primary series, moulting slowly and continuously during *c*. 1½ years until fully adult plumage attained.

**Measurements.** Skins (BMNH).

*leucogaster*

| | | | | | | | |
|---|---|---|---|---|---|---|---|
| WING | ♂ 402 | (13·3; 6) | 383–417 | ♀ 416 | (13·1; 8) | 393–431 |
| TAIL | 196 | (2·23; 7) | 193–199 | 193 | (9·83; 9) | 180–204 |
| BILL | 96·5 | (4·17; 8) | 92–105 | 98·6 | (6·82; 11) | 89–111 |
| TARSUS | 45·6 | (1·68; 8) | 43–48 | 47·7 | (1·95; 11) | 44–50 |
| TOE | 77·3 | (3·15; 7) | 73–83 | 83·9 | (4·29; 9) | 79–91 |

*plotus*

| | | | | | | |
|---|---|---|---|---|---|---|
| WING | 385 | (15·8; 7) | 363–404 | 413 | (11·5; 3) | 400–422 |
| BILL | 96·0 | (1·77; 8) | 92–98 | 96·7 | (5·82; 6) | 87–101 |

Sex difference in this small sample significant only for tarsus and toe of *leucogaster*, and wing of *plotus*. Vaurie (1965) gives average and range of wing of 20 ♂♂ *leucogaster*: 386 (370–415), of 20 ♂♂ *plotus* 398 (370 − 425). Ascension Island, wing 14 live ♂♂ ranges 370–410, 26 ♀♀ 390–420; bill 14 ♂♂ 87–91, 26 ♀♀ 91–99; differences significant (Dorward 1962*a*).

**Weights.** Nominate *leucogaster*: ♂♂ 724, 800, 1050, 1095; ♀♀ 1143, 1150, 1200, 1437. *S. l. plotus*: ♂♂ 870, 948; ♀♀ 960, 1130, 1245 (Voous 1964; BMNH, ZMA). Ascension Island: range 14 ♂♂ 850–1200, 26 ♀♀ 1100–1550; difference significant. Maximum weight of nestlings at *c*. 55–70 days: 1200–1400, weight variable during nestling period due to irregularities in food supply (Dorward 1962*a*).

**Structure.** Wing long and slender. 11 primaries: p10 longest, occasionally slightly shorter than p9, p9 0–13 shorter than p10, p8 22–34, p7 54–64, p6 84–100, p5 116–132, p1 210–240; p11 minute. Outer web of p10 narrow, inner of p8–p10 slightly emarginated. Tail long and strongly wedge-shaped: t7 *c*. 100 shorter than t1. Bill longer than head, high at base, gradually tapering. Cutting edges serrated. No external nostrils. Upper mandible with central and lateral hornplates and long, slightly curved, nail. Nail less strongly curved than in Red-footed Booby *S. sula*, which resembles *S. leucogaster* somewhat in juvenile plumage. Tarsus short and stout. Outer toe *c*. 95% of middle, inner *c*. 67%, hind *c*. 39%.

**Geographical variation.** Involves colour of bare parts and of plumage. Data on size not consistent, difference between sexes apparently greater in Indo-Pacific populations (see Measurements; Gibson-Hill 1947; Palmer 1962). *S. l. plotus* differs from nominate in being uniformly dark brown above, not paler on lower mantle, back, and tail. For recorded differences in bare parts see that section. In east Pacific subspecies, ♂ conspicuously pale headed.

## *Sula bassana* Gannet

PLATE 20
[facing page 183]

DU. Jan van Gent   FR. Fou de Bassan   GE. Basstölpel
RU. Северная олуша   SP. Alcatraz   SW. Havssula

*Pelecanus Bassanus* Linnaeus, 1758. Synonym *Morus bassanus*.

Monotypic

**Field characters.** 87–100 cm; wingspan 165–180 cm; ♂ slightly larger than ♀. Largest indigenous seabird in west Palearctic (only vagrant albatrosses *Diomedea* bigger). Mainly white, majestic, long-bodied and long-winged, spear-billed seabird with bewildering succession of mainly black-brown or brown and white plumages in juveniles and immatures. Sexes alike; no marked seasonal variation. Juvenile separable.

ADULT. Plumage white except for yellow-buff tinge on rear part of head (most intense on nape, especially in ♂, fading on hindneck) and contrasting black ends to wings. Bill stout, upper mandible with central and 2 lateral horny plates; bluish white, with dark lines. Bare skin of face and gular stripe dark blue-grey, eye blue-grey. Legs and fully webbed feet grey-black; mid-ridge of toes greenish, tending to yellow in ♂, blue in ♀. JUVENILE. Totally unlike adult; plumage essentially black-brown, inconspicuously speckled white everywhere except on wing-quills and tail.

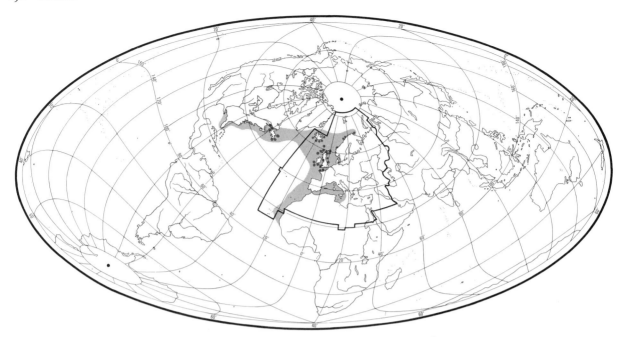

Bill dark horn. IMMATURES. 4 years taken to assume full adult plumage, erratic moult adding to complicated succession of plumages; relative uniformity of juvenile plumage increasingly invaded by white, starting on lower body, then head, neck and breast, and wing-coverts, and last on secondaries and tail (see, especially, Nelson 1964). Bill as adult by late 2nd or 3rd year.

At close range, adult unmistakable but at longer ranges or in poor light, identification (particularly of immatures) can be difficult even for experienced observers. At times recalls plumage and flight action of albatrosses *Diomedea* or larger shearwaters *Puffinus* and *Calonectris*. Thus close attention to structural characters essential: pointed, broad-based bill; heavy head; long, pointed tail; and striking configuration of long, pointed wings set across middle of long, oval body, giving essentially regular (even if angular) appearance to all flight attitudes. Possibility also of confusion with other Sulidae, especially in southern part of Atlantic range. Most boobies smaller with different facial characters (see under Sulidae). For differences from Brown Booby *S. leucogaster*, see latter. Adult Masked Booby *S. dactylatra* similar in plumage type to *S. bassana* but smaller, lacks buff colour on head, tail dark, largest scapulars with blackish tips, more black on wing (extending from primaries and primary coverts along rear edge of secondaries and tertials), and characteristic 'mask' of blue-black facial skin. Greater possibility of confusion between immature *S. bassana* and some ages of *S. dactylatra*, particularly as some sub-adults of former adult-like but may have yet to replace brown tail and secondaries. Thus care needed until transitional plumages of *S. dactylatra* better known, though immature *S. bassana* probably always distinguishable, when seen

well, by buffish colour on head (assumed as early as 2nd year). Adult of white morph Red-footed Booby *S. sula* also gannet-like in plumage, with golden tinge to feathers especially on head; but not yet recorded in west Palearctic, and much smaller (about size of *S. leucogaster*), usually with black along whole trailing edge of wing, and highly distinctive bare-part colours (on face as well as legs).

Flight action in light breezes alternates between series of deep, powerful beats and short glides, with occasional banking and soaring. In higher winds, adapts action, shortening wingspan and stroke and flying much more like large shearwater (particularly Cory's *C. diomedea*), often banking and side-slipping. Plunges for food, most dramatically in vertical descent with wings folded back, but if fish close to surface from other angles too. Swims well, with head up and tail held above surface; often unable to take off when stomach full. Short tarsi make walking appear difficult. Mainly silent at sea; colonies give far-carrying, gurgling roar.

**Habitat.** Marine, almost entirely within limits of continental shelf, mainly in temperate waters of North Atlantic up to arctic fringe; also for non-breeders, especially immatures, southward to fringe of tropics. Except in transit to and from breeding colonies, comes inshore freely only in pursuit of shoals of fish or in unusual weather conditions; even more infrequent as pelagic wanderer beyond *c.* 200 km from land. Flourishes both in cold Labrador Current and in warm North Atlantic Drift, wherever ample resources of fish concentrated within 10 m or at most 15 m of surface, and accessible by plunge-diving from cruising height of 10–40 m. Frequently flies low over waves, and soars to high altitudes, using unusual range of

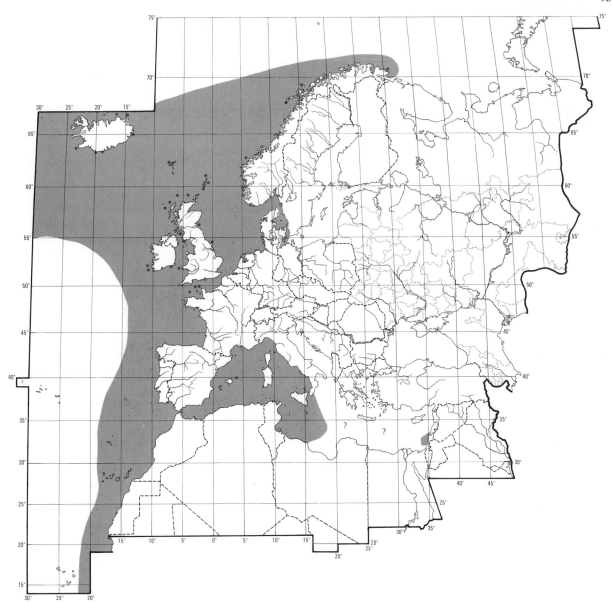

airspace compared with most marine species. Little affected by storms; avoids unproductive areas and adjusts rapidly to good opportunities, sometimes forming substantial local concentrations.

Breeding colonies on isolated stacks or small uninhabited islands, or less frequently on precipitous inaccessible cliffs of mainland or large islands, where human presence ignored unless persecuted. Nests usually on ledges of cliffs immediately above splash-zone, from *c*. 10 m up to *c*. 200 m or more; sometimes on flat tops, but normally on bedrock, or on shallow soil. Generally indifferent to man and artefacts, with local exceptions for shared exploitation of fish.

**Distribution.** Main colonies long established and re-markably stable, but declines in 19th century (see Population) led to extinction of some, mainly in western Atlantic, while increases in 20th century resulted in many new colonies. BRITAIN AND IRELAND: increased from 8 colonies in early 20th century to 16 in 1969–70; new colony *c*. 10 nests, Fair Isle, 1974 (R Broad). ICELAND: 4 new colonies and one extinct since 1943. FRANCE: colonized 1939. CHANNEL ISLANDS: first bred 1940. NORWAY: first bred 1946. Also new colonies in western Atlantic in Quebec and Newfoundland (Cramp *et al.* 1974).

Accidental. Spitsbergen, Bear Island (50 km off), Finland, USSR, Poland, Czechoslovakia, Austria, Yugoslavia, Greece, Turkey, Lebanon, Syria, Cyprus, Cape Verde Islands.

**Population.** Fisher and Vevers (1943–4) give history of colonies throughout range, together with estimates of changes in world population 1819–1939. Although for most colonies reliable figures available only in later years, their estimates suggest decline in total world population from *c.* 170000 pairs in 1829 to barely 53000 in 1894, followed by slow increase to nearly 83000 pairs in 1939. These changes were attributed primarily to predation by man and subsequent protection. Major predation was in western Atlantic where numbers declined from *c.* 115000 pairs in 1829 to *c.* 4000 in 1894, largely due to massacres at Bird Rocks colony, Gulf of St Lawrence, which was believed to have declined from *c.* 110000 pairs in 1834 (estimated at two-thirds of then world population, though based on limited historical data and best regarded as tentative); by 1939, estimated increased to *c.* 13000 pairs. In eastern Atlantic, much smaller decline, from *c.* 56000 pairs 1829 to *c.* 49000 pairs 1894, then rising to *c.* 70000 in 1939. More regular and accurate counts since show increase has continued and accelerated; most recent comprehensive figures, mainly for 1969–71, show world total of *c.* 198000 pairs, highest recorded, with *c.* 167000 pairs in eastern Atlantic and *c.* 31000 pairs in western Atlantic (Cramp *et al.* 1974); though decline of 16%, 1969–73 Bonaventure Island, Quebec, probably due to toxic chemicals and disturbance (Nettleship 1975). BRITAIN AND IRELAND: in early years of 20th century, *c.* 48000 pairs in 8 colonies; around 1930, *c.* 53000 pairs in 11 colonies; 1939, over 54000 pairs in 12 colonies; 1949, *c.* 64000 pairs in 15 colonies; between 1955–66, *c.* 110000 pairs in 15 colonies; 1968–70 (mainly 1969) *c.* 142000 pairs in 16 colonies, including *c.* 56000 pairs in St Kilda group, now world's largest. CHANNEL ISLANDS: Alderney, *c.* 3000 pairs in 2 colonies 1967. ICELAND: 1939, *c.* 14000 pairs in 3 colonies; one since extinct, probably due to cliff falls, and 4 new ones formed (1943, 1944, 1956, 1962); nearly 21000 pairs in 1962. FAEROES: 1972, estimated 1500–2550 occupied nests; apparently some increase, uncertain since differing method used for earlier counts of 1645 pairs in 1938; 1473 pairs in 1939, *c.* 1800 pairs in 1951, 1950 pairs in 1957, and 1081 occupied nests in 1966 (Olsen and Permin 1974). FRANCE: 3000 pairs at single colony, Brittany, 1970, and 3800 pairs, 1973 (J-Y M). NORWAY: 3 new colonies (all north of Arctic Circle) recorded 1961, 1965 and 1967; 569 pairs in 4 colonies, 1971, and 795 pairs, 1974 (E Brun). (See Cramp *et al.* 1974 for full details and references.)

Survival. Adult average annual mortality 6·1% (♂♂ 7·6%, ♀♀ 4·7%, difference not statistically significant) with mean life expectation 16·2 years; about 80% of fledged young die before breeding, mainly in 1st year of life (Nelson 1966c). Ringed birds controlled at Bonaventure Island colony, Quebec, up to 16 and 20 years old (Moisan and Scherrer 1973), and at Bass Rock, at least 21 years old (J B Nelson).

**Movements.** Partially migratory, otherwise dispersive. Following based largely on Thomson (1939) and Bannerman (1959). In west Palearctic some, mainly adults, remain in breeding range latitudes throughout winter, though rare Icelandic inshore waters late October to mid-December (Gudmundsson 1953). Many perform long southward migration, reaching sub-tropical and even tropical waters off west Africa; this tendency much stronger in young of year, but suggestion that migratory urge decreases year by year during lengthy immaturity (Witherby *et al.* 1940) not supported by ringing evidence. Main assembly of adults at most British and Irish breeding stations from early January, but month later at St Kilda, Shetland, and Faeroe Islands and 2 months later Iceland. Young leave nest sites before able to fly and begin dispersal by swimming; distances up to 72 km recorded (Kay 1949, 1950); adults tend to remain some weeks longer. Pre-breeders visit colonies during nesting season, arriving later and departing earlier than adults. Immatures widely dispersed in eastern Atlantic and North Sea spring to autumn, increased later by post-fledging dispersal of juveniles. British-ringed birds (adults and immatures, but mainly latter) recovered all maritime countries of north-west Europe north to Iceland and Norway. A few enter western Baltic; also recorded Latvia and one Scottish bird recovered Lithuania. Western pelagic limit for European breeders *c.* 33°W (Tuck 1968), perhaps representing dispersal from Iceland, but favours continental shelves to open ocean. Autumn passage south follows Atlantic seaboard of Europe and Africa; movements of young to Bay of Biscay and Portugal pronounced September, with further spread to north and west Africa November. Becomes numerous off Morocco and Canary Islands, smaller numbers regularly to Sénégal and off Cape Verde Islands, and occasionally as far as Guinea-Bissau where one recovery of Scottish juvenile. Some young remain south for 2nd calendar year. Return movements equally gradual; arrivals of pre-breeders back in north European waters from about March. Penetrates Mediterranean in autumn and winter, mainly western basin, but small numbers regularly to some eastern areas; single British birds recovered Gulf of Iskenderun (Turkey/Syria) and Israel. Inland occurrences rare, usually after gales. In west Atlantic, tendency to vacate seas around colonies in winter more marked; adults at breeding stations mainly April–October (Palmer 1962). Southward dispersal late autumn, young moving on average furthest south (as in east Atlantic); ranges south to Florida and Gulf of Mexico, and in latter 12:1 ratio of immatures to apparent adults; present there mainly December–March.

Not highly gregarious away from breeding stations, but considerable winter concentrations on good fishing grounds. Diurnal coastal passage commonly in small groups, sometimes forming almost continuous stream.

Young of year travel apart from parents. Many records of ringed birds returning to native or former colonies; but non-breeders sometimes recorded in summer near other colonies, even further north. Chick ringed Scotland (Ailsa Craig) 1966 discovered breeding 1970 in Norway (Nordmjele gannetry, founded 1967). ALT

**Food.** Chiefly fish, 2·5–30·5 cm in length, taken by plunging from heights of 10–40 m (Gurney 1913; Reinsch 1969) or sighted when swimming with head immersed and caught following dive from surface; prey usually swallowed during surfacing. Diving times 5–20 s. Fish and offal also taken whilst swimming, and will rob other birds feeding on surface. Observed once standing in shallow water looking for sand-eels *Ammodytes* (Hunt 1952).

Fish vary with locality, but mainly those within accessible depths, normally less than 25 m and living in large shoals. Brun (1972) suggested that breeding localities in Norway related to distributions of herring *Clupea harengus* and capelin *Mallotus villosus*. Other fish include (particularly) cod *Gadus morhua*, coal-fish (or saithe) *Pollachius virens*, pollack *P. pollachius*, whiting *Merlangius merlangus*, haddock *Melanogrammus aeglefinus*, poor cod *Trisopterus minutus*, sprat *Sprattus sprattus*, pilchard *Sardina pilchardus*, mackerel *Scomber scombrus*, sand-eel Ammodytidae, anchovy *Engraulis encrasicolus*, garfish *Belone bellone*, gurnard *Trigla*, salmon and sea-trout *Salmo* (Gurney 1913; Collinge 1924–7). Molluscs (cephalopods), crustaceans, and echinoderms recorded, but probably from fish eaten. In 18 stomachs: 84% fish, 14% miscellaneous animal material, and 2% molluscs (Collinge 1924–7). Fish also taken from fishing-nets near surface and include species normally out of diving range, including 'haddock' *Sebastes*, blue whiting *Micromesistius poutassou*, dolphin fish *Coryphaena*, blue ling *Molva dipterygia*, *Lycodes*, catfish *Anarhichas*, lemon-sole *Microstomus kitt*, dab *Limanda limanda*, long rough dab *Hippoglossoides platessoides*, and flounder *Pleuronectes platessa* (Reinsch 1969).

Young fed by incomplete regurgitation. At Bass Rock, Scotland, Nelson (1966*b*) found main food mackerel and, to lesser extent, herring, cod, saithe, and sand-eel. Fishing trips usually 7–13 hours, and estimated range over 100 miles and possibly up to 400 miles from breeding colony. Feeds throughout a 2-day watch averaged 2·7 per chick per day.

**Social pattern and behaviour.** Based mainly on material supplied by J B Nelson from studies at Bass Rock, Scotland (Nelson 1964*a*, 1965, 1966*a*, *b*, *c*, 1967*a*).

1. Highly gregarious. Usually gregarious at sea while fishing when birds already plunge-diving exert marked attraction on others. Newly fledged juveniles relatively solitary until able to fly from surface of sea; then often attach themselves to passing adults or join fishing flock. BONDS. Life-long monogamy the rule, pair-bond renewed each season. Unpaired ♀♀ often visit several receptive ♂♂ in quick succession. New pairs not infrequently break up after a few days or even weeks but, once fully formed,

usually remain intact from year to year, though 60% likelihood of change of mate during breeding life (usually 12 or more years) due to death or divorce; ♀♀ more likely to change than ♂♂. Pairs form first from 3rd year; first breeding usually in 5th or 6th. Both sexes tend single nestling until fledging when it leaves birth-site and becomes immediately independent. Chick does not stray from site during nestling period; neighbouring adults highly aggressive to straying young. Parents do not discriminate in favour of own young and will accept substitutes or additions even differing from own in age. BREEDING DISPERSION. Typically nests in large, dense colonies. Pairs spaced evenly within contact of neighbours at about 2·3 per m²; in large and growing colonies, intense competition for sites among or near already established pairs. Nest-site defended as territory by both sexes but especially ♂ who establishes it, usually in 4th or 5th year (occasionally 3rd), and normally retains without breeding for first season much of which spent attracting mate and establishing pair-bond Establishment preceded by extensive aerial reconnaissance followed by frequently transient attachment to relatively imprecise sites and gradually by hardening attachment to precise, strongly defended area. Once fully established, pair generally faithful to same site from year to year, or move negligible distance; breeding failure increases likelihood of change. ♀♀ somewhat less site-tenacious than ♂♂. New sites invariably taken up some weeks later than those of old-established pairs which, at Bass Rock, start to return January or even December. Site used for resting, sleeping, preening, etc., as well as for pair-formation, pair-interactions, copulation, incubation, and rearing of chick. Fully established sites rarely left unattended after return of most birds to colony, since nest-material quickly stolen if unguarded. Some adults and many immatures do not hold permanent sites but settle at 'clubs' on outskirts of breeding groups; site-owners do not congregate there even if mate, egg, or chick lost. ROOSTING. At breeding colony, established birds roost at nest or site, others at club. Sleep at intervals throughout day and almost totally inactive at night. Those arriving late from sea rarely, if ever, attempt to land at colony after dark, but sit on sea nearby until first light. Solitarily roosting individuals usually ill. Outside breeding season, or when temporarily absent from colony, roosts on open sea.

2. In gannetry, highly aggressive individuals (♂♂ especially) live within contact distance; this, together with marked similarity of sexes, favours well differentiated signal behaviour to control social interactions. Elements derived from aggressive behaviour have contributed markedly to most displays. Difficult for individual to acquire site in colony and unremitting, season-long attendance and display by both members of pair necessary to maintain it. ♂ and ♀ co-operate in defence by fighting and display, but ♀♀ do not defend site against ♂♂; this crucial in allowing pair relations to develop despite marked hostility shown by ♂ even to own mate (see below). ANTAGONISTIC BEHAVIOUR. Competition for socially adequate sites normally involves fighting, accompanied by loud calling; some clashes prolonged and damaging. Combatants grip one another on beak or face, trying to push other off site; fighting may continue in air or on sea. Bill-tip may be forced under rim of opponent's eye or far into mouth, but essentially fight consists of fierce gripping and pushing rather than stabbing; most damage results from stabs by neighbours as combatants violate their territories. Occasionally, challenger dives on opponent from air. Nest-biting, occurring e.g. after intruder repelled, may be interpreted as re-directed aggression; with wings held out, site-owner dips head and grips nest or ground while calling loudly. Highly ritualized version, Bowing, signals site-ownership, functioning as threat to repel intruders. Same downward dipping of head (see A), wings-out posture, loud

calling, and side to side Headshake between dips occur, but with no actual biting and sequence ends with Pelican-posture in which bill-tip pressed into breast. ♂ Bowing more frequent, of greater amplitude and longer duration than ♀'s. Ritualized Menacing another common, season-long pattern derived from biting: head, with mandibles gaping, thrust towards opponent, often twisted slightly and then withdrawn in single movement; used to defend site, sometimes developing into jabbing of rival's bill. Even well-established neighbours Menace one another frequently. Another important but relatively inconspicuous activity in site-establishment, of ♂ particularly, repetitive departure from and return to site in flight, each in-flight accompanied by individually distinctive calling and followed by Bowing. After pair-formation, ♀ behaves similarly. HETEROSEXUAL BEHAVIOUR. Pair-formation depends largely on ♀'s locating receptive site-owning ♂ by inspection of gannetry on wing, or where possible on foot. Such ♂♂ advertise status by inconspicuous display, Headshake-and-reach: exaggerated sideways Headshake with slight dips of head towards nest and reaching movements toward ♀; probably modified form of Bowing, with aggressive components (e.g. full downward Bow) suppressed and neutral Headshake emphasized. ♂ shows strong tendency to attack closely similar ♀, usually each time pair re-unite at site. ♂ vigorously Nape-bites approaching ♀ who immediately Faces-away by turning bill aside (see B); such behaviour, however, rather inconspicuous and not very success-ful in inhibiting ♂'s attack. ♀ does not Nape-bite ♂. Greeting ceremony of pair, Mutual-fencing, conspicuous and prolonged: partners stand breast to breast and, with wings out, shake heads from side to side, knocking bills together (see C) and intersper-sing forward and downward movements of head, sometimes over neck and back of mate, while calling loudly. This shares many movements with Bowing (though final Pelican-posture missing) and situations which elicit Bowing from solitary bird (e.g. intrusion) elicit Mutual-fencing from pair. Reciprocal and simultaneous Allopreening of head and neck frequently follows Mutual-fencing; similar fierce preening of intruding young or strange ♀ by ♂ often occurs, interspersed with attack. Sky-pointing occurs only when movement contemplated, typically preceding departure from nest, whether by flight or on foot: head and neck stretched vertically upwards, wings lifted at shoulder joint so wing tips raised, and tail depressed (see D); bird moves with special gait, slowly lifting feet in exaggerated manner, webs drooping, while eyes focus binocularly forwards and downwards, often swivelling. May be maintained until bird runs, jumps, or launches into flight when special call uttered. Lone bird will Sky-point before leaving, but most intense performances accompany change-over. Main function probably to synchronize nest-relief; when partners Sky-point simultaneously, one never leaves until other has stopped, avoiding leaving vulnerable egg or small chick

unattended. Sky-pointing not appeasement against neighbours whose territories must be transgressed; on contrary, elicits attack from neighbours. Before egg-laying, Mutual-fencing frequently followed by copulation: ♀ Headshakes vigorously; ♂ mounts, grips her nape and, with wings held out, patters webs on her back. Usually followed by brief Mutual-fencing and by ♀ touching nest material; ♂ then often leaves to collect more material, but no ritualized presentation. RELATIONS WITHIN FAMILY GROUP. Chick guarded throughout nestling period by one or both parents. Clearly recognizable food-begging occurs from c. 10 days. Later chick pesters adults, pointing bill up, swaying and then retracting head and touching adult's bill, usually at side. High-intensity begging includes repeated Yipping-call stimulating to adult and may elicit regurgitation before chick touches bill. ♂ rarely spontaneously attacks own young, but occasionally redirects attacks; then chick Faces-away, turning bill sharply down and to one side, in extreme form hiding bill below ventral surface and lying prone. Parents often preen chick, perhaps eliciting brief avoidance including Facing-away. Chick also preens parents. Once it can stand, and increasingly with age, chick begins to 'attack' parents in apparently playful way; attacks may be boisterous with excited Yapping-call, jabbing, and contorted head movements, and may elicit Facing-away and avoiding movements from adults, which occasionally retaliate briefly. Chick shows early interest in nest material and may seize from parent and add to nest, or help adult to do so. Rather rarely, fully grown chick performs short bout of Mutual-fencing with parent; also, rarely, bowing. Interactions between chicks and neighbour-ing adults rare. Rapid and intense Facing-away follows attack, if chick strays or falls from own site. Interactions with neighbour-ing chicks common, consisting of low intensity aggression: gripping bills, jabbing, and menacing. Strong hostility to intrusion by strange young but, if latter larger and retaliates, Faces-away in appeasement. Fully feathered chick's black plumage probably reduces likelihood of attack from parents, particularly ♂.

(Figs A–D after Nelson 1964*a*.)

**Voice.** Vocabulary limited (see Nelson 1965). Mainly silent away from breeding grounds, but utters rapid, excited 'urrah' calls when competing for fish offal, and when fishing communally, when often noisy.

Following uttered at colony. (1) Urrah-call. Loud, harsh 'urrah' or 'arrrrr' (see I), repeated several times, with individual variations sufficient to allow adults to recognize mates, and chicks their parents, by voice characteristics (White and White 1970). Given when flying in to nest,

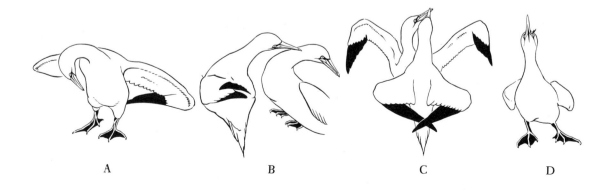

A                    B                    C                    D

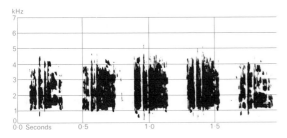

I P J Sellar Scotland July 1968

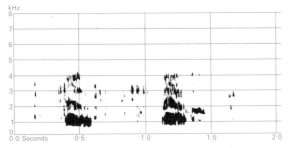

II P J Sellar Scotland July 1968

during site-ownership display, during pair greeting, often during fighting, and as threat or alarm call. (2) Groaning-call. Soft, protracted 'oo-ah', sometimes extremely high-pitched, typically given at take-off, usually preceded or accompanied by Sky-pointing posture. (3) Croaking-call. Soft, rapid 'crok-crok', rather like croaking of young Raven *Corvus corax*. Occasionally given in fast, level flight.

CALLS OF YOUNG, 'Cheeping' calls given from first day; later replaced by high intensity 'yip' Food-call. Young utter excited 'yapping' calls, like barking of small dog (see II), in aggression towards trespassing chicks from other nests, during 'attacks' on parents, and during Bowing-display by older young.

**Breeding.** Based mainly on Nelson (1964a, 1965, 1966a, b, c); Bass Rock, Scotland. SEASON. See diagram for British Isles; few data from elsewhere. SITE. On cliff ledges, steep slopes, and flatter ground. Colonial; from less than 10 pairs to many thousands (see Population). Nests evenly spaced and nearly touching; 72% of 408 nests had centres 60–80 cm apart, with maximum density 2·3 nests per m². Nests re-used in successive years by same birds. Nest: large compacted pile of seaweed, grass, and feathers, plus earth within reach, cemented together and on to substrate by excreta; usually 30–60 cm high, but built up to 2 m over several years. Firm, shallow cup lined with grass, seaweed, feathers, and finer material; lining results from removal of larger items. Occasionally slight structure or little more than depression in ground. Building: before laying, material collected by ♂ only; afterwards by both sexes, though mainly ♂. Material pushed into place with bill; cup formed by turning and pressing with chest and belly; earth scraped up sides of nest with bill. Material

gathered from nest area and elsewhere and stolen from other nests. Building continues during incubation. EGGS. Elongated elliptical to oval; pale, translucent blue turning white; covered with thick chalky layer, readily staining to brown or even black. 79 × 50 mm (62–87 × 41–54), sample 120 (Schönwetter 1967). Weight 105 g (81–130), sample 393; eggs from older ♀♀ tend to be heaviest. Clutch: 1. One brood. Replacement eggs laid after egg loss; once, twice, or possibly more. Eggs lost after up to 26 days incubation replaced; mean interval between loss and re-laying 19 days (6–32), sample 28. Greater tendency among experienced birds to re-lay: 26 of 34 cases, but inexperienced breeders 2 of 16. Eggs laid at any time of day, and possibly at night. INCUBATION. 44 days (42–46). By both sexes, ♂ taking first spell after laying; average spell ♂ 35·6 hours, ♀ 30·2 hours. No brood-patches, eggs incubated beneath webs. Egg retrieving behaviour poorly developed. Eggshells left in nest up to 4 days after hatching, then placed on nest rim, dropped over edge, or trampled underfoot. YOUNG. Altricial and nidicolous. Cared for and fed by both parents. Brooded continuously up to first 14 days, on top of or between parent's webs. Brooding spells average 22 hours (♂), 19 hours (♀). One parent in attendance throughout fledging period, and both present 15% of daylight hours. Fed by partial regurgitation, young putting head into parent's bill. Fed small amounts frequently at first, but from 28 days average 2·7 feeds per day. Parents defend young against other adults, avian predators, and humans by lunging with bill. FLEDGING TO MATURITY. Fledging period 90 days (84–97), sample 111. Become independent at fledging. Age of first breeding 5 or 6 years, rarely 4. BREEDING SUCCESS. Of 500 eggs laid, 82% hatched, 11% lost by predation or rolling out of nest, 7% infertile. 3rd time and more breeders hatched 86%, inexperienced birds 62·5%, difference due to eggs lost. 378 young fledged (92·3% of those hatched); range over 3 years, 89–94%. Artificial 'twinning' experiments showed that for twins of approximately same age, incubation period was 2 days longer, hatching success was as high as for one egg, while fledging success was somewhat lower at 83%.

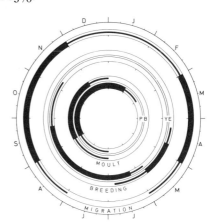

**Plumages.** ADULT. Almost entirely white. Hind crown, nape, sides of head, and hindneck tinged pale yellow to golden-brown, usually more intense in ♂♂; pale to nearly white in autumn, colour becoming stronger during winter and spring. Tail and secondaries white. Primaries, primary coverts, and bastard wing black, becoming brown by wear; shafts of primaries paler towards base; marginal upper and all under coverts of primaries white. NESTLING. 1st down sparse, short; white. 2nd down dense, long with only parts of face bare; white. JUVENILE. Head and upperparts, including upper wing-coverts, and marginal, lesser, and median under wing-coverts, brown-black with triangular white spots at tips of feathers. Spots minute and dense on head and neck, larger and further apart on mantle, still larger on rump; on upper wing progressively larger from marginal to greater coverts. Upper tail-coverts white on outer webs and at tips. Chest and flanks brown, feathers with pale bases and shaft-streaks and white spots at tips. Rest of underparts similar with more white (amount variable), giving mottled appearance. Under tail-coverts and tail-feathers brown-black with white spots at tips. Primaries and secondaries brown-black, primaries with basal half of shafts and of inner webs pale, secondaries with whole inner webs pale. Innermost secondaries with white spots at tips. Underside of flight-feathers and greater under wing-coverts grey with silver gloss. IMMATURE. Gradually developing, all-white plumage attained in 3–4 years; at any one time, birds of same age show much variation due to variable progress of moult. Following typical of late summer 2nd calendar year. Forehead still mainly dark brown with white spots (becoming white in autumn). Rest of head and neck white with occasional retained dark feathers. Upperparts dark brown, spotted white; some all-white feathers on upper mantle. Chest with mottled dark brown band, disappearing in autumn, rest of underparts mainly white, dark patch on thigh. Tail and wing mainly dark brown; white spots from retained juvenile wing-coverts worn off, smaller spots on new ones. Some marginal coverts white. Axillaries white. In late summer of 3rd calendar year, head, neck, upper mantle, and lateral upper tail-coverts white; head and neck with pale yellow-buff cast. Rest of upperparts dark brown, mixed with white feathers, becoming mainly white in autumn. Underparts white. Tail brown-black, 1–3 pairs of outer feathers white. Primaries and their coverts brown-black, secondaries brown-black with white tips, some white. Lesser and some median upper coverts white, great brown. Under wing-coverts white, marginal dark brown. Amount of white in plumage increases in autumn. In summer of 4th year, plumage mainly white like adult, but with dark central tail-feathers and some dark secondaries. Occasional dark tail-feathers or secondaries still present in 5th and 6th year (Nelson 1964b).

**Bare parts.** ADULT. Iris pale grey. Bill pale blue-grey, separate horn plates bordered by black lines, cutting edges of both mandibles black. Eyering blue, bare skin on face and narrow line down chin dark blue-grey. Leg and foot grey-black; greenish lines on mid-ridge of toes, continued and fusing on tarsus, tending to yellow in ♂, to blue in ♀. NESTLING. Bare skin at hatching black. Iris dark brown. Bill grey. Leg and foot dark grey. JUVENILE. Iris grey-brown. Bill dark horn-brown. Bare skin on face brown, leg and foot brown-black, lines on toes and tarsus

paler (Bauer and Glutz 1966). In 3rd calendar year, bill blue and leg and foot with conspicuous lines as adult (Nelson 1964b).

**Moults.** ADULT POST-BREEDING. Complete; primaries serially descendant, with 3 active centres in wing. July–December, occasionally as late as February (Witherby et al. 1940). Yellow colour on head and neck probably not acquired by separate partial pre-breeding moult as claimed by Witherby et al. (1940), but due rather to skin exudate; more information needed. POST-JUVENILE. Gradual body moult, usually from January of 2nd calendar year, but occasionally starting in November of 1st. Details of primary moult unknown, but in September of 2nd calendar year 1st series reaches outer primaries and 2nd series starts with inner. In subsequent years, timing of moult mainly as adult, so birds show greatest change in appearance in autumn; but more data needed.

**Measurements.** Adults, mainly autumn birds from Netherlands; skins (RMNH, ZMA).

| | ♂ | | | ♀ | | |
|---|---|---|---|---|---|---|
| WING | 491 | (11·0; 23) | 460–515 | 485 | (12·8; 23) | 460–520 |
| TAIL | 223 | (10·4; 20) | 210–250 | 220 | (9·16; 16) | 210–240 |
| BILL | 99·3 | (2·31; 24) | 95–103 | 97·4 | (3·10; 22) | 92–101 |
| TARSUS | 60·2 | (1·61; 21) | 58–64 | 60·3 | (1·61; 22) | 58–64 |
| TOE | 106 | (3·00; 19) | 100–110 | 104 | (4·22; 19) | 100–114 |

For live birds, wing: ♂ 513 (9) 500–535; ♀ 510 (14) 484–522; bill: ♂ 100·1 (66) 93·5–110; ♀ 99·2 (66) 92·5–104 (J B Nelson). ♂♂ average slightly larger than ♀♀, but difference not significant. In juveniles, tarsus and middle toe similar to adult, other measurements slightly smaller (skins RMNH, ZMA).

| | ♂ | | | ♀ | | |
|---|---|---|---|---|---|---|
| WING | 474 | (15·1; 18) | 445–505 | 479 | (16·6; 12) | 450–505 |
| BILL | 93·9 | (2·92; 19) | 90–100 | 96·0 | (2·07; 12) | 93–99 |

**Weights.** Adult breeding: Bass, ♂ 2932 (27) 2470–3470, ♀ 3067 (27) 2570–3610; Ailsa Craig, ♂ 3120 (17) 2400–3600, ♀ 2941 (18) 2300–3600 (J B Nelson). All ages: washed ashore Netherlands 1355–2760 (ZMA). Nestlings c. 90 at hatching, reaching maximum of c. 4250, but c. 3650 at fledging (Nelson 1964a).

**Structure.** Wings long and relatively narrow. 11 primaries: p9 or p10 longest, p9 0–5 shorter, p10 0–11, p8 20–27, p7 55–68, p6 95–110, p5 c. 140, p1 c. 270; p11 minute. Inner web of p10 strongly emarginated, of p8–p9 slightly; outer web of p8–p10 slightly. 26–31 secondaries (Snow 1967). Tail strongly wedge-shaped. 12 pointed feathers: t6 c. 110 shorter than t1. Bill robust, heavy at base, horny sheath of upper mandible divided in 1 central and 2 lateral plates with slightly curved nail at tip; no nostrils. Tarsus short and stout. All toes connected by webs, outer toe c. 95% of middle, inner c. 70%, hind c. 30%. Claws small, strongly curved.

**Geographical variation.** None in North Atlantic area. Close relatives in South Africa (S. capensis) and Bass Strait area and New Zealand (S. serrator) differ in pattern of black in wings and tail, in size, and in ecology. Sometimes considered conspecific, but under present conditions specific separation can neither be proved nor refuted, so probably best treated as forming superspecies.

# Family PHALACROCORACIDAE **cormorants, shags**

Medium-sized to large aquatic birds. About 30 species, all (apart from flightless *Nannopterum harrisi* of Galapagos) now placed in single genus *Phalacrocorax*; small cormorants sometimes separated as 3rd genus *Halietor*. 5 breeding species in west Palearctic. Body elongated, neck rather long. ♂ larger than ♀. Wings with long inner portion and short tip. 11 primaries, p8 and p9 longest; 17–23 secondaries, diastataxic (except *N. harrisi*, in which wings much reduced). Flight with regular wing-beats, low over water, rather higher over land. Tail long and strongly wedge-shaped, 12–14 pointed feathers. Bill strong, of medium length, laterally compressed, culmen rounded; hooked at tip; nostrils closed. Gular skin bare. Tarsus heavy; toes long, outer longest, totipalmate, nail of middle toe medially with comb. Tibia feathered. Legs set far back, stance upright; gait waddling. Able to perch in trees. Feet used for underwater propulsion. Oil gland feathered. Aftershaft absent. Plumage black, often with metallic sheen, or black above and white below; grey in some species. Bare face often brightly coloured. Sexes similar. Breeding plumage different from non-breeding, often by increase of white filoplumes. 2 moults per cycle, pre-breeding often involving relatively few feathers; primaries replaced in serially descendant order. Young altricial and nidicolous, naked at hatching, later growing single coat of dense white, brown, or black down. Contour feathers do not develop from same follicles as natal down. Juveniles differ from adult by being duller or paler; reach adult plumage in 3rd–4th calendar year.

Mainly tropical and temperate, many on marine coasts or islands but some inland. As specialized fish-eaters adapted to consume large meals, strictly linked with large or medium size aquatic habitats, both marine and freshwater, but require immediate access to suitable terrestrial resting places. Accordingly, frequent inshore waters, estuaries, lagoons, and lakes; also large, usually slow-flowing rivers. Although not highly mobile on land, swim for considerable distances, and dive deeply. Also make frequent, sometimes regular and prolonged aerial trips, often in parties, exploiting fishing waters far from breeding places. In west Palearctic, 2 of breeding species found only in south; of remaining 3, Cormorant *P. carbo* and Pygmy Cormorant *P. pygmeus* reduced, especially inland colonies, by drainage but also by human persecution, while marine Shag *P. aristotelis* has shown no marked changes in range. Only continental race (*sinensis*) of *P. carbo* and Caspian population of *P. pygmeus* truly migratory. Otherwise dispersive, in which a few individuals may make long (even overseas) movements. Migrations probably mainly diurnal.

Food basically fish, in some species entirely so; caught under water mainly by surface-diving and pursuit or benthic swimming. Tend to fish singly, but will gather when prey concentrated; co-operative fishing also reported from some species. Often return to colony or roost in flight formation. Monogamous pair-bonds, usually of seasonal duration; maintained more or less entirely at nest sites only. Typically colonial breeders, defending small nest-territories vigorously; often nest in close proximity with other species of similar ecological requirements, such as herons (Ardeidae). Leave water after feeding: rest during day on rocks, posts, trees, etc.; roost communally at night in similar but usually more protected situations. Factors influencing evolution of heterosexual signals much as in Sulidae, though considerable differences in forms of visual displays; advertising behaviour of ♂♂ particularly well developed, while marked ritualization also of movements by both sexes associated with take-off, landing, and locomotion on foot. Copulation only at nest-site. Voice relatively unspecialized in most species; calls more or less confined to breeding colony. Comfort behaviour similar to most other pelecaniform birds. Bathing spectacular at times, leading to misidentification as display (van Tets 1965). Other behavioural characters include: dissipation of heat by gular-fluttering; direct head-scratching; true yawning as well as jaw-stretching; and distinctive spread-wing posture, with wings extended at right-angles to body, adopted when loafing out of water. Significance of latter behaviour disputed, but widely thought mainly to relate to drying of wings; as in most other waterbirds, however, plumage otherwise thoroughly waterproof and frequent use made of preen-oil. Resting-posture as in Sulidae.

Breeding seasons extended, though more strictly seasonal at higher latitudes. Nests placed on ground, in open or in shelter; on cliff ledges; and in trees. Densely or loosely colonial, size and density of colonies usually correlated with distance travelled to obtain food (see Lack 1967). Nests re-used in successive years. Heaps of available vegetation, sometimes just scrapes; built by both sexes, with tendency to continue through incubation and nestling periods. Eggs elongate oval, pale blue or green, covered with chalky white coating; dark marks present in Socotra Cormorant *P. nigrogularis*. Clutches 2–4 (1–7). Single brood. Replacements laid after egg loss. Eggs laid at intervals of 2–3 days. Incubation periods comparatively short, 27–31 days. Eggs incubated on feet by both sexes in approximately equal spells, changing over once or twice a day; no brood-patches. Incubation starts with 1st egg; hatching asynchronous. Eggshells removed from nest. Young cared for by both parents. Brooded continuously while small. Fed by partial regurgitation; adults also bring water. Fledging periods up to *c.* 70 days, though mostly 48–53. Young fed for 2–3 months or more after fledging, then become independent.

# *Phalacrocorax carbo* Cormorant

PLATE 21
[after page 230]

Du. Aalscholver    Fr. Grand Cormoran    Ge. Kormoran
Ru. Большой баклан    Sp. Cormoran grande    Sw. Storskarv    N.Am. Great Cormorant

*Pelecanus Carbo* Linnaeus, 1758

Polytypic. Nominate *carbo* (Linnaeus, 1758), coasts North Atlantic; *sinensis* (Blumenbach, 1798), central and south Europe, Asia east to Japan and south to Ceylon; *maroccanus* Hartert, 1906, coasts north-west Africa; *lucidus* (Lichtenstein, 1823), coasts west and south Africa and inland east Africa. Extralimital: *novaehollandiae* Stephens, 1826, south New Guinea, Australia, New Zealand, and Chatham Islands.

**Field characters.** 80–100 cm; wing-span 130–160 cm; largest individuals larger than Shag *P. aristotelis* and much larger than Long-tailed Cormorant *P. africanus*. Large, heavy, hook-billed, marine and freshwater cormorant, with mainly black or dark plumages at all ages, and goose-like flight action. Sexes alike; obvious seasonal variation; juvenile separable. 4 races in west Palearctic, of which only northern and southern forms safely identifiable in field.

ADULT. North Atlantic race nominate *carbo*. In early breeding season, bare throat yellow, skin below eye orange, forecheeks and mid-throat white (forming obvious panel exaggerating depth of head); rest of head, neck, and body velvety black, with purple-blue gloss and relieved only by thin spray of long, white feathers on shaggy nape and oval white patch on thighs. Upper wing-coverts bronze-brown, each feather edged black (and whole area contrasting with body in even light), quills black, under wing-coverts black, contrasting with silvery cast on quills; tail black, faintly glossed bronze. Bill long, thick, and heavy, with upper mandible noticeably hooked at tip and obvious gape crease below eye; grey above, yellower below, with dark culmen and tip. Legs and large webbed feet black. Eyes green. Out

of breeding season, loses nape spray, high gloss to plumage and white on thighs. Other west Palearctic races. Eurasian race *sinensis*: plumage glossed dull green, throat whiter than in *carbo*, and hair plumes over crown to back and sides of neck complete and silvery-white; but old ♂♂ of *carbo* often show similar marks. Moroccan race *maroccanus*: plumage gloss greener still, whole throat and upper neck white. African race *lucidus*: foreneck completely white, extending to breast and even body. Winter adults of these 3 races duller, losing breeding plumage marks as *carbo*, but larger areas of white on necks of *maroccanus* and *lucidus* retained. JUVENILE. Bare face dull yellow; crown, hindneck, and upperparts dark yellowish-brown, with feathers edged dark brown or black; mid-face, sides of neck, lower foreneck, and breast dull buff or pale brown, with much mottling of same colour on flanks and lateral tail-coverts; rest of underparts dull white (purest in southern races); upper wing-coverts as back, quills brown-black, under-wing less black than in adult; tail brown-black. Bill grey with yellow tinge at base; legs and feet as adult. Extent of white on underparts variable in all races, but averages greatest in southern birds. IMMATURE. Plumage

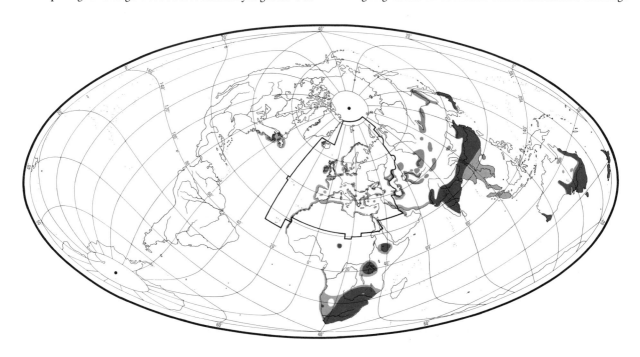

darkens with each moult and white on lower underparts disappears by 2nd autumn.

At long range, separation from *P. aristotelis* often difficult but combination of heavy bill, thick neck, prominent chest, bulky body, and long wings beaten slowly diagnostic in flight; first 2 characters remain obvious even in partly submerged bird. Plumage of adult appears blacker than in *P. aristotelis* but plumage of juveniles does overlap in spite of many comments to contrary. For distinctions from *P. africanus*, see that species.

Flight action and behaviour reminiscent of geese *Anser*. Wing-beats slow and powerful (in restricted horizontal plane) and noticeably shallow in high flight, with slight jerking movement at end of down-beat; and primaries rather squarely set and clearly separated on down-stroke (wing point of *Anser* more obvious, with primaries less separated); wing-beats more rapid and erratically angled in close-quarter flight, birds finding landings on cliffs in high winds difficult. Flocks on migration or parties moving to feeding grounds adopt echelon formations, more regularly shaped than in *P. aristotelis* but less so than in *Anser*. Dives for fish, submerging with or without obvious leap. Walk lumbering; perches in sentinel positions, often holding wings out in characteristic spread-wing posture. Usually silent at sea, but most characteristic call a low, guttural croak.

DIMW

**Habitat.** Mainly aquatic, in both salt and fresh waters, from Arctic to sub-tropical zone, principally in mid-latitudes; in west Europe, largely coastal while, in east Europe (and also in Netherlands), large populations breed colonially near extensive inland waters. Prefers fairly sheltered seas, avoiding deep water even close to land, and rarely extending far offshore, exceptionally to continental slope. Also on lakes, reservoirs, lagoons, floodlands, open water in swamps and other wetlands, deltas, estuaries, saltpans, and broad rivers, usually sluggish but sometimes on torrents. Avoids widespread floating vegetation or obstructive emergent growth. Spends long periods ashore, especially on exposed and slightly elevated positions for drying, oiling, and preening. These include rocks or reefs, shingle spits, sandbanks, trees (especially with bare branches), breakwaters, seawalls, piers, disused vessels (moored or wrecked), floating logs and similar convenient perches, either natural or artefacts, usually close to water and relatively immune from disturbance. Gregarious for roosting and breeding; requires secure cliff ledges, rocks, stacks or islets, or groves of trees, including tall forest trees several km from water; also uses dense reedbeds or bare patches in swamps. Given suitable habitat, will penetrate far inland, even through mountain gorges. Recorded up to 3450 m in Kashmir (Ali and Ripley 1968) and up to 2000 m in Armenia (Dementiev and Gladkov 1951). Although generally linked to large bodies of water, adaptable, catholic, and fairly mobile, fitting somewhat loosely into basically specialized ecological niche. Circles or soars up to considerable heights and flies overland, often for long distances, occasionally or regularly. Tenacious of breeding haunts even in face of continuing persecution, but where nesting in trees tends to destroy them by coating with guano. Generally avoids man or manned artefacts, keeping well out of gunshot, but has been locally domesticated for aiding fishermen. Shows unusual powers of survival and of adaptability to varying conditions.

**Distribution.** Considerable changes in colonies, especially inland, during last 100 years or so due to persecution by fishermen, habitat changes, and destruction of nesting trees by birds' droppings. Has bred at least once Tunisia (Heim de Balsac and Mayaud 1962) and perhaps formerly in Cape Verde Islands (Bannerman and Bannerman 1968).

Accidental. Jordan, Cyprus (irregular winter visitor), Azores, Madeira, Canary Islands, Cape Verde Islands.

**Population.** ICELAND. *c.* 2000–2500 pairs, 1975 (A Gardarsson). FAEROES: formerly common breeding, now almost extinct (J Dyck). BRITAIN AND IRELAND. 8136 pairs breeding 1969–70. Probably decreased in 19th century when tree-nesting colonies in Norfolk abandoned. More recently, marked increases (*c.* 4% a year since 1905)

in colonies Firth of Forth and Farne Islands, with probable increases in other areas, e.g. Yorkshire and north Scotland, but declines north-west Scotland (Cramp *et al.* 1974). NORWAY: some southern colonies deserted this century (Haftorn 1971). FRANCE: 237 pairs, 1968, increasing 1969 due to protection (Terrasse *et al.* 1969*b*); 300–330 pairs (Cruon and Vieillard 1975). DENMARK: bred until *c.* 1876 when became extinct; returned in 1938 to some of old sites, but now restricted to Vorsø and Funen (TD); *c.* 600 pairs, 1969–70 (W R P Bourne). SWEDEN: also became extinct 19th century, but has bred again since late 1940s, with 150–200 pairs in 1970 (SM). NETHERLANDS. Many changes in colonies and numbers due to persecution, land reclamation, and probably river pollution (Coomans de Ruiter 1966). In 1918–19, 3000 pairs in 10 colonies; 1940, 4000 pairs in 6 colonies; 1955, numbers controlled at 1200 pairs in 3 colonies; 1971, 1500 pairs in 2 colonies. Acute mortality from polychlorinated biphenyls in recent years (Rooth and Jonkers 1972). BELGIUM: various small colonies often destroyed; extinct since 1965, but re-introduced in 1970 when 3 nests (Lippens and Wille 1972). WEST AND EAST GERMANY. Formerly bred in northern areas with many lakes; full history of numerous small and some large colonies in Bauer and Glutz (1966). In West Germany, now single colony *c.* 25 pairs (H Ringleben). AUSTRIA: fluctuating colonies along Danube and occasional elsewhere (Bauer and Glutz 1966). CZECHOSLOVAKIA: breeds along Danube and irregularly elsewhere; one colony of nearly 2000 pairs reduced by shooting and extinct after 1962 (Bauer and Glutz 1966; KH). HUNGARY: only 1 regular colony, at Kisbalaton, found after 1962, and perhaps established by birds from Czechoslovakia; increased under protection to *c.* 180 pairs 1964, 298 pairs 1971, 149 pairs 1972; occasionally nests elsewhere (Bauer and Glutz 1966; KH, IS, and AK). Data inadequate to assess population trends in central Europe, but Bauer and Glutz (1966) considered total population apparently relatively stable. POLAND: now increasing; in 1955 estimated 950–1000 pairs in north-west Poland with additional 1000–1500 pairs in Olsztyn province (Tomialojc 1972). RUMANIA: numbers decreasing sharply owing to destruction of nests by fishermen (MT). GREECE: nearly extinct; in 1971, only 2 colonies with 540–570 pairs (WB). TURKEY: 4 known colonies, *c.* 800–900 pairs (OST). MAURITANIA: Banc d'Arguin, *c.* 1400 pairs on 2 islands, with 2 small colonies further north (de Naurois 1969*a*).

Survival. In Netherlands when population increasing *c.* 10% a year, annual ♀♀ mortality 36% 1st year, 22% 2nd, 16% 3rd, and 9–14% subsequent years; ♂♂ similar, except 7–12% after 3rd year (Kortlandt 1942). For one Scottish colony, mortality 70% 1st year, varying between 28% and 46% in subsequent years; large proportion shot, especially younger birds (Stuart 1948). In both samples, ring loss probably led to too high assessment of apparent mortality. Oldest ringed bird 19 years 8 months (Rydzewski 1973).

**Movements.** Migratory, partially migratory, and dispersive, according to population.

*P. c. carbo.* British and Irish populations non-migratory but show extensive dispersal with small proportion crossing English Channel; mainly coastwise, but also inland usually within 60 km of coast (Mills 1965). No significant differences in dispersal patterns for different age-classes, but dispersal from individual colonies (September–March) varies in direction, distances, and extent to which large water-bodies and land-masses crossed (Coulson and Brazendale 1968, on which paragraph based). Galloway, Northumberland, and Co. Wexford colonies alone show substantial northward movement; Welsh and Northumberland colonies provide most recoveries in south-east England; only Galloway birds quite often cross to opposite side of Britain; Co. Galway birds show little coastal movement but tend to move east across Ireland. Recoveries in France and Iberia (mainly Brittany to north Portugal) continuance of southward dispersal in Britain and Ireland, and originate mainly from Irish Sea and Galloway. North Sea effective barrier to eastward movement from Britain, recoveries Norway to north France being quite exceptional. Breeders of Iceland and from Kola Peninsula to west Norway disperse around respective coasts, and latter occur south to Skaggerak and Kattegat, small numbers entering west Baltic (Vaurie 1965).

*P. c. sinensis.* North Caspian and Baltic breeders mostly migratory (latter providing recoveries up to 2400 km); variable, often substantial, numbers overwinter within other parts of breeding range, to some extent dependent on degree of freezing. Many move south as far as Mediterranean basin; unlike nominate *carbo*, regularly migrates overland. West European breeders migrate chiefly south to south-west, those of east Europe chiefly south to SSE, but recoveries for both show longitudinal scatter. Those ringed Netherlands (where many overwinter) found abroad mainly France, Iberia, and Tunisia, but also from Britain and Morocco east to Hungary, Yugoslavia, Italy, and Sicily; those ringed Rügen (East Germany) found chiefly Denmark, Yugoslavia, Italy, and Tunisia, but also from Poland and Greece west to North Sea, Bay of Biscay, and Algeria (Heckenroth and Voncken 1970). Breeders of Balkans, Black Sea and Turkey perhaps just dispersive, but probably from these areas come those that winter in east Mediterranean all around Levant; many remain on Black Sea in mild winters. According to Meinertzhagen (1930) common winter visitor (October or November–April) in lower Egypt and a few ascend Nile to Sudan frontier, but none seen Egypt, October–November 1957 (Horváth 1959*b*). Migrates away from Volga delta but overwinters in large numbers on south Caspian; probably from this population come those, mainly immatures, which winter commonly through Iraq to northern half Persian Gulf (Ticehurst *et al.* 1922) Juveniles begin moving in all directions from June–July; adults also

disperse, often overland, from late July, prior to departure of migratory element, Reassembles in Dutch colonies in January or February, but in Baltic and USSR from mid-February to late March. Noted many parts of Europe during passage periods, also occasionally at other seasons since non-breeders wander even in summer. See also Bauer and Glutz (1966), Dementiev and Gladkov (1951).

*P. c. maroccanus* and *P. c. lucidus.* No information, but probably only locally dispersive within their respective north-west and west African ranges; over 3000 Banc d'Arguin, late December 1971 (Pététin and Trotignon 1972).

**Food.** Normally, entirely fish; obtained mainly during day by diving from surface with or (especially when well fed) without forward leap. Normally only feet used for underwater propulsion with wings held close to body. Seldom remains long below surface; dives range 15–60 s; and depths *c.* 3–9 m, with average 1–3 m (Dewar 1924; van Dobben 1952). In 461 dives, maximum depth 9·5 m, longest dive 71 s (Dewar 1924). Most prey, particularly larger and more resistant, brought to surface and often shaken and thrown in air before swallowing. When hunting, swims dipping head and eyes under surface. Typically solitary feeder, less often in loose flocks, and may travel up to 50 km to feeding area (Hachler 1959).

243 stomachs collected throughout year from estuaries of Dee, North and South Esks, and coasts of Angus and Kincardineshire, Britain, contained chiefly flatfish (especially flounder *Platichthys flesus*), gadoids (mainly saithe *Pollachius virens*, cod *Gadus morhua*, and whiting *Merlangius merlangus*), and shore and estuarine fish, particularly viviparous blenny *Zoarces viviparus* and sea-scorpions Cottidae; less frequently, in smaller quantities, clupeoids, common eel *Anguilla anguilla*, and salmonids (Rae 1969). Similar results from 27 stomachs from Cornwall, England, mainly estuaries: 40% flatfish, 50% shore and estuarine fish, and only 10% marketable fish i.e. herring *Clupea harengus* and sprat *Sprattus sprattus* (Steven 1933). Qualitative assessment based on regurgitations and pellets in June and July, Caithness, Scotland, again indicated flatfish (particularly dab *Limanda limanda*) most important, followed by shore and estuarine fish, gadoids, and sand-eels Ammodytidae (Mills 1969). Using similar techniques, Pearson (1968) on Farne Islands also found flatfish most frequent, followed by gadoids, sand-eel, common eel, and various shore species.

In 7 breeding colonies, Ireland, regurgitated food from birds feeding fresh, brackish, and saltwater areas contained mainly wrasse (Labridae) 10–36 cm (60% by weight), eel 10–65 cm (20%), flatfish (Pleuronectidae) 7–20 cm (10%), and salmonids (2%); differences between colonies related to fish availability (West *et al.* 1974). All studies from marine and brackish areas emphasize that most feeding on bottom-living fish (see *P. aristotelis* which feeds more often on mid-water fish and in different areas). From Denmark,

298 stomachs collected mainly April–October from Vorsø, contained common eel (38% by frequency; 10–60 cm long, average 20–25 cm), viviparous blenny (25%; average length 20 cm), herring 29%, cod 18%, and 9 other species (though these taken much less). Herring taken mainly in spring and summer when spawning schools occur on surface near coast (Madsen and Spärck 1950). In Holland, from regurgitations and pellets mainly from freshwater feeding areas, chiefly eel and roach *Rutilus rutilus* in breeding season, with ruffe *Gymnocephalus cernua*, pike-perch *Stizostedion lucioperca*, bream *Abramis brama*, white bream *Blicca bjoerkna*, perch *Perca fluviatilis*, rudd *Scardinius erythophthalmus*, and tench *Tinca tinca*. Results show that main prey taken always locally dominant species (van Dobben 1952). From Scotland: 21 stomachs from lochs contained mainly brown trout *Salmo trutta* (43% by frequency), perch (38%), and young salmon *S. salar* (19%); 93 stomachs from rivers mainly brown trout (23%), young salmon (17%), and eels (13%); 11 stomachs from estuaries mainly flounders (55%), sea trout (27%), and fewer gadoids and clupeoids (Mills 1965). From Co. Mayo, Ireland, 22 stomachs May–October main fresh-water prey brown trout and common eel; and from saltwater mainly herring (Piggins 1959). 9 stomachs from Windermere contained perch (2), trout (2), char *Salvelinus alpinus* (4), pike *Esox lucius*, and eel (Hartley 1948). From Norway, Collett (1921) found brown trout and perch in 3 stomachs. Food items other than fish usually form only small part of diet. Polychaetes and crustaceans, especially prawns, shrimps, and crabs often noted, but generally thought to be derived from fish stomachs (Madsen and Spärck 1950; Collinge 1924–7; Rae 1969). Occasionally birds recorded: waterfowl and Moorhen *Gallinula chloropus* (St John 1882), ducklings of Shelduck *Tadorna tadorna* (van Dobben 1952), Swallows *Hirundo rustica* flying low over water (Kortlandt 1940). Also water vole *Arvicola amphibia* (Kortlandt 1940), frogs *Rana temporaria* (Ponting 1967), and even kitten of 28 cm (Fisher and Lockley 1954).

Young fed by parents regurgitating fluid into pharyngeal pouch; later, more solid food taken from within parent's mouth and gullet (Madsen and Spärck 1950; van Dobben 1952).

Daily consumption varies between 425–700 g and averages between 15–17% of body weight (van Dobben 1952; Mills 1965; Skokova 1955; Macan and Worthington 1951; Meinertzhagen 1959; Rae 1969). Still much dispute on effects on fishing interests, conclusions varying with locality, time of year, and investigator.

**Social pattern and behaviour.** Based mainly on material supplied by G F van Tets; see especially Kortlandt (1938), also van Tets (1965), van Dobben (1952).

1. Often gregarious. Flocks formed as temporary aggregations, with no formal hierarchy for feeding, breeding and roosting; also for commuting between feeding, breeding and roosting areas. Individual nomadism the rule outside breeding season, with loose flocks forming during migration and at good feeding and roosting sites; no territories established, but some individuals may defend favourite roosting perch. Flock size depends on availability of food; often hundreds unite to feed on fish shoals, and on fish forced out of tidal inlets by receding water. BONDS. Pair-bond basically monogamous, of single season duration but may persist for more than one brood or season; mate occasionally changed early in breeding cycle. Both parents tend young until after fledging. In some colonies, especially on relatively flat ground, sometimes crèche phase lasting 2 weeks before chicks fully fledged and start to wander from colony. Juveniles fed by own parents away from nest and colony until contact lost. BREEDING DISPERSION. Colonial. Territory, restricted to nest-site, serves for courting, pair-formation, and raising young until at least 4 weeks after hatching. ROOSTING. Roosts are of two kinds. Diurnal haul-out spots in feeding areas serve mainly as places to rest and digest meal before flying to nocturnal roost or colony; usually floating logs, wave-washed rocks, buoys, and dead trees. Nocturnal roosts, similar to nesting colonies, usually located on small islands, steep cliffs and in groups of trees surrounded by water. Individuals arrive at and depart from roost singly or in loose, temporary flocks. Little regularity in arrival and departure and a few birds may be present most of day. Most, however, leave at first light and are back before dusk; some do not leave until late in day when others may have already returned.

2. Both ♂ and ♀ show similar behaviour patterns at or near nest which is centre of activity within breeding colony; only ♂, however, performs sexual advertising display. ANTAGONISTIC BEHAVIOUR. Nest-site claimed by ♂ and area within pecking distance of sitting bird defended continuously, members of pair taking turns from pair-formation until young strong enough to defend themselves and nest. Conflicts between ♂♂ occur over nest-sites, but ♀ will also defend ♂ and his site. Threat consists of lunging head forward with mouth wide open and irregular snake-like motion of neck; accompanying Threat-call loud and raucous in ♂, subdued and hoarse in ♀. Lesser forms of threat occur when conspecific wanders too closely but with no obvious aggressive intentions: closed bill pointed downwards with neck arched, or nest-material worried with bill in demonstrative and perfunctory manner. Both displays silent, and appear to indicate that site will be defended if necessary. Where 2 opponents close enough to fight, they grab with beaks and try to push and pull each other off contested site. HETEROSEXUAL BEHAVIOUR. ♂ advertises on nest-site for mate by Wing-waving (see A): uttering no call, he repeatedly raises wing-tips upwards and outwards with primaries folded behind secondaries, thus revealing and covering large

A

white rump patches about twice per second (white flashes readily seen even when bird barely visible against dark background). During Wing-waving, closed bill tilted up and forward, crest partly raised, neck S-shaped, body horizontal with rump up and breast down, and spread tail cocked up and backwards; head bobs up and down and throat and cloaca pulsate in time with wing movement. Wing-waving stops when ♀ comes near and does not occur when eggs present. During and after pair-formation bird on nest greets mate outside nest by Gaping. ♂ version consists of (1) pointing closed bill, head and neck straight upwards (see B); (2) swinging head and neck slowly backwards, bill opening widely while loud call made (see C); (3) touching base of tail with closed bill while head and bill rotated rapidly from side to side with gargling sound (see D); (4) returning head and neck rapidly forwards. ♀ version begins like that of ♂ but does not go beyond pointing open bill up and slightly backward before head and neck returned forward; subdued hoarse call given. In both sexes, body horizontal and wings drooped loosely alongside body; spread tail cocked up and slightly backwards in ♂ and drooped downwards in ♀. Pointing another greeting ceremony of both sexes: closed bill points upwards and head and neck stretched and slowly waved back and forth and from side to side; no call given, crest depressed, body horizontal with drooping wings and spread tail cocked up and slightly backwards. ♂ fetches most of nest material, presenting to ♀ to weave into nest structure. Before leaving mate on nest, either sex performs elaborate Pre-flight display: head, throat and upper neck distended and crest raised while closed bill tilted up and forwards. On returning, Post-landing display consists of giving head and neck discoid shape by flattening them laterally; crest raised, closed bill pointed forward and slightly downward and ♂ Roars. These 2 displays often combined into Hop-displays when bird moves in vicinity of nest. Kink-throat display also performed when nest approached, either while walking, swimming, or flying, with and without material in bill; also used on nest when bird weaves in new material. Kink-throat display characterised by angular appearance of throat due to forward protrusion of hyoid bones; accompanied by repetitive call (loud in ♂, hoarse and subdued in ♀). When no material held, mouth wide open in flight but closed on land or water. No distinctive pre- and post-copulatory displays, but all behaviour described may occur before and after copulation. During copulation ♂ closes wings and holds ♀'s head or neck with bill. In addition to more formal displays mentioned, members of pair Allopreen while on nest and rub and entwine their necks. RELATIONS WITHIN FAMILY GROUP. Young beg for food with Kink-throat display, with closed bill and persistent calling; in older chicks, wings spread and whole body wiggles from side to side. Begging for water occurs only during extremely hot and dry weather: chick waves mouth upwards, wide open. Off-duty parent then leaves to fetch water in throat; sometimes water also given by duty parent. Displays of young towards each other less coordinated versions of displays of adults. Sometimes put heads down each other's throats, presumably in search of food.

(Figs A–D after van Tets 1965.)

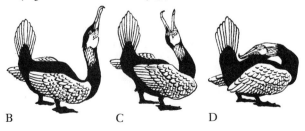

B          C          D

**Voice.** At nesting sites, vocal reportoire most complex and varied of pelecaniform birds studied (G F van Tets); otherwise usually silent. Various deep, guttural calls, with much individual variation. Voice of adult ♂ loud and raucous. Early in breeding-cycle, that of adult ♀ hoarse and subdued but becomes more masculine later in cycle (Kortlandt 1938).

Following calls described at Netherlands colony (G F van Tets). (1) Threat-calls. Those of ♂ 'agock-kock-kock' or 'tock-gock-gock' (recording, see I, shows variant 'a-gock a-gock a-gock'); of ♀ 'fhi-fhi-fhi'. (2) Calls associated with Kink-throat display. 'Gurr-gurr-gurr', 'curr-curr-curr', 'kirr-kirr-kirr', 'arre-arre-arre', 'urr-urr-urr', 'arr-roo-roo', 'gorr-rorr-rorr', or rahr-rahr-rahr' (♂) and 'ghi-hi-hi-hi' or 'for-for-for' (♀) given in Kink-throat display on land. 'Ghi-ghi-ghi, 'roh-roh-roh', or 'reh-reh-reh' (♂) and 'fhi-fhi-fhi', 'fee-hi-hi-hi-hi-hi', or 'fi-hi-hi-hi' (♀) uttered in Kink-throat display before landing. (3) Calls given with Hop-display. 'Ah-ah-ah' (♂); 'fi-fi-fi' or 'fhi-hi-hi' (♀). (4) Pre-flight and pre-hop calls. 'R-r-r-t' or 'rrrrr' (♂), 'rrrrr' (♀). (5) Post-landing and post-hop calls. Roaring 'rooo', 'rooj', or 'rooor' (♂); 'fheee' (♀). (6) Calls associated with Gaping-display. 'Arr', 'wi', or 'ah' (♂) and 'fee-hee-hee-hee', 'fee-fee-fee-fee', 'f-f-f-f-f-f', 'fee-he-he-he', or 'sh-hi-hi-hi' (♀) uttered when mouth open; 'ooorrr', 'r-r-r-r-r-r', or 'rooorr-rrr' (♂ only) given when mouth closed. (7) Call associated with rubbing and entwining of necks. 'Rrrrr' (both sexes).

CALLS OF YOUNG. Whining 'kra-wee kra-wee kra-wee' uttered when hungry (Witherby *et al.* 1940).

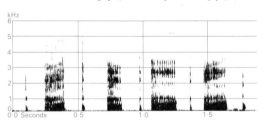

I Sture Palmér Norway June 1958

**Breeding.** SEASON. See diagram for Britain, Netherlands, and Baltic. In south-east Europe and USSR, laying early to mid-April, hatching from mid-May. SITE. On cliff ledges, from just above high water mark to over 100 m, usually fairly sheltered. Also in trees to 10 m above ground, or over water; in reedbeds; occasionally on bare ground. Colonial up to 2000 pairs; nests sometimes touching. Sites re-used in successive years. Nest: heap of twigs, seaweed, reeds, etc., lined with finer material. 75–100 cm high; diameter at base *c.* 1 m, of cup 30–40 cm (Dementiev and Gladkov 1951). Building: by both sexes, but mostly ♀ with ♂ supplying material; continues through incubation and fledging. EGGS. Elongate oval; pale blue or greenish, covered with chalky white deposit. 66 × 41 mm (58–74 × 38–43), sample 200; calculated weight 58 g (*P. c. carbo*). 63 × 40 mm (56–68 × 35–44), sample 250; calculated weight 53 g (*P. c. sinensis*). 63 × 40 mm (57–64 × 37–42), sample 6; calculated weight 55 g (*P. c. lucidus*) (Schönwetter 1967). Clutch: 3–4(–6). One brood; records of late broods may refer to 2nd, but more likely replacements after egg loss. Eggs laid at intervals of 2–3 days. INCUBATION. 28–31 days recorded for both *carbo* and *sinensis* (Witherby *et al.* 1940; Dementiev and Gladkov 1951), but also 23–24 days for *sinensis* (Portielje 1927). Begins with 1st egg; hatching asynchronous. Eggshells removed from nest. Both parents incubate, changing over twice a day. YOUNG. Altricial and nidicolous. Cared for and fed by both parents. Brooded continuously while small. Fed by partial regurgitation twice a day, once by each parent (van Dobben 1952). Young puts head into parent's bill. FLEDGING TO MATURITY. Fledging period *c*. 50 days. Young return to nest site to be fed for further 40–50 days (Kortlandt 1942); independent thereafter. Age of maturity 4–5 years, occasionally 3 (Kortlandt 1942). BREEDING SUCCESS. No data.

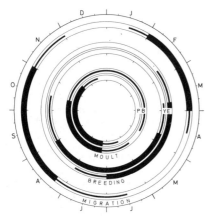

**Plumages** (nominate *carbo*). ADULT BREEDING. Head, neck, centre of mantle, back, rump, tail-coverts, and underparts black, glossed metallic blue. Crown, upper neck, and lower throat with varying number of white elongated feathers. Black feathers in centre of nape and hindneck elongated; form short crest. Chin, sides of head, and patch of loose feathers at base of leg white. White of chin and that of elongated feathers on head separated by dark band. Sides of mantle, scapulars, and upper wing-coverts bronze; feathers rounded, evenly bordered glossy blue-black. Tail-feathers black, shafts light horn. Flight-feathers black, secondaries with bronze outer webs. Underwing black with slight blue gloss. White feathers on crown, neck, and, to lesser extent, thigh lost during course of breeding season. ADULT NON-BREEDING. Like breeding, but blue-black of crown and hindneck tinged brown; white feathers of chin and sides of head tipped buff. Upper throat and sides of neck dark brown, mottled by grey or white of base of feathers. Underparts less glossy; no elongated white feathers on head. NESTLING. Naked at hatching; dense black down appears *c*. 6th day, leaving chin, lores, and area round eye bare; fully feathered *c*. 7 weeks. JUVENILE. Feathers of crown and nape pale brown with narrow blue-black tips, edged brown. Feathers of lower neck, centre of mantle, back, rump, tail-coverts and flanks brown, tipped more or less glossy blue or brown-black; sides of mantle, scapulars and upper wing-coverts dull bronze; feathers pointed (against more rounded in adults), evenly

edged brown-black and tipped pale brown. Small white hair-like filoplumes on nape and lower throat, few also on thigh; soon wear off. Feathers of chin, sides of neck, throat and chest dark brown, edged buff; palest on chin. Feathers of breast and belly basally white, tipped with variable amount of sooty black. Underparts variable from glossy sooty, with only a few light spots in centre of breast or belly, to white on chin, centre of throat, breast, belly, and vent; sometimes even under tail-coverts white. Tail and flight-feathers black, slight bronze gloss on edges of outer webs, tipped buff. FIRST IMMATURE WINTER. Like juvenile, but feathers of crown and hindneck broader, with darker, broader, glossy blue tips, contrasting with dirty white chin and cheeks and mottled brown throat. Underparts often whiter. FIRST IMMATURE SPRING AND SUMMER. Like first winter, but some filoplumes on head and neck early spring which soon disappear. Gradual moult of all body feathers to 2nd winter plumage follows, including replacement of juvenile mantle and wing-coverts; start of wing and tail moult during late summer. SECOND IMMATURE WINTER. Upperparts like adult non-breeding, but less glossy, and feathers of crown and hindneck edged brown. Underparts lack blue gloss, except sometimes on breast; feathers dark brown with black subterminal spot and white bases, giving mottled appearance on belly; chin feathers white, narrowly edged brown. SECOND IMMATURE SPRING AND SUMMER. Like 2nd winter, but more glossy on crown and neck, a few short filoplumes on crown, neck, lower throat, and thigh; breast and belly dull blue-black, white feather-bases partly visible. SUBSEQUENT PLUMAGES. Like adult. In breeding plumage, white elongated feathers in younger birds relatively short and narrow, larger and more numerous when older.

**Bare parts.** ADULT. Iris emerald green. Upper mandible and tip leaden or olive-grey, culmen black, lower mandible yellow at base. Skin round eye variable; dark grey and grey, yellow and black, or blue-black; cinnamon to orange triangular spot below eye. Lores yellow to orange; gular pouch black, with granular yellow spots. Leg and foot black. NESTLING. Skin brown-black. Bill-tip black; base of bill, skin round gape and pouch pink. Iris grey-brown. JUVENILE. Iris grey-brown, changing to green after 1 year. Bill yellow with black or dark horn culmen. Naked skin yellow. Leg and foot black.

**Moults.** ADULT POST-BREEDING. Complete; moult of primaries serially descendant each month July–December about one feather of each moulting series replaced; 2, rarely 3, active series in one wing. Tail moult irregular, feathers moulted more or less alternating, mainly June–December. Body plumage and apparently completely moulted same time, starting with head and neck. ADULT PRE-BREEDING. Feathers of head, neck, and part of body replaced January–April. YOUNG. From August of 1st to December of 2nd calendar year, body plumage moulted almost continuously. Various moults may theoretically be distinguished, but in practice hard to separate, as one moult may continue on some part of body while next already started elsewhere. Following may, for convenience, be recognized. POST-JUVENILE. Partial; August–December. Head, neck, upper mantle, part of breast and belly. Much individual variation. IMMATURE SECOND CALENDAR YEAR. Complete; body about February–December, wing and tail June–December. Descendant moult of primaries slow; in December, *c*. 6 juvenile outer primaries left and often also part of juvenile tail-feathers. In spring of next year, new series of primaries starts with innermost, and moult of juvenile primaries continues; juvenile p10 sometimes retained until spring of 4th calendar year. SUBSEQUENT MOULTS. Like adult. (Palmer 1962; Witherby *et al.* 1940; CSR; interpreted by reference to Potts 1971.)

**Measurements.** Bill length taken from feathers at side; depth is lowest depth of bill in middle. Skins (RMNH, MNHN, ZMA). *P. c. carbo.* Full-grown, Iceland, Norway, Finland, Britain, Brittany.

| | | | | | | |
|---|---|---|---|---|---|---|
| WING | ♂ 357 | (5·19; 6) | 350–363 | ♀ 339 | (9·27; 12) | 318–351 |
| TAIL | 154 | (6·44; 6) | 141–159 | 150 | (8·00; 12) | 135–163 |
| BILL | 69·6 | (2·52; 6) | 67–73 | 63·7 | (2·58; 12) | 59–68 |
| DEPTH | 16·3 | (0·92; 4) | 16–18 | 14·1 | (0·45; 8) | 13–15 |
| TARSUS | 74·0 | (3·18; 6) | 68–78 | 70·6 | (2·35; 12) | 67–74 |
| TOE | 91·0 | (3·79; 6) | 85–95 | 83·7 | (3·50; 12) | 78–91 |

*P. c. sinensis.* Breeding adults Netherlands.

| | | | | | | |
|---|---|---|---|---|---|---|
| WING | ♂ 347 | (8·04; 38) | 330–364 | ♀ 325 | (6·49; 18) | 311–337 |
| TAIL | 155 | (5·21; 38) | 145–165 | 144 | (5·54; 18) | 133–154 |
| BILL | 62·6 | (2·84; 34) | 58–67 | 55·7 | (3·00; 15) | 50–58 |
| DEPTH | 14·0 | (0·82; 34) | 13–16 | 11·8 | (0·59; 14) | 11–13 |
| TARSUS | 69·4 | (2·33; 38) | 66–73 | 66·1 | (1·84; 18) | 64–70 |
| TOE | 84·8 | (2·47; 38) | 80–90 | 79·7 | (2·50; 18) | 76–85 |

Sex differences significant, except tail and tarsus in nominate *carbo*. Juvenile measurements on average similar to adults, but some have shorter tarsus and toe, and juvenile *sinensis* has significantly longer tail and more slender bill:

| | | | | | | |
|---|---|---|---|---|---|---|
| TAIL | ♂ 160 | (6·64; 36) | 150–171 | ♀ 148 | (6·54; 19) | 135–161 |
| DEPTH | 13·2 | (0·97; 23) | 12–15 | 11·4 | (0·78; 17) | 10–12 |

*P. c. maroccanus* (1) and *P. c. lucidus* (2). ♂♂ (MNHN; Vaurie 1965).

(1) WING 335 (15·7; 4) 314–355  BILL 66·8 (3·31; 4) 62–70
(2)      334 (–  ; 7) 321–349       67·5 (–  ; 7) 64–70

**Weights.** Nominate *carbo*. ADULT. ♂♂: March, Iceland, 3600; April, France, 3170. JUVENILE. ♂: October, Norway, 3250. ♀♀: September, Scotland, 2127; October, Norway, 3490. *P. c. sinensis.* ADULT. East Germany, April (1) and June (2):

(1) ♂ 2423 (86) 2020–2810 ♀ 2085 (38) 1810–2555
(2) 2283 (36) 1975–2687 1936 (17) 1673–2174

Various others. ♂♂: April, Netherlands, 3042; October, East Germany, 3180; winter, Lower Rhine, 2345. ♀♀: February, Netherlands, 2123; January 3rd calendar year, Switzerland, 1860. JUVENILE. ♀♀: winter, Lower Rhine, 1730, 2270; January, Switzerland, 1732, 1757. Sex unknown: October, Netherlands, 6

average 2216, range 1570–2770. NESTLING. At hatching, 28–30; just before fledging, 1938–2515. (Bauer and Glutz 1966; Koeman *et al.* 1972; BTO; MNHN, ZMA.)

**Structure.** Inner half of wing relatively long and broad; tip short and pointed. 11 primaries: p9 and p8 longest and about equal, p10 6–12 shorter, p7 4–16, p6 35–50, p1 115–143; p11 minute, concealed by primary coverts. P7–9 emarginated outer web and p8–10 emarginated inner web. Tail rounded; 14 narrow feathers; t7 40–65 shorter than t1. Tips of flight- and tail-feathers rounded in adult, pointed in juvenile. Bill strong, heavy, sharply hooked; culmen straight, sides of bill somewhat scaly in adult, smooth in juvenile. Nostrils closed, present only in first days of life. Tarsus short, reticulate; toes long, 4th *c.* 16% longer than middle, 2nd *c.* 30% shorter, 1st *c.* 54%; all 4 connected by webs; 1st directed inwards; claws strong, short, curved; inner edge of 3rd pectinated.

**Geographical variation.** *P. c. sinensis* smaller than nominate *carbo*, especially bill; plumage glossed blue-green rather than blue-purple, but variable and some *carbo* have green-glossed chest, some west European *sinensis* slight purple gloss; difference in gloss between juveniles of both subspecies often visible on back and rump. *P. c. sinensis* has more white plumes on head and neck, but number dependent on age, and, rarely, old *carbo* have as many as typical *sinensis* (Stokoe 1958). East and south-east populations of *sinensis* (those of Japan sometimes named *hanedae* Kuroda, 1925) average smaller in size but much overlap. *P. c. lucidus* like *sinensis*, but smaller and greener; throat, breast, and sometimes belly white. *P. c. maroccanus* intermediate between *sinensis* and *lucidus*; throat and upper chest white (Mackworth-Praed and Grant 1970; Vaurie 1965). Populations of *lucidus* in east Africa dimorphic: some with neck and chest black, strongly resembling *sinensis* (described as separate subspecies *patrick:* by Williams, 1966); others with much white from throat to breast. Dark specimens most frequent in east Zaire and west Uganda, proportions decreasing to east and south (Urban and Jeffard 1974). In view of this variation, and occurrence of many intermediates, recognition of subspecies based on amount of white on neck in east Africa not warranted.          CSR

---

## *Phalacrocorax aristotelis* Shag

PLATE 22
[after page 230]

Du. Kuifaalscholver    Fr. Cormoran huppé    Ge. Krähenscharbe
Ru. Хохлатый баклан    Sp. Cormoran moñudo    Sw. Toppskarv

*Pelecanus aristotelis* Linnaeus, 1761

Polytypic. Nominate *aristotelis* (Linnaeus, 1761), north and west Europe; *riggenbachi* Hartert, 1923, north-west Africa; *desmarestii* (Payraudeau, 1826), Mediterranean and Black Sea.

**Field characters.** 65–80 cm, ♂ averaging larger; wing-span 90–105 cm. Medium-sized, slender-billed, marine cormorant with oily-green plumage and yellow gape. Sexes alike but noticeable seasonal differences; juvenile easily distinguishable.

ADULT BREEDING. Mainly oily black glossed dark green, but head and neck glossed dark blue-green and mantle, scapulars, and wing-coverts purplish, latter feathers bordered with velvet-black to give scaled effect across mantle and wings. Forward-curving crest usually con-

spicuous. Bill black with yellow base to lower mandible, gape orange-yellow; eyes bright green; legs and feet black. ADULT NON-BREEDING. Duller and browner; no crest. Chin white to brownish-white and throat brown. JUVENILE. Medium to dark brown above with inconspicuous green gloss; scaly effect on mantle and scapulars but pale tips to wing-coverts. Paler on sides of head and generally paler brown below, with small but variable areas of brownish-white on chin and in centre of throat, breast, and belly, though Mediterranean race *desmarestii* (and some individuals elsewhere) wholly whitish to brownish-white below. Base to lower mandible pink-brown, gape pale pink or tinged yellow; eyes pale yellow; legs black with brownish-white webs.

At close range adult unmistakable and immature differs from young Cormorant *P. carbo* in smaller size, slimmer build, much slenderer bill, and (except for *desmarestii*) much less white on brown breast, though sometimes white spot on chin. At longer ranges, sometimes difficulty in distinguishing these 2 species, particularly immatures but also adults if white patches on *P. carbo* cannot be seen. At all times, however, slimmer build, shorter neck, smaller head, and faster wing-beats of *P. aristotelis* help identification.

Although generally exclusively marine, in some areas immatures more often seen inland in winter than *P. carbo*. Behaviour much as latter, but perches almost exclusively on rocks, and seldom on posts, piers, buoys, and other man-made structures, though exceptionally, in some areas of Britain, on trees and buildings inland, notably on and inside cooling towers of power stations. Though prefers sheltered waters for feeding generally more at home in rough seas than *P. carbo* and when diving usually does so with more marked forward spring. Often perches in less upright attitude than other west Palearctic *Phalacrocorax*; frequently adopts spread-wing posture when loafing. Generally flies low over water with rapid wing-beats, sometimes interspersed with glides, and neck fully extended; less inclined than *P. carbo* to fly high or overland. Swims low in water with neck erect and bill lifted diagonally upwards. Breeds colonially and often roosts in large groups; flocks to exploit fish shoals, but otherwise often seen singly and less frequently than *P. carbo* in straggling parties or lines, except on migration when flying in distinctive, loose, goose-like flocks at fair height over sea.

**Habitat.** Essentially marine, not usually ranging far from coast; both offshore and inshore, especially along rocky coastlines or island groups fronting fairly deep water and, except in Mediterranean, under direct oceanic influence. Only exceptionally to or beyond continental slope. Stops well short of ice, but flourishes in cool, as well as fairly warm waters. Prefers less turbulent and more sheltered fishing grounds such as bays or channels. Usually however, avoids estuaries, shallow sandy or muddy inlets, and brackish or fresh water. Also, like Cormorant *P. carbo*, spends much time resting, drying, oiling, and preening on waterside vantage points such as rocks, including skerries, stacks, ledges near foot of cliffs, talus, and less frequently on artefacts, shingle spits, or sandbanks; not usually on trees. Although overlaps on some coasts, rarely shares nesting sites with *P. carbo*, preferring more sheltered and shady overhung ledges, fissures, or caves to open rock tables. Accomplished and sustained swimmer and fisher, diving to *c.* 4 m, but essentially land-based. Flies mostly

within lower airspace up to *c.* 100 m and normally nests no higher, although exceptionally to 200 m or more. Less given than *P. carbo* to daily or seasonal movements overland or to flying in formation, although commonly in loose skeins passing low over water to and from fishing grounds. Owing to less accessible habitat, less frequently in contact with man on land or water than *P. carbo* in most localities, and less in competition for same food fishes (see Lack 1945), but still often mistakenly persecuted; reacts tenaciously and proves difficult to extirpate.

**Distribution.** No recent changes in range except for probable extension east along Murman coast, USSR (Dementiev and Gladkov 1951).

Accidental. Belgium, Netherlands, Finland, Switzerland, Rumania, Israel, Iraq, Egypt, Malta. West Germany: almost annually Heligoland; rare elsewhere, including inland (Bauer and Glutz 1966).

**Population.** BRITAIN AND IRELAND. In 1969–70, *c.* 31 600 pairs, majority Scotland; recent declines in a few areas and increases, some marked, in many others, especially east coast of Britain (Cramp *et al.* 1974). Population between Firth of Tay (Scotland) and Humber (England), grew steadily 1905–65 from 10 to 1900 pairs; increase of 11% a year in eastern Britain attributed mainly to relaxation of human persecution (Potts 1969). NORWAY: has become much scarcer in some areas; for details of numbers at some colonies, including one of at least 1000 pairs, Rundöy, 1966, see Haftorn (1971). FRANCE: Brittany 1550–1600 pairs (Brien 1974); *c.* 300 pairs elsewhere (Y Brien).

Survival. 1st year survival rate Farne Islands 59 ± 12%, but higher on west and south coasts of Britain, late hatched chicks being at disadvantage; survival after 1st year averages 84 ± 5%, varying greatly from year to year, with average rate for ♂♂ higher than for ♀♀ (Potts 1969). Oldest ringed bird 15 years 10 months (Rydzewski 1973).

**Movements.** Dispersive, some adults perhaps resident. All populations of nominate *aristotelis* show coastwise dispersal, though this not normally so extensive as in sympatric Cormorant *P. carbo carbo* for *P. aristotelis* more closely associated with rocky coasts. Icelandic population believed resident. Those breeding Murmansk partially resident, but some move south-west around Norwegian coasts as far as 65°N (Dementiev and Gladkov 1951). Movements of Norwegian breeders better known as more intensively ringed (see Johansen 1975); pronounced autumn/winter dispersals, mainly southwards, and evidence of colony-specific wintering grounds without obvious age-class differences such as occur in Britain (see below); mean distances for January recoveries increase with latitude of colony, e.g. Lepsøy in Sunnmøre (62°37′N) 112 km, Røst in Nordland (67°26′N) 486 km, Bleik in Vesterålen (69°17′N) 600 km; Finnmark and Troms coasts may be largely vacated in winter—southward

recoveries up to 1300 km. Rudøy (62°24′N) birds exception, having shifted wintering grounds to north-east of colonies since 1960, corresponding with change in spawning grounds of herring *Clupea harengus* (Johansen 1975). No ringing evidence of emigration by Norwegian breeders, though probably stragglers to Swedish and south Finnish waters from Norway to Murmansk breeding range. Breeders of Brittany and Channel Islands also have restricted winter ranges, and seldom cross English Channel.

In Britain and Ireland, movements during first winter greater than subsequent winters, though displacements, north and south, mainly within breeding range; on east coast southward movement predominates in autumn and

winter, northward typical of non-breeders in summer; no true migration (Coulson 1961). Minority, mainly from Scotland, penetrate north to Norway; while rather more, mainly from Ireland, Irish Sea and Bristol Channel, move south in autumn and winter to Brittany, rarely to Spain; one atypical recovery of Welsh bird in Valencia on Spanish Mediterranean. North Sea appears to act as barrier to eastward movement, though very few recoveries from east Scotland and east England to Denmark, West Germany, Netherlands and Belgium where species rare visitor; possible exception Heligoland Bight where may be regular in small numbers (Bauer and Glutz 1966), origins uncertain. Degree of exodus from colonies variable, e.g. virtually absent from Lundy (Devon) October to late

February (Snow 1960); but adults present all year on Farne Islands (Northumberland). Juveniles begin dispersing soon after fledging and most gone from Farne Islands by mid-October when influx of juveniles from other colonies (Potts 1969). Sudden food shortages after periods of onshore winds can turn dispersal into eruption; then immatures travel further afield, and weakened birds sometimes in flocks occur on low-lying coasts and inland, mortality increasing (Potts 1969). Such eruptive behaviour apparently more typical of, but possibly just better documented for, British east-coast birds. Most juveniles return to natal colony to breed, but 8% of those reared Farne Islands surviving to adulthood breed elsewhere (Potts 1969).

*P. a. desmarestii* of warmer waters mainly resident, but some disperse; movement in Black Sea follows fish shoals (Dementiev and Gladkov 1951). Extent of dispersal little known, but one ringed Corsica moved 130 km to Sardinia and species occurs at intervals on coasts where does not breed in Mediterranean, Black Sea, and apparently Sea of Azov (Dementiev and Gladkov 1951; Bauer and Glutz 1966; Etchécopar and Hüe 1967).

**Food.** Chiefly and often entirely fish caught between mandibles under water and brought to surface, though occasionally swallowed below, especially if small. Typically surface-diver, but plunging from air, especially in rough water, also recorded (King 1972). May spring clear of water before diving or, particularly when laden with fish, slide under surface without preliminary leap. 155 dives: 5–100 s with mean of 40 s (Lumsden and Haddow 1946), maximum submergence, 3–4 mins (Dementiev and Gladkov 1951). Typically solitary feeder, though flocks of 300–500 noted following dense fish shoals (King 1972). Feeding may be some distance from roosting or nesting areas: 18 km (Pearson 1968), 13 km (Rees 1965).

Fish chiefly midwater and to lesser extent bottom-living species from coastal and estuarine areas (compare Cormorant *P. carbo*). Diet varies with season and locality but mainly sand-eel Ammodytidae, Clupeidae (herring family), and Gadidae (cod family); less often shore-fishes and crustaceans. 188 stomachs from Cornish waters, England, contained (by frequency) 51% sand-eel and 11% Clupeidae, especially sprat *Sprattus sprattus* (Steven 1933; Lack 1945). In 20 stomachs from Dee Estuary, England, preference found for small Gadidae, mainly whiting *Merlangius merlangus* and saithe *Pollachius virens* (Rae 1969). 24 regurgitations and 28 stomachs from Farne Islands, England, contained mainly sand-eels *A. tobianus*, *A. marinus* and less frequently Gadidae (*M. merlangus* and cod *G. morhua*) (Pearson 1968). In 176 from Angus coast, Scotland, 29% (by frequency) sand-eel, 14% Gadidae, and 5% Clupeidae; also a number of shore-living species, particularly viviparous blenny *Zoarces viviparus* (9%) and crustaceans (14%), mostly *Pandalus* shrimps (Rae 1969). In 12 stomachs from North and South Esk estuaries and

Montrose basin, Scotland, chiefly inshore fishes and mainly viviparous blennies (Rae 1969). Sand-eels again chief item (54·1% by volume) in 81 stomachs from southwest Scotland, with rather less Clupeidae, Blenniidae, and Crustacea (Lumsden and Haddow 1946), though marked differences in composition of diet depending on feeding locality, e.g. rocky sea floor or sandy sea floor. In 62 birds from Loch Ewe, Scotland, sand-eels not as frequently eaten as Gadidae (*G. morhua*, *P. virens*, and poor cod *Trisopterus minutus*) and herring *Clupea harengus* (Mills 1969). All recent work from British waters suggests that Collinge (1924–27) over-emphasized proportion of man-food fish in diet. Cod, herring, sand-eel, sculpin (Cottidae), and gobies *Gobius* recorded from Murman coast, USSR (Dementiev and Gladkov 1951); also capelin *Mallotus villosus* (Belopolskii 1957). Polar cod *Boreogadus saida* and cod recorded from Novaya Zemlya (Belopolskii 1957). In only freshwater analyses (Windermere, England), 5 stomachs contained 3 items: stickleback *Gasterosteus aculeatus*, trout *Salmo trutta*, and pike *Esox lucius* (Hartley 1948).

According to van Dobben (1952), only 1 meal a day taken, but probably more strictly follows tidal cycles and may feed twice a day. Mills (1969) estimated daily food consumption as 246 g, or 13·5% of body weight; Rae (1969) found maximum stomach content weight of 195 g, or 10·2% of body weight, but average of only 51 g.

**Social pattern and behaviour.** Based mainly on material supplied by B K Snow; see especially Snow (1960, 1963) for population breeding at Lundy, England.

1. Less gregarious than Cormorant *P. carbo*. Often solitary in winter and away from nesting colony, occasionally flocking at fish shoals with minimal social interaction. BONDS. Monogamous pair-bond normal but where shortage of good sites, as on Farne Island, simultaneous bigamy in 3–5% of breeders (G R Potts). May pair with same mate in successive seasons, especially if breeding successful. First-time breeders, generally aged 2 (♂♂) or 3 (♀♀), pair up later than older birds, in May instead of January–March. Some 1st year birds also pair and try to hold nest-sites. At some colonies, e.g. Farne Islands (G R Potts), some pair-bonds may be prolonged and maintained except during autumn moult and early winter; no evidence that any birds remained paired through non-breeding season at Lundy. BREEDING DISPERSION. Usually in small, loose colonies, though sometimes breeds densely. Defends only nest-site territory. Occupancy of sites extremely variable, e.g. none from mid-August to early March at Lundy, but site re-occupied from November at Farnes (G R Potts). ROOSTING. When nest-site re-established, first ♂, later pair, roost there, and one or both do so throughout breeding season. When colony deserted, roost on stack rocks or cliffs near sea; large numbers may use same rock, but some roost alone on narrow ledges.

2. Away from immediate vicinity of colony social interactions extremely uncommon. ANTAGONISTIC BEHAVIOUR. At colony, aggression involves pointing bill and looking directly at rival. Away from site, bird (usually ♂) thrusts head out with bill partly open, uttering Ark-call, while approaching object of aggression with wings held slightly away from body and head and neck feathers erected. Aggressive postures at site, when bird unwilling to move away, performed by both ♂ and ♀. Bird first bill-points

with head-shake and stretches neck out, then sits, raises tail and
periodically makes lateral head movements while holding nest
material (Nest-quivering). With increase of aggression, sitting
bird raises wings from body, draws head and neck back over body
with bill closed, then suddenly stretches head forward, opening
bill wide, so displaying yellow gape, and shaking head (Threat-
gape). Ark-call accompanies Threat-gape of ♂; ♀ gives hissing
sound. In all appeasing movements avoids pointing bill at other
bird. HETEROSEXUAL BEHAVIOUR. Appeasement between paired
birds achieved by more ritualized behaviour. In least ritualized
attitude of submission, by recently paired ♀♀ (and also by
juveniles), bill hidden by turning back of head towards aggressor.
Landing-gape (see A) main appeasing posture of mature birds of
both sexes; always associated with movement, either when
landing from a hop forwards or from flight. Body held in upright
position with feathers flattened, head thrown back and beak,
opened wide, pointed upwards. At start of breeding season, all
mature birds on rocks below colony land or hop about with
Landing-gape when near others, but primarily both members of
pair perform Landing-gapes when moving to and from site.
Advertising display, by ♂♂ only, consists of drawing head back
over body and rapidly and repeatedly darting it upwards and
forwards, showing gape (Dart-gape; see B). At closer approach,
this followed by Throw-back (see C) with neck laid backwards
along back, beak pointed skywards. When ♀ beside ♂, he moves
head from Throw-back to Bow (see D), a sign of acceptance of ♀,
and she preens his head and neck. Sequences of ♂ advertising
display done at nest-site or on rocks near sea below, though at
some colonies, e.g. Farne Islands (G R Potts), performed almost
exclusively at site. After courtship on rocks near sea, ♂ shows ♀
site, flying ahead and then hopping up to it with Landing-gape;
he again advertises while sitting at nest and ♀ approaches in same
manner as previously. In early stages of courtship, ♂ always sits at
site and several different ♀♀ may visit; more frequent visits by
one ♀ establish bond which culminates in ♂ allowing ♀ to sit at
nest; soon after, first copulation occurs. Greeting ceremony of
either sex on nest consists of Sitting-gape and Bow. In Sitting-
gape, bird displays gape upwards with lateral head movement
similar to Nest-quivering; in Bow, tail cocked acutely and bill
pressed into breast. Sitting-gape may be followed by Nest-
quivering instead of Bow, with beak in rim of nest. At nest-relief,
incoming bird often brings material which is usually taken by
partner, who may hold it upwards with a quiver followed by
actual Nest-quivering at rim of nest; this replaces Sitting-gape as
greeting. When incoming bird gains possession of nest, it
immediately does Sitting-gape. Greeting ceremony of approaching bird, continuing after it has arrived and stood by partner,
Throat-clicking (see E); performed with plumage flattened and
head, neck, body and tail in semi-horizontal line held over back or
neck of partner, Clicking made in throat with hyoid lowered, so
distending gular pouch. Throat-clicking also performed by ♀ in
early stages of courtship when first approaching ♂. Mounting and
copulation between paired birds occurs frequently, normally at
nest with ♂ usually in upper position. ♀ solicits with Sitting-gape
and Bow; ♂, who has been standing besides her, Throat-clicks,
hops on her back with Landing-gape, grips back of her neck with
beak, and copulates, uttering Ark-call from mounting stage
onwards. Allopreening frequent during courtship and fairly
common during incubation. Normally, standing bird preens
sitting bird, on head, neck or dorsal surface of body, so in early
stages of courtship most frequently ♀ that preens ♂. After pair-
formation complete, either sex may preen other. RELATIONS
WITHIN FAMILY GROUP. Food-begging by young consists of
holding head up and waving it from side to side while giving

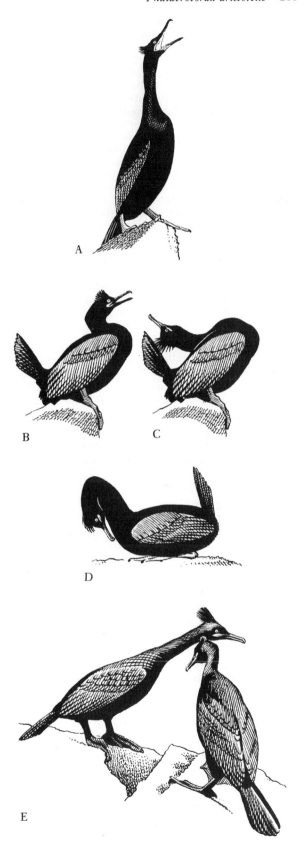

A

B     C

D

E

Food-call. When capable, chick always stands holding head at right angles to outstretched neck of parent, just below gular pouch. After young have left nest, they wing-wave when begging. Parent presents back of head towards full-grown, begging young just before feed. While still in nest, chick (from age of 15–20 days) will Threat-gape followed by Nest-quivering as aggressive response to intruders. Between 20–30 days, returning parent greeted with an 'Upward-gape', forerunner of Sitting-gape; also response to parent or sibling preening it. Will also take material brought by parent at nest-relief. While in nest, chicks frequently preen each other and parents. For last 10 days in nest and when first out of it, young curious; investigate anything that moves, mandibulating strange objects and snatching them from one another. When first at sea, they chase each other under water, pulling each other's tails and feet.

(Figs A–E from photographs and drawings in Snow 1963.)

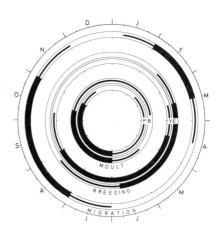

**Voice.** Vocabulary of adults, especially ♀, limited (Snow 1963, on which following account based). Both sexes produce clicking sounds from throat, mainly at nest where typically performed by standing partner while other sits; also given as greeting to mate and young.

CALLS OF MALE. Besides Throat-clicking, main vocalization grunting 'ar(k). . . .' given in long continuous series, interspersed with clicks—'ar(k)-ik -ar(k)- ik -ar(k)-ik. . . .' (see I); at loudest, audible for at least 300 m. Aggressive and self-assertive, uttered in various social situations in nesting area, both on ground and in flight; apparently confined to breeding season, initially frequent, then gradually decreasing, ceasing when young independent. Muted version given during copulation.

CALLS OF FEMALE. Except for Throat-clicking, and hissing in Threat-gape display, ♀ voiceless.

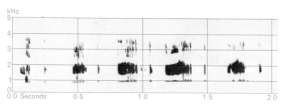

I C Fuller Wales July 1964

CALLS OF YOUNG. Calls of ♂ and ♀ indistinguishable until ♀♀ start losing voices at 35 days or more. 3 main calls. (1) Squeaking-call. 'Wee-ik' lasting c. 0·5 s; first uttered from piped egg. Basically complaint-call in tiny chicks, given with bill closed; later used as greeting with open bill. (2) Begging-call. Repeated 'weeeu . . .', lasting 0·75 s with 0·25 s intervals between notes, later becoming louder and longer 'wee-ee-ee. . . .'; in juvenile audible 50 m. (3) Threat-gape call. Frequently repeated, husky 'wee-aa', with emphasis on 2nd syllable, each note lasting c. 1 s. Uttered during chicks' versions of Threat-gape display. In ♀♀, at c. 30–35 days becomes huskier and more like hiss of adult ♀; in ♂, becomes louder.

**Breeding.** Based mainly on Snow (1960); Lundy Island, England. SEASON. See diagram for British Isles and southern Norway. In USSR, eggs laid early May (Dementiev and Gladkov 1951). Records of laying November and January in Britain thought to be early rather than late clutches (Tong and Potts 1967). SITE. On cliff ledges from just above high water mark to over 100 m. Normally sheltered; often just inside small caves. Ledges as small as 30 × 30 cm used. Also nests on boulder beaches, underneath large boulders. Colonial, but nests usually well spaced. Sites re-used in successive years by same birds. Nest: heap of vegetation with thicker items (bracken, seaweed) as base; lining of finer material (grasses, etc.). Average 45–55 cm diameter, 15 cm high (Dementiev and Gladkov 1951). Building: by both sexes. ♂ starts on arrival and selects site. Main building activity begins c. 36 days (but up to 70) before laying. Material added during incubation and rearing, with most just after laying of first egg. EGGS. Elongated oval; pale blue, overlaid with whitish chalky deposit, readily becoming stained during incubation. 63 × 39 mm (52–72 × 35–41), sample 100 (Schönwetter 1967). Weight 49 g (37–63), sample 272. Within clutch, 1st egg smallest, last egg next in size; size, measured by breadth and volume, increases with increasing ♀ age (Coulson et al. 1969). Clutch: 1–6. Mean of 447 clutches, Lundy, 3·07 (range 1–6), with 76% of 3. One brood. Replacement clutches laid after egg loss, especially if 1st clutch not complete. In 10 nests where eggs lost before clutch complete, interval between laying of 1st egg of 1st clutch and 1st egg of 2nd clutch 17·3 days (8–29). In 2 nests where eggs lost after 20 and 26 days incubation, interval between loss and re-laying 21 and 22 days. Eggs laid at 3-day intervals, occasionally 4–5 days, rarely more. Most eggs laid at night or early morning. INCUBATION. 30–31 days. By both sexes, changing about twice daily. Starts with 2nd egg, though 1st covered. Hatching asynchronous, c. 1 day apart. Eggshells removed. YOUNG. Altricial and nidicolous. Cared for and fed by both parents. Fed by partial regurgitation with young putting bill inside parents'. Parents defend young against avian predators. FLEDGING TO MATURITY. Fledging period 53 days (48–58), sample 35. After fledging young continue to be fed for c. 20 days, but sometimes up to 50 days with one record of 14 months; independent when feeding ceases. Age of first breeding 4 years, some at 3. BREEDING SUCCESS. Of 893 eggs laid in 4 seasons, Lundy, 71% hatched (range 69–73%), and 87% of these fledged (67–95%); mean young reared per nest 1·87 (1·32–2·25). Main egg losses due to infertility and predation. Losses of young due to falling from nest, starvation, disease, and predation. 47% of young lost in days 1–10, 18% in days 11–20 and 21–30, 13% in days 31–40, 5% in days 41–50. Higher loss from nests on ledges with vertical drop beneath. Marked correlation between size of nest structure and breeding success, as small, poorly built nests allowed eggs or young to fall out.

**Plumages.** Nominate *aristotelis*. ADULT BREEDING. All black, strongly glossed metallic green on body, blue-green on head and neck. Crest of upcurved feathers on forecrown. Feathers of mantle and scapulars and upper wing-coverts evenly and narrowly edged velvety black, centres slightly tinged purple. Tail black, glossed purple-green; flight-feathers dull black, slightly glossed green on outer webs; inner webs brown. Axillaries and under wing-coverts sooty black, slightly glossed green. In fresh plumage, numerous tiny white filoplumes between feathers of neck. ADULT NON-BREEDING. Like breeding, but duller black, with less green gloss; chin buff or white, throat mottled brown. No crest. NESTLING. Naked at hatching, skin brown. After a few days, dense, brown down with pale grey bases develops. Off-white down on throat later. Lores, area round eye, and area round base of mandibles bare. Fully feathered at 5–6 weeks. JUVENILE. Whole upperparts and flanks dark brown with faint green gloss; feathers of mantle, scapulars, and upper wing-coverts evenly edged dark brown (edges broader than in adult), tips buff when worn. Chin pale grey or off-white, throat buff-brown. Sides of head, neck, breast, and flanks dark brown, but vent and centre of breast and belly washed buff, palest on vent; feather-bases grey. Under tail-coverts grey-brown. Whole underparts (except chin) appear uniform buff-brown to pale buff, with slightly greyer central belly and vent, in contrast to juvenile Cormorant *P. carbo*, in which centre of breast, belly, and vent usually variably mottled sooty-black and white. Tail-feathers dull black, narrowly edged buff when fresh. Flight-feathers like adult, but duller, less glossy, narrowly tipped buff. Upper wing-coverts dark brown, slightly glossed dark green in centres and tipped buff; lack evenly black edging of adult. Underwing and axillaries sooty brown. IMMATURE. Gradually acquired from 1st winter to 2nd autumn; overlaps with juvenile and 1st adult plumage; see Moults. Upperparts resemble adult, but some brown mottling of feather-bases on head; edges on mantle, scapulars, and upper wing-coverts broader and less deep velvety black; scapulars and wing-coverts wide and rounded at tips, rather than narrow and pointed as juvenile. Underparts darker brown than juvenile, particularly on flanks; often faint green gloss on sides of neck, lower neck, and flanks. Chin pale buff, throat and foreneck dark brown, speckled by off-white feather edges and bases. In mid-summer of 2nd calendar year, belly, vent, and part of back and rump still juvenile, contrasting with darker immature feathers on rest of body. Immatures in worn plumage have underparts mottled by pale buff feather edges. FIRST ADULT PLUMAGE. Acquired gradually between 2nd and 3rd autumn; some specimens, however, like adult in March of 3rd calendar year. In mid-summer, typically black with green gloss from chin to lower breast, dark brown on belly and vent. When in adult plumage, some brown of feather-bases often visible on underparts; some

worn brown immature feathers on underparts or wing-coverts occasionally retained to spring of 4th calendar year. (Potts 1971; Witherby *et al.* 1940; Bauer and Glutz 1966; C S Roselaar.)

**Bare parts.** Nominate *aristotelis*. Iris bright emerald green. Bill black, cutting edges and small area at base of lower mandible yellow. Skin round eyes, inside mouth, and gape pale orange-yellow; skin of chin and base of lower mandible black, thickly spotted yellow. Foot black with paler webs. NESTLING. Iris light brown; bare skin brown. Foot dark brown. JUVENILE. Iris yellow-white. Bill pale pink-brown, culmen black. Bare skin round eyes, inside mouth, and at base of mandibles pale flesh with yellow tinge. Foot like adult, but webs and innerside of tarsus pale flesh-brown to yellow-brown. In 2nd summer, iris yellow-green and bare skin at face bright yellow; bill like juvenile; foot dark brown. Adult colour attained in next year. (Bauer and Glutz 1966; Dementiev and Gladkov 1951; Witherby *et al.* 1940; RMNH.)

**Moults.** ADULT POST-BREEDING. Complete; primaries serially descendant. Starts June–August, 1 primary replaced per month. Arrested from November; continued from February at slower rate than in autumn; speeds up after breeding, when also new series starts. Mostly 2 active series in wing, sometimes 3 when p10 replaced after 2 winters. Head and neck from June, followed by rest body from August; finished January. ADULT PRE-BREEDING. Partial; overlaps with post-breeding. November–February. Plumage of head and part of body renewed again. POST-JUVENILE. Complete; lasting over a year. Primaries serially descendant from May in 2nd calendar year, sometimes earlier; mostly finished December, then arrested. 30% of ♂♂ and 70% of ♀♀ retain p10 until next summer, 2% of ♂♂ and 4% of ♀♀ until summer of 4th calendar year. Head and neck start in 1st year during September, followed by mantle and some feathers elsewhere in late autumn; arrested during winter. Continues in spring, proceeding from head towards tail; mostly finished late autumn. Tail-feathers moult in irregular sequence during whole 2nd calendar year. Upper wing-coverts moulted with body, but greater moult at same time as corresponding flight-feathers. POST-IMMATURE. Complete; wing from summer of 3rd calendar year, timing like adult wing. Body from July; arrested in winter, finished by November 4th calendar year; in exceptional cases, ♂♂ already completed by March. A few brown feathers, especially on wing and underparts, sometimes retained until 5th calendar year. (Potts 1971; Witherby *et al*, 1940; J Wattel, C S Roselaar.)

**Measurements.** Nominate *aristotelis*. Britain, Brittany, and Netherlands. Skins (RMNH, MNHN, ZMA).

| | | | | | | |
|---|---|---|---|---|---|---|
| WING | ♂ 271 | (5·31; 12) | 261–278 | ♀258 | (4·78; 18) | 251–269 |
| TAIL AD | 129 | (5·40; 6) | 119–133 | 119 | (4·07; 10) | 114–125 |
| JUV | 137 | (5·83; 6) | 129–143 | 127 | (8·10; 7) | 114–138 |
| TARSUS | 64·7 | (2·07; 12) | 62–68 | 62·0 | (1·99; 18) | 58–65 |
| TOE | 78·1 | (2·71; 12) | 75–82 | 75·1 | (2·47; 19) | 70–81 |

Sex differences significant. Difference adult and juvenile tail significant. Other measurements of adult and juvenile similar; combined. Bill length (from feathering at side) and depth of bill (in middle) locally variable: (1) Norway, (2) north Scotland, (3) south-west England, Brittany (RSME, ZFMK, MNHN, RMNH, ZMA; R Murray, M van den Berg, C S Roselaar).

| | | | | | | |
|---|---|---|---|---|---|---|
| BILL (1) | ♂ | 55·0(1·85; 21) | 51–58 | ♀ | 55·3(2·42; 18) | 51–59 |
| (2) | | 58·9(1·65; 14) | 56–61 | | 59·3(2·08; 11) | 57–63 |
| (3) | | 55·7(1·68; 10) | 53–58 | | 56·2(1·18; 5) | 55–58 |
| DEPTH (1) | | 11·9(0·76; 19) | 10·6–12·5 | | 10·1(0·60; 18) | 9·3–11·3 |
| (2) | | 11·6(0·80; 15) | 10·4–12·1 | | 9·9(0·95; 10) | 9·1–12·1 |
| (3) | | 10·5(0·32; 5) | 10·2–11·0 | | 9·2(0·71; 14) | 8·3–10·4 |

Bills of Scottish birds significantly longer; those of south-west England and Brittany significantly more slender.

*P. a. desmarestii*. West Mediterranean. Skins, including type (ZFMK, MNHN, RMNH; Vaurie 1965 for wing).

| | | | | | | |
|---|---|---|---|---|---|---|
| WING | ♂ 258 | ( — ; 12) | 243–271 | ♀249 | ( — ; 11) | 240–265 |
| BILL | 60·9 (2·17; 6) | 58–65 | | 63·2 (1·68; 5) | 61–65 | |
| DEPTH | 10·0 (0·36; 5) | 9·7–10·6 | | 8·7 (0·46; 5) | 8·2–9·3 | |

**Weights.** Nominate *aristotelis*. ADULT. ♂♂: July, 1930 (Hagen 1942); December, 1760 (Bauer and Glutz 1966); February, 2154 (MNHN).

**Structure.** 11 primaries: p9 and p8 longest, about equal, p10 shorter 6–10, p7 1–5, p6 15–24, p1 72–89; p11 minute. Tail rounded, 12 feathers: t6 35–50 shorter than t1. Bill slender, smooth, less heavily hooked than *P. carbo*; bills of some juveniles, however, do tend to be relatively heavily hooked. Tuft of *c.* 5 cm long upcurved feathers on forecrown in adult. Rest of structure as in *P. carbo*.

**Geographical variation.** Slight in north-west Europe; see Measurements. Mediterranean subspecies *P. a desmarestii* slightly smaller than nominate *aristotelis*, but bill longer and more slender; bare skin at base of lower mandible in adult more extensive, paler yellow; bill usually yellow except black culmen and tip; foot brown with yellow webs; crest on average shorter, sometimes absent. Juvenile with much white on underparts; throat, breast, and belly at least white, sometimes chin to under tail-coverts (but not flanks). North-west African *P. a. riggenbachi* combines body size and colour of bare parts of *P. a. desmarestii* with bill dimensions of nominate *aristotelis* (Hartert 1912–21).

CSR

---

## *Phalacrocorax nigrogularis* Socotra Cormorant

PLATE 23
[between pages 230 and 231]

Du. Arabische Aalscholver   Fr. Cormoran de Socotra   Ge. Sokotra Kormoran
Ru. Черногорлый баклан   Sp. Cormoran de Socotra   Sw. Sokotraskarv

*Phalacrocorax nigrogularis* Ogilvie-Grant and Forbes, 1899

Monotypic

**Field characters.** 77–84 cm, ♂ averaging larger; wingspan 102–110 cm. Medium-sized, rather long-billed, marine cormorant with largely dark plumage and only minor white feathering in breeding season. Sexes alike with small seasonal differences; juvenile readily distinguishable.

ADULT. Entirely black with purple gloss on head and dark green gloss on rest of upperparts; still blacker tips to bronzed wing-coverts and scapulars give scaled appearance. At onset of breeding season, tuft of white hair-like plumes behind eye, fine white streaks on rump, and white flecks on neck, but only last more than ephemeral and sometimes apparently present throughout year. Bill grey-black with paler tip and green tinge at base of lower mandible; bare skin of face blackish; legs black. JUVENILE. Generally grey-brown above, mantle feathers with darker centres and scapulars with pale edges, and dirty white below flecked with grey-brown. Bare face dull yellow.

Unlikely to overlap in distribution with any west Palearctic congeners except Cormorant *P. carbo* (considerably larger, with relatively shorter, heavier bill and white on face) and Pygmy Cormorant *P. pygmeus* (much smaller, with short bill, rounded head, and found on fresh water). Most like Shag *P. aristotelis* in size and proportions, and juveniles rather similar in coloration (but adult *P. aristotelis* distinguished by yellow gape, oily-green plumage, and lack of white in breeding season).

Behaviour much as other *Phalacrocorax*. Apparently exclusively marine and generally described as shy and wary. Highly gregarious, breeding in vast colonies. Parties rest on sandbanks or fly from one feeding ground to another in close V-formation, sometimes at wave height and sometimes at up to 150 m. When fishing, remains submerged 1–3 min (Meinertzhagen 1954).

**Habitat.** Accurate data inadequate. Tropical, marine on rocky islands; breeding colonially Persian Gulf on boulder area at corner of smallish island *c.* 160 ha among jumbled heap of coral conglomerate boulders just above high water mark (D A Scott). On Arabian south-east coast during monsoon July–August frequents cool upwelling water (below 20°C), close inshore where highest nitrogen and oxygen concentrations and richest phytoplankton recorded (Bailey 1966).

**Distribution.** Only known breeding areas now on a few islands in Persian Gulf (Meinertzhagen 1954; D A Scott), though suspected breeding on islets off Dhufar, Aden and Socotra (Bailey 1966). In west Palearctic, according to Meinertzhagen (1954), breeding Umm al-Maradhia and Quaru, off Kuwait, but not included as nesting there by Ticehurst et al. (1925); no longer nests on former and unlikely on latter, which now inhabited, though seen Kuwait area May–October (SH).

**Movements.** Complex dispersals in Persian Gulf where present all year, but difficult to separate seasonal from regular feeding movements. Degree of emergence beyond Strait of Hormuz uncertain; occurs Arabian Sea and Gulf of Aden, sometimes in large numbers, but possibly breeds there (see Distribution). Once collected southern Red Sea at Assab, Eritrea (Moltoni 1942).

**Food.** No specific information; probably mainly fish.

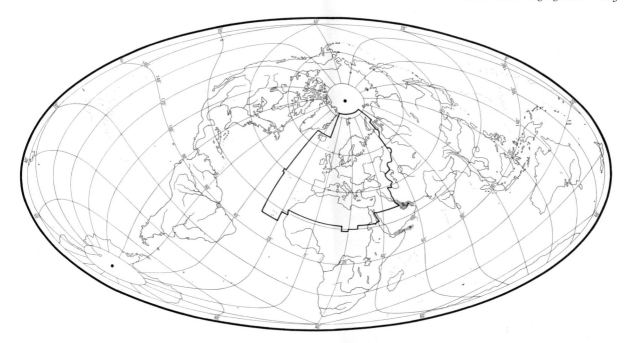

**Social pattern and behaviour.** Based on material supplied by R S Bailey.

1. Highly gregarious at all times. Flocks of up to 10 000 recorded at sea (Ticehurst *et al.* 1925). BONDS. Nothing known. BREEDING DISPERSION. Nests colonially in groups of up to at least 10 000 pairs. Territorial behaviour not studied but nests may be almost touching. On Arabi (Persian Gulf), 250 nests found in 1000–1500 m². Colonies may be confined to small parts of apparently suitable area (Ticehurst *et al.* 1925; Løppenthin 1951). ROOSTING. Communal outside breeding season, adults and immatures together; on coastal sand-spits in number of areas (Ogilvie-Grant and Forbes 1903; Bailey 1971).

2. Nothing recorded on behaviour.

**Voice.** Nothing known. No utterances so far reported, either at sea or at breeding colonies.

**Breeding.** Based mainly on Meinertzhagen (1954). SEASON. Last week January to mid-March. SITE. On ground on gravel and stony ridges of islands; among boulders. Colonial in tens of thousands; nests almost touching. Nest: scrape in gravel or slightly raised above surface as low heap of gravel. Building: no information. EGGS. Ovate; pale blue with chalky covering, dark umber-brown spots and blotches at large end. 57 × 39 mm (56–59 × 38–41), sample 6; calculated weight 48 g (Schönwetter 1967). Clutch: 2–3. No further information.

**Plumages.** ADULT. Head, neck, back, rump, and underparts velvet black with slight purple gloss. Mantle, scapulars, and upper wing-coverts glossy dark bronze-green with black spots to tips of feathers, except on greater upper wing-coverts. At some stage of annual cycle, probably before start of breeding, many white filoplumes among black feathers; dense on head and neck, sometimes forming tufts behind eyes, long on rump. During rest of year, filoplumes reduced to a few on neck. ♀ has apparently fewer filoplumes than ♂ (Bates 1938). Tail-feathers grey-black, becoming paler with wear. Flight-feathers bronze-green when fresh, secondaries with black shaft-streaks, primaries dark grey when worn, secondaries retaining some gloss. Under wing-coverts and axillaries brown-black. NESTLING. Naked at hatching. After a few days, sparse white down on back; later wholly covered in white down, except forehead, sides of face, chin and upper throat, underwing, and patch at side of breast. JUVENILE. Head and neck grey-brown paler on chin and throat, feathers edged off-white. Upperparts and upper wing-coverts grey-brown with pale feather margins. Feathers more pointed than in adult. Breast and belly white, breast sometimes slightly mottled grey-brown. Flanks grey-brown. Tail-feathers grey-brown with pale shafts and white tips. Flight-feathers dark grey, primaries with black shafts, secondaries with black shaft-streaks. Under wing-coverts and axillaries brown. IMMATURE. Head and neck dark brown, chin and throat paler. Feathers of mantle, scapulars, and upper wing-coverts dark grey with brown-black spot at tip and narrow white margin. Back, rump, and underparts dark brown-grey. Tail-feathers grey-black with black shafts. Flight-feathers grey-black without gloss. Age at which fully adult plumage reached unknown; probably 3rd calendar year.

**Bare parts.** ADULT. Iris green. Bill black or grey-black, paler towards tip; green band along basal half of mandible. Bare skin on face black with grey ridges; eyelid black with yellow warts; gular pouch black, sometimes brown or grey-green. Leg and foot black, webs browner; joints and toes sometimes tinged pink. Not known whether lighter colours on gular pouch and leg referable to non-breeding or not fully adult stage. NESTLING. Iris pale pink. Bill light brown, partly tinged blue. Leg and foot ivory-white. JUVENILE. Iris off-white. Bill light brown to blue. Gular pouch pink. Leg and foot pink-brown. IMMATURE. Iris steel-grey. Bill grey-green. Bare skin on face yellow tinged green; eyelid yellow-pink; gular pouch pink, tinged yellow near base of mandible. Leg and foot dusky; toes alternatively described as black or dusky with pink tinge. (Ogilvie-Grant and Forbes 1899, 1903; Ticehurst 1922, 1925; Ripley and Bond 1966; BMNH.)

**Moults.** Limited data. Presumably like other *Phalacrocorax*. Primaries serially descendant. Moult not wholly arrested during breeding. Timing of post-breeding moult and timing and extent of pre-breeding unknown. Juvenile in February moulting to immature (possibly aged *c*. 10 months). Immature in December moulting to adult. No data on other months. (BMNH.)

**Measurements.** BMNH.

|  | adult ♂♂ | immature ♀♀ | adult unsexed |
|---|---|---|---|
| WING | 288, 296 | 274 | 304 |
| TAIL | 85 | 83, 84 | – |
| BILL | 69, 73 | 74 | 77 |
| TARSUS | 73, 75 | 74, 73 | 76 |
| TOE | 76, 78 | 73, 73 | 86 |

Range of wing: ♂ 285–310, ♀ 275–298 (Ticehurst 1925; Bates 1938).

**Weights.** No data.

**Structure.** 11 primaries: p9 longest, p8 0–4 shorter, p10 3–7, p7 2–7, p6 34–37, p5 49–60, p1 99–109. Primaries pointed at tips; p8–p10 emarginated inner webs, p7–p9 outer. Secondaries long, reaching tips of primaries in closed wing. Tail strongly wedge-shaped, 14 feathers. Bill slender, slightly longer than head. Tarsus heavy, foot large; outer toe *c*. 120% of middle, inner *c*. 65%, hind *c*. 45%.

## *Phalacrocorax pygmeus* Pygmy Cormorant

PLATE 24
[between pages 230 and 231]

Du. Dwergaalscholver    Fr. Cormoran pygmée    Ge. Zwergscharbe
Ru. Малый баклан    Sp. Cormorán pigmeo    Sw. Dvärgskarv

*Pelecanus pygmeus* Pallas, 1773

Monotypic

**Field characters.** 45–55 cm, wing-span 80–90 cm, ♂ averaging larger. Small, short-necked, round-headed, fairly long-tailed, largely freshwater cormorant with dark plumage. Sexes alike but clear seasonal differences; browner juvenile with pale underparts readily distinguishable.

ADULT. In autumn and early winter, head, upper half of neck, and breast dark brown, sometimes looking black with brown tinge; some black on forepart of head, variable white on throat and around eyes, black scapulars and upper wing-coverts with grey tinge and darker edges giving scaled appearance; otherwise black with green sheen above and light brown below. Bill black-brown, bare facial skin tinged pink, feet blackish. In summer (moult starts in December), head and upper neck strongly tinged red-brown, becoming nearly black before breeding, with short crest and scattering of white filoplumes over head, neck, black underbody and upper tail-coverts; scapulars and upper wing-coverts greyer with scaled appearance more marked. JUVENILE. Dark brown crown and neck, whitish chin, grey-brown foreneck and breast, and brownish-white belly blotched with darker brown and orange-brown, blackish flanks and under tail-coverts black-brown, back with lighter feather edges, and grey-tinged scapulars and upper wing-coverts with dark edges; also yellowish bill.

Small size, short rounded, often brown head, short neck and, in summer, white flecks on head and underparts make species unmistakable; also found mainly by marshes and reedbeds, lakes, and rivers. Only Long-tailed Cormorant *P. africanus* comparable in size (and almost indistinguishable as juvenile), but these do not approach each other geographically; leaving aside possible vagrants, only much larger, longer-necked, shorter-tailed Cormorant *P. carbo* occurs in same areas.

Breeds colonially, but not otherwise markedly gregarious. Flight buoyant with rather rapid wing-beats occasionally interspersed with glides, usually rather low over water. If *P. africanus* loosely likened to duck *Anas* in flight silhouette, this species more recalls Coot *Fulica atra*. Perches easily on branches, reeds, and other waterside projections, sitting almost bolt upright with neck hunched and, particularly in reeds, tail sometimes acting as partial support. Croaking call seldom uttered except when nesting.

**Habitat.** Concentrated far from ocean in middle warm latitudes of continental climate. Lowland species, absent from mountainous, arid, and cooler regions; some shrinking of breeding range with drying-out and drainage of suitable wetlands. Usually on open standing or slow-flowing fresh waters, including oxbows and backwaters, ricefields, swamps, and floodlands, where coarse fish can readily be caught in shallow water by swimming and diving. Attached to densely vegetated places, including stands of trees or scrub in wetlands, reedbeds, herbage, and even small floating islets of dead plants. These heron-like habitats lead to frequent mingling with breeding herons (Ardeidae) of various species, both in reedbeds and in trees (on which also habitually rests and roosts). In winter, more often on brackish or salt water. Apart from such seasonal movements, makes fairly frequent diurnal trips, taking full advantage of warm airmasses for easy soaring flight. Appears indifferent to man, but makes little use of artefacts.

**Distribution.** Restricted range in south-east of west Palearctic area steadily decreasing, mainly owing to drainage. HUNGARY: has not bred for at least 100 years

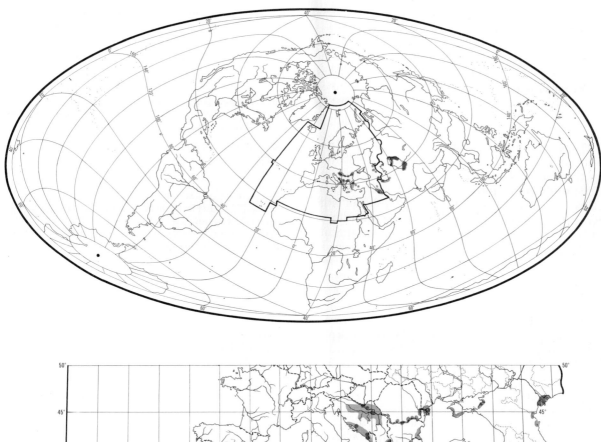

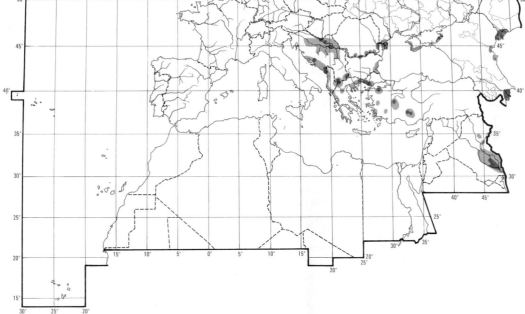

(Keve 1960). YUGOSLAVIA: formerly common in all large marshes of Vojvodina; many colonies here and elsewhere in the country abandoned due to drainage (VJV). TURKEY: formerly bred Amik Gölü, now drained (OST). ISRAEL: a few pairs formerly bred Lake Huleh; last nested between 1950 and 1955 (HM). ALGERIA: bred in numbers Lakes Fetzarah and Halloula in 19th century, but long extinct there (Heim de Balsac and Mayaud 1962).

Accidental. France, Sweden, West Germany, East Germany, Poland, Czechoslovakia, Hungary, Austria, Switzerland, Italy, Tunisia.

**Population.** YUGOSLAVIA. Montenegro: up to 200 pairs Lake Skadar, numbers variable; 100 pairs Lake Sasko in 1969. Former large colony Hutovoblato on Adriatic coast decreased to 30 pairs by 1967. Still relatively abundant in southern Macedonia, but no numerical data (VJV). GREECE: now 5 colonies with *c.* 550 pairs, majority at Lake

Mikra Prespa with up to 650 pairs 1971 and 400 pairs 1973 (WB). RUMANIA: far less numerous in last 30 years or so; 35 colonies in Danube delta with 8000 nests in 1962 (Vasiliu 1968). TURKEY: only 2 definite colonies with 170–200 pairs in 1967, but 3 further small colonies discovered since (OST). IRAQ: still common in marshes near Al Qurnah and large colonies reported in Hawr al Hammar, 1973 (PVGK).

**Movements.** Migratory, partially migratory, and resident; movements believed generally limited to relatively short distances, though few ringed. Balkans breeders winter partly inland, partly on or near Adriatic, Aegean, and north-east Mediterranean coasts; Yugoslavian birds recovered south-west Rumania (November) and in Albania (December). Evidently more definite seasonal movements by Black Sea breeders; largely deserts northern sectors in winter, when many present north Aegean (Hafner and Hoffman 1974). Birds ringed Danube delta (Rumania) recovered November–January in Greece (4) and central Yugoslavia (Spiess 1934). Turkish wintering flocks may include some Black Sea breeders, and unknown to what extent local breeders move. Deserts European colonies August–September, returning March–April (Bauer and Glutz 1966). Many known instances of straggling west and north of breeding range (see Distribution); occasionally irruptive movements in small flocks as autumns 1957 and 1958 on Ismaninger Reservoir, Bavaria, when numbers dwindling until following springs (Wüst 1958, 1959).

North Caspian breeders migratory; some Transcaucasian and Kazakhstan breeders overwinter locally, but numbers remaining within USSR vary markedly with winter temperatures (Dementiev and Gladkov 1951); common winter visitor (several thousands) to Iranian Caspian (D A Scott). Autumn exodus from USSR breeding areas may continue into November–December under influence of cold weather; colonies re-occupied March to early April. Irregular northward autumn movement in USSR also, wanderers having occurred Sea of Azov and inland in Ukraine and central Caucasus (Dementiev and Gladkov 1951). Vagrant to Pakistan (once). Iraq breeders believed mainly resident, and not recorded Kuwait; occur on winter floods of Tigris and Euphrates valleys and Kirkuk area (Ticehurst et al. 1922, 1926).

**Food.** Primarily fish, but data sparse with other prey only occasionally recorded; includes young water-voles *Arvicola*, crustaceans (e.g. shrimps), and (according to Dombrowski 1912) leeches. Mainly by diving; often first watches from perch just above surface (e.g. vertical reed stem on which presses tail for support). 8 dives in up to 2·4 m on Lake Antioch, Turkey, 9–42 s (Meinertzhagen 1959). Normally daytime feeder; singly, in pairs, less often in groups.

Fish include rudd *Scardinius erythrophthalmus*, pike *Esox lucius*, carp *Cyprinus*, and roach *Rutilus rutilus* up to 15 cm long (Bauer and Glutz 1966; Dementiev and Gladkov 1951). Andone et al. (1969) found 15 fish species in 130 birds collected in Danube delta: perch *Perca fluviatilis* (frequency 18·8%), *Rutilus rutilus* (14·8%), *Scardinius erythrophthalmus* (15·8%), Crucian carp *Carassius carassius* (10·8%), loach *Cobitis taenia* (9·7%), *Esox lucius* (5·6%), bitterling *Rhodeus sericeus amarus* (4·1%), bream *Abramis brama* (4·1%), weatherfish *Misgurnus fossilis* (3·6%), tench *Tinca tinca* (3·1%), and others in small quantity. Average weight of fish 15 g (max. 71, min. 7).

**Social pattern and behaviour.** Little known.

1. Mainly gregarious. In small flocks (usually family parties) in autumn (Dementiev and Gladkov 1951) and also during occasional irruptive movements, but often in larger flocks in winter (Bauer and Glutz 1966). Sometimes solitary. BONDS. No detailed information on pair-bond but said to be monogamous (Dementiev and Gladkov 1951), probably over at least one breeding season. Both parents tend young. BREEDING DISPERSION. Usually colonial; sometimes in 100s, often mixed with other species such as egrets *Egretta* and *Ardeola*, Night Heron *Nycticorax nycticorax*, and Glossy Ibis *Plegadis falcinellus*. ROOSTING. Communal and nocturnal, in reedbeds and trees but no detailed studies.

2. No information on behaviour.

**Voice.** Poorly studied. Apparently silent except during breeding season, when utters croaking calls in well-defined temporal pattern, alternating brief lower-pitched grunts with longer higher-pitched quacking notes (see I).

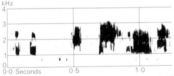

I C Chappuis/Sveriges Radio Rumania May 1967

**Breeding.** SEASON. South-eastern Europe and southern USSR: eggs laid from end April to end June or early July (Dementiev and Gladkov 1951; Bauer and Glutz 1966). Reported as prolonged within single colony (Aral Sea), but replacement layings may have been included (Dementiev and Gladkov 1951). SITE. In trees, or thick reedbeds when 1–1·5 m above water. Colonial; colonies of hundreds, usually with other Phalacrocoracidae, Ardeidae, etc. Old nests re-used and built up. Nest: hemispherical structure of twigs, lined grass and reeds; c. 25 cm diameter with deep cut (Dementiev and Gladkov 1951). Building: probably by both sexes, material both brought to site and gathered from within reach. EGGS. Ovate; smooth chalky coating on pale green surface. 47 × 30 mm (40–52 × 28–33), sample 100; calculated weight 23 g (Schönwetter 1967). Clutch: 4–6 (3–7). One brood. Replacements highly probable, but no data (see Season). INCUBATION. 27–30 days (Bauer and Glutz 1966). By both sexes. Starts with 1st egg; hatching asynchronous. YOUNG. Altricial and nidicolous. Cared for and fed by both parents.

Fed by partial regurgitation. Small young brooded by one parent while other obtains food. When a few weeks old, young leave nest and scramble round colony. FLEDGING TO MATURITY. Fledging period not accurately recorded, but *c*. 70 days. Age of independence and first breeding unknown. BREEDING SUCCESS. No data.

Plumages. ADULT BREEDING. Early in season head, neck, mantle, back, rump, and underparts black, speckled by many white filoplumes, densest on crown and sides of head. Small crest on forehead. Scapulars glossy grey with narrow black margins. Tail black. Flight-feathers grey-black; upper wing-coverts like scapulars, but marginal black; under coverts and axillaries black. ♀ slightly duller than ♂. Uncertain whether all individuals develop this plumage; pronounced variation at least in number of white filoplumes. Late in season similar, but head and neck velvety brown and far fewer filoplumes on crown and sides of head, those on rest of body gradually disappearing. ADULT NON-BREEDING. Crown dark brown, hindneck brown, chin white, throat and foreneck pale brown; plumage on head and neck not velvety. Rest of upperparts like breeding, but feathers of mantle, scapulars, and upper wing-coverts with narrow pale edges. Feathers of chest with white fringes, breast and belly variably mottled dark brown and pale buff. Some white filoplumes on back and flanks. Tail and wing like adult breeding. NESTLING. Naked at hatching, black; gradually grows short, dark brown down, developing last on head and neck. JUVENILE. Similar to adult non-breeding, but browner above; feathers of mantle black, with brown edges. Scapulars and upper wing-coverts dark grey with black-brown margins and paler fringe, more pointed than adult. Tail black. Flight-feathers dark grey tinged brown and with brown fringe at tips. SUBSEQUENT PLUMAGES. Juvenile probably succeeded by immature plumage resembling adult non-breeding (Hartert 1912–21) with retained juvenile tail and flight-feathers. Stresemann (1920), however, supposed that fully developed adult breeding immediately succeeds juvenile, of which flight-feathers and probably tail retained. More information needed. No information on 2nd winter plumage, but presumably like adult non-breeding.

Bare parts. ADULT BREEDING. Iris dark. Bill black; bare skin round eye and at corner of mouth black. Foot black. ADULT NON-BREEDING. Bill dirty brown, darkest on ridge; cutting edge reddish-yellow with dark spots; mandible light brown. NESTLING. Bill blue; bare skin on head fleshy. Foot fleshy, blue on outer

side. JUVENILE. Iris paler brown than adult. Bill dirty yellow, marbled brown at sides, dark brown along ridge; naked skin on head dirty reddish-yellow. Foot brown or brown-black (Naumann 1842; Hartert 1912–21; Bauer and Glutz 1966).

Moults. ADULT POST-BREEDING. Complete; primaries serially descendant. June to October or November, with much individual variation (Dementiev and Gladkov 1951). PRE-BREEDING. Partial; probably involves most body feathers, but at least head, neck, and underparts. December–March, finishing with development of white filoplumes. In another partial moult in late spring, head and neck lose black feathers and acquire brown; majority of filoplumes lost. POST-JUVENILE. Apparently partial moult in autumn of 1st calendar year leads to 'immature' plumage (but see Plumages). Timing of moult of flight-feathers unknown; may start shortly after body moult and proceed slowly through 2nd calendar year; details lacking. Further moults probably as adult. (Wüst 1958; Bauer and Glutz 1966.)

Measurements. Adult and juvenile. Skins (RMNH, ZFMK); Stresemann (1920), Makatsch (1950).

| | | | | | | |
|---|---|---|---|---|---|---|
| WING | ♂ 206 | (6·43; 11) | 195–217 | ♀ 201 | (4·74; 7) | 193–208 |
| TAIL | 142 | (4·67; 6) | 137–145 | 141 | (3·21; 5) | 139–147 |
| BILL | 30·5 | (1·33; 6) | 29–33 | 29·2 | (1·25; 5) | 27–31 |
| TARSUS | 38·0 | (1·26; 6) | 37–40 | 37·8 | (1·30; 5) | 36–39 |
| TOE | 56·2 | (2·05; 5) | 54–58 | 52·8 | (0·84; 5) | 52–54 |

♂ tends to be larger than ♀, but differences not significant in small sample, except for middle toe.

Weights. ♂♂ 650, 710, 870; ♀♀ 565, 640, 640 (ZMFK; Makatsch 1950; Schüz 1959).

Structure. 11 primaries: p8 longest, p9 0–4 shorter, p10 6–15, p7 2–4, p6 13–18, p1 54–61; p11 minute. P7–p9 slightly emarginated outer web, p8–p10 inner web. Tail long, markedly wedge-shaped. 12 feathers: t6 49–65 shorter than t1, t5 28–38. Bill short, straight, hooked at tip, no external nostrils. Leg short and stout. Outer toe *c*. 117% of middle toe, inner *c*. 74%, hind *c*. 52%. Small comb on nail of middle toe.

Recognition. Differs from Long-tailed Cormorant *P. africanus* by black instead of yellowish bill, and in adult plumage by absence of black subterminal spots on scapulars and upper wing-coverts. Museum specimens in juvenile plumage with discoloured bill probably indistinguishable, but tail of *P. africanus* averages longer. Oriental *P. niger* also closely similar.

## *Phalacrocorax africanus* Long-tailed Cormorant

PLATE 25
[facing page 231]

Du. Afrikaanse Dwergaalscholver    Fr. Cormoran africain    Ge. Gelbschnabelzwergscharbe
Ru. Длиннохвостый баклан    Sp. Cormorán africano    Sw. Långstärtad skarv

*Pelecanus africanus* Gmelin, 1789

Polytypic. Nominate *africanus* (Gmelin, 1789), Africa. Extralimital: *pictilis* Bangs, 1918, Madagascar.

Field characters. 50–55 cm, ♂ averaging larger; wingspan 80–90 cm. Small, long-tailed, short-necked, largely freshwater cormorant, mainly black in breeding season, otherwise browner above and whitish below. Sexes alike but marked seasonal differences; juvenile rather like non-breeding adult.

ADULT. In breeding plumage all velvet-black, with tuft

of feathers on forehead, apart from ephemeral white plumes behind eyes and silver-grey scapulars and wing-coverts roundly tipped and edged with black, giving striking pattern. Bill yellowish, bare face orange-yellow, legs and feet black. In non-breeding plumage mainly dark brown above and lacking crest, with less contrasted wing-coverts and scapulars (greyish blotched with black and

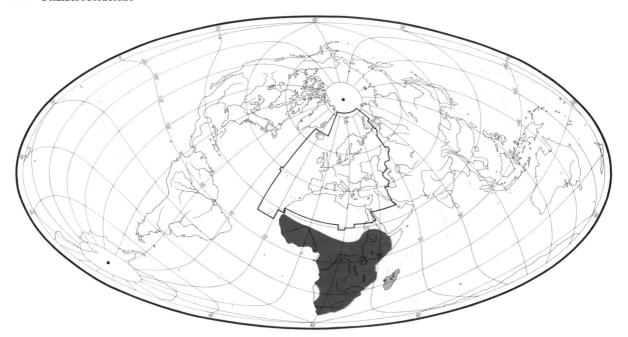

edged with buff) and pale brown breast shading into whitish throat and belly (but black flanks and undertail). JUVENILE. Resembles non-breeding adult, but lacks black blotches on scapulars and wing-coverts and has variable brown spots and streaks on breast (probably due to new feathers of next stage).

Combination of small size, short neck, and long tail give characteristic shape; much larger *P. carbo* only other cormorant with which it normally ever overlaps in west Palearctic. Long tail and similar habit of swimming low in water, often with only head and neck above surface, may cause confusion with Darter *Anhinga melanogaster* but latter considerably larger with much longer chestnut-and-white neck and long pointed bill. See also Pygmy Cormorant *P. pygmeus*.

Not as gregarious as marine cormorants and individuals often spend day singly, though in sight of each other, based on waterlogged stump, patch of reeds, or branch overhanging water, either swimming after fish or standing in wing-spread position. Numbers generally roost together and parties or flocks join in roosting flights, when can have surprisingly duck-like action in silhouette; groups usually compact rather than in lines or V-formation. Mainly freshwater areas, but also coastal lagoons, swamps, and river estuaries.

**Habitat.** Mainly tropical creeks, rivers, streams, lakes, and lagoons; also sea-coasts and inshore islands. Sometimes far inland at altitudes up to 1750 m or more. Swims deep in water often with only long neck visible, but, like congeners, spends much time drying outspread wings on posts or other suitable raised perches. Nests on Banc d'Arguin on bare rock and sand, on cliffs, and in mangroves (de Naurois 1969a; Gandrille and Trotignon 1973). On ground and on rocks, uses claws for climbing (Etchécopar and Hüe 1964). Takes-off easily and flies considerable distances between roosts and fishing grounds (Bannerman 1930).

**Distribution.** MAURITANIA: only proved breeding area in west Palearctic. RIO DE ORO: seen coast and may nest Legtoaa (Valverde 1957). EGYPT: formerly abundant resident Faiyum and probably Nile delta lakes, but not recorded since 1875 (Meinertzhagen 1930).

**Population.** MAURITANIA. Colonies on Banc d'Arguin: total *c.* 1500 pairs (de Naurois 1969a), 2585 pairs in 1973 (Duhautois *et al.* 1974).

**Movements.** Dispersive. Chapin (1932a) and Bannerman (1953) considered those seen northern tropics migrants from some central African breeding area. Now seems unlikely, for since found breeding, in part sporadically, in Mauritania, Sénégal, Nigeria, and Chad; probably nests opportunistically when water levels suitable. In Nigeria, common on waters throughout country, dispersing to pools away from larger rivers at peak flood when sandbanks submerged (Elgood *et al.* 1973). Degree of movement to or from Banc d'Arguin unknown.

**Food.** Mainly fish, crustaceans to much lesser extent. Feeds singly, catches prey under water by diving from swimming position and returning to surface where swallows prey head-first.

Of 7 collected coast Cape Province all had fish—*Clinus superciliosus* (7) and windtoy *Pterosmaris axillaris* (1)—and

2 crustaceans (isopods); 3 immatures collected July contained pipefish *Syngnathus* (2), sole *Heteromycteris* (1), shrimp *Palaemon pacificus* (3) (in one, only shrimp found, suggesting taken deliberately), and isopod *Paridotea ungulata* (Rand 1960). In 98 stomachs from Uganda lakes, main prey cichlids—*Haplochromis*, with much fewer *Haplochilichthys*, *Engraulicypris*, *Alestes*, *Synodontis*, *Tilapia*, and *Lates*; economically important *Tilapia* formed only 1·4% (Cott 1952, 1961). Birds from central African lakes similarly ate few of economically important cichlids. In 66 birds from Bangweulu swamps, main fish (23·8% by frequency) Mormyridae (*Gnathonemus macrolepidotus*, *G. monteiri*, *Marcusenius discorhyncus* and *M. strappersii*); also (17·5%) *Schilbe mystus*. In 17 birds from Lake Mweru, Mormyridae again most common prey (67%), followed by Clariidae and Cybrinodontidae (*Barbus*) (Bowmaker 1963, 1964). Tait (1967) from 34 observations on Kafue river, Zambia, found *Synodontis microstigma* taken on 27 occasions, *Schilbe mystus* on 3, and *Mastacembalus* once. At Lake Lifura, Katanga, *Barbus paludinosus* (14–15 cm) in birds (Ruwet 1963).

Average daily food intake estimated *c.* 14% of body weight (Bowmaker 1964), 6–16% (Rand 1960), and by others as much as 25%; 3 birds hand-reared on cyprinids and cichlids took *c.* 16% of body weight daily after reaching flying stage (Junor 1972).

**Social pattern and behaviour.** Based on material supplied by P L Britton.

1. Often solitary, but gregarious when nesting and when roosting outside breeding period. Usually feeds individually and loafs singly when digesting food; any assemblies then uncommon, though 8000 recorded in clusters in dead trees after feeding (Ruwet 1963), and due to local abundance of food or lack of dispersed resting sites. BONDS. Monogamous pair-bond of seasonal duration; not known if retained over successive broods in equatorial regions where 2 breeding seasons per year or prolonged season recorded, but likely. Both parents tend young; small family parties occasionally recorded after fledging, but probably disperse within month or two. BREEDING DISPERSION. Colonial; dozens or hundreds of pairs, often in mixed colonies with herons (Ardeidae), storks (Ciconiidae), etc. Defends nest-site territory of normally about 1 m cube, varying with proximity of neighbours. Even rudimentary nest hardly ever left undefended since material stolen by neighbours; while building, both sexes remain at nest when not fishing or collecting sticks and apparently digest food there. As Bowen *et al.* (1962) noted flightless young swimming from nesting site to adjoining islands, parents may establish distinct brood-territory in vicinity of some colonies just prior to fledging. ROOSTING. Communal. Outside breeding season, in dozens or hundreds with other species, especially Cattle Egret *Bubulcus ibis* and occasionally Cormorant *P. carbo*. Though Miller (1946) considered present species 'aloof' and Craufurd (1966) reported clear species stratification in shared roost, R K Brooke found it very tolerant of others and not segregated within roost. Mainly in trees, *Typha*, or *Phragmites* near or surrounded by water; inaccessible roost in trees down deep mining hole exceptional. Arrives hour or more before sunset, flying high in small parties and dropping down directly into roost or assembling nearby. Sometimes depart before sunrise but normally wait up to hour or more after, leaving for feeding areas in groups of *c.* 20, then dispersing to feed singly.

2. Little known; most information from P L Britton (Kenya). Decorative tuft of feathers on forehead of adults raised during all displays; also while incubating. ANTAGONISTIC BEHAVIOUR. Territorial defence shown against any species entering nest-site territory; bird stabs forwards towards intruder, thrusting upwards, downwards, or horizontally, while slightly raising wings and fanning and elevating tail. Off-bird sometimes stabs while standing on mate. Bird fleeing from boat usually flies but sometimes, after assuming upright Alert-posture, will swim or dive away or show intention of flying and then perform dipping and drinking sequence, so-called 'displacement-dive' (Bowmaker 1963). It first sharply dips head and neck under water with rotary action up to 3 times, next assumes Alert-posture again, and then arches neck quickly to 'drink' by touching water with open bill. Such behaviour interpreted as incipient or 'symbolic' dive but may more likely involve bathing movements. HETEROSEXUAL BEHAVIOUR. Apparent mate-seeking behaviour observed in unpaired adults of unknown sex in colony; neck arched, bill pointed forward and slightly up, and wings raised jerkily about three times every 2 s. In Kenya, initial pair-formation correlated with first heavy rain of season and, as copulation, building and laying occur very soon after, pair-bond itself fully established in brief period of up to a week or so after laying, often while one or both members of pair still completing body moult into breeding plumage. Subsequently, ♂ and ♀ meet less at site: early in incubation, change-over lasts 10–20 min with relieved ♂ returning once or twice with offering for nest (see below); later, takes as little as 1–2 min with minimal preening and nest arranging, but no offerings. Interactions between paired birds little developed even at start of nesting; include mutual Neck-rubbing, Bill-touching, and Allopreening; also some intertwining of necks and Stick-offering by ♂. Latter behaviour also follows copulation with ♂ returning some minutes later to present material with much Head-shaking; offering dropped on or placed in nest by ♀. Bill-rubbing and self-preening also follow copulation. RELATIONS WITHIN FAMILY GROUP. Little reported. In young, red skin at base of bill vivid and swollen before and during feeding.

**Voice.** Usually silent (Williams 1963), although utterances may have been overlooked in mixed colonies (P L Britton). Nesting birds hiss; young utter incessant throaty cackling (Mackworth-Praed and Grant 1962).

**Breeding.** SEASON. Eggs laid Banc d'Arguin from late May, most June, some July (de Naurois 1969a). SITE. In thick reed beds, 1–2 m above water, or in tops of low trees, 4–6 m above ground. Occasionally on rocky ledges (Malzy 1967). Colonial with nests 1–2 m apart. Nest: platform of twigs and other vegetation, *c.* 25 cm diameter (Ruwet 1963). Building: no data. EGGS. Ovate; pale blue with whitish chalky covering. 46 × 31 mm (41–50 × 29–34), sample 27; calculated weight 23 g (Schönwetter 1967). Clutch: 4 (3–5), Banc d'Arguin (de Naurois 1969a). One brood. No data on replacements. INCUBATION. 23–25 days. By both sexes. Starts with 1st egg; hatching asynchronous. YOUNG. Altricial and nidicolous. Cared for and fed by both parents. Fed by partial regurgitation. FLEDGING TO MATURITY. Fledging period not known. Mature at 4 years (Ruwet 1963). BREEDING SUCCESS. No data.

**Plumages** (nominate *africanus*). ADULT BREEDING. Lores and region round eyes bare, strewn with tiny black feathers; crown bordered with small, bristle-like feathers, black at tips, white at bases; small tufts of white filoplumes over ears. Rest of head, neck, centre of mantle, back, rump, and underside glossy black. Feathers of forehead elongated to short crest. Feathers of sides of mantle, scapulars, and lesser and median wing-coverts grey, glossed silver or bronze, with black spots at tips; longest scapulars and greater coverts with black margins but no spots. Tail and flight-feathers black with slight gloss, becoming paler with wear. Marginal coverts, under wing-coverts, and axillaries black. ADULT NON-BREEDING. Crown dark brown, feathers edged pale grey-brown, hindneck paler; no crest. Chin white, throat and foreneck pale brown, breast and belly off-white; flanks and under tail-coverts black. Upperparts as breeding but all feathers edged white or buff; tail and flight-feathers as breeding. NESTLING. Presumably naked at hatching; later covered in black or grey down (Böhm 1882); top of head and throat bare. JUVENILE. Upperparts browner than adult non-breeding; scapulars and upper wing-coverts without black spots, grey-brown with black-brown margin and pale brown fringe. Underparts dirty white, washed pale brown, particularly on foreneck and chest. Tail-feathers, primaries, and secondaries dark brown with narrow pale brown tips abrading with wear. Variation in immatures poorly understood. Juvenile plumage presumably followed by one resembling adult non-breeding as in *P. pygmeus*, although Verheyen (1953) claimed to have found 4 different immature plumages gradually approaching adult. Some birds with juvenile scapulars, mainly dark on underparts (Heuglin 1869; Ruwet 1963).

**Bare parts.** ADULT. Iris red. Bare skin of face described as yellow (Sclater 1906), yellowish-white (Chapin 1932a), yellowish-flesh (Reichenow 1900–1), or bright red (Bannerman 1930); probably more intensely coloured when breeding. Bill yellow, black along ridge, with dark bars on lower mandible. Foot black. NESTLING. Iris pale blue. Bare skin on head yellow, base of gular pouch blood-red (Böhm 1882). Bill and foot black, webs brown. JUVENILE. Iris brown, orange-red, or red. Bill flesh-yellow with black tip. Foot black.

**Moults.** ADULT POST-BREEDING. Complete; primaries serially descendant. Starts soon after fledging of young (Ruwet 1963). ADULT PRE-BREEDING. Partial; head, neck, and underparts. No definite moult season as breeding may take place in almost any month. POST-JUVENILE. Presumably parallel to Pygmy Cormorant *P. pygmeus* but no firm data.

**Measurements.** Adult and juvenile. Skins (RMNH, ZMA).

| | | | | | | |
|---|---|---|---|---|---|---|
| WING | ♂ 212 | (4·17; 6) | 206–219 | ♀207 | (7·23; 8) | 194–216 |
| TAIL | 149 | (4·16; 5) | 144–153 | 148 | (10·2; 5) | 139–164 |
| BILL | 30·7 | (1·58; 6) | 29–33 | 30·4 | (1·66; 8) | 26–32 |
| TARSUS | 37, 37 | | | 34, 36, 36, 36 | | |
| TOE | 51, 56 | | | 49, 51, 51, 53 | | |

Sex differences not significant. Tail given as 160–180 by Reichenow (1900–01), as 148–177 by Mackworth-Praed and Grant (1962), indicating great variation, even allowing for differences in method of measuring.

**Weights.** Adult ♂ Cape Province 685, ♀ 550 (ZMA). Immature ♂ Lake Naivasha *c.* 680 (Grant 1915).

**Structure.** 11 primaries: p8 longest, p9 0–3 shorter, p10 6–11, p7 1–3, p6 11–15, p1 *c.* 60; p11 minute. P7–p9 slightly emarginated outer web, p8–p10 inner web. At least 17 secondaries. Tail longer and more strongly wedge-shaped than in *P. pygmeus*, 12 feathers: t6 *c.* 65–70 shorter than t1, t5 *c.* 35–40. Structure of bill and foot as *P. pygmeus*.

**Geographical variation.** Birds from Madagascar (*pictilis*) differ from nominate *africanus* in being larger with bigger, less rounded black spots on upperparts (Bangs 1918); non-breeding and juvenile darker below (RMNH). Crowned Cormorant *P. coronatus* (Wahlberg, 1855), coasts of southern Africa from Benguela to east Cape Province, sometimes considered conspecific. Strongly resembles *africanus*, but tail on average shorter, upperparts darker with narrower black markings, and frontal crest in breeding longer. Marine, whereas nominate *africanus* lives on fresh water; but *pictilis* both coastal and inland (Meinertzhagen 1950).

# Family ANHINGIDAE darters

Large, aquatic birds, mainly on fresh water. 2 species forming 1 superspecies in single genus *Anhinga*. 1 species breeding in west Palearctic. Body elongated; neck extremely long and slender with special hinge mechanism at 8th and 9th vertebrae which aids in spearing fish (Stresemann 1927–34; Owre 1967), head small. ♂ and ♀ about equal in size, but ♂ on average longer billed. Wings long and broad, with relatively long upper arm and long primaries. 11 primaries; p8 longest, p11 minute. 17–18 secondaries; diastataxic. Wings adapted for powerful flapping flight, alternating with gliding, but also for soaring (Owre 1967). Tail long, strongly graduated; 12 broad, transversely corrugated feathers. Bill rather long and slender; sharply pointed, without hook, cutting edges serrated; nostrils reduced. Gular skin bare. Leg short, sturdy. Foot large, totipalmate; nail of middle toe medially with horny flange with comb-like incisions. Lower part of tibia bare. Stance upright; nearly always perch in trees, rarely walk. Swim deeply submerged; underwater propulsion by feet, wings often partly spread. Oil gland feathered. Aftershaft absent. Plumage black, vividly patterned white and silver-grey; head and neck brown or black with ornamental pattern on sides in breeding. ♀ slightly duller than ♂ in nominate race of Old World Darter *A. melanogaster*, but sexual difference more pronounced in Australian *A. m. novaehollandiae*, and New World Anhinga *A. anhinga*. Non-breeding plumage less bright than breeding. 2 moults per cycle. Flight-feathers moulted simultaneously, causing temporary flightlessness. Young altricial and nidicolous; naked at hatching, later single coat of thick buff down. Juveniles duller than adult, paler below. Adult plumage attained in 3rd calendar year.

Most of tropical and sub-tropical areas of world. In west Palearctic, *A. melanogaster* breeds only in extreme south-east. Comfort behaviour and general behavioural characters much as in Phalacrocoracidae, including wing-spread attitude. As Old and New World species closely similar, see species account for further information.

## *Anhinga melanogaster* **Darter**

PLATE 26
[facing page 231]

| | | |
|---|---|---|
| Du. Slangehalsvogel | Fr. Anhinga | Ge. Schlangenhalsvogel |
| Ru. Змеешейка | Sp. Marbella | Sw. Ormhalsfågel |

*Anhinga melanogaster* Pennant, 1769

Polytypic. *A. m. rufa* (Daudin, 1802), Africa and Middle East. Extralimital: nominate *melanogaster* Pennant, 1769, India east to Philippines and Celebes; *vulsini* Bangs, 1918, Madagascar; *novaehollandiae* (Gould, 1847), New Guinea and Australia.

**Field characters.** 85–97 cm, which head and neck form more than third; wing-span 116–128 cm. Mainly black and chestnut, streaked with white, recalling attenuated cormorant *Phalacrocorax* but characterized by extraordinary length of pointed bill, of snake-like neck, and of graduated tail. Sexes not dissimilar, but distinguishable, and only minor seasonal differences; juvenile clearly separable.

ADULT MALE. Crown and hindneck black with chestnut feather fringes; rest of neck chestnut with white stripe from gape down upper third of each side. Body and wings glossed black with whitish streaks on scapulars and wing-coverts, tail grey-black, In breeding season, feathers forming white stripe down side of neck lengthen into short and inconspicuous plumes. Bill greenish with yellower tip; bare skin on throat creamy and around yellow eyes greenish; legs brown to olive-green. ADULT FEMALE. Crown and hindneck brown instead of black; white neckstripe much less distinct; rest of neck appreciably paler (pink-buff rather than chestnut); and streaks on scapulars and wing-coverts whiter. JUVENILE. Resembles adult ♀, but lacks white stripes down sides of neck, and has whitish foreneck and otherwise buff underparts with black confined to flanks.

Easily distinguished from *Phalacrocorax* by different proportions, particularly pointed bill, snake-like neck, and long, conspicuous tail, and by contrast between mainly chestnut and white neck and black body; behaviour also markedly different.

Flies strongly with neck in distinctive kink, tail often spread in fan, and interspersing every few wing-beats with characteristic prolonged glide. Often swims with whole body and tail submerged and only head and neck above surface. Frequently associates by day, roosts by night, and rests with herons (Ardeidae) and *Phalacrocorax*, but seldom itself in any numbers. Holds out wings like *Phalacrocorax*, but with long tail hanging down more noticeably. Perches freely in trees, both in open and among dense foliage. Will soar in wide circles to considerable heights in thermals. Usually rather silent; harsh, rattling call, rapidly repeated, sometimes uttered both when perched and in flight.

**Habitat.** Tropical and sub-tropical open, fresh waters including lakes, pools, lagoons, reservoirs, slow-flowing rivers, streams, and swamps with open water; normally fringed with trees, and fairly shallow. Also on brackish waters of estuaries and sheltered inlets of sea. Avoids floating vegetation: dependent on overhanging, emergent, or floating bare branches, stumps, logs, posts, or similar resting places for drying out with open wings on emerging from water. Accomplished swimmer and diver, but to no great depth; otherwise spends time perching on usually wooden vantage points, and soars to considerable heights. Wary, but can exist close to human settlements on navigated waters, and even on those artificially created and significantly disturbed.

**Distribution.** In west Palearctic, now breeds only marshes of southern Iraq. Formerly bred Amik Gölü (Lake Antioch), Turkey, but extinct following draining in 1950s (RP).

**Population.** IRAQ: not uncommon in extensive marshes round Kut and Amara; large colonies reported 1973 near Qurna (PVGK). TURKEY: 55 pairs nested Amik Gölü, 1933 (Meinertzhagen 1935).

**Movements.** In Palearctic now restricted to southern Iraq where apparently resident; isolated winter records north to Babylon and Baghdad (Allouse 1953; Moore and Boswell 1956). Former population of Amik Gölü apparently partly resident since several February records (Kumerloeve 1963), but most moved to Israel winter quarters (Lake Huleh and Jordan valley south to River Yarmuk mouth), where present mid-September to March, sometimes April (Hardy 1946); no evidence this population migrated in part to Africa, as sometimes stated (Hovel 1947).

**Food.** Fish and possibly on occasions other aquatic animals, especially amphibians. Feeds singly, often stalking prey. Bill, which lacks terminal hook found in *Phalacrocorax*, becomes in effect triggered spear as long neck, carried in S-shaped bend when above water, shoots

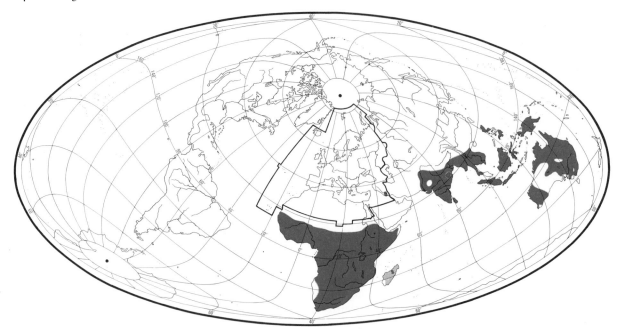

forward and impales prey. Smaller fish appear to be mostly pierced by upper mandible, and larger fish by both mandibles; latter held apart in act of stalking. Recent photographs disprove claim of some authors that prey merely caught between mandibles. Fish brought to surface, thrown in air, and swallowed head first.

In 50 stomachs from Lake Victoria and Lake Albert, Uganda, only fish, mainly *Haplochromis*; other species, which occurred much less frequently, included *Lates*, *Alestes*, *Bagrus*, *Haplochilichthys* and *Tilapia*—only latter being of economic importance (Cott 1952, 1961), one from near Lamu, Kenya, containing 8 small *Tilapia* (Jackson 1938). On Katanga, *Barbus paludinosus* main prey (Ruwet 1963).

Young fed predigested food from bill tip, but later take food from within parents' mouth and gullet. 4 hand-reared birds taken from Lake Kyle, Rhodesia, and fed twice daily on cyprinids (*Barbus*) and cichlids (*Tilapia*), initially took up to 50% of own body weight daily; this gradually declined to *c*. 16% at flying stage and remained constant thereafter (Junor 1972).

**Social pattern and behaviour.** Based mainly on unpublished observations on *A. m. novaehollandiae* in Australia by W J M Vestjens.

1. Moderately gregarious. Singly or in small flocks outside breeding season. BONDS. Sustained monogamy may occur as some pairs have bred in same nests in 2 successive years. Both parents tend young. BREEDING DISPERSION. Nests solitarily or in loose groups of 6–14 pairs in a number of trees, sometimes with cormorants *Phalacrocorax*. Aggression of nesting birds towards conspecifics perching in same tree suggests nest-area territories, though other species allowed to breed within these. ROOSTING. Starts just before and soon after dark, and lasts until first light. Breeding birds in nest trees; otherwise in pairs and small groups generally in dead trees near feeding area.

2. ANTAGONISTIC BEHAVIOUR. Occurs commonly between ♂ with nest and intruding ♂. Encounters start with threat behaviour; nest bird hops along branches of tree towards intruder with spread wings and open bill, calling as it flies from one branch to another. When near intruder, makes snapping movements of bill and intruder usually then leaves. If stays, they stab at each other on head and neck and sometimes both drop into water. Encounters between ♀♀ rare and then only between mated and unmated birds in nest tree. Threat behaviour as in ♂ but stabbing exceptional; generally unmated ♀ leaves trees. HETERO-SEXUAL BEHAVIOUR. ♂ selects nest-site, collecting a few green twigs and placing them on horizontal fork of branch or adding to existing nest. ♂ performs advertising displays. In Pointing towards ♀ flying over, sits on nest with wings closed, neck stretched up, and head and tail raised at 80° from horizontal; this changes to Wing-waving (see A) in which ♂ raises each wing in turn (with no constant speed), pointing bill up at 45°. Attracted ♀ greeted by ♂ with forward thrust of head with closed bill while uttering 1–2 calls. ♂ continues to Point and Wing-wave while ♀ in tree, then leaves and circles area in flight. ♀ moves from branch to branch and flies on nest, calling. When ♂ returns gives 3–15 loud calls just before alighting in tree and ♀ Points when ♂ comes to nest. Above may lead to copulation when birds lean against each other. In mounting, ♂ usually takes ♀'s bill crosswise in his or may hold green twig throughout copulation, with head alongside and below that of ♀. Mating occurs several times during first 2 days of pair-formation. Allopreening between paired birds performed unilaterally. Nest-material collected mainly by ♂ but after first copulation often by both birds. ♀ usually stays on nest, receives material from ♂, and adds to nest. Once nest established at least one bird always remains at site. Nest-relief display starts from first egg and lasts until one week after hatching. Relieving bird calls several times before alighting near nest. Bird on nest greets by lifting head to 80° and calling 3–4 times, keeping tail horizontal, then Wing-waves with tail vertical and presses head on to neck while pointing bill down at nest. Wing-waving followed by Snap-bowing (see B): with wings raised and tail elevated at 80°, bird pushes head forwards and holds sticks in nest with bill. Wings and tail held thus for a few seconds after Snap-

A

B

bow, then bird stands up and shakes feathers, flying away when mate enters nest. When non-breeders and other species land in nest-tree, incubating bird performs mild form of Wing-waving, lifting tips of primaries alternately 2–3 cm above rump while watching intently and moving restlessly around. RELATIONS WITH FAMILY GROUP. Begging signals for food similar for small and large chicks and fledglings. Young bird reaches up to full length with stretched neck and bill directed towards head of adult. Hyoid apparatus moves forwards in gular pouch and head moves up and down quickly, while upper part of body and neck move back and forth at slower pace and wings move in flying motion. Meanwhile, Food-calls produced which alternately increase and decrease in volume. Feeding order decided by relative size, larger chicks obtaining most of food. Aggressive pecking occurs between chicks during feeding. Food-begging occurs at times between 2 chicks and one will push bill and head into throat of other. During play, chick will grab bill of other, moving head and neck in all directions. Siblings seek each other's company, not only at nest but in branches near and also after leaving nest tree. No aggression between fledglings of different broods.

(Figs A and B after original drawings by W J M Vestjens.)

**Voice.** Usually silent except near nest. Variety of harsh, rattling, or grunting calls described at or near nest (W J M Vestjens).

(1) Rattling-call. Rapidly-repeated 'krrr', ending in 'kururah'; also uttered singly, and in flight. Used before or after next 2 calls. (2) Pre-landing call. Rolling 'kah', repeated 3–15 times, decreasing in intensity after first few repetitions. More rapidly repeated and harsher in ♂.

(3) Pre-mating call. ♂ utters 1–2 explosive 'chaah' notes; ♀ 1–4 explosive 'tjeeu' notes. (4) Hissing-call. Occasionally used by adults and immatures, apparently associated with threat displays.

CALLS OF YOUNG. Food-call a 'treu'-like squeak, rising and falling in volume, alternating with 2–3 clicks. Uttered by chicks of all ages.

**Breeding.** SEASON. Turkey: late March to late June (Aharoni 1930). SITE. In low trees, bushes, or occasionally reeds; from water level to 2 m above. Usually colonial; often mixed with cormorants (Phalacrocoracidae) and herons (Ardeidae). Nest: platform of branches, sticks, or sometimes reeds, average 45 cm diameter (Ruwet 1963); lined leaves and stems of water plants. Building: ♂ gathers material while ♀ builds, often finishing in 1 day. EGGS. Strongly elongated at one end; pale green or blue-white with chalky coat; proportion have many dark brown surface spots and deeper spots of violet-grey scattered irregularly, but in some gathered to one end; uncertain how much this is natural colour, and how much staining from nest material (Ottoson 1908; Aharoni 1930). 55 × 35 mm (53–61 × 34–37), sample 28; calculated weight 37 g (Schönwetter 1967). Clutch: 3–5. No data on number of broods or replacements. INCUBATION. Period unknown. 25–28 days in Anhinga *A. anhinga*, starting with 1st egg; by both sexes, though in 2 instances ♀ left after laying and ♂ did all incubating and rearing (Palmer 1962). YOUNG. Semi-altricial and nidicolous. Captive young able to climb about at 8 days (Aharoni 1930). Only recorded information from wild: will leave nest when less than 30 cm long and will swim away and dive when alarmed (Meinertzhagen 1935). Young of *A. anhinga* fed by complete and partial regurgitation by both parents, taking food dropped into nest and also putting head into parents' gullet (Palmer 1962). FLEDGING TO MATURITY. Fledging period in wild not known; nor time of independence. Attain flight at 6 weeks in captivity (Aharoni 1930). Said to breed at 1 year old while still in immature plumage (Meinertzhagen 1935). Age of first breeding of *A. anhinga* not less than 2 years (Palmer 1962). BREEDING SUCCESS. No information.

**Plumages** (*A. m. rufa*). ADULT BREEDING. Crown and hindneck dark brown to brown-black, blacker in ♂ than ♀; sides of neck paler. Streak of fluffy white feathers runs from gape halfway down side of neck, bordered on both sides by black streak of variable width, the ventral one sometimes absent. Chin and throat buff, gradually darkening to reddish-chestnut on lower foreneck, paler in ♀. Upper mantle brown-black with tiny black streaks on sides; lower mantle, back, rump, upper tail-coverts and underparts glossy black. Upper scapulars narrow lanceolate, black with white or pale brown shaft-streaks and black shafts. Lower scapulars broad, black with tiny white streaks at tips. Uppermost of lower scapulars with broad velvety brown to grey streak along shaft on inner web and corrugated outer web. Tail black, outer web of central feathers corrugated. Primaries and secondaries black; tertials black with white shaft-streaks, tips pointed. Greater upper wing-coverts velvety-brown to pale grey, outer web narrowly, inner broadly margined black. Rest of upper wing-coverts black with white shaft-streaks. Under wing-coverts and axillaries black. Black of scapulars and wings may become grey by bleaching and wear. ADULT NON-BREEDING. Like breeding, but head, neck, and upper mantle browner; white streak on side of head and neck lacks elongated feathers, not bordered black ventrally, and sometimes separated from dark crown and hindneck by pink-buff area. Variation in pattern of side of neck (see e.g. Friedmann 1930) probably dependent upon progress of moult. NESTLING. Down mostly white; crown, hindneck, and upper mantle buff in ♂, pale buff in ♀ (Meinertzhagen 1935). Some nestlings entirely brown (Aharoni 1930). JUVENILE. Crown and hindneck buff-grey, sides of head and neck pale buff without white streak. Mantle brown with buff feather edges; back, rump, and upper tail-coverts brown-black. Upper scapulars much less elongated than adult, brown-black with white shaft-streak and pale brown fringe. Underparts pale buff, darkest on chest, flanks and thighs dark brown. Tail black, tipped pale brown, central feathers not corrugated. Flight-feathers like adult, upper wing-coverts edged pale brown, most strongly on marginal coverts at bend of wing. In some young birds, white streak on side of head and neck vaguely indicated and breast, belly, and under tail-coverts brown-black with paler feather fringes; not known whether this sub-adult stage, or variant juvenile plumage of ♂ (as thought by Ogilvie-Grant 1898).

**Bare parts.** ADULT. Iris yellow with outer ring of brown (Granvik 1923); described as golden-yellow by many authors, but also as brown (in ♂, Meinertzhagen 1935), salmon-pink (Vincent 1934), or red (Reichenow 1900–01); evidently variable. Bill pale brown tinged yellow or green, with paler and more yellow tip; occasionally darker brown. Bare skin of face green; on throat cream, but said to be deep black in breeding ♂ (Aharoni 1930). Leg brown, paler and with yellow or green tinge in front, darker behind (Granvik 1923); webs brown-black in breeding ♂ (Aharoni 1930). NESTLING. Iris dark grey or blue-grey. Bill pale horn. Bare skin of face green, with dark line behind eye in ♂; on throat yellowish. Foot ivory-white to flesh (Hartert 1912–21; Meinertzhagen 1935). JUVENILE. Iris with brown outer ring, narrow white or yellow inner, and grey or yellow-grey in between. Bill grey-green. Bare skin of face dark green, on throat pale pink. Leg pale grey, tinged green, or brown-grey; darker at back of tarsus. (Granvik 1923; Chapin 1932; ZMA.)

**Moults.** Undescribed, except for simultaneous moult of flight-feathers and irregular moult of tail (Friedmann 1930; Middlemiss 1955). Anhinga *A. anhinga* has complete post-breeding moult and partial pre-breeding in which all body plumage replaced (Palmer 1962).

**Measurements.** Adult, skins (BMNH, ZMA).

| | ♂ | | | ♀ | | |
|---|---|---|---|---|---|---|
| WING | 349 | (12·8; 8) | 328–364 | 344 | (11·4; 7) | 331–360 |
| TAIL | 238 | (8·65; 8) | 229–253 | 239 | (7·35; 4) | 233–248 |
| BILL | 81·2 | (3·19; 12) | 75–87 | 75·5 | (2·43; 6) | 71–78 |
| TARSUS | 43·6 | (1·75; 11) | 41–46 | 42·4 | (1·40; 7) | 41–45 |
| TOE | 77·2 | (2·59; 9) | 72–80 | 77·0 | (2·16; 4) | 75–80 |

Difference between sexes not significant, except for bill.

**Weights.** 2 adult ♂♂, 1815, 1058; immature ♂ 1100 (ZMA).

**Structure.** Wing broad. 11 primaries: p8 longest, p9 1–3 shorter, p10 *c.* 20–25, p7 *c.* 5–10, p6 *c.* 30, p5 *c.* 70, p1 *c.* 125; p11 minute. P6–p9 emarginated outer webs. Tail long, strongly wedge-shaped. 12 stiff and broad feathers: t6 *c.* 50 shorter than t1. Bill slender and sharply pointed, cutting edges in distal half with minute backward-directed serrations. Neck long and thin. Tarsus short and heavy. Outer toe about equal to middle, inner *c.* 75% of middle, hind *c.* 40%. All toes joined by webs.

**Geographical variation.** Middle East population often considered separate race *chantrei* (Oustalet, 1882); said to have foreneck paler and greater upper wing-coverts greyer than *rufa*, but these characters variable in *rufa* (Ticehurst 1922) so recognition of *chantrei* not warranted. Nominate *melanogaster* differs from *rufa* in having neck greyer, white streak less pronounced; chin and throat in breeding plumage conspicuously chequered black and white; greater upper wing-coverts silver-white, never brown; juvenile brown below, not buff. *A. m. novaehollandiae* shows more pronounced sexual dimorphism in plumage than either *rufa* or nominate *melanogaster*. ♂ with head and neck darker than in *rufa*, ♀ with underparts buff-white. Nominate *melanogaster*, *rufa*, and *novaehollandiae* sometimes treated as 3 separate species (Vaurie 1965), or united with New World Anhinga *A. anhinga* into one. *A. anhinga* more strongly different from Old World forms than these from each other, so kept apart (Voous 1973), but considered to form superspecies with *A. melanogaster* (Mayr and Short 1970).

# Family PELECANIDAE pelicans

Huge aquatic birds. 7 species in single genus *Pelecanus*; 2 breeding in west Palearctic, 1 accidental. Body broad and heavy, neck long, head large. ♂ larger (particularly longer billed) than ♀, or sexes about equal in size. Wings long and broad, with long lower and much shorter upper arm; as in other Pelecaniformes do not fit closely to body when folded. 11 primaries; p8 longest, p11 minute; over 30 secondaries, diastataxic. Flight strong with deep wing-beats and occasional glides; often soar. Tail short and rounded, 20–24 feathers. Bill huge; long and broad. Large gular pouch, distensible by action of tongue muscle on rami of lower mandible. Upper mandible flat with strongly ridged culmen and powerful terminal hook. Nostrils tiny slits. Facial skin partly bare. Legs short and sturdy, lower part of tibia bare, totipalmate; claw of middle toe medially with horny flange, sometimes with comb-like incisions. Stance horizontal, gait slightly waddling; float high on surface when swimming. Oil gland feathered. Aftershaft absent. Plumage white or pale grey, more or less tinged pink, flight-feathers dark except for Brown Pelican *P. occidentalis*. Sexes similar, sometimes differing in bare parts which often more brightly coloured when breeding. 2 moults per cycle, pre-breeding restricted to development of crest on head or other ornamental feathers; some species then grow fleshy excrescences on front of bill. Primaries replaced in serially descendant order. Young altricial and nidicolous, naked at hatching, later covered in dense white or dark brown down. Contour feathers do not grow from same follicles as down. Juveniles differ from adults in being browner or duller. Fully adult plumage attained after several years.

Almost cosmopolitan; absent from north Holarctic, north-west Africa, south-east South America, and most oceanic islands. Aquatic, in warm low latitudes. Most species inhabit both inland lakes (fresh or brackish) and coast; only extralimital *P. occidentalis* true seabird. In west Palearctic, mainly in deltas, estuaries, lagoons, floodlands, and along sheltered waters, especially of more or less land-locked seas. Limited to places where biological productivity can sustain heavy biomass of large, active, sociable birds; restricted to shallow waters by poor diving capacity, and mainly to lowlands by aquatic requirements. Clumsy in take-off, but strong fliers up to considerable heights, with excellent capacity for sustained soaring by use of thermals; except on migration, rarely out of sight of land. Extremely sensitive to disturbance. Both west Palearctic species have declined markedly in range and numbers due to drainage, disturbance, and human persecution, but, in recent years, protection has led to stability or even some recovery. Migratory and dispersive. Travel in flocks, which may be large where common, in undulating lines or wedge-shaped formations; judging from predilection to soar, probably essentially diurnal. Generally migrate along river valleys and marine coasts. Surprisingly little modern information on routes or winter quarters of either European breeding species.

Food almost entirely fish. Most species surface-feeders, catching prey in bill while head-ducking or up-ending and then scooping with gular pouch acting as dip-net. Often hunt co-operatively in semi-circular groups. Only *P. occidentalis* surface or shallow plunge-diver. Typically gregarious at all times. Monogamous pair-bonds of seasonal duration; no evidence that birds associate as mates away from nest-site or colony. Usually colonial breeders, defending small nest-territories. Spend more time on water than either Phalacrocoracidae or Anhingidae, often loafing there as well as ashore or on sandbanks, etc. Roost communally at night, usually in shallow water or on shore. Some species loaf and roost in trees, at least at times. Heterosexual displays not intensively studied but seem decidedly simpler and less ritualized than in nearest pelecaniform families; include use of gular pouch, raising of bill, and bowing. Copulation only at nest-site. Voice relatively unspecialized; also non-vocal bill-clapping used chiefly at breeding colony where calls relating to threat, alarm, copulation, and display identified, suggesting a fair repertoire of social signals; also call when engaging in communal feeding. Comfort behaviour similar to most other pelecaniform birds; dissipate heat by gular-fluttering. General behavioural characters include direct head-scratching. Do not adopt wing-spread posture of Phalacrocoracidae and Anhingidae.

Seasonal breeders. Nest on ground on islands and in thick vegetation. Though usually colonial, can be solitary. Nest heaps of available vegetation; built by ♀♀ with ♂♂ bringing material in pouch. Eggs oval, rough, white. Clutches 1–3 (rarely 5–6); single brood. Replacements laid after egg loss. Eggs probably usually laid daily, or at 2-day intervals or longer. Incubation period fairly short, 30–36 days. Eggs incubated on feet by both sexes, in equal shares or mostly by ♀♀; no brood-patches. Incubation starts with 1st egg; hatching asynchronous. Young cared for by both parents; not left for first few days. Fed from pouch by partial or complete regurgitation. Young may leave nest at 20–30 days and form crèche. Fledge in 65–80 days, becoming independent then or later. Age of maturity 3 or 4 years.

## *Pelecanus onocrotalus* White Pelican

PLATES 27 and 28
[after page 254]

Du. Rose Pelikaan    Fr. Pélican blanc    Ge. Rosapelikan
Ru. Розовый пеликан    Sp. Pelícano vulgar    Sw. Pelikan

*Pelecanus Onocrotalus* Linnaeus, 1758

Monotypic

**Field characters.** 140–175 cm, bill 29–47 cm, wing-span 270–360 cm; huge, the second largest pelican, but some overlap with smaller individuals of Dalmatian *P. crispus*. Huge, pouch-billed, rather long-necked, apparently tail-less waterbird; like other pelicans markedly gregarious fish-eater. Sexes alike; seasonal variation in plumage and bare part colours. Juvenile and immatures separable.

ADULT. Whole body, tail, and all wing-coverts white (sometimes tinged buff). Uppersurface of secondaries and inner primaries whitish darkening towards wholly black outer primaries and primary coverts; undersurfaces of secondaries and primaries black. During breeding cycle, plumage acquires rosy tint (particularly on elongated wing-coverts) and both sexes grow short ragged crest on hind-crown, larger in female. Bill yellow-grey, brighter when breeding, with mandible edges pink or red. Eyes red. JUVENILE. Head, neck and upper plumage dull buff-brown, mottled overall but showing white only on rump, lower plumage dirty white, flight-feathers brown; immatures whiter, but dingier, more streaked and mottled than adults, flight-feathers black-brown. Bill grey, becoming yellower with age. Legs and feet flesh. Eyes grey-brown.

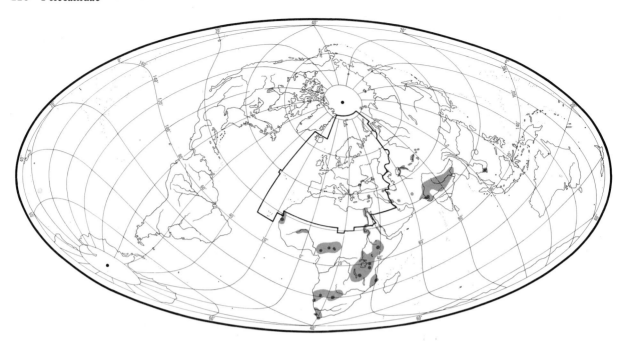

At distance on water or ground, difficult to separate from other west Palearctic pelicans, but, in flight, adult distinguished from *P. crispus* and Pink-backed Pelican *P. rufescens* by diagnostic wing pattern showing above only black outer primaries and coverts but below wholly black wing-quills contrasting sharply with white coverts. Such underwing pattern also marked in immatures, but indistinct in juveniles. At close range, adult separable from *P. crispus* by cleaner, whiter plumage, black wing-quills, leg colour, red eyes, and greater area of orbital skin which narrows forecrown feathering to sharp point just reaching centre of base of upper mandible; from *rufescens*, by longer, usually yellower bill and pouch, larger size, much whiter appearance (lacking grey tones), and again larger area of orbital skin and pointed forecrown feathering. Separation of young birds difficult (see other pelicans).

Flight pattern recalls stork *Ciconia*, but bill and head retracted over thick body cause different silhouette; action majestic with strong and powerful wing-beats, alternating with glides on slightly depressed wings; on migration or during feeding movements flocks travel in lines or V-formation, occasionally massing to soar and wheel. Usually needs long approach to land on water, but less space to take off. Gait a slow, clumsy waddle. Stands on bent tarsi; also perches freely on rocks and trees. Essentially gregarious, fishing in highly regimented troops, and loafing in clubs. Breeds colonially. Flight-call deep, quiet croak.   DIMW

**Habitat.** Surviving in scattered strongholds within middle and low latitudes in Mediterranean, steppe, desert, savanna, and dry-winter tropical climatic zones. Frequents low-lying, shallow, warm fresh waters, especially larger river deltas and wetlands; less often brackish or saline lagoons, estuaries or coastal waters, especially of land-locked seas, and inshore, rocky, or sandy islets. Nests among dense wetland vegetation, infrequently woody, and on bare soil or rock. For roosting sites see Social Pattern and Behaviour. Prefers waters near land and land near water; ample accessible supplies of medium-sized fish essential. Despite large size and clumsy build, remarkably aerial; flies and soars in thermals up to several hundred metres above breeding area, as well as on migration. Practice of mass fishing and requirements for undisturbed and inaccessible nesting quarters often involve regular journeys between separate areas, extending between 10 and 50 km or exceptionally up to 100 km. High biological productivity in habitat essential to sustain heavy biomass burden of concentrations of such large birds. Choice of suitable areas narrowed by lack of diving capability inherent in ultra-light structure required for soaring flight. Strong feet give high performance in swimming and fair mobility on ground, where rests freely. Does not seek or readily tolerate human proximity, and favours areas guarded against disturbance by such barriers as extensive reedbeds or difficult navigation. Conspicuous vulnerability and exacting habitat requirements of large social units make protection increasingly necessary for survival.

**Distribution.** Marked shrinkage in breeding range due to disturbance and persecution by man and drainage of nesting areas. HUNGARY: last bred 1858 (AK, IS). YUGO-SLAVIA: last bred in north-east (Slavonia and Vojvodina marshes), *c.* 1868 and Crna Reka, Macedonia, *c.* 1940 (VJV). BULGARIA: last bred 1932 although one pair reported with *P. crispus* at Lake Sreburna 1967 (SD, Terrasse *et al.* 1969a). RUMANIA: at end 19th century nested inland along rivers as far as Călăraşi but now confined Danube delta (Vasiliu 1968). GREECE: discovered nesting Lake Mikra Prespa 1968 (WB). Formerly bred Iraq (1884–1922) and Kuwait (near Bubyan Island 1922–23), but no recent records (Ticehurst *et al.* 1926; PVGK; SH).

In Ethiopian Africa, some colonies occupied irregularly and others probably undiscovered; wintering areas, there and in Asia, not fully known.

Accidental. Records in Europe from France, Finland, Poland, Czechoslovakia, Hungary, Austria, Switzerland, and Italy, but many old and most probably refer to escapes. Also Cyprus, Malta, Libya, Jordan, Algeria, Tunisia, Rio de Oro.

**Population.** RUMANIA. Formerly largely persecuted, but since early 1950s breeding areas given special protection. Location and size of colonies vary from year to year and population trends uncertain, but perhaps some recent decline after earlier recovery—1956, not exceeding 2000 pairs; 1961, 6000 pairs (of which 15–20% *P. crispus*); 1960–61, 4000 nests in 14 colonies; 1964–5, 1000 pairs in 2 colonies; 1969, *c.* 1700 pairs; 1971, *c.* 1200 pairs (Vasiliu 1968; Bauer and Glutz 1966; Terrasse *et al.* 1969a; SC). GREECE: Lake Mikra Prespa up to 40 pairs 1969 and, under protection, *c.* 150–180 pairs 1971; a few pairs may nest Macedonia (WB). USSR. Less numerous than *P. crispus*. Decreasing numbers, especially Aral Sea, for a decade, also in Volga delta, but increase after establishment of reserves (Dementiev and Gladkov 1951); no longer breeding Volga delta, 1972 (Yu A I). TURKEY: 2 colonies with *c.* 2500 pairs in 1969; birds summer in other areas also; formerly bred

Amik Gölü, now drained (OST). MAURITANIA: found breeding Île Arel, Banc d'Arguin, 1957; 1959–65, some 300–400 pairs frequently robbed (de Naurois 1969a).

**Movements.** Northern populations migratory, but perhaps only dispersive in tropics. Considering conspicuousness, extraordinarily little known of movements; no ringing recoveries. Few winter Balkans or Turkey, so most European and Turkish breeders must migrate. Substantial numbers winter Egypt—in Delta, Faiyum, Suez Canal, and Red Sea (G E Watson)—and others possibly continue south where presence might be masked by large resident African populations. Fairly common along Red Sea coasts in winter, less so on Arabian side than African (Meinertzhagen 1954); limited southward passage noted through Gulf of Aqaba November–January (Safriel 1968). Has occurred winter on Black Sea in Crimea and Georgia, but migrants passing through Turkey, Syria, and Israel probably include Danube delta and Sea of Azov breeders taking land route to Egypt or beyond; rarity on east Mediterranean islands may indicate avoidance of long sea crossings. Like storks *Ciconia*, not infrequently migrate by soaring in thermals, though more usually by direct flight in V-shaped formations at high altitudes. Winter ranges of Caspian and Lake Rezaiyeh breeders problematical, and unknown whether they depart south-west or south-east; most leave, for few winter among Dalmatian Pelicans *P. crispus* of south Caspian, and only small numbers elsewhere Iran (D A Scott). Migration noted along River Tigris, Iraq, in March and early April and September to early November (Marchant 1963a); these presumably moderate numbers that winter south Iraq

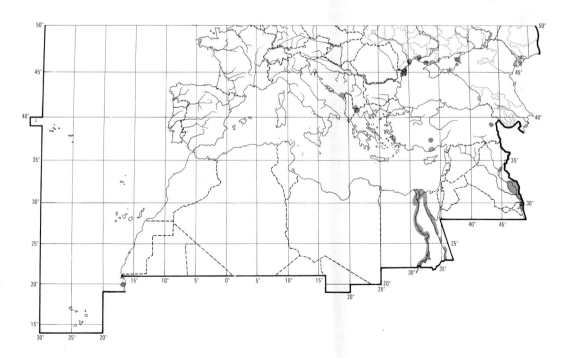

marshes (Allouse 1953; Moore and Boswell 1956). Immigrants reach Indian subcontinent, where winter numbers exceed size of Rann of Kutch colony (Ali and Ripley 1968); but perhaps these mostly from Kazakhstan.

Danube delta breeders depart September to early November, returning late March to April, both movements later than *P. crispus*. Main passages through Greece late September and October, and April (Zelenka 1963). USSR breeders depart with advent of frosts, continuing into November; colonies re-occupied March to early April (Dementiev and Gladkov 1951). Non-breeders do not all return to colonies, and flocks seen in summer in Balkans, Iraq, and Iran; puzzling record of 3000 flying south-east over Israel on late date 20 April (Harrison *et al.* 1962) probably also non-breeders. In Ethiopian Africa (as probably elsewhere), moves in response to water level changes, some sightings far from known colonies, and non-breeders wander at all seasons; 3000 present Banc d'Arguin December 1971 and 7000 August–September 1972, coinciding with low water levels in Sénégal rivers, aggravated by 2 years of severe drought (Pététin and Trotignon 1972; Gandrille and Trotignon 1973). Those recorded north to Denmark, Fenno-Scandia, Poland, Lithuania, and Leningrad, probably escapes. Few seen Europe away from south-east genuine vagrants, unlike several flocks recorded in Hungary, Italy, Austria, and Germany in 18th and 19th centuries when bred farther west than now (Bauer and Glutz 1966).

**Food.** Fish, caught in bill with gular sac held open in form of scoop. Often by co-operative fishing in groups forming semicircle with open side forward; fish scooped up with rapid dip of bill into water. Solitary feeding also common, usually within recognisable loose group, more especially when fish abundant. Species taken mainly gregarious, living in shallow water or near surface.

In Danube delta, 38 stomachs contained 22·9% (frequency) carp *Cyprinus carpio*, 15·2% bitterling *Rhodeus sericeus amarus*, 11·4% bream *Abramis brama*, 9·5% rudd *Scardinius erythrophthalamus*, 7·6% spined loach *Cobitis taenia* and, less frequently, 12 other fish species; average weight and size of each species relatively small (Andone *et al.* 1969). Adult feeding nestlings contained 3·95 kg fish, including a *Cyprinus carpio* of 1·85 kg (Dombrowski 1912). In Volga delta, mainly feeds on carp; also in spring *Rutilus caspicus*, perch *Perca fluviatilis*, and pike *Esox lucius*. In early spring, Lenkoran lowland rivers, *R. caspicus* and *R. frisii kutum* (Yu A Isakov). In Africa, main species probably cichlids *Tilapia*. In Lake Nakuru, Kenya, and Lake Natron, Tanzania, chiefly *T. grahami*; in Lake Abiata, Ethiopia, probably *T. nilotica* (Brown and Urban 1969). On Kafue river, Zambia, main prey in summer 1965 probably *Coptostomobarbus vittei* and *Aplocheilichthys johnstoni*; in one disgorged meal, however, 3 *Tilapia*, estimated 300–600 g each, and smaller unidentified cichlids found (Tait 1967). *Tilapia* main food

found Lake George, Uganda (Cott 1952). 65 stomachs from Lake Edward, Uganda, contained 83% (frequency) *Tilapia*, 39% *Haplochromis*, and 12% *Haplochromis* fish fry (Din and Eltringham 1974a). Mullet *Mugil* recorded as prey in south China (Caldwell and Caldwell 1931).

In early stages of life, chicks given frequent small amounts of liquid, pre-digested food, graduating later to whole fish up to 500 g taken from parent's gullet (Brown and Urban 1969; Dementiev and Gladkov 1951). Young at Kutch, India, disgorged fish (mainly *Cyprinodon dispar*) up to 25 cm long and weighing 500–600 g each (Ali and Ripley 1968).

Brown and Urban (1969) estimated intake of 900–1200 g of food per day for adults, Din and Eltringham (1974a) 1201 g, and Andone *et al.* (1969) 1600 g (captive birds).

**Social pattern and behaviour.** Based largely on material supplied by L H Brown from observations in Ethiopia; see especially Brown and Urban (1969).

1. Gregarious at all seasons. In flocks, often nomadic, outside breeding season, composed of both adults and immatures but no evidence of discrete family parties. When feeding, large flocks break up into smaller parties (6–20) which feed communally and often co-operatively. When moving to another feeding ground, or on migration, travel in large flocks (50–500) composed of smaller sub-units (20–100). BONDS. Probably forms monogamous pairs for single season; may re-mate in successive years, but no proof. Both parents tend young. BREEDING DISPERSION. Colonial; usually in large, dense groups of 300–40000 pairs, each colony with number of more or less discrete sub-units which lay synchronously. Flies from breeding to feeding grounds in small flocks (4–50), then feeds in normal feeding bands (see above). Only nest area defended and incubating birds may touch. No feeding territory defended at any time. Copulation, nesting, and feeding small young occur in nest-territory but pair-formation, self-feeding, and feeding larger young all away from it. When 2–3 weeks old, young collect in crèches ('pods') until 7th week, when walk about freely. ROOSTING. Communal. Roosting and day-time loafing areas the same. Normally in single or multiple flocks (100–2000). Main sites traditional, used by flocks year after year; at sandbanks (islands preferred), reedbeds, and rocky islands. Infrequently roosts in trees, e.g. when forced down by bad weather when travelling between feeding grounds. Sites usually completely open though on islands or in reedbeds may be screened by e.g. papyrus. Travelling flocks descend to temporary roost 1–2 hours before dark, depending on thermal activity, remaining there until thermal activity permits soaring flight, usually 1–3 hours after daylight. Flocks more permanently based on feeding lake split into small feeding parties early in morning; afterwards returning to loaf at roost area or sometimes another similar spot. May leave roost to feed at night, especially in clear moonlight. At breeding colony, bird incubating or with small young roosts on nest. Partner may roost beside it, but more often away near feeding grounds. Unmated adults and pre-breeders roost in flocks in breeding island. Parents with large young may roost there too, some with pods of young (see below), others in flock together. Earliest departures from roost on breeding island soon after dawn; main exodus when thermal activity permits.

2. ANTAGONISTIC BEHAVIOUR. Generally peaceful. Typical threat-posture consists of raising and stretching bill upwards (Head-up) and towards opponent (see A), sometimes lunging

PLATE 21. *Phalacrocorax carbo* Cormorant (p. 200). *P. c. carbo :* **1** ad breeding, **2** ad non-breeding, **3** 2nd imm winter, **4** 1st imm winter, **5** juv – light and dark types, **6** nestling. *P. c. sinensis :* **7** ad breeding. *P. c. lucidus :* **8** ad breeding, **9** juv. (CJFC)

PLATE 22. *Phalacrocorax aristotelis* Shag (p. 207): **1** ad breeding, **2** ad non-breeding, **3** 2nd imm winter, **4** 1st imm winter, **5** juv, **6** nestling. (CJFC)

PLATE 23. *Phalacrocorax nigrogularis* Socotra Cormorant (p. 214): **1** ad breeding, **2** ad non-breeding, **3** juv, **4** imm, **5** nestling. (CJFC)

PLATE 24. *Phalacrocorax pygmeus* Pygmy Cormorant (p. 216): **1** ad early breeding, **2** ad breeding, **3** ad non-breeding, **4** imm, **5** juv, **6** nestling. (CJFC)

PLATE 25. *Phalacrocorax africanus:* Long-tailed Cormorant (p. 219): **1** ad breeding, **2** ad non-breeding, **3–4** stages between juv and ad, **5** juv, **6** nestling. (CJFC)

PLATE 26. *Anhinga melanogaster* Darter (p. 223): **1** ad ♂ breeding, **2** ad ♀ breeding, **3** ad in faded plumage, **4** ad non-breeding, **5** juv, **6** nestling. (CJFC)

A

B

sharply; accompanied by repeated grunts or Moo-call. Rival may parry with similar lunge or raised bill, or turn and walk or swim away with bill lowered. Attacker seldom pursues. ♂♂ more aggressive than ♀♀. Normally no threat in feeding parties and only little at low intensity when loafing or roosting; threat intensifies, however, at pair-formation and mating. HETERO-SEXUAL BEHAVIOUR. Sexes develop differently coloured patch or 'knob' on bare skin of forehead early in breeding season. Pairs form at or away from breeding colony. Knobbed ♀♀ attract one or more knobbed ♂♂ from flocks of displaying or resting birds. When several ♂♂ follow same ♀, one ♂ quickly becomes dominant, repelling others with threat behaviour. Dominant ♂ swims after ♀ and may attempt copulation, forcing ♀ underwater. At colony, groups of unmated, knobbed ♂♂ perform Group-display, standing together, walking about, sometimes lunging with outstretched bills into centre of group, grabbing each other's bills, raising bills and Mooing; actions confused, irregular with no distinct pattern. Unmated ♀♀ join these groups at edge; then approached and courted by one or more ♂♂. Preferred ♂ apparently quickly selected and then repels other ♂♂. One or several pairs may then fly or swim out performing Setting-to-partners on water with ♂ swimming beside or behind ♀. 2 pairs may approach, with ♂♂ giving Head-up display or seizing each other's bills, then breaking away to swim beside ♀♀ again. Pairs may circle one another for some minutes, or soon move away and return to displaying group on land. On land, ♂ follows ♀ with Strutting-walk (see B): head and neck extended, bill lowered below horizontal, wings partly open, body swinging from side to side. Pair may then re-join displaying group, stop and stand about, or walk deliberately to future nesting area. Strutting-walk of ♂ indicates more advanced stage of pair-formation, but not necessarily followed quickly by site selection and copulation. When responsive, ♀ walks over to nesting area, followed by ♂, and ♀ selects site among other pairs; rarely resented or repelled. She squats on site and almost at once begins nesting activity, scraping ground with bill. At copulation, ♂ stands beside ♀, leans over and seizes her neck in bill, then mounts from side, pressing standing ♀ to ground. Successful copulation possible only on firm ground, not in water; lasts 3–5 s and may be repeated many times a day. Soon after copulating, ♂ walks away and fetches nest

material in pouch; may rob other nests. Usually ♂ drops materials in front of ♀ who incorporates them into rude nest around and beneath her. Nest material sometimes thrown into air by ♀ with swift jerk of bill; similarly earth or sand. At nest-relief, relieving bird alights near colony, walks in with head and neck raised, bill stretched upwards, wings partly open; sometimes lunged at by others and lunges back. Arriving at nest, walks up to sitting partner, raises bill and emits deep grunting calls, usually face to face, sometimes from side. Partner rises slowly, backs off nest and lowers bill to point downwards in Bow (see C). Pair may shuffle to and fro or walk round each other several times before relief complete. Once relieving bird well settled, partner stands for few moments beside it, wing-flaps vigorously, and then either takes-off or walks to bare ground to fly away, Little Allopreening occasionally recorded; also mutual billing or desultory sparring occurs between partners or adjacent birds. RELATIONS WITHIN FAMILY GROUP. Newly hatched young induced to take 1st feed by parent lowering bill between feet; bright red nail on bill stimulates pecking reaction from young. Later, naked young stimulated to feed by parent nibbling heads or bodies. Smaller young in pods do not recognise own parent but known by it. Adult arriving to feed walks up to pod, searches for own young, reaches over others, seizes it by neck or body, drags it out, and shakes it vigorously. Young thus stimulated to feed grabs parent's bill-tip and thrusts bill into pouch, sometimes into oesophagus. Larger young (28–42 days) in pods, and feathered young (42–70 days) moving freely about colony, recognize own parents and run to them on alighting. They solicit vigorously, weaving head and neck about, flapping wings, sometimes biting at own wings or even falling down as if stunned. Parent usually runs away with bill and neck fully extended upwards until suitable place reached; then stands or lies with bill level, neck fully extended for young to obtain food from pouch, oesophagus, or gullet. After feed, young may again pursue parent, soliciting; parent runs away and soon

C

flies. Parent may also forcibly disengage young by shaking head and neck. Small siblings in nest not aggressive; nevertheless brood usually reduced to 1 by pod stage. Little aggression noted in older young until after 42 days when free walking and large. Big young may attack smaller young which have been fed, piling over them, seizing them by neck and shaking them, apparently to make them disgorge. 2 hungry young may solicit one another and go through feeding motions, placing bills in each other's pouches. As feathers grow, young develop normal threat behaviour of adults: lunging at one another, Head-up display, etc.

(Figs A–C after Brown and Urban 1969.)

**Voice.** Mainly silent away from colonies. Continuous low-pitched roar or hum heard from large colony at rest; louder by day but never entirely silent, with sharp increase at dawn. Flight-call deep, quiet croak. Following vocalizations distinguished at colony (L H Brown). CALLS OF ADULTS. (1) Moo-call. Deep, resonant 'ha-ooogh', given in Head-up and Setting-to-partners display. (2) Threat-calls. Deep, nasal grunting 'huuh-huuh-huuh'; also probably deep grunting or growling 'orrrh-orrrh-orrrh'. (3) Food-call. Low, resonant 'huh-huh-huh', uttered by adults summoning young to be fed. (4) Rattling-call. Low rattling growl; not apparently associated with any particular activity. CALLS OF YOUNG. High-pitched and always clearly audible above deeper calls of adults. (1) Wailing-call. High-pitched, baby-like wailing; uttered by small nestlings. (2) Yelping-call. Continuous, yelping 'yewk-yewk-yewk', uttered by larger young being brooded or when grouped together in pods (see Social Pattern and Behaviour). (3) Yelling-call. Frantic wailing yells, 'eeee-yeeeee-eh eeee-yeeeee-eh', given by large, walking young soliciting parents for food.

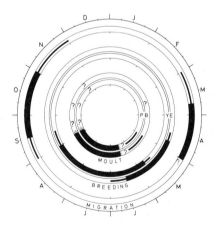

**Breeding.** SEASON. See diagram for Balkans. SITE. In thick vegetation, on islands, or in reedbeds. Occasionally on bare earth or rock outcrops. Colonial; nests can be touching. Nest: substantial pile of reeds, twigs, and other vegetation; sometimes lined with feathers (Ali 1960). Average height 30–60 cm, diameter at base 70–100 cm,

diameter of cup 30–50 cm (Terrasse *et al.* 1969*a*). Building: by ♀, ♂ bringing material in pouch. Takes 2–3 days (Dementiev and Gladkov 1951). EGGS. Oval, rough; white. 94 × 59 mm (80–104 × 52–64), sample 150 (Schön-wetter 1967). Weight 155–195 g, mean and sample not given (Dementiev and Gladkov 1951). Clutch: 2 (1–3). One brood. Replacement eggs laid if first eggs lost within 10 days of laying. Eggs probably laid daily, sometimes at 2-day intervals. INCUBATION. 29–30 days (Dementiev and Gladkov 1951). 35–36 days in Kenya (Brown *et al.* 1973). By both sexes, probably in equal shares. Starts with first egg; hatching asynchronous. YOUNG. Altricial and nid-icolous. Cared for and fed by both parents. Brooded continuously for first 2–3 days, and not left until about 14 days. Fed on partially regurgitated liquid until about 14 days, then feeds itself on solids, putting head into parent's pouch (Brown and Urban 1969). At *c*. 20–30 days young leave nest and form crèche (pod). Able to swim and fish at *c*. 55 days. FLEDGING TO MATURITY. Fledging period 65–70 days. Become independent at about the same time. Age of maturity probably 3 or 4 years. BREEDING SUCCESS. No data from west Palearctic. In Ethiopia, 0·8–0·9 young reared per pair, equalling *c*. 50% of eggs laid (Brown and Urban 1969).

**Plumages.** ADULT. Head, neck, and body white, tinged pink during breeding, possibly due to coloured secretion of oil gland; underparts, particularly lanceolate feathers on chest, often stained brown by iron oxide from water (Brown and Urban 1969; Baxter and Urban 1970). Crest of long, white feathers on nape during breeding (10–14 cm, Dombrowski 1912). Long scapulars sometimes with black streak along margin of outer web. Tail white. Primaries dark grey, tips and edges of inner webs silver-grey. Outer secondaries with dark grey centres, tips and edges of inner webs paler; edges of outer webs white, corrugated; inner secondaries increasingly paler and with more white. Under-surface of primaries and outer secondaries black. Upper and under wing-coverts and axillaries white. NESTLING. Naked at hatching. Dark brown down develops at 3–14 days, becoming paler after 28 days. Feathers on upperparts grow at 28–42 days, on underparts at 42–56 days. JUVENILE. Head, neck, and rest of upperparts, including upper wing-coverts grey-brown. Forehead and crown paler. Scapulars and wing-coverts with pale grey edges. Underparts and under wing-coverts off-white. Tail brown; flight-feathers dark brown. IMMATURE. Head, neck, centre of mantle, and back pale grey; centre of hindneck darker. Sides of mantle cinnamon-brown. Long scapulars dark grey with paler margins. Underparts white. Tail grey. Flight-feathers like adult but duller. Upper wing-coverts cinnamon-brown; marginal under wing-coverts mottled grey. Older immatures white like adult, but with some grey on long scapulars and greater upper wing-coverts. No ornamental feathers on nape or chest. Age at which immature plumages worn not exactly known; fully adult in 3rd (Koenig 1932) or 4th year (Dombrowski 1912).

**Bare parts.** ADULT. Iris red. Ridge and base of upper mandible and basal half of lower mandible grey-blue, cutting edges pink or red, nail cherry-red; rest of bill yellow-grey, brighter in breeding. Pouch and bare skin of face yellow; pouch bright yellow in breeding, bare skin in ♂ pink-yellow, in ♀ intense orange especially conspicuous prior to egg-laying due to swollen forehead or 'knob' (Brown and Urban 1969). Leg and foot pink or flesh, tinged crimson in breeding. NESTLING. Iris lead-grey (Dombrowski 1912). Bare skin at hatching, including bill and foot, deep pink; becomes dark slate-grey at 3 days (Brown and Urban 1969, Lehmann 1974). JUVENILE. Iris grey-brown. Bill grey, becoming yellower with age. Leg and foot flesh. Iris of older immatures dark brown.

**Moults.** Poorly known. ADULT POST-BREEDING. Complete; primaries serially descendant with up to 3 active centres in wing. Starts July with loss of occipital crest (Dombrowski 1912), but Bauer and Glutz (1966) supposed that body feathers already moulted during breeding as in *P. crispus*. ADULT PRE-BREEDING. Occipital crest acquired before breeding. Extent of moult of other feathers unknown. Serial moult of primaries apparently continues through most of non-breeding period. POST-JUVENILE. Probably starts in autumn of 2nd calendar year (Koenig 1932).

**Measurements.** Skins (RMNH, ZFMK): additional data on wing and bill from Chapin and Amadon (1950).

| | ♂ | | | ♀ | | |
|---|---|---|---|---|---|---|
| WING | 684 | (13·9; 24) | 665–772 | 620 | (17·1; 17) | 586–650 |
| TAIL | 176 | (13·0; 5) | 155–188 | 162 | (13·7; 7) | 138–178 |
| BILL | 409 | (25·1; 9) | 347–471 | 328 | (31·2; 14) | 289–400 |
| TARSUS | 145, 146, 149 | | | 132 | (7·42; 6) | 125–145 |
| TOE | 139, 146, 149 | | | 136 | (7·93; 7) | 125–145 |

Sex difference significant for wing and bill, though both show great individual variation.

**Weights.** ♂ 11 000 (Ali and Ripley 1968). ♂♂ up to 11 000, ♀♀ up to 10 000 (Dementiev and Gladkov 1951).

**Structure.** Wings long and broad, with long arm. 11 primaries: p8 longest, p9 0–10 shorter, p10 *c*. 40, p7 *c*. 10, p6 *c*. 30, p5 *c*. 80, p1 *c*. 250; p11 much shorter than primary coverts. P6–p9 emarginated on outer webs, emargination of p8 and p9 hidden under coverts; p6–p10 on inner webs, p6–p7 only slightly, emarginations just beyond coverts. Tail short and rounded; 24 feathers (sometimes 22), outer *c*. 50 shorter than central. Bill long and broad, with large hooked nail at tip, ridge over culmen, and enormously extensible pouch between rami of lower mandible. Feathers on forehead protrude in point towards base of upper mandible. Nostrils small and slit-like, concealed in nasal groove at extreme base of bill. Large area of bare skin on side of face. In early stage of breeding, fleshy knob develops at base of bill on forehead, shrinking after eggs laid (Brown and Urban 1969). Leg sturdy. Lower part of tibia bare. Outer toe slightly shorter than middle, inner *c*. 70% of middle, hind *c*. 50%.

**Geographical variation.** Strong individual variation, and existence of small subspecies *roseus* Gmelin, 1789, disproved. Name *roseus* synonym of *P. philippensis* (Chapin and Amadon 1950; Amadon 1955), but sometimes still used for African or east Asian populations of *P. onocrotalus* (e.g. Brown and Urban 1969).

## *Pelecanus crispus* Dalmatian Pelican

PLATES 27 and 28
[after page 254]

DU. Kroeskoppelikaan    FR. Pélican frisé    GE. Krauskopfpelikan
RU. Кудрявый пеликан    SP. Pelicano ceñudo    Sw. Krushuvad pelikan

*Pelecanus crispus* Bruch, 1832

Monotypic

**Field characters.** 160–180 cm, bill 37–45 cm, wing-span 310–345 cm; huge, the largest of all pelicans, though not markedly bigger than White Pelican *P. onocrotalus*, 25% larger than Pink-backed *P. rufescens*. Sexes alike; seasonal variations in plumage and bare part colours. Juvenile and immature separable.

ADULT. Whole head, neck, body, and folded wings greyish-white, tinged dirty blue below, with fringe of curly feathers on hindcrown and hindneck and large crop patch of straw-coloured feathers. During breeding cycle, plumage brighter, slightly silvery above, bluer below and head fringe grows into crest. Uppersurfaces of secondaries ash-grey with whitish edges, of primaries and primary coverts increasingly black, outer feathers completely so; undersurfaces of quills much paler with faintly dusky tips visible on inner primaries but black only on outer six. Orbital skin yellow; purple during breeding. JUVENILE. Upperparts dull, mottled grey-brown; underparts dirty white, quills brown. Immature becomes whiter, but dingier than adult with paler quills. Bill of adult pale yellow with orange nail in breeding cycle, greyer at other times; of young birds grey, with increasing cream or

yellow tone; pouch of adult in breeding cycle orange, otherwise pale yellow, of young greyish-yellow. Eyes pale off-white or yellow. Legs and feet at all ages and seasons lead-grey.

Separation from either congener difficult at distance, but adult's lack of any pink or reddish plumage tones and leg colour allow instant identification at close range. Flying adult readily separable from *P. onocrotalus* by uniform underwing (except for dark primary tips) and from *P. rufescens* by larger size, longer bill, and again uniform underwing. Young birds confusing, but greyer plumage tone and leg colour rule out *P. onocrotalus* and size and bill length *P. rufescens*. Flight silhouette similar to *P. onocrotalus* but action most powerful of all pelicans.

Gait a ponderous waddle, stance upright on bent tarsi. Essentially gregarious; breeds colonially. Usually silent outside breeding season.                DIMW

**Habitat.** Similar to and often shared with White Pelican *P. onocrotalus*, but less exclusively lowland and coastal. Tolerates undulating or hilly terrain with small open

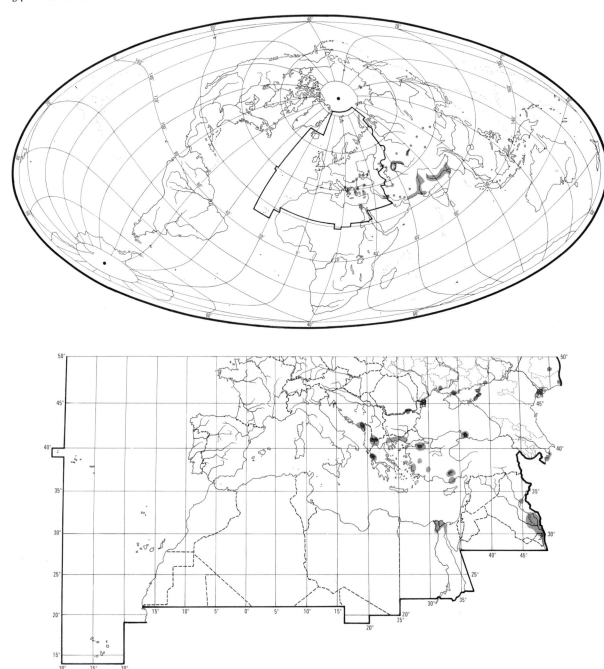

waters, and will accept breeding habitats only capable of carrying lower density, although nesting at high density favoured where conditions permit. Uses floodlands as well as shallow open waters; sometimes fishes inshore on sheltered coasts.

**Distribution.** Considerable decrease in range, wherever human population has increased, due to drainage, sensitivity to disturbance when nesting, and persecution by fishermen. According to Pliny, formerly bred western Europe in estuaries of Scheldt, Rhine, and Elbe; nested Hungary until middle 19th century (Voous 1960). RUMANIA: nested as far inland along Danube as Balafat in 1873; since 1904, has bred only in delta (Vasiliu 1968). In south-east Europe, several former colonies deserted, some recently (see Population).

Accidental. Hungary, Italy, Israel, Cyprus, Algeria, Rio de Oro.

**Population.** RUMANIA. In 1873, said to number millions of birds; still thousands 1896. 900 pairs in 1900, 200 pairs in 4 colonies in all Danube delta in 1909; 1300 pairs 1939. Apparently little change 1960–3, but only a few pairs after major floods 1964–5 (Vasiliu 1968). 300 pairs in 1969 (Terrasse *et al.* 1969*a*); *c.* 60 pairs in 1971 (SC). YUGOSLAVIA. Bred Vojvodina to 1870; near Adriatic coast 1858; in Macedonia to 1950s, with *c.* 20–30 pairs at Crna Reka in 1940 but drained by 1955. Only surviving colony at Lake Skadar where 30 pairs 1967, *c*, 25 pairs in 1973, 16 nests 1974 (VJV; Terrasse *et al.* 1969*a*). BULGARIA: bred for many years Lake Sreburna; *c.* 30 pairs 1960, destroyed by fishermen (Mountfort and Ferguson-Lees 1961). GREECE. Formerly bred many wetlands. Now nesting near Prewesa, *c.* 20 pairs in 1960–71, and Lake Mikra Prespa, *c.* 75 pairs 1969 and then under protection *c.* 90–100 pairs 1971, 120 pairs 1972, 150 pairs 1973 (WB). ALBANIA: some nests at Lake Mikra Prespa; colony at nearby Lake Malik in 1930 (Géroudet 1974). TURKEY. 3 known colonies: Lake Manyas, *c.* 30 pairs 1970; Kizil Irmak delta, 25 pairs 1961, *c.* 60 pairs 1971; near Eregli, a few pairs in 1970. In Thrace, colonies formerly River Ergene and Meric delta in 1962 (OST). USSR. Many scattered nesting areas, but little population data except for Volga delta, where abundant second half 19th century but numbers declined sharply early 20th century. Almost non-existent in 1920, *c.* 60 nests 1930, increasing to 300 pairs or more by 1949 under protection (Dementiev and Gladkov 1951).

**Movements.** Migratory and partially migratory. Rather little known as paucity of recent information on wintering areas and few ringed. Apparently European breeders winter east Mediterranean from Balkans (mainly Greece) to lower Egypt (where, however, Horváth (1959*b*) saw none though many White Pelicans *P. onocrotalus*). Moderate numbers pass through Turkey, and up to 600 winter (exceptionally 1000 in 1972); possibly include Sea of Azov breeders. Few recent reports from elsewhere in Levant. Migratory in north Caspian; but perhaps dispersive in south Caspian where several hundred winter Iran and Azerbaydzhan coasts; also small numbers in Fars, Seistan and Khuzestan (D A Scott), and in Tigris–Euphrates marshes of Iraq (Allouse 1953). Not uncommon winter visitor across Indian subcontinent from Baluchistan to Bengal, probably from Kazakhstan; possible that some Caspian birds reach western part of this area, or that some Kazakhstan breeders move west to Caspian; several ringed Kazakhstan showed WSW movements (Dementiev and Gladkov 1951), perhaps circuiting around Pamirs, but 2 from Lake Sassyk Kul (near Chinese border) moved 1300 km to lower Syr-Darya River (not far from Aral Sea) and 1500 km to Kagan in Uzbekistan, both rather far west.

Autumn desertions of and spring returns to colonies both earlier than in *P. onocrotalus*. Danube breeders leave August and return March (Bauer and Glutz 1966). In Volga delta, return March but depart about October; during southward exodus, a few linger on lakes while ice-free; first signs of return passage apparent Transcaucasia late February (Dementiev and Gladkov 1951). Present Greek winter quarters August–March (Zelenka 1963). Immatures seldom seen in colonies. Has occurred Kuwait and Gulf of Suez, but otherwise unrecorded Arabia (Meinertzhagen 1954); nor ever identified Ethiopian Africa. On average winters farther north and wanders less than *P. onocrotalus*; no European records north or west of Hungary can be accepted as genuine vagrants (Bauer and Glutz 1966). USSR records north to Tobolsk, Omsk, Krasnoyarsk, and Cherepovets assumed to be vagrants (Dementiev and Gladkov 1951), but perhaps also include escapes.

**Food.** Entirely fish. Caught in gular sac used as scoop in water, and often by group-fishing when form into semi-circle and drive fish forward.

10 stomachs from Danube delta, August, contained: carp *Cyprinus carpio* (8), perch *Perca fluviatilis* (6), asp *Aspius aspius* (4), roach *Rutilus rutilus* (4), pike *Esox lucius* (4), tench *Tinca tinca* (2), and (singly) 4 other species (Korodi-Gál 1961, 1964). In Volga delta, mainly carp, and also in spring perch, pike, and *Rutilus caspicus* (Yu A Isakov). In Lake Sreburna, Bulgaria, mainly Crucian carp *Carassius carassius*, carp, and tench; to lesser extent roach, orfe *Leuciscus idus*, and rudd *Scardinius erythrophthalamus* (Bauer and Glutz 1966). On hot days, pike up to 50 cm long taken as they swim near surface. In Nile winter quarters, mainly Siluridae (*Siluris auritus, Schilbe, Bagrus, Synodontis, Clarias, Heterobranchus*) and Mormyridae found in stomachs and crops (Heuglin 1873).

Estimated that 2 adults and 2 juveniles ate 1080 kg fish in 8 months, on average 1123 g daily per bird (Dementiev and Gladkov 1951). Average weight of stomach contents of 12 birds 1269 g (Korodi-Gál 1964).

**Social pattern and behaviour.** No detailed studies; based mainly on Bauer and Glutz (1966). See also Meischner (1958, 1962), van Tets (1965).

1. Gregarious throughout year. Feeds communally and co-operatively; also at times singly. Winter flocks often relatively large. BONDS. Monogamous pair-bond of at least seasonal duration. Both parents tend young, family group continuing to associate after fledging. BREEDING DISPERSION. Colonial, often densely with nests touching though looser groups reported; sometimes with White Pelicans *P. onocrotalus* and other waterbirds, but always discrete. Small area round couple defended during pair-formation, possibly largely by ♂; later restricted to eventual nest-site. After 2½ weeks, unfledged young collect in dense crèches in centre of colony, both during day and night; at 4–5 weeks move in flock to water close to colony. Adults may feed in area of colony, but mostly commute to distant feeding grounds. Non-breeders frequently spend summer in colony. Immatures remain on feeding grounds; usually visit colony for only a few days before being driven away by established birds. ROOSTING. Communal. Away from colony, flocks roost nocturnally in trees often with cormorants *Phalacrocorax*. At colony,

adults sleep at own nest-sites, both during height of day and at night, until cease guarding young; when young 4–5 weeks old, tend to roost on shore near colony or on water, moving entirely to water when young 7–8 weeks old. Young roost in colony, at first crowded together but later dispersed singly; older young visit colony to loaf during day and to roost at night, otherwise spending much time on water.

2. ANTAGONISTIC BEHAVIOUR. Threat-display to conspecifics and other birds consists of waving wide-open bill (van Tets 1965; see A). During pair-formation, courting ♂♂ threaten other ♂♂ that approach, and attack them with blows of bill; adults later similarly drive off visiting immatures. HETEROSEXUAL BEHAVIOUR. During pair-formation, ♂ turns laterally to ♀ and performs Bowing-display: droops carpal joints of closed wing and elevates primaries, spreads tail widely, and bows repeatedly while vigorously vibrating wing and tail feathers; simultaneously, loudly snaps and holds bill shut as tight as possible, inflates gular pouch like balloon, and rhythmically emits spitting and hissing sounds as air expelled. ♀ generally stands quietly throughout. Later, such behaviour occurs at nest-site chosen by ♀. Here pair copulate up to 10 times daily. ♀ frequently sits at site for progressively longer periods while ♂ stands beside her. ♂ may try to initiate copulation by attempting to grab ♀ by nape feathers, ♀ at first avoiding. Later, ♂ repeatedly grabs at back of ♀'s head, nape, and neck, then suddenly takes firm grip with bill, mounts, and copulates quickly while raising and vigorously beating wings; at same time, ♀ sinks bill obliquely forward and under body, repeatedly raising wings slightly or opening them sideways. ♂ dismounts with beating wings, slipping down at ♀'s side. ♀ remains at site for long periods before egg-laying, building nest with material brought by ♂ in pouch. Both sexes direct Bill-raising display at mate (and also at chicks and neighbours) from or near nest, lifting closed bill (see B) often with gular pouch expanded; may be low-intensity version of Bowing-display, but interpreted as 'recognition display' by van Tets (1965) and as threat by Meischner (1958, 1962). RELATIONS WITHIN FAMILY GROUP. Little information. Young call throughout day especially at feeding time when they may beg by moving bill up and down.

(Figs A–B after van Tets 1965.)

A

B

**Voice.** Generally silent away from colonies, although noisy when driving fish. Adults produce a few grunting, barking, hissing, spitting, and groaning sounds, generally similar to White Pelican *P. onocrotalus* but higher pitched. Recording at colony, Bulgaria, shows series of short repeated barks recalling Raven *Corvus corax*, possibly alarm-calls (see below). Utterances and their significance inadequately distinguished and not fully understood. Following account based mainly on data from L Güthert (in Bauer and Glutz 1966).

(1) Greeting-call. Soft, prolonged low-pitched hissing 'hchhchhchhchh'. (2) Display-calls. Series of hissing and spitting calls uttered by ♂ only. (3) Copulation-call. Muffled, grating 'ch!-ch!-ch!' (possibly ♂); softer 'chi!-chi!-chi!' (possibly ♀). (4) Alarm-call. Husky, baying 'wo-wo-wo', especially from alerted birds in colony. (5) Threat-call. Prolonged spitting 'hchschchschchsch-hhh'; similar but shorter and huskier 'hachchchhh' given during fighting. (6) Bill-clappering. Non-vocal clappering of mandibles produced during antagonistic behaviour in colony, and by both adults and juveniles on lake.

CALLS OF YOUNG. Prolonged muffled grunting or sheep-like bleating, resembling 'much-much-much' but varying according to intensity of excitement, becoming baying 'wawawawawawawaw' when highly excited; uttered throughout day, though especially when being fed.

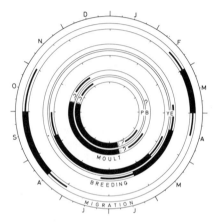

**Breeding.** SEASON. See diagram for Balkans and southern USSR. SITE. On islands or in thick aquatic vegetation; usually in cover but occasionally out in open. Normally loosely colonial, in places with White Pelican *P. ono-crotalus*; sometimes solitary. Nest: heap of grass, reeds, twigs, and small branches up to 1 m long and 5–7 cm wide; gradually sealed together with droppings. 1–1·5 m diameter and 1–1·5 m above water, though settles during season (Dementiev and Gladkov 1951). Building: by ♀, ♂ bringing material in pouch 25–40 times a day; takes 3–4 days to complete, occasionally 5 (Dementiev and Gladkov 1951). EGGS. Oval, rough; white. 95 × 60 mm (78–106 × 53–64), sample 150 (Schönwetter 1967). Weight 143–195 g, mean and sample not given (Dementiev and Gladkov 1951). Clutch: 2–3 (–5, rarely 6). One brood, Replacements probable after egg loss. INCUBATION. 30–32 days. By both sexes though mostly ♀. Starts with first egg, hatching asynchronous. YOUNG. Altricial and nidicolous. Cared for and fed by both parents bringing food and water in pouch. Food at first regurgitated into nest; later young inserts head into pouch. FLEDGING TO MATURITY. Fledging period c. 85 days. Become independent at c. 100–105 days (Dementiev and Gladkov 1951). Age of maturity probably 3 years. BREEDING SUCCESS. No data.

**Plumages.** ADULT. Head, neck, and body white, slightly powdered grey. Feathers of head and neck soft and curly. In breeding season, untidy crest of elongated feathers on nape. Feathers of upperparts elongate, those of mantle and scapulars with black shafts. Patch of lanceolate feathers on chest; often stained pale yellow, sometimes brown. Tail white, some feathers with black shafts. Primaries and their coverts dark brown, dusted grey. Secondaries dark grey near shaft, paler at inner edge, white or pale grey at outer. Undersurface of flight-feathers pale grey, darker at tips of primaries, shafts white. Upper and under wing-coverts and axillaries white; marginal upper coverts of primaries mottled grey; long upper coverts of secondaries with black shafts. NESTLING. Naked at hatching, skin vermilion-pink (Lehmann 1974). Down white; develops from 5th day. JUVENILE. Head and neck pale grey to dirty white, feathers shorter than adult; bushy crest on crown (Bauer and Glutz 1966). Mantle, scapulars, and upper wing-coverts mottled white and pale grey-brown; feathers less elongate than adult. Rest of body dirty white. Tail grey-brown. Primaries like adult, but without grey dusting; secondaries dark grey, edges grey. IMMATURE. Intermediate between juvenile and adult. Mantle mixture of white feathers and grey with white edges. Tail with white and grey feathers. Marginal and lesser upper wing-coverts like mantle; median and greater white faintly mottled brown-grey. Gradually approaches adult plumage, probably reached in 3rd year (Hartert 1912–21), but sequence of plumages not known.

**Bare parts.** ADULT. Iris pale yellow. Bill pale yellow, cutting edge distally tinged red, nail orange, ridge of culmen and base of upper mandible horn-grey. Gular pouch and bare skin of face yellow; pouch becomes deep orange or blood-red, orbital skin purple during breeding. Leg dark grey. NESTLING. Iris pale grey, later reddish grey (Reiser and von Führer 1896). Bill at hatching pink with pale tip, later dark grey. Eye-ring blue, later pale grey. Leg pink-flesh. JUVENILE. Iris pearl-grey. Bill yellow-grey. Gular pouch yellow. Leg lead-grey. IMMATURE. Bill yellow, ridge at base grey. (Bauer and Glutz 1966.)

**Moults.** ADULT POST-BREEDING. Complete; primaries serially descendant, up to 3 active centres in wing. Timing apparently variable: body and wing moult recorded during breeding, May–August (Bauer and Glutz 1966); or later, July–November (Dementiev and Gladkov 1951). ADULT PRE-BREEDING. Long feathers on nape develop prior to breeding; no further details known. POST-JUVENILE. Not known. Gradual change of plumage may indicate that young birds moult slowly during early years of life.

**Measurements.** Adults (Korodi-Gál 1964).

| | | | |
|---|---|---|---|
| WING | ♂ 710, 710, 720, 730 | ♀ | 710 |
| TAIL | 230, 230, 250, 250 | | 230 |
| BILL | 420, 430, 445, 450 | | 440 |
| TARSUS | 130, 130, 130, 135 | | 130 |

Average of 15 adult ♂♂: wing 745, bill 420. 15 ♀♀: wing 725, bill 390. Juveniles smaller than adults. Average of 3 ♂♂: wing 686, bill 397. 4 ♀♀: wing 670, bill 370 (Korodi-Gál 1964). Dombrowski (1912) gave following ranges: ♂ wing 690–800, bill 390–450; ♀ 670–780, bill 350–430. Difference between sexes apparently less marked than in White Pelican *P. onocrotalus*.

**Weights.** Adult ♂♂: 10500, 10500, 11500, 12000. ♀: 10000. Average of 3 juvenile ♂♂ 9250; 4 ♀♀ 9100 (Korodi-Gál 1964). Maximum up to 13000 (Bauer and Glutz 1966).

**Structure.** Wings long and broad, spread up to 3080 (Korodi-Gál 1964). 11 primaries: p8 longest, p9 *c.* 5–10 shorter, p10 *c.* 35–50, p7 *c.* 5–10, p6 *c.* 25–40, p5 *c.* 90–110, p1 *c.* 240–280. Tail rounded; 22–24 feathers, outer *c.* 50 shorter than central. Less bare skin on forehead and round eyes than in *P. onocrotalus*, feathers not ending in point but in slightly concave line over base of bill. Outer toe *c.* 95% of middle, inner *c.* 70%, hind *c.* 45%. Rest of structure as *P. onocrotalus*, but never fleshy knob at base of bill.

**Recognition.** In all plumages, distinguished from *P. onocrotalus* by feathers on forehead ending in slightly concave line over base of bill.

---

# *Pelecanus rufescens* Pink-backed Pelican

PLATES 27 and 28
[after page 254]

Du. Kleine Pelikaan   FR. Pélican gris   GE. Rötelpelikan
RU. Красноспинный пеликан   Sp. Pelicano roseo-gris   Sw. Rödryggad pelikan

*Pelecanus rufescens* Gmelin, 1789

Monotypic

**Field characters.** 125–132 cm, bill 29–38 cm, wing-span 265–290 cm; large, but distinctly smaller and shorter-billed than either congener in west Palearctic. General appearance essentially as other pelicans but slightly less grotesque and awkward on ground. Body plumage and wing-coverts of adult and underparts of younger birds composed of noticeably long, loose, plume-like feathers, whereas those of White Pelican *P. onocrotalus* and Dalmatian Pelican *P. crispus* of normal structure. Sexes alike; seasonal variation in plumage and bare part colours. Juvenile and immature separable.

ADULT BREEDING. Whole plumage basically greyish-white, with back, shoulders, and rump tinged vinaceous, inner wing-coverts and lower underparts all washed pink, with obvious occipital crest and patch of lance-like feathers on breast. Upper surfaces of secondaries ashy, of primary coverts and primaries dark grey; undersurfaces much paler, though primaries show increasing amount of grey (until completely so on outer ones). Bill yellowish-white with orange nail, pouch yellow-flesh or orange-yellow. ADULT NON-BREEDING. Plumage loses all vinaceous or pink tones, becoming generally drab ash-grey with brown wing-coverts and white rump. JUVENILE and IMMATURE. Brown, with rumps whiter than in adult and less contrast in wing pattern with quills brown (darkening with increasing age). Bills of non-breeding adult and younger birds greyer and pouches of latter flesh-coloured. Legs and feet cream or yellowish-white, but pale pink in eastern populations. Eyes yellow-brown.

At all distances, adult on water or ground appears darker and greyer than similarly aged *P. onocrotalus* and *P. crispus*. At close range, smaller size, crouched stance, and shorter bill and neck compared to *P. onocrotalus* obvious; markedly grey plumage tones and broader feathering on forecrown ending in concave line across base of upper mandible, rule out confusion. Separation from *P. crispus* less easy, though size difference even more pronounced; while forecrown feathering similarly shaped, pink plumage tone in breeding cycle and basically white legs at all times diagnostic. In flight, adult shows wing pattern recalling washed out *P. onocrotalus* and absence of isolated dark outer primaries on undersurface rules out *P. crispus*. For younger birds, bill length (only four times as long as head), wing pattern, and leg colour should allow identification.

Flight as other pelicans, though action slightly less heavy; flocks regularly adopt V-formation. Less gregarious than *P. onocrotalus*, but large assemblies frequent in Africa. Habitat as congeners but more frequent in coastal waters. DIMW

**Habitat.** Tropical, along northern fringe of Ethiopian Region, on islands and along coasts of Red Sea and West Africa, commonly ascending rivers and even smaller streams. Resorts to lakes, both freshwater and alkaline, and even at times to dry country, especially when locusts plentiful (Archer and Godman 1937). Concentrates in bays and harbours where abundant fish, shifting to roost on piers, walls, coral reefs, or ledges of cliffs and lines of sand-dunes. In West Africa, nests on tall trees, often with Marabou *Leptoptilos crumeniferus* and other species; elsewhere on sandy islands, in mangrove swamps, and sometimes in built-up areas (Bannerman 1930).

**Distribution.** Breeds much of Africa south of Sahara, Red Sea north to about 23°N, and Madagascar.

Accidental. Egypt: recorded May–October in 7 years between 1902–1920, and in 1959; sometimes in small flocks (Flower 1933; Horváth 1959*b*). Israel: 1 shot, probably near Tel-Aviv, between 1934 and 1936; juvenile ♂ Lake Huleh, May 1939, and juvenile ♀ Jordan valley, May 1962 (HM).

**Movements.** At least partially migratory; little known though position undoubtedly complex in view of variable breeding seasons, e.g. in north Nigeria September–March in Sokoto, but August–November in Kano (Bannerman 1951). Northward movement into sub-Saharan steppes in short summer wet season may be to undiscovered colonies, or involve non-breeders. Recorded as non-breeding visitor to Niger Inundation Zone, mainly June–August (Guichard 1947; Duhart and Deschamps 1963); common Lake Chad June–September, but no known colonies

(Hopson 1964). In Eritrea, wide dispersal, partially inland, from coastal colonies in non-breeding season (Smith 1957). Irregular occurrences in west Palearctic (see Distribution) fit this pattern. Movement in northern tropics not all north–south; flocks appear in local dry seasons around Gulf of Guinea coasts (Bannerman 1951), while small numbers visit south-west Arabia May–November (Browne 1950; Meinertzhagen 1954).

**Voice.** Generally silent away from breeding colonies.

**Plumages.** ADULT BREEDING. Crown pale grey tinged vinaceous; crest of elongated feathers on nape, most strongly developed during breeding. Sides of face white. Feathers of hindneck downy, grey; of foreneck longer, of normal structure, white. Mantle, scapulars, upper tail-coverts, and upper wing-coverts pale grey with slight vinaceous tinge. Feathers of upper mantle with white shafts and grey edges. Scapulars with black shafts, long ones also with black shaft-streaks. Back, rump, flanks, and under tail-coverts vinaceous-pink. Breast and belly white, patch of stiff, lanceolate feathers on chest slightly stained yellow. Tail grey with brown shaft-streaks. Primaries dark grey, secondaries dark brown along shafts, silver-grey at edges. Undersurface of flight-feathers pale grey with white shafts. Greater upper wing-coverts with brown shaft streaks, concealed under long, black-shafted median coverts. Under wing-coverts white, tinged pink. Axillaries pale vinaceous-pink. ADULT NON-BREEDING. Plumage loses all vinaceous or pink tones. JUVENILE. Like adult, but general colour of upperparts brown-grey, not pale vinaceous-grey. Crest on nape short. Feathers of mantle darker than adult, brown-grey with white edges. Pink tinge on back and elsewhere much fainter. Tail dark brown. Secondaries wholly dark grey, without silver-grey edges. Median upper wing-coverts not elongated, not covering greater coverts.

**Bare parts.** ADULT. Iris recorded as black, brown, red, or yellow-brown; nature of these differences not clear. Bill pale yellow with orange nail. Gular pouch flesh with narrow transverse yellow lines. Orbital skin mottled black. Leg grey, dull orange, or yellow. At beginning of breeding season, bare parts more vividly coloured: pouch deep yellow with brown-black lines, internally deep red; large black patch in front of eye; skin over eye pink (paler in ♂); below eye yellow; leg red (Burke and Brown 1970). At start of courtship, Din and Eltringham (1974*b*) found stripes on pouch pale yellow, orbital skin black, and upper eyelid in ♂ deep yellow, in ♀ dark orange. JUVENILE. Unrecorded.

**Moults.** Unrecorded.

**Measurements.** Wing 570–620, tail 176–195, tarsus 80–105, culmen 300–360 (McLachlan and Liversidge 1957). Maximum tarsus 107 (Bannerman 1930). Chapin (1923*a*) recorded bill of ♂ 378, of ♀ 292. Average of 83 adults: tail 190, culmen 320 (Din and Eltringham 1974*b*).

**Weights.** Average of 83 adults 5200 (Din and Eltringham 1974*b*).

**Structure.** Wings long and broad. 11 primaries: p8 longest, p7 equal to *c*. 10 shorter, p9 *c*. 25 shorter, p10 *c*. 50, p6 *c*. 15, p5 *c*. 70, p1 *c*. 200–220. Tail rounded; 20 feathers, outer *c*. 35 shorter than central. Bare orbital skin like Dalmatian Pelican *P. crispus* and feathers ending in concave line over base of bill. Outer toe *c*. 95% of middle, inner *c*. 72%, hind *c*. 45%. Rest of structure like White Pelican *P. onocrotalus*, but bill narrower and no fleshy knob at base of bill during breeding.

# Family FREGATIDAE frigatebirds

Large to very large, highly aerial seabirds. Adaptively conservative group of 5 closely related species in single genus. 1 species breeding in west Palearctic. Body slender, neck short. ♂ smaller than ♀. Wings long and narrow, area of wing large in relation to weight of bird. 11 primaries; p10 longest, p11 minute. About 23 secondaries; diastataxic. Excellently adapted for dynamic soaring flight with occasional deep wing-beats, constantly manoeuvring with long, deeply forked tail, hardly if ever alighting on surface. 12 narrow tail-feathers. Bill long, strongly hooked at tip. Nostrils reduced. Gular skin bare; inflatable like balloon in displaying ♂. Tarsus extremely short; foot rather small, totipalmate, nail of middle toe medially with comb. Hardly able to walk, often perching in trees. Oil gland feathered. Aftershaft absent. Plumage glossy black occasionally with white patches on underparts in ♂♂, mainly white below (except Ascension Frigatebird *F. aquila*) in ♀♀; orbital ring, gular skin, and feet often brightly coloured. 1 moult per cycle. Primary moult in serially descendant order. Young naked at hatching; later clad in dense white down. Contour feathers do not develop from same follicles. Juveniles differ from adults; duller above, pale underneath (often with dark breast band), head white or pale rufous. Attain adult plumage after several years.

Occur in all oceans, mainly tropical and sub-tropical. Essentially pelagic, extent of ocean range greatly underestimated until recently. Flight capability outstanding. Though obtain most of food from sea, not adapted to and reluctant to maintain more than momentary contact with water. Resort to land to roost and breed, often in bushes and trees. Employ powers of flight to take nesting material and food from other birds, and frequently soar for long periods in warm air. Readily make utmost practicable use of human artefacts and activities. Only tenuous foothold in west Palearctic where declining. Dispersive and sedentary.

Principal foods fish (especially flying-fish) and cephalopods (especially squid). Taken directly, or by harassing other seabirds, notably boobies *Sula*, in air or at surface of sea entirely by flight-feeding: mainly pursuit-flying, swooping, sky-plunging, and hovering. Food-piracy practised more by ♀♀ than smaller ♂♂. Also pick up stranded fish and hatchling sea-turtles on wing from beach, and eggs and small nestlings of other seabirds either from wing or when perched. Readily take offal and other edible scraps provided by man. Strictly diurnal. At sea, mostly solitary, soaring high; will form feeding congregations with own or other species. At breeding station or roost, often soar overhead in loose parties. Monogamous pair-bond, lasting for part or all of 1 cycle; re-pairing with same mate in successive cycles, after absence for moult, less likely than in many other seabirds. Colonial, often in small groups or more scattered. Site defence and tenacity much less marked than in other marine Pelecaniformes. Rest on wing when at sea or for long periods on any suitable perch, never settling on water; at all times of year, however, whether breeding or not, usually resort at night to communal sites in trees, bushes, or on cliffs. Heterosexual display relatively unspecialized: ♂♂, with highly inflated gular pouches, display in perched group to ♀♀ flying overhead. Voice also relatively unspecialized. Silent at sea and at colony except when coming in to land, displaying, fighting, or chasing (Diamond 1975). Calls of sexes dissimilar, especially Arrival-calls; variety of croaking, whistling, twittering, whinnying, warbling, drumming, and reeling noises. Main sound of ♂♂ produced during display by combination of vocal and non-vocal means. Comfort behaviour much as in other Pelecaniformes but preening and scratching often carried out high in air. Bathe on wing, repeatedly splashing momentarily on to surface of water without ceasing flight; also drink from air. Heat dissipated by gular-fluttering; spectacular sunning posture frequently adopted ashore—lie or sit back on tail with wings fully extended at right angles to body and rotated so undersurfaces facing upward. Unlike any other group of seabirds, so far as known, scratch heads by indirect method when perched but directly in flight.

Prolonged breeding season (*c*. 5–9 months), though most eggs in same colony often laid within period of 2 months. Actual periodicity of breeding uncertain but, in case of successful breeders, not annual as cycle takes over a year. Peak in laying annual, however, usually at similar time each year; due mainly to previously unsuccessful breeders (responding to same environmental factors, such as seasonal increase in food supply) with successful pairs nesting again either outside period of peak laying or missing an entire season, i.e. at intervals of *c*. 18 months to 2 years (see especially Diamond 1975). Nests normally in bushes or trees; flimsy structures of twigs built by ♀, ♂ bringing most of material. Also on ground with little or no material. Eggs ovate, white; not glossy. Clutch 1; single brood, but losses sometimes replaced. Incubation periods long, 44–55 days; by both sexes, each with single median brood-patch, equally and often in long spells of up to 18 days. Young cared for and fed, by incomplete regurgitation, by both sexes at variable but fairly long intervals. Guarded for as long as possible, up to 45 days. Fledging periods long, up to 6–7 months, followed by long period of post-fledging care—4–10 months or longer. Age of maturity unknown.

*Fregata magnificens* **Magnificent Frigatebird**

PLATE 19
[between pages 182 and 183]

Du. Amerikaanse Fregatvogel     Fr. Frégate superbe     Ge. Pracht-Fregattvogel
Ru. Блестящий фрегат     Sp. Rabihorcado grande     Sw. Praktfregattfågel

*Fregata minor magnificens* Mathews, 1914

Monotypic

**Field characters.** 95–110 cm, ♀ averaging larger; wingspan 215–245 cm. Large, black or black-and-white seabird, with long, hook-tipped bill, long, pointed wings, long, deeply forked tail, and short legs; adult ♂ has large, inflatable, red gular pouch. Sexes dissimilar, but no seasonal differences; succession of immature plumages.
ADULT MALE. Plumage entirely black, with metallic green gloss on head and wings, and purple on back and scapulars. Bill blue-grey to brown; gular pouch tinged pink to dull orange and becoming scarlet when inflated; legs blackish. ADULT FEMALE. Mainly black-brown, with slight purple and green gloss above, apart from greyish lesser coverts forming pale band on wing, indistinct lighter hind-collar, and anvil-shaped white patch on breast (white extending to sides of neck and belly, so that black-brown of foreneck and centre belly each come to a point). Bill grey-blue; no gular pouch; legs pink to red or crimson. JUVENILE. Head, neck, and most of underparts white, contrasting with black-brown back, rump, wings, tail, breast-band (usually incomplete), flanks, and under tail-coverts. Brown-grey to off-white lesser coverts form broad pale band on wing. Bill pale grey-blue, with pink tip in early stages; legs dull pinkish. IMMATURE STAGES. First, breast-band largely disappears, some dark streaks show on head, and blackish streaks and spots on belly, and back becomes blacker and glossier. Later, more streaks appear on head and hindneck, and blackish on belly becomes solid and more extensive. During next 2–3 years, both sexes darken until ♂ resembles adult ♀, but with throat and breast white streaked with blackish, while ♀ apparently retains largely white head and foreneck but mottled black-brown. By sub-adult stages, ♂ like adult ♂ but less glossy and with browner breast sometimes flecked white, while ♀ has dark areas duller and greyer-brown with light streaks on head; some individuals breed in what appears to be adult plumage but for white on belly. Full adult plumage may not be acquired until 4th year.
Frigatebirds unlikely to be mistaken for any other seabirds, but the 5 species are readily confused and 3 of other 4 could occur in west Palearctic waters as vagrants. Great Frigatebird *F. minor* has red or pinkish bill in all plumages; ♂ also distinguished by red feet and brownish wingband (like ♀ *F. magnificens*) and ♀ by square-cut, dirty white to greyish throat contrasting with dark cap; immature plumages have white of head tinged rufous and juvenile also has broad black breast-band. ♂ Lesser Frigatebird *F. ariel* has distinctive white axillary patch on each side, and ♀ has square-cut black throat and purple

tinge to bill; immature plumages closely resemble *F. minor*. ♂ Ascension Frigatebird *F. aquila* probably indistinguishable in field from *F. magnificens*, but ♀ also largely black (only all-dark ♀ *Fregata*) and distinguished from ♂ by absence of gular pouch and plumage iridescence, as well as by rusty-brown tones, paler brown hindneck (forming indistinct collar), and brownish upper breast; juvenile has white head and underparts with broad, brown-black breast-band, and pale blue feet.
Sails effortlessly and buoyantly with head sunk back between shoulders and forked tail closed; in active flight, wing-beats deep and loose. Never settles on water. Obtains fish prey and other food in flight, by circling lower and lower or by beating back and forth and, on spotting prey, either gliding down or dropping like stone from great height to check just above surface and snap it up with the bill. Flying-fish taken above surface, often after aerial pursuit, and others just below it with only bill, or at most bill and head, entering water. Floating food snatched delicately from surface. Also adept at picking up fish or young sea-turtles from beach in flight. Food also obtained, at least by ♀♀, by harassing other seabirds (particularly boobies *Sula*); ♂♂ also rob other species of nest material. Habitually preens and scratches in flight. Highly gregarious at breeding sites, and also congregates on bushy islets elsewhere and at other seasons, but at sea likely to be seen singly, or in small, scattered parties. Generally silent except when nesting. Perches and roosts in trees.

**Habitat.** Inshore and offshore warm waters; in west Palearctic only around Cape Verde Islands, where barely surviving, breeding on bare surface of islet. Elsewhere more commonly on tops of bushes or low trees, exceptionally up to 10 m. Despite great lifting capacity, has difficulty in taking off from flat surface. Clumsy and scarcely mobile on ground, but adept at alighting and perching on rigging of vessels, posts, buoys, hulks, quays, and other structures offering good opportunities for alighting or taking flight, and ample visibility for detecting approach of danger. Often uses mangroves and other trees, rocks, coral reefs, and sand or shingle spits (see also Roosting in Social Pattern and Behaviour). Where not molested becomes tolerant of humans and can be tamed by deliberate feeding; also frequently approaches fishing vessels and other places where offal is cast away, even including slaughterhouse on a river front. Highly aerial, using airspace at wide range of heights over land as well as sea and littoral, readily crossing from one coast to another

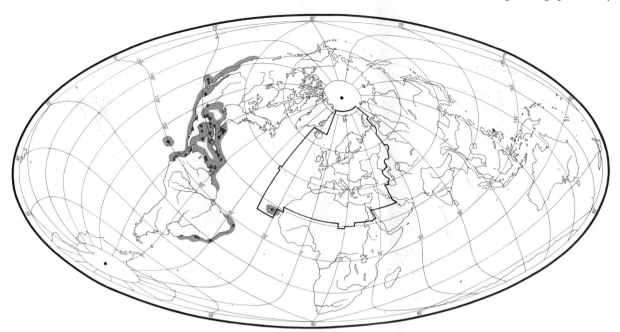

at Panama isthmus, and often soaring for long periods, up to 300 m and more. When fishing or hunting normally cruises well above surface, often in range 20–30 m. Among few birds able to ride out a hurricane in flight. Recorded in mid-ocean, and sometimes out of normal tropical range. Opportunist and parasitic feeding practices lead to degree of commensalism with man more comparable with gulls (Laridae) than with most pelagic species, especially around ports and shipping lanes. Despite powers of flight, spends long periods at rest on perch.

**Distribution.** In west Palearçtic, breeds only on 2 islets off Boa Vista, CAPE VERDE ISLANDS.

Accidental. France, October 1852 and March 1902 (CE); Azores, November, before 1903 (Bannerman and Bannerman 1966); Scotland, Inner Hebrides, July 1953 (British Ornithologists' Union 1971); Denmark, adult ♀ found dead, March 1968 (FS). Also single specifically unidentified *Fregata* off Aberdeen, August 1960 (British Ornithologists' Union 1971); near The Hague, August 1968 (*Limosa* 1970, **34**, 59–60), and off Co. Cork, Ireland, August 1973 (Smith *et al.* 1974).

**Population.** Has apparently decreased CAPE VERDE ISLANDS where now only 10–12 pairs (de Naurois 1969*b*).

**Movements.** Dispersive and sedentary. Resident within breeding range; young may disperse considerable distances, as do some adults outside breeding cycles, especially June–September in Caribbean (Palmer 1962). No recent records from waters around Cape Verde Islands; in past collected off Gambia (Bourne 1957). From major colonies in Brazil and Caribbean, occurs regularly Florida and Gulf of Mexico and has wandered north as far as Lake Michigan, New England, Quebec, Nova Scotia, and Newfoundland, and south to Argentina (Palmer 1962). Frigatebirds, presumed *F. magnificens*, reported present all across tropical North Atlantic during one raft voyage from west Africa to West Indies in November–December (Bombard 1953); this perhaps abnormal, though *Fregata* may cross open oceans more than has been supposed (Simmons 1969), as proved by Sibley and Clapp (1967) for immatures of Lesser Frigatebird *F. ariel* in Pacific. Single *F. magnificens* seen mid-Atlantic, 30°11′N 46°W, May 1962 (Bourne 1963), another 46°27′N 16°24′W, June 1968 (*Ardea* 1968, **56**, 285), and an unidentified *Fregata* at 51°04′N 13°05′W, May 1953 (*Fair Isle B.O. Bull.* 1953, **2**, 39); in view of these and small numbers now in Cape Verde Islands, exceptional records from European waters probably concern transatlantic vagrants.

**Food.** Chiefly fish; also squids, jellyfish, larger plankton (Audubon 1835; Howell 1932), young turtles (Lowe 1911), young Sooty Terns *Sterna fuscata* (Beard 1939), and eggs of Red-footed Boobies *Sula sula* (Brown 1947). Feeds from air with minimal or no contact with water. Takes marine prey from or near surface of water by swooping from flight and hovering; will also pursue flying-fish in air, and chase and harass other seabirds until food dropped or disgorged (see also under Field Characters). Also takes dead fish and offal from fishing boats, scraps from slaughterhouses, docks, and sewage outlets. Hatching turtles taken from flight on beach; nestling terns from flight, or when perched, on ground.

Fish include: unspecified flying-fish and mullet *Mugil*

(Audubon 1835; Wurdeman 1861; Watson 1908), herring *Clupea harengus* (Watson 1908), mackerel *Scomber*, and carangids (Murphy 1936). Of 25 stomachs from Florida, 13 contained menhaden *Brevoortia*; also pinfish *Lagodon rhomboides*, sea catfish *Galeicthyes*, and weakfish *Cynoscion regalis* in small quantities (Howell 1932). One captured inland Pennsylvania, October, contained freshwater fish: walleye pike *Stizostedion*, shad *Dorosoma*, and crappie *Pomoxis* (Palmer 1962).

**Social pattern and behaviour.** Virtually unknown in remnant population on Cape Verde Islands, but more information available from New World populations; see especially Murphy (1936) and Palmer (1962) for general observations, Diamond (1972, 1973) for more detailed study. In many respects, biology similar to that of better-studied oceanic frigatebirds—e.g. Great *F. minor* and Lesser *F. ariel* (Nelson 1967b, 1968; Diamond 1971)—with some important differences linked with more coastal feeding habits in areas rich in food.

1. Gregarious (often with other pelecaniform birds) when nesting and roosting; to lesser extent also when feeding. Probably largely absent from nesting areas when not breeding. At sea, spends much time soaring individually, often at considerable height; evidently largely solitary when searching for food, but will congregate at concentrated sources. BONDS. Monogamous pair-bond but, owing to nature of pair-formation and prolonged breeding cycle of ♀, unlikely to be sustained from one season to next, still less life-long as in many other seabirds. Unlike in other frigatebirds, pair-bond ends before completion of breeding cycle: while both sexes tend single chick until 3–4 months old, ♂ then departs to moult leaving ♀ to feed young for remaining 2–3 months of fledging period and also for at least 4 further months of post-fledging dependence (Diamond 1972). ♂ thus appears to breed annually, taking different mate, while successful ♀ cannot breed more than once in 2 years. If confirmed, this regime unique; apparently correlated with excess of ♀♀ in breeding population and with feeding on plentiful and dependable food supply close to colony, permitting rearing of more young than if both parents nested every 2 years. BREEDING DISPERSION. Colonial; often nesting densely but, like many coastal seabirds, colonies often small, usually from *c.* 40 pairs to 2500, while occasionally nests singly or in groups of 2–3 pairs. At start of breeding period, ♂ has no fixed territory, often displaying at more than one site, usually in company of other ♂♂. After joined by ♀, pair defend nest-site as territory; attachment even to nest not strong, with no site-fidelity in successive seasons (when site may still be occupied by juvenile of previous nesting). Pair-formation completed in territory, and copulation and nesting occur only there, as also most of protracted care of young. ROOSTING. Communal and largely nocturnal, in small to large, dense groups. Pairs roost mainly in colony area when breeding, guarding nest and small young; later in cycle, when juveniles free-flying though still dependent, hundreds of ♀♀ may continue to roost with young. Other adults, and possibly non-breeders, roost not far from colony. When absent from breeding station, probably sleeps on wing (for never voluntarily alights on sea); will also rest on posts (etc.) during day. Also regularly uses roosts elsewhere, even outside breeding range; sometimes in thousands. Uses tops of trees, mangroves, or shrubs, sometimes cliff-faces, mostly on small islands; also larger islands and mainland. Roost-sites more or less permanent if not seriously disturbed. Occupants often return to area well before sunset, but do not settle until just before dark, soaring overhead meanwhile. Said to appear drowsy at night, but in fact sits tight, loath to flush. Rises at dawn, soars high overhead in numbers, then later glides out over sea (birds some 15–60 m apart); similar flights noted during evening return.

2. At colony, in period prior to egg-laying, ♂ frequently inflates gular sac or throat-pouch into huge, scarlet balloon sometimes in air but usually while perching in tree or shrubs. Even when not actively displaying, ♂ may keep pouch inflated to some extent. ANTAGONISTIC BEHAVIOUR. Lack of true territory in colony during stage before pair-formation reflected in tolerance even of physical contact between ♂♂ during display (see below). Later, defence of nest-site relatively weak; though van Tets (1965) saw no interspecific fighting or threatening, ♂♂ attempting to land nearby may be repulsed with lunges and Bill-snapping, leading to gripping of bills. However, fights rare and never serious. Loud Bill-rattling also directed at neighbours attempting to steal nest-material (Palmer 1962), while ♀ seen waving bill at Red-footed Booby *Sula sula* (van Tets 1965). Both sexes give Arrival-call on descending to site (or roost perch), ♀ also while circling above (Diamond 1973)—but, like other frigatebirds, lacks any ritualized Post-landing or Take-off displays (van Tets 1965). ♂ has no true, distinctive, Site-ownership display but, unlike other frigatebirds, will at times direct low-intensity version of Advertising-display to incoming ♂♂ (see below). Aerial encounters between ♂♂ over colony reported as common; said to involve calling and fighting, but details lacking. HETEROSEXUAL BEHAVIOUR. Unusual for seabirds in that pair-formation precedes true territorial behaviour. ♂ gives Advertising-display (see A), often in group of 4–8 ♂♂, moving if unsuccessful in attracting ♀. With pouch fully inflated, sits back on spread tail, wings raised and outstretched sideways, head thrown back on mantle, bill vertical, and scapular feathers erected; posturing thus, swivels and rapidly vibrates wings back and forth, turns head from side to side, and alternately gives reeling and drumming sounds while Bill-rattling against pouch. Performs thus when ♀ overhead, especially if she hovers to inspect group, when every ♂ breaks into intense display. As ♀♀ of all frigatebirds show species-characteristic ventral patterning, this may be important recognition factor for ♂. ♀ of present species lacks ritualized wing or neck movements found in hovering *F. ariel* ♀ responding to ♂'s display. ♂ continues to display fully to ♀ after she lands beside him, usually in front, 1 or 2 (but rarely all) of group joining in. ♂ and ♀ also engage in mutual bouts of lateral Head-shaking, bills pointing forward and down, often with necks crossed. May also clash bills and snake heads at one another, ♀ giving twittering sounds; ♀ also Bill-clatters and bends neck round to nibble at ♂'s flanks, or stretches across to rub bill or head along side of ♂'s

A

pouch. ♂ stated to rub inflated pouch against ♀'s back. Contrary to Murphy (1936), ♀ does not have reciprocal Advertising-display similar to ♂'s. ♀ may quietly depart after first landing, or remain for hours during which initial pair-formation presumably occurs. Nest then built near spot, unsuccessful ♂♂ moving elsewhere. Only ♂ gathers nest-material, while ♀ guards site. ♂ delivers twigs to ♀, followed by Head-waggling and sometimes mutual nest-building; but no elaborate Nest-ceremony and no Allopreening. Once pair-formation started, repeated forays by ♂ for material, building by ♀, and intermittent copulation proceed for several or many days, probably without either feeding—while ♀ may go without food right up to egg-laying. Copulation probably as in other frigatebirds: does not usually follow initial bout of mutual display or any special soliciting, but immediately preceded by extensive Head-waggling (J B Nelson). Otherwise, pair-relations relatively low-key, and, after egg-laying, interaction between mates minimal, change-over occurring without contact or display. RELATIONS WITHIN FAMILY GROUP. Little information. Begging of young apparently similar to many other pelecaniform birds (van Tets 1965): bill kept closed and repetitive Food-call given. Tiny young raise and wave head from side to side; later direct it at parent, prodding at bill or gular area, and movements develop into more ritualized Head-waggling accompanied by Wing-waving, older chicks spreading tails. Diamond (1973) also refers to hunched Begging-posture, with half-open wings, accompanied by Bobbing of head and shoulders. In addition, hungry young said to Bill-clatter to attract parent (Palmer 1962). When older, young collected in groups beg to ♀ flying over but, as parent and own offspring apparently recognize one another individually, ♀ descends to feed only one of them (Diamond 1973).

(Fig A after Diamond 1973.)

**Voice.** Adults usually silent, especially in flight and at sea away from colony; also when at roost. Noise from breeding colony distinctive both by day and by night, sometimes carrying several km in daytime. Described variously (see Palmer 1962): curious chattering or muttering effect from rapidly-repeated, half-guttural, half-whistling sounds; continuous nasal cackling, with rather sharp, whining tone (more of a snorting after dark); and variety of more distinctive noises, probably mainly from parents and young. Accounts of specific calls inconsistent and incomplete. Main sounds of ♂, at least, appear basically non-vocal, being produced by vibration, rattling, or snapping of mandibles. Following sounds recognized by Diamond (1973).

(1) Arrival-call. Rapid rattle, descending in pitch, deeper and hoarser in ♀. Produced in flight as bird of either sex descends to alight; also by ♀ only while circling nest-site or roost, when slower and produced in disyllabic burst, accelerating into pre-landing rattle as ♀ starts to descend. Nature of sound not discussed by Diamond (1973), but probably at least partly bill-rattling. Term 'rattling' used by van Tets (1965) both for Advertising-display of ♂ and for sound given by both ♂ and ♀ when together at nest-site (see below). Repeated but not loud 'chuck', noted from birds flying in to roost, may be same—but uncertain whether vocal sound, or whether produced by bill or wings (Murphy 1936). Palmer (1962) also mentions: 'grating cry'

by ♂♂ fighting in sky, and series of monosyllabic 'wick' calls by parents in air over colony—same call apparently also uttered by ♀ when following calling young before alighting and parental feeding. (2) Drumming. By ♂ only during Advertising-display when pouch fully inflated. Deeper, more resonant sound than related 'reeling' and 'purring' (see below); all 3 sounds evidently largely non-vocal, being produced by rapid bill-vibration and resonated by inflated pouch acting as sounding chamber. Resonant 'gurgling' or 'chuckling' vocalizations by both sexes during courtship (Murphy 1936) do not correspond to any other descriptions of ♂ calls in this species; may refer to 'twittering'-call of ♀ (see below) or may arise from confusion with unique vocal 'whinny' of ♂ Great Frigatebird *F. minor* (see Nelson 1968; Diamond 1973). (3) Reeling. Resembles sound made by spokes of bicycle wheel. Alternated with drumming by ♂ in full Advertising-display to ♀; given with bill held at any angle, with pouch either inflated or deflated. (4) Purring. Heard rarely, from ♂ only, when pouch partly deflated; may be sound produced when bird attempts to drum with pouch only partially inflated. Directed at ♂♂, ♀♀, and juveniles, or given when ♂ has no audience. (5) Twittering. Rapid twittering, by ♀ only, accompanied by vibration of mandibles. Only ♀ sound heard in heterosexual context; given after ♀ has joined displaying ♂. 'Rattling' by ♀ mentioned by van Tets (1965) may be same sound: ♀ said to rattle with bill pointing downwards, ♂ with bill pointing up, but not confirmed by Diamond (1973).

CALLS OF YOUNG. Harsh, rhythmic, insistent, plaintive pig-like calls given when soliciting food. Series of drawn-out, nasal 'waaanuh' crying sounds, uttered on wing (Palmer 1962) may be same call, as may be squealing, chirping calls mentioned by Murphy (1936). Young may also bill-clatter to attract parents when hungry (Palmer 1962).

**Breeding.** SEASON. Cape Verde Islands: fresh eggs, small and large young found from beginning February to end March, so season clearly very prolonged if indeed annual (Bourne 1955; de Naurois 1969*b*). Lesser Antilles: ♂♂ capable of breeding every year, but successful ♀♀ not more than every other year (Diamond 1972, 1973). New World: at most colonies north of equator, laying usually between December and May, with peak of laying and hatching coming in dry season. Breeding may be timed to avoid rainy season (Palmer 1962). SITE. Normally in low bushes 0·6–1·5 m above ground, but on ground in Cape Verde Islands. Colonial; nests often touching. Nest: flimsy, open construction of small sticks and twigs, *c.* 25–35 cm diameter. Building: ♂ brings most material, ♀ guards nest and does most building (Diamond 1972); sticks gathered from surroundings, including other nests. EGGS. Elliptical ovate, not glossy; white. 68 × 47 mm (65–74 × 44–50), sample 50; calculated weight 84 g (Schönwetter 1967). Clutch: 1. One brood. No information on replacements.

INCUBATION. Not precisely known but *c.* 40–50 days (Bannerman 1959); estimated at 50 days (Diamond 1973). By both sexes. YOUNG. Altricial and nidicolous. Cared for and fed by both parents. Brooded more or less continuously while small as protection against predators including other adults (Bent 1922). Fed by incomplete regurgitation direct into bill of young. In Lesser Antilles, ♂ leaves breeding ground to moult when young *c.* 12 weeks while ♀ carries on rearing (Diamond 1972). FLEDGING TO MATURITY. Fledging period *c.* 20–24 weeks (Diamond 1972). Young fed by ♀ for at least further 16 weeks (Diamond 1972). Age of maturity not known, but full breeding plumage probably attained in 4th year. BREEDING SUCCESS. No data.

Plumages. ADULT MALE. Generally brown-black. Feathers on crown and wings glossed green, on mantle and breast purple. Tail black-brown with concealed white base. Primaries brown-black, with faint gloss when newly moulted; secondaries with brown-black outer webs and grey-brown inner, inner secondaries wholly brown-grey. ADULT FEMALE. Mainly brown-black, less glossy than ♂; with narrow collar on hindneck, breast, upper belly, and sides of chest and breast cream-white. Wings as ♂, but median and lesser upper coverts sandy, forming bar over inner wing, median with dark centres; inner lesser coverts brown-black. Wing bar paler in worn plumage. NESTLING. Naked at hatching. Single down; fluffy, white. Black juvenile feathers soon appear on mantle. Secondaries grow before primaries. JUVENILE. Head, neck, and most of underparts white, mottled brown band across lower breast; mantle, back, rump, flanks, vent, and under tail-coverts dark brown without gloss. Tail and wing as adult ♀. In worn plumage, dark breast band disappears, leaving some dusky shaft-streaks. IMMATURE MALE. Head and underparts mottled brown-black and white. Feathers of mantle elongate with slight gloss; otherwise like juvenile. Adult ♂ plumage apparently acquired gradually: some have developed throat pouch, but breast still mottled white; others black-brown, but less glossy than fully adult ♂. IMMATURE FEMALE. Undescribed; probably head and neck become gradually dark as in adult ♀. Adult plumage probably reached in 4th year but data lacking (Palmer 1962).

Bare parts. ADULT MALE. Iris dark brown, orbital ring black. Bill grey or dark horn. Gular sac bright red when inflated, orange when deflated. Foot black, pale yellow underneath. ADULT FEMALE. Iris dark brown, orbital ring blue. Bill pale blue-grey.

Naked skin at chin and throat blue-purple. Foot dark pink-red. NESTLING. Iris brown, orbital ring pale grey-blue. Bill pale grey-blue with pink-blue tip. JUVENILE AND SUBSEQUENT. Iris brown, orbital ring grey-blue. Bill light grey-blue. Foot flesh. In older ♂, foot becomes black, bill darker grey; gular sac develops before bird fully adult. Immature ♀ apparently like adult ♀. (Palmer 1962, Voous 1957.)

Moults. Poorly known; primaries serially descendant, 1, 2, or 3 series active at same time. No primary moult when breeding (Stresemann and Stresemann 1966), further details of timing unknown, apparently much individual variation. Juvenile p10 may be retained for about 2 years. Body moult continuous over most of year, even when no wing moult.

Measurements. Murphy (1936), Central America.

| | ♂ | | | ♀ | | |
|---|---|---|---|---|---|---|
| WING | 633 | (21) | 611–661 | 650 | (21) | 628–674 |
| TAIL | 431 | (21) | 339–472 | — | (21) | 404–506 |
| BILL | 112 | (21) | 105–118 | 121 | (21) | 109–130 |
| TARSUS | 22·4 | (21) | 21–25 | 22·9 | (21) | 21–25 |
| TOE | 71 | (21) | 69–74 | 75·6 | (21) | 71–80 |

Mean wing and bill west coast Central America ♂ 622, 109; ♀ 648, 123; Caribbean ♂ 610, 112; ♀ 636, 127 (Wetmore 1965). ♀ averages larger than ♂.

Weights. Curaçao: 6 ♂♂ mean 1340, range *c.* 1100–1580 (K H Voous). Peru: ♀ 1587 (Murphy 1936). British Honduras: ♂ 1061; ♀ 1419 (Russell 1964).

Structure. Wing very long and narrow. 11 primaries: p10 longest, p9 26–40 shorter, p8 73–90, p7 123–149, p6 173–197, p1 367–394; p11 shorter than primary coverts, lanceolate, pointed. About 23 secondaries, short, inner ones longer. Tail deeply forked, 12 feathers: t6 220–280 longer than t1. Feathers on head and mantle lanceolate in adult, more strongly in ♂ than ♀. Bare throat patch small in juvenile, larger in adult ♀, very large in adult ♂ with widely dispersed, short, black feathers at sides; strongly inflated during courtship. Bill long, compressed towards tip, with strong terminal hook; groove separates horn plate on culmen from lateral plate; nostrils narrow slits. Feet weak, tarsus short, webs deeply incised; outer toe *c.* 80% of middle, inner *c.* 60%, hind *c.* 40%. Inner edge of middle claw with comb.

Geographical variation. Cape Verde and Galapagos populations average larger than those of American coasts, but much overlap and division into subspecies not practicable (Bourne 1957; Wetmore 1965).

# Order CICONIIFORMES

Medium-sized to huge, long-legged wading birds. 5 families, of which 3 (Ardeidae, Ciconiidae, and Threskiornithidae) represented in west Palearctic; others —Balaenicipitidae (Whale-billed Stork) and Scopidae (Hammerhead)—monotypic and exclusively Ethiopian. Related to Phoenicopteriformes, which sometimes considered as belonging to same order, and, more distantly, to Anseriformes. Behavioural similarities suggest affinities also to Pelecaniformes (van Tets 1965; Meyerriecks 1966), but close relationship not supported by studies of egg-white proteins (Sibley and Ahlquist 1972). Suggested also, mainly on osteological and other anatomical characters, that Ardeidae should be placed in separate order from Ciconiidae and that Cathartidae (New World vultures) should be placed in same order as latter (Ligon 1967).

# Family ARDEIDAE bitterns, herons

Medium-sized to huge wading birds. Much difference of opinion over number of species and genera. 64 species in 15 genera recognized by Bock (1956), but recent tendency to combine many monotypic genera into *c.* 10 larger ones and to merge closely related, allopatric forms previously treated as separate species (see Curry-Lindahl 1971). Affinities of atypical, extralimital Boat-billed Heron *Cochlearius cochlearius* uncertain; sometimes considered separate family Cochleariidae. Remaining species (to which discussion now limited) currently placed in 2 subfamilies: Botaurinae and Ardeinae. 12 breeding species in west Palearctic; 5 others accidental. Body slim, neck long with kink at 6th vertebra. ♂ larger than ♀. Wings long and broad. Flight strong with regular wing beats, neck retracted. 11 primaries: p7–p10 longest, p11 minute. 15–20 secondaries; diastataxic. Tail short, square or slightly rounded; 8–12 feathers. Under tail-coverts nearly as long as tail-feathers. Bill mostly long, straight and sharply pointed; often serrated with notch near tip. Nostrils long slits. Lores bare. Legs long, lower part of tibia bare. Toes long; small web between middle and outer. Hind and inner toes broadened at base; claw of middle pectinated. Stance upright, gait striding. Oil gland small, often with short tuft (longer in night herons *Nycticorax*). Aftershaft well developed. Plumage loose, feather tracts narrow; down confined to apteria. 2–4 pairs of powder-down patches; down soft and friable, producing fine particles used in plumage care. Ornamental plumes on head, back, or chest in many species; usually more highly developed in breeding season. Bare parts yellow, brown, or black; usually become more colourful in season of display and pair-formation. Seasonal differences in plumage small. Moults poorly known; mostly 2 per cycle, but pre-breeding moult often restricted. Moult of primaries irregular. Young semi-altricial and nidicolous; single coat of sparse down white, grey, or pale brown. Except in *Nycticorax*, juveniles like adult or duller. Reach adult plumage in 2nd, 3rd, or 4th calendar year.

Cosmopolitan, with main area of adaptive radiation in tropics. Absent from arctic and antarctic areas and adjoining boreal zones. Adapted to catch medium-size prey moving in shallow water, and thus fairly restricted in habitat. Avoid areas far from fresh water or sea beyond shoreline. Otherwise, widely distributed from fairly high, temperate latitudes through sub-tropics and tropics wherever suitable feeding areas occur, even in forest, mountain, or agricultural surroundings. Usually found at water's edge, especially where gentle slopes and un-obstructed bottom makes for easy fishing, but some species may feed far from cover and taller, longer legged species (e.g. Goliath Heron *Ardea goliath*) in deeper water. Some smaller species, however, largely arboreal, while Cattle Egret *Bubulcus ibis* now mainly a terrestrial commensal of herd-living herbivores. Certain species (e.g. reef herons *Egretta sacra* and *E. gularis*) adapted to littoral habitats, while others (notably bitterns *Botaurus*) habitually cover-haunting, favouring tall, close vegetation such as reedbeds. Reedbeds, with trees and shrubs, main breeding and roosting sites. With ample flight endurance, often travel over wide areas; when conditions stable, tend to become sedentary and conservative in habitat and topographical attachments. In west Palearctic, most species suffered contraction in range and marked reduction in numbers due mainly to drainage and (especially in egrets *E. alba* and *E. garzetta*, and Squacco Heron *Ardeola ralloides*) plumage trade. In Grey Heron *Ardea cinerea*, whose numbers in some areas markedly affected by cold winters, some expansion to north; in Purple Heron *A. purpurea*, limited spread in central Europe since 1940. *B. ibis*, which has increased explosively in North America, South America, and Australia, has expanded only slightly in west Palearctic. In west Palearctic most species migratory, with pronounced post-fledging and post-breeding dispersals, even leading to transatlantic vagrancy. Migration mainly at night, in small flocks often in line formation. Migrants in west Europe prone to overshoot in spring.

Food basically fish, amphibians, and insects and their larvae; also, for some species, reptiles, small mammals, birds and their young, molluscs, and crustaceans. Indigestible material ejected as pellets. Prey grabbed by bill; sometimes speared. Basic feeding methods (1) stand and wait for prey, and (2) wade or walk slowly while stalking prey—in both cases, followed by striking out with neck and bill when within range. In Ardeinae at least, slow stalking sometimes accompanied by (3) movements serving to uncover or startle prey by foot-stirring (scraping, raking, or vibrating movement of single foot), foot-paddling (trampling movements of both feet), or wing-flicking (wings suddenly extended and retracted rapidly). (4) In disturb-and-chase technique, bird runs and dashes about in shallow water, flushing prey; may also stab repeatedly at surface, wing-flick, and, when prey sighted, extend wings sideways (open-wing feeding). In a few species, may then stop and stab in water below extended wing (underwing-feeding) or engage in full canopy-feeding by bringing wings forward over head, thus providing false refuge for fish previously disturbed. Other feeding methods include (5) swimming in deeper water and surface-diving; (6) hovering above water and plunge-diving; (7) plunge-diving from perch. For further details, see especially Meyerriecks (1960). Feeding largely diurnal or crepuscular, or both (e.g. *A. cinerea*); or crepuscular or nocturnal, or both (e.g. Night Heron *N. nycticorax*). Most species solitary feeders, some territorially. A few (e.g. *B. ibis*) sociable at all times, but majority gregarious mainly at roost and breeding colony, and during migration. With partial exception of some Botaurinae, monogamous pair-bond typical; usually of seasonal duration and not shown away from vicinity of nest-site, birds rarely if ever meeting as mates elsewhere. In a few strictly resident populations, however, pair-bonds permanent (see, e.g. Snow 1974). When breeding, both colonial and solitary species typically defend nest-site territory only; this established initially by ♂ as pairing territory, or ♂ frequents one or more temporary display sites before pair move later to permanent nest-territory. Most species roost communally, often conspicuously at traditional and protected sites; roosts mainly nocturnal but in some species diurnal. In otherwise nocturnally roosting species, diurnal standing grounds or day-roosts also reported (see *B. ibis* and *A. cinerea*). For further information on social behaviour, see under subfamily headings. Comfort behaviour generally similar to other marsh and waterbirds. Bathe while standing in shallow water. Liberal use made of powder-down patches and oil gland while preening, with frequent use of pectinated claw in scratching head, neck, and bill. In some species, underwing preened by extending wing at right-angle to body. Heat dissipated by gular-fluttering; characteristic sunning posture with upright stance and wings held, shieldlike, out at sides but not fully spread (so that primaries meet behind tarsi). Other behavioural characters include: direct head-scratching; frequent yawning; resting while standing upright, often on one foot, in hunched posture, at times with bill inserted into side of retracted neck (will also sit down).

In many and especially colonial species, onset of breeding protracted. Seasonal breeders in temperate zone but prolonged season further south. Nest in dense vegetation or in trees; sometimes on ground in open. Colonial, often with other Ciconiiformes or non-related species, or solitary. Nests piles of available vegetation, in tree-nesting species of interlocked twigs; built wholly or mainly by ♀ with material brought by ♂. Eggs blunt oval; light blue or green, smooth. Clutches 3–5 (1–10). Normally single brood, but 2 or even 3 possible in some species. Replacements laid after loss of eggs or even young. Eggs laid at intervals of 1–3 days. Incubation periods comparatively short, 22–30 days. Typically by both sexes in roughly equal spells. Single median brood-patch. Incubation starts with 1st or 2nd egg, so hatching asynchronous. Eggshells removed from nest. Young cared for and fed, typically by both parents (but see Botaurinae), by complete and partial regurgitation. Brooded continuously when small; then and later, sheltered from strong sun or rain by special spread-wing posture. Older young often guarded by parents in turn. May leave nest before fledging, though often return to be fed. Fledging periods vary from 30–55 days; young may become independent soon after, but prolonged periods of post-fledging semi-dependence probably more typical, especially in larger species. Age of first breeding usually 1 or 2 years.

# Subfamily BOTAURINAE bitterns

About 12 species in 2 main genera, *Botaurus* and *Ixobrychus*; both genera represented in west Palearctic. Body stouter and neck and legs relatively shorter than in Ardeinae. 8–10 tail-feathers. Inner toe longer than outer. Plumage cryptic; brown, buff, black, and yellow—in well-defined patches (*Ixobrychus*) or delicately mottled (*Botaurus*). Ornamental plumes on crown and lower neck not so elaborate as in many Ardeinae; in *Botaurus*, also whitish tufts in scapular region (used in display by at least 1 species). Sexes similar in *Botaurus*, strongly dimorphic in *Ixobrychus*. 2 pairs of powder-down patches, on breast and rump.

At most times, essentially cover-haunting and secretive. When alarmed, typically assume upright concealing posture (Bittern-stance). Social pattern and behaviour generally less well known than in Ardeinae, but at least 1 species of *Botaurus* polygynous at times, ♂ (whether with 1 mate or more) taking no part in rearing of young. In *Ixobrychus*, pair-bond and joint care of young by ♂ and ♀ as in other Ardeidae, so far as known. Visual displays, in both antagonistic and heterosexual situations, poorly known but, correlated with cryptic and mostly dispersed nesting pattern, probably less well developed than in Ardeinae, with even greater emphasis on frequent vocal displays (or song), especially in *Botaurus*.

# *Botaurus stellaris* **Bittern**

PLATE 29
[between pages 254 and 255]

Du. Roerdomp    Fr. Butor étoilé    Ge. Rohrdommel
Ru. Выпь    Sp. Avetoro común    Sw. Rördrom

*Ardea stellaris* Linnaeus, 1758

Polytypic. Nominate *stellaris* (Linnaeus, 1758), Palearctic. Extralimital: *capensis* (Schlegel, 1863), southern Africa.

**Field characters.** 70–80 cm, of which body half or less, ♂ averaging larger; wing-span 125–135 cm. Stocky, rather bulky, thick-necked heron, generally golden-brown with distinct black margins and black crown. Sexes alike and no seasonal differences; juvenile just distinguishable on close view.

ADULT. Golden-brown above, mottled and barred with black; more uniform brown on sides of head and neck; crown and moustache black. Pale yellow-buff below, striped with reddish-brown and blackish; feathers on foreneck thick and elongated. In flight, primaries and all except innermost secondaries show up as pale reddish-brown barred and spotted with black. Bill green-yellow; bare loral area green to pale blue; eyes yellow; feet pale green with yellow soles and back of tarsus. JUVENILE. Similar but generally paler, particularly below, with browner stripes and less well-defined moustache; indistinguishable by 1st winter.

Vagrant American Bittern *B. lentiginosus* smaller with chestnut (not black) crown, black mark down side of neck, less boldly marked plumage, and narrower-looking wings with black tips. Juvenile Night Heron *Nycticorax nycticorax* sometimes misidentified, but considerably smaller, greyer, speckled with whitish rather than black, and lacking dark crown and moustache; flight also different.

Generally associated with reeds. Skulking and secretive, and not often out in open during daylight; much of food obtained within reed cover, though sometimes seen at edge of reeds or feeding in shallows. Usually walks slowly and deliberately, but can run quite well; may stand motionless for some time before taking next step. Despite its bulk, clambers easily among reeds, grasping several stems at once; rarely perches in trees (unlike *Nycticorax*). Quite often flies over reeds during day when feeding young, but otherwise essentially crepuscular and, if flushed, travels only short distance before dropping back into reeds. Flies with distinctive, owl-like action on broad, round-ended wings low over reeds, with retracted neck hidden (appearing short and thick) and legs trailing; but neck remains extended until well on wing and may not be fully retracted over short distance. In breeding season, more often heard than seen on account of ♂'s characteristic booming.

**Habitat.** Closely restricted to lowland swamps and densely vegetated wetlands, in middle latitudes, mainly below 200 m with extensive shallow standing water, not unduly fluctuating in level. Favours tracts or fringes overgrown with tall emergent vegetation, especially reed *Phragmites*, giving dense cover close to sheltered open waters, including small pools and channels, natural or man-made. Sometimes emerges on to more open areas, e.g. lake or river banks or where small patches of reed have been cut, but not normally where visible from distance. Usually avoids even-aged and especially older, drier stands of vegetation, and acid waters (pH below *c.* 4·5), and scrub or tree growth unless sparse. Prefers reeds close enough for climbing to top by gripping several together to spread weight; normally moves through reedbed by walking on reed stems (Percy 1951). Tolerates brackish water, e.g. in estuarine or delta marshes and in areas where sea has broken in, but exceptional on sea coast. In frosty weather, hungry birds may occur anywhere near fresh water. Sometimes colonizes artificial wetlands, such as overgrown gravel-pits. Although more shy of disturbance or observation than most herons, occasionally becomes accustomed to regular and disinterested human activities close by. Lives within fairly compact territory, sometimes at high density where carrying capacity of habitat permits; thus uses airspace more sparingly than most of family, unless flushed or displaced by adverse weather, then usually flying at low or medium heights. Sometimes circles, soaring to great height. Generally moves at night.

**Distribution.** Many breeding areas abandoned, especially in 19th century, due to drainage and sometimes human persecution. Some recent recovery in a few countries due to protection and the establishment of reserves. BRITAIN AND IRELAND. Widespread decrease led to extinction in most areas by mid-19th century. Re-established Norfolk 1911 and has since increased and spread. Has bred in 11 counties since 1940; still absent Scotland and Ireland (Alexander and Lack 1944; British Ornithologists' Union 1971). Recent decline, Norfolk. SWEDEN: disappeared as breeding species after middle 19th century, perhaps due to hard winters; now breeding many areas (SM; Curry-Lindahl 1961). FINLAND: not reported regularly until 1930s, now breeds annually in small numbers (Merikallio 1958). WEST GERMANY: formerly bred in almost all parts; now nesting regularly only in north, though recently a few pairs found breeding again in Bavaria (EB). SWITZERLAND: almost annual in breeding season and may nest occasionally, but no proof (Bauer and Glutz 1966). ITALY: probably decreasing; no longer breeds Sicily (FS, GS). GREECE: formerly bred near Salonica (WB). PORTUGAL: formerly bred, but no certain records

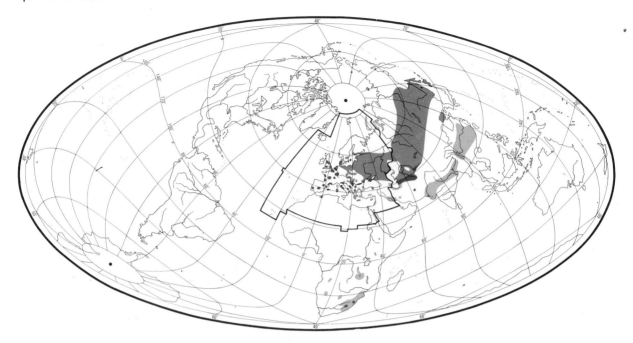

since 1930s (MDE). NORTH AFRICA: breeds near Oran, Algeria (EDHJ), and in past heard calling in spring in various areas north Tunisia, Algeria, and north-west Morocco (Heim de Balsac and Mayaud 1962).

Accidental. Iceland, Faeroes, Norway, Syria, Jordan, Cyprus, Azores, Madeira, Canary Islands.

**Population.** BRITAIN AND IRELAND: under 100 pairs (British Ornithologists' Union 1971). FRANCE: estimated 220–390 pairs in 1974 (Brosselin 1975). BELGIUM: some 20 pairs (Lippens and Wille 1972). NETHERLANDS: *c.* 320–350 pairs in 1950 and 1957 (Braaksma 1958), recent increase to *c.* 500 due to succession of mild winters (S Braaksma). DENMARK: maximum 25 pairs, decreasing recent years due to habitat destruction (TD). SWEDEN: 150 ♂♂ heard during census 1969 on lakes in centre and south; marked increase in only one area (Broberg 1971). WEST GERMANY: Nordrhein-Westfalen not more than 10 pairs in 1967 and 1971 (Mebs 1972). CZECHOSLOVAKIA: marked decrease 1950 (KH). POLAND: scarce, especially in south (Tomiałojć 1972). BULGARIA: apparently marked decline (Mountfort and Ferguson-Lees 1961). GREECE: 6–8 ♂♂ heard Evros delta 1973 (WB). In north of range (e.g. Britain, Netherlands, and Sweden), population drops after hard winters.

Survival. Oldest ringed bird 10 years 8 months (Rydzewski 1974).

**Movements.** Partially migratory, dispersive, and a resident. Uncertain whether has true migratory tendency, or whether movement dispersal in direct response to cold weather. In central and east Europe, will overwinter in breeding areas when marshes not frozen, in mild winters even in Baltic countries, but many (especially from north and east) move away. In North Sea countries, breeders mostly resident and many starve rather than emigrate in occasional severe winters (Bannerman 1957). Post-fledging dispersal less observed than in most European herons perhaps due to secretive habits, but some distant autumn recoveries of juveniles (e.g. Norfolk to Cheshire; Suffolk to Derbyshire and Co Longford; Silesia to Mecklenburg).

No evidence for any emigration by British breeders; those of Low Countries mainly resident, though a few recoveries south to Gironde (France) and to England; winter visitors to Britain (annually in small numbers) also include some from Germany (Saxony, Silesia) and Sweden. Late autumn and winter movements by continental breeders mainly between south and west; from Germany, where most ringed, recoveries in Netherlands, Belgium, France, Britain, Spain, Italy, and Czechoslovakia (Zink 1958). Longest ringing movements: Silesia to Landes (France) and Navarra (Spain), 1500 km; Lithuania to Morbihan and Bouches du Rhône (France), 1800 km; and from Sweden to Biscay provinces, Spain, 2000 km (Bernis 1966). Birds (origins unknown) winter regularly Iraq and both sides of Mediterranean and have straggled to some Atlantic islands. Winter ranges of those emigrating out of USSR uncertain in absence of recoveries; resident in Black Sea and Transcaucasia (Dementiev and Gladkov 1951), and fairly common winter visitor in north and west Iran (D A Scott), these presumably USSR breeders. In Egypt, reported as common in winter (Meinertzhagen 1930). A few of Palearctic race collected winter Nigeria, Sudan, Abyssinia, Eritrea, and north-east Congo; and individuals found

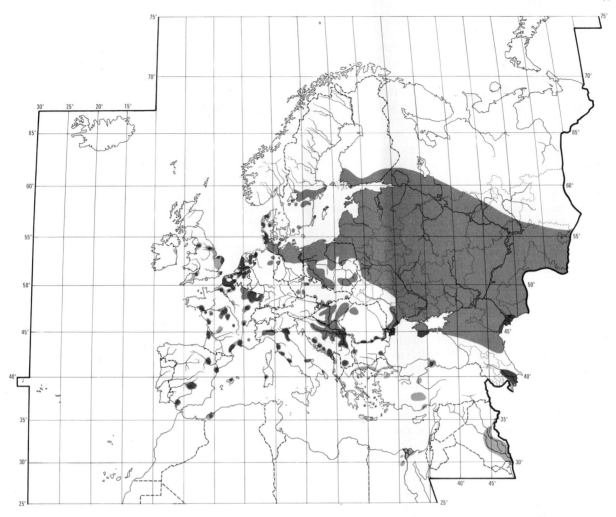

Saharan oases west to Colomb Bechar, mainly in spring (Moreau 1967).

Post-juvenile dispersal apparently begins July, but little evidence of adult movement until onset of night frosts, in east Europe from late September; movement protracted, often continuing into December. Return to breeding areas apparently February–April; some in USSR while still snow cover and night frosts. Migration mainly nocturnal, but sometimes by day in dull weather. Usually in ones and twos; rarely in parties of up to 10 in USSR with 'dozens' in small areas of suitable cover during passage, though these usually fly off individually at dusk (Dementiev and Gladkov 1951).

**Food.** Mainly fish, amphibians, and insects; also worms, leeches, molluscs, crustaceans, spiders, lizards (including *Lacerta*), small birds (including Wren *Troglodytes troglodytes* and Bearded Tit *Pānurus biarmicus*), and small mammals (including water-vole *Arvicola*, vole *Microtus*, and shrew *Neomys*). Takes prey while slowly wading or when stationary; fish held across neck (Lundevall 1953) or

around middle, and killed by shaking and biting before being swallowed head first (Gentz 1965); frogs and mice often impaled through back, and swallowed whole (Liebe 1892). Solitary, crepuscular, and daytime feeder, mainly in shallow water in or near cover.

Fish include pike *Esox*, carp *Cyprinus*, roach *Rutilus*, dace *Leuciscus*, eel *Anguilla*, miller's thumb *Cottus*, tench *Tinca*, perch *Perca*, stickleback Gasterostidae, trout *Salmo*, grayling *Thymallus*. Amphibians: *Rana*, Anura, *Triton*, *Bombina*, *Pelobates*. Insects: Hemiptera, particularly waterboatmen *Notonecta* and *Naucoris*; Coleoptera, especially waterbeetles (Dytiscidae and Hydrophilidae); Orthoptera; Odonata.

Diet constituents vary with locality and season (Gentz 1965). 9 stomachs from various parts of France, and survey of literature, indicated diet primarily fish (including loach *Cobitis* and *Leuciscus*) and amphibians (*Rana*), with small mammals and insects second, but rarely birds and crustaceans (Madon 1935). 16 Italian birds contained frogs *Rana esculenta* most frequently (26·6%), followed by insects (*Hydrous*, *Dytiscus*, *Notonecta*) in 5, fish (*Esox c.* 30 cm, sunfish *Eupomotis gibbosus c.* 20 cm), grass snake *Natrix natrix* (*c.* 73 cm), and water-vole *Arvicola amphibius* (Moltoni 1948). 51 stomachs from various areas in Hungary frequently contained amphibians (Anura 41%, *Triton* 15·6%), insects especially aquatic and terrestrial Coleoptera (50·9% and 29·4%) and aquatic Hemiptera (27·0%); and to a lesser extent, fish (19·6%) and voles *Microtus* (Vasvari 1929, 1938). Fish, however, said to be main food in USSR, especially *Cyprinus, Tinca, Perca*, and small *Esox* (occasionally up to 35 cm), though frogs, tadpoles, salamanders, worms, insects, and occasionally mammals also taken; near Lenkoran, in winter, water-beetles (*Macrodytes*, Hydrophilidae), and *Notonecta*, and possibly some vegetation recorded, while, in February, mole-crickets *Gryllotalpa*, waterbeetles, and salamanders taken; at other times, leaf-miners and fish (Dementiev and Gladkov 1951). Main prey from Franconian ponds, West Germany, fish (*Tinca, Cyprinus, Perca, Esox*) and frogs (Bauer and Glutz 1966).

Diet of juveniles principally tadpoles, though a nestling in August at Naurzum Reserve contained only water-voles (Dementiev and Gladkov 1951).

Lundevall (1953) calculated daily consumption as one-fifth body weight, though this based on captive and probably very hungry bird picked up in cold spell.

**Social pattern and behaviour.** Few detailed studies. Summaries in Gentz (1965), Bauer and Glutz (1966); see also Turner (1924), Portielje (1926), Zimmerman (e.g. 1929, 1931), Percy (1951), and especially Gauckler and Kraus (1965).

1. Essentially solitary at all times. During winter, not confined to such delimited areas as when breeding but ranges more widely. BONDS. Nature of pair-bond disputed. Polygamy by ♂ suggested by early writers (e.g. Percy 1951) but doubted by Gentz (1965); now confirmed that ♂ can have 1–5 mates (Gauckler and Kraus 1965). Probably no real pair-bond, ♂ and ♀ associating mainly for copulation. Only ♀ tends young, until shortly after fledging. BREEDING DISPERSION. ♂ strongly territorial from late winter until at least June–July, maintaining large area from which other ♂♂ excluded and within which nests of various ♀♀ situated, often near ♂'s calling place and seldom more than 50 m away (Gauckler and Kraus 1965), though up to 500 reported. Nests may be well separated from each other, though in some areas (e.g. Norfolk, England) fairly close at times and up to 5 nests side by side reported by Gauckler and Kraus (1965). Density of ♂♂ variable: from maximum of 1 every 2 ha in particularly favourable areas to 1 every 40–50 ha or more. ROOSTING. Little information, but evidently solitary; in dense cover and apparently off ground. Captive bird habitually slept in protective Bittern-stance (Lundevall 1953); see below. Active by day and at dusk, so main roosting possibly nocturnal; probable feeding flights around sunset and during dusk by 1–3 birds reported in spring, late summer, and autumn (Gauckler and Kraus 1965).

2. Intraspecific displays largely unknown, including those associated with pair-formation. Significance of posture in which pale shoulder plumes, usually hidden by folded wing, conspicuously exposed and spread crosswise over back, uncertain; in American Bittern *B. lentiginosus*, however, performed by ♂ (1) as threat-display preceding fighting with another ♂, (2) while persistently giving Advertising-call, and (3) prior to copulation

(Palmer 1962). Well-known Bittern-stance with neck and body fully stretched vertically, bill pointed skywards, and eyes swivelled forwards to watch intruder, assumed as behavioural component of camouflage when disturbed; will stay thus for hours, swaying to match movements of reeds, and can be approached closely. Birds frightened by predators, e.g. ♀ on nest, adopt Defensive-posture, crouching with feathers ruffled (especially elongated plumes of foreneck and crest), bill raised, and wings outspread. From both protective and defensive position, stares at face of enemy from below bill; liable to launch sudden, rapid attacks, stabbing with slightly open bill. ANTAGONISTIC BEHAVIOUR. Characteristic, far-carrying booming of ♂ seems largely territorial Advertising-call. Given by inflating oesophagus, with neck first stretched forward and downwards, then raised; sometimes preceded by clattering of mandibles. ♂ watched by Coward (1920) had plumes of neck and breast erected. Delivered frequently from regular calling place, both day and night but mostly in evening, throughout much of breeding season; neighbouring ♂♂ often answer one another, and calls of individuals distinctive. ♂♂ evidently highly aggressive in defence of territory; sometimes found mortally wounded or dead from stab wounds and 2 ♂♂ reported interlocked in ground combat. Will also engage in aerial combats, both circling in flight and trying to stab each other. More formal flight activity of 4–6 birds during middle of season may also be of territorial significance: said to involve mostly ♂♂ (some coming from territories up to 5 km distant) which circle in air, sometimes rising rapidly and gliding down; occasionally joined by 1–2 ♀♀. HETEROSEXUAL BEHAVIOUR. Little known, including displays associated with pair-formation, though Advertising-call of ♂ presumably attracts ♀♀ (as well as warning off rival ♂♂) for ♀ will sometimes answer booming ♂ with similar but much softer call, causing ♂ to intensify his calling. Copulation apparently not confined to nest as in most other Ardeidae; at least at times, has appearance of assault on ♀ who calls throughout repeated mountings by ♂ and for long period afterwards; ♂ spreads wings and neck plumes and pushes bill deep into ♀'s throat as if feeding her (Yeates 1940). (In *B. lentiginosus*, however, reported copulations more formal; preceded by ♂ giving Advertising-call, then approaching while displaying shoulder plumes.) Though ♂ thought by some to keep guard while ♀ incubating or tending young, does not normally visit nest; if he does, ♀ aggressive—rises from nest, ruffles neck plumes, shakes body from side to side, and glares. Feeding of ♂ by ♀ asserted, but not confirmed. RELATIONS WITHIN FAMILY GROUP. ♀ feeds tiny young without stimulation; later, young beg by grasping and tugging her bill. When 15–16 days old, wander at times from nest, clambering about surrounding vegetation; often seek shade if original nest in exposed site, ♀ sometimes building second nest for them. First adopt typical Bittern-stance if alarmed at *c.* 8 days. Give Food-calls whether ♀ present or not (Gauckler and Kraus 1965).

**Voice.** Distinctive, far-carrying booming of ♂♂ in breeding season; various other croaking and cackling sounds. Following based mainly on Gauckler and Kraus (1965).

(1) Booming of male. Deep, slow resonant booming, not loud but of great carrying power, sometimes audible up to 3–5 km. Preceded by 2–3 short coughs or grunts, 'up' (see I); also sometimes by bill-clappering. Varies in volume (sometimes quite soft) and frequency (most often at dusk in April–May). Repeated at 1–2 s intervals, usually 3–4

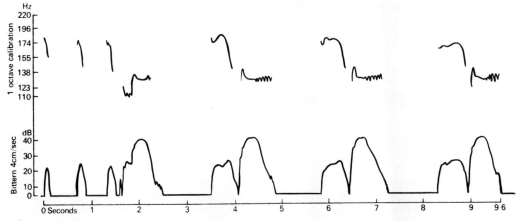

I L Shove England May 1966

times but sometimes 5–6 or more (Bauer and Glutz 1966; Percy 1951). Normally 'uprumb' or 'up-up-(up)-rumb' but also quiet 'umb üh üh üh ub' and 'hu hu umb ub' or 'üh ub'. Serves both to attract ♀♀ and defend territory; other birds flying over giving Flight-call early in breeding season cause ♂♂ to increase frequency of booming (see also Social pattern and Behaviour). ♀♀ sometimes respond with soft, booming, subdued 'wumph' (Witherby *et al.* 1939). (2) Flight-call. Nasal 'kau' (Gauckler and Kraus 1965) given by both sexes; also described as 'aark aark' or 'awk awk' (Witherby *et al.* 1939). Used in variety of circumstances, sometimes when perched, including by ♀♀ when foraging or flushed from nest. Common in autumn, when used as contact-call during circling flights at dusk. (3) Warning-call. Rapid, short, not loud 'ko ko ko' or 'kro, kro, kro'. Also thin, sibilant, high-pitched 'zeez' when rises in alarm (Turner 1911). (4) Excitement-call. By ♀ on nest. Rapid, trisyllabic 'tschätä-schätä-schätä' (Bauer and Glutz 1966). (5) Threat-call. By ♀ on ground or in flight when young picked up from nest: 'chra' or 'krä'.

CALLS OF YOUNG. Food-call incessant, cackling 'quaq-uaqua' or 'rah rah rah'. When older, give louder version, or whimper quietly when picked up; also utter bubbling noise (Turner 1911).

**Breeding.** SEASON. See diagram for central Europe (Gentz 1965); few local data but probably earlier in Mediterranean than in northern Europe. SITE. In dense reedbeds among previous year's growth. Nest of monogamous pairs well dispersed; females of polygamous male may nest 15–20 m apart (Gauckler and Kraus 1965). Nest: loose heap of dead reeds and other vegetation lined with a little finer material, forming circular platform 30–40 cm diameter and 10–15 cm high; material added until young well-grown and may finally be up to 90 cm diameter and 50 cm high (Dementiev and Gladkov 1951; Gentz 1965). Rests on matted roots and stems at water-level. Building: by ♀, using both material within reach and brought from a distance. EGGS. Blunt-ovoid; olive-brown, occasional fine

spotting at large end; matt. 53 × 39 mm (48–58 × 34–41), sample 100 (Schönwetter 1967). Weight 40 g (35–48), sample not given (Bauer and Glutz 1966). Clutch: 5–6 (3–7). Of 47 clutches, Germany: 3 eggs, 2%; 4, 34%; 5, 49%; 6, 15%; mean 4·76. Of 31 clutches, Denmark: 3 eggs, 3%; 4, 10%; 5, 48%; 5, 36%; 7, 3%; mean 5·25 (Hermansen 1972). One brood. Replacement clutches not recorded. Eggs laid at 2–3 day intervals. INCUBATION. 25–26 days. Normally by ♀; statement that ♂ may help (Dementiev and Gladkov 1951) presumably refers only to monogamous ♂. Starts with 1st egg; hatching asynchronous, spread over 12–13 days (Gentz 1965), or over 7–8 days (Gauckler and Kraus 1965). Eggshells ejected over side of nest or carried a few metres. YOUNG. Semi-altricial and nidicolous. Cared for and fed by ♀, though (monogamous) ♂ may bring food near to nest for ♀ to collect; food completely regurgitated into bottom of nest and young feed themselves. Disperse short distances from nest at 15–20 days and fed there. FLEDGING TO MATURITY. Fledging period 50–55 days; become independent soon after. Age of first breeding 1 year. BREEDING SUCCESS. Of 21 clutches, Germany, 62% successful, most losses due to man; of 66 eggs in 13 nests, 63% hatched and 56% reared to 14 days (Gauckler and Kraus 1965).

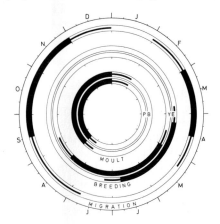

**Plumages.** ADULT. Crown and nape black, feathers of nape with wide buff margins, narrowly barred black; feathers of crown and nape elongated and loose. Supercilium pale buff, sides of head and neck buff, narrowly barred dark brown on neck. Moustachial streak dark brown to black. Chin and throat cream-buff with rufous-brown central streak. Mantle and scapulars black, boldly patterned by irregular buff or tawny-buff margins; back, rump, and upper tail-coverts paler, buff with grey-brown spots. Underparts pale buff with brown streaks, finely spotted black; streaks densest and widest on chest, scarcer and narrower towards belly, almost absent on vent and under tail-coverts. Tail-feathers tawny-buff, mainly black along shaft and speckled black on both webs. Flight-feathers and primary coverts black, irregularly barred rufous; inner secondaries with buff margin to outer web. Upper wing-coverts buff, strongly mottled and freckled tawny and black. Under wing-coverts pale buff with grey mottling, axillaries similar but barred instead of mottled. NESTLING. Down long and scanty on upperparts, shorter and even scantier on underparts; sides of head, hindneck, and sides of body bare. Upperparts cinnamon-tawny to rufous-brown, chin and throat white, rest of underparts rufous-buff. JUVENILE. Closely similar to adult but black of crown tinged brown and less extensive, usually not extending to nape. Moustachial streak brown, not black. Mantle and shorter scapulars less boldly patterned than adult, contrast of markings obscured by many fine mottlings and vermiculations. Shape of feathers different: body and tail feathers shorter, narrower near tip, and more loosely textured than those of adult; outer primaries tend to be narrower with more pointed tip (broader and rounder in adult), but this difference not always clear. In mid-winter, upper wing-coverts usually paler in juveniles than in adults. IMMATURE. Apparently not separable from adult.

**Bare parts.** ADULT AND JUVENILE. Iris yellow. Bill green-yellow, ridge of culmen brown becoming darker towards tip. Bare skin of face green to pale blue. Leg pale green; tarsal joint, back of tarsus and soles yellow. NESTLING. Bill and leg blue-green. (Witherby et al. 1939.)

**Moults.** ADULT POST-BREEDING. Complete; August to January (Witherby et al. 1939), but usually finished November. Sequence of primary replacement apparently irregular; poorly known as elusive during moult (Stresemann and Stresemann 1966). POST-JUVENILE. Partial; body and tail, July–November. Starts with head and neck soon after fledging. Head apparently moulted again mid-winter or early spring (needs confirmation). Occasionally some scapulars or tail-feathers retained during 1st winter, but other individuals (particularly ♂♂) may moult some flight-feathers and wing-coverts.

**Measurements.** Adults and juveniles (tail and tarsus adults only); Netherlands; skins (ZMA).

| | | | | | | | |
|---|---|---|---|---|---|---|---|
| WING | ♂ 346 | (6·67; 21) | 335–357 | ♀ 311 | (7·34; 16) | 296–327 |
| TAIL | 117 | (4·40; 11) | 112–126 | 104 | (4·87; 8) | 96–110 |
| BILL | 69·1 | (3·24; 20) | 61–74 | 64·3 | (2·10; 16) | 60–68 |
| TARSUS | 102 | (3·53; 11) | 97–109 | 91·0 | (2·63; 9) | 87–95 |
| TOE | 118 | (5·04; 20) | 110–132 | 107 | (5·02; 15) | 98–119 |

Sex differences significant. Bauer and Glutz (1966) gave maximum wing for adult ♂, measured in flesh, as 377, tail 131, bill 80, tarsus 125. Juvenile slightly smaller than adults, but difference significant only for tail and tarsus of ♂♂.

| | | | | | | | |
|---|---|---|---|---|---|---|---|
| TAIL | ♂ 113 | (2·75; 10) | 109–116 | ♀ 101 | (6·63; 7) | 92–110 |
| TARSUS | 98·6 (3·13; 10) | | 92–103 | 90·7 (4·54; 7) | | 81–94 |

**Weights.** Fat or in normal condition: ♂ 966–1940, ♀ 867–1150 (Bauer and Glutz 1966). Lean or emaciated: ♂ 639 (94·6; 12) 462–787, ♀ 556 (86·0; 16) 430–750 (ZMA).

**Structure.** Wings broad. 11 primaries: p9 usually longest, p10 and p8 0–5 shorter (though either occasionally longest), p7 0–10, p6 10–25, p5 25–45, p4 40–60; p11 minute. Webs of primaries not emarginated, p10 pointed, p9 slightly so. Tail short, slightly rounded; 10 feathers. Under tail-coverts reach nearly to tip of tail. Bill heavy and broad at base, rather short; ridge of upper mandible gently curved towards tip; distal part of cutting edges serrated. Tibia ¾ feathered. Outer toe c. 75% of middle, inner c. 82%, hind c. 65%, hind claw, c. 30%.

**Geographical variation.** B. stellaris capensis differs from nominate stellaris in being darker above, less strongly spotted buff, with primaries spotted rather than barred. Populations of east Palearctic slightly more contrastingly striped underneath (Bauer and Glutz 1966); sometimes considered separate race orientalis Buturlin, 1908. All species of Botaurus much alike and allopatric, so may be considered to form one cosmopolitan superspecies (Bock 1956).

## Botaurus lentiginosus American Bittern

PLATE 29
[between pages
254 and 255]

DU. Noordamerikaanse Roerdomp   FR. Butor d'Amérique   GE. Nordamerikanische Rohrdommel
RU. Американская выпь   SP. Avetoro lentiginoso   SW. Amerikansk rördrom

Ardea lentiginosa Montagu, 1813

Monotypic

**Field characters.** 60–85 cm, ♂ averaging larger; wingspan 105–125 cm. Stocky, medium-sized, brown heron with rusty crown, black neck patches, and black flight-feathers. Sexes similar and no seasonal differences; juvenile distinguishable on close view.

ADULT. Brown above, only finely freckled with black, apart from rusty brown crown, diagnostic black flight-feathers and characteristic elongated black patch from behind and below eye down side of neck; below, throat white with rufous streaks outlined in black, shading to buff on breast with more diffuse yellowish streaks. Bill dull yellow with dark-tipped upper mandible; bare lores greenish-yellow; eyes yellow; legs bright yellow-green. JUVENILE. Lacks black neck patches; these assumed in 1st winter.

Distinguished from Bittern B. stellaris by smaller size, black flight-feathers, black neck marks (except juvenile), rusty crown, and lack of bold black mottling and streaking on upperparts. Smallest individuals similar in size to juvenile Night Heron *Nycticorax nycticorax*, but latter greyer and without black on neck and flight-feathers.

Behaviour similar to *B. stellaris*, but seen much more often in open, habitually feeding in short sedges and grasses away from reed cover. Rarely perches in trees (unlike *Nycticorax*), though may stand a metre above ground on stump or in aquatic vegetation. Flight more like typical herons *Ardea*, wings being less broad and not so rounded and beats more rapid than in *B. stellaris*, giving impression of greater manoeuvrability. Flight note a rapid, throaty 'kok-kok-kok'.

**Habitat.** Like that of Bittern *B. stellaris*, allowing for secondary differences in vegetation and ecological conditions of North America, but apparently more often extends to meadows or pastures, hunting large insects. Perhaps also more frequent in tidal and saline situations, and on quaking morasses, but selects similar stands of tall emergent vegetation fronting shallow pools or watercourses. May differ from *B. stellaris* in lesser aversion to proximity of trees, and vagrants in Europe sometimes use atypical habitats for lack of choice.

**Distribution.** Breeds over large area of North America, north to British Columbia, Great Slave Lake, Alberta, Saskatchewan to Hudson Bay, and central Quebec to Newfoundland. Southern limits hard to define as apparently thins out, breeding locally from Texas to Florida.

Accidental. Britain and Ireland: *c.* 52, of which some three quarters before 1914, mainly October–November but also all months September–March (British Ornithologists' Union 1971), and 1 Channel Islands. Also Iceland 3; Faeroes 2; Norway 1; Spain 1; Azores 4; Canary Islands 1.

**Movements.** Migratory and dispersive. Considerable wandering late summer and early autumn, presumably post-fledging dispersal. Southward migration mainly September–October, stragglers into November. Though occasionally wintering north to Ontario, mainly withdraws from northern two-thirds of breeding range, except near Pacific coast where resident north to south-east British Columbia; winters chiefly from Nevada, Utah, New Mexico, Arkansas, Mississippi valley, and Virginia south through Mexico to Yucatan, with a few to Costa Rica and once Panama (Wetmore 1965); also visits some West Indian islands (Bond 1960). Early return, majority passing through USA March and first half April, reaching northern breeding range in April (Bent 1926; Godfrey 1966; Palmer 1962).

Has straggled to Canadian Northwest Territories, Labrador, west Greenland, Bermuda, and west Palearctic. Transatlantic vagrants in north-west Europe probably originate Gulf of St Lawrence region, where autumn

migrants most likely to be caught up in westerly storm tracks; but those in latitudes of Spain and Canary Islands perhaps blown out to sea further south in North America.

**Voice.** See Field Characters.

**Plumages.** ADULT. Crown rusty brown, bordered by cream-white supercilium. Hindneck olive-grey, speckled pale buff. Sides of head buff, line under eye olive or olive-brown, spot at base of lower mandible chestnut. Elongated black patch on side of neck, longer in ♂ than in ♀. Chin and upper throat white, streak of pale brown spots down centre. Mantle and scapulars dark brown or dark grey-brown, finely mottled and vermiculated buff and black; much less boldly patterned than Bittern *B. stellaris*; feathers of upper mantle edged buff. Back, rump, and upper tail-coverts duller and more delicately mottled than mantle. Underparts pale buff to cream-white, streaked brown; streaks finely speckled dark grey or black, densest on chest, absent on vent and under tail-coverts. Feathers of chest elongated; on sides of upper breast concealed tufts of cream-white feathers. Tail brown or grey-brown with paler vermiculations at edges of feathers. Flight-feathers dark grey, not barred as in *B. stellaris*; tips of inner primaries and outer secondaries brown, speckled dark grey, edged pale buff. Inner secondaries wholly brown and speckled. Upper wing-coverts vermiculated grey and pale buff, paler than mantle and scapulars. Under wing-coverts and axillaries buff, speckled dark grey. Whole plumage said to be warmer brown and brighter in autumn and winter (Roberts 1955), but variable at all seasons. JUVENILE. Like adult, but plumage looser and softer, no black patch on side of neck.

**Bare parts.** ADULT AND JUVENILE. Iris yellow. Bill dull yellow, ridge of culmen and tip of upper mandible dark brown. Bare skin of face green-yellow with olive-brown strip from eye to base of upper mandible. Leg bright yellow-green (Palmer 1962).

**Moults.** ADULT POST-BREEDING. Complete; August–November (Palmer 1962). Primaries irregular (Stresemann and Stresemann 1966). PRE-BREEDING. According to Palmer (1962) and Roberts (1955) body plumage renewed February–May, but Bent (1926) supposed only single moult per cycle; further information needed. POST-JUVENILE. Partial; body, but not wings and tail; August–November.

**Measurements.** Full-grown, eastern North America (Palmer 1962).

| | | | | | | |
|---|---|---|---|---|---|---|
| WING | ♂ | 280 | (14) | 267–291 | ♀ 247 | (17) | 238–255 |
| BILL | | 74 | (14) | 70–80 | 68·6 | (17) | 63–72 |

Wing ♂ 250–305, ♀ 245–270; tail ♂ 80–100; bill ♂ 68–86, ♀ 68–75; tarsus ♂ 86–97; middle toe 98–104 (Witherby *et al.* 1939).

**Weights.** ♂ 372; ♀ 482, 571, latter said to be extremely fat (Palmer 1962).

**Structure.** Wings broad. 11 primaries: p8 and p9 longest, either may be slightly shorter than the other; p10 2–10 shorter, p7 0–10, p6 13–25, p5 23–40, p1 66–86; p11 minute. Webs of primaries not emarginated, p10 sharply pointed at tip. Tail short, slightly rounded; 10 feathers. Under tail-coverts almost as long as tail. Bill heavy and broad at base, rather short; ridge of upper mandible gently curved towards tip. Tibia $\frac{3}{4}$ feathered. Outer toe *c.* 72% of middle, inner *c.* 80%, hind *c.* 65%, hind claw *c.* 30%.

## *Ixobrychus exilis* **Least Bittern**

PLATE 30
[between pages 254 and 255]

Du. Amerikaans Woudaapje  Fr. Blongios minute  Ge. Amerikanische Zwergdommel
Ru. Американский волчок  Sp. Avetorillo yanqui  Sw. Amerikansk dvärgrördrom

*Ardea exilis* Gmelin, 1789

Polytypic. Nominate *exilis* (Gmelin, 1789) North America, accidental. 3–4 other extralimital races.

**Field characters.** 28–36 cm, wing-span 40–45 cm, ♂ averaging larger. Tiny heron with blackish crown and back, chestnut to yellowish neck and underparts, and contrasting pale wing-patches. Sexes distinct, but only seasonal difference high gloss in spring plumage; juvenile also distinguishable.

Adult Male. Glossy green-black crown, back and tail, and slaty primaries, contrast with yellow-brown to chestnut neck, chestnut greater coverts and inner secondaries, and brown-yellow lesser and median coverts. These coverts form pale oval wing-patch conspicuous both in flight and on folded wing, which otherwise appears chestnut; light buff to whitish stripe along scapulars forms pair of pale braces on either side of back. Underparts brown-yellow with black-brown patch at side of breast and pale yellow wing-linings. Bill yellow with dark ridge; bare lores dull yellow to light green; eyes yellow; feet straw with greenish front to tarsus. Adult Female. Crown and mantle purple-chestnut instead of black, with 2 light stripes outlining scapulars, more buff-coloured greater coverts, and dark-streaked throat and foreneck. Juvenile. Differs from adult ♀ in paler, more brown crown and mantle, dusky streaks on lesser wing-coverts, and striped appearance of streaking on browner throat and breast. Colour Morphs. Rare dark form known as 'Cory's Least Bittern' normally has paler areas rich chestnut to chocolate (see also Geographical variation).

Likely to be confused only with slightly larger Little Bittern *I. minutus*, but ♂ of latter has more striking cream wing-patch and green-black secondaries and ♀ much less chestnut, while both lack light stripes on scapulars; juvenile darker and browner with less defined wing-patch. Green Heron *Butorides virescens* and Green-backed Heron *B. striatus* both slightly larger and rather chunkier with coverts same colour as back.

Usually in tall freshwater reeds and sedges. As New World counterpart of *I. minutus*, generally rather similar in behaviour, but even more secretive, burrowing through grass like rodent, and flight weaker, less butterfly-like: so might be momentarily mistaken for rail. Walks easily and runs rapidly, preferring to dash through or climb up aquatic vegetation rather than fly; if it has to fly, seldom does so for more than 25 m. Also 'freezes' like other bitterns with bill upward, eyes swivelled towards intruder, and neck elongated and laterally compressed. Usually solitary. Crepuscular peak of activity. Chattering call like Magpie *Pica pica* may be uttered if startled.

**Habitat.** Closely similar to that of Little Bittern *I. minutus*, but less arboreal and apparently more often in extensive wetlands with scattered shrubs and other woody plants among varied, dense stands of tall, emergent aquatic plants such as cat-tail *Typha*. In tropics, resorts to mangroves, and generally more tolerant of brackish water and saltmarsh than *I. minutus*. Fishes in series of small pools visited in definite order, never remaining at one after making a capture (Palmer 1962). Highly reliant on concealment afforded by habitat and related camouflage and behaviour; accordingly tolerant of human presence and disturbance in neighbourhood.

**Distribution.** *I. e. exilis* breeds eastern and central North America, from southern Canada (locally in southern Manitoba, Ontario, Quebec, and New Brunswick), to Gulf of Mexico and Caribbean, and reported breeding locally through eastern Mexico to El Salvador and southern Nicaragua.

Accidental. Azores: ♀ September 1951, ♀ November 1951, ♀ September 1952, ♂ October 1964 (Bannerman and Bannerman 1966). Iceland: one, September 1970 (FG).

**Movement.** Migratory and dispersive. Nominate race winters from Mexico, Texas, Florida, and Caribbean islands as far as Panama and Colombia, possibly further south, but winter distribution imperfectly known due to secretive habits and presence of other races in Neotropics. Autumn exodus mainly September from Atlantic states of USA, but later (up to mid-October) from Mississippi valley; only stragglers north of Gulf states from end of October. Return passage through USA most marked mid-April to late May; but infrequent March records thought

Plate 27. *(facing) Pelecanus onocrotalus* White Pelican (p. 227): 1 ad breeding, 2 ad non-breeding, 3 imm, 4 juv, 5 nestling. *Pelecanus crispus* Dalmatian Pelican (p. 233): 6 ad breeding, 7 ad non-breeding, 8 imm, 9 juv, 10 nestling. *Pelecanus rufescens* Pink-backed Pelican (p. 237): 11 ad breeding, 12 ad non-breeding, 13 imm, 14 juv. (PJH)

Plate 28. *(overleaf,* left) *Pelecanus onocrotalus* White Pelican (p. 227): 1 ad breeding, 2 ad non-breeding, 3a–c imm, showing progressive loss of brown under wing, 4 imm 2nd moult, 5 juv. *Pelecanus crispus* Dalmation Pelican (p. 233): 6 ad breeding (above and below), 7 imm, 8 juv. *Pelecanus rufescens* Pink-backed Pelican (p. 237): 9 ad breeding, 10 older imm, 11 imm, 12 juv or imm plumages from below, 13 juv. (PJH)

PLATE 29. *Botaurus stellaris* Bittern (p. 247): **1** ad, **2** juv, **3** nestling. *Botaurus lentiginosus* American Bittern (p. 252): **4** ad, **5** juv. (PJH)

PLATE 30. *Ixobrychus exilis* Least Bittern (p. 254): **1** ad ♂, **2** ad ♀, **3** juv, **4** ad dark morph. (PJH)

PLATE 31. *Ixobrychus minutus* Little Bittern (p. 255): **1** ad ♂, **2** ad ♀, **3** imm ♂ moulting into ad, **4** juv, **5** nestling. (PJH)

PLATE 32. *Ixobrychus eurhythmus* Schrenck's Little Bittern (p. 260): **1** ad ♂, **2** ad ♀, **3** imm ♂, **4** juv. (PJH)

to include early arrivals as well as exceptional overwintering birds. Seasonal withdrawal from much of breeding range shows that this supposedly weak flier travels considerable distances; has straggled to Nova Scotia, Newfoundland, and Bermuda in spring and autumn, as well as across Atlantic (see Distribution). (Bent 1926; Godfrey 1966; Palmer 1962; de Schauensee 1966.)

**Voice.** See Field Characters.

**Plumages** (nominate *exilis*). ADULT MALE. Crown, nape, mantle, scapulars, back, rump, and tail black, glossed green. Some scapulars with pale cream outer edges. Streak over eye and hindneck chestnut, sides of head and neck buff. Chin and throat white with pale buff gular stripe down centre. Rest of underparts pale buff to off-white. Concealed patch of dark brown feathers with buff edges on sides of chest. Flight-feathers black, inner primaries and outer secondaries with chestnut spot at tips; inner secondaries with chestnut outer webs. Primary coverts black, broadly tipped chestnut. Greater upper wing-coverts chestnut; median and lesser grey-buff with chestnut patch near bend of wing. Under wing-coverts pale grey-buff, greater grey tipped white. Axillaries dusky. ADULT FEMALE. Like adult ♂, but crown with dark chestnut edges. Mantle and scapulars dark brown, cream or buff outer edges to scapulars more prominent. Gular stripe mottled grey, more distinct. Chest darker buff with dark brown feather shafts. Lower breast, upper belly, and flanks streaked dark brown. Inner secondaries and greater upper wing-coverts less deep chestnut. JUVENILE. Like adult ♀, but crown with more chestnut, feathers of rest of upperparts with narrow buff tips which wear off in autumn (Bent 1926). Gular stripe and chest with more pronounced dark brown shaft-streaks, median and lesser upper wing-coverts with dark grey triangle at base extending into dark brown shaft-streak. ♂ probably recognizable by darker crown, ♀ by darker streaking on throat and chest (Bent 1926). FIRST WINTER AND BREEDING. Like adult but at least some dark-shafted wing-coverts retained. DARK MORPH. Melano-

erythristic variant, 'Cory's Least Bittern', dark all over. Adult: wing-coverts chestnut; sides of head and neck, throat and chest chestnut; rest of underparts dusky rufous with some white feathers at thighs. Juvenile also dark.

**Bare parts.** Iris yellow. Bill yellow with dark horn-brown culmen. Bare skin on sides of face dull yellow to light green. Leg green in front, yellow behind and on soles (Roberts 1955; Palmer 1962).

**Moults.** According to Palmer (1962), complete post-breeding moult July–August and partial pre-breeding late winter to early spring; 1st year birds should have partial post-juvenile moult in autumn and partial 1st pre-breeding in late winter to early spring. Confirmation needed.

**Measurements.** Adult (Dickerman 1973; middle toe ZMA).

| | | | | | | | |
|---|---|---|---|---|---|---|---|
| WING | ♂ 116 | (3·4; 29) | 109–122 | ♀ 114 | (3·4; 30) | 108–121 |
| TAIL | | 41·8 | (2·8; 20) | 37–47 | 40·1 | (1·8; 16) | 37–43 |
| BILL | | 45·3 | (1·9; 27) | 42–48 | 44·3 | 1·8; 29) | 40–48 |
| TARSUS | | 40·3 | (1·7; 12) | 38–42 | 39·5 | (1·4; 16) | 37–41 |
| TOE | | 43,44 | | | 43,45 | | |

Sex difference significant except for tarsus.

**Weights.** ♂♂ 46–85 (Palmer 1962).

**Structure.** 11 primaries: p9 longest, p10 1–4 shorter, p8 1–2, p7 3–5, p6 6–9, p5 10–12, p1 27–31; p11 minute. Primaries not emarginated. Tail square, 8 feathers. Bill relatively longer and more slender than in Little Bittern *I. minutus*. Lower *c.* 7 mm of tibia bare. Outer toe *c.* 80% of middle, inner *c.* 85%, hind *c.* 65%, hind claw *c.* 28%.

**Geographical variation.** Involves colour of hindneck, underparts, and wing-coverts (more richly coloured in tropical forms), and size (Palmer 1962; Dickerman 1973).

## *Ixobrychus minutus* Little Bittern

PLATE 31
[facing page 255]

| | | |
|---|---|---|
| Du. Woudaapje | Fr. Blongios nain | Ge. Zwergdommel |
| Ru. Волчок | Sp. Avetorillo común | Sw. Dvärgrördrom |

*Ardea minuta* Linnaeus, 1766

Polytypic. Nominate *minutus* (Linnaeus, 1766), Europe and west Asia; *payesii* (Hartlaub, 1858), Africa, accidental. 3 other subspecies extralimital.

**Field characters.** 33–38 cm, wing-span 52–58 cm; much smaller than any other west Palearctic heron except for vagrants, Least Bittern *I. exilis*, Schrenck's Little Bittern *I. eurhythmus*, and Dwarf Bittern *Ardeirallus sturmii*. Tiny, sharp-billed, thick-necked (in most postures), extremely skulking heron, with strongly patterned plumage in ♂ and characteristic pale forewing panels in both sexes. Sexes differ; some seasonal variation. Juvenile separable. ADULT MALE. Crown and slightly crested nape, back, scapulars and tail black (with faint green gloss). Neck, breast, and underparts pale buff, with longest of loose chest feathers tinged grey and patch on side of breast dark brown. Inner wing-coverts pale pink-buff, forming

striking oval panel on otherwise all black wing; underwing uniformly whitish. After breeding all buff areas become dirty and chest feathers more compact. Bill finely pointed, green-yellow with dark ridge to culmen. Bare lores yellow to dull red. Eyes yellow. Legs and feet green. ADULT FEMALE. Plumage pattern essentially as ♂, but duller, with crown glossy black, upperparts brown-black with pale feather-margins obvious at close range, underparts dirtier and lined with streaks of dark buff and brown and upperwing pattern less striking, with both coverts and quills browner. JUVENILE. Resembles adult ♀ but whole plumage duller and redder, with feather-margins of upperparts forming broad streaks, lines on underparts

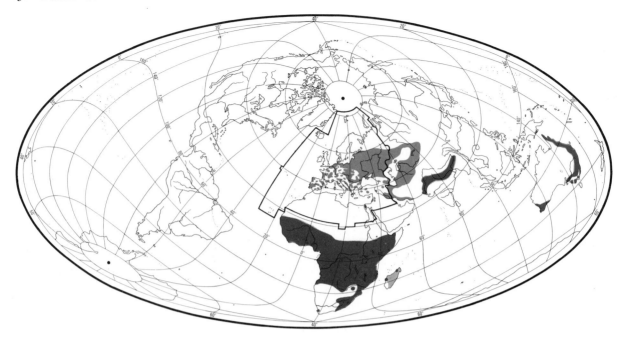

much stronger, and wing-coverts mottled and showing streaks on greater. Bare parts also duller.

Due to secretive behaviour and habitat preference, most often seen in flight. Combination of small size, plumage pattern and wing action can suggest small duck *Anas* or Corncrake *Crex crex* to inexperienced observer who only glimpses bird in distance. Separation from rare vagrant relatives varies from simple to difficult. *A. sturmii* has quite different plumage pattern with head, neck mane, upperparts, and entire wings dark slate (or dark brown in sub-adult) and underparts yellow- or grey-buff heavily streaked with black. Basic plumage pattern of *I. exilis* and *I. eurhythmus* as *I. minutus*. *I. exilis* identified by chestnut on sides of face, hindneck, and inner quills and (diagnostic) pair of thin white lines over back at all ages. *I. eurhythmus* distinguished by dark chestnut upperparts (brightly spotted in ♀ and young), red-brown on sides of face, sides and back of neck; much duller, ochre wing panels; and darker, heavier streaks on underparts (restricted to dark central line on foreneck of ♂). Great care needed with half-seen immature birds, with allowance made for effect of shadow on plumage tones.

Flight action rapid for family, with combination of plumage pattern and clipped wing-beats adding to impression of speed. Enters cover with characteristic glide followed by sudden bank and pitch; when rising, feet trail noticeably and looks to be in difficulty. Walks easily, but in crouching posture; climbs freely through dense cover (even up reeds), and adopts vertical concealing posture when discovered. Most active at dusk; feeds by snapping up small fish and wide variety of invertebrates. Not gregarious but migrants often form small groups. Flight-call low, short, sudden monosyllable, changing in alarm or excitement to a harder, more explosive cough. In spring, ♂ repeats a low, sighing croak at about 2-second intervals.

DIMW

**Habitat.** Like Bittern *Botaurus stellaris* avoids boreal or icy conditions, but extends to tropical as well as middle latitudes, and ranges above 500 m, although mostly in lowlands below 200 m. More adaptable than *B. stellaris*, occupying wide range of freshwater swamps, riverain zones, fringes of lakes or pools, and inundations carrying tall dense stands of reeds *Phragmites* or similar emergent plants, or trees and shrubs such as willows *Salix*, alders *Alnus*, and other deciduous species; rarely pines *Pinus* and creepers. Unlike *B. stellaris*, markedly arboreal, climbing freely and adroitly and favouring fairly dense cover fringing or overhanging still or gently flowing water. Also accepts artificial situations, such as fish-ponds, excavations, canals, ditches, and ornamental waters. Tolerates fairly close human presence provided adequate cover available, and unlike *B. stellaris* does not require large unbroken tracts of suitable habitat. Moves on ground in somewhat rail-like manner, avoiding open. Usually feeds in breeding area, but sometimes flights elsewhere. Flight normally low and direct, often involving difficult take-off; on migration sometimes higher. Outside breeding season, wanderings may lead to occurrence in abnormal habitat, but wintering birds seek conditions similar to those of breeding places, although sometimes on more open wetlands and waters, including even coastal regions, and also in clumps of trees or scrub away from water.

**Distribution.** Local as breeding species and precise distribution inadequately known in many areas.

ITALY: no longer breeds Sicily (GS). ALGERIA: breeds

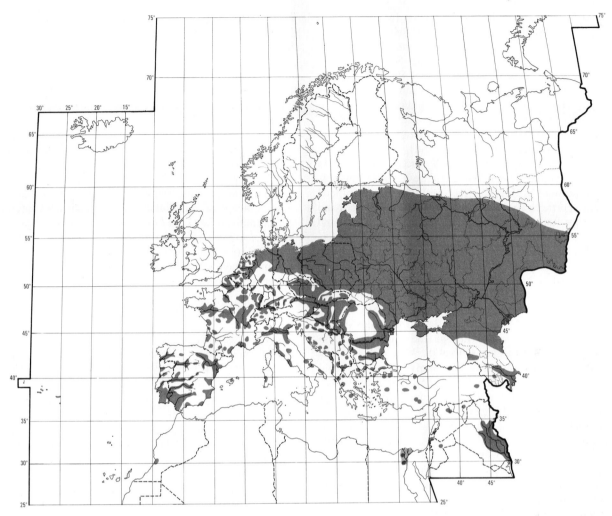

suitable localities in north (Heim de Balsac and Mayaud 1962); no recent proof (EDHJ). MOROCCO: bred once 19th century; recent records Tassela in 1965 and 1967, and Massa 1970–1 (Vernon 1973). TUNISIA: breeds suitable localities in north (Heim de Balsac and Mayaud 1962); no recent records (MS). BRITAIN: occasional breeding strongly suspected East Anglia 19th century and also southern England 1947, while apparent pairs have summered in recent years (British Ornithologists' Union 1971).

Accidental. Britain and Ireland: over 200, almost annually in recent years; chiefly southern half of England and Wales, occasionally north to Shetland and west to Ireland (British Ornithologists' Union 1971). Also Iceland, Faeroes, Norway, Denmark, Sweden, Finland, Azores, Madeira, Canary Islands.

**Population.** FRANCE: 1050–2000 pairs in 1974 (Brosselin 1975). BELGIUM: 100–200 pairs before 1960, now *c.* 60 pairs; decline due mainly to pollution and drainage (Lippens and Wille 1972). NETHERLANDS: some 170 pairs 1961–67, reaching 200–225 pairs in peak years; some local decreases due to loss of habitat (Braaksma 1968). WEST GERMANY. Baden-Württemberg: *c.* 250–300 pairs (Hölzinger 1970). Westfalen: under 20 pairs (Peitzmeier 1969). GREECE: Lake Mikra Prespa, probably hundreds of pairs (Terrasse *et al.* 1969a).

Oldest ringed bird 5 years (Rydzewski 1974).

**Movements.** Migratory and dispersive. Post-fledging dispersal late July to early September, random in direction; longest known movements by German birds up to 220 km to south-west, north-east, and east (Zink 1961). Merges into southward migration in August–September as adults depart, with few (mainly juveniles) left in Europe after October. Recorded exceptionally in winter from Mediterranean Basin north to Ireland, Britain, Low Countries, and Germany; also Sahara (Ahaggar Oasis in January). Principal winter quarters of west Palearctic breeders east Africa, from Sudan and Ethiopia west to Congo and south to Transvaal and eastern Cape Province (White 1965); present October to late April, though passage noted Egypt from mid-August (Meinertzhagen 1930). Of those ringed Netherlands and Belgium, minority south to south-west through France and Spain, most south-east through Italy; those ringed Switzerland and Germany, mainly SSE to south-east via Italy and Balkans (Zink 1961). Can cross Mediterranean and Sahara in single flight, and various oases records indicate broad-front crossings of latter; in North Africa, most in evidence on spring passage (Moreau 1967). Not infrequently appears in Azores, Madeira, and Canary Islands, while migration through Iberia and occurrences Moroccan and Algerian oases suggest that some winter in west Africa; but sole identifications of Palearctic race west of 9°E, one trapped Kano, northern Nigeria, April 1966 (Sharland 1966) and one shot Richard Toll, Sénégal, March 1961 (Morel and Roux 1962). Only 2 recoveries south of Sahara: Switzerland to Kasai (Congo) and Czechoslovakia to River Ubangi (Congo). No direct information on routes of USSR and Iran breeders but believed to winter in east Africa also (Dementiev and Gladkov 1951), thus accounting for passage over Arabia and through Iraq and Egypt.

Return across Mediterranean Basin from mid-March; breeding areas of central Europe and south Russia re-occupied in April and first week of May. ♂♂ generally precede ♀♀ by several days; immatures arrive considerably later and may form bulk of those after early May. Non-breeders wander in summer, mainly within and south of breeding range; one nestling ringed Loire Atlantique (France), July 1963, recovered following May near Rostov-on-Don (Russia).

Tropical African race *I. m. payesii* at least dispersive; little information available, but reported summer visitor Darfur (Lynes 1925) and has straggled once to Canary Islands (Bannerman and Bannerman 1965).

Migrates by night, spending daylight hours in reedbeds or swamps or, when preferred habitat not available, will alight in trees. Birds usually fly off individually at dusk; flock of 40–45 flying over Aboukir, Egypt, September 1944 (Bodenham 1945) quite exceptional.

**Food.** Primarily fish, amphibians, and insects taken while waiting in cover of reedbeds or at edge of open water, and by slowly stalking through water. Essentially crepuscular, apparently solitary.

Diet varies considerably with locality, availability, and time of year. Fish include especially bleak *Alburnus*, dace *Leuciscus*, carp *Cyprinus*, perch *Perca*, pike *Esox*, gudgeon *Gobio*, and sunfish *Eupomotis*. Amphibians include especially *Rana* and, insects particularly waterboatmen *Notonecta*, *Naucoris*, mole-cricket *Gryllotalpa*, waterbeetle *Dytiscus*, dragonflies *Libellula* and *Aeshna*. Also molluscs, crustaceans, spiders, worms, small mammals, and eggs and young of marsh-nesting birds.

In forest reservoirs of Lenkoran, USSR, chiefly toads and insects, and only occasionally fish, though elsewhere small fish constitute bulk of diet before autumn migration

and spring arrival. Main prey during autumn movements insects, especially orthopterans such as grasshoppers and field-crickets (Dementiev and Gladkov 1951). Similar food items found in 53 stomachs from various parts of Hungary and in 19 stomachs from various parts of Italy. Prey of Hungarian birds mainly insects and larvae: aquatic hemipterans, chiefly *Notonecta* and *Naucoris* (45·2% by frequency); terrestrial Coleoptera (adults 30·1%, larvae 26·4%); aquatic Coleoptera (11·3%); *Gryllotalpa* (11·3%); Odonata (adults 9·4%, larvae 7·5%); also 2·4% fish (especially *Esox*, *Alburnus*, *Cyprinus*, *Blicca*, *Leuciscus*) and 13·0% amphibians (Vasvari 1929, 1938). Italian birds contained insects most frequently, especially *Dytiscus* and other Coleoptera, Libellulidae, and aquatic Hemiptera; also amphibian *Rana esculenta*, and fish *Eupomotis gibbosus* and *Gobio gobio* (Moltoni 1948). Of 8 stomachs from central Africa, fish found in 4; also small frog, spiders, shrimps, mole-cricket, and a number of unidentified insects (Chapin 1932).

**Social pattern and behaviour.** No detailed studies. Summary in Bauer and Glutz (1966); see also Wackernagel (1950), Gentz (1959).

1. Mainly solitary or in pairs. Feeds alone, usually within or on edge of cover. Often in small, loose parties (not family groups) of 5–15 on migration. BONDS. In definite pairs during breeding season, unlike Bittern *Botaurus stellaris*; evidently monogamous, with pair-bond of at least seasonal duration. Both parents tend young until fledging or just after. BREEDING DISPERSION. Territory established by ♂ in late spring at start of breeding season. Size of territory and density of nests vary considerably: 68 nests close together in marsh, Yugoslavia, reduced to 8–10 pairs when reeds cut; 40 pairs in 100 ha in area in Saxony one year, but only 4 in others; elsewhere, 3 pairs on pit only 1800 m² (Bauer and Glutz 1966). Evidently least dense when nesting cover discontinuous, most dense in extensive, old stands of vegetation and round small ponds, though density also fluctuates from other causes. Essentially solitary (not colonial) nester, even though nests quite close at times in particularly favourable areas; these then said to be occupied in stages (Braschler *et al.* 1961). ROOSTING. Little information. In summer and winter quarters, probably mainly solitary amongst aquatic vegetation, at least sometimes in bushes or trees. Active both day and night but especially at dusk and dawn, so likely to roost at any time of day and night. Birds on passage in Egypt and elsewhere reported spending day in small parties high in trees (e.g. Meinertzhagen 1954).

2. Has Bittern-stance and Defensive-posture when disturbed much as in *B. stellaris*; also threat-display in which bird turns side-on to enemy with wings spread and, in characteristically stiff pose, lifts wing nearest it while lowering other. Uncertain whether any of such behaviour shown during intraspecific encounters. ANTAGONISTIC BEHAVIOUR. Established ♂ drives off ♂ intruders by chasing and flight-attack to border of territory, bill flushing even redder than usual and crest slightly raised. Advertising-call (see below), though said to be chiefly courtship call, continues into June–July, presumably after pairing, so evidently also territorial call. HETEROSEXUAL BEHAVIOUR. After first settling in territory, ♂ chooses site and starts to build nest to which he attracts ♀ by repeatedly uttering Advertising-call, both by day and night but especially in late afternoon and evening. Same or different nest may be used for breeding, so-called 'mock

nests' being abandoned if ♂ fails to entice ♀ to them. Displays associated with pair-formation still unknown. Copulation occurs on nest; not immediately preceded by any special displays though sometimes follows Greeting Ceremony when one of pairs joins other at site. Greeting Ceremony more characteristic of nest-relief; preceded by prolonged exchange of signals at approach of off-duty partner in which pair appear to threaten one another: open and close red lower mandible, ruffle feathers of back and breast, and repeatedly raise and lower crown feathers, sometimes calling. Change-over then follows silently: relieving bird ascends edge of nest (if ♂, often carries item of material which it deposits), steps over sitting mate, and touches its bill—or both cross bills before other departs. Either sex may, exceptionally, feed other at nest during incubation. RELATIONS WITHIN FAMILY GROUP. From 3rd day, young able to beg for food by seizing parent's bill and pulling it downward. If undisturbed, remain at or near nest for entire fledging period, though start to explore vicinity up to 10 m from early age. One substantiated case of adults removing young (aged 2–4 days) in bill from damaged nest to another 3 m away.

**Voice.** Rather silent except in breeding season. Utterances mostly hoarse or croaking, short but often repeated, frequently from dense cover.

(1) Advertising-call of male (so-called 'Spring Song'). Low-pitched, short, far-carrying 'hōgh', 'rru', or 'woof', described as croaking, froglike or resembling deep bark of dog, muffled and repressed; uttered repeatedly at roughly 2 s intervals, with only brief pauses (Voigt 1950; Bauer and Glutz 1966). (2) Flight-call. Low-pitched, abrupt 'quer' or throaty 'ker-ack' (Richardson and Hayman 1952); 'quer' sometimes preceded by a short higher-pitched 'quee' (Witherby *et al.* 1939). (3) Alarm-call. Short, hoarse 'gāt', 'gāck', or 'yick', also rendered as 'eke eke' and attributed also to excitement (Witherby *et al.* 1939). (4) Anxiety-call. 'Aark' (Bauer and Glutz 1966). (5) Nest-relief call. Loud 'gōōgōō', 'yiip yiip', or 'yĭrrp yĭrrp' (♀) (Witherby *et al.* 1939); similar note to latter sometimes used on brood dispersal probaby by ♀ (Bauer and Glutz 1966).

CALLS OF YOUNG. Whispered twitter 'tū-tū, tū-tū, tū-tū' in greeting parents (Witherby *et al.* 1939), possibly identical with soft mewing 'chet-chet-chet' or 'chiätt-chiätt' (Bauer and Glutz 1966); when being fed, harsh corvine note (Witherby *et al.* 1939).

**Breeding.** SEASON. See diagram for central and southern Europe; little data on variation within west Palearctic. SITE. Dense reedbeds, rushes, willow *Salix* thickets, or bushes, usually 5–15 m out from shore. When in reeds, close to water or up to 60 cm above, and over water 25–30 cm deep (Wackernagel 1950); in willows, up to 2 m above water (Dementiev and Gladkov 1951). Not colonial, but 2–3 pairs may breed on same pond with nests only 50 m apart, exceptionally 5–10 m apart (Braschler *et al.* 1961). Nest: compact pile 20–25 cm high, base diameter 25–35 cm; shallow cup 12–15 cm across and 2–4 cm deep; typically of 250–300 dead reed stems 10–20 cm long, exceptionally up to 50 cm, together with 300 reed leaves (Wackernagel 1950). Rushes and twigs used where no

reeds available; begun by bending down reeds and then breaking off and adding pieces to form platform (Ali and Ripley 1968). Completed in 2–4 days (Dementiev and Gladkov 1951). EGGS. Regular ovate, rough and matt; white, occasionally greenish. 36 × 26 mm (32–39 × 24–28), sample 250 (Schönwetter 1967); weight 10–14 g (Bauer and Glutz 1966). Clutch: 5–6 (4–9). One brood. No information on replacements. Eggs laid daily though sometimes gap of 2 days. INCUBATION. 17–19 days (16–21, exceptionally 24). ♀ begins incubation after 1st egg laid but sits intermittently; both sexes incubate when clutch complete; hatching asynchronous. ♀ sits at night and ♂ during middle of day (Groebbels 1935). Eggshells usually ejected from nest. YOUNG. Semi-altricial and nidicolous; develop rapidly, can stand at 1 week. Cared for and fed by both parents; by partial regurgitation. Young leave nest at 17–18 days and conceal themselves in surrounding vegetation. FLEDGING TO MATURITY. Fledging period 25–30 days; few records. Become independent at or soon after fledging and juveniles disperse from breeding area individually. Age of first breeding 1 year. BREEDING SUCCESS. No data.

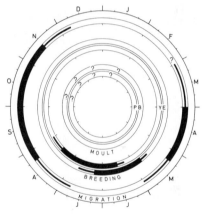

**Plumages** (nominate *minutus*). ADULT MALE. Crown, nape, mantle, scapulars, back, rump, and tail black with green gloss. Hindneck buff-grey washed vinaceous. Sides of head and neck, centre of chin and throat, foreneck, and rest of underparts buff, faintly streaked deeper buff with brown shafts on flanks and sides of breast and belly; paler towards vent, under tail-coverts cream-white. Sides of chin and throat white. Patch of feathers on sides of chest brown-black with pale buff margins, concealed by elongated buff feathers of chest. Primaries and outer secondaries black, outer webs faintly glossed green; p10 with narrow cream margin to outer web. Inner secondaries black glossed green like scapulars. Upper wing-coverts pale grey-buff, marginal darker, innermost greater covert with glossy black inner web, 2nd with inner web mottled brown. Under wing-coverts and axillaries cream-white. ADULT FEMALE. Crown and nape black, less glossy than ♂, with red-brown feather edges. Hindneck red-brown. Rest of upperparts brown, sometimes strongly tinged rufous, narrowly streaked pale buff. Sides of head grey-buff, sides of neck buff, sometimes with strong rufous-brown wash. Sides of chin and throat white, centre grey-brown. Foreneck and chest pale buff streaked pale brown. Rest of underparts buff, streaked dark brown. Patch of dark brown feathers on sides of chest as in ♂. Tail black with slight green gloss. Primaries and secondaries dark grey, p10 margined cream, inner secondaries brown like back. Upper wing-coverts pale grey-brown; marginal and patch near bend of wing red-brown; innermost greater covert with inner web brown. Under wing-coverts and axillaries cream-white. NESTLING. Densely covered with pink-buff down, chin and centre of underparts white; lores, round eyes, and hindneck bare. JUVENILE. Similar to adult ♀ but forehead and crown with wider and paler brown edges. Feathers of hindneck tipped buff, of rest of upperparts broadly edged buff or pale buff at sides and tips. Sides of neck, foreneck, and chest with darker brown streaks. Greater upper wing-coverts with pale brown-grey inner webs, outer webs mottled brown-grey with wide buff margins; median and lesser with dark brown-grey centres producing pattern of triangular spots. Narrow pale or brownish tips to tail, flight-feathers, and primary coverts. Colour of upperparts variable, some dark brown, others almost entirely chestnut on mantle and scapulars. FIRST WINTER AND BREEDING. Differs from adult in darker and more spotted wing-coverts; ♂ apparently less grey on sides of head and neck, more streaked underneath, and with brown rather than black inner secondaries (von Lucanus 1914; Witherby *et al.* 1939; Braschler *et al.* 1961).

**Bare parts.** ADULT. Irish yellow. Bill yellow or green-yellow, culmen dark brown. In breeding season, base of bill and bare skin on side of face may become temporarily suffused orange or red when excited; bare skin of face otherwise yellow with narrow dusky line from nostril to eye. Leg green or grey-green, back of tarsus and soles yellow. NESTLING. Iris dark brown. Bill pink. Leg pale yellow-pink. JUVENILE. Like adult; yellow of bill may be tinged flesh, green of leg olive (Wackernagel 1950; Bauer and Glutz 1966; specimens ZMA).

**Moults.** Poorly known; elusive during wing moult. ADULT POST-BREEDING. In *I. m. payesii* complete; p1–p7 descendant, others irregular, p9 may be last. Presumably similar in nominate *minutus* but data lacking. Timing of moult unknown, probably late autumn or early winter (Stresemann and Stresemann 1966; Herroelen 1973). POST-JUVENILE. Starts with body feathers late November (Chapin 1932) to January (von Lucanus 1914). Herroelen (1973) found no evidence for moult before arrival in Central African winter quarters. Not known whether flight-feathers moult during 1st year, but probably not.

**Measurements.** Adult breeding (RMNH, ZMA).

| | | | | | | |
|---|---|---|---|---|---|---|
| WING | ♂ 153 | (2·47; 15) | 149–157 | ♀ 148 | (3·42; 12) | 142–153 |
| TAIL | 49·6 | (1·65; 14) | 47–53 | 49·3 | (1·57; 10) | 47–51 |
| BILL | 48·6 | (1·91; 14) | 46–53 | 46·5 | (1·75; 11) | 44–49 |
| TARSUS | 46·2 | (2·79; 12) | 43–52 | 43·0 | (2·00; 8) | 41–47 |
| TOE | 51·8 | (1·48; 12) | 50–54 | 50·0 | (2·00; 8) | 48–53 |

Sex difference significant, except for tail. Extremes of range of wing: ♂♂ 139–162; ♀♀ 136–157.

**Weights.** Adult breeding ♂ 149 (11) 145–150; ♀ 146 (7) 140–150. Full-grown, extreme range 64–170 (Bauer and Glutz 1966). Nestling at hatching *c.* 10–11, at fledging *c.* 100–125; increase from 13 to 123 in 11 days (details in Wackernagel 1950).

**Structure.** 11 primaries: p9 and p10 longest, either occasionally 0–2 shorter than other; p8 2–6 shorter, p7 5–12, p6 10–16, p5 15–22, p1 43–49; p11 minute. Outer web of p10 narrow; of p8–p9 attenuated towards tip, but not emarginated. Tail slightly rounded; 10 feathers. Tibia wholly feathered. Outer toe *c.* 80% of middle, inner *c.* 85%, hind *c.* 62%, hind claw *c.* 25%.

**Geographical variation.** Involves colour of sides of head and neck, of upper wing-coverts, and size. Australian forms differ in many respects from Palearctic and African. It has been suggested these might best be considered separate species (Voous 1960). *I. m. payesii* differs from nominate *minutus* by chestnut sides of head and neck in adult ♂ and by shorter wing: p1 29–35 shorter than p9; range of wing in 7 ♂♂ 135–142, in 5 ♀♀ 141–150 (Chapin 1932); p10 distinctly shorter than p9.

---

## *Ixobrychus eurhythmus* Schrenck's Little Bittern

PLATE 32
[facing page 255]

Du. Mandsjoerijs Woudaapje  FR. Blongios de Schrenck  GE. Mandschurenzwergdommel
RU. Амурский волчок  SP. Avetorillo de Wagler  SW. Asiatisk dvärgrördrom

*Ardetta eurhythma* Swinhoe, 1873

Monotypic

**Field characters.** 35 cm; wing-span 55–59 cm; similar in size to Little Bittern *I. minutus* though rather shorter-winged. Tiny, skulking heron. Sexes differ; seasonal variation apparently not marked. Juvenile separable.

ADULT MALE. Crown, sides of face, hindneck, upperparts, tail, and tertials blackish-chestnut. Chin, throat, and lower neck to breast pale cream, darkening to buff, with area below side of face and along upper hindneck strikingly pale and all divided by central line of dark brown (spreading into streaks on throat and spreading out on upper breast). Rest of underparts cream, with shoulder patch dark rich brown. Inner wing-coverts ochre, darker, more olive in tone than in *I. minutus* but still forming obvious panel on dark brown wing in flight; underwing pale cream. Bill finely pointed, culmen black, rest of mandibles yellow. Bare lores purple-flesh, tinged green. Eyes yellow. Legs and feet green, yellowest around tibio-tarsal joint, along rear edges and on soles. ADULT FEMALE. Plumage less uniform in colour than ♂, with mantle and scapulars spotted with white and wing-coverts with buff; underparts sandier than ♂, with obvious brown streaks spreading out from dark neck line and over chest and flanks. Bill as ♂, but legs more olive. JUVENILE. Resembles ♀ but plumage pattern apparently less distinct and richly coloured; spots irregular in shape (see also Plumages). Bare parts duller.

If well seen, separation from *I. minutus* can be safely based on dark sides of face and definite line down centre of foreneck, chestnut mantle of ♂ and spotting, not streaking, of ♀ and juvenile.                                    DIMW

**Habitat.** Differs from Little Bittern *I. minutus* in less aquatic and arboreal tendencies, generally preferring drier, open meadows of long grass to swamps and thickets—although these also frequented, together with banks of ditches, open waters, lake sides, and, infrequently, margins of large rivers. Also uses irrigated gardens and watercourses in villages. Perches on lower branches of trees only exceptionally, and little seen by daylight. Mainly in lowland river valleys, becoming rare and sporadic in uplands. (Dementiev and Gladkov 1951.)

**Distribution.** Breeds south-east Transbaikalia and central Amur valley, east to Ussuriland and Sakhalin, south to Manchuria, Korea (probably), China, Japan, and perhaps northern Indo-China (Vaurie 1965). In west Palearctic, one obtained Mark Brandenburg or Mecklenburg towards end 19th century, now in Berlin Museum (*Orn. Mber.* 1954, **42**, 90); juvenile ♀ caught near Turin, Italy 12 November 1912, now in Zoological museum, Turin (*Riv. ital. Orn.* 1912–13, **2**, 86). German record poorly documented, but Italian specimen examined in flesh by competent ornithologist, and unlikely to have escaped from captivity.

**Movements.** Migratory. Timing little known, partly due to secretive habits; apparently absent from USSR late September to late May (Dementiev and Gladkov 1951). Winters south to south-west of breeding range, from south-east China and Indochina through Malaysia to Sumatra and Java, and also in Philippines and Celebes (Vaurie 1965). Migration routes unknown in absence of ringing, but thought by Deignan (1945) to be mainly coastal. Has wandered to Palau Islands (Micronesia), Thailand, and south Burma. Thus European records (see Distribution) quite extraordinary.

**Voice.** Taciturn. On take-off utters abrupt, coarse, muffled cry (Dementiev and Gladkov 1951).

**Plumages.** ADULT MALE. Crown brown-black, sides of head and rest of upperparts chestnut, centre of mantle tinged black. Chin and throat pale cream, central feathers buff with wide dark brown

shaft streaks, forming gular stripe extending to chest. Breast, belly, and flanks ochre; vent and under tail-coverts cream. Patch of dark brown feathers with ochre margins on side of breast. Tail black-brown. Flight-feathers dark brown, paler below, p10 with narrow cream edge to outer web. Inner primaries and outer secondaries with white terminal line. Upper wing-coverts pale olive-grey, tinged slightly ochraceous, innermost, marginal, and those at bend of wing chestnut. Under wing-coverts pale cream, axillaries pale yellow-grey. ADULT FEMALE. Crown black, hindneck chestnut, sides of head chestnut with white specks. Feathers of mantle and scapulars black with rows of white spots along margins, scapulars tinged chestnut near tips. Back and rump dark grey mottled white. Chin and throat off-white with dark brown gular stripe down centre. Rest of underparts buff, streaked brown. Tail dark brown. Flight-feathers like ♂. Upper wing-coverts chestnut, heavily spotted buff along margins. Under wing-coverts pale grey with dark centres, sometimes giving streaked appearance. JUVENILE MALE. Mantle and scapulars spotted like adult ♀, upper wing-coverts olive-ochre like adult ♂, probably slightly more strongly tinged ochraceous. JUVENILE FEMALE. Not known whether separable from adult ♀ with certainty, probably specimens with margins of upper wing-coverts strongly tinged ochraceous are juvenile. Also described as having mantle and scapulars browner than adult ♀ and spots pale ochre rather than white, particularly at tips of feathers (Dementiev and Gladkov 1951; La Touche 1931–4). More information needed. IMMATURE. ♂♂ recognizable in 1st autumn as long as some spotted or streaked feathers present on mantle and scapulars. Otherwise like adult.

**Bare parts.** No evidence of colour changes with age or season. Iris yellow. Bill yellow-green with dark brown culmen. Orbital skin purple-flesh tinged green in ♂, lemon-yellow tinged green in ♀. Leg olive-green, yellower around tibio-tarsal joint, along rear edge, and on sole.

**Moults.** No data. Young birds arrive in winter quarters mainly in adult plumage, so presumably post-juvenile moult early in autumn.

**Measurements.** Adults and juveniles, sexes combined; wing, bill, and toe from skins (RMNH); tail and tarsus from La Touche (1931–34). Wing 149 (3·42; 12) 142–153; tail 39–47; bill 47·2 (1·66; 11) 45–50; tarsus 45–52; toe 50, 51, 52, 54.

**Weights.** No data.

**Structure.** 11 primaries: p9 longest, p10 3–5 shorter, p8 0–2, p7 5–9, p6 10–15, p5 16–20, p1 36–41; p11 minute. Primaries not emarginated but outer web of p10 narrow. Tail short, rounded. 12 feathers: t6 4–6 shorter than t1. Bill relatively heavy at base, sharply pointed. Outer toe *c.* 80% of middle, inner *c.* 82%, hind *c.* 62%, hind claw *c.* 25%.

# *Ardeirallus sturmii* (Wagler, 1827) **Dwarf Bittern**

FR. Blongios de Sturm    GE. Graurückendommel

Ethiopian species, reported from freshwater habitats over much of tropical Africa from Sénégal, Nigeria, Chad, southern Sudan and Abyssinia south to Transvaal and eastern Cape Province; but breeding range not yet

properly defined, especially north of equator. Mainly summer visitor (November–April) to southern half of range, appearing with onset of rains, departing for long dry season (Benson and Irwin 1966). Breeds at least locally farther north though equatorial belt records may include migrants from south. Probably off-season movements towards equator in northern tropics also. Present Chad only July–September, breeding suspected (Salvan 1967). Nigerian dry-season records rare (Elgood *et al.* 1973); in Borgu region recorded only July–October (Walsh 1971), probably wet-season migrant breeder. Vagrant collected Tenerife, Canary Islands, 1889 or 1890 (Meade-Waldo 1890) sole Palearctic record.

# Subfamily ARDEINAE tiger-bitterns, egrets, herons

About 52 species in 7 main genera. 3 main groups (sometimes separated formally as tribes): (1) tiger-bitterns (or tiger-herons) *Tigrisoma*, etc.; (2) night-herons *Gorsachius* and *Nycticorax*; and (3) typical egrets and herons including *Butorides*, *Ardeola*, *Bubulcus*, *Egretta*, and *Ardea*. 9 breeding species and 5 accidental in west Palearctic. Usually 12 tail-feathers. Except in *Nycticorax*, legs relatively longer than in Botaurinae. Outer toe longer than inner. Ornamental plumes usually well developed; in *Egretta*, special aigrettes with long, free barbs drooping from shaft. Plumage white, grey, black, blue, purple, brown or greenish; often all-white, sometimes all-black, or with contrasting or cryptic pattern. Several species polymorphic. Sexes similar or nearly so. Usually 3 pairs of powder-down patches—on breast, rump, and thighs; 4th pair on back in a few species.

Many species live conspicuously when feeding, roosting, and breeding. Others more secretive, some (e.g. Purple Heron *A. purpurea*) resembling Botaurinae in many respects, including use of Bittern-stance (found also in other cryptic but tree-nesting genera such as *Butorides* and *Ardeola*). Terminology of displays follows Meyerriecks (1960) with modifications and additions. Nest-site centre of social activity; defended vigorously, in both colonial and solitary species, mainly by Upright and Forward-displays (in some cases, also by Snap-display). Interactions elsewhere limited and mainly hostile, but significance of activities at gathering grounds away from colony in some species not clear. Pair-formation initiated by ♂ who establishes site and attracts ♀♀ by calling and posturing; latter includes stationary Stretch and Snap-displays, Twig-shaking, and, especially in smaller species, aerial displays. 3 main pair-formation types: (1) succession of ♀♀ visit ♂ at site (usually existing nest) and same site used subsequently for breeding (e.g. Grey Heron *A. cinerea*); (2) several ♀♀ watch active ♂ who frequents one or more, often temporary display-sites (e.g. Cattle Egret *B. ibis*); (3) much as last, but both ♂♂ and ♀♀ attracted and often move as displaying group round colony (e.g. Little Egret *E. garzetta*). In all cases, ♂♂ initially show hostility to all comers regardless of sex. In most species, ♀ passive (may perform ritualized preening displays) and approaches ♂ cautiously over several days until his hostility subsides; in other, notably *B. ibis*, ♀ eventually becomes aggressively active and assaults ♂ in attempts to stay close to him. Subsequently, pair at site often engage in activities such as Billing and Allopreening; especially on return of mate, also perform Greeting Ceremony. Copulation at or near site. Except for Advertising-calls of ♂♂ at start of breeding cycle and calls accompanying pairing displays, not highly vocal even in species, such as *A. purpurea*, nesting in dense reedbeds. Rely more on visual display and contact, supplemented by harsh flight-calls during approach to site, when alarmed, or in dark. Some use made of non-vocal sound signals, especially Bill-rattling and Bill-snapping, but loud clattering of some Ciconiidae rare or absent. Young, especially in colonial species, maintain harsh clamour.

*Nycticorax nycticorax* **Night Heron**

PLATE 33
[after page 302]

Du. Kwak     Fr. Héron bihoreau     Ge. Nachtreiher
Ru. Кваква     Sp. Martinete     Sw. Natthäger     N.Am. Black-crowned Night Heron

*Ardea Nycticorax* Linnaeus, 1758

Polytypic. Nominate *nycticorax* (Linnaeus, 1758), Europe, Asia, Africa. Extralimital: *hoactli* (Gmelin, 1789), North America and northern South America and 2 races in southern South America.

**Field characters.** 58–65 cm, ♂ averaging larger; wingspan 105–112 cm. Small, stocky heron with stout bill, rather short legs, and distinctive black, grey, and white adult plumage. Sexes alike and only minor seasonal differences; juvenile easily distinguishable.

ADULT. Black crown and back, with green gloss which becomes stronger blue-green in breeding plumage; bluish-grey wings, rump, and tail; and white forehead, cheeks, and underparts. 2–3 long white plumes extend from nape right down back. Bill black; bare lores green-blue, becoming red during pairing and mating, often appearing greyish from loose scales and dried mud. JUVENILE AND FIRST-YEAR. Quite different, suggesting bittern *Botaurus*. Dark brown above, spotted with whitish-buff, and greyish

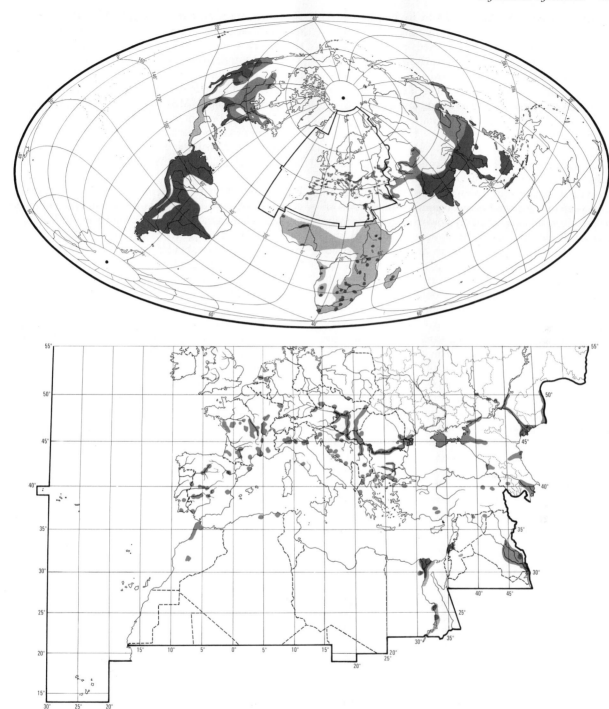

below, streaked with dark brown. Eyes brown; legs yellow-green. After 1st summer more like adult, but crest plumes shorter, black areas browner with dull green cast, grey areas tinged with brown, and white underparts washed with brown and grey.

Adult unmistakable, but juvenile needs to be distinguished from Bittern *B. stellaris*, which is larger, more

golden-brown, mottled and barred with blackish (see also American Bittern *B. lentiginosus*).

Mainly crepuscular, except when feeding young in nest. Roosts by day in thick cover, especially leafy trees, and then easily approachable; when disturbed, sometimes utters characteristic single croak, recalling Raven *Corvus corax*, and eventually flies out with clatter like Woodpigeon

*Columba palumbus*. At dusk, flight to feeding grounds; short, rounded wings and toes projecting only slightly beyond tail (sometimes feet retracted altogether) give distinctive outline and, though wings flapped in way characteristic of other herons (Ardeidae), beats rather faster than most; often calls in flight. Feeds in shallow water at edges of swamps and reedbeds, and along ditches, moving slowly and deliberately and, when necessary, clambering about easily on reeds.

**Habitat.** Within west Palearctic, concentrated in warm temperate and sub-tropical zones, especially in drier continental climates, but within these using wide range of habitats up to *c.* 2000 m. Forages commonly on margins of shallow fresh standing or gently flowing water, especially lakes, pools, ponds, lagoons, rivers, streams, swamps, and other wetlands; also sometimes on drier grasslands, on temporary inundations, and such man-made sites as reservoirs, canal banks, ditches, ricefields, and small ornamental waters. Markedly arboreal when resting, roosting, or nesting, using trees of many species, e.g. willows *Salix*, poplars *Populus*, and Alder *Alnus*, usually riverain but at times away from water and (outside west Palearctic) within populous human settlements. Where woody growth absent, uses reedbeds and other emergent vegetation. Not usually in exposed situations; intermediate between skulking bitterns *Botaurus* and *Ixobrychus* and open-ranging egrets *Egretta* in insistence on cover. On passage and in winter, occurs also along coasts and more freely in drier situations. Walks, wades, climbs, and sometimes swims; flies freely, usually at low altitudes, but fairly high on migration. Although often stationary for long periods, characteristically mobile and adaptable, making good use of artefacts and tolerating human neighbourhood and disturbance. Adaptability of behaviour may have assisted recovery after decline in 19th century.

**Distribution.** Breeding sometimes irregular, especially on edges of range. Disappeared from several countries on western edges in late 19th and early 20th century, but has returned to some since 1946. BELGIUM: doubtful breeding record 1885 (Lippens and Wille 1969). NETHERLANDS: formerly more common, bred until 1876; again, in small numbers, since 1946 (JW). FRANCE: considerable decline in 18th and 19th centuries, but spread since 1945 (Brosselin 1975). WEST GERMANY: breeding irregularly but increasingly since 1950 (EB). EAST GERMANY: bred occasionally early years 19th century (Bauer and Glutz 1966). POLAND: bred irregularly in past, especially Silesia; then from 1964 one small regular colony in south (Tomiałojć 1972; AD). CZECHOSLOVAKIA: probably bred Bohemia 19th century, definitely breeding since 1948 (Hudec and Černy 1972). AUSTRIA: extending range in lower Austria, and new colony lower Inn; irregular Neusiedlersee (HS). SWITZERLAND: first proved breeding 1967, then 1968–71, but none 1972 (WT). ITALY. Sicily: now only on passage (GS). ALGERIA: formerly bred at 2 sites shown, but present status uncertain (EDHJ). MOROCCO AND ALGERIA: proved breeding areas shown, but may not reflect current position (KDS, EDHJ). TUNISIA: formerly bred Lake Ichkeul (Heim de Balsac and Mayaud 1962), but no longer (MS).

Accidental. Britain and Ireland: over 200, chiefly May–June; some perhaps escapes (British Ornithologists' Union 1971). Also Iceland, Faeroes, Norway, Denmark, Sweden, Finland, Azores, Madeira.

**Population.** NETHERLANDS: in recent years only 2 regular colonies; not more than 20 pairs, decreasing (JW). FRANCE: recent decline; 2159 pairs in 31 colonies in 1968, 1550 in 28 colonies in 1974 (Brosselin 1975). SPAIN: total population unknown, but for details of colonies, with many counts, see Fernandez Cruz (1975). WEST GERMANY. Hessen: 1 pair 1968, 3 pairs 1969. Bodensee: 1 pair 1967. Bavaria: 1–2 pairs bred mouth of Isar 1950–68. Lower Inn on Austrian border: at least 5 pairs 1964, 27 pairs 1968, over 40 pairs 1971. (EB, HS). CZECHOSLOVAKIA. Bohemia: 4 colonies totalling 100–160 pairs (Hanzák 1962). HUNGARY: 652 pairs 1951 (Szijj 1954). SWITZERLAND: 6 pairs 1968, 8 pairs 1969, 2 pairs 1970, 2–3 pairs 1971, none 1972 (WT). ITALY: *c.* 8000 pairs in 1975 (SF). YUGOSLAVIA: Kopacevski area: 225–796 pairs, 1954–70, numbers fluctuating (Majić and Mikuska 1972). GREECE: *c.* 1500–1600 pairs in 1973 (WB). TURKEY. Meric delta: 600–700 pairs 1967. Manyas Gölü: *c.* 500 pairs 1967. About 50 pairs elsewhere 1969–72 (OST). USSR. Abundant, though rarer in northern and eastern limits of range; some colonies up to 2500 pairs (Dementiev and Gladkov 1951). Lower Volga delta: 720 pairs in 1951, 19 pairs 1958 (Lugovoy 1961). MOROCCO: Allal Tazi 1965–67, varying numbers from nil to *c.* 400 pairs (Haas 1969; Vernon 1973).

Survival. North American race *hoactli*. Mortality rates 1st year 69·0% (1926–45), 63·6% (1946–65); after 1st year 23·3% and 25·8% respectively, one recovered in 18th year (Henny 1972). Oldest ringed bird 21 years 1 month (Kennard 1975).

**Movements.** Migratory and dispersive. In July–August juveniles disperse in all directions, mostly north and west of colonies, usually within 800 km, but one 1200 km (France to Mecklenburg). Young ringed Hungary recovered east to Rostov and Kuybyshev (USSR) (Lippens and Wille 1969), though over 70% French recoveries at this time north of ringing locality. This dispersal merges into autumn migration which in Europe lasts through September and October; some linger into December in North Africa. Only stragglers in Europe November–February and some possibly escapes; 2 overwintered England (Essex) 1953–4, and other rare winter occurrences in North Sea countries. Substantial numbers reported occasionally in winter on south Caspian

(Dementiev and Gladkov 1951), but no dates given and possibly mainly late migrants. Tendency to leave Caspian a little later than Europe; occurs Kuwait October–May and a few winter Iran. Overwhelming majority of west Palearctic birds, however, trans-Saharan migrants, wintering tropical Africa where southern limits unknown as resident breeding population present. Some almost reach equator; one ringed Volga delta recovered Yaundé (Cameroons), indicating degree of south-westerly movement performed by some Russian birds. In Nigeria, occurs in much larger numbers in winter, with recoveries of singles from Hungary and Volga delta. Further west, Hungarian and Czechoslovakian birds recovered in Guinea; French in Mali, Gambia, and Sierra Leone (2); and one from Spain in Sénégal. One ringed Azerbydzhan recovered eastern Sudan—only firm evidence for presence of Russian birds in East Africa.

Broad front movements over Mediterranean suggested by recoveries of Hungarian birds in Malta, Italy, and Greece (spring and autumn); Spain, Tunisia, and Cyprus (May); Egypt (autumn); and Libya (December). French recoveries from Mediterranean show scatter from Iberia and Morocco east to Italy and Libya. Doubtless some use Nile valley and Atlantic coast routes, but broad-front Saharan crossings shown by oases records and spring recovery Ahaggar of one ringed Hungary (Moreau 1967).

Rather early return to west Palearctic colonies, from mid-March with most back by mid-April; scarce May migrants probably non-breeders. Tendency for returning migrants to overshoot natal areas and occur as vagrants further north; most British and Irish records occur March–May, especially April. Immatures seldom seen near colonies; may remain in wintering areas or wander. Recoveries of 2–8 year old, Hungarian-bred birds in Volga delta, and of French ones 2 years later in Rumania and 6 years later 200 km north of Moscow, suggest possibility of aberrant migration. Migrates singly or in small parties but flocks of hundreds, sometimes with Grey Herons *Ardea cinerea*, noted in south Russia (Dementiev and Gladkov 1951); chiefly by night, departing at dusk, but movements occasionally continue throughout day. (See also Bauer and Glutz 1966.)

**Food.** Chiefly amphibians, fish, and insects, mainly taken either by stalking in shallow water or from stationary position, but also reported hovering over water and diving (Meyerriecks 1960). Nocturnal and crepuscular feeder; normally solitary on individual feeding territories.

Amphibians include adult and young frogs *Rana* and salamanders *Triton*. Fish include carp *Cyprinus*, eel *Anguilla*, tench *Tinca*, mullet *Mugil*, chub *Leuciscus*, sandsmelt *Atherina*, *Gambusia*, loach *Cobitis*, sunfish *Eupomotis*. Insects include adult and larvae waterbeetles *Dytiscus*, Hydrophilidae, *Macrodytes*, *Colymbetes*; waterboatmen *Notonecta* and *Naucoris*; mole-cricket *Gryllotalpa*, dragonflies *Aeshna*, lace-wing (Neuroptera), and flies (Diptera). Also crustaceans *Triops*, small mammals (vole *Arvicola*, mole *Talpa*), lizard *Lacerta*, snake *Natrix natrix*, molluscs, spiders, leeches, and young birds (including other Ardeidae).

114 stomachs from Hungary contained, by frequency: fish 35%, amphibians 33%, terrestrial beetles 40·7%, large aquatic beetles 20·1%, waterboatmen *Notonecta* and *Naucoris* 16·6%, dragonfly adults and larvae 17·6%; also smaller numbers of spiders, lizards, salamanders, snails, and many insect species (Vasvari 1938). 148 stomachs from Italy: relatively more amphibians (51%), especially *Rana esculenta* (adult and young); rather less fish (22%), including *Eupomotis*, *Tinca*, *Cyprinus*, *Cobitis*; and insects (50%), including Hydrophilidae, Notonectidae, Libellulidae, and *Gryllotalpa*; similar components and proportions found in further 30 stomachs (Moltoni 1936, 1948). 120 stomachs from Danube delta contained, by volume: 31·7% fish (*Rutilus rutilus*, *Perca fluviatilis*, *Carassius carassius*, *Alburnus alburnus*, *Abramis brama*, *Scardinius erythrophthalmus*, and *Misgurnus fossilis*) and 68·3% frogs and invertebrates (Andone *et al.* 1969). From 4 colonies in Rhône delta, 35 stomachs contained, by frequency insects 74%, especially hydrophilids, *Dytiscus*, *Gryllotalpa*; fish 57%, including *Anguilla*, *Atherina*, *Gambusia*; crustaceans 43% (all *Triops*); and amphibians 32% (all *Rana*) (Valverde 1955–6). Similar diet found in south Bohemia, though fewer amphibians taken (Bauer and Glutz 1966). In USSR, Bessarabia: diet mainly fish, aquatic insects, and aquatic plant seeds; and, in Kharkov region, worms and larvae of horse botfly (Tabanidae) from cattle drinking areas. Winter foods near Lenkoran: fish (Cyprinidae), water insects (*Macrodytes*, *Colymbetes*), and plants (Dementiev and Gladkov 1951).

Nestlings fed at first on regurgitated, semi-digested fish, frogs, insects, and water-voles (Dementiev and Gladkov 1951).

**Social pattern and behaviour.** Except when breeding, nocturnal, elusive, and infrequently observed in west Palearctic (Voous 1960).

1. Solitary and gregarious. Although American race *hoactli* characterised by Meyerriecks (1960) as highly sociable throughout year in all activities, Palearctic race (nominate *nycticorax*) only loosely gregarious outside breeding season; then solitary except when roosting and migrating. At all times, feeds singly, establishing individual feeding territory (Lorenz 1938; Voisin 1970). BONDS. Monogamous pair-bond of season duration. Age of first pairing not known in nominate race; *hoactli* usually breeds first in 2nd or 3rd year, occasionally 1st (Palmer 1962). Both parents tend young until well after leaving nest; brood-bond between siblings ends soon after abandonment of nest (Lorenz 1935). BREEDING DISPERSION. Colonial, in small to large numbers on own or, more usually, with other herons and species such as Glossy Ibis *Plegadis falcinellus* and Pygmy Cormorant *Phalacrocorax pygmeus*. In mixed colonies, typically keeps in discrete, often dense units in higher part of trees. Each pair strictly maintains own nest-territory for as far as sitting bird can reach with bill (Voisin 1970). Defence of nest-territory by pair follows establishment of one or more pairing territories by ♂, not

necessarily at same site. Birds disperse to feed, sometimes as far as 10–20 km (Voisin 1970). New-fledged young wander freely in colony without evoking hostility of adults (Lorenz 1938); in *hoactli*, scatter widely among nesting trees, each defending own perching place (Noble *et al.* 1938). ROOSTING. Typically diurnal. Communal in small to large groups, from less than 10 to several hundreds; sometimes singly. Except when breeding, roosts by day and disperses at dusk to feed, sometimes for considerable distances (Witherby *et al.* 1939; Bannerman 1957; Voous 1960). Absent all night, returning before sun-up (Meinertzhagen 1954), though not equally active for whole of intervening time (see Steinfatt 1934). Defends own perch in roosting tree (see Palmer 1962). Roosts sited protectively in trees and other suitable cover often over or near water; also sometimes far from water, even in parks and gardens. Birds remain hidden in dense foliage, quiet and inactive. During pairing and mating phases, displays as much during night as day, but prolonged period of inactivity around mid-day (Allen and Mangels 1940). In Nigeria, flock of 100 seen resting in evening on open, grassy bank of river (see Bannerman 1957).

2. Earlier studies on nominate *nycticorax* by Lorenz (1934, 1935, 1938) and on *hoactli* by Noble *et al.* (1938) indicated considerable differences between 2 races, but this resolved by later study of Noble and Wurm (1942) and field work of Allen and Mangels (1940), Meyerriecks (1960, amplified in Palmer 1962), and Voisin (1970). Though pair-formation of heron-type 2 (see Ardeinae), and most displays appear strikingly different, main trends in social behaviour otherwise similar to Grey Heron *Ardea cinerea*. Use made of white display-plumes on head disputed: according to Lorenz (1938) and Noble *et al.* (1938) erected only in Greeting Ceremony, but Meyerriecks stressed use in threat-display and main pairing-display (see Palmer 1962) while Voisin (1970) commented on absence of plume-erection in all types of courtship. ANTAGONISTIC BEHAVIOUR. Threat-displays of both sexes include simple type of Upright-display and 2 versions of Forward-display shown during disputes throughout year. Sites in colony also defended by Snap-display (Meyerriecks 1960; Palmer 1962). In Upright-display, body and neck fully erected with slight forward inclination; crown, neck, and back feathers lightly or moderately erected but little or no use made of head plumes. In standard version of Forward-display, bird crouches with neck retracted and bill pointing at opponent; same feather areas as before moderately to fully erected, plus those of breast, with more use of head plumes. In full version of Forward, maximum erection of head, neck, and body feathers and frequent use of vertically raised head plumes. Usually silent during Upright-display, or gives infrequent Threat-calls; wings usually closed. During both versions of Forward-display, calling more frequent and varied, wings often opened and waved, and tail

flipped repeatedly up and down. Forward-displays typically lead to stalking rival from low crouch position, iris colour deepening during full Forward; then bird extends neck slightly and either Bill-snaps repeatedly or lashes out with open bill. Snap-display similar to that of *A. cinerea* but often culminates in brief seizure of twig (Twig-snap) rather than in Bill-snap only. Performed suddenly while ♂ moving slowly about about branches at site, with moderate erection of crown, neck, and back feathers, and head plumes depressed or slightly raised; as in Forward-displays, eyes bulge conspicuously. Aggressive birds also make supplanting attacks on others. When fighting, rivals strike at each other from crouched position, seizing bill or wing. ARRIVAL AND GATHERING-GROUND BEHAVIOUR. As in *A. cinerea*, both sexes return to breeding area at same time but typically in flocks (Noble *et al.* 1938; Allen and Mangels 1940; Voisin 1970). In pre-pairing phase, mostly inactive in trees. In 2 seasons, Long Island, soon after arrival, large flock (many birds already had red legs) suddenly rose silently from perches high in colony and flew to open ground nearby where birds milled about, threatening and fighting; later divided and continued encounters elsewhere (see Palmer 1962 for further details). Demonstration apparently confined to single day at start of breeding season. HETEROSEXUAL BEHAVIOUR. Though a few ♂♂ start displaying while still in non-breeding colour, pair-formation typically correlated with bare-part colour changes—especially reddening of legs—in both sexes. Sooner or later after arrival, ♂♂ start to leave flock; then give Advertising-calls and perform pairing displays from a series of temporary territories, often at sites of old nests, each attracting small group of other birds, presumably mainly ♀♀ (Meyerriecks 1960; Palmer 1962; Voisin 1970). Unpaired ♂ fetches twigs, though bulk of nest constructed by pair later. Occupant of pairing territory at first threatens and attacks any onlooker that comes near site. ♀'s approach to displaying ♂ cautious, due partly to his ambivalent behaviour. Pair-formation accomplished in 2–26 days, mean 16 (Allen and Mangels 1940), but, once ♀ permitted to stay at site, bond may be established quickly—within *c.* ½ hour (Voisin 1970). Watchers usually silent and inactive. May periodically supplant one another, or attempt to approach ♂, often Mock-preening (Voisin 1970); this takes form of Wing-touch display much as in *Bubulcus* and *Ardeola* (see A). Accounts of ♂'s pairing displays and mutual behaviour after pair-formation hard to reconcile. According to Meyerriecks (see Palmer 1962), ♂ performs separate Snap and Stretch-displays then, when ♀ allowed to approach and join ♂ on site, pair engage in mutual Billing and Allopreening; also Greeting Ceremony, copulation and associated activities. Main pairing display, which accompanies Advertising-call, termed Song-dance by Allen and Mangels (1940). ♂ (1) stands obliquely erect with neck extended in line with body but bill horizontal; (2) starts treading with feet,

A          B          C          D

lifting them alternately with toes flexed, arching back and bending head forward (see B); (3) bends legs and lowers body, starting to call when bill almost level with feet (see C); (4) crouches with neck over side of nest, completing call; (5) stands up and usually Mock-preens feathers of underparts (Belly-touch; see D) before repeating whole sequence. Feathers of back, crown, neck, and throat raised, head plumes erected vertically; pupils dilate, and eyes bulge. Song-dance obviously equivalent to Courting Dance of Lorenz (1938), Snap-hiss Ceremony of Noble *et al.* (1938), and Dance of Voisin (1970), though considerable variation in details particularly as to use of head plumes. Classified as reversed Stretch-display by Meyerriecks (see Palmer 1962) but more likely combined and modified Stretch-snap display. During crouching phase call sometimes replaced by typical Bill-snap (Noble *et al.* 1938), and bird may dance without making any noises or Twig-snap instead (Allen and Mangels 1940). Song-dance display repeated frequently by unpaired ♂, usually 3–4 times in succession (Voisin 1970) at rate of 8–10 per min. Advertising-call sometimes given without posturing. Other pairing displays of ♂ include Bowing and Wing-touch (Voisin 1970), but no flight displays recorded. In Bowing-display, often with twig in bill (Twig-bow); (1) stretches up with bill drawn in near neck (see E), (2) lowers head in front of body without bending legs or moving rest of body, (3) stretches up again. Wing-touch display (see A) performed independently of other pairing displays. Song-dances cease when ♀ joins ♂ at site (Allen and Mangels 1940). In approaching ♂, ♀ may Wing-touch, both doing so when first together at site though not after pair-formation completed (Voisin 1970); now also perform first mutual Greeting Ceremony (see below). ♀ may then attempt to take twig from ♂ but usually couple soon start Billing and Allopreening, with some Bill-rattling. At this stage, such behaviour tends to develop into hostile Bill-fencing, but soon changes back to Bill-rattling and Billing (Lorenz 1935). Allopreening performed simultaneously, on feathers of head and elsewhere, or unilaterally; often resembles Back-biting (see *Bubulcus*). Nest-building starts soon after pair-formation, ♀ staying on nest and ♂ bringing twigs (Twig-passing display). Pair also spend much of time quietly inactive at nest together, standing in hunched position. In absence of ♀, ♂ performs Song-dances. When either sex returns to nest, Greeting Ceremony follows. Preceded by loud calling as returning bird flies in and lands near nest (Voisin 1970). 1st bird then approaches and both then assume horizontal position with neck stretched out, most feathers raised, and head plumes erected, and touch bills, Bill-rattle, shake heads, and give Greeting-calls (Allen and Mangels 1940); if returning bird ♂, often carries twig (see F). Greeting Ceremony unusual in that no Stretch-display involved as, e.g., in *Ardea*; same as Overture-and-display of Noble *et al.* (1938) and

Appeasing Ceremony of Lorenz (1935, 1938), who stressed exhibition of white plumes. Copulation, at or near nest-site, does not normally occur on first day pair formed but usually from next day (Allen and Mangels 1940; Voisin 1970). Often follows Greeting Ceremony, especially after return of ♂ with twig, subsequent Twig-passing, and building by ♀ who then crouches (Allen and Mangels 1940); ♂ may first perform long bout of Twig-bowing (Voisin 1970; see E). May also follow period of quiescence and then prolonged Billing and Allopreening, including Back-biting. No formal post-copulatory display. At nest-relief, following Greeting Ceremony, relieving bird bends over other to perform building movements on far side, pressing down on its back to some extent and slowly driving it from eggs or small young (Lorenz 1935). RELATIONS WITHIN FAMILY GROUP. Young first give Food-call shortly after hatching (Allen and Mangels 1940); beg indirectly by calling and waving head in irregular manner (Noble and Wurm 1942). By 3rd week, indirect begging movements consist of series of sharp, upward angulations of head plus slight raising of wings; these movements and Food-call repeated for hours, even if parents absent. Young also beg directly by seizing bill of parent crosswise in own. Growing chicks hostile to any adult approaching nest; tend to give Forward-displays and stab with bill even at own parent. Parent approaches young in stealthy manner, walking to nest in crouched posture with head lowered, neck retracted; then habitually performs Greeting-display with call. Young reciprocate, then continue begging, but adult does not feed them until physically induced to do so by direct begging. Young tend to wander from nest at early age (by at least 12th day); parents will feed them elsewhere from *c.* 3rd week. Food-calling by young in absence of parent apparently reduced after they first start to perch away from nest, at *c.* 18th day; later suppressed entirely, at least during day.

(Figs A and E after Voisin 1970; rest after Allen and Mangels 1940.)

**Voice.** Largely hoarse and guttural. Away from breeding colony, generally silent apart from familiar coarse croak, rather like Raven *Corvus corax*, variously rendered 'quick', 'wok', 'quark', 'hwack', 'guok', etc. (Witherby *et al.* 1939; Peterson *et al.* 1954; Palmer 1962; Bauer and Glutz 1966). Sometimes uttered by day from roost but more usually as Flight-call at dusk or after dark.

Variety of raucous sounds at colony; also non-vocal Bill-snapping and Bill-rattling. Following main calls, those of ♂ and ♀ similar unless otherwise indicated. (1) Advertising-call. 2 distinct components: (a) clicking 'plup' like

E

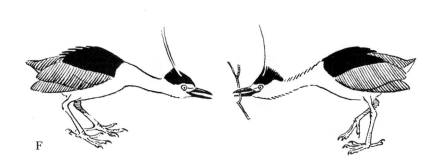

F

bursting bubble of mud, (b) twanging 'buzz'. (Allen and Mangels 1940). 1st component given in throat, not by snapping of bill; 2nd likened to prolonged hiss (Noble *et al.* 1935), deep and quite low like steam escaping through valve (Lorenz 1938). Given first from pairing territory at start of season by ♂ only. (2) Threat-calls. Highly variable, harsh and rasping: 'squaar', 'ok ok', 'rok-rok', 'scraak-scraak', etc. (see Palmer 1962); also 'kak kak kak' (Voisin 1970). (3) Alarm-call. More rapid and slightly higher pitched version of Flight-call (Palmer 1962); rendered 'squok' by Meyerriecks (1960). (4) Alerting and Mobbing call. Low 'wok-wok'; uttered by birds at sites on edge of colony upon intrusion by man, and by those flushed from nests while hovering over intruder (Palmer 1962). (5) Alighting-call. Sonorous 'kak kak'; given by incoming bird, sometimes answered by mate at nest (Voisin 1970). (6) Greeting-call. Series of soft 'wok wok' notes accompanied by Bill-rattling (Allen and Mangels 1940); disyllabic and low, with accent on 2nd note (Lorenz 1938), more guttural in ♂ (Noble *et al.* 1938). Also rendered 'quahck quahck' or 'goorock goorock', often with squealing 'va-va-va-va-va' (Steinfatt 1934). Uttered by both birds during Greeting Ceremony at nest. (7) Pre-copulatory call. Soft 'wok wok' (Palmer 1962).

CALLS OF YOUNG. Food-call a persistent, cackling 'kak-kak-kak', shrill and metallic during first 2 weeks, then gradually becoming harsher (Noble and Wurm 1942); also 'chip chip chip' like stones being scraped together (Allen and Mangels 1940). After leaving nest, while still flightless, utter continuous, chesty gurgle as Threat-call, high-pitched screech as Frightening-call, and low-pitched squawk as Fighting-call when disputing among themselves (Noble *et al.* 1938). Juveniles (aged *c.* 70 days and over) that formed premature pairs in captivity gave 'krwawrk-krwawrk-krwawrk' as Recognition-call, throaty and prolonged version of Flight-call as Greeting-call, and combination of Food and Flight-calls in recalling partner when latter moved off (Noble *et al.* 1938).

**Breeding.** SEASON. See diagram for western Europe. No information from elsewhere in west Palearctic. SITE. In trees or shrubs, less often in reedbeds; nests 2–50 m from ground. Colonial; often densely with 20–30 nests in one tree; frequently with other Ardeinae. Nest: platform of twigs and small branches, varying from frail to very solid; 30–45 cm diameter, 20–30 cm high; little or no cup; sticks up to 3 cm diameter and 60 cm long used; made of reeds in reedbed sites (Hanzàk 1949–50; Moltoni 1936). Building: usually ♀ with material brought by ♂, gathered from surrounding trees or stolen from others nests. Sticks and reeds often arranged pointing to centre. Continues throughout day, completed in 2–5 days (Hanzák 1949–50; Moltoni 1936). EGGS. Elongated oval; light blue-green, sometimes fading during incubation. 50 × 36 mm (44–56 × 32–38), sample 300 (Schönwetter 1967); weight 34 g (32–36), sample not given (Moltoni 1936). Clutch:

3–5 (1–8). Normally one brood, occasionally 2. Replacement clutches laid; generally only 2–3 eggs (Moltoni 1936). One egg laid every 48 hours. INCUBATION. 21–22 days. By both sexes but ♀ does larger share. Begins with 1st egg; hatching asynchronous at 48-hour intervals. Eggshells usually discarded over side of nest. YOUNG. Semi-altricial and nidicolous. Cared for and fed by both parents. Smaller young fed by partial regurgitation; later, food regurgitated completely into nest. Feeding stimulated by young tugging at parent's bill. Leave nest and perch in nearby branches at *c.* 20–25 days, returning to nest to be fed. Little or no nest sanitation. FLEDGING TO MATURITY. Fledging period 40–50 days. Young independent soon after. Mature at 1 year, but most do not breed until 2 years (Dementiev and Gladkov 1951). BREEDING SUCCESS. No data from west Palearctic. Of 24 eggs laid Georgia, USA, 22 hatched and 17 young fledged (Teal 1965).

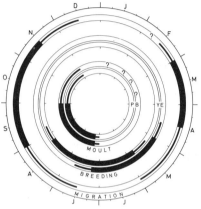

**Plumages.** ADULT. Forehead and narrow line over eye white. Crown, nape, mantle, and scapulars black, glossed green; longest scapulars grey. In breeding plumage, steel-blue gloss on black (Hartert 1912–21). On nape, 1–8 elongated narrow white feathers (up to 24 cm in ♂, to 17 in ♀); sometimes dark grey at base or tip. Hindneck and sides of head and neck light grey, grading into white of chin and throat. Back and upper tail-coverts light grey. Centre of underparts white, grading to pale grey on flanks. Tail, flight-feathers, and upper wing-coverts grey; outer webs of outer primaries darker; narrow white line along margin of outermost. Underside of wing paler grey, axillaries still paler. NESTLING. Down at hatching brown; darker (with long, hair-like, white tips) on crown, grey-brown on lower back, white on abdomen. Sparser by 6th day, mainly grey-brown, but abdomen pale grey. Feathers start to grow by 12th day; mainly feathered with down adhering to crown on 28th (McVaugh 1972). JUVENILE. Crown dark brown, narrowly streaked buff; sides of head with broader buff streaks. Hindneck and sides of neck streaked brown and pale buff in equal proportions. Chin and throat white or pale buff, central row of drop-shaped brown spots on throat. Mantle dark brown, feathers with triangular buff spots at tips; occasionally with wide cinnamon streaks. Scapulars paler, spots on longer ones progressively bigger. Back and rump grey-brown streaked white. Upper tail-coverts brown. Breast and belly white to pale buff, feathers with lateral brown streaks. Vent and under tail-coverts white. Flanks grey-brown, streaked white. Tail grey. Flight-feathers grey-brown with white tips, reduced on secondaries. Upper wing-coverts grey-brown with triangular buff spots at tips. Undersurface of flight-feathers pale brown-grey, rest of underwing pale brown with buff streaks. Axillaries pale grey-brown with pale buff streaks. IMMATURE. Worn from January of 2nd calendar year; may nest in this plumage. Crown dark brown with slight gloss. Sides of head and neck streaked buff and pale grey-brown, less contrasting than in juvenile. Mantle and scapulars dark grey-brown, slightly glossed purple. Centre of chin and throat white. Breast and belly white, streaked pale grey; flanks more grey-brown. Juvenile wings retained, but much worn and bleached, spots white, not buff (Gross 1923; Bent 1926; Heinroth and Heinroth 1926–7). SUBSEQUENT PLUMAGES. Juvenile wing starts to moult in spring of 2nd calendar year. In summer and autumn, immature replaced by plumage resembling adult, but crown and back browner, glossed oily green rather than metallic; grey parts also tinged brown. Both Gross (1923) and Bent (1926) supposed that this plumage moults to plumage resembling adult breeding, but with white forehead tinged brown. Alternatively, there may generally be only 1 moult per year (Heinroth and Heinroth 1926–7) and plumage acquired in autumn of 2nd calendar year may be worn during breeding season of 3rd (see also Hoogers and Kluyver 1959); more data needed. Subsequently like adult.

**Bare parts.** ADULT BREEDING. Iris crimson. Bill black. For short period at start of season, during pairing and mating phase, lores and bare skin round eye azure-blue or darker, nearly black, leg and foot red (claws black); later, base of lower mandible yellow, lores greenish, and legs pale yellow (Moltoni 1936). ADULT NON-BREEDING. Iris crimson or brown-red. Bill green-black, cutting edges and base of lower mandible yellow-green. Lores yellow-green or green-black. Leg and foot pale yellow. NESTLING. Iris at hatching pale yellow. Bare skin, bill, and leg off-white with yellow wash; small dark tip to culmen. Bare skin later greenish; bill and lores grey-green, flesh on culmen, darker at tip and base; leg pale green, foot pale yellow. JUVENILE. Iris at fledging yellow. Bill yellow-brown, darker at culmen, lower mandible yellow. Lores greenish-yellow. Leg yellow-green. Iris becomes orange in autumn (Belman 1974). IMMATURE. Iris brown-red. Upper mandible black with green or yellow tinge at cutting edge, lower mandible yellow-green with black tip. Lores and bare skin round eye black, eyelid azure. Leg pale yellow with green or red wash.

**Moults.** ADULT POST-BREEDING. Complete. August–November (Witherby *et al.* 1939); primaries irregularly (Stresemann and Stresemann 1966). No complete moult in spring, but alleged pre-breeding moult of body plumage (Palmer 1962; Bauer and Glutz 1966) needs corroboration. POST-JUVENILE. Partial; starts January of 2nd calendar year (Bent 1926) or even earlier in juveniles hatched early in season (Heinroth and Heinroth 1926–27). Moult of juvenile wing in spring of 2nd year (Stresemann and Stresemann 1966). Next partial moult almost continuous with post-juvenile, July to autumn of 2nd calendar year (Bent 1926). Subsequent moults like adult.

**Measurements.** Adult. Palearctic, Africa, Asia; skins (RMNH, ZMA). ♂ averages slightly larger than ♀, but difference not significant, sexes combined.

| | | | | | |
|---|---|---|---|---|---|
| WING | 291 | (7·79; 27) | 278–308 | TARSUS | 75·0 (4·67; 10) | 68–84 |
| TAIL | 108 | (3·71; 9) | 102–112 | TOE | 77·8 (5·00; 6) | 71–85 |
| BILL | 70·8 (3·54; 27) | | 64–78 | | | |

**Weights.** Adults, Italy: ♂ 682 (66·8; 12) 600–800, ♀ 590 (49·9; 11) 525–690 (Moltoni 1936). Older than 1st calendar year, Camargue, unsexed: 532 (125) 339–780; averages April 529, May 531, June 525, July 515, August 601. Juveniles, Camargue, unsexed: 514 (197) 375–776 (Bauer and Glutz 1966).

**Structure.** Wings broad and rounded. 11 primaries: p8 and p9 longest, p10 2–10 shorter, p7 0–5, p6 15–20, p5 25–30, p1 *c.* 80; p11 vestigial. P7–p9 slightly emarginated outer webs; p9 slightly inner web, p10 strongly. Tail square; 12 feathers. Bill not long, heavy and broad at base; ridge of culmen decurved towards tip, cutting edge slightly so. Nostrils slitlike, situated in deep nasal groove, extending nearly to tip of bill. Neck short. Legs short, lower fifth of tibia bare; outer and middle toe basally joined by short web; outer toe *c.* 78% of middle, inner *c.* 72%, hind *c.* 56%, hind claw *c.* 22%.

**Geographical variation.** Slight in Holarctic; *hoactli* larger than nominate *nycticorax* with shorter white stripe over eye in adult. Leg and foot brilliant salmon pink and lores deep blue-black until laying completed (Allen and Mangels 1940). South American races darker below.

## *Butorides striatus* Green-backed Heron

PLATE 34
[after page 302]

| | | |
|---|---|---|
| Du. Mangrovereiger | Fr. Héron vert | Ge. Streifenreiher |
| Ru. Зеленая кваква | Sp. Garzita azulada | Sw. Grönhäger |

*Ardea striata* Linnaeus, 1758

Polytypic. *B. s. brevipes* (Ehrenberg, 1833), Red Sea, Gulf of Aden, Somalia. Extralimital: nominate *striatus* (Linnaeus, 1758), northern South America; *atricapillus* (Afzelius, 1805), Africa; *amurensis* (Schrenck, 1860) eastern Asia; and *c.* 25 others.

**Field characters.** 40–48 cm, wing-span 52–60 cm, ♂ averaging slightly larger. Small, dark heron with paler underparts, rather short and heavy-looking neck, and large bill. Sexes alike and no seasonal differences; juvenile distinct.

ADULT. Dark glossy green upperparts, looking bluish on mantle, and edged buff on scapulars and wing-coverts, with dark grey primaries and secondaries. Nape feathers form crest and scapulars also elongated, but neither conspicuously; white patch behind eye edged below with black. Throat and elongated feathers of upper breast white, spotted and streaked with yellow-brown in middle and edged with grey on sides of neck; rest of underparts grey, darker on flanks. Appears blackish or green at distance, according to light; paler underparts often difficult to see unless head-on at close range. Bill heavy, blackish with yellow base to lower mandible; bare lores yellowish and black; eyes yellow; feet grey-brown with

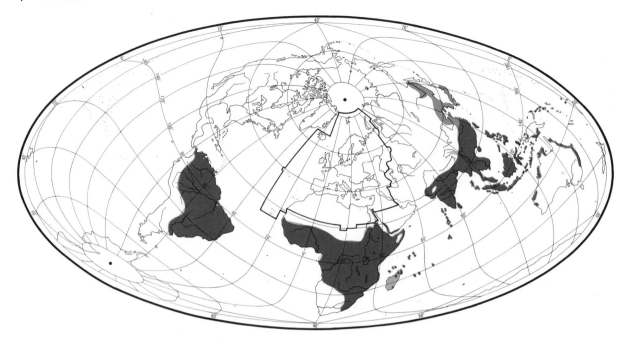

yellowish rear to tarsus and undersides of toes, becoming orange-red in breeding season. JUVENILE. Browner above, with white spots on wing-coverts; crest and scapular plumes less developed. Throat and upper breast more buff-white, speckled black on centre of throat and streaked dark olive-brown on breast; rest of underparts darker grey.

Quite different from other west Palearctic herons but needs to be distinguished, particularly at distance, from Green Heron *B. virescens* of North America, of similar proportions and much more likely as vagrant in Europe, which has rich rufous-chestnut sides of head and neck, with whitish stripe on foreneck, and yellow legs; juvenile also more heavily streaked below.

In some areas, rather secretive, tending to be crepuscular and slipping into tangled cover when disturbed; but in other parts much less shy, occurring even at urban ponds and feeding in open all through day, flying if disturbed and flighting between feeding places. Heavy-looking head and neck often stretched forward in flight. Wing action much like other small herons, but flies easily among trees and rank vegetation. Alights on elevated perches.

**Habitat.** Tropical fresh and marine waters including lowland rivers, streams swamps and, in the Red Sea and some other areas, seashore, mudbanks, harbours, estuaries, lagoons, and coral reefs. Unlike Little Bittern *Ixobrychus minutus*, ventures freely into open and exposed situations especially near sea, wading in tidal waters, but prefers cover for creeping away or skulking. Climbs or perches among mangroves or on rock ledges.

**Distribution.** Stragglers in west Palearctic on islands in northern Red Sea, 1919 and 1942 (Nicoll 1919; Bulman 1944), and coasts of Sinai, 1956 and 1967-71, presumed wanderers from breeding colonies further south in Red Sea, but breeding suspected near Nabq on eastern shore of Sinai 1972, and proved 1973 (HM).

**Population.** SINAI: 20 nests at Nabq, 1973 (HM).

**Movements.** Little known in absence of ringing. Those breeding in arid zone below Sahara move around in dry season, and probably summer (rainy season) migrant to Darfur, Sudan (Lynes 1925), but most coastal and island populations rarely identified outside their breeding ranges. Apparently limited dispersal in Red Sea area (see Distribution). High level of subspeciation (some 28 races in all) further indicates basically sedentary nature, though an Asiatic mainland subspecies *amurensis* migratory.

**Food.** Exclusively animal, taken while stalking along water's edge (often under overhanging branches or within reach of shady vegetation) and lunging forward, or when perched on low branches or artefact such as jetty or boat above water, craning neck down and jabbing forward or diving in (Mayo 1931; Hindwood 1933). May completely submerge when pursuing fish, or even dive from flight (Meinertzhagen 1954). Normally solitary, crepuscular and nocturnal feeder, occasionally daytime.

Chiefly fish including, in India, garfish *Belone strongylura* (16·5-20·5 cm long), mudfish *Periophthalmus*; in west Africa, mudskippers, crustaceans (crabs, shrimps, prawns), amphibians (frogs), molluscs, insects and larvae—of waterbeetles (Coleoptera), and grasshoppers (Orthoptera)—and annelids (Ussher 1874; Archer and

Godman 1937; Jackson 1938; Meinertzhagen 1954; Ali and Ripley 1968; Mackworth-Praed and Grant 1970). 26 stomachs, Sundarban, India, contained by weight 31·8% crustaceans (decapods only, including *Metapenneus, Scylla, Portunus, Varuna*), 29·0% fish (11 species from fresh and brackish water areas, on average 5–50 mm, including *Periophthalmus, Boleophthalmus, Puntius, Mystus*), 14·5% insects (including terrestrial and aquatic species), 13·8% amphibians (mostly *Rana* tadpoles), 3·6% annelids (Mukherjee 1971).

**Social pattern and behaviour.** Apart from general information (e.g. Dementiev and Gladkov 1951; Meinertzhagen 1954; Ali and Ripley 1968), little known in areas adjoining west Palearctic though observed in more detail, mainly at nest, in South Africa (Cowles 1930) and Australia (Hindwood 1933). Evidently closely similar to Green Heron *B. virescens* of North and Central America (see Palmer 1962 for summary of biology), the 2 forms considered by Curry-Lindahl (1971) and Payne (1974) to be probably conspecific.
1. Largely solitary, feeding alone and probably (like *B. virescens*) maintaining individual feeding territories at times. In India, birds frequent same secluded spots day after day (Ali and Ripley 1968). May stay in family parties for a while after breeding. BONDS. Monogamous pair-bond of seasonal duration indicated. Both sexes tend young until well after fledging (Hindwood 1935). BREEDING DISPERSION. Usually solitary and well-dispersed in concealing and protective sites. Nests sometimes nearer together and, rarely, loosely colonial; does not normally associate closely with other herons or marsh birds when breeding. Nothing known of territorial behaviour (for data on *B. virescens*, see Meyerriecks 1960). ROOSTING. Apparently mostly solitary and diurnal, but little information. Usually rests by day, flying to feeding grounds at sundown (Meinertzhagen 1954). Sometimes feeds during day, and follows tidal routine in some areas, feeding at ebb and resting during rise on tops of palms, sea wall, or coral outcrop (Ali and Ripley 1968), though day roosts otherwise normally concealed. In *B. virescens*, nocturnal roosting recorded in breeding area (Meyerriecks 1960): newly arrived birds land shortly after sunset in reeds and bushes close to ground; in early part of cycle, do not use same sites on successive nights but, after nest-territories established, roost only there.
2. Display and related behaviour not described, though said by Curry-Lindahl (1971) to be almost identical to *B. virescens* (see Meyerriecks 1960 for detailed study). Pair-formation of latter species of heron-type 1, with associated changes in colour of bill, lores, iris, and leg, and rich repertory of vocal and visual displays, both standing and aerial. Colour changes affecting legs noted during breeding season for *B. striatus* in Burma (Stanford 1936; Smythies 1953) and Surinam (Haverschmidt 1953). Erection of crest before landing observed by Glenister (1951) and Smythies (1953), suggesting possible Alighting-display (see Grey Heron *A. cinerea*). Raising of crest also recorded from incubating bird when mildly alarmed and when man near nest; accompanied respectively by rapid flipping up and down of tail and by threatening movements and calls (Cowles 1930). Tail-flipping develops in young at *c.* 6 days; also noticeable in adults away from nest, especially when hunting (Hindwood 1933). On sighting enemy, bird at nest usually adopts concealing Bittern-stance (Cowles 1930; Hindwood 1933) more typical of ground-nesting herons (see *Botaurus stellaris*), remaining motionless apart from almost imperceptible turning of head to follow intruder's movements. Same posture later shown by young from 6th day

but especially from 10th, when left alone by parents except for feeding; camouflage effect enhanced by green colour of skin (Hindwood 1933).

**Voice.** No information for Red Sea race *brevipes*, but some on other races.

Largely silent, but when flushed gives 'k-yow k-yow' or 'k-yek k-yek' alarm-call (Ali and Ripley 1968). Bird disturbed from nest during incubation uttered harsh croaks (Hindwood 1933) and threat-call of sitting bird to man near nest consisted of a few raucous squawks (Cowles 1930). 'Tewn-tewn-tewn' like call of Redshank *Tringa totanus* mentioned by Ali and Ripley (1968), but circumstances not given; nor for sharp squawk and high-pitched, cheerful 'chuck' of Mackworth-Praed and Grant (1962). Call of adult to another prior to selection of nest-site a harsh, sneezing 'tcha-aah', sometimes with an explosive 'hoo' (Hindwood 1933). Adult approaching nest when young present gave harsh call; changed to single, chicken-like 'chuck' on reaching nest prior to feeding (Cowles 1930); also on arrival at nest a long-drawn 'tchee-unk' (Mackworth-Praed and Grant 1962).

Calls of young from egg near hatching a soft croaking or cheeping; older young loafing between parental feeding visits occasionally uttered a soft, croaking 'toc-toc-toc-toc' (Hindwood 1933). Hiss when disturbed (S Cramp).

**Breeding.** SEASON. Sinai: eggs found May (H Mendelssohn). Sudan: June–September (Mackworth-Praed and Grant 1952). SITE. In low bushes, often standing in water, or on branch overhanging water; also in trees *c.* 3–9 m above ground (Hindwood 1933). In Sinai, most *c.* 50 cm above ground in mangroves (H Mendelssohn). Mostly single and well dispersed though small colonies recorded. Sometimes nests with other species, including Western Reef Heron *Egretta gularis* and Goliath Heron *Ardea goliath*. Generally well hidden and shaded. Nest: flimsy structure of twigs with shallow cup, *c.* 30 cm diameter; eggs often visible from beneath. Building: probably by both sexes. EGGS. Oval, almost elliptical; pale blue-green, dull and close-grained, 37 × 28 mm (35–40 × 26–30), sample 14 (Roberts 1965). Calculated weight 28·0 g (Schönwetter 1967). Clutch: 2–4. One brood. Replacement clutches not recorded. Eggs laid at 2-day intervals. INCUBATION. 21–25 days (Hindwood 1933). By both sexes. Begins with first egg, hatching asynchronous. YOUNG. Semi-altricial and nidicolous. Cared for and fed by both parents. Food regurgitated. FLEDGING TO MATURITY. Fledging period not recorded. Age of independence not known but young fed for 1 month or more after leaving nest tree (Hindwood 1933). Age of maturity not known. BREEDING SUCCESS. No data.

**Plumages** (*B. s. atricapillus*; material for west Palearctic *brevipes* inadequate). ADULT. Crown and crest on nape black, strongly glossed green. Hindneck and sides of neck ash. Short streak behind eye black, separated from black of crown by pale cream streak. Feathers of mantle and scapulars lanceolate, glossed green, tips tending to grey. Back dark grey. Upper tail-coverts glossed green. Chin and throat white with row of grey spots down centre. Foreneck and centre of upper breast rufous-brown, some white showing through; foreneck mottled dark brown-grey. Rest of underparts, including flanks, axillaries, and under wing-coverts ash, centre of belly slightly lighter. Tail dark grey, central feathers glossed green, paler below. Primaries and secondaries dark grey, outer webs slightly glossed green, paler below. Outer web of p10 margined cream. Inner primaries and secondaries tipped white, inner secondaries with outer webs strongly glossed green and margined cream to buff. Upper wing-coverts glossy green with white margin. Under wing-coverts and axillaries grey; leading edge of wing at base of primaries pale cream. NESTLING. Down light grey with white hair-like tips on head and white streak down centre of throat and underparts, skin blackish (Cowles 1930; Mackworth-Praed and Grant 1952). In Australian race *macrorhynchus*, body skin at first brown or flesh; for further details see Hindwood (1933). JUVENILE. Crown less glossy than adult, feathers with buff streaks along shafts, less elongated. Hindneck, sides of neck, foreneck, and chest streaked dark brown and buff. Chin and throat white with small dark brown spots. Feathers of mantle and scapulars not lanceolate, brown with slight green gloss, edges chestnut-buff. Some scapulars with buff spots at tips. Lower breast and belly streaked grey-brown and pale buff, long under tail-coverts off-white with dark brown streaks on outer webs. Flanks grey. Primaries, primary coverts, and secondaries with white spots at tips. Upper wing-coverts green, less glossy than adult, with buff spots at tips; greater with wide buff margins, others margined chestnut-buff. Under wing-coverts mottled grey and white, axillaries grey. In presumably older immatures, crown less streaked, mantle and scapulars mainly dull green, pattern of neck and chest more contrastingly white and black-brown, less suffused rufous; comparable plumage worn in 1st autumn by Green Heron *B. virescens* of North America (Palmer 1962).

**Bare parts.** ADULT. Iris bright yellow. Upper lores yellow-green, lower black. Upper mandible black, lower mandible yellow-green with darker cutting edges and tip. Leg grey-brown

in front, yellow at back, sole of foot yellow (Clancey 1964). Variant colour, apparently associated with display: iris yellow with red outer ring; base of upper mandible pink, rest black; alternatively, entire bill coral-red; lores pink; leg pink to orange-red (Haverschmidt 1953; Benson and Penny 1971). NESTLING. Iris greenish-yellow, later becoming more yellow (Hindwood 1933). Bill yellow, facial skin green-yellow (Cowles 1930). Legs brown or flesh colour; later green with some yellow on back of tarsi (Hindwood 1933). JUVENILE. Iris pale yellow. Bill dark horn with yellow-green lower mandible. Lores yellow-green. Front of leg grey-green, rest yellow.

**Moults.** Poorly known. ADULT POST-BREEDING. Complete; primaries irregular. ADULT PRE-BREEDING. Partial. Some feathers of crown and mantle; needs confirmation. POST-JUVENILE. Complete; 4–12 months after fledging (may be preceded by partial moult soon after fledging). Starts with crown and mantle, continues with remainder of body feathers; finishes with wing, moulting earlier than adult post-breeding.

**Measurements.** Skins *atricapillus* (no data for *brevipes*), sexes combined (RMNH, ZMA).

| | | | | | | |
|---|---|---|---|---|---|---|
| WING | 179 | (6·27; 16) | 167–190 | TARSUS | 45·5 (1·27; 10) | 44–47 |
| TAIL | 63·7 | (3·81; 16) | 57–70 | TOE | 49·8 (2·30; 10) | 46–54 |
| BILL | 60·7 | (2·52; 15) | 55–65 | | | |

**Weights.** No data for *atricapillus* or *brevipes*. Nominate *striatus* from northern South America 187 (28·0; 7) 135–214.

**Structure.** 11 primaries: p8 longest, p9 0–5 shorter, p10 1–10, p7 0–6, p6 6–12, p5 11–18, p1 37–43. Outer web of p10 narrow. P8–p9 slightly emarginated outer web, p9–p10 inner. Tail square; 12 feathers. Bill relatively heavy. Leg short. Inner toe *c.* 78% of middle, outer *c.* 80%, hind *c.* 65%, hind claw *c.* 25%.

**Geographical variation.** Involves colour of foreneck and chest (white in Asiatic populations) and size. *B. s. brevipes* said to be darker and less bright green on crown and back than *atricapillus* (White 1965). Green Heron *B. virescens*, North and Central America, and Lava Heron *B. sundevalli*, Galapagos Islands, closely related; sometimes considered conspecific (Payne 1974).

## *Butorides virescens* (Linnaeus, 1758) **Green Heron**

FR. Héron vert    GE. Mangrovereiher

New World species, occupying wide range of damp habitats (freshwater and marine) in North, Central and northern South America. Seven subspecies recognized: nominate *virescens*, the most migratory, breeds from southern New Brunswick, Quebec, and Ontario south to Gulf states and Mexico, wintering mainly south of breeding range, in Florida and from Gulf of Mexico to Colombia and Venezuela (Palmer 1962). Main passages

through USA in March–April and September–October. Has also straggled to Bermuda, Manitoba, Nova Scotia and Newfoundland, and once to west Greenland. One shot near St Austell, Cornwall, England, 27 October 1889, sole Palearctic record (Hudson 1972). Perhaps conspecific with Green-backed Heron *B. striatus* (Payne 1974) but possible overlap in Panama.

# *Ardeola ralloides* Squacco Heron

PLATE 35
[between pages 302 and 303]

Du. Ralreiger     Fr. Héron crabier     Ge. Rallenreiher
Ru. Желтая цапля     Sp. Garcilla cangrejera     Sw. Rallhäger

*Ardea ralloides* Scopoli, 1769

Monotypic

**Field characters.** 44–47 cm (body 20–23 cm), wing-span 80–92 cm; smaller and slighter than Cattle Egret *Bubulcus ibis*. Rather small, distinctively maned heron, shaped like Bittern *Botaurus stellaris*, with mainly brownish-buff appearance on ground but sudden explosion of white wings, rump and tail on take-off. Sexes alike; some seasonal variation; juvenile separable.

ADULT BREEDING. Bare lores yellow-green, crown and neck yellow-buff with overlay of black-edged and white-centred feathers forming long mane (reaching mantle). Mantle, scapulars, and tertials pink-brown, with longest mantle feathers more golden, drooping over white wings and tail. Chin white, foreneck and chest golden-buff, rump, tail, and rest of underparts white. Bill not long but sharply pointed, basal half mostly green or blue, distal half dark horn, even black. Rather short legs and feet warm pink early in breeding season, when lores bright blue. Eyes yellow. ADULT NON-BREEDING. Mane shorter and plumage duller, even brown above, and dirtier and streaked below. Bill duller; legs and feet yellow-green. JUVENILE. No mane or elongated body feathers; plumage darker, with mantle earthy in tone, breast more strongly striped, and wing tinged or mottled brown. Bill pattern less marked than in adult, though still dark tipped.

Can be confused with *Bubulcus ibis* which in breeding dress has buff on crown, chest, and lower back but always looks white at any distance and has diagnostic heavy jowl and less sharply pointed bill. Close similarity of Indian Pond Heron *A. grayii* potentially more troublesome; for distinctions see that species.

Flight recalls that of small *Egretta* but appears faster and more fluttering (as contrast of dark buff and white plumage exaggerated by wing-beats); walks freely and climbs well. Active throughout day but hunts most intently from or along edge of cover at dusk; more skulking than *Egretta* and *B. ibis*. Mostly solitary, but in winter some tens often within sight of each other in large wetlands; flocking occurs in winter and on migration. Unusually quiet, even for a heron, but harsh, high-pitched monosyllable often given at dusk or in alarm.                    DIMW

**Habitat.** Continental mainland species, of warm Mediterranean, sub-tropical, and tropical climates. Although recorded breeding on mountain lakes up to 2000 m, mainly in lowland valleys and floodplains, wetlands, deltas, and estuaries; infrequently on shallow sea coasts and inshore reefs or islets. Prefers still fresh waters, especially small pools or ponds, canals, and ditches flanked by dense aquatic vegetation, often including some woody shrubs, climbing plants, or scrub trees such as willows *Salix* and poplars *Populus*. Also feeds in open, e.g. on irrigated ricefields or floodlands, but more at home in or by dense cover, fishing from stem or branch or while wading in shallow water. Wades and walks easily and perches freely; does not swim or dive. For breeding, prefers to join mixed colonies of other herons, Pygmy Cormorant *Phalacrocorax pygmeus*, and Glossy Ibis *Plegadis falcinellus* in low trees (but up to 20 m), shrubs, or, failing these, in reedbeds. Flies considerable distances to and from nest and elsewhere, but uses airspace sparingly, not normally deviating or rising higher than essential. Avoids both arid areas and those of high rainfall. Will forage on irrigated land subject to disturbance; at Antanarive, Madagascar, closely-related *A. idae* even breeding on pond close to passers-by in botanic garden. In normal haunts, however, has little contact with man.

**Distribution.** Considerable changes in last 100 years due to plumage trade and habitat destruction affecting particularly areas in east and south of Palearctic range, with some expansion in limited areas in recent years. (See Population.) African breeding range imperfectly known; often apparently sporadic north of equator (Moreau 1972).

CZECHOSLOVAKIA: 1–2 pairs bred 1963 (KH).

Accidental. Iceland, Britain and Ireland, Belgium, Netherlands, Denmark, Sweden, Finland, East Germany, West Germany, Poland, Austria, Switzerland, Azores, Madeira, Canary Islands, Cape Verde Islands.

**Population.** Based mainly on exhaustive analysis of changes in colonies and numbers of Palearctic population by Józefik (1969–70). In last decades of 19th century, reaching peak *c.* 1900 and lasting until 1920, drastic decrease in breeding sites and numbers mainly due to human persecution, especially for plumage trade; larger sites often destroyed, with increase in smaller sites and sporadic nesting. From 1920–40, further small decreases, with renewed concentration at large sites caused mainly by habitat destruction elsewhere. From 1940–60, some increase due to protection, establishment of reserves, reduction of predators, and, in central and western parts of range, adaptation to transformed habitats. Accurate population estimates rarely possible due to fluctuating numbers, changes in sites, and limited information in more

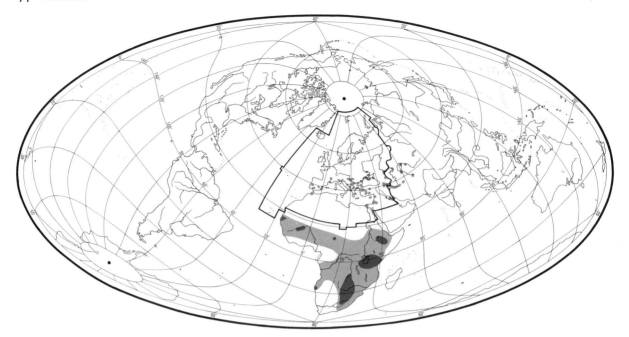

remote areas. Estimates by various methods suggested that overall numbers in Palearctic range varied from 16400 pairs at 165 sites in 1850–1900, to 6800 pairs at 115 sites in 1900–20, 6000 pairs at 80 sites in 1920–40, and 8200 pairs at 71 sites in 1940–60.

Changes since 1850, in the main regions in west Palearctic given by Józefik (1969–70), summarized below, with more recent data added where available.

A. Iberian peninsula. Increasing; 19th century one site *c.* 20 pairs; 20th century 3 sites 60–80 pairs. SPAIN: 4 colonies occupied since 1963 (FB); for details, see Fernandez Cruz (1975). PORTUGAL: one known colony; others suspected and believed increasing (MDE).

B. Southern France. 2 sites established 20th century. Camargue 49 pairs 1968, 115 pairs 1972, over 75 pairs 1974 (Hafner 1975). Also a few pairs Dombes (CJ).

C. Italian peninsula. In 19th century and first half 20th century, 10 sites with, on average, 100–130 pairs; after 1950, 4 sites with *c.* 30 pairs. No recent counts; never abundant, but common in some heronries (SF).

D. Middle reaches of Danube. 46 sites recorded second half 19th century, with 5500–6000 pairs, so area of most intense concentration in Palearctic. Decreased to 40 sites, with 1800–2000 pairs, early 20th century, and later to 20 sites, with 1000–1200 pairs; slight increase in sites and numbers since 1960. HUNGARY. 6 colonies; numbers fluctuating, but perhaps recent decline. Kisbalaton: 7–14 pairs 1968–72. Hortobágy: 30 pairs 1967 (AK, IS). YUGOSLAVIA. Kopačevski area: numbers fluctuated between 139 and 478 pairs 1954–70, with some tendency to decrease (Majic and Mikuska 1972).

E. Balkans. In 19th century, 10 sites with 450–500 pairs; first half 20th century, 14 sites, numbers similar; in 1950–60, 6 sites with 130–150 pairs; some increase since. GREECE: recent counts suggest *c.* 2050–2200 pairs in 9 main colonies; recent increase indicated, or perhaps some colonies unrecorded in past (WB).

F. North of Black Sea Region. In second half 19th century 31 sites with 3800–4200 pairs; decreased early 20th century when 37 sites but only 2100–2300 pairs; in 1950–60, only 26 sites, but population increased to 4500–4800 pairs. RUMANIA: increase apparently continuing after protection; now most common heron (MT).

G. Caucasian Region. In second half 19th century 22 sites, with 3800–4000 pairs; in second half 20th century, 24 sites with 2000–2200 pairs. Present number uncertain, but marked increase in some areas (e.g. Lenkoran).

H. Turkey and Eastern Mediterranean Coast. In middle 19th century, 12 sites with 450–500 pairs; in first half 20th century, 10 sites, 300–500 pairs; after 1950, 5 sites, 200–230 pairs. TURKEY. Recent counts in 4 colonies in different years suggest 800–1000 pairs; numbers in other colonies unknown. May also breed in eastern Turkey where present in summer (OST). ISRAEL: breeds in small numbers with frequent changes of colony sites (HM).

I. Persian Gulf Drainage Area. Probably bred sporadically at 5 sites in early 20th century with probably not more than 20–40 pairs; at present, breeding unlikely. IRAQ: rather common south and centre in summer 1973 but no record of colonies (PVGK).

J. North Africa. In 19th century, 9 sites with 6–700 pairs; in first half 20th century, number of sites about same but population halved; not more than 100 pairs 1950–60. ALGERIA: present status uncertain, but may breed (EDHJ). TUNISIA: has bred in past at least in favourable years; no recent breeding records (MS). MOROCCO: little recent data; at Allal Tazi *c*. 40–50 nests (Haas 1960). EGYPT: present all year, but no breeding records (GEW).

**Movements.** Migratory and dispersive. Some winter (perhaps only irregularly) western Morocco, Mediterranean, Iraq, Iran and Persian Gulf; most from west Palearctic migrate to northern tropics of Africa. Juvenile dispersal begins July; adults leave colonies August–early September, but southward movements, at times leisurely, continue well into October, especially in south Russia, or even later in Iran (D A Scott). Migrants occur throughout Mediterranean on both mainland and islands, and majority probably make broad-front crossing of Sahara, where recorded spring and autumn from oases. Also regularly across Arabia in September–October and March–April (Meinertzhagen 1954). Southern limit of winter range unknown, as another breeding population in tropical Africa, but probably equatorial forests form barrier (Bannerman 1957). Only 8 recoveries south of Sahara: Camargue birds in Guinea and Sierra Leone (2) and 5 (3 Yugoslavian, 2 Bulgarian) in Nigeria, where, in northern winter, numbers increase from small to very large (Moreau 1967). Spanish bird found on passage in Mauritania. Return migration begins March; normally European breeding colonies re-occupied late April–early May; in Hungary only short gap between vanguard and bulk of breeders (Bauer and Glutz 1966). Like other south European herons, tendency for individuals to overshoot well north of breeding range in spring; possibly now reach North Sea countries more rarely, thus *c*. 90 in Britain and Ireland to 1914, but only *c*. 10 between 1954–71.

**Food.** Relatively small (up to *c*. 10 cm) and mostly insects and their larvae, amphibians, and fish. Feeds singly or in small, spaced groups; mainly crepuscular, occasionally in daytime.

Insects recorded include waterbugs *Naucoris*, *Notonecta*, *Nepa*; waterbeetles, *Dytiscus*, *Hydrophilus*, *Cybister*, *Hydrous*; mole-cricket *Gryllotoalpa*; larvae of mayfly (Ephemeroptera) and dragonfly (Odonata, especially *Libellula* and *Aeshna*); tipulid and chironomid larvae. Amphibia include frogs (*Rana*, *Hyla*, and *Bombina*). Fish include bitterling *Rhodeus*, carp *Carassius*, bleak *Alburnus*, sunfish *Eupomotis*, tench *Tinca*, dace *Leuciscus*. Also recorded: Crustacea (*Gammarus*, *Apus*), spiders, lizards (*Lacerta*), molluscs, worms (*Lumbricus*), and small mammals (shrew *Sorex*).

In 108 stomachs from Hungary, mostly invertebrates, especially insects and their larvae: by frequency, *Naucoris*, *Notonecta* 67·5%; *Dytiscus*, *Cybister*, and *Hydrous* 55·5%; terrestrial beetles 37·9% (some possibly via frog stomachs); *Gryllotalpa* 36·1%; small fish, chiefly *Scardinius*, *Alburnus* and *Carassius* 31·4%; and Amphibia 26·8% (Vasvari 1938, 1939).

Similar prey eaten by birds feeding in naturally flooded areas of Tisza, Hungary; but in rice fields now frequented (particularly in early summer) relatively more insects and less fish taken (Sterbetz 1962, from observation and

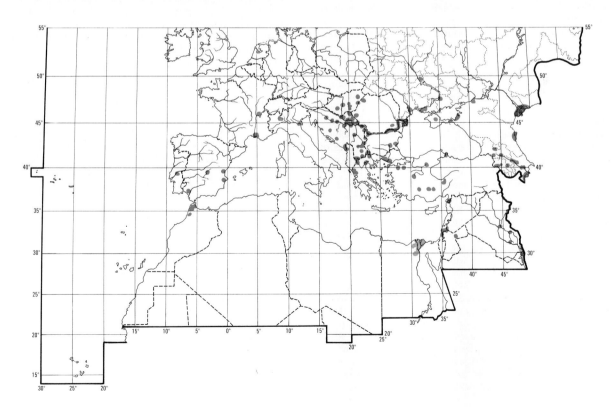

analyses of 34 stomachs). 92 Italian birds contained 83·3% insects, chiefly *Notonecta, Naucoris, Hydrous, Dytiscus,* and *Gryllotalpa*; 40% Amphibia *Rana esculenta*; and 5·5% fish (*Eupomotis gibbosus* and *Tinca tinca*) (Moltoni 1936); further 5 stomachs showed similar prey (Moltoni 1948). In Danube delta, Rumania, chiefly invertebrates, with fish only 22% by weight (Andone *et al.* 1969). In USSR, diet chiefly insects (especially grasshoppers), spiders, and small frogs (Dementiev and Gladkov 1951).

In west Tanzania, young fed mainly grasshoppers (Stronach 1968).

**Social pattern and behaviour.** Based mainly on material supplied by R K Murton.

1. Largely solitary except when nesting and roosting. Throughout year, mostly feeds individually, defending small feeding territory against conspecifics; in some areas, may feed gregariously among cattle like Cattle Egret *Bubulcus ibis*. BONDS. Monogamous pair-bond of seasonal duration. In Europe, pair-formation occurs after return from winter quarters. Both sexes tend young which become independent 1–2 weeks after leaving nest; family group may then break up, juveniles tending to congregate in colony as if deserted, or young may join parents on feeding grounds. BREEDING DISPERSION. Often gregarious when nesting, usually in small numbers in heronries of mixed species; also occasionally in hundreds, in small homogeneous groups, or singly. Nests usually 5–10 m apart in large colonies, but sometimes as close as 0·5 m. Pair defend territory in immediate vicinity of nest after lone ♂ initially holds larger area at same place (see below). ROOSTING. Communal, often at site shared with other Ardeinae; generally from dusk till dawn, but diurnal rhythm of activity needs better definition since often feeds at night. When breeding, pair and later duty birds roost nocturnally at or near nest site; off-duty birds and non-breeders also loaf during day in communal areas within colony, often with other species. Pairs also frequent favoured sites away from nest where they preen. Immediately before and after breeding, same trees in colony where nests built may be used for roosting. Outside breeding season separate roosts established by both resident and migrant populations. Usually in sheltered tree belt or wood, less frequently in extensive reed marshes or occasionally in reedbeds. Winter roosts usually larger than nesting colonies, drawing birds from radius of up to *c.* 80 km. Also, during times of year when food not limited, form small secondary roosts near feeding sources within area covered by larger roosts.

2. Little published information. Unpublished data (R K Murton) suggest general similarity to *Bubulcus, Egretta,* and *Ardea* with pair-formation of heron-type 1, associated with marked changes in bill and leg colour (Ferguson-Lees 1959). Use of protective Bittern-stance recorded both from adults (Riddell 1944) and new-fledged young (Sterbetz 1962). ANTAGONISTIC BEHAVIOUR. Aggressive Forward-display used as threat in defence of nest-site and feeding station: body inclined forward in low crouch and bill pointed at intruder. Full version, with feathers of head and back raised, occurs at higher intensity; accompanied by calls. Territory owner may also adopt Upright-display with body and neck at 45° and bill raised slightly above horizontal, all feathers sleeked; noted particularly when conspecific flies overhead or passes in foliage—and as prelude to Forward-display. Will also supplant rival, fights occurring if latter does not retreat, both adopting Forward-display and striking with beak. HETEROSEXUAL BEHAVIOUR. In migrant

populations birds occupy colony immediately after arrival (Sterbetz 1960–1). ♂ settles at site and builds small platform which he uses for pairing-displays to attract ♀ near. Observations on Indian Pond Heron. *A. grayii* and Chinese Pond Heron *A. bacchus* (R K Murton) supplemented by scattered reports and photographs of closely related *A. ralloides*, suggest behaviour closely similar in all (often considered conspecific). ♂ *A. bacchus* performs Flap-flight as advertisement at beginning of breeding season, flying out from site with curved neck and slow, exaggerated wing-beats (joint flights, with ♂ flying close behind ♀, occasionally noted after pair-formation). First stationary pairing-display, Snap-display much as in *A. cinerea*; as well as Bill-snapping, ♂ may also touch or hold twig. Alternates with Forward-display and leads to Stretch-display in which stretching and crouching phases evidently combined, as bird raises and lowers legs in dancing movements and pumps vertically pointed neck and bill up and down, with feathers of head and of scapular and pectoral areas erected. Any ♀♀ attracted to displaying ♂ at first repulsed, but usually one persists and eventually joins ♂ at site. Mutual Stretch-displays may now occur, accompanied by Billing, Allopreening, and Stick-passing; also by temporary increases in intensity of bare-part colours ('blushing'). When Billing, ♂ and ♀ hold and mandibulate each other's bills, sometimes with rapid nibbling movements; integrades with Stick-passing. When Allopreening, pair simultaneously nibble one another's plumage, chiefly back feathers; interspersed with bouts of unilateral Allopreening, mostly by ♀. At this stage of cycle, ritualized self-preening, involving cursory touching of primaries with bill (recalling Wing-touch display of *B. ibis*), and Forward-display alternate with Stretch-display. Later, full Stretch-display replaced by less ritualized version during Greeting Ceremony when one pair returns and joins other at site; legs remain bent in crouching position and plumes raised but body lowered less and head not pointed up; resembles Forward-display but bird less erect and bill pointed down. Copulation occurs on nest or nearby branch (see also Sterbetz 1962). Usually associated with Billing and Allopreening. ♂ may mount without obvious prelude or after brief Stretch-display by ♀. RELATIONS WITHIN FAMILY GROUP. No detailed information.

**Voice.** Mainly silent. Shrill harsh 'karr' (see Ib), sometimes shortened to 'kah' (see Ia), and also rendered 'charr', 'rrra', or 'rrra-ra' (Bauer and Glutz 1966), often heard at dusk, especially during breeding season, or when alarmed. Also uttered with bill open, in Forward-display, and in flight. Frog-like croak, followed by deep, descending gulping calls heard when bird landed (P A D Hollom). Rasping 'kek-kek-kek-kek' given in aggressive Forward-display.

Begging-call of young—'kri-kri kri-kri' (Bauer and Glutz 1966).

I C Chappuis/Sveriges Radio Kenya January 1967

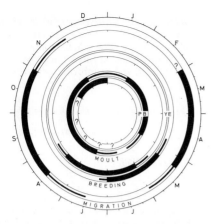

**Breeding.** SEASON. See diagram for south-east Europe. SITE. Dense thickets of willow *Salix*, or other bushes, reedbeds and wide variety of trees; from 2–20 m above ground or water. Colonial; but more often in scattered pairs or small groups in large, mixed colonies of other herons, though not usually in same tree as others which nest earlier and are dominant (Sterbetz 1962). Nest: platform of branches, twigs, or reeds, depending on site; 17–27 cm diameter, up to 20 cm high with cup 8–11 cm deep (Dementiev and Gladkov 1951; Sterbetz 1962); substantial or flimsy. Building: by both sexes, but mainly ♀ using material brought by ♂ from surrounding area. Completed in 6–8 days but material often added after laying (Sterbetz 1962). EGGS. Oval, pointed at both ends; green-blue. 39 × 28 mm (35–42 × 27–32), sample 160 (Schönwetter 1967). Weight 15–17 g (Moltoni 1936). Clutch: 4–6 (7). One brood. No information on replacements. Eggs laid normally at 24-hour intervals, occasionally 48. INCUBATION. 22–24 days (Sterbetz 1962). Mostly by ♀. Begins after clutch complete; hatching synchronous. YOUNG. Semi-altricial and nidicolous. Cared for and fed by both parents. Food partially regurgitated when young small, later completely into nest. Leave nest at *c.* 30–35 days and scramble about colony; freeze in undergrowth if disturbed (Sterbetz 1962). FLEDGING TO MATURITY. Fledging period *c.* 45 days. Become independent soon after. Age of first breeding not known. BREEDING SUCCESS. No data.

**Plumages.** ADULT BREEDING. Crown and hindneck yellow-buff with brown-black streaks; feathers elongated. Some lanceolate, long (up to 140 mm) feathers on nape cream-white with brown-black streak on each side. Sides of neck yellow-buff with a few dark grey streaks. Chin and centre of throat white. Mantle and inner scapulars pink-buff, washed vinous; outer scapulars and elongated ornamental feathers on chest golden-buff. Scapulars loosely structured, elongated, reaching to tip of tail or beyond, covering rest of upperparts. Rest of plumage white, inner secondaries and upper wing-coverts washed golden-buff. ADULT NON-BREEDING. Feathers of crown and hindneck shorter than breeding, buff, streaked brown-black; sides of neck more heavily streaked dark grey than breeding. Mantle and scapulars drab with slight vinous tinge; some outer scapulars with buff shaft-streaks at tips. Scapulars shorter than breeding. Chest golden-

buff streaked brown-grey. Rest of plumage white, inner secondaries drab, upper wing-coverts washed golden-buff. NESTLING. Skin yellow-green. Down of crown long, dark grey with light brown tips, becoming lighter with age; of nape brown with cream tips. Throat dark grey; mantle and scapulars grey-buff; back and underparts off-white. White spot on inner wing. (Valverde 1953; Moltoni 1936.) JUVENILE. Similar to adult non-breeding but feathers of head and neck shorter, scapulars not forming loosely structured coat. Head and neck paler buff, streaked dark brown. Mantle and scapulars duller brown, scapulars with wide buff shaft-streaks. Tail-feathers vaguely mottled brown. Inner secondaries brown. Shafts of primaries brown, outer primaries tinged dusky on outer webs and rear tips. Upper wing-coverts mottled brown. IMMATURE WINTER. Feathers of head and neck more elongated than juvenile, warmer buff. Feathers of mantle and scapulars of looser structure, yellow-brown, tinged rusty; scapulars with yellow-buff shaft-streaks. Wings like juvenile. FIRST BREEDING. Like adult breeding, but primaries with brown shafts, mottled brown at tips. Brown mottlings on rest of tail and wings mostly worn off.

**Bare Parts.** ADULT BREEDING. Iris chrome-yellow. Bill and bare skin at sides of face bright blue for short time at start of breeding season; about time of laying bill changes to turquoise-blue with black tip, lores to emerald green. Leg flesh-red or bright coral-red at onset of breeding, fading later through dull pink to yellow-brown (Ferguson-Lees 1959). ADULT NON-BREEDING. Iris yellow. Bill green-yellow, culmen and tip black. Lores yellow-green. Leg yellow, with green wash. NESTLING. Iris olive, tinged yellow. Bill yellow or orange-yellow with dark spot near tip. Lores green-yellow with blackish spots in front of and below eye and on forehead. Leg bright olive in front, pale yellow behind (Valverde 1953). JUVENILE. Iris yellow. Bill orange-yellow or grey-flesh, culmen and tip darker. Lores bright green, or dark brown. Leg yellow-green. (See also Moltoni 1936.)

**Moults.** ADULT POST-BREEDING. Complete; June–December. Data on sequence of primary moult lacking. Also not known whether moult of flight-feathers arrested during migration or starts on arrival in winter quarters. ADULT PRE-BREEDING. Partial; January–May. Body, wing-coverts, and inner secondaries, but not rest of wings or tail. POST-JUVENILE. Partial; July–December. Body, but not wing-coverts. FIRST PRE-BREEDING. Partial; at same time as adult pre-breeding. (Witherby *et al.* 1939.)

**Measurements.** Adults; skins (RMNH, ZMA).

| | | | | | | |
|---|---|---|---|---|---|---|
| WING | ♂ 225 | (7·95; 9) | 208–234 | ♀216 | (7·73; 9) | 209–228 |
| TAIL | | 81·1 | (4·42; 8) | 73–84 | 73·4 | (5·27; 9) | 66–84 |
| BILL | | 64·8 | (2·49; 8) | 62–70 | 61·8 | (2·49; 9) | 58–65 |
| TARSUS | | 58·0 | (3·57; 9) | 51–62 | 56·2 | (1·86; 9) | 54–59 |
| TOE | | 67·7 | (3·00; 9) | 62–72 | 66·8 | (1·79; 5) | 64–68 |

Sex difference significant for wing, tail, and bill.

**Weights.** (Moltoni 1936; Rokitansky and Schifter 1971).

| | | | | | | |
|---|---|---|---|---|---|---|
| ADULT | ♂ 285 | (37·2; 14) | 230–350 | ♀ 291 | (37·8; 9) | 250–370 |
| FLEDGLING | 225, 225, 280, 300 | | | 206, 250 | | |

Sex difference not significant.

**Structure.** Wings long and broad. 11 primaries: p9 longest, p10 0–5 shorter, p8 0–5 (usually longer than p10), p7 2–7, p6 10–15, p5 18–28, p1 *c.* 55; p11 small, *c.* 5–10 shorter than primary coverts. P7–p9 slightly emarginated outer webs. 15 secondaries (Stephan 1970). Tail slightly rounded; 12 feathers. Under tail-coverts reach tip of tail. Bill straight and slender. Leg slender with long toes, outer and inner *c.* 80% of middle, hind *c.* 55%.

**Geographical variation.** Populations of central and southern Africa described as separate race *paludivaga* Clancy, 1968; for the present not considered sufficiently distinct for separation as northern African populations poorly known and possibility of regular cline not excluded.

*Ardeola ralloides* may form superspecies with closely similar *A. grayii* (south-west Asia), *A. bacchus* (east Asia), *A. speciosa* (south-east Asia), and *A. idae* (Madagascar). These not conspecific as thought by Salomonsen (1929) as ranges show marginal overlap (Hartert and Steinbacher 1932–8).

## *Ardeola grayii* Indian Pond Heron

PLATE 36
[between pages 302 and 303]

Du. Indische Ralreiger     Fr. Héron crabier de Gray     Ge. Paddyreiher
Ru. Восточная желтая цапля     Sp. Garza cangrejera persa     Sw. Grays rallhäger

*Ardea Grayii* Sykes, 1832

Monotypic

**Field characters.** 42–45 cm (body 18–20 cm), wing-span 75–90 cm; small, almost identical in size to Squacco Heron *A. ralloides*. Closely related to latter with similar plumage pattern but dumpier silhouette, shorter wings (looking more rounded), and less skulking behaviour. Sexes alike; seasonal variation in plumage and bare part colours. Juvenile resembles non-breeding adult.

ADULT BREEDING. Head and neck unstreaked yellow-buff, latter overlaid by creamy plumes falling from browner crown. Mantle rich maroon, but duller, mottled dark brown when plumage worn, rest of upperparts pink-buff. Underparts white tinged buff; tail and wings (except brownish tips to outer primaries) pure white but, except in flight, mostly cloaked by loose scapulars. Bill basal two-thirds pale blue, yellow or orange, and tip horn-black. Orbital skin, legs, and feet green; eyes yellow. ADULT NON-BREEDING. Head and neck dark brown streaked buff, and lacking plumes, upperparts drab brown (feathers shorter when breeding); underparts white with streaked breast. Wings and tail pure white. JUVENILE. Similar to adult winter (see Plumages). Bill much darker, even black with colour restricted to base of lower mandible.

At distance, most birds indistinguishable from *A. ralloides* but darker, more vinaceous back of breeding adult surprisingly obvious in sunlight; at close range, breeding adult also separable by uniform head, mane and plumes (these streaked black in *A. ralloides*), and whitish underparts (chest strongly buff in *A. ralloides*). Differentiation of non-breeding adults more difficult, but lack of residual head mane again indicative of present species. Doubtful if juveniles separable in field, but again darker than *A. ralloides*.

Flight action and behaviour similar to *A. ralloides*, with marked agility in entering and leaving cover, and similar 'explosion' of white plumage on take-off. Gait also similar to *A. ralloides* but stance includes less upright postures related to habitat preferences. More gregarious than *A. ralloides*, roosting communally and occurring in markedly wider range of habitats. Far less secretive than *A. ralloides* and markedly tame in India. Call a high, harsh squawk.

DIMW

**Habitat.** Mainly sub-tropical and tropical, finding niche along edge of almost any kind or size of water, fresh or salt, but most commonly stagnant or slow-moving and often very small. Mainly in lowlands, but ranges up to *c.* 1500 m in India (Ali and Ripley 1968). Favours muddy margins, including mangrove swamps, and usually found near trees, on which roosts and nests. Uses jheels, flooded pits, ditches, ponds, and other artefacts freely; tolerant of human presence.

**Distribution.** Iran, southern end of Persian Gulf, eastwards to southern Baluchistan, India, south to Ceylon, Laccadives, Maldives, Andamans and Nicobars, and Burma. Meinertzhagen (1954) included Kuwait in its distribution, without details; has not been recorded there in last 17 years (SH).

**Movements.** Apparently mainly sedentary in nearest breeding area in Iran (D A Scott). In India and Pakistan, resident but shifting locally with drought and flood conditions (Ali and Ripley 1968).

**Voice.** See Field Characters.

**Plumages.** ADULT BREEDING. Head and neck yellow-buff, shading to white on chin and centre of throat. 3–4 lanceolate feathers on nape cream-white, shorter than in Squacco Heron *A. ralloides*. Mantle and inner scapulars deep purple-chestnut, much darker than *A. ralloides*, washed grey at tips; outer scapulars and elongated ornamental feathers on chest buff. Scapulars form loosely structured coat over rest of upperparts as in *A. ralloides*. Rest of plumage white, inner secondaries and upper wing-coverts washed pale buff; outer primary dusky with brown-grey shaft, 2–3 adjacent primaries faintly washed dusky at tips. ADULT NON-BREEDING. Similar to adult winter *A. ralloides*, but head, neck, and chest more heavily streaked dark brown; mantle and scapulars darker, earth-brown; outer scapulars and inner secondaries grey-buff. Rest of plumage like breeding. JUVENILE. Similar to adult non-breeding, but feathers of head, neck, and chest shorter; scapulars not forming loosely structured coat. Head and neck cream-white, streaked dark brown; chest cream-white, spotted pale brown rather than streaked dark brown. Mantle and scapulars dark grey-brown, some scapulars with dull cream shaft-streaks. Tail-feathers faintly mottled brown. Inner secondaries like mantle. Outer primaries with dark grey shafts, tinged dusky on outer webs and near tips. Upper wing-coverts mottled grey. Differs from juvenile *A. ralloides* in having the light parts paler and the dark parts duller and greyer. FIRST WINTER. Differs from juvenile in darker head and neck, streaked warm ochre not buff, streaked chest, and warmer brown mantle and scapulars. Differs from adult non-breeding in retained juvenile wing. FIRST BREEDING. Like adult breeding but with juvenile wing.

**Bare parts.** Iris yellow. Bill yellow with black tip, horny ridge of culmen, and blue base; also described as horn-black with dull orange or blue base to lower mandible (Meinertzhagen 1954); much darker in juvenile, with colour restricted to base of lower mandible. Bare skin on face duller or brighter green. Leg green; nuptial colour red or bright yellow (Ali and Ripley 1968, Scheer 1960). Data needed on any colour changes prior to breeding.

**Moults.** Presumably like *A. ralloides*; more information needed, ADULT POST-BREEDING. Complete; autumn. Primaries descendant, irregular in outer part of wing. ADULT PRE-BREEDING. Partial; body and inner secondaries, but not rest of wings. Starts March, but presumably earlier in tropical parts of range where breeding early. POST-JUVENILE. Partial; starts on head, neck and

mantle soon after fledging. Subsequent moults apparently parallel to those of adults.

**Measurements.** Mainland of Asia (Scheer 1960), adults and juveniles combined.

| | | | | | | |
|---|---|---|---|---|---|---|
| WING | ♂ 211 | (67) | 191–230 | ♀ 198 | (62) | 182–224 |
| BILL | 60·6 | (67) | 48–66 | 58·8 | (62) | 51–66 |
| TARSUS | 56·6 | (67) | 50–63 | 54·5 | (62) | 49–62 |

Baker (1929) gives range of tail as 73–84.

**Weights.** Adult ♂ Addu atoll, Maldives, 230 (Scheer 1960).

**Structure.** 11 primaries: p8 and p9 longest, often equal but either may be 1 or 2 shorter than other; p10 1–7 shorter, p7 0–5, p6 3–10, p5 10–19, p1 40–53; p11 minute. Portion of wing beyond tip of p1 relatively shorter than in *A. ralloides* (averaging 23·1% of wing length, against 28·4%). Outer web of p10 broad, p8–p10 only slightly emarginated inner webs, p7–p9 outer. Tail square; 12 feathers. Outer toe *c.* 79% of middle, inner *c.* 76%, hind *c.* 65%, hind claw *c.* 25%.

**Geographical variation.** Population of south Maldives described as separate race *phillipsi* Scheer, 1960; not recognized by Vaurie (1965).

## *Bubulcus ibis* Cattle Egret (Buff-backed Heron)

PLATE 37
[between pages 302 and 303]

DU. Koereiger    FR. Héron garde-boeufs    GE. Kuhreiher
RU. Египетская цапля    SP. Garcilla bueyera    SW. Kohäger

*Ardea Ibis* Linnaeus, 1758. Synonym: *Ardeola ibis*.

Polytypic. Nominate *ibis* (Linnaeus, 1758), Africa, south-west Asia, south Europe, North and South America. Extralimital *coromandus* (Boddaert, 1783), Asia east from Baluchistan to Australia.

**Field Characters.** 48–53 cm, wing-span 90–96 cm, ♂ averaging larger. Small, mainly white heron with distinctive heavy jowl, short neck, relatively short, pale legs and pale, stoutish bill. Sexes more or less alike, but distinct seasonal changes and ♀ separable in non-breeding plumage. Juvenile closely resembles non-breeding ♀.

ADULT BREEDING. Appears white at distance, but at close range elongated tufts of pink-buff to rich orange-buff on crown, lower throat, and mantle conspicuous. ADULT NON-BREEDING. Crown paler. Feathers of lower throat and mantle creamy-white and slightly elongated in ♂, white and hardly elongated in ♀. Bill varies from yellow or, sometimes, orange-yellow in winter to bright orange, or red at onset of breeding; bare loral area similarly changes from yellow to purple-pink; eyes from yellow to red; and legs from dark green-brown (yellowish on upper legs and soles) to dusky red. JUVENILE. Entirely white without elongated feathers. Bare parts as non-breeding adult. Immatures and non-breeding adults sometimes have grimy grey cast to plumage.

Unmistakable at close range, particularly in summer, but at distance may be confused with Little Egret *Egretta garzetta* and, in flight, with Squacco Heron *Ardeola ralloides*, with both of which it often associates. *E. garzetta*, however, has longer and slenderer neck, longer black bill,

and longer black legs with yellow feet; while *A. ralloides* has darker, thinner bill and thicker neck, and streaked head and darker body usually visible. Pale-billed white morph of Reef Heron *E. gularis* could cause confusion, but shape differences as in *E. garzetta* and behaviour usually different.

Often on dry grass, arable or even fresh-ploughed fields, rather than marsh, associating characteristically with cattle, walking between their legs and perching on their backs. Much less graceful than most other small herons: feathers hanging under chin give heavy jowled effect, and short, thick-looking neck usually hunched in to shoulder or extended awkwardly forward or diagonally upward, rather than in slender curve of *Egretta*; also, walks in search of insect prey with swaying, somewhat goose-like action. Less shy than most Ardeidae. Markedly gregarious, usually in small parties which may spread out while feeding but join up when taking flight. Wing-beats rather rapid and shallower than other Ardeidae. Perches readily in trees and moves about on twigs with some ease. Generally rather silent except when breeding.

**Habitat.** In lowland areas of warm tropical to Mediterranean climatic range; least restricted to aquatic habitats of west Palearctic herons. Walks more than wades; passes

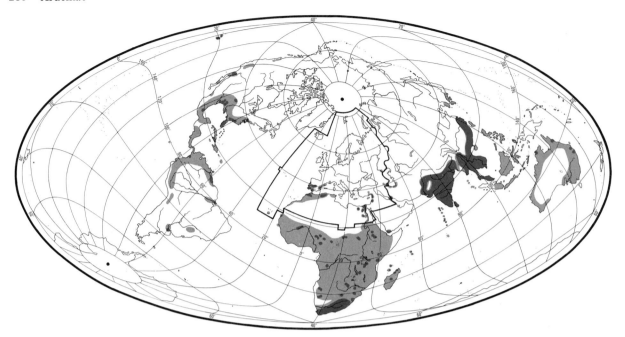

long daylight periods away from water on pastures, semi-arid steppes, arable fields, or open olive groves, often preferring fringing grasslands to swampy core of wetlands, but also freely resorting to banks of rivers, streams, ponds or lakes, ricefields, canals, tanks, lagoons, and reservoirs. Avoids coastal and marine habitats, cool or deep waters, mountains, and large unvegetated surfaces but ranges above 1700 m in East Africa. In forest areas, keeps to fringes and clearings. Generally indifferent to human disturbance, and eagerly follows plough. In some parts of range, on islets in public parks. Commonly close and continuous associate of cattle, and other livestock, and in some areas of wild game animals, up to rhinoceros and elephant. Particular herds often selected and followed. Daily cycle normally involves long morning and evening flights up to 60 km or more from feeding grounds (shifting with movement of herd) to evening watering-place and assembly area and eventual roost in trees or on ground. Flight direct, in close bands, and usually fairly low. Breeds usually below tops of medium or scrub, broad-leafed trees, or in tangle of climbing plants, or low thicket of bramble *Rubus*, treeheath *Erica arborea*, and similar shrubby vegetation.

**Distribution.** Marked extension of world range in recent years, especially from Africa to South America (probably established there early 1930s, but occasional earlier sight records), and later to Central and North America (Crosby 1972). Has also spread in eastern hemisphere, aided at least partly by introductions, to Australia and New Zealand; introduced Hawaiian Islands 1949 (Palmer 1962). Breeding distribution imperfectly known over large parts of range, especially where expanding. Considerable dispersive movements, particularly by immatures (see Move-

ments), have led to erroneous assumption of nesting in some areas in the past, while elsewhere (e.g. South America) colonies almost certainly exist which have not been recorded. Distribution outside breeding season varies greatly with water levels in many regions, and not known in detail for large areas, again especially in South America.

In west Palearctic, range expansion on much more limited scale but some, especially in west. SPAIN: expansion noted by 1961, though breeding colonies still restricted to south-west (Bernis 1961); now nesting further north, with recent colony in extreme north-east (FB). PORTUGAL: information limited, but apparently spreading, with colonies established north-east of Lisbon in 1940s and south of Oporto 1950–2 (MDE, H Wille). FRANCE: breeding attempted Camargue 1957, first successful 1969 (Hafner 1970). ALGERIA: formerly bred Lakes Fezzara and Halloula (Heim de Balsac and Mayaud 1962); little information on present position elsewhere. TUNISIA: not now breeding Lake Ichkeul (MS). CAPE VERDE ISLANDS: formerly considered accidental; 2 pairs bred 1966 and 1968 (Bannerman and Bannerman 1965; de Naurois 1969b). USSR: new colonies south of Lenkoran and in Volga delta (Dementiev and Gladkov 1951). ISRAEL: breeding from 1950, and increasing (HM).

Accidental. Iceland, England, Belgium, Netherlands, Denmark, Sweden, Hungary, Switzerland, Yugoslavia, Greece, Bulgaria, Rumania, Malta, Azores, Madeira, Canary Islands. Records in north and western Europe may refer to escapes; in Austria, various records of escapes from Vienna colony (Bauer and Glutz 1966).

**Population.** FRANCE: in Camargue, 2 pairs bred in 1968, increasing to 22 pairs 1970, 52 pairs 1973, and 98 pairs 1974

(Hafner 1975). SPAIN: no complete census, but details of colonies in Fernandez Cruz (1975) suggest over 6500 pairs; often marked fluctuations, but possibly increasing. MOROCCO: *c.* 1150 nests in 2 colonies near Tetouan; *c.* 1000 pairs near Rabat (Heim de Balsac and Mayaud 1962). TURKEY: largest known colony 30 pairs, but reports more frequent in recent years in south and may be other colonies not yet recorded (OST).

Mortality. Data available only for South Africa, where expanding rapidly. First year 37%, older birds 25%. Lower 1st-year mortality rate than in other herons studied; if typical, suggested may result in over-population in natural habitats, and so to dispersal and colonization elsewhere. Oldest ringed bird $13\frac{1}{2}$ years (Siegfried 1970, 1971*c*).

Movements. Migratory, partially migratory, and dispersive. Spanish breeders apparently disperse randomly, with no clear distinction between post-fledging and winter movements, Recoveries show internal displacements up to 265 km from colonies; also one south-east Portugal, and 6 north-west Morocco; all longer movements by immatures, with African recoveries October, December, and March–May (Fernandez Cruz 1970). Breeders of north Africa resident but not necessarily sedentary (one recovery showed 510 km displacement within Morocco); move according to feeding opportunities, and some southward movement from and along coastal fringe (Heim de Balsac and Mayaud 1962; Etchécopar and Hüe 1967); lack of oases records, and absence of winter influx to Sahel region, suggest Sahara seldom, if ever, crossed.

Movements of those breeding Near East to Caspian little known. Summer visitor to Transcaucasia and Volga delta, departing by October and reappearing late March to early April (Dementiev and Gladkov 1951); winter quarters unknown, but these birds may augment resident Iran population and account for those visiting Iraq and Kuwait. Small numbers seen December–January in Gulfs of Suez and Aqaba (Marchant 1941; Safriel 1968), and 2 small parties flew down Red Sea at *c.* 25°N in January (Elliott and Monk 1952), while small groups re-appear Gulf of Aqaba in spring and sometimes stay several weeks; but origins and directions of movement unknown. Greatest mobility by those breeding drier parts of Ethiopian Africa, with definite periodical movements related to rainfall and therefore to optimum feeding and nesting conditions (Moreau 1966; Elgood *et al.* 1973). That some transequatorial shown by recoveries Uganda and north Congo of young ringed up to 3500 km away in South Africa, though extreme cases may be examples of the erratic movements to which species owes its dynamic expansion. Almost certainly New World (where now migratory in USA) colonized from Ethiopian region rather than Palearctic, as indicated by occasional records from St Paul's Rocks, Ascension Island, St Helena, and probably Tristan da Cunha.

Food. Chiefly insects. Feeds in loosely associated groups or singly, on wet and dry land, in shallow water and on or with cattle and other large mammals. In dry season in Sierra Leone, also in vegetable gardens (Craufurd 1966), and in central Africa follows forest and savannah fires (Ruwet 1963). Commonest method of feeding: steady walk or run interspersed with downward stabs; other methods described by Meyerriecks (1960), Skead (1966), Siegfried (1971*a*).

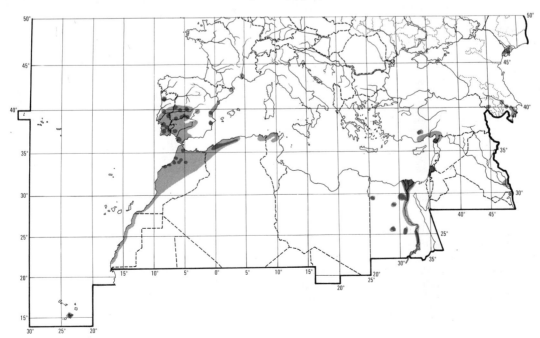

Insects especially grasshoppers (Orthoptera), with Coleoptera, Lepidoptera, Diptera, Hemiptera, and Odonata. Also arachnids (spiders), molluscs, amphibians, reptiles (lizards, snakes), and small mammals. Tick-eating records few, mainly from abdomen, flanks, and legs of mammal or from ground; back-riding probably gives vantage perch only (Bates 1937; North 1945; Beven 1946b; Holman 1946).

Grasshoppers main food in USSR (Dementiev and Gladkov 1951). In Egypt, 139 stomachs collected throughout year near Cairo (Kirkpatrick 1925) contained mainly Orthoptera especially Gryllus, mole-crickets Gryllotalpa, Liogryllus, Epacromia, and Chrotogonus; Lepidoptera larvae mainly Agrostis ypsilon; Coleoptera Pterostichus and Scleron orientale; Diptera Musca domestica; and toad Bufo regularis (both adults and tadpoles). In similar survey of 498 birds throughout Egypt, same species in approximately same proportions found (Kadry 1942). Nestling regurgitations from a colony in southern Spain contained 79·3% Orthoptera (grasshoppers, crickets, long-horned grasshoppers, mole-crickets), 15·0% other insects, and 5·7% varied, including crustaceans, Arachnida, Myriapoda, fish, frogs, skink Chalcides chalcides, lizard Psammodromus algirus, gecko Tarentola mauretanica and shrew Crocidura rullula (Herrera 1974). Grasshoppers main summer food in South Africa; nestlings fed grasshoppers and Lepidoptera (adults, pupae, and larvae) with occasionally Odonata, Diptera, Hymenoptera (wasps), spiders, amphibians (including Rana, Bufo, and Xenopus), snakes, mice, and birds (Skead 1966). Siegfried (1971b), from 250 pasture-feeding birds in South Africa, found similar wide range, most important being insects (in order of importance Lepidoptera, Orthoptera, Coleoptera, and Diptera) and earthworms Lumbricus, which together comprised 90% by weight and numbers. 98 nestlings fed mainly insects, primarily Orthoptera, adult Lepidoptera, caterpillars, and frogs; also relatively few earthworms.

**Social pattern and behaviour.** Based mainly on material supplied by D Blaker.

1. Most gregarious of west Palearctic herons. Throughout year, feeds in small, loose, nomadic flocks of up to c. 20 birds, or singly; up to several hundred may gather around locally abundant food. Flocks of mixed age and sex, size and composition varying continuously (Skead 1966; Blaker 1969a). Typically attend grazing mammals, both wild and domestic, thus significantly improving feeding efficiency (Heatwole 1965). Individuals reported defending immediate area of particular animal (Palmer 1962). BONDS. Monogamous pair-bond of seasonal duration though, arising in part from manner of pair-formation, temporary trios of 2 ♀♀ and 1 ♂ not infrequent at start of season (Lancaster 1970); multiple copulations by ♂ (Mountfort 1958) and rape or rape attempts (Lancaster 1970; Blaker 1969a) also reported. Age at which first pairs and breeds not known, but probably at end of first year (Siegfried 1966). Both sexes tend young until well after fledging. Brood-bond also established between siblings which continue to associate after leaving nest (Blaker 1969a). BREEDING DISPERSION. Highly

colonial, usually in groups of a few hundred pairs though also up to several thousand (W R Siegfried). Often in colonies with other herons, Spoonbills Platalea leucorodia, and cormorants Phalacrocorax. Early in season, ♂ frequents 1 or more pairing territories in colony area (Blaker 1969a; Lancaster 1970); defends site against own and other species (Palmer 1962). Later, pair defend nest territory of 2–3 m radius at same or different site; nests up to 6 m apart, though often more densely packed. Trespassers not tolerated closer than c. 80 cm from centre of nest (Blaker 1969a). Territories tend to become more diffuse when young start to wander from nest. ROOSTING. Communal and nocturnal; rarely solitary. Often with other herons and large waterbirds. Studied in most detail in South Africa (Vincent 1947; Skead 1966; Blaker 1969a; Siegfried 1971). Most roosts situated conspicuously in tall trees over or near water; some in reedbeds. May be in or near colony throughout year, though separate sites more often used outside breeding season. Each roost then most often occupied by 1000–2000 birds, though sometimes as few as c. 20 or, occasionally, over 10000. In area covered by Siegfried (1971e), roosts fairly evenly spaced at intervals of c. 16 km, each with own feeding area rarely more than 19 km distance. Flocks start moving towards roost c. 1 hour before sunset; often congregate at gathering points 1–10 km away (Blaker 1969a). Sometimes finally assemble at one of traditional standing grounds within 1 km from roost (Vincent 1947), then occupy roost all at once. Gathering grounds typically situated conspicuously in open; may be used chiefly where roost concealed in reeds for not common in vicinity of tree-roosts, birds flying in direct (see Siegfried 1971e). Roost occupied from just before sunset to just after sunrise. Within roost, individuals tend to occupy and defend own perches. At onset of breeding season, numbers at roost dwindle, though small numbers of non-breeders may return nightly throughout season. At all times of year, birds sometimes gather at mid-day, when cattle inactive, to loaf in flocks of up to 200 at standing grounds in or near feeding area (Vincent 1947; Blaker 1969a).

2. Virtually nothing reported from west Palearctic (Blair 1957), all substantial studies being extralimital: see Meyerriecks (1960, expanded in Palmer 1962), Skead (1966), and especially Blaker (1969a), Lancaster (1970). Apart from heterosexual interactions between pair, at or just near nest only, relationships elsewhere chiefly confined to simple antagonistic encounters, with no evidence of individual recognition even between mates. Though highly gregarious, flock integration loose; movements not highly synchronized and no true flight-calls (Blaker 1969a). Pair-formation of heron-type 2 (see Ardeinae) but behaviour of ♀ atypical; associated with conspicuous changes in colour of bare parts (Tucker 1936 and references above). Legs and bill also said to show short-term changes in colour intensity ('blushing') at moments of sexual excitement (Mountfort 1958). Pair-formation also preceded by changes in colour of plumes on crown (crest), breast, and upper back which persist throughout cycle. These plumes used in a number of displays but crest predominantly in antagonistic situations when motivational state indicated by position: fear inhibits erection of front part, aggression stimulates erection of posterior (Blaker 1969a). ANTAGONISTIC BEHAVIOUR. Away from colony, encounters relatively infrequent. Among feeding birds, mainly Supplanting-run with stiff-legged strides; if threatened bird does not give way, then both usually fly at one another and clash briefly, giving low-intensity Raa-calls. Supplanting-runs occur occasionally at standing grounds; also low-intensity Rick-rack calls from incoming birds. Little aggression at roost; perches defended by Raa-calls, Forward-displays (see below), and supplanting, especially at arrival of each new group, while Rick-rack calls common (Blaker 1969a).

Encounters most intense and complex at colony. Each bird returning to own site gives Rick-racks as Alighting-calls. Forward-displays of varying intensities most common threat in defence of site (Blaker 1969a). In full version, neck held horizontal, feathers of crest, neck, and other plumes fully erected, and bill opened and slightly lowered; usually stabs rapidly towards rival, giving Raa-call while making single forward and downward beat of partly spread wings. May then run along branches or fly and supplant opponent; among ♂♂ during pairing phase, can lead to aerial fighting. Bill-fighting following Forward-display probably most frequent between unpaired ♂♂; often vicious during pair-formation and appropriation of occupied nests, leading to wounding (Lancaster 1970). Especially during incubation, Forward-displays between birds at adjacent nests develop into formal stab-and-counter stab, see-saw ritual. Upright-display mainly of standard type (see Night Heron *N. nycticorax*); in variant, neck and head held at 45°, neck somewhat kinked, and bill pointed sharply up. Individual strongly threatened by Forward-display from immediately above may assume Withdrawn-crouch display instead of fleeing, squatting with body horizontal, neck retracted, bill obliquely raised, legs fully bent, scapular plumes slightly raised, and crest more or less depressed; then turns and leaves. Head-flick display, consisting of rapid, small-amplitude, lateral movements, more common than Withdrawn-crouch; may also be given as response to threat and in a variety of other, mainly antagonistic situations. HETEROSEXUAL BEHAVIOUR. During pairing phase, usually for period 3–4 days, ♂ becomes highly aggressive and parades up and down display site in colony, frequently performing pairing displays; may change site several times. Assumes hunched and sometimes almost horizontal posture (see A). Highly vocal, uttering modified Rick-rack and Chatter calls; also special Thonk-call when directing Forward-displays at unmated ♀♀. Latter strongly attracted, up to 10 gathering within 5-metre radius, typically perching 1–2 m above level of ♂. ♀♀ initially largely passive and silent, adopting long-necked posture while craning and peering at ♂ (see B), often perform Wing-touch display much as in *Nycticorax* and *Ardeola*. Will follow if ♂ changes site. Regardless of sex, any close-approaching birds, including watching ♀♀, threatened and supplanted by site ♂ at

C

this stage. In addition to hunched posture (A), ♂ frequently rocks from side to side while lowering almost fully spread wings (Wing-spread display). Often launches out from site in Flap-flight display; deep, exaggerated wing-beats produce loud thudding noise. Twig-shake display common: ♂ extends neck and grasps leaf or twig in bill (see C), shaking it for 1–3 s, sometimes giving Chatter call. Wing-touch display also frequent, especially when ♀♀ near. Particularly when alone, ♂ performs typical ardeid Stretch-display (much as in *A. cinerea*) with or without associated calls in stretching and crouching phases; if call given in 1st phase, sometimes repeated and each note then accompanied by slight downward bob. Modified Snap-display (termed Forward-snap) described by Lancaster (1970): with neck and crest feathers erected, ♂ extends neck and stabs, usually downwards, in deliberate and ritualized manner but without opening and snapping bill or producing any noise. Directed indiscriminately at birds both near and far; changes to Forward-threat with lunges if one flies or lands close. Eventually, in 24-hour period before pair-formation, ♀♀ themselves become more aggressive; they now approach displaying ♂ in Flap-flight and attempt to alight on his back and subdue him, only to be driven away. For several hours before pairing, such encounters reach peak as 1 or more ♀♀ in turn repeatedly alight on ♂ and attempt to Back-bite (see below); he later tends to crouch, eventually submitting and permitting one of ♀♀ to stay. Mutual Back-biting now common; form of Allopreening in which ♂ and ♀ typically run partly open bills through each others' back and neck feathers, with quivering motion, frequently while standing in opposite directions (see D) and often with necks crossed; accompanied by soft Chatter-calls. Other displays at this time include Stretch (mainly ♀), Flap-flight and Twig-shake (usually ♂). If ♀ not driven off again, pair-bond stabilizes within a few hours. Nest-building and copulation start, at same or more usually another site; move initiated by ♂, ♀ following. Throughout mating, incubating, and nestling phases, Greeting Ceremony with fully erected scapular plumes and flattened crest occurs whenever one of pair returns to nest and joins mate; often linked with Twig-passing by ♂. After brief

B

A

D

absences, and later in cycle, may consist only of a few Rick-rack calls, but more intense at other times. While still 2–10 m from nest, incoming bird typically gives Rick-rack calls in flight; answered by mate. Responses then variable but may include Roo-calling by nest-bird, while pointing head down, and Stretch-display by other. Afterwards, incomer usually crosses neck over that of mate while squatting beside it; in mating phase especially, they may next Back-bite, loud Rick-rack calls giving way to harsh Chatter-calls. Stretch-displays occurred in 73% of greetings by newly paired birds in colony studied by Lancaster (1970), both Stretch and Back-biting displays decreasing in frequency as hatching approached. Copulation at or near nest site, usually on nest if building started. Mostly little display by either sex, or preceded by Stretch-display in ♀. Soon after pairing, Stretch-display by ♂ sometimes elicits reversed mounting by ♀ (Lancaster 1970). Unlike many Ardeinae, ♀ crouches for copulation (Blaker 1969a); gapes with bill pointing diagonally up (Lancaster 1970). Mounted ♂ often seizes or nibbles ♀'s crown, nape, or neck feathers. After copulation, both usually shake body feathers and self-preen, taking no further notice of one another; in early part of mating phase, may nest-build or Back-bite together, and ♂ may Twig-shake. RELATIONS WITHIN FAMILY GROUP. Studied in detail by Blaker (1969a); see also Skead (1966), Siegfried (1973d). Young beg by giving Food-call almost incessantly. For first 5–8 days, also peck and grab weakly at any part of adult's bill, taking food mainly from floor of nest. Soon after, begging becomes increasingly stereotyped and violent: chick waves wings conspicuously, bobs head, neck, and body erratically up and down, quivers head with bill open, and attempts to seize base of parent's bill crosswise in own; later, often chases adult, buffeting it with wings. Adult tends to response by jerking head up beyond chick's reach, often giving Kok-call; passes food directly into mandibular pouch of young. Avoids older young as much as possible; after feed, if does not depart, waits out of reach until ready to feed again. Though younger wanderers still accepted into brood, parents discriminate against trespassing chicks older than 12–14 days, attacking and often killing them. From 3rd day, young defend nest against strange adults with Forward-threat, lunging and calling; also sometimes Head-flick. If attacked, show appeasing displays: Facing-away occasionally, Nest-crouch frequently. At first, also weakly threaten parents but rapidly less and less and not after c. 17th day; older young readily distinguish them, at times even before they alight (Skead 1966). Chicks start to wander around nest tree as soon as brooding over, each defending favourite perch 2–3 m from nest from strangers; when over c. 30 days, spend most of time away from nest, returning only to be fed. Threaten or peck trespassing chicks but associate with them elsewhere, gathering in flocks of up to 200 on ground below nests when c. 40 days old and still semi-dependent on parents. Siblings over age of c. 8 days peaceful and highly sociable when alone. Individual recognition marked after 15th day, and birds seek out each other's company; frequently give Chirp-call in greeting especially when re-uniting at nest. Competition for food intense at feeding time; then, from c. 13th day (more or less coinciding with end of parental guarding phase), one-sided pecking, fighting, and chasing frequent. Broods usually of 1–2 only, 3rd chick rarely surviving after 16th day. From day 10–15, chicks' bills become completely black until c. 30th; as young strongly induced to peck at yellow colour of adult's bill, black may help to inhibit this response between siblings. Many changes in behaviour, in both adults and young, coincide with period of maximum energy demand in latter (see Siegfried 1973d).

(Figs A–D after Blaker 1969a.)

Voice. Throaty, somewhat husky, and unmusical; calls mostly simple and subdued (Skead 1966). Relatively silent away from colony or roost; no flocking calls and no special flight-call though quiet, abbreviated version of Rick-rack call occasionally given by flying bird, especially when moving from colony (Blaker 1969a).

Most calls heard during breeding season at colony; often occur in variety of situations and undergo considerable modification during red-billed phase at beginning of breeding cycle so as to become almost unrecognizable, especially in unpaired ♂♂. In first few days after pairing, all calls muffled and husky in both sexes (Blaker 1969a). Both ♂ and ♀, especially in pairing phase, produce loud, thudding sounds with wings during Flap-flight display. No bill noises mentioned by Blaker (1969a) but clicking reported during Snap-display by A J Meyerriecks (see Palmer 1962). (1) Rick-rack call. Variable, but normally a harsh, double, croaking 'RICK-rack' with 1st syllable louder and higher pitched. Most common call, given throughout year. At colony, Alighting-call by bird flying in to site, whether mate present or not; typical of Greeting Ceremony between pair, bird at nest answering. In red-billed phase, quieter, low-pitched, muffled, hoarse version ('RUK-rok') in both sexes. Normal Rick-rack occasional at standing grounds, when uttered by arriving birds, but frequent at roost where subdued version provides continuous background of low gobbling notes. Abbreviated version also heard infrequently on feeding grounds when associated with incoming birds. (2) Raa-call. Brief, harsh 'raa'. Threat-call; characteristic of Forward-display in defence of nest-site; also occurs at roost in defence of perch. Subdued version given with Supplanting-run at feeding place. (3) Thonk-call. Snorting, gulping 'thonk', soft and muffled. Special version of Threat-call uttered only by red-billed ♂ in pairing phase, especially to approaching ♀♀. (4) Kraah-call. Harsh, drawn-out 'kraah'; similar to Raa-call but much louder and longer. Defensive-threat call, when predator close to nest, especially after young hatched. (5) Kok-call. Low-pitched 'kok'; uttered singly, or in series at up to 3 per s for minutes on end. Alarm-call, readily elicited by (e.g.) man, especially at colony; also given by parent in response to violent begging or larger young. Described as 'krok . . .' by Skead (1966) and said to be rare; intermittent, soft 'chok . . .' by birds collecting sticks and when alone at nest probably version of same. (6) Chatter-calls. 3 distinct versions. (a) Nasal Chatter: phrase of 5–13 notes, descending in volume and lasting 1–2 s. Given only by unpaired ♂ as Alighting-call or when returning on foot through branches to territory; also in response to presence of onlooking ♀♀, especially as accompaniment to Twig-shake display. Audible at least 40 m. (b) Soft Chatter: quieter and of uniform volume. In ♂, replaces nasal version after pair-formation; also given by both sexes during mutual Back-biting display, mainly in mating phase only. (c) Harsh Chatter. Loud, harsh, 'kakakakakaka', louder, slower, and longer than (b). Occurs in Greeting Ceremonies, following loud Rick-rack calls. (7) Ow-roo call. Typical Stretch-call of unpaired ♂: soft 'ow' given 1 or more times in 1st phase of Stretch-display, followed by soft, crooning 'rooo' in 2nd; either note may be weak or absent. Replaced by harsh 'aah' during Stretch-displays performed when red bare-part colour reverts to yellow at end of mating phase; similar in tone to Raa-call but more drawn out. Roo-call only occasionally uttered by paired ♀ during Stretch-display; harsh version also by bird of either sex at site with Rick-rack calls in answer to Rick-racks of incoming mate.

CALLS OF YOUNG. Combined calling from many nests produces remarkably subdued effect of continuous chattering (Skead 1966). Food-call of tiny young continuous, high-pitched 'ziz ZIT', sometimes also of 1 or 3 syllables (Skead 1966; Blaker 1969a); becomes harsher and more raucous in older young and, at high intensity, replaced by harsh squealing. Other calls include (1) raspy squeak as threat with Forward-display, developing into adult-like Raa-call at *c.* 50 days; (2) Fear-squeals when attacked; (3) 'eeeh' Discomfort-call (from just before hatching until 10th–12th day); (4) Chirp-call, probably serving in recognition and appeasement between siblings (Blaker 1969a). Chirp-call consists of 2 to *c.* 15 squealing and croaking notes; variable but usually a series of ascending notes followed by descending series. First given on 3rd day, becomes longer and more complex over next 20. Between 25th and 30th day, develops pattern of adult Rick-rack – Chatter sequence in Greeting Ceremony. Also high-pitched, repetitive Alarm-scream ('zeekle zeekle . . .') described from older young (Skead 1966).

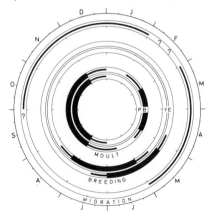

**Breeding.** SEASON. See diagram for Mediterranean area; no apparent variation within this. SITE. Reedbeds, bushes, low trees in water, and tall trees with nests to 20 m above ground; wide variety of trees used but those in leaf preferred. Colonial; from a few pairs, often with other species, to many thousands; nests sometimes close or touching, sometimes 6–8 on a branch, and up to 100 in a tree; colony may move short distance within same wood from one year to next. Nest: pile of reeds, twigs, or branches; can be flimsy or solid. 20–45 cm diameter,

12–25 cm high, shallow cup 7–12 cm deep. Twigs used up to 46 cm long in 32 tree nests just after laying, average 156 (range 110–280) sticks per nest (Siegfried 1971*f*); lower mass of nest tends to be of longer (up to 90 cm), more robust twigs, with thinner green twigs forming upper part (Skead 1966; Blaker 1969a); occasionally thin grass lining. Building: by both sexes, mainly ♀ using material brought, one item at a time, by ♂; first items placed haphazardly in tree fork, with *c.* 30% falling and not recovered; after 2 days, 20–40 sticks accumulate, thereafter builds by pushing twig into structure, withdrawing and re-inserting several times; nest ready for laying in 5–6 days but then only one-third final size and material added throughout incubation and long after hatching (Skead 1966; Blaker 1969a; Siegfried 1971*f*). EGGS. Broad oval, somewhat pointed towards one end; almost white, tinged pale green or blue. 45 × 34 mm (41–53 × 32–36), sample 80; calculated weight 28 g (Schönwetter 1967). Clutch: 4–5 (3–9). Determinate layer (Blaker 1969a). Normally one brood, though 2, perhaps 3, reported (Witherby *et al.* 1939; Rencurel 1972). Replacement clutch laid if first lost before complete (Siegfried 1971*f*). Eggs laid at 1-day intervals, sometimes 2. INCUBATION. 23·7 days (range 22–26), sample 20 eggs (Blaker 1969a). By both sexes changing over 3–4 times a day. Starts with 1st or 2nd egg; hatching asynchronous, at 1-day intervals in 12 nests, 2-day interval in 1 nest (Blaker 1969a). Eggshells dropped over side of nest. YOUNG. Semi-altricial and nidicolous. Cared for and fed by both parents. Not left unattended until *c.* 10 days old. Fed by regurgitation, into nestling's mouth while small, then into nest. While young under 10 days, each parent makes 1 food-collecting trip per day, though young fed several times per day. Thereafter single young fed average 1·5 times per day, twins 2·5 times per day (Blaker 1969a). Leave nest and stand on nearby branches at *c.* 20 days. FLEDGING TO MATURITY. Fledging period 30 days; become independent *c.* 15 days later. First breeding at 2 years, though 1 recorded (Rencurel 1972). BREEDING SUCCESS. No data from west Palearctic. In one colony (61 nests), Cape Province, South Africa, 17·6% egg loss mainly due to falling out of nest and attributed to laying in incompleted nests (Blaker 1969a). In 3 colonies (400–500 nests), Cape Province, South Africa, 34% eggs laid produced fledged young; main losses due to eggs falling out of nests and starvation in first 2 weeks after hatching (Siegfried 1972*b*).

**Plumages** (nominate *ibis*). ADULT BREEDING. Feathers of crown, chest, and centre of mantle loose, hair-like, greatly elongated, those of mantle nearly reaching tip of tail; ginger-buff. Rest of plumage, including wings and tail, white. ADULT NON-BREEDING. White, feathers of crown paler than breeding; in ♂, feathers of chest and centre of mantle tinged cream-buff, slightly elongated; in ♀, white and of normal structure. NESTLING. Skin olive-green at hatching, grey later; down white, long, shorter on wings, soon pushed out by developing feathers; nape, throat, and centre of back bare (Skead 1966). Completely feathered at 7 weeks, wings and tail full-grown at 8 weeks (Siegfried 1971*d*). JUVENILE. Like adult ♀ non-breeding, but crown also white, becoming slightly tinged cinnamon at 10 weeks (Siegfried 1971*d*).

**Bare parts.** ADULT. Most of year: iris, bill, and lores yellow; leg dark green, foot darkest, tibia and sole yellowish. 10–20 days before egg-laying: iris and bill red; lores ruby-magenta; leg dusky red. Red colours fade during pair-formation and iris gets yellow inner ring and red outer, bill becomes yellow at tip shading to vinous-pink or scarlet at base. 1–4 days before egg-laying: iris, bill, lores and leg yellow. Irregularities in seasonal shift of colours account for great variability. NESTLING. Iris pale yellow. Bill and lores at hatching pale flesh or horn (also described as green), darkening to black at 5–10 days, becoming gradually yellow during 2nd and 3rd month. Leg olive-green in front, pale yellow-green behind, becoming black before fledging. JUVENILE. Iris, bill, and lores yellow; leg black, dark green later. (Skead 1966, Blaker 1969*a*; see also Palmer 1962, Tucker 1936.)

**Moults.** ADULT POST-BREEDING. Complete; July–November, occasionally December. Either p1–p6 descendant, p7–p10 irregular (p10 often before p7), or sequence of primaries irregular from start. Ornamental feathers partially or entirely lost before start of wing moult. ADULT PRE-BREEDING. Partial. Probably restricted to development of ornamental plumes before breeding; February–May. Timing of moult reversed with the season in South Africa, undescribed for tropical populations. POST-JUVENILE. In South Africa, buff crown feathers appear at 5 months, moult of body starts at 6, mainly completed at 9, axillaries and lesser wing-coverts retained to *c*. 18, middle coverts moulting with body. Witherby *et al.* (1939) supposed moult at 2–6 months in Palearctic birds; confirmation needed. Ornamental plumes develop in spring of 2nd calendar year; at same time, wing and tail moult starts, probably arrested when birds breed. Sequence p1–p6 descendant, irregular afterwards. During next autumn and winter, wing apparently moulted again (1st post-breeding moult, sometimes overlapping later stages of post-juvenile primary moult). Full details in Stresemann and Stresemann (1966) and Siegfried (1971*d*).

**Measurements.** Data from Vaurie (1963), but from Witherby *et al.* (1939) for bill and middle toe.

| | ♂ | | | ♀ | | |
|---|---|---|---|---|---|---|
| WING | 253 | (20) | 241–266 | 248 | (20) | 240–258 |
| TAIL | 87·5 | (20) | 79–93 | 86 | (20) | 74–93 |
| BILL | – | (12) | 52–60 | – | (–) | 52–58 |
| TARSUS | 77 | (20) | 70–85 | 76 | (20) | 70–81 |
| TOE | | | 71–81 | | | 71–81 |

Wing averages shorter in Ethiopian populations than in Palearctic (Salomonsen 1934; Clancey 1968*b*): ♂♂ Africa 253 (4·92; 26) 245–265, Palearctic 261 (1·92; 5) 258–263; ♀♀ Africa 243 (8·66; 10) 234–260, Palearctic 251 (2·88; 5) 247–255. Sexes differ significantly in wing length.

**Weights.** America: ♂ 311; 2 ♀♀ 304; range of 4 unsexed 300–400 (Palmer 1962). South Africa: unsexed adult 345 (4) 325–387; nestlings reach *c*. 350 at 40 days (Siegfried 1973*d*).

**Structure.** Wing long and broad. 11 primaries: p8 and p9 longest, p10 0–8 shorter, p7 0–6, p6 10–20, p5 25–45, p1 *c*. 80. P7–p9 slightly emarginated outer webs, p9 to p10 inner. 15 secondaries (Siegfried 1971*d*). Tail square, 12 feathers; under coverts reach tip. Bill short, high at base; prominent nasal groove. Leg slender with long toes, inner *c*. 80% of middle, outer *c*. 85%, hind *c*. 65%, hind claw *c*. 30%. Claws long and heavy, especially hind claw.

**Geographical variation.** *B. i. coromandus* differs from nominate *ibis* in buff colour on head spreading to cheeks and throat, ornamental feathers more golden, bill and tarsus averaging longer, and more of tibia bare (Vaurie 1963). Colour of ornamental plumes somewhat variable in both forms and occasional specimens similar. Racial identification hardly possible in non-breeding plumage. African populations tend to be shorter-winged and have been named *ruficrista* Bonaparte, 1855, but variation probably clinal with apparently much overlap; therefore naming as separate subspecies not warranted. Similarly, slightly smaller population of Seychelles with slightly more golden plumes, sometimes named *seychellarum* (Salomonsen, 1934), considered to belong to nominate *ibis* (for discussion see Vaurie 1963; Clancey 1968*b*; Benson and Penny 1971).

## *Egretta gularis* Western Reef Heron

PLATE 38
[between pages 302 and 303]

DU. Westelijke Rifreiger    FR. Aigrette des Récifs    GE. Riffreiher
RU. Западная цапля рифа    SP. Garceta sombria    SW. Revhäger

*Ardea gularis* Bosc, 1792

Polytypic. Nominate *gularis* (Bosc, 1792), West Africa; *schistacea* (Hemprich and Ehrenberg, 1828), Red Sea east to India.

**Field characters.** 55–65 cm, wing-span 86–104 cm. Medium-sized, rather thick-billed, and slightly clumsy-looking egret, occurring in 2 colour morphs with many intermediates. Sexes alike; no obvious seasonal variation. Juvenile separable. 2 races, adults distinguishable in field.

WHITE MORPH. Wholly white or with cream wash on elongated feathers in adult; white, often with some mouse-brown feathers in juveniles. DARK MORPH. Wholly dark slate-grey, with white chin and throat in adult (bluer in *schistacea* and often faded in worn plumage); mouse-brown with chin and throat paler in juveniles, and pale brownish-slate in immatures. Bill of white morph brown in nominate *gularis*, yellow in adult *schistacea*, dull brown in younger birds; in dark morph, bill brown in nominate *gularis*, greenish-yellow (with dusky marks along grooves and mandible edges) in *schistacea*, dull grey-brown in younger birds. Legs of white morph dark olive-green, of dark morph black; feet respectively sulphur-yellow and dark yellow-horn with lemon soles; all parts duller, even brown in young birds. Eyes yellow. Intermediate birds variable, some apparently dark birds suddenly showing white quills in flight; a few individuals almost black.

At distance, impossible to separate from *E. garzetta*, which has rare dark morph; at close range, identification

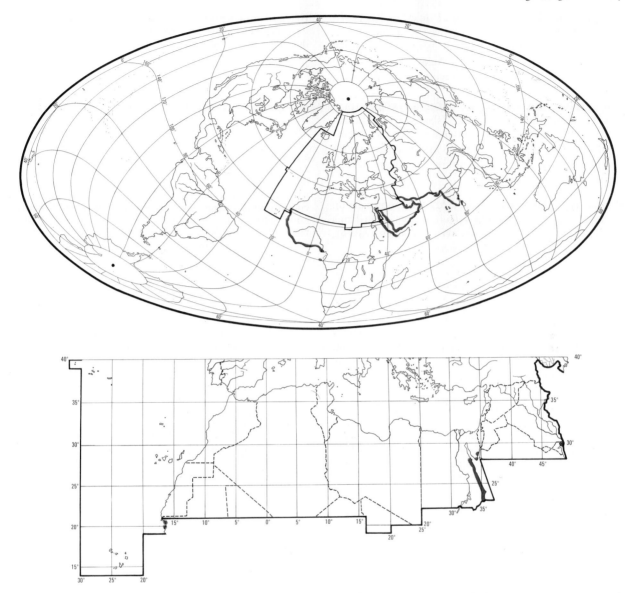

(especially when close comparison possible) can be safely based on thicker, never black bill, less graceful character, dull legs, and behaviour.

Flight similar to *E. garzetta* but action a little slower and less buoyant. Gait as *E. garzetta* in enclosed areas but can appear sluggish; in open habitats, moves more freely. Occurs in smaller groups than most smaller herons and frequently solitary along barren coasts. Food obtained by sudden lunging rushes through shallows or up ditches, sometimes associated with wing-screening as in Black Heron *E. ardesiaca* of Africa. Usually silent but gives a short harsh squawk when disturbed.                    DIMW

**Habitat.** Tropical and, over much of range, exclusively coastal, preferring rocky and sandy coasts and reefs, but also sometimes saltmarshes, mudflats, estuaries, tidal creeks, and lagoons. Wades in rock-pools or shallow surf; usually solitary. Breeds colonially in mangroves and other trees or scrub, including low growth of *Salicornia*, on inshore islands, sometimes in large trees on dry ground, and in some African areas inland behind mangrove belt (Bannerman 1930). Frequently away from cover and often trustful of man; recorded perching on masts of moored boats. Flight mostly low.

**Distribution.** Breeds southern extremities of west Palearctic. MAURITANIA: breeds on several islands, Banc d'Arguin (de Naurois 1969*a*). EGYPT: breeds Red Sea coast, including Sinai (Meinertzhagen 1954); few definite records but nested Nabq, on eastern coast in 1970 (HM).

IRAQ: breeding Fao (Ticehurst *et al.* 1922) and believed still resident (PVGK). KUWAIT: has bred on various islands, including Kubbar, Bubiyan, Warba (Ticehurst *et al.* 1925, 1926), and more recently Auha (Dickson 1942); still common Kuwait Bay, but no recent proof of nesting (SH).

Accidental. Azores: specimen 19th century (Bannerman and Bannerman 1966). Cape Verde Islands: 2 seen, March and April 1924 (Bannerman and Bannerman 1968). Spain: specimen, June 1970 (see *Ardeola* 1971, **15**, 103–10 for discussion of this and possible 1956 record). Israel: several seen, Gulf of Aqaba (HM).

**Population.** MAURITANIA. Banc d'Arguin: *c.* 900 pairs on 7 islands 1959–65 (de Naurois 1969a); counts on 4 islands suggested no marked change (Westernhagen 1970b); 562 pairs 1973 (Duhautois *et al.* 1974).

**Movements.** Dispersive; perhaps some resident. In Mauritania, degree of movement by Banc d'Arguin population of *E. g. gularis* unknown, but probably some dispersal since visits Port Etienne (Sarro and Pons Oliveras 1966). Outside breeding season, *E. g. schistacea* spreads along coasts of Persian Gulf, Arabian and Red Seas, and Gulfs of Suez and Aqaba, and immatures similarly scattered in all months (Meinertzhagen 1954; Bulman 1944), but, in absence of ringing, not known how far individuals disperse from colonies. Inland records in Africa of this coastal species need treating with caution due to possible confusion with dimorphic *E. garzetta*; such records from Kenya established by specimens (Owre 1967a), but Niger inundation zone occurrences (Guichard 1947) not thus confirmed.

**Food.** Probably chiefly fish, crustaceans and molluscs. Feeds mainly in shallow water, stalking slowly with sudden rush forward to stab prey. 2 from Port Said and one from Hurghada contained respectively 41, 25, and 30 fish approximately 10 cm long (Meinertzhagen 1954). In India, mudskipper *Periophthalmus* regularly taken by adults and nestlings (Ali and Ripley 1968).

**Social pattern and behaviour.** Based on material supplied by R K Murton.

1. Gregarious and solitary. At all times of year feeds alone, isolated or in scattered groups spaced along shore, each individual maintaining feeding territory. BONDS. Monogamous pair-bond, probably only of seasonal duration; light and dark morph birds often pair. Both parents tend young. BREEDING DISPERSION. Colonial, densely in small groups of 5–20 pairs, sometimes up to *c.* 100; occasionally solitary where population low. At start of season, ♂ establishes pairing territory which later becomes breeding territory extending in circle of *c.* 1 m from nest. ROOSTING. Communal. In tropics, where breeding season extended, often at colony, roost elsewhere apparently not required. Near coast, may roost in trees (e.g. mangrove swamps) or on rocky headland, cliff, or island. Regular nocturnal roosting, however, less evident than in other *Egretta* since often feeds according to tides and at night. In closely related Eastern Reef

Heron *E. sacra*, birds frequent resting areas near shore, waiting for falling tide to uncover feeding grounds; individuals leave at different times according to situation of feeding territory, last leaving up to 1 hour later than first (Recher and Recher 1972; Recher 1972).

2. Little published information. Display plumage like that of Little Egret *E. garzetta* and behaviour may be similar, though pair-formation evidently different (heron-type 2). ANTAGONISTIC BEHAVIOUR. Judging from scattered observations by R K Murton, that shown during encounters on feeding grounds similar to *E. sacra* (Recher 1972), including (1) typical Upright-display (also assumed by alert bird); (2) Forward-display; (3) supplanting attack by walking or running; (4) chasing; (5) fighting (opponents fly at each other, rising slightly into air, and jab with bills while raising crown feathers); (6) Withdrawn-crouch display in appeasement. Behaviour at colony basically same. HETEROSEXUAL BEHAVIOUR. Territory selected by ♂ who builds stick platform which first serves as display site and later forms basis of nest, defended by pair. Not known how ♂ initially advertises himself and site. At first highly aggressive, defending site with Forward-display; interspersed with Snap-display in which neck extended downwards to 45° and held still while single Bill-snap given before being briefly retracted and process repeated (R K Murton). In next phase, Stretch-display used to attract ♀♀ to site: closely similar to that of Snowy Egret *E. thula* (Meyerriecks 1960): all plumes erected, especially scapular and pectoral, and head pumped vigorously up and down. Eventually, one of attracted ♀♀ associates with ♂ and helps to repel others which approach closely. Once accepted, pair-bond strengthened by mutual Billing (in which ♂ and ♀ hold and fondle each others' bills, using movements which seem akin to stick mandibulation), Allopreening, Stick-passing, mutual Stretch-displays, and mutual Stroking-display (in which they face and rub crossed necks together, or stand side by side and rub heads together). Stretch-display also basis of Greeting Ceremony when pair meet at nest, including change-over during incubation and care of young. Copulation on nest; usually associated with Allopreening and Billing. RELATIONS WITHIN FAMILY GROUP. No information.

**Voice.** Normally silent, except for short, throaty croak when disturbed or supplanting rival (Ali and Ripley 1968). At breeding colonies little studied, but deep cries uttered in greeting (Dragesco 1961) and bill-snapping noted (R K Murton).

CALLS OF YOUNG. When small cheep and squeal; when older utter mournful cries (Dragesco 1961).

**Breeding.** SEASON. Red Sea: June–August; newly-hatched young found first week July (Meinertzhagen 1954). West Africa: prolonged, eggs from end April to end September, exceptionally October; young in down found once in early November; cycle may vary from year to year (de Naurois 1969a). SITE. In low shrubs and trees, occasionally among boulders; 2–3 m above ground in shrubs. Usually colonial; sometimes with other herons. Previous year's sites re-used. Nest: substantial platform of twigs with shallow cup; *c.* 30 cm diameter, 15–20 cm high, sticks up to 35 cm used, together with seaweed in shoreline sites (Meinertzhagen 1954). Building: by both sexes, but mainly ♀ with ♂ collecting material. EGGS. Oval, smooth; pale blue-green. Mean of 50 Indian eggs 45 × 34 mm (Ali and Ripley 1968); mean of 6 African eggs 47 × 34 mm (45–49 × 34); calculated weight 29 g (Schönwetter 1967). Clutch: 2–3(4), West Africa (de Naurois 1969*a*); 3, Red Sea (Meinertzhagen 1954). INCUBATION. 23–26 days (de Naurois 1969*a*). By both sexes. Begins with 1st egg; hatching asynchronous. YOUNG. Semi-altricial and nidicolous. Cared for and fed by both parents. Food regurgitated. FLEDGING TO MATURITY. Fledging period not known. Age of independence and maturity not known. BREEDING SUCCESS. No data.

**Plumages.** Polymorphic; dark and white morphs with intermediates in both races. ADULT BREEDING. WHITE MORPH. White with ornamental feathers as Little Egret *E. garzetta*: 2 long strap-like nape feathers, scapulars developed to coat of 'aigrettes' reaching beyond tail, and tuft of loose and lanceolate feathers on chest. DARK MORPH (subspecies *gularis*; for *schistacea* see Geographical Variation). Generally slate-black. Chin and upper throat white. Crown, nape, aigrettes, and plumes on chest dark blue-grey with green gloss in some lights. Lower breast and belly slightly tinged brown. Primaries dark grey on inner web, dark blue-grey on outer. Many have white patch in wing, consisting of white primary coverts or, more rarely, bastard wing, often differently developed in either wing; these wing patches also present in juvenile and immature plumages. Intermediates between white and dark morphs exist which are mainly white with scattered dark feathers or mainly dark with some white feathers, usually on crown. ADULT NON-BREEDING. Like breeding, but without elongated nape feathers; scapulars shorter and not aigrette-like. NESTLING. Not described for nominate *gularis*. In *schistacea* down of crest white, hair-like, 25 mm long; side of face and neck pale grey, extending to upperparts; wings, flanks, and underparts paler; skin grey (Meinertzhagen 1954). Nestlings of white morph probably white. Reichenow (1900–01) referred to a case of whitish nestlings which developed to dark adults in captivity. JUVENILE. Lacks ornamental feathers in both morphs. WHITE MORPH. White, at least in some cases (BMNH), so not generally pale mouse-brown as claimed by Meinertzhagen (1954). Probably more often mottled grey than white morph adults, as is regular in related Eastern Reef Heron *Egretta sacra* (Amadon 1953). DARK MORPH (nominate *gularis*). Pale brown-grey; chin and throat white; rest of underparts mottled off-white, centre of belly sometimes wholly white. Outer webs of primaries less blue than adult, margins tinged brown. IMMATURE. Feathers of chest slightly elongated in both morphs, aigrettes partially developed, but not as long as adult. WHITE MORPH. White. DARK MORPH. Crown and mantle slate-grey, paler than adult. Feathers of chest and aigrettes grey, not glossy dark blue-grey. Some wing-coverts and outer primaries retained from juvenile, brownish. Information needed on development of plumage in this species.

**Bare parts.** ADULT. Iris pale yellow to silver-grey. Bill horn-brown, lower mandible paler with yellow wash, sometimes yellow or orange-yellow; latter probably breeding colour but data lacking. In *schistacea*, more often yellow, particularly in white phase. Bare skin of face dark olive or yellow tinged olive; apparently lighter in white phase. Leg green-black, lower third often with yellow markings, toes yellow. NESTLING. Iris green-grey at hatching, later almost white. Bill pale yellow with black spots at tips of both mandibles. Leg brown-green tinged yellow. JUVENILE. Iris pale yellow. Upper mandible brown, lower yellow-flesh. Orbital skin yellow-green. Leg black, lower half of tarsus and toes green (Sharpe 1898).

**Moults.** Presumably as in *E. garzetta* but full data lacking. Juvenile starts moult on crown and mantle, primaries begin when plumes on chest growing. Inner primaries moult descendantly in adult and juvenile, outer irregularly.

**Measurements.** Skins, adult and juvenile, sexes combined (RMNH; Serle 1965; Hiraldo 1971; Bernis 1971*b*; Eisentraut 1973). Wing 265 (11·9; 23) 244–285; tail 91·4 (7·12; 11) 82–101; bill 83·8 (3·30; 21) 79–89; tarsus 89·8 (3·45; 19) 82–94; toe 66·0 (4·47; 5) 60–71. Strongly variable in all measurements. Vaurie (1965) gave slightly different value for wing, 10 ♂♂ averaging 277. *E. g. schistacea* larger: ♂♂ wing 288 (10) 272–311; bill 98 (10) 94–103; tarsus 104 (10) 92–116 (Vaurie 1965).

**Weights.** Adult ♂ nominate *gularis* 400, juvenile ♀ 350 (Eisentraut 1973).

**Structure.** 11 primaries: in *gularis* p8 and p9 longest, p9 occasionally slightly shorter, p10 1–11 shorter, p7 4–6, p6 17–20, p5 33–36, p1 76–89 (no data for *schistacea*); p11 minute. Outer webs of p8 and p9 and inner of p9 and p10 slightly emarginated. Tail square; 12 feathers, outer sometimes projecting beyond rest. Bill long, slightly heavier than *E. garzetta*, higher at base. Tarsus shorter, 106 (4·67; 19) 97–117% of bill against 122 (6·94; 37) 105–142% in *E. garzetta*. About half of tibia bare. Claws short.

**Geographical variation.** *E. gularis schistacea* averages larger than nominate *gularis* (see Measurements); dark morph adults paler, more ashy-blue, less slate than *gularis* (Vaurie 1965). Juveniles also paler, mouse-brown (Meinertzhagen 1954). Proportion of morphs varies geographically. Dark morph greatly predominant in *gularis* but white locally well represented, e.g. São Thomé (Amadon 1953), Cameroons (Good 1952). White morph generally more numerous in *schistacea*, predominant in north Red Sea, but dark more numerous in south. In Karachi, 80% dark, 14% white, 6% intermediate; in Bombay, 60% dark, 40% white (Meinertzhagen 1954).

**Recognition.** White phase closely similar to *E. garzetta garzetta* and it has been suggested that *E. garzetta* and *E. gularis* may be conspecific (Berlioz 1959). *E. gularis* usually identified by following combination of characters: bill brown or yellow, not jet-black (but juvenile *E. garzetta* has brown bill); bill rather heavy at base, quite stout towards tip; tarsus/bill ratio below 106 (Hiraldo 1971; Bernis 1971*b*). Melanistic *E. garzetta* equally hard to separate from dark morph *E. gularis*. More information on separation of species needed.

# *Egretta garzetta* Little Egret

PLATE 39
[facing page 303]

Du. Kleine Zilverreiger    Fr. Aigrette garzette    Ge. Seidenreiher
Ru. Малая белая цапля    Sp. Garceta común    Sw. Silkeshäger

*Ardea Garzetta* Linnaeus, 1766

Polytypic. Nominate *garzetta* (Linnaeus, 1766), south Europe, south Asia, north-west Africa, Cape Verde Islands, east and south Africa. Extralimital: *nigripes* (Temminck, 1840), Java and Philippines, east to New Guinea; *immaculata* (Gould, 1846), north and east Australia; *dimorpha* Hartert, 1914, Madagascar, Aldabra, and adjacent Assumption.

**Field characters.** 55–65 cm (of which body forms only half), ♂ averaging larger; wing-span 88–95 cm. Fairly small, white heron with long, thin neck, long, black bill and legs, and contrasting pale feet. Sexes alike and seasonal differences confined to crest and scapular plumes and bare-part colours; juvenile similar apart from plumes.

ADULT BREEDING. Immaculate white with elongated plumes on upper breast, mantle, and scapulars, last forming loose train; 2 long crest feathers up to 16 cm long hang down back of neck. ADULT NON-BREEDING. As breeding, but with fewer, rather shorter elongated plumes on mantle and scapulars, and no long crest feathers. Bill and legs always black and eyes yellow, but bare loral area and orbital ring change from green-grey in most of year to orange in breeding season, and feet similarly from yellow to orange (conspicuous in flight). JUVENILE. Like non-breeding adult but without elongated plumes, which start to appear on upper breast and scapulars in 1st winter.

Sometimes confused with Cattle Egret *Bubulcus ibis*, which looks all-white at distance, particularly in non-breeding plumages, and in flight with Squacco Heron *Ardeola ralloides*, which has all-white wings but tawny-buff body; but both these stockier, with shorter, thicker necks and shorter, coloured bills and legs. White morph of Reef Heron *E. gularis* much more difficult to distinguish, being of same size, with similar plumes (though shorter on head), all-white plumage, and dark legs with yellow feet; but has thicker bill (grey-brown in West African race *gularis* and yellow in Red Sea race *schistacea*), less graceful character, and duller, dark olive-green legs. Rare melanistic Little Egrets have grey to dark slate plumage, with or without white chin and throat, and are even more difficult to distinguish from normal dark *E. gularis*, except by thinner, black (not yellow or grey-brown) bill and paler, purer yellow feet. Also possibility of occurrence in Europe of Snowy Egret *E. thula*, American counterpart of *E. garzetta*; indistinguishable in non-breeding plumages and in summer differing only in having longer crest without 2 elongated plumes, and yellow on leg extending further up back (G J Oreel, P J Grant, D I M Wallace).

Rarely away from (usually fresh) water, frequenting marshes, floods, irrigated areas, and shallows at edges of lakes, lagoons and rivers. Commonly seen in parties or small flocks and frequently associates with other smaller gregarious herons. Breeds and roosts socially in trees and climbs agilely on twigs and in reeds. Active in feeding, rushing to and fro, sometimes running through shallow water with wings raised. Flight leisurely, interspersed with occasional glides, on rather rounded wings, with head retracted and feet projecting well beyond tail. Usually silent except at colonies.

**Habitat.** Margins of warm water in middle and low latitudes, both coastal and inland, usually in lowland but up to 2000 m in Armenia. Prefers shallow lakes, pools and lagoons, and gently flowing rivers or streams: also brackish estuarine and saline coastal waters, not always sheltered. Partial to floodlands and ephemeral or fluctuating water bodies, including saltpans, ricefields, and irrigated areas, either in open country or among scattered trees or forest. Favours generally open, unvegetated places or low herbage and open water with little floating vegetation, but for nesting chooses tall trees (deciduous or conifers), scrub such as willow *Salix* or tamarisk *Tamarix*, and sometimes reeds or other dense stands of swamp vegetation; occasionally on rocks or cliff ledges and cavities. Less often than most herons in close cover; sometimes in dry fields even following cattle like Cattle Egret *Bubulcus ibis*, and occasionally riding buffaloes (Dementiev and Gladkov 1951). Rarely far from water. Sometimes on inshore reefs or projecting rocks.

Walks and runs actively and wades to full depth; flies freely 20 km or more between nest-site and feeding area, usually at no great altitude. Habitat outside breeding season not significantly different. Tolerant of human presence except where persecuted.

**Distribution.** Marked declines and extinction in many areas in 19th century due to plumage trade. Some signs of increase in range following protection in 20th century.

FRANCE. Nested 19th century. Seen again from *c.* 1920, first colony discovered Camargue 1931. Has spread in last 30 years to Dombes 1938, Saône-et-Loire 1942, Loire estuary 1949, Forez 1957, Limagne 1958 (Valverde 1955–6, CJ). HUNGARY: extinct 1895, recolonized *c.* 1928 (Sterbetz 1961). CZECHOSLOVAKIA: has bred in small numbers in recent years; 1953, 1959, and 1963 (Hudec and Cerny 1972). ALGERIA: present status uncertain; probably breeds coastal zone (EDHJ). ISRAEL: began nesting after draining of Amik Gölü, Turkey (HM). IRAQ: suspected breeding, but no proof (PVGK).

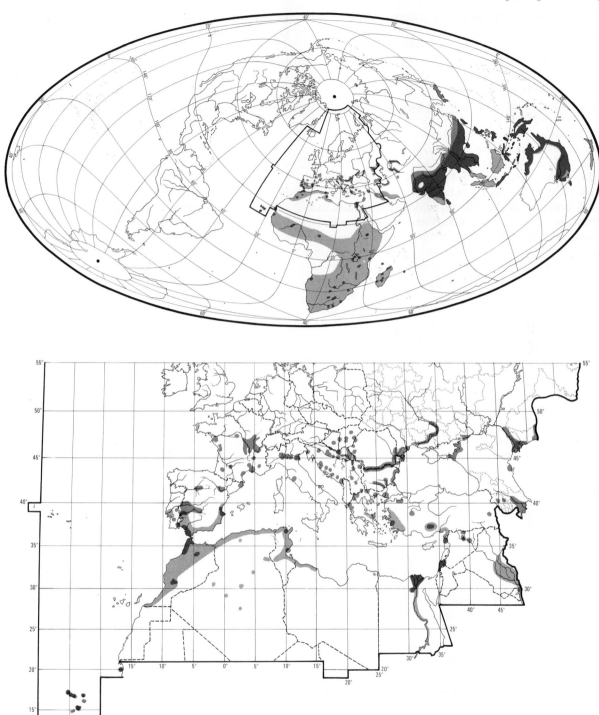

Accidental. Britain and Ireland, Belgium, Netherlands, Denmark, Norway, Sweden, West Germany, East Germany, Poland, Azores, Madeira, Canary Islands. More frequent in recent years and in some of these countries now recorded almost annually.

**Population.** FRANCE. 1549 pairs 1968, 1815 pairs 1974 (Brosselin 1975). Camargue: *c.* 1200–1500 pairs 1955 (Valverde 1955–6), 1430 pairs 1968, 1645 pairs 1973, 1700 pairs 1974 (Hafner 1975). HUNGARY. Numbers rose to 83 pairs in 1951 and 203 pairs in 1958 (Bauer and Glutz 1966).

From 1959–68, total numbers fluctuated between 150 and 214 pairs (Sterbetz and Szlivka 1972). At Kisbalaton, number declined from 16 pairs in 1968 to 6 pairs in 1972 (AK). YUGOSLAVIA. Vojvodina: incomplete counts showed 617 pairs in 1959 and 640 pairs 1968; numbers fluctuated greatly but no signs of major change (Sterbetz and Szlivka 1972). Kopačevski area 1949–70: regular counts show numbers varying from 88–243 pairs, with no clear trends (Majic and Mikuska 1972). GREECE: c. 1500 pairs in 9 large and several small colonies in 1972 (WB). RUMANIA: decreased last 50 years (Vasiliu 1968). ITALY: marked increase since 1935 in north Italian colonies, largest 400 pairs (Warncke 1960); estimated 4000–4500 pairs for whole country 1975 (SF). SPAIN: for details of colonies, see Fernandez Cruz (1975). USSR. Almost entirely exterminated Volga delta in early 20th century but numerous later in Astrakhan reserve, with c. 22000 birds autumn 1935 (Dementiev and Gladkov 1951). In lower delta 575 pairs in 1951, 100 pairs 1954, 400 pairs 1955, but only 40 pairs 1956 and 9 pairs 1957 (Lugovoy 1961). TURKEY: counts at 4 major colonies suggest total breeding numbers c. 900 pairs (OST). TUNISIA: 70–80 pairs breed annually Lake of Tunis; c. 100 pairs Kneiss 1971 (MS). MOROCCO: 150 nests in new colony Allal Tazi, 1967 (Haas 1969). MAURITANIA: breeds in small numbers on 4 islands Banc d'Arguin; c. 25–30 pairs 1959–65 (de Naurois 1969a). CAPE VERDE ISLANDS: colonies now small; probably less numerous than in 19th century (de Naurois 1969b).

Movements. Migratory and dispersive. After fledging, juveniles disperse more or less at random, such movement lasting from July–September, its extent apparently largely dependent on prevailing feeding conditions (Valverde 1955–6). Spanish juveniles may reach 400 km to north; movement recorded from Camargue to Iberia and Italy, and individuals of uncertain origins occur annually in west and west-central Europe. Occasional adults north of breeding range at this time have probably remained after spring overshooting (see below). Juvenile dispersal merges imperceptibly into true autumn migration, which becomes marked by adult departures in late August to early September, and may last into November or later, especially in Caspian region.

Except in severe weather, minority overwinters in Mediterranean basin (flocks up to 70 in Camargue), and others do more or less regularly in Transcaucasia, Azerbaydzhan, Iraq and, quite commonly, Iran. However, majority of west Palearctic population travels south to south-west (apparently seldom south-east) to become trans-Saharan migrants, some penetrating at least to equator (White 1965). Scale of trans-Saharan passage has become appreciated only recently (Moreau 1967). Southerly recoveries include: 2 from south Russia (Sea of Azov) in southern Nigeria, singles from France and Yugoslavia in Mali (Niger inundation zone), one each from Spain to Guinea, France to Gambia and Ghana, Hungary to Sierra Leone; also from Spain to Madeira and Canary Islands. Despite absence of recoveries, also winters east Africa, where common in Sudan; strong autumn passage through Eritrea (Moreau 1972). Volga delta bird recovered Iraq in November, when perhaps still migrating. Only clue to origin of those wintering Persian Gulf single recovery Kuwait, August, ringed 3 years before in Daghestan ASSR.

Return migration well under way in March; those back in Camargue colonies in February presumably had wintered locally; Hungarian and Volga delta colonies reoccupied from early April (Dementiev and Gladkov 1951). Prone to spring overshooting, resulting in singles or even small parties appearing well to north, more so than other south European herons. Annual spring visitor (chiefly April–June) to Britain and Ireland since c. 1950, though previously rare vagrant, while spring influxes, sometimes large, regular now in south Germany, Austria, and Switzerland (Bauer and Glutz 1966); birds not infrequently stay several weeks, sometimes months, and has exceptionally overwintered in British Isles.

3 records from New World: one Newfoundland, May 1954; 2 ringed as nestlings Spain (1956, 1962) recovered respectively Trinidad, January 1957, Martinique, October 1962 (Bernis 1966); latter at least attributable to juvenile dispersal rather than disorientated migration.

Food. Mainly small fish, small amphibians, terrestrial and aquatic adult and larval insects; also crustaceans, lizards, worms, snails, small mammals, and snakes. Prey from 1·2–15 cm, with preference for 3·8 cm (Valverde 1955–6; Sterbetz 1961). Prey taken as bird slowly wades, or when running and snapping; occasionally when standing motionless and, according to Moule (1953), by agitating the water and mud with one foot. Daytime and crepuscular feeder; chiefly in shallow fresh or brackish water, c. 16 cm deep in Camargue (Valverde 1955–6), flooded meadows, rice fields, and, in late summer, often on dry land.

Analyses show diet components, though broadly similar, vary with locality, availability, and time of year. Fish include tench Tinca, ruffe Acerina, carp Cyprinus, Crucian carp Carassius, loach Cobitis, sunfish Eupomotis, stickleback Gasterosteidae, eel Anguilla, roach Rutilus, bitterling Rhodeus, bleak Alburnus, mullet Mugil, sand-smelt Atherina and Gambusia. Amphibia include frogs (Rana, Pelobates, and Hyla) and toad Bombina. Insects include beetles Dytiscus, Cybister, Hydrous, and Hygrobia; waterboatmen Naucoris, Notonecta, and Corixa; water scorpion Nepa; dragonfly Libellula and Aeshna; mole-cricket Gryllotalpa; and cricket Gryllus. Crustaceans include shrimp Gammarus, prawn Palaemonetes, and phyllopod Triops (Apus).

58 stomachs from Rhône delta, chiefly in June, contained in order of frequency insects 72% (including larvae of Hydrophilidae, Hygrobia, Dytiscus, Cybister), fish 50% (including Gambusia, Gasterosteus, Cyprinus, Athe-

*rina*), crustaceans 24% (including *Triops*) and amphibians (including *Rana, Pelobates*) (Valverde 1955–6). 40 birds taken throughout year, Hungary and Yugoslavia, had similar diet, though no crustaceans and higher preponderance of fish, especially *Alburnus, Cobitis, Carassius, Cyprinus, Leuciscus*, and *Gymnocephalus* (Vasvari 1954). 77 samples, Danube delta, Rumania, contained 47% fish of 0·3–14·6 g average weight; *Gobius kessleri* 28·6% by frequency, *Carassius carassius* 12·6%, *Rutilus rutilus* 12·6%, *Scardinius erythrophthalmus* 9·8%, *Esox lucius* 9·8%, *Eupomotis gibbosus* 4·2%, *Cyprinus carpio* 2·8% (Andone *et al.* 1969). Nestlings, also from Danube delta, contained fewer fish (21·5%) including *C. carpio* (20–30 mm), *Gobius* (25–45 mm), *Atherina mochon* (25–35 mm), *Misgurnus fossilis* (45–95 mm); also *Rana* (tadpoles and juveniles), insects (especially *Corixa*), and larvae of Dytiscidae and *Hydrous* (Munteanu 1967).

From Saser area in Hungary, significant changes in feeding patterns took place 1948–53 correlated with development of rice culture, and by 1953 rice fields became main feeding areas from May–September (Sterbetz 1961). 50 stomachs from rice fields contained mainly insects (especially *Naucoris, Notonecta, Nepa*, and *Dytiscus*), frogs, crustaceans, and a few fish; from areas outside rice fields, mainly insects with more fish (especially *Rhodeus, Alburnus, Carassius*) but few human-food fish e.g. *Cyprinus* (Sterbetz 1961). From Italy, 36 stomachs contained mainly insects (88%), 29% fish, and 35% amphibians; a further 8 stomachs had similar prey (Moltoni 1936, 1948). In north Yugoslavia, chiefly small fish (*Cyprinus, Rutilus, Carassius*), frogs, lizards, locusts, and sparrows *Passer* taken (Sterbetz 1961). Prey similar in southern USSR, though water insects more important and, later in season, many locusts and other insects taken from dry land; in water, Kzyl-Agach reserve, prey mostly small *Gambusia* (Dementiev and Gladkov 1951). 80 stomachs from watered cultivated land, west Bengal, contained (by weight) mainly insects (74%), and a few fish (16%), while 58 stomachs from neighbouring mudflats contained chiefly fish (90%) and a few insects (2%) (Mukherjee 1971).

**Social pattern and behaviour.** Based mainly on material supplied by D Blaker.
1. Both gregarious and solitary. At all times of year, normally feeds singly, small groups of feeding birds probably temporary aggregations rather than true flocks (D Blaker). Sometimes associates with other Ardeinae, e.g. Great White Egret *E. alba* (R K Murton) and Cattle Egret *Bubulcus ibis* (Voous 1960); like latter species will attend grazing mammals, but not habitually. Disputes reported between feeding birds when one approaches another too closely (Mees 1950). BONDS. Monogamous pair-bond, probably of seasonal duration only and confined to nest-site and immediate vicinity, partners giving no sign of mutual recognition when meeting elsewhere. Both sexes tend young until well after fledging. Strong brood-bond between siblings which maintain contact with one another after fledging (see also Lorenz 1935). BREEDING DISPERSION. Colonial, often in large heronries,

alone or more usually with other species. At start of season, unpaired ♂ frequents series of small, temporary display territories in colony. Later, pair strongly defend nest-territory mainly against conspecifics but also against other species; territory may extend for diameter of 3–4 m round nest but sometimes nests less than 1 m apart. ROOSTING. Communal and nocturnal at traditional sites in trees; usually shared with other herons and, where population resident, may be on same site as colony or some distance away.

2. No substantial studies. Preliminary work by Blaker (1969*b*) suggests close similarity to extralimital Snowy Egret *E. thula* in pairing behaviour (see Meyerriecks 1960; Palmer 1962). According to Curry-Lindahl (1971), *E. garzetta* and *E. thula* conspecific (see also Delacour 1947), and behaviour largely identical. Pair-formation of heron-type 3, typical of *Egretta*; associated with striking changes in bare-part colour lasting until egg-laying (Blaker 1969*b*), but short-term 'blushing' of lores and orbital skin of *E. thula* during antagonistic encounters not reported in *E. garzetta*. Unlike *Bubulcus*, crest and other plumes used equally in both heterosexual and antagonistic situations, apparently functioning mainly to increase size and conspicuousness of displaying bird. In antagonistic encounters, extent of fear and aggression can be judged from relative positions of anterior and posterior parts of crest, much as in *Bubulcus*; long lanceolate plumes act together with rest of anterior portion of crest, not independently (Blaker 1969*b*). ANTAGONISTIC BEHAVIOUR. Little information away from colony, but chasing and threat calling observed during disputes between feeding birds (Mees 1950). At colony, particularly aggressive during pair-formation. Forward-display usual threat, both from lone birds (probably mostly ♂♂) in defence of display site during pairing phase and later from pair in defence of nest-territory. Similar to that of *Bubulcus* in most aspects, but S-curve of neck more marked; accompanied by Ggrow-call. Slight lateral Head-flicks (as in *Bubulcus*) common display in many antagonistic situations; may have appeasing function. Especially in early part of season, fights common if rival does not retreat; mainly aerial, birds flying upwards, pecking and buffeting each other with wings while hovering (Blaker 1969*b*). DANCING BEHAVIOUR. Collective dance performances mentioned but not described by Curry-Lindahl (1971). HETEROSEXUAL BEHAVIOUR. At start of season, unpaired ♂ highly mobile, defending and advertising from succession of small, ill-defined territories. ♂ spends much of time walking up and down site, repeatedly uttering loud Advertising-calls of at least 2 types. Frequently makes short Flap-flight displays with thudding sound of wings, much as in *Bubulcus*, and longer Circle-flights for 30–300 m in ordinary flight but with neck partly extended, returning to original perch or one close by. Distinctive Extended-neck Flight also observed from birds of uncertain sex: neck fully stretched at *c.* 20–30° above horizontal and scapular plumes raised for first 5–10 m after take-off. Both ♂♂ and ♀♀ strongly attracted by displaying ♂; all form highly vocal group and move about colony, many giving Advertising-calls. Several ♀♀ may perch in vicinity of displaying ♂♂, peering at them in long-necked manner; repeatedly fly towards them only to be threatened or supplanted. ♂ and any attendant ♀ occasionally perform Wing-touch display as (e.g.) in *Nycticorax*. Twig-shake display, as in *Bubulcus*, also characteristic of unpaired ♂; accompanied by Chatter-call. Stretch-display occurs at this stage, but said to be infrequent and confined to period just prior to pair-formation Stretching phase much as in *Bubulcus* but apparently silent while downward movement of head and neck in crouching phase (during which only scapular plumes erected and call given) short and rapid, unlike slow, large movement of

A

*Bubulcus*. After a few days, ♂ accepts a particular ♀ (exact manner of pair-formation not described); if ♀ not driven off again, bond stabilizes within a few hours. Common displays by both sexes at this time: Flap-flight and Bill-rattling. Nest-building by ♀, after twig-bringing and passing by ♂, and copulation also start within a few hours. Pair also remain perched together quietly on nest for long periods. Bill-rattling display, in which sound produced by rapid movements of mandibles, equivalent to Back-biting of *Bubulcus*; often performed above mate's back or across its neck; no contact made with feathers, and no Allopreening of any sort reported. Throughout mating phase, incubation and nestling periods, Greeting Ceremony occurs every time one of pair returns and meets partner at nest-site (never elsewhere). Alighting behaviour of incoming bird not described but utters Da-wah calls; other assumes Upright-display on nest (see A) and reciprocates calls, both then Chatter-calling. For several seconds after incomer lands, both birds adopt similar posture to medium-intensity Forward-display with all plumes erected (see B), then relax or Bill-rattle. Unlike in *Bubulcus*, crests not kept depressed, full Stretch-displays absent, and seldom any bodily contact. Usually little display precedes copulation, or ♂ may perform Stretch-display; ♀ does not crouch but stands with legs only slightly bent much as in *Ardea* (Blaker 1969*a, b*). RELATIONS WITHIN FAMILY GROUP. Broadly similar to *Bubulcus* and other colonial herons, so far as known. Unlike in *Nycticorax*, parents

B

said readily to feed young with little or no stimulation, pushing bill into that of chick; latter shows no pronounced attempt to seize adult's bill but usually begs with open bill pointing up (Lorenz 1935). Age of young, however, not indicated (see also Grey Heron *A. cinerea*). Siblings show strong mutual attachment; maintain cooperative defence even after leaving nest against attacker (usually strange young of same age).

(Figs A–B after Blaker 1969*b* and photograph by E Hosking.)

I  C Chappuis/Sveriges Radio  France June 1965

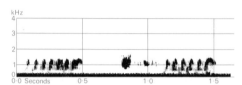

II  C Chappuis/Sveriges Radio  France June 1965

**Voice.** Little information from west Palearctic (see Witherby *et al.* 1939; Bauer and Glutz 1966). Mostly silent away from colony or roost, especially outside breeding season, but hoarse 'kgááár', somewhat less deep and rolling than corresponding call of Great White Egret *E. alba*, reported during disputes between feeding birds (Mees 1950). Highly vocal at colony, especially early in season, with variety of snarling, quacking, and croaking calls; including those rendered 'ärk', 'ork', 'kre-kre-kre', harsh 'rrhe' (Bauer and Glutz 1966), croaking 'kark', and bubbling 'wulla-wulla-wulla' (Peterson *et al.* 1954; see below).

8 calls (often with characteristic hollow inflexion) distinguished by Blaker (1969*b*) at South African colony, plus non-vocal Bill-rattling and thudding noise of wings (weaker than corresponding sound in Cattle Egret *Bubulcus ibis*) during Flap-flight display. (1) Aaah-call. Long, grating 'aaah' given at take-off (see I). Also heard at feeding grounds. (2) Da-wah call. Loud 'da-WAH' uttered in various situations, but especially when alighting at nest and during Greeting Ceremony. Homologous to Rick-rack call of *Bubulcus* and double croak of other herons. (3) Ggrow- call. Harsh, gargling 'ggrow'. Accompanies Forward-display (threat). Probably same as call described by Mees (1950); see above. (4) Gargling-call. Long, gargling sound. Advertising-call of ♂. Probably same as 'wulla-wulla-wulla' of Peterson *et al.* (1954) of which 'gargling' apt description (P J Sellar). Recording (see II) resembles 'la-la-la-la-ah-h-h-h . . .', an approximation of which obtained by vibrating finger in mouth while singing 'aaaaaah', suggesting bird may produce short 'la' by vibrating tongue while sounding long 'aaaaaah' (J Hall-Craggs). (5) Dow-call. Brief, loud, hollow 'dow'. 2nd

Advertising-call of unpaired ♂. (6) Po-call. Descending series of 5–9 hollow 'po . . .' notes. Given early in season by birds of uncertain sex without any visual display; presumably another Advertising-call. (7) Chatter-call. Brief, nasal stammering (no rendering) like chatter of *Bubulcus*. Accompanies Twig-shake display of ♂, and uttered by both sexes during Greeting Ceremony. (8) Ow-call. Brief, gulping 'ow', accompanying downward movement of Stretch-display.

CALLS OF YOUNG. Begging-call 'hé-hé-hé-hé-hé' (Bauer and Glutz 1966).

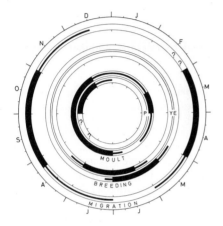

**Breeding.** SEASON. See diagram for general situation in southern Europe. Eggs laid from end April (south France), middle of May (Balkans). SITE. In reedbeds, alder thickets, bushes, and wide variety of trees; up to 20 m above ground or water. Colonial; nests at closest 1–2 m apart, up to 10 nests per tree. May form separate colonies but more often with other herons, Pygmy Cormorant *Phalacrocorax pygmaeus*, and Glossy Ibis *Plegadis falcinellus*. Nest: platform of sticks or reeds, varying from flimsy to solid; diameter 30–35 cm, height 10–15 cm; sticks usually *c.* 10–15 cm long (Sterbetz 1961). Building: by both sexes, but mostly by ♀ using material brought by ♂. EGGS. Elongated oval with slightly pointed ends, matt; green-blue, discoloured during incubation. 46 × 34 mm (42–54 × 31–38), sample 200 (Schönwetter 1967). Weight 22–27 gm (Géroudet 1967). Clutch: 3–5 (8). Of 199 Spanish clutches: 3, 3%; 4, 24%; 5, 61%; 6, 12%; 7, 0·5%; mean 4·8 (Valverde 1955–6). One brood. Replacement clutches after early egg loss. Eggs laid at 24-hour intervals, occasionally 48-hour. INCUBATION. 21–22 (25) days. By both sexes. Starts with 1st egg; hatching asynchronous. YOUNG. Semi-altricial and nidicolous. Cared for and fed by both parents. Fed by regurgitation into bill of nestling while small, later into bottom of nest. Leave nest at *c.* 30 days, and perch on nearby branches. FLEDGING TO MATURITY. Fledging period 40–45 days. Become independent soon after. Age of first breeding not certainly known but probably 1 year. BREEDING SUCCESS. No data.

**Plumages** (nominate *garzetta*). ADULT BREEDING. Wholly white with ornamental feathers on head, chest, and upperparts. Feathers of crown loose, 2–3 narrow, elongated (13–16 cm) plumes of firm structure on nape. Feathers of chest with loose bases and lanceolate tips, similar to nuchal plumes. Centre of mantle and scapulars with tuft of long (up to 24 cm) 'aigrettes' with long, loose, recurved rachis, covering upperparts and reaching to tip of tail. ADULT NON-BREEDING. Like breeding but without plumes on nape, while 'aigrettes' on upperparts fewer and shorter. NESTLING. Down white, soft, and hair-like. Feathers develop from *c.* 10 days. Remains of down, particularly on head, retained until after fledging. JUVENILE. White, without ornamental feathers. IMMATURE. White; feathers of chest slightly elongated with some lanceolate tips; imperfectly developed 'aigrettes' on mantle and scapulars (Witherby *et al.* 1939).

DARK MORPH. Exceptional in western Palearctic, but regular East Africa and Malagasy region (subspecies *dimorpha*). Specimen collected May 1869, Bulgaria, grey, including ornamental feathers, underparts lighter, head white with wavy grey bars, primaries mottled white, shafts black (Reiser 1894; von Boetticher 1949, 1952). Darker, paler, and also particoloured variants described from Africa. Closely similar to dark morph of *E. gularis*, recognizable only by pure black, more slender, bill and relatively longer tarsi—121% of bill (6·5; 53) 106–141, against 106% (4·67; 19) 97–116 in *E. gularis*. At least some of sight records of alleged melanistic *E. garzetta* in south Europe may have referred to *E. gularis*; specimen, Spain, 1970, identified as such (see *E. gularis*, Distribution).

**Bare parts.** ADULT. Iris yellow. Bill black. Bare skin of face grey, tinged green; orange at start of breeding. Leg black, foot yellow, tinged green; bright yellow or orange-yellow at start of breeding. NESTLING. Bill yellow-green; leg and foot olive. JUVENILE. Bill brown. Bare skin on face lead-grey. Leg black, foot grey-green (Witherby *et al.* 1939; Bauer and Glutz 1966; Sterbetz 1961).

**Moults.** ADULT POST-BREEDING. Complete; June–November or December. Starts with shedding of ornamental feathers on nape, followed by scapulars (Sterbetz 1961). Main moult of flight-feathers probably in winter-quarters. Inner primaries descendant, outer irregular. PRE-BREEDING. Partial; January–April occasionally May. Amount of moult of body plumage unknown, but nuchal plumes appear and 'aigrettes' become denser and longer. POST-JUVENILE. Partial. Body but not wings or tail; August–November, after departure from breeding area. Next moult is first complete one, earlier than adult post-breeding, spring of 2nd calendar year (Stresemann and Stresemann 1966).

**Measurements.** Adult and juvenile, skins (BMNH, RMNH, ZMA).

| | | | | | |
|---|---|---|---|---|---|
| WING | ♂ 280 | (13·2; 17) | 245–303 | ♀ 272 | (15·0; 17) | 251–297 |
| TAIL | 97·9 | (8·03; 16) | 84–113 | 93·6 | (5·91; 14) | 81–101 |
| BILL | 84·4 | (5·84; 17) | 67–93 | 79·5 | (6·67; 17) | 68–89 |
| TARSUS | 101 | (8·24; 17) | 78–112 | 97·3 | (6·36; 17) | 88–110 |
| TOE | 71·9 | (3·58; 15) | 65–79 | 71·1 | (8·33; 14) | 64–96 |

Sex difference significant for bill, but actual differences may be obscured by wrongly sexed specimens in museum collections.

**Weights.** Hungary: adult ♂♂ 496–614; adult ♀♀ 490–530 (Vasvari 1954). Italy: adult ♂♂ 450–460 (Moltoni 1936). Camargue: adults July 388 (6) 344–438; August 373 (5) 280–435; juveniles July 447 (8) 370–550; August 412 (24) 305–555 (Bauer and Glutz 1966). South Caspian: lean ♂ 396; fat ♂ 600; fat ♀ 482

(Schüz 1959). Juveniles from various sources: ♂♂ 400–540, once 300; ♀♀ 445–470, once 220.

**Structure.** 11 primaries: p8 and p9 longest, p10 0–10 shorter, p7 0–10, p6 10–25, p5 25–35, p1 *c*. 85; p11 minute. P8–p9 slightly emarginated outer webs, p9 and p10 inner. Tail square, 12 feathers. Bill slender, with deep nasal grooves; distinct notch at 3 mm from tip of upper mandible. Legs long and slender, *c*. half of tibia bare. Outer toe *c*. 85% of middle, inner *c*. 79%, hind *c*. 59%, hind-claw *c*. 22%.

**Geographical variation.** In *nigripes* and *immaculata* foot black, not yellow; *immaculata* said to be smaller and *dimorpha* larger than *garzetta* (Grant and Mackworth-Praed 1933).

## *Egretta intermedia* Intermediate Egret

PLATE 40
[facing page 303]

Du. Middelste Zilverreiger   Fr. Aigrette intermédiaire   Ge. Mittelreiher
Ru. Средняя беляя цапля   Sp. Garceta intermedia   Sw. Mellanägretthäger

*Ardea intermedia* Wagler, 1829. Synonym: *Mesophoyx intermedius*.

Polytypic. *E. i. brachyrhyncha* (Brehm, 1854), Africa south of Sahara, accidental. Extralimital: nominate *intermedia* (Wagler, 1829), east Asia; *plumifera* (Gould, 1848), Moluccas, New Guinea, Australia.

**Field characters.** 65–72 cm, ♂ averaging larger; wingspan 105–115 cm. Medium-sized white heron with yellow bill and mainly black legs. Sexes alike, except length of scapular plumes, and no certain seasonal differences. Juvenile similar, apart from plumes.

ADULT. White with slightly creamy tinge. Moderate crest on hind crown and nape, scapulars elongated to form train, and feathers of lower throat and upper breast also elongated. Rather short bill orange-yellow; bare lores yellowish; eyes pale yellow; legs yellow above tarsal joint and black on tarsi and feet. Scapular plumes shorter in ♀. JUVENILE. No elongated feathering.

Although nearer in size to Little Egret *E. garzetta*, much closer in appearance to Great White Egret *E. alba* which can have yellow bill and variable leg colours: confusion surprisingly easy where no size comparison available. *E. intermedia* has proportionately stubbier bill, shorter, thicker neck and shorter legs; and at close range black line of gape ends immediately below eye, whereas in *E. alba* it extends over 1 cm past eye. Stubby yellow bill and off-white plumage give some resemblance to Cattle Egret *Bubulcus ibis*, but that species easily distinguished by much smaller size, short pale legs, heavy jowl, and short neck, even if buff on crown, mantle and throat not visible.

Behaviour much as other egrets and associates with them, but generally more sluggish and occurs singly or in small, loose parties. Flight more like *E. alba* than *E. garzetta*.

**Habitat.** Margins of shallow standing or gently flowing water: rivers, streams, floodlands, pools and lakes, lagoons, and estuaries or coastal waters, including mangroves used for roosting (Nisbet 1968). Breeds in trees, bamboos, and other suitable vegetation near water.

**Distribution.** India, China, and Japan, south-east through Philippines and Sunda Archipelago to Australia; in Africa north to Sénégal, south-west Mauritania and to *c*. 19°N in Sudan. Visits Banc d'Arguin (R de N); specimen obtained Cape Verde Islands 1965 (Bannerman and Bannerman 1968); one Dead Sea, Jordan, April 1963 (IJF-L).

**Movements.** Little known. No evidence of true migration in Africa; like other aquatic species breeding in northern tropics, appears dispersive to some extent in dry season, though present throughout year on rivers of Chad and Sudan; only identified south Nigeria in dry season (Elgood *et al*. 1973). In Asia, migratory in Far East, wintering south to Micronesia and East Indies; but in Indian subcontinent no more than dispersive, shifting locally with water conditions (Ali and Ripley 1968).

**Voice.** Generally silent, but sometimes utters short 'kwark' (Williams 1963).

**Plumages.** ADULT. White. On mantle and scapulars tuft of loose ornamental feathers ('aigrettes'), reaching beyond tip of tail, shorter in ♀; similar but shorter plumes on chest. These plumes often said to be absent in non-breeding plumage (see Mackworth-Praed and Grant 1970), but Chapin (1932) mentioned non-breeding ♂ with long aigrettes. JUVENILE. White, without aigrettes.

**Bare parts** (subspecies *brachyrhyncha*). ADULT. Iris pale yellow. Bill yellow to orange-yellow. Bare skin of face pale yellow, yellow-green in breeding. Leg black, bare part of tibia yellow. JUVENILE. Not known.

**Moults.** Undescribed.

**Measurements.** Sexes combined; skins (BMNH, RMNH, ZMA).

| | E. i. brachyrhyncha | | | E. i. intermedia | | |
|---|---|---|---|---|---|---|
| WING | 311 | (4·66; 5) | 305–318 | 299 | (14·0; 13) | 275–327 |
| TAIL | 125 | (6·11; 5) | 118–132 | 118 | (11·1; 7) | 103–135 |
| BILL | 71·2 | (3·85; 8) | 66–78 | 72·8 | (2·94; 14) | 66–76 |
| TARSUS | 107 | (1·91; 7) | 104–110 | 106 | (6·37; 7) | 93–111 |
| TOE | 96, 97, 98, 104 | | | 92·9 | (4·63; 7) | 84–99 |

Nominate *intermedia* apparently much more variable than *brachyrhyncha*. Wing *brachyrhyncha*, 295–320 (Clancey 1964); tail 125–136; bill 70–76; tarsus 100–112 (McLachlan and Liversidge 1957).

**Weights.** No data.

**Structure.** 11 primaries: p9 longest, p10 6–10 shorter, p8 0–2, p7 0–5, p6 10–15, p5 28–34, p1 85–90; p11 minute. P10 slightly emarginated inner web, p7–p9 outer. Tail square; 12 feathers. Bill relatively short, heavy at base, rapidly tapering towards tip.

Outer toe *c.* 85% of middle, inner *c.* 80%, hind *c.* 57%, hind claw *c.* 23%.

**Geographical variation.** Slight, mainly in size, but overlap large. Nominate *intermedia* smaller than *brachyrhyncha*, *plumifera* still smaller. Bill of *intermedia* black at onset of breeding. (Hartert 1912–21.)

## *Egretta alba* Great White Egret

PLATE 40
[facing page 303]

Du. Grote Zilverreiger      Fr. Grande Aigrette      Ge. Silberreiher
Ru. Большая белая цапля      Sp. Garceta grande      Sw. Ägretthäger      N.Am. Great Egret

*Ardea alba* Linnaeus, 1758. Synonym: *Casmerodius albus*

Polytypic. Nominate *alba* (Linnaeus, 1758), Europe and temperate Asia. Extralimital: *modesta* (J E Gray, 1831), south and east Asia, Australia, and New Zealand; *melanorhynchos* (Wagler, 1827), Africa south of Sahara; *egretta* (Gmelin, 1789), Americas.

**Field characters.** 85–102 cm, of which body forms only 40–45 cm, ♂ averaging larger; wing-span 140–170 cm. Large white heron with long, thin neck, long, fairly deep bill, and long legs. Sexes alike and seasonal differences confined to scapular plumes and bare part colours; juvenile similar but just distinguishable.

ADULT. Immaculate white with long plumes on upper breast and base of neck and, in breeding dress, on scapulars forming loose train (but no head plumes). Bill varies from yellow during most of year to black with yellow base or even apparently all black (or all orange in extralimital races) at onset of breeding. Bare loral area and orbital ring green; eyes yellow; legs blackish with green tinge and sometimes yellowish at sides, but underside of thigh becomes yellow in breeding season and rarely red for short time. JUVENILE. Like non-breeding adult, but without plumes, which start to appear on upper breast and scapulars in 1st winter.

Sometimes confused with Little Egret *E. garzetta* at distance, but much larger and proportionately even slimmer with more snake-like neck and narrower wings, as well as black feet and, except at start of breeding, yellow bill. More difficult to distinguish from Intermediate Egret *E. intermedia*, whose smaller size and yellowish upper legs not always apparent, but has proportionately longer bill and longer, more kinked neck, while at close range black line of gape extends over 1 cm behind level of eye, whereas in *E. intermedia* it ends immediately below eye. Confusion also possible with albino Grey Heron *Ardea cinerea*, of similar size with yellow bill, but paler legs and less snake-like neck and head.

Commonly seen in parties, even flocks, but also singly or scattered along shores of lakes, rivers, and sea coast. Legs project well beyond tail in flight. Flight lighter and more buoyant than *A. cinerea* with wing-beats slower than *E. garzetta*. Feeds sedately in shallows, sometimes in deeper water. Although breeding in reeds, not secretive or reed-haunting and usually out in open. Seldom perches in trees,

except African race *E. a. melanorhynchos* which commonly nests in them.

**Habitat.** In warm middle and low latitudes, usually of continental climate, not occupying western sector of west Palearctic; sometimes for brief periods tolerates cool or even icy conditions, but avoids boreal and arctic zones. Largely restricted to extensive wetlands and margins of fresh water in lowland regions but recorded as nesting in mountainous terrain up to *c.* 1800 m in USSR (Dementiev and Gladkov 1951). Forages in meadows (moist or sometimes dry), marshes, depressions, swamps, flood-lands, drying pools, and on margins of rivers, oxbows, streams, channels, and lakes; sometimes also on irrigated land and ricefields and, mainly in winter, shallow estuarine or coastal waters. Breeds usually in large, dense, in-accessible reedbeds or other tall aquatic vegetation, but in some areas also frequently in osiers *Salix viminalis* and other willows, wild olive *Oleaster*, and other shrubs or low trees; nests often in contact with water. In parts of range, sometimes roosts in trees, including mangroves. Watchful and normally avoids situations not affording clear views of possible approaching danger; in flight rises at once on sighting hunter below (Dementiev and Gladkov 1951). Wades and walks well, but perches infrequently away from nest or roost; flies freely and uses airspace up to considerable altitude in movements which may involve daily trips of up to 10–15 km between feeding area and nest.

**Distribution.** Decreased in both range and numbers in late 19th and early 20th century due mainly to killing of breeding adults for millinery trade; some recovery since, following protection, but still threatened by drainage of suitable habitats, and position precarious in most of west Palearctic (see Population).

CZECHOSLOVAKIA: bred 1949, 1960 (Hudec 1972). POLAND: one pair bred 1863, Silesia (Tomiałojć 1972).

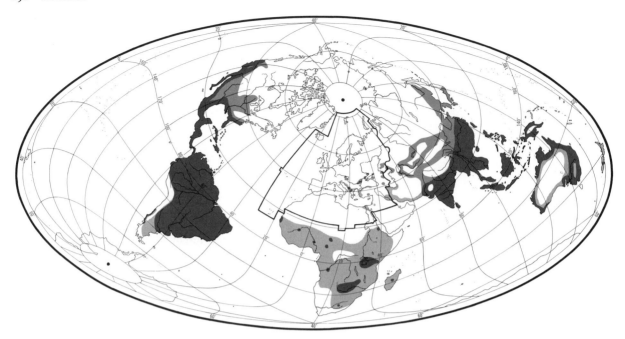

Accidental. Britain, Belgium, Netherlands, Norway, Denmark, Sweden, Finland, West Germany, East Germany, Poland, Czechoslovakia, Switzerland, Spain, Malta, Algeria, Morocco, Libya, Mauritania, Canary Islands. Azores record poorly documented (Bannerman and Bannerman 1966).

**Population.** AUSTRIA. Has bred Neusiedlersee from at least 1682 and probably most years since; increased from *c.* 100 pairs 1946 to *c.* 140 pairs 1951 and 200 pairs 1959 (Bauer and Glutz 1966). No recent counts (HS). HUNGARY. Main colony Kisbalaton: 100 pairs 1886; 7 pairs 1895; almost extinct 1915–30; fluctuating numbers to 1960, with maximum 40 pairs 1966 but only 4–11 pairs 1968–72. Lake Velence: first bred 1936; 12 pairs 1960. Hortobágy: one pair 1969 (Bauer and Glutz 1966; AK, IS). YUGOSLAVIA. Decreasing; now only *c.* 10 pairs. Ceased to breed Vojvodina after 1956 due to reclamation (VFV). Kopačevski: declined from 73 pairs 1954 to 5 pairs 1967; none since (Majić and Mikuska 1972). ALBANIA: bred Lake Skadar and possibly Lake Malik (Thorpe *et al.* 1936); present status unknown. GREECE: breeds only Lake Mikra Prespa; *c.* 12–15 pairs 1971 (WB). BULGARIA. Now rare. Lake Sreburna: 4–5 pairs 1972. One pair near Lake Burgas, 1957 (SD). RUMANIA. Formerly numerous Banat and Danube delta, but has not bred Banat since 1851. Considerable decline Danube area with *c.* 80 pairs 1939, none 1950; then, after protection, 120 pairs 1960–2 but now again decreasing, 30–40 pairs in all (Vasiliu 1968). USSR. Before 1917, almost exterminated Volga delta and Kazakhstan by hunting; increased Volga delta after protection, with 1070 adults and juveniles 1935, but more slowly Kazakhstan (Dementiev and Gladkov 1951); in

lower Volga delta 557 pairs 1951, but none in 1958 (Lugovoy 1961). Recently found breeding Crimea (Kostin 1974). TURKEY: present during nesting season in 6 localities, but breeding proved only one, with *c.* 50 pairs 1971 (OST).

Survival. USA: tentative results from ringing of *E. a. egretta* suggest 76% die in first year and 26% per annum thereafter; life expectancy at fledging 1·4 years, at start of 2nd year 3·3 years (Kahl 1963). Oldest ringed bird 22 years 10 months (Kennard 1975).

**Movements.** Partially migratory and dispersive. Post-fledging dispersal in all directions begins July; ringing recoveries of juveniles all within 400 km, but individuals occurring rarely north to Baltic and North Sea then may include extreme examples of this dispersal. Main departures from European and Volga delta colonies late September to early November, but in mild winters minority stay late or overwinter nearby (Bauer and Glutz 1966); thus absent from Kisbalaton, Hungary, only when frozen. In general, European breeders migrate shorter distances than other European herons, those of Austria and Hungary wintering south to Adriatic and central Mediterranean—north-east Italy, Yugoslavia, Albania, and Greece (Kuhk 1955). Small numbers wintering Tunisia (MS) perhaps have same origins, since one Austrian bird recovered Sardinia. Said to be not infrequent winter visitor, November–March, to Delta and Canal Zone of Egypt (Meinertzhagen 1930, 1954), and to winter in small numbers coastal Turkey (OST) and Israel (Hardy 1946); origins unknown. Some (possibly local birds) winter Black Sea and Transcaucasia, usually few except when mild (Dementiev and Gladkov 1951). Those wintering abundantly Iran presumably USSR breeders, as must be some wintering Iraq where one ringed Sea of Azov colony recovered in December. Scarce but regular visitor, September–April, to Persian Gulf, whence skins of Palearctic race nominate *alba* and Asiatic race *modesta*, but not known which predominates (Gallagher 1974).

Extent of penetration by west Palearctic breeders into Ethiopian Africa uncertain as another race (*melanorhynchos*) resident; one, presumed migrant, found dead central Sahara between Ahaggar and Bilma, and one ringed Russia recovered Oubangui (Central African Republic), but no indication of seasonal change in abundance in these northern tropics (Moreau 1967, 1972). Iran winter visitors arrive October–November, departing March (D A Scott).

Arrives in breeding areas late February to early April, mostly March (Bauer and Glutz 1966). Immatures rarely return to colonies; may account for late spring passage and early summer wandering, on rare occasions far north and west of breeding range.

**Food.** In wet season, mainly fish and aquatic insects; in dry, chiefly small mammals and terrestrial insects. Lizards, molluscs, and young birds also recorded. In shallow water, feeds by (1) slowly stalking with body held horizontally, stopping and stabbing down; (2) by standing with stiff neck and leaning forward waiting for prey to come within striking distance. When water too deep for wading (3) stands at edge, with head and neck 20–25° below horizontal peering into water, then takes off and in flight stabs down on to prey (Palmer 1962; Clarke 1965). Usually solitary, daytime feeder.

In Italy, 2 birds contained only sunfish *Eupomotis gibbosus*, carp *Cyprinus carpio*, and tench *Tinca tinca* (Moltoni 1948). In USSR, single stomachs from Syr Darya (July) contained many locusts and fish remains; from south of Lenkoran (January), a trout and 4 voles; near Lenkoran (March), a social vole *Microtus socialis*; near Dzaudzhikav (February), a trout and 2 water shrews; near Kizlyar (November), several voles (Dementiev and Gladkov 1951). In Hungary, some 20 stomachs contained mainly fish (Crucian carp *Carassius*, rudd *Scardinius*, bleak *Alburnus*, carp *Cyprinus*, dace *Leuciscus*, and bitterling *Rhodeus*), with fewer insects (waterbeetles *Dytiscus*, *Hydrophilus*, *Cybister*; mole-cricket *Gryllotalpa*; dragonfly *Aeshna*; and waterboatmen *Naucoris*) (Vasvari 1954). In Rumania, 16 stomachs contained chiefly fish, with 3 Great Reed Warblers *Acrocephalus arundinaceus*, 1 grass snake *Natrix natrix*, and various insects, especially locusts (Dombrowski 1912). 70 stomachs, west Bengal, contained by weight 80% fish, 5–175 mm long (chiefly *Mystus gulio*, *Periophthalmus*, *Mugil tade*, *Aplocheilus panchax*, and *Anguilla bengalensis*); 6% molluscs; 5% crustaceans; 4% aquatic insects; 2% water snakes; and 1% plant materials (Mukherjee 1971). For American studies see Palmer (1962).

Diet of nestlings probably mainly insects; dragonflies found in disgorged food in Syr Darya in May (Dementiev and Gladkov 1951).

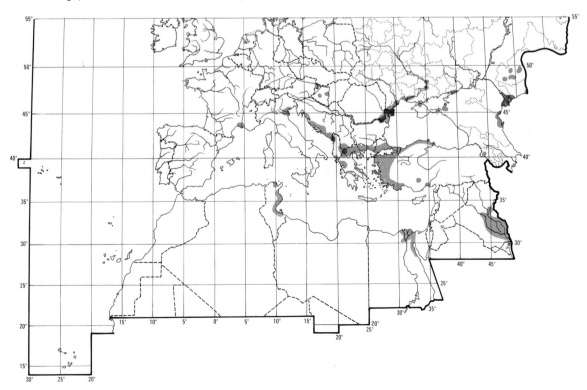

**Social pattern and behaviour.** Based mainly on material supplied by R K Murton.

1. Both solitary and gregarious, much as Grey Heron *Ardea cinerea*. At all times of year, often feeds solitarily (see, e.g. Bannerman 1957); also in small, scattered groups of 3–15. Nominate race does not appear to establish individual feeding territories (R K Murton) but defence of such adjacent to pairing territory in American birds (*egretta*) (see Palmer 1962). After breeding, in small family parties which may migrate together; where common, larger flocks encountered. In most areas, flocks never completely break-up in winter quarters; disperse to feed and gather to roost. BONDS. Monogamous pair-bond, probably mainly of seasonal duration only; in Europe, established in spring on return from migration. Both sexes tend young, family remaining together after fledging. BREEDING DISPERSION. Solitary or colonial. Nests of nominate race in reedbeds, often well dispersed (*c.* 300 m apart reported) but sometimes in small groups of 2–3 almost touching; sometimes associates with *A. cinerea* or Purple Heron *A. purpurea* (Witherby *et al.* 1939; Bannerman 1957). Other races, e.g. in Africa (*melanorhynchos*) and America, more densely colonial, particularly at protected sites in trees. Territory established by ♂ at start of season larger than area of *c.* 1 m round same site later defended vigorously by pair mostly against conspecifics; used for pair-formation, post-pairing courtship, copulation, and nesting (Meyerriecks 1960; R K Murton). ROOSTING. Communal and nocturnal, arriving and departing in flocks, but little information for nominate race in Europe. Elsewhere, often roosts in trees (Witherby *et al.* 1939). In Africa, parties return at dusk to traditional sites; sometimes shared with darters *Anhinga* or smaller egrets *Egretta* (Bannerman 1957). Roosting flights, often over considerable distances, noted in USA; arrive at sunset or just after dark, depart at first light (Palmer 1962). Primary nocturnal roost, usually shared with other species, may be in nesting colony for part of year, but, in post-breeding period, single larger roost serves birds from several colonies; usually occupied from about dusk to dawn from late autumn to early spring. Also uses secondary roosts within foraging area of primary roosts during lulls in feeding, or for short-term overnight resting. Secondary roost, and less often primary one, may be in extensive reedbeds, but mostly in trees (R K Murton). In breeding season, small parties may roost temporarily away from colony nearer feeding grounds. Gatherings of up to 20 reported during day at standing ground by side of lake (see Bannerman 1957).

2. Little known in west Palearctic. Early territorial and pairing behaviour studied in USA by Meyerriecks (1960, expanded in Palmer 1962) and by McCrimmon (1973). Similar in most features to that of herons of genus *Ardea*—to which assigned by Curry-Lindahl (1971), see also Voous (1973)—rather than to typical egrets of genus *Egretta*, of which behaviour of Intermediate Egret *E. intermedia* closest (see Blaker 1969*b*). Onset of breeding protracted. Pair-formation of heron-type 1 (see Ardeinae); associated with temporary colour changes in bare parts, particularly bill, which, in American race at least, revert to more usual colour during laying. Use made of scapular plumes (aigrettes) in many displays, both antagonistic and heterosexual; rest of plumage also erected, including short feathers of crest which lack ornamental plumes. ANTAGONISTIC BEHAVIOUR. At American colony, for several weeks before establishment of territory, some birds (probably ♂♂) spent increasing amount of day in breeding area; threatening, supplanting attacks, pursuit-flights common, with occasional fights (Meyerriecks 1960; Palmer 1962). Threat-displays by both sexes in colony include: (1) standard Upright-display with plumage sleeked similar to posture of alert bird; (2) Arched-neck Upright with plumes moderately erected; (3) Forward-display with body horizontal, head and neck retracted, flipping movement of tail, and extreme plumage erection, especially of scapular plumes. In early part of season, birds threatening each other closely in Forward-display may also engage in formal 'see-saw' pecking fights. DANCING GROUND BEHAVIOUR. Not certainly recorded but, as early as December, some birds in flock said by J J Audubon to strut round others while calling and raising plumes (see Bent 1926; Palmer 1962). During spring migration in Hong Kong, one bird also seen to advance on group of others with stiff-legged gait, neck outstretched, bill pointing up at 45°, neck and scapular feathers raised; apparent heterosexual behaviour with high degree of aggression (R K Murton) but of uncertain significance. HETEROSEXUAL BEHAVIOUR. After initial 'flying-round' period in colony area, individual ♂♂ begin to defend potential nest-sites vigorously (Meyerriecks 1960; Palmer 1962). Each gathers material to build small platform which serves as display site and from which, or close by, ♂ gives Advertising-calls and launches out and returns in simple Circle-flights. ♀♀ attracted by such behaviour but at first repulsed by threat-displays. Later courtship not studied in detail by Meyerriecks but post-pairing behaviour includes mutual Allopreening and Twig-passing by ♂. According to O Koenig, Snap-display usually directed diagonally upwards (see Baerends and van der Cingel 1962) but that of American race at least, observed from lone individuals (presumably ♂♂) during early pairing phase, closely similar to that of *A. cinerea* (McCrimmon 1974). Scapular plumes erected as neck extended and lowered so that bill well below level of feet, then neck and head feathers as single Bill-snap given; may be purely hostile. During same phase of cycle, individuals also perform Stretch-displays with scapular plumes raised: (1) neck and bill pointed skywards in stretching phase, and (2) head lowered towards back in crouching phase. Movements thus much as in *A. cinerea*, but no call mentioned and sequence usually immediately preceded by modified Snap-display: head lowered much as in typical Snap-display (but with less fully extended neck and no erection of head and neck feathers or snapping) and then twig grasped or trembling movements made with bill as if inserting stick in nest. Modified Snap-display sometimes omitted, or repeated several times with rapid flexion of tarsal joints. Wing-touch display, much as *Nycticorax*, often interspersed; most frequently between preliminary Snap-display and stretching phase of Stretch-display proper, or comes at end of crouching phase of latter. Snap-stretch display attracts other individuals, presumably ♀♀; twice, one of latter seen to fly on to back of displaying bird suggesting possibility of pairing assault by ♀ as in *Bubulcus*. Little information on post-pairing behaviour, but Greeting Ceremony at nest-relief described by Bent (1926); involves elevation of scapular plumes and rest of plumage, and raising of wings; one of pair may sometimes grasp head of mate in bill, probably as variant of Allopreening. Copulation not described. RELATIONS WITHIN FAMILY GROUP. No detailed information.

**Voice.** Mostly silent away from roost or colony. Normally silent in flight (Bauer and Glutz 1966) but 'kind of rolling grunt' once heard (Paige 1948), which may be same as loud, low-pitched croak of Palmer (1962), occasional croaking 'kraak' of Peterson *et al.* (1974), and strong, harsh 'grààar of Mees (1949). Last reported from birds both in flight and on ground. Occasional throaty croak, when one supplants another (Ali and Ripley 1968), also probably same.

I Anthony Walker/Sveriges Radio Rhodesia Spring 1968

More vocal at colony, where utters a number of deep, quite vehement, cawing sounds ('rha', 'rhä, 'arrp') of uncertain significance; also 'rrrooo rrrooo' as Greeting-call at nest (Bauer and Glutz 1966). Usually silent when disturbed but rapid 'cuk-cuk-cuk' reported, also harsh 'frawnk' from perched birds (see Palmer 1962). Cuk-call rapid and rattling (see I); may be repeated *c.* 20 times per s, resembling drumming of Lesser Spotted Woodpecker *Dendrocopus minor*. Advertising-call of ♂, rather soft 'fra-fra' or 'frawnk' (see Palmer 1962).

CALLS OF YOUNG. Food-call quick chatter (Witherby *et al.* 1939). Rendered 'ket ket ket' by Hanzák (1949–50), also described as soft, piping 'tchee tchee tchee' (J Hall-Craggs).

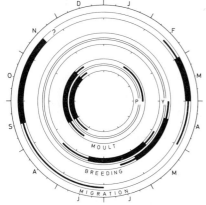

**Breeding.** SEASON. See diagram for central and south-east Europe; few data from elsewhere (Warga 1938). SITE. Usually in reedbeds in water up to 1 m deep; occasionally in low pines *Pinus* and willow *Salix* scrub over water and 4–5 m above. Normally colonial with up to 50 nests per tree (often with other Ardeidae) (Ali and Ripley 1968), though 4–5 more usual; nests sometimes touching but in reedbeds more spaced out. Nest: large pile of dead reeds or twigs lined with finer material, *c.* 100 cm (80–120) diameter and 20 cm (16–38) high, sample 20 (Warga 1938). Previous year's nests re-used. Building: by both sexes, with ♂ often starting. Additions made until young fledge. EGGS. Variable but usually fusiform; pale blue. 61 × 43 mm (54–68 × 40–46), sample 120 (Schönwetter 1967). Weight 44–53 g, sample 5 (Warga 1938). Clutch: 3–5 (2–6). One brood. Replacement clutches after egg loss. Eggs probably laid at 2-day intervals. INCUBATION. 25–26 days. By both sexes, Usually starts with 1st egg but may not be regular until *c.* 3rd egg. Hatching asynchronous.

YOUNG. Semi-altricial and nidicolous. Cared for and fed by both parents. Fed by regurgitation, stimulated by young grabbing and tugging parents' bill. In large colonies, may wander from nest to nest at *c.* 20 days and older. FLEDGING TO MATURITY. Fledging period *c.* 42 days (Warga 1938). Stay with parents until autumn migration. Age of first breeding probably 2 years. BREEDING SUCCESS. In colony of *c.* 80 nests, Hungary, young reared per nest varied from 1·13–3·06 over 6 years (Warga 1938).

**Plumages.** Wholly white in all races at all ages. ADULT BREEDING. Feathers of nape slightly elongated. Among scapulars 30–50 long aigrettes (up to 50 cm), reaching far beyond tip of tail. Feathers of chest loose and soft, but not aigrette-like. ADULT NON-BREEDING. Like breeding, but without aigrettes. Some scapulars with loose aigrette-like tips. NESTLING. Down white; elongated on crown, forming kind of crest; short and sparse on underparts; orbital skin, chin, and throat bare. Feathers start to emerge at 7th day; mostly feathered, except thighs, by 21st. Down adheres to feather tips, but by 37th day remains mainly on head and neck (McVaugh 1972). JUVENILE. Feathers of nape short, of chest not soft and loose; no scapulars with aigrette-like tips. IMMATURE WINTER. Some scapulars with loose tips but shorter than adult winter (Witherby *et al.* 1939), all primaries unmoulted and equally worn; difficult to recognize.

**Bare parts.** ADULT BREEDING. Iris yellow. Bill black with variable amount of yellow at base. Orbital skin pale green. Tarsus and toes black; tibia pinkish-yellow, this colour extending in narrowing stripe down sides of tarsus (Mountfort 1962). ADULT NON-BREEDING. Bill yellow. Tibia yellow, rest of leg dark grey-green. NESTLING. Not described for nominate *alba*. In subspecies *egretta*, iris off-white. Upper mandible grey with black culmen and tip, lower yellow-grey; bill later yellow. Orbital skin at first pale blue-grey, later yellow; throat pink. Leg pale pink-flesh with blue tinge, later grey-green (McVaugh 1972). JUVENILE. Iris yellow. Bill yellow. Orbital skin yellow tinged green. Leg brown-black; tibia yellow-green, sides of tarsus dark yellow, sole of foot yellow.

**Moults.** ADULT POST-BREEDING. Complete; primaries irregularly. August–November. ADULT PRE-BREEDING. Partial; extent of moult not certain, probably involving only development of aigrettes among scapulars. POST-JUVENILE. Partial; body but not wings or tail; August–November (Witherby *et al.* 1939). Juvenile flight-feathers moulted in spring of 2nd calendar year (Stresemann and Stresemann 1966).

**Measurements** (nominate *alba*). Ranges from Witherby *et al.* (1939), single values from skins RMNH, other data as indicated. Wing ♂ 438 (10) 410–485 (Vaurie 1965), average 451 (Dementiev and Gladkov 1951); ♀ 400–450, average 429 (Dementiev and Gladkov 1951). Tail ♂ 140–185, ♀ 166, 176. Bill ♂ 123 (10) 117–130 (Vaurie 1951), range 110–135; ♀ 110–132. Tarsus ♂ 190 (10) 170–215 (Vaurie 1951); ♀ 160, 172. Middle toe ♂ 117, ♀ 103, 110; range unsexed 109–126. ♂ rather larger than ♀.

**Weights.** Limited data. Lean winter ♂ 1030, ♀ 960; juvenile ♂ 1680 (Bauer and Glutz 1966).

**Structure.** 11 primaries: p8 longest, p7 and p9 equal or up to 10 mm shorter, occasionally p8 slightly shorter than p7; p10 5–10 shorter, p6 20–30, p5 38–60, p4 70–90, p1 c. 130; p11 minute. P10 strongly emarginated inner web, p8–p9 slightly; p7–p9 slightly outer web. Tail slightly rounded. 12 feathers: t6 c. 10 shorter than t1. Bill fairly stout, rather suddenly attenuated towards tip. Cutting edge hardly serrated on outer quarter of bill (more strongly in *Ardea*). Bare skin at gape projects beyond eye; in other *Egretta* ends just below eye (Stronach 1968). Leg long, relatively longer than in other *Egretta* and *Ardea*. Middle toe c. 62% of tarsus; outer toe c. 82% of middle, inner c. 73%, hind c. 50%, hind claw c. 22%.

**Geographical variation.** Involves size and colour of bare parts. Nominate *alba* largest. *E. a. modesta* smaller: average wing ♂♂ India 361, China 366 (Vaurie 1965); larger again in Australia and New Zealand. Latter populations sometimes separated as subspecies *maoriana* (Iredale and Mathews 1913). Population of Japan and adjacent areas probably intermediate between *alba* and *modesta* (Amadon and Woolfenden 1952). *E. a. melanorhynchos* (average wing 383, McLachlan and Liversidge 1957) and *egretta* (average wing ♂♂ 383, Palmer 1962), also small, bare part of tibia black; in *egretta*, bill in adult breeding orange-yellow not black.

## *Ardea melanocephala* Vigors and Children, 1826 Black-headed Heron

Fr. Héron mélanocéphale    Ge. Schwarzhalsreiher

Tropical African species, breeding dry and damp open areas from Cape Province north to Sénégal, Nigeria, Chad, Sudan and Ethiopia (White 1965). Migratory movements demonstrated for central Africa (Benson *et al.* 1970) and northern tropics. In Chad, moves south of c. 12°N in dry season (Salvan 1967). Nigerian colonies occupied April–October; a few remain north all year,

but most spread south after breeding (Elgood *et al.* 1973). Old records from Algeria, Spain, south France and Italy (Hartert 1912–21) never substantiated; adult claimed seen Camargue, France, 29 November 1971, but underwing pattern not noted and observers caution slight possibility of aberrant Grey Heron *A. cinerea* (Hovette and Kowalski 1972a).

## *Ardea cinerea* Grey Heron

PLATE 41
[after page 326]

Du. Blauwe Reiger    Fr. Héron cendré    Ge. Fischreiher
Ru. Серая цапля    Sp. Garza real    Sw. Gråhäger

*Ardea cinerea* Linnaeus, 1758

Polytypic. Nominate *cinerea* Linnaeus, 1758, Eurasia east to Sakhalin, Manchuria, and India, Africa and Comoro Islands; *monicae* Jouanin and Roux, 1963, islands of Banc d'Arguin. Extralimital: *jouyi* Clark, 1907, Japan, China, Indochina, Malaya, Sumatra, and Java; *firasa* Hartert, 1917, Madagascar.

**Field characters.** 90–98 cm (body 40–45 cm), wing-span 175–195 cm; of west Palearctic herons, second only in size to Goliath *A. goliath*. Large, heavy-billed, long-necked, deep-bodied, and long-legged, with whitish head and neck, contrasting with grey wings and body. Sexes alike; little seasonal variation. Juvenile separable.

ADULT. Head white, marked by black lines of loose feathers from over eye to rear crown (extending as plumes), nape and hindneck pale grey. Mantle and wing-coverts blue-grey, with scapulars elongated and greyish-white in breeding season, quills blue-black, tail grey. Chin and throat white, foreneck greyish-white, with double line of lengthening black streaks which intermingle with loose grey feathers falling from chest. Shoulder patch (noticeable at rest) and sides of belly black; flanks grey; rest of underparts white. Head-on in flight, wing shows whitish marks along leading edge, most noticeably on outer feathers of carpal joint; underwing uniformly blue-grey. Bill yellow; lores yellow, tinged green round eye. Eyes yellow. Legs and feet dull brown. Albinistic and dilute individuals occur. JUVENILE. Usually more compact than adult, with tighter and more uniformly grey plumage.

White restricted to face and sides of neck, and underparts and crown grey, lacking strong black plumes of adult. Foreneck marks brown-grey and broader, appearing more mottled. Marks on leading edge of wing indistinct. Some markedly brownish, even buff morphs occur; may be confused with immature Purple Heron *A. purpurea*. Subadults retain partly dull plumage until 2nd autumn and, even then, may lack white crown. Bill dark horn, becoming yellower in 2nd year; legs and feet greenish-grey.

Typical birds unlikely to be misidentified but albinistic or discoloured adults and immatures frequently puzzle inexperienced observers, being mistaken for Great White Egret *Egretta alba* and Purple Heron *A. purpurea*. Black-headed Heron *A. melanocephala* of tropical Africa may also be confused (though no certain record for west Palearctic). Like *A. cinerea*, essentially grey but adult has slate tone to plumage, with uniform black crown, hindneck, and lower foreneck contrasting with white cheeks and upper throat; juvenile also has generally darker plumage, with uniform dark grey crown and hindneck, deep buff lower throat, and unstreaked brownish underparts. Slighter build, smaller

PLATE 33. *Nycticorax nycticorax* Night Heron (p. 262): **1** ad breeding, **2** imm 3rd year, **3** imm 2nd year, **4** juv, **5** nestling. (PJH)

PLATE 34. *Butorides striatus* Green-backed Heron (p. 269): **1** ad breeding, **2** imm, **3** juv, **4** nestling. (PJH)

PLATE 35. *Ardeola ralloides* Squacco Heron (p. 273): **1** ad breeding, **2** ad non-breeding, **3** imm winter, **4** juv, **5** nestling. (PJH)

PLATE 36. *Ardeola grayii* Indian Pond Heron (p. 278): **1** ad breeding, **2** 1st breeding, **3** imm winter, **4** juv. (PJH)

PLATE 37. *Bubulcus ibis* Cattle Egret (p. 279): **1** ad breeding, **2** ad non-breeding, **3** juv, **4** nestling. (PJH)

PLATE 38. *Egretta gularis* Western Reef Heron (p. 286). Nominate *gularis*: **1** ad white morph, **2–4** ad intermediate morphs, **5** ad dark morph. *E. g. schistacea*: **6** ad white morph, **7** nestling. (PJH)

PLATE 39. *Egretta garzetta* Little Egret (p. 290): **1** ad breeding, **2** ad non-breeding, **3** juv, **4** nestling. (PJH)

PLATE 40. *Egretta intermedia* Intermediate Egret (p. 296): **1** ad breeding, **2** ad non-breeding. *Egretta alba* Great White Egret (p. 297): **3** ad breeding, **4** ad non-breeding, **5** juv, **6** nestling. (PJH)

size (by 10%) and more crooked neck of *A. melanocephala* obvious in comparison, while latter's distinctly white under wing-coverts contrasting with dark quills diagnostic at all ages.

As with all herons, flight silhouette characterized by retracted head and large neck (creating bulge), long, broad, rounded wings, and protruding feet. Flight action slow, with pronounced flaps of noticeably bowed wings, but when, for example, descending to nest site or confined feeding area, markedly agile, side-slipping and parachuting downwards, often with head and legs extended. Small groups in flight adopt pointed formation. Gait a careful walk; swims rarely across deep channels. Occasionally forms small flocks but usually rather solitary. Prey obtained by patient, deliberate tactics, either a slow walk or a motionless watch with neck muscles obviously tensed for spearing. Flight call a loud, harsh 'frarnk'; breeding adults give out variety of squawks, yelps and softer notes. DIMW

**Habitat.** Mainly in middle latitudes up to limits of frequent intense frosts and snow, becoming much less numerous and more scattered in Mediterranean, subtropical, and tropical zones. Mainly lowland, below *c.* 500 m; locally in mountain regions to 1000 m, exceptionally higher. Normally closely linked with distribution of suitable waters and trees, being more arboreal than other west Palearctic herons except Night Heron *Nycticorax nycticorax*. Prefers shallow fresh waters, standing or flowing, including broad rivers, narrow streams (not too rapid), oxbows, deltas, marshes, and estuaries; also lakes, pools, floodlands, muddy and sandy shores or flats, sandspits, islets and emerging rocks, and many artefacts including reservoirs and barrages, farm and ornamental ponds or fishponds, tanks, ricefields and other irrigated areas, ditches, dykes, canals, and sewage farms. Resorts to virtually all water margins not too deep, steeply shelving, of rough or broken material, closed by ice or dense vegetation, or so oligotrophic as to lack animal life for food.

Also often on grassland and other open ground for food or resting. Frequents both wide open and, less freely, enclosed surroundings. Tolerant of change in character or use of habitat except in immediate vicinity of breeding sites. Prefers nesting in trees, up to at least 40 m, either deciduous or coniferous, but most frequently in quiet and commanding situations; rarely in heart of extensive forest. In west Europe, occasionally on cliff ledges or lake islets; in east Europe especially, nests in reedbeds or rough herbage on ground near water. Nests also on marine islands e.g. Banc d'Arguin and atolls in Indian Ocean. Avoids sites where actively persecuted; exceptionally breeds in ornamental parks in city centres (Amsterdam, London). While hunting, extremely immobile, but otherwise walks and wades freely and readily takes flight to other feeding places, travelling 10–30 km when necessary between foraging area and nest. Flight normally around 50–150 m, but on passage or in thermals rises much higher, to at least 1500 m. Non-breeders and migrants show even wider habitat tolerance, including flattish coasts, inlets, lagoons, estuaries, and other suitable areas distant from breeding sites.

**Distribution.** In north of west Palearctic, has expanded north in Norway, spread in Sweden and now breeds regularly in Finland; possible decline in south, but both past and current data scanty. NORWAY: common around 1500 near Oslo, then disappeared in east due to persecution; since 1900, continued to expand north, and has also spread up valleys, with small numbers breeding in east (Haftorn 1971). FINLAND: only recently breeding regularly; few nests found and range may extend further north (Merikallio 1958, LvH). CZECHOSLOVAKIA: some westward extension since 1932 (Hudec and Černý 1972). IRAQ: has bred at head of Persian Gulf (Allouse 1953). ISRAEL: bred Lake Huleh after 1948; since lake drained only occasionally (HM). MOROCCO: perhaps nested in past (Heim de Balsac and Mayaud 1962); only modern record in 1962, when 2 nests, both failed (PR). ALGERIA: bred 19th century at Lake Fetzara (Heim de Balsac and Mayaud 1962); present status uncertain (EDHJ). TUNISIA: perhaps bred in past (Heim de Balsac and Mayaud 1962); several pairs building nests, Ichkeul 1972 (François 1975). LIBYA: bred 1943 (Stanford 1954).

Accidental. Spitsbergen, Azores.

**Population.** In west of range, populations apparently stable though with considerable fluctuation due to mortality in hard winters. In some countries elsewhere, persecution by fishermen and others still often of major importance. In north of range, climatic amelioration may have played some part in expansion in Fenno-Scandia though in Denmark other reasons suggested (Dybbro 1970). In central Europe, increase after 1880–90 with more recent decline beginning 1930–50; causes not stated though decrease after 1949 in Bavaria apparently due to increased shooting (Bauer and Glutz 1966). Water pollution may also be relevant in recent years, though in eastern England, where high organochlorine residues found (causing deaths in breeding season, thin eggshells and abnormal parental behaviour), no obvious effects on total population (Prestt 1970).

BRITAIN AND IRELAND. Following first national census in 1928, long series of estimates available for England and Wales (Nicholson 1929; Stafford 1971; Reynolds 1974). Totals have fluctuated around some 4000 nests, falling below after hard winters and rising above after mild winters (minimum 2250 nests in 1963, maximum 4925 in 1973), though trends differ considerably between regions. Total population normally recovers after 2–3 years (see Lack 1954*a*), but prolonged to 7 years after exceptionally severe winters in 1962 and 1963 (see graph). Scotland: 768 nests in 1928–9; 1086 in 1954, increase partly due to increased coverage (Garden 1958). FRANCE. Apparently

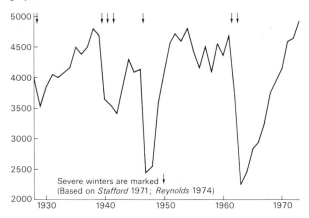

Fig II Nests of Grey Heron *Ardea cinerea* in England and Wales 1928–73.

increasing last 30 years especially in west; 2500–2700 pairs in 1962, 1900 in 1967, 2032 in 1968 and 3360 in 1974 (Brosselin 1975). First bred Camargue 1964 (CJ). BELGIUM: 450 nests in 1939, 652 in 1952, 49 in 1955, 180 in 1966, and 297 in 1969 (Lippens and Wille 1972). NETHERLANDS. 6000–7000 pairs in 1925, 8000–8500 in 1935, 4000–5000 in 1949 and 1956, 6000–7000 in 1962, then dropped after severe winter to 3500–4000 in 1963, 3250–3750 in 1964 (Rooth and Jonkers 1972). Recovery since, *c.* 8000 pairs in 1974, and over 10000 in 1975 (A A Blok). DENMARK: *c.* 1280 pairs in 1880, 1300 pairs in 1910, 1450 pairs in 1927, 2037 pairs in 1953, 1883 pairs in 1968; increasing in north and west since about 1940 but decrease in areas with dense human population (Jensen 1954; Dybbro 1970). SWEDEN: increased rapidly this century (Curry-Lindahl 1960); 1792 nests in incomplete census 1972 (probably 2000–2500 pairs in all) (S Svensson).

FINLAND: 5–10 pairs (Merikallio 1958). WEST GERMANY. Approximately 4625 nests in almost complete census in 1961 (Kramer 1962). Considerable decreases in recent years; e.g. Bayern *c.* 60% between 1960 and 1970; Baden-Württemberg 350–380 pairs in 1962, 363 in 1968, and *c.* 277 in 1972; Rheinland 197 pairs in 1961, 50 pairs in 1971; Hannover 282 pairs in 1958, 156 pairs in 1967; Hessen *c.* 220 pairs in 1950–4, 75 pairs in 1969 (EB, Hölzinger 1973). See also Bauer and Glutz (1966). EAST GERMANY: 2064–2135 pairs in 1960 (Creutz and Schlegel 1961). POLAND: in north-west, after decrease early 20th century, little change last 40 years; in Silesia, increase from 60 nests *c.* 1923 to 280 (Tomiałojć 1972). CZECHOSLOVAKIA. Decreased in last decade (Hudec and Černý 1972). Slovakia: 700–800 pairs in lowlands in 1956–8 (Hell and Sladek 1963). HUNGARY: incomplete census gave 982 pairs in 1951, population probably double this (Szijj 1954). AUSTRIA. Neusiedlersee: *c.* 180 pairs in early 1950s. Oberösterreich: 160 pairs around 1960 (Bauer and Glutz 1966). SWITZERLAND: by about 1900 reduced to *c.* 50 pairs by persecution, and extinction feared until protected in 1925; increased to *c.* 500–750 pairs in 1953, though only 350–400 pairs in 1970, reason unknown (Géroudet 1955; WT). ITALY: *c.* 2000 pairs in 1975 (SF). YUGOSLAVIA. No complete counts. Censuses at different heronries show considerable variations: e.g. Obedska Bara 10 pairs in 1955–6, 50 pairs in 1968 and 75 pairs in 1969; Kopačevski reserve 99 pairs in 1955, 9 in 1957, nil in 1958–68, 35 in 1970; Lake Skadar 100 pairs in 1967, 150 pairs in 1968, 80 pairs in 1970 (VFV). GREECE: *c.* 50–100 pairs Lake Mikra Prespa (Terrasse *et al.* 1969a). RUMANIA: numbers decreasing owing to persecution by fishermen (Vasiliu 1968). USSR. Density variable, but especially common lowland rivers flowing into Black Sea and Caspian, and

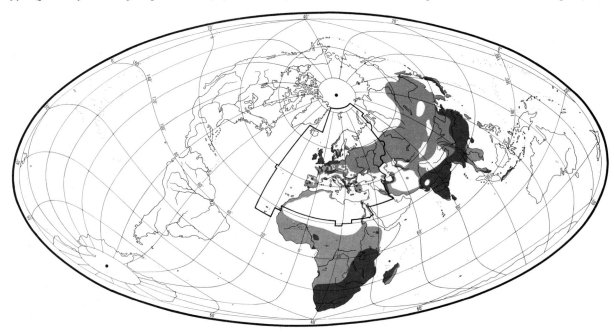

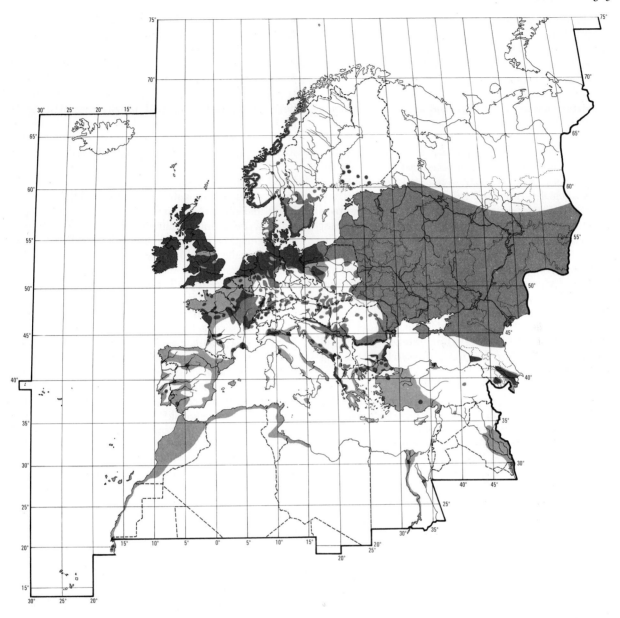

Caucasus. Astrakhan reserve: *c.* 18 500 in 1924 and 23 400 in 1935 (Dementiev and Gladkov 1951). Lower Volga delta: fluctuating numbers; 623 pairs in 1951, 1805 pairs in 1955, 403 pairs in 1958 (Lugovoy 1961). TURKEY: 350–500 pairs Manyas Gölü in 1966–7; also 7 smaller colonies of which 4 held some 120 pairs in recent years (OST). ITALY: enormous decrease in recent years (SF). SPAIN. For details of colonies, see Fernandez Cruz (1975). Guadalquivir: first bred 1951 (Valverde 1960); now *c.* 50–100 pairs (Ree 1973). MAURITANIA: 1100–1600 pairs on 2 islands, Banc d'Arguin (de Naurois 1969a); 450 pairs in 1973 (Duha-utois *et al.* 1974).

Survival. Britain: mortality of first-year birds 69%, of older birds 31%, life expectancy at fledging 1·5 years, at second year 2·7 years (Lack 1949). Belgium: mortality first year 78%; life expectancy at fledging 1·0 years (Verheyen and le Grelle 1952). Sweden: mortality first-year birds 67%, older 28%; life expectancy at fledging 1·7 years, at 2nd year 3·1 years (Olsson 1958). Oldest ringed bird 25 years 4 months (Rydzewski 1974).

Movements. Migratory, partially migratory and disper-sive. Post-fledging dispersal starts as soon as young can fly and lasts to mid-September; in almost any direction, though south-west predominates, recoveries to 300–400 km not uncommon in continental Europe where movement averages 150 km in June rising to 250 km in August; considered to reduce density near heronries for efficient use of food resources. At this time, open sea (even so narrow as English Channel) and mountains over 1000 m severely limit dispersal (Rydzewski 1956). Autumn migration proper from early September to late October; cold weather movements may occur later.

British and Irish breeders sole European non-migratory population, showing continuance of nomadic movements mainly within 150 km of natal heronries, though some move from Britain to Ireland and a few from southern England to Low Countries and France, rarely Spain; most starve rather than emigrate in occasional severe winters. Those Norwegian breeders which emigrate winter in Britain (chiefly Scotland) and Ireland, though a few recovered elsewhere in western maritime countries (including Iceland). Other European populations migrant in varying degrees but not to specific wintering areas, overwintering throughout all but most northern and eastern parts of breeding range except in very cold seasons. Average distances (km) for December recoveries: from Switzerland 250, south Germany 330, France 430, Netherlands 470, Poland 800, Denmark 920, Sweden 950, north Germany 980, and Soviet Baltic 1120 km (Rydzewski 1956), i.e. migratory trait weakest in central Europe, strongest in north and east. Nearly all autumn migration between south and south-west, with over 70% of Swedish and Danish breeders, 50–60% from other maritime countries, and 25–45% from central and eastern Europe moving south-west. Migrates on broad front, but some tendency to follow coasts and rivers; seas and mountains now crossed freely. Occurs throughout Near East (including Iraq) and Mediterranean basin in winter in increased numbers. Movements, if any, by few North African breeders unknown. Some European birds continue further south, mainly into steppe below Sahara where recoveries from 14 September. Recoveries in Sénégal, Guinea, and Sierra Leone from France, Netherlands, Switzerland, and Russia; in Mali and Upper Volta from Sweden, Netherlands, Hungary, Czechoslovakia, Poland, and Russia; in Togo and Nigeria from Germany, Hungary, and Russia; in Sudan and southern Egypt from Russia; and in Kenya from north Caspian (Volga delta) (Rydzewski 1956; Moreau 1967). Other recoveries of west European birds in Azores, Canary Islands, Madeira, and Cape Verde Islands, probably due to drift in easterly winds. Azores recovery of juvenile on 6 August, from France, clearly post-fledging dispersal as, perhaps, some other Atlantic islands recoveries, or even some of 5 ringed France recovered in west Atlantic (Martinique (2), Montserrat, Trinidad, and off Bermuda).

Return passage from winter quarters begins February, and colonies mostly re-occupied during March; but some young stay south during immaturity. Return movements predominantly reverse of autumn routes to north-east, but some evidence that slightly higher proportion moves through Italy in spring. Postulated that in response to climatic amelioration slow change towards nomadic rather than migratory habit, and that winter quarters in Africa below Sahara were in past of more importance (Rydzewski 1956). Many returns to natal heronries, but some displacements to other colonies to 490 km and one ringed Kaliningrad recovered July of seventh year in Komi ASSR (1800 km north-east) (Payevski 1971). Migration largely, but not exclusively, nocturnal; usually in small parties but flocks up to 200–250 noted in Caspian region (Dementiev and Gladkov 1951). May form mixed flocks with other Ardeidae, especially at night (Suchantke 1960).

**Food.** Chiefly fish, amphibians, small mammals, insects, and reptiles; occasionally crustaceans, molluscs, worms, birds, and (possibly as an aid in pellet formation) plant material. Feeds mostly during day, especially in morning and evening, on land or in shallow water where wades or stands, usually singly or in loosely associated groups. Prey caught by grabbing or stabbing and usually killed before swallowing. Exceptionally recorded swimming and diving for food (Lowe 1954).

Diet varies considerably with habitat and season. Numerous species recorded (see Witherby *et al.* 1939; Palmer 1962; Dementiev and Gladkov 1951), based on stomach analyses, regurgitations, pellets (limited value due to different rates of digestion), and observations. 200 stomachs from Hungary contained, by frequency, mainly fish (40·1%, chiefly carp *Cyprinus*, bleak *Alburnus*, Crucian carp *Carassius*, sunfish *Eupomotis*, pike *Esox*, rudd *Scardinius*), waterbeetles and their larvae (25·1%, chiefly *Dytiscus*, *Cybister*, *Hydrous*, *Hydrophilus*), frogs (20·3%, *Rana* and once *Pelobates*), and small mammals (14·5%, water-shrew *Neomys*, mole *Talpa*, water-vole *Arvicola*, field-voles *Microtus*) (Vasvari 1954). From northern Italy, 89 stomachs collected during summer months, contained by frequency 68·3% insects and their larvae (mainly mole-cricket *Gryllotalpa*, Hydrophilidae and Dytiscidae, and dragonflies), 26% fish (mainly *Eupomotis*), 21% amphibians (especially *Rana*), 24% reptiles (all grass snake *Natrix natrix*), 12·5% small mammals (including *Talpa* and *Arvicola*) (Moltoni 1936). Further 24 showed similar diet proportions (Moltoni 1948). 18 birds from France collected throughout 1921–26 showed relatively more fish (including perch *Perca*, tench *Tinca*, *Alburnus*, eel *Anguilla*), mammals (*Arvicola*), and amphibians (*Rana*), and rather less insects (including *Agabus*, *Helochares*, *Philhydrus*) (Madon 1935). In 72 stomachs collected at Oberlausitzer carp-ponds, fish again most important with 89% *Cyprinus carpio* (Schlegel 1964). 43 stomachs Danube delta, Rumania, contained by volume 87% fish (chiefly *Esox lucius*, *Carassius carassius*, *Perca fluviatilis*, *Cyprinus carpio*, sturgeon *Acipenser ruthenus*, *Tinca tinca*, and loach *Cobitus taenia*), weighing 1–125 g, with apparent preference for average of 70 g (Andone *et al.* 1969). Regurgitated food from 3 British heronries during several breeding seasons revealed diet differences related to different habitats (Owen 1955, 1960). At Wytham, where feed mainly in slow-moving waters, chief foods roach *Rutilus*, bleak *Alburnus*, perch *Perca*, gudgeon *Gobio*, three-spined sticklebacks *Gasterosteus*, and shrimps *Palaemonetes varians*. Fish caught selectively and those between 10–16 cm most frequently taken. Nestling regurgitations in Outer Hebrides showed mainly fed from coastal areas and mostly on fish (eel, wrasse *Labrus*, sea-scorpion Cottidae, 15-spined stickleback *Spinachia*), and crustaceans (Campbell 1949). Dementiev and Gladkov (1951) summarized data for USSR and emphasized wide variations in area and season: in Bessarabia, mice and susliks *Citellus*; in Kharkov region, frogs, tadpoles, toads, lizards, and mice; near Zaporozhe, fish, frogs, lizards, and insects; Astrakhan reserve, frogs, fish, water-voles and mice, and (for nestlings) grass snakes, insects and their larvae; near Lenkoran in summer, small fish, insects (including locusts, cicadas, blue darners), frogs, and in winter fish, especially *Rutilus frissi*, bream *Abramis*, carp, pike; on east Caspian shores in autumn fish (Gobidae) and, later on, crabs (*Potambus leptodactylus*, and *P. pachypus*).

Minimum daily quantity of food by adults *c.* 330 g up to 500 g (Creutz 1958, 1964; Heinroth and Heinroth 1926–7).

**Social pattern and behaviour.** Most intensively studied heron; see summaries in Witherby *et al.* (1939), Lowe (1954), Bauer and Glutz (1966), Milstein *et al.* (1970).

1. Both solitary and gregarious, At all times typically solitary feeder; at least some defend individual feeding territories to which they may return in successive years, especially outside breeding season (A A Blok). Congregations, sometimes large, also occur where available feeding areas restricted or at abundant but temporary food sources (see also Lowe 1954; Birkhead 1973b). Will join feeding aggregations of other species, including corvids and birds of prey, and form loosely mixed flocks with other long-legged marshbirds. Flocking tendency otherwise rather weak outside breeding season, temporary parties consisting usually of a few to some dozens though more may gather for roosting or migration. Juveniles associate in immediate post-fledging period; later small parties of yearlings sometimes keep together outside breeding season during dispersal, migration, and roosting (A A Blok; see also e.g. Lowe 1954, Olsson 1958). Yearlings also tend to stay together within flocks of mixed ages; individuals show interest in older birds when latter feeding (A A Blok), probably thus learning of good sites (see also below). BONDS. Monogamous pair-bond of season duration though in most cases ends earlier if breeding unsuccessful (A A Blok). In addition to some multiple associations during pair-formation, paired ♂♂ show tendency to promiscuity at times, mainly in form of rape attempts on ♀♀ other than mate (Verwey 1930; A A Blok). Attempts at bigamy also reported, with 2 ♀♀ at same or separate nests (A A Blok); in one, ♂ and 2 ♀♀ shared all duties and successfully raised 6 young (Spillner 1968). 1st-year birds frequent colony but most do not breed until 2nd year. Breeding by some yearlings well established, however (Holstein 1927; Verwey 1930; Milstein *et al.* 1970; etc.); these mostly ♀♀ paired to older ♂♂, but wholly yearling pairs also reported. Both parents tend young until after fledging, young returning for up to 3 weeks (see Milstein *et al.* 1970). Siblings show brood-bond, at least while still in colony; though initially attacked, fledglings sometimes join those of other broods and may even be fed by adults not their parents (Beetham 1910, and others). BREEDING DISPERSION. Normally colonial in exposed but protected sites; solitary nests also reported. Colonies typically spaced at intervals of several kms in suitable habitat, usually near feeding area though sometimes up to 20–30 km away. Size and spacing vary with food supply (Lack 1954b). Over much of range, colonies with other ciconiiform and pelecaniform birds quite common, but in Britain and more northerly parts usually nests on own or with totally unrelated birds such as crows (Corvidae). Formation of mixed colonies generally initiated by other, later nesting species. Although tends to nest higher than other birds in colony (Dementiev and Gladkov 1951), interspecific competition for sites often severe, especially with corvids and cormorants (see e.g. Nicholson 1929); when sites not limited, species tend to nest somewhat apart (A A Blok) or with groups of *A. cinerea* scattered between rest (Sterbetz 1961). Within colony, each ♂ initially defends display-site, usually existing nest, and adjoining area as pairing territory; later, breeding territory of pair generally confined to nest and immediate vicinity (Verwey 1930; Milstein *et al.* 1970; and others). Territory used for pair-formation, copulation, nesting, and most of raising of young. Later in cycle, wandering chicks and fledglings tolerated within other territories though not yearlings at any time. Some of latter usually present in colony, often initially at least in small visiting parties (A A Blok). Unpaired yearlings may occupy sites and build, but often perch above active nests, being especially attracted by social interactions of owners; frequently follow adults to and from colony in direction of feeding grounds (Owen 1959). ROOSTING. Both communal and solitary, nocturnal and diurnal, depending mainly on feeding routine. Active both by day and, perhaps to lesser extent, by night; often crepuscular. Information on roosting, especially at night, limited. At all times of year, single birds will sometimes roost solitarily, by day or night, mainly at or near feeding spot. In many areas, trees of colony probably used as communal nocturnal roost throughout year, especially by yearlings, though winter occupancy may be overlooked as birds arrive after sunset and leave before sunrise (A A Blok). Establishment of pairing territory may follow from roosting at particular site (Milstein *et al.* 1970). While breeding, pairs roost at or near nest, for periods both by day and night. Juveniles roost together in colony during post-fledging period (A A Blok). Outside breeding season, both yearlings and older birds roost nocturnally at times in groups in high trees or reeds (Bauer and Glutz 1966); also use cliffs, low rocks, or islets on treeless coasts (Witherby *et al.* 1939). At all times of year, tends to make use of communal standing grounds for day roosting; these often traditional and typically situated in isolated areas, usually open fields, but also sometimes under river banks or among vegetation at edge of field (T R Birkhead). Separate sites may be used at different seasons, i.e. in post-breeding period, during winter, and just before and during breeding season (when near colony). Birds arrive 2–3 hours after first light, depart in evening before dark, spending intervening period mostly inactive or preening. At winter day-roost, birds present throughout day but in breeding season, when attended by off-duty birds and unemployed yearlings, traffic between standing ground and nesting trees considerable (see especially Birkhead 1973a, b). Outside breeding season, numbers relatively small and adults often outnumber yearlings disproportionately, probably because latter need to spend more time feeding. In breeding season, numbers often larger, at any one time depending on size, composition, and activity of colony.

2. Key reference Verwey (1930). See also Krohn (1903), Stülcken (1943), Lowe (1954), Baerends and van der Cingel (1962), Strijbos (1962) and especially Milstein *et al.* (1970). Main

A      B

social interactions occur in colony, at or near nest-site away from which pair probably never meet as recognized mates, apart from aerial activities centred on nest. Pair-formation of heron-type 1 (see Ardeinae); associated with temporary colour changes involving iris, bill, and legs, especially in ♂ who assumes display colour just prior to site-establishment, ♀ somewhat later. Birds frequent colony and immediate vicinity well before nesting. Away from colony, interactions confined largely to antagonistic encounters while feeding and at roost, especially standing ground. During both antagonistic and heterosexual behaviour, variable use made of crest and other erectile plumes on lower neck and scapulars. ANTAGONISTIC BEHAVIOUR. Bird, especially ♂ in early part of cycle, defends site in colony from intruders by performing Arch-neck Upright display and Forward-display as threat. In Arch-neck Upright, stands erect facing in direction of opponent with body and neck near vertical but head and bill inclined downwards, and all plumes raised (see A); frequently accompanied by soft call (Baerends and van der Cingel 1962). Often precedes Forward-display in which bird moves body to horizontal position, curves neck back with crest and neck plumes erected, and vigorously thrusts head obliquely forward with bill open (see B)—typically while loud threat-call given though often Bill-snaps once instead. Will advance quickly to lunge at intruder perched in vicinity of nest, but prolonged combat rare. Threatened birds usually flee when often followed in pursuit-flight by territorial ♂ giving harsh call. Forward-display also directed at low-flying birds, and ♂ will leave site to fly at and supplant others nearby especially yearlings engaged in pairing displays (see Milstein et al. 1970). When returning to site, ♂ performs Alighting-display whether mate present or not (Milstein et al. 1970; Strijbos 1935; Thearle 1951): utters

characteristic call while flying in with powerful and somewhat laboured wing-beats, neck strongly arched, and feathers raised, particularly crest and neck plumes (see C). Similar behaviour shown by either sex later in cycle. According to Lowe (1954) and Baerends and van der Cingel (1962), ♂ at times directs distinctive Snap-display (see below) as threat at intruder or occupants of neighbouring nests, though Milstein et al. (1970) considered this common variant of Forward-display in which call replaced by Bill-snap. As shown by Baerends and van der Cingel (1962), however, although Snap and Forward-displays may share common initial posture with neck curved back, typically quite distinctive. They also recorded another threat version of Snap-display in March at standing ground in which one bird approaches other with neck stretched straight out then, when near, pecks near ground, sometimes seizing tuft of grass. Main threat-display at standing ground in spring, Arch-neck Upright: given mainly by incoming bird and, to lesser extent, by others in response to it; also at times by settled birds when one approaches or chases another (see Birkhead 1973b). In winter, little interaction between birds at standing site besides erection of dorsal plumes by alighting bird (Birkhead 1973a). GATHERING-GROUND BEHAVIOUR. Use of day-roost as 'dancing-grounds' at start of breeding cycle reported by Huxley (1924b), Lowe (1954), and others. Though birds mostly inactive, possible Dancing-display elicited by arrival of new bird, involves brief running or skipping with open wings, recalling dances of cranes Grus, during which short-term colour changes affect legs and bill (Lowe 1954). Such behaviour discounted as being no more than brief hostile chasing and dodging (Baerends and van der Cingel 1962; Milstein et al. 1970), but further observations needed. HETERO-SEXUAL BEHAVIOUR. Both sexes return to colony area at about same time early in year, but only ♂♂ take up sites in trees where pair-formation subsequently takes place. At site, each ♂ persistently utters far-carrying Advertising-call while perched conspicuously with head raised; continues at frequent intervals throughout day, less frequently at night, until mate obtained. In response to passing and in-coming birds, and calling or movement elsewhere in colony, lone ♂ also performs more visually dynamic pairing displays. In Stretch-display: (1) stretches body and neck up with bill pointing skyward (see D), (2) with neck and bill in same posture, lowers body by bending legs while inclining head backwards (see E); only neck plumes erected and posturing accompanied by distinctive double call. In Snap-display, ♂: (1) with body more or less horizontal, extends neck forward with bill pointing down; (2) lowers body by bending legs while depressing

C

D      E

F

neck and head until latter often at level of feet or lower (see F); (3) opens mandibles and Bill-snaps loudly once when head at lowest point. Crest and neck plumes erected during Snap-display; according to Selous (1927), crest shot upwards and forward at climax. ♂ may then also seize twig or Twig-shake more formally (Baerends and van der Cingel 1962), though neither behaviour seen by Verwey (1930) or Milstein *et al.* (1970). Advertising ♂ reacts at high intensity if ♀ approaches. Whereas ♂♂ treated in purely hostile manner, and ♂ continues to give Advertising-call, responses to ♀ more ambivalent, with reduction or cessation of calling. On sighting ♀, first performs Stretch-displays. When ♀ nearer, ♂ tends to respond with Arch-neck Upright and Forward-displays. If ♀ remains, ♂ gradually becomes less aggressive and does more and more Snap-displays, up to 40 in succession. Pair-formation follows quickly from initial acceptance of ♀ at site (Strijbos 1935), or more prolonged. Rôle of ♀ during pairing phase essentially passive, usually with little more than Bill-rattling (see below); typically accepted only if her approach cautious. After pair-formation, ♂ performs Snap-display infrequently; also then recorded rarely in ♀ (see Milstein *et al.* 1970). Although Stretch-display shown only by ♂ during pairing phase, also performed by ♀ afterwards, especially in meeting situation (see below). Once couple together at site, start mutual Billing and Allopreening; in earlier stages, often lead to brief Bill-sparring. When Billing, pair face and touch tips of mandibles briefly (Selous 1927), caress or prod each other's bills, or occasionally interlock them. Allopreening done mainly unilaterally by ♂ who also not infrequently initiates bouts of simultaneous Allopreening by pair, but ♀ rarely initiates or performs on own (Milstein *et al.* 1970). All parts of plumage treated, especially neck and mantle; also at times gently grips head, carpal-joint, or other parts (occasionally twigs in nest). Bill-rattling, produced by rapid opening and closing of mandibles, also performed by pair at nest (particularly ♂), especially in period immediately following acceptance of ♀ by ♂ (Hudson 1965). First shown by ♂ and ♀ during pairing phase (in ♀, especially when ♂ seems reluctant to accept her); between pair-formation and incubation (mating phase), typically associated with Allopreening. Bill-rattling bird holds head clear of mate's feathers, frequently with neck arched and lowered and bill pointing down; neck often over that of mate, over its back, or in front or behind it. Bill-rattling, Allopreening, and Billing typically associated with copulation which also follows soon after acceptance of ♀ at site, particularly after return of one of pair (Selous 1927)—often ♂ coming with twig. In early part of mating phase especially, ♂ frequently flies out in search of material after allowing ♀ to occupy nest; returns and passes twig to ♀. Both before and after pair-formation, ♂ will also engage in brief Circle-flights from site (Selous 1905, 1925, 1927): flies out with 3–4 loud, heavy flaps, recalling Flap-flight of smaller herons (see Cattle Egret *Bubulcus ibis*), and neck outstretched or somewhat bent. Circle-flights also observed later from ♀ during mating phase (Verwey 1930; Milstein *et al.* 1970), while joint flight by pair may occur after copulation (Selous 1927). Performance of

Alighting-display by twig-bringing ♂, or on return to nest by either sex at other times, initiates Greeting Ceremony if mate at site; especially intense during mating phase but persists more or less throughout cyclè. As mate flies in, bird on nest usually responds with Arch-neck Upright or Stretch-display, latter predominating as cycle progresses (see Milstein *et al.* 1970); one or both may also erect crest and neck plumes and wave open wings while both call loudly (Huxley 1924*b*). Alighting-display by returning bird and Stretch-display by bird incubating, brooding, or guarding young also typical later in cycle of nest-relief (Selous 1901); Allopreening and Billing also tend to occur then, especially at hatching (Milstein *et al.* 1970). Once pair-formation achieved, ♂ as well as ♀ largely inactive at nest when mate absent. Copulation occurs up to 3 times a day, lasting up to 16 s and continues until most or all eggs laid. Often initiated by Allopreening (and Bill-rattling), mostly by ♂ who tends to concentrate on ♀'s back and rump feathers before mounting standing ♀ with raised wings and grasping her neck in bill. ♀ may at times also stimulate ♂ to mount by deliberately pushing against him or adopting horizontal copulatory posture with tail raised; often calls during copulation. Afterwards, ♀ and sometimes ♂ fluff feathers and both frequently self-preen; ♀ also often Twig-shakes and sometimes does Stretch-display, while both may perform Snap-display. Allopreening, mostly by ♂, may also occur (Milstein *et al.* 1970). RELATIONSHIPS WITHIN FAMILY GROUP. Chick-raising phase divisible into 4 stages: (1) brooding, (2) guard, (3) post-guard, and (4) fledging, young gradually becoming independent during latter (Milstein *et al.* 1970). During first 2 stages, lasting 26–31 days, always at least one parent at nest; thereafter, adults usually visit nest briefly only to feed young. During guard stage, attendant parent perches near nest; will come to shelter young from rain, snow, or hot sun. Young beg by calling soon after hatching; from 7–10 days of age also stimulate parent by grasping and tugging its bill, pulling head down to floor of nest. Such direct soliciting becomes more and more vigorous as chicks grow, parents' feathers becoming soiled and damaged, or even plucked out; such treatment important factor in making adults avoid nest as much as possible after brooding stage over. Chicks grasp one another when hungry, and sometimes quarrel with erect necks and Bill-snapping. More intense fights usually confined to feeding sessions or just before or after. Start at 7th day and up to 15th consist mostly of harmless pecking exchanges. Can be more vicious later, sometimes resulting in bullying of smaller chicks and even their eviction; evidence of cannibalism by siblings or parents summarized by Milstein *et al.* (1970). Return of fledged sibling to nest may elicit brief threat from others followed by food-soliciting; after end of guard stage, strange adults and yearlings first greeted by begging, then by threat. Arrival of wandering chicks results in threat from resident young and fighting. Threat-displays of young include Forward-display, with Bill-snap more frequent than Threat-call, and true Snap-display, with bill directed away from intruder (Baerends and van der Cingel 1962); also aggressive Upright-display with slight snapping and pecking at opponent (for further details, see Milstein *et al.* 1970).

(Figs A, C, and F after Milstein *et al.* 1970; rest after Baerends and van der Cingel 1962.)

**Voice.** Away from colony, main utterance familiar Flight-call usually uttered by solitary bird: loud, harsh, rather cough-like 'frarnk' (see I) with variants; rendered 'kräick' by Bauer and Glutz (1966). Much more vocal at colony, with variety of calls, some harsh and croaking. Studied in

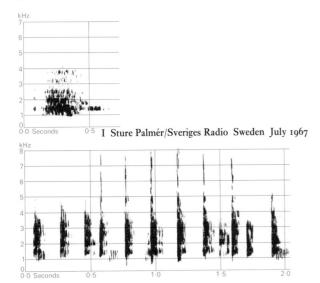

I Sture Palmér/Sveriges Radio Sweden July 1967

II Sture Palmér/Sveriges Radio Sweden July 1967

detail by Verwey (1930); see summaries in Witherby *et al.* (1939) and Bauer and Glutz (1966), and recent account by Milstein *et al.* (1970). In addition to calls, non-vocal sounds include: (1) single, formal Bill-snap producing sharp 'clop', given at climax of special Snap-display (mainly by ♂); (2) one or more Bill-snaps during antagonistic encounters and sometimes as apparent greeting between mates; (3) quieter Bill-rattling during pair-formation and later when pair together at nest (see Hudson 1965). Frequency of much louder Bill-clattering, much as in White Stork *Ciconia ciconia*, disputed—but evidently rare.

Following main calls at colony; uttered by both sexes unless otherwise indicated. (1) Advertising-call. Sharp, single, yelping 'rwo', loud and far-carrying. Song of unpaired ♂ only; given from nest-site at start of breeding season to attract ♀. (2) Stretch-call. Short, rather soft 'hoo' followed by longer, gargling 'oooo'; 'koop choo-oo-oor' of Selous (1927) probably same. Accompanies Stretch-display: 1st component at height of stretching phase, 2nd during crouching phase. (3) Threat-call. Loud, harsh, squawking 'qooo'; given during lunge of Forward-display. (4) Pursuit-call. Even louder, harsh 'schaah'; uttered in flight by ♂ chasing intruder from site (5) Anxiety-call. Ringing, but low and soft, nasal 'go-go-go', etc.; not confined to breeding season, given when mildly alarmed (e.g. when man below nest). Soft 'gog-gog-gog' while in Arch-neck Upright-display (Baerends and van der Cingel 1962) probably same. (6) Alighting-call. Series of grunting squawks, diminishing in volume and often ending in clucking; uttered in flight by incoming bird before landing at nest, whether mate present or not (see Milstein *et al.* 1970). Rendered 'arre-arre-ar-ar-ar-ar' by Verwey (1930). If mate present, first stage in Greeting Ceremony; may be answered by toned-down version or, more typically, by Stretch-call. (7) Copulation-call. Growling or wailing

sound, increasing in volume; difficult to describe (Milstein *et al.* 1970) but clearly same as soft, plaintive call of Holstein (1927) and perhaps croak of Huxley (1924*b*). Given at times by ♀ only; not certainly indicative of unsuccessful mating as claimed by Verwey (1930).

CALLS OF YOUNG. Chittering sound uttered as Food-call for first few days, later becoming harsher chacking which persists to after fledging; uttered intensely as young solicit food from parent, becoming a high-pitched squealing (like domestic pigs being fed) as large chicks gobble up meal (Milstein *et al.* 1970). Rapidly repeated call of older young rendered 'ek-ek-ek' (J Hall-Craggs; see II). For other calls, including development of adult-like ones, see Verwey (1930) summarized by Witherby *et al.* (1939).

**Breeding.** SEASON. See diagram for northern Europe. In Britain, first eggs in early February; normal start of laying early March, with peak end March, early April, and last eggs end April; repeat layings to June (Owen 1960). Few data from elsewhere in west Palearctic. SITE. Usually in tall trees with nests up to 25 m above ground, but also in low trees and bushes, especially on islands in lakes and sea; sometimes on cliff-ledges, in reedbeds, on ground among heather, exceptionally on buildings, bridges, or in open on shingle beach or bare ground. 1–3 nests per tree most common, but often up to 10, rarely up to 25. Normally colonial but single nests not uncommon. Same sites occupied for many decades. Nest: pile of sticks and twigs, usually lined with smaller twigs, leaves and grass; in reedbeds, built of reeds; generally substantial but can be frail (Milstein *et al.* 1970); ground nests may be reduced to slight scrape, ringed with a few small stones and debris. Previous year's nests often re-used with new lining or additions to structure. Building: by both sexes, ♂ mostly brings material while ♀ constructs. Twigs gathered from ground, from trees, or stolen from other nests. Foundation may take several days, then nest completed in further 3–5 days. EGGS. Blunt oval; pale blue, often becoming stained during incubation. 61 × 43 mm (53–70 × 39–50), sample 300 (Schönwetter 1967). Weight 60 g (48–69) sample not given (Bauer and Glutz 1966). Clutch: 4–5 (1–10). Of 222 clutches, England: 2 eggs, 2%; 3, 19%; 4, 61%; 5, 18%;

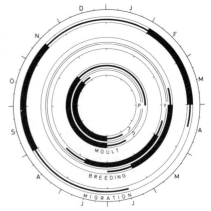

mean 3·95 (Owen 1960). Of 85 clutches, Switzerland: 1 egg, 2%; 2, 1%; 3, 20%; 4, 65%; 5, 11%; 6, 1%; mean 4·53 (Glutz von Blotzheim 1962). In France, 5-egg clutches more common than 4-egg; in Denmark, 4-egg clutches more common (Guichard 1949; Holstein 1927). Normally 1 brood; rarely 2. Replacement clutches common, up to 3 times, even after losing large young; usually same size as first clutch but may be larger or smaller (Owen 1960). One egg laid every 2 days on average; sometimes gap of 3–4 days. Most laid early morning. INCUBATION. 25–26 (–28) days, but up to 32 recorded. Both sexes incubate. Begins with 1st egg, occasionally later. Asynchronous hatching. Eggshells removed after hatching. YOUNG. Semi-altricial and nidicolous. Cared for and fed by both parents. Brooded continuously for *c*. 18 days (range 11–21), and parents stand guard for further 11 days (7–15) (Milstein *et al.* 1970). Food completely regurgitated on to floor of nest after bill tugging by young. Young leave nest and clamber on nearby branches at 20–30 days. FLEDGING TO MATURITY. Fledging period *c*. 50 days (42–55). Young return to nest for further 10–20 days; become independent soon after. Age at first breeding probably mostly 2 years, but 1 often recorded. BREEDING SUCCESS. Linked to brood size at hatching and to food supply. Average number of young reared, England: 2·0 from 29 broods of 2 at hatching; 2·6–2·9 from 63 broods of 3; 2·9–3·8 from 74 broods of 4; 3·1–3·5 from 9 broods of 5; all broods of 2 successful, broods of 3 lost none in years of good food supply, a few in bad years, broods of 4 and 5 lost more in bad years; smallest young always dies first in year of food shortage. Percentage of young dying varied from 2–46% over 3 years. Of young in 125 broods, 472 hatched and 367 reared (Owen 1960).

**Plumages.** ADULT. Forehead and crown white, often with slight admixture of grey. Sides of crown down to eyes and nape black, feathers loose and elongated; on nape 2 or 3 long (up to 20 cm) narrow black plumes. Sides of head, chin, and upper throat white. Neck pale grey, slightly tinged mauve; gular stripe of black and white spots down centre of foreneck. Upperparts blue-grey; ornamental feathers of lower mantle and upper scapulars paler, overlying much of upperparts and each split in several lanceolate points; shoulder patches at sides of mantle black. Ornamental feathers of chest long and loose like scapulars; pale grey at sides, black and white in centre. Rest of underparts mainly white with black wedge extending from centre of belly along sides of breast towards black shoulder patches. Flanks grey. Tail grey with grey-black tip. Primaries, primary coverts, bastard wing, and most of secondaries black; inner secondaries blue-grey. Underside of inner webs of primaries, underside of secondaries, and under wing-coverts grey. Marginal coverts at leading edge and some lesser coverts at carpal joint white; rest of upper wing-coverts grey, greater paler than median and lesser, those on outer part of wing palest, but concealed inner webs of greater coverts dark grey. Axillaries grey. NESTLING. Down soft and long, brown-grey on upperparts, paler on flanks, white on underparts. Tips of down on forehead, crown, and nape long, silver-white, forming conspicuous crest. Chin, throat, and orbital area bare. Pins of growing feathers appear at *c*. 1 week, fully feathered with down

adhering to tips at *c*. 4 weeks. Tips of down retained longest on feathers of crown and nape. JUVENILE. Forehead, crown, and streak over eye dark grey; nape dull black, feathers moderately elongated. Sides of head pale grey, chin white. Neck darker grey than adult, gular stripe mixed with buff. Upperparts darker grey without ornamental feathers. Underparts without black wedge, streaked brown-grey, particularly on chest; feathers of chest only slightly elongated. Upper wing-coverts darker grey than adult, often tinged brownish; paler, often white, at tips; white line along leading edge of wing strongly suffused cinnamon-buff. IMMATURE. Gradually acquired 1st autumn and winter; may nest in this plumage. More blue-grey above than juvenile, feathers of lower mantle and scapulars structured like adult, but shorter; black feathers on nape slightly glossy, longer; breast and belly with less brown-grey spots; gular stripe less buff; feathers of chest slightly more elongated. FIRST SUB-ADULT. Worn from autumn of 2nd calendar year. Differs from full adult by grey forehead and crown with only some white and by imperfectly developed black body wedge. Differs from immature by black streak over eye, long black crest plumes, black shoulder patches, and much paler grey neck. Upperparts variably darker or lighter grey. This and subsequent plumages highly variable, more like adult in some, more like juvenile in other individuals. Age determination on plumage characters therefore not precise after middle of 2nd calendar year. From autumn of 3rd calendar year another plumage worn, not always separable from preceding, but normally with more white on forehead and crown and with fully developed black body wedge. Grey on crown may apparently persist for several more years.

**Bare parts.** ADULT. Iris yellow. Bill yellow, darker along culmen; lower mandible yellow-brown. Bare skin of face yellow, round eye tinged green. Leg dull brown, tibia and inner side of tarsus often yellow. In sexually active but unpaired birds, iris, upper mandible, and to some extent lower mandible become deep orange or vermilion. Often tibia and tarsus also red. After pair-formation, iris and leg yellow, but bill remains partly red until after hatching of young. NESTLING. Iris yellow. Upper mandible dark grey, lower pale grey. Leg pale green-grey. JUVENILE. Iris yellow. Upper mandible dark brown, lower yellow. Leg dark grey, paler and with yellow-green tinge at joint and on bare part of tibia. In spring of 2nd calendar year, red tinges may appear as in adult, often more pronounced in ♂♂ than in ♀♀. (Heinroth and Heinroth 1926–27; Witherby *et al.* 1939; Milstein *et al.* 1970; A A Blok.)

**Moults.** ADULT POST-BREEDING. Complete; June–November. Begins before fledging of young. Inner primaries descendant, outer irregular, p10 often moulted before p8 and p9. Occasionally greater irregularities in older birds. Moult of body feathers may continue into winter; plumes on nape sometimes still growing March. Not known whether this may be considered partial pre-breeding moult. POST-JUVENILE. Partial; involving most of body, rarely some upper wing-coverts, not rest of wing. September–February, timing highly variable. IMMATURE. Parallel to adult post-breeding, but primaries may start earlier; some with most replaced June, presumably non-breeders. (Witherby *et al.* 1939.)

**Measurements.** Older than juvenile, Netherlands; skins (ZMA).

| | | | | | | |
|---|---|---|---|---|---|---|
| WING | ♂ | 457 (12·2; 20) | 440–485 | ♀ | 443 (9·77; 12) | 428–463 |
| TAIL | | 174 (7·21; 20) | 161–187 | | 166 (4·89; 12) | 157–174 |
| BILL | | 120 (5·76; 26) | 110–131 | | 112 (5·06; 19) | 101–123 |
| TARSUS | | 151 (8·14; 23) | 136–172 | | 141 (5·74; 16) | 132–153 |
| TOE | | 108 (5·44; 8) | 101–116 | | 100 (3·56; 5) | 96–104 |

Sex differences significant. Witherby *et al.* (1939) gave smaller minimum lengths for wing (♂ 430, ♀ 425) and bill (♂ 100, ♀ 100), probably pertaining to juveniles.

**Weights.** Adults, Holland; ♂♂ 1505 (300; 17) 1071–2073; ♀♀ 1361 (258; 13) 1020–1785; difference not significant (ZMA). Maximum, adult ♂, 2300; minimum, sex and age not specified, 810 (Bauer and Glutz 1966). Winter Caspian sea: ♂♂ 1350–1770; ♀♀ 1100–1470 (Dementiev and Gladkov 1951). Nestlings at hatching 45, reaching peak of *c.* 1200–1600 at 3½ weeks (Owen 1960).

**Structure.** Wings long and broad. 11 primaries: p8 longest, p7 and p9 often equal, may be 0–5 shorter, p10 5–20 shorter, p6 10–30, p5 29–65, p1 132–161; p11 minute. P8–p10 emarginated inner webs, p7 slightly, p7–p9 outer webs, p6 slightly. 19 secondaries (Stephan 1970). Tail square, 12 feathers; under tail-coverts about as long as tail. Bill heavy, broad and high at base, compressed and pointed at tip; nasal groove deep, hence culmen ridged; outer ⅓ of cutting edge serrated, slight notch near tip. About ⅓ of tibia bare. Middle toe *c.* 70% of tarsus, relatively shorter than in Purple Heron *Ardea purpurea*; outer toe *c.* 81% of middle, inner *c.* 75%, hind *c.* 51%, hind claw *c.* 17%.

**Geographical variation.** *A. c. jouyi* paler on neck and upper wing-coverts than nominate *cinerea*; this variation probably clinal, with wide area of intermediate populations in Asia (Vaurie 1965). *A. c. monicae* generally still paler than *jouyi*; sides of neck pure white; black on gular stripe reduced; upperparts pale grey with white tips to longest scapulars; upper wing-coverts grading from pale grey lesser to white greater; wing apparently shorter (Jouanin and Roux 1963). *A. c. firasa* with bill and tarsus longer, bill heavier than *cinerea* (Hartert 1912–21). Populations of Java and Sumatra also tend to heavy bill, especially lower mandible; named *altirostris* Mees, 1971, but difference slight and presumably not sufficient for subspecific recognition (Parkes 1974).

## *Ardea herodias* Linnaeus, 1758 **Great Blue Heron**

FR. Grand héron    GE. Nordamerika-Reiher

New World counterpart of Grey Heron *A. cinerea*, sometimes regarded as conspecific (Curry-Lindahl 1968). Separable in all plumages (except white colour phase) by rufous to cinnamon thighs; for other differences see Palmer (1962). Breeds British Columbia, across southern Canada to southern Quebec, and in New Brunswick and Nova Scotia; south through USA into Mexican lowlands; also in West Indies and Galapagos Islands. 9 subspecies recognized. Most migratory race nominate *herodias* of northern and eastern North America, which, after period of random post-breeding dispersal, mostly migrates south of breeding range and in winter occurs throughout Central America, Caribbean and on north coast of South America (Godfrey 1966; Palmer 1962). Notable ringing recoveries include: Minnesota to Mexico and Panama; Michigan to Jamaica, Cuba, Nicaragua and Honduras; Illinois to Honduras and Guatemala. Autumn migration mainly mid-September to late October, return early March to early May; on both passages notable concentrations along Atlantic coastline. Straggles north to Hudson and Ungava Bays, Newfoundland and south Greenland. One landed on east-bound ship in North Atlantic at 34°50′N 33°20′W on 29 October 1968, staying aboard in free state until captured when ship past Azores (King and Curber 1972).

## *Ardea purpurea* **Purple Heron**                                    PLATE 42
[after page 326]

DU. Purperreiger    FR. Héron pourpré    GE. Purpurreiher
RU. Рыжая цапля    SP. Garza imperial    SW. Purpurhäger

*Ardea purpurea* Linnaeus, 1766

Polytypic. Nominate *purpurea* Linnaeus, 1766, south-west Palearctic east to Russian Turkestan and Iran, east and south Africa; *bournei* de Naurois, 1966, Santiago, Cape Verde Islands. Extralimital: *madagascariensis* van Oort, 1910, Madagascar; *manilensis* Meyen, 1834, south and east Asia.

**Field characters.** 78–90 cm (body 38–45 cm), wing-span 120–150 cm; smaller, shorter winged, and slighter than Grey Heron *A. cinerea*. Quite large, thin-headed, long-necked, narrow-bodied, noticeably long-toed heron, with dark, rich plumage tones. Sexes alike; little seasonal variation. Juvenile separable.

ADULT. Bare lores yellow-green. Head and neck bright reddish-buff, strongly patterned by black crown, plumes, and stripes from below eye to nape and from upper throat along sides of neck. Mantle and wing-coverts slate, with scapulars elongated and pale chestnut in breeding season; quills black; rump and tail grey. Chin and cheeks white; foreneck pale buff and white, lower half overlaid with elongated black spots dispersing into loose chestnut and white feathers falling from chest. Shoulder area, breast, and belly vinaceous-chestnut (increasingly black below); flanks brown-grey; under tail-coverts black. Head-on in flight, leading edges of wings show buff on carpal area (as in *A. cinerea* but less bright); underwing intensely dark, with most coverts vinaceous-chestnut and quills dark grey. Bill as long as slender head (appearing rather fine), bright buff-yellow in breeding season, duller at other times. Eyes yellow. Legs and feet dark brown. JUVENILE. Plumage lacks dark intensity and sharply delineated head and neck patterns of adult, being generally paler and less slate, with buff edges to upperpart feathers and more uniform buff underparts. 2nd-year birds brighter, showing plumage pattern of adult but colours often sandy in tone. Bill duller than adult.

At long range or in poor light, liable to be confused with *A. cinerea* but, in addition to plumage differences, structure and character distinct. Flight silhouette characterized by coiled neck (drooping in obvious bulge), narrower body and wings than *A. cinerea*, and marked extension of long feet. On ground, more furtive than *A. cinerea*, rarely leaving cover, and gait affected by length of toes.

Flight action as *A. cinerea* but wing-beats freer, its lighter body sometimes lifting noticeably on down-stroke. Feeds mainly by slow stalking with long pauses with neck sharply coiled for instant strike. Flight-call similar to *A. cinerea* but less loud and thinner in tone.      DIMW

**Habitat.** Within west Palearctic, chiefly in warm, mid-latitude steppe and mixed forest zones; strongly attached to wetlands bearing extensive tall, dense stands of vegetation such as reeds *Phragmites*, with slight or no woody intrusions, usually in shallow, permanent fresh water. Sometimes in willows *Salix*, tamarisk *Tamarix*, and other shrub growth; also, in Africa, in mangroves growing in saline coastal waters. Atypical subspecies *A. p. bournei* in Cape Verde Islands nests in tall mango trees growing in deep ravine, foraging on neighbouring dry hillsides and flighting some 10 km to coastal inlet (Bannerman 1968). Nominate race prefers fairly shallow, eutrophic waters, with sandy, silty, muddy, or vegetated, fairly level bottom, free of rocks and other hard obstructions, standing or only gently flowing, and usually surrounded or closely flanked by dense cover of reeds, rushes, and similar tall vegetation. Nests in such tall vegetation, preferably on stems emerging from 0·4 to 1·3 m depth of water (Glutz von Blotzheim 1962). On passage and in winter, more often on open river banks or other waterside sites, including seashore and sandspits, or on short grassland. Normally fishes out to extreme wading depth, if possible near cover. Often finds ample food near nest, but readily undertakes long foraging trips when needed. Flies high, strongly, and for long distances, often thus giving sole indication of local presence. Intolerant of human disturbance (except *A. p. bournei*), and rarely accepts man-made substitutes for natural haunts; so vulnerable to drainage and readily deserts. Much less catholic, less mobile, and less arboreal than Grey Heron *A. cinerea*, rather resembling in habitat Bittern *Botaurus stellaris*, although less specialized and flies more frequently, both locally and seasonally.

**Distribution.** In central Europe, tendency to expand from *c*. 1940 leading to several, mostly temporary settlements in centre and west (Bauer and Glutz 1966). CZECHOSLOVAKIA: bred irregularly 19th century, and annually since 1947 (KH). WEST GERMANY: bred oc-

casionally in 19th century; nested 1947, and probably annually from 1955 in south (Bauer and Glutz 1966). SWITZERLAND: breeding since 1941 (Bauer and Glutz 1966). POLAND: bred Silesia 19th century; since 1956 probably only at Milicz and once elsewhere (AD, LT). BELGIUM: nested 1943 and possibly 1950, 1955 (Lippens and Wille 1972). ITALY: no longer breeding Sicily (GS). CAPE VERDE ISLANDS: not discovered breeding until 1951 (Bourne 1955).

Accidental. Faeroes, Britain and Ireland, Norway, Denmark, Sweden, Finland, East Germany, Azores, Madeira, Canary Islands.

**Population.** NETHERLANDS: increasing in last decade; *c*. 900 pairs 1971 (Rooth and Jonkers 1972). FRANCE. Marked recent decrease. Estimated 2100–2500 pairs in 1964–8, 1350–1500 pairs in 1974 (Brosselin 1975). WEST GERMANY. Probably 20–30 pairs in all. Rhine valley: first bred 1967, at least 9 pairs 1968, and 18–20 pairs 1970 (EB). SWITZERLAND. Fluctuating numbers. 3 pairs 1941, then marked increase, especially from 1949, to 54 pairs in 1955. Declined until 1960, 40–45 pairs 1961, and only 8–10 pairs 1972 (Bauer and Glutz 1966; WT). AUSTRIA: 240–350 pairs Neusiedlersee 1950–5; few elsewhere (Bauer and Glutz 1966). CZECHOSLOVAKIA: 4–20 pairs since 1947 at Lomnice, Bohemia (Bauer and Glutz 1966). HUNGARY: colony at Kisbalaton declined from 107 to 37 pairs 1968–72 (AK, IS). ITALY: 1500–2500 pairs 1975 (SF). SPAIN: for details of colonies, see Fernandez Cruz (1975). YUGOSLAVIA. Kopačevski area: number fluctuating between 91 and 287 pairs 1954–70 (Majić and Mikuska 1972). Lake Ludaško: *c*. 100–250 pairs 1959–68 (VFV). GREECE: at least 300 pairs, Lake Mikra Prespa (Terrasse *et al.* 1969). RUMANIA: common, but decreasing (MT). CAPE VERDE ISLANDS: only São Tiago; not more than 200 pairs (de Naurois 1969*b*).

Survival. Oldest ringed bird 23 years 2 months (*CRMMO Bull.* 1964, 18, 24).

**Movements.** Migratory and dispersive. As with most European herons, juveniles disperse in all directions after fledging, and remain nomadic at least through August. At this season appears relatively often outside breeding range, e.g. England, Belgium, northern West Germany, Denmark, exceptionally north to Fenno-Scandia and central Siberia (Tomsk). Some Swiss juveniles reach south-east France in post-fledging dispersal (Bauer and Glutz 1966). True migration begins August, lasting well into October, with stragglers to December; found exceptionally in mid-winter north to Belgium and England.

Though a few overwinter (perhaps only irregularly) Mediterranean basin, Egypt, and from south Caspian to Arabia (recorded mainly Bahrain), sporadically east to Baluchistan, great majority of west Palearctic breeders winter in Africa south of Sahara, mainly in adjoining steppe country, from early September to mid-April, but

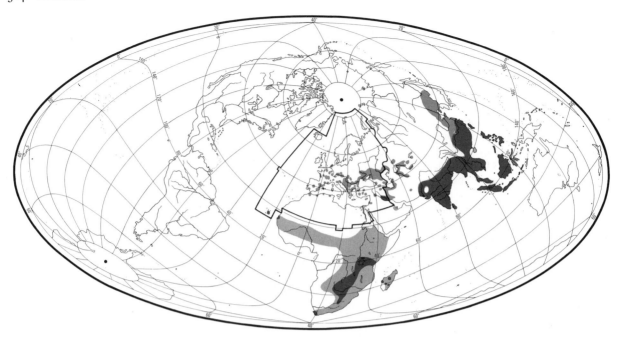

chiefly October–March (Moreau 1967). Most Dutch birds migrate SSW in autumn through France and Spain, but some SSE to south-east via Italy and Greece. Those from Switzerland and Hungary also move predominantly south-west, and growing evidence that USSR breeders do likewise. Standard south-west heading may apply to unringed populations as well (see Grey Heron *A. cinerea*). Occurs throughout Near East and Mediterranean basin on both migrations, and various oasis records demonstrate broad-front crossing of Sahara (Moreau 1967). African recoveries include Netherlands to Mauritania, Sénégal, Sierra Leone, Mali and Niger; France to Gambia, Sierra Leone and Mali; Sea of Azov to Nigeria and west Sudan; and Volga delta to Cameroons. Absence of recoveries from east Africa surprising, as common migrant through Egypt (Meinertzhagen 1930), and southward passage occurs late September to early October along coastal Eritrea (Smith 1957).

First migrants reach European breeding range early March, but majority during April and May. Then prone to overshoot nesting areas in warm anticyclonic weather with southerly winds, when annual in Britain (rare Ireland); recoveries of Dutch birds in May on Isles of Scilly and Fair Isle. Some Swiss birds perhaps use different route in spring: of recoveries up to 1965, 20 in autumn in Spain and south France and only one in Italy, while 9 in April included 5 from Italy (Bauer and Glutz 1966). One of French origin recovered Mali in July of 2nd calendar year indicates some remain well south during immaturity.

Migrates mainly at night, usually singly or in small groups; in USSR individuals have been seen in flocks of migrating Night Herons *Nycticorax nycticorax* (Dementiev and Gladkov 1961). *A. p. bournei* of Cape Verde Islands not identified elsewhere; may be sedentary (de Naurois 1966).

**Food.** Chiefly fish and insects (larvae and adults) taken generally early morning and evening, mainly from stationary position. Feeds singly, though often several in small area, either standing on floating vegetation or in shallow water among dense vegetation while stretching neck at *c.* 60° angle above horizontal with eyes looking down. Also fishes by wading slowly with beak horizontal and close to water: fish caught crossways with rapid strike of head and neck, out and down, and food then turned head towards gape before swallowing (Tomlinson 1974). To lesser extent, eats small mammals and amphibians; also snake (*Natrix*), lizard (*Lacerta*), and occasionally birds, crustaceans, molluscs, and spiders.

Fish include bream *Abramis*, carp *Cyprinus*, perch *Perca*, 10-spined sticklebacks *Pungitius*, roach *Rutilus*, pike *Esox*, rudd *Scardinius*, eel *Anguilla*, sunfish *Eupomotis*, carp *Carassius*, *Gambusia*, *Tilapia*. Insects include waterbeetles (*Dytiscus*, *Cybister*, *Acilius*, *Hydrophilus*, *Coelambus*); aquatic hemipterans (*Notonecta*, *Naucoris*); dragonflies *Libellula*; and mole-cricket *Gryllotalpa*. Mammals include vole *Microtus*, mole *Talpa*, water-vole *Arvicola*, and shrew *Sorex*. Amphibians include frogs *Rana* and *Pelobates*, and salamander *Triton*.

113 stomachs from Hungary contained primarily fish (52%) and insects (terrestrial beetles 46·0%, aquatic beetle larvae 38·9%, dragonflies 30·9%, aquatic Hemiptera 28·3%), with small mammals 24·7% (*Microtus arvalis* and occasionally *Sorex*) and amphibians 22·1% (Vasvari 1931, 1938). In Italy, 12 samples 91% fish, especially *Eupomotis gibbosus*, and *Cyprinus*; and 50% insects, chiefly Hy-

drophilidae, Dytiscidae, and Libellulidae (Moltoni 1936); similar prey in further 7 stomachs (Moltoni 1948). Madon (1935) concluded, mainly from 28 analyses in France, that chief prey fish and insects, followed by small mammals and amphibians. Fish main diet in USSR (Dementiev and Gladkov 1951). In Africa, Cott (1952) suggested exclusively fish, especially *Tilapia*, and Chapin (1932) found from 4 stomachs mainly fish, including small puffer *Tetraodon*, and some insect remains. 105 stomachs, Danube delta, Rumania, contained by number 56% fish (predominantly *Esox lucius, Rutilus rutilus, Carassius*, and *Perca fluviatilis*—1–102 g) and 44% undetermined animal remains (Andone *et al.* 1969). Also in Rumania takes suslik *Citellus citellus* (Dombrowski 1912; Lintia 1955). 70 stomachs, Sundarban, India, contained by weight 57% fish (32 species, mostly *Mystus gulio, Anguilla bengalensis*), 20·6% snakes (especially *Natrix stolata*, up to 190 mm long), 14·3% crustaceans (prawns, crabs), and 7·7% insects (Mukherjee 1971).

73 food samples from nestlings in Naardermeer and Nieuwkoop in Holland, late June, most frequently contained fish—mostly 110–200 mm, especially *Scardinius* (27% at Naardermeer)—and water-voles *Arvicola amphibius* (27% at Naardermeer), with a few water insects (Owen and Phillips 1956). In Camargue, 7 young and 17 pellets examined in late May and early June had many more insects (especially *Cybister* and *Gryllotalpa*) and smaller numbers of larger fish (Williams 1959); may

indicate that more insects fed to younger birds, and larger prey (such as fish and mammals) to older nestlings.

**Social pattern and behaviour.** Least known of European herons; for summaries, see Bannerman (1957); Bauer and Glutz (1966). For important recent observations in South Africa, see Tomlinson (1974*a, b*).

1. Largely solitary away from colony or roost. Usually feeds singly. In small flocks on migration, occasionally up to 100. BONDS. Monogamous pair-bond so far as known, of at least seasonal duration. Both parents tend young until after fledging. BREEDING DISPERSION. Mostly in small, loose colonies or singly, at well-concealed sites. Often only 2–3 pairs together (Smythies 1960). Will nest near other Ardeidae, including Grey Heron *A. cinerea*. 7–19 nests in 8 colonies located with tendency to group in twos and threes (Tomlinson 1974*a*); similar clumping of nests in dense cover recorded by Child (1971). Suitable cover in which young can hide important factor in siting of colonies. Each pair defends nest-site territory (Tomlinson 1974*b*), but no details on size. ROOSTING. Communal, at least at times. As more or less crepuscular feeder, rests both during day and night, either in open or in dense aquatic vegetation. In Africa, travels singly or in small parties to communal roost; over 40 seen to gather at dusk for night on bare expanse by lake, beginning to disperse *c.* 1 hour after sunrise (see Bannerman 1957).

2. Available data on European population suggest closely similar to *A. cinerea* in displays and other social behaviour (Bauer and Glutz 1966), but details lacking and similarity not confirmed by observations of Tomlinson (1974*b*). Though displays of two species, so far as known, clearly homologous, those of *A. purpurea* show a number of apparently unique features. Type and details of pair-formation unknown, though ♂ said to choose site (Bauer and

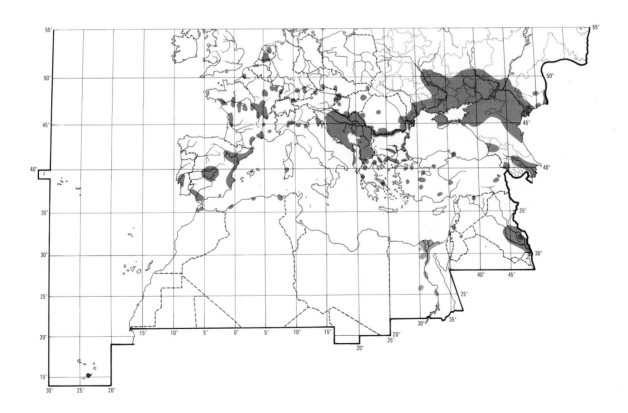

Glutz 1966). Unlike tree-nesting *Ardea*, both adults and young adopt concealing Bittern-stance (see *Botaurus stellaris*) when disturbed (Tomlinson 1974*a*, *b*). Prior to fleeing or adopting Bittern-stance, head and neck curved back over body and crest completely depressed. Bird then faces enemy, points bill, head, and neck skyward to almost vertical position with eyes fixed on intruder; blends well with reeds and, especially, bulrushes, remaining thus almost until touched. Similar skyward posture of neck and head adopted during feeding, and in Upright threat-display and Stretch-display. ANTAGONISTIC BEHAVIOUR. Use of crest feathers in threat-display discussed by Tomlinson (1974*b*): unlike Cattle Egret *Bubulcus ibis*, aggressive tendencies cause frontal part of crest to be raised. When responding aggressively to approaching rival, bird at nest first erects crest fully and slightly opens bill. Next, usually adopts Upright-display facing intruder: while resting on whole length of tarsi, assumes sky-pointing posture; unlike in Bittern-stance, crest, and pectoral and scapular plumes raised, gular region puffed out (exhibiting conspicuous white feathers), and wing half-spread at sides, coverts erected. Chin-puffing component unique to *A. purpurea*, so far as known. From Upright-display, bird moves into standard Forward-display, curving neck over back with bill open and held horizontally; wings and plumes as before, but crest temporarily depressed until lunges forward at intruder, usually while calling. While lunging, usually stays sitting (at least when defending eggs or young). Likely that full version of Forward-display follows extreme provocation, but as yet seen only from hand-reared birds towards dog: bird erects all head, neck, pectoral, and scapular feathers, including crest, advances slowly on stiff legs with bill pointing down, and, if necessary, flies directly towards intruder and supplants it. Aerial encounter over colony seen once: 2 of circling birds attacked one another, turning sideways, and thrusting out with legs before veering away. HETEROSEXUAL BEHAVIOUR. Only responses of already paired birds studied by Tomlinson (1974*b*), mainly in form of Greeting Ceremony at nest-relief; unlike in threat-display, birds do not usually face each other, while crest (especially frontal part) typically depressed. Returning partner of either sex flies in and lands near nest uttering characteristic call. As it approaches nest on foot, other bird orientates itself so as to face-away or be side-on to it, then rises slowly and assumes evident Stretch-display: (1) in stretch-phase proper, fully extends legs until body, neck, and head point nearly vertically with neck feathers erected, uttering sex-specific call; (2) after *c.* 1 s., suddenly performs crouch phase by bending legs, retracting neck (with bill at less acute angle), and lowers itself to nest while Bill-clattering 3 times ($\male$) or giving soft, 3-note call ($\female$) with gular area now flattened and scapular and pectoral plumes partly erected; (3) lowers head further with bill just above horizontal, same plumes fully raised, and feathers at rear of crest slightly up. Bill-clattering of $\male$, and sudden retraction of neck in crouch-phase by both sexes, among apparently unique features of Stretch-display in this species. Other bird, meanwhile, stops near nest as mate performs its Stretch-display, watching it while probably itself in partial and sustained Stretch-display with neck and head erect, bill pointing obliquely up, and neck feathers raised. Then moves on to nest, sometimes raising crest and partly opening wings, pushing latter backwards and forwards with short, jerky movements. Relieved bird then leaves nest, though sometimes reluctant to do so until made to go by pecks from other. Also seen to perform Sway-and-bob display after Stretch, bending forward in nest with head down and tail up as mate approached, swaying from side to side *c.* 6 times (bill pointing one way, then other) before rapidly lowering tail and raising head in bobbing motion; whole

performance then repeated 3 times. Occasionally, and apparently only when $\male$ relieved, Allopreening or secondary greeting may follow his Stretch-display; $\male$ main performer in Allopreening, reaching forward with scapular and pectoral plumes partly raised and Bill-clattering while nibbling $\female$'s chest and wing and often grasping her neck. In greeting, $\male$ bows to $\female$ and rises from nest, $\female$ then moving close and placing lowered head under his chest. Returning bird once seen to bring and present twig to mate, then left without nest-relief after mutual performances of Stretch-display. No Snap-display yet described, though almost certainly present earlier in cycle; according to O Koenig, usually directed diagonally upwards (Baerends and van der Cingel 1962), but possibility of confusion with Bill-clattering of $\male$ during crouch phase of Stretch-display. RELATIONS WITHIN FAMILY GROUP. From first, young beg by half-extending and slightly shaking wings while giving characteristic call. Later, sometimes mildly threaten parents when latter approach nest. Older chicks raise crests and vigorously grasp and pull down adult's bill. From 20th day, spend much time away from nest; after 30th, return there during day only to be fed. Intense rivalry between siblings lasts up to 20th day; largest chick dominates others by raising neck into stiff, vertical position then dropping it forward to deliver peck. Fighting most frequent at feeding; may continue up to 2 min at a time before one stops encounter by adopting crouched position, often with head over side of nest. Disputes noisy only if parent present. Upright-display, with expanded throat, shown to human intruder from 2nd day. In hand-reared birds, this display first common during 6th–12th day; Stretch-display as greeting to keeper on 25th day, full Forward-displays directed at dog on 32nd day (Tomlinson 1974*b*).

**Voice.** Similar to that of Grey Heron *A. cinerea* but used less frequently (Witherby *et al.* 1939; Bauer and Glutz 1966).

Main call of adults away from colony, given in flight, like that of *A. cinerea* but higher pitched and not so loud and deep; rendered 'frarnk' by Witherby *et al.* (1939), 'rrank' by Peterson *et al.* (1974) and 'krreck' or 'rrhe' by Bauer and Glutz (1966). Most vocal at colony, with rich vocabulary of harsh calls still not fully studied. Following described by Tomlinson (1974*b*): (1) high-pitched squawk during Forward-display threat; (2) low, guttural 'craak cr-ra-raak' by relieving bird just before alighting near nest; (3) musical 'whoop' ($\male$) or low 'craak' ($\female$) in response by relieved bird while performing 1st phase of Stretch-display; (4) loud 'clack-clack-clak' ($\male$) or soft 'crak-crak-crak' ($\female$) in 2nd phase of Stretch-display. Sounds made by $\male$ in (4) due to non-vocal Bill-clattering; also given during Allopreening. Greeting calls at nest: 'quärrärk quärrärk qwo qwo' answered by soft 'korr' (Bauer and Glutz 1966) probably equivalent to (2) and (3) above. Other calls mentioned by Bauer and Glutz: short, harsh 'quärr' (alarm); 'quäääät' (fear); and short, sharp 'quäck' (anger).

CALLS OF YOUNG. Characteristic Food-call of chicks up to 10 days, continuous 'chik . . .', later becoming raucous 'chak . . .' (Tomlinson 1974*b*). Young greet with 'görrör-rör' or 'grau-rau-rau', beg with 'kökökö' (see Bauer and Glutz 1966). When fledged, utter 'krrick' or 'kerkerker . . .' (see Witherby *et al.* 1939).

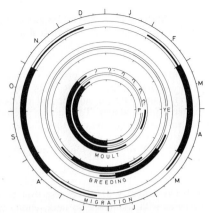

**Breeding.** SEASON. See diagram for central Europe; up to 3 weeks earlier in southern Europe. SITE. Selected by ♂. On clumps of dead reeds in dense reedbeds; 0·5–1 m above water level, always near open water, and up to 30 m from shore (Hanzák 1949–50). Sometimes in trees to 25 m above ground. Usually colonial with nests close together, but may be scattered and among those of other Ardeidae. Old nests not re-used. Nest: loosely constructed pile of dead reed stems, especially *Typha*; small branches and twigs used in tree nests. Average height 20–25 cm, diameter 50–70 cm, exceptionally 135 cm, shallow cup 5–10 cm deep (Hanzák 1949–50). Platforms of dead reed also constructed; used by off-duty adults, and later as resting and feeding sites for young. Building: by both sexes. Takes 7–12 days to complete (Steinfatt 1939). Material mainly gathered from near nest, but sometimes up to 100 m (Bauer and Glutz 1966). EGGS. Blunt oval; light blue-green, often splashed white or stained during incubation. 57 × 41 mm (50–61 × 37–44), sample 300 (Schönwetter 1967). Weight 50 g (37–60), sample 42 (Figala 1959). Clutch 4–5 (2–8). Possible cline in clutch size from 3·3 (2–5), sample 28, in Camargue to 5·1 (3–8), sample 96, in central France, but not enough data from same years (Ferry and Blondel 1960). Of 151 clutches from Switzerland, 2 eggs, 6%; 3, 21%; 4, 31%; 5, 36%; 6, 7% (Bauer and Glutz 1966). One brood. Replacement clutches laid after early egg loss. Eggs laid at 2–3 day intervals, but 5–7 days recorded. INCUBATION. 26 days (25–30). By both sexes; ♀ does larger share. Starts with 1st egg; hatching asynchronous. Eggshells discarded over side of nest. YOUNG. Semi-altricial and nidicolous. Cared for and fed by both parents. Brooded for first few days but at 8–10 days able to leave nest and clamber on reeds or branches; resort to platforms in preference to nests at *c.* 20 days. Food regurgitated by parents on to floor of nest or direct into bill of young. FLEDGING TO MATURITY. Fledging period 45–50 days. Independent at 55–65 days. Age of first breeding 1 year. BREEDING SUCCESS. No records.

**Plumages** (nominate *purpurea*). ADULT. Crown black, feathers rather long and loose, 2 lanceolate plumes (up to 15 cm) on nape; centre of hindneck black. Chin and upper throat white. Sides of head and neck warm buff to tawny with black line from gape joining black of nape and another black line running down side of neck. Centre of lower throat and foreneck with gular stripe of short black and white or pale buff streaks. Upperparts dark grey; mantle and upper scapulars with slight purple gloss in ♂, paler, more olive-grey in ♀. Feathers of lower mantle and upper scapulars each split into several lanceolate tips, longer and slightly broader in ♂. Long tips in ♂ cinnamon-chestnut, in ♀ buff. At sides of mantle, deep vinous-chestnut shoulder patches. Feathers of chest elongated with lanceolate tips, mainly cream-white with heavy black streaks. Breast deep chestnut streaked black; belly black with some chestnut; under tail-coverts black with white bases and some white streaks; flanks grey; tibia tawny-buff. Tail dark grey with slight gloss. Primaries, primary coverts, and bastard-wing brown-black, paler below; secondaries slightly paler, outer webs with grey-green gloss when fresh. Lower rows of lesser, median, and outer webs of greater upper wing-coverts grey, suffused tawny, more extensively in ♀♀ than in ♂♂; rest of lesser and inner webs of greater coverts darker; marginal with wide tawny margins. Cream-white patch at base of primaries. Lesser and median under wing-coverts chestnut; greater grey with tawny shaft-streaks or margins. Axillaries grey. NESTLING. Down soft and hair-like, sparser than Grey Heron *A. cinerea*, greenish skin showing through; dark brown on upperparts, white below. Tips of down on crown long and white, forming conspicuous crest. Sides of head, chin, throat, neck, centre of underparts, centre of mantle, and sides of back bare. At 6–9 days, pins of feathers appear over entire body; at 11–20, feathers break from sheaths; at 20, primaries begin to grow; at 30–35, able to fly; at 40–50, in full juvenile plumage but downy tips of crown feathers retained. (Witherby *et al.* 1939; Hofstetter *et al.* 1949; Tomlinson 1975.) JUVENILE. Forehead and part of crown black, rest of crown, nape, and hindneck chestnut. Short crest at nape, but no lanceolate feathers. Chin white; sides of head and neck, throat, and foreneck tawny buff, bases of feathers black, producing vaguely mottled appearance. Mantle, scapulars, and upper wing-coverts dark brown, slightly glossed, with wide tawny feather edges. Back, rump, and upper tail-coverts grey-black. Chest and breast pale buff, streaked dark brown; belly, vent, and under tail-coverts almost unmarked, but longest under tail-coverts with some brown streaks. Feathers of tibia dark brown with tawny edges; flanks dark grey-brown. Tail dark brown with slight gloss. Primaries brown-black, secondaries slightly paler and more glossy, inner edged tawny. Median and lesser under wing-coverts dark brown-grey, tipped tawny, greater paler, tipped off-white. Patch at base of primaries pale tawny. Axillaries dark brown-grey. IMMATURE. Gradually acquired 1st autumn and winter; may nest in this plumage. Differs from juvenile by presence of short ornamental scapulars and some lanceolate chest feathers. SUB-ADULT. Worn from autumn of 2nd calendar year. Like adult, but black of crown mixed with some chestnut, ornamental feathers on upperparts less bright, underparts not so deep chestnut, less black on belly, upper wing-coverts broadly edged pale tawny (Witherby *et al.* 1939). Not known whether always distinguishable.

**Bare parts.** ADULT. Iris yellow. Bill yellow with darker, brown, culmen. Bare skin of face yellow-green. Yellow tinges of iris, bill, and face brighter and more orange at start of breeding season. Leg dark brown, back of tarsus and soles yellow. NESTLING. Iris at first yellow-grey (Hofstetter *et al.* 1949), later yellow. Bill dirty yellow. Leg flesh. JUVENILE. Bill with more brown along culmen than adult. Leg pale yellow-green, darker towards foot (Naumann 1838; Bauer and Glutz 1966).

**Moults.** ADULT POST-BREEDING. Complete. Lengthy: June–November, or even to February. Sequence of primaries presumably as in other herons with inner descendant and outer irregular, but data needed. ADULT PRE-BREEDING. Partial; extent of this moult not known; probable that only some ornamental feathers develop, particularly nuchal plumes. POST-JUVENILE. Partial. Involves body and some wing-coverts; November–April. Subsequent moults like adult, but first complete moult tends to start early, May. (Witherby *et al.* 1939.)

**Measurements.** Adult; west Palearctic; skins (RMNH, ZMA).

| | ♂ | | | ♀ | | |
|---|---|---|---|---|---|---|
| WING | 371 | (6·80; 13) | 357–383 | 355 | (11·2; 9) | 337–372 |
| TAIL | 125 | (4·44; 13) | 118–136 | 119 | (4·98; 8) | 112–127 |
| BILL | 126 | (3·06; 13) | 120–131 | 116 | (3·87; 8) | 109–123 |
| TARSUS | 122 | (4·58; 13) | 113–131 | 118 | (4·07; 8) | 112–125 |
| TOE | 132 | (5·23; 13) | 121–139 | 126 | (4·46; 7) | 118–131 |

Sex difference significant. Juveniles smaller than adults.

| | ♂ | | | ♀ | | |
|---|---|---|---|---|---|---|
| WING | 367 | (6·03; 8) | 354–373 | 341 | (1·73; 3) | 339–342 |
| BILL | 119 | (5·56; 9) | 110–127 | 109, 111 | | |

Witherby *et al.* (1939) gave small minimum values for wing: ♂ 330, ♀ 290.

**Weights.** Netherlands: ♂♂ 617–1218, ♀♀ 525–1135 (ZMA). Chicks at hatching *c.* 35, reach 1100–1200 in 50 days (Tomlinson 1975).

**Structure.** 11 primaries: p8 longest, p9 about equal or slightly shorter, p10 5–10 shorter, p7 0–5, p6 5–25, p5 25–40, p1 *c.* 100–110; p11 minute. P7–p9 emarginated on outer webs, p8–p10 on inner. 18 secondaries. Tail square; 12 feather. Bill slightly more slender and evenly pointed that *A. cinerea*. About ⅓ of tibia bare. Foot large in relation to leg, middle toe *c.* 105% of tarsus. Outer toe *c.* 78% of middle, inner *c.* 76%, hind *c.* 60%, hind claw *c.* 25%.

**Geographical variation.** *A. p. bournei* described as paler than nominate *purpurea* with less black on neck, chest, and belly; centre of underparts white and pale chestnut (de Naurois 1966). *A. p. manilensis* paler and more grey above, gular stripe of black streaks poorly developed or absent, ornamental feathers on chest whiter, rest of underparts with more black (Vaurie 1965); juveniles less heavily streaked underneath. *A. p. madagascariensis* darker than nominate *purpurea* (Hartert 1912–21).

## *Ardea goliath* Goliath Heron

PLATE 43
[between pages 326 and 327]

DU. Goliathreiger   FR. Héron goliath   GE. Goliathreiher
RU. Цапля голиаф   SP. Garza goliat   SW. Goliathäger

*Ardea Goliath* Cretzschmar, 1826

Monotypic

**Field characters.** 135–150 cm; wing-span 210–230 cm. Huge, heavily built heron, standing almost as high as man, with large and unusually deep bill; head, hindneck, and belly reddish-chestnut, foreneck striped, and back and wings grey. Sexes alike and no seasonal differences; juvenile similar but distinguishable.

ADULT. Top and sides of head and rear and sides of neck reddish-chestnut, darkest on top of head where elongated feathers form crest; in contrast, chin and throat white and foreneck and upper breast white streaked with black, the breast feathers extended into plumes. Upperparts, wings, and tail dark grey with scapulars elongated; lower breast, belly, and underwing rich purplish-chestnut. Bill black above and horn below; bare loral area yellow tinged green; eyes yellow, rimmed red; legs black. JUVENILE. Like adult, but head and neck paler and duller, dark grey feathers of upperparts edged with rusty-buff, lower breast and belly buff-white streaked with dark brown, and scapulars and upper breast feathers not elongated.

Huge size makes species unmistakable, but colour pattern not unlike far smaller and slighter Purple Heron *A. purpurea* which, however, has comparatively thin bill, black crown and crest, and bold black lines down side of head and neck.

Frequents both inland and coastal waters. Generally solitary and rather inactive, often standing motionless for long periods in shallow water, usually in open and not secretive. Flight slow and ponderous, with wings bowed and legs sagging below horizontal. Utters characteristic deep, raucous 'arrk' when flushed.

**Habitat.** Tropical and sub-tropical margins of both coastal and inland waters, especially those permitting full use of exceptional wading capabilities. Favours margins of large rivers and lakes commanding long views of approaching enemies, especially man; also coastal sandbanks and mudflats, coral reefs, islets, and sand dunes. Penetrates estuaries, creeks, and mangrove fringes, or edges of tall stands of reedy vegetation, where usually nests, (although also on trees). Not highly aerial but can fly strongly at various altitudes. Normally avoids neighbourhood of man and artefacts (Archer and Godman 1937).

**Distribution.** Breeds marshy areas of southern Iraq where formerly common (Allouse 1953).

Accidental. Egypt: Red Sea February 1935, September and December 1942 (Jourdain and Lynes 1936; Bulman 1944). Israel: formerly occasional winter visitor to Lake Huleh, but not recorded since drained (HM). Syria: recorded Raqqa, 19th century (Kumerloeve 1967b).

**Population.** IRAQ: now much less common; no recent nesting records and few seen (PVGK).

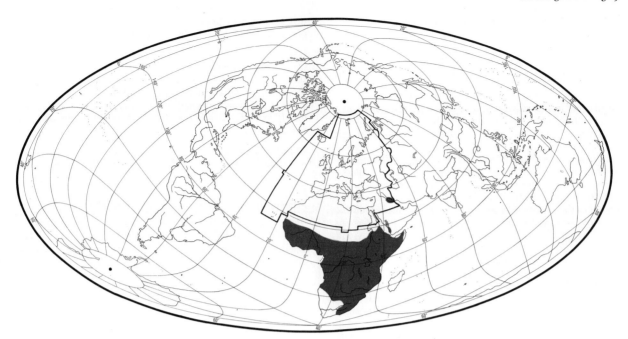

**Movements.** Resident throughout breeding range, but probably dispersive or nomadic. African birds disperse outside breeding season over suitable coastal and inland habitats, mainly within breeding range, in response to seasonal changes in water levels, but distances individuals move not known. In Chad, reported strictly sedentary (Salvan 1967), but none ringed. Some dispersal from Iraq marshes recorded along Tigris to Samarra and along Euphrates to Al Fallujah (Ticehurst *et al.* 1922), and at least once to Syria. Rare vagrant to Indian subcontinent (Ali and Ripley 1968) and straggled once to Aden (Sage 1959).

**Food.** Chiefly fish and, probably to lesser extent, frogs and other aquatic animals. Usually feeds alone or in pairs, in rather deep water (often up to abdomen) where stands or wades slowly and deliberately until prey sighted, then head and neck rotated until flush with surface, and prey speared with quick lunge (Bannerman 1953; Meinertzhagen 1954). Fish include mullet, eel, catfish (Pearson 1969; no specific identifications), and barbel *Barbus* (Jackson 1938).

**Social pattern and behaviour.** Based on material supplied by J Cooper, but little information.
1. Typically solitary or in single pairs throughout year. Apparently never flocks, always feeding individually. No information on relations within family group, though both parents tend young as in other Ardeinae. Usually solitary breeder, though occasionally in small colonies with other herons and pelecaniform birds (*Phalacrocorax* or *Anhinga*); in South Africa, nests of conspecifics often no closer than *c.* 100 m, but also recorded only *c.* 50 m apart and exceptionally 2 nests in same tree (J Cooper). No information on roosting, but probably not communal.

2. Making due allowance for solitary and terrestrial habits, what little known about displays suggest these similar to more gregarious and arboreal herons, e.g. Grey Heron *A. cinerea.* Apparently adult-like Forward-displays (both types) and Withdrawn-crouch reported from nestlings in response to human intruders (Cooper and Marshall 1970); full Forward-display also described from nestling by Audin (1963). In encounter away from nest (Schutte 1969), one bird extended neck on approach of other; latter then also stretched neck, with bill pointing down, and called, afterwards lowering extended neck until throat touched water, bill pointing at first bird, repeating call several times.

**Voice.** Call, usually uttered when flushed (Williams 1967), variously described as (1) gargling rattle; (2) like bellowing of calf or barking of baboon (Hauling 1953); (3) loud, deep 'arrk'. Courtship call (presumed ♂): 'mmmmmmmmmm haw-haw-haw', 1st note high-pitched, rest sharp and throaty (Schutte 1969).
CALLS OF YOUNG. No information.

**Breeding.** SEASON. No information from west Palearctic. Somaliland: breeds September, also November–December (Archer and Godman 1937). Guinea-Bissau: mid-August to mid-September (de Naurois 1969*a*). SITE. On ground, among trampled down reeds or sedges, or low bush or tree often standing in water; occasionally among boulders on shore; usually solitary, but sometimes with other herons or cormorants. Nest: large platform of sticks and reeds lined with smaller twigs; up to 1 m diameter. Building: by both sexes. EGGS. Elongated oval; pale blue. 75 × 52 mm (68–80 × 49–56), sample 30; calculated weight 109 g (Schönwetter 1967). Clutch: 2–3 (4). One brood. No data on replacements.

INCUBATION. *c.* 28 days (de Naurois 1969*a*). By both sexes. YOUNG. Semi-altricial and nidicolous. Cared for and fed by both parents. Fed by regurgitation. FLEDGING TO MATURITY. Fledging period *c.* 40 days (de Naurois 1969*a*). BREEDING SUCCESS. No data.

Plumages. ADULT. Crown vinous chestnut; feathers of hind crown and nape long and loose, forming bushy crest; some more elongated, projecting beyond rest. Hindneck and sides of head and neck vinous buff, occasionally paler. Chin and upper throat white. Lower throat and foreneck spotted black and white. Base of hindneck and rest of upperparts dark grey faintly glossed green. Ornamental feathers of lower mantle and upper scapulars each split into several lanceolate points, but less strongly developed than in Grey Heron *A. cinerea*. Tuft of black and white elongated ornamental feathers on chest. Rest of underparts, including axillaries and under wing-coverts deep vinous chestnut, paler on flanks and under wing-coverts. Base of tibia with some grey. Tail and flight-feathers dark grey. Upper wing-coverts dark grey, faintly glossed green, greater palest. NESTLING. Down grey-brown at hatching; grey when feather sheaths appear (Benson and Serventy 1956). JUVENILE. Forehead and crown mainly black. Upperparts including upper wing-coverts rusty-red, bases of feathers grey. Feathers of chest shorter than adult, with brownish webs and white shaft-streaks. Rest of underparts streaked, each feather having pale cinnamon-chestnut central streak and dark grey edges to both webs. Flanks and under wing-coverts dark grey, lesser ones mottled white and rusty red. IMMATURE. Differs from juvenile in lacking rufous tinges on upperparts and having underparts chestnut with pale shaft-streaks and grey inner webs of feathers. Differs from adult by

mainly dark grey forehead, pale streaking on underparts, dark grey flanks, and dark grey mottling on under wing-coverts.

Bare parts. ADULT AND JUVENILE. Iris yellow. Upper mandible black, lower blue-grey or pale flesh with darker cutting edge. Bare skin of face yellow tinged green. Leg and foot black. Any cyclical changes in colour apparently not recorded. NESTLING. Iris light green at hatching, becoming yellow later. Bill olive-grey, lower mandible sepia. Bare skin of lores olive-green (Benson and Serventy 1956).

Moults. Few data. Inner primaries of adult sometimes moulted descendantly (Milstein 1969), but skins in RMNH show irregular moult of primaries.

Measurements. Adult, skins (BMNH, RMNH).

| WING | ♂ 591 (19·2; 7) 570–630 | ♀ 575 (15·0; 8) 560–599 |
|------|------|------|
| BILL | 193 (8·86; 9) 183–208 | 177 (12·5; 10) 156–196 |

Sex difference significant for bill. Tail both sexes 200–235, mean 226; tarsus 225–238, mean 231 (McLachlan and Liversidge 1957). Middle toe 3 ♂♂ 158, 158, 159; 2 ♀♀ 145, 160 (skins BMNH, RMNH).

Weights. No data.

Structure. 11 primaries: p8 longest, occasionally 1–2 shorter than p7; p9 3–8 shorter, p10 19–40, p7 0–10, p6 10–25, p5 *c.* 50–60, p1 *c.* 160–180; p11 minute. Outer webs of p6–p9 slightly emarginated, inner webs p8–p10. Tail slightly rounded, 12 feathers: t6 *c.* 15–20 shorter than t1. Bill long, heavy at base; nostrils narrow slits, outer half of cutting edges serrated. Leg long, more than $\frac{1}{2}$ of tibia bare; outer toe *c.* 82% of middle, inner *c.* 77%, hind *c.* 58%, hind claw *c.* 19%.

# Family CICONIIDAE Storks

Large to huge wading and terrestrial birds. 17 species in 6 genera (Kahl 1971*a*, 1972*b*). 3 groups: (1) wood storks *Mycteria* (4 species) and openbill storks *Anastomus* (2 species); (2) typical storks *Ciconia* (5 species); (3) Saddle-bill and Black-necked Storks *Ephippiorhynchus* (2 species), Jabiru Stork *Jabiru* (monotypic), and Marabou and adjutant storks *Leptoptilos* (3 species). 2 breeding species and 2 accidental in west Palearctic. Body sturdy, elongate; neck long, head rather large. ♂ slightly bigger than ♀. Wings long and broad. Flight flapping, with neck outstretched (except in *Leptoptilos*); also engage in dynamic soaring. 12 primaries: p7–p9 longest; p12 minute (lacking in *Anastomus*). About 22 secondaries; diastataxic. Tail short, square or slightly rounded; 12 feathers. Under tail-coverts long and sometimes fluffy. Bill long and heavy, massive in some, tapering to blunt point. Straight or slightly upturned, moderately decurved in *Mycteria*, and with remarkable gap between cutting edges in *Anastomus*. Nostrils mainly slits at base of bill. Head partly or wholly bare. Large air-sacs under skin of neck; long, pendant, naked, distensible pouch in 2 species of *Leptoptilos*. Legs long, lower half or more of tibia bare. Toes relatively short with small webs at base, hind toe elevated; no claws

pectinated. Tarsus reticulate. Stance upright, gait striding. Oil gland feathered. Feathers with short aftershaft (lacking in some tropical species); down on feather tracts and apteria. No powder-down patches (reports of such in *Leptoptilos* probably refer to ornamental under tail-coverts). Plumage white, grey, brown, and black, often with metallic reflections. Usually boldly patterned, often black and white; most species with pale underparts. Elaborate plumes of Ardeidae lacking, but thick ruff of feathers on lower foreneck in *Ciconia*. Bare parts brightly coloured throughout year. Sexes generally similar. Breeding plumage like non-breeding, or slightly more colourful. Moults poorly known; mostly 1 per cycle. Primaries replaced in descendant or serially descendant order, or irregularly. Young semi-altricial and nidicolous. 2 coats of down, 1st growing from same follicles as later contour feathers, 2nd from same follicles as adult down. Juveniles reach full plumage in 2nd calendar year.

Cosmopolitan, in tropical, sub-tropical, and temperate regions; main centre of distribution in tropical Africa and tropical Asia. While superficially similar to Ardeidae in frequenting wetland habitats and water margins, more typically hunters or scavengers and less purely fishers;

often feed on dry land, at times far from water. Prefer warm, continental climatic types, ranging less far north than Ardeidae. In west Palearctic, more highly migratory, occurring in a greater variety of less-preferred habitats on passage. In west Palearctic, both breeding species (Black Stork *C. nigra* and White Stork *C. ciconia*) have shown marked decline in north and west Europe for reasons not clearly understood. Migrate almost always in flocks, often very large; mainly diurnal as thermals used extensively. Avoid long sea crossings, so passage of west Palearctic populations concentrated at either end of Mediterranean.

Food primarily insects, amphibians, and fish; also reptiles, small mammals, birds, etc. Bill of *Anastomus* adapted for dealing with molluscs, especially large water-snails (see Kahl 1971*b*). *Leptoptilos* largely scavengers. Prey taken in water, while walking or running. Food located visually in most species then grabbed with bill and swallowed whole with backward toss of head (Kahl 1971*b*). Feeding in *Mycteria*, however, largely tactile: often stir underwater vegetation with feet, then snap shut mandibles on prey that comes into contact; method used at times by other storks, including *C. ciconia* and *C. nigra* (see Kahl 1971*b*, 1972*a*). Most species forage singly, but some *Ciconia* especially often form large flocks when prey abundant and also gather at grass fires, while *Leptoptilos* congregate at carcases. Most species habitually either colonial or solitary when breeding; some, however, colonial or solitary depending on local conditions. Roosting both solitary or communal, but little detailed information; sustained resting largely nocturnal, but much time also spent loafing during day, especially in scavenging species. All storks strongly territorial when nesting, defending nest-site territory. Monogamous pair-bond, usually of seasonal duration though may pair up again in following years; in solitary nesting genera, some evidence that birds remain in pairs throughout year. In most species, nest-site territory initially established as pairing territory by ♂ who shows hostility to all intruders: interested ♀♀ approach repeatedly until ♂'s aggressiveness wanes (Kahl 1971*a*). Responses of site ♂ include: (1) Up-down display with calls or Bill-clattering, or both; (2) Aerial-clattering threat while flying after intruder; and, more specifically towards approaching ♀♀, (3) Display-preening (*Mycteria*); (4) Swaying Twig-grasping (*Mycteria*, *Leptoptilos*); (5) Advertising-sway (*Anastomus*); and Head-shaking Crouch (*Ciconia*)—terminology after Kahl (1971*a*). Responses of ♀ include: (1) Balancing-posture with wings extended, and (2) Gaping-display (neither shown by *Ciconia*). Main display after pair-formation mutual Up-down Ceremony; chiefly as greeting at site.

Copulation on nest; accompanied by Bill-clattering of ♂ in most genera. Away from site, individuals of most species largely ignore one another except (e.g.) for Forward-display at times by threatening bird and Upright-display by those threatened. Apart from hissing, storks often regarded as voiceless but this true only of certain species (e.g. *C. ciconia*). Others produce variety of vocal sounds, including melodious whistles as in most *Ciconia*, mostly during Up-down display in addition to characteristic Bill-clattering. Latter reduced or absent in some species (e.g. *C. nigra*); other non-vocal sounds include Bill-snapping. Comfort behaviour generally similar to other waterbirds but not intensively studied. Heat dissipated by panting with raised tongue (not by gular-fluttering as in Ardeidae), and by excreting dilute urine over legs (Kahl 1963). Wing-spread postures used in thermo-regulation, wing-drying, and to shelter nestlings (Kahl 1971*c*); those used in sunning either full-spread or shieldlike as in Ardeidae. Other behaviour characters include (1) direct head-scratching; (2) resting with neck retracted and bill resting against base of foreneck. In *Ciconia*, bill often hidden in erected ruff feathers; this, and habit of sometimes standing on one leg found in all storks, probably helps to conserve body heat.

Breed mainly seasonally, especially temperate, migratory species and those nesting in tropical areas with seasonal rains. Nest in trees and on cliff ledges; also on buildings and other man-made structures. Nests re-used in successive seasons. Large piles of twigs, branches, earth, etc. in most species, but rather flimsy at times in some (e.g. *Mycteria*); lined with finer materials. Built by both sexes. Eggs oval or blunt oval; white, becoming stained during incubation. Clutches usually 3–5 (1–7); single egg typical in Saddlebill Stork *E. senegalensis*. One brood. Replacement eggs possible but rarely recorded. Eggs laid at intervals of 1–4 days. Incubation periods 33–36 days in west Palearctic species, starting with 1st or 2nd egg—so hatching asynchronous. Both sexes incubate in roughly equal spells, changing 1–2 times a day; single median brood-patch. Young cared for by both parents. Fed, wholly or mainly by complete regurgitation on to floor of nest, several times daily; also brought water, which parents regurgitate over them, especially on warm days (e.g. Kahl 1971*b*, 1972*a*). Usually considerable size difference between siblings; competition between young at times of food shortage often results in death of smaller (see *C. ciconia*). Fledging period in west Palearctic species *c*. 60–70 days; young independent 7–20 days later. Age of first breeding 3–5 years.

# *Mycteria ibis* Yellow-billed Stork

PLATE 44
[between pages 326 and 327]

Du. Afrikaanse Nimmerzat    Fr. Tantale ibis    Ge. Nimmersatt
Ru. Африканский клювач    Sp. Tantalo africano    Sw. Afrikansk ibisstork

*Tantalus Ibis* Linnaeus, 1766. Synonym: *Ibis ibis*

Monotypic

**Field characters.** 95–105 cm, ♂ averaging larger; wing-span 150–165 cm. Predominantly white with black flight-feathers and tail, and long pale legs, but with bare red face and long, somewhat decurved yellow bill recalling ibises (Threskiornithidae). Sexes alike and no seasonal differences; juvenile quite different.

ADULT. Mainly dull white, with crimson-tinged back-feathers (when breeding), scapulars, upper wing-coverts, and axillaries, but primaries, secondaries and tail black glossed with purple and green. Conspicuous bare face red to reddish-orange contrasting with orange-tipped yellow bill; eyes brown; upper legs pale orange-red and lower brownish-pink with black toes. JUVENILE. Greyish-brown head, neck, and upperparts; white breast, belly, flanks, and under wing-coverts; black flight-feathers and tail. Bare face smaller in area than adult's and, like bill, duller. By end of first year more like adult, but greyer above without pink feather tips and with face and bill still dull.

Adult unmistakable on ground; only White Stork *Ciconia ciconia* at all similar, yet easily distinguished by straight red bill, white-feathered face, and bright red legs. High overhead 2 species look similar, and black (as against white) tail of *M. ibis* then best character; this also separates it from soaring White Pelican *Pelecanus onocrotalus* and Egyptian Vulture *Neophron percnopterus*. Juvenile also quite distinctive: juvenile Black Stork *C. nigra* bears some resemblance, but has shorter, straight bill, blacker mantle and wings, and brownish-black (as against white) underwing.

Gregarious and associates freely with herons (Ardeidae). Flight and behaviour much like other storks, but more aquatic than some: seldom far from water and spends much time standing in shallows (often feeding by submerging bare portion of head), or on adjacent mudbanks and sandbars. Largely silent, apart from bill-clattering and harsh guttural calls when nesting, but rarely utters long, loud, nasal sound like squeaking hinge.

**Habitat.** Fresh and saline waters, including broad, shallow rivers with low sandbanks, channels, islets, lake shores, floodlands, water-holes and coastal lagoons, salt-flats, and beaches. Wades to utmost practicable depth when feeding; often rests and breeds in trees, sometimes of great size, tolerating village and town locations, and perching well. Ordinarily flies low and for fairly short distances but, despite laboured appearance, rises high and covers long distances on occasion. (See Archer and Godman 1937.)

**Distribution.** Breeds Africa south of Sahara from Gambia to Sudan and south to South Africa, and Madagascar. Non-breeding visitor to Banc d'Arguin, Mauritania.

Accidental. Morocco and Tunisia: May 1959 and March 1959 respectively (Heim de Balsac and Mayaud 1962). Egypt: recorded regularly before 1913, May–September (Flower 1933); several October 1957 (Horvath 1959*b*). Israel: obtained 1944, 1968; also several sight records (HM).

**Movements.** Dispersive and partially migratory. Moreau (1966) believed that in northern and southern tropics post-breeding movements made to higher latitudes on own side of equator for moult. Elgood *et al.* (1973) found situation in west Africa more complex; in Nigeria, resident in northern savannah, and non-breeders common on coasts and great rivers in dry season (November–March) with northward withdrawal in wet season from humid regions having high rainfall. In northern tropics, breeds mainly dry season (September–March); though in Chad, where partially migratory between 12°N–14°N, numbers at Mare d'Agan rose from 40 in April to 200 in June, after onset of rains, with breeding June–September (Salvan 1967). Certainly some post-breeding season movement north into Sahel zone during short summer rains. Non-breeders present all year Eritrea, with major influx June–October (Smith 1957). Small numbers visit Niger inundation zone, Mali, where once seen in hundreds in October (Paludan 1936; Guichard 1947). Regular summer visitor to Banc d'Arguin, Mauritania, mainly immatures (Westernhagen 1970) with 130 August 1972 (Gandrille and Trotignon 1973); only vagrant further north (see Distribution).

**Voice.** See Field Characters.

**Plumages.** ADULT. Head mainly bare; back of head and neck white. Rest of upperparts white, tinged crimson when breeding (Kahl 1971, 1972*b*). Scapulars tinged crimson, long ones most strongly. Underparts white, under tail-coverts faintly tinged crimson. Tail, flight-feathers, and primary coverts black, glossed green. Tertials and upper wing-coverts white with crimson subterminal bar, darkest on greater coverts, absent on marginal. Median and lesser under wing-coverts pale crimson; greater deep crimson, especially subterminal; all with pure white tips. Axillaries faintly tinged crimson. JUVENILE. Feathering on head extending further forward than in adult; feathers of head and neck grey-brown, throat and foreneck paler. Mantle, scapulars, and upper wing-coverts grey-brown with pale feather edges. Back and rump off-white. Underparts dirty white. Tail and flight-feathers black, under wing-coverts brown. IMMATURE.

Like adult, but tinged grey, without crimson on scapulars and upper wing-coverts (Clancey 1964).

**Bare parts.** ADULT. Iris brown. Bill golden-yellow. Bare skin of face red, tinged orange near feathers. Leg red. JUVENILE. Iris grey-green. Bill grey-green, orange-yellow at base (Castan 1959). Bare skin of face yellow, tinged pink. Leg grey-green. IMMATURE. Bill and face duller than adult (Clancey 1964).

**Moults.** ADULT POST-BREEDING and JUVENILE. Complete; primaries descendant (Stresemann and Stresemann 1966). Further details not known.

**Measurements.** Wing 455–517, tail 170–190, bill 203–243, tarsus 195–218 (McLachlan and Liversidge 1957). Maximum length of tarsus 229 (Bannerman 1930). ♂ slightly larger than ♀

(Kahl 1972a). Juvenile apparently shorter winged: 415 (Castan 1959).

**Weights.** No data.

**Structure.** Wings broad. 12 primaries: p9 longest, p10 equal or slightly shorter, p11 *c.* 60 shorter, p8 *c.* 5–10, p7 *c.* 25, p6 *c.* 90, p5 *c.* 130, p1 *c.* 190; p12 minute. P7–p10 emarginated outer web, p8–11 inner. Tail square; 12 feathers. Bill long and heavy, high and broad at base, ridge of upper mandible rounded; decurved at tip. Nostrils narrow slits, high on sides of bill, near base. Feathering on head reaching to nape and over ear-coverts, but during display feathered skin retracted and nape becomes bare. Legs long, lower ⅔ of tibia bare. Middle toe *c.* 55% of tarsus; outer toe *c.* 85% of middle, inner *c.* 70%, hind *c.* 42%.

## *Ciconia nigra* Black Stork

PLATE 45
[between pages 326 and 327]

| | | |
|---|---|---|
| DU. Zwarte Ooievaar | FR. Cigogne noire | GE. Schwarzstorch |
| RU. Чёрный аист | SP. Cigüeña negra | SW. Svart stork |

*Ardea nigra* Linnaeus, 1758

Monotypic

**Field characters.** 95–100 cm, of which body only half, ♂ averaging larger; wing-span 145–155 cm. Large, long-necked, long-legged waterbird, flying with neck out-stretched; mainly glossy black with white underbody and red bill and legs. Sexes alike and no seasonal differences; juvenile distinguishable.

ADULT. Black with metallic purple and green gloss, except for white lower breast, belly, axillaries, and under tail-coverts. Black feathers of lower throat and foreneck somewhat elongated to form shaggy chest. Bill, bare loral area, and legs dull red to bright scarlet; eyes brown. JUVENILE. Similar, but head, neck, and upper breast rather browner and rest of upperparts much less glossy, with pale tips to scapulars and wing and upper tail-coverts. Bill, bare loral area, and legs grey-green.

On ground, adult unmistakable; for slight possibility of confusion with juvenile and immature Yellow-billed Stork *Mycteria ibis*, see that species. In air from below, dark head, neck, wings, and tail contrasting with white underbody characteristic. (Abdim's Stork *C. abdimii* of west and central Africa, which could occur as vagrant in west Palearctic, has similar black and white pattern, but is much smaller with white lower back and rump, green bill with red only at base, and dusky green legs with pink joints.)

Flight, gait and general habits much as White Stork *C. ciconia*, though feeds chiefly on fish and amphibians, so generally more closely associated with water. Much less gregarious than other storks, and chiefly seen singly; does not breed colonially, but as single pairs usually in undisturbed forest. Also much shyer than other storks, avoiding vicinity of man. More vocal on breeding grounds than *C. ciconia*, but silent on passage.

**Habitat.** In contrast to White Stork *C. ciconia*, mainly in old undisturbed forest areas, interspersed with streams, pools, swampy patches, damp meadows, and banks of rivers, or occasionally large bodies of fresh water. Only in certain regions, e.g. Caspian lowlands (Dementiev and Gladkov 1951), nests in isolated giant trees or groves in open cultivated country, approximating to habitat of *C. ciconia*, but even here much less sociable and usually well dispersed. Although equally a middle-latitude species in west Palearctic, linked with strongly continental climate and avoiding winter cold, with more easterly and northerly range (except for detached population in Iberia). In some mountainous regions, e.g. of USSR, Carpathians, Spain, and South Africa, nests not only in valleys but on steep crags, slopes, rocks, or in caves at altitudes up to 2000 m (exceptionally somewhat higher), and as far as 1 km from nearest water. Nesting more usually on upper parts of well-grown forest trees, in most areas deciduous but, in some, conifers freely used. Ample cover, wet or moist feeding areas, and in most regions segregation from human influences and contacts, leading criteria in habitat selection.

Much less dependent than *C. ciconia* on thermals, rising readily to considerable heights by slow flapping. On passage over Mediterranean and Sahara, apparently at most makes few and brief surface stops, as suitable moist forest habitats unavailable. Although closely associated with water, makes limited direct use of rivers and lakes, and still less of seas, frequenting mainly vegetated fringes and wetlands of all types and sizes. Regular and highly conservative in resort to routes and places, but sometimes re-occupies long abandoned haunts.

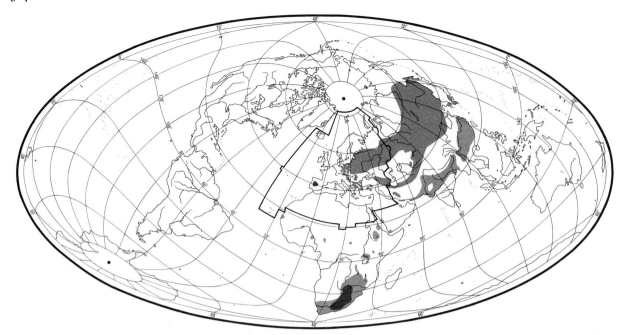

**Distribution.** Retreat in west of range in 20th century. DENMARK: formerly *c.* 50 pairs; last nested 1951 (TD). SWEDEN: last bred 1953 (SM). WEST GERMANY. Schleswig-Holstein: declined after 1909, only one possible breeding record since. In south-west, bred around middle 19th century, but then western breeding areas deserted one by one (for details see Bauer and Glutz 1966). BELGIUM: nested *c.* 1860, and 1892 (Lippens and Wille 1972).

Accidental. Britain, Belgium, Denmark, Norway, Sweden, Finland, Switzerland, Malta, Cyprus, Libya, Madeira.

**Population.** Despite shrinkage of western part of range, indications of increasing numbers this century in adjoining areas to east. POLAND. In former East Prussia, increased from 40 pairs in 1914 to 140 pairs 1935, and in Silesia increase noted from 1936 after 50 years decline (Bauer and Glutz 1966). Lowest population in 1920s (Tomiałojć 1972). In 1966, 480 occupied nests found; total population probably 500–530 pairs (Bednorz 1974). WEST GERMANY. Niedersachsen: probably *c.* 20 pairs about 1900, *c.* 6 pairs 1964 (Bauer and Glutz 1966). Bavaria: 1–2 pairs in recent years (EB). EAST GERMANY. Oberlausitz: first bred 1957; since 1960, 1–2 pairs regularly (Bauer and Glutz 1966). For whole country, 13 pairs in 1960; remaining near 18 pairs since 1963 (Schiemenz 1972). CZECHOSLOVAKIA: breeds occasionally in west; in Slovakia *c.* 100 pairs 1960 (Bauer and Glutz 1966), increasing since 1940 (KH). AUSTRIA: first bred a few years before 1939; at least 5 pairs 1951, probably slight increase since (Bauer and Glutz 1966). HUNGARY: *c.* 50–60 pairs around 1941 (Bauer and Glutz 1966). GREECE: under 20 pairs (WB). USSR. Considerable density only in east Transcaucasia, where in 1934 nests in suitable areas rarely more than 1 km apart

(Dementiev and Gladkov 1951). Estonia: *c.* 150 pairs 1963 (Bauer and Glutz 1966). SPAIN. Population *c.* 50–60 pairs fluctuating; occasionally nests outside area shown (Heydt 1972). Total Iberian population estimated a little over 150 pairs in normal years (Bernis 1974). YUGOSLAVIA: declining, especially in north (VFV).

Survival. Oldest ringed bird 18 years (Rydzewski 1974).

**Movements.** Migratory and dispersive. Apart from Spain, where may be partially resident (Bernis 1966), winter records from Europe exceptional, virtually all migrating south to winter tropical Africa, where few in west, more in east, but always thinly spread as usually solitary in winter (Bannerman 1957). Narrow-front passage around Mediterranean marked, but less so than in White Stork *C. ciconia*, and will apparently cross wider stretches of water; thus small spring passage Tunisia, autumn ringing recoveries Italy and Greece, and recorded Aegean Sea and Mediterranean islands fairly regularly (Schüz 1940a; Bauer and Glutz 1966). Having narrower wings, less dependent on soaring and gliding in thermals, though this still commonest form of migratory flight. Migratory divide in Europe not properly defined, but farther east than in *C. ciconia*, while more marked post-fledging dispersal may influence migratory direction. Thus from brood of 4 ringed Denmark, 2 recovered to south-west (Netherlands, north France) and 2 to south-east (Hungary, Rumania); while 2 young from an East German nest recovered France and Hungary.

As west European breeding population now small, not surprising now rarely seen west Africa. Up to 16 Sénégal delta January–March 1972 (Roux and Dupuy 1972); otherwise a few recent records, of 1–2 birds, mainly Nigeria and Chad (Moreau 1967) and no ringing re-

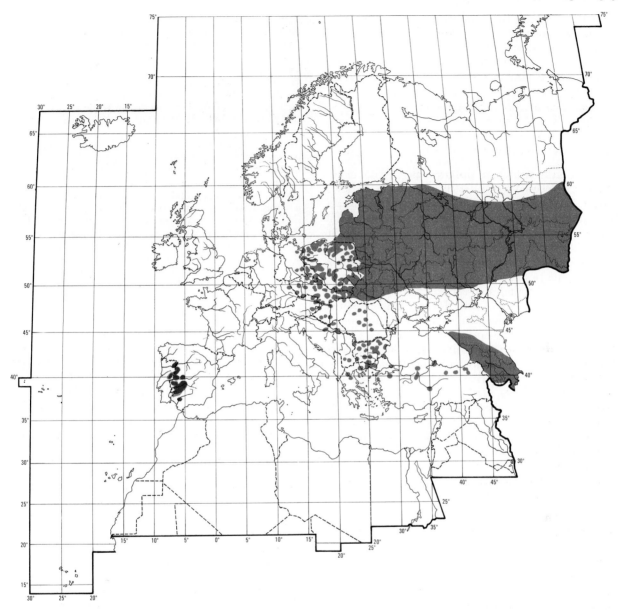

coveries; in earlier decades, recorded also Gambia, Guinea-Bissau and Ghana (Bannerman 1957). Those using Levant route concentrate for crossing of Bosporus; over 7400 counted 8 September to 3 October 1973 (OST); no comparable concentrations yet noted further south, and Israeli migrant flocks always under 100 (Safriel 1968). Travels via Near East and Sinai, then apparently Red Sea coast, as said to be rare Nile valley (Horváth 1959a), to winter quarters in east Africa where 3 recoveries: Kaliningrad to Ethiopia, and Poland to Ethiopia and north Uganda. Southern limits uncertain due to other breeders in South Africa; no ringing evidence that Palearctic birds cross equator, but occurs Kenya (regularly in small numbers) and Tanzania (fewer) only September to early

April, so probably of European origins (Backhurst *et al.* 1973). In Arabia, recorded Jiddah on Red Sea, a few October–December (Trott 1947), and Aden where 4 recent records of small flocks December–February (Smith 1956; Paige 1960; Ennion 1962). Migration routes of those breeding Caucasus, Caspian region, and Iran unknown. Rare Iraq (PVGK), and only summer visitor to Iran (D A Scott). Winter visitors to Pakistan and north India may include Caspian and west Asian breeders (Deméntiev and Gladkov 1951); but possibly some move south-west to Africa, and 4 seen to arrive Hurgada, Egypt, from across Red Sea in November (Horváth 1959a).

Autumn departures from Europe and western USSR early August to September, stragglers into October; peak

Bosporus crossings late September to early October, a month later than *C. ciconia* (Porter and Willis 1968). Return to breeding range about 2 weeks later than *C. ciconia*; noted late February in Azerbaydzhan, but major arrivals generally late March to mid-April, or late April in north (e.g. east Baltic). Sporadic passage in late May perhaps non-breeders. Straggles north and west of breeding range spring and autumn; now mainly early autumn (possibly post-fledging dispersal) in Low Countries, but exceptional British records mainly April–June.

**Food.** Chiefly fish from shallow waters. Also insects (including locusts), frogs, salamanders, and to lesser extent, small mammals, snakes, lizards, crustaceans, and passerine nestlings (Bauer and Glutz 1966; Dementiev and Gladkov 1951; Witherby *et al.* 1939). Feeds mainly by deliberate stalking, often singly or in small groups. In north-east Portugal reported catching fish (15–25 cm) by shading water with wings: standing or walking and every few moments raising and spreading wings wide and bringing them forward of normal flying position, bowed, and inbent, while darting head forward and into water to grasp prey; occasionally catch made after forward run of 2 or 3 steps (England 1974). In Barycz valley, Poland, travels to forage up to 9·6 km from nest (Cramp 1966).

Fish include loaches *Misgurnus* and *Cobitis*, pike *Esox lucius*, burbot *Lota lota*, rudd *Scardinius erythrophthalmus*, roach *Rutilus rutilus*, perch *Perca fluviatilis*, eel *Anguilla anguilla*, sticklebacks Gasterostidae (Cramp 1966, 1969; Bauer and Glutz 1966). In Rumania, from stomachs in summer and September, chiefly weatherfish *Misgurnus fossilis* (Dombrowski 1912). In India in January and February, observed feeding on snails *Vivipara bengalensis* and *Lymnaea acuminata* (Roberts 1969). At Oka sanctuary, USSR, young fed mostly (78–100% total number of food items) small (9–25 cm, 6–60 g each) *Esox lucius* and *Misgurnus fossilis*, with frog and other fish occasionally. Each bird fed 3–8 times daily with 180–600 g fish in one portion; daily consumption of each young 400–500 g (Priklonski 1958; Priklonski and Galushin 1959).

**Social pattern and behaviour.** Based mainly on material supplied by M P Kahl. See especially Siewert (1932, 1955), Kahl (1971, 1972*c*).

1. Mainly solitary; often in single pairs, especially when nesting. Singly or in small flocks up to 100 on migration, but in winter quarters typically in ones or twos. At all times tends to be solitary when feeding though often gathers into small, loose groups, probably at plentiful food sources. BONDS. Monogamous pair-bond, probably of long duration. Members of pair often return to nest-site together at beginning of season, so may sometimes associate on migration or in winter; otherwise pair-bond renewed seasonally. Age of first pairing and breeding unknown. Both parents tend young until after fledging. BREEDING DISPERSION. Usually nests in solitary pairs, with nests spaced 1 km or more apart even where numerous. Strictly territorial, defending nest-site and immediate area throughout breeding season. ROOSTING. At least one member of pair present on nest from first occupation until young 3–4 weeks old; mate

often roosts on adjacent branch or in nearby tree. Young assemble to roost on nest after fledging. Away from breeding area, no detailed information.

2. Nest-site centre of social activity during breeding season; away from nest, birds appear to ignore each other most of time. Pair often return to nest-site together; otherwise ♂ usually precedes ♀. ANTAGONISTIC BEHAVIOUR. Intraspecific threat and fighting at nest-site less common than in White Stork *C. ciconia*, probably owing to lower population density and concealed, less conspicuous nature of nest-site. Threat Up-down display given towards intruders near nest; similar to Up-down between mates (see below) but more vigorous and prolonged, and usually orientated towards intruder. Bill sometimes lifted to vertical during threat. Also said to Bill-clatter when aggressively excited (Witherby *et al.* 1939; Hudson 1965). Apparently lacks Nest-covering display found in *C. ciconia*. HETEROSEXUAL BEHAVIOUR. So far as known, generally similar to *C. ciconia*. ♂ gives Head-shaking Crouch on nest when mate approaches: drops into lying position as if incubating, erects neck-ruff and tail, lifts wing-tips slightly up from back, and shakes head vigorously 2–3 times per s, inscribing horizontal arc; display usually diminishes as season progresses. After return of ♂ or ♀ to nest, pair perform mutual Up-down display as greeting (see A). While standing more or less horizontal, with wing-tips slightly raised, each extends neck forward, with pronounced downward bend, and makes series of quick Head-tossing movements, bobbing up and down so that bill moves from approximately horizontal to 15–30° above horizontal; white under tail-coverts spread widely and conspicuously sideways, extending to both sides beyond tail, black feathers of latter being shut tight and both tail and undertail coverts alternately and repeatedly depressed to vertical and lifted to nearly horizontal. Accompanied by series of rather weak, melodious, bisyllabic whistles, synchronized with Head-tossing. Little similarity with homologous but more ritualized Up-down of *C. ciconia*; elaborate Throw-back and Bill-clattering of latter evidently rarely (if ever) performed by European populations (Heinroth and Heinroth 1926–7; Siewert 1932; Stoll 1934; Kahl 1971), though Kahl saw adult in South Africa throw bill up vertically during Up-down and heard it Bill-clatter a few times as bill lowered. Allopreening occurs regularly in both sexes. Copulation not usually preceded by any obvious display; initiated by ♂ walking round ♀ on nest, though sometimes ♀ will lean forward first, with back nearly horizontal, and press against ♂'s breast. ♂ steps on ♀'s back slowly from one side and hooks toes over ♀'s shoulders as ♀ opens wings slightly from sides of body. Then much as in *C. ciconia*, including Copulation-clattering of ♂. Copulation sometimes followed by mutual Up-down display, and both birds usually self-preen. Before egg-laying, ♂ and ♀ stay together on nest for much of time, engaging in frequent mutual displays; but also, as most storks, spend much time standing inactively with necks retracted and bill resting

A

PLATE 41. *Ardea cinerea* Grey Heron (p. 302): **1** ad breeding, **2** ad non-breeding, **3** imm, **4** juv, **5** juv dark type, **6** nestling. (PJH)

PLATE 42. *Ardea purpurea* Purple Heron (p. 312): **1** ad breeding, **2** sub-adult, **3** imm, **4** juv, **5** nestling. (PJH)

PLATE 43. *Ardea goliath* Goliath Heron (p. 318): **1** ad breeding, **2** imm, **3** juv, **4** nestling. (PJH)

PLATE 44. *Mycteria ibis* Yellow-billed Stork (p. 322): **1** ad breeding, **1a** ad, **2** juv. *Leptoptilos crumeniferus* Marabou (p. 336): **3** ad, **4** juv. (RG)

Robert Gillmor

PLATE 45. *Ciconia nigra* Black Stork (p. 323): **1** ad, **2** juv, **3** nestling. (RG)

PLATE 46. *Ciconia ciconia* White Stork (p. 328): **1** ad, **2** juv, **3** nestling. (RG)

PLATE 47. *Plegadis falcinellus* Glossy Ibis (p. 338): **1** ad breeding, **2** ad non-breeding, **3** imm non-breeding, **4** nestling. (RG)

PLATE 48. *Geronticus eremita* Bald Ibis (p. 343): **1** ad, **2** imm, **3** juv, **4** nestling. (RG)

B

against base of foreneck (see B). RELATIONS WITHIN FAMILY GROUP. Generally similar to *C. ciconia* so far as known. Begging of young resembles Up-down of adults in movements, timing, phrasing of voice, and lack of Bill-clattering; probably develops towards that display during ontogeny (M P Kahl). Like adults, young will Bill-clatter when aggressively excited, e.g. during ringing, though young of *C. ciconia* then silent (Cramp 1966).

(Figs A–B from photographs by M P Kahl.)

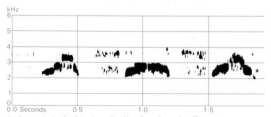

I Sture Palmér/Sveriges Radio Sweden April 1937

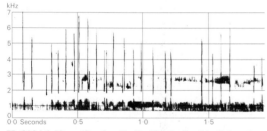

II Oldrich Unger/Sveriges Radio Czechoslovakia July 1967

**Voice.** In contrast to White Stork *C. ciconia*, has well-developed vocal calls. Bill-clattering much less developed and regular, although sometimes with vibrant or resonant quality, or with audible inspiration of breath (Voigt 1950); occasionally produced when angry, and Copulation-clattering, much as in *C. ciconia*, occurs (see II).

Most commonly heard call at nest rather soft 'chee lee, chee lee' (see I), 'chee' sounding like air being drawn into open mouth, 'lee' clear and prolonged; also rendered 'huji-ji' or 'chi-chu' (Bauer and Glutz 1966). Threat-call,

uttered on intrusion of another bird at nest-site, resembles a prolonged piping, hissing 'fiiich' (Bauer and Glutz 1966), 'fleeeee, he fleeee, he fleeeeehe' (Witherby *et al.* 1939). Series of weak, melodius disyllabic whistles given during head-tossing phase of Up-down display (Kahl 1971). Also utters high, thin 'hhiio ... hhiio', sometimes slightly resembling voice of Common Buzzard *Buteo buteo* (Priklonski and Galushin 1959); perhaps equivalent to melodious 'füo' of rarely heard Flight-call (Bauer and Glutz 1966).

CALLS OF YOUNG. Nestlings beg continually with 'ha-chi-chi . . .' or softer, deeper 'geék, gaák, gaók, goók', or occasionally 'päpäpäpä', a deeper 'gogogo' or soft 'pitjau' (Bauer and Glutz 1966). At *c.* 2 months, prolonged whining 'quieeeeee'; when fledged, hissing 'ool, ooi, ooi' (Witherby *et al.* 1939). Hoarse, gasping 'ziap' uttered when disturbed; well-grown young make deep growling 'uä-uä-uä' (Bauer and Glutz 1966). Young also clatter bills when excited (Cramp 1966).

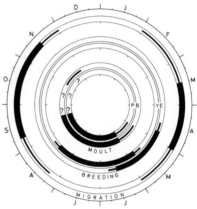

**Breeding.** SEASON. See diagram for Europe. Little apparent variation throughout range. SITE. In large old trees, rarely at top, most often in upper third, from 4–25 m above ground. Also on cliff ledges and in shallow caves. Not colonial but occasionally 2 nests in 1 tree and may be close to nests of Ardeidae (Dementiev and Gladkov 1951). Nest: large structure of sticks and twigs, on base of branches as large as bird can carry, re-inforced with earth and grass, lined typically with moss, but also grass, leaves, paper, and rags. Up to 1·5 m diameter, but smaller when new. Old nests re-used; also takes over old raptor nests. Building: by both sexes; materials brought from ground, or broken from nest tree. EGGS. Blunt oval; white, becoming soiled during incubation. $65 \times 49$ mm ($60–72 \times 44–53$), sample 80; calculated weight 86 g (Schönwetter 1967). Clutch 3–5 (2–6). Of 82 clutches, Poland 1955–72: 1 egg, 6%; 2, 15%; 3, 45%; 4, 24%; 5, 10%; mean 3·2 (some smaller clutches probably incomplete) (A Mrugasiewicz). One brood. Eggs laid at 2-day intervals. INCUBATION. 35–36 days (Cramp 1966), but 32–38 and 35–46 days reported (Witherby *et al.* 1939). By both sexes. Begins with 1st or 2nd egg; hatching asynchronous. YOUNG. Semi-altricial and nidicolous. Cared for and fed by both parents. One parent always on guard for first 10–15 days. Fed by regurgitation; at first direct into chick's throat, then on to floor of nest. FLEDGING TO MATURITY. Fledging period 63–71 days. Become independent soon after. Age of first breeding 3 years. BREEDING SUCCESS. In Poland, c. 80% of eggs produce fledged young, but annual range great, 31–92%; 3·4 young produced per successful pair (Cramp 1966).

**Plumages.** ADULT. Head, neck, upperparts, chest, tail, and wings black with strong purple and green gloss, most intense on head, neck, mantle, and lesser wing-coverts; tail tinged brown. Breast, belly, thighs, and under tail-coverts, and axillaries white. Feathers of chest elongated, but less than in White Stork *C. ciconia*. NESTLING. Down white; 1st sparse and short, 2nd longer and denser. First signs of developing dark feathers on ear-coverts at c. 20 days, on back at 22–24 days; fully feathered at c. 60 days (Schröder and Burmeister 1974). JUVENILE. Pattern like adult. Head, neck, and chest dark grey-brown with pale feather-tips giving speckled appearance. Rest of upperparts browner and less glossy than adult. Upper wing-coverts with narrow pale brown margins.

**Bare parts.** ADULT BREEDING. Iris brown. Bill and leg bright scarlet, orbital skin lighter. NON-BREEDING. Bill, leg, and orbital skin darker, crimson. NESTLING. Iris at hatching pale grey; bill yellow; leg pink. Later, bill orange-yellow at base, grading through pale olive to nearly colourless tip; orbital skin dark grey; leg pale olive. JUVENILE. Iris grey-brown. Bill olive-green,

orange-yellow at base. Orbital skin grey-green. Leg olive-green, bare part of tibia lighter, tinged yellow. Bill and leg become red in spring of 2nd calendar year. (Naumann 1838; Skovgaard 1920; Schröder and Burmeister 1974.)

**Moults.** ADULT. Single complete annual moult begins May–June, lasts until winter. Primaries descendant. POST-JUVENILE. Complete. Body moult starts February 2nd calendar year; wing moult follows from May onwards and lasts until winter. (Bauer and Glutz 1966.)

**Measurements.** Sexes combined. Wing 539 (8) 520–580 (Dementiev and Gladkov 1951); range of 9, 520–600; tail 190–240; bill 160–190; tarsus 180–200 (Witherby *et al.* 1939). Toe 85–97 (RMNH).

**Weights.** Adult c. 3000; 2 juveniles 2400 and 2500 (Bauer and Glutz 1966).

**Structure.** Wings long and broad. 12 primaries: p8 and p9 equal and longest, p10 9–21 shorter, p11 53–94, p7 0–15, p6 20–50, p5 80–133, p1 200–220; p12 shorter than primary coverts. P6–p10 emarginated outer webs, emarginations partly hidden by primary coverts; p7–p11 inner webs, just beyond coverts, emarginations strong and abrupt in p9–p11. 22 secondaries. Tail round. 12 feathers: t6 c. 35 shorter than t1. Bill long, more slender than *C. ciconia*, culmen straight, gonys sloping upward over half its length towards tip, giving bill upturned aspect. Lores partly feathered, orbital skin not continuous with bill as in *C. ciconia*. Leg long, lower half of tibia bare. Toes joined by webs at base; outer c. 80% of middle, inner c. 75%, hind c. 35%.

## *Ciconia ciconia* White Stork

PLATE 46
[between pages 326 and 327]

DU. Ooievaar   FR. Cigogne blanche   GE. Weissstorch
RU. Белый аист   SP. Cigüeña común   SW. Vit stork

*Ardea Ciconia* Linnaeus, 1758

Polytypic. Nominate *ciconia* (Linnaeus, 1758), Europe, North Africa, Middle East. Extralimital: *asiatica* Severtzov, 1872, central Asia; *boyciana* Swinhoe, 1873, east Asia.

**Field characters.** 100–115 cm, of which body only half; wing-span 155–165 cm, ♂ averaging larger. Large water-bird, mainly white with black flight-feathers, long red bill and legs, long neck, and short tail. Sexes alike and no seasonal differences; juvenile distinguishable.

ADULT. Entirely white except for black primaries, secondaries, greater coverts, and long scapulars; breast feathers long and shaggy. Bill, legs, and feet red; eyes grey with black surrounding patch. JUVENILE. Similar, but black of coverts and scapulars tinged brown, which extends on to smaller feathers, and bill and legs brownish-red (former with darker tip).

Unmistakable on ground. High overhead in flight species can look similar to adult Yellow-billed Stork *Mycteria ibis*, which, however, has black instead of white tail. Possibility of confusion also, when soaring high and details of bill and legs cannot be made out, with White Pelican *Pelecanus onocrotalus* and Egyptian Vulture *Neophron percnopterus*, which show similar patterns of

black and white from below.

Highly gregarious on migration, in winter quarters, and when congregating at special food source, flocks of hundreds or even thousands being seen; parties of up to 40 or 50 non-breeders also associate in breeding season. Flies with neck stretched out and drooping slightly, legs extending well beyond tail. Wing-beats slow and even; sometimes glides, particularly when coming to land, and soars in thermals on rigid wings. Walks sedately with neck upright or slightly forward, but rests with bill sunk between shoulders, often on one leg. Perches on buildings and trees, but does not clamber easily among branches. Often breeds in association with man, and non-breeding parties in summer can usually be approached fairly closely; at other times, often rather wary. Sound of bill-clattering in display draws attention to breeding pairs.

**Habitat.** In mid-latitude continental and Mediterranean climates, where open wetlands, savannas with scattered trees, steppes, floodlands, ricefields and irrigated lands, moist meadows and pastures, or arable fields offer extensive, easily accessible food supplies. Prefers shallow standing water in lagoons, pools, open ditches, and sluggish streams to large or fast rivers or margins of deep lakes and seas. Mainly in lowland regions down to or below sea-level; where similar habitats occur among mountains, breeding e.g. up to 2500 m in Morocco (Voous 1960) and up to 2000 m in Armenia, USSR (Dementiev and Gladkov 1951), but in central Europe records above 500 m exceptional. Avoids chilly and humid places, significant degrees of frost, and tracts of tall dense vegetation, from reedbeds to forests. For breeding uses sunny sites on tall trees, free-standing or in clumps, alive or dead; also on lofty artefacts such as church and other towers, chimneys, roofs, walls, ruins, stacks of hay and straw, and even specially erected platforms, big posts, or flat wheels. Nest-site usually adjoins terrain favourable for feeding.

Large size and mixed diet call for much flying; reliance on soaring and gliding rather than flapping flight over long distances or to higher altitudes put premium on habitats generating warm rising air currents. For same reason on migratory journeys prefers desert crossings to overflying thermally deficient areas such as forests or broad seas, where use of airspace makes undue demands on energy. Historical accident of convergence of habitat requirements with human settlement has long led to commensalism, reinforced by locally intense sentiment.

**Distribution.** Decrease in north and west of European range, with last regular breeding Belgium 1895, Switzerland 1949, and Sweden 1954.

ITALY: probably bred Po valley 14th century; has nested occasionally in north in recent years e.g. 1954, 1959, 1960 (SF). ISRAEL: bred unsuccessfully 1951, 1972; one pair successful 1973–4 (HM). BRITAIN: nested Scotland 1416 (British Ornithologists' Union 1971).

Accidental. Iceland, Britain and Ireland, Norway, Sweden, Finland, Malta, Azores, Madeira, Canary Islands, Cape Verde Islands.

**Population.** Counts of breeding birds more extensive than for any other species, covering considerable areas of European range, some of long duration. Culminated in international censuses 1934 (Schüz 1936) and 1958 (Schüz and Szijj 1962, 1971). Figures in summaries below from these sources (and more recent data from E Schüz) unless otherwise stated. See also Schüz and Szijj (1975) for fuller details, including breeding success for many areas in central Europe 1959–72.

SPAIN: 26000 pairs 1948, 18500 pairs 1957 (Bernis 1971*a*). PORTUGAL: 5471+ pairs 1958. FRANCE. Mainly Alsace, where 155 pairs 1932, 173 pairs 1948, 133 pairs 1958; sharp decline after 1960 to 23 pairs 1969, 18 pairs 1971, and 10 pairs 1973. A few nest in Saône and Ardennes, and occasionally elsewhere in recent years (CJ;

Parent 1973; Cruon and Vielliard 1975). BELGIUM: bred 1972, province of Luxemburg, first in wild since 1895, though full-winged birds nesting in Knokke reserve in recent years (Lippens and Wille 1972; Collin 1973). NETHERLANDS. Marked decline in last 30 years: 273 pairs 1934, 310 pairs 1939, 83 pairs 1950, 56 pairs 1958, 48 pairs 1960, 14 pairs 1970, 9 pairs 1972, 8 pairs 1974 (Schuilenburg 1974). DENMARK. Marked decline since 1934: 859 pairs 1934, 186 pairs 1958, 126 pairs 1963, 65 pairs 1969, 51 pairs 1972, 38 pairs 1973 and 40 pairs 1974 (TD). SWEDEN: 12 pairs 1934, 4 pairs 1948; last bred 1954. WEST GERMANY. Oldenburg: decrease; 108 pairs 1958 to 68 pairs 1972. Baden-Württemberg: sharp decrease; 143 pairs 1958 to 26 pairs 1972. Bayerisch Schwaben: decreased since 1967; 46 pairs 1958 to 18 pairs 1972, and 22 pairs 1973. Oberpfalz and Niederbayern: slight decrease since 1967; 59 pairs 1958 to 49–51 pairs 1968. EAST GERMANY. 2899 pairs in 1974, compared with estimated 2500 pairs in 1958 (Schildmacher 1975). POLAND. Lowest level of population 1928–31; present numbers similar to 1934 (Tomiałojć 1972). Milicz district: fluctuating, with 144 pairs 1959, 105 pairs 1969, 114 pairs 1970. CZECHOSLOVAKIA. 2413 pairs 1934. Increase since in Bohemia by 16% and in Moravia by 97%; decreased Slovakia by *c.* 23% (Bauer and Glutz 1966). Slovakia: 1203 pairs 1968 (Stollman 1969). Colonized higher areas in all 3 regions (Hudec and Černý 1972). HUNGARY: 7473 pairs 1958, 6017 pairs 1963; decline ascribed to drainage of wetlands, decline of buildings suitable for nesting, destruction of nests, and shooting (Marián 1970). AUSTRIA: small but increasing population; 118 pairs 1934, 276 pairs 1958, 384 pairs 1972, and 392 pairs 1974 (Aschenbrenner and Schifter 1975). SWITZERLAND. Some 150 pairs about 1900 when already decreasing; declined to 16 pairs 1930–40, 6 pairs 1948, last bred 1949 (Geroudet 1955). Attempts at reintroduction not completely successful so far (WT). YUGOSLAVIA. Vojvodina: 802 pairs 1957. Macedonia: 1490 pairs 1958. Not decreasing, but considerable fluctuations (VFV). GREECE: 9184+ pairs 1958, *c.* 2500 pairs 1968–70 (Schüz and Szijj 1975). RUMANIA: 729 pairs in part of west 1958; decreasing generally (MT). BULGARIA: 2504+ pairs 1958; considerable annual variations but not thought to be decreasing (SD). USSR. Commonest in south-west and western parts of range, and in some areas of Transcaucasia (Dementiev and Gladkov 1951). Estonia: 320 pairs 1939, 354 pairs 1958, 1060 pairs 1974 (Veroman 1975). Latvia: 6750+ pairs 1939, 6125+ pairs 1958. Byelorussia, Ukraine and Moldavia: 10319 pairs 1958 (Lebedeva 1960). ALGERIA: 6500 pairs 1935, 8800 pairs 1955 (Heim de Balsac and Mayaud 1962). MOROCCO. Varying estimates, from 12000 to 24000 pairs (Heim de Balsac and Mayaud 1962). Increasing and extending range (PR); estimates, based on sample census in 1974, 9619–*c.* 13500 pairs (Thiemann 1975). TUNISIA: 87 pairs 1938, 249 pairs 1954 (Heim de Balsac and Mayaud 1962); *c.* 180 pairs 1973, decreasing *c.* 15% a year (MS).

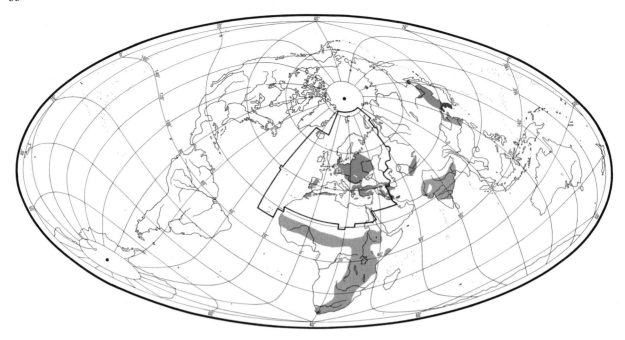

In international censuses, for areas counted in both 1934 and 1958, number of breeding pairs declined from 10 307 to 5087. Rough estimate of total European breeding population, in 1958, at least 93 000 pairs. Decline most marked in cool and wet areas of north-west Europe, and, in Germany, in hilly rather than low areas though in Austria and Czechoslovakia has recently spread to higher areas. Reasons not fully established—possible factors include climatic change, habitat changes, loss of nest sites, overhead wires, insecticides (by direct poisoning and by reducing food supply in breeding and wintering areas), less favourable human attitudes in some nesting areas, and shooting (especially on migration). In northern periphery of range, less young per pair fledged in recent years (Schüz and Szijj 1962). Lack (1966a), in extensive analysis of population dynamics, concluded that output of young raised per pair had no appreciable influence on subsequent changes in breeding population in region (Oldenburg) with largest source of data, but mortality critical, either of young betweeen attaining independence and breeding, or of adults, or of both. He argued that numbers probably regulated by mortality in African winter quarters, though this modified by at least one factor operating during breeding season—dispersive behaviour; latter presumably responsible for decline occurring faster in dry, hilly country than moist low-lying plains, and faster also in damper and cooler climate of north-west Europe than in drier eastern Europe.

Survival. Mortality of adults after 2 years old *c*. 21%, possibly 30% or higher in first 2 years of life (Schüz 1955). Oldest ringed bird 26 years (*Vogelwarte* 1972, **26**, 355–6).

**Movements.** Migratory. A few winter southern parts of breeding range, exceptionally north to Denmark and Kaliningrad, and in some numbers in Spanish marismas (Ree 1973); but overwhelming majority migrate to tropical Africa, Iran, or Indian subcontinent (Haverschmidt 1949; Bauer and Glutz 1966; Moreau 1972). Main departures during August, and peak Bosporus migration mid-August to early September; juveniles generally leave nest vicinity before parents, but young and adults migrate together in large flocks, of up to 11 000 over Bosporus (Porter and Willis 1968). At this time mostly soars in thermals and so avoids long stretches of open water; thus crosses Mediterranean on 2 unusually narrow fronts, across Straits of Gibraltar, and around coasts of Levant. Hence migratory divide in Europe, with those moving mainly south-west or south-east separated by band along *c*. 11°E, from Alps foothills to Harz, then west through Osnabrück to southern IJsselmeer. Division not rigid and in fairly broad belt breeders may use either route (Bernis 1959; Schüz 1963).

Those migrating south-west, including breeders of Iberia and Maghreb, winter mainly in steppe and savannah zones of northern tropics. Routes south of North African littoral still conjectural (see Moreau 1967); generally rare near Atlantic coast of north-west Africa, but perhaps broad-front movement across western Sahara, presumed norm-ally nonstop as oases records only occasional. Recoveries indicate Spanish breeders winter largely Niger inundation zone (Mali) to north Nigeria; those of Alsace from Sénégal to Mali; those of Maghreb mainly from north Nigeria and Niger eastwards (Bernis 1959; Moreau 1967). Thus evidence for tendency towards separate winter quarters for

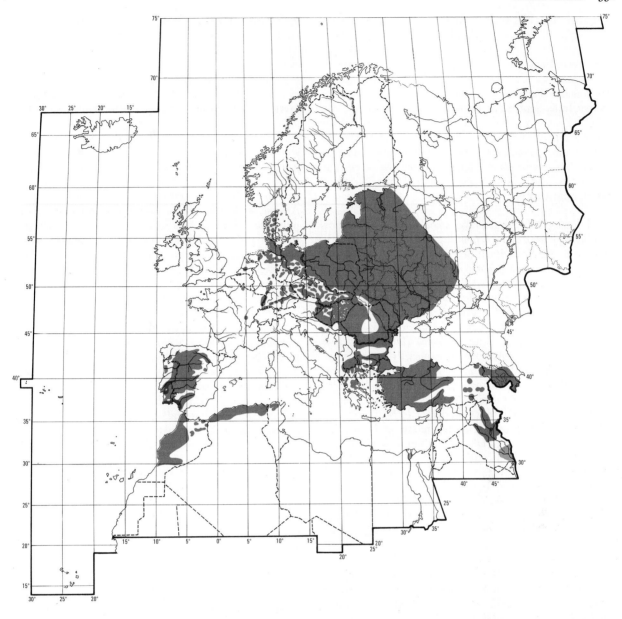

each population, but mixing occurs, even with central and eastern European breeders which winter chiefly in east and southern Africa. Other major route south-east through Europe, across Bosporus into Turkey, and thence western Syria, Jordan valley (also numbers east to Azraq), Dead Sea and east coast of Sinai, across southern Gulf of Suez at Tur, to enter Nile valley at Quena bend (26°N) (Meinertzhagen 1954; Phillips *et al.* 1959; Moreau 1972). In 1972, 339000 counted passing over Bosporus and Princes Islands, Turkey, between 5 August and 4 October (OST), compared with estimated 35000 in autumn over Gibraltar (Bernis 1975). 30000 recorded in 1½ hours passing over Tur (Mackenzie 1910). Probably begin fanning out soon afterwards; some winter Abyssinia and Sudan, especially

in locust years, but majority continue south to principal winter quarters in savanna from Kenya and Uganda to eastern Cape Province. In South Africa, where proportion of young higher than further north, present mainly November–February, though a few ringed juveniles recovered as early as September. Lost juveniles may continue south-east heading from Turkey (instead of turning south), with occasional recoveries in eastern Arabia and twice (from Germany) in India. No recoveries for population of Caucasus, Iraq, and west Iran; another migratory divide postulated from east coast Black Sea south to River Euphrates, with Turkish birds migrating mainly to Africa, but movement south-eastwards from Armenia and Azerbaydzhan (Schüz *et al.* 1971). Winters commonly Iraq and abundantly Khuzestan (Iran) though others apparently reach Pakistan and western India where fairly common winter visitor (Ali and Ripley 1968); but migrants to Indian subcontinent probably include some of Turkestan race *C. c. asiatica*, identified on passage in Afghanistan (Paludan 1959; Niethammer 1972). Species present India September–October to March–April.

Spring routes reverse of autumn. First arrivals back on upper Rhine late February, in Kaliningrad not until early April; spring passage more rapid than leisurely autumn exodus, but may be delayed (affecting breeding success) in some years. Many young remain south of breeding range during immaturity.

**Food.** Exclusively animal, taken mainly whilst walking or running with head and bill pointing downward, often with some wing-flapping. Small prey swallowed whole, larger prey (e.g. mice) usually killed and broken up beforehand. Normally daytime feeder, singly, in pairs, families, or large flocks, especially where prey (e.g. locusts) concentrated. In breeding areas, usually feeds near nesting site, but occasionally flights 3–5 km and, when prey densely concentrated, may fly long distances (Dementiev and Gladkov 1951). Feeds rapidly especially when prey abundant: one bird ate 44 mice, 2 hamsters, and 1 frog within an hour (Kálmán 1929–30); another caught 25–30 crickets every minute (Szijj and Szijj 1955); another 77 earthworms in 18 minutes (Hornberger 1967).

Eats wide variety of species depending on availability and locality; in dry years may be (e.g.) mostly insects and mice, in wet years mostly aquatic organisms. Items include: insects and larvae, especially Coleoptera (beetles) and Orthoptera (grasshoppers, crickets, and locusts); amphibians, particularly adult frogs (*Rana esculenta, R. temporaria*) and tadpoles; reptiles (lizards, snakes); small mammals, especially voles *Microtus* and *Arvicola*, mole *Talpa europaea*, shrews *Sorex*, young rats *Rattus*, and in Hungary young hamster *Cricetus cricetus* (Thorbias 1934); annelids (earthworms). Less often takes young and eggs of ground-nesting birds; molluscs and crustaceans, including in Germany introduced mitten-crab *Eriocheir chinensis* (Peters 1934; Berndt 1938); fish and scorpion. (Stammer 1934; Steinbacher 1936; Witherby *et al.* 1939; Schüz 1940*b*; Bouet 1950, 1956; Haverschmidt 1949; Dementiev and Gladkov 1951; Hornberger 1967.) Odd items include young goat, cat, weasel, domestic duck, and chicken (Skovgaard 1920*a*; Haverschmidt 1949; Kummer 1968).

Most data based on pellet analyses from spring and summer. In Hungary, April–July (mainly June), pellets and a few stomachs contained mostly insects, especially Orthoptera (particularly *Gryllotalpa* and *Gryllus*) and Coleoptera (especially Hydrophilidae and Carabidae), some mammals (*T. europaea* and fewer *Microtus arvalis*), and amphibians (*Rana* and *Bufo*) (Szijj and Szijj 1955). 2 stomachs from Hungary (spring) contained frogs, newts, lizards, field-mice, mole, small tortoise, small carp, snake,

and insects (Darázsi 1959). 50 pellets in June, Hungary, contained mostly insects, especially cricket *Gryllus campestris* (Szijj and Szijj 1955). 24 pellets, May–June, Alsace, contained mainly insects (chiefly Coleoptera, especially *Carabus*, and Orthoptera) and mammals *T. europaea* and *M. arvalis* (Schierer 1962). South-west Germany: in spring, mainly earthworms, snails (*Paludina*), and some insects (especially in years of abundance cockchafer *Melolontha* and grasshoppers), mice, and shrews; in summer, more Orthoptera and other insects (especially Carabidae) and, when following agricultural machinery, also spiders, snails, wood lice, mice (Hornberger 1967). Stomach contents from East Prussia, May–August, 90% by volume Coleoptera and other insects, plus mice, frogs, moles (Steinbacher 1935). In USSR: 1 stomach, Transcaucasia, contained 3 small tortoises *Clemmys caspica*, 4 grass snakes *Natrix natrix*, 6 frogs; nestlings near Kherson, Ukraine, fed at first on grasshoppers and lizards, later on frogs, snakes, and susliks (*Citellus*) (Dementiev and Gladkov 1951).

In North Africa, no essential difference between diet of adults and young: essentially insects, particularly (in years of abundance) locust *Schistocerca gregaria* and (in some years, chiefly in Algeria) locust *Dociostaurus maroccanus*. Also takes reptiles (snakes, lizards, skinks), small mammals, occasional scorpion *Scorpio maurus*, molluscs (in Morocco, mainly common *Helix lactea*), frogs *R. esculenta* and *Discoglossus pictus*, and small fish, especially introduced *Gambusia affinis* (Bouet 1950, 1956).

In south and central Africa, large numbers concentrate on migratory and plague invertebrates, e.g. locusts *Locustana pardalina*, armyworms *Spodoptera*, and caterpillars of *Laphygma exempta* and *Chloridea obsoleta* (Hale 1948; Milstein 1968; Schüz 1960).

**Social pattern and behaviour.** Based mainly on material supplied by F Haverschmidt and M P Kahl; see especially Schüz (1942), and Haverschmidt (1949).

1. Basically gregarious. Though often breeding in solitary pairs, commonly feeds in small parties even then and also nests colonially in parts of range. Outside breeding season, extremely gregarious, assembling in great flocks when migrating to winter-quarters in Africa where typically occurs in large, nomadic flocks that follow locust swarms. BONDS. Monogamous pair-bond often only of seasonal duration. Same pair sometimes breed together for more than one season at same nest-site; as ♂♂ and ♀♀ do not appear to winter in pairs, associate in pairs on migration, or arrive in breeding area at same time, faithfulness to nest-site and not to previous partner may be more important in determining re-pairings. ♂, usually returning first, often appears to accept first ♀ to come to site, whether previous mate or not. Most individuals start breeding at 3–5 years, but sub-adults occupy sites and form pairs without breeding for part of 1 or more seasons previously, sometimes moving sites between years. Both parents care for young until after fledging; latter usually depart singly, earlier than parents, and migratory flocks not made up of family parties. BREEDING DISPERSION. Loosely colonial or in solitary pairs; where numerous, numbers of nests normally on same tree, roof, or building; where populations small, in scattered and isolated

A                    B

pairs. Strictly territorial, defending nest-site through season and, at beginning of occupation, also surroundings and even nearby unoccupied nests. After fledging, young assemble at nest to be fed for some time before departure; then used for further period by adults for resting and roosting. ROOSTING. At least one of pair present on nest during breeding season until young $3\frac{1}{2}$–4 weeks old, while mate roosts at night near by on elevated point such as chimney, roof, or tree-top. During migration, flocks roost on similar elevated points.

2. Nest-site centre of social activity during breeding season; away from immediate vicinity of nest, e.g. on feeding grounds, birds appear to ignore each other most of time (Kahl 1972c). As a rule, ♂ returns first, taking possession of and defending nest. Stays on nest persistently, often displaying, and leaves infrequently and briefly to feed in immediate neighbourhood. Nest strongly defended against intruders and heavy battles fought with other ♂♂ and pairs—and also single ♀♀ (M P Kahl)—that try to alight. Disputes occur throughout breeding season as sub-adults, usually arriving later, try to occupy nests, showing strong predilection for occupied ones. Usually, established pairs drive off intruders. ♂ often accepts ♀ quickly; manner of sex recognition unknown and some prolonged encounters with ♀♀ may have been misinterpreted as fights between ♂♂. Rapid acceptance of ♀ in most cases, with occurrence of some functional copulations even on day of ♀'s arrival. CLATTERING AND UP-DOWN DISPLAY. Main sound signal mechanical Bill-clattering given during characteristic Up-down display; only true vocalization, Hissing, also given at times during Up-down. When performing Up-down, bird typically throws head rapidly upward and backward, then returns it more slowly forward and downward; can be repeated several times in succession, and at climax crown touches back (see A). Bill-clattering can occur throughout sequence, starting when head and bill initially point slightly downward. Start of sequence different in threat-version (Hiss-clattering); with very quick movements, head thrown back with bill widely open and only Hissing-call audible, Bill-clattering being delayed until after crown touches back, when head and neck move forward again. While performing Up-down displays, bird stands straddle-legged, often stepping to and fro, though may also stand on one leg, sit on tarsi, or even lie flat. Sequence nearly always ends by bird sinking bill down and pecking into nest-material. In threat version of Up-down, given to strange birds nearby, wings of ♂ somewhat open and pumped rhythmically and slowly up and down, but ♀ shows little if any

wing movement (see B). Up-down most common display, usually performed on nest by either sex in both antagonistic and heterosexual situations. Regular version of Up-down given first by ♂ on nest prior to arrival of ♀ when probably functions as Advertising-display to warn off rivals and attract mate. Also occurs as invariable post-landing behaviour, bird alighting on nest carrying out clattering, Up-down ritual whether second bird present or not. Regular Up-down also used as greeting between mates. ANTAGONISTIC BEHAVIOUR. Threat Up-down occurs as direct response to intruding or neighbouring storks flying over site or alighting in area immediately outside territory; ♂ and ♀ together, with drooping wings and cocked tails, will also Bill-clatter against more persistent intruders. When latter try to land at site, response changes to special threatening or defensive Nest-covering display (see C) in which site-bird stands motionless with legs somewhat bent, wings widely drooped, tail cocked, and head and neck extended forward, long feathers of foreneck ruffled. From this position, will sometimes sink down and even lie flat on nest with wings widely open. Fighting starts only when intruder manages to land on nest; consists of pecking with bill, beating with wings and attempts to push intruder over rim of nest. Fights can last for hours; blood may flow, and occasional deaths occur. In pursuing intruding bird, territory owner flies slightly above and behind, slowing when near and performing Aerial-clattering threat (Kahl 1972c): head lifted so that bill slightly above horizontal and oriented at opponent, while bill clattered loudly. HETEROSEXUAL BEHAVIOUR. Pair-formation not spectacular. ♂ reacts to strange ♀ by giving Head-shaking Crouch (Kahl

C

D

1972c), dropping to floor of nest as if incubating, erecting feathers of neck-ruff fully and shaking head vigorously from side to side. Head-shake may continue at rate of 2–3 per s for some minutes if ♀ continues to approach or remains nearby. After ♀ first alights on nest, received by ♂ with Up-down behaviour, in which she joins; copulations follow shortly, even within ¼ hr of ♀'s return. Later, mutual greeting, expressed by series of Up-downs, always follows arrival at nest; returning bird starts in air, few metres from nest, and guard or incubating bird joins in immediately, often while still lying down. Allopreening occurs regularly in both sexes, especially ♀ who nibbles crown feathers of ♂ as he sits or lies down. Up-down behaviour never used by ♂ as introduction to copulation. He walks slowly to ♀; both usually move for short time in circle or semicircle until ♀ eventually stops. When ♀ stands in suitable position, ♂ lays bill and neck sideways over her back and mounts by gently stepping on, adjusting balance with slowly flapping wings, and bending legs to lower body into position. ♀ supports ♂ by loosening wings slightly at sides and, drawing head backward until crown touches that of ♂, she continuously nibbles his foreneck feathers in bill. ♂, when in correct position, claps mandibles together among long feathers of ♀'s foreneck; this soft, rapid Copulation-clattering (Kahl 1972c; see D) sometimes lasts throughout, or precedes or occurs during cloacal contact; always indicates that copulation in progress. Up-down sometimes follows copulation, but often no post-copulatory displays, long period of self-preening by both sexes usually following. Copulation can also occur when ♀ sitting or lying down in nest. Reversed copulation occurs only rarely, and then mostly when ♂ lying flat in nest or on tarsi. RELATIONS WITHIN FAMILY GROUP. Up to 3–4 weeks, nestlings immediately flatten into nest with head and neck stretched out on sudden return of parent; similar response occurs when intruding stork alights, or with human intruder. Nestlings then limp, and can be handled. Small nestlings beg with mewing call and, from age of 1 day, by performing Up-down; Bill-clattering, though much weaker than adult's, similar and clearly audible. Nestlings also flap part-open wings and peck into nest-material when begging.

When almost fully fledged, threaten with feathers of foreneck ruffled when frightened by sudden arrival of parent, but soon start begging. Siblings do not fight among themselves (F Haverschmidt) unless hungry, when disputes often intense and lead to death through starvation of smallest (M P Kahl); latter also often ejected from nest before it dies, or is killed and even eaten by parent (Schüz 1957a). See Wynne-Edwards (1962) and Lack (1966a) for different views on significance of this. Siblings often Allopreen each other, and guarding parent will also often Allopreen nestlings for long periods.

(Fig C from photograph by F Haverschmidt; rest from photographs by M P Kahl.)

**Voice.** Adults voiceless apart from weak Hissing audible only at close range; main sound of both sexes non-vocal Bill-clattering (Schüz 1942; Haverschmidt 1949). Latter rapid, regular clapping or rattling of mandibles, with marked crescendo (see I), tongue-bone being extended and throat pouch acting as resonator; likened to distant machine-gun fire. Uttered in various social situations, mainly as component of characteristic Up-down Display when sometimes preceded by hissing in throw-back phase (see Social Pattern and Behaviour). Distinctive forms of slower clattering given by ♂ during copulation, and by both sexes in brief phrases when alarmed.

CALLS OF YOUNG. More vocal than adults, as soon as hatched uttering distinctive Food-call resembling prolonged mewing of cat; also cheeping sounds and harsh hissing. Young bill-clatter from hatching.

**Breeding.** Based mainly on Haverschmidt (1949), Netherlands. SEASON. See diagram for northern Europe. Commences up to 3 weeks earlier in southern part of range (Witherby et al. 1939). SITE. On trees and sometimes cliff-ledges and considerable use of man-made sites such as roofs of buildings, strawstacks, telegraph poles, and cartwheels on poles erected for them. Often solitary, but nests in colonies of up to 30 pairs in parts of range. Return to same site year after year, and repair and add to existing nest. Nest: large pile of branches (up to 3–4 cm thick) and sticks, often including earth, dung, and turf, lined with twigs, grass, paper, and rags. Average height 1–2 m and diameter 80–150 cm, but large nests up to 2·5 m high and 2 m in diameter. Building: by both sexes. Begun by first bird of pair to return, usually ♂. New nest can be built in 8 days. Material gathered from up to 500 m or more; added throughout nest period and even after young

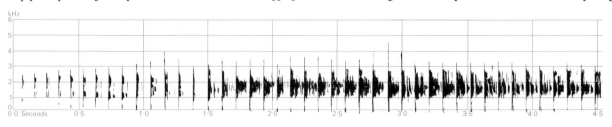

I Sture Palmér/Sveriges Radio  Denmark May 1967

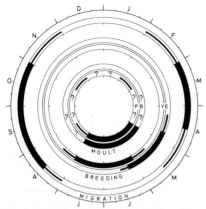

fledged. EGGS. Variable in shape; chalky white. 73 × 52 mm (65–82 × 47–56), sample 150 (Schönwetter 1967). Weight 111 g (96–129), sample not given (Bauer and Glutz 1966). Clutch: 4 (1–7). Of 80 clutches in Netherlands: 2 eggs, 6%; 3, 25%; 4, 48%; 5, 16%; 6, 5%. One brood. Replacement clutches only rarely recorded (Haas 1963) but ♀ may lay infertile eggs before arrival of ♂, then throw these out and lay new clutch. Eggs laid at intervals of 1–4 days, mostly 2. INCUBATION. 33–34 days. By both sexes; ♀ probably at night. Starts with 1st egg; hatching asynchronous. Eggshells discarded over side of nest. YOUNG. Semi-altricial and nidicolous. Cared for and fed by both parents. Rarely left for first 5–10 days. Fed by parents regurgitating on to floor of nest; every hour until day 10, every two hours to day 15. At *c.* 12 days, grab parents' bill to encourage feeding. When under 14 days, defecate into nest; thereafter over side. Commence wing-flapping at day 14, able to stand at day 22. FLEDGING TO MATURITY. Fledging period 58–64 days. Young leave area and become independent in 7–20 days after fledging, parents staying for further 15–30 days. Majority of young breed within 25 km of natal site. Age at first breeding normally 4 years, sometimes 2 or 5–7. In north-west Germany, 13% first bred at 2 or 3 years, and 87% at 4 to 7 years; on average more than 1 year later than in Baden and Alsace (Meybohm and Dahms 1975). BREEDING SUCCESS. Netherlands, 1934–45: brood size reared varied from 1·3–2·6 (mean 2·0), 12–20% of pairs reared no young in normal year, but up to 50% in bad year. Denmark, 1952–71: brood size for all pairs varied from 1·3–2·9 (mean 1·9), and for successful pairs from 2·4–3·2 (mean 2·8) (Dybbro 1972). Germany; birds breeding for first time at 5 years reared 2·3 young per successful pair compared with 1·9 and 1·8 young per successful pair for 4 and 3-year-old birds respectively (Schüz 1957a). Breeding success better in years when birds return earlier to breeding sites (Lack 1966a).

**Plumages.** ADULT. Mainly white; feathers of chest elongated, loose, forming ruff. Long scapulars, primaries and their coverts, secondaries, greater upper wing-coverts, and bastard wing black, glossed green or purple. Outer web of inner primaries and secondaries frosted grey when fresh. NESTLING. 1st down short and sparse, off white; bare patches on sides of body under wings, on centre of lower foreneck, centre of belly, inner sides of thighs, and round eyes and on throat. 2nd down dense, woolly, white, covering entire body; develops at 1 week (Bauer and Glutz 1966). By 3rd week, black scapulars and flight-feathers appear. JUVENILE. Like adult, but marginal coverts at base of primaries dark grey, mottled with some white (entirely white in adult); inner median upper wing-coverts and sometimes some short scapulars black; outer webs of secondaries evenly tinged grey (in adult mixed black and grey through differences in wear); and black parts less glossy, with some brown tinges at tips of feathers.

**Bare parts.** ADULT. Iris dark brown, sometimes grey (Bauer and Glutz 1966). Bill and leg bright red. Bare skin round eye and on lore black; gular skin black, but orange-red near feathers of chin, extending backwards in narrow streak on either side. NESTLING. Iris grey. Bill black with brown tip. Bare skin on head black. Leg at hatching pink, later grey-black. JUVENILE. Iris grey. Bill at first black, later grey-brown or pale red at base. Bare skin grey-black. Leg dull red.

**Moults.** ADULT. Single complete moult each year. Primaries replaced in irregular sequence during breeding season, some also in winter; details not known (Stresemann and Stresemann 1966). POST-JUVENILE. Complete; starting between December and May (Witherby *et al.* 1939). Group of almost flightless moulting birds in June may have been non-breeders in 2nd calendar year (Sterbetz 1968).

**Measurements.** Adults, sexes combined (RMNH, ZMA).

| | | | | | | |
|---|---|---|---|---|---|---|
| WING | 565 | (14·6; 10) | 542–580 | TARSUS | 220 | (4·07; 6) | 213–225 |
| TAIL | 227 | (10·6; 10) | 218–251 | TOE | 89·4 | (1·82; 5) | 87–91 |
| BILL | 176 | (9·61; 12) | 158–191 | | | |

Range wing ♂ 530–630, ♀ 530–590; bill ♂ 150–190, ♀ 140–170 (Witherby *et al.* 1939). Average wing 9 ♂♂ nominate *ciconia* 580, 14 ♂♂ *asiatica* 610 (Vaurie 1965). ♀ averages slightly smaller than ♂. Bill shorter in juvenile.

**Weights.** Non-breeding adults East Prussia: ♂ 3571 (41) 2610–4400, ♀ 3325 (27) 2275–3900. Increase in weight during summer: average of 14 ♂♂ June 3341, 14 ♀♀ 3150; of 12 ♂♂ July–August 3970, 12 ♀♀ 3521. (Bauer and Glutz 1966.)

**Structure.** Wings long and broad. 12 primaries: p8 longest, but sometimes p7 or p9; p9 0–11 shorter, p10 0–30, p11 43–90, p7 0–19, p6 15–68, p5 95–140, p1 202–264; p12 shorter than primary coverts. P7–p11 slightly emarginated on inner webs, just beyond coverts; p6–p10 on outer webs beneath coverts. 22 secondaries. Tail square; 12 feathers. Bill long, heavy at base; gonys allegedly more angled in ♂ than in ♀ (Schierer 1960), but not diagnostic. Nostrils slit-like. Leg long, lower half of tibia bare; toes webbed at base, outer toe *c.* 82% of middle, inner *c.* 73%, hind *c.* 39%.

**Geographical variation.** *C. c. asiatica* larger than nominate *ciconia*. *C. c. boyciana* much larger, with paler wings; bill black, bare skin of face red; bill heavier, slightly upturned, cutting edges not completely closing in middle; sometimes considered separate species (Vaurie 1965), but see Kahl (1972b).

*Leptoptilos crumeniferus* **Marabou**

PLATE 44
[between pages 326 and 327]

Du. Afrikaanse Maraboe    Fr. Marabout    Ge. Marabu
Ru. Африканский марабу    Sp. Marabú africano    Sw. Marabustork

*Ciconia crumenifera* Lesson, 1831

Monotypic

**Field characters.** 115–130 cm, standing about 120 cm high; wing-span 225–255 cm. Huge, hulking stork, with massive, pointed bill, bare head, neck, and gular pouch, and long legs; mainly dark grey above, and white below. Sexes alike, with small seasonal differences; juvenile distinguishable.

ADULT. Bill greyish-horn to yellow-white with darker base, bare head dull red, spotted black and scurfy-looking, with hindneck tinged blue and sides of neck and gular pouch bluish-pink shading into brick red at base of neck (though these colours often not striking in field, appearing generally as dirty flesh). Upper back and feathering on sides of neck and underbody white; rest of upperparts slate-grey with green sheen (sometimes appearing bluish-grey at distance), flight-feathers and tail blacker; under-wing dark. Breeding adult develops long, white fluffy undertail-coverts. Legs black, often stained by excreta and appearing dirty grey. JUVENILE. Similar but duller, without green sheen and with sparse down and scattered feathers on head and neck.

Unmistakable on ground and even in flight unlikely to be confused except at extreme heights. Black Stork *Ciconia nigra* has comparable but more contrasted black-and-white pattern, but is much less bulky with long neck. Vultures (Accipitridae) have similar outline, but lack projection of massive bill. Pelicans *Pelecanus* have much more white beneath wings.

Mainly carrion-eating stork with habits of vulture, locating carcases while cruising high in sky, but also feeds on locust swarms and wide variety of other animal prey from frogs to nestling birds. Sometimes singly but often in parties, particularly when at carrion or soaring; breeds colonially in trees or on cliffs. Not very aggressive, but can drive vultures from carcase. Walks and stands upright with head hunched between shoulders and gular pouch partly inflated. Perches easily in trees. Flies with neck drawn in, unlike most storks, but with bill still causing sizable projection; wing-beats slow and even, but once high in air soars effortlessly on outstretched wings. Generally silent, apart from bill-clappering and uproar of grunts and groans in breeding colony, but has croaking alarm note.

**Habitat.** Tropical, often along broad shallow rivers with ample sandbanks and on shores of lakes, but relatively independent of water and widespread over dry savanna as well as by swamps. Habitually perches, roosts, and breeds in commanding positions such as treetops, but avoids dense forests. Spends much time resting inactive, singly or gregariously, nature of diet not demanding or facilitating prolonged or vigorous movement. Often also circles or soars at considerable height, and undertakes diurnal journeys to and from roosts. Indifferent to man unless persecuted. (Archer and Godman 1937.)

**Distribution.** Breeds Africa south of Sahara, from Sénégal to Sudan and south to southern Africa.

Accidental. ISRAEL: one obtained Huleh Valley, May 1951; one shot Lake Tiberias, May 1957 (HM). MAURITANIA: one Banc d'Arguin, March 1966 (de Naurois 1969*a*).

**Movements.** Migratory or partially migratory, also dispersive. Breeds local dry seasons which vary regionally; north of equator nests mostly September–March. In East Africa, no more than 20% breed in any year, so large numbers range widely over suitable habitat; foraging non-breeders must move considerable distances (Kahl 1966*a*). In areas with heavy or prolonged wet season, most move away for peak of rains, and in northern hemisphere this apparently involves northward movements into higher, drier latitudes to moult (Chapin 1932; Moreau 1966), though Elgood *et al.* (1973) considered such movement ill-defined in Nigeria. Present Chad in flocks up to 500 June–September, north to Ennedi, though many fewer in dry season and no nesting colony known (Gillet 1960; Salvan 1967); common non-breeding visitor to Eritrea April–September or October (Smith 1967); and reaches 18°N in Mauritania (Heim de Balsac and Heim de Balsac 1954).

**Voice.** See Field Characters.

**Plumages.** ADULT. Head bare with some dark brown hairs; neck with sparse dirty grey downy filoplumes, bare skin shining through; down denser on nape and in narrow streak on upper hindneck. Collar of white and often some dark grey feathers at base of neck. Rest of upperparts, including median and lesser wing-coverts dark grey, glossed green on mantle, scapulars, and coverts. Underparts white; at start of breeding season under tail-coverts become long and fluffy, reaching to end of tail. Tail and flight-feathers dark grey with green gloss, paler below. Greater upper wing-coverts grey, with wide white margin on outer web and narrow on inner. Under wing-coverts dark grey; axillaries and some innermost coverts white. JUVENILE. Like adult. Down on head and neck denser. Upperparts without green gloss, slightly tinged brown. Greater upper wing-coverts with narrow brown margins.

**Bare parts.** ADULT. Iris brown; in ♀ lighter prior to egg-laying. Bill dirty yellow-brown. Bare skin of crown and sides of head red or orange with many black spots; forehead and lores more strongly mottled black; 2 short rows of black spots run backward from upper and lower rim of ear-opening. Chin and neck, including air pouch on foreneck pink. Early in breeding cycle, colour of head and neck more brilliant red with light blue patch on hindneck. Air pouch at base of hindneck usually concealed by white feathers of collar, but visible when inflated; deep crimson, conspicuousness individually variable. Leg black, often stained white by excreta; upper part of bare tibia red. (Kahl 1966b.) JUVENILE. Presumably like non-breeding adult.

**Moults.** Poorly known. Post-breeding complete; primaries probably serially descendant (Stresemann and Stresemann 1966). Pre-breeding apparently confined to development of long under tail-coverts. Post-juvenile with primaries normally descendant (Stresemann and Stresemann 1966).

**Measurements.** Sexes combined. Wing 673 (10) 620–720; tail 276 (10) 250–295; bill 244 (10) 226–280; tarsus 242 (10) 215–270

(McLachlan and Liversidge 1957). Bannerman (1930) gives higher values for 5 specimens: wing 650–755; tail 290–335; tarsus 225–295. Maximum bill length 316, Ethiopia (Friedmann 1930). ♂ larger than ♀ (Kahl 1966b).

**Weights.** Approximately 5000–6000 (Kahl 1966b).

**Structure.** Wings large and rounded. 12 primaries: p9 longest, p8 and p10 slightly shorter, p11 $c.$ 40, p7 $c.$ 15, p6 $c.$ 30, p5 $c.$ 100, p1 $c.$ 300; p12 minute. Tail square; 12 feathers. Bill long and heavy, depth at base up to 80 mm (Friedmann 1930); cutting edges straight, or slightly upturned in ♂ (Kahl 1966b). Nostrils slit-like, near base. Elongated extensible (up to 30 cm) subcutaneous air pouch at lower foreneck, normally partly inflated; smaller, rounded one at junction of hindneck and mantle (Akester et al. 1973). Leg long, $\frac{3}{4}$ of tibia bare. Middle toe $c.$ $\frac{1}{2}$ of tarsus, outer toe $c.$ 80% of middle, inner $c.$ 70%, hind $c.$ 45%.

# Family THRESKIORNITHIDAE  ibises, spoonbills

Medium-sized to large wading and terrestrial birds. About 30 species in $c.$ 15 genera, divided into 2 groups (sometimes treated as subfamilies): (1) typical ibises ($c.$ 14 genera) and (2) spoonbills *Platalea*. 4 breeding species, mainly peripheral, in west Palearctic. Body elongated, neck moderately long to long. ♂ larger than ♀. Wings rather long and broad. Flight with strong wing beats, sometimes soaring; neck and legs extended. 11 primaries: p8 and p9 longest, p11 minute. About 20 secondaries; diastataxic. Tail short, square, or slightly rounded (shallow fork in juveniles of some species); 12 feathers. Bill long: decurved in *Plegadis* and other ibises, straight with flattened, spatulate end in *Platalea*; nostrils slit-like. Variable extent of bare skin on head; neck also naked in *Threskiornis*. Legs rather long, lower half of tibia bare; toes of medium length, with small webs basally, hind toe slightly elevated, middle toe pectinated in *Plegadis*. Carriage of body upright, gait striding. Oil gland feathered. Feathers with aftershaft. Down on feather tracts and apteria; no powder-down patches. Plumage usually white, brown, or black (dark colours often glossy) but scarlet in Scarlet Ibis *Eudocimus ruber* and pink in Roseate Spoonbill *P. ajaja*. Sexes alike or nearly so. In some species, breeding plumage differs from non-breeding by presence of ornamental feathers. Bare parts, especially of face, brightly coloured in some; colour may intensify during period of pair-formation. 2 moults per cycle; pre-breeding may involve only small part of plumage. Moult of primaries descendant or serially descendant. Young semi-altricial and nidicolous. 2 downs: white, grey, or black; 1st sparse, growing from follicles of later contour feathers and soon overgrown by dense 2nd, growing from follicles of later down. Juveniles usually similar to adults, but often darker with bare areas of head more restricted; occasionally, markedly different from adults (e.g. in *E. ruber*).

Cosmopolitan in tropical, sub-tropical, and warm temperate areas. Mainly non-marine, aquatic birds of warm continental climates, preferring standing or slow-flowing fresh water, marshes, floodlands, etc. Some species, however, including Bald Ibis *Geronticus eremita*, often feed in dry habitats. Powers of resistance to adverse pressures and of adapting to fresh opportunities often apparently limited and, in west Palearctic, all 4 breeding species have shown reductions in ranges and numbers, often marked; *G. eremita* now in danger of extinction. West Palearctic species migratory, with much post-fledging and post-breeding dispersal. Migrate diurnally and probably also nocturnally, at high altitudes at times; fly in transverse-line formation, but often soar.

Eat wide variety of invertebrates (especially insects and their larvae, molluscs, and crustaceans) and small vertebrates (particularly fish, reptiles, and amphibians). Feed mostly in shallow wet areas where typically probe in soft mud (*Plegadis* and most other ibises), or sweep bill from side to side in water (*Platalea*). Mostly gregarious when foraging, and at other times—e.g. when loafing during day and when roosting at night. Typically colonial breeders, each pair defending nest-site territory only; a minority of species solitary or colonial, depending on local conditions. Monogamous pair-bond, mostly of seasonal duration so far as known (at least in migratory species—but see *G. eremita*). Pair-formation appears to be much as in other Ciconiiformes, but not widely studied. Similarly, displays not so extensively studied as in Ardeidae but include at least some similar elements including Stretch and Snap-displays, Twig-grasping, and Stick-passing (see Palmer 1962). Voice mainly harsh, guttural, wheezing, or grunting; among west Palearctic species, Spoonbill *P. leucorodia* also produces non-vocal, bill-snapping sounds. Away from colony or roost, generally silent except when flocks alarmed. Nestlings highly vocal, often producing shriller sounds than adults. Comfort behaviour apparently generally similar to other waterbirds; bathe in shallow water, rapidly beating wings against water and simultaneously sweeping face and bill along surface (see Skead 1951). Heat dissipated by gaping and gular-fluttering, while unfledged young often stand with one wing lowered; eggs and nestlings sheltered by drooped wings of adult (see Urban 1974). Other behaviour characters include (1) direct head-scratching; (2) resting with neck retracted.

Annual, seasonal breeders in temperate parts of range, but in tropics nesting often initiated by local factors such as rain or rise in water level. Nest in trees, dense vegetation, or sometimes on ground or cliffs. Colonies often of mixed-species composition. Nests large piles of available vegetation, lined grasses, rushes, etc.; roughly interwoven in case of sticks. Built largely by ♀ with material brought by ♂. Eggs oval, white, pale green, or blue, unmarked or spotted, with chalky coat and often rough surface.

Clutches 2–5 (1–7). One brood. Replacement clutches possible after egg loss but, in west Palearctic, common in only one species. Eggs laid at intervals of 1–3 days. Incubation periods 21–29 days in west Palearctic species, beginning with 1st egg—so hatching asynchronous. Both sexes incubate, but ♀ probably takes larger share, changing over at least once in 24 hours; single median brood-patch. Eggshells discarded over side of nest. Young cared for by both sexes; nestlings brooded continuously when small. Fed mainly by partial regurgitation. May leave nest at *c.* 2 weeks, returning to be fed. Fledging periods variable: 28–50 days in west Palearctic species, young becoming independent 1–4 weeks later. Age of maturity, where known, 3–4 years.

## *Plegadis falcinellus* Glossy Ibis

PLATE 47
[facing page 327]

Du. Zwarte Ibis    Fr. Ibis falcinelle    Ge. Sichler
Ru. Каравайка    Sp. Morito    Sw. Svart ibis

*Tantalus Falcinellus* Linnaeus, 1766

Polytypic. Nominate *falcinellus* (Linnaeus, 1766), southern Europe, north Africa, central Asia, south-east USA, and Greater Antilles. Extralimital: *peregrinus* (Bonaparte, 1855), Madagascar, Indonesia, and Australia. Subspecific status of populations of India, south-east Asia, east and south Africa not known.

**Field characters.** 55–65 cm, of which body only half, ♂ generally larger; wing-span 80–95 cm. Fairly small, all-dark ibis with decurved bill, long neck, and long legs. Sexes alike, but marked seasonal differences; juvenile distinguishable on close view.

ADULT BREEDING. Looks black at distance, but neck, upper mantle, shorter scapulars, shoulders, and underparts rich chestnut. Head, rest of upperparts, wings, and tail blackish glossed with purple and green. Long, thin, evenly downcurved bill dark olive-brown; bare loral area purplish-black (but briefly pale blue at onset of breeding); legs variable from green-brown to dark brown. ADULT NON-BREEDING. Head, throat, and neck brown-black streaked with white. Underparts brown-black with chestnut tinge and some purple gloss. Otherwise as breeding. JUVENILE. Like adult non-breeding, but head and neck browner, sometimes with large white spots on foreneck and crown; rest of upperparts duller, glossed green rather than purple, and underparts paler and browner. By 1st autumn (September–December), head and neck streaked with white, but less distinctly than adult non-breeding.

Often likened at distance to black Curlew *Numenius arquata*, which has comparable size and bill length, but *P. falcinellus* has more upright stance, heron-like gait, more rounded wings, longer legs, and longer, more curved neck. In flight, rapid wing-beats interspersed with glides and elongated appearance, with decurved bill and legs drooping slightly, give characteristic, curiously flattened outline. Can hardly be confused with Bald Ibis *Geronticus eremita*, which is noticeably larger and more heavily built, with thinner-looking bill, shorter neck and legs, and red, not blackish, bare parts.

Generally associated with shallow lakes, rivers, and floods, but also sometimes by coastal lagoons; often roosts in trees far from water. Feeds by probing in mud and shallows, and will swim for short distances in deeper water. Walks sedately like Spoonbill *Platalea leucorodia*, more compact than egret *Egretta*, though with neck extended. Breeds in reeds or trees, and perches easily on thin twigs, though clambers about rather slowly. Markedly gregarious; usually seen in small flocks and breeds in large colonies, often with various small herons (Ardeidae). Flight direct and purposeful, as fast as *P. leucorodia*, with wing-beats much quicker than those of herons of comparable size. Flocks may fly in compact groups, but more often in long lines (sometimes single file, in line abreast or as V, more usually diagonally) which undulate in snake-like fashion; individuals occasionally glide for long distances. Single birds or parties sometimes circle higher and higher, then plunge down with partly closed wings and dangling legs. Usually rather silent, but has harsh crow-like croak and flocks sometimes utter subdued chattering or grunting sounds for long periods.

**Habitat.** Generally in lowlands through warmer middle and low latitudes, preferring shallow waters of lakes and lagoons, floodlands, deltas, rivers, estuaries, and some-

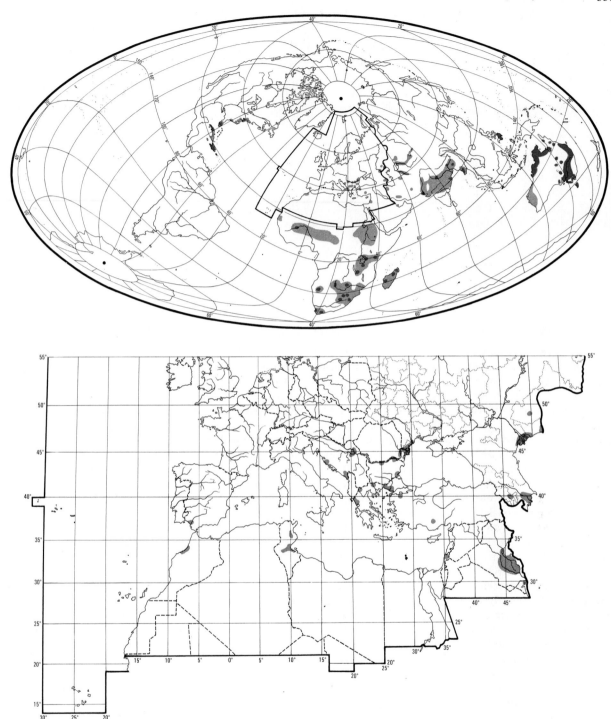

times shallow coastal beaches; also irrigated farmland and ricefields. In breeding season, favours wetlands such as swamps with stands of tall dense reeds *Phragmites* and rushes *Juncus*: fairly often nests in low trees such as willows *Salix*. Locally up to nearly 2000 m (Armenia). Avoids deep, fast-flowing, or turbulent water. Makes long

daily flights from feeding to breeding or roosting site; also wanders far, often in groups, circling widely for exploration or flying high in close formation. Intolerant of human presence, especially while breeding. Increasingly circumscribed through drainage and spread of settlement.

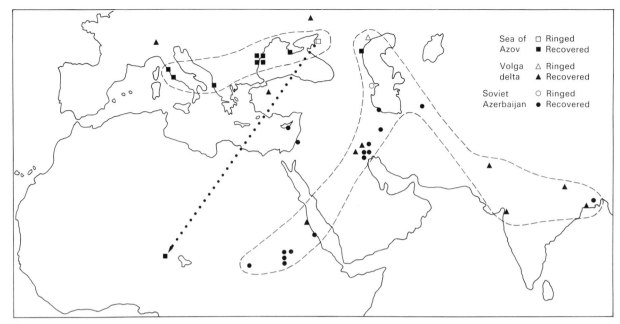

Fig III Recoveries of Glossy Ibis *Plegadis falcinellus* ringed in USSR (after Sapetin 1968)

**Distribution.** Marked contraction of range in western central Europe and North Africa in last 100 years. SPAIN: bred from at least 1774 to early 20th century, then attempts made in 1909, 1930, and 1935 (Valverde 1960). FRANCE: bred Garde (Rhône) 1844, and Dombes 1961 (CE). ITALY: a few pairs breed frequently, but not every year; last Piedmont 1971 (SF). AUSTRIA. Neusiedlersee: probably breeding from 1890s; bred 1920s to early 1930s with unsuccessful attempts 1938 and 1958 (Bauer and Glutz 1966; HS). HUNGARY: formerly irregular but numerous breeder, now occasional (see Population). CZECHOSLOV-AKIA: possibly bred 19th century (KH). IRAQ: common, breeding in suitable localities in marshes (Ticehurst *et al.* 1922); probably still breeding (Allouse 1953); present status unknown (PVGK). EGYPT: no longer breeding, but possibly did so fairly recently (Meinertzhagen 1930); now passage visitor, mostly late April to June (GEW). MOROCCO: said to have bred commonly in north in 19th century (Heim de Balsac and Mayaud 1962). ALGERIA: bred Lakes Halloula and Fetzara 19th century (Heim de Balsac and Mayaud 1962); present status unknown (EDHJ).

Accidental. Iceland, Faeroes, Britain and Ireland, Belgium, Netherlands, Denmark, Norway, Sweden, Finland, West Germany, East Germany, Poland, Czechoslovakia, Austria, Switzerland, Azores, Madeira, Cape Verde Islands.

**Population.** HUNGARY: earliest record end 17th century; breeding irregularly Kisbalaton 1912–53, with consider-able fluctuations, reaching 1000 pairs between 1922 and 1928 (Keve 1968); none there 1968–72 and only nesting

occasionally elsewhere (AK, IS). YUGOSLAVIA. 4500 pairs 1869; 600 pairs 1931 (Bauer and Glutz 1966). No current total figures; Kopačevski area, bred 1954 (27 pairs) and 1968 (Majić and Mikuska 1972). GREECE. 6 colonies in recent years, fluctuating, but perhaps 1100–1500 pairs. Largest Evros delta with 1000–1200 pairs 1971, 400–500 in 1973; and Lake of Kerkini with 400 pairs 1972 (WB). USSR. Overall numbers fairly considerable. Volga delta: 8780 birds 1934, 12620 birds 1935 (Dementiev and Gladkov 1951), but declined to 704 pairs 1951 and 1 pair 1958 (Lugovoy 1961). TURKEY. 3 known colonies: Manyas Gölü, maximum 105 pairs between 1960 and 1973; Meric delta, 500–600 pairs 1967; and Eregli, 450 pairs 1970 (OST). ISRAEL: one colony, with 16–20 pairs 1968–71 (HM).

Survival. Oldest ringed bird 19 years 10 months (Rydzewski 1973).

**Movements.** Migratory and dispersive, with considerable nomadic element. Flocks form after breeding, adults and juveniles often separated, and latter begin dispersing in all directions. Records north of breeding range, some of flocks, mainly August–November, and rare December–March occurrences in west and west-central Europe perhaps young which remained following dispersal (Bauer and Glutz 1966). From Hungary, young dispersed south-west and south-east, mainly within 300 km, but re-coveries also July to early September in Algeria, south France, Italy, Yugoslavia, Rumania and Bulgaria; in marked northwards irruption, autumn 1926, one juvenile travelled 2600 km to Kuybyshev, USSR, in 23 days, and others recovered Netherlands (2) and west Norway

(Schenk 1929). Other recoveries of Hungarian birds Spain, Sicily, Malta, Greece, Egypt, and Rostov (USSR).

Southerly aspects of post-fledging dispersals merge into autumn migration, which becomes dominant September, as adults and young withdraw south of breeding range in Europe and west USSR. Winters in small numbers Mediterranean basin west to Morocco (Smith 1965), and apparently Iraq, but not Iran (D A Scott); only straggler to Arabia. Most European breeders probably trans-Saharan migrants; many spring and autumn records from oases of Algeria, Libya, Niger, and Chad, and common October–March in steppe zone southern edge of western desert, while one ringed Sea of Azov recovered 75 days later Niger (Elgood *et al.* 1966; Moreau 1967). Extent of wintering east Africa uncertain; migrants pass through Nile valley, especially spring, with Hungarian ones recovered upper Egypt and Caspian ones Sudan, while small numbers winter Eritrea (Smith 1957). Southern limits in Africa of Palearctic migrants unknown, since other breeding populations there. Migrations of those breeding around Caspian more complex (see map); some move south-west towards Africa (recoveries Iraq, Levant, Sudan), especially from Azerbaydzhan colonies, others south-east (recoveries Kazakhstan, India, Bangladesh), notably from Volga delta colonies, though birds from both regions may take either route (Sapetin 1968).

Return movement through Mediterranean basin and Black Sea late March, some still moving early May, though colonies re-occupied April. Some spring overshooting, with rare April–May appearances north to Britain and Germany, but, in contrast to autumn, records of 1–2 birds only. Migrates dawn to dusk, perhaps also nocturnally; generally in transverse-line formation, with flocks of 500–1000 in USSR, especially Turkmenia (Dementiev and Gladkov 1951).

**Food.** Mainly insects and their larvae, including Diptera, Coleoptera (waterbeetles *Hydrophilus*), Orthoptera (grasshoppers, crickets), Odonata (dragonflies), and Trichoptera (caddisflies); also Hirudinea (leeches), Mollusca (*Planorbis, Ampullaria*), worms, crustaceans, and possibly small amphibians, reptiles, and fish. Feeds usually in small flocks in or near wetlands.

♂ in Scotland, October, contained larvae of fly *Eristalis tenax*, beetles, and molluscs *Mytilus, Cyclas*, and *Lymnaea* (Sim 1881). Of 2 juveniles from France, September (Madon 1935), one had gastropod snails, a bivalve, *Hydrous* debris, and unidentified insect remains; other, marine molluscs and crabs. Juvenile ♀ examined October contained mainly snail *Lymnaea minuta* (Rey 1907).

**Social pattern and behaviour.** No detailed studies; following based on scattered observations, summarized mainly in Bent (1926), Witherby *et al.* (1939), Dementiev and Gladkov (1951), Bannerman (1957), Palmer (1962), Bauer and Glutz (1966).

1. Gregarious throughout year. Often in large flocks; single birds seldom seen. Usually feeds in smaller parties. BONDS. Pair-bond evidently monogamous and of at least seasonal duration, but details lacking. Both parents tend young, continuing for a while after fledging. BREEDING DISPERSION. Colonial, often densely; almost always in association with other species (see Breeding). ROOSTING. Communal. During breeding season, in colony on or near nest or in branches of nearby trees. After fledging of young, family may at first return to sleep on nest at night. Outside breeding season, uses communal roost nocturnally, often with herons (Ardeidae), at such sites as waterside trees; information scanty. Recorded roosting in trees on passage in Egypt and in winter quarters in Kenya (Meinertzhagen 1954).

2. Flock forms compact group in air or nearly straight line, commute thus between colony or roost and feeding grounds. Flocks sometimes circle to great heights above colony at start of breeding season, but also elsewhere and at other times of year so reproductive significance (if any) not established. ANTAGONISTIC BEHAVIOUR. Apparently nothing recorded, possibly because overt threat posturing and fighting infrequent. HETEROSEXUAL BEHAVIOUR. Display said to begin at nest-site, where mutual Bowing and Allopreening occur (Hoogerwerf 1953), but details lacking. No details of copulation (which occurs on nest immediately after construction, on branches near, or on heaps of rushes) or associated displays (if any). Allopreening and mutual Billing frequently performed at nest-relief, pair standing erect and giving cooing calls. This ritual said to occur for as long as nest in use, but only when ♀ relieves ♂ as when ♀ at nest she leaves before ♂ alights on hearing his Flight-call (Baynard 1913); not clear, however, how sexes distinguished, or stage of breeding when observations made. Relieving bird often brings twig or leaf, adding it to nest. RELATIONS WITHIN FAMILY GROUP. Young beg by stroking parent's bill with own; also said to place bill sideways into parent's bill before thrusting bill deep into throat. Fed thus by incomplete regurgitation though, at least at times, adult may regurgitate fully into nest where food picked up directly by young. When older, still unfledged young tend to wander in colony, sometimes collecting in groups especially where nests on ground; there said to be fed collectively by any adult. Evidence conflicting, however, and no marked birds studied. More likely that each pair feeds only own young, as in Sacred Ibis *Threskiornis aethiopicus*, especially as family groups continue to associate after fledging, with persistence of parental feeding away from colony.

**Voice.** Generally silent away from nest, though harsh, low-pitched, crow-like notes resembling 'graa graa graa' (see I) sometimes given in flight by single birds, pairs, and flocks, with variants 'rha', 'rraa', and 'gra-ak' and ranging from long guttural croaks to subdued grunts. Grunting sounds also uttered at nest, followed by one or more notes like bleating of young calf. Guttural, cooing Contact-calls uttered between mates and between parents and young. (See Witherby *et al.* 1939; Bauer and Glutz 1966.)

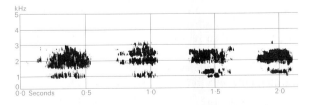

I C Chappuis Bulgaria May 1967

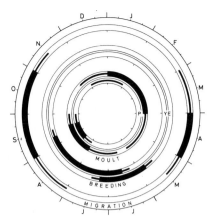

**Breeding.** SEASON. See diagram for Black Sea area; data lacking elsewhere. SITE. In trees or bushes growing in water, at heights to 5–7 m, or in dense reeds or rushes. Colonial; from a few to thousands of pairs, usually with other Ciconiiformes, also with Pygmy Cormorant *Phalacrocorax pygmeus*. Nest: compact pile of interwoven twigs (in trees) or reeds (in reedbeds), lined with green vegetation; diameter 30 cm, cup depth 5–8 cm; twigs up to 80 cm long used (Dementiev and Gladkov 1951). Building: by both sexes, though ♂ may bring material for ♀ to build; nest can be ready for eggs within 2 days, but material added throughout nesting cycle (Baynard 1913). EGGS. Regular oviform; deep green-blue, occasionally light blue, with less discolouring during incubation than related species (Dementiev and Gladkov 1951). 52 × 37 mm (46–59 × 33–40), sample 100 (Schönwetter 1967). Weight 31–39 g (Dementiev and Gladkov 1951). Clutch: 4 (3–6). One brood. No data on replacements. Eggs laid at 24-hour intervals. INCUBATION. 21 days. By both sexes, but ♀ does larger part, probably incubating through night while ♂ does spell during day (Baynard 1913). Begins on completion of clutch though ♀ may sit for periods before this; hatching synchronous. Eggshells discarded over side of nest. YOUNG. Semi-altricial and nidicolous. Cared for and fed by both parents. One parent always present for first 5 days. Early feeding by partial regurgitation but later food disgorged into bottom of nest. For a few days, ♂ brings food and passes to ♀ who feeds young. In first 5 days, fed 8–12 times per day; days 6–10, 6–10 times, then 5–8 times per day thereafter (Baynard 1913; Dementiev and Gladkov 1951). Young remain in nest for first 2 weeks, then move to nearby branches and reeds, returning to be fed. FLEDGING TO MATURITY. Fledging period *c*. 28 days. Young stay in or around nest until *c*. 50 days when leave with parents for feeding grounds; may then get some food for themselves but also fed by parents (Baynard 1913; Dementiev and Gladkov 1951). Time of independence not known. Age of first breeding unknown. BREEDING SUCCESS. No data.

**Plumages.** Nominate *falcinellus*. ADULT BREEDING. Head, neck, mantle, scapulars, and chest to vent and thighs deep chestnut (paler on underparts), tinged black with purple gloss on crown, foreneck, and chin; faintly glossed purple on upperparts, tips of lower scapulars purple-blue. Back, rump, tail-coverts, tail, and flanks black with strong purple gloss, more metallic green towards edges of feathers. Lesser upper wing-coverts deep chestnut; rest of upper wing-coverts, flight-feathers, tertials, underwing, and axillaries glossed dark metallic green, strongly tinged purple on centres of median and greater upper wing-coverts, inner secondaries, and tertials. ADULT NON-BREEDING. Head and neck sooty-brown, distinctly streaked by white feather-margins; sooty-black with narrow streaking on crown, paler with rather wide streaking on throat and foreneck. Mantle and scapulars black with purple and green gloss; chest to vent dark brown with faint purple gloss to base of feathers. In some, feathers of underparts variably tipped red-brown, giving faint rufous tinge to underparts; thighs usually red-brown. Tail, wing, and rest of body like adult breeding. NESTLING. 1st down sparse, sooty-black with variable amount of white on crown and throat. 2nd down apparently similar. JUVENILE. Head and neck sooty-brown, faintly mottled white on face; variable amount of white on crown, throat, and foreneck: usually small white patch of white above eye and some bars or spots on foreneck, but some with distinct white bar over crown and mostly white throat and foreneck; others completely sooty-brown. White feathers wear off relatively quickly. Upperparts, tail, flanks, and wing-feathers dark sepia with faint bronze green tinge, especially on wing; no strong green and purple gloss like adult, no chestnut on forewing. Underparts brown, feathers only faintly tinged green to base, no rufous tinge to underparts or thighs. IMMATURE NON-BREEDING. Like juvenile, but new feathering of head and neck streaked like adult non-breeding (but streaking less distinct and brown of head slightly paler); new feathers of upperparts glossed green or purple (less strong than adult); no rufous on underparts. During winter, more feathers like adult non-breeding gradually acquired; advanced immature specimens in spring like adult non-breeding, but wing without chestnut on lesser coverts (rarely a few coverts variably tinged chestnut), part of juvenile outer primaries or coverts retained, and underparts brown without rufous tinge. IMMATURE BREEDING (does not nest in this plumage). Variable. Feathers of head and neck with chestnut tips, brown bases, and variable off-white margins, giving head and neck streaked or mottled rufous appearance; in some, mainly like adult non-breeding. Upperparts like adult non-breeding, but feathers of mantle and scapulars variably tinged chestnut; underparts chestnut like adult breeding, but usually some brown of feather-bases visible. Wing more strongly glossed purple and green than juvenile, but less than adult; usually no chestnut on forewing, but a few coverts tinged rufous in some. SECOND IMMATURE NON-BREEDING. Like adult non-breeding, but wing, when not moulted, like immature breeding.

**Bare parts.** ADULT. Iris dark hazel. Bill dark olive-brown. Foot varies from pale olive-grey to dark brown with faint red tinge. Bare skin round eye, on lores, corner of mouth, and lower mandible black, tinged purple on lores; during courting period, bare skin of face cobalt-blue, bordered by narrow white or pale blue line along feathering of forehead and cheeks. NESTLING. Iris grey-brown. Bill pink with base, tip, and band round middle portion black. Foot pink. Bare skin round eye, on lores, and of corner of mouth pink. JUVENILE. Iris dark brown. Bill at first as in nestling, but gradually darkens to adult colour from ridge and tip of culmen onwards, last on basal portion of lower mandible. Foot dark olive-grey. Bare skin like adult non-breeding. (Meinertzhagen 1930; Palmer 1962; Witherby *et al.* 1939; RMNH.)

**Moults** (in west Palearctic). ADULT POST-BREEDING. Complete; flight-feathers descendant. Starts on head and neck when feeding young, followed from July or early August by body, tail, and inner flight-feathers with corresponding coverts. Moult finished, except for some primaries when autumn migration or dispersal starts: adults visiting Netherlands, September–November often retain 3–4 old outer primaries, usually without showing active moult; primary moult finished in winter quarters. ADULT PRE-BREEDING. Partial. Head and body, probably not always back, rump, tail, or wing-coverts. Sometimes starts November, but usually later; finished April–May. Underparts first, followed by scapulars, mantle, head, and neck. POST-JUVENILE. Partial. Head and neck August–October, followed by scapulars, chest, back, rump, underparts, and mantle, together with variable amount of inner lesser and median coverts. Tail, flight-feathers, and remaining coverts from mid-winter, usually finished before spring migration. Some juvenile coverts, outer primaries, and feathers of mantle or tail sometimes retained. IMMATURE PRE-BREEDING. Partial; head and body. January–May, but usually not complete and part of immature non-breeding plumage retained until mid-summer when immature post-breeding moult follows like adult.

**Measurements.** Nominate *falcinellus*. Mainly Netherlands; skins (RMNH, ZMA).

| | | | | | |
|---|---|---|---|---|---|
| WING AD ♂ | 297 (11·0; 7) | 280–306 | ♀ 273 (5·73; 7) | 267–281 |
| JUV | 292 (5·89; 11) | 283–303 | 267 (5·32; 6) | 257–272 |
| TAIL AD | 106 (6·03; 6) | 96–111 | 94 (3·24; 7) | 90–99 |
| BILL AD | 132 (4·31; 11) | 126–141 | 110 (2·55; 8) | 106–114 |
| TARSUS | 107 (3·58; 18) | 101–113 | 86 (2·38; 13) | 82–90 |
| TOE | 81 (1·73; 18) | 78–84 | 72 (3·23; 12) | 67–77 |

Sex differences significant. Wing and tail of juvenile slightly (not significantly) shorter. Bill of juvenile not full-grown until *c.* 3–6 months old, e.g. ranges: August–September, ♂ 98–113, ♀ 86–102; October–November, ♂ 108–132, ♀ 95–103; December–March, ♂ 120–134, ♀ 101–111.

**Weights.** Adults USSR, probably summer, ♂♂ 557–768, ♀♀ 530–680; juvenile at fledging 500 (Dementiev and Gladkov 1951). Juvenile ♂ October, Netherlands, 560 (RMNH).

**Structure.** Wing rather long and broad, slightly rounded. 11 primaries: p9 and p8 longest, about equal; p10 2–10 shorter, p7 7–20, p6 22–38, p1 85–110; p11 reduced, concealed by primary coverts. Inner web of p10–9 and outer p9–8 emarginated. Tail rather short; 12 feathers. Bill long and slender, higher than broad at base, evenly curved downwards. Upper mandible ridged and lower grooved, tip of bill slightly flattened. Nostrils small and slit-like. Feathers of head and neck pointed and somewhat elongated in breeding plumage, shorter and rounder in other plumages. Small area round eye, lores, area at corner of mouth, and skin between lower mandible bare in all plumages. Foot long and slender, lower half of tibia bare; small web between middle and outer toe. Claws long, slightly curved, that of middle toe slightly incised on inner edge. Outer toe *c.* 83% middle, inner *c.* 75%, hind *c.* 47%.

**Geographical variation.** Wing of nominate *falcinellus* of south Europe distinctly longer than *peregrinus* from Java and Celebes; bill of both similar; tarsus of *peregrinus* slightly shorter. Related White-faced Ibis *P. chihi* of west USA, Mexico, and South America (sometimes considered conspecific) differs in breeding plumage by rim of white feathers bordering bare reddish facial skin and by paler upperparts of juvenile.           CSR

## *Geronticus eremita* Bald Ibis

PLATE 48
[facing page 327]

Du. Heremietibis     Fr. Ibis chauve     Ge. Waldrapp
Ru. Краснощекий ибис     Sp. Ibis calvo     Sw. Waldrapp

*Upupa Eremita* Linnaeus, 1758

Monotypic

**Field characters.** 70–80 cm; wing-span 125–135 cm, ♂ averaging larger. All-dark ibis with metallic sheen, wispy hind-ruff of long pointed feathers, and dull red decurved bill, head, and legs. Sexes alike and no seasonal differences; juvenile easily distinguishable at reasonable range.

ADULT. Neck, body, wings, and tail dark metallic bronze-green shot with purple, except for upper wing-coverts tinted with copper and violet iridescence. Elongated hackles at rear and sides of neck form wispy ruff giving head tufted, shaggy appearance. Long, heavy, decurved bill, bare head and throat, and shortish legs and feet all dull crimson. JUVENILE. Duller, much less iridescent, and lacking copper-tinted shoulder patch. Bare head also duller with thin covering of greyish hairs.

Considerably larger and more heavily built than Glossy Ibis *Plegadis falcinellus* and easily distinguished even at distance or in flight, when colour of bill, head, and legs not apparent, by shorter neck and legs, less bulbous head, and less rounded wings with 3–4 short, well separated 'fingers'; *P. falcinellus* has characteristic look in flight, seeming flattened with hunched back.

Different habitat from other ibises, not lowland marshes but breeding on cliffs by rivers, coasts, or cultivated ground. Often seen on arable fields or grassy or stony ground both near and far from rivers. Feeds by pecking at or into ground while walking forward and making quite rapid progress. Flies with rather shallow wing-beats, often interspersed with glides on slightly bowed wings; legs do not extend to end of tail. Gregarious, though now scarce (see Population).

**Habitat.** In contrast to other ibises, mainly between tropical and temperate zones, in somewhat arid regions with precipitous rocky escarpments and fissures, ledges or caverns. Range of tolerance for climate, diet, disturbance and other factors not yet clear, but apparently unadapt-

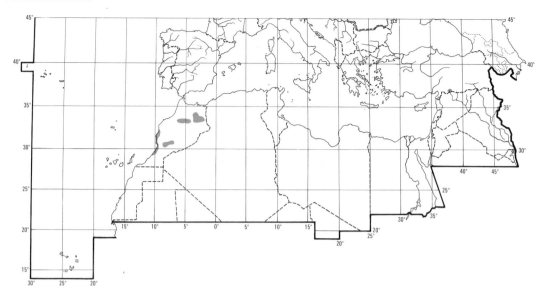

able, and now seriously threatened. Most often on dry wadi beds, rocky slopes, or semi-desert, but also, especially in non-breeding season, frequents high meadows, pastures, swamps, and valleys of mountain streams and seashore. On migration up to 2000 m or more (Archer and Godman 1937). Sometimes associates freely with man, nesting, especially in past, on castles, ruins and walls, even in towns. Most colonies now in caves, holes, on ledges, among boulders, or on inaccessible crags inland or on seacoast in remote, thinly inhabited regions. Still subject to some human interference and to damage by falling rock. Capable of walking on rugged terrain and flies freely, circling when prospecting and attaining considerable heights on migration.

**Distribution.** As far as known, now breeds only Morocco (only general areas mapped) and Birecik, Turkey. No proof of breeding Yemen, Ethiopia, hills of Red Sea, or Sudan as previously suggested by various authors (Smith 1970). For winter distribution, see Movements. CENTRAL EUROPE. Formerly bred until 16th or 17th century in southern Germany (Kelheim, Passau), Austria (Salzburg, Graz), and Switzerland, and perhaps elsewhere, though firm details lacking. Causes of extinction unknown, but some shot or taken for food (Géroudet 1965; Bauer and Glutz 1966). SYRIA: formerly bred, extinct c. 1910–16 (HK). IRAQ: no proof of breeding. EGYPT: may have bred in distant past (Meinertzhagen 1930). ALGERIA: has nested at Boghari, but irregularly; none present 1972 (Heim de Balsac and Mayaud 1962; PR).

Accidental. Spain, Iraq, Israel, Egypt, Azores, Cape Verde Islands.

**Population.** Continues to decline: now seriously threatened, with known world breeding population c. 275 pairs. Recent marked decline Birecik colony: e.g. c. 1000 birds 1911; 600–800 pairs 1954; c. 65 pairs 1964; 45–48 pairs 1967; 37–39 nests 1969; 35–36 nests 1970; c. 30 pairs 1971; 23 pairs 1973; 25 pairs 1974 and 1975. Another colony c. 20 km from Birecik deserted c. 1940 (Kumerloeve 1969; U Hirsch; SC). Reasons for marked decline unknown, but pesticides strongly suspected, over 600 found dead 1959–60 (U Hirsch; Parslow 1973b). MOROCCO. Colonies scattered and generally smaller (3–40 pairs); also some isolated nests. Some occupied irregularly and breeding may not occur in dry years (Robin 1973). No complete census; details of colonies in Heim de Balsac and Mayaud (1962), Géroudet (1965), Vernon (1973), Rencurel (1974) and Hamel (1975). Evidence suggests steady decline in recent years, e.g. in east 3 of 7 known colonies abandoned since 1966, and in centre and west in 14 colonies numbers fell from c. 229 pairs 1970–1 to c. 188 pairs 1972–3 and c. 173 pairs 1974–5. Breeding population estimated c. 250 pairs 1975, with perhaps 100–150 non-breeding birds, mostly immatures (U Hirsch). See also Hamel (1975), for discussion of decline and possible causes.

**Movements.** Migratory in east, perhaps only dispersive in west. Following notes based largely on Smith (1970). Colonies vacated late June to early July, but emigration not immediate; in Morocco, apparently most disappear August though small flocks noted November or even later, while upper Euphrates birds do not reach Ethiopian winter quarters before December. Both populations return to colonies February and March.

Turkish breeders winter north-east Africa; specimens from Asia Minor and Ethiopia similar, yet separable from Moroccan birds (Siegfried 1972a). Only modern reports of flocks from Ethiopia, mainly Massawa/Asmara region, where numbers December–February 1951–4 consistent with size of Birecik colony then; other scattered winter records out to east Sudan and Addis Ababa region. Absence of definite reports between Euphrates and winter quarters, suggests eastern population migrates over Near East deserts, perhaps in non-stop flight (Bourne 1959). Has, exceptionally, wandered to Egypt, Saudi Arabia, Aden, and Somalia; also old January–February records for north Iraq (Ticehurst *et al.* 1922, 1926).

Even less known about movements of Moroccan breeders. Small parties identified in Mauritanian deserts at 25°N in February *c.* 1950 and 18°N in February 1960; while a few seen Rio de Oro in August, October, and May, last presumably immatures. Winter range of this population unknown, but presumed inland rather than coastal. Rencurel (1974) believed there were enough winter reports from Morocco (notably Moyen-Atlas) to question whether truly migratory (as assumed by Smith 1970); suggested pronounced dispersive and erratic movements outside breeding season, rather than trans-Saharan migration.

**Food.** Mainly animal material, especially invertebrates. Daytime feeder, solitary or in groups, using bill to probe under stones, in crevices, and in tufts of vegetation.

Invertebrates include: Orthoptera (crickets, grasshoppers, locusts), Dermaptera (earwigs), Coleoptera (beetles), Formicidae (ants and eggs), Isopoda (woodlice), Araneae (spiders), Scorpionidae (scorpions), and molluscs. Vertebrates include: amphibians (adult frogs and tadpoles), reptiles (snakes and lizards), fish, rodents, and birds (including dead ones). Plant materials include rhizomes of aquatic species, duckweed *Lemna*, berries and shoots. (Heim de Balsac and Mayaud 1962; Meinertzhagen 1954; Parslow 1973*b*; Porter 1973; Rencurel 1974; Siegfried 1966; Smith 1970.)

Adult, Birecik, Turkey, in June, contained 70% locusts *Calliptamus italicus*, 25% bush-crickets *Dociostaurus maroccanus*, 5% Tettigoniidae, and 1 spider, while in chick 10–15 days old, May, 15 insect species including mole-cricket *Gryllotalpa gryllotalpa*, ants *Messor* and *Camponotus*, nocturnal ground-beetles (Tenebrioniidae), ground-beetles (Carabidae), jewel beetles (Buprestridae), chafer and dung beetle (Scaraboidea), weevils (Curculionidae), and indeterminate crustacean claws (Pala 1971).

**Social pattern and behaviour.** Compiled mainly from material supplied by H Kumerloeve, U Hirsch, and D T Holyoak. Studied at Birecik colony, Turkey (Kumerloeve 1962, 1965, 1967*a*; U Hirsch), in Morocco (Rencurel 1974), and at Zoological Garden, Basel (Wackernagel 1964).

1. Usually gregarious throughout year. Winter flocks vary from a few to over 100. Feeds in loose, usually small groups with birds well apart. Commutes in flocks between colony or roost and feeding grounds, flying in V-formation. BONDS. Monogamous pair-bond; probably only of seasonal duration as frequent changes of mates before egg-laying and promiscuous copulations recorded frequently at Birecik, while some birds (both wild and captive) known to find new mates between breeding seasons. Both parents tend young until after fledging, juvenile returning to nest for food at least 1 week after first flight. BREEDING

DISPERSION. Normally colonial, though some single pairs in North Africa. Nests often quite dense, but at least 15 cm between rims of adjacent nests even then. Each pair defends territory of up to 5 m round nest. Pairs breeding late unable to occupy spaces between existing nests even when these 5 m apart. Age of first pairing unknown in wild, but 3 reared in captivity bred first when 3 years old. During breeding season, Birecik adults feed up to 60 km from colony, although within 5 km when incubating or brooding young. ROOSTING. Communal and largely nocturnal. During breeding season, on ledges of nesting cliffs, mainly on or besides occupied nests; also, to smaller extent, in fields where forage. Wintering flock of 59 in Eritrea roosted at night in trees of village garden (Smith 1970); may also roost on cliffs during winter.

2. ANTAGONISTIC BEHAVIOUR. Little known outside breeding season. In Eritrean winter quarters, pre-roosting birds in trees pecked vigorously at those landing, giving croaking calls with raised plumes (K D Smith). At colony, encounters usually silent and most common when birds crowd on to ledge near nests, or when one enters territory of another. Though fighting not frequent, can be severe; rivals peck at each other with open bills, usually in short bursts. More often, one simply touches other with tip of bill; or gives threat-display with erected feathers, especially those of mantle—first gaping with head retracted and bill lifted (see A), then quickly lowering bill to ground with neck extended. Such threatening also performed by neighbours, thus defining border between territories. Birds, including both members of pair standing side by side, also adopt static appeasement posture with crests and neck hackles raised and bill lowered. Incubating birds will assume defensive Nest-covering display at approach of stranger, half-rising, ruffling body feathers, partly lowering wings, and stretching neck. HETEROSEXUAL BEHAVIOUR. Most birds pair after return to colony. After choosing site and cleaning it, unmated ♂ stands on threshold for long periods. Sight and activity of conspecifics in vicinity causes him to become restless and utter various calls. Points bill up with neck extended and throat enlarged but, in delivering main (disyllabic) call, lowers head with bill touching breast during 1st note and then raises it sharply for 2nd. Intensity of calling increases at approach of another bird: ♂ crouches, vigorously grasps and bites suitable object in vicinity (e.g. twig), and, with crest waving, makes far-carrying rumbling sound; swaying from one foot to other, passes bill under flanks several times as if inserting nest-material beneath body—sometimes actually so doing. Other bird, if ♀, watches inactively; lowers head and thus identifies her sex when ♂ finally approaches more closely and noisily on wing. ♂ renews intense, disyllabic calling; then Allopreens ♀'s back, neck, and head feathers. Couple next engage in mutual Billing, ♂ repeatedly entwining neck over ♀'s

A

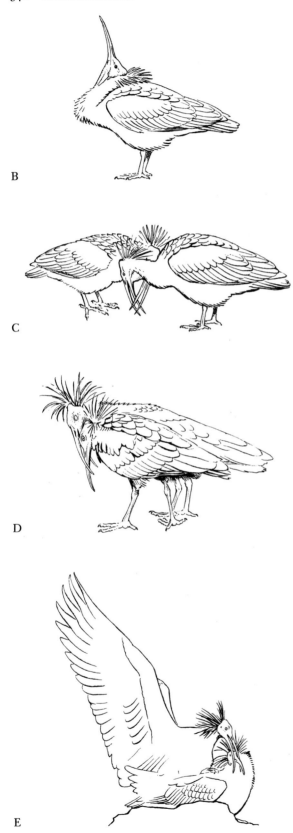

B

C

D

E

while moving round her. Pair-formation may follow encounter. Paired bird, recognizing mate flying towards cliff, gives 'jum' call and faces it. When latter lands by nest, mutual Bowing-display follows: ♂ and ♀ often face, then, synchronously, each first throws head back to shoulder (see B) and next lowers it to near ground (see C). Sequence repeated 2–3 times more in quick succession; crest and breast feathers raised throughout and 'jum' call given once by each bird during every downward movement. Slower, deeper bowing action, often without calling and followed by return to upright position also recorded. Movements of both versions of Bowing-display measured and formal; often followed by perfunctory self-preening. Bowing-displays occur regularly while pair at nest, but infrequent elsewhere. Mutual Allopreening (see above) also common at nest-site, particularly when one of pair returns and other remains on nest; one may also grasp other's bill (see D). Bird returning to site with nest material may leave it close to nest if mate asleep; often, however, passes it to mate and one may deposit it over other's neck. Copulations performed at colony many times each day, from a few days before laying until hatching, often following soliciting by ♀, who usually lies prone with raised head (typically on nest). ♂ Allopreens ♀'s head, etc., making gentle pecking movements, mounts, grasping base of her bill, and remains in position for *c.* 2 mins, rapidly shaking ♀'s bill in his (producing characteristic clappering noise) while extending and beating wings (see E). Nest-relief rarely completed without physical contact; often involves returning bird pushing other from nest. RELATIONS WITHIN FAMILY GROUP. Adult helps hatchling, loosening eggshell from around it. Young fed within 10–15 mins of hatching, parent grasping and guiding their bills; later, beg by simultaneously swinging head from side to side and back and forth, with upward pointed bill and giving Food-call. Parent returning from foraging trip gives 'jum' call on arrival at nest and when regurgitating food or water for young; shakes head forward, then lowers it, and nestling inserts own head into adult's throat. Body of nestling aligned with that of parent during feeding, and nictitating membrane drawn across eye immediately before it enters gape of adult. When about 14 days old, young begin to direct begging at parent's bill-tip and make rapid back and forward, vibrating movements of bill simultaneously with slower swing of larger amplitude. At 20 days, change to begging at corner of adult's gape; at about same time, parent starts grasping base of chick's bill. Soon after young reach directly into parent's throat with bill-tip. When parent arrives with food, eldest nestling pecks at heads of smaller siblings so it can beg first; when eldest satiated, next eldest pecks at other young and thus fed next. Older young often unguarded during day; those from 2 or more nests may gather together, running back to own nests when parents come with food. Neighbouring young not tolerated near nest by adults. During renewal of adult heterosexual behaviour in mid-May, well-grown young often adopt submissive posture with neck bent and head hidden near chest or belly.

(Figs B–C after Wackernagel 1964; others from photographs by U Hirsch.)

**Voice.** Normally silent when feeding, but vocal at breeding colonies and especially vociferous at roosts. At colony, monosyllabic, guttural 'jum' or 'jupe' call uttered in various situations, e.g. when greeting mate and in Bowing-display (Wackernagel 1964). Raucous, low and stifled 'horr . . . horr' given by unmated ♂ at entrance to nest-site often followed by disyllabic 'ouahh yooohhh'; also loud rumbling noise while displaying to approaching bird (Rencurel 1974). Gobbling sounds uttered at nest-relief (U Hirsch). Non-vocal, clappering sound made with bills during copulation. When alarmed, higher-pitched, harsh, monosyllabic 'gru' call given, with variants 'ga', 'gr', or 'gu' (Bauer and Glutz 1966). Guttural, repeated 'kyow' croaks uttered when congregating at winter roosts (K D Smith).

CALLS OF YOUNG. Small nestlings give clear 'lib, lib, lib' Food-calls (Kumerloeve 1967).

**Breeding.** SEASON. Eggs laid March, Morocco (Heim de Balsac and Mayaud 1962); end March and early April, Turkey (Kumerloeve 1967a). SITE. On cliff ledges and among boulders on steep slopes up to 100 m above ground; cliffs by sea or inland, when usually near river. On top of old buildings in one area of Morocco (P Robin). Will use artificial ledges provided by conservationists, Turkey. Normally colonial; from a few pairs to (formerly) several hundreds (Kumerloeve 1962). Nest: loose platform of small branches, lined softer materials such as grass and straw; tufts of vegetation and pieces of paper sometimes incorporated (Kumerloeve 1967a). Building: by both sexes. Material brought from surrounding area, occasionally stolen from another nest. EGGS. Oval, rough and pitted; blue-white with small brown markings and spots, becoming stained brown during incubation. $63 \times 44$ mm ($61-69 \times 42-46$), sample 46; calculated weight 68 g (Schönwetter 1967). Weight of captive laid eggs 61 g (54–67), sample 13 (Wackernagel 1964). Clutch: 2–4 (1–7). One brood. No information on replacements. Eggs laid at 1–3 day intervals. INCUBATION. 24–25 days (Heim de Balsac and Mayaud 1962); 27–28 days in captivity (Wackernagel 1964). By both sexes. Begins with 1st egg; hatching asynchronous. YOUNG. Semi-altricial and nidicolous. Cared for and fed by both parents. Fed by partial regurgitation. FLEDGING TO MATURITY. Fledging period 40–50 days. Time of independence not known. Age of maturity in Basel Zoo 3 years (Wackernagel 1964). BREEDING SUCCESS. In Morocco in favourable years, from clutches of 2–4 eggs, 20% pairs rear on average no young, 60%, 1 young; 15%, 2 young; 5%, 3 young; in dry years, 50% rear no young, nearly all the rest 1 (P Robin).

**Plumages.** ADULT. Head naked. Nape, neck, and rest of plumage black with green metallic gloss; feathers becoming browner with wear. Median and lesser upper wing-coverts glossed purple-red. Crest of narrow, elongated feathers on nape, glossed blue or purple. NESTLING. Down grey-brown, paler on underparts; forehead, lores, and orbital skin bare. JUVENILE. Like adult, but head not bare, feathers of head black-brown with white margins; median upper wing-coverts glossed green like rest of plumage; feathers of nape wider and less elongated. In 2nd calendar year, upper wing-coverts purple-red, nuchal crest present but head still mainly feathered.

**Bare parts.** ADULT. Iris orange-red, yellow-grey before 3rd year. Bill red. Bare crown black with orange central streak, sides of head and throat red. Leg and foot dirty red. NESTLING. Bill pink, mixed with black. JUVENILE. Iris grey. (Bauer and Glutz 1966.)

**Moults.** ADULT POST-BREEDING. Complete; primaries probably serially descendant (Stresemann and Stresemann 1966). Timing unknown, ♀ November shows mixture of old and new feathers, p1–p8 new, p9 and p10 old and worn (ZMA). POST-JUVENILE. Captive birds moult primaries descendantly in spring of 2nd calendar year; at same time purple-red upper wing-coverts and nuchal crest develop, but head still feathered.

**Measurements.** Wing ♂♂ 403–420, ♀♀ 390–408; tail ♂♂ and ♀♀ 196–220; bill ♂♂ 133–147, ♀♀ 115–131; tarsus ♂♂ and ♀♀ 68–72 (Hartert 1912–21). Bill eastern populations ♂♂ 129 (2·3; 8) 126–132, ♀♀ 124 (2·4; 5) 121–127; western populations ♂♂ 141 (5·5; 16) 132–146, ♀♀ 134 (4·6; 12) 128–145 (Siegfried 1972a). 3 ♀♀ western populations: tail 176, 178, 179; tarsus 67, 68, 71; toe 60, 60, 63 (ZMA).

**Weights.** Only data from captive birds. Just-fledged juveniles 1080–1230; maximum weight nestlings 1150–1440 (Bauer and Glutz 1966).

**Structure.** Wings long and broad. 11 primaries: p9 longest, p8 slightly shorter, p10 $c.$ 25 shorter, p7 $c.$ 15, p6 $c.$ 50, p5 $c.$ 75, p1 $c.$ 165; p11 minute. P7–p9 emarginated outer web, p8 and p10 slightly inner web. Tail rounded; 12 feathers: outer $c.$ 20–25 shorter than central. Bill long, narrow, but rather broad at base, moderately down-curved; nostrils small slits at base; nasal groove pronounced along entire length to tip. Leg short; outer toe $c.$ 85% of middle, inner $c.$ 75%, hind $c.$ 50%.

**Geographical variation.** Siegfried (1972a) argued that eastern and western populations completely separated; bills of eastern population significantly shorter.

---

## *Threskiornis aethiopicus* **Sacred Ibis**

PLATE 49
[after page 374]

DU. Heilige Ibis    FR. Ibis sacré    GE. Heiliger Ibis
RU. Священный ибис    SP. Ibis sagrado    SW. Helig ibis

*Tantalus aethiopicus* Latham, 1790

Monotypic

**Field characters.** 65–75 cm, ♂ averaging larger; wingspan 112–124 cm. Mainly white, with dark, long, heavy, decurved bill, dark, wholly bare head and neck, black scapular plumes, and black tips to flight-feathers. Sexes alike and no seasonal differences; juvenile easily distinguishable.

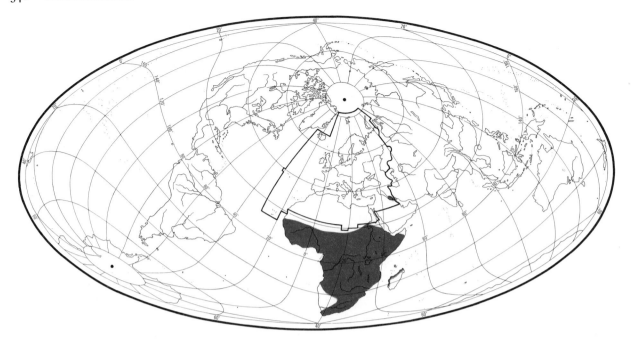

ADULT. Bare skin on head and neck dull grey-black. All feathers pure or dirty white apart from long, metallic blue-black scapular plumes, forming loose tuft falling over closed wing and tail, and black tips to primaries and secondaries. Bare skin at sides of breast and underwing scarlet. Bill black; eyes brown with dark red orbital ring; legs black, tinged red. JUVENILE. Head and neck feathered, white mottled and streaked with black. Less developed scapulars greenish-brown; more black on outer primaries and primary coverts. Under wing-coverts dark streaked; tail white with brown corners.

Unmistakable on ground. In flight, narrow, black rear outline to broad, white wing and dark head, neck, and curved bill characteristic; at dusk or in strong sunlight may show pinkish glow on underparts. Wing-beats faster than herons (Ardeidae) and storks (Ciconiidae) and often interspersed with glides, sometimes for long distances.

Gregarious, feeding in groups in wide variety of habitats, sometimes even probing carcases like kites *Milvus*. Nests colonially. Perches easily in trees, but moves about in them rather slowly. Walks slowly with body nearly horizontal and neck more or less straight, but often stands with neck retracted. Flies with neck outstretched, usually in flocks in V formation or long lines extended diagonally. Generally rather silent, but harsh croak sometimes uttered in flight.

**Habitat.** Sub-tropical as well as tropical, adapted to fairly wide range of mainly inland habitats by lakes and rivers and in cultivated areas. Breeding sites range from high trees to bare surface of rocky marine islands. Often roosts and breeds on islets in rivers or floodland, and not infrequently in town and villages. Feeds in mud of large rivers, resting on sand-banks, or on wetlands and tidal pools along coast. Walks and flies well; flight especially at low and medium altitudes.

**Distribution.** In west Palearctic, now breeds only Iraq where fairly common in southern part Amara to Fao, especially in winter, though definite nesting records few (Allouse 1953). Formerly bred Egypt where common in Dynastic times (many mummified specimens and murals of breeding birds and young) and apparently until 1800, but almost completely disappeared by 1850 (Meinertzhagen 1930).

Accidental. USSR: formerly vagrant to Caspian and Black Seas; small flocks near Baku 1944 (Dementiev and Gladkov 1951). Kuwait.

**Population.** No information. Oldest ringed bird 21 years 1 month (Rydzewski 1974).

**Movements.** Little information on movements of Iraq population. Present all year, though not necessarily sedentary, in marshes of lower Tigris and Euphrates; this presumed origin of Kuwait stragglers and of small numbers wintering Khuzestan wetlands, Iran, where maximum 42 in January 1972 (Carp 1972). Just possibly breeds Khuzestan, where summer records (none of over 8 birds).

In northern tropics of Africa, timings of movements ill-defined due to geographically variable nesting periods; mostly in dry season, involving dispersals to permanent fresh waters and coasts, and south–north migration (Lynes 1925; Moreau 1966; Elgood *et al.* 1973). Comparable movements in southern hemisphere, where recoveries

Angola and Zambia of birds ringed 1000–1500 km away in South Africa. Stragglers noted Aden in January, March, April, and June (Browne 1950).

**Food.** Invertebrates mainly, including insects (particularly locusts and grasshoppers), spiders, annelids, crustaceans, and molluscs. Also frogs, reptiles, fish, young birds and eggs, carrion, and offal (Hartert 1912–21; Clancey 1964; Bannerman 1930; Mackworth-Praed and Grant 1952, 1970). Feeds during day mainly in flocks, mostly by wading in shallow wet areas, and occasionally on dry land close to water. Many fly some distance from breeding colony to feeding grounds: over 10 km in Ethiopia (Urban 1974).

On Lake Shala, Ethiopia, takes fish scraps and contents of eggs of White Pelican *Pelecanus onocrotalus* broken by Egyptian Vultures *Neophron percnopterus* (Brown and Urban 1969; Urban 1974). In South Africa, forces Cape Cormorants *Phalacrocorax capensis* from nests to eat young and eggs (Bolster 1931). Also eats 'wildfowl' young (Clancey 1964) and, on Lake Rudolph, crocodile eggs exposed by monitor lizards (Modha 1967).

Regurgitations of 2 young, Lake Shala, contained chiefly beetle larvae, Lepidoptera larvae, and beetles (Urban 1974).

**Social pattern and behaviour.** Based on material supplied by E K Urban mainly from observations at Lake Shala, Ethiopia (see Urban 1974).

1. Gregarious throughout year. Outside breeding season in flocks. BONDS. Seasonally monogamous; duration of pair-bond unknown. Pair-formation normally in colony just before nesting. Both sexes tend young until after fledging. BREEDING DISPERSION. Colonial, often in association with but segregated from herons (Ardeidae), storks (Ciconiidae), or cormorants (Phalacrocoracidae). ♂ first establishes tiny pairing territory among as many as 100 ♂♂ packed in *c.* 10 m² on top of small trees or on ground. Later, pair defend separate nest-site territory in colony; one colony held 27 nests within 5 m². Nest-territory abandoned when young, 2–3 weeks old, leave nest and wander in vicinity. During breeding season, feeds away from colony, normally in flocks of 5–200 or more, sometimes in twos or threes, or singly. ROOSTING. Communal. Only information from Lake Victoria (Kenya) where many hundreds roost at night January–February along waterfront (L Brown), and from Lake Shala, where during breeding season 200–300 birds occasionally arrive near colonies at sunset and remain until first light on rocky, steep-sided islands in trees.

2. Breeding activity begins when parties of 10–50 birds, apparently mainly or wholly ♂♂, fly about islands, land briefly, then fly and land again, often at new spot. In 1–2 days, settle on one place where ♂♂ establish pairing territories; in following 1–2 days additional birds, both ♂♂ and ♀♀, arrive, new ♂♂ eventually going to site of pairing territories where much chasing and fighting. Soon after arrival, ♀♀ fly to territories where pair-formation occurs. Once pair formed, move to breeding area close by where ♀, with ♂ following, apparently selects nest-site; within few hours, usually same day, 100 or more pairs may be present. Vacated pairing territories soon claimed by other unmated ♂♂. Eventually all pairs move to colony and each pair then remains on own nest-territory. ANTAGONISTIC BEHAVIOUR. When occupying

A

pairing territory, ♂ stands with wings held slightly downwards with rectrices spread and shaken and tertiary plumes erected. Sometimes two ♂♂ face and repeatedly bill-feint, even striking each other's bills, at same time producing squeals, squeaks, and series of sounds resulting from inhaling and exhaling air. Similar bill-feinting squabbles between neighbours of same or different sex occasionally occur at nest-site territories. ♂ at pairing territory also performs Forward-threat (see A), adopting horizontal stance, extending head and neck towards opponent, then repeatedly feinting with open bill (no snapping of mandibles), while raises and opens wings, showing bright red skin outlining bones of underside of wing and adjoining area of breast. Modified form of Forward-threat occurs also when 2 or more birds, standing side by side and usually facing same direction, open wings but Bill-point skyward, nearly touching bills and calling as during bill-feinting. Other threat behaviour occurring at pairing territory includes pursuit-flight, in which ♂ flies towards and chases opponent, and supplanting attacks, in which bird displaying in Forward-threat flies towards perched opponent, forcing it to leave. Known Advertising-displays are modified Snap-display of ♂ and Stretch-display of both sexes. In rather infrequent Snap-display, ♂ stands in low horizontal position and suddenly extends neck forward and down, keeping mandibles open but not snapping them. In Stretch-display, bird in one rapid motion moves neck up and back and, with bill pointed skyward (see B), flicks neck backward causing crown to

B

touch mantle either on right or left side, after which neck and head quickly brought back to normal position. Especially in pairing territory, some ♂♂ and ♀♀ inflate and show sac in lower part of neck. HETEROSEXUAL BEHAVIOUR. If ♂ does not chase ♀ away when she first joins him in pairing territory, they face each other; then, both Bow with neck and head extended forward and downward, each afterwards returning to normal standing position. Next, ♂ and ♀ may intertwine necks and bills, bow further, preen extensively (no Allopreening, however) and call as during bill-feinting (see above). Rarely, either bird may grasp or nibble twigs. Some pairs continue such behaviour elsewhere on island but most settle in nest-site territory, ♂ and ♀ standing side by side with bodies often pressed together. Copulation begins only after nest-territory established. ♀ crouches at nest-site, ♂ straddles her, placing feet under her wings, then ♀ clamps wings together and ♂ grabs and shakes her bill, moving partly open mandibles up and down it, ♀ meanwhile sometimes also shaking bill. After copulation, both assume normal standing position at nest and preen extensively. Only after copulation period started will ♂ leave to collect material; ♀ remains on site and builds nest. ♂ occasionally flies back to pairing territories to display with other ♂♂. When ♂ returns to nest, sexes face, ♂ standing, ♀ crouching; they angle bills skyward and produce, especially, inhaling-exhaling vocalizations (see above), copulation normally following. If ♂ at nest, especially during incubation, and ♀ returns, ♂ normally crouches while ♀ stands. Once nest near completion, ♀ sometimes emits 1–3 sharp calls; ♂, standing near, moves to her and they may copulate. RELATIONS WITHIN FAMILY GROUP. Adults always recognize and feed only own offspring; young usually recognize parents. On returning to colony with food, adult often lands on one side of colony, and emits single, short nasal 'keerooh' call. Young, which may be several metres away, in group of up to 50 young, recognize parent's call and respond by running, jumping, or even flying to adult for food. Other young occasionally rush up, only to be chased away. No aggression seen from parent to own young. Once young able to fly, may chase returning parent, even flying round colony area before adult lands and feeds. No parental warning behaviour, elaborate begging behaviour, or appeasement on part of adult and young recorded. No obvious communication between siblings; young typically inactive, usually standing about colony until parents return with food, or wander about.

(Figs A–B from photographs by E K Urban.)

**Voice.** During breeding season, utters variety of noises but mainly silent at other times. Following vocalizations described by Urban (1974).

In antagonistic or heterosexual confrontations, both sexes utter various squeals, squeaks, moans, and inhaling/exhaling noises resembling 'whoot-whoot-whoot-whooeeoh', 'pyuk-pyuk-peuk-pek-peuk', or wheezing 'hnhh-hnhh'. During final period of nest building, at least ♀ utters sharp 'whaank' 1–3 times; attracts ♂ and copulation may follow. Adults utter 'turrooh' or 'keerooh', falling in pitch, to call young which have left nest. Sometimes utters harsh croak in flight.

CALLS OF YOUNG. Food-calls when small resemble repeated, high-pitched screaming 'chrreeeee-chree-ah-chrreeeee . . .'; those of older young similar but louder.

**Breeding.** Based mainly on Urban (1974), Ethiopia. SEASON. Iraq: eggs laid April–May. Ethiopia: most laying between 14 March and 24 April, but extremes 1 March and 20 August. SITE. Most often in trees, especially flat-topped thorns; but on islands, also in low scrub, under 1 m from ground, and on ground, among rocks; sometimes in rushes in swamps. Colonial; nests close but rarely touching. Nest: large pile of branches and sticks, lined grass, rushes, and leaves, and occasionally shells and coloured beads. Sticks up to 40 cm in length used, with largest outside. Average diameter of 10 nests 34 cm (range 28–43); maximum height 20 cm. Building: ♂ collects most material, while ♀ builds. EGGS. Oval or slightly rounded, rough; dull white with faint blue tinge, a few dark red spots sometimes visible. 63–43 mm (58–68 × 39–46), sample 42; calculated weight 62 g (Schönwetter 1967). Clutch: 2–4 (1–5). Of 34 clutches, Ethiopia: 1 egg, 2; 2, 23; 3, 8; 4, 1; mean 2·24. One brood. Replacement clutch possible after egg loss but not proven. INCUBATION. 28–29 days (Ethiopia), though reported as c. 21 days (Roberts 1965). Starts on completion of clutch; hatching synchronous. By both sexes, changing at least once every 24 hours. YOUNG. Semi-altricial and nidicolous. Cared for and fed by both parents. One parent always present for first 7–10 days, each parent staying for 24 hours. Young fed at intervals through day by partial regurgitation, putting head and upper neck into parent's mouth. Young leave nest between 14 and 21 days old and congregate in loose groups near colony. Fed once daily after leaving nest. FLEDGING TO MATURITY. Fledging period 35–40 days. Leave colony at 44–48 days, and probably independent soon after. Age of first breeding not known. BREEDING SUCCESS. At Lake Shala, Ethiopia, in 1966–70 and 1972, between 89 and 298 pairs attempted to breed and average 35% of pairs reared young (range 0–81%). In most successful year, 1968, 253 pairs attempted to breed, and 190 pairs (81%) successfully reared 201 young, or 1·06 young per pair. At 21 days old, average brood size was 1·46 young per pair; at 39 days, 1·13 young per pair. Main reasons for losses of eggs and young were adverse weather, in particular heavy rain storms, and possibly low-flying aircraft.

**Plumages.** ADULT. Bare head and neck black, plumage of body white; wide outer edges and tips of somewhat elongated lower scapulars and tertials decomposed, loose in texture, mainly black with purple-blue gloss and some pale grey towards white base of feathers. Tips of flight-feathers metallic blue-green (c. 5 cm on outer primaries, less on other remiges). Underparts and greater coverts often partly stained cinnamon or brown. Outside breeding season, at least some sub-adults have head and neck partly feathered white with variable black mottling, and scapulars and tertials narrower and shorter with narrower dark tips and outer edges, less detached rami, and more extensive white and pale grey. NESTLING. 1st down of crown, sides of head, hindneck, and sides of neck dense, black with white patch of sparse down on hindcrown; of foreneck and body white, rather short and sparse. Throat and skin round base of bill and eye bare. 2nd down similar; white spot on crown may soon wear away. JUVENILE. Head and neck feathered, black, slightly mottled by white bases of short feathers; throat and foreneck white, variably mottled black in some. Body white, tertials and longer scapulars with narrow black tip and outer edge, bordered pale grey inside and with white towards base of inner web; rami of tip and outer edge hardly detached. Flight-feathers more extensively tipped black with faint olive gloss, black not so sharply demarcated from white as adult (where tip appears to have been dipped in ink). Tip of outer primaries black for *c.* 7 cm, some with much black on outer webs; shafts of flight-feathers black. Bastard wing and tips and shafts of greater primary coverts black. Tips or shafts of some longer body or tail-feathers sometimes also black. IMMATURE NON-BREEDING. Like juvenile but feathers of head and neck relatively longer, more pointed, black with more white on bases and sides. Longer scapulars and tertials as for non-breeding sub-adult. Wing juvenile, but when moulted like immature breeding, although some worn juvenile outer primaries, primary coverts, or bastard wing often retained. IMMATURE BREEDING (does not nest in this plumage). Like adult, but head and neck feathered as in immature non-breeding. Rami of longer scapulars and tertials detached like adults, but somewhat more white and pale grey on bases and centres. Wing like adult, but flight-feathers with slightly more black on tip (less than juvenile and more strongly glossed), shafts partly dark brown. SUBSEQUENT PLUMAGES. Like adult. Completely bare head and neck and flight-feathers with only limited black on tips and without dark shafts apparently not attained until 3–4 years old, although ornamental scapulars and tertials strongly developed.

**Bare parts.** ADULT. Iris brown with crimson outer ring (latter most obvious at onset of breeding). Bill glossy black with grey corrugations at side of mandibles. Foot black, dark red on tarsus and tibia. Skin of head and neck dull black, lower eyelid pale pink. Bare spot at sides of breast, extending in strip over underwing along leading edge, dull grey-flash, but blood-red at start of breeding cycle (Holyoak 1970). NESTLING. Iris pale grey. Bill and foot pale pink or ivory-white. JUVENILE. Iris dark brown, at first with narrow white outer ring. Bill ivory or brownish-flesh; when older, dark brown to black with smooth sides. Foot dull grey. Bare skin at base of mandibles and round eye dull black, strip on underwing near leading edge dull grey-flesh (Sharpe 1898; Hartert 1912–21; Reichenow 1900–01; Urban 1974).

**Moults.** ADULT POST-BREEDING. Complete; flight-feathers descendant. Flight-feather moult during non-breeding season, may stop when birds migrate; then either finished later, or, if breeding season started in meantime, terminated; such complications apparently produce serial descendant pattern of replacement in some. Body moulted shortly after completion of breeding cycle, but tertials and scapulars apparently not until most flight-feathers moulted. At least part of body (upper scapulars, some feathers of underparts) also moulted shortly before breeding cycle starts again. POST-JUVENILE. Insufficiently known (much of material examined undated). Head, neck, and part of body moulted soon after fledging. Flight-feathers moulted in regular descendant sequence at end of non-breeding season; head, body (including lower scapulars and tertials), and tail at about same time. SUBSEQUENT MOULTS. Poorly known. Probably like adults, but for 2–3 years new feathering on head and neck acquired at same time as body feathers.

**Measurements.** ADULT. Sudan; skins and mounted specimens (RMNH, ZMA). Wing: ♂♂ 380, 385, ♀♀ 361, 361, 362. Tail: ♂♂ 129, 132, 140; ♀♀ 131 (4) 127–135. Bill: ♂♂ 173, 174, 182; ♀♀ 145 (5) 138–148. Tarsus: ♂♂ 100, 107, 116; ♀♀ 98·0 (5) 94–103. Toe: ♂♂ 88, 96, 97; ♀♀: 84·6 (5) 81–87. Ethiopia, bill: ♂♂ 170·4 (7) 162–183; ♀♀ 145·5 (8) 135–157 (Urban 1974). JUVENILE. Sudan. Wing: ♂ 362; ♀♀ 336, 337. Tail: ♀ 111.

**Weights.** Kenya, April: ♂ 1530 (Britton 1970).

**Structure.** Wing rather long and narrow, slightly rounded. 11 primaries: p9 longest, p8 about equal or 0–6 shorter, p10 18–28 shorter, p7 6–14, p6 20–32, p1 105–125; p11 reduced, narrow, concealed by primary coverts. Inner web of p10–9 and outer p9–8 emarginated. Tail rather short, nearly square (slightly forked in juvenile); 12 broad feathers with smoothly rounded tips. Bill long, rather heavy, decurved, high at base, laterally compressed. Nostrils slit-like, situated in grooves running over upper mandible at sides of sharply ridged culmen. Sides of mandibles rugged and scaly in adults, smooth in juveniles. Head and neck bare in adult; feathered (except for skin round mandibles, lores, and area round eye) in juvenile and sub-adult. Strip of skin over underside of wing bones, extending to sides of breast, bare. Lower scapulars and tertials elongated and wide, rami of tip and outer webs detached and curved, forming loose plumes which cover wing tip and tail at rest. Lower half of tibia bare. Toes rather long with long and pointed slightly decurved nails; nail of middle toe with a few dentations. Outer toe *c.* 81% middle, inner *c.* 74%, hind *c.* 48%.

**Geographical variation.** Several closely related forms elsewhere in tropical Old World; sometimes all considered to be subspecies of *T. aethiopicus* (e.g. Holyoak 1970), but geographical distribution of certain plumage characteristics rather patchy, probably indicating long isolation of different populations; considered to be separate species, comprising superspecies with *T. aethiopicus*. They are: *T. bernieri* on Madagascar and Aldabra, *T. melanocephalus* in south and east Asia, and *T. molucca* in Moluccas, New Guinea, Solomons, and Australia.    CSR

# *Platalea leucorodia* Spoonbill

PLATE 50
[after page 374]

Du. Lepelaar    Fr. Spatule blanche    Ge. Löffler
Ru. Колпица    Sp. Espátula    Sw. Skedstork

*Platalea Leucorodia* Linnaeus, 1758

Polytypic. Nominate *leucorodia* Linnaeus, 1758, Palearctic north to 55°N, India, and Ceylon; *balsaci* de Naurois and Roux, 1974, Mauritania. Extralimital: *archeri* Neumann, 1928, islands southern Red Sea and north Somalia.

**Field characters.** 80–90 cm, of which body less than half, ♂ generally larger; wing-span 115–130 cm. Large, white, and rather graceful heron-like bird with long, broad, spatulate bill, long neck, and long legs. Sexes similar (but larger bills and longer legs of ♂♂ usually obvious in flocks) and only small seasonal differences; juvenile easily distinguished.

ADULT BREEDING. All white except for variable gorget of yellow-buff at base of foreneck, and sometimes partial retention of juvenile's black tips to primaries; conspicuous crest of long, narrow feathers. Bill black with orange-yellow tip; bare skin of chin, throat, lores, and round eyes mainly yellow; eyes red; legs black. ADULT NON-BREEDING. Lacks crest and yellowish foreneck. White plumage often sullied. JUVENILE. Similar to non-breeding adult, but with black tips to primaries and black shafts to secondaries. Bill pink, becoming dark during 1st winter; legs pinkish-flesh changing to slate.

Combination of white plumage, long spoon-shaped bill, and neck extended in flight unmistakable. Egrets *Egretta* have retracted necks and much shorter, thinner bills; slight possibility of confusion with flying swans *Cygnus* at long range, but *P. leucorodia* flaps and glides, and long legs extending beyond tail usually visible. African Spoonbill *P.*

*alba* distinguished by bare red face, usually some red on upper mandible, and bright pink legs; also by absence of yellowish foreneck.

Gregarious, but usually in small parties, though flocks of up to 100 may be seen in some areas; nests colonially. Walks like *Egretta*, but has loping run after prey; often rests on one leg. Perches easily in trees, but more usually seen on ground or in shallow water where feeds by characteristically sweeping tip of bill from side to side; will swim for short distances in deeper water. Flies with slower and more measured wing-beats than *Plegadis*, but not as slow as larger *Ardea*, often gliding for short distances; parties usually fly in single file, or in wandering V on migration; sometimes soars in thermals to considerable heights. Generally silent.

**Habitat.** Basically within warm climates, locally penetrating deep into temperate zone. Normally in coastal lowlands or alluvial river basins but breeds exceptionally to nearly 2000 m on Lake Sevan, Armenia (Dementiev and Gladkov 1951). Highly specialized bill ties it while feeding strictly to shallow, usually extensive waters of fairly even depth with bottoms of mud, clay, or fine sand, preferably with gentle tidal changes or slow currents, or newly flooded,

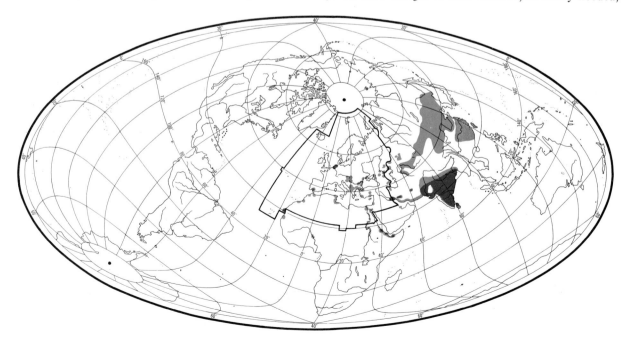

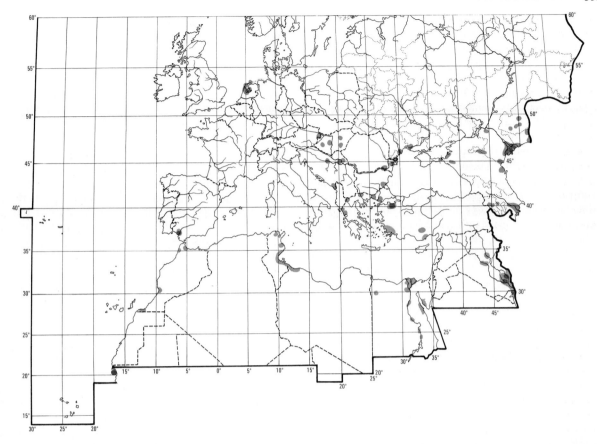

whether fresh, salt, or brackish. Avoids both stagnant and turbulent waters, and copious vegetation whether submerged, floating, or emergent. For breeding prefers dense reedbeds and similar masses of emergent plants, often with scattered shrubs or trees, such as willows *Salix*, poplars *Populus*, or *Oleaster*, which used for nesting, normally up to 2–5 m. Breeding colonies usually front pools or channels; high degree of security against disturbance and predation essential, and, given this, nests may be placed on ground on islands in lakes, rivers, or even sea. When necessary, makes regular flights to feed, rarely extending beyond 15 km. Uses airspace freely, both circling low and soaring to considerable altitudes. In winter quarters, more often marine, but confined to deltas, estuaries, lagoons, inlets, and sheltered shallow coasts. Wades and walks about widely. Limited distribution of suitable habitats and intolerance of disturbance demand strict reserves for breeding.

**Distribution.** Breeding distribution always scattered but overall range, especially in western Europe, much reduced due to drainage of nesting areas and, in earlier years, human exploitation of eggs and nestlings for food (Brouwer 1964). Now seriously endangered in most areas.

BRITAIN: bred until 17th century (British Ornithologists' Union 1971). FRANCE: bred until middle of 16th century and probably again in 1943–4, Loire (CE). ITALY: said to have nested near Ravenna in 1500 (SF). PORTUGAL: bred *c.* 1616 (Tait 1924). More recently has occasionally nested in several countries. DENMARK: in 1900, 1919, 1921–5, 1928–9, 1942–4, 1948–9 and 1962–9 usually in small numbers; in some cases probably linked with disturbance at Netherlands colonies (Kortegaard 1973). WEST GERMANY: unsuccessful attempt by one pair, Memmert, 1962 (*J. Orn.* 1963, **104**, 97–100). CZECHOSLOVAKIA: a few pairs have nested occasionally since 1949 in south Moravia (KH). PORTUGAL: 1966 (MDE). ISRAEL: 1951, unsuccessful (HM). NORTH AFRICA: one egg found Lake Fetzara, Algeria, in 19th century only certain record (Heim de Balsac and Mayaud 1962).

Accidental. Iceland, Faeroes, Norway, Sweden, East Germany, Poland, Switzerland, Jordan, Cyprus, Azores, Madeira, Canary Islands, Cape Verde Islands.

**Population.** NETHERLANDS. Some 300 pairs breeding around 1900; increased under protection to 450 pairs 1925 and *c.* 500 pairs in 1950s. Declined from 1959, falling to well below 200 pairs 1964; fluctuating since between 150–200 pairs. Pollution of feeding areas and poisoning by organochlorines main cause of recent decline (Brouwer 1964; Rooth and Jonkers 1972). SPAIN. Bred irregularly Guadalquivir area, usually in small numbers for many years

(Brouwer 1964; Valverde 1960). Increased in recent years, now 300–500 pairs, Coto Doñana (Ree 1973). AUSTRIA: average 200–250 pairs Neusiedlersee after 1945 (Koenig 1952). HUNGARY. 220 pairs in 3 main areas 1951, but considerable fluctuations. Kisbalaton: varied from nil to 120 pairs, averaging c. 50 pairs; 38–57 pairs 1968–72. Lake Velence: since 1930, seldom over 25 pairs but 150 pairs 1951; 17 pairs 1962. Hortobágy: 110 pairs 1969. (Szijj 1954; Brouwer 1964; Sterbetz 1972; AK). YUGOSLAVIA. Once large population, now small and still decreasing. Formerly numerous Vojvodina, where several colonies, now mostly abandoned. Obedska Bara: between 300 and 1000 pairs in various years from 1838–1908; c. 60 pairs 1930; increased from 1955 under protection to 100 pairs 1961, but only 10 pairs 1972. Bred Kopačevski area until 1954, when 11 pairs, and Banat in 1957. Also nested Posavina (Croatia), 3–32 pairs 1962–68; Crna River (south Macedonia), c. 200 pairs 1939, now largely drained; Lake Sasko (Montenegro), c. 50 pairs 1969. (Brouwer 1964; Majić and Mikuska 1972; VFV.) GREECE. Although no colonies listed by Brouwer (1964), 5 now known with total of 200–240 pairs; largest Lake of Kerkini and Lake Mikra Prespa, with 100 pairs and 60–80 pairs respectively in 1971. Evros delta colony extinct 1965 (WB). RUMANIA. Now restricted to Danube delta, Dobrogea, and one site Danube floodland (last nested Banat, 1950). Total population c. 400 pairs 1909, over 250 pairs before 1940. Marked decrease after 1945 and fluctuating numbers from under 20 to over 220 pairs 1953–63 (Vespremeanu 1968). USSR. Rare on estuaries of Black Sea rivers; more common north Caucasia, Transcaucasia, and Volga delta (Dementiev and Gladkov 1951). Lower Volga delta: decrease from 690 pairs 1951 to 19 pairs 1958 (Lugovoy 1961). TURKEY. Manyas Gölü: perhaps increasing, c. 700 pairs 1968. Recently discovered colony Eregli area: c. 70 pairs 1969, c. 380 pairs 1970. Meric delta: may still nest, perhaps 40 pairs (OST). IRAQ: breeds in south, but no recent data (PVGK). MAURITANIA. Banc d'Arguin: c. 1430 pairs on 5 islands 1959–65 (de Naurois 1969a); 562 pairs in 1973 (Trotignon 1974a).

Survival. Oldest ringed bird 28 years 2 months (Rydzewski 1973).

**Movements.** Migratory and dispersive. Such post-fledging dispersal as occurs (July–August) normally for relatively short distances; Volga delta birds up to 90 km north to north-east (Skokova 1959), Dutch birds reaching north-west Germany and (fewer) south-east England; but 3 exceptional early autumn recoveries of juveniles (Yugoslavia to Scotland, Hungary to Camargue, Netherlands to Azores) indicate some long-distance wandering. Main departures from European colonies August–September; few remaining October, though appreciable numbers still present Volga delta (Skokova 1959). Occasional Netherlands in mild winters, when a few fairly regularly south-west England and south Ireland. Most Dutch birds (see Brouwer 1964) migrate along Atlantic coast, alighting at major estuaries; winter recoveries and sightings of colour-ringed birds Canary Islands (2), Banc d'Arguin (2), and Sénégal delta (1) (Roux 1973), while Smith (1965) noted migrant flocks on Atlantic coast Morocco in September and March; thus believed that Dutch birds winter among much larger Banc d'Arguin population. Latter dispersive, some moving south to Sénégal, and apparently in small numbers to Gambia (Cawkell and Moreau 1963); but many overwinter Banc d'Arguin where 4000 late December 1972 (Pététin and Trotignon 1972). 2 recoveries of Spanish birds: one December 45 km from colony, other Morocco September; 200 birds in marismas January 1972 (Ree 1973). Breeders of south-east Europe winter partly Mediterranean basin, partly northern tropics of Africa; ringed ones from Austria, Hungary, and Yugoslavia recovered October–March in Italy, Sicily, Greece, Bulgaria, Turkey, Tunisia, Libya, and lower Egypt, with notable concentration Tunisia; also single recoveries Austria to Algerian Sahara 400 km ESE of Ouargla (October), Hungary to north Sudan (November) and Niamey in Niger (February); absence of recoveries Morocco or coastal Algeria suggest no contact with western colonies (Brouwer 1964; Bauer and Glutz 1966).

Since European populations not large, and partially wintering Mediterranean, numbers migrating to northern tropics must be relatively small; occurs Mali east to Sudan, but in Nigeria known only as vagrant to Lake Chad (Elgood *et al.* 1966; Moreau 1967, 1972). Palearctic subspecies identified south to north-east Congo and 14°N in Sudan; records of vagrants to inland lakes Kenya and Uganda possibly northern birds, since *P. l. archeri* of Red Sea appears to be coastal (Backhurst *et al.* 1973). Some numbers winter Iraq (some north of limited breeding area) and Iran. Also found Kuwait in February. Birds ringed Manyas Gölü colony, west Turkey, moved south to south-east, with recoveries lower Egypt (February), Sudan (October), Israel (February, March), Iraq (Euphrates delta in January, March, and May), and Pakistan (November) (Schüz 1957b). USSR populations appear to migrate south-east, since one ringed Sea of Azov recovered Pakistan, and from Caspian colonies one found Iranian Baluchistan and 4 in north India (Sapetin 1968); large flocks arrive Indian subcontinent in October (Ali and Ripley 1968).

Early arrivals back in Netherlands and Azerbaydzhan in second half February, but most return to European colonies late March and April. Later arrivals non-breeders which wander during summer; recoveries of Turkish bird Iraq in May, and one from Caspian in India July, suggest some immatures remain in winter quarters. Migration often diurnal, small parties or large flocks; usually in transverse line formation, or soaring and circling at considerable heights. See also Dementiev and Gladkov (1951).

**Food.** Insects and their larvae including Coleoptera (especially waterbeetles), Odonata (dragonflies), Trichoptera (caddisflies), Orthoptera (locusts), Diptera, and Hemiptera; small fish, molluscs (*Tellina*, aquatic snails), crustaceans, frogs, tadpoles, worms, leeches, reptiles, and some plant material. Feeds in small flocks usually in shallow water and marshes by wading, often in oblique lines. Tip of bill, with mandibles slightly open, immersed and swept from side to side.

In Netherlands, main food small fish, up to 15 cm long and 4 cm high, especially sticklebacks (*Gasterosteus aculeatus, Pungitius pungitius*); also crustaceans (e.g. *Crangon vulgaris*) and aquatic insects and larvae (E P R Poorter). In USSR, adults and juveniles contained blue darner larvae, water-scavengers, waterbeetles, occasionally fish and, more rarely, young frogs. In one from Turkmenia, insects, other invertebrates, and tadpoles found. In Astrakhan reserve and Kzytorda region, main foods aquatic insect larvae, small fish, molluscs, and frogs (Dementiev and Gladkov 1951). In Rumania, diet includes small fish (e.g. weatherfish *Misgurnus fossilis*), worms, larvae of aquatic insects, frogs, tadpoles, and gastropod molluscs (Dombrowski 1912).

**Social pattern and behaviour.** Little studied; following based mainly on observations by E P R Poorter in Netherlands.
1. Gregarious throughout year. In flocks on migration and during flights between roost or colony and feeding grounds. Does not usually associate with other long-legged marshbirds. Feeding groups tend to be small, and solitary feeding not uncommon; no feeding territories maintained, but individuals often frequent same feeding site regularly. BONDS. Though mated ♂♂ regularly show promiscuous copulatory behaviour, pair-bond monogamous; possibly only of seasonal duration. Both parents tend young until well after fledging. BREEDING DISPERSION. Typically colonial; not usually in association with other species in west Palearctic. Solitary nests occasionally reported. In colonies on ground (e.g. reedbeds), many nests in discrete groups or 'neighbourhoods', while others more scattered. Nests in same neighbourhood may be packed closely, often connected by bridges of reed stems and branches resulting from mutual robbing of nest material. Each pair defends own nest and immediate area in which only close neighbours tolerated. Bands of neighbours often co-operate to drive off strange adults; so neighbourhood may represent group territory. Egg-laying tends to be synchronized within each neighbourhood, and late-comers not allowed to build within group, at least not until it breaks up when young nearly fledged. Unfledged young wander freely about colony from *c.* 3 weeks; tolerated everywhere except on strange nests. Nest-site and area usually less strongly defended when young more mobile, but used as rendezvous for feeding up to at least 5 weeks, after which young fed mainly away from nest; territory may shift if whole brood settles in another locality within colony. Non-breeders associate in bands; keep away from colony and from feeding grounds of breeders. ROOSTING. Communal. Sleeps at any time but, as largely nocturnal feeder, mostly during daylight. When breeding, adults with eggs or small young roost in colony; when unfledged young older, parents usually absent during night feeding. After fledging, when colony deserted, mixed flocks of adults and young roost together in groups for long periods during day, preferring shallow water,

meadow land surrounded by ditches, or shrubs; leave at dusk to feed but return in morning. Same communal roost may be used for several consecutive years. At feedng grounds, flocks may at first rest together after arrival, but resting after feeding less synchronized.
2. Not studied in detail. ANTAGONISTIC BEHAVIOUR. Away from colony, e.g. on feeding grounds, generally peaceable—though chasing, threat, and pecking observed. At colony, during establishment and maintenance of territory, billfighting regular: rivals slash and peck, producing snapping sound by contact between bills and shutting tight of mandibles. In such skirmishes, pecking may alternate quickly with Facing-away, thus avoiding bill-slashes of opponent and showing yellow and red skin of throat. Will also attack opponent, including persistently intruding young on nest, by vigorous pecking and slashing at both bill and legs. Intruders from other neighbourhoods, and non-aggressive strangers, chased on foot or briefly in air, sometimes by several birds together (see above). When pursued, intruders flee, usually with feathers sleeked. In threat-display, head feathers raised (jutting from head in uneven fan) and, at higher intensity, bill opened; performed during chasing (sometimes also by pursued bird) and on nest, either while sitting or standing. Sitting threats directed at conspecifics or other species flying low over, or at other species near nest; standing threats given when defending nest-site and area. Similar behaviour occurs at nest-relief. At some colonies (e.g. Zwanenwater, Netherlands), where adults rather tame, young defended against human intrusion by Nest-covering display: wings spread slightly over young, head feathers raised, and bill pointed down. Such behaviour not recorded against intruders of own species, but also performed, with or without erection of head feathers, during showers and strong sunshine—so protective rather than threat. HETEROSEXUAL BEHAVIOUR. Behaviour during early stages of pair-formation and nest-site selection not fully observed in Netherlands to avoid disturbance. Display described in which head feathers raised and bill pointed up exposing conspicuous bare skin on throat (Thijsse 1906; Binsbergen 1959), but significance uncertain. Allopreening, by members of pair simultaneously (see A), performed regularly on nest, and elsewhere in colony when young nearly fledged; also frequent, however, between casual associates at feeding grounds, and no evidence to suggest behaviour then other than for plumage care.

A

B

No details of copulatory behaviour between mated birds. Paired ♂♂, however, regularly try to rape neighbouring ♀♀ sitting on eggs, sometimes mounting and copulating. One ♂, for several mornings in succession, seen to copulate with same ♀ neighbour before relieving own mate after returning from feeding grounds. Paired ♂♂ also copulate, or attempt to copulate, with non-breeding ♀♀. One ♂ observed to land back at nest with strange ♀ and copulate with her; after he dismounted, another ♂ left eggs and also copulated with same ♀—no introductory display recorded in any instance. At nest-relief, bird on nest spreads head feathers and makes groaning call with bill open (see B); relieving partner sometimes does same. RELATIONS WITHIN FAMILY GROUP. Young beg by bobbing head up and down, giving calls; nearly fledged chick also throws one wing over parent's neck at feeding. Mobile young react to nearly every adult landing near by begging, but usually stop when own parent not involved. When parent does not feed begging chick after arrival, latter may suddenly turn away and walk off; adult then often follows hurriedly but, in turn, walks away when chick stops and turns back; may be repeated several times before young fed. At feeding, 2 young sometimes simultaneously insert bills into parent's throat. Young highly sociable, with own siblings and other broods; engage in much mutual billing and, when older, sometimes insert bill into mouth of another which may make reciprocal regurgitating movements. When mobile, often visit other nests and spend night there in absence of adults. Strange young often tolerated for considerable time after return of adults next day, but persistent visitors eventually driven away. No fighting occurs between young themselves. After fledging and departure from colony (at 5–6 weeks), bond between parents and young, and to lesser degree between different young, may continue up to at least 10 weeks; parental feeding continues throughout though young increasingly find own food.

(Figs A–B from photographs by E P R Poorter.)

**Voice.** Generally silent; voice little used except for soft murmurs audible only at close range (Bauer and Glutz 1966), and at breeding places, where adults occasionally utter deep grunting 'huh-huh-huh' (see I*b*), also rendered 'hou-hou-hourour-houm' (Lippens and Wille 1972), and probably equivalent to groaning calls uttered at nest-relief

(E R P Poorter). When excited, e.g. during fighting, various snapping sounds produced at shutting and contact of bills.

CALLS OF YOUNG. Produce shrill high-pitched calls when begging (E R P Poorter): continuous, rasping 'cheeerr' (see I*a*) or 'chrie' (smaller young), 'tschibb-tschibb-tschirr-tschirr' (older young); also utter 'trara' or 'tritt' pleasure-calls, and, at fledging, 'hürück' calls (Bauer and Glutz 1966).

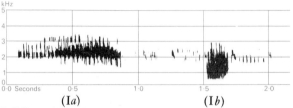

(I*a*)        (I*b*)

I C Chappuis/Sveriges Radio  Netherlands June 1967

**Breeding.** SEASON. Netherlands: see diagram. Austria: laying begins end April. Rumania: laying from mid-April to mid-May, depending on spring temperature and water levels. Older birds begin before first-time breeders (Brouwer 1964; Dementiev and Gladkov 1951; Seitz 1937; Vespremeanu 1968). SITE. On ground or mat of old reeds in dense reedbeds; in willow *Salix* thickets close to water, and up to 5 m above ground. Occasionally in grass tussocks where no other vegetation. Colonial; nests mostly 1–2 m apart, but sometimes touching. Colony site used in successive years. Sometimes with other species, particularly Glossy Ibis *Plegadis falcinellus* in Rumania; rarely with herons (Ardeidae). Nest: large pile of reeds, twigs, and grass stems, lined grass and leaves. Average height 30–60 cm, diameter 50–70 cm. Building: by both sexes with material gathered within reach and brought in to site. EGGS. Slightly elongated oval; chalky white with small reddish-brown spots and lines. 67 × 46 mm (56–75 × 40–50), sample 250 (Schönwetter 1967); 61 × 42 mm (52–77 × 36–48), sample 105, Rumania; large clutches have smaller eggs (Vespremeanu 1968). Calculated weight 76 g (Schönwetter 1967). Clutch 3–4 (6, rarely 7). Of 54 clutches, Macedonia: 3 eggs, 56%; 4, 37%; 5, 6%; 6, 1% (Makatsch 1952). Of 415 clutches, Rumania: 2 eggs, 23%; 3, 54%; 4, 23%; mean 3·5 (Vespremeanu 1968). One brood. Replacement clutches laid after egg loss. Eggs laid 2–3 day intervals. INCUBATION. 24–25 days (Dementiev and Gladkov 1951); 21 days (Holstein 1929). By both sexes. Begins with 1st egg; hatching asynchronous. YOUNG. Semi-altricial and nidicolous. Cared for and fed by both parents. Fed by partial regurgitation. At *c.* day 30, young leave nest for trampled down space close by. FLEDGING TO MATURITY. Fledging period 45–50 days. Age of independence not known. Age of maturity probably 3–4 years (Bauer and Glutz 1966). BREEDING SUCCESS. Of 1234 eggs laid, Rumania, 1958–63, 96% hatched; of 1188 hatched young, 6·4% died in first 2 days, 24·8% by day 10, 28·9% by day 40, and 29·7% by day 50. Mortality of chicks

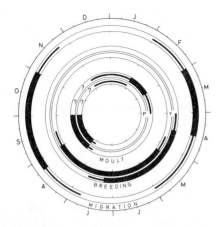

in nest varied annually from 17·2%–100%. Total loss caused by rise in water levels; other factors include food supply and weather. (From data in Vespremeanu 1968.)

**Plumages.** ADULT BREEDING. Entirely white, except yellow-buff collar of variable extent round base of neck. Loose crest of narrow, elongated, and pointed feathers on hindcrown, sometimes partially tinged yellow-buff. Rami of elongated feathers of hindcrown soon detached by wear; later also those of tertials and lower scapulars. During breeding season, elongated feathers and yellow-buff colour of base of neck may disappear completely. ADULT NON-BREEDING. Like adult breeding, but no crest on hindcrown and yellow-buff tinges absent. NESTLING. 1st down sparse, white with silky tips; 2nd down short and dense, cream-white; soon covered by growing juvenile feathers. JUVENILE. Like adult non-breeding, but shafts of flight-feathers distinctly horn-black; tips of *c.* 4 outer primaries and outer webs of *c.* 2 outer black. Shafts of larger wing-coverts and scapulars basally horn-black; also to variable extent those of feathers elsewhere on body, tail, or wing. IMMATURE WINTER. Much variation, apparently due to great differences in hatching dates (from early May to early August). Advanced birds like non-breeding adults, but new flight-feathers still with some black at tips (less than juvenile); retarded ones similar, but retain many black-shafted juvenile feathers. IMMATURE SUMMER. Much variation. Advanced birds like adult breeding, but outer primaries each usually with black tip and shaft; only exceptionally short crest or yellow patch on base of lower neck. SUBSEQUENT PLUMAGES. Like adult, but some younger birds in breeding plumage apparently still without crest or yellow-buff marks. However, some specimens in otherwise full breeding plumage retain black wing-tips (Bauer and Glutz 1966; Witherby *et al.* 1939; C S Roselaar).

**Bare parts.** ADULT. Iris carmine. Bill black, barred black and horn-grey on base and middle portion; terminal part of wide tip orange-yellow, edged black. Amount, pattern, and tinge of yellow on tip of bill variable. Foot black. Bare skin at base of upper mandible and narrow line on lores to eye black, bordered by white or pale yellow; bare skin round eye, at corner of mouth, chin, and throat yellow, tinged orange and often with red band bordering feathers of lower throat when breeding. NESTLING. Iris grey. Bill and foot orange at hatching, pink-flesh after a few days. Bare skin round eye blue; on chin, throat, and base of upper mandible pink. JUVENILE. Iris grey-brown. Bill pink. Foot pink-flesh. Bare skin as nestling. During 1st winter and 2nd summer, adult colours gradually attained: bill gradually darker from base, middle and tip change to yellow (in 2nd summer, bill usually mainly yellow); foot changes to olive-grey and slate (first on bare tibia and tarsus);

bare skin on face yellow. (Bauer and Glutz 1966; E P R Poorter; RMNH.)

**Moults.** ADULT POST-BREEDING. Complete; flight-feathers probably regularly descendant (Stresemann and Stresemann 1966). August–March; head and neck first, followed by underparts, tail, and upperparts. Wing starts with inner primaries August–September; moult probably stops during migration, and finished in winter quarters. ADULT PRE-BREEDING. Partial; January–March. Head and neck, probably also underparts or some feathers on rest of body. POST-JUVENILE. Complete; December–May. Timing of start in winter probably dependent on age (see Plumages). Some advanced specimens apparently start post-juvenile moult in October, and have partial pre-breeding moult in early spring like adult. SUBSEQUENT MOULTS. Probably like adults. (Bauer and Glutz 1966; C S Roselaar.)

**Measurements.** Netherlands; skins (RMNH, ZMA).

| | | | | | | |
|---|---|---|---|---|---|---|
| WING AD ♂ | 394 | ( 8·2; 13) | 386–412 | ♀ 370 | (5·3; 10) | 360–377 |
| JUV | 382 | ( 5·6; 12) | 374–393 | 355 | (6·2; 12) | 345–364 |
| TAIL | 117 | ( 5·1; 25) | 108–126 | 113 | (3·1; 24) | 108–118 |
| BILL AD | 213 | (10·2; 15) | 195–231 | 182 | (5·7; 14) | 168–191 |
| TARSUS | 149 | ( 5·7; 24) | 140–163 | 131 | (4·2; 19) | 123–141 |
| TOE | 93 | ( 3·5; 21) | 90–104 | 85 | (2·8; 22) | 79–90 |

Sex differences significant, except tail. Wing of juvenile significantly shorter than adults. Tail, tarsus, and toe similar; combined (but wing, tail, tarsus, and toe often not full-grown until *c.* 1 month after fledging). Full bill-length not attained until 3–6 months: June–July, 4 ♂♂ 142 (125–171) 6 ♀♀ 111 (87–129); August–September, 9 ♂♂ 172 (151–213) 9 ♀♀ 143 (103–164); October–February, 4 ♂♂ 210 (186–230).

**Weights.** Data scanty. Adult ♂, Hungary, 1960 (Bauer and Glutz 1966). Following from Netherlands (ZMA). Breeding adults poisoned by chemical residues, June: ♂♂ 1323, 1463; ♀ 1130. ♂ killed by frost, February: 1200. Recently fledged birds, June: ♀♀ 795, 838 (condition poor); ♂ 1656 (condition fair). Juvenile ♀, died from botulism: September, 1055.

**Structure.** Wing rather long, pointed. 11 primaries: p9 longest, p10 10–17 shorter, p8 0–5, p7 7–14, p6 35–50, p1 122–145; p11 reduced, concealed by primary coverts. Inner webs of p10–8 and outer p9–7 emarginated. Tail short, square in adults, slightly forked in juveniles; 12 broad feathers. Bill long, flat, spoon-shaped: 30–34 mm broad at base, 17–20 in middle, 43–50 near rounded tip. Tip very flat, outer tip slightly decurved. Inner side of mandibles with some rows of blunted knobs basally. Upper surface of basal and central part of upper mandible and basal sides of lower corrugated in adult, smooth in juvenile (first faint corrugations appear basally at sides of mandibles from 1st winter). Narrow, elliptical nostrils in groove at sides of culmen near base. Skin round eye and on lores, chin, and central throat bare (throat less so in juveniles). Loose crest of elongated feathers on hindcrown in adult breeding plumage, up to 120–150 mm long in ♂, 105–135 in ♀. Lower half of tibia bare. Outer toe *c.* 86% middle, inner *c.* 76%, hind *c.* 46%. Outer, middle, and inner toe basally connected by webs.

**Geographical variation.** Asiatic birds sometimes considered larger (Hartert 1912–21). In *balsaci*, bill completely black and virtually no yellow-buff on chest. Size smaller: 9 ♂♂ and 8 ♀♀ respectively, wing 372 (364–390) and 352 (340–362), bill 184 (157–200) and 167 (151–180), tarsus 130 (123–144) and 120 (115–129) (de Naurois and Roux 1974). *P. l. archeri* like *balsaci*, but still smaller: wing 325–360, bill 145–170, tarsus 102–118 (Neumann 1928; RMNH).

# Order PHOENICOPTERIFORMES

## Family PHOENICOPTERIDAE flamingoes

Large wading birds with highly specialized feeding habits. Sole family in order. Intermediate between Ciconiiformes and Anseriformes in many characters, but considered by some authorities to be closer to Ciconiiformes and sometimes placed in that order (see, especially, Sibley and Ahlquist 1972). 5 species in single genus *Phoenicopterus*, though some authors recognize 2 further genera for smaller species—*Phoeniconaias* for Old World form *minor* and *Phoenicoparrus* for New World forms *andinus* and *jamesi* (see Kear and Duplaix-Hall 1975). 2 species in west Palearctic, one breeding. Body ovate, neck long, head small. ♂ larger than ♀. Wings long and broad. Flight strong and rapid, with neck and legs extended. 12 primaries: p10 longest, p12 minute. About 27 secondaries; diastataxic. Tail short and rounded; 12–16 feathers. Bill large, sharply decurved in middle; adapted for filter-feeding (see Jenkin 1957). Upper mandible smaller than lower, lid-like; either shallow-keeled (2 larger species) or deep-keeled (3 smaller species). Lower mandible larger and trough-like. Inner surfaces of both mandibles lined with filtering lamellae—more extensively in deep-keeled species. Tongue thick and fleshy, with spines. Legs extremely long, tibia almost wholly bare, feet small; front toes webbed, hind toe reduced and slightly elevated or absent. Walk and swim well. Plumages largely pink and red; flight-feathers black. Bare parts often brightly coloured: pink, red, or orange. Sexes nearly alike. Presumably single moult per cycle; primaries shed irregularly or simultaneously (occasionally producing period of flightlessness). Young precocial and semi-nidifugous. 2 downs, grey; bill straight, becoming curved later. Juveniles mottled brown, attaining adult plumage over several years.

Local and discontinuous in tropical and sub-tropical regions of Europe, Africa, Asia, and New World; at places (e.g. South America), high in mountains in cold climate zones. Invariably associated with brackish, salt-water, or alkaline lakes and lagoons. Well illustrate ecological principle that abandoning diversity involves risks of instability; tenure of many haunts precarious and intermittent, with strong likelihood of more or less complete failure to reproduce in successive years and, from time to time, mass mortality. In west Palearctic, sole nesting species, Greater Flamingo *P. ruber*, breeds somewhat irregularly in a few scattered localities; overall population trends unknown. Some species make considerable migratory movements, chiefly at night. Local nomadism in search of nest sites, food, and fresh water widespread. In west Palearctic, long-distance movements, especially those north of colonies, appear to be of an irregular nature. Because of long immaturity and in-

termittent breeding, summer range more extensive than actual breeding range.

Food of smaller, deep-keeled species (e.g. Lesser Flamingo *P. minor*) chiefly blue-green algae and diatoms obtained at or near surface of water. That of larger, shallow-keeled species (e.g. *P. ruber*) more varied: as well as algae and diatoms, chiefly invertebrates such as small molluscs, crustaceans, and chironomid larvae obtained, together with organic particles, from mud. Both groups feed by day or night. Water and mud filtered through bill by action of lower mandible and tongue, mainly as bird wades with head submerged and inverted; in 3 smaller species, especially, head often swept from side to side. Often trample with feet to stir up food items. Surface-feeders, e.g. *P. minor*, will also feed while swimming. Highly gregarious at all times. Typically form huge, dense colonies, each pair defending nest-site territory only. Monogamous pair-bond, strong and often sustained. Display mostly in large, highly vocal groups; not confined to breeding season or eventual breeding area. Both sexes participate, though display bouts usually initiated by ♂♂ (Kahl 1975b). Group display may be interpreted as adaptation to special conditions of unstable breeding habitat, intense display serving to form groups of birds in similar physiological state which later breed as unit with highly synchronized egg-laying (A Studer-Thiersch). Thus breeding groups form independently of favourable conditions; if birds then encounter such conditions, soon able to start nesting. Separate pair-courtship, with copulation away from site, mainly while feeding, facilitates pair-formation and maintains pair-bond. Calls chiefly resonant, nasal honking and gabbling, often goose-like; used continually in flock at all times, especially during group display. Silent when alone. Noisy at colony, especially at hatching and during parental feeding. Comfort behaviour generally similar to that of other waterbirds. Bathing frequent; in shallow fresh water when possible. Foot-shaking used to dry feet prior to taking turn at nest. Manner of heat dissipation uncertain; does not include gular-fluttering (A Studer-Thiersch). Wings spread slightly to deflect rain from nest during incubation (Studer-Thiersch 1975a). Other behavioural characters include: (1) direct head-scratching, (2) resting on one leg with bill inserted into scapulars—on same side as standing leg, unlike in Anatidae; (3) pre-flight signals, with head stretched up and slightly forward while intermittently bobbing head backwards to rub on back and calling (A Studer-Thiersch).

In many areas, breeding irregular and non-seasonal; may not nest at all in some years. Even where breeding seasonal, laying period extended. When breeding does occur, laying usually highly synchronized within colony, often in discrete groups. Dense colonies typically situated on areas of wet mud, less frequently on islands or vegetated ground, surrounded by water. Nests truncated mounds of mud, with shallow depression in top, built by both sexes. At island sites, nest may be slight one of feathers, stones, and debris. Clutch typically 1 large egg, rarely 2; pale blue or whitish with chalky covering. One brood. Replacements sometimes laid after egg loss. Incubation by both sexes, with legs doubled under and projecting behind; no brood-patches. Eggs not recovered if displaced. Incubation

period 27–31 days. Parents may assist young to hatch. Eggshells left in nest or discarded over side, but fragments eaten by chick. Downy young cared for by both parents. Fed by regurgitation on glandular secretion from crop. Young leave nest 7–12 days after hatching and move into herds (crèches) attended by only a few adults; these groups often mobile and may reach immense size in *P. minor* (Brown and Root 1971; Brown 1975). Young in herds fed by own parents, mainly at dusk or at night. Fledged at 65–85 days, when bills adult-like in shape, and only then independent though may stay in crèche for further month. Mature at 2–3 years.

## *Phoenicopterus ruber* Greater Flamingo

PLATE 2
[after page 38]

Du. Flamingo     Fr. Flamant rose     Ge. Flamingo
Ru. Фламинго     Sp. Flamenco     Sw. Flamingo

*Phoenicopterus ruber* Linnaeus, 1758

Polytypic. *P. r. roseus* Pallas, 1811, south Europe, south-west Asia, Africa. Nominate *ruber* Linnaeus, 1758, extralimital, West Indies, Galapagos.

**Field characters.** 125–145 cm, of which body only about ⅔ths, ♂ averaging larger; stands 105–155 cm high; wing-span 140–165 cm. Large, grotesque, yet graceful waterbird with very long neck and legs, pinkish-white plumage, crimson and black wings markedly contrasting in flight, and bent bill. Sexes alike and no rigid seasonal differences, but some individual variation in pink suffusion of plumage. Juvenile distinguishable.

ADULT. Head, neck, body, and tail white with tinge of pink, strongest in fresh plumage, but variable individually and according to diet. Wing-coverts bright crimson and flight-feathers black, partly hidden at rest by elongated scapulars (of which lowermost pale pink), but contrasting conspicuously with body in flight; axillaries and under wing-coverts pink. Bill with black tip noticeably bent downwards and otherwise pink extending to bare lores; eyes yellow; legs pink with darker joints and feet. JUVENILE. Shorter neck and legs, grey-brown head and neck speckled with white, streaky brown mantle and scapulars, brown and white wing-coverts, pink axillaries, blackish-brown flight-feathers; otherwise dirty white body and tail. Black-tipped grey bill and grey to brown legs and feet. Plumage becomes gradually whiter and then pinker over succeeding 3–4 years.

Often distinguishable from Lesser Flamingo *P. minor*, when together, by whiter coloration and larger size, but diagnostic field marks are mainly pink bill, and brighter, redder, more contrasting wing-coverts. Flamingoes in south Europe, Middle East, and North Africa virtually certain to be *P. ruber* (endemic race *roseus*) though *P. minor* has occurred. Farther north in Europe, however, strong likelihood of escapes of other forms, particularly Chilean Flamingo *P. chilensis* from South America and, to lesser

extent, *P. r. ruber* from Caribbean area, both of which similar in size to *P. r. roseus*. *P. r. ruber* has even brighter pink body than *P. minor*, which distinguishes it from *P. r. roseus*, and orange-pink bill with white base and black tip, which separates it from *P. minor*. *P. chilensis* has body and bill like *P. r. roseus*, but readily distinguished at reasonable range by yellowish-grey to grey legs with characteristic bright pink joints and webs.

Rarely seen singly, being markedly gregarious in flocks of hundreds or thousands. On ground, walks sedately; flocks group together and move unhurriedly away from danger, not taking flight unless pressed; but can run when necessary (e.g. when moulting). Taking to air normally preceded by short run with flapping wings, and on alighting runs several metres after touching down. Flies with rapid wing beats, neck and legs outstretched; does not usually glide for more than beat or two, except when about to alight. Flocks fly in variety of formations, from loosely massed to V, long diagonal, single file, or wavy line. Never perches off ground and spends most time wading in shallow water, feeding with head partly or wholly submerged and bill inverted, frequently treading or marking time in one place or slowly moving backwards to stir up food organisms from mud; in deeper water will immerse head and whole neck. Swims easily, sometimes in tight flocks, even on sea well off shore; as on ground, neck may be straight or curved. Frequently stands on one leg: rests with neck curved, or folded along back, and bill tucked in feathers. Babbling noise on ground and deep, resonant honking in flight both remarkably goose-like.

**Habitat.** Exceptionally restricted and well-defined, within tropical to Mediterranean zones. Frequents open,

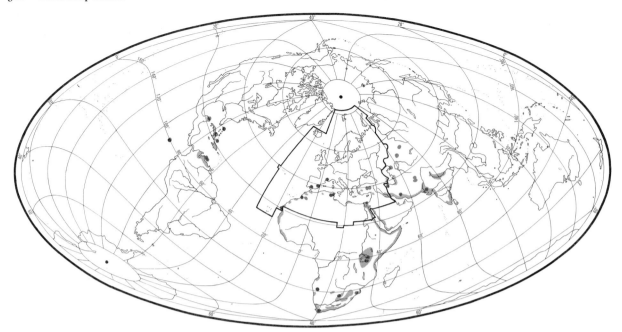

shallow lagoons, muddy, extensive lakes or deltas, coastal or less often inland, of highly saline or alkaline water (up to *c*. pH 11), not normally exceeding *c*. 1 m deep over substantial tracts. Sometimes on tidal mudflats (Ali and Ripley 1968); exceptionally on vegetation, e.g. on luxuriant fresh growth of grass in east Africa (Brown 1958). Some east African breeding haunts on lakes up to *c*. 2000 m. As bottom feeder, using mud layer, has ecological niche distinct from surface algae and diatom food resource of Lesser Flamingo *P. minor* and accordingly more limited in numbers. Apparently also has greater need of seasonal access to fresh drinking water (Brown 1958). Requires large, undisturbed expanses, shifting restlessly from one part to another, normally in large flocks. Prefers not to mix with birds of other families.

Successful mass breeding possible only in optimal conditions of water levels and biological productivity, failing which will remain on traditional area through successive seasons without breeding. In Kenya, brings food from some 20 km distant from breeding place when local resources insufficient; flies freely to considerable heights, ranging widely around chosen centres. Vigorous wader and walker. Highly intolerant of human disturbance or close approach (including low-flying aircraft), but resorts to artificial salt pans and at times content to live on extensive waters surrounded by and in view of dense human settlement.

**Distribution.** Breeding often irregular even at favoured localities, and infrequent at other sites. Some evidence that one or more of group of sites (e.g. Camargue, southern Spain, and perhaps Tunisia) used according to suitability in different years. Map shows all known breeding sites used in last 30 years.

Formerly bred in several other countries. EGYPT. Once nested Lakes Menzaleh and Bardawil, eggs taken 1894; date of extinction unknown (Meinertzhagen 1930). For recent colony Sinai, see Population. KUWAIT: bred Rubiyan Island 1922 (Ticehurst *et al.* 1926); now present in summer in small numbers, but no evidence of nesting (SH). CAPE VERDE ISLANDS: probably nested irregularly and in small numbers in past (de Naurois 1969*b*). ALGERIA: said to have bred formerly; no recent observations (Johnson 1975).

Accidental. Belgium, Netherlands, Denmark, Norway, West Germany, East Germany, Poland, Czechoslovakia, Austria, Switzerland, Hungary, Yugoslavia, Rumania, Greece, Bulgaria, Israel, Malta, Canary Islands. Records from northern and central Europe probably often escapes but see Movements for influxes. Portugal: irregular visitor (MDE).

**Population.** FRANCE. Bred Camargue since 1914 and probably earlier. Breeding attempted in 36 years since 1914, annually 1947–63, none 1964–8, annually since 1969. In years when breeding attempted, number of pairs varied between under 2000 and *c*. 8000, and young reared from 0–6000. Nesting attempts, and success, influenced by spring weather, disturbance, predation, and erosion of nesting islands. For discussion of numbers and factors involved, see Johnson (1975). SPAIN. Colonies in marismas of Guadalquivir 1935, 1944, 1945; largely unsuccessful. Bred elsewhere Andalucia, 1963–7, 1969, and 1973; *c*. 3500 pairs 1963, *c*. 2000 birds 1967 (Valverde 1963; Studer-Thiersch 1972; Johnson 1975). In 1973, new colony Alicante; 20–30 young reared (*Ardeola* 1974, **20**, 328–30). MOROCCO. Breeds irregularly Iriki; *c*. 1500 nests 1956, over 500 young 1968 (Robin 1968). Breeding suspected on

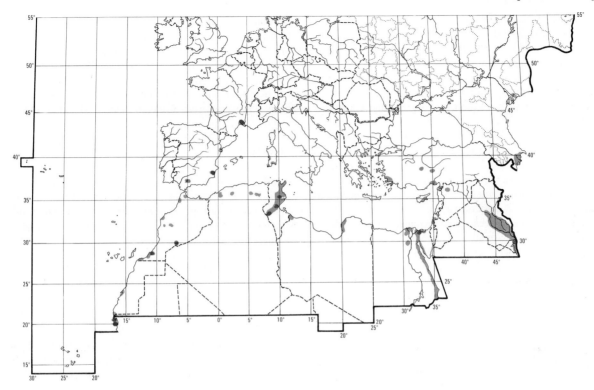

coast near Puerto Cansado (de Naurois 1969*a*) and said to nest Merdja Zerga (Vernon 1973). TUNISIA. Bred 1926, 1948–9, and regularly in recent years (Bauer and Glutz 1966). Breeds in Schott Djerid region, at least in favourable years; *c.* 2000 young in 1959 (Heim de Balsac and Mayaud 1962). In 1972, *c.* 10000 pairs bred lake in south. In summer 1973, 14000 birds present Tunisia, but no colonies (Johnson 1975). EGYPT: colony discovered north-west shores Sinai 1970, 500–600 pairs, with *c.* 500 young; probably nested successfully since (HM). MAURI-TANIA: breeds intermittently 2 islands Banc d'Arguin; variable numbers, from 10000 to 16500 pairs 1959–65 (de Naurois 1969*a*). TURKEY. Breeding recently proved Tuz Gölü; at least 2000 pairs 1969, at least 5000 nests 1970, at

least 300 pairs 1972. Kurbağa Gölü: at least 1500 pairs 1970. Suspected Acigol, where over 2000 birds in recent summers (OST).

Mortality. Little known, but see discussion by Rooth (1965) and Johnson (1974) of factors involved. Oldest ringed bird 13 years 4 months (*CRMMO Bull.* 1966, **20**, 18).

**Movements.** Migratory, partially migratory, and disper-sive; erratic at times. Intensive ringing only in Camargue; foreign recoveries shown in Table. Population resident in coastal lowlands of west Mediterranean; minority, num-bers varying with weather, winter Camargue, most Andalucia and north-west Africa; 119000 in Tunisia and

Recoveries outside France of *Phoenicopterus ruber* ringed Camargue (based on Johnson 1975)

| | Jan | Feb | Mar | Apr | May | June | July | Aug | Sept | Oct | Nov | Dec | Undated | TOTAL |
|---|---|---|---|---|---|---|---|---|---|---|---|---|---|---|
| Portugal | — | — | — | — | — | — | — | — | 1 | 5 | 2 | — | — | 8 |
| Spain | 11 | 16 | 6 | 9 | 1 | 2 | 2 | 5 | 19 | 40 | 18 | 19 | 11 | 159 |
| Balearic Islands | — | 1 | — | — | — | — | — | 1 | — | — | — | 1 | — | 3 |
| Sardinia | 2 | — | 2 | — | — | — | 2 | 2 | 2 | 3 | 4 | 2 | 2 | 21 |
| Italy and Sicily | 1 | — | — | — | — | — | — | 1 | — | 2 | — | — | — | 4 |
| Turkey | — | — | 1 | — | — | — | — | — | — | — | — | — | — | 1 |
| Libya | — | — | 1 | — | — | — | — | — | — | — | 1 | — | — | 2 |
| Tunisia | 9 | 5 | 5 | 1 | 5 | 1 | — | — | 6 | 6 | 9 | 7 | 4 | 58 |
| Algeria | 1 | 1 | — | — | — | — | — | — | 1 | 6 | 2 | — | — | 11 |
| Morocco | 1 | 1 | 2 | 1 | — | — | 2 | 1 | 4 | 3 | 2 | 2 | 1 | 20 |
| Mauritania and Sénégal | — | — | 1 | — | — | — | 2 | — | — | 1 | — | — | — | 4 |
| TOTAL | 25 | 24 | 18 | 11 | 6 | 3 | 8 | 10 | 33 | 66 | 38 | 31 | 18 | 291 |

Algeria alone November 1971 (Johnson 1975) will have included Tunisian breeders also. Camargue birds rarely recovered east of Italy and Sicily, while ringing indicates limited coastal movement south from Morocco into range of Banc d'Arguin population. One Libyan recovery was in Kufra Oasis, Sahara, 2700 km south-east of Camargue. Partial emigration from Camargue begins July, continues to October or later in cold weather; return movement apparent March, but non-breeders abroad all months.

Banc d'Arguin population believed to disperse round coasts of Mauritania and Senegambia, and perhaps these which occur occasionally south to Liberia; but many winter locally, and c. 50000 Banc d'Arguin late December 1971 (Pététin and Trotignon 1972).

Those wintering in east Mediterranean basin, chiefly Turkey, Cyprus, Cyrenaica (Libya) and lower Egypt, presumably of Asiatic origins. Migrants pass through Levant, including large December flocks Lake Djabboul, Syria, probably to north-east Africa; also limited passage through Gulf of Aqaba, November (Safriel 1968). At Kurbaga Gölü colony site in Turkey, 18000 in November 1969 but down to 1700 in following January (Koning and Dijksen 1970); but no ringing there. Recent ringing at Lake Rezaiyeh colony, Iran, showed wide scatter of juveniles, most of which appear to leave Iranian region during their first winter (Scott 1975); distinct groupings of recoveries in Cyrenaica and in Indus–Ganges plains of Pakistan and India, which emerge as most important wintering areas for Lake Rezaiyeh immatures; but also recoveries in Ethiopia, Turkey, Cyprus, Saudi Arabia, and Persian Gulf States (Qatar and Muscat). Thus valuable confirmation that some from Palearctic reach Ethiopian Africa and Indian subcontinent. Though most Lake Rezaiyeh birds leave area in autumn, 5000–12000 winter regularly in Fars, Iran (where recovery of Rezaiyeh adult), and 52500 in February 1972 after large numbers of waterfowl driven south from Caspian by bad weather; while up to 7000 normally winter Gorgan Bay, south Caspian, and small numbers Khuzestan and Iranian Baluchistan (Carp 1972; D A Scott). These must include some USSR breeders, and there are 16 Iranian recoveries (south Caspian 9, Azerbaydzhan 3, Fars 4) of Greater Flamingoes ringed Lake Tengis (Kazakhstan) and Gasan Kuli Reserve (Turkmeniya) (Scott 1975). However, large numbers of USSR birds winter Soviet Azerbaydzhan and Turkmenian regions of Caspian unless driven south by cold weather; first winter visitors reach Azerbaydzhan late November, and begin departing March, reappearing north Caspian early May (Dementiev and Gladkov 1951). Moderate numbers winter Iraq, Kuwait, and coastal Arabia; recoveries from Lake Rezaiyeh mentioned above, but this may not be only source.

Non-breeders present all year in wintering areas, while throughout range prone to irregular movements in response to temperature and water-level changes. Most occurrences in central and north Europe probably escapes, especially since majority adults, whereas immatures wander most (Lippens and Wille 1972). Occasional erratic influxes occur, e.g. May 1924, when noted Rhône valley, France, and flock of c. 60 immatures Switzerland (Poncy 1926); autumn 1935, when small flocks of juveniles reached Germany—these thought, on basis of parasites, to have originated from Caspian region (Bauer and Glutz 1966); autumns of 1906 and 1907, when large number visited central Siberia east to Lake Baikal (Johansen 1908; Allen 1956).

**Food.** Chiefly small invertebrates, insects, crustaceans, molluscs, and annelids, with some protozoa, diatoms, algae, seeds and plant fragments; also 'organic ooze' (Gallet 1950; Guichard 1951; Allen 1956) and possibly occasionally fish (Ticehurst 1923; Ali and Ripley 1968). Gregarious, feeding by day (particularly at dawn and dusk) and by night, walking in water and mud (up to 70–80 cm deep), or more rarely swimming (up to 120–130 cm); head faces backwards between legs and immersed (unlike Lesser Flamingo *P. minor*) so that hinged upper mandible at bottom. Most common feeding method: moving slowly forward with scything action of head, sucking in and expelling water by pumping action of tongue, food particles being trapped by lamellae of bill (Jenkin 1957). Other methods, including up-ending and 'treading', summarized by Allen (1956) and Rooth (1965).

For *P. r. roseus*, insects include fly larvae (Chironomidae and Ephydridae); waterboatmen *Sigaria*, *Micronecta*, *Notonecta*; waterbeetles (Coleoptera); and ants (Formicoidae). Molluscs include *Paludestrina*, *Neritina*, *Cerithium*, *Cardium*, *Venus*, *Mytilus*, *Tapes*, and *Tympanotomus*; crustaceans, *Sphaeroma*, *Gammarus*, *Asellus*, and *Paradiaptomus*; algae, *Spirulina*, *Arthrospira*, and *Oscillatoria*; also seeds of sedges *Cyperus* and *Scirpus*, tassel pondweed *Ruppia*, and rush *Juncus* (from Allen 1956; Ridley *et al.* 1955; Palmer 1962). 7 stomachs from eastern shore of Caspian Sea, USSR, May 1938, contained the unicellular water plant *Apanothece*, seeds of *Ruppia maritima*; near Lenkoran, in winter, mainly cockle *Cardium edule*; and at Kara-Bogaz Bay mainly brineshrimp *Artemia salina* (Dementiev and Gladkov 1951). In East Africa, 6 stomachs from Lake Elmenteita (November, December, and April) and 2 from Lake Hannington (July) contained in order of frequency: chironomid larvae, corixids (*Sigaria*, *Micronecta*), seeds of *Cyperus*, copepods, and plant fragments (Ridley *et al.* 1951).

Parents feed young initially, and to some extent until specialized bill structure formed, on regurgitated liquid secretion of glands in upper digestive tract (Lang *et al.* 1962; Studer-Thiersch 1966).

**Social pattern and behaviour.** Based mainly on material supplied by A Studer-Thiersch.

1. Highly gregarious at all times. Will concentrate up to many thousands in areas rich in food or at freshwater sources.

Flightless young form herds (crèches) on breeding ground. BONDS. Monogamous pair-bond, probably life-long. Both parents tend young. BREEDING DISPERSION. Normally nests in large dense colony up to many thousand pairs, with number of discrete groups, each having highly synchronized egg-laying (Lomont 1954; Uys *et al.* 1963; and others). Defends nest-site territory only (Brown 1958), but hostility between members of same nest-group seems less than between those of different groups (Studer-Thiersch 1974). ROOSTING. Gathers to sleep in shallow water, often in dense groups, both during day and night.

2. In most parts of range, nests wherever favourable at time, though often returning to same general area, e.g. large lake. Activities preceding nesting, such as intense group display and copulation, not always confined to eventual breeding ground; even preliminary nest-building not restricted to final site (Brown 1958; Brown *et al.* 1973). When begins to visit breeding area, goes on land in separate groups which keep distinct at nests (Studer-Thiersch 1974). FLOCK BEHAVIOUR. Activities of flock more or less synchronized, especially within sub-groups during breeding season (Studer-Thiersch 1972); not known whether same degree of group cohesion sustained in flocks after breeding. At feeding grounds, scatters rather evenly over wide area. At freshwater source, will crowd in enormous masses, individuals touching one another. Just before nesting, when often much group display, behaviour within flock highly synchronized; individual distance between sleeping birds so small that one losing balance will touch several others. ANTAGONISTIC BEHAVIOUR. Some squabbling frequent in group, but real attacks and fights uncommon (A Studer-Thiersch). At low level, Raised-scapulars Threat performed (see Ali 1945; Brown 1958) while uttering deep, short call repeatedly; with increasing intensity, performs Neck-Swaying Threat (terminology of displays follows Kahl 1975*b*), stretching head and neck horizontally forward and moving from side to side, occasionally pecking at rival. During fights, 2 peck haphazardly and tear at one another with bill, but without causing serious injury (Brown 1958). Such fights limited; occur only during site-selection, when ♀ (supported by ♂) extremely aggressive (Studer-Thiersch 1967). Moderate aggression also shown at start of period of intense group display (see below), during nest-building, and when young hatch. Aggression related mainly to maintenance of individual-distance. Mates support each other during fights. Escape behaviour little pronounced, though sometimes bird walks, runs or flutters away after quarrel, and may then be chased. On feeding grounds, Raised-scapulars Threat by one bird often deters another from approaching. In sleeping group even bill-fights do not necessarily cause bird to change position. If one passes between others close together, often lowers neck and head in form of submissive behaviour (Studer-Thiersch 1967). GROUP DISPLAY. See Allen (1956), Brown (1958), Studer-Thiersch (1974, 1975*b*). 8–10 weeks before nesting, group display—which occurs irregularly and at low intensity throughout year—becomes frequent, some groups displaying all day long, usually near islands where eventually breed. Individuals appear to stay in constant groups; when more than one group performs simultaneously, no obvious mingling. Many birds (often 15–30 or 40) may take part, standing together in loosely formed party, ♂♂ and ♀♀ equally, but ritualized movements not orientated towards any particular bird or sex. With necks stretched upwards, they repeatedly Head-flag (see A) by moving heads from side to side jerkily in relatively fixed rhythm, while giving short, harsh disyllabic calls. Then, several birds simultaneously begin to perform ritualized comfort-movements, whereupon calling immediately stops. Sequence of displays more or less fixed: Wing-salute, Twist-preen, Inverted

Wing-salute, and Twist-preen (to other side). In Wing-salute (see B) open wings for *c.* 2 s with neck and body stretched upwards; in Inverted Wing-salute (see C), bows forward with stretched neck and opens wings beyond body; in Twist-preen (see D), lowers head, extends wrist of wing and touches inner

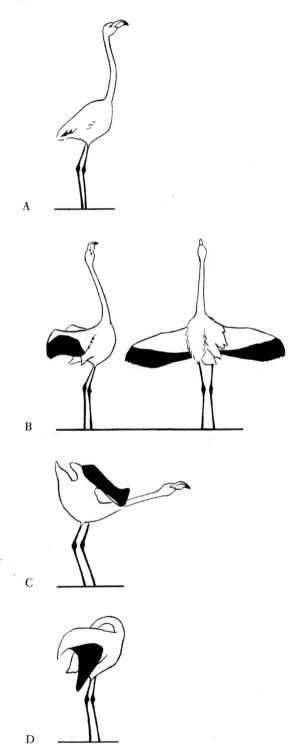

E

wing-coverts with bill. Wing-leg Stretch, stretching of one wing and leg, also occurs, but only seldom during intense group ceremonies. Sequence of movements often ends with head-scratching or foot-stretching, after which birds begin to look upwards again. Sequences repeated most quickly in groups of between 20–30. In larger groups (of up to 100 or more), birds often display at lower intensity or just walk about; aggressive behaviour then frequent, and threatening or pecking often answered by Group-hooking when several dart in same direction side by side in Hooking-posture (see E), with necks obliquely stretched forward and head bent down, bill pointing back towards underside of neck. Several may False-feed after outburst of Group-hooking. Though period of intense display can last for weeks or months, individuals generally take part only for a few days up to 2–3 weeks before beginning to copulate and show interest in nests. PAIR-COURTSHIP. See Studer-Thiersch (1974). Pair-formation unobtrusive; possibly occurs during feeding, when ♂ and ♀: (1) stand close together and, from time to time, look up with fully stretched necks, uttering repeated soft calls; (2) False-feed while walking quickly side by side, or one after another, with necks stretched down, one or both sometimes self-preening at outer wing-coverts. When paired, help each other during fights, have short mutual individual-distances, and synchronize routine activities to some extent; also call together with necks stretched down—irregularly after fights, as kind of Triumph Ceremony, after copulations and during nest-relief. Copulation (see Brown 1958; Suchantke 1959; Studer-Thiersch 1967, 1972) usually begins some days after pairs cease taking part in group display. Both may invite copulation, but typically ♀ walks around and ♂ follows with neck stretched forward, sometimes touching her back. When ♀ stops, often partly opens wings and ♂ mounts with flapping wings, then lowers neck and stretches head obliquely down. After ejaculation, ♂ stands up on ♀'s back and then jumps down in front of her. Sequel varied, but can include soft mutual calling, self-preening while standing side by side, feeding, and threatening other birds—Hooking often

occurring before and after such squabbles. ♂ will sometimes try to mount feeding ♀, pre-copulatory and feeding postures being closely similar. Occasional copulations with strangers do not alter pair-bond. RELATIONS WITHIN FAMILY GROUP. See Gallet (1950), Brown (1958), Studer-Thiersch (1967). Parents care for chick in nest, then when latter mobile follow, brood, feed and guide it, defending small area round it until it joins crèche at c. 8–10 days. Chick then unguarded; parents at first return frequently to feed when chick calls persistently, but later less and less until parental care ceases. Parents recognize own chick, and feed it exclusively. Hatchlings answer all loud calls in neighbourhood, but soon after leaving nest chick seems to recognize parents; older young sometimes search for parent by pushing several sleeping adults, thus provoking them to call. Parent offers food to tiny chick by touching its bill with own several times; later nestling begs by calling and sometimes pecking neck feathers of brooding bird. Older chick begs by approaching parent while calling softly, legs straddled, neck lowered to ground, and tries to place itself under adult's breast; then stretches neck and bill upwards and calls more loudly and frequently. Should adult run away, chick pursues it, then crouches, or attempts to crouch, in front. Stretching down of head and neck by begging young seems to have appeasing function; in small chicks causes brooding behaviour by adult, in older often inhibits adult from running away. Threat and pecking between young in crèche rare.

(Figs A–E after sketches by A Studer-Thiersch.)

**Voice.** Based mainly on information supplied by A Studer-Thiersch for *P. r. roseus*. See also Social Pattern and Behaviour.

Flocks utter continual brassy, nasal 'ka-ha' or 'a-ha' (see I), resembling honking of Greylag Goose *Anser anser* but more rapid (Brown 1958); call also rendered 'kawuck' and 'aahonk aahonk' (Witherby *et al.* 1939). Conversational babble given on ground includes continuous 'ke-kuk-kuk-kuk-kuk' or 'wuh-wuh-wuh-wuh' (see III) during feeding, and deep grunting 'murrt-murrt-murrt' in threat, which becomes louder, harsher, and sometimes higher-pitched as aggressiveness increases (Brown 1958). Low nasal wheeze, 'pmaaa' or 'kngaaa' (Brown 1958), bassoon-like (see II), indicates intention to fly. After copulation, pair may call softly.

CALLS OF YOUNG. Small young call persistently with rhythmically repeated, high-pitched 'kwick'; older chicks give harsher croaking calls (Bauer and Glutz 1966). Calls individually distinctive and likely to be major factor in identification of own young by parents, especially in crèche stage.

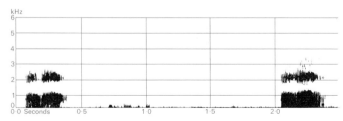

I J-C Roché France

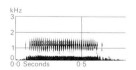

II J-C Roché France

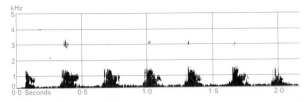

kHz

III J-C Roché France

**Breeding.** SEASON. See diagram for Mediterranean region. In Camargue, France, dependent on spring rainfall, temperature, and water level, though precise relationships not fully known (Johnson 1974). Mauritania: late April to late May, 1959–60 (de Naurois 1969*a*); 3–4-week-old young seen, 20 July 1965 (Trotignon 1975). Kenya: laying in almost every month of year, but main peak April, lesser peaks July and October–December, coinciding with rainy reasons (Brown *et al.* 1973). SITE. On damp, usually muddy ground. Sometimes in vegetated area (e.g. *Salicornia*), first cleared by trampling, tearing, and digging out of roots (Gallet 1950). Colonial; distance between nests *c.* 35 cm (range 12–72) with most 20–50 cm (Uys *et al.* 1963). Colony site may shift from year to year. Nest: conical mound of sun-baked mud with shallow cup in top. When built of clay, diameter 30–5 cm, height 30–40 cm, surrounded by circular trench up to 20 cm deep, excavated for material; when of sand, may be only 10–15 cm high, with small or no trench; exceptionally only slightly raised bed of mud pellets. No lining (Gallet 1950). Building: by both sexes. Bird stands in centre of nest and scrapes up mud pellets from immediate surround and rolls them to top; also daubs on semi-liquid mud. Building sporadic for many days prior to egg-laying. Sometimes eggs laid in rudimentary nest which is then quickly built up, from 10–30 cm, in 48 hours. EGGS. Oval; off-white overlain with chalky deposit, becoming muddy and discoloured during incubation. 90 × 55 mm (77–103 × 48–60), sample 133; calculated weight 140 g (Schönwetter 1967). Gallet (1950) reports 2 types of egg: (1) 89 × 58 mm (short, regular, more common); (2) 94 × 54 (long, narrow, less usual). Clutch: 1, rarely 2. 2 eggs in 40 of 2000 clutches, Camargue (Gallet 1950); in 5 in 900 clutches, Kenya (Brown 1958); 1 in 400, South Africa (Uys *et al.* 1963). One brood. Replacement clutches laid within 7 days of early egg loss, but not recorded after loss of young (Gallet 1950). Eggs of breeding group laid almost simultaneously (Gallet 1950). INCUBATION. 28–31 days (27–36). By both sexes. Eggshells discarded over side of nest, or sometimes left in nest and consumed by chick. YOUNG. Precocial and semi-nidifugous. Cared for and fed by both parents. Fed on regurgitated liquid secretion of glands in upper digestive tract ('crop-milk'), direct into bill. Can feed themselves at *c.* 30 days but still fed by parents at least until fledging. Stay on or close to nest until *c.* day 10, then move into crèche of young mostly similar age and size. FLEDGING TO MATURITY. Fledging period

70–75 days (65–90). Most stay in crèche until at least day 100; move away rapidly and become independent on break up of crèche. Age of first breeding 2–3 years. BREEDING SUCCESS. 47% of breeding pairs on average reared young at Camargue, in 20 breeding years from 1946–72, range 0–95%. Main losses of eggs and young to Herring Gulls *Larus argentatus*; heavy rain and wave action destroys nests in some years. Egg-collecting formerly serious; other human disturbance (e.g. tourists, low-flying aircraft) still a threat (Johnson 1974). Nesting colonies, Banc d'Arguin, Mauritania, destroyed by high tides, 1959 (Trotignon 1975). At one colony, Kenya, 1951–71, 19% young hatched reared, range 0–66% in 11 breeding years; at 2nd colony, 48%, range 42–56% in 4 breeding years. Main losses of eggs due to rising water level, competition for nest sites with White Pelicans *Pelecanus onocrotalus*, human disturbance, and predation and chasing of birds by Maribou Storks *Leptoptilos crumeniferus*; main losses of young due to predation and chasing by Maribous, disease, and rising water level (Brown *et al.* 1973).

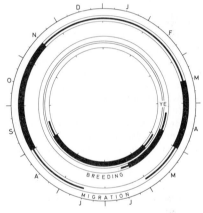

**Plumages.** ADULT. Head, neck, and body white with delicate pink hue, especially in fresh plumage; variable, sometimes much brighter pink. Tail-feathers pink on outer web; paler, almost white, on inner. Primaries and *c.* 18 outer secondaries black; inner secondaries pink, primary coverts pink, paler on inner webs; rest of upper wing-coverts crimson-pink, marginal coverts paler; greater under wing-coverts black, median and lesser crimson-pink. Axillaries bright crimson. NESTLING. 1st down short, woolly, pale ash-grey; lighter, almost white on underparts and at sides of nape (Gallet 1950). 2nd down darker grey. JUVENILE. Head and neck pale brown-grey. Feathers of mantle brown, shafts darker, edges pale grey. Upper scapulars lanceolate, pale grey with brown shaft-streaks; lower scapulars broader, brown with pale edges, white at base of inner webs, grading into mainly white feathers, with faint pink hue and brown tips. Feathers of back, rump, and upper tail-coverts white, with dark shafts and often little dark wedges at tips. Tail-feathers white with slight pink wash, sometimes with brown streaks at edges of outer webs. Primaries and secondaries dark brown, paler below; inner secondaries with white inner webs. Primary coverts white with dusky tips and inner webs. Greater coverts white with dark brown tips; rest of upper wing-coverts brown with white bases, marginal mainly white, tinged pink. Underwing white tinged pink, greater coverts pale brown like under surface of secondaries. Axillaries crimson-pink, often with some brown at tips. IMMATURE. Feathers of entire body white, wings like juvenile. Older birds apparently variable, body and most lesser and median upper wing-coverts white or pale pink. Primary coverts with dusky and greater coverts with dark brown tips. Still older birds like adult breeding with crimson wings, but dusky tips to primary coverts. Exact determination of age in immatures impossible (Bauer and Glutz 1966).

**Bare parts.** ADULT. Iris yellow. Bill black at tip, pink at base; black does not reach upwards to bend of culmen. Bare skin at side of face pale pink. Leg pink. NESTLING. Iris black. Bill pink at hatching, later pinkish-grey. Bare skin of face red. Leg pink; at 5 days dark brown with tinge of pink, at 7 dark brown or black (Brown 1958). JUVENILE. Iris brown. Bill lead-grey with black tip. Bare skin of face lead-grey. Leg lead-grey to brown or pale pink. (Witherby *et al.* 1939; Gallet 1950.)

**Moults.** ADULT. Extraordinarily variable and poorly known. Captive birds often shed flight-feathers simultaneously just before breeding, after failure of nesting attempt, or in non-breeding years; on average every other year, but sometimes moult at longer intervals or in 2 successive years (Stresemann and Stresemann 1966). In wild, moult of flight-feathers irregular, probably almost continuous; relation to breeding not clear (Uys *et al.* 1963). Regular descendant primary moult in summering birds claimed by Middlemiss (1961). Simultaneous post-breeding moult of primaries in separate moult locality, involving moult migration, described for nominate *ruber* (Palmer 1962). Body feathers probably moult slowly during most of year, suggesting separate post-breeding and pre-breeding moult (Witherby *et al.* 1939), but unknown whether individual feathers renewed twice a year. Sometimes at least, plumage of head and neck moulted twice in autumn (Bauer and Glutz 1966). JUVENILE. Partial; involving most body feathers but not wings and tail, September–March (Witherby *et al.* 1939). Flight-feathers renewed in 2nd calendar year (captive birds, Bauer and Glutz 1966).

**Measurements.** ADULT. Skins and mounted specimens (RMNH, ZMA) and 1 ♂ in flesh (Ali and Ripley 1968). Sex of some old specimens determined by size.

| | | | | | | |
|---|---|---|---|---|---|---|
| WING | ♂ 428 | (17·9; 8) | 406–464 | ♀ 380 | (11·8; 6) | 360–396 |
| TAIL | 156 | (7·52; 8) | 144–167 | 135 | (3·43; 6) | 130–138 |
| BILL | 121 | (2·85; 7) | 117–125 | 116 | (3·34; 7) | 112–121 |
| TARSUS | 323 | (23·4; 8) | 291–373 | 259 | (18·4; 7) | 232–280 |
| TOE | 86·7 | (4·07; 7) | 80–91 | 78·4 | (4·76; 7) | 72–86 |

Sex differences significant. Witherby *et al.* (1939) gave rather different ranges: 12 ♂♂ wing 360–445, tail 125–57, bill 120–135, tarsus 240–365; ♀♀ (no sample given) wing 355–425.

**Weights.** ADULT (Gallet 1950).

♂ 3579 (340; 13) 3000–4100 ♀ 2525 (413; 12) 2100–3300

**Structure.** Wing long and narrow. 12 primaries: p10 longest, p11 0–11 shorter, p9 3–10, p8 31–41, p7 49–65, p6 81–91, p1 156–192; p12 rudimentary. P10 and p11 emarginated inner webs, p9 and p10 outer; outer web of p11 narrow. 27 secondaries, 4 behind elbow (Stephan 1970). 14 tail-feathers, outer and central longer than those in between: t7 *c.* 15–25 shorter than t1, t6 *c.* 5–12 shorter than t7. Under tail-coverts as long as tail or longer. Bill heavy, sharply bent in middle, small elongate nail at tip, many lamellae along cutting edges; upper mandible flat (details in Jenkin 1957). Nostrils slit-like. Bill of nestling straight. Neck long and slender. Leg long, tibia bare for $\frac{3}{4}$ of length. Toes webbed, outer $\frac{1}{3}$ of middle toe projecting beyond webs. Hind toe small, elevated. Leg of nestling short and swollen.

**Geographical variation.** 2 clearly distinct forms, formerly considered separate species. American *ruber* differs from Old World *roseus* by bright vermilion plumage, especially on head, neck, and upper breast; wing-coverts deep vermilion, not crimson; crimson patch on vent and lower flanks; yellow base of bill; and black at tip of bill extending to curvature. *P. chilensis* Molina, 1782, southern South America, sometimes considered conspecific.

---

## *Phoenicopterus minor* Lesser Flamingo

PLATE 2
[after page 38]

DU. Kleine Flamingo  FR. Petit Flamant  GE. Afrika-Flamingo
RU. Малый фламинго  SP. Flamenco enano  SW. Mindre flamingo

*Phoenicopterus minor* Geoffroy, 1798. Synonym: *Phoeniconaias minor*

Monotypic

**Field characters.** 80–90 cm, of which body only $\frac{2}{5}$, ♂ averaging larger; stands 90–105 cm high; wing-span 95–100 cm. Small, often rose-pink flamingo with characteristic dark, heavy-looking, short-based, and sharply down-turned bill, dark face, and blotched crimson and black wings. Sexes alike and no rigid seasonal differences, but considerable individual variation in pink suffusion. Juvenile and other immature stages distinct.

ADULT. Key point: black-tipped bill and bare area of lores and around eyes dark red, looking almost black at distance and combining with high-ridged shape of bill to give rather fierce 'black-eyed' look. Head, neck, body, and tail typically pale rose-pink, blotched deeper crimson-pink on breast, but sometimes extremely pale or even white when plumage worn and after rearing young. Wing-coverts pink, lesser and median with some crimson centres, producing blotched or scaled effect, and flight-feathers black, these colours conspicuous in flight, but largely hidden at rest by elongated scapulars (some of which have dark crimson centres); axillaries pinkish-crimson contrasting with mainly pink under wing-coverts. Eyes red to yellow-red; legs greyish-pink, with slightly darker joints and feet, to pale vermilion. JUVENILE. Upperparts pale brown-buff with darker head, neck, wing-coverts, and especially mantle and scapulars, all with browner streaking, and flight-feathers blackish-brown; underparts brownish-white. Pinkish axillaries only touch of colour. Bill black at tip and purplish-brown at base; eyes brown; legs dark grey. Plumage grows gradually whiter and then pinker over succeeding 3–4 years, while bill turns redder, and legs greyer before becoming tinged with pink.

Confusable only with Greater Flamingo *P. ruber* (apart from escapes of other Phoenicopteridae, for which see *P. ruber*). Best distinctions much darker colour and different profile of bill, which combine to give more ill-disposed expression than comparatively severe appearance of *P. ruber*. More extensive markings on upperparts another good point. Otherwise, typically smaller, shorter-necked, and pinker, but size often difficult to estimate, and Old World race of *P. ruber* surprisingly variable in this respect. Degree of pinkness (dependent on strength of keto-carotenoid pigments derived from food) also variable, some *P. minor* being pinkish-white and actually paler than pinkest *P. ruber*. Juveniles still more difficult to separate on size because young *P. ruber*, with shorter necks and legs, are noticeably smaller than adults, but *P. minor* still has darker and different-shaped bill, as well as blacker legs, darker brown head, neck, and wing-coverts, and less white underparts. In flight, *P. minor* shows more compact silhouette with thicker neck, shorter legs, and more blotched wing-coverts.

Gait, flight, and general habits, including marked gregariousness, much as *P. ruber*, but usually feeds nearer surface level with side-to-side, scythe-like movements of only partially immersed bill, while *P. ruber* (essentially a bottom feeder) makes deeper sweepings with head partly or wholly submerged; nevertheless, inverted position of bill much the same in both and *P. minor* will sometimes immerse head. Goose-like honking higher-pitched, rather more yelping, than that of *P. ruber*, but not reliably distinguishable except to practised ear.

**Habitat.** Often shared, unlike diet, with Greater Flamingo *P. ruber*, especially on high-lying inland lakes of strongly alkaline character productive of masses of diatoms and blue-green algae. Differs from *P. ruber* in feeding by walking or swimming, in water usually *c.* 30–45 cm deep, preferably smooth but if need be exposed to wave action. However, in shallow brine only few cm deep, as in salt pans and Little Rann of Kutch, India, feeds like *P. ruber* (Ali and Ripley 1968; Brown 1960; Brown and Root 1971).

**Distribution.** Main breeding areas Rift Valley lakes of east Africa and pans north of Kalahari, southern Africa. Erratic breeder elsewhere and in 1965 found nesting Aftout-es-Saheli, Mauritania (de Naurois 1969a).

Accidental. Spain. Malaga: one, May 1966, and at least 4, July 1972 (*Ardeola* 1968, 12, 229, and 1973, 19, 14–15). Mauritania: for occurrences at Banc d'Arguin, see Movements.

**Movements.** Dispersive; often erratic and over long distances, and may occur any month. Mainly centred on east Africa, wandering south and north in Rift Valley. Presumably from here originate irregular movements to Madagascar, Mascarene Islands, Aden, and lower Persian Gulf, while irregular appearances Pakistan and north-west

India, sometimes in thousands, like to be mainly African birds since local breeding only once detected. Rather few records in west Africa, despite recent discovery of breeding Mauritania, and there may not be a permanent population; reported south to Cameroon coast, where 2000 February–March 1932 (Bannerman 1932), and north to Banc d'Arguin, where 600 August 1972 and 3100 August 1973 (Gandrille and Trotignon 1973; Duhautois *et al.* 1974). Vielliard (1972) reported recent November–February records from Lake Chad basin, and with this evidence suggested that west African occurrences stem from irregular incursions out of Rift Valley, with opportunistic nesting.

**Voice.** See Field Characters.

**Plumages.** ADULT. Head, neck, and body crimson-pink, fading to white with pink hue in worn plumage; small feathers bordering bare skin of face crimson; in some ♂♂ feathers of mantle, scapulars, and breast with crimson centres. Tail-feathers crimson-pink, paler on inner webs. Primaries and *c.* 20 outer secondaries black, inner secondaries pink. Primary coverts, greater upper wing-coverts, and marginal coverts pink; median and lesser strongly contrasting, fiery crimson with pink fringe. Greater under wing-coverts black, rest pink; axillaries crimson, inner webs paler. JUVENILE. Head, neck, mantle, and scapulars brown, darker than Greater Flamingo *P. ruber*; feathers of mantle and scapulars with dark brown shaft-streaks. Lower scapulars basally white; distally dark brown with paler margins. Back, rump, and upper tail-coverts white, mottled and streaked brown. Underparts pale brown, vent and under tail-coverts white. Tail-feathers with white inner webs and shafts, dark brown outer webs. Primaries dark brown. Primary coverts white with *c.* 2 cm wide terminal bar light brown. Secondaries dark brown, slightly paler on margins of inner webs, inner with white margins to inner webs. Upper wing-coverts brown with dark brown shaft-streaks, lesser and marginal edged paler. Under wing-coverts white with dark brown shafts, greater pale brown. Axillaries white, slightly tinged pink, with dark brown shafts and wide pale brown margins to outer webs. This plumage becomes strongly bleached, much of lighter brown off-white, resulting in more boldly patterned upperparts. Greater upper wing-coverts white with wide terminal brown bars. IMMATURE. Head and neck grey-brown faintly suffused pink. Body feathers pale pink with white shafts. Upper wing-coverts pink, median coverts more deeply tinged than others, some lesser and marginal with dark brown spots in centre. Primary coverts pink with brown tips. Axillaries bright pink. This plumage also becomes paler when worn. At next moult, dark tipped primary coverts lost and crimson median and lesser coverts acquired. Pale brown head last sign of immaturity, but known to breed in that plumage (Berry 1972).

**Bare parts.** ADULT. Iris yellow-red to vermilion. Bill dark crimson with black tip, less extensive than in *P. ruber roseus*, red lightest behind black tip, forming conspicuous pale band in dried skins. Bare skin of face dark red. Leg pale vermilion, deeper tinged at joint and foot. JUVENILE. Iris brown. Bill black at tip, purple-brown at base. Leg dark grey.

**Moults.** Limited data. Simultaneous primary moult involving flightlessness apparently more frequent than in *P. ruber*; may occur before, during, or just after breeding cycle, or also when no breeding takes place. Flightlessness lasts *c.* 3 weeks in individuals, 6–8 weeks in population (Brown and Root 1971; Berry 1972). In other cases, wing moult irregular as shown by skins with primaries in various stages of development and wear. Moult of immatures undescribed.

**Measurements.** ADULT. ♂♂: wing 321–354, tail 120–142, bill 100–118, tarsus 190–242. ♀♀: wing 310–325, bill 93–104 (Kear and Duplaix-Hall 1975). Adult ♂♂: wing 340; tail 99, 105; bill 95, 96; tarsus 217, 245; toe 70, 76 (ZMA). Adult ♀: bill 93 (ZMA).

**Weights.** Mean 1900 (Kear and Duplaix-Hall 1975).

**Structure.** Wing large. 12 primaries: p10 longest. Other primaries in adult ♂: p11 7 shorter, p9 12, p8 32, p7 48, p6 73; p1 144. P12 vestigial. Approximately 27 secondaries. Tail short, 14 feathers; outer longest, central pair slightly projecting beyond rest. Bill decurved; lower mandible relatively heavier than in *P. ruber*, more strongly arched. Upper mandible very narrow, deeply keeled, sinks between rami of lower in closed bill; lamellae much finer than in *P. ruber*; nasal groove short and broad (Gray 1896; Jenkin 1957). Structure of leg like *P. ruber*.

# Order ANSERIFORMES

Medium-sized to large aquatic, marine, and terrestrial birds. 2 families: (1) Anhimidae (screamers), and (2) Anatidae; first confined to South America, other cosmopolitan. Seem distantly related to Phoenicopteriformes and Ciconiiformes (see Sibley and Ahlquist 1972), though some anatomical similarities with gamebirds such as Cracidae (chachalacas, etc.) may suggest distant affinity with Galliformes via anseriform groups Anhimidae and Anseranatinae (Simonetta 1963; Johnsgard 1968). For recent survey of order, see Matthews (1974).

## Family ANATIDAE wildfowl

Waterbirds (some more or less terrestrial) with relatively short legs and front toes connected by webs. Though considerable adaptive diversity in outward appearance, size, plumage colours, behaviour, and ecology, homogeneous in many characters—as attested by numerous, often fertile, interspecific hybrids reported, chiefly in captivity (see Gray 1958). About 140 species in 3 subfamilies: (1) extralimital Anseranatinae (single species, Magpie Goose *Anseranas semipalmata* of Australia); (2) Anserinae; and (3) Anatinae—last two subdivided into several tribes (see further below). Classification and systematic order adopted here follows Delacour and Mayr (1945, 1946) and Delacour (1954–64), largely as modified by Johnsgard (e.g. 1965, 1968).

Body broad and rather elongated in many, though more rotund in some (especially diving species). Plumage thick and waterproof; contour feathers distributed over distinct feather-tracts with underlying coat of down. Neck medium to long. Wings generally relatively small; mostly pointed, fairly broad in many, but narrower in some highly migratory species. 11 primaries; p9 nearly always longest, p11 minute. Wide range in number of secondaries, from *c.* 12 to *c.* 24, innermost (tertials) often long and brightly coloured; diastataxic. Majority fast fliers with large and highly keeled sternum. Tail short and square or slightly rounded in most; long in some diving species (serving as rudder), pointed or with elongated central feathers in some others. 14–24 tail-feathers, but variable even within single species. Bills show much adaptive variation but typically of medium length, broad, often flattened centrally and distally but high at base, and rounded at tip with horny nail at tip—producing slight terminal hook; covered with soft skin. Edges of mandibles with rows of lamellae, showing differential development in various ecological types and taxonomic groups; most highly specialized in surface plankton feeders, least so in species (such as scoters *Melanitta*) that swallow molluscs whole. Tongue thick and fleshy; epithelium covered with papillae and horny spines. Lower part of tibia and tarsus bare; front toes connected by webs (reduced in *Anseranas* and some Anatini), hind toe elevated. Gait striding or waddling. Oil gland feathered. Aftershaft reduced or absent. Special intromittent copulatory organ present in ♂♂; vascularized sac everted from wall of cloaca, protruded by muscular action; facilitates sexing by examination, even of small young. Salt-secreting nasal glands subject to adaptive variation in size, even within same species; enlarged in forms inhabiting saltwater or brackish habitats, modifying profile of head considerably. In many species, ♂♂ have remarkably lengthened, bent, or locally widened trachea forming resonating tubes; also syringo-bronchial sound-boxes (bullae), either fully ossified or with membraneous fenestrae. These vocal structures highly characteristic of species or larger taxonomic units (see, especially, Johnsgard 1961*d*, 1971). Considerable diversity of plumage types. ♂ and ♀ similar, nearly similar, or show extreme sexual dimorphism. Except in *Anseranas*, flight-feathers moulted simultaneously, producing period of flightlessness lasting 3–4 weeks. 2 moults per cycle. Young precocial and nidifugous, covered with thick down (pattern often cryptic and characteristic of taxonomic groups within subfamilies); able to swim soon after hatching.

Cosmopolitan, but absent from continental Antarctica

and some islands. Usually on or close to water. Highly vulnerable to human pressures on wetland habitats. 1 species (Labrador Duck *Camptorhynchus labradorius*) extinct during last century, and 3 more (Crested Shelduck *Tadorna cristata*, Pink-headed Duck *Rhodonessa caryophyllacea*, Auckland Island Merganser *Mergus australis*) probably so in this. A few species domesticated: Swan Goose *Anser cygnoides*, Greylag Goose *A. anser*, Muscovy Duck *Cairina moschata*, and Mallard *Anas platyrhynchos* (see Goodwin 1965); some populations of a few more (Mute Swan *Cygnus olor*, Canada Goose *Branta canadensis*, Egyptian Goose *Alopochen aegyptiacus*) kept in semi-domesticated or feral conditions. For most wildfowl, censuses of breeding numbers extremely difficult and, within west Palearctic, even estimates available only for a few species. Extensive series of winter counts organized by International Waterfowl Research Bureau, therefore, often provide best indications of total populations for many species, though still rarely possible to make any accurate assessments of population trends. For winter counts, main sources used: Isakov (1970c), Szijj (1972), Atkinson-Willes (1974, 1975). In interpreting figures from these, it should be noted that (1) 'north-west Europe' as defined by Szijj (1972) includes Spain and Portugal, whereas Atkinson-Willes (1974) excludes Iberia except for Cantabria and Galicia; (2) region 'Europe–Black Sea–Mediterranean' of Szijj (1972) covers central and south-east Europe, and overlaps slightly with western part of USSR (for which see map in Isakov 1970c). Northern forms, especially, often highly migratory; autumn movements preceded by marked moult-migrations by ♂♂ of many species to special areas for period of flightlessness. More sedentary in warmer latitudes, but local dispersal often initiated by factors such as drought.

Wide range in diet, from totally vegetable to totally animal, and in feeding habits, from terrestrial grazing to bottom diving; correlated with conspicuous adaptations in structure of bill, head musculature, length of neck, and in general body proportions. Terminology of feeding methods in species accounts mainly after Szijj (1965b) and Bauer and Glutz (1968, 1969); see also Olney (1963b). Most species gregarious—feeding, loafing, roosting, and travelling in cohesive flocks, integrated by calls, special pre-flight signals, etc. Generally solitary breeders, however, with cryptic nests in concealed sites, though some species colonial—either habitually or, more often, as alternative to dispersed nesting—usually at protected sites such as islands. Degree of territorialism when breeding and relation between territory and nest-site vary between species and larger taxa; some strictly territorial while others occupy wholly or largely undefended home-ranges. Monogamous pair-bond in most species but much variation between taxonomic groups in duration of bond and degree of ♂ promiscuity (if any). Social systems and displays correlated with formation and maintenance of pairs complex and largely dissimilar in 2 main subfamilies

(see below). Copulation on water in vast majority of species (exceptions in *Anseranas*, and some Anserini and Tadornini), typically with ♂ grasping ♀'s nape in bill. Vocalizations varied but generally simple (mainly honks, grunts, quacks, coos, and whistles); often different between sexes when linked with anatomical differences in vocal apparatuses. Non-vocal sound signals produced in some cases. Calls of downy young include (1) Contact-call or Greeting-call (also termed Pleasure and Contentment-call) and (2) Distress-call (see Kear 1968). Comfort behaviour well known from study of McKinney (1965b). Bathing frequent and elaborate. Typically performed while swimming in water too deep for standing; involves head-dipping, wing-thrashing, somersaulting, and diving. Followed by oiling (with use of both bill and head) and preening. No elaborate heat-responses besides feather erection. Other behaviour characters include: (1) direct head-scratching; (2) resting, often on 1 leg, with head turned back and bill inserted in scapulars on same side as lifted leg (Heinroth and Heinroth 1954), latter being characteristically stowed away in waterproof flank 'pocket'.

Breeding strictly seasonal in more northern, migratory species and populations, less so or largely opportunistic in warmer latitudes. Although breeding habitat and nest-sites show considerable diversity, nests usually placed over water or on or near ground. Well hidden in vegetation or sometimes concealed in other dark places such as burrows and tree holes (or nest-boxes); some species also use old nests of other birds or cliff ledges. Distance from water often small; however, much adaptive variation with some species moving, at least at times, well away. Nests made only of vegetation, etc., within reach of sitting bird, using sideways-building movements (see Harrison 1967); usually lined with down plucked from ♀'s belly (often cryptic and grown specially for this purpose). Eggs large, unmarked; surfaces greasy. Clutches often large, especially in species with reduced parental care. Covered by down in most species in absence of attendant. Tendency in some species to lay odd eggs in nests of other anatids; though such nest-parasitism may reach quite serious proportions in some populations, especially of pochards (Aythyini) and stiff-tails (Oxyurini), only 1 species (Black-headed Duck *Heteronetta atricapilla*) obligate parasite. In some species also, 2 or more ♀♀ may lay at same site, at extreme producing 'dump' of eggs without incubating them. Single clutch though most species will re-nest if eggs lost. With exception of *Anseranas*, Dendrocygnini, and some species in Anserini, incubation by ♀; starts with last egg so hatching synchronous. No true brood-patches. Displaced eggs retrieved if within reach of sitting bird, using bill. Eggshells left in nest. Downy young typically led to water after leaving nest. Self-feeding in all species except *Anseranas* and Musk Duck *Biziura lobata*, but with indirect provision of some food in earlier stages by minority in both Anserinae and Anatinae (see Kear 1970). Establish recognition of own species by special learning process of imprinting upon parent during brief critical period; in exceptional (e.g. experimental) circumstances, may become imprinted on wrong species or even inanimate objects (see especially Heinroth 1911; Lorenz 1935; Hess 1957; Boyd and Fabricius 1965; Schutz 1965). Variable incubation and fledging periods, correlated with latitude at which breeding takes place—i.e. proportionally shorter in migratory species nesting in high latitudes with short summer season.

# Subfamily ANSERINAE whistling ducks, swans, true geese, freckled ducks

Wildfowl mostly with long necks and somewhat elongated bodies, fairly long tarsi with reticulated fronts, and feet placed in middle of body. No important sexual dimorphism; visual display and voice also closely similar in ♂ and ♀. Juveniles like adults, slightly duller, or (when adults white) brown or dark grey. About 33 species in 3 tribes: (1) wholly extralimital Dendrocygnini (whistling ducks)—9 species in genus *Dendrocygna*, tropics and sub-tropics of all continents, plus White-backed Duck *Thalassornis leuconotus* of Africa and Madagascar (an aberrant diving species often placed in tribe Oxyurini of Anatinae); (2) Anserini (see below); and (3) extralimital Stictonettini—single genus *Stictonetta* (1 species, aberrant Freckled Duck *S. naevosa* of Australia, often placed in tribe Anatini of Anatinae).

# Tribe DENDROCYGNINI whistling ducks

*Dendrocygna bicolor* (Vieillot, 1816) **Fulvous Whistling Duck**

FR. Dendrocygne fauve    GE. Fahlpfeifgans

Resident in widely separated regions: California and Nevada to Louisiana and Mexico; discontinuously in South America east of Andes and south to Uruguay and central Argentina; tropical Africa from Sénégal, Chad, and Ethiopia southwards, most in eastern half of continent; Madagascar; Indian sub-continent (rare in west) and Burma. Limited migrations in cooler regions, while vagrants not unusual beyond normal distribution (Delacour 1954). Recorded Panama, Guatemala, and Honduras; in recent years occurring with increasing frequency in eastern USA, West Indies, and Bermuda, and straggling north as far as Ontario and New Brunswick, while recorded October 1964 at 23°N 60°W in Sargasso Sea (Godfrey 1966; Munro 1967; Watson 1967). In northern tropics of Africa, numbers occurring Niger inundation zone and Chad apparently variable (Bannerman 1951; Malzy 1962), perhaps indicating erratic movement (Salvan 1967). Thus singles killed Camargue, France, September 1970 (Hovette 1972) and Guadalquivir marismas, Spain, September 1971 (Castroviejo 1972) possible vagrants, but caution needed as species commonly kept in captivity.

# Tribe ANSERINI swans, true geese

Largest wildfowl. 20–23 species. 2 genera of swans: *Cygnus*, and extralimital *Coscoroba* (South America); most with all-white plumage. *Cygnus* composed of 2 groups: (1) Mute Swan *C. olor* and extralimital Black Swan *C. atratus* (Australia) and Black-necked Swan *C. melanocoryphus* (South America); (2) 4 largely allopatric forms of northern swans. Latter sometimes separated in genus *Olor*; here treated as 1 polytypic and 2 monotypic species (following Voous 1973), but also often considered as comprising 4 monotypic species, or 2 polytypic species—usually in combination *cygnus* and *buccinator*, *columbianus* and *bewickii*—or sometimes single polytypic species (see Johnsgard 1974). 2 main genera of geese: 'grey' geese *Anser* (9 species); 'black' geese *Branta* (5 species). Aberrant *Cereopsis* (Cape Barren Goose, Australia), included by Johnsgard (e.g. 1965) in Anserini (see also Kear and Murton 1973) but often placed in Anatinae by other taxonomists (Delacour and Mayr 1945; Veselovsky 1970); not considered further here. In west Palearctic, 3 breeding *Cygnus*; 5 breeding and 1 accidental *Anser*; and 2 breeding, 1 introduced, and 1 wholly migrant *Branta*.

Bills of both swans and geese strong; adapted for grazing, especially in more terrestrial geese in which lamellae take form of varying number of horny 'teeth' along edges of upper mandible. No iridescent coloration, pied pattern on wing, or contrastingly coloured tertials. Plumages of *Anser* and *Branta* combine mostly grey, brown, or black with white. Especially in *Anser*, neck feathers of geese arranged in vertical furrows. Vocal apparatus in both sexes a simple tympaniform membrane where bronchi join trachea; in northern swans, trachea convoluted inside sternum. Lores naked in adult *Cygnus*.

Bill and foot usually bright pink or orange-yellow in *Coscoroba* and *Anser*, dark slate or black in *Branta*; in *Cygnus*, bill usually black with orange or yellow, feet dark. Webs between front toes reduced in terrestrial Hawaiian Goose *B. sandvicensis*. During post-breeding moult, ♂ and ♀ of mated pair normally shed flight-feathers and become flightless at different times. In at least some *Cygnus*, ♀ first to moult, followed by ♂ when ♀ flying again or nearly so (see Kear 1972); in Bewick's Swan *C. columbianus bewickii*, wing moult of pair said to be simultaneous (Dementiev and Gladkov 1952). In geese *Anser* and *Branta*, gap of 1–2 weeks between wing moult of sexes; according to Dementiev and Gladkov (1952) and Bauer and Glutz (1968), ♂ moults first, but in at least some species established that ♀ first (Hanson 1965; M A Ogilvie). Downy young simply patterned with varying shades of white, grey, olive-yellow, or brown.

Largely Holarctic. Most prefer cool or cold regions but largely stop short at ice or deep snow. Large aquatic and terrestrial herbivores; no more than marginally marine (with notable exception of Brent Goose *B. bernicla*) and avoid most deep or fast-flowing waters. Many attached to grasslands and other areas of low, non-woody vegetation in high to mid latitudes, from tundra to steppe, stopping short at deserts and mountains and most avoiding dense, tall vegetation. Vigilant and wary; when breeding, favour sites which are inaccessible—e.g. islands and cliff ledges— or eminences commanding wide views over open country. Strong fliers. In west Palearctic, most marked distributional changes in *C. olor*, where spread of feral populations in 20th century, especially in recent years; considerable decline of Whooper Swan *C. cygnus* in USSR during 19th century but recent spread in Fenno-Scandia and increase Iceland; contraction and splitting of range in Greylag Goose *A. anser* with some recent increases assisted by introductions; decline in *B. bernicla* during 1930s with marked recovery recently helped by protection; decline in Barnacle Geese *B. leucopsis* in USSR with probably some recent recovery; and spread and increase in north-west Europe of introduced Canada Goose *B. canadensis* assisted by further introductions. Majority of both swans and geese in northern hemisphere highly migratory (in west Palearctic, *C. olor* partial exception). Moult migration restricted to non-breeders (i.e. mainly immatures) as breeding ♂♂ remain with mates and families, moulting during breeding cycle. In *Cygnus*, non-breeders tend to unite near breeding areas; in *Anser* and *Branta*, move northwards—most European species to tundra and forest tundra. Normal migration often at high altitudes, day and night; traditional halting places used on way (see especially Hochbaum 1955).

Essentially vegetarian, feeding in shallow water and on land, mainly on grasses (including grain in some species) and aquatic and marsh plants. *Cygnus* mainly underwater grazers, neck-dipping and up-ending with frequent foot-paddling (*C. cygnus*); will also graze on land—in case of *C.*

*columbianus bewickii*, by digging with bill and sometimes feet. *Anser* and *Branta* mainly specialized terrestrial grazers while walking, also probing and digging —sometimes in soft mud; will also feed in water by up-ending, etc. With partial exception of *C. olor*, often highly gregarious at all times when not breeding, typically in flocks comprised of pairs and family parties. Pre-flight signals largely Chin-lifting (*C. olor*), Head-bobbing (northern *Cygnus*), lateral Head-shaking (*Anser*), or Head-tossing (some *Branta*); in most cases reinforced by vociferous calls. When breeding, often loosely colonial (at times with small territories) at protective sites, especially in *Anser* and *Branta*; *Cygnus* mostly with well-dispersed nests (in large territories), though *C. atratus* usually and *C. olor* sometimes colonial. Strong, strictly monogamous pair-bonds of long-term, non-seasonal duration. Also strong family bonds, both between parents and young and between siblings. No communal courtship. Most important display in formation and maintenance of pair-bond mutual Triumph Ceremony with characteristic calls (except in *C. olor* and allies); especially in geese (see *A. anser*), often initiated by ♂ after attack on rival. Also performed at times by members of same family group. Unlike in most Anatinae, little ritualization of comfort behaviour especially in heterosexual situations though some movements (e.g. Body-shake, Wing-flap) used in threat by some species. Copulation typically while swimming on water except in *Coscoroba* (in shallows) and *B. sandvicensis* (on land). Pre-copulatory display consists of mutual Head-dipping, ♀ eventually assuming Prone-posture. In mutual post-copulatory display, pair rise in water to greater or lesser extent in most species; posture and movements variable in *Cygnus*, similar in all *Anser* and *Branta*—terminated by bathing and wing-flapping. Elaborate nest-relief ceremony recorded only in *C. atratus*. Though varying degrees of reliance on visual displays, vocalizations generally play key rôle in most species for individual recognition and flock cohesion. Voice considerably reduced in *C. olor* and allies; used largely at close quarters and not, e.g., for territorial advertisement as in northern *Cygnus*—while far-carrying, non-vocal throbbing sound from wings replace flight-calls of others. Voice loud but relatively unspecialized in remainder; quite powerful, sonorous, and often musical in northern *Cygnus* (in which sometimes used in pair duet), and honking in *Coscoroba*, *Anser*, and *Branta*. Apparent greater noisiness of last two genera in part related to almost continuous vocal activity of larger, close-knit flocks outside breeding season; but vocabulary of calls also more extensive than in *Cygnus*, especially in *Anser*. Calls closely similar in both sexes, though sometimes differ in pitch. In addition to usual calls of most Anatidae, downy young also have distinctive Sleepy-calls, given when nestling down before sleeping—also at times while feeding (see Kear 1970). When threatened at close quarters, all species hiss freely. Comfort behaviour and other behaviour characters much as in other Anatidae, but bathing often spectacular with somersaulting and kick-diving.

Seasonal breeding in most with, in arctic species, highly synchronized laying periods. Nests on ground in open or in vegetation, usually near water but can be distant. Lined with down, though considerably less in *Cygnus* than in *Anser* and *Branta* (and most other Anatidae). Built by both sexes in *Cygnus* though ♀ does larger share. Eggs white, creamy-white, or pale green; smooth or with chalky covering. Clutches usually 4–7 (1–14) in *Cygnus*, 4–7 (2–11) in *Anser* and *Branta*; smaller in high-latitude forms (which do not lay replacements). Eggs laid at intervals of 1–2 days. Incubation by both sexes in *C. atratus*; in others, ♂ may cover eggs only during laying or when ♀ off feeding. ♂ often mounts guard at various distances from nest, especially in *Anser* and *Branta*. Incubation periods 29–36 days (swans), 24–30

(geese) (Kear 1970). Downy young tended by both parents, but brooded only by ♀. In *Cygnus*, adults indirectly provide some food in early stages (plucking underwater vegetation and foot-paddling), young taking it from surface; in *C. olor* and allies, small young also habitually carried on back. Brood aggressively defended from predators; deferment of wing moult by 1 of pair in most or all species (see above) ensures that one parent always able to protect young. Fledging periods relatively short in high arctic breeders, long in temperate species. Distraction-display by both sexes, in form of 'injury-feigning' or 'injury-flight' also recorded in some *Anser* and *Branta*, but apparently lacking in *Cygnus* (see Hebard 1960). Young stay with parents after fledging at least through 1st autumn, in most species through 1st winter, and, in some, spring migration; may reunite with them at end of one or more subsequent breeding seasons. Mature at 2–3 years.

## *Cygnus olor* Mute Swan

PLATE 51
[facing page 375]

Du. Knobbelzwaan   Fr. Cygne tuberculé (muet)   Ge. Höckerschwan
Ru. Лебедь-шипун   Sp. Cisne vulgar   Sw. Knölsvan

*Anas Olor* Gmelin, 1789

Monotypic

**Field characters.** 145–160 cm, of which body only half, ♂ averaging larger; wing-span 208–238 cm. All-white, large and heavy swan with rounded head, graceful curve to long neck, wings often arched, long, pointed tail, and orange bill with black knob and base. Sexes alike and no seasonal difference, except that ♂ usually has larger bill knob, further increased in size at onset of breeding. Juvenile different, but becomes more or less like adult after 1st year.

ADULT. Entirely white, sometimes stained (see Plumages). Bill red-orange outlined in black from tip along cutting edge, around base to swollen knob, and around nostrils; feet black, but grey-flesh in morph known as 'Polish Swan'. JUVENILE. Generally but unevenly dingy brown above and whitish below, with some white extending to cheeks, lower scapulars, and centre of mantle to rump, but 'Polish Swan' morph all-white from 1st down. Bill (lacking knob) grey gradually becoming suffused pink and then dull orange, feet grey becoming black. During 1st winter and spring, plumage grows increasingly white, though sometimes still partly brownish above until 2nd winter.

Except at long range, unlikely to be confused with other Palearctic swans, which have black and yellow bills, straighter necks, wings not arched, less erect tails, and comparatively silent wing-beats. Juvenile darker and browner above than those of the other two species without pink or livid white on bill. Looks more graceful on water, with curved neck and down-pointed bill, but on land moves more slowly and clumsily with waddling gait.

Seen mainly on or by water, but will graze in flooded meadows and even dry fields. Feeds by dipping head and neck below surface and also by up-ending like duck in shallow water, but full-grown bird seldom dives (though downy young will do so). Takes off from land or water rather awkwardly, with heavy flapping of wings and feet pattering along surface. Like other swans, flies with neck extended and slow, regular wing-beats, but characteristic musical throbbing noise produced by the flight-feathers peculiar to this species and audible for 1–2 km. Parties generally fly in diagonal lines. Markedly social in winter, often in concentrations of 100 or more, and non-breeders may gather in numbers throughout summer, but in many areas breeding pairs usually isolated. Largely silent at all times, but for snorting and hissing sounds. Aggressive towards other birds and, when breeding, towards man; aggressive posture characteristic with arched wings, neck drawn back, and jerky surging advance across water as both feet paddled simultaneously. Small young often carried on parent's back.

**Habitat.** Patchily distributed through mainly temperate middle latitudes in both continental and oceanic climates, enabling occupancy of habitat throughout year. Has switched extensively from natural to humanly influenced habitats and patterns of living; transition from unquestionably wild stocks in east Europe to feral, introduced, and domesticated stocks in west. In east, breeds typically in widely scattered pairs on large open lakes with

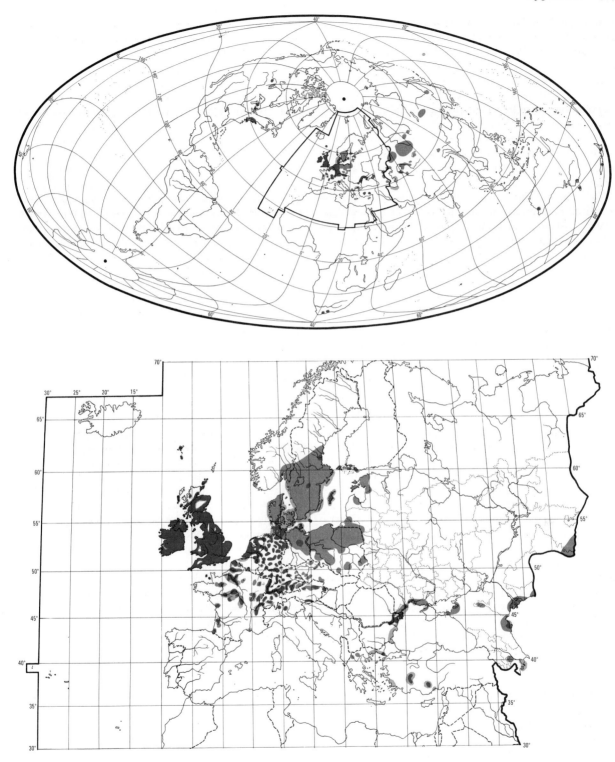

extensive shallows and much floating, bottom, and emergent vegetation; also in deltas of southern rivers (e.g. Volga). Sluggish rivers, brackish lagoons, saltwater inlets, and quiet coasts also frequented, notably by moulting concentrations and non-breeders. Seasonal shifts also involve floodlands and tidal waters. In contrast to east European stocks, which avoid proximity to man, west European birds largely conditioned to more or less close human presence, and frequently become dependent on artificial habitats and food. Uniformly deep, unvegetated, steep-margined, or oligotrophic standing waters little visited; prefers even regulated rivers, streams, canals, ditches, channels, ponds created by subsidence or for ornament, gravel-pits, reservoirs, and wetlands. Tolerant of pollution except by oil and toxic chemicals, benefitting from widespread eutrophication and disposal of waste products in fresh and coastal waters. Locally effective in controlling aquatic weeds, and takes readily to direct feeding. Although relatively sedentary in most parts of range, and even less mobile on land than other swans, makes frequent wide-ranging excursions on wing, usually at altitudes below 200 m and often below 50 m, suffering many accidents, e.g. through collisions with power-lines (Beer and Ogilvie 1972).

**Distribution.** Introduced in many countries of western and central Europe, often from 16th and 17th centuries, but major spread in 20th, especially recently, after declines between 1939–45. Not always possible to distinguish between wild and semi-feral stocks. Elsewhere, introduced into North America, South Africa, Australia and New Zealand. FAEROES: introduced 2 islands from c. 1940 (Williamson 1970). BRITAIN: long domesticated and used for food, especially in Middle Ages, but probably always some wild stocks (Ticehurst 1957). NETHERLANDS: spread since 1947 over whole country, mainly derived from feral stocks. DENMARK: almost extinct 1920s, since colonized all suitable areas. NORWAY: recently established, now breeding in several places in south (Haftorn 1971). SWEDEN. Formerly in 2 discrete areas in south and central Sweden; these have joined since 1950s (Mathiasson 1973a). Now colonizing western areas, but heavy mortality in severe winters militates against northward expansion (Curry-Lindahl et al. 1970). FINLAND: first bred 1934 after introduction Åland Islands; considerable expansion since 1950, probably assisted by influx from Sweden (LvH, Grenquist 1970). POLAND: bred only northern provinces c. 1900; since 1930 has spread to many lowland areas in south (Tomiałojć 1972). CZECHOSLOVAKIA: bred regularly since 1955 in small numbers (Hudec and Černý 1972). SWITZERLAND: first introduced 1690, but mainly from middle of 19th century (WT). ITALY: feral populations on Lakes Como and Garda (SF). YUGOSLAVIA: bred 1954, 1965 (VFV). BULGARIA: many nesting areas in Danube and on coast deserted due to drainage etc., but slight recent increase (Dontchev 1973). USSR: first bred Lithuania 1927, Latvia 1935, and Estonia 1959 (Mourashka and Valius 1961; Viksne 1968; Jogi 1968).

Accidental. Spain, Hungary, Iraq, Jordan, Egypt, Cyprus, Malta, Azores.

**Population.** Breeding. BRITAIN. 3500–4000 pairs with 11000 non-breeders in 1955–6; apparently then increasing fast, but partial census 1961 suggested return to 1955 levels

(Campbell 1960; Eltringham 1963). Decrease of 25% by 1965, due mainly to hard winters (Ogilvie 1967); no substantial change since (Ogilvie 1972). Counts made on Thames on annual swan-upping expeditions from 1823–1972 showed that for much of period total numbers fluctuated between 400 and 600, with marked increase to over 1100 in 1955–6 but decline to 399 by 1972 (Cramp 1972). IRELAND: estimated 5000–6000 birds (Ogilvie 1972). FRANCE: c. 50–100 pairs (Jouanin 1970); 200–300 pairs, increasing last 20 years (FR). NETHERLANDS: steady increase to c. 2500 pairs (Ogilvie 1972). DENMARK: reduced to 3–4 pairs 1925; increased following protection to 30–40 pairs in 1935, 385 pairs in 1950, 758 in 1954, and 2740 pairs in 1966 (Paludan and Fog 1956; Bloch 1971). NORWAY: 15–20 pairs (Haftorn 1971). SWEDEN. About 300 pairs in 1920, 2000–2500 pairs 1971 (SM). In 2 areas, recent decreases: Scania 268 pairs in 1967, 226 in 1972; Blekinge 182 in 1960, 141 in 1972 (Nilsson 1973). FINLAND: c. 20 pairs (Merikallio 1958). WEST GERMANY: 478 semi-wild pairs plus 2054 non-breeding birds in south (Hölzinger 1973); at least 1042 pairs in north, giving calculated total of 9480 birds, including non-breeders, for whole country (Scherner 1974). EAST GERMANY: increased from c. 350 pairs 1957–9 to nearly 1000 pairs (Schiemenz 1972). POLAND: 400 pairs in 1939, 1000 in 1956, and 3600 with 3000 non-breeders in 1958 (Sokołowski 1960); c. 7420 birds in 1965 (Tomiałojć 1972). SWITZERLAND: c. 245 pairs in 1962 (Szijj 1965a); 175 pairs in north-east in 1968 (Forster and Wagner 1973). GREECE: 5–8 pairs Evros delta; 40–50 pairs introduced 1967 on Lakes Agras and Kastorias (WB). USSR. In 1974, total population estimated c. 4600 pairs, with 29300 non-breeding birds. Main breeding populations around Caspian Sea (3000 pairs), Baltic Sea (500 pairs), Kazakhstan (500 pairs) and Black Sea (300 pairs), with recent increases in most areas (G A Krivonosov). See also Nedzinkas (1975), Viksne (1968), and Kumari (1974) for increases in Baltic.

Winter. Estimated 120000 birds north-west Europe and 10000 Europe–Black Sea–Mediterranean region (Szijj 1972). USSR: 5200 birds in western part in not so severe winters (Isakov 1970c).

Survival. Britain: annual mortality 3–12 months 32·1%, 1–2 years 35·4%, 2–3 years and 3–4 years 25·0%, breeders over 4 years 18–20%; life expectancy at 4 years 4·8 years (Beer and Ogilvie 1972). Denmark: annual mortality 26% (Bloch 1971). Sweden: annual mortality 1965–70, 28·5% but only 21·0% at most if severe winter of 1969–70 excluded (Mathiasson 1973a). Oldest ringed birds: semi-feral, 19 years 5 months (*The Ring* 1965, **42**, 104); apparently wild, 15 years 4 months (BTO).

**Movements.** Wholly migratory in some parts, mainly sedentary in others; also partial migrant. In Britain, Ireland, Low Countries, and France, where populations feral, little or no migration or even much local movement; 94% of British-ringed birds moved less than 50 km, most

PLATE 49. *Threskiornis aethiopicus* Sacred Ibis (p. 347): **1** ad, **2** juv, **3** nestling. (RG)

PLATE 50. *Platalea leucorodia* Spoonbill (p. 352): **1** ad breeding, **2** ad non-breeding, **3** 1st imm winter, **4** nestling. (RG)

Peter Scott.

following watercourses (Ogilvie 1967). Some breeding birds remain on territory all year, others move short distances to and from winter flocks; young move to latter in late autumn and winter, and pre-breeders often remain in flocks all summer (Minton 1971). Small moult movements in Britain suspected from ringing: Midlands to Lancashire coast (Minton 1971), Durham and Northumberland to Loch Leven, Kinross (Ogilvie 1972).

In north Europe (Scandinavia, north Germany to Estonia), largely migratory. Major moult migration by non-breeders to Öresund area; formerly many to Danish islands but now mainly Swedish side (Malmöhus) with smaller numbers north to Kungsbacka Fjord in Halland; up to 15 000 present July–September, probably accounting for most pre-breeders of Denmark and Baltic countries (Mathiasson 1963a, 1964, 1972). In post-moult dispersal, joined by breeders which leave freezing lakes; winter west Baltic coasts of Danish archipelago, south Sweden and north Germany (especially Mecklenburg). In colder conditions, many move west into Denmark and Schleswig-Holstein, and in severe weather some reach areas where northern birds normally only stragglers, including Netherlands, Belgium, north France and south-east England (Harrison and Ogilvie 1967; Hilprecht 1970), and occasionally overland to Switzerland where single recoveries of Polish and Lithuanian birds. Nearly 4000 Bodensee in early 1963 must have included many immigrants. In hard winters, other Lithuanian birds moved nearly 1700 km to Solway Firth (Scotland), Belgium, north France, and Adriatic coast, Italy (Hilprecht 1970), so some cold winter stragglers to Mediterranean evidently from Baltic. Return to breeding lakes dependent on thaw, though apparently some Polish breeders return from January while lakes still frozen (Bauer and Glutz 1968).

Small, rather isolated populations of south-central Europe show some winter movement, e.g. south Germany birds to Switzerland, and Swiss birds from small waters to large. In south USSR (Ukraine to Kazakhstan), migratory populations leave inland breeding areas to winter on coasts of Black and Caspian Seas; some pass through Dardanelles to Greece and Turkey, where winter recoveries of Ukraine birds (Hilprecht 1970), and others occasionally reach Egypt, Persian Gulf, Afghanistan, and north India. Many immatures said to remain on Black and Caspian Seas until old enough to take up inland territories (Dementiev and Gladkov 1952).

◄ PLATE 51. (*facing*) *Cygnus olor* Mute Swan (p. 372): 1 ad ♂, 2 ad ♀, 3 1st winter 4–9 months, 4 juv up to 4 months, 5 downy young, 6 ad ('Polish' morph), 7 juv ('Polish' morph), 8 downy young ('Polish' morph). *Cygnus cygnus* Whooper Swan (p. 385): 9 ad, 10 ad showing characteristic staining, 11 2nd autumn, 12 juv at 7 months, 13 downy young. *Cygnus columbianus bewickii* Bewick's Swan (p. 379): 14 ad, 15 ad showing staining, 16 2nd autumn, 17 juv at 7 months, 18 downy young. (PS)

**Food.** Based mainly on information supplied by M Owen. Mainly aquatic vegetation obtained from water up to 1 m deep by full up-ending or by immersing head and neck only (dipping); also emergent plants and seeds by grazing and dabbling at water's edge, and grasses and herbs by grazing on land. Dipping occurs in depths 20–45 cm, average submersion under 10 s; up-ending when over 45 cm, with average submersion 13 s (Dewar 1942). During winter on Ouse Washes, England, when undisturbed, feeding activity increases during day, with plateau some 3 hours after sunrise, and continues a short while after dark (Owen and Cadbury 1975). Small animals also occasionally but regularly taken, including frogs, toads, tadpoles, molluscs, worms, insects and larvae (Witherby *et al.* 1939).

45 stomachs from Sweden, spring and summer, contained over 95% submerged vegetation; mainly stonewort *Chara* (26 stomachs), some green algae *Enteromorpha* (11) and *Cladophora* (9), and, occasionally, seeds of naiad *Najas marina* and sea clubrush *Scirpus maritimus* (5); also, in March and April especially, wigeon grass *Ruppia maritima* (11) and brown alga *Pylaiella rupurcola* (9). Animal items (23), in small quantities and probably incidental, included crustaceans, insects, molluscs (Berglund 1963). Other submerged plants, particularly milfoil *Myriophyllum*, green alga *Vaucheria*, and *Potamogeton*, important in other Swedish areas as determined by faecal analysis (Olsson 1963). On Swedish west coast, nearly exclusively eelgrass *Zostera marina* during May–August; in September-October also fed, sometimes exclusively, on green alga *Ulva lactuca* (Mathiasson 1973b). In parts of England in spring, grazes on saltmarsh turf, on arrow-grass *Triglochin maritima*, plantain *Plantago*, sea-aster *Aster*, milkwort *Glaux*, spurrey *Spergularis*, and saltmarsh grass *Puccinellia maritima*; during late summer and autumn, feeds on *Zostera* and *Enteromorpha* (Gillham 1956). In autumn, will also readily take leaves (plus galls) of waterside willow *Salix fragilis*, mainly after these have fallen but also directly from accessible branches (B King).

In Denmark, of 67 stomachs in winter, 39 contained *Ruppia spiralis*, 16 *Potamogeton pectinatus*, 14 *Chara*, 11 *Zostera*, and 7 horned pondweed *Zannichellia pedunculata*; seeds and invertebrates present in small quantities (Spärck 1957). Based on 8 stomachs and observations from winter on Ouse Washes, preference shown for soft grasses; marsh foxtail *Alopecurus geniculatus*, creeping bent *Agrostis stolonifera*, and floating sweet grass *Glyceria fluitans*. Starchy roots of marsh yellow-cress *Rorippa palustre* prominent in diet in 2 winters. In contrast to Bewick's Swan *C. columbianus bewickii*, larger proportion of aquatic plants, such as starwort *Callitriche*, hornwort *Ceratophyllum demersum*, and water crowfoot *Ranunculus*, taken (Owen and Cadbury 1975).

Regularly uses agricultural pastures for feeding, and occasionally cereals (Harle 1951). Takes advantage of any readily available foods such as waste grain from mills (Atkinson-Willes 1963) or fish refuse (Boase 1965). Also concentrates round rubbish barges on River Thames, England (Cramp 1957). Among wide range of items rarely taken; whole fish (Watson 1931; Hulme 1948), *Prunus* blossoms (King 1960), and raw meat (King 1962). Many fed in parks, etc., on bread and other titbits.

Cygnets eat aquatic vegetation torn off by parents (Dementiev and Gladkov 1952); probably also insects and aquatic invertebrates.

Daily food requirement of moulting adults between 3·6 and 4 kg of wet vegetable food (Mathiasson 1973*b*).

**Social pattern and behaviour.** Based largely on material supplied by J Kear; see references in Kear (1972).

1. Pairs territorial during breeding season or longer; non-breeders and pre-breeders mostly gregarious throughout year. If nesting unsuccessful, pairs often leave territory in June, before wing moult, and join hundreds, sometimes thousands, of other unemployed birds on traditional waters for flightless period. Later some parents and young leave territory and continue to associate in large feeding and roosting flocks on open water having adequate food, or on rivers in centre of towns. A few pairs remain on territory and defend it throughout winter. Differences in degree of gregarious behaviour outside breeding season presumably due in part to climate, since in freezing conditions most move to open water. BONDS. Long-term monogamy normal, ♂ and ♀ remaining paired outside breeding season. Over 6 years, divorce rate in one English population 3% of pairs that had already bred and 9% among pairs that had failed or not yet begun to breed (Minton 1968). Pair-formation slow, starting among grey-brown juveniles which engage in courtship greeting in winter herds. Incidence of courtship increases into 2nd summer when birds capable of breeding, but majority form pairs and go through engagement period of at least one season before laying; characteristic age of first breeding 3–4 years, with ♀♀ slightly more precocious than ♂♂ (Minton 1968). Both sexes tend young. Family party remains intact usually until well after fledging of young; in some cases, young driven from territory in autumn, in others, parents and juveniles continue to associate until January or February, occasionally into next summer (S Cramp), either in territory or winter flock. In regions where truly migratory, family departs as unit. Siblings continue to associate during 1st year and into 2nd winter; pairing between siblings, however, apparently not usual. BREEDING DISPERSION. Normally pairs territorial from about February to October in Britain, defending area size of which depends on local features of habitat. Average distance between nests on rivers near Oxford (England) 2·4–3·2 km, with 90 m minimum (Perrins and Reynolds 1967). On Alster (Hamburg), a shallow weedy lake, average territory 300 × 150 m (Hilprecht 1970). All maintenance and breeding activities, except pair-formation, confined to territory, and family usually remain there until cygnets fledge. Lack of territory can prevent nesting in otherwise mature pair, and, even when territory gained, sometimes held initially only during spring. Minority of pairs breed colonially, defending only nest-site and feeding gregariously (although as family parties). e.g. at Abbotsbury, England, and in Poland and Denmark, where in some cases nests less than 2 m apart. Danish colonies, up to 100 pairs, usually on coastal islands where families (as at Abbotsbury) spread out after hatching to feed on intertidal vegetation (Kear 1972). ROOSTING. In winter, and non-breeders in summer, on water, usually fresh; family parties stay together. Arrive before dark, leave just before sunrise. Family often sleeps on water with cygnets on ♀'s back when small or, if nest-site close to water, may return there.

2. Adults, especially ♂♂, highly aggressive. Threat behaviour sometimes occurs at nest-site but more especially during establishment and maintenance of territory; defence of latter evoked by other adults, but not usually by grey-brown juveniles. Conspicuous white plumage of adult has been termed 'continuous advertisement'; may play important rôle in social situations. Perhaps because of this, other forms of recognition signalling absent and birds tend to be silent except at close quarters (Kear 1972). FLOCK BEHAVIOUR. In winter flocks, threat behaviour in both sexes often related to maintenance of individual-distance; in general, however, herds seem less well integrated than those of other Palearctic swans. Typical *Cygnus* Head-bobbing pre-flight movements, for instance, appear absent, even in family groups, though Chin-lifting with some Head-flicks and occasional lateral Head-shakes recorded (McKinney 1965*b*). ANTAGONISTIC BEHAVIOUR. Main threat-display, of both sexes, well-known Busking (or 'Swanning') in which secondary feathers raised over back, neck fluffed, and head laid back with neck well arched (see A); often accompanied by characteristic jerky swimming through water, using both feet simultaneously. Wing-flapping also used as threat-display. In territorial disputes, 2 ♂♂ will parade boundary between territories while Busking. One bird will sometimes attack and other withdraw, but if aggressor goes too far beyond own territory likely to be counter-attacked and forced to retreat to boundary; snorts can be given as birds chase one another. Rarely, fighting develops which territory owner normally wins; such fighting involves both beak and wrists of wings, and nearly 3% of ringed birds found dead thought to have been killed in territorial disputes (Ogilvie 1967). Nests and young also defended with beak and wings. Some pecking seen between birds in winter herds, but serious fights infrequent and usually one bird turns and flees. Submission indicated by sleeked neck feathers, wings held flat against back, gently curving neck, and bill-tip pointing down; bird beaten in fight will extend neck forward along water or ground and remain immobile, victor soon ceasing to peck (K E L Simmons). HETEROSEXUAL BEHAVIOUR. Chief displays between pair include courtship greeting and Triumph Ceremony but copulatory behaviour also important in maintaining pair-bond, most pairs mating more often and over longer period of late winter and spring than necessary to fertilize clutch. Courtship during pair-formation consists largely of mutual Head-turning (see B) as birds face each other with breasts almost touching, upper neck feathers fluffed, and secondaries sometimes raised. Same behaviour also between mates as greeting. Triumph Ceremony between pair occurs when rival has been repulsed; similar to Threat-busking, but posture assumed by both and accompanied by calling (a kind of snoring) and by mutual Chin-lifting. So-called nest-relief ceremonies, containing elements of Triumph Ceremony, described but probably not common since

A

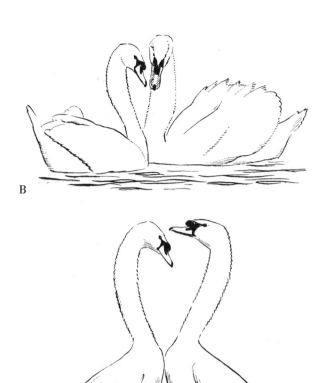

B

C

cygnets. Greeting within family usually simple Chin-lifting with erect neck; later, in winter flocks, Triumph Ceremony can involve both juveniles and parents.

(Figs A–C after illustrations in Scott *et al.* 1972.)

**Voice.** Not nearly so far carrying or persistent as in other west Palearctic *Cygnus*; used relatively infrequently in specific situations, only at fairly close quarters and especially during the breeding season. Calls often hoarse, explosively snorting, snoring, or grunting, usually with little of crooning, trumpeting, or honking quality of more vocal *Cygnus*. Hisses when cornered (see I) much more readily than others. Though will utter other calls when flying (see Witherby *et al.* 1939 and below), true contact Flight-call lacking; wing-beats produce loud, penetrating, rhythmic singing sound resembling 'vaou-vaou-vaou . . .' (see II) audible 1–2 km which may help birds keep in touch (J Kear). Take-off accompanied by loud pattering of feet on water.

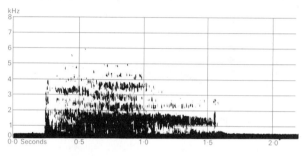

I V Maxse  Ireland June 1965

II Sture Palmér/Sveriges Radio  Sweden April 1959

normally only ♀ incubates. In pre-copulatory display, both birds alternately dip head beneath water, preen or rub back or flanks, and occasionally up-end; gradually activities become synchronized and, between Head-dips, necks held upright and close together for moment or two. Wings held low, often dragging on water, and ♂ pushes neck and body over ♀'s until mounted on her back, grasping her neck feathers in bill. Post-copulatory behaviour also stereotyped: ♂ slips off, both birds often utter hoarse, muted calls, and rise half out of water breast to breast with necks extended (see C) and bills pointing first up, then down, and finally from side to side; both then usually bathe, preen, and tail-wag (Johnsgard 1965). RELATIONS WITHIN FAMILY GROUP. ♀ often, and ♂ occasionally, carries cygnets on back during first 10 days and intermittently for 6 weeks. Cygnets, unlike those of other Palearctic swans, show strong urge to climb; they board parent between tail and tips of short, folded wings, but not actively helped, though adult stays still and may form step with 'heel' of foot. Parents pull up underwater vegetation, and material from shore and overhanging branches, and drop round young—using sideways-passing action typical of nest-building (Kear 1972). Foot-paddling by parents serves to raise food particles for downy young, movement itself attracting them to adult's side. Distress-calls heard when cygnet isolated, cold, or hungry; deserted cygnet holds itself erect, neck stretched upwards, down on head erected, mouth open, uttering 1 cry to each expiration and making itself as conspicuous and noisy as possible (Kear 1968). For other calls, see Voice. In winter flock, family and area round defended by all members, especially ♂. Parental hostility checked partly by grey-brown plumage of young—as suggested by attacks recorded on white ('Polish')

CALLS OF ADULTS. Little detailed information. Following distinguished by J Kear: (1) snoring calls by both sexes during Triumph Ceremony and when greeting at nest; (2) hoarse, muted trumpeting by both sexes at times during post-copulatory ceremony; (3) threat snorts by ♂, especially when chasing rival; (4) quiet, breathy bubbling sounds by ♀ to young, even before hatching. Display snores probably fairly variable: described as 'kgiyurr' (♂), 'kiorr' (♀) by Dementiev and Gladkov (1952) and when uttered by several birds, e.g. in Triumph Ceremony, produces group chorus of grunts, gurgles, or growls (as well as snores) combined with crude whistling in tempo somewhat resembling Wigeon *Anas penelope* (Sture

Palmér recording). Other calls described include following by Söderberg (1950); (5) loud, resonant cry like Crane *Grus grus* (reported in east and north of range but not from feral or captive birds in west); (6) explosive, snorting 'o-hi-ang' or 'arung-goa' (as Alarm-call); (7) slow 'glock glock' (soliciting call by ♀). Also (6) piping 'wia', sometimes becoming gurgling 'wiachrr' (Lost-call); begins with high-pitched note, followed by deep sighing, and ends with whinnying sound.

CALLS OF YOUNG. Cygnets have 4 main calls (J Kear; see also Kear 1968). (1) Contact-call (Pleasure-call): soft sounds, uttered in series of 2–5 notes, given when warm and wide awake, feeding or preening together or with parents; softer and more goose-like than in Whooper Swan *C. cygnus*. (2) Greeting-call: similar to last, but louder and with rising and falling cadence. (3) Sleepy-call: low, soft trills, like quiet, lazy sighing, uttered at long intervals when tired or resting. (4) Distress-call (Lost-call): regular and high-pitched, noisy peeping, given more slowly and over much longer period than Contact-calls. Also (5) defensive, adult-like hiss by 10th day.

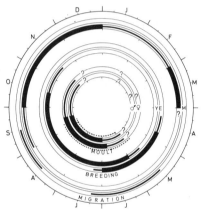

**Breeding**. Main reference: Kear (1972). SEASON. See diagram for north-west Europe. In England, early eggs in mid-March. SITE. Near water, within 100 m, generally less; on bank, island, in reedbed or other aquatic vegetation, sometimes on shore of sheltered sea-inlet. Normally far apart but colonial in some areas, e.g. Denmark, with closest nests 2 m apart (Bloch 1970). Nests may be re-used in subsequent years after rebuilding. Nest: large mound of rushes, reeds, and other vegetation; 1–2 m diameter at base but up to 4 m when standing in water; height 60–80 cm, shallow depression in top 5–15 cm deep. Lined softer vegetation and a little down. Building: by both sexes; ♂ sometimes selects site and lays foundations. Typically ♂ stands near nest and passes material over shoulder, dropping it for ♀ on nest who produces cup by pressing with breast, bill, and feet. Build any time of day. Completion time varies; average 10 days. ♀ adds small quantities of material within reach during incubation. EGGS. Rounded at both ends; pale green with chalky covering when fresh, becoming scratched and stained

brown or yellow during incubation. 113 × 74 mm (100–22 × 70–80), sample 88 (Schönwetter 1967). Weight 345 g (294–385), sample 80 (Valius 1959). Clutch: 5–8 (1–11). Of 102 English non-colonial nests: 1 egg, 1%; 2, 8%; 3, 8%; 4, 10%; 5, 12%; 6, 19%; 7, 19%; 8, 14%; 9, 7%; 10, 2%; 11, 1%; mean 6·0 (Perrins and Reynolds 1967). Of 181 Danish colonial nests: 1, 7%; 2, 8%; 3, 8%; 4, 14%; 5, 24%; 6, 18%; 7, 11%; 8, 8%; 9, 2%; 10, 1%; mean 4·9 (Jensen 1967). One brood. Replacement clutches 2–4 weeks after loss of eggs, and always smaller; not recorded after loss of young. Eggs laid at 48-hour intervals, normally early morning. INCUBATION. 36 days (35–41). Only ♀ incubates, but ♂ will sit during laying, and while ♀ feeding; ♂ incubated successfully after death of ♀ (Stevenson 1966). Eggs covered with nest material and down when ♀ off. Begins with last egg; hatching synchronous, within 24–36 hours. Eggshells left in nest. YOUNG. Precocial and nidifugous. Cared for by both parents. Often ride on parents' back, more usually on ♀, while small; ♀ broods young at night when small, on nest or on ground. Both sexes, especially ♂, aggressively defend nest and young against predators and humans. Self-feeding, but with some indirect provision by parents which pull vegetation from bank and beneath water surface for small young, which also take insects so disturbed. FLEDGING TO MATURITY. Fledging period 120–150 days. Young of some pairs driven off breeding territory as soon as plumage predominantly white (during late autumn or winter); other broods accompany parents to wintering area, usually joining large flock in which they remain when parents leave for breeding territory. Exceptionally, young may remain with parents on territory or in flock throughout next summer if latter do not breed. First pairing at 2 years, rarely 1; first breeding of ♂ commonest at 4 years, sometimes 3, rarely 2; first breeding of ♀ at 3 years, occasionally 2. BREEDING SUCCESS. England (compiled from Reynolds 1965; Minton 1968; Beer and Ogilvie 1972): 42% eggs laid fail to hatch; 9% mortality first week after hatching; 22% mortality 2nd week; 12% in weeks 3 and 4; 8% in weeks 5 and 6; 2–4% mortality in each subsequent 2 weeks equalling 50% in first 3 months after hatching. Vandalism an important cause of egg loss in some areas. Starvation most important cause of young mortality in first 2 weeks after hatching.

**Plumages**. ADULT. Entirely white (sometimes stained rusty-red on crown, and to lesser extent on rest of head, neck, and belly, by ferrous content of water). DOWNY YOUNG. Upperparts pale grey, underparts white; in domesticated and semi-domesticated stock, all-white morph ('Polish Swan', *Cygnus immutabilis* Yarrell, 1838) commonly occurs. JUVENILE. Head and neck grey-brown, paler on chin and foreneck. Feathers of rest of upperparts, sides, and flanks tipped brown, white bases giving mottled appearance on lower mantle, back, and rump. Underparts pale grey. Scapulars, tail, flight-feathers, and upper wing-coverts pale grey, suffused brown on tips and on outer webs. Underwing and axillaries white. FIRST WINTER AND SUMMER. New feathers attained in autumn on crown and nape dark brown, narrowly tipped white; later on, wholly white. Rest of head and foreneck white. Juvenile feathers at base of upper mandible not replaced; wear off during winter. From autumn to spring, white feathers develop on mantle, scapulars, and flanks, and brown edges of most remaining feathers wear off, producing off-white plumage. Juvenile flight-feathers and most of wing-coverts and tail retained. SECOND WINTER. At least until January, some 2nd year birds recognized by brown spots on shafts near tips of primaries or brown-tipped feathers or shaft-streaks on head, rump, scapulars, or outer median coverts.

**Bare parts.** ADULT. Iris hazel. Bill orange-red in spring, pink rest of year. Knob on upper mandible, bare triangle in front of eye, skin round nostril, nail of bill, cutting edges, and base of lower mandible black. Tip of lower mandible flesh-coloured. Knob of adult ♂ enlarged in spring, smaller outside breeding season and in ♀; mostly feathered in young. Foot black, but pink, yellow-grey, or pale grey in some semi-domesticated birds and all swans of '*immutabilis*' type. DOWNY YOUNG. Bill black, nail pale horn; skin in front of eye feathered. Foot dark grey. JUVENILE. Like adult, but bill dark grey, becoming pink during 1st winter.

**Moults.** ADULT. POST-BREEDING. Complete; flight-feathers simultaneous. Flightless *c*. 6–8 weeks. Breeding pair moult at different times: ♀ when nestlings small, ♂ when ♀ almost able to fly again (Heinroth and Heinroth 1928; Dementiev and Gladkov 1952). After completion of wing, moult continues with tail and tail-coverts, upperparts, underparts, neck, and head (Hilprecht 1970). POST-JUVENILE. Starts shortly after fledging; feathers of nape and crown first, followed by rest of neck and head, upper mantle, shorter scapulars, and some feathers of sides of body in autumn. Most of mantle, scapulars, sides of body, breast, and some central tail-feathers with their coverts renewed when 1st complete summer moult starts in June of 2nd calendar year, with sequence like adult.

**Measurements.** Netherlands, breeding. Skins (RMNH, ZMA). Bill measured to base of knob over nostrils. ADULT.

| | | | | | | | |
|---|---|---|---|---|---|---|---|
| WING | ♂ 606 | (12·1; 12) | 580–623 | ♀ 562 | (16·3; 10) | 533–589 |
| TAIL | 224 | (12·3; 6) | 205–246 | 211 | (12·1; 10) | 190–232 |
| BILL | 80·6 | (3·88; 12) | 74–88 | 74·2 | (2·89; 13) | 69–79 |
| TARSUS | 114 | (3·49; 12) | 107–118 | 104 | (4·69; 10) | 99–114 |
| TOE | 158 | (5·15; 11) | 147–166 | 142 | (5·00; 10) | 134–150 |

Sex differences significant. JUVENILE. Like adult, but wing and tail shorter, e.g.:

| | | | | | | |
|---|---|---|---|---|---|---|
| WING | ♂ 582 | (18·2; 4) | 552–598 | ♀ 556 | (13·1; 7) | 540–572 |

**Weights.** All in kg. Upper Thames, England (Reynolds 1972). (1) 1st winter (September to March or April), (2) 1st summer, (3) adult (over 3 years) winter. (4) adult summer.

| | | | | | |
|---|---|---|---|---|---|
| (1) | ♂ | 9·7(0·73; 159) | 8·1–12·1 | ♀ | 7·8(0·79; 221) | 5·5–9·5 |
| (2) | | 10·9(0·98; 42) | 9·3–13·5 | | 8·3(0·72; 36) | 6·4–9·7 |
| (3) | | 11·8(0·89; 59) | 9·2–14·3 | | 9·7(0·64; 35) | 7·6–10·6 |
| (4) | | 11·9(0·83; 21) | 10·6–13·5 | | 9·6( – ; 6) | 8·3–10·8 |

Adult weight constant all year round, except ♀ in early April prior to laying, when heavier: 10·6 (8) 9·7–11·7. Weight depends on local availability of food. Weights elsewhere may be much heavier, e.g. north-east Poland adults in autumn up to 22·5; juveniles up to 20 (von Sanden 1935).

**Structure.** Wing long, pointed. 11 primaries: p9 and p8 longest, about equal; p10 14–25 shorter, p7 4–14, p6 50–65, p1 210–240; p11 reduced. Outer web of p9–7 and inner p10–8 emarginated. Tail rather long, wedge-shaped; 20–24 feathers. Bill fairly long, with conspicuous bare knob at base of culmen in adult (merely indicated in juvenile). Outer toe *c*. 97%, inner *c*. 80%, hind *c*. 24% of middle toe. Other structure as in Whooper Swan *C. cygnus*, but lacks concave depression near base of culmen, and nail of bill narrow. CSR

## *Cygnus columbianus* Bewick's Swan and Whistling Swan

PLATE 51
[facing page 375]

DU. Kleine Zwaan    FR. Cygne de Bewick    GE. Zwergschwan
RU. Малый лебедь    SP. Cisne chico    SW. Mindre sångsvan

*Anas Columbianus* Ord, 1815

Polytypic. *C. c. bewickii* Yarrell, 1830 (Bewick's Swan), tundra north-east Europe and Siberia. Nominate *columbianus* (Ord, 1815) (Whistling Swan), extralimital, tundra North America.

**Field characters.** Bewick's Swan *C. c. bewickii*. 115–127 cm, of which body only half, ♂ averaging larger; wing-span 180–211 cm. All-white and medium-sized swan (i.e. smaller than other 2 Palearctic species) with proportionately shorter, straight neck, rounded head, and black and yellow bill. Sexes alike and no season differences. Juvenile distinctive, but becomes more like adult after 1st year. ADULT. Entirely white, though head, neck, and underparts may be stained rusty. Bill black with yellow base which comes to rounded, jagged, or squared end well behind nostril; feet black. Shape and extent of yellow on bill varies so much that probably all may be recognized individually (see Scott 1966). JUVENILE. Generally still greyer and paler above than juvenile Whooper Swan *C. cygnus*. Bill again highly variable, generally having dark tip and dirty pink to reddish-pink base sometimes extending in wedge to below or beyond nostril as in *C. cygnus*. During 1st spring and summer, black extends progressively over forward pink areas and base becomes livid white, then yellow, though amount of pink remaining varies greatly and culmen may be black, yellow-white, or any combination of these colours. Feet dark grey, becoming blacker. Also during 1st winter and spring, plumage grows increasingly pale and many white feathers acquired on upperparts.

Most likely to be confused with *C. cygnus*, particularly in flight when latter's larger size and distinctively shaped yellow bill-patch not easy to judge, but has more goose-like flight, proportions, and shape with rather shorter-looking neck, stockier body and, above all, more rounded head,

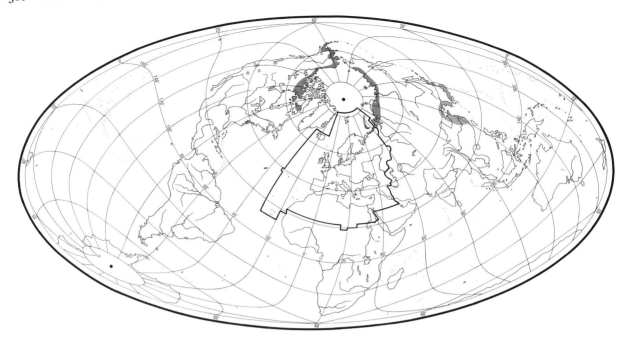

these distinctions also applying to juveniles. Like *C. cygnus*, typically carries neck erect, both on water and on land, but also often curved when feeding or sitting around. Walks easily.

Feeds by day or night, both on water and by grazing. Flight powerful like that of other swans, but wing-beats slightly faster than those of larger species. Parties often fly in no definite formation, but sometimes in V or diagonal lines. Markedly gregarious in winter, in even larger herds than *C. cygnus*, sometimes of several hundreds or even thousands. Noisier than *C. cygnus* on water or ground, producing musical babble of shorter, less trumpeting notes, while flight-calls also softer and lower pitched. Typically less wary of man than *C. cygnus*.

**Habitat.** In arctic breeding quarters, Bewick's Swan *C. c. bewickii* aptly known also to Russians as 'Tundrovi lebed'—tundra swan. During stay, limited by ice-free period within *c.* 130 days, lives mainly on low, grassy, swampy, and fairly level tundra, with many pools and lakes; sometimes also along broad, slow-flowing rivers and backwaters or on islands and coasts. Less commonly within treeline; avoids mountain regions. In breeding season, freedom from disturbance by man apparently essential, and humanly modified habitats not frequented. Normally widely dispersed over breeding grounds, but in flocks at other times.

On migration to winter quarters, pauses briefly on suitable lakes, pools, and rivers; sometimes also on intermediate tundra or on shallow saltwater lagoons and coastal waters. In west European winter quarters strongly attached to regular inland localities, not necessarily used for whole season, but can drastically change routes, staging points, and areas of seasonal residence over period of decade or so. Prefers lowlands below 100 m, free from persecution, using lakes, reservoirs, pools, and rivers accompanied by suitable grazing areas, especially flooded grasslands. Occasionally visits arable fields with spring wheat and has become accustomed to feeding on provided grain in shallow water and close to humans (Slimbridge and Welney, England). Short-distance movements normally within *c.* 100 m of ground, but much higher on migration. Some evidence of growing response to conservation measures; relatively small population not readily explicable in terms of habitat limitations. See Ogilvie (1972).

**Distribution.** Bewick's Swan *C. c. bewickii*. Breeding range. No information on changes.

Winter. Formerly reported wintering commonly Caspian and Aral Seas and some lakes to east (e.g. Dementiev and Gladkov 1952; Delacour 1954), but recent surveys have failed to record any in USSR (Isakov 1970*b*), while only small numbers normally in Iran (D A Scott). SCOTLAND: in 19th and early 20th centuries common, especially Tiree and South Uist, but by 1930s scarce visitor only. ENGLAND: began wintering only in late 1930s. WEST GERMANY: much reduced due to drainage in last 30 years.

Accidental. Spitsbergen, Bear Island, Poland, Czechoslovakia, Hungary, Austria, Switzerland, Yugoslavia, Greece, Israel, Iraq, Spain, Italy, Libya, Algeria.

POPULATION. Bewick's Swan *C. c. bewickii*. No breeding counts.

Winter. North-west Europe. Now 6000–7000 birds of

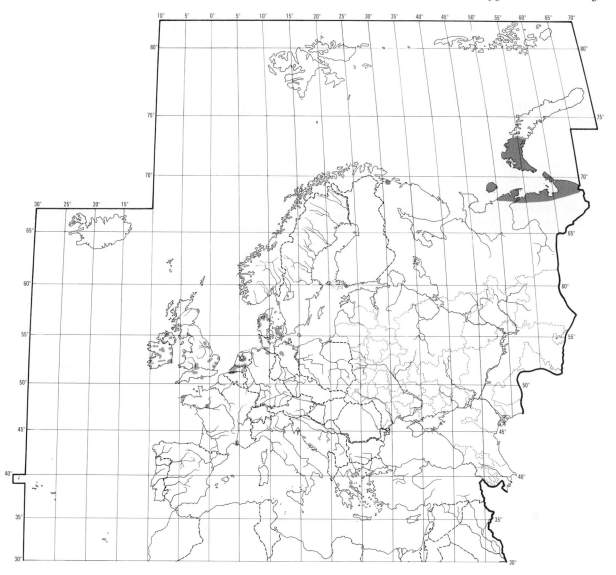

which 2500–3000 Netherlands, 2000–2500 Britain and Ireland, 700 Denmark, and 300 West Germany. Total fluctuates according to breeding success, while distribution dependent to some extent on local weather (for January counts 1967–73, see Atkinson-Willes 1975). England: great increase in last 30 years; following influxes in 1938–9 and 1955–6, now established in several localities and increasing, from a few hundreds in late 1950s to over 1500 by 1970 (Nisbet 1959; Ogilvie 1969b, 1972). Ireland: some decrease; probably 1500 in late 1950s and only 700–1000 now, due to drainage and possibly to stopping short in England (Ogilvie 1972).

Survival. Birds over 3 years old: mean mortality 15%, life expectancy 6·2 years (Beer and Ogilvie 1972). Oldest ringed bird 10 years 11 months (Wildfowl Trust).

**Movements.** Bewick's Swans *C. c. bewickii* wintering west Europe from single wholly migratory population breeding east to Taymyr. Movements begin early September from tundra breeding grounds, which abandoned by mid-October. Narrow-front migration route along arctic coasts Russia to White Sea, overland via Lake Ladoga to Gulf of Finland, thence over north Estonia, Baltic, Gotland and extreme south Sweden to winter quarters in Denmark and north-west Germany (few), Netherlands, Britain, and Ireland, with some to Norway, Belgium, and France (including small numbers Camargue since mid 1960s). Passage through Baltic early October to mid-November. Some reach winter quarters mid-October, but arrivals continue for many more weeks as birds pushed further west by approach of cold weather. Peak numbers

Netherlands in December, exceptionally November, but Britain and Ireland in January. Movements between winter haunts Britain and Ireland occur in either direction throughout winter, influenced by food and weather; similarly between Netherlands and Britain.

Spring return begins early February in mild winters, but normally early March, with bulk leaving Britain and Ireland by mid-March and Netherlands by end March. Return passage through Baltic, but passes further north than in autumn with more recorded central Sweden and south Finland; movements there until third week May. Early record Leningrad first half April, and on breeding grounds mid-May, but majority of arrivals there in latter half May continuing into early June. Like Whooper Swan *C. cygnus*, small parties may occur outside main wintering areas, especially in cold weather, e.g. France and central Europe, with rare stragglers further south. Isolated regular wintering area Iran (south Caspian and Lake Rezaiyeh); usually under 100 though 843 in severe winter 1968–9 (D A Scott); routes used unknown. (See also Bauer and Glutz 1968; Dementiev and Gladkov 1952; Högström and Wiss 1968; Hilprecht 1970.)

**Food.** Based mainly on information supplied by M Owen. *C. c. bewickii*. Almost entirely leaves, shoots, roots, rhizomes, and tubers, obtained by up-ending—for 20–30 s (Acland 1923)—in water less than 1 m deep or by immersing head and neck. Frequently grazes and digs for roots and stolons in partly flooded pastures; sometimes using feet to aid digging (Brouwer and Tinbergen 1939). Seeds occasionally, but animal food probably incidental. During winter on Ouse Washes, England, when undisturbed, increased feeding activity during day, with plateau 6 hours after sunrise, and continued to feed for at least one hour after dark; on adjoining farmland fed intensively in morning and late afternoon, with period of relatively low activity mid-day (Owen and Cadbury 1975).

In Netherlands in winter, mainly rhizomes of pondweeds *Potamogeton pectinatus* and *P. perfoliatus*, with some leaves of hornwort *Ceratophyllum*, horned pondweed *Zannichellia*, milfoil *Myriophyllum*, and stonewort *Chara*. In tidal basin areas, mainly rhizomes of eelgrass *Zostera* (Brouwer and Tinbergen 1939). Similarly, of 8 stomachs from Denmark, 5 contained only *Potamogeton* and 3 only *Zostera* rhizomes and roots (Spärck 1957).

Based on 20 stomachs and observations in winter on Ouse Washes, preference for soft grasses (floating sweet grass *Glyceria fluitans*, creeping bent *Agrostis stolonifera* and marsh foxtail *Alopecurus geniculatus*) though coarser reed grass *G. maxima* taken later in winter, and in deep flooding. Starchy roots of marsh yellow-cress *Rorippa palustre* prominent in diet in 2 winters. From 1972–3 also fed on waste crops, especially potatoes. In contrast to Mute Swan *C. olor* in same area, also fed on flood banks taking white clover *Trifolium repens*, stolons and leaves of grasses such as rye grass *Lolium perenne*, and cocksfoot *Dactylis glomerata* (Owen and Cadbury 1975). At Slimbridge, England, hand-fed on grain, but also feeds on neighbouring pastures, droppings containing mostly *L. perenne*, *A. stolonifera*, *Poa trivialis*, *Alopecurus*, and *G. fluitans*. In mild periods, feeds extensively on clover *T. repens* stolons (M Owen). At Wexford Slobs, Ireland, waste potatoes taken intensively, with winter wheat and carrots (Merne 1972).

Only information from breeding grounds indicates that families feed in and on margins of lakes, and occasionally graze tundra (Dementiev and Gladkov 1952).

**Social pattern and behaviour.** Based largely on material supplied by J Kear for *C. c. bewickii*; see references in Kear (1972).

1. Gregarious except when nesting. Family leaves breeding territory and travels as unit to wintering area, where flocks with other families, pairs, and individuals. BONDS. Monogamous pair-bond of long-term duration. No positive case of divorce amongst population wintering at Slimbridge (England); if mate lost, bird often takes another within 12 months, though may remain unpaired as long as 3 years (Kear 1972). Pair-formation slow and, although courtship seen in winter, firm pairs formed elsewhere, probably in herds in summer area. Pairs form by 3rd or 4th winter, typically breed by 5th or 6th (M E Evans); delay in breeding may be due to lack of nesting territory; have bred during 3rd summer in captivity (J Kear). Both parents tend young; family remain together throughout winter and depart together; mechanism of break-up unknown, and only adult pair seen in nesting territory. Siblings often associate through 2nd and 3rd winters, and may rejoin parents in winter area (one associated with parents and subsequent young for 5 winters). No record at Slimbridge of pairing between siblings. BREEDING DISPERSION. Territorial while nesting. In USSR, average 1 pair per 2000 ha of wet tundra, although may nest more closely on some islands (Dementiev and Gladkov 1952). All breeding activities except pair-formation occur in territory, and family remain until cygnets fledge. ROOSTING. As Mute Swan *C. olor*, but owing to winter movements temporary sites more frequent.

2. Displays similar to Whooper Swan *C. cygnus* except performed with more rapid movements and more frequent spreading wide of wings. Although territories probably large, voice not so strong as *C. cygnus*. Pre-flight signals, which integrate family for take-off, consist mainly of conspicuous Head-bobbing movements, in which neck alternately bent and stretched, while calling; Wing-flaps and Body-shakes also recorded (McKinney 1965b). ANTAGONISTIC BEHAVIOUR. Threat-display common in winter flocks, especially upon arrival of newcomers, but members of family parties do not quarrel among

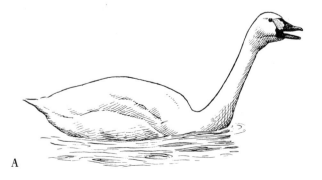

A

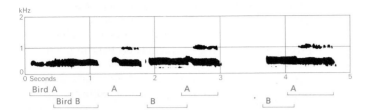

I D S McChesney England March 1962

themselves. In most frequent posture, bird faces opponent with neck extended at 45° calling loudly (see A); may also quiver partly or wholly spread wings, and repeatedly bend and extend neck while calling. Aggressive encounters typically initiated by adult pair, often by ♂, but cygnets and others may join circle of displaying birds. Fights, using beak and wings, not uncommon in winter and, having pecked and driven off another, victor usually Tail-wags; also both may Wing-flip and Body-shake. Neck feathers fluffed in aggression but, being short, often inconspicuous. General alarm indicated by erect, sleeked neck and, sometimes gaping bill. Drinking or Bill-dipping seem occasionally to act as appeasement. Little reported on territorial behaviour, but probably similar to *C. cygnus*. Stare-down posture, much as in latter, reported from captive ♂ when young in danger (Evans 1975). HETEROSEXUAL BEHAVIOUR. Courtship consists of mutual Head-turning, neck feathers erected, as ♂ and ♀ face. Triumph Ceremony by pair occurs after aggressive encounters; also used as greeting by pair and, as in threat, wings quivered in half-open and lifted position while loud calls given. When either of pair calls, mate usually joins in. Pre-copulatory display limited to Head-dipping movements; often initiated by ♀, with lowered wings and ruffled head and neck feathers, ♂ joining in with wings slightly raised; when movements synchronized, ♂ mounts (Evans 1975). In post-copulatory display, ♂ and sometimes ♀ spread wings and both call, rather more softly than usual; often also rise together in water and turn partial circle before settling back and starting to bathe, tail-wag, and preen (Johnsgard 1965). For details of nest-relief ceremony of captive birds, see Evans (1975). RELATIONS WITHIN FAMILY GROUP. Cygnets at first helped to find food by parents; not carried. Calls similar to *C. olor*. Aggression within family not reported. Greeting within family normally consists of Head-bobbing. In winter flock, family and area around defended by all, particularly adults; act as unit in threatening other birds, and join in subsequent Triumph Ceremony. For observation of captive broods, see Evans (1975).

(Fig A from photograph in Scott *et al.* 1972.)

**Voice.** Bewick's Swan *C. c. bewickii*. Strong in both sexes. Of similar range and significance to that of Whooper Swan *C. cygnus* but of a less trumpeting and more honking, crooning, or barking character; also quicker and less noisy (J Kear) with fundamental frequency somewhat higher than in *C. cygnus* (see Johnsgard 1971) though frequency range similar. Used extensively, year-round, individually and in chorus, especially on water and in flight, for wide-ranging territorial and contact purposes as well as at close range. Unlike Mute Swan *C. olor*, wing-beats do not produce audible sounds except on take-off, landing, or abrupt changes of direction, when brief swishing (probably of no signal value) produced. Resting flocks make

gentle, musical murmuring (see below); more integrated than in other west Palearctic swans. Immatures have more wheezy calls, and probably do not develop adult voice until 2nd summer (J Kear). Calls of Whistling Swan *C. c. columbianus* appear generally similar (Scott 1966).

CALLS OF ADULTS. Basically similar in most contexts (J Kear). (1) Threat-call. Particularly loud, consisting of repeated phrases of 3 or more syllables; accompanies special aggressive postures. (2) Triumph-call. Similar to last; given by pair during Triumph Ceremony. Sometimes performed as duet (see I) much as in *C. cygnus* but less precisely timed: 5 distinct components identified, but normally call consists of 1–3 syllables, resembling baying 'couc' or 'hougoucouc' (Lippens and Wille 1972). (3) Flight-call. Repeated, monosyllabic, honking 'bong', 'ong', or 'bung' (see II); louder and sharper than *C. cygnus* (Dementiev and Gladkov 1952). Also serves as individual Contact-call. (4) Lost-call. Single, fairly loud note, repeated at rather infrequent intervals by alert bird out of sight of mate and young. (5) Pre-flight call. Series of single notes like Lost-call but softer and more frequent, with increase in rate to take-off; resembles stiff, jerky 'hob hog hog', intensifying on increased excitement to 'hugquggug' or 'gu-uh-e-lu-ok'. (Bauer and Glutz 1968.) (6) Greeting-call. Quiet and shorter version of Triumph-call; 1–2 syllables, repeated. Given by mates to each other and young. From resting flock, produces pleasant babble of musical sounds, ranging from short 'guhk', longer 'ho' or 'wah' to higher pitched 'kuyhk' of immatures. Nocturnal vocalizations from winter flock probably same; gives effect of quiet crooning, inaudible beyond 50 m, and consists of variety of low notes of ascending and descending pitch. (7) Hiss. Given much as in *C. cygnus*.

CALLS OF YOUNG. Like those of *C. olor* but gruffer and lower pitched (J Kear).

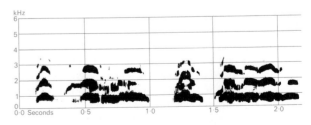

II Sture Palmér/Sveriges Radio Sweden October 1965

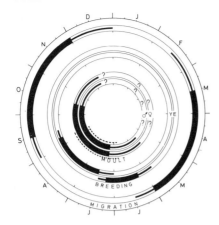

**Breeding.** Bewick's Swan *C. c. bewickii*. SEASON. See diagram for whole breeding range. SITE. On dry elevation, bank or hummock, selected as snow-free before thaw, sometimes in standing water after thaw. Nests well dispersed; may be re-used in subsequent years with new material added. Nest: large mound of vegetation, including moss, sedges, and grasses. Base diameter about 1 m, height 50–60 cm, small depression in top, lined softer grasses and some down. Building: by both sexes in captivity, ♂ passes material to ♀ standing in nest; cup produced by pressing with breast, bill and feet (Johnstone 1957), but in wild state by ♀ alone (Dementiev and Gladkov 1952). A little material added by ♀ during incubation. EGGS. Almost equally rounded at both ends; white or slightly yellow, often slight lustre, becoming scratched and stained brown or yellow during incubation. 35 eggs from west of River Lena 103 × 67 mm (96–111 × 65–71), calculated weight 260 g; 6 eggs from east of River Lena 108 × 71 mm (104–112 × 69–73), calculated weight 306 g (Schönwetter 1967). Weight of captive laid eggs 290 g (252–326), sample 44 (J Kear). Clutch: 3–5 (2–6). One brood. Replacement clutches not recorded. Eggs usually laid at 48-hour intervals in captivity (Evans 1975). INCUBATION. 29–30 days, few records. Only ♀ incubates, covering eggs with nest material and down when off feeding. ♂ sits during egg-laying, and sometimes during incubation when ♀ off feeding. Begins on completion of clutch; hatching synchronous. Eggshells left in nest. YOUNG. Precocial and nidifugous. Cared for by both parents, brooded by ♀ at night when small. ♂ will drive off mammalian predator by flying towards it while ♀ leads young into hiding on edge of water or among vegetation (Dementiev and Gladkov 1952). Self-feeding, but small young take vegetation pulled from banks or beneath water by parents. FLEDGING TO MATURITY. Fledging period 40–45 days; few records. Young stay with parents through first autumn and winter, leaving wintering areas with them on spring migration but generally remain behind at some stopping place while they fly to breeding grounds. May rejoin parents and new cygnets at traditional wintering haunt in 2nd, 3rd, or even 4th winter, again leaving with them in spring; autumn migration, however, completed separately (Ogilvie 1972). First pairing occasionally in 2nd winter, more commonly in 3rd; first breeding at 3–4 years, but often later. BREEDING SUCCESS. No data from breeding grounds. Winter counts in Britain 1953–72 show considerable annual fluctuation in proportion of young, 7%–44%, and in mean brood size, 1·5–3·1 (Evans 1971; Ogilvie 1969b). Short available breeding season in Arctic dictates almost synchronous breeding over wide area, so eggs and young vulnerable to adverse weather; losses on migration from shooting also a factor, as 12% of 53 first winter birds X-rayed carried lead shot (Evans *et al.* 1973).

**Plumages.** Similar to Whooper Swan *C. cygnus* in all stages, but juvenile slightly paler grey on upperparts, especially in spring.

**Bare parts.** *C. c. bewickii.* ADULT. Iris brown, eye-ring occasionally yellow. Base of bill and bare skin in front of eye yellow, tip of bill black. Yellow less extensive than in *C. cygnus*, usually not extending beyond vertical line at two-thirds of bill from tip, though sometimes protruding beyond this in irregular shape along culmen to nostril, but never in wedge along cutting-edge to below nostril as in *C. cygnus*. Basal third of culmen and cutting-edges variably black or yellow; for individual recognition based on variable pattern in bill colour see Scott (1966). Leg and foot black, in some with grey or olive tinge. For bare part colours in leucistic birds, see Evans and Lebret (1973). DOWNY YOUNG AND JUVENILE. Shape of bill as adult; otherwise as in *C. cygnus*; characteristic pattern visible from early spring.

**Moults.** ADULT POST-BREEDING. Complete; flight-feathers simultaneous. Flightless for *c.* 29 days between late July and early September; ♂ and ♀ at same time. Starts body moult when wings in quill; finishes December (Dementiev and Gladkov 1952; Heinroth and Heinroth 1927–8). POST-JUVENILE. Partial. Sequence apparently as in *C. cygnus*, but quicker; March specimens of *C. c. bewickii* on average with more white on mantle and scapulars than *C. cygnus*.

**Measurements.** Netherlands, winter; skins (RMNH, ZMA).

| | | | | | | |
|---|---|---|---|---|---|---|
| WING AD ♂ | 519 | (17·0; 12) | 493–548 | ♀504 | (9·85; 17) | 469–525 |
| JUV | 512 | (13·6; 7) | 494–534 | 484 | (18·6; 6) | 466–517 |
| TAIL AD | 151 | (7·64; 9) | 141–164 | 152 | (7·85; 15) | 139–164 |
| JUV | 134 | (12·0; 6) | 125–155 | 131 | (4·46; 6) | 125–137 |
| BILL AD | 91·0 | (5·31; 12) | 82–102 | 90·9 | (3·24; 18) | 86–97 |
| TARSUS | 109 | (3·57; 16) | 100–116 | 101 | (3·73; 21) | 92–106 |
| TOE | 128 | (6·03; 16) | 118–141 | 121 | (4·94; 21) | 110–130 |

Sex differences significant, except tail and bill. Difference between adult and juvenile not significant, except tail and ♀ wing. Bill of juvenile as adult, but difficult to measure due to variable feathering on forehead. Adult and juvenile combined for tarsus and toe.

England, winter; live (Scott *et al.* 1972).

| | | | | | | |
|---|---|---|---|---|---|---|
| WING AD ♂ | 531 | (93) | 485–573 | ♀510 | (92) | 478–543 |
| JUV | 505 | (17) | 458–552 | 487 | (31) | 445–525 |
| BILL AD | 94·7 | (94) | 81–108 | 90·9 | (94) | 75–100 |
| TARSUS | 106 | (110) | 93–119 | 102 | (125) | 85–111 |

**Weights.** England, winter (Scott *et al.* 1972).

| | | | | | | |
|---|---|---|---|---|---|---|
| AD | ♂ 6400 | (96) | 4900–7800 | ♀5700 | (95) | 3400–7200 |
| JUV | 5400 | (15) | 4500–6700 | 5000 | (29) | 3300–6400 |

**Structure.** 11 primaries: p9 longest, p10 10–22 shorter, p8 0–3, p7 11–24, p6 60–75, p1 212–225. Bill relatively short. Outer toe *c.* 95% middle, inner *c.* 82%, hind *c.* 25%. Other structure as in *C. cygnus*.

**Geographical variation.** Nominate *columbianus* slightly larger; bill mostly black, yellow absent or confined to small spot before eye. East Siberian population with larger bill, higher at base and wider near tip: maximum width (adults) 30·5–35, average 31·6, compared with 28–32, average 28·5, in *bewickii*. Increase in bill size clinal, overlap large, and size extremes may occur in all populations; so no justification for recognizing *C. c. jankowskii* Alpheraky, 1904 (Siberia east from Lena delta) as separate subspecies. (Delacour 1954; Dementiev and Gladkov 1952; Vaurie 1965.)                                                    CSR

# *Cygnus cygnus* Whooper Swan

PLATE 51
[facing page 375]

Du. Wilde Zwaan          Fr. Cygne sauvage          Ge. Singschwan
Ru. Лебедь-кликун        Sp. Cisne cantor           Sw. Sångsvan

*Anas Cygnus* Linnaeus, 1758

Monotypic

**Field characters.** 145–160 cm, of which body only half, ♂ averaging larger; wing-span 218–243 cm. All white and large swan, with flattened forehead, long erect neck, and black and yellow bill. Sexes alike and no seasonal differences. Juvenile distinctive, becoming like adult during 1st summer and 2nd winter.

ADULT. Entirely white, though head, neck, and underparts may become stained rusty. Bill black with variably yellow base extending in wedge usually to below nostril or beyond. Legs and feet black. JUVENILE. Head, neck, and upperparts generally grey-brown (darkest on crown), except for whitish cheeks, throat, and foreneck; underbody white with grey-brown flanks. Bill blackish with dirty pink base which, during 1st winter, starts to show pattern of adult and becomes livid white, then tinged yellow; feet dark grey, becoming blacker. Also during 1st winter and spring, plumage gradually grows whiter, with much individual variation, until by 15–20 months old predominantly white.

Most likely to be confused with Bewick's Swan *C. columbianus bewickii*, particularly in flight when latter's smaller size and truncated yellow bill-patch not easy to judge, but characterized by flat profile of bill and head and usually somewhat lower-pitched, bugle-like clanging, uttered more loudly and regularly on wing. Mute Swan *C. olor* more easily distinguished by orange bill with black knob, curved and thicker-looking neck, often arched wings and, in flight, mechanical throbbing of primaries. Juveniles closely similar in colour of plumage and bill to those of *C. c. bewickii*, but head-shape and, where comparisons possible, size remain valid distinctions; juvenile *C. olor* browner and darker above and has bill grey without dirty pink.

Swims with straight, stiff neck, and walks well (not with awkward, waddling action of *C. olor*), with neck either erect or kinked back over body and then straight up (though occasionally more curved). Feeds by day or night, both in water and grazing on land. Flight powerful with neck extended and slow, regular wing-beats that produce swishing noise but not loud throb of *C. olor*. Parties sometimes fly in diagonal lines or V formation, particularly when flying high and far, but often in no order. Markedly gregarious in winter, in flocks of up to 100 or more, and non-breeding immatures remain in smaller parties through summer, but pairs always breed in isolation. Generally wary of man but not usually as aggressive towards other species as *C. olor*.

**Habitat.** Unlike Bewick's Swan *C. columbianus bewickii*, prefers breeding habitats south of tundra, except in Iceland and parts of arctic Europe, but equally demands immunity from disturbance. Diverse types include large, shallow, reed-fringed lakes in steppe region; swampy wetlands and pools, or even small ponds set among low-lying or upland grasslands and heaths, or surrounded by forests, including taiga; or reedbeds. Also rivers, estuaries, lagoons, and arms of sea. Exceptionally on wetlands or hill lakes up to *c.* 800 m. Although using deeper waters when resting, requires waters of maximum depth *c.* 1 m for underwater grazing, with rich bottom vegetation but free from floating and emergent plants; will also graze on land. Nest-sites normally in well-vegetated surroundings and widely dispersed, often on islets.

On migration, occurs more widely than *C. c. bewickii*, following sea coasts, chains of lakes, and river systems, with brief stop-overs to and from winter quarters. More ready than *C. c. bewickii* to use marine habitats, especially in land-locked seas and sheltered inlets, when more suitable inland habitats not available. In regular winter quarters, when preferred aquatic plants lacking, readily forages on stubble fields and arable crops, even potatoes. Increasingly grazes floodlands and other wetlands in late winter and early spring. Flies at great height on migration, but at low or moderate heights on local journeys. Unlike Mute Swan *C. olor*, accepts human proximity only where non-interference fully established. Extensive losses of tenable sites within now humanly settled zones have led to fragmented breeding distribution but recent spread in Fenno-Scandia. In past, shooting pressures and disturbance also restricted winter quarters. (See Matthews 1972.)

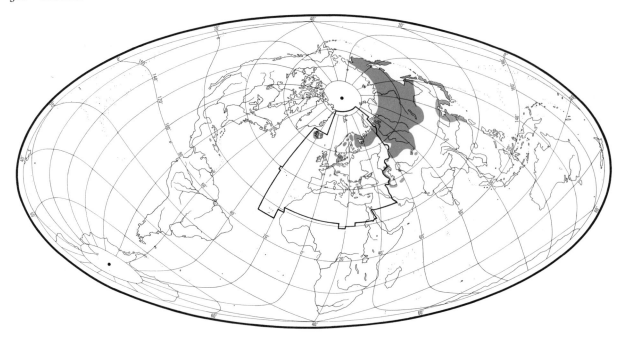

**Distribution.** Marked recent spread Fenno-Scandia following protection after earlier decline. FAEROES: bred until 17th century (Williamson 1970). NORWAY: in last 15 years settled northern coasts in small numbers (Haftorn 1971). SWEDEN: from *c*. 1800, southern limits retreated due to human persecution from *c*. 60°N to *c*. 63°N by late 19th century and to 67°N by about 1920; in last 15 years, protection and introduction has led to new breeding sites in south (Fjeldså 1972). FINLAND: marked decline, perhaps due to illegal hunting and plundering of eggs and young; close to extinction by early 1950s, but considerable increase since due to protection, with much of old range re-occupied and some extension further south (Merikallio 1958; Haapanen *et al.* 1973). POLAND: nested Biebra marshes 1973 (P Koźniewski and K Machata). BRITAIN: became extinct Orkney 18th century; now breeds occasionally Scotland, most recent 1939, 1947, 1963 (British Ornithologists' Union 1971; Parslow 1973a). USSR: during 19th century, major decline with many breeding sites in south abandoned due to hunting and loss of nesting areas (Dementiev and Gladkov 1952); bred once Udmurt ASSR in last 10 years (Priklonsky *et al.* 1970).

Accidental. Spitsbergen, Bear Island, Jan Mayen Island, Czechoslovakia, Hungary, Cyprus, Egypt, Spain, Algeria, Tunisia.

**Population.** Breeding. ICELAND: probably 5000–7000 birds, of which only a small proportion breed; no increase in last 25 years (Gardarsson 1975). SWEDEN: *c*. 200 birds, of which only a few breeding; increased Lapland in last 15 years (Curry-Lindahl *et al.* 1970); 15–20 pairs in south (Fjeldså 1972). FINLAND: *c*. 80 pairs with many non-breeders; marked increase 1950–70 (Haapanen *et al.* 1973).

Winter. In west Palearctic, 3 main wintering pop-ulations. (1) Icelandic: up to 5000 birds (but see above), of which *c*. 2500 Britain, *c*. 2000 Ireland, and *c*. 500 Iceland, with Irish population increased in severe winters. Has increased and spread in Ireland since *c*. 1943 (Ruttledge 1974), but decreased Iceland, where formerly 1000–1500 birds (FG; Ogilvie 1972). (2) North-west Europe: *c*. 14000 birds, mainly Baltic, moving further south in severe winters; marked increase Norway recent years (Haftorn 1971). (3) Black and Caspian Seas: at least 25000 birds; also recent counts of 400, Greece. (Ogilvie 1972.)

Survival. Iceland: mean annual mortality of birds over 4 months 17% (Beer and Ogilvie 1972). Finland: adult mortality May–August 2% (Haapanen *et al.* 1973). Oldest ringed bird 7 years 6 months (Reykjavik Museum).

**Movements.** 2 discrete breeding populations in west Palearctic, one wholly, other partially migratory. USSR and Fenno-Scandian population leaves breeding grounds at start of freeze, generally mid to late-September in north and east, through to mid-October in west. Passage through White Sea and east Baltic to winter on coasts of north Germany, Denmark, and Sweden, arriving October–November; movement further west to Low Countries in cold weather. Baltic winterers from at least as far east as River Ob (74½°E). Uncertain whether any of these reach Britain regularly: only sure evidence one ringed Sussex recovered Sweden, February; but small numbers in winter east and south England seem more likely continental than Icelandic. Winterers in east Mediterranean, Black, Caspian, and Aral Seas, and further east thought to come from more eastern parts of breeding range, passing through east Europe, west and south-west USSR. Return movements from both these major wintering regions begin mid-March, continuing through

April and early May, arriving on breeding grounds in mid to late May; see Bauer and Glutz (1968); Hilprecht (1970); Ogilvie (1972).

Iceland population partially migratory, with *c.* 75% wintering Britain (mainly Scotland) and Ireland; proportions vary according to weather and food in Iceland, and further emigration can occur at any time in winter. Icelandic birds leave when breeding waters freeze; non-migrants gather on coastal brackish lagoons, interior lakes kept open by thermal springs, and larger rivers. Such areas also used by non-breeding summer moulting flocks; some short movements to moulting areas probable. Main arrivals north Britain late September to November. Some fly direct to Ireland, others move there later as severe weather displaces them from Scotland. Movements in either direction between Scotland and Ireland occur throughout winter. Departures March–April, with stragglers into May.

Occasionally summers Britain and Baltic countries. Singles or small parties appear not infrequently, chiefly in cold weather, outside main wintering areas; one ringed Switzerland shot same winter north Italy. Has straggled to USA (New England) and Greenland; also Persian Gulf, Pakistan, and north India.

**Food.** Based mainly on information supplied by M Owen. Almost entirely aquatic vegetation (leaves, stems and roots) from fresh and saline waters, by up-ending or dipping (immersing head and neck). Up-ends for 1–20 s at time (Airey 1955). Feeding actions often accompanied by frequent foot-paddling movements which loosen under-water plants (Venables 1950). Forages on land, increasingly in recent years.

Of 9 stomachs from Denmark, 3 contained only eelgrass *Zostera*, 2 wigeon grass *Ruppia*, 1 pondweed *Potamogeton*, and 3 roots of dicotyledons obtained from land (Spärck

1957). *Zostera* eaten extensively in the Hebrides and Ireland (Freme 1931; Kennedy *et al.* 1954). *Potamogeton natans*, Canadian pondweed *Elodea canadensis*, and stonewort *Chara* also recorded (Hewson 1964). One stomach, East Anglia, England, contained 90% by weight roots and crowns of marsh yellow-cress *Rorippa palustre*, with shoots and leaves of horsetail *Equisetum* and *Glyceria* (M Owen). Frequents farmland extensively in Scotland, mainly taking grain from stubbles, potatoes, and grass (Kear 1963*b*), but also growing and cut turnips from stock fields (Downie 1961). Rarely scavenges on refuse dumps, eating waste potatoes and other refuse (Baxter and Rintoul 1953; Ruttledge 1963) and, exceptionally, acorns and plums (Hilprecht 1970).

In breeding season, roots and shoots of aquatic plants and small aquatic animals (Dementiev and Gladkov 1952). One stomach, Iceland, contained mainly crowfoot *Ranunculus trichophyllus* leaves and cottongrass *Eriophorum* seeds. Non-breeders in summer took algae from lake and crowberry *Empetrum* berries from surrounding slopes (C G Hancock).

**Social pattern and behaviour.** Based largely on material supplied by J Kear; see references in Kear (1972).

1. Gregarious except when nesting. If breeding unsuccessful, pairs may leave nesting grounds before moult and join other non-breeders on traditional, often coastal, waters. Family leaves territory and moves to wintering area as unit. There flocks with other families and non-breeders. BONDS. Long-term monogamy seems typical, but not known whether divorce ever occurs. Pair-formation slow. Courtship occurs in winter flocks, but more often in non-breeding herds where most pairs formed, with engagement period lasting at least until following season. Probably capable of breeding at 3 years, but factors such as lack of suitable territory prevent many from nesting until 4–6 years. Both parents tend young; family remain together on migration and in winter, departing for summer area together, though only adults arrive at nesting territory (Kear 1972). BREEDING DISPERSION. Territorial while nesting, defending large area possibly averaging *c.* 100 ha of scrub zone of taiga, containing number of shallow, well-vegetated pools. All breeding activities, except pair-formation, occur in territory and family usually remain there until cygnets fledge. ROOSTING. As Bewick's Swan *C. columbianus bewickii*, but sometimes on sea.

2. As in other white swans, conspicuous plumage of adults may play important part in advertising ownership of territory, and in keeping others at distance during breeding season. Territory size much larger than in Mute Swan *C. olor*, and loud voice added to repertory of displays. Pre-flight signals, which integrate family take-off, consist of conspicuous Head-bobbing movements while calling. ANTAGONISTIC BEHAVIOUR. Threat-display takes several forms frequently seen in winter flocks. Commonly bird faces opponent, neck extended at angle of *c.* 45°, while calling; at same time may quiver partially, or more rarely, fully open wings, and repeatedly bend and extend neck as calls given. Body-shaking, with hissing and wing-rustling noises, also used in threat, sometimes terminating in Staring-down (see A) in which bird holds fluffed neck rigid, bill pointing down, wings usually open and still (on water, head may be completely submerged). Neck feathers always fluffed in aggression, but inconspicuous. General alarm indicated by erect, sleeked neck and slightly gaping bill. Group display described by Ekström

A

(1973) involving 3 pairs probably threat-display followed by Triumph Ceremonies (J Kear). In territorial defence, most common during nest-building and early incubation, one or both members of resident pair may give Warning-display by half-opening and quivering wings, stretching neck and calling. If intruder still approaches, defender flies up and then chases it over boundary. Fights on land, water, and in air sometimes occur. HETEROSEXUAL BEHAVIOUR. Head-bobbing usual greeting. Courtship consists of mutual Head-turning, neck feathers erected, while ♂ and ♀ face. Triumph Ceremony between pair occurs after aggressive encounters; also used in greeting by pair. As in threat-display, wings quivered in half-open, lifted position, while neck repeatedly bent and extended as birds call loudly (see B). When either gives loud calls, mate usually joins in; notes at first alternate but finally synchronized, and display may terminate with necks rigidly outstretched. Pre-copulatory display limited to Head-dipping movements closely resembling those of bathing, except birds do not thrash wings on water, and synchronized movements last only a few seconds before ♂ mounts. In post-copulatory display, ♂ and sometimes ♀ spread wings and both call, rather more softly than usual; then rise together on water and turn partial circle before settling back and starting to bathe, tail-wag, and preen (Johnsgard 1965). RELATIONS WITHIN FAMILY GROUP. Small cygnets helped to find food by parents; not carried. Calls similar to *C. olor*. No record of aggression within family. All members may act as unit in threatening other birds in wintering flocks, and join in subsequent Triumph Ceremony.

(Figs A–B from photographs in Scott *et al.* 1972.)

B

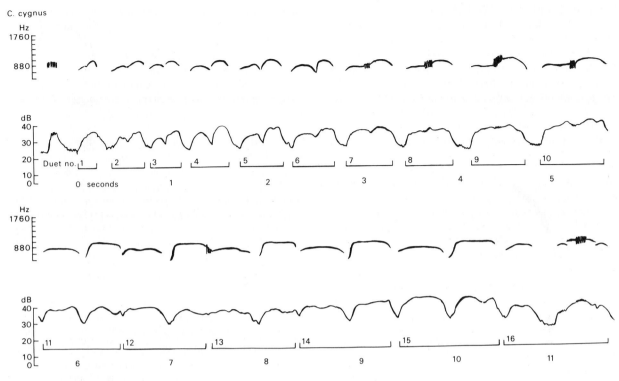

C. cygnus

I P J Sellar Iceland May 1967

**Voice.** Based mainly on information supplied by J Kear. Strong in both sexes, sometimes lower pitched in ♂ (Scott 1950; Scott *et al.* 1972). Loudest calls with musical trumpet or bugle-like quality, deeper and stronger than Bewick's Swan *C. columbianus bewickii*; difference probably linked with extra loop of trachea within sternum (Johnsgard 1971), full tracheal length and adult voice developing in 2nd summer (J Kear). Prolonged exhalation of air from trachea in dying bird can produce a series of musical notes, probably accounting for legend of 'Swan-song' (see also Witherby *et al.* 1939). Individual utterances brief, averaging *c.* 0·2 s. Duration and effect of performances often enhanced in flock choruses, giving trumpet-like massed sound. Use made of voice, and factors influencing calling, much as in *C. c. bewickii*. Calls most frequently heard from flocks on water and in flight (J Kear). On take-off or landing, or when abruptly changing direction, wings produce brief swishing sound as in *C. c. bewickii* but lacking strong persistent throb of Mute Swan *C. olor*.

CALLS OF ADULTS. Vocabulary probably much as in *C. c. bewickii* (J Kear). 'Song', attributed to ♂ on breeding grounds, said to consist of low, musical series of *c.* 7 distinct notes, rising slightly then falling and rising again (Kirkman and Jourdain 1930); of uncertain significance. (1) and (2) Threat and Triumph-calls. Basically same. Composed of brief, melodious notes; though loud, probably more subdued and lower pitched than Flight-call, resembling deep nasal 'ang' (Voigt 1950). Given on

water or land. When uttered by members of pair together during Triumph Ceremony, may take on character of integrated duet (Armstrong 1940; Johnsgard 1965), at times developing into controlled antiphonal calling, with gradual and matched increase in duration of constituent calls, i.e. when initiating bird alters tempo other follows suit; calls precisely pitched, at *c.* 880 Hz and 980 Hz respectively (Hall-Craggs 1974; see I). (3) Flight-call. Double 'whoop whoop', with 2nd syllable slightly higher pitched (see II); 'whoop-a', 'ahng-ha' (Witherby *et al.* 1939), etc. Apparently whole call pitched 1 octave higher than Triumph-call on water. Repeated several times with brief pauses. Short, weak, high-pitched, monosyllabic version—'ang', 'huck', or 'wick'—also given as individual Contact-call on land or water as well as in flight (Witherby *et al.* 1939; Bauer and Glutz 1968). (4) Alarm-calls. Harsh, excited 'krow' (Witherby *et al.* 1939), audible at distance. Weak, nasal, goose-like version used in breeding area. (5) Lost-call. Loud, high-pitched, goose-like 'wak'; uttered

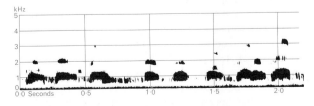

II Sture Palmér/Sveriges Radio Sweden April 1966

repeatedly, with straight neck, in situations causing anxiety, including temporary separation from mate or young (Bauer and Glutz 1968). (6) Pre-flight call. Apparently much as last. (7) Greeting-call. Probably subdued version of Triumph-call; given between members of pair and family (J Kear), in flock producing varied but harmoniously blended group chorus. (8) Hiss. Given as threat to conspecifics, and when cornered by enemy.

CALLS OF YOUNG. As in *C. olor*, but with similar lower and gruffer character as *C. c. bewickii* (J Kear).

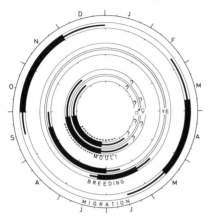

Breeding. SEASON. See diagram for Fenno-Scandia and USSR. In Iceland, similar except first eggs late April in favourable years (Hilprecht 1970). SITE. Normally close to water, on bank of pool, or commonly on islet or promontory; snow-free and dry before thaw, but sometimes in shallow water afterwards. Territorial. Nests may be re-used in subsequent years after rebuilding. Nest: large mound of reeds, sedges, and other plants; 1–2 m across at base, exceptionally 3 m, and 50–70 cm high; depression in top 5–20 cm deep, lined softer grasses and small amounts down. Building: by both sexes. ♂ standing by nest passes material over shoulder to ♀; cup produced by pressing with breast, bill, and feet. ♀ adds a little material during incubation. EGGS. Almost equally rounded at both ends; white or tinged yellow, often with slight lustre, becoming scratched, and stained brown or yellow during incubation. 113 × 73 mm (105–126 × 68–77), sample 232; calculated weight 330 g, sample 109 (Schönwetter 1967); no significant differences between eggs from Iceland or continental Europe and Asia. Weight of captive laid eggs 329 g (293–356), sample 15 (J Kear). Clutch: 3–5 Iceland, 4–7 USSR (Kear 1972; Dementiev and Gladkov 1952). Of 31 clutches, Finland: 2 eggs, 10%; 3, 13%; 4, 29%; 5, 29%; 6, 13%; 7, 6%; mean 4·4 (Haapanen et al. 1973). One brood. Replacement clutches not recorded. Eggs usually laid at 48-hour intervals. INCUBATION. 35 days (31–42); few records. Only ♀ incubates, though ♂ may sit in her absences; eggs otherwise deeply covered with nest material and down when ♀ feeding. Begins with last egg; hatching synchronous. Eggshells left in nest. YOUNG. Precocial and nidifugous. Cared for by both parents. Brooded by ♀ at night when small; both parents highly aggressive in defence of nest or young, displaying at low-flying aircraft and threatening mammalian predators, though some leave when human intruder approaches within 1 km (Kear 1972). Self-feeding, but small young take vegetation pulled from banks or beneath water by parents. FLEDGING TO MATURITY. Fledging period 87 days (78–96), sample 9 (Haapanen et al. 1973). Young stay with parents through first autumn and winter, leave wintering area together in spring but separate from them at or before reaching breeding site. First pairing at 2 years, first breeding generally at 4. BREEDING SUCCESS. Of 86 eggs in 18 clutches, Finland, 74% hatched. Of 110 cygnets in 29 broods at hatching, 90% fledged. Average brood size at hatching 3·8, from average clutch size 4·4; average brood size at fledging 3·3. Mortality of young stable at *c.* 10% per month from hatching to autumn migration; significantly higher (10–14%) in north Finland than in south Finland (3–6%). Only 60% of summers in north Finland long enough for cygnets to reach fledging stage before waters freeze (Haapanen et al. 1973). Counts of Icelandic breeding birds in Britain in winter show marked annual variation in proportion of young, from under 10% to over 40%, and in mean brood size from *c.* 1 to 3–4 (M A Ogilvie).

Plumages. ADULT. Entirely white; sometimes stained rusty on head, neck, or belly. DOWNY YOUNG. Head and upperparts silver-grey, below white. JUVENILE. Head and neck pale brown-grey, slightly darker on crown, paler round base of bill, on chin, and throat. Feathers of upperparts, sides, and flanks white, tipped grey (browner when worn), those of underparts white. Tail-feathers, flight-feathers, and upper wing-coverts pale grey, suffused brown on tips and outer webs. Underwing and axillaries white. FIRST WINTER AND SUMMER. New feathers on crown and nape grey-brown, narrowly tipped or edged white; rest of head and neck like juvenile. Pure white feathers develop on mantle, scapulars, chest, and flanks; on rest of body dark edges partly wear off. Juvenile tail and wing retained. SECOND WINTER. Sometimes a few brown feathers retained on outer upper wing-coverts, a few brown feathers on head, or dark shaft-streaks on scapulars and primaries.

Bare parts. ADULT. Iris brown, rarely blue or with blue ring (Koch 1940; Kinlen 1963). Base of upper mandible and bare skin in front of eye yellow; tip of upper mandible, cutting edges, and lower half of culmen (rarely whole culmen) black. Yellow extends in wedge along edge of upper mandible to below nostrils. In some, narrow black rim along forehead. Lower mandible black, yellow near base and on bare skin at base. Leg and foot black. DOWNY YOUNG. Bill flesh-pink, grey at tip and along sides. Feet pale orange in small young, flesh in older (Boyd 1972). JUVENILE. Bill basally pink, tip dark horn. During 1st winter and spring, tip blackens, and base first turns white and later yellow, showing characteristic pattern. Feathers on lores and upper culmen wear off during 1st winter, but small pores remain at least until 2nd winter.

Moults. ADULT POST-BREEDING. Complete; flight-feathers simultaneous. Flightless for *c.* 5–6 weeks, June to late August (Dementiev and Gladkov 1952; Heinroth and Heinroth 1928). As in *C. olor*, ♀ moults wing before ♂ (Kear 1972). Body moult starts when wing in quill, sequence apparently as *C. olor*; finished December (Hilprecht 1970). POST-JUVENILE. Partial. Crown, nape, and sides of head about October–January, followed by feathers on part of neck, mantle, scapulars, flanks, chest, and sides of body about January–March; sometimes also a few tertials, inner wing-coverts, and tail-feathers. Other juvenile feathers replaced completely in summer, sequence as adult post-breeding; some juvenile wing-coverts sometimes retained.

**Measurements.** Netherlands (eastern origin), winter. Skins (RMNH, ZMA).

| | | | | | |
|---|---|---|---|---|---|
| WING AD ♂ | 610 (13·0; 17) | 587–635 | ♀ | 583 (14·2; 14) | 562–615 |
| JUV | 611 (17·9; 6) | 586–632 | | 553 (23·1; 7) | 530–597 |
| TAIL AD | 169 (9·0; 16) | 151–182 | | 167 (8·4; 15) | 151–181 |
| JUV | 149 (5·3; 6) | 141–155 | | 139 (3·4; 7) | 134–144 |
| BILL AD | 106 (4·7; 19) | 98–116 | | 102 (4·4; 15) | 92–111 |
| TARSUS | 123 (3·9; 21) | 116–130 | | 113 (3·8; 22) | 104–119 |
| TOE | 156 (6·6; 21) | 142–172 | | 141 (6·8; 22) | 129–155 |

Sex differences significant, except adult tail. Tail and wing (♀ only) of juvenile significantly shorter than adult, other measurements similar (but juvenile bill difficult to measure as base of culmen variably feathered); tarsus and toe combined. Difference in wing between adult and juvenile ♂ more pronounced in live birds England (Wildfowl Trust): adult 616 (10) 590–640, juvenile 597 (6) 581–609. Maximum wing USSR, ♂ 660, ♀ 635 (Dementiev and Gladkov 1952). Birds from Greenland smaller: 7 adults wing 562 (536–573), bill 94 (78–104), tarsus 114 (108–119),

toe 135 (131–141) (Schiøler 1925). When sex taken into account, probably little overlap with Netherlands specimens.

**Weight.** All in kg. ADULT. ♂♂: January, Netherlands, 8·7; March, Netherlands, 10·4, Denmark, 14; winter, Norway, 10; May, Iceland, 8·5. ♀: March, Denmark, 8·75. Average 3 breeding and 3 non-breeding adults, August, Iceland: 9·4 and 7·9 respectively, range 7·4–10·3. JUVENILE. ♂: March, Netherlands, 7·8. Minimum of frost-killed birds Netherlands: adult ♂ 5·1, adult ♀ 5·0, juvenile ♀ 3·7. (Hagen 1942; Kinlen 1963; Schiøler 1925; ZMA.)

**Structure.** Wing long, pointed. 11 primaries: p9 longest, p10 3–20 shorter, p8 0–3, p7 12–24, p6 60–82, p1 215–260; p11 reduced, pointed, concealed by primary coverts, Outer web of p9–7 and inner of p10–8 emarginated. Tail short, rounded; 20 feathers. Bill long, straight; base of culmen with broad concave depression. Sides of bill parallel, nail broad. Lores and skin between eyes and base of bill bare. Neck long. Tarsus reticulated; outer toe *c.* 95% middle, inner *c.* 80%, hind *c.* 21%.

**Geographical variation.** Slight; in size only. Recently extinct Greenland population small, Icelandic intermediate. Few data on sexed specimens; until more known of variation in breeding stocks, *C. c. islandicus* Brehm, 1831 included in nominate. Sometimes considered conspecific with Trumpeter Swan *C. buccinator* of North America.　　　　　CSR

# *Anser fabalis* Bean Goose

PLATES 52 and 61 [after page 398 and between pages 494 and 495]

DU. Rietgans　　FR. Oie des moissons　　GE. Saatgans
RU. Гуменник　　SP. Ánsar campestre　　SW. Sädgås

*Anas Fabalis* Latham, 1787

Polytypic. Nominate *fabalis* (Latham, 1787), taiga of north Europe to Ural; *rossicus* Buturlin, 1933, tundra of USSR from Kanin to Taymyr Peninsula; *johanseni* Delacour, 1951, west Siberian taiga from Urals to Yenisey, perhaps also Altay mountains to Lake Baikal. Extralimital: *middendorffii* Severtsov, 1872, Siberian highland taiga east of Yenisey; *serrirostris* Swinhoe, 1871, tundra of east Siberia from Lena delta to Anadyr. (Delacour 1951; Vaurie 1965; C S Roselaar.)

**Field characters.** 66–84 cm (body 43–51 cm); wing-span 142–175 cm; averaging only slightly smaller than Greylag Goose *A. anser* but not so bulky, with less weight in the rear half of body. Large, tall, rather long-billed and long-necked, essentially brown goose, with very dark head and neck obvious in flight. Sexes alike though ♂ tends to be larger; no seasonal variation. Juvenile separable at close range. 3 races in west Palearctic, distinction often difficult.

ADULT. (1) Western race nominate *fabalis*. Head and neck dark brown, almost black on forehead. Entire upper-parts rather uniform dull brown, feathers with buff tips forming transverse pattern; upper tail-coverts white contrasting with ash-brown back and tail, latter with white rim. Chest buff-brown, merging into brown (indistinctly banded darker brown) flanks and belly; upper flanks edged white forming obvious line. Vent and under tail-coverts white. Some birds have white feathering across top and down sides of base of upper mandible. Surface of upper wing lacks distinct pattern, with coverts similar in tone to upperparts and contrasting only slightly with dull black quills. Bill rather heavy and long, with upper mandible almost straight-edged, orange-yellow with varying amount of black extending from base; legs and feet orange-yellow. Eyes dark brown, with no ring. (2) Western tundra race *rossicus*. Smaller than nominate *fabalis*, with typically shorter neck, higher base and finer tip to bill which has orange-yellow nearly always restricted to small patch behind nail; distinction requires considerable experience of species. (3) West Siberian race *johanseni*. Almost as large as nominate *fabalis*, with longer bill, but variation in latter race and *rossicus* prevents certain identification in field (see Geographical Variation). JUVENILE. Subspecific distinctions doubtful though bill pattern and size differences essentially as adult. Whole plumage duller: pattern of upperparts diffuse and of underparts mottled without white edges to flanks. Some birds almost as pale as juvenile Pink-footed Goose *A. brachyrhynchus* so that close attention to size, bulk, and bare parts crucial to identification. Yellow on bill, legs, and feet duller than in adult, often tinged grey.

At long range and in poor light, difficult to separate from other grey geese but noticeable length of dark head and

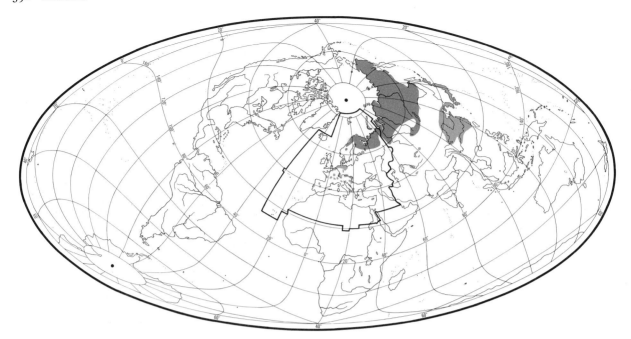

neck, upright stance, and uniform upperwing in flight, characteristic. High head carriage, bulk, uniform dark plumage tone, lack of forewing contrast, rather long, deep bill, and orange legs all diagnostic at closer ranges. Confusion with large, dark individuals of *A. brachyrhynchus* possible, but *A. fabalis* never shows grey upperparts and differences in bare-part colours afford certain distinction of typical birds.

Flight action and behaviour as *A. anser*, but with no hint of labouring, and extended neck a constant character of silhouette. Gait freer than *A. anser* but walk slightly more twisting; feeding rate slower than *A. brachyrhynchus*. Swims well but less frequently than other grey geese. Breeding populations occur in small scattered groups, taiga races choosing forest waters and tundra races more open coastal habitat. Migrant and wintering populations gregarious, banding into flocks of hundreds (rarely thousands) and exploiting traditional areas of grass or farmland; winter flocks break down into family groups in some areas e.g. Netherlands. Mixes freely with other grey geese. Shyer than *A. brachyrhynchus*, particularly when in small numbers; also more prone to roost on feeding grounds or on small waters nearby. Flight-notes loud: resonant, rather bassoon-like cackle most characteristic; lower and fuller than *A. brachyrhynchus*, flock chorus lower pitched than that of *A. brachyrhynchus* but sharper, more reedy, and more penetrating than that of *A. anser*.

DIMW

**Habitat.** Conspicuous differences in habitat characterize the various subspecies. More southerly breeders within west Palearctic (mostly nominate *fabalis*), extending in places to south of sub-arctic, unique among indigenous geese in nesting within dense coniferous forest or birch scrub. At times also on upland stony tracts with only scattered trees, usually near stream, pool or lake. Subspecies *rossicus* prefers more conventional *Anser* habitats on wet low tundra and Arctic Ocean islands, but habitat overlaps in some areas with nominate *fabalis*. While some northern birds spend no more than about 100 days on breeding grounds, others, and most forest breeders, remain up to 140, occasionally 180 days. Many forest breeders after rearing young move north for period to arctic tundra. Winter movements extend less far west than in other *Anser*; in normal winters, only small proportion reach grassy wetlands in oceanic climates. Fields under crops or fallow, steppe, floodlands, rivers, and coastal shallows used on passage and in winter. Flocks of different subspecies appear to remain mostly segregated, but whether through habitat factors not known. In use of fresh and coastal waters resembles other *Anser* species, but perhaps more often in areas less readily accessible to open water.

As forest breeder, can rise steeply from confined spaces, but frequency and patterns of aerial activity otherwise resemble other geese. Generally winters on areas where approaching danger visible at distance, and survival possible in face of intense hunting pressures.

**Distribution.** SWEDEN: marked contraction of range (SM).

Accidental. Iceland, Ireland, Greece, Turkey, Malta, Cyprus, Egypt, Algeria, Morocco, Portugal, Azores, Madeira.

**Population.** Breeding. NORWAY: declined in south of range since early 20th century, and in main stronghold Finnmark in last 20–50 years; mainly due to persecution, especially killing of moulting birds (Haftorn 1971).

SWEDEN: decline due to changes in forest habitat and human persecution (Curry-Lindahl *et al.* 1970). USSR: numerous, but main population outside west Palearctic, on Taymyr and Yamal peninsulas (Uspenski 1970).

Winter. BRITAIN: 400–500 in 1950s, now only 100–200. IBERIA: *c.* 4000. BALTIC–NORTH SEA: probably at least 40000 birds. CENTRAL EUROPE: 100000 to 150000 birds (Mörzer Bruyns *et al.* 1973; Ogilvie 1970). USSR: 1500 birds in western part in not so severe winters (Isakov 1970*c*).

Oldest ringed bird 7 years 10 months (Rydzewski 1974).

**Movements.** Migratory. Several more or less discrete wintering areas, but not yet related to specific parts of breeding range, while timing and migration routes imperfectly known. In west Europe, winters Britain (few), Denmark, south Sweden, East and West Germany, Netherlands, Belgium, France and Spain. Departures from breeding grounds usually early to mid-September, stragglers remaining longer; thus for those breeding USSR, passage Leningrad area still in progress late October. Arrivals winter quarters from late September to early October; departures middle to late March for rapid return migration. Small British group (nominate *fabalis*) exception; this rarely appears before early January, suggesting secondary movement from another wintering area, possibly Denmark, from which dispersal made south and west in cold weather; perhaps relevant that birds ringed Swedish Lapland recovered Denmark (October, November) and England (January).

Large-scale ringing only in Netherlands, from which breeding season recoveries in tundra and forest zones of USSR east to River Ob, thus involving at least two races (*rossicus* and nominate *fabalis*). Passage recoveries in

Denmark, north Germany, Poland, Gulf of Finland overland to White Sea and eastwards; also winter recoveries south-west to Spain, south-east to Hungary and Italy, and 2 in east England (where *rossicus* or intergrades occur occasionally). Thus some link with another major wintering group: Austria, Hungary, Italy, Yugoslavia, (irregularly to Greece and Turkey), present mainly October–March; little known about migration routes though recoveries from east to Yenisey delta (Sterbetz 1971); apparently *rossicus* and intergrades predominate, but a few *johanseni* (from west Siberia) identified (Johansen 1962). Only small numbers occur between these 2 major areas, e.g. in Switzerland, south-west Germany, east France. Further east, reported migrating through European Russia and Ukraine in autumn; these formerly wintered on Black Sea coasts but no recent information and may have shifted winter quarters. Breeders of central and east Siberia (3 races) winter from Turkestan eastwards; only occasional visitor to Caspian Sea. Major moult gathering of non-breeders reported south Novaya Zemlya, central Yamal, and west Taymyr (Uspenski 1965). See also Bauer and Glutz (1968); Dementiev and Gladkov (1952); Mathiasson (1963b); Markgren (1963); Mörzer Bruyns *et al.* (1973).

**Food.** Based mainly on information supplied by M Owen. Grasses, cereal grains, and other agricultural crops; mainly by grazing on arable and pastureland in winter. Little difference between races in winter food.

In southern Sweden, mainly found on pasture and clover fields (77% by observation) taking, from wetter areas, grasses *Alopecurus geniculatus*, *Glyceria fluitans*, *Poa trivialis*, *P. pratensis*, *Deschampsia caespitosa*, and *Holcus lanatus* as well as clover *Trifolium repens*. Also used ploughed fields (10%), fields of winter rye *Triticum* (8%), and, mainly autumn, stubbles (5%). Leaves of grasses (*Poa*, *Phleum*, *Alopecurus*, *Holcus*, and *Deschampsia*) predominated in 10 stomachs, with barley and roots, including rhizomes of *Agropyron repens* (Markgren 1963).

In Germany, analysis of 132 stomachs autumn and winter showed rhizomes of *Agropyron repens* important, with leaves of wild and cultivated grasses, potatoes *Solanum tuberosum*, stubble grain, and winter rye. *Equisetum* shoots, winter wheat *Triticum*, and clover also taken (Schröder 1969).

In Hungary, grain and seeds main constituents of 47 stomachs in autumn; winter wheat and grasses also frequent. Seeds mainly wheat, maize, and rice, but also barley *Hordeum*, sunflower *Helianthus*, and weeds (*Echinochloa*, *Polygonum*, *Datura*, *Suaeda*, *Bolboschoenus*, and *Atriplex*). 53 stomachs in winter and spring contained mostly winter wheat, grasses and other pasture plants, *Chara*, and *Suaeda* leaves, with grain and seeds of secondary importance (Sterbetz 1971).

In USSR, on spring migration, Upper Volga lakes, takes 75% by weight green parts (including 65%

*Deschampsia caespitosa* leaves) and 25% berries and seeds (including 20% cranberry *Oxycoccus*). In same period on Mologa flood plains, 50% vegetative parts (including 40% stands and roots of winter rye), 7% *D. caespitosa*, and 50% seeds and berries (including 30% rye, 10% oats, 4% *Polygonum*, 5% cranberry) (Isakov and Raspopov 1949; Yu A Isakov). In cold weather, Netherlands, Brussels sprouts *Brassica* taken (Philippona 1966).

In north European (USSR) summer, feeds on green parts of plants, and fruits of crowberry *Empetrum nigrum* (Yu A Isakov). In east Siberia, preferred foods in summer, leaves, flowers, and seeds of *Polemonium acutiflorum*; leaves, seeds, and rootstocks of *Polygonum viviparum*; and flowers and seeds of *Saxifraga cernua*. Grasses *Alopecurus alpinus* and *Poa* also important, occasionally with other tundra herbs (Bauer and Glutz 1968).

Goslings probably eat leaves, buds, and flowers of arctic vegetation and various aquatic and terrestrial insects, molluscs, crustaceans, and even fish roe. *Equisetum* shoots found in stomachs of older juveniles from Kolyma river (Dementiev and Gladkov 1952).

**Social pattern and behaviour.** Based mainly on material supplied by H Boyd.

1. Gregarious except when nesting; typically in smaller, less dense flocks than most other geese even when abundant (for sizes of migrating and wintering flocks, see e.g. Mathiasson 1963b). Composition of flocks much as in other geese (see Pink-footed *A. brachyrhynchus*). At first, consist chiefly of family units (see e.g. Elder and Elder 1949); these joined later by non-breeders which earlier gathered into separate flocks of several hundred prior to moult and made lengthy moult migrations northward (Salomonsen 1968). By late autumn and winter, flocks well mixed. BONDS. Monogamous pair-bond; forms life-long pairs in captivity and presumably in wild, though no studies of marked birds. Age of initial pair-formation not known, but some time in advance of first breeding at 2–3 years. Both parents tend young until fledging and family remains united until next breeding season. BREEDING DISPERSION. Solitary, nests well dispersed. No details of territorial behaviour in wild but in captivity ♂ defends area round ♀ and nest-site until young hatch. ROOSTING. Communal, often in multiple flocks, except when nesting. In winter quarters, typically in open on lakes or floodwater, comparatively infrequently on estuaries, marshes or other localities; usually near or within easy travelling distance of grazing grounds. Large area of free water required. Flocks using same roost either gather all together in fairly limited area or split up into 2 or more groups; in either case, individuals usually not close, though family parties stay loosely together. Flocks keep well away from shore, though in rough weather move closer to windward side. Usually nocturnal, but flocks not infrequently also visit roost area at midday for bathing and loafing (though in mid-winter, especially, will rest on grazing grounds during day); in some areas, occasionally spend night at grazing grounds. In southern Sweden, most sites used regularly and clearly traditional; when sites frozen, flocks will sleep on ice—though may use alternative site. Flocks arrive at roosts after sunset and leave before dawn, adjusting activity by changing intensity of feeding effort as well as responding to day-length; feeding flights usually 5–10 km from roost (see Markgren 1963; Mathiasson 1963b). Elsewhere, will commute to 15 km or beyond, but flights tend to be shorter than

those of White-fronted Geese *A. albifrons* in comparable situations. Small flocks in Britain show great attachment to traditional roosts and feeding areas (Watson 1972).

2. No detailed studies, but behaviour evidently closely similar to other *Anser* (see Greylag Goose *A. anser*). FLOCK BEHAVIOUR. When grazing, tends to separate into small groups, probably often of 1 or more families. When moving to and from roost and at roost, gather in larger flocks; similarly, migrant flocks tend to be small (usually less than 20) though larger when arriving at and departing from stopping places (Mathiasson 1963*b*). Behaviour serving to integrate flock much as in other geese, including frequent use of various 'gaggling' calls both on ground or water (when feeding or loafing) and in air, by day and by night. These include comparatively quiet Contact-calls; at times, such vocalization heard almost continuously from (e.g.) birds roosting on water (Markgren 1963). Calls much louder when flock takes wing, especially after disturbance or when individuals lose total contact with family or flock companions. Flocks fly in V-formations or oblique lines; when alighting, often drop down steeply from good height, tipping from side to side with wings outstretched stiffly or bent, flight-feathers vibrating while calling (Mathiasson 1963*b*). On approach of predator or whenever danger threatens, flock members become silent, stretch necks erect, remain still, and concentrate attention on it before showing appropriate response; before taking wing, give pre-flight signals by rapidly shaking heads (Markgren 1963). ANTAGONISTIC BEHAVIOUR. Threat includes Erect-posture, Forward-display, and Diagonal-neck posture (Johnsgard 1965). For most part, winter flocks peaceful, but encounters occur especially between strangers (individuals and families); birds call, assume intense Forward-displays, and repeatedly bow and stretch necks. Also, especially in spring, ♂ drives off rival ♂♂, then returns to mate and performs Triumph Ceremony (Markgren 1963); after which momentarily assumes Bent-neck posture (Johnsgard 1965). Group flights frequently noted immediately after arrival on breeding area; 3 birds follow each other, now descending, now rising; uncertain whether this chase of ♀ by 2 ♂♂ during pair-formation or attempts by unmated ♂♂ to steal ♀ from her mate (Dementiev and Gladkov 1952), or wholly antagonistic encounter. HETEROSEXUAL BEHAVIOUR. In Triumph Ceremony, ♂ returns to ♀ with head lifted, wings spread, while Cackling (Markgren 1963); full details lacking, however, but ♀ presumably joins ♂ in mutual display, similar behaviour also occurring during courtship and at other times as greeting ceremony (see *A. anser* and Canada Goose *Branta canadensis*). Copulatory behaviour similar to other *Anser* (Johnsgard 1965). RELATIONS WITHIN FAMILY GROUP. Apparently much as in other geese, but not studied in any detail.

**Voice.** Not studied in detail. Least vocal of grey geese; calls disyllabic, reedy, and bassoon-like, resembling lower notes of Pink-footed Goose *A. brachyrhynchus* (Scott and Boyd 1957).

Flight-calls of flock resemble 'ung-ank', or evenly-spaced 'bow-wow' (see I); lone birds utter 'ow ow-ow aw' calls (see II), also 'gock' (Bauer and Glutz 1968). Conversational Contact-calls quiet, rendered as rather deep 'ah-ah-ah ...' (Markgren 1963), though intensity rises under excitement, and louder 'rat rat' call uttered on flying away (Bauer and Glutz 1968); also rendered 'gack gack gack' (Markgren 1963). 'Treng' call given as warning and Hiss uttered in aggression (Bauer and Glutz 1968).

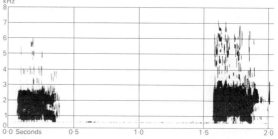

I Sture Palmér/Sveriges Radio Sweden October 1966

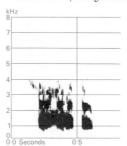

II S Wahlstrom Sweden

CALLS OF YOUNG. Contact-calls uttered in groups of 2–4 notes. Distress-calls fairly high and regular, 8 notes uttered in $2\frac{1}{2}$ s. Sleepy-call lasts *c*. 1 s; trill lower in pitch than in other calls (J Kear).

**Breeding.** SEASON. See diagram for western Palearctic. Onset of laying may be delayed by adverse weather. SITE. Low hummocks and banks free from snow and post-thaw flooding; at base of tree or among bushes. Usually close to water but up to 1 km away. Normally well dispersed, though sometimes forming loose colonies (Dementiev and Gladkov 1952). Previous year's nests re-used with new lining. Nest: low mound of grasses, dead leaves, moss, and other vegetation, with shallow cup lined with down, particularly after laying. Building: mostly by ♀, though ♂ may help, using material within reach of nest. EGGS. Oval, more rounded at one end, rough granular texture, only slightly glossy; pale straw colour, becoming stained brown during incubation. 84 × 56 mm (74–90 × 53–59), sample

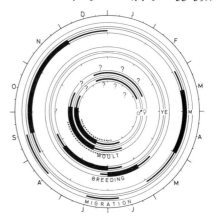

75; calculated weight 146 g (Schönwetter 1967). Weight of captive laid eggs 148 g (133–177), sample 31 (*fabalis*); 145 g (122–164), sample 53 (*rossicus*) (J Kear). Clutch: 4–6 (3–8). One brood. No information on replacement clutches. Eggs probably laid at 24-hour intervals. IN-CUBATION. 27–29 days (25–30). Only ♀ incubates, covering eggs with down when off nest. Begins with last egg; hatching synchronous. Eggshells left in nest. YOUNG. Precocial and nidifugous. Self-feeding. Both parents care for young, defending them against predators. Brooded by ♀ at night when small. FLEDGING TO MATURITY. Fledging period *c*. 40 days, but little data (Scott and Boyd 1957). Young stay with parents through first autumn and winter, migrating with them in spring but leaving before they reach breeding grounds. Age of first breeding normally 3 years, but likely that a few breed in 2nd spring (P Scott). BREEDING SUCCESS. No data.

**Plumages.** Nominate *fabalis*. ADULT. Head and neck dark brown, paler on nape and sides of head, brown-grey on chin. Mantle, scapulars, breast, and sides of body dark brown to brown-black; feathers broadly edged buff on mantle, tipped brown shading to white on scapulars and sides of body and broadly tipped off-white on chest. Often white rim at base of bill, varying from a few white feathers to broad white band, up to 12 mm, on forehead. Back and rump black-brown, longest rump feathers tipped white. Upper tail-coverts white, some central with grey-brown centre and shaft at times. Belly and thighs grey-brown, feathers of belly broadly tipped white. Vent, sides of rump, and under tail-coverts white. Tail-feathers brown-black, narrowly margined and broadly tipped white. Flight-feathers black; primaries with white shafts and grey bases of outer webs; inner secondaries and tertials narrowly edged white on tips and outer webs. Upper wing-coverts dark grey, lighter on primary coverts and outer greater secondary coverts, tinged brown on inner median and lesser coverts. Greater and median coverts broadly tipped white, narrow white margins on outer webs. Under wing-coverts and axillaries dark grey. Pale feather-edges abraded by wear in summer. DOWNY YOUNG. Upperparts olive-brown, tinged yellow on nape, mantle, and sides of head. Streak before eye dark brown. Faint spot below wing and bar across wing yellow; sides of body dusky brown, underparts pale yellow, bleaching when older. JUVENILE. Like adult, but edges of feathers of upperparts and upper wing narrower, browner. Usually no white near base of bill. Sides of body paler brown, with less contrasting edges. Grey-brown feathers on breast and belly less white on tips, giving darker, more scaly appearance. When worn, dark brown subterminal blotches along shafts of lower breast feathers often visible. From mid-winter, upper tail-coverts and tail appear darker than adult, as white edges more abraded. Feathers of neck square (pointed in adult); wing-coverts, scapulars, and feathers of mantle, breast, belly, and sides of body narrow and rounded (broad and square in adult). FIRST WINTER AND SUMMER. Like adult, but retaining worn juvenile wing-coverts and usually some body or tail-feathers.

**Bare parts.** ADULT. Nominate *fabalis*. Iris dark brown. Bill orange-red to orange-yellow; usually with at least some black on culmen above nostrils, at base of lower mandible and at tip; nail black. Foot orange-yellow, nails dark horn. *A. f. rossicus* and other subspecies: like nominate *fabalis*, but bill black, orange

confined to band round bill behind nail. Much variation in bill pattern in all subspecies, especially in zones of contact between nominate *fabalis* and others. Nominate *fabalis* regularly has black round nostrils and at base of upper mandible, orange confined to upper bill below nostrils and band behind nail; other subspecies often have streak of orange extending along cutting-edge of upper mandible to base of bill. In contact zones, a few nominate *fabalis* have as much black as typical *rossicus* or *johanseni*, but latter two only exceptionally with as much orange as typical nominate *fabalis*. Ground colour of bill in part of some populations sometimes pink or flesh, probably due to mutation causing loss of yellow carotenoid pigment; such birds, named *neglectus* Sushkin, 1895 (also foot pink), or *carneirostris* Buturlin, 1901 (foot normal orange), occur in all subspecies. Birds with intermediate, orange-pink bill-colour also occur fairly regularly. DOWNY YOUNG. Bill and feet dark grey. JUVENILE. Like adult, but orange of bill and feet duller, washed grey, becoming like adult in 1st winter. Inside of mouth pale flesh or pinkish white (grey in adults), becoming dark grey or black on central palate when older.

**Moults.** ADULT POST-BREEDING. Complete; flight-feathers simultaneous. Flightless for *c*. 1 month, July–August. Main moult of tail and body starts when wings half-grown, last feathers replaced late autumn. Feathers of head and neck apparently replaced again about mid-winter. POST-JUVENILE. Partial; involves most or all tail- and body feathers and some wing-coverts. Starts about October with head and neck, followed mid-winter by some scapulars, longer flank feathers, and chest. From late January, some have renewed all body feathers, at least central tail-feathers, and some median wing-coverts; others retain part of juvenile body feathers until at least April.

**Measurements.** Wing and culmen lengths of typical nominate *fabalis* (slender and mostly orange bill), and typical *rossicus* (heavy and mostly black bill) from sample of *c*. 1000 live birds, measured December–March 1960-9, Netherlands (L Hartsuyker, RIN; K H Voous); additional data from skins of typical specimens of both races (RMNH, ZMA). Depth of bill (D) is maximum visible depth of lower mandible near base when bill closed.

ADULT. Nominate *fabalis*.

| | | | | | |
|---|---|---|---|---|---|
| WING | ♂ 481 | (14·0; 87) | 452–520 | ♀460 (13·7; 73) | 434–488* |
| TAIL | 132 | (5·84; 5) | 126–139 | 132 ( – ; 2) | 127–137 |
| BILL | 63·6 | (2·96; 93) | 57–70 | 60·0 (2·73; 75) | 55–66 |
| BILL D. | 6·4 | (0·60; 13) | 5·5–7·3 | 6·2 (0·61; 6) | 5·3–7·0 |
| TARSUS | 82·2 | (3·61; 21) | 76–90 | 76·7 (3·57; 11) | 73–80* |
| TOE | 88·6 | (3·38; 12) | 82–94 | 82·7 (5·09; 11) | 78–84* |

ADULT. *A. f. rossicus*.

| | | | | | |
|---|---|---|---|---|---|
| WING | ♂ 454 | (9·97; 144) | 430–478 | ♀43² (10·8; 133) | 405–458* |
| TAIL | 125 | (8·51; 4) | 113–136 | 119 (8·13; 6) | 106–130 |
| BILL | 57·7 | (2·32; 142) | 52–63 | 54·6 (2·24; 134) | 49–60 |
| BILL D. | 7·9 | (0·81; 13) | 7·0–10·0 | 7·5 (0·58; 10) | 6·9–8·6 |
| TARSUS | 75·2 | (3·84; 13) | 70–81 | 73·9 (3·22; 13) | 69–79 |
| TOE | 76·4 | (3·90; 12) | 67–81 | 74·7 (3·12; 9) | 70–81 |

ADULT. *A. f. johanseni*. Shensi (China), Kazakhstan, and Lake Baikal (USSR) (J Delacour; Dementiev 1936; RMNH).

| | | | | | |
|---|---|---|---|---|---|
| WING | ♂ 479 | (16·7; 11) | 454–520 | ♀451 (15·5; 5) | 425–468* |
| BILL | 66·6 | (3·49; 11) | 62–72 | 62·3 (4·07; 5) | 58–70 |
| BILL D. | 7·1 | (0·33; 6) | 6·5–7·5 | 6·7 ( – ; 1) | – |

ADULT. *A. f. serrirostris*. Lena delta to Anadyr, breeding; China, Japan, wintering (J Delacour; Dementiev 1936; RMNH).

| | | | | |
|---|---|---|---|---|
| WING | ♂ 474 (21·4; 28) | 440–524 | ♀449 (21·5; 20) | 420–491* |
| BILL | 65·9 (3·43; 30) | 59–72 | 63·3 (3·06; 17) | 58–69 |
| BILL D. | 10·4 (1·05; 26) | 8–12·5 | 9·3 (0·83; 10) | 8·1–11·3* |

ADULT. *A. f. middendorffii*: whole breeding area, and migrants south-east Siberia, China, Japan (J Delacour; Dementiev 1936; RMNH).

| | | | | |
|---|---|---|---|---|
| WING | ♂ 492 (33·7; 16) | 440–558 | ♀488 (18·6; 9) | 465–524 |
| BILL | 73·3 (5·05; 15) | 64–81 | 72·7 (5·29; 9) | 63–80 |
| BILL D. | 8·6 (0·95; 15) | 7–10·5 | 8·6 (0·82; 8) | 7–9·8 |

JUVENILE. Like adult, but wing and tail shorter. Wing of (1) nominate *fabalis* and (2) *rossicus* (origin as adults).

| | | | | |
|---|---|---|---|---|
| (1) | ♂ 461 (13·3; 38) | 436–487 | ♀ 442 (13·8; 48) | 418–476* |
| (2) | 429 (12·7; 31) | 390–451 | 417 (12·8; 36) | 378–443 |

*Sex differences significant.

**Weights.** ADULT. (1) Typical nominate *fabalis* and (2) *rossicus*, Netherlands, December–March (see Measurements).

| | | | | |
|---|---|---|---|---|
| (1) | ♂ 3198 (302; 68) 2690–4060 | | ♀2843 (274; 58) 2220–3470 | |
| (2) | 2668 (233; 126) 1970–3390 | | 2374 (203; 117) 2000–2800 | |

JUVENILE. Average and range of both sexes *c.* 6–15% lower than adult.

**Structure.** Wing long, rather narrow. 11 primaries: p9 longest, p10 7–18 shorter, p8 6–12, p7 30–42, p6 65–78, p1 175–240; p11 narrow, pointed, concealed by primary coverts. P10 with narrow outer web; p10 and p9 emarginated on inner web, p9 and p8 on outer. Tail short, rounded; 18 feathers. Feathers of neck lanceolate in adult, projecting, giving 'furrowed' appearance. Bill heavy, high at base, culmen gradually sloping to oval nail in tundra-inhabiting *rossicus* and *serrirostris*; bill more slender, culmen concave between nostrils and tip, and rounded nail in taiga-inhabiting nominate *fabalis*, *johanseni*, and *middendorffii*. From above, sides of bill about parallel in taiga birds, converge towards tip in tundra. Cutting-edge of upper mandible arched, serrated; *c.* 26–27 horney 'teeth' visible *fabalis*, *c.* 23–24 *rossicus*. Lower mandible straight and slender in taiga subspecies, heavy and convex near base in tundra. Tarsus and toe reticulated. Outer toe *c.* 90%, inner *c.* 75%, hind *c.* 25% middle toe.

**Geographical variation.** 2 main groups of subspecies can be recognized: (1) those inhabiting tundra, with heavy bill (especially lower mandible) and shorter neck, tarsus, and toes, and (2) those inhabiting taiga, with slender bill and longer extremities. In both groups, size increase clinal from west to east: on tundra, from smaller *rossicus* to larger *serrirostris*, all with relatively dark bills; on taiga, from smaller *fabalis* with mainly orange bill to large, dark-billed *middendorffii*. *A. f. johanseni* intermediate in size between latter 2; bill mainly dark-coloured. Broad zone of intergradation between main groups in forested tundra, especially in western part of distribution. In this zone, height of lower bill best identifying character, irrespective of amount of black on bill. Pink-footed Goose *A. brachyrhynchus* closely related to *A. fabalis*; represents insular population of tundra birds, and fits in cline of decreasing body size— *serrirostris*—*rossicus*—*brachyrhynchus*, but kept separate on ground of difference in behaviour, proportions, and colour of plumage. Difference in colour of bill and foot not always consistent, as a few *A. brachyrhynchus* have bill and foot orange, and some *A. fabalis* pink. CSR

## *Anser brachyrhynchus* Pink-footed Goose

PLATES 52 and 61
[after page 398
and between pages 494 and 495]

DU. Kleine Rietgans    FR. Oie à bec court    GE. Kurzschnabelgans
RU. Короткоклювый гуменник    SP. Ánsar piquicorto    SW. Spetsbergsgås

*Anser Brachyrhynchus* Baillon, 1833

Monotypic

**Field characters.** 60–75 cm; wing-span 135–70 cm; smaller than Greylag Goose *A. anser*, and averaging slighter, smaller, and shorter necked than Bean Goose *A. fabalis*. Medium-sized, rather compact, rather short-billed and short-necked, essentially pinkish-grey goose, with dark, round head and foreneck and pale forewing obvious in flight. Sexes alike though ♂ larger; no seasonal variation. Juvenile separable.

ADULT. Head and upper neck dark brown, paler brown on lower neck; upperparts grey with pink tone and increasingly separated lines of feathers tipped brownish-white, forming regular transverse pattern; tertials and inner secondaries brown-grey, edged pale grey. Upper tail-coverts white contrasting with ash centre to back and dark grey, white-rimmed tail. Chest pink-brown, merging with faintly and irregularly banded brown flanks and belly; upper flanks edged white, forming obvious line. Some adults have white feathering across top and down sides of base of upper mandible. Upperwing shows distinct pattern, with pale ashy coverts contrasting with brown-black flight-feathers. Bill light and short, usually with upper mandible slightly concave in outline; pink with varying amount of dark brown extending from base; pink sometimes restricted to small patch behind black nail. Eye dark brown, with no ring. Legs and feet pink, sometimes almost purple. JUVENILE. Darker with less pink or grey tone, more mottled due to incomplete pattern of feather-tip colours and often with ochre legs and feet; in good light, quite distinct from bright adult.

At long range or in poor light, difficult to separate from other geese but compactness, daintier carriage, contrast of dark head with greyer body and grey forewing in flight characteristic. Shorter neck (looking waisted in most stances) and bare-part colours important in distinguishing from *A. fabalis*.

Flight action and behaviour generally as other grey

geese, but wing-beats faster and more fluid than larger species and silhouette daintier and more compact with small, round head closer to more rounded body; 'whiffling' aerobatics highly developed; flight formations less broken than *A. anser*. Gait freer than large geese; swims well. Breeding populations often occur as large associated communities. Migrant and wintering populations highly gregarious, exploiting traditional area of wetlands and pasture, roosting on coastal flats or undisturbed water. Tolerant of other geese but, in mixed groupings, flocks tend to feed apart; association with farms has accustomed it to man, but shyer than *A. anser*. Flight-notes not individually loud but shriller than *A. anser* and *A. fabalis* (counter-tenor rather than tenor); most characteristic a disyllabic, high-pitched, excited yelp of ♂, a trisyllabic, excited cackle, and a squeak (in alarm). Flock chorus distinctly higher pitched, more urgent, and more penetrating than *A. anser* and *A. fabalis*.                    DIMW

**Habitat.** Concentrated largely in Atlantic sector of west Palearctic, but presence in breeding habitat (up to 700 m above sea) limited to brief and uncertain ice-free period. Strict attachment, for reasons not understood, to parts of apparently suitable terrain, at mean densities above 130 nests per km² (Kerbes *et al.* 1971). Preference in Iceland for inaccessible nest-sites in river gorges suggests safety from ground predators primary requirement; apparent inconsistency of wide-spread grouping of oasis nests on low heathy mounds or ridges perhaps due to relative failure of such predators to reach these seasonally uninhabitable uplands. Evaluation complicated by rapid transition from snow and ice cover to shallow water after thaw and then to boggy grassland with abundant sedge *Carex* and moss *Rhacomitrium*, affording ample food for young. Drier conditions and overgrazing by sheep suggested as possible factors inhibiting spread over other areas. In Spitsbergen, where Arctic Fox *Alopex lagopus* apparently not a menace, nest on flat ground or grassy slopes when snow-free at laying time, as well as low cliffs and rock outcrops (Norderhaug *et al.* 1965). Scale of moult migration by non-breeders to east Greenland in June–July may also be linked with inadequacy of suitable habitat in Iceland; for breeding birds, this inadequacy appears even more marked in Greenland.

In winter quarters, roosts on estuarine flats and sandbanks, or on freshwater lakes and sometimes heather moors up to 30 km from feeding places; latter mostly on arable land. Stubble, especially barley and oats, and potato fields more often favoured than grassland. Highly gregarious, mobile, and alert, critically prospecting opportunities and hazards of local habitats in flight at wide range of heights. Few species walk so far; up to 24 km appears not unusual for families with flightless young (Bulstrode and Hardy 1970). Precariously poised for survival, although currently successful in maintaining recently expanded population strength.

**Distribution.** No recent changes in breeding range reported. Recorded Franz Josef Land in 1914, but breeding not fully established (Dementiev and Gladkov 1952). Contraction of winter range in Britain and Ireland during last few decades, with almost total withdrawal from Ireland and fewer England, possibly due to increased available food supply in Scotland (Boyd and Ogilvie 1969).

Accidental. Finland, USSR, Poland, Czechoslovakia, Hungary, Austria, Switzerland, Yugoslavia, Rumania, Italy, Spain, Azores, Madeira, Canary Islands.

**Population.** 2 discrete breeding populations. (1) ICELAND AND GREENLAND. Iceland: main colony Thjórsárver, *c.* 2500–4000 pairs, 1951–3, 10 700 pairs, 1970 (Kerbes *et al.* 1971), *c.* 8000 pairs, 1974 (Gardarsson 1975); elsewhere *c.* 2000 scattered pairs. Greenland: *c.* 1000 pairs in east. Winter in Britain, where increased from *c.* 30 000 birds in 1950 to 76 000 in 1966, since fluctuating around 70 000, but increased to 89 000 in 1974. Possible factors for increase include lower adult mortality rate, increased winter food supply, and more statutory refuges on wintering grounds (Boyd and Ogilvie 1969). (2) SPITSBERGEN. No complete breeding counts, but estimated total of 12 000–13 000 birds made from partial counts (Norderhaug 1970*b*). Counts on autumn passage, Denmark, 12 000–15 000 (Fog 1971) compared with 7000–10 000 in late 1950s (Mörzer Bruyns *et al.* 1969).

Survival. Mean annual mortality of adults 26%, and of 4–16 months old 42% (from ringing). Mean annual mortality of all over 4 months 21·5% (from censuses), with evidence of decline in rate 1950–72 (Boyd and Ogilvie 1969). Life expectancy of adults 3·3 years. Oldest ringed bird 21 years 5 months (BTO).

**Movements.** Migratory. Breeding population of Iceland and east Greenland winters Scotland and England, with a few in Ireland. Main departures from Greenland breeding and moulting grounds during first half September, though southward dispersal begins last half August; crosses to Iceland, mostly into interior, where joins bulk of breeding stock. Main emigration to Britain first half October, but occasionally late September when early snow covers Icelandic feeding grounds; passage almost complete by late October. Several major arrival points in Britain, from which birds disperse later, though some go direct to traditional haunts; movements due to weather, local

PLATE 52. (*facing*) *Anser fabalis* Bean Goose (p. 391). *A. f. fabalis:* 1 ad, 2 ad showing extreme bill colouring, 3 juv up to 5–7 months, 4 downy young. *A. f. rossicus:* 5 ad. *Anser brachyrhynchus* Pinkfooted Goose (p. 397): 6 ad, 7 ad white feathering round bill, 8 juv up to 5–7 months, 9 downy young. *Anser anser* Greylag Goose (p. 413). *A. a. anser:* 10 ad, 11 juv up to 5–7 months, 12 downy young. *A. a. rubrirostris:* 13 ad, 14 downy young. (PS)

Peter Scott

feeding conditions, and disturbance occur throughout winter (Wildfowl Trust). Return passage to Iceland in April, with some prior gathering, particularly in northern haunts. Peak passage second half April, continuing first half May. Whereas major arrivals over north and north-east coasts Scotland, spring departures include major contingent up west coast and over Hebrides (Williamson 1968). Massive moult migration from Iceland early June to east Greenland, involving thousands of non-breeders (Taylor 1953; Christensen 1967). Exceptional recoveries of Icelandic birds (associated mostly with unusual weather patterns) in west Norway, south Sweden, Denmark, Netherlands, Azores, and Canary Islands.

Spitsbergen breeding population winters Denmark, West Germany, and Low Countries, a few reaching east Britain, Belgium, and north France, especially in severe winters (Holgersen 1960). Begins flocking on breeding grounds in last half August, and moves southward at end of month, continuing through September (Løvenskiold 1964). Main route probably west coast Norway, though minority may go overland from Finnmark to Gulf of Bothnia, since recorded west Russia (Novgorod), Baltic states, Poland, East Germany, and Sweden. Main influx first half October; at this time practically whole population in Denmark (Fog 1971). Onward movement south to West Germany and Netherlands starts almost immediately and continues through November; in severe winters few remain in Denmark. Peak numbers in West Germany and Netherlands December–February. Return movements commence late February to give spring build-up in Denmark, in some years approaching three-quarters of population. Departs Denmark during May. Early arrivals Spitsbergen mid-May; main immigration end May to early June. No moult migration recorded.

**Food.** Based mainly on material supplied by I Newton. Vegetable material, including parts of plants both above and below ground. Feeds like Greylag Goose *Anser anser*, though much less commonly in water, and smaller bill and gizzard tend to restrict it to softer material. Gullet holds up to 150 cc of food.

In summer quarters, eats green parts, roots, and fruits of wide variety of tundra plants. In Iceland, main foods include rhizomes of alpine bistort *Polygonum viviparum*, shoots of horsetail *Equisetum variegatum* and cotton-grass *Eriophorum*, and, in autumn, seed-heads of sedge *Carex*. In summer, adults feed at first largely on leaves and catkins of willow *Salix glauca*, switching gradually to graminoids (*Carex, Calamagrostis stricta*), which form nearly whole diet in late July and early August. At first, goslings take more herbs and *Equisetum* than adults (Gardarsson 1972). 12 viscera and 20 faecal samples collected August, central Iceland, contained predominantly leaves of sedge *Carex bigelowii*, and to lesser extent, mosses; from August, leaves and ripened fruits of crowberries *Empetrum nigrum* and *E. hermaphroditum* increasingly important, coinciding with movement of geese from marshes to higher and drier areas, while other food-plants recorded include *Cerastium nigrescens, Saxifraga, Cochlearia, Selaginella*, various mosses, and lichens (Hardy 1967). Graminoids, *Salix, Polygonum viviparum*, and *Equisetum variegatum* also important in Spitsbergen (Nyholm 1965). 25 stomachs, Spitsbergen, contained *Cerastium, Saxifraga, Equisetum* and various grasses (Roi 1911); in July–August, cropped grass and leaves of mountain sorrel *Oxyria digyna* and scurvy-grass *Cochlearia officinalis* taken (Løvenskiold 1964). 3 stomachs from north-east Greenland, May, contained mainly tubers and bulbils of *P. viviparum* and leaves of *E. variegatum*, with smaller quantities of *Carex misandra* fruits, *Oxyria digyna* leaves and *Festuca vivipara* bulbils (Bird 1941). As arctic summer progresses, apparent change of emphasis in diet to *Poa* and later to *Carex*. Summer foods generally contain high levels of protein, phosphorus and other minerals, while autumn foods rich in protein and carbohydrate. Concentrated grazing changes composition of tundra vegetation over the years, reducing or locally eliminating favoured *Polygonum, Equisetum*, and *Eriophorum* (Cottam and Hanson 1938).

In winter quarters, now feeds mainly on farmland, including grassland, but exact composition of diet differs according to local, seasonal, and annual variations in crop-plant availability. Icelandic and Greenland birds, during stay in Britain, feed on grass throughout, but mainly in spring; on cereal grains (mainly barley *Hordeum*, also oats *Avena* and wheat *Triticum*) mostly in autumn; potatoes (*Solanum tuberosum*) mostly from late autumn to early spring; and growing cereals mostly in late spring (Kear 1964; Newton and Campbell 1970). Also take roots of carrots (*Daucus*) and leaves of rape (*Brassica napus*) and other brassicas in some areas. On pasture, common agricultural grasses, such as *Lolium, Phleum, Festuca*, and *Poa* eaten; also leaves and stolons of *Trifolium* and other plants. Rarely, seeds and leaves of *Polygonum, Atriplex, Stellaria*, and other weeds taken. Spitsbergen birds, on migration through Denmark, feed mainly from stubble, also from coastal and other grassland; and in winter in Holland from grass fields alone.

In 4 winters, near Loch Leven, Scotland, about half total feeding was on grass fields, a third on stubbles (mainly barley), and rest mainly on potatoes. *A. brachyrhynchus* took fewer root-crops than did *A. anser* in same area; difference probably linked with small bill of

◀ PLATE 53. (*facing*) *Anser albifrons* White-fronted Goose (p. 403). *A. a. flavirostris:* 1 ad, 2 ad showing heavy barring, 3 juv up to 5–7 months, 4 downy young. *Anser a. albifrons:* 5 ad, 6 ad showing heavy barring, 7 juv up to 5–7 months, 8 downy young. *Anser erythropus* Lesser White-fronted Goose (p. 409): 9 ad showing normal barring, 10 ad showing heavy barring, 11 juv up to 5–7 months, 12 downy young. (PS)

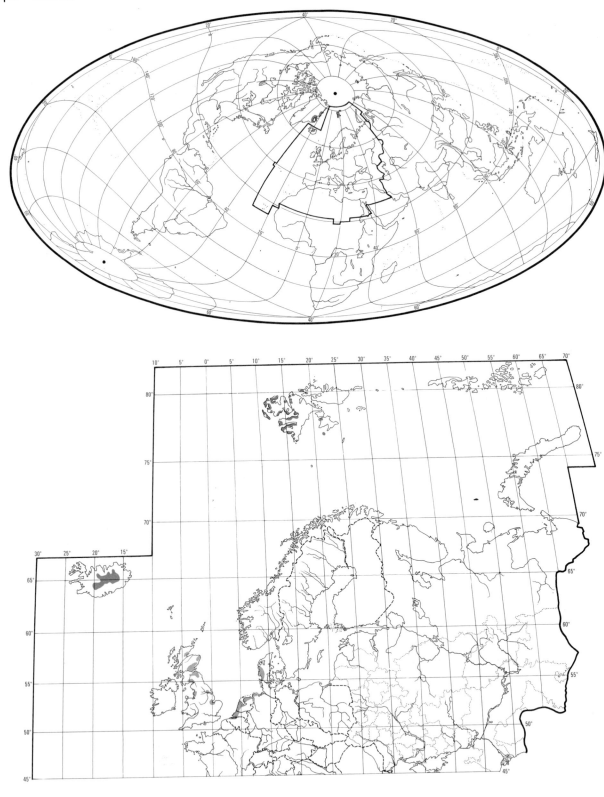

former. Main conflict with agriculture was eating of young grass in spring (Allison *et al.* 1974).

**Social pattern and behaviour.** Based mainly on material supplied by H Boyd.

1. Gregarious throughout year, except pairs for short period while nesting. In flocks outside breeding season, composed of family parties, pre-breeders, and mature adult pairs without young; autumn flocks large, commonly up to 5000 in Britain (and 7000–9000 recorded), decreasing somewhat in winter and spring as birds leave stubble for grass fields. Organization loose, large aggregations at roosts dispersing into smaller feeding groups which fluctuate widely in size from day to day and show non-territorial attachment to feeding sites. BONDS. Monogamous pair-bond of life-long duration; after loss of mate, some (but perhaps not all) form new attachments. Pairs first form in 2nd year but some inconstancy until first nesting, probably usually in 3rd summer. Both parents tend young; bonds maintained throughout 1st year, families breaking up at start of next nesting, young moving off voluntarily to better feeding area rather than being driven off by parents (I Inglis). BREEDING DISPERSION. Largely colonial over most of range in small groups of often not more than *c*. 10 pairs; nests only a few metres apart. At main Icelandic breeding area (Thjórsárver) nests much more dispersed on flatter, hummocky ground: sometimes as little as 5 m apart, usually 75 or more; mean density in 1970, 130 nests per km², maximum 540 (Kerbes *et al.* 1971). ♂ strongly territorial at start of season and throughout incubation. Earliest occupants of nesting territories usually retain them, though occasionally displaced by incomers even after laying started (I Inglis). Defence of ♀ and nest wanes during incubation, reviving at start of hatching. Parental defence of brood strong, but families aggregate into larger groups, often of several hundred, when young 10–20 days old, at about time adults become flightless. Remain gregarious thereafter, often moving long distances on foot from nesting area. Non-breeders separate from nesters, forming loose flocks. ROOSTING. Communal at all seasons, apart from nesting pairs. No fixed roost sites apparent during summer; pattern in early autumn prior to migration not recorded. In main British wintering areas, multiple flocks roost nocturnally (as a rule) at safe sites in estuaries and on inland waters such as pools, lakes, reservoirs, moorland pools, and (occasionally) floodwater; mainly on water but also, e.g., on sandbanks. Often in huge, dense groups, e.g. up to 28 500 on tree-fringed pool of 40 ha near Perth, Scotland; usually in lee of mainland shore or island. Flocks return to roost around sunset and leave soon after dawn (later and earlier in mid-winter), flying relatively long distances, up to 30 km, to feeding areas. Often arrive from and depart in several different directions, taking more than one hour to assemble and leave. Sites traditional, especially those few used immediately after autumn arrival from breeding area. Though same roosts (and feeding areas) frequented from year to year, not self-contained; birds continually shift between sites throughout wintering range, though more usually only locally. Occasionally return at mid-day to drink and bathe, but roost mainly deserted during day and use made of special 'rest stations' on grasslands and secluded pools in marshlands (mosses), particularly in longer days of late winter and spring. These stations often used as stopping places on way to main night roost. On moonlit nights, especially in mid-winter, sometimes feed at night, either remaining on feeding grounds or leaving roost in night. Roosts sometimes shared with Greylag Geese *A. anser*, 2 species keeping apart. For further details, see Brotherston (1964); Newton *et al.* (1973).

2. No detailed studies, but behaviour evidently closely similar to other *Anser*, especially Bean Goose *A. fabalis*. FLOCK

BEHAVIOUR. See *A. fabalis* and *A. anser*. When roosting, flocks approach site in level flight, rising somewhat when nearing roost, then gliding or side-slipping and tumbling down to water; first flocks more hesitant than later ones (Newton *et al.* 1973). Call loudly at roost, especially on arrival of each new group; noise declines after return of last flock, but maintained at low level throughout most of night, increasing again before dawn. ANTAGONISTIC AND HETEROSEXUAL BEHAVIOUR. As far as known, similar to *A. fabalis*. RELATIONS WITHIN FAMILY GROUP. No detailed studies. In family party, ♂ dominant over ♀ and both adults over young, but apparently no sex-determined ranking among siblings.

**Voice.** Resembles Greylag Goose *A. anser* and Bean Goose *A. fabalis* but more urgent and not so deep; lower pitched than White-fronted Goose *A. albifrons*, and lacks laughing quality. Calls usually disyllabic, sometimes trisyllabic (see I), resembling 'ung-unk' or 'ang-ank' (Scott 1951). Softer 'wink-wink' or 'king-wink' also uttered (Scott 1951). Alarm-call single, high-pitched, sharp note. Conversation-calls (see II), given while grazing or sometimes in flight, not unlike drumming of Snipe *Gallinago gallinago* (Scott 1951). Aggressive hiss uttered in anger. As in all geese, 1st autumn birds utter higher pitched calls than adults. Non-vocal wing-creaking sounds produced, as in other geese, by vibrating wing-tips on take-off and landing, and during manoeuvres in flight (see III).

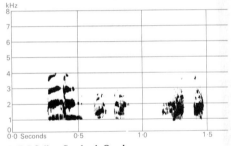

I P J Sellar Scotland October 1970

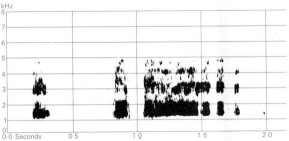

II Sture Palmér/Sveriges Radio Iceland June 1966

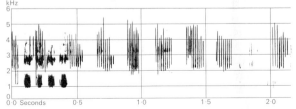

III Sture Palmér/Sveriges Radio Iceland June 1966

CALLS OF YOUNG. Contact-calls in groups of 2–3 notes. Distress-call fairly high, but less rapid though more regular than Contact-call (J Kear).

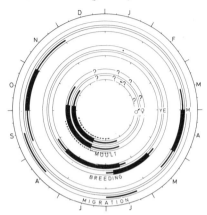

**Breeding.** SEASON. Iceland: see diagram. East Greenland: few data but laying probably commences last week May. Spitsbergen: laying commences last half May and completed first half June; main hatching period first half July with fledging from early August (Løvenskiold 1964). SITE. Low hummocks and banks snow-free at time of building, and above post-thaw floods; also tops of rock outcrops, ledges on river gorge cliffs, and tops of rock pinnacles in gorges. Colonial in some areas with nests as close as 5 m, but can be widely dispersed. Sites frequently re-used, becoming built up with permanent rim; some sites central Iceland estimated at least 40 years old (J B Sigurdsson). Nest: low mound of grasses, sedges and other vegetation, and sometimes droppings with shallow cup; average dimensions of 43 nests, diameter 30–40 cm, height 8–10 cm, diameter of cup 22 cm, depth of cup 8 cm (Nyholm 1965). Large amounts of down added during and after laying. Building: ♀ only, though ♂ may help, using material within reach. Droppings incorporated by ♀ during incubation. EGGS. Oval, rough texture, only slightly glossy; white or pale straw colour, becoming stained brown or yellow during incubation. 78 × 52 mm (70–90 × 48–58), sample 300 (Schönwetter 1967). Weight of captive laid eggs 132 g (111–149), sample 82 (J Kear). Clutch: 3–5 (1–9). Of 344 central Iceland clutches: 2 eggs, 3%; 3, 8%; 4, 45%; 5, 35%; 6, 8%; 7, 0·5%; 8, 0·5%; mean 4·4 (J B Sigurdsson). One brood. No replacement clutches recorded. Eggs laid at 24-hour intervals. INCUBATION. 26–27 days (25–28). By ♀ only, ♂ usually standing very close. Eggs covered with down when ♀ off. Starts on completion of clutch; hatching synchronous. Eggshells left in nest, and eaten by ♀ on return in spring (A Gardarsson). YOUNG. Precocial and nidifugous. Self-feeding. Cared for by both parents who defend them against predators; many pairs at nests made threat postures towards low-flying helicopter (Kerbes *et al.* 1971). ♀ broods young while small. FLEDGING TO MATURITY. Fledging period *c.* 56 days. Young stay with

parents through first autumn and winter, migrating together in spring, but leave parents at or before reaching breeding grounds. First breeding probably normally at 3 years, sometimes 2. BREEDING SUCCESS. Proportion of young of Icelandic breeding birds on British wintering grounds 1950–69 varied annually from 11%–49% (average 25·6%), with mean brood size 1·3–4·0 (average 2·5) (Boyd and Ogilvie 1969). Factors affecting breeding success include weather on breeding grounds at time of laying and hatching.

**Plumages.** ADULT. Head and neck dark brown, paler, more rufous on nape. Often narrow white line along base of upper mandible. Feathers of mantle and scapulars brown with marked blue-grey tinge and broad edges buff shading to white; sometimes blue-grey edged white. Back and rump grey, feathers on lower rump tipped white. Sides of back and rump and upper tail-coverts white. Feathers of breast and belly grey-brown, broadly tipped buff-grey; sides of body and flanks brown, feathers broadly tipped white. Vent and under tail-coverts white. Tail-feathers dark grey, with broad tips and narrow margins white. Flight-feathers black, primaries with white shafts, basally tinged grey, secondaries narrowly edged white. Primary coverts ash-grey, greater and median coverts grey broadly tipped white, inner and lesser-coverts tinged brown. Underwing and axillaries grey. DOWNY YOUNG. Mostly yellow-olive, yellow-brown on crown and back, buff on sides of head and body; down bleaches when older. Brown streak before eye. Bar on wing and spot below wing yellow-olive, belly pale yellow. Much individual variation. JUVENILE. Like adult, but light edges of mantle, scapulars, and sides of body narrower, pale grey to white, less contrasting; centres of feathers grey-brown or blue-grey. Feathers of breast and belly grey-brown, edged white, giving scaly appearance; dark shaft-streaks or blotches below visible when worn. Light edges of feathers much abraded by wear, especially upperparts and tail, dark shaft-streaks in central upper tail-coverts often visible. Neck feathers square (lanceolate in adult), feathers of mantle, scapulars, underparts, and wing-coverts narrow and rounded (broad and square in adult). FIRST WINTER AND SUMMER. Like adult, but some of juvenile feathers retained.

**Bare parts.** ADULT. Iris dark-brown. Bill pink, often with black base, black line on basal part of culmen, and black nail; in some black, pink only in narrow band behind nail. Leg and foot pink when breeding ♂ brighter than ♀), sometimes almost purple, exceptionally orange (Scott 1956). DOWNY YOUNG. Bill and foot black, nail of bill pale horn. JUVENILE. Like adult, but foot pink-grey, often tinged yellow ochre, changing to pink during winter (Witherby *et al.* 1939).

**Moults.** ADULT POST-BREEDING. Complete; flight-feathers simultaneous. Flightless for *c.* 25 days July–August. Body and tail moult starts mainly after renewal of flight-feathers; finished about October. POST-JUVENILE. Partial. Head, neck, lower scapulars, and lower flank feathers renewed about October–November, followed by mantle and breast about November–December, and by belly, back, rump, some or all tail-feathers, and some wing-coverts about January–February. Some juvenile feathers of mantle and belly may be retained until summer, but usually only most wing-coverts. Juvenile starts moult of wing somewhat earlier than adult, followed by complete body moult.

**Measurements.** Winter, Netherlands (Spitsbergen population) (ZMA, RMNH). ADULT.

| | | | | | | |
|---|---|---|---|---|---|---|
| WING | ♂ 443 | (10·8; 14) | 430–460 | ♀420 | (9·05; 11) | 405–435 |
| TAIL | 135 | (6·96; 14) | 125–149 | 127 | (7·87; 11) | 111–140 |
| BILL | 47·2 | (2·06; 16) | 43–52 | 43·2 | (1·41; 15) | 40–46 |
| TARSUS | 75·4 | (2·31; 16) | 72–80 | 70·3 | (2·66; 16) | 65–74 |
| TOE | 75·6 | (1·77; 16) | 72–78 | 70·1 | (2·03; 16) | 66–73 |

Sex difference significant. Adult wing and tail longer than juvenile. Other measurements similar; combined in above table. JUVENILE.

| | | | | |
|---|---|---|---|---|
| WING | ♂ 397,419 | | ♀401 | (6·37; 5)　394–412 |

**Weights.** ADULT. (1) Spitsbergen, moulting, July; (2) Britain, mainly October (Beer and Boyd 1962); (3) Netherlands,

December–March (L Hartsuyker, J J Smit, RIN).

| | | | | |
|---|---|---|---|---|
| (1) | ♂ 2600 (180; 162) | 1990–3000 | ♀ 2340 (400; 116) | 1850–2800 |
| (2) | 2770 (310; 750) | 1900–3350 | 2520 (270; 796) | 1810–3150 |
| (3) | 2490 (165; 46) | 2180–2880 | 2197 (197; 42) | 1790–2630 |

JUVENILE. Britain, mainly October (Beer and Boyd 1962).

♂ 2390 (240; 671) 1410–3080 ♀ 2170 (240; 627) 1450–2800

**Structure.** 11 primaries: p9 longest, p10 5–12 shorter, p7 3–12, p7 35–44, p6 65–75, p1 175–210. Bill short, but high at base. Lower mandible slender and straight; cutting edge slightly arched. Upper mandible with *c.* 20–23 horny 'teeth'. Rest of structure like Bean Goose *A. fabalis*.

**Geographical Variation.** See *A. fabalis*.　　　　CSR

## *Anser albifrons* White-fronted Goose

PLATES 53 and 61
[facing page 399
and between pages 494 and 495]

DU. Kolgans　　FR. Oie rieuse　　GE. Blässgans
RU. Белолобый гусь　　SP. Ánsar careto grande　　SW. Bläsgås

*Branta albifrons* Scopoli, 1769

Polytypic. Nominate *albifrons* (Scopoli, 1769), tundra Kanin Peninsula to Kolyma, USSR; *flavirostris* Dalgety and Scott, 1948, south-west Greenland, migrant west Europe. Extralimital: *frontalis* Baird, 1858, tundra Kolyma (Siberia) to Queen Maud Gulf (Canada); *gambelli* Hartlaub, 1852, willow taiga of Mackenzie Basin area, Canada; *elgasi* Delacour and Ripley, 1975, probably taiga Alaska.

**Field characters.** 65–78 cm (body 43–50 cm); wing-span 130–165 cm; smaller than Greylag Goose *A. anser* but averages larger and longer than Pink-footed Goose *A. brachyrhynchus*, and all but a few individuals bigger than Lesser White-fronted Goose *A. erythropus*. Medium-sized, rather angular, deep-chested, square-headed, essentially grey-brown goose, with white forehead and black bars on underparts. Plumage most variegated of grey geese. Sexes alike though ♂ larger; no seasonal variation. Juvenile separable. 2 races in west Palearctic distinguishable in field.

ADULT. (1) Eurasian race, nominate *albifrons*. Large area surrounding base of bill (and rarely forecrown) white, head and neck grey-brown (with crown and hindneck slightly darker). Rest of upperparts grey-brown, with increasingly separated lines of feathers finely tipped pale brown forming transverse pattern; upper tail-coverts white contrasting with dusky centre to back and dusky, white-rimmed tail. Chest buff-brown, often looking noticeably pale and contrasting with grey-brown flanks and belly, strongly but variably barred and blotched with black. Upper flanks edged white, forming obvious line; vent and under tail-coverts white. Upperwing lacks obvious pattern, as coverts dusky grey and do not contrast with greyish-black quills. Bill not heavy but quite long (with slightly concave outline to upper mandible); pink with white nail but some show orange tinge near nail. Eyes dark hazel, with thin, buff-grey (sometimes yellow) orbital ring, usually invisible in field. Legs and feet orange. (2) Greenland race *flavirostris*. Averages 5% larger than

nominate *albifrons*, with bill longer, heavier, and orange-yellow, while plumage generally darker, with marked olive tone to brown plumages, duller chest, and greater extent of black barring underneath (some nominate *albifrons*, however, are as dark and separable only on bill colour). JUVENILE. (1) Nominate *albifrons*. Distinct from adult, lacking white forehead and marked variegation, and having browner, more mottled plumage without any black bars underneath; can be confused with *A. anser*. (2) Race *flavirostris*. Darker, duskier, or more olive than nominate *albifrons*; bill colour as adult. SUB-ADULT. Both races. White forehead appears during 1st winter but black bars of underparts lacking or incomplete until 2nd autumn.

At long range or in poor light, difficult to separate from other grey geese but variegated plumage added to angular appearance always indicative. At close range, adult plumage unmistakable; single juveniles of both races can be puzzling, but close attention to bare-part colours, plumage tones, forewing colour, and shape allows identification. For separation of birds of all ages from *A. erythropus*, see that species.

Flight behaviour and formations as other *Anser* though flocks often bunch more; flight action markedly fluid, giving noticeably more agility in air, with particularly fast climb or turn over suddenly obvious danger. Flight silhouette characterized by square head, rather straight neck, deep chest, and noticeably narrower wings than other *Anser*. Gait more free than *A. anser* but slower than *A. erythropus*; swims well. Breeding populations occur as scattered communities in tundra and swamps, joining

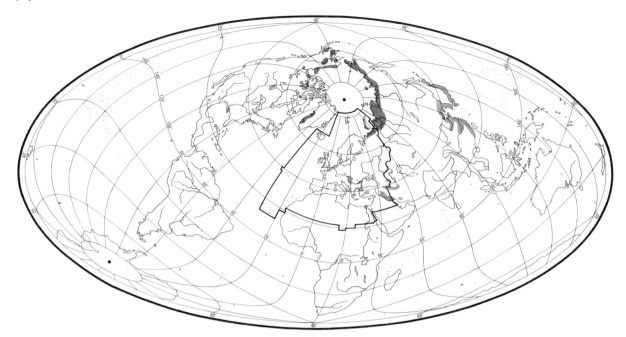

together on migration to form flocks numbering up to thousands in winter. Generally shyer than *A. anser* and *A. brachyrhynchus*. Flight-notes quite loud, higher pitched and almost merry in tone compared to other *Anser*. Musical, disyllabic bark and fast, laughing shout most characteristic. Flock chorus essentially contralto, more querulous than other large *Anser*.                    DIMW

**Habitat.** Tundra-based, staying rather longer than 100 days required for raising young in arctic breeding quarters. Depends chiefly on low-lying, shrubby tundra close to rivers, lakes, or pools, where sufficiently dry slopes, banks, mounds, hummocks, or patches of sand or clay provide suitable nest-sites commanding good view. At home on islands of Arctic Ocean, as well as on lakes and wetlands, but hardly extends into wooded taiga zone. Tolerates icy maritime conditions while awaiting thaw of breeding sites. Follows set patterns of daily local movements between feeding, resting, bathing, and nesting places, alternating between dry and wet ground and open water. Strongly prefers fresh waters. Greenland race *A. a. flavirostris*, breeding on plateaux with lakes at *c*. 700 m, favours more acid bogs and wet heaths even on uplands, where buried rather than emergent parts of plants eaten. Will rest or hide under clumps of dwarf willow, but less cautious than several other northern breeding species and consequently more vulnerable.

On passage and in winter quarters uses large lowland pastures and meadows, but also arable fields under clover and cereals, fallows, and rough grasslands including wetlands as well as steppes with halophyte or arid vegetation. *A. a. flavirostris* also frequents bogs in Irish and Welsh winter quarters. Catholicity of winter habitat linked with extent and variety of areas of winter residence, embracing also some Mediterranean coastal and even sub-tropical wetlands. On Anatolian plateau, winters at altitudes of up to and over 1000 m. Walks and swims well. Flies freely when disturbed or excited, circling widely at low or medium heights. Although conservative, climatic or human pressures can lead to sudden switches of habitat and range.

**Distribution.** No information on changes in breeding distribution.

Accidental. Portugal, Spain, Switzerland, Israel, Cyprus, Malta, Tunisia, Libya, Azores, Madeira.

**Population.** Breeding. USSR: estimated 100 000–150 000 adults west of River Khatanga (Uspenski 1970).

Winter. Counts and estimates for main sub-populations (see Movements) by Mörzer Bruyns *et al.* (1973) and Ogilvie (in Sedgwick *et al.* 1970 and unpublished). (1) Atlantic group. Counts at 2 major sites (Wexford, Ireland, and Islay, Scotland) and surveys of many lesser haunts suggest *c*. 12 000 birds. No marked trends, but probably slight decline last 10 years. (2) Baltic-North Sea group. Complete counts in mid-winter show increase from *c*. 30 000 in early 1960s to *c*. 70 000 in 1969, 100 000 in 1970, and 130 000 in 1973 (J Philippona); may represent shift from further east, but probably also reflects increased protection in Netherlands. France: decline from 3000–5000 before 1960 to tens in recent years (FR). (3) Pannonic group. Incomplete counts suggest 65 000–100 000 birds; thought to be more numerous in late 1950s particularly in Yugoslavia and north and central Greece. Italy: in Apulia up to 5000 birds before 1960; now

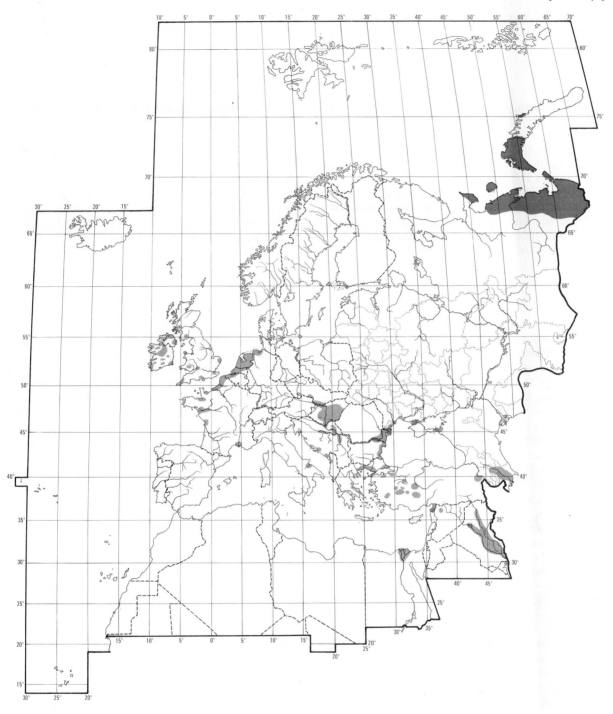

*c.* 1000 (SF). (4) Pontic group. Normally, highest estimate *c.* 100000 birds, but exceptional count *c.* 500000 in Rumania, 1968. Probably some increase in last decade, but may involve shift from Pannonic group (Johnson and Hafner 1970). (5) Anatolian group. Estimated 100000 birds (OST); recently increased (Johnson and Hafner 1970). (6) Caspian group. Numbers unknown, but thought to be 10000–40000 birds (Mörzer Bruyns *et al.* 1973).

Survival. Adults. *A. a. albifrons:* mean annual mortality (birds wintering Britain) 28% with birds wintering Netherlands, 30·9% (Doude van Troostwijk 1974). *A. a. flavirostris:* mean annual mortality 34%; life expectancy 2·4 years (Boyd 1962). Oldest ringed bird 17 years 8 months (Rydzewski 1973).

**Movements.** Migratory. 6 winter sub-populations in west Palearctic, fairly well separated; except for 2 of these, of which many ringed, not possible to relate to specific parts of breeding range. Similarly, migration routes mostly little known. (1) Baltic-North Sea group. Winters Netherlands, Belgium, England and Wales, and France in severe weather. Known from ringing to breed Novaya Zemlya and Kolguev, northern Arkhangel, and Kanin Peninsula; a few recoveries east to Taymyr. Major moult gatherings Novaya Zemlya, central Yamal, and west Taymyr (Uspenski 1965). Leaves these areas September, moving south-west to Gulf of Finland, reaching north Mecklenburg by early October where stays in large numbers to December, though may move to wintering grounds from early October. Return passage begins March in Britain and most have left Netherlands by early April; some halt in Estonia late April before moving on May. From recoveries, some move due east to Moscow region by early May, before turning north to reach breeding grounds by end May; others take same route in autumn. Several ringed England and Netherlands recovered south-east Europe in later winters indicate limited interchange between winter groups. (2) Pannonic group. Winters Yugoslavia, north Italy, Albania, and north Greece. Migrates through Pannonia (plains of Austria and Hungary), where peak passage early October to late November; return through same area late February to early March. (3) Pontic group. Winters on or near Black Sea, including Danube delta, European Turkey, the west coast of Turkey and recently north-east Greece (Thrace). (4) Anatolian group. Winters interior of Turkey (Anatolia), probably separate from Pontic group but likely to include flocks wintering Bay of Iskenderun. Small numbers move in some years to Israel and Egypt. Timing and route not clearly known but latter certainly lies through south-west USSR. (5) Caspian group. Winters round south coasts Caspian Sea and almost certainly Tigris-Euphrates basin, Iraq. Migration through south USSR. (6) Atlantic group, *A. a. flavirostris* breeds west Greenland; winters Ireland, west Scotland, and Wales (now only a few); vagrant to England and continental Europe. Southward movements Greenland from late August, main migration September to early October. Some move east over inland ice-cap, appearing on Greenland east coast, thence crossing to west and south coasts Iceland, where may stay until late October before moving to wintering grounds. Probably others continue south down west coast Greenland, thence across Atlantic direct to winter quarters where immigration late September to early November. Return movement mid-April to 3rd week May. Probably most fly first to Iceland, where recorded throughout May, then over Greenland ice-cap to breeding areas, though some may fly round Cape Farewell and up west coast. Arrive west Greenland 2nd week May to early June. Immatures remain on breeding ground to moult (Salomonsen 1967, 1968). Some *flavirostris* occur eastern North America; status uncertain.

See also Dementiev and Gladkov (1952); Bauer and Glutz (1968); Mörzer Bruyns *et al.* (1973).

**Food.** Based mainly on information supplied by M Owen. Vegetarian, chiefly leaves, stems, stolons, rhizomes, tubers, and seeds. Severs vegetation while grazing by pulling head sharply backwards with scissor-like action of side of bill while tongue moves backwards to carry food to oesophagus. Pecking rapid (up to 130 per minute), and performed while walking (Owen 1972a). Also probes with bill, especially for stolons of clover *Trifolium* (Owen 1972a). Gregarious, normally diurnal feeder, though will feed at night especially if disturbed (Markgren 1963; Owen 1972a, b). In shorter winter days in Gloucestershire, England, spent over 90% of time feeding, in longer days slightly less (Owen 1972a). Usually feeds close to roost, or will roost on feeding grounds if undisturbed: in Hungary, flew less than 10 km to feed (Sterbetz 1967), in Netherlands 10–15 km maximum (Philippona and Mulder 1960); in England, fed within 1–2 km of roost (Owen 1972b).

Little data from breeding grounds: heath berries in Alaska (Kortright 1942), and shoots of horsetail *Equisetum* recorded (Alpheraky 1905).

On autumn migration in Iceland, Greenland race *flavirostris* takes mainly rootstocks of bistort *Polygonum viviparum*, seed-heads of sedge *Carex*, and shoots of horsetail *Equisetum*, and cowberry *Vaccinium* (Gardarsson and Sigurdsson 1972). 110 stomachs of nominate *albifrons* from Hungary, collected mainly autumn, contained grasses (in over 50%), chiefly *Festuca pseudovina*, less rice, wheat and stubble weed seeds, with some snails probably taken accidentally (Sterbetz 1967). On migration in USSR, nominate *albifrons* feeds on oats from stubbles, pondweed *Ruppia maritima*, speedwell *Veronica*, *Frinbristylus*, peas *Pisum*, and the seeds of sedge *Bolboschoenus maritimus* (Dementiev and Gladkov 1952). 87 samples from East Germany contained 45% grass, 9% potatoes, 16% winter corn, 7% rhizomes of couch grass *Agropyron repens*, 10% stubble corn, and small amounts of *Equisetum*, rape *Brassica napus*, and scabious *Scabiosa* (Schröder 1969).

Analysis of 100 stomachs, numerous droppings, and observations at main British winter haunt of nominate *albifrons* showed that grasses, mainly *Lolium perenne*, *Agrostis stolonifera*, *Poa trivialis*, and *Phleum pratense* taken on inland pastures and mainly *Puccinellia maritima*, *Hordeum secalinum*, and other species taken from saltings, with stolons of *Trifolium repens* important mainly in midwinter; at same haunt, selective where range of vegetation types available, choosing shorter, younger herbage which contains less dead matter, e.g. heavily grazed pasture (Owen 1971, 1972a; M Owen). Selection of feeding sites and food in part determined by nutritional quality of food, which may be increased by proper summer grazing management (Owen 1973a). Broadly similar

winter foods taken by *flavirostris*, though in places tends to frequent wetter areas, and indications of differences in diet on some sites: on Wexford Slobs, Ireland, feeds in autumn mainly on stubble, then moves on to permanent grassland, winter wheat, potatoes, and sugarbeet fields, and, later in spring, feeds briefly on seed grain not properly covered, sprouting spring cereals, and spring flush of grass (O J Merne). 3 birds Solway pasture, Britain, each contained more than 90% stolons of *T. repens* (M Owen); 2 stomachs North Uist, Scotland, contained bulbils of deer-grass *Scirpus caespitosus*, and one clover stems and leaves (Campbell 1947). Stomachs from peat-bog, Wales, in winter contained roots of cotton-grass *Eriophorum* and bulbil of white beak-sedge *Rhynchospora alba* (Cadman 1953, 1956). Droppings from same site showed *Rhynchospora* predominant food, with grasses *Agrostis tenuis*, *Glyceria*, and *D. caespitosa* important in spring (Pollard and Walters-Davies 1968).

Between 650 and 800 g fresh food eaten per day, i.e. over 25% body weight (Owen 1972a).

**Social pattern and behaviour.** Based mainly on material supplied by H Boyd.

1. Gregarious for most of year, except pairs when nesting and during early phase of brood rearing. In flocks on migration and in winter quarters, varying from a few birds to several thousand; larger flocks composed of family groups, pre-breeders, and unsuccessful breeding pairs. Some large flocks persist and roost together, but others split up by day into smaller, ephemeral groups, while small isolated flocks often remain as closed groups for weeks. BONDS. Monogamous pair-bond of lifelong duration. Adults losing mates may obtain new ones rapidly, particularly in late winter and spring. Pairs first form, sometimes only temporarily, in 2nd year but in wild successful first breeding unknown before 3rd year, though some may attempt to nest in 2nd year. Nests in 2nd year in captivity (J Kear). Both parents tend young up to fledging, then family remains united until following spring when 1st-year birds leave parents at start of nesting; these sometimes re-join parents after succeeding breeding season, occasionally even when new young present (H Boyd). BREEDING DISPERSION. Over most of range, breeding pairs widely dispersed, not colonial; little recorded about territorial or other social behaviour before and during nesting. While young and moulting adults still flightless, families aggregate together with non-breeders though each remains as close-knit unit. Non-breeders form separate flocks and may migrate north to moulting areas (Salomonsen 1968), some apparently remaining apart from family groups until reaching winter quarters. ROOSTING. Communal. Larger flocks chiefly on estuaries, inland floodwater, or large lakes; small flocks sometimes on peatlands and other remote places without open water. Sometimes fly considerable distances to feed, though more like Greylag Goose *A. anser* than Pink-footed Goose *A. brachyrhynchus* in feeding on grass close to roost. In winter, feed all day; sometimes also at night, especially when moonlight strong.

2. No detailed studies, but behaviour virtually identical to closely-related *A. anser* though, arising from dichotomy between dispersed life of nesting pairs and gregariousness in all other situations and seasons, altogether more subdued. As in *A. anser*, re-inforcement of pair and family bonds characteristic of most social behaviour, even when sexually quiescent. FLOCK BE-

HAVIOUR. Evidently much as in other *Anser*, though interactions within wintering flocks of wild birds more closely studied (see Boyd 1953). In disputes, members of other family groups threatened by both parents and young; larger families dominate over small, families over pairs, and all groups over singles. Variation in shape of white forehead patch and belly markings suggested as possible basis of individual recognition (Johnsgard 1965). Activity within flocks generally well synchronized, especially in response to disturbance and early in each feeding session. Pre-flight signals as in other grey geese: repeated lateral Head-shakes accompanied by low calls (Johnsgard 1965); but Wing-flaps also recorded (McKinney 1965b). Synchrony in moving to drinking and bathing places less often extends from a family to neighbouring group. ANTAGONISTIC BEHAVIOUR. As other *Anser*, threat-displays such as Diagonal-neck posture and Forward-display occur during disputes, use being made of furrowed neck plumage which vibrates in lateral posturing (Johnsgard 1965). During hostile encounters in wintering flocks, 4 types of threat posturing described by Boyd (1953): (1) neck stretched forward with head at or rather above shoulder-level, bill more or less horizontal, and at times waved vertically; (2) neck stretched upward and only slightly forward, head carried high and waved backward, forward, and to lesser extent laterally, often vigorously; (3) head held low with neck close to ground, chin lifted, and bill pointed up, while wing tips sometimes lifted, with or without corresponding raising of carpal joints (4) head pointed down, neck bowed with bill pointing straight down. In 1st posture, head persistently directed toward opponent, but 2nd and 3rd often accompanied by lateral movements of body. All 4 postures performed while aggressor makes slight approaching movements or walks towards rival; 1st, 3rd, and 4th—but not 2nd—also while running, in case of 3rd culminating in violent twitching and opening of wings. Only 2nd posture, apparently, accompanied by calling. These postures evidently mostly versions of Forward-display, but 2nd seems to correspond to Erect-posture. In winter flocks in England, threats rarely lead to physical contact, and vigorous fighting involving use of wings rare; only 4% of 2129 observed encounters involved hits by one bird on another and many of those related to encounters within family. Initiators of hostile behaviour nearly always successful in causing withdrawal of target group. HETEROSEXUAL BEHAVIOUR. Much as in *A. anser*, including form and use of Triumph Ceremony in pair-formation and heterosexual and family greeting; pre-copulatory and post-copulatory displays closely similar to those of *A. anser*. RELATIONS WITHIN FAMILY GROUP. No detailed studies on behaviour of adults and downy young. Well-developed hierarchy established within each family, ♂ dominating over ♀, and adults (together or singly) over young; in most families, no linear rank-order discernible among siblings.

**Voice.** No detailed studies. Musical, higher pitched than most other *Anser*, with laughing quality. Calls disyllabic, sometimes trisyllabic, with metallic ring (see I—showing

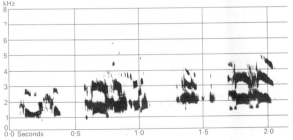

I Sture Palmér/Sveriges Radio Sweden October 1966

flight-calls of flock), resembling 'kow-lyow' or 'lyo-lyck' (Scott and Boyd 1957); also rendered 'ki-lick' or 'klä-lick' (Bauer and Glutz 1968), as shrill, jerky laughing 'klick-klick ... kleck-kleck ... kling' (Géroudet 1965), and as laughing, rapidly repeated 'wah' (Bent 1925). Voice of ♂ higher pitched than ♀ (Boyd 1953). Feeding flocks utter low buzzing. Alarm-call higher-pitched, more metallic version of normal call (M A Ogilvie).

CALLS OF YOUNG. Contact-calls uttered in groups of 2–3 notes. Distress-calls higher and louder, 8 calls in 2 s. Sleepy-call, lower in pitch than last; trill lasting ½–1 s. (Kear 1968).

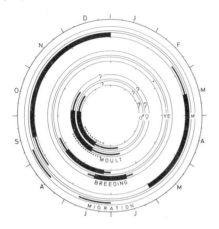

**Breeding.** SEASON. USSR: see diagram. West Greenland: main egg-laying period early June; fledging begins early August (Salomonsen 1967). SITE. Hummocks and low mounds, snow-free at time of building and above water-level following thaw. Old nests frequently re-used. Usually widely dispersed, though loose colonies in some areas. Nest: low circular rim with shallow saucer-shaped depression within, made of vegetation within reach of nest, lined grass and large amounts down, latter during and after laying. Building: by both sexes, though ♀ probably does most. EGGS. Oval, smooth, slightly glossy; white with pale straw tinge, becoming stained brown or yellow during incubation. 79 × 53 mm (72–89 × 47–59), sample 120; (Schönwetter 1967); no significant differences between eggs of races *flavirostris* and *albifrons*. Weight of captive laid eggs 114 g (97–126), sample 51 (J Kear). Clutch: 5–6 (3–7). One brood. Replacement clutches not recorded. Eggs usually laid at 24-hour intervals. INCUBATION. 27–28 days. By ♀ only, covering eggs with down when off nest. Starts on completion of clutch; hatching synchronous. Eggshells, left in nest. YOUNG. Precocial and nidifugous. Self-feeding. Cared for by both parents who defend nest and young against predators. Brooded by ♀ while small. FLEDGING TO MATURITY. Fledging period 40–43 days. Young stay with parents through first autumn and winter, depart with them in spring but remain at some stopping place while parents continue to breeding grounds. First breeding usually at 3 years. BREEDING SUCCESS. Russian

breeders wintering Britain, 1947–72, showed annual variation in proportion of young from 8%–45% (mean 28%), mean brood size 1·8–3·8 (Ogilvie 1966 and unpublished). Adverse weather on breeding grounds at time of laying and hatching among main factors affecting success. On wintering grounds, some evidence that 3-year old birds have smaller families than 4–6 year olds, and possibly smaller also in older birds, but data limited (Boyd 1965).

**Plumages.** Nominate *albifrons*. ADULT. Head, neck, and chest brown, darker on crown and nape. Forehead and area round base of bill white, bordered by black-brown band. White of forehead does not usually reach crown above eye. Feathers of mantle, scapulars, sides of body, and flanks dark brown with grey tinge on centres; edges buff, fading to white at tip. Back and rump dark grey, longest feathers tipped white; upper tail-coverts white, some with brown shaft-streak basally. Feathers of lower breast and belly grey-brown, broadly tipped buff to white. Belly and sides of body with varying amount of black feathers (edged white when fresh), forming irregular patches or transverse bars. Vent, sides of back and rump, and under tail-coverts white. Tail-feathers black with grey bloom; narrow white margins, becoming broad at tips. Flight-feathers black, primaries grey basally, shafts white; inner secondaries narrowly fringed white. Upper wing-coverts grey, primary coverts paler, lesser coverts and inner median and greater ones darker, tinged brown. Greater and median coverts broadly tipped white, lesser edged brown shading to grey. Underwing and axillaries dark grey. DOWNY YOUNG. Crown, back, and sides of body olive-brown. Dark brown streak on lores. Underparts pale grey, sides of head, throat, upper breast, and flanks tinged yellow-brown. Forehead, chin, and bar on wing yellow. JUVENILE. Like adult, but no white on head, forehead black-brown (feathers sometimes with white base, becoming visible by wear); edges of feathers of mantle, scapulars, and sides of body narrow, buff, less contrasting. Shape of feathers different: those of neck square and short (instead of lanceolate and projecting in adult), those of mantle, scapulars, breast, belly, and wing-coverts narrow and rounded (broad and square in adult). Rounded feathers of breast and belly grey-brown edged pale grey, giving scaly appearance; sub-terminal dark brown shaft-spots on feathers of breast or belly sometimes visible when worn. FIRST WINTER AND SUMMER. Like adult. Early moulting birds acquire black forehead, but feathers replaced by white ones during late winter; late moulting birds acquire white feathers directly, although often mixed with black. Longitudinal white streak on chin, often extending to upper throat. Shape of moulted feathers of mantle and breast square, but not as broad as adult; virtually no black feathers on belly, but those of sides of body with contrasting edges like adult. Sometimes median coverts also moulted, but usually many juvenile wing-coverts retained.

**Bare parts.** Nominate *albifrons*. ADULT. Iris dark brown. Bill pink, tinged yellow on middle of culmen and basal half of under mandible; nail white. Foot yellow-orange. Eyelids dull black but not uncommonly yellow especially in adult ♂, e.g. in c. 20% of Dutch adults (ZMA) and c. 18% of adults in Mecklenburg, Germany (Schröder 1966)—but less swollen and not so conspicuous as in Lesser White-fronted Goose *A. erythropus*. DOWNY YOUNG. Bill dark grey horn, nail pale horn; foot dark olive-grey. JUVENILE. As adult, but bill duller, sometimes tinged yellow; nail dark horn, becoming white during 1st winter. Foot dull yellow, sometimes tinged olive or grey. Eyelids blackish (in some yellow at end of winter). Bare parts of *flavirostris* like nominate *albifrons*, but bill of adult and juvenile cadmium-orange to cadmium-yellow, pink shade only just behind nail of bill. (Dalgety and Scott 1948; Witherby *et al.* 1939; C S Roselaar.)

**Moults.** ADULT POST-BREEDING. Complete; flight-feathers simultaneous. Flightless for *c.* 25 days; starts *c.* 25 July, finishes *c.* 20 August; non-breeders slightly earlier and ♀♀ caring young *c.* 1 week later, ready to take wing at same time as accompanying juveniles (Dementiev and Gladkov 1952). Body moult starts mainly when flight-feathers full-grown, mostly finished before reaching winter quarters. ADULT PRE-BREEDING. Head, neck, and some body feathers apparently replaced again mid-winter to spring. POST-JUVENILE. Partial. Plumage of head and neck renewed mainly November–December, forehead somewhat later; old feathers sometimes retained until February. Often new black forehead December–January, white feathers appearing in black from January (sometimes December); some specimens moult forehead directly to white. Some moult mantle, scapulars, chest, and longer flank feathers November–January and all body feathers (and sometimes tail and lesser and median wing-covers) renewed by February; others retain much of juvenile belly, back, tail, and wing-coverts until at least March.

**Measurements.** Nominate *albifrons*. Netherlands, winter; skins (RMNH, ZMA).

| | | | | | | | |
|---|---|---|---|---|---|---|---|
| WING AD ♂ | 428 | (12·3; 20) | 399–444 | ♀ 404 | (5·78; 14) | 393–415 |
| JUV | 407 | (6·21; 6) | 398–417 | 384 | (9·64; 12) | 369–397 |
| TAIL AD | 119 | (7·13; 15) | 106–217 | 113 | (7·04; 12) | 102–127 |
| BILL | 46·4 | (1·49; 32) | 43– 50 | 43·3 (1·86; 31) | | 39– 47 |
| TARSUS | 73·5 | (2·80; 26) | 69– 80 | 68·6 (2·83; 26) | | 63– 74 |
| TOE | 74·2 | (3·76; 26) | 69– 82 | 68·4 (3·43; 26) | | 63– 75 |

Sex differences significant, except for tail. Wing and tail of adult significantly longer than juvenile. Other measurements similar; combined.

*A. a. frontalis.* ADULT. Mainly USA, Japan; skins (Todd 1950; RMNH).

| | | | | | | | |
|---|---|---|---|---|---|---|---|
| WING | ♂ 425 | (12·1; 17) | 398–445 | ♀ 408 | (11·2; 9) | 389–432 |
| BILL | 52·2 (3·00; 18) | | 47–57 | 49·8 (3·09; 10) | | 45–56 |

*A. a. flavirostris.* Wexford Slobs, Ireland, winter (O J Merne).

| | | | | | | | |
|---|---|---|---|---|---|---|---|
| WING AD ♂ | 426 | (14·4; 232) | 389–463 | ♀ 423 | (14·5; 261) | 389–461 |
| JUV | 410 | (13·4; 94) | 370–441 | 405 | (15·6; 149) | 361–439 |
| BILL | 52·7(2·61; 326) | | 46–60 | 52·0(2·84; 410) | | 44–60 |
| TARSUS | 74·9(4·27; 326) | | 65–84 | 74·4 (3·91; 410) | | 63–83 |

Sex differences not significant. Adult wing significantly longer than juvenile. Other measurements similar; combined.

**Weights.** Nominate *albifrons*. (1) England, winter (Beer and Boyd 1963). (2) Netherlands, 6 winters (L Hartsuyker, RIN).

| | | | | |
|---|---|---|---|---|
| AD (1) | ♂ | 2450 (271; 87) | ♀ | 2180 (260; 92) 1720–3120 |
| | | 1790–3340 | | |
| (2) | | 2130 (183; 238) 1757–2650 | | 1905 (142; 287) 1430–2240 |
| JUV (1) | | 2150 (181; 63) 1670–2490 | | 1990 (187; 67) 1490–2400 |
| (2) | | 1901 (201; 148) 1440–2460 | | 1738 (205; 172 1150–2375 |

*A. a. flavirostris* Wexford Slobs, Ireland, winter (O J Merne).

| | | | | |
|---|---|---|---|---|
| AD | ♂ | 2543 (243; 232) 1815–3230 | ♀ | 2526 (283; 261) 1815–3290 |
| JUV | | 2405 (261; 94 1615–3035 | | 2293 (266; 149) 1445–2950 |

**Structure.** Wing long, rather narrow, pointed 11 primaries: p9 longest, p10 6–14 shorter, p8 2–11, p7 28–42, p6 60–74, p1 160–204; p11 minute, pointed, concealed by primary coverts. Emarginations of flight-feathers as in Bean Goose *A. fabalis*. Tail short, rounded; 16, rarely 18, feathers. Bill rather short, less heavy than *A. fabalis*, although high at base. Lower mandible slender, straight. Cutting edge of upper mandible arched; *c.* 25–29 horny 'teeth', partly visible when bill closed. Outer toe *c.* 92%, inner *c.* 80% and hind *c.* 28% middle toe.

**Geographical variation.** *A. a. frontalis* similar to nominate *albifrons*, but bill longer, and colour of head, upperparts, and wing-coverts slightly darker brown. Taiga subspecies *gambelli* and *elgasi* still larger and darker with relatively longer bill, tarsus, and toes; yellow eye-ring often present in *elgasi*, larger and darker of two (Delacour and Ripley 1975). Isolated *flavirostris* best characterized by long orange-yellow bill; mantle and scapulars black-brown with narrower white edges than in nominate; back and rump grey-black; throat sometimes dark brown, contrasting with paler brown sides of head and neck; breast darker brown; belly and vent more heavily barred black, sometimes all black (a few nominate *albifrons* can have as much black as *flavirostris*). Short and rounded belly feathers of juvenile *flavirostris* often dark brown to brown-black, edged white. CSR

# *Anser erythropus* Lesser White-fronted Goose

PLATES 53 and 61
[facing page 399
and between pages 494 and 495]

| | | | |
|---|---|---|---|
| DU. Dwerggans | FR. Oie naine | GE. Zwerggans | |
| RU. Пискулька | SP. Ánsar careto chico | Sw. Fjällgås | |

*Anas erythropus* Linnaeus, 1758

Monotypic

**Field characters.** 53–66 cm (body 40–42 cm); wing-span 120–135 cm; distinctly smaller, shorter-necked and smaller-headed than White-fronted Goose *A. albifrons* though a few overlap. Small, rounded, dainty, small-billed and small-headed, essentially dusky-brown goose, with extensive white forehead and forecrown. Sexes alike though ♂ larger; no seasonal variation. Juvenile separable.

ADULT. Wide surround to base of bill and on most of forecrown (to above eye) white, rest of head and body uniform dusky-brown, with upperparts showing only dull transverse lines. Underparts with delicate banding of brown paler than *A. albifrons*, and usually only a few black bars or blotches on lower flanks and belly centre; rest of plumage as *A. albifrons*. Bill light and small, almost triangular in shape, bright pink with white nail. Eye dark brown, set off by swollen ring of yellow orbital skin. Legs

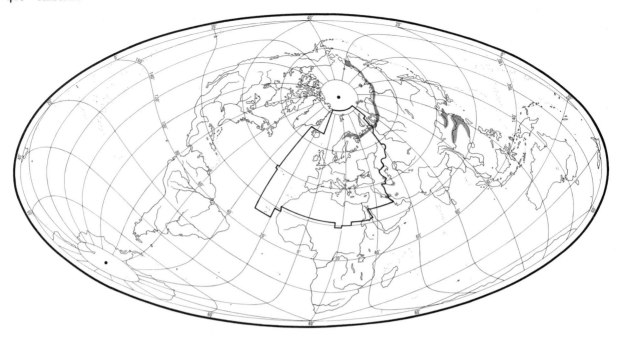

and feet orange. JUVENILE. Lacks white front to head, black bars on underparts and transverse lines on upperparts and has brown nail on bill; already shows yellow eye-ring.

At all ages, uniform, almost immaculate appearance of plumage striking and, unlike other *Anser* except some *A. albifrons*, tips of primaries often extend past tail. At long range, or in poor light, almost impossible to separate from small *A. albifrons* but, in prolonged observation, uniformity of plumage tone, daintiness and faster feeding rate (in both walking and pecking) discernible at all ages. At close range, identification confirmed by bright swollen eye-ring, though some (*c.* 20% or less) *A. albifrons* have narrow, dull eye-rings. Separation based solely on extent of white front to head or area of underbarring dubious, as some *A. albifrons* show extensive white or limited barring.

Flight similar to *A. albifrons*, but wing-beats faster and even more agile; silhouette more compact, with less neck extension and more rounded body. Gait remarkably free, with noticeably faster walk than *A. albifrons* and trotting run. Flight-notes not individually loud but much higher pitched and squeakier than *A. albifrons*; most characteristic a disyllabic or trisyllabic yelp (calls of single birds in flocks of *A. albifrons* distinctive to experienced ear); flock chorus essentially treble.                    DIMW

**Habitat.** Mostly breeds well south of Arctic Ocean, and relatively little associated with islands, shorelines, estuaries, and low-lying tundra. Biogeographically, falls between high arctic breeders and those overlapping temperate zone. Spends correspondingly long period in breeding area, especially where oceanic influences prevail. Prefers fringing zones of wooded tundra or forest edge, slopes by streams, and craggy faces of fells or mountain foothills up to at least 700 m. Breeding habitat thus intermediate between that of low tundra nesters and that of species using precipitous sites. Nests among dwarf shrubs, rough grasses or patches of stones, or in boggy hollows.

Shifts for summer moult to large open waters fringed with sedge *Carex*, etc. Adept at hiding, especially during moult, in such cover as dwarf willows *Salix*. On migration and in winter almost entirely within continental climatic zone, on open treeless tracts, including semi-arid salt steppes, meadows, pastures, and sometimes field crops. Roosts in reedbeds, rushes, or on water or banks of lakes and rivers, overlapping here more than on breeding grounds with other *Anser*. Rarely visits marine waters, and seems more terrestrial than most of relatives.

Markedly aerial in habits, apparently making more frequent use than other geese of higher airspace, diving down from great heights and making lengthy reconnaissances or flight demonstrations against ground predators.

**Distribution.** Data limited, but apparently some contraction in west of range. NORWAY: disappeared from Vadsö area after 1950; pair nested as far south as Dovrefjell 1962–3 (Haftorn 1971). SWEDEN: several breeding localities abandoned in last decade (Curry-Lindahl *et al.* 1970).

Accidental. Britain (now almost annual), Ireland, France, Belgium, Netherlands, Denmark, West Germany, Czechoslovakia (regularly in spring before 1960), Austria, Switzerland, Turkey, Egypt.

**Population.** Breeding. NORWAY: decline in Börgerfjell (Haftorn 1971). SWEDEN: always sparse, but decreased last decade (Curry-Lindahl *et al.* 1970). FINLAND: *c.* 200 pairs (Merikallio 1958); abrupt decline since *c.* 1950 (Soikkeli 1973).

Winter. Some indications of decrease in numbers wintering south-east Europe (Johnson and Hafner 1970); few recorded there and in north Turkey and south Iraq in recent years.

**Movements.** Migratory, with several separate wintering areas. Following based on Bauer and Glutz (1968);
Dementiev and Gladkov (1952). Depart breeding grounds north Scandinavia and arctic Russia late August and first half September; main migration routes south to south-east across western USSR, and from Fenno-Scandia through Baltic states, late September to early October. Some reach major wintering grounds on south coast Caspian Sea and Transcaucasia by mid-October, but farther west others still passing Sea of Azov in December. Movement from east Baltic across Poland towards River Danube; regular autumn migrant through Hungary (some overwinter in mild seasons), probably to winter in Rumania, Bulgaria, Yugoslavia, Greece (Evros delta), and north Turkey (Johnson and Hafner 1970). Others winter Khuzestan (Iran), Iranian–Afghanistan border, Iraq, and Kazakhstan; sporadic to India and Pakistan. Little ringing evidence, but 3 from Swedish Lapland recovered Stavropol (Caucasus), Greece, and central France; latter clearly a straggler.

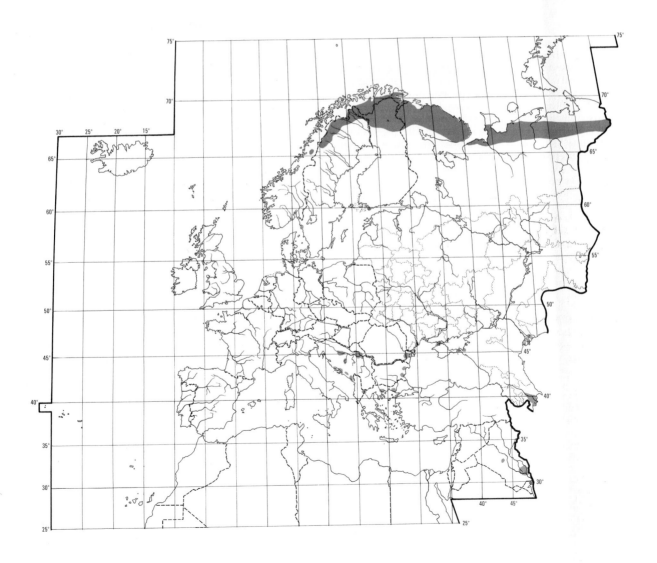

Return passage noted Caspian area from February, with most having left south coast by mid-March; reappear Sea of Azov early April. Little data on remainder of migrations to summer quarters, where arrive mainly May. South-east autumn migration routes in Europe cross south-west routes of White-fronted Goose *A. albifrons* and Bean Goose *A. fabalis*; stragglers to west Europe, almost annually Low Countries and Britain, usually in flocks of these species, having presumably been caught up with them.

**Food.** Little known. On Lapland breeding grounds, least willow *Salix herbacea* important (Scott *et al.* 1953). In USSR, 3 stomachs Taymyr peninsula contained only *Eriophorum angustifolium* and *E. scheuchzeri* (Zharkova and Borzhonov 1972).

On wintering grounds, mainly shoots, leaves and stems of grasses and green plants (Dementiev and Gladkov 1952). One shot February 1955, Dovey Estuary, Wales, contained only white beak-sedge *Rhynchospora alba* (Cadman 1955).

**Social pattern and behaviour.** Based mainly on material supplied by H Boyd.

1. Gregarious for much of year, though nesting pairs isolated. Within normal range, including south-east part of west Palearctic, often in large flocks outside breeding season. BONDS. Wild birds not studied intensively. Monogamous, life-long pair-bond in captivity, pairs first forming in 2nd year before first breeding, usually at 3 years (but occasionally at 2). Both parents tend young, family remaining together until next nesting season. BREEDING DISPERSION. Nests in solitary pairs. No information on territory in wild; in captivity, ♂ becomes territorial in 3rd year, then and in subsequent years defending mate and nest until eggs hatch. Non-breeders form moulting flocks and, in Scandinavia, move upwards in alpine zone (see Salomonsen 1968), though not known to undertake long moult migrations. ROOSTING. Communal and basically nocturnal outside breeding season, generally in dense flocks on large lakes. In south-east Caspian area, flocks habitually spend night on deep, open water among reeds in flooded complex of marsh, river, and irrigation channels, sometimes on banks, arriving during early evening and departing at first light; same area (apparently) visited during early afternoon for drinking (see Bannerman 1957).

2. Not significantly studied in wild, but evidently closely similar to other *Anser*, though display and other movements performed more quickly; Diagonal-neck Posture and Forward-display noted during hostile encounters among captive birds; copulation closely similar to that of (e.g.) Greylag Goose *A. anser* (Johnsgard 1965). Aerial evolutions recorded from wintering flock prior to alighting at drinking place (see Bannerman 1957). On migration, flocks fly in V or sloping-line formation like other geese; at other times, including flighting to and from roost, mostly fly in compact but disorderly-looking crowd, calling. Social hierarchy established within family group as in White-fronted Goose *A. albifrons*.

**Voice.** Noticeably higher pitched, more squeaky, and above all more rapid (Bauer and Glutz 1968) than White-fronted Goose *A. albifrons*. Calls not loud, ordinarily disyllabic or trisyllabic but sometimes in longer series

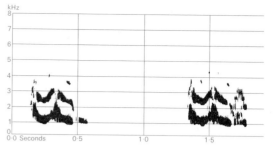

I Sture Palmér/Sveriges Radio Sweden June 1963

(Bauer and Glutz 1968), resembling 'kow-yow' in ♀ and high-pitched 'kyu-yu' (see I) or 'kyu-yu-yu' in ♂ (Scott and Boyd 1957). Call of ♂ to mate thin, high-pitched piping 'yi-yi-yi', almost inaudible to human ear (Witherby *et al.* 1939). In excitement, voice strangely thin, almost piercing, with clearly penetrating 'i' (Bauer and Glutz 1968). Contact-call loud 'kah', unlike other *Anser* (Witherby *et al.* 1939).

CALLS OF YOUNG. Contact-calls uttered in groups of 2–5 notes. Distress-calls: 7 notes uttered each $2\frac{1}{2}$ s. Sleepy-call fairly low and lasting *c.* $\frac{1}{2}$ s (J Kear).

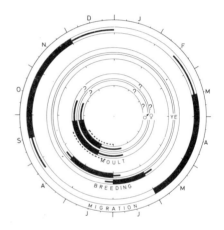

**Breeding.** SEASON. See diagram for the west Palearctic range. SITE. Snow-free rock outcrops or hummocks, usually close to water but up to 150 m away; sometimes in thick birch scrub but more usually in open (Dementiev and Gladkov 1952). Previous years' sites frequently used. Nests widely dispersed. Nest: slight construction of grass and other vegetation flattened at top with shallow depression. Old nest cups lined with fresh grass. Much down added after laying. Building: probably by ♀ only. EGGS. Oval, smooth, slightly glossy; light yellowish-white, becoming stained unevenly with ochre during incubation. 76 × 49 mm (69–85 × 43–52), sample 100 (Schönwetter 1967). Weight of captive laid eggs 104 g (82–125), sample 84 (J Kear). Clutch: 4–6 (2–8). One brood. No information on replacements. Eggs usually laid at 48-hour intervals. INCUBATION. 25–28 days (Witherby *et al.* 1939); 25 days in captivity (J Kear). By ♀ only, covering eggs with down when off nest. Starts with last egg; hatching synchronous. Eggshells left in nest. YOUNG. Precocial and nidifugous. Self-feeding. Both parents care for young; brooded by ♀ when small. FLEDGING TO MATURITY. Fledging period 35–40 days (Witherby *et al.* 1939); young stay with parents through first autumn and winter, leaving wintering grounds together in spring but parents return to breeding grounds alone. First breeding at 3 years, occasionally at 2. BREEDING SUCCESS. No data.

**Plumages.** ADULT. Similar to nominate race of White-fronted Goose *A. albifrons*, but white normally extends higher on forehead, posterior margin forming wedge on crown above hind-corner of eye, instead of being rounded and before eye; some variation, however, in both species. Head, neck, upperparts, and wing darker, with more earth-brown shade. Amount of black on underparts on average smaller. DOWNY YOUNG. Crown, streak before and through eye, and back dark brown, tinged olive on mantle; greyer on sides of body. Sides of head mottled yellow-brown and olive; chin pale yellow; forehead, throat, and nape golden-yellow; belly, bar across wing, and spot below yellow. Much individual variation. JUVENILE. Like *A. a. albifrons*, but head and neck darker brown, and feathers of belly darker grey-brown, edged grey or white, appearing darker when worn. FIRST WINTER AND SUMMER. Differs from adult in same way as 1st winter and summer *A. a. albifrons*.

**Bare parts.** ADULT, DOWNY YOUNG, AND JUVENILE, Iris, bill, and foot like *A. a. albifrons*; bill, without yellow hue, and usually slightly deeper pink. Eye-ring distinctly swollen, yellow to orange-yellow, more conspicuous than narrow yellow eye-ring of some adult *A. a. albifrons*; present in nestling when greyish.

**Moults.** Like *A. a. albifrons* but flight-feathers apparently moulted *c.* 1–2 weeks earlier (Dementiev and Gladkov 1952).

**Measurements.** Netherlands; some additional Denmark, France, Hungary, north Caspian (winter), and Lapland (June)

(Schiøler 1925; RMNH, ZMA).

| | | | | | | | |
|---|---|---|---|---|---|---|---|
| WING | AD ♂ | 378 | (6·23; 8) | 370–388 | ♀ 373 | (9·22; 7) | 361–387 |
| | JUV | 364 | (3·12; 5) | 360–369 | 345 | (9·50; 10) | 329–356 |
| TAIL | AD | 104 | (3·77; 4) | 98–108 | 102 | (3·74; 7) | 97–110 |
| | JUV | 88 | (2·62; 3) | 86–92 | 85 | (2·98; 6) | 81–90 |
| BILL | | 33·6 | (1·56; 13) | 31–37 | 31·3 | (1·28; 17) | 29–34 |
| TARSUS | | 63·7 | (2·20; 13) | 59–68 | 61·0 | (2·49; 15) | 57–65 |
| TOE | | 62·5 | (2·24; 13) | 59–67 | 59·7 | (1·91; 19) | 56–63 |

Sex differences significant, except adult wing and tail. Juvenile wing and tail significantly shorter than adult. Other measurements similar; combined. Wing ♂ USSR apparently larger, 392 (5) 381–412 (Dementiev and Gladkov 1952), but whether due to different method of measuring or to possibly larger size of eastern populations not clear.

**Weights.** ADULT. ♂♂: Norway, June, 1800, 2000. ♀♀: former East Prussia, September, 1400; Wales, February, 1843. JUVENILE. ♂♂: former East Prussia, September, 1445; north Germany, December, 1440; Belgium, January, 1850. ♀♀: Sweden, September, 1310; Netherlands, October, 1380; Austria, November, 1520 Bauer and Glutz 1968; ZMA). 4 ♂♂ Yakutskaya ASSR, winter, 1950–2300; ♀♀ 2100, 2150 (Vorobiev 1963). Range USSR 1600–2500 (Dementiev and Gladkov 1952).

**Structure.** Wing long, pointed, rather narrow. 11 primaries: p9 longest, p10 3–11 shorter, p8 2–9, p7 25–37, p6 50–68, p1 155–180; p11 pointed, short, concealed by primary coverts. Tail short, rounded; 16 feathers. Bill short, relatively high at base. Culmen almost straight, rather steeply sloping towards nail. Cutting edge of upper mandible straight, with *c.* 18–23 horny 'teeth', not visible when bill closed. Outer toe *c.* 93% inner *c.* 77%, hind *c.* 27%.

**Recognition.** Small size (especially bill), swollen eye-ring, more white on forehead, straight cutting edge of upper mandible lacking conspicuous visible 'teeth' when bill closed, and smaller number of 'teeth' (18–23, against 25–29) separate from *A. albifrons*. CSR

## *Anser anser* Greylag Goose

PLATES 52 and 61
[after page 398
and between pages 494 and 495]

DU. Grauwe Gans    FR. Oie cendrée    GE. Graugans
RU. Серый гусь    SP. Ánsar común    SW. Grågås

*Anas Anser* Linnaeus, 1758

Polytypic. Nominate *anser* (Linnaeus, 1758), west and north-west Europe; *rubrirostris* Swinhoe, 1871, south-east and east Europe, west and central Asia.

**Field characters.** 75–90 cm; wing-span 147–180 cm; except for some introduced races of Canada Goose *Branta canadensis*, averages larger than any other west Palearctic goose, with largest bill. Large, big-headed, rather thick-necked, heavy, grey goose, with strikingly pale forewing obvious in flight. Sexes alike, though ♂ larger; no seasonal variation. Juvenile separable at close range. 2 races distinguishable in field.

ADULT. (1) Western race, nominate *anser*. Head and neck pale buff-grey, darker on crown and hindneck. Upperparts grey-brown, with increasingly separated lines of feathers finely but noticeably tipped buffish-white forming regular transverse pattern; tertials and inner secondaries also grey-brown broadly edged buffish-white. Upper tail-coverts white contrasting with grey centre to back and grey, white-rimmed tail. Chest pale buff-grey, merging with grey (mottled white and buff) flanks and belly, latter with varying area of black spots or blotches, not forming bars. Upper flanks increasingly edged white towards vent, forming obvious line, and rear flank feathers with blackish centres, appearing as irregular barring on well-marked birds; vent and under tail-coverts white. Upperwing shows striking pattern, with pale blue-grey coverts contrasting with greyish-black quills. Bill heavy and quite long, essentially triangular in shape (with upper mandible lacking concave outline); pale orange, with pink

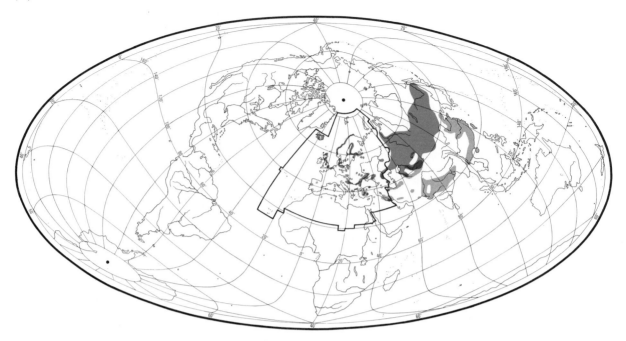

tinge showing behind white nail. Eyes brown, surrounded by spectacles of orange orbital skin. Legs flesh-coloured. (2) Eastern race *rubrirostris*. Slightly larger and generally paler than nominate *anser* with whitish tones replacing buff, particularly on head and neck (both appearing uniform), also upperparts (feather-edges greyish-white form stronger transverse pattern) and flanks. Forewing even paler, appearing whitish-grey in some lights. Bill longer than nominate *anser*, wholly pink except for white nail. Legs cold pink. JUVENILE. Young birds of nominate *anser* distinguishable from adults by more mottled plumage lacking in particular sharply defined transverse lines of upperparts, obvious flank edges and belly marks. Young birds of eastern race *rubrirostris* similarly distinguishable from adult but again paler than nominate *anser*, with pink bill.

At long range, or in difficult light, grey geese are difficult to separate but *A. anser* usually most readily identifiable due to its combination of bulk, heaviness (on ground or in flight), uniform pale plumage tone, lighter forewing, large bill and head, and pink legs. At close range on ground, confusion possible with juvenile or 1st winter White-fronted Goose *A. albifrons*, but *A. anser* larger (by 15%), always greyer (with less contrast between fore- and hind-parts) and less mobile (see also *A. albifrons*).

Flight powerful and fast (though appears a little laborious), with regular beats of rather broad wings alternating with gliding on outstretched wings or planing with angled wings when coming in to land. Lands easily with characteristic rotary flutter of wings; on take-off in calm conditions runs farther than other species and rises from water with more difficulty. Flocks indulge in dramatic, plunging aerobatics characterized by sudden banking and sideslipping. Walk slower and more rolling than other grey geese, and runs less easily. Swims well; dives to avoid sudden danger. Breeding populations occur as scattered, small communities. Migrant and wintering populations much more gregarious, combining into flocks numbering hundreds or thousands. Usually tamest of the grey geese and sometimes content to walk away from man. Flight-notes loud; most characteristic a trisyllabic, rather deep, sonorous cackle and a multisyllabic buzzing call. Flock chorus distinctly more bass, less penetrating and less far-carrying than that of other *Anser*.               DIMW

**Habitat.** Unlike relatives, has clung to spectrum of boreal and temperate habitats from arctic tundra (sparingly), through wetlands wherever inaccessible swamps, reed-beds, and lake islets offer security, to mountain regions (to 2300 m in Mongolia) and wetlands in steppe or semi-desert. Dependence on combination of secure aquatic and open grassland habitats results in patchy distribution (due largely to human impacts) over much of west Palearctic, in contrast to more concentrated and continuous habitat of related forms. Most breeding areas now based on extensive, open fresh waters with dense emergent vegetation, such as reedbeds, and ready access to suitable grazing pastures, meadows, and wetlands. Floating vegetation freely used. General preference for eutrophic sites. Infrequently nests on ledges of steep rocky slopes or escarpments, and in Scotland and elsewhere in tall heather (*Calluna*, *Erica*). Exceptionally in or near low trees, on flooded areas, especially in USSR; also on islets in salt water. Spends relatively long periods, and in some regions entire year, in breeding habitats, but tends to winter in milder climates than most geese; habitats then include

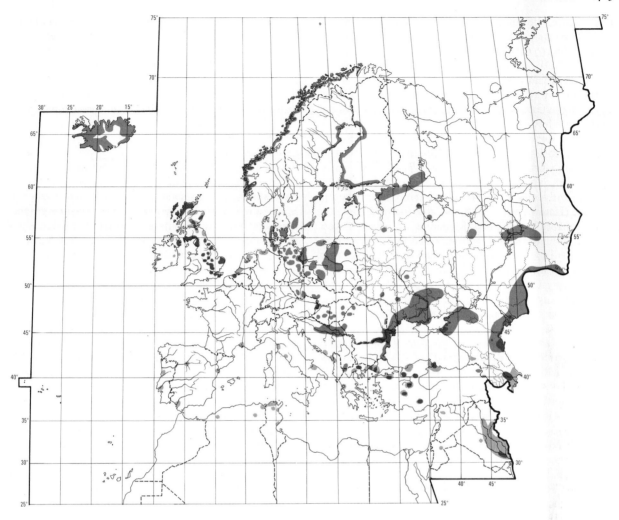

estuaries, floodlands, reservoirs, and lakes round Mediterranean. Forages over croplands, stubble, grasslands, and swamps, as well as in shallow water, either fresh or saline. Compared with other *Anser*, commonly finds less need for aerial movements, these tending to keep within narrow radius and at low or medium altitudes except on migration. Loss of breeding habitat through drainage, settlement, and excessive toll of birds and eggs have long been serious, but readily responds to manipulation of introduced or feral stocks, sometimes merging with wild populations or repopulating former habitats after local extinction.

**Distribution.** Originally probably bred over whole of Europe but now disjunct distribution due to drainage, etc. (Rooth 1971). In USSR, range probably continuous until early 20th century, but since, and especially in last 20 years, split into numerous small areas, due to hunting and habitat destruction (Isakov 1972). More recently introduced in many areas in Britain and to lesser extent elsewhere. ICELAND: some spread in last 30–40 years

(Rooth 1971). FAEROES: fairly common breeder until *c.* 1840; since bred 1939, 1942, and 1943 (Williamson 1970). BRITAIN AND IRELAND. Small natural population restricted to Scotland; extinct England since *c.* 1831 and Ireland since 18th century (Rooth 1971). In recent years, introduced on large scale, especially Britain. FRANCE: bred 1934 (FR). BELGIUM: introduced Zwin from 1956 (Lippens and Wille 1972). NETHERLANDS: breeding until 1909, then again since early 1950s, with some introductions 1962 (Rooth 1971). SWEDEN: more widespread in 19th century; much reduced by persecution, drainage, and disturbance, and, not long ago, in danger of extinction but has made partial recovery aided by local introductions (Curry-Lindahl *et al.* 1970; SM). ALGERIA: breeding reported Lake Fezara 19th century, but quite exceptional (Heim de Balsac and Mayaud 1962).

Accidental. Spitsbergen, Bear Island, Jan Mayen, Israel, Kuwait, Cyprus, Malta, Libya, Morocco, Azores, Madeira, Canary Islands.

**Population.** Breeding. ICELAND: increasing; estimated 3500 pairs 1960, 12000 pairs 1966, 18900 pairs 1969, 15000 pairs 1971 and 18600 pairs 1973 (Boyd and Ogilvie 1972; MAO). BRITAIN. Scotland: north and west, 1000–2000 mainly wild birds resident (Sedgwick *et al.* 1970); Galloway, up to 1000 feral birds by 1971 (Young 1972). WEST GERMANY: 410 pairs (Szijj 1973); increased Schleswig-Holstein (EB). EAST GERMANY: 400–500 pairs (Rutschke 1970); 500–1000 pairs, increasing (Rooth 1971); *c.* 2000 pairs 1973 (E Rutschke). DENMARK: 1000–2000 pairs, increasing (Rooth 1971); *c.* 2000 pairs 1972 (TD). NORWAY: some thousands of pairs (Rooth 1971); greatly reduced 20th century, but recent tendency to increase (Haftorn 1971). SWEDEN: 200–300 pairs 1953–4, now probably more (Curry-Lindahl *et al.* 1970); *c.* 500 pairs (Rooth 1971). FINLAND: increased from early 1950s (Grenquist 1970); *c.* 150–250 pairs (Rooth 1971). POLAND: *c.* 460 pairs in 3 main areas, with other sites elsewhere (Rooth 1971). CZECHOSLOVAKIA: *c.* 220 pairs (Hudec 1970). HUNGARY: 250–300 pairs; slight increase following protection in 1949 (Szabo 1970); 18 pairs, Kisbalaton, in 1968, 46 pairs 1972 (IS). YUGOSLAVIA: 60–80 pairs (Mikuska 1973). BULGARIA: *c.* 50 pairs (Rooth 1971). GREECE: 10–20 pairs (Rooth 1971). RUMANIA: *c.* 1000 pairs (Hudec and Rooth 1970). USSR. Major decline in all areas, with total of 50000–60000 pairs in 1967, including *c.* 700 pairs in north-west (mainly Estonia), *c.* 650 pairs south Ukraine, *c.* 3000 pairs Sea of Azov, and not more than 500 pairs Azerbaydzhan (Isakov 1972). Volga delta: 11 300 pairs in 1970 (Krivenko and Krivonosov 1972). Estonia: rapid increase on maritime islands; *c.* 300 pairs in 1936, 750 pairs 1972 (Kumari 1974).

Winter. BRITAIN AND IRELAND. Increased from 26000 birds in 1960 to 65000 in 1970, 76000 in 1973, and 68000 in 1974 but major decrease Ireland and some decline north England (Boyd and Ogilvie 1972; Ruttledge 1970; MAO). NORTH-WEST EUROPE: *c.* 30000 birds (Rooth 1971). RUMANIA, GREECE, AND TURKEY: *c.* 10000 birds (Rooth 1971). NORTH AFRICA: normally 5000–7000 birds in Tunisia (MS); few recent records Morocco (KDS). USSR: 69400 birds in western part in not so severe winters (Isakov 1970*c*).

Survival. Icelandic population: mean annual mortality adults 23% (ringing), over 4 months old 22% (censuses), perhaps declining from 1960 to 1971; adult life expectancy 3·8 years (Boyd and Ogilvie 1972). Mean annual mortality of birds ringed Denmark as young and adults 33%; further life expectancy of full-grown young 2·3 years and of adults 2·6 years (Paludan 1973). Oldest ringed bird 17 years 4 months (Rydzewski 1974).

**Movements.** Several populations in Europe, some discrete, some overlapping; most wholly or partially migratory. Icelandic breeders migrate second half October for British (mainly Scottish) wintering grounds; arrivals Britain occasionally continuing into November; in large concentrations late October or first half November before dispersal in smaller flocks to traditional winter haunts (Boyd 1957*b*; Wildfowl Trust). Some movement around country in winter, when a few recoveries Ireland, but none England south of Cumberland and Northumberland. Return passage early March to April, stragglers into May; major route out over Hebrides (Williamson 1968). Some gatherings of moulting non-breeders in Iceland but no moult migration detected. Small Scottish population basically resident.

Breeding population of Norway, Sweden, Denmark, and Germany mainly migrates south to south-west through Netherlands (where some overwinter) and France to Spain, a few reaching Portugal; leaves breeding grounds late September, rapid passage through Netherlands and France mainly October–November. Many follow French coast, but others take broad-front overland routes, thence over Pyrenees and across Spain to Guadalquivir delta where peak numbers December–January. Return movement begins late January; passage more leisurely than in autumn but completed by end April (Roux 1962). Moult migration still developing; formerly only in Danish population which moulted Limfjorden, numbers there increasing from 500 in 1934 to 3000 in 1955 as joined by birds from other areas known from ringing to include north Germany, Poland, and Austria (Paludan 1965; Salomonsen 1968). Since 1959, bulk have moved to other sites, and in 1957 began moulting Netherlands (mainly IJsselmeer and Haringvliet) where now several thousand June (Lebret and Timmerman 1968; Ouweneel 1970). Other moulting areas now known Gotland (Sweden) and Sjaelland (Denmark) (Haack and Ringleben 1972); also Vega and Vikna Islands off west Norway, where 2500 summer 1971, probably used only by Norwegian birds (Lund 1971). Moulting flocks assemble second half May, disperse towards natal areas as soon as moult completed. Little information on Gulf of Bothnia and east Baltic breeders, but may join west Baltic population in autumn and winter. Relatively small numbers overwinter around North Sea or in central Europe, but rather more in mild seasons; few cross North Sea to Britain.

Central European breeders absent from nesting areas September to mid-March; some move south-west to Spain, others south through or from Czechoslovakia, Hungary, Austria, and Switzerland as far as Tunisia and sometimes Algeria. Winter quarters of south-east Europe breeders uncertain; perhaps among those wintering Greece and north Africa. Breeders of Russia and west Siberia migrate south to south-west in October to Black Sea, Turkey, and Caspian Sea; major moulting area for non-breeders in Volga delta (Dementiev and Gladkov 1952). Non-breeders from all migratory populations not infrequently present in small flocks well south until May, or later.

Birds resembling Siberian race *rubrirostris* occasionally reported west Europe, but pink-billed Greylags of uncertain taxonomic status breed west through south

Russia to Hungary and Austria. Small feral populations introduced into parts of Austria, Germany, Belgium, Britain, and probably elsewhere; generally fairly sedentary, but some individuals ringed Belgium (where introduced *rubrirostris*-type) have reached Sweden, Denmark, Netherlands, and north France, perhaps as abmigrants (Schneidauer 1968). See also Bauer and Glutz (1968); Mörzer Bruyns *et al.* (1973).

**Food.** Based mainly on information supplied by I Newton. Plant-material, accessible from ground or water surface and not too hard, including roots and tubers, green leaves and stems, flower-heads, and fruits. Green leaves and other soft materials clipped off with side of bill, but pieces from large roots and tubers scraped off with terminal nail on upper mandible. Feeds mainly by grazing on land, but also while floating on water, occasionally up-ending to pull up submerged material; also probes in soft mud and pulls at stems to expose roots. In general, prefers in summer to eat young green parts, and in winter underground storage organs. Gullet holds up to 200 cc of food.

In summer quarters, feeds from marshes and lake margins and farmland, including pasture. In marshes, roots of *Scirpus* favoured—though not in European part of USSR (Yu A Isakov); other aquatic food-plants include *Lemna*, *Potamogeton*, *Sparganium*, *Glyceria*, *Equisetum*, *Phragmites*, *Phalaris*, and *Leersia* (Dementiev and Gladkov 1952; Hudec and Rooth 1970; Kvet and Hudec 1971). On farmland, eats various agricultural grasses, including *Lolium*, *Phleum*, *Poa*, *Festuca*, and *Bromus*, and less frequently leaves, roots, or seed-heads of *Polygonum*, *Stellaria*, *Chenopodium*, *Capsella*, *Trifolium*, *Chrysosplenium*, *Cochlearia*, *Potentilla*, *Sonchus*, *Taraxacum*, and other weeds. In Iceland, also potatoes (*Solanum tuberosum*) by pulling at tops to expose roots, turnips (*Brassica rapa* and *B. rutabaga*), kale, and seed-heads from ripening cereal crops (Kear 1967). Other summer foods recorded occasionally include fruits of *Vaccinium* and *Rubus*, and leaves and roots of *Eriophorum*. At Neusiedler See, leaves and young shoots of reed *Phragmites communis* taken by adults and chicks (Koenig 1952) and, particularly in March–April and August–September, sedge *Bolboschoenus maritimus* (see Bauer and Glutz 1968). Similarly in southern Moravia, *B. maritimus* tubers eaten in autumn and spring, and leaves of *P. communis* May–July (Hudec 1973). In Volga delta in summer, eats young stems of *P. communis*, *Butomus* and grasses, and seeds of cockspur *Echinochloa crus-galli*. In autumn, mainly nodules of arrowhead *Sagittaria sagittifolia*, stems of *Typha*, club-rush *Scirpus michellianus*, and willow *Salix triandra*, and seeds of *Sparganium* (Yu A Isakov).

Over most of winter range, feeds on grassland, and extent to which grass and leaf diet varied depends on local agriculture. Icelandic birds, during stay in Britain, eat largely grass, but also cereal grains (barley *Hordeum*, oats *Avena*, and wheat *Triticum*) mostly in autumn; potatoes mainly late autumn to early spring; and growing cereals mostly late spring (Kear 1964; Newton and Campbell 1970). Roots of carrots (*Daucus*), turnips (*Brassica*), sugar beet (*Beta*), and leaves of rape (*B. napus*) and other brassicas also taken in restricted areas, and occasionally leaves and seeds of various weeds, and rhizomes of couch (*Agropyron repens*). Main conflict with agriculture includes eating of turnips in winter and young grass in spring. Elsewhere in winter range, feeds on peas (*Pisum*), lentils, buckwheat (*Fagopyrum*), and other crops. In southern Moravia, green parts of cereals, *Triticum*, *Avena*, *Hordeum*, main foods in spring and less so in autumn, and cereal grains summer to autumn (Hudec 1973). In Upper Volga in spring, mainly green stands of winter rye, hairgrass *Deschampsia caespitosa*, fescue *Festuca rubra*, brome *Bromus inermis*, and rye and oat seeds. In Azerbaydzhan in winter, mainly grasses, winter wheat, and tubers of *B. maritimus* (Yu A Isakov).

Some populations (e.g. in Holland and Spain) winter primarily on inland or coastal marshes, eating mainly roots of *Scirpus* also leaves of *Carex*, *Hordeum marinum*, *Phalaris* and *Bolboschoenus maritimus* (Bernis 1964). In south-west Netherlands January–March, exclusively root pieces *Scirpus lacustris* and tubers *S. maritimus*; $185 \pm 25$ root pieces (0·8 kg) eaten per goose per 24 hours when foraging more than $5\frac{1}{2}$ hours (Loosjes 1974).

**Social pattern and behaviour.** Based mainly on material supplied by H Fischer; see especially Fischer (1965) and, for classic study, Heinroth (1911).

1. Gregarious except when nesting, in flocks made up mainly of families, pairs, and unpaired birds (mostly immature). After breeding, family parties may remain segregated, or join non-breeders. Several flocks often congregate to moult in well-sheltered areas with good feeding (Bauer and Glutz 1968; Hudec and Rooth 1970). From August onwards, flocks collect for migration. BONDS. Life-long monogamy the rule, ♂ and ♀ associating all year even when sexually inactive. Deviations occur, but mainly in semi-natural or captive conditions; include simultaneous association of 2 ♂♂ with ♀ or 2 ♀♀ with ♂ ('trios') and successive bigamy by both sexes (see also Harrison 1967). Stable pair-bonds not usually established until 3rd or 4th year. Pairings between siblings observed only in experimental conditions, but those between father–daughter and mother–son recorded in nature upon death of mate. Both parents tend young, family remaining in group until next breeding season. Manner of break-up of group not fully understood; not due only to parental aggression (see Pink-footed Goose *A. brachyrhynchus*). Single goslings stay longer with parents than larger broods. After leaving, *c.* 75% of young form casual groups comprising siblings, other juveniles, or unpaired adults; rest solitary until form pairs. About 50% resume more or less sporadic contact with parents in 2nd year, often rejoining them if no succeeding young. BREEDING DISPERSION. Variable, according to habitat. Loosely colonial where sites protective: in feral population south-west Scotland, 87% of nests on wooded islands, 2–21 m apart (mean 11), with 1 nest per *c.* 7 m² (Young 1972). At less safe sites, more widely dispersed. In all cases only nest defended as territory, especially by ♂. In colony, ♂ remains for long periods on guard near nest during incubation, especially near end (Young 1972); elsewhere,

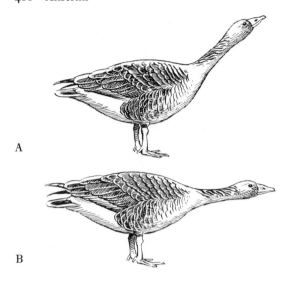

A

B

♂ stays within 1 km, returning if ♀ raises alarm and when young hatch, nest then being abandoned. Non-breeders remain in nesting area or in discrete loafing places (Haack and Ringleben 1972). ROOSTING. Typically communal (except when nesting) and nocturnal at traditional sites, grouped on and beside water (on bank or island). Sleep sometimes interrupted by foraging on water, or exceptionally flies to feed in fields. On average, 3 loafing breaks for drinking, bathing, and preening during day; duration depends on temperature (see Leisler 1969). Flocks wintering in south-east Scotland establish roosting and feeding routine much as in *A. brachyrhynchus* (Newton *et al.* 1973), but differ in certain respects: (1) while similarly using extensive estuarine mudflats, large lochs and reservoirs, and remote 'mosses', also frequent many smaller waters and rivers; (2) make less use of separate 'rest stations' by day (sited on damp riverside fields, islands, and 'mosses'); (3) in mixed roosts, tend to leave later and return earlier than *A. brachyrhynchus*. Like latter, sometimes visit roost to bathe and drink around mid-day, but then often accumulate steadily during rest of day for nocturnal roosting. At mixed roosts, keep separate and travel independently. When ♀ incubating, ♂ roosts near nest or elsewhere; some ♂♂ join non-breeders. In south-west Scotland, incubating ♀♀ spend most of night on nest, but will leave to defaecate and sometimes to feed, even joining other ♀♀ and ♂♂ elsewhere on nest island or on water (Young 1972).

2. Most intensely studied *Anser*, but mainly in semi-captivity or under feral conditions. Highly gregarious with strong pair and family bonds, type or absence of bond strongly influencing individual's whole behaviour (see especially Fischer 1965). FLOCK BEHAVIOUR. Each bird in flock maintains own individual-distance. Stable but not linear rank-order established between members of same flock, at least when latter relatively small and

permanent. While others feed, top-ranking ♂♂ (mainly) often assume Alert-posture with head held high. When strange flocks meet, on moulting grounds or prior to migration, may threaten by giving outcry of Rolling-calls as in mobbing (see below). ♂ of one flock often leaves to charge, bite, or wing-beat one of other flock. After encounter, members of same flock may unite in greeting ceremonial usually reserved for mate or family (see below). When different flocks combine, rank-order in each disturbed and skirmishes occur until new order established. Activities in same flock (e.g. feeding, bathing, preening) largely synchronized. Whole flock mobs predators, advancing with heads held high uttering Rolling-calls. If unfledged young present, mobbing can then consist of Hissing with wings raised high and develop into attack; adults of several families will also encircle and protect young of all. Flock cohesion aided by several types of calls: Pre-flight call, Flight-call, Alarm-call, and Distance-call. Pre-flight signals consist of lateral Head-shaking movements (Johnsgard 1965). ANTAGONISTIC BEHAVIOUR. In wild, flocks spend much of time peacefully feeding, with periodic signs of antagonism related mainly to individual-distance, rank-order, and defence of mate or family. In whole system of silent threat gestures (involving intention-movements of pecking), neck directed at opponent (Fischer 1965): varies progressively from version with neck straight and slanting obliquely forwards (Diagonal-neck posture of Johnsgard 1965; see A), by way of others with neck stretched horizontally forward or obliquely downwards, with or without lifted chin so that head and bill point up (Forward-posture; see B), to one with neck more or less horizontal and kinked to greater or lesser extent (Bent-neck posture). Diagonal-neck and Forward-posture indicative of strong aggression; assumed (e.g.) by top-ranking members of flock, those of intermediate or lower rank tending to adopt Bent-neck. Furrowed feathers of neck often vibrated during threat-display. Intense fighting occurs occasionally during disputes over nest-sites but rare in flock, though birds will then fight less severely at times when feeding, mostly just pecking. During 2nd stage of pair-formation, ♂ behaves generally in aggressively demonstrative, self-assertive manner otherwise confined to period when accompanying small young: holds himself upright, puffs up plumage, sticks out chest, and spreads tail; walks in exaggerated way, flies readily, and keeps wings outspread for longer than usual afterwards. Frequently makes attacks on rival ♂♂; flies towards them and charges silently before returning to ♀ initiating Triumph Ceremony (see below). Reaction to threat and attack variable: bird may turn slowly away; escape by running or flying off, or by diving; redirect threat or attack on another individual; or resist. One attacked persistently assumes Submissive-posture, depressing all body feathers and tail, hanging head, and drawing bill close to breast; stranger in flock adopts this posture almost continuously. HETEROSEXUAL BEHAVIOUR. Highly ritualised Triumph Ceremony, between actual or potential pair, marked feature at all times of year; see, e.g. Heinroth (1911), Fischer (1965), Lorenz (1966), Radesäter (1974). In most complete form, ♂ leaves ♀ and carries out 'attack' on real or imaginary rival, then

C

D

(1) returns to ♀ uttering loud Rolling-calls in the Diagonal-neck posture, often with wings at first still outstretched (Rolling-phase), and (2) joins ♀, both giving Cackling-calls in Forward-posture with heads low and close together (greeting or Cackling-phase). In latter, both sexes show aversion to meeting frontally, with bills aimed at each other in hostile manner; sooner or later orientate themselves laterally in some way (see C). In 1st stage of pair-formation, courting ♂ repeatedly approaches other unpaired birds in Neck-angled posture (see D); another ♂ retreats or reacts aggressively, but ♀ stands ground and ♂ able to come closer. In 2nd stage, ♂ demonstrates in manner already described, also uttering Alarm-calls at slightest provocation. After chasing other birds, especially if 2 ♂♂ court same ♀, returns to ♀ in intense display, attempting to initiate Triumph Ceremony. ♀ at first reluctant to participate and pair-formation not completed until she does so fully. Pair subsequently greet one another after every break in contact. Also during 2nd stage of courtship, ♂ attempts to initiate copulation: while still several metres from ♀, assumes high-riding Haughty-posture on water, with tail raised showing white under tail-coverts (see E), then approaches and swims parallel to ♀, repeatedly Head-dipping (see Canada Goose *B. canadensis*). At this stage, sequence progresses no further than mutual Head-dipping, especially exaggerated in ♂, which ends with bathing. Later in cycle, Head-dipping leads to copulation, usually first near nest during building. Mounting and copulation as in other geese (see *B. canadensis*). In post-copulatory display, ♂ (especially) stretches neck vertically with bill and wing-tips raised, then both bathe and preen. Similar cycle of behaviour repeated by mated birds each spring but courtship more perfunctory and can be omitted in mates of many years standing, though all pairs continue to perform Triumph Ceremony in some form as greeting, especially after hostile encounters. Incomplete, uncoordinated courtship may be shown by immatures in 2nd year, often initiated by older birds. RELATIONS WITHIN FAMILY GROUP. Only ♀ normally broods young, while ♂ defends family. Newly hatched goslings respond to parents with 'vi-call' and outstretched neck in greeting, recognizing parents by sight and sound almost from start. Young follow parents from 2nd day, running to them on hearing Alarm-call. When several days old, follow ♂ when he attacks others, thus learning rank-order of flock. When lost, run about with head held high, uttering Distress-call; parents wait or search, giving Distance-call. Family group remains constantly together, both adults and young frequently Contact-calling; whole behaviour highly

synchronized. After separation, all members of group perform Cackling-phase of Triumph Ceremony together, though this not fully developed in young until sexually mature. Sons tend to greet father preferentially; daughters, mother. Aggression rare between parents and young until break-up of family group; then either ♂ or ♀ pecks juveniles, driving them away. All behaviour shown by young to parents also shown to siblings. Aggression between siblings, chiefly pecking and wing-beating (also, more rarely, Hissing), occurs from *c.* 3rd day both spontaneously and during Triumph Ceremony. In latter case, goslings (unlike adults) greet one another initially with bills directed frontally, this leading to fighting. Soon, however, develop special appeasing movement not found in adults: Face-away (as in *B. canadensis*) with head and neck turned, inhibiting further aggression (Radesäter 1974). Such fighting ceases after 10th day, rank-order among siblings being established in which only subordinate birds Face-away (see also Stahlberg 1974). This rank-order apparently clear-cut and linear only in broods raised in artificial conditions; otherwise, rank-order (if any) loose. From *c.* 20th day, both rank-order and Facing-away disappear, all birds (of whatever status) having learned to orientate laterally in peaceful manner when Cackling, as in adult.
(Figs A–E after photographs and drawings supplied by H Fischer; see also Fischer 1965.)

**Voice.** Complex, with wide variety of often loud and deep cackling and honking calls. Following distinguished by H Fischer (see also Heinroth 1911; Fischer 1965); excludes intermediates.
  (1) Contact-call. Soft, low, conversational cackling (see I); 2–7 syllables, nasal in character. Uttered in flock or family group when grazing or preening, seldom by solitary birds. (2) Distance-call. Loud, high-pitched, sonorous cackling (see II); 2–4 syllables with stress on 1st. Rendered 'aahng-ung-ung' or 'gnong-ong-ong' (Witherby *et al.* 1939). Given by bird seeking lost partner, latter answering with same call; utterances highly variable, facilitating individual recognition. Also given by flocks in flight, answered by groups on ground; thus main flight-call—likened to pack of hounds in cry (Witherby *et al.* 1939). (3) Locomotion-call. Loud, staccato cackling which can develop from Contact-call; 7–20 syllables, with different rhythm. Uttered when change of location on foot anticipated. (4) Pre-flight call. Short, sharp notes, uttered in long series (with distinct pause between each note) by all members of flock when inclination to fly strong. (5) Alarm-call. Short, cut-off, high-pitched, monosyllabic 'gang'. Given when danger threatens or anything unusual occurs. (6) Distress-call. Drawn-out, high-pitched vibrating note. Given, rarely, by defeated bird after violent combat; also by geese attacked by other species (including man). (7) Copulation-call. Prolonged call, not adequately described and rarely heard. (8) Nest-alarm-call. Combination of Alarm, Distance, Locomotion, and Contact-calls, of variable duration and sequence, and irregular rhythm. Uttered by incubating ♀ when approached by birds other than mate or by small predators, or when nest left temporarily; each ♀ has own characteristic version to which only own mate responds. (9) Rolling-call. Complex

E

of different calls, producing resonant sound: loud notes related to both Alarm and Distance-calls; short, irregular, staccato notes related to Contact and Cackling-calls. Uttered by ♂ in 1st phase of full Triumph Ceremony as rejoins mate; by both sexes (particularly ♂) as they approach one another; and by many birds together, e.g. during encounter between different flocks and when flock mobs predator. (10) Cackling-call. Fast, polysyllabic cackling, like intense version of Contact-call. Given by both sexes during last phase of Triumph Ceremony, but more intensely by ♂. (11) Hissing. Occurs under extreme stress when bird dare not attack, mostly by incubating ♀♀ defending nest; otherwise rare.

CALLS OF YOUNG. Contact-call, consisting of 1 or more 'vi' notes; uttered from moment of hatching and later during feeding and preening; develops into repeated double or polysyllabic Greeting-call. When tired and in bodily contact with parents or siblings, give soft, trilling Sleepy-calls. When lost or cold (etc.), attract attention by loud, sharp, high-pitched, 1-syllable Distress-calls. Alarm-calls, similar to adult, first given at 7–8 days but not common until 40th day. Both Alarm and Distance-calls develop from Distress-call. Will occasionally hiss aggressively from 6–8 days.

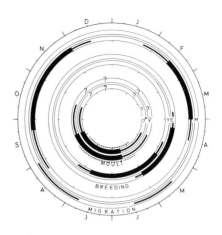

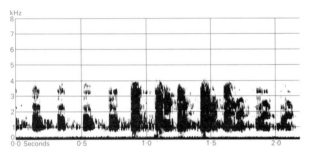

I Sture Palmér/Sveriges Radio Sweden June 1959

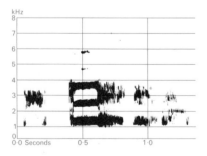

II Sture Palmér/Sveriges Radio Sweden June 1959

**Breeding.** SEASON. Continental Europe: see diagram. Iceland: laying begins end April, main laying period mid-May. Britain: first eggs laid end March or early April, main laying period mid-April. Fledging from mid-July. In Czechoslovakia, cold spring temperatures delay onset of laying (Hudec 1971). SITE. On ground, often sheltered hollow, or at base of tree, under bush, or in reedbeds; also on rafts of vegetation in rivers. Of 463 nests, Czechoslovakia, 59·3% in reedbeds, 21% on ground, 19·7% in pollard willows *Salix* (Hudec 1971). Rarely more than 10 m from water and then only on islands (Hudec 1971). Often colonial with nests as close as 2 m. Site frequently re-used in subsequent years. Nest: pile of vegetation, foundation often of twigs, up to 26 cm long, built up with heather, grass, reeds, and other material. 80–110 cm diameter, exceptionally 200 cm, height 13–60 cm, rarely 120 cm, cup diameter 25 cm, depth of cup 5–15 cm (Hudec and Rooth 1970); lined grass and small amounts down. Building: by ♀, though ♂ helps choose site and often stays near during building. Uses material within reach of nest. 3–6 days to complete, material sometimes added after laying, some nests gaining 6–8 cm in height from laying to hatching (Young 1972). EGGS. Ovate, sometimes elongated; creamy-white, staining pale to dark brown or yellow during incubation; smooth texture. British eggs: 85 × 58 mm (77–94 × 50–63), sample 100 (Witherby *et al.* 1939); weight 149 g (122–172), sample 60 (Young 1965). Czechoslovakia: 86 × 58 mm (77–97 × 53–66), sample 230; weight 164 g (151–179), sample 113 (Kux 1963). Clutch: 4–6 (3–12), exceptionally 14 or more, but then 2 or more ♀♀ involved. Of 476 British clutches: 3 eggs, 4%; 4, 13%; 5, 21%; 6, 35%; 7, 17%; 8, 4%; 9, 2%; 10, 2%; 11, 1%; 12, 1%; mean 5·9 (Young 1972). Of 453 Czechoslovakian clutches: 2 eggs, 5%; 3, 5%; 4, 16%; 5, 30%; 6, 16%; 7, 11%; 8, 7%; 9, 6%; 10, 2%; over 10, 3%; mean 5·75. Dump nests of 20–36 eggs recorded (Hudec and Kux 1971). One brood. Replacement clutches laid after egg loss, up to 7–8 eggs recorded. Eggs laid at 24-hour intervals, occasionally 48-hour; egg-laying at any time of day, most in morning. INCUBATION. 27–28 (30) days. By ♀ only. Starts with last egg; hatching synchronous. Eggshells left in nest. YOUNG. Precocial and nidifugous. Self-feeding. Cared for by both parents who defend nest and young against predators. Brooded by ♀ when small and may return to nest at night. FLEDGING TO MATURITY. Fledging period 50–60 days. Young stay with parents through first autumn and winter, leaving wintering areas with them in spring but moving elsewhere when parents return to breeding grounds. Mature at 3 years, occasionally 2. BREEDING SUCCESS. July brood size of British breeders, 1963–71, varied from 3·7–4·6 (mean 4·1), September brood size 2·8–4·0 (mean 3·2.); sample 365 broods (Young 1972). Autumn counts in Britain of Icelandic breeding birds 1960–71 showed proportion of young varying from 9%–45% (mean 26·5%) and mean brood size 1·5–3·5 (mean 2·3) (Boyd and Ogilvie 1972).

**Plumages.** Nominate *anser*. ADULT. Head, neck, breast, and thighs grey-brown; chin and throat paler. Narrow white line round base of upper mandible. Mantle brown with blue-grey tinge, feathers edged buff shading to white. Scapulars black-brown, edged white. Back and rump pale blue-grey, feathers edged light grey, longest edged white. Sides of body and flanks grey-brown, broadly barred white. Sides of back and rump, upper and under tail-coverts, belly, and vent white. Underparts with varying number of black or black-tipped feathers, not forming bars like White-fronted Goose *A. albifrons*. Tail-feathers grey-brown, narrowly margined and broadly tipped white. Flight-feathers black; primaries with white shafts, basally tinged grey. Outer webs and tips of secondaries narrowly bordered white. Upper wing-coverts mainly pale blue-grey, outer webs and tips of greater and median coverts tipped white; varying number of greater and median coverts tinged brown, at least inner ones black-brown like scapulars. Underwing and axillaries pale grey. Feathers of neck lanceolate; project, giving 'furrowed' appearance. DOWNY YOUNG. Upperparts, sides of body, and thighs olive-brown; streak on lores brown; sides of head olive; forehead, underparts, and bar across wing yellow. Much variation in colour; yellow fading soon. JUVENILE. Like adult, but white line round base of upper mandible absent; rarely white on chin; feather edges on upperparts and sides of body narrower, buff, not contrasting with brown centres of feathers. Breast grey-brown, feathers edged white, belly white; both, when worn, with small dark brown subterminal streaks or blotches. Shape of feathers different from adult; those of neck short, round (not long and projecting), those of mantle, scapulars, underparts, greater and median coverts narrow, rounded at tip (not broad and square). Worn tail-feathers pointed; in worn central upper tail-coverts grey-brown shaft-streak often visible. FIRST WINTER AND SUMMER. Like adult, but usually white longitudinal streak on chin. Feathers of mantle, although square, not as broad; narrowly edged buff to grey. No white round base of upper mandible. Belly white; rarely few black or dark brown spots. Most juvenile wing-coverts often retained. In subsequent plumages, increasing number of black feathers on underparts.

**Bare parts.** ADULT. Iris dark-brown. Eyelids pink or yellow. Bill orange in nominate *anser*, with some pink only behind nail and along cutting-edges; wholly pink in *rubrirostris*; intermediate types not uncommon. Nail of bill white-horn. Foot flesh-coloured. DOWNY YOUNG. Bill and foot dark olive-grey. JUVENILE. Like adult, but eyelids yellow-white, and bill and foot tinged grey; adult colour acquired during winter. In juvenile *rubrirostris*, often pale yellow tinge on base of bill and round nostrils. (Bauer and Glutz 1968; Witherby *et al.* 1939; Hudec and Rooth 1970.)

**Moults.** ADULT POST-BREEDING. Complete; flight-feathers simultaneous. Flightless for *c.* 1 month, mid-May to mid-August. Earliest southern part of range, non-breeders earlier than breeders; parents able to fly at same time as young. Wing-coverts

moulted simultaneously with flight-feathers; rest of body and tail start when flight-feathers complete, underparts first, followed by back, mantle, neck, and head. Tail moult prolonged; feathers alternating. Domestic geese said to moult body twice, partial moult involving most body-feathers starting shortly after complete post-breeding moult; details in wild geese unknown (Hudec and Rooth 1970). POST-JUVENILE. Partial. Head and neck, sides of body, and flanks renewed September–November, followed by mantle, breast, scapulars, belly, back, rump, tail-coverts, some central tail-feathers, and inner median coverts. Some specimens have most body feathers (but not all tail-feathers and no wing-coverts) new by late October; others retain juvenile belly, back, etc. until January or later.

**Measurements.** ADULT. Nominate *anser*. Denmark, April–October; mainly Danish to Bothnian population (Schiøler 1925), supplemented by tail measurements of wintering birds Netherlands; skins (ZMA, RMNH).

| | | | | | | |
|---|---|---|---|---|---|---|
| WING | ♂ 465 | (10·6; 7) | 448–480 | ♀442 | (14·4; 16) | 412–465 |
| TAIL | 139 | (7·71; 14) | 130–149 | 129 | (5·39; 10) | 120–139 |
| BILL | 66·6 | (3·16; 32) | 59–74 | 61·5 | (2·10; 24) | 58–65 |
| TARSUS | 84·7 | (3·80; 32) | 78–93 | 78·8 | (3·51; 24) | 71–87 |
| TOE | 91·6 | (3·84; 32) | 84–102 | 85·7 | (3·93; 24) | 80–97 |

Sex differences significant. Icelandic population (Scotland, winter) with significantly shorter bill, but otherwise similar (Matthews and Campbell 1969).

| | | | | | | |
|---|---|---|---|---|---|---|
| WING | ♂ 467 | (11·5; 191) | 436–500 | ♀447 | (13·3; 157) | 417–480 |
| BILL | 60·0 | ( 2·7; 125) | 54–66 | 56·2 | ( 2·6; 117) | 47–62 |

*A. a. rubrirostris.* USSR (Dementiev and Gladkov 1952).

| | | | | | | |
|---|---|---|---|---|---|---|
| WING | ♂ 468 | (16) | 435–513 | ♀448 | (7) | 395–470 |
| BILL | 68·8 | (19) | 59–78 | 63·8 | (8) | 47–73 |

JUVENILE. Bill, tarsus, and toe not different from adult, except in early autumn when not yet full-grown; wing and tail significantly shorter. Denmark and Netherlands combined for wing (1) and tail (Schiøler 1925; ZMA, RMNH); Scotland for wing (2) (Matthews and Campbell 1969).

| | | | | | | |
|---|---|---|---|---|---|---|
| WING (1) | ♂ 430 | (20·6; 12) | 379–455 | ♀ 416 | ( – ; 3) | 400–425 |
| WING (2) | 450 | (13·3; 122) | 418–482 | 433 | (15·2; 119) | 390–466 |
| TAIL | 109 | (4·95; 4) | 104–117 | 106 | ( – ; 2) | 105–108 |

**Weights.** (1)–(3) Scotland (Icelandic population): (1) arrive lean October, (2) gain weight until December; (3) then lighter January–March (less food available), but probably gain weight prior to departure April. (4) Galloway, Scotland (resident population), June, moulting (Matthews and Campbell 1969). (5) Kazakhstan, March–May (Dolgushin 1960). See Hudec and Rooth (1970) for additional weights.

ADULT

| | | | | | | |
|---|---|---|---|---|---|---|
| (1) | ♂ 3454 | (246; 42) | 3030–3790 | ♀ 3039 | (197; 45) | 2540–3470 |
| (2) | 3793 | (343; 52) | 2740–4250 | 3170 | (369; 25) | 2070–3960 |
| (3) | 3509 | (321; 94) | 2600–4560 | 3108 | (274; 75) | 2160–3800 |
| (4) | 3692 | (249; 36) | 3200–4300 | 3237 | (173; 30) | 3000–3600 |
| (5) | 3455 | (390; 10) | 2800–4100 | 2921 | (408; 7) | 2450–3600 |

JUVENILE

| | | | | | | |
|---|---|---|---|---|---|---|
| (1) | ♂ 2900 | (180; 9) | 2730–3170 | ♀2722 | (180; 8) | 2430–2990 |
| (2) | 3297 | (268; 21) | 2450–4250 | 3174 | (209; 22) | 2810–3540 |
| (3) | 3083 | (339; 82) | 2160–4160 | 2726 | (288; 52) | 1980–3220 |

**Structure.** Wing long, rather narrow, pointed. 11 primaries: p9 longest, p10 6–20 shorter, p8 0–8, p7 28–35, p6 54–70, p1

175–215; p11 narrow, pointed, concealed by primary coverts. P9–8 emarginated outer web, p10–9 inner. Tail short, rounded; 18 (rarely 20) feathers. Bill heavy, about length of head, high at base; cutting-edge of upper mandible distinctly arched, with *c.* 18–22 horny 'teeth'. Lower mandible rather heavy, but straight. Tarsus reticulate; outer toe *c.* 93% middle, inner *c.* 84%, hind *c.* 30%.

**Geographical variation.** *A. a. rubrirostris* has pale grey-brown upperparts; broad light edges to scapulars, feathers of mantle and wing-coverts; and pink bill. Much local variation, and birds with intermediate amount of pink on bill in large area between Ural and Czechoslovakia/East Germany. Populations of Iceland, Scotland, and coastal Norway (sometimes separated as *A. a. sylvestris*) have shorter bill and narrower intra-orbital space (larger salt glands) than others of nominate race. Probably 2 isolated populations in late Pleistocene: one coastal south-west Europe and one inland south-east Europe and Asia, developing into 2 groups, '*sylvestris*' and *rubrirostris* (compare Cormorant *Phalacrocorax c. carbo* and *P. c. sinensis*). Nominate *anser* from central Europe may be considered as somewhat variable intermediate but, as much overlap in bill-length with '*sylvestris*', north-west and central European populations treated as single subspecies. Geographical variation obscured recently by introductions where original populations extinct or threatened.

CSR

## *Anser indicus* (Latham, 1790) **Bar-headed Goose**

FR. Oie à tête barrée    GE. Streifengans

Breeds in high central Asia: south-east Russian Altai to west Manchuria, south to Ladakh, Tibet, and Inner Mongolia. Overwinters on southern edge of breeding range (e.g. valleys of south Tibet), but majority migrate over Himalayas into Pakistan, north India and Bangladesh. No reliable indications of westward vagrancy. Identified various European countries north to Finland and Sweden, south to Spain, but such birds so obviously escapes that many must go unrecorded; common in ornamental waterfowl collections, free-flying at some (Delacour 1954). A feral population Kalmarsund, south-east Sweden, from *c.* 1930 (Bàuer and Glutz 1968), but probably failed to become established and perhaps now dispersed.

## *Anser caerulescens* **Snow Goose**

Du. Sneeuwgans      FR. Oie des neiges     GE. Schneegans
Ru. Белый гусь     Sp. Ánsar hiperbóreo   Sw. Snögås

*Anas caerulescens* Linnaeus, 1758

PLATES 56 and 61
[after page 470
and between pages 494 and 495]

Polytypic. Nominate *caerulescens* (Linnaeus, 1758), north-east Siberia and North America from north Alaska to south Baffin Island; *atlanticus* (Kennard, 1927), islands surrounding north Baffin Bay, and north-west Greenland. Both accidental Europe.

**Field characters.** 65–80 cm (body 43–48 cm), wingspan 132–165 cm; close to Pink-footed Goose *A. brachyrhynchus* in size, but juveniles of both races and nominate *caerulescens* of all ages 5–10% smaller. Medium-sized, oval-headed, rather thin-necked, white or blue-grey goose, with stout bill, black primaries obvious in flight, and upright carriage on ground. Sexes alike though ♂ larger; no seasonal variation. Juvenile separable. 2 races in west Palearctic, one with 2 colour phases.

ADULT. (1) White morph. Western race, nominate *caerulescens* ('Lesser Snow Goose'): entirely white except for grey primary coverts and black primaries. Eastern race *atlanticus* ('Greater Snow Goose'): as white morph *caerulescens*, but with noticeably bulkier body, longer and heavier bill, and longer legs. (2) Dark morph ('Blue Snow Goose'). Head and foreneck white, most of hindneck and all upperparts grey, over-marked with black and tinged blue on mantle and scapulars, but pale on rump; upper tail-coverts white, contrasting with dark grey tail with white rim. Upper breast, chest, flanks, and belly grey, tinged brown and with many feathers edged buff; rest of underparts white. Upperwing surface shows contrast between coverts (coloured as mantle) and black quills; underwing grey, with white axillaries contrasting with flanks. Hybrids between morphs variable, most common having entirely white underparts except for upper flanks and chest sides. JUVENILE. (1) White morph. Pale grey, washed or speckled brown on crown, hindneck, mantle, wing-coverts, and flanks; longer wing-coverts and tertials dark centred, quills brown-black. Eastern race *atlanticus* larger. (2) Dark morph. Like adult, but head and neck dusky, grey plumage browner above and paler below, and tertials duller and of normal length. Bills of adults in all forms fairly heavy but not long, crimson edged black, with white nail. Eyes dark brown; no ring. Legs and feet deep reddish-pink. All bare parts of juveniles initially grey in tone.

White-morph birds among most unmistakable of wildfowl but Ross's Goose *A. rossii*, often kept in wildfowl collections, also white, though 40% smaller, dainty, with small triangular blue-based bill and greater agility in flight and on ground. Frequent escapes of both species, so all occurrences other than those within flocks of partly sympatric White-fronted Geese of Greenland race *A. albifrons flavirostris* now suspect.

Flight action and behaviour as *A. brachyrhynchus*, with considerable agility in close quarters. Gait rather nimble, particularly in nominate *caerulescens*; swims well, frequently roosting on water. Flight-notes loud; most characteristic a rather hard, clarion cackle, lacking singing tone of most *Anser*; flock chorus of tenor type but with many sharp and honking notes giving wider range than in *A. brachyrhynchus*. DIMW

**Habitat.** Arctic tundra, usually lowland but sometimes using rocky ledges and often close to water, fresh or marine. In winter quarters, grasslands and wetlands, frequently coastal, suitable for mass grazing or uprooting of food plants, which often leads to denudation and erosion (Soper 1941). Choice of habitat, whether for breeding, wintering, or staging on passage, exceptionally compact and intensive in use. Migratory flight at high altitude.

**Distribution.** Breeds north-east Siberia (Wrangel Island, and in past elsewhere, but now possibly only sporadically), eastward across arctic North America to north-west Greenland.

Accidental. Britain and Ireland: occurs almost annually Scotland and Ireland, less often England, but impossible to determine true status as kept in many British and European waterfowl collections and escapes quite frequent (British Ornithologists' Union 1971). Similar reservations apply to smaller numbers of records of vagrants elsewhere. Most frequent Iceland, Netherlands, Finland, Norway, with fewer (1–3) records for France, Belgium, Denmark, Czechoslovakia, Poland, West Germany, East Germany, Finland, Faeroes, and Azores.

**Movements.** Migratory. *A. c. caerulescens* divisible into sub-populations. A few east Siberian breeders winter Japan and east China, but most cross Bering Sea to Puget Sound (Washington) and central valleys of California; latter also used by breeders from west Canadian arctic. Those wintering inland in north-central Mexico believed to come (via Central flyway) from central Canadian arctic, though many from latter join breeders of Hudson Bay region and south Baffin Island which use Mississippi flyway to winter on Gulf of Mexico. In eastern North America, many stopping places used during leisurely spring passage, whereas in autumn many (notably blue morph) migrate direct from James Bay to Gulf coast. Absent from arctic breeding grounds chiefly late August–May or early June, but flocks make lengthy stays in sub-arctic Canada, especially James Bay region in east; main passage through Ontario and upper St Lawrence in October and April. Occurs Atlantic coast USA (few), and has straggled to West Indies and Greenland.

*A. c. atlanticus* has rather restricted range; from colonies around Baffin Bay, congregates in autumn at Cap Tourmente on St Lawrence River (Quebec), remaining 8–10 weeks before onward passage to winter quarters on Atlantic coast from Maryland to North Carolina; similar gathering there (for *c.* 6 weeks) in March–April. See Cooch (1957); Linduska (1964); Godfrey (1966).

**Voice.** See Field Characters.

**Plumages.** 2 plumage types. ADULT. (1) White morph (nominate *caerulescens* and *atlanticus*). White, except black primaries (outer grey at base), blue-grey primary coverts (outer webs rarely mottled white), and bastard-wing (rarely white). (2) Dark morph (nominate *caerulescens* only). Head and upper-neck white; lower hindneck, mantle, and scapulars black-brown, feathers tinged blue-grey in centre, edged dark grey or brown. Back and rump light grey; upper tail-coverts white, finely speckled grey. Lower neck, breast, belly, sides, and flanks grey-brown, feathers edged grey or buff. Under tail-coverts, vent, often centre of belly white; sometimes all belly and breast white in heterozygous hybrids between morphs. Tail-feathers dark grey, edged white. Flight-feathers black, outer primaries basally grey. Upper wing-coverts grey tinged blue; inner median coverts, with black-brown shaft-streaks, elongated greater coverts, tertials, scapulars, and inner secondaries brown-black, broadly edged white. Underwing pale grey, axillaries white. JUVENILE. Shape of feathers as other *Anser*, tertials and inner coverts not elongated as in adults. (1) White morph. Forehead and sides of head pale grey; crown, streak before and spot behind eye, nape, hindneck, and sides of neck grey-brown, feathers fringed white. Mantle and scapulars, grey-brown, broadly tipped and edged buff shading to white; back, rump, and upper tail-coverts, white, finely speckled grey on back and rump. Underparts white, speckled or suffused grey-brown on breast, sides of body, and flanks. Longer scapulars and tertials dark-brown, edged white. Tail grey-brown, broadly tipped white. Primaries and their coverts as adult, but coverts sometimes white with grey-brown freckling or clouding. Secondaries and greater coverts white, mottled or clouded grey-brown, innermost broadly edged white. Median coverts pale grey-brown, variably tipped and freckled white; lesser coverts grey-white, mottled grey-brown. Underwing and axillaries white. (2) Dark morph. Like adult, but head and neck slaty-brown, except white chin-spot; mantle and scapulars browner; brown underparts paler, more white on belly; tertials and inner wing-coverts less contrastingly edged, or plain brown-black. During winter head becomes mottled white. FIRST WINTER AND SUMMER. Like adult, but most juvenile wing-coverts and sometimes part of body plumage retained. (Kortright 1942; Witherby *et al.* 1939.)

**Bare parts.** ADULT. Iris dark brown. Bill crimson, cutting-edge of both mandibles black, nail white. Foot deep reddish-pink. JUVENILE. Like adult, but foot dark olive-grey to purple-grey, and bill dark grey, attaining adult colour during winter, but some dark spots sometimes retained to 2nd winter.

**Moults.** ADULT POST-BREEDING. Complete; flight-feathers simultaneous. Flightless *c.* 3 weeks, late July to late August; non-breeders *c.* 2 weeks earlier. JUVENILE able to fly at *c.* 6 weeks, at same time as parents (Cooch 1961). Sequence of moult of body feathers probably as in other tundra *Anser* (see White-fronted Goose *A. albifrons*).

**Measurements.** Canada and USA (Kennard 1927). ADULT. (1) nominate *caerulescens*, (2) *A. c. atlanticus*.

| WING | (1) ♂ | 430 | (45) | 395–460 | ♀ 420 | (43) | 380–440 |
|------|-------|-----|------|---------|-------|------|---------|
|      | (2) | 450 | (20) | 430–485 | 445 | (10) | 425–475 |
| BILL | (1) | 58 | (45) | 51–62 | 56 | (40) | 50–61 |
|      | (2) | 67 | (20) | 59–73 | 62 | (10) | 57–68 |
| TARSUS | (1) | 84 | (35) | 78–91 | 82 | (37) | 75–89 |
|      | (2) | 92 | (20) | 86–97 | 86 | (10) | 80–92 |

JUVENILE. Wing of both races slightly shorter than adult, other measurements similar.

Weights. ADULT. (1) Nominate *caerulescens*. James Bay, Canada; September–October (Cooch *et al.* 1960). (2) *A. c. atlanticus* (Kennard 1927).

| (1) | ♂ 2744 (235; 467) 2155–3402 | ♀ 2517 (217; 422) 1814–3175 |
|-----|------|------|
| (2) | 3626 ( – ; 13) 3175–4375 | 3065 ( – ; 5) 2835–3175 |

JUVENILE. Slightly less heavy than adult, especially in autumn.

Structure. Wing long, rather narow, pointed. 11 primaries: p9 longest, p10 5–15 shorter, p8 4–10, p7 28–40, p6 55–75, p1 170–215; p11 short, concealed by primary coverts. Longer scapulars, tertials, inner median and greater wing-coverts elongated, curved down. Tail rounded, short; 16 (rarely 18) feathers. Bill heavy, as long as head, cutting-edge of upper mandible strongly arched and lower bill curved like Bean Goose *A. fabalis rossicus* or *A. f. serrirostris*; *c.* 21–24 heavy 'teeth', clearly visible when bill closed. Outer toe *c.* 85% middle, inner *c.* 75%, hind toe *c.* 30%.

Geographical variation. *A. c. atlanticus* larger than nominate *caerulescens*, with heavier, larger bill; dark morph rare. In nominate *caerulescens*, 2 phases; formerly almost segregated, breeding in different areas and for long time thought to represent 2 species, *A. caerulescens* ('Blue Snow Goose') and *A. hyperboreus* Pallas, 1769 ('Lesser Snow Goose'). Dark morph formerly restricted to south Baffin Island, with only low percentage white birds, but in last 60 years or so (with amelioration of climate) blue has invaded white morph colonies in Hudson Bay area, and now extending westwards. Breeding birds prefer partner of same morph, but mixed pairs not rare and their offspring predominantly dark. Reasons for range extension of dark morph birds discussed by Cooch (1961, 1963).

# *Branta canadensis* Canada Goose

PLATES 54 and 61
[after page 446
and between pages 494 and 495]

Du. Canadese Gans     Fr. Bernache du Canada     Ge. Kanadagans
Ru. Канадская казарка     Sp. Barnacla canadiense     Sw. Kanadagås

*Anas canadensis* Linnaeus, 1758

Polytypic. Nominate *canadensis* (Linnaeus, 1758), south-east Canada and north-east USA, introduced various parts north-west Europe. About 11 other subspecies North America, Commander Islands, and Kuriles; for possible vagrants in west Palearctic, see Movements.

Field characters. Size extremely variable between different races: 56–110 cm, wing-span 122–183 cm but feral European populations mostly 90–100 cm, wing-span 160–175 cm, of which body less than two-thirds; ♂ averaging larger and much heavier. Generally large (but down to small) goose with long neck and large bill and feet; mainly grey-brown with black head and neck, white throat-patch, white lower belly and upper and under tail-coverts, and dark flight-feathers and tail. Sexes similar and no seasonal differences; juvenile rather similar and indistinguishable in field by 1st winter.

ADULT. Whole head and neck black (but not breast like other black geese *Branta*), apart from sharply demarcated white throat-patch extending up behind eye (sometimes divided by black line down centre of throat, and sometimes spotted with black); infrequently, white ring at base of neck. Dark brown upperparts and wings with black-brown flight-feathers, paler brown sides and flanks with light feather edgings, whitish-brown breast; pure white upper and under tail-coverts, and black tail. Bill and legs black. JUVENILE. Similar apart from greyer throat-patch at first, but by 1st autumn indistinguishable in field except by lighter build. Some small Canada Geese likely to be genuine vagrants (see Movements).

Unlikely to be confused with any other species. Barnacle Goose *B. leucopsis* closest in appearance, but generally much smaller and shorter-necked (though small northern vagrant Canada Geese may be of comparable size) with wholly white face, black breast and grey instead of brown plumage.

Walks sedately but easily; feeds by day on grassland and crops. Swims frequently, particularly during breeding season, and also feeds on aquatic plants by dipping head or up-ending. Markedly gregarious, occurring in feral flocks of up to several hundred, chiefly on or by fresh water but sometimes on sea or estuaries, and often associated with grey geese *Anser*. Flies fast with deep, regular wing-beats, sometimes in loose flocks, but often like grey geese *Anser* in V-formation, diagonals, or single file. Not usually noisy when settled (unlike wild flocks in North America), but characteristic deep resonant honking in flight.

Habitat. As introduced species in west Palearctic, adopts habitats differing from native Nearctic haunts. Acclimatized stock limited to north temperate, oceanic, middle latitudes where climate permits year-round residence, in contrast to migratory role of most races in North America. Favours lowland pools and ornamental waters, not too grown up with emergent vegetation and sheltered by groups or even compact, tall stands of forest trees, but amply fringed by open, flattish tracts of herbage suitable for close grazing; exceptionally breeds on open moorland.

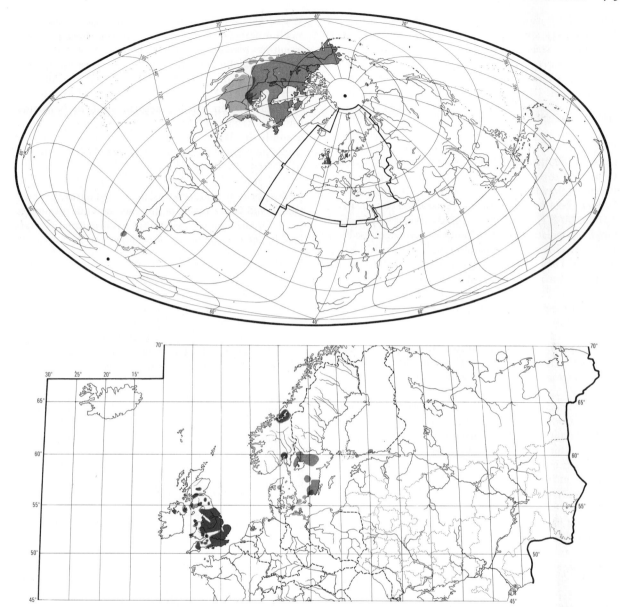

Grazes near water and infrequently forages on cropland, or in hilly terrain. Large or deep lakes, oligotrophic waters and those with rocky rather than well-drained sand or earth banks, reservoirs with sparse vegetation and rivers commonly avoided unless near area of settlement or moult gatherings or in weather severe enough to force temporary local displacement when may resort to estuaries or marine inlets.

Highly malleable to human influence. Learns safe limits quickly and soon settles within them. Tame or feral stocks thus readily planted in heart of metropolis such as London, as well as on ornamental waters in country parks. Damage to habitat caused by erosion and compaction, and excess deposit of droppings. Local introduced stocks. may fail

to survive on naturally available food, but show little enterprise in spontaneous relocation. Local, prospecting flight excursions frequent, usually within fairly narrow radius and at no great height.

**Distribution.** Introduced west Palearctic. Originally introduced Britain nearly 300 years ago; spreading elsewhere in north-west Europe, especially in recent years, due to new introductions. BRITAIN AND IRELAND: introduced England 17th century; spread to Wales, Scotland, and Ireland aided by further introductions or artificial re-inforcement (British Ornithologists' Union 1971; Ogilvie 1969a). NORWAY: introduced Oslo 1936 and Tröndelag 1958; expanding, at least in area of

Trondheimsfjord (SH). SWEDEN: introduced 1929, first breeding 1933; spread greatly since (B Nilsson). FINLAND: breeding in small numbers Porvoo since 1966 (LvH). WEST GERMANY: free-flying populations in some areas, but not yet fully feral (E Bezzel).

Accidental. Iceland, Faeroes, Belgium, Poland, Czechoslovakia, Switzerland, Spain. For races other than *B. c. canadensis*, see Movements.

**Population.** BRITAIN. Generally increasing, though considerable Scottish population (over 2000 birds) virtually eliminated 1939–45 (British Ornithologists' Union 1971). Breeding population 1000–2000 pairs (Atkinson-Willes 1970a). Post-breeding totals: 2500–4000 birds 1953; c. 10 500 birds 1967–9 (Ogilvie 1969a). SWEDEN. Increasing rapidly; population Blekinge (south-east Sweden) 800 birds 1956, 1200 birds 1957, 1500 birds 1965 (Curry-Lindahl *et al.* 1970). Total population 739 breeding pairs (3643 birds) 1966 and 1609 pairs (8990 birds) 1971 (Ahlbom *et al.* 1973).

Survival. England: mean annual mortality of adults 22%; life expectancy 3.9 years (Boyd 1962). Oldest ringed bird 23 years 5 months (Kennard 1975).

**Movements.** Different populations migratory or sedentary. Native to North America, where most populations highly migratory, with northward moult movements and southward autumn passage over continent, some reaching central Mexico; vagrant to Anadyrland, Hawaii, Bahamas, and Jamaica. 2 small eastern races *hutchinsii* and *parvipes*, occur autumn in west Greenland (former has bred), often among White-fronted Geese *Anser albifrons*. In most recent winters, 1–10 apparently wild *B. canadensis* seen Ireland and west Scotland in flocks of *A. albifrons flavirostris* and Barnacle Geese *B. leucopsis*. These suspected to include *parvipes*, *hutchinsii*, and nominate *canadensis*, and possibly *minima* and *interior*; but field identifications of any subspecies of *B. canadensis* questionably reliable especially since taxonomy still controversial, and only nominate *canadensis* certainly identified by Irish specimen of possible immigrant (Boyd 1961a; Merne 1970).

Species introduced to Europe, where sedentary in some areas, migratory in others; well-marked moult migration recorded. British breeders largely sedentary, and divided into more or less discrete groups, though limited interchanges between adjacent sub-populations occur; in most areas movements shown by ringing seldom over 50 km (Ogilvie 1969a); flocks wander locally depending on food supply and disturbance, some spreading out to breed. Long-distance moult migration recorded from Yorkshire to Beauly Firth, Inverness-shire: developed since c. 1950, non-breeders departing Yorkshire late April to mid-May, some June, returning late August to early September (Walker 1970). This moult movement may be still developing; Derbyshire and Nottinghamshire birds recaptured Beauly Firth 1973 and 1974 and one recovery

Warwickshire to Caithness, September 1972. Since 1969, small numbers from English Midlands recaptured in moulting periods in Yorkshire (Walker 1970). A few British (mainly Yorkshire) birds moved south as far as north France in 1963 cold weather; and 2 Hampshire birds recovered north France September 1964 and March 1965.

Breeders from Sweden migrate south in November, some wintering south Sweden, others moving to wintering grounds in West and East Germany, and Netherlands; one exceptional recovery Sweden to Spain (Cadiz) in January. Return movements begin February in some years, but usually March, continuing into April. Moulting birds appear on Hiddensee Island, East Germany, in July (Bauer and Glutz 1968); origin uncertain but presumably from Sweden. Some, apparently wild, reached Switzerland in 1963 severe winter (Leuzinger 1963).

**Food.** Primarily plant materials including roots, rhizomes, tubers, stems, leaves, fruits, and seeds obtained (much as in *Anser*) mostly by grazing. Occasionally feeds in water, dipping head and up-ending. Gregarious, especially outside breeding season; daytime feeding normal though will feed at night particularly if disturbed.

Little data available for Europe. Chiefly grasses, aquatic plants, clover *Trifolium*, winter wheat, spring cereals, grain, beans *Vicia*, kale *Brassica*, and on occasions mangolds *Beta* (Witherby *et al.* 1939; Rogerson and Tunnicliffe 1947; Kear 1963a, 1964). From Kent, 3 stomachs contained leaves, stems, and roots of white clover *Trifolium repens*, and less grasses. One other contained only barley grains (Olney 1967). In North America, winter foods mainly various grasses and marsh and aquatic plants including wigeon grass *Ruppia*, eel-grass *Zostera*, bulrush *Scirpus*, sedge *Carex*, spike-rush *Eleocharis*, pondweed *Potamogeton*, wild rice *Zizania*, glasswort *Salicornia*, skunk cabbage *Symphocarpus*, and sea lettuce *Ulva*. Cultivated grain and legumes also form large part of diet, especially corn, wheat, rice, oats, barley, rye, sorghum, peas, soybeans, peanuts, and sugarbeet (Martin *et al.* 1951; Glazener 1946; Yocom 1951; Hanson and Griffith 1952; Hanson and Smith 1950; Bossenmaier and Marshall 1958; Yelverton and Quay 1959; Hanson 1967; Ellis and Frye 1966).

**Social pattern and behaviour.** Based mainly on material supplied by E Fabricius from observations in Sweden.

1. Gregarious much of year. Outside breeding season, flocks composed of family parties, broodless adults, and yearlings (i.e. immatures no longer in family group and now in later part of first full year); both latter categories often in pairs. Flock uses home-range consisting of feeding grounds, roosts, and bathing places. BONDS. Monogamous, life-long pair-bond; will re-mate if partner lost. Pair-formation starts in yearlings after break-up of family group, but first bonds unstable and more durable pairs formed in 2nd year. Both parents tend young (sometimes adopting those from other broods); family-bond maintained until onset of new breeding cycle when broken by parental hostility. BREEDING DISPERSION. Often nests in loose colony of highly

territorial pairs at protective sites on islands (Blurton Jones 1951; Collias and Jahn 1959; Hanson 1965). In English colony: (1) territories established gradually in late winter and early spring, starting before family group finally disbanded; (2) maintained until young left nest; (3) each territory consisted of nest island and, almost invariably, adjacent area for grazing on mainland bank opposite, mean size 1870 m²; (4) pair used same territory each year (Blurton Jones 1951). Territories in Swedish colonies *c*. 60-200 m in diameter, mean 92·2 m; smallest where topography provided visual isolation. Interspecific territorialism recorded in mixed colonies with other geese, but latter tolerated closer than conspecifics (Fabricius *et al.* 1974). In Scandinavia, as also at times in North America, nests less densely on tarns where food scarce and suitable nest-sites few (Fabricius 1970; Hanson 1965). Feeding, courtship, and copulation mainly within territory; in Sweden, young first feed there though family tends to move out later. Several family groups may form loose flock, often joined by immatures. Non-breeders, mainly yearlings and 2nd-year birds, roam home-range of local population in pairs or small bands; pairs may inspect prospective nest-sites and occasionally show territorial behaviour. ROOSTING. Communal and nocturnal for much of year (breeders roost in territory); in single or multiple flocks on lake or other large and open body of water, but often uses mudbanks (Raveling 1969a). Regular flights to roost begin after post-breeding moult of adults and fledging of young; end at onset of nesting (pairs) or moulting (non-breeders). Parties fly to roost shortly after sunset, leave after sunrise; where distances short, walk—especially in Britain (Ogilvie 1969a), less so in Scandinavia. Roost sites and other waters visited for bathing and loafing at midday.

2. Best studied of *Branta* geese; mainly in feral or captive populations, though some American work on wild birds (e.g. Raveling 1970). FLOCK BEHAVIOUR. Winter flocks highly integrated. Spend much time in antagonistic encounters; most intense and numerous when flock foraging though also occur at roosting lake, especially just after return from morning feeding and during 1–2 hours before afternoon return to feeding grounds. Result in flock organization in which larger families dominate smaller; families, pair; both pairs and families, single birds (see especially Raveling 1970). Paired ♂♂ also defend area round mates and young (Hanson 1953; Raveling 1970). Flock becomes increasingly active in hour before morning or afternoon departure from roosting lake. Prior to leaving (also from grazing fields), incidence of pre-flight behaviour increases, including calling (Raveling 1969b). Birds assume pre-flight Erect-posture, with neck extended vertically, and periodically Head-toss, suddenly lifting chin and pointing bill obliquely up, at same time rapidly waving head from side to side up to 3 times, producing conspicuous flickering effect of white cheek patches. Also periodically Head-shake laterally (Johnsgard 1965; Klopman 1968), and Wing-flap, waving fully open wings up to 7 times. Sudden appearance of white rump pattern after birds take wing, together with Contact-calls, induces flight in other members of flock. ANTAGONISTIC BEHAVIOUR. See especially Collias and Jahn (1959); Blurton Jones (1960); Klopman (1968); Raveling (1970). Antagonistic encounters occur especially in winter flock, during pair-formation, and soon after establishment of territory. Complex system of threat-posturing involved. Main components: Erect-posture, Head-pumping, and Coil-down (tendencies to attack and escape in balance); Advertising-posture, Diagonal-neck posture, Bent-neck posture, and Forward-posture (tendency to attack stronger)—though transitional stages often occur. Erect-posture much as that assumed before flight but neck feathers raised, accompanied by loud Alarm-calls and sometimes

A

B

by Hissing and trampling movements of feet. In Head-pumping, neck repeatedly bent (see A) and straightened, with vertical bobbing movement. In Coil-down, bent neck retracted above level of carpal joints, bill pointed obliquely forward and downward; often accompanied by Wing-shaking, producing rattling or clapping noise, and (on water) sometimes by Bill-dipping, when air blown through nostrils. Advertising-posture of ♂ (see B), most pronounced when paired, shown from beginning of pair-formation through breeding season: body erect, neck stretched vertically with slight backward tilt, head horizontal, and cervical and inter-clavicular air-sacs inflated, with feathers of crop and breast erected. Often leads to Diagonal-neck posture at appearance of alien geese or any other disturbance, most typically as attack initiated, when accompanied by vigorous Head-tossing and loud Rolling-calls; in spring, latter given by pairs even in flight. At approach of opponent, Bent-neck posture evoked: in High-coil version, neck retracted well above level of carpal joints and directed forward, head horizontal and bill open; in Low-coil version (most common of threat-displays), neck held at or below level of carpal joint and head close to ground with bill usually open (see C). Finally, full Forward-posture often immediate

C

precursor of attack: neck fully extended and pointed at rival, bill open, tongue raised. Attacks launched by swimming or running, but territorial ♂♂ also fly. Rival grasped by bill and struck by powerful blows of carpus. Birds escape by running or swimming away, usually in Erect-posture, taking wing at close approach; will Hiss, threaten, and fight back if cornered. Those retreating often adopt Submissive-posture with neck curled, bill lowered almost to touch breast, neck feathers erected, and head often turned away. When flightless, young and moulting adults likely to dive, swim in half-submerged posture, or squat motionless on land. HETEROSEXUAL BEHAVIOUR. Characteristic Triumph Ceremony performed by pair much as in Greylag Goose *Anser anser* (Klopman 1961; Fabricius and Radesäter 1972; Radesäter 1974), though less highly ritualized and less emancipated, with more primitive features. Complete sequence consists similarly of Rolling-phase followed by Cackling-phase, pair often rapidly alternating as they swim, walk, or run forward while keeping parallel. In initial Rolling-phase, ♂ returns to ♀ much as in *A. anser* but tends also to perform elaborate neck-movements up, down, and back in vertical plane (Radesäter 1974). Cackling-phase similar but calls distinctly different and, on meeting, ♂ directs head straight at ♀ and may bite her slightly on nape, ♀ immediately turning head and neck from ♂ (Facing-away) or moving off. Thus, unlike *A. anser*, ♂ has no inhibition against 'threatening' ♀, so pair tend to move in concentric circles or semi-circles with ♂ on outer path. While important in pair-formation, Triumph Ceremony also occurs in some form between pair throughout year, ♂ and ♀ greeting each other thus after temporary separation and also, typically, after antagonistic encounters with others. Mostly outside breeding season, Triumph Ceremony may consist largely of terminal Cackling-phase only, Rolling-phase being often incomplete; also ♀ may not always play fully active role, often assuming Submissive-posture while Facing-away. Pair-formation initiated by ♂ who assumes Advertising-posture and generally behaves in demonstrative manner, making real or sham attacks on other geese, then moves away while performing Rolling-phase of Triumph Ceremony. Responsive, unpaired ♀ tends to stay near such a ♂, also approaching, and Rolling, then joining in mutual Cackling-phase. Either sex (but most often ♂) initiates pre-copulatory display, often several metres from partner, by rhythmically Head-dipping deep below surface: head held erect with neck arched backwards and then swept down and under water, before being returned in same manner to starting position; after brief pause, sequence repeated again and again (Klopman 1962). If partner responsive, also starts Head-dipping and the 2 slowly approach, until side by side, usually head to head with tails further apart. Later in cycle copulation follows. Tempo of mutual Head-dipping increases, especially in ♂ (Collias and Jahn 1959), ♀ turning rear-end further away until pair more or less face; ♂ then approaches obliquely from front, touches her back with bill, and mounts at shoulder grasping ♀ by back of neck with bill and forcing her into Prone-posture. Neck-grip maintained during copulation when ♀'s head, neck, and often entire body submerged. In post-copulatory display, ♂ stretches neck high and slightly back, with bill pointing up and breast out of water, arches partly closed wings laterally from body, and sometimes calls. Incomplete version of same behaviour sometimes shown by ♀ who usually turns away and begins vigorous bathing instead; followed by preening. Before also bathing and preening, ♂ does Both-wings-stretch and series of Wing-flaps. RELATIONS WITHIN FAMILY GROUP. Much as in other geese (see *A. anser*). Parents elicit following response of young by Head-tossing, neck-waving, and giving Contact-calls; goslings reply with Contact-calls. Lost young give Distress-call, parents reacting by approaching and giving their Contact-calls at great intensity (Würdinger 1970). On rejoining one another, both young and parents perform partial Triumph Ceremony. Siblings communicate by Contact-calls and even after temporary separation perform Cackling-phase of Triumph Ceremony together, greeting behaviour being shown from 1st day (though not fully at start). Fighting between siblings breaks out from time to time during first 2 weeks, often during greeting. Later reduced by development of Facing-away and formation (at least in experimental conditions) of firm, linear rank-order. Unlike in *A. anser*, however, Facing-away persists during all ontogenetic stages, being found also in Triumph Ceremony of adults (Fabricius and Radesäter 1972; Radesäter 1974).

(Fig A after Blurton Jones 1960; others from photographs by E Fabricius.)

**Voice.** Variety of deeply resonant, often honking calls. Also non-vocal sounds including (1) rattling or clapping produced by flicking secondaries against body in Wing-shaking display, (2) droning of primaries when territorial ♂ flies at intruder. Following calls distinguished by E Fabricius.

(1) Contact-call. Soft, grunting 'rorr-rorr, rorr-rott' (etc.); uttered by birds in group, when moving close to mate, young, or other members of flock; also given by parents to elicit following response in young. Characteristic of feeding flock. Conversational honking or gabbling, softer and lower pitched than flight-call, probably same (see Bent 1925). (2) Distance-call. Loud, disyllabic 'gorr-rrack' repeated at intervals of some seconds. Uttered in much the same situations as in Greylag Goose *Anser anser*. Thus main flight-call, producing clear and trumpet-like honking, 'ah-honk', etc. (see I); 2nd syllable higher than 1st (Witherby *et al.* 1939). (3) Pre-flight call. Resembles Contact-call but louder and more rapidly repeated. Coordinates take-off, producing characteristic background humming sound in flock; particularly loud just before take-off. (4) Alarm-call. Loud 'gorrack' or 'hucka', repeated at short intervals. Loud version accompanies hostile Erect-posture. (5) Distress-call. Loud 'oh-oo, oh-oo'; uttered when seized by enemy or bitten by opponent in combat. (6) Post-copulation call. Soft, wheezy groan (Klopman 1962) or brief, muted snore; uttered by ♂ only. (7) Rolling-call. In ♂, loud, deep, resonant, prolonged, and honking 'rack-rook, rack-rook-rook, rack-rook-rook' (etc.); in ♀, loud, staccato, yipping 'rack-rack-rack-

I Sture Palmér/Sveriges Radio Sweden April 1965

rack . . .'. Uttered by aggressive birds, particularly when sexually active, and also as introduction to Triumph Ceremony; often preceded by Alarm-call. (8) Cackling-call. So-called only by analogy with equivalent, but quite different call in *A. anser*. Snoring 'arrr' or 'orrr'; uttered by both sexes in terminal phase of Triumph Ceremony. (9) Hiss. Given by cornered birds and sometimes when threatening other species of geese or other waterfowl; in particular, by adults with eggs or young when closely approached by predator.

CALLS OF YOUNG. Contact-call: soft, twittering 'wheeo, wheeo, wheeo', rapidly repeated; uttered when close to parents or siblings, often in response to their Contact-calls and in greeting (Triumph Ceremony). Distress-call: disyllabic piping 'pew-vip'. Sleepy-call: soft, prolonged trill.

**Breeding.** Based mainly on unpublished information, England (M A Ogilvie). SEASON. Britain: egg-laying starts second half of March, main period first half April; hatching from mid-April; fledging from late June. Sweden: egg-laying starts mid-April. SITE. On ground close to water, usually within 30 m, often in shelter of bush or at base of tree, occasionally in open. Invariably on islands if these present at breeding water; if not, usually under 5 m from water but exceptionally up to 3 km on open moorland where well concealed by heather or bracken. Normally colonial though isolated nests occur; closest nests *c.* 5 m apart. Previous year's nests re-used. Nest: low pile of leaves, grass, and reeds, sometimes with base of twigs or small branches up to 1 m long, if available. Average diameter 50–75 cm, height 20 cm. Lined with down during and after laying. Building: mostly by ♀, using material within reach of nest. EGGS. Oval, matt; white or cream, becoming stained light brown during incubation. 86 × 58 mm (79–99 × 54–65), sample 100 (Schönwetter 1967). Weight of British eggs 220 g (173–240), sample 50 (J Kear). Clutch: 5–6 (3–11). Of 75 British clutches: 3 eggs, 3%; 4, 10%; 5, 28%; 6, 27%; 7, 24%; 8, 4%; 9, 1%; 10, 3%; mean 5·9. One brood. Replacement clutches laid after early egg loss. Eggs laid at 24 or 48-hour intervals. INCUBATION. 28–30 days. By ♀ only. Starts on completion of clutch; hatching synchronous. Eggshells left in nest. YOUNG. Precocial and nidifugous. Self-feeding. Cared for by both parents, brooded by ♀ when small. Both birds defend nest and young against predators and sometimes against human intruders. FLEDGING TO MATURITY. Fledging period 40–48 days. Young stay with parents through first autumn and winter, becoming independent at start of next breeding season. Age of first breeding usually 3 years, occasionally 2 or 4. BREEDING SUCCESS. No detailed information for west Palearctic.

**Plumages.** Nominate *canadensis*. ADULT. Head and neck black, broad band on upper throat over cheeks (narrowing on ear-coverts) white; feathers on centre of throat sometimes tipped black, splitting white patch. In fresh plumage, white feathers of

cheek patch finely tipped dusky. Base of neck, upper mantle, chest, and breast buff, feathers variably tipped pale buff shading to white; tips sometimes conceal buff bases, especially at base of neck. Lower mantle, scapulars, and upper wing-coverts dark grey-brown, feathers edged buff shading to white; sides of breast and flanks similar, but paler brown. Back dark brown; rump and shorter central upper tail-coverts black, rest of tail-coverts white. Belly, vent, and under tail-coverts white. Tail black. Flight-feathers dark sepia, primaries edged black at tips. Under wing-coverts and axillaries brown. DOWNY YOUNG. Centre of crown and hindneck, thin line lores through eye, and upperparts sepia-olive, longer tufts of down bright golden-olive. Forehead, sides of head and neck, patches on front and rear edge of wing and on body below wing, and underparts citron to olive-yellow. Gradually become greyer and darker when older; yellow fading, at *c.* 3 weeks some left only on head, chest, and flanks (Yocom and Harris 1965). JUVENILE. Like adult, but structure of feathers different: short and narrow, tapering to rounded tip, rather than long, broad, and square. Difference most marked on scapulars, tail and underparts. White edges of feathers of upperparts and flanks more intensely washed buff, plumage appearing more irregularly or indistinctly barred. Black of head and neck duller and browner, feather tips of cheek patch washed dark brown. Some brown on centres and tips of white upper tail-coverts. (Marquardt 1962). FIRST WINTER. Like adult; new feathers longer and more square-tipped than juvenile, but narrower than adult. Variable amount of juvenile plumage retained; fades when older, appearing buff.

**Bare parts.** ADULT. Iris dark brown. Bill black. Leg and foot leaden-black. DOWNY YOUNG. Iris pale blue-grey. Bill dark blue-grey. Leg and foot blue-grey. JUVENILE. Like adult (Witherby *et al.* 1939).

**Moults.** Timing markedly dependent on latitude and climate. ADULT POST-BREEDING. Complete; flight-feathers simultaneously. Like other geese, starts with some feathers of body shortly after hatching of young. Tail and wing soon follow; flightless for *c.* 3–4 weeks. When wing full-grown, rest of head and body completed. ADULT PRE-BREEDING. Partial. Head and neck, probably not body; spring (Delacour 1954). POST-JUVENILE. Partial; head, body, and tail, from early autumn onwards. Timing and amount of plumage involved probably strongly dependent on size, latitude, and migration.

**Measurements.** Range: (1) nominate *canadensis*, (2) *interior*, (3) *maxima*, (4) *parvipes*, (5) *hutchinsii*, (6) *minima* (Delacour 1954; some additional Aldrich 1946; Taverner 1931; RMNH).

|        | (1)     | (2)     | (3)     | (4)     | (5)     | (6)     |
|--------|---------|---------|---------|---------|---------|---------|
| WING   | 450–550 | 410–549 | 480–560 | 410–442 | 345–408 | 325–390 |
| TAIL   | 120–168 | 105–165 | 147–170 | 120–145 | 110–130 | 110–118 |
| BILL   | 48–65   | 43–64   | 60–68   | 36–49   | 30–39   | 25–32   |
| TARSUS | 78–97   | 71–92   | 90–106  | 73–88   | 65–75   | 60–75   |
| TOE    | 85–102  | 77–99   | 95–107  | 58–73   | 52–63   | 77–65   |

When sex and age taken into account, ranges less wide, and somewhat less overlap between subspecies in specimens of similar sex and age.

*B. c. interior.* Illinois (USA), winter (Hanson 1951).

|        |     |              |         |             |                |         |
|--------|-----|--------------|---------|-------------|----------------|---------|
| WING AD| ♂ 506 | (13·6; 110) | 463–546 | ♀481 (13·3; 92) | | 453–524 |
| JUV    | 486 | (13·1; 114)  | 430–520 | 460 (14·7; 98) | | 410–490 |
| BILL   | 53·6( 2·7; 224) | | 47–61 | 50·0( 2·5; 190) | | 43–56 |

Sex differences significant. Wing (and tail) of juveniles significantly shorter than adult. Bill similar; combined.

Nominate *canadensis*. Adults. Scotland, June (Dennis 1964).

|      |         |      |       |         |      |       |
|------|---------|------|-------|---------|------|-------|
| BILL | ♂ 56·5  | (36) | 50–63 | ♀ 53·7  | (41) | 49–58 |

**Weights.** Britain, in wing moult. June: 36 ♂♂ 4880 (4170–5410), 36 ♀♀ 4390 (3670–4950) (Dennis 1964). July: unsexed 3784 (524; 34) 2900–4000 (BTO). Netherlands: adult ♀♀, January, 3550, 3585; juvenile ♂, December, 2845 (J J Smit; RIN).

**Structure.** Nominate *canadensis*. Wing relatively long and pointed. 11 primaries: p9 longest, p10 5–12 shorter, p8 4–11, p7 22–37, p6 55–76, p1 190–220; p11 minute, reduced, concealed by primary coverts. Outer web of 9–7 emarginated, inner p10 and slightly p9–8. Tail short, rounded; 16–18 feathers. Bill about as long as head, reather heavy, culmen slightly concave. Lower mandible and cutting edges straight; *c.* 25–32 horny 'teeth', not visible when bill closed. Toes longer than tarsus; outer *c.* 93% middle, inner *c.* 78%, hind *c.* 29%.

**Geographical variation.** Considerable; gives rise to different opinions on subspecific delimitations and even on specific distinctions within this group (up to 4 species sometimes recognized). General and more conservative view of just one species adhered to here. Geographical variation concerns relative body size; proportions of neck, bill, tarsus, and foot; darkness and barring of plumage; amount of white on lower neck; and pattern of cheek patch. Size of subspecies south of *c.* 55°N large (e.g. nominate *canadensis*, *interior*, and *maxima*), those of taiga or inland tundra (e.g. *parvipes*) and north Pacific islands and coast intermediate, and of high arctic coastal tundra (*hutchinsii*,

*minima*) small. Eastern populations palest, gradually darkening towards Pacific coast (but southern subspecies inhabiting dry Great Plains region and those of Pacific islands pale). Southern subspecies with longer extremities (bill, neck, toes) than northern; middle toe of southern birds longer than tarsus, of northern shorter. *B. c. interior* differs from nominate in darker, less barred upperparts, chest, and flanks; less distinct pale collar round lower neck, hindneck as dark as mantle. *B. c. maxima* as pale as nominate, but barring less distinct; no pale buff collar, hindneck like mantle, but often narrow incomplete white ring bordering black of neck below. White cheek patch more extensive than nominate, often divided in two by black streak over throat. *B. c. parvipes* like *maxima*, but much smaller; usually no white neck-ring or black throat-streak, but incidence of these and colour of body variable. *B. c. hutchinsii* as pale as nominate; chest especially pale, but lower hindneck dark, similar in colour to indistinctly barred upperparts. Size small, rarely with white neck-ring. *B. c. minima* tiny and dark: upperparts dark grey, barred buff; underparts deep chocolate; incidence of white ring round lower neck and black streak on throat variable. Revision based on adequate sample of specimens of breeding grounds urgently needed (Delacour 1954). Introduced birds Great Britain and Sweden probably mainly nominate *canadensis*, but size and colour of some specimens may indicate that at least some *maxima* originally involved (Harrison and Harrison 1966; Merne 1970; RMNH).                                                          CSR

## *Branta leucopsis* Barnacle Goose

PLATES 54 and 61
[after page 446
and between pages 494 and 495]

Du. Brandgans    Fr. Bernache nonnette    Ge. Weisswangengans
Ru. Белощекая казарка    Sp. Barnacla cariblanca    Sw. Vitkindad gås

*Anas leucopsis* Bechstein, 1803

Monotypic

**Field characters.** 58–70 cm, of which body two-thirds; wing-span 132–145 cm. Medium-sized goose with fairly long neck and short, small bill; creamy face contrasts boldly with black hind-crown, neck, and breast. Sexes similar and no seasonal differences; juvenile and immature distinguishable.

ADULT. Whole face and forehead creamy or pale buff except for irregular black line from bill to eye and, sometimes, dusky marks on cheeks. Hind crown, neck, breast, and upper mantle glossy black, sharply demarcated from silvery-white lower breast and belly, but merging into blue-grey mantle and wing-coverts barred with black bands and finer white lines. Flanks strongly barred brown-grey; upper and under tail-coverts white; flight-feathers and tail black. Bill and legs black. JUVENILE. Black line through eye joining crown; white of upper face more obscured by dusky mottling. Rest of plumage duller, greyer black, and browner. Mantle, scapulars, and wing-coverts with brown subterminal bands and narrow, brownish-white tips; flanks more buff and less strongly barred. FIRST WINTER. Retains juvenile coverts and many other juvenile feathers, so that same distinctions apply to some extent through 1st year, though close views often necessary for recognition.

White face and forehead immediately distinguishes species from other black geese *Branta*. Smaller Brent Goose *B. bernicla* has white patch only on side of neck and lacks strikingly barred upperparts. Generally larger Canada Goose *B. canadensis* has white throat-patch extending to ear-coverts (but otherwise black face) and black extending only to base of neck.

Swims on fresh water and occasionally on calm sea, but generally seen on land not far from sea, being more terrestrial than *B. bernicla*. Breeds on cliff ledges and steep slopes or on flat ground on offshore islets. Flight as other geese, though wings generally look rather more pointed. Highly gregarious; breeding in colonies and wintering in some areas in huge flocks which do not usually associate with other geese. Flocks quarrelsome, with much more bickering than other geese; also noisy, sounding like yelping dogs. Generally less wary of man than grey geese *Anser*.

**Habitat.** In contrast to equally extreme arctic breeding habitats of Brent Goose *B. bernicla*, prefers precipices overlooking fjords or slopes and small islands or skerries to low tundra. Oceanic rather than continental in climatic preference; not so far found breeding in Eurasia on any

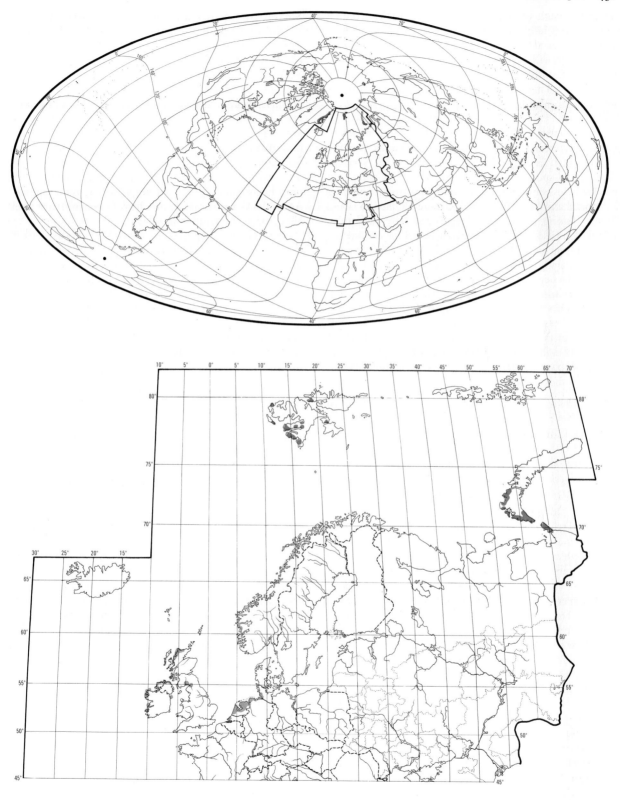

mainland south of Arctic Ocean. Inaccessible breeding sites on crags, hilltops or islands require shift for feeding to valleys or coastal fringes and banks of rivers and lakes with plentiful growth of grasses, herbage, or willow scrub. Nests usually within 1 km of sea or other open water, but often 200–300 m or more above it, sometimes along fjords reaching far inland.

Although 3 main breeding populations migrate by separate routes to distinct winter quarters, all maintain similar preference for terrestrial grazing, making little use of intertidal feeding grounds favoured by *B. bernicla*. Appears to spend more time than latter on wing, but markedly less on water. Greenland population, wintering in Ireland and Scottish Hebrides, more inclined to island living. During winter, Spitsbergen and Novaya Zemlya populations use coastal lowlands of British or continental mainland, including floodlands, and merselands periodically inundated by high tides although carrying terrestrial grasses. Dependent, in places, on improved grassland grazed by farm animals or autumn stubbles. Such habitat presents wider range of choice than narrow intertidal zones, and drastic re-deployment can occur in winter quarters, as recently, in favour of polder habitats in Netherlands. 3 distinct breeding and wintering populations show variations in habitat adaptability and response to management, of special significance for conservation research.

**Distribution.** ICELAND: pairs breeding successfully in recent years probably escapes (FG). SPITSBERGEN: proved breeding in north-west since 1960, and south-east since 1969 (MAO).

Accidental. Bear Island, France, Spain, East Germany, Poland, Czechoslovakia, Hungary, Austria, Switzerland, Yugoslavia, Rumania, Egypt, Morocco, Azores (some occurrences may refer to escaped birds).

**Population.** Breeding. SPITSBERGEN: considerable recent increases at most sites (MAO). USSR: recent and catastrophic decreases (Dementiev and Gladkov 1952); winter counts (below) suggest subsequent recovery.

Winter. (1) GREENLAND population. 5 censuses 1957–66 showed increase from 11800 to 19920 (Boyd 1968); 24100 in 1973; Ireland no change, increase entirely in Scotland (Cabot and West 1973; Ogilvie and Boyd 1975). (2) SPITSBERGEN population. Solway: several thousands wintered 1930s, but massive reduction 1939–45 due to disturbance; 400 in late 1940s, 1500 in 1954–5, 2500 in 1961, and 4250 in 1963–4 (Roberts 1966); then fluctuating between 3000 and 3500 with increase to 4350 in 1972–3, 5100 in 1973–4, and 5200 in 1974–5 (MAO). (3) USSR population. Now mainly wintering Netherlands, due to increased habitat and protection there as well as draining of former haunts in West Germany; *c.* 19700 in 1959 (Boyd 1961*b*), increasing slowly to 25000 in 1968, 31000 in 1969–70, 40000 in 1972–3, and 45000 in 1973–4 (Mörzer Bruyns *et al.* 1973).

Survival. Oldest ringed bird 18 years 5 months (MAO).

**Movements.** 3 discrete populations, all wholly migratory.

East Greenland population migrates via Iceland to winter west Scotland and Ireland (Salomonsen 1967). Flocks gather late August, start leaving end August to early September; majority gone from Greenland by mid-September. Most halt Iceland, particularly valleys of northern fjords; departures begin early October, with main arrival Scotland and Ireland 2nd half October, or occasionally delayed till early November. Limited winter movements between various Scottish haunts and between Scotland and Ireland; then in late March to early April some northward movement within British Isles; departures begin mid-April, continuing for about 1 month. Again flocks halt Iceland; first back Greenland 2nd week May; main immigration late May.

Spitsbergen population winters Solway Firth (Scotland–England). Departures from Spitsbergen early to late September: move down west coast Norway and over North Sea, crossing overland north England and south Scotland. First arrivals Solway Firth from 3rd week September to 1st week October in different years; build-up to near peak numbers usually within 3 weeks. Main departures from Solway in 2nd half April, some occasionally remaining into 1st week May; arrivals in Spitsbergen not until last 10 days of May (Løvenskiold 1964).

Novaya Zemlya and Vaigach Island population migrates south-west through White Sea, overland to Gulf of Finland, across Baltic, north Estonia, Gotland, south Sweden, north Germany, and south Denmark to principal winter quarters Netherlands; one ringed Netherlands recovered Norfolk suggests possible origin of rare but regular visitors to south-east England. Departures from breeding grounds end August to early September; migration over Leningrad area late September to late October and Estonia mid-September to mid-November, heaviest passage October (Dementiev and Gladkov 1952; Kumari 1971). Stops made Danish archipelago and north Germany; peak counts Netherlands generally January, though cold weather can bring birds through earlier from east. Scarce visitor Belgium and north-west France, usually in severe weather. Spring passage starts late March when many move to North Friesian Islands and stay to late April; regular overland migration across base of Jutland peninsula during April. Few spring records from Swedish mainland but major stopping place on Gotland, where about half population present in 1st half May, and remainder in Estonia; in both areas peak numbers at rather few sites in mid-May followed by sudden departure to north-east. Arrives back on breeding grounds last 10 days of May. See also Uspenski (1965); Bauer and Glutz (1968).

**Food.** Based mainly on information supplied by M Owen. Vegetation, chiefly leaves and stems of grasses and other herbs, stolons, and seeds. Molluscs and crustaceans recorded but probably taken accidentally (Alpheraky 1905;

Dementiev and Gladkov 1952). Grazes like other *Branta*; also uses bill to probe and pull up vegetation (e.g. stolons), and to strip seeds from stalks. Gregarious, day-time and night-time feeder.

In Greenland in summer, mainly grasses and sedges (e.g. *Scirpus*), catkins, leaves, and shoots of willow *Salix herbacea*, and leaves of mountain avens *Dryas octopetala*, alpine bistort *Polygonum viviparum*, mountain sorrel *Oxyria digyna* (Manniche 1910; Bauer and Glutz 1968).

Autumn and winter habitat mainly coastal pastures, marshes, and small islands, but increasingly uses agricultural pastures in most main haunts. Seeds of saltmarsh rush *Juncus gerardii* formed up to 44% autumn diet in Solway, Scotland, with 14% stolons of clover *Trifolium repens*, and 42% leaves, mainly grass *Puccinellia maritima* (Owen and Kerbes 1971). Analysis of droppings collected throughout season at this site showed arrow-grass *Triglochin maritima* also important in autumn; stolons and grasses from sown pastures, mainly *Lolium* and *Agrostis*, throughout winter; and leaves of *Festuca rubra*, *J. gerardii*, *Triglochin*, and hawkbit *Leontodon autumnalis* in spring. Other species regularly taken include grasses *Poa pratensis*, *Agrostis stolonifera*, and *Holcus lanatus*, and plantain *Plantago maritima*, thrift *Armeria maritima*, and samphire *Salicornia*. Seeds of *J. gerardii*, *F. rubra*, sedges *Carex flacca* and *C. nigra*, spike-rush *Eleocharis uniglumis*, *P. maritima*, and *Salicornia* also recorded (M Owen). Up to 60% winter diet in Solway may be stolons of *T. repens* (Owen 1973a). In Inishkea Islands, Ireland, droppings contained mainly stolons of *Festuca rubra* and *T. repens*, leaves of *P. pratensis*, *Agrostis*, *H. lanatus*, and other plants, though *Plantago maritima* (a typical constituent of the swards) only occasionally taken (M Owen). 27 stomachs from Uist, Scotland, in January and February, contained mainly grasses *Festuca* (especially *F. rubra*), *Poa*, and *Puccinellia maritima*. A wide range of other plants also eaten, including leaves and stems of clover *T. repens* and *T. dubium*, fool's watercress *Apium nodiflorum*, buttercup *Ranunculus repens*, woodrush *Luzula pilosa*, daisy *Bellis perennis*, scurvy-grass *Cochlearia*, and plantain *Plantago coronopus*, with mosses, liverworts, and seeds in small quantities (Campbell 1936, 1946b); in Iceland, autumn foods mainly rhizomes of *P. viviparum*, *Carex* seeds and horsetail *Equisetum variegatum* (Gardarsson 1975). In Gotland, Sweden, in autumn and spring, grazes on leaves and stems from saltings, mainly *Festuca* and *Agrostis* (Bjärvall and Samuelsson 1970). From Friesland, Netherlands, important constituent of droppings *Puccinellia maritima* (M Owen). Apart from using agricultural pastures, flocks occasionally eat grain from stubbles in autumn; also use, rarely, waste potatoes (Berry 1939; Kennedy *et al.* 1954).

**Social pattern and behaviour.** Based mainly on material supplied by H Boyd.

1. Gregarious throughout year. Outside breeding season, flocks include family parties, adults without broods (nearly all in pairs), and 2nd year birds (some in pairs). Those wintering on small islands form persistent groups of one or a few family parties, and may gather into flocks of several hundred birds only shortly before or after migratory flights (Darling 1940); on migration, flocks usually under 150. BONDS. Monogamous, life-long pair-bond, but will find new mate if first lost. No precise information on initial pair-formation; starts in yearlings (see Canada Goose *Branta canadensis*), probably most frequently in late winter, but pairs may then be impermanent. Both parents tend young. Family group maintained until start of next breeding season; manner of break-up in wild not reported. BREEDING DISPERSION. Usually colonial at highly protective sites (Jourdain 1922); in small groups (5–50 pairs), occasionally up to 150 pairs or singly (Salomonsen 1967; Uspenski 1965). Territorial in captivity, during nesting period only; no information from arctic breeding places on territory size, but nests sometimes only 2 m apart (Ferns and Green 1975). In Novaya Zemlya, where few breed, occurs in seabird colonies and in association with nests of Gyr Falcon *Falco rusticolus* (Uspenski 1965); no similar associations noted elsewhere. Broods led from nest soon after hatching and usually reared in nomadic groups at some distance, up to 40 km, from birth place. Family parties gradually aggregate during rearing period, especially after parents flightless due to wing moult. Non-breeders, including birds old enough to breed as well as pre-breeders, sometimes found in separate flocks; during flightless period and after, less often mingle with families, but mix almost completely in staging and wintering areas. Whereabouts and behaviour of non-breeders at time of territory establishment and during nesting not recorded. ROOSTING. Communal at all seasons, except nesting pairs. In arctic summer, probably no synchronized cessation of feeding activity, but diurnal rhythm develops on estuarine wintering grounds, flocks roosting—chiefly at night—on sandbanks or open water. Small groups living on offshore islands without predators probably do not resort regularly to coasts for roosting in winter or gather in large flocks (Darling 1940). At Icelandic staging areas, some flocks roost on inland lakes, but this unusual in Britain and Netherlands in winter. Flocks usually fly to roost, rarely more than 10 km in Britain, though may walk from water or mud to saltmarsh and back when circumstances permit. Roost sites and other open water visited for bathing and loafing during day, frequency and duration depending on season, disturbance at feeding places, and (presumably) availability of food.

2. No detailed studies, but evidently closely similar to *B. canadensis*. FLOCK BEHAVIOUR. Much as in other geese. Pre-flight signals almost identical to those of *B. canadensis* (Johnsgard 1965). Flock wintering on Solway Firth, Scotland, spent 87–97% of time feeding, depending on habitat, but in all cases less than 1% spent in hostile encounters—at least up to late February, when flock departed. Though silent at roost after all assembled, often kept up continuous low chatter when feeding during day, usually breaking into loud clamour on taking wing. When disturbed by Peregrine *Falco peregrinus*, low-flying aircraft, etc., flock rose in body, often to considerable height, then wheeled and dived at high speed like waders (Scolopacidae), sometimes in complete silence. When moving between sites (e.g. feeding and roosting areas) flew in compact but formless flocks, not in formation as on migration (Roberts 1966). ANTAGONISTIC BEHAVIOUR. Threat-postures include Bent-neck posture and Forward-display, much as in *B. canadensis*. In Solway winter flock, aggression never long sustained and usually consisted of a few Wing-flaps together with low-intensity pecks or short, darting run at near neighbour. HETEROSEXUAL BEHAVIOUR.

A

Triumph Ceremony similar to that of *B. canadensis* (Johnsgard 1965); accompanied by alternate wing-flicking (see A). Pre-copulatory display consists of mutual Head-dipping as in *B. canadensis* but, in otherwise identical post-copulatory display, ♂ shows more exaggerated lifting of wings (see B). RELATIONS WITHIN FAMILY GROUP. Adult reported to carry young on back or in bill away from precipitous nesting sites (Madsen 1925; Pedersen 1934), though goslings probably usually descend on foot with parents or jump. Bond between parents and goslings obviously strong for during flocking of family groups mixing of broods or transfer of young from one pair to another unusual in undisturbed conditions. Later, though family persists when in large winter flocks, members tend to straggle more than in other goose species, so temporarily detached young common; this perhaps due to high speed with which such flocks travel across open places when feeding and to characteristic 'leap-frogging', in which groups falling behind fly up and over flock to alight near front (H Boyd).

(Figs A–B after Johnsgard 1965.)

B

**Voice.** Not studied in detail. Calls freely, but repertoire limited. Shrill monosyllabic bark, 'gnuk' (see I), of varying pitch, sometimes rapidly repeated and resembling shrill yapping of small dogs, uttered in flight and on ground. In flight, barks accompanied by creaking noise of wings (see

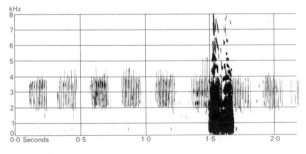

I Sture Palmér/Sveriges Radio Sweden 1965

I—vertical lines) made as birds bank steeply. Flocks also utter muffled, lower pitched, conversational feeding calls rendered 'hoog', 'hogoog', or 'wräck', rarely grazing for long in silence (Bauer and Glutz 1968). Loud 'ark' described at breeding grounds (Witherby *et al.* 1939) probably alarm-call; also uttered in winter. Harsh, sharp 'wok' uttered by captive ♀ leading small young when humans near (S Cramp). Triumph-call high pitched, shrill, quite unlike other geese (Witherby *et al.* 1939).

CALLS OF YOUNG. Contact-calls uttered in groups of 2–5 notes. Distress-call higher, more widely spaced and regular than Contact-call. (J Kear.)

**Breeding.** SEASON. Spitsbergen: see diagram. USSR: few data but probably similar. SITE. Ledges on steep cliffs and rock pinnacles; tops of rock outcrops, on ground on low hummocks and snow-free places, particularly on islands in river channels and up to 3 km from shore. In Spitsbergen, ground nests on offshore islands now predominate, though cliff breeding sites formerly favoured (Norderhaug 1970a). In east Greenland, used cliff ledges *c*. 75 cm deep by 1–2 m wide (Ferns and Green 1975). Colonial, though isolated nests occur; closest about 2 m apart. Old nests re-used. Nest: shallow mound of grass and other vegetation, often built up with own droppings, shallow depression in top; average of 4 nests, diameter 43 cm, diameter of cup 18 cm, depth of cup 7 cm (Nyholm 1965). Re-use of sites gradually produces lasting saucer shape with rim. Lined with much down during and after laying. Building: by both sexes using material within reach. Droppings added during incubation. EGGS. Oval; grey-white, staining yellow during incubation. 77 × 50 mm (68–82 × 46–54), sample 75 (Schönwetter 1967). Weight 103 g (94–114), sample not given (Dementiev and Gladkov 1952). Weight of captive laid eggs 105 g (77–125), sample 100 (J Kear). No apparent variation between eggs of 3 breeding populations of Greenland, Spitsbergen, and USSR. Clutch: 4–5 (2–9); no differences between 3 populations. One brood. No data on replacements, though unlikely. Eggs probably laid at 24-hour intervals. INCUBATION. 24–25 (28) days. By ♀ only, with ♂ standing near. Eggs covered with down when ♀ off. Starts with last egg; hatching synchronous. Eggshells left in nest. YOUNG. Precocial and nidifugous. Self-feeding. Cared for by both parents, who defend nest and young against predators.

FLEDGING TO MATURITY. Fledging period 40–45 days. Young stay with parents through first autumn and winter, leaving wintering grounds with them in spring, but becoming independent when parents reach breeding grounds. Age of first breeding probably 3 years, occasionally 2. BREEDING SUCCESS. Counts on wintering grounds of all 3 breeding populations show annual variation in proportion of young from 7·5% to 47·2% with mean brood size from 1·4–3·1 (Boyd 1968; M A Ogilvie).

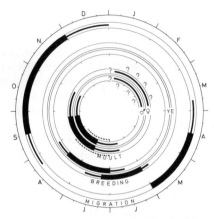

**Plumages.** ADULT. Forehead, sides of head, chin, and throat cream-white, usually with some black at sides of upper mandible, round eyes, or from lores through eye to hindcrown. Occasionally crown to below eye and lores black, pattern of head then resembling Canada Goose *B. canadensis*, but always white triangular spot on forecrown, and white on cheeks more extensive. Hindcrown, neck, mantle, chest, upper breast, back, and rump deep black; dark grey feather-bases sometimes visible on lower mantle and back. Scapulars and upper wing-coverts ash-blue, tips broadly black with white edges; outer wing-coverts with fainter and narrower black bar and white tip. Considerable variation in colour of scapulars and wing-coverts, possibly related to sex: some appear dark without white edges and only limited grey of feather-bases visible, others mainly blue-grey, each feather with broad white tip and faint black subterminal bar. Feathers of sides of body and flanks pale grey, broadly edged white. Thighs and sides of back black, blotched white; sides of rump, tail-coverts, belly, and vent white. Tail black. Primaries black, outer webs broadly edged grey; primary coverts, secondaries, and tertials pale grey with sepia or black inner webs and tips; some inner occasionally fringed white at tip. Under wing-coverts and axillaries grey, narrowly tipped white. In fresh plumage, white feathers on head narrowly edged black; when worn, white edges of scapulars and wing-coverts abraded and sides of body and flanks browner, more contrastingly barred white. DOWNY YOUNG. Centre of crown, upperparts, chest, and thighs medium-grey to pale grey-brown; loral streak dark grey or sepia. Head suffused pale yellow at hatching; grey colours duller and browner after some weeks. Forehead, sides of head, spots on wing, sides of back and rump, and underparts white. JUVENILE. Like adult, but white of head greyer, partially mottled brown. Black of neck, mantle, and chest duller, tinged brown; sometimes dark grey with tips of feathers suffused dull black. Scapulars and wing-coverts grey-brown, gradually darker towards dark brown or black tip, faintly margined buff or off-white, not deep black with contrasting white edges. Sides of body, flanks, and thighs grey-brown to off-white, feathers without contrasting pale edges.

Feathers short and narrow with somewhat rounded tips, against broad and square in adults; most marked on scapulars, wing-coverts, tail, flanks, and underparts. FIRST WINTER AND SUMMER. Plumage resembling adult gradually acquired, but never completely; at least abraded juvenile wing-coverts retained. New feathers like adult, but duller black on head, neck, mantle, and chest; slightly narrower (although square) on scapulars and flanks, with rather less well-defined white edges.

**Bare parts.** ADULT AND JUVENILE. Iris dark brown. Bill, leg, and foot black.

**Moults.** ADULT POST-BREEDING. Complete; flight-feathers simultaneously. Flightless for 3–4 weeks: breeding adults between mid-July and mid-August, non-breeders slightly earlier. Moult of head, body, and tail starts when wing almost full-grown; completed about October–November. ADULT PRE-BREEDING. Apparently head and neck moulted again late winter and spring, but not rest of body. POST-JUVENILE. Partial; from about October. Head, neck, upper mantle, and chest first; followed November–January by flanks, upper scapulars, and sometimes central tail-feathers. In spring, rest of body moulted except wing-coverts, part of back and rump, tail-coverts, and usually underparts and outer tail-feathers.

**Measurements.** Netherlands, winter (Novaya Zemlya population); skins (RMNH, ZMA).

| | | | | | | |
|---|---|---|---|---|---|---|
| WING AD ♂ | 410 | (10·8; 23) | 388–429 | ♀392 | (8·60; 19) | 376–410 |
| JUV | 389 | (8·79; 9) | 374–399 | 372 | (9·76; 9) | 362–393 |
| TAIL AD | 126 | (5·50; 22) | 116–134 | 119 | (4·62; 18) | 113–130 |
| JUV | 111 | (6·24; 9) | 103–120 | 106 | (5·38; 8) | 98–117 |
| BILL | 29·6 | (1·43; 32) | 28–33 | 28·6 | (1·28; 28) | 27–32 |
| TARSUS | 72·4 | (2·77; 31) | 67–80 | 67·8 | (2·46; 28) | 64–72 |
| TOE | 62·6 | (2·45; 24) | 57–68 | 58·2 | (2·39; 22) | 54–63 |

Sex differences significant, except juvenile tail. Wing and tail of juveniles significantly shorter than those of adults. Other measurements similar; combined. Measurements Spitsbergen and Greenland populations similar.

**Weights.** ADULT. (1) Spitsbergen, July, in wing moult (Jackson *et al.* 1975). (2) East Greenland, mid-July to early August, in wing moult (Marris and Ogilvie 1962). (3) Solway, Scotland, February (Boyd 1964). (4) Netherlands, winter (Bauer and Glutz 1968).

| | | | | | | |
|---|---|---|---|---|---|---|
| (1) | ♂ 1756 | (175) | 1410–2170 | ♀1505 | (177) | 1310–1820 |
| (2) | 2010 | (255) | 1500–2400 | 1772 | (229) | 1360–2230 |
| (3) | 1870 | ( 20) | 1590–2100 | 1702 | ( 13) | 1420–1870 |
| (4) | 1672 | ( 33) | 1370–2010 | 1499 | ( 35) | 1290–1785 |

Arrive heavy in breeding area. Steady decrease in weight during breeding, but increase during wing moult: maximum for adult breeding, ♂, east Greenland, 2560 (J de Korte). Rather fat when autumn migration starts. Minimum weight adult ♂, dying during cold spell, Netherlands: 925 (ZMA). JUVENILE. Netherlands, January–February (Bauer and Glutz 1968): ♂ 1504 (14) 1180–1660; ♀ 1359 (11) 1220–1690.

**Structure.** Wing rather long and narrow, pointed. 11 primaries: p9 longest, p10 and p8 both 3–12 shorter, p7 27–45, p6 62–80, p1 165–195; p11 minute, pointed, concealed by primary coverts. Inner web of p10–9 and outer p9–8 emarginated. Tail rather long, slightly rounded; 16 feathers. Bill short, higher than broad at base, sides slightly tapering towards large and broadly rounded nail. Culmen straight, but slightly concave behind nail. Cutting edges almost straight, lower mandible slender. About 30 horny 'teeth' in upper mandible, not visible when bill closed. Outer toe *c.* 90% middle, inner *c.* 79%, hind *c.* 25%. CSR

## *Branta bernicla* Brent Goose

Du. Rotgans     Fr. Bernache cravant     Ge. Ringelgans

Ru. Черная казарка     Sp. Barnacla carinegra     Sw. Prutgås     N.Am. Brant

*Anas Bernicla* Linnaeus, 1758

Polytypic. Nominate *bernicla* (Linnaeus, 1758), tundra USSR from Kolguev to Taymyr Peninsula; *hrota* (O F Müller, 1776), arctic Canada east from Melville Island; north Greenland, Spitsbergen, and Franz Josef Land. Straggler: *nigricans* (Lawrence, 1846), tundra USSR east from Taymyr; north Alaska, and Canada east to about Perry River and west arctic islands.

**Field characters.** 56–61 cm, of which body two-thirds; wing-span 110–120 cm. Small, stocky, and dark goose, with short neck and legs, rather long wings, and short tail; whole head to upper breast black except for whitish neck-patch, and upperparts dark. Sexes similar and no seasonal differences; juvenile and immature distinguishable. 3 races in west Palearctic distinguishable in field.

ADULT. (1) Dark-bellied race, nominate *bernicla*. Head and neck, upper mantle, and upper breast sooty-black except for small irregular whitish patch on side of neck; rest of upperparts dark brown-grey to slate-grey, contrasting with white upper tail-coverts. Rest of underparts slate-grey (often hardly paler than upperparts) with paler bars on flanks, contrasting with white vent and under tail-coverts. Flight-feathers and tail black, but latter so short that appears as no more than narrow black wedge at end of conspicuous white stern. Bill and legs black. (2) Pale-bellied race *hrota*. Similar except for lower breast, belly, and flanks being generally much paler grey-brown shading to whitish in centre and towards rear. (3) West North American and east Siberian race *nigricans*. Differs from dark-bellied, nominate race in generally browner tinge, both above and below, with broader pale barring on flanks, and especially much deeper and more conspicuous white collar extending to foreneck. (NB. Intermediates between races not uncommon and some stragglers difficult to distinguish. In particular, underparts colour variable in both adult and 1st winter birds of dark-bellied, nominate race, with all shades from normal slate-grey to virtually as pale as *B. b. hrota*; extent of pale flank barring also highly variable, with some individuals showing little and others as much as *nigricans*.) JUVENILES. In all races, resemble respective adults, but lack white patch on neck and show more prominent barring on lower mantle, scapulars, and wings due to grey-brown to grey-white tips to many of these feathers and particularly white tips to secondaries and greater coverts. FIRST WINTER. Starts to show neck patch, but still distinguishable by white bars on secondaries and greater coverts.

Unlikely to be confused with any other species: all-black head, dark body, and prominent white stern distinguish it from white-faced Barnacle Goose *B. leucopsis* and generally much larger, white-throated Canada Goose *B. canadensis*.

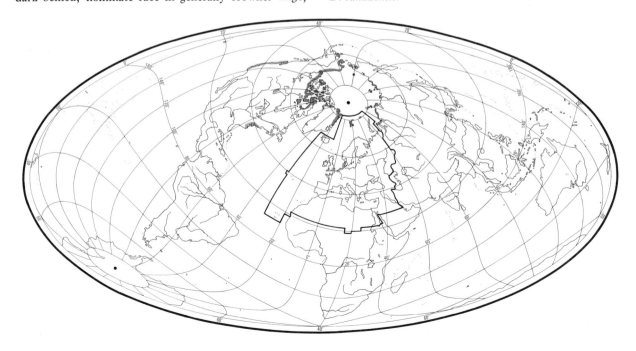

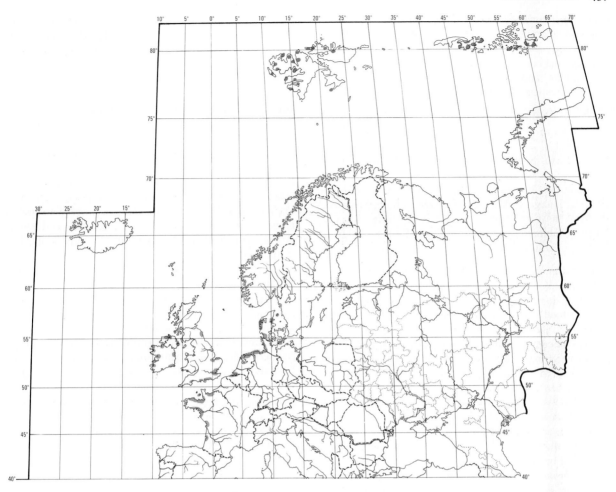

Strictly coastal and more aquatic than other geese. Swims readily, particularly on sea, where rests at high tide and roosts at night; also dips head or up-ends in shallows to feed on eel-grass *Zostera*, and dives readily when flightless. When swimming, stern-high posture accentuates white undertail. Walks daintily and precisely, with quick movements. Highly gregarious, particularly outside breeding season when occurs in flocks of several hundreds. Flocks feed on mudflats in rather dispersed manner, but bunch more tightly in fields, often moving together and hardly spreading out as, for example, *B. leucopsis*. Flocks generally active and rather quarrelsome, but not usually noisy except in flight. Flies fast, with quicker wing-beats than larger geese. Flocks usually in irregular masses or oblique or diagonal lines, seldom in V formation, wavering and undulating in distinctive and graceful manner, often quite low over mudflats. All movements quicker, less deliberate than those of *B. leucopsis* and *B. canadensis*.

**Habitat.** Restricted largely to low tundra with many pools in high arctic breeding quarters, where nesting must begin before snow and ice melt, and juveniles must leave with parents within 100 days. Nominate race favours grassy tundra along river valleys or near seacoasts; *B. b. hrota* often breeds also on islands or islets off coast. Unlike Barnacle Goose *B. leucopsis*, seems always to choose flattish situations, rarely far from sea.

On leaving breeding quarters, resorts to shallow seacoasts and estuaries, especially with extensive mudflats rich in *Zostera, Enteromorpha*, and, less frequently, other green littoral plants. Each year, regularly uses same routes, stopping places, and habitats on migration. Strongly attached to intertidal feeding zones, and in contrast to *B. leucopsis* has until *c.* 1970 only exceptionally been recorded grazing on land in winter or spring. When not feeding, prefers to rest or sleep on sea surface, although rarely far from land or in rough waters. Occurrences on fresh water unusual except in, or on way to or from Arctic. Although strong flier, takes wing with some reluctance, and flies no higher than essential. Being conspicuous, easily located, and predictable in routine movements, and lacking suitable alternatives to clearly defined existing habitats, can flourish only in areas unfrequented by man or under strict protection. Increasing pressures for reclamation and competing demands for land-use intensify problems of maintaining ample stock following traditional way of life. Has recently shown positive although limited response to international conservation but fluctuations of food plants and climate, and growing threats to remaining habitat indicate prospect of great difficulty (Ogilvie and Matthews 1969).

**Distribution.** No recorded changes in breeding range. Up to 1930s, 4000 birds *B. b. hrota* wintered regularly Moray Firth, Scotland, but none now (Atkinson-Willes 1963).

Accidental. Poland (formerly winter visitor), Czecho-slovakia, Hungary, Austria, Switzerland, Italy, Spain and Portugal (now irregular in winter), Yugoslavia, Bulgaria, Greece, Rumania, Cyprus, Egypt, Tunisia, Algeria, Morocco, Azores.

**Population.** No counts of breeding populations.

Winter. *B. b. bernicla.* Severe decrease, probably of at least 75%, in north-west Europe in 1930s following disease of main food plant *Zostera marina* (Atkinson-Willes and Matthews 1960). Estimated 16 500 in 1955–7; fairly steady increase since (fluctuations mainly due to great variation in annual breeding success) with 22 000 in 1960–1, 25 000 in 1964–5, 30 500 in 1966–7, 39 000 in 1970–1, 34 000 in 1971–2, then sharp increase to 48 000 in 1972–3, and to *c.* 80 000 in 1973–4, following two highly successful breeding seasons (Ogilvie and Matthews 1969*a*; M A Ogilvie). Increase due to improved protection and perhaps recovery of *Zostera*; also, from 1972–4, mild winters and successful breeding seasons.

*B. b. hrota.* (1) North-east Greenland population (and some from arctic Canada) wintering in Ireland. Fluctuat-ing numbers, with 12 000 birds in 1960–1 and 1961–2, 7000–8000 in 1965–6 and 1968–9, 12 000 in 1970–1 (*Irish Bird Reports* 1961–70). (2) Spitsbergen and Franz Josef Land population wintering in Denmark and north-east England. Estimated 4000 birds from 1946–7 to mid-1950s (Salomonsen 1958), then decline to 2750 in 1965–6, 2200 1969–70, and 1600 in 1970–1 (Norderhaug 1970*c*; Fog 1972). Shooting in Denmark and disturbance in breeding areas may be contributory factors (Norderhaug 1970*c*).

Survival. Mean adult annual mortality of *B. b. bernicla* wintering Britain 14%, and *B. b. hrota* from Spitsbergen 17% (Boyd 1962).

Oldest ringed bird over 13 years (M Fog).

**Movements.** (1) *B. b. bernicla.* Single wholly migratory population. Depart USSR tundras (Taymyr westwards) mid-August to 1st week September. Main route west along arctic coasts USSR to White Sea, then overland to Gulfs of Finland and Bothnia. Passage through Baltic mid-September to early October; first arrivals Denmark late September but bulk October. Some stay Denmark and West Germany through November before moving on in colder weather; others go straight to winter quarters in

Netherlands, south-east England, and west France, where peak numbers December–February, with movements between England and France and within each influenced by food supply and weather.

Return passage begins early March; most leave England and France by mid-April. Spring gatherings in Nether-lands, Denmark and West Germany until main departures in mid-May; only stragglers after mid-June. Spring route reverse of autumn, except fewer reach White Sea via Gulf of Bothnia: most via Gulf of Finland and Lake Ladoga; main passage through these areas in second half May and breeding grounds reached first half June. One ringed Denmark 11 May found 7 days later near Pechora, USSR (Fog 1967). Some non-breeders stop on Kola and Kanin Peninsulas where summer and moult; others turn north to Novaya Zemlya and moult there (Uspenski 1960).

(2) *B. b. hrota.* 2 populations, discrete and wholly migratory. Breeders of north-east and north Greenland, plus others from Canadian arctic islands, winter Ireland (Salomonsen 1967). 2 main routes: down east coast Greenland and then via Iceland; down west coast Greenland, some crossing inland ice-cap at about latitude of Arctic Circle, others continuing down and round Cape Farewell—latter probably bypassing Iceland. Substantial numbers from Canadian high arctic islands join Greenland breeders in migrating via Iceland to Irish winter quarters, scale only recently realized (Maltby-Prevett *et al.* 1975; *Irish Bird Rep.* 1975); *hrota* from Ellesmere, Axel Heiburg and Bathurst Islands, and eastern Melville Island, certainly involved; thus originate west to at least 108°W, where intergrades with *nigricans*. Also one Ellesmere Island bird recovered north-west France, where regular occurrence in small numbers of *hrota* only recently recognized. Other Canadian birds winter on Pacific (*nigricans*) and Atlantic (*hrota*) coasts of USA. Passage mid-September to third week October; main arrivals Ireland from second half October. Some movement between haunts during winter. Migration north starts early April, peak in second half. Many spend late April to mid-May on west coast of Iceland, continuing north to reach breeding grounds by first week June; some again cross inland ice.

Breeders of Spitsbergen and Franz Josef Land flock from second half August and migrate south first half of September (Løvenskiold 1964), following outer skerries down west coast Norway. Arrive Denmark end September and early October, rapidly reaching peak numbers. First arrivals Northumberland winter haunts mid-October with normal level of few hundreds reached by mid-November. Hard weather in Denmark can cause further movement to Northumberland giving peaks in January or February. Depart Northumberland by early April, for Denmark where spring gathering in north-west before main depart-ure in second half May. Arrivals in Spitsbergen late May and early June.

Wintering grounds in Europe of nominate *bernicla* and *hrota* now restricted following declines (see Population); formerly more widespread on western seaboard.

(3) *B. b. nigricans*. Migrates from tundras of Siberia (east of Taymyr), Alaska, and west Canadian arctic islands, to winter on both sides of North Pacific south to Yellow Sea and Baja California. Vagrant to England (2), Netherlands (1), Hawaii, Texas, and Fukien.

See also Dementiev and Gladkov (1952); Uspenski (1960); Fog (1967); Bauer and Glutz (1968); Ogilvie and Matthews (1969).

**Food.** Vegetation by grazing and, in shallow water, by pulling up and tearing underwater plants. Also feeds on drifting eel-grass *Zostera*. Trampling movements and grubbing may expose and loosen from mud plant systems (Fog 1967; A St Joseph). Rarely dives. When plants covered by tide, will up-end or swim with head and neck below surface. Gregarious; mainly diurnal though will feed by moonlight. Daily feeding rhythm related to tidal cycle.

In winter quarters, most commonly *Zostera*, marine green algae of shallow waters, especially *Enteromorpha, Ulva* and less often *Cladophora*. Also saltmarsh plants: glasswort *Salicornia;* grasses *Puccinellia, Festuca*, and *Spartina*; arrow-grass *Triglochin maritima*; and sea aster *Aster tripolium*. Brown and red algae and sponges (Collet 1894*b*) occasionally. Animal materials taken occasionally but rarely, if ever, deliberately: molluscs and crustaceans (Gätke 1900; Campbell 1946*a*), and lugworms (Burton 1961). On breeding grounds, mosses and lichens and, as thaw progresses, grasses and other tundra plants.

Diet changes with season and availability. At Terschelling, Netherlands, September–December, mainly *Z. noltii* roots and leaves. In December and January, flocks move increasingly to *Enteromorpha*, and from end of January to mid-March primarily *Enteromorpha*. In April and May, after moving into saltings during March, feed almost exclusively on *Salicornia*, especially seedlings *S. herbacea*, and *Puccinellia* (Mörzer Bruyns and Tanis 1955). In south-west Netherlands, close correlation shown between distribution of flocks and their main foods *Z. noltii, Z. marina, Enteromorpha*, and *Ulva lactuca* (Wolff *et al.* 1967). Also feed at times on grassland and wheat (Mörzer Bruyns and Timmerman 1968). At Scolt Head, England, mainly *Enteromorpha* and other green algae through winter, especially in mid-winter, and *Z. noltii* most prominent in early autumn. Higher saltmarsh plants, especially *Puccinellia maritima* and *A. tripolium*, taken in late winter and early spring. Most species preferred at their most active growth phase. Similar diets in Hampshire and Essex (Ranwell and Downing 1959). In Essex, majority changed during winter from exclusive diet of *Z. noltii* to mixture of *Z. noltii* and *Enteromorpha* (*E. clathrata, E. intestinalis, E. prolifera*, and *E. ramulosa*) and *Ulva* (*U. lactuca* and *U. latissima*); with spring tides, some birds

moved on to higher saltmarsh plants (Burton 1961). Recent reports from same area show, from December onwards, feeding on wheat, barley, and grassland (A K M St. Joseph). 28 stomachs mainly from Essex contained chiefly *Zostera* and *Enteromorpha*. One stomach contained over 40 laver spire shells *Hydrobia ulvae* (Campbell 1946). In Gulf of Morbihan, France, almost entirely *Zostera* throughout winter (Voisin 1968); similarly in Denmark, where also may feed on ejected grain-filled pellets of Herring Gulls *Larus argentatus* (Fog 1967).

On breeding grounds in Spitsbergen, adult *B. bernicla hrota* take algae *Fucus*, moss, grass stalks, leaves of *Oxyria digynia*, and young sprouting *Saxifraga* (le Roi 1911). Young also take grasses, freshwater algae, and *Oxyria* (Walter 1890). In Greenland, adults take leaves of *Ranunculus nivalis* and *R. sulphureus, Eriophorum scheuchzeri*, and *Cerastium alpinum* (Salomonsen 1950*a*; Nyholm 1965). On breeding grounds of *B. b. bernicla*, mosses and lichens and, as season progresses, young grass shoots (Dementiev and Gladkov 1952).

Population decrease from 1930s has been correlated with dramatic decline of main food *Zostera marina* in north Atlantic; not totally proven and other factors, such as shooting pressure and breeding failure, perhaps significant (Olney 1958; Ranwell and Downing 1959; Uspenski 1961; Voisin 1968). Evidence suggests green algae important food at least since 1930s, and in places *Z. noltii* now more important than *Z. marina*. Stomach analyses of American *B. b. hrota* collected before 1932 had *Zostera* as main food, after 1932 algae (Cottam *et al.* 1944).

**Social pattern and behaviour.** Based mainly on material supplied by H Boyd.

1. Gregarious throughout year. Outside breeding season, usually in flocks varying from a few birds to several thousand, including family parties, pre-breeders, and non-breeding adults. In Alaska, flocks of *B. b. nigricans* arriving in autumn 1965 composed either of relatively small flocks of family groups or much larger ones of non- (or failed) breeders; these kept distinct at first but later integrated (Jones and Jones 1966). In winter quarters, much shifting of haunts as food supplies depleted ensures that few flocks persist as isolated groups for long, but some evidence of persistent small groups larger than a single family (H Boyd). BONDS. Monogamous, life-long pair-bond believed general. Age at first breeding in wild not known, but some paired by 12 months (H Boyd). Both parents tend young before fledging and family group usually persists until next breeding season. In *B. b. nigricans* at autumn staging area in Alaska in 1965, all families appeared to have disintegrated by October, due mainly to hostility shown by adults to juveniles (Jones and Jones 1966); perhaps abnormal because of social chaos arising out of great concentration of birds (L S Maltby-Prevett). On Melville Island, North West Territories, some yearlings remained as part of family group if adults did not nest, or rejoined parents after nest failure or loss of young (L S Maltby-Prevett). As most feeding takes place in dense flocks moving rapidly over mud or swimming in shallow water, cohesion of family parties in winter quarters sometimes less obvious than in *Anser* and other *Branta* (even Barnacle Goose *B. leucopsis*), but families easily detected in local flights and when flocks feed in fields (H Boyd).

BREEDING DISPERSION. Colonial in many parts of range, in small groups nesting protectively on small islands close to sea shore or in lakes; elsewhere, nests may be dispersed, typically within a few hundred metres of tideline though some up to nearly 10 km inland (L S Maltby-Prevett). Territorial behaviour prior to and during nesting like that of other geese. Most non-breeders move away from nesting areas soon after occupation and migrate before moulting gregariously. ROOSTING. Communal. Daily feeding and resting rhythms determined by state of tide; thus roost at any time of day or night close to where they feed, on water, sandbanks, or on fields in some areas.

2. Little information except on captive birds. Though probably closely related to *B. leucopsis*, behaviour rather dissimilar (Johnsgard 1965). FLOCK BEHAVIOUR. No detailed studies. Pre-flight signals consist of Head-shaking closely similar to Snow Goose *A. caerulescens* (L S Maltby-Prevett). Exaggerated Head-tossing of *B. leucopsis* and Canada Goose *B. canadensis* lacking; resembles Red-breasted Goose *B. ruficollis* in this respect (Johnsgard 1965). In Alaska in autumn, flocks of non-breeders placid, with little social interaction and no cohesion, but flocks comprising 4 or more family groups (each of 2 adults plus 1–5 juveniles) excitable, quarrelsome, and much better integrated. Hostile encounters frequent in latter flocks; occur when one family comes too close to another, members of each family defending area round them all. Also, in such flocks, odd birds occasionally make unprovoked attacks on others which usually just avoid them (Jones and Jones 1966). ANTAGONISTIC BEHAVIOUR. Unlike *B. leucopsis* and *B. canadensis*, shows little ritualization of threat-postures, though white neck-patch vibrated in hostile situations in same way as neck furrowing of other geese and Forward-display usually precedes overt attack; atypically, has Diagonal-neck posture resembling that found in *Anser* but, like *B. ruficollis*, seems to lack characteristic Bent-neck posture of other *Branta* (Johnsgard 1965). In Alaska, hostile encounters in autumn flocks of family groups occur on both water and land (but not in flight); attacking bird thrusts head and neck forward and, with bill open, rushes at opponent (Jones and Jones 1966). HETEROSEXUAL BEHAVIOUR. Pre-copulatory display comprises mutual Head-dipping, as in other geese, but at times almost develops into up-ending; unlike *B. canadensis* and particularly, *B. leucopsis*, does not raise wings during post-copulatory display (Johnsgard 1965). In Alaska, separated individuals reciprocally assume posture resembling hostile Forward-display, then join up (Jones and Jones 1966); this probably partial description of first phase of typical goose Triumph Ceremony. RELATIONS WITHIN FAMILY GROUP. No detailed information, particularly on behaviour of parents and downy young. In autumn flocks in Alaska, family integrity and movement maintained by apparent difference in roles by adults, one taking lead while other brings up rear (Jones and Jones 1966).

**Voice.** Often silent, but voice when used far-carrying, guttural, and metallic. Flocks can produce deafening clamour in flight and characteristically noisy on water. Distinctions between various calls inadequately studied, and comparisons with vocabularies of other *Branta* not yet practicable. Insufficient evidence available to attribute calls to either or both sexes, or to distinguish significance of particular calls. Confused babbling audible at great distances (Kortright 1942) represents indistinctly audible calls described under (1) below.

(1) Trumpeting. 'Rott rott rott' (Söderberg 1950), hard and brief, or monosyllabic 'rronk' (see I), soft in quality and

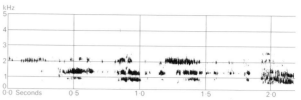

I Sture Palmér/Sveriges Radio Netherlands April 1961

less loud than in other geese (Scott 1950). (2) Alarm call. Nasal 'wauk' (Witherby *et al.* 1939). (3) Hiss. Aggressive; uttered at close quarters, as in other geese.

CALLS OF YOUNG. No information.

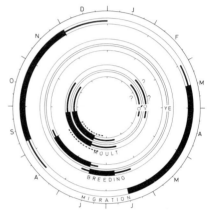

**Breeding.** SEASON. See diagram for breeding range of *bernicla*. Periods for *hrota* in Spitsbergen similar, though few data (Løvenskiold 1964). SITE. Dry hummock above thaw flood level, often on small islands in rivers, or offshore. Colonial; closest nests *c.* 25 m apart. Previous year's nests re-used. Nest: shallow depression lined grasses and moss, with much down added during and after laying. Average dimensions of 9 nests: diameter 32 cm, cup diameter 20 cm, cup depth 6 cm (Nyholm 1965). Building: by both sexes, using material within reach of nest. EGGS. Oval; creamy-white or yellow-white (*hrota*), green-white or light olive (*bernicla*), becoming stained during incubation. 71 × 47 mm (64–80 × 42–52), sample 280 (*hrota*); 75 × 47 (73–77 × 46–48), sample 9 (*bernicla*) (Schönwetter 1967). Weight 79 g (69–83), sample 12 (*hrota*) (Nyholm 1965); calculated weight 91 g (*bernicla*) (Schönwetter 1967). Clutch: 3–5 (1–8). One brood. No data on replacements, though unlikely. Eggs probably laid at 24-hour intervals. INCUBATION. 24–26 (21–28) days. By ♀ only, covering eggs with down when off nest. Starts with last egg; hatching synchronous. Eggshells left in nest. YOUNG. Precocial and nidifugous. Self-feeding. Cared for by both parents, and defended against predators. FLEDGING TO MATURITY. Fledging period not recorded, but probably *c.* 40 days (compare other small high Arctic breeding geese, e.g. Barnacle Goose *Branta leucopsis*). Young stay with parents through first autumn and winter, and depart with them on spring migration, but leave them at or before reaching breeding grounds. Age of first breeding 2 or 3 years. BREEDING SUCCESS. Siberian *bernicla* have most variable breeding performance of all Palearctic geese, with annual proportion of young in winter flocks in Britain 1954–68 varying from under 1% to over 45%, with mean brood size 1–4 (Ogilvie and Matthews 1969). Greenland and Spitsbergen breeding *hrota* also highly variable with similar range of figures. Adverse weather, affecting laying and hatching (which are synchronous over wide area) main factor influencing brood size.

**Plumages** (nominate *bernicla*). ADULT. Head, neck, upper mantle, chest, and upper breast black. Triangular white patches at sides of neck, which do not meet in front or behind neck. Rest of mantle grey-black, narrowly edged pale grey-brown. Scapulars, back, rump, and wing-coverts grey-black. Sides of back and rump, and upper tail-coverts white. Lower breast, belly, and underwing dark grey, but paler than back, with faint grey-brown edges on belly. Feathers of sides of body and flanks darker grey with brown tinge and contrasting white edges. Vent and under tail-coverts white. Tail and flight-feathers black. Grey edges of underparts fade to brown during winter. DOWNY YOUNG. Down of crown, streak before eye, nape, back, breast, and thighs grey-brown; sides of head, bar across wing, and underparts pale grey. JUVENILE. Like adult, but head, neck, chest, upper breast, and upper mantle dull brown-black, without white markings; rest of upperparts tinged brown, edged grey-buff on mantle and white on scapulars. Lower breast and belly ash-grey, feathers of sides of body and flanks lack contrasting white edges of adult. Tail tipped white. Greater and median coverts broadly edged white, other wing-coverts and secondaries narrowly edged white. Feathers of mantle and underparts, scapulars, and wing-coverts narrow and rounded (broad, almost square in adult). FIRST WINTER AND SUMMER. Like adult, but head and neck dull black, white neck patches often less conspicuous. Most of juvenile white-tipped wing-coverts retained.

**Bare parts.** ADULT AND JUVENILE. Iris dark brown; bill and foot black; foot sometimes tinged olive on front of tarsus and upper surface of toes. DOWNY YOUNG. Bill dark slate with paler nail. Leg and foot dark slate, slightly paler on front of tarsus and on toes.

**Moults.** ADULT POST-BREEDING. Complete; flight-feathers simultaneous. Flightless for *c.* 3 weeks, mid-July to mid-August. Body moult starts when flight-feathers complete, first on underparts, later elsewhere; tail feathers last. Sometimes entire plumage replaced before autumn migration (Dementiev and Gladkov 1952). Feathers of head and neck apparently renewed again in spring. POST-JUVENILE. Partial; feathers of neck, head, sides of body, and flanks mainly replaced October–December (white patches on neck and barred flanks sometimes already present on arrival winter quarters, or appear soon after). During winter, moult of mantle, breast, scapulars, and belly; followed about March–April, by tail-coverts, tail, and some inner wing-coverts in some birds; others retain most of these until 1st complete summer moult.

**Measurements.** Netherlands, winter. Skins (RMNH, ZMA). Nominate *bernicla*. ADULT.

| | ♂ | | | ♀ | | |
|---|---|---|---|---|---|---|
| WING | 340 | (5·93; 18) | 330–353 | 324 | (4·60; 13) | 317–335 |
| TAIL | 94·6 | (2·90; 17) | 90–101 | 90·8 | (2·59; 14) | 86–95 |
| BILL | 34·9 | (1·80; 17) | 32–38 | 31·7 | (1·11; 14) | 29–33 |
| TARSUS | 63·7 | (2·27; 17) | 60–67 | 58·1 | (1·34; 13) | 56–61 |
| TOE | 58·4 | (1·94; 16) | 55–62 | 54·3 | (1·42; 13) | 51–57 |

One adult ♀ wing only 303 (not included). Sex differences significant.
JUVENILE. Like adult, except for significantly shorter wing and tail, e.g.

| WING | ♂ | 321 | (8·05; 6) | 313–336 | ♀ | 309 | (7·24; 10) | 294–318 |
|---|---|---|---|---|---|---|---|---|

*B. b. hrota.* Size similar to nominate *bernicla*, except slightly longer wing:

| AD | ♂ | 344 | (–; 3) | 338–348 | ♀ | 329 | (3·22; 7) | 323–333 |
|---|---|---|---|---|---|---|---|---|
| JUV | | 326 | (4·33; 7) | 321–332 | | 316 | (6·52; 4) | 308–325 |

Other subspecies. Sizes similar to nominate *bernicla*.

**Weights.** Nominate *bernicla*. Averages and sample sizes (A) Denmark (Fog 1967); (B) live, Essex, Norfolk, and Lincolnshire, 1972–3 and 1973–4 (A K M St Joseph; Wildfowl Trust):

| (A) | Oct. | Nov. | Dec. | Spring |
|---|---|---|---|---|
| AD | 1385 (36) | 1501 (25) | 1558 (6) | 1411 (75) |
| JUV | 1225 (61) | 1395 (52) | 1244 (4) | 1336 (6) |
| (B) | Nov. | Dec. | Jan. | Feb. | Mar. |
| AD | 1392 (43) | 1420 (27) | 1377 (22) | 1327 (40) | 1328 (39) |
| JUV | 1276 (27) | 1285 (32) | 1336 (8) | 1211 (112) | 1219 (116) |

*B. b. hrota.* Denmark, autumn; 19 adults, average 1489; 12 juveniles, 1364 (Fog 1967). Adults Southampton Island, Canada. Average and sample size (Barry 1962):

| | On arrival (early June) | Pre-nesting (mid-June) | Breeders (end June to mid-July) | Non-breeders | Wing-moult (end July to August) |
|---|---|---|---|---|---|
| ♂♂ | 1575 (4) | 1530 (10) | 1395 (10) | 1270 (8) | 1275 (8) |
| ♀♀ | 1395 (4) | 1260 (3) | 1090 (10) | 1225 (11) | 1230 (5) |

**Structure.** Wing long, rather narrow, pointed. 11 primaries: p9 longest, p10 2–9 shorter, p8 9–13, p7 25–38, p6 46–64, p1 149–176. P10–9 emarginated inner web, p9–8 outer. Tail short, rounded; 16 feathers. Tail-coverts reach tip of tail. Bill short, high at base. Cutting edges of upper mandible straight, with *c.* 29 horny 'teeth', not visible when bill closed. Feathering of head extends far on bill. Outer toe *c.* 90% middle, inner *c.* 80%, hind *c.* 25%.

**Geographical variation.** *B. b. hrota* has belly white or pale grey, feathers of sides of body and flanks pale brown, broadly edged white; edges less contrasting than nominate *bernicla*. In fresh plumage, white edges of flanks wider than in nominate *bernicla*, overlapping, so little brown visible on flanks. Upperparts browner than nominate; feathers of mantle and scapulars grey, broadly edged buff. Adults in worn plumage and juveniles may have belly light brown; then often difficult to separate from worn juvenile nominate, but upperparts diagnostic. Downy young paler. Adult *nigricans* has broader white patches on neck than nominate *bernicla* meeting in front and forming nearly complete white collar; upperparts darker grey; lower breast and belly dark slate, hardly differing from black upper breast, but more contrastingly defined from white tips of feathers of flanks. Birds of east Siberia sometimes separated as *orientalis* Tugarinov, 1941 (e.g. Uspenski 1960), underparts probably slightly paler on average than in arctic west America, but individual variation in colour considerable in all subspecies. Nominate *bernicla* may hybridize with *nigricans* in central Taymyr (Uspenski 1960), while *nigricans* from Melville Island, Canada, shows some intermediate characters with pale-breasted race *hrota* (Maltby-Prevett *et al.* 1975). CSR

## *Branta ruficollis* Red-breasted Goose

Du. Roodhalsgans    Fr. Bernache à cou roux    Ge. Rothalsgans
Ru. Краснозобая казарка    Sp. Barnacla cuellirroja    Sw. Rödhalsad gås

*Anser ruficollis* Pallas, 1769

Monotypic

**Field characters.** 53–56 cm, of which body over two-thirds; wing-span 116–135 cm. Small goose with rather thick-looking neck and small, extremely short bill; elaborate and unmistakable pattern of black and white with chestnut-red. Sexes similar and no seasonal differences; juvenile distinguishable.

ADULT. Face, crown, and hindneck black except for white in front of eye; cheek patch, foreneck, and upper breast rich chestnut, cheek patch edged white and foreneck separated from black areas by white line down neck and white band (bordered black in front) round upper mantle and breast. Rest of body, wings, and tail black except for broad, shaggy white line along sides and flanks, 2 fine white lines across wing-coverts, and white lower belly and upper and under tail-coverts. Bill and legs black. JUVENILE. Duller and browner. Smaller, paler reddish patch on side of head and cinnamon-brown foreneck and breast mottled with black-brown and whitish. White circle round upper mantle and breast and white line along sides and flanks less well defined, tips to wing-coverts forming 3–4 brownish or brownish-white lines. Usually noticeable white tip to tail. FIRST WINTER. Retains many juvenile feathers and, in particular, wing-coverts remain as juvenile, but probably not distinguishable from adult after January–February.

Unlikely to be confused with any other species, but individuals can easily be overlooked in flocks of grey geese *Anser*. At distance, broad white flank-stripe is best field mark.

Swims readily on both fresh and salt water, and in winter quarters also roosts at night on water. Feeds terrestrially on grasses and crops. Highly gregarious, breeding in small colonies (though sometimes singly) on river banks and congregating on migration and in winter quarters in large flocks, often associating with White-fronted Geese *Anser albifrons* and Lesser White-fronted Geese *A. erythropus*. Flies fast, with noticeably agile action, in irregular groups or occasionally in diagonal lines, but seldom in V formation or single file. Like Barnacle Geese *B. leucopsis*, flocks quarrelsome and noisy, uttering high-pitched double staccato cackles and series of squeaks. All movements quick and dainty, and head moves remarkably fast when grazing.

**Habitat.** Although less northerly breeder than Brent Goose *B. bernicla*, must also await ice melting, and completes breeding cycle in little over 100 days. Choice of nesting site in tundra or open parts of northern wooded tundra zones intermediate between *B. bernicla* and

Barnacle Goose *B. leucopsis*, favouring high and dry situations on steep river banks, low rocky crags or gulleys, where cover afforded by such plants as dwarf birch *Betula* or willow *Salix*, or dead grass. Nest may be open and even visible from afar, chief peculiarity being preference for sharing habitat with predators such as Peregrine *Falco peregrinus*, and placing nests close to theirs. Alternative choice of inaccessible islets for nest security also recorded. Usually breeds within easy reach of water adequate to provide refuge for young and sentry position for ♂. Proximity of fresh growth of grasses or other edible green plants such as cotton grass *Eriophorum* also essential. Association with trees not recorded.

Outside breeding season, steppe habitats freely used, not only on migration but also for wintering. Salt-steppes clothed with *Salicornia* and pastures or semi-arid grazing range usually chosen, at low altitudes and below sea-level by Caspian. Crops of onions, garlic, and shoots of winter wheat and barley attractive. In Azerbaydzhan, usually spends nights near shallow water of steppe lakes or sea. Proximity to water, however, less essential in winter than for other *Branta*, and often content with regular brief visits to waterholes. Small outlying bands resort to wetlands farther south and east. Odd birds sometimes caught up in migrant flocks of other geese and survive with them on other types of grazing land, including winter wheat in major Rumanian haunts. Markedly less aquatic and more terrestrial and aerial than others of genus. In urgent need of improved conservation.

**Distribution.** Formerly Caspian and Iraq often given as sole wintering areas (Dementiev and Gladkov 1952; Vaurie 1965), but has wintered Rumania since 1940, and more recently Bulgaria and Greece in smaller numbers (Johnson and Hafner 1970). Numbers in Caspian area much smaller now suggesting distribution has changed (see also Population and Movements).

Accidental. Britain, Belgium, Netherlands, Norway, Sweden, Finland, West Germany, East Germany, Poland, Czechoslovakia, Austria, Hungary, Italy, Spain, Yugoslavia, Israel, Cyprus, and Egypt; some of these records almost certainly refer to escaped birds. No substantiated record for Iceland (FG) and old records for Algeria (Heim de Balsac and Mayaud 1962) must be regarded as uncertain.

**Population.** Breeding. USSR: estimate of Taymyr breeding birds (nearly whole of Siberian population), based on

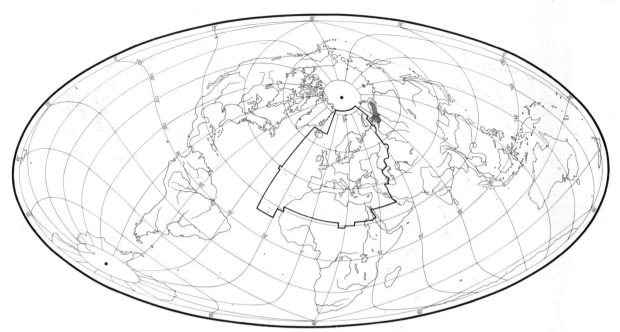

surveys late summer 1972 and 1973, 25 000 including young, with not more than 15 000 breeding adults (A A Kistchinski). Decreasing, population apparently halved since 1965 (Holloway 1970); threatened by excessive hunting, disturbance and habitat loss (Isakov 1972), and possibly affected by decline in Peregrines *Falco peregrinus* nesting in same region (J Kear).

Winter. RUMANIA: 25 000 December 1968, but only 3750 same area 1969–70 (Johnson and Hafner 1970); 9300 in 1970–1 and 6000 in 1971–2 (IWRB). BULGARIA: 300 in winter 1969–70 (Johnson and Hafner 1970). GREECE: up to 75 in recent winters (Johnson and Hafner 1970). USSR. Caspian: formerly many thousands, with highest count 60 000 Azerbaydzhan 1956, but only 1000–2000 located since 1968 (Yu A Isakov).

**Movements.** Single migratory population. Leaves west Siberian tundras mid to late September. Narrow-front passage SSW across central USSR through October, following Rivers Tas and Ob southwards, then Ob tributaries (Irtysh and Tobol) south and south-west towards north Caspian Sea. Then some via Aral Sea to River Syr-Darya in Kazakhstan; others down west coast Caspian to Azerbaydzhan, with small numbers continuing Iraq (Euphrates marshes), while rest move west to Black Sea and Greece, with a few to Turkey. In some recent years, majority seem to have followed latter course, though formerly numbers much smaller (Mörzer Bruyns *et al.* 1969) (see also Population). Wintering grounds reached November. In hard weather movements, large numbers may occur temporarily Iran, though not recently (D A Scott), and coast of Thrace. Departures begin March; noted over Emba River (north-east of Caspian) and in

Turgai River basin further east in late April to early May. Back on breeding grounds early June, so clearly stops on way, though localities unknown. Straggles to west Europe, where rather more frequent North Sea countries (especially England and Netherlands) since *c.* 1960, perhaps connected with westward shift in winter quarters; generally with White-fronted Geese *Anser albifrons*, and may in part have travelled by more northern route with White-front flocks. One ringed Netherlands January 1972 shot Yamal Peninsula June 1974. Based on Bauer and Glutz (1968); Dementiev and Gladkov (1952); Uspenski (1965); and Sterbetz and Szijj (1968).

**Food.** Feeds only by grazing, cropping with relatively rapid head movements. In breeding areas, mostly grasses Gramineae and 'sedges' Cyperaceae, particularly cottongrass *Eriophorum angustifolium* (Uspenski 1965). From Taymyr Peninsula 5 stomachs contained *Eriophorum angustifolium* and *E. scheuchzeri* (Zharkova and Borzhonov 1972). In winter areas, mainly coastal grasses; to lesser extent pondweed *Potamogeton fluitans*, and seeds of *Galium* and *Bolboschoenus maritimus*. On occasions quantities of glasswort *Salicornia europaea*, and frequently winter wheat and barley shoots (Dementiev and Gladkov 1952; Uspenski and Kishko 1967). In Rumania, feeds regularly in fields of sprouting winter wheat and maize stubble (Scott 1970). On passage, Kazakhstan, sometimes supplements tender green shoots with wild garlic tubers *Allium* and tubers and rhizomes of steppe plants (Grote 1939).

**Voice.** Principal call, uttered in various situations (e.g. in flight and by feeding flocks), shrill, loud, double staccato

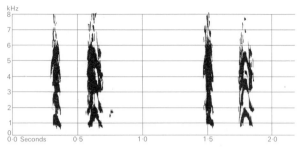

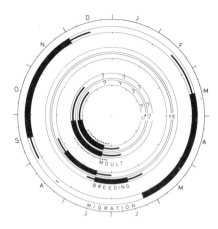

I P J Sellar England (captive) February 1975

shriek 'kee-kwa' (see I), 'ki-kwi', 'ti-che', 'ka-ak', or 'kik-wik'; with 2 disjointed syllables, 2nd somewhat broken (Witherby *et al.* 1939). Aggressive hiss frequent in feeding flocks (Bannerman 1957).

**Plumages.** ADULT. Head and neck with bold pattern of black, russet, and white. Upper mantle and upper breast russet, bordered caudally by narrow black and wider white line round body. Rest of upperparts glossy black except white upper tail-coverts, dusky tips to scapulars, and broad white tips to greater and median upper wing-coverts. Lower breast and upper belly black, sides of body white, flanks boldly barred black and white. Lower belly, vent, and under tail-coverts white. Tail and flight-feathers black. Under wing-coverts and axillaries black. JUVENILE. Like adult, but black duller and browner, russet paler, white lines and patches less clearly defined. Russet feathers of breast tipped black. Pale grey bases of black feathers of underparts shine through. Tail more broadly tipped white; wing-coverts tipped brown, shading to white on median and greater coverts. Shape of feathers like juveniles of other *Branta*. FIRST WINTER AND SUMMER. Like adult, but black duller, and juvenile wing-coverts and some juvenile body feathers (especially below) retained.

**Bare parts.** ADULT AND JUVENILE. Iris dark chestnut, bill and foot black.

**Moults.** ADULT POST-BREEDING. Complete; flight-feathers simultaneous. Flightless for *c.* 15–20 days between mid-July and late August; non-breeders *c.* 2 weeks earlier than breeders. Body moult starts when flight-feathers fully grown. Occurrence of partial autumn–winter moult not established (Dementiev and

Gladkov 1952; Uspenski 1965). POST-JUVENILE. Partial; starts on breeding grounds with neck and head, followed by some of body feathers and tail in winter quarters.

**Measurements.** ADULT. Mainly Netherlands, some Rumania and France, winter; skins (RMNH, ZMA).

| | | | | | | |
|---|---|---|---|---|---|---|
| WING | ♂ 367 | (7·12; 8) | 355–379 | ♀ 343 | (6·09; 7) | 332–352 |
| TAIL | 109 | (7·39; 8) | 98–121 | 102 | (3·45; 8) | 96–107 |
| BILL | 24·9 | (1·02; 9) | 23–27 | 24·2 | (0·95; 8) | 22–26 |
| TARSUS | 61·3 | (2·50; 8) | 58–65 | 57·1 | (2·03; 8) | 54–61 |
| TOE | 54·2 | (2·36; 9) | 50–59 | 51·4 | (1·11; 8) | 49–53 |

Sex differences significant, except bill. JUVENILES. Wing and tail *c.* 7% shorter than adults; other measurements similar.

**Weights.** Few published data. ADULTS AND JUVENILES combined; Netherlands, East Germany, Rumania, Kazakhstan. ♂♂: November 1200, 1315, 1465; January 1270; February 1625. ♀♀: December 1058, 1130. Sex unknown: November 1445. (Bauer and Glutz 1968; Dolgushin 1960; ZMA.)

**Structure.** Wing long, rather narrow, pointed. 11 primaries: p9 longest, p10 1–10 shorter, p8 8–14, p7 30–42, p6 55–71, p1 157–192; p11 reduced, pointed, concealed by primary coverts. P10–9 emarginated inner web, p9–8 outer. Tail rather long, slightly rounded; 16 feathers, upper tail-coverts not reaching tip of tail. Bill short, high at base, cutting edges of upper mandible straight, with *c.* 16 horny 'teeth', not visible when bill closed. Feathering of head extends far on bill. Outer toe *c.* 91% middle, inner *c.* 81%, hind *c.* 20%. CSR

# Subfamily ANATINAE ducks

Medium-sized to fairly large wildfowl. Tarsi scutellated in front. Marked sexual dimorphism in plumage and syrinx structure in majority of species; correlated with sexual differences in visual displays and voice. About 110 species, comprising 6–9 tribes, all but 2nd and 4th represented in west Palearctic though 3rd only by introductions: (1) Tadornini; (2) Tachyerini (steamer ducks)—3 species (2 flightless) in genus *Tachyeres*, South America and Falkland Islands; (3) Cairinini; (4) Merganettini (torrent ducks)—1 polytypic species in genus *Merganetta*, South

America; (5) Anatini; (6) Aythyini; (7) Somateriini; (8) Mergini; (9) Oxyurini.

Trachea of ♂ usually with bony, asymmetrical bulla on left side of syrinx. Double annual moult in both sexes, resulting in breeding and non-breeding plumages. These usually closely similar and cryptic in ♀♀, though non-breeding plumage usually duller. ♂ breeding plumage of many species in temperate regions elaborate and colourful ('bright'), contrasting with sombre and cryptic non-breeding plumage (eclipse) worn for short period while birds flightless during post-breeding moult and resembling plumages of ♀♀ and juveniles. Wing typically brightly coloured in both sexes, often with metallic speculum on greater coverts and secondaries which contrasts with colourful median and lesser wing-coverts or tertials; this pattern maintained all year as wing moulted only once. As a rule, juvenile plumage resembles non-breeding plumage of both adult sexes, but juveniles separable by tail-feathers—notched tip with bare shaft protruding—and by narrower, shorter, and more pointed body feathers and wing-coverts. In many Holarctic ducks, juveniles develop immature non-breeding plumage within a few months of hatching. In some species breeding in 1st year, this plumage involves growth of only a few new feathers and is quickly replaced by breeding plumage; in others which defer breeding until 2nd year, immature plumage more complete and retained longer, being only gradually replaced by immature breeding plumage during whole 1st year of life. In all cases, juvenile wing retained until 1st complete moult in summer of 2nd calendar year, although tertials often and some wing-coverts sometimes replaced earlier.

TERMINOLOGY OF PLUMAGES. Terms 'breeding' and 'non-breeding' used for this group so as to maintain homology with other taxa treated in this volume—even though bright (breeding) ♂ plumage of most duck species (often termed 'nuptial' in ornithological literature) usually worn for much of year when birds not actually breeding, including autumn and winter when pair-bonds initiated and maintained until nesting in spring (see below). Thus, ♂♂ often attain non-breeding plumage soon after start of nesting when their reproductive activities (but not those of ♀♀) are over. In ♀♀, while timing of both moults tend to correspond roughly with those of ♂♂, also subject to adaptive variation. In many species, post-breeding moult of ♀♀ more protracted, with greater individual variation in timing—particularly in successfully breeding ♀♀; moult usually inhibited during nesting, starting 1–2 months later than in ♂♂. ♀♀ of some species (e.g. some Anatini) start moult shortly before nesting and finish after eggs hatch, thus combining breeding and non-breeding plumages while incubating. ♀♀ of some other species (e.g. some Aythyini) take advantage of slightly more cryptic plumage while nesting by moulting in early spring. Although such ♀♀ in fact nest in 'non-breeding' plumage, terminology maintained for reasons of homology and because no other system of naming plumages and moults seems to offer any better solution to what is often a most complex situation. Due to different moulting patterns in ♂♂ and ♀♀, annual cycle diagrams modified from standard version in case of Anatinae: 2 inner circles (usually showing moult of wing and body respectively) now used for body moult of ♂ and ♀ respectively, wing moult for each sex being represented as interrupted line beside corresponding circle.

# Tribe TADORNINI sheldgeese, shelducks

Fairly large, often semi-terrestrial, goose-like waterfowl; comprised mainly of relatively long-necked grazing birds with short toes and rather long tarsi inserted well forward. Intermediate in many characters between Anserini and remaining Anatinae. 14 living species in 5 genera: 2 typical sheldgeese (*Cyanochen, Chloephaga*); 2 intermediate (*Neochen, Alopochen*); and shelducks (*Tadorna*). Latter consists of 6 species in 2 groups (1) typical *Tadorna* (*tadorna, radjah*); (2) '*Casarca*' (remainder including *ferruginea*). Steamer ducks *Tachyeres* sometimes included in Tadornini but here considered as constituting separate tribe Tachyerini, following Johnsgard (e.g. 1965). 3 species of Tadornini in west Palearctic, all breeding.

Wings with bony, spur-like knob on metacarpal joint. Tails fairly long. Bills comparatively short and thick (sheldgeese), or depressed (*Alopochen*, shelducks) with distinct lamellae (shelducks generally) and turned slightly upwards (typical *Tadorna*). Sexes differ in tracheal structure: that of ♀♀ simple and goose-like, only ♂♂ with enlarged bullae—usually on left of trachea but in Shelduck *T. tadorna* enlarged on both sides; bullae much reduced in '*Casarca*' group. Sexes dimorphic in some *Chloephaga* and *Tadorna*; similar or nearly so in remainder. Plumages of both ♂ and ♀ usually bright; no discernible non-breeding plumage in most species. True eclipse plumages rare. Large metallic green speculum; lesser and median wing-coverts usually plain (often white). Juveniles like adults but colours duller. Downy young boldly patterned black and white.

Largely cosmopolitan but most species sub-tropical and temperate southern hemisphere, and all absent from North America. Basically birds of low or lower middle latitudes, or (as *Cyanochen* and some *Chloephaga*) of mountains in tropical zone. As a group, best characterized by continental warm or mild climatic requirements, acceptance of high altitudes, and preference for inland waters or, except in case of maritime Kelp Goose *Chloephaga hybrida*, at most for sheltered coastal waters. In case of *Cyanochen* and *Chloephaga*, also characterized by use of grassland in terrestrial grazing; and, in *Alopochen* and *Tadorna*, by hole-nesting propensities and liking for unvegetated sand, silt, or mud margins, and by adaptability to semi-domestication. Orinoco Goose *Neochen jubatus*, atypically, a bird of dense, tropical forests where (like *Alopochen*) it perches freely. In west Palearctic, *Alopochen* contrasts with *Tadorna* in being a partly tropical breeding species much more tolerant of forest country. During flightless period of post-breeding moult, often frequent open areas of land or water where can observe and avoid terrestrial predators. In west Palearctic, Egyptian Goose *A. aegyptiacus* now confined to Egypt (where decreasing) and possibly Chad, with small feral population in England. *T. tadorna* has re-colonized Finland and spread elsewhere, especially inland; numbers increased in parts of western Europe and Black Sea. Ruddy Shelduck *T. ferruginea* has declined in south-east Europe (but slight recent increase under protection) and USSR, and now extinct Tunisia. Little detailed information of movements of extralimital sheldgeese and most other southern Tadornini but many populations resident to greater or lesser extent. In west Palearctic, *A. aegyptiacus* mainly sedentary though partially migratory within Ethiopian range. Of 2 shelducks, many *T. tadorna* migratory or partially so, some *T. ferruginea* migratory; both also dispersive at times and *T. ferruginea* also nomadic in some areas. In case of *T. tadorna*, major moult migration at end of breeding season to traditional moulting areas within western Europe involves immatures and most adults—except those attending crèches of young which moult *in situ*. Little comparable information for *T. ferruginea*, but moult gatherings occur. Movements in most species mainly nocturnal and in flocks; do not hesitate to cross land-masses.

Some (e.g. *Chloephaga*, *Alopochen*) chiefly plant feeders, mainly by terrestrial grazing; some shelducks (e.g. *T. ferruginea*) omnivorous, feeding by grazing, dabbling, and up-ending; others (e.g. *T. tadorna*) primarily animal feeders, mainly by dabbling in mud or shallow water, swimming with head submerged, and up-ending. Often feed and otherwise associate in pairs and family parties or in flocks. Pre-flight signals consist largely of lateral Head-shaking and repeated Chin-lifting. Most species highly aggressive. Maintain territories at least while breeding, nesting within, but territories of some *Tadorna* (e.g. *T. tadorna*) mainly for feeding and meeting of pairs with nest-sites elsewhere, sometimes in groups. Long-term, mono-gamous pair-bonds much as in Anserini—though thought by Johnsgard (1965) to be less strong in shelducks than in sheldgeese. In some cases, pair occupy territory together throughout year. Courtship often terrestrial; more elaborate than in Anserini but less so than in other Anatinae, with no true communal displays. Not fully studied in majority of species, especially in wild; among west Palearctic species, behaviour of *T. tadorna* far from typical of rest of tribe. In all species, ♀♀ play important or major role in pair-formation; mainly by use of characterist-ic, aggressive Inciting-display typical of most Anatinae but finding most complete expression in Tadornini —where often directly functional in causing chosen ♂ to attack others. ♂ pairing and other heterosexual displays include Puffing, Bowing, and High-and-Erect, often with wing-raising and strutting gait (see Johnsgard 1965), but often difficult to distinguish from antagonistic

behaviour, while displays of different genera or even of different species-groups within *Tadorna* often divergent. In some sheldgeese (including intermediate *Alopochen*) and some *Tadorna*, mutual Triumph Ceremony much as in Anserini. Displays more typical of other Anatinae (see Anatini) found in some species, mainly *Tadorna*; include unilateral and mutual Bill-dip, Ceremonial-drinking, Mock-preening in form of Preen-behind-Wing display, and, in *T. tadorna*, vocal version of Upward-shake (♂ only). Pursuit-flights (see Anatini) also reported from at least 1 species. Copulation typically on water, sometimes in shallows or on land. Pre-copulatory behaviour resembles Anserini with mutual Head-dipping. Post-copulatory behaviour distinctive; includes High-and-erect display by ♂, usually with wing lifted on side furthest from ♀. Voices often loud and sexually well differentiated. ♀♀ of all species with low-pitched, quacking calls like those of some other Anatinae (Johnsgard 1968). ♂♂ of most sheldgeese have whistling calls as also those of 2 species of typical *Tadorna*; in '*Casarca*' group of shelducks, however, voice of ♂♂ loud and honking, while that of ♂ *Alopochen* different again. Call frequently in flight as well as on water, land, or perch. In most species, ♂'s vocal response to Inciting-call of ♀ of 2 types—aggressive (to other ♂♂), friendly (to ♀); see Johnsgard (1965).

Seasonal breeders, at least in west Palearctic. Nests on ground in open (*Chloephaga*, etc); in burrows and holes in ground, trees, or buildings (*Tadorna*); on ground in thick cover, cliff ledges, or in holes (*Alopochen*). Old nests of other species also sometimes used (*Alopochen*, *T. fer-ruginea*). Sites sometimes far from water. Usually solitary nesters but sometimes close in hole-nesting *T. tadorna*. Amount of nest material variable, from mound of vegetation on ground to little or nothing in holes; lined down. Building by ♀ only. Eggs rounded, creamy-white and smooth. Clutches 3–12, averaging larger (8·4) in hole-nesting species than in open nesting ones (6·1)—Lack (1968). Multiple laying by ♀♀ in 1 nest not uncommon in some species (see *T. tadorna*). Replacement clutches produced after early egg-loss. Eggs laid at intervals of 1 day. Incubated by ♀ only, leaves nest 1 or more times per day when usually joins ♂; latter may stand guard in open-nesting species. Incubation periods 28–30 days (Kear 1970), with no significant difference between hole-nesters and others (Lack 1968). Young attended by both parents but brooded by ♀ only. Both parents aggressively defend young at times in most or all species. Distraction-display

PLATE 54. (*facing*) *Branta canadensis* Canada Goose (p. 424). *B. c. canadensis*: 1 ad, 2 juv up to 4–7 months, 3 downy young. *B. c. maxima*: 4 ad. *B. c. interior*: 5 ad. *B. c. taverneri*: 6 ad, 7 ad showing white neck collar. *B. c. minima*: 8 ad, 9 ad showing white neck collar. *B. c. hutchinsii*: 10 ad.
*Branta leucopsis* Barnacle Goose (p. 430): 11 ad showing white face, 12 ad showing cream face, 13 juv up to 7 months, 14 downy young. (PS)

Peter Scott.

by ♀ or both sexes, in form of 'injury-feigning', reported in *Chloephaga* and *Tadorna*, but evidently lacking in *Neochen* and *Alopochen* (see Hebard 1960). Other anti-predator reactions by parents include 'tolling' (Sowls 1955)—i.e. moving or flying conspicuously, often while calling, away from or in vicinity of predator. In most species, young not independent until fledging or after, remaining with parents for up to 6 months. In *T. tadorna*, broods may amalgamate into crèches, some parents then deserting own young. Mature usually at 2 years.

## *Alopochen aegyptiacus* Egyptian Goose

Du. Nijlgans   Fr. Oie d'Egypte   Ge. Nilgans
Ru. Нильский гусь   Sp. Oca egipcia   Sw. Nilgås

*Anas aegyptiaca* Linnaeus, 1766

Monotypic

PLATES 57 and 61
[facing page 471
and between pages 494 and 495]

**Field characters.** 63–73 cm, of which body two-thirds; wing-span 134–154 cm. Closer to shelducks *Tadorna* than true geese *Anser* and *Branta*, but larger than former, with heavier bill, coarse-looking head, longer legs, and more upright stance; grey to red-brown above, with paler head and underparts, chestnut spectacles, collar, and breast patch, and white wing-coverts striking in flight. Sexes similar and no seasonal differences apart from swelling at base of bill; juvenile and immature distinctive.

ADULT. Pale grey head mottled brown, with reddish hindneck, chestnut patches round eyes, and chestnut collar. Mantle and scapulars vary from red-brown through rusty grey to pale grey; underparts dirty white or pale grey, tinged rusty, with chestnut patch on lower breast (this and reddish on hindneck sometimes lacking in pale individuals). Rump, tail, and primaries black; secondaries metallic green forming bright green speculum; upper and under wing-coverts white, former with black line along greater coverts. In flight, bold white forewings, partly broken by black line, contrast with black flight-feathers. Bill pink with narrow black tip, cutting edge and base; eyes yellow to red-brown; legs pink to purple-flesh. JUVENILE AND FIRST WINTER. Duller and generally paler, lacking eye and breast patches, but with darker head and hindneck from area of eye backwards. Bill and legs yellow-grey; eyes yellow-brown.

Adult only ever likely to be confused with Ruddy Shelduck *T. ferruginea*, which also has paler head and similar black-and-white wing pattern, but is slightly smaller and shorter legged, with conspicuously orange-chestnut body and no eye-patches. Juvenile bears slight resemblance to young Shelduck *T. tadorna*, because of dark hood and pale face, but latter much blacker above and whiter below, with white rump and upper tail-coverts.

Like *T. ferruginea*, occurs in wide variety of lowland and montane habitats where water present, but more commonly in forested areas and by small streams; seldom near sea. Generally wary, but little hunted and so approachable with care. Swims well with stern-high shape, and will also dive readily, but lives mainly on land, and walks easily if sedately. Often perches and roosts on trees and other prominences. Flight strong and fast, though with relatively slow wing-beats, recalling goose rather than duck. Generally in pairs or family parties, but flocks of up to 100 or more in Africa during flightless post-breeding period. Noisy and demonstrative in breeding season.

**Habitat.** Contrasts biogeographically with other Anatidae of west Palearctic in representing overflow from Africa of marginal colonists. Mainly sub-tropical, almost entirely on inland fresh water, distributed along rivers and near lakes and pools, occurring up to fairly high altitudes in mountainous regions. Successfully introduced feral stock in England demonstrates adaptability to new lowland habitats in temperate zone, around ornamental water in parkland and in managed aquatic and riparian habitats such as alder-willow swamp woodlands, pastures, and meadows of Norfolk Broads. At home amid luxuriant vegetation, terrestrial and aquatic, even perching and nesting in trees as well as on ground under thick plant cover and in various kinds of holes, but avoids dense forest and not averse to rocky or other unvegetated places (see Breeding). Unlike true geese (*Anser* and *Branta*), flies freely within and over forests.

**Distribution.** In west Palearctic, now breeds only England (where introduced 18th century and long found in feral state in several parts), Egypt, and perhaps Chad, where seen Tibesti in July, probably breeding (Salvan 1967). See Ringleben (1975) for escapes in Netherlands; not yet established in feral state (JW).

Former breeding status uncertain. According to Vaurie (1965) breeding also Syria and Palestine, and formerly Algeria and Tunisia, while occurred in Danube valley until

◀ PLATE 55. *(facing)* *Branta bernicla* Brent Goose (p. 436). *B. b. hrota*: 1 ad, 2 1st winter and summer, 3 juv, 4 downy young. *B. b. bernicla*: 5 ad, 6 1st winter and summer, 7 juv, 8 downy young. *B. b. nigricans*: 9 ad. *Branta ruficollis* Red-breasted Goose (p. 442): 10 ad, 11 1st winter and summer, 12 1st winter and summer showing lack of chestnut cheek-patch on some individuals, 13 downy young. (PS)

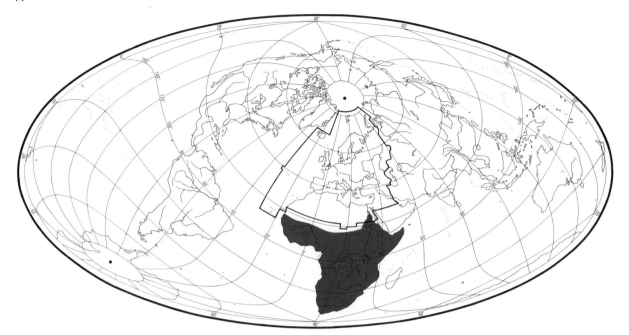

early 18th century. No evidence, however, for breeding Syria (Kumerloeve 1967), or Algeria and Tunisia (Heim de Balsac and Mayaud 1962). Still nesting Palestine, but becoming scarce (Meinertzhagen 1930); not seen in wild in Israel since 1933, but nested Tel Aviv Zoological Gardens, where *c*. 30 full-winged, but infertile, birds driven away 1955–68 (HM). In Hungary, observed late 17th century but no later records (IS).

Accidental. Algeria: occasionally winters. Tunisia: occasional in winter. Malta: once. Cyprus: formerly (mainly 1900–20) scarce but regular late November to March, now rare; perhaps some from Tel-Aviv (Stewart and Christensen 1971). Occasional records western Europe (e.g. Sweden, France, Spain) probably escapes.

**Population.** EGYPT: formerly very abundant throughout Nile valley, but becoming scarce in Lower Egypt (Meinertzhagen 1930); still common Nile north of Aswan (Ripley 1963), but no recent reports. ENGLAND: in numbers only East Anglia, mainly Norfolk, where largest colony Holkham Park (*c*. 25 pairs), with smaller groups elsewhere (British Ornithologists' Union 1971; Wildfowl Trust).

**Movements.** English introduced population resident. In Egypt, resident upper Nile valley and no evidence of long-distance movements from there; not recorded Sinai or Red Sea (Meinertzhagen 1930). Occasional winter occurrences oases of Algeria and Tunisia (Heim de Balsac and Mayaud 1962) more likely trans-Saharan stragglers than westward wanderers from Nile. Seemingly partially migratory in Ethiopian Africa, penetrating southern Sahara in wet season. In Chad, some move north into Sahel zone, even to

Tibesti, with summer rains, and may breed (Salvan 1967), though resident Niger Inundation Zone of Mali (Guichard 1947). In Nigeria, where breeds only northern savanna, big influx to Lake Chad in latter half dry season as inland pools dry out (Golding 1934), and occurs sporadically south to Lagos (Elgood *et al.* 1973). Ringing shows movements up to 1100 km in South Africa.

**Food.** Mainly vegetation, chiefly grass, plant leaves, seeds, and growing crops (maize stubble, wheatlands, groundnuts, sweet potatoes); and possibly to some extent animal material (Bannerman 1930; Clancey 1964; Eltringham 1973; Mackworth-Praed and Grant 1962; Pitman 1965). Feeds mainly terrestrially, by grazing in pairs, family parties, and seasonally in large flocks. Near inland waters in eastern Cape Province, South Africa, much damage reported to young wheat; 1 stomach contained mainly corms of *Cyperus esculentus* (Taylor 1957).

**Social pattern and behaviour.** Based mainly on material supplied by W R Siegfried (South Africa).

1. Usually gregarious except when nesting. Outside breeding season, often in flocks; these commonly of hundreds and occasionally thousands. Small part of population dispersed in pairs throughout year, each occupying small permanent water. BONDS. Sustained, monogamous pair-bond. Post-breeders in pairs in winter flocks; rest of adult population also probably then paired, but confirmation needed. Both parents tend young; family bonds maintained throughout pre-fledging period and afterwards for at least 6 weeks. Juveniles tend to associate in groups within non-breeding summer flocks. BREEDING DISPERSION. Pairs well dispersed, each maintaining relatively large, discrete territory of variable size (in open water situation *c*. 1 ha). Used for courtship, mating, and nesting, occasionally for pair-formation which, however, normally occurs in flock. When territory embraces all isolated water, as is common, young reared

entirely within. On large waters shared by families, each tends to remain within own territory. In more arid parts of range, whole family may leave territory and travel overland when water dries up. Adults mostly feed within territory. In many areas of South Africa in most years, large non-breeding population, consisting partly of pairs, gathers in flocks. ROOSTING. Flock spends most of daylight bathing, preening, and loafing on favoured flat shore of permanent lake. Towards sunset leaves to visit communal grazing grounds, apparently returning after few hours to roost and not spending whole night in fields (Shewell 1959).

2. In areas with seasonally dependable and regular rainfall, antagonistic and courtship activities generally renewed in flocks once moult completed, up to 4 months before dispersal to breeding territories. FLOCK BEHAVIOUR. Flocks well integrated and synchronized. Individual-distance maintained, but mated birds tend to stay closer together, so spacing within flock uneven. Alarm registered by erect posture (with neck straight up, feathers raised) and Alarm-call. Make for water if disturbed, swimming well out and giving Alarm-call constantly if pressed; seldom fly unless taken by surprise, or persecuted by hunters. Pre-flight signals include Chin-lifting (Johnsgard 1965; McKinney 1965b). Regular evening and morning feeding flights; increased calling and activity precede successive departures of pairs and parties. ANTAGONISTIC BEHAVIOUR. Particularly aggressive—in winter flock (where fights, between ♀♀ as well as ♂♂, mostly occur), during courtship, and, especially, during establishment and maintenance of breeding territory. ♂ of pair assumes major role as attacker and defender in territorial disputes. Pairing and paired ♀♀ threaten and attack unpaired ♀♀ which attempt to solicit chosen ♂ (see below). Threat-display of ♂ usually in 3 stages: (1) in Wing-showing display, folded wings lifted slightly away from body, exposing dorsal surface of white wing-coverts and bony protuberances on wrists, and held rigid and half-dragging or frequently flicked; (2) in Bent-necked posture, feathers on back and sides of neck erected and neck drawn in—with head lowered against breast and bill pointing down; (3) in Forward-posture, wings slightly opened, neck lowered, head and neck stretched horizontally forward, and bill opened and directed at rival. ♂ then often utters Threat-call in preparing to pursue rival. ♀ may threaten similarly, sometimes giving ♀-version of Threat-call, but tends to go straight into chase and attack. Territorial pairs assume Wing-spread display, calling while facing and threatening overflying conspecifics, body erect, chest out, neck and head high, and wings lifted and fully open. As this changes into full attack, aggressor rears up in Wing-spread display just before body contact made if rival does not flee. Rivals facing thus may move uplifted heads from side to side. If fighting follows, they stand or swim breast to breast and try to hold on to and bite each other's backs, near base of neck, while beating with wings and at times striking with feet. In both sexes, appeasement signalled by erect posture with neck stretched vertically but bent slightly backwards, head high, and bill horizontal; feathers sleeked to present thin neck, and wings hidden in flank pockets masking any white. When fleeing, neck erect and feathers raised, head high, bill horizontal, and wings slightly open. HETEROSEXUAL BEHAVIOUR. Contains many hostile elements; distinction from antagonistic behaviour mostly not clear-cut. Pairing and paired ♂♂, in flock or nesting territory, commonly perform Advertising-display by strutting with wings held out from flank pockets, neck bent with feathers erected, and bill pointing down. ♂ adopts same posture when returning to mate after attacking rival. Both ♂♂ and ♀♀ at times perform Sky-call, drawing erect, stretching neck and raising head and bill vertically uttering loud, single-syllabled, sex-specific calls; not mutual display, sometimes response to overflying birds. Mate-selection and courtship behaviour not adequately studied. ♀ often takes initiative and courts ♂, but ♂ may also make approach (see below). ♀♀ cause potential and actual mates to attack other ♀♀ as well as ♂♂ by Inciting-display. Form partly determined by position of rival; ♀ displaying from side or front of chosen ♂ makes threatening bowing movements of head and attacking movements towards rival, normally pointing bill towards it, while in Bent-neck or Forward-posture and giving ♀-version of Threat-call (Inciting-call). In courting and newly paired birds, ♂ will also approach ♀ and greet her by performing Ceremonial-drinking; may swim high in water and turn sideways to show yellow under tail-coverts. If not threatened or chased away, tries to initiate Triumph Ceremony, placing himself in front of ♀ giving Triumph-call while rhythmically craning neck back and forth, wings held out from flank pockets and flicked, neck feathers erected. If ♀ joins in, adopts similar posture and utters Inciting-calls. ♂ and ♀ now both call while facing or standing side by side in erect posture with necks extended almost vertically and bills pointing up, exposing chin. Voices of both then quicken and rise, while wing-flicking and bowing movements increase, at times leading to crossing of necks. If rival nearby, ♂ will display further by suddenly Wing-spreading. Usually, both sexes also perform mutual Wing-spread display as climax to Triumph Ceremony. Though courting ♀♀ may initiate elements of Triumph Ceremony, in paired birds normally initiated by ♂ usually following aggressive encounters with rivals. Often preceded by mutual Ceremonial-drinking. Copulation normally while birds swimming; also in shallow water or on land. Pre-copulatory display consists mainly of mutual Head-dipping, on water ♂ swimming rather more erect than ♀. Alternate Head-dipping increases in tempo, with neck crossing, before ♀ assumes Prone-posture and ♂ mounts. ♀ may initiate copulation, going straight into Prone-posture without mutual Head-dipping. After dismounting, ♂ retains hold on ♀'s nape as both call before he releases her and adopts post-copulatory High-and-erect posture while often lifting far wing, both birds then usually bathing (Heinroth 1911; Johnsgard 1961b, 1965; W R Siegfried). RELATIONS WITHIN FAMILY GROUP. Little detailed information.

**Voice.** Normally silent when loafing and feeding, but calls loudly and persistently in social situations. Both sexes hiss when annoyed but otherwise calls totally dissimilar. (Delacour and Mayr 1945; Delacour 1954; Johnsgard 1965, 1968.)

CALLS OF MALE. No whistled notes, unlike typical sheldgeese *Chloephaga*. All variants of husky, asthmatic, breathing sound, strong and gusty, if rather laboured at times, recalling steam engine. Intensity and rapidity of notes depend on degree of excitement; given as Alarm-call, Threat-call, Triumph-call, and (special loud, 1-syllable version) as Sky-call; also after copulation.

CALLS OF FEMALE. Harsh quacking or trumpeting, often loud, rapid and strident (see also Ripley 1963; Williams 1963); rendered as guttural, cackling 'kek kek' (Etchécopar and Hüe 1967), and 'honk-haah-haah-haah'. One version used as ♀ threat-call, Inciting-call, and Triumph-call—when given almost incessantly. As in ♂, special Sky-call loud and of single syllable.

CALLS OF YOUNG. Contact-calls rapid and high, uttered in groups of 6–7 notes. Distress-calls rather higher and more rapid (12 in $2\frac{1}{2}$ s) than in geese Anserini. Apparently no Sleepy-call. (J Kear).

**Breeding.** SEASON. Egypt: few data but breeds from March. Britain: most eggs laid late March and April (E Forster). SITE. Highly variable: on ground in thick vegetation or under bushes probably commonest; also among rocks, in holes in banks, caves, on cliff ledges up to 60 m, occasionally on buildings; often in tree using old nests of other species including Goliath Heron *Ardea goliath* and Hammerkop *Scopus umbretta* or in hollow or crown of pollard willow *Salix*. Nest: on ground, mound of grass or reeds; cliff and hole nests usually depression without material; in trees, adds lining of twigs or leaves. All nests lined down. Building: by ♀ only. EGGS. Rounded at both ends; creamy white. 69 × 50 mm (62–74 × 47–54), sample 90 (Schönwetter 1967). Weight of captive laid eggs 97 g (79–110), sample 81 (J Kear). Clutch: 8–9 (6–12). One brood. No information on replacements. Laid at 24-hour intervals, occasionally 48-hour. INCUBATION. 28–30 days. By ♀ only. Starts after completion of clutch; hatching synchronous. Eggshells left in nest. YOUNG. Precocial and nidifugous. Self-feeding. Cared for and defended by both parents; brooded by ♀ while small. FLEDGING TO MATURITY. Fledging period 70–75 days (Siegfried 1965a). Young stay with parents for several weeks or months after fledging but precise time of independence unknown. Age of first breeding probably 1 year. BREEDING SUCCESS. Population of *c*. 25 pairs, Norfolk, England, rear on average only 2 young per pair to fledging; predation by Crows *Corvus corone*, and competition with Canada Geese *Branta canadensis* and Greylag Geese *Anser anser* thought to be main reasons (E Forster).

**Plumages.** ADULT. Band round base of bill (not always reaching chin), lores, patch round eye, hindneck, collar round lower neck, and sometimes some spots on throat and foreneck chestnut; colour deepest round eye, feathers with grey or off-white bases elsewhere. Extent of chestnut variable; chestnut feather-tips may disappear by wear. Forecrown, indistinct line bordering eye patch behind, lower sides of head, chin, throat, foreneck, and sides of neck white; ill-concealed grey-brown feather-bases, especially on neck. Hindcrown and ear-coverts dark buff-brown to pale cinnamon-grey. Upper mantle finely vermiculated dark grey and yellow-buff or warm cinnamon. Lower mantle and upper scapulars dark olive-grey, edged and finely speckled pale buff or cinnamon; in some, cinnamon tinge prevalent but others mainly dark grey, with limited speckling. Lower scapulars cinnamon, tips often finely speckled pale buff; inner webs often tinged olive-grey or ash, sometimes completely ash. Back, rump, and upper tail-coverts black, finely vermiculated off-white on back and sides of rump. Chest like upper mantle, but sometimes as pale as flanks. Flanks white or pale cream-yellow, vermiculated grey. Chestnut spot of variable size on centre of breast. Belly and vent white, under tail-coverts pale cinnamon or yellow-buff. Tail, flight-feathers, bastard wing, and primary coverts black, outer webs of secondaries glossed dark metallic green or purple, except at tip. Upper wing-coverts white, greater with black subterminal bar. Outer webs of outer tertials chestnut, of inner cinnamon; inner webs ash, olive-grey, or cinnamon-grey. Tertial-coverts white (often partly tinged pale grey), with freckled tip or subterminal bar black. Under wing-coverts of primaries black, rest of under wing-coverts and axillaries white. Sexes similar; no non-breeding plumage discernible. DOWNY YOUNG. Crown, hindneck, streak behind eye to nape, upperparts, thighs, and sometimes indistinct spot on cheek dark cinnamon-brown. Forehead, wide streak over eye, sides of head, large spots on wing and at sides of back, and underparts white or pale grey. JUVENILE. Like adult, but no chestnut marks on head, neck, and breast. Crown, sides of head, upperparts, and flanks duller grey-brown; face, throat, and underparts pale grey-buff. Tips of shafts of tail-feathers bare, projecting. Greater upper wing-coverts pale grey-brown, tips beyond black subterminal bar white with some terminal black mottling. Tips of median and lesser coverts variably suffused pale grey-brown. Tips of outer webs and whole inner webs of secondaries variably edged white; gloss of secondaries usually less extensive. Tertials duller, shorter, and narrower. FIRST ADULT. Like adult, but juvenile flight-feathers and usually at least outer lesser and median upper wing-coverts and variable number of greater upper wing-coverts retained. (Delacour 1954; C S Roselaar.)

**Bare parts.** ADULT. Iris golden or orange-yellow, sometimes light brown or red-brown. Bill flesh-pink; base, edges of upper mandible, and nail black, area round nostrils sometimes dusky or (at least in some nesting ♀♀) whole side of upper mandible spotted black. Foot light pink to pale purple-flesh. DOWNY YOUNG. Iris yellow-brown to brown, bill and foot dull olive-grey. JUVENILE. Iris yellow-brown. Bill and foot yellow-grey. (Delacour 1954; RMNH, ZMA.)

**Moults.** Little known; moult of wing related to breeding and local rainy seasons, both variable geographically; detailed studies only South Africa (Siegfried 1965b, 1967). Moult of body plumage and tail starts shortly after breeding (ZMA). POST-JUVENILE. Partial. Head, body, tail, tertials, and variable number of upper wing-coverts. Moult starts soon after fledging, completed within a few months. Juveniles moult wing at same time as adults (Siegfried 1967).

**Measurements.** East and south Africa; skins (RMNH, ZMA).

| | | | | | | | |
|---|---|---|---|---|---|---|---|
| WING AD ♂ | 392 | (12·1; 6) | 378–406 | ♀ 375 | (14·5; 7) | 352–390 |
| TAIL AD | 123 | (5·16; 6) | 116–130 | 121 | (7·35; 6) | 111–131 |
| BILL | 49·6 | (3·18; 7) | 46–54 | 47·9 | (2·87; 6) | 43–52 |
| TARSUS | 85·5 | (5·14; 7) | 80–95 | 80·2 | (4·39; 10) | 73–85 |
| TOE | 76·9 | (3·44; 7) | 72–82 | 72·4 | (3·66; 10) | 67–76 |

Sex differences of wing, tarsus, and toe significant. Wing and tail of juvenile average *c*. 10 mm shorter than adult. Other measurements similar; combined. More extensive ranges in Delacour (1954) and Bannerman (1930; no bill given): wing ♂ 360–420, ♀ 330–400; tail ♂ 125–155, ♀ 110–145; tarsus ♂ 76–92, ♀ 67–89; bill ♂ 45–55, ♀ 40–52.

**Weights.** Average ♂ *c*. 2500, ♀ *c*. 2040 (Mackworth-Praed and Grant 1962). ♀♀ *c*. 680 less than ♂♂ (Bannerman 1930). ♂, 1900–2250; ♀, 1500–1800 (Kolbe 1972).

**Structure.** Wing rather long and pointed. 11 primaries: p9 and p8 longest, about equal, p10 and p7 each 4–16 shorter, p6 18–32, p5 38–62, p1 110–145. Inner web of p10 and p9 and outer p9 and p8 emarginated. Tail rather long, slightly rounded; 14 feathers, each tip broadly rounded. Bill slightly shorter than head, higher than broad at base, straight and heavy, nail large. Small swelling at base of culmen during breeding. Legs long and slender. Outer toe *c*. 92% middle, inner *c*. 81%, hind *c*. 37%. CSR

## *Tadorna ferruginea* **Ruddy Shelduck**

Du. Casarca    Fr. Tadorne casarca    Ge. Rostgans
Ru. Огарь    Sp. Tarro canelo    Sw. Rostand

*Anas ferruginea* Pallas, 1764.    Synonym: *Casarca ferruginea*

Monotypic

PLATES 58 and 61
[facing page 471
and between pages 494 and 495]

**Field characters.** 61–67 cm, of which body two-thirds, ♂ averaging slightly larger; wing-span 121–145 cm. Rather goose-like duck with small bill, orange-chestnut body, paler head, black primaries and tail, and white wing-coverts conspicuous in flight. Sexes rather similar but distinguishable, and only minor seasonal differences in ♂ non-breeding plumage. Juvenile and 1st year resemble adults, but separable on colour of upperparts.

ADULT MALE. Much individual variation. Head cinnamon-buff with variable creamy forehead and face, sometimes with dusky patch at rear of crown, separated from orange-brown to chestnut body by black collar, which may be narrow and indistinct or quite broad. Rump, tail, and primaries black with green gloss; black-brown secondaries with bright green speculum; white wing-coverts, often washed pink-buff, largely concealed at rest, but usually show as white line. Bill and legs black; eyes brown. In flight, white upperwing and underwing contrast sharply with black flight-feathers. MALE NON-BREEDING. Loses black collar, or nearly so, but otherwise no field difference. ADULT FEMALE. Similar to adult ♂, but less orange, more chestnut and so darker, with generally whiter head and wing-coverts, and no collar. JUVENILE. Resembles adult ♀, but duller with browner back. 1st year like adult (♂ acquires collar about February), but with grey wash on upper wing-coverts and scapulars.

Cape Shelduck *T. cana* (South Africa), which may occur as escape, rather similar, but has grey head (♀ with whiter face), chestnut back and belly contrasting with yellow-buff neck and breast, yellow eyes, and no collar. Otherwise, in flight only likely ever to be confused with Egyptian Goose *Alopochen aegyptiacus*, which has shelduck build and similar black-and-white wing pattern but it larger, thinner-necked, and longer-legged, mainly brown above and grey below, with chestnut eye-patches, neck ring, and (usually) breast patch, and pink bill and legs.

Generally by fresh water, seldom on sea coast. May be wary or tame, depending on degree of persecution, but rarely shot for food. Feeds largely at night. Swims well with neck erect and stern-high shape, but spends much time on land, where walks easily, even runs, and grazes like geese *Anser* and *Branta*. Perches on rocks and also trees. Flight powerful and fast, though with rather slow, measured wing-beats. Characteristically in pairs or family parties in most areas, but elsewhere (e.g. steppes) and on migration large flocks may be seen. Rather noisy with high-pitched whooping calls.

**Habitat.** Unlike Shelduck *T. tadorna*, has not extended range to maritime coastal habitats but occupies wider spectrum of inland biotopes, from coasts of inland seas and lakes, salt lagoons and marshes, rivers, streams, and pools, up to hills, high plateaux, and mountainous regions, reaching nearly 5000 m in Pamirs and Tibet. Occupancy here only in warmer seasons, altitudinal and northward limits buffering from regular or severe cold. Boreal and taiga zones overlapped only marginally in summer. Dependence on water for resting and feeding less than for most Anatinae, although needs access to some, shallow and preferably saline. Often on unvegetated banks and in steppe or semi-desert, with only small pools or streams accessible. Breeding recorded several kilometres from water. Avoids tall or dense herbage, and emergent or floating aquatic plants. Uses holes in cliffs, banks or trees for breeding; sometimes old nests of raptors and crevices in walls or ruined buildings. Preferred terrain on passage and in winter differs little from breeding habitat. Not assiduous user of airspace, although capable of strong flight rising to considerable heights on migration and on local movements.

**Distribution.** Breeding often local in areas shown while, in some desert regions such as southern Morocco and Algeria, nests only in wet years (Vielliard 1970). Major decreases in breeding range in south-east Europe, while more eastern areas in North Africa abandoned. BULGARIA: considerable decrease, no longer nesting in north and east (Boev 1965); now protected and increasing (SD). RUMANIA: no longer breeding north of Danube (Vasiliu 1968). USSR: many western areas abandoned in fairly recent times, e.g. Voronezh Region (after 1855), central Volga (early 20th century), Azov area, and Kharkov region (Dementiev and Gladkov 1952). TUNISIA: now extinct as breeding species (MS). ALGERIA: no recent breeding records from eastern areas (Vielliard 1970). SPAIN: breeding not proved, though possibly nested (Valverde 1960).

Accidental. Iceland, Britain and Ireland, Portugal, France, Belgium, Netherlands, Denmark, West Germany, East Germany, Norway, Sweden, Finland, Poland, Czechoslovakia, Hungary, Austria, Switzerland, Italy, Yugoslavia, Malta, Libya, Madeira.

**Population.** Breeding. GREECE: under 200 pairs (WB). RUMANIA. About 150–160 pairs 1896, then major decline.

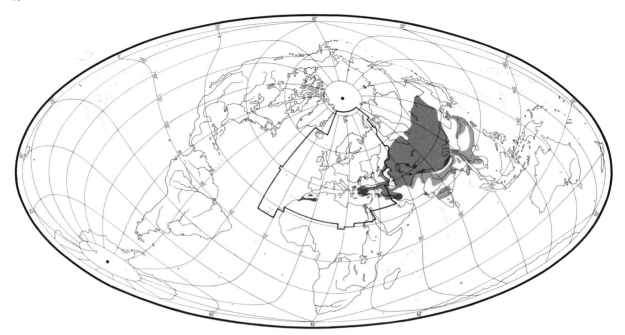

Some recent increase under protection: 2–11 pairs 1955–8, 12–15 pairs 1960, *c.* 20–25 pairs now (Vasiliu 1968). NORTH AFRICA: in Atlas region, up to 1000 pairs, and in southern desert region several hundred pairs (Vielliard 1970). MOROCCO. Iriki: estimated 50 pairs 1965, 30 pairs 1966, none in dry season 1967, 50 pairs 1968 (Robin 1966, 1968).

Winter. Estimated, 20000 birds in Europe—Black Sea—Mediterranean region (Szijj 1972). USSR: 16 700 birds in western part in not so severe winters (Isakov 1970*c*).

**Movements.** Mainly dispersive or nomadic, but some migratory. In north-west Africa, many moult and over-winter on or near nesting grounds as far inland as water found, but move away from drought-affected areas; temporary wetlands important in dispersal as for breeding (Vielliard 1970, on which this account based). Some cold weather exodus from higher parts Atlas ranges to coastal plains, especially the Rharb; but July–September concentrations Moulaya delta presumably moulting, since scarce there in winter. Formerly up to 200 crossed to Spanish marismas (Guadalquivir delta) from mid-August to early October, remaining to February–March, but no more than 10 seen in winters 1968–72 (Ree 1973). Occasional elsewhere on Mediterranean coast Spain, Portugal, and south France during post-breeding dispersal. This only instance of species nesting Africa migrating to Europe.

Southward dispersal by some breeders of Balkans and Soviet Black Sea countries; 1800 on Sea of Azov January 1968 possibly exceptional for season. Small numbers winter Greece (Hafner and Walmsley 1971); 150 Nestos delta January 1973 (Johnson and Carp 1973). In Turkey, many resident though dispersive, and degree of emigration obscured by movements of Black Sea breeders; one ringed Ukraine recovered Smyrna in January; generally 12 000–13 000 present Turkey November–December, but 9700 highest January count (Dijksen *et al.* 1972). Possibly small numbers which pass through Syria and Israel, and reputedly winter north-east Africa come from above areas; few reach Ethiopia, but has straggled to Aden. Though reported fairly common November–April in Lower and Upper Egypt and Suez Canal zone (Mein-ertzhagen 1930), and common October–March in Nile valley of north Sudan (Cave and Macdonald 1955), little recent information (Vielliard 1970).

Further east, movements little known (one ringing recovery); probably most resident or nomadic, concentrating on saline lakes, though withdraws from northern parts of breeding range in USSR. Estimated 60000 winter between Anatolia and Afghanistan, with up to 40000 on Lake Rezaiyeh (north-west Iran) and smaller concentrations on lakes of east Turkey and other parts of Iran, Iraq, Turkmenia, and, more erratically, around Caspian. Small numbers reach Persian Gulf south to Oman (Seton-Browne and Harrison 1968). Longest regular migration apparently by those wintering north India and Pakistan; one recovered latter ringed 42°N 75°E in Kirgiz SSR.

Post-breeding flocking begins early August; USSR nesting grounds largely deserted (except extreme south) by mid-September, though migration through Caspian continues into November. Rapid return early March to early April, non-breeders following later; some arrivals on Soviet breeding areas before general thaw. Not infrequently recorded spring and autumn in non-

Mediterranean Europe; probably most recent extra-limitals escapes, but several small influxes recorded May–June; major invasion in summer 1892 when obtained north and west to Ireland and Britain (flocks up to 20), Iceland, Greenland, Norway, and Sweden (Ogilvie 1892; Bauer and Glutz 1968).

**Food.** Omnivorous; plant material probably predominates, but proportions vary with locality and season. On land, where feeds more often than Shelduck *T. tadorna*, plucks vegetation and picks up seeds and invertebrates. Also grubs in wet soil, and dabbles and up-ends in shallow water. Gregarious feeder in flocks or family groups.

In USSR (Kara-Kum, Barsa-Kelmes), chiefly young green shoots of grasses and thistles in spring (Dementiev and Gladkov 1952): in July–August, insects (especially locusts) and autumn seeds from land plants (Bannikov and Tarasov 1957); in other areas of USSR (Taman, Irgiz, Kurgaldzhin, Achinsk), juveniles and adults in autumn feed on grain fields, especially on millet *Panicum* (gullets containing 40–65 g). In Volga delta, stomachs from migrating birds contained littoral crustaceans, and tender parts of plants (Dementiev and Gladkov 1952). In Askaniya-Nova, diet wholly wheat *Triticum* from harvested fields. In winter in Kakhetian valleys, sprouting greens; spring broods eat mainly insects, especially locusts (north-west Kazakhstan), and brine shrimps *Artemia salina* (Pamirs) (Dementiev and Gladkov 1952). In Tibet,

shrimps *Gammarus* and aquatic molluscs main foods, and in Iraq chiefly grain (Meinertzhagen 1954). Many also feed on rubbish near houses and possibly on carrion, though may merely be feeding on invertebrates associated with carrion (Dementiev and Gladkov 1952; Jourdain 1940; Meinertzhagen 1954; Ali and Ripley 1968). Others items include fish, frogs, spawn, and worms (Witherby *et al.* 1939).

**Social pattern and behaviour.** Based mainly on material supplied by J Vielliard.

1. Mainly gregarious, except when nesting. Outside breeding season, usually in family parties or loose flocks, sometimes in many thousands (Vielliard 1970). Flocks not nomadic but based on more reliable feeding grounds. Where migratory, travels in family parties or small flocks. Some live in pairs on streams in nesting area throughout winter, often defending territory. BONDS. Monogamous pair-bond, perhaps life-long (Delacour 1954). No information on duration in wild, but often in pairs outside breeding season. Both parents tend young, family remaining together after fledging; bonds sustained into winter, lasting longest in families which do not join others. Where feeding area restricted families leave nesting area and young gather in crèches guarded by a few adults (Vielliard 1970). BREEDING DISPERSION. Where nesting holes close to one another, breeds somewhat gregariously; otherwise well dispersed. In all cases, strongly territorial but probably defends nest-site only. When nests far from wetland feeding area, many may feed at same place; dispersed breeders tend to feed solitarily. Non-breeders gather in small flocks. ROOSTING. On water at night. Isolated pairs or families often roost together. Communal roosts of winter (and perhaps passage) flocks on large open waters used

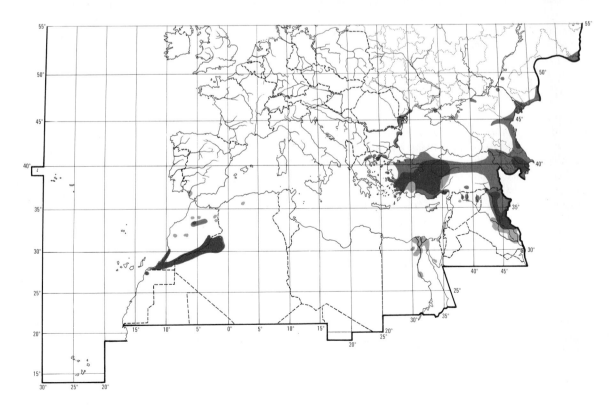

for feeding by day and sometimes at night, and by non-breeding birds in summer; no special movements or pre-roost gatherings.

2. Little known of behaviour in wild; most information from captive birds (Heinroth 1911; Johnsgard 1965). Pre-flight signals consist of Chin-lifting together with lateral Head-shaking. ANTAGONISTIC BEHAVIOUR. Disputes rare in flock, individual-distance maintained largely by mutual avoidance. Aggression strongest in defence of territory (J Vielliard). In threat-display, wings partly spread to show white markings, tail fanned, and head and bill lifted while threat-call given. When fighting, mostly strikes with wings, also pecking (P A Johnsgard). Calling and Neck-up posture reported in both sexes as advertisement while breeding or in winter territory (J Vielliard). HETEROSEXUAL BEHAVIOUR. Initiative in pair-formation taken by ♀ who attaches herself to ♂ and performs Inciting-display with outstretched neck and head low, giving repeated Inciting-calls and making frequent lateral Bill-pointing movements, usually over shoulder, towards other ♂♂. Chosen ♂'s first response to Inciting of ♀ is to give 1-syllable threat-call; latter often replaced by 2-syllable call when ♂ also tends to assume High-and-erect posture, jerking head back and sometimes lifting tail. ♀ said to select ♂ who responds strongest to her display (Heinroth 1911). Pre-copulatory display consists of mutual Head-dipping in water, while ♂ frequently gives 2-syllable call. After ♂ dismounts, ♀ calls, then ♂ while still holding ♀'s nape with bill, afterwards slightly lifting wing opposite ♀ in High-and-erect posture; both then bathe. RELATIONS WITHIN FAMILY GROUP. Parental warning-calls well in evidence almost throughout time young present; small young sometimes carried on back (J Vielliard). Otherwise no relevant information.

Voice. Rather whining, nasal trumpeting or whooping; loud and penetrating. Used frequently, including in flight—when rendered (e.g.) 'ang'. Calls of sexes distinguishable: those of ♂ with predominantly 'o' sound; those of ♀ generally louder, deeper, and harsher with predominantly 'a' sound (Heinroth 1911).

CALLS OF MALE. Main calls: (1) loud, prolonged, rolling 'chorr' or 'kor' (see I), uttered as threat-call especially as initial response to ♀'s Inciting-display; (2) 2-syllable 'cho-HOO' which often replaces 'chorr' call as response to ♀'s Inciting-display—also given before and after copulation (Johnsgard 1965). Other calls include: (3) prolonged, deep 'ho' (alarm or warning-call). Following also described by Delacour (1954) but significance uncertain: (4) prolonged, rather high-pitched, whooping 'ah-onk'; (5) repeated, ascending 'ka-ha-ha'.

CALLS OF FEMALE. (1) Inciting-call: repeated 'gaaa' (Johnsgard 1965); shown in I at end of ♂ threat-call. (2) Alarm-call: 'hä-hä' (Heinroth 1911).

CALLS OF YOUNG. No detailed information but similar to Shelduck T. tadorna (J Kear), at least in downy stage.

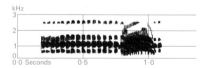

I C Chappuis/Sveriges Radio Morocco April 1966

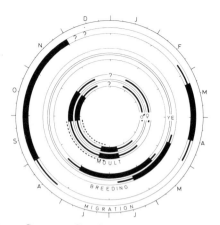

Breeding. SEASON. South-east Europe and southern USSR: see diagram. North Africa: eggs laid from mid-March to mid-April (Heim de Balsac and Mayaud 1962). Iraq: eggs found end April (Bannerman 1957); eggs laid from mid-March, well incubated eggs found end April and early May, young under one week old seen 1 May (Marchant 1963b). SITE. Holes in sand or clay banks, either natural or excavated by another animal; in abandoned buildings and farm sheds; in hollow trees up to 10 m above ground; crevices in rocks and cliffs, rarely on cliff ledge in open. Will use nest-boxes. Nest: shallow depression lined with down, occasionally a little grass incorporated. Building: shallow nest cup formed by ♀ turning body. Reports that burrows excavated (Dementiev and Gladkov 1952) not confirmed. EGGS. Rounded oval; white, dully glossy. 68 × 47 mm (62–72 × 45–50), sample 110 (Schönwetter 1967). Weight of captive-laid eggs 83 (69–99), sample 70 (J Kear). Clutch: 8–9 (6–12); up to 16 almost certainly 2 ♀♀. One brood. Replacement clutches probable but no data. One egg laid per day. INCUBATION. 28–29 days (27–30). By ♀ only. Starts on completion of clutch; hatching synchronous. Eggshells left in nest. YOUNG. Precocial and nidifugous; self-feeding. Cared for by both parents, who defend them against predators. Some amalgamation of broods may take place where several in one area, but these not deserted by own parents as in Shelduck T. tadorna. FLEDGING TO MATURITY. Fledging period c. 55 days. Young become independent soon after. Age of first breeding probably 2 years (Dementiev and Gladkov 1952). BREEDING SUCCESS. No data.

Plumages. ADULT MALE BREEDING. Head and upper neck cinnamon-buff, darkest towards neck, paler cream-buff on lores and round eye; sometimes grey patch at rear of crown. Ring round lower neck black. Lower neck, mantle, scapulars, and underparts rufous-cinnamon; deep chestnut chest, centre of belly, vent, and sometimes under tail-coverts. Inner scapulars finely vermiculated dark grey and cinnamon-buff; back vermiculated black and buff. Rump, upper tail-coverts, and tail black, slightly glossed green. Sides of tail-base black, or cinnamon vermiculated black. Flight-feathers black, strongly glossed metallic green and slightly purple on outer web of secondaries. Primary coverts black, lesser mottled white near base; rest of upper wing-coverts white, often strongly washed pink-buff or cream-yellow on tips when fresh. Outer webs of outer tertials deep chestnut, inner webs grey; inner tertials rufous-cinnamon, inner webs pale buff, grey, or vermiculated buff and grey. Underwing and axillaries white, greater primary coverts grey. In worn plumage, cinnamon areas may fade and bleach to pale yellow, especially on head and upperparts. ADULT FEMALE BREEDING. Like adult ♂ breeding, but feathers near base of upper mandible, round and behind eye white, contrasting with cinnamon-buff of rest of head. No black collar round neck. Inner web of inner tertials usually darker grey than ♂, contrasting more with outer web but rather variable in both sexes. In worn plumage, cinnamon of head fades to yellow or pale buff, no longer contrasting with white sides of head. ADULT MALE NON-BREEDING. Like breeding, but black collar narrow or absent. ADULT FEMALE NON-BREEDING. Apparently like adult ♀ breeding. DOWNY YOUNG. Like Shelduck *T. tadorna*, but down of crown and upperparts dark olive-grey rather than dark brown; patches on mantle and sides of back and rump generally smaller, and small white spot above eye (present in most *T. tadorna*) usually absent. JUVENILE. Like adult ♀ but crown and hindneck dark grey-brown; rest of head pale grey, ♀♀ often with whiter face than ♂♂; feathers of lower mantle and scapulars dark grey-brown, tinged cinnamon-buff at tips. Back, rump, tail-coverts, and tail dull black, feathers of rump and some outer tail-feathers tipped buff. Underparts cinnamon-pink, some grey-brown feather-bases visible on flanks and belly. Wing like adult, but flight-feathers duller; upper wing-coverts virtually without buff or yellow tinge, greater ash, median and lesser white, usually variably tipped grey (which may wear off). FIRST IMMATURE NON-BREEDING and BREEDING. Like adult, but with juvenile wing, except sometimes tertials and some inner coverts. (Bauer and Glutz 1968; Dementiev and Gladkov 1952; C S Roselaar.)

**Bare parts.** ADULT and JUVENILE. Iris dark brown, somewhat paler in juvenile. Bill black, in some ♀♀ slightly spotted flesh near tip. Foot dark grey to dull black. DOWNY YOUNG. Iris brown. Bill grey-black. Foot like adult. (Bauer and Glutz 1969; RMNH.)

**Moults.** ADULT POST-BREEDING. Complete. Starts, when birds leave broods, with scapulars and flanks, followed by head and rest of body late June to August; often not finished when flight-feathers shed. Part of body plumage moulted only once a year. Flight-feathers simultaneously; flightless for *c.* 4 weeks mid-July

to September. Tail between July and November. ADULT PRE-BREEDING. Partial; most of head and body, and tertials. Head and part of mantle, scapulars, and flanks start when wing full grown, rest of body later. Moult often arrested December–February; full breeding plumage attained March–April. POST-JUVENILE. Partial; body, and tertials, from July onwards. Usually complete immature non-breeding plumage attained before December, except juvenile wing and sometimes part of tail. FIRST PRE-BREEDING. Sequence of head and body like adult pre-breeding; mainly in spring, but full breeding plumage on head and neck sometimes from November. Juvenile wing (except tertials and often some inner wing-coverts) retained. SUBSEQUENT MOULTS. Like those of adult. (Dementiev and Gladkov 1952; Witherby *et al.* 1939; C S Roselaar.)

**Measurements.** ADULT. Whole range, all year; skins (RMNH, ZMA).

| | | | | | | |
|---|---|---|---|---|---|---|
| WING | ♂ 366 | (10·3; 9) | 354–383 | ♀339 | (17·1; 6) | 321–369 |
| TAIL | 125 | (6·23; 8) | 116–135 | 118 | (3·81; 6) | 112–122 |
| BILL | 44·4 | (2·24; 11) | 41–49 | 39·5 | (2·63; 9) | 35–43 |
| TARSUS | 61·5 | (1·69; 11) | 59–64 | 54·6 | (1·51; 10) | 52–57 |
| TOE | 64·5 | (2·98; 11) | 60–69 | 57·2 | (1·59; 10) | 55–59 |

Sex differences significant, except tail. JUVENILE. Wing and tail *c.* 20 mm shorter than adult. Other measurements similar; combined above.

**Weights.** Full-grown. Volga delta, central Kazakhstan, and Balkhash area, USSR. March–April: 4 ♂♂, 1465 (1360–1600); 2 ♀♀, 1150, 1280. May–June: 5 ♂♂, 1354 (1260–1500); 4 ♀♀, 1088 (925–1500). August–September: 5 ♂♂, 1362 (1210–1600); 2 ♀♀, 1125, 1250. October: 15 ♂♂, 1360 (1200–1600); ♀, 1100 (Dolgushin 1960).

**Structure.** Wing rather long, pointed. 11 primaries: p9 longest, p10 2–13 shorter, p8 7–13, p7 21–35, p6 38–54, p1 125–160. Tail rather long. Bill slightly shorter than head, higher than broad at base, sides parallel. Culmen slightly concave, cutting edges slightly upcurved towards tip of bill. Nail of bill relatively broader than *T. tadorna*. No knob at base of culmen. Feathers of head and neck dense and elongated in breeding plumage, especially on hindcrown and hindneck. Outer toe *c.* 93% middle, inner *c.* 79%, hind *c.* 30%. Rest of structure as in *T. tadorna*.

CSR

## *Tadorna tadorna* Shelduck

PLATES 59 and 61
[after page 494
and between pages 494 and 495]

DU. Bergeend   FR. Tadorne de Belon   GE. Brandgans
RU. Пеганка   SP. Tarro blanco   SW. Gravand

*Anas Tadorna* Linnaeus, 1758

Monotypic

**Field characters.** 58–67 cm, of which body two-thirds, ♂ averaging larger; wing-span 110–133 cm. Large, somewhat goose-like duck with conspicuous bill, elongated head, and boldly patterned plumage; mainly white, apart from dark green head, broad chestnut belt round forepart of body, black scapulars, flight-feathers, tail tip, and belly stripe, and red bill and pink legs. Sexes similar but distinguishable, and only slight seasonal differences. Juvenile quite different.

ADULT MALE. Head and upper neck blackish-green, lower neck white; chestnut band round breast; broad black line along scapulars; blackish stripe from chestnut band on breast to vent, varying individually in width and widening out on belly; rest of body white except for yellow-brown under tail-coverts. Wings white with black flight-feathers; tail white tipped black. Bill bright red, with prominent fleshy knob in early spring; eyes brown; legs red-pink. At distance and in flight, looks white and black with head,

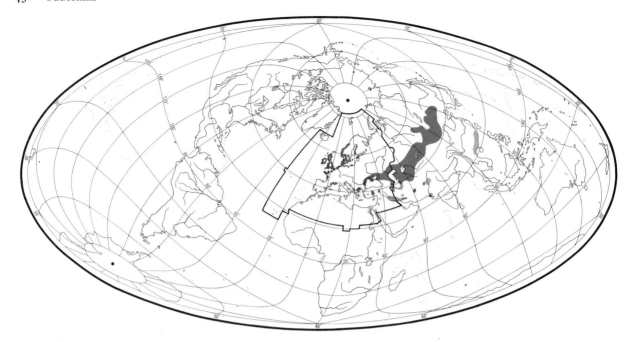

breast band, braces, and belly stripe standing out as dark. MALE ECLIPSE. Duller and less clear-cut pattern: head brownish-black with hoary white about face and throat; chestnut belt paler, uneven, and mixed with white and blackish; belly stripe reduced and mottled white and grey; upper tail-coverts flecked with black, and under tail-coverts much paler. Frontal knob disappears and bill becomes paler. When flightless, looks much whiter owing to lack of black primaries. ADULT FEMALE. Closely resembles adult ♂, but head and neck usually duller. Chestnut band generally narrower and paler with some black vermiculation; black belly stripe also narrower, and less clear-cut; under tail-coverts paler and yellowish-brown. Bill duller than ♂, with black around tip. FEMALE NON-BREEDING. Usually like ♂ eclipse, but head browner; chestnut belt largely mixed with grey and whitish above and broadly mixed with white at sides and below; and belly stripe no more than black mottling. Bill dull red, tinged with grey. Some have completely white underparts and so much white on face and throat that they resemble juveniles, except for wings. JUVENILE. Quite different from adults, apart from ♀♀ in extreme non-breeding. Top of head, hind neck, and whole upperparts dark grey-brown, contrasting with white forehead, face, spectacles, foreneck and underparts, the last tinged pale grey and lacking any trace of breast band or belly stripe. Bill pink-grey; legs pale grey. In flight, forewing pale greyish rather than white, all flight-feathers except outermost primaries broadly tipped white, and secondaries with white inner webs. By 1st winter, more or less resembles adult, but ♂♂ usually have head and neck duller, chestnut belt mixed with black and white, belly stripe narrower and more broken (sometimes as black as adult ♂), and paler under tail-coverts; ♀♀ have

browner heads with more white on face and throat, and chestnut band and dark belly stripe narrower and more broken; both sexes also readily distinguished by retention of juvenile wing with white tips to primaries and secondaries, and grey greater coverts.

Adults unmistakable. Juvenile only ever likely to be confused with immature Egyptian Goose *Alopochen aegyptiacus*, which is similarly dark above and pale below with pinkish bill and legs, but more brown and buff with more clearly indicated dark eye-patch.

Largely associated with marine habitats, especially sandy and muddy coasts and estuaries, but in Asia breeds by inland salt lakes, and in some areas nests along rivers or in meadows and small woods far from open water. Less aquatic than other ducks, and swims more in muddy channels than open sea, but buoyant on water. Young dive freely, but adults only when wounded or frightened. Walks and even runs easily and in breeding season regularly perches on cliff-ledges, sheds, and haystacks. Rises easily from either water or land, sometimes with preliminary run. Flight much more goose-like than other ducks, not particularly fast and with slow, powerful wing-beats. Often in pairs or small parties, but sometimes in much larger flocks; non-breeding birds congregate in nesting season. Flocks flying to feeding grounds tend to travel in V-formation or diagonal lines, but on migration usually in straight lines at considerable heights. Generally silent outside breeding season, but in summer harsh, chattering or growling quacks of ♀♀, in series 'ark-ark-ark . . .', characteristic and audible over considerable distances.

**Habitat.** Both in habitat and biogeographical range, displays consistent narrow preferences contrasting with

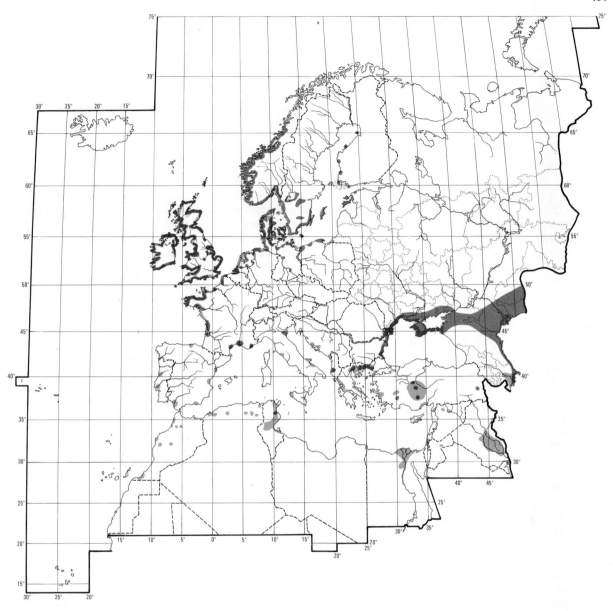

marked discontinuities, suggesting basically relict status offset by capability to adapt and survive. Favours warm, semi-arid, and mild maritime climate, only sparingly entering Mediterranean zone and almost wholly avoiding boreal and sub-arctic. Choice of habitat, especially for breeding, governed by attachment to salt or brackish water, either on shallow coasts and estuaries or inland seas and lakes. As coastal species in north-west Europe, typically ranges only 1–2 km out to sea, and little farther inland. Wholly freshwater habitats distant from sea coast or brackish lakes used only sporadically and by odd pairs or single birds, but access to some fresh water favoured. No marked preference for islands except in moulting period. Needs foraging areas of fairly high biological productivity,

especially sands and mud flats over which shallow water alternates with drying out through tides or evaporation. Central Asian range, extending over south-eastern west Palearctic, consists mainly of widely scattered salt lakes and marshes in steppe and semi-desert, far from marine waters in most cases even from large inland seas.

Suitable nesting holes, essential for breeding. Sand dunes and light soils favoured, but also ground covered by dense prickly or thorny shrubs, weed plants or reeds, old grain stacks, and other artefacts such as walls, tombs, deserted buildings, or nest-boxes, holes in trees or rock fissures, and burrows of various mammals. Overhead shelter and concealment common factor. On migration, crosses mountainous and other unsuitable terrain, between stopovers of suitable type. Moulting concentrations, some immense, on open, shallow saline waters, inland and marine. Winter habitat similar; in USSR mostly on large salt-flats covered with water. Moves readily and fast on land, and flies strongly, rising on passage to 1000 m or more above ground.

**Distribution.** In north-western parts of range, recently recolonized Finland and spread elsewhere, especially to inland sites; in part due to protection. In south-west, retreating in North Africa and Italy.

FINLAND: scattered pairs breeding south-west since 1962 after absence since 1888 (LvH, Merikallio 1958). ITALY: breeding Sardinia (Bezzel 1957) and said to have bred Apulia up to 1930–40, but no recent records (SF). IRAQ: probably bred near Baghdad (Ticehurst et al. 1922). ALGERIA: bred 19th century (Heim de Balsac and Mayaud 1962); no recent records. TUNISIA: bred several areas 19th century (Heim de Balsac and Mayaud 1962); a few pairs now probably attempt to breed in central areas in wet seasons (MS).

Accidental. Iceland, Faeroes, Czechoslovakia, Hungary, Austria, Switzerland, Kuwait, Libya.

**Population.** Breeding. BRITAIN AND IRELAND: despite local decreases has continued to increase in Britain, especially eastern and southern England, with more frequent inland nesting (Parslow 1967); estimated 2500–5000 pairs in Britain (Atkinson-Willes 1970a), c.12 000 pairs in Britain and Ireland (Yarker and Atkinson-Willes 1971). FRANCE: increase and spread last 30 years, helped by protection since 1962; 120–200 pairs, mostly Brittany and Camargue (FR). BELGIUM: c. 115 pairs; formerly rare, now increasing (Lippens and Wille 1972). NETHERLANDS: increased inland nesting; 2500–3000 pairs 1957 (Timmerman 1960). WEST GERMANY: increased inland nesting; 1730 pairs (Szijj 1973). DENMARK: continued increase; many new inland localities occupied recent decades (TD). NORWAY: some recovery after marked reduction 1940–5 due to persecution (Haftorn 1971). SWEDEN: increasing breeding population and pronounced increase in non-breeding summer visitors (SM). EAST GERMANY: c. 400 pairs (Bauer and Glutz 1968). POLAND: extremely scarce breeder (Tomiałojć 1972). GREECE: 200–300 pairs (WB). USSR. Estonia: 150 pairs, no increase (Onno 1970). Ukraine: marked increase last 10 years on islands Tendrovsky inlet, Black Sea coast (Ardamatskaya 1970).

Winter. Estimated 130 000 birds in north-west Europe and c. 75000 in distinct southern population, mainly USSR, Turkey, Iran, Greece and North Africa (Atkinson-Willes 1975).

Survival. Mean annual mortality adults 20%; life expectancy 4·5 years (Boyd 1962). Oldest ringed bird 14 years 6 months (Rydzewski 1974). At Aberlady, Scotland, where population stable, numbers of breeders and non-breeders regulated primarily by competition for feeding space in late winter, with territorial behaviour secondary (Jenkins et al. 1975).

**Movements.** Migratory, partially migratory, and dispersive. In north-west Europe, major moult migration considerably affects picture of normal migration. Breeders and immatures leave nesting grounds in Britain, Ireland, Belgium, Netherlands, Denmark, Scandinavia, north Germany, north Poland, and Baltic States, for annual moult in German Waddenzee area, leaving ducklings in care of small number of adults which moult locally (Goethe 1961a). Flock prior to moult migration, which probably undertaken in single flight, passing across land in their path, e.g. England or Jutland. First move June, probably mostly immatures, adults following July; some movement as late as September, especially from east. Waddenzee (chiefly Grosser Knechtsand) moulting place for virtually all Shelduck in north-west Europe, c. 100 000 or slightly more (Salomonsen 1968), except for 3000–4000 (perhaps of Irish origin) which gather Bridgwater Bay, south-west England.

When moult completed, autumn migration begins in rather leisurely fashion. For breeders of Netherlands, Britain, and Ireland, this entails return to breeding areas. Timing varies considerably; some years almost complete by late October, in others delayed until December. Breeding populations from areas to east of moulting grounds winter around coasts of southern North Sea, west France, and to some extent Britain and Ireland; many thousands remain on moulting grounds (Goethe 1961a). Those wintering outside their breeding areas begin return March. After reaching flying age, young of year dispersive to large extent for 1st autumn and winter, when some British move west to south-west to Irish Sea and south-west England, others south at least to Bay of Biscay (recoveries to Basses Pyrénées). Only scattered individuals in midwinter in Baltic and Scandinavia.

Breeding populations of south-east Europe (Greece, Black Sea, Turkey) mainly sedentary, flocking in winter, moving only if bad weather. This basically true for breeders of area from Iran to Afghanistan though some winter Kuwait (where rare) and Pakistan. Those breeding central to southern USSR (Volga and Ural steppes east to Irtysh-Ob basin) migrate to Caspian, where join resident breeders (Dementiev and Gladkov 1952); small local moulting flocks recorded but no evidence of moult migration. South of main wintering areas, small numbers occur fairly regularly Iberia, Mediterranean basin east to Nile valley, Middle East, and Persian Gulf; include juveniles, older birds displaced from further north by cold weather, and some local (isolated) breeders.

Singles or small parties not rare on inland waters during migration, even in west Europe where more strictly coastal. See also Bauer and Glutz (1968).

**Food.** Mainly invertebrates, especially molluscs, insects, and crustaceans. Feeds in groups of varying size by wading mainly in shallow water or on wet mud; less often on dryer, inland areas. Feeding methods: (a) surface digging in exposed mud (Olney 1965), (b) scything action or dabbling on exposed mud with moist surface (Swennen and van der Baan 1959), (c) dabbling and scything in shallow water 1–10 cm, (d) head-dipping in water 10–25 cm, (e) up-ending in deeper water 25–40 cm (Bryant and Leng 1975). Rarely dives (Jung 1968 recorded submergences of 5–15 s in 5–10 m), though young do so freely. Foot-trampling (mud-pattering) movements to bring prey to surface well described. Frequency of use of each method related to tide state, to type, size, availability and abundance of prey, and to substratum (Swennen and van der Baan 1959; Olney 1965; Bauer and Glutz 1968; Oelke 1970; Bryant and Leng 1975).

Varies considerably throughout range: in north and west Europe, mainly small molluscs; in south Europe and central and south USSR, small crustaceans and insect larvae predominate. Molluscs include: *Hydrobia, Cardium, Macoma, Mytilus, Montacula, Cingula, Buccinum, Littorina, Skenea, Paludina, Tellina, Nucula, Mya,* and *Theodoxus.* Crustaceans: small crabs, shrimps and prawns, sandhoppers (species not specified), *Artemia,* and *Corophium.* Insects: grasshoppers (Orthoptera), beetles (Coleoptera—*Carabus nitens*), and fly larvae (Diptera—Chironomidae). Other items include small fish and spawn, annelid worms (Nereidae and *Arenicola*), and plant materials (chiefly algae, grasses and seeds of various aquatic species).

All 46 birds collected each month (except July) from English coastal waters contained snail *Hydrobia ulvae.* In 18 from one Kent locality, 89·5% of volume occupied by *H. ulvae,* one containing over 3000 individuals. Another contained *H. jenkinsi,* and a third traces of amphipod *Corophium volutator.* 6·4% volume plant material, chiefly green seaweed *Enteromorpha* and, less often, *Vaucheria;* also seeds of sea club-rush *Scirpus maritimus* and herbaceous seablite *Suaeda maritima.* In 7 from moult area in Bridgwater Bay, England, August–October, 82·4% by volume consisted of *H. ulvae,* 7·3% baltic tellin *Macoma balthica,* and 7·7% (1 bird) *C. volutator;* also small amounts of *Enteromorpha* and nereid 'jaws'. Cycle of feeding correlated to cyclic behaviour of main prey, and thus closely related to ebb and flow of tide (Olney 1965a, 1965b). However, at Firth of Forth, Scotland, association between intensity of feeding and distribution and abundance of main food, *H. ulvae,* observed and this influenced by tidal movements, weather conditions, and possibly by human disturbance. Behaviour of *Hydrobia* in relation to water depth apparently strongly influenced both spatial distribution of feeding flocks and feeding method used (Bryant and Leng 1975). One bird from North Uist contained mainly *H. ulvae* and rough periwinkle *Littorina saxatilis* (Campbell 1947). 2 stomachs from Knechtsand,

Heligoland Bight, October, had chiefly *H. ulvae,* juvenile sand-gaper *Mya arenaria* and *M. balthica,* foraminifera shells, and algal chlorophyll. Faecal samples contained *M. balthica, Nucula, Littorina littorea, Mytilus edulis,* and *Cardium edule.* 2 from Mellum contained *H. ulvae* and mussel *Mytilus* remains (Goethe 1961a, 1961b). In USSR winter quarters on Kirov Bay and Mugan steppe, mainly plant materials; one February stomach contained 8·2% by volume *H. ulvae* and *Theodoxus;* 16·1% seeds, mainly tassel pondweed *Ruppia* and reeds, and 75·7% decomposed algae and seeds. Spring foods on Ili delta also mainly plant, especially Chenopodiaceae seeds. Summer diet chiefly brine shrimp *Artemia salina* and eggs, chironomid larvae, and single-celled algae, especially *Aphanoteca* (Lake of Gor'koye group, central Kazakhstan, and Naurzum, also Volga-Ural steppe and Sivash river—from latter average number per stomach of *Artemia* and chironomid larvae 1800 and 5970, and astounding maxima of 19 800 and 63 880). On salt lakes of Uxboi, mainly crustaceans: also swarming ants (Formicidae). On Naurzum lakes area, locusts *Calliptamus italicus* (Dementiev and Gladkov 1952).

8 large nestlings in July contained chironomid larvae, and 5 in July from Naurzum lakes filled with *C. italicus.*

**Social pattern and behaviour.** Based mainly on material supplied by I J Patterson. See especially Hori (1964a, 1969), Young (1970a), Williams (1974).

1. Gregarious except in spring and early summer. Congregates in large numbers when flightless during post-breeding moult. Later, major wintering grounds contain flocks of several thousands which disperse during winter into smaller units in breeding areas. Usually scattered when feeding, but may gather densely on concentrated food sources. Sub-adults, with some non-territorial adults, remain in loose flocks during summer; joined later by failed breeders and those whose ducklings have joined other broods, then by remaining breeders after fledging of young. BONDS. Monogamous pair-bond, most pairs persisting from year to year. May separate on moult migration, but pairs often distinguishable within winter flock. Pair-formation occurs in flock. Some ♀♀ pair when 1 year old, but most birds unpaired at least until 2nd year. ♂ attends ♀ throughout incubation, and may both tend young until fledging. In some areas, complex family and crèche organization. At first, ducklings spaced out in family groups; at this stage, any large party of tiny young of same age in charge of single pair probably from multiple nest. When pairs with broods interact, young may mingle and 1 pair sometimes left without brood. Amalgamation of broods to form crèches frequent, especially where many broods occupy same area; crèches (up to 100 or more strong) mostly attended by some of original parents. Crèches form during first 2–3 weeks at nursery area, in many cases initial brood-tie lasting less than 1 week. Family bond often broken, apparently by aggression of parents, usually after encounters with other families when attacks re-directed against own young. For first few days, crèche (like brood) compact group closely following adults. Young join nearest brood or crèche when alarmed, size and composition of crèches frequently changing. Become increasingly self-reliant; by *c.* 16–20 days, many largely independent, brooding each other at night. BREEDING DISPERSION. Winter flocks begin to break up late March as pairs disperse to territories where remain for most

of day. Territory usually on muddy shore; used only for feeding. In approximately same place each year; if mate changed, ♀ tends to return to previous territory while ♂ goes to that of new mate. If ♂ dies during breeding season, ♀ usually retains territory until new ♂ joins her; if ♀ dies, ♂ usually returns to flock. Adjacent territories not always mutually exclusive, and territorial defence not always consistently of same area. Concurrently with establishment of feeding territory, pairs prospect for burrows in nesting area, which may be up to several miles away. Visits, made mainly early morning, may last several hours; individuals known to visit same localities day after day, and in successive seasons. Only part of time spent actively searching for burrows, and birds commonly form small groups to sleep and preen. These persistent groupings of adults in nesting areas distinct from other types of flock; termed 'parliaments' (Young 1970b), or 'communes' (Hori 1964a). No territory round nests which may be close together, even in same burrow. While ♀ incubating, ♂ remains on feeding territory where joined by ♀ 3–4 times a day for feeding. When young emerge from burrow, taken to nursery area usually quite separate from feeding territory which now abandoned. Both parents defend area round mobile young, though ranges of different broods overlap widely; parents with brood can displace birds still defending feeding territories. ROOSTING. No special night roosting places reported, birds probably feeding throughout. At high tide in winter, roost communally on islands, sandspits, saltmarsh, or on water. Territorial pairs roost singly near feeding territory.

2. FLOCK BEHAVIOUR. Little information. Pre-flight signals include lateral Head-shaking and probably Chin-lifting (Johnsgard 1965); see also below. ANTAGONISTIC BEHAVIOUR. Territorial defence often vigorous, involving complex interactions between pairs (Young 1970a). In commune, aggressive behaviour occasional, mainly between ♂♂. Both parents attack adults coming near young. Postures and associated calls described by several authors, often using different names. Bowing (Boase 1935) or Head-throwing (Hori 1964a) main threat-display of ♂: rotary pumping movement of head, usually several times in succession, accompanied by whistling; at higher intensity, whole body may be moved rhythmically up and down. Bowing also occasionally seen in ♀♀, particularly when groups prospecting for nest-sites. Inciting-display main threat of ♀: head and neck held low and pointing movements (often lateral) made, usually at opponent; accompanied by special call. When ♀ Incites close to mate (see A), ♂ usually responds with Bowing, often then attacking opponent. Attack behaviour shown by both ♂ and ♀:

B

C

head and neck lowered and orientated towards opponent (Head-down), sometimes with bill slightly open and usually with back feathers and wings raised. Bird may remain stationary in Head-down, or walk, run, or fly at opponent, biting if contact made. Object of attack usually flees with plumage sleeked and head held high; when does not, fight may occur, especially between ♂♂, with mutual jabbing with open beaks and beating with partly extended wings. PURSUIT-FLIGHTS. Of 2 main types (Hori 1965). (1) Courtship-flights most common in May though occur at other times, especially February–March: up to 8 or more ♂♂, often from assembly on water, follow single ♀ closely in wide circles, each ♂ in turn trying to fly alongside; rapid flighting followed or interspersed with gliding and calling. (2) Sexual pursuits occur at same seasons when 1 or more unpaired ♂♂ chase mated ♀; latter defended vigorously by own mate. HETEROSEXUAL BEHAVIOUR. Still inadequately studied, but includes following postures, some of which, however, may be threat. In Rest-intent of ♂ (Boase 1935), erect posture assumed, nape feathers raised; High-and-erect of Johnsgard (1965) may be same (see B)—often accompanied by whistling while head motionless. In Whistle-shake of ♂ (Johnsgard 1965), exaggerated Upward-shake involves rearing of body and throwing up of head while whistle given; often occurs in mildly alarming situations. In Salute (Boase 1935), whistles while head flicked quickly upwards, bill in line with erect neck (see C); used in pre-flight situations by both sexes, though chiefly ♂, may be flight-intention signal between mates. Inciting-display of unpaired ♀ important in pair-formation; may be directed at several ♂♂. Whether ♂ responds by Bowing or attacking others (see above) may be important in selection of mate by ♀. Both sexes perform exaggerated Preen-behind-Wing display on side towards mate, usually in sequence Bill-dip—Head-shake—preen (McKinney 1965b), particularly after ♂ returns from making attack. Copulation on water, preceded by vigorous Head-dipping or shallow dives by both sexes while swimming rapidly (Johnsgard 1965). ♀ may adopt Prone-posture, lying motionless in water, ♂ usually Preening-behind-Wing before mounting. As ♂ dismounts, ♀ calls and ♂ pulls her head back, both rotating

A

slightly in water; ♂ assumes post-copulatory High-and-erect display with wing opposite ♀ slightly raised. Both sexes then bathe and self-preen. RELATIONS WITHIN FAMILY GROUP. Based mainly on Hori (1964a, 1969). Ducklings cheep as eggs chip, without apparent reciprocal notes from ♀. Before young leave nest, ♀ utters variety of soft calls for a few hours; then induces brood to leave nest-hole with series of further quiet calls. Brood led quickly to nursery area by pair or by ♀ only via territory; family sometimes accompanied by one or more other pairs (parasitic, or failed breeders). Family may remain together for *c.* 9 weeks. Young range widely, returning quickly to parents in response to alarm-call, especially of ♀; young give loud piping call if separated from parents or siblings, otherwise continual, soft contact-calls between adults and ducklings. Parents, especially ♂, attack strange ducklings in brood; extend this to own young when strangers of similar size (C M Young, M J Williams, I J Patterson).

(Fig A after Lorenz 1966; Figs B–C after Johnsgard 1965.)

**Voice.** Often loud and far-carrying. Calls uttered in flight and on ground and water. Except for loud hiss when cornered, calls of ♂ and ♀ differ greatly—though, according to Boase (1935) and Witherby *et al.* (1939), sexes have some alarm-calls in common. Non-vocal, rasping noise produced during Preen-behind-Wing display, at least

by ♂ (McKinney 1965b). Following based on outline by I J Patterson.

CALLS OF MALE. Variety of melodious whistling notes, difficult to distinguish and describe. Often rather passerine in quality, in particular recalling cardueline finches; initially rising in pitch, then dropping back more gradually (Johnsgard 1965, 1971). Often used in conjunction with threat postures and courtship displays. Descriptions include: (1) sibilant 'sos-thieu', given in flight (Kirkman and Jourdain 1930); (2) 'piu-pu' (from tape, see I); (3) 'whee-o' or 'whee-ee' (Witherby *et al.* 1939); (4) 'djudjuju' (Voigt 1950). Especially during antagonistic encounters, occurs as rapid, rhythmic series—'wheesp-wheesp-wheesp . . .'; more distinctive, long, whirring or trilling version given with Whistle-shake display. During gliding phase of courtship-flights, ♂♂ produce infrequent chorus of soft, clear whisles, recalling Wigeon *Anas penelope* (Hori 1965). When attempting to attract attention of predator in vicinity of young by tolling (see under Tadornini), ♂ uses variety of whistles: soft and clear when danger not severe; increasing in volume or number, or both, when danger greater—e.g. 'whee-chew whee-chew', with occasional long trilling phrases (Hori 1964b). Whistles also given as alarm-call.

CALLS OF FEMALE. Lower and deeper than ♂'s, lacking whistling quality. (1) Long, nasal, fast, pulsating 'ak-ak-ak-ak . . .' (Witherby *et al.* 1939; see II) or 'gagagaga' (Voigt 1950); loud and insistent, often with notes decreasing in duration and loudness towards end. Sometimes accelerated to *c.* 12 notes per s (see III). Given in variety of situations, in flight and when settled, but often used as contact-call by bird separated from mate; incubating ♀ flying from nest to join ♂ on territory gives

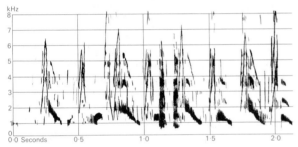

I Sture Palmér/Sveriges Radio Sweden March 1965

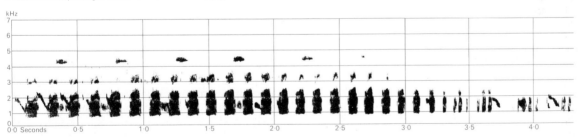

II L Shove England April 1964

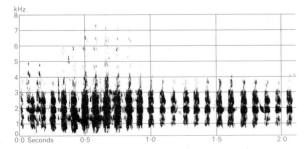

III Sture Palmér/Sveriges Radio Sweden April 1963

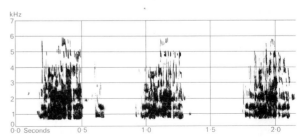

IV Sture Palmér/Sveriges Radio Sweden April 1963

this call at intervals. Flight-call of birds in late summer and autumn described as deep, nasal 'ga-ak' by Voigt (1950). (2) Longer, most distinctive, harsh 'aaark aaark', also rendered 'ro-ow ro-ow' (see IV) (Witherby *et al.* 1939) and characterized as growling by Hori (1964*b*). Typical threat and Inciting-call. Said by Boase (1935) also to be given by ♂ as intense threat-call. (3) Softer, high-pitched 'aank', usually uttered singly. Given, e.g., when nest or brood threatened, young responding by grouping round ♀. Same monosyllabic, nasal 'aank' repeated slowly and regularly while tolling (Hori 1964*b*). (4) *Anas*-like quacking notes, also given as alarm-call (Hori 1964*b*); may be harsh 'quack-wack-wack-wack' of Kirkman and Jourdain (1930). Said also to be given by ♂. (5) Various soft calls uttered as time for brood to leave nest approaches; last only a few hours (Hori 1969). Soft 'arrnk' or 'arrk', with varied amplitude and frequency, used as contact-call, call of reassurance, and alarm-call. This and, less frequently, soft 'ugg ugg ugg' given to call young from nest; when first young appear, quiet but hard and bubbling version of first call, 'ak ak ak' uttered. Brood led with gentle 'arruk arruk'.

CALLS OF YOUNG. Hatchlings start cheeping as eggs chip; later keep up continuous peeping while still in nest (Hori 1969). High-pitched piping, with rhythm closely similar to 'ak-ak' call of ♀, given when out of nest, usually if separated from parents; soft trill uttered when in close contact.

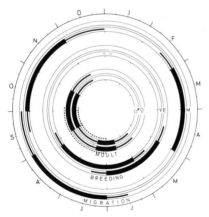

Breeding. SEASON. See diagram for North Sea countries. Southern USSR: laying begins first half May (Dementiev and Gladkov 1952). SITE. In holes of various kinds, on ground or up to 8 m above it, occasionally in thick vegetation. Generally near water but may be up to 1 km away. Of 100 sites in Kent, 29 in hollow trees (14 in base, 15 above ground), 28 in hay or straw stacks, 27 in burrows of rabbit *Oryctolagus cuniculus*, 23 under buildings and other objects, and 3 in open (Hori 1964*b*). Nearly all in rabbit burrows in Aberdeen, in area where these plentiful (Patterson *et al.* 1974). In hollow trees up to 5 m above ground, in stacks up to 8 m and 5 m inside (Hori 1964). Territorial, but in well-populated areas nests may be only 1 m apart. Nest: shallow depression. Little or no material

used in hole sites, grass and other vegetation in open sites, hay and straw in stack sites; all lined large amounts down. Building: site chosen by ♀ with ♂ in attendance; nest cup formed by ♀. Reports that makes own burrows (Dementiev and Gladkov 1952) not confirmed. EGGS. Rounded oval, smooth with little gloss; creamy white. $66 \times 47$ mm ($61-71 \times 43-40$), sample 175 (Schönwetter 1967). Weight 78 g (66–93), sample 57 (J Kear). Clutch: 8–10 (3–12). Of 95 clutches, Kent, by single ♀♀: 3 eggs, 1%; 4, 2%; 5, 1%; 6, 3%; 7, 3%; 8, 28%; 9, 23%; 10, 24%; 11, 10%; 12, 5%; mean 9·0. Of 35 clutches laid by 2 or more ♀♀: 13 eggs, 11%; 14, 20%; 15, 17%; 16, 9%; 17, 9%; 18, 5%; 19, 5%; 20, 6%; 21, 3%; 22, 3%; 23, 6%; 25, 3%; 32, 3%; mean 17·0 (Hori 1969). Of 30 clutches Aberdeen, mean 8·1 (5–11) (Patterson *et al.* 1974). One brood. Replacement clutches laid only when eggs lost early in incubation. One egg laid per day, usually in early morning, occasionally a day missed. INCUBATION. 29–31 days. By ♀ only. Starts with last egg; hatching synchronous. ♀ covers eggs with down when off nest and recorded covering eggs with nest material during laying (Hori 1964*b*). Eggshells left in nest. YOUNG. Precocial and nidifugous. Self-feeding. Both parents lead brood from site to feeding area (up to 3 km). Remain in brood for 15–20 days or until moult migration urge in parents strong, then broods amalgamate into crèches of up to 100. Some broods stay with parents until fledging, though this uncommon. Crèche adults sometimes successful breeders with own broods in crèche, but others failed or non-breeders (Hori 1964*b*). FLEDGING TO MATURITY. Fledging period 45–50 days. Some broods largely independent at 15–20 days, brooding themselves and in groups without adults (Hori 1969). Fully independent at or soon after fledging. Mature at 2 years; most ♀♀ breed at this age but few ♂♂ not until 4 or even 5 (Hori 1964*b*). BREEDING SUCCESS. Average of 27% eggs laid produced fledged young in Kent 1960–63 (range 16–49%), sample 665–733 eggs per year (Hori 1964*b*). Of 25 clutches, Aberdeen, 1962–64, 76% hatched; of 458 young in 67 broods aged 1–5 days, 65% survived to day 12, 43% to day 18, 37% to day 25, 29% to day 33, 25% to day 43, 24% to day 50 and fledging. Average number of young reared per pair to fledging, 1962–64, 1·5 (range 1·2–1·9). Survival of young to fledging varied according to brood size: 11·9% of young in broods 1–5, 38·6% in broods 6–10, 26·9% in broods 11–15 (Patterson *et al.* 1974).

**Plumages.** ADULT MALE BREEDING. Head and upper neck black, glossed green; often tiny white spot on chin and below eye. Lower neck, upper mantle, chest, and upper breast white. Broad band round body over lower mantle and lower breast tawny chestnut; narrowly bordered black in front on breast, interrupted by black in centre of breast. Chestnut feathers on lower mantle bordering back sometimes finely vermiculated black. Shorter and outer scapulars glossy black, inner white, or with inner web white and outer web vermiculated buff and black. Back, rump, upper tail-coverts, sides of body, and flanks white. Centre of belly and vent black, feathers with broad white bases, in fresh plumage with narrow buff tips. Under tail-coverts cinnamon, pale yellow towards sides and tips. Tail white, broadly tipped black. Primaries, their coverts, and tip of bastard wing black. Secondaries black, strongly glossed bronze-green on outer webs, white towards base of inner webs. Outer webs of outer tertials deep chestnut, inner webs white, separated by dark grey shaft-streak; inner tertials white, longest with black outer edge. Rest of upperwing, under wing-coverts, and axillaries white. All black and chestnut feathers of head and body may have narrow white tips when fresh. ADULT FEMALE BREEDING. Like adult ♂ breeding, but feathers of head shorter and duller, tinged brown on face and throat. In worn plumage, white bases of feathers visible round base of bill. Lower mantle more extensively vermiculated black, sometimes narrowly tipped black and white. Black of scapulars duller, often minutely speckled buff or white. Black band over centre of breast and belly narrower, white of feather-bases exposed. Cinnamon of under tail-coverts usually paler and less extensive, bordered white at sides and tip. Dark longitudinal streak on tertials broader and less clearly defined; white of tips and outer webs of tertials sometimes washed grey. ADULT MALE NON-BREEDING (Eclipse). Feathers of head and neck short, black, edged brown, white bases more or less exposed round base of bill. Feathers of mantle finely mottled grey and cinnamon, with dull black subterminal bar and narrow white edge at tip. Scapulars dull black. Underparts white; cinnamon suffusion and some black mottling on breast and central belly, where white of feather-bases visible. Narrow dusky tips to white feathers of sides of body and upper tail-coverts. Under tail-coverts pale cinnamon. Wing as in breeding. Dark feathers show narrow white tips when new. ADULT FEMALE NON-BREEDING. Like adult ♂ eclipse, but head duller and paler brown-black. Much white round base of bill, on cheeks, and chin; mantle grey-buff, finely mottled and barred dark grey. Underparts white, except sides of breast, where feathers white, suffused cinnamon, in some with black subterminal bar. DOWNY YOUNG. Crown, streak through eye, and hindneck dark brown. Mantle (except white patch in centre), centre of back and rump, flank, thigh, and broad bar over upper wing dark grey-brown. Forehead, small streak or spot above eye (absent in some), cheeks, rear of wing, large spots at sides of back behind wing and at sides of rump, and underparts white. JUVENILE. Crown, sides of head behind eye, and sides of back of neck sooty-black, feathers narrowly tipped grey or buff on crown and white elsewhere. Forehead, spots above and below eye, sides of head in front and below eye, chin, throat, and foreneck white. Mantle grey, feathers dark grey to black towards tip, finely speckled and narrowly tipped buff or grey. Scapulars dark brown, paler towards grey-buff edges. Feathers of sides of breast and body, flanks, and upper tail-coverts white, sub-terminally mottled or washed grey and buff, on flank with dark brown shaft-streak. Rest of body white. Tail-feathers sooty-brown, paler buff at sides and tip, bare shaft projecting. Wing like adult, but flight-feathers broadly tipped white (except *c.* 4 outer primaries), secondaries with white inner web; greater upper wing-coverts pale grey, median and marginal coverts white tipped grey; tertials with inner web and tip pale grey and outer web cinnamon. FIRST IMMATURE NON-BREEDING. When fully developed, like adult non-breeding, but juvenile wing and usually tertials and tail retained. Often not all juvenile body feathers changed for 1st immature non-breeding, and ♀♀ especially moult much of juvenile direct into 1st immature breeding. FIRST IMMATURE BREEDING. Like adult breeding, but white-tipped flight-feathers and grey-tinged upper wing-coverts of juvenile plumage retained. Black on belly of ♂ shows slight white mottling, and buff-brown feather edges, instead of uniform

black of adult ♂; centre of breast and belly of ♀ white mottled buff and sooty-brown, in contrast to adult ♀, where belly-streak uniform black or white boldly spotted black. SUBSEQUENT PLUMAGES. Like adult. Specimens with fully adult wings except for some grey spots on tips of greater and median upper wing-coverts may be 2nd year, but this probably not regular.

**Bare parts.** ADULT MALE. Iris brown or red-brown. Bill flesh or pale red, in spring bright carmine or blood-red, including knob. Nail of bill and often spot at nostrils black. Foot red-flesh. ADULT FEMALE. Like adult ♂, but bill duller, flesh-pink, usually with some black round tip and on nail. DOWNY YOUNG. Iris brown. Bill and foot grey. JUVENILE. During growth, bill (except tip) and foot fade to pink-white or pale flesh; attain adult colour in winter. Bill of 1st year not usually as bright red as adult. (Witherby *et al.* 1939; RMNH.)

**Moults.** ADULT POST-BREEDING. Complete; flight-feathers simultaneous. Starts with body late June, shortly after hatching of young; non-breeders somewhat earlier, a few others much later. Most feathers of underparts and head first, followed by some feathers of mantle and scapulars (starting during moult migration). Apparently no more body feathers lost when flight-feather moult starts; part of body plumage moulted only once a year. Flightless for 25–31 days between early July and mid-October. Non-breeders and failed breeders first, those adults which have looked after flocks of young last. Tail at same time as wing. ADULT PRE-BREEDING. Partial; head and body. Starts when full-grown; completed August–December. POST-JUVENILE AND SUBSEQUENT MOULTS. Post-juvenile starts soon after fledging. Between late July and early October, most of head, breast, and belly, and variable amount of chest, flanks, scapulars, and mantle attain 1st immature non-breeding, but most of this soon lost. From late September, 1st immature breeding feathers appear, and often most feathers of head, mantle, scapulars, and underparts in 1st breeding before moult slows or stops during mid-winter. A few juvenile or 1st non-breeding feathers remain on mantle and scapulars, and rarely all juvenile belly; these sometimes retained until spring. Back, rump, tail-coverts, and tail sometimes moulted September–December, but mostly start later and part of juvenile plumage here retained during winter; rarely completely juvenile until spring. Some juvenile inner tertials with coverts, and more rarely some inner wing-coverts, moulted from December, but usually from spring, sometimes retained until summer. 1st post-breeding and following pre-breeding like adults. (Bauer and Glutz 1968; Schiøler 1925; Witherby *et al.* 1939; C S Roselaar.)

**Measurements.** Netherlands, mainly spring. Skins (RMNH, ZMA).

| | | | | | |
|---|---|---|---|---|---|
| WING AD ♂ | 334 | (8·75; 33) | 312–350 | ♀ 303 (7·83; 28) | 284–316 |
| JUV | 315 | (12·2; 13) | 291–334 | 290 (9·18; 13) | 277–307 |
| TAIL AD | 108 | (4·35; 27) | 96–115 | 96·9 (4·79; 27) | 89–106 |
| JUV | 83·1 | (2·30; 8) | 80–87 | 75·0 (2·74; 9) | 72–79 |
| BILL | 53·0 | (1·82; 37) | 50–58 | 47·3 (1·70; 36) | 44–50 |
| TARSUS | 55·8 | (1·64; 34) | 52–60 | 50·1 (1·86; 36) | 46–54 |
| TOE | 60·1 | (1·85; 34) | 57–64 | 54·2 (2·34; 36) | 51–59 |

All sex differences significant. Juvenile wing and tail significantly shorter than those of adult. Other measurements similar; combined. Tail in 1st breeding *c.* 5 mm shorter than adults.

**Weights.** ADULT. (1) February, south-west Caspian, USSR (Dementiev and Gladkov 1952). (2) April–May, Netherlands (ZMA), Denmark (Schiøler 1925), and Kazakhstan, USSR (Dolgushin 1960); combined. (3) June–August, as in (2).

(1) ♂     1180 (—; —)   830–1500 ♀   813 (—;—)   562–1085
(2)       1261 (110; 11)  1100–1450    1043 (138; 5)   926–1250
(3)       1167 (129; 7)  1000–1350     952 (88; 5)   850–1075

Highest weight ♂ USSR 1650 (Dementiev and Gladkov 1952). Lowest weights, emaciated birds killed by February frosts, Netherlands (ZMA): ♂♂ 770 (adult), 720 (juvenile); ♀♀ 580 (adult), 540 (juvenile).

**Structure.** Wing rather long, pointed. 11 primaries: p9 longest, p10 2–10 shorter, p8 8–17, p7 20–33, p6 43–55, p1 130–160; p11 minute. Inner web of p10 and outer p9 emarginated. Tail rather short, square; 14 feathers. Upper tail-coverts almost reach tip of tail. Bill about as long as head, higher than broad at base; tip slightly broader than base. Bill strongly upcurved towards flat tip, culmen concave; nail of bill small and narrow. In spring, adult ♂ with large fleshy knob at base of culmen; reduced during rest of year; 1st-year ♂ and adult ♀ virtually without knob. Feathers of head and upper neck dense and elongated in breeding ♂, up to c. 35 mm long, especially on hindcrown and hindneck; shorter or absent in ♀ and non-breeding. Outer toe c. 94% middle, inner c. 82%, hind c. 32%.      CSR

# Tribe CAIRININI perching ducks

Small to fairly large wildfowl, usually living in well-wooded areas, freely perching in trees, and nesting in holes high above ground. Some semi-terrestrial. Highly diversified group of 13 species in 9 mainly monotypic genera, often showing striking convergences with other Anatidae. 2 groups (Johnsgard 1965): (1) more generalized genera *Plectropterus* (Spur-winged Goose—Ethiopian Africa), *Cairina* (Neotropical America, south-east Asia), *Pteronetta* (central Africa), and *Sarkidiornis* (Comb Duck—South America, Ethiopian Africa, south Asia); (2) more specialized genera *Nettapus* (pygmy geese—central Africa to Australia), *Callonetta* (Ringed Teal—South America), *Aix* (North America and east Asia), *Chenonetta* (Maned Goose—Australia), and *Amazonetta* (Brazilian Teal). With possible exception of *Plectropterus*, none occur naturally in west Palearctic—though Mandarin *Aix galericulata* introduced (see species' account for further information on habitat, status, and numbers).

Wings often wide and rounded; bony, spur-like knob on metacarpal joint. Tails fairly broad and elongated; slightly graduated but not pointed. Bills rather thick and goose-like, not depressed, often heavy; large nail. Hind toe well developed, not lobed, and claws strong and sharp at all ages; legs set far forward, tarsus usually short (especially in *Nettapus*), but longer in some (especially semi-terrestrial *Plectropterus*). Diving ability usually restricted. ♂ noticeably larger than ♀ in some species. Sexes differ in tracheal structure to varying degrees; except in *Nettapus*, ♂♂ with bony, enlarged bullae—in *Aix*, rather large and rounded, somewhat resembling Anatini. Plumage bright in many; often iridescent, especially in more generalized genera. Patterns more complex in other genera, particularly *Aix*. No real speculum in most species but tertials and wing-coverts often bright and metallic. Sexual dimorphism slight in some, considerable in others—especially *Aix*. Eclipse plumage only in *Aix* and 1 species of *Nettapus*. Juveniles like adult ♀♀. Downy young, patterned dark brown and white or yellow, resemble most those of Anatini; remarkable for long, stiff tails and capacity for climbing.

Cosmopolitan but most species tropical or sub-tropical. Most species surface-feeders though some (notably *Plectropterus*) terrestrial grazers. Often in flocks; pre-flight signals diverse—include Neck-craning, Chin-lifting, and Head-thrusting movements, also lateral Head-shaking. Social pattern and behaviour of *Aix* most resembles that of typical Anatini in many respects, including type of pair-bond and communal courtship. Latter includes versions of such *Anas*-like ♂ displays as Upward-shake, Water-flick, Turn-back-of-Head, Burp, and Bridling; correlated with striking wing-patterns, Preen-behind-Wing displays highly developed. Inciting-display of ♀ also much as in *Anas*. In more generalized genera, however, pair-bonds weak or absent (Johnsgard 1965). Pre-copulatory behaviour variable; includes Head-pumping (as in *Anas*), Head-dipping, and Bill-dipping. Copulation by forced rape normal in *Sarkidiornis* and Muscovy Duck *Cairina moschata*. Post-copulatory behaviour also variable, but little studied (for example, see *A. galericulata*). Voice characteristics variable; sexually differentiated to greater or lesser extent. ♂ calls mostly whistles; ♀ calls honking, quacking, or squeaking (characteristic Decrescendo-calls of *Anas* lacking). Some species more or less silent.

Apart from hole-nesting, breeding biology of most species much as in Anatini—including incubation and care of young by ♀ only and short-term family-bonds. See further under *A. galericulata*.

## *Plectropterus gambensis* (Linnaeus, 1766) **Spur-winged Goose**

FR. Oie armée    GE. Sporngans

African species, breeding rivers and swamps from Cape Province north to latitudes 13°N–15°N from Gambia to Sudan. Generally considered resident and dispersive; ringing recoveries up to 570 km in South Africa. In northern tropics, recorded wandering in dry season, e.g. influx to Lake Chad in February as inland pools dry out, dispersing again in May after rains (Golding 1934). Some pre-1919 records from Egypt perhaps escapes as species often then imported from Sudan (Nicoll 1919); this less likely to apply to 4 singles seen between Abu Simbel and Kertassi (area now under Lake Nasser), 2–7 March 1962 (Ripley 1963). Since breeds Nile valley north to Khartoum, Upper Egypt records may include wild birds; occasional dry season vagrancy indicated by flock of 5 seen, Oman, Arabia, January 1963 (Seton-Browne and Harrison 1968).

## *Aix sponsa* (Linnaeus, 1758) **Wood Duck**

FR. Canard huppé    GE. Brautente

North American species, breeding secluded streams, ponds, and marshes. Apart from isolated Pacific coast population, occurs eastern parts of continent north to Great Lakes, Quebec and New Brunswick, south to Gulf of Mexico and Cuba; now probably most numerous nesting duck in eastern USA (Linduska 1964). Migratory in northern half of range, wintering freshwater swamps and marshes from South Carolina and Arkansas to central Mexico (Stewart and Robbins 1958; Grice and Rogers 1965; Godfrey 1966). Commonly kept in captivity in Europe, where most wanderers certainly escapes. Free-flying colony in Berlin Zoological Gardens dwindled from 1930, and wild-state nesting apparently attempted in 1930s on Bodensee (Bauer and Glutz 1968). In Britain, feral populations Lake District, East Anglia, and south-east England, small and perhaps unestablished; not fully admitted to national list (Sharrock 1971). One shot Furnas, Azores, December 1963 more likely transatlantic vagrant (Bannerman and Bannerman 1966).

## *Aix galericulata* **Mandarin**

PLATES 60 and 74
[after page 494
and facing page 543]

DU. Mandarijneend    FR. Canard mandarin    GE. Mandarinente
RU. Мандаринка    SP. Pato mandarín    SW. Mandarinand

*Anas galericulata* Linnaeus, 1758

Monotypic

**Field characters.** 41–49 cm, of which body over two-thirds; wing-span 68–74 cm. Rather small, compact, but long-tailed perching duck with short bill, large head and thick-looking neck; exotic-looking ♂ has red bill, multi-coloured crown and crest, broad white band from bill to nape, orange-chestnut 'side-whiskers' and wing 'sails', and maroon breast edged with black and white stripes. Sexes dissimilar, but no seasonal differences apart from eclipse. Juvenile resembles adult ♀, but distinguishable.

ADULT MALE. Forehead, crown, and long crest falling down to back all metallic green with purple and copper shades; broad band of white on side of head and crest takes in eye; cheeks buff, shading into ruff of pointed, whitish-shafted, orange-chestnut feathers at sides and front of neck. Upperparts generally dark olive; scapulars white, edged blue-black; tail brown, glossed green. Remarkable orange-chestnut sails, bordered in front with black and steel-blue, on either side of back, are greatly enlarged central tertials; olive-brown primaries have whitish edgings which show up as series of streaks behind these sails. Upper breast rich maroon, edged with 3 black and 2 white vertical stripes; flanks pale orange-brown vermiculated with black and white; lower breast, belly, and under tail-coverts white. Bill red with pink nail; eyes brown; legs yellow to orange-buff. In flight, wing-coverts and flight-feathers largely olive-brown, but speculum bright green edged behind with white tips to secondaries; contrary to many illustrations, sails not erect, though they may flutter up and down. MALE ECLIPSE. Closely resembles adult ♀, but distinguished by reddish bill, yellower legs, less well-defined white marks at base of bill and around eyes, thicker crest, and glossier plumage with browner mottling below. ADULT FEMALE. Oval head mainly grey, darker slate on top, with white spectacles extending back as line nearly to nape; white edge at base of bill runs into white chin and throat. Entire upperparts from mantle to tail, including wings, olive-brown, except for bright green speculum and white-edged primaries. Upper breast and flanks brown

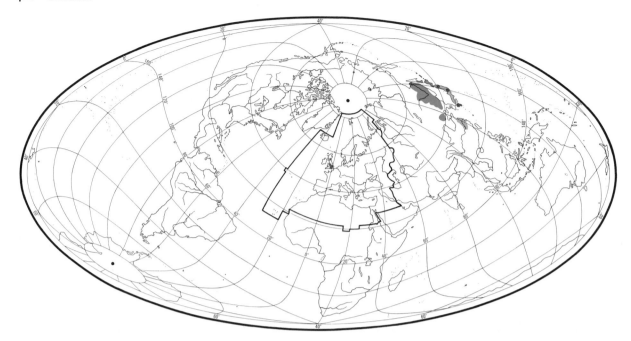

with whitish oval spots arranged in rows and increasing in size from foreneck to rear edge of flanks; lower breast, belly and under tail-coverts white. Bill usually grey-black to grey-brown with orange nail, but sometimes pinkish; eyes brown; legs dirty yellow. JUVENILE. Closely resembles adult ♀, but more uniformly grey-brown with less distinct face markings (often little or no white around and behind eyes) and less clearly dappled upper breast and flanks.

Adult ♂ unmistakable. ♀ and immature quite unlike any other west Palearctic duck, but need to be distinguished from corresponding plumages of Wood Duck *A. sponsa* (North America, but becoming established ferally in western Europe), which is generally darker and glossier green-brown above, with more pronounced white patch around and behind eye (less spectacled effect), as well as looking bigger and longer with larger, flatter bill, less peaked forehead, and shorter legs.

Essentially found on fresh water in wooded areas, including streams and rivers, small ponds, and large lakes. Generally wary, but becomes tame where fed regularly. May be seen out in open water, but feeds along wooded banks, both in and out of water, taking wide variety of vegetable matter. Swims well, with body high on water and tail often raised, but seldom up-ends and does not normally dive for food, though will readily dive if alarmed or wounded. Walks easily and perches freely on fallen trunks or low branches. Rises easily from either water or land and flies fast with rapid beats and considerable agility. Usually singly, in pairs, or small parties, but sometimes flocks. Often quite noisy: ♂ has variety of whistles; ♀ also has fair vocabulary of soft, plaintive, clucking and rolling notes and, in particular, a loud 'eck' sound recalling sharp 'kewk' of Coot *Fulica atra*.

**Habitat.** In contrast with nearly all other Palearctic Anatidae, essentially belongs to middle-latitude temperate broad-leaf forest zone, and its standing or slow-flowing fresh waters, fringed by dense growth of trees and shrubs—preferably those overhanging to provide cover down to water, often with some reed, sedge, or other emergent vegetation. Primarily in lowland, but occupies valleys and forested uplands to 1500 m, using small ponds as well as lakes and rivers with wooded islets. Nests in hollow trees, stumps, fallen logs and roots. Needs ready access to arboreal or shrub cover for concealment, and infrequently resorts to open ground or waters, especially marine, although visiting ricefields and cornfields accessible from nearby cover. Although normally secretive, adapts readily to close human presence when undisturbed. Able to rise steeply and fast out of confined spaces, and uses powers of flight freely, mostly at fairly low level. Not much given to movement over land except at water's edge. Deforestation and disturbance over much of natural range, and lack of detailed knowledge of essentials for survival give special value to establishment of feral colonies in England.

**Distribution.** Natural range: eastern Palearctic. BRITAIN: established 20th century as breeding resident due to escapes and deliberate releases; mainly south-east England, notably Surrey and Berkshire, and also locally elsewhere England, and Perth, Scotland (British Ornithologists' Union 1971).

Accidental. Faeroes, Finland, Czechoslovakia, Yugoslavia, Italy, Spain; all probably escapes (see Movements).

**Population.** BRITAIN: estimated rather more than 500 birds (Savage 1952); little recent change (Parslow 1967).

**Movements.** Migratory and dispersive in Asia. Breeders of USSR Maritime Territories and Manchuria (and presumably Hopei) migrate, some via Korea, to winter quarters in east China (mainly south of Yangtze Kiang); passage late August to November, returning, already paired, March to early May (Dementiev and Gladkov 1952; Cheng Tso-hsin 1963). Resident in Japan, though some southward shift (Savage 1952). Has occurred Ryu Kyu Islands, Taiwan, Mongolia, Upper Assam and Manipur.

Introduced British population resident. Individuals seen occasionally far from nesting areas may be strays from ornamental collections, which capable of long-distance movement. One left St James's Park, London, summer 1930, found next April in Hungary (Witherby 1934); 2 left Oslo, Norway, 8 November 1962, shot together next day 900 km away in Northumberland, England (Holgersen 1963).

**Food.** Omnivorous; mainly vegetable, especially seeds and nuts, with some animal at times, especially land snails and insects. Nocturnal and daytime feeder, on land and in water; mainly from surface, dabbling, up-ending, and head-dipping, but rarely diving.

Diet changes throughout year, but detailed information scant. USSR: in spring, insects especially beetles, land snails, seeds of wild grape *Vitis*, horsetail *Equisetum*, acorns *Quercus*, rice *Oryza sativa*, grasses (Gramineae, probably cereals), and fish. Latter include minnow *Phoxinus phoxinus*, lamprey young (Petromyzonidae), and dead fish spawn. In August and September, flocks visit fields of rice and buckwheat *Fagopyrum esculentum*, in early autumn land snails particularly important, and as family parties join together in flying groups, acorns become main diet (Dementiev and Gladkov 1952). In England, autumn and winter diet mainly acorns, chestnuts *Castanea* or *Aesculus*, and beechmast *Fagus sylvaticum* and in summer chiefly insects (especially Diptera) off water and aquatic plants (Savage 1952). In China, weed seeds and plant remains recorded (Shaw 1936).

Young take chiefly insects and larvae, as well as small fish and plant materials.

**Social pattern and behaviour.** Based mainly on material supplied by R L Bruggers. Information available only from feral and captive birds in Europe and North America (Savage 1952; Bruggers 1974).

1. Gregarious except when nesting. Congregates for summer moults: ♂♂ in June when most pairs separate; joined by unsuccessful ♀♀ that have not re-nested, and later by successful ♀♀. Flocks relatively large early in autumn, Virginia Water, England; smaller in winter, at least for most of day while not feeding (Savage 1952). BONDS. Pair-bond primarily mono-gamous, sometimes over several breeding seasons; ♂ sometimes also promiscuous while mate incubating, or bigamous, forming loose pair-bond with 2nd ♀ (R L Bruggers). Pair-formation usually begins September (adults) or later (1st-winter birds), continuing through winter; most paired by February. Although ♂ keeps vigilance during incubation period, accompanying ♀ while she feeds, pair-bond usually temporarily discontinued before hatching; some ♂♂ remain with mates afterwards (Savage 1952), but contact between most former mates re-established in moulting flocks. Only ♀ tends young. BREEDING DISPERSION. No detailed studies in wild, but nests probably well scattered and limited by availability of natural tree-cavities. Most pairs maintain territory while near nest-hole during laying or early incubation; size in wild uncertain (R L Bruggers). ROOSTING. Feeds at dusk, preens, then sleeps—but activity after dark uncertain. Outside breeding season, apparently spends night in flocks on rocks or logs in water at concealed sites, e.g. along shore under overhanging trees. During breeding season, ♂♂ apparently roost singly near nest; ♀♀ feed just before sunset then spend night on nest, leaving again to feed just before dawn—during early incubation, often take 3rd break in afternoon.

2. Detailed descriptions of main elements of social behaviour given by Heinroth (1911); Savage (1952); Lorenz (1953); Johnsgard (1965); Bruggers (1974). Crest and unique, ornamen-

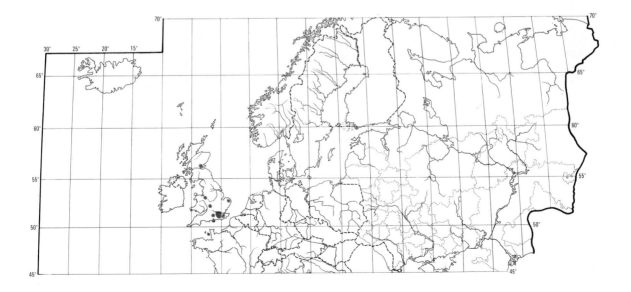

A

tal tertial 'sails' of ♂ erected during communal courtship and most other social situations, including copulation. FLOCK BEHAVIOUR. Little information. Pre-flight signals consist of Neck-craning movements (Johnsgard 1965) in which head and bill stretched diagonally up while birds sometimes call. ANTAGONISTIC BEHAVIOUR. Not highly aggressive, but several types of reaction often initiated and accompanied by ♀ Inciting-display, in which ♀ rapidly jerks head repeatedly at rejected ♂♂ while calling. During communal courtship, ♂♂ engage in quick, formal chases, darting wildly side by side; frequently threaten by Bill-jerking—rapidly flicking bill upwards. During nest-site inspection by pair, Open-bill threat (♂♂) and Bill-up display (both sexes) predominate. During and following pair-formation, and in territorial defence, ♂♂ threaten approaching individuals of either sex by Bill-pointing and Open-bill threat, also making short rushes while pair pursue intruders continuously, ♂ in lead. COMMUNAL COURTSHIP. Highly developed, recalling in many features analogous behaviour in *Anas* ducks. Occurs mainly in poor light (morning, evening, and on dark days) or shade, involving groups of ♂♂ and 1 or more ♀♀. Begins September, continuing through winter; peak February–March. Consists of intermittent chases and threats between ♂♂ and short bursts of display, mainly on water. Mate selection said to be entirely by ♀. Before and during bouts of courtship, ♂♂ assume Full-sail or Courtship-intent posture and frequently utter basic Courtship-call from Alert-posture with neck stretched vertically while Bill-jerking. Secondary displays include ordinary Upward-shake (evidently infrequent), but 2 more types of ritualized shake performed as major displays without raising body from water. In Display-shake, ♂ lowers then rapidly thrusts head upwards (see A) while quickly shaking head and tail and uttering special call. In Double Display-shake, recalling Water-flick of *Anas*, (1) dips bill

B

in water, shakes head while giving another special call, and pulls head back to normal position; (2) rapidly and sharply dips and shakes again, then thrusts neck upwards while again calling and vigorously wagging tail. When performing Burp-display, ♂ utters prolonged nasal call while stretching neck vertically, crest and 'beard' conspicuously spread. Main secondary activities consist of various Mock-preen displays, including Preen-dorsally. Of these, Ceremonial-drink and Preen-behind-Wing sequence most elaborate. More obviously than other displays, orientated towards particular ♀: ♂ positions lateral to ♀, raises crest, dips bill in water, pauses, then quickly passes bill behind erected sail-feather on wing (see B), usually on side nearest ♀. Main activities of ♀: Inciting-display and Coquette-display. In latter, special Coquette-call given with sudden dip of bill and forward and downward thrust of head; functional equivalent of Nod-swimming and other activities of *Anas* ♀♀, provoking ♂♂ to display. PURSUIT-FLIGHTS. Courtship-flights (see Anatini) observed infrequently during communal courtship (Savage 1952). PAIR BEHAVIOUR. Pair-formation indicated by (1) ♀ directing Inciting-display at rejected ♂♂ while near preferred ♂; (2) unilateral and reciprocal allopreening by pair; (3) ♂ directing 'caressing' movements toward ♀, gently dipping bill and head towards her without touching, then passing bill down own chest or side (Bruggers 1974); (4) following of ♂ by ♀, occasionally while ♂ ceremonially Leads her with Turn-back-of-Head display (see Anatini); (5) ♂ following ♀, especially in spring; (6) other ritualized behaviour which often follows antagonistic behaviour in spring. Last category includes Ceremonial-drinking as pair stand side by side; Ceremonial-drink—Preen-behind-Wing sequences by ♂; occasional Preen-behind-Wing displays by ♀ only; and Bridling-display of ♂ ('rebuff' of Savage 1952). Later display initiated on land in winter (occasionally spring) when ♀♀ start Inciting as pairs leave loafing spots and meet at water's edge, but also seen on water during communal courtship in autumn on close approach of strange ♂: ♂ assumes position with chest puffed out, bill resting on foreneck, giving Jibbering-call; then lifts head back and performs Bill-jerks repeatedly. Bill-jerking also used as greeting display to ♀, e.g. when latter joins ♂ during incubating recess. COPULATION. Always occurs on quiet section of water. Often initiated when ♀ swims while performing Lateral-dabbling display (see Shoveler *Anas clypeata*) followed by ♂; or ♀ may Head-nod, swinging head and neck back and forwards while changing location. Then stops, turns laterally in front of ♂, performs Head-pumping movements, and gradually assumes Prone-posture. ♂ responds quickly by also Head-pumping, in which ♀ may join briefly (in both sexes, display differs from *Anas* Head-pumping in that head and neck moved repeatedly forward and up). ♂ may then Bill-dip near or over ♀'s neck and allopreen her side and back, while both revolve in tight circle, before mounting. After mating, ♂ swims rapidly *c.* 3 m from ♀ while Turning-back-of-Head as ♀ bathes (Johnsgard 1965); may then perform Display-shake before also bathing, pair then leaving water to preen (Bruggers 1974). RELATIONS WITHIN FAMILY GROUP. No detailed information.

(Figs A–B after Bruggers 1974.)

**Voice.** Used infrequently, except during communal courtship and when disturbed (Bruggers 1974, on which this account mainly based). Information derived mainly from study of feral or captive birds (see also Savage 1952; Lorenz 1953; Johnsgard 1965). Calls mostly brief; differ between sexes.

CALLS OF MALE. Mostly variations of sharp, rising whistle accompanied by lower, snorting, nasal component.

4 main calls given during communal courtship, all involving whistles. (1) Basic Courtship-call: short 'prfruib' (see I), reminiscent of 'whip' call of Spotted Crake *P. porzana*, uttered up to 50 times per min, especially at start of communal courtship; also given by paired ♂ in response to Coquette-call of mate. (2) Display-shake call: soft, melodious, whirring whistle like 'fwwwww' and 'rrrrr' pronounced simultaneously. (3) Double Display-shake call: 'gnk-zit' like half-suppressed sneeze, followed by short, sneezy whistle. (4) Burp-call: long, drawn-out 'pffrrruuiehb'; same or similar call also given as contact-call when separated from mate. Following among other calls described. (5) Jibbering-call: soft, repeated twittering, audible only at close range; uttered (at up to 4 notes per 0·5 s) during Bridling-display and when disturbed (often preceding flight). (6) Quiet-call: soft, whispering 'ppffhhtt', given to ♀ prior to her entering nest-hole. (7) Alarm-call: shrill, whistling 'uib', on taking wing (Savage 1952). (8) Flight-call: brief, sharp, plaintive, whistling 'wriick' or 'hwick' (Savage 1952).

CALLS OF FEMALE. (1) Coquette-call: loud, sharp, single 'kett' or 'ke' (easily mistaken for 'kewk' call of Coot *Fulica atra*); main component of Coquette-display during communal courtship, also given by ♀ separated from mate. (2) Inciting-call: repeated, plaintive 'ack', uttered up to 53 per min; given during Inciting-display, sometimes in flight (see II), and shorter version when nervous or disturbed. (3) Clucking-call: soft, repeated, barely audible, twittering 'cluca-cluca-cluca'; response to Bridling-display of ♂, also uttered when disturbed (much as equivalent Jibbering-call of ♂)—probably same as 'Go-away' call of Lorenz (1953). (4) Hissing: uttered when disturbed late in incubation. (5) Exodus-call: soft, melodious note used to call young from nest-hole (Savage

1952). (6) Distraction-call: deep, throaty, rolling 'rrr-r-ruck' used to warn young and divert predators (Savage 1952).

CALLS OF YOUNG. Contact-calls uttered in groups of 2–3 notes. Distress-call clear and bell-like, with single peak. (J Kear.)

**Breeding.** SEASON. Britain: first eggs mid-April, main laying period end-April and early May; fledging period July. SITE. Nearly always in hole in tree, ground level to 10 m, rarely to 15 m; in bole, trunk, or branch. Rarely on ground in thick vegetation, under bush or fallen branch (Savage 1952). Nest: shallow depression; virtually no material in hole nests, grass and leaves in open sites; lined down. Building: site chosen by ♀ with ♂ in attendance. Depression or nest formed by ♀, turning body and using material within reach of nest. EGGS. Oval; pure white. $51 \times 37$ mm ($46$–$55 \times 34$–$41$), sample 30 (Schönwetter 1967). Weight of captive laid eggs 44 g (36–52), sample 100 (J Kear). Clutch: 9–12 (rarely 14). One brood. No information on replacements. One egg laid per day. INCUBATION. 28–30 days. By ♀. Starts with last egg; hatching synchronous. Eggshells left in nest. YOUNG. Precocial and nidifugous. Self-feeding. Cared for by ♀ who may lead them up to 2 km from nest to first feeding area; brooded by ♀ at night when small (Savage 1952). FLEDGING TO MATURITY. Fledging period 40–45 days. Become independent at or just after fledging. Age of first breeding 1 year. BREEDING SUCCESS. No data.

**Plumages.** ADULT MALE BREEDING. Tuft of narrow elongated feathers on forehead and crown black, glossed blue-green in front, purple-red in middle, and purple-blue behind; bordered by white at upper sides of head. Lores and area above and below eye warm cinnamon, grading to white behind eye and to chestnut on chin and throat. Narrow elongated feathers on nape blue-green, on lower cheeks and upper sides of neck chestnut with pale buff shaft-streak, variably tinged purple-red towards tip. Lower neck olive-grey, finely barred buff. Upperparts, tail, and upper wing-coverts dark grey-brown with faint olive gloss, but inner scapulars strongly glossed purple-blue, outer scapulars white with black border on outer web, spot at side of tail-base purple-red, and upper tail-coverts black with green gloss. Chest and sides of breast and upper mantle chestnut with strong purple-blue or purple-red gloss, separated from flanks by double white bar bordered black in front of closed wing. Flanks pale cinnamon with fine black vermiculations; paler towards tail, longest feathers tipped black and white. Flight-feathers sepia; outer webs of primaries edged silvery-white, tips glossed blue-green; *c.* 5 inner secondaries strongly glossed blue-green on outer webs; secondaries broadly tipped white with subterminal black border. Tertials olive-sepia; 2 outer with strong blue-green gloss on outer web and occasionally narrow white tips. Central tertial ('sail') with extremely wide warm cinnamon inner web and narrow purple-blue outer; tip of inner web bordered black, border at base and narrow shaft-streak white. Underwing and axillaries dark grey or sepia, longer inner coverts white barred sepia. ADULT FEMALE BREEDING. Forehead, crown, and tuft on hindcrown and nape grey-brown with distinct lead-grey tinge. Chin, throat,

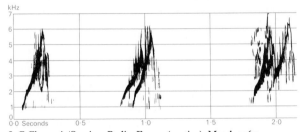

I C Chappuis/Sveriges Radio France (captive) March 1965

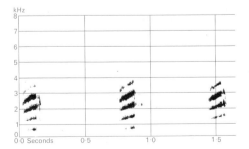

II C Chappuis/Sveriges Radio France (captive) March 1965

narrow ring round eye, and streaks along sides of base of upper mandible and behind eye to nape white, sharply defined from grey lores and cheeks, and also from sometimes considerably elongated feathers of upper neck (latter sometimes with narrow white shaft-streak). Upperparts, tail, and upper wing-coverts olive-brown, somewhat darker on upper tail-coverts. Chest, sides of breast, and flanks olive-brown to lead-grey, each feather with large pale cream oval spot towards tip. Sides of under tail-coverts and vent buff-brown, rest of underparts white. Wing like adult ♂, but less gloss on outer webs of secondaries, inner one with white bordered by black bar or spot towards base on outer web. Outer tertials sepia, outer webs strongly glossed purple-blue and narrowly edged black (sometimes with thin white border); central tertial uniform cinnamon or olive-brown, fairly wide. ADULT MALE NON-BREEDING. (Eclipse). Like adult ♀ breeding, but white parts of head less clearly defined, no white streak along base of upper mandible, and crown in some faintly glossed purple. Pattern on feathers of chest, sides of breast, and flanks less distinct, feathers brown with buff central streak instead of oval cream spot; chest and flank feathers sometimes partially vermiculated grey and white towards tips; upperparts occasionally faintly glossed green. Wing like adult ♂ breeding, but tertials cinnamon tinged bronze, outer webs of outer 2 glossed blue-green; central one rather narrow. ADULT FEMALE NON-BREEDING. Like adult ♀ breeding, but head paler grey with less distinct white pattern and no white along base of upper mandible; crest shorter, upperparts duller. Cream spots on chest and flanks indistinct, pattern like adult ♂ non-breeding. Wing like ♀ breeding. DOWNY YOUNG. Crown, upperparts, and sides dark olive-brown, longer filaments tipped olive-yellow. Spots on wing and on sides of back and rump yellow. 2 streaks behind eye sepia; sides of head, throat, and chest yellow-buff, underparts pale yellow. JUVENILE. Like adult ♀ non-breeding, but structure of feathers different; shorter and narrower with rounded tips. White underparts often with sepia shaft-streaks or dusky spot on feathers, appearing streaked or spotted; upperparts sometimes faintly streaked sepia. Tail-feathers notched at tip, bare shaft projecting. Wing like adult, distinguishing sexes, but less grey on outer webs of primaries, tips of inner webs of secondaries sepia, tertials narrower and shorter, dull olive-brown, outer partially glossed blue-green on outer web. FIRST IMMATURE NON-BREEDING. Like adult non-breeding, and difficult to distinguish when no juvenile tertials, outer tail-feathers, or feathers of underparts, back, or rump retained. Wing juvenile. FIRST IMMATURE BREEDING. New feathers like adult breeding, but 1st non-breeding feathers usually retained for considerable time; some new feathers show intermediate character between non-breeding and breeding. In ♂, feathers of crown and neck less elongated; central tertial narrower; bright colours of head and body duller, partly suffused brown. In ♀, elongated feathers of crown and neck short; head greyer, crown less deep lead-grey.

**Bare parts.** ADULT MALE. Iris dark brown with pale yellow outer ring. Bill bright red with pink or white nail, foot orange-yellow or orange-pink; both duller during eclipse, but still brighter than ♀. ADULT FEMALE. Iris dark brown. Bill brown with pink tinge (sometimes completely pink); nail and upper culmen orange-

yellow or pale grey, cutting edges grey. Foot dirty yellow. DOWNY YOUNG. Iris brown. Bill brown, nail and lower mandible pale yellow. Foot dark grey with black webs and olive-yellow streaks on front of tarsus and along sides of toes. JUVENILE. Like adult ♀. (Delacour 1959; Dementiev and Gladkov 1952; Hartert 1912–21; C S Roselaar.)

**Moults.** ADULT POST-BREEDING. Complete; flight-feathers simultaneously. In ♂, tertials first followed by head and body mid-May to July; tail about July. Flightless for *c.* 1 month between May and early August. ♀♀ probably as ♂♂, but wing starts *c.* 1 month later. ADULT PRE-BREEDING. Partial; head, body, and tertials. Starts late August in ♂, completed September–October; ♀♀ later. POST-JUVENILE. Partial; head, body, and tail. Attains 1st immature non-breeding plumage from August; often completed by late September when first feathers of 1st immature breeding appear, but tertials and some feathers of tail, underparts, back, and rump sometimes juvenile until winter. FIRST PRE-BREEDING. Partial; head, body, and tertials. In ♂♂, gradually acquired from late September, first on chest, sides of breast, scapulars, flanks, and sides of tail-base, but feathers often show some non-breeding characters, e.g. partial olive-brown suffusion. Part of 1st non-breeding plumage usually retained until spring or later. No information for ♀♀. SUBSEQUENT MOULTS. Probably like adult. (Dementiev and Gladkov 1952; C S Roselaar.)

**Measurements.** ADULT. China, Japan, and captive birds west Europe; skins (RMNH, ZMA).

| | | | | | | |
|---|---|---|---|---|---|---|
| WING | ♂ 235 | (5·30; 11) | 226–242 | ♀ 226 | (7·21; 8) | 215–234 |
| TAIL | 101 | (5·41; 13) | 94–111 | 99·3 | (3·70; 8) | 94–104 |
| BILL | 28·6 | (0·97; 15) | 27–31 | 28·2 | (0·62; 8) | 27–30 |
| TARSUS | 38·1 | (1·47; 14) | 36–40 | 36·7 | (0·94; 8) | 35–38 |
| TOE | 48·6 | (1·55; 15) | 46–52 | 47·0 | (1·25; 8) | 45–49 |

Sex difference in wing significant. JUVENILE. Wing on average *c.* 10 mm shorter, tail *c.* 20; other measurements as adults.

**Weights.** ADULT. December, USSR: ♂, 628 (571–693); ♀, 512 (428–608) (Dementiev and Gladkov 1952). ♂♂ in semi-captivity, Netherlands, up to 725 in winter (ZMA).

**Structure.** Wing rather long and narrow, pointed. 11 primaries: p9 longest, p10 0–4 shorter, p8 6–11, p7 17–22, p6 29–35, p1 90–104. Inner web of p10 and outer p9 emarginated. Inner web of longest tertial in breeding ♂ extremely widened towards tip, curved upwards to form 'sail'; slightly broadened but with sides parallel in non-breeding ♂ and breeding ♀, normal in other plumages. Tail long, slightly rounded; 16 feathers. Bill short, rather high and narrow at base, gradually narrowing to broad and rounded nail. Culmen slightly concave; cutting edges nearly straight. In breeding ♂, feathers of crown and hindneck long, narrow, and hair-like, forming dense tuft; those of lower cheeks and upper sides of neck long and pointed. In adult ♀ breeding and ♂ non-breeding and 1st breeding, feathers elongated to varying extent; shorter in other plumages. Outer toe *c.* 92% middle, inner *c.* 72%, hind *c.* 28%. CSR

PLATE 56. (*facing*) *Anser caerulescens* Snow Goose (p. 422). *A. c. atlanticus*: **1** ad, **2** juv up to 6 months. *A. c. caerulescens*: **3** ad white morph, **4** juv white morph up to 6 months, **5** ad blue morph, **6** juv blue morph up to 6 months. (PS)

Peter Scott

PLATE 57. *Alopochen aegyptiacus* Egyptian Goose (p. 447): **1** ad.rufous morph, **2** ad grey morph, **3** juv, **4** downy young. (NWC)

PLATE 58. *Tadorna ferruginea* Ruddy Shelduck (p. 451): **1** ad ♂ breeding, **2** ad ♀, **3** ad ♂ non-breeding, **4** juv, **5** downy young. (NWC)

# Tribe ANATINI dabbling ducks

(Known also as surface-feeding, puddle, or river ducks.) Fairly small to medium-sized wildfowl. About 40 species in 4 genera, all but 2nd monotypic: *Hymenolaimus* (aberrant Blue Duck, New Zealand); *Anas*; *Malacorhynchus* (aberrant Pink-eared Duck, Australia); *Marmaronetta* (Marbled Teal). South American torrent ducks *Merganetta* also often included, but here treated as belonging to separate tribe Merganettini. 36 species in *Anas*, including following main species-groups in Holarctic, some or all formerly treated as separate genera: (1) wigeons, including *A. penelope*; (2) gadwalls, including *A. strepera*; (3) true teals, including *A. crecca*; (4) mallards, including *A. platyrhynchos*; (5) pintails, including *A. acuta*; and (6) blue-winged ducks, including Shoveler *A. clypeata*. Term 'teal' used loosely in ornithological literature to indicate small ducks generally, not only in different species-groups of *Anas* but also in Cairinini. 13 species in west Palearctic: 7 breeding and 5 accidental *Anas*; plus *Marmaronetta* (breeding). Bodies fairly slender. No marked difference in size between sexes (♂♂ somewhat larger). Wings long and pointed; in flight, wing-beats less rapid than in Aythyini and other diving ducks. Tails usually fairly short, pointed; central feathers elongated in some species. Bills fairly long in most species; flattened, with distinct lamellae. Legs quite short and inserted centrally giving horizontal stance; hind toe much reduced, not lobed. Take off from water and land with facility. Walk easily but with waddling gait; able to perch well, though only a few species regularly perch in trees. Diving ability rather limited, submerging briefly with use of wings. Sexes differ in tracheal anatomy, ♂♂ having enlarged, rounded, bony bullae on left side of syrinx. Plumage of both sexes usually with bright speculum. In many species, sexes alike also in other plumage characters; majority of these relatively sombre or wholly cryptic but some quite bright—in both types, non-breeding plumage differs little from breeding. In 14 species of *Anas*, mainly migrants within temperate parts of northern hemisphere, ♂♂ only with bright plumage worn for much of year; alternates with eclipse-type non-breeding plumage during flightless period of post-breeding moult. ♀♀ of these species highly cryptic at all times. Colour of bill or foot, or both, sometimes bright. Juveniles resemble adults in non-breeding plumage. Downy young typically brown and buff or yellow, often with dark and light streaks on sides of head and light spot on each wing and on each side of back or rump.

(Remaining outline confined to *Anas*; see species' account for *Marmaronetta*.) Cosmopolitan and predominantly continental in distribution, though some island forms. Adapted for living in shallow, biologically productive waters with plenty of vegetation—marginal, submerged, and often emergent and floating. Range widely through mid-latitudes, making only limited penetration into Arctic tundra or even taiga zones and, in west Palearctic, becoming markedly localized southwards. Faster streams and unsheltered or offshore marine waters normally avoided. While some species penetrate among trees (especially flooded or swamp forests) and others tolerate wide open spaces, most prefer more or less dense fringing vegetation to chosen waters, latter being either standing or slow flowing with ready access to secure and sheltered resting and breeding places. Need for concealment when breeding or in flightless stage of post-breeding moult may require penetration, more or less deeply, within dense marginal or emergent vegetation and swamps with little open water; some species nest, at least at times, far from water. As main habitats unstable in many areas, or affected by varying human persecution, exceptional flight mobility enables reconnaissance of wide range of alternative waters and rapid shift when necessary. Vulnerable to reclamation of wetlands, especially when these few and scattered, but readily accept artificially created waters. In west Palearctic, breeding distribution complicated in some species because nesting often sporadic, especially on edges of range. *A. strepera* and *A. clypeata* have expanded in several areas in north-west Europe (former also in south Germany), perhaps linked with higher summer temperatures during period of climatic amelioration, as may also have been temporary expansions of Garganey *A. querquedula* from 1920–30. Information on breeding numbers limited because of impossibility of accurate nest counts, but large-scale winter counts organized by International Waterfowl Research Bureau now sufficiently comprehensive to provide reasonable estimates of wintering numbers and main locations, and, in some cases, tentative indication of trends in certain areas (Atkinson-Willes 1974). Some species migrate over considerable distances, especially in northern hemisphere. Moult migrations, by adult ♂♂ only, often extensive; except in *A. penelope* among west Palearctic species, no significant numbers of non-breeders involved as most birds breed in 1st year. ♀♀ moult during late summer and early autumn on or near breeding grounds. All migration mainly nocturnal, sometimes at high altitudes, often in irregular, wavy lines.

Essentially surface feeders, though will dive for food under certain conditions. Some primarily vegetarian, on land and in shallow water. Many omnivorous, however, taking chiefly seeds and invertebrates mainly from shallow water by dabbling at surface (pumping water and mud through bill, using lamellae to sieve out food), dipping head and neck below water, and up-ending; some highly specialized filter feeders, others also forage on land. Will feed singly, but more often in pairs and flocks—and otherwise usually gregarious when not nesting. Main pre-flight signals lateral Head-shaking and repeated vertical

Head-thrusting. Prior to and in initial stages of nesting, each pair typically occupies home-range which overlaps with those of other pairs. Within home-range, 1 or more smaller areas frequented for feeding, loafing, etc.; variously named—'core area', 'activity centre', 'waiting area' (where ♂ stays while ♀ at nest and where pair meet at times during laying and incubation)—defended as territories, to great or lesser extent, in some species (mainly by ♂). Monogamous pair-bonds, long-term in monomorphic resident or nomadic, often tropical species (see Siegfried 1974) but more usually of seasonal duration, especially in northern migratory ones. In latter, pair-formation typically starts in flock during autumn and winter following assumption of breeding ('nuptial') plumage, though initial pairings often temporary; final pair-bond terminated at some stage during incubation when ♂♂ flock again. In addition to maintaining firm bond with eventual mate, ♂♂ of many species also show promiscuous tendencies, displaying to other ♀♀ and also copulating with them—mainly by rape; extent of such promiscuity subject to ecological factors which affect intensity of defence of own mate and territory. Same factors also influence types and frequency of pursuit-flights of single ♀♀ characteristic of ♂ Anatini. Of 3 main types: (1) courtship-flights—chase by several ♂♂ originating from displaying party on water and initiated by ♀; (2) 3-bird flights—chase of ♀ of intruding pair by single ♂ based on own activity centre; (3) rape-intent flights—chase by several ♂♂ often ending in rape attempts. 2nd and 3rd types connected by intermediates; much controversy over details and interpretation, especially role of such pursuits in dispersing pairs. Courtship, typically on water but sometimes on land or even in flight (during pursuits), of 2 main types: (1) communal courtship (also termed 'social display') and (2) pair-courtship ('directed courtship' of von de Wall 1965). In communal courtship, often starting in autumn or winter, group of several ♂♂ typically display to 1 or more ♀♀, both unpaired and (increasingly as season advances) paired birds of both sexes taking part. Courting party develops progressively in many species, as more and more ♂♂ join in; in some, however, notably *A. platyrhynchos*, group typically assembles before display starts. ♂ displays often elaborate, consisting of secondary and major forms, ♂♂ tending first to assume special Courtship-intent posture indicative of impending display. Marked tendency for each ♂ to align body parallel to chosen ♀ prior to displaying; components of some displays also show marked directional bias towards ♀. Secondary displays, mainly derived from comfort behaviour and closely similar to latter in form, usually silent; often precede one or other of major displays. Include Upward-shake and Wing-flap (both involving brief rise as bird treads water), lateral Head-shake (with bill inclined down), and Head-flick (with vertical component most marked). Major displays often more elaborate; usually with vocal components produced by contortion of tracheal tubes which determines posture of neck. Include Water-flick (or Grunt-whistle), Head-up-Tail-up, Burp-display, and Down-up (last not usually addressed to ♀). Each species has own repertoire of displays, some of which may be combined in different sequences; may include silent Nod-swimming and Turn-back-of-Head components, latter performed as ♂ swims in front of ♀, inducing her to follow (ceremonial Leading-display)—though these also performed independently of other displays or each other. In some species, notably *A. platyrhynchos*, major displays of ♂♂ often synchronized in bursts. ♀♀ noticeably less active than ♂♂. Displays include Nod-swimming (silent) and Inciting (with characteristic calls), either of which may induce ♂♂ to display according to species. Inciting-display, though often directed at specific rejected ♂♂, of type not functional in causing preferred ♂ to attack them (unlike in Tadornini). Considerable controversy over nature of communal courtship, but now little doubt of importance in formation and maintenance of pair-bond and extra-pair relationships. Strong competition between ♂♂, arising both from often marked preponderance of that sex and from need to secure favourable positions for display relative to preferred ♀. In most species, pair-courtship distinct from communal, though elements of it often occur during latter as bonds start to form. As well as ♂ Turn-back-of-head and ♀ Inciting, includes Bill-dip, full Ceremonial-drinking, and various Mock-preen displays—notably highly ritualized Preen-behind-Wing (in which species-characteristic speculum pattern briefly exposed); other areas preened less formally include back (Preen-dorsally, Preen-Back-behind-Wing), and underparts (e.g. Preen-Belly). Other displays include Bridling-display on land (see *A. crecca*); also copulatory display and behaviour, initiated well before need to inseminate ♀ in many species and thus also associated with maintenance of pair-bond. Except sometimes in case of rapes, copulation on water. Pre-copulatory displays consist typically of mutual Head-pumping; post-copulatory displays of ♂♂ more variable but include Burp-display, Bridling, and Nod-swimming. Marked sexual differences in voice. Calls of ♂♂ variable; often weak nasal, rasping, wheezing, clucking, or rattling sounds but also include penetrating whistles (sometimes followed by grunts) in many species—uttered chiefly during display, when disturbed, aggressive, or separated from mate or flock companions. Calls of ♀♀ typically louder and coarser, often quacking; most characteristic vocalizations include Decrescendo-call (pattern of which tends to be individually constant, facilitating identification) and Inciting-call. In some species, pair call simultaneously while posturing during and after antagonistic encounters (Pair-palaver); when mates separated, often call—Decrescendo-calls from ♀♀, e.g. Burp-calls from ♂♂. Non-vocal sound signals produced in some species. Behaviour characters include mass dashing-and-diving during bathing (see *A. platyrhynchos*).

Breeding strictly seasonal in non-tropical species; most have limited breeding periods, especially those forms nesting in arctic, but more prolonged in others (especially *A. platyrhynchos*). Sites mainly on ground, concealed in thick cover often well away from water; less often in open or above ground, but cavities in trees, artificial nest-boxes, and old nests of other species used by at least 1 species. Nests usually well dispersed but sometimes grouped even quite densely, at protected sites. Shallow depressions with rim of vegetation, lined copiously with down plucked by ♀. Building by ♀ only. Eggs oval, yellowish or pinkish-white, grey-green, buff, rarely bluish; smooth. Clutches usually 6–12, averaging smaller in forms on remote islands (see Lack 1968); multiple layings sometimes occur. Replacements laid after egg loss and some species double brooded.

Eggs laid daily. Incubation by ♀ only, leaving nest 2 or more times per day when usually joins ♂ (if still present). Incubation periods usually 21–28 days (Johnsgard 1968). Young cared for by ♀ only in most species with ♂ occasionally in attendance at start; in minority (mainly southern or tropical irregular migrants with long-term pair-bonds), ♂ as well as ♀ accompanies young though only ♀ broods them (see Kear 1970; Siegfried 1974). Young aggressively defended by both sexes in latter species, but main anti-predator reaction otherwise distraction-display of ♀ in form of 'injury-feigning', parent flapping awkwardly over water or land with wings open, exposing speculum, and giving Distraction-calls. Young become independent just before or at fledging. Mature at 1 year.

## *Anas penelope* Wigeon

PLATES 62 and 74
[facing pages 495 and 543]

Du. Smient    Fr. Canard siffleur    Ge. Pfeifente
Ru. Свиязь    Sp. Ánade silbón    Sw. Bläsand    N.Am. European Wigeon

*Anas Penelope* Linnaeus, 1758

Monotypic

**Field characters.** 45–51 cm, of which body two-thirds; wing-span 75–86 cm. Medium-sized, short-necked, and compact dabbling duck with small bill, peaked forehead, pointed tail, and narrow wings; ♂ has creamy-yellow forehead and crown contrasting with reddish-chestnut head and neck, pinkish breast, grey back and flanks, white forewing and belly, and black undertail. Sexes dissimilar, with some seasonal differences. Juvenile closely resembles rather variable adult ♀.

ADULT MALE. Dominant features are yellowish forehead and crown (looking creamy-white in strong sunlight), contrasting with chestnut head and neck, and conspicuous white forewing showing as white band along side at rest. Chestnut of head, with green-glossed patch or speckling behind eye and blackish throat, appears noticeably darker than pink-brown breast and vermiculated grey upperparts and sides; white lower breast and belly, extending up into white patch at rear of flanks, contrast with black around whitish-grey tail. Elongated scapulars vermiculated grey; speculum dark green, broadly edged with black, gives strong contrast to white forewing and grey-brown primaries; axillaries and underwing typically dusky grey, but on some can appear greyish-white. In flight, large white areas on forewings, white belly, and black stern striking. Bill grey-blue with black tip; eyes brown; legs blue-grey to yellow-brown with dusky webs. MALE ECLIPSE. Superficially resembles adult ♀, but upperparts darker, flanks richer rufous, and easily identified by white forewings still showing as white band along side at rest. ADULT FEMALE. Small bill, peaked forehead, pointed tail, and more rufous and unstreaked plumage give characteris-

tic outline and pattern. 4 morphs, with tone rufous or grey and each plain or barred. Head and neck pink-brown, barred and spotted with black (but sometimes appearing grey); mantle and scapulars brown with pink-buff bars (variable in number and completeness) and edges (or latter may be deep grey). Upper breast and flanks brown with extensive pink-white or pink-buff fringes; rest of underparts white with dark marks on under tail-coverts. Axillaries greyish; underwing brown with paler markings. Wing-coverts grey-brown; speculum blackish with little green gloss (occasionally as bright as ♂) enclosed between wing bars. Bill and legs duller grey-blue than ♂. JUVENILE. Closely resembles adult ♀ and, with variability of plumage, not certainly distinguishable in field. By 1st winter much like adults, but ♂ does not usually assume white forewings until 2nd winter.

Combination of pale forehead and white forewings preclude confusion of ♂ with any other west Palearctic species except rare American Wigeon *A. americana* (which see). Also possibility of confusing ♀ or juvenile with immature Gadwall *A. strepera* which, however, is patterned more like Mallard *A. platyrhynchos* with larger and darker bill, paler and differently shaped head, and white speculum.

Breeding habitat rather specialized, but outside nesting season found on both fresh and salt water, particularly favouring larger open sheets, floods, estuaries, and muddy coasts. Rather shy. Swims fairly low in water, keeping neck retracted. Will up-end when necessary, but not habitually like some *Anas*; dives only when wounded or frightened. Walks and runs easily; frequently loafs on land and is only duck to graze in compact flocks on short grass, salt-marsh,

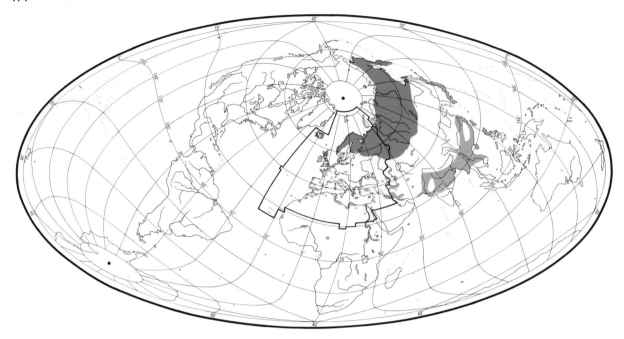

and water margins. Rises straight from water and flies with rapid beats, not twisting and plunging to same extent as Teal *A. crecca*. Long, narrow wings more depressed below body at downstroke than in Mallard *A. platyrhynchos*. Flocks overhead produce fluttering sound by braking in air. Highly gregarious in flocks or parties of up to several hundred. Flocks usually fly in tight formation, but sometimes in long straggling lines. ♂ has distinctive whistle, frequently uttered, carrying considerable distances.

**Habitat.** When breeding, concentrated in boreal and subarctic zones with slight overlap into temperate and fairly numerous occurrences, often sporadic, in steppe zone. Prefers shallow, open, broad, fresh waters, of medium quality, neither strongly eutrophic nor oligotrophic, with ample submerged or floating vegetation but without dense, emergent or marginal stands. Avoids fast-flowing streams and rivers, but tolerates saline or alkaline steppe lakes and wetlands. Predominantly lowland, within continental climatic zone; tolerant of open woodland and preferring wooded to open tundra, but thinning out towards dense forest and mountains. Uses good nesting cover in coniferous or deciduous wooded areas, as well as steppes, both near and fairly distant from water. In treeless areas, requires dense and tall stands of heather *Calluna*, bracken *Pteris*, or similar cover. Avoidance of extreme arctic climates permits early summer occupancy of breeding grounds and early departure, with rapid return movement in early autumn.

Migration to winter quarters not significantly delayed by moves, sometimes over long distances, to moulting assemblies on lakes, reservoirs, rivers, and deltas. In-

dividuals not infrequently summer in winter quarters. Winter habitat mainly in oceanic climates, lowland and largely maritime, especially along coasts where shallow, fairly sheltered waters and extensive tracts of tidal mud, sand, or salt-marsh offer sustenance and security for gatherings. Freshwater and brackish lagoons, and tracts of flooded grassland also attractive, and may be used in preference to coastal waters. Space and visibility essential; closed landscapes usually avoided. Freely uses suitable man-made lakes and regulated floodlands. Makes full use of powers of flight for daily movements in light or darkness, usually at no great height.

**Distribution.** BRITAIN. First bred Scotland 1834; marked increase and spread to northern England, with sporadic nesting elsewhere (Alexander and Lack 1944). Perhaps some recent contraction of Scottish range and apparently never permanently established elsewhere (Parslow 1967). POLAND: formerly bred Silesia (Tomiałojć 1972). EAST GERMANY: first proved breeding 1908 (Bauer and Glutz 1968).

Occasional breeding: Faeroes (*c.* 1872, 1948–50), Ireland (1933, 1953), Netherlands (1914, 1919, 1948, 1952–3, 1963), West Germany (1964), Denmark, Czechoslovakia (1967), Rumania (1967), Italy (last 1966). May have bred Belgium, and pairs seen nesting season Turkey, but breeding not proved.

Accidental. Jan Mayen, Bear Island, Spitsbergen, Kuwait, Rio de Oro, Azores, Madeira.

**Population.** Breeding. ICELAND: major decline 1961–74 at Lake Mývatn (Fjeldså 1975). BRITAIN: *c.* 350 pairs (Yarker and Atkinson-Willes 1971). FINLAND: *c.* 40000

pairs (Merikallio 1958); increased recently, especially in south (Grenquist 1970). NORWAY: decline in some areas (Haftorn 1971).

Winter. About $1\frac{1}{2}$ million birds in Europe and western Asia (Atkinson-Willes 1975).

Survival. Mean annual mortality adults ringed north-west Europe 47%; life expectancy 1·6 years (Boyd 1962). Oldest ringed bird 18 years 3 months (Rydzewski 1973).

**Movements.** Most highly migratory, though local resident populations in west Europe. A few resident Iceland, but most winter Ireland and Britain, with smaller numbers reaching other North Sea countries, France, and Iberia; a

few ringing recoveries east Canada and USA (one Barbuda, West Indies), and probably most vagrants to east North America (where rare but fairly regular) and Greenland from Iceland; several exceptional recoveries of Icelandic birds in USSR in subsequent summers indicate abmigration. British breeders apparently mostly resident or make short south-west movements (e.g. Scotland to Ireland); but rather few ringed, of which 3 recovered as juveniles Netherlands and south-west France, and 2 as abmigrants in Iceland and Russia (Novgorod). Breeding populations of Fenno-Scandia and USSR east to lower Yenisey basin and south to 50°N migrate to winter quarters in west and south-west Europe, especially West Germany, Netherlands, Britain, Ireland, France, and to lesser extent Iberia; Dutch-ringed bird recovered 108°E in Irkutsk exceptional. Autumn passage mainly along Baltic and North Sea coasts, but many overland across Europe in spring. Rather few overwinter Baltic and Denmark, when also scarce central Europe.

Those breeding west and central Siberia (east to Irkutsk) winter on Caspian and Black Seas, west to Mediterranean, especially Turkey, Greece, north Italy, and south France. Those reaching north Africa (notably Tunisia) probably also from west-central Siberia (at least one recovery). Some winter east Africa (exceptionally south to Tanzania), Arabia (rare) and Iraq (presumably also Siberian). Origins of small numbers wintering Canary Islands, Sénégal, Nigeria, and Chad unknown; maximum in this region 300 on Lake Chad winter 1969–70 (Roux 1970b).

Moult migration reported from widely separated areas; ♂♂ leave breeding grounds early and move to moulting localities where join immature non-breeders. European moult gatherings noted Estonia, south Sweden, Denmark, and Netherlands (Salomonsen 1968). In USSR, moulting flocks at many localities within and to south of breeding range, especially Volga delta, lakes in Urals, Trans-Urals and Kazakhstan, and on upper Pechora. Of ringed birds which moulted Volga delta, 20% recovered subsequently North Sea countries, 56% Mediterranean and Balkans, 24% between Levant and Caspian Sea (Shevareva 1970). Leave moulting areas on autumn migration, often moving due west or even north-west to European wintering grounds. Arrivals at and departures from nesting grounds (other than moult migration) vary to some extent with latitude and weather. Typically, flocks form August, mass departures from breeding areas September (some earlier), main arrivals winter quarters October–November. Depart from North and Black Seas mid-March to early April, earlier in mild winters, and this probably general for start of return passage; major NNE movements in USSR during April, but northern tundras not re-occupied until thaw in 2nd half May. See also Bauer and Glutz (1968); Dementiev and Gladkov (1952); Donker (1959).

**Food.** Almost entirely vegetarian, mainly leaves, stems, stolons, bulbils, and rhizomes; also some seeds and occasionally animal materials. Obtained (1) on land, by grazing while walking; (2) on water, from surface; (3) less often, under water by immersing head and neck. Gregarious, night and day-time feeder (depending in particular on disturbance and tides). May fly up to 16 km to feed (Owen 1973b). Often associates with and takes food brought to surface by other species, e.g. Coot Fulica atra, swans Cygnus, White-fronted Goose Anser albifrons, Brent Goose Branta bernicla (Ussher and Warren 1900; Millais 1902; Boase 1950; Soding 1950; Szijj 1965b).

Spring diet in USSR mainly leaves and shoots (Dementiev and Gladkov 1952). In March and April, Volga delta, chiefly aquatic vegetation: 23% tape-grass Vallisneria spiralis, 25% fringed waterlily Nymphoides pellatum, 10% water crowfoot Ranunculus aquatilus, 7·5% flowering rush Butomus umbellatus, and 10% filamentous algae. In late April, Mologa river: 70% duckweed Lemna minor, 10% Canadian pondweed Elodea, and 20% hairgrass Deschampsia caespitosa. Near Astrakhan and in steppe belt (Baraba, Ili delta), takes young orache Atriplex and thistles. On northern rivers as thaw sets in, feeds on floating rhizomes of pondweeds Potamogeton. In May, southern Yamal, eats buds of willow Salix. In Iceland, spring, large numbers of flies taken (Millais 1902). Summer diets USSR: June and July, Rybinsk reservoir, 99·7% foliage and shoots, including 35% bent grass Agrostis stolonifera, 25% bur-reed Sparganium simplex, and 25% lady's smock Cardium pratensis; in northern Aktybinsk region and Naurzum lakes (Kazakhstan), June and July, swarming locusts Calliptamus italicus (Dementiev and Gladkov 1952).

Autumn and winter diets studied by Millais (1902); Coward (1910); Phillips (1923); Madon (1935); Witherby et al. (1939); Glegg (1943); Olney (1958, 1967); Owen (1973b). On coast, mainly eel-grass Zostera, tassel pondweed Ruppia, and various algae (e.g. Enteromorpha, Ulva, Cladophora, Ulothrix). Zostera probably chief food prior to decline in 1930s (Thompson 1851; Millais 1902; Phillips 1923), though wide range of other foods previously recorded (Glegg 1943) with considerable feeding inland and on saltings (Yarrell 1843; Thompson 1851; Cordeaux 1896). Since 1930s, apparent increase in inland feeding. Items include grasses (especially Puccinellia, Poa, Festuca, Glyceria, Agrostis, Lolium, Alopecurus, Holcus); other herbs (buttercup Ranunculus, pondweeds Potamogeton, clover Trifolium, fool's watercress Apium, horsetail Equisetum, samphire Salicornia, duckweed Lemna, bistort Polygonum); occasionally seeds of sedges (Carex, Scirpus, and Eleocharis), plantain Plantago, and blackberries Rubus; and some animal materials (molluscs, crustaceans, amphibians, fish spawn). Animal items probably chiefly taken accidentally, some (e.g. cockles Cardium in winter Dornoch, Scotland) occasionally deliberately.

344 stomachs collected in 12 areas (though mainly coastal North Uist, Scotland), August–February, contained chiefly algae (especially Enteromorpha, and less Ulva, Cladophora, and others), Ruppia, Zostera, and less grass leaves, Potamogeton, Salicornia, and seeds; results influenced by method of collecting, as numbers taken at one time by punt gunning on mudflats (Campbell 1936, 1946). 112 stomachs from salt and brackish water areas, Kent, England, September to February, contained chiefly stems and leaves of grass and Potamogeton pectinatus, and algae, mainly Enteromorpha and Ulva (Olney 1965b). In Bridgwater Bay, England, September–February, diet (based on gut analyses) consisted largely of grass leaves, with roots, stolons, bulbils, and seeds of secondary importance. In early winter, Puccinellia maritima selected in preference to Agrostis stolonifera and Festuca rubra,

though often disturbance masks real food preferences (Owen 1973*b*). In Denmark, 80 birds collected throughout year, though mainly winter, contained by frequency chiefly grass and leaves (88%) and seeds (30%), and 13% animal materials (Spärck 1957). From south-east Finland, 12 stomachs, August–November, all contained seeds (mostly *Potamogeton natans* and *Eleocharis palustris*), 10 also plant shoots (especially *Equisetum fluviatile*), and 6 insects, though significance small (Tiussa 1972). Autumn foods studied in USSR (Isakov and Vorobiev 1940; Dementiev and Gladkov 1952). At Rybinsk reservoir, 98% leaves of aquatic plants (mainly *Potamogeton lucens*, 65%) and filamentous algae. In Volga delta, when moulting, chiefly green parts of plants, especially *Vallisneria*, and shoots and seeds of *Najas marina*. In north, berries (*Vaccinium myrtillus* and *Empetrum nigrum*). Depending on harshness of winter, grasses (*Poa bulbosa* and *Puccinellia*), sedges, and *Butomus umbellatus* form 39–76% of diet in Azerbaydzhan, with seeds 24–61%, mainly ricefield weeds. On east Caspian Sea coast, alga *Cladophora* and amphipods taken.

**Social pattern and behaviour.** Compiled mainly from material supplied by D A Saunders.

1. Highly gregarious except when nesting, often feeding in large flocks, at times with other *Anas*. ♂♂ form post-breeding flocks after pair-bonds ended (mainly June), aggregating for wing moult June–July; joined later by ♀♀ and fledged young. Move back to breeding area in small flocks of 25–30 (Millais 1902). BONDS. Monogamous pair-bond, prolonged but probably essentially seasonal. Pairing starts late autumn (D A Saunders), continuing through winter (Bezzel 1959). Pair-bond strong (Heinroth 1911), not usually ending until later stage of incubation, with little evidence (even in captivity) of promiscuous matings or of rapes. ♀ probably always tends young alone, however, staying with brood until about time of fledging; no recent evidence that ♂ accompanies brood other than rarely, but parental role suspected by Lorenz (1953) and earlier observers claimed that ♂ re-joined ♀ and brood immediately after hatching (see Bannerman 1968). BREEDING DISPERSION. Nesting pairs well dispersed in typical semi-wooded, freshwater habitats (Hildén 1964), but nests on island, Loch Leven, Scotland 8–34 m apart (I Newton). Size of home-range variable, some pairs making extensive foraging trips while others more sedentary; availability of local food supply probably major factor. Extent of true territorialism uncertain but ♂ readily defends ♀ against other ♂♂, especially during aerial pursuits; while ♀ at nest-site, remains at special 'waiting area' in vicinity, defending this against other ♂♂ (S-A Bengtson). Small flocks of 2–6 ♂♂ also observed in breeding area (Hildén 1964) but status not known. ROOSTING. Wintering flocks, when not feeding, often rest communally in large numbers (at times up to several thousand) at undisturbed sites such as open sea and sandbars, along coasts and estuaries, and on inland lakes. When feeding flocks subjected to persistent disturbance, such roosting largely diurnal; otherwise, both diurnal and nocturnal, often depending on tide cycles and local weather conditions which modify grazing routine (Bannerman 1968; Owen 1973*b*). Little information from breeding area, but reported as largely crepuscular, probably with bimodal peaks of feeding activity during early morning and late evening (D A Saunders).

2. Distinct from that of other *Anas* (except Chiloë Wigeon *A. sibilatrix* and, especially, American Wigeon *A. americana*), with louder and more frequent ♂ vocalizations and largely different repertoire of visual displays. Studied in captivity (Lorenz 1953; Johnsgard 1965) and, more recently, in wild (D A Saunders). FLOCK BEHAVIOUR. Large flocks often densely packed and well co-ordinated; extremely wary of avian predators. Characterized by noisy calling of ♂♂ which utter disyllabic Whistle-call when alarmed. Pre-flight signals consist of lateral Head-shakes and rapid, vertical Head-thrusting (McKinney 1965*b*). ANTAGONISTIC BEHAVIOUR. Disputes frequent in winter flocks when both sexes combine Head-forward threat-display, with open bill and quick pecks, with repeated Chin-lifting and much threat calling (multi-syllable in ♂ and continuous in ♀); similar calls given in flight. In disputes during communal courtship, and when pairs approach after pair-formation, ♂♂ threaten each other at high intensity by calling and adopting Wings-up display with wingtips crossed and folded wings raised high above back, exposing conspicuous wing-patches (see A); chases and attacks, in which opponent grabbed with bill and beaten with wings, also occur. Paired ♀♀ also extremely aggressive in similar situations; perform Inciting-display (see below), threatening component of which may develop into chase and attack. COMMUNAL COURTSHIP. General features described by Millais (1902). Absence of many typical *Anas* displays, combined with frequent hostility between courting ♂♂, gives communal courtship of this and other wigeons less highly ritualized character than that of Mallard *A. platyrhynchos* and similar species, thus in some ways resembling blue-winged ducks (Shoveler *A. clypeata* and allies). Occurs frequently in wintering area when ♂♂ compete for mates, several crowding round ♀ on water or land, displaying to her while thwarting each other's approach to ♀. At start of courting bout, ♂♂ perform Burp-displays—either intermittently or in long series. As head raised, with nape and crest feathers erect, characteristic Whistle-call uttered with bill slightly open while body aligned laterally to ♀. Secondary displays at this time include Upward-shakes and Wing-flaps (also laterally to ♀), Preen-dorsally, and Bill-dip. ♂ may also attempt to ceremonially Lead ♀, moving in front of her while in Turn-back-of-Head display—though this not so well differentiated as (e.g.) in *A. platyrhynchos* and Pintail *A. acuta*, nor typically associated with ♀ Inciting as in those species. As ♂♂ manoeuvre for position near ♀, threaten one another vigorously (see above) while orientating themselves laterally to each other on moving towards ♀. If one ♂ does not become dominant, grappling attacks break out. Eventually, dominant ♂ in group rushes towards ♀ performing unique Forehead-turning display: faces her with head erect and held slightly back, pale crown feathers erect, and bill pointed down and moved slightly from side to side. ♀'s acceptance of particular ♂ indicated by Inciting-display when other ♂♂ intrude: Chin-lifts towards favoured ♂, alternating with vigorous

A

Head-forward threats towards others while giving continuous calls. During initial pairing encounters, ♂ performs Forehead-turning display repeatedly to ♀ until they touch. PURSUIT-FLIGHTS. Courtship-flight initiated when ♀ sometimes takes wing if ♂♂ pack near her on water, group making rapid twists and turns in air as some ♂♂ drop out and others join. Typical 3-bird-flights also reported on breeding grounds but not rape-intent-flights. PAIR BEHAVIOUR. When separated from and out of sight of mate, ♂ performs long series of Burp-displays (with Whistle-call); corresponding Decrescendo-calling of ♀ brief and infrequent. In early winter, when pair-bond still recently established, ♂ greets ♀ with Forehead-turning display when rejoining her after separation. In presence of other birds, ♀ performs Inciting-display close to mate; especially in period following pair-formation, may develop into Pair-palaver if ♂ reciprocates with multi-syllable calls and weak Chin-lifting (Lorenz 1953). May be preceded by ♂'s swimming to ♀, calling, and then Head-shaking briefly before repeatedly performing series of Preen-behind-Wing displays—each time lifting wing on side nearest ♀—to which ♀ sometimes reciprocates. In following palaver, behaviour of ♂ and ♀ rigidly co-ordinated so that calling has character of duet. Preen-behind-Wing display also likely to occur after antagonistic encounters, e.g. between 2 pairs, each pair withdrawing and mates then directing display to one another (McKinney 1965b); paired ♂ also sometimes does Bill-dip followed by Preen-Shoulder display in similar circumstances, touching bill against white outer wing-coverts to side nearest ♀. Pair also orientate Preen-behind-Wing and other Mock-preen towards each other at other times, often reciprocally. Include, as well as Preen-Shoulder by ♂ only, Preen-dorsally and Preen-Back-behind-Wing. COPULATION. Lack of records in wintering area, suggest not used frequently as form of pair-courtship (as in many other *Anas*) but occurs mainly in breeding area immediately prior to and during egg-laying. Preceded by typical *Anas* Head-pumping by both sexes. In single copulation described by Adams (1947), ♀ then turned away from ♂ and assumed Prone-posture with head low over water. ♂ mounted for several seconds; afterwards, both bathed before dashing-and-diving and then flying a few metres and self-preening. Chin-lifting and calling by ♂ reported as post-copulatory display (M Schommer). RELATIONS WITHIN FAMILY GROUP. No detailed information. ♀ and brood probably most active at night (S-A Bengtson).

(Fig A after Johnsgard 1965.)

**Voice.** Radical differences between sexes. ♂ highly vocal, though vocabulary sounds limited owing to basic similarity of whistling calls. Pair may call together, giving effect of single utterance (Witherby *et al.* 1939); see below.

CALLS OF MALE. Loud, musical, 2-syllable, whistling 'whee-OO' most characteristic (see I), with both 1 and 2-syllable variants. Following distinctive (D A Saunders). (1) Typical Whistle-call: 'whee-OO' uttered singly or in long series as courtship-call, as pair contact-call (much of whistling by ♂♂ in flying flocks being of this nature), and alarm-call (often just after taking wing). (2) Multi-syllable calls: 'wip . . . wee . . . wip-weu', uttered as threat-call during disputes—presumably same as 'wheeye-irr' of Witherby *et al.* (1939); also given in flight, especially when flock changes direction, and by ♂ during Pair-palaver while ♀ utters Inciting-call (this simultaneous calling referred to above—see II). Peculiar cheeping call said to be given when frightened (Millais 1902).

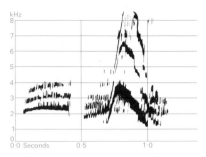

III L Ferdinand Iceland June 1967

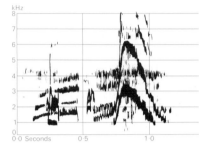

II P Bondesen Denmark June 1963

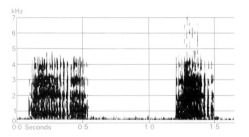

I P Bondesen Denmark June 1963

CALLS OF FEMALE. Main vocalization a distinctive purring or grating growl (see III), continually repeated and varying in loudness. Described as 'rerr' (Lorenz 1953) and 'krrr- krrr . . .' (D A Saunders); see also Delacour (1956), Johnsgard (1965). Such 'continuous calls' (D A Saunders) uttered in antagonistic encounters, often in flight, as (1) threat-call, and (2) Inciting-call. Alarm-call said to be similar, but quite distinct and harsher (Millais 1902). Decrescendo-call also similar, of 1–3 syllables of decreasing intensity (e.g. KRRR-kw'); lacks quacking quality of most other *Anas*. Further study needed on these and remaining ♀ calls.

CALLS OF YOUNG. Little information. Distress-calls of small, downy young figured in Kear (1968).

**Breeding.** SEASON. See diagram for northern Europe. Timing of thaw influences onset of breeding in far north. Britain: laying commences mid-April, main period May (Campbell and Ferguson-Lees 1972). Iceland: mean date first laying 1961–70 from 19–30 May, average 24 May; temperature April and May had some influence (Bengtson 1972). SITE. On ground, in thick cover, well concealed under overhanging vegetation, grass tussock, or scrub. Usually near water but also up to 250 m away. Not colonial, but nests as close as 5 m. At Scottish locality where 35–40 pairs breed, nests randomly spaced from 8–34 m (I Newton). Nest: shallow depression, lined grass, leaves and sometimes small twigs; much down added after laying. Diameter 20–30 cm, height 5–7 cm. Building: by ♀, using material within reach. Cup formed by turning and pressing with body. EGGS. Ovate, smooth; cream or pale buff. 55 × 39 mm (49–60 × 35–42), sample 200 (Schönwetter 1967). Weight of captive laid eggs 42 g (34–48), sample 79 (J Kear). Clutch: 8–9 (6–12). Mean of 23 Finnish clutches 8·7; range 6–12 (Linkola 1962). Mean of 357 Icelandic clutches 9·0 (Bengtson 1972). One brood. Replacement clutches laid after egg loss, with 2nd re-nesting possible; mean of 59 Iceland renests 7·2 (Bengtson 1972). Occasional dump nests occur, mean of 10 Icelandic 15·2 (Bengtson 1972). One egg laid per day. INCUBATION. 24–25 days. By ♀. Starts on completion of clutch; hatching synchronous. Eggshells left in nest. YOUNG. Precocial and nidifugous. Self-feeding. Cared for by ♀, and brooded while small. FLEDGING TO MATURITY. Fledging period 40–45 days. Independent at or just before fledging. Age of first breeding 1 year, sometimes not until 2. BREEDING SUCCESS. Of 301 eggs laid Finland, 78% hatched and 34% young reared to fledging; survival per brood in 3 years, 4·3, 1·1, and 3·3, with losses of 46%, 86%, and 59% (Hildén 1964). Hatching success of 551 nests, Iceland 1961–70, averaged 67·9% (range 40·4–80·6) with Raven *Corvus corax*, mink *Mustela vison*, and desertion main reasons for failure (Bengtson 1972). Of 148 Scottish nests, 55% hatched, 44% predated, and 1% deserted (I Newton).

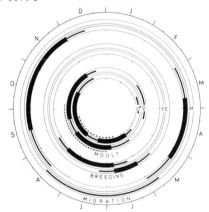

**Plumages.** ADULT MALE BREEDING. No morphs. Forehead and crown cream-yellow, fading to pale cream by wear. Chin, throat, and foreneck dull black, feathers variably chestnut at bases. Rest of head and neck chestnut, feathers of hindcrown and hindneck mottled black at tip; streak behind eye to nape with green specks, rarely all green. Upperparts, flanks, and sometimes narrow collar round lower neck finely vermiculated dark grey and white; centres of lower scapulars and feathers of rump with variable amount of grey. Central upper tail-coverts white, finely vermiculated grey at centres, lateral upper and all under tail-coverts black. Chest and sides of breast vinaceous pink. Breast, belly, vent, and sides of tail base white. Central tail-feathers black with grey bloom; others pale brown-grey, edged white. Primaries and primary coverts dark grey, browner on outer web; lesser primary coverts and tips of greater finely speckled white. Outer webs of secondaries black, basal half glossed metallic green, inner webs dark grey, edged white at tips. Outer web of innermost secondary white, narrowly margined black; rarely velvet black with dark grey streak. Greater upper wing-coverts white, tipped black (widest on outer), inner webs grey. Other wing-coverts white; innermost and marginal dark mouse-grey, latter finely speckled white. Outer webs of tertials black, inner dark grey; narrow outer edge and shaft-streak white; outer edge of inner tertials often partly suffused grey. Tertial-coverts dark grey, edged white. Under wing-coverts and axillaries white, heavily mottled and clouded pale brown-grey. ADULT FEMALE BREEDING. 4 morphs of 2 colours (grey and rufous), each with 2 types of pattern (plain and barred). In all, head and neck pink-buff to rufous-buff, boldly spotted and barred black on crown, hindneck, and behind eye, and with smaller spots elsewhere; throat and foreneck paler, sometimes uniform. Upperparts of grey barred morph olive-sepia on mantle, with feathers broadly edged olive-grey and barred pink-buff subterminally; scapulars dark sepia with buff edges and incomplete bars; back, rump, and upper tail-coverts dark grey or sepia, feathers narrowly edged pink-buff or white and similarly barred or mottled subterminally. In rufous barred morph, feathers of mantle, back, and rump sepia, with many pink-buff bars and narrow buff or white edges. In plain morphs, upperparts with fewer and usually incomplete subterminal bars, edges of feathers pink-buff or olive-grey. Underparts variable to same extent. In rufous morphs, chest and sides of breast sepia, feathers broadly tipped cinnamon-buff and barred pale pink-buff or white subterminally; feathers of chest sometimes cinnamon-buff with faint sepia subterminal bar. Flanks cinnamon-buff, usually uniform but occasionally heavily barred sepia. When fresh, feathers of chest, sides of breast, and flanks fringed white. In grey morphs, chest pale pink-buff, feathers variably tipped vinaceous-grey; flank feathers pale cinnamon, tips washed white. In all morphs, belly, vent, and sides of tail-base white, sometimes finely barred dark grey partially. Under tail-coverts black, variably fringed cinnamon, broadly edged and irregularly marked white. Tail like adult ♂ breeding, but sometimes incompletely barred buff, and feathers narrowly edged off-white to pale buff. Flight-feathers like adult ♂, but outer webs of outer 2–4 secondaries dark grey-brown with narrow white tips. Secondaries usually less glossed green; white of inner secondary often suffused grey; white of tips of black inner secondaries sometimes continued over outer web in grey line. Upper wing-coverts dark grey-brown, contrastingly edged white at tips, occasionally faintly barred and laterally suffused buff. Outer web of greater coverts tipped black (narrower than adult ♂), narrowly bordered white subterminally, often edged white at tip. Tertials and tertial-coverts dark sepia, narrowly edged and irregularly streaked buff. Under wing-coverts grey-brown, edged white; axillaries white with much fine grey speckling. ADULT MALE NON-BREEDING (Eclipse). Like adult ♀ breeding, and also with several morphs: feathers of mantle, chest, and flanks edged cinnamon, buff, or olive-grey, subterminally uniform or barred and vermiculated cream or buff. In all morphs, head, chest, flanks, and edges and bars of upperparts rich rufous-brown, rather than cinnamon as ♀ breeding; wing like adult ♂ breeding; forecrown with large round black dots, less barred than ♀; hindcrown and hindneck darker, more richly glossed green. Back and rump dull dark grey, faintly edged and vermiculated off-white. Flanks sometimes partially barred grey and white. Central tail-feathers shorter than breeding ♂, less sharply pointed, dark grey with narrow white edges. Tertials like adult ♂ breeding, but shorter and tips rounder. ADULT FEMALE NON-BREEDING. Like adult ♀ breeding, but head heavily spotted; mantle, chest, flanks, back and rump usually with only a few bars; scapulars more broadly edged cinnamon. Tail dark grey, narrowly edged off-white. Wing like adult ♀ breeding, but tertials dark olive-brown, narrowly edged off-white or partially cinnamon; outer webs often tinged black. DOWNY YOUNG. Crown, hindneck, upperparts, and sides of body dark sepia, tips of longer down filaments, bar at rear edge of wing, and often faint spots at side of back and at rump cinnamon-buff. Chest and sides of head and neck pale cinnamon. No distinct line through eye or over cheek, but sometimes dark spot on hindcheek or line from lores to below eye. Chin, throat, sides of body below wing, and belly cream-buff. JUVENILE MALE. Superficially like adult ♀ non-breeding, but mantle grey-black, feathers narrowly edged grey or pale buff, usually with narrow subterminal pink-buff bar on centre of mantle or fine white vermiculation at sides. Scapulars narrower than adult, narrowly edged buff. Back and rump dull dark grey, feathers faintly edged buff-grey; upper tail-coverts dull black, margined pale buff. Feathers of underparts narrow, on chest and flanks pale grey with faint subterminal white bars and yellow-buff tips; white feathers on belly or vent sometimes mottled dusky grey. Tail-feathers narrow, dark grey-brown, edged off-white; notched at tip, bare shaft projecting. Wing variable; secondaries like adult ♂, but outer webs usually with only slight green gloss towards base, often partly tinged dark grey, sometimes tipped white. Greater upper wing-coverts grey, outer webs with tip black, often narrowly bordered white. Median and lesser coverts grey-brown, edges and centres suffused pale buff. Wing sometimes almost like adult ♂, but at least lesser and median coverts pale grey with faint grey-buff edges and bars. Wing-coverts never contrastingly edged white like adult ♀, and secondaries more extensively black with more green gloss than juvenile ♀. Feathers of 1st immature non-breeding (resembling adult ♂ non-breeding) appear early in autumn on head and body; sides of rump become grey with white vermiculations. JUVENILE FEMALE. Like juvenile ♂, but ground colour of head slightly paler, yellow-buff; mantle usually without bars or vermiculations. Wing like adult ♀, but secondaries virtually without green gloss, sometimes sepia instead of black and usually partly marbled or streaked dark grey or sometimes mottled white on outer web. Tips of secondaries often broadly white. Greater upper wing-coverts usually without dark blotch on white tip of outer web, but sometimes some black freckling. Median and lesser coverts less contrastingly edged: tips pale buff or grey rather than white, sometimes with faint pink-buff subterminal bar; coverts darker, without pink wash on centres and edges like juvenile ♂. Inner secondary sepia, bordered by dark grey band and narrow white margin on outer web; often not as white as adult ♀ or juvenile ♂. FIRST BREEDING MALE. Like adult ♂ breeding, but juvenile plumage of back and rump often partly retained until summer or some feathers of mantle, belly, vent, or outer tail until spring. Green speckling behind eye limited; about half of birds have chest variably speckled or barred black. Wing as juvenile ♂, but tertials and often some upper wing-coverts like adult ♂ from spring. FIRST BREEDING FEMALE. Many juveile feathers retained until spring, but some birds with considerable amount of 1st immature non-breeding plumage (resembling adult ♀ non-breeding) on head, scapulars, flanks, and chest from early autumn. In spring (sometimes December), like adult ♀ breeding, but often at least juvenile back, rump, and outer tail-feathers retained. Wing like juvenile ♀; tertials and

some upper wing-coverts sometimes changed for breeding plumage (resembling feathers of mantle) in spring.

**Bare parts.** ADULT MALE. Iris yellow-brown to dark brown. Bill bright slate to leaden blue; tip, edge of upper mandible near tip, and lower mandible black. Foot light blue-grey, slate or olive-grey, or yellow-brown; webs dusky. ADULT FEMALE. Like adult ♂, but bill darker, dull slate to leaden black, foot leaden or blue-grey. DOWNY YOUNG. Iris dark brown, bill and foot leaden or blackish-grey. JUVENILE. Iris dark grey-brown or dark brown, bill and food like adult ♀; ♂ attains adult colour with development of breeding plumage (Schiøler 1925; Witherby et al. 1939; RMNH; ZMA).

**Moults.** ADULT POST-BREEDING. Complete; flight-feathers simultaneous. Head, underparts, upper mantle, and scapulars first, late May to July in ♂ and unsuccessful nesting ♀, followed by wing between late June and early September. Back, rump, and tail prolonged; start before wing, finish later. Successful ♀ similar, but later: body July to early September; wing between late July and September, sometimes October. In both sexes, but especially ♀, part of back, rump, and tail moulted only once a year. ADULT PRE-BREEDING. Partial. Starts when wing completed, involving head, body, tertials (rarely innermost secondary), and central pair of tail-feathers (sometimes others). Starts August in ♂, with back, rump, and part of mantle and scapulars; followed by head and rest of body September to early November, and tertials, central tail-feathers, and some tail-coverts October–November. About 15% of ♂♂ retain non-breeding tertials, central tail-feathers, and odd feathers elsewhere until March–April. Breeding ♀ like ♂, but starts later; most in breeding plumage by December. Relatively more non-breeding body feathers than ♂, and usually all tertials, retained until spring when tail also moulted. POST-JUVENILE and FIRST PRE-BREEDING MALE. Both partial. Limited amount of juvenile plumage changed for 1st immature non-breeding September–October. Mainly restricted to head, flanks, scapulars, chest, and sides of breast; usually replaced by breeding plumage before December. Breeding plumage appears from October (sometimes September); sides of rump, chest, flanks, and scapulars first, soon followed by head and rest of body. Marked variation in amount of breeding plumage acquired in autumn. Some in breeding by November–December, except back, rump, or some feathers on underparts; others mainly juvenile until spring, and at least back and rump until summer. Juvenile tail-feathers changed for breeding October–May (central pair first); some outer often not until July–August. Occasionally breeding tertials and some upper wing-coverts acquired from September, but often not before spring. POST-JUVENILE and PRE-BREEDING FEMALE. 1st. immature non-breeding plumage more strongly developed than ♂, but much juvenile plumage retained on underparts, back, and rump. Breeding plumage acquired from December. Lower mantle, tail-coverts, and underparts often juvenile until March, back and rump until summer. Advanced specimens grow 1st immature non-breeding plumage from September and breeding from late October; complete breeding plumage from December (except back, rump, and some tail-coverts). Retarded specimens with only limited 1st immature non-breeding plumage until February, moulting into breeding in spring. Tail-feathers as in juvenile ♂. Tertials changed for breeding in spring in advanced birds; retarded ones sometimes change a few, May–August. Occasionally, some wing-coverts changed for breeding ones in spring. (Dementiev and Gladkov 1952; Witherby et al. 1939; C S Roselaar.)

**Measurements.** Netherlands, whole year. Skins (ZMA, RMNH).

| | | | | | | |
|---|---|---|---|---|---|---|
| WING AD ♂ | 267 | (6·37; 45) | 252–281 | ♀ 250 | (4·54; 19) | 242–262 |
| JUV | 257 | (4·92; 42) | 246–266 | 244 | (6·54; 32) | 228–261 |
| TAIL AD | 106 | (3·84; 19) | 102–119 | 90·7 (2·50; 13) | | 86–95 |
| JUV | 81·7 (2·84; 15) | | 77–87 | 79·4 (3·62; 17) | | 74–86 |
| BILL | 34·7 (1·29; 84) | | 32–38 | 33·8 (1·48; 51) | | 31–37 |
| TARSUS | 39·5 (1·47; 49) | | 37–44 | 38·6 (1·42; 40) | | 35–41 |
| TOE | 49·4 (1·58; 48) | | 46–52 | 48·5 (1·50; 38) | | 45–51 |

Sex differences significant, but tarsus, toe, and juvenile tail only marginally. Tail and wing of juvenile significantly shorter than adult. Other measurements similar; combined. Tail of adult ♂ in non-breeding plumage and juvenile ♂ in 1st breeding both average 92·4, range 84–102. Tail of 1st breeding ♀ like juvenile.

**Weights.** ADULT. Various parts USSR. Average or range of averages from several areas (sample size, when known, in brackets) and total range (Dementiev and Gladkov 1952; Dolgushin 1960).

| | | | | |
|---|---|---|---|---|
| (1) | ♂ 647, 697, 725; | 400–820 | ♀ 465–650; | 400–780 |
| (2) | 701–775, 750 (4); | 620–850 | 650–750; | 620–770 |
| (3) | 735 (4); | 681–800 | 550 (4); | 530–590 |
| (4) | 695 (8), 838; | 600–1000 | 590, 615; | 530–715 |
| (5) | 872 (6), 925; | 670–1090 | 710, 748, | |
| | | | 777 (7); | 600–910 |

(1) Weight rather low in winter, November to early March (south-west Caspian, including juveniles). (2) Prior to nesting, May. (3) Decrease when nesting June–July, especially ♀♀. (4) Birds in varying stages of wing moult. (5) Peak in October. High level maintained until adverse weather. JUVENILE. On average less heavy than adult. Rybinsk reservoir, USSR, November: ♂♂ 831 (760–990), ♀♀ 712 (600–800) (Dementiev and Gladkov 1952).

**Structure.** Wing rather long, sharply pointed. 11 primaries: p9 longest, p10 0–4 shorter, p8 9–14, p7 20–30, p6 36–47, p1 114–140. Inner web of p10 and outer p9 emarginated. Tail short, rounded; 14–16 feathers, central pair elongated and pointed in ♂ breeding, short and rounded in non-breeding plumage. Bill shorter than head, rather narrow; sides almost parallel, but converging near rounded tip. Culmen slightly concave, cutting edges almost straight, nail fairly broad and rounded. Tertials long and tapering in ♂ breeding, shorter and broader at tip in ♀ and non-breeding ♂; still shorter, narrower and with more pointed tip in juvenile. Outer toe *c.* 95% middle, inner *c.* 79%, hind *c.* 29%.                                         CSR

## *Anas americana* American Wigeon

PLATES 63 and 74 [facing pages 495 and 543]

DU. Amerikaanse Smient          FR. Canard siffleur d'Amérique          GE. Nordamerikanische Pfeifente
RU. Американская свиязь          SP. Ánade silbón americano          SW. Amerikansk bläsand

*Anas americana* Gmelin, 1789

Monotypic

**Field characters.** 45–56 cm, of which body two-thirds; wing-span 76–89 cm. Medium-sized, short-necked, and compact dabbling duck with short, small bill, peaked forehead, pointed tail, and narrow wings; ♂ has white forehead and crown, glossy green patch on side of otherwise speckled grey head and neck, pink-brown back, breast, and flanks, white forewing and belly, and black undertail. Sexes dissimilar, but no seasonal differences apart from eclipse. Juvenile closely resembles adult ♀.

ADULT MALE. Dominant features white forehead and crown, and grey-buff head peppered with black, contrasting with mainly pink-brown body, and conspicuous white forewing showing as band along side at rest. Head thus paler than body, with glossy green patch from around eye to nape surprisingly inconspicuous in anything but good light. White lower breast and belly, extending up into white patch at rear of flanks, contrast with black around whitish-grey tail. Mantle and scapulars vermiculated pink and grey; back, rump, and elongated scapulars finely vermiculated grey; speculum largely black (with only narrow strip of dark green), giving strong contrast to white forewing. Axillaries and median under wing-coverts white, though this not easy to see in flight when large white areas

on forewings and white belly and black stern more striking. Bill grey-blue with black tip; eyes brown; legs blue-grey to greenish-brown with blackish webs. MALE ECLIPSE. Resembles adult ♀, but head, sides of breast, and flanks much richer and more chestnut, upperparts also rather richer, and easily identified by white forewings still showing as white band along side at rest. ADULT FEMALE. Small bill, peaked forehead, pointed tail, and more rufous and unstreaked plumage give characteristic wigeon-outline and pattern. Head and neck creamy-buff heavily spotted with dusky brown, thus appearing grey in contrast to more pink-grey breast. Axillaries and under wing-coverts white as in ♂. Otherwise similar to Wigeon *A. penelope*: although upperparts generally more strongly barred, speculum largely or entirely black, and greater wing-coverts showing slightly more black-and-white contrast, all these points within range of variation of ♀ and immature ♂ *A. penelope*. JUVENILE. Similar to adult ♀, but more heavily streaked and barred on forehead, crown, and nape. By 1st winter, much like adults, but ♂ does not assume white forewing until 2nd winter.

Combination of pale forehead and white forewing precludes confusion of ♂ with any other west Palearctic species except *A. penelope* which has yellowish forehead, reddish-chestnut head and mainly grey body. Thus head of *A. penelope* darker than body, whereas in *A. americana* position reversed. Possibility also of confusion with escapes of third closely related species, Chiloë Wigeon *A. sibilatrix* of South America, which has white face, black crown and hind part of head, dark green band from eye to nape, white patch on ear-coverts, black and white upperparts, barred black-and-white breast, rusty chestnut flanks, and white undertail (in addition, sexes nearly alike). ♀♀ and other plumages of *A. americana* and *A. penelope* much more difficult to distinguish, but head of latter usually pink-brown (never so greyish or finely marked), sides and flanks duller, and axillaries and under wing-coverts typically dusky greyish; although last difficult to see in field, *A. americana* gives impression of more white underneath. ♀ or immature Gadwall *A. strepera* has larger, orange-sided bill and white speculum.

Prefers fresh water but otherwise habits and flight closely resemble those of *A. penelope*. Generally swims higher on water. Call of ♂ usually 3 mellow whistles, with none of ringing strength of *A. penelope*.

**Habitat.** Chiefly inland low-arctic and boreal as breeding species, wintering mainly in warm temperate, sub-tropical, and some tropical coastal regions. Frequents shallow, sheltered inshore waters, estuaries, and deltas, as well as freshwater lakes, pools and rivers, and neighbouring grasslands or sometimes crops. Breeds on dry ground, often far from water, sometimes in woodland, and frequently on islands. For feeding, requires shallow water unless grazing on land. Generally similar to Wigeon *A. penelope* in habitat.

**Distribution.** Breeds across North America from Alaska, Mackenzie basin, Great Slave Lake, east to Hudson Bay, locally Ontario and Quebec, and recently as far east as Nova Scotia; south to north-east California, northern Nevada, Colorada, and Nebraska, and rarely in Michigan and Pennsylvania.

Accidental. Britain and Ireland: about 75 (including 2 recoveries); mostly single ♂♂ but occasionally ♀ with ♂, and flock of 13 in October 1968, Kerry (British Ornithologists' Union 1971; Smith *et al.* 1975). Iceland: 12; mainly May–July, others November and December (Gardarsson 1968). Also Azores, 1956 and 2 of unknown date (Bannerman and Bannerman 1966); West Germany, 1960 (Bauer and Glutz 1968); Norway, 1967 (Haftorn 1971), and Spain, 1971 (*Ardeola* 1973, **19**, 15–16).

**Movements.** Migratory. Winters mainly south of breeding range, principally coastal areas of Atlantic (south from Massachusetts), Gulf of Mexico, Pacific (south from southern British Columbia), and Neotropics south to Colombia; sporadically West Indies and Venezuela. Relatively few winter inland. Those wintering New England to Florida include many which have migrated east from prairie region, via Manitoba and Great Lakes, to Atlantic, thence southwards along coast; reverse passage in spring. ♂♂ may begin dispersing in September after wing-moult, but main movement southwards October and early November; present Colombia October–April (de Schauensee 1966); strong northward passage through USA in April but northernmost breeding grounds (Alaska, Mackenzie) not re-occupied until late May.

Regular vagrant to Europe. 8 Icelandic records May–July indicate aberrant migration (Bruun 1971). 2 trans-Atlantic recoveries of birds ringed New Brunswick: duckling on 6 August 1966 shot 2 months later, Shetland, Scotland; juvenile on 29 August 1968 shot 6 weeks later Co. Kerry, Ireland, as part of small influx. Has also straggled to Greenland, Newfoundland, Bermuda, Aleutian Islands, Hawaii, and Japan. See also Bent (1923); Linduska (1964); Godfrey (1966).

**Voice.** Resembles Wigeon *A. penelope*, but less noisy. Mellow, throaty whistle of ♂ weaker, more wheezy or lisping; uttered in groups of 3 notes with 2nd loudest—'whew-WHEW-whew' or 'whee-WHEE-whew'. ♀ relatively silent, but vocabulary closely similar to that of *A. penelope*. (Kortright 1942; Pough 1951; Delacour 1956; Johnsgard 1965.)

**Plumages.** Closely resembles Wigeon *A. penelope*, but differs in all plumages by mostly white axillaries and off-white ground colour of head and neck. ADULT MALE BREEDING. Forehead and crown pale cream, patch round and behind eye to hindneck glossy metallic-green, usually more extensive than patch sometimes seen in *A. penelope*. Rest of head and neck pale cream or buff, heavily barred and spotted black. Body like adult ♂ *A. penelope*, but vermiculations on mantle, scapulars, and flanks vinaceous-pink and dark grey rather than white and dark grey. Wing like adult ♂ of *A. penelope* but less green in speculum (usually confined to narrow band bordering greater coverts); inner secondary more extensively tinged grey, black tips of outer greater coverts wider; median under wing-coverts and axillaries white, often slightly instead of extensively, speckled dusky at tip. ADULT FEMALE BREEDING. Like breeding plumage of *A. penelope*, but ground colour of head pale buff or off-white (heads of some grey morph *A. penelope* closely similar, however), heavily streaked dark sepia; hindcrown, hindneck, and feathers behind eye to nape black with some pink-buff bars. Forehead and forecrown usually not barred as in *A. penelope*. Wing like adult ♀ *A. penelope*, but outer web of greater coverts white or pale grey with black tip, not dark grey-brown with black tip and narrow white edges and subterminal line. Inner secondary pale grey, narrowly edged white, sometimes (when, together with tertials, changed for breeding plumage) dark grey-brown with pale grey marbling and white edge on outer web. Median under wing-coverts and axillaries like ♂. ADULT MALE NON-BREEDING (Eclipse). Like adult ♀ breeding but chest and flanks rich cinnamon-brown (like eclipse ♂ *A. penelope*), and wing like adult ♂ breeding. ADULT FEMALE NON-BREEDING. Head and wing as adult ♀ breeding, body like non-breeding ♀ of *A. penelope*. JUVENILE MALE. Head and neck like adult ♀, but head more heavily streaked and crown with narrow V-bars. Body and tail like juvenile ♂ *A. penelope*. Wing like adult ♂, but at least outer and sometimes inner median and lesser upper wing-coverts pale grey-brown with pale buff or off-white centres and broad edges, and dark shaft-streaks; tertials short, sepia with off-white edges. Upper wing-coverts usually show more white than juvenile ♂ *A. penelope*, especially on greater coverts and inner median and lesser coverts. JUVENILE FEMALE. Head, body, and tertials like juvenile ♂, wing like adult ♀, but edges of median coverts narrower, pale buff, less clearly defined; outer webs of greater coverts show much sepia at bases and white at tips; outer webs of secondaries partly sepia and dark grey, virtually without green gloss, tips white. Differs from juvenile ♀ *A. penelope* in white ground colour of head and neck, and usually in more extensive black dots on tips of greater coverts. FIRST BREEDING. Like adult breeding, but juvenile wing and (especially in ♀) also many juvenile body and tail feathers retained. (Carney 1964; Kortright 1942; Witherby *et al.* 1939; C S Roselaar.)

**Bare parts.** As in *A. penelope*. Often narrow indistinct black line at base of upper mandible.

**Moults.** Apparently do not differ from those of *A. penelope*.

**Measurements.** Like *A. penelope*, but adult tail significantly longer (RMNH, ZMA).

TAIL AD   ♂  116 (7·23; 6)   101–126   ♀ 96·3   (–; 3)   96–97

**Weights.** North America; ♂ 765 (12) 650–1135; ♀ 710 (73) 510–825 (Kortright 1942).

**Structure.** Similar to *A. penelope*, including wing formula, but tail more acutely pointed in breeding plumage. Outer toe *c.* 94% middle, inner *c.* 76%, hind *c.* 26%.

## *Anas falcata* Falcated Duck

PLATES 64 and 74
[after page 518 and facing page 543]

Du. Bronskopeend    Fr. Canard à faucilles    Ge. Sichelente
Ru. Касатка    Sp. Cerceta de alfanjes    Sw. Praktand

*Anas falcata* Georgi, 1775

Monotypic

**Field characters.** 48–54 cm, of which body two-thirds, ♂ appearing bulkier; wing-span 76–82 cm. Medium-sized, large-headed, thickset dabbling duck with peaked forehead, mane on nape, unusually long inner secondaries, and straight bill; ♂ has purple-chestnut and bright green head, white throat and foreneck crossed by green collar, grey body with profuse black crescents on breast, grey and black sickle-shaped secondaries covering black and yellow-buff stern, and pale grey forewing. Sexes dissimilar, but no seasonal differences apart from eclipse. Juvenile resembles adult ♀, but distinguishable.

ADULT MALE. Perhaps most exotic of genus, with head shape recalling *Aix*. Head metallic purple-chestnut shading to almost black on crown and to brilliant bronzed green on sides of head and neck, including long, bushy, silky crest falling on to back and giving impression of extremely thick neck; in contrast, white spot above base of bill and large area of white on foreneck, latter crossed by narrow, dark green half-collar on lower part. Most of body varying shades of finely patterned silver-grey, streaked sooty on back, heavily pencilled with crescentic black on breast (recalling ♂ Gadwall *A. strepera*, but more clearly defined), and finely vermiculated on flanks and belly. Velvet-black upper and under tail-coverts with black-bordered, yellow-buff patch at sides (recalling Teal *A. crecca*), and strip of white in front; tail grey. Black on outer scapulars, forming narrow, dark braces; sickle-shaped (falcated) inner secondaries blue-black, pale grey and white; speculum black and green; forewing light grey contrasting with grey-brown primaries. Axillaries and underwing white. In flight, grey forewing adds to generally grey effect apart from darker head, neck ring, and braces;

head still looks disproportionately large. Bill black; eyes brown; legs grey with darker webs. MALE ECLIPSE. Resembles adult ♀, but head often with some green gloss, cheeks darker, upperparts less boldly patterned, inner secondaries not elongated, and forewing clearer grey. ADULT FEMALE. Superficially like ♀ *A. strepera*, but browner and darker. Short mane at back of head gives slightly large-headed appearance. Generally dark brown above, with pale streaking on head and neck, and lighter, reddish edgings to mantle, back, and scapulars; underparts yellow-brown with darker spots. Inner secondaries slightly elongated; speculum and forewing as in ♂, but duller. Bill and legs as in ♂. JUVENILE. Resembles adult ♀, but crown blacker and upperparts darker with narrower feather-edgings and more uniform nape and scapulars. ♂♂ do not become like adults until February or later.

Extraordinarily large-looking head of bronzed green and reddish-purple, with long mane and green-banded white throat, makes ♂ unmistakable. ♀ in size and proportions most like ♀ *A. strepera*, but easily distinguished by larger head, grey forewing, and green-and-black speculum, while *A. strepera* has orange-sided bill, white speculum, and white belly. Only other ♀ dabbling duck with grey forewing is much smaller Garganey *A. querquedula*.

Usually on fresh water, including lakes, rivers, and floods, but in winter quarters sometimes on sea. Extremely shy and little known. Regularly associates in winter quarters in south-east Asia with *A. strepera*, which species it most closely resembles in general character, habits, and flight. Large-headed appearance gives squat look on water, but leaps easily into air like smaller ducks, with swishing wings. Usually in ones, twos, or small parties, but occurs in large flocks on migration. Generally silent away from breeding grounds.

**Habitat.** Breeding grounds in cool, northerly, middle latitudes of eastern Asia within forest limits, chiefly in river basins and by large or small lakes, in both open and wooded terrain. In winter, also on coastline and on floodlands and rice fields, as well as lakes and rivers.

**Distribution.** Breeds eastern Siberia, north-east Mongolia, central Manchuria, Amurland, Ussuriland, Sakhalin, Kuriles, and Hokkaido (Vaurie 1965).

Accidental. Isolated records from Europe (e.g. Austria, Czechoslovakia, Sweden) probably refer to escaped birds, but following likely to be genuine vagrants (see Movements): Turkey, ♂ November 1968, Lake Van (JV); Jordan, ♀ January 1969, Azraq (Savage 1969); Iraq, ♂ March 1916 (Allouse 1953) and recorded (no details) 1968–9 winter (Savage 1969).

**Movements.** Migratory. South to south-east passage from breeding areas through Lakes Baykal and Khanka mainly during September; arrivals in Japanese, Korean, and Chinese winter quarters from mid-September. Spring departures March and early April; passage through Maritime Territories mid-March to mid-April, reaching Yakutsk late April and early May, furthest north (River Yana, Verkhoyansk) late May (Dementiev and Gladkov 1952). Not uncommon winter visitor to Upper Assam (Savage 1966); at that season not infrequently straggles to Manipur, and vagrant in various localities across north India to Kutch, Sind, and Pakistan (Ali and Ripley 1968). Thus highly probable that Middle East occurrences (see Distribution) genuine vagrants, especially since 1968–9 records there and India (near Delhi) were during season of severe drought (Savage 1969).

**Voice.** Whistle of ♂ variously described as shrill, piercing, or low and trilling; rendered as 'tyu-vit . . . tyu-tyu-vit'. Burp-call a vibrating 'rruh-urr' (see Johnsgard 1965). Quacking call of ♀ like ♀ Gadwall *A. strepera* (Ali and Ripley 1968), but Inciting-call resembles Wigeon *A. penelope* (Johnsgard 1965).

**Plumages.** ADULT MALE BREEDING. Forehead, crown, and elongated feathers on hindcrown chestnut, except for small white spot near culmen. Sides of head behind eye and elongated feathers of nape and central hindneck metallic bronze-green. Lores and cheeks glossy bronze. Collar round upper neck black, glossed green; chin, throat, sides of upper neck, and collar round lower neck white. Mantle with dense pattern of fine black and white arcs. Upper scapulars vermiculated dark grey and white. Lower scapulars pale grey finely vermiculated white, some inner ones darker grey; central edged white, some outer with large black spot on tip of outer web. Back and rump dull black, finely vermiculated grey, central upper tail-coverts finely barred dark

grey and white. Feathers of chest and upper breast white with broad black crescents (tipped cream-buff when fresh), lower breast dotted black. Flanks finely vermiculated black and white; belly and vent grey and white. Square patch at sides of tail-base white; under tail-coverts and lateral upper tail-coverts black with green gloss, former with triangular cream-buff patch. Tail dark grey (outer feathers paler). Primaries dark grey with blackish outer webs and tips; primary coverts dark grey with pale grey outer web. Secondaries black with faint blue-green gloss, some narrowly tipped white (sometimes cinnamon); outer 1–2 dark grey with narrow white edges and fine white vermiculation, outer webs of 2 innermost secondaries dark metallic-green with white shafts. Narrow, elongated, and falcated tertials black with green gloss; narrow outer edge and shaft-streak pale grey or white, broad inner edge dark grey with fine white vermiculations. Upper wing-coverts pale grey, greater with white over much of outer webs and with narrow cinnamon tips. Variable number of median, marginal, and inner lesser coverts finely vermiculated white. Greater tertial coverts pale grey. Underwing pale grey; median coverts and axillaries white, marginal coverts or tips of axillaries vermiculated or mottled grey or sepia. ADULT FEMALE BREEDING. Forehead, crown, hindneck and streak through eye, dull black, narrowly streaked buff. Sides of head and neck pale buff, rather heavily streaked dark sepia, throat and foreneck off-white, finely spotted grey-brown. Mantle dark sepia with cinnamon-pink tips and V or U-shaped marks; scapulars similar but edges broader and warmer cinnamon. Back and rump dull black, feathers narrowly barred and tipped buff; upper tail-coverts sepia with broad irregular marks and edges cinnamon-pink. Feathers of chest and flanks warm buff, each with broad dark sepia crescent and sepia central spot at base. Breast, belly, and vent pale buff; in some, spotted sepia on breast; in others, faintly spotted or streaked pale grey. Under tail-coverts buff, irregularly streaked or spotted sepia. Tail pale grey, feathers edged white; some feathers dark sepia streaked and edged buff. Wing like adult ♂, but upper wing-coverts pale grey-brown; lesser narrowly and median broadly tipped pale buff or off-white, some inner with subterminal pink-cinnamon mark; tips of greater coverts tinged pale buff, with cinnamon margin to tip; greater tertial coverts dark grey-brown, narrowly edged white. Secondaries like adult ♂, but inner 2 only faintly glossed green. Tertials straight, sepia with black tinge at sides of white shaft, outer webs tinged ash-blue basally; narrow margin and tip white. In spring, a few new tertials sometimes resemble scapulars, and some new inner median coverts like feathers of mantle. Underwing and axillaries like ♂, but centres of marginal coverts sepia. ADULT MALE NON-BREEDING (Eclipse). Like adult ♀ breeding, but forehead, crown, and hindneck black, only faintly barred buff; sides of head and neck pale buff, finely streaked sepia; mantle and scapulars dull black with olive-grey edges and fine white vermiculation or speckling towards tips of feathers. Back, rump, and upper tail-coverts dull black, narrowly marked buff; chest and flanks buff with sepia crescents (less distinct than ♀ breeding); underparts buff, indistinctly barred or blotched sepia. Wing like adult ♂ breeding, but tertials shorter (hardly curved); black with shaft-streak, narrow outer edge, and broader inner dark grey (not vermiculated). ADULT FEMALE NON-BREEDING. Like adult ♀ breeding, but head darker, heavily streaked black at sides; upperparts and flanks dark sepia with narrower warm buff edges and marks; chest warm buff, streaked or mottled black; underparts buff, indistinctly streaked grey-brown. JUVENILE MALE. Like adult ♀ breeding, but forehead and crown brown, streaked buff; rest of head and neck pale grey, finely speckled brown on sides of head; feathers of upperparts narrowly edged buff, only some scapulars with narrow subterminal buff mark; underparts pale buff (warmer buff on chest) streaked grey-brown or sepia; feathers of body short and narrow. Tail-feathers sepia; notched at tip, bare shaft projecting. Wing like adult ♂, but upper wing-coverts browner grey, without white vermiculation; white terminal part of greater coverts narrower, greyer, cinnamon margin faint or absent; up to 5 outer secondaries dark grey with white edges, not vermiculated white. Greater coverts narrower with more pointed tip, rather than broad and smooth as in adult; tertials short and straight, sepia with narrow white edges. JUVENILE FEMALE. Like juvenile ♂, but no subterminal marks on scapulars, and wing mainly like adult ♀, although upper wing-coverts somewhat darker grey-brown; median coverts narrower and less contrastingly tipped white on outer web; no subterminal marks on inner coverts; greater coverts narrower, tips pointed; black on outer secondaries restricted. Tertials like juvenile ♂. FIRST BREEDING MALE. Like adult ♂ breeding, but crest slightly shorter, belly and vent paler and less distinctly barred. Wing juvenile, but tertials with coverts like adult ♂; tertials less strongly decurved. FIRST BREEDING FEMALE. Like adult ♀ breeding, but subterminal marks on feathers of upperparts narrower and chest and flanks with less distinct arcs (body plumage intermediate in character between adult ♀ breeding and non-breeding). Wing juvenile, but tertials like adult ♀ (often some juvenile tertials retained; sometimes some inner upper wing-coverts changed for sepia ones with edges and subterminal marks buff).

**Bare parts.** ADULT MALE. Iris dark brown. Bill black. Foot dark grey, webs black. ADULT FEMALE. Iris brown. Bill dark grey, culmen and patches at sides of upper mandible black. Foot grey-brown or olive grey. JUVENILE. No information. (Delacour 1956; Dementiev and Gladkov 1952; Hartert 1912–21.)

**Moults.** ADULT POST-BREEDING. Complete; flight-feathers simultaneous. ♂♂ start head and body mid-June, followed by flight-feathers between early July and late August. No infor-

mation for ♀♀. ADULT PRE-BREEDING. Partial. Head, body, and tertials; in ♀♀, apparently sometimes also some upper wing-coverts and tail-feathers in spring. Starts when flight-feathers full-grown; complete breeding plumage apparently not attained before mid-winter (Hartert 1912–21). POST-JUVENILE AND FIRST PRE-BREEDING. Small number examined indicate that partial 1st immature non-breeding (in which juveniles resemble adult non-breeding) may be acquired in early autumn. ♂♂ attain complete breeding plumage (except wing, and sometimes back and rump) only slightly later than adult ♂♂; ♀♀ not before spring and then sometimes not completely. Moults apparently comparable with those of Gadwall *A. strepera*. (Dementiev and Gladkov 1952; C S Roselaar.)

**Measurements.** ADULT. Extreme south-east USSR, Japan; skins (RMNH, ZMA).

| | | | | | | |
|---|---|---|---|---|---|---|
| WING | ♂ 259 | (4·28; 5) | 253–264 | ♀ 242 | (— ; 3) | 237–249 |
| BILL | | 43·3 | (1·86; 9) | 40–46 | 40·5 (2·70; 5) | 38–44 |
| TARSUS | | 39·4 | (1·54; 7) | 37–41 | 38·2 (1·10; 5) | 37–40 |

Sex differences significant for bill. Wing of juveniles on average *c*. 10 mm shorter.

**Weights.** Hopeh, China: ♂♂, 713 (4) 590–770; ♀♀, 585 (5) 422–700 (Shaw 1936).

**Structure.** Wing rather long and narrow, pointed. 11 primaries: p9 longest, p10 0–4 shorter, p8 6–10, p7 16–25, p6 27–40, p1 102–121. Inner web of p10 and outer p9 emarginated. Tail short, straight; 14–16 bluntly pointed feathers. Tertials long and narrow in breeding ♂, curved in semi-circle over closed wing. Bill rather long and narrow, higher than broad at base; sides parallel, tip oval with narrow nail. Cutting edges straight, culmen slightly concave in front of nostril. Feathers of crown and hindneck elongated in ♂ breeding, up to 9 cm long, forming dense, drooping crest; slightly elongated and dense in adult ♀ breeding and adult ♂ eclipse. Tail-coverts reach tip of tail in ♂ breeding. Outer toe *c*. 94% middle, inner *c*. 77%, hind *c*. 26%.    CSR

## *Anas strepera* Gadwall

PLATES 65 and 74 [after page 518 and facing page 543]

Du. Krakeend    Fr. Canard chipeau    Ge. Schnatterente
Ru. Серая утка    Sp. Ánade friso    Sw. Snatterand

*Anas strepera* Linnaeus, 1758

Polytypic. Nominate *strepera* Linnaeus, 1758, Holarctic. Extralimital: *couesi* Streets, 1876, Fanning Islands, Pacific; extinct.

**Field characters.** 46–56 cm, of which body about two-thirds; wing-span 84–95 cm. Medium-sized, slightly built dabbling duck of uniform appearance, with rather abrupt forehead and medium-sized bill; ♂ mainly dark grey with black (and no white) around tail, white speculum bordered black in front, and chestnut wing-coverts. Sexes dissimilar (though differences less striking than in many other ducks), with some seasonal differences. Juvenile resembles adult ♀, but distinguishable.

ADULT MALE. Head and neck recall ♀♀ of *Anas* generally, with dark crown and nape and pale buff or grey sides, usually finely spotted with black (thus increasing grey effect and at close quarters showing as distinctive

peppering), though some ♂♂ have this spotting absent, crown and nape grey-brown, and forepart of head tinged cinnamon. Close views also reveal detail of delicate black and whitish vermiculations on mantle, back, and flanks, and grey scapulars, primaries and outer secondaries; upper breast barred or freckled black with white crescentic markings; lower breast and belly white with only faint darker freckling. Upper and under tail-coverts black contrasting with grey body and grey-brown tail. White speculum, often visible at rest, forms bold white rectangle at rear edge of inner wing in flight, bordered in front by broad black band and chestnut wing-coverts (though these colours usually concealed on water); in flight also, clear-

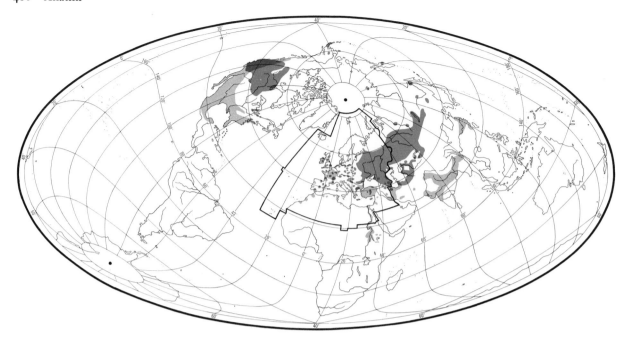

cut white lower breast and belly contrast with grey flanks and breast. Bill dark grey with touch of orange on cutting edge; eyes brown; legs dull yellow or orange with blackish webs. MALE ECLIPSE. Resembles adult ♀, but greyer with darker and less heavily marked upperparts, and chestnut retained on wing-coverts. Sides of bill become dull orange. ADULT FEMALE. Much like delicately built ♀ Mallard *A. platyrhynchos*, but greyer. Upperparts, breast, and sides brown with dark feather centres and buff edges and bars; head and neck patterned like ♂, but spotting coarser and tone somewhat browner (though still greyish in contrast to body). General effect duller and less strongly marked than ♀ *A. platyrhynchos*, with grey-brown instead of whitish tail. Best characters are clearly defined, dull orange side-panels to bill (♀ *A. platyrhynchos* only rarely shows comparable pattern), white speculum usually forming wing-patch like ♂ in flight (but some ♀♀ have only 1 white feather, as in ♀ Wigeon *A. penelope*), similarly bordered black in front, though wing-coverts show much less chestnut; clear-cut white lower breast and belly also contrast with brown flanks in flight (producing pattern like *A. penelope*). JUVENILE. Resembles adult ♀, but with boldly streaked and spotted underparts and generally rather darker upperparts. After moult, young ♂ becomes like eclipse ♂, and young ♀ like adult ♀, except that black on wing-coverts much reduced and chestnut virtually absent.

Easily overlooked among *A. platyrhynchos*, with which it sometimes associates, but uniformly grey-bodied ♂ quite distinctive and black stern not bordered in front by white as in ♂♂ of *A. platyrhynchos*, Pintail *A. acuta*, *A. penelope*, Shoveler *A. clypeata*, and Blue-winged Teal *A. discors*. Square white patch on inner secondaries unique among surface-feeding ducks, but when reduced in ♀ (see above)

possibility of confusion in flight with ♀ *A. penelope* which, however, has different shape with small bill, narrower wings, and pointed tail. ♀ best distinguished from rather similar ♀♀ of *A. platyrhynchos* and *A. acuta* by white on secondaries and by orange side-panels on bill; rather abrupt forehead, short grey-brown tail and, in flight, contrasting white belly are additional points.

Largely confined to freshwater lakes and pools, often with extensive reed cover; seldom on sea. Habits much as *A. platyrhynchos*, but tends to be rather shy and rarely in large flocks. Swims higher on water than *A. platyrhynchos*, often with elevated stern like *A. acuta*. Flies with rather pointed wings and rapid beats, recalling *A. penelope*, wings producing whistling sound. ♂ has short, deep, rasping or chuckling croak and ♀ a quack like *A. platyrhynchos* but less loud and slightly higher-pitched.

**Habitat.** Basically lowland, continental, middle-latitude species, of open rather than forest or hilly terrain. Coming from regions subject to strong climatic fluctuations, appears to be gradually colonizing more oceanic bioclimatic zones within west Palearctic (see Distribution), and so modifying habitat demands, though still averse to marine or brackish and turbulent waters. Strongly prefers fairly shallow, eutrophic, standing or slow-flowing, open water, offering plenty of cover from patches or fringes of emergent vegetation and dry banks or islands. Breeds both in open places and under thorny bushes; in USSR, on floodlands, in haystacks of previous year. Readily forages and nests some way from water, on natural grassland or heath, pasture, or crop fields.

In winter, tends towards local concentration in suitable shallow sheltered parts of large wetlands, lakes, deltas,

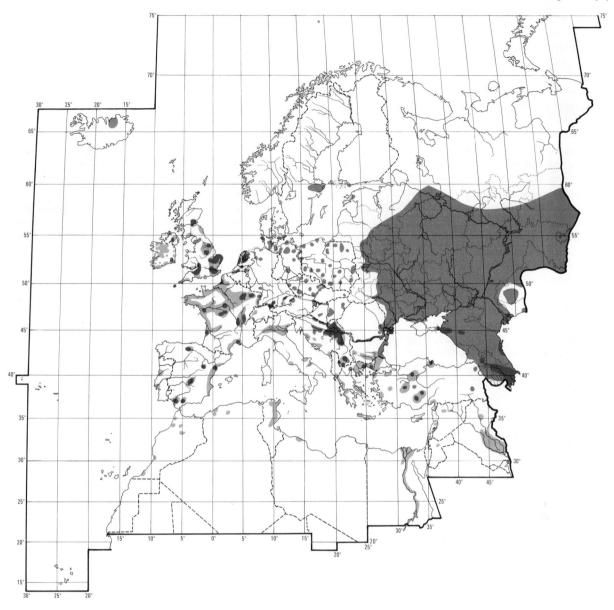

estuaries, or lagoons. Conservative in living patterns, and reluctant to shift farther than essential from regular haunts. Shows less initiative than Mallard *A. platy-rhynchos* in taking advantage of human artefacts or assistance, but relatively easy to introduce in suitable unused habitats. In England, stocks on ornamental waters and reservoirs tend to originate from introductions, though further build-up possibly assisted by settlement of wild birds. Apparently less tolerant of extreme cold, wet, or wind than others of genus. Mobile on ground as well as swimming and flying. Takes off less often in steep ascent from confined spaces than *A. platyrhynchos* or Teal *A. crecca*, but equally prone to make circling reconnaissance flights at low or medium heights, and frequent local

movements between resting, foraging, and breeding haunts.

**Distribution.** Signs of spread and increases especially in west of Palearctic range, mainly in 20th century; compared with eastward spread in North America (Henny and Holgersen 1974). ICELAND: rare 19th century; since increased greatly and spread slightly from Lake Mývatn (Gudmundsson 1951). BRITAIN AND IRELAND: first nested Scotland 1906 (where increased, at least initially) and Ireland 1937 (where still sporadic breeder); introduced England (Norfolk) *c.* 1850 and has since spread, with small groups established elsewhere, often by introductions or escapes (Parslow 1967). FRANCE: no breeding record before 1920; first nested Argonne 1925, Camargue 1927, Dombes 1936, Brenne 1960s (FR). WEST GERMANY: first bred Bavaria 1930, east Holstein 1931, Württemberg 1934, Lower Saxony c. 1935 (Bauer and Glutz 1968). DENMARK: few breeding records (Fog and Joensen 1970); regular in recent years in small numbers (TD). USSR: nested Estonia from 1960 (Onno 1970); sporadic other Baltic republics (Kumari et al. 1970); elsewhere precise distribution often not known, but sporadic or rare in several areas (Dementiev and Gladkov 1952).

Occasional breeding. NORWAY: 1965, 1966 (Haftorn 1971). FINLAND: may have bred 1927, 1954, and 1967 (Merikallio 1958; Grenquist 1970). SWITZERLAND: 1959, 1961–3, and probably in other years (Bauer and Glutz 1968). ITALY: nests occasionally in north, most recent 1973 (SF). NORTH AFRICA: only definite breeding record Lake Fetzara, Algeria (Heim de Balsac and Mayaud 1962), but may occasionally nest Morocco (JDRV).

Accidental. Faeroes, Finland, Malta, Kuwait.

**Population.** Breeding. ICELAND: estimated 200–300 pairs, 1975 (A Gardarsson). BRITAIN: 100–200 pairs (Atkinson-Willes 1970a). FRANCE: 300–500 pairs, with over half in Dombes (FR). NETHERLANDS: c. 80 pairs 1970; considerable increase since, perhaps 200–240 pairs 1973 (JW). WEST GERMANY. About 730 pairs (Szijj 1973). Bavaria: population in south increased from 2 pairs 1930 to c. 230 pairs 1965 (Bauer and Glutz 1968). EAST GERMANY. About 210 pairs (Isakov 1970a). Mecklenburg: increased since 1939. Oberlausitz: c. 30–50 pairs, highest numbers around 1930 (Bauer and Glutz 1968). DENMARK: now c. 10 pairs annually (TD). SWEDEN: c. 30–75 pairs (Curry-Lindahl et al. 1970). POLAND: local and scarce (Tomiałojć 1972). CZECHOSLOVAKIA: several hundred pairs, apparently increased and spread last decade (Hudec 1970; Hudec and Černý 1972). USSR: about 163 000 pairs with largest concentrations west Palearctic in Volga–Kana region (10000 pairs), Dniester–Dnieper (32000 pairs), Lower Volga (12 000 pairs), and North Caspian (15 000 pairs) (Isakov 1970a, b). GREECE: 100–200 pairs (WB). SPAIN. Increased last 15 years, due to new reservoirs (Bernis 1972). Marismas: 20–50 pairs in 1950s (Valverde 1960); somewhat increased since (Ree 1973).

Winter. Estimated 10 000 birds in north-west Europe, and 50 000 in Europe–Black Sea–Mediterranean region (Szijj 1972). USSR: estimated 109 200 birds in not so severe winters (Isakov 1970c).

Survival. Oldest ringed bird 13 years (Wildfowl Trust).

**Movements.** Migratory in north and east parts of west Palearctic range, largely resident elsewhere. Icelandic breeders winter Ireland and Britain. Breeders from Scotland move to Ireland (chiefly) and England; but those of England, believed mostly descended from captive stock, apparently rather sedentary, though a few young ringed East Anglia recovered Spain, Italy, Mediterranean and Atlantic coasts of France, Netherlands, and Denmark, some possibly due to abmigration. Those breeding north Germany, Poland, south Sweden, and west-central Russia winter North Sea, chiefly Netherlands and Britain, with some as far as Mediterranean. One autumn-ringed Netherlands exceptionally recovered Kemerovo, USSR (86°E), May. Breeders of Netherlands and France perhaps resident. Larger populations breeding central Europe (Austria, Hungary), Balkans, and south-central Russia to west Siberia, move south to Caspian and Black Seas, Greece, Turkey, Nile valley (Egypt and Sudan), rarely east Africa south to Kenya and Tanzania (Backhurst et al. 1973); some through Iran and Iraq to Persian Gulf. Black Sea and east Mediterranean now main European winter haunts. Few penetrate west and central African tropics (recorded Upper Volta, Nigeria, Chad), origins unknown.

Small moulting flocks of ♂♂ reported from several localities, but extent of movement little known. Up to 400 on Ismaninger reservoir, Bavaria (Bezzel 1964), include ♂♂ from south Germany, Austria, and Czechoslovakia; other European moulting grounds include IJsselmeer and Bodensee (Salomonsen 1968). In USSR, important Volga delta moulting area draws ♂♂ from valleys of Volga, Tobol and Irtysh, up to 2400 km distant; other moulting flocks on Rybinsk reservoir and Kazakhstan lakes. Moult migration begins early June while ♀♀ incubating; flocks reach peak numbers early July, dispersing late July and early August. ♀♀ and young leave nesting areas from early August; autumn passage leisurely, at peak west Europe October–November; majority in winter quarters by mid-December. Timing of return migration little known, but passage through central Europe chiefly March–April. See also Bauer and Glutz (1968); Dementiev and Gladkov (1952).

**Food.** Chiefly vegetative part of plants, obtained mainly while swimming with head under water, less often by up-ending; rarely from surface (Szijj 1965b). Also, occasionally grazes and takes cereal grains on land (Macgillivray 1852; Dementiev and Gladkov 1952; Szijj 1965b; Bauer and Glutz 1968). Obtains some food by parasitising other species: Red-crested Pochard Netta rufina, Goldeneye Bucephala clangula, Coot Fulica atra (Jauch 1952; Eggenberger 1953; Hudson et al. 1960). On Bodensee, Germany in autumn when water depths increased, c. 300 completely dependent for food on F. atra bringing stonewort Chara to surface (Berthold 1961). Normally daytime feeder in small to large flocks.

Plant materials include roots, leaves, tubers, buds, and seeds of pondweeds Potamogeton, sedges Carex and Scirpus, rushes Juncus, hornwort Ceratophyllum, bur-reed Sparganium, wigeon-grass Ruppia, grasses (including Glyceria), and stoneworts Chara; also algae. Animal material (probably mainly accidental) includes insects, molluscs, annelids, small amphibians and spawn, and small fish (Macgillivray 1852; Witherby et al. 1939;

Dementiev and Gladkov 1952).

Palearctic data sparse. 2 stomachs, North Uist, Scotland, January and February, contained *Ruppia*, alga *Cladophora*, and seeds of marestail *Hippuris vulgaris* (Campbell 1947). 9 stomachs, western Europe, all contained aquatic plant parts, and 2 seeds (Madon 1934). 4 stomachs, flooded plains Tisza river, Hungary, contained *Chara* spores, and seeds of *Carex*, *Potamogeton*, and orache *Atriplex* (Sterbetz 1969–70). From USSR, seasonal and area differences apparent. At Rybinsk reservoir, vegetative parts predominate (over 99%) throughout season; chiefly duckweed *Lemna* in May, *Sparganium* leaves June–August, and *Potamogeton* leaves October. In Volga delta, March and April, vegetative parts 88·4% by volume (25·3% rhizomes and leaves wild celery *Apium*, 16·3% *Ceratophyllum*, and 15·6% filamentous algae), seeds 11·3%, and animal materials 0·3%; during moult, exclusively vegetative parts, mainly rockcress *Najas* and *Apium* leaves. In summer, north Kazakhstan, large numbers Orthoptera (locusts) eaten from water surface and by foraging 1·5–2 km into steppes; later, many flight to cereal (wheat, buckwheat, millet, rice) fields. In early autumn, Baraba steppes, chiefly fennel-leaved pondweed *P. pectinatus* and arrowhead *Sagittaria* tubers (Dementiev and Gladkov 1952). In winter, southern Caspian, when green parts scarce, largely seeds (81·4% by weight), especially cereals, *Ruppia*, *Potamogeton* and *Scirpus* (Isakov and Vorobiev 1940).

American data suggest similar diet pattern and similar species (e.g. *Potamogeton*, *Ruppia*, *Carex*, *Ceratophyllum*, *Zannichellia*) with vegetative parts predominating (Mabbott 1920; Gates 1957; Anderson 1959; Keith 1961; Serie and Swanson 1972).

Ducklings north Kazakhstan, USSR, contained locusts, chiefly *Calliptamus italicus* (Dementiev and Gladkov 1952). In southern Alberta, Canada, ducklings ate chiefly surface invertebrates during first few days, gradually changing to aquatic invertebrates and plants, until essentially herbivorous by 3 weeks old. Change of diet parallels changes in feeding behaviour and sites, i.e. surface-feeding replaced by subsurface-feeding (Sugden 1973, based on 167 stomachs).

**Social pattern and behaviour.** Compiled mainly from material supplied by F McKinney and M Schommer.

1. Gregarious except when nesting. Flocks mostly small in winter (up to *c*. 30). ♂♂ first to form post-breeding flocks, after pair-bonds ended (see Movements). Migrant populations often form large flocks prior to autumn departure, and again before return to breeding area; migrate mostly in small parties (10–40), in spring consisting largely of pairs. BONDS. Monogamous pair-bond of seasonal duration typical, but with some promiscuity. Slight surplus of ♂♂ in breeding populations studied by Gates (1962) and Duebbert (1966). Extent of promiscuous copulations not finally established though paired ♂ at times shows strong interest in other ♀♀ (see below); some evidence also that ♀ will pair with different ♂ for replacements, some 50% of first nestings being unsuccessful (Gates 1962). Pair-formation occurs in flock,

starting late July and August, continuing until May; in Bavaria, 90% of ♀♀ paired by late October (Bezzel 1959). Pair-bond ends soon after laying completed (Hochbaum 1944) and usually before mid-incubation, but sometimes persists to hatching; timing related to date of laying, mates of late-nesting ♀♀ leaving earliest in incubation phase (Gates 1962; Duebbert 1966; Oring 1969). Only ♀ tends young. Broods remain distinct (Gates 1962). BREEDING DISPERSION. ♀♀ home to previous breeding areas (Sowls 1955; Gates 1962). Pairs arrive on breeding grounds at least 3–4 weeks before egg-laying. Unpaired ♂♂, usually in groups of 2–3, often attempt to associate with pairs (Dwyer 1974). In Utah, USA, 17 days between arrival and establishment of breeding home-range, followed by 11-day pre-nesting period during growth of suitable nesting cover (Gates 1958, 1962). In typical habitats, pairs well dispersed, each occupying own home-range often centred on ♀'s nest-site of previous year and within which located 1 or more preferred feeding and loafing spots ('activity centres'); these defended as territories, much as in Shoveler *A. clypeata* (Dwyer 1975). During laying and incubation, ♀ uses only centre nearest nest, ♂ becoming less and less faithful to original home-range and finally deserting it (and mate) (Gates 1962; Dwyer 1974). Nests usually well dispersed on mainland but on islands sometimes densely aggregated (see e.g., Hammond and Mann 1956; Duebbert 1966). Home-ranges generally smaller than those of Mallard *A. platyrhynchos* and Pintail *A. acuta*, but larger than in *A. clypeata* (Gates 1962). In one mainland population, mostly 14–35 ha (mean 27), decreasing as season progressed; position reversed in one island population, with small, discrete territories initially and eventually home-ranges in order of 'several hundred acres'—in both cases size of home-ranges dependent on distance between suitable activity centres and nest-sites (see Gates 1962 and Duebbert 1966). ROOSTING. Little information. Before incubation, pairs on breeding grounds use loafing areas in channels and ditches (Gates 1962) or on open water (Duebbert 1966) during day when not feeding, and presumably at night. At Delta, Manitoba, in July, moulting ♂♂ spend most of day sitting along shore preening and sleeping; feed mostly in early morning and late evening (Oring 1969). During winter, communal roosting probably both diurnal and nocturnal depending on local feeding regime. Like other *Anas*, prefers to rest out of water (on bank, etc.) when safe to do so. Nocturnal courtship not uncommon.

2. In some respects, intermediate between wigeon-group (*A. penelope* and allies) and mallard-group (*A. platyrhynchos* and allies). Studied in wild (Gates 1958, 1962; Dwyer 1974, 1975) and in captivity (Lorenz 1953; von de Wall 1963; Johnsgard 1965; M Schommer). FLOCK BEHAVIOUR. Little information. Judging from spring observations in breeding area (Dwyer 1974), spatial pattern of pairs suggest maintenance of individual-distance in flocks. Head-shaking and Head-thrusting main pre-flight signals (McKinney 1965b). ANTAGONISTIC BEHAVIOUR. During disputes in flock, main display of both sexes Open-bill threat with neck extended and mandibles apart, bird approaching rival thus as if to attack. ♂ also makes lateral Head-jerking movements similar to more ritualized Inciting-display of ♀; in latter, neck retracted and bill repeatedly pointed sideways towards adversary while Inciting-call given. Chin-lifting by both sexes also frequent, especially during close-quarter disputes between pairs and when pair reject advances of unpaired ♂. Often takes form of Pair-palaver involving Chin-lifting by both and Inciting by ♀. When Chin-lifting, neck stretched up with bill raised obliquely, opening and shutting as multi-syllable calls given. Fighting of ♂♂, both on land and water, generally similar to *A. platyrhynchos*. After encounter, ♂ reported to perform Mock-preens: Preen-

dorsally and Preen-Back-behind-Wing (McKinney 1965*b*). COMMUNAL COURTSHIP. Begins early, being already intense by mid-August (Lebret 1961), and continues until May or June; main peak August–October, with smaller one March–April (Bezzel 1959). In courting party, ♂♂ (calling noisily) tend to group centrally with ♀♀ on periphery. Major courtship displays include Water-flick (Grunt-whistle), Head-up-Tail-up, and Down-up much as in *A. platyrhynchos*, plus Burp-display, Head-high display, and Chin-lifting; all accompanied by calls. Secondary displays include Upward-shake and Head-flick; Tail-wag also frequent as independent activity as well as being integral part of some displays, e.g. Water-flick. Other display activities of ♂ less well known; include Preen-dorsally and other Mock-preen displays, Bill-dip, and probably Wing-flap. Antagonistic encounters between ♂♂ also frequent. Courtship bouts appear less tightly organized than in *A. platyrhynchos*, each ♂ displaying independently. Most displays performed with body parallel to ♀; feathers of forehead and crown erected and those on nape flattened laterally to produce darkened, triangular disc-shaped patch. ♂♂ also adopt rounded-looking Courtship-intent posture at times with necks retracted and head feathers uniformly erected. In Burp-display, ♂ stretches neck moderately erect, gives Burp-call, and immediately lowers neck. In Head-high display, normally facing ♀, same posture much more sustained and Flute-whistle given (or ♂ remains silent). In Water-flick display, movements slower than in *A. platyrhynchos* and terminal Tail-wag often accompanied by Head-shake or Head-flick, usually with neck held obliquely forward. In both Head-up-Tail-up (see A) and Down-up displays (see C), movements less exaggerated than in *A. platyrhynchos* but effective use made of exposed white speculum and conspicuous black area at rear-end of body. Though Head-up-Tail-up often and Down-up occasionally performed independently, usually linked by transitional Turn-Head-to-♀ component of Head-up-Tail-up display (see B). Unlike in *A. platyrhynchos*, no Nod-swimming associated with Head-up-Tail-up complex; Turn-back-of-Head display largely performed independently. Most conspicuous activities of ♀ during communal courtship, sequence of Chin-lifting directed at favoured ♂ (who may reciprocate) and Inciting movements away from him towards 1 or more rejected ♂♂. PURSUIT-FLIGHTS. (1) Courtship-flights. Originate from courting party on water in winter and spring if paired ♀ takes wing when pressed too closely by unmated ♂♂, these and mate following in swift, twisting flight typically lasting several minutes (Lebret 1955; Wüst 1960; Gates 1958; Dwyer 1974). In air, ♂♂ in turn manoeuvre to get near ♀, each then performing aerial version of Burp-display; much fighting between ♂♂ as ♂ paired to ♀ attempts to drive others from mate while she Incites. (2) 3-bird-flights. Of usual type: ♀ of one pair chased by ♂ of another followed by mate of ♀. First common in spring, especially after pairs dispersed in home-ranges where resident ♂♂ initiate such flights from activity centre against intruding pairs; frequency, intensity, and duration increase as time of laying approaches, then decrease after clutch half-completed (Dwyer 1974). If intruders on water, resident ♂ either alights close to other ♀ while assuming Chin-lift posture and calling or, later in pre-nesting period, sometimes lands directly on her. In ensuing flight, intruding ♂ persistently tries to come between mate and pursuer, frequently attacking latter as ♀ Incites; just before and during laying, however, chase so vigorous that intruding ♂ seldom catches up. Resident ♂ returns to activity centre afterwards, typically giving 1 or more Burp-calls as he glides in. Makes fewer and shorter pursuits if mate absent. (3) Harrying-flights. Later in laying period and during incubation, ♀♀ subjected to chasing, by

up to several ♂♂ together, resembling rape-attempt-flights of certain other *Anas* (Hochbaum 1944; Gates 1962; Duebbert 1966; Dwyer 1974); especially common after mates desert sitting ♀♀, latter being particularly vulnerable when returning to activity centres from nests. Flights often prolonged with much fighting between ♂♂ and intense escape manoeuvres by ♀; forcing down of ♀ and attempted rape apparently rare, however, and Repulsion-behaviour of ♀ infrequent or absent. Rape attempts probably more frequent where birds artificially crowded (see Wüst 1960). PAIR BEHAVIOUR. After returning to activity centre from 3-bird-flight, ♂ rejoins ♀ while performing Burp-displays (Dwyer 1974). Contact between separated mates achieved by repeated Burp-calling by ♂ and frequent Decrescendo-calling by ♀; as in other *Anas*, ♀ also engages in Persistent-quacking in spring (see *A. platyrhynchos*). Waiting ♂

A

B

C

alone at activity centre while ♀ laying spends over 50% of time Burp-calling (Dwyer 1974). Highly ritualized Pair-palaver most common feature of pair-courtship, particularly in circumstances outlined earlier. Described by Lorenz (1953) who stressed general similarity to homologous ceremony in *A. penelope*. Pair-palavers on meeting of 2 or more pairs at times initiate communal courtship (in which other birds join); during ensuing bout, paired ♂♂ direct displays to own mates. During pair-formation and subsequently, Mock-preening displays also frequent, especially Preen-behind-Wing by ♂ and, to lesser extent, by ♀; often performed as greeting after separation, especially during antagonistic encounters. Preceded by Bill-dip and Head-shake. Bill-dip and full Ceremonial-drinking also occur independently, again mainly when mates meet after separation. Other Mock-preen displays include Preen-dorsally, Preen-Back-behind-Wing (McKinney 1965*b*), and, on land, Preen-Belly (M Schommer). Pair also touch each other briefly with bill at times. From initial stages of pairing, not infrequently engage in pre-copulatory display together; often leads to full copulatory sequences. COPULATION. Pre-copulatory display and copulation much as in other *Anas*. Afterwards, ♂ (1) usually Bridles (see *A. platyrhynchos*), (2) swings round to face ♀ as she starts post-copulatory bathing, (3) performs Burp or Head-high display, (4) bathes. RELATIONS WITHIN FAMILY GROUP. Little information, but calls and behaviour of ♀ and young said to be closely similar to *A. platyrhynchos*. Chin-lift when excited, at times apparently in form of Family-palaver (Lorenz 1953).

(Figs A–C after Johnsgard 1965.)

**Voice.** Distinctive, especially in ♂. As with other *Anas*, pronounced dichotomy between calls of ♂ and ♀. Those of ♂ variously described (see, e.g. Witherby *et al.* 1939) mainly as often loud, deep, croaking, raucous, grunting, and reedy; often combined with whistles, latter also being given on own. Calls of ♀ mainly quacking and chattering. Non-vocal sounds not mentioned in literature.

CALLS OF MALE. Following distinguished by M Schommer, consisting mostly of various whistles and grunts. First 5 calls typical of communal courtship, though some also uttered at other times. (1) Flute-whistle: series of 2 (1–4) low whistles separately uttered; given during Head-high display. Rendered 'ööii' (von de Wall 1963). (2) Burp-call: 2 syllables, consisting of short, low whistle immediately followed by loud, short grunt (see I); component of Burp-display, given also in variety of situations other than communal courtship, e.g. after copulation and by lone ♂ in absence of mate. (3) Call accompanying Water-flick display (typical Grunt-whistle call): 2 syllables, consisting of long, high whistle followed by long grunt (see II). Described by Johnsgard (1965) as 'zee-raeb'. (4) Call accompanying Head-up-Tail-up —Down-up sequence: multi-syllabic, consisting of grunts and whistles. Described by Johnsgard (1965) as 'raeb-zee-zee-raeb-raeb', first 'raeb' component associated with Head-up-Tail-up display, rest with Down-up (M Schommer). (5) Chattering-call: complex and con-tinuous, consisting of long, quickly repeated series of whistles of various types and short grunts (see III); accompanies Chin-lifting display in Pair-palaver (see below) and lateral Head-jerking threats. Described by

Lorenz (1953) as 'oeh oe-ee-oe oeh'. (6) Alerting-call: separate repetitions of faint, very short, throaty clicks, rather like 'took' call of ♂ Shoveler *A. clypeata*.

CALLS OF FEMALE. Vocabulary not fully described. (1) Decrescendo-call (see IV): similar to *A. platyrhynchos* but with fewer syllables (Lorenz 1953), more rapid sequence of notes (Johnsgard 1965), and more nasal tone (M Schommer). Given in much the same situations as *A. platyrhynchos*, e.g. by lone birds in absence of mate. This or Alarm-call, or both, evidently loud quacking often mentioned in literature. (2) Alarm-call: similar to last but more uniform in cadence and loudness (Johnsgard 1971). (3) Inciting-call: rapid, multisyllabic chattering with no clear pattern (M Schommer). In Pair-palaver, ♂ and ♀ often give respective Chattering and Inciting-calls simul-taneously (Lorenz 1953). (4) Maternal-calls: vocabulary said by Lorenz (1953) to be similar to that of *A. platyrhynchos*, including 2-syllable Contact-calls (with Chin-lifting).

CALLS OF YOUNG. Contact-calls uttered in groups of 2–3 notes. Distress-call similar to *A. platyrhynchos* (J Kear).

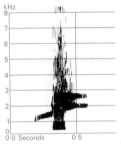 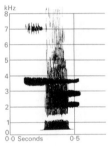

I M Schommer England March 1974     II M Schommer England March 1974

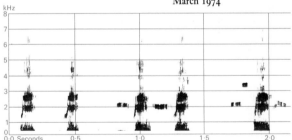

III M Schommer England April 1974

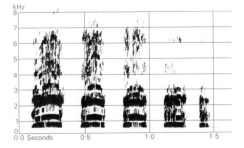

IV M Schommer England December 1975

**Breeding.** SEASON. See diagram for north and central Europe. Onset of laying in southern Germany probably governed by spring temperature (Bezzel 1962). SITE. On ground, usually close to water (less than 20 m) but up to 100 m on islands. In dense vegetation, under overhanging tussocks, thick bushes, or in tall grass or rushes. Occasionally in open but then normally in colonies of gulls *Larus* or terns *Sterna*. Not colonial, but nests as close as 5 m. At Scottish locality with 40–45 pairs, nests evenly spaced 15–25 m apart (I Newton). Nest: slight·hollow, lined grass and leaves to form low rim; lined down. External diameter 20–30 cm, internal 18–20 cm, cup up to 7 cm deep. Building: by ♀, gathering material within reach of nest and shaping cup with turning movements of body. EGGS. Blunt oval; pale pink. 55 × 39 mm (51–59 × 35–44), sample 200 (Schönwetter 1967). Weight of captive laid eggs 44 g (35–55), sample 100 (J Kear). Clutch: 8–12 (6–15). Of 73 Czechoslovakian clutches: 6 eggs, 3%; 7, 7%; 8, 8%; 9, 23%; 10, 27%; 11, 12%; 12, 12%; 13, 4%; 14, 3%; 15, 1%; mean 9·96 (Balat and Folk 1968). One brood. Replacement clutches if eggs lost. INCUBATION. 24–26 days. By ♀. Commences on completion of clutch; hatching synchronous. Eggshells left in nest. YOUNG. Precocial and nidifugous. Self-feeding. Cared for by ♀ and brooded while small. FLEDGING TO MATURITY. Fledging period 45–50 days. Independent at or just before fledging. Age of first breeding 1 year. BREEDING SUCCESS. Of 499 eggs laid in 54 Czechoslovakian nests, 59·9% hatched; in 34 successful nests 92·6% of eggs hatched (Balat and Folk 1968). In Germany, mean brood size of 105 7-day old young 8·97, of 88 14-day 7·09, of 78 21-day 6·84, and of 54 aged 28–49 days 6·08 (Bauer and Glutz 1968). Of 168 Scottish nests, 43% hatched, 55% predated, and 2% deserted (I Newton).

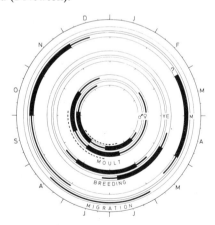

**Plumages.** ADULT MALE BREEDING. Crown and hindneck black, barred cinnamon-buff. Rest of head and neck pale buff or grey specked black (heaviest eye to nape, limited on chin and throat). Some ♂♂ have crown and hindneck uniform grey-brown with strong cinnamon tinge on former; ground colour of rest of head buff or pink-cinnamon. Mantle, upper and outer scapulars, and flanks dull black or dark grey with narrow white or pale buff

vermiculations; lower inner scapulars buff with dark grey centres, outer web sometimes vermiculated. Back and upper rump black with olive tinge, finely vermiculated or speckled off-white. Lower rump and tail-coverts black. Feathers of chest and sides of breast black with narrow white crescents. Upper breast white with wide black crescents; lower breast, belly, and vent white, faintly vermiculated grey on lower belly and vent. Tail dark grey, some feathers margined black; outer narrowly edged off-white towards tip. Primaries with coverts, inner webs of all secondaries and outer webs of outer 2–3 dark mouse-grey; primaries with paler grey inner webs and bronze tinged tips. Outer webs of inner 2 secondaries white, 3rd white, usually with broad black outer edge, others black. Greater upper wing-coverts black, few outer chestnut or more often mouse-grey with variable black edges and white or chestnut tips. Outer median coverts chestnut, inner black; lesser coverts dark mouse-grey, marked and tipped chestnut; marginal coverts dark mouse-grey, finely vermiculated or freckled cream. Tertials pale ash, narrowly edged white on outer web; greater tertial-coverts with inner webs mouse-grey and outer webs black. Under wing-coverts and axillaries white. ADULT FEMALE BREEDING. Crown and indistinct streak from base of upper mandible through eye to nape black, feathers broadly edged warm buff; rest of head and neck pale buff, finely spotted or streaked dark grey-brown; chin and throat white. Upperparts and flanks sepia, feathers with broad rufous-buff edges and subterminal bars; outer mantle, scapulars, and flanks with broad irregular V-shaped marks. Feathers of chest warm buff with broad sepia subterminal crescents or U-shaped marks; under tail-coverts similar, sometimes fringed white. Breast, belly, and vent white, feathers often spotted sepia subterminally on upper breast, vent, and sometimes sides of belly; vent tinged pale buff or cream. Tail like adult ♂, but edges broader and buff, partly marked buff subterminally. Wing like adult ♂, but middle secondaries with outer web grey, bordered black; black on greater coverts less extensive, outer mouse-grey, tipped white. Black and chestnut on median and lesser coverts variable, usually black restricted to inner median coverts and chestnut to large subterminal marks on central median; outer median coverts and all lesser olive-grey with pale-grey edges and often narrow subterminal cream bars or arcs, usually no chestnut tinge and fine vermiculations as in ♂. In spring, at least some inner greater or median coverts, changed for somewhat longer and narrower sepia feathers with broad buff edges or marks like feathers of upperparts of body. Black-and-chestnut pattern on wing-coverts may disappear then. Tertials olive-brown, edged and marked buff like scapulars; tertial coverts similar, but sepia instead of olive-brown. ADULT MALE NON-BREEDING (Eclipse). Superficially like adult ♂ or juvenile ♂, but wing like adult ♂ breeding. Head and neck like adult ♀ breeding, but edges on crown narrower and whole head more heavily streaked. Upperparts and flanks sepia, with olive-buff edges (narrower and less distinct than adult ♀ breeding); parts of mantle, upper tail-coverts, and flanks with cream subterminal bars (wider than juvenile ♂); upper scapulars and flanks with small V-shaped marks. Feathers of chest white with broad sepia marks and buff tips; under tail-coverts white, irregularly streaked or spotted; rest of underparts white, sometimes with round black spots. Tail and wing like adult ♂ breeding, but tertials shorter, olive-brown, faintly edged cream at tip. ADULT FEMALE NON-BREEDING. Like adult ♀ breeding, but crown darker, rest of head and neck more heavily spotted, edges of feathers of upperparts and flanks narrower, buff; subterminal marks faint or absent. Underparts with variable number of large round subterminal dusky spots. Wing and tail like adult ♀ breeding, but tertials shorter, dark olive-grey, edged cream at tip; tertial-coverts with outer web black and inner dark olive-brown, tipped white. Wing-coverts partly as in breeding plumage until start of wing moult. DOWNY YOUNG. Like Mallard *A. platyrhynchos*, but upperparts sepia, without olive tinge; light patches on wing and at sides of back and rump more extensive, cream-buff instead of yellow; underparts cream-buff, in some tinged cinnamon on chest. At hatching, light parts tinged yellow, but this fades within 1–2 days. JUVENILE MALE. Crown black; rest of head and neck off-white, heavily spotted dark grey-brown at sides of head and hindneck; spotting finer on chin, throat, and foreneck. Upperparts dull black, feathers narrowly edged grey-buff on mantle, scapulars, and upper tail-coverts. Narrow, pale buff, subterminal bars on sides or whole of mantle and V-shaped marks on part of scapulars and upper tail-coverts. Chest and flanks dull black or sepia, feathers margined buff and irregularly barred white; rest of underparts white, tips of feathers usually heavily spotted or streaked dusky brown; spots narrower and duller than round black spots of some adults. Tail dark grey-brown; feathers edged white, notched at tip, bare shaft projecting. Wing like adult ♀, but more black on outer web of middle secondaries; bars and edges on lesser and inner median coverts narrower, more wavy and irregular, instead of distinct bars and arcs. Amount of chestnut and black variable, but always less than adult ♂. Juvenile ♂ differs from adult ♂ by light edges and bars on lesser coverts, instead of fine freckling and vermiculation without light edges. Tertials with coverts like adult ♀ non-breeding, but shorter and duller brown. No development of breeding wing-coverts in spring like ♀♀. In early autumn, part of body plumage (sides of mantle and scapulars, chest, flanks, sometimes head or upper tail-coverts) replaced for 1st immature non-breeding (resembling adult non-breeding) or sometimes showing characters intermediate between breeding and non-breeding (e.g. coarse barring on chest or flanks); this plumage soon replaced by breeding. JUVENILE FEMALE. Like juvenile ♂, but less barring on mantle, scapulars, and sides of breast; dark sepia feathers of back and rump edged buff; tail dark brown, less grey than ♂. Wing like adult ♀, but outer webs of mouse-grey middle secondaries with narrower black margin, or no black; 2 inner secondaries often tinged pale ash instead of white, sometimes hardly differing from other secondaries; greater and median coverts grey-brown, edged white at tip, inner with black outer web; usually only few central median coverts with some chestnut at tip or on outer web. Tertials like juvenile ♂. Wing differs from juvenile ♂ by much less black and chestnut, broader edges, and more regular bars on median and lesser coverts. Development of 1st immature non-breeding plumage like juvenile ♂. FIRST BREEDING MALE. Like adult ♂ breeding, but juvenile wing (except tertials with coverts), and sometimes some feathers elsewhere (especially on back and rump) retained until 2nd summer. Feathers of lower breast and belly sometimes with subterminal dusky bars or spots (this rarely also in adult ♂). FIRST BREEDING FEMALE. Like adult ♀, but juvenile wing retained until 2nd summer. During moult in 1st autumn, feathers growing first have characters of adult non-breeding plumage, new feathers later gradually acquiring more adult breeding characteristics. Feathers resembling adult non-breeding, juvenile tertials, and often juvenile feathers of back and upper rump, or some elsewhere, retained until 2nd spring. In spring, some upper wing-coverts changed for coverts resembling adult ♀ breeding; one white innermost secondary sometimes changed for dark grey new one, exceptionally both. SUBSEQUENT PLUMAGES. Like those of adults, but ♀♀ sometimes change part of 1st breeding plumage direct to 2nd breeding. (Oring 1968; Schiøler 1925; Witherby *et al.* 1939; C S Roselaar.)

**Bare parts.** ADULT MALE. Iris dark brown. Bill leaden-grey; when in non-breeding plumage, sides orange-yellow, like ♀. Foot dull orange or yellow, webs dusky. ADULT FEMALE. Iris dark brown. Bill dark brown or dark olive-grey on culmen, dull flesh-orange to ochre-yellow with some dark brown dots on sides. In summer, sides brighter orange-yellow, and dots more distinct. Size and number of dots tends to increase with age. Foot yellow-brown to orange-yellow, webs dusky grey. DOWNY YOUNG. Iris brown. Bill grey, edges of upper mandible flesh (in contrast to more uniform bill of *A. platyrhynchos*). Foot dark grey or olive-black, sides of tarsus and toes yellow-buff. JUVENILE. Iris dark brown. Bill dark olive-grey, sides of upper mandible yellow; ♀♀ often with some small dark dots along edges. Foot dull yellow, webs dusky grey or blackish. All bare parts change to adult colour when breeding plumage attained (Bauer and Glutz 1968; Oring 1968; Witherby *et al.* 1939; RMNH).

**Moults.** ADULT POST-BREEDING. Complete; flight-feathers simultaneous. ♂♂ head and body May–July, ♀♀ June–August—in North America, ♀♀ already in moult March–April (Oring 1968), but this not apparent in Netherlands' sample. When head and body finished, wing follows; tail and tertials moulted with wing but start slightly earlier. Flightless for *c.* 4 weeks: ♂♂ late June to late August; ♀♀ later, sometimes not until October. ADULT PRE-BREEDING. Partial; head, body, tail, and tertials. Underparts and central pair of tail-feathers first, followed by head and upperparts; tertials and rest of tail (not always completely) last. ♂♂ in almost complete breeding plumage by September; ♀♀ by October, but tertials with coverts and variable amount of tail-feathers and greater and median (sometimes lesser) coverts not attained until spring. POST-JUVENILE. Part of juvenile plumage replaced by 1st immature non-breeding August–October, mainly on head, flanks, chest, sides of mantle, and scapulars, sometimes also rest of mantle or upper tail-coverts. Breeding feathers appear from September, central tail-feathers and underparts first. Most 1st non-breeding and all remaining juvenile feathers changed for 1st breeding before December (sometimes October), but retarded specimens retain much of it until spring. ♂♂ retain some juvenile outer tail-feathers or tertials during winter only occasionally, but ♀♀ often retain these (and also juvenile feathers of back and rump and 1st autumn plumage on upperparts) until spring, when also number of wing-coverts changed as in adult ♀. SUBSEQUENT MOULTS. Like those of adults, but some ♀♀ do not acquire complete 2nd non-breeding on upperparts. (Bauer and Glutz 1968; Dementiev and Gladkov 1952; Oring 1968; C S Roselaar.)

**Measurements.** Netherlands, whole year. Skins (RMNH, ZMA).

| | | | | | |
|---|---|---|---|---|---|
| WING AD ♂ | 269 | (5·85; 24) | 261–282 | ♀ 252 (5·24; 14) | 243–261 |
| JUV | 264 | (5·52; 37) | 251–274 | 246 (6·58; 32) | 233–262 |
| TAIL AD | 86·0 | (3·24; 13) | 81–92 | 80·2 (2·33; 9) | 77–84 |
| JUV | 77·5 | (2·68; 12) | 75–82 | 73·7 (2·61; 11) | 69–77 |
| BILL | 42·4 | (1·63; 60) | 39–46 | 39·8 (1·60; 48) | 37–43 |
| TARSUS | 40·3 | (1·01; 42) | 38–42 | 38·8 (1·24; 34) | 36–42 |
| TOE | 53·4 | (1·80; 42) | 50–57 | 51·2 (1·77; 34) | 47–55 |

Sex differences significant. Wing and tail of juveniles significantly shorter than those of adults. Other measurements similar; combined. Tail in both adult non-breeding and in 1st breeding *c.* 80 (75–88) in ♂, *c.* 77 (73–83) in ♀.

**Weights.** USSR; mainly Volga delta and Kazakhstan (Dementiev and Gladkov 1952; Dolgushin 1960). ADULT. General trend like *A. platyrhynchos*, though data limited. Various averages (sample size, when known, in brackets), and total range.

| | | | | | |
|---|---|---|---|---|---|
| (1) | ♂ 771 (22); | 605–950 | ♀ 716, 737 (9), 750, 770; | 650–820 | |
| (2) | 808 (6); | 660–1100 | 777 (3); | 750–830 | |
| (3) | 763 (18), 766 (7); | 650–900 | 500 (1), 744 (8); | 500–800 | |
| (4) | 913; | 800–1000 | 854, 900 (1); | 800–1000 | |
| (5) | 780 (5), 807; | 605–1000 | 630, 707 (13), 716, 780; | 540–840 | |
| (6) | 958, 959 (11); | 800–1300 | 869; | 800–1000 | |
| (7) | 500, 600, 707, 830; | 470–1050 | 562, 663; | 470–680 | |

(1) Weight rather low just after winter (♂ March, ♀ March–April). (2) Slight increase during maximal development of gonads (♂ April, ♀ May). (3) Return to spring level after reduction of gonads and breeding of ♀ (♂ May–June, ♀ June). (4) Increase just before wing moult (♂ August, ♀ late June to August). (5) Low when wing full-grown (August–September).

(6) Increase to peak in October. (7) Weight variable in winter (January to early March); high October level maintained in some until January or later in mild weather (ZMA). JUVENILE. Like adult from October–November.

**Structure.** Wing rather long, pointed. 11 primaries: p9 longest, p10 1–7 shorter, p8 7–15, p7 18–27, p6 31–44, p1 103–130. Inner web of p10 and outer p9 emarginated. Tail short, rounded; 16 feathers, central pair pointed in breeding, somewhat shorter and rounder in non-breeding plumage. Bill shorter than head, rather narrow, about as high as broad at base, sides parallel, straight. Lamellae strong and long. Culmen slightly concave in middle, straight and flat at tip. Nail rather narrow, rounded. Scapulars elongated and pointed in breeding plumage, broader and rounder in non-breeding. Feathers of hindcrown and hindneck form short shaggy crest. Outer toe *c.* 96% middle, inner *c.* 79%, hind *c.* 26%.

CSR

## *Anas formosa* Georgi, 1775 **Baikal Teal**
FR. Sarcelle élégante    GE. Gluckente

Breeds within forest zone of north to north-east Siberia, from west Anadyrland and Kamchatka west through Yakutsk and Krasnoyarsk to Yenisey basin (to *c.* 70°N), south to Lake Baykal and southern Sea of Okhotsk. Migrates to winter quarters in Japan and east and south-east China, being absent from breeding range chiefly mid-September to early May. Spring passage overland through Korea, Manchuria, and Soviet Maritime Territories, but autumn route(s) still conjectural (Dementiev and Gladkov 1952). Straggles south-west from breeding range to Tomsk region and Baraba Steppe, and occasional (mainly winter) records across Assam and north India (Ali and Ripley 1968)

indicate vagrancy potential. In Europe, recorded Finland, Sweden, Germany, Netherlands, Belgium, Britain, Ireland, France, Switzerland, Italy, Malta (Bauer and Glutz 1968); but origins of a good many suspect since species commonly kept in waterfowl collections. However, 5 obtained Saône valley, France, November 1836 (Mayaud 1936) pre-dated any known introduction into Europe (Delacour 1956), while one Fair Isle, Scotland, 30 September 1954, occurred simultaneously with 3 other Siberian species (Harrison 1958) though in neither case admitted to national list concerned.

## *Anas crecca* **Teal**

DU. Wintertaling    FR. Sarcelle d'hiver    GE. Krickente
RU. Чирок-свистунок    SP. Cerceta común    SW. Kricka    N.AM. Green-winged Teal

PLATES 66 and 74
[between pages 518 and 519 and facing page 543]

*Anas Crecca* Linnaeus, 1758

Polytypic. Nominate *crecca* Linnaeus, 1758, Europe and north Asia; *carolinensis* Gmelin, 1789, North America, accidental. Extralimital: *nimia* Friedmann, 1948, Aleutian Islands.

**Field characters.** 34–38 cm, of which body over two-thirds; wing-span 58–64 cm. Small, compact, and short-necked dabbling duck with somewhat oval head and narrow, pointed wings; ♂ *A. c. crecca* has chestnut head with metallic green eye-patch outlined whitish, spotted breast, grey body with horizontal white stripe above wing, and black and yellow-buff stern. Sexes dissimilar, with some seasonal differences. Juvenile resembles adult ♀, but just distinguishable. ♂ (but not ♀) of North American race (*carolinensis*) distinct.

ADULT MALE. Head bright chestnut, with broad green patch, curving back from eye to nape and surrounded by narrow line of buff-white which also extends forward and down to base of bill. At distance, head appears uniformly darker than grey body, but prominent horizontal white stripe along scapulars and black-bordered yellow-buff patch on either side of black undertail remain conspicuous.

Rich cream breast with blackish spots, vermiculations of dark grey and cream on back and flanks, and black line below scapular stripe visible only at close ranges; speculum metallic green and black, broadly edged with whitish to buff in front and only narrowly so behind. In flight, medium grey-brown forewing makes whole wing appear dark apart from 2 light bars on either side of speculum; dark head and white and black braces not so conspicuous as on ground; from below, grey flanks stand out against white belly, axillaries, and underwing. Bill dark grey; eyes dark brown; legs green-grey with blacker webs. *A. c. carolinensis* (Green-winged Teal) easily recognized by vertical white line at side of breast, down from bend of wing, and absence of horizontal white stripe on scapulars; breast generally darker and more rosy-brown, and whitish outline to green eye-patch much restricted, especially above—but these points noticeable only at close range.

PLATE 59. *Tadorna tadorna* Shelduck (p. 455): **1** ad ♂ breeding, **2** ad ♀ breeding, **3** ad ♂ eclipse, **4** 1st imm breeding ♂, **5** juv, **6** downy young. (NWC)

PLATE 60. *Aix galericulata* Mandarin (p. 465): **1** ad ♂ breeding, **2** ad ♀ breeding, **3** ad ♂ eclipse, **4** juv, **5** downy young. (NWC)

PLATE 61. (*facing*) Geese and shelducks in flight
*Anser fabalis* Bean Goose: **1** adults.
*A. brachyrhynchus* Pink-footed Goose: **2** adults.
*A. anser* Greylag Goose: **3** adults, **4** juv (*A. a. anser*); **5** adults (*A. a. rubrirostris*).
*A. albifrons* White-fronted Goose: **6** adults, **7** juv (*A. a. flavirostris*); **8** adults, **9** juv (*A. a. albifrons*).
*A. erythropus* Lesser White-fronted Goose: **10** adults, **11** juv.
*A. caerulescens* Snow Goose: **12** ad white morph, **13** juv white morph, **14** ad intermediate, **15** ad blue morph, **16** juv blue morph (all *A. c. caerulescens*).
*Branta bernicla* Brent Goose: **17** adults, **18** juv (*B. b. hrota*); **19** adults, **20** juv (*B. b. bernicla*).
*B. leucopsis* Barnacle Goose: **21** adults.
*B. canadensis* Canada Goose: **22** adults (*B. c. canadensis*); **23** adults, **24** ad with white collar (*B. c. minima*); **25** adults, **26** ad with white collar (*B. c. hutchinsii*)
*B. ruficollis* Red-breasted Goose: **27** adults, **28** juv.
*Alopochen aegyptiacus* Egyptian Goose: **29** adults, **30** juv.
*Tadorna tadorna* Shelduck: **31** adults, **32** juv.
*T. ferruginea* Ruddy Shelduck: **33** adults. (PS)

Peter Scott.

PLATE 62. *Anas penelope* Wigeon (p. 473): **1** ad ♂ breeding, **2** ad ♀ breeding, **3** ad ♂ eclipse, **4** juv, **5** downy young. (NWC)

PLATE 63. *Anas americana* American Wigeon (p. 481): **1** ad ♂ breeding, **2** ad ♀ breeding, **3** 1st breeding ♂, **4** ad ♀ breeding. (NWC)

MALE ECLIPSE. Resembles adult ♀, but upperparts darker and more uniform, with less distinct dark line through eye. In flight, forewing more uniform grey-brown, not mottled, and less contrast from buff (rather than white) wing-bars. Period of eclipse short, largely August–September. ADULT FEMALE. Head and body with dark crown and stripe through eye, streaked cheeks, and neck paler than dark brown upperparts with lighter feather edges; foreneck, breast, and flanks variably marked with dark spots, crescents, and U-shapes; centre of breast and belly white. Speculum metallic black and green as in ♂, but narrowly and distinctly bordered white both in front and behind. In breeding plumage, breast and flanks become more strongly spotted and upperparts fall into 2 colour morphs: one darker and plainer (but not as uniform as ♂ eclipse and juvenile), other paler with more conspicuous feather edgings and barrings. Bill grey with pinkish-orange on upper mandible. JUVENILE. Closely resembles adult ♀, but darker and more uniform above (resembling ♂ eclipse) and distinguished on land by spotted lower breast and belly. Bill pink-horn. Moults quickly and full plumage assumed by November.

Minute size (smallest west Palearctic duck, apart from vagrant ♀ Bufflehead *Bucephala albeola*) distinguishes from all other dabbling ducks of area except summer Garganey *A. querquedula* and vagrant Blue-winged Teal *A. discors*. Combination of dark head, grey body, and black-edged, yellow-buff stern diagnostic of ♂ at all ranges. (♂ *A. c. carolinensis* easily separated by vertical, not horizontal, white mark on side.) ♀ and other plumages distinguished in flight by darker forewing and well-defined speculum bordered with white, as well as shorter neck and more compact shape than small ducks mentioned above; from below, underwing less contrasted than ♀ *A. querquedula*, which has distinctive dark leading edge. On water, ♀♀ of *A. querquedula* and *A. discors* both paler, with whiter throat and whiter spot at base of bill, though these differences often difficult to see.

Chiefly associated with wide variety of freshwater habitats, less often on sea. Swims daintily; usually dives only when wounded or frightened. Walks more clumsily than size and shape might suggest; habitually rests out of water, even on dead trees and branches. When flushed, shoots almost vertically up from water or land, and flies rapidly in tight packs, usually low and erratically. On migration, flies high and in lines or V-formation. Commonly in pairs or small parties, but often in sizeable flocks and sometimes in vast concentrations. ♂ frequently utters short, high, soft but far-carrying, double bell-like whistle; ♀ has sharp, high 'queck', used most often when alarmed.

**Habitat.** Breeds throughout middle latitudes of west Palearctic, from cold coastal tundra southward through steppe forest and steppe to desert fringes, but infrequent in semi-desert and outnumbered by Garganey *A. querquedula* on steppe waters. Normally well scattered over small outlying or isolated pools, ponds, lagoons, oxbows, slow-flowing streams, and other diminutive water bodies, often forming part of larger wetland, lake, or river system, especially in valleys of small forest rivers, but frequently far from other waters. Eutrophic waters normally favoured, but given plentiful food will tolerate neutral or even somewhat acid conditions. Bordering plants must form some kind of dense herb layer, over which tree canopy fully acceptable. Nests fairly freely well away from water, on heathlands or under scrub; up to 2000 m or higher in mountains in south of range. At home on any kind of water except deep and lifeless, fast-running, exposed, or wave-troubled. Will adapt to such open habitats as shallow tidal coasts, large estuaries, salt-marshes, and lagoons, brackish or saline, mainly on passage or in winter, when artificial waters such as reservoirs devoid of vegetation also frequented. Fringe zones between open shallow water and emergent floating or overhanging vegetation most favoured generally. Relatively independent of reliance on human artefacts, and averse to frequent or intensive human disturbance. Makes full use of flight, mainly at medium and low heights, for reconnoitring secure places and for daily movements between feeding and resting areas. Where wintering areas long-established, settled patterns of resting areas suitable for large concentrations to satisfy their daytime needs each related to own nocturnal feeding grounds at a mean distance (in the Camargue) of *c.* 15 km. For feeding require fine mud under a few centimetres of water with abundant small seeds, and often also animal food, and in resting areas enough open water for hundreds or thousands of birds, often with well-vegetated borders fringing fine-particled marshes and a slightly sloping shore of less than 1/1000, preferably exposed to sun. Suitable perches up to 2 m above the water also used. Habitat types outlined above must therefore be understood as normally forming part of a closely linked series together satisfying complex of diurnal and seasonal behaviour patterns (Tamisier 1974).

**Distribution.** Breeding often sporadic on edges of range. SPITSBERGEN: bred 1938 (Løvenskiold 1963). FRANCE: also bred Landes 1940, and Pyrénées 1927 (FR). SWITZERLAND: regular breeding Lake Neuchâtel, sporadic elsewhere (Bauer and Glutz 1968). ITALY: breeds, at least occasionally, further south e.g. Apulia and Tuscany (SF). SPAIN: bred 1929 (*Ibis* 1932, (13) 2, 167). YUGOSLAVIA: bred 1889, 1968 (VFV). RUMANIA: bred 1915 (MT).

Accidental. *A. c. crecca*: Jan Mayen Island, Bear Island, Spitsbergen, Cape Verde Islands. *A. c. carolinensis*: Britain and Ireland (*c.* 90, annually in recent years), Netherlands (4), Belgium (2), Morocco (1).

**Population.** Breeding. BRITAIN AND IRELAND: *c.* 1000–1500 pairs (Atkinson-Willes 1970a); perhaps slight decrease Britain (Parslow 1967), and marked decrease,

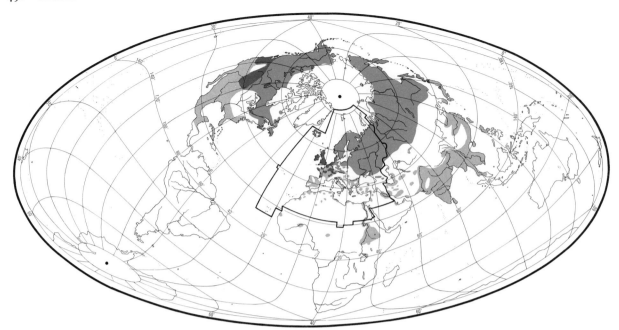

mainly due to drainage, Ireland (Ruttledge 1970). FRANCE: perhaps 500–1000 pairs (FR). BELGIUM: *c*. 400 pairs (Lippens and Wille 1972). NETHERLANDS: *c*. 750 pairs (JW). WEST GERMANY: *c*. 1020 pairs (Szijj 1973). DENMARK: probably not over 200 pairs (Joensen 1974). CZECHOSLOVAKIA: some hundreds of pairs (Hudec 1971). FINLAND: *c*. 80 000 pairs (Merikallio 1958). ESTONIA: slightly over 2000 pairs (Onno 1970). POLAND: decreasing in south (Tomiałojć 1972).

Winter. In north-west Europe, *c*. 150 000 birds; in Black Sea—Mediterranean region, *c*. 750 000 birds (Atkinson-Willes 1975). USSR: estimated 609 200 birds in not so severe winters in western part (Isakov 1970*c*).

Survival. Annual mortality. Britain 1949–55, ♂ 49%, ♀ 57%; Pembrokeshire, Wales, both sexes, 64% in 1934–8, 49% in 1945–8, 65% in 1949–53, but in war years 1941–5 only 39%; about three-fifths of ♂ losses and half of ♀ probably attributable to man (Boyd 1957*a*). For both sexes, 55% France, 58% Italy and Spain; in both areas, mortality of mainly 1st-year birds higher than that for mainly 1–2 year olds; for mainly 1–2 year olds, mortality Europe 47%, USSR 51% (Tamisier 1972*c*). Oldest ringed bird 16 years 10 months (*Ornitologiya* 1960, 3, 385).

**Movements.** Largely migratory, small numbers resident some areas. Extensive ringing has shown number of flyways but with considerable overlap and evidence of change from year to year. All but a few Icelandic breeders migrate to Britain (chiefly Scotland) and Ireland. Most or all breeders from north Russia, Baltic states, Fenno-Scandia, north Poland, north Germany, and Denmark fly south-west in autumn to North Sea wintering grounds, chiefly Netherlands and Britain. Winter distribution

between Netherlands, England, Wales, and Ireland greatly dependent on weather; cold spells cause immediate westward movement, and in severe winters many move to France and Iberia; minority reach north Africa (chiefly Morocco). Breeders of Netherlands, Britain, France, and south Europe largely resident, moving only in severe weather. Roughly parallel to above flyway, another runs south-west from north USSR (Nenets, Komi, Yamal-Nenets, Taimyr); some overlap in breeding areas, but routes diverge as latter flyway on more southerly course through central Europe to Spain, south France and Italy; some cross to north Africa. This part of Mediterranean also wintering area for breeders of central Europe. Ringing recoveries for both major flyways show change from one to another in different season not uncommon; thus many ringed Britain and Netherlands recovered subsequent seasons in south and south-east Europe and Black Sea; exceptional recoveries of North Sea birds in Kazakhstan, Krasnodarsk, and Iran probably abmigrants. Unlike Mallard *A. platyrhynchos*, only relatively small numbers winter north and north-central Europe (Atkinson-Willes 1969).

Further east, flyway from west-central Russia and Ukraine runs through Black Sea and Balkans to Greece and west Turkey; another from central and south USSR (Volga-Ural, Ob-Irtysh basins, and north Kazakhstan) to south Caspian and on to Iran and Iraq (Shevareva 1970). Some moulting Volga delta disperse southwards through Turkey and east Mediterranean into Egypt where many overwinter Nile valley, and some winter-ringed there recovered north-east to Komi, Sverdlovsk, and Omsk; those wintering Arabia and Persian Gulf perhaps from these populations. Yet many from Komi winter Europe,

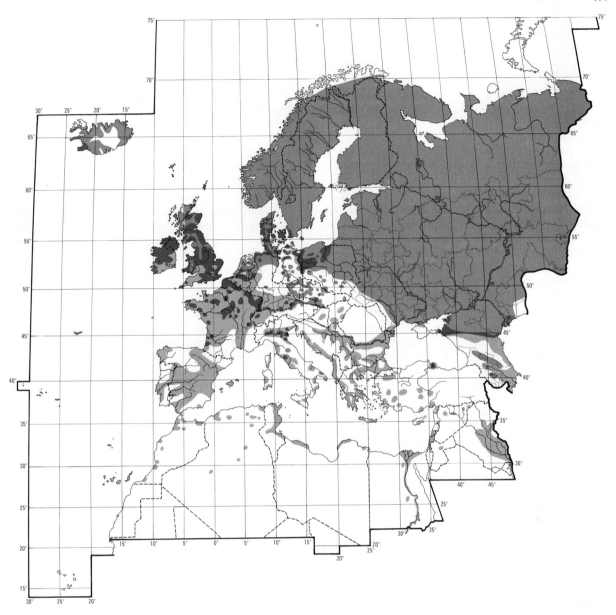

while others from Omsk winter India and Pakistan. Birds of uncertain origin winter in relatively small numbers northern tropics of Africa, south to Nigeria, north Congo and north Tanzania; Dutch migrant recovered Upper Niger; largest aggregates 500 Sénégal delta January 1972 (Roux 1972) and 1200 Lake Chad basin 1970–1 winter (Vielliard 1972). Summer wanderers to Novaya Zemlya, Bear Island, Spitsbergen (has bred), and Jan Mayen; vagrants to Greenland and Atlantic North America, including one ringed England, November, shot Newfoundland 4 weeks later.

Moult migration in some areas, scale very variable, ♂♂ leaving ♀♀ when incubation begins (Salomonsen 1968). Large concentrations reported USSR whilst, in other areas, birds moult in small numbers or singly on or near nesting grounds. ♂♂ moulting Volga delta from as far away as Karelia, Arkhangel, and Ob watershed. Movement of ♂♂ through Denmark in June apparently involves fairly local breeders as well as more northern birds, while small moult gatherings IJsselmeer, Netherlands, include some from Poland (Salomonsen 1968). Autumn movement may begin July (failed breeders), but peak passage across Europe October–November, ♂♂ tending to precede ♀♀; cold-weather movements at any time during winter. Return migration from late February, mostly March–April, and tundras not re-occupied until late May.

New World subspecies, *A. c. carolinensis*, withdraws south after breeding, wintering British Columbia, west and south USA, West Indies, and Central America (Bruun 1971); stragglers to Greenland, Bermuda, Colombia, Hawaii, Polynesia (Palmyra), and Japan. Almost annual vagrant to Europe (notably Britain and Ireland); juvenile ringed New Brunswick, Canada, August 1970, shot Isles of Scilly, England, January 1971. See also Bauer and Glutz (1968); Dementiev and Gladkov (1952); Hoffman (1960); Wolff (1966).

**Food.** Omnivorous, seeds predominating in winter. Considerable variation in frequency of particular feeding method depending on habitat, time of day, season, and sex (Szijj 1965b; Willi 1970; Tamisier 1972a, b). (1) Walks slowly, filtering mud with bill, in water of a few centimetres—70% of night spent feeding, Camargue, France (Tamisier 1972a). (2) Swims with head, or head and neck under water; most frequent method Bodensee (Szijj 1965b), Klingnauer reservoir (Willi 1970), and near Berne, Switzerland (Mazzuchi 1971). (3) Up-ends. (4) Picks items off surface of water, or vegetation. (5) Occasionally feeds by skimming with bill held flat on water, usually only when food abundant (Willi 1970), and, rarely, (6) dives (Bennett 1965). At Klingnauer, ♂♂ feed in deeper water on average than ♀♀, so feeding grounds and methods differ: e.g. feeding by walking on silt-banks mainly by ♀♀, by up-ending mainly by ♂♂ (Willi 1970). In Camargue, feeds essentially by night, though when feeding time (8–12·7 hours per day) exceeds night length, c. 30% feeding occurs in daytime: variation in feeding time depends on physiological condition, and on climatic factors (Tamisier 1972a, b, 1974). In other areas, mostly diurnal feeding with occasional night feeding depending on moonlight (Szijj 1965b; Leuzinger 1968; Willi 1970). At Klingnauer in autumn, 2 feeding peaks in early morning and late afternoon; in winter, 1 peak mid-afternoon; in spring, 2 peaks (Willi 1970). On tidal wintering grounds, activity rhythms determined by variation in water levels and by moon phases (Lebret 1970; Tamisier 1974). Gregarious during day, scattered at night (Tamisier 1970).

Food varies with locality and season, but basically seed-eater in autumn and winter, with relatively more animal materials in summer. Adapted to take seeds of certain size, mainly 1–2·6 mm; larger seeds (e.g. grain) taken rarely and, like smaller seeds (e.g. oospores), only when abundant. Animal materials often larger, up to 11 mm, involving differing feeding behaviour (Olney 1963b, 1965b; Tamisier 1972b, 1974). Plant materials chiefly seeds of aquatics; sedges and bulrushes (Carex, Scirpus, Schoenoplectus, Eleocharis), pondweeds Potamogeton, wigeon-grass Ruppia, milfoil Myriophyllum; buttercup Ranunculus, dock Rumex, bistort Polygonum, birch Betula, blackberry Rubus, alder Alnus, samphire Salicornia, and sea aster Aster. Also grasses Festuca, Panicum; rice Oryza sativa; oospores of stonewort Characeae; and occasionally algae (Entermorpha), eel-grass Zostera, and duckweed Lemna. Animal materials include molluscs (Hydrobiidae,

Physidae, Limnaeidae), Diptera larvae (Chironomidae, Eristalis), caddisfly larvae (Phryganeidae), waterbeetles (Agabus, Hydrobius), waterbugs (Plea), crustaceans (Ostracods), and annelids. (Thompson 1851; Witherby et al. 1939; Campbell 1947; Lynn-Allen 1954; Dementiev and Gladkov 1952; Olney 1963b, 1967; Tamisier 1971; Mazzuchi 1971.)

All 313 stomachs, Camargue, August–March, contained seeds (Tamisier 1971, 1974). As percentage of total dry weight: 25% oospores of Characeae; 25% seeds of Cyperaceae (mainly Scirpus litoralis, S. lacustris); 25% rice grains and cockspur Panicum crus-galli (from harvested rice fields). Rest seeds of Salsolaceae (chiefly seablite Suaeda); seeds of Potamogeton (chiefly P. pusillus), Ruppiaceae (chiefly R. maritima), and Myriophyllum. Animal materials present in 50%: mainly molluscs, chironomid larvae, and ostracods. Oospores and seeds of Cyperaceae important August–September, and again February–March, and rice and cockspur seeds October–December. Seeds of Cyperaceae and Potamogetonaceae most frequent, 60% and 85% respectively; average size 1·2–2·6 mm. In England, analyses of 191 stomachs north Kent marshes, and 82 stomachs from other salt-marsh and brackish water areas September–February, showed: (1) on salt-marsh, chiefly seeds of Salicornia, rather less seeds from other plants, and, in certain areas, Enteromorpha and mollusc Hydrobia ulvae; (2) in brackish water, mainly seeds of Scirpus maritimus, S. tabernaemontani, and Eleocharis; and rather less molluscs H. jenkinsi (Olney 1963b). 9 stomachs, Terschelling, Netherlands, December and February, contained similar salt-marsh and brackish water seeds—e.g. S. maritimus, Salicornia (de Vries 1939). In 96 stomachs, freshwater areas, England, September–January, 66 plant species identified (83·6% total volume), chiefly seeds of Eleocharis palustris and Ranunculus repens; also less chironomid larvae (Olney 1963b). 16 stomachs from inland area contained basically same main items (Olney 1967). Similar results also from 31 stomachs south-east Finland, August–November, when all contained seeds (mostly Carex rostrata, C. nigra, E. palustris, Potamogeton natans), 69% other parts of aquatic plants, 41% beetles and larvae and some other invertebrates (Tiussa 1972). 32 stomachs, September–March, from Berne area, all contained seeds (80% Cyperaceae, Rosaceae, Betulaceae, and Urticaceae), but here 87·5% also contained animal materials, mainly Diptera (especially chironomid larvae) and ants (Formicidae) (Mazzuchi 1971). From Hungary, 23 stomachs contained mainly seeds (especially bristle-grass Setaria glauca, Carex, Scirpus and Chara) and less chironomid larvae and waterbeetles (Sterbetz 1969–70). In USSR, data available throughout year, and seasonal and regional differences more obvious (Isakov and Vorobiev 1940; Dementiev and Gladkov 1952). In general, mostly seeds during autumn and winter, animal material during spring and summer. Volga delta, moult period: mostly seeds

(70%, chiefly unripe rockcress *Najas marina*); less plant green parts (28%) and chironomid larvae. In autumn, migrants mainly take seeds (53–86%, August–November); in early spring, again chiefly seeds (69% of volume), rockcress, pondweeds, bur-reeds *Sparganium*, with a few molluscs *Planorbis* and waterbeetles. At Rybinsk reservoir, however, migrants take relatively more animal foods: 67%, chiefly chironomids, in autumn; 84%, mainly chironomids, caddisfly larvae *Phryganea grandis*, and waterboatmen *Corixa*, in April; 74%, almost exclusively chironomids, in May and June. In Mologa river area, September–October, animal foods (beetles, fly larvae, molluscs) only 15%, with 15% arrowhead *Sagittaria* bulbs and 70% seeds. In April and May, animal-food consumption rose to 77% (35% volume) and in breeding season 98% (70% volume), with wide range of species, including dragonflies and chironomid larvae, molluscs, caddisfly larvae, phyllopod crustaceans, and aquatic oligochaetes. In October and November near Krasnovodsk, 45% marine molluscs *Theodoxus pallasi*, but, in December–February on Azerbaydzhan lowlands, animal food only 1% of total contents. Winter diets, as in other Palearctic areas, generally show change to plant materials, principally seeds: *Ruppia* and *Najas* on coasts; *Panicum*, *Echinochloa crus-galli*, other grasses, and *Scirpus* in ricefields, and goosefoot *Atriplex* and thistles at steppe ponds.

One food source utilized if abundant, often for long periods, e.g. in Switzerland fed almost exclusively on seeds of purple loosestrife *Lythrum salicaria* during time waters frozen (Leuzinger 1968).

Daily food requirements estimated 20–30 g fresh weight, i.e. 7–8% body weight (Tamisier 1971).

Diet of downy and feathered ducklings Mologa river, USSR, 85% and 75% by volume animal materials; chiefly small molluscs pecked from vegetation, larvae of dragonflies, chironomids and mosquitoes, and adult waterbeetles (Dementiev and Gladkov 1952). 8 young, Pripyat marshes, near Minsk, USSR, contained only animal materials, of which 83% molluscs, mainly *Planorbis* and *Limnaea* (Dolbik 1959).

**Social pattern and behaviour.** Compiled mainly from material supplied by F McKinney. 1. Gregarious except when nesting, though will disperse to feed. Outside breeding season, most often in small flocks of *c.* 30–40, but also in small groups or up to 100 or more. In Camargue, scatter widely when feeding nocturnally (Tamisier 1974). As in other *Anas*, sex-ratio often shows local excess of ♂♂, especially in migrant populations: ♂♂ largely predominate in Netherlands, November–February (Lebret 1950, 1961); in Camargue, ♂♂ on average 70% of wintering population (Tamisier 1974). Winter flocks composed increasingly of paired birds, these later migrating together to breeding grounds. ♂♂ first to form post-breeding flocks after pair-bonds ended. Moulting aggregations occur in late summer, after pre-moult movement by ♂♂ in some populations. BONDS. Monogamous pair-bond of seasonal duration. Formed in flock mainly during winter: in

Bavaria, up to 50% paired November–January, 95% by end March (Bezzel 1959); in Netherlands, pairs observed from late October and numerous from November, but many ♂♂ remain unpaired until spring when immigrant ♀♀ arrive (Lebret 1961). ♂♂ also promiscuous, probably to large extent. In North American race *carolinensis*, captive ♂♂ maintained strong pair-bonds but at times courted other ♀♀ and, during May–June, pursued them in rape attempts (F McKinney). Pair-bond ends when ♂ leaves ♀ during incubation, apparently almost immediately it begins (Hildén 1964). Only ♀ tends young; family-bonds loosen when *c.* 1-month old (Hildén 1964). BREEDING DISPERSION. Pairs usually well dispersed on small ponds, usually surrounded by trees (Hildén 1964). Pair-spacing probably much influenced by ♀'s need for seclusion from assaults of other ♂♂. Little known about nature and size of home-ranges, but no indication that ♂ ever attempts to defend activity centre or waiting area as territory. In *carolinensis*, wild ♂♂ associate peacefully even during laying period (Munro 1949), while captive ♂♂ show more interest in courting and raping other ♀♀ than in defending territories and, consequently, mates—unlike Shoveler *A. clypeata* in similar circumstances. ROOSTING. Outside breeding season, rest by day in tight-packed groups on open water, ice, or edge of mudbanks; at dusk, progressively move into dense vegetation (Bauer and Glutz 1969). In winter quarters, Camargue, roosting communal and diurnal (Tamisier 1966, 1970, 1974), birds spending 6–10 hours per day sleeping on water or shore on bare ground or in clumps of vegetation, or perching in trees, with little effective feeding. Formation of permanent day-time flocks, reduction in diurnal feeding, restriction of preening and sleeping to certain times of day, and separation of loafing areas from nocturnal feeding grounds thought to be determined by pressure from diurnal bird-predators. In same locality, flighting to and from predator-free feeding grounds occurs crepuscularly. Nocturnal courtship common.

2. Studied mainly in captivity: nominate *crecca* by Lorenz (1953); von de Wall (1963); Johnsgard (1965); *carolinensis* by McKinney (1965c). Displays closely similar in both races. Studies of wild birds restricted, but see Boase (1925), Tamisier (e.g., 1974). FLOCK BEHAVIOUR. Little detailed information. Main pre-flight signals as in Mallard *A. platyrhynchos*. Wintering flocks, Camargue, react to flying predators with loud whistling by ♂♂. If on land, birds fly and settle on water where re-group and swim away in tight pack; eventually fly in close formation on roughly circular course, remaining on wing over water until predator stops hunting (Tamisier 1970). ANTAGONISTIC BEHAVIOUR. Hostile encounters between ♂♂ associated mainly with pair-formation. More quarrelsome than most *Anas* during communal courtship, bouts being interspersed with periods when chasing and fighting between ♂♂ predominate (McKinney 1965c). At close quarters, threaten with Chin-lift (Bill-up) display in which head sunk into shoulders and bill raised; accompanied by threat-calls, including whistles at higher intensity when also perform rapid, repeated, lateral Head-shakes. Open-bill threats also occur during pairing disputes. ♀♀ threaten ♂♂ during pairing, often in form of Inciting-display in which sideways, neck-jerking movements of head made repeatedly at rejected ♂♂ while Inciting-call given. In *carolinensis*, ♂ adopts special Bill-down posture while holding ground in front of aggressive ♀ in courting group, with crown pointing towards her (F McKinney). In *carolinensis*, aggression in breeding area, just before and during nesting, associated with rape attempts rather than defence of territory, paired ♂ attacking others trying to rape mate; in absence of ♀♀, ♂♂ mostly peaceful. ♀'s Repulsion-display during laying and incubation similar to other

A

*Anas.* COMMUNAL COURTSHIP. Occasionally begins August (Lebret 1961) but not regular until October–November, continuing until June (Bezzel 1959). Usually 5–7 ♂♂ in group and 1 ♀, but up to 25 ♂♂ in spring (McKinney 1965c); additional ♂♂ strongly attracted by displaying group, particularly by whistling of ♂♂. In autumn and early winter, participating ♂♂ largely unpaired but, in *carolinensis*, paired ♂♂ increasingly take part, leaving mates to join courting groups. ♂♂ rapidly swim round ♀, jockeying for positions from which to display to her while making aggressive moves towards rivals or avoiding them. Short Jump-flights over a few meters also fairly common; less obviously ritualized than in *A. clypeata* (Lebret 1958a). Major displays, all accompanied by Courtship-calls, include Burp, Water-flick, Head-up-Tail-up, and Down-up; except in case of Burp-display, and sometimes Down-up, synchronization of such displays in sudden bursts not highly developed. ♂♂ also utter rapid series of multi-syllable calls without special posturing. Secondary displays include Upward-shake, Head-shake, Head-flick, Wing-flap, and other movements more typical of pair-courtship. At beginning of courting bout, and at intervals later, several ♂♂ perform series of Burp-displays lateral to ♀; with each brief lifting of head, feathers of crown and nape raised, bill opened wide, and loud Burp-call uttered, while dorsal feathers and wings vibrated (von de Wall 1963; McKinney 1965c). Upward-shakes, lateral to ♀, also performed as introduction to courting bout and intermittently later, when often immediately precede major displays; similarly, Head-shakes and Head-flicks can occur independently of other displays or linked with them. Water-flick (Grunt-whistle) most frequent of major displays: much as in *A. platyrhynchos*, but quicker; frequently preceded by several Head-shakes (until bird in suitable position relevant to ♀) and often terminated by integral Head-flick while Tail-wagging. Immediately followed (in 50% of cases) by Head-up-Tail-up display with body broadside to ♀; more exaggerated even than in *A. platyrhynchos*, exhibiting yellow and black pattern of under tail-coverts (see A), culminates in swing of body to face ♀ with head erect while single whistle given. This Turn-Body-to-♀ component held for several seconds; then ♂ sometimes Nod-swims away (see below) or, more frequently, performs Turn-back-of-Head display and attempts to ceremonially Lead ♀ while swimming in front of her. Head-up-Tail-up display apparently never performed on own, only as sequel to Water-flick. Both Leading and Nod-swimming may occur independently of Head-up-Tail-up, often during lulls in courtship as ♂♂ manoeuvre for positions in case of Nod-swimming. Latter display lacks extended neck posture found in *A. platyrhynchos*, ♂ just jerking head forward and backwards. Down-up display usually performed broadside to ♀, nearer than Water-flick and only when another ♂ close (see below). Differs considerably from Down-up of *A. platyrhynchos*, superficially

resembling Head-up-Tail-up. Preceded by Head-shakes in Chin-lift posture, then bill sharply lowered while 3 rapid whistles given as rear of body swung high to meet head and immediately lowered, ♂ again assuming Chin-lift posture (McKinney 1965c). When courting group close to shore, ♂♂ may land and perform other displays lateral to ♀; in Bridling-display (see B), ♂ pushes retracted head back over mantle while pouting chest and giving single whistle; in Preen-Belly display, lowers head to touch belly with bill. Both displays usually preceded by Body-shake which also occurs on own. Inciting-display main pairing activity of ♀, both on land and water; performed beside preferred ♂ and directed at rejected ♂♂. Also occurs in flight (Lebret 1958b), especially during Rape-intent pursuits. ♀ also Nod-swims like ♂ but this stimulates ♂ displays less than equivalent behaviour of ♀ *A. platyrhynchos*. At times ♀♀ give loud Single-quacks in response to major courtship displays of ♂. PURSUIT-FLIGHTS. Common. (1) Courtship-flights. Courting group often takes wing, from water or land, and flies round for short periods; ♂♂ call in flight, though no special aerial displays noted (McKinney 1965c). Seen from December in Netherlands (Lebret 1961). (2) 3-bird flights. In Netherlands, common during February–April; similar to those of *A. platyrhynchos* (Lebret 1961), but probably overlap more with pursuits of 3rd type if indeed distinguishable. (3) Rape-intent flights. Later in spring, paired ♂♂ leave mates to pursue other ♀♀ vigorously, on water and land and in air; frequently culminate in rape, each ♂ afterwards returning to own mate (F McKinney). Such pursuits continue while ♀ laying and incubating. PAIR BEHAVIOUR. In *carolinensis*, paired ♂ attending courting group later directs displays to mate should she join in (F McKinney); if another ♂ approaches paired ♀, her mate often does Down-up display near both, all 3 tending to align themselves parallel. ♀ will also then Incite next to mate. When pair separated, Burp-calling by ♂ and Decrescendo-calling by ♀ re-establish contact. A number of displays, first performed by ♂♂ during communal courtship also occur when pair alone, ♂ positioned lateral to ♀. As well as Bridling and Preen-Belly displays on land, include 2 further Mock-preen displays (Preen-dorsally and Preen-behind-Wing), brief Bill-dips (often preceding Mock-preen displays), and full Ceremonial-drinking. Bridling, Ceremonial-drinking, and (sometimes also by ♀) Preen-behind-Wing display in particular likely to occur when mates meet after separation; also, less frequently, Inciting by ♀—but no well-developed Pair-palaver, as (e.g.) in Wigeon *A. penelope*, recorded. In *carolinensis*, Persistent-quacking by ♀ associated with exploratory-flights by pair at time of nest-site selection; Decrescendo-calls sometimes incorporated. Exact role of copulatory behaviour in pair-courtship uncertain. In Netherlands, copulation never observed in winter and pre-copulatory display rarely (Lebret 1961); elsewhere, copulation recorded as

B

early as November (Bezzel 1959). COPULATION. Pre-copulatory display consists of mutual Head-pumping and ♂'s post-copulatory display of Bridling (accompanied by whistle)—only situation where this performed on water. In case of rape, no formal preliminaries, 1 or more ♂♂ assaulting ♀ vigorously—at times, including own mate if unable to drive others off (F McKinney). In captivity, ♀♀ sometimes drowned by rape assaults. RELATIONSHIPS WITHIN FAMILY GROUP. Little detailed information, but probably much as in *A. platyrhynchos*. Broods remain hidden in vegetation much of day, moving more freely in open water at night (Hildén 1964).

(Fig A after Lorenz 1953 and Johnsgard 1965; Fig B after McKinney 1965*a*.)

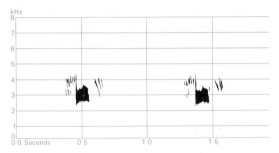

I Sture Palmér/Sveriges Radio Sweden April 1968

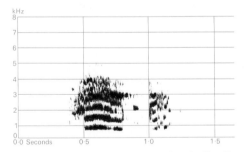

II Sture Palmér/Sveriges Radio Sweden April 1968

**Voice.** Calls—especially those of ♂—often much in evidence during winter and spring displays. Main note of ♂ penetrating, melodious, and far-carrying whistle, just discernibly of 2 syllables. Rendered, e.g.: 'prip-prip' (Kirkman and Jourdain 1930), 'krick' or 'kedick' (Lorenz 1953), 'krick-et' (Johnsgard 1965), 'thu' (McKinney 1965*c*); recording (see I) shows that each note lasts almost 0·1 s and ascends a semitone from about D to E flat (J Hall-Craggs). Calls of ♀ quite different, consisting of various forms of quacking, higher pitched and more rapid than corresponding calls of ♀ Mallard *A. platyrhynchos*; many have harsh, grating, and, at times, squeaking quality (F McKinney). Statements that ♀♀ seldom call (see Millais 1902; Witherby *et al.* 1939) or have restricted vocabulary not warranted; that flight-calls of sexes similar (see Géroudet 1965) highly unlikely. Whistling noise produced by wings in flight. Only other reported non-vocal sound unusual noise like water thrown in fine jet (see Bent 1923); almost certainly made by water directed at ♀ during Water-flick display of ♂.

CALLS OF MALE. Following listed by F McKinney (see also McKinney 1965*c*). (1) Burp-call: series of even-spaced whistles delivered from Burp-posture, one each time head raised (see I). During communal courtship, often given in bursts as several ♂♂ perform Burp-display simultaneously; also uttered in situations of mild alarm, and when ♂ separated from mate. Equivalent of Repeated-calls of ♂ Shoveler *A. clypeata*; heard from autumn until late spring or early summer. Functions mainly in attracting ♀'s attention during pairing, warning her of danger, and maintaining contact between mates. (2) Other Courtship-calls: single loud whistle given at peak of Water-flick display, followed by quiet grunt (Grunt-whistle); single quieter whistle during Turn-body-to-♀ phase of Head-up-Tail-up Display; 3 rapid whistles, rendered 'zee-zee-zeet' by Johnsgard (1965), during Down-up display; single whistle during Bridling-display on land (also in water after copulation). (3) Multi-syllable calls: variety of staccato whistles given during communal courtship but not linked to any distinctive visual display; often rapid series of 3–4 notes—e.g. 'te tiu te' or 'te tui tu te-te'. (4) Threat-calls: quiet, wheezy notes, rapidly repeated, given with Chin-up display; at high intensity, merge into rapid peeping whistles. (5) Other calls: include very quiet, peeping notes when pair together; audible only at close range.

CALLS OF FEMALE. Vocabulary identical to that of ♀ *A. platyrhynchos* (Lorenz 1953). Series of loud quacks (see II) uttered in a variety of situations, e.g. in flight with mate, at times during communal courtship in response to ♂ displays (Single-quacks), and immediately before egg-laying (Persistent-quacking); in most cases, may play role in maintaining pair-bond (F McKinney). Other main calls listed by F McKinney include following. (1) Decrescendo-call: series of usually 4–7 loud quacks, 1st longer and higher pitched; usually 4 notes and frequently given (Johnsgard 1965). Used to re-establish contact when mates separated. (2) Inciting-call: series of harsh, rapid, rattling notes, higher pitched than corresponding call of ♀ *A. platyrhynchos*. (3) Repulsion-call: loud, often squeaky quacks given when harassed by ♂♂ intent on rape; recording, of ♀ pursued by ♂, suggests repeated nasal cackling—'gack ki-ki-ki-ki . . .' (4) Distraction-call: loud, excited-sounding quacks, especially when disturbed while leading young; rendered 'whelp whelp' (Coward 1920). Bird flushed from nest gave low, continuous, frog-like croaking (see Witherby *et al.* 1939).

CALLS OF YOUNG. Little detailed information.

**Breeding.** SEASON. See diagram for north-west Europe. Egg-laying starts central Europe mid-April; in south Finland, early to mid-May; in USSR, late May to early June (Bauer and Glutz 1968; Dementiev and Gladkov 1952; Hildén 1964). SITE. On ground in thick cover sheltered by overhanging tussocks or bushes; exceptionally in crown of tussock up to 50 cm off ground. Not colonial but sometimes nests only 1 m apart. Never far from water. Nest: slight hollow, with lining and rim of leaves, grass, and other vegetation; also much down. Outside diameter 21 cm, inner 13 cm, depth of cup 6 cm (Hui and Hui 1968). Building: by ♀, shaping cup with body, and gathering material within reach of nest. EGGS. Blunt ovate, smooth; yellowish-white. 45 × 33 mm (42–50 × 31–36), sample 250 (Schönwetter 1967). Weight 29 g (25–31), sample 19 (Verheyen 1967). Clutch: 8–11 (7–15). One brood. Replacement clutches after egg loss. INCUBATION. 21–23 days. By ♀ only, covering eggs with down when off nest. Starts with last egg; hatching synchronous. Eggshells left in nest. YOUNG. Precocial and nidifugous; self-feeding. Cared for by ♀, brooded at night when small. May return to nest for first few nights. FLEDGING TO MATURITY. Fledging period 25–30 days. Independent at or just before fledging. Age of first breeding 1 year. BREEDING SUCCESS. Little data. Average size of 9 young broods, south Finland, 7·6, and of 15 well-grown broods 4·8 (Hildén 1964).

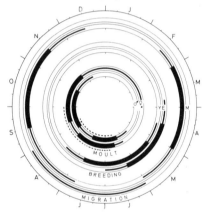

**Plumages.** Nominate *crecca*. ADULT MALE BREEDING. Forehead, crown, and narrow strip along base of upper mandible chestnut, bordered by narrow cream-buff line. Broad patch round eye extending to nape glossy green, bordered in front and on cheek by narrow cream line. Chin black, rest of head chestnut. Narrow ring round lower neck, mantle, inner scapulars, outer webs of lower scapulars, sides of body, and flanks finely vermiculated black and white. Outer scapulars white, broadly edged black on outer web. Inner web of lower scapulars dark grey; back and rump dark grey, often finely vermiculated white, feathers edged olive-grey. Upper tail-coverts black with broad buff edges. Chest and breast white, dotted black; in fresh plumage clouded with buff. Belly and vent white, latter narrowly barred grey. Central under tail-coverts black, longer edged pale buff; those at sides cream-buff. Tail dark grey, narrowly edged off-white. Primaries and primary coverts sepia, inner webs paler, greater coverts margined grey. Secondaries grey-brown, narrowly tipped white or pink-buff; outer webs of 3–6 inner metallic green, of others velvety-black (partly glossed green towards base). Tertials dark olive-grey, outer edges black with narrow white margin (in some freckled black); outer tertial, bordering speculum, with velvety black outer edge, narrowly bordered pink-buff outside and white or silvery-grey inside. Greater upper wing-coverts white, tinged cinnamon at tips, inner with much grey at base and some grey freckling at tip. Median and lesser wing coverts dark ash-grey

with black shafts, in some with olive-brown tinge; tertial coverts similar, narrowly tipped buff. Under wing-coverts dark grey, tipped white; greater and axillaries white, latter freckled dusky at tip. ADULT FEMALE BREEDING. Forehead, crown, hindneck, and streak through eye to nape dark grey-brown, feathers broadly edged pink-buff and grey; rest of head pale buff to white with numerous dark spots, fewest on chin and throat. Mantle and upper scapulars sepia, feathers with olive-grey edges and pink-buff wavy bars. Lower scapulars sepia, narrowly edged pale buff and with irregularly U-shaped pale cinnamon marks and bars. Feathers of back and rump black, edged grey and subterminally marked buff; upper tail-coverts same, with narrow off-white U-shaped marks. Feathers of chest and sides of breast white with dark subterminal crescents, tips broadly edged pink-buff shading to white or grey-buff. Feathers of flanks sepia with broad pale buff edges and streaks. Breast, belly, and vent white with variable subterminal pale grey mottling, under tail-coverts white, spotted or streaked dusky-brown. Tail-feathers dark grey, narrowly edged white, variable number of feathers marked buff. Wing like adult ♂, but white tips of secondaries wider, outer secondaries often mottled white, and green gloss less extensive. Lesser and median coverts darker grey, narrowly tipped grey to white. Tertials sepia, narrowly edged and irregularly streaked buff, those bordering speculum with broad black streak and narrow white margin on olive-grey outer web. Another ♀ morph has crown, upperparts, and chest black with rufous-buff marks and edges; feathers of mantle, scapulars, back, and rump with cinnamon edges and U-shaped marks instead of pink-buff bars. ADULT MALE NON-BREEDING (Eclipse). Like adult ♀ breeding, but crown and hindneck darker, dull black, feathers only narrowly edged buff; sides of head with more numerous deeper but smaller black spots, light streak over eye indistinct. Scapulars sepia with olive tinge and buff edges and subterminal spots, sometimes vermiculated white. Underparts like ♀, but sides of breast with more extensive dark centres and often white wavy bars, sides of body and flanks often partly vermiculated; breast and vent often with round dark subterminal spots. Wing like adult ♂ breeding, but tertials olive-grey with broad brown-black edge to outer web; outer tertial like adult ♂ breeding. Several morphs occur, varying in pattern and coloration of mantle: feathers sepia with olive-grey edges (plain morph), often with variable amount of black and white subterminal vermiculation or freckling (freckled morph); or with buff edges and subterminal pink-buff crescents or wavy bars (barred morph: marks usually narrower and less extensive as in adult ♀ breeding). ADULT FEMALE NON-BREEDING. Head and neck like adult ♂ non-breeding mantle and scapulars dark sepia with narrow pink-buff or grey feather-edges and small buff subterminal spots or crescents. Differs from plain morph of adult ♂ non-breeding in wing (like adult ♀ breeding) and by absence of bars or vermiculations at sides of breast and flanks. Tail dark grey-brown, narrowly edged off-white. Rest of body like ♀ breeding, but back and rump usually unmarked, feathers of flanks shorter and duller, underparts often more heavily mottled grey. Tertials like adult ♂ non-breeding, but edge of outer more olive-black with broader pale buff margin; rest of outer web ash or olive-grey, rather than white or silver-grey. DOWNY YOUNG. Like that of Mallard *A. platyrhynchos*, except for smaller size, better developed dark line over cheek, and (sometimes) cinnamon-tipped down on upper-parts. JUVENILE MALE. Like barred morph of adult ♂ non-breeding or adult ♀ breeding, but underparts usually with narrow dusky shaft-streaks or spots, tail-feathers with notched tip, bare shaft projecting, tertials short and narrow. Head like adult non-breeding, but crown tinged dark grey-brown, feather edges olive-grey or pale rather than warm buff. Mantle dark grey-brown, not as dark as adult, feathers narrowly edged grey, marked with narrow incomplete cream bar or V. Scapulars narrow and with pointed tip, rather than long and wide with square tip as adult non-breeding. Back and rump dark grey, often edged and marked like mantle. Underparts like adult ♀, but marks on flanks often incomplete and paler buff, chest and sides of breast sometimes without white marks, sepia centres smaller; feathers of breast, belly, and vent narrow, often with narrow dark streak at tips, breast and vent sometimes with many small dusky spots, smaller than those of non-breeding adult plumage. Tail dark grey-brown, narrowly edged off-white or buff. Wing like adult ♂, but grey of upper wing-coverts somewhat duller and browner; tertials shorter and narrower as in adult non-breeding, dark sepia with outer web slightly greyer and narrowly edged white; outer tertial, bordering speculum, with outer web ash with broad black streak along narrow pale buff margin. JUVENILE FEMALE. Like juvenile ♂, but creamy marks on mantle usually absent or restricted to small spots at sides of feathers; back and rump sepia, feathers edged pale grey, often unmarked subterminally. Wing like adult ♀. Tertials like juvenile ♂, but outer browner, dark streak over outer web sepia rather than velvety black. Wing differs from juvenile ♂ in contrasting grey edges to lesser and median coverts, and in less extensive green on speculum. FIRST BREEDING MALE. Like adult ♂ breeding, but sometimes some juvenile tail-feathers, tertials, or feathers on upperparts retained until spring. Wing juvenile; often difficult to separate from adult. Inner side of tip of outer greater primary coverts sometimes narrowly edged white in juvenile; median coverts frayed, narrow, and trapezoidal, against more broad and smoothly rounded in adult (Carney 1964). FIRST BREEDING FEMALE. Like adult ♀ breeding, but part of juvenile plumage on body, tail, and tertials retained until spring. Median upper wing-coverts edged buff in juvenile (but also in some adults), shape and wear like 1st breeding ♂.

**Bare parts.** ADULT MALE. Iris dark hazel. Bill dark slate; when in eclipse, sides of upper mandible orange-yellow or olive-green. Foot olive-grey to brown-grey, webs dark grey to dull black. ADULT FEMALE. Iris light brown to dark hazel. Bill dark slate or olive-grey on culmen and tip; pink-yellow or orange-yellow with some black dots at sides of upper mandible. Foot like adult ♂. DOWNY YOUNG. Bill dark slate, edges of upper mandible and lower mandible basally yellow-brown. Foot blue-grey or grey-brown with darker webs. JUVENILE. Iris and foot like adult. Bill pink-horn on culmen and tip, paler pink-yellow or flesh on lower mandible and at edges of upper; ♀ usually with small dots at lower edge of upper mandible. Adult colour attained from September onwards. (Bauer and Glutz 1968; Schiøler 1925; Witherby *et al.* 1939; RMNH.)

**Moults.** Timing and sequence mainly as in *A. platyrhynchos*. ADULT POST-BREEDING. Complete. Head and body in ♂ between early June and late July; in ♀ later; part of breeding plumage, especially on back and rump, sometimes retained until September. Flight-feathers simultaneous, flightless for *c.* 4 weeks; ♂♂ between early July and late August, ♀♀ mid-July to late September. Tertials and tail-feathers moult at same time, but somewhat earlier start. ADULT PRE-BREEDING. Partial. In ♂, head and body mainly September–October, tertials and central and variable number of other tail-feathers October–November. In ♀, head, body, and central tail-feathers late September to November, but sometimes part of body plumage and usually tertials and part or all other tail-feathers retained until February–March. POST-JUVENILE AND FIRST PRE-BREEDING MALE. Partial. Some 1st

immature non-breeding August–September resembling adult non-breeding, but juvenile and 1st non-breeding usually completely replaced by breeding September–November. Underparts moulted first, followed by head and rest of body; tertials mainly October–November (sometimes as late as February); central tail-feathers September–November, rest of tail completed from October to spring. POST-JUVENILE AND FIRST PRE-BREEDING FEMALE. Partial. 1st immature non-breeding (resembling adult ♀ non-breeding) August–October; partly retained until spring, but usually much of it, and of remaining juvenile, changed for breeding September–December. Some individuals in complete breeding plumage on body by December, others usually not before February–March, after moult-stop in winter. Tertials moulted in spring, rarely earlier. Juvenile tail sometimes retained until March, but usually at least central pair changed for adult from September; sometimes whole tail like adult November. SUBSEQUENT MOULTS. Like adults. (Dementiev and Gladkov 1952; Schiøler 1925; C S Roselaar.)

**Measurements.** Netherlands, whole year. Skins (RMNH, ZMA).

| | | | | | | |
|---|---|---|---|---|---|---|
| WING AD ♂ | 187 | (3·26; 34) | 181–196 | ♀ 180 | (2·67; 22) | 175–184 |
| JUV | 184 | (4·20; 63) | 176–192 | 177 | (4·48; 30) | 166–185 |
| TAIL AD | 66·9 | (2·13; 14) | 64–71 | 64·7 | (2·71; 10) | 62–69 |
| JUV | 60·9 | (2·56; 18) | 56–66 | 59·8 | (1·96; 20) | 56–63 |
| BILL | 36·4 | (1·37; 85) | 34–40 | 34·9 | (1·29; 50) | 32–38 |
| TARSUS | 30·4 | (0·95; 38) | 29–32 | 29·8 | (0·98; 32) | 28–31 |
| TOE | 39·1 | (1·37; 38) | 37–42 | 38·6 | (1·67; 32) | 36–40 |

Sex differences significant for wing and bill. Juvenile tail significantly shorter than adult breeding tail; differences of other measurements not significant, so bill, tarsus, and toe combined. Tail of adult non-breeding and 1st breeding about as long as juvenile tail.

**Weights.** Averages several areas (sample size, when known, in brackets), and total range. USSR: Dementiev and Gladkov (1952), Dolgushin (1960). France: Bauer and Glutz (1968). Netherlands: ZMA, RIN.

ADULT

| | | | | |
|---|---|---|---|---|
| (1) | ♂ *c.* 275, 341 (300); | 200–371 | ♀ *c.* 240, 308 (285); | 185–400 |
| (2) | 347, 359 (100); | 240–375 | –, 324 (48); | 250–430 |
| (3) | 360, | 320–400 | 341, 410; | 250–450 |
| (4) | 284 (11); | 230–440 | 289 (11); | 250–330 |
| (5) | 334, 334; | 250–450 | 323, 325; | 300–375 |
| (6) | 350 (13), 305 (23); | 163–500 | 280 (4), 274 (16); | 260–410 |
| (7) | 335 (54); | 250–425 | 294 (48); | 200–430 |

(1) Rather variable January–March (Azerbaydzhan, USSR, and Camargue, France, respectively). (2) Increase before departure April, same localities. (3) Summer quarters, prior to breeding, April–June (north USSR: ♂ Pechora River, ♀ Kamchatka and south Yamal). (4) Low during breeding June–July (Kazakhstan, USSR). (5) Increase prior to wing moult (Rybinsk reservoir, shortly after start moult, and Volga delta, USSR). Decrease during moult, followed by increase prior to departure from breeding or moult area, after which birds arrive lean in winter quarters. (6) September–October (moult area and winter quarters, Netherlands and Camargue respectively). (7) November–December (Camargue).

JUVENILE. Weight of some like adults from August, but average generally lower until November, or later when conditions adverse. Breeding area September (Rybinsk reservoir and Netherlands, respectively): ♂ 378, 322 (19); 290–420 ♀ 325, 289 (4); 275–390.

**Structure.** Wing rather long, narrow, pointed. 11 primaries: p9 longest, p10 0–3 shorter, p8 4–9, p7 12–18, p6 20–30, p1 75–92. Inner web of p10 and outer p9 emarginated. Tail short, rounded; 16 pointed feathers, central pair projecting. Bill slightly shorter than head, somewhat higher than broad at base, narrow. Culmen slightly concave, decurved at tip. Sides of bill almost parallel; nail small and narrow. Tertials and scapulars long and pointed in breeding plumage, shorter and rounder in non-breeding. Feathers of hindcrown and hind-neck slightly elongated. Outer toe *c.* 92% middle, inner *c.* 78%, hind *c.* 28%.

**Geographical variation.** Slight clinal variation in size, becoming larger towards east. North American *carolinensis* similar in size to nominate *crecca*, but plumage different. Outer scapulars of breeding ♂ dark olive-grey, outer webs vermiculated like mantle, inner variably vermiculated white; outer edges of scapulars sometimes black (no white streak on scapulars as in nominate). Broad white crescent at sides of breast in front of closed wing. Cream lines on sides of head restricted, especially those bordering forehead and crown. Vermiculation of upperparts and flanks somewhat finer. Non-breeding, ♀, and juvenile plumages indistinguishable from nominate *crecca*, although tips of greater upper wing-coverts tend to have more extensive and deeper cinnamon tinge, especially on outer, but variable and with much overlap.

CSR

## *Anas capensis* Cape Teal

PLATES 67 and 74
[between pages 518 and 519
and facing page 543]

Du. Kaapse Taling     Fr. Sarcelle du Cap     Ge. Fahlente
Ru. Капский чирок     Sp. Pardilla del Cabo     Sw. Kapbläsand

*Anas capensis* Gmelin, 1789

Monotypic

**Field characters.** 44–48 cm, of which body two-thirds; wing-span 78–82 cm. Medium-sized dabbling duck with upturned pink bill, high rounded forehead, bulbous nape, and generally pale dappled appearance; green and black speculum bordered with white. Sexes similar and no seasonal differences; juvenile difficult to distinguish.

ADULT MALE. Whole head and underparts silvery-grey to buff-white, indistinctly flecked with black-brown on head and boldly spotted light brown on breast and darker brown on flanks; white throat, and faint pale collar; upperparts brown with pale red-buff feather edgings. Tail grey-brown edged with white; wings dark brown-grey with white outer secondaries and green and black speculum broadly bordered with white. Bill depressed near base, broad and markedly upturned near tip, pink shading to lilac at tip, with black edges and black surround to base; eyes variable from brown through yellow to red and deep orange. Feet and legs yellow to yellow-brown with dusky webs. ADULT FEMALE. Virtually indistinguishable, though slightly smaller and described as paler and less speckled (see Winterbottom 1974). JUVENILE. Similar to adult but slightly duller and less speckled below.

Easily distinguished by generally pale, spotted colouration and pink bill; conspicuous U-shaped white edges to speculum diagnostic. Marbled Teal *Marmaronetta angustirostris*, only other pale dappled duck in west Palearctic, paler on upperparts and otherwise much less grey, more cream and brown, with dark bill, shaggy head, dark smudge around eye, and no well-marked speculum; Red-billed Teal *Anas erythrorhyncha*, with similar range to *A. capensis* in Africa, much darker with dark brown cap.

Usually in pairs or small parties, but moulting flocks of several hundreds occur. Swims well and dives more than most *Anas*, swimming under water with wings apparently closed. Flight not fast and usually rises in long, gradual ascent; does not normally fly far or high when disturbed, though can jump and change direction rapidly like Teal *A. crecca*. Spends much time on shore, where walks with not ungainly waddle. Moulting birds swim with head and neck outstretched and body half-submerged, diving if approached closely. Feeds mainly by swimming with beak submerged or by up-ending. Generally rather silent.

**Habitat.** Tropical and warm temperate; on fresh and brackish waters and soda or salt lakes or salt-pans, including major lakes, reservoirs, and marshes, from lowlands up to over 1700 m in East African Rift Valley. Usually in open or savanna country, nesting sometimes on islets in open situations, or under thick bushes. Evidence conflicting as to frequency of flying to different waters. Full extent of habitat preferences not yet recorded.

**Distribution.** Breeds over large area of southern and eastern Africa, north to Ethiopia and Sudan (Winterbottom 1974), and, in northern tropics, perhaps west to Lake Chad basin (but see Movements).

Accidental. Libya: Kufra oasis, April 1961, and one found dead 250 km north-east earlier; 2 pairs seen daily 31 March–5 April 1968, display and territorial behaviour suggested intention to nest (Cramp and Conder 1970).

**Movements.** Nature uncertain; extended breeding season (e.g. Urban and Brown 1971) must preclude any simple pattern. Apparently scarce wet season migrant breeder to Darfur, west Sudan, arriving late April (Lynes 1925). No proof yet of nesting in northern tropics west of Darfur; those seen irregularly December–April in north-east Nigeria, south-east Niger, and in Chad perhaps non-breeding visitors, though may be small resident population more conspicuous in dry season when concentrated on the relatively few permanent water-bodies. 300 once reported Lake Chad (Vielliard 1972), otherwise no records over 80 birds. In Chad found as far north as Ounianga Kébir (200 km from Libyan border), where one shot April 1954 and 50–60 present December 1963, but unknown whether resident or migratory there (Salvan 1967). Dry season dates, in conjunction with Darfur situation and records from Libyan desert (see Distribution), suggest seasonal movement in response to arid conditions.

**Voice.** Generally rather silent (Delacour 1956; Winterbottom 1974). In Burp-display ♂ utters whistle, described as clear and of 3 syllables, 'oo-WHEE-oo' (Johnsgard 1965), also as husky and like nasal squeak of Marbled Teal *Marmaronetta angustirostris* (Delacour 1956). Calls of ♀ low and quacking but Inciting-call rather harsh 'rrak'.

**Plumages.** Apparently no non-breeding plumages discernible. ADULT MALE. Crown and sides of head pink-buff, rest of head and neck white, with longitudinal dark brown streaks or spots except chin and throat. Mantle dark grey-brown to dull black, each feather with broad pink-buff edge and subterminal bar, encircling large roundish black subterminal spot. Scapulars dark grey-brown broadly edged pink-buff, some shorter ones barred pink-buff towards base. Back, rump, and upper tail-coverts dull black, barred buff. Chest and upper breast narrowly barred pale buff or white and dark grey-brown; flanks and under tail-coverts same, but more broadly barred or spotted. Feathers of belly and vent pale grey-brown, broadly edged white. Tail-feathers dark grey, narrowly edged pale buff. Primaries black with grey bloom; inner narrowly edged pale buff on outer web. Outer secondaries white (bases dark grey), others with dark grey inner webs and glossy green outer webs; those bordering white ones velvety black. All secondaries broadly tipped white; tips narrowly bordered black subterminally. Tertials like scapulars; outer 2 (bordering green speculum) ash, with broad and sharp black edges to outer web. Upper wing-coverts dark grey-brown, more or less tinged dark grey at tip, greater broadly tipped white. Under wing-coverts dark grey, edged white; axillaries white. When feathers worn, pink-buff colour fades to pale grey or white. ADULT FEMALE. Like adult ♂, but fewer and broader bars on chest and upper breast which appears spotted instead of narrowly barred. Black on outer 2 tertials usually more restricted, and gradually shading into grey-brown of rest of webs. JUVENILE. Like adult ♀, but edges and bars of upperparts narrower and

paler buff; feathers narrower, black subterminal spots smaller. White bars and edges of feathers of underparts contrast less with grey-brown rest of feathers, underparts appearing paler and only faintly spotted. Tertials shorter, outer 2 duller, less ash. Tail-feathers notched at tip, bare shaft projecting. FIRST ADULT PLUMAGE. Like adult.

**Bare parts.** Iris variable: light brown, yellow, red, or deep orange; in at least some, ♂♂ yellow and ♀♀ orange-brown. Bill pink, shading to lilac and light-blue at tip; base, culmen from nostrils to forehead, and edges along bill-tip black (in young, black sometimes absent). Foot yellow-ochre to dull yellow-brown, webs dull dark grey. (Delacour 1956; Winterbottom 1974; C S Roselaar.)

**Moults.** ADULT. Limited data on timing available; probably varies locally, and related to rain periods. Only one complete moult a year; flight-feathers simultaneous. Flightless for 23–24 days. JUVENILE. Fully fledged at *c.* 7 weeks. Head, body, and sometimes tertials and tail moulted at *c.* 3 months, wing at *c.* 6. Next moult when about 1 year old. (Winterbottom 1974; C S Roselaar.)

**Measurements.** Sex and age unknown: (1) 12 skins South African Museum; (2) 52 live specimens Rondevlei, South Africa (Winterbottom 1974). Average and range.

|     | WING | TAIL | BILL | TARSUS |
|-----|------|------|------|--------|
| (1) | 196 (180–206) | 64·6 (54–75) | 48·7 (45–55) | 34·7 (30–41) |
| (2) | 194 (168–206) | 64·3 (58–74) | 39·6 (36–44) | 37·0 (32–40) |

Bill in (1) apparently to skull, in (2) exposed culmen. Data for few additional South African skins RMNH and ZMA give slight indication of sex differences: wing adult ♂ 201, 204, 209; ♀ 194, 197. Bill adult ♂ 40, 42, 46; ♀ 39, 40. Juveniles slightly shorter wing.

**Weights.** Full-grown, dates not given. Cape Province, South Africa: 402 (63) 342–590 (Winterbottom 1974). Lake Nakuru, Kenya: 419 (35; 24) 382–505 (Britton 1970).

**Structure.** Wing rather long, pointed. 11 primaries: p10 and p9 longest, about equal, p8 3–8 shorter, p7 8–14, p6 17–24, p1 65–84. Inner web of p10 emarginated, and also, slightly, inner and outer p9. Tail short, slightly rounded; 14 feathers. Bill as long as head, higher than broad at base; upcurved, depressed at tip; tip slightly wider than base. Culmen concave. Nail small and narrow. Forehead high and rounded. Outer toe *c.* 93% middle, inner *c.* 76%, hind *c.* 28%. CSR

## *Anas platyrhynchos* Mallard

PLATES 68 and 74
[between pages 518 and 519 and facing page 543]

Du. Wilde Eend    Fr. Canard colvert    Ge. Stockente
Ru. Кряква    Sp. Ánade real    Sw. Gräsand

*Anas platyrhynchos* Linnaeus, 1758

Polytypic. Nominate *platyrhynchos* Linnaeus, 1758, Iceland, Europe, Asia, and North America. Extralimital: *conboschas* C L Brehm, 1831, Greenland; 3 subspecies in south USA and Mexico, and 2 Hawaiian Islands.

**Field characters.** 50–65 cm, of which body just over two-thirds, ♂ averaging larger; wing-span 81–98 cm. Large, heavily built dabbling duck with rather long head and bill; ♂ has greenish-yellow bill, green head, white neck ring,

rusty breast, mainly grey body, and black and white stern. Sexes dissimilar, with some seasonal differences. Juvenile closely resembles adult ♀, but just distinguishable.

ADULT MALE. Head bottle-green, separated from

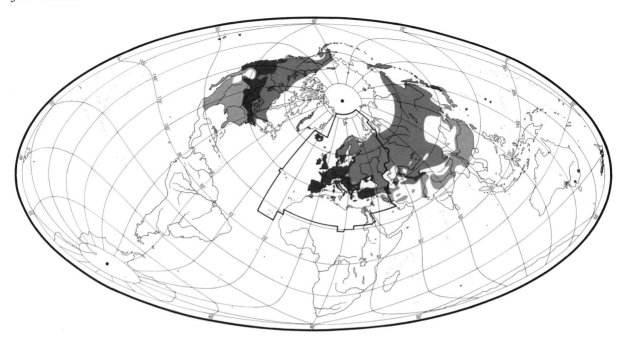

purple-brown breast by narrow, white collar. Body mostly vermiculated grey (darker above than below), contrasting with green-glossed, black upper and under tail-coverts bordered in front by white band; tail grey and white with 2 black central feathers curling up. Speculum blue or purple according to light, edged with black and more broadly with white at both front and rear, showing in flight as bands of these colours running whole length of secondaries. Also in flight, wings mainly grey-brown above contrasting with wedges of dark brown on back and of black on rump; tail appears black-centred and broadly outlined white (pattern otherwise found only in ♂ Shoveler A. clypeata); below, cream under wing-coverts contrast with brown flight-feathers, but look only marginally lighter than underbody. At distance on water, head, breast, and stern appear black and, in sunlight, rest of body looks palest grey. Bill usually greenish-yellow; eyes brown; feet orange-red. MALE ECLIPSE. Superficially like ♀, but easily distinguished by heavier build, yellower bill, blacker crown, paler and greyer face and neck, greyer and more uniform upperparts and darker or warmer brown breast with less streaking. ADULT FEMALE. Mainly brown, mottled, spotted and streaked with blackish. Crown, nape, and line through eye darker than sides of head and neck, leaving narrow supercilium and cheeks of pale buff finely streaked with blackish; throat and foreneck usually unmarked buff. Rest of upperparts blackish with pale feather-edges and thus darker than underparts which are usually buff with blackish streaks, drops, or U-shapes on centres of feathers. Much individual variation in depth of colour and strength of markings. Features in flight are speculum and spread wing much as in ♂; tail broadly outlined greyish-white (pattern otherwise found only in ♀ A. clypeata); and creamy wing-linings

contrasting with brown body. Bill usually dull orange to olive, with blackish ridge and tip; eyes and feet as ♂. JUVENILE. Closely similar to adult ♀, but duller and with underparts more narrowly streaked, without U-shaped marks; young ♂ has crown, nape and upperparts darker. Bill reddish-brown; feet orange. Young ♂ often distinguishable by uniform greenish bill before any noticeable changes in plumage.

♂ unmistakable. Blue or purple speculum bordered with black and white diagnostic of A. platyrhynchos in all plumages. Apart from rather similar Black Duck A. rubripes (which see for differences), ♀ needs to be distinguished from ♀♀ of Gadwall A. strepera (smaller with white belly, clear orange panels on sides of much finer bill, more abrupt forehead, more pointed wings, and black and white speculum producing white square on inner secondaries in flight), Pintail A. acuta (slender build with small head, long thin neck, pointed tail, narrower wings, greyer plumage with more delicate crescentic markings on flanks, blue-grey bill, and indistinct brownish speculum bordered with white behind), Wigeon A. penelope (smaller with much shorter bill, more rounded head, short neck, pointed tail, more rufous colouration, whitish belly, and obscure green and black speculum), and Shoveler A. clypeata (slightly smaller but heavier build, much longer bill, shorter neck, wings set far back in flight, blue forewing, and green speculum with white bar in front and only thin white line behind). Note, however, that both sexes also need to be distinguished from domesticated individuals of A. platyrhynchos which are usually larger and often have plumage differences, such as blackish bodies or excess white on neck and wings.

Catholic in choice of habitat, and found on all types of

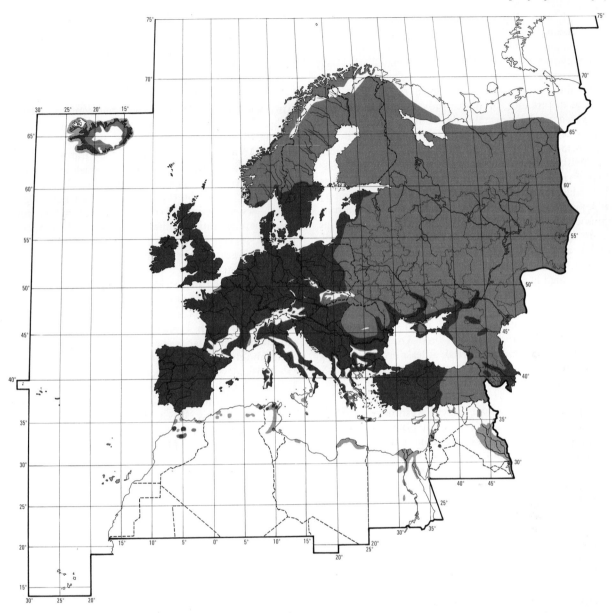

fresh and salt water from small ponds to open sea. Swims high on water with tail cocked; young dive freely and adults occasionally. Rises straight off water, leaping into air, and flies with shallow beats (wings not depressed much below level of body) which are not as fast as many other ducks, but produce whistling sound. Readily comes to land and walks or runs easily with nearly horizontal carriage. Markedly gregarious. Only ♀ produces the familiar deep quacking, while ♂ has soft, nasal, single or double 'raehb', as well as high-pitched whistle and deep grunt during courtship.

**Habitat.** Adaptability to extremely wide range of habitat outstanding characteristic. Throughout middle latitudes of west Palearctic, from Arctic tundra to sub-tropical zone, on or by standing or flowing fresh water, brackish estuaries, and lagoons, or coasts where saline water shallow, fairly sheltered, and within sight of land. Essentially still and shallow-water bird, limited to less than 1 m for foraging, and shunning depths beyond few metres even for resting. Ample plant growth (submerged, floating, emergent, and riparian) attractive; even dense stands of reeds or overhanging branches, flooded swampy woodlands, or seasonal floods. Open waters with mudflats, banks, or spits, and reservoirs, ornamental waters, canals, irrigation networks, and sewage farms also favoured. Hardly any pond or pool too small, at least in breeding season, when also resorts to sites some distance from water.

Although mainly lowland, will extend to 2000 m or higher on occasion (Bauer and Glutz 1968). Generally avoids deep, exposed, rough, fast-flowing, oligotrophic, or rockbound waters, and hard or dry unvegetated areas such as rocks, sand dunes, and artificial surfacing.

No duck more tolerant of human presence or even intense disturbance, or more ready to exploit artificial supplies of food, incidental or deliberately provided. None adapts better to living in cities, and few more readily use man-made structures, even occupied dwellings and special nest-baskets. Also occupies hollows or holes in old trees. Freely visits crop stubbles and watercress beds. Being so adaptable, not wholly tied to aquatic existence, promptly adjusting local numbers to changing food resources. Although highly palatable to man and predators, maintains stocks in face of heavy toll. Can rise steeply up and away even in awkwardly confined spaces; much given to circling around, usually at heights below c. 500 m. Walks readily, even through fairly dense herbage, but underwater use of habitat by adults usually confined to up-ending.

**Distribution.** NORWAY: marked expansion northward since 1870s (Haftorn 1971). Has bred Azores (1903) and Cyprus (1970–1), and may breed, at least occasionally, in Syria, Iraq, and Egypt, though proof lacking.

Accidental. Spitsbergen, Bear Island.

**Population.** Breeding. Only tentative estimates available but despite some local increases, all main geographical populations thought to be declining due to hunting pressure (Isakov 1970a). ICELAND: at least 5000 pairs (Gardarsson 1975). BRITAIN: no evidence of marked change (Parslow 1967); c. 40 000 pairs (Atkinson-Willes 1970a). IRELAND: marked decrease due mainly to drainage (Ruttledge 1970). FRANCE: decreasing last 20 years most areas; 30 000–60 000 pairs (Jouanin 1970), now probably 20 000–40 000 pairs (FR). BELGIUM: c. 10 000 pairs (Lippens and Wille 1972). NETHERLANDS: c. 150 000 pairs (Isakov 1970a). SWEDEN: slight decrease last decade (Curry-Lindahl et al. 1970). FINLAND: c. 160 000 pairs (Merikallio 1958); population doubled one area in south 1951–60 (Grenquist 1970). POLAND: c. 90 000–100 000 pairs (Isakov 1970a). USSR: total population c. 882 000 pairs (Isakov 1970a); for details of numbers in main regions of Soviet west Palearctic (of which highest Baltic with 142 000 pairs), see Isakov (1970b).

Winter. Estimated $1\frac{1}{2}$ million birds in north-west Europe, $1\frac{1}{2}$ million Black Sea–Mediterranean and 1 million Middle East, with total west Palearctic population of some 4–5 million birds (Atkinson-Willes 1975). In western USSR, in not so severe winters, c. 1 100 000 birds (Isakov 1970c).

Survival. From ringing in north-west Europe, mean annual mortality adults 48%, life expectancy 1·6 years (Boyd 1962). Finland: mortality 64% first year, 55% in succeeding years (Grenquist 1970). Sweden: mortality

76% juveniles, 64% adults (Curry-Lindahl et al. 1970). For mortality in North America, where adult mortality lower for ♂♂ than ♀♀, and discussion of possible bias, see Anderson and Henny (1972). Oldest ringed bird 29 years 1 month (Kennard 1975).

**Movements.** Mostly migratory, but some populations rather sedentary. Extensive ringing has demonstrated various flyways, but overlapping wintering areas for different populations. In south and west Europe, resident, apart from abmigration; even in occasional severe winter, movement only to nearest open water, usually coasts, and hardly any emigration (for Britain, see Boyd and Ogilvie 1961). In north, east, and central Europe, some remain all year where conditions suitable, e.g. waters kept open by thermal springs in Iceland, or by power stations and sewage effluent near coastal towns, as in Baltic; most, however, migrate.

Many Icelandic birds winter Britain and Ireland. Those from north-west Russia, Fenno-Scandia, Baltic states, north Poland, north Germany, and Denmark winter from Denmark to north France and in Britain, with slight onward passage as far as north Spain. Some breeders from Baltic and upper Volga basin join those of central Europe (Switzerland, south Germany, south Poland, Austria, Hungary, Czechoslovakia); latter partially resident in mild winters but otherwise move south as far as north Mediterranean and Adriatic coasts (France, Italy, Balkans), some to north African coast. Also movement south-east down River Danube to winter in delta and on Black Sea. Latter area also important for those from south-west Russia (Byelorussia, Ukraine); some of which continue south to south-west to Balkans (especially Greece), and presumably north-east Africa (where common winter), Egypt, and Gulf of Suez, with 2 Egyptian recoveries from Volga delta moult area. Breeders from further east and north in USSR (Volga–Ural and Ob–Irtysh basins as far as Tomsk and north Kazakhstan) move south to winter Azerbaydzhan and Caspian regions, and further into Iran, Turkey, and Iraq, irregularly along Persian Gulf. Juveniles independent from mid-July and begin wandering all directions, often NNW to north-east. Migrations protracted; movements to winter quarters from August, peak numbers south and west Europe November–December; breeding grounds re-occupied early February to early May, depending in north and east on inland thaw.

Moult movements reported from various areas beginning mid-May; probably fairly common on small scale, but few large concentrations, and most marked in populations having long migrations to winter quarters (Salomonsen 1968). Volga delta, Astrakhan, long known as important moulting area where thousands of ♂♂ (from Ukraine to west Siberia) gather June, not finally dispersing until well into September, when migrant ♀♀ and juveniles already arriving. Other major moult assembly areas in

USSR on Lake Ilmen (Novgorod), including many from Pechora basin, on Rybinsk reservoir (upper Volga), and various Trans-Ural and Kazakhstan lakes. Up to 10 000 moult in Matsalu Bay, Estonia (Kumari 1962). Other significant European concentrations include Lake Tåkern (Sweden), IJsselmeer (Netherlands) and Bodensee (Germany). Ringing studies indicate even smaller moult gatherings may include birds that have travelled some distance, e.g. 2 sites in east England used by ♂♂ from Netherlands and Denmark.

Abmigration frequently reported for Mallard in north-west Europe, most obvious from artificially-reared stock released by sportsmen; interchange between winter quarters in different years shown by ringing, e.g. from North Sea countries to south-east Europe and Black Sea, from Volga delta and Camargue (France) to Britain, doubtless in part due to aberrant migration. Birds of unknown origin have occurred Azores, Canaries, Madeira, and tropical Africa south to Gambia, Eritrea, and north Sudan; but exceptional south of Sahara, and Kenya–Uganda reports not verified (White 1965). Exceptionally, one ringed England, September, recovered 3 years later in Alberta, Canada. See also Bauer and Glutz (1968); Dementiev and Gladkov (1952); Shevareva (1970).

**Food.** Omnivorous and opportunistic, with wide range of food and feeding methods. Frequency of methods related to habitat, season, food availability, and feeding mechanisms. Food obtained from water while floating or swimming by (1) pecking and sieving (i.e. pumping in and expelling water) at surface with bill ('dabbling'), (2) submerging head and neck, (3) up-ending, (4) diving. Up-ends in depths up to 48 cm for average of 4·8 s, maximum 5–8 s (Szijj 1965*b*), and from age of 6 days (Weidmann 1956). Young dive regularly for food at 4–7 weeks (Weidmann 1956), adults only occasionally, e.g. for submerged acorns autumn and winter, and then in depths 1–2 m for average of 6–10 s (Heinroth 1910; Witherby *et al.* 1939; Mylne 1954; Kumerloeve 1960; Ringleben and Creutz 1961; Ern 1970); submerge with or without forward leap, with wings held against body, using feet alone for propulsion (Dabelow 1925). Methods 1–3 above may be preceded and accompanied by foot-paddling to bring items to surface (Millais 1902; Willi 1970); while engaged in method 2, may swim persistently backwards, feeding in turbulence (K E L Simmons). Methods 2 and 3 used most often on Bodensee, Germany (Szijj 1965*b*), method 2 on Klingnauer reservoir, Switzerland (Willi 1970). Food also taken on land, where (5) grazes like geese *Anser* and Wigeon *A. penelope*, plucking leaves and shoots; (6) picks up items with tip of beak; (7) 'grubs' with beak at base of plants pulling out anything that can be grasped (Goodman and Fisher 1962); (8) bites off pieces from larger food items, e.g. potatoes (Weidmann 1956); (9) strips seeds from plants and shakes plants to loosen seeds and invertebrates (Weidmann 1956). Young, and oc-

casionally adults, snatch insects from air (Spalding 1973; Weidmann 1956; Chura 1961). Gregarious feeder, though in breeding season in pairs and family parties; daytime and sometimes, especially when disturbed, nocturnal. On Klingnauer reservoir, daily two-peak feeding activity, in early morning and late evening, often after dark, depending on light intensity (Willi 1970).

Diversity of feeding behaviour allows wide use of different habitats. This reflected in long list of recorded food items: mainly immobile with size range up to 5·0 cm but chiefly 1·0 mm–1·0 cm (Olney 1964), or, in case of more flexible items such as eels *Anguilla anguilla*, up to 10·2 cm (Thompson 1851). Plant materials include seeds, buds, and leaves of aquatic and terrestrial species of many families; animal materials insects, molluscs, crustaceans, annelids, amphibians, fish, and occasionally even birds and mammals. For specific identifications see Thompson (1851); Morris (1891); Millais (1902); Naumann (1905); Phillips (1923); Niethammer (1938); Witherby *et al.* (1939); Campbell (1947); Salomonsen (1950*a*); Dementiev and Gladkov (1952); Bannerman (1958); Gillham (1961); Olney (1962, 1964, 1967); Toufar (1965); Pirkola (1966); Bauer and Glutz (1968); Tiussa (1972).

Considerable variation in diet with locality and season, though most data based on samples collected during shooting season, and rarely relate to ecological factors which affect availability and productivity of food items. In general terms: in early part of year, chiefly seeds and overwintering green parts; as spring leads into summer, animal materials taken more frequently—Diptera emergences, e.g. of midges (Chironomidae), stoneflies (Plecoptera), mayflies (Ephemeroptera), especially important—until may predominate in midsummer; in early autumn, plant intake increases, especially as seeds ripen (including cereals: wheat *Triticum*, barley *Hordeum*, oat *Avena*, rye *Secale*, millet *Panicum*, maize *Zea*, rice *Oryza* and waste potatoes), until, in late autumn and in winter, such materials may be exclusive diet.

USSR studies show changes with area and season (Isakov and Vorobiev 1940; Dementiev and Gladkov 1952). In early spring before thaw, feeds on overwintered plants, e.g. duckweed *Lemna*, hornwort *Ceratophyllum*, sedges *Carex*. On Baraba steppe lakes as thaw begins, chiefly seeds and shoots of orache *Atriplex pendunculata*, glasswort *Salicornia*, seablite *Suaeda corniculata*; later, as waters rise, debris from edges, such as seeds, molluscs, adult insects and pupae. Where spring cereals (especially oats) present, these may constitute bulk of diet. In flood plain shallows and pot-holes as temperatures rise, animal materials predominate (60–97%), particularly phyllopod crustaceans, beetles (Coleoptera), dragonfly (Odonata) and caddisfly (Trichoptera) larvae, chironomids, and molluscs. On Baraba steppe, animal materials (65·6% frequency) primarily insects, mainly chironomid larvae, less pulmonate molluscs, caddisflies, and beetles —especially adult and larval leaf-beetle *Donacia* (also important on Mologa river flood plains and Volga delta). In northern Kazakhstan lake areas in summer, forage for locusts (Orthoptera); plant materials also form significant part of summer diet, particularly *Ceratophyllum* tops, *Lemna*, water plantain *Alisma* leaves, and flowering rush *Butomus*. During flightless moult period, almost exclusively plant materials; in Volga delta, shoots and seeds of rockcress *Najas*, tapegrass *Vallisneria*, and pondweed *Potamogeton*; in Baraba steppe, chiefly *Potamogeton*. In late July and August in central belt, and in September in south, plant materials begin to predominate, especially buds of arrowhead *Sagittaria*, tubers of fennel-leaved pondweed *P. pectinatus*, buds of frogbit *Hydrocharis*, and seeds of *Potamogeton*, *Scirpus*, and *Carex*. As cereals ripen, flight into fields especially on to rye, wheat, barley, millet, and, in south, rice. Winter diets chiefly seeds (about 70% by weight) and aquatic plant shoots; in steppe flood waters, mainly thistles; at lakes and coastal lagoons, mainly seeds of Japanese millet *Echinochloa*, wigeon-grass *Ruppia*, *Ceratophyllum*, buckwheat *Polygonum*, bulrushes *Scirpus*, and bur-reed *Sparganium*.

45 crops and gizzards from south and south-west Finland, August (42) and October (3), contained chiefly seeds, especially *Carex rostrata*, *C. vesicaria*, *C. nigra*, *Scirpus lacustris*, *Potamogeton natans*, *P. perfoliatus*, and *Sparganium simplex*; and, from one barren area, berries of *Vaccinium myrtillus* and *Empetrum nigrum*. Animal materials varied from 0–75% frequency; mainly Trichoptera larvae, chironomid and simulid larvae, pupae and eggs; less crustaceans *Asellus aquaticus* and molluscs *Pisidium* (Pirkola 1966). 77 artificially reared birds released on to inland lake Kent, England, where mainly sedentary, had, over 4 shooting seasons, yearly and seasonal differences in diet correlated to availability of main foods: though cereal (wheat and barley) fed throughout season, principal item only in September; from October–January, chiefly seeds of oak *Quercus robur*, bur-reed *Sparganium erectum*, and hornbeam *Carpinus betulus*; in one year, no *Quercus* seeds produced and then comparatively more cereal eaten; though recorded as food elsewhere (Spencer 1960; Thönen 1968), hazel *Corylus avellana* seeds available but not taken (Olney 1964). Observations and results of 210 stomach analyses over 8 shooting seasons, September–January, near Sevenoaks, Kent (Olney 1967), showed variations in diet and feeding habits correlated to changes in habitat (e.g. river clearance, grazing) and to effects of differing weather conditions (e.g. flooding, dry summers). Main foods: in river, leaf and stems of water crowfoot *Ranunculus aquatilis*, mollusc *Hydrobia jenkinsi*, caddisfly larvae *Hydropsyche angustipennis*, and less seeds of flotegrass *Glyceria fluitans*; from riverbanks, mainly seeds of bur-reed *S. erectum* and water-pepper *Polygonum hydropiper*; from wet meadows, chiefly seeds of creeping buttercup *Ranunculus repens*, persicaria *Polygonum persicaria*, hammer sedge *Carex hirta*, and dock *Rumex conglomeratus*; in and around gravel-pits, wide variety but

chiefly seeds of alder *Alnus glutinosa*, *Polygonum*, *S. erectum*, parts of horsetail *Equisetum*, and chironomid and *Hydropsyche* larvae; acorns of *Quercus robur* eaten in years when produced. Changes in feeding areas throughout 8 years related to factors affecting production and availability of food items.

Analyses of 177 birds collected, shooting season September–February, from coastal and estuarine areas, England (Olney 1964), showed food mainly seeds taken from nearby brackish-water areas and, at beginning of season, from inland fields of cereal (chiefly *Hordeum*). From salt-marshes, mainly seeds of *Salicornia*, *Atriplex*, and *Suaeda*, and less molluscs *Hydrobia ulvae*; also crustaceans (crab *Carcinus maenas*) and shrimps *Crangon vulgaris* and *Corophium*. From brackish-water areas, chiefly seeds of sea clubrush *Scirpus maritimus*, less *S. tabernaemontani*, *P. pectinatus*, wigeon-grass *Ruppia* and *Ceratophyllum submersum*, and less mollusc *Hydrobia jenkinsi* and shrimp *Palaemonetes varians*. Some common constituents of salt-marshes not taken, e.g. cord grass *Spartina townsendii*, sea beet *Beta vulgaris* (Olney 1967). Yet in Greenland, where in winter restricted to sea coast, of necessity marine feeder, almost entirely on molluscs including *Margarita helicina*, *Modiolaria discors*, *Macoma calcarea* and *Tellina*, and rather less on amphipod crustaceans (Winge 1898; Salomonsen 1950*a*).

Ducklings, in Utah, USA, as they matured, showed gradual change from almost entirely animal diet to almost entirely plant diet. Correlates to changes in feeding behaviour: up to 19–25 days, ducklings feed on land or on surface of water without immersing nares, and mainly on insects; from 25 days, feed with head submerged and by up-ending, and insects gradually replaced by seeds (Chura 1961). 16 ducklings collected Lake Mývatn, Iceland, contained mainly seeds (80% by wet weight), and chiefly those of *Sparganium* (older birds), *Carex*, and marestail *Hippuris* (younger birds), with animal materials (20%) mainly Diptera and chiefly adult chironomids (Bengtson 1975). In USSR, 83·4% of animal foods taken from overhanging vegetation, and from surface by downy ducklings (Dementiev and Gladkov 1952).

For North American studies see bibliography in Anderson *et al.* (1974).

**Social pattern and behaviour.** Based on material supplied by U Weidmann and co-workers (J A Darley, K E L Simmons).

1. Highly gregarious. For most of year, in flocks of up to several hundreds or even thousands. Where migratory, travels in flocks, those in spring consisting mainly of pairs. Resident and migrant flocks break up in early spring. In most areas, flocks seldom dissolve for long: tend to be small and scattered in summer, consisting mainly of unattached birds; with break up of pairs, in March–April onwards (see Eygenraam 1957; Lebret 1961), flocks at first wholly or mainly ♂♂, but joined later by unsuccessfully breeding ♀♀. ♀♀ leading broods tend to avoid flock until break up of family groups. Winter flocks composed of pairs, trios (see below), and unpaired birds. Although sex-composition of flocks in some areas more or less equal (see

Witherby *et al.* 1939), in general unbalanced in favour of ♂♂. Excess of ♂♂ already evident at hatching (Hochbaum 1944) and later increased by greater mortality of breeding ♀♀; in migrant populations, sex-ratio also affected by tendency for ♀♀ to move further south. During winter (November–February), excess of ♂♂ 130–110: 100, Finland (Raitasuo 1964); 112–107: 100, south Sweden, with higher proportion in urban areas (Hansson 1966); 114–108: 100, Netherlands (Eygenraam 1957). BONDS. Monogamous pair-bond; basically of seasonal duration, though old mates in resident populations sometimes pair up again in later years (Lebret 1961; Dwyer *et al.* 1973). Pairs form from end August, exceptionally earlier. In migrant populations especially, pair-formation for many delayed until spring. In some resident populations, e.g. in Netherlands (Lebret 1961) and Bavaria (Bezzel 1959), most paired by end October, including some 1st-year birds, and rest by February; in others, e.g. in Finland (Raitasuo 1964), most pairing February. Trios, i.e. simultaneous association of 2 ♂♂ and 1 ♀ (or, less frequently, 2 ♀♀ and 1 ♂), occasionally recorded (Weidmann 1956; Lebret 1961; Raitasuo 1964). ♂♂ also show marked promiscuous tendency; besides maintaining firm pair-bonds, often attempt to form liaisons with additional ♀♀ during communal courtship and will copulate, usually by rape, with as many other ♀♀ as possible. Pair-bond ends soon after ♀ starts incubating (see McKinney 1965*a*), and usually by mid-incubation. Exceptionally, ♂ stays even after young hatch; this more likely if breeding early (McKinney 1965*a*). Occasionally, ♂ pairs with another ♀ while original mate incubating (Sowls 1955; Weidmann 1956; Lebret 1961), and original pair may also re-unite for replacement clutch or second brood—or ♀ pairs to another (U Weidmann and co-workers). Only ♀ tends young. No brood amalgamations or crèching. BREEDING DISPERSION. See, especially, Hochbaum (1944), Sowls (1955), Dzubin (1955, 1969), McKinney (1965*a*). Break up of resident flocks starts February, especially in mild winters, as pairs prospect for sites for increasingly long periods daily, before dispersing finally in March. Pairs in migrant flocks tend to disperse as quickly as possible after arrival. Each pair then frequents large home-range, size depending on distance between feeding and nesting areas. Within home-range, pair use 1 or more activity centres where feed and loaf; ♂ defends area round ♀ but extent to which activity centres defended as true territories depends on local population density and features of habitat, as home-ranges of different pairs overlap (sometimes considerably). ♀♀ home to previous breeding areas, sometimes using same nest-site (Sowls 1955). When site chosen and laying begun, ♂ frequents waiting area usually not far from nest. During laying and incubation, home-range of ♀ restricted to vicinity of nest, including waiting area of ♂. ♂ ranges increasingly widely once incubation started; at first, continues to frequent 1 or more waiting areas fairly near nest where joined by ♀ during her recesses from eggs, but present less and less as incubation progresses, until moves right away and pair-bond broken. Over same period, ♂ also associates to increasing extent with other ♂♂. In typical mainland habitats, establishment of home-ranges based on activity centres, plus associated pursuit-flights, result in wide dispersion of nests at concealed sites, sometimes far from water. In some areas (e.g. North American prairies), stations of ♂♂ (and nests) at considerable distance from one another. Will also at times form dense, protected colonies on islands—e.g. at Loch Leven, Scotland, where ♂♂ establish small loafing and waiting territories round mainland bank (I Newton). ROOSTING. Communal (or, when breeding, in pairs and family parties), both diurnally and nocturnally. Time and site of roosting flexible and readily modified by local and seasonal factors, with wide variation

in routine of different populations. In some areas, affected by tidal pattern. Except where severely disturbed, moves from water to land for preening, oiling, and resting, following bathe. Uses, e.g., bare edge of bank, suitable shorelines generally, and waterside vegetation—often on islands; also on sandbanks, mudflats, logs in water, etc. Seeks shelter on windy days. Often roosts in feeding area but, where necessary, will fly between loafing and feeding places, often for considerable distances. In autumn and winter, many flocks make evening and morning flights to and from loafing area, especially around sunset and sunrise, also visiting fields to feed by day (see McKinney 1969); such flights shown abortively even by strictly sedentary birds (Weidmann 1956), indicating intrinsic rhythm. Especially where persecuted, may rest mainly by day in safe places, e.g. town moats (Lebret 1961), and feed by night. Otherwise may intersperse bouts of sleeping throughout 24-hour activity cycle (Weidmann 1956), or throughout day while roosting all night. Especially in autumn and winter flocks, rest periods tend to be synchronized, but, in breeding season, each pair follows own routine. In Finland, day divided into many activity periods lasting 45–75 mins, with intervening rest periods of 30–35 mins (Raitasuo 1964).

2. Extensive literature of which papers by Lorenz (1953), Weidmann (1956), Johnsgard (1960), Lebret (1961), Raitasuo (1964), and McKinney (1969) most comprehensive or recent; unpublished data from U Weidmann and co-workers also used—see also Weidmann and Darley (1971*a*, *b*), Simmons and Weidmann (1973). FLOCK BEHAVIOUR. Synchronization of activities within flock achieved by certain calls and movements, and by social facilitation of bathing, other social comfort behaviour (including preening and stretching), feeding, and resting. 'Diving-play' (see Lebret 1948) particularly characteristic: mass bathing accompanied by dashing-and-diving escape movements (McKinney 1965*b*)—rushes across water (often with abrupt changes in direction), short flights, and sudden dives. Main pre-flight signals: Head-shakes and Head-thrusts, usually while facing into wind with neck erect; accompanied by calls from ♀♀. In presence of danger, flock members give sex-specific warning-calls in alarm-posture with neck sleeked and head lifted high. Dog on shore, or smaller ground predator, mobbed: group aligns in wide circle round enemy, calling incessantly. Birds temporarily separated, or newly arrived, show strong tendency to join flock, calling when alone. Though flock closely integrated, not highly organized, e.g. no peck-order observed in wild. Large flocks, at least, of 'open-ended' composition, i.e. many members do not identify each other individually and birds able to join and leave at will. ANTAGONISTIC BEHAVIOUR. Hostilities often arise at feeding place or elsewhere when birds meet. Particularly prevalent between ♂♂ in spring when engaged in communal courtship or otherwise competing for mates. One ♂ may chase another over water, with neck extended and bill often open; on land, goose-like charging with head lowered forward. Fighting not uncommon. In simple encounter, bill thrust over rival's neck with jabbing and pecking. More serious disputes sometimes begin with birds moving rapidly side by side (parallel-swimming or running) before facing and chest-pushing with bills sharply lowered as they spar at each other's bills and grab at chest feathers; may then resort to vigorous wing-beating while holding on with bill, often rotating repeatedly on spot with much splashing (Circular-fights) until one ♂ flees. On both land and water, especially in presence of mate, ♂ may threaten silently with Head-jerking movements directed repeatedly, often over shoulder, at rival. Short-range threat, again often when close to mate, consists of Rabrab-calling with erect neck and bill slightly

raised (Chin-lifting or Bill-up display). 2 or more ♂♂ frequently group together thus and engage in characteristic Rabrab-palavers during which one of group may suddenly perform Down-up display—though latter much more typical of communal courtship. When 2 ♂♂ meet, often perform Ceremonial-drinking; may have appeasing function. ♀♀ occasionally fight together, or with ♂♂, by chest-pushing but this rarely proceeds to wing-beating, etc. Attacks by ♀♀ on ♂♂ occasional, however; more usually threaten, sometimes with ♂-like Head-jerking but more often with ♀-specific Inciting-display (with every gradation between the two). In typical Inciting, ♀ retracts neck and repeatedly jerks bill over shoulder in direction of rejected ♂, giving loud Inciting-calls. Incubating ♀♀ coming off nest, and ♀♀ leading young, also repel conspecifics approaching too near (especially ♂♂ intent on rape) with Repulsion-display; ruffle feathers (especially on head and back), fan and depress tail, draw head back on mantle, open bill (with upper mandible stretched upwards), and utter Repulsion-calls. COMMUNAL COURTSHIP. From August to early May, exceptionally in July when ♂♂ still in eclipse; peaks October–November, February–March (but dates vary somewhat from place to place). Typically in flocks of both sexes; many at first unpaired but courtship continues even when majority paired. Bouts last up to 15 mins or longer and most displays of ♂♂ usually highly synchronized. At start, ♂♂ assume Courtship-intent posture with head drawn down between shoulders and head feathers erected; often group together, with ♀♀ spaced out on periphery. Activity consists mainly of persistent swimming manoeuvres, secondary displays (particularly repeated Tail-wagging, Head-flicks, and Upward-shakes), and major displays (Water-flick, Head-up-Tail-up complex, and Down-up), together with antagonistic behaviour between ♂♂ and elements of pair-courtship. Most ♂ displays performed lateral to preferred ♀, while some show marked directional bias of movements towards her (see Simmons and Weidmann 1973). Much manoeuvring by ♂♂ consist of attempts to reach suitable positions relative to ♀ for displaying to best advantage, competition for ♀♀ and for places in which to display appearing to be major cause of hostility often accompanying communal courtship, particularly in spring (U Weidmann and co-workers). In Upward-shake, ♂ Tail-wags and rises briefly in water while stretching neck and rotating neck and head rapidly; in Head-flick, head movements similar but no integral Tail-wagging or rise in water. Burst of more or less simultaneous major displays by 2 or more ♂♂ often follows series of Upward-shakes and Head-flicks. Such bursts tend to consist of Head-up-tail-up and Down-up, while Water-flick more often given on own. Bursts in which all ♂♂ perform same display more common than expected by chance. Gasping-call (without any obvious visual display) sometimes given by non-displaying ♂♂ when others perform major displays. In Water-flick display (Grunt-whistle), ♂ (1) Bill-dips, (2) rises in water; (as in Upward-shake; see A), (3) jerks spray of water droplets sideways with bill, (4) subsides and Tail-wags; water directed at preferred ♀ (von de Wall 1965) and movements accompanied by characteristic Whistle-and-Grunt combination of calls. Head-up-Tail-up complex usually consists of 4 linked displays following in quick succession. ♂ (1) abruptly raises head, tail, and closed wings (Head-up-Tail-up proper); (2) holds posture briefly while formally facing preferred ♀ and giving Whistle-call (Turn-Head-to-♀); (3) lowers head and neck and swims away with neck stretched forward along water (Nod-swimming); (4) moves in front of ♀, attempting to Lead her by swimming ahead with erect head while pointing bill away and presenting rear view of head with feathers flattened laterally to form darkened patch (Turn-back-of-Head). If ♂ not in

favourable position, Turn-back-of-Head, Nod-swimming, or both omitted. Turn-back-of-Head also often occurs as independent activity, but Nod-swimming rarely (in ♂). In Down-up display, ♂ (1) suddenly erects head (this sometimes omitted), (2) abruptly tilts forward and downwards and Bill-dips in water (see B); (3) raises bill while dragging up column of water or sending spray forward, (4) stretches up again and assumes Chin-lift position while lowering tail; gives 1 or more Whistle-calls during Down-up proper followed by Rabrab-calling in terminal phase. Unlike other major displays, Down-up not obviously addressed to ♀♀ and may primarily be response between ♂♂. Further secondary displays include lateral Head-shakes, mostly during lulls when ♀ unresponsive (Simmons and Weidmann 1973), Bill-dip on own, and Preen-dorsally. ♂♂ also sometimes make short, ritualized Jump-flights (Lebret 1958a), often one after the other in quick succession or even more or less simultaneously. Each (1) rises steeply in characteristic posture (see C), (2) flies 3–8 m over ♀, (3) drops ahead of her, (4) Ceremonially-drinks in lateral position. Through synchronization of various ♂ displays depends partly on interactions between ♂♂ themselves (Weidman and Darley 1971a), ♀ usually plays key role: periodically Nod-swims among and around ♂♂ (see D); also Head-nods by jerking head up and down. Most displays by ♂♂ occur when 1 or more ♀♀ Nod-swim or Head-nod (Weidmann and Darley 1971b). ♀ often Incites during communal courtship, usually while following preferred ♂ as he Leads her: directs Inciting movements from latter towards 1 or more other ♂♂. PURSUIT-FLIGHTS. Extensively studied (see, e.g., Geyr 1924, 1953; Hochbaum 1944; Dzubin 1955, 1957; Lebret 1958b, 1961; Hori 1963; McKinney 1965a; Titman 1973), but much controversy over details and interpretation. 3 or 4 main types, with many intermediates. (1) Courtship-flights. During communal courtship on water, whole group may rise, fly round briefly, and resettle near (Dzubin

A

B

C

1957). More extended flights, lasting 5–10 mins, also occur April to mid-May: 5–15 ♂♂ chase ♀, group flying about erratically and eventually settling elsewhere. Initiated by ♀ who appears to be only loosely paired when pressed too closely by unpaired ♂♂ in courting party on water. No antagonism between ♂♂ reported. ♀ occasionally Incites (with characteristic call) during such flights. (2) 3-bird flights. Typical of period when pairs dispersed, March–April—after start of Persistant-calling by ♀ (Dzubin 1957)—lasting until ♂ deserts ♀ during incubation. ♀ (only occasionally ♂) of intruding pair flying over occupied activity centre of another pair chased away by ♂ of latter while mate of pursued ♀ follows at some distance. Pursuing ♂ returns to starting point afterwards, chases being longer (up to half a kilometre) during incubation than earlier (up to 200 m). During flight, ♀ may perform Inciting and, less frequently, Repulsion-displays (with calls); seldom if ever defended by mate. Such pursuits may be main method by which pairs—and thus nests— spaced out because ♀♀ tend to avoid areas from which chased. (3) Rape-intent flights occur later in season, after break-up of pair when ♂♂ in small flocks. Up to 20 ♂♂ pursue ♀—typically when she takes recess from nest during incubation—over long distances, trying to force her down and rape her. ♀ performs Repulsion-display in flight; may make short, hopping flights, drop to ground, or even dive into thick cover. Rapes also initiated in flock earlier in season when mated ♂ harasses another ♀ on water; ♀ attempts to escape while further ♂♂ join in—often including own mate who may also rape her (Barratt 1973). (4) Pairing-flights. Possible further type (see Dzubin 1957) showing some of feature of rape-intent-flight but apparently initiated by ♀ after losing eggs late in incubation—i.e. after desertion by original mate and when new mate needed for replacement clutch. Up to 30 ♂♂ slowly follow ♀; latter does not perform Repulsion-

display and occasionally hovers low over ground. PAIR BEHAVIOUR. Process of pair-formation gradual and early stages still imperfectly understood. Crucial phase, however—often seen during bouts of communal courtship, and at other times in presence of other birds—consists of ♂ Leading and ♀ following, often while Inciting. Sometimes, ♂ will stealthily approach and hang round ♀, adopting stationary Head-high posture (U Weidmann and co-workers)—then, at opportune moment, try to Lead her. Inciting by paired ♀ occurs at times also when no 2nd ♂ present, e.g. when pair meet after temporary separation. From autumn onwards (long before functional copulation), pairs also engage in copulatory activities. Other elements of pair-courtship, all of which tend to occur most as greeting behaviour when mates re-join and when others come close, include (1) Pair-palaver resembling Triumph Ceremony of geese *Anser* and *Branta*, especially after chasing away rival ♂, in which ♂ Rabrab-calls in Chin-lift posture while ♀ typically Incites; (2) mutual Ceremonial-drinking; (3) various forms of Mock-preen displays, including Preen-behind-Wing, Preen-dorsally, Preen-Belly, and Preen-Breast. Preen-behind-Wing by ♂ to ♀, or (less often) ♀ to ♂, most differentiated of these: bird usually first Bill-dips, then turns head back sharply over shoulder on side nearest mate, passes bill rapidly along inside of wing (producing characteristic sound), quickly returns head while closing wing, and often then Bill-dips or Ceremonially-drinks. Mates call when separated from one another: Slow Rab-calls by ♂, loud Decrescendo-calls by ♀♀ in which calls vary individually, enabling ♂♂ to identify own ♀♀ by voice (Lockner and Phillips 1969). Persistent-calling by ♀, in spring before eggs laid, also likely to inform mate of her location—as well as initiating dispersion of pairs and aggressive defence of activity centres by ♂ (see Dzubin 1957). COPULATION. Occurs on water fairly regularly from late September to May, and several times daily February–March. Unlike rapes, usually preceded by mutual Head-pumping: head rhythmically and repeatedly jerked right down and stretched up; initiated by either sex. ♀ then adopts Prone-posture: stretches out neck and head along water, slightly droops wings, and eventually raises tail (may occasionally display thus without initial Head-pumping). After copulating, ♂ dismounts and, still holding ♀'s head, quickly Bridles, by jerking head back on mantle while raising chest out of water and whistles; then Nod-swims round ♀, sometimes adding brief spell of Leading, and finally bathes and Wing-flaps. ♀ also performs last two activities immediately after release; occasionally Nod-swims first. RELATIONSHIPS WITHIN FAMILY GROUP. Especially during first few days, ♀ broods young periodically, leads them to water and food, and protects them. When leading brood, ♀ tends to move slowly; at intervals, utters quiet Maternal-calls which become louder on appearance of danger or if ducklings give Distress-calls. If brood split, ♀ dashes back and forth until re-united. Young follow and one another—closely during first days, at increasing distances later. Naive ducklings appear not to recognize ♀ visually but quickly become imprinted to visual and vocal features of mother; initial tendency to approach and follow sound source emitting Maternal-call in preference to sources giving other sounds (Gottlieb 1971). When cold, hot, wet, hungry, or lost, young give loud Distress-call; lost ducklings rejoining family utter Greeting-calls, similar calls often heard when brood bathing or feeding intensely (Weidmann 1968; Kear 1968; Abraham 1974). Fighting between siblings rare, though young from *c.* 10 days sometimes attack other broods. Ducklings that have become wet, and presumably, unidentifiable, sometimes killed by own mother—♀♀ tending to kill all small, strange young they encounter closely.

(Figs A, B, and D from photographs by U Weidmann; Fig C after Lebret 1958*a*.)

D

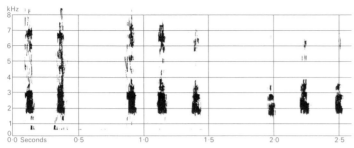

I M Schommer England December 1975

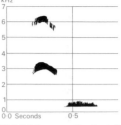

II M Schommer England December 1975

**Voice.** Marked differences between calls of ♂ and ♀: familiar loud, deep, harsh quacking peculiar to ♀ while main utterances of ♂ weak-sounding, somewhat rasping, nasal 'raehb' (also grunts and whistles during heterosexual displays). Both sexes also produce brief, non-vocal, rasping sound 'rrrr' during Preen-behind-Wing display; quite loud but not far-carrying. For description of calls, see especially Abraham (1974), where most analysed in detail; also Lorenz (1953), Weidmann (1956), McKinney (1969). Except where otherwise indicated, calls given on water, land, and in flight at all times of year; amount and variety of calling much reduced, however, during incubation (♀) and post-breeding moult. Marked tendency for most calling to occur during morning and evening; at other times of day, often polyphasic—periods of vocal activity for 30–45 min alternating with periods of inactivity for 20 min to 2–3 hours.

CALLS OF MALE. 'Raehb' calls (abbreviated to 'rab' in forming name of calls) fall into 2 main groups. (1) Slow Rab-call: series of quiet, spaced, long-drawn notes, 'raehb' or 'yaarb' (see I); given in many situations mostly as alarm-call (Warning-call of Weidmann 1956), and contact-call (Summoning-call of Lorenz 1953), when separated from mate and in response to Decrescendo-call of latter or calls of other flock members. Unlike corresponding call of ♀, not used as pre-flight call, but sometimes uttered after taking wing—also while alighting among flock. Alerts or attracts conspecifics and indicates location to mate. (2) Rabrab-call (Conversation-call of Lorenz 1953): faster series of shorter 'raehb' notes, typically in repeated 2-note phrase with emphasis on 2nd syllable but also in phrases of 1 or 3 syllables. Given in many situations, mostly as threat-call or greeting-call, mainly October–June. Uncertain if ever given in flight. Characteristic of antagonistic encounters between ♂♂, including Rabrab-palavers of 2 or more birds, and when pair re-unite after separation and engage in Pair-palavers; also given after sudden scare, often by all ♂♂ in flock, and in terminal phase of Down-up display—in all cases, Chin-lifting posture adopted while calling. Quieter version, intermediate between Rabrab and Slow Rab, given without Chin-lifting in other situations, e.g. before changing location and before and after flight; also at times before, during, and after copulation. (3) Whistle-call: loud, high-pitched, single whistle;

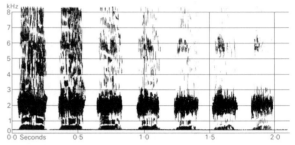

III M Schommer England December 1975

uttered mainly on water (not in flight), chiefly September–June, during all 3 major displays of communal courtship and during post-copulatory Bridling-display—particularly flute-like during Bridling, Water-flick, and Head-up-Tail-up displays but of lower amplitude during Down-up. (4) Grunt: deep, hollow-sounding; given, mainly on water (never in air), during Water-flick display following Whistle-call (see II). Lorenz (1953) suggested that sound produced by release of compressed air after delivery of associated whistle. Similar grunt heard at close range after post-copulatory Whistle-call (Abraham 1974). (5) Gasping-call: wheezy, 3-note 'chachacha'; uttered at times on water during communal courtship (see also von de Wall 1963).

CALLS OF FEMALE. More varied than ♂'s. (1) Decrescendo-call: series of fast, loud, far-carrying 'quack' notes (see III), delivered in characteristic phrase of 2–10 notes (1–20) in which emphasis placed on 1st or 2nd note with decreasing amplitude, frequency range, and duration of succeeding ones together with increase in intervals between them. Note duration and number of calls vary greatly between birds. Uttered by unpaired ♀♀ at any time of year, except when moulting, and by paired ♀♀ in absence of mate (especially in response to his Slow Rab-calls) chiefly September–April; rarely in flight. Shows most obvious diurnal periodicity of all calls; particularly common shortly before and after sunrise and sunset—but at all times strongly induced by sight of flying conspecifics and sound of other ♀♀ giving same call. Decrescendo-like calls, of 1–20 descending notes with nasal quality of Repulsion-call (see below), also heard from ♀♀ with broods for 1–2 weeks after hatching; significance un-

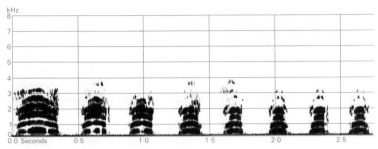

IV J-C Roché

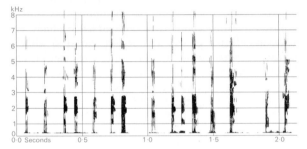

V M Schommer England December 1975

certain. (2) Single-quacks: given on own or in short series in variety of other situations (see especially Weidmann 1956); attracts or indicates presence to others, e.g. in response to calling birds of either sex and when joining flock. May sometimes have same function as Decrescendo-call, and latter occasionally preceded by series of 8–9 Single-quacks. Loud quack sometimes given at end of copulation (K E L Simmons). (3) Persistent-quacking (Advertising-call): series of even-spaced 'quacks' (see IV) produced monotonously, often for minutes on end, at rate of 1 or 2 per s, rate varying individually. Given, in presence or absence of mate, throughout period in spring when pair settled on home-range until first egg laid (Dzubin 1957), often during twilight—especially just before and after dawn (Hori 1963). Frequently uttered in flight; also before and after inspection-flights by pair over nesting area. Advertises presence of ♀ preparing to nest and informs mate of ♀'s location. Also associated with appearance of ground predator. (4) Pre-flight call (Going-away call of Lorenz 1953): repetition of short, sharp, evenly spaced notes—typically a harsh 'gack' with bill open (but whimpering sound when bill closed)—becoming louder and more frequent just before take-off. (5) Alarm-call (Warning-call): single 'quaaack' in which sound energy spread more evenly throughout frequency spectrum and duration than in normal 'quack'; alerts conspecifics in face of danger, but used less often than corresponding call of ♂. (6) Quegqueg-call (Conversation-call): 2-syllable 'queg-queg' notes, equivalent to Rabrab-call of ♂, occasionally given in Chin-lift posture during Pair-palaver but largely replaced by Inciting-call with head lowered (see Weidmann 1956). (7) Inciting-call:

rapid succession of loud, rhythmic querulous, stuttering 'queg' and 'geg' notes (see V), each beginning abruptly and ending faintly; uttered persistently, typically during communal and pair-courtship while directing Inciting-movements away from chosen ♂ towards others. Mainly September–June. Occurs also, e.g. with Rabrab-calling of mate during Pair-palavers, especially after antagonistic encounters or when pair unites after separation. Though primarily indicative of preference for certain ♂ and rejection of others in heterosexual situations, also at times given as pure threat-call to intruders or rivals of either sex. Frequently uttered in flight. (8) Repulsion-call: series of loud, harsh 'gaeck' notes, each starting and ending abruptly; phrase often of 3–4 notes, with 1st quietest and 2nd loudest. Given, usually in characteristic Repulsion-posture, shortly before start of and during incubation—especially on approach of ♂ intending rape; infrequent from ♀♀ leading broods but persists in ♀♀ that have lost clutches. (9) Maternal-calls: series of quiet low-amplitude notes uttered at infrequent intervals; start 2 days before hatching, becoming louder and more frequent as hatching nears and reaching peak just before brood leaves nest—during this period changes from 'gu-gu-gu . . .' (bill closed) to 'quai-quai-quai . . .' (bill open). This Broody-call develops into Leading-call (Weidmann 1956) as ♀ induces brood to follow her from nest and subsequently; more variable in pattern and intensity than Broody-call, amplitude and duration increasing if ♀ alarmed or young utter Distress-calls. In early stages, Maternal-calls facilitate imprinting of ducklings on mother. (10) Distraction-call: series of soft, nasal notes uttered in irregular pattern; closely resembles Repulsion-call but not so harsh. Given only by ♀ with young in presence of predator, mainly on water. (11) Other quacking-like calls: include whimpering or high, quick mumbling 'kn' sounds mixed with more intense 'quai' notes during pre-copulatory Head-pumping, movement over short distances, and feeding; quiet, rapid chuckling after copulation (K E L Simmons) probably same. Faint, drawn-out 'gnn-gnn . . .', etc. different from above, also uttered during feeding (Weidmann 1956), but repeated 'tuckata' notes from feeding birds (Hochbaum 1944) probably Inciting-call.

CALLS OF YOUNG. Include following (see especially

Weidmann 1956; Kear 1970). (1) Pre-hatching calls: series of cries and irregular, tapping-like clicks, often in response to Broody-calls of mother; start just over 3 days before hatching (Gottlieb and Vandenbergh 1968), cries corresponding to later Contact and Distress-calls of new-hatched young (see Abraham 1974). (2) Contact-call: soft, fast, high-pitched 'pipi' notes, irregularly spaced; uttered when feeding, bathing, preening, dozing, and being brooded; also when siblings and mother gather together and after alarm or other distress. (3) Distress-call: loud, high-pitched 'peeea' notes, harsher and more even-spaced than Contact-call; given when wet, cold, hungry, alarmed, or separated. (4) Air-alarm call: described as 'piii' (Weidmann 1956) or shrill whistle (Starkey and Starkey 1973); so far heard only from ducklings, 1-week or older, in response to aeroplane or hawk-models overhead.

Gradual voice changes in young during weeks 4–12 (Abraham 1974), and sex differences already evident during weeks 4–6 when some ♂♂ utter rasping 'raehb' and some ♀♀ 'quack', alternately with clear, loud 'piiii'; see also Abs (1969) for correlation between voice change in ♂ and growth of trachea.

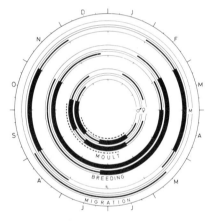

**Breeding.** SEASON. See diagram for north-west Europe. Iceland: mean date of first laying, 1961–70, May 18, range May 12–23 (Bengtson 1972a). USSR: first laying early April in south to mid-May in north (Dementiev and Gladkov 1952). Finland: first laying mid-April to early May (Hildén 1964). Bavaria: first laying end-March to early April (Bezzel 1962). Season greatly prolonged in some areas, perhaps due to presence of domestic strains selected for long breeding season. Start of laying in southern England and Scotland dependent, at least in part, on February temperatures (Ogilvie 1964; I Newton). Late breeders may include birds under 1 year old (see below). SITE. Variable. Of 224 nests in wood, south-west England, 77 in thick ground cover (bramble *Rubus*, nettles *Urtica dioica*), 58 in thin ground cover (nettles, rushes *Juncus*), 38 at foot of tree or post, 26 inside hollow tree, 14 in crown of pollard willow *Salix*, and 11 in open (Ogilvie 1964). Of 62 nests on islets, Finland, 45% in thick cover, 32% in thin cover, 5% under boulders and 18% in open (Hildén 1964).

Up to 10 m in trees, occasionally higher. Readily uses artificial nest-boxes and baskets. Not colonial, but nests sometimes only 1 m apart. On Scottish island, where 500 nests, most 5–10 m apart, rarely as close as 2 m; in uniform habitat, ♀♀ spaced their nests and tolerated nests of other species of Anatidae closer than those of their own species, but none nearer than 2 m (I Newton). Usually close to water, though up to 3 km. Nest: shallow depression with low rim formed from grass, leaves, and sometimes small twigs; lined much down. Average internal diameter 17·5 cm (17–22), depth 8·7 cm (6–14) (Bauer and Glutz 1968). Building: by ♀, using material within reach of nest. Material often added during laying, sometimes during incubation. EGGS. Blunt oval; grey-green or buff, occasionally bluish. 57 × 41 mm (50–65 × 37–46), sample 500 (Schönwetter 1967). Weight 51 g (42–59), sample 200 (J Kear). Clutch: 9–13 (4–18); over 18 probably always 2 ♀♀. Of 84 clutches, Denmark: 5 eggs, 4%; 6, 2%; 7, 8%; 8, 15%; 9, 15%; 10, 20%; 11, 24%; 12, 5%; 13, 5%; 14, 1%; 17, 1%; mean 9·6 (Fog 1965). Of 33 clutches, Finland: 6 eggs, 15%; 7, 15%; 8, 34%; 9, 12%; 10, 15%; 11, 9%; mean 8·2 (Hildén 1964). Of 95 1st clutches, south-west England: 6 eggs, 1%; 7, 1%; 8, 3%; 10, 7%; 11, 18%; 12, 23%- 13, 16%; 14, 13%; 15, 10%; 16, 6%; 17, 1%; 18, 1%; mean 12·6 (Ogilvie 1964). Larger clutch size in England may be due to domestic strains in wild stock. Mean of 68 1st clutches, Iceland, 8·5 (Bengtson 1972a). One brood. Replacement clutches laid after egg loss, or even after early loss of ducklings. Interval from loss of eggs to re-lay 20–38 days, sample 3; interval from loss of ducklings to re-laying 15–30 days, sample 6 (I Newton). Of 114 clutches (all or nearly all 2nd clutches), south-west England: 4 eggs, 1%; 5, 2%; 6, 1%; 7, 6%; 8, 9%; 9, 24%; 10, 18%; 11, 19%; 12, 10%; 13, 4%; 14, 2%; 15, 2%; 16, 1%; 17, 1%; mean 9·9 (Ogilvie 1964). Mean of 31 2nd clutches, Iceland, 6·9 (Bengtson 1972a). One egg laid per day, usually in morning, though sometimes gap of 2 or even 3 days. INCUBATION. 27–28 days (23–32); mean 27·6 days (24–32), sample 51 (Ogilvie 1964). By ♀, ♂ sometimes nearby at least in early stages. Eggs covered with down when ♀ off. Starts when clutch complete; hatching synchronous, usually within 24 hours, occasionally to 36 hours. Some evidence for most hatching during daytime (Bjärvall 1968). Eggshells left in nest. YOUNG. Precocial and nidifugous. Self-feeding. Leave nest at mean age of 14–21 hours after hatching (range 4–36), sample 103, for ground nests; and 20–25 hours (13–46), sample 53, for hole nests (Bjärvall 1968). Cared for by ♀ and brooded at night when small. ♀ will defend nest and young against predators, and sometimes against human intruders. FLEDGING TO MATURITY. Fledging period 50–60 days. Become independent at or just before fledging. Age of first breeding normally 1 year, but 6 or 7 months recorded for ♀ and 7 months for ♂ (Kear 1961). BREEDING SUCCESS. Finland: average brood size at fledging 7·0 (63 broods) from average clutch size 8·1 (119 clutches) (Linkola 1962).

Iceland: mean hatching success of 131 nests, 1960–71, 68·4%, range 52·7%–83·1%; of 34 failures, 52% predated, 28% deserted, 9% flooded and 11% unknown causes; average young reared per ♀ 3·5. (Bengtson 1972a). Denmark: average brood size reared to fledging 5·0–7·5 (Fog 1965). South-west England: hatching success of 2292 eggs, 82·4%; 88·7% of 203 nests hatched at least 1 egg (Ogilvie 1964). Mean brood size from 80–140 pairs, 1957–62, 6·9 (4·9–7·6); and young reared per successful ♀, 4·7 (3·6–6·4) (Boyd and King 1964). Scotland: of 1717 nests, 1966–71, 55% hatched, 39% predated, 6% deserted; some evidence that predation, mainly by crows Corvidae, increased by visits of researchers (I Newton).

**Plumages.** Nominate *platyrhynchos*. ADULT MALE BREEDING. Head and neck metallic green, glossed purple in some lights. Narrow white ring round lower neck. Mantle and scapulars finely vermiculated grey and white; feathers of lower mantle suffused cinnamon-buff; outer edges of scapulars vermiculated black and chestnut, inner webs of scapulars bordering back black, often vermiculated or speckled pale grey. Back black, finely vermiculated pale grey; rump and tail-coverts black, glossed blue-green. Chest and upper breast rich mahogany, feathers tipped white when fresh; flanks and rest of underparts white, finely vermiculated pale grey. Central 2 pairs of tail-feathers black, others with centres sepia or grey, broad edges and mottling white; outer feathers mainly white. Primaries with coverts and bastard-wing dark brown-grey, inner webs paler, outer edges of outer webs pale grey. Secondaries brown-grey, broadly tipped white; outer webs strongly glossed purple-blue or blue-green, with black bar bordering white tip. Greater upper wing-coverts white with black tips and mouse-grey bases. Median and lesser coverts mouse-grey, tinged olive-buff at tips. Tertials pale grey, inner slightly vermiculated white at tip; outer webs of outer tertials and tertial-coverts suffused deep chestnut, edged black near base. Under wing-coverts and axillaries white. ADULT FEMALE BREEDING. Head and neck pale buff, heavily streaked black with green gloss on crown and streak through eye, narrower and duller on hindneck and sides of head and neck; chin, throat, and foreneck warm buff, usually unstreaked. Feathers of mantle, scapulars, and upper tail-coverts sepia or dull black, broad edges and V-shaped subterminal marks, crescents, or streaks buff. Back and rump black, edges and irregular bars and spots on feathers buff. Chest and sides of breast cinnamon to buff, feathers subterminally with sepia crescents, central streaks, or blotches. Flanks like scapulars, but light edges and irregular U-shaped marks wider. Breast, belly, and vent pale buff, some feathers fringed white; under tail-coverts white. Underparts variably spotted or streaked, but clearer and less heavy than in juvenile and feathers relatively wider. Tail buff, irregularly streaked, barred, and spotted sepia or black; outer feathers edged white. Wing like adult ♂, but tips of greater tertial coverts white: white line formed by subterminal bars on greater coverts reaching nearer body. Speculum narrower; median and lesser upper wing-coverts slightly more olive-brown, often narrowly tipped grey-buff. Tertials olive-brown, edged buff or white, longer often with faint buff streaks; outer edge of outer tertial chestnut-black. ADULT MALE NON-BREEDING (Eclipse). Superficially like ♀ or juvenile, but upperparts more uniform than ♀ and underparts less heavily streaked than juvenile. Crown and streak through eye uniform black with slight green gloss; rest of head off-white, heavily streaked dark grey-brown; chin and throat usually

uniform off-white. Upperparts dark sepia, feathers narrowly edged buff or grey on scapulars, upper tail-coverts, and part of mantle, glossed olive on back and rump. Underparts like adult ♀ breeding, but spots on under tail-coverts larger, chest sometimes tinged chestnut, and flanks vermiculated cream. Breast and belly pale buff, sometimes with broad sepia spots or streaks. Tail grey and white like adult ♂ breeding but central pairs sepia, edged pale buff. Wing like adult ♂ breeding (tertial coverts without white tips), but tertials buff-brown rather than grey, inner webs tinged olive. ADULT FEMALE NON-BREEDING. Like adult ♀ breeding, but marks and edges on upperparts and tail narrower and paler; on mantle, back, and rump usually only few incomplete buff bars or crescents, on scapulars, upper tail-coverts, and tail incomplete narrow U-shaped marks. Head like adult ♂ eclipse, but usually some faint buff streaks on crown. Wing like adult ♀ breeding (white tips to greater tertial coverts), but tertials duller, greyer, without broad buff streaks and edges, more like those of adult ♂ eclipse. DOWNY YOUNG. Forehead, crown, nape, lores, streak through eye to nape, and short streak on hindcheek to hindneck dark sepia with slight olive tinge. Superciliary streak, cheeks, chin, sides of neck, and foreneck yellow-buff, paler yellow on throat. Upper side of body and wing, sides of breast, patch on side of body, and thighs sepia, darkest on back, longer down tipped olive-brown. Short streaks at side of back and over rear edge of wing, and smaller patch at side of rump pale yellow. Underparts pale yellow-buff, tinged cream-buff on chest and sides of body. JUVENILE MALE. Head like adult non-breeding, usually with faint buff streaks on crown. Upperparts like adult ♂ eclipse, dark sepia, narrowly edged buff on mantle, scapulars, lower rump, and upper tail-coverts; upper tail-coverts variably glossed green near tips. Feathers of flanks sepia with pale buff margins. Feathers of underparts narrow, warm buff on chest and sides of breast, paler buff or off-white elsewhere, fringed with white on belly; dark central dots or streaks at tips of feathers. Tail-feathers gradually narrow towards notched tip, bare shaft projecting; dark grey-brown, edged pale buff. Wing like adult ♂, but tertial-coverts and median and lesser upper wing-coverts often tinged brown and edged buff-grey (edges soon wearing off). Tertials shorter and narrower than adult ♂ eclipse, dull olive-brown, less grey and cinnamon than adult ♂ breeding. In late summer and early autumn, part of juvenile plumage (mainly flanks, scapulars, chest, sides of mantle, or head) replaced by 1st immature non-breeding, resembling adult ♂ eclipse, but soon changed for breeding. JUVENILE FEMALE. Like juvenile ♂, but feathers of back and upper rump edged and sometimes barred buff. Wing like adult ♀, lesser and median coverts narrowly tipped buff (edges soon wearing off). Greater tertial coverts tipped white (sometimes faintly grey only) in contrast to juvenile ♂; tertials like juvenile ♂, but slightly narrower and darker. Development of 1st immature non-breeding (resembling adult ♀ non-breeding) as in juvenile ♂. FIRST BREEDING MALE. Like adult ♂ breeding, but juvenile wing (except tertials) and sometimes some feathers of back and rump retained. Juvenile wing often similar to adult, as buff edges may wear off; then only separable by more frayed and narrower median coverts, which gradually narrow towards rounded tip, instead of unfrayed and broad median coverts with squarer tips. Sometimes pale buff or off-white edges on tips of inner webs of outer greater primary coverts (Carney 1964). FIRST BREEDING FEMALE. Like adult ♀ breeding but juvenile tertials and part of back and rump often retained until spring; wing juvenile, often only separable from adult by pale buff edges on tips of inner webs of outer greater primary coverts, and by more frayed and narrower coverts.

**Bare parts.** ADULT MALE. Iris dark brown. Bill olive-green to bright yellow, occasionally pale blue or green-blue (dullest in eclipse), culmen and tip dull black. Foot orange-red. ADULT FEMALE. Iris and foot like adult ♂. Bill dull yellow-orange to olive-brown, culmen, upper sides, and tip dark horn; some dark dots along edge of upper mandible. DOWNY YOUNG. Iris brown. Bill flesh, spotted black; nail pink-white. Foot dark olive-grey. JUVENILE. Iris dark brown. Bill red-horn, becoming darker when older; in ♀ sides paler, often small dots along edge of upper mandible. Foot dull flesh or yellow to yellow-orange, webs darker. (Bauer and Glutz 1968; Lebret 1961; Schiøler 1925; Witherby *et al.* 1939; RMNH, ZMA.)

**Moults.** ADULT POST-BREEDING. Complete; flight-feathers simultaneous. Sides of breast, flanks, and 1–2 central pairs of tail-feathers with coverts first, followed by head, rest of body and tail. Moult of tail prolonged, feathers in alternate sequence, often not finished before can fly again. Moult of head and body lasts *c.* 1 month, mainly between late May and mid-July in ♂♂ and failed breeding ♀♀; later in successful ♀♀, starting when young *c.* 2 weeks old, finishing between early July and mid-August. When head and body almost complete, tertials shed, followed by flight-feathers, wing-coverts, and axillaries. Flightless for *c.* 4 weeks, mainly between early June and late August (peak mid-July) in ♂, and between late June (starting when young independent) and late September in ♀. ADULT PRE-BREEDING. Partial; head, body, tail, and tertials. Starts shortly after end of wing moult. Sides of body and underparts first, head and upperparts later, part of rump and mantle last. Tail prolonged, not always complete; central tail-feathers and tertials with coverts moult when head and body mainly in breeding. ♂♂ moult between August and early October (some feathers of upperparts, central tail-feathers, and tertials occasionally later), ♀♀ usually finished before November (but tail, tertials, and some other feathers sometimes not until spring). POST-JUVENILE. Part of juvenile plumage changed for 1st immature non-breeding between July and September, mainly on chest, sides of body, flanks, scapulars, and outer mantle, sometimes (especially in ♀♀) head, rest of mantle, or some other feathers. Occasionally starts before wing full-grown. Moult to 1st immature non-breeding usually incomplete, number of feathers involved variable. In ♀♀, feathers growing earlier in autumn like adult ♀ non-breeding, later gradually more like adult ♀ breeding. 1st immature and non-breeding in both sexes usually soon replaced by 1st breeding. FIRST PRE-BREEDING. ♂♂ at about same time as adult ♂♂, but slower on body; belly and parts of back and rump last. Moult slows or stops in winter, but only limited amount of juvenile or 1st immature non-breeding retained until spring. Tail and tertials mainly before December, occasionally partly retained until April. ♀♀ like ♂♂, but post-juvenile and first pre-breeding moults generally slower (sometimes, however, in breeding early September, except tertials); more often part of 1st immature non-breeding or juvenile tertials retained until end of winter pause. (Bauer and Glutz 1968; Dementiev and Gladkov 1952; Schiøler 1925; Witherby *et al.* 1939; C S Roselaar.)

**Measurements.** Netherlands, mainly late summer and autumn. Skins (RMNH, ZMA).

| | | | | | | | |
|---|---|---|---|---|---|---|---|
| WING AD | ♂ 279 | (4·08; 13) | 272–285 | ♀ 265 | (4·66; 13) | 257–273 |
| JUV | 272 | (8·06; 27) | 258–287 | 257 | (7·61; 23) | 245–272 |
| TAIL AD | 85·8 | (3·64; 14) | 80–91 | 84·5 | (2·84; 12) | 81–90 |
| JUV | 75·2 | (4·43; 5) | 72–83 | 72·7 | (3·56; 6) | 68–77 |
| BILL | 55·4 | (2·44; 58) | 51–61 | 51·8 | (2·21; 48) | 47–56 |
| TARSUS | 45·3 | (1·51; 45) | 42–48 | 43·4 | (1·38; 37) | 41–46 |
| TOE | 59·2 | (2·93; 46) | 54–66 | 56·5 | (2·40; 36) | 52–60 |

Sex differences significant, except tail. Tail of adult non-breeding and 1st breeding on average *c.* 3 mm shorter than adult breeding. Juvenile tail significantly shorter than adult, difference in wing not significant. Other measurements of adults and full-grown juveniles similar; combined.

Adults (1) Iceland and (2) Greenland (*conboschas*) (Schiøler 1925).

WING

| | | | | | | | |
|---|---|---|---|---|---|---|---|
| (1) | ♂ 282 | (29) | 275–298 | ♀ 263 | (24) | 252–277 |
| (2) | 292 | (69) | 275–306 | 272 | (41) | 261–285 |

BILL

| | | | | | | | |
|---|---|---|---|---|---|---|---|
| (1) | 53·4 | (29) | 49–59 | 50·8 | (24) | 47–55 |
| (2) | 46·6 | (69) | 44–51 | 48·1 | (41) | 45–52 |

**Weights.** ADULT. Czechoslovakia, based on 527 ♂♂ and 277 ♀♀ (Folk *et al.* 1960).

| | | | | | | | |
|---|---|---|---|---|---|---|---|
| (1) | ♂ 1088 | (155; 26) | 859–1572 | ♀ 944 | (120; 11) | 750–1140 |
| (2) | 1142 | (132; 26) | 873–1458 | 1096 | (89; 11) | 1003–1270 |
| (3) | 1100 | (100; 31) | 943–1300 | 957 | (94; 19) | 788–1279 |
| (4) | 1218 | (114; 79) | 850–1450 | 1020 | (74; 29) | 840–1150 |
| (5) | 1101 | (90; 91) | 880–1410 | 959 | (108; 30) | 750–1200 |
| (6) | 1216 | (139; 15) | 1017–1442 | 1084 | (102; 14) | 921–1320 |

(1) Rather low in mid-winter, e.g. January. (2) Increase to peak during maximal enlargement of gonads (♂♂ March, ♀♀ April). (3) Low in May (♂♂) or June (♀♀) after reduction of gonads, and when ♀♀ breeding. (4) Both sexes gain weight prior to wing moult (♂♂ June, ♀♀ July). (5) Loss in weight, as much energy required during growth of wings. Minimum when able to fly again (♂♂ August, ♀♀ September). (6) Gradual increase up to December, while fat stored for winter (at onset of winter, weight decreases quickly). Data include breeding, migrating, and wintering populations, as these factors apparently without influence here (Folk *et al.* 1960). In USSR, cycle similar, but weights prior to departure late October and November high: ♂♂ 1330 (1000–1700), ♀♀ 1240 (1000–1400) (Dementiev and Gladkov 1952). Annual cycle of specimens from Netherlands (ZMA) and Kazakhstan (Dolgushin 1960) in agreement with Czechoslovakian data.

JUVENILE. Average weight reaches that of adult from October–November (included in adult data above from then); some singles reach adult level July–August. In adverse weather, averages for juveniles tend to be lower than adults (Folk *et al.* 1960; Dementiev and Gladkov 1952).

**Structure.** Wing rather narrow, pointed. 11 primaries: p9 longest, p10 0–6 shorter, p8 5–13, p7 17–27, p6 27–45, p1 105–130. Inner web of p10 and outer p9 emarginated, inner p9 and outer p8 sometimes slightly. Tertials broad and long in breeding, shorter and narrower in non-breeding and juvenile. Scapulars tapering towards pointed tip in breeding, tip round in other plumages. Tail short, rounded: 18–20 feathers, 1 or 2 central pairs curled to semi-circle at tip in ♂ breeding, straight in non-breeding and ♀. Bill broad, about as long as head, higher than broad at base. Middle part of culmen slightly concave, but convex near wide, flat, and broadly rounded tip of bill. Nail rather narrow, small. Outer toe *c.* 93% middle, inner *c.* 78%, hind *c.* 29%.

**Geographical variation.** Nominate in North America slightly larger: average 10 ♂♂—wing 286, bill 56, tarsus 47·7, toe 62·2; 5 ♀♀—wing 272, bill 55, tarsus 45·5, toe 60. Greenland subspecies *A. p. conboschas* still larger, but bill smaller; plumage paler and greyer. Icelandic birds somewhat intermediate in size and colour between *A. p. conboschas* and nominate *platyrhynchos*, but, as

PLATE 64. *Anas falcata* Falcated Duck (p. 483): **1** ad ♂ breeding, **2** ad ♀ breeding, **3** juv. (NWC)

PLATE 65. *Anas strepera* Gadwall (p. 485): **1** ad ♂ breeding, **2** ad ♀ breeding, **3** ad ♂ eclipse, **4** ♂ 1st pre-breeding moult, **5** juv, **6** downy young. (NWC)

PLATE 66. *Anas crecca* Teal (p. 494). *A. c. crecca:* **1** ad ♂ breeding, **2** ad ♀ breeding, **3** ad ♀ non-breeding, **4** 1st winter, **5** downy young. *A. c. carolinensis:* **6** ad ♂. (NWC)

PLATE 67. *Anas capensis* Cape Teal (p. 504): **1** ad, **2** juv.
*Marmaronetta angustirostris* Marbled Teal (p. 548): **3** ad ♂ breeding, **4** ad ♀, **5** juv, **6** downy young. (NWC)

PLATE 68. *Anas platyrhynchos* Mallard (p. 505): **1** ad ♂ breeding, **2** ad ♀ breeding, **3** ad ♂ eclipse, **4** juv, **5** downy young. (NWC)

PLATE 69. *Anas rubripes* Black Duck (p. 519): **1** ad ♂, **2** ad ♀, **3** juv. (NWC)

PLATE 70. *Anas acuta* Pintail (p. 521): **1** ad ♂ breeding, **2** ad ♀ breeding, **3** ad ♂ eclipse, **4** juv ♂, **5** juv ♀, **6** downy young. (NWC)

PLATE 71. *Anas querquedula* Garganey (p. 529): **1** ad ♂ breeding, **2** ad ♀ non-breeding, **3** ad ♂ eclipse, **4** 1st imm non-breeding, **5** downy young. (NWC)

most specimens not separable from nominate, included in latter. In extralimital, low latitude subspecies, breeding plumage of ♂ greatly reduced, ♂♂ wearing female-like dress whole year. Subspecies in south USA and Mexico (*fulvigula, maculosa, diazi*) have plumage of both sexes in varying degree intermediate between ♀ nominate *platyrhynchos* and Black Duck *A. rubripes*, but in view of local hybridization, currently included in former (Delacour 1956; Johnsgard 1961*c*) although sometimes treated as separate species. Pacific Ocean forms (*wyvilliana, laysanensis*) small, those of Laysan with also variable amount of white on head; these often treated as separate species.      CSR

## *Anas rubripes* Black Duck

PLATES 69 and 74
[between pages 518 and 519
and facing page 543]

Du. Zwarte Eend      Fr. Canard obscur      Ge. Dunkelente
Ru. Американская чёрная кряква      Sp. Ánade sombrio americano      Sw. Svartand

*Anas rubripes* Brewster, 1902

Monotypic

**Field characters.** 53–61 cm, of which body just over two-thirds, ♂ generally larger; wing-span 85–96 cm. Large dabbling duck with proportions of Mallard *A. platyrhynchos*; both sexes like extremely dark sooty-brown ♀ *A. platyrhynchos* with strikingly contrasted silvery-white wing-linings, no white in tail, and little or no white in speculum. Sexes similar, but distinguishable, and no seasonal differences apart from slight eclipse. Juvenile closely resembles adult ♀.

ADULT MALE. General pattern similar to ♀ *A. platyrhynchos*, but dark sooty-brown body gives much blacker appearance. Crown and line through eye black; supercilium and rest of head and neck light brown finely streaked with black, contrasting sharply at base of neck with rest of body. Pale brown edges to black-brown body feathers produce distinctive pattern; feathers at sides of breast show U-shaped buff marks; tail sooty-brown. Speculum purple, edged broadly with black at both edges, but with no white in front and at most a thin border behind, so hardly stands out from otherwise sooty-brown wings in flight; wing-linings gleaming silvery-white rather than cream, producing distinctive flashing white and black effect in flight against sooty-brown underbody and flight-feathers. Bill yellow to olive-yellow; eyes brown; feet orange-red to coral-red. MALE ECLIPSE. Little different but head and neck greyer with less fine streaking and bill olive; thus difficult to separate from adult ♀, except by greener bill, redder feet, and U-shaped buff marks at sides of breast. ADULT FEMALE. Similar to ♂, but not quite so dark. Sides of head and neck greyer and more coarsely streaked; feather edgings produce less well-defined pattern; feathers at sides of breast with V-shaped rather than U-shaped buff marks. Bill olive-green variably blotched blackish; legs red-brown to dull orange or greenish-yellow, with blackish webs. JUVENILE. Similar to adult ♀, but duller with more streaked underparts, no V or U marks at sides of breast. Bill dull olive with black blotches; legs dull orange-yellow to green-yellow. Becomes like adult after 1st autumn.

Needs to be distinguished from ♀ *A. platyrhynchos* which is much paler brown with less contrasted head and neck, more creamy and much less contrasted wing-linings, whitish tail, and well-defined blue or purple speculum with clear white borders in front and behind (*A. rubripes* shows virtually no white on upperwings or tail as it flies away). Unfortunately, they commonly hybridize and many intermediates occur in America. At distance might be confused with ♀ Common Scoter *Melanitta nigra*, which has smaller black bill, heavier head, no supercilium, shorter neck, pointed tail, and blacker wings.

Habitat, general habits, and even voice virtually identical to *A. platyrhynchos*, though less often on small ponds and showing preference for brackish and salt water.

**Habitat.** Generally as Mallard *A. platyrhynchos*, which it replaces over most of range in lowlands of north-east North America where, despite hardiness in cold weather, most compelled by icing of inland waters to shift to coastal waters and marshes during much of winter. Relatively more attached than *A. platyrhynchos* to salt water, tolerating even fairly rough seas and favouring salt-marshes. Readiness to rely on larger proportion of animal matter in diet than *A. platyrhynchos* and wary vigilance assist towards maintenance of large population in face of increasing encroachment on habitat by man. Flexibility and resourcefulness also manifest in wide choice of nest-sites, on ground under cover of herbage, on islets, on old stumps surrounded by water, and even up to *c.* 15 m in old nests in large trees, no single type being clearly preferred. Walks and swims strongly and flies swiftly often at considerable height.

**Distribution.** Breeds eastern North America from northern Manitoba east to Labrador and Newfoundland, south to north Minnesota, Wisconsin, north Illinois, Indiana, Pennsylvania, Maryland, and Virginia, then in smaller numbers to North Carolina.

Accidental. Azores: November 1968 (*Bull. Br. Orn. Cl.* 1969, **89**, 86–88). Ireland: February 1954, February 1961 and November 1966. England: March 1967, September 1969 (British Ornithologists' Union 1971). Sweden: November 1973 (*Vår Fågelvärld* 1975, **34**, 53–5).

**Movements.** Partially migratory and dispersive. Part of northern breeding population migrates south in late autumn (from September, peak passage early November); those from Atlantic provinces and states chiefly to tidal and freshwater marshes from New Jersey to South Carolina, though some winter as far north as open water, a few even in Newfoundland; only dabbling duck likely to be present in numbers then in New England and Maritime Provinces. Return movement begins mid-February, probably complete by end April. West of line Allegheny Mountains to James Bay, southward movements on average shorter, *c.* 85% wintering north from Tennessee, mainly in Ohio Indiana, but some from Atlantic and Mississippi flyways migrate farther south to Gulf of Mexico and Florida. Has wandered west to Alberta, Washington State, and California, north to Baffin Island, south to Puerto Rico, and across Atlantic. (Based on Bruun 1971; Linduska 1964; Stewart and Robbins 1958.)

**Voice.** Closely similar to Mallard *A. platyrhynchos*.

**Plumages.** ADULT MALE BREEDING. Crown, streak through eye to nape, and hindneck black, feathers narrowly margined buff; rest of head and neck pale buff or cream, streaked sepia. Body sooty-black, darkest on back and rump; feathers narrowly edged buff or off-white on mantle, scapulars, and upper tail-coverts, somewhat broader on underparts. Feathers of chest and sides of breast each with narrow pale buff or white U-shaped subterminal mark, on upper flanks with cinnamon-buff subterminal streak or incomplete U-shaped mark. Tail black. Flight-feathers dark sepia, paler on inner webs; outer webs of secondaries (except 1–2 outer) strongly glossed purple-blue; tips of secondaries black, occasionally with narrow white edge. Upper wing-coverts sooty-black, sometimes slightly tinged olive or with narrow dark grey or grey-buff edges; tips of greater coverts black, with pale grey tinge subterminally in some. Tertials with coverts greyish-sepia; outer webs of outer black, in some glossed purple; tertial-coverts and some longer tertials narrowly edged dark grey or grey-buff. Under wing-coverts and axillaries white, latter sometimes with dusky subterminal mottling. ADULT FEMALE BREEDING. Like adult ♂ breeding, but subterminal marks at chest and sides of breast usually V-shaped, not U-shaped. Outer webs of *c.* 3 outer secondaries lack purple gloss; tertials slightly shorter and narrower, more extensively edged grey-buff; greater tertial-coverts in some with pale grey tinge subterminally. ADULT MALE NON-BREEDING (Eclipse). Like adult ♂ breeding, but edges of body feathers narrower, head more heavily streaked; feathers of chest and sides of breast without distinct U-shaped marks, flanks without buff streaks or marks; tertials somewhat shorter and narrower, duller sepia. ADULT FEMALE NON-BREEDING. Like adult ♂ eclipse, but wing like adult ♀ breeding. JUVENILE. Head and upperparts as in adult non-breeding; underparts sepia,

feathers margined buff, except at tips, giving streaked appearance, especially on belly. No subterminal marks on flanks, chest, or sides of breast (but ♂♂ sometimes small pale crescents). Body feathers and tertials shorter and narrower than adults, not wide with slightly rounded tips. Tail-feathers dark sepia; notched at tip, bare shaft projecting. Wing like adults, but inner web of 4 outer primary coverts with light edges, median coverts shorter and narrower, trapezoidal with frayed tip instead of wide and smoothly rounded. Secondaries and greater coverts narrower, former with less distinct dusky tips. Development of 1st immature non-breeding (resembling adult non-breeding) probably as in Mallard *A. platyrhynchos*. FIRST BREEDING. Like adult breeding, but wing (except tertials) juvenile. (Carney 1964; Kortright 1942; C S Roselaar.)

**Bare parts.** ADULT MALE. Iris dark brown. Bill olive-yellow to orange-yellow, rarely with dusky spot on culmen; duller olive during eclipse. Foot orange-red to coral-red, duller in eclipse. ADULT FEMALE. Iris brown. Bill more green or olive than adult ♂, with dark streak over culmen and black dots on sides of upper mandible. Foot brown, usually with salmon or orange tinge. JUVENILE MALE. Bill light olive, foot orange-brown, gradually changing to adult colour; in early spring, bill yellow, foot salmon or orange-red. JUVENILE FEMALE. Bill dusky or olive-grey, usually with some black dots at sides of upper mandible. Foot brown, sometimes with olive tinge. During autumn and winter, sides of bill change to yellow-olive and foot acquires some orange or salmon tinge. (Kortright 1942.)

**Moults.** ADULT POST-BREEDING. Complete. Head and body first, in ♂ from May or June, in ♀ already started March or April; followed by complete wing in summer. ADULT PRE-BREEDING. Partial; head, body, tail and tertials, mainly September–October. POST-JUVENILE. Partial; like adult pre-breeding, but slower; usually finished before December (Kortright 1942).

**Measurements.** ADULT. 5 ♂♂ RMNH; additional data on range from Delacour (1956). Wing 290 (265–301); range given by Delacour apparently included both adults and juveniles, as wing of adult ♂ usually over 282 (Carney 1964) and juveniles generally smaller. Tail 93–105, bill 55·4 (52–60), tarsus 48·5 (44–50), toe 64 (61–66). Range ♀♀ (Delacour 1956), wing 245–275, bill 45–53.

**Weights.** North America: 35 ♂♂, 1245 (905–1730); 33 ♀♀, 1135 (850–1330) (Kortright 1942).

**Structure.** 11 primaries: p10 longest, p9 0–4 shorter, p8 6–12, p7 18–27, p6 32–44, p1 102–130. Central tail-feathers not curled at tip. Rest of structure as in *A. platyrhynchos*.

**Geographical variation.** None. In view of widespread hybridization with Mallard *A. p. platyrhynchos* in parts of east North America, sometimes considered as conspecific with that species. For relationship with subspecies of *A. platyrhynchos* in south USA and Mexico, see Johnsgard (1961*c*).   CSR

# Anas acuta Pintail

Du. Pijlstaart    Fr. Canard pilet    Ge. Spiessente
Ru. Шилохвость    Sp. Ánade rabudo    Sw. Stjärtand

*Anas acuta* Linnaeus, 1758

Polytypic. Nominate *acuta* Linnaeus, 1758, Holarctic. 2 extralimital, south Indian Ocean: *eatoni* (Sharpe, 1875), Kerguelen; *drygalskii* Reichenow, 1904, Crozet Islands.

**Field characters.** 51–66 cm, of which body two-thirds, excluding central tail-feathers of generally larger ♂, which project up to 10 cm; wing-span 80–95 cm. Large, slender, thin-necked dabbling duck with long bill, rounded head, and pointed tail; ♂ has chocolate head and upper neck, white neck stripe and underparts, black undertail with creamy area in front, mainly grey upperparts, and largely black tail. Sexes dissimilar, with some seasonal differences. Juvenile closely resembles adult ♀, but distinguishable.

ADULT MALE. Chocolate-brown head and upper fore-neck, with black nape and hindneck, contrast sharply with white stripe from behind ear down side of neck broadening out into white lower foreneck, breast, and belly. Back and sides vermiculated grey; scapulars produce black line above grey flanks, while some feathers farther back strikingly elongated with cream edges. Black under tail-coverts have broad, cream-buff band in front; tail mainly black with whitish edges. Speculum metallic green glossed with bronze, shading to black at rear, with rich buff border in front and white behind; these colours show in flight as bands running whole length of secondaries, but speculum itself rather obscure, the buff and white areas being much more noticeable. Otherwise, in flight, black braces on scapulars contrast with grey forewings and upperparts, while underparts stand out as gleaming white against dark head and greyish and mottled flanks and wing-linings. At long range on water, looks mainly grey with dark head and conspicuous white breast. Bill grey-blue with black base, ridge and nail; eyes yellow-brown; feet grey with black webs. MALE ECLIPSE. Resembles adult ♀, but upperparts generally greyer and more uniform without broad buff edges to feathers. ADULT FEMALE. General pattern like other ♀ *Anas*, but rather paler and greyer. Main characters apart from different shape, bold crescent-shaped markings on flanks, whitish belly, obscure speculum (browner than ♂'s with buff bar in front much reduced), blue-grey bill, and green-grey feet. In flight, light border on rear of secondaries conspicuous; wing-linings mottled brown and white and appearing darker than underbody, while pointed tail shows mainly brown. JUVENILE. Resembles adult ♀, but darker above, without pale feather edgings, and much more heavily streaked and spotted below; cheeks and sides of neck paler. Bill dark grey.

♂ unmistakable. ♀ can be separated from ♀♀ of Mallard *A. platyrhynchos* and Gadwall *A. strepera* by blue-grey bill, thin neck, slender shape, more elegant patterning, narrow pointed wings, pointed tail, and obscure speculum, as well as differences in colour of belly, wing linings, and tail. ♀ Wigeon *A. penelope* shows some of these characters but distinguished by rounded head, short neck, more compact shape, much more rufous coloration, and more uniform appearance on water.

In breeding season, found on moorland pools, lowland lochs, and brackish and freshwater lagoons and marshes, but in winter chiefly estuaries and sea coasts, or fenland washes. Generally shy and wary. Swims high on water with neck straight or curved and tail raised (though lowered if uneasy); up-ends with tail depressed, often in relatively deep water; rarely dives except when wounded or flightless. Generally tends to be rather inactive by day and to feed nocturnally. Walks easily and gracefully with body and tail nearly horizontal. Flies with rapid wing-beats and action more like *A. penelope* than *A. platyrhynchos*; wings produce low hissing or rustling sound. Often in pairs or parties up to half a dozen, but sometimes large loose flocks, particularly on sea or stubble. Flocks of up to 100 fly in long lines or, sometimes, V-formation, giving different appearance to more haphazard grouping of other *Anas*. ♂ generally quiet, though has weak, nasal warning call and, in display, double whistle recalling Teal *A. crecca*, but lower; ♀ also rather quiet, but utters thin, hoarse quack or guttural croak.

**Habitat.** Within mainly continental breeding distribution in west Palearctic, ranging into northern tundra and including outliers through steppe to some isolated Mediterranean settlements. Shows consistent preference for shallow, aquatic habitats of open, fairly spacious, usually lowland grassland or prairie, most often eutrophic, and at least moderately productive biologically. Such preferences carry risk of habitat modification through drought, disappearance of temporary pools, or flooding, requiring shift to new breeding sites. Therefore character-istically mobile, accepting equivalent vegetation or topo-graphy within different zones. Narrow waters dominated by flanking forest or overgrown with dense aquatic vege-tation, deep lakes or pools with naked rocky shores, and fast-running rivers not favoured. Sometimes uses bare ground at some distance from water although normally nesting near it, in fairly dense but not tall herbage or shrub cover. Breeding among or under trees normally avoided. For summer moult usually shifts habitat to undisturbed

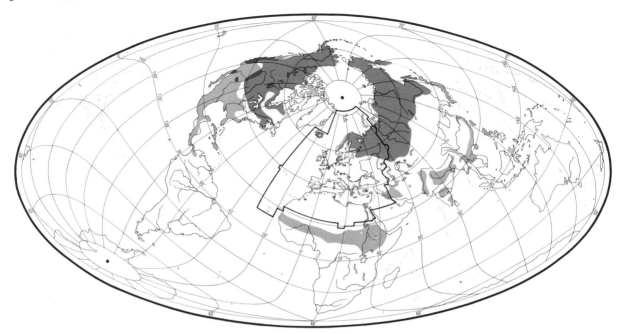

lakes, deltas, or other estuaries, salt-marshes, and some-times reservoirs with good vegetation cover, often as first stage on return migration. In winter, prefers sheltered coastal habitats, especially on estuaries, floodlands and inland waters nearby. Also feeds on farmland, including stubble. Relatively little use made of reservoirs and other artefacts, except managed floodlands. Use of airspace even freer than for other *Anas* and often at considerable heights.

**Distribution.** Nesting occurs sporadically over large areas of western and southern edges of range, with occasional breeding, and summering pairs well to south (see below); map shows only areas of regular breeding, but even within these, localities frequently change.

FAEROES. Bred 1871, 1926 (KW). SPITSBERGEN: bred 1935 (Løvenskiold 1964). BRITAIN AND IRELAND. Scotland: first bred 1869; since then frequent changes in areas occupied. England: sporadic nesting since *c.* 1910; now regular southern half. Ireland: in north, bred sporadically 1917–38 then regularly for a time Lough Neagh; in south, sporadic since 1917 (Parslow 1967; RFR). NETHERLANDS: first bred 1923 and has nested on 74 occasions, mainly since 1952, though some may have involved decoy ducks (CSR). WEST GERMANY: frequent sporadic breeder, mainly in north, but a few as far south as Bavaria (Bauer and Glutz 1968). POLAND: scarce; almost extinct Silesia (Tomiałojć 1972). CZECHOSLOVAKIA: breeds rarely and irregularly (Hudec 1970). HUNGARY: breeds sporadically over wide area; most frequent on alkaline soils east of River Tisza (Szabo 1970). YUGOSLAVIA: bred 1964 (VFV). RUMANIA: has nested occasionally since 1961, especially Danube delta (MT; Vasiliu 1968). ITALY: breeds oc-casionally (SF). PORTUGAL: bred 1943 (MDE). SPAIN:

breeds occasionally (FB); formerly quite common Ma-rismas but has probably not nested recently (Ree 1973). TURKEY: bred 1971 (OST). USSR: nests sporadically as far south as steppes and islands of Black Sea (Dementiev and Gladkov 1952).

Accidental. Bear Island, Kuwait, Azores, Madeira.

**Population.** Breeding. ICELAND: probably n.e. 500 pairs (Gardarsson 1975). BRITAIN: *c.* 50 pairs (Atkinson-Willes 1970a). WEST GERMANY: *c.* 30 pairs (Szijj 1973). EAST GERMANY: *c.* 100 pairs (Isakov 1970a). DENMARK: *c.* 200 pairs (Joensen 1974). NORWAY: decline in some areas and fluctuating numbers in others (Haftorn 1971). FIN-LAND: *c.* 20 000 pairs; considerbale decline some areas but elsewhere recent increases (Merikallio 1958). USSR: *c.*316 000 pairs in western part, despite great reduction in numbers (Isakov 1970a, b). AUSTRIA: regular only east of Neusiedlersee, where marked fluctuations with maximum 10–15 pairs 1966–7 (Bauer and Glutz 1968). MOROCCO: breeds Iriki, when water levels high, with perhaps several hundred pairs in favourable years (Robin 1968).

Winter. Some 50 000 birds north-west Europe and at least 250 000 Black Sea–Mediterranean region (Atkinson-Willes 1975). USSR: *c.* 184 600 birds in western part, mainly Black Sea and Caspian (Isakov 1970c).

Survival. Mean annual adult mortality, based on USSR recoveries, 48% (Boyd 1962). Oldest ringed bird 26 years 6 months (Kennard 1975).

**Movements.** Mainly migratory. Breeders from Iceland winter Britain and Ireland though a few recovered further south-east in Europe and single recoveries west Greenland and Quebec (Bruun 1971). Degree of movement by small British population unknown. Breeding populations of

north Russia east to north-west Siberia (Omsk), Fenno-Scandia, and Baltic migrate south-west to winter Netherlands and British Isles—movement from former to latter in hard weather. Some ringed Novgorod recovered south-west to Portugal (Shevareva 1970). Small numbers ringed Britain autumn-winter moved south to Iberia and north-west Africa, and singles to Sénégal and Ghana. Breeding areas uncertain of large numbers wintering west Africa (notably Sénégal, Nigeria, Mali, Niger, Chad); some from USSR since 7 recoveries of Volga delta birds. 80 000 Sénégal delta January 1971, outnumbered only by Garganey *A. querquedula* (Roux 1971).

In east Europe and south USSR, flyways not well defined. Ringing of Trans-Ural population produced recoveries in west Mediterranean (31%); Balkans, east Mediterranean, and Black Sea (39%); Caspian (24%) and other 6% east to India. Birds ringed Ob-Irtysh basin recovered Indian subcontinent (61%), Caspian (17%), Black Sea (17%), and 5% elsewhere (Shevareva 1970). Thus large numbers wintering Greece, Turkey, and to lesser extent on Black Sea, originate USSR from Byelorussia and Moscow to Ob-Irtysh basin of west Siberia; and wintering flocks around south Caspian into Iran and Iraq originate vast area from southern Urals to Omsk. Numbers migrate up Nile to winter east Africa; some ringed Egypt recovered such widely separated areas as Finland (28°E) and Novosibirsk (84°E) but most Pechora, Ob, and Irtysh basins. Some interchange between flyways in different years indicated by recoveries of birds ringed North Sea countries in south-east Europe, Turkey, and Kazakhstan, and of Baltic birds on Black and Azov Seas, though some aberrant migration probably involved. 2 juveniles ringed Labrador, Canada, recovered England (Devon, Hampshire) same autumns, but uncertain to what extent birds from Canada cross Atlantic (Bruun 1971). North American population highly migratory, including extraordinary passage over Pacific to Hawaii, and south in smaller numbers into central Polynesia (Cooper and Lysaght 1956; Clapp 1968).

Moult migrations frequent, in large and small numbers (Salomonsen 1968). ♂♂ desert ♀♀ late May and early June (latest on tundras), as soon as clutch complete; peak of moult in July when some ♀♀ begin migrating. Flocks of several hundred ♂♂ moulting IJsselmeer, Netherlands, not local birds. In USSR moulting areas, recorded lower Ob and lower Yamal, also many localities to south, including Kazakhstan where up to 30000 in adjacent Lakes Tengiz and Kurgaldzhino (*Oiseau* 1967, **36**, 292), and Volga delta which draws ♂♂ from as far off as Arkhangel, Ryazan, Volgograd, and Ob basin; these later migrate west or south-west to Mediterranean and Africa (northern tropics), a few reaching Spain and even North Sea. Major movements away from moult areas and breeding grounds mid-August to early September, and by October some ringed Volga delta already as far west as Italy and Netherlands. Early passage through Europe in August, peak movements mid-September to November, ♂♂ preceding ♀♀ due to earlier moult (Salomonsen 1968). Further movements under weather influence at any time during winter.

Departures from west Africa begin February, from west Europe late February or March; reach Leningrad April, tundras late May. Major flyways tend to follow coasts; normally relatively small numbers inland central Europe, but in spring birds which have wintered to south-west likely to cross overland in some numbers, suggesting loop migration. See also Bauer and Glutz (1968); Dementiev and Gladkov (1952).

Food. Wide variety of plant and animal materials; obtained mainly in depths 10–31 cm inland (Willi 1970; Sugden 1973), 20–30 cm on coast (Spitz 1964), up to 53 cm (Bauer and Glutz 1968). Chiefly from mud bottom by up-ending—for 4·8–6·2 s (Szijj 1965)—with tail depressed; also when swimming with head and neck submerged. Long neck may be adaptation for bottom feeding providing advantage over other *Anas* in same habitat (Olney 1965b). Occasionally dives for food: bringing head back over shoulders, plunges forward with slightly spread wings, submerging up to 10 s or more (Ryall 1943; Chapman *et al.* 1959). Also feeds on land, picking up grain and digging out rhizomes and tubers with beak. Usually feeds in pairs and small parties to large flocks, associating freely with other *Anas*. Feeds by day and night on Klingnauer reservoir,

Switzerland, February; 2 peaks of maximum activity in late morning and late afternoon (Willi 1970).

Plant materials chiefly seeds, tubers, and rhizomes of following species: pondweeds *Potamogeton, Elodea, Vallisneria*; sedges *Carex*; docks *Rumex*; bistort *Polygonum*; grasses including *Glyceria*; scurvey-grass *Cochlearia*, pillwort *Pilularia*, spurrey *Spergularia*, hornwort *Ceratophyllum*, cereals, rice, and potatoes. Also takes eel-grass *Zostera*, stoneworts *Chara*, and various algae. Animal materials mainly insects, especially Coleoptera (waterbeetles *Dytiscus*), Diptera (chironomid midges), Trichoptera (caddisfly larvae *Phryganea*), Odonata (dragonfly larvae), Orthoptera (grasshoppers); also molluscs (*Hydrobia, Planorbis*), annelids (leeches), crustaceans (brine shrimp *Artemia*, copepods, phyllopods, and ostracods), amphibians (tadpoles), and rarely fish. (Thompson 1851; Macgillivray 1852; Millais 1902; Madon 1935; Witherby *et al.* 1939; Gregory 1947; Lynn-Allen 1954; Dementiev and Gladkov 1952; Bauer and Glutz 1968.)

Considerable variation with locality and season though, in general, plant materials predominate in autumn and winter and, in some areas, animal materials in summer and spring. In early spring, Camargue, France, mainly feeds in *Artemia* ponds, later moving to other ponds, and takes mainly snails (*Lymnaea, Physa*) and insect larvae; from October–March, chiefly seeds, especially rice *Oryza sativa* (Bauer and Glutz 1968). In river Medway, Kent, December–February, almost entirely *Hydrobia ulvae* and a few seeds of samphire *Salicornia* and seablite *Suaeda maritima* (Olney 1965b), but during same months on Terschelling, Netherlands, almost entirely plant seeds, particularly *Carex, Rumex*, and *Polygonum* (de Vries 1939). In USSR, local and seasonal differences pronounced and markedly different food reserves exploited (Dementiev and Gladkov 1952). On Mologa floodplains in March, feeds exclusively on vegetation but, as soon as rivers thaw, takes many Chironomidae larvae. On Volga, February–March chiefly green vegetation (41·5%) and seeds (40·6%). Yet, in April, Rybinsk reservoir, 60% total volume animal (chiefly caddisfly larvae *Phryganea grandis* and chironomids); in May–June, animal food increases to 70·5% (particularly insect larvae); in June–August, animals still predominate (especially molluscs). In summer on Mologa flood plains, animal proportion of diet rises to 70% (molluscs 30%, dragonflies 20%, chironomids 15%, other insects 5%) with vegetative parts 25%, seeds only 5%; but moulting birds on Volga, feeding at dusk on flooded meadows, take 50% green vegetative parts (chiefly *Najas marina*) and 48% seeds and, later in October, still chiefly vegetative parts (especially bulbs of arrowhead *Sagittaria* and leaves of eel-grass *Zostera*) and seeds. In winter on Caspian coast, mostly small molluscs and algae; in Azerbaydzhan, mainly seeds (74·6% by weight), some green parts of plants (23·7%), and only 17% animal materials. 16 moulting birds Kurgaldzhino lake, USSR, contained seeds (mainly *Potamogeton* 58·7% and *Ruppia*

25%), and leaves, roots, and stems of other water plants (43·7%), and 12·5% molluscs and insects (Gavrin 1964).

Ducklings from Mologa, USSR, contained 80% invertebrates, chiefly molluscs, dragonfly larvae, chironomids, and small crustaceans (Dementiev and Gladkov 1952). During pre-fledging period (i.e. first 50 days), 144 ducklings collected Alberta, Canada, ate *c.* 67% (dry weight) invertebrates, mostly gastropod molluscs and dipterous larvae (Sugden 1973). Surface invertebrates composed 70% diet first 5 days, gradually being replaced by aquatic invertebrates and plant foods (mostly seeds of grasses and sedges). During first 1–12 days, pecking from surface predominated, gradually being replaced by sub-surface feeding either by bill-dipping, head-ducking, or up-ending; these in turn give way to mainly bottom feeding by head-ducking or up-ending.

For American studies see Mabbott (1920); Munro (1944); Anderson (1959); Keith and Stanislawski (1960); Keith (1961); Glasgow and Bardwell (1962).

**Social pattern and behaviour.** Based mainly on material supplied by R I Smith and S R Derrickson; see especially Smith (1968).

1. Highly gregarious for much of year. Flocks often fairly small, especially inland, but large numbers often occur on areas of extensive water. ♂♂ first to form post-breeding flocks, when pairs break up, leaving nesting area for larger, more permanent wet areas where undergo moult. In late summer and autumn, sexes still tend to flock separately, merging only during winter; by mid-winter, however, flocks consist of many pairs plus small groups of unpaired ♂♂. Sex-ratio often locally unbalanced, e.g. 70% ♂♂ November–February, Netherlands (Lebret 1961). BONDS. Monogamous pair-bond, of seasonal duration. Pair-formation starts quite early in winter, but most attachments resulting in strong bonds probably develop gradually during spring migration; in Netherlands, usually no pairs seen before December (Lebret 1961). ♂♂ also strongly promiscuous in breeding area, spending much time pursuing and attempting to rape other ♀♀ (see also McKinney 1973). Pair-bond often ends during first week or so of incubation (see McKinney 1965a). Only ♀ tends young; ♂ occasionally seen with ♀ and brood, but relationship casual and no evidence that original mates involved. BREEDING DISPERSION. As breeding habitat often temporary, ♀'s fidelity to previous nest-site not likely to be strong. Nests usually well spaced. Dispersion of pairs variable depending on distribution of suitable feeding waters; latter often widely scattered and each pair characteristically occupies large home-range with much overlap with other pairs and frequent joint use of restricted feeding areas, making defence of clear-cut territories impractical (McKinney 1973). Spacing of nesting pairs begins with dispersal of ♀♀ from communal feeding areas; where latter widely distributed, ♂♂ disperse with their mates but, in some situations, ♂♂ remain at communal areas throughout nesting period. These areas also visited by pairs and lone ♀♀ for feeding. ROOSTING. Little detailed information, but clearly much as most other *Anas*, i.e. communal on water or, where undisturbed, shore, both by day and at night depending on local feeding regime. Roost on sea as well as lakes and rivers, and roosting often largely diurnal because of crepuscular and nocturnal feeding habits. In early spring and autumn, flocks tend to use open shorelines, flooded fields, and mats of emergent vegetation, often in association with

Mallard *A. platyrhynchos*; in late winter, small flocks observed roosting in similar situations with Teal *A. crecca*, Shoveler *A. clypeata*, and Gadwall *A. strepera*.

2. Studied both in wild and captivity (Lorenz 1953; Johnsgard 1965; Smith 1968; S R Derrickson). Among west Palearctic *Anas*, closest to *A. platyrhynchos* in behaviour. FLOCK BEHAVIOUR. Much as in *A. platyrhynchos*, though calls distinctive. Pre-flight signals chiefly Head-shakes and Head-thrusts while soft calls given from alert position. ANTAGONISTIC BEHAVIOUR. Threat-displays include Chin-lifting in which bill rapidly raised and lowered while neck stretched up; closely linked with minor threat posture in which bill lowered to centre of breast. In 2 further such postures, bill directed slightly forward or placed horizontally on neck. All such threat accompanied by loud calls. Rest of aggressive behaviour similar to *A. platyrhynchos*, but ♂♂ more inclined to use wings while fighting, without holding with bill (Lorenz 1953) and no equivalent to Rabrab-palaver of that species. Preen-Back-behind-Wing display occurs, at least occasionally, after hostile encounters (McKinney 1965b). ♂ becomes increasingly aggressive in March–April, mainly in vicinity of mate, chasing away others (especially ♂♂) approaching within 3–5 m; ♀ will also pursue intruding ♀♀ tolerated by mate. However, in contrast to more territorial ducks (e.g. *A. clypeata*), ♂♂ strikingly less aggressive and more sociable at water areas frequented by many pairs in breeding area, especially during laying period, lacking distant threat-displays and rarely fighting (see McKinney 1973). Inciting-display of ♀ consists of lateral threatening movements of head directed against rejected ♂♂, with head held close to body and breast uplifted while repeated Inciting-calls given. Repulsion-display of broody ♀ more intense and frequent than in *A. platyrhynchos*. When ♀ harassed by intruding ♂ intent on rape, often combines head-rolling movements (as when oiling) with Repulsion-posture; even in absence of such a ♂, performs Repulsion-display in response to other kinds of disturbance—e.g. by man or other predator. COMMUNAL COURTSHIP. Gatherings of several ♂♂ closely round a ♀ occur on water or land, mostly between mid-December and early April. ♂♂ assume Courtship-intent posture much as in *A. platyrhynchos* (see Lorenz 1953). Major displays include Chin-lifting (see above), Burp, Water-flick, and Head-up-Tail-up—but Down-up of (e.g.) *A. platyrhynchos* absent. Secondary displays include Upward-shake which acts as introduction to major displays, though partly replaced in this rôle by repeated Burp-displays; also Preen-behind-Wing, though this more typical of pair-courtship. In Burp-display (see A), often alternating repeatedly with Chin-lifting, neck extended vertically as head moved up and somewhat back, bill down, while characteristic composite call given. Water-flick display similar to *A. platyrhynchos* but accompanied only by Whistle (not Grunt). Head-up-Tail-up display, which often follows Water-flick, also similar to *A. platyrhynchos* but elbows not raised and Turn-

A

B

Head-to-♀ component apparently not usually followed by Nod-swimming—latter, which may also occur independently, takes form of somewhat undifferentiated Head-nodding and may have been largely overlooked (see Smith 1968). Turn-back-of-Head display, which exhibits striking pattern on nape and neck (see B), much as in *A. platyrhynchos*; often follows Burp-display, also occurring when ♂ separately tries to ceremonially Lead ♀. Preferred ♂ in courting group usually performs Chin-lifting and Burp-display more than other ♂♂ and frequently Turns-back-of-Head; others mostly do Water-flick and Head-up-Tail-up displays. ♀ lacks true Nod-swimming of *A. platyrhynchos*; main pairing display Inciting, performed while moving towards or following preferred ♂ (often mate), frequently in response to his Leading. PURSUIT-FLIGHTS. (1) Courtship-flights. In January–February and especially March–April, courting parties frequently take wing, often on initiation of ♀. ♂♂ constantly move near and away from ♀, giving aerial version of Burp-display and attempting to perform Turn-back-of-Head. ♀ shows aerial versions of Inciting-display, 3 minor threat postures (see above), and, occasionally, special Head-up display with head raised slightly above horizontal axis of body while neck kept outstretched (Smith 1968). Apparently similar flights also develop from exploratory-flights of pair (see below) when birds joined in air by other members of flock. (2) and (3) 3-bird flights and rape-intent flights. These less easily distinguished from each other than in other *Anas*, even *A. platyrhynchos*. Though often involving 2 ♂♂ and a ♀, majority of flights of rape-intent type—and true 3-bird flights probably rare or absent, as also aerial chases between ♂♂ only. Flights initiated mainly when paired ♀ approaches or approached by paired ♂ other than own mate, resulting in close pursuit of ♀ while her mate (if present) follows at greater distance, other ♂♂ (also mostly paired) often then joining in. Last for a few seconds to over 1 hour; may continue for many kilometres from original point, often high in sky. Pursuing ♂ usually approaches ♀ from rear and below, often repeatedly, grabbing at or touching her flank or breast feathers with bill; ♀ performs Head-up display and minor threat postures frequently. Longer flights ended by ♀ descending rapidly and often landing far from water; rape or attempted rape by 1 or more ♂♂ follows (♀'s own mate not seen to take part). Such flights most characteristic of 12-day period before egg-laying (see Smith 1968). ♂♂ also harass ♀♀ on ground. If ♀♀ reluctant to leave area, may make series of short flights (10–30 m) to evade ♂♂, often taking refuge in cover. Broody ♀♀ often thus involved; perform Repulsion-display and give other distinctive calls; in flight, adopt special Head-down display as well as usual threat postures. PAIR BEHAVIOUR. Distinction between communal courtship and pair-courtship not clear-cut; during latter, ♂ performs Chin-lifting and Burp-display frequently, often after ♀ has directed Inciting-display towards intruding ♂, Burp-call often attracting other ♂♂ which form courting group. Paired birds also feed, drink, Bill-dip

and Head-shake, and carry out other comfort behaviour when together; of these, Preen-behind-Wing display by both sexes probably most ritualized. Burp-display of ♂ also used frequently in temporary absence of mate, with corresponding Decrescendo-call from ♀. Exploratory-flights by pair over nesting cover occur, in morning and evening, starting shortly after arrival in breeding area; initiated by ♀ who often performs aerial Head-up display. COPULATION. Copulations by pair not seen before March and then rarely; most frequent during final stages of spring migration, evidently as part of pair-courtship. Later in cycle, pair-copulations seem less important in achieving fertilization than promiscuous rapes of ♀, largely by paired ♂♂ other than own mate, round time of egg-laying—both for first clutches and replacements (Smith 1968). Rapes not preceded by displays and occur frequently on land, often following pursuit-flight. In pair copulations, ♀ usually first assumes Prone-posture on water and ♂ responds with Head-pumping—mutual Head-pumping, initiated by ♀, being less frequent. Post-copulatory display of ♂ variable, but usually involves Bridling as in *A. platyrhynchos* but not true Nod-swimming. RELATIONSHIPS WITHIN FAMILY GROUP. Much as in *A. platyrhynchos* so far as known.

(Figs A–B after Lorenz 1953.)

**Voice.** Often said to be used infrequently except during display, but vocabulary probably basically similar to Mallard *Anas platyrhynchos*, though sounds species-characteristic and mostly quieter. Calls of sexes distinct (Millais 1902). Pre-flight calls, however, differ mainly in tone only (R I Smith, S R Derrickson); described as soft, but, according to Lorenz (1953), considerably deeper and hoarser than similar call of *A. platyrhynchos*, and rich in rolling 'r' sounds. Main note of ♂ described by Lorenz (1953) as thin, nasal, passerine-like 'geeeee', also written 'whee' (Smith 1968). ♂ also whistles rather like Teal *A. crecca* (Millais 1902) but lower pitched. ♀ utters variety of quacking notes; said, e.g., to be short and sharp (Kirkman and Jourdain 1930), and higher pitched and more staccato than in *A. platyrhynchos* (Géroudet 1965). Unlike *A. platyrhynchos*, calls rarely given by either sex in flight following disturbance (R I Smith, S R Derrickson). Wing noise in flight described as low hissing (Witherby *et al.* 1939). Similar non-vocal sound signal to that of *A. platyrhynchos* produced during Preen-behind-Wing display (McKinney 1965b).

CALLS OF MALE. (1) Slow-gee call: monosyllabic, drawn-out 'geeeee' corresponding to Rabrab-call of ♂ *A. platyrhynchos* (Lorenz 1953). (2) Burp-call (see I):

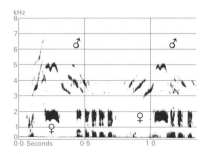

I J-C Roché

extended 'geee . . . geeeee', rising and falling as neck raised and lowered during Burp-display, usually with soft flute-like whistle ('preep') uttered simultaneously at peak of movement and ascending 'geeeee' note (Lorenz 1953). Given during both pair and communal courtship, whistling component attracting other ♂♂ (Smith 1968); also by ♂ separated from mate, and in all other situations in which ♂ *A. platyrhynchos* produces Slow-rab call (Lorenz 1953). (3) Other whistles: given during Water-flick display, Head-up-Tail-up display, and Bridling-display. (4) Calls during disputes in flock: loud notes, accompanying Chin-lifting and other threat gestures, presumably 'geeeee' or variants, but not adequately described; appear to induce others to join group from afar (R I Smith, S R Derrickson). (5) Other calls: not closely studied, but may include peculiar, rather faint wheezing note 'while flighting' (Kirkman and Jourdain 1930), and, according to Millais (1902), peevish cheeping like ♂ Wigeon *A. penelope*.

CALLS OF FEMALE. (1) Decrescendo-call: series of deep, loud quacks—1st unique in form and tone (R I Smith, S R Derrickson)—with decreasing amplitude and pitch; not so loud as in *A. platyrhynchos* and usually less than 6 syllables (Johnsgard 1965), giving impression of 2-syllable 'quahrr-quack' though uncertain whether 2 notes or many uttered rapidly (Lorenz 1953). Heard most frequently in fading twilight, but less often than in *A. platyrhynchos*. (2) Inciting-call: rather wooden, rattling or chuckling sounds (see I), softer than in *A. platyrhynchos*, much faster and run together in almost continuous 'arrrrrrrr' but with similar querulous and scolding, undulating intonation (Lorenz 1953). Also described as soft 'RARRerrerr' (see Witherby *et al.* 1939), and as succession of 'kuk' notes, similar to squeal of juvenile ♀ but lower pitched—'kuk-kuk-kuk-kuk-kuk . . . kuk-kuk-kuk-kuk' (Smith 1968). (3) Calls during disputes in flock: 'kuk' notes accompany threat posturing, on water or land (see under ♂ above)—also during pursuit-flights. (4) Repulsion-call: similar cackling to corresponding call of *A. platyrhynchos* but harsher and deeper (Lorenz 1953), a series of distinctive 'kak' notes—'kak-kak-kak-kak' (Smith 1968). (5) Distraction-call: variable 'gaak', 'gaaak . . . kek', or 'gaaag . . . keek-keek' (sometimes softer 'kee-kee'); given in response to predator approaching brood—also by broody ♀ harassed by ♂ intending rape when may alternate or intergrade with Repulsion-call, though latter uttered when ♀ reluctant to fly whereas Distraction-call more associated with flight (Smith 1968). (6) Other calls: note like growling croak of ♀ *A. penelope* mentioned by Millais (1902); equivalents of Persistent-quacking, alarm-call, maternal-calls, etc., of *A. platyrhynchos* still not fully described.

CALLS OF YOUNG. Calls of downy young similar to *A. platyrhynchos* (Lorenz 1953). During weeks 7–16, ♂ develops slightly modified calls: 'peep' of downy young becomes 'whee' or 'kwee' (corresponding to main 'geeeee'

note of adult ♂), given singly or in series; during weeks 18–20, whistle added, coinciding with first displays to ♀♀. At 7 weeks, calls of ♀ have tonal quality of adult ♀'s, 'peep' of downy young becoming 'kuk', given singly or in series; also utter high-pitched squealing—'ke-ke-ke-ke-ke-ke'. Calls of both sexes associated with minor threat postures and Chin-lifting (see Smith 1968 for further details).

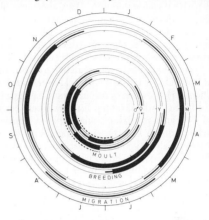

**Breeding.** SEASON. See diagram for north and west Europe. Little apparent variation over considerable area from central Europe to British Isles (Hudec and Touskova 1969). In far north onset of laying dependent upon thaw. SITE. On ground, in short cover, grass, rushes, and sometimes under scrub; often on bare ground as 9 of 18 nests, Czechoslovakia (Hudec and Touskova 1969). Can be close to water but up to 200 m away usual, occasionally 1–2 km. Not colonial, but nests sometimes as close as 2–3 m. Nest: slight hollow; lined grass, leaves and other vegetation, but sometimes little or none, especially in open site. Always lined down. Building: by ♀, material reached from nest; shaped by turning movements of body. EGGS. Ovate; yellowish-white, sometimes yellowish-green. 55 × 39 mm (48–60 × 36–42), sample 75 (Hudec and Touskova 1969). Weight of captive laid eggs 43 g (37–50), sample 28 (J Kear). Clutch: 7–9 (6–12). Of 39 Finnish clutches: 6 eggs, 13%; 7, 10%; 8, 31%; 9, 26%; 10, 17%; 11, 3%; mean 8·3 (Hildén 1964). One brood. Replacement clutches laid after egg loss. One egg laid per day. INCUBATION. 22–24 days. By ♀, but ♂ often close by. Starts on completion of clutch; hatching synchronous. Eggshells left in nest. YOUNG. Precocial and nidifugous. Self-feeding. Cared for by ♀, ♂ said frequently to be in attendance at least in early stages (Bent 1923; Witherby *et al.* 1939; Bannerman 1959), but more usually deserts ♀ after egg-laying (Hochbaum 1944; Sowls 1955). Brooded by ♀ while small, and nest and young defended by ♀ using distraction-displays. FLEDGING TO MATURITY. Fledging period 40–45 days. Independent around fledging time. Age of first breeding 1 year, though some not until 2. BREEDING SUCCESS. 76% of eggs in 33 Finnish nests hatched. 29 broods averaged 7·1 young at or soon after hatching, and 4·7 young in latter half of fledging period (Hildén 1964).

**Plumages.** Nominate *acuta*. ADULT MALE BREEDING. Head and throat chocolate, darkest on crown; often with patch from eye to nape glossed bronze-green or purple. Centre of nape and hindneck black, separated from brown sides of head and throat by white band. Mantle, upper scapulars, sides of breast, and flanks finely vermiculated black and white. Outer scapulars black, streaked cream or buff-grey on inner web and along shaft; elongated lower scapulars black broadly edged cream. Lower scapulars partly vermiculated grey. Back and rump like mantle, but feathers with more black at shafts and less vermiculation towards lower rump. Upper tail-coverts ash with black shaft-streak or black inner web, broadly edged pale cream. Patch at sides of rump cream-buff. Foreneck, sides of neck, chest, breast, belly, and vent white; finely vermiculated grey-brown on lower belly and vent. Under tail-coverts black, outer bordered white. Tail-feathers dark brown-grey on outer web, paler grey on inner; narrowly edged off-white, central feathers black, conspicuously elongated. Primaries and bastard-wing sepia on outer webs, paler ash on inner; primary coverts sepia, tinged grey on outer web. Secondaries with glossy green or bronze-pink outer webs, narrow black subterminal bars, and broad white tips (often faintly tinged buff); *c*. 4 outer ones dull black. Inner webs of secondaries dark grey. Upper wing-coverts mouse-grey with brown shafts; greater broadly tipped cinnamon, marginal finely speckled white or pale buff. Tertials light mouse-grey with broad black shaft-streaks; outer one, bordering glossy secondaries, velvety black on outer web, with broad pale grey (sometimes tinged pink) streak beside shaft. Tertial-coverts mouse-grey, outer edged cinnamon. Greater under wing-coverts light grey with white tips and specked edges; rest of underwing dark grey, heavily speckled and vermiculated white; axillaries white freckled dusky at tips and along shafts. ADULT FEMALE BREEDING. Crown and nape cinnamon, broadly streaked black; sides of head and neck paler, more finely streaked or spotted black; chin, throat and foreneck uniform pale buff or white. Mantle, scapulars, sides of breast, and flanks sepia, tinged olive, feathers with broad U-shaped pale cinnamon subterminal marks and broad buff edges shading to pale buff and white at tips. Back and rump dark sepia broadly edged pale buff or white; edges often finely speckled grey. Upper tail-coverts like back and rump, but irregularly marked pink-buff subterminally. Chest pale buff, feathers marked sepia towards base. Feathers of breast, belly, and vent buff with broad white fringes; vent often spotted grey-brown, and spots or streaks sometimes elsewhere on underparts. Under tail-coverts white or pale buff, irregularly marked dusky at centres. Tail-feathers grey brown, central black, all narrowly edged pale buff and with broad, irregular, pale cinnamon marks. Primaries like ♂, but outer webs of secondaries buff-brown, slightly glossed green in some lights, and mottled black towards base, especially on outer secondaries. White tips and black subterminal bars narrower than in ♂. Upper wing-coverts dark grey-brown, greater narrowly tipped white, pink-buff, or cinnamon, others narrowly margined off-white and sometimes with buff subterminal marks. Tertials sepia, broadly edged and irregularly streaked pale cinnamon. Under wing-coverts and axillaries dark grey broadly edged and barred white. ADULT MALE NON-BREEDING (Eclipse). Crown and hindneck black, feathers narrowly edged buff, sides of head behind eye grey-brown; chin, throat, and foreneck white, rest of head and neck buff, all finely speckled dusky-brown. Mantle, back, and rump dark grey-brown irregularly vermiculated, barred, or freckled white, feathers tinged olive and grey towards tips. Scapulars dull black bordered grey, tips narrowly edged off-white; those bordering mantle heavily vermiculated white at base. Chest, breast, belly, vent, and under tail-coverts

like adult ♀ breeding, but lower breast and belly more often spotted grey-brown and rest of underparts often partly vermiculated. Sides of breast, flanks, and upper tail-coverts sepia broadly edged and barred pale buff and white, often with much white vermiculation towards base of feathers, sometimes wholly vermiculated. Tail like adult ♂ breeding, but central pair shorter and duller. Wing like adult ♂ breeding, but tertials shorter, with more rounded tips, and often edged off-white. ADULT FEMALE NON-BREEDING. Like adult ♀ breeding, but cinnamon and buff of head paler, edges and bars of upperparts narrower and paler, off-white; underparts sometimes more heavily spotted; tertials sepia, narrowly edged white, tinged black along shaft without buff marks. DOWNY YOUNG. Crown, streak through eye, irregular streak over cheek, upperparts, and thighs light sepia, tips of longer down pale buff. Streaks over and below eye, on rear of wing, and on sides of back and rump white; those on head often partly tinged buff. Underparts white, tinged buff or grey on foreneck, chest, and vent. JUVENILE MALE. Like adult ♀ breeding, but head darker: crown dark grey-brown, slightly mottled buff; rest of head and neck off-white, heavily spotted and streaked dark brown. Upperparts, sides of breast, and flanks dark brown with narrow irregular white bars and edges. Underparts pale buff heavily spotted or streaked with grey-brown marks; no broad white fringes as adult. Tail-feathers notched at tip, bare shaft projecting; sepia with narrow, pale buff edges and bars. Wing like adult ♂, but buff or white tips of secondaries and greater upper wing-coverts narrower; median and lesser coverts duller grey, often suffused grey-brown towards tip and narrowly edged white. Tertials short and narrow, duller and browner than adult ♂ non-breeding, black centres less clearly defined. Greater tertial-coverts olive-grey, narrowly edged white. Before first pre-breeding moult, feathers on part of body partially changed for 1st immature non-breeding plumage, sometimes starting when wings not yet full-grown; resembles adult ♂ eclipse. JUVENILE FEMALE. Like juvenile ♂, but usually differing from it and adult ♀ by unbarred mantle. Wings like adult ♀, but outer web of secondaries often dark grey-brown with irregular patches, bars, or streaks dull buff speckled brown-grey, rather than uniform buff-brown with limited number of specks and slight green gloss as most adult ♀♀. White tips of secondaries and greater coverts narrower. Tertials short and narrow, sepia, narrowly edged and irregularly marked off-white. Development of 1st immature non-breeding (resembling adult ♀ non-breeding) like juvenile ♂. FIRST BREEDING MALE. Like adult ♂ breeding, but sometimes part of juvenile plumage on back, rump, underparts, or tertials retained. Wing juvenile, but often light and dark edges of coverts worn away (often showing more wear than adult ♂ at same season). Light tips of secondaries and greater coverts narrower; tertials like adult ♂ breeding. FIRST BREEDING FEMALE. Like adult ♀ breeding, including tertials; juvenile wing and sometimes some juvenile or 1st non-breeding plumage retained.

**Bare parts.** ADULT MALE. Iris yellow-brown to yellow. Bill pale blue-grey; nail, base, and strip over culmen black. Foot grey with darker joints and black webs. ADULT FEMALE. Iris brown to yellow-brown. Bill dark blue-grey, black on top of culmen; lower edge of upper mandible with some dark dots. Foot olive-grey or leaden-grey; webs and joints dull black. DOWNY YOUNG. Iris brown. Bill dark grey-horn, nail and base of lower mandible flesh. Foot olive-grey. JUVENILE. Iris brown or dark cinnamon. Bill dark grey, ♀ with dots at lower edge of upper mandible. Foot dark olive-grey or yellow-grey; webs and joints dull black. (Bauer and Glutz 1968; Dementiev and Gladkov 1952; Delacour 1956; Witherby *et al.* 1939; RMNH.)

**Moults.** Like Mallard *A. platyrhynchos*, but post-breeding, wing, and pre-breeding moults start 2–3 weeks later. In autumn, moult often not finished before migration. Adult ♂ usually in full breeding plumage by November (but central tail-feathers sometimes then still growing), adult ♀ by December (but sometimes some non-breeding on upperparts, and often tertials retained until April). Post-juvenile and 1st pre-breeding moult also later. Juvenile ♂♂ in breeding mainly by December, but some retain juvenile back, rump, belly, vent, some tail-feathers, tertials, and sometimes some other juvenile or 1st non-breeding plumage, until spring. Juvenile ♀♀ attain much 1st non-breeding plumage in September–November on head, mantle, scapulars, chest, flanks, and sometimes belly and vent; some breeding plumage develops here October–January, but remaining juvenile plumage retained during winter. In spring, much of breeding plumage attained, but some 1st non-breeding usually retained on upperparts, chest, and flanks, and some juvenile on back, rump, tertials, or sometimes belly and vent. (Dementiev and Gladkov 1952; C S Roselaar.)

**Measurements.** Netherlands, mainly July–December. Skins (RMNH, ZMA).

| | | ♂ | | ♀ | | |
|---|---|---|---|---|---|---|
| WING AD | 275 | (4·00; 20) | 267–282 | 260 | (3·80; 12) | 254–267 |
| JUV | 266 | (5·96; 35) | 254–279 | 248 | (6·70; 20) | 236–260 |
| TAIL AD | 179 | (6·42; 8) | 172–189 | 104 | (6·30; 7) | 95–113 |
| JUV | 92·3 | (5·24; 16) | 80–99 | 88·3 | (3·66; 13) | 83–95 |
| BILL | 50·9 | (2·07; 55) | 47–56 | 46·7 | (1·91; 31) | 44–51 |
| TARSUS | 42·6 | (1·26; 40) | 40–45 | 41·0 | (0·98; 30) | 39–43 |
| TOE | 55·4 | (1·92; 34) | 42–59 | 52·9 | (1·79; 24) | 50–57 |

Sex differences significant. Wing and tail of juvenile significantly shorter than adult. Other measurements similar; combined. Tail of ♂ in 1st-breeding plumage 168 (13·9; 12) 148–195 (not significantly different from adult). Average of adult ♂ non-breeding tail 108, of adult ♀ non-breeding 93, and of immature ♀ in 1st breeding 97.

**Weights.** ADULT. Various parts USSR (Bauer and Glutz 1968; Dementiev and Gladkov 1952; Dolgushin 1960). Various

averages (sample size, when known, in brackets), and total range.

| | | | | | |
|---|---|---|---|---|---|
| (1) | ♂ 851 (183), 923; | 680–1150 | ♀ 700, 735 (68), 773; | 550–900 |
| (2) | 917, 930, 950; | 800–1060 | 860, 910, 935; | — |
| (3) | 856, 865 (13), 888; | 750–1035 | 693 (13), 700; | 600–850 |
| (4) | 1080; | 900–1120 | 742 (12), 800; | 650–1050 |
| (5) | 807 (36); | 600–1000 | 708 (19); | 540–870 |
| (6) | 1057, 1081; | 900–1300 | 870; | 800–900 |
| (7) | 732, 829, 1000; | 550–1150 | 586, 650, 728; | 400–860 |

(1) Weight generally low in spring (♂ March, ♀ March–April; Caspian Sea). (2) Increases when gonads develop (♂ April to early May, ♀ April to late May; north USSR). (3) Loss when gonads reduced and ♀ breeding (♂ May to early June, ♀ June to early July; north USSR and Kazakhstan). (4) Heavy prior to loss of wing-feathers (♂ 1st half June, ♀ July; Volga delta, Kazakhstan). (5) Low when wing moult just finished (♂ July–August, ♀ August–September; Kazakhstan). (6) Rises to peak in autumn (♂ September–October, ♀ October; Kazakhstan, Volga delta). (7) In winter, highly variable (November–February; Caspian Sea). JUVENILE. Some individuals reach adult level from September. Average generally like adult from November, except in adverse conditions.

**Structure.** Wing long, pointed. 11 primaries: p9 longest, p10 0–5 shorter, p8 6–13, p7 18–28, p6 32–46, p1 106–138. Inner web of p10 and outer p9 emarginated. Tail rather long, pointed; 16 pointed feathers, central 2 pairs elongated in breeding plumage, about equal to others in non-breeding. Bill longer than head, narrow; sides parallel; culmen slightly concave over nostril, otherwise straight; nail fairly small, narrow. Tertials and scapulars elongated, tapering to point in breeding plumage, shorter and rounder in non-breeding and juvenile. Body and neck long and slender. Outer toe *c.* 94% middle, inner *c.* 78%, hind *c.* 30%.

**Geographical variation.** None in northern hemisphere. Southern subspecies much smaller, and adult ♂♂ never reach full breeding plumage. (Delacour 1956.)                                  CSR

---

## *Anas querquedula* Garganey

PLATES 71 and 74
[facing pages 519 and 543]

DU. Zomertaling    FR. Sarcelle d'été    GE. Knäkente
RU. Чирок-трескунок    SP. Cerceta carretona    SW. Årta

*Anas Querquedula* Linnaeus, 1758

Monotypic

**Field characters.** 37–41 cm, of which body slightly over two-thirds; wing-span 60–63 cm. Small, slightly built, and slender-necked dabbling duck with rather flat crown and straight bill; ♂ has broad white supercilium, otherwise mottled brown head, breast, upperparts, and stern, contrasting with greyish flanks and white belly, long black and white scapulars, and pale blue-grey forewing. Sexes dissimilar, but no seasonal differences apart from eclipse which is retained unusually long (5 or 6 months). Juvenile closely resembles adult ♀, but just distinguishable when standing.

ADULT MALE. Broad white stripe, curving from just in front of and above eye to well down nape, dominant character, contrasting with black-brown forehead and crown and golden-brown cheeks and foreneck flecked with white; black patch on chin and throat. Mantle to upper tail-coverts black-brown with lighter feather edges and bars; breast and sides of upper mantle pink-brown barred with black (crescentic markings on upper breast), sharply demarcated from vermiculated greyish flanks and white belly; undertail white with dark brown bars and spots; tail brown with white edgings. Elongated and drooping scapulars striped grey, dark green, black and white; forewing pale blue-grey; speculum dull green edged in front and more narrowly behind with white. Bill lead-grey; eyes brown; legs dull grey. In flight, blue-grey forewing distinctive; also, from below, sharp demarcation between brown breast and white belly, and prominent dark leading edge to wing. MALE ECLIPSE. Closely resembles adult ♀, but brighter, with whiter throat and belly, and easily distinguished in flight by much brighter blue-grey forewing and green speculum. ADULT FEMALE. Plumage much like ♀♀ of other dabbling ducks, being brown with lighter feather-edges and white lower breast and belly, and closely resembles ♀ Teal A. crecca, but rather paler, with whiter throat, pale patch at base of longer bill, and more distinct dark eyestripe and light supercilium. In flight, greyish (but not blue-grey) forewing and indistinct greenish-brown speculum characteristic; also, from below, dark leading edge to wing, as in ♂, prominent especially against light. Bill greenish-grey. JUVENILE. Closely resembles adult ♀, but lower breast and belly mottled and streaked with brown.

Broad white supercilium and sharp demarcation of brown breast from greyish flanks and white belly readily identify ♂ at considerable distances and even in flight, when blue-grey forewing also characteristic. Only other west Palearctic ducks with blue forewings are widespread Shoveler A. clypeata and vagrant Blue-winged Teal A. discors, but in their cases colour bright blue, not grey-blue, and A. clypeata also larger, with unmistakable bill and plumage pattern. Other plumages likely to be confused only with A. crecca, but latter looks darker with brownish throat, almost no supercilium, and distinct green speculum, as well as being shorter-necked and more compactly built. Dark leading edge to forewing of A. querquedula also diagnostic from below.

Found mainly on shallow fresh water with low but extensive cover; seldom on sea except on migration. Quiet and unobtrusive. Swims low on water and, despite much smaller bill, dabbles more like A. clypeata than A. crecca. Up-ends rather seldom and dives only when wounded or frightened. Rises easily from water and flies with rapid wing-beats and great agility, but without twists and plunges of A. crecca. Usually in pairs or small parties, but concentrates into flocks of several hundreds on migration and in tens of thousands in African and Asiatic winter quarters. Generally rather quiet, but ♂ has dry, crackling or rattling call like fishing reel and ♀ a short, sharp quack.

Habitat. Distinguished among genus as fully migratory, whole population changing habitat seasonally throughout west Palearctic. Breeds mainly within Mediterranean, steppe, and temperate climatic zones, with some overspill north into boreal and south into desert fringes. Averse to arctic or oceanic influences reflected in persistently cool, windy, or rainy conditions; also tends to avoid high upland or mountain regions and dense forest. In west Palearctic, makes only brief localized use of marine or even of tidal estuarine habitats. Also avoids extensive deep and lifeless, oligotrophic waters without ample aquatic vegetation, especially where margins naked and rocky. Favours narrow or well compartmented, sheltered, and shallow standing fresh waters, merging into grassland, floodland, or other wetland, with plenty of floating and emergent vegetation, but not too tall or dense, unbroken, fringing cover. Sites of high biological productivity and benign microclimate most congenial, some being permanent but others arising or declining through climatic fluctuations or human influences. Like other Anas, keeps informed of such changes by constant and wide-ranging use of airspace, but tends to settle in small area, not shifting much daily between feeding, resting, and breeding places. Spends little time on land except immediate margin. Makes little use of artefacts. Relatively indifferent to man unless persecuted or badly disturbed. Habitat outside breeding season similar, but where suitable equivalents unavailable temporarily uses more exposed and poorer waters, or small ponds, ditches, and irrigation pools.

Distribution. Considerable fluctuations on edges of range, with much sporadic nesting. Precise data often limited, but Bauer and Glutz (1968) suggested expansion 1850–80, followed by decline, then some increase around 1900, further marked increase in north of range 1920–30, and decline since 1950s; changes often linked with climatic conditions, e.g. warm springs stimulating northern spring migration, or dry years in south forcing birds further north than usual.

BRITAIN AND IRELAND. Gradual spread and increase in southern England until c. 1952, but decline since; considerable annual variations dependent on size of spring arrivals, e.g. marked influx 1959 with temporary nesting in new areas, notably south-west England. Has bred Scotland 1928, Wales 1936, and Ireland 1956 and 1959 (Parslow 1973a; British Ornithologists' Union 1971). CHANNEL ISLES: bred Jersey 1952 (La Société Jersiaise 1972). NORWAY: first recorded 1862; since invasion in 1947, probably breeds almost annually in small numbers (Haftorn 1971). SWEDEN: expansion in mid 19th century; during last two decades, colonized lakes along Baltic coast in north (Bauer and Glutz 1968; Curry-Lindahl 1964). FINLAND. Fairly common some areas mid 19th century with weak expansion 1850–80, then apparently vanished; reappeared 1930s, with some expansion 1940s (Merikallio 1958; LvH). Decreased since early 1950s (Grenquist 1970). PORTUGAL: bred 1938 and for some years after (MDE). SPAIN: formerly bred marismas of Guadalquivir in small numbers when water levels high (Ree 1973); sporadic breeding elsewhere (FB). ITALY: in Sicily, now only on passage (GS). CYPRUS: bred 1910 (Stewart and Christensen 1971). ICELAND: record of nesting 1860 not accepted (FG). IRAQ: may have bred, but no proof (Allouse 1953).

Accidental. Iceland, Faeroes, Azores.

Population. Breeding. BRITAIN: 50–100 pairs (Atkinson-Willes 1970a). FRANCE: 1000–2000 pairs (Jouanin 1970).

BELGIUM. Some 20 pairs, apparently more numerous in warm summers (Lippens and Wille 1972). NETHERLANDS: *c.* 5000 pairs (CSR). WEST GERMANY: *c.* 1100 pairs (Szijj 1973). DENMARK: *c.* 200 pairs (Joensen 1974). FINLAND: 1000–2000 pairs, but considerable fluctuations (Merikallio 1958). CZECHOSLOVAKIA: thousands breed in lowlands up to 600 m (Hudec 1970). USSR: 2000–2500 pairs Estonia (Onno 1970).

Winter. Tentative estimate of 250000 birds in Europe–Black Sea–Mediterranean and West Africa (Szijj 1972).

Survival. Oldest ringed bird 10 years (Rydzewski 1974).

**Movements.** Migratory. Summer visitor to north and west Europe, where only stragglers November–February. Quite small numbers winter Mediterranean basin (Atkinson-Willes 1970*b*), but majority of west Palearctic breeders trans-Saharan migrants. Large numbers winter all across northern tropics, especially in west Africa; some dispersal south to Transvaal. By far commonest Palearctic duck in Sénégal and Gambia, especially Sénégal delta where 200000 January 1971, 140000 January 1972; in latter month, 93000 seen in incomplete aerial count of Niger inundation zone (Mali) (Roux 1971, 1972). Also common north Nigeria, though no figures available. Thus west African winter population may exceed size of west Europe breeding numbers, so major involvement by USSR breeders (Moreau 1967). Birds ringed Sénégal delta recovered later Netherlands, France, Italy, Yugoslavia, Poland, and USSR (Roux 1971); and 2 ringed Nigeria recovered west Siberia (BTO).

Breeders of Britain, Netherlands, France, and West Germany have 2 migration routes: (1) south through France and Iberia, across Mediterranean to Morocco and Algeria, thence south to wintering areas; (2) south-east to Italy and Balkans, then probably direct across Sahara. No evidence that west Europe breeders reach east Africa. Breeders of north-west and north USSR (some from east to 80°E), Baltic countries, and central Europe, migrate south-west to south France and Italy, then to tropical Africa. Few sightings anywhere coastal north Africa indicates most cross Mediterranean and Sahara in single flight (Moreau 1967). From west-central and south USSR (Kazakhstan, Volga, Ob and Irtysh basins), breeders migrate south-west over Caspian, some then via Ukraine to east Mediterranean and down Nile, while south-west flights across Red Sea, September, indicate passage across Arabia, probably continued as diagonal movement across Sahara to major west African winter haunts (Moreau 1967); another part of this west Siberian population migrates south-east to winter in Indian subcontinent (Shevareva 1970). Loop migration apparent as returns from west and east Africa tend to be via Italy and Balkans, smaller proportions occurring west and east ends of Mediterranean on spring passage (Bauer and Glutz 1968; Shevareva 1970). Recoveries in subsequent seasons of

several from west Europe across west Siberia to Irkutsk (98°E), also singles Leningrad to India and India to Ukraine and Italy, presumably aberrant migration.

Moult gatherings of ♂♂ known (assembling from late May), but not most of catchment areas; Volga delta particularly important, drawing birds from at least as far away as Leningrad, Komi ASSR, and Tomsk. Migration to winter quarters mainly late July to October, peak through Europe August and early September. Early arrivals reach Sénégal early September, Niger late September, and most by mid-October; perhaps somewhat earlier in east Africa since migration through Egypt mainly August and early September (Meinertzhagen 1930). Return movement begins February, with early arrivals North Sea countries at end of month; nearly all gone from Sénégal by end March; main northward passage through Europe March–April, arriving further north about mid-May. See also Dementiev and Gladkov (1952), Impekoven (1964).

**Food.** Animal and plant materials collected mainly while swimming with head under water, somewhat less often upending briefly, and from surface. Often also snaps at individual items on or flying above surface (Szijj 1965*b*). Normally daytime feeder, in pairs, small parties, or large flocks.

Items include: insects and larvae, especially Hemiptera (waterbugs *Notonecta*, *Naucoris*), Trichoptera (caddisfly Phryganeidae), Coleoptera (waterbeetles *Haliplus*, *Hydrobius*, *Helophorus*, *Cyclonotum*, *Dryops*), and Diptera (midge Chironomidae); molluscs (*Planorbis*, *Anodonta*, *Bythinia*, *Physa*, *Viviparus*, *Lymnaea*); crustaceans (amphipod *Talitrus*, ostracods, phyllopods); annelids (worms, leeches Hirundinea); young and spawn of frogs; and fish. Plant materials include buds, leaves, roots, tubers, and seeds of aquatics: *Potamogeton* (pondweeds), *Sparganium* (bur-reeds), *Scirpus* and *Carex* (sedges), *Glyceria* (grasses), *Juncus* (rushes), *Nymphaea* (waterlilies), *Rumex* (docks), *Polygonum* (bistort), *Ranunculus* (water crowfoot), *Phragmites* (common reed), *Lemna* (duckweed), *Chara* (stoneworts); also rice and seaweeds. (Naumann 1896–1906; Madon 1935; Witherby *et al.* 1939.)

Specific data and detail of seasonal and local differences limited. 29 stomachs from France, Germany, and Italy all contained plant materials (24 contained seeds, mainly grasses Gramineae, and *Phragmites*), 7 had molluscs, 4 insects, and 2 crustaceans (Madon 1935). 23 stomachs, flood plains Tisza river, Hungary, contained *Lemna* (9 stomachs), *Carex* (8), *Chara* spores (4), bristle-grass *Setaria glauca* (2), cockspur *Echinochloa crus-galli* (1), and insects *Notonecta glauca* (2) and *Helophorus* (2) (Sterbetz 1969–70). 1 stomach, Bodensee, Germany, April, contained 100 young snails (*Bythinia*, *Lymnaea*) *c.* 1 mm, and 13 Trichoptera larvae (Szijj 1965*b*). On occasions, one food exclusively utilized: in Austria, 27 August–3 October, 20–30 feed on crustacean *Triops crancriformis* (Bauer and

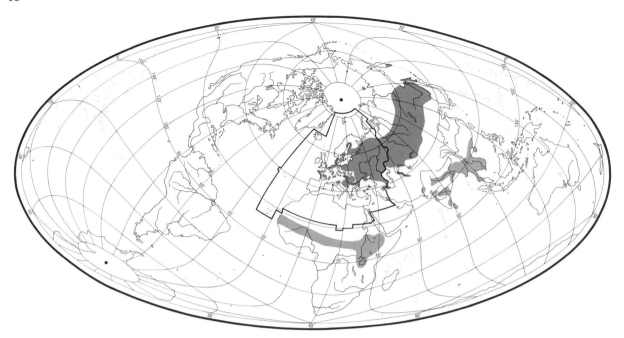

Glutz 1968); in Sénégal, when migrants first arrive, essentially only *Bulinus*, though later (as valleys flooded) chiefly grass seeds (Morel and Roux 1966).

In USSR, seasonal and area differences more apparent (data mainly from Dementiev and Gladkov 1952). In migrants, Volga delta, March and early April, 26·7% by volume animal (18·3% molluscs, 6·7% waterboatmen *Corixa*, 17% crustacean *Corophium*), 50% seeds, 23·3% leaves and roots. In late spring, Mologa river, seeds *c.* 50% frequency, though volume small; animal materials *c.* 95% volume, mainly small crustaceans (Branchiopoda: *Triops, Branchipus, Estheria, Limnadia*), molluscs (*Viviparus, Bythinia, Lymnaea, Planorbis, Physa*), and insects mainly chironomids (Isakov and Raspopov 1949). 12 stomachs, Pripyat marshes, spring and summer, 84·6% by volume animal, mostly molluscs (*Lymnaea, Planorbis*); 15·4% plants, mainly seeds *Rumex, Sparganium, Polygonum* (Dolbik 1959). At Rybinsk reservoir in spring, animal 68% (molluscs 34%, chironomid larvae 7%, waterbeetles *Dytiscus* and *Corixa* 27%), seeds 30% (chiefly *Sparganium*), fennel-leaved pondweed *P. pectinatus* 2%; in July and August, animal foods up to 90% (molluscs 54%, aquatic insect larvae chiefly *Clyptotendipes gripekoveni* 26%, other insects 10%). Similarly, on Mologa river in summer, animal foods up to 90%, but here 32% Diptera larvae (mainly chironomids), 27% molluscs, 25% larvae and adult Odonata (dragonflies). During moult period, Volga delta, animal foods only 8·1%, but seeds 67% and vegetative parts (mostly rockcress *Najas marina*) 24·9%; in August and September, however, animal foods 41·2% (molluscs, primarily *Valvata piscinalis* 34·6%; insects, chiefly larvae of soldier-fly Stratiomyidae, and aphids 6·6%), vegetative parts 19·1%, seeds 39·7%; while

in October, animal food decreased to 27·5%, and green parts increased to 37·5%. In autumn, Mologa river, animal foods 60% (mainly molluscs 40% and insects 19·5%), vegetative parts 10% (mostly filamentous algae and bulbs of arrowhead *Sagittaria*), seeds 30%. In September, Rybinsk reservoir, animal foods rose to 77% (molluscs 35·5%, chironomid larvae 21%, caddisfly larvae 17%, adult insects 3·5%), with seeds 23%.

From Pripyat marshes, USSR, 10 stomachs of young contained 90% animal food (Dolbik 1959).

**Social pattern and behaviour.** Based mainly on material supplied by F McKinney.

1. Gregarious except when nesting. Within European summering range, mostly in pairs and small parties. Often in much larger numbers on passage and in winter quarters. Winter flocks increasingly consist of pairs which migrate to breeding grounds in groups. ♂♂ first to form post-breeding flocks after pair-bond ended (May–June), moulting aggregations occurring in late summer. BONDS. Monogamous pair-bond, probably usually of seasonal duration. Promiscuity, including rape attempts, not recorded and pair-bond apparently strong—reported to last throughout incubation (but see Witherby *et al.* 1939) though only ♀ tends young (Guichard 1957). BREEDING DISPERSION. Similar to Shoveler *A. clypeata* in many ways (Guichard 1957) though not so closely studied: pairs dispersed on small ponds and appear to maintain more clear-cut territories than (e.g.) in Teal *A. crecca*. Behaviour recalls most that of Blue-winged Teal *A. discors* (see Dzubin 1955): during laying and incubation, ♂ frequents waiting area on water in vicinity of nest, intolerant towards approach of other ♂♂. ROOSTING. Little information. Probably much as in *A. crecca* as both crepuscular feeding and nocturnal courtship reported.

2. Least well known of west Palearctic *Anas*; distinctive in a number of respects but nearest *A. clypeata*. Studied mainly in captivity (Lorenz 1953; Johnsgard 1965). FLOCK BEHAVIOUR.

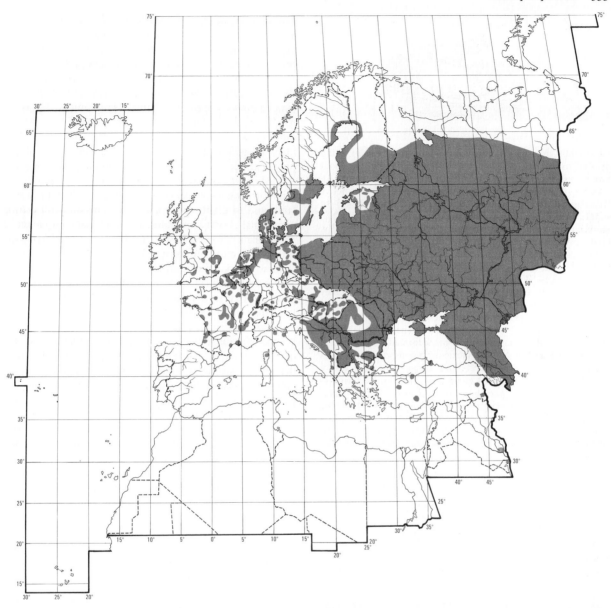

Flocks often dense and well co-ordinated, especially in flight when manoeuvre with speed and agility, often close to water. Pre-flight signals said to be lacking (Johnsgard 1965), but according to Lorenz (1953) similar to those of Mallard *A. platyrhynchos* though vertical thrusting (or pumping) movements of head even more pronounced. ANTAGONISTIC BEHAVIOUR. Vertical pumping movements of head stated to occur also 'at any general excitement' (Lorenz 1953); possibility, therefore, of confusion between true Head-thrusting (pre-flight) movements and Hostile-pumping such as found as threat-display in *A. clypeata* and other blue-winged ducks (see McKinney 1970). ♂ also said to give 1-syllable calls in same situation as Rabrab-palavers between ♂♂ in *A. platyrhynchos* (Lorenz 1953) but no details of posture given. ♀ Inciting-display consists of usual *Anas* head-jerking and bill-pointing threat movements towards rejected ♂; alternate with pronounced Chin-lift, each movement being accompanied

by single Inciting-call. COMMUNAL COURTSHIP. Not recorded during autumn passage, but frequent in winter quarters (where majority of pairs formed) and continues into April–May in breeding area, though 97% of ♀♀ paired by spring (Bezzel 1959). Display repertoire of ♂♂ includes elements resembling displays both of extralimital spotted-teal group (Hottentot Teal *A. versicolor* and Silver Teal *A. punctata*) and of blue-winged ducks (Johnsgard 1965). At start of courting bout on water, ♂♂ mill about in Courtship-intent posture with windpipe bulging in throat while performing secondary displays. These include Ceremonial-drinking, Mock-preening, Upward-shakes, and, as excitement increases, Wing-flaps (Lorenz 1953)—though Johns-gard (1965) found Upward-shakes infrequent and not restricted to beginning of bout. Mock-preen display, in which folded wing on side nearest ♀ raised briefly showing conspicuous blue patch, said by Lorenz (1953) to take unique form of Preen-outside-

A

Wing—but Johnsgard (1965b) and McKinney (1965b) found display to consist of typical Preen-behind-Wing. As bout progresses, major displays become increasingly frequent. All variants of a Burp-drink display accompanied by distinctive Rattling-call and terminating in Ceremonial-drinking. (1) In simple Burp-drink, ♂ raises head slightly, calls, and drinks. (2) In more intense version, first stretches neck up and back rapidly, holding bill horizontal. (3) In most elaborate form, head flung right back so that front of crown touches lower back, bill pointing up (see A), then returned rapidly to starting position immediately after call given. This movement—also termed Laying-Head-back display—unique among *Anas* ducks and superficially similar to Head-throw of pochards *Aythya* and goldeneyes *Bucephala*. ♂ also performs Turn-back-of-Head display while swimming ahead of ♀ (characteristically as she Incites); during such ceremonial Leading, white V-mark formed by convergent eye-stripes at back of ♂'s head particularly conspicuous to ♀. As well as Inciting, Preen-behind-Wing frequent in ♀. When ♀ particularly responsive during communal courtship, performs ♂-like laying-Head-back version of Burp-display (Lorenz 1953; Géroudet 1965) with special 2-note call. PURSUIT-FLIGHTS. Little information. Those reported, April to mid-May, on breeding grounds (Guichard 1957) probably include courtship-flights and 3-bird flights but not rape-intent flights; much calling from both sexes. Courtship-flights initiated when ♀ takes wing on being pressed too intensely by displaying ♂♂ who follow her, calling; party rises high in air, performing erratic evolutions (Géroudet 1965). PAIR BEHAVIOUR. Few details, but some of displays mentioned earlier, including Preen-behind-Wing and Ceremonial-drinking, probably more typical of pair-courtship than communal courtship. Preen-behind-Wing, by both sexes, more common than in other *Anas* with possible exception of Wigeon *A. penelope* and allies (Johnsgard 1965). Decrescendo-call of unpaired ♀ or ♀ separated from mate rare and brief (Lorenz 1953); corresponding call of ♂ apparently not described. COPULATION. Not known if forms part of pair-courtship and seems not to have been described from wild. In captivity, preceded by mutual Head-pumping but other behaviour undescribed. RELATIONS WITHIN FAMILY GROUP. Little information, Guichard (1957) reports that brood lives in aquatic vegetation; when disturbed, ♀ gives alarm-call.

(Fig A after Lorenz 1953; see also Johnsgard 1965.)

**Voice.** Rather limited in both sexes. Characterized especially by peculiar call of ♂ who lacks pure whistles in his repertoire (Johnsgard 1971), as do most species of blue-winged ducks. Calls of ♀ said to be infrequent and more like those of ♀ Mallard *A. platyrhynchos* than more rasping ones of Pintail *A. acuta* (Lorenz 1953). Some utterances likened also to quack of ♀ Teal *A. crecca* (Witherby *et al.* 1939), but longer and not so sharp—like 'knek' (Kirkman and Jourdain 1930). Sexes said to have some calls in

common, including quiet chattering (Witherby *et al.* 1939) and 'love notes' anglicized from German sources as 'yeck yeck yeck' or 'djü djü djü' (Kirkman and Jourdain 1930), but no adequate details of circumstances given.

CALLS OF MALE. Main utterance, equivalent of Repeated-calls of Shoveler *A. clypeata* and other blue-winged ducks, unique, mechanical-sounding rattle (see I); recalls crackling noise of breaking ice (P J Sellar), and quite closely resembles drumming of woodpeckers (Picidae) in acoustical structure and spacing; if taped call played at quarter speed sounds like pleasing tune of even-spaced notes tapped out on wooden xylophone or miniature wood-blocks (J Hall-Craggs). Given on water, land, and in flight—but mainly as Courtship-call during communal activity on water, especially as component of Burp-drink display. Used throughout period of pair-formation and courtship. Variously described, e.g.: strident clacking or rattling 'klerrep' like stick drawn across iron railings or grinding of teeth (Kirkman and Jourdain 1930); short 'rrrrp' with quality of wooden rattle, hard to locate (Lorenz 1953); rolling rattle 'rrar . . . rrar . . . rrar . . . rrar' (McKinney 1970). Produced in broken bursts, with pulse rate of *c.* 50 per s, somewhat like rattling of Mute Swan *C. olor* (Johnsgard 1971). Few other calls described; include 1-syllable 'gegg . . .', aggressive and conversational calls equivalent to Slow Rab of *A. platyrhynchos* (Lorenz 1953).

CALLS OF FEMALE. (1) Decrescendo-call: like that of *A. clypeata*, especially in 'swallowed' character—i.e. last notes sound as if bill suddenly dipped in water—and rapid rise and fall producing peculiar roaring emphasis; consists of only 2–3 notes and said to be rare (Lorenz 1953). (2) Pre-flight call: similar to that of *A. platyrhynchos* (Lorenz 1953). (3) Alarm-calls: 'krrrt' (Géroudet 1965); 'quake' when flushed from nest (see Bannerman 1958). (4) Inciting-call: disjointed series of 1-syllable 'gaeg' notes, single note accompanying each movement of Inciting-display. (4) Call given during ♀ version of Laying-Head-back display: 'QUAEH-geg' (Lorenz 1953); 'ghaeghaeghaeg' (Géroudet 1965).

CALLS OF YOUNG. Contact-calls uttered in groups of 2–5 notes. Distress-calls similar to *A. platyrhynchos*, rather slower than in Marbled Teal *Marmaronetta angustirostris*. (J Kear.)

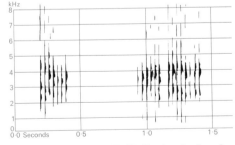

I Sture Palmér/Sveriges Radio Sweden April 1958

**Breeding.** SEASON. See diagram for central and western Europe. In south and west Russia, main laying period early May (Dementiev and Gladkov 1952). SITE. On ground in thick vegetation, grass, or rush tussocks. Usually close to water, within 20 m, rarely over 100 m. Not colonial. Nest: shallow depression lined leaves and grass, *c.* 20 cm diameter and up to 10 cm deep. Lined down. Building: by ♀, using material within reach and shaping depression with body. EGGS. Ovate; light straw colour or light brown. 46 × 33 mm (40–50 × 30–36), sample 170 (Schönwetter 1967). Weight 28 g (26–29), sample 9 (Verheyen 1967). Clutch: 8–9 (6–14). One brood. No data on replacement clutches, though probably occur. Eggs laid daily. IN-CUBATION. 21–23 days. By ♀, covering eggs with down when off nest. Starts on completion of clutch; hatching synchronous. Shells left in nest. YOUNG. Precocial and nidifugous. Self-feeding. Cared for by ♀, who broods them while small. FLEDGING TO MATURITY. Fledging period 35–40 days. Independent at about same time. Age of first breeding 1 year. BREEDING SUCCESS. No data.

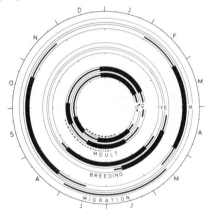

**Plumages.** ADULT MALE BREEDING. Forehead and crown black, feathers on forehead streaked buff, on crown narrowly tipped brown. Broad white streak over eye to lower nape. Hindneck dark grey-brown, chin and upper throat black; sides of head and neck auburn, narrowly streaked white. Feathers of sides of upper mantle black with pale buff U-shaped marks; chest similar, but marks broader and more yellow-buff, grading into dark grey-brown and pink-buff barred feathers on breast. Centre of mantle and back brown-black, feathers edged olive-grey. Outer scapulars pale blue-grey narrowly edged white, upper with much olive-brown at bases; inner scapulars elongated, upper ones black, tinged olive with narrow white shaft-streak, lower ones with outer webs blue-grey and inner webs black with broad, white shaft-streak. Rump and upper tail-coverts sepia-black, irregular wavy bars and edges pale buff. Flanks white, finely vermiculated black, longer feathers broadly tipped grey; rest of underparts white, finely marked grey on lower belly and vent, boldly spotted dark brown on under tail-coverts. Tail black with grey bloom, feathers narrowly edged white; outer feathers often suffused white. Primaries dark grey-brown on outer webs, paler on inner; shafts white, tips and edges of inner feathers narrowly fringed white. Primary coverts pale grey-brown, edged white on outer webs. Secondaries broadly tipped white; outer webs green (blue-grey on some outer feathers), inner webs sepia edged white.

Greater upper wing-coverts pale brown-grey broadly tipped white, rest of upper wing-coverts pale blue-grey, some median narrowly edged white. Tertials sepia with narrow white edges; outer tinged grey on outer web, inner with narrow white shaft-streak. Greater tertial-coverts blue-grey, edged white. Lesser under wing-coverts dark grey, rest of under wing and axillaries white. ADULT FEMALE BREEDING. Crown black, hindneck dark grey-brown, feathers of both broadly edged buff. Sides of head and neck pink-buff or white, narrowly streaked dark brown, heaviest on lores and behind eye and beneath faint streak below eye; chin and throat white. Feathers of upperparts dull black with olive tinge, broadly edged cinnamon or pink-buff (paler buff towards tip) on mantle and scapulars, and pale buff or white on back and rump. Edges and irregular marks on upper tail-coverts pink-buff. Slightly elongated inner scapulars often with narrow cinnamon shaft-streaks. Feathers of chest and sides of breast sepia, contrastingly edged pink-buff to white (giving scaled appearance when worn); sides of body and flanks sepia-brown, broadly edged and irregularly marked pale buff or white. Under tail-coverts white boldly marked or streaked dark brown along shafts; rest of underparts white, often with some subterminal sepia to pink-buff marks. Tail sepia, feathers narrowly edged white and irregularly marked pink-buff. Wing like adult ♂, but secondaries dark grey with less green gloss; often narrowly edged white on outer web. Greater upper wing-coverts with much narrower white tips, outer webs narrowly edged white; lesser and median coverts pale grey-brown, suffused grey towards tips and often edged white. Tertials sepia with olive tinge, outer webs and tips edged pink-buff to white, often with dull cinnamon shaft-streaks. ADULT MALE NON-BREEDING (Eclipse). Like adult ♀ breeding, but sides of head more heavily streaked; crown with narrower feather edges, appearing darker. Upperparts sepia, feathers more narrowly edged pale grey or buff, some sub-terminal buff V-marks in rump. Feather edges at chest and sides of breast buff, feathers of flanks sepia, edged white, subterminally with white shaft-streaks and bars or white vermiculations. Lower belly and vent mottled grey. Tail and wing like adult ♂ breeding, but tertials like adult ♀ breeding. In autumn, feathers of non-breeding plumage on part of body sometimes replaced by a 3rd plumage, preceding breeding: feathers on chest and sides of breast dark grey-brown with white crescentic marks; like non-breeding on sides of mantle, scapulars, and flanks, but with some coarse white vermiculation. ADULT FEMALE NON-BREEDING. Like adult ♂ non-breeding, but all feather edges of upperparts pale buff, usually no V-marks in rump. Feathers of flanks usually unmarked subterminally; wing like adult ♀ breeding. Differ from latter by dark crown, more heavily streaked sides of head and neck, absence of shaft-streaks or marks on scapulars, tertials, and upper tail-coverts. DOWNY YOUNG. Crown, hindneck, and 2 distinct lines from base of bill through eye to nape and from base of bill over cheeks to hindneck dark olive-brown; also upperparts, sides of body, and thighs. Superciliary streak, small spot in lores, indistinct line below eye, chin, throat, and small spots on wing and sides of back and rump pale yellow. Chest grey, underparts grey-yellow. Differs from downy young of Teal *A. crecca* by more distinct dark lines through and below eye (latter reaching to base of bill); also by yellow spots on lores, absence of ring round eye, and grey-yellow rather than cream-yellow underparts (Delacour 1956; Dementiev and Gladkov 1952; Witherby *et al.* 1939). JUVENILE MALE. Like ♀ breeding, but crown black with narrow buff feather edges, streaks at sides of off-white head and neck more numerous (but narrower than in similar adult non-breeding); feather edges of upperparts narrower, grey-buff on upper mantle and blue-grey on lower mantle, back, and rump; scapulars dull black, narrowly edged buff. All body feathers narrower, tips less square than adults. Dark centres of chest feathers more pointed towards tip; feathers of breast, belly, and vent strongly suffused grey towards bases, often small dusky spot on tips, giving spotted or streaked appearance to underparts, especially on vent. Underparts often tinged yellow-buff. Tail like adult, but notched at tip, and bare shaft projecting. Wing like adult ♂, but white tips of secondaries and greater upper wing-coverts narrower; tips of lesser and median coverts more or less suffused grey-brown; tips of inner webs of inner greater coverts with grey-brown spot, or at least with some mottling. Tertials and greater tertial-coverts shorter and narrower than those of adults; dark sepia narrow white edges, tips of inner broadly edged white. JUVENILE FEMALE. Like juvenile ♂, but feather edges of upperparts pale buff, sometimes off-white on back and rump, rarely blue-grey as ♂. Wing like adult ♀, but speculum dark grey, usually without green gloss; white tips of secondaries and greater coverts narrower; white tips on inner webs of inner greater coverts usually narrowing to white line or absent. Median and lesser coverts sometimes uniform dark grey-brown, virtually without ash-grey tinge. FIRST IMMATURE NON-BREEDING. Like adult non-breeding, but juvenile wing and tertials, and some of juvenile upperparts, belly, vent, and outer tail-feathers retained. Sexes separable by wing. FIRST BREEDING. Like adult breeding and often difficult to separate when no juvenile tertials, tail, or body feathers retained. In ♂, dark tinge of juvenile median and lesser upper wing-coverts often worn away; white tips of secondaries and greater coverts narrow (but wear may make those of adults also appear narrow). Dark spot or mottling on tip of inner web of inner greater coverts, characteristic of juvenile ♂, often worn away in spring. In ♀, wing-coverts duller brown and green of secondaries virtually absent. In both sexes, wing-coverts usually show more wear than adults, and contrast with relatively fresh tertial-coverts (if latter not still juvenile).

**Bare parts.** ADULT MALE. Iris dark brown. Bill lead-grey, almost black. Foot dull leaden-grey to ash-grey. ADULT FEMALE. Iris umber-brown. Bill dark olive-grey, almost black on top of culmen; some black dots near base of upper mandible, apparently larger and more numerous when older. Foot grey or olive-grey, webs darker grey. DOWNY YOUNG AND JUVENILE. Iris grey-brown to brown until about September, later like adults. Bill flesh-grey, darkening to dark olive-grey after *c.* 2 months, first on culmen, last at base of lower mandible. In ♀, some small black dots on lower edge of upper mandible. Foot yellow-grey or brown-grey, webs dark grey; attains adult colour in autumn. (Dementiev and Gladkov 1952; Witherby *et al.* 1939; RMNH.)

**Moults.** ADULT POST-BREEDING. Complete; flight-feathers simultaneous. Head and body in ♂ late May to July; in ♀, mostly later, starting when young almost independent, finished as late as end August. Flightless for 3–4 weeks: ♂♂ between mid-June and mid-August; ♀♀ up to *c.* 1 month later. Tail about same time, but often more prolonged. ADULT PRE-BREEDING. Partial; head, body, tail, and tertials. Starts shortly after renewal of flight-feathers, but moult slow prior to autumn migration. Shortly before or after migration, some feathers of 3rd plumage sometimes develop; see Plumages (adult ♂ breeding). Breeding plumage attained in winter quarters, November to early March, first on underparts, last on scapulars, underside of head and neck, and tertials; elongated scapulars or tertials sometimes not full-grown upon arrival in breeding area. POST-JUVENILE AND SUBSEQUENT MOULTS. Variable number of body feathers on chest, breast, flanks, vent, and scapulars and sometimes whole head and body (except back and rump) replaced for 1st immature non-breeding plumage from August or shortly after arrival in winter area, but these and remaining juvenile feathers (except wing) changed for 1st breeding from December onwards. Tail (central feathers first), and sometimes few tertials with coverts from September. Full breeding plumage from March, but occasionally some 1st immature non-breeding or juvenile feathers retained on mantle, back, rump, or tertials until post-breeding moult in midsummer. (Dementiev and Gladkov 1952; Witherby *et al.* 1939; C S Roselaar.)

**Measurements.** Netherlands. March–December. Skins (RMNH, ZMA).

| | | | | | |
|---|---|---|---|---|---|
| WING AD ♂ | 198 | (4·36; 34) | 190–211 | ♀189 (3·14; 16) | 184–196 |
| JUV | 194 | (4·22; 35) | 187–201 | 186 (3·26; 17) | 182–194 |
| TAIL AD | 66·1 | (3·13; 28) | 60–73 | 62·6 (3·37; 17) | 58–69 |
| JUV | 62·9 | (3·00; 34) | 57–71 | 59·6 (2·13; 16) | 54–64 |
| BILL | 39·6 | (1·18; 70) | 38–43 | 38·0 (1·06; 34) | 36–40 |
| TARSUS | 31·3 | (1·03; 38) | 29–33 | 30·1 (1·00; 20) | 28–32 |
| TOE | 42·5 | (1·22; 38) | 40–45 | 41·8 (1·57; 20) | 39–45 |

Sex differences significant, except toe. Wing and tail of juveniles significantly shorter than those of adults. Other measurements similar; combined.

**Weights.** Data from Camargue, France (Bauer and Glutz 1969); Netherlands (ZMA); Kazakhstan, USSR (Dolgushin 1960); Rybinsk reservoir and Volga delta, USSR (Dementiev and Gladkov 1952). Mean; sample size and range given where known.
ADULT.

| | | | | | | |
|---|---|---|---|---|---|---|
| (1) | ♂ 342 | (100) | – | ♀ 310 | (100) | – |
| (2) | 371 | (9) | 340–440 | 351 | (8) | 320–420 |
| (3) | 333 | (13) | 320–430 | 320 | (6) | 290–350 |
| (4) | 367 | (–) | 250–450 | 360 | (–) | 250–550 |
| (5) | 463 | (–) | 370–500 | 441 | (–) | 427–460 |
| (6) | 500 | (–) | 400–600 | 450 | (–) | – |

(1) Arrive lean from winter quarters (March, Camargue). (2) Gain weight in breeding area (May, Kazakhstan). (3) Decrease during breeding (June–July, Kazakhstan). (4) Various stages of regrowing remiges August, Volga delta. (5) Increase when wing full-grown (August, Volga delta). (6) Heavy prior to autumn migration (October, Volga delta). Limited winter data indicate level of about (1) or (3).
JUVENILE. Generally reach adult level in early autumn. August, Netherlands: ♂ 416 (9) 360–464; ♀ 400 (6) 334–472.

**Structure.** Wing rather long and narrow, pointed. 11 primaries: p9 longest, p10 0–3 shorter, p8 3–8, p7 12–18, p6 21–29, p1 76–98; p11 minute. Inner web of p10 and outer p9 emarginated. Scapulars and tertials elongated and pointed in ♂ breeding, shorter and with rounder tip in other plumages. Tail short, rounded; 14 feathers. Bill slightly longer than head, higher than broad at base; tip only slightly wider than base. Culmen somewhat concave, edges of mandibles straight; nail small and narrow. Outer toe *c.* 92% middle, inner *c.* 74%, hind *c.* 28%.

**Geographical variation.** East Asian populations slightly smaller, especially bill.　　　　　　　　　　　　CSR

# *Anas discors* Blue-winged Teal

PLATES 72 and 74
[after page 542 and facing page 543]

Du. Blauwvleugeltaling    Fr. Sarcelle soucrourou    Ge. Blauflügelente
Ru. Синекрылый чирок    Sp. Cerceta aliazul    Sw. Amerikansk årta

*Anas discors* Linnaeus, 1766

Monotypic

**Field characters.** 37–41 cm, of which body just over two-thirds; wing-span 60–64 cm. Small, rather straight-billed dabbling duck of similar proportions to Garganey *A. querquedula*; ♂ has purplish-grey head with bold white crescent in front of eye, pale blue forewing, green speculum, mainly brown body, and black undertail bordered in front by white patch. Sexes dissimilar, but no seasonal differences apart from eclipse which is retained unusually long (4 or 5 months). Juveniles closely resemble adult ♀, but ♂♂ just distinguishable.

ADULT MALE. Head pattern diagnostic: bold white crescent at front of face, extending back over eye and down to throat, contrasts with black-brown crown, forehead and throat, and otherwise lead-grey head and upper neck with purple tinge. Upperparts dark brown with green tinge and buff feather edges; elongated scapulars green-black with buff shaft-streak. Tail and flight-feathers dark brown; wing-coverts form bright pale blue forewing (as Shoveler *A. clypeata*, not dull blue-grey as *A. querquedula*) separated by white band from bright green speculum. Underparts pink-buff closely and profusely spotted with black; under tail-coverts black with bold white patch in front on each side. In flight, white axillaries and underwing contrast with dark-looking underbody. MALE ECLIPSE. Closely resembles adult ♀, though with darker crown, more uniform upperparts and more defined dark line through eye; only clearly distinguishable in field by retention of brighter blue coverts, brighter and greener speculum, and broader and whiter band between. ADULT FEMALE. Resembles ♀ *A. querquedula*, but distinguished by pale blue (not blue-grey) forewing only slightly duller than ♂; speculum much duller than ♂ with only slight green gloss and bordered in front by inconspicuous whitish band. Forehead and crown blackish with buff edgings; rest of head and neck light buff narrowly streaked with dusky, apart from marked white or whitish loral spot and dark line behind eye. Upperparts dark brown with buff feather edgings. Upper breast and flanks brown with buff edgings; lower breast, belly, and undertail whitish with brown mottling. Axillaries and underwing as ♂. JUVENILE. Closely resembles adult ♀, but young ♂ has noticeably richer green speculum and iridescence on rear upperparts. Moult begins in August, but full plumage (particularly head pattern) not acquired until January onwards.

Apart from distinctive head of ♂, size and blue forewing preclude confusion with any other west Palearctic duck except *A. querquedula*, which is paler, with greyer forewing and shorter bill. Note, however, that ♀ difficult to distinguish in field from escaped ♀ Cinnamon Teal *A. cyanoptera* (south-western North America and South America), though latter has longer bill, less well-developed loral spot and eyestripe, and tendency to browner plumage. ♀♀ or eclipse ♂♂ also closely resemble corresponding plumages of Teal *A. crecca* at rest on water with blue forewing hidden and only green speculum showing, though *A. crecca* paler with smaller bill and lacking distinctive loral spot.

Behaviour much as *A. querquedula*, of which it is North American ecological counterpart. Found chiefly on freshwater ponds, lakes, and marshes, seldom on sea. Generally rather shy. Feeds mainly from surface, with or without submerging head, but only rarely up-ends and hardly ever dives unless injured or in grave danger. Walks easily and perches readily on thick branches or roots overhanging water. Flies fast in small, tight parties, but without sudden changes of direction of *A. crecca*. Generally rather silent.

**Habitat.** Like Garganey *A. querquedula* favours warm, fairly luxuriant, and standing or sluggish fresh waters. Feeds on surface and tolerates small open pools amid dense floating and emergent vegetation; also narrow channels or ditches, floodlands, and mudbanks, but infrequent on brackish or saline waters.

**Distribution.** Breeds across North America, mainly in prairie region, from British Columbia (rarely southern Yukon), Great Slave Lake, central Manitoba, Ontario, and southern Quebec to Cape Breton and Newfoundland in north, and in south mainly from southern California, New Mexico, and central Texas to Tennessee and North Carolina, with isolated area in Louisiana.

Accidental. Although undoubtedly wild birds occur in west Palearctic, with 3 recoveries of juveniles (see Movements), some records may refer to escapes; this possibility thought to have increased markedly in Britain since *c.* 1970 (Smith *et al.* 1973). Britain and Ireland: over 70, vast majority since 1949, mainly September–January, but recently more February–May in Britain (British Ornithologists' Union 1971; Smith *et al.* 1972, 1974). France: December 1962, November 1964, December 1965, early 1969 (CE). Belgium: March 1966 (Lippens and Wille 1972). Denmark: May 1886 (Salomonsen 1963). Netherlands: October 1899, June 1943, January 1956 (Commissie voor de Nederlandse Avifauna 1970), May 1974 (CSR). West Germany: March 1952 (*Orn. Mitt.* 1952, **4**, 186). Sweden: May 1970 (*Vår Fågelvärld* 1973, **32**:

32–3). Spain: January 1974 (*Ardeola* 1974, **20**, 336). Azores: November 1935, 2 in September 1962 (Bannerman and Bannerman 1966). Morocco: October 1970 (RH).

**Movements.** Migratory. Withdraws south of breeding range in autumn to winter quarters in Gulf States (Florida to Texas), Mexico and West Indies, south into Neotropics as far as Peru and Brazil; stragglers to Galapagos, Chile, Uruguay, and Argentina; only small numbers trans-equatorial migrants. Also summer wanderer to Canadian tundras and west Greenland. As consequence of breeding concentration in prairie region, more migrants use inland Mississippi and Central than Atlantic and Pacific flyways. Early migrant: main southward passage mid-August to early October; main spring return March to early May, though few through USA before April. Many ringing recoveries show rapid autumn migration; one from north Alberta to Venezuela in 4 weeks, another from Manitoba to Colombia in 16 days.

Fewer vagrants to eastern Atlantic (see Distribution) than in case of Green-winged Teal *Anas crecca carolinensis*, presumably because scarcer in coastal North America. 3 trans-Atlantic recoveries of juveniles: New Brunswick to Suffolk, England (September 1971); Prince Edward Island to Tetuan, Morocco (October 1970) and Ebro delta, Spain (January 1974). Thus likely that Canadian Maritime Provinces origin of trans-Atlantic vagrants, as apparently also in Green-winged Teal and American Wigeon *A. americana*. See Bent (1923); Haverschmidt (1970); Linduska (1964).

**Voice.** Mainly silent (Bent 1923) except in late winter and spring. Calls generally similar to those of other blue-winged ducks with exception of various whistling calls of ♂. These latter loud, nasal, and high pitched, or squeaky and low, depending on type of call given (see Shoveler *A. clypeata*); uttered singly or in phrases, on water and in flight, and rendered (e.g.) 'peew', 'pew', 'peep', or 'tseel'— in some individuals, replaced by deeper, nasal 'paay' or 'pay' like ♂ *A. clypeata* (see McKinney 1970; Johnsgard 1965; Delacour 1956). Calls of ♀ high pitched and quacking, closely resembling ♀ *A. clypeata*.

**Plumages.** ADULT MALE BREEDING. Forehead, crown, lores, and chin black. Large white crescent at sides of face from forehead to throat, narrowly bordered black behind. Rest of head and neck deep lead-grey, faintly glossed magenta-pink at sides in some lights. Mantle and upper scapulars sepia, feathers with broad olive-grey edges and cream or pink-buff U-shaped marks, sometimes with pink-buff centres and large round sepia spots. Some lower outer scapulars with outer web blue and inner web sepia, longer with broad pink-buff shaft-streak; inner scapulars and tertials black with blue-green gloss and off-white shaft-streak. Back and rump dark sepia with faint green gloss, feathers edged blue-grey; upper tail-coverts same, edged buff. Chest, upper breast, and flanks pink-buff, heavily dotted black; belly and vent same, but marking more irregular, forming narrow bars on vent. Under tail-coverts black, large patches at sides of vent white. Tail black with grey bloom, feathers narrowly edged off-white. Flight-feathers and primary coverts sepia; inner webs of primaries paler; outer edges of greater primary coverts grey, of lesser blue. Outer webs of secondaries metallic green (except some outer), sometimes narrowly tipped white. Greater upper wing-coverts black; inner narrowly tipped white (gradually becoming more extensive towards wing-tip). Some inner greater partly tinged blue towards tip. Median and lesser upper wing-coverts blue (like Shoveler *A. clypeata*; in Garganey *A. querquedula* paler and greyer, blue less brilliant). Lesser under wing-coverts dark grey, marginal tinged blue; rest of under wing-coverts and axillaries white. ADULT FEMALE BREEDING. Forehead, crown, and faint streak through eye sepia, feathers broadly edged buff on crown. Hindneck dark grey-brown, sides of head and neck off-white, finely speckled sepia. Chin, throat, foreneck, and faint spot on lores white. Upperparts sepia, feathers edged olive-grey on mantle and buff on scapulars, both with irregular buff subterminal crescent or U-shaped marks; lower scapulars often with pale buff shaft-streaks. Feathers of back, rump, and upper tail-coverts edged and irregularly barred or marked buff. Feathers of chest sepia, broadly edged buff; flanks sepia with broad pale buff U-shaped marks and edges. Rest of underparts off-white, often with dark grey-brown spots. Tail sepia, broadly edged and irregularly marked buff. Wing like adult ♂, but speculum duller, usually black with dark green tinge; greater coverts blue with white edges and usually symmetrical sharply angled V-shaped marks; tertials sepia, edged and streaked buff. ADULT MALE NON-BREEDING (Eclipse). Like adult ♀ breeding, but edges on crown narrower, sides of head and neck more heavily spotted sepia; chin, throat, and foreneck finely spotted. Feather edges of mantle, scapulars, and upper-tail-coverts narrower, marks smaller, less distinct; back and rump like adult ♂ breeding. Chest and sides of breast sepia, feathers with pale buff edges and U-shaped marks; on flanks same, with some irregular pale buff bars. Rest of underparts like adult ♀ breeding. Tail and wing like adult ♂ breeding, but tertials slightly shorter, with round tips, dark sepia with green tinge and faint pale shaft-streak, narrowly edged off-white on outer web and tip. ADULT FEMALE NON-BREEDING. Like adult ♀ breeding, but feathers of crown more narrowly edged buff, rest of head more heavily spotted sepia; feathers of upperparts sepia with narrow grey or pale buff edges and virtually without subterminal marks, except some on sides of mantle or upper tail-coverts. Tail sepia narrowly edged white. Wings like adult ♀ breeding, but tertials with coverts like adult ♂ eclipse. JUVENILE MALE. Head and body like adult ♀ non-breeding, but feathers narrower and less square, on upperparts without subterminal marks except few on upper tail-coverts; on chest, sides of breast, and flanks sepia with yellow-buff edges and faint subterminal pale buff crescent; belly feathers narrow, off-white, centres mottled pale grey-brown. Tail-feathers notched at tip, bare shaft projecting; sepia with narrow off-white edges. Wing like adult ♂, but usually small dusky dots on white tips of greater upper wing-coverts; tertials like adult ♂ eclipse, but shorter and narrower, edged and broadly tipped pale buff or off-white. In autumn, variable amount of 1st immature non-breeding develops on head and body; resembles adult ♂ eclipse, but often part of juvenile plumage and always some tail-feathers and all wing retained. JUVENILE FEMALE. Like juvenile ♂, development of 1st immature non-breeding (resembling adult ♀ non-breeding) similar. Wing like adult ♀, but white edges of greater coverts wider, subterminal mark of greater coverts usually straight bar, almost never acutely angled V like adult ♀ (Carney 1964; Dane 1968); tertials with coverts narrower, relatively more frayed at tip than adults. FIRST BREEDING. Like adult breeding. Juvenile wing retained until 2nd summer.

**Bare parts.** ADULT MALE. Iris brown. Bill black, tinged blue near base; tinged olive in eclipse, when lower mandible flesh, spotted brown. Foot dull yellow to orange, webs dull black. ADULT FEMALE. Iris dark grey-brown to brown. Upper mandible olive-black, sides yellow-flesh with large black dots (fading in summer). Size of spots at sides of upper mandible increases with age; in 2nd autumn, spots usually over 5 mm, often larger than 10 mm; ♀♀ with some spots over 15 mm at least 2 years old (Dane 1968). Lower mandible horn-yellow. Foot yellow-brown to dull yellow, webs grey-brown. JUVENILE. Iris brown. Bill dusky olive-black, edges and under mandible yellow-flesh; becoming darker in late autumn in ♂ and with some fine dots at sides of upper mandible in ♀. Foot dirty olive-grey, joints and webs dark olive-brown; attain adult colour with development of 1st breeding plumage. (Kortright 1942; Witherby *et al.* 1939; ZMA, RMNH.)

**Moults.** Apparently like *A. querquedula*, but adult ♂♂ seem to attain breeding plumage somewhat earlier in winter.

**Measurements.** Central USA, Mexico, Antilles, and Surinam. Skins (RMNH, ZMA).

| | | | | | | |
|---|---|---|---|---|---|---|
| WING AD ♂ | 191 | (3·52; 8) | 186–195 | ♀ 183 | (4·45; 5) | 176–188 |
| BILL | 40·8 | (1·52; 18) | 38–44 | 40·0 | (1·30; 5) | 37–41 |
| TARSUS | 31·9 | (1·06; 18) | 30–34 | 31·1 | (0·89; 5) | 30–32 |

Sex differences wing and bill significant. Juvenile wing on average *c.* 7 shorter.

**Weights.** (1) Mexico, December–January (Bauer and Glutz 1968). (2) Netherlands Antilles, April, just prior to spring migration (ZMA, H J Koelers). (3) Iowa, USA, October–November (Bauer and Glutz 1968).

| | | | | | | |
|---|---|---|---|---|---|---|
| (1) | ♂ 370 | (7) | 300–400 | ♀ 346 | (6) | 325–360 |
| (2) | 478 | (3) | 439–499 | 451 | (5) | 419–492 |
| (3) | 353 | (6) | 290–415 | 315 | (6) | 280–360 |

**Structure.** Wing rather long and narrow, pointed. 11 primaries: p9 longest, p10 0–3 shorter, p8 3–7, p7 10–16, p6 19–28, p1 74–90. Inner web of p10 and outer p9 emarginated. Tail rather short, pointed; 14 tapering feathers. Bill slightly longer than head, higher than broad at base. Culmen slightly concave in front of nostril, convex near tip. Cutting edge of upper mandible slightly curved upwards near base, exposing lamellae, but curved downward near slightly distended tip of otherwise rather narrow bill. Lower inner scapulars and tertials long and tapering to point in breeding ♂, shorter and with rounder tips in other plumages. Outer toe *c.* 92% middle, inner *c.* 77%, hind *c.* 31%.

**Geographical variation.** Slight. In breeding populations of salt and brackish tidal marshes along Atlantic seaboard of north-east USA and south-east Canada (sometimes separated as *A.d. orphna* Stewart and Aldrich, 1956) dark parts of plumage more intensely black in both sexes and grey of head and neck and pink-buff of underparts in breeding ♂♂ deeper in tone. Broad intergradation zone between paler and darker birds in Great Lakes region and central Canada; overlap too large and differences too slight to warrant recognition of subspecies. CSR

## *Anas clypeata* Shoveler

PLATES 73 and 74
[after page 542 and
facing page 543]

DU. Slobeend    FR. Canard souchet    GE. Löffelente
RU. Широконоска    SP. Pato cuchara    SW. Skedand    N.AM. Northern Shoveler

*Anas clypeata* Linnaeus, 1758

Monotypic

**Field characters.** 44–52 cm, of which body about three-fifths; wing-span 70–84 cm. Medium-sized, heavy-bodied, and fairly short-necked dabbling duck with flattened head and huge spatulate bill; ♂ has dark green head, white neck, breast, and scapulars, chestnut belly and flanks, and bright, pale blue forewing. Sexes dissimilar; ♂ has supplementary plumage as well as eclipse, and ♀ distinctive breeding and non-breeding. Juvenile resembles adult ♀ non-breeding, but distinguishable.

ADULT MALE. Dark green head, appearing black at distance, slaty-black centre to back and rump, and chestnut belly and black undertail contrast markedly with white of neck and breast, scapulars, axillaries, and underwing, and white strip between belly and undertail. Elongated scapulars pale blue, dull green, black and white; forewing bright pale blue; speculum green with white bar in front. Bill grey-black; eyes yellow to orange; legs orange-red. In flight, blue forewing conspicuous against brown primaries, while black back contrasts with broad white braces; below, from head to tail, distinctive pattern of green–white–red–white–black. MALE ECLIPSE AND SUPPLEMENTARY. Resembles adult ♀ non-breeding, but

head and upperparts darker, redder, and more uniform, and easily distinguished in flight by much brighter blue forewing and green speculum. Bill browner. ADULT FEMALE. In breeding plumage, whole head and body patterned much like ♀♀ of other dabbling ducks, but all paler areas and feather edges cinnamon-pink to pink-buff. In non-breeding plumage, brown with paler feather edges, greyer head and neck, and blacker forehead and crown. Tail whitish like ♀ Mallard *A. platyrhynchos*. Quick plumage distinctions in flight are green speculum and bluish forewing, even though latter much duller than in ♂. Bill olive-grey or dark brown with yellow to orange cutting edges and base; eyes brown or yellow; legs orange-red. JUVENILE. Resembles adult ♀ non-breeding, but upperparts more uniform and underparts more streaked, while wings of ♂♂ brighter (particularly median and lesser coverts) and those of ♀♀ duller (median and lesser coverts grey-brown and speculum sometimes absent). During moult from August to March or April, gradually becomes like adult non-breeding, then adult supplementary and finally full adult.

Adult ♂ unmistakable. Other plumages most closely

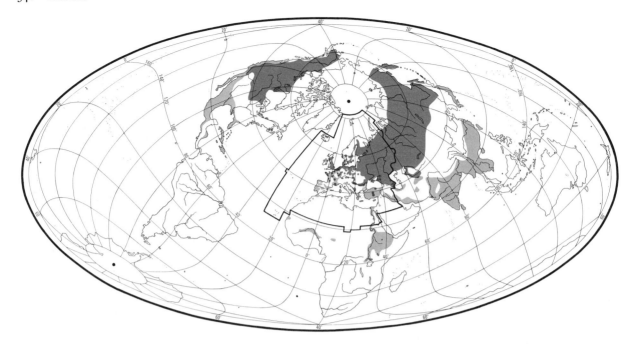

resemble *A. platyrhynchos*, but readily distinguished by pale blue forewing and green speculum. Proportions and carriage quite different, and enormous spatulate bill, coupled with distinctive carriage as bird swims with front end low in water, is characteristic. Only other west Palearctic ducks with blue forewings, Garganey *A. querquedula* and Blue-winged Teal *A. discors*, much smaller with normal bills; also wing-coverts of *A. querquedula* greyer.

Found mainly on fresh water, particularly with extensive cover and muddy shallows, and seldom on sea except on migration. Fairly tame. Swims rather low with bill dabbling in water; also paddles in shallows or mud with similarly dabbling action, but up-ends or dives comparatively infrequently. Walks in ungainly manner. Rises easily, with characteristic drumming rattle of wings; quite agile in air with rapid wing-beats making loud whistling noise and flight resembling Wigeon *A. penelope*, but small, pointed wings appearing set far back. Active and spends much time on wing, particularly in spring. Generally rather quiet.

**Habitat.** Although penetrating in small numbers through taiga into tundra zone, main haunts are in temperate open woodland, grassland, and steppe regions, extending marginally into semi-desert. Accommodates to milder, oceanic climates better than some of steppe-based relatives. Everywhere only in lowlands, on exceptionally shallow but permanent and productive waters usually of no great extent, and most often free of overhanging trees or fringing forest, although normally fringed by dense stands of reeds or other emergent vegetation. Copious floating vegetation tolerated if enough patches of open water and abundant surface planktonic population. Frequently resorts to mown grass or pastures adjoining clumps of denser and taller herbage or protective shrubs or low trees screening suitable nest-site. Attracted by sewage farms, ricefields, and other shallow artificial waters bordered by lush grassland. Differentiation between resting and feeding habitats not pronounced, preferring those serving both purposes. Specialized feeding mechanism (see Food) compels choice of equivalent conditions on passage and in winter quarters; so avoids marine waters, except briefly, although freely using inland seas and brackish or saline inland waters.

Like relatives, quick to take advantage of extension of suitable habitat. Circles like other *Anas* and usually flies below 100 m; shares generic capacity for quick and ready movement between sites. Where immune from persecution, tolerates distant human presence fairly well, using artificial nest-sites where natural ones lacking.

**Distribution.** Some expansion in north and west of European range, possibly linked with increase in summer temperatures (Voous 1960), but breeding also sporadic and fluctuating in many areas on edges of range. ICELAND: first bred 1931; now nesting in a few localities in north and south-west (Gardarsson 1967). BRITAIN AND IRELAND. Major increase and spread in early years of 20th century (in Scotland, at height 1900–20, with lower rate of spread continuing probably until early 1950s) but present trends not clear, with local fluctuations and declines due to habitat loss (Parslow 1967). Ireland: increase over last 10 years (Ruttledge 1970). NORWAY: gradual but irregular increase since mid-19th century (Haftorn 1971). SWEDEN: increasing in south, with recent spread to lakes in north

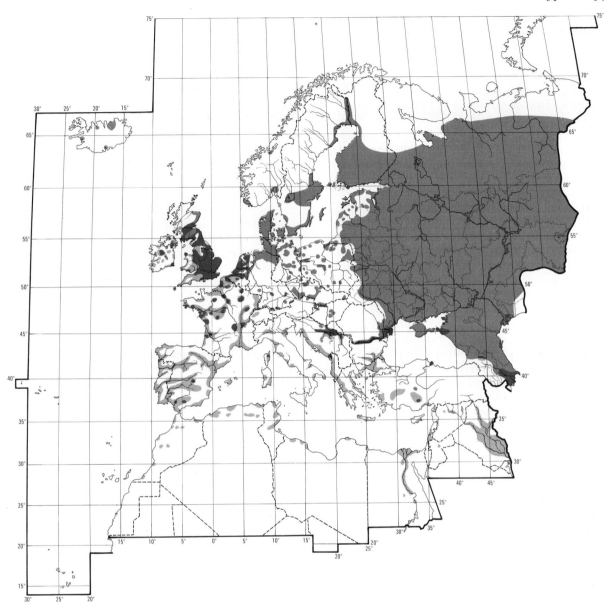

from Finland (Curry-Lindahl *et al.* 1970). FINLAND: fluctuating and sporadic in north (Merikallio 1958). USSR. Estonia: spread in last 30 years over whole country (Kumari *et al.* 1970). SWITZERLAND: bred 1917, then 1929–55 more or less regularly; sporadic since (WT). GREECE: formerly bred Macedonia and Thessaly; lakes now drained (WB). CYPRUS: bred 1910 (Stewart and Christensen 1971). SPAIN: sporadic or scarce nester (FB); formerly bred marismas of Guadalquivir when water level high (Ree 1973).

Accidental. Bear Island, Spitzbergen, Azores, Madeira, Canary Islands.

**Population.** Breeding. ICELAND: 30–60 pairs (Gudmundsson 1951); under 100 pairs (Gardarsson 1975). BRITAIN: *c.* 250 pairs (Atkinson-Willes 1970*a*). FRANCE: numbers unknown, but perhaps 600–1000 pairs (FR). BELGIUM: *c.* 165 pairs, increasing (Lippens and Wille 1972). NETHERLANDS: *c.* 9000 pairs (CSR). DENMARK: scarce most parts, but locally common; perhaps slight increase (Fog and Joensen 1970); probably not exceeding 500 pairs (Joensen 1974). FINLAND: *c.* 4000 pairs (Merikallio 1958); most abundant in west and south-west (Grenquist 1970). EAST GERMANY: decline in parts of Saxony (Bauer and Glutz 1968). WEST GERMANY:

considerable increase lower Rhine in recent years; fluctuating in south Bavaria (for other details see Bauer and Glutz 1968). POLAND: widespread, but scarce (Tomiałojć 1972). AUSTRIA: considerable fluctuations in east (Bauer and Glutz 1968). USSR: 2700–3000 pairs in Baltic region, mainly in Estonia (Kumari et al. 1970).

Winter. Total population west Palearctic probably in order of 1½ million birds but data incomplete (Atkinson-Willes 1975).

Survival. Mean annual mortality adults ringed Britain 44%; life expectancy 1·8 years (Boyd 1962). Oldest ringed bird 20 years 5 months (Rydzewski 1973).

**Movements.** Mostly migratory. Breeders of Iceland all migrate, probably to Ireland or Britain though confirmation needed. Most British breeders move southwards to south France, south Spain, north and central Italy, a few to north Africa; bulk have left Britain by end October, before main arrivals of Continental birds. Breeders from south Fenno-Scandia and USSR east to c. 60°E (rarely from 68°E) and south to c. 55°N migrate west and south-west to western seaboard, chiefly Netherlands, Britain, and Ireland, some going further to west and south France and north Spain. Degree of movement by breeders of west France unknown. Breeders of central and south-east Europe probably join USSR breeders in Mediterranean Basin. Populations of east Russia, Trans-Urals, and west Siberia migrate south through Volga region, then to south Caspian, Azov and Black Seas, and to Mediterranean, particularly Turkey, Greece, Italy, and north Africa, where overlap with north European breeders. One from Netherlands (autumn) to Kazakhstan (April), and another from Volga delta (summer moult) to Ireland (November), presumably aberrant migrants.

Those wintering from Egypt south to east Africa, and Persian Gulf, presumably also from USSR, for several ringed Volga delta recovered Egypt and Kenya. Considerable numbers in west and north-central Africa (notably Senegambia, Niger, Chad, Nigeria) more likely to involve European breeders, and autumn-ringed Dutch birds recovered Sénégal and Canary Islands. Commonest Palearctic duck Lake Chad basin in recent winters, 10 500 in December 1970 (Roux 1971). Also common Sénégal delta where 6600 January 1973 (Morel and Roux 1973).

Tendency for ♂♂ to stay late with ♀♀ in nesting season, but small-scale ♂♂ moult gatherings reported in late summer from several European localities, e.g. Matsalu Bay (Estonia), IJsselmeer (Netherlands), and Bodensee (Germany/Switzerland) (Salomonsen 1968); bigger moult concentrations in USSR, notably Volga delta, Trans-Ural and Baraba steppes, and Kazakhstan lakes.

Main autumn migration rather earlier than other Palearctic ducks, except Garganey A. querquedula. Principal passage across Europe in September–October, with major passage through Britain in November, while first arrivals Sénégal in October. Departs tropical Africa in February, peak movement through Europe mid-March to mid-April, and virtually all breeders returned by early May. Continental birds pass through Britain in March–April, at same time as returning local breeders; some abmigration results, with British-bred birds recovered in USSR. Some suggestion of partial loop migration, for substantial numbers ringed Camargue and Netherlands recovered Italy in March–April, indicating return passage by more direct route.

Highly migratory North American population has element passing over Pacific via Hawaii into central Polynesia, as in Pintail A. acuta. See also Bauer and Glutz (1968); Dementiev and Gladkov (1952); Ogilvie (1962).

**Food.** Omnivorous, but particularly planktonic crustaceans, small molluscs, insects and larvae, seeds, and plant debris. Collected by: (1) surface-feeding; (2) swiming with head and neck immersed, a method which predominates in places (Szijj 1965b); (3) up-ending less often than other Anas, but possibly for longer periods (Bauer and Glutz 1969) and on occasions as method of bringing food to surface (Reuver 1959); (4) diving, possibly more often than other Anas, but still not frequently. Normally singly, in pairs, and small groups. When surface-feeding, swims with neck stretched forward, sweeping surface with side-to-side movements of bill and filtering out food particles. Shows extreme specialization of Anas filtering mechanism: proportionally long bill with wide distal end which increases amount of water which can be sucked in; fine, hair-like serrations of upper and lower jaws which intermesh to form most efficient filtering device, especially for plankton (Olney 1965b; McKinney 1973). Pairs, or 3–4 birds, often swim together surface-feeding in circles, head to tail, presumably on stirred-up food; one pair may get food up to 1½ hours in same spot (Millais 1902; Heinroth 1910). Occasionally, single bird swims in circle creating a whirlpool to bring food to surface (Lippens 1959). Dives with no forward leap, using wings underwater, in depths up to 80 cm, and rarely for more than 5 s (Dean 1950; Steinbacher 1963). Day-time and, when disturbed, probably night-time feeder.

Wide variety of foods recorded, though diet constituents rarely accurately known. Often undue prominence given to certain items retained for longer periods and easily identified (e.g. seeds), in contrast to items rapidly digested and often, because of feeding methods and environment, eaten already broken up (e.g. plankton, inorganic debris). Particularly important: small crustaceans (amphipods, copepods, cladocerans, ostracods); small molluscs (Planorbis, Hydrobia, Rissoa, Viviparus, Valvata, Littorina); insects and larvae—caddisflies (Trichoptera), waterbugs (Hemiptera), dragonflies (Odonata), flies (Diptera), beetles (Coleoptera); seeds of aquatic plants, including sedges Scirpus, Eleocharis, and Carex, pondweeds Potamogeton, grasses Glyceria. Less annelids; amphibian spawn and tadpoles; spiders; fish and vegetative plant

PLATE 72. *Anas discors* Blue-winged Teal (p. 537): **1** ad ♂ breeding, **2** ad ♀ breeding, **3** juv ♂. (NWC)

PLATE 73. *Anas clypeata* Shoveler (p. 539): **1** ad ♂ breeding, **2** ad ♀ breeding, **3** ad ♂ eclipse, **4** ad ♂ supplementary, **5** juv ♂, **6** downy young. (NWC)

1                2                3                4                5

6                7                8                9

10               11               12               13

14               15               16               17

18               19               20               21               22

23               24               25        N.W.CUSA.        26

parts, including buds and shoots. (Yarrell 1843; Thompson 1851; Millais 1902; Witherby *et al.* 1939; Dementiev and Gladkov 1952; Bauer and Glutz 1968.)

USSR data show seasonal and locality changes, and emphasize importance of molluscs. Rybinsk reservoir: in May, molluscs 47% of volume (especially *Planorbis contortus*), ostracod crustaceans 17%, insect larvae 25% (chironomids, Odonata, occasionally Trichoptera), adult insects 6%, and seeds 5%; in July and August, molluscs consumption increased to 75% (chiefly *Viviparus viviparus*), seeds to 13%, but less crustaceans, insects and larvae; in September and October, 62% molluscs, and insect larvae increased to 31% (mainly Trichoptera), and seeds 6·5%. On Mologa river, early spring, animal foods (molluscs, chironomids) only 30%; rest green vegetation (duckweed *Spirodela*, Canadian pondweed *Elodea*, and hornwort *Ceratophyllum*) while autumn migrants ate mainly molluscs (89·5%, especially *Viviparus*). Spring foods, Volga, mainly molluscs (particularly *Valvata*); also taken in August and September, with crustaceans, insects and their larvae, and small amounts (0·6%) fish larvae (bleak *Alburnus* and catfish *Silurus*). On Baraba steppe, molluscs again important, 71·4% frequency; Coleoptera and larvae 35·7%, undetermined insects 35·7%, and spiders 7·1%. In winter, Azerbaydzhan (Lenkoran), animal foods decrease in importance—chiefly molluscs (*Dreissena, Planorbis, Hydrobia, Lymnaea*) and water-beetles—and seeds predominate (Isakov and Vorobiev 1940; Isakov and Raspopov 1949; Dementiev and Gladkov 1952).

Few data from elsewhere: 3 stomachs, October and November, Benbecula, Scotland, contained only seeds (*Ruppia, Hippuris, Galeopsis, Carex, Eleocharis*) and plant debris (Campbell 1947); 13, September–February, north Kent, England, contained, in those feeding brackish and fresh water, mainly adult waterboatmen *Corixa* (*C. nigrolineata, C. concinna,* and *C. stagnalis*) and less chironomid larvae, waterfleas *Daphnia*, and seeds (*Ruppia spiralis* and *Scirpus maritimus*); in those from saltwater areas, chiefly mollusc *Hydrobia ulvae*, and less *Salicornia* (Olney 1965*b*). 2, Tisza river plains, Hungary, contained only seeds of *Polygonum*, (Sterbetz 1969–70). USA studies show preponderance of plant debris, seeds, molluscs, and insects (McAtee 1922; Anderson 1959).

In USSR, Mologa river area, stomachs of downy young contained 75% by volume planktonic crustaceans (*Daph-*

*nia, Cyclops, Diaptomus, Estheria, Limnadia* and Ostracoda) (Isakov and Raspopov 1949). Incubator-hatched ducklings readily ate *Daphnia*; this thought to be one of main foods Lake Manitoba, Canada (Collias and Collias 1963).

**Social pattern and behaviour.** Based mainly on material supplied by F McKinney.

1. Gregarious except when nesting. Outside breeding season, often in fairly small flocks of *c.* 20–30, but up to several hundred not rare in particularly favourable areas. Most likely to congregate densely when feeding. Flocks often show unbalanced sex-ratio; e.g. often excess of ♂♂ in Netherlands, rising to over 80% in autumn and winter (Lebret 1950); increasingly consist of pairs as winter progresses, these migrating to breeding grounds in flocks. ♂♂ first to form post-breeding flocks after end of pair-bond, and moulting aggregations occur in late summer. BONDS. Monogamous pair-bond, probably largely of seasonal duration. Pair-formation occurs in flock during winter, e.g. from mid-December in Louisiana, USA (McKinney 1970). Earliest bonds often unstable. Final pair-bond strong, not ending until near hatching. Promiscuous tendencies on part of ♂♂ weak and rape of other ♀♀ rare (McKinney 1970, 1973), though some harassment suffered by ♀ when off nest from territorial and unpaired ♂♂ (see Seymour 1974*a*). Polyandry reported as common (Millais 1902; Kortright 1942), but not substantiated by recent studies. Only ♀ tends young. BREEDING DISPERSION. ♀♀ home to previous breeding areas (Sowls 1955). Pairs frequent relatively small home-ranges (Poston 1969): 6 such, including peripheral areas visited occasionally by ♂♂, measured 6–36 ha (mean 20). Within home-range, highly territorial (Hochbaum 1944; Sowls 1955; Seymour 1974*a*), defending core area where most activity (including feeding) occurs and from which ♂ drives away intruders of both sexes (Poston 1969). Territories more clear-cut than in most other ducks, with little if any overlap (McKinney 1970) and boundaries well defined where interactions frequent (Seymour 1974*b*). In one area, along roadside ditch, size of territory *c.* 2·9 ha in pre-laying period and *c.* 0·9 ha during laying and incubation; defence by ♂ mostly confined to water, but overflying birds also threatened and sometimes chased (Seymour 1974*b*). ♀, especially, spends increasing amount of time within territory until there all daylight period towards end of laying and during incubation. Territorial system, as well as promoting nest dispersion (McKinney 1965*a*), maintains exclusive feeding place for ♀ during pre-laying, laying, and incubation periods; localization of activities within small area associated with time-consuming feeding by plankton straining, and intensive use of locally rich food source (McKinney 1973). ♀ with brood sometimes uses same feeding area for many days, but others leave territory soon after hatching (Poston 1969). ROOSTING. Little information, but outside breeding season, probably much as in other ducks that feed diurnally—i.e. communal and nocturnal with periods of loafing during day. Though mainly on fresh water inland, will sometimes roost in estuaries.

2. Studied chiefly in captivity: Lorenz (1953), Johnsgard (1965), McKinney (1967, 1970). As with other blue-winged ducks (e.g. Garganey *A. querquedula*), behaviour differs appreciably from most other *Anas*, particularly in communal courtship displays; also considerably influenced by strong pair-bond and marked territorialism. FLOCK BEHAVIOUR. Little studied. Lateral Head-shaking and vertical Head-thrusting, in erect posture while facing into wind, most important pre-flight signals. Flocks often pack densely, birds swimming in circle, dabbling close to tail of another (Heinroth 1911); such behaviour—and similar behaviour by pair—associated with

◀ PLATE 74 (*facing*) Surface-feeding ducks in flight (adult breeding plumage). *Aix galericulata* Mandarin: 1 ♂, 2 ♀. *Marmaronetta angustirostris* Marbled Teal: 3. *Anas falcata* Falcated Duck: 4 ♂, 5 ♀. *Anas platyrhynchos* Mallard: 6 ♂, 7 ♀. *Anas rubripes* Black Duck: 8 ♂, 9 ♀. *Anas strepera* Gadwall: 10 ♂, 11 ♀. *Anas acuta* Pintail: 12 ♂, 13 ♀. *Anas penelope* Wigeon: 14 ♂, 15 ♀. *Anas americana* American Wigeon: 16 ♂, 17 ♀. *Anas crecca* Teal: 18 ♂, 19 ♀. *Anas capensis* Cape Teal: 20. *Anas querquedula* Garganey: 21 ♂, 22 ♀. *Anas discors* Blue-winged Teal: 23 ♂, 24 ♀. *Anas clypeata* Shoveler: 25 ♂, 26 ♀. (NWC)

A

feeding, not display. ANTAGONISTIC BEHAVIOUR. Quite peaceable throughout much of year, especially in flock (even when feeding in close-packed groups) up to period of spring migration. ♂ becomes increasingly and strongly aggressive in defence of mate and territory, continuing to protect ♀ from attention of other ♂♂ throughout much of incubation period. Hostile-pumping most distinctive threat-display: head (with crown feathers depressed) repeatedly raised high, bill pointing slightly up (see A); accompanied by 'took' calls uttered regularly and out of phase with head movements. Though most common during pairing and territorial disputes, used at all seasons at close range and as long-distance threat. At close quarters, ♂♂ also threaten with sleeked crown feathers while sometimes freezing with heads low. Open-bill threats, rushes, and highly ritualized Circular-fights occur on water or land, usually with Hostile-pumping; in Circular-fight, mostly at territorial boundaries, 2 ♂♂ repeatedly jump past each other, flailing wings as they spin round but seldom making contact. In territorial disputes, ♂♂, after Hostile-pumping, also chase intruders; will pursue passing birds of either sex in flight, often pecking other ♂♂ or clashing with wings. Loud Wing-rattling noise of ♂ rising from territory aids long-range territorial advertisement. After returning to territory, often in absence of other birds, ♂ tends to perform Hostile-pumping, sometimes for 10 mins or longer (Seymour 1974b). ♂♂ hardly ever fight with ♀♀, ♀♀ rarely with one another. While threatening, ♀ performs Hostile-pumping movements; when alternated with sideways threatening movements of lowered head and neck (which may also occur on own), form typical Inciting-display, especially in presence of preferred ♂ or mate, pumping being accompanied by Inciting-calls. ♀ particularly aggressive when nesting and with young; may perform lateral Inciting movements when threatening others near brood even though mate now absent. Repulsion-display of ♀ similar to that of other Anas ♀♀, but rare since incubating ♀♀ usually protected from other ♂♂ by mate. COMMUNAL COURTSHIP. From mid-winter, ♂♂ assemble at times in courting parties, typically composed of up to c. 12 ♂♂ and 1 ♀; particularly noticeable in April–May when some ♂♂ still unpaired and display to paired ♀♀, even after laying begun (Hochbaum 1944; McKinney 1970). Displays much simpler than in most other Anas outside blue-winged duck group, and easily overlooked. Same displays performed by lone ♂ to ♀, others often then joining in. During courtship bout, ♂♂ swim slowly but persistently after ♀. First attract ♀'s attention with several lateral displays with body sideways to her. Of these, 3 ritualized from feeding movements—all with back and flank feathers erected and feet rapidly paddling; occur in similar situations, and probably with similar function, to Grunt-whistle, Head-up-tail-up, and Down-up of other Anas. In Lateral-dabbling display, ♂, with slightly exposed speculum, dips bill in water and dabbles much as when feeding—but position held stiffly for several seconds and Fast-calls often given. In Head-dip display, head and neck briefly submerged; in Up-end display, whole front of body immersed—both distinguishable from true feeding mainly by

lateral orientation to ♀. In lateral display of a different type, ♂ assumes Head-high posture with head moderately erect and utters series of Repeated-calls. Other lateral displays consist largely of movements somewhat ritualized from comfort behaviour, including: (1) Bathe-and-Wing-flap sequence; (2) Body-shake-and-Belly-preen sequence; (3) Preen-dorsally and Preen-behind-Wing; (4) Ceremonial-drinking; and (5) rather infrequent Upward-shakes (not performed as introduction to major displays). In addition to lateral displays, Turn-back-of-Head one of most common displays associated with pair-formation: ♂ moves in front of ♀ holding head stiffly with bill pointing away as he ceremonially Leads her. Short, ritualized Jump-flights also frequent (see especially Lebret 1958a; Hori 1962, 1963); effectiveness in attracting ♀'s attention increased by Wing-rattling noise and striking pattern of ♂'s wings (McKinney 1970). Preceded by Head-shakes and Fast-calls in erect posture; ♂ then rises steeply and, with neck horizontal, flies 5–15 m, alights ahead of group, and swims back. Inciting, given close beside preferred ♂ while discouraging other ♂♂, most important ♀ pairing display. PURSUIT-FLIGHTS. (1) Courtship-flights: displaying groups often soon take to air and fly round, twisting here and there as ♂♂ follow ♀ with much calling (McKinney 1970). (2) 3-bird flights: established ♂ frequently pursues pairs short distances, then typically returns to territory (McKinney 1965a); as in other Anas, pursuit often directed at ♀ of pair, but, in contrast to A. platyrhynchos, her mate always defends her (Hori 1963). Attacks also frequently made on ♂ of intruding pair, much as in simple territorial chasing of single birds. (3) Rape-intent flights: rare or absent in most populations. PAIR BEHAVIOUR. In autumn and early winter, ♀♀ give loud Decrescendo-calls, mostly near dawn and dusk. At same seasons, ♂♂ also often utter Single-calls and ♂ Decrescendo-calls. During communal courtship, both ♂♂ and ♀♀ orientate respective displays to mate. Turn-back-of-Head display by ♂, and Inciting by ♀, continue after pair-bond formed—♂ often performing Lateral-dabbling display when mate Incites against another ♂. Pair make exploratory flights together over nesting cover prior to egg-laying; during same period, ♀ frequently gives loud Persistent-quacking calls. When ♀ feeds during pre-laying and laying periods, or during off-nest spells when incubating, ♂ remains close, alert for intruders. Copulatory behaviour not a feature of earlier pair-courtship, occurring regularly only from c. 3 weeks before egg-laying; throughout this period, however, and during laying, copulate 1–2 times daily (McKinney 1970). COPULATION. Pre-copulatory behaviour much as in other Anas; during Head-pumping movements, ♂'s crown feathers not depressed and bill usually down-pointed at c. 30°—unlike in superficially similar Hostile-pumping. While mounted, ♂ utters quiet calls. In post-copulatory Bill-down display, closely lateral to ♀ (characteristic of blue-winged ducks among Anas), ♂ lowers bill sharply, raises wing-tips slightly, and erects head and neck feathers while giving ♂ Decrescendo-call. Other displays by ♂ occasionally follow, e.g. Turn-back-of-Head, but bathing only activity noted from ♀. RELATIONSHIPS WITHIN FAMILY GROUP. No detailed information. Hostile-pumping performed even by small downy young.

(Fig A after photograph in McKinney 1970.)

Voice. Reputedly little used and weak in both sexes (see, e.g., Bent 1923; Witherby et al. 1939), due largely to predominantly rather quiet calls of ♂, but probably used as freely as in most Anas. Calls of ♀ mostly variants of louder quacking. Dichotomy between calls of ♂ and ♀ well established and no evidence that sexes share any quacking

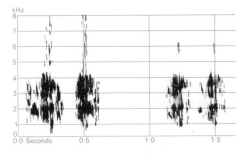

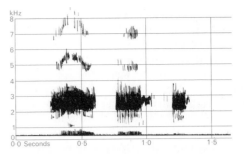

I Sture Palmér/Sveriges Radio Denmark March 1964

II L Ferdinand/Sveriges Radio Denmark August 1963

notes (see Coward 1920). ♂ also produces loud, non-vocal sound signal—unique among west Palearctic Anatinae—in form of Wing-noises of 2 similar (perhaps not wholly distinguishable) types: (1) Wing-rattling on taking wing; (2) Wing-clapping with final beat of wings in Wing-flap display (McKinney 1970).

CALLS OF MALE. Generally rather quiet, hoarse 'took' most common note; also louder, nasal 'paay' and wheezy 'whe' or 'thic' notes (F McKinney). 'Took' notes also described as hoarse 'chutt' (Lorenz 1953); recording (see I) suggests somewhat wooden-sounding 'chuck', 'tuck', or 'sluck'. Following calls distinguished by F McKinney (see also McKinney 1970). (1) Single-call: loud, nasal 'paay'. Typically uttered in autumn and early winter before pairs formed, rarely later (except after copulation—see 6), standing with head somewhat raised. (2) Decrescendo-call: Single-call followed by series of Repeated-calls (see 3 and 6). Given at same time of year and in same posture as Single-call on own, in both cases without any obvious inducement. (3) Repeated-call: series of even-spaced, 1 or 2-syllable, 'took' notes—'took-took . . . took-took . . .' (see I) or 'took . . . took . . . took . . .'; intergrading at times without pause and, at beginning or end of series, sometimes replaced by quiet, wheezy sounds with same rhythm. Uttered, from beginning of pair-formation to desertion of ♀, initially by stationary ♂♂ on land or water while still unpaired; also in mild alarm, after pair take wing, when separated from mate or returning to territory after chase, and as one of courtship displays. Functional equivalent, e.g. of Burp-call of ♂ Teal *A. crecca* and Slow Rab-call of Mallard *A. platyrhynchos*. A few quiet 'took' notes, or wheezy equivalent, accompany Wing-clapping noise of Wing-flap display. (4) Threat-call: louder 'took' notes, fairly regularly spaced. Given during Hostile-pumping display; also during communal courtship and courtship-flights. (5) Fast-calls: rapid series of 'took' notes in short bursts. Typical of pair-formation period, e.g. during Lateral-dabbling display, immediately before Jump-flight display, when 2 pairs approach on water, and—especially when flushed—immediately after take-off by pair (changing abruptly to Repeated-calls after rising). (6) Calls associated with copulation: series of quiet, wheezy 'whe' or 'thic' notes occasionally heard during pre-copulatory Head-pumping but more usually after ♂ mounts, oc-

casionally then changing to quiet, single 'took' notes. After dismounting, ♂ Decrescendo-call given, e.g. 'paay took-took took-took . . .'; sometimes followed by Repeated-calls.

CALLS OF FEMALE. (1) Decrescendo-call: usually 1–4 long notes (see II), each with downward inflection, in descending pitch and followed immediately by 3–9 shorter Single-quacks of approximately same pitch. Deeper and longer than in Blue-winged Teal *A. discors* and Garganey *A. querquedula*. Rendered by Johnsgard (1965) as 'gack-gack-gack-ga-ga' with last 1–2 notes muffled, as in other blue-winged ducks. Considerable variation between birds in number of notes and pattern of complete call. Given throughout year, except when incubating or raising young; frequent in juvenile from *c*. 2 months. Especially common at period of morning and evening twilight; appears to announce individual identity and establish contact with mate. (2) Single-quacks: as well as forming terminal part of Decrescendo-call, given in short series of loud, even-spaced notes mostly at dusk when in flying mood during period of pair-formation, and also (in phrase of 3–7 notes) after flying up when flushed in spring. (3) Persistent-quacking: long series of loud, repeated, 1-syllable 'quacks' similar to Single-quacks. Given, with only brief pauses between bursts, for many minutes, e.g. while pair swimming or standing together, especially in period immediately before egg-laying; (4) Inciting-call: rapid, rippling chatter, rising in pitch; uttered as head raised. Higher pitched and quieter than in *A. platyrhynchos*. (5) Repulsion-call: single, loud, querulous 'quack' followed by series of quieter, rapid 'gack' notes. (6) Maternal-call: rapid and extremely quiet 'bub bub bub bub bubbubub-ubub' notes, with occasionally trills. Given both as Broody-call and later Leading-call (see *A. platyrhynchos*). (7) Distraction call: loud quacking (McKinney 1967); long series of 1 or 2-syllable notes—varying in pitch, volume, and speed depending on degree of disturbance—uttered when flushed from nest, or when disturbed with brood.

CALLS OF YOUNG. Contact-calls much as in other *Anas*. Distress-calls high and loud, swooping from high to low frequency (J Kear); according to Lorenz (1953) of 2 distinct types (a) rather quick 'tit tit tit' (lower intensity) changing abruptly to (b) long-drawn 'teet teet teet' (higher intensity).

**Breeding.** SEASON. See diagram for north-west and central Europe. Laying begins early to mid-May in northern USSR (Dementiev and Gladkov 1952). SITE. On ground in grass or rushes, sometimes in heather or low scrub; often more or less in open. Usually close to water. Not colonial, but nests as close as 5 m. Nest: shallow hollow, lined grass, leaves, down. Average diameter 20–30 cm, depth of cup 7–10 cm. Building: by ♀, using material within reach of nest; cup formed by body turning. EGGS. Ovate; buff to pale olive. 52 × 37 mm (48–57 × 35–40), sample 275 (Schönwetter 1967). Weight 40 g (35–43), sample 27 (Niethammer 1938); weight of captive laid eggs 39 g (34–44), sample 82 (J Kear). Clutch: 9–11 (6–14). Of 43 Finnish clutches: 6 eggs, 2%; 7, 5%; 8, 12%; 9, 44%; 10, 28%; 11, 9%; mean 9·2 (Hildén 1964). One brood. Replacement clutches laid after egg loss. INCUBATION. 22–23 days (21–25). By ♀, some ♂♂ stay near during early stages. Begins on completion of clutch; hatching synchronous. Eggshells left in nest. Eggs covered with down when ♀ off nest. YOUNG. Precocial and nidifugous. Self-feeding. Cared for by ♀, who defends nest and young against predators; broods young while small. FLEDGING TO MATURITY. Fledging period 40–45 days. Become independent at or just before fledging. Age of first breeding 1 year. BREEDING SUCCESS. Of 451 eggs laid Finland in 3 years, 74% hatched and 17·5% were reared to fledging. Survival of young per pair 4·7, 0·2, and 2·0, so losses of 49%, 97%, and 78% (Hildén 1964). Of 26 Scottish clutches, 54% hatched, 42% predated, and 4% deserted (I Newton).

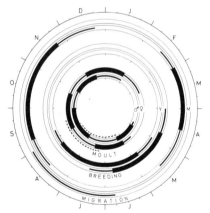

**Plumages.** ADULT MALE BREEDING. Head and upper neck black, glossed metallic blue-green on nape and sides, slightly mottled brown on crown, chin, and throat. Lower neck, sides of upper mantle and breast, and chest white. Centre of mantle dark sepia, feathers edged white on upper mantle and olive-grey on lower. Outer scapulars white, some longer with broad pale blue outer web and narrow blue or black edge on inner web, or partly freckled blue; inner scapulars dark sepia or black with slight green gloss, longer ones with broad white shaft-streak. Back, centre of rump, and tail-coverts black, strongly glossed metallic green in some lights; sides of rump and tail-base white partly vermiculated grey or cinnamon. Upper sides of flank pale cinnamon, finely vermiculated black; lower flanks, lower breast, belly, and vent chestnut, sometimes partly barred or spotted black on vent. Central tail-feathers dark olive-sepia, narrowly edged off-white and buff, other feathers suffused and broadly edged white. Flight-feathers dark sepia; inner webs of primaries dark grey, outer webs of secondaries (except one outer) strongly glossed metallic-green, but tips dark sepia. Tertials black with faint blue-green tinge; white shaft-streak towards tip of outer. Greater upper wing-coverts dark grey, broadly tipped white on outer, gradually narrowing on inner. Inner greater coverts and tips of black tertial coverts tinged blue. Greater primary coverts and bastard-wing dark sepia, tinged blue-grey on outer webs. Median and lesser upper wing-coverts pale blue. Smaller under wing-coverts and axillaries white, greater and marginal coverts grey. ADULT FEMALE BREEDING. Head and neck cinnamon-pink, broadly streaked black on forehead, crown, and line through eye, narrowly on sides of head and on neck; chin and throat uniform white or buff. Feathers of mantle and scapulars dark sepia with broad pink-buff edges and U-shaped subterminal marks, or pink-buff with sepia shaft-streak and some subterminal sepia dots. Back, rump, and upper tail-coverts dark sepia, irregularly marked with wavy pink-buff bars or streaks. Chest, sides of breast, and flanks pink-cinnamon or buff with U-shaped sepia marks. Breast, belly, and vent pink-buff, feathers with partially concealed sepia blotches; sometimes uniform pale cinnamon-pink or white. Under tail-coverts pale buff with sepia crescents and dots. Tail sepia, broadly barred pink-buff; central feathers narrowly, outer broadly edged pale buff or off-white. Wing like adult ♂, but green gloss on secondaries usually fainter (occasionally absent); secondaries, inner greater coverts and tertial coverts narrowly edged white; pale blue of lesser and median coverts greyer, less brilliant, often partly edged pale buff or cream; white tips of greater coverts less sharply defined. Tertials sepia, widely edged and centrally suffused pale buff. ADULT MALE NON-BREEDING (Eclipse). Superficially like adult ♀ breeding, but much darker and feathers less marked subterminally. Crown black, faintly marked pale buff; neck and sides of head grey-brown, narrowly barred and mottled pink-buff and pale grey; chin and throat pink-buff or off-white, finely speckled brown. Mantle, scapulars, upper tail-coverts, and sides of breast dark sepia; feathers narrowly edged grey on mantle and upper tail-coverts and pale buff on scapulars and sides of breast (often also with variable narrow buff crescents subterminally). Chest and flanks pink-buff, feathers with U-shaped sepia mark subterminally. Breast, belly, and vent grey-brown, feathers tipped buff and blotched sepia subterminally, often suffused chestnut. Under tail-coverts heavily barred sepia and pale buff. Sides of rump barred pale buff and olive-sepia, rest of rump, back, wing, and tail like adult ♂ breeding, but tertials somewhat shorter, with more rounded tip and duller shaft-streak. ADULT FEMALE NON-BREEDING. Head and neck like adult ♀ breeding, but sepia streaks on crown and sides broader, although head not as dark as adult ♂ non-breeding. Mantle, scapulars, and upper tail-coverts like adult ♂ non-breeding, but fringes of feathers buff and somewhat broader; often with buff U-shaped mark on scapulars and tail-coverts. Back and rump dull black, tips of feathers often narrowly marked buff. Feathers of chest and flanks sepia, edged buff, with narrow buff subterminal mark. Rest of plumage, including wing, like adult ♀ breeding, but tertials uniform dark olive-sepia, narrowly edged off-white, each with faint grey shaft-streak. ADULT SUPPLEMENTARY. In ♂, non-breeding plumage on part of head and body replaced in early autumn by plumage with some breeding characters. Variable number of feathers involved; on parts of body some non-breeding feathers changed for supplementary, while others moult directly to breeding. Feathers at side of head black with much white at base, or with pale buff barring subterminally; when these shed, large white crescent on fore cheek may be left for short period. Chest, upper breast, and sides of breast and mantle white, feathers with black crescent near tip and terminally suffused buff or cinnamon-pink. Shorter scapulars white, with black and buff barring or black vermiculation at tips, longer scapulars sepia with white edges and blue tinge on outer webs. Flanks coarsely barred black and chestnut. Not certainly known if ♀♀ attain supplementary plumage. DOWNY YOUNG. Like Mallard *A. platyrhynchos*, but upperparts darker, with light patches ill-defined and absent on wing. Longer down on upperparts, streak over eye, cheeks, chin, and throat cinnamon-buff, not yellow. Chest greybuff; underparts pale grey, sometimes suffused cream-yellow. When older, shape of bill characteristic. JUVENILE MALE. Forehead and crown black, feathers narrowly edged buff; rest of head grey-brown or pale buff, feathers streaked brown. Upperparts sepia, underparts grey-brown; feathers fringed off-white on belly, buff elsewhere; incomplete buff bars on sides of breast, outer scapulars, and lower flanks. Feathers short and narrow, those of belly to under tail with small dark spots and streaks. Tail-feathers brown, barred pale buff; notched at tip, bare shaft projecting. Wing like adult ♂, but gloss on secondaries more restricted, bronze-green rather than metallic blue-green, tips of outer webs narrowly edged white; white wing bar formed by tips of greater coverts usually narrower; tips of greater coverts mostly with distinct round spot; lesser and median coverts greyer at base; tips of upper wing-coverts slightly broader and rounder; tertials as in adult ♂ non-breeding, but shorter. JUVENILE FEMALE. Head, body, and tail like juvenile ♂. Wing like adult ♀, but secondaries only faintly glossed green; sometimes dull black or sepia without gloss; off-white edges of tips of secondaries broader; white wing bar formed by tips of greater coverts narrower; white of inner greater coverts often with brown dot at tip; lesser and median coverts duller, usually with only faint blue or grey tinge, sometimes dark brown; cream edges to lesser and median coverts as in adult ♀. FIRST IMMATURE NON-BREEDING AND SUBSEQUENT PLUMAGES. 1st immature non-breeding, acquired in early autumn, resembles adult non-breeding. Appears first on head and sides of tail-base; on latter, ♂♂ barred or vermiculated white and sepia, ♀♀ pink-buff dotted sepia. Variable amount of juvenile plumage sometimes retained on underparts, back, rump, tertials, and tail until spring; wing juvenile. Immature ♂♂, and probably also ♀♀, attain supplementary plumage later in autumn, and both sexes 1st breeding plumage from late autumn or early spring. Supplementary and 1st breeding like adults; when no juvenile body or tail-feathers retained, ageing possibly only by juvenile wing. Some 1st breeding ♂♂ have black dots on chestnut of belly. Some supplementary or 1st non-breeding plumage often retained until summer.

**Bare parts.** ADULT MALE. Iris golden yellow or orange. Bill leaden black, lower mandible spotted paler; in eclipse (until about November), yellow-brown or orange-brown, gradually changing to black-brown or olive-brown. Foot orange-red. ADULT FEMALE. Iris brown, yellow-brown, or straw-yellow. Bill olive-grey or dark brown with yellow-brown sides; edges and lower mandible pale orange to orange-red. Edges of upper mandible with black dots basally. Foot orange-red. DOWNY YOUNG. Iris brown. Bill red-brown with grey culmen. Foot dark olive-grey. JUVENILE MALE. Iris grey-brown or light sepia. Bill dark olive-brown, edges and lower mandible yellow-orange. Foot yellow-orange, webs dark grey-brown. In early autumn, iris

changes to yellow-brown and yellow, and foot to orange; in spring, bare parts as adult ♂. JUVENILE FEMALE. Iris brown; sometimes yellow-brown from early autumn. Bill olive or brown-grey; base, edges, and lower mandible orange-brown to pale orange. Usually some small black dots along edge of upper mandible. Foot ochre-yellow to orange, webs dull yellow-grey. Adult colour attained in winter. (Bauer and Glutz 1968, Kortright 1942; Schiøler 1925; Witherby *et al.* 1939; RMNH, ZMA.)

**Moults.** ADULT POST-BREEDING. Complete; flight-feathers simultaneously. In ♂ between early May and early June; starts with head, underparts, mantle, scapulars and central pair of tail-feathers; back, rump, wing, and other tail feathers start *c.* 1 month later. Moult in ♀ *c.* 1 month later than ♂. Flightless for 3–4 weeks, ♂♂ mid-June to mid-August, ♀♀ late July to early September. ADULT PRE-BREEDING. In ♂, non-breeding plumage on chest, sides of breast and mantle, on part of head flanks, scapulars, and sometimes tail coverts, changed for supplementary plumage August–September; shed for breeding November–December, but some feathers on chest or sides of breast occasionally not before April. On rest of body, non-breeding plumage completely replaced by breeding September–November. Central and some other tail-feathers and tertials moult about September. Moult in ♀♀ difficult to follow; sequence apparently as in ♂♂. Breeding plumage appears on underparts from September; completed by February–March with tertials and scapulars. POST-JUVENILE AND SUBSEQUENT MOULTS. In ♂♂, juvenile shed for 1st immature non-breeding August–September, but some juvenile outer tail-feathers, part of back and rump, and tertials sometimes retained until November. In September–October, juvenile belly and vent changed for breeding plumage, and part of 1st non-breeding plumage on head, chest, flanks, outer mantle, scapulars, and breast changed for supplementary; remaining 1st non-breeding plumage changed for breeding late October to December. Supplementary plumage replaced November to early spring, first on breast; often some 1st non-breeding and supplementary feathers retained until 1st post-breeding moult in summer. In ♀♀, juvenile changed for 1st non-breeding September–October, but juvenile tertials, and often juvenile back, rump, and some outer tailfeathers retained until March; many juvenile feathers occasionally retained elsewhere on body. In October, plumage of head, chest, flanks, and part of mantle and scapulars replaced by one probably equivalent to supplementary plumage of ♂♂. This and remaining 1st non-breeding plumage changed for breeding February–April, but sometimes partially retained until 1st post-breeding moult. (Dementiev and Gladkov 1952; C S Roselaar.)

**Measurements.** Netherlands, whole year. Skins (RMNH, ZMA).

| | | | | | | |
|---|---|---|---|---|---|---|
| WING AD ♂ | 244 | (3·77; 27) | 239–249 | ♀230 | (4·52; 18) | 222–237 |
| JUV | 235 | (5·04; 41) | 227–251 | 222 | (4·45; 29) | 213–229 |
| TAIL AD | 81·7 | (2·62; 26) | 76–86 | 75·8 | (2·60; 15) | 72–80 |
| JUV | 74·9 | (3·88; 11) | 70–81 | 68·6 | (3·74; 15) | 63–74 |
| BILL | 66·1 | (1·91; 61) | 62–72 | 60·7 | (2·12; 47) | 56–64 |
| TARSUS | 37·2 | (1·14; 48) | 35–40 | 36·0 | (0·86; 39) | 35–38 |
| TOE | 51·5 | (1·77; 47) | 48–55 | 48·6 | (1·75; 39) | 46–52 |

Sex differences significant. Wing and tail of juvenile significantly shorter than adult; rest similar, combined.

**Weights.** ADULT. Mainly USSR; some additional Netherlands. Averages for several areas (sample size, when known, in brackets) and total range (Dementiev and Gladkov 1952; Dolgushin 1960; ZMA).

| | | | | | |
|---|---|---|---|---|---|
| (1) | ♂ 418–510; | 300–620 | ♀ 387–478; | 300–c. 570 | |
| (2) | 568–650, 569 (8); | 490–750 | 550, 575 (7); | 480–715 | |
| (3) | 600, 638 (8); | 530–700 | 692; | 611–774 | |
| (4) | 600 (1), 635; | 600–650 | 527 (3), 576; | 470–700 | |
| (5) | 850; | 650–1000 | 642; | 600–700 | |
| (6) | 634 (8), 717; | 475–800 | 600, 612 (4); | 525–750 | |
| (7) | 692, 860 (2); | 560–950 | 664, 680 (3); | 600–800 | |

(1) Low January–February (south-west Caspian). (2) Heavier on arrival breeding area March–May (USSR, some Netherlands). (3) June and (4) July: ♀♀ heavy at start of breeding, but lose weight during it (USSR). (5) August: strongly variable (Volga delta). (6) Rather low August–September when flight-feathers new (USSR, Netherlands). (7) Heavy October–November (Volga delta, Netherlands).

JUVENILE. Netherlands, October (ZMA): ♂ 602 (16), 500–700; ♀ 599 (5), 500–670.

Structure. Wing rather long and narrow, pointed. 11 primaries: p9 longest, p10 0–4 shorter, p8 6–11, p7 16–24, p6 28–40, p1 96–127. Inner web of p10 and outer p9 emarginated. Tail short, rounded; 14 pointed feathers. Bill longer than head, much higher than broad at base, strongly widened towards rounded tip. Culmen nearly straight, sides of upper mandible curved upwards in basal half of bill (exposing long and thin lamellae), outwards near tip. Nail of bill small and narrow, decurved. Scapulars and tertials long and tapering to point in adult ♂ breeding, shorter and broader at tip in other plumages. Outer toe c. 92% middle, inner c. 76%, hind c. 29%.                      CSR

## *Marmaronetta angustirostris* **Marbled Teal**

PLATES 67 and 74
[between pages 518 and 519
and facing page 543]

Du. Marmereend    Fr. Sarcelle marbrée    Ge. Marmelente
Ru. Мраморный чирок    Sp. Cerceta pardilla    Sw. Marmorand

*Anas angustirostris* Ménétries, 1832

Monotypic

**Field characters.** 39–42 cm, of which body under two-thirds; wing-span 63–67 cm. Small dabbling duck with long, narrow bill, relatively long neck, big shaggy head, dark patch around eye and generally pale spotted, grey-brown plumage; no speculum. Sexes similar and no seasonal differences; juvenile distinguishable.

ADULT MALE. Generally grey-brown, noticeably spotted with pale cream feather centres and contrasting dark grey-brown edges. Slightly darker above than below, with broad dark stripe from eye to nape; nape feathers elongated to form pendant shaggy crest. Primaries pale silver-grey, secondaries pale brown fading into grey-brown tips; tail partly tipped white. In flight, no distinct pattern apart from paler secondaries and sometimes visible white tips on tail. Bill blackish with dull grey-green base and whitish transverse line near tip; legs green-brown with blackish webs. MALE ECLIPSE. None. ADULT FEMALE. Similar to adult ♂, but slightly smaller, crest smaller or absent and slight differences in colour of bill. JUVENILE. Resembles adult, though duller and greyer above without creamy spots on back, and almost uniformly cream below with much less distinct markings on flanks.

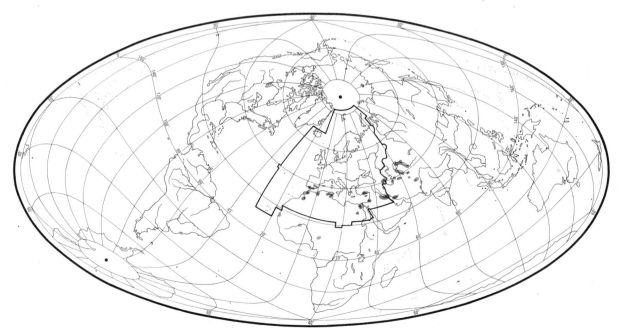

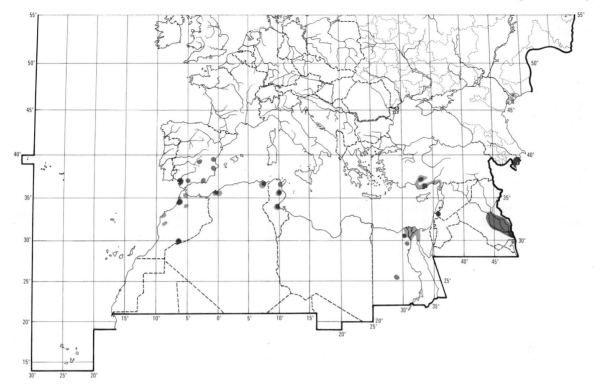

Looks uniformly pale grey-brown or mortar-coloured, with dark eye-patch and shaggy head. This combination, and lack of coloured speculum, distinguish it from all other west Palearctic ducks. Accidental Cape Teal *Anas capensis* larger and greyer with darker back, reverse pattern of spotting (dark centres and light edges), well-marked speculum, high rounded head, and pink bill.

Usually shy and unobtrusive, spending much time on reed-covered lakes and slow-moving rivers, usually in pairs or small parties. Swims well, but not usually very far. Flies with slower wing-beats than Teal *A. crecca*, with much less springing take-off, and generally for only short distances into fresh cover. Generally quiet.

**Habitat.** Restricted and fragmented range due to limited suitable habitats, through warm steppe and Mediterranean zones of continental climate. Mainly lowland, but breeds to *c.* 1500 m Baluchistan (Ali and Ripley 1968). Reluctance to shuttle between distinct resting and feeding places, and preference for instant access to shelter and deep shade in dense stands of fringing aquatic vegetation, also limiting factors. Rests on low, overhanging branches of tamarisks and other bushes, but not otherwise concerned with trees or woody plants. Prefers small or medium-sized, fairly shallow freshwater pools, with sometimes floating and normally much fringing and submerged vegetation; also floodlands and shallow, brackish saline or alkaline lagoons or pools; artificial waters such as reservoirs and irrigation networks; slow-flowing rivers or their oxbows and backwaters, and inlets of inland seas, but only excep-

tionally tidal waters. Foraging on banks and even in cornfields recorded, but frequency not clear. Seasonal and periodic displacements enforced by instability of water regime in characteristic habitats, but otherwise less restless and fond of flying around than *Anas*. Vulnerable to development and persecution through expanding human population near habitats.

**Distribution.** In south of range, breeding and wintering areas affected by water levels, so areas mapped not always occupied annually. TUNISIA: breeds at various sites in wet seasons e.g. 1957, 1962, 1964, 1966, and attempted 1968; most eggs taken (Jarry 1969; MS). FRANCE: bred Camargue in 1890s and perhaps later, but not proved since at least 1940 (FR). ITALY: bred 1892 (SF). YUGOSLAVIA: nested Skopje (Macedonia), 1950–60 (VFV). GREECE: Crete, 1925 (WB). CYPRUS: bred 1888, 1910, and 1913 (Stewart and Christensen 1971). CANARY ISLANDS: bred *c.* 1857 or earlier, and probably up to 1914; site largely drained by 1948 (Bannerman 1963). CAPE VERDE ISLANDS: perhaps bred Boa Vista late 19th century, but no recent records (Bannerman and Bannerman 1968).

Accidental. West Germany, Czechoslovakia, Hungary, Greece, Rumania, Italy, Portugal, Lebanon, Cyprus, Malta, Madeira.

**Population.** SPAIN. Main breeding area marismas of Guadalquivir, where marked decline. One of commonest ducks end 19th century with several thousand pairs; only 100–200 pairs in early 1960s (Valverde 1964); under 100 pairs (Ree 1973). USSR: in west Palearctic, now breeding only in Lenkoran lowlands in extremely small numbers (YuAI). TURKEY. Ereğli area: 20 pairs in 1971. Göksu delta: 2 pairs in 1971, up to 70 birds in April 1972. Possibly breeding elsewhere (OST). IRAQ: colonies small and scattered, probably tens of pairs (PVGK). MOROCCO: Iriki: *c.* 500 pairs in 1965, 1966, and 1968 (Robin 1968). ALGERIA: apparently declined since 19th century, though present position inadequately known; may also breed Hauts Plateaux (Heim de Balsac and Mayaud 1962; EDHJ).

**Movements.** Migratory and dispersive, but little understood in virtual absence of ringing. Apparently irregular at times, according to availability of shallow waters. Spanish breeders leave marismas as these dry out in late summer; many records September to north-east near coasts Alicante and Valencia, and in Ebro delta. 3 ringed Coto Doñana July recovered north-west Morocco November (2) and January (Thévenot 1972), but some present Spain all year, though no major return to marismas until late January and early February (Valverde 1964). Apparently largely deserts Tunisia in winter though staying within breeding range Morocco and Algeria, numbers highly irregular: in January–February 1972, 125 found Morocco, 200 Algeria, none Tunisia (Hovette and Kowalski 1972*b*), though at least 1500 maritime Morocco, chiefly Rharb, January 1964 (Blondel and Blondel 1964). Southward movement occurs; 1500 on temporary desert flood south of Tafilalet 1–2 May 1970 (*Alauda* 1970, **38**, 119), while small numbers evidently cross Sahara, since encountered recent winters in Sahel zone of Sénégal, Mali, and Chad (maximum 45 Lake Chad 1969–70 winter) (Roux 1970, 1972; Morel and Roux 1973).

May be partially resident Turkey but, again, winter numbers variable: 2000 Adana region January 1968 (Johnson and Hafner 1970), but only 600 in Anatolia, January 1971, and none found January 1972 or 1973 (Koning 1973). A few winter Israel and lower Egypt. By far largest winter concentration known is in Khuzestan, Iran, where 12600 February 1971 and *c.* 10000 January 1972 (Carp 1972; Scott and Carp 1972); these assumed to include breeders of Iraq and Iran (where only summer visitor to Azerbaydzhan), and also some from USSR (where migratory except in south Caspian), since small-scale passage occurs Iranian Caspian (D A Scott). Present rarity in former winter quarters, Pakistan and north India, linked to decline in breeding numbers in Turkestan and Uzbekistan (e.g. Savage 1970*b*).

Little information on timing of movements. Autumn passage on River Uzboi, Turkestan, late September and early October, and on River Atrek, Turkestan/Iran border, November; spring return rather late in USSR, with movement through Turkestan/Afghanistan border in April, and around Caspian early May (Dementiev and Gladkov 1952). No moult migration recorded; ♂♂ flock while ♀♀ incubating, moulting nearby. Now only oc-

casional in Europe outside Iberia; past influxes include numbers Italy and stragglers north to Bavaria and Bohemia summer 1892, when also major influx to Europe of Ruddy Shelduck *Tadorna ferruginea*, and small flock Hungary in September 1896 (Bauer and Glutz 1968).

**Food.** Little known. Feeds mainly in shallow water, dabbling at surface and amongst vegetation; up-ends, occasionally dives (Johnsgard 1961*a*), will also 'grub' on shore in mud. According to Ali and Ripley (1968) chiefly vegetarian taking seeds, tubers and shoots of aquatic plants, but worms, molluscs, aquatic insects and their larvae also taken (Dementiev and Gladkov 1952).

**Social pattern and behaviour.** Based mainly on material supplied by P A Johnsgard.

1. Gregarious, at times even when nesting. Outside breeding season flocks often small, though large numbers reported in winter in some areas. BONDS. Little known in wild. Monogamous pair-bond seems strongly developed in captivity, re-formed each autumn and winter (P A Johnsgard). Timing of pair-formation in wild populations largely unknown, but January and March indicated in Spain (Valverde 1964), and in USSR spring migrant flocks consist of paired birds (Dementiev and Gladkov 1952). In pairs from some time prior to nesting until egg-laying begun; ♂♂ then reported to desert mates and gather into small flocks (Dementiev and Gladkov 1952). Only ♀ tends young. BREEDING DISPERSION. Probably solitary at times, but colonial nesting reported in south-west Spain under favourable conditions, with nests only 1 m apart (Valverde 1964; see also Hawkes 1970). ROOSTING. No detailed information.

2. Studied only in captivity (Jones 1951; von de Wall 1962; Johnsgard 1961*a*, 1965). Behaviour nearer pochards (Aythyini) than typical dabbling ducks *Anas*. Only pre-flight signal recorded repeated, pochard-like Chin-lifting (McKinney 1953). ANTAGONISTIC BEHAVIOUR. Captive birds show little aggression in flock; fights not reported, von de Wall (1962) noting only hostile stretching of neck and pointing of bill. Threat behaviour of ♀ takes main form of Inciting-display consisting of slightly ritualized alternation of attack and retreat movements (see below). COMMUNAL COURTSHIP. Occurs in groups of several ♂♂ and 1 or more ♀♀; in captive birds, southern England and northern USA, from late autumn to spring (P A Johnsgard). ♂♂ swim round and near ♀, generally alternating 3 major displays (Neck-stretch-Head-jerk, Sneak, Turn-back-of-Head) with either or both 2 secondary ones (Upward-shake, Neck-stretch-Head-shake) from position lateral to ♀. Upward-shake like that of *Anas* but more frequent after major display than before. In Neck-stretch-Head-shake display, head raised and lowered while simultaneously shaken laterally, finally resting on back (von de Wall 1962). Most ritualized activity, Neck-stretch-Head-jerk display, somewhat similar to last but lacks Head-shaking movement: with crest erected, head (1) suddenly raised vertically; (2) held still briefly at height of Neck-stretch (see A); (3) suddenly drawn back and down with jerk, bill opening and call uttered as head touches back (see B). Display sometimes consists only of Neck-stretch phase; then no call given. In pochard-like Sneak-display, head and neck extended forward, bill pointing in direction of another bird (♂ or ♀); performed frequently, appears to represent ritualized threat. In Turn-back-of-Head display, ♂ swims ahead of Incting ♀, ceremonially Leading her. ♂ displays in general appear to be stimulated by Inciting-display of ♀ in which (1) ♀ makes rushes towards rejected ♂, (2) swims rapidly

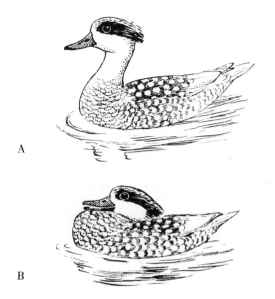

A

B

back to favoured ♂ while nodding head back and forth with short jerks. ♀ also performs ♂-like Neck-stretch-Head-jerk display (with call), but this less vigorous and frequent. PURSUIT-FLIGHTS AND PAIR BEHAVIOUR. No detailed information. Only Mock-preen display recorded, Preen-dorsally (see McKinney 1965*b*). COPULATION. Pre-copulatory behaviour much closer to Aythyini than to other Anatini. Head-pumping of *Anas* replaced by fairly long periods of mutual Bill-dipping, Ceremonial-drinking, and Preen-dorsally—♂ mounting after ♀ assumes Prone-posture. After copulation, ♂ calls once with neck fully extended diagonally, then assumes pochard-like Bill-down posture with head drawn back and swims in partial circle round ♀. Latter bathes immediately after ♂ dismounts. RELATIONS WITHIN FAMILY GROUP. No detailed information.

(Figs A–B after Johnsgard 1965.)

**Voice.** Relatively weak in both sexes and used infrequently, though no detailed studies. Tracheal bulla of ♂ unique, with some features common to both Anatini and Aythyini (see Johnsgard 1961*a*). Main call of ♂ (see I) normally heard only during Neck-stretch-Head-jerk display of courtship (Johnsgard 1961*a*) and after copulation; described as nasal squeak—'eeeeep' (Jones 1951; Johnsgard 1965). ♀ lacks strong quacking calls typical of *Anas*, including Decrescendo-call. Inciting-display performed silently or, possibly, with barely audible sounds. Call given during ♀-version of Neck-stretch-Head-jerk display similar to ♂'s. Weak, high-pitched, whistling

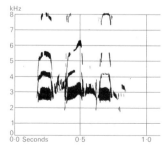

I C Chappuis/Sveriges Radio
Morocco April 1966

'pleep-pleep' call also ascribed to ♀ by Jones (1951); as such, unique within Anatini, but nature unknown.

CALLS OF YOUNG. Contact-calls much as *Anas*, uttered in groups of 2–4 notes. Distress-call like Mallard *A. platyrhynchos* but at higher frequency and faster. (J Kear.)

**Breeding.** SEASON. Spain and North Africa: eggs laid occasionally late April, most in May and first half June (Valverde 1964). Southern USSR: season protracted, eggs laid May–July (Dementiev and Gladkov 1952). SITE. Usually on ground under low bushes and in thick vegetation; in Spain sometimes concealed in roofs of grass and reed huts (Hawkes 1970). Never far from water. Nest: slight depression, lined grass and down. Building: by ♀, using material within reach. EGGS. Ovate; pale straw. 46 × 34 mm (42–51 × 32–36), sample 100 (Schönwetter 1967). Weight of captive laid eggs 30 g (26–35), sample 62 (J Kear). Clutch: 7–14 (up to 24 recorded, presumed by 2 ♀♀). One brood. No data on replacements, but probable. One egg laid per day. INCUBATION. 25–27 days. By ♀. Starts on completion of clutch; hatching synchronous. Eggshells left in nest. YOUNG. Precocial and nidifugous. Self-feeding. Cared for by ♀; brooded while small. FLEDGING TO MATURITY. Fledging period not recorded. Age of first breeding 1 year. BREEDING SUCCESS. No data.

**Plumages.** ADULT MALE BREEDING. Forehead, crown, and hindneck barred buff and black (feathers buff with black subterminal bars and pale grey tips). Ill-defined patch round and behind eye dark brown; also lores in some. Cheeks, throat, sides and front of neck white, feathers narrowly margined brown, giving finely streaked appearance. Mantle, back, rump, and upper tail-coverts dark grey-brown, feathers broadly tipped cream-buff or pink-buff, paler grey towards base; scapulars similar, each with large round grey or cream buff spot near tip. Feathers of chest, sides of breast, flanks, lower vent, and under tail-coverts pale grey or pale buff, each with 2 dark grey-brown subterminal bars; bar near tips darkest, often margined buff; tips of feathers of chest, lower flanks, vent, and under tail-coverts often tinged buff. Rest of underparts white, feathers with faint buff-grey subterminal bar or faint buff tinge towards tip. Central tail-feathers dark grey-brown, tipped white, rest of tail pale grey-brown. Flight-feathers and tertials pale grey-brown, palest, almost white, towards tips of secondaries, darker near tips of inner webs of primaries. Outer webs of primaries silver-grey towards tip. Upper wing-coverts pale grey-brown, edged pale grey; shafts dark brown. Under wing-coverts white; marginal coverts and axillaries white with some dusky blotches near tips. Pale tips of feathers may either wear off, causing darker and more heavily barred plumage, or dark parts of feathers may fade, giving paler appearance. ADULT FEMALE AND ADULT MALE NON-BREEDING. Like adult ♂ breeding, but crest short or absent. DOWNY YOUNG. Crown, narrow streak through eye, hindneck, upperparts, and thighs dark fawn-brown; streak over eye, rest of head, underparts, and ill-defined patches on wing and at sides of back and rump pale buff-yellow. JUVENILE. Like adult ♀, but markings less distinct and duller, spots on upperparts less creamy and ill-defined; underparts uniform dull grey, less buff. SUBSEQUENT PLUMAGES. Like adult. (Bauer and Glutz 1968; Dementiev and Gladkov 1952; Delacour 1956; Hartert 1912–21; C S Roselaar.)

**Bare parts.** ADULT MALE. Iris brown. Bill dull black with narrow white band behind nail, and faint blue-grey band along cutting edges and lateral base of upper mandible. Foot olive-green or pale yellow, webs dull black. ADULT FEMALE. Iris and foot as ♂. Bill dull black with dull olive-green triangular patch basally on side of upper mandible. DOWNY YOUNG. Iris brown. Bill and foot green-grey. (Bauer and Glutz 1968; Delacour 1956; C S Roselaar.)

**Moults.** ADULT. Apparently twice a year, but few details known. ♂ in pre-breeding body and tail moult early October, wings freshly moulted (Dementiev and Gladkov 1952). Post-breeding probably in late summer.

**Measurements.** Captive birds, Wildfowl Trust, England (M A Ogilvie) and skins from west Mediterranean (RMNH, ZMA); combined.

| | | | | | | |
|---|---|---|---|---|---|---|
| WING AD ♂ | 207 | (5·93; 9) | 195–215 | ♀ 198 | (6·84; 10) | 186–206 |
| JUV | 196 | (11·4; 6) | 180–208 | 190 | (12·3; 4) | 174–202 |
| TAIL AD | 69·8 | (— ; 4) | 67–73 | 67·7 | (— ; 3) | 67–68 |
| BILL | 44·2 | (1·86; 9) | 42–47 | 42·3 | (1·80; 11) | 39–45 |
| TARSUS | 38·1 | (1·00; 8) | 36–40 | 37·4 | (2·01; 11) | 35–40 |
| TOE | 47·0 | (— ; 4) | 46–48 | 46·0 | (— ; 3) | 45–47 |

Sex differences significant for adult wing and bill. Tail measurements by Hartert (1912–21) and Delacour (1956) somewhat larger (up to 105).

**Weights.** Turkmeniya, USSR: ♀, October, 400; unsexed, December, 477 (425–500) (Dementiev and Gladkov 1952). India: ♂ 535–590, ♀ 450–535 (Salim Ali and Ripley 1968). Maxima of adults died in captivity (Wildfowl Trust): ♂ 425, ♀ 470; minima: adult ♂ 270, adult ♀ 255; juvenile ♂ and ♀ 205 (M A Ogilvie).

**Structure.** Wing rather long, pointed. 11 primaries: p9 longest, p10 1–2 shorter, p8 1–7, p7 9–14, p6 17–23, p1 75–90. Inner web of p10 and outer p9 emarginated. Tail rather long, slightly rounded; 14 feathers with pointed tips. Bill long and narrow, higher than broad at base, but depressed beyond nostrils; cutting edges parallel, lamellae short, nail of bill small. Feathers of nape elongated in adult ♂ breeding, forming loose thin crest. Outer toe *c.* 91% middle, inner *c.* 77%, hind *c.* 28%.          CSR

# Tribe AYTHYINI pochards

Medium-sized, mainly freshwater diving ducks. (N.B. Designation 'diving duck' used, not as taxonomic term, but as ecological characterization for all ducks in tribes Tachyerini, Aythyini, Somateriini, Mergini, and Oxyurini, plus a few aberrant members of other, predominantly surface-feeding tribes.) 15 species in 2 genera: *Netta* (3 species) and *Aythya* (12 species); monotypic *Rhodonessa* (Pink-headed Duck, India and Nepal) recently extinct. *Netta* intermediate in some characters between Anatini and *Aythya*. Latter composed of 3 species-groups: (1) typical pochards, including *A. ferina*; (2) white-eyed pochards, including Ferruginous Duck *A. nyroca*; (3) scaups, including *A. marila* and Tufted Duck *A. fuligula*. 1 species of *Netta* and 4 of *Aythya* breeding in west Palearctic; 1 *Aythya* accidental.

Bodies in *Aythya* short and heavy, heads big; wings broader and less pointed than in typical Anatini, necessitating faster wing-beats often producing whistling sound; tails short; bills rather heavy (less so in white-eyed pochards), about as long as head, flattened and, in some, wider at tip; legs short, with large toes and broadly lobed hind toe, and set well apart far back on body. *Netta* similar but bodies longer and narrower, bills narrower, legs longer and more slender. All take off from water with some difficulty. *Aythya* clumsy on land; *Netta* much less awkward, with even more upright stance. Though *Netta* somewhat less well adapted for diving than *Aythya* (Delacour and Mayr 1945), all dive with considerable facility, typically without using wings. Sexes differ in tracheal anatomy; as well as showing 1–2 enlargements of tracheal tubes, ♂♂ have large, rather angular bullae, with several fenestrae, not rounded and evenly ossified as in

*Anas* ♂♂. ♂♂ mainly patterned simply—black, brown, or chestnut and white; unstreaked ♀♀ varying shades of brown. Broad pale (often white) panel on rear half of upper wing; no metallic speculum. ♂ non-breeding plumage in most species of eclipse type; ♀♀ often nest in plumage homologous to non-breeding plumage of other Anatinae. Bills usually slate or bluish but red in 2 *Netta*; eyes red (most pochards of both genera), white (♂ white-eyed pochards), brown or yellow (♀♀ of last group), or yellow (scaups). Juveniles resemble ♀♀. Downy young most like those of Anatini but head stripes faint or absent; young of scaups dark.

Cosmopolitan, but most species Holarctic. Concentrated both as breeders and in winter on standing fresh water of moderate depth, usually 1–15 m; Scaup *A. marila*, marine in winter, partial exception. Tolerate fairly restricted open waters with dense marginal vegetation, even in forest setting. In most areas, suitable sites restricted and vulnerable to desiccation, drainage, and other adverse factors, leading to some instability in distribution and population. Some colonize modern artefacts such as reservoirs, gravel-pits, and ornamental waters. In Red-crested Pochard *N. rufina*, evidence of decline in west Europe in 19th century and definite spread more recently. Both Pochard *A. ferina* and *A. fuligula* have spread markedly in north and west Europe since mid-19th century. In central Europe, *A. nyroca*, after increase in 19th century, showed marked decline (recently halted) while declined also further south. In north-west Europe, *A. marila* retreated north in 19th century, but recent extension south. In winter, oil pollution major hazard for both *A. marila* and *A. fuligula*, especially in Baltic. All west

Palearctic species migratory to greater or lesser extent. Movements include major moult migrations, mainly by adult ♂♂; some adult ♀♀ join moult assemblies in late summer. Tendency towards winter segregation of sexes perhaps due in part to different migration timing (see also below). Migrate at considerable heights, in long wavering lines, blunt wedges, or compact groups.

Range from chiefly vegetarian (e.g. *Netta*) to omnivorous; in some species (e.g. *A. marila*), animal food predominates. Food obtained in water, mainly by diving from surface to bottom. Usually submerge for shorter periods than Somateriini and Mergini. Difference between sexes in preferred diving depths, and hence in mean duration of dives, recorded in some species and probably widespread; may be contributory factor in partial winter segregation of sexes. Most species (especially in *Netta*) also dabble on surface at times, head-dip, and up-end. Feed mainly in pairs and flocks. Largely gregarious at most other times also (but see *A. nyroca*). Repeated Chin-lifting main pre-flight signal, but Head-flicks also frequent in some *Aythya* while other shaking and stretching comfort-movements recorded (see McKinney 1965*b*). Prior to nesting, each pair frequents home-range overlapping those of other pairs but flocking tends to persist and activity centres and waiting areas (see Anatini) often shared by ♂♂ and pairs; not defended as territories so far as known, though each ♂ defends small area round mate when latter present (see, e.g., Dzubin 1955). Monogamous pair-bonds of limited seasonal duration typical in Holarctic species. In *Aythya*, pairs form later than in most northern Anatini, mainly late winter and spring (see, e.g. Weller 1965); earlier in *N. rufina*. Pair-bond ends early in incubation period (most *Aythya*), but later in *N. rufina*. Promiscuous tendencies of ♂♂ much less marked than in Anatini; except in *Netta*, rape attempts rare, and pursuit-flights largely of courtship type. Communal courtship on water much as in Anatini though most major displays different—with differences also between *Netta* (particularly *N. rufina*) and *Aythya*; often nocturnal as well as diurnal. Secondary displays of ♂♂ include Head-flick and Upward-shake, though latter infrequent in some species. Typical major displays, usually accompanied by calls, include Head-throw, Kinked-neck, and Sneak-display. Latter takes 2 main forms: full version with head along

water; incomplete version (or Crouch-display) with head inclined forward. Other displays include Turn-back-of-Head, Neck-stretch, and Coughing-display—though some confusion in literature whether Neck-stretch and Coughing-displays different or partially same. In some species, ♀♀ perform ♂-like major displays at times; Inciting-display of same functional type as in Anatini but differs largely in form. In most species, some of displays used by ♂ in communal courtship also used in pair-courtship; others distinct, including unique Courtship-feeding of *N. rufina*. Displays performed by both ♂ and ♀, sometimes mutually, include Ceremonial-drinking and Mock-preening. Copulation also part of pair-courtship. Pre-copulatory displays include Bill-dipping and Preen-dorsally; in *Netta*, also *Anas*-like Head-pumping. Prone-posture of ♀ differs from that of *Anas* in that neck stretched diagonally forward not flat on water. Post-copulatory displays include characteristic Bill-down posture by ♂ or both sexes. Calls of ♂♂ often whirring or cooing and not far-carrying, but some (notably scaups) also whistle. Used chiefly in courtship, of 2 main types given during (1) Head-throw and Kinked-neck displays (2) Coughing-display. ♀♀ usually not highly vocal; calls mostly growling and harsh, louder than those of ♂♂—include Inciting-calls but Decrescendo-calls lacking in most species. Non-vocal rattling sound produced in Preen-behind-Wing display in all or most species.

Holarctic species strictly seasonal breeders. Nests sited over shallow water or on ground never far from water; usually in thick cover. Well dispersed or grouped, sometimes close together, at protective sites such as islands (sometimes associated with gulls *Larus*). Shallow depressions with rim of available material, lined with down plucked by ♀. Building by ♀ only. Eggs oval, green-grey or pale buff; smooth. Clutches usually 5–12; multiple laying not uncommon in some species. One brood; replacements laid after egg loss. Eggs laid daily. Incubation by ♀ only. Incubation periods 24–28 days (Kear 1970). Young cared for by ♀ only. Distraction-display, in form of 'injury feigning' occurs (at least in *Aythya*) but less common than in Anatini. No true crèching but broods sometimes amalgamated. Young independent at or before fledging in most species. Mature in 1st year.

## *Netta rufina* Red-crested Pochard

PLATES 75 and 99
[after pages 566 and 686]

Du. Krooneend   Fr. Nette rousse   Ge. Kolbenente
Ru. Красноносый нырок   Sp. Pato colorado   Sw. Rödhuvad dykand

*Anas rufina* Pallas, 1773

Monotypic

**Field characters.** 53–57 cm, of which body less than two-thirds; wing-span 84–88 cm. Large, plump-bodied, and rather long-necked diving duck with tapering bill and

large, rounded head increased in size by short, bushy, erectile crest; ♂ has red bill, rich golden chestnut head, mainly brown upperparts, and black neck and underparts

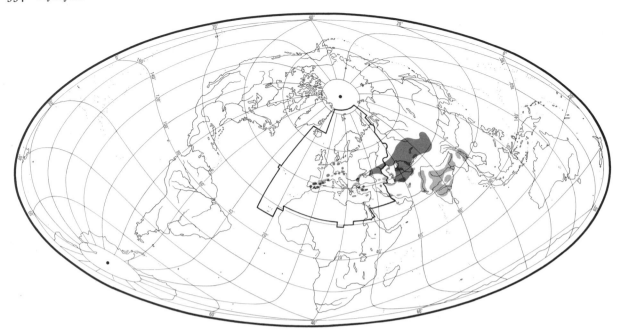

except for white side-patches. Sexes dissimilar, but no seasonal differences apart from eclipse. Juvenile resembles adult ♀, but just distinguishable.

ADULT MALE. Bill bright red; whole head and upper foreneck rich chestnut, shading to paler and more golden on crown. Mainly white flanks contrast with jet-black lower neck, breast, and stern; upper flanks barred dark and light brown. Upperparts and wings dull brown to grey-brown with white crescent (sometimes concealed) on shoulders, and white flight-feathers forming white patch; tail grey. Eyes red; legs orange-red to vermilion. In flight, exceptionally broad white stripe, often tinged pink, runs almost whole length of upperwing; from below, oval white flank patches contrast with black breast and central belly stripe, separated from white axillaries and underwings by brownish line along upper flanks. MALE ECLIPSE. Resembles adult ♀, but readily distinguished by all-red bill and redder eye; also has larger-looking head and usually more pronounced bushy crest, darker body, greyer forewing, and whiter wing-stripe. ADULT FEMALE. Sharp contrast between dark brown forehead, crown to eye-level, and nape, and grey-white lores, cheeks, and foreneck. Rest of upperparts dull, light brown shading to buff-brown on scapulars, and olive-brown on rump. Underparts also mainly light brown marked with yellowish on breast and barred with whitish on flanks, but belly greyer. Bill grey-black with pink-red edges and patch towards tip; eyes brown, becoming red-brown or dull orange-red in spring; legs pinkish with darker webs. In flight, wing-stripe nearly as conspicuous as ♂, but faintly tinged brown; forewing

also browner; white axillaries and underwing stand out against pale brown and greyish underbody; contrasting head pattern still conspicuous. JUVENILE. Closely resembles adult ♀, but darker with more mottled underparts which, however, are lost early in autumn; young ♂♂ have larger-headed appearance due to rudimentary erectile crest. Bill black-brown, becoming red during 1st winter; eyes brown, becoming red-brown or red during 1st winter; legs dull orange. ♂ assumes duller version of full plumage by April, with paler bill, duller eyes, shorter crest, and browner underparts.

♂ unmistakable: apart from bright bill and large, rounded head, and black, white, and brown body, breadth as well as length of upperwing stripe, and contrast of oval flank patches and black belly make flight identification easy. Red on ♀'s bill not always easy to see, but sharply contrasting pale cheeks and otherwise generally brown body preclude confusion with any other species except ♀ Common Scoter *Melanitta nigra*, which has stouter and shorter-looking bill, darker upperparts and breast, and no white in wings. ♀ Smew *Mergus albellus* also has sharply contrasting pale cheeks, and white in wings, but much smaller, with top of head chestnut-red, cheeks pure white, and upperparts and breast grey.

Found chiefly on fresh water with extensive reed cover, but also on open, brackish lagoons, and some coasts and inland seas. Not skulking, but usually wary. Swims characteristically higher on water than other diving ducks. Adapted to less essentially aquatic existence than *Aythya*. Dives easily, but more often simply submerges head or up-ends like dabbling duck *Anas*. Walks more easily than other diving ducks, with less rolling and more horizontal carriage; sometimes even grazes on land. Rises with difficulty, pattering along surface of water, but in air flies strongly with rapid wing-beats. Usually in pairs and small parties, but sometimes in larger flocks. Rather silent, but ♂ utters harsh, rasping wheeze in spring, and ♀ has hard, grating churr.

**Habitat.** In warm, central Asian main haunts primarily uses fairly large, moderately deep, reed-fringed eutrophic lakes, with ample open water and few or no flanking trees. Also occupies saline or alkaline lagoons and pools; not averse to reaches of slow-flowing rivers. Mainly lowland, but up to *c*. 500 m in Bavaria (Bauer and Glutz 1966). In west Europe, where some increase and spread in last 80 years, has adopted atypical and sometimes ephemeral habitats; often prefers for breeding quite small pools with ample submerged vegetation and open surfaces flanked by tall, dense stands of emergent aquatic plants and fringing trees. Sometimes uses same water for foraging, resting, and breeding, diving to *c*. 2–4 m. Shifts seasonally for moulting gatherings and to winter quarters, resorting mainly to

similar habitats, but some using inland seas (e.g. Caspian) and coasts of Mediterranean. Moves easily on land, but rarely far from water's edge; not so much on wing as many Anatinae. Not highly sensitive to human disturbance, but makes limited use of artificial waters.

**Distribution.** In western Europe, breeding often irregular and data incomplete, but some signs of decline 19th century and definite spread more recently. FRANCE: marked spread; first bred Camargue 1894, Forez 1896, Dombes 1910, Corsica 1972; irregular nesting elsewhere (FR). NETHERLANDS: first bred 1942 (Voous 1943). WEST GERMANY: bred Schleswig-Holstein since 1920 and Bavaria since 1957, with irregular nesting elsewhere (see Bauer and Glutz 1969). EAST GERMANY. Krakower Obersee: bred 1847 to *c*. 1898 and again from 1961. Lewitz: bred until at least 1924, now absent (Bauer and Glutz 1969). DENMARK: first bred *c*. 1940; spread to 10 localities, but since 1965 only one area occupied (TD). RUMANIA: first proved breeding 1958, but perhaps nested earlier (MT). GREECE: formerly bred Lake Artzon, drained 1920–2; introduced and bred Lake Aarus 1967–9; perhaps nests occasionally elsewhere (WB). ITALY.

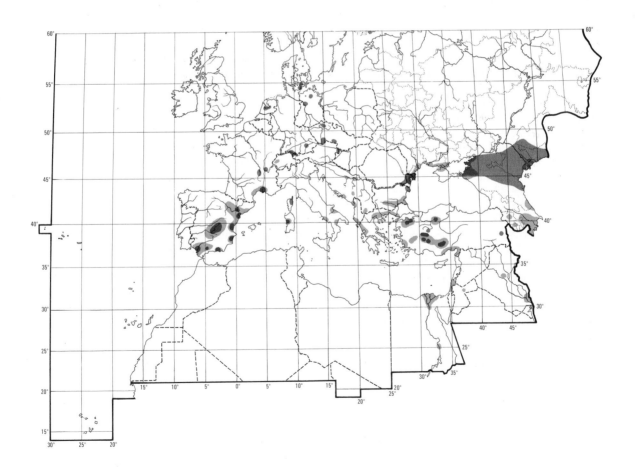

Occasional breeding since 1950 on mainland (SF). Sardinia: breeding (Bezzel 1957); bred 1969 and present 1972 (Demartis 1973). Sicily: now only accidental visitor (GS). SPAIN: spread to marismas of Andalucia 1935–40 (Valverde 1960). ALGERIA: bred 19th century and early 20th century, present status uncertain (Heim de Balsac and Mayaud 1962; EDHJ). MOROCCO: suspected breeding, but no proof (Heim de Balsac and Mayaud 1962).

Occasional breeding. BRITAIN: 1937, 1958, and most years since 1968; probably all escaped birds (British Ornithologists' Union 1971; Wildfowl Trust). POLAND: 1968–70 (AD, LT). BELGIUM: probably bred 1903, 1905 (Lippens and Wille 1972).

Accidental. Norway, Finland, Hungary, Portugal, Jordan (could be regular in winter), Cyprus, Malta, Libya, Tunisia. Britain and Ireland: see Movements.

**Population.** Breeding. FRANCE: 500–600 pairs, mostly Camargue, Dombes, and Forez (FR). NETHERLANDS: at least 25 pairs annually (JW). WEST GERMANY. About 100 pairs (Szijj 1973). Schleswig-Holstein: at least 25 pairs, Fehmarn, 1950; 56 pairs 1966. Bavaria: at least 11 pairs, Ismaning 1967; 30–40 pairs, Bodensee (Bauer and Glutz 1969). EAST GERMANY: c. 15 pairs (Isakov 1970a). DENMARK: c. 50 pairs (Isakov 1970a); maximum 5 pairs 1972 (TD). CZECHOSLOVAKIA: 25–45 pairs (Isakov 1970a). AUSTRIA. c. 10 pairs, delta of upper Rhine (Bauer and Glutz 1969). RUMANIA: increasing (MT). SPAIN. 3000–6000 pairs Ciudad Real; 200–400 pairs Valencia (Bernis 1972). Up to 100 pairs marismas, Andalucia, in wet seasons in 1950s; only small numbers 1971 (Ree 1973).

Winter. Estimated 10 000 birds in north-west European region and 50 000 birds in Europe–Black Sea–Mediterranean region (Szijj 1972). USSR: 409 800 birds in western parts in not so severe winters (Isakov 1970c).

Oldest ringed bird 7 years 2 months (Rydzewski 1974).

**Movements.** Migratory and partially migratory. Most or all leave breeding range north of 46°N; in west and central Europe, northernmost regular wintering places in Switzerland (usually 30 or less); infrequent winter records north and north-west Europe likely to include escapes. Winters north Mediterranean basin. Main western concentrations Spain, notably Ebro delta and Lake Albufera (Bernis 1966), and south France, especially Camargue (Jouanin 1970). In Balkans and east Mediterranean basin, thousands reported in recent winters in Rumania and Danube delta, north Greece, and Turkey (Johnson and Hafner 1970); no recoveries, but assumed to include Balkans and Asia Minor breeders, and presumably some Russian birds. Many winter south Russia, especially northern Black Sea, Sea of Azov, and Caspian; in latter, main concentrations eastern shores with 214 000 in January 1967 and 262 000 in January 1968 (Isakov 1970c), far more than occur anywhere to west. Further south, small to moderate numbers winter lower Egypt, Jordan, Iraq, and especially

Iran—where c. 12 500 in 1971–2, not restricted to south Caspian (D A Scott); a few to Maghreb. A few November–December records Oman and Bahrain (Seton-Browne and Harrison 1968; Gallagher 1971).

Rare but regular autumn migrant to south-east England c. 1952–62, occurrences apparently correlated with moult flocking in Netherlands (Pyman 1959; British Ornithologists' Union 1971); reversal to vagrant status coincided with decline of Danish breeding population (see Distribution). Danish and north-west German populations migrate south and south-west to Mediterranean France and Spain; those of west-central Europe appear to use same wintering areas since birds from Untersee recovered there and Italy. Czechoslovakian birds recovered south-east France and Yugoslavia (Hudec 1968). Populations of south France and Spain partially migratory (perhaps resident in latter); 2 young ringed Camargue moved to east Spain for winter.

Moult migration of adult ♂♂ and immatures begins early June, flocks of several hundred gathering in late summer. In west, moult Fehmarn Island or Kleiner Binnensee in north-west Germany, Zwarte Meer or Veluwemeer (now mainly latter) in Netherlands, and in south Germany and Switzerland in variable numbers on Ismaninger reservoir, Bodensee, and Untersee. Alternative sites and numerical fluctuations perhaps indicative of developing moult-migration tradition. Distances travelled to moult areas uncertain, though one Untersee bird found Netherlands in September, while recoveries indicate Camargue birds moult on Bodensee (Mayaud 1966). New influx of moulted birds to Bodensee area October, when up to 3000–4000 (7000 in 1951) though less in recent years, departing November (Szijj 1963); such numbers exceed size of north and central European breeding populations, and some must originate further east (Salomonsen 1968).

Main autumn migration in west and east late October and early November, though ringed Danish and south German birds have reached Spain and south France by 2nd half September; by December, most in winter quarters. Return movement February–March; most northern and eastern breeding areas re-occupied April and early May. See also Bauer and Glutz (1969); Dementiev and Gladkov (1952).

**Food.** Mainly vegetation, obtained by diving and by dabbling on surface—on Untersee, West Germany, approximately equal time spent using each method (Szijj 1965c)—up-ending, and foraging with head immersed. Normally only one feeding method used at a time within flock. Daytime feeding generally, especially early morning and evening, in pairs, small parties, and large flocks. Dives, usually preceded by leap, mainly in 2–4 m (Dementiev and Gladkov 1952). On Untersee, dives 6–10 s, maximum 13·7 (Szijj 1965c).

Specific data on food items sparse. Mainly stems, leaves,

roots, seeds, buds of aquatic plants (especially Characeae), and pondweed *Potamogeton*, milfoil *Myriophyllum*, and hornwort *Ceratophyllum*. Occasionally aquatic insects and larvae, small fish, frogs (adults, tadpoles, and spawn); also crustaceans, and molluscs, but importance as food not known as possibly taken accidentally or as grinding agents. Characeae most frequently recorded food; *Chara aspera*, *C. ceratophylla*, and *C. foetida* identified (Naumann 1905; Millais 1913; Poncy 1924; Witherby *et al.* 1939; Bauer and Glutz 1969).

2 stomachs Untersee in September and November contained only *Chara* (Szijj 1965c). Similarly, wintering birds from Caspian Sea, almost exclusively eat *Chara* in Krasnovodsk Bay; in Lenkoran, chiefly *Chara* (89·7%), less hornwort, and seeds (10·3%) mainly *Bolboschoenus* (Isakov and Vorobiev 1940; Dementiev and Gladkov 1952). Summer foods even less known. On Baraba steppe lakes, USSR, mainly pondweed leaves and tops of hornwort and milfoil (Dementiev and Gladkov 1952). In London parks in April, semi-wild ♂♂ brought alga *Rhizoclonium hieroglyphicum* to surface for ♀♀, and in June ♀♀ brought same alga to surface for young (Gillham 1955).

**Social pattern and behaviour.** For general information, see especially Witherby *et al.* (1939), Bannerman (1950), Géroudet (1965), Bauer and Glutz (1969).

1. Highly gregarious most of year. Outside breeding season, in main wintering areas, often in large flocks (see Movements). ♂♂ often greatly outnumber ♀♀ in both autumn and winter flocks; ♂♂ first to flock (late May, June). BONDS. Monogamous pair-bond of seasonal duration; stronger and more prolonged than in *Aythya* ducks. Some indication also of promiscuous tendencies as in many *Anas* ducks, though this may mostly involve unmated ♂♂. Pair-formation starts in autumn in flock (see Lind 1961), but many birds do not obtain mates until late winter or spring because of unbalanced sex-ratios in winter. ♂ deserts ♀ at later stage than *Aythya* ♂♂, patrolling nest area when ♀ laying and incubating, accompanying her when she leaves eggs (Jauch 1954; Schifferli 1952). Usually departs before hatching and ♀ typically rears young alone though ♂ sometimes stays near ♀ and brood (Wüst 1963). Parent-young bond stronger and more prolonged than in *Aythya*, lasting at least 11 weeks; indications that it may persist in some cases after fledging. BREEDING DISPERSION. Little information. Nests usually well dispersed but sometimes as close as 30 m. ♂ defends ♀ in presence of other ♂♂; otherwise ♂♂ peaceful in waiting areas in vicinity of nests. ROOSTING. Communal for much of year, usually on water out in open but sometimes rests on shore. Roosts both diurnally and nocturnally, depending in part on local conditions. Frequently engages in nocturnal courtship, especially at dusk and by moonlight, when grey parts of plumage conspicuous (Steinbacher 1960).

2. Studied in some detail but mainly in captivity. See Lind (1958, 1962); also Steinbacher (1960); Johnsgard (1965), Veselovský (1966). Shows features of both *Anas* and *Aythya*, especially in copulatory displays. FLOCK BEHAVIOUR. When feeding in tight-knit group, all birds tend to use same feeding method (Szijj 1965b). Main pre-flight signals rapid, repeated Chin-lifting (involving head only); additional Head-flicks of Pochard *A. ferina* lacking (McKinney 1965b). Head-pumping

movements said to be indicative of intention to dive (Lind 1965b). When alarmed, birds stretch necks fully vertical, calling continuously. ANTAGONISTIC BEHAVIOUR. Generally peaceable, though hostility may break out when birds (particularly ♂♂) in close proximity, especially during or after communal courtship in spring. Threat takes 3 main forms (see Lind 1962); in all cases, feathers depressed. In Forward-display, both sexes silently stretch neck and head horizontally above water, bill usually wide open. In Downward-sneeze, ♂ draws head into neck and points bill obliquely down, uttering Sneeze-call usually with bill in water; unlike Sideways-Sneeze of courtship (sometimes also directed at other ♂♂ in apparent threat), never preceded by Head-flick. In Attack-intent display, ♂ holds head low and somewhat thrust forward as Threat-call given. Latter also uttered irregularly throughout antagonistic encounters. Threat sometimes followed by rush across water at rival; then often bathes (Lind 1959b). Inciting-display of ♀ involves alternation of (1) threat and forward tossing movement of head at rejected ♂ and (2) component directed towards favoured ♂ (see below). ♀ also performs *Anas*-like Repulsion-behaviour when chased by ♂♂ attempting rape (Johnsgard 1965). Rape attacks by 1 or more ♂♂ on ♀♀ other than mate common in spring well after start of normal pair-copulations and at time when overall aggressiveness of ♂♂ increases. COMMUNAL COURTSHIP. From autumn to spring—though Courtship-calls of ♂♂ given occasionally at other times, even during moult (Platz 1963). Usually on water. Typically, 5–20 ♂♂ crowd round and follow 1 ♀, though more ♀♀ present at times; can also involve single ♂ and ♀, other ♂♂ often then joining in. Bill-dipping, full Ceremonial-drinking, and other activities by ♂♂ often precede courtship proper; in April, Head-pumping also quite frequent (Lind 1962). Courtship-intent posture of ♂ much as in Mallard *Anas platyrhynchos*; ♂ also often erects crest during courtship. Secondary displays include Upward-shake (with Tail-wag), Head-flick, and less frequent Wing-flaps. According to Johnsgard (1965), 3 of the 4 major displays each accompanied by its own Courtship-call. In Sideways-sneeze (see A), ♂ withdraws neck, lowers head and turns it in direction of ♀; with quick lateral movement of head, then jerks crown toward ♀ while uttering Sneeze-call with mandibles further apart. Sideways-sneeze probably homologous to Burp-displays of *Anas* and Kinked-neck display of *Aythya*. In incomplete Sneak-display (Crouch-display), ♂ lowers crest, points bill towards ♀ while stretching neck out over water, and gives nasal call; thus resembles behaviour described by Lind (1962) as Attack-intent threat directed by one ♂ at another (see above). In Neck-stretch display, ♂ elevates head and raises crest while calling; this resembles behaviour when alarmed. In final major display, ♂ ceremonially Leads ♀ while performing Turn-back-of-Head, exhibiting darkened patch on nape formed by full erection of crest. Bouts of courtship tend to proceed in 3 phases (Lind 1962): (1) repeated series of secondary displays; (2) these become less frequent as major displays, particulary Sideways-sneeze and ceremonial Leading, predominate; (3) ♂♂ Ceremonially-drink and sometimes perform Neck-stretch display. Secondary displays also immediately introductory to major displays; in particular, Sideways-sneeze almost invariably preceded by Head-flick. Sideways-sneeze frequently leads to attempts at ceremonial Leading; while swimming after ♀, ♂♂ often form group side by side with heads slightly raised and inclined forward, at times taking on character of excited chase across water after ♀. During courtship, ♀ may show ambivalent behaviour, attacking and fleeing, tendency to escape taking form mainly of flight-intent movements (Chin-lifting) or dive-intent movements (Head-pumping). Incites by alternately threatening

1 or more rejected ♂♂ and repeatedly turning in Neck-stretch posture away from them and towards favoured ♂; accompanied by calling and exaggerated Chin-lifting. ♂♂ often respond to attack of ♀ by Turning-back-of-Head. PURSUIT-FLIGHTS. Aerial chases of ♀ by 2–7 (occasionally up to 15) ♂♂ frequent from 2nd half April, lasting until mid-July with peak mid-May to mid-June (Jauch 1953; Lebret 1952; Lind 1962). Commonly follow intensive swimming chases during which ♂♂ perform Sideways-sneeze (Lind 1962); also initiated when ♂ swims towards strange ♀ causing her to fly away, ♂ following and other ♂♂ joining in. All pursue ♀ in flight, sometimes performing aerial version of Sideways-sneeze with wings held stiffly down while lowering head and neck (Lebret 1955). In air, ♀ calls repeatedly; tries to evade ♂♂ and will peck at nearest. ♀ and pursuers alight close together afterwards; then usually peaceable (Lind 1962). Though usually interpreted as rape-intent flights, best considered as form of courtship-flight. PAIR BEHAVIOUR. During communal courtship, ♂ displays to mate in presence of other ♂♂. Otherwise, pair-courtship usually consists of quite different activities, including Courtship-feeding (Gillham 1955; Buxton 1962; Platz 1964). Typically, ♀ waits on surface while ♂ dives and brings up food or occasionally inedible objects such as twigs. ♂ either then waits for ♀ or swims to her; may place stuff on water or present it to ♀, not eating offering himself (if at all) until ♀ does. Not associated with other displays. Evidently confined to well-established pair; mainly March–June (Jauch 1949; Platz 1963) but recorded from February (Buxton 1962) and even November (King and Prytherch 1963). Other pair-displays include Preen-behind-Wing. Unlike in *Anas*, as frequent in ♀♀ as in ♂♂; often performed by ♀ towards mate or potential mate (Johnsgard 1965). Especially after antagonistic encounters with other birds, pair often greet one another by Ceremonial-drinking and Preen-behind-Wing (Lind 1962). ♀ will Incite next to mate when approached by strange ♂♂. ♂ separated from mate may perform Lure-sneezing: utters Sneeze-call with head stretched slightly forwards or sideways; if ♀ in view, ♂ directs call at her until she joins him. Corresponding Decrescendo-call of ♀, as in *Anas*, reported but details lacking (see Johnsgard 1965). Pre-copulatory displays and full copulatory sequences occur from November (Lebret 1961; Lind 1962), so clearly also part of pair-courtship. COPULATION. Pre-copulatory displays include *Anas*-like Head-pumping by one or both sexes. According to Johnsgard (1965), ♂

initiates sequence by Bill-dipping and Preening-dorsally, interspersing these with Preen-behind-Wing, preening elsewhere, lateral Head-shaking, and Head-pumping. Characteristic sequence of displays: series of *c.* 3 Bill-dips→Head-flick→Preen-dorsally (McKinney 1965*b*). If ♀ unresponsive, ♂ sometimes performs Lure-sneezing. ♀ usually also Head-pumps, gradually assuming Prone-posture with body and tail low in water but neck stretched forward diagonally (Steinbacher 1960; see B). Post-copulatory displays of ♂ consist of single Sideways-sneeze (see A) directed at ♀ followed by swimming round in Bill-down posture. ♀ appears typically to bathe after being released by ♂, but Nod-swimming once observed by Lind (1958). Rapes not preceded by display. RELATIONS WITHIN FAMILY GROUP. Little information. ♀ brings food to surface for young (Gillham 1955).

(Fig A after Johnsgard 1965; Fig B after Steinbacher 1960.)

**Voice.** Generally silent during summer outside period of courtship display. Calls of ♀ especially require further study. Non-vocal sound produced during Preen-behind-Wing display by rapid drawing of bill over feather-quills (McKinney 1965*b*).

CALLS OF MALE. 3 main calls, given during courtship as well as at other times: (1) hoarse, relatively loud, repeated 'bät', uttered on water and in flight throughout year; (2) quiet 'geng', usually uttered in phases of several notes one after the other, occasionally in August but most frequently from October–May; (3) monosyllabic, quite far-carrying 'baix', uttered during same period as last (Steinbacher 1960). Call 1 given during any kind of excitement, e.g. as contact-call, on appearance of enemy, during attack on another ♂, and during low-intensity courtship. Evidently also call given during Attack-intent and Sneak displays; described as weak 'chrü' (Lind 1958, 1962) and as nasal note (Johnsgard 1965), may also be described as quick, harsh, resonant 'chik' (see I) often repeated 4–5 times. Call 2 given during courtship and whenever birds disturbed; thus warning or anxiety note, and typically associated with Neck-stretch display. Also described as soft, continuously repeated 'chrüüü' (Lind 1962). Call 3, Sneeze-call, given during various Sneeze displays. Described as 'loud, hard, rasping wheeze' (see Witherby *et al.* 1939), also rendered as loud, di-syllabic 'chi-zick' (see II), 'chriib' (Lind 1962), and likened to half-repressed sneeze (Johnsgard 1965).

CALLS OF FEMALE. 2 main calls, subject to considerable variation, recognized by Steinbacher (1960): (1) 'gock', and (2) 'kurr'. Call 1 used in full threat and as Inciting-call; also described as soft 'guk-guk' (Lind 1958), and 'quack' (Heinroth 1911). Call 2 usually given during quarrels with own or other species; more of an anxiety note than call 1. Seems to be same as hoarse, grating 'kur-r-r' of Witherby *et al.* (1939). When given while escaping from ♂♂ in rape-intent flights, described as hoarse, trisyllabic, almost grating croak (see Bannerman 1958), repeated 'rro' (Lind 1962), and 'wrah-wrah-wrah' or 'wu-wu-wu' (Bauer and Glutz 1969) like distant barking of dog. ♀ may also have distinctive Decrescendo-call as in *Anas* (see Behaviour). ♀ tending small young calls them to her with soft 'kock' (Steinbacher 1960).

A

B

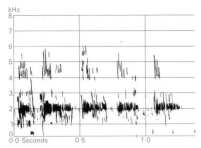

I Sture Palmér/Sveriges Radio

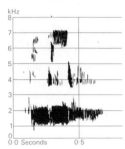

II Sture Palmér/Sveriges Radio

CALLS OF YOUNG. Pleasure-call: weak, low-pitched trilling repeated 2–5 times; also uttered as greeting. Anxiety-call: shrill piping 'pii-pii-pii' (Kear 1968).

Breeding. SEASON. See diagram for central and southern Europe. In southern USSR, laying starts mid-May (Dementiev and Gladkov 1952). SITE. On ground in dense vegetation, often deep in bush; in dense reed and rush beds. Never far from water, occasionally on matted reeds in water. Nest: depression, lined grass, leaves, rushes, and down. Average diameter 28–45 cm, depth of cup 10–20 cm (Kux 1963). Building: by ♀, using materials within reach of nest. Cup formed by turning movements of body. EGGS. Broad, rounded; light stone, occasionally tinged green. 58 × 42 mm (53–62 × 39–45), sample 150 (Schönwetter 1967). Weight of captive laid eggs 56 g (47–69), sample 101 (J Kear). Clutch: 8–10 (6–14). Dump nests by 2 or more ♀♀ not infrequent, up to 39 eggs reported in one (Hellebrekers and Voous 1964). Nest parasitism often recorded, with eggs (singles up to full clutch) laid in nests of Mallard *Anas platyrhynchos*, Gadwall *A. strepera*, and others. Some dump and parasitic laying takes place even after incubation has been started on nest by another ♀ (Hellebrekers and Voous 1964). One brood. Replacement clutches laid after egg loss. INCUBATION. 26–28 days. By ♀. Starts with last egg; hatching synchronous. Eggs covered with down when ♀ off. Eggshells left in nest. YOUNG. Precocial and nidifugous. Self-feeding. Cared for by ♀, brooded when small. FLEDGING TO MATURITY. Fledging period 45–50 days. Independent at about same time. First breeding at 1 year, though some probably not until 2. BREEDING SUCCESS. Of 13 German broods: mean size at 1 week old 6·3; at 2 weeks 5·5; at 3 weeks 4·8; and near fledging 4·4 (Bauer and Glutz 1969).

Plumages. ADULT MALE BREEDING. Head and upper neck vinous-chestnut; crown cinnamon, fading to pale yellow when worn. Sides of head often slightly tinged blue-grey. Hindneck, lower neck, sides of mantle, back, rump, tail-coverts, chest, breast, centre of belly, and vent black, glossed green on rump and tail-coverts, slightly purple elsewhere. Centre of mantle dull grey-brown, scapulars drab, broad crescent across upper scapulars white. Flanks and sides of belly white, bordered above and behind by cinnamon. Tail-feathers dark grey, tinged silver-grey. Primaries white, tips of all and outer edges of outer 4–5 black. Secondaries white, suffused grey-brown to black near tips, especially outer; tips narrowly edged white. Some inner secondaries and all tertials dark ash; inner tertials tinged cinnamon-olive. White of flight-feathers often suffused pale cinnamon-pink on outer webs. Marginal upper wing-coverts white, rest of upperwing dark ash or dark olive-grey with cinnamon tinge. Underwing and axillaries white. ADULT FEMALE BREEDING. Forehead and crown down to eye cinnamon-brown to red-brown. Rest of head and foreneck pale grey somewhat mottled brown on lores. Hindneck and mantle cinnamon-brown, scapulars, sides of breast and flanks buff-brown. Back dark brown, rump and upper tail-coverts pale buff-brown. Chest, breast, belly, and vent pale grey-brown, feathers tipped buff on chest, pale grey elsewhere. Under tail-coverts grey-brown, tipped white; longest white. Tail and wing like adult ♂, but white in marginal wing-coverts restricted, upper wing-coverts paler, cinnamon-grey; wing-bar suffused pale grey. In worn plumage, tips of feathers of scapulars and underparts fade to pale grey or white; cheeks and throat white. ADULT MALE NON-BREEDING (Eclipse). Like adult ♀ breeding, but upperparts and flanks darker brown; feathers of underparts darker grey-brown, more contrastingly tipped buff or white on chest and breast. Wing like ♂ breeding. (See also Bare Parts). ADULT FEMALE NON-BREEDING. Like adult ♀ breeding, but upperparts and flanks darker brown and feathers of breast and belly broadly tipped white. DOWNY YOUNG. Crown, hindneck, upperparts, and sides of body olive-brown; longer tufts of mantle paler yellow-olive; tinged grey on flanks. Spots on wing, at sides of back behind wing, at sides of rump, and on flanks yellow. Narrow streak behind eye brown; streak over eye, rest of head, and underparts pale yellow; in some, sides of head tinged buff; centre of belly and vent white. JUVENILE. Like adult ♀ breeding, but feathers of crown and cheeks short, crown dull cinnamon-grey, cheeks mottled grey-brown; upperparts and sides of body duller and greyer, less cinnamon-buff; feathers of underparts short and narrow, tipped buff on chest and sides of breast, and narrowly tipped white on belly and vent. Tail-feathers with bare shaft

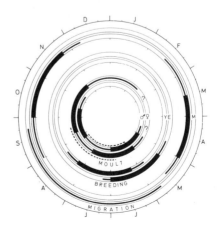

projecting. Wing like adult, but greater upper wing-coverts narrow and with rounded tips, instead of broad and square. Sexes similar, but ♂ more white on leading edge. SUBSEQUENT PLUMAGES. Like respective adults. During early autumn, limited part of juvenile plumage changed for 1st non-breeding (feathers like adult non-breeding). FIRST BREEDING. Soon develops, often complete by December; juvenile feathers on belly and vent moulted last. ♂ in 1st breeding resembles adult ♂ breeding, but feathers of belly and vent duller back with more extensive grey-brown bases and narrow white tips, appearing mottled; occasionally part of 1st non-breeding plumage retained on underparts until late spring. ♀ in 1st breeding like adult ♀ breeding, but feathers on crown and cheek shorter; crown and upperparts less cinnamon, greyer brown.

**Bare parts.** ADULT MALE. Iris bright red. Bill deep pink to bright vermilion; nail pink. Foot orange-red to vermilion, joints and webs dull black. ADULT FEMALE. Iris red-brown, darker brown in late summer to winter. Bill dark horn-brown, band behind nail deep pink to purple-red, cutting edges and tip of lower mandible orange-pink. Foot dull pink to orange-flesh, webs and joints dark slate-grey or brown. DOWNY YOUNG. Bill dark brown, tip pink, lower mandible pink or orange. Foot dark grey, dull black at joints; front of tarsus and streaks on webs along toes yellow or pink. JUVENILE. Iris yellow-brown; in early autumn, red-brown in ♂, brown in ♀; later in autumn, like adult but 1st-year ♂ not so bright red. Bill dusky brown with pink tip and cutting-edges; attains adult colour during autumn (but ♂ paler pink). Lower mandible of ♂♂ usually with base and skin between rami black; dusky skin often persists until 2nd year, and at least sometimes until 5th. Foot dull pink or flesh; brighter (pale orange-red) in ♂ from autumn; webs and joints dull black. (Dementiev and Gladkov 1952; Witherby *et al.* 1939; RMNH, ZMA, C S Roselaar.)

**Moults.** ADULT POST-BREEDING. Complete; flight-feathers simultaneous. Head and body first, from May in ♂, shortly after nesting in ♀; underparts, back, and rump often not yet completed when wing starts. Flightless for *c.* 4 weeks. Moult of wing and tail June–August in ♂, *c.* 1 month later in ♀. ADULT PRE-BREEDING. Partial; head and body. Underparts, back, and rump first; starts late August, when wing full-grown, overlapping with post-breeding moult of underparts and back. Followed by head and upperparts; completed mostly before November. POST-JUVENILE MALE. Partial; head, body, tail, and tertials. Between late July and October, head, flanks, scapulars, mantle, and chest attain variable amount of 1st non-breeding plumage; sometimes also part of breast, vent, and tail-coverts. From September, 1st breeding develops; usually all remaining juvenile (except part of belly and back in some) changed for breeding before November, and much of 1st non-breeding before December. Tail moulted

August–December, sometimes completed September. POST-JUVENILE FEMALE. Like post-juvenile ♂, but generally slower; some juvenile feathers retained on body or tail until December or even spring. Probably only limited amount of 1st non-breeding develops (difficult to separate from 1st breeding). SUBSEQUENT MOULTS. Like those of adult, but 1st post-breeding more prolonged, less complete; overlaps greatly with prolonged pre-breeding. (Dementiev and Gladkov 1952; C S Roselaar.)

**Measurements.** Whole year, Netherlands. Skins (RMNH, ZMA).

| | | | | | | |
|---|---|---|---|---|---|---|
| WING AD | ♂ 264 | (5.21; 16) | 255–273 | ♀ 260 | (7.28; 14) | 251–275 |
| JUV | 257 | (5.16; 7) | 250–264 | 248 | (7.11; 10) | 237–259 |
| TAIL AD | 70.6 | (3.16; 15) | 67–76 | 68.4 | (3.27; 15) | 62–74 |
| BILL | 48.2 | (2.04; 25) | 45–52 | 46.6 | (2.14; 26) | 42–50 |
| TARSUS | 44.1 | (1.37; 25) | 42–47 | 42.2 | (1.46; 26) | 40–45 |
| TOE | 66.0 | (2.18; 25) | 62–70 | 63.5 | (2.19; 25) | 58–67 |

Sex differences significant, except adult wing and tail. Juvenile tail on average 9 mm shorter than adult. Juvenile wing and tail significantly shorter than adult. Other measurements similar; combined.

**Weights.** ADULT. (1) February: south-west Caspian, USSR. (2) March: Kazakhstan, USSR. (3) May–September: various parts USSR and few data Netherlands; combined. (4) October: Kazakhstan.

| | | | | | | |
|---|---|---|---|---|---|---|
| (1) | ♂ 1135 | (—) | 900–1170 | ♀ 967 | (—) | 830–1320 |
| (2) | 1130 | (—; 17) | 990–1300 | 1100 | (—; 12) | 1000–1200 |
| (3) | 1167 | (112; 8) | 1020–1347 | 1146 | (122; 9) | 1020–1300 |
| (4) | 1320 | (—; 11) | 1200–1420 | 1220 | (—; 6) | 1100–1400 |

Range November–December, USSR: ♂♂ 1500–1550, ♀♀ 1350–1500; limited data Netherlands similar. JUVENILE. October, Kazakhstan: 18 unsexed 1170 (950–1350). November–December, USSR: ♂♂ 1350–1500, ♀♀ 1000–1200. (Netherlands: ZMA. USSR: Dementiev and Gladkov 1952; Dolgushin 1960). FULL-GROWN. Numerous data Camargue, France, indicate generally lower weights: ♂♂ 721–1195, ♀♀ 694–1205 (details in Bauer and Glutz 1969).

**Structure.** Wing rather long, pointed. 11 primaries: p10 and p9 longest (about equal), p8 7–12 shorter, p7 17–24, p6 31–42, p1 100–115. Inner web of p10 and outer p9 emarginated. Tail short, slightly rounded; 16 feathers. Bill slightly longer than head, about as high as broad at base, slightly narrowing towards flat tip. Culmen slightly concave. Cutting edges straight; nail rather broad, oval. Feathers of crown, nape, and, to lesser extent, cheeks elongated, forming dense crest in breeding ♂, somewhat shorter in adult breeding ♀ and adult eclipse ♂. Toes relatively short compared to other diving ducks but, like these, hind toe broadly lobed. Outer toe *c.* 98% middle, inner *c.* 79%, hind *c.* 30%.

CSR

# *Aythya ferina* Pochard

PLATES 76 and 99
[after pages 566 and 686]

Du. Tafeleend    Fr. Fuligule milouin    Ge. Tafelente
Ru. Красноголовый нырок    Sp. Porrón común    Sw. Brunand

*Anas ferina* Linnaeus, 1758

Monotypic

**Field characters.** 42–49 cm, of which body two-thirds; wing-span 72–82 cm. Medium-sized, short-necked, stocky diving duck with high crown, long sloping forehead, and long, broad bill; ♂ pale grey with chestnut-red head and neck, and black breast and stern. Sexes dissimilar, but only small seasonal differences apart from eclipse. Juvenile closely resembles adult ♀, but distinguishable.

ADULT MALE. Uniform' chestnut head and neck demarcated from black forepart of mantle, lower neck and breast, which in turn contrast with vermiculated pale grey body; black wedge round tail also stands out. In flight, broad, pale grey stripe running whole length of wing, though less well-defined on primaries, is same colour as back and contrasts with darker grey forewing; white under wing-coverts and light underbody stand out against black breast and under tail-coverts. Bill dark grey with broad, pale blue-grey band across middle; eyes reddish-orange becoming bright red in spring; feet dull grey with blackish webs. MALE ECLIPSE. Superficially like adult ♀, but fairly easily separated by more uniform golden-brown head and neck, and greyer back. ADULT FEMALE. Head, neck, forepart of mantle, and breast mainly yellowish-brown, but rather variable in depth of colour; dark brown on crown and distinctive face pattern with hoary areas round base of bill, cheeks, and throat, and (sometimes absent) grey-white streak behind eye; breast colour variably broken up by broad white feather tips. In winter, back, scapulars, and sides coarsely vermiculated grey-brown; centre of lower breast whiter; upper tail-coverts dark brown with white flecks; under tail-coverts greyer and more plentifully marked with white. In summer, back and scapulars rich dark brown with grey tinge; sides yellower brown with indistinct vermiculations. In flight, wing-pattern as ♂, but back not paler than wing-coverts and underparts not so clearly defined. Bill similar to ♂, but duller; eyes brown with reddish tint in summer; feet green-grey with dark grey webs. JUVENILE. Closely resembles winter adult ♀, but cheeks and sides of neck paler, no white streak behind eye, upperparts and sides less vermiculated grey, upper tail-coverts blacker, and lower breast and belly barred and mottled grey and white or buff. Juvenile ♀ even less grey than juvenile ♂. Some immature ♂♂ nearly in full plumage, but rather duller than adults, by October; others not until January or February.

Chestnut, black, and grey ♂ unmistakable among west Palearctic duck, but note that 2 American species have similar pattern: ♂ Canvasback *A. valisineria* larger and paler with blackish wash on head, even flatter forehead, and much longer, plain black bill; ♂ Redhead *A. americana* intermediate in size, with much darker grey back, shorter bill, rounded head, and yellow eyes. Eclipse ♂ *A. ferina* distinctive on uniform head colour, brown front, greyish back, and head shape. ♀'s combination of sloping forehead, grey-banded bill, hoary front to face, and (when present) line behind eye preclude confusion with any other west Palearctic duck, as do brown foreparts, mainly greyish body, darker undertail, and grey wing-stripe. ♀ Ring-necked Duck *A. collaris* has not dissimilar head pattern and grey wing stripe, but different head shape, shorter grey bill with white band behind nail, more clearly defined white round eyes, and darker back and sides. ♀♀ *A. valisineria* and *A. americana* show differences in size, head shape, bill, and back colour comparable to those of ♂♂. Some *A. ferina* × Tufted Duck *A. fuligula* hybrids closely resemble *A. ferina* (see Gillham *et al.* 1966).

In breeding season, mainly on medium-sized sheets of water with extensive reeds and similar vegetation, but in winter also on large lakes and reservoirs without emergent cover, and sometimes on tidal estuaries, but seldom on open sea. Swims rather low in water with tail trailing surface. Often dives with marked jump and, unlike other west Palearctic *Aythya*, not uncommonly paddles feet before diving; occasionally up-ends in shallow water. If disturbed, usually swims rather than flies away. When taking flight, patters along surface for some distance with more laboured action than other *Aythya*, but full flight strong and fast with rapid wing-beats producing whistling sound; heavy body and rather short wings give clumsy appearance. Markedly gregarious and flocks usually fly as compact groups, but over longer distances may straggle in long lines or even V formation. In summer, not infrequently stands or sits on shore and walks fairly easily for short distances with more horizontal carriage than most other *Aythya*. Generally rather quiet, but ♂ has soft whistle and ♀ harsh growl.

**Habitat.** Basically a steppe-dweller, stopping short of tundra and only thinly inhabiting taiga and boreal forest zones. In Europe has colonized new temperate habitats, extending beyond continental into oceanic climates, without substantial relaxation of strict requirements for breeding habitat. As predominantly bottom-feeding diver, normally at depths of 1–2·5 m, requires several hectares of open water uncluttered with floating vegetation but prolific in submerged plant and animal food. Prefers belts of tall, dense, fringing, aquatic vegetation and especially

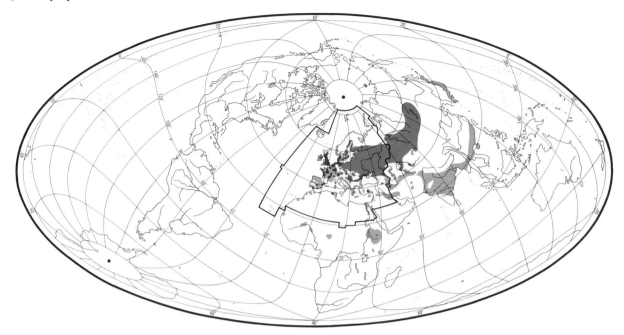

overgrown islets, inaccessible to most predators, affording choice of sheltered nest-sites. Ideal breeding habitat scarce, and this constraint apparent in pattern of breeding distribution. In high density situations, frequent instances of overflow on to small pools, as well as on to such man-made water bodies as reservoirs, ornamental ponds, and fishponds. Nests only exceptionally beyond 5 m from water's edge, making minimal movement on land at all seasons. In steppe situations, saline, brackish, or soda lakes, and inundations used fairly freely. Towards Atlantic, acid oligotrophic waters sometimes tolerated; also sluggish reaches of rivers and estuaries. Avoids typical mountain habitats, but in central Europe occupies pools up to altitudes of 500–800 m, and in Tibet even to 2600 m.

Although at all seasons strongly preferring fresh water, will shift to coastal and inshore maritime habitats when driven by frost or other compelling factors, and will at times even breed in sheltered marine bays. Freely uses excellent powers of flight, and will rise on occasion to considerable altitudes. Towards man, markedly opportunist; especially wary at moulting places but often readily approachable even tame where confidence built up by familiarity with human presence.

**Distribution.** Considerable expansions of range in north and west Europe in last 130 years. Although situation confused by sporadic and irregular breeding (while retreats as well as expansions have occurred), 3 main phases may perhaps be distinguished. (1) In 19th century colonized Swedish islands from 1849, spread to Swedish mainland and Finland in next 20 years or so; then to Denmark, and, at the end of 19th century, Netherlands and southern Bavaria. (2) First 20–30 years of 20th

century: further spread and increases in Sweden, and also Britain, central Europe, north USSR, and France. (3) Since 1945: further extensions, especially in East and West Germany, Belgium, and France. (See Bezzel 1969 for summary and map; Bauer and Glutz 1969.) Expansions of range originally ascribed to drying up of shallow breeding lakes in central Asia (Lönnberg 1924; Kalela 1940), or other aspects of climatic amelioration, but more recently to growth in artificial waters (reservoirs, fishponds, etc.) suitable for breeding, increasing eutrophication, and reduced disturbances (Bezzel 1969; Bauer and Glutz 1969; von Haartman 1973).

ICELAND: first bred 1954 (Gardarsson 1975). BRITAIN AND IRELAND. Marked increase and spread inland Scotland and Ireland still continuing (Alexander and Lack 1944). Continued increase England, particularly Kent, but probably decreased Scotland (Parslow 1967). Ireland: first bred 1907, regularly since 1930 Roscommon; recent further spread suggested by BTO Atlas enquiry (1968–72). FRANCE: probably first colonized early 20th century; increasing constantly last 20 years or so, and expanding to west and north (FR). BELGIUM: breeding regularly since 1956; expansion since 1965 (Lippens and Wille 1972). NETHERLANDS: first recorded breeding early 19th century, widespread but local early 20th century, then reduction due to habitat deterioration; marked spread and increase since 1950 (JW). DENMARK: first bred Jutland 1862 (Bauer and Glutz 1969). SWEDEN: first bred Gotland 1849 and Öland 1855, extending to mainland in 1860s (Bauer and Glutz 1969). FINLAND: first bred 1867, with further wave from 1889, continuing to increase until 1950 (von Haartman 1973). WEST AND EAST GERMANY: considerable changes and fluctuations, with many new or

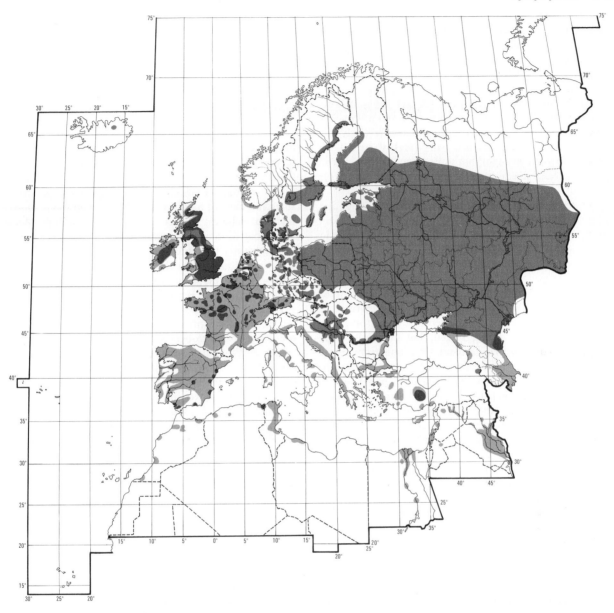

re-occupied sites since 1957 (Bezzel 1969); see also Bauer and Glutz (1969). POLAND: in 19th century, known in many regions only as migrant; now breeding, fairly numerous in all lowland areas (Tomiałojć 1972). SWITZER-LAND: first nested 1952; breeding regularly since 1970 and perhaps earlier (WT). AUSTRIA: first proved breeding 1957; spreading and increasing (Bauer and Glutz 1969). USSR. Expansion in some northern areas e.g. Moscow region, after 1900 (Dementiev and Gladkov 1952). Estonia: range expanded in 1930s (Kumari *et al.* 1970). ITALY: now probably breeding regularly in north, mainly Po valley (SF). SPAIN: breeding sporadic in most areas shown (FB). GREECE: bred Crete 1953; perhaps nesting Macedonia and Thrace (WB). TUNISIA: formerly bred in

north, but no recent evidence of nesting (MS).

Accidental. Faeroes, Azores, Canary Islands, Madeira, Cape Verde Islands, Rio de Oro, Kuwait.

**Population.** Breeding. ICELAND: only a few pairs (FG). BRITAIN AND IRELAND. *c.* 250 pairs (Parslow 1973). Britain: *c.* 200 pairs (Yarker and Atkinson-Willes 1971). FRANCE: 1700–2300 pairs (Jouanin 1970). BELGIUM: *c.* 80 pairs, increasing (Lippens and Wille 1972). NETHERLANDS: 110 pairs (Bezzel 1969); current estimate 650–750 pairs (CSR). DENMARK: 350–700 pairs (Fog and Joensen 1970). SWEDEN: estimated 1000 pairs (Isakov 1970*a*); decreased 1954–64, apparently stable since (Curry-Lindahl *et al.* 1970). FINLAND: *c.* 5000 pairs (Merikallio 1958). WEST

GERMANY: 2000 pairs (Szijj 1973); nearly half in Bavaria (Bezzel 1969). EAST GERMANY: 1500 pairs (Isakov 1970a); strong increase last 5–10 years (Rutschke 1970). CZECHO-SLOVAKIA: 2500 pairs (Bezzel 1969); only sparse breeder mid-19th century (Bauer and Glutz 1969). AUSTRIA: under 100 pairs (Bezzel 1969). SWITZERLAND: 5–6 pairs 1972 (WT). YUGOSLAVIA: c. 500 pairs (Bezzel 1969). SPAIN: 200–250 pairs (Bezzel 1969); mainly marismas of Guadalquivir, where perhaps some decline recently (Valverde 1960). USSR. About 200 000 pairs (Isakov 1970a). In Estonia, increase in 1930s, but some recent decline, but in Latvia increase due to protection (Onno 1970; Kumari et al. 1970).

Winter. Estimated 250 000 birds north-west Europe and 750 000 Europe–Black Sea–Mediterranean region (Szijj 1972; Atkinson-Willes 1975). USSR: c. 383 000 birds in not so severe winters (Isakov 1970c).

Survival. Oldest ringed bird 9 years 7 months (BTO).

**Movements.** Mainly migratory, but some resident or partially migratory. Few British or Irish breeders resident on breeding waters; extent of movements uncertain as few known breeders or young ringed, but evidently some emigrate from southern England since one from Essex found south-east France; summer adults found south to Iberia, but these may have been ringed as non-breeders from continental populations. Those breeding Netherlands, France, and Spain perhaps no more than partially migratory; winter movement, especially in hard weather, from Netherlands to Britain probably mainly eastern breeders. Also winter recoveries (apparently atypical) in Britain and Ireland of young ringed France, Bavaria, and Czechoslovakia. Breeding populations of Denmark, Fenno-Scandia, north Germany, Poland, Baltic States, and USSR between about 50°N and 60°N (east to 70°E) migrate west and south-west to winter West Germany, Switzerland, Netherlands, Britain, Ireland, France, Iberia, and north-west Africa (Boyd 1959). Recovery in Wales of juvenile ringed Barabinsk (78°E) exceptional. West Mediterranean basin, with Italy, also receives birds from central Europe, Balkans, and south-central USSR; but main wintering areas for all but western elements in Tunisia, Greece, Turkey, Danube delta and elsewhere on Black Sea, Sea of Azov, and to lesser extent Levant to Egypt and Jordan (Azraq).

Main winter distribution more southerly than Tufted Duck A. fuligula; 166000 in Tunisia and Algeria, November 1971, but under 600 of latter species (Johnson and Hafner 1972). Rather small numbers of uncertain origins trans-Saharan and Nile valley migrants, wintering northern tropics from Sénégal to Ethiopia, straggling south to Gambia, Uganda, and Tanzania (Roux 1971; Backhurst et al. 1973); largest counts 250 Chad 1970–1 winter, while in January 1972, 400 Lake Faguibine, Mali, and 420 Sénégal delta (Roux 1972). Substantial flocks winter Caspian Sea, Iraq, and Iran, and small numbers in

Persian Gulf south to Oman, probably from west Siberia. Not all individuals of a population follow its normal pattern; complexity of movement shown by recoveries of central European breeders in North Sea countries, and winter-ringed birds from latter found later Kazakhstan; aberrant migration doubtless occurs. One ringed winter England recovered 2 years later Magadan, Siberia (150°E), far east of breeding range.

Moult gatherings of ♂♂ frequent, in west Europe involving considerable numbers, e.g. on Abberton Reservoir, England (2000–3000), Ismaninger reservoir, Bavaria (up to 20000), and IJsselmeer, Netherlands (up to 50 000). In USSR, many sites carry comparatively small numbers of moulting ♂♂. Moult migrations shorter than in dabbling ducks Anas, many moulting flocks remaining near breeding waters. Flocks of ♂♂ begin to assemble early June and may be joined by some ♀♀ (under 10% of moulting flocks) (Salomonsen 1968); peak numbers mid-July, dispersing late August and September, early Russian birds reaching Italy by 10 September.

Peak autumn migration, ♂♂ generally preceding ♀♀ by c. 2 weeks, late September and October through east and south Europe, October–November in western maritime countries. Pronounced tendency for winter segregation of sexes perhaps due to different migration timing, ♂♂ occupying nearest wintering waters, while ♀♀, arriving later, have to move further to reach suitable habitat (Salomonsen 1968). Cold weather movements occur as late as January–February, e.g. mass exodus of Bavarian and Swiss flocks south as far as Camargue (♀♀ react more to cold than ♂♂: Bauer and Glutz 1969), returning as temperatures rise. Spring passage may begin February in mild seasons, but most leave winter quarters March and early April; return movement rapid, ♂♂ predominating early on. Breeding waters re-occupied from early March in temperate regions, from early May in Siberia. See also Dementiev and Gladkov (1952).

**Food.** Plant and animal; proportions vary with season and locality, though in many areas primarily seeds and vegetative parts. Obtained chiefly by diving, either preceded by jump—deeper the dive bigger the jump (Willi 1970)—or not. Also, unlike other Aythya so far as known, at times foot-paddles before diving, apparently to stir up food (Simmons 1968a). Also up-ends, feeds with head and neck immersed, and dabbles on surface, in driftlines, and shoreline mud (Szijj 1965b; Olney 1968). Most dives in depths of 1 m–2·5 m (Büttiker 1952; Bezzel 1959, 1969; Pehrsson 1965; Szijj 1965b; Reichholf 1966; Olney 1968; Willi 1970). ♀♀ dive less deeply than ♂♂, and prefer shallower areas (Bezzel 1959; Willi 1970). Average dive 20 s, maximum 25 s in 0·8–1·4 m (Höhn 1943); average 15·4 s in 0·9 m (Dewar 1924); average 24 and 28·6 s in 3·0–3·4 m and 3·7–4·0 m respectively (Ingram and Salmon 1941); average 13·6 s in 80–120 cm, range 3–24 s (Klima 1966); average 14·5 s in 0·3–1·8 m (Olney 1968). Diving in feeding groups not synchronized. Diurnal and nocturnal feeder (Klima 1966; Willi 1970). During 24 hours (April–June), 28·9% spent feeding (Klima 1966).

Plant materials include seeds, rhizomes, buds, shoots, leaves, tubers. Most frequently recorded include stoneworts *Chara*, pondweeds *Potamogeton*, milfoil *Myriophyllum*, hornworts *Ceratophyllum*, sedges *Carex* and *Scirpus*, persicarias *Polygonum*, and grasses. Animal materials include crustaceans, molluscs, annelids, insects and larvae, amphibians (frogs and tadpoles), and small fish. For reviews see Olney (1968) and Bezzel (1969). Size ranges from microscopic *Chara* spores to 35 mm molluscs and 8 cm shoots (Madsen 1954).

*Chara* particularly important throughout Palearctic region. Of 19 stomachs, France and Germany, 16 contained plants, 13 *Chara* (Szijj 1965*b*). From England and northern Ireland, inland September to January, 43 stomachs contained mainly plants (86·1% total volume), and chiefly stoneworts *Chara* and *Nitella* and seeds, especially *Potamogeton* (*P. pectinatus*, *P. berchtoldii*, *P. natans*, *P. gramineus*, *P. crispus*) and *Polygonum* (particularly *P. amphibium*). Animal materials 13·9% total volume, mainly chironomid larvae (Olney 1968). In 118 Danish winter stomachs, mainly from brackish water areas, *Chara* again most important, with seeds of *Potamogeton* (mostly *P. pectinatus*), *Scirpus*, horned pondweed *Zannichellia*, eel-grass *Zostera*, and wigeon-grass *Ruppia*. Animal materials occurred in 61%, and in 11% entire contents; mainly molluscs, especially blue mussel *Mytilus edulis*, laver shells *Hydrobia*, periwinkles *Littorina*, cockles *Cardium edule*, and Baltic clam *Macoma balthica* (Madsen 1954). Similar results from USSR, though differences with locality and season: in winter in Kirov Bay and near Sar Island, entirely molluscs, *C. edule* and fewer *Theodoxus pallasi*, but in Krasnovodsk Bay chiefly *Chara* (60%), *Ruppia* seeds (8·5%), and molluscs (28%), mainly *C. edule*. From Rybinsk reservoir in September, chiefly rhizomes of *Potamogeton lucens* (88·6%), but in May chiefly chironomid larvae *Tendipes plumosus* and caddisfly larvae *Phryganea grandis*. Migrants on arrival Baraba steppe lakes take mainly shoots of *Salicornia* and *Atriplex pedunculatus*, and some, mainly chironomid larvae. During spring passage on Mologa river, mainly seeds especially *Potamogeton*, also some green parts, particularly *Elodea* and duckweed *Lemna* (Dementiev and Gladkov 1952). Where *Chara*, *Potamogeton*, and other plants absent or infrequent, animal foods predominate: 39 stomachs from Klingnauer reservoir, Switzerland, examined in autumn contained mainly chironomid larvae and *Tubifex* and, as these depleted, apparently then fed on organic debris (Willi 1970). On Ismaninger reservoir, West Germany, moulting birds also said to feed on decomposing organic matter (Bezzel 1969). 31 stomachs from Hungary, February–November contained mainly *Chara* and seeds of *Polygonum*, *Potamogeton*, *Carex*, bristle-grass *Setaria*; also some insects (waterbeetles and waterbugs), and zebra-

mussel *Dreissena polymorpha* (see Willi 1970).

Young take insects, especially flies, and seeds floating on surface (Millais 1913). Young in August on Rybinsk reservoir, USSR, eat chironomid larvae almost exclusively (Dementiev and Gladkov 1952).

**Social pattern and behaviour.** For general information, see especially Bezzel (1969), Bauer and Glutz (1969); also Witherby *et al.* (1939), Bannerman (1958), Géroudet (1965).

1. Highly gregarious for much of year. Outside breeding season, in relatively small parties' on passage but frequently in large flocks of up to many thousands during post-breeding moult and winter. Sex composition of winter flocks often locally unbalanced. ♂♂ first to flock, from late May or early June when ♀♀ still nesting. BONDS. Monogamous pair-bond mostly of brief seasonal duration; occasional bigamy suspected (Bezzel 1969). Though some pair-formation evidently starts in winter in most parts of range, many pairs not formed until spring. In north Kent, pairs form in flock from mid-March onwards (Hori 1966); in Bavaria, peak of pair-formation in May (Bezzel 1959). Most pairs separate during incubation, in 1st or 2nd week (Hori 1966), though some ♂♂ occasionally accompany ♀♀ with broods until early July (Bezzel 1959). Only ♀ tends young. Although young said to remain in family group until fledged at *c.* 6½ weeks (Hori 1966), parent-young bond itself ends well before fledging in many cases (see also Bezzel and von Krosigk 1971). Some young leave even before this, not infrequently joining another ♀ or brood even when quite small. In general, however, ♀ seems to desert young when wing moult imminent (Gillham 1957). ♀♀ without broods recorded attracting young of other ♀♀ by calls until these attach themselves to foster-mother. Occasionally, and usually towards end of season, presumed brood accompanied by 2 ♀♀; although different broods may crowd together when in danger, no evidence of crèche formation (Hori 1966). BREEDING DISPERSION. In some areas, spring flock disperses from beginning of April (Hori 1966). In other areas, however, flock never totally disbands and, where common, pairs stay in flock until time of nesting. Little exact information on home-ranges of breeding pairs. No evidence that feeding, loafing, or waiting areas within home-range defended as real territory but, in presence of conspecifics, ♂ defends small area round mate. Nests concealed and usually well dispersed; some tendency to associate with colonies of gulls *Larus*. ♂ waits for ♀ during laying period and accompanies her at first during her recesses while incubating. In some areas, broods show marked tendency to frequent same nursery area during growth period of young (Hori 1966). ROOSTING. Communal for most of year; largely diurnal. Mostly on open water. As main feeding period starts after dark, rests mostly in middle of night and periodically during much of day. Nocturnal courtship common. For further information on activity rhythms, see Bauer and Glutz (1969).

2. Observed both in wild and captivity (e.g. Höhn 1943; Boase 1950; Steinbacher 1960; Johnsgard 1965; Klíma 1966; Bezzel 1968). Differs in some respects from other west Palearctic *Aythya*, being closer to North American Canvasback *A. valisineria* and allies (Redhead *A. americana*, Ring-necked Duck *A. collaris*). FLOCK BEHAVIOUR. Forms quite compact groups at times, e.g. when loafing, but more dispersed while feeding. When alerted, birds stretch necks vertically; tend to swim away and dive from danger rather than fly. Pre-flight signals, however, well developed: usually consist not only of characteristic Chin-lifting movements (Johnsgard 1965) but also of Head-flicks and certain other comfort-movements (McKinney 1965*b*). ANTAGONISTIC BEHAVIOUR. Relatively little information. ♂'s Neck-stretch

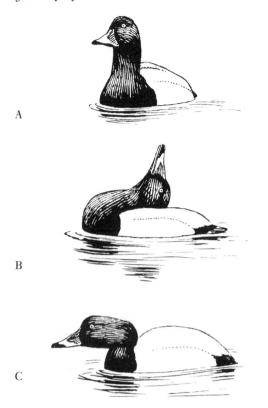

A

B

C

display (see A), typical of courtship, evidently also directed at other ♂♂ as threat as in *A. valisineria* (see Johnsgard 1965). Also when threatening, both sexes point head at rival with neck extended or bent, feathers ruffled, and bill sometimes open; accompanied by calls, especially from ♀ (Bauer and Glutz 1969). Hostile component of ♀'s Inciting-display basically similar; bird calls while performing lateral threatening movements with neck stretched horizontally, alternating these with Neck-stretch posture near favoured ♂; lacks Chin-lifting movements of *Netta* and some other *Aythya* (Veselovský 1966). COMMUNAL COURT-SHIP. Occurs on water from mid-February to mid-July, Bavaria, with May peak (Bezzel 1968); in Netherlands, first seen end October but not common until December (Lebret 1961). Often several ♂♂ and 1 ♀, but presence of additional ♀♀ not uncommon, ♂♂ tending to centre attention on particular ♀ in group; also follow any active ♀, all swimming about rapidly. During courtship, neck of ♂ often appears thickened (see A); according to Millais (1913), due to inflation but Höhn (1943) thought probably caused solely by erection of feathers. Millais (1913) also described blazing red appearance of eye during Sneak-display due to constriction of pupil. In Courtship-intent posture, ♂ swims with head only slightly raised (Steinbacher 1960). Secondary displays, performed especially during preliminary stages of courting session, include Head-shake, Head-flick, and Upward-shake; ♂♂ also said to engage in accelerated swimming at start of courtship, (Klíma 1966). 3 major displays (Head-throw, Kinked-neck, and Sneak) accompanied by Courtship-calls of at least 2 types, one of which also being given without special posturing, but information conflicting (see Voice). In full Head-throw, ♂ whips head back while calling until nape touches mid-back area (see B), then rapidly brings head forward again (Johnsgard 1965). No record of any initial Head-

shake or Head-flick (as in scaup-like ducks) though terminal Bill-dipping and Preen-behind-Wing (as in Tufted Duck *A. fuligula*) absent (Bezzel 1968). Kinked-neck display like that of *A. valisineria* in which ♂ calls with neck arched over breast and hyoid lowered, producing bulge in throat (see Johnsgard 1965). In full Sneak-display, ♂ calls while lying partly submerged with neck, head, and bill stretched flat on water in direction of ♀; usually floats inactively, but may pivot or even swim about thus. Incomplete version of Sneak-display (Crouch-display), with neck withdrawn and bill pointing obliquely down (see C), also common; frequently prelude to full Sneak. In Neck-stretch display, head well erected and neck enlarged (see A). This found to be frequent by Johnsgard (1965) who mentioned no accompanying call or associated movements; often performed lateral to ♀ or other ♂♂ (Klíma 1966), and usually in front of ♀ (Bezzel 1968). Possible that Coughing-call may also be given from Neck-stretch posture to produce Coughing-display; then apparently sometimes accompanied by Head-nodding movements (see Bezzel 1968; Bauer and Glutz 1969). Evident Head-nodding, in which head jerkily lowered and then raised so that neck held rigidly erect while bill remains in horizontal plane, also described by Höhn (1943). Ceremonial Leading and Turn-back-of-Head display not seen by Johnsgard (1965), but probably much as in other *Aythya*. ♀♀ often take little active part in communal courtship, but will at times swim about erratically, threaten ♂♂, rush aggressively at them over water, or perform Inciting-display. Also frequently Bill-dip (Millais 1913). Copulation, initiated by ♀'s adoption of Prone-posture, also recorded during communal courtship, starting well before establishment of definite pairs (Höhn 1943). PURSUIT-FLIGHTS. Courtship-flights occur mainly from late spring onwards and originate from courting group on water (see Bauer and Glutz 1969). ♂♂ follow ♀ in straight line or wide arc. ♀ may call in flight, but shows no Inciting or Repulsion behaviour. ♂♂ may break away during prolonged flights but whole party often returns to water, either continuing communal courtship or preening. 3-bird flight once recorded by Hori (1966). Rape-intent flights during incubation period not persistent, while actual rapes not recorded. PAIR BEHAVIOUR. Pair-courtship said to consist of much the same behaviour shown in communal courtship (Bezzel 1968) plus copulatory behaviour, but confirmation needed; also further observations on frequency of displays such as Preen-behind-Wing. COPULATION. Poorly documented. Bill-dipping by both sexes occurs at times both before and after copulation: bill lowered into water close to front of breast, then evidently twitched rapidly forward and backward (see Bauer and Glutz 1969). At other times, copulatory display may be partly or wholly absent. Though Preen-dorsally by ♂ recorded in unspecified context by McKinney (1965*b*), said to be totally absent in pre-copulatory situation (see Bauer and Glutz 1969). ♀ assumes Prone-posture characteristic of *Aythya* (Höhn 1943). After dismounting, ♂ performs Bill-down display while ♀ bathes. RELATIONS WITHIN FAMILY GROUP. ♀ and brood usually frequent open water where ♀ leads young away from danger by swimming with body low in water, calling softly (see Bauer and Glutz 1969). For general account, see Hori (1966).

(Figs A–C after Johnsgard 1965; see also Veselovský 1966.)

**Voice.** Mostly weak, low-pitched, and inaudible at a distance. Largely confined to courtship but incompletely studied. Calls of ♂ differ from most other *Aythya* in remarkable harmonic development with changing pitch and loudness (Johnsgard 1971).

PLATE 75. *Netta rufina* Red–crested Pochard (p. 553): **1** ad ♂ breeding, **2** ad ♀, **3** ad ♂ eclipse, **4** juv, **5** downy young. (NWC)

PLATE 76. *Aythya ferina* Pochard (p. 561): **1** ad ♂ breeding, **2** ad ♀, **3** ad ♂ eclipse, **4** juv, **5** downy young. (NWC)

PLATE 77. *Aythya collaris* Ring-necked Duck (p. 569): **1** ad ♂ breeding, **2** ad ♀, **3** juv. (NWC)

PLATE 78. *Aythya nyroca* Ferruginous Duck (p. 571): **1** ad ♂ breeding, **2** ad ♀ breeding, **3** ad ♂ eclipse, **4** 1st imm non-breeding, **5** downy young. (NWC) ·

CALLS OF MALE. 2 main Courtship-calls. (1) Soft, whispered, nasal 'wiwierr' (Steinbacher 1960); also described as 'hip-sierr' or 'meeyo-oo-oo' (Bezzel 1968). Asthmatical wheeze resembling 'ng' sound produced while breathing in through nose (Höhn 1943) probably same; tape-recording (see I) resembles 'weee ke-hooo.' (2) Louder, triple 'kil-kil-kil' (sometimes 1–2 syllables), descending slightly in pitch (Steinbacher 1960). Call 1 said by Steinbacher (1960) to accompany Head-throw and Sneak-displays, as well as being given in normal swimming posture; also said to be uttered in Kinked-neck posture (Johnsgard 1965). Call 2 may be main component of Coughing-display (Johnsgard 1965) but both calls linked with Sneak-display by Bezzel (1968). Soft, low whistle of Millais (1913), frequent soft piping 'quee-week' of Höhn (1943), and soft, whistling, half-piping, half-cheeping 'djudjn' or 'djudjidji' like conversation of young geese *Anser* (Voigt 1950) may represent a 3rd call.

CALLS OF FEMALE. 2 calls during courtship mentioned by Steinbacher (1960): (1) 'pack' or 'back'; (2) 'brerr' or 'err'. Call 1 probably same as explosive 'pwook' of Höhn (1943). Call 2 identified by Johnsgard (1965) as Inciting-call; probably same as harsh, repeated 'graa' of Höhn (1943). Tape-recording (see II) resembles 'karr . . .' (usually in series of 3–4 notes, sometimes only 2). See also Tufted Duck *A. fuligula*. Rasping 'girrr' during pursuit-flights probably same, also continual 'gagagagag grA grA . . .', etc., on water in presence of group of ♂♂ (Voigt 1950); see also Söderberg (1950). Other calls include: (3) conversational 'wuk-uk-uk' (Witherby *et al.* 1939); (4) 'gurr' given as warning call to young (Höhn 1943). For comments on flight-call, see Ferruginous Duck *A. nyroca* and *A. fuligula*.

CALLS OF YOUNG. Contact-calls uttered in groups of 2–4 notes. Distress-calls higher and more rapid than in young *Anas* of similar size. (J Kear.)

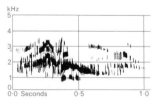

I  J-C Roché

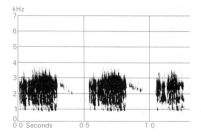

II  J-C Roché

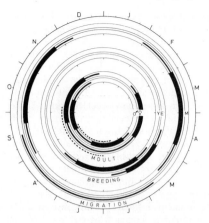

**Breeding.** Based mainly on Bezzel (1969), southern Germany. SEASON. See diagram for north and west Europe. In southern Germany, egg-laying begins early May, reaches peak June, and ends late July; in USSR, early to mid-May (Dementiev and Gladkov 1952). SITE. On ground close to water (usually within 10 m), in thick cover; or in water, built up above surface, in dense reeds and rushes. Of 380 nests, Czechoslovakia: 76·6% in water, 16·3% on islets, 7·1% on dry land (Harlin 1966b). Nest: platform, with shallow cup of reed stems, leaves, and other vegetation; lined down. Diameter 25–35 cm, external height up to 35 cm for nests in wet sites, but 10–15 cm in dry sites. Building: by ♀, using material within reach of nest. Cup shaped by movements of body. EGGS. Broad oval; green-grey. 62 × 44 mm (56–68 × 39–47), sample 300 (Schönwetter 1967). Weight 68 g (61–73), sample 12 (Germany); 65 g (55–74), sample 57, captive laid (J Kear). Clutch: 8–10 (4–22, over 15 probably always 2 ♀♀). Of 142 1st clutches, Germany: 6 eggs, 2%; 7, 13%; 8, 21%; 9, 18%; 10, 17%; 11, 15%; 12, 8%; 13, 5%; 14, 1%; mean 9·5. Of 307 clutches, Czechoslovakia: 3 eggs, 0·3%; 4, 2%; 5, 7%; 6, 16%; 7, 17%; 8, 17%; 9, 16%; 10, 10%; 11, 7%; 12, 6%; 13, 2%; mean 8·09; clutch size decreases through season (Harlin 1966a). One brood. Replacement clutches laid after egg loss. Of 74 replacement clutches, Germany: 5 eggs, 7%; 6, 16%; 7, 18%; 8, 28%; 9, 24%; 10, 5%; 11, 1%; 12, 1%; mean 6·7. One egg laid per day. INCUBATION. 25 days (24–28). By ♀. Starts on completion of clutch; hatching synchronous. Eggs covered with down when ♀ off. Eggshells left in nest. YOUNG. Precocial and nidifugous. Self-feeding. Cared for by ♀; brooded at night when small. FLEDGING TO MATURITY. Fledging period 50–55 days. Become independent at or before fledging, possibly at 3 weeks in some cases (Gillham 1957). Breed at 1 year, though some possibly not until 2. BREEDING SUCCESS. In Germany, average size of 1st broods at hatching 7·14 young, and of repeat broods 6·0 (45% ♀♀ lost 1st clutch before hatching). Average number of young reared to flying stage 4·42 per successful pair, and 1·83 for all pairs. Of 1151 eggs, Czechoslovakia, 1968–70, 56% hatched (Harlin 1972).

**Plumages.** ADULT MALE BREEDING. Head and neck deep chestnut. Small white spot on chin. Feathers of upper mantle black, central sometimes tipped chestnut. Chest black. Lower mantle, scapulars, sides of body, and flanks with many fine white and dark grey-brown vermiculations. Outer webs of tertials pale grey, inner brown-grey; both strongly vermiculated white. Back dark grey, vermiculated or speckled white; rump and upper tail-coverts black, in some finely speckled white. Lower breast and belly white, finely vermiculated grey-brown; vent white with heavy vermiculation (sometimes all black). Under tail-coverts black. Tail dark grey. Primaries pale ash, tips of all and outer edges of outer 3–4 sooty-black; coverts sooty, finely speckled white at tips. Secondaries pale ash, darker near tip, few inner blue-grey with narrow black outer edge; all narrowly tipped white, especially on outer webs of central secondaries, and with much fine white vermiculation and speckling near tip and on outer web of inner secondaries. Upper wing-coverts dark ash with much white vermiculation. Underwing and axillaries white, but marginal coverts and centres of greater grey-brown. ADULT FEMALE BREEDING (Does not nest in this plumage). Feathers of crown brown-black, tipped cinnamon; cheeks, hindneck and sides of neck earth-brown to cinnamon. Lores, streak behind eye, chin, and throat pale grey or buff. Feathers of upper mantle and chest cinnamon-brown with dark grey bases; tipped buff to white on chest. Lower mantle and scapulars vermiculated and speckled pale grey and dark grey-brown; tertials dark olive-grey with a few pale grey specks along edges. Back black with much white speckling; rump and upper tail-coverts same, but fewer specks. Flank feathers finely vermiculated grey-brown and white, occasionally with much olive-brown at bases. Breast and belly white, but some pale grey-brown of feather-bases visible. Vent and under tail-coverts grey-brown, sometimes speckled white, longer under tail-coverts broadly tipped white. Tail dark grey. Wing like adult ♂, but only tips of upper wing-coverts finely speckled; secondaries without speckling, except some near tips of central ones. ADULT MALE NON-BREEDING (Eclipse). Like adult ♂ breeding, but head and neck duller brown, somewhat mottled on throat and neck; crown sooty-brown. Upper mantle and chest dull black or dark grey-brown, finely speckled buff or white subterminally; broadly tipped rufous-brown on upper mantle, buff to white on chest. Lower mantle, scapulars, and tertials dark grey-brown, more coarsely vermiculated white. Feathers of flanks and vent grey-brown, vermiculated white at tips; those of belly grey-brown broadly edged white. May be confused with adult ♀ breeding, but head plain rufous-brown, upper mantle and chest darker brown, more vermiculations on underparts, and wing like adult ♂ breeding. ADULT FEMALE NON-BREEDING (nests in this plumage). Like breeding, but virtually no vermiculation on body; feathers of lower mantle, scapulars, tertials, and flanks dark olive-brown with buff edges; chest and sides of breast dark grey-brown, strongly tinged chestnut-brown near tips of feathers, and broadly edged white on chest. Breast and belly white, but rufous-grey of feather-bases visible. DOWNY YOUNG. Crown, hindneck, upperparts, thighs, sides of breast, and vent brown; longer tufts on forecrown, mantle, and sides tipped yellow-olive. Streak through eye and at rear of cheek brown; streak over eye, lores, sides of head and neck, patches on rear edge of wing, below wing, and at sides of back and rump yellow. Underparts yellow, grading to white on vent and under tail-coverts. JUVENILE MALE. Like adult ♀ breeding, but head and neck duller and greyer brown, without light streak behind eye; lower mantle and scapulars brown-grey, vermiculated or speckled white at tips of feathers. Upper mantle, chest, and flanks dark grey (mantle sometimes rufous-brown), feathers edged yellow-buff or grey. Rump and upper tail-coverts usually uniform dull black. Inner web of tertials sepia, outer dark olive-grey faintly speckled white. Feathers of breast and belly narrow; grey-brown, tipped grey or buff-white, giving underparts mottled appearance. Vent and under tail-coverts grey-brown with narrow buff-brown edges. Tail dark grey-brown; feathers narrow, tapering, notched at tip, bare shaft projecting. Wing like adult ♀, but wing-coverts slightly darker grey; differs from adult ♂ in reduced amount of speckling on wing-coverts and on outer webs of secondaries (only faintly spotted near tips). JUVENILE FEMALE. Like juvenile ♂, but mantle, scapulars, and tertials plain grey-brown with olive-brown feather edges, usually without any white speckling. Wing-coverts uniform dark-grey, tips of secondaries without speckling. FIRST IMMATURE NON-BREEDING. Like adult non-breeding, but underparts and varying amount of plumage on upperparts still juvenile. FIRST BREEDING. Variable; like adult breeding, but juvenile tertials and some feathers of tail or elsewhere often retained. Wing juvenile, but sometimes a few tertials or wing-coverts changed for adult ones. Some have 1st-breeding plumage on belly and vent by December, but others not until spring.

**Bare parts.** ADULT MALE. Iris orange-yellow to red, brilliant red in spring. Bill dark slate; broad band between nail and nostrils pale blue-grey. Leg and foot pale brown-grey to dark grey, webs and joints dark slate. ADULT FEMALE. Iris grey-brown or yellow-brown to deep brown; reddish hazel in summer. Bill dark slate to dull black, with pale grey band between nail and nostrils (indistinct in summer). Leg and foot pale green-grey to grey, webs and joints slaty-black. DOWNY YOUNG. Iris pale blue-grey. Bill horn-black with reddish nail and flesh lower mandible. Foot slaty-black with yellow streak at sides of toes. JUVENILE. Iris yellow-olive; grades after a few months to orange-yellow or orange-red in ♂, and grey-brown or yellow-brown in ♀. Bill olive-black to dark slate; ♂ develops grey band from autumn onwards. Leg and foot grey to grey-blue with dull black webs and joints. (Bauer and Glutz 1969; Schiøler 1926; Witherby *et al.* 1939; RMNH.)

**Moults.** ADULT POST-BREEDING. Complete; flight-feathers simultaneous. ♀♀ moult more or less whole body, and possibly also part of tail, late February or March to June. Body moult of ♂♂, June–July, less complete on back and belly; immediately followed by wing. Flightless for *c.* 3–4 weeks; ♂♂ late June to early September (mainly late July); ♀♀ early July to late October (mainly late August). Tail late July to mid-October. ADULT PRE-BREEDING. Partial. Head and body, mainly September–October. ♂♂ in full breeding plumage from about late October, ♀♀ *c.* 1 month later. POST-JUVENILE MALE. Starts July. Feathers replaced July–August may be considered as belonging to 1st immature non-breeding; from September, new feathers like adult ♂ breeding. Head and neck moult first, followed by scapulars, flanks, and part of mantle and chest. These parts of body may attain 1st non-breeding plumage, but most of it and part of remaining juvenile replaced for 1st breeding before December, or (after a moult-stop in winter) during spring. Tail moults from late August, mostly October–November; a few specimens acquire only new central feathers in autumn, completing moult in spring. Belly and vent moult from September; in some already finished October, others December, but quite a few retain all juvenile breast and belly until spring (although head and rest of body then in 1st breeding). Tertials replaced in spring (rarely October–December); usually also some inner greater coverts or more rarely inner secondaries, contrasting with worn juvenile feathers. POST-JUVENILE FEMALE.

Autumn moult less complete: often juvenile or 1st immature non-breeding on scapulars and flanks retained during winter, and usually many on breast and belly; otherwise moults like ♂. FIRST POST-BREEDING. Like adult in spring (♀) and summer (♂), but often less complete, and part of 1st non-breeding and breeding retained; juvenile wing retained until summer.

**Measurements.** Netherlands, whole year; skins (RMNH, ZMA).

| | | | | | |
|---|---|---|---|---|---|
| WING AD ♂ | 217 | (3·22; 19) | 212–223 | ♀206 (4·61; 22) | 200–216 |
| JUV | 213 | (4·47; 41) | 202–220 | 206 (7·85; 23) | 185–215 |
| TAIL AD | 54·2 | (1·87; 19) | 51–58 | 53·0 (1·52; 23) | 51–57 |
| JUV | 49·1 | (3·03; 10) | 45–54 | 50·1 (2·47; 9) | 46–54 |
| BILL | 47·1 | (1·80; 62) | 43–52 | 44·9 (1·48; 47) | 42–48 |
| TARSUS | 39·5 | (1·02; 52) | 37–42 | 38·8 (1·19; 47) | 36–41 |
| TOE | 67·4 | (2·14; 50) | 63–72 | 65·3 (2·80; 47) | 60–71 |

Sex differences significant, except tail. Wing and tail of juvenile significantly shorter than adult (except ♀ wing). Other measurements similar; combined in table. 1st breeding tail similar to adult tail.

**Weights.** FULL-GROWN. Winter. Much local variation. (1) December–February: Camargue, France (Bauer and Glutz 1969). (2) March: Switzerland (Bauer and Glutz 1969). (3) January–February: south-west Caspian, USSR (Dementiev and Gladkov 1952).

| | | | | | |
|---|---|---|---|---|---|
| (1) | ♂ 849 | (–; 119) | 585–1240 | ♀807 (–; 202) | 467–1090 |
| (2) | 882 | (–; 21) | 725–990 | 832 (–; 35) | 710–1110 |
| (3) | 1095 | | 900–1300 | 905 | 850–920 |

Spring to autumn. (4) April–May: Kazakhstan, USSR (Dolgushin 1960). (5) June–August: Denmark (Schiøler 1926) and Kazakhstan; combined (adults). (6) September–November: Netherlands (ZMA), Denmark, and Kazakhstan; combined (adults).

| | | | | | |
|---|---|---|---|---|---|
| (4) | ♂ 931 | ( 95; 8) | 800–1100 | ♀1056 ( – ; 6) | 975–1120 |
| (5) | 943 | (102; 8) | 787–1108 | 795 ( 79; 8) | 630–900 |
| (6) | 1012 | ( 92; 19) | 850–1228 | 940 (110; 10) | 760–1120 |

JUVENILE. From September as adult. September–October: Denmark.

| | | | | | |
|---|---|---|---|---|---|
| | ♂ 1040 | (158; 5) | 870–1254 | ♀980 (146; 5) | 786–1190 |

**Structure.** Wing rather long and narrow. 11 primaries: p9 longest, p10 0–3 shorter, p8 4–8, p7 14–21, p6 25–36, p1 86–102. Inner web of p10 and outer of p9 emarginated. Tail short, rounded; 14 feathers. Bill as long as head, higher than broad at base; sides parallel, but slightly wider near round and flat tip. Culmen slightly concave in front of nostrils. Nail of bill narrow, oval. Outer toe *c.* 98% middle, inner *c.* 78%, hind *c.* 29%. CSR

---

## *Aythya collaris* **Ring-necked Duck**

PLATES 77 and 99
[facing page 567 and after page 686]

DU. Amerikaanse Kuifeend    FR. Morillon à collier    GE. Ringschnabelente
RU. Кольчатая чернеть    SP. Porrón de collar    SW. Ringand

*Anas collaris* Donovan, 1809

Monotypic

**Field characters.** 37–46 cm, of which body almost two-thirds; wing-span 61–75 cm. Rather small, short-necked, compact diving duck with peaked triangular head (high dome far back on crown produces flat, even concave, back of head) and fairly large, broad, slightly upcurved bill with white ring near tip; ♂ black with pale grey side-panel outlined in white extending up in point at front. Sexes dissimilar, but no seasonal differences apart from eclipse; juvenile resembles ♀, but just distinguishable.

ADULT MALE. Entirely black above, glossed with purple on peaked head and elsewhere with green, apart from inconspicuous chestnut ring round lower neck. Breast, tail, and undertail also black; sides and flanks white largely covered by fine blackish lines producing pale grey effect, but outlined in pure white with a broader area of white at front extending up as diagnostic 'spur' (conspicuous at long ranges) in front of closed wing; belly white shading to grey on hind part. Wings blackish with obscure grey speculum and broad, pale grey stripe along whole length in flight; under wing-coverts white and grey. Bill dark grey-blue with conspicuous, sharply defined, broad, bluish-white band immediately behind black nail and narrow white band around base, producing markedly pied pattern; eyes orange-yellow; feet grey-blue to grey or yellow-grey. MALE ECLIPSE. Resembles adult ♀, but darker and lacking face markings. ADULT FEMALE. Mainly brown to grey-brown, darkest on crown and upperparts, greyest on sides of head and neck, and mottled by pale feather edgings on breast and sides; belly mottled brown and white. Characteristic features: 'spectacles' formed by distinct white eye-ring and narrow line behind, and whitish area behind base of bill (not forehead) and chin extending to mottling on cheeks. Wings brown with broad grey stripe in flight, similar to that of ♂. Bill dark grey with band behind nail less conspicuous than in ♂ and no white around base; eyes yellow-brown; feet grey or green-grey. JUVENILE. Closely resembles adult ♀, but darker above with less well-defined bill and head markings, and more mottled below; by September or October, immature ♂♂ distinguishable by darker breast and blackish feathers appearing among brown of head and back.

♂ at first sight like ♂ Tufted Duck *A. fuligula*, but that species lacks distinctive bill markings (having at most narrow white line behind nail) and has rounded head with drooping crest, pure white sides without white spur at front, and broad white stripe on wings in flight; ♂ Scaup *A. marila* has pale grey back and white wing stripe, no bill markings or white spur. Eclipse ♂ *A. collaris* distinguished by combination of (reduced) bill markings, head shape, dark back, and grey wing stripe. ♀'s spectacles and paler areas behind bill give appearance and surprised look of ♀ Pochard *A. ferina*, which also has pale grey wing stripe but longer black bill crossed with blue-grey band, less well defined white round eyes, different head and body shape, lighter back, and paler sides. Otherwise ♀ only likely to be confused with *A. fuligula* and *A. marila* which, however, often have sharply defined white area round base of bill; even when this absent, both are less delicate-looking, lack spectacles, white markings on lower part of face and conspicuous band on bill, and have white wing stripes; ♀ *A. fuligula* also slightly darker and less grey.

Found more on fresh water of marshes and ponds, or at edges of lakes, rather than out on middle of large expanses of deep water. Occasionally on rivers, particularly with marshy edges, or tidal estuaries, but seldom on sea. Swims high on water and dives extremely well; also dabbles and even up-ends in shallow water. Generally rather lively. Rises much more easily than other *Aythya* and flies fast on rapid wing-beats producing whistling noise. Usually silent.

**Habitat.** In Nearctic mid-latitudes, mainly in inland, lowland wetlands and on pools and ponds; less frequently on large or open lakes. Avoids very shallow waters, diving freely to *c.* 13 m. Breeds in fringing vegetation on shore or islands. On passage and in winter, still largely restricted to inland habitats of similar type but including some artefacts such as rice fields.

**Distribution.** Breeds North America from interior of British Columbia, Mackenzie, and Labrador to Quebec, Maritime Provinces, and Newfoundland south to Washington, Michigan, New York, and southern Maine, with isolated or sporadic breeding further south in western and eastern USA. Has greatly extended range east of the Great Lakes in last 30 years or so (Mendall 1958; Godfrey 1966).

Accidental. Although first described from specimen, said taken Lincolnshire, England, March 1801, not acceptably recorded Europe until 1955. Since then, records in various western European countries, mainly ♂♂ and from October–June; at first linked with eastwards expansion in North America but occurrences since 1969 in Britain and Ireland increasingly open to suspicion as escapes (Smith *et al.* 1970). Numbers of individuals involved uncertain, as tendency to reappear same sites in successive winters (possibly after summering with other *Aythya*) and some may move during winters. For details of occurrences to 1968, see Bruun (1971). Britain and Ireland: over 20 from 1955–74. France: 1966, 1967. Belgium: 1960, 1961, 1962. Netherlands: 1959, 1961, 1962. Norway: 1966–7; Denmark: 1963. West Germany: 1969. Switzerland: 1966–7, 1967–8, 1969. Also Iceland: ♀ May 1967 (Bauer and Glutz 1968), and Azores: at least 2 ♂♂, 2 ♀♀ February–March 1962 (Bannerman and Bannerman 1966).

**Movements.** Migratory. Withdraws south of breeding range to winter mainly in western, eastern, and southern parts USA, West Indies, Mexico and Guatemala; stragglers south to Venezuela and Trinidad. Most important wintering areas are (1) south-eastern USA from River Mississippi (Tennessee–Louisiana) to South Carolina and Florida, which draws birds from at least as far north-west as Saskatchewan (McIlhenny 1934); and (2) west of Rocky Mountains in Oregon and California. Moderate numbers Mexico, mainly Gulf coast. Migrates along all 4 major flyways; largest numbers use Mississippi route, though eastern fringe of continental migration passes through Ontario and Great Lakes states into Atlantic flyway. Autumn passage late September to November. Spring return begins late February in Florida; first arrivals Maine in late March, but principal passage through USA in third week April. Has wandered north in summer to Alaska, Ungava, and Labrador, while vagrants to Azores and various European countries notable for paucity of autumn dates (see Bruun 1971). See Kortright (1942); Mendall (1958); Linduska (1964).

**Voice.** Generally similar to that of other *Aythya* in same species-group. Calls mostly quiet and uttered infrequently. Include soft, breathy notes and louder, clearer whistling by ♂ (given only on water); soft, rolling notes by ♀ (see e.g., Johnsgard 1965).

**Plumages.** ADULT MALE BREEDING. Head and neck black, glossed green on crown and purple at sides of head and neck. Narrow collar round lower neck chestnut, small spot on chin white. Mantle, scapulars, back, rump, tail-coverts, chest and upper breast black. Scapulars finely speckled white, feathers of upper breast narrowly tipped white. Sides of body in front of closed wing, lower breast, and belly white; rest of sides, flanks, and vent white with grey vermiculations. Tail brownish-black. Primaries pale grey-brown, outer edges of outer and tips of all sooty black. Secondaries pearl-grey, inner webs and in some tips of outer webs sooty brown; tips often narrowly edged white. Outer webs of some inner secondaries blue-grey, narrow outer edge black, greenish-black towards base; tertials glossy dark green. Upper wing coverts sooty black, lesser coverts with narrow dark grey tips, others slightly glossed green. Greater and marginal under wing-coverts light grey with white tips, rest of underwing and axillaries white. ADULT FEMALE BREEDING (Does not nest in this plumage). Crown brownish-black, finely speckled or tipped brown. Chin, narrow line round base of upper mandible, ring round eye, and narrow, sometimes indistinct line behind eye white. Sides of head and upper neck pale grey, feathers mottled and vermiculated dark grey or black; feathers of throat and foreneck white narrowly tipped grey, of lower hindneck white broadly tipped olive-brown. Mantle, scapulars, and upper tail-coverts dark brown, feathers broadly tipped rufous to olive-brown; back black. Feathers of chest, breast, and belly dark grey-brown, broadly tipped rufous-brown on chest, buff on breast, and white on belly; feathers of flanks grey-brown, broadly tipped buff to white. Vent buff-grey; under tail-coverts black vermiculated and speckled white at tips. Tail and wing like adult ♂, but tertials and wing-coverts with less gloss, white edges to tips of secondaries narrow or absent, and smaller upper wing-coverts browner, edged cinnamon-brown rather than dark grey. ADULT

MALE NON-BREEDING (Eclipse). Like adult ♀ breeding, but head darker, with less white in lores and round eyes; wing like adult ♂ breeding. ADULT FEMALE NON-BREEDING. Like breeding, but paler brown. JUVENILE. Like adult ♀ breeding, but head and upperparts dull brown, feather edges narrower and paler; feathers of underparts narrow, less white on tips, giving belly spotted appearance. Tail-feathers narrow, notched at tip, bare shaft projecting. In 1st non-breeding (autumn), feathers of chest in ♂ dusky black, in ♀ grey-brown, both with buff edges. Wing duller brown, upper wing-coverts narrower and with rounder tips, instead of somewhat broader and squarer in adult. FIRST BREEDING. Like adult breeding, but wing juvenile, with more abraded coverts in spring; part of juvenile feathers of underparts sometimes also retained. Chestnut collar on lower neck of ♂ less complete or absent. (Kortright 1942; Delacour 1952; C S Roselaar.)

**Bare parts.** ADULT MALE. Iris orange-yellow. Bill dark grey-blue with broad bluish-white band behind black tip, and narrow white band at base of upper mandible; cutting edges also often narrowly edged white; colours fade in eclipse. Foot dark blue-grey, joints and webs dull black. ADULT FEMALE. Iris brown to yellow-brown. Bill dark grey with pale grey band over upper mandible behind black tip. Foot grey or olive-grey. JUVENILE. Iris brown. Bill dark grey with slight olive tinge; when reaching 1st-breeding plumage, as in adult. Foot as adult, tinged olive in some. (Kortright 1942; Bauer and Glutz 1969; Delacour 1959.)

**Moults.** Apparently little different from Tufted Duck *A. fuligula* but only specimens examined in post-juvenile and pre-breeding moult.

**Measurements.** ADULT. East USA, skins (ZMA, RMNH); additional data on range from Delacour (1959) and Godfrey (1966).

| | | | | | | | |
|---|---|---|---|---|---|---|---|
| WING | ♂ | 201 | (4·13; 6) | 194–206 | ♀ 195 | (–; 3) | 185–201 |
| BILL | | 47·0 | (1·87; 6) | 44–54 | 44·9 | (1·00; 4) | 43–47 |

Tail slightly longer than *A. fuligula*; tarsus and toe similar.

**Weights.** Maine, USA (Bauer and Glutz 1969).

| | | | |
|---|---|---|---|
| AD Spring | ♂ 737 (22) 708–822 | ♀ 653 (10) 596–708 |
| AD Autumn | 767 (7) 681–937 | 681 (6) 511–879 |
| JUV August | 631 (5) 567–738 | 653 (4) 496–767 |

**Structure.** Wing rather long, pointed. 11 primaries: p9 longest, p10 equal or slightly shorter, p8 3–7 shorter, p7 13–18, p6 25–31, p1 74–88. Inner web of p10 and outer of p9 emarginated. Tail short, rounded; 14 feathers. Bill as long as head, higher than broad at base, sides almost parallel; slightly elevated at tip. Nail narrow. Feathers on centre of crown and hindcrown dense and somewhat elongated in breeding plumage, especially in ♂♂; give head short and high appearance. Outer toe about equal to middle, inner *c.* 78% middle, hind *c.* 30%.                    CSR

---

## *Aythya nyroca* Ferruginous Duck (White-eyed Pochard)

PLATES 78 and 99
[facing page 567 and after page 686]

DU. Witoogeend     FR. Fuligule nyroca     GE. Moorente
RU. Белоглазый нырок     SP. Porrón pardo     SW. Vitögd dykand

*Anas nyroca* Güldenstädt, 1770

Monotypic

**Field characters.** 38–42 cm, of which body almost two-thirds; wing-span 63–67 cm. Small, neat diving duck with shape suggesting *Anas* as much as *Aythya*, flat forehead, and rather high crown; ♂ mainly reddish-chestnut, with white eyes, greenish-black back, and conspicuous white undertail. Sexes dissimilar, but no seasonal differences apart from eclipse. Juvenile closely resembles adult ♀, but distinguishable.

ADULT MALE. Reddish-chestnut head, neck, and breast, looking rich mahogany in some lights; yellower sides and flanks; black-brown upperparts and wings glossed dull green; and well-defined white undertail conspicuous at considerable ranges. Silky white belly, shading into grey towards rear, not visible when swimming; white spot on chin, black-brown collar round lower neck, and white wing-mark all inconspicuous. In flight, reddish-brown colour often still obvious; above, looks mainly dark with conspicuous, broad, somewhat curved stripe of white along whole length of wing (i.e. whiter on primaries than in Tufted Duck *A. fuligula*); below, white belly and staring white undertail framed by dark flanks and retrices contrasting with mainly white underwings, thus producing more variegated pattern than other *Aythya*. Bill slate-grey with grey tip and edges; eyes white; feet black marked greenish-yellow. MALE ECLIPSE. Like adult ♀, but redder-brown on top of head and breast, and paler brown on sides of face, neck and mantle; more important, eyes still white and white undertail still sharply defined. ADULT FEMALE. Duller and browner than adult ♂ with red-brown crown but more golden-brown sides of head and neck shading to dull red-brown on foreneck and breast, and brown on flanks and sides. Rest of upper and underparts much as ♂, but white at base of bill like some other *Aythya*. Flight pattern as ♂, but looks more chocolate-brown. Bill duller black than ♂, but still with grey tip and edges; eyes brown; feet greenish-grey with blacker webs. JUVENILE. Similar to adult ♀, but more uniform above. Head, neck, and breast dark brown with only slight reddish tinge, belly mottled brown and white, and undertail partly obscured with brown markings. After moult during 1st winter, juveniles become similar to adults; eyes of immature ♂ turn grey by March.

Adult ♂ quite unlike any other west Palearctic duck. Duller, brown-eyed female looks not unlike ♀ *A. fuligula*,

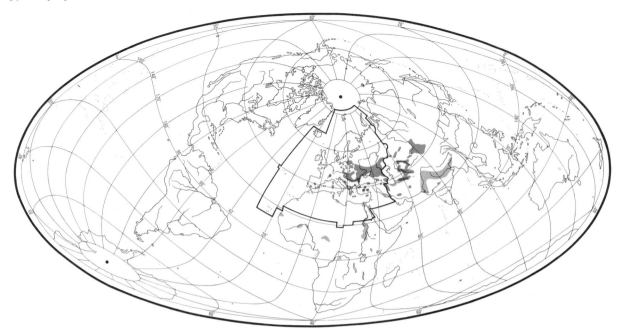

but high crown, more slender shape and white undertail should preclude confusion. (Some *A. fuligula* also show white on undertail, but this seldom so extensive and never so sharply defined.) Immatures, still browner and more uniform, might be confused with ♀ Pochard *A. ferina* which, however, has paler and greyer back, no rufous tinge, hoary face, and grey wing stripe. Broad, curved white wing-stripe of *A. nyroca* characteristic in all plumages. When identifying *A. nyroca* outside normal range, possibility of hybrids resembling one of parents or even a third species must be borne in mind: in particular, ♀ hybrids of *A. ferina* × *A. nyroca* at any distance (see details in Gillham *et al.* 1966).

Usually found in overgrown freshwater pools, lagoons, and marshes, but in some areas regularly on sea in winter. Generally rather secretive and tending to skulk among reeds, but not easily alarmed. Sits high on water with tail held less flat than *A. fuligula*. Dives extremely well, usually in shallow water, and will often swim under water away from danger; will also dabble or up-end for food in shallows. Rises from surface more easily than other European *Aythya*, sometimes almost springing into air. Flight like *A. fuligula*, but not so strong, usually fairly low, and only for short distances. Though sometimes in very large flocks in winter quarters, usually in pairs or small parties of up to 10–15 at most; seldom mixes with other species. Generally rather silent; ♂ has low grating wheeze, while ♀ has quiet version of harsh growl common to most *Aythya*.

**Habitat.** Concentrated in lowland, continental middle latitudes with no oceanic or arctic fringe, extending to high altitudes only marginally in central Asia. Among most characteristically steppe-based Anatinae, experiencing corresponding fluctuations in numbers and range but remaining fairly constant in choice of habitat. Prefers fairly shallow expanses of water, rich in submerged and floating vegetation, fringed by dense stands of emergent plants such as reeds *Phragmites*, often with willows *Salix*, alders *Alnus*, and other trees. Failing suitable fresh waters, saline and alkaline pools or wetlands with ponds and channels utilized. Coastal waters, inland seas, and large open lagoons sometimes frequented on passage or in winter. Flowing rivers or streams, oligotrophic or deep lakes, and open or exposed waters of all kinds normally avoided. Largely non-terrestrial; nests on anchored floating vegetation or on islands and banks with immediate access to water. Almost unaffected by character of wider environment, provided free from serious disturbance or pollution. Colonizes some regulated waters, including fishponds. Although some open water essential, this can be in smaller patches and more confined among heavy vegetation growth than would be tolerated by most Anatinae, in this resembling smaller grebes (*Podiceps* and *Tachybaptus*). Being adapted to relatively closed habitats, more given to swimming, diving, and resting on vegetation than to flying unless essential, when readily rises to some height.

**Distribution.** Range fluctuating and erratic even in main areas, due to changing water levels (Dementiev and Gladkov 1952), and on edges breeding often sporadic. In central Europe, expansion in mid-19th century and some local increases around 1900, though subsequent marked downward trend only recently halted (Bauer and Glutz 1969). Contraction and decline also in south of range, e.g.

Spain, Italy, and north Africa. BELGIUM: probably bred 1909 (Lippens and Wille 1972). FRANCE: now breeds regularly only Dombes; formerly nested occasionally elsewhere (FR). NETHERLANDS: breeds occasionally, first 1856, with 11 records in 20th century (Bauer and Glutz 1969); small numbers of escaped birds now nesting regularly (CSR). WEST GERMANY: breeding several sites 19th century and early 20th century; one occupied until 1963 (Bauer and Glutz 1969). EAST GERMANY: decrease in range with some areas abandoned as recently as 1950s (Bauer and Glutz 1969). POLAND: scarce, but perhaps breeds more widely than shown (Tomiałojć 1972a, d). USSR: sporadic nesting north of main range (Dementiev and Gladkov 1952). ITALY: now apparently only sporadic; no recent breeding records from south, Sicily, or Sardinia (SF; GS; Demartis 1973). BULGARIA: range decreased (Pateff 1950). GREECE: bred Crete 1943; may nest elsewhere (WB). ISRAEL: last bred *c.* 1956, when Lake Huleh drained (HM). MOROCCO: formerly bred north-west, last record 1942 (Heim de Balsac and Mayaud 1962; KDS). ALGERIA: formerly bred north-east (Heim de Balsac and Mayaud 1962); present status uncertain (EDHJ). TUNISIA: not known to breed (MS).

Accidental. Britain (now annual small numbers England though some refer to escapes), Ireland, Norway, Denmark, Sweden, Finland, Belgium, Canary Islands, Cape Verde Islands.

**Population.** Breeding. FRANCE: 1–5 pairs, probably decreased (FR). CZECHOSLOVAKIA: *c.* 100 pairs (Isakov 1970a); considerable fluctuations (Hudec and Černý 1972). HUNGARY: *c.* 80–100 pairs (Isakov 1970a); recent increase Hortobágy (Bauer and Glutz 1969). EAST GERMANY: *c.* 25 pairs (Isakov 1970a); considerable decrease last 50 years (Rutschke 1970). WEST GERMANY: 3 pairs; irregular (Szijj 1973). SPAIN: in marismas of Guadalquivir, formerly *c.* 500 pairs (Valverde 1960); great decrease, now few pairs (Ree 1973). RUMANIA: most common duck in lowlands (MT). USSR: *c.* 140 000 pairs, which 65 000 nest Dnestr-Dnepr and 10 000 Kuban, with remainder mainly outside west Palearctic in Kazakhstan and Aral Sea region (Isakov 1970b).

Winter. Estimated 75 000 birds Europe–Black Sea–Mediterranean region (Szijj 1972). USSR. *c.* 45 000 birds in western part in not so severe winters (Isakov 1970c).

**Movements.** Migratory, some perhaps only partially. Marked southward displacements away from northern half of breeding range, few wintering north of 46°N; routes

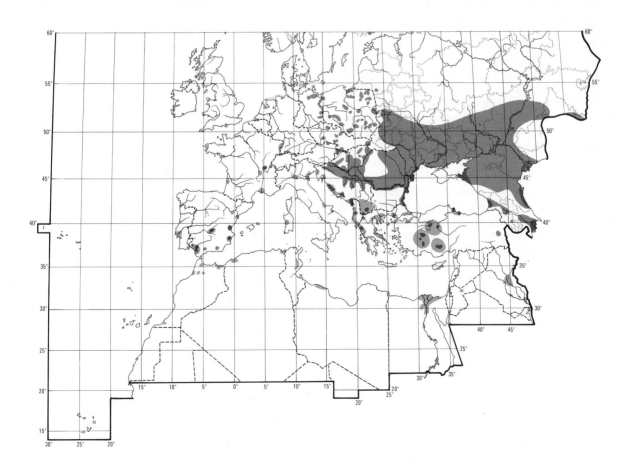

little known, as few ringed. Solitary and secretive behaviour persists in winter, e.g. not seen Lake Agoulinitsa, Greece, January 1970 though commonest duck in hunters' bags (Johnson and Hafner 1970). Does not form close flocks like other *Aythya* and so censusing difficult. Though winters throughout Mediterranean basin, Asia Minor, and south USSR, four-figure winter concentrations in west Palearctic reported only from Black Sea, Sea of Azov, and southern half Caspian Sea. New canal and reservoir wetlands in south Turkmenia hold many that probably wintered formerly on Caspian; link suggested by 6000 Caspian and 9400 south Turkmenia early 1967, but 3700 Caspian and 12200 south Turkmenia early 1968 (Isakov 1970c). Winter range extends south to Iraq and Iran, and has occurred November–December in Oman on Persian Gulf (Seton-Browne and Harrison 1968). Small to moderate numbers migrate (via Nile valley, and with direct Sahara crossings further west) to northern savannahs of tropical Africa, where commonest Palearctic diving duck, numbers decreasing east to west. Consistently common only Sudan, numbers irregular elsewhere, e.g. flock 1500 Kano, Nigeria, 1963–4 winter, but no other records there over 300 and apparently absent some seasons (Elgood *et al.* 1966; Moreau 1967; Roux 1970a). In January 1972, 230 in Sénégal delta where previously thought only straggler (Roux 1972); has wandered south to Sierra Leone, Cameroon, and Kenya (Roux 1970; Backhurst *et al.* 1973).

Few long-distance recoveries; 2 winter-ringed Netherlands and Camargue recovered north-west Germany (Niedersachsen) in March and west Poland in September respectively; Czechoslovakian juveniles found Italy (December, March) and Brittany (March), and one from Hungary recovered Yugoslavia in August. Apparent absence of moult movements unusual among Palearctic ducks, but not unique (see Marbled Teal *Marmaronetta angustirostris*). Departures from northern breeding localities begin September, peak October. Vanguard arrives southern wintering areas, even below Sahara, by late October, while from more southerly parts of breeding range departures delayed until onset of cold weather; migration on lower Dnepr and Dnestr September to mid-October, with large numbers gathering then at mouths, dispersing as estuaries and shallower lakes freeze. Thus in Hungary, 1967, 5800 September, 900–1000 October–November, but only 5 in December (Szabo 1970); 13 000 in Danube delta November–December 1968 but only 1000 January 1969 (Johnson and Hafner 1970).

Return movement from early March apparent Mediterranean basin, Black Sea, and south-west Caspian. First reach central Europe breeding places mid-March to early April, north Ukraine and Semipalatinsk around mid-April, and Tomsk early May. Passage continues, even in north-west Africa, well into May, but such late movement likely to involve non-breeders. See also Bauer and Glutz (1969); Dementiev and Gladkov (1952).

**Food.** Varied, plant materials predominating. Taken from surface, or when swimming with head submerged or up-ending, and by diving. In Hungary and Greece, each method occurs approximately equally often (Szijj 1965c; Sterbetz 1969). Mainly in shallow water, usually with rich littoral vegetation. In Hungary, in natural waters, artificial ponds, and rice fields mostly in depths 20–70 cm, and up to 1·6 m (Sterbetz 1969). Diving times up to 40–50 s, but probably normally less (Witherby *et al.* 1939). Mainly daytime feeding, evening and morning.

Data limited. Mainly plants apparently, but seasonal and regional differences not studied. Chiefly seeds and other parts of aquatic plants; pondweed *Potamogeton*, sedge *Carex*, hornwort *Ceratophyllum*, frog-bit *Hydrocharis*, *Polygonum*, *Bolboschoenus maritimus*, cockspur *Echinochloa*, water-lily *Nymphaea alba*, stonewort *Chara*, and duckweed *Lemna*. Animal materials include small fish and spawn, tadpoles and frogs up to 3 cm long, annelids, molluscs, and crustaceans; also insects, especially dragonflies (*Libellula*, *Agrion*), waterbugs, caddisflies (Phryganeidae), waterbeetles, and flies (Millais 1913; Witherby *et al.* 1939; Dementiev and Gladkov 1952; Bauer and Glutz 1969).

100 stomachs collected throughout year in Hungary from rice fields, flooded river areas, small ponds, and large lakes contained mainly plant materials; in order of frequency, seeds of bristle-grass *Setaria*, *Potamogeton*, *Polygonum*, *Carex*, *Bolboschoenus*, *Echinochloa*, barley *Hordeum*, rice *Oryza*, saltwort *Salsola*, and maize *Zea*; also leaves and shoots of *Chara*, *Lemna*, and various grasses, with animal materials less frequent, mainly insects, especially *Chironomus* larvae, waterbeetles (particularly Hydrophilidae), waterbugs (*Notonecta*, *Sigara*, *Corixa*), and dragonfly larvae; also occasional crabs *Branchinecta ferox*, molluscs, fish, and frog remains (Sterbetz 1969). 31 stomachs from flood area of River Tisza, Hungary, contained mainly seeds of *Setaria glauca* and *Potamogeton*, whole *Lemna* and *Chara*, chironomid larvae, and mollusc *Dreissena polymorpha* (Sterbetz 1969–70). Also from Hungary, in September, dodder *Cuscuta lupuliformis*, persicaria *Polygonum lapathifolium*, *Chenopodium album*, *Scirpus*, and *Amaranthus retroflexus* (Thaisz 1899). From Lake Geneva, Switzerland, mainly seeds, and fish and insect remains (Madon 1935). In USSR from Lenkoran wintering quarters, basically seeds (chiefly *B. maritimus* and *E. crus-galli*), with less green parts and insects (Dementiev and Gladkov 1952).

**Social pattern and behaviour.** For general information, see Witherby *et al.* (1939), Bannerman (1958), Bauer and Glutz (1969); also Géroudet (1965).

1. For most of time, less gregarious than other west Palearctic *Aythya*. Outside breeding season, mostly singly and in small, loose parties of 2–5 birds. Where common, found in larger numbers (100–300) on a few waters only during period between end of post-breeding moult and departure for wintering area. BONDS. Monogamous pair-bond of seasonal duration. Pairs

formed late, from January onwards (Steinbacher 1960); most birds arrive on breeding ground in pairs (Dementiev and Gladkov 1952), with ♂ leaving ♀ sometime during incubation. Only ♀ tends young. BREEDING DISPERSION. Where common will often nest in small colonies at protected sites, sometimes in association with *Larus* gulls; otherwise, nests well dispersed in concealed sites. ♂ waits for ♀ during early part of incubation period, accompanying her back to nest after each recess. ROOSTING. Communal. Rests periodically both at night and especially during day; particularly active at dusk, feeding mainly at night. Sleeps both on water, especially in middle of day, and ashore, on small islands or hidden among shore vegetation. Roosts on sea in India (Ali and Ripley 1968).

2. Little studied; for observations in captivity, see Steinbacher (1960), Johnsgard (1965). Resembles scaup-like ducks more than typical pochards. FLOCK BEHAVIOUR. Repeated Chin-lifting only pre-flight movement mentioned. ANTAGONISTIC BEHAVIOUR. Inciting of ♀ consists of (1) attacking sallies directed frequently at rejected ♂♂, followed by (2) swimming back to preferred ♂ in Neck-stretch posture while calling and Chin-lifting. COMMUNAL COURTSHIP. From mid-January to end May in captive birds (Steinbacher 1960), but as early as mid-October in wild (Lebret 1961). Apart from Head-shake (see below), Upward-shake only secondary display of ♂ mentioned and that infrequent (McKinney 1965b). ♂ has Courtship-intent posture with head moderately raised and crown feathers erected (Steinbacher 1960); also frequently depresses tail sharply into water so white under tail-coverts show as triangular patch on each side (Johnsgard 1965; see A). Of major displays, Head-throw, Kinked-neck, and Sneak accompanied by main Courtship-call (Johnsgard 1965); this also given without special posturing, but *A. nyroca* less vocal than either Pochard *A. ferina* or Tufted Duck *A. fuligula* during courtship. Frequently displays while swimming forward. Head-throw quicker than in *A. ferina*; preceded by Head-shake and probably also by Head-flick. In Kinked-neck display, ♂ stretches neck forward and up, then bends it back as call given (see A), upper part of neck suddenly swelling backwards enlarging apparent size of head. Sneak-display distinctive and common; usually of incomplete type (Crouch-display) with head inclined forward and bill pointing obliquely, but full version with neck stretched on surface of water also performed (see Millais 1913). Crouch-display immediately followed by Neck-stretch in which neck feathers usually depressed tightly while those of crown, side of head, and chin erected; at times, neck enlarged both during full Sneak-display and Neck-stretch, while bill may be elevated to 75° (Millais 1913). Coughing-display conspicuous and fairly frequent; ♂ rapidly flicks up tips of wing while uttering Coughing-call (Johnsgard 1965); said to assume Neck-stretch posture while so doing (see Bauer and Glutz 1969). In final group of silent displays, ♂ (1) often Head-nods while swimming, jerking neck

forward and backward, or up and down; (2) swims in front of ♀, e.g. when latter Inciting, ceremonially Leading her while Turning-back-of-Head. Activities of responsive ♀ during communal courtship varied: as well as Inciting not infrequently performs ♂-like Head-throws and Kinked-neck displays. Courtship-flights not recorded. PAIR BEHAVIOUR. Pair-courtship presumably, as in other *Aythya*, consists partly of same displays used in communal courtship plus copulatory and other behaviour. Role of Mock-preening outside context of pre-copulatory display uncertain, however, though Preen-behind-Wing and Preen-dorsally by ♂ (McKinney 1965b) and Preen-behind-Wing occasionally by ♀ (Johnsgard 1965) occur at times in unknown circumstances. COPULATION. Imperfectly known. Pre-copulatory display of ♂ consists of repeated Bill-dipping and Preen-dorsally (Johnsgard 1965); ♂ may also Preen-behind-Wing (McKinney 1965b). Displays of type occurring during communal courtship may also at times lead to copulation; ♀ noted once to dive after copulation, then both sexes drank (see Bauer and Glutz 1969). RELATIONS WITHIN FAMILY GROUP. No details.

(Fig A after Johnsgard 1965.)

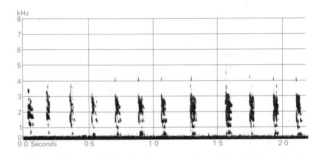

I C Chappuis France March 1965

**Voice.** Generally similar to other *Aythya*, with calls of ♀ louder and harsher than ♂'s. Mainly associated with courtship but information confused and fragmentary.

CALLS OF MALE. 3 Courtship-calls recognized by Steinbacher (1960): (1) quiet, soft 'weck'; (2) louder, hoarse 'wück' or 'wückwück'; (3) loud and much rarer 'witt'. Call 2 given during Kinked-neck and Sneak-displays; also accompanies Head-throw according to Johnsgard (1965) who rendered it 'wheeoooo' and also distinguished (4) rather high-pitched 'WEE-whew' as Coughing-call. Recording (see I) does not fit any of above closely: rapid chatter of brief, harsh, slightly metallic but muffled 'chk' notes, audible only at fairly close range and delivered at *c.* 6 notes per s for up to 3 s; occasionally given singly. Probably closest to call 2. May also be same as wheezing groan of Millais (1913) and low, rather grating wheeze of H Wormold (see Witherby *et al.* 1939).

CALLS OF FEMALE. More pleasant than in ♀ Scaup *A. marila* (Steinbacher 1960): (1) 'gek' or, when uttered quietly, 'dü'; (2) 'err'. Latter given with ♀-version of Kinked-neck display; further characterized as 'snoring' by Bauer and Glutz (1969). Call 1 presumably Inciting-call; this described as repeated 'gak-gak-gak' by Johnsgard (1965) who also distinguished (4) 'gaaak' associated with ♀-version of Head-throw display. Inciting-call higher

A

pitched and less gruff than in ♀ Pochard *A. ferina* but otherwise similar (Bauer and Glutz 1969). Identity of call given in flight uncertain; from recording (see II) loud, harsh, growling or croaking 'aark' repeated at *c.* 2 notes per s. Higher pitched and less hoarse than somewhat similar flight-call of *A. ferina* and usually uttered in much longer series. Also rendered 'grrr-grrr' (Voigt 1950) or 'kirr-kere, kirr' (Ali and Ripley 1968).

CALLS OF YOUNG. Described as rasping 'ürrr', with 'ü' emphasized, by Voigt (1950). Contact-calls uttered in groups of 2–4 notes. Distress-calls even more rapid than in *A. ferina*, but at same frequency (J Kear).

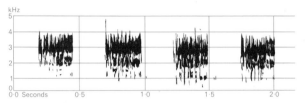

I C Chappuis France March 1965

**Breeding.** SEASON. See diagram for central Europe. Little apparent variation in most of range, but few records from Mediterranean. SITE. On ground close to water, or in water, in dense reeds and other water vegetation; in centre of clump, or sometimes on top of large grass tussock. Nest: low platform of grass, reed or rush stems, and leaves, with shallow cup; lined down. Diameter 20–22 cm, height 8–10 cm (Kux 1963). Building: by ♀, using material within reach of nest. EGGS. Blunt ovate; pale buff or pale brown. 53 × 38 mm (48–60 × 35–43), sample 160 (Schönwetter 1967). Weight of captive laid eggs 36 g (31–41), sample 63 (J Kear). Clutch: 8–10 (6–14). One brood. Replacement clutches laid after egg loss. One egg laid per day. INCUBATION. 25–27 days. By ♀. Starts on completion of clutch; hatching synchronous. Eggs covered with down when ♀ off nest. Eggshells left in nest. YOUNG. Precocial and nidifugous. Self-feeding. Cared for by ♀, brooded at night when small. FLEDGING TO MATURITY. Fledging period 55–60 days. Become independent at or before fledging. Age of first breeding 1 year. BREEDING SUCCESS. No data.

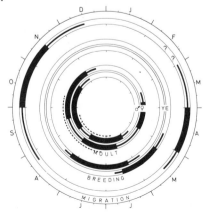

**Plumages.** ADULT MALE BREEDING. Head and neck rich chestnut-red, tinged purple on crown and hindneck. Small spot on chin white, band round lower neck black. Centre of upper mantle, lower mantle, and scapulars black, finely speckled chestnut, especially at tips of feathers. Back, rump, and upper tail-coverts black. Chest, upper breast, sides of breast, and upper flanks rich chestnut grading to chestnut and cinnamon on lower flanks. Lower breast, belly, and under tail-coverts white. Centre of vent white mottled grey-brown, sides of vent grey-brown freckled buff or grey. Tail dark grey-brown. Primaries white; edges of outer webs of outer 3–5 and all tips black. Secondaries white, tipped black with narrow white terminal edges. Tertials glossy dark green; narrow black edge on outer web. Greater upper wing-coverts dark olive-grey, rest of upperwing olive-brown. Underwing (except grey-brown marginal coverts) and axillaries white. ADULT FEMALE BREEDING. Like adult ♂, but generally duller; less white in primaries. Feathers of crown glossy purple-black with chestnut tips, hindneck sepia-black, rest of head chestnut-brown; some white mottling on throat. Mantle and scapulars brown-black, in some with slight cinnamon speckling and rufous-olive edges at tips of feathers. Underparts like adult ♂ breeding, but chestnut of chest, breast, and flanks paler, and some dark grey-brown of feather-bases visible, especially on belly. Centre of vent dark grey-brown, feathers tipped cinnamon or pale buff. Wing like adult ♂, but more primaries washed grey near tip, e.g. p6 grey over much of its length (showing little contrast with black-brown tip and outer edge), instead of white with faint grey hue (strongly contrasting with black tip and edge), as adult ♂. When feathers worn, sides of head, face, and throat paler cinnamon-buff, and feather edges of mantle and scapulars bleached buff. ADULT MALE NON-BREEDING (Eclipse). Like adult ♂ breeding, but head and neck duller chestnut; crown, sides of head, and hindneck dull black; upperparts black without much speckling, feathers edged olive-brown. Feathers of chest, upper breast, and flanks paler chestnut, tipped buff or rufous-cinnamon. On underparts, some grey-brown of feather-bases visible. ADULT FEMALE NON-BREEDING. Like adult ♀ breeding, but head paler brown, feathers of throat and foreneck white narrowly tipped chestnut; much grey-brown mottling on underparts. DOWNY YOUNG. Crown, hindneck, upperparts, sides of body, and sides of vent dark grey-brown; darker, brownish-black on lower back; longer filaments tipped olive-yellow on crown, mantle, and sides of body. Patches on wing, at sides of back, and tiny spots at sides of rump yellow. Sides of head (except narrow brown line through eye in some), chest, and breast golden-yellow; throat, neck, and centre of belly and vent pale yellow. JUVENILE. Like adult ♀ breeding, but head and neck duller and paler; sides of head cinnamon-brown, throat buff, and hindneck grey-brown. Feathers of underparts narrow and short, with pale grey-brown bases and buff tips on chest, upper breast, and flanks, and narrower white tips on lower breast, belly, and centre of vent. Under tail-coverts white with some grey-brown spots and bars. Tail-feathers narrow, notched at tip, with bare shaft projecting. Wing like adult, but slightly less green gloss. About 95% of ♂♂ distinguishable from ♀♀ by more extensive white near tips of outer primaries (see under Adult Female Breeding). FIRST IMMATURE NON-BREEDING. Part of plumage on head, neck, chest, flanks, and scapulars like adult non-breeding; remaining plumage juvenile. FIRST BREEDING. Like adult breeding, but amount of new feathers acquired highly variable; usually at least some of narrow juvenile belly feathers retained until 2nd summer, and juvenile wing-coverts narrow and frayed.

**Bare parts.** ADULT MALE. Iris white. Bill slate-grey, paler near tip and sides of upper mandible; nail black, lower mandible darker grey. Leg and foot olive grey, joints and webs black. ADULT FEMALE. Iris brown. Bill leaden-black, greyer near tip and at sides. Leg and foot like ♂, but duller. DOWNY YOUNG. Iris brown. Upper mandible dark olive-brown; cutting edges, nail, and lower mandible pale horn or pale flesh. Leg and foot dark horn-brown to black, front of tarsus and streaks on web at sides of toes olive-yellow. JUVENILE. Iris grey-brown, changing in course of winter to brown in ♀ and to pale grey in ♂. Bill leaden-blue to grey-black, with paler base. Leg and foot grey-blue with dull black webs and joints. (Bauer and Glutz 1969; Witherby *et al.* 1939; RMNH.)

**Moults.** ADULT MALE POST-BREEDING. Complete. Head, neck, scapulars, mantle, chest, and flanks June–July; followed by back, rump, and rest of underparts, but moult interrupted here during simultaneous moult of flight-feathers July–August. Tail at same time as wing. ADULT MALE PRE-BREEDING. Partial. Starts when wing full-grown with underparts, back, and rump August–October (probably partly moulted once a year only here), followed by scapulars, head, neck, and mantle late September to November. ADULT FEMALE. Like adult ♂, but post-breeding 1–1½ months later, and much overlap between post-breeding and pre-breeding; pre-breeding on mantle and probably elsewhere often not completed before winter, remainder replaced in spring. Head and neck, and possibly other parts of body, moult late March to May, but due to lack of specimens unsure whether this early post-breeding or late pre-breeding. POST-JUVENILE. Sequence of plumages and timing as in Pochard *A. ferina*, and equally highly variable. Advanced specimens have attained much of 1st-breeding plumage (including tertials, but rarely all mantle, back, breast, belly, and tail) by late November; retarded ones have only head, neck, chest, sides of breast, and flanks in 1st breeding by then, and rest of body sometimes still in 1st non-breeding or

juvenile plumage until spring. (Bauer and Glutz 1969; Dementiev and Gladkov 1952; C S Roselaar.)

**Measurements.** Mainly Netherlands, autumn. Skins (RMNH, ZMA).

| | | | | | | | |
|---|---|---|---|---|---|---|---|
| WING AD ♂ | 188 | (3·87; 31) | 180–196 | ♀ 182 | (2·39; 8) | 178–185 |
| JUV | 185 | (3·65; 27) | 177–192 | 177 | (2·98; 21) | 171–183 |
| TAIL AD | 54·0 | (2·36; 31) | 50–60 | 52·7 | (1·83; 8) | 50–55 |
| JUV | 51·4 | (2·37; 26) | 47–55 | 49·4 | (3·42; 18) | 46–52 |
| BILL | 40·3 | (1·27; 58) | 38–43 | 38·2 | (1·11; 28) | 36–40 |
| TARSUS | 32·7 | (0·93; 58) | 31–35 | 32·2 | (1·07; 29) | 30–34 |
| TOE | 54·2 | (1·68; 59) | 50–57 | 52·8 | (1·82; 29) | 50–57 |

Sex differences not significant, except bill and juvenile wing. Wing and tail of juveniles significantly shorter than adult (except wing ♂♂). Other measurements similar; combined.

**Weights.** Full-grown. (1) January–February: south-west Caspian, USSR. (2) March: River Ili, USSR, and Camargue, France; combined. (3) May–July: Kazakhstan (USSR), Denmark and Netherlands; combined. (4) September: River Ili. (5) October–December: Netherlands and Camargue; combined. (USSR: Dementiev and Gladkov 1952; Dolgushin 1960. Camargue: Bauer and Glutz 1969. Netherlands: ZMA. Denmark: Schiøler 1926.)

| | | | | | | | |
|---|---|---|---|---|---|---|---|
| (1) | ♂ | 583 | | 500–650 | ♀ 520 | | 410–600 |
| (2) | | 587 | ( – ; 9) | 480–740 | 545 | (54·5; 4) | 470–600 |
| (3) | | 589 | (59·2; 5) | 519–670 | 551 | ( – ; 2) | 535–567 |
| (4) | | 498 | ( – ; 5) | 440–540 | 513 | ( – ; 3) | 470–540 |
| (5) | | 589 | (100; 5) | 470–730 | 558 | (99·5; 5) | 464–727 |

Juveniles as heavy as adults from about September.

**Structure.** 11 primaries: p9 longest, p10 0–2 shorter, p8 3–8, p7 12–16, p6 21–28, p1 70–84. Bill slender, almost as long as head, cutting edges not upcurved towards tip. Base of bill slightly narrower than tip, culmen slightly concave. Outer toe *c.* 96% middle, inner *c.* 78%, hind *c.* 30%. Rest of structure as *A. ferina*.

CSR

## *Aythya fuligula* Tufted Duck

PLATES 79 and 99
[after pages 638 and 686]

DU. Kuifeend    FR. Fuligule morillon    GE. Reiherente
RU. Хохлатая чернеть    SP. Porrón moñudo    SW. Vigg

*Anas Fuligula* Linnaeus, 1758

Monotypic

**Field characters.** 40–47 cm, of which body almost two-thirds; wing-span 67–73 cm. Rather small, short-necked, compact diving duck with rounded head, drooping crest, and fairly large, broad bill; ♂ black with white side-panel. Sexes dissimilar, but only small seasonal differences apart from eclipse. Juvenile closely resembles adult ♀, but ♂ just distinguishable.
ADULT MALE. Mainly black, glossed purple on head and neck and green elsewhere, with boldly contrasting, pure white flanks and belly which stand out at long ranges. Long, black, drooping crest at back of head diagnostic when visible, but often inconspicuous. In flight, wings black with broad white stripe along whole length, becoming dingier and less well defined on primaries; mostly white underwing and white belly and flanks

contrast with otherwise black body. Bill dark grey-blue, usually with inconspicuous white band behind black nail; eyes golden-yellow; feet grey-blue. MALE ECLIPSE. Resembles adult ♀, but dark parts blacker, never any white round base of bill, and underparts whiter. Upperparts and breast dull brown-black with little gloss contrasting much less with sides and flanks (now pale brown) and belly (mixed with greyish-white and white); little or no crest. ADULT FEMALE. Rather variable but generally dark brown to almost black on upperparts, redder on cheeks and neck. In summer, breast typically dark brown flecked with yellow-brown, sides and flanks yellow-brown, belly red-brown, and under tail-coverts mottled white, but some more or less white on sides and belly. In winter, breast dark brown flecked with whitish, sides and flanks dull brown

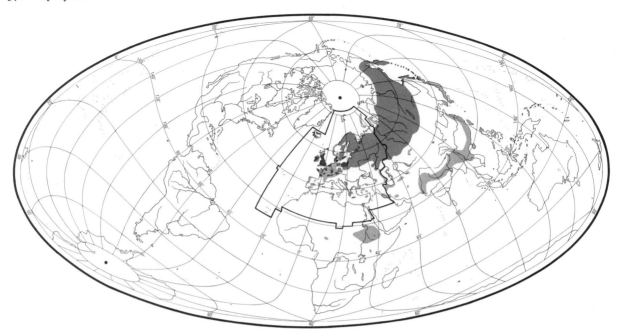

mixed varyingly with yellowish-white to white (showing some mottled whitish above waterline), belly white, and undertail whitish with variable dark brown markings. Many show a little white at base of bill and in winter this can be extensive and well defined; others develop nearly pure white undertails in autumn. Crest, much shorter than in ♂, diagnostic when visible. In flight, looks nearly black with wing stripe as in ♂. Bill dark grey with black nail; eyes yellow; legs pale greenish-grey. JUVENILE. Closely resembles adult ♀ but generally less white at base of bill (particularly juvenile ♀) and undertail, while ♂♂ have blacker sides of head and foreneck and white flecking on mantle and scapulars. Juvenile ♂ gradually becomes like adult ♂ by December onwards, when distinguished by short crest, barred undertail, and grey-brown vermiculations on sides and flanks.

Black and white adult ♂ unmistakable: ♂ Scaup *A. marila* has silvery grey back and no crest; see Ring-necked Duck *A. collaris* for differences from that species. 1st winter ♂ with short crest and grey-barred sides might be mistaken for *A. collaris*, which still distinguished by bill markings, head shape, spur of white in front of wing, and grey wing stripe. Dark brown ♀ and juvenile also distinctive with or without whitish sides, but need to be distinguished from ♀ and immature *A. marila* which have heavier-looking head with broader and usually more sharply defined area of white round larger bill, paler upperparts, and no crest. ♀ *A. collaris* has distinctive bill and facial markings, head shape, and grey wing stripe. ♀ Ferruginous Duck *A. nyroca* distinguished by sharply contrasted white undertail (though some ♀ *A. fuligula* do have conspicuous white undertail) and, in flight, by broader, more curved, white wing stripe and, from below,

white underparts restricted by dark flanks. (See *A. marila* and *A. nyroca* for problems of hybrids.)

Generally on freshwater lakes and rivers of fair size, but in summer on wide variety of reed-fringed or completely open waters, including large reservoirs, pools, slow rivers, or even marshland ditches; seldom on sea. Often becomes tame, particularly in town parks. Swims buoyantly but rather low in water, with tail on surface. Dives usually with distinct jump. Usually swims rather than flies away from intruders and, though rising more easily than Pochard *A. ferina*, still has to patter along surface. Flight straight and fast with rapid wing-beats producing whistling noise. Non-breeders usually in scattered parties up to 20–40, but often in much larger concentrations. Not infrequently stands or sits on shore and walks fairly easily for short distances with rather upright carriage. ♂ generally silent, apart from occasional low whistle, but ♀ has harsh growling call often heard in flight.

**Habitat.** Concentrated within upper middle latitudes, avoiding extremes of both heat and cold, and overlapping only marginally into tundra and steppe-desert zones. Increased greatly during recent decades in oceanic regions of west Palearctic, and no longer predominantly continental in distribution. Almost universally a lowland species. Capability for sustained diving to depths of 3–14 m enables use of deeper, more open fresh waters, frequently free of floating vegetation and not encroached on by emergent stands of reed *Phragmites* and similar marginal or shallow growth. Some preference for eutrophic waters where open, and in breeding season for moderate depths of 3–5 m. Avoids deep lakes with bottoms mainly below 15 m unless possess extensive shallower

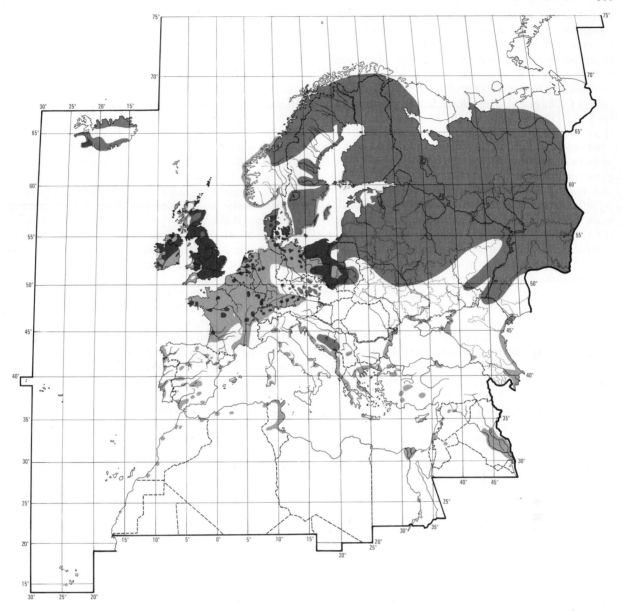

bays. Large exposed expanses also avoided, main preference being for areas between *c*. 100 and 0·4 ha. Quiet reaches of rivers and some sheltered coastal waters of more or less land-locked seas sometimes occupied; also nearly all types of man-made waters from barrages and large reservoirs to gravel-pits, ornamental ponds, even within city parks and fishponds. Strongly favours islets for breeding, normally close to water's edge, activity on land being slight. In winter, resorts to inlets and sheltered sea coasts, but in west Europe prefers lakes, reservoirs, and tidal estuaries, avoiding strong wave action and exposed maritime conditions unless frozen out from fresh waters. Vigorous flier making full use of surrounding airspace over fairly wide radius and to considerable heights. After

Mallard *Anas platyrhynchos*, no duck takes such full advantage of humanly provided habitats, or proves more tolerant of close human presence. See Olney (1963*a*).

**Distribution.** Marked spread and increase in west of range since late 19th century; according to Voous (1960), perhaps best regarded as distribution fluctuation in which climatic changes causing desiccation of lakes in southwest Asia played some part. Recent increases helped by ability to make use of man-made habitats (reservoirs, gravel-pits, and even lakes in city parks), while in Norway, Sweden, and Estonia, has apparently benefited from increase of Black-headed Gull *Larus ridibundus* and Common Gull *L. canus*, with which it often nests (Curry-

Lindahl *et al.* 1970; Onno 1970; Haftorn 1971). In Baltic, vulnerable to severe winters when marked decline with recovery after 3–4 years (Curry-Lindahl *et al.* 1970; von Haartman 1971). ICELAND: first recorded 1895; steady increase in range since and may be still increasing (Gudmundsson 1951; Gardarsson 1967). FAEROES: bred 1966–7 (Williamson 1970). BEAR ISLAND: pair may have bred 1932 (Løvenskiold 1964). BRITAIN AND IRELAND. Huge increase and spread late 19th and early 20th century (Alexander and Lack 1944). Continued England until late 1930s with further increases due to new reservoirs and gravel-pits; in Ireland has continued to expand since 1945, and also in Scotland, where may have slowed down recently (Parslow 1967). FRANCE: bred occasionally before 1940, regularly since 1964, colonizing new areas since 1966 (FR). BELGIUM: bred 1947, 1956; big increase 1969–70 (Lippens and Wille 1972). NETHERLANDS; recolonized 1904 after breeding in early 19th century, but sporadic until 1941; since spread and increased, especially in 1950s (Bauer and Glutz 1969; van Tol 1974). DENMARK: first proved breeding 1904; has since spread (TD). NORWAY: spread in south from about 1940 (Haftorn 1971). SWEDEN: before 1850, practically unknown outside Swedish Lapland, then bred southern Baltic; continued to spread along Baltic coast and inland, especially since 1925, with huge increase in south during 1950s (Curry-Lindahl *et al.* 1970). FINLAND: spread south over whole country mainly during 20th century (Grenquist 1970). WEST AND EAST GERMANY. Data for 19th century and early 20th century sparse, but some spread in northern areas more recently, and marked expansion Bavaria with first nesting in south in 1920 and in north 1952; much sporadic nesting elsewhere. For details and maps see Bauer and Glutz (1969). CZECHOSLOVAKIA: first bred 1914; marked extension in 1930s (Bauer and Glutz 1969). HUNGARY: bred 1965 and 1973, but doubtful if regular (AK, IS). AUSTRIA: first bred 1960 and has since spread (Bauer and Glutz 1969). SWITZERLAND: bred occasionally from 1958 and regularly since 1967. YUGO-SLAVIA: breeding for many years, apparently in small numbers, within mountain areas shown (Matvejev and Vasić 1973). ALBANIA: said to breed Lake Scutari (Ticehurst and Whistler 1932). RUMANIA: bred sporadically early 19th century, but no longer does so (Talpeanu 1964; Vasiliu 1968). BULGARIA: seen in nesting season in marshes along Danube and Black sea but no proof of breeding (SD). TURKEY: pairs seen in breeding season in 3 years since 1970 but nesting not proved (OST). CYPRUS: reported to have bred 3 areas in past, with eggs obtained 1910, but records have been disputed (Bannerman and Bannerman 1971; Stewart and Christensen 1971).

Accidental. Spitsbergen, Malta, Kuwait, Azores, Madeira, Cape Verde Islands.

**Population.** Breeding. ICELAND: *c.* 10000 pairs; probably now most numerous breeding freshwater duck (A Gardarsson). BRITAIN: *c.* 1500–2000 pairs (Atkinson–Willes 1970*a*). FRANCE: *c.* 50–100 pairs (FR). BELGIUM: *c.* 25 pairs (Atkinson-Willes 1972). NETHERLANDS: still increasing, with *c.* 2000–2500 pairs, 1973 (JW). WEST GERMANY: 2720 pairs (Szijj 1973). DENMARK: probably n.e. 500 pairs (Joensen 1974). FINLAND: *c.* 40000 pairs (Merikallio 1958). USSR. Baltic: at least 3500–4000 pairs, increasing (Kumari *et al.* 1970); Estonia: *c.* 5000 pairs (Onno 1970).

Wintering. Estimated 525000 birds in north-west Europe and 300000 in Europe–Black Sea–Mediterranean region (Szijj 1972). USSR: *c.* 618000 birds in western parts in not so severe winter but confusion with Scaup *A. marila* possible in large flocks on sea (Isakov 1970*c*). Oil pollution major hazard for birds wintering in western Baltic (*c.* 250000) and Black Sea (*c.* 200000) (Atkinson–Willes 1975).

Survival. Mean annual mortality adults ringed north-west Europe 46%, life expectancy 1·7 years (Boyd 1962). Oldest ringed bird *c.* 14½ years (BTO).

**Movements.** Predominantly migratory though largely resident in a few areas. Some Icelandic breeders winter on coast and other ice-free waters there, but most move to Ireland (chiefly) and Britain; minority continue to France and Iberia, with single recoveries Morocco, Norway, and Komi ASSR (latter abmigrant). Breeders of northern Britain mainly move south-west to Ireland, though Scottish juveniles recovered exceptionally north to Norway and south to Portugal; those of south Britain believed largely resident. Populations breeding north France and Netherlands perhaps also mostly resident, though some move west to Britain, especially in hard weather, while a few Dutch breeders recovered France and Switzerland and as abmigrants in Finland and Estonia. Breeders from Fenno-Scandia, north Germany, Poland, Baltic States, and USSR north of 55°N and east to *c.* 65°E (exceptionally to 73°E in Tyumen) mostly migrate south-west to Netherlands, Britain, and Ireland, many remaining on Baltic; limited onward movement as far as Iberia. A few recoveries of Baltic States birds in Switzerland suggest some passage over central Europe. Parallel flyway extending further north and east taken by birds from overlapping area of north USSR as far as France, Iberia, and north-west Africa. Those breeding central Europe mainly winter there while shifting up to 500 km south-west to WSW, but some from Czechoslovakia recovered WNW to England and north France (Hudec 1968). Breeders from central and south USSR, including west Siberia, migrate to east Mediterranean and Black Sea, with major concentrations Danube delta, Greece, and Turkey. Others reach northern Africa, some following Nile to Sudan and Ethiopia; recorded south to Tanzania and Malawi. Further west in Africa, very small numbers of unknown origins trans-Saharan migrants, reaching Chad and north Nigeria and exceptionally Mauritania, Sénégal, and Gambia; largest flock 60 Lake Chad, February 1963 (Elgood *et al.* 1966; Morel and Roux 1973). In Europe outside USSR, winters further north on average than Pochard *A. ferina*; in normal weather conditions over 70% winter north of line from English Channel to Adriatic, compared with only *c*. 10% of *A. ferina* (Bauer and Glutz 1969). Further wintering grounds on south Caspian, in Iraq and Iran, and a few down Persian Gulf to Oman; these birds come from western Siberia, east of south Urals, from breeding areas overlapping those of birds wintering Mediterranean.

Moult gatherings of ♂♂ more frequent than in *A. ferina*, but generally smaller; in Europe, tends to moult in areas where breeding local, indicating long movement involved, e.g. Matsalu Bay (Estonia), Krakower See (East Germany), Ismaninger reservoir (Bavaria), and IJsselmeer (Netherlands). Moult gatherings in USSR more numerous, but still small. Tendency for winter segregation of sexes results from sexual variation in timing of autumn migration due to different moult cycles, as in *A. ferina* (Salomonsen 1968). ♂♂ begin leaving breeding areas in June, moult migration reaching peak late July.

Autumn migration of both sexes begins September, and northern parts of breeding range vacated late October–early November. Begins arriving winter quarters early October, though peak numbers often not until January–February in central and west Europe as birds driven out of more northern parts by cold. Spring return movement more rapid, ♂♂ preceding ♀♀; begins late February in mild seasons, and in west and central Europe almost complete mid-April. Early arrivals on Fenno-Scandian breeding grounds late March, in north Russia and west Siberia about mid-May. See also Bauer and Glutz (1969), Dementiev and Gladkov (1952).

**Food.** Omnivorous: generally stationary or slow-moving items collected mainly from bottom and chiefly by diving, less often from emergent plants and from on and over water surface. Occasionally up-ends, and forages by wading in shallows, driftline, and ashore, especially for cereal grain. Gregarious and diurnal. In Czechoslovakia, 61% of daylight hours spent foraging, of which 54·5% by diving and 6·2% by skimming (Folk 1971). Mostly in depths 0·6–3 m, occasionally up to 4·9 m (Dewar 1924), 5 m (von Haartman 1945), or rarely over 7 m (Kennedy *et al.* 1954; Willi 1970). In depths 0·75–1·1 m, diving times 6–24 s, most frequently 14–17 s (Veselovský 1952). Maximum 20·1 s (Szijj 1965*c*) and up to 40 s (Dewar 1924; Groebbels 1942; Veřščagin 1950). Foraging ♂♂ remain underwater longer than ♀♀ (Willi 1970; Folk 1971). Food usually swallowed under water, though plant material in particular may be brought to surface. Occasionally takes dead or moribund fish (Gillham 1958), and regularly bread and scraps in some town parks.

Considerable variation in diet according to locality, year, and season, depending on availability and abundance of food. Obviously capable of adapting to existing trophic conditions. In coastal wintering areas, chiefly molluscs,

and inland often mainly animal materials including molluscs, crustaceans, and insects, though plants (especially seeds) may form major part of diet. Spring and summer diet more variable: in some areas mainly seeds, in others mainly insects or molluscs.

Winter marine areas: 193 stomachs from Danish coastal waters, mainly October–February, contained chiefly molluscs (90% of total), especially blue mussel *Mytilus edulis*, cockles *Cardium*, periwinkles *Littorina*, and laver shells *Hydrobia*, and only 5% crustaceans, 2% fish, and 3% plants, but, in 30 stomachs from Danish brackish water areas, 40% plants (mainly seeds of bulrush *Scirpus*, wigeon-grass *Ruppia*, and pondweeds *Potamogeton*), and 60% animal materials, chiefly molluscs, and fewer insects, crustaceans, and fish (Madsen 1954). 63 stomachs, south Swedish coasts, September–March, had similar contents, though some seasonal variation. In autumn and spring, mainly *Cardium lamarckii* and *Hydrobia* but in December *Mytilus edulis* more common and in March *Hydrobia* becomes dominant (Nilsson 1969). From south-eastern Caspian Sea, almost exclusively *Cardium edule*, and more recently from south-western shores *Mytilaster lineatus* (Isakov and Vorobiev 1940; Veřščagin 1950).

Winter inland areas: 54 stomachs, September–January from gravel-pit near London contained mainly zebra-mussel *Dreissena polymorpha* (82% by frequency, 81·1% by volume). 20 stomachs, September–January, from Northern Ireland (where *Dreissena* absent), contained mainly molluscs (39·4% by volume), especially *Hydrobia jenkinsi*; crustaceans (20·1%), particularly freshwater shrimps *Gammarus* and *Asellus*; and plants (32·3%), mainly seeds of *Potamogeton*, *Scirpus lacustris*, milfoil *Myriophyllum*, and spike-rush *Eleocharis palustris*; 8 stomachs from English inland waters had similar diet proportions, though more crustaceans taken and fewer molluscs (Olney 1963*a*). 6 stomachs from gravel-pit Kent, England, contained mainly molluscs, *Hydrobia jenkinsi*, *Pisidium*, *Anodonta* and *Lymnaea*, while 3 from nearby river all contained caddisfly larvae *Hydropsyche angustipennis*, and one also crayfish *Astacus pallipes* (Olney 1967). In contrast, 38 stomachs October–February from Klingnauer reservoir, Switzerland, had 84·6% plant detritus, 9% chironomid larvae, 4% *Gammarus*, 2% other animals, and only 0·4% molluscs. Some seasonal variation apparent, with more *Gammarus* and caddisfly larvae eaten beginning of winter, then, as these supplies depleted, more plant detritus taken (Willi 1970). Rybinsk reservoir, USSR, September–October, stomachs contained 44% molluscs (chiefly *Valvata*), 26% caddisfly larvae, 10% chironomid larvae, 5% dragonfly larvae and water beetles, and 10% seeds, 5% green plants; at this time obtained mainly in flooded shallows by pecking items from plants. Elsewhere in USSR inland lakes, only 14% molluscs, 26% crustaceans, and 46·6% seeds (Isakov and Vorobiev 1940). 8 stomachs October–November from Mologa river estuary, contained 40% animal material (mainly molluscs) and 60% plant materials (Isakov and Raspopov 1949).

Spring and summer inland: 98 stomachs southern Czechoslovakia, March–October, contained mainly seeds (especially *Potamogeton* and *Polygonum*) and fewer insects, particularly waterbug *Notonecta*, beetles, and molluscs; proportion of animal food increased with season (Folk 1971). 33 stomachs, Mologa river, USSR, June, contained only 5% seeds, 10% green parts of plants, and 85% animal materials (Isakov and Raspopov 1949). Yet, 11 stomachs from upper reaches Lena river, August and September, contained predominantly plant materials, mainly green parts and fewer seeds of buckbean *Menyanthes trifoliata*, sedges *Carex*, and fish (Verzuckij 1965). In spring and autumn in Pechora, chiefly molluscs (especially *Lymnaea*), some caddisfly larvae and beetles, and fish; over 50% stomachs also contained plants, especially leaves and roots of pondweeds. On Rybinsk reservoir in spring, molluscs occupied 32% of stomach content volume (mainly *Unio*, and some *Pisidium* and *Valvata*); caddisfly *Phryganea grandis* also 32%, fish 16%, and seeds 10% (Dementiev and Gladkov 1952). 72 summer birds from Lake Mývatn, Iceland, contained mainly molluscs, chiefly *Lymnaea*, chironomid larvae and some eggs of stickleback *Gasterosteus* (though only early in season), and less seeds of marestail *Hippuris vulgaris* and *Potamogeton*; no marked seasonal changes noted (Bengtson 1971).

Ducklings take more food from surface than adults, and on Lake Mývatn newly hatched and small ducklings fed almost entirely on seeds and adult insects; when about half grown, *Lymnaea* and chironomid larvae became important (Bengtson 1971). Downy young from Mologa river, USSR, took mainly molluscs (70% by frequency), especially *Viviparus duboisiana* and *Physa* picked from plants, and chironomid larvae. Older ducklings from Rybinsk ate molluscs (44%, by frequency), caddisfly larvae (20%), and chironomids (33%) (Isakov and Raspopov 1949). In Finland, young fed mainly on molluscs (*Theodoxus*, *Bithynià*, and *Lymnaea*), chironomids, and seeds of *Potamogeton* (Fabricius 1959). Small ducklings feed mainly on insects and molluscs seized in beak, but, as beak grows longer and broader, more by sieving on surface (Veselovský 1951).

**Social pattern and behaviour.** For general information, see Witherby *et al.* (1939), Bannerman (1958), Géroudet (1965), Bauer and Glutz (1969).

1. Highly gregarious for much of year. Outside breeding season, in large flocks (up to several thousand) in many areas, especially during winter. Sex composition of winter flocks unbalanced throughout range; probably due in part to differing feeding requirements of ♂♂ and ♀♀ (see, e.g., Folk 1971). ♂♂ first to flock (from June) after termination of pair-bond. BONDS. Monogamous pair-bond of seasonal duration; maintained from late winter in some cases, usually until just after ♀ starts incubation. Due at least partly to unbalanced sex-ratios in winter, pair-formation often delayed until spring. In south Germany,

most pairs formed March–April (Bezzel 1959). On Lake Mývatn, Iceland, some ♀♀ remain unpaired until late May or early June despite surplus of ♂♂ (Bengtson 1968). Simultaneous bigamy on part of ♂ occasionally recorded (Bezzel 1964; Reichholf 1965). Only ♀ tends young; ♂ rarely seen with brood, even where pairs scattered and ♂♂ tend to leave ♀♀ later. Parent-young bond ends before young can fly, most ♀♀ deserting brood mainly just prior to onset of wing moult (Gillham 1957, 1958). Even when tiny, young capable of surviving on own if weather favourable. Permanent amalgamation of broods under care of 1 ♀ rare; however, flocking of several ♀♀ and their young may occur (Bezzel and von Krosigk 1971) and some ♀♀ without young adopt lost or deserted young at times. BREEDING DISPERSION. Though solitary nests not uncommon, especially when situated far from water, often highly colonial, mainly on islands; frequent association with gulls *Larus* and terns *Sterna*, giving added protection to nests, feature of breeding biology (see Durango 1947; Bauer and Glutz 1969). Little information on home-ranges or territories of breeding pairs prior to incubation; reference to pair territories by Gillham (1958) not substantiated by details. Non-breeding adults, mostly ♂♂, present in some populations at least during breeding season (Gillham 1947, 1958). ROOSTING. Communal for most of year out on still, open water; said to be both diurnal and nocturnal, and though more active diurnally will feed nocturnally even on dark nights (Bauer and Glutz 1969). In Moravia, during 1967–70, active only by day, spending 23% of time preening, bathing, resting, and sleeping and remainder feeding and swimming, with main roosting period at night (Folk 1971).

2. Closely similar to Scaup *A. marila*. Published observations limited; see Boase (1926, 1950), Bezzel (1968), and, for studies on captive birds, Steinbacher (1960), Johnsgard (1965). FLOCK BEHAVIOUR. Feeding parties more integrated than Pochard *A. ferina*, birds often diving quickly one after the other or simultaneously (Bauer and Glutz 1969). Rapid and repeated Chin-lifting main pre-flight signal (Johnsgard 1965). ANTAGONISTIC BEHAVIOUR. ♂♂ in general peaceable. Aggressive rushes over water with lowered head and extended neck recorded from both sexes; also re-directed rush on mate by ♂ after encounter with another pair (Boase 1926, 1950). Inciting-display of ♀ consists of calling while alternating between (1) lateral threatening movements towards rejected ♂, as head executes quarter-turn with bill pointing obliquely down (Bezzel 1968), and (2) assumption of Neck-stretch posture with Chin-lifting (Veselovský 1966). Rape-attempts on water twice observed by Boase (1950). COMMUNAL COURTSHIP. Starts beginning November, Netherlands, but not common before December while Head-throw display of ♂ not seen before March (Lebret 1961); in Bavaria, between mid-January and mid-June with March peak (Bezzel 1968). Frequent calling of ♂♂ and excited swiming about give display parties appearance of greater activity than in *A. ferina*. ♂♂ adopt Courtship-intent posture with neck sunk in shoulders (Steinbacher 1960). Little use made of crest during courtship, though at times neck enlarged much as in *A. ferina* and eye blazes yellow due to contraction of pupil (Millais 1913). Secondary displays of ♂ include Head-shakes, frequent Head-flicks, and Upward-shakes. Of major displays, Head-throw and Kinked-neck display accompanied by 1st of 2 Courtship-calls (Johnsgard 1965). Head-throw relatively fast, lasting *c*. $\frac{1}{4}$ s. Usually preceded by Head-shake and Head-flick (McKinney 1965b) and often followed by a Bill-dip and Preen-behind Wing (Bezzel 1968). Bill-dipping and Preening-behind-Wing also sometimes performed on own, but probably essentially pair-courtship. In Kinked-neck display, neck not strongly bent as call

given. Full Sneak-display absent (Bezzel 1968) but inconspicuous version of Crouch-display, as in other scaup-like ducks, recorded (Johnsgard 1965; Bengtson 1968). Typical *Aythya* Neck-stretch display with thickened neck and raised crest often performed silently in front of ♀ (Bezzel 1968); Coughing-display version also frequent, ♂ giving Coughing-call while flicking tail and wings (Johnsgard 1965) with neck erect (see A). In final group of silent displays, ♂ (1) Head-nods while swimming; (2) ceremonially Leads ♀ (e.g. while latter Incites) by swimming in front of her in Turn-back-of-Head display with head feathers strongly depressed (Johnsgard 1965). Like Ferruginous Duck *A. nyroca*, ♀ will perform ♂-like Head-throw display at times, as well as Inciting (see above). According to Boase (1950), will also Head-nod, sometimes mutually with ♂. Frequently Bill-dips (Millais 1913) and often dives during communal courtship, followed underwater by some or all ♂♂; in some cases diving initiated by ♂♂ (Bezzel 1968). PURSUIT-FLIGHTS. Not seen in any form by Lebret (1961) or Bezzel (1968), but courtship-flights found to be frequent (Bengtson 1968) as also brief 3-bird flights (see Witherby *et al.* 1939). Rape-intent flights not recorded, though ♂♂ will, at least occasionally, attempt forced copulation with ♀ on water. PAIR BEHAVIOUR. Components of pair-courtship evidently much as in other *Aythya*; Head-throw by ♂ to own mate outside context of communal courtship probably important (see Bezzel 1968). ♂ accompanying ♀ courted by other ♂♂ assumes Courtship-intent posture while swimming close to ♀ on side furthest from rivals (Boase 1950). ♀ directs Preen-behind-Wing, often repeatedly, at ♂, producing non-vocal noise much as in Mallard *Anas platyrhynchos*; noted as greeting when joined by ♂ (McKinney 1965*b*). COPULATION. Initiated by ♂ (Johnsgard 1965). Pre-copulatory display consists mainly of repeated Preen-dorsally by ♂ with some Preening-behind-Wing and occasional Bill-dips; ♀ may respond with some of same movements before assuming typical *Aythya* Prone-posture, sometimes suddenly. In post-copulatory display, ♂ immediately performs Kinked-neck display (with call) then swims away in characterstic Bill-down posture (see B) as ♀ bathes. Copulation said at times also to be

preceded by Head-throw—Bill-dip—Preen-behind-Wing sequence by both sexes alternately, or by no display; similarly, post-copulatory activities may be confined to preening or diving by ♂ and ♀ (see Bauer and Glutz 1969; Bezzel 1968). RELATIONS WITHIN FAMILY GROUP. For general accounts, see Gillham (1957, 1958) and Hori (1966).

(Figs A–B after Johnsgard 1965.)

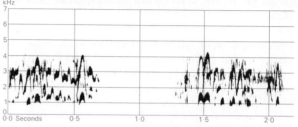

I  Sture Palmér/Sveriges Radio  Sweden  April 1960

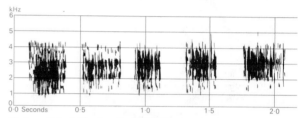

II  Sture Palmér/Sveriges Radio  Sweden  April 1960

**Voice.** Usually neither conspicuously loud nor so low as to be audible only at close range; largely confined to courtship period. Calls of ♂ mainly a musical whistling with drawn-out cadence; those of ♀ softer and more rolling than in other ♀ *Aythya*, though at times croaking and crow-like when distressed (E M Nicholson). Vocabulary still not fully studied. Non-vocal sounds include: (1) whistling of wings in flight (much less pronounced than, e.g., in goldeneyes *Bucephala*); (2) rattling of bill against quills of flight-feathers in Preen-behind-Wing display.

CALLS OF MALE. Only 1 main Courtship-call recognized by Steinbacher (1960): 'bückbückbück', often quite loud but at times uttered quietly; often introduced by single 'bück'. 2 Courtship-calls described by Johnsgard (1965): (1) mellow 'WHEE oo' during Head-throw and Kinked-neck displays; (2) rapid 3-note whistle 'WHA-wa-whew', 1st syllable highest and loudest, last more prolonged, during Coughing-display. This Coughing-call of Johnsgard (1965) same as call described by Steinbacher (1960); also used as flight-call (see I). In tempo, resembles whistle of Wigeon *Anas penelope*, although more creaky; may also be characterized as 'bubbling' and 'bell-like' and likened in rhythm to distant gobbling of Turkey *Meleagris gallopavo* by Bezzel (1968). Soft, ♀-like 'kack' and 'rr' sounds also mentioned by Steinbacher (1960).

CALLS OF FEMALE. 3 calls during courtship recognized by Steinbacher (1960): (1) 'quäck'; (2) 'gack'; (3) 'karr'. Of these, latter evidently Inciting-call; this described as soft 'kärr' by Johnsgard (1965). On recording (see II) sounds

A

B

like repeated 'quaark quaark quark', diminishing in intensity. Also rendered as growling 'körr' or 'krr' by Bezzel (1968). Guttural 'bre-bre-bre' described as flight-call (Bauer and Glutz 1969); firmer in timbre, quicker, higher, and more vigorous than corresponding utterance of Pochard *A. ferina*. ♀ leading young may give accented 'gŕr gŕr gŕr oŕr aŕr rŕr' when excited; otherwise call quite soft (Bauer and Glutz 1969).

CALLS OF YOUNG. Contact-call in groups of 2–5 notes. Distress-call rapid and high, faster than in *A. ferina* and similar to that of Ferruginous Duck *A. nyroca* (J Kear). Latter call described as persistent, piping 'risisisi' (Bauer and Glutz 1969).

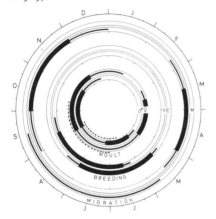

**Breeding.** SEASON. See diagram for northern Europe. Western and central Europe similar. Rather later in northern USSR where timing dependent in part on thaw. Mean date of first laying, Iceland 1961–70, 25 May, range 21 May–1 June (Bengtson 1972*a*). At Loch Leven, Scotland, 80% of clutches started in six week period; egg-laying coincides with greatest mass of chironomid larvae on loch floor, young hatching when greatest number of newly emerged adult chironomids on water surface (I Newton). SITE. On ground or in water, in rush or grass tussocks, under bushes; also in open, but then usually in colonies of gulls *Larus* or terns *Sterna* (a not uncommon site). Of 158 nests, Czechoslovakia, 89.9% beside or in water, 5.7% on islets, 4.4% on dry land (Harlin 1966*b*). Recorded on old wall 30 m from ground (Bannerman 1958). Not colonial but sometimes hundreds of nests at one locality; at Loch Leven nests regularly dispersed between 7–11 m apart (I Newton). Usually within 20 m of water, except on islands, where up to 150 m. Strong preference for islands where available. Nest: depression; lined grass, rushes, reeds, and down. Diameter 20–25 cm, cup depth 7–10 cm. Building: by ♀, using material within reach of nest, and shaping cup by turning movements of body. EGGS. Ovate; greenish-grey. 59 × 41 mm (53–66 × 38–46), sample 300 (Schönwetter 1967). Weight 53 g (46–65), sample 58, captive laid (J Kear). Clutch: 8–11 (3–22), but over 14 probably always 2 ♀♀. Dump nests occur, mean of 100 Icelandic 17.3 eggs (Bengtson 1972*a*). Of 72 1st

clutches, Finland: 7 eggs, 3%; 8, 6%; 9, 22%; 10, 27%; 11, 15%; 12, 8%; 13, 8%; 14, 3%; 15, 1%; 17, 6%; 19, 1%; mean 10.9 (Hildén 1964). Of 211 clutches, Czechoslovakia: 3 eggs, 1%; 4, 1%; 5, 7%; 6, 10%; 7, 16%; 8, 23%; 9, 17%; 10, 12%; 11, 4%; 12, 6%; 13, 3%; mean 8.25, with clutch size decreasing through season (Harlin 1966*a*). Mean of 636 Icelandic 1st clutches 10.1 (Bengtson 1972*a*). One brood. Replacement clutches laid after egg loss: 45% of 51 ♀♀ re-nested after loss of 1st clutch, Iceland; mean of 98 2nd clutches 8.0 (Bengtson 1972*a*). Of 20 Finnish 2nd clutches: 6 eggs, 15%; 7, 15%; 8, 25%; 9, 30%; 10, 5%; 11, 5%; 12, 5%; mean 8.3 (Hildén 1964). INCUBATION. 25 days (23–28). By ♀. Starts on completion of clutch; hatching synchronous. Eggs covered with down when ♀ off. Eggshells left in nest. YOUNG. Precocial and nidifugous. Self-feeding. Cared for by ♀; brooded while small. FLEDGING TO MATURITY. Fledging period 45–50 days. Become independent at or before this. In late seasons, Germany, some broods deserted at *c.* 4 weeks when moult of ♀ becomes imminent (Bezzel and von Krosigk 1971). In England, deserted from a few days old (rare) to 43 days; most 29–42 days, sample 34 (Gillham 1957, 1958). Can breed at 1 year, but some not until 2. BREEDING SUCCESS. Of 4342 eggs laid Finland in 3 years, 78% hatched and 11.4% young reared to fledging; annual survival per pair 2.2, 1.0, and 0.5, so losses of 76%, 89%, and 95% (Hildén 1964). Hatching success of 1115 nests, Iceland, 1961–70, averaged 71.8%, range 53.3–80.6% (Bengtson 1972*a*). Of 1692 Scottish nests, 57% hatched, 38% predated, and 5% deserted (I Newton). Of 1373 eggs laid Czechoslovakia, 1968–70, 59% hatched (Harlin 1972).

**Plumages.** ADULT MALE BREEDING. Head and neck black, glossed violet-purple. Sometimes small white spot on chin. Upperparts, chest, under tail-coverts, and lower vent black, finely speckled off-white on lower mantle and scapulars. Upper breast black, feathers tipped white; vent and sometimes lower flanks white, finely vermiculated grey; rest of underparts and flanks white. Tail-feathers dull black. Primaries grey, outer edge of outer 4–5 and all tips black, outer web of inner primaries white except tips and bases. Secondaries white, broadly tipped black on outer web, dark grey on inner. Upper wing-coverts and tertials black; median and lesser coverts often minutely speckled near tips. Underwing white; marginal coverts grey-brown, narrowly edged white; axillaries white, freckled grey-brown at tip and outer web. ADULT FEMALE BREEDING (Does not nest in this plumage). Head and upper neck black. Usually some white feathers round base of bill, especially on lores and chin, sometimes form clear white spot near sides of upper mandible, but always smaller and less clear-cut than in adult ♀ Scaup *A. marila*. Dark brown edges and bases of some head feathers may give brown tinge. Feathers of lower neck, upper mantle, and chest dark brown, tipped cinnamon; lower mantle and scapulars similar, but edges narrower and greyer, often with some white speckling. Back, rump, and upper tail-coverts black. Breast dark grey-brown, feathers broadly edged white; sides of body and flanks olive-brown or rufous-olive, partly vermiculated or suffused white. Vent pale grey-brown, feathers often tipped or vermiculated white. Under tail-coverts dark grey-brown, tipped and suffused white, sometimes all white. Rest of underparts white. Tail and wing like adult ♂, but minute specks on wing-coverts usually lacking; dark tips to secondaries narrower on average. ADULT MALE NON-BREEDING (Eclipse). Like adult ♂ breeding, but crest abraded; mantle and breast dull black with much white speckling; feathers of flanks grey-brown coarsely vermiculated black; off-white feathers sometimes near base of bill. Differs from adult ♀ breeding in longer crest; black mantle and breast with many white specks; greyer, more coarsely vermiculated flanks; more black on lower vent and under tail-coverts; speckled wing. ADULT FEMALE NON-BREEDING (Nests in this plumage). Like adult ♀ breeding, but tips of black feathers more broadly edged brown; mantle, scapulars, and flanks without white specks and vermiculations; feathers of lower breast and belly dark grey-brown, variably tipped white. In worn plumage, head dark brown with pale buff or white lateral spot near base of upper mandible; golden brown edges to upperparts, chest, breast, and sides of body; underparts more or less grey-brown or buff-brown, mottled white. DOWNY YOUNG. Crown, hindneck, upperparts, chest, and sides of body sooty brown; sides of head and chin dull yellow, with indistinct dark brown lines through and below eye. Centre of breast and belly pale buff or yellow. JUVENILE. Feathers of head, neck, and upperparts dark grey-brown, narrowly edged buff; darkest on crown, back, rump, and upper tail-coverts. Feathers on lores, near base of bill, and on chin pale buff. Chest, sides of breast and body, and flanks pale grey-brown, feathers edged buff to white. Feathers of underparts short and narrow; off-white, distinctly mottled grey-brown on shafts. Under tail-coverts white, subterminally blotched or barred grey-brown, giving spotted appearance. Tail-feathers narrow, notched at tip, bare shaft projecting. Wing like adult ♀, but often very finely speckled pale grey on median and lesser upper wing-coverts in juvenile ♂. Sexes otherwise similar; feathers of lower mantle and scapulars, however, moulted early, sometimes when wing not yet full-grown, resembling adult non-breeding; dull black speckled white in ♂; mostly plain dusky black edged paler brown in ♀. FIRST IMMATURE NON-BREEDING. In autumn, head dark brown (darkest in ♂; often with some white near base of upper mandible in ♀); mantle, scapulars and flanks like adult non-breeding; some juvenile feathers usually retained. FIRST BREEDING. Like adult breeding on head and neck from October, somewhat later on rest of body; sometimes resulting in complete breeding as early as December, but wing juvenile, and head often duller and slightly browner, especially in ♀♀. Some 1st non-breeding usually retained on lower mantle, outer scapulars, and flanks (♀♀ only rarely attain as strongly vermiculated flanks as adult breeding ♀♀). Juvenile lower breast, belly, vent and some tail-feathers sometimes retained. Much variation in timing (see Moults); retarded specimens may retain much juvenile or 1st autumn plumage until summer, even on head. SECOND NON-BREEDING. Like adult non-breeding, but sometimes juvenile plumage of underparts, and often some 1st breeding or 1st non-breeding feathers retained.

**Bare parts.** ADULT MALE. Iris golden-yellow. Bill light slate-blue; tip and nail black, usually separated from slate-blue by narrow white band. Leg and foot lead-blue with darker joints and black webs. ADULT FEMALE. Like ♂, but bill slightly darker slate; in some, dark horn with narrow slate band behind black tip of bill; leg and foot often greyer, with faint green tinge. DOWNY YOUNG. Iris brown-grey. Bill dark olive-brown, nail pink or white, lower mandible yellow. Leg and foot dark brown to olive-brown; sides of tarsus and edges of webs along toes yellow-green or grey. JUVENILE. After *c.* 1 month, iris changes to light yellow in ♂ and brown-yellow in ♀; tip and nail of bill darken to black and

rest of bill changes to slate. Adult colours attained during autumn, but iris of ♀ apparently not bright yellow before 2nd year and sometimes never. (Bauer and Glutz 1969; Witherby *et al.* 1939; RMNH.)

**Moults.** ADULT MALE POST-BREEDING. Complete; flight-feathers simultaneously. Body and tertials first (except head tuft), late May to early July; followed by back, belly, and vent, but moult interrupted here when flight-feathers shed, so some of these feathers thus replaced once a year only. Wing and tail late June to early September, flightless for 3–4 weeks. ADULT FEMALE POST-BREEDING. Complete. Body and tertials March–May, wing as in ♂, but 1–2 months later. Tail April–September; not known if replaced once or partially twice in this period. Head moulted twice a year, but timing uncertain; ♀♀ with fresh breeding plumage on head April (otherwise in full moult to non-breeding) may not moult until summer; same may apply to back or underparts. ADULT MALE PRE-BREEDING. Partial; head and body, starts when flight-feathers full grown. Back, underparts, head, and tail-coverts (September–October); followed by head tufts, mantle, scapulars, and flanks (about October to early November); some non-breeding feathers sometimes retained until mid-winter. Head tuft moulted only once a year; October–November. ADULT FEMALE PRE-BREEDING. Partial. Back, underparts, rump, tail-coverts, and chest about August–October; mantle, scapulars, and flanks September–October, sometimes still partly in non–breeding February. Head and neck variable; start August–November, finish October to early February. POST-JUVENILE AND FIRST PRE-BREEDING. Partial. Body (except sometimes part of belly and vent), tertials, and tail, from August. Much variation due to physical condition, local differences in hatching time, and variable occurrence of arrested moult in winter. Feathers replaced August–November resemble adult non-breeding (1st immature non-breeding plumage), November–April adult breeding. Some ♀♀ show little moult into breeding and change 1st non-breeding plumage and rest of juvenile directly for 2nd non-breeding in spring. 1st non-breeding develops August–November on head, neck, mantle, and scapulars; October–January on flanks; late October–November occasionally on chest, sides of breast, and vent. Breeding feathers replace juvenile and 1st non-breeding feathers on head in ♂ October to late December, in ♀ October to early February; on mantle and scapulars in both ♂ and ♀ November–March, but some ♀♀ do not start before moult-stop in winter and moult to 2nd non-breeding from March. Some juvenile and 1st non-breeding flank feathers changed for 1st breeding from December. Juvenile belly and vent retained until about June. Some in complete 1st breeding October–December. Tail highly variable; may start October, often before February, some much later. Tertials moult from November. FIRST POST-BREEDING. Like adult post-breeding, but usually part of fresh 1st breeding plumage, and often belly and vent, retained. (Bauer and Glutz 1969; Dementiev and Gladkov 1952; Witherby *et al.* 1939; C S Roselaar.)

**Measurements.** Netherlands, whole year. Skins (RMNH, ZMA).

| | | | | | | |
|---|---|---|---|---|---|---|
| WING AD ♂ | 206 | (3·78; 46) | 198–215 | ♀ 199 | (3·38; 40) | 193–205 |
| JUV | 202 | (4·20; 37) | 194–210 | 196 | (3·55; 36) | 185–203 |
| TAIL AD | 53·7 | (2·10; 44) | 49–58 | 52·7 | (2·72; 39) | 48–57 |
| JUV | 52·1 | (2·10; 14) | 49–56 | 49·9 | (2·51; 24) | 46–53 |
| BILL | 39·9 | (1·41; 66) | 37–44 | 38·6 | (1·27; 73) | 36–41 |
| TARSUS | 35·5 | (0·80; 40) | 34–37 | 34·7 | (1·16; 40) | 32–37 |
| TOE | 58·3 | (2·07; 38) | 55–63 | 57·1 | (1·96; 35) | 52–61 |

Sex differences significant, except adult tail and toe. Juvenile wing and tail significantly shorter than adult (except tail ♂). Other measurements similar; combined.

**Weights.** Full-grown. (1) and (2) November and December–March respectively, Camargue, France. (3) December–March, Switzerland (Bauer and Glutz 1969). (4) December–March, Netherlands (ZMA) and Denmark (Schiøler 1926); combined. (5) April–May, Kazakhstan, USSR (Dolgushin 1960). (6) May, Baraba Steppe, USSR (Dementiev and Gladkov 1952). (7) June–September, adults only, various areas USSR combined (Dementiev and Gladkov 1952; Dolgushin 1960).

| | ♂ | | | ♀ | | |
|---|---|---|---|---|---|---|
| (1) | 654 | (20) | 475–762 | 597 | (30) | 335–733 |
| (2) | 723 | (343) | 400–950 | 680 | (700) | 450–920 |
| (3) | 813 | (92) | 600–1020 | 718 | (58) | 560–930 |
| (4) | 925 | (11) | 835–1028 | 867 | (8) | 738–948 |
| (5) | 723 | (10) | 670–810 | 590 | (4) | 525–640 |
| (6) | 742 | (–) | 595–900 | 840 | (–) | 795–995 |
| (7) | 767 | (9) | 660–970 | 684 | (14) | 540–970 |

Juveniles reach adult weight from about September. In February, Azerbaydzhan, USSR, ♂♂ may reach 1400, ♀♀, 1150.

**Structure.** Wing rather long and narrow. 11 primaries: p10 and p9 about equal, longest; p8 4–10 shorter, p7 13–22, p6 25–37, p1 80–100; p11 minute. Inner web of p10 and outer p9 emarginated. Tail short, rounded; 14 feathers. Bill about as long as head, slightly higher than broad at base, wide and flat near tip. Culmen slightly concave, cutting edges much upcurved near tip of bill; nail narrow, rounded. Feathers of centre of crown narrow, elongated, forming long, curled tuft, 69·7 (55–80) mm long in 25 adult ♂♂, 58·5 (40–75) in 12 1st breeding ♂♂, 36·4 (25–45) in 12 adult and 1st breeding ♀♀; absent in juveniles. Outer toe slightly longer than middle, inner *c.* 80% middle, hind *c.* 32%.    CSR

*Aythya marila* **Scaup**                      PLATES 80 and 99
                                              [after pages 638 and 686]

Du. Toppereend    Fr. Fuligule milouinan    Ge. Bergente
Ru. Морская чернеть    Sp. Porrón bastardo    Sw. Bergand    N.Am. Greater Scaup

*Anas Marila* Linnaeus, 1761

Polytypic. Nominate *marila* (Linnaeus, 1761), north Europe and Asia east to about Lena River. Extralimital: *mariloides* (Vigors, 1839), further east in northern Asia, and western North America.

**Field characters.** 42–51 cm, of which body two-thirds; wing-span 72–84. Medium-sized, compact, broad-bodied diving duck with large, wide bill and large, round head, lacking any crest; ♂ has black head, breast, and stern, pale grey back, and white flanks and belly. Sexes dissimilar and seasonal differences in both. Juvenile closely resembles adult ♀, but just distinguishable.

ADULT MALE. Head and neck black, glossed green or, in some lights, purple; breast also black. Mantle, back and scapulars vermiculated white and brownish-black, giving pale grey effect; sides and belly white, with slight vermiculations on flanks and lower belly. Rump and upper and under tail-coverts black. Bill bright grey-blue; eyes golden-yellow; legs grey-blue. In flight, wings brownish-black with broad white wing-bar running almost whole length, but fading to grey on outer primaries; underwing and axillaries largely white, increasing contrast between black front and stern. MALE ECLIPSE. More easily distinguished from adult ♀, and less different from full plumage, than in most ducks. Head and neck browner, and only little white or whitish around base of bill; breast also brownish-black, with some white fringes. Upperparts still greyish with coarse vermiculations of white and black. Flanks vermiculated grey-brown; belly pale brown; under tail-coverts vermiculated black and white. Bill darker than in full plumage. ADULT FEMALE. In summer, head, neck, and breast dark brown, usually with cinnamon tinge; except in fresh plumage, increasingly clear whitish patch on ear-coverts; otherwise much as winter, but mantle and scapulars warmer yellowish-brown and not vermiculated. In winter broad band of white around base of bill. Head, neck, and breast brownish-black, and no pale ear-patch except in really worn plumage. Mantle and scapulars mainly brownish-black with slight grey effect produced by white vermiculations. Flanks light yellowish-brown; lower breast and belly white; under tail-coverts white with dark brown markings. Bill darker grey-blue than in ♂ and legs duller, but eyes still golden-yellow. In flight, wings as in ♂, brownish-black above with broad wing-bar, and largely white below. JUVENILE. Closely resembles adult ♀ winter, but white band around base of bill narrower, and often absent on forehead, or mixed with brown feathers; more white speckling on lower neck and upper mantle, and mantle and scapulars otherwise less vermiculated in juvenile ♂ and hardly at all in juvenile ♀; cheeks, sides, and front of neck paler yellowish-brown, particularly in juvenile ♀; juvenile ♂ and ♀ further distinguished by latter's lack of vermiculations on flanks. By late October to early December, vermiculations on upperparts become more marked, particularly in ♂♂, which assume duller version of adult plumage by March or April.

Adult ♂'s combination of black front and stern, grey back, and white sides distinctive enough and, at reasonable range, precludes confusion with all other species except similar Lesser Scaup *A. affinis* (North America, not yet recorded in Europe) which is slightly smaller and, more important, less broad-bodied and more buoyant; ♂ of latter species also has peaked rear to crown, more purplish head, greyer flanks, less white in wing (wing-bar hardly extends to primaries), and, diagnostic at close range, much smaller nail on bill. Note, however, that *Aythya* hybrids can resemble both scaup species (see Gillham *et al.* 1966). At longer ranges, ♂♂ of Pochard *A. ferina*, Tufted Duck *A. fuligula*, and even Mallard *Anas platyrhynchos* can cause confusion; all having dark fronts and sterns and lighter bodies, but see those species for differences. ♀ and immature *A. marila* most likely to be confused with corresponding stages of *A. fuligula* (apart again from ♀ *A. affinis* which has similar differences in proportions, shape, and wing-bar to ♂). In summer plumage, except when fresh, whitish patch on ear-coverts of ♀ *A. marila* diagnostic and in winter white band round base of bill broader and more clearly defined than in any *A. fuligula*, but juveniles and immature ♀♀ can be difficult to separate (immature ♂ *A. marila* distinguished by greyer back), although *A. fuligula* has smaller bill, less steep forehead, suggestion of crest, and much slighter build.

Breeds by moorland and tundra lakes and rivers, but for rest of year usually almost exclusively maritime, particularly in coast bays and estuaries, though in some areas winter concentrations normal on large expanses of fresh water; little affected by storms. Swims rather low in water with tail trailing; dives easily, even in deep and rough water. Walk ungainly and comes to land rather less than other *Aythya*. Rises with difficulty, but flies rapidly with quick beats. Generally silent except during courtship, but ♂ sometimes utters rapid series of soft, low whistles and ♀ has muffled growl varying in loudness and harshness.

**Habitat.** Most northerly of genus in breeding and wintering; also most marine, showing relatively little overlap in habitat. Even less continental and more lowland in range; summer habitats mainly low-arctic or sub-arctic, in tundra and wooded tundra zones, with some overlap into more open coniferous forest and, in Scandinavia only, upland birch country. Even small pools may be chosen where open water of adequate depth carries vegetation yielding food and some marginal cover. Such waters normally intermediate between eutrophic and oligotrophic, and often fed by springs. Larger waters freely used however. Nests placed close to water, often in association with breeding colonies of gulls (Laridae) and terns (Sternidae), especially in Baltic. Apparently feeds at shallower depths than Tufted Duck *A. fuligula*, and not proved to possess equal diving capability. Hundreds, often thousands, concentrate out of breeding season where large food resources, such as mussel beds or waste grain from sewage outfalls, occur. Such areas may be tidal, and exposed to rough seas, or otherwise inland or partly landlocked seas, including those of low salinity, e.g. Baltic. Brackish and fresh waters, large and small, less regularly or massively used; fresh waters on coast (including a few artificial pools on promenades of seaside resorts), and unvegetated reservoirs not infrequently visited. Resort to rivers uncommon, except locally on passage. Sometimes new artificial waters, such as IJsselmeer (Netherlands) quickly generate large new food resources at shallow depth resulting in spectacular concentrations.

Close contact with man infrequent, but not avoided. Rests freely on banks or spits, but moves little on land. Take-off somewhat laboured, and use of airspace relatively infrequent.

**Distribution.** Southern boundary of range in north-west Europe shifted northwards in 19th century (Voous 1960) but recently extending south again in Sweden, Finland, and Estonia. FAEROES: bred 1894 (Williamson 1970). BRITAIN. Has bred sporadically north Scotland since end of 19th century; perhaps now slightly more frequently, but not annually. Also bred England 1944 and infertile clutches laid 1967–71 (British Ornithologists' Union 1971; Parslow 1973a). SWEDEN: range decreased, especially in south, in 20th century; new colonization south-east coast from 1955 (Curry-Lindahl *et al.* 1970; SM). DENMARK: bred 1963; 1970, 1971, and 1974 (TD). FINLAND: new areas colonized in south-west archipelago and Gulf of Finland (Grenquist 1970). USSR. Estonia: probably colonized outlying islands *c.* 20 years ago; increased and occasionally nesting elsewhere (Onno 1970). SPITSBERGEN: ♀ seen with ducklings 1948 (Løvenskiold 1964). Records of occasional nesting in central Europe (e.g. Czechoslovakia, West Germany, East Germany) regarded as not fully authenticated by Bauer and Glutz (1966).

Accidental. Bear Island, Jan Mayen, Portugal, Spain (perhaps regular in winter), Greece, Iraq, Malta, Cyprus, Morocco, Tunisia, Azores.

**Population.** Breeding. ICELAND: in 1956, most common duck Lake Mývatn with 10000–15000 pairs (FG), but decline from late 1950s to 3000–4000 pairs in 1974; probably declined elsewhere, with perhaps *c.* 10000 pairs for whole of Iceland (A Gardarsson). NORWAY: probably more numerous in past (Haftorn 1971). SWEDEN: main population in mountain areas Lapland; decline in numbers 20th century, especially in south, marked in 1940s due to cold winters, but increased since 1955 in south-east (Curry-Lindahl 1970). FINLAND: *c.* 1000 pairs but main population, formerly in Lapland, now along coast; considerable fluctuations in numbers on coast, with increase in archipelago in 1950s and 1960s, but irregular south of 60°N (Merikallio 1958; Grenquist 1970; LvH). USSR: 115000 pairs in western part, of which *c.* 50000 in Europe; population rather stable, but breeding numbers lower in cold springs.

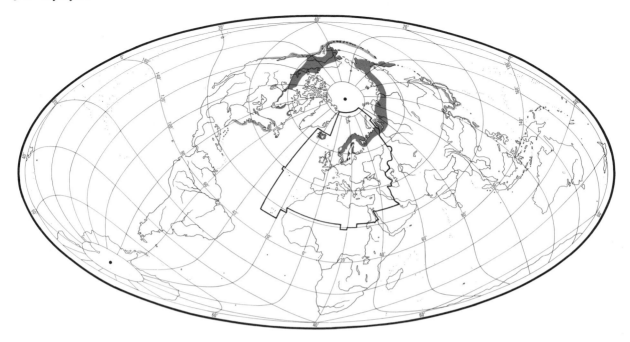

Winter. Estimated 150000 birds in north-west Europe and 50 000 Europe–Black Sea–Mediterranean region (Szijj 1972). Because of dense concentrations, e.g. 50000 birds in Danish waters, especially vulnerable to oil pollution (Joensen 1972). USSR: estimated 93 200 birds (though confusion with Tufted Duck *A. fuligula* possible in large flocks at sea) in western part in not so severe winters, mainly Black Sea, with some Caspian Sea (Isakov 1970*c*).

Survival. Mean annual mortality adults ringed Iceland 52% (Boyd 1962). Oldest ringed bird 13 years (Rydzewski 1973).

**Movements**. Migratory; perhaps only partially in a few areas. In Iceland, ringing of young showed migration chiefly to Ireland, Britain, and Netherlands (Boyd 1959), with smaller numbers reaching north France, Denmark, and north Germany; exceptionally (probable abmigrants) Swedish Lapland and Poland; only small numbers overwinter Icelandic waters. Small number breeding south Norway and Sweden may not migrate, though presumably move at least to coasts for winter, as certainly do breeders from central and north Norway. Breeders of Finland and arctic Russia, and perhaps north-west Siberia (of those ringed North Sea in winter, a few recovered later east to 73°E in Tyumen) migrate WSW, with overland passage from White Sea to Leningrad region, to winter west Baltic, especially coasts north Germany and Denmark; from Baltic, westward movement (especially in cold winters) to Netherlands, east Britain, and France, with one Finnish bird recovered Ireland. This population meets Icelandic birds in North Sea area (Boyd 1959),

where important winter concentrations in Waddensee and IJsselmeer (Netherlands), and Firth of Forth (Scotland). Relatively few reach Mediterranean basin, especially west; winters regularly Dalmatian coast (Yugoslavia) and locally in western Turkey (origins uncertain though a Finnish bird found November in Dalmatia). Further major wintering area in Black Sea of birds apparently from west Siberia; migratory divide between these and those moving to west and north-west seaboard unknown; main wintering area on Ukraine coast extending into Sea of Azov, but large flocks no longer winter coasts of Rumania and Bulgaria (Johnson and Hafner 1970). Birds of uncertain origins (probably Siberian) winter also on Caspian Sea. Generally scarce inland west and central Europe, chiefly at passage periods and in cold weather influxes; but large-scale overland migration across USSR to Baltic, Black and Caspian Seas. One ringed Switzerland found later 68°N Tyumen, USSR.

Large flocks of moulting ♂♂ reported from east Siberia, but in west Palearctic only small moulting groups (June–July) infrequently encountered except on IJsselmeer (Netherlands) where up to 1000 in late July. Most ♀♀ moult wings on breeding grounds. Slow autumn departure from north USSR mid-August to early September. Migration peaks mid to late September in arctic Russia, early October east Baltic, mid-October north Germany, late October in Netherlands and Britain; yet first arrivals on North and Black Seas during September, while others remain as far north and east as temperatures permit. Return movement begins late February in mild winters but normally about mid-March; main passage through Baltic in April, reaching Timan tundra (lower

Pechora) in 2nd half May, Yamal in early June. Immatures do not all return to breeding areas, some remaining in winter and passage areas. See also Bauer and Glutz (1969); Dementiev and Gladkov (1952).

**Food.** Omnivorous, though molluscs predominate in many Palearctic wintering areas. Chiefly obtained by diving, using feet only for propulsion, occasionally by up-ending and by dabbling in shallows and driftlines. If large, prey normally brought to surface. Gregarious, often in large flocks, feeding during day and, especially if disturbed, at night (Nilsson 1969*c*). Feeding pattern may be affected by tidal cycles. Most dives in less than 6 m, and majority probably 0·5–3·5 m. Rarely up to 7 m (Cronan

1957) and 10 m (Madsen 1954). 69 dives, mainly in less than 1·5 m, 9–33 s (Dewar 1924). Takes scraps, and scavenges, including dead fish (Christmas 1960), and in Scotland feeds on grain at sewage outfalls fed by breweries or distilleries (Milne and Campbell 1973; M A Ogilvie).

Annual and seasonal variation in diet associated with differences in abundance and availability of food, and possibly with breeding requirements. Molluscs predominate in most marine localities, rather fewer crustaceans, insects, annelids, and occasionally small fish. Plants, especially seeds, taken regularly and in some areas and seasons, form major part (Millais 1913; Naumann 1905; Witherby *et al.* 1939; Bauer and Glutz 1969). Most data on winter foods. In Danish saltwater areas, 111 stomachs

September–February contained primarily molluscs (80–95% of total volume), chiefly blue mussel *Mytilus edulis*, and to lesser extent cockles *Cardium*, periwinkles *Littorina*, dogwhelk *Nassa reticulata*, and laver shells *Hydrobia*. Of less importance: crustaceans, fish, annelids, and plants (mainly seeds and vegetative growth). In 11 stomachs from Danish brackish-water areas, molluscs mainly *Hydrobia* again predominate, but crustaceans (*Gammarus*), insects, and especially plants (mainly seeds of pondweeds *Potamogeton*, bulrush *Scirpus*, and wigeongrass *Ruppia*) relatively more important (Madsen 1954). Similar findings from 24 stomachs March–April, south Swedish coast: mainly *M. edulis* (87% by frequency) and *Cardium lamarckii* and Baltic clam *Macoma balthica* (Nilsson 1969c). From northern Ireland, 6 stomachs, September–January, contained chiefly *Hydrobia*, and seeds especially milfoil *Myriophyllum*, *Potamogeton*, and *Polygonum* while 2 also contained *Gammarus*, and over 1200 statoblasts of polyzoan *Cristatella mucedo*. 3 stomachs, October and January, from north Kent, England contained *Hydrobia*, *Cardium*, and seeds of *Potamogeton*, *Scirpus*, and *Myriophyllum* (P J S Olney). 3 from North Uist, Scotland, October and November, all contained *Mytilus* and plant remains (Campbell 1947). USSR studies show variation with area and season. October stomachs, Rybinsk reservoir, contained: 31·5% by volume molluscs, chiefly valve snail *Valvata*, and mussels *Anodonta* and *Unio*; 24% aquatic insects, mainly caddisfly larvae *Phryganea grandis* and chironomids *Tendipes plumosus*; 33% *Potamogeton lucens* roots (or, possibly, tubers) in 3–5 cm pieces; and 10·5% seeds. Also inland, October–November stomachs, Ukraine, contained mainly molluscs; also insects and some plants. In Caspian winter quarters, Lenkoran area, entirely cockles *Cardium edule*, and further north on Azerbaydzhan coasts mussel *Mytilaster lineatus* (Dementiev and Gladkov 1952).

Summer food data sparse. From Sweden, stomachs contained pea-shell *Pisidium*, insect larvae, and *Gammarus* (Bauer and Glutz 1969). Of 10 stomachs from Pechora, USSR, all contained leaves, mainly *Potamogeton*, 3 dragonfly larvae, and in one each molluscs and fish (Dementiev and Gladkov 1952). From Lake Mývatn, Iceland; in summer, 90 ♂♂ contained mainly chironomid larvae, also some eggs of stickleback *Gasterosteus* and molluscs *Lymnaea* and *Pisidium*; 119 ♀♀ mainly chironomid larvae, molluscs (chiefly *Lymnaea*), and relatively more stickleback eggs and seeds, chiefly *Carex*. Fish eggs and molluscs eaten mainly in summer. Importance of chironomids increased with season, with seeds relatively more important in May and June, In 1970 when chironomid larvae scarce, molluscs, stickleback eggs, and seeds eaten relatively more frequently than in preceding years. Fish eggs more important in 1968 and 1970, but absent in 1969 (Bengtson 1971).

Ducklings take more food from surface than adults. On Lake Mývatn, newly hatched and tiny ducklings fed mainly on chironomid larvae and seeds, less on water-fleas (Cladocera) and adult chironomids. Seeds and adult insects became less important as young grew, and Cladocera, molluscs (chiefly *Lymnaea*), and chironomid larvae more often taken (Bengtson 1971).

For North American studies, see Cottam (1939); Cronan (1957), Foley and Taber (1952), Kubichek (1933). Munro (1941a).

**Social pattern and behaviour.** For general information, see Witherby *et al.* (1939), Bannerman (1958), Géroudet (1965), Bauer and Glutz (1969).

1. Highly gregarious for much of year. Outside breeding season, often in large flocks of several hundreds or even thousands on sea. Sex composition of winter flocks often unbalanced; ♂♂ tending to predominate in north, ♀♀ further south. ♂♂ first to flock, from June after termination of pair-bond. BONDS. Monogamous pair-bond of seasonal duration; lasts, in some cases, from late winter until ♂ deserts ♀ during incubation, though many pairs not formed until spring. On Lake Mývatn, Iceland, 12% of ♀♀ still unpaired in second half May and 8% in first half July. Little evidence of sexual promiscuity in spite of preponderance of ♂♂ on some breeding waters, e.g. 120:100 on Lake Mývatn (Bengtson 1968). Though ♂ at times recorded accompanying ♀ and brood from nest to water (Pearson and Pearson 1895), usually gone by mid-incubation (Hildén 1964) and only ♀ tends young. ♀ apparently stays with brood much longer on average than other *Aythya* (Hildén 1964), and more ♀♀ undergo wing moult on breeding waters. Flocking of several ♀♀ and their young less common than in Tufted Duck *A. fuligula* but amalgamation of broods of up to 30 ducklings under care of single ♀ more frequent. BREEDING DISPERSION. Both solitary and colonial nesting recorded, latter sometimes densely on islands. Nesting associated with gulls (Laridae) and terns (Sternidae) pronounced in some areas (see Hildén 1964). No information on home-range. ROOSTING. Communal for much of year, on water both diurnally and nocturnally, depending largely on tidal rhythms which affect feeding.

2. Even less well studied than other west Palearctic *Aythya*; little published on wild birds but for observations in captivity see Steinbacher (1960), Johnsgard (1965). FLOCK BEHAVIOUR. Flocks often closely packed. Pre-flight signals more complex than in *A. fuligula*: besides similar Chin-lifting movements (Johnsgard 1965), lateral Head-shakes also frequent with some Head-flicks and other comfort-movements (McKinney 1965b). ANTAGONISTIC BEHAVIOUR. Little information. When ♂♂ meet during hostile encounters, Preen-behind-Wing sometimes frequent (McKinney 1965b). When Inciting, ♀ assumes Neck-stretch posture (see A) and makes occasional lateral threatening movements of head with lowered bill (see B) at rejected ♂, while calling (Johnsgard 1965); at times rushes over water at such ♂♂ between spells of Inciting (Bengtson 1968). COMMUNAL COURTSHIP. Mainly in late winter and spring, but no precise information; peak later than in *A. fuligula* nesting in same locality (see Bengtson 1968). Often several ♂♂ and one ♀, but few data. In Courtship-intent posture, ♂ swims in rather hunched-looking posture with head feathers smooth and rounded (Steinbacher 1960). Secondary displays include Head-shake, Head-flick, and Upward-shake—though latter infrequent (McKinney 1965b). Of major displays, Head-throw and Kinked-neck accompanied by crooning Courtship-call (Johnsgard 1965) with neck enlarged (Millais 1913). Head-throw much as in *A. fuligula* but faster; similarly preceded by Head-shake—Head-flick (McKinney 1965b), but confirmation needed whether followed by Bill-

A

B

dip—Preen-behind-Wing. In Kinked-neck display, stretches up neck, lowering hyoid as call given. Sneak-display inconspicuous (Johnsgard 1965); of incomplete type (Crouch-display) but involves little more than slight movement of bill obliquely downwards. 4th major display, Coughing, frequent and conspicuous: ♂ utters 2nd Courtship-call (whistling Coughing-call) with flick of wings and tail (Johnsgard 1965); posture not described by latter author but same call said to be given from Neck-stretch posture while head pushed upward (see Bauer and Glutz 1969). Head-nodding while swimming (see *A. fuligula*) not recorded, but ♂ frequently performs Turn-back-of-Head display with head feathers fully depressed while ceremonially Leading ♀ (Johnsgard 1965; see A–B). Preen-behind-Wing also commonly directed by ♂ at ♀ but role in communal courtship uncertain. Inciting-display main activity of ♀, lateral threatening component being orientated away from preferred ♂ (see B). ♀ also frequently Bill-dips (Millais 1913) and Preens-behind-Wing, but this probably essentially pair-courtship. No records of typical ♂-like displays from ♀. PURSUIT-FLIGHTS. Courtship-flights common (Bengtson 1968). PAIR BEHAVIOUR. As in other *Aythya*, pair-courtship probably consists partly of displays also shown in communal courtship plus copulatory activities. In addition, ♂ often directs Preen-behind-Wing at ♀, and ♀ repeatedly at ♂ (McKinney 1965b), but not clear whether this ever fully reciprocal. COPULATION. Initiated by ♂ and pre-copulatory displays generally much as in *A. fuligula* (Johnsgard 1965), though Bill-dipping more frequent. ♀ responds with similar behaviour before adopting Prone-posture. In post-copulatory display, ♂ performs Kinked-neck display (with call) and swims away in Bill-down posture; ♀ often swims briefly in same manner, then bathes (etc.). RELATIONS WITHIN FAMILY GROUP. No detailed information.

(Figs A–B after Johnsgard 1965.)

**Voice.** Calls of ♂, especially, so low as to be heard only over a few metres (Millais 1913); include soft, musical, accented cooing or crooning, gentler and more staccato than in ♂ Eider *Somateria mollissima*. Calls of ♀ typically harsh as in other ♀ *Aythya*, but distinctly pulsed and slowed, and include gentler, more crooning notes. Non-vocal sounds include: (1) rustling of wings in flight like goldeneyes *Bucephala* but less loud; (2) rattling of bill against quills of flight-feathers in Preen-behind-Wing display (at least by ♂).

CALLS OF MALE. 2 main Courtship-calls. (1) Soft, or sometimes louder, pleasant dove-like 'kucku', given on own or linked with Head-throw display (Steinbacher 1960); also described as soft, rapid 'WA hooo' (see I) during Head-throw and Kinked-neck displays (Johnsgard 1965), and as gentle 'pa-whoo' (Millais 1913). (2) Soft, rapid, whistling 'week-week-whew' during Coughing-display (Johnsgard 1965).

CALLS OF FEMALE. 3 calls during courtship recognized by Steinbacher (1960), all louder and coarser than corresponding calls of Pochard *A. ferina*: (1) 'kack'; (2) 'gock'; (3) 'querr'. Of these, latter may be Inciting-call; this described as low 'arrrr' by Johnsgard (1965)—see II. Similar call also given in flight and when alarmed: slow, slurred, muffled, growling 'kar-r-r-r'. Other calls during courtship described by Millais (1913) seem distinct from above: (4) crooning 'tuc-tuc-turra-tuc'; (5) gentle 'chup-chup' or 'chup-chup-cherr-err'.

CALLS OF YOUNG. No detailed information.

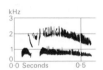

I J-C Roché Finland June 1968

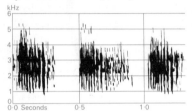

II J-C Roché Finland June 1968

**Breeding.** SEASON. See diagram for Fenno-Scandia and northern USSR. In Arctic, start of laying dependent on thaw. Mean date of 1st laying, Iceland 1961–70, 28 May, range 22 May–3 June, with temperature of preceding weeks important (Bengtson 1972a). SITE. On ground, less often in shallow water; usually well concealed in tussock, reeds, or under overhanging scrub, but sometimes in open. Never far from water. Not colonial, but in dense breeding area nests within 1 m. Nest: depression lined grass or rushes, and down. Average diameter 25–30 cm, height 17 cm (Dementiev and Gladkov 1952). Building: by ♀, using material within reach of nest. Cup formed by turning movements of body. EGGS. Blunt ovate; olive-grey. 62 × 43 mm (55–68 × 41–48), sample 170 (Schönwetter 1967). Weight of captive laid 61 g (53–71), sample 31 (J Kear). Clutch: 8–11 (6–15). More than 15 probably always 2 ♀♀; mean of 112 Icelandic dump nests 16·5 (Bengtson 1972a). Of 59 Finnish 1st clutches: 7 eggs, 2%; 8, 7%; 9, 27%; 10, 39%; 11, 14%; 12, 5%; 13, 3%; 14, 3%; mean 10·0 (Hildén 1964). Mean of 734 Icelandic 1st clutches 9·7 (Bengtson 1972a). One brood. Replacement clutches laid after egg loss; 31% of 45 ♀♀ re-nested after loss of 1st clutch in Iceland; mean of 113 2nd clutches 7·0 (Bengtson

1972a). Of 34 late clutches. Finland: 8 eggs, 35%; 9, 35%; 10, 15%; 11, 12%; 14, 3%; mean 9·2 (Hildén 1964). INCUBATION. 26–28 days. By ♀. Starts on completion of clutch; hatching synchronous. One egg laid per day. Eggs covered with down when ♀ off nest. Eggshells left in nest. YOUNG. Precocial and nidifugous. Self-feeding. Cared for by ♀ though ♂ occasionally in attendance at least at start. Brooded at night by ♀ while small. FLEDGING TO MATURITY. Fledging period 40–45 days. Become independent at or before fledging. Age of first breeding uncertain; at 1 year (Delacour 1959) or at 2 (Dementiev and Gladkov 1952). BREEDING SUCCESS. Of 2802 eggs laid Finland in 3 years, 77% hatched and 6·5% young reared to fledging; survival per pair 0·9, 0·9, and 0·2 young, giving losses of 91%, 90%, and 98% (Hildén 1964). Hatching success of 1323 Icelandic nests, 1961–70, averaged 67·9%, range 47·6–84·3% (Bengtson 1972a).

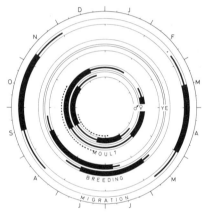

**Plumages.** Nominate *marila*. ADULT MALE BREEDING. Head, neck, upper mantle, rump, tail-coverts, chest, upper breast, and lower vent black, glossed green on head. Lower mantle and scapulars white, finely vermiculated dark grey to black on outer scapulars, somewhat coarser on lower mantle and inner scapulars. Back and tertials black, coarsely vermiculated white; outer tertials black, glossed green with white-speckled tips. Underparts white; vermiculated black where bordering black of upper breast and vent. Lower flanks sometimes finely vermiculated grey. Primaries dark grey, outer edges of outer 4 and tips of all black, outer web of inner 6 white, shading to grey near black tip. Secondaries white, broadly tipped black, few inner with inner web dark grey and outer web black glossed green; black tips of secondaries finely speckled white terminally. Primary coverts, greater upper wing-coverts, and bastard wing black with fine white specks at tips; rest of upper wing black vermiculated white. Marginal coverts of underwing grey, vermiculated white, rest of underwing and axillaries white. ADULT FEMALE BREEDING (Does not nest in this plumage). Head and upper neck black; crown feathers often edged cinnamon. Broad white band on face round base of mandibles (in some, slightly mottled black on lores, forehead, and chin); broader than white spot found on some ♀♀ Tufted Duck *A. fuligula*. White bases of feathers of hindcheek sometimes not concealed. Lower neck, upper mantle, chest, and upper breast grey-brown to dull black, feathers tipped cinnamon or olive-brown on lower neck and upper mantle, paler cinnamon to buff on chest and sides of breast, and white on upper breast.

Feathers of lower mantle, scapulars, and back black, coarsely vermiculated white at tips. Tertials and rump black; upper tail-coverts black, tipped cinnamon. Flanks light cinnamon, broadly vermiculated white at tips of feathers. Lower breast and belly white; vent grey-brown, often vermiculated white. Short under tail-coverts dark grey-brown, vermiculated white; longer, white mottled grey. Tail black with grey bloom. Wing like adult ♂, but secondaries and greater coverts almost without spots or vermiculations; median and lesser coverts dull dark grey, finely speckled white, not vermiculated. ADULT MALE NON-BREEDING (Eclipse). Like adult ♂ breeding, but black parts dull, not glossy. Feathers of head with much white or pale buff at bases, producing light patches on face round bill and mottling on hindcheek when worn. Feathers of upper mantle speckled white near bases and cinnamon at tips; those of chest edged buff and speckled white near bases, of upper breast narrowly edged white. Lower mantle and scapulars more coarsely vermiculated black and white, flanks and under tail-coverts coarsely vermiculated white and dark grey-brown. ADULT FEMALE NON-BREEDING (Nests in this plumage). Like adult ♀ breeding, but head dull cinnamon; much white at face round bill. White of feather-bases visible on hindcheeks, sides of neck, and throat in worn plumage. Lower mantle, scapulars, and flanks cinnamon or grey-brown, less vermiculated white at tips of feathers; when worn, tips bleached buff. Chest and breast dull black, feathers narrowly edged white rather than cinnamon or buff. DOWNY YOUNG. Like *A. fuligula*, but paler. Upperparts olive-brown, palest on mantle, darker on crown and back. Sometimes ill-defined yellow-brown patches at wing and sides of rump. Cheeks yellow-buff, occasionally with faint brown streak. Chest, sides, and vent olive-brown, rest of underparts pale yellow-buff. JUVENILE. Like adult ♀ breeding, but head and neck dark grey-brown; white usually confined to large lateral patch on lores and to chin. Mantle and scapulars grey-brown, sometimes with a few white specks; feathers edged olive-cinnamon. Feathers of underparts short and narrow: those of chest, sides of breast, and flanks grey-brown (sometimes tinged cinnamon) tipped white to yellow-buff; those of breast, belly, vent, and under tail-coverts grey-brown, narrowly tipped white, giving mottled appearance. Tail-feathers narrow, notched at tip, bare shaft projecting. Inner secondaries and greater upper wing-coverts without green gloss, black of wing duller and greyer. Vermiculations and specks of median and lesser coverts of juvenile ♂ like adult ♀, but coverts of juvenile ♀ uniform grey, only faintly speckled white on marginal coverts. FIRST IMMATURE NON-BREEDING MALE. In early autumn new feathers on head, neck, mantle, scapulars, chest, and flanks resembling adult non-breeding but head mottled brown, and flanks grey-brown with white vermiculated tips; rest of plumage juvenile. FIRST BREEDING MALE. New feathers, resembling adult breeding, from October. At times, in breeding plumage (except wing) before winter, but usually combination of juvenile (e.g. most underparts, part of tail), 1st non-breeding (e.g. some upperparts, chest, and flanks), and 1st breeding plumage (e.g. head, some of upperparts, chest, flanks, and tail) during winter. Juvenile wing and often part of juvenile back, rump, belly, and vent retained until summer. FIRST IMMATURE NON-BREEDING FEMALE. Starts on head and scapulars September; elsewhere October. In non-breeding before mid-winter, but juvenile wing, most underparts, and, in about half, tail with coverts retained. Head like adult ♀ breeding, but cinnamon instead of black (sometimes black tinge on feathers bordering white bill-patch); upperparts and flanks as in adult ♀ non-breeding; less vermiculated than in breeding plumage. Apparently only one moult in autumn, in contrast to juvenile. ♂. SECOND NON-BREEDING FEMALE. Attained in spring;

resembles adult ♀ non-breeding; juvenile wing, and usually part of belly, vent, and sometimes a few tail-feathers retained until pre-breeding moult.

**Bare parts.** ADULT MALE. Iris golden yellow. Bill blue-grey, nail black. Foot olive-grey to lead-blue; webs, joints, and back of tarsus black. ADULT FEMALE. Iris golden yellow but, at least during nesting, bright pale grey or pale yellow in some ♀♀. Bill leaden-grey, darker on tip and along culmen; dull black in some ♀♀, at least during nesting. Foot pale or greenish-grey to dull lead-blue; webs, joints and back of tarsus black. DOWNY YOUNG. Iris yellow-brown or greenish-yellow. Bill dark slate, tinged olive at sides; nail pale brown, base of lower mandible yellow-pink. Foot blue-grey to dull black, darker on web, joints, and back of tarsus, with paler streaks along sides of toes. JUVENILE. Iris grey-yellow to yellow-brown. Bill lead to slate, irregularly marked black at base and tip in ♀, less marked in ♂. Foot olive-grey to grey-blue, webs, joints, and back of tarsus dull black. Bare parts of juveniles attain adult colour during 1st winter, but adult colour of iris in ♀ probably *c.* 1 year later (Trauger 1974). (Bauer and Glutz 1969; Schiøler 1926; Witherby *et al.* 1939; RMNH.)

**Moults.** Like *A. fuligula*, but some ♀♀ of *A. marila* moult wing later, after arriving in winter quarters, September–November (Joensen 1973; ZMA). Juvenile ♀ apparently moults only once in autumn, in contrast to *A. fuligula*. For details see Billard and Humphrey (1972).

**Measurements.** Netherlands, whole year. Skins (RMNH, ZMA).

| | | | | | | |
|---|---|---|---|---|---|---|
| WING AD ♂ | 227 | (4·09; 45) | 219–237 | ♀ 217 | (3·78; 35) | 211–225 |
| JUV | 220 | (5·33; 56) | 208–229 | 213 | (4·55; 34) | 202–222 |
| TAIL AD | 56·3 | (2·00; 45) | 52–61 | 55·7 | (2·00; 34) | 51–60 |
| JUV | 54·1 | (2·27; 38) | 49–59 | 53·8 | (2·33; 20) | 51–58 |
| BILL | 44·0 | (1·31; 46) | 41–47 | 42·9 | (1·31; 36) | 40–45 |
| TARSUS | 39·9 | (1·26; 56) | 38–42 | 38·6 | (1·08; 45) | 37–41 |
| TOE | 65·9 | (2·64; 58) | 61–72 | 64·3 | (1·93; 40) | 61–68 |

Sex differences significant, except tail. Juvenile wing significantly shorter than adult; other differences between adult and juvenile not significant (tarsus and toe of both combined in table).

**Weights.** ADULT. (1) December–March, Scandinavia (Schiøler 1926; Hagen 1942) and Netherlands (ZMA) combined. (2) June–September, various areas USSR (Dementiev and Gladkov 1952; Dolgushin 1960) combined. JUVENILE. (3) October–November, Scandinavia and Netherlands. (4) December–March, Scandinavia.

| | | | | | |
|---|---|---|---|---|---|
| (1) | ♂ 1219 | (106; 17) | 972–1372 | ♀ 1183 | ( 75; 12) 1037–1312 |
| (2) | 936 | ( 87; 7) | 820–1050 | 883 | (132; 10) 690–1100 |
| (3) | 992 | (178; 6) | 744–1171 | 1110 | ( 75; 4) 1020–1190 |
| (4) | 1105 | (137; 10) | 850–1330 | 1024 | (148; 9) 791–1273 |

Heavy in winter, but lean after spring migration. ♀♀ store fat prior to nesting, but weight decreases during breeding: Yamal, USSR, range June 910–1100, July 825–880.

**Structure.** 11 primaries: p10 and p9 about equal, longest; p8 4–9 shorter, p7 15–23, p6 28–38, p1 90–107. Bill heavier at base and broader at tip than *A. fuligula*. No tuft on crown. Rest of structure as *A. fuligula*.

**Geographical variation.** Pigmentation of lower mantle and scapulars of adult ♂♂ increases towards east: Pacific *A. m. mariloides* more boldly vermiculated black than nominate *marila*. Birds from east Asia tend to be somewhat smaller than west Palearctic and North American ones. American population sometimes separated from typical *mariloides* as *A. m. nearctica* Stejneger, 1885, but overlap in size considerable, and recognition not warranted. CSR

# Tribe SOMATERIINI eiders

Medium-sized to fairly large, essentially marine diving ducks. 4 species in 2 genera: *Somateria* (typical eiders), monotypic *Polysticta* (Steller's Eider). Often included with other sea-ducks in tribe Mergini (Delacour and Mayr 1945; Johnsgard 1960, 1964) but differ in a number of characters and more conveniently placed in tribe on own, following Delacour (1956), Humphrey (1958), and others. 3 breeding species and 1 accidental in west Palearctic.

Bodies in *Somateria* broad and heavy, but more elongated than in *Aythya*, heads large; wings relatively short and flight heavy, usually low over water when strong and direct; tails fairly short, rounded; bills high at base (with conspicuous, shield-like knob in King Eider *S. spectabilis*), almost straight, about as long as head, with large nail; legs not so short as in *Aythya* and set fairly well back, lobe on hind toe large. Smaller and more slender *Polysticta* rather *Anas*-like, with longer, more pointed tail and totally different bill with reduced nail and flange of soft skin on upper mandible; also swift and manoeuvrable in flight. *Somateria* take off from water with difficulty, but *Polysticta* rise easily and steeply. Come out on land more readily than most other sea-ducks, walking with steady, rolling gait. Dive with great facility; unlike *Aythya*, however, partially open wings to submerge and some or all also use mainly folded wings for underwater propulsion, beating them with short, jerking movements and bastard-wing extended. Sexes differ in tracheal anatomy; ♂♂ with *Anas*-like, asymmetrical bullae (small in *Polysticta*), completely and uniformly ossified, and tracheal tubes of uniform diameter; bronchi considerably enlarged. Marked sexual dimorphism in both genera. ♂♂ boldly patterned in black and white, with dark belly in *Somateria* and chestnut underparts in *Polysticta*; distinctive green colour on head found only in this tribe. Forewing white and secondaries ornamentally modified. ♀♀ brown and highly cryptic; barred in *Somateria*, more uniform in *Polysticta*. Glossy speculum only in *Polysticta*. ♂ non-breeding plumage of eclipse type. Juveniles resemble adult ♀♀. Downy young distinctive: brown above with pale supercilliary stripe (reduced in *Polysticta*) and dark line through eye, paler below; lack capped appearance and bold patterning of most Mergini.

Entirely northern Holarctic. Ecology similar in many features to that of other sea-ducks (see Mergini). Mainly frequent coasts, inlets, and estuaries, often rocky; also adjacent fresh water on tundra, etc., for breeding. In west Palearctic, Eider *S. mollissima* declined markedly in 19th century owing to human persecution but great increase and slight spread in western Europe during 20th, especially in recent years under protection. All migratory or partially migratory, but movements relatively restricted. As in other sea-ducks (see Mergini), moult migration particularly well developed. In *S. mollissima*, crèche system releases many ♀♀ which join main moulting assemblies late July or August.

Omnivorous, but take mainly animal food (chiefly molluscs and crustaceans) obtained by diving from surface to bottom. Will also dabble, head-dip, and up-end at times; use of feet to scratch up substrate in shallow water also recorded. Feed mostly by day, but *S. mollissima* follows tidal rhythm in some areas. Usually highly gregarious when feeding and at most other times. Lateral Head-shaking main pre-flight signal, but variety of comfort behaviour (shake, wing-flap, bathing, both-wings stretch) also performed before flight by some species. Information on home-ranges and territories limited but ♂ defends vicinity of ♀, wherever she may be, in period immediately prior to egg-laying, thus permitting her to spend maximum time feeding. Monogamous pair-bonds of seasonal duration; not known if same pairs form again in successive years. Pair-formation in flock, starting in autumn (*S. mollissima*) but often not until late winter or spring; bond usually ends during incubation. Communal courtship well developed. In *Somateria*, ♂-displays include various distinctive Cooing-movements; those of *Polysticta* also distinctive (see species account), with those of Spectacled Eider *S. fischeri* intermediate between *Polysticta* and other eiders. Displays common to both genera include Neck-stretch, Head-toss, and lateral Head-turning. True Turn-back-of-Head display (see Anatini and Aythyini) absent in *Somateria*. Secondary displays (which also precede copulation) include Upward-shake, Wing-flap, Ritual-bathe, and Head-roll. In *Somateria*, ♀♀

at times perform ♂-like displays during communal courtship. Inciting-displays distinctive in both genera (most ritualized in *Polysticta*); of same functional type as in Anatini. Courtship-flights occur in *Polysticta* but no pursuit-flights of any type in *Somateria*, though engage in underwater chases at times during communal courtship. Pair-courtship confined largely to copulatory activity which first occurs well in advance of nesting. ♀♀ display little if at all before copulation but usually soon adopt prolonged Prone-posture on water, often during or after Inciting at other ♂♂. In addition to pre-copulatory displays noted above, ♂ also Preens-dorsally; after dismounting, does single Reaching-display (*Somateria*) or Rearing-display (*Polysticta*) then usually retreats rapidly from ♀ while Head-turning. Calls of ♂ *Somateria* mostly soft, dove-like cooing sounds linked with inflation of throat, though whether caused by inhalation or, as in most birds, by expiration not certain (see Johnsgard 1971); used mainly in courtship. ♂ *Polysticta* largely silent. Calls of ♀♀ mainly throaty or wooden (*Somateria*) and raucous (*Polysticta*); include well-developed Inciting-calls but Decrescendo-calls lacking.

Strictly seasonal breeders, especially in high Arctic. Ground nesters, often in open; near water or sometimes far away. Nests usually well dispersed but *S. mollissima* often colonial at protected sites; nesting association with Arctic Tern *Sterna paradisaea* frequent. Nests shallow depressions lined with available material and copious down. Building by ♀ only. Old nests often re-used. Eggs blunt or elongate oval, pale green, bluish, olive, or buff; smooth (slight gloss in *Polysticta*). Clutches usually 4–7 (*Somateria*), 6–8 (*Polysticta*). Multiple laying recorded in colonies of *S. mollissima*. One brood; replacements laid after egg loss except in high Arctic. Eggs laid at 1-day intervals. Incubation by ♀ only, leaving nest rarely. Incubation periods 22–28 days. Young cared for by ♀ only. Distraction-display, in form of 'injury-feigning', reported in *Somateria*; ♀♀ will also aggressively defend young from avian predators. Supernumerary ♀♀ present at times during hatching and nest-exodus in *S. mollissima*; these fly off readily if approached by predator and may provide diversion from mother and brood. Crèching also particularly well developed in *S. mollissima*. Young independent by time of fledging. Mature in 3rd year, sometimes 2nd.

*Somateria mollissima* **Eider**

PLATES 81 and 100
[between pages 638 and 639
and facing page 687]

Du. Eidereend    Fr. Eider à duvet    Ge. Eiderente
Ru. Обыкновенная гага    Sp. Eider    Sw. Ejder    N.Am. Common Eider

*Anas mollissima* Linnaeus, 1758

Polytypic. Nominate *mollissima* (Linnaeus, 1758), coasts north-west Europe from France and Britain to Baltic and Novaya Zemlya; *faeroeensis* C L Brehm, 1831, Faeroes; *borealis* (C L Brehm, 1824), arctic North Atlantic, from Franz Josef Land and Spitsbergen to Iceland, Greenland, and Baffin Island region of Canada. 3 extralimital: *dresseri* Sharpe, 1871, Atlantic coast of North America from Maine to Labrador; *sedentaria*, Snyder, 1941, Hudson Bay; *v-nigrum* Bonaparte, 1855, arctic coast of east Siberia and north-west North America.

**Field characters.** 50–71 cm (body 35–45 cm); wing-span 80–108 cm. Large, long-headed, rather short-necked, bulky sea-duck; bill with long triangular profile, eye set far back. At close range, extension of loral feathering forward to below nostril and beyond limit of frontal feathering diagnostic. Adult ♂ looks black below and white above on sea, black behind and white in front in flight. ♀ and juvenile of both sexes essentially dark brown, mottled above and barred below. Sub-adult ♂ shows bewildering variety of piebald plumages. Sexes dissimilar; marked seasonal variation in ♂. Juvenile and sub-adult separable.

ADULT MALE. Nominate *mollissima*. Crown black, rest of head and neck white but with panel down rear of cheeks and large patch on nape pale green. Mantle, scapulars, tertials, round patches on sides of rump, and wing-coverts white; rest of upperparts, most of primary coverts, and wing-quills black. Chest cream-pink, rest of underparts black. Bill olive-grey, with yellow or green tinge to base and strongly hooked, pale nail; legs and feet dull yellow or green. ♂♂ of other races vary little in field appearance from above, but ♂ *borealis* has bright orange-yellow bill. MALE ECLIPSE. Essentially uniform sooty brown with whitish supercilium and base to chest usually obvious; wing-coverts remain white. ADULT FEMALE. All plumage rather dull but warm brown, darker on back than head and body and closely marked overall with straight bars. No obvious pattern though in good light at close range pale stripe through eye contrasting with darker crown and cheeks, 1–2 white wing-bars (bordering sometimes purple, usually dark brown secondaries) and pale under wing-coverts may show. All bare parts duskier than ♂. SUB-ADULT. ♂ assumes breeding plumage slowly with periodic eclipse plumages and erratic or continuous moult contributing to strange, irregular piebald appearance during first year and a half. White appears first on neck (giving chance of confusion with Goldeneye *Bucephala clangula*) then on scapulars, breast, and later on rest of upperparts; clean face and pink chest not attained until 4th winter. Rest of plumage at first dark brown, then increasingly sooty (and less barred than ♀) until finally intensely black. JUVENILE. Resembles adult ♀ but duller, lacking any marked speculum. 1st autumn ♂ has more sooty appearance with less pronounced barring above.

Differentiation of breeding ♂♂ of 3 *Somateria* species

normally not difficult but at other ages often hard and even impossible, unless near. Details of head shape, bill feathering, and pattern of barring only certain bases of diagnosis.

As with all *Somateria*, flight silhouette noticeably chunky, with wings appearing both broad and short and head held low. Flight action strong and regular, after laboured take-off, birds flying low over water, usually in single file. Gait slow and rolling; swims well but not fast with head often retracted on to mantle; dives for food in shallow water, also up-ends and dips head. Highly sociable, breeding in small or large coastal colonies and wintering in large groups scattered along coasts and in inshore waters. Spends much time on sea but groups haul out to preen and loaf on sandbanks, reefs, and cliff bottoms. ♂ courts ♀ with pleasing croon, which in chorus carries far; ♀ utters grating growl. Rarely vocal in flight but conversations of winter or spring groups often long and varied.                                                                  DIMW

**Habitat.** In west Palearctic, wholly coastal and marine, extending south of Arctic Circle only in White Sea, Baltic, and Atlantic waters. Successful adaptation to arctic conditions enables wintering displacement mostly within breeding range. High dependence on molluscs and crustaceans, and daily food requirement of some 300 gm, prevent viable breeding colonies surviving where biomass of benthos in accessible coastal waters falls below 25 gm per cubic m; so, for example, nearly all nesting on Novaya Zemlya confined to west coast. Forages chiefly on ebbing tide, at depths normally up to 3 m, and avoids strong winds and waves. Further restricted in habitat by unsuitability, for landing or breeding, of steep rocky or spray-swept coasts. Prefers islands, skerries, or reefs with sheltered approaches and protective rocks, stones, herbage, or even trees, to safeguard nest. In high Arctic, where such shelter unavailable, open sites exposed to weather and predators must be accepted, and nests often closely packed. Association with other colonial birds active in driving off predators, or even with such predators as Snowy Owls *Nyctea scandiaca*, favoured for defence. Non-conducting properties of eiderdown assist successful hatching even on frozen terrain. Uses land only for nesting and resting as near sea as possible, and over most of range hardly visits

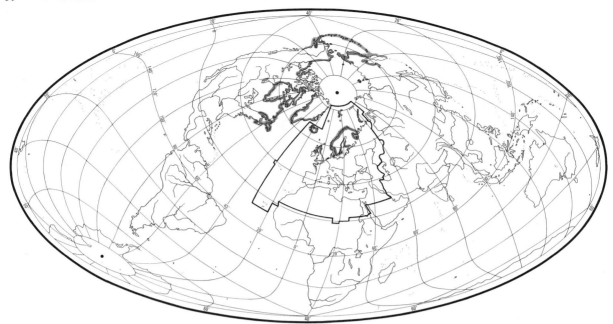

fresh water, although ducklings may be led up lowest reach of streams, and some winter on inland water.

Although flying freely, normally keeps to lowest few metres of airspace, mostly over water. Has little occasion to use man-made habitats or artefacts, except artificial nest-sites locally. Desire to maximize gathering of eiderdown has led to conservation for centuries in England, Iceland, and elsewhere; breeding ♀♀ tolerate close approach and handling in such conditions. See Uspenski (1972) on which this account mainly based.

**Distribution.** Some southward extension of range this century in western Europe. BRITAIN AND IRELAND. In Scotland, spread to mainland *c*. 1850 with main phase of expansion towards end of 19th century following protection. Colonized Ireland 1912 and north-west England 1949 (Parslow 1967). FRANCE: has bred occasionally Brittany since 1905 and probably regularly since 1964; occasionally elsewhere (FR). NETHERLANDS: first bred 1906 on Vlieland, spreading to all West Friesian islands in 1940s (Swennen 1972). WEST GERMANY: first occupied Sylt in North Friesian islands between 1785 and 1805; has spread since to other islands and bred sporadically on mainland (Bauer and Glutz 1969). DENMARK: long established; *c*. 12 localities 19th century, with increases mainly in these 1900–50, but many new areas colonized 1950–70 (Joensen 1973*b*). SWEDEN: spreading inwards in archipelagos in last decade (Curry-Lindahl *et al.* 1970). FINLAND: similar tendency to spread inland in parts of archipelagos but has disappeared from head of Gulf of Bothnia (Merikallio 1958; Grenquist 1970).

Accidental. Czechoslovakia, Hungary, Austria, Switzerland (but increasing in winter since late 1950s), Yugoslavia, Greece, Bulgaria, Rumania, Spain, Azores.

**Population.** Breeding. Declined markedly in many areas in 19th century due to human exploitation. Increased western Europe 20th century and especially in recent years, mainly due to protection, but local setbacks when exploitation increased (as during 1939–45 war), or by pollution (as in Netherlands and Denmark). Increases in some areas, e.g. Scotland and West Germany, began before any general protection. In Spitsbergen and USSR, still inadequately protected. ICELAND: estimated 200 000 to 300 000 pairs (Gardarsson 1975). SPITSBERGEN: population much reduced everywhere due to ruthless plundering by humans (Løvenskiold 1964). BRITAIN: about 10 000 pairs (Atkinson-Willes 1970*a*); still increasing in some areas e.g. Sands of Forvie, Aberdeenshire, where *c*. 1200 pairs 1961–3, *c*. 1800 pairs 1968, and 2000 pairs 1970 (Milne 1974). FRANCE: 1–10 pairs (FR). NETHERLANDS. 10 pairs in 1925, *c*. 300 in 1936, *c*. 1080 in 1948, *c*. 2350 in 1954, reaching peak of 5756 nests in 1960; then declined to 1768 nests in 1966 and 1329 in 1968. This decline correlated with pollution by chlorinated hydrocarbons from Rhine, and, after control measures, increased to 1919 nests in 1970 (Swennen 1972). WEST GERMANY. Most now on Amrum where increasing, but has declined Sylt. Total population *c*. 400 pairs (Bauer and Glutz 1969); *c*. 700 pairs (Szijj 1973). DENMARK: *c*. 1500 pairs 1935, *c*. 3500 pairs 1960, *c*. 7500 pairs 1970; some 28 000 killed by oil pollution in Kattegat 1969–71 (Joensen 1973*b*). NORWAY: 103 000 pairs (Kumari 1968). SWEDEN. Over 60 000 pairs (Kumari 1968); increasing recently due to abolition of spring hunting (Curry-Lindahl *et al.* 1970). Census of post-breeding ♂♂ suggested 297 000 pairs breeding in Baltic areas of Sweden, Finland and Estonia (Almkvist *et al.* 1974). FINLAND: *c*. 25 000 pairs (Merikallio 1958); *c*. 90 000 pairs (Alerstam *et al.* 1974). USSR. Barents and

White Seas: *c.* 50 000 nests some 12–25 years ago, with about half on Novaya Zemlya; now considerably reduced due to inadequate protection (Uspenski 1968). White Sea and Murmansk areas: *c.* 5500–6000 nests 1960 and 1961 (Karpovitch and Kester 1970). Estonia: *c.* 4000 pairs, considerably reduced by human activity last 70 years but now protected and increasing (Kumari *et al.* 1970).

Winter. Estimated some 2 million birds—mainly in six areas. (1) White Sea—Barents Sea: *c.* 107 000, including other eiders, but mainly *S. mollissima*. (2) Norway: at least 250 000, probably many more. (3) Baltic–Kattegat–west Jutland: *c.* 750 000. (4) Waddensee (Netherlands and West Germany): *c.* 200 000. (5) Britain: 50 000–60 000. (6) Iceland *c.* 50 000. Faeroes and Svalbard: no data. (Atkinson-Willes 1975.)

Survival. Netherlands: annual mortality of birds ringed as fledglings 1965–70 averaged 17%; ♀ annual mortality ranged from 15%–61% in 1964–68 reflecting poisoning by chlorinated hydrocarbons, and 2%–8% from 1969–71 after control measures (Swennen 1972). Denmark: annual mortality 20% (Paludan 1962). Adults ringed north-west Europe: mean annual mortality 39%, life expectancy 2·1 years (Boyd 1962). Mass deaths caused by parasitic

infections of breeding grounds after hard winter reported from many areas (Grenquist 1970). Oldest ringed bird 15 years 7 months (Rydzewski 1973).

Movements. Partially migratory and dispersive. British and Irish breeders show some small movement, rarely over 200 km, e.g. Aberdeenshire birds winter mainly Firth of Tay (110 km south-west), with 2 moulting areas between (Milne 1965*a*). Exceptional recoveries from Aberdeenshire in Denmark (5), Sweden, and Finland but no regular emigration by any British or Irish birds. Variable numbers occurring, mainly winter, south and east England coasts almost certainly from Netherlands (Taverner 1959, 1967), and recoveries of Dutch birds in Kent, Yorkshire, Northumberland, and Fife. Dutch population partially migratory; other recoveries Denmark to Normandy, and infrequently Finland and Sweden (abmigrants) to northern Bay of Biscay, but most Netherlands birds probably moult and winter there.

Extensive moult migration, mainly adult ♂♂ and immatures, profoundly affects picture of autumn migration in Baltic (see Salomonsen 1968; Joensen 1973*a*). Large moulting flocks assemble off south-west Jutland, in

German Waddensee and on Danish side of Kattegat. To reach former areas, many migrate nocturnally overland across south Jutland in late June and July; followed in early August by smaller numbers of ♀♀ (most ♀♀ moult in Baltic), these apparently originating mostly from southern Sweden and Finland. After moulting, some disperse (September–November) southwards to winter in Waddensee, mixing with Netherlands birds; at this time, more arrive to winter west Baltic, including adult ♀♀, juveniles, and those which moulted off Gotland (Sweden), Estonia, and Latvia; these tend to concentrate among Danish islands and (in smaller numbers) off north Germany, making only short movements according to temperatures. Some from south Norway winter Danish waters; extent of this movement unknown.

Faeroese and Icelandic populations mainly, if not wholly, resident, though in latter, northern breeders may move to south coast for winter. Breeders of Svalbard and Novaya Zemlya moult on nearby coasts and then most migrate south in September towards (respectively) Norwegian Sea and White Sea to Murmansk; birds ringed Spitsbergen recovered Tromsø (November) and north Iceland (October, May). Some winter as far north as Spitsbergen, appearing inshore when breaks occur in sea ice (Løvenskiold 1964). Populations of arctic Fenno-Scandia and USSR winter in open water of White Sea, off Murmansk, and north Norway; apparently partially migratory, some White Sea birds moving to Murmansk, Murmansk birds to Norwegian coasts. Westward movement may continue at least to January (Belopol'skii 1957), presumably linked with temperatures. Important moulting areas in White Sea used at least by breeders of Kandalaksha Gulf and east Murmansk (Karpovitch and Kester 1970).

Irregular visitor to inland waters; usually singles, though records from central Europe increasing since c. 1957 (following increase of mussel Dreissena polymorpha), and now almost annual inland Germany and Switzerland, mainly mid-September to mid-April; exceptionally over 300 Bodensee, winter 1971–2 (Leuzinger and Schuster 1973). One ringed Sweden recovered Switzerland, November, and one ringed Switzerland, December, found 6 weeks later Bay of Biscay (France).

Return movements commence late February in mild winters. Though first may reach Estonia in March, peak passage through Kalmar Sound (major route to Gulf of Bothnia) April and breeders reach Finland 2nd half April. Apparently recross Jutland in spring since no marked passage through Skagerrak (Salomonsen 1968). Main return to Svalbard and USSR colonies mid-April to May, depending on ice; in Kandalaksha Gulf first arrivals coincide with cracking of sea ice, main arrivals with appearance of wide leads. Immatures do not visit colonies, many remaining near wintering areas or in intermediate maritime localities.

**Food.** Chiefly immobile or slow-moving, bottom-living species, primarily molluscs and to lesser extent crustaceans and echinoderms; obtained benthically by surface-diving and, in shallow water, by head-dipping and up-ending. Usually gregarious daytime and occasionally night-time feeder with feeding cycle controlled by tide, time, and possibly day-length: when tidal range small, diurnal feeding with morning and evening peaks; when large, peak feeding normally at low tide (Bent 1925; Dunthorn 1971; Pethon 1967; Player 1971; Cantin et al. 1974). In arctic USSR, may feed 4 times daily (Dementiev and Gladkov 1952). Dives without forward leap, putting head into water and half opening wings just before slipping under surface using feet to propel (Millais 1913; Player 1971). Dives last 6–78 s: 47 s maximum in 333 dives (Dewar 1924); 30–40 s in 16 m (Pethon 1967); 78 s maximum in 12–15 m (Bergman 1939). Depths 2–4 m (48 of 50 dives in less than 3 m (Pethon 1967), with authenticated maximum depths of 15–20 m (Millais 1913; Bergman 1939; Pethon 1967). In shallow water, also scratches out small craters in sea bed with feet and explores with beak (Player 1971). Items torn off underwater rocks, taken off bottom, or out of surface mud or sand, swallowed under water, or brought to surface and shaken violently until separate e.g. mussel clumps, crabs (Picozzi 1958; Dunthorn 1971).

From marine areas, chiefly molluscs, especially bivalves and mainly blue mussel Mytilus edulis (related bivalves taken where mussel does not occur); also gastropods, particularly periwinkles Littorina. In general, rather fewer crustaceans (mainly crabs) and echinoderms (mainly common starfish Asterias rubens, sea-urchins Echinoidea, and holothurians). In individual meals, following of occasional importance: sea-anemones, actinarians, alcyonarians (Evans 1909; Krabbe 1907), cuttlefish (Cottam 1939); insects (Collinge 1924–7; Cottam 1939; Belopol'skii 1957; Scot-Ryen 1941; Pethon 1967), polychaetes (Montague 1925; Dementiev and Gladkov 1952; Pethon 1967). Fish often recorded, usually in small quantities (Collinge 1924–7; Cottam 1939; Salomonsen 1950a; Dementiev and Gladkov 1952; Madsen 1954; Belopol'skii 1957), probably when normal foods not available and where fish dead or immobilized e.g. in nets. Herring Clupea harengus eggs important in spring in eastern Canada (Cantin et al. 1974). Plants of importance only on breeding grounds when berries (especially crowberries Empetrum nigrum), green algae, leaves, and seeds taken by incubating ♀.

Diet changes seasonally within areas due to different requirements and restrictions imposed by breeding and, between areas according to fauna present. From Danish marine areas, October–April, 261 stomachs showed 85·1% (by frequency) molluscs, of which blue mussel 68·6%, periwinkles 22·2%, dogwhelks Nassa 17·3%, common whelk Buccinum undatum 8·4%, clams Mya 5·4%, and Spisula 4·9%; crustaceans 38·7%, of which shorecrab Carcinus maenas 30·7%, barnacles Balanus 9·2%; echinoderms 29·1%; common starfish 26·8%; fish 2·7%. From Danish brackish areas, food in 35 stomachs had similar composition though far fewer crustaceans and echinoderms, and more fish, mainly 3-spined stickleback *Gasterosteus aculeatus* (Madsen 1954). Studies in USSR show seasonal differences related to different local climatic conditions and breeding season. In Kandalaksha Bay, when littoral zone frozen, feeds in sub-littoral zone mainly on molluscs (87% by frequency). On Novaya Zemlya, where severe conditions limit littoral fauna diversity, mainly mussels, plus other molluscs *Pecten, Acmaea, Buccinum*, and *Saxicava artica*. In east Murmansk, blue mussel (84·1% of total) and far fewer periwinkle, crustaceans, and echinoderms; in west Murmansk, molluscs 94·7% of total. While incubating, ♀♀ take crowberries and green parts of plants; in some areas, may still feed in water though in others take little or no food. When accompanying young, ♀♀ take small prey, especially periwinkles (Dementiev and Gladkov 1952; Belopol'skii 1957; Geramisova and Baranova 1960). From south-east Norway, spring and summer stomachs of 97 adults and 76 young contained mainly isopods (*Idotea pelagica* and *I. granulosa*), amphipods (*Gammarelus angulosus* and *Parajassa pelagica*), and molluscs (especially blue mussel and *Thais lapillus*). Adults showed significant change from mainly blue mussel in spring and early summer to mainly small crustaceans in August–October; not related to corresponding changes in availability so possible that increased amounts of small crustaceans in adults and young important in feather maintenance during growth and moult (Pethon 1967). From Firth of Forth, Scotland, 94% of 50 stomachs, November–March, contained blue mussel, 24% shorecrab, and 10% periwinkle (Player 1971). From Argyll, Scotland, blue mussel occurred in all 13 stomachs collected, with blue mussel taken mostly in morning, and shorecrab, periwinkle, and echinoderms in evening from opposite end of loch (Dunthorn 1971). From Orkneys, 41 stomachs in February contained chiefly molluscs of 12 species, mainly limpet *Helcion pellucidus* (25) and *Lacuna divaricata* (10); also crabs (3 species in 17) (Evans 1909). From Greenland waters, Salomonsen (1950a) noted molluscs most frequent, and occasional crustaceans, fish spawn, and fish (especially sea scorpion Cottidae). Off south Spitsbergen, mainly sea slugs (holothurians) (Løvenskiold 1954); off west Spitsbergen in 10 stomachs, mainly crustaceans and relatively few molluscs (Hartley and Fisher 1936), 1 stomach in August containing whole sea-urchin *Echinus* (Kristofferson 1926). 63 stomachs from south-western archipelago, Finland in spring contained 98% by volume blue mussel (Bagge *et al.* 1970). In south Germany and Switzerland zebra mussel *Dreissenapolymorpha* taken, as well as fish and fish offal (Leuzinger and Schuster 1973). In St Lawrence estuary, Canada, pre-nesting birds ate mainly herring eggs and *Nereis virens*. Later, when accompanying ducklings, ♀♀ ate mostly *Littorina* and amphipods. Main diet in winter probably blue mussel (Cantin *et al.* 1974; H Milne).

Size of normal foods ranges from 1–80 mm, with possible selection for blue mussel under 40 mm (Madsen 1954; Belopol'skii 1957; Player 1971; Dunthorn 1971), shorecrab 20–52 mm (Madsen 1954; Player 1971), periwinkle 12–15 mm (Player 1971), and common starfish 30–40 mm (Madsen 1954).

From south Norway, 76 ducklings 0–10 weeks old, showed youngest ate mostly small crustaceans; as young grew, these (especially amphipods) decreased and blue mussel predominated (Pethon 1967). Scot-Ryen (1941) found crab larvae in downy young, as well as molluscs. In White Sea and Barents Sea, young up to 14 days old take small molluscs (mainly periwinkles); ducklings aged 10–12 days also small crustaceans, mainly *Gammarus*. After 15 days, blue mussel becomes more important until finally predominates (Geramisova and Baranova 1960). In 245 ducklings from St Lawrence estuary, Canada, *Littorina* made up 30–87% of diet, importance growing with age of birds. In first few weeks diet more diversified than in older ducklings, with insects forming surprisingly high part (Cantin *et al.* 1974).

**Social pattern and behaviour.** Based mainly on material supplied by H Milne and F McKinney.

1. Highly gregarious. In flocks on sea for much of year, composition varying according to season (see, e.g. Milne 1965a). From October–March, winter flocks at times of many thousands. 1st winter birds mostly flock separately, often in discrete wintering areas, or sometimes keep to perimeter of adult flocks. ♂♂ first to flock, June or early July, gathering locally to moult off coast; may later assemble elsewhere for flightless period far out at sea where joined by ♀♀ 3–4 weeks later (Milne 1965b; Salomonsen 1968). BONDS. Monogamous pair-bond of seasonal duration. Some evidence also, in minority of birds, for ♂ promiscuity in form of normal copulations (not rapes) with 2–3 ♀♀ in short space of time (Gross 1938; McKinney 1961). Pair-formation begins late September or October, continuing through winter; correlated with peaks in courtship display, and ♀♀ of one Scottish population at least become paired in 2 phases—up to 50% in autumn, rest following spring (Gorman 1974). On Farne Islands, England, some pairs still forming in May, shortly before nesting (McKinney 1961). ♂ usually deserts ♀ after start of incubation (but see below). 1st-year ♂♂ display but do not form pairs. Only ♀ tends young. Once family leaves nest, parent-young bond often weak partly because of presence of other ♀♀—first of so-called 'aunties' (which sometimes attend new-hatched young) and then of crèche 'guards'—and partly because mother tends to leave brood in early stages in order to go off to replenish reserves lost during incubation. In feeding area during first 2–3 days, individual young or whole broods tend to join others to form creches attended by more than one ♀ (see especially Gorman and Milne 1972). Size of crèche varies according to site and density of young, but 10–100 (with up to 10 guard-♀♀) common and up to 500 observed. Crèches intermingle and sub-divide throughout rearing period; young frequently of varying ages. Crèche ♀♀ mostly successful breeders. BREEDING DISPERSION. Though isolated nests not rare, typically colonial throughout much of range. Where area of colony site restricted and protected, density high: frequently up to 250 per ha in optimum habitat; nests elsewhere usually more loosely grouped. Until time of nest-

selection, pairs remain in flock on sea; then leave flock and come ashore. On Farne Islands, starting end April, pairs land in early morning and come further and further from shore on succeeding days (McKinney 1961; Tinbergen 1958). At colonies elsewhere, pairs may first space well out evenly on shore (H Milne). In all cases, ♂ accompanies ♀ during site-selection. No true territory defended, but pairs occupy relatively confined zones of shore during egg-laying, often in immediate vicinity of nests, and each ♂ defends area around ♀ against intrusion. ♂ accompanies ♀ to nest during laying; thereafter, at most colonies, ♀ at nest alone when ♂ departs, though at some colonies in Iceland and Spitsbergen pairs associate throughout incubation. ♀ leads young to water in early morning *c.* 24 hours after hatching; then taken to suitable feeding area close to shore, perhaps up to 34 km from nest. ROOSTING. Communal for most of year, often in rafts on sea but also ashore (e.g. on rocks) where beaches afford shelter from prevailing wind in rough weather (H Milne). Said to be diurnal feeder and to roost on rocks mainly at night (Bent 1925) but, where pattern of feeding tidal, will feed on night tides as well as on day tides. On Ythan estuary, Scotland, tidal pattern of roosting afloat at high water and feeding at low August–January (Gorman 1970), and probably typical of much of year. In Firth of Forth during winter, flocks prefer to feed at low tide but loaf at mid-day regardless of tide, flying out to sea at dusk to roost (Player 1971). During summer, Norway, where little tidal influence, flocks spend night ashore, starting to feed at dawn and many later loafing between 09.00 and 16.00 with smaller feeding peak at dusk; in winter, day-loafing much reduced (Pethon 1967). During egg-laying, pair often roost on own ashore just above high-water mark; when ♀ starts incubating, ♂ usually joins other ♂♂ and may roost on land or water until departure (H Milne).

2. Studied extensively in wild and captivity; see especially Hoogerheide (1950), McKinney (1961), Johnsgard (1964, 1965). FLOCK BEHAVIOUR. Winter flocks often dense, but tend to split up into smaller, looser units when feeding and synchronized diving not so marked as in Steller's Eider *Polysticta stelleri*; also much less ready to take wing, apparently preferring to dive in face of danger (McKinney 1973). Head-shaking only pre-flight movements observed. ANTAGONISTIC BEHAVIOUR. Most hostile interactions in flocks between ♂♂ in courting parties and between pairs, especially April–May. Paired ♂ highly aggressive when near mate prior to egg-laying, threatening with repeated ♀-like but silent Chin-lifting (see below); also swims towards rival with head and bill projecting forward, and frequently chases it over water, ending with severe pecking. In occasional fights, ♂♂ struggle with much wing-beating and splashing. Paired ♀ intolerant of ♂♂ other than mate; performs Inciting-display, either threatening rejected ♂ by stretching lowered neck and pointing bill or alternating this with repeated Chin-lifting towards mate while Gog-gog-calls uttered with chin swollen, giving head characteristic shape. ♀ leading young also hostile to other adults, especially ♂♂. COMMUNAL COURTSHIP. Can show 2 peaks: October–November and April–May (Gorman 1974). Typical courting party on water consists of ♀ and 3–10 ♂♂; groups larger and more active during spring peak when up to 20 ♂♂. Main displays of ♂, various Cooing-movements; these accompanied by Cooing-calls and often combined in fixed sequences (see Voice) and preceded by repeated lateral Head-turning through 90° arc while head held vertically in Neck-stretch posture. In Bill-toss display (1) head thrown back, (2) bill held vertically with short pause (see A), (3) head flicked back to starting position. In Neck-jerk display, head thrust forward and up while bill held down (see B). In Reaching-display, head (1) held stiffly erect (Neck-stretch phase), (2) brought forward until bill nearly touches water (Reaching-display proper), (3) finally

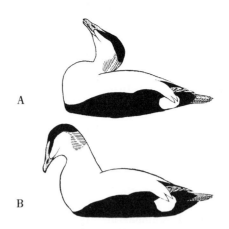

returned rapidly to normal position while breast puffed out as call given. ♂ also utters Cooing-calls without special movement or posture (Roo-calling of McKinney 1961). All displays performed while ♂♂ attempt to hold position close to ♀, constantly jostling, threatening, and chasing. Movement of party on water may be rapid, and, especially in spring, ♂♂ press ♀♀ so closely at times that she dives to evade them; most active ♂♂ follow and underwater pursuit results. ♂♂ also perform variety of secondary displays more characteristic of pre-copulatory situation, including (1) Neck-stretch on own; (2) Preen-dorsally; (3) Upward-shake (see C); (4) Wing-flap; (5) Head-roll (brief oiling-like movements of side of head over back); (6) Head-shake; and (7) Bill-dip. Inciting (see above) most frequent ♀ display; Gog-gog-calls also given on own and infrequently ♀ performs Reaching-display, Preen-dorsally, Neck-stretch, Ritual-bathe, Wing-flap, Head-roll, and Bill-dip much as in ♂. PAIR BEHAVIOUR. Same displays as in communal courtship occur in encounters between pairs and associated with copulation, but relative frequencies vary. Copulations begin last week February, over 2 months before first eggs. ♂ seeking to re-establish contact with mate swims about with head erect and bill pointing up (as in low-intensity Bill-tossing), calling 'wah' (McKinney 1961). COPU-LATION. Pre-copulatory displays most intense at dawn and dusk, with superimposed pattern of greater activity on flood tide, and least around high or low water; most frequent when flood tide coincides with dawn (Gorman 1970). Initiated by ♂; ♀ does not reciprocate but gradually assumes Prone-posture on water often after or while Inciting, as ♂ swims round her for several minutes performing continuous series of displays. Of these, Upward-shake, Ritual-bathe + Wing-flap, Neck-stretch, Preen-dorsally especially characteristic; other displays include Preen-behind-Wing and Head-roll, but Cooing-movements relatively infrequent. Copulation immediately preceded by any one of these actions; afterwards ♂ usually performs single Rearing-display, then swims away while Head-turning in Neck-stretch posture as ♀ (usually) just bathes and Wing-flaps. RELATIONS WITHIN FAMILY GROUP. See Driver (1960, 1974), Milne (1963), Gorman and Milne (1972). Young give variety of calls, some starting in

egg, which ♀ and siblings reciprocate. Cohesion-calls uttered as greeting between siblings, with Chin-lifting; or latter performed alone. During hatching or when ♀ leads brood to water, family sometimes attended by other ♀♀ (aunties). Role of latter puzzling but may be protective to some extent. Role of later guard-♀♀ in crèches mainly to watch for predators. Give alarm by calling, and ward off any predator overhead by leaping out of water; also brood smaller young at night or when weather inclement. When in danger, young crowd round nearest guard.

(Figs A–C after McKinney 1961, Johnsgard 1965.)

**Voice.** Exceptionally vocal, especially during courtship in flock in autumn and spring. Flocks often more silent at other times. Quality, power, and tempo of utterances assist audibility above sound of waves and wind. Basic calls show much minor variation but remain clearly recognizable through strong cooing or crooning quality in ♂ (recalling doves *Columba*, but more powerful) and hoarse, guttural, or raucous and rattling tones of ♀.

CALLS OF MALE. Largely various low Cooing-calls and given mostly during display, both when courting and in encounters between pairs; see especially Hoogerheide (1950), McKinney (1961), Johnsgard (1964, 1965), and, for technical analysis, Johnsgard (1971). In flocks on water, notable for volume and fusion of utterance with fluctuating surges and strong rising and falling rhythm (see I). Main calls include: (1) 'a-ooo' or 'ah-hOOO', given during Bill-toss and Neck-jerk displays (in latter case, slightly lower-pitched, quieter, and with less emphasis on 2nd syllable); (2) quiet 'whoo-hooo' or 'hoo-ooo' (sounding like 'cuckoo' at close range), during Reaching-display; and (3) soft 'roo' or 'rhoe' (Roo-calling). When displays combined, these calls become 'AH-oo ... a-oo (Bill-toss + Neck-jerk), 'wooh ... woohoo' (Neck-jerk + Bill-toss), and 'wooh ... woohoo ... wooh' (Neck-jerk + Bill-toss + Neck-jerk). Other descriptions of Cooing-calls include 'ah-oo, ah-ee-oo' or 'coo-roo-uh' (Witherby *et al.* 1939), and 'uhu uhu' or 'a-o-wah-ah-o-wah' (Kortright 1942). In addition 'anger call' described as 'wh-r-r-r-r' and call given immediately before copulation as 'aw, aw, aw' (Kortright 1942), while rarer calls include monosyllabic 'wah' and 'woof'—former given during low-intensity Bill-toss display (when ♂ appears to be seeking own mate and occasionally at other times) and later occasionally with Neck-stretch display (McKinney 1961).

CALLS OF FEMALE. ♀ does not coo, but when performing Reaching-display substitutes loud, throaty cawing version of ♂'s 'whoo-hoo' (McKinney 1961); this possibly 'coo-roo' mentioned in Witherby *et al.* (1939). Main calls, otherwise, mostly variants of deep, throaty 'gog'-like notes, often loud and far-carrying, given in many situations and probably basically hostile (McKinney 1961); rendered 'kok' by Hoogerheide (1950). Can also be described as staccato quacking 'qua-qua-qua' or 'kwar-kwar-kwar' (see II), hollower and less high-pitched than call of ♀ Mallard *Anas platyrhynchos*. During Inciting-display, typically delivered in continuous series of *c*. 6 or more notes: 'gogogog ...' or 'kokkokkok ...'; wooden quality mentioned by Johnsgard (1964). Other calls include: 'oh-oh-oh' when flying from nest (Bauer and Glutz 1969), and various calls given in presence of young—soft, purring 'whirr' when brooding or attending young, or responding to their Distress-calls; 'kuk-kuk-kuk' (like Inciting-call but louder and harsher) when danger imminent, and in response to Complaint-calls of young; and loud, deep, croaking 'krrok-krrock' when defending young against predators (H Milne). Descending, throaty, rattling 'whirr' also given as alarm-call causing young to cluster round (Driver 1960, 1974); alternatively rendered 'gurr-urr' (Salomonsen 1950a) or 'gang-gang-gang' (Bauer and Glutz 1969). Cohesion-call of ♀ with young, low, rather hoarse 'cow-cow-cow' or 'how-cow-cow-cow' (Driver 1974).

CALLS OF YOUNG. Mainly variants of repeated, high, shrill monotonous 'peep', 'niep', or 'twee'. Following distinguished by Driver (1960, 1974). (1) Brooding-call: monosyllabic and rich-sounding, rapidly rising or falling. (2) Cohesion-call: rising twitter of usually 4 notes, uttered at frequent intervals. (3) Contentment-call: phrase of 4 rapidly repeated notes; similar to last but faster, higher pitched, and more monotonous. (4) Complaint-call: disyllabic, with 2nd note higher pitched, insistent, and often repeated. (5) Distress-call: high-pitched, piercing, far-carrying, monosyllabic 'peep', also often repeated persistently (see Fig. 10 in Kear 1968). (6) Alerting-call: thin, rising, rather faint, single call, seldom heard more than once at any one time, given when puzzled by potentially dangerous, but indefinite stimulus; will alert ♀ and siblings. Imperfect versions of (2), (4), and (5) given in egg 2–3 days before hatching, together with much non-descript vocalizing.

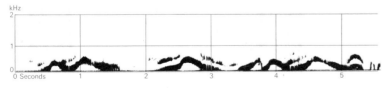

I S Wahlstrom Sweden April 1968

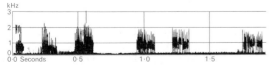

II S Wahlstrom Sweden April 1968

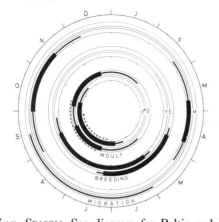

**Breeding.** SEASON. See diagram for Baltic and Fenno-Scandia. First eggs early April Britain and Netherlands, but last week April or 1st week May Scotland (Milne 1965). 2nd half May Iceland; late May and early June western USSR; mid-June Spitsbergen and arctic USSR (Bauer and Glutz 1969; Dementiev and Gladkov 1952; Løvenskiold 1964). In Scotland, onset of laying apparently independent of spring temperatures (Milne 1965b). In Spitsbergen, and other arctic areas, onset of laying highly correlated with appearance of snow-free ground, first eggs laid within 3 days of nest-site becoming exposed (Ogilvie and Taylor 1967; Ahlén and Anderson 1970). Laying in colonies highly synchronized; in Spitsbergen all laying within 1–2 weeks, in Scotland all within 3 weeks (Ahlén and Anderson 1970; Milne 1965b). SITE. On ground, often in shelter of rock or vegetation, but also completely exposed. Usually near seashore and islands much favoured, but up to 3 km inland (Jenkins 1971). Often colonial, with nests in densities up to 2 per m², and over 3000 nests in some colonies. Often in colonies of Arctic Tern *Sterna paradisaea*. Nest: slight hollow, lined with available material, including grass, droppings, and then down. Re-use of old nests leads to formation of permanent cup with rim. Average diameter 20–25 cm, depth 10 cm (Dementiev and Gladkov 1952). Building: by ♀ using material within reach. EGGS. Blunt ovate; greenish-grey, sometimes buff, green, or, rarely, blue. 77 × 52 mm (68–88 × 47–57), sample 400 (Schönwetter 1967). Weight 109 g (87–127), sample not given (Dementiev and Gladkov 1952). Weight of captive laid eggs 111 g (89–137), sample 100 (J Kear). Clutch: 4–6 (1–8, with more than 8 usually 2 ♀♀). Of 193 Finnish clutches: 3 eggs, 14%; 4, 27%; 5, 45%; 6, 12%; 7, 1%; 8, 1%; mean 4·6 (Hildén 1964). Of 2661 Spitsbergen clutches: 1 egg, 12%; 2, 24%; 3, 34%; 4, 21%; 5, 5%; 6, 2%; 7, 1%; 8, 1%; mean 2·95. In Spitsbergen, clutch size varied with nest density from 2·57 where nests 1 per 4 m² to 4·00 where nests 9 per 4 m². Also higher in scattered nests in Arctic Tern colonies (3·34) than away from them (2·58). Spitsbergen birds generally subject to heavy predation; clutch size in sites where no predation estimated at 4·5. Average number of eggs laid by each ♀ estimated at 6·2, though this may have included

some re-layings (Ahlén and Anderson 1970). One brood. Replacement clutches laid after egg loss. One egg laid per day. INCUBATION. 25–28 days. By ♀, ♂ often in attendance near nest at least in early stages. Starts with last egg, though ♀ sits for increasing period as laying proceeds; hatching synchronous. Eggs covered with down when ♀ off nest, though absences less frequent than in most other Anatidae. Eggshells left in nest. YOUNG. Precocial and nidifugous. Self-feeding. Cared for by ♀, and brooded at night while small. In dense breeding areas, broods frequently combine into gatherings of 100 or more young plus ♀♀; much interchange between such groups and adult–young bond slight or non-existent (Milne 1965b). FLEDGING TO MATURITY. Fledging period 65–75 days. Become independent at 55–60 days, sometimes earlier. Age of first breeding usually 3 years, some ♀♀ at 2. BREEDING SUCCESS. Of 1026 eggs laid Finland in 3 years, 75% hatched and 19% young reared; survival per pair 2·5, 0·5, and 0·1 young, giving losses of 46%, 90%, and 97% (Hildén 1964). In Spitsbergen, clutch size at start of incubation averaged 3·34, at hatching 2·8, and number of eggs hatched 1·9; main egg predator Glaucous Gull *Larus hyperboreus* (Ahlén and Anderson 1970). In Scotland, hatching success in 4 years from 417 nests averaged 61·7% (56·1–70·2%); annual chick survival 5·8%, 1·9%, 40·0%, and 0·1%. Shortage of food thought to be primary mortality factor in years of poor chick production (Milne 1965b).

**Plumages.** Nominate *mollissima*. ADULT MALE BREEDING. Forehead, centre of crown down to eye, and feathering along sides of frontal processes of upper mandible black. Streak from centre of crown to nape white. Dense, stiff, and somewhat elongated feathers at upper nape green, separated from green patch at upper sides of neck by white streak. Cheeks, chin, throat, neck, mantle, and scapulars white, like back, sides of rump, and patch at sides of tail-base; chest and upper breast also white, but these tinged pink-cream when fresh. Centre of rump, tail-coverts, flanks, lower breast, belly, and vent black. Tail sooty black. Primaries and primary coverts dark grey, outer webs black and part of coverts speckled white at tips; inner primaries sometimes tipped white. Secondaries, greater coverts, and bastard-wing black, varying number of outer coverts sometimes tipped white. Inner secondary bordering tertials black with inner web and tip white. Tertials and lesser and median upper wing-coverts white. Greater under wing-coverts grey, tipped white; other under wing-coverts and axillaries white, sometimes suffused grey at base. ADULT FEMALE BREEDING. Head and neck cinnamon-buff, forehead and crown narrowly streaked black, elsewhere finely spotted black. Usually pale buff streak through lores and behind eye. Feathers of mantle, chest, breast, and under tail-coverts black, broadly barred and tipped cinnamon-buff when fresh (tips often concealing black bars of sides of breast and chest). Scapulars, back, rump, and upper tail-coverts similar, but tips and bars warmer cinnamon and tips of scapulars broader. Lower breast, belly, and vent variable; either sepia with distinct buff bars or plain greyish sepia, only faintly barred. Tail, primaries, and primary coverts like adult ♂, but coverts not speckled. Secondaries and greater upper wing-coverts sooty-black, in some with cinnamon or purple tinge or mottling on outer web; distinctly tipped white. Tertials black, broadly edged warm cinnamon or chestnut on outer webs. Median and lesser coverts black with cinnamon-buff edges and grey base. Under wing-coverts sepia, tipped buff; greater coverts and axillaries off-white, variably suffused dusky grey. Broad cinnamon of buff feather-tips on plumage abrade and fade during spring and early summer, forehead and crown become darker, rest of head more heavily mottled black, upperparts, chest, and sides more contrastingly barred pale buff and black. ADULT MALE NON-BREEDING (Eclipse). Head, neck, mantle, scapulars, chest, breast, flanks, and sides of tail-base sooty-brown or sepia, darkest on crown and sides of head. Feathers of crown, mantle, and chest variably tipped buff; on mantle, scapulars, chest, and breast often with broad imperfectly concealed white base or white central mark towards tip. Irregularly mottled white or buff streak through eye. Some white breeding feathers sometimes retained. Back, rump, tail-coverts, breast to vent, tail, and wing like adult ♂ breeding. Differs from adult ♀ by darker, unbarred plumage. ADULT FEMALE NON-BREEDING. Like adult ♀ breeding, but head and neck darker, sooty-black, faintly edged buff on forehead and crown, thickly streaked sepia elsewhere. Feather edges on upperparts, chest, and flanks narrower and paler, buff sub-terminal bars narrower and less clearly defined, often incomplete. Sides of body, belly, and vent dark ash-brown, without distinct bars. DOWNY YOUNG. Crown, hindneck, cheeks, upperparts, and vent dark grey, paler grey on upper mantle. Streak over eye to hindneck, sides of chin, and belly pale grey, centre of chin, foreneck, and chest darker grey. Feathering of bill as adult. JUVENILE. Resembles adult ♀, but differs by absence of distinct bars on upperparts; underparts conspicuously marked with narrow and dense bars of sepia and pale buff or off-white; usually no white tips to secondaries and greater upper wing-coverts. Crown sepia, narrowly streaked buff on centre, mottled white on forehead and sides, forming indistinct white streak over eye. Cheeks and throat pale buff, finely streaked sepia; neck and upper mantle pale sepia or dark grey, feathers narrowly tipped buff. Lower mantle, scapulars, back, rump, and upper tail-coverts sepia, feathers narrowly edged buff with dull black bar subterminally, instead of straight and broad warm buff bars like adult ♀ breeding. Flank feathers grey with narrow pale buff tip and sepia subterminal bar. Tail sepia, bare feather-shafts projecting. Wing like adult ♀, but no white to tips of secondaries and greater coverts; secondaries, tertials, and greater coverts dark sepia, faintly edged pale buff; other upper wing-coverts like mantle. Juvenile ♂ as juvenile ♀, but ground colour of upperparts, upper wing-coverts, and secondaries of ♂ somewhat darker, dull black rather than sepia brown; tertials of ♂ longer, faintly sickle-shaped; buff feather edges of back and rump slightly narrower in ♂. FIRST IMMATURE NON-BREEDING MALE. Like juvenile ♂, but head, neck, and sometimes a few feathers on mantle, scapulars, and flanks new: feathers of forehead, crown, and upper nape buff, narrowly barred black; centre of crown, cheeks, and neck black, indistinctly barred buff; distinct streak through eye to nape white, finely speckled black; new feathers elsewhere like adult ♀ or ♂ eclipse. FIRST IMMATURE BREEDING MALE. Highly variable. Variable number of new feathers like adult ♂ breeding acquired gradually late autumn to spring. Chest, sides of breast (left side of body often much earlier than right), upper mantle, flanks, and lower scapulars first, later elsewhere on body, when 1st immature non-breeding plumage of head also becomes mixed with some white, green, and black. White feathers often tipped or suffused dull black (scapulars especially appear black), black feathers with brown tinge. Wing, back, and rump, usually underparts and tail, and variable amount

of plumage elsewhere juvenile until summer. FIRST IMMATURE NON-BREEDING AND BREEDING FEMALE. No distinctive 1st immature non-breeding and breeding plumage discernible; juvenile feathers replaced first resemble adult ♀ non-breeding (hardly barred), later on gradually intermediate in character between adult ♀ non-breeding and adult ♀ breeding. As in juvenile ♂, wing, tail, back, rump, and underparts (sometimes almost whole body) juvenile until summer. SECOND IMMATURE NON-BREEDING. Like adult non-breeding, but often distinguishable by retained abraded juvenile or 1st immature breeding feathers, intermixed with fresh non-breeding; wing juvenile, when not yet moulted. SECOND IMMATURE BREEDING MALE. Like adult ♂ breeding, but feathers on head, neck, mantle, scapulars variably tipped or suffused dull black or buff (dark tinge may disappear by abrasion); centre of back and rump more heavily suffused black. Wing like adult ♂, but secondaries and greater coverts sometimes narrowly tipped white. Lesser coverts and tips of median suffused sooty; tertials less sickle-shaped, usually broadly tipped and edged black. SECOND IMMATURE BREEDING FEMALE. Like adult ♀ breeding, but white tips of secondaries and greater coverts narrower; edges and subterminal bars of body feathers narrower, less straight and less sharply defined from black on feathers. THIRD IMMATURE NON-BREEDING. Like adult non-breeding, but wing like 2nd immature breeding when not yet moulted or, when moulted, as 3rd immature breeding. THIRD IMMATURE BREEDING MALE. Like adult ♂ breeding, but centre of back and rump still rather extensively black; tertials and some inner greater and median coverts usually mottled dull black at tip. THIRD IMMATURE BREEDING FEMALE. Like adult ♀ breeding, but brown bars and edges of body plumage somewhat less sharply defined from black. (Schiøler 1926; Witherby *et al.* 1939; C S Roselaar.)

**Bare parts.** Nominate *mollissima*, Netherlands. ADULT MALE. Iris brown. Frontal processes and upper base of bill pale yellow, rest of bill olive-grey, nail pale pink or yellow. During autumn, bill, including frontal processes, dull olive-grey. Foot olive-yellow or grey-green, webs and soles dull black. ADULT FEMALE. Iris brown. Bill dull olive-grey with paler nail. Foot dark olive-grey or dusky-yellow, webs dull black. DOWNY YOUNG. Iris dark brown. Bill and foot dark plumbeous olive. JUVENILE. Iris brown. Bill and foot dull olive-grey, somewhat paler in ♂ from mid-winter; in some, olive-yellow from January.

**Moults.** ADULT POST-BREEDING. Complete; flight-feathers simultaneous. ♂♂ from mid-June or mid-July, first on head, neck, chest, upper breast, flanks, mantle, scapulars, and sides of tail-base; followed, between mid-July and late August, by lower breast, belly, vent, back, centre of rump, wing, and tail. ♀♀ like ♂♂, but *c*. 1 month later. ADULT PRE-BREEDING. Partial. Head and body, except back, centre of rump, lower breast, belly, and vent. ♂♂ between mid-August and November; ♀♀ October–March; ♀♀ sometimes moult underparts, as late as May. POST-JUVENILE AND FIRST IMMATURE PRE-BREEDING. Partial; from September. Highly variable in amount of plumage involved. In ♂, head, neck, and some feathers of upperparts, sides, and flanks attain 1st immature non-breeding; from October, new feathers black with increasing amount of white. In ♀, new feathers intermediate between adult ♀ non-breeding and breeding, appearing from September–October on same parts of body as ♂. Juvenile plumage of wing, back, rump, usually underparts and tail and some elsewhere, retained until summer. FIRST IMMATURE POST-BREEDING. Like adults, but less complete and more prolonged. Some feathers of 1st immature breeding plumage, and occasionally juvenile back, rump, and underparts retained. SECOND

IMMATURE PRE-BREEDING. Complete. Like adult, but finished somewhat later, and number of non-breeding feathers retained for considerable time in some. (Dementiev and Gladkov 1952; Schiøler 1926; C S Roselaar.)

**Measurements.** ADULT. Nominate *mollissima*. Netherlands, April–July; skins (RMNH, ZMA).

| | | | | | | |
|---|---|---|---|---|---|---|
| WING | ♂ 304 | (7·37; 20) | 289–315 | ♀301 | (5·74; 21) | 286–312 |
| TAIL | 96·0 | (3·69; 17) | 90–104 | 94·7 | (2·43; 21) | 90–98 |
| BILL | 57·2 | (2·07; 22) | 53–61 | 54·4 | (2·10; 23) | 51–59 |
| TARSUS | 54·2 | (1·47; 21) | 52–57 | 52·8 | (1·64; 23) | 50–56 |
| TOE | 79·3 | (2·73; 21) | 75–86 | 78·9 | (2·59; 17) | 76–86 |

Sex differences significant for bill and tarsus only.
*S. m. borealis.* Greenland, Iceland; skins (RMNH, ZMA).

| | | | | | | |
|---|---|---|---|---|---|---|
| WING | ♂ 291 | (5·91; 15) | 284–302 | ♀282 | (3·36; 5) | 278–287 |
| BILL | 51·1 | (2·80; 17) | 47–56 | 47·6 | (2·16; 5) | 46–52 |
| TARSUS | 50·2 | (1·50; 17) | 48–53 | 49·1 | (2·01; 5) | 47–51 |

Sex differences significant for wing and bill. All measurements significantly smaller than nominate.
*S. m. faeroeensis.* Faeroes (Schiøler 1926).

| | | | | | | |
|---|---|---|---|---|---|---|
| WING | ♂ 270 | (22) | 260–284 | ♀264 | (7) | 257–271 |
| BILL | 49·4 | (22) | 48–56 | 48·0 | (7) | 45–50 |

JUVENILE. Nominate *mollissima*. Netherlands, mainly winter; skins (RMNH, ZMA).

| | | | | | | |
|---|---|---|---|---|---|---|
| WING | ♂ 285 | (7·46; 35) | 275–297 | ♀278 | (6·84; 33) | 265–292 |
| TAIL | 80·9 | (4·03; 22) | 74–88 | 77·2 | (3·37; 21) | 70–84 |
| BILL | 55·2 | (2·57; 37) | 50–61 | 53·1 | (2·46; 34) | 48–58 |

Sex differences significant. Wing and tail significantly shorter than adults; other measurements similar or only slightly shorter. Wing and tail in 2nd and 3rd year intermediate between adult and juvenile.

**Weights.** ADULT. Nominate *mollissima*. (1) Murmansk, USSR, summer (Bauer and Glutz 1969). (2) Denmark, mainly winter (Schiøler 1926).

| | | | | | | |
|---|---|---|---|---|---|---|
| (1) | ♂ 2218 | (22) | 1384–2800 | ♀1915 | (32) | 1192–2895 |
| (2) | 2315 | (22) | 1965–2875 | 2142 | (18) | 1864–2595 |

Averages of ♂♂, Murmansk: May, 2370; June, 2110. Ranges ♀♀, Murmansk, just prior to breeding 2365–2895, at end of breeding 1555–1637; averages, May 2370, June 2081, July 1629, August 1928 (Dementiev and Gladkov 1952).
*S. m. borealis.* East Greenland, June–July (J de Korte).

| | | | | | | |
|---|---|---|---|---|---|---|
| | ♂ 2000 | (12) | 1560–2710 | ♀1810 | (11) | 1575–2165 |

*S. m. faeroeensis.* ♀♀ Faeroes, mostly April–June, 1847 (6) 1703–2223.
JUVENILE. Nominate *mollissima*. Denmark, mainly autumn and winter (Schiøler 1926).

| | | | | | | |
|---|---|---|---|---|---|---|
| | ♂ 2080 | (37) | 1562–2567 | ♀1770 | (10) | 1552–2009 |

**Structure.** Nominate *mollissima*. Wing rather short, broad at base; pointed. 11 primaries: p9 longest, p10 0–8 shorter, p8 6–11,

p7 19–28, p6 34–44, p1 103–132. Inner web of p10 and outer p9 emarginated. Tertials falcate, curved outwards; slightly elongated, tapering to point. Inner lower scapulars stiff, broad at tip, with emarginated tip of outer web in adults; this tip sometimes erected actively for short periods in ♂ but not more or less permanently as in adult ♂ King Eider *S. spectabilis* (Palmer 1973). Tail short and rounded; 14 feathers. Bill about as long as head, higher than broad at base. Sides of bill slightly narrowing towards large, elliptical nail. Culmen straight, except for slight concave depression behind nail; cutting edges of upper mandible straight, only slightly curved upwards near nail. Feathering of forehead extends into sharp point on culmen, about halfway between nostrils and narrow, rounded tip of bare frontal processes. Feathering at sides of bill extends much further than frontal feathering, approximately to below middle of nostril. Outer toe slightly larger than middle, inner *c.* 79% middle, hind *c.* 34%.

**Geographical variation.** Considerable. Several distinctly defined subspecies in arctic Pacific and in west Atlantic and Hudson Bay, but variation in east and arctic Atlantic apparently more gradual and therefore more complex. Variation concerns body size, development of frontal processes of bill, extent of feathering on sides of bill, colour of bill of ♂, amount of green on sides of head of ♂, presence of black V on throat of ♂, and colour of breeding plumage of ♀. Birds of Faeroes (*faeroeensis*) like nominate, but smaller bill and narrow, short frontal processes dark olive-grey, and ♀ barred much darker brown; those of Shetlands and Orkneys slightly larger, but probably best included in *faeroeensis*. Birds elsewhere in Britain and in south-west. Norway intermediate between *faeroeensis* and nominate *mollissima* in size, bill-colour, and tone of ♀ plumage. Birds inhabiting north Norway, north-west USSR, and Novaya Zemlya like nominate *mollissima*, but tend to be slightly larger and darker. *S. m. borealis* of west Greenland and Baffin Island region smaller than nominate *mollissima*; feathering of sides of bill usually not extending to below nostril; bill of ♂, including rather short and narrow frontal processes, orange-yellow; tertials longer and more strongly decurved; and colour of ♀ more rufous. *S. m. borealis* of Spitsbergen slightly larger, of east and south Greenland and Iceland like those of west Greenland, but bill colour and extent of feathering at sides of bill tends somewhat to nominate *mollissima*. In west Atlantic subspecies *dresseri*, frontal processes of bill long and broadly rounded, bill of ♂ orange-yellow, green on sides of head extending to below eye, and ♀ barred rufous brown; size like nominate *mollissima*. *S. m. sedentaria* from Hudson Bay like *dresseri*, but frontal processes of bill smaller and ♀ paler and greyer. Frontal processes in *v-nigrum* of arctic Pacific short and narrow, feathering on sides of bill rounded instead of pointed, not reaching to below nostril; bill of adult ♂ orange-red, green on sides of head extends to below eye, often black V on throat (apex to chin), and ♀ plumage pale; large. Black V on throat exceptionally also in other subspecies. (Schiøler 1926; Roi 1912; Vaurie 1965; C S Roselaar.) CSR

605

PLATES 82 and 100
[between pages 638 and 639
and facing page 687]

# *Somateria spectabilis* King Eider

Du. Koningseider    Fr. Eider à tête grise    Ge. Prachteiderente
Ru. Гага-гребенушка    Sp. Eider real    Sw. Praktejder

*Anas spectabilis* Linnaeus, 1758

Monotypic

**Field characters.** 47–63 cm (body 34–40 cm); wing-span 86–102 cm; slightly smaller and shorter billed than Eider *S. mollissima*. Large, rather round-headed eider; more compact and less ungainly than *S. mollissima*, and in ♀ less extension of upper mandible towards eye. At close range, extension of frontal feathering forward to above nostril (and beyond limit of loral feathering) diagnostic, except in adult ♂. Adult ♂ looks white in front and black behind on sea, but shows white wing-coverts in flight: uniquely combines orange-red bill and bare shield on forehead with pearl-grey crown and nape of pendant feathers. ♀ and juveniles of both sexes essentially dark red-brown looking rustier and less marked than typical *S. mollissima* of similar ages. Sub-adult ♂ shows rather less variety of plumage than *S. mollissima*, with pale chest and upper mantle. Sexes dissimilar; marked seasonal variation in ♂. Juveniles and sub-adults separable.

ADULT MALE. Narrow surrounds to shield and V-shaped mark on throat black; crown and nape of broad pendant feathers pearl-grey, darkening at rear and outlined with white stripe; upper cheeks green, fading down sides of nape; rest of head, neck, and upper mantle white. Rest of upperparts, including 2 sickle-shaped plumes rising from base of tertials, black; round patches on sides rump white. Wing-quills black contrasting with large white patch on wing-coverts. Chest buff-pink; rest of underparts black. Bill with orange-flesh tip, deepening to orange-red at base and on shield; legs and feet dull orange, latter with dusky webs. MALE ECLIPSE. Dark, with uniform sooty plumage relieved on most by noticeably pale lower cheeks, throat, and chest; wing-coverts remain white. Shield above bill shrinks to insignificant lobe and becomes dull red. ADULT FEMALE. Nape with pendant feathers as in ♂ and separated from cheeks by thin dark line. All plumage uniform, warm, rusty-brown, more delicately marked than in *S. mollissima* with finer streaks on head, and noticeably more crescentic barring on flanks; further distinguished from that species by noticeably pale chin and throat, and boldly spotted scapulars. Bill, legs, and feet grey. JUVENILE. Resembles adult ♀ but with more uniform and duller brown upperparts, lacking rustiness. Legs and feet of ♂ dull olive. SUB-ADULT. ♂ assumes breeding plumage over 3 years, with erratic progression as in *S. mollissima*, Appears less piebald than young ♂ *S. mollissima*, with white cheeks and chest more obvious and completely sooty or black scapulars already noticeable in 1st summer. Shield above bill obvious in 2nd summer but full size and colour not attained before full maturity.

Adult ♂ unmistakable; ♀♀ and sub-adults difficult to distinguish from *S. mollissima* but warmer plumage tones and crescentic feather patterns give useful clues. Nape pattern and pale chin of ♀, different limit to bill feathering, and less hooked nail on bill diagnostic at close range, but brighter edges to ♀'s secondaries not safe distinction from *S. mollissima*. For distinction from Spectacled Eider *S. fischeri*, see that species.

Flight silhouette similar to *S. mollissima* but a little more compact, with extension head and neck less apparent in ♂. Flight action and formation as *S. mollissima* but a little lighter and quicker. Less markedly gregarious and marine when breeding than *S. mollissima*. On migration and in winter, large groups form and associate with *S. mollissima* and other sea-ducks but remain farther off shore often feeding in deep water. ♂ has more vibrant cooing-calls than *S. mollissima*; ♀ utters grunting croak, more akin to *Anas* than that of *S. mollissima*.                DIMW

**Habitat.** Even more arctic species than Eider *S. mollissima*. In west Palearctic, only to minor extent south of Arctic Circle, largely keeping clear of milder influence of North Atlantic Current. Consequently does not share oceanic extension of *S. mollissima*; even wintering birds mainly content to find open waters in Arctic Ocean rather than to migrate south into boreal areas. Farther offshore than *S. mollissima*, especially in winter. In summer, reverses roles, by spreading far inland across tundra instead of breeding largely on coast. Accordingly, less firmly linked to particular coastline or archipelago, and more inclined to face abrupt seasonal switching of habitats, diet, and pattern of life. Also more attracted towards less rocky and even muddy shores of bays or estuaries, and in summer makes more use of freshwater pools, lagoons, and even rivers. Such waters in parts of northern tundra ice-free as few as 60 days in year, against estimated minimum 80 days for breeding cycle, requiring evacuation to coast at early stage in growth of young (Uspenski 1972). Despite apparently precarious basis, species seems to maintain high and stable population; densities up to 1 breeding pair per 10 ha of tundra recorded in NE Yakutsk, but in west Palearctic no more than 0·2 on Bolshesemelskaja tundra, down to 0·02 in Novaya Zemlya (Uspenski 1972). Tundra habitats varied, including dry patches of mosses, lichens, herbage, grassy places, and peat-hummocks up to 50, exceptionally 100 km inland; may be some distance from nearest water. Sites less often used include wet, swampy hollows and sand or shingle spits. Shrubby or tree-grown

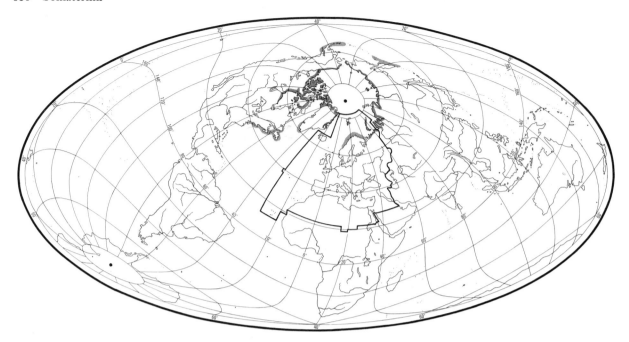

areas avoided. Readiness to go farther out from shore than *S. mollissima* indicative of deeper diving capacity. Lives right up to ice edge. Flight normally so low that birds used to be brought down by primitive missiles, but heights as high as 1000 m recorded (Wynne-Edwards 1952*a*).

**Distribution.** NORWAY: bred Trondheimsfjord, well south of normal range, in 1957, 1960, and 1961; perhaps same pair (Haftorn 1971). FINLAND: egg taken Åland Islands, May 1883, considered by R Kreuger to belong to this species (*Orn. Fenn.* 1967, **44**, 111). Summering birds (♂♂ usually identified) occur in many places outside breeding area, and have paired with ♀ Eider *S. mollissima* (especially Iceland), but apparently few progeny produced.

Accidental. Britain and Ireland: *c.* 100 all months, chiefly Scotland, but also east coast England and Ireland (Smith *et al.* 1975). Finland: *c.* 60 (Merikallio 1958). Sweden: over 50 (Bauer and Glutz 1969). Also Faeroes, Jan Mayen, Bear Island, Netherlands, Belgium, France, Denmark, West Germany, East Germany, Poland, Hungary, Italy.

**Population.** USSR Total of perhaps 1–1½ million adults (Uspenski 1972). Shot on large scale only in certain parts of range, e.g. 12 000–14 000 yearly on Ostrov Kolguyev (Dementiev and Gladkov 1952).

Survival. Oldest ringed bird 5 years 9 months (Rydzewski 1973).

**Movements.** Migratory, perhaps only partially in some cases. Following based on Dementiev and Gladkov (1952), Salomonsen (1968), and Bauer and Glutz (1969). Over-

winters high latitudes in 3 main areas. Breeders of east Siberia, Alaska, and arctic Canada west of Victoria Island (110°W) winter Bering Sea. Those breeding rest of arctic Canada and north-west Greenland winter from south-west Greenland to Labrador and Newfoundland, small numbers passing up Gulf of St Lawrence to Great Lakes and down Atlantic coast to New England. Breeders of north USSR (eastern limit unknown) and, presumably, some from Spitsbergen and Novaya Zemlya, have winter area extending from White Sea west to arctic Norway. From this region, small numbers move down Norwegian coast and sporadically into North Sea, while Baltic records (over 60 Finland) perhaps indicate exceptional overland movement between White Sea and Gulf of Finland, as in Steller's Eider *Polysticta stelleri*. Though sea-ice closes most of Polar Basin, a few generally manage to winter Barents Sea and Kara Sea. Others reach Iceland, probably from east Greenland (one recovery).

Moult migration sometimes spectacular, with *c.* 100 000 ♂♂ and immatures in Davis Strait (maximum density Disco Bay) coming from north-west Greenland and especially arctic Canada west to 110°W; from latter, major moult movement up to 2500 km, passing overland across Baffin Island. Some immatures go direct to moult gatherings from winter quarters, while others move towards breeding range but interrupt this to return to moulting ground (Salomonsen 1968); this appears to be origin of 'summering areas' mapped by Uspenski (1972). Small parties moult coasts of Finnmark, Kola peninsula, Vaigach, and south Spitsbergen; four-figure moulting flocks located south-west Novaya Zemlya, and Ostrov Belyy in Kara Sea, while 20 000–40 000 immatures present all summer off Kolguyev evidently moult there alongside the many moulting adults (Dementiev and Gladkov 1952). ♂♂ desert breeding grounds while ♀♀ incubating and first reach even distant moult areas by early July, peak movements in second half July. Autumn migration proper late August to October; last birds leave nesting grounds September. Spring passage largely determined by ice condition, e.g. first inshore reports Novaya Zemlya variably 30 March–21 May. First reach Spitsbergen and arctic Canada in April; but away from influence of warm currents return delayed; vanguard may wait offshore for weeks before inland waters thaw, and northernmost breeders may not be established before end of June. Present off Murmansk and Finnmark mainly October to mid-April.

**Food.** Almost entirely animal, particularly molluscs, crustaceans, and echinoderms. Gregarious, daytime feeder, mainly by surface-diving with partially opened wings and often in deep water, disputably up to 54·9 m (see Preble and McAtee 1923). 8 dives in 12·8 m ranged from 54–81 s (Meinertzhagen 1959). In shallows, will probably up-end and head-dip like Eider *S. mollissima*.

In breeding areas, crustaceans and insect larvae may predominate, with some plant material—chiefly leaves and stalks of grasses (Gramineae) and other tundra plants (Manniche 1910; Roi 1911; Summerhayes and Elton 1923). Stomachs filled with Chironomidae larvae reported from Taymyr and Novosibirskiye islands in May and June (Dementiev and Gladkov 1952), and these taken ex-

clusively in shallow lakes in north Greenland (Just 1967). When sea still ice-covered in June in north Greenland, stomach contents entirely recently thawed phyllopod crustaceans (*Branchinecta paludosa*) frozen in since previous autumn (Røen 1965). ♂♂ may eat little at beginning of breeding season, but marked changes can occur when they, and later ♀♀ and grown young, move to sea. Off Spitsbergen, flocks of ♂♂ fed on sea-slugs (holothurians) in shallow, coastal waters (Løvenskiold 1954, 1964). On Novaya Zemlya, molluscs especially mussel *Mytilus* and *Saxicava*, and some crabs, main summer foods, though a July stomach from Bezymyan-naya Bay had almost entirely isopod crustaceans *Idotea entomon*. In August, flocks of ♂♂ in same area 15–20 km from coast fed on swarming pteropods *Clio borealis* and planktonic molluscs *Limacina limacina* (Dementiev and Gladkov 1952). In northern Norway, birds in late winter contained molluscs (*Buccinum undatum*, *Trophon clathrus*, *Boreochiton marmaoreus*, and *Astarte*), sea-urchin *Strongly-locentrotus droebachiensis*, and crustaceans *Hyas coarctatis* and *Hippolyte turgida*. One, May from Tromsø contained only echinoderm pieces (Collett 1894b). 3 ♂♂ from Pechenga, USSR, in February and April, had mainly sea-urchins *S. droebachiensis*, and fewer unidentified crabs and mussels *M. edulis* (Siivonen 1941). Other marine items recorded include molluscs (*Modiolaria discors*, *M. laevigata*, *M. faba*, *Leda minuta*, *Pecten islandicus*, *Turritella erosa*, *Volutomitra groenlandica*, *Nucula*, *Cyprina*, *Mactra*, and *Mya*), and ophiurids, actinians, and fish spawn

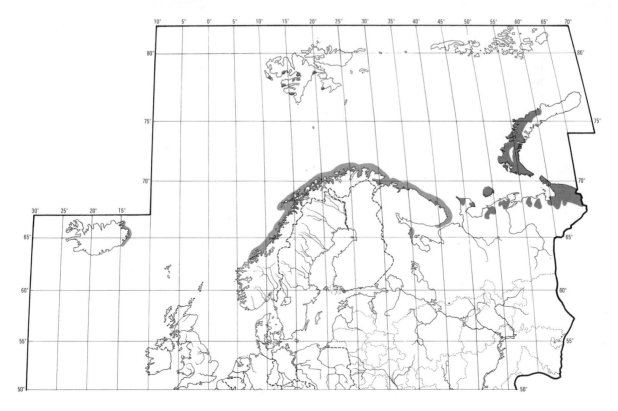

(Salomonsen 1950a; Witherby *et al.* 1939). 85 stomachs mainly from Alaska (January, February, March, May, June, and November) contained by volume 45·8% molluscs (chiefly *Mytilus* and *Modiolaria*), 18·6% crustaceans (mainly King crab *Dermaturus mandtii* and to lesser extent other crabs *Cancer*), amphipods, and sessile barnacles (Balanidae), 17·2% echinoderms (chiefly sand-dollar *Echinarachnius parma* and sea-urchin *S. droebachiensis*), 5·2% insects, 2·4% sea anemone *Aulactinia capitata*, and 5·5% plant materials. Insects, more prominent in June and November, mainly caddisfly larvae (Trichoptera), and fewer beetles and waterbugs. Most plant material probably taken accidentally, but eel-grass *Zostera marina* in some cases deliberately (Cottam 1939). This, and earlier American work, suggests wide variety of prey.

Young, which first feed exclusively in fresh water, take crustaceans, insect larvae, and probably some plant materials (Manniche 1910; Summerhayes and Elton 1923).

**Social pattern and behaviour.** 1. Highly gregarious except when breeding. Outside breeding season, in main area of distribution, winters in large flocks, of mixed sex and ages, on sea, often in association with other eiders. Migrates in smaller flocks of a few to *c.* 1000, mean 105 (Thompson and Person 1963), and parties in autumn smaller than in spring (see Bent 1925). Early spring migrants mainly ♂♂, but later birds often in pairs (Murdock 1885; see also Bent 1925). Within breeding range, ♂♂ first to flock in July when ♀♀ still nesting, returning to sea for body moult, often in company of immatures, and after making moult migration to undergo wing moult in large numbers at traditional areas (Salomonsen 1950a, 1968). ♀♀ flock later and moult elsewhere. Large flocks of immatures found throughout year on sea in some areas. See also Steller's Eider *Polysticta stelleri*. BONDS. Evidence suggests monogamous pair-bond of seasonal duration, as in other *Somateria*. Pair-formation mostly occurs late, about end April or early May; on sea, either near breeding area, during spring passage, or (in case of later migrants) in wintering area (Bauer and Glutz 1969). ♂ deserts ♀ early in incubation period. Mixed pairs of ♂ *S. spectabilis* and ♀ Eider *S. mollissima* occur in small numbers among colonies of latter species in Iceland (Pettingill 1959, 1962; Palmer 1973). Only ♀ tends young. Parent-young bond lasts till brood fledged; family stays on fresh water or ♀ leads young to sea shortly before they can fly (Løvenskiold 1964), sometimes earlier. 2 broods may combine under care of 1 ♀ (Bauer and Glutz 1969), but unfledged young also form large crèches of up to 300–400 (see Bent 1925; Witherby *et al.* 1939, 1944), with several ♀♀ in attendance. BREEDING DISPERSION. Unlike *S. mollissima*, essentially solitary nester. Nests and pairs well scattered but no information on home-ranges, or territories, though pairs evidently leave flocks on sea and first disperse on to inland waters before seeking nest-sites. ♂ remains with ♀ during egg-laying, and often stays on guard near nest for first few days of incubation. Immatures do not visit inland nesting areas (Bent 1925). ROOSTING. Evidently communal on sea for much of year, but little information.

2. Least well known of west Palearctic eiders (but see Myres 1959b). FLOCK BEHAVIOUR. Evidently much as in closely related *S. mollissima*. Head-shakes only pre-flight signals observed (McKinney 1965b). ANTAGONISTIC BEHAVIOUR. Includes silent

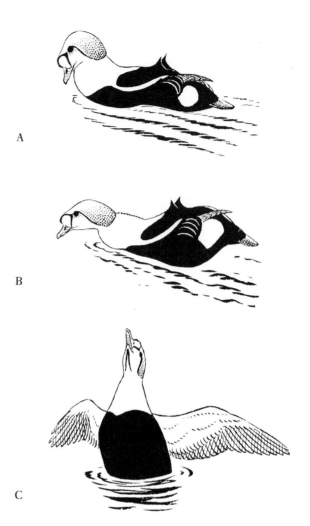

A

B

C

Chin-lifting in ♂ and Inciting-display in ♀ (see *S. mollissima*). COMMUNAL COURTSHIP. See Myres (1959b), Johnsgard (1965). As in *S. mollissima*, main displays of ♂ involve Cooing-movements accompanied by Cooing-calls, but these of only 2 types, with no compound sequences, followed (not preceded) by lateral Head-turning in Neck-stretch posture; latter display more deliberate and exaggerated, with bill moving through 180°. In conspicuous Pushing-display: head, from position with neck retracted, (1) deliberately pushed forward and up while bill points sharply down (see A), (2) brought back to starting position; performed smoothly, usually in series of 3–4, each push accompanied by one of Cooing-calls as white area of chest highly inflated. Resembles several Neck-jerks of *S. mollissima* but thought by Johnsgard (1964) not to be homologous. Reaching-display (see B for 2nd stage) more or less identical to homologous display in *S. mollissima*, but call quite different. Secondary displays, more characteristic of pre-copulatory sequences, also not uncommon; include Ritual-bathe, Wing-flap, Upward-shake, Head-roll, and Preen-dorsally. Of these, Wing-flap more ritualized than in *S. mollissima*: 2 flaps as in latter, but ♂ usually faces ♀ and holds head in almost rigid line, showing inverted-V mark on chin (see C). Main ♀ displays, Gog-gog-calling and Inciting, similar to those of *S. mollissima*. ♀♀ also sometimes perform incomplete version of ♂ displays, including Reaching and Pushing; also Ritual-bathe, Wing-flap, and Upward-shake. PAIR BEHAVIOUR. Probably much as in *S. mollissima*. COPULATION. ♂ swims round prone ♀ performing long series of precopulatory displays including Pushing, Reaching, Ritual-bathe and Wing-flap sequence, Upward-shake, and Head-roll. Mounting often immediately preceded by Wing-flaps and nearly always followed by single Reaching-display, after which ♂ swims away fairly quickly while Head-turning as ♀ bathes. RELATIONSHIP WITHIN FAMILY GROUP. Little information.

(Figs A–C after Johnsgard 1965.)

**Voice.** Apparently less vocal than Eider *S. mollissima*. Similarity of calls (especially of ♂) to those of *S. mollissima* overstressed. Most earlier descriptions, nearly all relating to spring and early summer, casual and incomplete.

CALLS OF MALE. As in *S. mollissima*, consist largely of Cooing-calls given during courtship display; evidence from tape recordings (of captive birds) suggests these coarser-grained, hesitant, gruffer, less loud and dove-like than in *S. mollissima*, with conspicuous tremulous quality. Main calls: (1) repeated 'hoo', accompanying each movement in Pushing-displays; (2) fading 'hoo-oo-oo-oo', in Reaching-display (Johnsgard 1965). Other descriptions apparently referring to these Cooing-calls include: loud, rolling 'urr-urr-URRR' increasing in volume on last note, and soft cooing sound generally heard at night (see Kortright 1942); 'broo rroo rroo' (see I), like 'rookooing' of Black Grouse *Tetrao tetrix*, soft and dove-like in quality and likewise heard continuously from many birds (Höhn 1957); triple loud quacking, 'arr . . . arr . . . arr' (Dementiev and Gladkov 1952); short 'huuu' of fluctuating pitch (Bauer and Glutz 1969); and tremulous 'arr' or 'huu' usually repeated 3–4 times with last syllable more strongly accented (Uspenski 1972). Flight-calls of large flock described as 'grating croaks' (see Gabrielson and Lincoln 1959); ascribed to ♂♂ but more probably given by ♀♀.

CALLS OF FEMALE. Main calls, variants of wooden 'gog'-like notes as in *S. mollissima* but more extended and slurred, voice being generally deeper, hoarser, and more clucking. 'Gogogog . . .' notes of Inciting-display 'like hammer hitting a wooden wall' (Johnsgard 1967); tape-recordings also suggest 'gnark-k-k-k' as alternative rendering (see II). When alarmed, gives 'angry grunt'

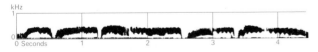

I D S McChesney England (captive) March 1962

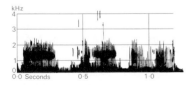

II D S McChesney England (captive) March 1962

followed by 'hissing and growling' (see Kortright 1942); these vocalizations also described as hoarse, grunting quacking (Godfrey 1966).

CALLS OF YOUNG. Much as in *S. mollissima*. Contentment-call (contact-call) given in groups of 1–4 notes (see Fig. 3 in Kear 1968).

**Breeding.** SEASON. Eggs laid second half June Spitsbergen and arctic USSR (Dementiev and Gladkov 1952; Løvenskiold 1964). Timing dependent on thaw. SITE. On ground, usually exposed, sometimes in shelter of rock or hummock. Usually near water, often fresh, showing preference for small islands. Nest: slight hollow, with little material other than down. Average diameter 25 cm. Building: by ♀, shaping nest by turning body. EGGS. Elongate oval; pale olive. 67 × 45 mm (61–78 × 41–49), sample 200 (Schönwetter 1967). Weight of captive laid 64 g (59–69), sample 14 (J Kear). Clutch: 4–5 (3–7); rare nests with more, by 2 ♀♀. One brood. No data on replacements but probable after egg loss. One egg laid per day. INCUBATION. 22–24 days. By ♀; ♂ often in attendance at start. Starts with last egg; hatching synchronous. Eggs covered with down when ♀ off nest. Eggshells left in nest. YOUNG. Precocial and nidifugous. Self-feeding. Cared for by ♀, brooded at night while small. In areas of high density, broods amalgamate. FLEDGING TO MATURITY. Fledging period not recorded. Age of first breeding 3 years. BREEDING SUCCESS. No data.

**Plumages.** ADULT MALE BREEDING. Crown and dense, soft feathers on nape and sides of nape pale blue-grey (palest on forecrown), separated from emerald green cheeks by narrow white line from eye to sides of neck. Feathers bordering bare frontal processes of upper mandible, on forehead, on small spot below and behind eye, and broad V on throat (apex to chin) black. Neck, mantle, chest, upper breast, and large spot at sides of tail-base white, mottled black on lower mantle in some, strongly tinged cream or pink-buff on lower breast, chest, and sides of mantle. Rest of body and tail black. Upperwing dull black, tinged sepia on inner webs of flight-feathers; deep black on strongly curved tertials and on outer webs of secondaries and greater coverts. Variable number of inner secondaries narrowly tipped white; tertials faintly streaked dark grey along shafts. Upper wing-coverts black but median and longer lesser coverts white, some shorter lesser coverts tipped white, forming large white patch on forewing. Underwing pale grey, marginal coverts brown, median and lesser coverts and axillaries white. ADULT FEMALE BREEDING. Head and neck rufous cinnamon, finely streaked black on forehead, crown, and nape; in some, lighter buff spot round eye and streak behind eye to sides of neck. Upperparts, chest, and flanks cinnamon-chestnut, feathers with black U-shaped subterminal marks (compared with straight bars in Eider *S. mollissima*). When fresh, subterminal marks almost concealed by broad rufous feather-margins, giving only slightly black-mottled appearance, but dark marks conspicuous in spring, when plumage abraded and faded, and chest and flanks especially appear pale buff or cinnamon with distinct sepia arcs. In worn plumage, head and neck pale buff with dense narrow sepia streaks on crown and neck, less dense on sides of head (appearing paler than *S. mollissima* in corresponding plumage). Back and rump black, edged rufous cinnamon; lower breast, belly, and vent uniform sooty brown. Flight-feathers and greater coverts black, slightly tinged cinnamon on outer webs of secondaries and greater coverts. Tips of outer secondaries and inner greater coverts white; broad outer edge of tertials chestnut, in some tinged purple. Lesser and median upper wing-coverts black, broadly edged and subterminally marked warm cinnamon; in worn plumage, edges sometimes fade to pale buff or off-white, so forewing appears light. Underwing and axillaries as in adult ♂; marginal coverts sooty, edged brown. ADULT MALE NON-BREEDING (Eclipse). Grey, green, and white head and body feathers replaced by dull black feathers with variable white marks on bases, but sometimes breeding feathers on parts of head, mantle, and chest retained. Wing, tail, and rest of body resemble adult ♂ breeding. Feathers of head and neck sooty-brown, much buff on bases visible on lores, cheeks, and sides of neck (appearing paler than eclipse ♂ S. mollissima); chin and throat grey with brown mottling. Mantle sooty-brown with variable white suffusing, scapulars uniform sooty. Feathers of chest and upper breast sepia with off-white subterminal spot, variable dull black bars, and buff tip; a few feathers sooty-black with broad white tips. ADULT FEMALE NON-BREEDING. Like adult ♀ breeding, but crown and hindneck black, narrowly streaked buff; ring round eye and streak behind it pale buff; cheeks, throat, and foreneck buff, finely streaked black. Upperparts black, feathers edged cinnamon (edges narrower and paler than breeding). Feathers of sides of breast, chest, and flanks grey-brown with pale cinnamon tip and black subterminal arcs (latter broader and less sharply defined than in breeding). Rest of body, wing, and tail like adult ♀ breeding. DOWNY YOUNG. Like S. m. mollissima, but upperparts, cheeks, chest, and sides of body paler; chin, throat, and underparts off-white; sides of head buff-yellow, dark line below eye to nape; loral and frontal feathering of bill as in full-grown birds. JUVENILE. Head and neck pale buff, heavily streaked black on crown and hindneck, narrower elsewhere; chin, throat, foreneck, and sides of neck pale grey, hardly streaked. Feathers of upperparts, chest, and flanks sepia (pale grey towards base), broadly edged cinnamon. Rest of underparts pale grey, feathers shading to sepia near buff tip, faintly barred sepia in some. Tail sepia, edged warm buff; feathers notched at tip, bare shaft projecting. Wing like adult ♀, but duller, sepia; secondaries and greater coverts edged cinnamon-buff instead of white; median and lesser coverts like upperparts, less contrastingly edged and marked as adult ♀. Juvenile ♀ differs from juvenile ♂ by slightly paler speculum and somewhat broader buff edges on feathers of back and rump. Differs from juvenile S. mollissima by paler and greyer head and neck; broader cinnamon margins to feathers of upperparts and flanks, dark feather-centres appearing rounder near tip; paler and less distinctly barred underparts. FIRST IMMATURE NON-BREEDING MALE. In autumn, newly acquired feathers of head and neck cinnamon-buff, finely barred black on crown and streaked sepia on sides of head and neck. Often distinct pale buff streak round eye to sides of neck. New feathers of scapulars and flanks sooty-black; tipped brown; rest of head and body, tail, and wing juvenile. FIRST IMMATURE BREEDING MALE. During winter and spring, acquires variable number of feathers resembling adult ♂ breeding. Advanced specimens in spring have head similar to 1st immature non-breeding, but variable number of grey feathers on crown and nape, green on lores, and white on sides of head and throat (all usually suffused dusky on tips); spot below eye and V on throat black. Scapulars, flanks, and tail coverts black; mantle, chest, upper breast, and sides of tail-base white with black or buff suffusion. Back, rump, and underparts usually juvenile; tail (except sometimes central pair) and wing always so. Retarded specimens attain only limited number of breeding-like feathers on head, scapulars, flanks, and chest; most of head 1st immature non-breeding; most of body, all wing and tail juvenile. FIRST IMMATURE NON-BREEDING and BREEDING FEMALE. Unlike ♂, no separate 1st immature non-breeding and breeding plumages discernible. During winter and spring, new plumage gradually develops, showing intermediate character between non-breeding and breeding of adult ♀. Juvenile back, rump, and underparts, tail (except sometimes central pair), wing, and at times many feathers on rest of body or head retained. SECOND IMMATURE NON-BREEDING. Like adult non-breeding, but some 1st immature breeding and juvenile feathers (especially on underparts) retained. Wing juvenile when not moulted. Head and neck of ♂ as in 1st immature non-breeding, paler brown than adult. SECOND IMMATURE BREEDING MALE. Like adult ♂ breeding, but grey, green, and white feathers of head, neck, and front part of body variably tipped dusky (tips may wear off); white patch on forewing less distinct and smaller, sometimes nearly absent: lesser coverts dull black, white median coverts tipped or suffused black. SECOND IMMATURE BREEDING FEMALE. Like adult ♀ breeding, but black arcs of upperparts, chest, and flanks somewhat broader and less clearly defined; tips of greater coverts and secondaries slightly narrower, in some pale grey instead of white. (Schiøler 1926; Witherby et al. 1939; C S Roselaar.)

**Bare parts.** ADULT MALE. Iris bright yellow. Bill violet-red, paler towards pale pink-blue nail; frontal lobes orange-yellow. Bill and lobes duller in eclipse. Foot dull orange, yellow, or pale yellow-brown; pale yellow on toes; webs and joints dusky grey. ADULT FEMALE. Iris dull yellow. Bill olive-grey, nail pale olive-grey. Foot dull olive-grey. DOWNY YOUNG. Iris brown. Bill pale slate-grey. Foot dull olive-grey. JUVENILE. Iris brown. Bill slate grey to dull black. Foot yellow-brown. In 1st winter, iris of ♂ dull yellow, bill and lobes pale olive-grey with orange tinge from December in some; in following summer, bill pale flesh with pale yellow lobes and bluish-black sides and tip, foot dull olive-ochre with dark webs and joints; during 2nd winter, bill of ♂ like adult ♂, but not as bright. (Schiøler 1926; Witherby et al. 1939.)

**Moults.** ADULT POST-BREEDING. Complete; flight-feathers simultaneous. From mid-July, ♂♂ moult head, neck, mantle, scapulars, chest, flanks, and sides of tail-base, followed by wing and tail between late July and late September. Underparts, back, and rump at about same time as wing; moulted once a year only (directly replaced by new breeding), sometimes also mantle and chest. In ♀♀, head, neck, mantle, scapulars, chest, flanks, and variable number of feathers of back and rump and tail-coverts August–September; followed by wing, tail, and rest of body September to late December. ADULT PRE-BREEDING. Partial; head, neck, and part of body; ♂♂ September–November; ♀♀ late November to January. POST-JUVENILE AND SUBSEQUENT MOULTS. Like S. mollissima; much overlap between 1st immature post-breeding and 2nd immature pre-breeding. (Dementiev and Gladkov 1952; Schiøler 1926; Witherby et al. 1939.)

**Measurements.** ADULT. Greenland and Spitsbergen; skins (RMNH, ZMA).

| | | | | | | |
|---|---|---|---|---|---|---|
| WING | ♂ 277 | (7·75; 15) | 266–293 | ♀ 270 | (7·32; 7) | 256–276 |
| TAIL | | 82·8 (2·55; 12) | 79–87 | | 79·2 (2·56; 7) | 76–83 |
| BILL | | 30·9 (1·37; 17) | 27–34 | | 32·6 (1·34; 10) | 31–35 |
| TARSUS | | 46·5 (1·92; 17) | 44–50 | | 45·5 (1·41; 10) | 44–48 |
| TOE | | 67·3 (2·90; 17) | 63–72 | | 63·0 (1·79; 10) | 60–66 |

Sex differences significant, except tarsus.

JUVENILE. Wing and tail shorter than adult; other measurements similar; e.g. Greenland (Schiøler 1926):

WING    ♂   256   (50)    242–270   ♀ 249   (30)    230–275

Wing in 2nd year intermediate in length between adult and juvenile.

**Weights.** ADULT (1) Spring migration, 1st half May: Hooper Bay region, Alaska (Conover 1926). (2) June: north-east Yakutsk, USSR (Uspenski *et al.* 1962), Adelaide Peninsula, Canada (Bauer and Glutz 1969), and east Greenland (J de Korte); combined. (3) Autumn migration, 1st half August (♂♂) or mid-August (♀♀): Point Barrow, Alaska (Thompson and Person 1963) (range calculated). (4) September: USSR (Dementiev and Gladkov 1952).

| (1) | ♂ 1852 | (12) | 1673–2013 | ♀ 1762 | (6) | 1502–1871 |
|-----|--------|------|-----------|--------|-----|-----------|
| (2) | 1642 | (6) | 1560–1720 | 1763 | (5) | 1557–1830 |
| (3) | 1655 | (39) | 1367–1954 | 1569 | (139) | 1213–1923 |
| (4) | 1748 | (–) | 1500–1870 | 1400 | (1) | — |

♀♀ after laying, July, Adelaide Peninsula: 1198, 1470 (Bauer and Glutz 1969). ♂♂ north Norway, March, 1877; May, 1932, 1975, 2140 (Schiøler 1926). JUVENILE. Winter, Lake Michigan, USA; ♂♂ 1180, 1205; ♀♀ 1071, 1162 (Mumford 1962). February, Denmark: ♀ 1390; February–March of 3rd calendar year, north

Norway: ♂♂ 1657, 1808, 1998; ♀ 1637 (Schiøler 1926).

**Structure.** Wing rather long, pointed. 11 primaries: p9 longest, p10 1–7 shorter, p8 3–11, p7 17–29, p6 32–45, p1 100–130. Inner web of p10–9 and outer of p9 emarginated. Tertials stiff, sickle-shaped, curved downwards over folded wing. Tail short, slightly rounded; 14 pointed feathers. Bill rather short and narrow, high at base, sides slightly narrowing towards large rounded nail. Feathering of forehead extends in narrow strip over culmen to above nostril, feathering at sides of bill only to about half-way between gape and nostrils (in *S. mollissima*, lateral feathering projecting much farther than frontal). Bare frontal processes of upper mandible wide and short. In adult ♂, conspicuous knob on upper culmen in spring, covered at sides by broad and swollen processes up to 45 mm long and 30 mm wide; knob and processes shrink during summer. In adult ♀, slight knob in spring and processes up to 12 mm wide; in juvenile ♂, swelling developing from 2nd autumn. Feathering of cheeks in adult ♂ plush-like, of crown, nape, and sides of nape soft, dense, and somewhat elongated; slightly elongated and dense feathers also on nape of adult ♀ breeding. Feathers of upperparts, chest, and flanks of adult ♀ rounded at tip, not square like adult ♀ *S. mollissima*. Some longer inner scapulars wide in adult; outer webs near tips drawn to point (this actively raised by ♂ in spring, forming short 'horn' on back). Outer toe *c.* 102% middle, inner *c.* 82%, hind *c.* 34%.

CSR

## *Somateria fischeri* Spectacled Eider

PLATES 83 and 100
[between pages 638 and 639
and facing page 687]

DU. Brileider      FR. Eider de Fischer      GE. Plüschkopfente
RU. Очковая гага      SP. Eider de Fischer      SW. Glasögonejder

*Fuligula Fischeri* Brandt, 1847

Monotypic

**Field characters.** 52–57 cm (body 36–40 cm), wing-span 85–93 cm; noticeably smaller than Eider *S. mollissima* (by 15%); overlaps with smaller King Eider *S. spectabilis*. Clumsy-looking, long, and angular-headed eider, with extraordinary facial expression caused by large, goggle-like eye-patches in both sexes; base of bill fully cloaked by unified loral and frontal feathering and bill tip rather thin. Adult ♂ recalls *S. mollissima* but has black chest instead of black crown. Crown and nape feathers markedly pendant. ♀ and juveniles paler than those of other *Somateria*; all show eye-patches and full feather cloak to base of bill. Sexes dissimilar; marked seasonal variation in ♂. Juvenile and sub-adult separable.

ADULT MALE. Upper head green, looking faded, contrasting with large, oval, satin-white eye-patch; latter with narrow black border. Lower edge of forehead, throat, neck, upper breast, mantle, scapulars, tertials, upper wing-coverts and round patches on sides of rump all white, dustier than in other ♂♂ *Somateria*. Chest and underparts grey-black, duller than in other ♂♂ *Somateria*. Wing-quills and tail brown-black. Bill orange; legs and feet dull yellow-brown. MALE ECLIPSE. Generally dark grey-brown (more uniform than adult ♀) but paler than other ♂♂ *Somateria*; eye-patches less obvious than in breeding

plumage. ADULT FEMALE. All plumage dull and dusty, noticeably grey on pale head and (in most lights) with yellower tone and less distinct barring than other ♀ *Somateria*; eye-patches buff and rounder in shape than in ♂. Bill dull blue; legs and feet duskier than ♂. SUB-ADULT MALE. Assumes breeding plumage slowly over 3 years, with similar erratic moult and piebald appearance to those of other ♂ *Somateria*. JUVENILE. Resembles adult ♀, but head pattern even more indistinct; less barred overall. Bare parts duller than adult.

Clumsiest and most grotesque of all eiders, looking dusty and faded in all plumages. ♂ unmistakable but ♀ shows similar head profile to *S. mollissima* so concentration on face (eye-patches and full feather cloak to base of bill) essential. ♀ more easily distinguished from ♀ *S. spectabilis* by duller, less warm plumage, long evenly angled forehead and again eye-patches and different bill feathering.

Flight silhouette similar to *S. mollissima* though smaller size evident in lighter flight action. Habitat and habits essentially as other *Somateria*, but generally much more silent.

DIMW

**Habitat.** Continental arctic coastal tundra, mainly on freshwater pools or lakes, well beyond treeline, breeding in low fringing herbage beside or near standing water including coastal lagoons; also on deltas or shallow muddy coastal inlets. Up to 40–50 km from sea (Kistchinski and Flint 1974); rarely away from lowlands. After breeding, mainly at sea, favouring icy waters, not necessarily inshore, still in or near arctic region.

**Distribution.** Coast of north-east Siberia from mouth of Yana east to Kolyuchin Bay, probably St Lawrence Island, and coast of northern Alaska south to Baird Inlet and east to Demarcation Point. Believed to breed sporadically in Lena delta and occurs summer, but does not breed New Siberian archipelago (Vaurie 1965).

Accidental. Norway: ♂ shot Vardö December 1933, and single ♂♂ seen at different places on arctic and Atlantic coasts, May, June, and September 1970 (SH). USSR: 4 (2 ♂♂) seen near Pechenga, Murmansk, March 1938 (*Ibis*, (14) 2, 758).

**Movements.** Migratory. Apparently withdraws from breeding range as northern seas become ice-bound, traversing arctic coasts and Bering Strait to winter Bering Sea at southern edge of ice. Though main winter areas said to be around Aleutian Islands and, more sparingly, Alaskan Peninsula to Kodiak Island (e.g. Kortright 1942; Delacour 1959), not encountered by Aleutian National Wildlife Refuge staff, so Siberian coast perhaps more important wintering area than hitherto recognized (Johnsgard 1964). Apparently makes moult migrations like other *Somateria* since ♂♂ desert nesting areas during breeding cycle, and big passage, mainly ♂♂, noted north-east Siberia on 11 July (Bailey 1948). Autumn migration proper begins September; latest recorded departure from Point Barrow, Alaska, 17 September (Bent 1923).

Early returns to Alaskan breeding range noted 4 May onwards when open leads in inshore ice; but main passage through Bering Strait 16 May–23 June (Bailey 1948). In Siberia, first arrivals noted New Siberian Islands 10 June and Indigirka delta 8 June (Dementiev and Gladkov 1952; Kistchinski and Flint 1974), though in Igiak Bay, Alaska, eggs already being incubated on 7 June (Johnsgard 1964). Rarely seen outside normal range and only 2 Canadian records (Godfrey 1966).

**Voice.** Calling relatively infrequent (Kistchinski and Flint 1974). Main Cooing-call of courting ♂: weak 'ah-HOO' (Johnsgard 1965); also described as 'gog-goo' (Uspenski 1972). ♀ Inciting-call: wooden or throaty 'gog-gog-gog . . .' as in other *Somateria* (Johnsgard 1964, 1965). Quiet 'cro cro ko ko . . . cro cro ko ko . . .' (perhaps same call) given while swimming in flock and in reply to display of ♂; also a rarely heard, hoarse 'krro' like distant croaking of Raven *Corvus corax* (Kistchinski and Flint 1974).

**Plumages.** ADULT MALE BREEDING. Forehead, crown, cheeks, and upper nape green; yellow-green on centre of forehead, crown, and nape; deep sea-green on dense stiff feathers at sides of nape, and on cheeks below and behind eye-patch or 'spectacle'. Latter composed of *c*. 3 cm wide circle of velvety silver-white feathers round eye; narrowly bordered by black rim above, in front, and behind. Feathers bordering upper mandible, on chin, throat, and neck, white. Body, tail, and wing like Eider *S. mollissima*, but dark colour of underparts extends to chest; flanks, chest, and underparts leaden-grey rather than black; black of back, rump, tail-coverts, and tail with grey bloom; under wing-coverts and axillaries white with grey-brown mottling; white of upperparts and upper wing variably tinged cream. ADULT FEMALE BREEDING. Like adult ♀ breeding of *S. mollissima*, but distinct pale cream circle with fine sepia streaks round eye, bordered in front by dark sepia feathers, indistinctly separated from pale grey-brown throat. Plush-like feathers on centre of forehead pale buff. Rest of head and neck cinnamon-buff, finely streaked sepia; body barred cinnamon and sooty-black. ADULT MALE NON-BREEDING (Eclipse). White parts of breeding plumage on head replaced by dark grey and on body by dark grey-brown; some white of feather-bases sometimes visible. Green on head changed for grey-brown with white mottling. Dark grey circle round eye sometimes contrasts with white mottling on crown and cheeks. Resembles adult ♀, but darker, without cinnamon bars, and with some white on crown, cheeks, or body. ADULT FEMALE NON-BREEDING. Differs from adult ♀ breeding in same way as adult ♀♀ of other *Somateria*. JUVENILE. Superficially like adult ♀, but head more heavily streaked; feather edges of upperparts narrower, more diffuse, and with much dark sepia of feather bases visible; barring faint or incomplete. Underparts pale grey-brown, narrowly and faintly barred sepia on breast and flanks, streaked sepia on chest. Wing dark sepia, lesser and median coverts narrowly tipped chestnut-buff. Circle round eye less distinct. Development of 1st immature non-breeding and breeding plumage apparently as in other *Somateria*. FIRST IMMATURE BREEDING MALE. In 1st spring, head and neck like adult ♂ breeding, but white on head mottled brown, circle round eye more faintly bordered black, streak below eye green-grey. Upper chest, mantle, and scapulars white, partly suffused sepia; breast, flanks, and tail-coverts brown, sometimes tinged plumbeous. Juvenile feathers on back, rump, belly, vent, and wing. FIRST IMMATURE BREEDING FEMALE. In 1st spring like adult ♀ breeding, but upperparts, chest, and flanks darker, pale brown edges less distinctly defined from dark centres of feathers. Wing, back, rump, and underparts juvenile. SECOND IMMATURE NON-BREEDING. Probably like adult non-breeding but part of 1st breeding and juvenile plumage retained until late summer. SECOND IMMATURE BREEDING MALE. Like adult ♂ breeding, but white of lesser and median upper wing-coverts variably suffused grey. SECOND IMMATURE BREEDING FEMALE. Presumably like adult ♀ breeding. (Dementiev and Gladkov 1952; Kortright 1942; Uspenski 1972.)

**Bare Parts.** ADULT MALE. Iris white with pale blue ring. Bill orange, nail paler; duller in eclipse. Foot dull yellow or olive-brown, webs dusky grey. ADULT FEMALE. Iris brown. Bill dull blue-grey. Foot yellow-brown. JUVENILE. Bill olive-brown, foot dull yellow. (Dementiev and Gladkov 1952; Kortright 1942; Uspenski 1972.)

**Moults.** ADULT POST-BREEDING. Complete; flight-feathers simultaneous. Head, neck, mantle, scapulars, chest, sides of breast, and flanks from late June to August in ♂, from mid-August in ♀. Back, rump, underparts, tail, and wing follow August–September in ♂, probably later in ♀. ADULT PRE-BREEDING. Partial; head and body, except back, rump, breast, belly, and vent. September–December in ♂; ♀♀ finish somewhat later. POST-JUVENILE AND SUBSEQUENT IMMATURE MOULTS. Apparently as in other *Somateria* (Dementiev and Gladkov 1952; Kortright 1942; Uspenski 1972.)

**Measurements.** Summer adults north-east Siberia (Uspenski *et al.* 1962). 7 ♂♂: wing 272 (265–280), bill 25·5 (24–28). 5 ♀♀: wing 263 (233–280), bill 24·6 (23–27). Ranges, USSR: wing ♂ 225–267, ♀ 240–250; bill ♂ 21–23, ♀ 25–28; tarsus ♂ 45–50, ♀ 46–48 (Dementiev and Gladkov 1952).

**Weights.** ADULTS. ♂♂. Lower Indigirka River, USSR: summer 1650 (7) 1500–1850 (Uspenski 1972). Alaska: late May to early

June 1645 (8) 1445–1700 (Conover 1926). ♀♀. Mid-June, prior to laying, 1667 (3) 1400–1850. Late July to early August, breeding just finished, 1277 (3) 1200–1300 (Kistchinski and Flint 1974; Uspenski 1972).

**Structure.** Wing, tail, and foot like other *Somateria*. Structure of bill different: feathering extends far on bill to about straight line from corner of mouth to culmen over nostrils; no bare frontal processes, no feathers extending far on sides of bill. Large round patch of velvety feathers round eye. Feathers of forehead short, plush-like; those on crown, nape, and sides of nape slightly elongated and stiff, forming short, broad crest.          CSR

## *Polysticta stelleri* Steller's Eider

PLATES 84 and 100
[between pages 638 and 639
and facing page 687]

DU. Stellers Eider    FR. Eider de Steller    GE. Scheckente
RU. Сибирская гага    SP. Eider de Steller    SW. Alförrädare

*Anas Stelleri* Pallas, 1769

Monotypic

**Field characters.** 43–47 cm (body *c.* 35 cm); wing-span 70–76 cm; much smaller than Eider *Somateria mollissima*, about size of Goldeneye *Bucephala clangula*. Rather small, compact sea-duck, with small, thick-based and slightly drooping bill, and steep forehead and nape; rather long, pointed tail usually carried clear of water and noticeably long tertials. May suggest *Anas* at first sight, but relationship with *Somateria* increasingly evident in prolonged observation. Breeding ♂ uniquely combines white head and buff underparts; ♀ shows no obvious plumage mark except speculum, and best identified by shape. Sexes dissimilar; marked seasonal variation in ♂. Juvenile and sub-adult separable.

ADULT MALE. Head white, with pale green feathers on lores and forehead, black eye-patch and small green tufts on sides of crown, stemming from black spots. Chin, throat and neck-collar, velvety black, connecting with black upper mantle and centre of back. Lower sides of mantle and forward scapulars white, rear scapulars and tertials marked with purple and white, all long and forming spray over folded wing; rest of upperparts black. Quills brown-black, secondaries with purple speculum and increasingly tipped and edged white, contrasting with white inner wing-coverts. Neck below collar and upper flank edges white, merging with warm buff chest, forebelly, and flanks; black spot on side of chest (not always visible); rest of underparts dusky, becoming black-brown under tail. Bill, legs, and feet blue-grey. MALE ECLIPSE. Resembles ♀ but pale head and barred chest show hint of breeding pattern and upper wing-coverts remain white. ADULT FEMALE. Noticeably dark and uniform, less barred, more mottled than other ♀ eiders *Somateria*. Head with rusty tinge and whole body rather matt, almost purplish-black with dark feather centres obvious only on back. Tertials elongated, marked purple and black and combining with similarly coloured secondaries to form speculum, recalling

that of Mallard *Anas platyrhynchos* and noticeably bordered white in front, less so behind; pattern distinctive at close range, even in flight. Rest of wing uniform brown-black. Underwing strikingly white, quite distinct from all ♀ *Somateria*. Bare parts duskier than ♂. JUVENILE. Somewhat resembles adult ♀ but plumage tone warmer and paler, particularly underneath. Feathers of back paler tipped than in adults and chest and underparts heavily barred. Bare parts as ♀. SUB-ADULT. ♂ assumes full breeding plumage in 2 years, with white head, scapular and tertial plumage and buff underparts incomplete in first.

♂ unmistakable, with plumage pattern different from ♂ *Somateria* in distribution of black (none below, but much on upper back). ♀ unlikely to be confused with any ♀ *Somateria* and more likely to recall ♀ Goldeneye *Bucephala clangula* or scoter *Melanitta*; at long range, pointed tail carried clear of water often only character indicating true identity. At close range, uniform plumage, speculum, and long tertials diagnostic.

Flight silhouette, showing rather narrow wings unlike *Somateria*, recalls *Melanitta*, even *Anas*, as does light and rapid flight action, allowing easy take-off. In flight, wing-beats produce singing tone, louder than in *B. clangula*. Gait freer than in *Somateria*; swims expertly. Stands out on rocks to preen or rest, often with other eiders.          DIMW

**Habitat.** Regular arctic habitats lie towards meeting of Arctic and Pacific oceans; even in winter rarely south of Arctic Circle. Unlike Eider *Somateria mollissima*, prefers moving several km inland for nesting, normally among many pools of different shapes and sizes which characterize flat coastal belt within tundra zone. Nests on low hillocks or ridges of peat, surrounded by swampy vegetation such as mosses and sedges, or on banks of standing waters overgrown with grass and sedge; at this season least tied to

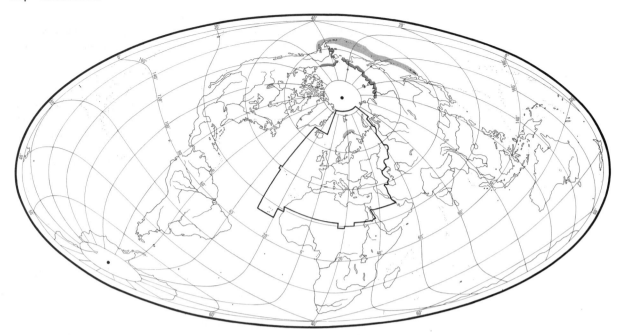

sea and seacoasts of all eiders (Uspenski 1972). Skilful swimmer and diver, requiring clear water, off clean rocky shores, or in bays where fresh water enters, or among or alongside sea ice. More mobile and restless than *S. mollissima*, moving easily on land, and flying higher with many changes of direction carried out in unison.

**Distribution.** Breeds occasionally well outside regular range, including central and west Siberia and northern Europe. In west Palearctic, definite records few though may nest more frequently than these indicate. USSR: southern island of Novaya Zemlya, pair seen in breeding plumage June, and brood of flying young September (Dementiev and Gladkov 1952). NORWAY: probably bred Varangerfjord 1924 when ♀♀ seen with young (Haftorn 1971; see also Bannerman 1958).

Accidental. Mainly Baltic and adjacent North Sea (Sweden, Finland, Poland, East Germany, Denmark) where now much less frequent than in 19th century (Bauer and Glutz 1969; SM), though a few may still winter Åland Islands when ice conditions permit (LvH). Britain and Ireland: 12, mainly Scotland (Smith *et al.* 1975). France: ♂ 1855 (CE). Spitsbergen: ♂ 1968 (MAO).

**Population.** In east Siberia, at least 100000 adults (Uspenski 1972); considerable decline since end 19th century suggested by fall in numbers wintering Commander Islands (Dementiev and Gladkov 1952) though latter may be due to warming of arctic seas (see Movements).

**Movements.** Migratory. Main wintering area southern Bering Sea, where all Alaskan breeders and most from north-east Siberia winter; birds ringed in flightless moult Izembek Bay, Alaska, recovered west to river Lena delta (127°E). During 1957–62, nearly 200000 moulted Izembek Bay, those from Lena delta having moved some 3200 km to moult within winter range. In 1963 and 1964, however, only *c.* 10000 appeared in early part of season; some evidence that in those years Siberian birds stopped short at Cape Vankarem, Chukotski Peninsula, to moult prior to migration to Bering Sea (Jones 1965). No firm evidence yet on origins of much smaller numbers which winter regularly (and sometimes summer) on coasts of Murmansk and Finnmark, but evidently from Siberia since no known regular breeding sites west of Khatanga Gulf. Considerable annual variation in numbers reaching Finnmark, where maximum 400 Varangerfjord, April 1967 (Haftorn 1971); degree of westerly movement probably influenced by pack-ice conditions in Polar Basin (Johnsen 1937). Not infrequent records 19th century from Baltic Sea and Gulfs of Bothnia and Finland suggest then scarce but probably regular winter visitor, and their distribution (concentrated in east) indicates overland passage between White Sea and Gulf of Finland, as still in some other sea-duck. Cessation of such movement probably connected with climatic amelioration reducing area of sea-ice in Barents Sea region (Belopol'skii 1957). Rarely straggles to temperate latitudes in east Atlantic or Pacific.

Mass arrivals Alaskan moult areas August, though rafts of up to 100 moulting birds seen Varangerfjord, 1st half July. In 1964, when most Bering Sea winterers moulted elsewhere, these arrived suddenly Alaskan Peninsula 6–9 November. Departures from Point Barrow, Alaska, breeding area from mid-July to mid-October (Bailey

1948). Return migration begins April with mass assemblies Commander Islands and northern Alaskan Peninsula; early appearances Point Barrow 5 June, lower Anadyr 19 May–7 June, New Siberian Islands 15 June, and east Taymyr 18 June. Arrivals and departures from Finnmark and Murmansk appear to fit same pattern, though picture complicated by summering non-breeders; disappearances of larger flocks from Porsanger and Varanger fjords reported late May (Bannerman 1958). Immatures also encountered all year in Bering Sea. See also Bauer and Glutz (1969), Dementiev and Gladkov (1952), Haftorn (1971).

**Food.** Almost entirely aquatic, saltwater invertebrates, particularly molluscs and crustaceans, but also polychaete worms, echinoderms, small fish, gephyrean worms, brachiopods; in fresh water, also insects and their larvae. Mainly daytime feeder; dives from surface with wings open in compact flocks, individuals often submerging simultaneously. Apparently rarely brings items to surface, unlike *S. mollissima* (McKinney 1965*d*). In 7·3 m, one bird dived 37, 34 and 42 s (Meinertzhagen 1959). In spring and summer, found more often in shallower water where wades or swims, dabbling on surface, up-ending, or feeding with head submerged (Blair 1958; McKinney 1965*d*).

8 taken January and February near Vadsø and Vardo at mouth of Varangerfjord contained molluscs *Littorina obtusata, Lacuna vincta, Trophon truncatus, Margarita helicina, M. groenlandica,* and *Buccinum groenlandicum*; and amphipod crustaceans *Amphithoe, Gammarus, Pleustes panopla, Anonyx lagena, A. gularis,* and *Podocerus anguipes* (Collett 1894*b*). One, April, had fed on molluscs (*L. obtusata, L. vincta, M. helicina,* and *Pyrene rosacea*), crustaceans (*G. locusta* and *Idotea baltica*), and polyzoan *Crisia* (Schiøler 1926). 4 also taken January and February, from Pechenga, USSR, contained by volume 73·3% molluscs, 21·6% crustaceans, and 5·1% echinoderms. Molluscs included small (2·3 mm) *M. helicina, Cingula interrupta, Onoba oculeus, Skeneopsis,* and *Planorbis,* and larger *L. vincta, Purpura lapillus, B. groenlandicum,* and *Mytilus edulis*; crustaceans mainly Gammaroideae, Idoteidae, and Caprellidae; echinoderms *Ophiopholis aculeata* and *Stronglyocentrotus droebachiensis.* Most prey small, largest being a 20·4 mm *O. aculeata.* Another February bird contained small pieces of scallop *Chlamys islandica* (Siivonen 1941). From Alaska in December, one contained chiefly amphipods; in February, another mainly univalves *Polinices recluziana* and *Melanella* and to a lesser extent amphipods and bivalve *Cardium ciliatum.* 66 from north-west Alaska and north-east Siberia during breeding season (May, June, July) had 45·2% by volume crustaceans, mostly amphipods; 19·3% molluscs, mostly bivalves (especially *M. edulis, Siliqua, Tellina, Macoma,* and *Saxicava artica*); 13·0% insects, those commonly found on tundra freshwater ponds particularly midge larvae (Chironomidae), and caddisfly larvae (Trichoptera)

and their cases); 3·3% annelid worms; 2·8% scutellid (sand dollars; 2·3% small fish; and 12·9% plant material; most of which probably taken accidentally (Cottam 1939). ♂ from Anadyr, north-east Siberia, in June contained stonefly larvae (Plecoptera), waterbeetles, and waterboatmen *Corixa* (Dementiev and Gladkov 1952).

4 juveniles had 40·3% by volume plant material, particularly *Potamogeton* and crowberry *Empetrum nigrum*; and 59·7% animal material, especially caddisfly larvae and midge larvae *Chironomus* (Cottam 1939). A juvenile from Gulf of Finland in November contained 20 shrimps *Gammarus locusta* (Siivonen 1941).

**Social pattern and behaviour.** Based mainly on material supplied by F McKinney.

1. Highly gregarious except when breeding. In main wintering area, Bering Sea, flocks of up to many thousands, including both adult and 1st winter birds. Pairs form in winter flocks, departing for breeding grounds together (McKinney 1965*d*). ♂♂ first to flock, from end June; may join immatures (Blair 1958). At Izembek Bay, Alaska, adults often return from more northerly breeding grounds to moult from August (Jones 1965); flocks of 1st year birds, with a few adult ♂♂, reported there in May (McKinney 1959) and many summer. Summering flocks of non-breeders feature both of present species and King Eider *S. spectabilis*. BONDS. Monogamous pair-bond of seasonal duration appears usual. At Izembek Bay, many paired by end March, but courting parties still frequent during April. ♂♂ desert ♀♀ soon after incubation begins (Blair 1958). 1st year ♂♂ show heterosexual behaviour but not known to form pairs. Only ♀ tends young. Tendency to form crèches, at least temporarily (Blair 1958). BREEDING DISPERSION. Little information. Apparently solitary nester, with well-scattered, concealed nests; not in colonies as often in Eider *S. mollissima*. ♂ accompanies ♀ to and from nest during laying period (see Blair 1958). ROOSTING. Communal for most of year. Essentially diurnal feeder, so main roosting time at night, though will also loaf and rest during day. Wintering flocks sleep on water, sandbars, or beds of exposed eelgrass (McKinney 1965*d*).

2. Studied both in wild, mainly on sea in wintering area off Alaska (McKinney 1965*d*), and in captivity (Johnsgard 1964, 1965); see also Uspenski (1972). Differs in a number of important features from typical eiders *Somateria*. FLOCK BEHAVIOUR. In wintering and moulting areas, extremely sociable in most activities, detached individuals and pairs seldom remaining long away from rest of flock except for copulation. Forms densely packed rafts when feeding with precisely synchronized diving (see also McKinney 1973; Blair 1958). Flocks take wing easily and frequently, especially when disturbed by birds of prey or gulls *Larus*. Pre-flight signals consist of lateral Head-shakes in Neck-stretch posture and certain other forms of comfort behaviour. Take-off by one flock can cause other flocks to fly; several often join in air to alight together in single massed group, birds then swimming restlessly close together with heads erect while repeatedly performing comfort behaviour. Though ♂♂ generally silent, ♀♀ particularly vocal and constant chatter emerges from winter flock. ANTAGONISTIC BEHAVIOUR. ♂♂ aggressive in courting parties, encounters between pairs, and situations involving pair and unmated ♂, especially when competing for ♀ or defending mate; rush at one another across water and sometimes fight briefly. Quiet calls may be heard at such times, apparently as threat. Also, displays characteristic of communal

A

courtship often closely associated with overt hostility; include Head-toss similar to Chin-lift of *Somateria* eiders but particularly rapid, exposing and then hiding black throat-patch as head jerked up and back (see A). Threat and Inciting-display of ♀ much as in *Somateria*; involves similar hostile movements of mainly lowered head towards rejected ♂, but alternates with ♂-like Head-toss in which head moved rapidly and smoothly, with loud call, from threatening posture to bill-up position with breast raised slightly out of water, and then forward to normal swimming posture. COMMUNAL COURTSHIP. Courting parties, usually including ♀ and 3–7 ♂♂, move rapidly over water and frequently fly up. ♂ displays not accompanied by vocalizations. Main displays on water include (1) Rearing-display, in which head and body suddenly swing back and up until body almost vertical, exposing breast (see B); (2) Head-toss; (3) exaggerated Upward-shake, in which head first held characteristically forward (see C), instead of in line with body as in other 2 west Palearctic breeding eiders, and then jerked back rapidly; (4) Head-turn, in which bill swung rapidly from side to side; and (5) Neck-stretch. These displays at times combined into compound series; see McKinney (1965d), and Johnsgard (1964, 1965) for examples. Frequent assumption of Neck-stretch posture and performance of comfort behaviour probably indicative of ♂'s willingness to fly during courtship: if ♀ does not initiate courtship-flight (see below), ♂ from time to time performs ritualized Short-flight instead, rising steeply with rapid wing-beats, flying a few metres, and alighting with conspicuous splash. Inciting-display main activity of ♀ in courting party; often performed while swimming behind preferred ♂ as he cere-

monially Leads her by Head-turning or performing Turn-back-of-Head display. COURTSHIP-FLIGHTS. Frequent during communal courtship on water, starting when ♀ flies up. Most flights under 30 s, but some last several minutes, group circling and swerving with many changes in direction. In flight, one ♂ usually stays close to ♀ who, judging from calls, directs Inciting-display at other ♂♂. After group re-settled on water, preferred ♂ frequently chases others and rushes after ♀, while several birds perform Upward-shakes, Head-tossing, and Head-turning. PAIR BEHAVIOUR. Displays mainly those associated with copulation. COPULATION. At least in wintering area, occurs either after pair drift away from flock to feed or, more often, after pair deliberately fly several hundred metres away from flock and begin pre-copulatory behaviour immediately. Although ♀ occasionally Preens-dorsally and Ritual-bathes, usually remains in Prone-posture while ♂ performs sequence of displays beside her. Most common of these Preen-dorsally, Bill-dip, and Ritual-bathe (latter consisting of long series of up to 17 head-dips), while Head-shake, Head-roll, and Head-turning much less frequent and Wing-flap entirely omitted. Before mounting, ♂ does single Upward-shake then rushes rapidly towards ♀ in Neck-stretch posture while Head-turning. Copulation brief and pair rotate in water momentarily afterwards before ♂ releases ♀. Post-copulatory display of ♂: single Rearing-display followed by rushing away in Neck-stretch posture, at times with Head-turning, and 1 or more Short-flights from ♀ (or pair then takes off and returns to flock). Many features of mating behaviour suggest pair vulnerable to attacks of flying predators when away from flock; sequence briefer and displays silent and less conspicuous than in *S. mollissima* (McKinney 1973). RELATIONSHIPS WITHIN FAMILY GROUP. No detailed information.

(Figs A–C after McKinney 1965a.)

**Voice.** Limited data emphasize difference from other eiders *Somateria* and relatively silent habits, at least in ♂. At take-off, large flock makes loud roar of wings audible a mile away (McKinney 1965d); loud, far-carrying whistling noise also produced by wings in flight (Percy 1958). Growls from flying birds of unstated sex compared to call of ♀ Wigeon *Anas penelope* (see Bannerman 1958); calls of ♀ also said to resemble those of *A. penelope* in Witherby *et al.* (1939).

CALLS OF MALE. Soft, growling threat note only vocalization heard by Johnsgard (1965). Evidently lacks any Cooing-calls comparable to those of *Somateria*, at least in winter flock. According to McKinney (1965d) and Johnsgard (1964, 1965), courtship not accompanied by any audible sounds, even when observed at close range in captivity, though 'courtship call' described as soft and deep by Uspenski (1972) and low and guttural by Dementiev and Gladkov (1952), while Blair (1958) referred to 'a murmuring croon' harder and considerably less in volume than coo of Eider *S. mollissima*.

CALLS OF FEMALE. Descriptions somewhat conflicting, but no support for statement of Blair (1958) that ♀ more silent than ♂. In contrast to ♂♂, ♀♀ noisy in winter flocks, uttering (1) rapid, rippling, guttural calls, and, when Inciting, (2) loud 'qua-haaa' or 'cooay' (McKinney 1965d). First may be equivalent to hurried, stuttering and somewhat rising 'ä ä ä ä är' described as common call in

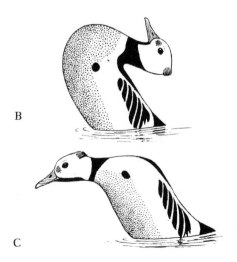

B

C

swimming and flying flocks (see Bauer and Glutz 1969), and to 'brief humming trill' of Uspenski (1972). Johnsgard (1965) pointed out absence of any 'gog'-like calls similar to those of *Somateria*. Alarm-call consists of loud, cawing 'guähää' (see Bauer and Glutz 1969), which may be same as distinctive 'growling bark' like noise of young dog (see Gabrielson and Lincoln 1959) and cry reminiscent of Teal *Anas crecca* but harsher (Bent 1925). Softer, clucking call noted from ♀ driven from eggs about to hatch (see Kortright 1942).

CALLS OF YOUNG. No information.

**Breeding.** SEASON. Little information. Main laying, arctic USSR, late June and early July (Dementiev and Gladkov 1952; Bannerman 1958). SITE. On ground, exposed on open tundra, or among scrub of willow *Salix* and birch *Betula*. Not far from water. Nest: hollow, lined grass, moss, lichens, and down, Building: by ♀. EGGS. Blunt ovate; light olive or olive-buff. 61 × 45 mm (56–68 × 38–48), sample 100 (Schönwetter 1967). Weight 55 g (52–59), sample 8 (Uspenski *et al.* 1962). Clutch: 6–8 (5–10). One brood. No data on replacements. Eggs probably laid one per day. INCUBATION. Period unknown. By ♀. Starts with last egg; hatching synchronous. Eggs covered with down when ♀ off nest. Eggshells left in nest. YOUNG. Precocial and nidifugous. Self-feeding. Cared for by ♀. Broods may amalgamate. FLEDGING TO MATURITY. Fledging period not recorded. Age of first breeding probably 3 years. BREEDING SUCCESS. No data.

**Plumages.** ADULT MALE BREEDING. Head and upper neck silver-white. Small plush-like tuft on hindcrown light green; white of forehead, lores, and below eye variably tinged light green. Small spots round eye and at sides of tuft on hindcrown black. Chin and throat velvety black, separated from broad black collar round lower neck by white ring; latter sometimes incomplete. Centre of mantle, back, rump, and upper tail-coverts black. Shorter scapulars white; inner partially dusky. Longer outer scapulars white, outer webs broadly bordered purple-blue; longer inner scapulars purple-blue with narrow white shaft-streak. Sides of mantle and collar below black of neck white; chest and upper sides of flanks cream-buff, gradually darkening to cinnamon and chestnut on breast and sides of belly and dull black on centre of belly, vent, sides of rump, and under tail-coverts. Small spot at side of breast in front of closed wing black; occasionally small white spot at sides of tail-base. Tail black. Primaries black, inner webs and bases sepia, sides and tips of innermost often edged white. Secondaries broadly tipped white; outer webs strongly glossed purple-blue, inner webs sepia. Tertials curved outwards, white, outer webs broadly edged purple-blue. Primary coverts and bastard-wing sepia, variably edged and tinged white, rest of upper wing-coverts white. Under wing-coverts and axillaries white. ADULT FEMALE BREEDING. Crown and hindneck sooty black, narrowly barred cinnamon; rest of head cinnamon, palest on chin, throat, and foreneck, narrowly barred sepia. Rim round eye often uniform cinnamon or pink-buff. Upperparts dark brown to black, variably barred with narrow pink-cinnamon marks; feathers of mantle tipped pink-cinnamon, those of scapulars and upper tail-coverts tipped warm cinnamon. Chest, sides of breast, and flanks warm cinnamon; feathers broadly

marked with sepia subterminal Vs on chest and sides of breast and with bars on flanks; flanks sometimes barred pink-buff. Breast, belly, vent, and under tail-coverts chestnut, feathers barred black subterminally; paler pink to base. Tail sepia. Wing like adult ♂, but upper wing-coverts dark sepia, faintly edged paler brown; inner greater coverts (not tertial coverts) broadly tipped white, outer more narrowly; speculum less extensive, outer secondaries sepia without gloss. Outer web of tertials purple-blue, inner web black, with silver-white shaft-streak or spot near tip. Underwing like ♂, but marginal coverts blotched sepia. ADULT MALE NON-BREEDING (Eclipse). Head, neck, and upperparts dark brown, feathers narrowly edged cinnamon-brown on head and neck, more broadly on scapulars. Chest, breast, and flank dark brown, feathers edged rufous brown; other parts of body and wing like adult ♂ breeding, upper wing-coverts white. ADULT FEMALE NON-BREEDING. Like adult ♀ breeding, but cinnamon subterminal bars on head, upperparts, and flanks virtually absent; feathers sepia with relatively narrow buff-brown tips and grey bases. Breast, belly, and vent uniform dark sepia. DOWNY YOUNG. Head, neck, upperparts, chest, and sides of body sepia. Small lines above and below eye and narrow streak behind eye cream-buff. Sides of head tinged grey, chin pale grey, underparts dark grey. JUVENILE. Head and neck grey-brown, darkest and with narrow buff feather edges on crown, paler on chin, throat, and foreneck. Sides of head and neck narrowly streaked sepia. Upperparts and flanks dark grey-brown, feathers narrowly tipped buff. Feathers of underparts sepia with rufous-brown tips (paler in ♀) and imperfectly concealed pale grey bases. Tail sepia; feathers notched at tip, bare shaft projecting. Wing like adult ♀, but white tips to greater coverts and secondaries narrow; only some inner secondaries with slight purple-blue gloss (♂♂) or no gloss at all (♀♀); tertials straight (♀♀) or slightly curved (♂♂), without white shaft-streak or spot, and outer web glossed purple-blue in ♂, dull black in ♀. FIRST IMMATURE NON-BREEDING MALE. Head and neck like adult non-breeding, with broader and warmer brown feather edges than juvenile and often pale patch round eye and pale streak behind it. New feathers on scapulars, flanks, sides of breast, or tail-coverts like adult ♂ eclipse. Juvenile throat, foreneck, wing, tail (except sometimes central pair) and many feathers elsewhere retained. Feathers grown later in autumn variably intermediate between 1st non-breeding and breeding. FIRST IMMATURE BREEDING MALE. Like 1st non-breeding, but chin, throat, and lower neck dull black, usually divided in middle of foreneck by off-white or grey-brown collar. New scapulars dull black or rufous-brown, centres and bases tinged white. During winter and spring, short black tuft on hindcrown, dark brown feathers with off-white or cinnamon suffusion on flanks, white feathers with buff tips on sides of head and with cinnamon tips and sometimes sepia subterminal bar on chest and sides of breast may develop. Some outer scapulars with purple-blue outer webs; underparts dull black or chestnut, usually with exposed grey feather bases and sometimes barred sepia. Occasionally some curved tertials acquired like adult ♂, but suffused dark grey. Wing juvenile. Often part of juvenile or 1st non-breeding on body retained. FIRST IMMATURE NON-BREEDING AND BREEDING FEMALE. Plumage development probably as in 1st immature non-breeding and breeding ♂. In spring resembles adult ♀ breeding, but subterminal bars to feathers of upperparts and flanks less developed, underparts uniform brown (if not still juvenile). Wing juvenile, frayed. SECOND IMMATURE NON-BREEDING (MALE AND FEMALE). Probably like respective adults, but some feathers of 1st breeding plumage retained. SECOND IMMATURE BREEDING MALE. Like adult ♂ breeding, but marginal and variable number of lesser and median upper wing-coverts suffused grey-brown; sometimes brown tips to white feathers of head, upperparts, and chest. SECOND IMMATURE BREEDING FEMALE. Apparently like adult ♀ breeding. (Delacour 1959; Dementiev and Gladkov 1952; Schiøler 1926; Uspenski 1972; Kortright 1942; Witherby *et al.* 1939; C S Roselaar.)

**Bare parts.** ADULT MALE. Iris red-brown. Bill blue-grey, duller towards base. Foot dark blue-grey, webs dusky, claws yellow-grey. ADULT FEMALE. Iris dark brown. Bill and foot like adult ♂, but darker. DOWNY YOUNG. Iris dark brown. Bill black with brown nail. Foot dark grey. JUVENILE. Iris brown or grey-brown. Bill and foot dark grey. (Delacour 1959; Kortright 1942; Schiøler 1926; Stübs 1960; Uspenski 1972.)

**Moults.** ADULT POST-BREEDING. Complete; flight-feathers simultaneous. Head and body first; in ♂, starts between late June and late July. Longer scapulars and head first, underparts and tail last, with wing *c.* 4 weeks after start of moult. Moult of back, rump, underparts, and tail not finished when wing starts; slows up during growth of flight-feathers, finished late September. Timing in ♀ not known; body probably starts when young independent, followed by wing *c.* 1 month later. In ♂, back, rump, belly, vent, and some scapulars moulted only once a year. ADULT PRE-BREEDING. Partial. Head and part of body start when flight-feathers full-grown. In ♂, breeding plumage attained on flanks and short scapulars from September, on chest, sides of breast, and mantle from October, on head, neck, and longer scapulars from mid-October to November. Moult of ♀♀ probably as in ♂♂, but later in autumn and probably into early winter; some moult on underparts also in spring. POST-JUVENILE AND FIRST PRE-BREEDING. Partial. Head and body, often including belly in contrast to *Somateria*, and apparently often complete tail. Feathers acquired first (on crown, hindneck, sides of head and body, on parts of scapulars, chest, and sides of breast) probably best considered as belonging to 1st immature non-breeding plumage; resemble adult non-breeding. Feathers attained from about October resemble adult breeding or show intermediate characters. Juvenile wing, occasionally some feathers on body, and much 1st non-breeding plumage usually retained. FIRST POST-BREEDING. Like adult, but less complete: feathers of breeding plumage acquired later in spring retained. Overlaps with following pre-breeding moult. (Bauer and Glutz 1969; Dementiev and Gladkov 1952; Uspenski 1972; C S Roselaar.)

**Measurements.** Skins; north Europe, Greenland, and Kamchatka (RMNH; Schiøler 1926).

| | | | | | | | |
|---|---|---|---|---|---|---|---|
| WING AD ♂ | 216 | (2·87; 10) | 208–225 | ♀ 207 | (2·06; 4) | 205–210 |
| JUV | 208 | (4·67; 7) | 199–213 | 206 | (—; 3) | 203–209 |
| BILL | 38·3 | (1·34; 15) | 36–41 | 39·7 | (1·52; 7) | 38–42 |
| TARSUS | 38·3 | (1·33; 15) | 36–40 | 37·8 | (1·78; 7) | 36–40 |
| TOE | 57·6 | (1·85; 15) | 53–60 | 57·1 | (1·37; 7) | 55–59 |

Tail: adult ♂♂ 84, 84, 85; juvenile ♂♂, 72, 75. Sex differences significant for adult wing only. Wing of juvenile ♂ significantly shorter than adult. Bill, tarsus, and toe of juvenile similar to adult; combined.

**Weights.** ADULT. Heavy at arrival breeding grounds in spring; in summer, weight decreases to a minimum in August–September (Uspenski 1972). ♂♂. North-east USSR: May, 850; June to early July, 794 (19) 670–900; October, 500–800; November, 826 (5) 680–1000 (Dementiev and Gladkov 1952; Portenko 1972; Uspenski *et al.* 1962). Alaska: late May to early July, 880 (5) 850–907 (Conover 1926). ♀♀. North-east USSR: June to early July 853 (14) 750–1000; July, 720; September, 720; November, 820, 865. Alaska: late May to early July, 890 (5) 850–964. JUVENILE. ♂♂. Germany: November, 675 (Stübs 1960). North Norway: May 660, 776 (Schiøler 1926). Alaska: June, 787 (4) 765–850.

**Structure.** Wing pointed, rather long and narrow. 11 primaries: p10 longest, p9 about equal or 0–4 shorter, p8 6–12 shorter, p7 16–26, p6 26–39, p1 80–105. Inner web of p10 and outer p9 emarginated. Tail rather long, wedge-shaped; 14 pointed feathers. Tertials in adults strongly curved outwards, relatively short, tips rounded; straight or nearly straight in juveniles. Outer scapulars in adult ♂ breeding long, pointed, curved down over wing at rest; inner webs narrow. Bill narrow, about as long as head, higher than broad at base, slightly narrowing towards tip. Culmen straight, indistinctly defined from broad curved nail. Sides of bill and base of culmen not covered with short plush-like feathers as in *Somateria*. Cutting edge of upper mandible with narrow soft strip of skin, projecting over lower mandible when bill closed (sometimes folded inwards in skins). Strip of skin widest near tip of bill, where cutting edge of upper mandible slightly upcurved. Outer toe almost equal to middle, inner *c.* 71% middle, hind *c.* 32%. CSR

# Tribe MERGINI scoters, sawbills, other sea-ducks

Medium-sized to fairly large, mainly marine diving ducks. 14 species in 5 genera, first 2 monotypic: *Histrionicus* (Harlequin), *Clangula* (Long-tailed Duck), *Melanitta* (scoters, 3 species), *Bucephala* (Bufflehead *B. albeola* and 2 species of goldeneyes), *Mergus* (sawbills, 6 species). Monotypic *Camptorhynchus* (Labrador Duck) recently extinct; also Auckland Island Merganser *M. australis*. 9 breeding species and 3 accidental in west Palearctic.

Bodies rather short and heavy in *Melanitta* and, to lesser extent, *Bucephala*; more slender in others, particularly larger *Mergus*. Tail fairly long in *Histrionicus* and with enlarged central feathers in ♂ *Clangula*; otherwise relatively short. Bill generally high at base, rather narrow, tapering towards tip, with broad, rounded nail; rather massive in *Melanitta*, narrow and often long in *Mergus* with serrated edges along mandibles. Other physical characters otherwise much as in Somateriini. Flight usually rapid and low; whistling or ringing sound often produced by wings. Some rise from water with difficulty but others much more easily. Gait on land varies from awkward to accomplished; some species rarely come ashore except when nesting or, in some cases, when loafing or roosting—though may then perch above ground especially ♀♀ of hole-nesting species. All dive expertly, with use of wings when submerging and also under water in *Histrionicus*, *Clangula*, and *Melanitta*; diving habits of

*Bucephala* and *Mergus* more like Aythyini. Sexes differ in tracheal structure; in ♂♂ more variable than in any other tribe of Anatinae. Similar to Somateriini in *Histrionicus*. Uniformly structured in *Clangula* with small, asymmetrical bulla and 2 fenestrae. Bulla absent or reduced and symmetrical in *Melanitta*, with greatly enlarged bronchi in some. Bulla large in goldeneyes *Bucephala* with 2 fenestrae and trachea with distinct swellings in middle area; bulla small in *B. albeola*, with no tracheal swellings. Most variable in *Mergus*: among west Palearctic breeding species, enlarged bulla of Smew *M. albellus* biassed to left with series of fenestrae and uniform enlargement of trachea; bulla of Red-breasted Merganser *M. serrator* with both chambers much enlarged with large, oval fenestrae and central area of tracheal tube expanded; bulla of Goosander *M. merganser* huge, fenestrated, and asymmetrical, and tracheal tube with 2 enlargements. Except in Brazilian Merganser *M. octosetaceus* (and extinct *M. australis*), sexual dimorphism extreme. Plumage of ♂ *Histrionicus* unique; in *Melanitta*, ♂♂ wholly or largely black with partial iridescent sheen. ♂♂ of other genera mostly patterned black and white, usually with some gloss. ♀♀ mostly brown and unstreaked in *Histrionicus* and *Melanitta*; mainly mottled brown and white in *Clangula*; mainly grey and white with brown heads in *Bucephala* and *Mergus*. White on secondaries in most species, but wing plain in *Clangula* and 2 *Melanitta*; only *Histrionicus* with metallic speculum. Secondaries ornamentally modified in some species. Non-breeding plumage of ♂♂ true eclipse in all but *Melanitta* and most extralimital *Mergus*. Bill and legs sometimes brightly coloured. Juveniles similar to adult ♀♀ non-breeding. Downy young variable but mostly with dark cap, pale cheeks, and pale underparts; brown or grey in *Histrionicus*, *Clangula*, 1 *Melanitta*; black and white in *Bucephala* and other *Melanitta*; streaked or spotted brown and white in *Mergus*.

Holarctic except 1 living and 1 recently extinct *Mergus* in southern hemisphere. Sturdy and specialized for living in rugged habitats, especially marine and, in case of *Histrionicus*, on turbulent running water; have outstanding powers of diving. Though strong fliers, less given to long daily or seasonal displacement than many other Anatinae. Concentrated mostly in high and upper mid-latitudes. Favoured coastlines more exposed and often more rocky than those frequented by other Anatinae with exception of Somateriini. Although no change in range of *C. hyemalis* within west Palearctic, recent declines in Iceland, Fenno-Scandia, and possibly USSR. Common Scoter *M. nigra* first recorded breeding Scotland 1885, Ireland 1905; some decreases Norway and Finland. Velvet Scoter *M. fusca* has declined on south-east coast Sweden and inland in north Finland though increasing in Finnish archipelago. Expansion by Goldeneye *B. clangula* in central Europe since mid-19th century but recent retreat north in USSR due to destruction of forests, and no longer breeds west Norway; increases in some areas helped by provision of nest-boxes. Range of *M. albellus* has shrunk markedly in USSR due to habitat destruction, but breeding Norway since 1925. Both *M. serrator* and *M. merganser* have spread south in Britain and Ireland though elsewhere isolated populations of latter species south of main range have tended to disappear in recent years. For winter counts, see Atkinson-Willes (1975). All Holarctic species migratory to greater or lesser extent. Moult migrations highly developed; moult assemblies include immatures as well as adult ♂♂, also some adult ♀♀. In most species, majority of ♀♀ moult on breeding grounds, or in winter quarters. Migrate over any landmass under direct route, almost entirely nocturnally; generally fly quite high on passage, in irregular groups or wavy lines.

Chiefly take animal food; in *Mergus*, entirely so. Obtained by diving from surface, mainly to bottom, but *Mergus* unique within Anatidae in pursuing swift-moving prey under water. Will feed singly, in pairs, or family parties, but usually in flocks. Gregarious also at most other times. Lateral Head-shaking main pre-flight signal, but certain other comfort-movements and bathing also regular in most species. On breeding grounds, ♂ tends to defend ♀ wherever she may be. Breeding dispersion of several species poorly known. Territories sometimes well defined (e.g. *Histrionicus*, *Clangula*) but some or all *Mergus* sociable and non-territorial when breeding. In territorial species, nest not always in ♂'s territory. Monogamous pair-bonds of seasonal duration, ♂♂ deserting ♀♀ during incubation. Pair-formation in flock, starting mostly in late winter. Communal courtship (mostly on water) well developed and ♂ displays often elaborate, especially in *Bucephala* and *Mergus*. Neck-stretch posture common to most species but otherwise major displays tend to differ according to genus. Include Head-throw and Kick-displays with calls; also Rushing or Steaming towards ♀ and Turn-back-of-Head or Head-turning. Secondary displays include Upward-shake, Wing-flap, various Mock-preens, and lateral Head-shake but some, especially last, elevated to major displays in some species. ♀♀ at times perform ♂-like displays or have special courtship displays as well as variable Inciting-displays of same functional type as in Anatini. Short-flight displays by ♂♂ frequent in a few species during communal courtship. Courtship-flights in several species (in some replaced by underwater pursuits of ♀); 3-bird flights reported in *Clangula*. Pair-courtship may involve displays used in communal courtship, notably in *Melanitta*; in all species, copulatory behaviour also involved. Pre-copulatory displays include Mock-preens or Ceremonial-drinking which may be mutual, or ♀ assumes Prone-posture more or less immediately—in some species, during or after Inciting. Other ♂ displays include Water-twitch, Wing-flap, Head-shake, and Upward-shake; immediately before mounting, may Steam or Rush towards ♀, sometimes after performing single major display. While mounted, ♂ of some species Wing-shake. After copulation, ♂ may perform one of major displays; swims or Steams away, usually calling and sometimes does Turn-back-of-Head or Head-turning display. Range of vocalizations broad and more or less strongly differentiated between sexes; concentrated mainly at courtship and pre-nesting periods. Calls include piping, whistling, or fluting (♂♂) and croaking, rattling, and grunting (♀♀). Nearly all utterances given on water or wing. ♀ vocalizations include Inciting-calls, but no known equivalent of Decrescendo-calls of ♀ Anatini.

Seasonal breeders in Holarctic. Mostly ground nesters in cover, but some nest in holes in ground or trees, close to or away from water. Nests often well dispersed but sometimes more concentrated at protected sites (often in association with gulls *Larus* or terns *Sterna* in some species); some species regularly loosely colonial. Nests shallow depressions lined with available material and down. Building by ♀ only. Old nests often re-used. Eggs elliptical or oval, buff to olive, green-grey, or bluish; smooth, slightly glossy in some. Clutches usually 5–10, in hole nesters up to 14. Multiple laying occurs. One brood; replacements laid after egg loss except in high Arctic. Eggs laid at 1–2 day intervals. Incubation by ♀ only. Incubation periods 22–32 days (Johnsgard 1968). Young cared for by ♀ only. Distraction-display in form of 'injury-feigning' reported in all genera (Hebard 1960). Amalgamation of broods at times in some species; young then become independent well before fledging, otherwise at or just before. Mature in 2nd or 3rd year.

## *Histrionicus histrionicus* Harlequin

PLATES 85 and 99
[facing page 639 and after page 686]

Du. Harlekijneend   Fr. Garrot arlequin   Ge. Kragenente
Ru. Каменушка   Sp. Pato arlequín   Sw. Strömand

*Anas histrionica* Linnaeus, 1758

Monotypic

**Field characters.** 38–45 cm, of which body just over two-thirds; wing-span 63–69 cm. Small, buoyant, short-necked sea-duck with short, narrow bill, high forehead, and pointed tail; ♂ mainly dark grey-blue with chestnut flanks and harlequin pattern of bold white marks on head, neck, breast, and scapulars. Sexes dissimilar, but no seasonal differences apart from eclipse. Juvenile indistinguishable in field from adult ♀.

ADULT MALE. Generally dark grey-blue and black, looking all black at distance, with pattern of white marks, mostly bordered with black, on head, base of neck, breast, and scapulars. Rear part of streak at side of black crown rufous; flanks chestnut; whole stern black with small white patch on side. Wings dark grey-blue with sooty flight-feathers; short white wing-bar; purplish-blue speculum; and white on innermost secondaries. In flight, looks mainly dark with white patches on head and white braces; underwing brown. Bill lead-blue; eyes reddish-brown; legs pale blue with blackish webs. MALE ECLIPSE. At first sight like adult ♀, but easily distinguished by blacker plumage, tinged grey-blue on back, with more distinct white on face, white retained on inner secondaries, and dark brown lower breast and belly. ADULT FEMALE. Generally dull blackish-brown, paler on throat and foreneck but looking all dark at distance, with usually 2 light patches in front of eye, circular white patch on ear-coverts, and mottled whitish lower breast and belly. In flight, wings show as uniform blackish-brown. Bill and legs duller than in ♂; eyes browner. JUVENILE. Closely resembles adult ♀, though upperparts generally paler and more olive-brown. ♂♂ start to assume adult pattern in October and particularly December onwards, general tone becoming dark blue-grey with chestnut suffusion on flanks, but white marks reduced or absent, and even in 1st summer still duller above with whitish breast and belly.

Both sexes recognized by combination of small size, tiny bill, pointed tail often cocked, generally dark body, and great buoyancy on water. ♂ at long range looks all black and might be confused with ♂ Common Scoter *Melanitta nigra*, which is noticeably larger with heavier bill. Head pattern of ♀ and immature might be mistaken for ♀ Velvet Scoter *M. fusca* or Surf Scoter *M. perspicillata*, which have only 2 pale, less well-defined patches on side of head and are much bigger with heavier bills, while former also has white patch in wing. ♀ often likened to ♀ or immature Bufflehead *Bucephala albeola*, but that has only single, elongated head patch behind eye and is noticeably smaller and much paler, more grey-brown above and whiter below, with white wing patches. At distance, young Long-tailed Duck *Clangula hyemalis* most likely confused, since it is similar shade of brown and its light sides of head look like those of *H. histrionicus* when individual patches cannot be made out, but *C. hyemalis* looks generally paler with white at sides of neck and grey and white underparts. Flight silhouette of *H. histrionicus* recalls Goldeneye *B. clangula* without white wing patches and belly.

In breeding season, on fast, rocky rivers; in winter, in rough waters off cliffs and headlands. Often relatively easy to approach, particularly in broken water. Rarely associates with other ducks, though occasionally with scoters *Melanitta*. Swims high and buoyantly with characteristic habit of bobbing or jerking head at each leg stroke; pointed tail frequently cocked or slowly raised and lowered. Dives habitually in roughest water; sometimes plunges in from

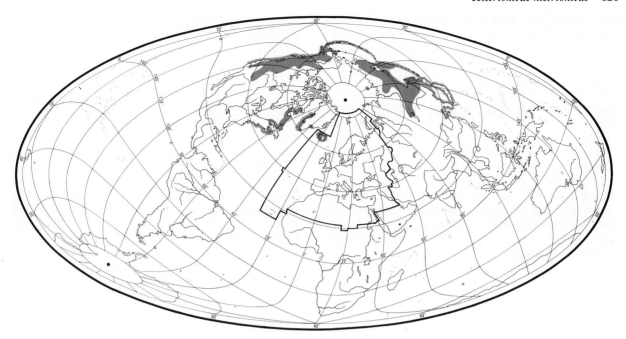

rock and frequently from wing; also walks on bottoms of streams, facing current with wings closed and head held down. In shallow streams, also up-ends or wades about and simply immerses head. Flies rapidly with quick wing-beats, usually in small, compact parties. Usually silent, but ♂ has low, piping whistle beginning with 2 separate notes and ending in descending trill, and ♀ harsh croak.

**Habitat.** At home only in continuously turbulent waters, at low temperature: even rapids and waterfalls acceptable. Not normally in contact with ice, and avoids rivers of glacial source. Immediately south of Arctic Circle in Iceland, ascends to highest streams; reaches much higher altitudes in American Rocky Mountains. Yearly cycle comparable to salmon *Salmo salar* with same movement into fjords and bays, and same lingering around river mouths and slow, hesitant upstream penetration (Gudmundsson 1971). Even nesting areas on river frequently coincide with salmon spawning grounds. Apparently rarely takes short cut by flying overland. Direct run-off rivers, being poor in plant and animal life, support only widely spaced pairs, concentration being confined to lake outlet and spring-fed rivers. Explores gravel stream-beds at depths of 1–2 m against strong current, and dives directly from flight or flies directly from below water. Moves adroitly even on steep slippery rock surfaces out of water. Flies strongly, following each turn of stream and passing under rather than over bridges. For nest-sites, requires well-concealed hollows beneath rocks or shrub and herb cover up to 2 m tall, especially on islets in rivers, sometimes behind waterfalls. ♀♀ passing downstream with ducklings to sea readily traverse torrents and intervening lakes. Wintering always on coast, keeps within equally restricted and rugged habitat, rarely more than some 300 m from most exposed rocky headlands and lava shores under steep cliffs, shunning sheltered bays and sandy or shingly stretches. Rests on rocks on shore. Virtually independent of man-made habitats or artefacts.

**Distribution.** ICELAND: only breeding area in west Palearctic.

Accidental. Britain and Ireland 7, Spitsbergen 1, Norway 1, Sweden 3, USSR several (including records from Baltic Sea and White Sea), West Germany 4, East Germany 6, Poland 2, Czechoslovakia 1, Austria 1 (possibly 2), Switzerland 1, Italy 1. These vagrants possibly of differing origins (see Movements).

**Population.** Iceland. Unlikely to exceed 3000 pairs. Eggs exported for aviculture after 1945, but now fully protected (Gudmundsson 1971; Gardarsson 1975).

**Movements.** Migratory and partially migratory; movements often short, only to nearest coasts. Icelandic population leaves inland waters to winter all around coasts, birds of north and north-east tending to move to west and south-west, especially to Reykjanes Peninsula (Gudmundsson 1971); no evidence of emigration, though European vagrants may include Icelandic birds. However, high proportion of such vagrancy records from Baltic Sea, with others north to Spitsbergen and White Sea, and one in Tomsk, suggest these may include some movements from east Siberia (Gudmundsson 1971). Siberian breeders more typically migratory than those of Iceland, those wintering Bering, Okhotsk, and Japan Seas evidently including breeders from west to River Lena basin and Lake Baykal (Dementiev and Gladkov 1952). Also, transatlantic vagrancy may occur, for east Canadian birds mostly withdraw from breeding range to winter from south-east Labrador and Newfoundland south to Maine and, less consistently, to Long Island, New York (Godfrey 1966). Greenland breeders winter around southern coasts there.

Accidental inland Canada, USA (south to Florida), Mongolia, and Ryu Kyu Islands. Leaves northernmost breeding places early September to early October, returning middle to late May. Immatures mostly remain on coasts; adult ♂♂ return there to moult 2nd half June and 1st half July; ♀♀ also moult on coasts, moving with juveniles about September. In Maine, recorded between 10 October and 30 April, peak numbers present November–March. Icelandic birds pass from exposed coasts to bays and fjords in spring, assembling at river mouths, then in late April–May move up rivers in short stages of alternate swimming and flying; in spring at least always follow rivers, not taking direct overland routes (Gudmundsson 1971).

**Food.** Almost entirely animal, obtained by surface-diving with partially opened wings, bird normally emerging almost exactly where submerged. In shallow places, also skims food off surface and head-dips but rarely up-ends. Uses bill under water to prise items off rocks. At sea, dives average 15–25 s (Alford 1920; Bengtson 1966a), 5–35 s (Palmer 1949); in depths not usually exceeding 3–4 m (Bengtson 1966a). Inland, in shallower waters—in Iceland 80–130 cm (Pool 1962)—1210 dives normally ranged from 15–18 s, maximum 39 s (Bengtson 1966a). In Iceland in summer, peaks in feeding frequency at *c.* 06.00 and 17.00–18.00 hrs, with smaller peak around midnight (Bengtson 1966a). In winter at sea, highly social feeder, flock often diving simultaneously; on breeding grounds, feeds mostly in pairs, rarely in small parties.

Specific data limited: winter foods mainly molluscs (periwinkles *Littorina*, nerite *Neritina*, blue mussel *Mytilus edulis*) and crustaceans, with fewer annelids, fish, and spawn (Millais 1913; Witherby *et al.* 1939). From Icelandic coasts, chiefly crustaceans (amphipods, isopods, and small crabs) and some molluscs (Gudmundsson 1961). Spring and summer foods, Iceland: insect larvae, chiefly of caddisfly (Trichoptera) and mayfly (Ephemeroptera), and small crustaceans (Witherby *et al.* 1939). From northern Iceland rivers, mainly larvae, pupae, and (from surface) adults of blackfly (Simuliidae) (Bengtson 1966a; Bengtson and Ulfstrand 1971). 12 stomachs, Iceland, contained predominantly Simuliidae (especially *Simulium vittatum*) and fewer midge (Chironomidae) and caddisfly larvae (Gudmundsson 1961). In USSR, early June on Anadyr, stomachs contained caddisfly larvae, also larvae of stonefly (Perlidae), waterbeetles, bugs (Hemiptera), and small fish (Dementiev and Gladkov 1952).

In Iceland, young feed mainly on Simuliidae larvae (Bengtson 1972b), but also take some plant materials (Bengtson and Denward 1966). Stomachs of ducklings from Kamchatka, USSR, in September, contained only insect remains (Dementiev and Gladkov 1952).

In North America, January–March and June–September, 63 stomachs contained 57·1% by volume crustaceans (mainly small crabs, especially *Hemigrapsus* and hermit crabs *Pagurus*, and amphipods), 24·7% molluscs (mainly *Chiton*, *Lacuna*, *Littorina*, *Margarites*, *Acmaea*, *Mytilus*), 10·2% insects (mainly stoneflies, mostly during summer and early autumn), 2·4% echinoderms, and 2·4% fish (Cottam 1939). 9 stomachs, December, Maine, contained chiefly crustaceans *Gammarellus angulosus*, and molluscs *Nucella lapillus* and *Lacuna vincta* (Palmer 1949).

**Social pattern and behaviour.** 1. Gregarious, except (usually) when nesting. Outside breeding season, mostly in small, compact flocks scattered on sea: from September to early spring in Iceland, where 5–30 birds in feeding parties (Bengtson 1966a). In Icelandic waters, by December, a few pairs may already be keeping apart from each main flock of unpaired birds, either in small, loose groups or singly. At end April, pairs start gradually to move up rivers for breeding (Bengtson 1972b). ♂♂ first to flock again from mid-June, gathering briefly in 'clubs' to start moult before leaving, together with some non-breeding and unsuccessful ♀♀, to complete it and winter on sea where later joined by majority of ♀♀ (after moult started inland) and fledged young (Gudmundsson 1961; Bengtson 1972b). Flocks of 5–12 non-breeding ♀♀ present in breeding area from end June. In Greenland and Iceland, immatures stay mostly on sea throughout year (Gudmundsson 1961; Salomonsen 1950a). BONDS. Monogamous pair-bond, usually of seasonal duration. Pair-formation starts in winter flock; mostly completed before leaving sea though continues on rivers with courtship display until mid-June (Bengtson 1966a). On breeding waters in May, Iceland, surplus of ♂♂ (mean ratio 125:100) (Bengtson 1972b); little evidence of promiscuity, however, though some pair copulations resemble rape (see Bengtson 1966a). ♂ deserts ♀ when incubation begins, sometimes remaining 1–2 days and accompanying ♀ during her recesses from nest. Exceptionally, ♂ seen with ♀ and brood (Harrison 1967); typically only ♀ tends young at least until fledged—in Iceland into September or October (Gudmundsson 1961). No amalgamation of broods. BREEDING DISPERSION. Mainly solitary and well-scattered, most concentrated where rivers rich in invertebrate food (see Bannerman 1968). Thus, *c.* 50 pairs on 5 km of River Laxá (Bengtson 1966a); 1–5 pairs using same nesting island (Gudmundsson 1961). In some cases, ♀♀ here known to use same sites each year. Where well dispersed, or on periphery of nesting concentrations, maintains clear-cut territory along small section of river (Bengtson 1966a, but see Bengtson 1972b). In more densely populated areas, though pairs keep well apart for much of time, territories much less distinct or absent; ♂ defends area around ♀, and maintains exclusive loafing spot. Also on River Laxá, both paired and unpaired birds spend first few days after spring arrival communally at clubs; these form again in 2nd week June when frequented by ♂♂ and non-breeding ♀♀. ♀ leads young eventually to sea. ROOSTING. Communal for much of year. Little winter information, but roosting mainly nocturnal, most of day and possibly some of night being spent feeding (Bengtson 1966a). Diurnal loafing reported on shore rocks when sea fairly calm, otherwise on water (Gudmundsson 1961). During summer on River Laxá, main rest periods 08.00–14.00 and each side of midnight, but birds also loaf between spells of feeding (Bengtson 1966a). Each pair frequents own loafing spot prior to nesting. Also up to 40–50 birds use communal loafing spots (clubs) early and late in season; these often situated in shallow but rough water in middle of river, with many obstacles protruding above water, but also on river bank or island (Bengtson 1966a). Often rests on stones or submerged logs, with feet in water (Gudmundsson 1961).

2. Studied in wild, especially by Myres (1959b) and Bengtson (1966a); see also Johnsgard (1965). Though full display repertory probably not yet known, striking visual displays absent, in contrast with most other sea-ducks, with greater emphasis apparently on vocalizations. Main display, shown in many situations by both sexes, variable Head-nodding similar to Rotary-pumping of Barrow's Goldeneye *Bucephala islandica*: at full intensity consists of repeated elliptical movements of head parallel to surface of water while bill held horizontally; in ♂ at least, often accompanied by high-pitched calling. FLOCK BEHAVIOUR. On sea and in flight, flocks often dense, and members of same feeding group will dive in close synchrony (Bengtson 1966a). On arrival in breeding area, flocks often highly vocal and restless (Millais 1913). No pre-flight movements recorded (Johnsgard 1965) but may call before taking wing (Alford 1920) though seldom escapes by flight. When alarmed, holds neck at angle of *c.* 70° to water and intensifies unritualized nodding movements of head that normally accompany locomotion, performing rapid sequence of short, elliptical jerks; when predator gets nearer, alarm-posture assumed in which neck, with feathers slightly ruffled, stretched out—then escapes, usually by swimming or diving (Bengtson 1966a). ANTAGONISTIC BEHAVIOUR. In breeding area, hostile encounters occur at clubs and in courting parties (Bengtson 1966a). ♂♂ especially then perform version of Head-nodding display in which movements, though rapid, less complete and stiffer than at other times. This interpreted as 'greeting' by Myres (1959b), but Bengtson (1966a) pointed out that it usually induces withdrawal and, if not, further intimidation follows in form of (1) threatening with horizontal lowering of head and neck, and (2) hostile approach, sometimes with bill open as hissing sound given (see Bent 1925). May lead to attack in which aggressor leaps or rushes at opponent; no underwater attacks recorded, though combatants often dive. Fights rare, and none observed in winter flocks on sea. Apparent Mock-preens also occur during hostilities. In both winter and summer, paired ♀♀ occasionally perform Inciting-display when approached by strange ♂♂ (Bengtson 1966a): (1) lower head and neck, often touching water with throat; (2) either point bill at rejected ♂ or, more usually, make distinctive Head-turns from side to side similar to those of goldeneyes *Bucephala*. Sometimes accompanied by call and followed by hostile rush over water at ♂, then more Inciting. Head-nodding usual response of threatened ♂; also directs forward threatening movements at ♀. COMMUNAL COURTSHIP. Occurs on sea in winter and, apparently to lesser extent, on breeding waters in spring. Off Iceland in December, courting parties of ♀ and 3–8 ♂♂ detach themselves from larger flocks (Bengtson 1966a). With necks stretched up and tails elevated, ♂♂ then frequently perform Head-nodding displays. Periodically, one ♂ will rush after ♀ for *c.* 10 m, causing her to flee and dive, though no underwater pursuits recorded; other ♂♂ intensify their Head-nodding but do not follow. Other ♂ activities include: (1) Wing-flap (much as in eiders *Somateria*, i.e. often with only *c.* 2 flaps), associated with Head-nodding and frequently repeated; (2) calling with bill open without any special

posture; (3) peering under water with head submerged; (4) forward threat-like posture; (5) Mock-preens, including Preen-dorsally and Preen-behind-Wing. More elaborate display, involving backwards and forwards head-throwing movements, wing raising and lowering, and integrated calls described by Bretherton (1896); not confirmed by recent studies, though account of Yeates (1951) somewhat similar. ♀♀ frequently call, sometimes also Bill-dipping, when ♂♂ perform Wing-flaps (see Johnsgard 1965). COURTSHIP-FLIGHTS. ♂♂ in courting parties not infrequently chase ♀ in flight (Bengtson 1966a). PAIR BEHAVIOUR. Paired birds said to call to one another on meeting (Millais 1913). Bouts of mutual Head-nodding by ♂ and ♀, mostly initiated by ♂, mentioned by Bengtson (1966a) though context not clear; movements of ♀ usually less intense but even slight movements by ♀ increase intensity of those of ♂. ♂ may also perform Water-twitch between Head-nods, shaking bill in water much as (e.g.) in *Bucephala*. COPULATION. Not a feature of pair-courtship, but occurs mostly just prior to egg-laying. In Iceland, usual around 10.00 hrs and at 15.00–18.00 hrs; preliminaries may occur in presence of other pairs but complete sequences seen only at secluded spots (Bengtson 1966a). Pre-copulatory display prolonged and mutual; usually started by ♂, movements of ♀ being briefer and less regular. Begins with Head-nodding lasting 5–30 mins, then Water-twitching during which bill dipped in water 3–5 times rapidly. Finally, ♂ rushes over surface at ♀ with bill open, calling; eventually mounts, but sometimes not until after series of 5–20 rushes usually over 10–30 mins. Activity resembling Rearing-display of Steller's Eider *Polysticta stelleri* (but possibly just Upward-shake) also occasionally performed up to 5 times prior to rush. ♀ does not usually assume Prone-posture until immediately before ♂ mounts, copulation lasting 2–6 s. Afterwards, activities of both variable, but ♂ may dive and then direct series of further, but lower intensity rushes at ♀ who sometimes dive-bathes and preens. Mating behaviour of ♂ shows many aggressive features and ♀ may at times fly or run away, pursued by ♂. ♀ may also not permit ♂ to mount, grabbing at his tail as he tries to grip her nape, both then self-preening instead. Rotation of pair on water after copulation may be due largely to ♀'s struggle to release herself. RELATIONS WITHIN FAMILY GROUP. Little information (see Bengtson 1966a for some general notes).

**Voice.** Imperfectly described. Most accounts relate to Pacific populations which may differ in voice from Atlantic; tape recordings from Canada and Iceland not fully reconcilable with field descriptions. Vocal element in display appears more important than visual (Myres 1959b; Bengtson 1966a); open bills indicating vocal activity noted when background noise made any utterances inaudible. Need for audibility against rushing torrents or breaking surf accounts for high-pitched squealing or whistling voice with little in common with most Anatinae. On Pacific coast of North America, local names of 'squealer' and 'sea mouse' attest vocal activity (Bent 1925), despite reputation as a generally silent bird (see Witherby *et al.* 1939).

CALLS OF MALE. Basically quiet and low-pitched but, under excitement, become much faster, shriller, louder, and higher pitched—almost defying description and classification. (1) Main call during communal courtship given without special posturing or accompanies Head-nodding display, singly or in trills: high-pitched, mouse-

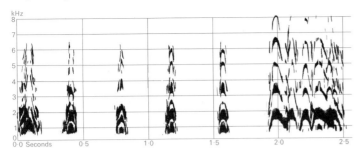

I  Grace Bell/Sveriges Radio  Canada  October 1962

like whistling or squealing (Bent 1925; Myres 1959b) — (see I); also uttered during pre-copulatory rush towards ♀ (Bengtson 1966a). Loud 'gi-ek' of Dementiev and Gladkov (1952) may refer to this call. Call accompanying rare Head-throw display (Bretherton 1896) may also be similar, at least in part: shrill whistle starting with 2 descending long notes as head thrown back and running off in long trill as head brought forward; quiet whistle mentioned in connection with same type of display by Yeates (1951). (2) Low-pitched piping whistle much like Common Sand-piper *Actitis hypoleucos* (Alford 1920): uttered before take-off from water and, during winter, while feeding; significance uncertain. (3) Soft 'drut' (Salomonsen 1950a): warning-call, evidently corresponding to fine 'dü' or soft 'da' of Witherby *et al.* (1939). (4) Low or harsh 'hu' or 'hey-hey': from flock in flight after arrival in breeding area in response to equivalent call of ♀ (Millais 1913); probably contact-call. (5) Rapidly repeated (7–8 times) 'oy . . .': in flight, usually when pursuing ♀ (see Bent 1925). (6) Hissing: uttered with open bill during aggressive en-counters (Bent 1925).

CALLS OF FEMALE. (1) High-pitched, staccato call like Little Tern *Sterna albifrons*: given in response to Wing-flap of ♂♂ during communal courtship (see Johnsgard 1965). A number of other high-pitched calls during courtship mentioned but not described by Johnsgard (1965). (2) 'Giak': response to 'gi-ek' of ♂ (Dementiev and Gladkov 1952). (3) Inciting-call (heard only once) described merely as 'harsh' (Bengtson 1966). (4) Persistent 'ek-ek-ek-ek': uttered frequently from flock in flight after arrival in breeding area (Millais 1913); apparently contact-call. (5) Low, somewhat raven-like call when brood endangered (Nicholson 1930); also described as jarring sound (Bengtson 1966a). (6) Low-pitched, smooth quacking: uttered in a variety of situations, significance uncertain.

CALLS OF YOUNG. No information.

Breeding. SEASON. Earliest eggs, Iceland, mid-May, most laid last week May and 1st half June; late eggs in nest mid-July. Hatching from mid-June onwards (Bengtson 1966a, 1972b; Gudmundsson 1971). Timing probably affected by temperature (Bengtson 1972b). SITE. On ground, in dense vegetation, usually well concealed, often under low scrub, occasionally among rocks or in rock cavity. Rarely more than 5 m from water. Nest: depression, with small amount grass and twigs; lined down. Building: by ♀, using material within reach of nest. EGGS. Blunt ovate; creamy-yellow. 58 × 41 mm (50–62 × 39–43), sample 105 (Gud-mundsson 1961). Calculated weight 53 g (Schönwetter 1967). Clutch: 5–7 (3–10); occasionally up to 12, probably 2 ♀♀ (Gudmundsson 1961). Of 77 Icelandic clutches: 3 eggs, 5%; 4, 14%; 5, 23%; 6, 32%; 7, 16%; 8, 9%; 9, 1%; mean 5·7 (Bengtson 1927b). One brood. No data on replacements but late eggs (July) indicate strong like-lihood. Eggs laid at 1, occasionally 2-day intervals (Gudmundsson 1971; Bengtson 1972b), not 2–4 days (Bengtson 1966a). INCUBATION. 27–29 days. By ♀. Starts with last egg; hatching synchronous. Eggs covered with down when ♀ off. Eggshells left in nest. YOUNG. Precocial and nidifugous. Self-feeding. Cared for by ♀, brooded at night while small. FLEDGING TO MATURITY. Fledging period probably 60–70 days (Gudmundsson 1971) and certainly longer than 40 days given by Bengtson (1966a). In Iceland, become independent after post-fledging move-ment with ♀ down river to sea (Gudmundsson 1971), though some broods apparently abandoned by ♀ when only half grown (Bengtson 1972b). Age of first breeding 2 years. BREEDING SUCCESS. Of 504 eggs laid, Iceland 1966–70, 81% hatched and between 50–70% of these survived to fledging (Bengtson 1972b).

Plumages. ADULT MALE BREEDING. Centre of forehead and crown black glossed blue, bordered by white bands at sides which gradually change to cinnamon-rufous above and behind eyes. Large semi-circle on lores, spot on ear-coverts, and narrow line at sides of neck white, margined bluish-black. Rest of head and neck dark lead-blue, extending to white line bordered black round body over upper chest and upper mantle. Mantle, upper back, scapulars, chest, and sides of breast blue-grey, outer web of longer inner scapulars and inner web of outer scapulars white. Lower back, rump, and upper tail-coverts black, glossed blue. At sides of breast in front of wing, white streak bordered black. Sides of body and flanks chestnut; feathers of breast and belly dark brown, edged blue-grey; lower vent and under tail-coverts black, glossed blue. White spot at sides of vent near base of tail. Tail black. Flight-feathers sooty-black; outer webs of most second-aries glossed steel-blue. Wing-coverts dark blue-grey, inner greater coverts with steel-blue outer web and broad white tip; some central median coverts sometimes with large white subterminal spot. Tertials dark grey, longest one with white shaft-streak, outer 2 with outer webs white, edged black. Underwing and axillaries dark brown, coverts tipped blue-grey or white. ADULT FEMALE BREEDING. Head and neck dark slate, darkest on crown, cheek, and hindneck, more dusky brown on forehead, lores, chin, and throat. Small white patches on ear-coverts and between forehead and eye, larger patch between gape and eye; those near eye mottled dark brown when fresh. In worn plumage, throat and sides of head in front of and below eye off-white, except dusky brown mottling on forehead and lores. Upperparts sooty-brown, feathers narrowly edged dark blue-grey; scapulars tinged and upper tail-coverts tipped olive-brown. Chest, upper breast, flanks, and vent dark slate-grey with pale grey tips; feathers of rest of underparts grey-brown, broadly tipped white, often with small terminal dusky spot. Tail sepia or dull black. Flight-feathers sooty-black, secondaries with purple gloss on outer web; upper wing-coverts like upperparts. Underwing and axillaries dark grey-brown, coverts variable tipped white. ADULT MALE NON-BREEDING (Eclipse). Like adult ♀ breeding, but crown, sides of head, and upperparts darker, more slate-blue; at times with some white in feathers on sides of breast and scapulars, and some chestnut in flanks. Breast and belly dark brown instead of mottled white as adult ♀. Wing like adult ♂ breeding, but new tertials narrower, outer with white confined to broad shaft-streak. ADULT FEMALE NON-BREEDING. Like adult ♀ breeding, but edges of feathers of upperparts narrower, and paler grey. DOWNY YOUNG. Crown to below eye, hindneck, upperparts, sides of body, and thighs dark brown; small streak above dark line in lores white; indistinct patches of white on wing, at sides of back behind wing, and at sides of rump. Cheeks and underparts white. JUVENILE. Like adult ♀ breeding, but head and upperparts dark brown, less greyish; edges of feathers of mantle and scapulars olive-brown instead of grey; feathers of breast and belly short and narrow, grey-brown with rather narrow white tips. Tail-feathers notched at tip, bare shaft projecting. Wings duller and browner; secondaries not or only slightly glossed purple. FIRST IMMATURE NON-BREEDING. In early autumn, both sexes apparently attain variable amount of feathers resembling adult non-breeding. FIRST IMMATURE BREEDING. Resembles adult breeding. Gradually acquired from autumn: first on head, later on upper mantle, flanks, scapulars, sides of tail, and often tail and tail-coverts. Marked individual variation in amount of plumage involved. In ♂, white patches less extensive than adult, especially on scapulars, and head duller blue. Breast, belly, and wing at least retain juvenile feathers. In spring, new feathers of ♂ like adult ♂ breeding, of ♀ like adult ♀ non-breeding; juvenile wing and usually belly retained. (Schiøler 1926; Witherby *et al.* 1939; C S Roselaar.)

**Bare parts.** ADULT MALE. Iris red-brown-Bill lead-blue, nail pale horn. Foot pale blue-grey, web and back of tarsus black. ADULT FEMALE AND JUVENILE. Iris brown. Bill dark slate-blue with paler tip. Foot grey-green or olive-green, web and back of tarsus dull black. (Bauer and Glutz 1969; Schiøler 1926; Witherby *et al.* 1939.)

**Moults.** ADULT POST-BREEDING. Complete. Body and tertials about April in ♀♀, June–July in ♂♂. ♀♀ stop moult when incubating. Flight-feathers simultaneous; flightless between late

July and September in ♂♂, *c.* 1 month later in ♀♀. ADULT PRE-BREEDING. Partial; head and body. Start immediately when wing completed, partly overlapping with post-breeding body in ♂. Finished before October in ♂♂, before December in ♀♀. POST-JUVENILE. Variable amount of 1st immature non-breeding attained in early autumn on head, chest, flanks, and scapulars. FIRST IMMATURE PRE-BREEDING. Marked variation in number of feathers of breeding plumage attained September–December. In ♂, involves head, neck, and part of mantle, scapulars, rump, flanks, and tail-coverts, often central or whole tail; in ♀ usually less extensive. SUBSEQUENT MOULTS OF IMMATURES. In spring, rest of juvenile feathers (including tertials, but not breast, belly, and wing) changed for 1st immature breeding feathers in ♂, mostly for 2nd immature non-breeding in ♀. 1st immature post-breeding of ♂ like adult ♂, but not all breeding feathers acquired in spring and not always all juvenile belly moulted. Following pre-breeding like adult but body usually finished later; flightless between August and early October. (Bauer and Glutz 1969; Schiøler 1926; Witherby *et al.* 1939; C S Roselaar.)

**Measurements.** ADULT. Iceland, Greenland, and Labrador; skins (RMNH, ZMA).

| | | | | | | |
|---|---|---|---|---|---|---|
| WING | ♂ 205 | (5·58; 13) | 197–214 | ♀ 198 | (3·77; 4) | 194–201 |
| TAIL | | 93·0 (5·85; 12) | 87–105 | 79·0 | (5·03; 4) | 74–86 |
| BILL | | 25·8 (0·73; 13) | 24–28 | 25·3 | (0·80; 4) | 24–26 |
| TARSUS | | 37·5 (1·17; 14) | 36–40 | 36·0 | (1·22; 4) | 34–37 |
| TOE | | 56·7 (2·41; 14) | 53–62 | 53·5 | (3·42; 4) | 50–58 |

Sex differences significant, except bill. Adults of east Siberian population similar, except larger bill: range 27–30 in ♂♂, 23–30 in ♀♀ (Dementiev and Gladkov 1952). Bill of 10 Canadian ♂♂: 25·8 (25–27) Godfrey 1966).

JUVENILE. Greenland and Iceland (Schiøler 1926). Wing ♂ 190 (17) 181–199; ♀ 183 (19) 177–191. Tarsus and toe about similar to adult; tail much shorter, *c.* 65–75 mm (RMNH).

**Weights.** ADULT. ♂♂. Alaska: spring migration, 736; mid-May 710; late May to early June, 582, 618, 639 (Conover 1926; Irving 1960; Johnston 1963). Iceland: June, 750 (Bauer and Glutz 1969). Kamchatka: May, 714; October, 650; November, 670; January, 675 (Dementiev and Gladkov 1952). ♀♀. Alaska: early June, 562; summer, probably breeding, 520, 553. Kamchatka: September, 510; October, 500.

**Structure.** 11 primaries: p9 longest, p10 0–4 shorter, p8 4–8, p7 13–20, p6 24–35, p1 72–91. Tail long, graduated, and pointed; 14 stiff, pointed feathers. Bill short and narrow, as high as broad at base, sides gradually narrowing towards large oval nail. Small lobe at base of upper mandible in corner of mouth. Feathers of hindcrown and hindneck elongated in adult ♂. Outer toe *c.* 97% middle, inner *c.* 76%, hind *c.* 26%.

**Geographical variation.** Slight. *H. h. pacificus* Brooks, 1915, north-east Siberia and north-west America, not recognized, as differences between Atlantic and Pacific populations small and overlap considerable.                                                    CSR

*Clangula hyemalis* **Long-tailed Duck**

PLATES 86 and 100
[facing pages 639 and 687]

Du. IJseend     Fr. Harelde de Miquelon     Ge. Eisente
Ru. Морянка     Sp. Havelda     Sw. Alfågel     N.Am. Oldsquaw

*Anas hyemalis* Linnaeus, 1758

Monotypic

**Field characters.** 40–47 cm, of which body two-thirds, plus up to 13 cm of elongated central tail-feathers in ♂; wing-span 73–79 cm. Rather small, neat, and delicately proportioned sea-duck with short, high-based bill, steep forehead, small head, narrow wings, and, in ♂, long and pointed tail; ♂ predominantly dark brown in summer, with whitish face patch, flanks, belly, and undertail, and predominantly white in winter, with dark brown cheek patch, pectoral band, back, tail, and wings. Sexes dissimilar and marked seasonal differences. Juvenile only superficially like adult ♀ and easily distinguished.

As complex and controversial plumage changes fully discussed under Plumages, not described in detail here. Adult ♂ unmistakable. Larger ♂ Pintail *Anas acuta* only other duck with long tail, but found mainly on fresh water, has all-dark head, white breast, grey flanks, and black stern. ♀ and, more particularly, juvenile sometimes confused with ♀♀ or immatures of local Harlequin *Histrionicus histrionicus* or vagrant Bufflehead *Bucephala albeola*, but both smaller with different patterns of white on head and, in addition, former much darker, especially on belly, and latter shows white in wings. Indeed, in flight, *C. hyemalis* only white-headed or extensively white-bodied duck with all-dark wings.

In breeding season, on tundra waters but otherwise largely marine and usually well out to sea, though not infrequently close inshore, occasionally on coastal fresh waters, and in some areas regular on large inland lakes. Generally not at all shy, but lively and restless, frequently diving or taking to wing. Swims buoyantly, ♂ with long tail often trailed on surface, but raised to 45° when alert, or almost vertical and partly expanded when agitated or excited. Dives with great agility, even in rough sea, throwing head back and plunging forward with spread tail and often partly opened wings. Except on migration, flies rapidly low over surface, characteristically swinging from side to side so that dark upperparts and white underparts show alternately. Wings held more downcurved than in other ducks and have peculiar beat, being brought hardly above body on upstroke and much lower on downstroke. Seldom walks on land except to and from nest, but moves fairly easily with erect gait. When alighting on water, drops down abruptly breast first, without gliding like most other ducks, and sometimes also dives from wing. Singly in southern parts of winter range, but generally highly gregarious. Flies in irregular flocks or straggling lines. Seldom associates with other sea-ducks. Far more vocal than most other ducks.

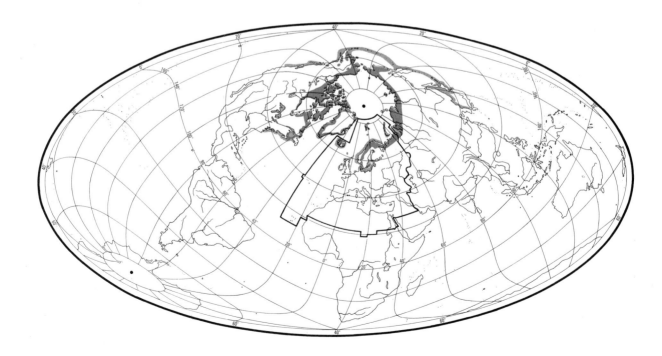

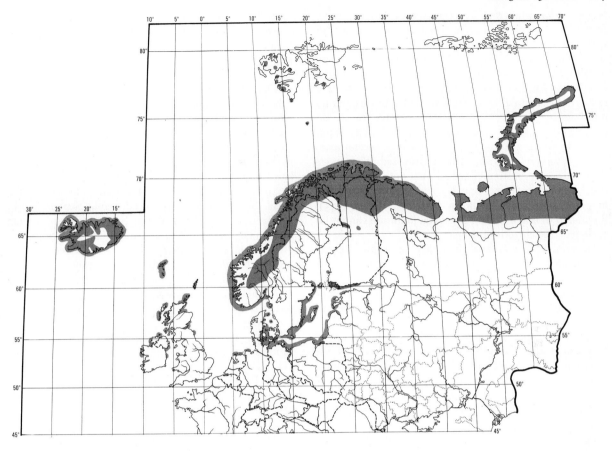

**Habitat.** Utilizes most ice-free habitats especially of high arctic, where often most abundant bird, tolerating proximity of pack-ice while awaiting summer thaw. Breeds on islands and skerries off mainland coasts, and on larger islands of Arctic Ocean, using low promontories, deltas, and fringing tundra as soon as ice melts. Up fjords, as in Greenland, also breeds on islets in salt water, but mainly flights a little way inland to dry spots in tundra vegetation, especially where mosaic of habitats formed by hummocks and ridges together with moist depressions and pools of standing water. Flowing water, and even lakes through which rivers flow, perhaps less favoured. Readily inhabits uplands, extending south in Scandinavia to arctic-alpine zone among willows *Salix* and dwarf birch *Betula*, but generally avoids wooded tundra and fringes of coniferous forest. Associated with other gregarious breeders, e.g. Arctic Tern *Sterna paradisaea*. Outstandingly capable as diver. Moves little on land, but makes maximum use of airspace, readily flying between coast and inland sites at up to 200–1200 m over land, although usually low over water. Winters predominantly off coasts and some distance out to sea, but also among skerries and to some extent up fjords and occasionally on fresh water. Compact flocking and restless mobility render species unusually susceptible to oil pollution. Also exposed locally to overshooting, and vulnerable to heavy losses through being trapped in submerged fishing-nets. See Salomonsen (1950*a*) and Kortright (1942).

**Distribution.** SWEDEN: occasionally nests further south (Curry-Lindahl *et al.* 1970). SCOTLAND: believed to have bred 3 occasions, Shetland, in 19th century; bred Orkney 1911 and probably 1912 and 1926 (British Ornithologists' Union 1971).

Accidental. Spain, Belgium, France, Czechoslovakia (perhaps regularly in winter), Hungary, Austria, Switzerland, Italy, Yugoslavia, Rumania, Greece, Azores.

**Population.** Breeding. USSR. Holds vast bulk of west Palearctic population; most numerous duck in far north. Estimates vary, e.g. 740 000 pairs in western part (Isakov 1970*a*); 1 038 000 pairs, of which 15 000 on Arctic islands; 500 000 European tundra and forest tundra; and, outside western Palearctic, 123 000 Taymyr and 400 000 Yamal-Gyda (Isakov 1970*b*); 5 million adults in western USSR (Uspenski 1970). ICELAND: decreased greatly since early 20th century (Gudmundsson 1951); total population perhaps 100 000 to 300 000 pairs (Gardarsson 1975).

NORWAY: less numerous in recent years (Haftorn 1971). SWEDEN: pronounced decline Swedish Lapland last 18 years but local and temporary increases on a few mountain lakes (Curry-Lindahl 1964). FINLAND: rough estimate 500 pairs (Merikallio 1958); decreased (Curry-Lindahl 1964).

Winter. Censuses attempted 1961–2 and 1962–3 winters in North Sea and Baltic (where many killed by oil), but no total figure possible, and earlier estimates based on numbers of passage birds varied greatly (Mathiasson 1970). In western Europe, highest count in recent years only 113000 and provisional total estimated at no more than 500000 suggesting either estimates of breeding and passage numbers at fault, or massive winter concentrations in places at present unknown (Atkinson-Willes 1975). Possibility of major decline due to oil pollution cannot be excluded. See also Movements.

Survival. Mean annual mortality of adults ringed Iceland 28%; life expectancy 3·1 years (Boyd 1962).

**Movements.** Migratory and partially migratory. Winters mainly offshore between 55°N and 75°N and in Baltic; no clear picture of movements from small-scale ringing to date. Icelandic breeders partial migrants, some remaining to winter around coasts, others moving WSW to southern Greenland. No evidence of Icelandic birds reaching continental Europe; yet ones ringed west Greenland moved south-east to Iceland and Denmark, and to Atlantic seaboard of North America, while others wintered in Greenland seas.

Baltic appears to be most important wintering area in west Palearctic, and densities up to 250000 per 100 km² reported offshore around Öland and Gotland (Mathiasson 1970). Moulting birds ringed Yamal peninsula, USSR, recovered there west to Denmark, and Baltic apparently used by most Russian and west Siberian breeders; heavy passage noted between White Sea and Gulf of Finland overland across Leningrad region and Karelia. Marked westerly movement may occur within Baltic in severe weather. Relatively small numbers, perhaps 1000–2000 (Isakov 1970c), winter in White and Barents Seas. Also overwinters along Norwegian coast, much more numerous in north, with over 60% north of Arctic Circle and over 18000 from Finnmark to Nordland in 1961–2 (Lund 1962). As main Norwegian and Baltic wintering areas well separated, may originate from different breeding areas, former perhaps mainly Fenno-Scandian breeders in view of absence of passage Gulf of Bothnia (Mathiasson 1970); but 2 winter recoveries of Yamal birds in west Norway (2 others Lapland summer perhaps abmigrants) indicate some from USSR use northern route round Scandinavia. Uncertain to what extent Spitsbergen, Bear Island, and Novaya Zemlya breeders winter north Norway, Barents and White Seas; most emigrate, though some resident when seas ice-free. Origins of British winterers unknown. On western seaboard, penetrates in small numbers to southern North Sea. Occurs almost annually on lakes of central Europe (Bauer and Glutz 1969); on Black Sea, where recovery of one ringed Swedish Lapland; also on Caspian (more rarely) and lakes of Kazakhstan, after migrating overland across USSR along Irtysh, Tobol, and Ural rivers (Dementiev and Gladkov 1952).

Though extensive moult migration defined in east Siberia (Anadyrland birds to Wrangel Island), and moulting area known Kurgan (53°30′N 66°E), 1000 km south of breeding range, in west Palearctic ♂♂ moult on coasts and lakes close to breeding areas either solitarily or in small flocks (Salomonsen 1968), movements beginning late June to early July. Large flocks build up August–September as ♀♀ and young desert breeding areas. Some reach south Sweden mid-September, but overland passage from White Sea to Gulf of Finland mostly 1st half October, and main influx to west Baltic November or December. Depart from Greenland and Iceland breeding areas as freeze September or October. Return movement North and Baltic Seas from mid-March, with major overland passage towards White Sea in May; return to breeding areas dependent on thaw, late April or early May in Iceland and west Greenland, mid-May to mid-June in east Greenland, Spitsbergen, and USSR tundras. See also Boyd (1959).

**Food.** Predominantly animal, especially crustaceans and molluscs. Obtained by surface-diving, often with jump as head thrown back and then forward (Millais 1913; Salomonsen 1950a). As submerges, tail fanned; wings usually partially opened under water (Oldham 1928). Depths normally 3–10 m (Hørring 1919; Phillips 1925–26), though many authors (see Cottam 1939; Schorger 1947, 1951) recorded gill-net casualties at greater depths, even down to 55 m. Submergence times 30–40 s (Alford 1920), 30–60 s (Phillips 1925), 22–61 s in 1·2–8·9 m (Ingram and Salmon 1941). Normally daytime feeder, though may be active at night (Millais 1913). Members of flock frequently dive successively in long line, or synchronize both submergences and surfacing (Millais 1913; Stewart 1967).

Both mobile and immobile food items taken. From salt and brackish water areas: molluscs include bivalves (especially blue mussel *Mytilus edulis*, cockles *Cardium*, clams *Spisula*, *Mya*, *Macoma*, and *Modiolaria*, *Margarita*, *Nucula*, *Scrobicularia*, *Astarte*, *Tellina*, *Tapes*, *Patella*), and gastropods (especially periwinkles *Littorina*, dogwhelk *Nassa reticulata*, and *Hydrobia*, *Lacuna*, *Rissoa*, *Buccinum*, *Neretina*); crustaceans include amphipods (*Gammarus*, *Crangon*), isopods (*Idotea*), mysids, small crabs (*Carcinus*); fish include gobies Gobiidae, sticklebacks Gasterosteidae, cod *Gadus morhua*, and flatfish Pleuronectidae; annelids (polychates); echinoderms; and plant materials (probably most accidentally), algae, eel-grass *Zostera*, wigeon-grass *Ruppia*, and (occasionally) grain from loading ships (Harrison and Jourdain 1919). From freshwater areas: insects and larvae including waterboatmen *Corixa*, caddisfly (Trichoptera), craneflies (Tipulidae), and midges (Chironomidae); crustaceans (*Gammarus*, waterfleas Cladocera); molluscs (Hydrobiidae); plant materials, including seeds and fruiting berries, tubers, roots, leaves, and moss. (Collet 1877; Cottam 1939; Hølboll 1843; Morris 1891; Naumann 1896–1905; Millais 1913; Witherby *et al.* 1939; Bird and Bird 1941; Dementiev and Gladkov 1952; Madsen 1954; Løvenskiold 1964.)

Winter and autumn foods. Denmark, November–April (Madsen 1954): 113 stomachs from saltwater areas contained chiefly molluscs (93·8% frequency) especially cockles (mainly *C. nodosum*, fewer *C. edule*), *Mytilus edulis*, fewer clams *Spisula* and *Mya*, *Littorina*, and *Nasa*; crustaceans (54·9%), particularly amphipods *Gammarus*, and isopods (*Idotea baltica*, *I. granulosa*, *I. viridis*); fish (14·4%) mainly Gobiidae; polychaetes (9·7%); and echinoderms (7·2%). Contents of 59 stomachs from brackish-water areas similar; mainly molluscs (94·9%), especially *M. edulis* and Baltic clam *Macoma baltica*, and fewer crustaceans (49·2%) chiefly *Gammarus* and *Idotea*. One stomach near Petsamo, USSR, contained 1709 molluscs, mainly *Cingula interrupta*, and fewer *Margarita helicina*, *Lacuna divaricata*, *Onoba oculeus*, and *M. edulis*; also a few crustaceans (Siivonen 1941). From Greenland coastal waters (Salomonsen 1950*a*), mainly crustaceans (including amphipods and schizopods) and molluscs (*Modiolaria faba*, *Mya truncata*, *Macoma calcarea*, *M. helicina*, *Saxicava arctica*). In USSR, on Pechora river during autumn passage, chiefly insects; at Rybinsk reservoir, 73% small fish (roach *Rutilus* 52%, perch *Perca* 21%), 20% insect larvae, (chiefly chironomids *Tendipes plumosus*, *T. reductus*, *Glyptotendipes gripekoveni*), and less caddisfly larvae and molluscs. In winter quarters, east Murman, stomachs contained amphipods; on Kamchatka, chiefly molluscs (Dementiev and Gladkov 1952).

Spring and summer foods. South-west Finnish archipelago: 31 stomachs contained chiefly *Macoma baltica* (*c.* 65% volume), less *M. edulis* (*c.* 21%), and crustaceans (*c.* 11%), mainly *Gammarus* (Bagge *et al.* 1970). In Greenland, mainly insect larvae (particularly gnats, chironomids, waterbeetles) and adult waterbeetle *Colymbetes dolabatus*, blackfly *Simulium*, gnat *Sciara*, and waterfleas, though just after arrival mostly plant materials (Manniche 1910; Salomonsen 1950*a*). In USSR: chiefly crustaceans and insect larvae on tundra; in early spring, Taymyr, stomachs contained chironomid larvae; on Yamal, larvae of caddisflies, and crustaceans (*Trops*, *branchipus*) which swarm in shallow waters; on Anadyr, mainly stonefly (Perlidae) and caddisfly larvae, small fish, roe, and beetles, also thawed dead fish from previous year's spawning (Dementiev and Gladkov 1952). 105 stomachs, Lake Mývatn, Iceland, in breeding season contained primarily chironomid larvae (importance of which increased towards end of season); fish eggs (chiefly of sticklebacks) relatively important in June and July; and waterfleas (Cladocera) in

August. Chironomid larvae main food 1958–70, though in terms of percentage weight large quantities of fish eggs, and, in 1969, more Cladocera taken; sticklebacks eaten only in 1970 (Bengtson 1971).

Ducklings ate gnat larvae and small crustaceans in Greenland (Salomonsen 1950*a*). 33 ducklings from Lake Mývatn, Iceland: those newly hatched to half-grown ate almost exclusively Cladocera; later mainly chironomid larvae, with some Cladocera and a few molluscs (Bengtson 1971).

Social pattern and behaviour. 1. Gregarious for much of year, except pairs in breeding area. In winter quarters, in small to large flocks on sea, adult ♂♂ predominating in some areas, adult ♀♀ and immatures in others. Migrant flocks in spring usually several hundred strong (Bergman and Donner 1964); after passage, wait on coast or within fjords until start of inland thaw then disperse, usually in small parties (see Schmidt 1966), over tundra pools and lakes during June; in northern Greenland, ♂♂ said to predominate at first (see Bent 1925) though not elsewhere (Dementiev and Gladkov 1952), birds in one area, Canada, invariably arriving in pairs (Alison 1975). ♂♂ and non-breeding ♀♀ first to flock, starting late June or July while breeding ♀♀ still sitting; gather on fresh water, soon leaving for sea. Largest flocks form in late summer and autumn after moult migration; otherwise, moults singly or in small flocks (150–200) near breeding area (see, e.g., Salomonsen 1965). Immatures mainly on sea throughout year, but sometimes present in breeding area in small numbers, flocking apart from breeders (see Alison 1975). BONDS. Strong, monogamous pair-bond of seasonal duration; possible that some pairs re-form in successive seasons (Alison 1975). Pair-formation starts November in wintering area (Schmidt 1966) and, though some birds still unpaired early May, mostly completed January–February before arrival in breeding area early June, taking *c.* 30 days (Alison 1975). During and after pair-formation, ♂ defends vicinity of ♀ wherever she may be. ♂♂ often in excess of ♀♀ on breeding grounds (Salomonsen 1950*a*; Drury 1961), but little evidence of ♂ promiscuity. ♂ usually deserts ♀ during first half of incubation period. Only ♀ tends young, usually until fledging but some ♀♀ leave after hatching and amalgamation of broods in care of 1 or 2 ♀♀ common (see Bent 1925; Alison 1975); up to 19 young reported in such groups by Dementiev and Gladkov (1952), 32 by Alison (1975). BREEDING DISPERSION. Mainly in pairs in nesting area. Freshwater pools occupied by single pairs but sometimes 2–3 pairs on larger lakes, ♂♂ and ♀♀ returning at times to waters and sites occupied in previous year (Alison 1975). Nests either well dispersed or in loose colonies on (e.g.) islands where often in association with terns *Sterna*. In case of both isolated and clustered nests, sites often traditional (Alison 1975). ♂ strongly territorial; mean of 5 territories 0·5 ha (Alison 1975). Same territories occupied each year. On loss of mate, ♂ abandons territory within 12 hours but ♀ stays in territory even though this soon occupied by new pair. ♀ may nest outside territory, frequently within territory of neighbouring ♂, but always near enough to hear own ♂ calling and always returns to mate's territory to feed. ROOSTING. Communal for much of year; mainly on water. Active by day and night (Millais 1913; Bauer and Glutz 1969) but also said to be mainly diurnal feeder (Witherby *et al.* 1939). Winter flocks off Nantucket Island, North America, started leaving feeding area for roost at *c.* 15·00 hrs, continuing until after dark and returning next morning (see Bent 1925).

2. See Millais (1913), Myres (1959*b*), Drury (1961), and

especially Alison (1975). Social interactions characterized by highly vocal behaviour of ♂. FLOCK BEHAVIOUR. Bill-dipping, in which bill lowered towards surface with or without touching it, important display of both sexes in flock (Alison 1975); also performed by lone birds but significance uncertain in either case. Flocks readily take wing, but Head-shaking only pre-flight signal recorded (McKinney 1965*b*); also part of ♂ courtship. During aerial evolutions, especially autumn and spring, tower in circles, then descend rapidly, some birds scattering and zigzagging (see Bent 1925). ANTAGONISTIC BEHAVIOUR. ♂♂ highly aggressive to one another at times in flock, especially when courting. Encounters involve much restless swimming and flying, persistent calling, rushes and chases over water, aerial and underwater pursuits, and pecking attempts. Fights, both on water and in air, relatively rare. Paired ♂ also highly aggressive to other ♂♂ approaching mate. ♂ defending territory in breeding area approaches intruders while performing Head-toss display with associated call (see below), then returns to mate; if pair intrudes, ♂ directs threat at ♀ of pair, often initiating 3-bird flight (see below). Repeated Chin-lifting, usually as head nodded rapidly through angle of 80° and accompanied by characteristic calls, main aggressive response of ♀; used both as threat and Inciting-display but full details lacking. ♀♀ also lunge at times towards both ♂♂ and other ♀♀, but particularly hostile to courting ♂♂ (see Alison 1975). COURTSHIP. Starts late October, soon after arrival in wintering area, but of low intensity and peak not reached until February (Alison 1975); intensity declines in March, but display continues until June in breeding area. Courting parties characterized by constant activity and frequent calling of swimming ♂♂ as 10–15 crowd round ♀; additional ♂♂ often attracted and fly in (Millais 1913). However, communal courtship less frequent than courtship of ♀ by single ♂ (Alison 1975), and much group display before February appears to be between ♂♂ themselves. Displays by individual ♂♂ at first directed towards any ♀ near but, towards end January, each ♂ starts persistently courting a particular ♀, though still met by hostility. When courtship reaches peak, and ♀'s aggression to preferred ♂ wanes, other ♂♂ may attempt to join in but driven away by both. After main peak over, paired ♂ directs displays to other nearby ♀♀ rather than mate, except before and after copulation. During courtship, ♂ adopts Courtship-intent posture by raising feathers of head (Johnsgard 1965); also frequently adopts Neck-stretch posture with head erect and tail cocked, occasionally with head inclined slightly backwards. Utters main Courtship-call repeatedly, either without accompanying movements or during Head-shake and Head-toss displays. Head-shake most frequent display (Alison 1975): rapid lateral movements which alternately flash black and white parts of head plumage; visible up to 150 m, especially in strong sunlight, and often followed by flight. Head-toss next most common display: (1) head lifted (first 2 notes of call given); (2) immediately thrown back (first 2 syllables of rest of call given), remaining briefly about half-way down back (see A); (3) head returned to starting point (final syllable given). Unique Rear-end display less frequent: ♂ (1) raises head, (2) suddenly swings it rapidly forward and down as tail erected vertically and feet kicked backwards, raising end of body out of water (see B); accompanied by own distinctive call. Other (silent) swimming displays include Wing-flap (♂ rises almost vertically in water, showing black breast band), Ritual-bathe, Upward-shake (occasional and mainly associated with Ritual-bathe), and brief Bridling (much as in Anatini). Turn-back-of-Head also occurs but appears rare in winter, though possibly overlooked; more commonly performed by paired ♂♂. During courting manoeuvres, ♂ also makes Short-flights (Drury

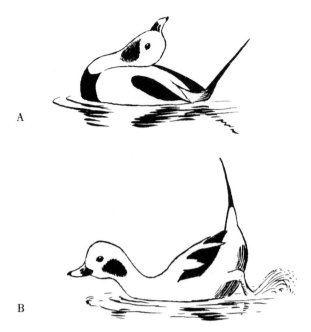

1961) or Parachute-displays (Alison 1975): (1) rises for 3–10 m, (2) drops vertically on stiff wings, (3) alights with heavy splash, always turning black breast band towards ♀; may be accompanied by repeated Courtship-calls of main type. Besides hostility and Chin-lifting (with calls), responses of ♀ to courting ♂ includes Hunched display with tail raised, back and scapular feathers ruffled, and head drawn tightly into shoulders; appears to be appeasing posture to preferred ♂. PURSUIT-FLIGHTS. In courtship-flights, 3–6 ♂♂ typically follow ♀ into air, group then zigzagging low over water, both sexes repeatedly calling—♂♂ while performing modified Head-toss displays (Phillips 1925; Alison 1975). Courting pursuits of ♀ by single ♂ also common for 2 or more days during main period of pair-formation, ♂ chasing ♀ over and under water and in air; induces ♀ to fly by assuming Neck-stretch posture and Head-shaking repeatedly. If other ♂♂ join in (see above), ♀ eludes all but original ♂ who continues courting her on water afterwards. Territorial 3-bird flights also common in nesting area, established ♂ pursuing ♀ of intruding pair whose mate then defends her in air, chase ending soon after ♀ leaves territory. Intruding ♂♂ may also threaten and attack ♀ of territorial pair, making her fly; in ensuing chase, ♀'s mate follows others. PAIR BEHAVIOUR. During peak of water courtship and pairing chases, ♀ no longer tries to evade ♂ and responds with Chin-lifting when he displays, latter ceasing his pursuits. When pair-formation completed, ♂ displays less frequently with changes in relative frequencies of certain displays (see Alison 1975). Pair spend much of time close together and engage in displays associated with copulation, displaying little otherwise. When separated from ♀, ♂ gives repeated long-range contact-calls while swimming persistently in Neck-stretch posture with tail and rump submerged; ♀ behaves similarly (Alison 1975), but full details lacking. Calls of ♂ show considerable individual variation, facilitating identification. COPULATION. Seen infrequently, mainly in wintering area, usually after pair moves away from main flock (Alison 1975). Lacks aggressive features found in Harlequin *Histrionicus histrionicus*. ♂ performs 1–10 pre-copulatory displays at any one time; include Head-toss, Head-shake, Ritual-bathe, Bridling, and Neck-stretch. Responses of ♀ include Chin-lifting, Hunch-display, Ritual-bathe, Neck-stretch, and, eventually, Prone-posture—♀ faces courting ♂ and swims thus towards him, turning when within 1–2 m and passing, repeating performance when ♂ turns towards her again. No Wing-shake by ♂ during copulation and no pair-rotation after. ♂ performs 1–4 post-copulatory displays; include Head-toss, Neck-stretch, Ritual-bathe—Head-shake—Wing-flap sequence, and Turn-back-of-Head. ♀ Ritual-bathes. RELATIONS WITHIN FAMILY GROUP. No detailed information.

(Figs A–B after Johnsgard 1965.)

**Voice.** Outstanding among Anatinae for frequency, richness, and melodious resonance of ♂'s calls. These loud, yodelling, and far-carrying with harmonic content similar to those of some swans and geese (Anserini). See Johnsgard (1971) for further analysis. Given in flock as well as by solitary birds, both on water and in flight throughout much of year, by night as well as day. From displaying flock on water, produce effect like distant sound of bagpipes (Witherby *et al.* 1939) or baying of hounds (see Bent 1925). Difficulty in describing them verbally and marked individual variation explain plethora of renderings. Noisy, garrulous, and sonorous qualities reflected in names such as Old Squaw, Old Injun, Old Wife, Noisy Duck, Hound Duck, and Organ Duck. Likely that 2 main calls of ♂ confused and that individual or any regional variants swamped by numerous alternative renderings. ♀ often as vocal as ♂ but calls less musical and penetrating and much less well known. In flight, especially during spring evolutions, flocks produce far-carrying, wing noise (see Bent 1925).

CALLS OF MALE. 2 similar Courtship-calls, differing mainly in slight cadence characteristics, distinguished by Johnsgard (1965, 1971). (1) Main Courtship-call: preliminary 'ugh ugh' (often inaudible) followed by highly characteristic 'ah-oo-GAH' (see I); also rendered 'ahr-ahr-ahroulit' or 'urk-urk- . . . urk' by Alison (1975) and main (terminal) component as 'ah-har-lik' by Myres (1959*b*)—but this, like other calls of ♂, subject to great individual variation, as shown by analysis of sonagrams (Alison 1975). Given on own and as component of Head-shake and Head-toss displays; also during courtship-flights and some Parachute-displays (Short-flights). Most frequent ♂ vocalization and little doubt that most descriptions in literature refer to it, all or part of phase variously being described as shrill, loud, resonant, and bugle-like (see, e.g., Bent 1925; Witherby *et al.* 1939;

Söderberg 1950; Murie 1963). (2) Other Courtship-call: 'a-oo a-oo A-ooo-gah'; given only during Rear-end display. (3) Close-range contact-call (Conversation-call): soft 'gut' or 'gut-gut', similar to equivalent call of ♀ but lower pitched and subject to much individual variation in duration and pitch; given especially during feeding (Alison 1975). (4) Long-range contact-call: 'urk-ow-ow' or 'ow-ow', uttered repeatedly by swimming ♂ in special posture when separated from mate (Alison 1975). (5) Alarm-calls: monosyllabic 'urk' (Alison 1975)—tensely uttered 'og' or 'ak' (Voigt 1950), probably same; low, growling 'utt-utt-utt', merging into call 1 as bird rises (Salomonsen 1950*a*).

CALLS OF FEMALE. Mainly low and quacking in character (see Bannerman 1958). (1) Threat and Inciting-call, accompanying Chin-lifting display: 'gut-gut-GOO ah-GOO ah', 'urk', or 'urk urk ang ang goo' (Drury 1961); also 'rurk urk urk urk ong ong goo' (Alison 1975) or 'ar-ar-arc-èng-èng' (Lippens and Wille 1972). (2) Call during courtship-flights: 'kak-kak-kak-kak' (see Alison 1975). (3) Close-range contact-call (Conversation-call): see under ♂. (4) Alarm-call: monosyllabic 'urk' (Alison 1975); compared to low or distant bark of dog (Nicholson 1930). (5) Maternal-calls: bird in charge of brood uttered subdued 'whob', both on water and in flight (Salomonsen 1950); series of grating notes given when Arctic Skua *Stercorarius parasiticus* attacked young (see Bent 1925). (6) Other calls, of uncertain significance, include soft quack (Delacour 1959) and frequent, soft 'ved', 'vahd', or 'vood' (see Witherby *et al.* 1939).

CALLS OF YOUNG. No detailed information.

**Breeding.** SEASON. See diagram for Lapland and Arctic USSR. Iceland: mean date of first laying, 1961–70, varied from 20–30 May, average 24 May (Bengtson 1972). Spitsbergen: first eggs mid-June; first young from mid-July (Løvenskiold 1964). Start of laying dependent on timing of thaw. SITE. On ground, in thick vegetation, scrub, sometimes in open, and occasionally in rock crevice. Rarely far from water. Nest: small depression, lined small amounts vegetation and down. Average diameter 20 cm, depth 8 cm. Building: by ♀. EGGS. Ovate; olive-buff. 54 × 38 mm (47–58 × 35–41), sample 200 (Schönwetter 1967). Weight 39 g (38–40), sample not given (Dementiev and Gladkov 1952). Clutch: 6–9 (5–11). Up to 17 recorded but by 2 ♀♀; mean of 10 Icelandic dump nests 15·1 (Bengtson 1972). Mean size 150 1st clutches, Iceland, 7·9 (Bengtson 1972). Mean clutch-size, Churchill, Manitoba 6·8; joint layings rare (Alison 1975). One brood. Replacement eggs laid after egg loss, sometimes twice; 37% of 27 Icelandic failed breeders re-nested. Mean of 20 2nd clutches, Iceland, 6·0 (Bengtson 1972). No replacements laid, Manitoba, 1968–71 (Alison 1975). One egg laid per day. INCUBATION. 24–29 days, mean 26; sample 106 eggs (Alison 1975). By ♀. Starts with last egg; hatching synchronous. Eggs covered with down when ♀ off; usually 2 recesses per day, sometimes prolonged (Alison 1975).

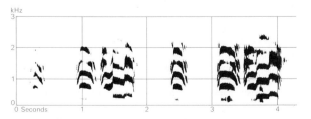

I L Koch/BBC Iceland June 1953

Eggshells left in nest. YOUNG. Precocial and nidifugous. Self-feeding, but in early stages will feed on material dislodged to surface by diving of ♀ (Alison 1975). Cared for by ♀; brooded at night when small. Broods sometimes amalgamate. FLEDGING TO MATURITY. Fledging period 35–40 days (Alison 1975). Become independent at or soon after fledging. Age of first breeding 2 years. BREEDING SUCCESS. Hatching success Iceland, 1961–70, mean 64·6% (Bengtson 1972). In Manitoba, 19% eggs non-viable; hatching success, 1968–71, mean 58·9% with predation responsible for 26·4% of losses (Alison 1975).

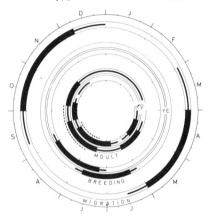

**Plumages.** Particularly complex and controversial (see Salomonsen 1941, 1949; Stresemann 1948), and difficult to fit into most terminologies of plumages and moults. Salomonsen (1941) has been mainly followed in recognizing 4 plumages in ♂: (1) transitional autumn plumage (worn about September–November), (2) winter plumage (about November to early April), (3) transitional summer plumage (about May to June), (4) full non-breeding plumage (about July to early September). Plumages 1 and 2 (differing only slightly) both 'bright' and together equivalent to breeding plumage of most other Anatinae, being similarly worn during period of pair-formation and most courtship display; mainly replaced, however, just before breeding season, by special summer or 'semi-eclipse' plumage. Incorrect to call latter 'nuptial' or 'breeding' plumage as these terms widely used for ornamental or bright (winter) plumage of Anatinae generally (Stresemann 1948); largely cryptic, and transitional between winter plumage and full eclipse assumed later, correlated with need for camouflage due to changes in habitat. Plumages 3 and 4 together equivalent to non-breeding plumages of most other Anatinae. Significance of brief autumn plumage uncertain. Convenient also to recognize 4 plumages in ♀, using much the same terminology. ADULT MALE AUTUMN. As winter, but head and neck white (slight orange tinge on crown) with grey-brown mottled patch at lower side of head and upper neck. ADULT MALE WINTER. Crown, hindneck, upper mantle, small spot round eye, chin, throat, chest, and often narrow line at base of upper mandible white, in some slightly suffused pale grey. Forehead cream-grey, sides of head mouse-grey. Sides of neck black-brown, grading to chestnut on lower sides of neck. Scapulars white, tinged variably pale blue-grey on centres. Lower mantle, and centre of back, rump, and upper tail-coverts black. Breast black-brown, sharply contrasting with white chest and rest of underparts. Sides of body and flanks pale blue-grey, feathers tinged white on tips. Sides of back, rump, and

upper tail-coverts white. Outer tail-feathers white, central pair black, those next central pair black with white sides. Primaries and primary coverts sepia. Secondaries and greater upper wing-coverts dark brown; outer web of outer secondaries tinged olive, of inner secondaries warm brown, of greater coverts sooty. Outer web of tertials warm brown or dark chestnut, inner web black. Median and lesser upper wing-coverts black. Underwing grey, axillaries and marginal and lesser coverts buff-brown. ADULT MALE SUMMER. Feathers of upper mantle and scapulars with black triangular centres and broad red-brown to yellow-buff sides. Sides of forehead, lores, and front part of cheeks mouse-grey; spot round and behind eye white. Rest of head and neck, chest, and upper mantle brown-black. From about mid-April to mid-May head with somewhat pied appearance through arrested moult: some white on hindcrown, chin, and base of upper mandible; also some on upper mantle or chest. Rest as winter. ADULT MALE FULL NON-BREEDING (Eclipse). Head, neck, upper mantle, and chest as in summer plumage, but somewhat faded. Scapulars short with rounded tips, dark sepia with buff edges. Sides of body and flanks grey-brown. Rest of plumage on body, wing, and tail still as in winter plumage; combines worn and faded feathers with fresh ones when in full moult. ADULT FEMALE AUTUMN. (About November–February.) Body, wing, and tail like adult ♀ winter, head and neck like adult ♀ non-breeding. ADULT FEMALE WINTER. (About December–April.) Centre of forehead, crown, nape, hindcheek and sides of upper neck dull black with faint olive tinge on nape and cinnamon tinge on sides of neck. Lores, sides of head, and forecheek cinnamon-buff with narrow white line along side of upper mandible and small white circle round eye, projecting in point to nape. Chin, throat, and centre of foreneck grey-brown. Lower neck white; where merging into buff of mantle and chest grey-brown. Feathers of mantle and scapulars broadly edged warm cinnamon (rarely white), in some with off-white tips or sides; centres of feathers of mantle sepia (especially much sepia on lower mantle), scapulars with distinct black triangular mark on centre. Back and centre of rump and upper tail-coverts black, feathers broadly edged olive-buff (except sometimes on back and rump). Chest and sides of breast pale cinnamon, rest of underparts and sides of rump and upper tail-coverts white; some grey-brown of feather-bases occasionally visible on chest and breast. Tail sepia (darkest on central feathers), sides margined white; outer feathers mainly white. Wing like adult ♂, but secondaries and tertials olive-cinnamon rather than warm brown or dark chestnut, some outer secondaries with grey outer webs sometimes; lesser and median coverts sepia, broadly edged cinnamon (like mantle); greater sepia tipped buff-white. ADULT FEMALE SUMMER. (About May–August; nests in this plumage.) Head and neck sooty black (in some with olive tinge), but sides of head in front of eye dark olive-grey, indistinctly defined from crown and hindcheek; spot round eye, streak behind eye down to nape, and lower sides of neck white; chin and throat mottled pale and dark grey. Old winter feathers of upper wing and upperparts abraded and faded on tips, showing grey or buff edges, but mantle, scapulars, and sometimes back and rump with variable number of new black feathers with narrower cinnamon or olive-grey edges than in winter plumage. Underparts like winter, but a number of feathers of chest or flanks sometimes new, dark grey-brown, fringed off white. ADULT FEMALE NON-BREEDING. (About August–November.) Head and neck white, with crown and large patch at lower cheek and upperside of neck dull black; sometimes forehead also black; indistinct band over chin and throat pale grey-brown. Scapulars and centre of mantle olive-brown, feathers tinged cinnamon at sides and with somewhat darker dusky centres; some scapulars tipped grey. Lower mantle, back, rump, and upper tail-coverts black, feathers variably fringed olive-brown. Chest and flanks grey-brown, feathers faintly tipped olive-brown on chest and off-white on flanks; rest of underparts white. DOWNY YOUNG. Sides of head in front of eye, crown, hindneck, chest, and upperparts dark grey-brown; long tufts of down olive-buff. Small off-white spots in lores, below eye, and sometimes above eye; often small white streak from eye to nape. Chin, throat, hindcheeks, and underparts pale grey, darker grey on sides of body. JUVENILE. Forehead, crown, nape, and patch on hindneck and upper side of neck dark grey-brown or dull black, rest of head pale grey, slightly darker on hindneck, chin, and throat; pale feather bases partly visible. Indistinct streak from eye to nape white. Mantle, back, rump, and upper tail-coverts dark sepia or sooty-black, feathers faintly edged olive-grey or buff on upper mantle; scapulars olive-brown with somewhat paler olive edges and dark sepia shaft-streaks. Chest and sides of breast dark grey-brown with faint buff feather edges, upper breast and flanks pale brown-grey; rest of underparts off-white. Tail-feathers sepia with narrow off-white margin; outer ones paler, grey; bare shaft projecting beyond abraded square tips. Wing like adult ♂, but duller, lesser and median coverts like upper mantle; outer webs of secondaries and tertials sepia without distinct warm brown tinge, tinged slightly cinnamon in juvenile ♂, more olive in ♀, but difference slight. Tertials short and rather narrow, tips rounded instead of pointed. FIRST IMMATURE NON-BREEDING, AUTUMN, AND WINTER MALE. Strongly variable. In 1st autumn, new feathers on head and neck tend to resemble those of adult ♂ autumn plumage while new scapulars white with grey centres (sometimes tinged chestnut). Later in autumn, some attain variable amount of plumage resembling adult ♂ winter; include central tail-feathers on occasion. Juvenile wing, outer tail-feathers, and underparts (without dark breast) retained. FIRST IMMATURE SUMMER MALE. Tends to resemble adult ♂ summer, but abraded wing and outer tail-feathers juvenile, and underparts (even when new) white, without dark breast; usually also variable number of juvenile 1st immature non-breeding, or 1st immature winter feathers retained on upperparts. SECOND IMMATURE NON-BREEDING MALE. Like full non-breeding adult ♂, but wing and outer tail (when not yet moulted) juvenile; variable number of feathers of preceding plumages retained. SECOND IMMATURE AUTUMN AND WINTER MALE. Like adult ♂ autumn and winter, but tail on average slightly shorter; occasionally some non-breeding or feathers of 1st immature spring plumage retained. FIRST IMMATURE NON-BREEDING, AUTUMN, AND WINTER FEMALE. Tends to resemble adult ♀ autumn, but wing, tail, back, rump, underparts, and variable number of feathers elsewhere juvenile. Head and neck like adult ♀ autumn, but plumage colours less distinctly defined; new scapulars and mantle feathers edged grey (only rarely warm cinnamon); new feathers of flanks, chest, and sides of breast white or pale grey, on chest and sides with sepia centres. Some advanced ♀♀ occasionally attain head pattern like adult ♀ winter. FIRST IMMATURE SUMMER FEMALE. Head, neck, and upperparts like adult ♀ summer, but variable amount of preceding plumages retained; new scapulars and mantle feathers dark olive-brown with pale grey edges. Wing and tail worn; juvenile. SECOND IMMATURE NON-BREEDING FEMALE. Part of preceding plumage retained; wing and tail juvenile until moult. SECOND IMMATURE AUTUMN AND WINTER FEMALE. Like adult ♀ autumn and winter, but scapulars usually broadly edged grey instead of cinnamon; sometimes part of immature non-breeding plumage retained. SUBSEQUENT PLUMAGES. Like adult. (Salomonsen 1941; Schiøler 1926; C S Roselaar.)

**Bare parts.** ADULT MALE. Iris yellow-brown to orange-red. Base of bill to nostril black, nail blue-black, broad band round mandibles behind nail yellow-orange to rose-red. Paired ♂♂, July, with bills entirely black (Alison 1975). Foot blue-grey, webs dull black. ADULT FEMALE. Iris light brown to yellow-brown. Bill slate-grey; in unpaired birds, June–July, green-blue (Alison 1975). Foot olive or blue-grey, webs dark grey. DOWNY YOUNG. Iris brown. Bill blue-black. Foot olive-slate. JUVENILE. Iris dark brown to brown. Bill blue-grey; in ♂♂, pink band develops from October. Foot pale blue-grey, webs dark grey. (Bauer and Glutz 1969; Kortright 1942; Salomonsen 1941; Schiøler 1926; Witherby *et al.* 1939; RMNH; ZMA.)

**Moults.** Highly peculiar and more complex than in other Anatinae (Salomonsen 1941); as with plumages, difficult to describe using accepted terminologies. Unique in that 4 ornamental scapulars, some smaller scapulars, and sides of head and neck moulted 3 times a year. Plumage anterior to breast line (head, neck, upper mantle, chest, and some smaller scapulars) and flanks moulted twice a year; plumage posterior to breast line (rest of back, rump, vent, belly, tail-coverts, wing, and tail) once a year (see also Stresemann 1948). As in other Anatinae, really 2 moults but both interrupted, producing series of 4 partial moults. In ♂, 1st two stages of these moults together equivalent to single post-breeding moult of most other Anatinae. The 1st stage, advanced to period before breeding, produces summer, semi-eclipse plumage ('spring' moult); 2nd stage follows *c.* 3 months later, producing full non-breeding, eclipse plumage ('summer' moult). Remaining stages together equivalent to single pre-breeding moult of most other Anatinae: in 1st stage ('autumn' moult), autumn plumage assumed; in 2nd ('winter' moult), winter plumage assumed. Moults of adult ♀ mainly like adult ♂'s; timing and amount of plumage involved variable, with marked difference in start of summer moult between non-breeders and late breeders. ADULT MALE SPRING MOULT. April–May, mainly late April to early May. Involves scapulars, head, neck, upper mantle, and chest. On head, black feathers on crown appear first; rest of head, neck, and forepart of body follow soon, but white on hindcrown, chin, upper throat and in some also on lores and upper mantle retained until late May (occasionally a few white feathers on hindcrown until July). ADULT MALE SUMMER MOULT. Late June to early September. Includes wing, tail, and those body feathers not replaced during spring moult; also scapulars for 2nd time. Flanks first, late June to mid-July, then longer scapulars lost (longest full-grown again by late August); wing, wing-coverts, axillaries, and tail simultaneously shed late July, and moult intensive on body in August, with simultaneous growth of flight-feathers and tail. Completed late August or early September, but central tail-feathers often not before October. ADULT MALE AUTUMN MOULT. 2–4 weeks after end of summer moult, flanks, scapulars, head, neck, upper mantle, and chest moulted mid-September to early October. Flanks and scapulars first, completed late September (elongated late October); then head, neck, chest, and upper mantle; throat last. ADULT MALE WINTER MOULT. Involves only sides of head and neck (for 2nd time), late October to mid-November. ADULT FEMALE SPRING MOULT. Late April to early June. Like adult ♂, but moult of scapulars, mantle, and chest not always complete; sometimes also parts of upper breast, lower mantle, back, rump, or lower breast to vent not moulted. ADULT FEMALE SUMMER MOULT. Variable; between July and September. Wing and tail early August to early October. Scapulars and flanks, and those feathers not replaced during spring moult: mantle, chest, and breast, apparently back and rump, and occasionally lower breast. ADULT FEMALE

AUTUMN MOULT. Lasts *c.* 1 month between mid-September and early December. Involves head, neck (throat last), scapulars, flanks, chest, and probably mantle, back, rump, and breast to vent. ADULT FEMALE WINTER MOULT. Between December and February. Head and neck only. (Possibly not in all birds, as some seem to retain autumn plumage on head until moult to summer plumage.) POST-JUVENILE AND IMMATURE AUTUMN AND WINTER MOULTS. Partial; September–December. Amount of plumage involved markedly variable; tends to follow equivalent moults of adult. Involves flanks, head, and neck, and part of chest, upper mantle, and scapulars. Retarded specimens retain almost complete juvenile plumage until spring; advanced attain winter plumage, except for breast, belly, vent, and outer tail-feathers. IMMATURE SPRING MOULT. Mid-April to late May in ♂, April to early May in ♀. Involves same plumage as adults, but usually less complete; part of immature winter plumage and juvenile wing, underparts, and outer or all tail retained. IMMATURE SUMMER MOULT. From mid-July onwards; same parts of plumage involved as in adult. Last juvenile feathers (including tail and wing) lost August. Often some immature winter plumage retained; Moult less complete than in adult. SUBSEQUENT IMMATURE MOULTS. Like adult. (Salomonsen 1941; C S Roselaar.)

**Measurements.** Netherlands, winter; Greenland and Scandinavia, summer. Skins (RMNH, ZMA).

| WING | AD | ♂ 228 | (4·60; 45) | 218–241 | ♀ 212 | (4·35; 20) | 204–220 |
|------|-----|-------|-----------|---------|-------|-----------|---------|
| | JUV | 214 | (6·71; 14) | 205–227 | 202 | (5·07; 16) | 192–211 |
| TAIL | AD | 215 | (17·7; 43) | 188–254 | 70·4 (3·74; 17) | | 64–78 |
| | JUV | 66·1 (3·57; 14) | | 61–74 | 56·8 (3·83; 17) | | 51–64 |
| BILL | AD | 27·2 (1·03; 56) | | 24–30 | 25·8 (0·99; 20) | | 24–27 |
| TARSUS | | 35·8 (1·23; 70) | | 34–38 | 34·1 (1·18; 36) | | 32–37 |
| TOE | | 54·8 (2·06; 70) | | 51–60 | 51·8 (1·41; 36) | | 49–56 |

Sex differences significant. Wing and tail of juvenile significantly shorter than adult. Bill of juvenile ♀ like adult ♀, but juvenile ♂ somewhat larger than adult ♂: 28·4 (1·78; 14) 26–32. Juvenile tarsus and toe similar to adult; combined. Wing in 2nd year intermediate between adult and juvenile; tail 2nd year ♂ 183 (26·3; 14) 140–236.

**Weights.** ADULT AND SECOND YEAR. (1) November–March, Denmark (Schiøler 1926) and some Netherlands (ZMA). (2) April–May, south Baltic (Bauer and Glutz 1969). (3) Late May and June, arctic Alaska (Conover 1926; Irving 1960) and east Greenland (J de Korte). (4) July, Yamal Peninsula, USSR (Dementiev and Gladkov 1952). (5) August–September, arctic Canada (Bauer and Glutz 1969) and east Greenland (J de Korte).

| (1) | ♂ 748 | (10) | 680–910 | ♀ 705 | (20) | 575–792 |
|-----|-------|------|---------|-------|------|---------|
| (2) | 788 | (8) | 616–955 | 735 | (2) | 590–879 |
| (3) | 788 | (20) | 621–880 | 657 | (5) | 510–794 |
| (4) | 781 | (–) | 666–800 | 624 | (–) | 516–730 |
| (5) | 883 | (1) | – | 702 | (8) | 650–800 |

♀♀ with young in August, Yamal, 550–600; without young at start of wing-moult, 700–800 (Dementiev and Gladkov 1952). JUVENILE. Rybinsk Reservoir, USSR: October, ♂ 616 (500–660), ♀ 603 (550–650); November, ♂ 670, ♀ 805 (Dementiev and Gladkov 1952).

**Structure.** Wing rather long and narrow, pointed. 11 primaries: p10 and p9 about equal, longest; p8 5–13 shorter, p7 18–28, p6 28–45, p1 80–110. Inner web of p10 and outer of p9 slightly emarginated. 14 tail-feathers with pointed tips; outer 5 short, inner elongated, especially central pair; pair next innermost *c.* 80–110 shorter. Bill short, rather narrow, about as high as broad at base; gradually depressed and narrowed to large round nail. Cutting edges of upper mandible curved upwards near tip. Lower scapulars elongated and pointed in ♂ breeding plumage. Feathers of hindcrown and hindneck somewhat elongated and dense in ♂ winter. Outer toe *c.* 102% middle, inner *c.* 82%, hind *c.* 32%. CSR

## *Melanitta nigra* Common Scoter

PLATES 87 and 100
[after pages 662 and facing page 687]

DU. Zwarte Zeeëend FR. Macreuse noire GE. Trauerente
RU. Синьга SP. Negrón común SW. Sjöorre N.AM. Black Scoter

*Anas nigra* Linnaeus, 1758

Polytypic. Nominate *nigra* (Linnaeus, 1758), north Europe and north Asia east to Olenek River; *americana* (Swainson, 1832), north Asia east from Yana River, and North America, accidental.

**Field characters.** 44–54 cm, of which body two-thirds; wing-span 79–90 cm. Medium-sized, compact sea-duck with squat shape, deep bill, thick neck, and rather long, pointed tail; ♂ all black apart from coloured bill. Sexes dissimilar, but no obvious seasonal differences; juveniles and 1st winter ♀ resemble adult ♀, but 1st year ♂ variably distinctive.

ADULT MALE. Entirely glossy black, except for undersides of flight-feathers and duller belly. Bill, swollen by knob at base of upper mandible, also mainly black, but relieved by conspicuous patch of yellow or orange in front of and around nostrils and extending back on to knob (♂ of American race *M. n. americana* has flatter knob, and it and whole base of bill yellow); eyes brown; legs brown-black.

In flight, underwing pattern noticeably 2-toned through grey inner webs of flight-feathers contrasting with black coverts. ADULT MALE NON-BREEDING. Hardly distinguishable, but head and neck duller and underparts browner. ADULT FEMALE. Dark brown, contrasting with well-defined and conspicuous area of brownish-white on cheeks, sides of upper neck, and throat; lower breast and belly paler, often whitish, and flanks and undertail more barred. Bill greenish-black, sometimes with narrow strip of orange-yellow between nostrils and down almost to nail. In flight, underwing pattern shows similar contrast to ♂, flight-feathers being distinctly greyish and coverts dark brown. JUVENILES AND FIRST-WINTER FEMALE. Closely resemble adult ♀, but paler brown above and whiter below, with small patch of orange-yellow around nostrils in ♂. FIRST-WINTER MALE. From December onwards, variably mottled with black on head, upperparts, upper breast, flanks, and undertail; adult plumage not assumed until 2nd autumn.

All-black plumage (except for undersides of flight-feathers) and markedly pointed tail distinguish ♂ from all other west Palearctic waterfowl at reasonable ranges: ♂ Velvet Scoter *M. fusca* and ♂ Surf Scoter *M. perspicillata* (which see for details) have white patches on wings, head, or both; coots *Fulica* smaller with stubby tails and white frontal shields and bills. In ♀ and immature, combination of brown plumage and conspicuous whitish cheeks precludes confusion with any other ducks except ♀ Red-crested Pochard *Netta rufina*, which has greyish cheeks more sharply demarcated, paler brown upperparts, orange-brown tip to bill, and white in wing, and is seldom seen on sea; and immature *M. perspicillata*, which has profile recalling Eider *Somateria mollissima*, variable line across cheeks below eye, and darker foreneck. At long range and in flight, however, distinction of all plumages from *M. perspicillata* often difficult, head shape and underwing pattern then being most important characters (contrasting flight-feathers of *M. nigra* show as silvery-grey in sunlight at distances of well over 1 km).

Outside breeding season, except temporarily on migration or when oiled, almost exclusively marine, but avoids really rough water as far as possible and prefers open sea just off coast, or broad estuaries, to broken surfaces among rocks and islands. Despite heavy build, swims buoyantly with pointed tail raised, unless alarmed when partly sinks body and watches with neck erect. Generally rather wary, more so than *M. fusca*. Except when nesting, rarely seen on land, where gait upright and awkward. Rises from water with less difficulty than other scoters. Rapid wing-beats accompanied by slight whistling sound. Usually flies in long wavering lines or in irregular groups, generally low over surface, but sometimes (more often than eiders *Somateria*) at several hundred metres. Flocks sometimes shoot down on closed wings, with loud rushing sound, from considerable height. Gregarious outside breeding season, often gathering in tight or loose flocks of several hundreds or even thousands. Some flocks in July consist largely of ♂♂ gathered to moult. Noisier than other *Melanitta*: ♂ has plaintive polysyllabic piping and other sounds which can be taken up by several birds of flock; ♀ has hoarse growl like other diving ducks.

**Habitat.** Within low-arctic and boreal limits, tolerates fairly wide range of habitat conditions. Occupies nest-sites well inland, and even some way from fresh water, amid tundra vegetation or dwarf heath, preferably sheltered by tall herbaceous plants or shrubs. Although occupying upland slopes and even arctic-alpine terrain, avoids steep rocky situations and those shut in by trees. Sometimes accepts wetter and more open sites than most arctic species, but shows preference for islets and low promontories. Also uses open banks of slow-flowing rivers, preferring wide pair dispersal and ample opportunities for flight from danger. During flightless stage of moult, however, flocks closely on open water, often at distance from breeding area, either on large lakes or at sea.

Outside breeding period, predominantly marine, resting and feeding in flocks in shallow, inshore waters, generally 500 m to *c*. 2 km from land, where depth not more than 10–20 m and animal food abundantly accessible. In such conditions, accepts exposure to strong wave action or rapid currents but more rugged coastlines rarely fulfil its requirements. Uses airspace often in long formations, and normally at low altitude, but flies fairly high overland; generally flies only when need pressing.

**Distribution.** BRITAIN AND IRELAND. First recorded breeding Scotland 1855, perhaps overlooked earlier (Bannerman 1958). First bred Ireland 1905, in Fermanagh, spread to Mayo 1948 (Ruttledge 1966). Breeds irregularly outside areas shown (British Ornithologists' Union 1971). FINLAND: southern breeding limit retreated somewhat to north (LvH). FAEROES: has bred Sandoy (Williamson 1970). SPITSBERGEN: bred 1905, 1963, 1965, and possibly other years (Løvenskiold 1964; M A Ogilvie).

Accidental. Hungary, Austria, Switzerland, Italy, Yugoslavia, Greece, Rumania, Bulgaria, Turkey, Kuwait, Cyprus, Malta, Libya, Algeria (no recent records), Azores, Madeira, Canary Islands.

**Population.** Breeding. ICELAND: estimated *c*. 500 pairs, 1975 (A Gardarsson). BRITAIN. After increase north Scotland, probably little recent change (Parslow 1967); *c*. 30 pairs, 1974 (Sharrock *et al*. 1975). IRELAND. Continued gradual increase (Parslow 1967). Fermanagh: over 50 pairs (Ruttledge 1966); 137–167 pairs 1967, then some decline to *c*. 127 pairs 1970, 104 pairs 1971, 78 pairs 1972, 122 pairs 1973, 110 pairs 1974 (J Cadbury). Mayo: 20–30 pairs (Ruttledge 1966). NORWAY: declined some areas, especially in north (Haftorn 1971). SWEDEN: population apparently rather stable (Curry-Lindahl *et al*. 1970). FINLAND. Declining, *c*. 500 pairs (Merikallio 1958). Sparsely distributed and rare many areas; evident decrease last few decades (Grenquist 1970). Markedly affected by oil pollution in Baltic and North Sea (Joensen 1973).

Winter. Highest January count (1967–73) 200 000 birds, but total west Palearctic population in winter probably 400 000–500 000 (Atkinson-Willes 1975), much lower than 1½ million estimated passing Finland in spring or 1 million on West German coast in summer 1952 (see Movements).

Survival. Birds ringed Iceland: mean annual mortality of adults 23%; life expectancy 3·8 years (Boyd 1962). Oldest ringed bird 15 years 11 months (Rydzewski 1973).

**Movements.** Migratory, though some small populations may not move far. Icelandic breeders winter around Ireland, Britain, and Atlantic coasts France, Spain, and

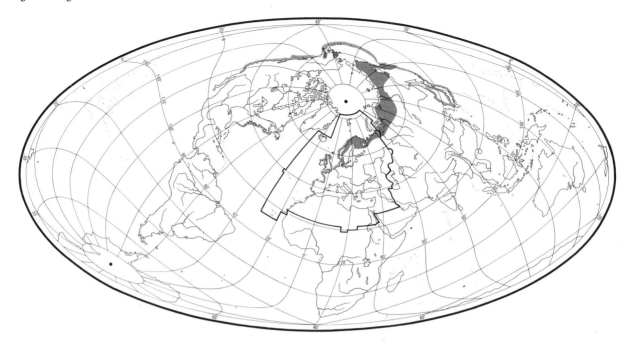

Portugal; one recovery from Azores; uncertain if any overwinter regularly in home waters. Movement by small British and Irish population largely unknown, but breeding lakes deserted. Birds wintering Norwegian coast (north to Lofoten Islands) presumably mainly local breeders. Breeders from arctic USSR, east to about River Lena, migrate west to WSW along Arctic Ocean to White Sea and overland across north Russia; majority crossing Karelia and Leningrad to Gulf of Finland to winter in Baltic (east to Estonia in mild seasons), on both sides of North Sea, west coasts of Britain (perhaps Ireland), France, and north and west Iberia. These wintering areas also used by breeders of Sweden and Finland. Common migrant Atlantic coast Morocco late September to early November, up to 3900 per day flying south. Winter quarters problematical, for no reported Moroccan winter flocks over 500 (Smith 1965); only small numbers reach Banc d'Arguin, Mauritania (e.g. Pététin and Trotignon 1972), and unrecorded farther south. Generally scarce Mediterranean; only straggler east of Balearic Islands. Apart from major passage across north Russia and Denmark (see below), and small-scale June–August moult movement across northern England (Spencer 1969), occurs inland outside breeding season only irregularly and in small numbers, including Black and Caspian Seas (Dementiev and Gladkov 1952). Occasional visitor Greenland and Persian Gulf.

Considerable moult migration, of which knowledge incomplete (see Joensen 1973). Many immatures stay in Danish waters through spring and summer (moult begins June); and from late June to mid-August are joined by adult ♂♂ which pass west through Baltic and overland by night across southern Denmark to major moulting areas off west coast Jutland, whence some spread south to Friesian Islands. Main assemblies off Fanø, Mandø and Rømø, and in July 1963 up to 150 000 in that area, under 10% ♀♀. Smaller numbers (20000–30000) moult in Kattegat. Danish moult concentrations the largest recorded, but the numbers, as known, small in comparison with scale of moult migration. Several hundred thousand pass Estonia July–August, mainly nocturnally (Jögi 1971); and 1½ millions estimated to pass Gulf of Finland in May, when (of course) ♀♀ also involved (Bergman and Donner 1964). Heavy westerly passage past Wangerooge (East Friesian Islands) occurs June–August, and in 1952 one million (c. 75% ♂♂) estimated to have passed (Bauer and Glutz 1969). Thus certain that major moulting areas in North Sea or Atlantic yet to be discovered, for many more birds not accounted for than present in other (small) moult gatherings known Scotland and Ireland. Irish breeders moult locally in Donegal Bay (A Ferguson). Tends to moult farther out to sea and mainly in deeper water than most other sea-ducks, up to 13 km offshore in Danish waters. Most Icelandic breeders moult on breeding lakes (Bannerman 1958), though some immatures may join Scottish moult flocks. After moulting off Denmark, immatures and adult ♂♂ disperse west and south, mainly September, with little movement back to Baltic; apparently Baltic winterers predominantly ♀♀ and juveniles, which leave breeding range September and early October, though by then several thousands of ♀♀ already moulting in Danish waters.

Peak autumn migration (after moult completed) early November in Baltic, and November to early December in North Sea. Proportions of ♀♀ and juveniles in western seaboard flocks rise steadily until passage ends December.

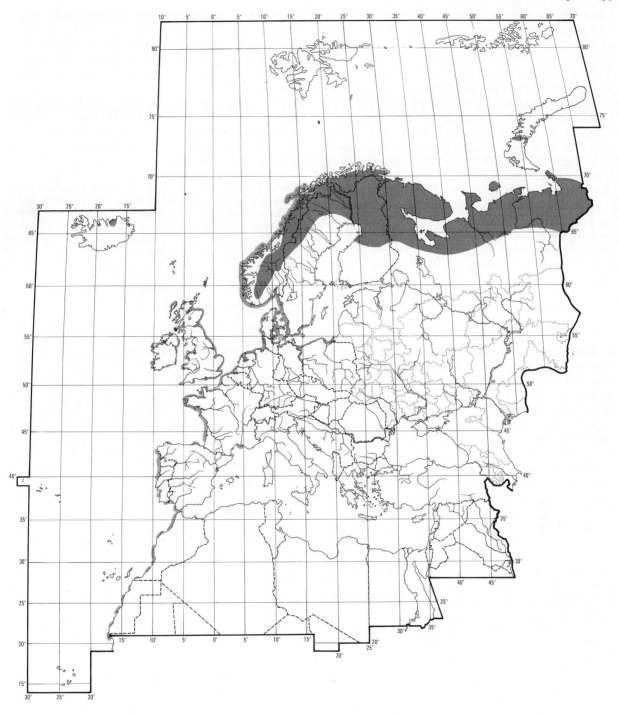

Icelandic breeders emigrate early September to early October. Return movements late February to April in Atlantic and North Sea, April–May Baltic, and northern breeding grounds re-occupied mid-May to early June. Spring migrations through Skagerrak and Kattegat, not across Jutland (Salomonsen 1968). Marked spring passage through Gulf of Finland, with overland movement in direction of White Sea. Small number of non-breeders remain in Atlantic and North Sea, more in Baltic.

East Siberian and North American race *M. n. americana* winters in North Pacific and, in western North Atlantic, from Newfoundland to South Carolina. Accidental Netherlands; 2 adult ♂♂ December 1954 and November 1967 (Bruun 1971).

**Food.** Mainly molluscs, obtained by diving, sometimes with partially spread wings (Townsend 1909; Witherby *et al.* 1939); in smooth water, wings usually closed (Marples 1929), in rougher conditions opened slightly (Brady 1951). Tail distinctly fanned (Oldham 1928). Chiefly daytime feeder, often in closely grouped flocks with regular massed dives (Brady 1951). 81% dives in 2·2–3·7 m, greatest depth 6·4 m; maximum submergence 49 s, most 18–30 s (Dewar 1924). 17 dives at unknown depths averaged 40 s (Brady 1951). In some areas, usual depths 10–20 m, maximum 30 m (Madsen 1954) but probably normally 1–3 m (Dementiev and Gladkov 1952); in Iceland, preferred depths 1·5–3·5 m (Bengtson 1971).

In marine and brackish-water areas, predominantly molluscs, especially blue mussel *Mytilus edulis* (up to 40 mm), fewer cockles *Cardium* (up to 40 mm), clams *Mya* and *Spisula* and other bivalves (*Venus, Tellina, Macoma, Solen, Venerupis, Cyprina, Nucula, Saxicava*), and gastropods, dogwhelk *Nassa recitulata*, periwinkles *Littorina*, and laver snails *Hydrobia*. Occasionally crustaceans, particularly isopods (*Idotea*), amphipods (shrimps *Gammarus*), and small crabs (*Carcinus*); annelids (polychaetes); and echinoderms. In freshwater areas: molluscs, especially mussel *Anodonta* and pondsnail *Lymnaea*; insects and larvae; annelids; small fish and their eggs; and roots, tubers, and seeds (Morris 1891; Naumann 1896–1905; Millais 1913; Witherby *et al.* 1939; Madsen 1954; Bengtson 1971). Availability of items may be prime factor in diet composition, e.g. wrecked cargo of 'horse beans', Heligoland, Germany, fed *c.* 1000 birds for over 4 weeks (Gätke 1900).

From saltwater areas, Denmark, October–April (mainly January), 219 stomachs contained by frequency 95·9% molluscs, especially blue mussel (50·7%), cockles (42·5%) particularly *Cardium edule* and *C. nodosum*, and dogwhelk (10·9%); crustaceans (10·9%), mainly amphipods, isopods and barnacles *Balanus* taken accidentally with mussels; and annelids (12·8%), mostly tube-worms *Pectinaria*. Contents of 8 stomachs from brackish areas similar, and all contained blue mussel (Madsen 1954). From Spitsbergen, mainly mollusc *Margarita helicina* (Roi 1911). In USSR in winter at sea, chiefly molluscs, especially *Mytilus* (Dementiev and Gladkov 1952).

From fresh water, Denmark, in November, 1 stomach contained mainly larvae of caddisfly *Phryganea* (Madsen 1954). Migratory birds, Upper Pechora, USSR, fed mainly on caddisfly and dragonfly larvae, less on molluscs, and still less on minnow *Phoxinus phoxinus*. Autumn birds, Rybinsk reservoir, USSR, chiefly bivalve molluscs (50% in volume), chironomid larvae (2%), and caddisfly larvae (16·3%) (Dementiev and Gladkov 1952). On Lake Mývatn, Iceland, in summer, 81 ♀♀ contained mainly chironomid larvae, with some fish eggs, mollusc *Lymnaea*, and seeds (chiefly pondweeds *Potamogeton*); 12 ♂♂ mainly fish eggs, and some chironomid larvae, Cladocera, and seeds. Cladocera and chironomid larvae increased in importance with advancing season; seeds mostly eaten in summer and molluscs and fish eggs in July. In 3 years 1968–70, chironomid larvae eaten extensively, though in 1968 and 1970 exceeded in weight by fish eggs; no molluscs recorded 1970 but relatively more seed (Bengtson 1971).

Ducklings, Lake Mývatn, less than half-grown ate mainly adult insects (obtained from or above surface) and seeds (mainly *Potamogeton*) but later mainly chironomid larvae and some Cladocera (Bengtson 1971, from analysis of 157 stomachs).

**Social pattern and behaviour.** 1. Highly gregarious except when nesting. Outside breeding season, often in large, mobile flocks on sea; usually several 100 and up to 1000 or more. Winter flocks may consist of both sexes and all ages, though ♂♂ tend to predominate in north and ♀♀ and immatures farther south (see Witherby *et al.* 1939; also Boase 1926; Brady 1951). During spring and autumn passage, flocks sometimes many thousands strong. ♂♂ first to flock again in June, usually forming small parties on fresh water before moult migration to sea where often gather in huge, dense flocks (Salomonsen 1968). BONDS. Monogamous pair-bond of seasonal duration. ♂♂ also show promiscuous tendencies, sometimes attempting to rape ♀♀ other than mate (Bengtson 1966b). Pair-formation starts in winter flock but final pairing often delayed until spring when sex-ratio less unbalanced. ♂ usually deserts ♀ a few days after start of incubation (Bengtson 1966b; see also Bent 1925; Bannerman 1958). ♀ tends young alone until fledging; amalgamation of 2 or more broods not uncommon (see Bauer and Glutz 1969). BREEDING DISPERSION. Nests well dispersed at concealed sites mostly near water. Little known about home-ranges or territories (if any) but forms small flocks on breeding waters at least for while after spring arrival. ♂ defends vicinity of ♀ wherever she may be (Bengtson 1966b). ROOSTING. Communal on sea for much of year. Mainly diurnal feeder so roosts nocturnally as well as loafing periodically during day. Will rest ashore, on islet (Bannerman 1958) or sandbanks (Witherby *et al.* 1939), but rarely.

2. Studied on sea, mostly in spring (Gunn 1927; Boase 1949; Schmidt 1952, 1957; Humphrey 1957; McKinney 1959; Myres 1959b); also Lake Mývatn, Iceland, in breeding season (Bengtson 1966b) and captivity (Johnsgard 1965). FLOCK BEHAVIOUR. Forms dense rafts at times on sea in winter, but tends to feed in smaller, more scattered groups and make synchronized mass dives (Brady 1951). Readily takes wing, and as in other mobile sea-ducks (such as Long-tailed Duck *Clangula hyemalis* and Steller's Eider *Polysticta stelleri*), pre-flight signals not infrequent (Bengtson 1966b), consisting mostly of assumption of Neck-stretch posture with head raised plus occasional lateral Head-shakes (Johnsgard 1965). Unlike other *Melanitta*, holds bill in raised position when alert (Bauer and Glutz 1969), indicating readiness to fly rather than to dive. ANTAGONISTIC BEHAVIOUR. On sea, aggression between courting ♂♂ mostly takes form of hostile skating rapidly over water with head and neck stretched out along surface; paddling of feet not vigorous (Boase 1949). Such rushes relatively infrequent (Gunn 1927; McKinney 1959), intention-movements of rushing often being sufficient as threat. ♀, however, frequently makes low, swift rushes towards rejected ♂♂; when Inciting, also alternately threatens them and turns head towards particular ♂, pointing bill up—slight Chin-lifting (see Myres 1959b; Johnsgard 1965). On Mývatn, May–June, ♂♂ disputing in courting parties adopt threat-display with head and neck lowered;

PLATE 79. *Aythya fuligula* Tufted Duck (p. 577): **1** ad ♂ breeding, **2** ad ♀ summer, **3** ad ♀ winter, **4** ad ♀ winter 'Scaup-type', **5** ad ♂ eclipse, **6** 1st imm ♂, **7** 1st imm non-breeding ♀, **8** downy young. (NWC)

PLATE 80. *Aythya marila* Scaup (p. 586): **1** ad ♂ breeding, **2** ad ♀ summer, **3** ad ♀ winter, **4** ad ♂ eclipse, **5** juv ♂, **6** juv ♀, **7** downy young. (NWC)

PLATE 81. *Somateria mollissima* Eider (p. 595): **1** ad ♂ breeding, **2** ad ♀ breeding, **3** ad ♂ eclipse, **4** 2nd imm breeding ♂, **5** 1st imm breeding ♂, **6** ad ♀ non-breeding, **7** downy young. (NWC)

PLATE 82. *Somateria spectabilis* King Eider (p. 605): **1** ad ♂ breeding, **2** ad ♀ breeding, **3** ad ♂ eclipse, **4** 2nd imm breeding ♂, **5** juv, **6** downy young. (NWC)

PLATE 83. *Somateria fischeri* Spectacled Eider (p. 611): **1** ad ♂ breeding, **2** ad ♀ breeding, **3** ad ♂ eclipse, **4** downy young. (NWC)

PLATE 84. *Polysticta stelleri* Steller's Eider (p. 613): **1** ad ♂ breeding, **2** ad ♀ breeding, **3** ad ♂ eclipse, **4** juv, **5** downy young. (NWC)

PLATE 85. *Histrionicus histrionicus* Harlequin (p. 620): **1** ad ♂ breeding, **2** ad ♀ breeding, **3** ad ♂ eclipse, **4** 1st imm breeding ♂, **5** juv, **6** downy young. (NWC)

PLATE 86. *Clangula hyemalis* Long-tailed Duck (p. 626): **1** ad ♂ summer, **2** ad ♂ eclipse, **3** ad ♂ winter, **4** ad ♀ winter, **5** ad ♀ autumn, **6** transitional 1st autumn ♂, **7** juv, **8** downy young. (NWC)

sometimes followed by fast rush over surface, but violent combats rare and some ♂♂ non-aggressive (Bengtson 1966*b*). Threatening behaviour of ♀ similar, but rushes rare. Both sexes also Head-shake frequently (see below) while calling in Neck-stretch posture with tail cocked. In defending mate, ♂ frequently chases off intruding ♂♂ and pairs, less often intruding ♀♀. Paired ♀ hostile to intruding ♂♂ and ♀♀, less often to intruding pairs; gives characteristic calls probably as part of Inciting-display. In encounters between pairs, all birds assume Neck-stretch posture. COMMUNAL COURTSHIP. Small groups of displaying ♂♂ gather round and closely follow single ♀. Usually 5–8 ♂♂ on sea, sometimes only 2 (McKinney 1959), but 3–12 typical inland, numbers increasing as season progresses and more ♂♂ available after mates start sitting (Bengtson 1966*b*); parties containing more than 1 ♀ rare. On Mývatn, 3 types of courting parties distinguished: (1) those in which ♂♂ attend active, unpaired ♀♀ which show hostility towards them; (2) those in which ♂♂ attempt to court ♀♀ accompanied by own mates; (3) those in which ♂♂ harrass immature, non-breeding, or incubating ♀♀ (Bengtson 1966*b*). These categories represent gradual change in nature of parties as season progresses. Courting group characterized by frequent calling (Gunn 1927), and almost continuous activity on part of all members. Each ♂ adopts Neck-stretch posture, with rear-end lowered and tail raised up to 45°, and periodically utters Courtship-whistle while following ♀; bill opens for several seconds during call, with conspicuous drop of lower mandible (McKinney 1959), and head often lowered obliquely forwards in jerky fashion while calling (Bowing-display; see Humphrey 1957, Bengtson 1966*b*). Deliberate, exaggerated, lateral Head-shake most frequent ♂ display; often follows Courtship-whistle (Johnsgard 1965). Other common displays observed by McKinney (1959) include Low-rush, Upward-shake, Wing-flap, and Short-flight. In Low-rush, head lowered as ♂ rushes for *c.* 1 m over surface then stops abruptly amid shower of spray, usually with terminal Head-shake. Differs from unritualized, purely hostile rushes in that bird moves more slowly, splashes more with feet, and adopts a more crouched posture with neck drawn in and back arched (see A); at least at times, apparently accompanied by rattling calls (Brooks 1920). Upward-shake much as in other Anatinae but with striking forward twitch of head (apparently peculiar to *Melanitta*). Wing-flap also ends in similar twitch. In Short-flight, ♂ (1) takes off and flies low over water for *c.* 1–3 m with *c.* 8–10 fluttering wing-beats, producing clearly audible wing noises; (2) alights with splash while suddenly extending feet forward, pulling head back, and assuming upright position as splash starts; (3) extends neck

B

and continues forward in fast Low-rush (see also Gunn 1927). Bengtson (1966*b*) found following additional displays also common: Tail-snap, in which tail fanned and then suddenly raised vertically over back (see B) or brought forward further still; and Body-up, in which body raised in water much as in Upward-shake but without shaking movements and accompanied by Courtship-whistle. Other displays include various Mock-preens and Steaming. Mock-preens mostly Preen-Breast and Preen-dorsally (McKinney 1959) while Preen-Breast and Preen-under-Wing described by Bengtson (1966*b*). When Steaming, ♂ swims towards ♀ with body half-erect, breast prominent, head held high, and neck drawn back slightly (Bengtson 1966*b*); probably same as High-rush of Gunn (1927) in which ♂ 'foams' past ♀ in half-upright position with breast protruding. 2 further displays mentioned (without details) and figured by Johnsgard (1965): Water-flick and Forward-stretch; former probably equivalent of Water-twitch of other *Melanitta* and *Bucephala*. For details of display sequences, see Johnsgard (1965), Bengtson (1966*b*). Actual frequency of various displays depends on whether ♀ responsive or not and on degree of hostility among ♂♂. During courtship, ♀ frequently shows aggression to ♂♂ (see above). When interested in particular ♂, adopts Neck-stretch posture and calls in unison; Courtship-whistle of ♂ followed by Tail-snap also tends to induce Preen-behind-Wing in ♀ which in turn often elicits Low-rush from ♂ (Johnsgard 1965). ♀ may also follow preferred ♂ closely, showing hostility only to other ♂♂ (Schmidt 1952), and, presumably, Inciting. Tail-snap and Low rush also reported from ♀♀ (Myres 1959*b*), other authors mentioning Head-shaking, Wing-flap, Upward-stretch, Preen-Breast, and further forms of Mock-preening. COURTSHIP-FLIGHTS. Courting parties frequently take wing, ♂♂ following ♀♀ in flight (Bengtson 1966*b*); exact details lacking. PAIR BEHAVIOUR. No fundamental differences between displays of ♂ in group and those of ♂ courting own ♀ away from group. Humphrey (1957) noted ♂ of isolated pair direct Courtship-whistles with Bowing, Tail-snap, and Low-rush at mate. Bengtson (1966*b*) showed that communal and pair-courtship shared many movements, including Tail-snap and Bowing, but paired ♂♂ perform no Body-ups to mate, fewer Short-flights, and less Steaming while Upward-shake and Wing-flap more common. At least occasionally, pair will Head-shake mutually. Copulatory behaviour also figures in pair-courtship, occurring on sea from end December, though no real evidence of strong pair-bond between birds copulating in winter quarters (Myres 1959*b*). COPULATION. See Myres (1959*b*), McKinney (1959), and Johnsgard (1965), but description incomplete. Preceded by Mock-preening of various types by both sexes. Then, as ♂ performs single Upward-shake, ♀ assumes Prone-posture and ♂ mounts immediately. Unlike in Velvet

A

Scoter *M. fusca*, copulation not accompanied by Wing-shake by ♂ who releases ♀ immediately and swims away in Neck-stretch posture, giving Courtship-whistle, as ♀ bathes. ♂ also seen to perform Tail-snap, Low-rush, and Upward-shake after dismounting. RELATIONS WITHIN FAMILY GROUP. No detailed information.

(Figs A–B after Johnsgard 1965.)

**Voice.** Limited in repertory, differing for each sex, but used more freely than by other scoters. Most characteristic utterances in flock when a number of birds call simultaneously with monosyllabic whistling, piping, or hooting notes, apparently of varying pitch and at varying intervals; speeded up under excitement. Heard at distance as pleasing murmur, these notes higher pitched and more metallic in ♂♂ (described as bell-like, plaintive, plover-like, or curlew-like) and hoarser or harsher in ♀♀. Difficulty of distinguishing utterances of individuals and sexes, and of separating these from non-vocal wing sounds in distant flocks has led to confusion, still not fully resolved. Typical wing noise in flight a quiet regular whistling 'vi-vi-vi' but, on take-off, pinions of ♂ make remarkably loud whirring likened to trilling of Alpine Swift *Apus melba* (see Bauer and Glutz 1969); probably same sound made during Short-flight in courtship.

CALLS OF MALE. Main Courtship-call of ♂ (also given after copulation) low, slurred, hooting pipe (see I), often audible over considerable distance. Described as brief 'pju' (Bengtson 1966*b*) or more prolonged, mournful 'wheeuu' (McKinney 1959). Other renderings, representing call of 1 ♂ or several in chorus, include 'cour-loo' or 'coor coor corLEE courlee' (Brooks 1920), 'tu-luk' (see Kirkman and Jourdain 1930), and 'tu-tu-tu' (see Bannerman 1958). Piping or whistling 'gyv gyv gyv gyv' (Söderberg 1950), given in flight, including when circling breeding area, may be same or distinctive—or possibly non-vocal (see above). Other calls include rattling 'tucka-tucka-tucka' of Brooks (1920), given during courtship but apparently different from main Courtship-call, and various tittering sounds (which could be wing noise).

I P J Sellar Iceland May 1967

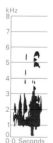

II Sture Palmér/Sveriges Radio Iceland June 1967

CALLS OF FEMALE. Not all separated from those of ♂. Frequently repeated, whistling call given at times in unison with main Courtship-call of ♂. Resembles latter but grating rather than mellow, like door with rusty hinges (Johnsgard 1965); tape recording (see II) also suggests it more croaking, lower pitched, and coot-like. Threat and Inciting-call, described as metallic 'ghä-ghä-ghä-ghä' (Bengtson 1966*b*) or hard 'ö' or 'ö-e' (Bauer and Glutz 1969), may be same; also hoarse, grating, whistling 're-re-re-re-re-re' of Bannerman (1958) and 'eu' or 'kerg' of Lippens and Wille (1972). Alarm-call described as wheezy, disyllabic 'kre-enk' (Gillham 1951). Call to young a bell-like 'gaga-gaga' or 'wah wah' (Bauer and Glutz 1969). Other calls or variants of those already described include harsh, creaking rattle, 'kark kark'; hoarse 'kr-r-r-rr' of Witherby *et al.* (1939) probably same.

CALLS OF YOUNG. No detailed information.

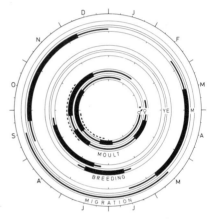

**Breeding.** SEASON. See diagram for Finland and USSR. Iceland: mean date of first egg 1961–70 varied from 21 May–1 June, average 25 May (Bengtson 1972). Ireland: eggs laid end May to end June (S White, J Cadbury). Laying controlled by timing of thaw in arctic. SITE. On ground, usually well concealed in thick vegetation, or under overhanging scrub; sometimes in natural hollow. Usually near water. Nest: hollow, lined grass, moss, lichens, and down. Building: by ♀, using material within reach. EGGS. Ovate; cream to buff. 66 × 45 mm (59–72 × 42–47), sample 150 (Schönwetter 1967). Weight 60–74 g, sample not given (Dementiev and Gladkov 1952). Clutch: 6–8 (5–11); more than 11 probably 2 ♀♀. Dementiev and Gladkov (1952) stated 6–8 in European USSR and 9–10 in Iceland. Of 35 Irish clutches: 4 eggs, 3%; 5, 9%; 6, 25%; 7, 37%; 8, 17%; 9, 9%; mean 6·8 (S White, J Cadbury). Mean of 101 1st clutches, Iceland, 8·7 (Bengtson 1972). One brood. Replacement clutches laid after egg loss, by 3 of a sample of 10 failed breeders in Iceland where mean of 30 2nd clutches 6·1 (Bengtson 1972). Eggs laid at 1–2 day intervals. INCUBATION. 30–31 days (Dementiev and Gladkov 1952), though 27–28 days reported in captivity (Witherby *et al.* 1939). By ♀. Starts with last egg; hatching synchronous. Eggs covered with down when ♀ off nest. Eggshells left in nest. YOUNG. Precocial and nidifugous. Self-feeding. Cared for by ♀ and brooded at night while small. Broods may merge. FLEDGING TO MATURITY. Fledging period 45–50 days. Independent at same time. Age of first breeding 2–3 years. BREEDING SUCCESS. Average hatching success of 159 nests, Iceland 1961–70, 81·8%, range 57·2–88·9% (Bengtson 1972). Only 16 of 38 Irish nests hatched; egg predation by Hooded Crows *Corvus corone cornix* and Magpies *Pica pica* main factor (S White; J Cadbury).

**Plumages.** ADULT MALE BREEDING. In fresh plumage, black, glossed violet-blue on sides of head, neck, upperparts of body, and chest; glossed green on underparts. Flight-feathers sooty-black, inner webs paler, pale grey near base of primaries. Rest of upperwing black. Underwing dark grey, lesser coverts and axillaries sooty-black. Less glossy in worn plumage with much brown of feather bases visible on head and underparts; feathers of upperparts and chest more square-tipped, with slight grey edges, or tips bleached buff. ADULT FEMALE BREEDING. Crown to eye sooty-black, dark brown round base of upper mandible and on hindneck, rest of head and neck pale grey with grey-brown mottling. Mantle, scapulars, sides of breast, and lower flanks dark brown, feathers broadly edged olive-brown (sometimes cinnamon); back, rump, and upper tail-coverts same, but edges narrower. Underparts glossy dark brown, feathers on chest and sides of body tipped olive. Flight-feathers black-brown with paler inner webs, secondaries tinged olive on outer web. Greater coverts as secondaries, rest of upper wing-coverts grey-brown, inner with olive tinge, tips buff or grey. Under wing-coverts and axillaries grey-brown, lesser coverts tipped pale grey. In worn plumage, crown dark brown, rest of head pale grey-brown; feathers of upperparts, flanks, and chest with square buff or grey tips, giving barred appearance; breast and belly grey-brown, tips of feathers grey or buff. ADULT NON-BREEDING. Combines new non-breeding feathers (resembling breeding) with old, un-moulted feathers. DOWNY YOUNG. Crown, lores, hindneck, upperparts, breast, and sides of body dark brown, palest on upper mantle. Sides of head, chin, throat, lower breast and belly pale grey. JUVENILE. Crown to eye brown; forehead, hindneck, and lower neck buff-brown, rest of head and neck pale grey. Upperparts, chest, sides of breast, flanks, tail, and tail-coverts dark grey-brown, darkest on back and rump; feathers faintly edged buff or grey on mantle, scapulars, chest, and flanks (plumage not black and olive like adult ♀). Breast, belly, and vent white, with small brown spots on belly and bars on vent (not uniform dark brown as adult ♀). Wing as adult ♀. Sexes similar but ♂ with darker tertials and, on average, darker brown centres to feathers on mantle and broader subterminal bars on belly, appearing darker, but much overlap in colour. FIRST IMMATURE NON-BREEDING. In early autumn of 1st calendar year variable (usually restricted) number of new feathers on body like juvenile, but longer and broader; more olive near tip; on head like adult ♀, but crown not so black. FIRST IMMATURE BREEDING. Later in autumn and in early spring juvenile feathers replaced by feathers resembling adult breeding; conspicuous in ♂ by appearance of black feathers. Most juvenile plumage and also immature autumn replaced on head, neck, upperparts, flanks and tail. In late spring, some ♂♂ acquire dark brown feathers on underparts and black feathers with olive tips on mantle (possibly homologous to adult non-breeding). In summer of 2nd calendar year, immature ♂♂ resemble adults but duller black with much brown visible at bases of feathers, some combining black feathers with newly moulted brown ones; wing juvenile until about July; variable number of juvenile feathers on belly until late summer. ♀♀ like adult, but with juvenile wing until July and usually many juvenile feathers on belly. SECOND IMMATURE BREEDING. From 2nd autumn like adult breeding, but ♂♂ with black of head and belly less glossy and more brown visible; ♀♀ may retain some juvenile belly feathers and worn 1st immature feathers on upperparts until mid-winter.

**Bare parts.** Nominate *nigra*. ADULT MALE. Iris dark brown, narrow eye-ring dark brown or orange-yellow. Bill black, spot on centre of upper mandible from nail to base of knob yellow, tinged orange at nostrils; yellow often extends in thin line over centre of knob to forehead. In some (mainly 2nd year, but also a few older), line over knob 3–6 mm wide. Foot olive-brown to dull black, webs black. ADULT FEMALE. Iris dark yellow-brown to dark brown. Bill olive-brown to black, inner side of nostrils yellow; often olive-yellow to yellow streak on culmen and spots round or between nostrils. Foot olive-brown, webs black. DOWNY YOUNG. Iris dark grey. Bill dark horn with red nail and yellow nostrils. Foot dark green, webs black bordered yellow-green. JUVENILE. Iris grey-brown to brown. Bill dark olive, innerside of nostril pink or yellow. From January, ♂ acquires orange-yellow on bill; in spring bill like adult, but yellow usually extends in 3–6 mm wide band towards forehead over less developed knob. Foot olive-grey to yellow-brown, webs dark grey; in spring like adult. (Bauer and Glutz 1969; Schiøler 1926; Witherby *et al.* 1939; C S Roselaar.)

**Moults.** ADULT POST-BREEDING. Complete. Tail late March to early May; scapulars, flanks, varying area of mantle and chest, and a few feathers elsewhere April–May; rest of body later, overlapping with and interrupted by simultaneous moult of flight-feathers. Flightless for *c.* 3–4 weeks: ♂♂ mid-July to mid-September; ♀♀ September–October, some later. ADULT PRE-BREEDING. Partial. Chest, sides of body, flanks, scapulars, tail, and part of mantle; partly overlaps with post-breeding body moult. ♂♂ September–December (mostly October–November, tail October), ♀♀ *c.* 1 month later. Belly in some ♀♀ not before February. Head and neck of ♀ moulted twice, about October and April–May, but 3 periods of active moult on head in ♂♂—September–October, December–January, and April–May. POST-JUVENILE. Partial; sexes alike. Slow moult from September onwards into 1st immature non-breeding plumage on head and lower flanks; rarely also a few tertials and scapulars. FIRST IMMATURE PRE-BREEDING. New feathers of breeding plumage appear on head, neck, scapulars, and flanks from late November (exceptionally early October); moult slow or arrested mid-winter, all feathers in these areas replaced by April–May. Upper mantle, chest, sides of breast, and tail-coverts moulted from January, but mainly April–May. Lower mantle, breast, and some feathers of back and rump moult from March; in some ♂♂, juvenile replaced by 2nd immature non-breeding from about April. Tail usually moulted April–May, but in some already complete January. Flight-feathers late June–October, mainly late July. In 2nd autumn, slow moult into 2nd immature breeding (Joensen 1973*a*; C S Roselaar).

**Measurements.** Netherlands, whole year; skins (RMNH, ZMA).

| | | | | | | |
|---|---|---|---|---|---|---|
| WING AD ♂ | 234 | (4·73; 91) | 224–247 | ♀ 226 | (4·78; 31) | 216–239 |
| JUV | 226 | (5·68; 30) | 217–241 | 218 | (4·25; 30) | 206–226 |
| TAIL AD | 92·3 | (5·30; 39) | 82–103 | 74·7 | (4·42; 24) | 68–84 |
| JUV | 74·7 | (3·01; 19) | 71–82 | 68·5 | (3·17; 22) | 62–73 |
| BILL AD | 47·5 | (1·68; 47) | 43–51 | 43·4 | (1·19; 32) | 41–46 |
| TARSUS | 45·4 | (1·28; 69) | 43–48 | 43·5 | (1·35; 55) | 41–46 |
| TOE | 71·5 | (2·73; 67) | 67–77 | 67·7 | (2·33; 54) | 63–73 |

Sex differences significant. Tarsus and toe of adult and juvenile similar; combined. Adult wing and tail significantly larger than juvenile of same sex; bill of juvenile ♂ on average 2 mm shorter than adults.

**Weights.** ADULT. (1) October–April, Netherlands and Denmark combined. (2) April–May, Greifswalder Oie, East Germany. (3) Netherlands coast, oiled and frost killed, emaciated December–March. JUVENILE. (4) October, Rybinsk Reservoir, USSR. (5) December–April, Netherlands and Denmark.

| | | | | | | |
|---|---|---|---|---|---|---|
| (1) | ♂ | 1165 (107; 14) | 964–1339 | ♀ 1059 (84·5; 10) | 973–1233 | |
| (2) | | 1363 (—; 4) | 1304–1450 | 1250 (—; 2) | 1231–1268 | |
| (3) | | 752 (62·5; 21) | 642–851 | 703 (45·5; 9) | 636–778 | |
| (4) | | 778 (—; —) | 710–818 | 735 (—; —) | 600–850 | |
| (5) | | 1126 (123; 17) | 878–1380 | 979 (159; 13) | 622–1227 | |

Various others (summer; breeding areas). ADULT. USSR. Lake Onega: May, ♂♂ 1306 (1215–1610). Arkhangel: May, ♀ 990; June, ♂♂ 960–1150. South Yamal: June, ♀ 1000; July, ♂♂ 975–1100; July and August, 2 ♀♀ each 915. Starvation weight outside winter, ♂♂, Netherlands, as low as 510. (Denmark: Schiøler 1926. East Germany: Bauer and Glutz 1969. Netherlands: ZMA. USSR: Dementiev and Gladkov 1952.)

**Structure.** Nominate *nigra*. 11 primaries: p9 longest; in adult, p10 4–10 shorter, p8 0–5, p7 10–16, p6 23–31, p1 84–97; in juvenile, p10 slightly longer to 5 shorter than p9, p8 3–9 shorter, p7 12–23, p6 25–38, p1 88–104. In adult ♂, p10 strongly emarginated, only 6–9 mm wide for 65–80 mm from tip; in adult ♀, narrow but usually only faint notch 35–55 mm from tip; in juveniles, gradually tapering towards tip, in some (mainly ♂♂) faint notch 33–55 mm from tip. Tail acuminated; 16 stiff and pointed feathers. Bill higher than broad at base, sides of bill parallel, nail large with rounded tip. In adult ♂, swollen knob with slight groove in middle at base of upper mandible behind nostril; in adult ♀, knob less obvious, base of upper mandible in some only slightly swollen. Knob absent in juveniles, but develops from spring onwards in ♂. Feathers at base of bill in adult ♂ short, plush-like, not extending over bill; feathers of rest of head and neck (except chin, throat, and nape) with pointed tips. Outer toe slightly longer than middle, inner *c.* 80% middle, hind *c.* 30%.

**Geographical variation.** East Asiatic and North American populations (*M.n. americana*) like those of Europe and West Asia (nominate *nigra*) in size and plumage, but bill different in form and pattern and relatively shorter: 43·7 (42·0–45·5) in adult ♂ (Godfrey 1966). Upper mandible of adult ♂ yellow with black tip and edges, knob on bill completely yellow. Knob wider and longer than in nominate, reaching to middle of nostrils; less swollen, more gradually sloping to flatter tip of bill. In both sexes and all ages, nostrils nearer tip of bill than in nominate; nail somewhat elevated, more strongly arched. Subspecies apparently replace each other abruptly at lower Lena, USSR, but not known whether ranges in contact. No certain intergrades known: specimens with much yellow on centre of knob probably immature ♂♂ of nominate *nigra*.                    CSR

## *Melanitta perspicillata* Surf Scoter

PLATES 88 and 100
[after page 662 and facing page 687]

Du. Brilzeeëend     Fr. Macreuse à lunettes     Ge. Brillenente
Ru. Пестроносый турпан     Sp. Negrón careto     Sw. Vitnackad svärta

*Anas perspicillata* Linnaeus, 1758

Monotypic

**Field characters.** 45–56 cm, of which body two-thirds; wing-span 78–92 cm. Medium-sized sea-duck with massive bill, short pointed tail, and mainly dark plumage without white in wings; ♂ has multi-coloured bill, all-black plumage apart from white on forehead and nape, whitish eyes, and reddish feet. Sexes dissimilar; only slight seasonal differences. Juveniles and 1st-winter ♀ not unlike adult ♀ unless underparts seen, but 1st-year ♂ variably distinctive.

ADULT MALE. Velvety black, less glossy than other scoters, with white patch on forehead and larger white triangle on nape. Bill large, slightly swollen on top and more so at sides (with black feathering extending on culmen to near nostrils); red-orange on top, yellow towards tip, and white on sides apart from square black patch near base bordered above and behind by orange to red. Eyes grey-white to yellow-white; orange-red feet with dusky webs visible when diving. MALE NON-BREEDING. Similar, though browner, except that white patch on nape disappears at one stage during moult. ADULT FEMALE. Black-brown above and slightly paler brown below, uniform but for some whitish edgings, with 2 whitish patches on side of head—small in front of eye and larger behind (both may be indistinct or even absent, particularly former)—and sometimes variably distinct whitish patch on nape corresponding to ♂'s. Legs yellow-orange with dusky webs; bill green-black. JUVENILES AND FIRST-YEAR FEMALE. Resemble adult ♀, but generally paler and browner with whitish or almost white breast and belly, while, in contrast to Velvet Scoter *M. fusca*, whitish patches on sides of head often less well-defined or run together; white patch on nape never present. FIRST-YEAR MALE. From November onwards, variable number of black feathers on head, neck, upperparts, upper breast, flanks, and undertail, these areas sometimes becoming wholly black early in new year, with small white patch on nape but none on forehead.

Adult ♂ easily distinguished from other scoters at close range by bill pattern and white patches on head, but, unless absence of white in wing can be established, adult ♀ distinguishable from ♀ *M. fusca* only by whitish patch on nape and bill shape; and immatures only by bill shape and whiter cheeks and underparts. Importantly, in all plumages, has more flattened head profile and heavier bill than other scoters with shape recalling Eider *Somateria mollissima*. Underwing lacks marked two-toned pattern of Common Scoter *M. nigra* and immatures further distinguished by leg colour.

Habits much like other scoters, but seldom flies in lines and generally more agile; flight less heavy than *M. fusca*, more like *M. nigra*. Mostly silent.

**Habitat.** Breeds on margins of lakes or ponds, in bogs and other wetlands, within or beyond northern tree limit, in dense but not tall vegetation. Winters almost entirely in inshore marine waters, rarely beyond *c.* 9 m depth for feeding, often within zone of breaking waves; rests in flocks farther out, but commonly within sight of shore. Sometimes visits neighbouring freshwater lagoons and ponds. Flies low over sea on everyday movements, but to considerable heights on migration. Comes little to land except when nesting.

**Distribution.** Breeds North America from western Alaska east into Canada, from Mackenzie to Hudson Bay, James Bay, Quebec, and Labrador, and south to northern British Columbia, Great Slave Lake, and Lake Athabasca.

Accidental. Most frequently recorded Britain and Ireland (over 100, widely spread, but chiefly Shetland and Orkney), France (16, including one Mediterranean coast), Finland (11), and Sweden (5); also Faeroes, Denmark, Netherlands, and Belgium (each 2), and Iceland, Norway, and Czechoslovakia (1). In Britain and Ireland, and most other countries, concentration of autumn and winter records; but in Finland and Sweden most May and June (Baltic, end April and May; Lapland, summer), suggesting stragglers may migrate with Velvet Scoters *M. fusca*, with which they often associate. (British Ornithologists' Union 1971; Bauer and Glutz 1969; Bruun 1971.)

**Movements.** Migratory. Winters Pacific coasts from Aleutian Islands and south-east Alaska to Baja California, and in Atlantic from south Newfoundland and Gulf of St Lawrence to Florida. Migration mainly coastwise, but common migrant Lake Mistassini, Quebec interior, suggesting breeders from James Bay at least migrate overland to and from Gulf of St Lawrence (Godfrey 1966). Apparently commonest of three scoters off Atlantic coasts North America (Bruun 1971); in autumn 1969 passage past Manomet, Massachusetts, more *M. perspicillata* identified than Common Scoter *M. nigra* and Velvet Scoter *M. fusca* combined (Petersen 1971). Rare in interior south of breeding range. In North Pacific, ranges in small numbers to Commander, Pribilof, and Bering Islands. In Atlantic has straggled to Bermuda and Greenland; and most frequent Nearctic duck in Europe (see Distribution). Many moulting ♂♂ assemble from late June on east coast Labrador and in Gulf of St Lawrence (Todd 1963); also off Vancouver Island in Pacific. Autumn migration along Atlantic coasts late September to early December, and

return movements March to early May; peak passages New England mid-October and first half April (Stewart and Robbins 1958; Petersen 1971); northern breeding places reached mid-May to mid-June (Gabrielson and Lincoln 1959).

**Voice.** Poorly known and evidently little used, except in courtship. ♂ has liquid, gurgling call (Myres 1959*b*), described as low and 'like water dropping in a cavern' by Brooks (1920). Call of ♀ crow-like (Myres 1959*b*).

**Plumages.** ADULT MALE BREEDING. Black, except for roughly triangular white patch on forecrown and large white patch of elongated hair-like feathers projecting in point downwards on nape. Black feathers on chin, throat, and belly with partly visible brown bases. Flight-feathers sooty with brown inner webs, rest of upperwing black; underwing black, greater coverts and axillaries sooty. Shortly before moult of head, short black feathers on nape exposed through wear of long white ones. ADULT FEMALE BREEDING. Crown to eye and hindneck sooty, feathers on hindcrown and nape elongated, with white shaft streaks of varying width, forming more or less distinct spot on nape. Sides of head from base of upper mandible to ear-coverts white finely speckled brown, interrupted by dark brown band from eye downwards. Lower cheeks, chin, throat, and neck dark brown. Upperparts, tertials, tail, and flanks sooty-brown, some feathers on rump, flanks, and upper tail-coverts tipped black. Underparts dark brown. In worn plumage, sides of head paler, upperparts barred by abraded square feathers with faded edges; feathers of underparts edged grey. Flight-feathers and upper wing-coverts dark brown, outer webs of primaries brown-black. Underwing and axillaries dark brown, part of greater under wing-coverts tipped white. ADULT NON-BREEDING. Part of plumage new, resembling fresh breeding; worn breeding plumage partly retained. JUVENILE. Like adult ♀, but no white on nape; pale buff or off-white patches at sides of head less distinct; crown brown, rest of head and neck grey-brown. Feathers of upperparts, chest, and flanks greyer brown, narrowly edged buff or grey; feathers of underparts short and narrow, white with grey-brown centres so belly appears spotted. Tail-feathers notched at tip, bare shaft projecting. FIRST IMMATURE NON-BREEDING AND SUBSEQUENT PLUMAGES. Development of new feathers (resembling those of adult) as in Common Scoter *M. nigra*. In ♂, white patch on nape develops in 1st winter, but no white on forecrown before 2nd winter; in ♀, some white in neck from 2nd winter.

**Bare parts.** ADULT MALE. Iris pale blue or yellow-white. Bill with large black patch basally at side of upper mandible; margined behind red, above orange, below white. Rest of side pale blue; ridge and area round nostril red, nail pale yellow; lower mandible basally yellow, tip flesh. Leg and foot bright orange, webs black. ADULT FEMALE. Iris brown, pale grey or yellow. Bill green-black; indistinct black patch with pale grey margins basally at side of upper mandible. Leg and foot dull orange-red, webs dull black. JUVENILE. Iris brown; bill and foot as adult ♀, but duller, bill greyer, feet more yellow. Attains adult colour from early spring. (Dwight 1914; Kortright 1942; Witherby *et al.* 1939.)

**Moults.** Apparently as in *M. nigra*.

**Measurements.** ADULT. Average 5 ♂♂ east USA (RMNH, ZMA); additional data on range from Witherby *et al.* (1939): wing 244 (238–56); bill 37 (34–41). Range ♀♀: wing 223–35; bill 35–8.

**Weights.** Full-grown, North America (Kortright 1942). ♂ 992 (10) 652–1134; ♀ 907 (7) 680–992.

**Structure.** 11 primaries: p10 and p9 about equal, longest; p8 6–12 shorter, p7 19–26, p6 33–40, p1 87–108. Flight-feathers not emarginated. Tail short, pointed; 14 feathers. In adult ♂, feathering of forehead and chin projects on bill to about hind-corner of nostril but feathers do not extend on sides of bill. Sides of bill basally parallel, narrowing towards tip; sides of upper mandible swollen. Nostril round, large. Bill as high as broad at base, culmen steeply declining above nostril, nail large and rounded. In ♀ adult and juveniles, bill less high and swollen at base, feathering on forehead projects less far on culmen. Outer toe equal to middle, inner *c.* 81% middle, hind *c.* 31%.   CSR

## *Melanitta fusca* Velvet Scoter

Du. Grote Zeeëend       Fr. Macreuse brune       Ge. Samtente
Ru. Турпан       Sp. Negrón especulado       Sw. Svärta       N.Am. White-winged Scoter

PLATES 89 and 100
[between pages 662 and 663
and facing page 687]

*Anas fusca* Linnaeus, 1758

Polytypic. Nominate *fusca* (Linnaeus, 1758), Europe and Asia east to Yenisey. Extralimital: *stejnegeri* (Ridgway, 1887), Asia east of Yenisey Basin; *deglandi* (Bonaparte, 1850), North America.

**Field characters.** 51–58 cm, of which body nearly two-thirds; wing-span 90–99 cm. Rather large sea-duck with long broad bill, heavy head, thick neck, and short, pointed tail; ♂ all black, except for patches of white in wing and below eye, and partly orange bill. Sexes dissimilar, but no obvious seasonal differences. Juveniles and 1st-winter ♀ barely separable from adult ♀, but 1st-year ♂ variably distinctive.

ADULT MALE. Glossy black, with white secondaries and small white mark immediately under eye. Bill has only small knob at base and black centre with orange sides and redder tip; eyes grey-white; feet red with black webs. MALE NON-BREEDING. Similar, but duller and browner, particularly on head and neck where no gloss at all. ADULT FEMALE. Dark brown above and slightly paler brown below, uniform but for some whitish mottling, with white secondaries, as in ♂, and 2 variable whitish patches on sides of head (indistinct in fresh plumage and sometimes absent). Bill olive-black; eyes brown; legs duller red than ♂. JUVENILES AND FIRST-YEAR FEMALE. Resemble adult ♀, but generally paler and browner, more mottled with white below, and with more pronounced white patches on side of head. FIRST-YEAR MALE. Variably mottled with black on head, neck, upperparts, upper breast, flanks, and undertail from December onwards. Adult plumage not assumed until 2nd autumn.

Distinguished from other scoters in all plumages by white wing patch, conspicuous in flight or when wings flapped on water as white rectangle covering whole secondaries, but seldom visible at rest. ♂'s small white eye-patch, coloured bill, and red feet usually seen only at closer ranges. Whitish patches in front of and behind eye of ♀ and

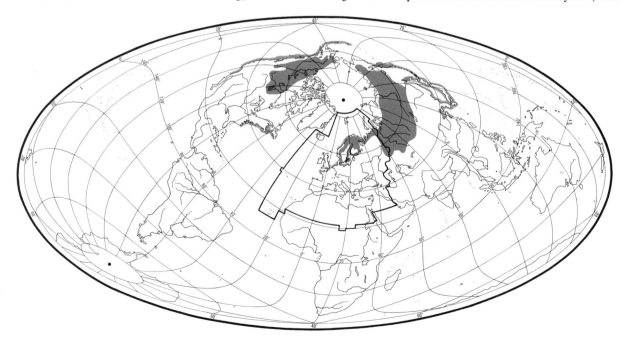

immatures also usually inconspicuous at distance.

General behaviour similar to Common Scoter *M. nigra*, but usually in much smaller parties of 20 or less, often swimming in single file, and seen more often feeding in broken water among rocks and islands. Less buoyant on water and rises from surface with more difficulty. In flight, heavy head and thick neck noticeable. Often flies low over waves, but also at heights of up to 20 metres. Generally rather tamer than other scoters, and mostly silent.

**Habitat.** Ranges across continental interior through higher middle latitudes; more boreal, montane, and varied than Common Scoter *Melanitta nigra*. Frequent contrast

also through association with trees, which provides link between northerly inland habitats, by lakes, pools and rivers within wooded tundra and taiga zones, and western habitats on wooded shores and skerries of Baltic. More generally breeds close to fresh or brackish waters than *M. nigra*. On migration, common on inland waters. Always highly aquatic, performing with skill on and under water to normal foraging depth of *c*. 5 m; occasionally much more. Clumsy on land. Has difficulty in becoming airborne and commonly flies low, especially over water. Great majority winter at sea, often along exposed coasts where food accessible on seabed; sometimes attracted by large inshore mussel-beds in estuaries or inlets. Single birds somewhat more liable to appear in winter on inland waters, natural or artificial, than *M. nigra*. Not sensitive to human presence, and correspondingly vulnerable to encroachment by development and to damage by oil pollution.

**Distribution.** BRITAIN: breeding suspected north Scotland, notably Shetland 1945, but no confirmed records (British Ornithologists' Union 1971). SPITSBERGEN: possibly bred 1882 (Løvenskiold 1964).

Accidental. Spitsbergen, Bear Island, Iceland, Faeroes, Hungary, Yugoslavia, Greece, Bulgaria, Rumania, Israel, Italy, Portugal, Algeria, Morocco, Azores.

**Population.** Breeding. SWEDEN: marked decrease last decade parts of south-east coast (Curry-Lindahl *et al.* 1970). FINLAND. Some 5000–8000 pairs, mainly in archipelago. Northern inland population markedly reduced due to hunting, but marine population of archipelago increased, especially in outer islands from 1920s to late 1950s, though now stabilized or slightly decreasing (Merikallio 1958; Grenquist 1970). USSR. Numbers everywhere low (Dementiev and Gladkov 1952). Estonia: *c.* 1100 pairs, steadily diminishing (Onno 1965).

Winter. Least numerous of sea-ducks. Highest January count in Europe 26 400 birds 1968 but as *c.* 60 000 moulting Danish waters (Joensen 1973*a*) perhaps total at least 150 000–200 000 birds (Atkinson-Willes 1975). USSR: in western parts *c.* 300 in not so severe winters (Isakov 1970*c*).

Oldest ringed bird 12 years 6 months (Rydzewski 1974).

**Movements.** Migratory. Winters mainly coastal waters of north and north-west Europe, generally closer to breeding range than Common Scoter *M. nigra*; no large numbers south of English Channel. Tends to keep offshore, and may be overlooked among mass movements of far more numerous *M. nigra*. Small numbers winter off Murmansk and north Norway, more down west coast Norway, in Baltic (mainly west, but some Gulf of Finland in mild winters), around Britain (chiefly east coast), and on east side of North Sea south to Normandy. Only small numbers reach Ireland, Brittany, Bay of Biscay, and Iberia. In January 1968, more in Danish waters than rest of Europe combined (Bauer and Glutz 1969).

Most from USSR east to Tyumen probably migrate to north-west Europe; no recoveries but no other major wintering areas of nominate *fusca* known. Recoveries of Fenno-Scandian breeders mostly from Danish archipelago; but also 3 from Finland to English Channel (March, November), and singles from Sweden to West Germany (April) and Norway to west Scotland (October). More frequent inland than *M. nigra* (mainly passage birds), though numbers insignificant compared with coast; on south German and Swiss lakes, sometimes flocks of 30 or more, and may stay weeks or even months (Bauer and Glutz 1969). Occasional north Mediterranean records suggest a few cross Alps. Small numbers winter south and west Black Sea (occasionally north Greece), and small

parties (possibly moulting) noted Black Sea coast of Turkey in July (*Orn. Soc. Turkey Bird Rep. 1966–7*; Johnson and Hafner 1970).

Moult movements variable. 'Summer' flocks of ♂♂ and immatures reported Kolguyev, Vaigach, and southern Novaya Zemlya (Dementiev and Gladkov 1952). In Lapland and west Siberia, ♂♂ congregate on lakes within breeding range; while some move south to steppe lakes of Chelyabinsk and Kurgan (Isakov 1970*c*). Important moulting grounds in Limfjorden and Kattegat, Denmark, where usually *c.* 45 000 in late July (Joensen 1973*a*), more than midwinter population; adult ♂♂ comprise 80% in July, with smaller numbers of immatures, while many adult ♀♀ arrive August–September; some of latter still flightless in October (Salomonsen 1968). Danish moulting birds include many from Fenno-Scandia, but perhaps some from farther east. Small numbers join moult migration of *M. nigra* to and through North Sea (some overland crossings of Jutland observed—Joensen 1973*a*), and irregular summer movements off Scottish coast may be continuation of this.

In north Russia, major departures of ♀♀ and juveniles September and October; those of Finnish skerries in September, when migration apparent west to North Sea. Pronounced passage Baltic region October–November, and numbers increase in Danish waters before continued westerly movement; in North Sea and English Channel, peak numbers often not until January. Return movement from early March, but spring passage later than in *M. nigra* and Eider *Somateria mollissima*, and large flocks in Danish waters until mid-May when peak passage occurs Sweden and Finland. In USSR, one of last ducks to return, reaching Tobolsk late May and Yamal mid-June (Dementiev and Gladkov 1952).

**Food.** Chiefly molluscs, obtained by surface-diving with partially opened wings. In Finland, depths 2–5 m, more rarely up to 7 m (von Haartman 1945; Koskimies and Routamo 1953), though in Danish waters frequently 14–30 m (Hørring 1919; Madsen 1954). Submergence times: south-west Finland, mostly 20–40 s; maximum 46 s ♀♀, 56 s ♂♂ (Koskimies and Routamo 1953); southern Scotland, maximum 51 s in 296 dives (Dewar 1924); in Bodensee, Switzerland, up to 60–65 s (Bauer and Glutz 1969). Also in shallower water, and occasionally will dabble in driftlines (Naumann 1896–1905). Normally daytime feeder; often gregarious with synchronized diving.

Diet similar to Common Scoter *M. nigra*, but more varied probably because more often feeds near coast. In salt and brackish-water areas: mainly molluscs, especially blue mussel *Mytilus edulis* (usually 5–20 mm), cockles *Cardium* (up to 20 mm), dogwhelks *Nassa* (up to 25 mm), and less often *Mya* (up to 40 mm), *Macoma, Spisula, Mactra, Venus, Nucula, Astarte, Cyprina, Modiolaria, Leda, Solen, Tellina, Donax, Littorina*, and *Buccinum* (up to 60 mm); crustaceans, including small crabs (shore-crab *Carcinus maenas*, hermit-crab *Eupagurus bernhardus*), isopods (*Idotea*), and amphipods (*Gammarus*, *Pallasea*); echinoderms, including heart-urchin *Echinocardium* and starfish *Asterias*; annelids, including Polychaetes (*Pectinaria*, *Arenicola* up to 120 mm long, *Nereis*); rarely small fish. In freshwater areas: molluscs, including *Bithynia*, *Valvata*, *Anodonta*, *Unio*, *Dreissena*, and *Pectunculus*; insects; annelids; small fish; and seeds, roots, tubers, buds, and leaves (Macgillivray 1852; Collett 1877; Mitchell 1892; Naumann 1896–1905; Witherby *et al.* 1939).

In winter, 144 stomachs, saltwater areas Denmark, mainly October–February, contained chiefly molluscs (97·2% frequency, 83% by volume), and less crustaceans (16%, 6%), echinoderms (9·7%, 6%), annelids (8·3%, 4%), and fish (4·2%, 2%); main items dogwhelk, cockles, and blue mussel. Contents of 13 stomachs, brackish-water areas, similar (Madsen 1954). 7 stomachs, Sempach, Switzerland (when food scarce), contained only roe of white fish *Coregonus* (Huber 1956; Hofer 1968); and from Lower Saxony gravel-pits (also when food scarce), frogs (probably *Rana ridibundus*) (Rettig 1961). On Rhine and Bagger, Germany, birds fed on molluscs *Dreissena polymorpha*, *Unio*, *Anodonta*, *Viviparus viviparus*, and *Bulimus tentaculata* (Bauer and Glutz 1969). In USSR: autumn birds in Pechora took mainly molluscs, and fewer caddisfly larvae and plants (chiefly leaves of pondweed *Potamogeton*); in October, Rybinsk reservoir, animal food predominated (99·5% volume) of which 62·5% molluscs (mainly *Unio* and *Anodonta*), 30% caddisfly larvae, 7% small fish (Dementiev and Gladkov 1952).

Spring and summer data sparse. 38 stomachs from south-west Finnish archipelago in spring contained 71% by volume blue mussel, and less baltic mussel *Macoma baltica*, laver snails *Hydrobia*, and shrimps *Gammarus* (Bagge *et al.* 1970). A stomach from Norway contained amphipod *Pallasea* (Collett 1877). Summer diet for *M. fusca deglandi* mainly insects, especially caddisfly larvae; some crustaceans; fish; and plant materials (Cottam 1939).

Ducklings, when small, mostly take insects from or just above surface (Koskimies 1957). In southern Norway, mainly insects (Collett 1894); and southern Finland, mainly molluscs *Theodoxus*, *Bithynia*, and *Lymnaea* (Fabricius 1959).

**Social pattern and behaviour.** 1. Moderately gregarious for much of year. Outside breeding season, birds of nominate race mostly in small, scattered parties on sea; sometimes up to *c.* 100, rarely more. In North American race (*deglandi*), flocks up to several thousands (Bent 1925). Among breeders, ♂♂ first to flock again from late June, soon leaving for flightless period mainly at sea where assemble, mostly in small groups. These may join flocks of 1st-year birds that have spent summer at sea (Salomonsen 1968). ♂♂ and immatures joined later by ♀♀ and juveniles, but winter flocks often of unbalanced composition, ♂♂ tending to predominate to north (e.g. up to 30:1 in Orkneys) and ♀♀ and young to south (see Witherby *et al.* 1939). BONDS.

Monogamous pair-bond of seasonal duration. Pairing starts in winter flock and completed on sea after spring passage while flocks wait for breeding areas to thaw, though unmated ♂♂ continue to seek mates later. ♂ usually deserts ♀ during incubation, apparently fairly late (Millais 1913); particularly in case of isolated pairs, ♂ occasionally stays to accompany (and defend) ♀ and brood (see Bauer and Glutz 1969). ♀ usually tends young alone, showing hostilities to other young and mothers. Parent-young bond otherwise not strong, at times ending while young fairly small, and groups of mixed young tend to form under certain conditions (see Koskimies 1957; Koskimies and Lahti 1964); these often several dozen strong and may be as many as 115 (Witherby *et al.* 1939; Hildén 1951; see also Bauer and Glutz 1969). BREEDING DISPERSION. Nests well dispersed at concealed sites near water; in some areas, especially on coast, associates with colonies of gulls (Laridae) and terns (Sternidae) (see Bauer and Glutz 1969). After arrival on nesting waters, pairs at first in loose flocks; then and later, whenever pair comes near single ♂♂ or pairs, ♂ defends moving area round ♀. After nest-site chosen, pairs disperse for part of day; each maintains small territory restricted mainly to water in front of site from which ♂ excludes conspecifics when these approach within 50–100 m, though boundary continually changing (Koskimies and Routamo 1953). Pairs continue to associate away from territories—e.g. at communal feeding, gathering, or loafing places—until ♀♀ start sitting. ♂ also defends small area in front of pair's night-roosting site. ♂ constantly attends mate during laying period, except when ♀ at nest, and later accompanies ♀ during her recesses from incubation. When not with ♀, and after pair-bond ended, ♂ associates with other ♂♂, small groups of which often persist throughout breeding season. ROOSTING. Communally on sea for most of year. At all times, essentially diurnal feeder, roosting nocturnally with periods of loafing during day. Though sometimes comes ashore in winter (see Witherby *et al.* 1939), this unusual. In breeding area, frequently roosts ashore at night; depending on wind, pair sleep together as near nest-site as possible until ♀ starts sitting (Koskimies and Routamo 1953). In south-west Finland, pair move to roost from communal gathering area about sunset, ♀ going to site ashore first while ♂ remains on water for further 30 min or longer. ♂ first into water during night (usually between 01·30–02·30 hrs), considerably before ♀. At about sunrise, some pairs go off to feed on own, but most assemble at communal area on water where at first loaf together. Eventually, at *c.* 04·00, begin 'morning flights' (see below) which continue for further 1–2 hrs before birds move off to feed; then from 09·00–18·00 hrs feed and loaf communally on water. Between 17·00 and 20·00 hrs, assemble again before dispersing in first part of night. ♀ broods young ashore during day. Family also loafs ashore during day, same spot often used in turn by succession of different broods. Family roosts ashore at night at similar sites, but often use same one repeatedly.

2. Studied in both winter quarters and breeding area (see Boase 1949, 1950; Koskimies and Routamo 1953; Myres 1959*a*, *b*; and, for summaries, Johnsgard 1965, Bauer and Glutz 1969). FLOCK BEHAVIOUR. Little information, but see part 1. Generally reluctant to take wing while on sea in winter. No pre-flight behaviour described, and assumption of Neck-stretch posture with neck obliquely forward not vertical (see below) suggests tendency to dive rather than to fly, unlike in Common Scoter *M. nigra*. This supported by somewhat more deeply sunk carriage in water and frequent holding of bill slightly down not up, while diving clearly usual response to predators. Synchronized diving for food reported. ANTAGONISTIC BEHAVIOUR. That of ♂♂ much as in *M. nigra*, but swimming attacks and pursuits, in which rush

across water with beating wings, also recorded and underwater chases seem more common (Koskimies and Routamo 1953). ♀ pecks at rejected ♂♂ in courting parties, while Inciting-display involves calling and rapid Chin-lifting directed towards chosen ♂ (Myres 1959*b*). Paired ♂ in breeding area defends ♀, especially just before egg-laying; often leads to intense hostile encounters especially with unpaired ♂♂ which try repeatedly to join pairs in their courtship. ♂♂ of other pairs, however, tend to be tolerated or merely elicit Wing-flaps (Koskimies and Routamo 1953). COMMUNAL COURTSHIP. Small groups of ♂♂ form round single ♀ much as in *M. nigra* but fewer ♂♂ really active in display at any one time and exact role of vocalizations uncertain; in nominate race at least, ♂♂ seem to lack conspicuous piping of *M. nigra*, though whistles reported during courtship in American *deglandi*. Following behaviour similar to *M. nigra*: swimming rapidly after ♀ in forward variant of Neck-stretch posture, Low-rush, Upward-shake, and Wing-flap; but Head-shaking (at least in frequent and exaggerated form), Bowing, Tail-snap, Short-flight, Body-up, and Forward-stretch absent, while Ceremonial-drinking common. Water-twitch and Mock-preens also recorded, but these probably mainly pre-copulatory displays. Unlike in *M. nigra*, courtship-flights appear to be totally absent, but these replaced by underwater pursuits lasting 15–20 s; initiated by ♀, who may dive several times in succession followed by most or all ♂♂ (see Boase 1949, 1950). PAIR BEHAVIOUR. As in other *Melanitta*, pair-courtship seems similar in most features to communal courtship with addition of behaviour associated with copulation. Chin-lifting by ♀ (with calling) performed next to ♂ in apparent Triumph Ceremony (Johnsgard 1965) or Pair-palaver. Morning-flights by pair prior to incubation striking phenomenon; though probably linked with nest-site selection (see Stoll 1931), said also to be important part of pair-courtship in nominate *fusca* (Koskimies and Routamo 1953). For similar behaviour in American race *deglandi*, see Bent (1925). After roosting alone at night, pairs assemble before all engage in persistent flights over land, taking off one after the other repeatedly. Each pair, with ♀ usually leading and calling persistently, circle low inland 1–3 times then return to starting point. Between each outburst of flighting, pairs swim around excitedly. Morning-flights continue into egg-laying period, ♀ breaking off her flight to visit nest. COPULATION. See especially Myres (1959*a, b*). As in *M. nigra*, copulation preceded by frequent Mock-preening (mostly Preen-behind-Wing, exposing white speculum, and Preen-dorsally)—but by ♂ only. Mock-preening occurs on own or immediately follows Water-twitch in which ♂ bill-dips with slight shake of head. ♂ also performs Ceremonial-drinking, often mutually with ♀. ♀ does not assume Prone-posture until ♂ mounts. ♂ may perform any one of above displays immediately before mounting; unlike *M. nigra*, invariably does 1–2 vigorous, double Wing-shakes about time of intromission. No marked pair-rotation after ♂ dismounts, or any elaborate post-copulatory activities; both birds re-arrange wing feathers, then ♂ moves slowly away from ♀ in hunched position. No bathing reported. RELATIONS WITHIN FAMILY GROUP. Studied by Koskimies and Routamo (1953) and Koskimies (1957). Although bond between ♀ and young tends to be unstable, survival of latter (at least in earlier stages) dependent on membership of group, whether original brood or not. ♀ guards smaller young while they dive for food. Tiny young stay close to ♀ and brooded on land if weather cold or wet; later, scatter more to feed, followed by ♀. Young follow ♀, in tight-knit group or line when she leads them off in Neck-stretch posture (e.g. in face of danger). ♀ manoeuvres any straying duckling back into group by swimming towards it in threat-like posture with neck drawn in

and bill held sharply down; then, when *c.* 50–100 cm away, turns quickly away through 180° and slowly swims back, repeating process if it does not follow. ♀♀ without young similarly entice unaccompanied ducklings to follow them. Only when accompanied by ♀ do smaller young learn to alternate between periods of feeding and loafing; in absence of ♀, young do not dry, preen, and oil themselves adequately. Lone lost young panic, call, and wander about, soon dying if they do not join a brood. Hostility by brood ♀ recorded towards own and other young during disputes between broods or mixing of young.

**Voice.** Much less used than in Common Scoter *M. nigra*, with little if any of chorus effect from active flocks, and largely confined to brief, crude calls ranging from harsh croaking and growling through softer purring to higher pitched piping or whistling. Information fragmentary and uncertainty whether descriptions of American race *deglandi* applicable to nominate *fusca*. As in *M. nigra*, confusion between non-vocal wing noise and possible flight-call also unresolved, at least in *deglandi*; described as like tinkling ice or intermittent series of low, bell-like sounds in Bent (1925) where whistling wing noise during courtship likened to that of goldeneyes *Bucephala* but louder and audible for some distance.

CALLS OF MALE. Apparent Courtship-call of nominate *fusca* ringing or piping 'skryck-lyck' or 'vak-vak' (Söderberg 1950), higher pitched than usual flight-call; also rendered 'kju' (Bauer and Glutz 1969). Whistled, double 'whur-er' mentioned for *deglandi* in Witherby *et al.* (1939) probably same as whistling note uttered during courtship of Johnsgard (1965) but this call appears to have no counterpart in vocabulary of nominate race; same applies to curious, low purring note of Brooks (1920) and peculiar rattling or cackling call of Phillips (1926). One flight-call in nominate *fusca* hoarse, slow, growling or croaking 'koarr', 'karr', or 'kraa' (see e.g. Dementiev and Gladkov 1952; Bannerman 1958), usually twice repeated. Recording resembles mournful groaning 'AW-e-AW AW-e-AW', the 'e' sound representing momentary rise in pitch (see I). In winter quarters, flight-call said to be more growling but less grating and rattling than in *M. nigra* (Voigt 1950).

CALLS OF FEMALE. In *deglandi* at least, Inciting-call while Chin-lifting described as 'very thin whistle' (Johnsgard 1965; Myres 1959*a*). In nominate *fusca*, call during 'morning-flight' of pair excited, vibrating 'braaa-braaa-braaa' (Koskimies and Routamo 1953); sharp-

I  Sture Palmér/Sveriges Radio  Sweden  May 1950

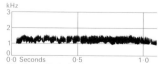

II  Sture Palmér/Sveriges Radio  Sweden  May 1950

sounding 'kerr' (see Bannerman 1958) and harsh 'kerr-kerr' or 'koARRR kew' (see Witherby *et al.* 1939) may be same, but 'kjuorrr kjuorrr' of Bauer and Glutz (1969) apparently different. Call enticing young into water a prolonged, vibrant trill (see II) recalling quavering call of Tawny Owl *Strix aluco*.

CALLS OF YOUNG. No detailed information.

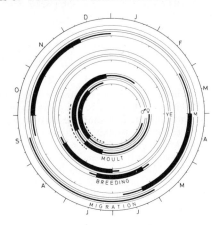

**Breeding.** SEASON. See diagram for Finland and north-west USSR. From mid-June, in arctic USSR. SITE. On ground, in thick vegetation, grass, low scrub; well concealed, rarely in open. Sometimes in woods, at foot of tree. Usually within 100 m of water, can be up to 2–3 km. Will use nest-boxes. Not colonial but often several nests on one islet, as close as 3 m. Nest: shallow cup with rim of small twigs, leaves, and grass; lined down. Building: by ♀, using material within reach of nest. EGGS. Ovate; creamy to buff. 72 × 48 mm (64–78 × 43–52), sample 250; calculated weight 92 g (Schönwetter 1967). Clutch: 7–9 (5–12), more than 12 usually 2 ♀♀. Of 187 Finnish clutches: 6 eggs, 2%; 7, 16%; 8, 33%; 9, 40%; 10, 7%; 11, 1%; 12, 1%; mean 8·43 (Hildén 1964). One brood. Replacement clutches laid after egg loss. Mean size of 19 late clutches, Finland, 8·2 compared with 8·5 for 30 early clutches. Eggs laid 1–2 day intervals (mean 1·69), usually in early morning (Koskimies and Routamo 1953). INCUBATION. 27–28 days (26–29), 27·5 days (range 26–29), sample 23 (Koskimies and Routamo 1953). By ♀. Starts on completion of clutch; hatching synchronous. Eggs covered with down when ♀ off nest. Eggshells left in nest. YOUNG. Precocial and nidifugous. Self-feeding. Cared for by ♀ and brooded while small, but parent-young bond loose and even small young will leave and live independently, though with low survival; other broods amalgamate under 1 ♀ (Koskimies 1955). FLEDGING TO MATURITY. Fledging period *c.* 50–55 days, but not accurately recorded. Independent normally at *c.* 30–40 days. Age of first breeding 2 years, possibly 3. BREEDING SUCCESS. Of 7418 eggs laid, Finland, in 3 years, 84% hatched but only 3·2% reared to fledging; survival per ♀ in 3 years was 0·02, 0·6, and 0·1 young, giving losses of 99·8%, 92%, and 99% (Hildén 1964). Low survival

believed due to (1) tendency to overcrowding in brood rearing areas, (2) extremely loose parent-young bonds, and (3) sensitivity of young to bad weather, including low water temperatures (Koskimies 1955; Hildén 1964). Survival of young to at least 40 days on inland lakes, Finland: 4·3 young per ♀, 1966 (12 broods); 6·4, 1967 (10 broods) (Waaramaki 1968).

**Plumages.** Nominate *fusca*. ADULT MALE BREEDING. Entirely black, except small white crescent below and behind eye, white secondaries and tips of upper greater coverts (sometimes narrowly fringed black) and white-tipped axillaries, median and greater under wing-coverts. Lesser under wing-coverts black, median and axillaries sooty-black tipped white, greater more or less white. In worn plumage, much brown visible at bases of feathers; tips of feathers worn and faded brown, edges grey on upperparts. ADULT FEMALE BREEDING. Head and neck dark brown, darkest on hindcrown; rest of upperparts, chest, and flanks, dark brown with dark olive-brown edges. Feathers of lores and ear-coverts basally white, showing 2 white patches at side of head when dark tips wear off. Lower breast and belly dark brown, feathers often with white margins and bases, giving streaked appearance. Tail dark brown. Wing as ♂, but upper wing-coverts dark brown edged olive-brown; white tips to greater coverts narrower. In worn plumage, 2 distinct white patches on head; upperparts faded, feathers show buff or grey edges, giving barred appearance; much mottling on belly. ADULT MALE NON-BREEDING. Like breeding, but less glossy; in some, a few feathers of scapulars, mantle, and flanks dark brown tipped olive-brown; unmoulted breeding feathers show much fading and wear. ADULT FEMALE NON-BREEDING. Newly moulted feathers resemble fresh breeding; rest of plumage worn breeding, with large white spots at sides of head. DOWNY YOUNG. Like Common Scoter *M. nigra*, but cheeks, sides of neck, and throat purer white, often small white spots on lores; underparts white except for grey-brown band on chest. JUVENILE. Like adult ♀, but dark grey-brown where ♀ black-brown; feathers of upperparts edged grey-brown, chest, sides of breast, and flanks edged buff. Mantle and tertials in juvenile ♂ darker brown, contrasting with grey-edged scapulars, upperparts of juvenile ♀ paler and uniform brown. Feathers of lores with conspicuous buff-white bases; ear-coverts with white bases, tips soon wearing off producing larger pale patches than worn adult ♀ in winter. Feathers of underparts short and narrow; those of belly variable, in ♂ mostly grey-brown with white edges, in ♀ white with grey-brown subterminal spot. Tail-feathers notched at tip, bare shaft projecting. Wing as adult ♀, but coverts paler brown, and white tips narrow in outer greater coverts, almost absent in inner. FIRST IMMATURE BREEDING MALE. From autumn onwards, black develops on sides of head, throat, and lores, first reducing large white spot behind eye (December), later that on lores. In spring, head and neck all black like adult ♂, but no white crescent at eye. From March onwards, some black feathers develop in scapulars, flanks, and chest, few elsewhere; part of belly and usually tail retain juvenile feathers. FIRST IMMATURE BREEDING FEMALE. Acquires new feathers like those of adult ♀ on head, neck, scapulars, and flanks; timing as ♂, but usually fewer feathers moulted. SECOND IMMATURE BREEDING MALE. Like adult ♂ breeding, but less glossy on head, belly browner; in 2nd autumn some juvenile feathers often still present. SECOND IMMATURE BREEDING FEMALE. Like adult ♀, but hindcrown less glossy, and some retain part of juvenile plumage on belly until 2nd winter. (Schiøler 1926; Witherby *et al.* 1939; C S Roselaar.)

**Bare parts.** Nominate *fusca*. ADULT MALE. Iris pale grey to white, with dusky outer ring. Knob on bill black; also central culmen, cutting edges, and often line on each side of nail. Sides of upper mandible orange-yellow, nail orange-red to deep pink. Foot deep red, more orange on inner side tarsus, darker at joints; webs black. ADULT FEMALE. Iris brown. Bill olive-black. Foot dull red, webs black. JUVENILE. Iris grey-brown to brown. Bill grey-brown or green-black, becoming darker during 1st winter; sides paler in ♂ from January onwards, yellow in spring, but nail still dusky. Foot grey-flesh to greyish yellow-brown, webs dull black; foot dull red from late spring. (Schiøler 1926; Witherby *et al.* 1939; RMNH.)

**Moults.** ADULT POST-BREEDING. Body apparently partial; April–May. In ♂, involves part of flanks, scapulars, mantle; also tail (not always complete). In ♀, limited number of new feathers acquired on same parts of body. Flight-feathers simultaneous. In Denmark, flightless for *c.* 3–4 weeks between late July and late August in ♂, late August and early October in ♀ (Joensen 1973*a*). ADULT PRE-BREEDING. Partial; head, body, and tail. In ♂, body October–November, head and neck last; tail before November. In ♀, somewhat later. POST-JUVENILE AND SUBSEQUENT MOULTS. Some new feathers on sides of head November, more January; head and neck (rarely also tail) renewed by spring, when also parts of scapulars, flanks, tail-coverts, chest, thighs, sides of breast or, occasionally, all body moulted. Rest of juvenile feathers, including wing, moulted in late summer like adult; some feathers of belly sometimes not replaced before 2nd autumn.

**Measurements.** Netherlands, winter; skins (RMNH, ZMA).

| | | | | | | |
|---|---|---|---|---|---|---|
| WING AD ♂ | 280 | (4·81; 31) | 269–286 | ♀ 263 | (4·80; 7) | 255–271 |
| JUV | 268 | (5·68; 17) | 260–282 | 251 | (7·00; 22) | 232–262 |
| TAIL AD | 81·2 | (2·81; 26) | 75–89 | 73·4 | (3·55; 7) | 67–78 |
| JUV | 71·9 | (2·84; 15) | 68–78 | 68·0 | (3·43; 21) | 61–73 |
| BILL | 44·9 | (1·71; 47) | 41–51 | 40·8 | (1·87; 27) | 37–44 |
| TARSUS | 48·8 | (1·56; 43) | 46–53 | 45·8 | (1·71; 27) | 43–49 |
| TOE | 78·7 | (2·74; 42) | 74–84 | 71·4 | (3·00; 27) | 66–79 |

All sex differences significant. Wing and tail of juveniles significantly shorter than adult. Tarsus, toe, and bill similar; combined in table.

**Weights.** Autumn and winter. ADULT. Denmark, (1) September–November; (2) December–January (Schiøler 1926).

JUVENILE. (3) Denmark and Netherlands, October–January (Schiøler 1926, ZMA).

| | | | | | |
|---|---|---|---|---|---|
| (1) | ♂ 1642 | (131; 4) | 1517–1793 | ♀ 1533 | (148; 4) 1360–1710 |
| (2) | | 1794 | (171; 5) | 1589–1980 | 1730 (142; 7) 1519–1895 |
| (3) | | 1685 | (301; 6) | 1173–2104 | 1214 (66; 4) 1140–1271 |

Other times of year. ADULT. Greifswalder Oie, East Germany, spring: ♂♂ 1631 and 2034, ♀ 1255. ♂♂ South Baltic: June, 1272; September 1300 (Bauer and Glutz 1969). ♀♀ USSR: Yamal, July, 1100; Arkhangelsk, August, 1230 (Dementiev and Gladkov 1952). Minimum weight, emaciated winter birds Netherlands, ♂ 1024, ♀ 894 (ZMA). JUVENILE. USSR: ♀♀ Rybinsk reservoir, September, 995 (860–1210); ♂ Karelia, May, 1090 (Dementiev and Gladkov 1952).

**Structure.** Nominate *fusca*. Wing pointed. 11 primaries: p10 longest, p9 0–7 shorter, p8 6–19, p7 20–35, p6 35–51, p1 95–124. Primaries not emarginated. Tail short, graduated; 14 pointed feathers. Bill large, broad, swollen at sides of base of upper mandible; slight knob behind slit-like nostrils; culmen gently sloping from knob via nostrils to flat and wide tip of bill. Feathers of lores and forehead extend in straight line from corner of mouth to point *c.* 6–8 mm behind nostril, and then vertically over top of culmen. Outer toe slightly longer than middle, inner *c.* 81% middle, hind *c.* 29%.

**Geographical variation.** Mainly involves structure and colour of bill. Base of upper mandible of East Asiatic (*stejnegeri*) and North American (*deglandi*) populations higher, knob extending over nostrils; feathering at sides of bill extends in both to *c.* 1–4 mm behind nostril, that on culmen as far as at sides of bill in *deglandi*, but somewhat less in *stejnegeri*. In adult ♂♂ *deglandi* and *stejnegeri*, knob distinctly swollen, particularly in latter. Black at sides of upper mandible in both more extensive basally, sides purple-red shading to orange at base, rather than yellow-orange as in nominate *fusca*. Size of white crescent below and behind eye variable in all subspecies, but on average smallest in nominate *fusca*. Sides of body and flanks olive-brown in adult ♂ *deglandi*, black in *stejnegeri*. As nominate *fusca* and *stejnegeri* not known to meet in Yenisey Basin, we prefer to treat *fusca* and *stejnegeri*/*deglandi* as conspecific until sympatric breeding (if any) proved, although bill and trachea structure different. (Dwight 1914; Kortright 1942; Vaurie 1965; C S Roselaar.)   CSR

---

## *Bucephala albeola* Bufflehead

PLATES 90 and 99
[between pages 662 and 663
and after page 686]

DU. Buffelkopeend   FR. Garrot albéole   GE. Büffelkopfente
RU. Малый гоголь   SP. Porrón albeola   SW. Buffelhuvud

*Anas Albeola* Linnaeus, 1758

Monotypic

**Field characters.** 32–39 cm, of which body two-thirds; wing-span 54–61 cm; ♂ generally slightly larger than ♀. Tiny, compact sea-duck with low profile on water, large puffy head and small bill; general pattern of ♂ like small Goldeneye *B. clangula* but with broad band of white round back of head. Sexes dissimilar, with some seasonal differences. Juvenile and immature resemble ♀, but young ♂ distinguishable.

ADULT MALE. Strikingly black and white, with dark part of head black glossed green, purple, and bronze, and broad wedge of white beginning below eye and widening out round back of head; black back and rump paling to grey tail, white outer scapulars, and white neck and underparts shading to grey undertail. In flight, wings black with conspicuous square of white on inner half extending almost to front edge and broad 'braces' of white separating wings from body. Bill small, fine and triangular, blue-grey usually with yellowish line along edge of upper mandible; eyes brown; feet pink. MALE ECLIPSE. Resembles adult ♀ but has blacker head, usually with some purple-glossed feathers, and larger white patch behind eye. ADULT FEMALE. Dark grey-brown head with narrow white oblong patch extending backward from behind eye; sooty-brown back, rump, and scapulars shading to grey tail; grey flanks, whitish breast and belly, and white patch sometimes showing on secondaries. In flight, wing blackish with white rectangle formed by inner secondaries, smaller in area than those of other *Bucephala*. Bill dark grey; eyes brown; feet grey with pink or blue tinge. JUVENILE. Like ♀, but ♂ generally already larger than ♀, with bigger white patch on head and more white in wings. In 1st winter, some ♂♂ start to show adult plumage before end of year, but many not until February or later and full plumage not assumed until 2nd winter.

In flight, looks slightly longer-necked than other *Bucephala*. Size of Teal *Anas crecca* but with white in wings. Large head and general body pattern recalls *B. clangula* but readily distinguished by broad white wedge on back of head of ♂ and white patch behind eye of ♀ (also grey instead of brown head). Head pattern of ♂ recalls Hooded Merganser *Mergus cucullatus*, but that species has cinnamon flanks with 2 black lines down front edge, black neck, and longer saw-bill. ♀ likely to be confused only at considerable distances with ♀ Harlequin *Histrionicus histrionicus* (larger, with more rounded patch behind eye and 2 indistinct patches in front) and ♀ Long-tailed Duck *Clangula hyemalis* in summer plumage (larger still, with whitish streak behind eye but also whitish area round eye and whitish patch on side of lower neck) but neither shows any white on wings in flight.

General behaviour much as *B. clangula* but even more restless and active. Swims buoyantly, dives easily, flies restlessly for short distances, springing quickly from water, without running along surface like other sea-ducks, and usually flying low. Flight swift and dashing, but wings do not produce singing sound of other *Bucephala*. Seldom comes to land. Seen on lakes and rivers, or in estuaries and on sea.

**Habitat.** Mainly boreal and cool temperate waters, standing or slow-flowing, such as pools or lakes and connecting channels, set among wooded lowlands or uplands affording suitable nestholes in trees; infrequently in treeless areas where nest-sites available only in earth banks. Winters both on inland waters and along shallow, sheltered coasts, in bays, frequently in breaking surf, normally within *c.* 5 m depth limit. Outstanding as diver, but not often on land where movement clumsy and restricted. Mostly flies fairly low above water with easy take-off, sometimes directly from below surface.

**Distribution.** Breeds in forested areas of North America from central Alaska to southern Yukon, western Macken-zie, Saskatchewan, middle and southern Manitoba, and sparsely Ontario south to British Columbia, northern Montana, and locally Oregon and north-east California. Formerly bred Wyoming, north Iowa, and south-east Wisconsin.

Accidental. Britain: ♂ *c.* 1830, one winter 1864–5, ♀ January 1920, ♀ February 1932, and ♂ February–March 1961; all England (British Ornithologists' Union 1971). Iceland: ♂ November 1956 (FG). Czechoslovakia: adult ♀ shot March 1885 (K Hudec).

**Movements.** Migratory. Most withdraw from breeding range, wintering from Commander and Aleutian Islands, Alaskan Peninsula, British Columbia, northern Montana, Great Lakes, and Canadian Maritime Provinces south to Baja California, Mexico, Gulf States, and north Florida; main concentrations in coastal states from southern British Columbia to California, and from New England to maritime Texas. Ringing shows Alaska and British Columbia populations migrate to western areas; those of Saskatchewan and Manitoba south and south-east (major routes Mississippi valley and over Great Lakes) towards Gulf of Mexico and Atlantic. Ill-defined migratory divide in Alberta, with large area from which birds migrate south-west or south-east. ♂♂ make moult movements within breeding range, beginning June or July. Autumn migration starts when ♀♀ also moulted, beginning September, but main departures October; first reach wintering areas (except extreme south-east) by end of month, passage continuing into first half December as more northern areas closed by ice. Return passage begins mid-February in south (♂♂ on average earliest); under way everywhere by mid-March; in strength southern parts breeding range April, later spread north dependent on thaw.

Vagrants, mainly autumn and winter, recorded Kamchatka, Kurile Islands, Japan, Hawaii, Greenland, West Indies, Bermuda, and Europe. See Erskine (1972) for comprehensive analysis.

**Voice.** Generally silent, but rapid, guttural 'ik-ik-ik-ik' given at times by ♀; only sound reported from ♂ a brief growling (Myres 1959*b*; Erskine 1972).

**Plumages.** ADULT MALE BREEDING. Head and upper neck black, glossed violet, purple, bronze, and green; large wedge starting below eye and widening to hindcrown white. Lower neck, upper mantle, and underparts white, flanks narrowly bordered black above and behind; vent and under tail-coverts washed pale grey. Lower mantle, inner and longer scapulars, tertials, back, and centre of rump black; outer scapulars white with black edge on outer webs. Sides of rump and upper tail-coverts light grey. Tail dark grey, usually tipped white. Primaries and some outer secondaries sooty black on outer webs, sepia on inner. Leading edge of wing and inner lesser coverts black, rest of upperwing, including most secondaries, white. Underwing and axillaries mottled dusky and white. ADULT FEMALE BREEDING. Head and neck dark slate, darkest and slightly glossed purple on crown. Patch on cheek behind eye white. Upperparts dark grey-brown, almost black on centre of mantle and back, mantle feathers occasionally edged paler. Chest, flanks, vent, and under tail-coverts grey, feathers tipped white when fresh; rest of underparts white. Tail and underwing like ♂. Primaries and some outer secondaries sooty with paler inner web, rest of secondaries white, sometimes tipped or edged black. Upper wing-coverts dark grey-brown; greater coverts usually spotted white on outer web, or whole tip white with black margin. ADULT MALE NON-BREEDING (Eclipse). Like adult ♀ breeding, but head and upperparts darker, some feathers at sides of head glossed purple, white patch behind eye slightly larger, tail-coverts greyer; wing like adult ♂ breeding. ADULT FEMALE NON-BREEDING. Like ♀ breeding, but browner, less greyish. JUVENILE. Like adult ♀ breeding, but head, upperparts, chest, and flanks paler brown, feathers narrowly edged buff on upperparts, chest, and flanks. White cheek-patch smaller. Feathers of belly and vent narrow, short; grey-brown, narrowly tipped white, giving mottled appearance to underparts. Tail-feathers narrow, notched at tip, bare shaft projecting. FIRST IMMATURE NON-BREEDING. In early autumn, feathers of head, upperparts, sides of body, and tail changed for 1st non-breeding; later in autumn, also underparts. Head darker than juvenile, light cheek-patch more extensive, body greyish instead of brownish. Resembles adult non-breeding, but wing juvenile, with narrow and somewhat abraded coverts and short, straight tertials. Upper tail-coverts in 1st non-breeding ♂ pale grey-brown, contrasting with darker back and tail; in ♀ uniformly dark. FIRST IMMATURE BREEDING. From late autumn onwards, ♂ new scapulars and flank feathers like 1st non-breeding but with some white centres in ♂; later, mainly in spring, new feathers like adult breeding. Breeding plumage also develops on head and elsewhere. In early summer, immature ♂♂ combine worn 1st immature non-breeding and fresh 1st immature breeding feathers. During summer, part of 1st breeding usually retained; no full non-breeding like adult. SECOND IMMATURE BREEDING. Like adult. (Erskine 1972; Kortright 1942; Witherby *et al.* 1939; C S Roselaar.)

**Bare parts.** ADULT MALE. Iris dark brown. Bill blue-grey, cutting-edge sometimes pink or yellow; nail, tip, and base dusky. Leg and foot pink or pale flesh. ADULT FEMALE. Iris dark brown. Bill dark grey, tinged blue behind nail. Leg and foot purple-grey or lilac-pink, darker on webs and joints. (Kortright 1942; Witherby *et al.* 1939.)

**Moults.** ADULT MALE POST-BREEDING. Complete. Head and body early May to mid-July, followed by wing mid-July to mid-August; tail prolonged. Flight-feathers simultaneous; flightless

for *c.* 3–4 weeks. ADULT MALE PRE-BREEDING. Partial; head and most of body. Starts when wing full-grown, mostly finished before late September. ADULT FEMALE POST-BREEDING. Complete. Head and body start before ♀♀ leave broods, about July to early August; followed by and overlapping with wing moult August–September. ADULT FEMALE PRE-BREEDING. Partial. Head and body; immediately follows post-breeding body moult. Part of body plumage probably moulted only once a year. POST-JUVENILE AND SUBSEQUENT MOULTS. Whole juvenile plumage except wing replaced by 1st immature non-breeding August–September. From about October, some ♂♂ replace number of 1st non-breeding feathers for 1st immature breeding on head, scapulars, flanks, and (after moult-stop during winter) elsewhere. Full breeding sometimes attained March–June, but juvenile wing and usually part of 1st non-breeding plumage retained. Not known whether ♀♀ attain 1st immature breeding plumage. During summer, old 1st non-breeding and breeding feathers changed for 2nd non-breeding, but fresh 1st breeding feathers often retained; probably replaced directly by 2nd breeding in autumn. (Erskine 1972.)

**Measurements.** ADULT. Ranges only, combining 12 of each sex from Erskine (1972) and 4 of each sex from RMNH.

|   | WING | TAIL | BILL | TARSUS | TOE |
|---|------|------|------|--------|-----|
| ♂ | 169–179 | 67–68 | 27–31 | 31–35 | 54–60 |
| ♀ | 151–161 | 59–70 | 23–27 | 28–31 | 48–55 |

**Weights.** Full-grown. North America (Erskine 1972).

| (1) | ♂ | 433 | (21) | 270–570 | ♀ | 313 | (13) | 260–365 |
|-----|---|-----|------|---------|---|-----|------|---------|
| (2) |   | 457 | (22) | 400–560 |   | 336 | (12) | 250–425 |
| (3) |   | 429 | (9)  | 390–480 |   | 301 | (61) | 260–400 |
| (4) |   | 473 | (55) | 340–600 |   | 350 | (33) | 230–470 |

(1) December–March (winter); (2) April–May (spring migration, optimal development of gonads, ♀ laying); (3) June–August (incubation, wing moult); (4) September–November (autumn migration).

**Structure.** 11 primaries: p10 and p9 about equal, longest; p8 4–8 shorter, p7 8–16, p6 21–26, p1 63–76. P10 slightly emarginated on inner web, p9 on outer. Tail short, rounded; 16 feathers. Bill like other *Bucephala*, but much smaller, with nostrils relatively nearer base of bill. Feathers of crown and nape, and to lesser extent those on sides of head, elongated in adult ♂ breeding; form short dense crest. Outer toe about equal to middle, inner *c.* 76% middle, hind *c.* 30%.

CSR

---

## *Bucephala islandica* Barrow's Goldeneye

PLATES 91 and 99
[between pages 662 and 663
and after page 686]

Du. IJslandse Brilduiker     Fr. Garrot d'Islande     Ge. Spatelente
Ru. Исландский гоголь     Sp. Porrón islándico     Sw. Islandsknipa

*Anas islandica* Gmelin, 1789

Monotypic

**Field characters.** 42–53 cm, of which body two-thirds; wing-span 67–84 cm; ♂ rather larger than ♀. Medium-sized, stocky sea-duck with flat-crowned, bulbous, oval head and short, stubby bill; ♂ has purple-glossed head with white crescent behind bill and black and white body. Sexes dissimilar, with some seasonal differences. Juvenile

and immature resemble ♀, but distinguishable.

ADULT MALE. Strikingly black and white, with black head glossed purple and violet, bold white crescent between eye and bill, black back and rump shading to grey tail, row of 6–7 large white blobs on otherwise black scapulars, and white neck and underparts. In flight, wing

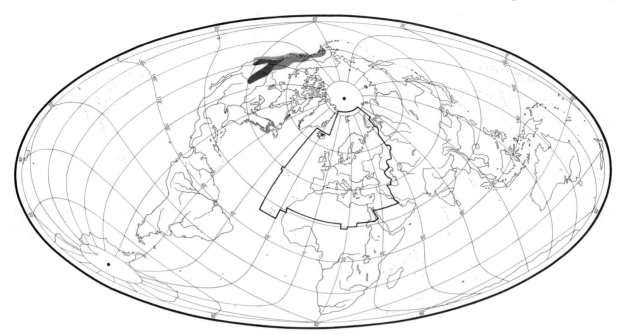

black with conspicuous square of white on inner half covering rear two-thirds and broken by transverse black line near front. Bill black; eyes bright yellow; feet yellow-orange with grey-brown webs. ADULT MALE NON-BREEDING. Resembles adult ♀ but has greyer head, and all-black bill. ADULT FEMALE. Less bulbous, but still puffy, chocolate-brown head (no white patch), often inconspicuous white collar, mottled blue-grey upperparts extending in band across upper breast, black rump and grey-brown tail, white underparts with grey flanks and undertail, and white showing on coverts and secondaries of otherwise black wings. In flight, wing has conspicuous square of white as ♂, but extending in greyish shade nearly to front edge and broken by 2 black lines. Bill black, usually with large yellow patch near tip (some north American birds have all-yellow bill); eyes pale yellow; feet dull orange with grey-brown webs. ADULT FEMALE NON-BREEDING. Duller head, no collar, and browner breast, sides and flanks. JUVENILE. Like non-breeding ♀ (no collar), but ♂ already larger than ♀. By 1st winter, both sexes resemble adult ♀ (though 1st-winter ♂ often noticeably larger), but with only ill-defined whitish collar; ♂♂ extremely variable in assuming adult plumage and partial loral patch may start to appear in 1st autumn.

In flight, looks large-headed and short-necked. Confusion likely only with Goldeneye *B. clangula*, but looks larger and bulkier (though much overlap in size), and has shorter, deeper, stubbier bill which tapers markedly towards tip, and vertical forehead, low rounded crown, and distinct mane at back of head producing oval shape (instead of peaked crown and 'triangular' shape). ♂ has purple (rather than green) gloss on head with large loral crescent of white (instead of round blob), but care needed

with partly moulted *B. clangula* where developing patch may appear as small crescent; while much less white on scapulars (row of white blobs instead of close, white streaking forming solid patch) gives blacker appearance to upperparts and black extends forward in pointed spur in front of wing. ♀♀ difficult to distinguish, but shape, and darker, richer brown colour of head and deeper, stubbier bill with more yellow at tip useful guides. In flight, both sexes of *B. islandica* show smaller area of white on wing, not extending so far forward in ♂ and greyer at front in ♀.

General habits similar to *B. clangula*. Wings produce similar singing noise. More frequently seen in fast-flowing streams, but this applies chiefly to breeding season when they replace each other geographically in west Palearctic.

**Habitat.** Mainly cold but ice-free, inland waters of subarctic and arctic-alpine zones, ranging in Rocky Mountains to *c*. 3000 m, but mainly below 1000 m and mostly 200–500 m. Requirements for habitat appear inflexible compared with most Anatinae. Favours both large and small, clear, productive lakes and pools with little floating or emergent vegetation and of medium depth. Strongly alkaline waters often preferred. On rivers, rapids and torrents frequented, although to less extent than Harlequin *Histrionicus histrionicus*. Nests in tree-holes where available, up to *c*. 16 m above ground, but contrasts with Goldeneye *B. clangula* in being able to flourish in treeless regions where uses dense dwarf scrub, and even cavities in abandoned artefacts such as walls and sheep-folds, as well as nest-boxes. Although normally by water, such sites may be 1 km or more distant. At times will nest in open.

Differs from most northern breeders among Anatinae in

shifting for non-breeding season only to minimum necessary to reach ice-free waters. Corresponding change of habitat involves use of lowland, standing, or running waters, and to some extent of sheltered, marine bays and inlets. Resort to land minimal, but local airspace freely used at low and intermediate levels. Not preferred quarry species for hunters (Kortright 1942). Apparently stable current status may be misleading in view of indications of possible relict stage after earlier wider distribution (Voous 1960).

**Distribution.** ICELAND: only breeding area in west Palearctic. Breeding range in eastern Canada inadequately known (Godfrey 1966); has bred south-west Greenland, but true status uncertain (Vaurie 1965).

Accidental. Faeroes: 1944, 1945, 1949. Norway: 1848, 1851. Spitsbergen: 1965. USSR: 2 specimens on western lakes, in second decade 20th century and 1935. Finland: 1958, 1964. West Germany: 1955, 1956. East Germany: 1853, 1957. Poland: 1957. France: 1829, 1834.

**Population.** ICELAND: *c*. 800–1000 pairs, Lake Mývatn (Bengtson 1972c); population stable since 1960, *c*. 800 breeding pairs, 1975 (A Gardarsson).

**Movements.** Migratory, partially migratory, or resident in different regions. Icelandic breeders resident, over-wintering mainly in breeding areas or on other large ice-free inland waters (Boyd 1959). Breeders of Pacific coast North America partially migratory, wintering southern portion of breeding range (south British Columbia to north California); small numbers reach San Francisco Bay. Population of east Canada migratory, wintering from northern Gulf of St Lawrence along coasts to Maine, a few to upper River St Lawrence and Great Lakes and south to New York; has straggled further south and to Newfoundland (Godfrey 1966). Those wintering north-east USA arrive later and depart earlier than Goldeneye *B. clangula*; extreme dates 18 September and 23 April, but normally present in strength only late December to February (Bauer and Glutz 1969). Small numbers recorded all seasons, but mainly winter, in south-west Greenland may represent separate breeding population or eastward movement from Labrador (Vaurie 1965; Salomonsen 1967). Vagrants to west and north Europe (see Distribution) thus possibly of New World rather than Icelandic origin.

**Food.** Mainly insects, molluscs, and crustaceans, obtained by similar methods to Goldeneye *B. clangula*, chiefly by surface-diving and rarely by up-ending and dabbling. Inland, often feeds at depths 0·5–3·5 m, in running water and in shallow water close to shore where vegetation sparse or absent.

On Lake Mývatn, Iceland, in summer: 24 adult ♀♀ had fed mostly on chironomid larvae (75% by frequency) and molluscs *Lymnaea* (25%), and to less extent on Simuliid

larvae; 24 adult ♂♂ obtained over same period had fed chiefly on chironomid larvae (42%), *Lymnaea* (33%), and fish eggs (17%), especially of 3-spined stickleback *Gasterosteus aculeatus*. In both groups, seeds (mostly pondweeds *Potamogeton*), though frequent not taken in any appreciable quantity. Fish eggs taken only June–July; slight increase as summer progresses in use of chironomid larvae and decrease in *Lymnaea* (Bengtson 1971). In Greenland, mainly molluscs (*Margarita helicina* and *Modiolaria faba*) and small crustaceans (Salomonsen 1950a). 71 stomachs from different seasons and areas of North America, had chiefly insects (36·4% of total volume), especially dragonfly nymphs (Odonata), caddisfly larvae (Trichoptera), and waterboatmen (*Corixa* and *Notonecta*); also molluscs (19·2%), especially blue mussel *Mytilus edulis*, and crustaceans (17·7%), particularly amphipods. Plant materials 22·3% of total, consisted mainly of pondweeds, especially *Potamogeton*. Molluscs and crustaceans taken more in winter, insects and plants more in summer (Cottam 1939).

Of 34 ducklings, newly-hatched and small ducklings ate mainly *Lymnaea*, with some chironomid larvae and adults; older ducklings ate more chironomid larvae until fully grown when these formed nearly entire diet (Bengtson 1971).

**Social pattern and behaviour.** 1. Apparently similar to Goldeneye *B. clangula* in most respects. Gregarious except when nesting. Flocks seem mainly small and scattered in winter, but may be larger at times during and after movement between breeding area and sea; up to 200 reported in North America (see Bent 1925). ♂♂ first to flock, while ♀♀ incubating, initially in small parties; leave for sea in company of immatures. BONDS. Monogamous pair-bond of seasonal duration indicated. Pair-formation starts in winter flocks and continues on inland breeding waters in spring. ♂ deserts ♀ during incubation. Only ♀ tends young. BREEDING DISPERSION. Nests usually solitary and pair probably territorial after break-up of spring flock. Immatures occur in breeding area, but on Lake Mývatn, Iceland, numbers small; 1st year ♀♀ spend much of first spring seeking nest-sites (Bengtson 1966a). ROOSTING. Little information; probably much as in *B. clangula*.

2. Studied in captivity by Johnsgard (1965) and in wild chiefly by Myres (1957, 1959b); see also Sawyer (1928), Munro (1939). Behaviour in west Palearctic little known. In many respects, closely similar to *B. clangula* but with some significant differences in heterosexual displays associated with pair-formation. FLOCK BEHAVIOUR. Probably much as in *B. clangula*; pre-flight behaviour similar though wing-stretching and bathing movements not reported (see McKinney 1965b). ANTAGONISTIC BEHAVIOUR. Like all *Bucephala*, highly aggressive at times to conspecifics; will also threaten and chase other waterfowl (Myres 1957; McKinney 1965a). In Head-forward display (or Threat-crouch), ♂ floats or swims with neck stretched low along water; often immediately precedes underwater attack (McKinney 1965a). Hostile encounters between pairs in breeding area frequent (Munro 1939): ♂♂ direct Head-forward posture and underwater attacks at other ♂♂, approaching closer by wing and flying short distances. ♂♂ also recorded directing splashing attacks across surface at one another, then rising breast to breast and hitting with wings (see Bent 1925). ♀♀ at times show similar

A

B

C

threat and underwater attack behaviour as ♂♂; Inciting-display (Jiving of Myres 1957) more frequent than in *B. clangula* and different in form, consisting solely of silent pointing movements of head alternately from side to side. COMMUNAL COURTSHIP. Masthead, Bowsprit, and simple Head-throw of *B. clangula* absent. Main display, Rotary-pumping, homologous with somewhat similar Bowsprit-pumping of *B. clangula*, but differs in that bill held almost horizontally, movements of head rotary, and soft call given. Head-throw-kick display much as in *B. clangula*, but of fast type only and, as head flung back and call given, bill turned slightly sideways towards ♀. This frequently preceded by Crouch-display; closely similar to Head-forward threat, but ♂ makes clicking noises with bill open. Upward-shakes, Wing-flaps, and Short-flights towards ♀ also common. As in *B. clangula*, responses of ♀ in courting parties more complex than in many ducks; include Rotary-pumping like ♂, swimming in Head-up posture, and Inciting-display. Latter often performed while ♀ follows favoured ♂ but lacks component orientated towards that ♂; Neck-dip-kick display of ♀ *B. clangula* rare or absent. In response to Inciting, ♂ often swims ahead of ♀ while performing lateral Head-turning movements or assumes Head-up posture, while opening and closing bill and calling, and periodically performs Neck-withdrawing display by pulling head back and down near shoulder (see A). PAIR BEHAVIOUR. Pair-courtship consists mainly of pre-copulatory displays and fuller copulatory sequences (but see *B. clangula*). During site-seeking by ♀ in breeding area, pair fly overland; in Iceland, usually when slight drizzle (Gudmundsson 1961; Bengtson 1966a). COPULATION. Almost identical with that of *B. clangula* (Sawyer 1923; Myres 1951, 1959a; Johnsgard 1965; McKinney 1965b). Copulatory sequences may start with mutual Ceremonial-drinking (less exaggerated than in *B. clangula*), but ♀ usually soon adopts Prone-posture on water, often maintaining it for several minutes (see B–C). ♂ sometimes performs Rotary-pumping but this gives way to 1st stage of true pre-copulatory displays, i.e. repeated Ceremonial-drinking (Water-flip of Myres 1959a; see B) and Wing-Leg-stretches, though Ritual-bathe, slight Water-twitching, and Both-Wings-stretch also recorded. Just prior to mounting, ♂ performs vigorous Water-twitching (or Jabbing), splashing water by series of head-shakes with bill in water; then suddenly does single Preen-Back-between-Wings on side nearest ♀ before swimming rapidly towards her with head erect (Steaming; see C). While mounted, ♂ does single Wing-shake at intromission; after dismounting, retains grip on ♀ while pair rotate on water for several seconds, then Steams away while performing lateral Head-turns and calling. Afterwards, ♂ may do Wing-Leg-stretch but more usually only bathes and completes sequence with body-shake and Wing-flap. ♀, meanwhile, also bathes. RELATIONS WITHIN FAMILY GROUP. Not studied in any detail.

(Figs A–C after Johnsgard 1965.)

**Voice.** Little used except by ♂ during courtship. Incompletely studied, but evidently not similar to

Goldeneye *B. clangula*. Shares with that species distinctive, clearly audible wing-whistling in flight (see I), but less loud and less metallic (Millais 1913) and possibly slightly higher pitched, with wing-beat rate of *c.* 7–8 per s compared to *c.* 9 per s in shorter winged *B. clangula*.

CALLS OF MALE. 2 main Courtship-calls described by Johnsgard (1965): (1) weak, grunting 'ka-KAA' accompanying Head-throw-kick display (quite different from corresponding call of *B. clangula*), with similar but softer version given during Rotary-pumping display; (2) clicking call during Crouch and Neck-withdrawing displays (for analysis, see Johnsgard 1971). Call 2 nasal, rather wooden, resembling vibrations of teeth of comb (see II). Repeated grunting call uttered during post-copulatory display (Johnsgard 1965).

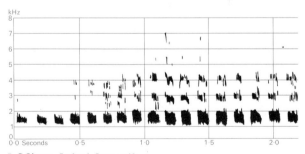

I S Jónson Iceland June 1968

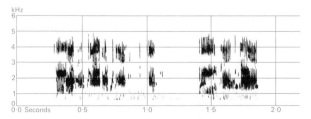

II P J Sellar Iceland May 1967

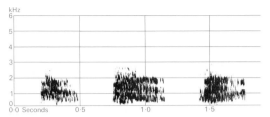

I S Jónson Iceland June 1968

CALLS OF FEMALE. No calls during courtship mentioned by Johnsgard (1965) and Inciting-display apparently performed silently. Flight-call (probably uttered in other situations) emphatic, coarsely slurred, growling 'arrr arrr' (see III); also rendered 'gä-gä-gä-gair' or 'grrr gärr' and described as a rough, high-pitched croak (Larrison and Sonnenberg 1969).

CALLS OF YOUNG. No information.

**Breeding.** SEASON. Mean date of first laying, Lake Mývatn, Iceland, 1961–70, varied from 8–12 May; average 10 May (Bengtson 1972a). SITE. Most often in hole or deep crevice; in Iceland, especially in lava and rock, but in North America in holes in trees. Also found in thick vegetation always well concealed under overhanging low scrub. Will use nest-boxes, even on sides of houses in Iceland. Rarely far from water. Nest: depression, with little or no material in hole, or with grass and leaves when in vegetation; lined down. Building: by ♀. EGGS. Ovate; bluish-green. 62 × 45 mm (57–65 × 42–48), sample 153 (Gudmundsson 1961). Calculated weight 70 g (Schönwetter 1967). Weight of captive laid eggs 64 g (58–66), sample 6 (J Kear). Clutch: 8–11 (6–14); larger clutches by 2 ♀♀. Mean of 49 dump nests, Iceland, 17·4; mean of 179 clutches, 10·4 (Bengtson 1972a). One brood. Replacement clutches laid after egg loss; mean of 39, Iceland, 7·5 (Bengtson 1972a). INCUBATION. 28–30 days, By ♀. Starts on completion of clutch; hatching synchronous. Eggs covered with down when ♀ off. Eggshells left in nest. YOUNG. Precocial and nidifugous. Self-feeding. Cared for by ♀. Brooded while small. FLEDGING TO MATURITY. Fledging period not recorded. Age of first breeding 2 years. BREEDING SUCCESS. Average hatching success of 246 nests, Iceland 1961–70, 75/0%; range 31·5–85·0% (Bengtson 1972a).

**Plumages.** ADULT MALE BREEDING. Differs from Goldeneye B. clangula in purple gloss on head and neck; large white crescent along most of base of upper mandible, extending into point on side of forehead; black of mantle extends down to side of breast in front of wing; white only on outer row of scapulars (which are shorter and broader), with broad black edges elongated in point beyond white centre of feather; longer flank feathers broadly tipped black; dark bases to white greater upper wing-coverts not covered by short white median coverts, so form broad black wing-bar; lesser upper wing-coverts black. ADULT FEMALE BREEDING. Like B. clangula, but head darker (dark chestnut with

slight purple gloss in some); brown extends lower on neck, white neck-ring nearer body; ragged crest from hindcrown to hindneck; grey extending farther down on chest and sides of breast; black on bases of greater upper wing-coverts more extensive; upper row of median coverts darker grey, broadly tipped white; lesser coverts black, tipped dark grey. ADULT NON-BREEDING. Like adult ♀ breeding, but head duller and greyer chestnut, neck-band grey; wing of ♂ like ♂ breeding. DOWNY YOUNG. Like B. clangula, except for shape and colour of bill. JUVENILE. Like B. clangula, but head darker brown; grey on chest and sides of breast more extensive; upperwing darker, with all median and lesser coverts dull black—some median coverts of lower row tipped pale grey in ♂, but only a few rarely in ♀.

**Moults.** Few specimens examined indicate no difference from B. clangula.

**Bare parts.** Like Goldeneye B. clangula, but adult ♀ has broader yellow band behind nail of bill in late winter and spring; yellow sometimes extending below nostrils along sides of upper mandible. In population of west North America (where sympatric with B. clangula), whole upper mandible, except nail, yellow, with a few dusky markings. Bill of downy young dull black, with flesh tip and reddish-grey nail; foot grey-green. (Bauer and Glutz 1969; Delacour 1959; Kortright 1942; Schiøler 1926.)

**Measurements.** ADULT. Iceland; skins (RMNH, ZMA).

| | | | | | | |
|---|---|---|---|---|---|---|
| WING | ♂ 237 | (6·10; 8) | 229–248 | ♀ 218 | (4·15; 5) | 211–221 |
| TAIL | 88·9 | (2·10; 8) | 86–91 | 85·0 | (3·08; 5) | 81–88 |
| BILL | 33·0 | (0·83; 8) | 32–34 | 30·7 | (1·10; 5) | 29–32 |
| TARSUS | 43·0 | (2·27; 8) | 38–46 | 38·0 | (1·00; 5) | 37–39 |
| TOE | 69·4 | (2·45; 8) | 67–73 | 60·8 | (3·11; 5) | 58–64 |

Sex differences significant. Wing, tarsus, and toe significantly longer than those of B. clangula.

JUVENILE. Wing and tail on average c. 10 mm shorter, other measurements similar to adults; e.g. Iceland, wing (Schiøler 1926) ♂ 227 (5) 220–233; ♀ 209 (11) 200–217.

**Weights.** Few published data suggest somewhat heavier weight than B. clangula. Winter range adult ♂♂ 1191–1304; ♀♀ 737–907; juvenile ♂ 1077 (Bauer and Glutz 1969; Kortright 1942).

**Structure.** 11 primaries: p9 longest, p10 2–4 shorter, p8 3–9, p7 14–21, p6 21–36, p1 80–105. Inner web of p10 and outer of p9 emarginated. Bill relatively higher, shorter, and more massive than Goldeneye B. clangula; seen from above, bill narrows gradually to pointed tip, instead of sides parallel and tip rounded as in B. clangula. Cutting edge of upper mandible somewhat upcurved behind nail, and tip of bill slightly elevated. Nail longer and broader than B. clangula: breadth of B. islandica at least 6 in ♂, 5 in ♀ (Dementiev and Gladkov 1952). Elongated feathers form dense crest, especially in hindneck; in adult ♂ from forecrown to hindneck, in adult ♀ from hindcrown downwards. Outer scapulars short, broad, and bifid in adult ♂ breeding (long, narrow, and square-tipped in B. clangula). High forehead in both sexes due to protuberance on skull (Kortright 1942). Outer toe c. 98% middle, inner c. 78%, hind c. 30%. Rest of structure as B. clangula.

**Geographical variation.** None except colour of bill of ♀ (see Bare parts). CSR

## *Bucephala clangula* **Goldeneye**

PLATES 92 and 99
[between pages 662 and 663
and after page 686]

Du. Brilduiker     Fr. Garrot à oeil d'or     Ge. Schellente
Ru. Гоголь     Sp. Porrón osculado     Sw. Knipa     N.Am. Common Goldeneye

*Anas Clangula* Linnaeus, 1758

Polytypic. Nominate *clangula* (Linnaeus, 1758), Europe, north Asia. Extralimital: *americana* (Bonaparte, 1838), North America.

**Field characters.** 42–50 cm, of which body two-thirds; wing-span 65–80 cm ♂ rather larger than ♀. Medium-sized, stocky sea-duck with high-crowned 'triangular' head and short bill. Sexes dissimilar, with some seasonal differences. Juvenile and immature resemble ♀, but distinguishable.

ADULT MALE. Strikingly white and black, with black head glossed green, bold circular white patch between eye and bill, black back and rump shading to grey tail, broad outer scapulars predominantly white with black edges, and white neck and underparts. In flight, wing black with conspicuous square of white on inner half extending almost to front edge and partly broken by black line extending inwards from primary coverts. Bill black; eyes bright yellow; feet pale orange with blackish webs. MALE ECLIPSE. Resembles adult ♀, but has darker brown head with blacker lores and usually some green-glossed feathers, and more white on scapulars and wing-coverts. ADULT FEMALE. Less peaked, chocolate-brown head (no white patch), often inconspicuous white collar, mottled blue-grey upperparts extending in band across upper breast, black rump and grey-brown tail, white underparts with grey flanks and undertail, and white showing on coverts and secondaries of otherwise black wings. In flight, wing black with conspicuous square of white as ♂, but partly or completely broken by two black lines. Bill black with yellow band near tip; eyes pale yellow to white; feet dull yellow-orange with blackish webs. FEMALE NON-BREEDING. Paler brown head, no collar, and browner breast, sides and flanks. JUVENILE. Like non-breeding ♀ (no collar), but ♂ already larger than ♀ and usually with larger whitish patch on folded wing. By 1st winter, both resemble adult ♀ (though 1st-winter ♂ often noticeably larger), but with little or no white on scapulars (none in 1st-winter ♀) and only ill-defined whitish collar. Some ♂♂ start to show adult plumage on head as early as October, but many not until January or later and green-glossed black head and partial or complete loral patch may appear at any time from October to March or April; full adult plumage not assumed until 2nd winter.

In flight, looks large-headed and short-necked. Combination of head shape, white loral patch, and white and black pattern of ♂, and chocolate and grey pattern and white neck ring of ♀, make confusion with any other species unlikely, except for similar Barrow's Goldeneye *B. islandica* (which see for differences). Unlike pochards *Aythya*, *Bucephala* ♀♀ show good deal of white on closed wing.

In summer, on lakes and rivers in wooded areas, but during rest of year may be seen on large, open sheets of fresh water or on sea, where eschews shelter even in rough weather. Shy and usually keeps apart from other ducks. When swimming purposefully, holds head forward with neck extended, but at other times head is sunk on shoulders. Swims buoyantly and dives continuously. Active and restless, springing easily from water and generally flying rather low. Wing-beats rapid and produce distinctive singing sound, loudest in adult ♂. Walks easily with erect carriage, but seldom comes to land. In breeding season, ♀ often perches on large branches (but flies directly in and out of nest hole). Generally rather silent, except during display.

**Habitat.** For breeding requires tall forest growth (with hollow trees), close to lake, pool or river of adequate productivity, neither very shallow nor very deep, and having largish areas of rather cold open water, free of emergent, floating, and even much submerged vegetation, and at least seasonally free of ice. Biogeographic and climatic limitations restrict such conditions to fairly narrow belt in upper middle latitudes, overlapping wooded steppe to south and wooded tundra and arctic-alpine belts above Arctic Circle, with main concentration in coniferous forest zone. These habitat requirements so critical that evidence frequent of competition for relatively few suitable breeding holes in appropriately located trees. Man has in some regions intensified this pressure by deforestation and by eliminating hollow trees. In other regions pressure eased by systematic provision of artificial nest-boxes—for centuries in some areas. Climatic fluctuations also affect situation by extending or shrinking tracts of suitable forest and bodies of water, with result that colonisation and recessions occur, especially along southern fringes of regular breeding range. So far as known, arboreal-aquatic habitat requirements remain firmly defined, and allow little flexibility.

Apart from surprising arboreal capability, only minimally terrestrial, but performance both on and beneath surface outstanding in still, flowing, and even turbulent water, turning over stones on bottom to depths of *c.* 5–8 m. On passage, use made of almost any water, standing or flowing, natural or artificial, inland or marine. In winter

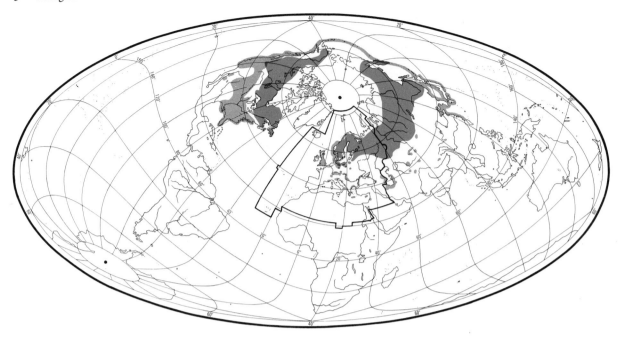

also, both less gregarious and more opportunist than most relatives, resorting indiscriminately to fresh and salt water with, however, some apparent bias towards estuaries and marine bays, sheltered shallow coasts, and sewage outfalls. Airspace over water and land freely used at wide range of altitudes.

**Distribution.** In northern central Europe appears to have bred west of Vistula only for last 130 years, and west of Oder only for last 100 years. Recently, still slight tendency to expansion in north (west Mecklenburg and Holstein) but further south (Moritzburg and western Lausitz) expansion halted and even reversed (Bauer and Glutz 1969). CZECHOSLOVAKIA: first bred southern Bohemia 1960 (Hudec 1970). NORWAY: no longer breeding in western Norway, but slight, recent spread in south (Haftorn 1971). USSR: marked contraction in south in historic times due to destruction of forests (Dementiev and Gladkov 1952).

Nests irregularly on edges of range and sporadically well outside normal range. BRITAIN: said to have bred Cheshire, England, in 1931 and 1932 (British Ornithologists' Union 1971); breeding in small numbers, Inverness, Scotland 1970–4 (Sharrock *et al.* 1975). BULGARIA and RUMANIA: bred in small numbers around 1900 in Dobruja, lower reaches of Danube and delta (Vasiliu 1968; Bauer and Glutz 1969). YUGOSLAVIA: has bred Montenegro; last record 1956 (Matvejev and Vasić 1973). SWITZERLAND: unsuccessful breeding 1955 (WT).

Accidental. Faeroes, Portugal, Spain, Algeria, Morocco, Malta, Cyprus, Israel, Iraq (formerly fairly common in winter), Jordan, Azores.

**Population.** Breeding. FINLAND: commonest diving duck, *c.* 50 000 pairs (Merikallio 1958); increased rapidly in 1950s due to provision of nest-boxes (Grenquist 1970). SWEDEN: most frequent freshwater diving duck, commoner in past, but no significant change last 10 years; nesting sites reduced by modern forestry management, but offset in some areas by erection of nest-boxes (Curry-Lindahl *et al.* 1970). WEST GERMANY: *c.* 100 pairs (Szijj 1973). EAST GERMANY: *c.* 70 pairs (Isakov 1970a). POLAND: more numerous than formerly (Tomiałojć 1972). USSR: *c.* 120 000 pairs in Europe and west Asia, with 13 500 pairs Karelia and Kola peninsula, 60 000 northern Dvina and Pechora, 1000 upper Volga, 5000 Baltic, and 1500 Volga–Kama (Isakov 1970a, b).

Winter. Estimated total population in northern and central Europe *c.* 200 000 birds, with 170 000 in Denmark and west Baltic, 20 000–30 000 in Netherlands, Scotland and Ireland, and 20 000–30 000 on larger lakes of central Europe (Atkinson-Willes 1975). USSR: 52 200 in western part in not so severe winters (Isakov 1970c).

Survival. Annual mortality (breeding ♀♀) Sweden: *c.* 37%; adult life expectancy *c.* 2 years (Nilsson 1971). Oldest ringed bird 8 years (Helsinki).

**Movements.** Migratory, though some movements short. Breeding populations of Fenno-Scandia, north Germany, Poland, Baltic States and north-west Russia mostly winter Baltic (east to Estonia in mild seasons), Denmark, Netherlands, Britain and Ireland, and in progressively decreasing numbers from Belgium south to Mediterranean France. Some, however, move only to nearest coasts, e.g.

in south Sweden (Nilsson 1969*a*). A few now winter south-west Iceland (FG). Those from Swedish Lapland apparently migrate south-west across Norway, bypassing Baltic (Nilsson 1969*a*). Small flocks, perhaps of local origin, winter Varanger Fjord–Murmansk. Small to moderate numbers wintering on chain of lakes and reservoirs along northern side of Alps, and east to Danube valley, apparently mainly European breeders; recoveries from Finland to south France, Switzerland, south Germany, Italy, Balkans, and Black Sea; from Oberlausitz (East Germany) to Italy; from Saxony to Bavaria; and from Kola peninsula (Russia) to Switzerland. Degree of movement by small Czechoslovakian population unknown, but one winter recovery Italy (570 km south-west). Larger winter populations south Russia, apparently breeders of central and west Russia and west Siberia, though migration routes little known. In Black Sea, winters chiefly on north-east coasts where 8000 in 1967 (Isakov 1970*c*); also smaller numbers in west and south Black Sea, reaching Greece and Turkey. On Caspian Sea, 37000 wintered 1968, mainly in west and south-west parts (Isakov 1970); in Iran, common south Caspian and straggles to Khuzestan (D A Scott).

Knowledge of moult migration fragmentary (Salomonsen 1968). Assemblies of ♂♂ and immatures reported from several areas of tundra and steppe in north Russia and west Siberia, also one of thousands in Kandalaksha Gulf, White Sea (Dementiev and Gladkov 1952). Further west, moult gatherings occur Estonia (Matsalu Bay), south Sweden (Lake Tåkern), and Denmark (mainly Limfjorden). Extent of migration to these assemblies unknown for most. 12–14000 moulting Denmark evidently mainly from breeding populations of Sweden and south-west Finland; ♂♂ arrive from early June, but peak numbers August with arrivals then of many adult ♀♀; while totals include many immatures which probably remained in Danish waters from previous spring, concentrating July–August for moult (Jepsen 1973). Those moulting Transural (Kurgan) lakes, USSR, evidently come from basins of Irtysh, Ob, Yenesey, and Pechora; on dispersing, some return to north-west Russia, others go south to Caspian (Dementiev and Gladkov 1952).

Autumn migration begins late August, reaching peak in Baltic and North Seas in November; virtually completed

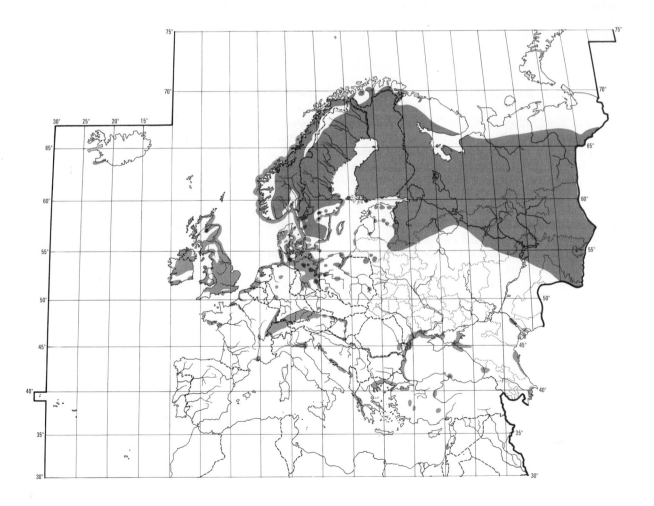

by early December, though may move later west and south in hard weather. On average, adult ♂♂ migrate less far than others, predominating in Baltic and northern wintering areas generally; but proportions become progressively smaller to west and south, while juveniles apparently move furthest of all (Bauer and Glutz 1969; Nilsson 1969a). Return early, passing through southern Baltic normally from mid-February, and most have left wintering areas by end March; even in Lapland and west Siberia main arrivals late April and first half May.

Food. Primarily molluscs, crustaceans, and insect larvae obtained during daytime, mainly by surface-diving, sometimes synchronously, rarely by dabbling and up-ending. Mobile and immobile items taken from bottom and mid-water at depths of up to 4 m (Dewar 1924; Ingram and Salmon 1941; Kennedy et al. 1954; Nilsson 1969), though chiefly 1–3 m (Willi 1970; Leuzinger 1972) and probably rarely above 9 m (Olney and Mills 1963). Diving times 15–20 s (Emeis 1936), 9–23·8 (Szijj 1965); maximum, in 385 dives, 36 s (Dewar 1924). ♂♂ have significantly superior diving capacity (von Haartman 1945). May turn over stones with bill. Generally swallows food under water, but occasionally brings large and mobile items to surface (Olney and Mills 1963).

Diet varies considerably with habitat and availability. Mainly animal, including molluscs, small crustaceans, insects and larvae, and small fish; occasionally earthworms, frogs, tadpoles, water-shrews, and leeches. Plant material, more prominent in autumn; includes seeds, tubers, roots, and leaves of aquatic species, also algae (Naumann 1905; Millais 1913; Phillips 1925; Witherby et al. 1940).

Following data from saltwater and brackish areas. 90 stomachs from saltwater sites, Denmark, October–February, contained by frequency 76% crustaceans (chiefly shrimps Crangon vulgaris, Gammarus, Idotea, and shore crab Carcinus maenas), 70% molluscs (mostly small periwinkles Littorina, spire shells Hydrobia, blue mussel Mytilus edulis, and cockles Cardium), 22% fish (mainly gobies Gobiidae, and three-spined stickleback Gasterosteus aculeatus), and small amounts of plant materials; 109 stomachs from brackish sites showed much variation in species related to habitat, but mainly molluscs (especially M. edulis and Hydrobia) and crustaceans (particularly Gammarus and Idotea), some insect larvae, occasional fish, and, in some areas, considerable amounts of plant materials (mainly seeds and leaves of wigeon-grass Ruppia, and pondweeds Potamogeton) (Madsen 1954). Similar diet shown from Baltic coasts with M. edulis dominating, and fewer Littorina, Gammarus, Idotea baltica, and fish. 1 stomach from south Sweden contained Gammarus, Idotea viridis, and Hydrobia (Nilsson 1969). From Kent estuaries, England, 14 stomachs November–February contained chiefly crustaceans, particularly C. maenas (65·9% by volume) and fewer laver-spire

shell Hydrobia ulvae, and fifteen-spined stickleback Spinachia spinachia (Olney and Mills 1963). USSR studies also emphasize variation of diet with habitat: on south Caspian wintering grounds on Turkmen shores, mainly crustaceans (Gammaridae) and Potamobius; on opposite shores in Transcaucasia, mainly molluscs; in Gulf of Kizil-Agach, Cardium edule; on rocky reefs of north Azerbaydzhan and Dagestan, Mytilaster lineatus (Isakov and Raspopov 1949; Dementiev and Gladkov 1952).

Following data from inland areas. 18 stomachs, Bodensee, December–February, showed main food until 1968–9 larvae of caddisfly Hydropsyche and fewer Limnophilus, but subsequently, (correlated with great increase of species), chiefly zebra mussel Dreissena polymorpha. 1 stomach from upper end of Bodensee, January, contained mainly stoneflies (Plecoptera) and midges (Chironomidae) (Leuzinger 1972). 17 stomachs from Scotland showed caddisfly larvae, especially Hydropsyche angustipennis, molluscs, stonefly, and mayfly. Ephemeroptera nymphs main diet from fast-flowing rivers; chironomids and waterbeetle larvae principal diet from lakes. In 12 stomachs Lough Beg, Ireland, September–December, caddisfly larvae again predominated, with rather fewer Gammarus duebeni, waterbug Corixa, and common spire shell Hydrobia jenkinsi; plant material, mainly seeds of Potamogeton, only 3·9% total volume. 5 stomachs from England contained roach Rutilus rutilus, chironomids, corixids, crustacean Asellus, and (in 2) plant materials including, surprisingly, hairy seeds of reed Phragmites communis, marsh thistle Cirsium palustre, and narrow-leaved dandelion Taraxacum palustre (Olney and Mills 1963). From south-east Finland, August–October, 36 stomachs contained chiefly Corixa, some waterbeetles, insect larvae, molluscs, fresh-water sponges (Spongillidae), and, as autumn progressed, more plant materials, chiefly seeds of sedge Carex lasiocarpa, bur-reed Sparganium, and water-lily Nuphar luteum (Tiussa 1966). From Danube, 4 stomachs had mainly Gammaridae and stoneflies (Festetics and Leisler 1970). In USSR, type of food depends on area and time of year. At Rybinsk reservoir, caddisfly larvae occupied 75% of total volume September–October, but only 47% in April and chironomid larvae 32%. In September, 10 stomachs from River Mologa valley contained dragonfly nymphs (Odonata), occupying 45% of volume, and plant materials (chiefly tubers of arrowhead Sagittaria sagittifolia and fennel-pondweed Potamogeton pectinatus), 50%. In early spring, 15 stomachs from Mologa contained 60% fish, mainly R. rutilus, silver bream Blicca bjoerkna, and perch Perca fluviatilus; 20% chironomids; and 10% snails Lymnaea stagnalis. On upper Pechora, mainly caddisfly larvae, terrestrial insects, some fish, and pondweed leaves (Isakov and Raspopov 1949; Dementiev and Gladkov 1952).

9 stomachs and some faeces of downy young from Mologa contained mainly dragonfly nymphs (particularly Coenagrion), waterbeetle larvae (especially Dytiscus mar-

*ginalis*), and fewer chironomids, waterlice *Asellus*, and caddisfly larvae (Isakov and Raspopov 1949).

**Social pattern and behaviour.** Based partly on material supplied by L Nilsson.

1. Gregarious except when breeding. For most of year, in small flocks, but sometimes up to several hundred, especially when roosting. During winter, sex and age composition of flocks usually grossly unequal; over much of wintering range, adult ♂♂ greatly in minority, e.g. on Untersee, south Germany generally 10–20% (Leuzinger 1972), predominating only in north, e.g. in Sweden north of Scania (Nilsson 1969a). Larger winter flocks disband at start of migration, but pairs remain in groups until breeding grounds reached. After break-up of pairs, ♂♂ first to gather into flocks for moult; joined by unsuccessful ♀♀. ♀♀ with young remain apart, but sometimes gather later for moulting. BONDS. Monogramous pair-bond of seasonal duration. Pair-formation starts in winter flock, in late January, south Germany (Bezzel 1959); often however, continues during spring migration and in breeding area, with many birds still unpaired in April (south Sweden). ♂ deserts ♀ during incubation. Only ♀ tends young; latter partly independent after a few days and brood starts to separate rather early. BREEDING DISPERSION. Nests usually well dispersed, though comparatively dense breeding populations build up in some areas because of provision of nest-boxes (McKinney 1965a). Same ♀ may use same hole in successive years (Sirén 1957). Pairs territorial when nesting, defending (♂♂ mainly) well-defined area; this generally in small bay or at shore near nest tree, though sometimes more distant from nest (latter often situated outside territory). Immatures remain in flock over summer. ROOSTING. Chiefly diurnal feeder. In winter quarters, loafs at times during day and roosts nocturnally, often well away from feeding area. Flocks in some areas gather in large, communal roost at traditional site; in other areas, each flock roosts separately—though early in season, when numbers low, often go to neighbouring roost, frequenting own roost later (Nilsson 1970). Sites generally on open water some way from shore. In some areas, flights to and from roost correlated with sunset and sunrise. On Untersee, flights to roost made in both large and small flocks, timing dependent on variety of factors (see Leuzinger 1972). Pre-dusk gatherings in large numbers, after wide dispersion during day, noted at wintering sites (see e.g. Linsell 1969), birds flying or swimming, singly or in small parties, to roost area and forming dense raft for night. On breeding grounds, local breeders roost solitarily in pairs, others in flock.

2. Nominate race studied by Boase (1924, 1950), Gurín (1939), Lind (1959a), Johnsgard (1965), Nilsson (1969b), and others; work on American race, e.g. by Townsend (1910), Brewster (1911), Myres (1957, 1959b), Dane *et al.* (1959), and Dane and van der Kloot (1964), indicates no important differences. FLOCK BEHAVIOUR. Composition of flock not fixed, individuals and groups arriving and leaving without provoking aggression in most cases (L Nilsson). When alarmed, neck sleeked and head lifted into full Neck-stretch posture, with bill horizontal and head feathers somewhat depressed. Flocks often take wing. Pre-flight behaviour complex, consisting of rapid lateral Head-shakes (while facing into wind in Neck-stretch posture) interspersed between Ritual-bathing, Upward-shakes, and Both-Wings-stretches; Wing-and-Leg-stretches also occur (Lind 1959a; McKinney 1965b). Exaggerated Neck-stretch posture, with head and neck feathers tightly depressed and bill raised at angle of *c.* 15°, also assumed at times when about to dive. Accompanied by Head-shaking and peculiar Head-jerking movements which take neck backward to *c.* 45°, so that bill points up vertically (Lind

1959a). When feeding, members of flock often dive simultaneously or in rapid succession (Géroudet 1965); preening and other comfort behaviour also often rather synchronized in flock (Nilsson 1965b). ANTAGONISTIC BEHAVIOUR. Like other *Bucephala*, often highly aggressive. Hostility generally rare at sea but common inland, probably because of higher frequency of pairs; shown mainly by ♂♂, in courting parties in spring and in territory. Paired ♂ also hostile towards other ♂♂ and pairs when defending mate; normally directs attack at other ♂ of pair. ♀♀ with young shows marked aggression to conspecifics (including other young) coming too close. Antagonistic encounters between ♂♂ often consist only of static performances of Head-forward display (Threat-crouch; see A). 3 types of attack (Nilsson 1969b): (1) in swimming version, ♂ approaches opponent in Head-forward threat; often followed (2) by diving and underwater attack, and (3) by 'wild' chase, both rushing over water with beating wings until exhausted (occasional leads to fighting). As in Barrow's Goldeneye *B. islandica*, ♂ at times makes brief flights to approach adversary closer (McKinney 1965a). During communal courtship, ♂♂ may chase one another in version of Head-forward in which neck completely stretched forward and bill open, or patter across water with forepart of body raised (Lind 1959a). Certain other displays performed in courting parties also seem basically hostile, including Head-throw and those (such as Bowsprit-pumping) associated with Head-forward. According to Dane and van der Kloot (1964), Head forward, Head-up-pumping, Head-back, and Head-back-bowsprit displays all associated with fighting, last 2 being involved in territorial defence in breeding area. ♀ sometimes briefly attacks rejected ♂♂ during communal courtship, and frequently adopts Head-forward posture. Inciting-display (Jiving of Myres 1957) of ♀ performed silently; hostile component consists of stretching neck sideways or backwards as necessary, from Head-forward, in direction of rejected ♂ then slowly returning head to starting position. COMMUNAL COURTSHIP. At low intensity from September but true display parties not formed until December; most intense February–March, while still in wintering area, continuing into May in breeding area. Groups often of several ♂♂ and single ♀ (sometimes 1–2 more), both paired and unpaired; also a pair and single ♂ (Nilsson 1969b), while parties of up to 30 ♂♂ and 8 ♀♀ reported (Dane *et al.* 1959). Bouts of display typically initiated by meeting of pair or pairs and single ♂; when courtship starts, other ♂♂ (paired and unpaired) join in. Activities of ♂♂ in courting parties varied and complex as they crowd round or follow ♀, swim ahead, or move towards or away from core of group. Each ♂ usually assumes initial Courtship-intent posture with head drawn down almost to shoulders; when swimming at speed, neck stretched slightly upwards and forepart of body raised out of water by vigorous paddling (Lind 1959a). Especially when swimming beside or ahead of ♀, also frequently performs Head-turning display, pointing bill from side to side with head feathers erected and tail sometimes cocked; if ♀ Inciting, ♂ may also interpose Neck-withdrawing display but this not so highly ritualized or conspicuous as in *B. islandica* (Johnsgard 1965). At times,

A

individuals make Short-flights towards or away from group: (1) ♂ first assumes Neck-stretch-posture; (2) takes off with longer run than usual; (3) flies with fluttering wing beats, while head (with feathers raised) held slightly lower than usual; (4) alights conspicuously, skidding to stop along water (Dane *et al.* 1959). Secondary displays include Upward-shake (not confined to start of bout nor preliminary to other displays), Wing-flap, and Head-shake; latter also invariably given in terminal phase of many of major displays. Head-forward (Threat-crouch) frequently performed on own, when may be followed by diving, and also as first stage of certain major displays. Latter fall into 3 groups. In 1st group (Head-up, Masthead, Bowsprit, Bowsprit-pumping, Head-up-pumping), head erected vertically or obliquely, with or without neck movements; usually performed lateral to ♀, some accompanied by soft calls. In Head-up display, ♂ assumes partial Neck-stretch posture; sometimes this changes to Head-up-pumping when head thrust rapidly up and down. In Masthead-display, ♂ (1) stiffly jerks neck from Head-forward into vertical position; (2) remains thus for brief moment; (3) snaps head and neck stiffly back into initial posture, usually while paddling feet vigorously. Bill remains horizontal except briefly at height of upward movement when it points vertically; accompanied by call and often followed by dive. In Bowsprit-display, ♂ (1) goes into Head-forward posture, (2) jerks head rigidly up into Oblique-posture (see B) while calling. When Bowsprit-pumping, assumes Oblique-posture and swims while drawing head back into Neck-stretch posture and forward into Oblique again, often repeatedly and either silently or with faint call. In 2nd group of major displays (Head-throw, Head-throw-bowsprit, Head-back, Head-back-bowsprit), head flung or 'snaked' over back; performed while facing or lateral to ♀, all with terminal Head-shake, and at least some with buzzing or snarling calls. Head-throw consists (1) of rigid erection of neck, as breast partly lifted out water, (2) followed by backward fling of head until nape touches rump, bill pointing vertically; call uttered just after, with bill open (see C), and (3) head brought rapidly back to Courtship-intent posture while Head-shake given. Head-throw-bowsprit similar but head brought forward into Oblique-posture before terminal Head-shake. Other 2 displays in this group recorded mainly in breeding area (Dane and van der Kloot 1964): in Head-back, ♂ quickly 'snakes' head over mantle from Courtship-intent posture into mid-back position, then completes movement as in full Head-throw; in Head-back-bowsprit, however, head brought up into Oblique-posture before sequence terminated by Head-shake. In 3rd group of major displays (Fast Head-back-kick, Slow Head-throw-Kick), main head movements of 2nd group followed by backward kicking and splashing of both feet, exhibiting their bright colour, and erection of head much as in 1st group; performed facing or with back to ♀, all with terminal Head-shake and relatively loud calls. ♂ first assumes more or less same posture with head on rump as in climax of Head-throw, either (in Fast Head-back-kick) by moving head quickly along back from Courtship-intent posture or (in Slow Head-throw-kick) by slowly raising head into Neck-stretch posture and then lowering it backwards; in both cases, Kick-display proper follows as ♂ strikes back with feet, raising rear part of body and forcing front under water, at same time thrusting head forward until neck and bill vertical (see D). Call then given as neck stretched even higher and kick completed, bird finally resuming Courtship-intent posture while performing terminal Head-shake. Which display occurs at any given time partly depends on position of particular ♂ in relation to ♀ and other ♂♂ (Lind 1959*a*). For further analyses of display durations and sequences, and factors influencing

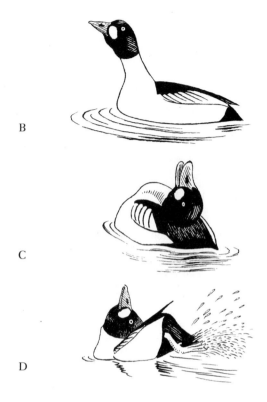

B

C

D

performances of specific displays, see Dane *et al.* (1959) and Dane and van der Kloot (1964). Behaviour of ♀ during communal display more complex than in many Anatinae. Often swims in Neck-stretch posture, mostly with head feathers depressed and bill slightly raised, often while Bowsprit-pumping. Will also from time to time assume Head-forward posture. Frequently makes dives of *c.* 5 s; then followed below surface by ♂♂ (Lind 1959*a*), resulting in underwater pursuits. ♀ Incites by (1) following particular ♂ in Neck-stretch posture and (2) periodically directing hostile movements of head at 1 or more other ♂♂ (see above). Will also then (and at other times) orientate Neck-dip-kick display at favoured ♂: (1) from Neck-stretch posture, goes into Head-forward facing ♂ then (2) thrusts neck (and sometimes head) below water while pointing bill up, elevating rear-end of body (with tail depressed) by kick back of feet, and giving faint call; (3) as body subsides, keeps head pointing up, sometimes looking from side to side, then (4) slowly lowers it. Like ♂, ♀ also Head-shakes from time to time. See Dane and van der Kloot (1964) for discussion on stimulus-response relationships between displays of ♂♂ and ♀♀. PAIR BEHAVIOUR. At least some of displays more usually seen as part of communal courtship performed at times by single ♂ to ♀ away from group, but such birds probably still unpaired. Paired ♂ will, however, direct many of courtship displays to own ♀ when latter courted by 2nd ♂ or part of displaying group. Reciprocal Head-turning by ♂ and Inciting (including Neck-dip-and-kick display) by ♀ as she follows mate then probably part of pair-courtship, but main activities in this category involve pre-copulatory displays and full copulation sequences. These first occur from late December, becoming particularly common in March–April; not usually associated directly with communal courtship, though afterwards pair occasionally withdraw from group and perform on own.

PLATE 87. *Melanitta nigra* Common Scoter (p. 634): **1** ad ♂, **2** ad ♀, **3** 1st imm breeding ♂, **4** 1st imm non-breeding, **5** downy young. (NWC)

PLATE 88. *Melanitta perspicillata* Surf Scoter (p. 642): **1** ad ♂, **2** ad ♀, **3** 1st imm breeding ♂, **4** juv. (NWC)

PLATE 89. *Melanitta fusca* Velvet Scoter (p. 644): **1** ad ♂, **2** ad ♀, **3** advanced 1st imm breeding ♂, **4** juv, **5** downy young. (NWC)

PLATE 90. *Bucephala albeola* Bufflehead (p. 650): **1** ad ♂ breeding, **2** ad ♀ breeding, **3** ad ♂ eclipse, **4** 1st imm non-breeding ♂, **5** ad ♀ non-breeding. (NWC)

PLATE 91. *Bucephala islandica* Barrow's Goldeneye (p. 652): **1** ad ♂ breeding, **2** ad ♀ breeding, **3** ad ♂ eclipse, **4** 1st imm breeding ♂, **5** juv, **6** downy young. (NWC)

PLATE 92. *Bucephala clangula* Goldeneye (p. 657): **1** ad ♂ breeding, **2** ad ♀ breeding, **3** ad ♂ eclipse, **4** 1st imm breeding ♂, **5** juv, **6** downy young. (NWC)

PLATE 93. *Mergus cucullatus* Hooded Merganser (p. 666): **1** ad ♂ breeding, **2** ad ♀ breeding, **3** 1st imm breeding ♂. (NWC)

PLATE 94. *Mergus albellus* Smew (p. 668): **1** ad ♂ breeding, **2** ad ♀ breeding, **3** ad ♂ eclipse, **4** ad ♀ breeding, **5** juv, **6** downy young. (NWC)

PLATE 95. *Mergus serrator* Red-breasted Merganser (p. 673): **1** ad ♂ breeding, **2** ad ♀ breeding, **3** ad ♂ eclipse, **4** 1st imm breeding ♀, **5** downy young. (NWC)

PLATE 96. *Mergus merganser* Goosander (p. 680): **1** ad ♂ breeding, **2** ad ♀ breeding, **3** ad ♂ eclipse, **4** juv, **5** downy young. (NWC)

PLATE 97. *Oxyura jamaicensis* Ruddy Duck (p. 689): **1** ad ♂ breeding, **2** ad ♂ non-breeding, **3** ad ♀ breeding, **4** juv, **5** downy young. (NWC)

PLATE 98. *Oxyura leucocephala* White-headed Duck (p. 694): **1** ad ♂ breeding, **2** ad ♀ breeding, **3** ad ♂ eclipse, **4** juv, **5** downy young. (NWC)

E

COPULATION. Closely similar to *B. islandica*; for full details, see Myres (1957, 1959*a*), Lind (1959*a*), Dane *et al.* (1959). Sequence often initiated by ♀ assuming Prone-posture (see E), though there may be some mutual Ceremonial-drinking first; ♀ usually remains still while ♂ displays around her (generally within 30–60 cm). Initial phase of ♂'s pre-copulatory display may last several minutes: as well as highly exaggerated Ceremonial-drinking and frequent Wing-Leg-stretches, less often performs Both-Wings-stretch, Preen-Breast, and Head-rubbing on side of body. Towards end of this phase, Water-twitches increase in frequency as ♂ backs slowly from ♀. Followed by single Preen-Back-between-Wings (on side nearest ♀; see E) and then ♂ Steams back towards ♀, calling. Copulation lasts 5–15 s; see *B. islandica* for details of this and subsequent behaviour. RELATIONS WITHIN FAMILY GROUP. Before young leave nest-hole, ♀ surveys neighbourhood for long time and makes exploratory flights; then induces young out of hole with special call. Brood follows ♀ to water, during first days staying close to her, contact facilitated by Maternal-calls; Warning-call of ♀ makes young collect round her. Generally no aggression between siblings, but sometimes hostility between broods.

(Figs A–E after Johnsgard 1965.)

**Voice.** Usually silent when not displaying, i.e. before or after main period of courtship. Vocabulary of sound signals supplemented by vibrant, rhythmic whistling of wings in flight (see I); recognizable at a distance even from equivalent noise of Barrow's Goldeneye *B. islandica*. Audible at least 1 km; louder in ♂. Likened to ringing noise of flat stone made to ricochet over thin ice, with syllables 'pjyb jybjybjub' (Söderberg 1950); also to distant tinkling of bell (Géroudet 1965).

CALLS OF MALE. Courtship-calls most recently studied by Lind (1959*a*), Dane *et al.* (1959), and Johnsgard (1965). Last recognized 3 main types: (1) loud 'zeee-ZEEE' (see II) uttered during both types of Head-throw-Kick displays; (2) quieter, rattling 'rrrt' during ordinary Head-throw display; and (3) soft 'rrrrrrrt' during Bowsprit and Masthead displays. Lind (1959*a*) pointed out that Courtship-call 1 accompanies displays most often performed on periphery or at a distance from courting party, whereas quieter calls usually given within group. Courtship-call 1 also described as a fairly loud 'rretsch-ree' (Lind 1959*a*); sharp, nasal call (Dane *et al.* 1959) identified by them as flat, vibrant 'paaap' of Brewster (1911); harsh, vibrant rasping 'zzee-at' (Townsend 1910). Penetrating, far-carrying 'speer speer' (see Bent 1925); vibrating

'quirrick ... crirric' (Géroudet 1965); and 'quee-reek' (Boase 1924) probably also same. For analysis of structure, see Johnsgard (1971). Courtship-call 2 (see III) described also as a faint snarling 'rrrrrr' (Lind 1959*a*) and as a buzzing, insect-like sound (Dane *et al.* 1959); calls like winding up watch (Géroudet 1965) and 'ur' of Boase (1924) probably same. Courtship-call 3 requires further study; said to resemble fainter and shorter version of call 2, with different timbre, or terminal part of call 1, and also to accompany Bowsprit-pumping display as well as acting as an apparently hostile call among ♂♂ (Lind 1959*a*). Other calls include a faint 'bzzzzt' and low grunts ('uig-uig-uig') uttered, respectively, during pre-copulatory and post-copulatory Steaming-display (Johnsgard 1965); and short, deep 'arrr' on flying off (see Bauer and Glutz 1969).

CALLS OF FEMALE. Only call mentioned by Lind (1959*a*) and Johnsgard (1965) uttered during Neck-dip-kick display: faint 'eeuu' or 'weak screeching cry'; evidently equivalent of Inciting-call of other ♀ ducks. Whistling 'peep peep' of Kortright (1942), identified as Courtship-call, though this doubtful. Loud, hoarse note like that of *Aythya* ♀♀ also mentioned by Witherby *et al.* (1939); this probably call, best rendered 'ah-ah-ah', recorded from flying bird (see IV). Variants of same seem to be given

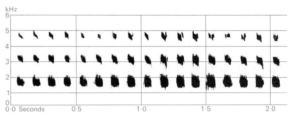

I P J Sellar Sweden May 1970

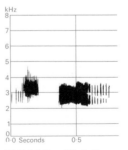

II P J Sellar Slimbridge February 1975

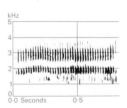

III Sture Palmér/Sveriges Radio Sweden March 1960

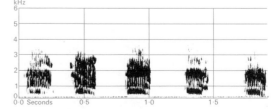

IV P J Sellar Sweden May 1970

when with young, e.g. low 'wah-wah' and squeaking 'ach' or 'heck'. Other calls include 'cuk-cuk-cuk' (Witherby *et al.* 1939) or 'gärk-gärk-gärk' (Bauer and Glutz 1969) when searching for nest-site; excited, repeated 'err-kiörr-kiörr' used to entice young to jump from hole.

CALLS OF YOUNG. 3 calls mentioned by Bauer and Glutz (1969): (1) trilling 'bibibi' (Pleasure-call); (2) harder 'pipipi' (when lost), (3) notably louder, high 'trii'.

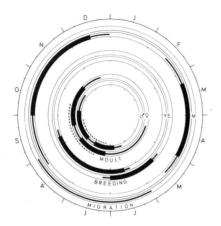

**Breeding.** SEASON. See diagram for Fenno-Scandia and western USSR. In Germany, earliest eggs 1st half April, with main laying 2nd half April and early May (Bernhardt 1940). SITE. Holes, in trees or artificial sites, especially nest-boxes; rarely in burrows of rabbit *Oryctolagus cuniculus*. Minima for diameter of hole 9 cm, inside depth 45 cm; height above ground 1·4–5 m (Bernhardt 1940; Rajala and Ormio 1970). Same nest-site used in successive years. Nest: shallow depression; lined available material (often little) plus down. Building: no excavating; depression shaped by ♀. EGGS. Blunt ovate; bluish-green. 59 × 43 mm (52–67 × 39–45), sample 200 (Schönwetter 1967). Weight 68 g (61–77), sample not given, upper Volga (Dementiev and Gladkov 1952); 60 g (55–65), sample 5, Germany (Heinroth 1927–28); 59 g (58–70), sample 38, captive laid (J Kear). Clutch: 8–11 (5–13); up to 22 recorded, but by 2 or more ♀♀. Mean of 63 clutches, Finland, 9·3; range 5–17 (Linkola 1962). One brood. Replacement clutches laid after egg loss, but only by minority (Linkola 1962). INCUBATION. 29–30 days (27–32). By ♀. Starts on completion of clutch; hatching synchronous. Eggs covered with down when ♀ off nest. ♀♀ studied Finland left nest-box on average 4·5 (3–6) times per day with absences averaging 79 minutes (Sirén 1952). Eggshells left in nest. YOUNG. Precocial and nidifugous. Self-feeding. Cared for by ♀ and brooded at night while small. On hatching, stayed in nest 24–37 hours before departure; broods in 5 nests all left in morning; departure complete in 40–150 s (Sirén 1952). On leaving nest, brood frequently taken by ♀ considerable distance to rearing area; 1·5–2 km recorded (Sirén 1952). FLEDGING TO MATURITY. Fledging period 57–66 days. Independent at

*c.* 50 days. Age of first breeding 2 years. BREEDING SUCCESS. Finland: average clutch size at hatching 9·3, average number of young reared 4·7 (Linkola 1962). Finland (another area): of 1552 eggs, 50·6% hatched; mean clutch size at start of incubation 10·3, mean brood size at hatching 9·6 (Grenquist 1963).

**Plumages.** ADULT MALE BREEDING. Head and upper neck black, glossed green; large round spot between eye and upper mandible white. Lower neck, upper mantle, sides of breast, and underparts white, mottled brown on thighs. Outer webs of elongated feathers on lower flanks broadly edged black. Lower mantle, inner scapulars, back, rump, and upper tail-coverts black. Outer scapulars white; shorter with black outer webs, longer ribbon-shaped with both webs edged black. Tail black with dark grey bloom. Primaries and primary coverts black; outer 4 secondaries black with white tips, others white; tertials black. Marginal coverts black with narrow white tips, rest of upper wing-coverts white. Greater coverts basally black. Axillaries and under wing-coverts dark grey, tipped white. ADULT FEMALE BREEDING. Head and upper neck dark chestnut. Lower neck white, mottled grey on hindneck. Upperparts brown-black, broadly edged blue-grey on mantle and inner scapulars, white on tips of outer scapulars, and pale grey on rump. Chest and sides of breast dark grey, flanks grey-brown; feathers tipped white. Thighs and lower vent dark grey, mixed with white. Rest of underparts white. Tail and wing as ♂, but tips of greater coverts black; longer median coverts black with white tips; marginal and small lesser coverts more extensively dark grey. Large white patch on forewing. ADULT NON-BREEDING. Like adult ♀ breeding, but head and upper neck dark brown with some black tinge at sides in ♂, dull brown in ♀; lower neck grey-brown; mantle, inner scapulars, and flanks dark brown with narrow ash-brown or off-white feather edges. Wing as breeding, distinguishing sexes. DOWNY YOUNG. Crown to lores, centres of hindneck, upperparts, wing, and sides of body dark sepia, tinged grey on mantle and chest. White patches on rear edge of wing, flanks, and sides of back and rump. Sides of neck, cheeks, and underparts white. JUVENILE. Like adult ♀ breeding, but head and upper neck grey-brown, lower neck grey; upperparts brown, narrowly edged dark grey on mantle and grey-brown on scapulars. Chest sepia, feathers narrowly edged buff. Feathers of flanks and lower vent narrow, grey-brown. Tail-feathers narrow, notched at tip, bare shaft protruding. Sexes differ in pattern of wing. In ♂, greater coverts mainly white, lesser and upper row of median coverts varying between grey-brown with white or pale grey tips and all whitish, forming wing patch as adult ♀ but duller. Other upper wing-coverts dark grey-brown or dull black. In ♀, median and lesser coverts darker on average, dark grey to grey-brown; only a few outer coverts in upper row of median tipped white or pale grey; rarely all tipped white. Juvenile differs from adult non-breeding in pattern of wing, and narrower and shorter feathers of underparts and tail. FIRST IMMATURE NON-BREEDING. In autumn, variable number of new feathers attained, resembling adult non-breeding. Sometimes complete in 1st immature non-breeding by October (except wing), but often juvenile belly, vent, and back retained; occasionally only head, flanks, or scapulars changed. Head and neck warmer brown than juvenile, sides of head in ♂ often tinged black. Chest grey-brown, feathers tipped white; flanks grey-brown, often mottled pale grey in ♂. Juvenile outer tail-feathers often retained until spring. Later in autumn, feathers intermediate between 1st non-breeding and breeding plumage develop; in ♂, scapulars with much white at base of outer web, and sides of breast white with black subterminal bar. FIRST IMMATURE BREEDING. Acquired by some from October onwards; others in spring. ♂♂ with white spot at base of upper mandible; feathers of head dark brown with black tips; outer scapulars white, but black edges broader than adult ♂ breeding, and tips mottled grey. In spring, sometimes like adult breeding, but wing juvenile, and often some juvenile or 1st non-breeding feathers on underparts, back, or outer tail. ♀♀ especially retain much juvenile and 1st non-breeding plumage until summer. SECOND NON-BREEDING. Like adult non-breeding, but feathers of breeding plumage acquired late in spring retained. Wing juvenile until moult, distinguishing sexes. SECOND BREEDING MALE. Like adult ♂ breeding, but feathers of crown less elongated, brown of bases of head feathers showing through. Wing apparently variable: sometimes new wing as in juvenile or like adult, but some black median coverts; advanced specimens probably inseparable from adult. Longer scapulars and flank feathers with broader black edges; tips occasionally mottled grey. Sides of breast sometimes in part faintly barred grey-brown subterminally. SECOND BREEDING FEMALE. Like adult ♀ breeding, but, at least in some, new wing as in juvenile ♀.

**Bare parts.** Nominate *clangula*. ADULT MALE. Iris bright yellow. Bill blue-black. Foot yellow-orange, webs olive-black. ADULT FEMALE. Iris pale yellow to white. Bill black, with narrow orange-yellow band between nail and nostrils (slight or absent in non-breeding). Foot yellow-ochre to pale orange, webs dull black. DOWNY YOUNG. Iris brown, paler after *c.* 3 weeks. Bill grey. Foot yellow-grey. JUVENILE MALE. Iris yellow-brown to yellow. Bill dark olive-brown; yellow-ochre band behind nail, darkening during 1st winter. Foot dull yellow-brown to yellow-ochre, webs dark olive-grey. JUVENILE FEMALE. Iris light brown, later pale yellow. Bill and foot as juvenile ♂, but band on bill behind nail brightens to yellow in winter. (Bauer and Glutz 1969; Dementiev and Gladkov 1952; Schiøler 1926; Witherby *et al.* 1939; RMNH, ZMA.)

**Moults.** ADULT POST-BREEDING. Complete; flight-feathers simultaneous. Starts with head and body June or early July (♀♀ sometimes in May); back, rump, belly, and vent last, interrupted by wing moult. Flightless for 3–4 weeks mid-July to mid-September in ♂♂, *c.* 3 weeks later in ♀♀; tail at about same time, but more prolonged. ADULT PRE-BREEDING. Partial; head and body. Starts when flight-feathers full-grown. Back, rump, belly, and vent moulted first, overlap with post-breeding; in part, moulted once a year only. ♂♂ mainly in full breeding plumage by late October, ♀♀ by early December. POST-JUVENILE. Partial. Feathers belonging to 1st immature non-breeding acquired early in autumn, then immature breeding later on. Head, mantle, chest, and flanks often before November, mantle and sides of breast before mid-winter, scapulars and tail prolonged. FIRST IMMATURE PRE-BREEDING. Partial. Head and body; sometimes from October, but mostly from December; varying greatly in amount of plumage involved. Usually variable amount of juvenile and 1st non-breeding plumage retained. FIRST IMMATURE POST-BREEDING AND SECOND PRE-BREEDING. Like adults, but more

prolonged and more variable in timing. ♂♂ start wing moult late June to late August (mostly mid-July), finish mid-July to late September; ♀♀ at times some weeks later. Some 2nd non-breeding body feathers sometimes retained until December. (Dementiev and Gladkov 1952; Jepsen 1973; Witherby *et al.* 1939; C S Roselaar.)

**Measurements.** Netherlands, winter; skins (RMNH, ZMA).

| | | | | | | | |
|---|---|---|---|---|---|---|---|
| WING AD ♂ | 220 | (5·33; 31) | 209–231 | ♀ 203 | (2·57; 24) | 197–207 |
| JUV | 214 | (6·06; 21) | 202–224 | 195 | (3·62; 16) | 186–200 |
| TAIL AD | 84·6 | (3·60; 20) | 78–91 | 77·5 | (3·51; 25) | 71–82 |
| 1st W | 76·6 | (2·97; 15) | 73–82 | 69·5 | (3·69; 13) | 65–76 |
| BILL | 33·3 | (1·27; 57) | 30–36 | 29·4 | (0·98; 48) | 28–31 |
| TARSUS | 38·8 | (0·95; 40) | 37–41 | 35·1 | (1·06; 46) | 33–37 |
| TOE | 65·0 | (1·80; 41) | 62–70 | 57·6 | (2·30; 45) | 54–62 |

Sex differences significant. Wing and tail of adult significantly longer than juvenile. Bill, tarsus, and toe similar; combined. Wing and tail 2nd winter intermediate between adult and juvenile. Juvenile tail somewhat shorter than 1st winter.

**Weights.** ADULT. (1) Winter: Denmark (Schiøler 1926), Netherlands (ZMA). (2) April–May: north-west USSR (Dementiev and Gladkov 1952), averages of several samples and total range. (3) June–August: Kazakhstan. (4) September–October: north-west USSR. (5) September–October: Kazakhstan.

| | | | | | | |
|---|---|---|---|---|---|---|
| (1) | ♂ | 1136 | (9) | 996–1245 | ♀ 787 | (9) | 707–860 |
| (2) | | 951; 956; 986 | | 820–1150 | 635; 708; 806 | 605–882 |
| (3) | | 982 | (4) | 888–1180 | 583 | (3) | 500–650 |
| (4) | | 912 | (–) | 850–1010 | 731 | (–) | 650–810 |
| (5) | | 1010 | (12) | 750–1100 | 720 | |

Maximum adult ♂ winter, south Caspian, USSR, 1300 (Dementiev and Gladkov 1952.)

JUVENILE. Denmark. October–November: ♂, 934 (7) 727–1035; ♀, 496, 650. December–February: ♂, 1040 (11) 905–1140; ♀, 740 (7) 710–796.

**Structure.** Wing pointed, rather narrow. 11 primaries: p10 and p9 about equal, longest; p8 6–14 shorter, p7 19–26, p6 30–39, p1 80–105. Inner web of p10 and outer p9 emarginated. Tail short, rounded; 16 feathers. Bill rather short and heavy, higher than broad at base, gradually depressed. Sides of bill parallel, tip rounded; nail small and narrow, decurved. Feathers on centre of crown elongated in ♂ breeding, forming dense, broad crest. Feathers of hindcrown and upper nape somewhat elongated in adults of both sexes. Feathers not elongated in juvenile, and only slightly in non-breeding and 1st and 2nd immature breeding. Outer scapulars ribbon-like and elongated in ♂ breeding. Outer toe equal to middle, inner *c.* 81% middle, outer *c.* 32%.

**Geographical variation.** Slight; ♂♂ of North American race *americana* somewhat larger, with longer and broader bill, but much overlap. Range of bill length of adult ♂♂ Netherlands (30–36; ZMA, RMNH), USSR (32–41; Dementiev and Gladkov 1952), and North America (33–42; Vaurie 1965) suggesting clinal increase in bill length from Europe eastward.  CSR

## *Mergus cucullatus* Hooded Merganser

PLATES 93 and 100
[between pages 662 and 663
and facing page 687]

Du. Kokardezaagbek  Fr. Harle couronné  Ge. Haubensäger
Ru. Хохлатый крохаль  Sp. Serreta cabezona  Sw. Kamskrake

*Mergus cucullatus* Linnaeus, 1758

Monotypic

**Field characters.** 42–50 cm, of which body two-thirds; wing-span 56–70 cm. Small to medium-sized, large-headed, and comparatively short-bodied sawbill duck with relatively short, thin beak and rather high-domed or oblong head; ♂ noticeably black and white, with large, black-bordered patch of white towards rear of otherwise black head, white breast edged at sides with 2 vertical black stripes, black back, and vermiculated brown flanks. Sexes dissimilar but seasonal differences in ♂. Juvenile and immatures resemble adult ♀, but just distinguishable.

ADULT MALE. Head and neck black with conspicuous fan-shaped crest of white, broadly tipped with black. When crest raised, head large and rounded with much white apart from black forepart and edges; when depressed, oblong with broad and widening strip of white behind eye. Rest of upperparts black, shading to dark grey-brown rump and tail; inner secondaries and elongated tertials black striped with white, forming bold streaks towards rear of folded wing; white speculum, divided by black bar, sometimes visible at rest. Breast and belly strikingly white; at sides of breast 2 vertical lines of black project down from mantle; yellow-buff to reddish-brown flanks provide obvious patch of colour. Bill black; eyes bright yellow; legs yellowish-brown. In flight, thin bill and attenuated horizontal shape typical of *Mergus*, but lacks largely white inner wing characteristic of other species; instead, forewing grey, rear edge of secondaries striped white and black, and only clear white is elongated speculum divided by narrow black line; from below, coloured flanks stand out against white underwings and belly. MALE ECLIPSE. Closely resembles adult ♀, but distinguished by all-black bill, yellow eyes, yellow-brown head, and more white on secondaries; variable white patch behind eye often present. ADULT FEMALE. Head and neck mainly brown, darkest on crown; crest shorter than in ♂, but of comparable shape, strongly tinged yellowish-brown and paler, glowing whitish, at tips; chin and throat white, or whitish speckled with brown. Rest of upperparts dark brown; wing as ♂ but with less white, particularly on secondaries and tertials. Upper breast and sides grey, tinged yellowish-brown on flanks; lower breast and belly white; under tail-coverts mottled brown. Bill black, edged distinctly with orange; eyes light brown; legs dusky brown. JUVENILE. Closely resembles adult ♀, but just distinguishable by shorter crest, paler brown upperparts with light edgings, brown edgings on flanks, and still less white in wings, particularly on tertials.

Pattern of adult ♂ almost unmistakable, though black and white appearance, and particularly erectile fan-shaped white crest, suggest slight possibility of confusion with Bufflehead *Bucephala albeola*, which is much smaller and whiter, with white scapulars and flanks and no black outer border to white on head. Adult ♀ and immatures bear some resemblance to corresponding plumages of Goosander *M. merganser* and, especially, Red-breasted Merganser *M. serrator*, which are larger with noticeably longer, reddish bills, usually paler faces, less high-domed and bushy crests, paler and greyer upperparts and flanks, and much more white on secondaries.

Chiefly found on shallow fresh water, particularly on lakes, pools, and slow rivers in wooded areas; rarely on salt-water. Secretive, but not especially shy. Floats buoyantly, but sinks low in water when alarmed or when swimming purposefully; dives from surface by simply submerging or with forward jump. Walks more easily than larger *Mergus*, and with surprisingly horizontal carriage. Readily perches on fallen trees and on branches overhanging water. Like other *Mergus*, rises from water with some difficulty, after splashing along surface, but spends more time on wing than them. Flies rapidly with fast wing-beats and sharp changes of direction, but not usually high above water. Usually singly or in pairs, and seldom more than small parties. Does not associate much with other species.

**Habitat.** In cool to warm temperate climatic zones frequenting inland waters, usually small, standing, or gently flowing (except when displaced by freezing to more rapid watercourses), surrounded more or less closely by trees, essential for breeding. Ecologically, North American equivalent of Smew *M. albellus*, but much less ready to use marine or wide open waters and more adapted to small pools or channels in forests or swampy wetlands.

**Distribution.** Breeds wooded areas of North America from southern Alaska, British Columbia, Alberta, probably Saskatchewan, and Manitoba, east to southern Quebec, New Brunswick, and Nova Scotia, south to Oregon, Wyoming, Iowa, western Tennessee, and locally further south.

Accidental. Britain: North Wales, immature ♂, winter 1830–1. Ireland: Cork, pair December 1878; Kerry, ♀ January 1881; Armagh, ♀ or immature December 1957 (Witherby *et al.* 1939; British Ornithologists' Union 1971).

West Germany: Lower Saxony, ♂ shot 1906 (Bauer and Glutz 1969).

**Movements.** Migratory. Winters from southern edge of breeding range southwards. Autumn passage protracted, with tendency to linger on inland waters as far north as these unfrozen; some may remain all winter in central USA, a few even north to southern Great Lakes (Wood 1951). Most concentrate in winter in coastal states from southern British Columbia to California, and from Massachusetts to Florida and Gulf states of USA. Has straggled to Baja California, eastern Mexico, West Indies, and Bermuda. Autumn migration mid-September to mid-December and spring return late February to mid-May; peak movements October to mid-November and mid-March to mid-April. Has wandered in summer north to southern Mackenzie, southern Keewatin, northern Quebec, and Newfoundland (several). See Kortright (1942), Godfrey (1966), Robbins *et al.* (1966).

**Voice.** Not highly vocal but quite loud, rolling, frog-like 'crrrooooo' reported from ♂ and hoarse 'gak' from ♀ (Johnsgard 1965).

**Plumages.** ADULT MALE BREEDING. Head and neck black, except white ear-coverts, white feathers with black tips on crown and nape, and olive-tinged forehead. Feathers of centre of crown and nape elongated; can be raised to semicircular crest. Mantle, back, and scapulars black; rump and upper tail-coverts brown-black edged rufous. Tail black, tinged grey. Broad black crescent extends from mantle down white side of chest, narrower one in front of wing. Sides of body yellow-buff grading to deep cinnamon on lower flanks, vermiculated black. Sides of vent and under tail-coverts white, mottled grey-brown and buff; rest of underparts white. Primaries and outer secondaries with their coverts brown-black to black; inner secondaries broadly edged white on outer web, except base. Greater upper wing-coverts black, inner broadly tipped white; median coverts pale ash or brown-grey; inner median and lesser wing-coverts brown-black, tipped blue-grey at leading edge. Tertials black; broad white shaft-streak on outer, faint narrow grey one on inner. Centre of underwing and axillaries white, lesser coverts grey-black, rest of underwing grey. ADULT FEMALE BREEDING. Head and neck dark grey-brown, tinged rufous on crest at hindcrown and nape. Chin white, mottled grey. Upperparts brown-black, darkest on rump, feathers edged olive-brown on mantle. Feathers on chest grey, tipped white, on sides of body grey-brown, tipped olive-brown. Rest of underparts white, barred brown-black at sides of vent and under tail-coverts. Wing as adult ♂, but median coverts brown

instead of pale grey, outer tertials with shaft-streaks narrower and pale grey, inner all black. ADULT NON-BREEDING (Eclipse). Both sexes like adult ♀ breeding, but wing and eye colour of ♂ as in ♂ breeding. JUVENILE. Like adult ♀, but crest short; upperparts lighter, dark sepia edged ash-grey on mantle and brown on scapulars; greater wing-coverts black with white subterminal spot on outer web (this subterminal spot larger in ♂♂ than in ♀♀; although strongly variable in both, in ♀♀ sometimes mottled dusky); tertials shorter, only few inner with pale grey shaft-streak. FIRST IMMATURE NON-BREEDING. New feathers attained in autumn like adult non-breeding or ♀ breeding. FIRST IMMATURE BREEDING. Feathers resembling adult breeding appear on head, neck, chest, sides of breast, flanks, and scapulars from late autumn or winter, but usually some juvenile feathers retained on back, underparts, or tail, and wing juvenile. SECOND IMMATURE NON-BREEDING. Like adult non-breeding, but breeding feathers attained latest in spring retained, and wing juvenile; until 1st wing moult. (Kortright 1942; Witherby *et al.* 1939; C S Roselaar.)

**Bare parts.** ADULT MALE. Iris bright yellow. Bill black. Foot yellow-olive to light brown, webs dusky. ADULT FEMALE. Iris yellow-brown to brown. Upper mandible black, edges and lower mandible orange. Foot brown, olive-grey or dusky yellow, webs darker. JUVENILE. Iris brown. Upper mandible leaden grey, edges and lower mandible flesh; nail brown. Foot as adult ♀. (Kortright 1942; Witherby *et al.* 1939.)

**Measurements.** ADULT. North America: skins (RMNH, ZMA); range includes additional data from Witherby *et al.* (1939).

| | ♂ | | | ♀ | | |
|---|---|---|---|---|---|---|
| WING | 198 | (1·21; 6) | 193–202 | 190 | ( – ; 3) | 184–198 |
| TAIL | 88·5 | (4·00; 6) | 80–96 | 85·7 | ( – ; 3) | 82–90 |
| BILL | 39·1 | (0·95; 8) | 38–41 | 36·8 | (2·15; 4) | 34–40 |
| TARSUS | 31·5 | (0·66; 8) | 29–33 | 30·6 | (0·85; 4) | 29–32 |
| TOE | 54·1 | (1·13; 8) | 52–55 | 51·1 | (2·17; 4) | 48–53 |

Sex differences significant for wing, bill, and toe. Wing and tail of juvenile both on average *c.* 8 mm shorter; other measurements similar to adult.

**Weights.** North America (Kortright 1942).

| ♂ | 680 | (19) | 595–879 | ♀ 554 | (12) | 453–652 |
|---|---|---|---|---|---|---|

**Structure.** Wing long and pointed. 11 primaries: p9 longest, p10 0–4 shorter, p8 5–9, p7 14–22, p6 23–34, p1 77–92. Inner web of p10 and outer of p9 emarginated. Tertials narrow, elongated; often curved downward over wing in adult ♂. Tail long, rounded; 18 feathers. Bill rather short, slender, nearly cylindrical, with blunt horny 'teeth' not inclined backwards as in other *Mergus* species. Long, narrow, fan-shaped crest of hair-like elongated feathers on head in adult. Outer toe *c.* 97% middle, inner *c.* 81%, hind *c.* 30%.                                    CSR

## *Mergus albellus* Smew

PLATES 94 and 100
[between pages 662 and 663
and facing page 687]

DU. Nonnetje    FR. Harle piette    GE. Zwergsäger
RU. Луток    SP. Serreta chica    SW. Salskrake

*Mergus Albellus* Linnaeus, 1758

Monotypic

**Field characters.** 38–44 cm (24–28 cm); wing-span 55–69 cm; much smaller than other sawbills, being only slightly larger than Teal *Anas crecca*. Small-billed, slightly crested, compact sawbill, with steep forehead; striking plumage patterns in both sexes and at all ages. Predominantly white plumage of adult ♂ beautifully patterned by black face mask and other dark markings. All other plumages exhibit unique head pattern, dark rufous crown and nape contrasting with white lower face. Sexes highly dissimilar in breeding plumage, and ♂ larger; seasonal variation in ♂. Juvenile separable.

ADULT MALE. Head white, with frontal mask and loose feathers on sides of nape black. Neck and upper body mostly white (appearing almost wholly so at distance), showing inverted black V's on sides of chest, which stem from black panel down centre of back. Rump and tail grey-black. Underparts white, with flanks behind inverted V vermiculated dark grey (appearing uniform pale grey at distance). Wing also shows contrasting black and white pattern. Primaries all black, but secondaries broadly edged white on trailing edge. Contrast between black inner tertials and white outer ones marked in flight. Greater wing-coverts black with bold white tips (completing white frame to secondaries), median wing-coverts white, lesser ones black, as primary coverts. In distant flight, display of white less striking; body and wing appear more piebald. Bill atypical of genus, short and with profile recalling small *Anas*, grey; legs and feet grey. Eyes red. MALE ECLIPSE. Resembles ♀, but back blacker and large white patch on median wing-coverts retained. ADULT FEMALE. Face mask blackish but much less distinct than in ♂. Crown, nape, and upper hindneck deep rufous-brown, forming distinct head-cap and contrasting sharply with white lower face and throat. Rest of upperparts and tail cold grey. Breast dusky-grey (further exaggerating head pattern), fading into greyish-white underparts. Wings patterned as ♂ but with all white areas reduced. Bare parts as ♂ but eyes browner. JUVENILE. As adult ♀, but white wing-coverts occluded by grey-brown tips, and lores not black.

At all times and ranges, most distinctive duck; head pattern of ♀, immatures and eclipse ♂ as diagnostic as general plumage pattern of breeding ♂. Flight silhouette rather compact (more reminiscent of goldeneyes *Bucephala* than other *Mergus*) but action typical of genus, though with marked ease at take-off and when manoeuvring in limited space. Groups fly in oblique lines and V's, more rarely (over short distances) in bunches, like Teal *Anas crecca*. Swims with notable buoyancy, diving easily and quickly; walks well, with characteristic upright carriage.

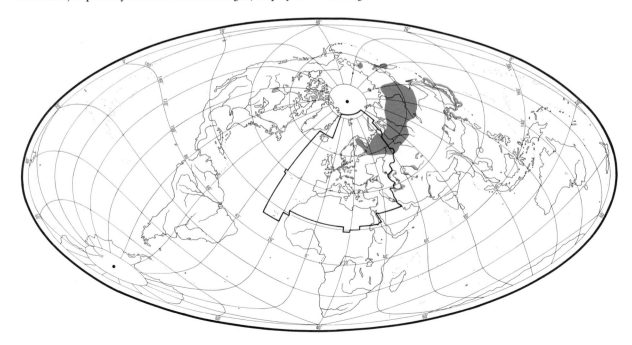

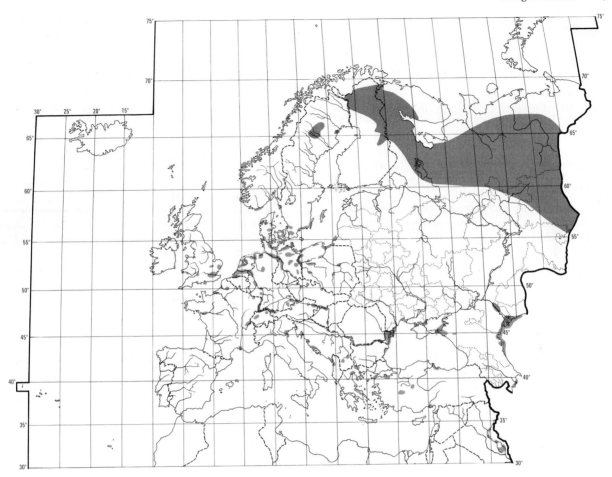

Found mainly on slow-running waters in taiga when breeding, resorting on migration and in winter mostly to southern fresh waters; more tolerant of shallower water than other 2 west Palearctic *Mergus*. Normally silent, though ♂ gives occasional hoarse, grating call.    DIMW

**Habitat.** Like Goldeneye *Bucephala clangula*, intimately linked with well-grown trees, affording holes for nesting, often made by Black Woodpecker *Dryocopus martius*, close to cool, boreal, fresh waters, either still or gently flowing. Scattered mainly south of Arctic Circle and wooded tundra, through coniferous and mixed forest zones. Likes drowned woodlands with many dead trees, and oxbows or other backwaters of large rivers, but avoids fast-flowing streams and mountain torrents. Access to quite small ponds suffices in otherwise suitable habitats. As acceptable holes in trees scarce, takes readily to nest-boxes intended for *B. clangula* in inhabited localities.

On passage, exploits varied opportunities afforded by small size and powers of take-off in confined spaces to rest and feed on miniature bodies of water or small streams. In winter, mainly on fresh waters; resorts to salt water for fishing only in sheltered and fairly shallow bays, estuaries, and inlets, normally up to 4 or at most 6 m deep. Reluctant to leave in hard weather until compelled by complete freezing, and continues to fish below surface ice as long as unfrozen patches give access. Highly mobile and restless, frequently shifts feeding and resting areas, and explores on wing at varying heights surrounding opportunities in diverse habitats. Tends to occur in company with *B. clangula*, but only moderately gregarious. Reservoirs, barrages, and other man-made habitats often used in winter and on passage.

**Distribution.** USSR. Boundaries of range not clarified, especially in east. Breeds, or has bred sporadically, further south, e.g. Ilmen, lower Dnepr, Voronezh, and Volga delta. Range appears to have shrunk in recent times due to disappearance of suitable habitat (Dementiev and Gladkov 1952). NORWAY: breeding first proved 1925, south Varanger; increase to 1930, but no great expansion since (Haftorn 1971). SWEDEN: supposed to have invaded during 19th century; now extremely rare, but recently discovered Norrbotten archipelago (Curry-Lindahl *et al.* 1970). RUMANIA: formerly bred small numbers Danube flood area and, more rarely, delta; several clutches obtained

1902–9 (Vasiliu 1968; Dombrowski and Lintia 1955).

Accidental. Spain, Algeria, Libya, Malta, Cyprus, Egypt, Jordan.

**Population.** Breeding. No numerical data, but sparse or rare in Fenno-Scandia.

Winter. Total west Palearctic population estimated 75 000 birds, with recent increases in west (Atkinson-Willes 1975).

**Movements.** Migratory. Routes little known since few ringed. Small to moderate numbers wintering northern parts East and West Germany, also Denmark, Netherlands, and Britain, must account for breeders of Fenno-Scandia; but also some from north-west USSR. Birds, winter-ringed Netherlands and West Germany, recovered in breeding range in Komi (52°E) and Tyumen (76°E), and as autumn migrants in Kalinin and Tambov; Kalinin bird probably on passage back towards Baltic, but Tambov recovery so far south that apparently then migrating to Black or Caspian Seas. Dividing line in USSR between breeders migrating west to Baltic and North Seas, and much larger numbers going south to Balkans and south Russia, not known. Winters commonly Sea of Azov, where up to 25 000 (Isakov 1970c), with smaller numbers in western Black Sea, Hungary (2000–3000, Szabo 1970), northern Greece, and Turkey; on Caspian Sea largest flocks in west, and up to 20000 winter Volga delta alone (Krivonosov 1970). Small numbers pass south from Caspian into Iraq and Iran. Additionally, occurs in winter irregularly or in very small numbers in southern Norway and Sweden, Ireland, Belgium, and (especially in hard weather) central and western European countries south to Mediterranean; 1000–2000 in France in severe weather of early 1963 though normally under 50 (Jouanin 1970). Much rarer in western Mediterranean basin than in east.

Groups of moulting ♂♂ reported, dates not stated, from Siberia (Baraba Steppe, Ob-Irtysh confluence, and north Yakutsk) (Dementiev and Gladkov 1952); but not west Palearctic. Otherwise autumn departures from breeding areas begin September, completed early October; main passage through Swedish hinterland and Baltic countries mid-October to November; early records North Sea countries October but main arrivals not until December or January following cold weather further east. Proportion of adult ♂♂ highest northern Germany (48% Hamburg area), falling to south and west; in England ♀♀ and juveniles far outnumber adult ♂♂. Return movement conspicuous by March, and most wintering waters vacated then; stragglers in April and even May, and immatures often summer south of breeding ranges (especially Finland and Russia). Reaches Arkhangel early to mid-May, Kola peninsula May to early June. See also Bauer and Glutz (1969).

**Food.** In winter and early spring mainly fish, at other times mainly insects. Obtained chiefly by surface-diving.

Depending on prey, dives nearly vertically or at long slant. Short submergence time, usually less than 30 s but occasionally up to 45 s, normally in depths 1–4 m. During 59 dives in 3–4 m, maximum submergence 30 s (Ingram and Salmon 1941). 114 dives in depths less than 2 m, 8–31 s, mean 18·2 s (Nilsson 1974). Prey usually brought to surface. Diurnal, most commonly in small flocks, often diving synchronously. In Netherlands, winter mass-fishing reported in flocks of c. 750 with birds in front continuously diving and those at back making short flights to keep up with leaders; compare co-operative feeding in Goosander *M. merganser* and Red-breasted Merganser *M. serrator* (Källander et al. 1970).

Freshwater fish include: salmon and trout *Salmo*, gudgeon *Gobio gobio*, roach *Rutilus rutilus*, bleak *Alburnus alburnus*, loach Cobitidae, sticklebacks Gasterosteidae, pike *Esox lucius*, minnow *Phoxinus phoxinus*, burbot *Lota lota*, eel *Anguilla anguilla*, perch *Perca fluviatilis*, and carp *Cyprinus carpio*. Marine fish include: plaice *Pleuronectes platessa*, sand-eel Ammodytidae, sandsmelt *Atherina*, sticklebacks *Gasterosteus* and *Spinachia spinachia*, blenny *Zoarces viviparus*, herring *Clupea harengus*, and bream *Abramis brama*. Fish include pelagic and bottom-living species; mainly small, 3–6 cm long but occasionally up to 10 (carp) or 11 cm (perch), and exceptionally up to 29 cm (eel). Insects mainly aquatic, both adults and larvae, especially waterbeetles, dragonflies, and caddisflies. Occasionally crustaceans, molluscs, marine polychaetes, frogs, and plant materials including seeds, leaves, and roots. (Dementiev and Gladkov 1952; Madsen 1957; Millais 1913; Naumann 1905; Witherby et al. 1939.)

16 stomachs from Danish waters, October–February contained mainly fish, up to three-quarters of total. 11 from coastal fjord areas contained fish (in 8), especially gobies (*Gobius niger, Chaparrudo flavescens*), sticklebacks (*Gasterosteus aculeatus, Pungitius pungitius, S. spinachia*), and fewer crustaceans (*Palaemon, Mysis*), nereid worms, and small quantities of plant material (*Zostera, Ruppia, Potamogeton*). 5 from fresh water also contained chiefly fish (in 4) (*R. rutilus, P. fluviatilis, E. lucius, P. pungitius, Abramis*), some insect remains (caddisfly larvae, waterbeetles, waterbugs *Corixa*, chironomid larvae), and plant materials (duckweed *Lemna* and seeds of *Potamogeton* and *Scirpus*); in one also a frog (Madsen 1957). 2 ♀♀ January from Kent contained trout *Salmo trutta* including one 14 cm long (Olney 1967). 1 from Scotland contained 18 minnows, another an eel (Macpherson 1887). 1 from Sweden filled with caddisfly larvae (Nilsson 1974). USSR studies show changes in diet with season. In early spring on way to breeding grounds, chiefly fish; in March on Volga delta, 76% by volume fish, mainly carp, rest insects; in April on Mologa river, 60% fish, mainly roach, 40% insects and larvae. As lakes thaw and birds move from rivers to open waters, fish become minor part of diet and insects increase: 65% waterbeetles and waterbugs, 35% dragonfly larvae. On Rybinsk reservoir in April, 68%

caddisfly larvae (*Phryganea grandis*) and 26% fish, chiefly ruffe *Gymnocephalus* with fewer young perch and roach; in May, 100% waterbeetle *Dytiscus marginalis*. Similar pattern in Anadyr, where in June mostly caddisfly larvae and waterbeetles, and in Ufa region, where summer diet chiefly dragonfly larvae. Insect diet predominates through summer into autumn. On Rybinsk reservoir, September and October, 77% caddisfly, 16% fish, but in November as water cools and insects die or move to bottom abrupt change to fish (87%), chiefly young ruffe. Further south, winter diet chiefly fish; in Lenkoran, mainly carp and bleak; at mouth of Sulak river, mainly flounder (Dementiev and Gladkov 1952).

**Social pattern and behaviour.** Based mainly on material supplied by L Nilsson.

1. Gregarious most of the year. Outside breeding season, mainly in small flocks; larger flocks (50 or more), usually of rather short duration, occur where common. Sex-ratios vary considerably with range, season, and habitat (Nilsson 1974). In small flocks or pairs on spring migration. BONDS. Monogamous pair-bond of seasonal duration. Pairs formed mainly in winter quarters and on migration: none seen before February, south Sweden, and about a third of population in pairs by March–April (Nilsson 1974); commonly in pairs from February, Schleswig-Holstein, with a few in December–January and peak in March (Sudhaus 1966); *c.* 30% ♀♀ paired in February, south Germany, *c.* 40% March–April (Bezzel 1965). Pair-bond ends during incubation. Only ♀ tends young. BREEDING DISPERSION. No detailed information. Discrete colonies reported (Dementiev and Gladkov 1952). ROOSTING. Gregarious and nocturnal outside breeding season. In south Sweden, spends night in small flocks in daytime feeding area on water or edge of ice; may also establish large roost elsewhere in sheltered bay, arriving at sunset and leaving before or at sunrise (Nilsson 1974). Also rests during day but for short periods (4–18% of total time). Where possible, prefers to rest out of water, e.g. on stones, tree-trunks, and branches over water.

2. FLOCK BEHAVIOUR. Flock composition, pre-flight signals, alarm behaviour, etc., much as in other sawbills *Mergus* so far as known. Communal hunting for fish recorded (see Källander *et al.* 1970). ANTAGONISTIC BEHAVIOUR. Aggression occurs mostly in courting parties, rarely between pairs or individuals though paired ♂♂ will defend mates against intruding ♂♂; disputes over food rare. Courting ♂♂ frequently jab or attack other ♂♂, sometimes ♀♀; swimming attack most common form of hostility, ♂ rushing at opponent over water with head well forward; diving attacks rare (see Red-breasted Merganser *M. serrator*). In flock, ♀♀ show much aggression towards ♂♂, jabbing at them with bill. COMMUNAL COURTSHIP. In flock from late December but main peak late February and March; most intense in afternoon and when gather to roost (Nilsson 1974). Usually 2–7 ♂♂ group around 1–2 active ♀♀; larger flocks generally divide into several groups. Display often initiated when bird of either sex flies in to join inactive flock, ♂ swims to a flock, 2 pairs or a pair and a single ♂ meet, or when birds disturbed; once courting party formed, others (both paired and unpaired) join group. Displays described by Hollom (1937), Lebret (1958c), Johnsgard (1965). ♂♂ adopt Courtship-intent posture with neck slightly curved backwards, feathers of forehead erected (see A), and usually circle ♀ while performing displays. Secondary displays include frequent Upward-shakes in which ♂ raises body to 40–60° then shakes

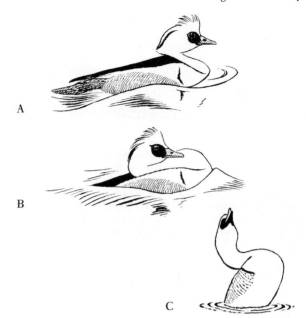

head backwards and forward before re-settling; lateral Head-shakes and Wing-flaps also frequent. Pouting-display most common of major displays; bill held horizontal throughout and quiet call given. Starts from Courtship-intent posture, then ♂ smoothly pumps back of head and neck backwards over mantle (see B) before returning to starting position. More elaborate movement, Head-fling, sometimes follows Pouting: ♂ suddenly throws head back while rising in water to *c.* 45° (see C), calling, then brings head rapidly forward while re-settling on water with bill pointing sharply down so that black V-pattern on nape exhibited frontally. In Neck-stretch display, ♂ suddenly extends neck fully upwards while calling then brings it down. ♀ taking active part in courtship swims in Courtship-intent posture with extended neck; frequently engages energetically in highly ritualized and conspicuous Inciting-display, often several times in succession, towards particular ♂ swimming away from her, typically lunging forward violently with body, bill pointing sharply down. Lacks lateral pointing movement of most other *Mergus*; accompanied by harsh call, also sometimes heard on own. Especially when followed by Inciting ♀, ♂ tends to perform Turn-back-of-Head display, showing black lines on nape. When courting activity intense, ♂♂ will also swim rapidly and persistently after inactive ♀, displaying behind her (Nilsson 1974); birds at rear of group may perform Short-flights, suddenly rising, passing over rest of group, and alighting in front, showing white of wings; then turn back to join others, often Pouting. ♀ may escape by diving, followed by pursuing ♂ and some of rest of ♂♂. PURSUIT-FLIGHTS. Courtship-flights frequently arise from swimming pursuits when ♀ takes wing instead of diving followed by ♂♂ from courting group; may continue, with short pauses, for up to 30 mins, ♀ and 1–2 ♂♂ sometimes diving from wing and continuing chase under water. Such pursuits most common when most of ♀♀ paired (Nilsson 1974). PAIR-BEHAVIOUR. No precise information on pair-courtship, but includes copulation and associated displays. COPULATION. See especially Johnsgard (1965). Occurs from January (Bauer and Glutz 1969), often when pair stops diving activity or sometimes after withdrawing from courting party (Nilsson 1974). Pre-copulatory behaviour initiated by mutual Ceremonial-drinking or by ♀ adopting Prone-posture immediately, sometimes after first directing

D

Inciting movements at ♂. Prone-posture differs from that of other *Mergus*: ♀ floats with neck extended, head near surface, and tail elevated to *c*. 30°; faces ♂ throughout, frequently Tail-quivering and sometimes Inciting. ♂ circles ♀, repeatedly approaching and retreating while displaying; displays include Ceremonial-drinking, Head-shakes, Upward-shakes, and Preen-dorsally; also Water-twitch (L Nilsson). Eventually mounts without special preliminaries, Wing-shaking 2–5 times during copulation (see D); afterwards, ♂ may release ♀ immediately or pair may rotate, then ♂ performs Single Head-fling before moving rapidly away while Turning-back-of-Head to ♀, latter sometimes following and Inciting. Both sexes generally then bathe. RELATIONS WITHIN FAMILY GROUP. No detailed information.

(Figs A–D after Johnsgard 1965.)

Voice. Little studied. Calls mostly quiet and infrequent except during courtship or when anxious or alarmed. Those of ♂ prolonged creaking, grunting, or rattling; of ♀, hoarse and grating or rattling.

CALLS OF MALE. Courtship-call soft, mechanical-sounding rattle (Johnsgard 1965); like noise of wrist watch being wound-up (Lebret 1958*c*). Quite high-pitched initially but slowing and hestitating towards end. Also likened to sound of fingernail drawn along teeth of comb (see I), with renderings 'kur-rik' or 'krrr-eck' (Bauer and Glutz 1969). Accompanies Pouting and Neck-stretch displays with somewhat louder version during Head-fling display (Johnsgard 1965). Significance of half-growling, half-humming, ventiloquial 'rroh' call (Bauer and Glutz 1969) uncertain.

CALLS OF FEMALE. Inciting-call harsh, rattling 'krrrrr krrrrr'; louder version associated with more energetic threatening movements (Johnsgard 1965). Low, hollow, rapid 'wok' or 'quok', uttered singly or in series (also recorded by C Chappuis), apparently different but significance uncertain.

CALLS OF YOUNG. Contact-trills given in groups of 2–4 notes. Distress-call rather swooping, starting at a low frequency and rapidly moving to higher; performed rather slowly. (J Kear.)

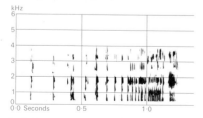

I C Chappuis/Sveriges Radio France (captive) March 1965

Breeding. SEASON. See diagram for Finland and USSR; based on rather little data. SITE. Holes in trees in natural state, but takes readily to nest-boxes. Nest: slight depression lined with available material (often little or none) plus down. Building: no excavating; depression shaped by ♀. EGGS. Ovate; cream to pale buff. 52 × 38 mm (48–58 × 34–40), sample 215 (Schönwetter 1967). Weight of captive laid eggs 39 g (35–45), sample 67 (J Kear). Clutch: 7–9 (5–11); up to 14 recorded, presumably 2 ♀♀. One brood. No information on replacements. INCUBATION. 26–28 (30) days. By ♀. Starts on completion of clutch; hatching synchronous. Eggs covered with down when ♀ off nest. Eggshells left in nest. YOUNG. Precocial and nidifugous. Self-feeding. Cared for by ♀. FLEDGING TO MATURITY. Fledging period not recorded. Age of first breeding probably 2 years. BREEDING SUCCESS. No data.

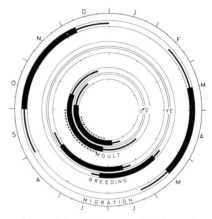

Plumages. ADULT MALE BREEDING. Head, neck, and upper mantle white, except black patch round and below eye to base of bill and black streak with green gloss at side of hind crown. Lower mantle black; black lines extending across side of breast and in front of wing. Most scapulars white, outer webs edged black. Lower scapulars and tertials light ash. Back black, grading into dark ash of upper tail-coverts and tail on rump; latter finely speckled white at sides. Underparts white, flanks finely vermiculated dark grey. Median coverts white, all other feathers on wing black (tinged brown on primaries and inner webs); secondaries broadly, inner primaries, greater coverts, and lesser coverts near bend of wing all narrowly tipped white; outer web of outer tertial white, narrowly edged black. Lesser under wing-coverts dark brown, tipped dark grey, rest of under wing-coverts and axillaries white. ADULT FEMALE BREEDING. Crown and hindneck deep chestnut. Spot round and under eye to base of bill black, grading to chestnut on ear-coverts. Chin, throat, and sides of neck white, sharply contrasting with dark parts of head, neck and chest. Feathers of mantle sooty black, broadly edged blue-grey, scapulars olive-brown, tipped grey. Back and centre of rump black; sides of rump and upper tail-coverts dark grey, tipped white. Tail dark grey. Feathers of chest grey, broadly edged white; sides of chest dark brown, vermiculated white on tips of feathers; flanks white, suffused grey-brown, caudally olive-grey. Rest of underparts white. Wing as adult ♂, but area of white on median coverts smaller, white edges to greater coverts and secondaries narrower, tertials dark brown; outer web of outer tertial ash edged black. ADULT MALE NON-BREEDING (Eclipse). As adult ♀ breeding, but top of head down to eye orange-brown;

no black in front of eye. Centre of mantle sooty-black, edged grey; scapulars dark grey-brown or sooty-black, edged pale grey, without olive tinge. Some white vermiculation on grey-brown flanks. Wing like adult ♂ breeding, but tertials darker ADULT FEMALE NON-BREEDING. Head and neck as adult ♂ eclipse. Rest of plumage like adult ♀ breeding, but upperparts browner, less olive, feathers edged grey. DOWNY YOUNG. Crown to below eye, nape, upper part of body and thighs sooty-black; small spot below eye and large patches at rear of wing, below wing, and at sides of back and rump white; upper breast and sides of body grey-brown or sooty-brown, rest of under parts white. Much like Goldeneye *Bucephala clangula*, but bill narrower. JUVENILE. Superficially like adult ♀. White median coverts variably suffused grey-brown. Crown to eye, and nape buff-orange. Upperparts sooty or grey-brown, edged grey; chest grey-brown, feathers tipped white; flanks with narrow grey-brown feathers, suffused white near bases and tips. Vent and under tail-coverts suffused or barred grey-brown. Secondaries and greater coverts more broadly tipped white than adult ♀. Tertials in juvenile ♂ longer and paler than juvenile ♀; outer tertial next to speculum tinged grey in ♂, mostly dark brown in ♀. FIRST IMMATURE NON-BREEDING. In autumn variable amount of plumage resembling adult non-breeding attained on head, scapulars, part of mantle, flanks, and chest, sometimes elsewhere. At least juvenile wing, belly, vent, back, and some tail-feathers retained. In ♂, crown and nape brownish-chestnut, sometimes some black in front of eye; mantle black, feathers edged blue-grey (rarely with olive tinge); scapulars pale ash-grey; chest grey-brown, feathers broadly tipped white and vermiculated near tip, sometimes with black subterminal bar; flanks pale ash, some feathers partly vermiculated grey. New feathers of ♀ like adult ♀ non-breeding; feathers of mantle tinged olive (not blue-grey as in 1st immature non-breeding ♂). FIRST IMMATURE BREEDING. From winter onwards number of feathers resembling adult breeding gradually develop; during winter, new feathers gradually whiter on bases; in late spring like adult breeding. ♂♂ mid-winter usually with black in front of eye, crown and scapulars intermixed with white, and indistinct black lines at sides of breast; ♀♀ with some black in front of eye. Tail like adult, wing juvenile. SECOND NON-BREEDING. Probably mostly as adult non-breeding but breeding plumage acquired latest in spring retained, and wing juvenile until mid-summer. SECOND BREEDING. Like adult, but moult out of non-breeding later in autumn; some ♂♂ have white scapulars partly suffused grey at tip.

**Bare parts.** ADULT MALE. Iris red-brown, in older ♂♂ pearl-grey. Bill blue-grey, nail dark-horn. Leg and foot slate, webs dull black. ADULT FEMALE. Like adult ♂, but iris red-brown, never grey, bill darker grey, sometimes tinged green laterally; foot paler, tinged green. DOWNY YOUNG. Iris grey-brown, leg and foot dark greyish-horn. JUVENILE. Iris dark brown. Bill dark grey.

Leg and foot grey, webs dark grey. (Bauer and Glutz 1969; Witherby *et al.* 1939.)

**Moults.** ADULT. Examined skins and limited published data suggest no differences from Goosander *M. merganser*. POST-JUVENILE AND FIRST PRE-BREEDING. Chest and sides of body mostly renewed before November, followed by head, neck, mantle, and scapulars before January (a few juvenile feathers retained in some). Moult of tail, tail-coverts, and flanks variable, some finished late November, others still in progress March. Feathers changed early in autumn like adult non-breeding; later on gradually more like adult breeding, especially in spring. SUBSEQUENT MOULTS. Like adult, but more prolonged.

**Measurements.** Netherlands, winter. Skins (ZMA, RMNH).

| | | | | | |
|---|---|---|---|---|---|
| WING AD ♂ | 202 | (2·68; 25) | 197–208 | ♀184 (2·63; 10) | 181–189 |
| JUV | 196 | (3·50; 20) | 188–202 | 177 (3·78; 24) | 171–184 |
| TAIL AD | 74·8 | (1·59; 24) | 72–78 | 69·7 (2·63; 10) | 65–73 |
| 1ST W | 69·3 | (3·57; 19) | 65–75 | 65·4 (3·53; 18) | 59–73 |
| BILL | 29·6 | (1·11; 46) | 27–32 | 26·8 (1·01; 33) | 25–29 |
| TARSUS | 34·0 | (1·02; 46) | 31–36 | 30·6 (0·72; 33) | 29–32 |
| TOE | 55·5 | (2·17; 46) | 51–62 | 49·6 (1·54; 34) | 46–55 |

Sex differences significant, except 1st winter tail. Juvenile wing (both ♂♂ and ♀♀) and 1st winter tail (♂♂ only) significantly shorter than adults. Bill, tarsus, and toe of juvenile and adult similar; combined in table. Juvenile tail tends to be somewhat shorter than 1st winter.

**Weights.** ADULT. ♂♂ Rybinsk reservoir USSR: April, 700 (598–795); May, 622 (565–650); October, 652 (540–825); November 814 (720–935). Arkhangel, USSR: July, 655; Denmark and Netherlands, January, 950; March, 718, 800. ♀♀ Rybinsk reservoir: April, – (510–670); October, 568 (515–630); November, 572 (550–650). Netherlands: December, 530; March, 600. JUVENILE. ♂♂ Rybinsk reservoir: October, 630 (500–760); November, 822 (645–920). Netherlands: January, 900. ♀♀ Rybinsk reservoir: October, 556 (500–680); November 588 (535–670). Netherlands: January, 606. (USSR: Dementiev and Gladkov 1952.) Netherlands: RMNH, ZMA. Denmark: Schiøler 1926.) Little difference between adult and juvenile in autumn, and apparent similarity with annual curve of *Mergus merganser*, with tendency for heavy weight in winter.

**Structure.** 11 primaries: p10 and p9 longest (about equal), p8 7–11 shorter, p7 17–23, p6 36–38, p1 77–98. Inner web p10 and outer p9 emarginated, slightly also inner p9 and outer p8. Tail 16 or 18 feathers. Bill short, higher than broad at base; nostrils slightly behind middle of bill. Frontal and occipital crest in adult ♂ breeding, only occipital in ♀; crest short in non-breeding, rudimentary in juvenile. Outer toe *c.* 96% middle, inner *c.* 81%, hind *c.* 27%. Other structure as *Mergus merganser*. CSR

## *Mergus serrator* Red-breasted Merganser

DU. Middelste Zaagbek    FR. Harle huppé    GE. Mittelsäger
RU. Длинносый крохаль    SP. Serreta mediana    SW. Småskrake

PLATES 95 and 100
[between pages 662 and 663
and facing page 687]

*Mergus Serrator* Linnaeus, 1758

Monotypic

**Field characters.** 52–58 cm (body 33–36 cm); wing-span 70–86 cm; smaller and slighter than Goosander *M. merganser*. Medium-large sawbill, with noticeably ragged head (elongated by thin bill and wispy, divided crest) and rather dark plumage tones. Only breeding ♂ has breast with definite pattern, appearing not as red but as brown

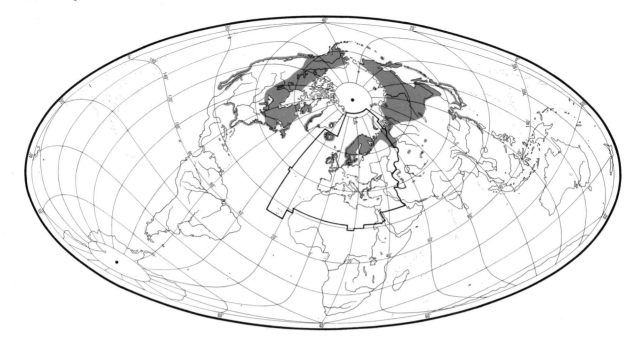

with black spots. ♂'s combination of wholly dark head, white neck collar, and dark chest unique amongst *Mergus*. ♀ and immature much less distinctive, inviting confusion with both *M. merganser* and Hooded Merganser *M. cucullatus*. Marked seasonal variation in ♂. Juvenile not normally separable.

ADULT MALE. Head and upper and centre of hindneck greenish-black; large patch of feathers at shoulder black with bold white centres, merging with reddish-brown breast, boldly spotted black. Mantle and inner scapulars black, outer scapulars mostly white; rump and tail-coverts vermiculated black and white, tail dark grey-brown. Flanks vermiculated black and white, rest of underparts white. Wing pattern essentially dark on outer half, white on inner; secondary coverts show dark leading edge and 2 black bars along centre of wing; tertials mostly black. Underwing mostly white. Thin, almost upturned bill and legs red. MALE ECLIPSE. Resembles adult ♀ but back darker and wing pattern unchanged. ADULT FEMALE. Much less definitely patterned than ♂ (and ♀ *M. merganser*). Head (with crest a little shorter than ♂) dull ash-brown (reddest on cheeks) with indistinct white centre to throat and (usually) black spots or mask before eyes. Head and upper neck uniform but not sharply demarcated from chest or rest of body. Sides of chest mottled grey-brown or grey (centre of chest usually paler but not noticeably so), merging into mottled pale grey flanks. Upperparts dirty grey, with more obvious mottling than *M. merganser*. Inner-wing pattern darker than in ♂, with secondary coverts as back. Bare parts duller than ♂. JUVENILE. Resembles adult ♀ but crest shorter and plumage generally darker grey, especially on chest. Sub-adult ♂ in first spring shows pattern of adult, particularly

on chest, but wings retain dull wing-coverts. Bill orange, legs and feet yellower than adult.

Breeding ♂ unmistakable but at long range or in poor light distinction of other plumages from *M. merganser* often difficult. Very thin, almost upturned bill, slimmer shape, ragged head, and more pronounced wing-bars characteristic however, and allow identification even when plumage detail invisible. Much longer, red bill, larger size, and different head shape make distinction from *M. cucullatus* easier. Flight silhouette and action as *M. merganser*, but gives less impression of bulk, and wing noise reduced. Swims easily, often moving head at each push of feet and lower in water than other *Mergus*.

Resorts for breeding to northern rivers and waters, particularly near coast; moves to sea in winter, coming inland only in hard weather. Noticeable flocking off river mouths in late summer and on brackish waters in winter, otherwise occurs in small parties. ♂ courts ♀ with loud, rough purr; ♀ utters harsh, grating note.     DIMW

Habitat. Although fairly adaptable, seeks unusual combination of habitat conditions. Mainly boreal, but also oceanic climatic distribution leads marginally into tundra as well as temperate forest zones; not limited to eutrophic waters. Salt water commonly favoured, in sheltered, shallow, clear bays, inlets, straits or estuaries, with sandy rather than muddy bed, and with convenient spits, projecting rocks, or grassy banks for resting just out of the water. Islands or islets also preferred, and channels, rather than large open expanses, not infrequently utilized; also rivers of suitable depth and of moderate current. Although ground nester, requires substantial surrounding cover, usually woody, so indirectly linked with forest habitat.

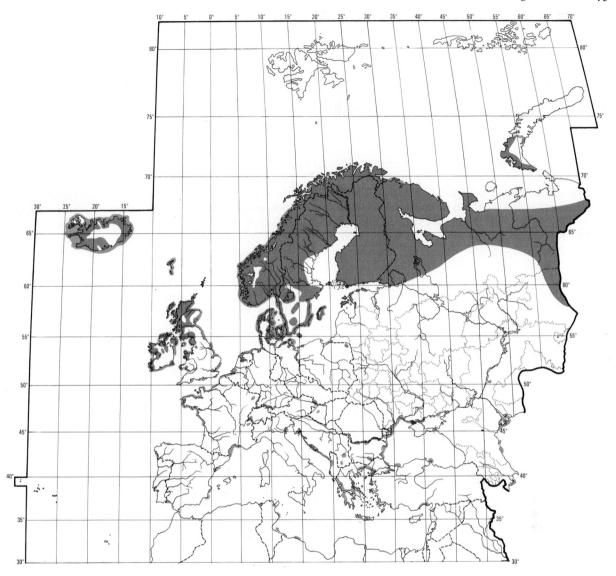

Wintering birds predominantly marine in habitat, but still favour similar enclosed waters. Reservoirs and barrages visited on passage, and occasionally artefacts used for nest-sites, but in general independent of man-made habitat features. Although not well adapted to walking, comes to land oftener and nests farther from water than most Anatinae. Swims and dives with great skill and speed, and spends much time in aerial excursions, usually at fairly low altitude. Locally much persecuted by man as competitor for fish.

**Distribution.** BRITAIN: marked expansion Scotland from *c*. 1885–1920; colonized northern England 1950 and north Wales 1953, spreading since in both areas. IRELAND: more widespread since early 20th century, but little recent change (Parslow 1967). NETHERLANDS: has nested 6 times since 1916 (JW). USSR: has bred rarely south Lithuania

(Ivanauskas 1964); no longer breeds Lake Sevan, Armenia (YuAI).

Accidental. Spitsbergen, Jan Mayen, Israel, Jordan, Iraq, Cyprus, Malta, Morocco, Azores, Madeira.

**Population.** Breeding. FAEROES: apparently increasing (Williamson 1970). BRITAIN AND IRELAND. 1000–2000 pairs (Atkinson-Willes 1970*a*); probably still increasing some areas though at much reduced rate (Parslow 1967). Many shot Scotland where not protected, but little overall change (Mills 1962*b*). WEST GERMANY: at least 66 pairs 1965 (Bauer and Glutz 1969); *c*. 60 pairs (Szijj 1973). EAST GERMANY: at least 280 pairs 1965, considerable annual fluctuations (Bauer and Glutz 1969); *c*. 200 pairs, increasing (Rutschke 1970). SWEDEN: possibly slight increase south-west coast from 1950s (Curry-Lindahl *et al.* 1970). FINLAND: estimated 10000 pairs (Merikallio 1958).

USSR. Baltic region: 450–500 pairs (Kumari *et al.* 1970). Estonia: 600 pairs, increasing rapidly under protection (Onno 1970).

Winter. Estimated 40 000 birds north-west Europe and 50 000 Europe—Black Sea—Mediterranean region (Szijj 1972). USSR: nearly 48 000 pairs in western part in not so severe winters (Isakov 1970c).

Survival. Oldest ringed bird 9 years 4 months (Reykjavik).

**Movements.** Migratory and partially migratory. Breeding population of Iceland partially migratory; some resident, others migrate to Britain and Ireland, and single recoveries Netherlands and east Greenland. Movements of British breeders not fully known, but at least disperse from inland breeding waters. Some breeders from Denmark, Norway, Sweden, and northern Germany do not move far, wintering Baltic, coastal Norway north to Arctic Circle, and Sweden north almost to zone of total icing; others join those migrating from Finland, Poland, Baltic States, and north-west USSR, which winter in force in Baltic and further WSW to Netherlands and Britain, smaller numbers reaching west France (Finnish and German birds recovered Normandy). Denmark and Baltic Germany constitute main north and west European wintering area. Small numbers winter on Murmansk coast. Some passage through continental Europe; birds ringed Baltic area recovered north Spain, Czechoslovakia, Italy, and Bulgaria, while exceptional recoveries of Finnish birds in Smolensk and Sea of Azov. Thus Baltic and north Russian origins likely for small numbers wintering Iberia and north Mediterranean. Only straggler further south. Further regular wintering areas on Black Sea, mainly west Ukraine coast, with smaller numbers in Rumania, Greece, Yugoslavia, and on south Caspian Sea (stragglers south to Iranian Baluchistan); these presumably from north Russia and west Siberia, but no confirmatory recoveries. Outside breeding season, always scarcer inland than Goosander *M. merganser* or Smew *M. albellus*, especially in west and central Europe.

♂♂ leave nesting areas early June, and, with immatures, moult in small coastal or near-coastal groups, sometimes at considerable distances from breeding places, reaching peak numbers mid-July. Biggest known moulting concentrations in Europe in Denmark, where *c.* 12 000 of which *c.* 8000 in Limfjorden, but these probably mostly relatively local birds; some evidence that larger flocks (up to 700–800) mainly adult ♂♂, smaller ones mainly immatures (Joensen 1973a).

Autumn migration may begin September, but final departures from most northern breeding areas not until mid- or late October when peak movement through Baltic and to Black Sea region. As with other diving ducks, tendency for ♀♀ and young to move earlier and further than ♂♂; adult ♂♂ often over 50% of flocks in Baltic, though probably less than 20% among Swiss migrants and British winter visitors. Spring return may begin late February; arrives April in Baltic breeding range, later (dependent on thaw) further north and east. Not all immatures return to breeding range, though all but stragglers withdraw from southern wintering areas. See also Dementiev and Gladkov (1952), Bauer and Glutz (1969).

**Food.** Diet and method of feeding similar to Goosander *M. merganser*, with no significant difference in diet when feeding together in same area (Aass 1956; Munro and Clemens 1939; White 1937, 1939). Primarily fish obtained by foraging from surface with head and eyes immersed and subsequent diving. Unlike *M. merganser*, wings as well as legs used under water (Dementiev and Gladkov 1952; Saxby 1874; Seebohm 1885; Venables and Venables 1955). Prey brought to surface or, if water deep or prey small, swallowed under water. Hunts in pairs, small groups or large flocks—often co-operatively, line of moving birds driving fish forward into shallower water where diving activity increased (Dementiev and Gladkov 1952; Des Lauriers and Brattstrom 1965; Emlen and Harrison 1970). Behaviour may be modified by pursuing gulls *Larus* (Munro and Clemens 1939). Most active early morning and evening. Prefers shallow waters, mostly below 3·5 m, occasionally up to 5·6 m. Submergence times up to 2 min, but normally below $\frac{1}{2}$ minute. Of 22 dives in 3–3·4 m, maximum 27 s, minimum 15 s (Ingram and Salmon 1941). 333 dives in 1·8–3·7 m, maximum 45 s (Dewar 1924). Occasionally takes food from or at surface (Curth 1954).

Prey mostly small, less than 8–10 cm long, and often shoaling. Fish (Palearctic only) include: salmon and trout *Salmo*, eel *Anguilla anguilla*, perch *Perca fluviatilis*, brook lamprey *Lampetra planeri*, river lamprey *L. fluviatilis*, minnow *Phoxinus*, stickleback Gasterosteidae, grayling *Thymallus thymallus*, carp *Cyprinus*, pike *Esox lucius*, chub *Leuciscus cephalus*, dace *L. leuciscus*, and roach *Rutilus rutilus* (all inland); founder *Platichthys flesus*, gobies Gobiidae, butterfish *Pholis gunnellus*, coalfish *Pollachius virens*, cod *Gadus morhua*, sculpins Cottidae, sand-eel Ammodytidae, herring *Clupea harengus*, sprat *Sprattus sprattus*, smelt *Atherina*, blennies *Blennius* and *Zoarces viviparus*, goldsinny *Ctenolabrus rupestris*, sticklebacks Gasterosteidae, pipefish Syngnathidae, and hake *Merluccius merluccius* (all coastal and estuarine). Crustaceans (prawns and shrimps *Crangon*, *Palaemon*, *Gammarus*; shore-crab *Carcinus maenas*; also *Idotea*, *Asellus*, mysids), annelids (nereids, Lumbricidae), molluscs (*Hydrobia*, *Mytilus*, *Littorina*, *Cardium*, *Mya*), insects and larvae (Coleoptera, Trichoptera, Odonata, Lepidoptera, chironomids), plants (leaves, roots, seeds of aquatic species) also taken. Some smaller invertebrates and plant materials probably taken accidentally or in fish stomachs. (Bauer and Glutz 1969; Collinge 1924–7; Dementiev and Gladkov 1952; D'Urban and Mathew 1892; Millais 1913; Robinson 1909; Witherby *et al.* 1939.)

From inland and estuarine waters of Scotland, 114 stomachs showed following. From fresh water, salmon most important (42·5% by frequency), with fewer brook lamprey (6·2%, taken at spawning time as also river lamprey in Sweden), eels (4·4%) (Mills 1962a), and minnows (1·8%). Eels and sticklebacks preferred food in some Scottish areas (Berry 1936). From saltwater, mainly crustaceans (mysids, common shrimp *Crangon vulgaris*, and shore-crab), flounders, gobies, butterfish, and coalfish (Mills 1962a). Similarly, in 6 stomachs in February from Outer Hebrides, chiefly crustaceans (common shrimp and *Gammarus*) and flounder. One from Essex, December, contained common shrimp and sand goby *Pomatoschistus minutus* (Campbell 1947). From Firth of Tay, Scotland, mainly young herring, sprats, and smelt (Berry 1939). 158 stomachs from Danish marine and brackish waters contained mostly fish (75–80% of total diet) and crustaceans: fish chiefly sticklebacks (*G. aculeatus*, *P. pungitius*, *S. spinachia*), 61% by frequency; gobies (*Chaparrudo flavescens*, *P. minutus*, *Gobius niger*), 43%; blenny (*Zoarces viviparus*), 13%; and 8 other fish species in small numbers; crustaceans (mainly Crangonidae and Palaemonidae), 37%, and number of mostly fish-derived items, e.g. nereids, molluscs, insect larvae, and plant materials (Madsen 1957). Sticklebacks, mainly *G. aculeatus*, again most important item in 21 stomachs from Finnish archipelago in spring (Bagge *et al.* 1970). Similarly, in 64 birds during breeding season, Lake Mývatn, Iceland, predominantly sticklebacks, with no seasonal or annual variation in diet (Bengtson 1971). In Sweden, however, mainly salmon parr and miller's thumb *Cottus gobio* (Lindroth 1955). USSR studies show some variation with season and area, though still chiefly fish with insects to lesser extent. In winter quarters, Caspian Sea (Lenkoran Bay), exclusively fish, mainly young carp (5–15 cm long) and *Alburnoides*. On Black Sea from April–December, chiefly *Neogobius fluviatilis*. Inland on autumn passage on Rybinsk reservoir, 3 stomachs contained only caddisfly larvae *Phryganea grandis*. On Pechora, 50% of stomachs contained caddisfly larvae and waterbeetles, 63% fish (chiefly minnows, some Siberian sculpins *Cottus sibiricus*, and a few grayling). One stomach contained a frog (Dementiev and Gladkov 1952; Bauer and Glutz 1969).

From Lake Mývatn, 65 newly hatched and small ducklings mainly ate sticklebacks (63% by frequency, 84% by weight), adult insects (94%, 8%), and seeds (72%, 7%). 5 half-grown had only sticklebacks, 9 full-grown juveniles almost entirely sticklebacks (Bengtson 1971). In Scotland, Trichopteran larvae (mostly limnophilids) found in 1st-year birds, and in Norway young ate Ephemeroptera, Trichoptera, Plecoptera, and chironomids (Mills 1962a).

**Social pattern and behaviour.** Based mainly on material supplied by L Nilsson.

1. Gregarious throughout year. For much of year in flocks, mostly of moderate size. Flocks of up to a few hundred or more in suitable areas during autumn migration, but in spring travel in rather small flocks. After break-up of pairs, ♂♂ and unsuccessful ♀♀ gather into flocks for moult. Immatures stay in separate flocks during summer, or frequent area of breeding colony. BONDS. Monogamous pair-bond of seasonal duration; polyandry and polygyny seem also to occur rarely. Pairs sometimes seen in November, but mainly formed later in winter and during spring migration. Pair-bond ends during incubation, though ♂♂ sometimes seen with ♀♀ and small young. Only ♀ tends young, often leaving brood after a few weeks; young of 2–3 weeks then join large, mixed broods led by 1 ♀. BREEDING DISPERSION. Gregarious during breeding season, Pairs stay in flock during spring migration and behave sociably on breeding grounds, often assembling at communal gathering place on shore when not incubating or feeding. Colonial in areas of high density, with many pairs nesting on same island. Not obviously territorial, though may defend small area round nest. ROOSTING. During breeding season, spends night communally. In some localities, on shore at gathering place or on sandbanks; from sunset until *c.* 1 hour after sunrise. In others, collects at gathering place in evening, but flock leaves to sleep on sea, returning at dawn. Outside breeding season, seems to spend night at sea in small flocks, but detailed observations lacking. Comparatively little time during day spent resting (Nilsson 1974).

2. During breeding season, gathering place (see above) of central importance in social life in many areas of dense population. FLOCK BEHAVIOUR. Composition much as in Goosander *M. merganser*, especially outside breeding season. Later, attendance and communal activities at gathering place seem important in maintaining flock cohesion. Pre-flight signals consist of fast Head-shakes with neck erect and bill slightly uplifted (Johnsgard 1965), facing into wind. Alarm-posture, with erect neck and depressed head feathers, assumed from stationary position while facing suspect object. ANTAGONISTIC BEHAVIOUR. See Nilsson (1965). Most disputes occur within flock, mainly during communal courtship and feeding; seem mainly related to maintenance of individual-distance and to sexual competition, disputes over food being rare (unlike in *M. merganser*). Attacks, mostly between ♂♂, of 2 main types: (1) swimming attack, much as in *M. merganser*; and, less frequently, (2) diving attack launched from under water. ♂♂ also threaten in Crouch-posture with head low over water. ♀♀ highly aggressive against ♂♂ in courting parties, often jabbing with bill. COMMUNAL COURTSHIP. See especially Johnsgard (1965). Usually occurs in flocks of several ♂♂ and 1 or a few ♀♀ from late November to June or early July; peak reached in breeding area. Initiated much as in *M. merganser*; most intense morning and evening, especially when birds arrive at or leave roost or gathering place. During intense display, ♂♂ swim about in Courtship-intent posture with head withdrawn into shoulders, bill pointing slightly up, head feathers raised a little. Secondary displays include Upward-shake, Wing-flap, and Head-shake. Most elaborate of major displays Salute-curtsey sequence (for recent analysis, see van der Kloot and Morse 1975), often performed by several ♂♂ simultaneously. Starts from Courtship-intent posture, or posture with bill lowered, and usually preceded by 1 or more rapid Head-shakes. Crest then depressed and neck suddenly jerked forward into Salute pose with bill, head, and neck extended diagonally (see A) as soft call given from position lateral to ♀. After brief pause, Curtsey follows: neck lowered into water, rear-end of body raised above surface, and tail sharply depressed; bill opened widely and pointed towards ♀ (or another ♂) as call given (see B). Finally, neck withdrawn to varying degree while bird back-paddles with

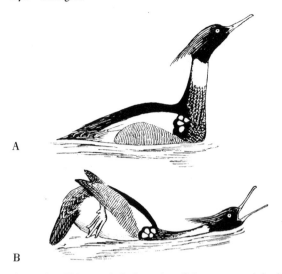

A

B

feet, tail still lowered. Independent Salute seen mainly during low-intensity display. ♂ assumes diagonal posture with jerk of head from normal swimming posture, then jerks neck slightly down, uttering call. Salute or full Salute-curtsey sequence often performed in modified form by ♂♂ on shore. Most frequent activity of courting ♂♂ Display-skating; almost indistinguishable from swimming attack, but less prolonged, as birds race over surface. Used both in overt attacks on other ♂♂ and to approach or attract ♀; often followed by Turn-back-of-Head to ♀. Apart from aggression towards ♂♂ (see above), ♀ not highly active in courting parties. Utters harsh calls, either without special movement or as component of Inciting-display: from swimming posture, ♀ suddenly stretches head and lowers neck until *c.* 45° to water with bill pointing down, repeating movement 2–3 times in rapid bobbing fashion (much as in Smew *M. albellus*), often towards particular ♂. Will sometimes also jab at ♂♂ while Inciting. Also makes Head-nodding actions (Ringleben 1951; Curth 1954) and may perform ♂-like Salute-curtsey sequence during active courtship. ♀ likely to dive in escape when approached by many ♂♂ Display-skating simultaneously; ♂♂ then follow ♀ under water. PURSUIT-FLIGHTS. During intense display, Display-skating by ♂♂ can develop into courtship-flight with ♀ leading party of ♂♂; after a while, all settle on water and display continues. PAIR BEHAVIOUR. Little information on pair-courtship, but includes copulation and associated displays. COPULATION. See Adams (1947), Johnsgard (1965). Occurs from December, but not frequent until spring. Pre-copulatory behaviour initiated by Ceremonial-drinking by both birds, or ♀ first adopts Prone-posture, lying deep in water with neck extended and head and tail somewhat raised. ♂ responds by circling ♀ (who turns towards him throughout), repeatedly drinking, Preening-dorsally, Wing-flapping, Upward-shaking, Head-shaking, and Bill-dipping; sometimes also Water-twitches (L Nilsson). Frequently performs partial Salute-curtsey immediately before mounting. No Wing-shaking reported during copulation; afterwards, ♂ retains hold of ♀ briefly as pair rotate on water, then performs Salute-curtsey sequence before sometimes Display-skating away. Both ♂ and ♀ bathe and Wing-flap, sometimes also self-preening or dashing-and-diving (see Mallard *Anas platyrhynchos*). RELATIONS WITHIN FAMILY GROUP. Young led from nest by ♀. Largely independent while feeding from early age, many ♀♀ spending much time away from brood. ♀ warns young against intruders by call.

(Figs A–B after Johnsgard 1965.)

**Voice.** Not closely studied, with some confusion in literature between calls of ♂ and ♀ though these appear to be distinct as in most Anatinae. Largely linked with courtship or disturbance. Wing noise at take-off but not in full flight (Townsend 1911).

CALLS OF MALE. Courtship-call complex; drawn-out, wheezy, metallic rattling or purring uttered during Salute and Salute-curtsey displays, mainly in spring, sometimes in autumn or winter. Accounts hard to reconcile, so may vary considerably. Described (e.g.) as loud, rough, purring 'da-ah' (Townsend 1911); single, soft, cat-like 'yeow' during Salute with additional 2 faint 'yeow' notes during Salute-curtsey (Johnsgard 1965); 'juiw' (L Nilsson). Recording (see I) resembles 'chit-up . . . pititee'—short, disyllable phrase followed by prolonged, harsh, slurred, trisyllabic phase, both showing wide frequency range. Alarm-call described as deep, gruff 'gra' or 'gro' sometimes rising to somewhat quicker 'gragrag' (Voigt 1950); also as hoarse 'ra' or barking 'rap rap' (Söderberg 1950)—but possible confusion in both cases with calls of ♀.

CALLS OF FEMALE. Little understood, and different utterances probably confused; all harsh—croaking or rasping. Inciting-call a harsh 'krrrr-krrrr' (Johnsgard 1965; L Nilsson); also uttered without special posturing during communal courtship. Single note given during ♀-version of Salute-curtsey display may be special call or version of Inciting-call: louder than equivalent call of ♂ (Townsend 1911). Other descriptions of vocalizations given in response to ♂ (either same as or different from above) include: loud, harsh 'karr-r-r' (Witherby *et al.* 1939); rasping croak like 'garrr' or deeper, muffled 'gorr' (Voigt 1950); gruff 'räg' (Bauer and Glutz 1969); hoarse 'harr' (Scott 1954). Rough croaks heard from birds flying to and from nests (Townsend 1911); soft, monotonous, gruff, and deliberate 'rokrok-rokrok' or 'rab-rab-rak' also

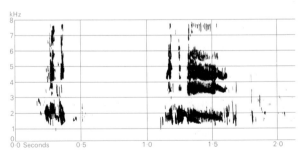

I Sture Palmér/Sveriges Radio Iceland June 1967

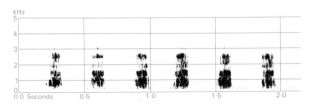

II Sture Palmér/Sveriges Radio Iceland June 1961

given in flight, especially on landing or take-off (see Bauer and Glutz 1969)—recording (see II) resembles 'uck-uck-uck-uck'. Warning-call, 'garr' or 'prrack prack', said to be similar to flight-call but more urgent (Söderberg 1950); also described as single or repeated 'wak' or 'rok'—or, when intensely anxious, 'wark-wark-wark' (see Bauer and Glutz 1969). Alarm-call, if different from last, a harsh 'quack' (see Witherby *et al.* 1939). Call to young a low, husky but distinctive 'kha-kha-kha' (Kortright 1942).

CALLS OF YOUNG. No detailed information.

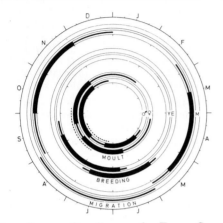

Breeding. SEASON. See diagram for Fenno-Scandia and USSR. Britain: first eggs end April, main period May and early June (Campbell and Ferguson-Lees 1972). Denmark: first eggs in last third May, peak laying 1–20 June, ceases end July (Kortegaard 1968). Iceland: mean date of first laying, 1961–70, varied from 1–8 June; average 3 June (Bengtson 1972a). SITE. On ground among tree and scrub roots, in hollows and crevices in banks and cliffs, in shallow burrows, or concealed in thick vegetation. Never far from water. Nest: shallow depression lined grass and leaves, plus down. Average diameter 27 cm, height 8 cm (Curth 1954). Building: by ♀, using materials within reach of nest. EGGS. Blunt ovate; stone to buff-olive. $65 \times 45$ mm ($57–70 \times 41–8$), sample 280 (Schönwetter 1967). Weight 73 g (68–77), sample 102 (Curth 1954). Clutch: 8–10 (6–14); larger numbers by 2 ♀♀; mean of 9 dump nests, Iceland, 16·9 (Bengtson 1972a). Of 144 clutches, Finland: 6 eggs, 1%; 7, 11%; 8, 26%; 9, 28%; 10, 19%; 11, 6%; 12, 5%; 13, 1%; 14, 1%; 15, 1%; 17, 1%; mean 9·2 (Hildén 1964). Mean of 100 first clutches, Iceland, 9·5 (Bengtson 1972a). One brood. Replacement clutches laid after egg loss. Mean of 30 late, probably 2nd clutches, Finland 8·4 (6–14), against 9·7 for 20 early clutches (Hildén 1964). Mean of 27 2nd clutches, Iceland, 6·2 (Bengtson 1972a). Eggs laid every 1–2 days, mean 0·7 eggs per day (Curth 1954). INCUBATION. 31–32 days (29–35). By ♀. Starts on completion of clutch; hatching synchronous. Eggs covered with down when ♀ off nest. Eggshells left in nest. YOUNG. Precocial and nidifugous. Self-feeding. Cared for by ♀, brooded at night while small. Tendency for broods to amalgamate where numerous, often in charge

of only 1 ♀ (Hildén 1964). FLEDGING TO MATURITY. Fledging period 60–65 days. Independent before this, *c.* 50 days. Age of first breeding 2 years. BREEDING SUCCESS. Of 3110 eggs laid, Finland in 3 years, 77% hatched and 14% young reared; average young per pair in 3 years was 1·5, 2·0, and 0·7, so losses of 84%, 78%, and 92% (Hildén 1964). In Iceland, 1961–70, average hatching success of 131 nests 63·8%; range 33·1–73·5% (Bengtson 1972a). Of 26 nests, Denmark, 23% hatched, 42% deserted and 35% predated (mainly by otters *Lutra lutra* and gulls *Larus*). 61% of 74 incubated eggs hatched, 31% were addled, 8% lost from nests (Kortegaard 1968).

Plumages. ADULT MALE BREEDING. Head, crest, upper neck, and central hindneck, black; glossed green on sides of crown, faintly purple elsewhere. Lower neck white. Mantle, inner and longer scapulars, and sides of chest black, large central spots on feathers of chest in front of wing white. Outer scapulars white, faintly vermiculated dark grey on outer web. Back, rump, and flanks vermiculated dark grey and white. Upper tail-coverts dark ash, tail dark slate, both sometimes with tips and edges of feathers vermiculated like rump. Lower neck and chest cinnamon-buff; in centre of chest, feathers narrowly bordered black laterally, on sides of chest more broadly. Rest of underparts white. Innermost tertials, lesser coverts, primaries, and primary coverts sooty-black; lesser coverts tipped dark grey. Median coverts white. A few outer secondaries and coverts black, greyer near tip and on inner web; other secondaries and greater coverts white with broad black bases (forming 2 wing-bars); innermost secondary and outer tertials white, bordered with black on outer web and suffused grey on inner. Under coverts of primaries grey, rest of underwing and axillaries white. ADULT FEMALE BREEDING. Crown down to eye, crest, and nape dark olive-brown. Lores and stripe under eye white (stripe occasionally indistinct). Sides of head and neck, and streak below lores to ear, orange-brown. Chin and throat white, not sharply defined from orange-brown on head. Spot around or behind eye black; variably often also black near gape and on forehead, or on chin and streak from gape to below eye; rarely whole face and throat black. Upperparts and tail dark grey, broadly edged olive-grey on mantle and scapulars and ash on back, rump, and upper tail-coverts. Feathers of chest white, suffused orange-buff near tip and sometimes grey-brown subterminally. Rarely, edges of mantle, scapulars, and chest suffused richly orange-buff. Sides of breast and flanks grey-brown, feathers broadly tipped olive-grey or white; rest of underparts white. Wing like adult ♂, but lesser and median coverts grey-brown, often broadly tipped ash, greater coverts narrowly edged black; outer tertials and sometimes inner secondary ash (where ♂ white). ADULT MALE NON-BREEDING (Eclipse). Like adult ♀, but wing as adult ♂ breeding; crest shorter, no black round eye or on face, no white loral streak. Mantle and scapulars sooty-black, narrowly edged olive-grey; back, rump, and tail-coverts grey-brown, feathers tipped grey, or vermiculated grey near tip. Sides of breast and flanks grey-brown, sometimes tinged buff on sides of breast and partly vermiculated white on flanks. ADULT FEMALE NON-BREEDING. Like adult ♀ breeding, but sides of head and neck paler orange-brown; forehead, crown, short crest, and streak through eye grey-brown. Chin and throat pale orange. No black on face; loral streak buff. Centres of feathers of chest dark grey-brown. DOWNY YOUNG. Like Goosander *M. merganser*, but white streak from lores to eye less defined suffused tawny; dark line below loral streak narrower and paler brown. Bill structure different (see Structure). JUVENILE. Like adult ♀ breeding, but tail-feathers narrow, pointed, with notch at tip. Sides of head paler, buff-brown, crown and streak through eye grey-brown, streak in lores buff. No black on face. Feathers of mantle and scapulars edged grey instead of olive; feathers on sides of breast and flanks narrow, white, suffused grey-brown. Part of vent and some under tail-coverts grey-brown. Wing of juvenile ♂ like adult ♀, but juvenile ♀ has tertials shorter, dark grey-brown with black edge on outer webs, and little or no grey bloom. FIRST IMMATURE NON-BREEDING. New feathers acquired in autumn on head, flanks, scapulars, or sometimes elsewhere resemble adult non-breeding. In ♂, flanks sometimes partly vermiculated grey-brown and white, mantle and upper scapulars black, edged olive-grey. FIRST IMMATURE BREEDING. Feathers resembling adult breeding gradually develop from late autumn. Most obvious in ♂: sides of breast in front of wing first (black feathers with large white spots), then scapulars, tertials, part of head, neck, chest, and flanks during winter. When no juvenile feathers of tail, tertials, back, or belly retained, 1st immature breeding of plumage strongly resembles adults, but in ♀♀ usually no black on face, and wing relatively more abraded. SECOND NON-BREEDING. Like adult non-breeding, but often that part of breeding plumage acquired latest in spring, retained.

**Bare parts.** ADULT MALE. Iris carmine. Bill deep carmine, culmen and nail black. Foot deep vermilion. ADULT FEMALE. Iris pale brown to dull red. Bill brown-red to orange-red, culmen dark brown. Foot dull red to orange-red, webs pale brown. DOWNY YOUNG. Iris grey-brown. Bill dark horn. Foot olive-brown. JUVENILE. Iris pale brown, in ♂ changing to yellow or orange-red during winter. Bill reddish horn, culmen brown, changing to carmine (♂) or orange-red (♀) in spring. Foot yellow-brown, webs dull brown; change to pale vermilion (♂) or yellow-red (♀) during winter. (Witherby *et al.* 1939; ZMA, RMNH.)

**Moults.** ADULT POST-BREEDING. Complete. Head and body between May and August in ♂♂ (latest in north of range), some weeks later in ♀♀. Belly, vent, and back last, interrupted by simultaneous moult of flight-feathers. Flightless for *c.* 1 month mid-July to late August in ♂♂, *c.* 1 month later in ♀♀; belly, vent, and back completed later (often moulted once a year only); moult of tail prolonged. ADULT PRE-BREEDING. Partial. Head and body. Flanks, mantle, chest, and scapulars first, followed by head, neck, and sides of breast. Tail completed about November, rest of head and body mostly before December in ♂♂, *c.* 1 month later in ♀♀. POST-JUVENILE. Partial; amount of plumage changed for 1st immature non-breeding variable. Head, neck, flanks, and scapulars October–January, sometimes some feathers elsewhere

on body. FIRST IMMATURE PRE-BREEDING. Partial. Head, body, tail, and tertials, but highly variable, and sometimes much of juvenile or 1st non-breeding retained. Starts about December with sides of breast and tail; more breeding plumage acquired in spring, but usually not completely, and at least juvenile wing retained. SUBSEQUENT MOULTS. Like adult, but those in 2nd year usually more prolonged and less complete. (Dementiev and Gladkov 1952; Joensen 1973a; C S Roselaar.)

**Measurements.** Netherlands, mainly winter. Skins (RMNH, ZMA).

| | | | | | | | |
|---|---|---|---|---|---|---|---|
| WING AD | ♂ 247 | (5·18; 32) | 235–255 | ♀ 228 | (6·54; 14) | 216–239 |
| JUV | 236 | (6·13; 12) | 226–245 | 217 | (4·18; 14) | 208–221 |
| TAIL AD | 81·2 | (3·14; 34) | 76–87 | 76·4 | (2·82; 14) | 73–81 |
| IMM | 70·8 | (2·64; 10) | 67–75 | 64·0 | (4·47; 12) | 57–70 |
| BILL | 59·2 | (2·16; 46) | 56–64 | 52·1 | (2·21; 28) | 48–55 |
| TARSUS | 47·0 | (1·47; 45) | 44–50 | 42·7 | (1·51; 28) | 40–45 |
| TOE | 65·5 | (1·90; 46) | 62–69 | 60·2 | (2·10; 28) | 57–64 |

Sex differences significant. Wing and tail of juveniles significantly shorter than adults. Other measurements similar; combined in table. Tail given for immatures is juvenile for ♀, 1st winter for ♂ (1st winter tail *c.* 5 mm longer than juvenile tail).

**Weights.** ADULT. November–February, Netherlands and Denmark. ♂♂, 1197 (118; 11) 947–1350; ♀♀, 984 (84·3; 5) 900–1100. ♂♂, north-west USSR: May, 1082 (970–1120); June, 1265; July, 1100 (1000–1200). ♀♀, August–October: Denmark, 882, 900, 959, 993; Rybinsk reservoir, USSR, 780–930. JUVENILE. Denmark, winter: ♂♂ 1231, 1251; ♀♀ 828, 860, 1055. (USSR: Dementiev and Gladkov 1952. Denmark: Schiøler 1926. Netherlands: RMNH, ZMA.)

**Structure.** 11 primaries: p10 and p9 longest, about equal; p8 9–16 shorter, p7 22–31, p6 36–45, p1 98–116. Bill long, more slender than *M. merganser*, lower at base; nostrils nearer base, about a quarter of bill-length from base. Nail of bill small. Sides of lower mandible compressed, intercrural space small with only a little extension of feathering of chin. Feathers of central crown and nape elongated in adult breeding, forming ragged double crest; shorter in non-breeding and juvenile. Outer toe about equal to middle, inner *c.* 83% middle, hind *c.* 26%. Rest of structure as *M. merganser*.

**Geographical variation.** Differences too slight to warrant subspecific distinction of *M. s. schiøleri* (see Salomonsen 1949), for Greenland birds, which have wing and bill slightly longer, (bill also said to be wider at base through data conflicting), but much overlap (Vaurie 1965).                   CSR

## *Mergus merganser* Goosander

PLATES 96 and 100
[between pages 662 and 663
and facing page 687]

DU. Grote Zaagbek      FR. Harle bièvre      GE. Gänsesäger
RU. Большой крохаль      SP. Serreta grande      SW. Storskrake      N.AM. Common Merganser

*Mergus Merganser* Linnaeus, 1758

Polytypic. Nominate *merganser* Linnaeus, 1758, Europe and north Asia. Extralimital: *comatus* (Salvadori, 1895), highlands of Pamir and Tibet; *americanus* Cassin, 1852, North America.

**Field characters.** 58–66 cm (body 40–46 cm); wing-span 82–97 cm; largest of saw-billed ducks, noticeably bigger in head and body than Red-breasted Merganser *M.*

*serrator*. Bulky, lengthy sawbill, with bulbous head, long thin bill, and rather clean tones in all plumages. ♂ uniquely combines long, red bill and dark green head with elongated, creamy-white underbody; ♀ and immatures less distinctive but chestnut heads contrast strikingly with pale clean chests. Marked seasonal variation in ♂. Juveniles separable at close range.

ADULT MALE. Head (with slight, even crest) and ruff round upper neck green-black; centre of mantle black, rump grey and tail grey-black. Outer scapulars and underparts white, tinged cream or pink and with grey sometimes visible on lower flanks. Wing basically black on outer half, white on inner, with base to greater secondary coverts and edges of tertials black. Underwing mostly white, so when overhead looks mostly white with black head. Bill, legs, and feet red. MALE NON-BREEDING. Somewhat resembles adult ♀ but wing pattern unchanged, body sides with much white and upperparts blacker. ADULT FEMALE. Although plumage pattern less contrasting than ♂, more striking and cleaner toned than other ♀ *Mergus*. Head (with obvious crest) and upper neck red-brown, throat with noticeable white centre; upper neck sharply demarcated from white lower neck and bright greyish-white chest. Upperparts rather blue-grey (cleaner than *M. serrator*) with rump and tail darker. Flanks grey (forepart mottled white), exaggerating cleanness of chest; underbody as ♂. Inner-wing pattern darker than in ♂, with blue-grey coverts contrasting with dark grey tertials and outer secondaries and white inner secondaries. Legs and feet orange. JUVENILE. Resembles adult ♀ but head colour less intense (and crest shorter), white throat less distinct, chest and upperparts duller grey (without blue tone). Sub-adult ♂ distinguishable in 1st spring by retention of dull upperwing pattern, blackish collar below chestnut upper neck and black marks on chin and throat. Bare parts lack full red of adults.

Breeding ♂ unmistakable, but distinction of ♀ (at longer ranges or in poor light) and immatures from *M. serrator* needs care. In addition to plumage differences, *M. merganser* has regular head shape and is larger. Basic plumage patterns also resemble those of goldeneyes *Bucephala* but long bill and size rule out confusion. Flight silhouette not typically duck-like, recalling diver *Gavia* (or even cormorant *Phalacrocorax*) since length of bill, head, and neck balanced by prominent tail at end of cigar-shaped body. Take-off laboured but flight action rapid, with powerful, even wing-beats (associated with notably straight body plane) producing humming whistle. Flying groups adopt no regular formation. When afloat, bulkiest of genus, resting more buoyantly on water than *M. serrator*; walks with surprising ease.

Resorts to lakes and wider rivers for breeding, and retains marked preference for deep, fresh waters in winter, commonly exploiting reservoirs and canals; not uncommon on brackish water in winter but frequent at sea only during migration. Usually silent, except in display. DIMW

**Habitat.** Despite substantial geographical overlap with Red-breasted Merganser *M. serrator* in boreal and north temperate west Palearctic range, largely separated in habitat, favouring upper basins of rivers, large clear inland lakes in forest or mountainous regions, and landlocked or non-tidal seas. Avoids warm waters and, although in places reaching north of Arctic Circle, keeps well clear of any ice, and equally of floating or fringing aquatic vegetation or luxuriant submerged aquatic plants. Demands fairly high productivity of fishes but, not being primarily bottom feeder, tolerant of deep waters and of fast-flowing streams, extending to narrow headwaters of smallest which support sufficient fish. In Ladakh, to at least 4000 m (Ali and Ripley 1968). In common with other large hole-nesters, tied to breeding habitat which includes within easy reach of water mature trees (often broad-leaved) with suitable natural holes or those of Black Woodpeckers *Dryocopus martius*. Failing these, will use rock crevices, cavities in steep slopes or artefacts such as nest-boxes, holes in walls, chimneys, and inside abandoned structures. Although made shy by intense persecution from human fishing interests, freely uses barrages and reservoirs, and occasionally nests near man, for example on ornamental waters. In winter, resorts to large open waters, including marine inlets, but only incidentally and to minor degree a sea-duck. Apart from resting by water margins, comes little to land, but makes much use of airspace, adroitly following courses of streams round sharp bends among fringing trees or rock faces. Shows remarkable tenacity and power to maintain numbers in face of sustained campaigns for extermination.

**Distribution.** BRITAIN. First recorded breeding Scotland 1871, where increased after large winter influx in 1876, and by end of 19th century distributed over much of north; then slower spread, not breeding regularly in south until 1940. Spread to England 1941, when bred Northumberland, and in 1950 nested Cumberland (Mills 1962b; Parslow 1967); further spread since in northern England and breeding one site in Wales. IRELAND: bred Co. Donegal since 1969 (RFR).

South of main range, some isolated populations have disappeared in recent decades, e.g. in West Germany, Austria, and USSR where no longer breeds Lake Sevan, Armenia (Bauer and Glutz 1969; Dementiev and Gladkov 1952; YuAI), while other nesting sporadic. YUGOSLAVIA: nested Bosnia (up to 1908), Montenegro (1902), and Macedonia (1939 and recently) (Matvejev and Vasić 1973); may breed more regularly than records suggest (VFV). GREECE: a few pairs bred in north, 1939, 1968–71, and 1973 (WB). RUMANIA: old nesting record Dobrogea not confirmed (Vasiliu 1968).

Accidental. Faeroes, Spitsbergen, Bear Island, Spain, Italy, Greece, Turkey, Israel, Iraq, Cyprus, Malta, Tunisia, Morocco.

**Population.** Breeding. BRITAIN: 500–1000 pairs (Atkinson-Willes 1970a). FRANCE: *c.* 35 pairs (FR).

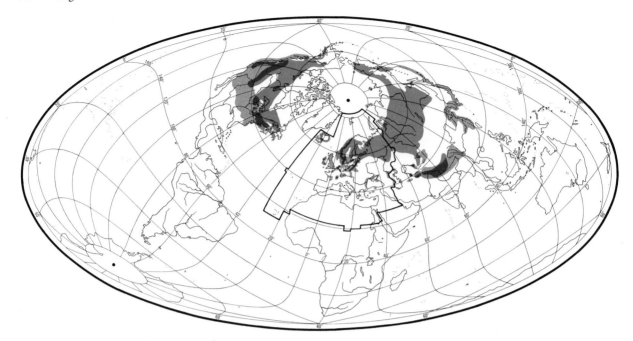

DENMARK: 15–20 pairs, decreasing (TD; Ferdinand 1975). NORWAY: less plentiful than formerly, especially in south (Haftorn 1971). SWEDEN: apparently stable (Curry-Lindahl *et al*. 1970). FINLAND: estimated 4000 pairs (Merikallio 1958); stable or increasing, helped by nest-boxes (Grenquist 1970). WEST GERMANY. 75–85 pairs (Bauer and Glutz 1969); *c*. 100 pairs (Szijj 1973). Bavaria: 52–56 pairs (Bauer and Zintl 1974). EAST GERMANY: *c*. 50 pairs; marked decrease last 30 years (Rutschke 1970). POLAND: apparently more numerous than in 19th century (Tomiałojć 1972). SWITZERLAND: *c*. 140 pairs, with considerable number of non-breeders (Bauer and Glutz 1969). USSR. 1300–1400 pairs Baltic region (Kumari *et al*. 1970). Estonia: 1500 pairs, increased recently since protected (Onno 1970).

Winter. Estimated 75 000 birds north-west Europe and 10 000 Europe–Black Sea–Mediterranean region (Szijj 1972). USSR: estimated 26 000 birds west of 90°E (mainly Black Sea and Caspian) in not so severe winters (Isakov 1970*c*).

Survival. Adult mean annual mortality 40%, life expectancy 2·0 years (Boyd 1962). Oldest ringed bird 7 years 4 months (Helsinki).

**Movements.** Migratory and partially migratory. No evidence yet that any Icelandic breeders emigrate. Similarly, British breeders almost entirely resident, moving short distances (mainly within 150 km) from breeding waters to lakes and sheltered estuaries; recoveries of Northumberland young west to Lake District, Solway Firth, and Ayrshire, and exceptionally northward to Aberdeenshire, Ross-shire, and (once) Norway. No evidence that breeders of southernmost Scandinavia,

north Germany, and Poland move further than western Baltic, but those breeding central and northern Scandinavia, Finland, Baltic States, and USSR east to Pechora migrate west to Baltic and beyond to Netherlands and Britain, in smaller numbers to west France and north Spain; recovery of Finnish bird in Yugoslavia probably atypical. Some evidence of westward movement in Lapland: over 3000 passed west along Barents Sea coast (69°45′N) 4–26 September 1962 (Fjeldså 1969). Small population breeding northern side of Alps thought to winter on lakes of Switzerland and south Germany, but some move south (notably in cold weather) as far as south France and north Italy. Populations presumably originating further east (into west Siberia) migrate to Balkans and south Russia. Small numbers occur Rumania, Yugoslavia, and Greece, while larger flocks occur north-east Black Sea. Major wintering area in west Caspian, up to 15 000 in Volga delta alone (Krivonosov 1970). A few penetrate into Iran and Iraq.

No marked moult migration or moult assemblies of ♂♂ known from west Palearctic, apart from reports of several hundred gathering on Novaya Zemlya (Dementiev and Gladkov 1952); but local moult movements often apparent, e.g. river breeders moving to lakes, those of Baltic skerries to outer, treeless ones (Hildén 1964). In late August and early September, moulting and breeding waters often deserted as flocks build up on estuaries and shallow parts of some inland lakes. Mass departures not until advent of freezing; thus major movements through USSR and Baltic October and early November, some lingering Rybinsk reservoir until late November. In Baltic area, movement from inland waters to coasts marked from mid-October, and former practically deserted by end of year. Early arrivals in North Sea countries late October and early November, but no large numbers until December, while numbers build up on Black Sea and Sea of Azov from mid-October to mid-December.

Return migration from early March and, apart from stragglers, non-breeding range vacated by mid-April. Returns to Finnish archipelagos earlier than other breeding Anatidae, and while shores still icebound (Hildén 1964). British and Swiss breeders begin returning March, but further north and east return dependent on thaw; Lapland and northern USSR reached late April or May. See also Bauer and Glutz (1969).

**Food.** Primarily fish obtained mainly by surface-diving, using legs only for propulsion. Forages with head (including eyes) submerged, pursuing prey once sighted; in shallow water, may catch prey without diving but in deeper water will always dive. Chases and catches prey with head and neck held in front of body in straight line. Returns to surface to swallow or, in deep water (especially if prey small), may swallow underneath. Beak with its backward projecting 'teeth' used to seize and hold prey across middle, then quickly turned and swallowed head

first. Bill also used to probe around stones. Occasionally may up-end. (Goodman and Fisher 1962; Lindroth and Bergstrom 1959; Mills 1962b; Nilsson 1966; Sayler et al. 1940; White 1957.) Can be opportunistic feeder, taking dead or dying fish caught in turbines (Timken and Anderson 1969) or from gulls (Hofer 1968). Hunts in pairs, small groups, or large flocks, often in apparently organized way in which birds form moving line to drive fish forward (Dementiev and Gladkov 1952; Mills 1962b; Tebbutt 1961; White 1957). Most active early morning and evening. Prefers shallow waters up to 4 m. Submergence times up to 2 min, but normally not more than $\frac{1}{2}$ min. Of 97 dives in 2–3 m, maximum submergence 52 s (Nilsson 1966); of 96 dives, most in 18–37 m, maximum 37 s (Dewar 1924). In deep and turbid water, possibly locates prey by movement rather than by sight (Heard and Curd 1959), but most fishing in clear water (White 1957).

Diet constituents depend on habitat and availability. Possibly some preference for salmonids (Aass 1956; Lindroth and Bergstrom 1959; Sayler et al. 1940; White 1957), though where other species abundant these may be taken more frequently (Munro and Clemens 1937). Prey mostly small, less than 10 cm, with upper limit determined

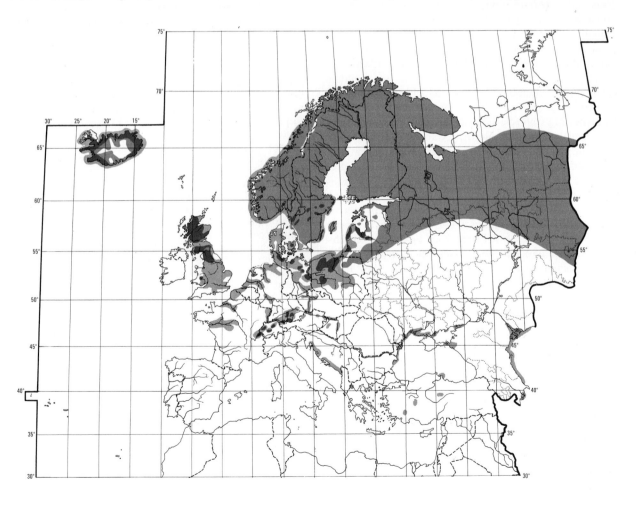

by girth rather than length (Latta and Sharkey 1966). Freshwater fish (Palearctic only) include (with maximum length in brackets): salmon and trout *Salmo* (36 cm), eel *Anguilla anguilla* (46 cm), perch *Perca fluviatilis*, minnows *Phoxinus*, grayling *Thymallus thymallus*, pike *Esox lucius* (31 cm), miller's thumb *Cottus*, dace *Leuciscus leuciscus*, rudd *Scardinius erythophthalmus*, roach *Rutilus rutilus*, barbel *Barbus* (30 cm), bleak *Alburnus alburnus*, carp *Cyprinus* (31 cm). Coastal fish include: gunnel *Pholis gunnellus*, herring *Clupea harengus*, cod *Gadus morhua*, plaice *Pleuronectes platessa*, sand-eel Ammodytidae, gold-sinny *Ctenolabrus rupestris*, and blenny *Zoarces viviparus*. Insects and larvae (Coleoptera, Diptera, Trichoptera, Lepidoptera), crustaceans (*Neomysis*, *Gammarus*, *Idotea*, *Palaemon*, *Crangon*), annelids (nereids), molluscs (*Mytilus*, *Macoma*), frogs (*Rana*), mammals (water shrew *Neomys fodiens*), and birds (unidentified) all less frequently recorded. Plant materials often present, probably taken accidentally or within fish (Bolam 1912; Collinge 1924–7; Dementiev and Gladkov 1952; Lindroth and Bergstrom 1959; Naumann 1905; Madsen 1957; Mills 1962a).

Many studies in America and Palearctic have tried to determine effects on salmon and trout waters; in certain areas and at certain times, detrimental influence highly probable, but many complicating factors—e.g. also eats salmon and trout predators (Aass 1956; Elson 1962; Lindroth 1955; Lindroth and Bergstrom 1959; Mills 1962a; Sayler *et al.* 1940; White 1957). On Indalsälven, Sweden, mainly parr of salmon and sea trout, and to minor extent miller's thumb; in late summer, chiefly grayling fry (Lindroth 1954; Lindroth and Bergstrom 1959). From Scotland, 124 stomachs contained mainly salmon (57·2% by frequency), brown trout (9·7%), and eels (14·5%); less frequently, 6 other fish species, insects, water-shrew, and birds (Mills 1962a). In Canada, eels can be total diet (Coldwell 1939). From Finnish archipelago, 21 stomachs in spring contained mainly fish, especially stickleback *Gasterosteus aculeatus* (10), and rather fewer blenny, perch, pike, roach, herring, molluscs (*Mytilus edulis*, *Macoma baltica*), and crustaceans (Bagge *et al.* 1970). 48 stomachs from Danish coastal waters October–February (1 in March) contained large number of eels (54% by frequency), sticklebacks (35%), blenny (17%), gobies (17%), 4 other fish species, and crustaceans (*Crangon*, *Gammarus*). 5 stomachs from Danish freshwaters had virtually only fish: dace, rudd, bleak, roach, and ten-spined stickleback *Pungitius pungitius* (Madsen 1957). In USSR, chiefly fish with species varying with area and season. On spring passage on Mologa river, 75% fish (miller's thumb and young perch), 25% mollusc *Pisidium amnicum*. On Rybinsk reservoir, entirely fish (burbot *Lota lota* and roach) in spring and, in autumn, 97% by volume fish (mainly roach 68%, ruffe 18%, and some perch and smelt *Osmerus eperlanus*), and 3% caddisfly larvae *Phryganea grandis*. In Semiryeche, Turkestan, chiefly barbels *Diptychus* and water snakes. On Pechora breeding grounds, 70% by frequency fish (mainly minnow, grayling, salmon, roach, and Siberian sculpin *Cottus sibiricus*), 16·4% aquatic insects (waterbeetles, larvae of dragonflies, and caddisflies), 10·9% frogs *Rana*, 2·7% molluscs, and once a water shrew. In Volga delta on autumn passage exclusively fish (roach, carp, pike). In winter on coasts of Turkmenia, chiefly gobies and sandsmelt *Atherina* (Dementiev and Gladkov 1952).

Ducklings at first take mainly insects, graduating after 10–12 days to fish which eventually predominates (White 1957). In Sweden, ducklings ate caddisfly adults and crustacean *Neomysis vulgaris* (Fabricius 1959).

Percentage of body weight eaten per day by captive birds variable: 38·5% (Sayler *et al.* 1940), 17·9–26·8% (Latta and Sharkey 1966).

**Social pattern and behaviour.** Based mainly on material supplied by L Nilsson.

1. Gregarious most of year. Outside breeding season in flocks; flocks of 1000–10000 or more on suitable waters during autumn migration and in winter, but generally smaller on spring migration. Pairs stay in flock until reach breeding grounds. After pair break up, ♂♂ gather for moult; joined by unsuccessful ♀♀. Immatures stay in flocks over summer. BONDS. Monogamous pair-bond of seasonal duration, but occasional polygamy suspected. First pairs seen November (south Germany), but most form later in winter and during spring migration. Pair-bond ends during incubation, sometimes at start, but ♂♂ have been seen with ♀♀ and young. Only ♀ tends young; often leaves before young able to fly. BREEDING DISPERSION. No territorial behaviour. ♀♀ rather gregarious on breeding grounds, sometimes searching for nest-sites together; up to 3–4 (occasionally to 10) may nest in same tree. ROOSTING. Communal. Prefers stones, trees, or other perches at traditional sites protected by surrounding water, but large flocks on water in some localities; wintering birds sometimes on edge of ice. Returns to roost well before sunset and leaves rather late; but generally keeps close to roost site during whole day, leaving only for short feeding expeditions and spending much more of day resting than other sawbills (Nilsson 1974).

2. FLOCK BEHAVIOUR. Flocks of no fixed composition. Smaller groups often gather into larger ones for cooperative fishing (Mills 1962b). Preening and bathing often performed by many simultaneously. Pre-flight signals consist of bathing movements with some lateral Head-shaking (McKinney 1965b); alarm behaviour much as in Red-breasted Merganser *M. serrator*. ANTAGONISTIC BEHAVIOUR. Most disputes arise during communal courtship and among feeding birds in hard weather (Nilsson 1966); rare between single birds or pairs. Attacks between ♂♂ of 2 main types: (1) in swimming attack, most common, bird races against opponent with head slightly forward; (2) in 'wild' chase (see also Goldeneye *Bucephala clangula*), wings used to increase speed over surface. (3) Diving attack much as in *M. serrator* but rare. In food disputes, both ♂♂ and ♀♀ pursue successful birds, mostly in swimming attacks and wild chases. In courting parties, ♀♀ show hostility to ♂♂ by jabbing them with bill. COMMUNAL COURTSHIP. See Johnsgard (1965), Nilsson (1966). Usually in flocks of several ♂♂ and 1 or a few ♀♀, many participants already paired; from December until start of breeding season with peak late winter and spring. Often initiated when pairs meet or new individuals join flock; also when gathering to roost. Once courting party formed, neighbouring ♂♂ (both paired and unpaired) join in. ♂♂ often adopt Courtship-

A

B

C

intent posture with neck forward and head feathers erected. Secondary displays include frequent Wing-flaps and Upward-shakes; ♂♂ also perform Ceremonial-drinking during communal courtship, but this generally pre-copulatory. Salute (see A) only elaborate major display of ♂; from normal position or Courtship-intent posture, stretches neck until bill points straight-up and utters faint bell-like call. Also frequently adopts partial Neck-stretch display with head feathers erected (see B) while repeatedly uttering twanging call. Head-throw, recalling *B. clangula*, seen rarely: ♂ throws head over back but only at angle of *c.* 45°. ♂ also rarely performs Kick-display, as in *B. clangula* but with no head movements, and display similar to Curtsey of *M. serrator*. Most frequent activities of courting ♂♂ include: (1) circling ♀ in Courtship-intent posture while calling; (2) making rapid, often short rushes over water with neck slightly forward. This Display-skating aggressive in nature and often directed against other ♂♂, leading to full attacks (see above); preferred ♂♂, especially, also Display-skates or swims ahead of Inciting ♀ in Turn-back-of-Head display with crown feathers erected. ♀ adopts Courtship-intent posture like ♂. Often jabs at displaying ♂♂, giving harsh call; same call used during slightly ritualized Inciting-display in which ♀ makes sideways pointing movements of head towards rejected ♂♂. ♀ sometimes escapes by swimming or diving when several ♂♂ Skate towards her; often results in underwater pursuit. Pursuit-flights. Courtship-flights with

♀ leading (as in *M. serrator*) occur rarely when ♀ takes wing after surface chasing during communal courtship. Pair Behaviour. Little information, but includes copulation and associated displays. Copulation. See especially Johnsgard (1965). Occurs from December; pair sometimes leave courting party and start pre-copulatory behaviour immediately (Nilsson 1966). Initiated by either sex Ceremonial-drinking (see C); continues mutually until ♀ assumes full Prone-posture or ♀ may go prone without previous drinking. ♂ responds by circling ♀ who turns in his direction as he displays; includes Ceremonial-drinking, Bill-dip (often with Water-twitch as in *B. clangula*), Preen-dorsally, Preen-behind-Wing, and Upward-shake. Sooner or later, ♂ swims towards ♀ in normal posture, or with slightly extended neck, and mounts. Copulation lasts 5–10 s (with no Wing-shaking); ♂ retains grip afterwards as pair rotate once or twice, then moves away while Turning-back-of-Head to ♀ and calling. One or both may then bathe and Wing-flap, generally then preening. Relations within Family Group. After 1–2 days in nest-hole, young climb out, jump to ground, and follow ♀ to water; ♀ may carry them on back when danger threatens (Erskine 1971*b*). Young stay close to ♀ during first days; brooded by night for *c.* 2 weeks. Rather independent during food-seeking, and often left by ♀ for while. Lost young join other broods. ♀ warns young against danger by call; sometimes dives with them on back.

(Figs A–C after Johnsgard 1965.)

**Voice.** Not closely studied and little used except in courtship and when alarmed. As with Red-breasted Merganser *M. serrator*, calls of ♂ and ♀ probably confused at times in literature, especially as sexes will call simultaneously during courtship. Peculiar hollow, rushing sound produced by wings when birds plunge from a height with necks extended and wings half-closed (see Witherby *et al.* 1939).

Calls of Male. Courtship-calls main vocalization. Perhaps highly variable (see Witherby *et al.* 1939). 2 types described by Johnsgard (1965) and confirmed by L Nilsson: (1) faint, twanging 'uig-a' given in Neck-stretch posture, throat enlarging with each utterance; (2) faint, high-pitched, bell-like note during Salute-display. Other descriptions, which may refer to either of above, include 'kerr kerr kerr', etc. (Witherby *et al.* 1939); repeated, purring 'door-door' (Townsend 1916); recording suggests strikingly musical 'pa-poor', 1st note higher than 2nd, with vibrant quality suggestive of frog croaking (P J Sellar)—peculiar, wooden croaking ascribed to ♀ by Witherby *et al.* (1939) perhaps same. During courtship-flights, ♂♂ call 'gagaga', 'rrga', or combinations like 'kragagagagaga' (Voigt 1950). Warning or alarm-call, given at take-off or when disturbed, a hoarse 'grrr' or 'wak' (Voigt 1950), rough 'karrr' (Söderberg 1950), or hoarse croak (Kortright 1942). Other, possibly distinct calls include soft, prolonged 'bab-o-bab' (Niethammer 1938) and raucous, muffled clucking (Géroudet 1965).

Calls of Female. Mostly harsh. Threat and Inciting-call a loud 'karr karr' (Johnsgard 1965). May develop into quick, excited-sounding, variable, cackling — 'kokokokokok' or 'eck-eck-eck-eck' (see I)—alter-

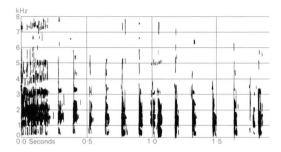

I Sture Palmér/Sveriges Radio Sweden May 1961

nating with one of ♂'s Courtship-calls. Loud, snorting 'karr-r-r' given in flight; apparently also serves as warning or alarm-call. Call summoning young from nest-hole an urgent 'cro . . . cro . . . cro . . .' (Géroudet 1965); other calls to young 'krrra' or 'rarrk rarrk' (Bauer and Glutz 1969). When disturbed with brood, ♀ hisses (Witherby *et al.* 1939).

CALLS OF YOUNG. 'Tit . . . tit . . . tit' (Niethammer 1938); 'ik-ik-ik', or 'fir-fir' when disturbed at roost (Bauer and Glutz 1969).

Breeding. SEASON. See diagram for Fenno-Scandia and USSR. In central and western Europe, earliest eggs laid end March to early April; main laying period 2nd half April and early May. SITE. Usually hole in tree, less often in banks or among rocks. Readily uses nest-boxes, as well as holes beneath buildings and roof spaces of houses. Generally close to water, but can be up to 1 km away. Nest: slight hollow with lining of available material (sometimes little) plus down. Building: no excavating; depression shaped by ♀. EGGS. Broad ovate; creamy-white. 68 × 47 mm (62/74 × 42–49), sample 300 (Schönwetter 1967). Weight 84 g (69–98), sample not given (Dementiev and Gladkov 1952). Weight of captive laid eggs 85 g (80–89), sample 9 (J Kear). Clutch: 8–11 (4–22, but over *c.* 13 probably 2 ♀♀). Of 35 clutches, Finland: 6 eggs, 6%; 7, 6%; 8, 11%; 9, 34%; 10, 17%; 11, 17%; 12, 9%; mean 9·4 (Hildén 1964). One brood. Replacement clutches laid after egg loss. INCUBATION. 30–32 days (28–35). By ♀. Starts on completion of clutch; hatching synchronous. Eggs covered with down when ♀ off. Eggshells left in nest. YOUNG. Precocial and nidifugous. Self-feeding. Cared for by ♀, brooded at night while small. May be led long distances to rearing area after leaving nest. Broods may amalgamate where numerous. FLEDGING TO MATURITY. Fledging period 60–70 days. Become independent at about same time. Age of first breeding 2 years. BREEDING SUCCESS. In Finland, average clutch size at hatching 10·8, average young reared 6·8 (Linkola 1962).

Plumages. Nominate *merganser*. ADULT MALE BREEDING. Head and neck black, glossed green on crown and hindneck. Lower neck, upper mantle, sides of body and rump, breast, belly, and under tail-coverts white. Underparts often tinged pale salmon or deep cream. Lower mantle and inner scapulars black, outer scapulars white. Back, centre of rump, and upper tail-coverts grey with black shaft-streaks, rump with white edges. Lower flank grey, feathers banded and freckled white near tips. Tail dark grey. Primaries, primary coverts, and bastard wing sooty. Inner tertials and 4 outer secondaries with their coverts dark grey, suffused white on tips; rest of secondaries white, outer tertials white edged black on outer webs. Inner lesser and median, and marginal coverts dark grey with white tips; rest of coverts white (greater sooty at base). Greater coverts of underwing grey, rest of under wing-coverts and axillaries white. ADULT FEMALE BREEDING. Crown, crest, sides of head, and upper neck red-brown; feathers of forehead and lores slightly darker. Chin and upper throat white, contrasting sharply with sides of head; merges into red-brown of lower throat. Feathers of lower neck, chest, and sides of breast grey, broadly tipped white. Mantle, scapulars, back, rump, upper tail-coverts and tail blue-grey with dark shaft-streaks; feathers of mantle and scapulars with grey-brown centres, of sides of rump with white tips. Feathers of sides of body and flanks white with grey arcs. Lower breast, belly, vent, and under tail-coverts white or cream. Primaries and outer secondaries with their coverts sooty, rest of secondaries white. Tertials pale grey, narrowly edged dark brown. Greater coverts white, often with narrow grey-brown tips. Rest of upper wing blue-grey. Underwing as ♂. ADULT NON-BREEDING. As adult ♀ breeding; but head and neck paler, crown suffused olive; crest shorter; variable white or buff streak from side of upper mandible to eye (sometimes absent); white streak down centre of throat. In ♂, centres of feathers of mantle and scapulars black or dark brown, flanks white with some grey bars. Wings like breeding, in ♂ with white coverts, in ♀ grey. DOWNY YOUNG. Crown to below eye and centre of nape tawny; rest of upperparts, sides of body, and thighs grey-brown, darkest on back and rump, tips of down

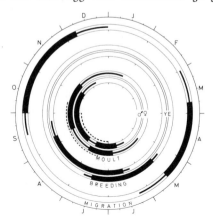

PLATE 99. (*facing*) Diving ducks in flight (adult breeding plumage), ►
A.
*Netta rufina* Red-crested Pochard: 1 ♂, 2 ♀.
*Aythya ferina* Pochard: 3 ♂, 4 ♀.
*Aythya collaris* Ring-necked Duck: 5 ♂, 6 ♀.
*Aythya nyroca* Ferruginous Duck: 7 ♂, 8 ♀.
*Aythya marila* Scaup: 9 ♂, 10 ♀.
*Aythya fuligula* Tufted Duck: 11 ♂, 12 ♀.
*Histrionicus histrionicus* Harlequin: 13 ♂, 14 ♀.
*Bucephala albeola* Bufflehead: 15 ♂, 16 ♀.
*Bucephala islandica* Barrow's Goldeneye: 17 ♂, 18 ♀.
*Bucephala clangula* Goldeneye: 19 ♂, 20 ♀.
*Oxyura leucocephala* White-headed Duck: 21 ♂, 22 ♀.
*Oxyura jamaicensis* Ruddy Duck: 23 ♂, 24 ♀. (NWC)

1

2

3

4

5

6

7

8

9

10

11

12

13

14

15

16

17

18

19

20

21

22

23

N.W. CUSA.

24

N.W.CUSA.

buff. White patches on wing, on sides of back above thighs, and on sides of rump; sometimes also below wing. Dark brown streaks from base of culmen to eye and from gape to below eye (latter more defined that in Red-breasted Merganser *M. serrator*), separated by narrow interrupted creamy band on lores. Rest of cheeks, nape, and sides of neck suffused tawny (but cheeks less tawny than *M. serrator*), underparts from chin to undertail white. Able to fly after *c.* 65 days, but flight-feathers not full-grown until 80–85 days. JUVENILE. As adult ♀ breeding, but crown, crest, and nape buff-brown, suffused olive-grey; crest short and less dense; streak from lower base of upper mandible to eye pale buff, or mottled white; white on chin not sharply defined, extending in narrow streak down centre of throat. Mantle and scapulars browner. Grey-brown tips to white greater coverts absent or faintly indicated. Juvenile ♂ differs from ♀ in pale grey or off-white outer lesser and median coverts, and pale grey outer tertials. FIRST IMMATURE NON-BREEDING. Usually many new feathers, resembling adult ♀ non-breeding, attained on head, mantle, scapulars, chest, and flanks; sometimes on whole body, but not wing. Shorter and narrower juvenile feathers variably retained on back, rump, underparts, or tail. FIRST IMMATURE BREEDING. Feathers resembling adult breeding gradually acquired from late autumn onwards. Typical mid-winter ♂♂ combine 1st non-breeding on most of head and body with black feathers on chin, throat, and semicircular ring bordering red-brown neck; outer scapulars partly pale grey. In spring, advanced birds like adult breeding, but wing juvenile: in ♂♂, brown mottling on head, some grey at sides of breast, grey tips to outer scapulars; in ♀♀, crest usually shorter than adult ♀ breeding. Retarded birds still partly in 1st non-breeding or juvenile plumage. SECOND NON-BREEDING. Like adult non-breeding, but some 1st immature breeding feathers retained, and wing juvenile until moult. (Anderson and Timken 1971; Erskine 1871*a*; Witherby *et al.* 1939; C S Roselaar.)

**Bare parts.** ADULT. Iris dark brown, often with narrow dull outer ring in ♂. Bill ♂ bright red, ridge of culmen and nail black; ♀ duller. Leg and foot vermilion (♂) or orange (♀); webs duller. DOWNY YOUNG. Iris grey-brown. JUVENILE. Iris yellow, brown round pupil. Bill yellow-brown, ridge of culmen dark brown; yellow becomes more orange-red near base during winter. Leg and foot dull yellow-brown, changing to orange on tarsus during winter. (Bauer and Glutz 1969; Witherby *et al.* 1939.)

**Moults.** ADULT POST-BREEDING. Complete; flight-feathers simultaneous. Flightless for *c.* 1 month, ♂♂ between mid-July and late September, ♀♀ somewhat later. Scapulars, sides of body, and flanks mid-June to July; head, neck, back, rump, and underparts July to August. Back, rump and underparts arrested

◀ PLATE 100 (*facing*) Diving ducks in flight (adult breeding plumage), B.
*Somateria mollissima* Eider: 1 ♂, 2 ♀.
*Somateria spectabilis* King Eider: 3 ♂, 4 ♀.
*Somateria fischeri* Spectacled Eider: 5 ♂, 6 ♀.
*Polysticta stelleri* Steller's Eider: 7 ♂, 8 ♀.
*Melanitta perspicillata* Surf Scoter: 9 ♂, 10 ♀.
*Clangula hyemalis* Long-tailed Duck: 11 ♂, 12 ♀.
*Melanitta nigra* Common Scoter: 13 ♂, 14 ♀.
*Melanitta fusca* Velvet Scoter: 15 ♂, 16 ♀.
*Mergus albellus* Smew: 17 ♂, 18 ♀.
*Mergus cucullatus* Hooded Merganser: 19 ♂, 20 ♀.
*Mergus merganser* Goosander: 21 ♂, 22 ♀.
*Mergus serrator* Red-breasted Merganser: 23 ♂, 24 ♀. (NWC)

during wing moult, so partly moulted once a year only, especially in ♀♀. ADULT PRE-BREEDING. Partial; starts when wing full-grown. First underparts and back, overlapping with post-breeding, followed by tail, scapulars, sides, and flanks September–December, and by tertials, head and neck late October to November; ♀♀ sometimes later. POST-JUVENILE. Partial; head, body, tail, and tertials. Variable amount of 1st immature non-breeding plumage acquired from September, but this and variable number of juvenile feathers changed for 1st immature breeding from November or March. Moult sometimes arrested December–March. Tail often moulted before December, tertials not until spring. (Anderson and Timken 1971; Dementiev and Gladkov 1952; Erskine 1971*a*; C S Roselaar.)

**Measurements.** Netherlands, winter; skins (RMNH, ZMA).

| | | | | | | | |
|---|---|---|---|---|---|---|---|
| WING AD ♂ | 285 | (5·23; 30) | 275–295 | ♀ 262 | (4·04; 23) | 255–270 |
| JUV | 275 | (5·76; 27) | 263–291 | 252 | (4·96; 20) | 242–260 |
| TAIL AD | 105 | (2·73; 24) | 100–111 | 100 | (3·27; 23) | 95–106 |
| 1st W | 96·4 | (4·72; 25) | 89–109 | 90·8 | (2·51; 19) | 86–96 |
| BILL | 55·8 | (1·92; 58) | 52–60 | 48·7 | (1·83; 43) | 44–52 |
| TARSUS | 51·7 | (1·40; 58) | 49–55 | 47·4 | (1·76; 43) | 44–51 |
| TOE | 72·9 | (2·44; 57) | 69–80 | 66·4 | (2·07; 43) | 63–71 |

Sex differences significant. Wing and tail of juvenile significantly shorter than adult. Other measurements similar; combined. Juvenile tail 10–20 mm shorter than 1st winter tail.

**Weights.** Limited data on nominate *merganser*, but apparently tendency same as similar-sized *M. m. americanus*. Nova Scotia, Canada (breeding area): (1) March–April, (2) May, (3) June–September, (4) November (Erskine 1971*a*). South Dakota, USA (winter area): (5) November to early December, (6) January to early February, (7) March (Anderson and Timken 1972). Average and sample size.

| | | ADULT | | | JUVENILE | | |
|---|---|---|---|---|---|---|---|
| (1) | ♂ | 1604 (16) | ♀ 1106 (2) | ♂ | 1401 (5) | ♀ 989 (8) |
| (2) | | 1651 (16) | 1220 (7) | | 1632 (6) | 1228 (6) |
| (3) | | 1581 (10) | 1071 (26) | | 1436 (15) | 1140 (29) |
| (4) | | 1709 (13) | 1220 (5) | | 1585 (14) | 1223 (15) |
| (5) | | 1600 (21) | 1160 (13) | | 1411 (26) | 1095 (22) |
| (6) | | 1890 (62) | 1390 (21) | | 1741 (8) | 1469 (4) |
| (7) | | 1660 (22) | 1270 (8) | | 1537 (8) | 1172 (9) |

Ranges: adult ♂♂ 1264–2160, adult ♀♀ 898–1770; juvenile ♂♂ 1162–2119, juvenile ♀♀ 860–1544.

**Structure.** Wing pointed. 11 primaries: p9 longest, p10 0–6 shorter, p8 6–15, p7 25–33, p6 41–52, p1 110–130. Tail short, rounded; 18 feathers. Bill long, narrow, serrated, higher than broad at base; culmen slightly concave; nail large, strongly decurved. Feathering of chin extends to about half way along lower mandible. In contrast to *M. serrator*, nostrils slightly behind middle of bill (already diagnostic in downy young). Adult ♂ breeding has feathers of crown, upper sides of head, and nape dense, elongated; crest of juvenile short, of ♂ non-breeding and ♀ adult long, drooping, on hindcrown only. Outer toe *c.* 98% middle, inner *c.* 83%, hind *c.* 24%.

**Geographical variation.** *M. m. americanus* similar to nominate *merganser*, but in adult ♂ black on bases of white greater upper wing-coverts less concealed by median coverts, forming broad bar on speculum. *M. m. comatus* differs from nominate by shorter, narrower, and more slender bill, and longer wing: wing 25 ♂♂, 295 (286–305); bill 51·5 (48–56) (Vaurie 1965); adult ♀ also differs by paler head and mantle. CSR

# Tribe OXYURINI stiff-tailed ducks

Fairly small to medium-sized, mainly freshwater diving ducks. 8 species in 3 genera, first and last monotypic: *Heteronetta* (Black-headed Duck, South America), *Oxyura* (typical stiff-tailed ducks), *Biziura* (Musk Duck, Australia). 3 species-groups in *Oxyura*: (1) Masked Duck *O. dominica*, Neotropical America; (2) Ruddy Duck *O. jamaicensis* and White-headed Duck *O. leucocephala*; (3) remaining 3 species. Highly distinctive tribe with no close affinity to other Anatinae. *Heteronetta* obligate brood-parasite and aberrant in other features; not considered further here. Monotypic genus *Thalassornis* (White-backed Duck) often included but here considered aberrant member of Dendrocygnini (Anserinae), following Johnsgard (1967). 1 *Oxyura* native breeding species in west Palearctic, another introduced.

Bodies broad and short; necks short and thick, with loose, expansible skin. Sexes nearly similar in size except in *Biziura*. Wings broad and short; flight rapid and jerky. Tails long, narrow, and stiff; tail-coverts very short. Bills about as long as head; broad, widened and flattened at tip, swollen at base in some; nail narrow and sharp. In *Biziura*, ♂ with wattle-like fold of skin hanging below bill. Legs short and thick, placed far back on body; feet large, hind toe lobed. Except in *O. dominica*, take off from water with long run. Walking ability restricted. Dive with facility without using wings, steering under water with tail. Tracheal structure of ♂♂ simple; no bulla. In most species, trachea connected to inflatable air-sac, or oesophagus itself enlarged and presumably inflatable; in *Biziura*, inflatable pouch below tongue. Plumage grebe-like, dense and shiny. Sexual dimorphism marked except in *Biziura*. ♂♂ of *Oxyura* mainly chestnut, white below with black and white or wholly dark heads and cobalt-blue bills; ♀♀ cryptic—brown, finely vermiculated, freckled, and barred on body, and with dark cheek and eye stripes in most species. Wing uniform brown except in *O. dominica*. Both sexes barred brown in *Biziura*. Non-breeding plumage present, but not much different in pattern from breeding in some forms—though definite ♂ eclipse in *O. leucocephala* and North American race of *O. jamaicensis*. Wing moulted twice a year in at least some species. Juveniles resemble adult ♀♀. Downy young of *Oxyura* typically with dark crown and cheek markings like adult ♀♀.

Mainly southern hemisphere. Largely tropical, subtropical, and warm temperate, but some extend to boreal or high-latitude wetlands free of ice. Habitat and habits somewhat convergent with smaller Podicipedidae (grebes) including use of fairly shallow, restricted, and heavily vegetated waters unacceptable to other diving ducks. Normally avoid turbulent, flowing, or exposed waters. In winter, readily resort to shallow, inshore, marine waters, as well as brackish or saline lagoons and estuaries. In west Palearctic, range of *O. leucocephala* fragmented and breeding often sporadic, especially on periphery. Disap-peared from many western and southern parts of range. For movements of northern *Oxyura*, see species' accounts.

*Oxyura* omnivorous: primarily invertebrates and seeds and leaves of aquatic plants obtained by diving from surface and foraging on bottom; occasionally dabble on surface. *Biziura* essentially animal feeder (Johnsgard 1965). *Oxyura* typically gregarious for much of year, but *Biziura* less often in flocks. Pre-flight signals may be lacking (but see species' accounts). Territorial behaviour little studied; in some, e.g. Maccoa Duck *O. maccoa*, ♂ establishes pairing territory to which ♀♀ attracted by display. Exact nature of pair-bond uncertain in most *Oxyura*; may involve short-term monogamy, successive polygyny, or promiscuity. Evidently no true pair-bond in *Biziura*. Communal courtship well developed in some species; in *Biziura*, almost in form of 'lek' as (e.g.) in some grouse (Tetraonidae)—see Johnsgard (1966, 1967). ♂ displays, particularly bizarre in *Biziura*, differ appreciably from those of other Anatinae. In *Oxyura*, involve vertical stretching of neck and cocking of tail, inflation of neck, and splashing—either with bill and neck (Sousing-display) or with feet or wings; in some, also drumming of bill against neck (see *O. jamaicensis*). Apart from Wing-flapping, usual secondary displays of most other Anatinae appear to be lacking; also Mock-preening. ♀♀ during courtship lack Inciting-displays and other activities (e.g. Nod-swimming) while induce ♂ display in other tribes. Copulatory behaviour, so far as known, differs in most respects from that of other Anatinae (see *O. jamaicensis*) and appears to involve a varying degree of aggression from both sexes. ♂ vocalizations simple but extremely variable; modified by inflatable throat-sacs, etc. and extended by instrumental sounds (such as drumming and rattling with bill) and non-vocal noises (such as splashing). ♀ mostly silent but produce rattling, wheezing, or purring calls; lack Inciting-calls, Decrescendo-calls, and any harsh quacking or croaking sounds.

Northern *Oxyura* seasonal breeders. Nests well dispersed; in thick cover over shallow water. Platforms of vegetation with shallow cup; sometimes lined down. Built by ♀. Old nests of other species sometimes used. Eggs dull white, particularly large in relation to size of ♀. Clutches usually 5–10. Casually parasitic at times, depositing some eggs in nests of own species and other waterbirds. One brood; replacements probable after egg loss. Eggs laid at 1½-day intervals. Incubation by ♀ only. Incubation periods 21–27 days (Johnsgard 1968), 23–26 days (Kear 1970). Young cared for by ♀—but ♂ may play some role, probably only temporary, in some—see *O. jamaicensis*. Distraction-display not recorded (Hebard 1960). Although young self-feeding in *Oxyura*, parental feeding highly developed in *Biziura*. Except in latter, young independent well before fledging. Mature in 1st year (*Oxyura*), 2nd year (*Biziura*).

# *Oxyura jamaicensis* Ruddy Duck

Du. Rosse Stekelstaart    Fr. Erismature roux    Ge. Schwarzkopf Ruderente
Ru. Американская савка    Sp. Malvasía canelo    Sw. Amerikansk kopparand

*Anas jamaicensis* Gmelin, 1789

Polytypic. Nominate *jamaicensis* (Gmelin, 1789), North and Central America, introduced England. Extralimital: *andina* Lehmann, 1946, Andes of Colombia; *ferruginea* (Eyton, 1838), Andes south from Ecuador.

**Field characters.** 35–43 cm, of which body and tail (6–8 cm) nearly three-quarters; wing-span 53–62 cm. Small, dumpy, and short-necked stiff-tailed duck with broad, concave-ridged bill, prominent head, and relatively long, stiffly projecting tail often cocked; ♂ has blue bill, black crown, white cheeks, and chestnut body. Sexes dissimilar and with seasonal differences; juvenile resembles ♀ but distinguishable.

ADULT MALE. Bright blue bill, black cap and nape contrasting with white sides of head and chin. Rich chestnut upperparts, upper breast, and flanks; silvery-white lower breast and belly finely barred with dusky brown, white under tail-coverts, brown tail, and plain brown wings (mixed with white on under wing-coverts). Legs bluish-grey with dusky webs. MALE ECLIPSE. Retained through winter; quite distinct with bill becoming greyish, crown browner, white cheeks speckled with black, and all chestnut areas dark brown and grey like winter ♀, from which distinguished by white cheeks and whiter undertail. ADULT FEMALE. Blue-grey bill, dark brown cap and nape, and buff-white cheeks bisected by often poorly defined dark brown line from base of bill to nape. Pale brown neck, red-brown upperparts and upper breast, silvery-white underparts barred and mottled with brown, particularly on flanks, and white under tail-coverts variably barred with brown. Wings, tail, and feet as ♂. FEMALE NON-BREEDING. Similar, but with greyer bill and red-brown areas replaced by brown. JUVENILE. Like winter ♀, but duller and paler with poorly defined cheek line and less barring on flanks.

Often swims with tail cocked and this, coupled with dumpy shape, blue to blue-grey bill, contrasted head pattern, and (in summer) reddish body rules out all other ducks except related White-headed Duck *O. leucocephala*; latter, however, larger with swollen base to bill, more white on head of ♂, which also has greyer upperparts and lacks white undertail, and usually more clearly defined cheek stripe on ♀. ♀ with faint cheek stripe might be confused at distance with larger ♀ Common Scoter *Melanitta nigra* or Red-crested Pochard *Netta rufina* on account of contrasted pale cheeks, but shape and other differences establish identity.

Almost exclusively found on fresh water—lakes, ponds, and gravel-pits—with reed fringes. Swims buoyantly, ♂ often with throat and breast puffed out, and dives quickly or, remarkably, sinks below surface without diving. Seldom leaves water, but though sometimes stated to be almost helpless on land, because legs set far back, stands and walks quite adequately and ♀ and brood recorded walking over 200 m overland (H Hays); but at water's edge sometimes moves by pushing with both feet together while resting on breast. Also disinclined to take wing except when travelling from one water to another. Flight whirring, with rapid beats. Generally silent except in courtship.

**Habitat.** In North America, lowland wetlands with lush emergent vegetation and suitable patches of open water in zone of temperate continental climate. Favours pools with fairly shallow bottoms and rich in floating and submerged aquatic plants, appearing more grebe-like than duck-like in habitat and habits. Avoids flowing fresh water, unless very sluggish, but makes limited use of saline estuaries and sheltered coastal waters. Avoids rocky, steep, or artificial margins and beds, but can adapt to changing patterns and levels of water due to climatic or other fluctuations, other than drainage. Becomes airborne only with difficulty, normally flying close to surface and not rising to much height even on migration.

**Distribution.** Nearctic and Neotropical. BRITAIN. Introduced western England and now firmly established. Escaped captive birds began breeding regularly Avon 1960 and Stafford 1961; their descendants have since spread to Gloucester, Hereford and Worcester, Warwick, Shropshire, Cheshire, Derby, and Leicester, with the main centres in the west Midlands; also bred Hertford 1965–8 (Hudson 1976). IRELAND: bred Lough Neagh, 1974 (J T R Sharrock).

**Population.** BRITAIN: increased by 25% per annum during 1965–75; in latter year, 50–60 breeding pairs and post-breeding population of 300–350 birds (Hudson 1976).

**Movements.** Small English introduced population resident; but most make regular, short (up to 70 km), seasonal movements, deserting small breeding pools and meres to winter on some large reservoirs. Flocks develop September–December and disperse March–April. Occasional records elsewhere may refer to escapes from collections but mainly wandering feral birds; as does not normally breed until 2 years old, some wanderers perhaps immature birds prospecting for future nesting sites.

Migratory in North America, withdrawing from

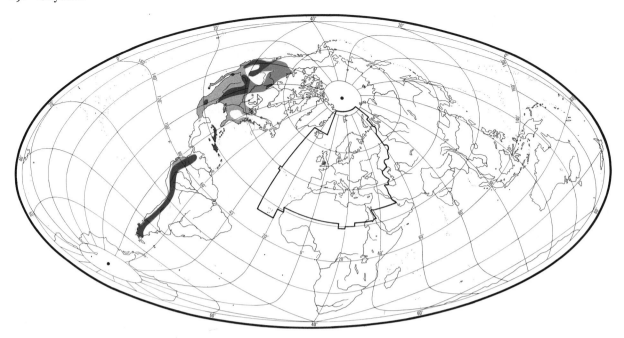

breeding range (except in California) to winter in Pacific, Atlantic and Gulf coastal states of USA and south into Mexico, British Honduras, and Guatemala (Eisenmann 1955; Robbins *et al.* 1966); no evidence of transatlantic vagrancy.

**Food.** Omnivorous; primarily insect larvae and aquatic plant seeds obtained by surface-diving and then straining ooze overlying bottom. Swims along bottom with head extended in front, bill opening and closing in rapid straining action as head and neck move in lateral sweeping movement. Occasionally skims surface, swimming with head and neck stretched out and head moving in arc with beak half immersed (Goodman and Fisher 1962; Siegfried 1973*b*). Daytime feeder, in shallow water. In 1 m depth, mean of 46 dives (♀♀) 20·6 s, 61 dives (♂♂) 18·6 s (Siegfried 1973*b*).

From various localities in North America over 9 months of year, 163 stomachs contained 72·4% by volume plant materials, chiefly seeds and tubers, and occasionally leaves and stems of pondweeds *Potamogeton* (74% by frequency), seeds of sedges *Scirpus*, and, in lesser quantities, parts of at least 24 other species. Animal material ranged from 20% by volume in September and November to 39% in June, with Diptera predominating, especially midges Chironomidae (Cottam 1939). However, in 41 adults from pothole region of Manitoba collected June–August, animal materials predominated, constituting 90% (23 ♂♂) and 95·3% (19 ♀♀) by volume. Chironomidae, especially *Chironomus* larvae and pupae, easily chief items: in ♂♂, 75% by frequency and 82·1% by volume, in ♀♀ 84% and 62·9%. Other invertebrates—insects (Hemiptera, Coleoptera, Odonata, Trichoptera), crustaceans, molluscs,

annelids, and nematodes—taken less frequently and in smaller quantities. Plant materials, mainly seeds, only 4–9% by volume (Siegfried 1973*b*).

Ducklings of all ages forage by surface-diving, though dives shallower and shorter than adults, and occasionally by surface-straining (Collias and Collias 1963; Siegfried 1973*b*). 14 juveniles, July and August, various North American localities, contained mainly insect larvae, especially chironomids (Cottam 1939). 18 ducklings, 1–5 weeks old, from Manitoba, had similar diet but with chironomids even more frequent (100%) and in greater percentage volume (73·4%) (Siegfried 1973*b*). 4 ducklings from Delta area, Lake Manitoba, contained only chironomids (Collias and Collias 1963).

**Social pattern and behaviour.** Compiled mainly from material supplied by H Hays and D E Ladhams from observations made respectively in Manitoba, Canada (migrant population, Roseneath pothole country), and in Avon, England (resident and wintering populations, Chew Valley and Blagdon Lakes); supplementary information from M W Weller and G V T Matthews.

1. Gregarious for much of year. In flocks outside breeding season; usually of mixed sex and age, though some autumn groups mostly juveniles. At Delta Marsh, Manitoba, ♂♂ outnumber ♀♀ 9:1 in spring; though unbalanced ratio probably influenced by earlier arrival of ♂♂, also noticeable to some extent throughout breeding season (Hochbaum 1944); elsewhere, less marked preponderance of ♂♂ (see Low 1941). Sex-ratio at Avon lakes more or less even. Flocks often fairly small (50 or less) even in North America, though up to several thousand reported there at some undisturbed sites. ♂♂ first to flock while ♀♀ incubating, from mid-June at Delta. BONDS. Relationships between sexes, and between ♂ and young, require further study. Extent and duration of pair-bond disputed: some association of ♂♂ and ♀♀ in separable couples before and even during breeding clearly

indicated, but uncertain whether seasonal monogamy involved or ♂ polygamous, engaging in temporary liaisons with series of ♀♀. As definite couples hard to recognize in spring flocks in Roseneath area, heterosexual associations evidently start mainly after arrival in May; later, 'pairs' apparent in courting parties and when birds dispersed near likely nesting areas. At Chew, in loose pairs June–August; ♂ and ♀ seek nest-site together and ♂ associates with ♀ when latter off eggs; later, same or different ♂ (sometimes 2) may attend ♀ with brood for 2–3 weeks. In North America, association after egg-laying uncertain; in some cases, ♂ appears to maintain contact with ♀ and brood but no evidence same individual involved throughout. Only ♀ incubates and effectively tends young. Though ♂ said commonly to lead and help protect brood until full-grown (Bent 1925), this apparently not general (see Chura 1962) and interest of ♂ probably primarily in ♀, not young. At Delta, records of ♂♂ even temporarily accompanying young exceptional; however, at Chew, odd ducklings often accompanied and protected by a ♂ (King 1976). Bond between ♀ and brood often ended well before highly precocial young can fly. Broods of various ages often mingle, but no large group reported even where population large. At Chew, siblings stay as feeding group until winter, not necessarily with ♀. BREEDING DISPERSION. Nests usually well scattered. In Iowa, USA, usually density 1 per 8·5 ha (Low 1941), but here and elsewhere sometimes higher. Density at Chew low: up to 10 pairs over 480 ha. No evidence of well-defined territories, though ♂♂ at Chew will patrol near nest and show aggression to intruders. At Chew, ♀ with brood later keeps to restricted area of *c.* 1 ha for feeding until winter, but this not defended as territory. Surplus ♂♂ at Delta remain in flock when others still attached to nesting ♀♀; later, joined by majority of breeding ♂♂. ROOSTING. Largely diurnal and communal. Rests and sleeps on open water for long periods during day; usually in flocks, especially in winter. No information on nocturnal roosting; probably amongst thick cover. At Delta in spring, when water still cold, ♀♀ and (to somewhat lesser extent) ♂♂ built special individual roosting platforms spending *c.* 65% of 24 hours resting or sleeping there (Siegfried 1973a); these isolated or in small groups, maximum 8 together (averaging 8 m apart).

2. Studied in wild by H Hays and D E Ladhams, but few published observations; see especially Wetmore (1920), Johnsgard (1965, 1967). Unlike most ducks, little social interaction in winter flock, pair-formation occurring in spring in breeding area.

FLOCK BEHAVIOUR. When alarmed, birds assume Head-high posture with neck stretched vertically. Seldom fly, preferring to dive; said to lack pre-flight signals (Johnsgard 1965), but rapid lateral Head-shaking preceded or replaced by Head-rolling while in Head-high posture recorded by H Hays. When Head-rolling, side of face merely touched quickly on back, then apparently moved lightly and briefly across preen-gland before neck erected with bill slightly lifted; may be repeated several times before bird flies after giving Head-shake. ♀ also reported Head-jerking (see below) before flying short distance. In winter flock, Avon lakes, communal Patter-rushes of uncertain significance initiated by 1 bird scuttling 50–100 m over surface, 20–30 others doing same in succession; unlike in more ritualized rushes (see below), only slight sound of splashing. ANTAGONISTIC BEHAVIOUR. In Roseneath area, intense hostility shown at times by both sexes during breeding season. ♂♂ successful in obtaining 'mates' each aggressively defends area of *c.* 3 m around own ♀. ♀♀ mostly aggressive against any approaching ♂, from arrival until just prior to pairing, then again when incubation starts. During antagonistic encounters, ♂ persistently assumes Head-high-Tail-cock posture with neck and tail erected more or less vertically; may also fan tail and erect ear-tufts while depressing feathers on crown. When threatening, both sexes assume Hunched-posture with scapular feathers slightly raised and head withdrawn into them; then neck thrust forward and head raised while bird gapes at opponent (Open-Bill-threat), ♀ at least also erecting ear-tufts and hissing. In Hunched-rush, lower head and extend neck, erect scapulars (often to 90°), and swim rapidly after rival; latter usually withdraws, then aggressor immediately stops. In dispute between 2 ♂♂, when neither gives way, birds may face bill to bill for several seconds in rigid posture with neck and tail extended flat on surface; then either one withdraws, one attacks other or both jump at each other more or less simultaneously with necks extended. Combats often brief as each ♂ nips throat of other just below bill (pecking fight) or uses feet to claw at its scapulars and back of nape or to strike its head (foot-fight). 2 reactions shown at times by ♂ when passing or being passed by other ♂♂ probably appeasing: turns head aside to present rear aspect (Facing-away) or also places bill in scapulars (Pseudo-sleeping). Both sexes escape from rivals by swimming rapidly away in crouched position or diving, usually into cover; fly if attacked persistently. Withdrawing ♂ sometimes performs Bubbling-display; this used frequently in courtship but evidently also as threat to other ♂♂

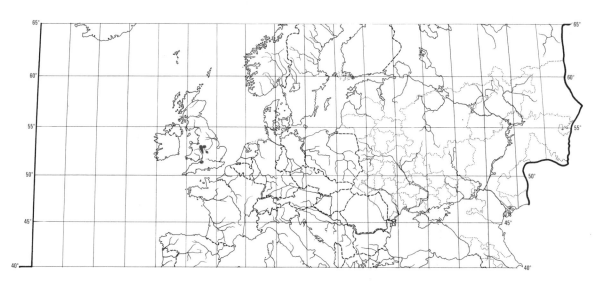

and in response to disturbance. Certain other displays of ♂ (Tail-flash, Ring-rush, and Surfboarding) may also be basically aggressive as only performed when other ♂♂ present, sometimes in apparent absence of ♀. COMMUNAL COURTSHIP. In Roseneath area, groups of ♂♂ form round 1 or more ♀♀ and perform 3 major displays (Bubbling, Ring-rush, and Tail-flash) from arrival until start of nesting. At Chew, display groups of up to 8 ♂♂ noted March–August, before, during, and after nesting. Courting ♂♂ assume Head-high-Tail-cock posture, other displays following from this. Gatherings often initiated when Bubbling ♂, or even one just in extreme Head-high-Tail-cock near ♀, joined progressively by other birds. Highly ritualized Bubbling-display main ♂ activity. Starting from stationary position, sometimes facing ♀, ♂ (1) erects ear-tufts into deep V and fully inflates tracheal air-sac (see A); (2) starts Chest-beating, drawing in head and lowering bill, and then jerking head rapidly up and down so that bill strikes side of breast above air-sac repeatedly (see B); (3) ends by partially deflating air-sac while suddenly extending head and calling (see C)—folded wings briefly lifted, possibly due to pressure of tail. During stages 1–2, tail raised steadily past vertical until tip 4–5 cm from head. When ♂ Chest-beating, profusion of air-bubbles form on water round breast (see B); at same time, inflated sac acts as tympanum, blows producing sound. Now clear that ♂ first taps chest rapidly 5 times, then slaps it once or twice to press against feathers, expelling air and causing bubbles to appear, thus producing 2 different sounds; in 50% of cases, no slaps or bubbles (D E Ladhams). Next major display, Tail-flash, often alternates with Bubbling; in extreme Head-high-Tail-cock, with tail held over back at *c.* 135° and head canted close to tip, ♂ places himself in front of ♀ then swims ahead while exhibiting white under tail-coverts and manoeuvring body from side to side. Tail-flash functional equivalent of Turn-back-of-Head of *Anas* ducks, inducing ♀ to follow particular ♂. In final major display, hostile-looking Ring-rush, ♂ sits low on water then (1) spreads cocked tail, at same time drawing head back to within 5–8 cm of it; (2) suddenly snaps tail down, and extends neck; (3) rushes forward *c.* 3 m while beating wings against water *c.* 8 times. During stages 2–3, anterior part of body raised out of water; during stage 3, tail submerged, scapulars raised, and peculiar, far-carrying ringing sound produced by wing-beats. ♂ often Ring-rushes thus to within 1 m of ♀, then stops, turns round, and either Bubbles, assumes intense Head-high-Tail-cock, or pursues ♀ in Hunched-rush. Little aggression between ♂♂ in courting parties until one obtains interest of particular ♀, then hostile to other ♂♂. For much of time, ♀♀ show no overt heterosexual response to displaying ♂♂, but tend to react aggressively with Open-Bill-threat and Hunched-rush, particularly before pairing. Dive when pressed too closely. PAIR BEHAVIOUR. When willing to associate with ♂, ♀ (1) ceases to be aggressive and follows passively as he performs Tail-flash. May also (2) Head-jerk by bobbing bill down against breast, then quickly stretch into Head-high posture; (3) Chest-beat in less-ritualized, slower, and more jerky manner than ♂, usually producing no bubbles. When paired, ♂ seems no longer to direct Ring-rush at own ♀; after pursuing rival in Hunched-rush, however, often turns quickly toward ♀ (when other ♂ dives) and then rapidly Surfboards back to her, producing conspicuous wake, in quite different posture with tail and lower back almost submerged, chest lifted high and puffed out, and head held back, bill largely hidden on breast. On reaching ♀, ♂ turns aside and performs Tail-flash or Bubbling-display. At times, ♂ intersperses Bubbling-display with Bill-dip and Water-flick while facing ♀, first submerging bill and then raising head high before flicking water laterally while moving bill in figure-of-eight movement. Can occur when pair feeding, but more typically pre-copulatory display. COPULATION. See especially Johnsgard (1965). ♂ cautiously approaches ♀ while Head-nodding slightly in Head-high posture, frequently performing Bill-dip and Water-flick. Usually copulates only when able to get close enough to mount ♀ without warning. ♀ shows no reciprocal displays; often dives away and may threaten, though will sometimes eventually adopt partly submerged Prone-posture. ♀ almost totally submerged during brief copulation, after which ♂ faces ♀ and does Bubbling-display several times in rapid succession; pair then self-preen. RELATIONS WITHIN FAMILY GROUP. Little information. At one nest, Roseneath area, young led away by ♀ *c.* 20 hours after hatching, then brooded at night in nest of American Coot *Fulica americana*. For first 2 weeks, young dive for food only when ♀ dives, latter timing duration of own dives to match that of young, but after *c.* 3 weeks scatter to feed on own; may then join other broods, adults showing no hostility to strange young (D E Ladhams). If ♂ accompanies brood, stays near while ♀ and young feed, threatening ♂♂ and other waterfowl that approach; reported performing Bubbling-display on seeing young taking to water for first time. Family can walk overland when necessary, up to 225 m (H Hays). Young noted doing ♀-like Chest-beating from *c.* 4 days old, mostly when disturbed but also on encountering other broods.

(Figs A–C after Johnsgard 1965.)

**Voice.** Calls few and unspecialized. Several sounds earlier assumed by some authors to be produced vocally, now found to be wholly or partly non-vocal.

MALE. Most sounds non-vocal. During Chest-beating phase of Bubbling-display, rattling noise made by open bill, while series of 5 'tic' sounds followed by 1–2 'croak'

A

B

C

noises produced as bill first taps then slaps inflated tracheal air-sac (D E Ladhams); also described as hollow tapping sound (Johnsgard 1965), and chittering sound (G V T Matthews). In terminal phase of Bubbling-display, produces soft, low, belching noise (Johnsgard 1965) or bumping sound (G V T Matthews). Possibly due to partial deflation of air-sac, but described as low, grunting vocal note by Wetmore (1920) in which 2 sounds uttered simultaneously, producing terminal 'quek' sound: (1) harsh, frog-like 'quok', (2) slightly drawn-out, almost disyllabic, reedy note (like that of ♂ Gadwall *Anas strepera*). ♂ also makes far-carrying, ringing noise during Ring-rush display: series of rapid, staccato notes, increasing then decreasing in speed and caused by impact of wings on water (H Hays).

FEMALE. Makes rattling noise with bill when Chest-beating; quite different from that produced by ♂ (Wetmore 1920). Other sounds truly vocal. Falsetto 'qua-er' uttered at end of Head-jerking display (Wetmore 1920). Hisses when threatening ♂♂ (H Hays). Other calls given when disturbed (alone or with young) include: (1) high-pitched, sharp 'e' note, (2) low-pitched, grating, nasal 'raanh' (H Hays); these may be same as high-pitched 1-syllable 'keow', flat-tone 'wharp' or 'quep' of Wetmore (1920). Only vocalization heard by Johnsgard (1965), high-pitched squeak; uttered occasionally by ♀ when threatening or fleeing from conspecifics.

CALLS OF YOUNG. Little information in wild. Incubator-hatched ducklings up to 7 weeks old gave 2 types of call (H Hays): (1) succession of sharp, high-pitched peeps ('e-e-e-e') when chased or suddenly picked up, (2) soft, polysyllabic trill when crowding away from intruder; both calls also heard from pipped eggs, and 1st call from day-old duckling in wild when lost. From 8th week, only known call a ♀-like 'raanh' when frightened.

**Breeding.** Little data from west Palearctic. SEASON. Britain: prolonged, with egg-laying from mid-April; first brood recorded (Avon) 18 May, most broods in June and July, but small duckling 31 October (D E Ladhams; M A Ogilvie). SITE. In thick vegetation, in water of average depth 35 cm, but up to 100 cm; averaging 80 m from shore (0–250 m) and 30 m from open water (0–200 m) (Low 1941). Nest: platform of reed or rush stems and leaves, with shallow cup, interwoven among upright stems. Most often of previous year's dead vegetation; sometimes green material added, or solely used. Average diameter 30 cm, depth of cup 7 cm, height above water 20–25 cm; built up if level rises. 43 of 71 nests had overhead 'cupola' formed by bending down vegetation; 25 had access ramps from water (Low 1941). Little or no down. Building: by ♀, though both sexes build resting platforms before nesting season (Siegfried 1973a). EGGS. Ovate; dull or creamy white, often becoming stained during incubation. 62 × 46 mm (60–68 × 43–48), sample 100 (Schönwetter 1967). Weight of captive laid eggs 76 g (64–88), sample 100

(J Kear). Clutch: 6–10; up to 20 recorded, but by more than 1 ♀. Mean of 71 clutches, Iowa, 8·1 (Low 1941). One brood. No data on replacement clutches. Eggs laid at 1½ day intervals (J Kear). INCUBATION. 25–26 days. By ♀. Begins on completion of clutch; hatching synchronous. Eggs covered with nest material when ♀ off nest. Eggshells left in nest. YOUNG. Precocial and nidifugous. Self-feeding. Many records of ♂ accompanying brood in early stages or throughout fledging period (Bent 1925; Kortright 1942), but not proven always parent of brood (Delacour 1964). FLEDGING TO MATURITY. Fledging period 50–55 days. Independent before fledging, often well before. Age of first breeding mostly 2 years, but some probably at 1. BREEDING SUCCESS. Of 546 eggs laid, Iowa, 69·4% hatched (Low 1941). No data from west Palearctic.

**Plumages.** ADULT MALE BREEDING. Crown to below eye and nape glossy black, sharply contrasting with white cheek and chin. Neck, mantle, scapulars, chest, sides of body, and upper tail-coverts rich chestnut, tinged purple on chest and tail-coverts. Back and rump dark brown, vermiculated buff or grey. Lower breast, belly, and vent grey-brown, feathers broadly tipped silver-white; often tinged yellow on breast. Under tail-coverts white. Tail dark brown. Flight-feathers and upper wing-coverts dark brown; coverts, secondaries, and tertials finely speckled buff. Often red tinge on inner wing-coverts. Underwing and axillaries grey-brown, centre of underwing and edges of axillaries white. ADULT FEMALE BREEDING. Crown to below eye and hindneck dark brown, feathers tipped chestnut or buff. Sides of head off-white, with narrow indistinct streak of dark brown feathers with white tips on cheek from gape to nape. Chin white, sides of neck and throat brown-grey. Mantle and scapulars dark brown with rufous tinge, finely speckled and vermiculated buff on feather tips. Sides of chest and upper tail-coverts buff or off-white, feathers glossy black basally. Chest and sides of body dark brown, indistinctly barred buff-brown. Breast, belly, and under tail-coverts as adult ♂ (brown feather bases sometimes partly visible); tail and wing as ♂, but underwing and axillaries white. ADULT MALE NON-BREEDING (Eclipse). Body as in adult ♀, but mantle and scapulars sometimes intermixed with chestnut; head like adult ♂ breeding, cheeks white, but crown and nape dull black, feathers tipped buff; neck and throat pale grey. ADULT FEMALE NON-BREEDING. As breeding, but cheeks greyer, dark streak ill-defined; mantle, scapulars, and upper tail-coverts brown, not reddish, more conspicuously barred. DOWNY YOUNG. Mostly sooty-brown, tinged grey on mantle and chest; sides of head, chin, and throat pale grey, narrow dark streak on cheek. Lower breast, belly, vent, and patch at each side of back pale grey. JUVENILE. Like adult ♀, but dark brown greyer and duller, crown dull black; feathers of sides of chest grey, tipped grey-brown; less speckled on wing-coverts, more barred on mantle and scapulars. Juvenile tail-feathers narrow, with notch at tip, and bare shaft protruding. Feathers of underparts and flanks narrow, rounded, grey-brown with narrow grey-white tips, giving scaly appearance to breast and belly and distinct bars to flanks and under tail-coverts. FIRST IMMATURE NON-BREEDING. Like adult ♀, but tail and often belly still juvenile. FIRST IMMATURE BREEDING. Gradually attained in spring by both sexes; late May like adult. (Kortright 1942; C S Roselaar.)

**Bare parts.** Iris yellow-brown to brown-red, eyelids blue-grey. Bill dusky blue-grey, in breeding ♂ bright blue. Leg and foot blue-grey, webs dark grey. (Kortright 1942.)

**Moults.** ADULT POST-BREEDING. Complete. Flight-feathers first, simultaneously, late July and August in ♂, from late August in ♀; followed by head, body, and tail. ADULT PRE-BREEDING. Complete; flight-feathers simultaneous. About March–April, before spring migration (Siegfried 1973c). POST-JUVENILE. Partial, head and body; not tail and wing. August–October, belly and vent later. FIRST IMMATURE PRE-BREEDING. Probably as in adult; March–May.

**Measurements.** Full-grown. North America; skins (RMNH, ZMA). Range includes additional data on wing (Delacour 1959) and tail, bill, and tarsus of ♂ (Godfrey 1966).

| | | | | | | |
|---|---|---|---|---|---|---|
| WING | ♂ 149 | (1·97; 7) | 142–154 | ♀143 | (3·61; 7) | 135–149 |
| TAIL | 68·8 | (3·56; 5) | 64–79 | 68·2 | (2·56; 6) | 65–71 |
| BILL | 40·4 | (0·64; 8) | 38–41 | 38·4 | (1·00; 7) | 37–40 |
| TARSUS | 33·5 | (1·29; 8) | 32–38 | 31·5 | (1·38; 7) | 30–33 |
| TOE | 63·5 | (0·93; 8) | 62–72 | 59·4 | (1·13; 7) | 58–61 |

Sex differences significant, except tail.

**Weights.** North America (Kortright 1942). ♂ 610 (8) 540–795; ♀ 510 (13) 310–650.

**Structure.** 11 primaries: p9 longest, p10 0–3 shorter, p8 2–5, p7 8–13, p6 16–22, p1 48–60. Bill rather slender, slightly upcurved, broad and flat at tip, not high and swollen at base as White-headed Duck *O. leucocephala*. Outer toe *c.* 98% middle, inner *c.* 77%, hind *c.* 27%. Rest of structure as *O. leucocephala*.

**Geographical variation.** None in North and Central America—*O. j. rubida* (Wilson, 1841), for continental North American populations, not recognized. South American *O. j. ferruginea* larger and darker, adult ♂ with head black, except white chin; *O. j. andina* intermediate between *ferruginea* and nominate *jamaicensis*. (Delacour 1959). Both forms sometimes considered separate species. CSR

## *Oxyura leucocephala* White-headed Duck

PLATES 98 and 99
[facing page 663 and after page 686]

DU. Witkopeend     FR. Erismature à tête blanche     GE. Ruderente
RU. Савка     SP. Malvasía     SW. Kopparand

*Anas leucocephala* Scopoli, 1769

Monotypic

**Field characters.** 43–48 cm, of which body and tail (8–10 cm) nearly three-quarters; wing-span 62–70 cm. Medium-sized, short-necked and dumpy stiff-tailed duck, with broad, swollen bill, large and prominent head, and relatively long and graduated, stiffly projecting tail often characteristically cocked; ♂ has blue bill, largely white head, black neck, and grey and reddish body. Sexes dissimilar and with clear seasonal differences in ♂; juveniles resemble ♀, but young ♂ slightly different.
    ADULT MALE. Bright blue bill, conspicuously swollen at base. Black crown patch, and otherwise all-white head and nape. Black neck shading into chestnut breast and grey and red-brown upperparts. Silvery-white lower breast and belly mottled with grey, black-brown wings and grey-brown tail. Legs bluish-grey with dusky webs. MALE ECLIPSE. More black on crown, sometimes extending right down nape and with occasional spots of black on the white cheeks. Rest of plumage paler and greyer with more prominent fine barring. Bill slate-grey. ADULT FEMALE. Blue-grey bill; dark brown head and nape, except for buff-white chin, throat, lower cheeks, and narrow line below eye; grey-brown neck and red-brown upper breast. Rest of plumage much as summer ♂, but more grey-brown with stronger barring on upperparts, breast, and flanks. FEMALE NON-BREEDING. Similar, but with slate-grey bill and dusky grey-brown plumage, rather than grey and red-brown. JUVENILE. Somewhat variable, but like duller and paler winter ♀; young ♂ said to be redder on back (Delacour 1959).
    Adult ♂ unmistakable, but in eclipse might be confused with eclipse ♂ Ruddy Duck *O. jamaicensis*, when distinguished by swollen base to bill, larger-looking head and, even in individuals whose black cap extends down nape, by more extensive white extending above eyes and across forehead. ♀♀ of these species much more similar but, apart from smaller size, ♀ *O. jamaicensis* lacks swollen base to bill and usually has less well-defined cheek stripe, though this variable in both.
    Largely confined in breeding season to brackish lakes and lagoons fringed with reeds and with good growth of pondweeds; in winter typically in large areas of shallow water and sometimes on sea, usually in rocky bays. General characters similar to *O. jamaicensis*, but less dainty and buoyant. Swims with pointed tail depressed or cocked, dives easily or sinks below surface, and sometimes swims with only head above water. Hardly ever seen on land, but can stand and walk without apparent strain. Flies seldom, with rapid beats producing whirring action. Generally silent.

**Habitat.** Confined mainly to steppe zone of Palearctic continental middle latitudes. Accomplished diver, remaining submerged over long periods and travelling 30 m or more under water, but remarkable for preferring shallow pools or parts of lakes, often less than 1 m deep. Consequently extremely local: needs ample tracts of open water, flanked by dense stands of emergent aquatic plants providing cover and nest-sites with immediate access to clear paths for underwater escape. Rarely recorded voluntarily moving on land. Breeds often on small pool

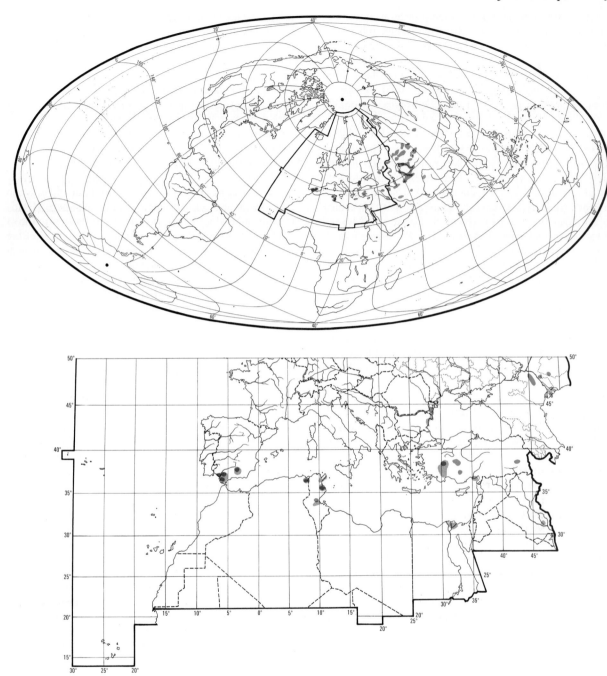

(preferably fresh but sometimes saline, alkaline, or otherwise chemically changed); mainly natural, but at times artificial waters such as fishponds.

Remains well out of gunshot on large shallow waters or marine lagoons (Lake of Tunis), sometimes without vegetation cover. Severe wave action avoided and does not appear to venture out of sight of land or on to deep or icy waters, but rivers with strong current sometimes used on passage. Uses airspace reluctantly, only when necessary,

requiring long take-off, and generally keeping low. Fragmented distribution and decline in former breeding areas suggests biologically precarious state.

**Distribution.** Range much fragmented and nesting often sporadic, especially on periphery. Has disappeared from many western and south-eastern parts. Widest distribution in south-east Europe probably *c.* 1850–1930 (Bauer and Glutz 1969). HUNGARY: sporadic breeding from 1853, perhaps some local increase *c.* 1900, but now occasional; last confirmed nesting 1958 (Bauer and Glutz 1969; AK; IS). YUGOSLAVIA: occasionally breeds Vojvodina; recent records 1967–9 (VJV). ALBANIA: may have bred formerly but present status unknown. GREECE: may have bred 1957–8; now only rare winter visitor (WB). RUMANIA. Bred Transylvania 1872–1925. Dobrogea: probably nested annually since 1958 but in small numbers (MT). FRANCE: nested Corsica until early 1960s (FR). ITALY. Formerly widely reported, but breeding proved only Apulia, Sardinia, and Sicily. Now rare; no longer nesting Sicily (G Sorci), recent breeding Sardinia unconfirmed, but at least 2 pairs bred Apulia 1957 (SF). NORTH AFRICA: in 19th century, apparently common with breeding records from several areas (Heim de Balsac and Mayaud 1962); seems to have declined markedly since, though position inadequately known. ALGERIA: present status ill-defined, but almost certainly nests regularly Lake Tonga (EDHJ, MS). MOROCCO: no recent breeding records (KDS). TUNISIA: breeds in small numbers, perhaps only sporadically (MS). USSR: northern limits of range vary widely from year to year; increases sharply at steppe lakes during drought periods, with attendant salinification, declining markedly in years with high water (Dementiev and Gladkov 1952); no longer breeding Lake Sevan, Armenia (YuAI).

Accidental. Portugal, Belgium, Netherlands, West Germany, East Germany, Poland, Austria, Switzerland, Israel, Cyprus, Malta, Libya. Records from Czechoslovakia not fully documented (KH).

**Population.** Breeding. SPAIN: now only a few dozen pairs at most (FB); no longer breeds marismas of Guadalquivir (Ree 1973). TURKEY: 80 pairs located Central Plateau, 1971; probably 100–150 pairs attempt to breed annually (OST). NORTH AFRICA: position unclear; few definite recent breeding records, yet Blondel (1964) suggested that, with Marbled Teal *Marmaronetta angustirostris*, numbered several hundred pairs and winter numbers there of over 1000 birds (see Movements). USSR. Nowhere regarded as numerous breeder (Dementiev and Gladkov 1951). Kazakhstan probably main headquarters, with estimated 25 pairs per 100 ha in south and 0·3 pairs per 100 ha in north (Gavrin 1970).

Winter. Estimated world population some 15000 birds (Matthews and Evans 1974).

**Movements.** Migratory and partially migratory, some perhaps resident. Little known in absence of ringing; withdraws from most of USSR breeding range, but degree of movement from Mediterranean countries mostly uncertain. Winters locally Mediterranean basin (west to Spain and Morocco), Asia Minor, southern USSR, and Pakistan, but generally scarce. In west Palearctic, only recent reports of 4-figure aggregates from Turkey, where numbers rise in late autumn to 3000–4000 November and

early December and 6000–9000 January, *c.* 90% on Burdur Gölü which too saline to freeze in winter (Dijksen *et al.* 1972; Koning 1973); presumably many immigrants from USSR. Few winter Rumania, where 108 on one lake November 1968 dropped to 29 by mid-January 1969 (Johnson and Hafner 1970). 1200 estimated total wintering USSR, of which up to 800 on east Caspian, remainder Turkmeniya and Kazakhstan (Isakov 1970c); and small numbers sometimes winter south Caspian (Firouz and Ferguson 1970; D A Scott). In recent years, *c.* 1000 have taken to wintering Pakistan (Savage 1970a); assumed to be USSR breeders. In Mediterranean basin, only important regular wintering area Tunisia where, in January 1973, 670 Lake Tunis and 350 in small groups elsewhere (Goldschmidt and Hafner 1973); may be mostly concentration of Maghreb breeders despite paucity of recent breeding records North Africa. Former small population of Spanish marismas mainly resident, leaving during summer drought and reappearing December or earlier (Bernis 1966). Formerly wintered lower Egypt (mainly Nile delta), but no recent information, and few recent sightings Iraq.

In Kazakhstan, ♂♂ recorded leaving mates during breeding cycle and gathering on large lakes, but apparently do not undertake moult migration (Bauer and Glutz 1969). Autumn departures begin late September and northern breeding areas deserted by mid-October; concentrations on east Caspian and in Turkey grow during 2nd half October and November; first appear Pakistan, November. Return begins February, and last noted Pakistan and Iraq late April; main spring passage Kazakhstan around 29 April–5 May, and by mid-May return movement finished even for west Siberian breeders (Dementiev and Gladkov 1952; Johansen 1959).

**Food.** Omnivorous, with vegetable matter possibly predominating, though data limited. Occasionally dabbles at floating matter or at water's edge, but most food obtained by surface-diving, by quietly lowering into water or by rolling forward head first. Up to 40 s submergence in pool no more than 2 m deep, and as many as 50 successive dives recorded with only brief surface pauses of 5–8 s (Matthews and Evans 1974).

Plant materials include leaves of pondweeds (*Potamogeton* and *Vallisneria*) and eel-grass *Zostera*, seeds of bulrush *Scirpus* and wigeon-grass *Ruppia*. Animal matter includes Chironomidae larvae, molluscs, and crustaceans. Wintering birds on Caspian Sea contained snail *Hydrobia*, red seaweed *Polysiphonia*, and stonewort *Chara*; also seeds of *Ruppia maritima* (Dementiev and Gladkov 1952).

Bird from Punjab Salt Range Lakes in West Pakistan in December contained seeds, possibly of *R. maritima* and *Melilotus indica* (Savage 1965). ♀♀ from central Kazakhstan, USSR, July, contained seeds of *Potamogeton* and *Najas*, and waterboatmen *Corixa* and *Micronecta*. Young caught at same time had only insects (Dolgushin 1960). Young hatched by parents in captivity dived within 7 min of reaching water, and by 3rd day dived for up to 15 s reaching adult performance levels by 6th week; unlike adults, young had to leap into air to make dive (Matthews and Evans 1974).

**Social pattern and behaviour.** Little known in wild; compiled largely from information supplied by G V T Matthews, mainly from captive birds at Slimbridge, England (see Matthews and Evans 1974).

1. Gregarious, except while nesting. Flock size variable outside breeding season, but 400–500 not unusual on moulting and wintering grounds. ♂♂ congregate to moult while ♀♀ sitting. Migrates in small groups. BONDS. Type of pair-bond uncertain; ♂ displays to more than one ♀ throughout breeding season and any association in separate pairs must be brief. Only ♀ incubates and tends young; cases in wild of ♂ remaining with ♀ and young may refer to passing ♂ attracted, but not paired to, brood ♀. BREEDING DISPERSION. Gregarious until shortly before nesting. Not colonial; territorial activity, by ♀ only, seems limited to areas round nest and (for first 2 weeks) brood. ROOSTING. Mostly in flocks on water; both diurnal and nocturnal. Breeding ♀ and young roost off water on old nest-site, for up to 3 weeks after hatching. Non-breeding adults observed building platforms, but not known whether used for roosting.

2. Scattered accounts in literature (summarized by Johnsgard 1965, 1966) inaccurate in some respects. FLOCK BEHAVIOUR. Lateral Head-flicks presumably pre-flight signals (but see Ruddy Duck *O. jamaicensis* and below). ANTAGONISTIC BEHAVIOUR. Hunched-posture of ♂, with scapulars raised, assumed as threat; often accompanies Open-Bill-threat with head thrust low over water, or may develop into Hunched-rush at opponent (see A), after which attacker often Head-shakes, sometimes with bill in water. Opponent usually escapes by diving; actual physical contact rare, lasting a few seconds. ♂ attacks on other ♂♂ seem concerned with driving them away from desired ♀ rather than from fixed area. Other behaviour shown by ♂♂ in courting parties (see below) probably also mainly hostile. Simple Open-Bill-threat given by ♀♀ to ♂♂ (see B). COMMUNAL COURTSHIP. In initial stages, flock of ♂♂ and ♀♀ engage in Flotilla-swimming, moving steadily at 3·4 km per hour, leading bird (♂ or ♀) in Head-high posture with neck erect, bill horizontal, and tail raised 45° or flat on water. All generally silent, but ♀♀ may give soft calls and ♂♂ harsh grunts. Lateral Head-shakes sometimes lead to 'scutter-rushes' across water, at least in pinioned birds. Ritualized comfort behaviour of several kinds appears during Flotilla-swimming (and in other contexts), often after scutter-rush, including: Dip-diving—body briefly immersed with down-

C

D

A

B

up movement; Head-rolling—head rubbed on shoulder; Wing-shuffling—folded wings slightly raised and lowered, alternately or together; and Dab-preening—bill rapidly and repeatedly jabbed into breast feathers. May be pre-flight movements. More elaborate group displays appear next. ♂♂ swim about rapidly giving Tickering-purr calls. After Hunched-rush at another, ♂ rushes similarly back towards ♀, swinging broadside on into Sideways-hunch while calling (see C); then swims away and, if ♀ does not follow, repeats sequence. Sometimes ♂ between 2 ♀♀ may Sideways-hunch first on one side, then other, until one responds. Next, Sideways-hunching may be succeeded by Head-high-Tail-cock (see D): ♂ assumes Head-high posture with closed tail erected just over 90°; head and tail movements slow and synchronized, except tail flicked up more rapidly when first in water; posture held for a few seconds, then both head and tail lowered. Once ♂ seen to have penis protruding stiffly as tail flicked up. Although Head-high-Tail-cock posture looks like that assumed by *O. jamaicensis* at start of its Bubbling-display, and may be homologous, ♂ *O. leucocephala* does not apparently inflate neck nor beat bill on breast to produce non-vocal sound. Separate Dab-preening probably cause of erroneous reports of Bubbling-display in present species. ♂ not so insistent as *O. jamaicensis* in positioning in front of ♀ to expose white under tail-coverts. Head-high-Tail-cock often develops when ♂ not previously engaged in display; then series of up to 9, with variable interval of several seconds between each. Also often precedes or follows more intense display activity when, after series of Sideways-hunches, ♂ suddenly Kick-flaps, thrusting bill into water and jerking up folded wings while legs strike vigorously backwards. Then resumes Sideways-hunching at high intensity with bill open, making series of double Pipe-calls; between each call, Tail-vibrates—twisting fully spread tail sideways while strongly vibrating it, folded wings slightly raised and lowered together. Up to 6 sequences of calling and Tail-vibration may follow before ♂ subsides, usually giving slow Head-high-Tail-cock. ♀'s reaction minimal, even to most intense ♂ display: may perform Open-Bill-threat; if follows ♂ as he swims away after Sideways-hunch, may erect feathers on top of head. COPULATION. Not yet observed in captive birds, though fertile clutches produced.

RELATIONS WITHIN FAMILY GROUP. Young emerge from nest within 12 hours of hatching, diving within 7 min of reaching water. Give contact-call; also double Whit-call, both when separated from and near ♀—who produces low Purr-call, especially in situations of mild alarm. In first few hours on water, less frequently later, ♀ will swim first to one duckling, then to another, with head low, bill slightly open, and scapulars smooth, briefly touching them lightly on back of head; such Head-touching strengthens family-bond. When ♂ remains on same water, usually ignores young; but once seen to approach and make Sideways-hunch towards duckling on first appearance on water (may parallel Bubbling-display of *O. jamaicensis* towards newly emerged young). Whit-call of young not heard after 4th week when Open-Bill-threat against others evident. When frightened, duckling elevates stiff retrices and erects head, posture being reminiscent of Head-high-Tail-cock of adult ♂— but held while swimming rapidly away. One duckling (32–34 days old) observed to direct adult-like Sideways-hunches, Kick-flaps, and Head-high-Tail-cocks at mother during bouts of communal display among moulting adults; at point when Sideways-hunch reached intensity where Piping would occur in adult, only wheezy sound emerged. ♀ responded with whole repertory of ♂ movements not seen from ♀ in normal heterosexual context.

(Figs A–D after photographs in Matthews and Evans 1974.)

**Voice.** Silent most of year; no utterance heard in winter quarters (Ali and Ripley 1968; Etchécopar and Hüe 1967; M Smart). Other utterances little studied except in captivity (Matthews and Evans 1974), on which following account largely based.

CALLS OF MALE. Harsh grunt sometimes uttered during Flotilla-swimming phase of communal courtship, but main vocalizations: (1) Tickering-purr rather like rattling of ♂ Garganey *Anas querquedula*, though softer and more continuous, or bouncing of ping-pong ball on table; (2) Pipe-call, series of high, double, reedy notes resembling 'pipe-pipe', in timbre similar to sound made by ♂ Common Scoter *Melanitta nigra*. Also makes non-vocal splashing noises in Kick-splash display.

CALLS OF FEMALE. Soft 'gek' sometimes uttered during Flotilla-swimming phase of communal courtship; otherwise normally silent, though wheezy sound made with open bill during post-breeding courtship by ♀♀ behaving like ♂♂. Warning-call, 'huu-rugru-u', given before diving at breeding place to escape bird of prey (Bauer and Glutz 1969). Call to missing or inattentive duckling, a low rattling purr.

CALLS OF YOUNG. (1) Buzzing 'whit whit', softer than piping of other Anatinae; not heard after 4th week. (2) Low twitter; contact-call between young. (3) Wheezy sound, like equivalent ♀ call, uttered by precocious duckling displaying to own mother.

**Breeding.** SEASON. Little data but eggs laid late May and June in Spain, North Africa, and southern USSR (Castan 1958; Dementiev and Gladkov 1952; Bauer and Glutz 1969). SITE. In water in thick vegetation. Nest: platform of dead reed stems and leaves woven into upright stems; also often uses old nests of Coot *Fulica atra* and Tufted Duck *Aythya fuligula*, adding further material. Vegetation often bent down to form roof. Little or no down. Building: by ♀. EGGS. Broad ovate; dull white. 67 × 51 mm (63–73 × 48–54), sample 100; calculated weight 96 g (Schönwetter 1967); weight of captive laid eggs 97 g (91–101), sample 10 (J Kear). Clutch: 5–10; up to 15 recorded but by 2 or more ♀♀. One brood. No information on replacement clutches. Eggs laid at 1½ day intervals (J Kear). INCUBATION. 25–26 days in captivity (Matthews and Evans 1974). By ♀, and in captivity incubated throughout with only normal periods off for feeding (Matthews and Evans 1974), not intermittently as recorded (Dementiev and Gladkov 1952). YOUNG. Precocial and nidifugous. Self-feeding. Cared for by ♀. FLEDGING TO MATURITY. Fledging period not known. Young independent before fledging. Age of first breeding not known. BREEDING SUCCESS. No data.

**Plumages.** ADULT MALE BREEDING. Head mostly white; crown, narrow ring round eye, and collar round upper neck black; in some, nape and lower chin also black. Lower neck, chest, and upper mantle deep chestnut-red, sometimes tinged violet. Lower mantle, scapulars, and flanks cinnamon-pink, yellow-grey when worn. Upperparts finely speckled and vermiculated grey-brown; feathers of chest and flanks finely barred black subterminally. Back and rump grey-brown, barred grey or buff, upper tail-coverts deep copper-chestnut. Feathers of belly, vent, and under tail-coverts grey-brown, tipped yellow-ochre. Tail black or dark brown. Primaries and inner webs of secondaries and inner tertials grey-brown, rest of upperwing similar, but finely speckled or vermiculated buff or pale grey. Underwing pale grey, suffused white on larger coverts; axillaries white. ADULT FEMALE BREEDING. Forehead, crown, hindneck, and sides of head dark brown, feathers marked chestnut subterminally. Narrow streak from base of bill below eye to occiput off-white, speckled dark brown. Sides of neck and throat grey-brown, barred pale grey or buff; chin white. Lower neck, chest, mantle, scapulars, flanks, and upper tail-coverts deep chestnut-red, feathers subterminally barred sooty-black; lower scapulars paler, yellow or grey-brown. Back and rump grey-brown, barred grey or buff. Breast, belly, and vent ochre, mottled grey-brown on belly and barred on vent and under tail. Wing and tail as adult ♂. ADULT MALE NON-BREEDING (Eclipse). As breeding (including white sides of head), but sides and front of neck pale grey, barred dark grey, instead of black collar. Black of crown more extensive, sometimes contiguous with black hindneck and eye-ring; occasionally, black spot behind ear-coverts. Chest orange-yellow, mantle and scapulars cinnamon to buff, more heavily vermiculated black, upper tail-coverts chestnut; underparts more heavily spotted grey-brown. ADULT FEMALE NON-BREEDING. As breeding, but dark brown replaces deep chestnut; streak below eye and sides of neck often whiter. DOWNY YOUNG. Head and neck dark brown, distinct buff-white streak from base of bill under eye to ear; chin, throat, and sides of neck pale grey. Chest, upperparts, and flanks dark brown, mantle and centre of chest greyer; indistinct grey-brown spot on side of back behind wing. Belly and vent white. Down of tail well developed, with long, stiff shafts. JUVENILE. As adult ♀, but without chestnut tinge, except a little on barred upper tail-coverts. Mantle and scapulars dark grey-brown, finely vermiculated and speckled pink-cinnamon. Head and hindneck browner, white line below eye broader and paler. Chin, upper throat, and sides of upper neck pale grey; lower neck mottled pale grey and sooty. Feathers of chest yellow-brown, those of breast, belly, and vent grey-brown, edged buff. Under tail-coverts white. Shafts of tail feathers bare and broadened near tip (down still adhering in late summer). Sexes difficult to distinguish but ♂ with rusty cinnamon cast on chest and ♀ more uniform pattern on head and neck (Dementiev and Gladkov 1952). FIRST IMMATURE NON-BREEDING AND BREEDING. Not described. Probably gradually change from juvenile into immature non-breeding (resembling adult ♀ non-breeding), in winter, and then into breeding in spring. Some ♂♂ with almost completely black head, recorded locally in east of range, perhaps rare black-headed morph of adult ♂. (♂♂ of closely related *O. maccoa* and *O. australis* always black-headed.) (Dementiev and Gladkov 1952; Savage 1965; C S Roselaar.)

**Bare parts.** ADULT MALE. Iris bright yellow to orange-yellow. Bill in breeding plumage bright sky blue, in eclipse slaty. Foot reddish or plumbeous grey, webs and joints black. ADULT FEMALE. Iris light yellow. Bill dark leaden. Foot leaden grey, tinged red on tarsus and toes. DOWNY YOUNG. Iris brown. JUVENILE. Iris pale grey, light-brown, or yellow-brown. Bill light leaden. Foot pale cinnamon. (Dementiev and Gladkov 1952; Bauer and Glutz 1969; Ali and Ripley 1968.)

**Moults.** ADULT POST-BREEDING. Body and tail August–September; tail-feathers almost simultaneous. ADULT PRE-BREEDING. Body and tail April–May. Time of wing moult not precisely known. Possibly both sexes replace flight-feathers simultaneously July–August when young growing; at least some ♂♂ in wing moult December–January (no data on ♀♀ available), showing new-grown wings February, or flightless January (Dementiev and Gladkov 1952). In related *O. maccoa* and *O. jamaicensis*, ♂ and ♀ moult wing twice a year while in *O. australis* ♂ seems to moult twice, ♀ once (Palmer 1972). POST-JUVENILE. Not described. Timing of 1st pre-breeding moult apparently as in adult.

**Measurements.** ADULT. Skins, various parts summer and winter range (RMNH, ZMA). Additional data for bill and tarsus from full-grown birds, Pakistan (M A Ogilvie; Wildfowl Trust); wing range includes data of apparently partly juvenile birds from Pakistan (Ali and Ripley 1968; Wildfowl Trust).

| | | | | | | |
|---|---|---|---|---|---|---|
| WING | ♂ 162 | (3·85; 10) | 157–172 | ♀159 | (2·19; 6) | 148–167 |
| TAIL | 92·4 | (4·85; 9) | 85–100 | 85·5 | (7·23; 6) | 75–93 |
| BILL | 45·5 | (1·28; 17) | 43–48 | 44·5 | (0·85; 16) | 43–46 |
| TARSUS | 35·9 | (1·12; 16) | 35–38 | 34·9 | (1·06; 16) | 33–37 |
| TOE | 68·6 | (2·82; 7) | 66–73 | 67·5 | (1·73; 4) | 65–69 |

Sex differences not significant. Tail of ♂♂, USSR, 110–120 (Dementiev and Gladkov 1952).

JUVENILE. Average of wing and tail *c.* 10 mm shorter; other measurements similar to adult.

**Weights.** ADULT. Kazakhstan, USSR (Dementiev and Gladkov 1952; Dolgushin 1960). ♂♂: April, 755 (720–800); June–July, 600, 750, 760, ♀♀: April–May, 720, 890, 900; July, 620; September, range 510–820. Spain (ZMA), ♂, September, 705. Pakistan, December (Savage 1965). ♂♂ 553, 794, 865; ♀♀ 539, 610, 631.

JUVENILE (including some of doubtful age). Kazakhstan and Turkmeniya, USSR (Dementiev and Gladkov 1952; Dolgushin 1960). ♂♂: October, 500, 650. ♀♀: October, range 400–500; November, range 600–650.

**Structure.** Wing short, pointed. 11 primaries: p9 longest, p10 0–4 shorter, p8 2–7, p7 11–17, p6 19–26, p1 56–67. Inner web of p10 and outer p9 emarginated. Tail long, stiff, strongly rounded. 18 narrow, pointed feathers; t9 40–60 shorter than t1. Tail-coverts short. Bill high at base, suddenly sloping above nostril to broad and flat tip (tip narrower in juvenile). Nail narrow, sharply hooked. Sides of bill near base strongly swollen in adult, less in juvenile. Outer toe *c.* 97% middle, inner *c.* 77%, hind *c.* 25%. In all plumages, feathers stiff and dense, with glossy tips on underparts.

**Geographical variation.** None known. Black-headed ♂♂ observed only in east part of range (see Plumages).    CSR

# REFERENCES

AASS, P (1956) *Norg. Jeger og Fiskerforb* **85**, 8–14. ABRAHAM, R L (1974) *Condor* **76**, 401–20. ABS, M (1970) *Verh. dt. zool. Ges.* Suppl. **33**, 298–301. ACLAND, C M (1923) *Br. Birds* **17**, 63–4. ADAMS, R G (1947) *Br. Birds* **40**, 186–7. AHARONI, J (1930) *Beitr. Fortpfl. Biol. Vögel* **6**, 145–51. AHLBOM, B, VON ESSEN, L, and FABRICIUS, E (1973) *Svensk Jakt* **111**, 223–30. AHLÉN, I (1961) *Vår Fågelvärld* **20**, 296–302; (1966) *Vår Fågelvärld* suppl. **4**, 1–45. AHLÉN, I and ANDERSSON, Å (1970) *Ornis scand.* **1**, 83–106. AINSLIE, J A and ATKINSON, R (1937) *Br. Birds* **30**, 234–48. AIREY, A F (1955) *Bird Study* **2**, 143–50. AKESTER, A R, POMEROY, D E, and PURTON, M D (1973) *J. Zool. Lond.* **170**, 493–9. ALDRICH, J W (1946) *Wilson Bull.* **58**, 94–103. ALERSTAM, T, BAUER, C-A, and ROOS, G (1974) *Ibis* **116**, 194–210. ALEXANDER, W B (1955) *Birds of the Ocean*. London; (1965) *Ibis* **107**, 401–5. ALEXANDER, W B and LACK, D (1944) *Br. Birds* **38**, 42–5, 62–9, 82–8. ALEXANDER, B (1898) *Ibis* (7) **4**, 74–118, 277–85. ALFORD, C E (1920) *Br. Birds* **14**, 106–10. ALI, S (1945) *J. Bombay nat. Hist. Soc.* **45**, 586–93; (1960) *J. Bombay nat. Hist. Soc.* **57**, 412–5. ALI, S and RIPLEY, S D (1968) *Handbook of the birds of India and Pakistan* **1**. Bombay. ALISON, R M (1975) *Orn. Monogr.* **18**. ALLAN, R G (1962) *Ibis* **103b**, 274–95. ALLEN, R P (1956) *The Flamingos. Nat. Audubon Soc. Res. Rep.* **5**. ALLEN, R P and MANGELS, F P (1940) *Proc. Linn. Soc. N.Y.* **50–1**, 1–28. ALLISON, A, NEWTON, I, and CAMPBELL, C (1974) *Loch Leven National Nature Reserve*. WAGBI Cons. Pub. ALLOUSE, B E (1953) *The avifauna of Iraq. Iraq nat. Hist. Mus. Publ.* **3**. ALMKVIST, B, ANDERSSON, Å, JOGI, A, PIRKOLA, M K, SOIKKELI, M, and VIRTANEN, J. (1974) *Wildfowl* **25**, 89–94. ALPHERAKY, S (1905) *The Geese of Europe and Asia*. London. AMADON, D (1953) *Bull. Am. Mus. nat. Hist.* **100**, 393–451; (1955) *Bull. Br. Orn. Club* **75**, 21–3. AMADON, D and WOOLFENDEN, G (1952) *Am. Mus. Novit.* **1564**. ANDERSON, B W and TIMKEN, R L (1971) *J. Wildl. Mgmt* **35**, 388–93; (1972) *J. Wildl. Mgmt.* **36**, 1127–33. ANDERSON, D R and HENNY, C J (1972) *Population ecology of the Mallard* **1**. US Dept. Int. Fish and Wildl. Service Res. Pub. **105**. Washington. ANDERSON, D R, SKAPTASON, P A, FAHEY, K G, and HENNY, C J (1974) *Population ecology of the Mallard* **2**. US Dept. Int. Fish and Wildl. Service Res. Pub. **119**. ANDERSON, H G (1959) *Bull. Ill. nat. Hist. Surv.* **27**, 289–344. ANDERSSON, T (1954) *Vår Fågelvärld* **13**, 133–42; (1972) The Fulmar, in *Encycl. of the Animal World*. London. ANDONE, G. ALMĂSAN, H, RADU, D, ANDONE, L, CHIRIAC, E, and SCĂRLĂTESCU, G (1969) *ICSPS, Stud. si Cerc.* **27**, 133–83. ANTHONY, A W (1898) *Auk* **15**, 140–4. ARCHER, G E and GODMAN, E M (1937) *The Birds of British Somaliland and the Gulf of Aden* **1**. Edinburgh. ARDAMATSKAYA, T (1970) *Proc. Int. Reg. Meet. Cons. Wildfowl Resources Leningrad 1968*, 67–9. ARMSTRONG, E A (1940) *Birds of the Grey Wind*. Oxford; (1947) *Bird display and behaviour*. London; (1951) *Ibis* **93**, 245–51. ARMSTRONG, T E and ROBERTS, B B (1956 and 1958) *Illustrated Ice Glossary*. Scott Polar Res. Inst. ASCHENBRENNER, L and SCHIFTER, H (1975) *Egretta* **18**, 8–17. ASHMOLE, N P (1971) In Farner *et al.* (eds) (1971), 223–86. ASHMOLE, N P and ASHMOLE, M J (1967) *Bull. Peabody Mus. nat. Hist.* **24**, 1–131. ASHMOLE, N P, BROWN, R G B, and TINBERGEN, N (1956) *Br. Birds* **49**, 501. ATKINSON-WILLES, G L (1963) *Wildfowl in Great Britain*. London; (1969) *Wildfowl* **20**, 98–111; (1970a) *Proc. Int. Reg. Meet. Cons. Wildfowl Resources, Leningrad 1968*, 101–7; (1970b) *Proc. Int. Reg. Meet. Cons. Wildfowl Resources, Leningrad, 1968*, 221–38; (1972) *Proc. Int. Conf. Wetlands and Waterfowl, Ramsar 1971*, 87–110; (1975) *Aves* **12**: 177–253, 254–283. ATKINSON-WILLES, G L and MATTHEWS, G V T (1960) *Br. Birds* **53**, 352–7. AUDIN, H (1963) *Honeyguide* **40**, 4–5. AUDUBON, J J (1835) *Orn. Biogr.* **3**. Edinburgh. AUSTIN, O L Jr. (1952) *Bull. Mus. comp. Zool. Harv.* **107**, 389–407. AUSTIN, O L Jr. and KURODA, N (1953) *Bull. Mus. comp. Zool. Harv.* **109**, 279–637.

BACKHURST, G C, BRITTON, P L, and MANN, C F (1973) *J. E. Afr. nat. Hist. Soc. natn. Mus.* **140**. BAERENDS, G P and VAN DER CINGEL, N. A (1962) *Symp. zool. Soc. Lond.* **8**, 7–24. BAGGE, P R, LEMMETYINEN, R, and RAITIS, T (1970) *Suom. Riista* **22**, 35–45. BAILEY, A M (1922) *Condor* **24**, 204–5; (1948) *Birds of Arctic Alaska*. Colorado Mus. nat. Hist. pop. Ser. **8**. BAILEY, R S (1966) *Ibis* **108**, 224–64; (1968) *Ibis* **110**, 493–519; (1971) *Ibis* **113**, 29–41. BAILEY, R S, POCKLINGTON, R, and WILLIS, P R (1968) *Ibis* **110**, 27–34. BAIRD, S F (1887) *Auk* **4**, 71–2. BAKER, E C S (1929) *The Fauna of British India. Birds* **6**. London. BALAT, F and FOLK, C (1968) *Zool. Listy* **17**, 327–40. BANDORF, H (1968) *Vogelwelt Beihefte* **1**, 7–61; (1970) *Der Zwergtaucher*. Wittenberg Lutherstadt. BANGS, O (1918) *Bull. Mus. comp. Zool. Harv.* **61**, 489–511. BANNERMAN, D A (1914) *Ibis* (10) **2**, 438–94; (1930) *The birds of tropical West Africa* **1**. London; (1932) *Ibis* (13) **2**, 689; (1951) *The birds of tropical West Africa* **8**. London; (1953) *The birds of West and Equatorial Africa*. London; (1957) *The birds of the British Isles* **6**; (1958) **7**; (1959) **8**. Edinburgh; (1963) *Birds of the Atlantic Islands* **1**. Edinburgh; (1969) *Bull. Br. Orn. Club* **89**, 86–8. BANNERMAN, D A and BANNERMAN, W M (1965) *Birds of the Atlantic Islands* **2**. Edinburgh; (1968) *History of the birds of the Cape Verde Islands*. Edinburgh; (1971) *Handbook of the birds of Cyprus and migrants of the Middle East*. Edinburgh. BANNIKOV, A G and TARASOV, P P (1957) *Trudy Byuro Kol'tsev.* **9**, 208–14. BARNHILL, M V and DUMONT, P G (1973) *Am. Birds* **27**, 17. BARRATT, J (1973) MSc Thesis. Minnesota Univ., USA. BARRY, T W (1962) *J. Wildl. Mgmt.* **26**, 19–26. BARTH, H (1857–8) *Reisen und Entdeckungen in Nord- und Central Afrika in den Jahren 1849–55* **1–5**. Gotha. BARTONEK, J C (1968) Ph D dissertation. University Wisconsin, Madison, Wisconsin. BARTONEK, J C and HICKEY, J J (1969) *Condor* **71**, 280–90. BATES, G L (1937) *Ibis* (14) **1**, 47–65; (1938) *Ibis* (14) **2**, 437–62. BAUER, K M and GLUTZ VON BLOTZHEIM, U N (1966) *Handbuch der Vögel Mitteleuropas* **1**; (1968) **2**; (1969) **3**. Frankfurt am Main. BAUER K M and ZINTL, H (1974) *Anz. orn. Ges. Bayern* **13**, 71–8. BAXTER, E V and RINTOUL, L J (1953) *The Birds of Scotland*. Edinburgh. BAXTER, G P (1970) *Flor. Nat.* **43**, 66. BAXTER, R M and URBAN, E K (1970) *Ibis* **112**, 336–9. BAYNARD, O E (1913) *Wilson Bull.* **25**, 103–17. BEAMISH, H V (1945) *Field* **186**, 325. BEARD, D B (1939) *Auk* **56**, 327–9. BECK, J R and BROWN, D W (1972) *Br. Antarctic Survey Sci. Rep.* **69**. London. BEDNORZ, J (1974) *Ochr. Przyr.* **39**, 201–43. BEER, J V and BOYD, H (1962) *Bird Study* **9**, 91–9; (1963) *Wildfowl Trust Ann. Rep.* **14**, 114–19. BEER, J V and OGILVIE, M A (1972) In Scott P *et al.* (1972), 125–42. BELLROSE, F C and HAWKINS, A S (1947) *Auk* **64**, 422–30. BELMAN, P J (1974) *Br. Birds* **67**, 120. BELOPOL'SKII, L O (1957) *Ecology of the Sea Colony Birds of the Barents Sea*. Moscow. BENGTSON, S-A (1966a) *Wildfowl Trust Ann. Rep.* **17**, 79–94; (1966b) *Vår Fågelvärld* **25**, 202–26; (1968) *Wildfowl* **19**, 61–3; (1971) *Ornis fenn.* **48**, 77–92; (1972a) *Oikos* **23**, 35–58; (1972b) *Ornis scand.* **3**, 1–19; (1972c) *Bull. Br. Orn. Club* **92**, 100–1; (1975) *Ornis fenn.* **52**, 1–4. BENGTSON, S-A and DENWARD, C (1966) *Fauna Flora, Upps.* **177**–84. BENGTSON, S-A and ULFSTRAND, S (1971) *Oikos* **22**,

235–9. BENNETT, C G (1965) *Br. Birds* **58**, 190–1. BENSON, C W and IRWIN, M P S (1964) *Bull. Br. Orn. Club* **84**, 134–7; (1966) *Puku* **4**, 49–56. BENSON, C W, BROOKE, R K, DOWSETT, R J, and IRWIN, M P S (1970) *Arnoldia* **4** (40), 1–59. BENSON, C W and PENNY, M J (1971) *Phil. Trans. R. Soc. B* **260**, 417–527. BENSON, C W and SERVENTY, D (1956) *Ostrich* **27**, 171–2. BENT, A C (1919) *Life histories of North American birds. Bull. US natn. Mus.* **107**; (1922) **121**; (1923) **126**; (1925) **130**; (1926) **135**. BERGLUND, B (1963) *Acta vertebr.* **2**, 241–2. BERGMAN, G (1939) *Acta zool. fenn.* **23**, 1–134. BERGMAN, G and DONNER, K O (1964) *Acta zool. fenn.* **105**, 1–59. BERGMAN, S (1936) *Fauna Flora Upps.* **31**, 133–5. BERG-SCHLOSSER, G (1968) *Die Vögel Hessens.* Ergänzungsband. Frankfurt am Main. BERLIOZ, J (1959) *Ostrich* Suppl. **3**, 415–7. BERNDT, R (1938) *Beitr. Fortpfl. Biol. Vögel* **14**, 95–9. BERNDT, R, GOETHE, F, and RAHNE, U (1966) *Bonn. zool. Beitr.* **17**, 241–56. BERNHARDT, P (1940) *J. Orn.* **88**, 488–97. BERNIS, F (1959) *Ardeola* **5**, 9–80; (1961) *Ardeola* **7**, 204–17; (1964) *Ardeola* **9**, 67–109; (1966) *Aves Migradoras Ibericas* **1**, Sociedad Española de Ornitologia. Madrid; (1971a) *Adena* **1**, 11–5; (1971b) *Ardeola* **15**, 107–10; (1972) *Proc. Int. Conf. Wetlands and Waterfowl Ramsar 1971*, 239–45; (1974) *Ardeola* **19**, 151–224; (1975) *Ardeola* **21**, 489–580. BERRY, H H (1972) *Madoqua*, (I) **5**, 5–31. BERRY, J (1936) *Rep. Avon biol. Res.* 1934–5, 31–64; (1939) *The status and distribution of Wild Geese and Wild Duck in Scotland.* Cambridge. BERTHOLD, P (1961) *Vogelwarte* **21**, 142–4. BERTRAM, G C and LACK, D (1933) *Ibis* (13) **3**, 283–301. BEVEN, G (1946a) *Br. Birds* **39**, 122–3; (1946b) *Ibis* **88**, 133; (1972) *World of Birds* **1** (3), 3–7. BEZZEL, E (1957) *Anz. orn. Ges. Bayern* **4**, 589–707; (1959) *Anz. orn. Ges. Bayern* **5**, 269–355; (1962) *Anz. orn. Ges. Bayern* **6**, 218–33; (1964a) *Anz. orn. Ges. Bayern* **7**, 43–79; (1964b) *Vogelwelt* **85**, 39–43; (1965) *Vogelwelt* **86**, 89–91; (1968) *Vogelwelt* **89**, 102–11; (1969) *Die Tafelente.* Wittenberg Lutherstadt. BEZZEL, E and VON KROSIGK, E (1971) *J. Orn.* **112**, 411–37. BIBER, O and HOFFMAN, L. (1973) *Bull. Int. Waterfowl Res. Bur.* **35**, 32–38. BIERMAN, W H and VOOUS, K H (1950) *Ardea* **37**, suppl. 1–123. BILLARD, R S and HUMPHREY, P S (1972) *J. Wildl. Mgmt* **36**, 765–74. BILLINGS, S M (1968) *Auk* **85**, 36–43. BINSBERGEN, A F (1959) *Vogeljaar* **7**, 190–197, 220–228, 233–35. BIRD, G G and BIRD, E G (1941) *Ibis* (14) **5**, 118–161. BIRKHEAD, T R (1973a) *Naturalist* **924**, 13–19; (1973b) *Br. Birds* **66**, 147–56. BJÄRVALL, A. (1968) *Wildfowl* **19**, 70–80. BJÄRVALL, A and SAMUELSSON, A (1970) *Zool. Revy* **32**, 26–33. BLAIR, H M S (1957) In Bannerman (1957) 92–6; (1958) In Bannerman (1958) 164–7. BLAKER, D (1969a) *Ostrich* **40**, 75–129; (1969b) *Ostrich* **40**, 150–5. BLOCH, D (1970) *Dansk orn. Foren. Tiddsskr.* **64**, 152–62; (1971) *Danske Vildtundersøgelser* **16**, 1–47. BLONDEL, J and BLONDEL, C (1964) *Alauda* **32**, 250–79. BLURTON-JONES, N G (1951) *Reading Orn. Club Rep.* **5**, 19–22; (1960) *Wildfowl Trust Ann. Rep.* **11**, 46–52. BOASE, H (1924) *Br. Birds* **18**, 69–71; (1925) *Br. Birds* **19**, 162–4; (1926) *Br. Birds* **19**, 226–30; (1935) *Br. Birds* **28**, 218–24; (1949) *Scott. Nat.* **61**, 10–8; (1950) *Scott. Nat.* **62**, 1–16; (1965) *Scott. Birds* **3**, 301–10. BOCHENSKI, Z (1961) *Bird Study* **8**, 6–15. BOCK, W J (1956) *Amer. Mus. Novit.* **1779**; (1963) *Proc. Int. Orn. Congr.* **13**, 39–54. BODENHAM, K L (1945) *Bull. zool. Soc. Egypt* **7**, 14–47. BODENSTEIN, G and SCHÜZ, E (1944) *Orn. Mber.* **52**, 98–106. BOEV, N (1965) *Wildfowl Trust Ann. Rep.* **16**, 58–63. BÖHM, R (1882) *Orn. Cbl.* **7**, 113–16. BOLAM, G (1912) *Birds of Northumberland and the Eastern Borders.* Alnwick. BOLSTER, R C (1931) *Ostrich* **2**, 18–19. BOMBARD, A (1953) *The Bombard Story.* London. BOND, J (1960) *Birds of the West Indies.* London BORMAN, F W (1929) *Ibis* (12) **5**, 639–50. BORODULINA, T L (1971) *Byull. Mosk. Obch. Isp. Prir. Biol.* **76**, 47–53. BOSWALL, J (1960) *Br. Birds* **63**, 212–15. BOSSENMAIER, E F and MARSHALL, W H (1958) *J. Wildl. Mgmt.* **1**, 1–32. BOUET, G (1950) *La Vie des*

*Cigognes.* Paris; (1955) *Faune Union franc.* **16**, 63; (1956) *L'Oiseau* **26**, 59–61. BOURNE, W R P (1953) *Bull. Br. Orn. Club* **73**, 79–82; (1955a) *Ibis* **97**, 145–9; (1955b) *Ibis* **97**, 508–56; (1957) *Ibis* **99**, 182–90; (1960) *Sea Swallow* **13**, 26–39; (1961) *Sea Swallow* **14**, 7–27; (1962) *Sea Swallow* **15**, 9–27; (1963) *Sea Swallow* **16**, 9–40; (1964) *Bull. Br. Orn. Club* **84**, 114–16; (1965) *Sea Swallow* **17**, 10–39; (1966a) *Sea Swallow* **18**, 9–39; (1966b) *Ibis* **108**, 425–9; (1967a) *Ibis* **109**, 141–67; (1967b) *Sea Swallow* **19**, 51–76; (1970) *Sea Swallow* **20**, 50. BOURNE, W R P and DIXON, T J (1973) *Sea Swallow* **22**, 29–60. BOURNE, W R P and WARHAM, J (1966) *Ardea* **54**, 45–67. BOWEN, W, GARDINER, N, HARRIS, B J, and THOMAS, J D (1962) *Ibis* **104**, 246–7. BOWMAKER, A P (1963) *Ostrich* **34**, 2–26; (1964) *Rep. jt. Fish. Res. Org. N. Rhodesia 1961* **11**, 14–18. BOYD, H (1953) *Behaviour* **5**, 85–130; (1954) *Br. Birds* **47**, 137–63; (1957a) *Ibis* **99**, 157–77; (1957b) *Wildfowl Trust Ann. Rep.* **8**, 51–54; (1959) *Wildfowl Trust Ann. Rep.* **10**, 59–70; (1961a) *Wildfowl Trust Ann. Rep.* **12**, 153–6; (1961b) *Wildfowl Trust Ann. Rep.* **12**, 116–24; (1962) In *The Exploitation of Natural Animal Populations* (ed. Le Cren and Holdgate) Oxford, 85–95; (1964) *Wildfowl Trust Ann. Rep.* **15**, 75–76; (1965) *Wildfowl Trust Ann. Rep.* **16**, 34–40; (1968) *Wildfowl* **19**, 96–107; (1972) In Scott P et al. (1972) 17–27. BOYD, H and FABRICIUS, E (1965) *Behaviour* **25**, 1–15. BOYD, H and KING, B (1964) *Wildfowl Trust Ann. Rep.* **15**, 47–50. BOYD, H and OGILVIE, M A (1961) *Wildfowl Trust Ann. Rep.* **12**, 125–36; (1969) *Wildfowl* **20**, 33–46; (1972) *Wildfowl* **23**, 64–82. BRAAKSMA, S (1958) *Ardea* **46**, 158–66; (1968) *Limosa* **41**, 41–61. BRADY, F (1951) *Br. Birds* **44**, 413–14. BRANDT, M (1940) *Beitr. Fortpfl. Biol. Vögel* **16**, 135–9. BRASCHLER, K, LENGWEILER, O, FELDMANN, G, and EGLI, V (1961) *Orn. Beob.* **58**, 59–75. BRATTSTROM, B H and HOWELL, T R (1956) *Condor* **58**, 107–20. BRETHERTON, B J (1896) *Oreg. Nat.* **3**, 45–9, 61–4, 77–9, 100–2. BREWSTER, W (1911) *Condor* **18**, 69–71. BRIEN, Y (1974) *Avifaune de Bretagne.* Brest. BRINCKMANN, A and HAEFELFINGER, H-R (1957) *Orn. Beob.* **51**, 182–95. BRITISH ORNITHOLOGISTS' UNION (1971) *The Status of Birds in Britain and Ireland.* Oxford. BRITTON, P L (1970) *Bull. Br. Orn. Club* **90**, 142–4. BROBERG, L (1971) *Vår Fågelvärld* **30**, 91–8. BROCK, S E (1914) *Scott. Nat.* **79**, 78–86. BROEKHUYSEN, G J and FROST, P G H (1968a) *Bonn. zool. Beitr.* **19**, 350–61; (1968b) *Ostrich* **39**, 242–52. BROOKS, A (1920) *Auk* **37**, 353–67. BROOKS, W S (1915) *Bull. Mus. comp. Zool. Harv.* **59**, 393. BROSSELIN, M (1975) (ed) *Hérons Arboricoles de France.* Paris. BROTHERSTON, W (1965) *Wildfowl Trust Ann. Rep.* **15**, 57–70. BROUWER, G A (1964) *Zoöl. Meded. Leiden* **39**, 481–521. BROUWER, G A and TINBERGEN, L (1939) *Limosa* **12**, 1–18. BROWN, L (1947) *Birds and I.* London. BROWN, L H (1958) *Ibis* **100**, 388–420; (1960) *The Mystery of the Flamingos.* London; (1975) In Kear and Duplaix-Hall (1975), 34–8. BROWN, L H POWELL-COTTON, D, and HOPCRAFT, J B D (1973) *Ibis* **115**, 352–74. BROWN, L H and ROOT, A (1971) *Ibis* **113**, 147–72. BROWN, L H and URBAN, E K (1969) *Ibis* **111**, 199–237. BROWN, R G B (1970) *Ibis* **112**, 44–51. BROWNE, P W P (1950) *Ibis* **92**, 52–65. BRUGGERS, R L (1974) Ph D Thesis. Bowling Green State University. BRUN, E (1972) *Ornis scand.* **3**, 27–38. BRUUN, B (1971) *Br. Birds* **64**, 385–408. BRYANT, D M and LENG, J (1975) *Wildfowl* **26**, 20–30. BUCHANAN, A (1921) *Out of the wild north of Nigeria.* London. BUCKLEY, P A and WURSTER, C F (1970) *Bull. Br. Orn. Club* **90**, 35–8. BULMAN, J and BEALE, S (1944) *Ibis* **86**, 480–92. BULSTRODE, C J K and HARDY, D E (1970) *Wildfowl* **21**, 18–21. BURKE, V E M and BROWN, L H (1970) *Ibis* **112**, 499–512. BURKHARD, S (1961) *Falke* **8**, 97. BURN, D M and MATHER, J R (1974) *Br. Birds* **67**, 257–96. BURTON, P J K (1961) *Wildfowl Trust Ann. Rep.* **12**, 104–12. BÜTTIKER, W (1952) *Biol. Abh.* **1**, 1–40. BUXTON, E J M (1962) *Wildfowl Trust Ann. Rep.* **13**, 170. BUXTON, J (1952) *Orn. Beob.* **49**, 53. BYLIN, K (1971) *Vår*

*Fågelvärld* **30** (2), 79–83.
CABANNE, F and FERRY, C (1950) *L'Oiseau* **20**, 84–5. CABOT, D and WEST, B (1973) *Proc. R. Irish Acad.* **73**, 415–43. CADMAN, W A (1953) *Br. Birds* **46**, 374–5; (1955) *Br. Birds* **48**, 325; (1956) *Nature in Wales* **2**, 348–9. CALDWELL, H R and CALDWELL, J C (1931) *South China Birds*. Shanghai. CAMPBELL, A C (1933) *Emu* **33**, 86–92. CAMPBELL, A G and MATTINGLEY, A H E (1906) *Emu* **6**, 185. CAMPBELL, B (1960) *Bird Study* **7**, 708–23. CAMPBELL, B and FERGUSON-LEES, I J (1972) *A Field Guide to Birds' Nests*. London. CAMPBELL, J W (1936) *Br. Birds* **30**, 209–18; (1946a) *Br. Birds* **39**, 194–200, 226–32; (1946b) *Br. Birds* **39**, 371–3; (1947) *Ibis* **89**, 429–32; (1949) *Scott. Nat.* **61**, 73–100. CANTIN, M, BEDARD, J, and MILNE, H (1974) *Can. J. Zool.* **52**, 319–34. CARNEY, S M (1964) *US Fish and Wildl. Serv. Spec. Sci. Rep. Wildl.* **82**. CARP, E (1972) *Bull. Int. Waterfowl Res. Bur.* **33**, 37–9. CARRICK, R and DUNNET, G M (1954) *Ibis* **96**, 356–70. CARRICK, R and INGHAM, S E (1970) In Holdgate, M W (ed) *Antarct. Ecology* **1**, 505–25. London. CASTAN, R (1958) *Alauda* **26**, 52–62; (1959) *Alauda* **27**, 148–150. CASTROVIEJO, J (1972) *Alauda* **40**, 287. CAVE, F O and MACDONALD, J D (1955) *Birds of the Sudan*. Edinburgh. CAWKELL, E M and MOREAU, R E (1963) *Ibis* **105**, 156–78. CHAPIN, J P (1932) *The Birds of the Belgian Congo* **1**. *Bull Amer. Mus. nat. Hist.* **65**. CHAPIN, J P and AMADON, D (1950) *Ostrich* **21**, 15–18. CHAPMAN, F M (1905) *Bull. Amer. Mus. nat. Hist.* **21**, 53–77; (1926) *Bull. Am. Mus. nat. Hist.* **55**. CHAPMAN, S, KING, B, and WEBB, N (1959) *Br. Birds* **52**, 60. CHASEN, F M (1933) *Bull. Raffles Mus.* **8**, 55–87. CHENG TSO-HSIN (1963) *China's Economic Fauna: Birds*. Washington D.C. CHILD, G (1971) *Ostrich* **43**, 60–2. CHRISTENSEN, N H (1967) *Dansk orn. Foren. Tidsskr.* **61**, 56–64. CHRISTMAS, J Y (1960) *Auk* **77**, 346–7. CHRISTOLEIT, E (1927) *J. Orn.* **75**, 385–404. CHURA, N J (1961) *Trans. N. Am. Wildl. and Nat. Resources Conf.* **26**, 121–34. CLANCEY, P A (1964) *The birds of Natal and Zululand*. Edinburgh; (1965) *Ostrich* **36**, 36; (1968a) *Arnoldia* **3** (37), 1–5; (1968b) *Ostrich* **39**, 193–4. CLAPHAM, C S (1964) *Ibis* **106**, 376–88. CLAPP, R B (1968) *Ibis* **110**, 573–5. CLARK, A H (1903) *Auk* **20**, 285–93. CLARKE, G. (1965) *S. Aust. Orn.* **24**, 51–2. CLASE, H F, COOKE, F, HILL, T A, and ROFF, W J (1960) *Bird Study* **7**, 76–81. COLDWELL, C (1939) *Can. Fld. Nat.* **53**, 55. COLLETT, R (1877) *Nyt. Mag. Naturvid.* **23**, 85–225; (1894a) *Ibis* (6) **6**, 269–83; (1894b) *Nyt. Mag. Naturvid.* **35**, 1–376; (1921) *Norges Fugle* **3**. Oslo. COLLIAS, N E and COLLIAS, E C (1963) *Wilson Bull.* **75**, 6–14. COLLIAS, N E and JAHN, R J (1959) *Auk* **76**, 479–509. COLLIN, A (1973) *Aves* **10**, 29–69. COLLINGE, W E (1924–7) *The food of some British wild birds*. York. COLLINS, J W (1884) *US Commission of Fish and Fisheries. Rep. for 1882*. COLMAN, H R and BOASE, H (1925) *Br. Birds* **18**, 313–16. COMMISSIE VOOR DE NEDERLANDSE AVIFAUNA (1970) *Avifauna van Nederland*. Leiden. CONDER, P J (1949) *Ibis* **91**, 649–55. CONDON, H T (1936) *S. Austr. Orn.* **13**, 141–61. CONOVER, H B (1926) *Auk* **43**, 162–80. CONROY, J W H (1972) *Scient. Rep. Br. Antarct. Surv.* **75**, CONROY, J W H, BRUCE, G, and FURSE, J R (1976) *Ardea* **63**, 87–92. COOCH, F G (1957) *Wildfowl Trust Ann. Rep.* **8**, 58–67; (1961) *Auk* **78**, 72–89; (1963) *Proc. Int. Orn. Congr.* **13**, 1182–94. COOCH, F G, STIRRETT, G M, and BOYER, G F (1960) *Auk* **77**, 460–5. COOK, F and MILLS, E L (1972) *Ibis* **114**, 245–51. COOMANS DE RUITER, L (1966) *Limosa* **39**, 187–212. COOMBS, C J F (1960) *Ibis* **102**, 394–419. COOPER, J and LYSAGHT, A (1956) *Ibis* **98**, 316–19. COOPER, J and MARSHALL, B E (1970) *Bull. Br. Orn. Club* **90**, 148–52. CORDEAUX, J (1896) *British birds and their nests and eggs* **4**. London. CORKHILL, P (1973) *Br. Birds* **66**, 136–43. COTT, H B (1952) *E. Afr. Fisheries Res. Organ. Ann. Rep.*, 23–5; (1961) *Trans. zool. Soc. London* **91**, 211–356. COTTAM, C (1936) *J. Wash. Acad. Sci.* **26**, 165–177; (1939) *US Dept. Agric. Tech. Bull.* **643**, 88–93. COTTAM, C and HANSON,

H C (1938) *Fld. Mus. Pub. nat. Hist.* **20**, 405–26. COTTAM, C and KNAPPEN, P (1939) *Auk* **56**, 138–9. COTTAM, C, LYNCH, J, and NELSON, A I (1944) *J. Wildl. Mgmt* **8**, 36–44. COULSON, J C (1961) *Br. Birds* **54**, 225–35. COULSON, J C and BRAZENDALE, M G (1968) *Br. Birds* **61**, 1–21. COULSON, J C POTTS, G R, and HOROBIN, J (1969) *Auk* **86**, 232–45. COWARD, T A (1910) *Fauna of Cheshire* **1**. London; (1920) *The birds of the British Isles and their eggs* **1**. London. COWLES, R B (1930) *Auk* **47**, 465–70. CRACRAFT, J (1974) *Ibis* **116**, 494–521. CRAMP, S (1957) *London Bird Rep.* **21**, 58–62; (1966) *Br. Birds* **59**, 147–50; (1969) In *Birds of the World* **1**. London; (1972) *Wildfowl* **23**, 119–24. CRAMP, S, BOURNE, W R P, and SAUNDERS, D (1974) *The Seabirds of Britain and Ireland*. London. CRAMP, S and CONDER, P J (1970) *Ibis* **112**, 261–3. CRAUFURD, R Q (1966) *Ibis* **108**, 411–18. CREUTZ, G (1958) *Falke* **5**, 98–101, 116–19; (1964) *Zool. Abh. Mus. Tierk. Dresden* **27**, 29–64. CREUTZ, G and SCHLEGEL, R (1961) *Falke* **8**, 377–86. CRONAN, J M (1957) *Auk* **74**, 459–68. CROOK, J H (1964) *Behaviour* Suppl. **10**; (1965) *Symp. zool. Soc. Lond.* **14**, 181–218; (1970) *Anim. Behav.* **18**, 197–209. CROSBY, G T (1972) *Bird-Banding* **43**, 205–11. CROWE, R W (1951) *Br. Birds* **44**, 391. CRUON, R and VIELLIARD, J (1975) *Alauda* **43**, 1–21. CULLEN, J M (1963) *Ibis* **105**, 121. CURRY-LINDAHL, K (1957) In *Atlas över Sverige*. Stockholm; (1959) *Våra Fåglar i Norden*. Stockholm; (1960) *Terre et Vie* 1960 (4), 169–93; (1961) *Bijdragen tot de Dierkunde* **31**, 27–44; (1964) *Proc. 1st Eur. Meeting Wildfowl Cons., St Andrews 1963*, 3–13; (1968) *Vår Fågelvärld* **27**, 289–308; (1971) *Ostrich* Suppl. **9**, 53–70. CURRY-LINDAHL, K, ESPING, L-E, and HÖJER, J (1970) *Proc. Int. Reg. Meet. Cons. Wildfowl Res, Leningrad 1968*, 88–96. CURTH, P (1954) *Der Mittelsäger*. Wittenberg Lutherstadt.

DAANJE, A (1950) *Behaviour* **3**, 48–99. DABBENE, R (1926) *Hornero* **3**, 311–48. DABELOW, A (1925) *Morph. Jb.* **54**, 288–321. DALGETY, C T and SCOTT, P (1948) *Bull. Br. Orn. Club* **68**, 109–21. DANE, B, DRURY, W, and WALCOTT, C (1959) *Behaviour* **14**, 265–81. DANE, B, and VAN DE KLOOT, W G (1964) *Behaviour* **22**, 282–328. DANE, C W (1968) *J. Wildl. Mgmt.* **32**, 267–74. DARAZSI, J (1959) *Aquila* **65**, 339–40. DARLING, F F (1940) *Island Years*. London. DAVIS, P (1957) *Br. Birds* **50**, 85–101, 371–84. DEACON, G E R (ed) (1962) *Oceans: An Atlas-History of Man's Exploration of the Deep*. London. DEAN, M (1950) *Br. Birds* **43**, 19. DE CHAVIGNY, J and MAYAUD, N (1932) *Alauda* **4**, 304–48. DEICHMANN, H (1909) *Medd. om Grøn.* **29**, 141–56. DEIGNAN, H G (1945) *Bull. US natn. Mus.* **186**. DEKEYSER, P L (1947) *Nature, Paris,* **3127**, 23; (1957) *Bull. Inst. franc. Afr. noire* **19**, sér. A, No. 1, 620–40. DE KORTE, J (1972) *Beaufortia* **19**, 113–50. DELACOUR, J (1947) *Birds of Malaysia*. New York; (1951) *Ardea* **39**, 135–42; (1954) *The Waterfowl of the World* **1**; (1956) **2**; (1959) **3**; (1964) **4**. London. DELACOUR, J and MAYR, E (1945) *Wilson Bull.* **57**, 3–55; (1946) *Wilson Bull.* **58**, 104–10. DELACOUR, J and RIPLEY, S D (1975) *Am. Mus. Novit.* **2565**, 1–4. DE LANNOY, J (1967) *Z. Tierpsychol.* **24**, 162–200. DEMARTIS, A M (1973) *Alauda* **41**, 35–62. DEMENTIEV, G P (1936) *Alauda* **8**, 169–93. DEMENTIEV, G P and GLADKOV, N A (1951) *Ptitsy Sovietskogo Soyuza* **1**, **2**; (1952) **4** Moscow. DE NAUROIS, R (1969a) *Mém. Mus. nat. d'Hist. Nat.* (A) **56**, 1–293; (1969b) *Bull. Inst. fond. Afr. noire* **31**, ser. A, 143–218; (1966) *L'Oiseau* **36**, 89–94. DENNIS, R H (1964) *Wildfowl Trust Ann. Rep.* **15**, 71–4; (1973) *Scott. Birds* **7**, 307–8. DE SCHAUENSEE, R M (1966) *The Species of Birds of South America and their Distribution*. Philadelphia. DES LAURIERS, J R and BRATTSTROM, B H (1965) *Auk* **82**, 639. DE VRIES, V (1939) *Limosa* **12**, 87–98. DE WAARD, S (1937) *Levende Nat.* **41**, 159–60. DEWAR, J M (1924) *The Bird as a Diver*. London; (1942) *Br. Birds* **35**, 224–6. DIAMOND, A W (1971) Ph D Thesis Aberdeen Univ.; (1972) *Ibis*

114, 395–8; (1973) *Condor* **75**, 200–9; (1975) *Ibis* **117**, 302–23. DICKERMAN, R W (1973) *Auk* **90**, 689–91. DICKSON, V. (1942) *J. Bombay nat. Hist. Soc.* **43**, 258–64. DIJKSEN, L J, KONING, F J, and WALMSLEY, J G (1972) *Bull. Int. Waterfowl Res. Bur.* **33**, 17–34. DILLERY, D G (1965) *Auk* **82**, 281. DILLON, O W (1959) *N. Am. Wildl. Conf. Trans.* **24**, 374–82. DIN, N A and ELTRINGHAM, S K (1974a) *Ibis* **116**, 28–43; (1974b) *Ibis* **116**, 477–93. DINSMORE, J J and FFRENCH, R P (1969) *Wilson Bull.* **81**, 460–3. DIXON, C C (1933) *Trans. R. Can. Inst.* **19**, 111–39. DOLBIK, M S (1959) *The Birds of White Russian Polesie*. Minsk. DOLGUSHIN, I A (1960) *Ptitsy Kazakhstana* **1**. Alma–Ata. DOMBROWSKI, R VON (1912) *Ornis Romaniae*. Bucharest. DOMBROWSKI, R VON and LINTIA, D (1955) *Păsările din R. P. R.* **3**. Bucharest. DONKER, J K (1949) *Ardea* **47**, 1–27. DONTCHEV, S (1973) *Proc. Int. Waterfowl Res. Bur. Symposium 1972*, 72–4. DORWARD, D F (1962a) *Ibis* **103b**, 174–220; (1962b) *Ibis* **103b**, 221–34. DOTT, H E M (1973) *Bird Study* **20**, 221–5. DOUDE VAN TROOSTWIJK, W J (1974) *Ardea* **62**, 98–110. DOWNIE, W B (1961) Unpubl. Rep. to Dept. of Agric. and Fish. for Scotland. DRAGESCO, J (1961) *Science et Nature* **47**, 1–4. DRIVER, P M (1960) *J. Arct. Inst. N. America* **13**, 201–4; (1974) *In search of the Eider*. London. DRURY, W H Jr (1961) *Can. Fld. Nat.* **75**, 46–53. DUBOIS, A D (1919) *Auk* **36**, 170–80. DUEBBERT, H F (1966) *Wilson Bull.* **78**, 12–25. DUFFEY, E (1951) *Ibis* **93**, 237–45. DUFFEY, E and SERGEANT, D E (1950) *Ibis* **92**, 554–63. DUHART, F and DESCAMPS, M (1963) *L'Oiseau* **33**, Suppl. 1–106. DUHAUTOIS, L (1974) with CHARMOY, M-C, CHARMOY, F, REYJAL, D, and TROTIGNON, J *Alauda* **42**, 313–27. DUMONT, P G (1973) *Amer. Birds* **27**, 739–40. DUNBAR, M J (1955) *Ice Terminology* in Kimble, G H T and Good, D *Geography of the Northlands*. New York. DUNKER, H (1974) *Norw. J. Zool.* **22**, 15–29; (1975) *Norw. J. Zool.* **23**, 149–64. DUNTHORN, A A (1971) *Bird Study* **18**, 107–12. DUNNET, G M and ANDERSON, A (1961) *Bird Study* **8**, 119–26; (1965) *Scott. Birds* **3**, 219–35. DUNNET, G M, ANDERSON, A, and CORMACK, R M (1963) *Br. Birds* **56**, 2–18. DURANGO, S (1947) *Fauna Flora Upps.* **42**, 185–205, 249–59; (1954) *Ornis fenn.* **31**, 1–18. D'URBAN, W S M and MATHEW, M A (1892) *The Birds of Devon*. London. DURY, G H (1959) *The Face of the Earth*. London. DWIGHT, J (1914) *Auk* **31**, 293–308. DWYER, T J (1974) *Auk* **91**, 375–86; (1975) *Wilson Bull.* 335–43. DWYER, T J, DERRICKSON, S R, and GILMER, D S (1973) *Auk* **90**, 687. DYBBRO, T (1970) *Dansk orn. Foren. Tidsskr.* **64**, 45–69. DZUBIN, A. (1955) *Trans. N. Am. Wildl. and Nat. Resources Conf.* **20**, 278–98; (1957) *Blue Jay* **15**, 10–13; (1969) *Saskatoon Wetlands Seminar. Can. Wildl. Serv. Rep. Series* **6**, 138–60.

EATON, E H (1910) *Birds of New York* **1**, 95. *Mem. N. York State Mus.* **12**. ECKSTEIN, K (1907) *Deutsche Fischereiz*. EDBERG, R (1957) *Vår Fågelvärld* **16**, 137–8. EGGENBERGER, H. (1953) *Vögel der Heimat* **24**, 21–3. EISENMANN, E. (1955) *Trans. Linn. Soc. New York* **7**, 1–128. EISENMANN, E and SERVENTY, D L (1962) *Emu* **62**, 199–201. EISENTRAUT, M (1973) *Bonn. zool. Monogr.* **3**. ELDER, W H and ELDER, N L (1949) *Wilson Bull.* **61**, 113–40. ELGOOD, J H, FRY, C H, and DOWSETT, R J (1973) *Ibis* **115**, 1–45. ELGOOD, J H, SHARLAND, R E, and WARD, P (1966) *Ibis* **108**, 84–116. ELLIOTT, C C H (1970) *Ostrich* Suppl. **8**, 385–96. ELLIOTT, H F I (1957) *Ibis* **99**, 545–86. ELLIOTT, H F I and MONK, J F (1952) *Ibis* **94**, 528–9. ELLIS, J W and FRYE, J R (1966) *J. Wildl. Mgmt* **29**, 396–7. ELSON, P F (1962) *Bull. Fish Res. Bd. Can.* **133**, 1–87. ELTRINGHAM, S K (1973) *Wildfowl* **24**, 81–7. ELTRINGHAM, S K and BOYD, H (1960) *Wildfowl Trust Ann. Rep.* **11**, 107–17; (1963) *Br. Birds* **56**, 433–44. ELY, C A and CLAPP, R B (1973) *Atoll Res. Bull.* **171**. EMEIS, W. (1936) *Schrift d. Naturw. Ver. f. Schleswig-Holstein* **21**, 365–402. EMLEN, J T

(1957) *Ibis* **99**, 352. EMLEN, S T and AMBROSE, H W (1970) *Auk* **87**, 164–5. ENGLAND, M D (1957) *Br. Birds* **50**, 439; (1974) *Br. Birds* **67**, 236–7. ENNION, H E (1962) *Ibis* **104**, 560–2. ERN, H (1970) *Vogelwarte* **25**, 334–6. ERSKINE, A J (1971a) *Ibis* **113**, 42–58; (1971b) *Wildfowl* **22**, 60; (1972) *Buffleheads*. Can. Wildl. Serv. Mon. Ser. **4**. ETCHÉCOPAR, R D and HÜE, F (1967) *The Birds of North Africa*. Edinburgh. EVANS, M E (1971) *Wildfowl* **22**, 140–3; (1975) *Wildfowl* **26**, 117–30. EVANS, M E and LEBRET, T (1973) *Wildfowl* **24**, 61–2. EVANS, M E, WOOD, N A, and KEAR, J (1973) *Wildfowl* **24**, 56–60. EVANS, P R (1966) *J. Zool. Lond.* **150**, 319–69; (1968) *Br. Birds* **61**, 281–303. EVANS, W (1892) *Ann. Scott. Nat. Hist.* **1892**, 74–6; (1909) *Br. Birds* **3**, 165–6. EYGENRAAM, J A (1957) *Ardea* **45**, 117–43.

FABRICIUS, E. (1959) *Wildfowl Trust Ann. Rep.* **10**, 105–113; (1970) *Zool. Revy* **32**, 19–25. FABRICIUS, E, BYLIN, A, FERNÖ, A, and RADESATER, T (1974) *Ornis scand.* **5**, 25–35. FABRICIUS, E and RADESÄTER, T (1972) *Zool. Revy.* **33**, 60–9. FALLA, R A (1934) *Rec. Auckland Inst. Mus.* **1**, 139–54; (1937) *Rep. B.A.N.Z. antarctic res. Exped.* Ser. B. **2**. FARNER, D S, KING, J R, and PARKES, K C (eds) (1971) *Avian biology* **1**; (1972) **2** New York. FERDINAND, L (1975) *Ornis fenn.* **52**, 45–8. FERGUSON-LEES, I J (1959) *Br. Birds* **52**, 185–9. FERNANDEZ CRUZ, M (1970) *Ardeola* **16**, 36–9; (1975) *Ardeola* **21**, 65–126. FERNS, P H and GREEN, G H (1975) *Wildfowl* **26**, 131–8. FERRY, C and BLONDEL, J (1960) *Alauda* **28**, 62–4. FESTETICS, A and LEISLER, B (1970) *Wildfowl* **21**, 42–60. FFRENCH, R (1963) *Auk* **80**, 379. FIALA, V (1974) *Anz. orn. Ges. Bayern* **13**, 198–218. FIGALA, J. (1959) *Sylvia* **16**, 105–12. FIROUZ, E and FERGUSON, D (1970) *Proc. Int. Reg. Meet. Cons. Wildfowl Resources, Leningrad 1968*, 185–8. FISCHER, H (1965) *Z. Tierpsychol.* **22**, 247–304. FISCHER, H I (1965) *Condor* **67**, 355–7. FISHER, J (1952) *The Fulmar*. London; (1966) *Bird Study* **13**, 5–76. FISHER, J and LOCKLEY, R M (1954) *Sea-Birds*. London. FJELDSÅ, J. (1969) *Sterna* **8**, 275–82; (1972) *Sterna* **11**, 145–63; (1973a) *Ornis scand.* **4**, 55–86; (1973b) *Sterna* **12**, 161–217; (1973c) *Vidensk. Medd. dansk. natur. Foren. Kbh.* **136**, 57–95; (1973d) *Vidensk. Medd. dansk. natur. Foren. Kbh.* **136**, 117–89; (1973e) *Feltornithologen* **15**, 85–90. FLEET, R R (1974) *AOU Orn. Monograph* **16**; (1975) *Dansk orn. foren. Tidsskr.* **69**, 89–102. FLOWER, S S (1933) *Ibis* (13) **3**, 34–46. FOCHLER-HAUKE, G. (1959) *Allgemeine Geographie; Das Fischer Lexikon*. Munich. FOG, J (1965) *Dan. Rev. Game Biol.* **4**, 61–95. FOG, J, and JOENSEN, A H (1970) *Proc. Int. Reg. Meet. Cons. Wildfowl Resources, Leningrad 1968*, 112–13. FOG. M (1967) *Dan. Rev. Game Biol.* **5**, 1–40; (1971) *Dan. Rev. Game Biol.* **6** (3) 3–12; (1972) *Rep. Vildtbiologisk Station*, Kalø. FOLEY, D D and TABER, W R (1952) P. R. Proj. 52-R. Final Rep. N.Y. Cons. Dept. FOLK, C (1971) *Acta. Sc. Nat. Brno* **5**, 1–39. FOLK, C, HUDEC, K, and TOUFAR, J (1966) *Zool. Listy* **15**, 249–60. FORMÁNEK, J (1970) *Sylvia* **18**, 135–213. FORSTER, R and WAGNER, G (1973) *Orn. Beob.* **70**, 67–80. FRANÇOIS, J (1975) *Alauda* **43**, 279–93. FRANK, F (1952) *J. Orn.* **93**, 142–3. FRANKE, H (1969) *J. Orn.* **110**, 286–90. FREME, S W P (1931) *Br. Birds* **24**, 369–71. FRIEDMANN, H (1927) *Bull. Mus. comp. Zool. Harv.* **68**, 139–236; (1930) *Bull. US natn. Mus.* **153**. FRIELING, H (1933) *Zoogeographica* **1**, 485–550.

GABRIELSON, I N and LINCOLN, F C (1959) *The Birds of Alaska*. Harrisburg. GALLAGHER, M D (1971) *Gulf Birdwatchers' Newsletter* **19**, 8; (1974) *Bull. Br. Orn. Club* **94**, 122–6. GALLET, E (1950) *The Flamingos of the Camargue*. Oxford. GANDRILLE, G and TROTIGNON, J (1973) *Alauda* **41**, 129–59. GARDARSSON, A (1967) *Proc. 2nd Eur. Meeting Wildfowl Cons. 1966*, 78–80; (1968) *Náttúrufraedingurinn* **38**, 165–75; (1972) *Ibis* **114**, 581; (1975) *Rit Landvernder* **4**. Votlendi, 100–34. GARDARSSON, A and SIGURDSSON, J B (1972) Unpublished Report. Museum of Nat. Hist. Iceland, 269–80. GARDEN, E A (1958) *Bird Study* **5**, 90–109.

GARRIDO, O H and MONTAÑA, F G (1968) *Torreia* 4, 1–13.
GARTLAN, J S (1968) *Folia Primat.* 8, 89–120. GATES, J M (1957)
*Utah Acad. Proc.* 34, 69–71; (1958) MS Thesis. Utah State
Univ.; (1962) *Wilson Bull.* 74, 43–67. GATKE, H (1900) *Vogelwarte
Helgoland.* Braunschweig. GAUCKLER, A and KRAUS, M (1965)
*Vogelwelt* 86, 129–46; (1968) *Anz. orn. Ges. Bayern* 8, 349–64.
GAVRIN, V F (1964) *Trudy Inst. Zool. Acad. Nauk Kazakh. SSR*
24, 5–58; (1970) *Proc. Int. Reg. Meeting Cons. Wildfowl
Resources, Leningrad 1968*, 77–80. GEIGER, W (1957) *Orn. Beob.*
54, 97–133. GENTZ, K (1959) *Falke* 6, 39–47; (1965) *Die Grosse
Dommel.* Wittenberg Lutherstadt. GERAMISOVA, T D and
BARANOVA, Z M (1960) *Proc. Kandalaksha Game Reserve* 3,
55–68. GÉROUDET, P. (1955) *Proc. Int. Orn. Congr.* 11, 72–80;
(1965) *Waterbirds with webbed feet.* London; (1967) *Les
Échassiers.* Neuchâtel; (1974) *Nos Oiseaux* 32, 188–201. GEYR
VON SCHWEPPENBURG, H (1924) *J. Orn.* 72, 472–6; (1953) *J. Orn*
94, 117–27. GIBSON, J D (1967) *Notornis* 14, 47–57. GIBSON-
HILL, C A (1947) *Bull. Raffles Mus.* 18, 87–165; (1951) *J.
Bombay nat. Hist. Soc.* 48, 214–35. GIFFORD, E W (1913) *Proc.
Calif. Acad. Sci* (4) 2, 1–132. GILL, F B (1967) *Proc. US natn.
Mus.* 123 (3605), 1–33. GILLET, H. (1960) *L'Oiseau* 30, 45–82.
GILLHAM, E H (1951) *Br. Birds* 44, 67; (1955) *Br. Birds* 48,
322–3; (1956) *Bird Study* 3, 205–12; (1957) *Br. Birds* 50, 2–10;
(1958) *Br. Birds* 51, 413–26; (1961) *Br. Birds* 54, 375–9.
GILLHAM, E, HARRISON, J M, and HARRISON, J G (1966)
*Wildfowl Trust Ann. Rep.* 17, 49–65. GIOL, A (1957) *Riv. ital.
Orn.* 27, 118–21. GLASGOW, L L and BARDWELL, J L (1962)
*Conf. S. E. Ass. Game and Fish Comm. Proc.* 16, 175–84.
GLAZENER, W C (1946) *J. Wildl. Mgmt* 10, 322–9. GLAUERT, L
(1946) *Emu* 46, 187–92. GLEGG, W E (1943) *Ibis* 85, 82–7.
GLENISTER, A G (1951) *The birds of the Malay Peninsula.*
London. GLUTZ VON BLOTZHEIM, U N (1962) *Die Brutvögel der
Schweiz.* Aarau. GODFREY, W E (1966) *The birds of Canada.*
Ottawa. GODMAN, F DU CANE (1907–10) *A Monograph of the
Petrels.* London. GOETHE, F (1961a) *Br. Birds* 54, 106–14;
(1961b) *Br. Birds* 54, 145–61. GOLDING, F D (1934) *Ibis* (13) 4,
738–57. GOLDSCHMIDT, B DE, and HAFNER, H (1973) *Bull. Int.
Waterfowl Res. Bureau* 35, 38–46. GOOD, A-I (1952) *Mém. Inst.
fr. Afr. noire Cent. Cameroun.* Sér. Sc. nat. 2. GOODMAN, D C
and FISHER, H I (1962) *Functional Anatomy of the Feeding
Apparatus of Waterfowl.* Carbondale. GOODWIN, D (1965)
*Domestic Birds.* London. GORDON, S and GORDON, A S (1928)
*Br. Birds* 22, 2–5. GORMAN, M L (1970) *Wildfowl* 21, 105–7;
(1974) *Ibis* 116, 451–67. GORMAN, M L and MILNE, H (1972)
*Ornis scand.* 3, 21–6. GOTTLIEB, G (1971) *Development of species
identification in birds.* Chicago. GOTTLIEB, G and VANDENBERGH,
J G (1968) *J. Exp. Zool.* 168, 307–26. GOTZMAN, J (1965) *Ekol.
pol.* Ser. A 13, 289–302. GRANT, C H B (1915) *Ibis* (10) 3, 1–76.
GRANT, C H B and MACKWORTH-PRAED, C W (1933) *Bull. Br.
Orn. Club* 53, 189–96. GRANVIK, H (1923) *J. Orn.* 71, Sonderheft.
GRAY, A P (1958) *Bird Hybrids.* London. GRAY, G R (1869) *Ibis*
(2) 5, 438–43. GREENWAY, J C (1958) *Extinct and vanishing birds
of the world.* New York. GREGORY, T C (1947) *Br. Birds* 40, 186.
GRENQUIST, P (1963) *Proc. Int. Orn. Congr.* 13, 685–9; (1970)
*Proc. Int. Reg. Meet. Cons. Wildfowl Resources Leningrad, 1968,*
83–7. GRICE, D and ROGERS, J P (1965) *The Wood Duck in
Massachusetts.* Mass. Division of Fisheries and Game. GRIFFIN,
D R (1965) *Bird Migration.* London. GROEBBELS, F (1935) *J.
Orn.* 83, 525–31; (1942) *Verh. orn. Ges. Bayern* 22, 223–54. GROSS,
A C (1912) *Auk* 29, 49–71. GROSS, A O (1923) *Auk* 40, 191–214;
(1938) *Auk* 55, 387–400; (1947) *Bird-Banding* 18, 117–26. GROSS,
W A O (1935) *Auk* 52, 382–99. GROTE, H (1939) *Orn. Mber.* 47,
170–6; (1950) *Zool. Gart.* 17, 87–90. GRUBB, T C Jr (1970) *Auk*
87, 587–8; (1973) *Auk* 90, 78–82; (1974) *Animal Behaviour* 22,
192–202. GUDMUNDSSON, F (1951) *Proc. Int. Orn. Congr.* 10,

502–4; (1952a) *Náttúrufraedingurinn* 22, 44–5; (1952b) *Náttúrufraedingurinn* 22, 76–7; (1961) In BLAEDEL, N (1961),
*Nordens Fugle* 5. Copenhagen, 220–6, 253–60; (1971) *Náttúrufraedingurinn* 41, 1–28, 64–98; (1972) *Ibis* 114, 582.
GUICHARD, G (1949) *L'Oiseau* 19, 85–91; (1951) *L'Oiseau* 21,
48–54; (1957) *L'Oiseau* 27, 270–6. GUICHARD, K M (1947) *Ibis*
89, 450–89. GUNN, D (1927) *Br. Birds* 20, 193–7; (1939) *Br. Birds*
33, 48–50. GURNEY, J H (1913) *The Gannet.* London. GYLLIN, R
(1965) *Fauna Flora Upps.* 60, 147–58.

HAACK, W and RINGLEBEN, H (1972) *Vogelwarte* 26, 257–76.
HAAPANEN, A, HELMINEN, M, and SUOMALAINEN, H K (1973)
*Paper Finn. Game Res.* 33, 39–60. HAAS, G (1963) *Vogelwarte* 22,
100–9. HAAS, W (1969) *Alauda* 37, 28–36. HACHLER, E M (1959)
*Sylvia* 16, 282–3. HAFNER, H (1975) *Ardeola* 21, 819–25. HAFNER,
H and HOFFMAN, L (1974) *Bull. Int. Waterfowl Res. Bur.* 37, 95–7.
HAFNER, H and WALMSLEY, J G (1971) *Bull. Int. Wildfowl Res.
Bur.* 32, 41–51. HAFTORN, S (1971) *Norges Fugler.* Oslo. HAGEN, Y
(1942) *Arch. Naturgeschichte, Neue Folge* 11, 1–174; (1952) *Results
Norw. scient. Exped. Tristan da Cunha* 20. Oslo. HAGERUP, O
(1926) *Vidensk. Medd. dansk naturh. Foren.* 82, 127–56. HALE,
P E (1948) *Ostrich* 19, 232–4. HALL, A B and ARNOLD, G P (1966)
*Dansk orn. Foren. Tidsskr.* 60, 141–5. HALL-CRAGGS, J H (1974)
*Ibis* 116, 228–31. HAMEL, H D (1975) *Vogelwelt* 96, 213–21.
HAMMOND, M C and MANN, G E (1956) *J. Wildl. Mgmt.* 20,
345–52. HANCOCK, C G and BACON, P J (1970) *Br. Birds* 63,
299–30. HANSON, H C (1951) *Auk* 68, 164–73; (1953) *Auk* 70,
11–16; (1965) *The Giant Canada Goose.* Carbondale. HANSON,
H C and GRIFFITH, R E (1952) *Bird-Banding* 23, 1–22. HANSON,
H C and SMITH, R H (1950) *Bull. Illinois nat. Hist. Surv.* 25,
67–210. HANSSON, L (1966) *Vår Fågelvärld*, Suppl. 4, 95–140.
HANTZSCH, B (1905) *Beitrag zur Kenntnis der Vogelwelt Islands.*
Berlin. HANZÁK, J (1949–50) *Sylvia* 11–12, 85–97; (1952) *Acta.
Mus. nat. Prag.* 8B 1, 1–37; (1962) *Sborník Předásek* (eds Černý,
W and Urbanck, B), 77–88. Prague. HARDY, D E (1967) *Wildfowl
Trust Ann. Rep.* 18, 134–41. HARDY, E (1946) *A Handbook of the
Birds of Palestine.* Palestine. HARLE, D F (1951) *Br. Birds* 44,
287–8. HARLIN, J (1966a) *Zool. Listy* 15, 175–89; (1966b) *Zool.
Listy* 15, 333–44; (1972) *Zool. Listy* 21, 85–95. HARRIS, M P
(1966a) *Ibis* 108, 17–33; (1966b) *Bird Study* 13, 84–95; (1969a)
*Proc. Calif. Acad. Sci.* 37, 95–166; (1969b) *Ardea* 57, 149–57.
HARRIS, M P and HANSEN, L (1974) *Dansk orn. Foren. Tidsskr.* 68,
117–37. HARRISON, C J O (1965) *Behaviour* 24, 161–209; (1967)
*Ibis* 109, 539–51. HARRISON, J G and OGILVIE, M A (1967)
*Wildfowl Trust Ann. Rep.* 18, 85–7. HARRISON, J M (1958) *Bull.
Br. Orn. Club* 78, 105–7. HARRISON, J M and HARRISON, J G
(1966) *Br. Birds* 59, 547–50. HARRISON, J M, HOVEL, H, and
HARRISON, D L (1962) *Bull. Br. Orn. Club* 82, 76. HARRISON, J M
and JOURDAIN, F C R (1919) *Br. Birds* 13, 85–6. HARRISSON, T H
and HOLLOM, P A D (1932) *Br. Birds* 26, 62–92, 102–31, 142–55,
174–95. HARTERT, E (1903–10) *Die Vögel der paläarktischen
Fauna* 1; (1912–21) 2; (1921–2) 3. HARTERT, E and STEINBACHER,
F (1932–8) *Die Vögel der paläarktischen Fauna.* Ergänzungsband.
Berlin. HARTLEY, C H (1934) *Geog. J.* 84, 127–8. HARTLEY, C H
and FISHER, J (1936) *J. Anim. Ecol.* 5, 370–89. HARTLEY, P H T
(1933) *Br. Birds* 27, 82–6; (1937) *Br. Birds* 30, 266–75; (1948) *Ibis*
90, 361–81. HAULING, S R (1953) *Ostrich* 24, 10. HAVERSCHMIDT,
F (1949) *The Life of the White Stork.* Leiden; (1953) *Ibis* 95, 699;
(1970) *Vogelwarte* 25, 229–33. HAWKES, B (1970) *Wildfowl* 21,
87–8. HEARD, W R and CURD, M R (1959) *Proc. Okla. Acad. Sci.*
39, 197–200. HEATH, H (1915) *Condor* 17, 20–41. HEATWOLE, H
(1965) *Anim. Behav.* 13, 79–82. HEBARD, F V (1960) *Wildfowl
Trust Ann. Rep.* 11, 53–4. HECKENROTH, H and VONCKEN, I.
(1970) *Auspicium* 4 (2), 81–99. HEIM DE BALSAC, H and HEIM DE
BALSAC, T (1954) *Alauda* 22, 145–205. HEIM DE BALSAC, H and

MAYAUD, N (1962) *Les oiseaux du Nord-Ouest de l'Afrique*. Paris. HEINROTH, O (1910) *J. Orn.* **58**, 101–56; (1911) *Proc. Int. Orn. Congr.* **5**, 589–702. HEINROTH, O and HEINROTH, K (1954) *Aus dem Leben der Vögel*. Berlin. HEINROTH, O and HEINROTH, M (1926–7) *Die Vögel Mitteleuropas* **2**; (1927–8) **3**. Berlin-Lichterfelde. HEINTZELMAN, D S and NEWBURY, C J (1964) *Wilson Bull.* **76**, 291. HELL, P and SLÁDEK, J (1963) *Sbornik vych. muz.* **4**, 77–91. HELLEBREKERS, W P J and VOOUS, K H (1964) *Limosa* **37**, 5–11. HELLMAYR, C E and CONOVER, B (1948) *Publs. Field Mus. nat. Hist., Zool. Ser.*, **13** I (2). HELMUTH, W T (1929) *Auk* **37**, 255–61. HEMMING, J E (1968) *Wilson Bull.* **80**, 326–7. HENNY, C J (1972) *Wildlife Res. Rep.* **1**, US Fish and Wildlife Service. HENNY, C J and HOLGERSEN, N E (1974) *Wildfowl* **25**, 95–101. HERMANSEN, P (1972) *Dansk orn. Foren. Tidsskr.* **66**, 57–63. HERRERA, C M (1974) *Ardeola* **20**, 287–306. HERROELEN, P (1973) *Ornis Brabant* **57**, 1–4. HESS, E H (1957) *Ann. N.Y. Acad. Sci.* **67**, 724–32. HEUGLIN, T von (1873) *Ornithologie Nordost-Afrika's* **2**. Cassel. HEUSCHER, J (1907) *Schweiz. Fischerei* **15**, 257–62. HEWSON, R (1964) *Br. Birds* **57**, 26–31. HEYDT, J G (1972) *Adena* **2**, 3–6. HILDÉN, O (1964) *Ann. zool. fenn.* **1**, 153–279. HILPRECHT, A (1970) *Höckerschwan, Singschwan, Zwergschwan* (2nd edn). Wittenberg Lutherstadt. HINDE, R A (1956) *Ibis* **98**, 340–69. HINDWOOD, K A (1933) *Emu* **33**, 27–43, 97–102. HIRALDO, F (1971) *Ardeola* **15**, 103–7. HOCHBAUM, H A (1944) *The Canvasback on a Prairie Marsh*. Harrisburg; (1955) *Travels and Traditions of Waterfowl*. Minneapolis. HOFER, J (1915) *Schweiz. Fischerei* **23**, 154–6, 330–2; (1968) *Orn. Beob.* **65**, 129–30; (1969) *Orn. Beob.* **66**, 1–6. HOFFMAN, L (1960) *Orn. Beob.* **57**, 37–50. HOFSTETTER, J, CHMETZ, I, BAUMANN, R, and CROUSAZ, G de (1949) *Nos Oiseaux* **20**, 81–5. HÖGSTRÖM, S and WISS, L-E (1968) *Vår Fågelvärld* **27**, 14–42. HÖHN, E O (1943) *Br. Birds* **37**, 102–7; (1957) *Auk* **74**, 205–6. HOLBØLL, C (1843) *Nat. Tidsskr.* **4**, 361–457. HOLD, T (1970) *Ibis* **112**, 151–72. HOLGERSEN, H (1945) *Scient. Results Norw. Antarct. Exped.* **23**, 1–100; (1957) *Publner Kommander Chr. Christensens Hvalfang-stirius.* **21**; (1960) *Proc. Int. Orn. Congr.* **12**, 310–16; (1963) *Sterna* **5**, 229–75. HOLLOM, P A D (1937) *Br. Birds* **31**, 106–11. HOLLOWAY, C W (1970) *Proc. Conf. on Productivity and Cons. in Northern Circumpolar Lands*, Edmonton, 175–92. HOLMAN, F C (1946) *Ibis* **88**, 232–3. HOLSTEIN, V (1927) *Fiskehejren*. Copenhagen; (1929) *Dansk orn. Foren. Tidsskr.* **22**, 111–18. HOLYOAK, D (1970) *Bull. Br. Orn. Club* **90**, 67–74. HÖLZINGER, J. (1969) *Anz. orn. Ges. Bayern* **8**, 473–509; (1970) *Anz. orn. Ges. Bayern* **9**; 155–69; (1973) *Anz. orn. Ges. Bayern* **12**, 10–4. HOOGERHEIDE, C (1950) *Ardea* **37**, 139–60. HOOGERS, B J and KLUYVER, H N (1959) *Limosa* **32**, 8–13. HOOGERWERF, A (1953) *Limosa* **26**, 20–30. HOPSON, A J (1964) *Bull. Nigeria orn. Soc.* **1** (4), 7–15. HORI, J (1962) *Wildfowl Trust Ann. Rep.* **13**, 173–4; (1963) *Wildfowl Trust Ann. Rep.* **14**, 124–32; (1964a) *Ibis* **106**, 333–60; (1964b) *Wildfowl Trust Ann. Rep.* **15**, 100–3; (1965) *Wildfowl Trust Ann. Rep.* **16**, 58; (1966) *Bird Study* **13**, 297–305; (1969) *Wildfowl* **20**, 5–22. HORNBERGER, F (1967) *Der Weiss-Storch*. Wittenberg Lutherstadt. HØRRING, R (1919) *Danmarks Fauna* **23**, *Fugle* **1**. Copenhagen. HORVÁTH, L (1959a) *Acta zool. Acad. Sci. Hungaricae* **5**, 353–67; (1959b) *Ann. Hist. nat. Mus. nat. hungarici* **51**, 451–81. HOSKING, E J (1939) *Br. Birds* **33**, 170–3. HOVEL, H (1957) *Aquila* **63–4**, 365–6. HOVETTE, C (1972) *Alauda* **40**, 343–52. HOVETTE, C and KOWALSKI, H (1972a) *Alauda* **40**, 397; (1972b) *Bull. Int. Waterfowl Res. Bur.* **34**, 42–58. HOWELL, A H (1932) *Florida Bird Life*. Flor. Dept. of Game and Freshwater Fish. N.Y. HUBER, J (1956) *Orn. Beob.* **53**, 5–9. HUDEC, K (1970) *Proc. Int. Reg. Meet. Cons. Wildfowl Resources, Leningrad 1968*, 135–8; (1971) *Zool. Listy* **20**, 177–94; (1973) *Zool. Listy* **22**, 41–58. HUDEC, K and ČERNÝ, W (1972) *Fauna ČSSR—Ptáci—Aves* **1**. Prague. HUDEC, K and KUX, Z (1971)

*Listy* **20**, 365–76. HUDEC, K and ROOTH, J (1970) *Die Graugans*. Wittenberg Lutherstadt. HUDEC, K and TOUSKOVA, I (1969) *Zool. Listy* **18**, 253–62. HUDSON, M J (1965) *Ibis* **107**, 460–5. HUDSON, M T, PIERCE, G, and TAVERNER, J H (1960) *Br. Birds* **53**, 271–2. HUDSON, R (1972) *Br. Birds* **65**, 424–7; (1976) *Br. Birds* **69**, 132–43. HUDSON, R and PYMAN, G A (1968) *A Guide to the Birds of Essex*. Chelmsford. HUI, P and HUI, M (1968) *Orn. Beob.* **65**, 5–10. HULME, D C (1948) *Br. Birds* **41**, 121. HUMPHREY, P S (1957) *Condor* **59**, 139–40; (1958) *Condor* **60**, 129–35. HUMPHREY, P S and PARKES, K C (1959) *Auk* **76**, 1–31. HUNT, G H (1952) *Br. Birds* **45**, 420. HUNTINGTON, C E (1963) *Proc. Int. Orn. Congr.* **13**, 701–5; (1966) *Proc. Int. Orn. Congr.* **14**, Abstracts, 71–2. HUNTINGTON, C E and BURTT, E H (1972) *Proc. Int. Orn. Congr.* **15**, 653. HUXLEY, J S (1914) *Proc. zool. Soc. Lond.* **1914**, 491–562; (1919) *Br. Birds* **13**, 155–8; (1923) *J. Linn. Soc.* **35**, 253–92; (1924a) *Br. Birds* **18**, 129–34; (1924b) *Br. Birds* **18**, 155–63.

IMPEKOVEN, M (1964) *Orn. Beob.* **61**, 1–34. INGRAM, G C S and SALMON, H M (1941) *Br. Birds* **35**, 22–8. INT. COMM. ZOOL. NOMENCL. (1956a) *Opin. Decl. int. Commn. zool. Nom.* **13**, 1–64 (Opinion 401); (1956b) *Opin. Decl. int. Commn. zool. Nom.* **13**, 119–30 (Opinion 406). IREDALE, T (1914) *Ibis* (10) **2**, 423–36. IREDALE, T and MATHEWS, G M (1913) *Ibis* (10) **1**, 402–52. IRVING, L (1960) *Bull. US natn. Mus.* **217**, 1–409. ISAKOV, Yu A (1970a) *Proc. Int. Reg. Meet. Cons. Wildfowl Resources, Leningrad 1968*, 19–23; (1970b) *ibid.*, 24–45; (1970c) *ibid.*, 239–54; (1972) In Kumari, E (1972) (ed), 127–31. ISAKOV, Yu A and RASPOPOV, M P (1949) *Trudy Darvinsk. Gosud. Zapowednikl.* ISAKOV, Yu A and VOROBIEV, K A (1940) *Taus. Hassan-Kuli Orn. State Reserve*, **1**, 5–159. IVANAUSKAS, T (1964) *Lietovos Paukščiai* **3**. Vilnius.

JACKSON, F J (1938) *The birds of Kenya Colony and the Uganda Protectorate*. London. JACOBY, H, KNOTZSCH, G, and SCHUSTER, S (1970) *Orn. Beob.* **67**, Suppl. 1–260. JACOBY, H and SCHUSTER, S (1972) *Anz. orn. Ges. Bayern* **11**, 176–80. JAMESON, W (1958a) *The Wandering Albatross*. London; (1958b) *Sea Swallow* **11**, 29–30. JARRY, G (1969) *L'Oiseau* **39**, 112–20. JAUCH, G A (1953) *Vögel der Heimat* **23**, 209–13. JAUCH, W A (1952) *Vögel der Heimat* **23**, 69–70. JEHL, J R JR (1971) *Trans. S. Diego Soc. nat. Hist.* **16**, 291–302. JENKIN, P M (1957) *Phil. Trans. R. Soc. London* Ser. B, **240**, 401–93. JENKINS, D (1971) *Scott. Birds* **6**, 251–5. JENKINS, D, MURRAY, M G, and HALL, P (1975) *J. anim. Ecol.* **44**, 201–31. JENSEN, L L (1954) *Dansk orn. Foren. Tidsskr.* **48**, 189–218. JEPSEN, P U (1973) *Dan. Rev. Game Biol.* **8** (6), 1–23. JESPERSEN, P (1930) *Danish 'Dana' Exped. 1920–22*, *Oceanogr. Rep.* **7**, 1–36. JOENSEN, A H (1964) *Dansk orn. Foren. Tidsskr.* **58**, 127–36; (1972) *Dan. Rev. Game Biol.* **6** (8), 1–24; (1973a) *Dan. Rev. Game Biol.* **8** (4), 1–42; (1973b) *Danske Vildtundersøgelser* **20**, 1–36; (1974) *Dan. Rev. Game Biol.* **9**, 1–206. JÖGI, A (1971) *Orn. Mitt.* **23**, 65–7. JOHANSEN, H (1959) *J. Orn.* **100**, 313–36; (1962) *Aquila* **67–68**, 33–38; (1968) *Orn. Jahrb.* **19**, 215–25. JOHANSEN, O (1975) *Sterna* **14**, 1–21. JOHNSEN, S (1937) *Bergens Mus. Arbok* **3**, 1–18. JOHNSGARD, P A (1960) *Wilson Bull.* **72**, 133–55; (1961a) *Bull. Br. Orn. Club* **81** (3), 37–43; (1961b) *Ibis* **103a**, 71–85; (1961c) *Auk* **78**, 3–43; (1964) *Wildfowl Trust Ann. Rep.* **15**, 104–7; (1965) *Handbook of waterfowl behavior*. New York; (1966) *Auk* **83**, 98–110; (1967) *Wildfowl Trust Ann. Rep.* **18**, 98–107; (1968) *Waterfowl. Their Biology and Natural History*. Lincoln, Neb.; (1971) *Wildfowl* **22**, 46–59; (1974) *Wildfowl* **25**, 155–61. JOHNSON, A R (1975) In Kear and Duplaix-Hall (1975), 17–25. JOHNSON, A R and CARP, E (1973) *Bull. Int. Waterfowl Res. Bur.* **35**, 47–57. JOHNSON, A R and HAFNER, H. (1969) *Bull. Int. Wildfowl Res. Bur.* **27–8**, 50–3; (1970) *Wildfowl* **21**, 22–36; (1972) *Bull. Int.*

*Waterfowl Res. Bur.* 33, 51–62. JOHNSON, R A and JOHNSON, H S (1935) *Wilson Bull.* 47, 97–103. JOHNSTON, D W (1963) *Wilson Bull.* 75, 435–46. JOHNSTON, W B (1953) *Can. Fld Nat.* 67, 181. JOHNSTONE, G W (1974) *Emu* 74, 209–18. JOHNSTONE, S T (1957) *Avic. Mag.* 63, 27–8. JONES, R (1951) *Avic. Mag.* 57, 183–4. JONES, R D (1965) *Wildfowl Trust Ann. Rep.* 16, 83–5. JONES, R D and JONES, D M (1966) *Wildfowl Trust Ann. Rep.* 17, 75–8. JOUANIN, C. (1955) *L'Oiseau* 25, 155–61; (1964) *Bolm Mus. munic. Funchal* 18, 142–57; (1970) *Proc. Int. Reg. Meeting Cons. Wildfowl Resources, Leningrad 1968*, 154–60. JOUANIN, C and ROUX, F (1963) *L'Oiseau* 33, 103–6; (1965) *Bolm Mus. munic. Funchal* 19, 16–30; (1966) *Bolm Mus. munic. Funchal* 20, 14–27. JOUANIN, C, ROUX, F, and ZINO, A (1969) *L'Oiseau* 39, 161–75. JOURDAIN, F C R (1922) *Auk* 39, 166–7; (1940) *Ibis* (14) 4, 181. JOURDAIN, F C R (1922) *Auk* 39, 166–7; (1940) *Ibis* (14) 4, 181. JOZEFIK, M (1969–70) *Acta orn. Warsz.* 11, 103–262; 12, 57–102, 394–504. JUNG, K (1968) *J. Orn.* 109, 22–4. JUNOR, F J R (1972) *Ostrich* 43, 193–205. JUST, J (1967) *Dansk orn. Foren. Tidsskr.* 61, 133–7.

KADRY, I (1942) *Bull. zool. Soc. Egypt* 4, 20–6. KAHL, M P (1963) *Auk* 80, 295–300; (1966a) *J. Zool. Lond.* 148, 289–311; (1966b) *Behaviour* 27, 76–106; (1971) *Living Bird* 10, 151–70; (1972a) *Ibis* 114; 15–29; (1972b) *J. Zool. Lond.* 167, 451–61; (1972c) *Z. Tierpsychol.* 30, 225–52; (1975a) In Kear and Duplaix-Hall (1975), 93–102; (1975b) In Kear and Duplaix-Hall (1975), 142–9. KALELA, O. (1940) *Orn. fenn.* 23, 77–98; (1949) *Bird-Banding* 20, 77–103. KÄLLANDER, H, MAWDSLEY, T, NILSSON, L, and WADEN, K (1970) *Br. Birds* 63, 32–3. KÁLMÁN, B (1929–30) *Aquila* 26–7, 211–312. KARPOVITCH, V and KESTER, B (1970) *Proc. Int. Reg. Meet. Cons. Wildfowl Resources, Leningrad 1968*, 58–9. KEAR, J (1961) *Br. Birds* 54, 427–8; (1963a) *Wildfowl Trust Ann. Rep.* 14, 54–65; (1963b) In Atkinson-Willes (1963), 318–38; (1964) *IUCN Publ.* 3, 321–31; (1967) *IUGB Trans.* 7, 615–22; (1968) *Beihefte der Vogelwelt* 1, 93–113; (1970) In Crook J H (ed) *Social Behaviour in Birds and Mammals*, 357–92. London; (1972) In Scott P et al. (1972), 79–124. KEAR, J and DUPLAIX-HALL, N (1975) *Flamingos*. Berkhamsted. KEAR, J and MURTON, R K (1973) *Wildfowl* 24, 141–3. KEITH, D B (1937) *Br. Birds* 31, 66–81. KEITH, L B (1961) *Wildl. Monogr.* 6, 1–88. KEITH, L B and STANISLAWSKI, R P (1960) *J. Wildl. Mgmt.* 24, 95–6. KENDEIGH, S C (1961) *Animal Ecology*. London. KENNARD, F G (1927) *Proc. New Engl. Zool. Club.* 9, 85–93. KENNARD, J H (1975) *Bird-Banding* 46, 55–73. KENNEDY, P G, RUTTLEDGE, R F, and SCROOPE, C F (1954) *The Birds of Ireland*. London. KERBES, R H, OGILVIE, M A, and BOYD, H (1971) *Wildfowl* 22, 5–17. KEVE, A (1960) *Nomenclator Avium Hungariae*. Budapest; (1968) *Zool. Abhandl.* 29, 159–75. KING, B (1960) *Wildfowl Trust Ann. Rep.* 11, 154; (1962) *Wildfowl Trust Ann. Rep.* 13, 171; (1963) *Wildfowl Trust Ann. Rep.* 14, 172; (1967) *Br. Birds* 60, 300; (1971) *Br. Birds* 64, 372; (1972) *Br. Birds* 65, 480–1; (1976) *Br. Birds* 69, 34. KING, B and CURBER, R M (1972) *Br. Birds* 65, 442–3. KING, B and PRYTHERCH, R (1963) *Wildfowl Trust Ann. Rep.* 14, 172. KING, W B (1967) *Seabirds of the Tropical Pacific Ocean*. Washington D.C.; (1974) *Pelagic Studies of Seabirds in the Central and Eastern Pacific Ocean*. Washington D.C. KINLEN, L (1963) *Wildfowl Trust Ann. Rep.* 14, 107–14. KINSKY, F C and FOWLER, J A (1973) *Notornis* 20, 14–20. KIRKMAN, F B and JOURDAIN, F C R (1930) *British Birds*. London. KIRKPATRICK T W (1925) *Bull. Min. Agric. Egypt* 56, 1–28. KISTCHINSKI, A A and FLINT, V E (1974) *Wildfowl* 25, 5–15. KLÍMA, M (1966) *Zool. Listy* 15, 317–32. KLOPMAN, R G (1961) *Mag. Ducks and Geese* 12, 6–9; (1962) *Living Bird* 1, 123–9; (1968) *Behaviour* 30, 287–319. KNOPFLI, W (1935) *Orn. Beob.* 32, 93; (1956) *Die Vögel der Schweiz* 16. Bern. KOCH, J C (1940) *Limosa* 13, 101–20.

KOEMAN, J H, BOTHOF, T, DE VRIES, R, VAN VELZEN-BLAD, H, and VIS, J G (1972) *TNO-nieuws* 27, 561–9. KOENIG, A (1932) *J. Orn.* 80, Sonderheft. KOENIG, O (1952) *J. Orn.* 93, 207–89. KOERSVELD, E VAN (1951) *Proc. Int. Orn. Congr.* 10, 592–4. KOLBE, H (1972) *Die Entenvögel der Welt*. Neudamm. KONING, F J (1973) *Bonn. zool. Beitr.* 24, 219–26. KONING, F J and DIJKSEN, L J (1970) *Bull. Int. Wildfowl Res. Bur.* 29, 28–9; (1971) *Bull. Int. Wildfowl Res. Bur.* 32, 51–67. KOP, P P A M (1971a) *Ardea* 59, 56–60; (1971b) *Levende Nat.* 74, 39–41. KORODI GÁL, J (1961) *Bul. Inst. Cercet. Pisc.* (2), 91–5; (1964) *Aquila* 69–70, 71–82. KORTEGAARD, L (1968) *Dansk orn. Foren. Tidsskr.* 62, 38–67; (1973) *Dansk orn. Foren. Tidsskr.* 67, 3–14. KORTLANDT, A (1938) *Ardea* 27, 1–40; (1940) *Arch. neerl. Zool.* 4, 401–42; (1942) *Ardea* 31, 175–280. KORTRIGHT, F H (1942) *The Ducks, Geese and Swans of North America*. Harrisburg. KOSKIMIES, J (1955) *Proc. Int. Orn. Congr.* 11, 176–9; (1957) *Ann. zool. Soc. Vanamo* 18/9, 1–69. KOSKIMIES, J and LAHTI, L (1964) *Auk* 81, 281–307. KOSKIMIES, J and ROUTAMO, E (1953) *Papers Finn. Game Res.* 10, 1–105. KOSTIN, Y V (1974) *Vest. Zool.* 1, 83–4. KRABBE, T N (1907) *Dansk orn. Foren. Tidsskr.* 1, 98–112. KRAMER, H (1962) *J. Orn.* 103, 401–17. KRETCHMAR, A V and LEONOVITCH, V V (1965) *Falke* 12, 268–72. KRISTOFFERSON, S (1926) *Norsk orn. Tidsskr.* 7, 186. KRITZLER, H (1948) *Condor* 50, 5–15. KRIVENKO, V and KRIVONOSOV, G (1972) In Kumari (1972), 57–63. KRIVONOSOV, G (1970) *Proc. Int. Reg. Meet. Cons. Wildfowl Resources Leningrad. 1968*, 70–2. KROHN, H (1903) *Der Fischreiher und seine Verbreitung in Deutschland*. Leipzig. KUBICHEK, W F (1933) *Iowa St. Coll. J. Sci.* 8, 107–26. KUENEN, P H (1955) *Realms of Water*. London. KUHK, R (1955) *Orn. Beob.* 52, 2–5. KUMARI, E V (1962) *Wildfowl Trust Ann. Rep.* 13, 109–16; (1968) (ed.) *The Common Eider in the USSR*. Tallinn. 23–33; (1971) *Wildfowl* 22, 35–43; (1972) (ed.) *Geese in the USSR*. Tartu; (1974) (ed.) *Estonian Wetlands and their Wildlife*. Tallinn. KUMARI, E V, MICHELSONS, H A, and IVANAUSKAS, T L (1970) *Proc. Int. Reg. Meeting Cons. Wildfowl Resources, Leningrad 1968*, 60–2. KUMERLOEVE, H (1960) *Orn. Mitt.* 12, 221; (1962) *J. Orn.* 103, 389–98; (1963) *Alauda* 31, 110–36, 161–211; (1965) *Vogelwelt* 86, 42–8; (1967a) *Alauda* 35, 194–202; (1967b) *Alauda* 35, 243–66; (1969) *Alauda* 37, 260–1. KUMMER, J (1968) *Beitr. Vogelk.* 14, 180. KUNKEL, P (1974) *Z. Tierpsychol.* 34, 265–307. KUX, Z (1963) *Acta Mus. Moraviae* 48, 167–208. KVET, J and HUDEC, K (1971) *Hydrobiologia* 12, 351–9.

LACK, D (1940) *Condor* 42, 269–86; (1945) *J. anim. Ecol.* 14, 12–16; (1949) *Br. Birds* 42, 74–9; (1954a) *Br. Birds* 47, 111–21; (1954b) *The Natural Regulation of Animal Numbers*. Oxford; (1966a) *Population Studies of Birds*. Oxford; (1966b) *Ibis* 108, 141–3; (1967) *Proc. Int. Orn. Congr.* 14, 3–42; (1968) *Ecological Adaptations for Breeding in Birds*. London. LACK, D and LACK, L (1933) *Br. Birds* 27, 179–99. LADHAMS, D E (1968) *Br. Birds* 61, 27–30. LADHAMS, D E, PRYTHERCH, R J, and SIMMONS, K E L (1967) *Br. Birds* 60, 295–9. LAMBERT, K (1971) *Beitr. Vogelkund* 17, 1–32. LANCASTER, D A (1970) *Living Bird* 9, 167–94. LANG, E M., STUDER-THIERSCH, A, THOMMEN, H, and WACKERNAGEL, H (1962) *Orn. Beob.* 59, 173–6. LAPTHORN, J, GRIFFITHS, R G, and BOURNE, W R P (1970) *Ibis* 112, 260–1. LARRISON, E J and SONNENBERG, K G (1969) *Washington Birds*. Seattle. LA SOCIÉTÉ JERSIAISE (1972) *Birds in Jersey*. Jersey. LA TOUCHE, J D D (1931–4) *A handbook of the birds of eastern China* 2. London. LATTA, W C and SHARKEY, R F (1966) *J. Wildl. Mgmt.* 30, 17–23. LEBEDEVA, M I (1960) *Vogelwarte* 21, 229. LEBRET, T (1948) *Br. Birds* 41, 247; (1950) *Ardea* 38, 1–18; (1952) *Orn. Beob.* 49, 30–1; (1955) *J. Orn.* 96, 43–9; (1958a) *Ardea* 46, 68–72; (1958b) *Ardea* 46, 73–5; (1958c) *Ardea* 46, 75–9; (1961) *Ardea*, 49, 97–158; (1970) *Limosa* 43, 11–30.

LEBRET, T and TIMMERMAN, A (1968) *Limosa* **41**, 2–17. LEHMANN, H (1974) *Jber. naturw. Ver. Wuppertal* **27**, 80–104. LEHTONEN, L (1970) *Ann. zool. fenn.* **7**, 25–60. LEISLER, B (1969) *Egretta* **12**, 1–52. LENSINK, C J (1967) *Murrelet* **48**, 41. LEUZINGER, H (1963) *Orn. Beob.* **60**, 223–6; (1968) *Orn. Beob.* **65**, 1–5; (1972) *Orn. Beob.* **69**, 207–35. LEUZINGER, H and SCHUSTER, S (1973) *Orn. Beob.* **70**, 189–202. LEWINSOHN, C H and FISHELSON, L (1967) *Israel J. Zool.* **16**, 59–68. LEYS, H N, MARBUS, J, and DE WILDE, J J F E (1969) *Levende Nat.* **72**, 11–8. LEYS, H N and DE WILDE, J J F E (1968) *Levende Nat.* **71**, 265–72; (1971) *Limosa* **44**, 133–83. LIEBE, K (1892) *Orn. Monatsschr.* **17**, 321–8. LIGON, J D (1967) *Occas. Pap. Mus. Zool. Univ. Mich.* **651**. LIND, H (1958) *Z. Tierpsychol.* **15**, 99–111; (1959a) *Dansk orn. Foren. Tidsskr.* **53**, 177–219; (1959b) *Behaviour* **14**, 123–35; (1962) *Z. Tierpsychol.* **19**, 607–25. LINDBERG. P (1968) *Zool. Revy* **30**, 83–8. LINDROTH, A (1955) *Rep. Inst. Freshw. Res. Drottning.* **36**, 126–32. LINDROTH, A and BERGSTROM, E (1959) *Rep. Inst. Freshw. Res. Drottning.* **40**, 165–75. LINDUSKA, J P (1964) (ed.) *Waterfowl Tomorrow.* Washington. LINKOLA, P (1962) *Suom. Riista* **15**, 157–74. LINSELL, S E (1969) *Wildfowl* **20**, 75–77. LIPPENS, L (1959) *Gerfaut* **49**, 357–61. LIPPENS, L and WILLE, H (1953) *Gerfaut* **43**, 165–8; (1969) *Gerfaut* **59**, 123–56; (1972) *Atlas des Oiseaux de Belgique et d'Europe occidentale.* Tielt. LISTER, M (1962) *A Glossary for Bird Watchers.* London. LITTLER, F M (1910) *A handbook of the birds of Tasmania.* Launceston. LIVERSIDGE, R (1959) *Ostrich* suppl. **3**, 47–67. LOCKLEY, R M (1936) *Ibis* (13) **6**, 712–18; (1942) *Shearwaters.* London; (1952) *Ibis* **94**, 144–57; (1953) *Br. Birds* **46**, suppl. 1–48; (1959) In Bannerman (1959), 92–100. LOCKNER, F R and PHILLIPS, R E (1969) *Behaviour* **35**, 281–7. LOMONT, H (1954) *Terre et Vie* **101**, 28–38. LÖNNBERG, E (1924) *Fauna Flora Upps.* **19**, 97–119. LOOSJES, M (1974) *Limosa* **74**, 121–43. LØPPENTHIN, B (1951) *Proc. Int. Orn. Congr.* **10**, 603–10. LORENZ, K (1934) *J. Orn.* **82**, 160–1; (1935) *J. Orn.* **83**, 137–213, 289–413; (1938) *Proc. Int. Orn. Congr.* **8**, 207–18; (1941) *J. Orn.* **194**–293; (1953) *Comparative studies on the behaviour of the Anatinae.* London; (1966) *On aggression.* London. LOUW, G N (1972) *Symp. zool. Soc. Lond.* **31**, 297–314. LOVEGROVE, R (1971) *Ibis* **113**, 269–72. LØVENSKIOLD, H L (1954) *Norsk. Polarinstit. Skrifter,* **103**; (1964) *Avifauna Svaldbardensis, Norsk. Polarinstit. Skrifter,* **129**. LOW, J B (1941) *Auk* **58**, 506–17. LOWE, F A (1954) *The Heron.* London. LOWE, P R (1911) *A naturalist on desert islands.* London. LOWE, P R and KINNEAR, N B (1930) *Nat. Hist. Rep. Br. antarct. Terra Nova Exped.,* Zool. **4** (5), 103–93. LUCANUS, F VON (1914) *J. Orn.* **62**, 49–56. LUGOVOY, A E (1961) *Trudy Astrakh. zapov.* **5**, 211–9. LUMSDEN, W H R and HADDOW, A J (1946) *J. anim. Ecol.* **15**, 35–42. LUND, H. M-K (1962) *Norg. Jeger og Fiskerforb* **91**, 440–2; (1971) *Sterna* **10**, 247–50. LUNDEVALL, C-F (1953) *Vår Fågelvärld* **12**, 1–8. LUTHER, H and RZOSKA, J (1971) *Project Aqua. IBP Handbook* **21**. Oxford. LYNES, H (1925) *Ibis* (12) **1**, 71–131, 344–416, 541–90, 757–97. LYNN-ALLEN, E (1954) *Bird Notes* **26**, 11–16.

MAAG, A (1917) *Schweiz Fischerei* **25**, 72–4. MABBOT, D C (1920) *US Dept. Agric. Bull.* **862**. MCALLISTER, N M (1958) *Auk* **75**, 290–311. MACAN, T T and WORTHINGTON, E B (1959) *Life in Lakes and Rivers.* London. MCATEE, W L (1922) *Auk* **39**, 380–6. MCATEE, W L and BEAL, F E L (1912) *US Dept. Agric. Farmers Bull.* **497**. MCCARTAN, L and SIMMONS, K E L (1956) *Ibis* **98**, 370–8. MCCRIMMON, D A (1974) *Wilson Bull.* **86**, 165–7. MCGILL, W (1970) *Bokmakierie* **22**, 62. MACGILLIVRAY, W (1852) *A History of British Birds* **4, 5**. London. MCILHENNY, E A (1934) *Auk* **51**, 328–37. MACKENZIE, R H (1910) *Ibis* (9) **4**, 566. MCKINNEY, F (1953) Ph D Thesis. Bristol; (1959) *Wildfowl Trust Ann. Rep.* **10**, 133–40; (1961) *Behaviour Suppl.* **7**; (1965a)

*Wildfowl Trust Ann. Rep.* **16**, 92–106; (1965b) *Behaviour* **25**, 120–220; (1965c) *Wilson Bull.* **77**, 112–21; (1965d) *Condor* **67**, 273–90; (1967) *Wildfowl Trust Ann. Rep.* **18**, 108–21; (1969) In Hafez, E S (ed.) *The behaviour of domestic animals* 593–626. London. (1970) *Living Bird* **9**, 29–64, (1973) In Farner, D S (ed.) *Breeding Biology of Birds,* 6–21. Washington. MACKWORTH-PRAED, C W and GRANT, C H B (1952) *Birds of eastern and north eastern Africa.* **1**. London; (1962) *Birds of the southern third of Africa* **1**. London; (1970) *Birds of west central and western Africa* **1**. London. MCLACHLAN, G R and LIVERSIDGE, R (1957) *Roberts' Birds of South Africa.* Johannesburg. MACLAREN, P I R (1946) *J. Bombay nat. Hist. Soc.* **46**, 543–5. MACMILLAN, A T (1970) *Scott. Birds* **6**, 62–128. MCNEIL, R and BURTON, J (1971) *Auk* **88**, 671–2. MACPHERSON, H A (1887) *Zoologist* (3) **11**, 271; (1892) *A Vertebrate Fauna of Lakeland.* Edinburgh. MCVAUGH, W JR (1972) *Living Bird* **11**, 155–73. MADON, P (1926) *Rev. française Orn.* **18**, 108–30; (1931) *Alauda* **3**, 264–310; (1935) *Alauda* **7**, 177–97, 546–68. MADSEN, C (1925) *Dansk orn. Foren. Tidsskr.* **19**, 33–41. MADSEN, F J (1954) *Dan. Rev. Game Biol.* **2**, 157–266; (1957) *Dan. Rev. Game Biol.* **3**, 19–83. MADSEN, F J and SPÄRCK, R (1950) *Dan. Rev. Game Biol.* **1**, 45–70. MAJIĆ, J and MIKUSKA, J (1972) *Larus* **24**, 65–77. MAKATSCH, W (1950) *Die Vogelwelt Macedoniens.* Leipzig; (1952) *Die Vögel der Seen und Teiche.* Radebeul. MALBRANT, R (1954) *L'Oiseau* **24**, 1–47. MALLETT, G E and COGHLAN, L J (1964) *Ibis* **106**, 123–5. MALTBY-PREVETT, L S, BOYD, H, and HEYLAND, J D (1975) *Bird-Banding* **46**, 155–61. MALZY, P (1962) *L'Oiseau* **32**, suppl. 1–81; (1967) *L'Oiseau* **37**, 122–42. MANNICHE, A L V (1910) *Meddr Grøn.* **45**, 1–200. MARCHANT, S (1941) *Ibis* **83**, 265–95; (1963a) *Ibis* **105**, 369–98; (1963b) *Ibis* **105**, 516–57. MARKGREN, G (1963) *Acta vertebr.* **2**, 295–418. MARKUZE, V K (1965) *Ornithologia* **7**, 244–57. MARIÁN, M (1970) *Vogelwarte,* **25**, 255–7. MARPLES, G (1929) *Br. Birds* **23**, 45. MARQUARDT, R E (1962) *J. Wildl. Mgmt.* **26**, 96–7. MARRIS, R and OGILVIE, M A (1962) *Wildfowl Trust Ann. Rep.* **13**, 53–64. MARSHALL, A J (1952) *Ibis* **94**, 310–33. MARTIN, A C, ZIM, H S, and NELSON, A L (1951) *American Wildlife and Plants.* New York. MASSA, B (1974) *Riv. ital. Orn.* **44**, 210–12. MATHEWS, G M (1912) *The Birds of Australia* **2**. London. MATHIASSON, S (1963a) *Acta Univ. Lund N F Avd. 2.* **58** (13), 1–19; (1963b) *Acta vertebr.* **2** (3), 419–533; (1963c) *Vår Fågelvärld* **22**, 271–89; (1964) *Göteborgs Naturhist. Mus. Årstryck,* 15–19; (1970) *Br. Birds* **63**, 414–24; (1972) *Proc. Int. Orn. Congr.* **15**, 667; (1973a) *Viltrevy* **8**, 400–52; (1973b) *Wildfowl* **24**, 43–93. MATTHEWS, G V T (1954) *Ibis* **96**, 432–40; (1968) *Bird Navigation.* Cambridge; (1972) In Scott P et al. (1972), 181–95; (1974) In *Encyclopaedia Britannica* 938–47. MATTHEWS, G V T and CAMPBELL, C R G (1969) *Wildfowl Trust Ann. Rep.* **20**, 86–93. MATTHEWS, G V T and EVANS, M E (1974) *Wildfowl* **25**, 56–66. MATVEJEV, S D and VASIĆ, V F (1973) *Catalogus Faunae Jugoslaviae—Aves.* Lubljana. MAYAUD, N (1931) *Alauda* **3**, 230–49; (1932) *Alauda* **4**, 41–78; (1936) *Inventaire des Oiseaux de France.* Paris; (1941) *L'Oiseau* **11**, No. spéc. 44–6; (1949–50) *Alauda* **17–18**, 144–55, 222–33; (1966) *Alauda* **34**, 191–9. MAYO, A L W (1948) *Ibis* **90**, 22–5. MAYO, L M (1931) *Emu* **31**, 73. MAYR, E (1969) *Principles of systematic zoology.* New York. MAYR, E and SHORT, L L (1970) *Species taxa of North American birds.* Publs Nuttall orn. Club **9**. Cambridge, Mass. MAZZEO, R (1953) *Auk* **70**, 200. MAZZUCHI, L (1971) *Orn. Beob.* **68**, 161–78. MEADE-WALDO, E G (1890) *Ibis* (6) **2**, 429–38. MEBS, T (1972) *Anthus* **9**, 16–18. MEES, G F (1949) *Br. Birds* **42**, 249; (1950) *Br. Birds* **43**, 302; (1971) *Zoöl. Meded. Leiden* **45**, 225–44. MEINERTZHAGEN, R (1930) *Nicoll's Birds of Egypt.* London; (1935) *Ibis* (13) **5**, 110–51; (1941) *Ibis* (14) **5**, 105–17; (1950) *Bull. Br. Orn. Club* **70**, 8; (1954) *Birds of Arabia.* Edinburgh; (1959) *Pirates and Predators.* Edinburgh. MEISCHNER, I (1958) *Zoolo-*

gische Garten (NF) 25, 104–26; (1962) Falke 4, 34–9. MELDE, M (1973) Der Haubentaucher. Wittenberg Lutherstadt. MENDALL, H L (1958) Univ. Maine Bull. 60 (16), 1–317. MERIKALLIO, E (1958) Fauna Fenn. 5, 1–181. MERNE, O J (1970) Irish Bird Rep. 1969, 12–7; (1972) Br. Birds 65, 394–5. METCALF, W G (1966) Ibis 108, 138–40. MEYBOHM, E and DAHMS, G (1975) Vogelwarte 28, 44–60. MEYERRIECKS, A J (1960) Comparative breeding behavior of four species of North American herons. Publs Nuttall orn. Club 2; (1966) Auk 83, 683–4. MIDDLEMISS, E (1955) Ostrich 26, 40–1; (1961) Bokmakierie 13, 9–14. MIKUSKA, J (1973) Larus 25, 55–60. MILLAIS, J G (1902) The natural history of British surface-feeding ducks. London. (1913); The natural history of British diving ducks 1, 2. London. MILLER, L (1937) Condor 39, 44. MILLER, R C (1966) The Sea. London. MILLER, W T (1946) Ostrich 17, 181–7. MILLS, D H (1962a) Sci. Invest. Freshw. Fish Scot. 29; (1962b) Wildfowl Trust Ann. Rep. 13, 79–92; (1965) Freshwater and Salmon Fish Res. 35: 1–16; (1969) Scott. Birds 5, 264–8, 268–76. MILNE, H (1965a) Bird Study 12, 170–80; (1965b) Scott. Birds 3, 221–5; (1974) Ibis 116, 135–54. MILNE H and CAMPBELL, L H (1973) Bird Study 20, 153–72. MILSTEIN, P LE S (1968) Ostrich Suppl. 6, 197–215; (1969) Ostrich 40, 215. MILSTEIN, P LE S, PRESTT, I, and BELL, A A (1970) Ardea 58, 171–257. MINTON, C D T (1968) Wildfowl 19, 41–60; (1971) Wildfowl 22, 71–88. MITCHELL, F S (1892) Birds of Lancashire. London. MODHA, M L (1967) E. Afr. Wildl. J. 5, 74–95. MOISAN G and SCHERRER, B (1973) Terre et Vie 27, 414–34. MOLTONI, E (1936) Riv. ital. Orn. 6, 109–48, 211–69; (1942) Riv. ital Orn. 12, 46–61; (1948) Riv. ital. Orn. 18, 87–93. MONTAGUE, F A (1925) Br. Birds 19, 139. MOORE, H J AND BOSWELL, C (1956) Field Observations on the birds of Iraq. Iraq Nat. Hist. Mus. Publ. 9. MOORE, W G (1974) A Dictionary of Geography. London. MOREAU, R E (1950) Ibis 92, 419–33; (1966) The Bird Faunas of Africa and its Islands. London; (1967) Ibis 109, 232–59; (1972) The Palaearctic-African Bird Migration Systems. London. MOREL, G and ROUX, F (1962) L'Oiseau 32, 28–56; (1966) Terre et Vie 20, 19–72, 143–172; (1973) Terre et Vie 27, 523–50. MORRIS, B R (1891) British Game Birds and Wildfowl. London. MORRIS, D (1954) Behaviour 6, 271–322; (1956) Behaviour 9, 75–113. MÖRZER BRUYNS, M F, PHILLIPONA, J, and TIMMERMAN, A. (1969) Unpublished Rep. of Goose Res. Grp of Int. Waterfowl Res. Bur.; (1973) Rev. Rep. of Goose Res. Grp of Int. Waterfowl Res. Bur. MÖRZER BRUYNS, M F and TANIS, J (1955) Ardea 43, 261–71. MÖRZER BRUYNS, M F and TIMMERMAN, A (1968) Limosa 41, 90–106. MÖRZER BRUYNS, W F J (1967) Ardea 55, 144–5. MOUGIN, J-L (1967) L'Oiseau 37, 57–103; (1968) L'Oiseau 38, no. spécial, 1–52. MOULE, G W H (1953) Br. Birds 46, 258. MOUNTFORT, G (1958) Portrait of a Wilderness. London; (1962) Br. Birds 55, 475–8. MOUNTFORT, G and FERGUSON-LEES, I J (1961) Ibis 103a, 443–71. MOURASHKA, I P and VALIUS, M I (1961) Proc. Baltic orn. Congr. 4, 71–80. MUKHERJEE, A K (1971) J. Bombay nat. Hist. Soc. 68, 37–64, 691–716. MUMFORD, R E (1962) Wilson Bull. 74, 288–9. MUNRO, J A (1939) Trans. R. Can. Inst. 22, 259–318; (1941a) Can. J. Res. 19D, 113–38; (1941b) The Grebes. Studies of waterfowl in British Columbia. Occ. Pap. Br. Columb. prov. Mus. 3, 1–71; (1944) Trans. R. Can. Inst. 22D, 60–86; (1945) Auk 62, 38–49; (1949) Can. J. Res. 27D, 149–78. MUNRO, J A and CLEMENS, W A (1937) Bull. Fish. Res. Bd. Can. 55, 1–50; (1939) J. Wildl. Mgmt 3, 46–53. MUNRO, R T (1967) Can. Fld. Nat. 81, 151–2. MUNTEANU, D (1967) Ocrot. Nat. 11, 235–40. MURIE, A (1963) Birds of Mount McKinley. Nature Park Alaska 1963. MURPHY, R C (1924) Bull. Am. Mus. nat. Hist. 50, 211–78; (1936) Oceanic birds of South America. New York; (1952) Am. Mus. Novit. 1586; (1958) US Fish and Wildl. Serv. Spec. Sci. Rep.—Fisheries 279; (1967) Distribution of North Atlantic Pelagic Birds. Ser. Atlas Mar. Env. 14, Amer. Geog. Soc. New York. MURPHY, R C and

CHAPIN, J P (1929) Am. Mus. Novit. 384. MURPHY, R C and IRVING, S (1951) Am. Mus. Novit. 1506, 1–17. MURPHY, R C and SNYDER, J P (1952) Am. Mus. Novit. 1596, 1–16. MURRAY, B G (1971) Ecology 52, 414–23. MYLNE, C K (1954) Br. Birds 47, 395. MYRBERGET, S, JOHANSEN, V and STORJORD, O (1969) Fauna 22, 15–26. MYRES, M T (1957) MA thesis, Univ. British Columbia; (1959a) Wilson Bull. 71, 159–68; (2959b) PhD thesis, Univ. British Columbia.

NAUMANN, J A (1838) Naturgeschichte der Vögel Deutschlands 9; (1842) 11 Leipzig; (1903) Naturgeschichte der Vögel Mitteleuropas 12; (1905) 10. Gera. NEDZINSKAS, V (1973) Zool. Zhurn. 52, 1360–6. NELSON, E W (1899) Birds of the Tres Marias Islands. North American Fauna 14, 21. NELSON, J B (1964a) Ibis 106, 63–77; (1964b) Scott. Birds 3, 99–137; (1965) Br. Birds 58, 233–88, 313–36; (1966a) Br. Birds 59, 393–419; (1966b) Ibis 108, 584–626; (1966c) J. anim. Ecol. 35, 443–70; (1967a) Ardea 55, 60–90; (1967b) Nature, London 214, 318; (1968) Galapagos. Islands of birds. London; (1970) Oceanogr. Mar. Biol. Ann. Rev. 8, 501–74; (1971) Ibis 113, 429–67. NETHERSOLE-THOMPSON, C and NETHERSOLE-THOMPSON, D (1942) Br. Birds 35, 160–9, 190–200, 214–23, 241–50. NETTLESHIP, D N (1975) Can. Fld. Nat. 89, 125–33. NEUMANN, O (1928) J. Orn. 76, 783–4. NEWELL, R G (1968) Br. Birds 61, 145–59. NEWSTEAD, R A and COWARD, T A (1908) Br. Birds 2, 14–7. NEWTON, A (1852) Zoologist 10, 3691–8. NEWTON, I and CAMPBELL, C R G (1970) Scott. Birds 6, 5–18. NEWTON, I, THOM, V M, and BROTHERSTON, W (1973) Wildfowl 24, 111–21. NICE, M M (1941) Amer. Midland Nat. 26, 441–87. NICHOLSON, E M (1929) Br. Birds 22, 269–323, 333–72; (1930) Ibis (12) 6, 280–313, 395–428; (1952) Br. Birds 45, 41–55; (1973) In Lovejoy, D (ed.) Land Use and Landscape Planning, 287–97. London. NICHOLSON, E M and DOUGLAS, G L (1971) Int. Biol. Prog. CT Progr. Rep. 1971. London. NICOLL, M J (1919) Handlist of the Birds of Egypt. Cairo. NIETHAMMER, G (1938) Handbuch der Deutschen Vogelkunde 2. Leipzig. NIETHAMMER, J (1964) J. Orn. 105, 389–426. NILSSON, L (1965a) Vår Fågelvärld 24, 244–256; (1965b) Vår Fågelvärld 24, 301–9; (1966) Vår Fågelvärld 25, 148–60; (1969a) Wildfowl 20, 112–18; (1969b) Vår Fågelvärld 28, 199–210; (1969c) Oikos 20, 128–35; (1970) Oikos 21, 145–54; (1971) Vår Fågelvärld 30, 180–4; (1973) Vår Fågelvärld 32, 115–19; (1974) Wildfowl 25, 84–8. NISBET, I C T (1959) Br. Birds 52, 393–416; (1968) Ibis 110, 348–52. NISBET, I C T and SMOUT, T C (1957) Br. Birds 50, 201–4. NOBLE, G K (1939) Auk 56, 263–73. NOBLE, G K, WURM, M, and SCHMIDT, A (1938) Auk 55, 7–40; (1942) Auk 59, 205–24. NOLL, H and SCHMALZ, J (1935) Orn. Beob. 32, 102–5. NORDERHAUG, M (1970a) Årbok norsk Polarinst. 1968, 24–35; (1970b) Årbok norsk Polarinst 1969, 55–69; (1970c) Årbok norsk Polarinst. 1968, 7–23. NORDERHAUG, M, OGILVIE, M A, and TAYLOR, R J F (1965) Wildfowl Trust Ann. Rep. 16, 106–10. NORTH, M E W (1945) Ibis 87, 469–70; (1950) Ibis 92, 99–114. NYHOLM, E S (1965) Ann. zool. fenn. 2, 197–207.

OELKE, H (1970) Vogelwelt 91, 107–11. OGILVIE, F M (1892) Zoologist (3) 16, 392–98. OGILVIE, M A (1962) Wildfowl Trust Ann. Rep. 13, 65–9; (1964) Wildfowl Trust Ann. Rep. 15, 84–8; (1966) Wildfowl Trust Ann. Rep. 17, 27–9; (1967) Wildfowl Trust Ann. Rep. 18, 64–73; (1969a) Wildfowl 20, 79–85; (1969b) Br. Birds 62, 505–22; (1970) Proc. Int. Reg. Meet. Cons. Wildfowl Resources, Leningrad 1968, 108–9; (1972) In Scott, P et al. (1972) 29–55. OGILVIE, M A and BOYD, H (1975) Wildfowl 26, 139–47. OGILVIE, M A and MATTHEWS, G V T (1969) Wildfowl 20, 119–25. OGILVIE, M A and TAYLOR, R J F (1969) Ibis 109, 299–309. OGILVIE-GRANT, W R (1896) Ibis (7) 2, 41–55; (1898) Cat. Birds Br. Mus. 26, 329–653. London; (1899) Bull. Liverpool

*Mus.* **2**, 2–3. OGILVIE-GRANT, W R and FORBES, H O (1903) *The Natural History of Sokotra and Abd-al-Kuri–Aves.* Liverpool. OLDHAM, C (1928) *Br. Birds* **22**, 214–15. OLIVER, P J (1971) *Br. Birds* **64**, 56–60, 322–6. OLNEY, P J S (1958) *Wildfowl Trust Ann. Rep.* **9**, 47–51; (1962) *Wildfowl Trust Ann. Rep.* **13**, 119–25; (1963a) *Ibis* **105**, 55–62; (1963b) *Proc. zool. Soc. Lond.* **140**, 169–210; (1964) *Proc. zool. Soc. Lond.* **142**, 397–418; (1965a) *Ibis* **107**, 527–32; (1965b) *Int. Union Game Biologists Trans.* **6**, 309–22; (1967) *Wildfowl Trust Ann. Rep.* **18**, 47–55; (1968) *Biol. Cons.* **1**, 71–6. OLNEY, P J S and MILLS, D H (1963) *Ibis* **105**, 293–300. OLROG, C C (1963) *Op. lilloana* **9**. OLSEN, B. and PERMIN, M (1974) *Dansk orn. Foren. Tidsskr.* **68**, 39–42. OLSON, S T and MARSHALL, W H (1952) *Occ. Pap. Univ. Minn. Mus. nat. Hist.* **5**. OLSSON, V (1958) *Acta vertebr.* **1**, 91–189; (1963) *Acta vertebr.* **2**, 256–64. ONNO, S (1958) Thesis (Akad. Nauk Estn. SSR, Inst. Zool. 1. Bot Tartu); (1960) *Proc. Int. Orn. Congr.* **12**, 577–82; (1965) *Wildfowl Trust Ann. Rep.* **16**, 110–14; (1970) *Waterfowl in Estonia.* Tallinn. 18–46; (1971) *Beitr. Vogelk.* **17**, 339–48. OREEL, G J and VOOUS, K H (1974) *Ardea* **62**, 130–2. ORIANS, G H (1969) *Amer. Nat.* **103**, 589–603; (1971) In Farner and King **1**, 513–46. ORIANS, G H and WILLSON, M F (1964) *Ecology* **45**, 159–66. ORING, L W (1968) *Auk* **85**, 355–80; (1969) *Wilson Bull.* **81**, 44–54. ORLANDO, C (1958) *Riv. ital. Orn.* **28**, 101–13. OTTOSSON, O (1908) *Arkiv. for Zool.* **4** (9), 1–4. OUSTALET, E (1891) *Mission Scientifique du Cap Horn 1882–83*, Zool. **6**, 1–341. Paris. OUWENEEL, G L (1970) *Limosa* **43**, 159. OWEN, D F (1955) *Ibis* **97**, 276–95; (1959) *Ardea* **47**, 187–91; (1960) *Proc. zool. Soc. Lond.* **133**, 597–617. OWEN, D F and PHILLIPS, G C (1956) *Br. Birds* **49**, 494–9. OWEN, M (1971) *J. appl. Ecol.* **8**, 905–17; (1972a) *J. anim. Ecol.* **41**, 79–92; (1972b) *J. appl. Ecol.* **9**, 385–98; (1973a) *Wildfowl* **24**, 123–30; (1973b) *Ibis* **115**, 227–43. OWEN, M and CADBURY, C J (1975) *Wildfowl* **26**, 31–42. OWEN, M and KEAR, J (1972) In Scott P *et al.* (1972), 57–77. OWEN, M and KERBES, R H (1971) *Wildfowl* **22**, 114–19. OWRE, O T (1967a) *J. E. Afr. nat. Hist. Soc. natn. Mus.* **26**, 61–3; (1967b) *Orn. Monogr.* **6**.

PAIGE, J P (1948) *Br. Birds* **41**, 350; (1960) *Ibis* **102**, 520–25. PALA, S (1971) *Tabiat ve Insan* **5**, 14–22. PALMER, R S (1949) *Bull. Mus. comp. Zool. Harv* **102**; (1962) *Handbook of North American birds* **1**. New Haven; (1972) In Farner *et al.* **2**, 65–102; (1973) *Wildfowl* **24**, 154–7; (1974) *Notornis* **21**, 121–3. PALUDAN, K (1936) *Vidensk. Medd. dansk naturh. Foren.* **100**, 247–346; (1962) *Danske Vildt.* **10**, 1–86; (1965) *Danske Vildt.* **12**, 1–54; (1973) *Vidensk. Meddr. dansk naturh. Foren.* **136**, 217–32. PALUDAN, K and FOG, J (1956) *Danske Vildt.* **5**, 1–46. PARENT, G N (1973) *Aves* **10**, 70–122. PARKES, K C (1952) *Condor* **54**, 314–15; (1974) *Notornis* **21**, 121–3. PARSLOW, J L F (1967) *Br. Birds* **60**, 2–47; (1973a) *Breeding Birds of Britain and Ireland.* Berkhamsted; (1973b) *Bull. Br. Orn. Club.* **93**, 163–6. PASHBY, B S and CUDWORTH, J (1969) *Br. Birds* **62**, 97–109. PATEFF, P (1950) *Ptitzite w Bulgarija.* Sofia. PATTERSON, I J, YOUNG, C M, and TOMPA, F S (1974) *Wildfowl* **25**, 16–28. PAULSEN, D R (1969) *Auk* **86**, 759. PAYEVSKI, V A (1971) *Trudy Zool. Inst. Leningrad* **50**, 1–110. PAYNE, R B (1974) *Bull. Br. Orn. Club.* **94**, 81–8. PEARSON, D (1969) In *Birds of the World* I. London. PEARSON, H J and PEARSON, C E (1895) *Ibis* (7) **1**, 237–49. PEARSON, T H (1968) *J. anim. Ecol.* **37**, 521–52. PEDERSEN, A (1934) *Polardyr.* Copenhagen; (1942) *Medd. om. Grøn.* **128** (2), 51–116. PEHRSSON, O (1965) *Vår Fågelvärld* **24**, 107–32. PEITZMEIER, J (1969) *Abh. Landesmus. Naturk. Münster* **31** (3), 1–480. PENNYCUICK, C J and WEBBE, D (1959) *Br. Birds* **52**, 321–32. PERCY, W (1951) *Three Studies in Bird Character.* London; (1958) In Bannerman (1958) **7**, 168–70. PERRINS, C M (1967) *Skokholm Bird Obs. Rep.* 1967, 23–5. PERRINS, C M, HARRIS, M P, and BRITTON, C K (1973) *Ibis* **115**, 535–48. PERRINS, C M and REYNOLDS, C M (1967) *Wildfowl*

*Trust Ann. Rep.* **18**, 74–84. PETERKEN, G F (1967) *Guide to the Checksheet for IBP areas.* Oxford. PETERS, J L (1931) *Check-list of birds of the world* **1**. Cambridge, Mass. PETERS, N (1931) *Abh. naturw. Ver. Hamburg* **23**; (1934) *Orn. Monatsberichte* **42**, 174. PETERSEN, W R (1971) *Manomet Bird Obs. Res. Rep.* **1**. PETERSON, H (1955) *Svensk Fisk-Tidsskr.* **65**, 99–101. PETERSON, R T (1947) *A Field Guide to the Birds.* Boston. PETERSON, R T, MOUNTFORT, G, and HOLLOM, P A D (1954) *A Field Guide to the Birds of Britain and Europe.* London. PÉTÉTIN, M and TROTIGNON, J (1972) *Alauda* **40**, 195–213. PETHON, P (1967) *Nytt. Mag. Zool.* **15**, 97–111. PETTINGILL, O S Jr. (1959) *Wilson Bull.* **71**, 205–7; (1962) *Wilson Bull.* **74**, 100–1. PHILIPPONA, J (1966). *Wildfowl Trust Ann. Rep.* **17**, 95–7. PHILIPPONA, J and MULDER, T (1960) *Limosa* **33**, 90–127. PHILIPSON, W R and DONCASTER, C C (1951) *Br. Birds* **45**, 11–13. PHILLIPS, J C (1923) *A Natural History of the Ducks* **2**; (1925) **3**; (1926) **4**. Boston. PHILLIPS, J H (1963a) *Ibis* **105**, 340–53; (1963b) *Br. Birds* **56**, 197–203. PHILLIPS, W W A (1947) *J. Bombay nat. Hist. Soc.* **46**, 593–613; (1955) *J. Bombay nat. Hist. Soc.* **53**, 132–3. PHILLIPS, W W A, ROSENBERG, H, SCHÜZ, E, and SEILKOPF, H (1959) *Vogelwarte* **20**, 116–21. PICOZZI, N (1958) *Br. Birds* **51**, 308. PIGGINS, D J (1959) *Rep. Statem. Accts. Salm. Res. Trst. Irel.* Inc. Append. **1**. PIRKOLA, M K (1966) *Suom. Riista* **18**, 67–81. PITMAN, C R S (1965) *Wildfowl Trust Ann. Rep.* **16**, 115–21. PLATZ, F (1963) *J. Orn.* **104**, 291; (1964) *J. Orn.* **105**, 190–6. PLAYER, P V (1971) *Wildfowl* **22**, 100–6. POLLARD, D F W and DAVIES, P W (1968) *Wildfowl* **19**, 108–16. PONCY, R (1924) *Bull. Soc. zool. Genève* **3**, 21–8; (1926) *Bull Soc. zool. Genève* **3**, 433; (1941) *Orn. Beob.* **38**, 139; (1943) *Orn. Beob.* **40**, 22. PONTING, E D (1967) *Br. Birds* **60**, 482. POOL, W (1962) *Wildfowl Trust Ann. Rep.* **13**, 126–9. PORTENKO, L A (1959) *Aquila* **66**, 119–34; (1972) *Psitsy Chukotskogo Poluostrova i Ostrova Vrangelya* **1**. Leningrad. PORTER, R (1973) *Birds* **4**, 227–8. PORTIELJE, A F J (1926) *Ardea* **15**, 1–15; (1927) *Ardea* **16**, 107–23. POST, P W (1967) *Bird-Banding* **38**, 278–305. POSTON H J (1969) *Can. Wildl. Serv. Rep. Ser.* **6**, 132–7. POTTS, G R (1969) *J. anim. Ecol.* **38**, 53–102; (1971) *Ibis* **113**, 298–305. POUGH, R H (1951) *Audubon Water Bird Guide.* New York. PREBLE, E A and McATEE, W L (1923) *US Dept. Agric. N. Amer. Fauna* **46**, 1–128. PRESTT, I (1970) *Proc. Int. Union Cons. Nat. 11 Tech. Meeting*, **1**, 95–102. PRESTT, I and JEFFERIES, D J (1969) *Bird Study* **16**, 168–185. PRESTT, I and MILLS, D H (1966) *Bird Study* **13**, 163–203. PREUSS, N O (1969) *Dansk orn. Foren. Tidsskr.* **63**, 174–83. PRIKLONSKI, S G (1958) *Trudy Okskogo zapov.* **2**, 102–15. PRIKLONSKI, S G and GALUSHIN, V M (1959) *Trudy 3 Prib. Conf.* **3**. Vilnus 1959. PRIKLONSKY, S, PAVLOV, M, and PANCHENKO, V (1970) *Proc. Int. Reg. Meet. Cons. Wildfowl Resources, Leningrad 1968.* 63–6. PRINZINGER, R. (1974) *Anz. orn. Ges. Bayern* **13**, 1–34. PRYTHERCH, R J (1965) *Br. Birds* **58**, 305–9. PYMAN, G A (1959) *Br. Birds* **52**, 42–56.

RADESÄTER, T (1974) *Behaviour* **50**, 1–15. RAE, B B (1969) *Dept. Agric. and Fish. Scot. Marine Res.* **1**, 1–16. RAITASUO, K (1964) *Paper Finn. Game Res.* **24**, 1–72. RAJALA, P and ORMIO, T (1970) *Paper Finn. Game Res.* **31**, 3–9. RALPHS, P (1955) *Skokholm Bird Obs. Rep.* **1955**, 20–22; (1956) *Skokholm Bird Obs. Rep.* **1956**, 19–22. RAND, R W (1960) *Invest. Rep. Div. Fish. S. Afr.* **42**, 1–32. RANKIN, M N and DUFFEY, E A G (1948) *Br. Birds* **41** suppl., 1–42. RANKIN, N (1947) *Haunts of British divers.* London. RANWELL, D S and DOWNING, B M (1959) *Anim. Behav.* **7**, 42–56. RAVELING, D G (1969a) *J. Wildl. Mgmt.* **33**, 319–30; (1969b) *Auk* **86**, 67–681; (1970) *Behaviour* **37**, 291–319. RECHER, H F (1972) *Emu* **72**, 126–30; RECHER, H F and RECHER, J A (1972) *Emu* **72**, 85–90. REE, V (1973) *Sterna* **12**, 225–68. REES, E I S (1961) *Sea Swallow* **14**, 54–5; (1964) *Ibis* **106**, 118–19; (1965) *Br. Birds* **58**, 508–9. REICHENOW, A (1900–1) *Die Vögel*

*Afrikas* 1. Neudamm. REICHHOLF, J (1965) *Anz. orn. Ges. Bayern* 7, 339–40; (1966) *Anz. orn. Ges. Bayern* 7, 536–604. REINSCH, H H (1969) *Der Basstölpel.* Wittenberg Lutherstadt. REISER, O (1894) *Ornis balcanica* 2. *Bulgarien.* Wien. REISER, O and VON FÜHRER, L (1896) *Ornis balcanica* 4. *Montenegro.* Wien. RENCUREL, P (1972) *Alauda* 40, 278–86; (1974) *Alauda* 42, 143–58. RETTIG, K (1961) *Orn. Mitt.* 13, 174. REUVER, H J A DE (1959) *Limosa* 32, 169. REY, E (1907) *Orn. Mschr.* 32, 259–71. REYNOLDS, C M (1965) *Bird Study* 12, 128–9; (1972) *Wildfowl Trust Ann. Rep.* 23, 111–18; (1974) *Bird Study* 21, 129–34. RICHARDSON, R A and HAYMAN, P J (1952) *Br. Birds* 46, 450–1. RICHDALE, L E (1943) *Trans. R. Soc. NZ* 73, 217–32; (1951) *Sexual Behavior in Penguins.* Lawrence (Kansas); (1963) *Proc. zool. Soc. Lond.* 141, 1–117; (1965) *Trans. zool. Soc. Lond.* 31, 1–86. RICHTER, A, SCHWARZ, M, and WINKLER, R (1970) *Orn. Beob.* 67, 133–8. RIDDELL, W H (1944) *Ibis* 86, 503–11. RIDLEY, M W, MOSS, B L, and PERCY, R C (1955) *J. E. Afr. nat. Hist. Soc.* 22, 147–58. RINGLEBEN, H (1951) *Vogelwelt* 72, 43–4; (1975) *Falke* 22, 230–3; RINGLEBEN, H and CREUTZ, G (1961) *Orn. Mitt.* 13, 186–7. RIPLEY, S D (1963) *Ibis* 105, 108–9. RIPLEY, S D and BOND, G M (1966) *Smithson. misc. Collns* 151 (7), 1–37. RITCHIE, G S and BOURNE, W R P (1966) *Sea Swallow* 18, 64–5. ROBBINS, C S, BRUUN, B, and ZIM, H S (1966) *Birds of North America.* New York. ROBERTS, A (1965) *The Birds of South Africa.* Cape Town. ROBERTS, B (1940) *Scient. Rep. Graham Ld Exped.* 1, 141–94. ROBERTS, E L (1954) *The Birds of Malta.* Malta; (1966) *Wildfowl Trust Ann. Rep.* 17, 36–45. ROBERTS, T J (1969) *J. Bombay nat. Hist. Soc.* 66, 616–9. ROBERTS, T S (1932) *The Birds of Minnesota.* 1. Minneapolis; (1955) *Manual for the Identification of the birds of Minnesota and neighboring states.* Minneapolis. ROBERTSON, C J R and KINSKY, F C (1972) *Notornis* 19, 289–301. ROBIN, P (1966) *Alauda* 34, 81–101; (1968) *Alauda* 36, 237–53; (1973) *Bonn. Zool. Beitr.* 24, 317–21. ROBINSON, H W (1909) *Br. Birds* 2, 311. ROEN, U (1965) *Dansk orn. Foren. Tidsskr.* 59, 85–91. ROGERSON, S and TUNNICLIFFE, C (1947) *Our bird book.* London. ROI, O LE (1911) In *Avifauna Spitzbergensis.* Bonn; (1912) *Orn. Monatsber.* 20, 65–6. ROKITANSKY, G and SCHIFTER, H (1971) *Annln. naturh. Mus. Wien* 75, 495–538. ROOTH, J (1965) *Uitg. natuurw. StudKring Suriname* 41, 1–151; (1971) *Ardea* 59, 17–27. ROOTH, J and JONKERS, D A (1972) *TNO—nieuws* 27 551–5. ROTHSCHILD, Lord (1919) *Bull. Br. Orn. Club* 39, 81–3. ROUX, F (1962) *Wildfowl Trust Ann. Rep.* 13, 74–8; (1969) *Bull. Int. Wildfowl Res. Bur.* 27–28, 23–4; (1970a) *Proc. Int. Reg. Meet. Cons. Wildfowl Resources, Leningrad 1968.* 265–73; (1970b) *Bull. Int. Wildfowl Res. Bur.* 30, 18–21; (1971) *Bull. Int. Waterfowl Res. Bur.* 32, 37–9; (1972) *Bull. Int. Waterfowl Res. Bur.* 34, 33–6; (1973) *L'Oiseau* 43, 1–15. ROUX, F and DUPUY, A (1972) *L'Oiseau* 42, 61–5. ROUX, F and JOUANIN, C (1968) *Br. Birds* 61, 163–9. ROWAN, M K (1952) *Ibis* 94, 97–121. ROYAMA, T (1966) *Bird Study* 13, 116–29. RUMMEL, L and GOETZINGER, C (1975) *Auk* 92, 333–46. RUSSELL, S M (1964) *Orn. Monogr.* 1. RUTHKE, P (1935) *Beitr. Fortpfl. Biol. Vögel* 11, 218–9; (1938a) *Beitr. Fortpfl. Biol. Vögel* 14, 13; (1938b) *Orn. Fenn.* 15, 70–3. RUTSCHKE, E (1970) *Proc. Int. Reg. Meet. Cons. Wildfowl Resources, Leningrad 1968,* 114–26. RUTTLEDGE, R F (1963) *Br. Birds* 56, 340; (1966) *Ireland's Birds.* London; (1970) *Proc. Int. Reg. Meet. Cons. Wildfowl Resources, Leningrad 1968.* 110–11; (1974) *Bird Study* 21, 141–5. RUWET, J-C (1963) *Revue Zool. Bot. afr.* 68, 1–60. RYALL, R H (1943) *Br. Birds* 36, 201. RYDZEWSKI, W (1956) *Ardea* 44, 71–188; (1973) *Ring* 7, 63–70, 91–6; (1974) *Ring* 112–17, 141–5, 169–71. RYVES, B H (1948) *Bird Life in Cornwall.* London.

SAFRIEL, U (1968) *Ibis* 110, 283–320. SAGE, B L (1959) *Ibis* 101, 252–3; (1971) *Br. Birds* 64, 519–28; (1973) *Br. Birds* 66, 24–30. ST JOHN, C (1882) *Sketches of the Wild Sports and Natural History of the Highlands.* London. SALOMONSEN, F (1929) *Bull. Mus. natn. Hist. nat. Paris,* 2ᵉ Ser. 1, 347–57; (1934) *Proc. zool. Soc. Lond.* 219–24; (1935) *Zoology of the Faeroes: Aves.* Copenhagen; (1941) *J. Orn.* 89, 282–337; (1949) *Dansk orn. Foren. Tidsskr.* 43, 186; (1950a) *Grønlands Fugle.* Copenhagen; (1950b) *Dansk orn. Foren. Tidsskr.* 44, 100–5; (1958) *Vidensk. Meddr. dansk naturhist. Foren.* 120, 43–80; (1963) *Oversigt over Danmarks Fugle.* Copenhagen; (1965) *Auk* 82, 327–55; (1967) *Fuglene på Grønland.* Copenhagen; (1968) *Wildfowl* 19, 5–24; (1971) *Dansk orn. Foren. Tidsskr.* 66, 11–19. SALT, G and HOLLICK, F S J (1946) *J. Exp. Biol.* 23 1–46. SALVADORI, T (1895) *Cat. Birds Br. Mus.* 27. London. SALVAN, J (1967) *L'Oiseau* 37, 255–84. SALVIN, O (1891) *Ibis* (6) 3, 411–14. SAPETIN, Y V (1968) *Migr. Zhivot.* 5, 94–112. SARRO, A and PONS OLIVERAS, J R (1966) *Ardeola* 12, 121–141. SAUER, E G F (1971) *Z. Kölner Zoo,* 14, 43–64; (1972a) *Auk* 89, 717–37; (1972b) *Bonn. zool Beitr.* 23, 3–48, SAUER, E G F and ROTHE, P (1972) *Science* 176, 43–5. SAUER, E G F and SAUER, E M (1959) *Bonn. zool. Beitr.* 10, 266–85; (1966a) *Living Bird* 5, 45–75; (1966b) *Ostrich* suppl. 6, 183–91. SAUNDERS, A A (1951) *Guide of Bird Songs.* New York. SAUNDERS, H (1889) *Manual of British Birds.* London. SAVAGE, C D W (1952) *The Mandarin Duck.* London; (1965) *Wildfowl Trust Ann. Rep.* 16, 121–3; (1966) *Wildfowl Trust Ann. Rep.* 17, 45–8; (1969) *Wildfowl* 20, 144–7; (1970a) *Proc. Int. Reg. Meeting Cons. Wildfowl Resources, Leningrad 1968,* 189–91; (1970b) *Wildfowl* 21, 88. SAWYER, E J (1928) *Wilson Bull.* 40, 1–17. SAXBY, H L (1874) *The Birds of Shetland.* Edinburgh. SAYLER, J, CLARK, I I, and LAGLER, K F (1940) *J. Wildl. Mgmt* 4, 186–219. SCHEER, G (1960) *Senckenberg. biol.* 41, 143–7. SCHENK, H (1970) *Vogelwarte* 25, 268–9. SCHENK, J (1929) *Aquila* 34–5, 16–85. SCHERNER, E R (1974) *Vogelwelt* 95, 161–9. SCHIEMENZ, H (1972) *Falke* 19, 44–5. SCHIERER, A (1960) *L'Oiseau* 30, 169–72; (1962) *L'Oiseau* 32, 265–8. SCHILDMACHER, H (1975) *Falke* 22 366–71. SCHIØLER, E L (1925) *Danmarks Fugle* 1; (1926) 2. Copenhagen. SCHLEGEL, R (1964) *Abh. Mus. Tierk., Dresden Zool.* 27, 65–7. SCHMIDT, G (1952) *Vogelwelt* 73, 123–5; (1957) *Vogelwelt* 78, 125–6; (1966) *Corax* 1, 216–50; (1967) *Corax* 2, Beiheft 1, 2–9. SCHMIDT, TH (1970) *Naturens Verd.* 1970. 363–74. SCHMIDT-NIELSEN, K, KANWISHER, J, LASIEWSKI, R C, COHN, J E, and BRETZ, W L (1969) *Condor* 71, 341–52. SCHNEIDAUER, T R DE (1968) *Ardea* 56, 228–47. SCHÖNWETTER, M (1927) *Orn. Mber.* 35, 13–17; (1967) *Handbuch der Oologie* 1. Berlin. SCHORGER, A W (1947) *Wilson Bull.* 59, 151–9; (1951) *Wilson Bull.* 63, 112. SCHREIBER, R W and ASHMOLE, P (1970) *Ibis* 112, 363–94. SCHRÖDER, H (1966) *Beitr. Vögelk.* 12, 189–93; (1969) *Beitr. Vögelk.* 14, 269–80. SCHRÖDER, P and BURMEISTER, G (1974) *Der Schwarzstorch.* Wittenberg Lutherstadt. SCHUILENBURG, H L (1974) *Het. Vogelj.* 22, 913–9. SCHUTTE, G W (1969) *Lammergeyer* 10, 100. SCHUTZ, F (1965) *Z. Tierpsychol.* 22, 50–103. SCHÜZ, E (1936) *Orn. Mber.* 44, 65–71; (1940a) *Vogelzug* 11, 23–31; (1940b) *Beitr. Fortpfl. Biol. Vögel* 16, 203–5; (1942) *Z. Tierpsychol.* 5, 1–37; (1954) *Vogelwarte* 17, 65–80; (1955) *Proc. Int. Orn. Congr.* 11, 55–28; (1957a) *Vogelwarte* 19, 1–15; (1957b) *Vogelwarte* 19, 41–44; (1957c) *Vogelwarte* 19, 132–5; (1959) *Die Vogelwelt des Südkaspischen Tieflandes.* Stuttgart; (1960) *Vogelwelt* 20, 205–22; (1963) *Proc. Int. Orn. Congr.* 13, 475–80. SCHÜZ, E, BERTHOLD, P, GWINNER, E, and OELKE, H (1971) *Grundriss der Vogelzugskunde.* Berlin. SCHÜZ, E and SZIJJ, J (1962) *Bull. Int. Council Bird Pres.* 8, 86–98; (1971) *Bull. Int. Council Bird Pres.* 11, 141–5; (1975) *Vogelwarte* 28, 61–93. SCLATER, W L (1906) *The birds of South Africa* 4. London. SCOTT, D A (1970) PhD Thesis. Univ. of Oxford. SCOTT, D A and CARP, E (1972) *Proc. Int. Conf. Cons. Wetlands and Waterfowl, Ramsar 1971,* 291–9. SCOTT, P (1950) *Wildfowl Trust Ann. Rep.* 3, 121–9; (1956) *Br. Birds* 49, 172–3; (1966)

*Wildfowl Trust Ann. Rep.* **17**, 20–6; (1970) *Wildfowl* **21**, 37–41. SCOTT, P and BOYD, H (1957) *Wildfowl of the British Isles.* London. SCOTT, P, FISHER, J, and GUDMUNDSSON, F (1953) *Wildfowl Trust Ann. Rep.* **5**, 78–115. SCOTT, P and THE WILDFOWL TRUST (1972) *The Swans.* London. SEDDON, B (1971) *Introduction to Biogeography.* London. SEDGWICK, N M, WHITAKER, P, and HARRISON, J G (1970) *The New Wildfowler in the 1970s.* London. SEEBOHM, H (1885) *A History of British Birds* 3. London. SEGONZAC, M (1972) *L'Oiseau* **42**, no. spéc., 3–68. SEITZ, A (1937) *Beitr. Fortpfl. Biol. Vögel* **13**, 35–45. SELANDER, R K (1965) *Amer. Nat.* **99**, 129–41. SELOUS, E (1901) *Zoologist* (4) **5**, 161–83, 338–50; (1905) *Bird Life Glimpses.* London; (1915) *Wildlife* 1915, 29–35, 38–42, 98–99, 137–41, 175–78, 219–30; (1920–1) *Naturalist* 1920, 97–102, 195–8, 325–8; 1921, 173–6, 197–200, 301–5; (1925) *Naturalist* 1925, 179–82; (1927) *Realities of Bird Life.* London. SERIE, J R and SWANSON, G A (1972) *34th Midwest Wildl. Conf. Des Moines, Iowa,* 10–13. SERLE, W (1965) *Ibis* **107**, 60–94. SERVENTY, D L, SERVENTY, V N, and WARHAM, J (1971) *The handbook of Australian sea-birds.* Sydney. SETON-BROWNE, C and HARRISON, J (1968) *Bull. Br. Orn. Club* **88**, 59–73. SEYMOUR, N (1974a) *Auk* **91**, 423–7; (1974b) *Wildfowl* **25**, 49–55. SHARLAND, R E (1966) *Bull. Niger. orn. Soc.* **3** (10), 45. SHARPE, R B (1898) *Cat. Birds Br. Mus.* **26**, 1–328. London. SHARROCK, J T R (1968) *Br. Birds* **61**, 423–5; (1971) *Ibis* **113**, 430–3. SHARROCK, J T R and Rare Breeding Birds Panel (1975) *Br. Birds* **68**, 489–506. SHAUGHNESSY, P D (1971) *Aust. J. Zool.* **19**, 77–83. SHAW, T H (1936) *The birds of Hopei Province.* Peking. SHELLEY, L O (1930) *Auk* **47**, 238–40. SHERWOOD, M P (1957) *Flor. Nat.* **30**, 11–8. SHEVAREVA, T (1970) *Proc. Int. Reg. Meet. Cons. Wildfowl Resources, Leningrad 1968,* 46–55. SHEWELL, E L (1959) *Ostrich* Suppl. **3**, 168. SHORROCK, H (1962) *Wildfowl Trust Ann. Rep.* **13**, 168. SIBLEY, C G and AHLQUIST, J E (1972) *Bull. Peabody Mus. nat. Hist.* **39**. SIBLEY, C G and FRELIN, C (1972) *Ibis* **114**, 377–87. SIBLEY, F C and CLAPP, R B (1967) *Ibis* **109**, 328–37. SIEGFRIED, W R (1965a) *Bokmakierie* **17**, 57; (1965b) *Ostrich* **36**, 94; (1966a) *Bokmakierie* **18**, 54–7; (1966b) *Ostrich* **37**, 198; (1967) *Ostrich* **38**, 173–8; (1970) *Ostrich* **41**, 122–35; (1971a) *J. appl. Ecol.* **8** (2), 447–68; (1971b) *Ardea* **59**, 38–46; (1971c) *Zool. Afric.* **6**, 289–92; (1971d) *Ostrich* Suppl. **9**, 153–64; (1971e) *Trans. R. Soc. South Africa* **39**, 419–43; (1971f) *Ostrich* **42**, 193–7; (1972a) *Bull. Br. Orn. Club.* **92**, 102–3; (1972b) *Ostrich* **43**, 43–55; (1973a) *Wildfowl* **24**, 150–3; (1973b) *Can. J. Zool.* **51**, 1293–7; (1973c) *Bull. Br. Orn. Club.* **93**, 98–9; (1973d) *Living Bird* **11**, 193–206. SIEWERT, H (1932) *J. Orn.* **80**, 533–41; (1955) *Störche.* Gütersloh. SIIVONEN, L (1941) *Orn. fenn.* **18**, 35–43. SIM, G (1881) *Zoologist* (3) **5**, 26. SIM, R J (1904) *Wilson Bull.* **16**, 67–74. SIMMONS, K E L (1951) *Ibis* **93**, 407–13; (1954) *Bird Study* **1**, 53–6; (1955) *Studies on Great Crested Grebes.* London; (1956) *Br. Birds* **49**, 432–5; (1959) In Bannerman (1959) 215–23; (1962) *Bull. Br. Orn. Club* **82**, 109–16; (1965) *Bull. Br. Orn. Club* **85**, 161–8; (1967a) *Living Bird* **6**, 187–212; (1967b) MSc Thesis. Bristol Univ.; (1968a) *Br. Birds* **61**, 308–9; (1968b) *Br. Birds* **61**, 556–8; (1968c) *Birds* **1**, 122–5; (1969a) *Wilson Bull.* **81** 113–16; (1969b) *Bristol Orn.* **2**, 71–2; (1970a) In Crook, J H (ed.) *Social Behaviour in Birds and Mammals,* 37–77; (1970b) *Br. Birds* **63**, 300–2; (1970c) PhD Thesis. Bristol Univ.; (1972) *Br. Birds* **65**, 465–79, 510–21; (1973) *Br. Birds* **66**, 30–1; (1974a) *Br. Birds* **67**, 413–37; (1974b) *Avic. Mag.* **80**, 143–6; (1975a) *Bristol Orn.* **8**, 89–107; (1975b) *Wildfowl* **26**, 58–63. SIMMONS, K E L and WEIDMANN, U (1973) *J. Zool. Lond.* **170**, 49–62. SIMONETTA, A M (1963) *Arch. Zool. Ital.* (Toringo) **48**, 53–135. SIMPSON, K N G (1972) *Birds in Bass Strait.* Sydney. SIRÉN, M (1952) *Paper Finn. Game Res.* **8**, 101–11; SJÖLANDER, S (1957) *Suom. Rüsta* **11**, 130–33; (1968) *Zool. Revy* **30**, 89–93. SJÖLANDER, S and ÅGREN, G (1972) *Wilson*

*Bull.* **84**, 296–308. SKEAD, C J (1951) *Ibis* **93**, 360–82; (1966) *Ostrich* Suppl. **6**, 109–39. SKEAD, D M (1974) *Ostrich* **45**, 189–92. SKOKOVA, N N (1955) *Vop. Ikht.* **5**, 170–85; (1959) *Migratsii Zhivotnykh* **1**, 67–94. SKOVGAARD, P (1920a) *Dansk Fugle* **1**, 2–9; (1920b) *Den sorte Stork, saerlig i Danmark.* Viborg. SLADEN, W J L, WOOD, R C, and MONOGHAN, E P (1968) *Antarctic Res. Ser. Washington* **12**, 213–62. SMITH, F R and the Rarities Committee (1973) *Br. Birds* **66**, 329–60; (1974) *Br. Birds* **67**, 310–48; (1975) *Br. Birds* **68**, 306–38. SMITH, K D (1956) *Ibis* **98**, 303–7; (1957) *Ibis* **99**, 1–26, 307–17; (1965) *Ibis* **107**, 493–526; (1970) *Bull. Br. Orn. Club* **90**, 18–24. SMITH, P A (1969) *Bull. Br. Orn. Club* **89**, 52–60. SMITH, R I (1968) *Auk* **85**, 381–96. SMITH, R W J (1974) *Scott. Birds* **8**, 151–9. SMYTHIES, B E (1953) *The birds of Burma.* Edinburgh; (1960) *The birds of Borneo.* Edinburgh. SNOW, B K (1960) *Ibis* **102**, 554–75; (1963) *Br. Birds* **56**, 77–103, 164–86. SNOW, D W (1965) *Condor* **67**, 210–14; (1967) *A guide to moult in British birds.* Br. Trust Orn. Tring. SÖDERBERG, R (1950) *Våra Fåglar.* Stockholm. SODING, K (1950) *Orn. Mitt.* **2**, 146–7. SOIKKELI, M (1973) *Paper Finn. Game Res.* **33**, 28–30. SOKAL, R R and ROHLF, F J (1969) *Biometry.* San Francisco. SOKOŁOWSKI, J (1960) State Council Nat. Cons. Warsaw **1**, 1–28. SOOT-RYEN, T (1941) *Tromsø Mus. Aaish.* **59**, 1–42. SOPER, J D (1941) *The Blue Goose.* Dept of Interior, Ottawa. SORCI, G, MASSA, B, and CANGIALOSI, G (1972) *Riv. Ital. Orn.* **42**, 232–47. SOWLS, L K (1955) *Prairie Ducks.* Harrisburg, Penn. SPALDING, D (1973) *J. Anim. Behav.* **2**, 2–11. SPÄRCK, R (1957) *Int. Union Game Biol. Trans.* **3**, 45–7. SPENCER, K G (1969) *Br. Birds* **62**, 332–3. SPIESS, S VON (1934) *Vogelzug* **5**, 191–2. SPILLNER, W (1968) *Beitr. Vogelk.* **14**, 29–74. SPITZ, F (1964) *Terre et Vie* **18**, 452–88. STAFFORD, J (1971) *Bird Study* **18**, 218–21. STAHLBERG, B-M (1974) *Wildfowl* **25**, 67–73. STAMMER, J (1937) *Berichte Vereins Schles. Orn.* **22**, 20–28. STAMP, D (ed) (1966) *A Glossary of Geographical Terms.* London. STANFORD, J K (1936) *Br. Birds* **30**, 136. STANFORD, W P (1953) *Ostrich* **24**, 17–26; (1954) *Ibis* **96**, 449–73, 606–24. STARKEY, E E and STARKEY, J F (1973) *Condor* **75**, 364–66. STEENHUIZEN, P L (1934) *Ardea* **23**, 127–31. STEINBACHER, G (1960) *Vogelwelt* **81**, 1–16; (1963) *Orn. Mitt.* **15**, 106–8. STEINBACHER, J (1935) *Naturf.* **12**, 95–6; (1936) *Schr. phys.-ökon Ges. Königsberg* **69**, 23–36. STEINFATT, O (1934) *Beitr. Fortpfl. Biol. Vögel* **10**, 85–96; (1939) *Beitr. Fortpfl. Biol. Vögel* **15**, 191–8, 240–51. STEJNEGER, L (1885) *Bull. US natn. Mus.* **29**, 161. STEPHAN, B (1970) *Mitt. zool. Mus. Berl.* **46**, 339–437. STERBETZ, I (1961) *Der Seidenreiher.* Wittenberg Lutherstadt; (1962) *Aquila* **67–68**, 39–70; (1967) *Aquila* **73–74**, 43–49; (1968) *Aquila* **75**, 295; (1969–70) *Aquila* **76–7**, 141–63; (1969) *Falke* **16**, 292–5; (1971) *Limosa* **44**, 54–60; (1972) *Derc. Muz. Evk.* 33–52. STERBETZ, I and SLIVKA, L (1972) *Larus* **24**, 141–8. STERBETZ, I and SZIJJ, J (1968) *Vogelwarte* **24**, 266–77. STEVEN, G A (1933) *J. Marine Biol. Assoc.* **19**, 277–92. STEVENSON, H (1866) *Birds of Norfolk.* London. STEWART, P A (1967) *Auk* **84**, 122–3. STEWART, P F and CHRISTENSEN, S J (1971) *A Check-list of birds of Cyprus.* STEWART, R E and ALDRICH, J W (1956) *Proc. biol. Soc. Wash.* **69**, 29–36. STEWART, R E and ROBBINS, C S (1958) *Birds of Maryland and the District of Columbia.* N. Amer. Fauna 62. STOKOE, R (1958) *Br. Birds* **51**, 165–79. STOLL. F E (1931) *J. Orn.* **79**, 541–7; (1934) *Ardea* **23**, 51–6. STOLLMAN, A (1969) *Vogelwarte* **25**, 65–6. STOLPE, M (1935) *J. Orn.* **83**, 115–28. STONEHOUSE, B (1958) *Ibis* **100**, 204–8; (1960) *Wideawake Island.* London; (1962a) *Ibis* **103b**, 107–23; (1962b) *Ibis* **103b**, 124–61; (1963) *Ibis* **103b** 474–9. STORER, R W (1956) *Condor* **58**, 413–26; (1960) *Proc. Int. Orn. Congr.* **12**, 694–707; (1961) *Auk* **78**, 90–2; (1963) *Proc. Int. Orn. Congr.* **13**, 562–9; (1967) *Condor* **69**, 469–78; (1969) *Condor* **71**, 180–205; (1971) In Farner *et al.* (eds) (1971) **1**, 1–18. STRESEMANN, E (1920) *Avifauna Macedonica.* München; (1926) *Orn. Mber.* **34**, 131–9;

(1927–34) *Handbuch der Zoologie* **7** (2). *Sauropsida : Aves.* Berlin; (1936) *Orn. Mber.* **44**, 100–2; (1963*a*) *Auk* **80**, 1–8; (1963*b*) *Condor* **65**, 449–59. STRESEMANN, E and STRESEMANN, V (1966) *J. Orn.* **107**, Sonderheft; (1970) *J. Orn.* **111**, 378–93. STRESEMANN, V (1948) *Avic. Mag.* **54**, 188–194. STRIJBOS, J (1962) *De Blauwe Reiger.* Amsterdam. STRONACH, B W H (1968) *Ibis* **110**, 345–8. STUART, D (1948) *Br. Birds* **41** 194–9. STÜBS, J (1960) *J. Orn.* **101**, 499; (1972) *Mitt. zool. Mus. Berl.* **48**, 325–92. STUDER-THIERSCH, A (1966) *Orn. Beob.* **63**, 85–9; (1967) *Zool. Garten* **34**, 159–229; (1972) *Orn. Beob.* **69**, 237–52; (1974) *Z. Tierpsychol.* **36**, 212–66; (1975*a*) In Kear and Duplaix-Hall (1975), 121–30; (1975*b*) In Kear and Duplaix-Hall (1975), 150–8. STÜLKEN, K (1943) *Beizwild der Könige; ein Reiher biologie.* Wedel in Holstein. SUCHANTKE, A (1959) *Orn. Beob.* **56**, 94–7; (1960) *Vogelwelt* **81**, 33–46. SUDHAUS, W (1966) *Vogelwelt* **87**, 89–91. SUGDEN, L G (1973) *Can. Wildl. Service Rep.* **24**. SUMMERHAYES, V S and ELTON, C S (1923) *J. Ecol.* **2**, 258. SWALES, M K (1965*a*) *Ibis* **107**, 17–42; (1965*b*) *Ibis* **107**, 215–29. SWANSON, G A and BARTONEK, J C (1970) *J. Wildl. Mgmt* **34**, 739–46. SWENNEN, C (1972) *TNO-nieuws* **27**, 556–60. SWENNEN, C and van der BAAN, G (1959) *Br. Birds* **52**, 15–18. SZABO, F (1970) *Proc. Int. Reg. Meet. Cons. Wildfowl Resources, Leningrad 1968*, 139–42. SZIJJ, J (1954) *Aquila* **55–58**, 81–87; (1963) *Vogelwarte* **22**, 1–17; (1965*a*) *Orn. Beob.* **62**, 61; (1965*b*) *Vogelwarte* **23**, 24–71; (1965*c*) *Vogelwelt* **86**, 98–104; (1972) *Proc. Int. Conf. Cons. Wetlands and Waterfowl, Ramsar 1971*, 111–19; (1973) *Bull. Int. Waterfowl Res. Bur.* **35**, 14–15. SZIJJ, J and SZIJJ, L (1955) *Aquila* **59–62**, 91–94.

TÄCKLIND, E (1954) *Vaør Faøgelvärld* **31**, 40. TAIT, C C (1967) *Fisheries Res. Bull. Zambia*, **3**, 31–2. TAIT, W C (1924) *The Birds of Portugal.* London. TALPEANU, M (1964) *1st Eur. Meeting Wildfowl Cons., St. Andrews 1963*, 45–48. TAMISIER, A (1966) *Terre et Vie* **20**, 316–37; (1970) *Terre et Vie* **24**, 511–62; (1971) *Alauda* **39**, 261–311; (1972*a*) *Alauda* **40**, 107–35, 235–56; (1972*b*) Thesis. Univ. Montpellier; (1972*c*) *Chasse et mortalité chez les sarcelles d'hiver.* Le Sambuc; (1974) *Wildfowl* **25**, 123–33. TANSLEY, A G (1968) *Britain's Green Mantle.* London. TATE, J D (1969) *Proc. Nebraska Acad. Sci.* **79**, 50. TAVERNER, J H (1959) *Br. Birds* **52**, 245–58; (1967) *Br. Birds* **60**, 509–15. TAVERNER, P A (1931) *Ann. Rep. natn. Mus. Canada 1929*, 29–40. TAYLOR, J S (1957) *Ostrich* **28**, 1–80. TAYLOR, R J F (1953) *Ibis* **95**, 638–47. TEAL, J M (1965) *Wilson Bull.* **77**, 257–63. TEBBUTT, C F (1961) *Br. Birds* **54**, 284. TERRASSE, J-F, TERRASSE, M, and BROSSELIN, M (1969*a*) *L'Oiseau* **39**, 185–201. TERRASSE, J-F, TERRASSE, M, and CHAPPUIS, C (1969*b*) *L'Oiseau* **39**, 252–60. THAISZ, L (1899) *Aquila* **6**, 139–41. THEARLE, R F (1951) *Middle Thames Nat.* **4**, 10–11. THEVENOT, M (1972) *Bull. Soc. Sci. Nat. and Phys. Maroc* **52**, 243–88. THIEMANN, H (1975) *Orn. Mitt.* **27**, 109–10. THIJSSE, J P (1906) *Het intieme leven der vogels.* Haarlem. THOBIAS, J (1934) *Kocsag.* **7**, 65. THOMASSON, K (1953) *Zool. Bidr. Från Uppsala* **30**, 157–66. THOMPSON, D Q and PERSON, R A (1963) *J. Wildl. Mgmt* **27**, 348–56. THOMPSON, W (1851) *The Nat. Hist. of Ireland* **3**. London. THOMSON, A L (1965) *L'Oiseau* **35**, suppl. 130–40. THÖNEN, W (1968) *Orn. Beob.* **65**, 187. THORPE, W H, COTTON, P T, and HOLMES, P F (1936) *Ibis* (13) **6**, 557–80. TICEHURST, C B (1922) *Bull. Br. Orn. Club* **42**, 120–1; (1923) *Ibis* (11) **5**, 438–74; (1925) *J. Bombay nat. Hist. Soc.* **30**, 725–33. TICEHURST, C B, BUXTON, P A, and CHEESMAN, R E (1922) *J. Bombay nat. Hist. Soc.* **28**, 650–74. TICEHURST, C B and CHEESMAN, R E (1925) *Ibis* (12) **1**, 1–31. TICEHURST, C B, COX, P, and CHEESMAN, R E (1925) *J. Bombay nat. Hist. Soc.* **30**, 725–33; (1926) *J. Bombay nat. Hist. Soc.* **31**, 91–119. TICEHURST, C B and WHISTLER, H (1932) *Ibis* (13) **2**, 40–93. TICEHURST, N F (1957) *The Mute Swan in England.* London. TICKELL, W L N (1967) *Emu* **66**, 357–67; (1968) *Antarctic Res. Ser. Washington* **12**,

1–55. TICKELL, W L N and GIBSON, J D (1968) *Emu* **68**, 6–20. TICKELL, W L N and SCOTLAND, C D (1961) *Ibis* **103a**, 260–66. TIMKEN, R L and ANDERSON, B W (1969) *J. Wildl. Mgmt.* **33**, 87–91. TIMMERMAN, A (1960) *Limosa* **33**, 159–73. TINBERGEN, N (1952) *Quarterly Rev. Biol.* **27**, 1–32; (1953) *Social Behaviour in Animals.* London; (1957) *Bird Study* **4**, 14–27; (1958) *Curious Naturalists.* London. TITMAN, R D (1973) PhD Thesis. University of New Brunswick, Canada. TIUSSA, J (1966) *Suom.Riista* **18**, 42–9; (1972) *Suom. Riista* **24**, 40–6. TODD, W E C (1950) *Condor* **52**, 63–8; (1963) *Birds of the Labrador Peninsula and adjacent areas.* Toronto. TOMIAŁOJĆ, L (1972) *Ptaki Polski Wykaz Gatunkówi rozmieszcenie.* Warsaw. TOMLINSON, D N S (1974*a*) *Ostrich* **45**, 175–81; (1974*b*) *Ostrich* **45**, 209–23; (1975) *Ostrich* **46**, 157–65. TONG, M and POTTS, G R (1967) *Br. BIRDS* **60**, 214–15. TOSCHI, A (1969) *Introduzione alla ornitologia della Libia.* Bologna. TOUFAR, J (1965) *Inst. Vert. zool. Czech. Acad. Sci. Brno.* 215–29. TOWNSEND, C W (1905) *The birds of Essex County Massachusetts.* Mem. Nuttall orn. Club **3**; (1909) *Auk* **26**, 234–48; (1910) *Auk* **27**, 177–81; (1911) *Auk* **28**, 341–5; (1916) *Auk* **33**, 9–17. TRAUGER, D L (1974) *Auk* **91**, 243–54. TROTIGNON, J (1975) In Kear and Duplaix-Hall (1975) 35–7. TROTT, A C (1947) *Ibis* **89**, 77–98. TUCK, G S (1967) *Sea Swallow* **19**, 11–14. TUCKER, B W (1936) *Br. Birds* **30**, 70–3. TURNER, E A (1911) *Br. Birds* **5**, 90–7; (1924) *Broadland Birds.* London. TUTIN, T G, HEYWOOD, V H, BURGES, N A, VALENTINE, D H, WALTERS, S M, and WEBB, D A (eds) (1964) *Flora Europaea.* Cambridge.

URBAN, E K (1974) *Ibis* **116**, 263–77. URBAN, E K and BROWN, L H (1971) *A Checklist of the Birds of Ethiopia.* Addis Ababa. URBAN, E K and JEFFORD, T G (1974) *Bull. Br. Orn. Club* **94**, 104–7. USPENSKI, S M (1959) *Bull. Moscow Soc. Nat. Res. Biol. Sec.* **64** (2), 39–52; (1960) *Wildfowl Trust Ann. Rep.* **11**, 80–93; (1965) *Die Wildgänse Nordeurasiens.* Wittenberg Lutherstadt; (1968) In Kumari, E (1968) 7–14; (1970) *Proc. Int. Reg. Meet. Cons. Wildfowl Resources, Leningrad 1968*, 56–7; (1972) *Die Eiderenten.* Wittenberg Lutherstadt. USPENSKI, S M, BĔME, R L, PRIKLONSKI, S G, and VEKHOV, V N (1962) *Ornitologiya* **5**, 49–67. USPENSKI, S M and KISHKO, Y I (1967) *Problemy Severa* **11**, 235–43. USSHER, H T (1874) *Ibis* (3) **4**, 43–75. USSHER, R J and WARREN, R (1900) *The Birds of Ireland.* London. UYS, C J, BROEKHUYSEN, G J, MARTIN, J, and MACLEOD, J G (1963) *Ostrich* **34**, 129–54.

VALIUS, M I (1959) *Ornitologiya* **2**, 221–7. VALVERDE, J A (1953) *Alauda* **21**, 250–2; (1955–56) *Alauda* **23**, 145–71, 254–79; **24**, 1–36; (1957) *Aves del Sahara Español.* Madrid; (1960) *Arch. Inst. acclim. Almeira* **9**, 1–168; (1963) *Ardeola* **9**, 55–65; (1964) *Ardeola* **9**, 121–32. VAN DAMME, B (1968) *Gerfaut* **58**, 162. VAN DOBBEN, W H (1952) *Ardea* **40**, 1–63. VAN HERP, J V and DEKKER, D (1966) *Natur. Amst.* **64**, 59–60. VAN KAMMEN, I I (1916) *Oologist* **33**, 172. VAN IJZENDOORN, A L J (1944) *Limosa* **17**, 8–13. VAN LYNDEN, A J H (1962) *Het Vogeljaar* **10**, 291. VAN OORDT, G J and HUXLEY, J S (1922) *Br. Birds* **16**, 34–46. VAN OORDT, G J and KRUIJT, J P (1953) *Ibis* **95**, 615–37. VAN TETS, G F (1965) *Orn. Monogr.* **2**. VAN TOL, J (1974) *Het Vogeljaar* **22**, 781–3. VAN ZINDEREN BAKKER, E M Jr (1971) *Rep. S. Afr. biol. geol. Exped. 1965–6*, 161–71. VASILIU, G D (1968) *Systema Avium Romaniae.* Paris. VASVARI, M (1929) *Aquila* **34–5**, 342–74; (1931) *Aquila* **36–37**, 231–93; (1938) *Proc. Int. Orn. Congr.* **9**, 414–22; (1939) *Aquila* **42–5**, 556–613; (1954) *Aquila* **55–8**, 23–38. VAURIE, C (1963) *Bull. Br. Orn. Club* **83**, 164–6; (1965) *The birds of the Palearctic fauna. Non-Passeriformes.* London. VENABLES, L S V (1936) *Br. Birds* **30**, 60–9. VENABLES, L S V and LACK, D (1934) *Br. Birds* **28**, 191–8. VENABLES, L S V and VENABLES, U M (1950) *Scott. Nat.*

62, 142–52; (1955) *Birds and Mammals of Shetland*. Edinburgh. VEREŠČAGIN, N K (1950) *Trudy Inst. Zool. Azerb.* **4**, 133–214. VERHEYEN, R. (1953) *Explor. Parc. natn. Upemba Miss. G F de Witte* **19**; (1967) *Oologica Belgica*. Brussels. VERHEYEN, R and GRELLE, G le (1952) *Gerfaut* **42**, 214–22. VERNON, J D R (1973) *Alauda* **41**, 101–9. VEROMAN, H (1975) *Ardeola* **21**, 827–37. VERWEY, J (1930) *Zool. Jb. Allg. Zool. Physiol.* **48**, 1–120. VERZUCKIJ, B N (1965) *Vses. orn. konf. Alma Ata* 62–4. VESELOVSKÝ, Z. (1951) *Sylvia* **13**, 1–19; (1952) *Věst Cs. zool. společnosti*, **16**, 354–76; (1966) *Acta Soc. Zool. Bohemoslov.* **30**, 77–90; (1970) *Z. Tierpsychol.* **27**, 915–45. VESPREMEANU, E E (1968) *Ardea* **56**, 160–77. VIELLIARD, J (1970) *Alauda* **38**, 87–125; (1972) *Alauda* **40**, 63–92. VIKSNE, J A (1968) *Baltic Comm. for study of Bird Migration* **5**, 80–107. VINCENT, J (1934) *Ibis* (13) **4**, 305–40; (1947) *Ibis* **89**, 489–91. VLUG, J J (1974) *Limosa* **47**, 16–22. VOIGT, A (1950) *Excursionsbuch zum Studium der Vogelstimmen*. Heidelberg. VOISIN, C (1968) *L'Oiseau* **38**, 151–74; 225–48; (1970) *L'Oiseau* **40**, 307–39. VOISIN, J-F (1968) *L'Oiseau* **38**, no. spéc., 95–122; (1969) *L'Oiseau* **39**, no. spéc., 82–106. VON BOETTICHER, H (1949) *L'Oiseau* **19**, 29–30; (1952) *L'Oiseau* **22**, 61–2. VON BRAUN, C, HESSLE, A-C, and SJÖLANDER, S (1968) *Zool. Revy* **30**, 94–5. VON DE WALL, W (1962) *J. Orn.* **103**, 150–2; (1963) *J. Orn.* **104**, 1–15; (1965) *J. Orn.* **106**, 65–80. VON HAARTMAN, L (1945) *Acta zool. fenn.* **44**, 1–120; (1971) In Farner *et al.* (1971) 448–81; (1973) In Farner D S (ed.) *Breeding biology of birds*. Washington, 448–81. VON SANDEN, W (1935) *Orn. Mber.* **43**, 82–5. VON TORNE, H (1940) *Beitr. Fortpfl. Biol. Vögel* **16**, 173–80. VOOUS, K H (1948) *Ardea* **36**, 265–8; (1949) *Ardea* **37**, 113–22; (1954) *Ardea* **42**, 217–8; (1957) *Stud. Fauna. Curaçao*, **7**; (1960) *Atlas of European Birds*. London; (1961) *Limosa* **34** 185–88; (1964) *Nytt Mag. Zool.* **12**, 38–47; (1965) *Ardea* **53**, 237; (1967) *Ardea* **55**, 268–9; (1969) *Ibis* **111**, 106–12; (1970) *Ardea* **58**, 265–6; (1972) *Ardea* **60**, 128–9; (1973) *Ibis* **115**, 612–38. VOOUS, K H and PAYNE, H A W (1965) *Ardea* **53**, 9–31. VOOUS, K H and WATTEL, J (1963) *Ardea* **51**, 143–57. VOROBIEV, K A (1963) *Ptitsy Yakutii*. Moscow.

WAARAMAKI, T (1968) *Suom. Riista* **20**, 87–93. WACKERNAGEL, H (1950) *Orn. Beob.* **47**, 41–56; (1964) *Orn. Beob.* **61**, 49–56. WALKER, A F G (1970) *Wildfowl* **21**, 99–107. WALLACE, D I M (1973) *Ibis* **115**, 559–71. WALSH, F (1971) *Bull. Nigerian Orn. Soc.* **8** (30), 25–7. WALTER, A (1890) *J. Orn.* **38**, 243. WARD, P and ZAHAVI, A (1973) *Ibis* **115**, 517–34. WARENTIG, W (1953) *Vögel Heimat* **23**, 97–9. WARGA, K (1938) *Proc. Int. Orn. Congr.* **8**, 655–63. WARHAM, J (1955) *West. Aust. Nat.* **5**, 31–9, (1957) *West. Aust. Nat.* **7**, 235–6; (1958a) *Br. Birds* **51**, 269–72; (1958b) *Br. Birds* **51**, 393–97; (1962) *Auk* **79**, 139–60; WARHAM, J and BOURNE, W R P (1974) *American Birds* **28** (3) 598–603. WARHAM, J, BOURNE, W R P, and ELLIOTT, H F I (1966) *Br. Birds* **59**, 376–84. WARNCKE, K (1960) *Vogelwelt* **81**, 129–41. WATERS, E (1964) *Br. Birds* **57**, 309–15. WATERS, W E (1964) *Scott. Birds* **3**, 73–81. WATSON, D (1972) *Birds of Moor and Mountain*. Edinburgh. WATSON, G E (1966) *Seabirds of the tropical Atlantic Ocean*. Washington; (1967) *Auk* **84**, 424; (1971) *Auk* **88**, 440–2. WATSON, G E, ANGLE, J P, HARPER, P C, BRIDGE, M A, SCHLATTER, R P, TICKELL, W L N, BOYD, J C, and BOYD, M M (1971) *Birds of the Antarctic and Subantarctic*. Am. Geogr. Soc. New York. WATSON, J B (1908) *Tortugas Lab. of Carnegie Inst. Washington*, 2. no. 103; (1931) *Br. Birds* **24**, 367–8. WEIDMANN, U (1956) *Z. Tierpsychol.* **13**, 208–71; (1958) *Z. Tierpsychol.* **15**, 277–300. WEIDMANN, U and DARLEY, J A (1971a) *Animal Behav.* **19**, 287–98; (1971b) *Rev. Comp. Animal* **5**, 131–5. WELLER, M W (1965) *Auk* **82**, 227–35. WEST, B, CABOT, D,

and GREER-WALKER, M (1974) *Proc. R. Irish Acad.* **75**, 285–304. WESTERNHAGEN, W V (1970a) *Vogelwarte* **25**, 185–93; (1970b) *J. Orn.* **111**, 206–26. WETMORE, A (1920) *Auk* **37**, 221–47; (1924) *US Dept. Agric. Bull.* 1196; (1965) *The birds of the Republic of Panama*. Part 1. Smithson. misc. Collns 150. WHITE, C M N (1965) *A revised check list of African non-passerine birds*. Lusaka. WHITE, F B (1931) *Auk* **48**, 559–63. WHITE, H C (1937) *J. Biol. Bd. Canada* **3**, 323–38; (1939) *J. Fish Res. Bd. Canada* **4**, 309–11; (1957) *Bull. Fish. Res. Bd. Canada.* **111**, 1–63. WHITE, S J and WHITE, R E C (1970) *Behaviour* **37**, 40–54. WIDMER, E (1928) *Schweiz. Fischerei* **36**, 269–73. WILBUR, H M (1969) *Auk* **86**, 433–42. WILHELM, O (1938) *Dansk orn. Foren. Tidsskr.* **32**, 101–53. WILLI, P (1970) *Orn. Beob.* **67**, 141–217. WILLIAMS, A J (1971) *Astarte* **4**, 31–6. WILLIAMS, G (1959) *Terre et Vie* **106**, 104–120. WILLIAMS, J G (1963) *A field guide to the birds of east and central Africa*. London; (1966) *Bull. Br. Orn. Club.* **86**, 48–50. WILLIAMSON, K (1952) *Scott. Nat.* **64**, 138–47; (1965) *Fair Isle and its birds*. Edinburgh; (1968) *Scott. Birds* **5**, 71–89; (1970) *The Atlantic Islands*. London. WILSON, E A (1970) *Nat. Antarct. Exped. 1901–1904, Nat. Hist.* **2**, Vertebrata 2, 1–121. WINGATE, D B (1964) *Auk* **81**, 147–59. WINGE, H (1898) *Grønlands Fugle. Medd. om Grøn.* 21. WINTERBOTTOM, J M (1974) *Ostrich* **45**, 110–32. WITHERBY, H F (1922) *A practical handbook of British birds* **13**, London; (1934) *Aquila* **38–41**, 443. WITHERBY, H F, JOURDAIN, F C R, TICEHURST, N F, and TUCKER, B W (1939) *The handbook of British birds* **3**; (1940) **4**; (1941) **5**; (1944) **5**, additions and corrections. London. WITTGEN, A B (1962) *Levende Nat.* **65**, 64–72. WOBUS, U (1964) *Der Rothalstaucher*. Wittenberg Lutherstadt. WOLFF, W J (1966) *Ardea* **54**, 230–70. WOLFF, W J, KOEIJER, P, SANDEE, A J J, and WOLF, L (1967) *Limosa* **40**, 163–74. WOLTON, A W, RANISTHORNE, J R, and ROGERS, M J (1950) *Br. Birds* **43**, 91–2. WOOD, N A (1951) *The Birds of Michigan*. Misc. Pub. Mus. Zool. Univ. Mich. 75. WOODS, R W (1970) *Ibis* **112**, 259–60. WRIGHT, C A (1864) *Ibis* (1) **6**, 135–57. WRIGHT, P A, SHARROCK, J T R and DOBINSON, H M (1964) *Br. Birds* **57**, 200–2. WURDEMAN, G (1861) *Ann. Reps. Smiths. Inst. for 1860*, 426–33. WÜRDINGER, I (1970) *Z. Tierpsychol.* **27**, 257–302. WÜST, W (1958) *Anz. orn. Ges. Bayern* **5**, 89–93; (1959) *Anz. orn. Ges. Bayern* **5**, 167–80; (1960) *Proc. Int. Orn. Congr.* **12** (2), 795–800; (1963) *J. Orn.* **104**, 291. WYNNE-EDWARDS, V C (1935) *Proc. Boston Soc. Nat. Hist.* **40**, 233–346; (1952a) *Auk* **69**, 353–91; (1952b) *Scott. Nat.* **64**, 84–101; (1962) *Animal dispersion in relation to social behaviour*. Edinburgh.

YARKER, B and ATKINSON-WILLES, G L (1971) *Wildfowl* **22**, 63–70. YARRELL, W (1843) *History of British Birds* **3**. London. YEATES, G K (1940) *Br. Birds* **34**, 98–9; (1950) *Br. Birds* **43**, 5–8; (1951) *The land of the loon*. London. YEATMAN, L J (1971) *Histoire des Oiseaux d'Europe*. Paris. YELVERTON, C S and QUAY, T L (1959) *Game Div. N. Carolina. Wildl. Res. Comm*, 1–44. YOCOM, C F (1951) *Waterfowl and their food plants in Washington*. Seattle. YOCOM, C F and HARRIS, S W (1965) *J. Wildl. Mgmt* **29**, 874–7. YOUNG, C M (1970a) *Ibis* **112**, 330–5; (1970b) *Ardea* **58**, 125–30. YOUNG, J G (1965) *Wildfowl Trust Ann. Rep.* **16**, 54–5; (1972) *Wildfowl* **23**, 83–7.

ZEDLITZ, O (1913) *J. Orn.* **61**, 179–88. ZELENKA, G (1963) *Alauda* **31**, 212–7. ZHARKOVA, Y and BORZHONOV, B (1972) In Kumari (1972), 117–26. ZIMMERMAN, R (1929) *J. Orn.* **77**, 249–66; (1931) *J. Orn.* **79**, 324–32. ZINK, G (1958) *Vogelwarte* **19**, 243–7; (1961) *Vogelwarte* **21**, 113–18. ZINO, P A (1971) *Ibis* **113**, 212–17. ZUSI, R L and STORER, R W (1969) *Misc. Publs. Mus. Zool. Univ. Mich.*, **139**.

716

PLATE 101 EGGS
*Gavia immer* Great Northern Diver, 2 eggs

*Gavia arctica* Black-throated Diver, 3 eggs

*Gavia stellata* Red-throated Diver, 2 eggs

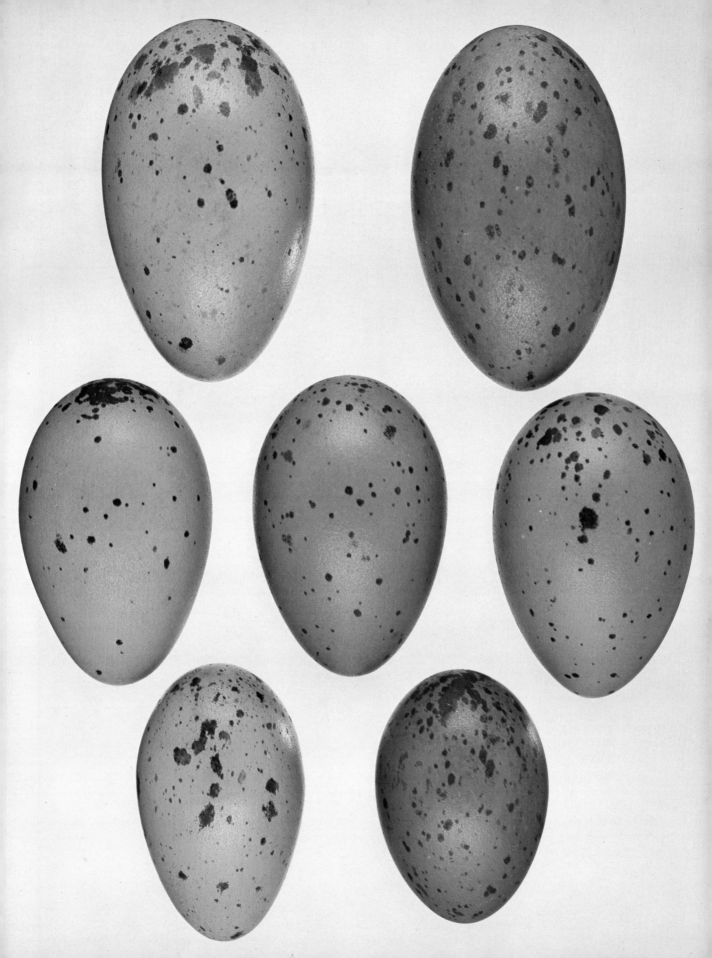

PLATE 102 EGGS

*Tachybaptus ruficollis* Little Grebe, 2 eggs

*Podiceps nigricollis* Black-necked Grebe, 3 eggs

*Podiceps auritus* Slavonian Grebe, 3 eggs

*Podiceps grisegena* Red-necked Grebe, 2 eggs

*Podiceps cristatus* Great Crested Grebe, 3 eggs

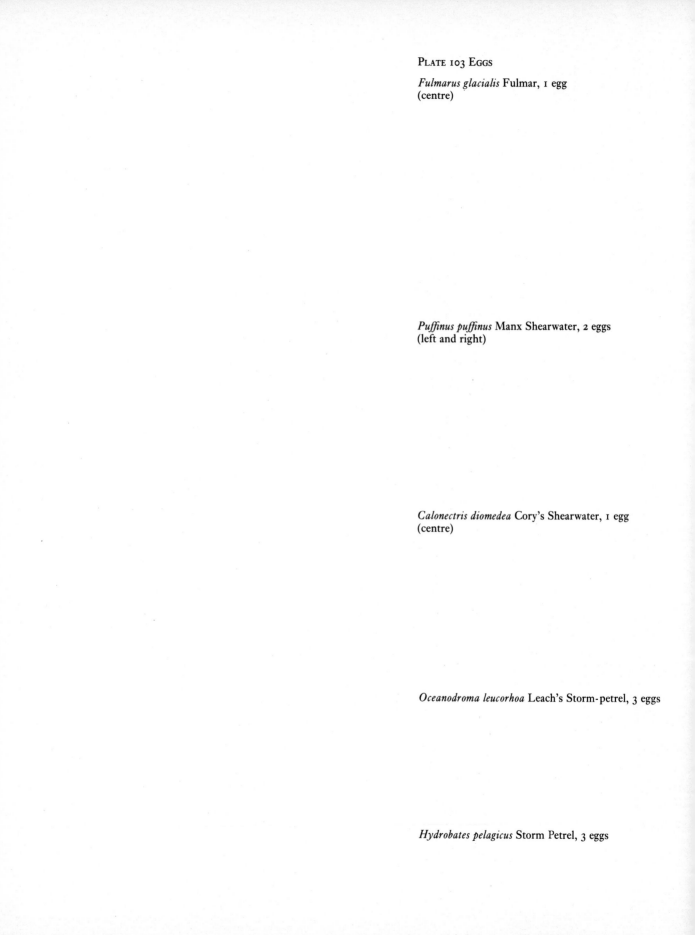

PLATE 103 EGGS

*Fulmarus glacialis* Fulmar, 1 egg
(centre)

*Puffinus puffinus* Manx Shearwater, 2 eggs
(left and right)

*Calonectris diomedea* Cory's Shearwater, 1 egg
(centre)

*Oceanodroma leucorhoa* Leach's Storm-petrel, 3 eggs

*Hydrobates pelagicus* Storm Petrel, 3 eggs

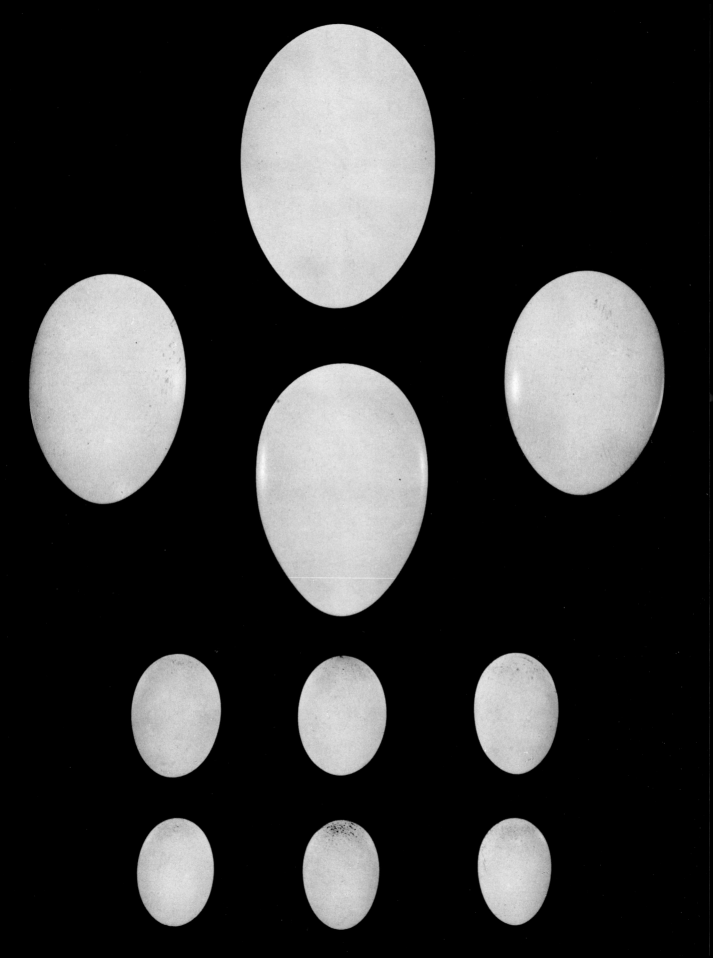

PLATE 104 EGGS

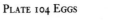

*Pelecanus crispus* Dalmatian Pelican, 1 egg
(left)

*Pelecanus onocrotalus* White Pelican, 1 egg
(right)

*Sula bassana* Gannet, 1 egg
(centre)

*Phalacrocorax aristotelis* Shag, 1 egg
(left)

*Phalacrocorax carbo* Cormorant, 1 egg
(right)

*Phalacrocorax pygmeus* Pygmy Cormorant, 2 eggs

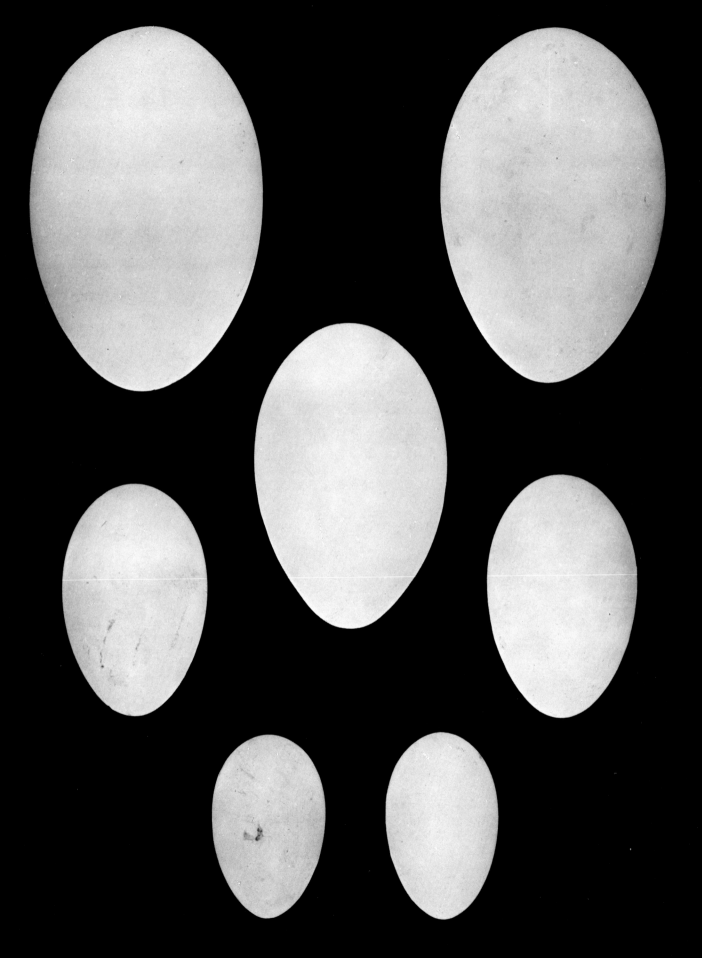

PLATE 106 NEST DOWN

*Cygnus olor* Mute Swan

*Cygnus cygnus* Whooper Swan

*Anser fabalis* Bean Goose

*Anser brachyrhynchus* Pink-footed
Goose

*Anser albifrons* White-fronted Goose

*Anser erythropus* Lesser White-fronted
Goose

*Anser anser* Greylag Goose

*Branta canadensis* Canada Goose

*Branta leucopsis* Barnacle Goose

*Branta bernicla hrota*
Brent Goose (pale-bellied race)

*Alopochen aegyptiacus* Egyptian Goose

*Tadorna tadorna* Shelduck

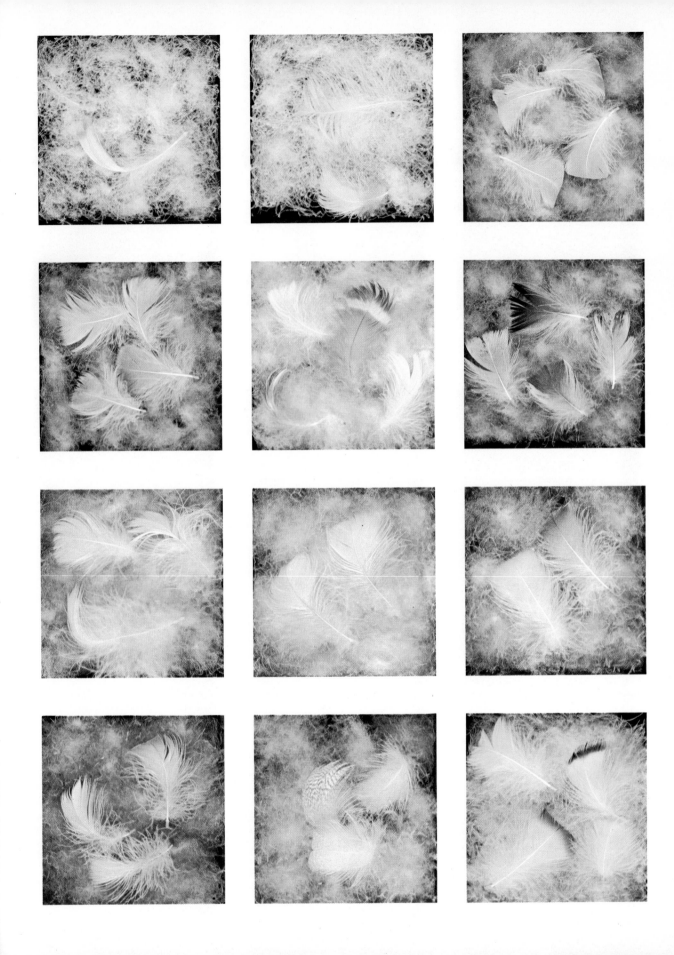

PLATE 107 NEST DOWN

*Anas platyrhynchos* Mallard

*Anas crecca* Teal

*Anas strepera* Gadwall

*Anas penelope* Wigeon

*Anas acuta* Pintail

*Anas querquedula* Garganey

*Anas clypeata* Shoveler

*Marmaronetta angustirostris* Marbled Teal

*Netta rufina* Red-crested Pochard

*Aythya ferina* Pochard

*Aythya nyroca* Ferruginous Duck

*Aythya fuligula* Tufted Duck

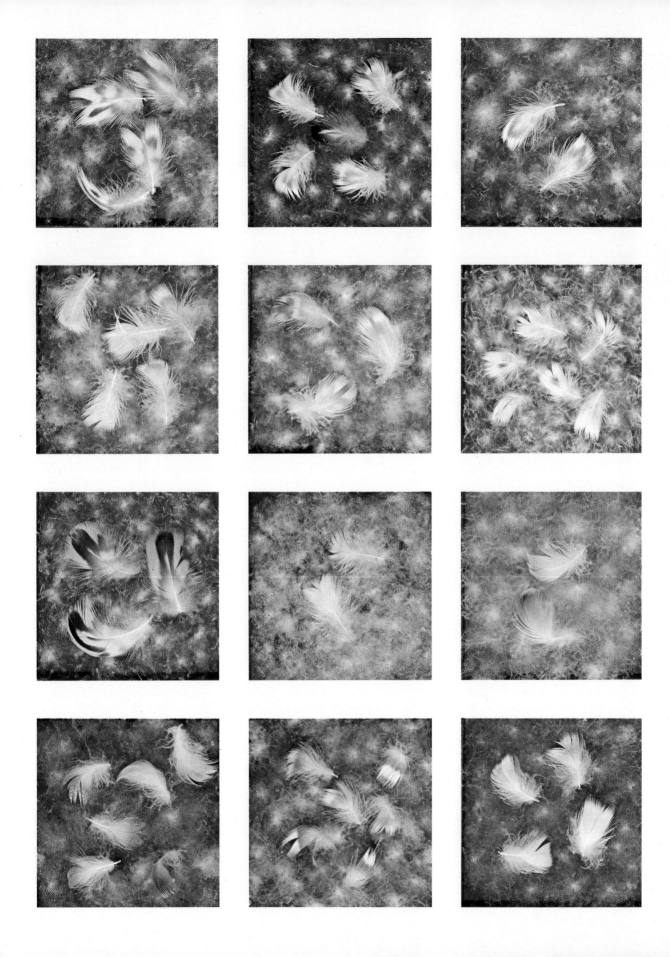

PLATE 108 NEST DOWN

*Aythya marila* Scaup

*Somateria mollissima* Eider

*Somateria spectabilis* King Eider

*Histrionicus histrionicus* Harlequin

*Clangula hyemalis* Long-tailed Duck

*Melanitta nigra* Common Scoter

*Melanitta fusca* Velvet Scoter

*Bucephala islandica* Barrow's Goldeneye

*Bucephala clangula* Goldeneye

*Mergus albellus* Smew

*Mergus serrator* Red-breasted Merganser

*Mergus merganser* Goosander

# INDEXES

Figures in **bold type** refer to plates

## SCIENTIFIC NAMES

## FRENCH NAMES

## GERMAN NAMES